GEOMETRIC FORMULAS

A = area, P = perimeter, C = circumference, SA = surface area, V = volume

Triangle

$$A = \frac{1}{2}bh$$

Heron's Formula:

$$A = \sqrt{s(s-a)(s-b)(s-c)}$$

where $s = \dfrac{a+b+c}{2}$

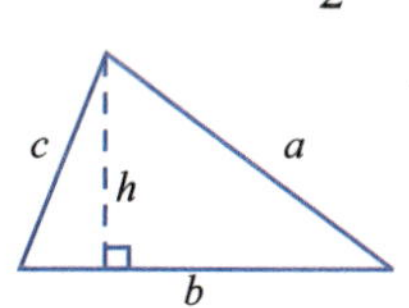

Rectangle

$$A = lw$$
$$P = 2l + 2w$$

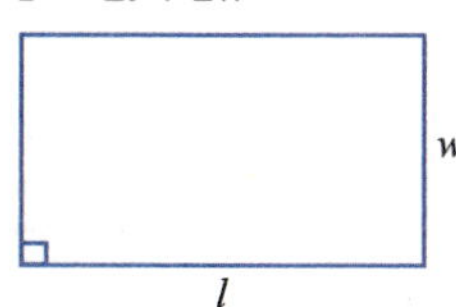

Parallelogram

$$A = bh$$

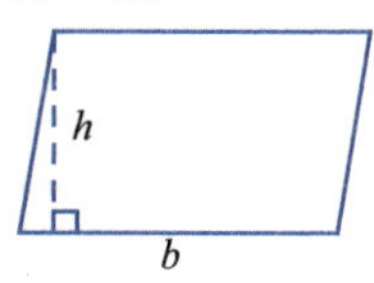

Trapezoid

$$A = \frac{1}{2}h(b+c)$$

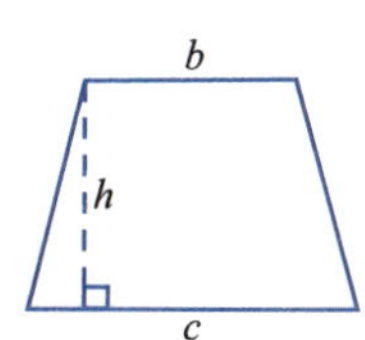

Circle

$$A = \pi r^2$$
$$C = 2\pi r$$

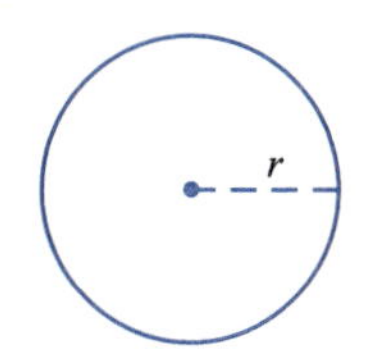

Ellipse

$$A = \pi ab$$

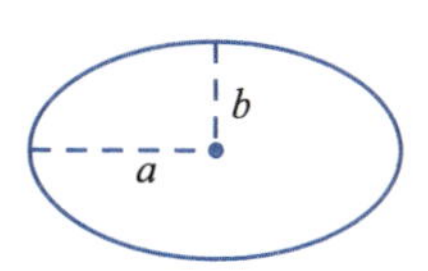

Rectangular Prism

$$V = lwh$$
$$SA = 2lh + 2wh + 2lw$$

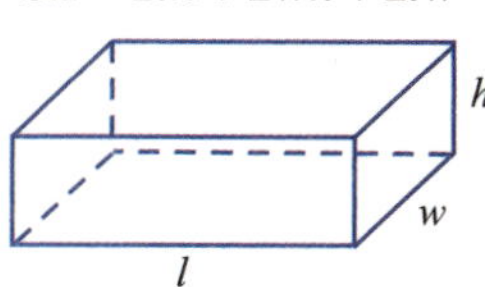

Circular Cone

$$V = \frac{1}{3}\pi r^2 h$$

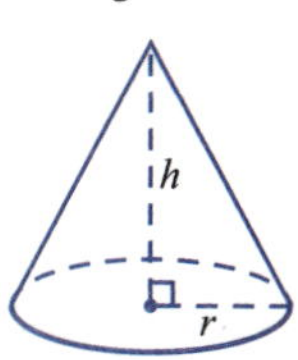

Rectangular Pyramid

$$V = \frac{1}{3}lwh$$

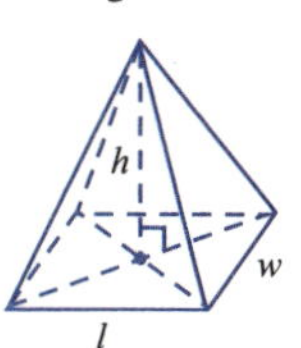

Right Circular Cylinder

$$V = \pi r^2 h$$
$$SA = 2\pi r^2 + 2\pi rh$$

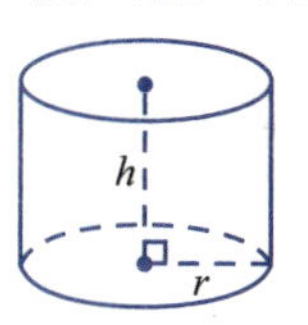

Right Cylinder

$$V = (Area\ of\ Base)h$$

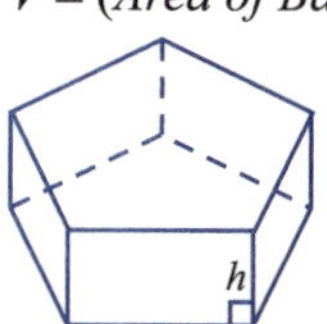

Sphere

$$V = \frac{4}{3}\pi r^3$$
$$SA = 4\pi r^2$$

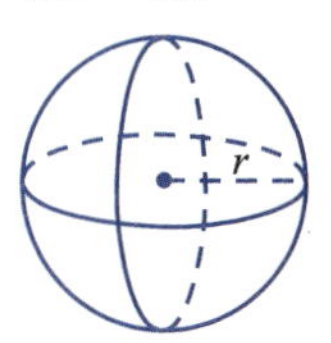

PROPERTIES OF ABSOLUTE VALUE 1.1

For all real numbers a and b:

$|a| \ge 0$ $\quad |-a| = |a|$

$a \le |a|$ $\quad |ab| = |a||b|$

$\left|\dfrac{a}{b}\right| = \dfrac{|a|}{|b|},\ b \ne 0$

$|a+b| \le |a| + |b|$ (the triangle inequa

PROPERTIES OF EXPONEN RADICALS 1.2

$a^n \cdot a^m = a^{n+m}$ $\quad (a^n)^m = a^{nm}$ $\quad (ab)^n = a^n b^n$

$\dfrac{a^n}{a^m} = a^{n-m}$ $\quad a^{-n} = \dfrac{1}{a^n}$ $\quad \left(\dfrac{a}{b}\right)^n = \dfrac{a^n}{b^n}$

$a^{\frac{1}{n}} = \sqrt[n]{a}$ $\quad \sqrt[n]{ab} = \sqrt[n]{a}\sqrt[n]{b}$ $\quad a^{\frac{m}{n}} = \sqrt[n]{a^m} = \left(\sqrt[n]{a}\right)^m$

$\sqrt[n]{\dfrac{a}{b}} = \dfrac{\sqrt[n]{a}}{\sqrt[n]{b}}$ $\quad \sqrt[m]{\sqrt[n]{a}} = \sqrt[mn]{a}$

SPECIAL PRODUCT FORMULAS 1.3

$$(A-B)(A+B) = A^2 - B^2$$
$$(A+B)^2 = A^2 + 2AB + B^2$$
$$(A-B)^2 = A^2 - 2AB + B^2$$

FACTORING SPECIAL BINOMIALS 1.3

$$A^2 - B^2 = (A-B)(A+B)$$
$$A^3 - B^3 = (A-B)(A^2 + AB + B^2)$$
$$A^3 + B^3 = (A+B)(A^2 - AB + B^2)$$

QUADRATIC FORMULA 1.8

The solutions of the equation $ax^2 + bx + c = 0$ are

$$x = \frac{-b \pm \sqrt{b^2 - 4ac}}{2a}.$$

PYTHAGOREAN THEOREM 2.1

Given a right triangle with legs a and b and hypotenuse c,

$$a^2 + b^2 = c^2.$$

DISTANCE FORMULA 2.1

$$d = \sqrt{(x_2 - x_1)^2 + (y_2 - y_1)^2}$$

MIDPOINT FORMULA 2.1

$$\left(\frac{x_1 + x_2}{2}, \frac{y_1 + y_2}{2}\right)$$

STANDARD FORM OF THE EQUATION OF A CIRCLE 2.2

The standard form of the equation of a circle with radius r and center (h,k) is

$$(x-h)^2 + (y-k)^2 = r^2.$$

SLOPE OF A LINE 2.4

$$m = \frac{y_2 - y_1}{x_2 - x_1}$$

Horizontal lines $y = c$ have a slope of 0.

Vertical lines $x = c$ have an undefined slope.

FORMS OF LINEAR EQUATIONS 2.4

Standard form: $ax + by = c$

Slope-intercept form: $y = mx + b$

Point-slope form: $y - y_1 = m(x - x_1)$

PARALLEL AND PERPENDICULAR LINES 2.5

Given a line with slope m:

slope of parallel line $= m$

slope of perpendicular line $= -\frac{1}{m}$

TRANSFORMATIONS OF FUNCTIONS 4.1

Horizontal Shifting

The graph of $g(x) = f(x - h)$ has the same shape as the graph of f, but shifted h units to the right if $h > 0$ and shifted h units to the left if $h < 0$.

Vertical Shifting

The graph of $g(x) = f(x) + k$ has the same shape as the graph of f, but shifted upward k units if $k > 0$ and downward k units if $k < 0$.

Reflecting with Respect to Axes

- The graph of the function $g(x) = -f(x)$ is the reflection of the graph of f with respect to the x-axis.
- The graph of the function $g(x) = f(-x)$ is the reflection of the graph of f with respect to the y-axis.

Stretching and Compressing

- The graph of the function $g(x) = af(x)$ is stretched vertically compared to the graph of f by a factor of a if $a > 1$.
- The graph of the function $g(x) = af(x)$ is compressed vertically compared to the graph of f by a factor of a if $0 < a < 1$.
- The graph of the function $g(x) = f(ax)$ is stretched horizontally compared to the graph of f by a factor of $\frac{1}{a}$ if $0 < a < 1$.
- The graph of the function $g(x) = f(ax)$ is compressed horizontally compared to the graph of f by a factor of $\frac{1}{a}$ if $a > 1$.

OPERATIONS WITH FUNCTIONS 4.3

Let f and g be functions.

$$(f + g)(x) = f(x) + g(x)$$

$$(f - g)(x) = f(x) - g(x)$$

$$(fg)(x) = f(x)g(x)$$

$$\left(\frac{f}{g}\right)(x) = \frac{f(x)}{g(x)}, \text{ when } g(x) \neq 0$$

$$(f \circ g)(x) = f(g(x))$$

RATIONAL ZERO THEOREM 5.3

If $f(x) = a_n x^n + a_{n-1}x^{n-1} + \cdots + a_1 x + a_0$ is a polynomial with integer coefficients with $a_n \neq 0$, then any rational zero of f must be of the form $\frac{p}{q}$, where p is a factor of the constant term a_0 and q is a factor of the leading coefficient a_n.

FUNDAMENTAL THEOREM OF ALGEBRA 5.4

If p is a polynomial of degree n, with $n \geq 1$, then p has at least one zero. That is, the equation $p(x) = 0$ has at least one solution. (**Note:** The solution may be a nonreal complex number.)

COMPOUND INTEREST 6.2

An investment of P dollars, compounded n times per year at an annual interest rate of r, has a value after t years of

$$A(t) = P\left(1 + \frac{r}{n}\right)^{nt}.$$

An investment compounded continuously has an accumulated value of $A(t) = Pe^{rt}$.

PROPERTIES OF LOGARITHMS 6.3, 6.4

For $a, x, y > 0$, $a \neq 1$, and any real number r:

$\log_a x = y$ and $x = a^y$ are equivalent

$\log_a 1 = 0$ $\log_a a = 1$

$\log_a(a^x) = x$ $a^{\log_a x} = x$

$\log_a(xy) = \log_a x + \log_a y$

$\log_a\left(\frac{x}{y}\right) = \log_a x - \log_a y$

$\log_a(x^r) = r\log_a x$

CHANGE OF BASE FORMULA 6.4

For $a, b, x > 0$ and $a, b \neq 1$:

$$\log_b x = \frac{\log_a x}{\log_a b}$$

THIRD EDITION

PRECALCULUS

PAUL SISSON

Editors:
Danielle C. Bess,
Daniel Breuer,
S. Rebecca Johnson,
Claudia Vance

Assistant Editors:
Sarah L. Allen,
Allison Conger,
Marvin Glover,
Lisa Hinton

Index:
Barbara Miller

Creative Services Manager:
Trudy Gove

Contributing Designers:
Patrick Thompson,
Joshua A. Walker

Composition and Answer Key Assistance:
Quant Systems India Pvt. Ltd.

Courseware Developers:
Vince Cellini,
Adam Flaherty

Technology Assistant:
Kyle Gilstrap

Cover Design:
Patrick Thompson

Cover Artwork:
George Hart

George Hart is a sculptor and applied mathematician who demonstrates how mathematics is cool and creative in ways you might not have expected.

Whether he is slicing a bagel into two linked halves or leading hundreds of participants in an intricate geometric sculpture barn raising, he always finds original ways to share the beauty of mathematical thinking.

Hart's career includes eight years as a professor at Columbia University, fifteen years as a Research Professor at Stony Brook University, and five years cofounding the Museum of Mathematics in New York City. Now a full-time artist and consultant, he also makes videos that show the fun and creative sides of mathematics. See http://georgehart.com for examples of his work.

Manager of Math Content Development: Chelsey Cooke

Project Manager: Emily Christian

A division of Quant Systems, Inc.

546 Long Point Road
Mount Pleasant, SC 29464

Library of Congress Control Number 2020906521

Printed in the United States of America

10 9 8 7 6 5 4 3 2 1

ISBN: 978-1-64277-171-8

Table of Contents

2 Equations and Inequalities in Two Variables

3 Relations, Functions, and Their Graphs

4 Working with Functions

5 Polynomial and Rational Functions

6 Exponential and Logarithmic Functions

7 Trigonometric Functions

8 Trigonometric Identities and Equations

9 Additional Topics in Trigonometry

10 Conic Sections

11 Systems of Equations and Inequalities

12 Sequences, Series, Combinatorics, and Probability

13 An Introduction to Limits, Continuity, and the Derivative

Preface

Introduction

Goals of this book: Of the many types of people who take precalculus, a fortunate few are firmly set on their academic path and already know how the skills and knowledge gained will open the door to later studies. But for a far larger number of students, precalculus is a step in a long and tremendously important process of discovery. My goals in writing are to provide readers the means to acquire needed skills and knowledge, but even more importantly, to provide the inspiration to do so.

I do this because if I were forced to give just one reason that people should learn mathematics, I would say, "Life is simply easier for those who know how to think mathematically." Whether the issue at hand is one of family budgeting, constructing a business model, making an investment decision, selecting a major to pursue, or any of the other innumerable choices we are all faced with every day, such choices are made with less stress and more confidence by those who possess the right mathematical tools and experience in critical thinking.

Style of this book: My style of writing arises from my own experiences as a student and as an educator. I firmly believe that mathematics is easiest to learn and remember when it's tied to history and context, and the human element of mathematics is stressed throughout this book. Mathematicians know that math is a creative art form calling for intuition, experimentation, and imagination, all of which I call upon the reader to use. The habit of thinking mathematically comes through understanding, not rote memorization, so underlying principles and motivation are consistently emphasized. Finally, I have written a text that is meant to be read, comprised of a linearly presented narrative as opposed to a scattered collection of bullet points and rigid cookie-cutter procedures.

To the student: If you are a student taking precalculus, the old saying that "math is not a spectator sport" should be taken to heart. Mathematics is not something you can learn by watching someone else do it. Chances are you've already had the unhappy experience of watching a teacher solve a problem and marveling at how easy it seems, only to discover later that a nearly identical problem is much harder to solve at home or on a test. The lesson to be learned is that you have to practice math in order to learn math.

Another key point is that very, very few people will fully grasp a mathematical concept or master a mathematical skill the first time they are presented with it. Most math is learned in a cyclic process of plowing ahead until lost, backing up and rereading, and then plowing ahead a bit further. A math book is not like a novel—you shouldn't expect to be able to read it straight through from cover to cover.

Finally, be sure to take full advantage of your instructor and fellow students. Pay attention to what your instructor emphasizes and learn from his or her unique insight into the material. Work with fellow students when possible, and ask others for help if they understand something you haven't gotten yet. And in turn, when you have the opportunity to explain some math to someone else, take advantage of that too. Teaching mathematics to others is an amazingly effective way to deepen your own understanding.

Features

Application Previews

Each chapter begins with a list of sections and an engaging preview of an application appearing in the chapter.

CHAPTER 8

Trigonometric Identities and Equations

SECTIONS

8.1 Fundamental Trigonometric Identities
8.2 Sum and Difference Identities
8.3 Product-Sum Identities
8.4 Trigonometric Equations

WHAT IF ...

What if you were walking up to bat in a baseball game? At what angle from the ground should you hit the ball to make sure that it clears the back fence, giving you a home run?

By the end of this chapter, you will have been introduced to and know how to solve equations using several fundamental trigonometric identities. These identities are very useful for simplifying trigonometric expressions. On page 621, you will find a baseball problem similar to the one given above. You will need to use identities such as the double-angle identities on page 604 and inverse trigonometric functions to find the angle to hit the ball in order to achieve the needed range.

Historical Contexts

Each chapter includes a brief introduction to the historical context of the math that follows. Mathematics is a human endeavor, and knowledge of how and why a particular idea developed is of great help in understanding it. Too often, math is presented in cold, abstract chunks completely divorced from the rest of reality. While a (very) few students may be able to master material this way, most benefit from an explanation of how math ties into the rest of what people were doing at the time it was created.

Pythagoras

Introduction

In this chapter, we review the terminology, notation, and properties of the real number system frequently encountered in algebra; the extension of the real number system to the larger set of complex numbers; and the basic algebraic methods we use to solve equations and inequalities.

We begin with a discussion of common subsets of the set of real numbers. Certain types of numbers are important from both a historical and a mathematical perspective. There is archaeological evidence that people used the simplest sort of numbers, the counting or natural numbers, as far back as 50,000 years ago. Over time, many cultures discovered needs for various refinements to the number system, resulting in the development of such classes of numbers as the integers, the rational numbers, the irrational numbers, and ultimately, the complex numbers, a number system that contains the real numbers.

Many of the ideas in this chapter have a history dating as far back as Egyptian and Babylonian civilizations of around 3000 BC, with later developments and

Topics

Each section begins with a list of topics. These concise objectives are a helpful guide for both reference and class preparation.

4.3 COMBINING FUNCTIONS

TOPICS

1. Combining functions arithmetically
2. Composing functions
3. Decomposing functions
4. Recursive graphics

TOPIC 1: Combining Functions Arithmetically

In Section 4.1, we gained experience in building new functions from old ones by shifting, reflecting, and stretching the old functions. In this section, we will explore more ways of building functions.

We begin with four arithmetic ways of combining two or more functions to obtain new functions. The basic operations are very familiar to you: addition, subtraction, multiplication, and division. The difference is that we are applying these operations to functions, but as we will see, the arithmetic combination of functions is based entirely on the arithmetic combination of numbers.

DEFINITION: Addition, Subtraction, Multiplication, and Division of Functions

Let f and g be two functions. The **sum** $f+g$, **difference** $f-g$, **product** fg, and **quotient** $\frac{f}{g}$ are four new functions defined as follows.

1. $(f+g)(x)=f(x)+g(x)$
2. $(f-g)(x)=f(x)-g(x)$
3. $(fg)(x)=f(x)g(x)$
4. $\left(\frac{f}{g}\right)(x)=\frac{f(x)}{g(x)}$, provided that $g(x)\neq 0$

The domain of each of these new functions consists of the common elements (or the intersection of elements) of the domains of f and g individually, with the added condition that in the quotient function we have to omit those elements for which $g(x)=0$.

Definitions, Theorems, Formulas, Properties, Identities, and Procedures

All definitions are clearly identified and set apart in highly visible green boxes for easy reference, and all theorems, formulas, properties, identities, and procedures are similarly set apart in distinctive blue and purple boxes. All formally identified terms appear in bold print when first defined, and other useful terms appear in italic font.

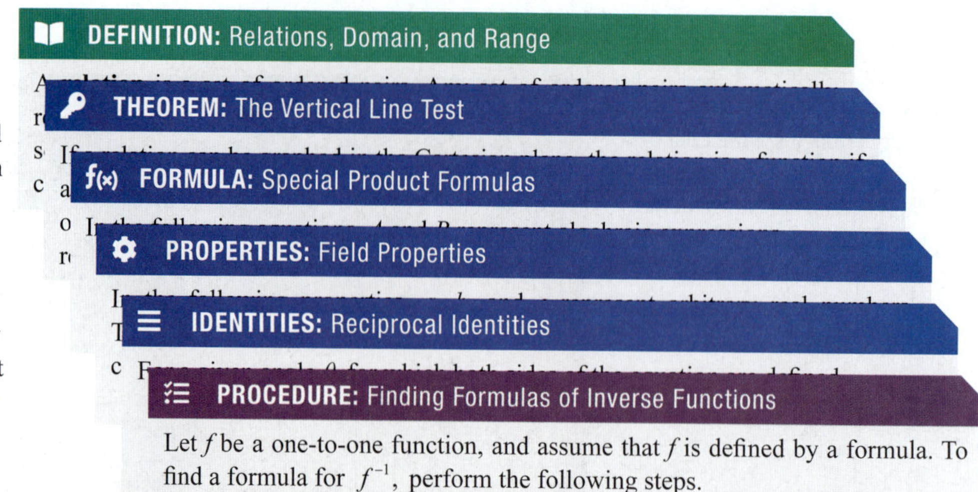

DEFINITION: Relations, Domain, and Range
THEOREM: The Vertical Line Test
FORMULA: Special Product Formulas
PROPERTIES: Field Properties
IDENTITIES: Reciprocal Identities
PROCEDURE: Finding Formulas of Inverse Functions
Let f be a one-to-one function, and assume that f is defined by a formula. To find a formula for f⁻¹, perform the following steps.

Cautions

Many common errors are pointed out, along with how to correct them. These are set apart in red boxes.

CAUTION

We are faced with another example of the reuse of notation. Note that f^{-1} does *not* stand for $\frac{1}{f}$ when f is a function! We use an exponent of -1 to indicate the reciprocal of a number or an algebraic expression, but when applied to a function or a relation it stands for the inverse relation.

Sidebars

Historical and biographical anecdotes appear in sidebar format, providing additional background to the main body of the text.

Technology Instructions

Technology notes and screenshots are included throughout the text to highlight ways that graphing utilities can help solve problems or explain concepts. Step-by-step instructions for using a TI-84 Plus are given in many cases.

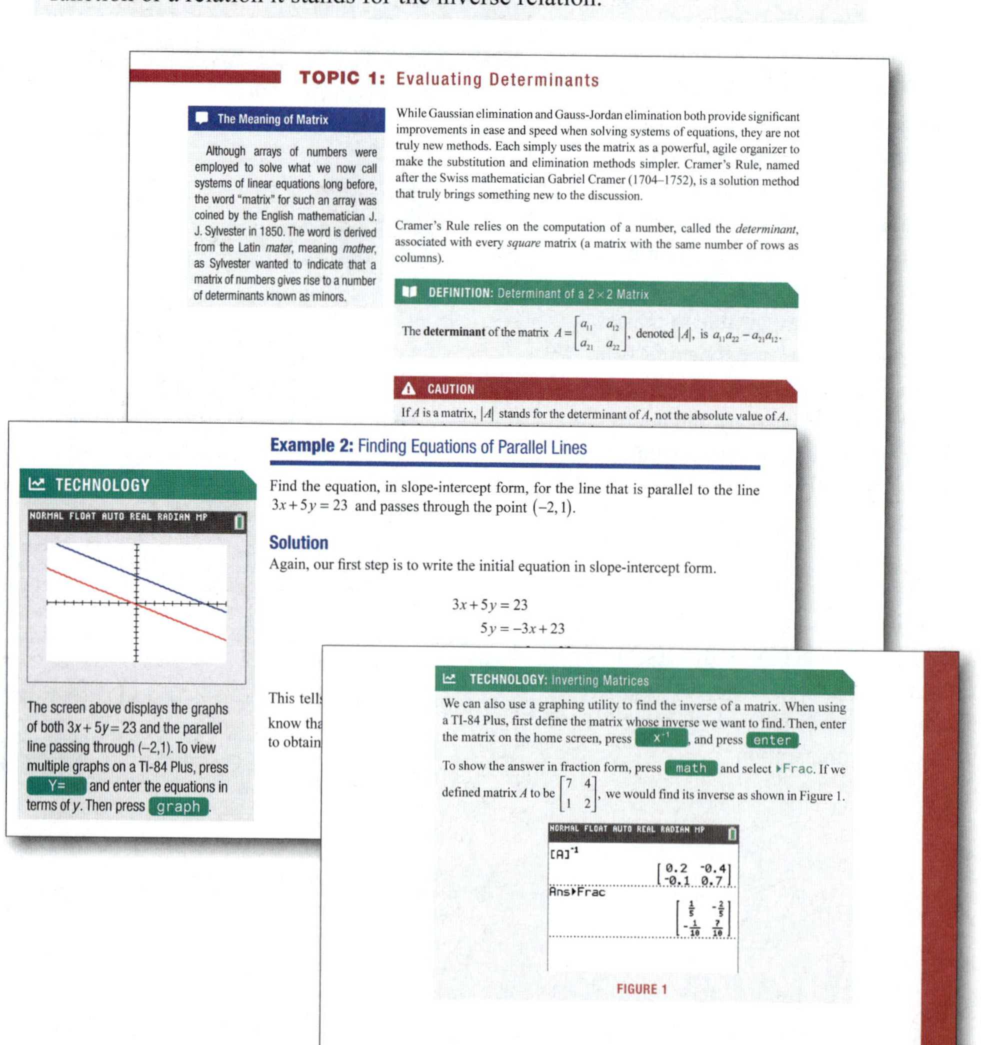

TOPIC 1: Evaluating Determinants
The Meaning of Matrix
Although arrays of numbers were employed to solve what we now call systems of linear equations long before, the word "matrix" for such an array was coined by the English mathematician J. J. Sylvester in 1850. The word is derived from the Latin mater, meaning mother, as Sylvester wanted to indicate that a matrix of numbers gives rise to a number of determinants known as minors.
While Gaussian elimination and Gauss-Jordan elimination both provide significant improvements in ease and speed when solving systems of equations, they are not truly new methods. Each simply uses the matrix as a powerful, agile organizer to make the substitution and elimination methods simpler. Cramer's Rule, named after the Swiss mathematician Gabriel Cramer (1704–1752), is a solution method that truly brings something new to the discussion.
Cramer's Rule relies on the computation of a number, called the determinant, associated with every square matrix (a matrix with the same number of rows as columns).
DEFINITION: Determinant of a 2 × 2 Matrix
The determinant of the matrix A = [a11 a12; a21 a22], denoted |A|, is a11a22 − a21a12.
CAUTION
If A is a matrix, |A| stands for the determinant of A, not the absolute value of A.
Example 2: Finding Equations of Parallel Lines
Find the equation, in slope-intercept form, for the line that is parallel to the line 3x + 5y = 23 and passes through the point (−2, 1).
Solution
Again, our first step is to write the initial equation in slope-intercept form.
3x + 5y = 23
5y = −3x + 23
TECHNOLOGY
NORMAL FLOAT AUTO REAL RADIAN MP
The screen above displays the graphs of both 3x + 5y = 23 and the parallel line passing through (−2,1). To view multiple graphs on a TI-84 Plus, press Y= and enter the equations in terms of y. Then press graph.
TECHNOLOGY: Inverting Matrices
We can also use a graphing utility to find the inverse of a matrix. When using a TI-84 Plus, first define the matrix whose inverse we want to find. Then, enter the matrix on the home screen, press x⁻¹ and press enter.
To show the answer in fraction form, press math and select ▸Frac. If we defined matrix A to be [7 4; 1 2], we would find its inverse as shown in Figure 1.
NORMAL FLOAT AUTO REAL RADIAN MP
[A]⁻¹
Ans▸Frac
FIGURE 1

Applications

Many exercises and examples illustrate practical applications, keeping students engaged.

Example 4: Using Trigonometric Functions

Before cutting down a dead tree in your yard, you very sensibly decide to determine its height. Backing up 40 feet from the tree (which rises straight up from level ground), you use a *theodolite* (a surveyor's instrument that accurately measures angles) and note that the angle between the ground and the top of the tree is 61°55′39″. How tall is the tree?

opp

61°55′39″

40 feet

FIGURE 5

Solution

As in so many problems, a picture is of great help. Note that this problem indeed involves solving a right triangle: we know the measure of an angle and the length of the angle's adjacent leg, and we want the length of the opposite leg. This observation gives us the best clue as to which trigonometric function to use; since tangent and cotangent are the two that don't depend on the length of the hypotenuse, chances are one of these is a good choice.

Note also that we have to convert the theodolite's reading into decimal form before using a calculator.

$$61°55'39'' = 61 + \frac{55}{60} + \frac{39}{3600} = 61.9275°$$

Now we can use Figure 5 to see that

APPLICATIONS

66. The three sides of a triangle are related as follows: the perimeter is 43 feet, the second side is 5 feet more than twice the first side, and the third side is 3 feet less than the sum of the other two sides. Find the lengths of the three sides of the triangle.

67. Eric's favorite candy bar and ice cream flavor have fat and calorie contents as follows: each candy bar has 5 grams of fat and 280 calories; each serving of ice cream has 10 grams of fat and 150 calories. How many candy bars and servings of ice cream did he eat during the weekend he consumed 85 grams of fat and 2300 calories from these two treats?

68. A farmer plants soybeans, corn, and wheat and rotates the planting each year on her 500-acre farm. In a particular year, the profits from her crops were $120 per acre of soybeans, $100 per acre of corn, and $80 per acre of wheat. She planted twice as many acres of corn as soybeans. How many acres did she plant with each crop that year if she made a total profit of $51,800?

Categorized Exercises

Each section concludes with a selection of exercises designed to allow the student to practice skills and master concepts. References to relevant examples from the section are clearly labeled where applicable. Many levels of difficulty exist within each exercise set, providing instructors with flexibility in assigning exercises and allowing students to practice elementary skills or stretch themselves, as appropriate. Exercise sets are organized into the four categories of Practice, Applications, Writing & Thinking, and Technology.

PRACTICE

Find equations for the vertical asymptotes, if any, for each of the following rational functions. See Example 1.

1. $f(x) = \frac{5}{x-1}$
2. $f(x) = \frac{x^2+3}{x+3}$
3. $f(x) = \frac{x^2-4}{x+2}$
4. $f(x) = \frac{-3x+5}{x-2}$

APPLICATIONS

27. The distance that an object falls from rest, when air resistance is negligible, varies directly as the square of the time. A stone dropped from rest travels 144 feet in the first 3 seconds. How far does it travel in the first 4 seconds?

28. A record store manager observes that the number of records sold seems to vary inversely as the price per record. If the store sells 840 records per week when the price per record is $15.99, how many does he expect to sell if he lowers the price to $14.99?

29. A person's Body Mass Index (BMI) is used by physicians to determine if a patient's

WRITING & THINKING

61. Given two odd functions *f* and *g*, show that $f \circ g$ is also odd. Verify this fact with the particular functions $f(x) = \sqrt[3]{x}$ and $g(x) = \frac{-x^3}{3x^2-9}$. Recall that a function is odd if $f(-x) = -f(x)$ for all *x* in the domain of *f*.

62. Given two even functions *f* and *g*, show that the product is also even. Verify this fact with the particular functions $f(x) = 2x^4 - x^2$ and $g(x) = \frac{1}{x^2}$. Recall that a function is even $f(-x) = f(x)$ for all *x* in the domain of *f*.

TECHNOLOGY

Use a graphing utility to graph the following functions. Experiment with different viewing windows until you obtain a sketch that seems to capture the meaningful parts of the graph.

47. $f(x) = 10x^5 - x^3$
48. $g(x) = x^5 + x^2$
49. $f(x) = x^3 - 5x^2 + x$
50. $g(x) = \sqrt{x} - x^2$
51. $f(x) = \sqrt{x} + 3x - 1$
52. $g(x) = x^4 - 3x^3 + 2$

Chapter Projects

Each project describes a plausible scenario related to the concepts of the chapter and is suitable for an individual or group assignment.

CHAPTER 11 PROJECT

Market Share Matrix

Assume you are the sales and marketing director for Joe's Java, a coffee shop located on a crowded city street corner. There are two competing coffee shops on this block—Buck's Café and Tweak's Coffee. The management has asked you to develop a marketing campaign to increase your market share from 25% to at least 35% within 6 months. With the resulting plan to meet this goal, you predict that each month

a. you will retain 93% of your customers, 4% will go to Buck's Café, and 3% will go to Tweak's Coffee;

b. Buck's Café will retain 91% of their customers, 6% will come to Joe's Java, and 3% will go to Tweak's Coffee; and

Chapter Review Exercises

Immediately following each Chapter Project is a Chapter Review, which presents several more exercises pertaining to the major ideas of the chapter.

CHAPTER 4 REVIEW EXERCISES

Section 4.1

Sketch the graphs of the following functions by first identifying the more basic functions that have been shifted, reflected, stretched, or compressed. Then determine the domain and range of each function.

1. $f(x)=(x-1)^3+2$ **2.** $G(x)=4|x+3|$

3. $m(x)=\dfrac{1}{(x+2)^2}$ **4.** $g(x)=-\sqrt[3]{x}+4$

5. $r(x)=\dfrac{1}{x-2}-3$ **6.** $f(x)=\sqrt{x-1}+3$

Formula Pages

Three pages in the front of the text and additional pages in the back of the text detail the most important formulas, theorems, graphs, and identities covered in precalculus.

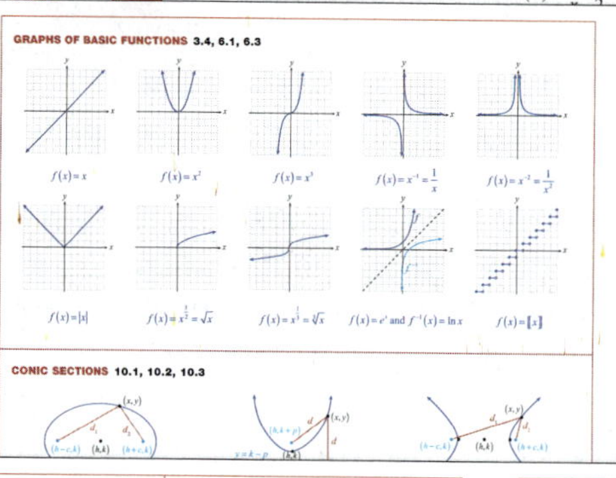
GRAPHS OF BASIC FUNCTIONS 3.4, 6.1, 6.3

CONIC SECTIONS 10.1, 10.2, 10.3

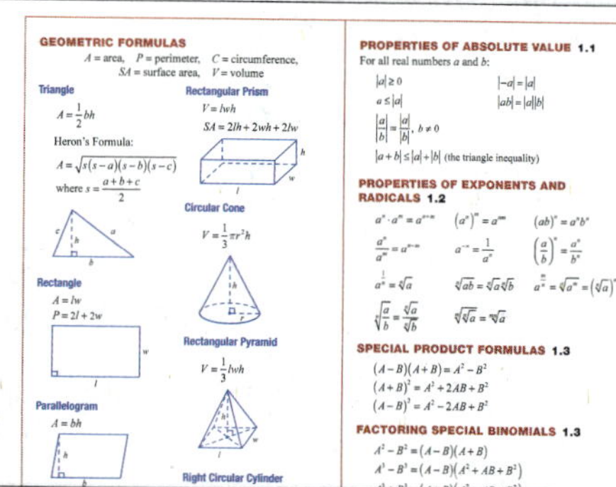
GEOMETRIC FORMULAS

PROPERTIES OF ABSOLUTE VALUE 1.1

PROPERTIES OF EXPONENTS AND RADICALS 1.2

SPECIAL PRODUCT FORMULAS 1.3

FACTORING SPECIAL BINOMIALS 1.3

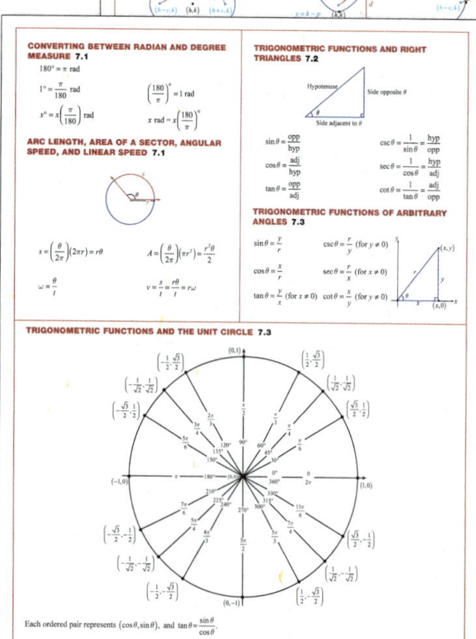
CONVERTING BETWEEN RADIAN AND DEGREE MEASURE 7.1

ARC LENGTH, AREA OF A SECTOR, ANGULAR SPEED, AND LINEAR SPEED 7.1

TRIGONOMETRIC FUNCTIONS AND RIGHT TRIANGLES 7.2

TRIGONOMETRIC FUNCTIONS OF ARBITRARY ANGLES 7.3

TRIGONOMETRIC FUNCTIONS AND THE UNIT CIRCLE 7.3

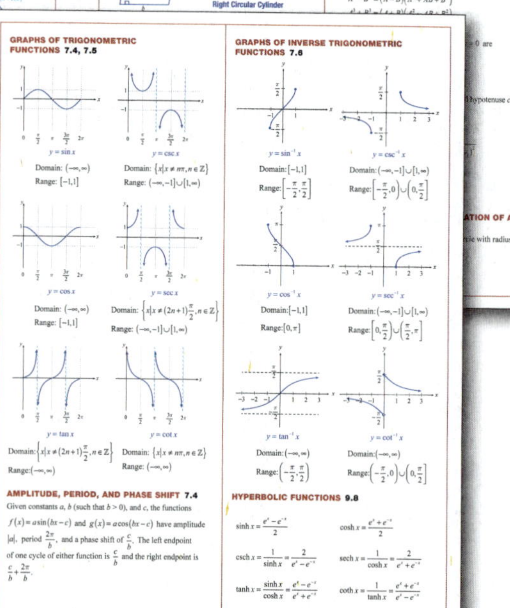
GRAPHS OF TRIGONOMETRIC FUNCTIONS 7.4, 7.5

GRAPHS OF INVERSE TRIGONOMETRIC FUNCTIONS 7.6

AMPLITUDE, PERIOD, AND PHASE SHIFT 7.4

HYPERBOLIC FUNCTIONS 9.8

Major Changes in the Third Edition

- **New and Reorganized Chapters**
  A new chapter titled "An Introduction to Limits, Continuity, and the Derivative" has been included; and the chapter introducing functions has been expanded and split into two chapters, the first titled "Relations, Functions, and their Graphs" and the second titled "Working with Functions."

- **New Sections**
  A new section devoted to the development and use of mathematical models has been written, as has a new section introducing hyperbolic functions. The section titled "Mathematical Models" contains a broad array of interesting and realistic examples, including an extensive discussion of cost and revenue functions and the interpolation and extrapolation of data. The section "Hyperbolic Functions" provides students experience with these functions prior to their usual abrupt appearance in calculus.

- **Expanded and Split Sections**
  The section titled "Linear and Quadratic Functions" has been expanded and split into two sections dealing with these classes individually. The section titled "Transformations of Functions" has been expanded and split into two sections, the first with the same name and the second titled "Properties of Functions." The section titled "Graphs of Trigonometric Functions" has been expanded and split into two sections. The first is "Graphs of Sine and Cosine Functions" and the second is "Graphs of Other Trigonometric Functions."

- **New and Expanded Topics**
  The discussion and use of many existing topics have been expanded, and new topics have been added. These include
  - regression (linear, quadratic, exponential, logistic, and logarithmic),
  - constructing mathematical models,
  - interpolation and extrapolation,
  - intervals of monotonicity,
  - local extrema,
  - average rate of change,
  - systems of linear inequalities,
  - transition matrices, and
  - the Multinomial Theorem.

- **New Methodology**
  The extension of trigonometric functions to the real line now includes a comprehensive discussion and application of the unit circle approach.

Acknowledgements

I am very grateful to all at Hawkes Learning for their hard work and dedication to this project, as well as for their indefatigable good cheer. In particular, many thanks to

Dr. James Hawkes
Marcel Prevuznak
Chelsey Cooke
Emily Christian
Daniel Breuer
Danielle C. Bess
Claudia Vance
S. Rebecca Johnson
Marvin Glover
Lisa Hinton
Sarah L. Allen
Allison Conger
Barbara Miller
Quant Systems India Pvt. Ltd.
Adam Flaherty
Vince Cellini
Kyle Gilstrap
Trudy Gove
Joshua A. Walker
Patrick Thompson
George Hart

I am also very grateful to all the readers of the first and second editions who have taken the time to offer helpful comments and suggest alterations and additions. I've benefitted greatly from your collective wisdom, and many of the improvements in this edition are due to your valued feedback.

Many, many thanks are due to the reviewers of this edition and previous editions for their careful and thorough work!

Reviewers of the third edition:

Mike Azlin *University of Mississippi*
Rick Bailey *Midlands Technical College*
Brent Bollich *South Louisiana Community College*
Allison Elowson *University of North Carolina at Charlotte*
Fenecia Foster *Southeast Technical Institute*
Heidi Griffin *Arkansas State University*
Mark Pelfrey *Southwestern Michigan College*
Becky Pierson *Savannah Technical College*
Brenda Reed *Navarro College*
Charlotte Schmitz *New Mexico Junior College*
Chris Schroeder *Morehead State University*
Theresa Thomas *Blue Ridge Community College (VA)*
Veronica Young (formerly Respress) *Truett McConnell University*

Reviewers of previous editions:

Dhruba Adhikari *Mississippi University for Women*
Froozan Afiat *College of Southern Nevada*
Donna Ahlrich *Holmes Community College*
Dora Ahmadi *Morehead State University*
Eva Allen *Indian River State College*
Anna Pat Alpert *Navarro College*
Larry Anderson *Louisiana State University Shreveport*
Lisa Anglin *Holmes Community College*
Marchetta Atkins *Alcorn State University*
Russ Baker *Howard Community College*
Madelaine Bates *Bronx Community College of the City University of New York*

Shari Beck *Navarro College*
Sage Bentley *Navarro College*
Richard Alan Blanton *Morehead State University*
Stephanie Blue *Holmes Community College*
Brent Bollich *South Louisiana Community College*
Stephanie Burton *Holmes Community College*
Candace Carter-Stevens *Mississippi Valley State University*
Brenda Cates *Mount Olive College*
Deanna Caveny *College of Charleston*
Brenda Chapman *Trident Technical College*
Michelle DeDeo *University of North Florida*
Gilbert Eyabi *Anderson University*
Hamidullah Farhat *Hampton University*
Mary Ellen Foley *Louisiana State University Shreveport*
Terry Fung *Kean University*
Nathan Gastineau *Arkansas State University*
Mark Goldstein *West Virginia Northern Community College–New Martinsville*
Leslie Gomes *University of Arkansas Community College at Morrilton*
Heidi Griffin *Arkansas State University*
Joshua Hanes *Mississippi University for Women*
Bobbie Jo Hill *Coastal Bend College*
Leslie Horton *Delta State University*
Christopher Imm *Johnson County Community College*
Abdusamad Kabir *Bowie State Univeristy*
Bathi Kasturiarachi *Kent State University at Stark*
Richard LeBorne *Tennessee Technological University*
Bill Lepowsky *Laney College*
Alice Lou *Columbia College*
Heidi Lyman *South Seattle Community College*
Richard Mabry *Louisiana State University Shreveport*
Katherine Malone *Fort Scott Community College*
Monica Meissen *University of Dubuque*
Virginia Metcalf *Somerset Community College*
Angela Miles *Holmes Community College*
Mike Miller *Minnesota State University Moorhead*
Cailin Mistrille *University of Arkansas Community College at Morrilton*
Charles Naffziger *Central Oregon Community College*
Paula Norris *Delta State University*
Carol Okigbo *Minnesota State University Moorhead*
Bonnie Oppenheimer *Mississippi University for Women*
Ron Palcic *Johnson County Community College*
Nancy Parkerson *Holmes Community College*
Jennie Pegg *Holmes Community College*
Stan Perrine *Charleston Southern University*
Kimberly Potters *Eastern New Mexico University*
Brenda Reed *Navarro College*
Harriette Roadman *Community College of Allegheny County*
David Rule *Holmes Community College*
Joan Sallenger *Midlands Technical College–Beltline*
Mike Schramm *Indian River State College*
Christopher Schroeder *Morehead State University*
Elizabeth Schubert *Saddleback College*
Barbara Sehr *Indiana University Kokomo*
Mack Smith *Delta State University*
Carlos Spaht *Louisiana State University Shreveport*
Mary Jane Sterling *Bradley University*
Gloria Stone *State University of New York at Oswego*
Gail Stringer *Somerset Community College*
Preety Tripathi *State University of New York at Oswego*
Al Vekovius *Louisiana State University Shreveport*
Vance Waggener *Trident Technical College*
Danae Watson *University of Arkansas Community College at Morrilton*
Bill Weber *University of Wyoming*
Elizabeth White *Trident Technical College*
Ralph L. Wildy Jr. *Georgia Military College Augusta Campus*
Mary Beth Williams *Eastern New Mexico University*
Raymond Williams *Mississippi Valley State University*
Clifton Wingard *Delta State University*
Shaochen Yang *Mississippi University for Women*
Lixin Yu *Alcorn State University*

Finally, I am deeply grateful to my wife Cindy for her unstinting love and support, and for putting up with thousands of hours of textbook writing over the years.

Hawkes Learning: A Clear Path to Mastery

Hawkes' software employs an adaptive, competency-based approach to knowledge mastery supported by a user-friendly interface. The student-centric platform promotes positive active learning by adapting to each student's needs through algorithmically generated questions based on an individual learner's pace, skill, and knowledge level. The real-time adaptive feedback addresses errors immediately, so that students learn from their mistakes when they make them. For each topic, the Hawkes Learning path to content mastery engages students through three simple modes: Learn, Practice, and Certify.

Competency-based learning made simple in three steps:

LEARN offers a multimedia-rich presentation of the lesson content. It includes instructional videos, interactive examples, and more.

PRACTICE engages students with algorithmically generated questions and intelligent tutoring in an ungraded, penalty-free environment.

CERTIFY requires students to demonstrate mastery of the material at a defined proficiency level without access to tutoring aids.

Support

If you have questions or comments we can be contacted as follows:

24/7 Chat: chat.hawkeslearning.com

Phone: 1-800-426-9538

Email: support@hawkeslearning.com

Web: support.hawkeslearning.com

THIRD EDITION

PRECALCULUS

PAUL SISSON

CHAPTER 1

Algebraic Expressions, Equations, and Inequalities

SECTIONS

WHAT IF ...

What if while you were picking strawberries and dropping them into a bucket, your brother was sneaking strawberries out of the bucket to eat? How would this affect the rate at which you can fill the bucket?

By the end of this chapter, you'll be able to work with algebraic expressions, exponents, and polynomials, solve equations and inequalities in one variable, operate with rational expressions, and solve formulas involving radicals. You'll encounter the answer to the berry-picking question on page 127. You'll master this type of problem using the techniques for solving rational equations, found on page 117.

Pythagoras

Introduction

In this chapter, we review the terminology, notation, and properties of the real number system frequently encountered in algebra; the extension of the real number system to the larger set of complex numbers; and the basic algebraic methods we use to solve equations and inequalities.

We begin with a discussion of common subsets of the set of real numbers. Certain types of numbers are important from both a historical and a mathematical perspective. There is archaeological evidence that people used the simplest sort of numbers, the counting or natural numbers, as far back as 50,000 years ago. Over time, many cultures discovered needs for various refinements to the number system, resulting in the development of such classes of numbers as the integers, the rational numbers, the irrational numbers, and ultimately, the complex numbers, a number system that contains the real numbers.

Many of the ideas in this chapter have a history dating as far back as Egyptian and Babylonian civilizations of around 3000 BC, with later developments and additions due to Greek, Hindu, and Arabic mathematicians. Much of the material was also developed independently by Chinese mathematicians. It is a tribute to the necessity, utility, and objectivity of mathematics that so many civilizations adopted so much mathematics from abroad and that different cultures operating independently developed identical mathematical concepts so frequently.

As an example of the historical development of just one concept, consider the notion of an irrational number. The very idea that a real number could *be* irrational (which simply means not rational, or not the ratio of two integers) is fairly sophisticated, and it took some time for mathematicians to come to this realization. The Pythagoreans, members of a school founded by the Greek philosopher Pythagoras in southern Italy around 540 BC, discovered that the square root of 2 was such a number, and there is evidence that for a long time $\sqrt{2}$ was the only known irrational number. A member of the Pythagorean school, Theodorus of Cyrene, later showed (c. 425 BC) that $\sqrt{3}, \sqrt{5}, \sqrt{6}, \sqrt{7}, \sqrt{8}, \sqrt{10}, \sqrt{11}, \sqrt{12}, \sqrt{13}, \sqrt{14}, \sqrt{15}$, and $\sqrt{17}$ also are irrational. It wasn't until 1767 that European mathematician Johann Lambert showed that the famous number π is irrational, and the modern rigorous mathematical description of irrational numbers is due to work by Richard Dedekind in 1872.

As you review the concepts in Chapter 1, keep the larger picture firmly in mind. All of the material presented in this chapter was developed over long periods of time by many different cultures with the aim of solving problems important to them.

1.1 REAL NUMBERS AND ALGEBRAIC EXPRESSIONS

TOPICS

1. Common subsets of real numbers
2. The real number line
3. Order on the real number line
4. Set-builder notation and interval notation
5. Basic set operations and Venn diagrams
6. Absolute value and distance on the real number line
7. Components and terminology of algebraic expressions
8. The field properties and their use in algebra

TOPIC 1: Common Subsets of Real Numbers

Some types of numbers occur so frequently in mathematics that they have been given special names and symbols. These names will be used throughout this book and in later math classes when referring to members of the following sets.

TECHNOLOGY

```
NORMAL FLOAT AUTO REAL RADIAN MP
1/3
                    0.3333333333
4/9
                    0.4444444444
3/8
                           0.375
```

Examples of Rational Numbers

```
NORMAL FLOAT AUTO REAL RADIAN MP
π
                     3.141592654
√2
                     1.414213562
```

Examples of Irrational Numbers

DEFINITION: Types of Real Numbers

The Natural (or Counting) Numbers: This is the set of numbers $\mathbb{N} = \{1, 2, 3, 4, 5, \ldots\}$.

The Whole Numbers: This is the set of natural numbers and 0: $\{0, 1, 2, 3, 4, 5, \ldots\}$.

The Integers: This is the set of natural numbers, their negatives, and 0. As a list, this is the set $\mathbb{Z} = \{\ldots, -4, -3, -2, -1, 0, 1, 2, 3, 4, \ldots\}$.

The Rational Numbers: This is the set, with symbol $\mathbb{Q}$ (for quotient), of ratios of integers (hence the name). That is, any rational number can be written in the form $\frac{p}{q}$, where p and q are both integers and $q \neq 0$. When written in decimal form, rational numbers either terminate or have a repeating pattern of digits past some point.

The Irrational Numbers: Every real number that is not rational is, by definition, irrational. In decimal form, irrational numbers are nonterminating and nonrepeating.

The Real Numbers: Every set above is a subset of the set of real numbers, which is denoted $\mathbb{R}$. Every real number is either rational or irrational, and no real number is both.

Figure 1 shows the relationships among the subsets of $\mathbb{R}$. This figure indicates, for example, that every natural number is automatically a whole number, and also an integer, and also a rational number.

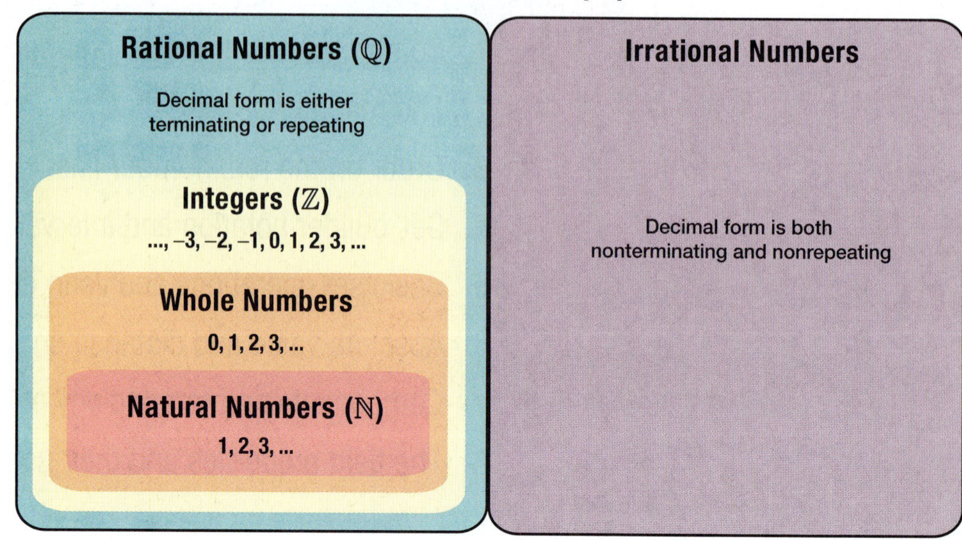

FIGURE 1: The Real Numbers

Example 1: Types of Real Numbers

Consider the set $S = \left\{-15, -7.5, -\frac{7}{3}, 0, \sqrt{2}, 1.\bar{6}, \sqrt{9}, \pi, 10^{17}\right\}$.

a. The natural numbers in S are $\sqrt{9}$ and 10^{17}. Note that $\sqrt{9}$ is a natural number since $\sqrt{9} = 3$.

b. The whole numbers in S are 0, $\sqrt{9}$, and 10^{17}.

c. The integers in S are -15, 0, $\sqrt{9}$, and 10^{17}.

d. The rational numbers in S are $-15, -7.5, -\frac{7}{3}, 0, 1.\bar{6}, \sqrt{9}$, and 10^{17}. The numbers -7.5 and $1.\bar{6}$ are both rational numbers since $-7.5 = \frac{-15}{2}$ and $1.\bar{6} = \frac{5}{3}$ (the bar over the last digit indicates that the digit repeats indefinitely).

e. The only irrational numbers in S are $\sqrt{2}$ and π.

NOTE

Any integer p is also a rational number, since it can be written as $\frac{p}{1}$.

TOPIC 2: The Real Number Line

Mathematicians often depict the set of real numbers as a horizontal line, with each point on the line representing a unique real number (so each real number is associated with a unique point on the line). The real number corresponding to a given point is called the **coordinate** of that point. Thus one (and only one) point on the real number line represents the number 0, and this point is called the **origin**. Points to the right of the origin represent positive real numbers, while points to the left of the origin represent negative real numbers. Figure 2 is an illustration of the real number line with several points plotted. Note that two irrational numbers are plotted, though their locations on the line are approximations.

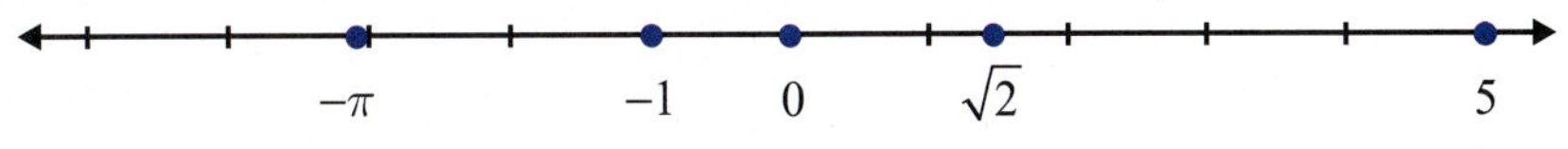

FIGURE 2: The Real Number Line

Example 2: Drawing the Real Number Line

We choose which portion of the real number line to show and the physical length that represents one unit based on the numbers that we wish to plot.

a. If we want to plot the numbers 101, 106, and 107, we might construct the graph below.

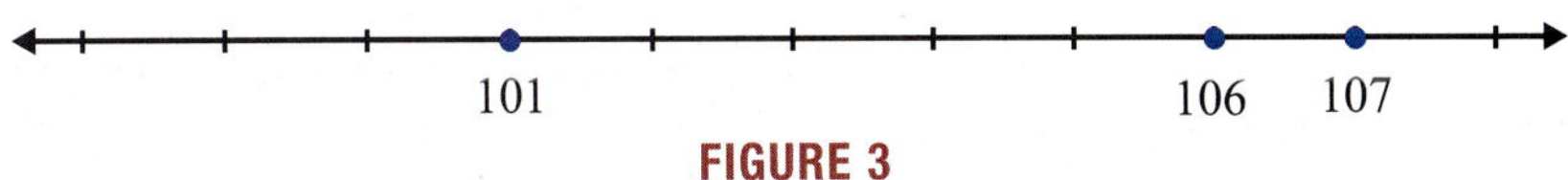

FIGURE 3

b. If we want to plot the numbers $-\frac{3}{4}$, $-\frac{1}{2}$, and $\frac{1}{4}$, we might make the unit interval longer.

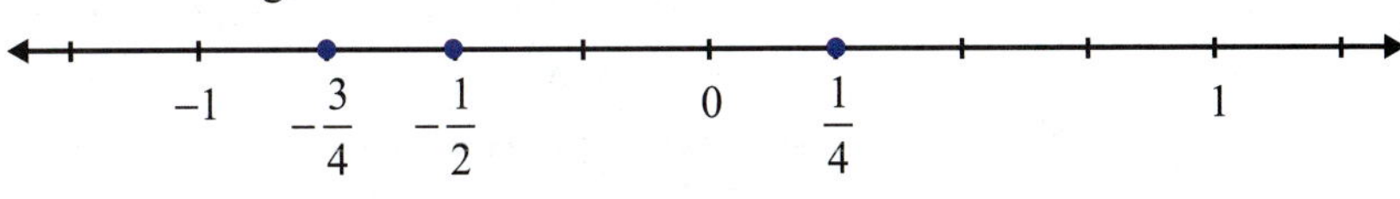

FIGURE 4

TOPIC 3: Order on the Real Number Line

Symbol History

The inequality symbols $<$ and $>$ first appeared in a book by the English mathematician Thomas Harriot published in 1621. The modified symbols $\leq$ and $\geq$ were later introduced by the French mathematician Pierre Bouguer in 1734. The equality symbol $=$ was invented by the Welsh mathematician Robert Recorde in 1557.

Representing the real numbers as a line leads naturally to the idea of *ordering* the real numbers.

DEFINITION: Inequality Symbols (Order)

Symbol	Reading	Meaning
$a < b$	"a is **less than** b"	a lies to the left of b on the number line
$a \leq b$	"a is **less than or equal to** b"	a lies to the left of b or is equal to b
$b > a$	"b is **greater than** a"	b lies to the right of a on the number line
$b \geq a$	"b is **greater than or equal to** a"	b lies to the right of a or is equal to a

The two symbols $<$ and $>$ are called *strict* inequality signs, while the symbols $\leq$ and $\geq$ are *nonstrict* inequality signs.

CAUTION

Remember that order is defined by the placement of real numbers on the number line, *not* by magnitude (its distance from zero). For instance, $-36 < 5$ because -36 lies to the left of 5 on the number line.

Example 3: Working with Order

a. $5 \le 9$ since 5 lies to the left of 9.

b. $5 \le 5$, since 5 is equal to 5. Note that for every real number a, we have $a \le a$ and $a \ge a$.

c. $-7 > -163$, since -7 lies to the right of -163.

d. The statement "5 is greater than -2" can be written $5 > -2$.

e. The statement "a is less than or equal to $b + c$" can be written $a \le b + c$.

f. The statement "x is strictly less than y" can be written $x < y$.

g. The negation of the statement $a \le b$ is the statement $a > b$.

h. If $a \le b$ and $a \ge b$, then it must be the case that $a = b$.

TOPIC 4: Set-Builder Notation and Interval Notation

To describe the solutions to equations and inequalities, we need a precise, consistent way of expressing sets of real numbers. **Set-builder notation** is a general method of describing the elements that belong to a given set. **Interval notation** is a way of describing certain subsets of the real line.

DEFINITION: Set-Builder Notation

The notation $\{x \mid x \text{ has property } P\}$ is used to describe a set of real numbers, all of which have the property P. This can be read "the set of all real numbers x having property P."

The symbol $\mid$ is also read as "such that," so the above notation can also be read "the set of all real numbers x, *such that* x has property P."

Example 4: Set-Builder Notation

a. $\{x \mid x \text{ is an even integer}\}$ is another way of describing the set $\{\ldots,-4,-2,0,2,4,\ldots\}$. We could also describe this set as $\{2n \mid n \text{ is an integer}\}$, since every even integer is a multiple of 2.

b. $\{x \mid x \text{ is an integer and } -3 \le x < 2\}$ describes the set $\{-3,-2,-1,0,1\}$.

DEFINITION: The Empty Set

A set with no elements is called the **empty set** or the **null set**, and is denoted by the symbol $\varnothing$ or $\{\ \}$.

The empty set can arise from a set defined using set-builder notation; for example, the set $\{y \mid y > 1 \text{ and } y \leq -4\}$ is equivalent to the empty set, since no real number y satisfies the stated property.

Sets that consist of all real numbers bounded by two endpoints, possibly including those endpoints, are called **intervals**. Intervals can also consist of a portion of the real line extending indefinitely in either direction from just one endpoint.

We can describe such sets with set-builder notation, but intervals occur frequently enough that special notation has been devised to define them succinctly.

DEFINITION: Interval Notation

Interval Notation	Set-Builder Notation	Meaning
(a,b)	$\{x \mid a < x < b\}$	all real numbers strictly between a and b
$[a,b]$	$\{x \mid a \leq x \leq b\}$	all real numbers between a and b, including both a and b
$(a,b]$	$\{x \mid a < x \leq b\}$	all real numbers between a and b, including b but not a
$(-\infty,b)$	$\{x \mid x < b\}$	all real numbers less than b
$[a,\infty)$	$\{x \mid x \geq a\}$	all real numbers greater than or equal to a

Intervals of the form (a,b) are called **open** intervals, while those of the form $[a,b]$ are **closed** intervals. The interval $(a,b]$ is **half-open** (or **half-closed**). A half-open interval may be open at either endpoint, as long as it is closed at the other. The symbols $-\infty$ and ∞ indicate that the interval extends indefinitely in the left and the right directions, respectively. Note that $(-\infty,b)$ excludes the endpoint b, while $[a,\infty)$ includes the endpoint a.

CAUTION

The symbols $-\infty$ and ∞ are just that: symbols! They are not real numbers, so they cannot be solutions to a given equation. The fact that they are symbols, and not numbers, means that they can never be included in a set of real numbers. For this reason, a parenthesis always appears next to either $-\infty$ or ∞; a bracket should never appear next to either infinity symbol.

Example 5: Intervals of Real Numbers

a. The interval $(2,8)$ represents the set $\{x \mid 2 < x < 8\}$. This interval is open at both endpoints, so neither 2 nor 8 is included in the set.

b. The interval $[-5,-1]$ is another way to write the set $\{x \mid -5 \le x \le -1\}$. This interval is closed at both endpoints, so both –5 and –1 are included in the set.

c. The interval $[-3,10)$ stands for the set $\{x \mid -3 \le x < 10\}$. This interval is closed at the left endpoint, –3, and open at the right endpoint, 10.

d. The interval $(4,\infty)$ stands for the set $\{x \mid x > 4\}$. Since the interval is open on the left endpoint, it is the set of numbers greater than (but not equal to) 4.

e. The interval $(-\infty,\infty)$ is just another way of describing the entire set of real numbers.

TOPIC 5: Basic Set Operations and Venn Diagrams

The sets that arise most frequently in algebra are sets of real numbers, and these sets are often the solutions of equations or inequalities. We will need to combine two or more such sets through the set operations of *union* and *intersection*. These operations are defined on sets in general, not just sets of real numbers, and can be illustrated by means of Venn diagrams.

A **Venn diagram** is a pictorial representation of a set or sets, and it indicates, through shading, the outcome of set operations such as union and intersection. In the following definition, these two operations are first defined with set-builder notation and then demonstrated with a Venn diagram. The symbol $\in$ is read "is an element of."

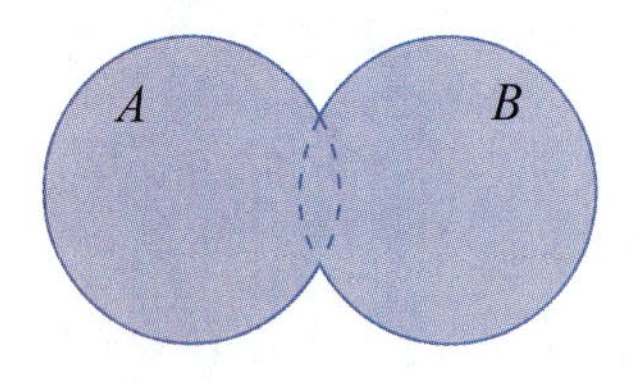

FIGURE 5: $A \cup B$

DEFINITION: Union

In this definition, A and B denote two sets, which are represented in the Venn diagram by circles. The operation of union is depicted in Figure 5 by shading.

The **union** of A and B, denoted $A \cup B$, is the set $\{x \mid x \in A \text{ or } x \in B\}$. That is, an element x is in $A \cup B$ if it is in the set A, the set B, or both. Note that the union of A and B contains both individual sets.

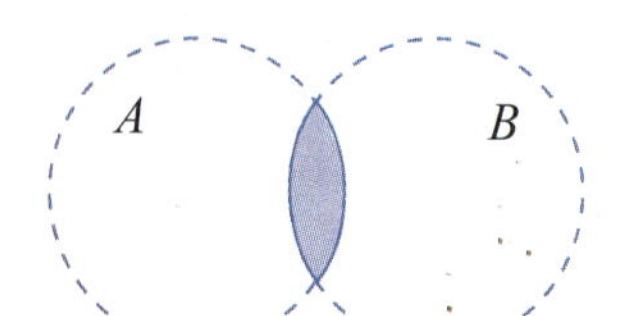

FIGURE 6: $A \cap B$

DEFINITION: Intersection

In this definition, A and B denote two sets, which are represented in the Venn diagram by circles. The operation of intersection is depicted in Figure 6 by shading.

The **intersection** of A and B, denoted $A \cap B$, is the set $\{x \mid x \in A \text{ and } x \in B\}$. That is, an element x is in $A \cap B$ if it is in both A and B. Note that the intersection of A and B is contained in each individual set.

Example 6 applies the set operations of union and intersection to intervals. Recall that intervals are sets of real numbers.

Example 6: Union and Intersection of Intervals

Simplify the following set expressions.

a. $(-2,4]\cup[0,9]$

b. $(-2,4]\cap[0,9]$

c. $[3,4)\cap(4,9)$

d. $(-\infty,4]\cup(-1,\infty)$

Solution

a. $(-2,4]\cup[0,9]=(-2,9]$

Since these two intervals overlap, their union is described with a single interval.

b. $(-2,4]\cap[0,9]=[0,4]$

This intersection of two intervals can also be described with a single interval.

c. $[3,4)\cap(4,9)=\varnothing$

These two intervals have no elements in common, so their intersection is the empty set.

d. $(-\infty,4]\cup(-1,\infty)=(-\infty,\infty)$

The union of these two intervals is the entire set of real numbers.

Example 7: Union and Intersection

Simplify the following set expressions.

a. $\{1,2\}\cup\{0,3\}$

b. $\{x,y,z\}\cap\{w,x\}$

c. $\mathbb{Z}\cup\mathbb{R}$

d. $\mathbb{Z}\cap\mathbb{R}$

Solution

a. The union of the two sets consists of all elements in either set: $\{0,1,2,3\}$.

b. The intersection consists only of elements in both sets: $\{x\}$.

c. Since the integers are all also real numbers, the union of these two sets is simply the set of real numbers $\mathbb{R}$. We say that $\mathbb{Z}$ is contained in $\mathbb{R}$.

d. Similarly, since all integers are also real numbers, the integers are the elements contained in both sets. Thus, the intersection is $\mathbb{Z}$.

TOPIC 6: Absolute Value and Distance on the Real Number Line

In addition to order, the depiction of the set of real numbers as a line leads to the notion of *distance*. Physically, distance is a well-understood concept: the distance between two objects is a nonnegative number, dependent on a choice of measuring system, indicating how close the objects are to one another. The idea of *absolute value* gives us a means of defining distance in a mathematical setting.

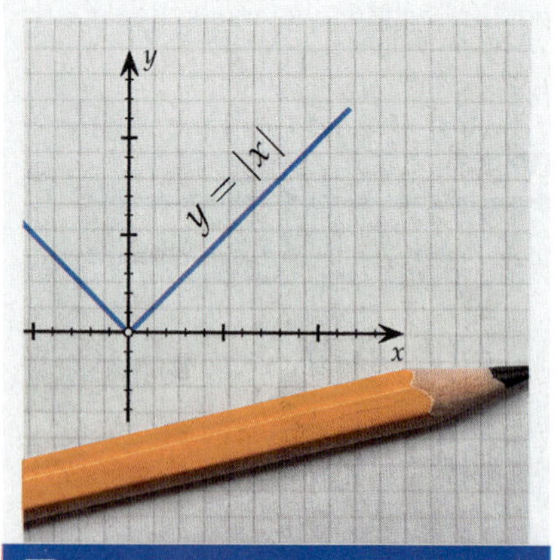

Early History of Magnitude

The Swiss-born amateur mathematician Jean Robert Argand used the idea of magnitude in 1806 while discussing the then-new complex plane (we will study complex numbers and the complex plane in later sections). The German mathematician Karl Weierstrass introduced the vertical bar symbols for the concept in 1841. Absolute value is closely related to the notions of magnitude, distance, and norm in various mathematical and physical contexts.

DEFINITION: Absolute Value

The **absolute value** of a real number a, denoted as $|a|$, is defined as follows.

$$|a| = \begin{cases} a & \text{if } a \geq 0 \\ -a & \text{if } a < 0 \end{cases}$$

The absolute value of a number is also referred to as its **magnitude**; it is the nonnegative number corresponding to its distance from the origin.

Note that this definition implicitly gives us a system of measurement; 1 and -1 are the two real numbers which have a magnitude of 1, and so the distance between 0 and 1 (or between -1 and 0) is one unit. Note also that 0 is the only real number whose absolute value is 0.

DEFINITION: Distance on the Real Number Line

Given two real numbers a and b, the **distance** between them is defined to be $|a-b|$. In particular, the distance between a and 0 is $|a-0|$ or just $|a|$.

Distance should be symmetric. That is, the distance from a to b should be the same as the distance from b to a. Also, no mention was made in the above definition of which of the two numbers a and b is smaller, and our intuition suggests that this is immaterial as far as distance is concerned. These two concerns are really the same, and happily (not by chance) our mathematical definition of distance coincides with our intuition and is indeed symmetric. Given two distinct real numbers a and b, exactly one of the two differences $a-b$ and $b-a$ will be negative, and since $b-a=-(a-b)$ the definition of absolute value makes it clear that these two differences have the same magnitude. That is, $|a-b|=|b-a|$.

Example 8: Absolute Value

TECHNOLOGY

```
NORMAL FLOAT AUTO REAL RADIAN MP
|17-3|
                               14
|3-17|
                               14
```

a. $|17-3|=|3-17|=14$. 17 and 3 are 14 units apart.

b. $|-\pi|=|\pi|=\pi$. Both $-\pi$ and π are π units from 0.

c. $\frac{|7|}{7}=\frac{7}{7}=1$

d. $\frac{|-7|}{-7}=\frac{7}{-7}=-1$

e. $-|-5|=-5$. Note that the negative sign outside the absolute value symbol is not affected by the absolute value. Compare this with the fact that $-(-5)=5$.

f. $|\sqrt{7}-2|=\sqrt{7}-2$. Even without a calculator, we know $\sqrt{7}$ is larger than 2 (since $2=\sqrt{4}$), so $\sqrt{7}-2$ is positive and hence $|\sqrt{7}-2|=\sqrt{7}-2$.

g. $|\sqrt{7}-19|=19-\sqrt{7}$. In contrast to part f., we know $\sqrt{7}-19$ is negative, so its absolute value is $-(\sqrt{7}-19)=19-\sqrt{7}$.

Example 8 illustrated some of the basic properties of absolute value. The following properties can all be derived from the definition of absolute value.

PROPERTIES: Properties of Absolute Value

In the following properties, a and b represent arbitrary real numbers.

1. $|a| \geq 0$
2. $|-a| = |a|$
3. $a \leq |a|$
4. $|ab| = |a||b|$
5. $\left|\frac{a}{b}\right| = \frac{|a|}{|b|},\ b \neq 0$
6. $|a+b| \leq |a| + |b|$ (This is called the **triangle inequality**, as it is a reflection of the fact that one side of a triangle is never longer than the sum of the other two sides.)

TECHNOLOGY

```
NORMAL FLOAT AUTO REAL RADIAN MP
|(-3)(5)|
                              15
|(-15)|
                              15
|(-3)|*|5|
                              15
```

Example 9: Using Absolute Value Properties

a. $|(-3)(5)| = |-15| = 15 = |-3||5|$

b. $1 = |-3+4| \leq |-3| + |4| = 7$

c. $7 = |-3-4| \leq |-3| + |-4| = 7$

d. $\left|\frac{-3}{7}\right| = \frac{|-3|}{|7|} = \frac{3}{7}$

TOPIC 7: Components and Terminology of Algebraic Expressions

Algebraic expressions are made up of constants and variables, combined by the operations of addition, subtraction, multiplication, division, exponentiation, and the taking of roots. **Constants** (like -3, 5.9, and π) are fixed numbers, while **variables** (like x, y, and ω) are usually letters that represent unspecified numbers. To **evaluate** a given expression means to replace the variables (if there are any) with specific numbers, perform the indicated mathematical operations, and simplify the result.

The **terms** of an algebraic expression are those parts joined by addition (or subtraction), while the **factors** of a term are the individual parts of the term that are joined by multiplication (or division). The **coefficient** of a term is the constant factor of the term, while the remaining part of the term is the **variable factor**.

Example 10: Terminology of Algebraic Expressions

Consider the algebraic expression $-17x(x^2+4y)+5\sqrt{x}-13$.

a. This expression contains three terms: $-17x(x^2+4y)$, $5\sqrt{x}$, and -13. The terms are combined by addition and subtraction to form the whole expression.

b. The factors of the term $-17x(x^2+4y)$ are -17, x, and (x^2+4y). The factors are combined by multiplication to form the whole term. The coefficient of $-17x(x^2+4y)$ is -17, and the variable factor is $x(x^2+4y)$.

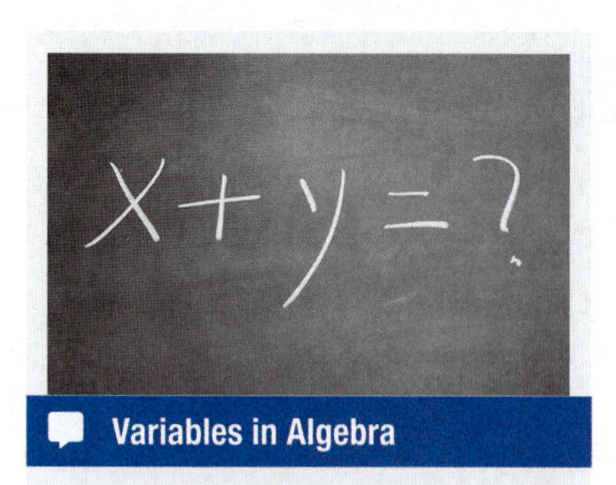

Variables in Algebra

The use of lowercase letters like *x* and *y* for variables and our exponent notation for powers, along with much other modern notation, was popularized in the book *La Géométrie* by the French mathematician René Descartes, published in 1637. Descartes' inspiration for such usage likely dates back far earlier to the work of the Arabic mathematician Mohammed al-Khowârizmî and his influential 9th-century book *Hisâb al-jabr w'al-muqâ-balah*, part of the title of which gave rise to our word *algebra*.

Example 11: Evaluating Algebraic Expressions

Evaluate the following algebraic expressions.

a. $5x^3-16$ for $x=4$

b. $-3x^2-2(x+y)$ for $x=-2$ and $y=3$

Solution

In both cases, we simply "plug in" the given values for each variable, then simplify.

a.
$$\begin{aligned} 5(4)^3-16 &= 5(64)-16 \\ &= 320-16 \\ &= 304 \end{aligned}$$

b.
$$\begin{aligned} -3(-2)^2-2(-2+3) &= -3(4)-2(1) \\ &= -12-2 \\ &= -14 \end{aligned}$$

TOPIC 8: The Field Properties and Their Use in Algebra

The following properties of addition and multiplication on the set of real numbers are probably familiar. You have likely used many of them in the past, though you may not have known the technical names of the properties. These properties and the few that follow them form the basis of the logical steps we use to solve equations and inequalities in algebra.

The set of real numbers forms what is known mathematically as a *field*, and consequently, the properties below are called *field properties*. These properties also apply to the set of complex numbers, which is a larger field containing the real numbers. Complex numbers will be discussed in Section 1.5.

PROPERTIES: Field Properties

In the following properties, a, b, and c represent arbitrary real numbers. The first five properties apply to addition and multiplication, while the last combines the two.

Name of Property	Additive Version	Multiplicative Version
Closure	$a+b$ is a real number	ab is a real number
Commutative	$a+b=b+a$	$ab=ba$
Associative	$a+(b+c)=(a+b)+c$	$a(bc)=(ab)c$
Identity	$a+0=0+a=a$	$a\cdot 1=1\cdot a=a$
Inverse	$a+(-a)=0$	$a\cdot\frac{1}{a}=1$ (for $a\neq 0$)
Distributive	$a(b+c)=ab+ac$	

It is important to remember that variables represent numbers, so the field properties apply to algebraic expressions as well as single variables.

Example 12: Using the Field Properties

a. $3[2+(-8)]=3\cdot(-6)=-18$ and $3\cdot 2+3\cdot(-8)=6-24=-18$

This demonstrates the distributive property: $3[2+(-8)]=3\cdot 2+3\cdot(-8)$.

TECHNOLOGY

```
NORMAL FLOAT AUTO REAL RADIAN MP
3-4
                                -1
3+-4
                                -1
-4+3
                                -1
```

b. $3-4=3+(-4)=-4+3=-1$

While subtraction is not commutative, we can rewrite any difference as a sum (with the sign changed on the second term) and then apply the commutative property.

c. $(-x)\left(\frac{1}{y}\right)=\frac{-x}{y}=\left(\frac{1}{y}\right)(-x)$

Division can be restated as multiplication by the reciprocal of the denominator, and multiplication is commutative.

d. $(x^2+y)\left(\frac{1}{x^2+y}\right)=1,$ provided that $x^2+y\neq 0.$

Any nonzero expression, when multiplied by its reciprocal, yields the multiplicative identity 1. The expression $\frac{1}{x^2+y}$ is the multiplicative inverse of x^2+y.

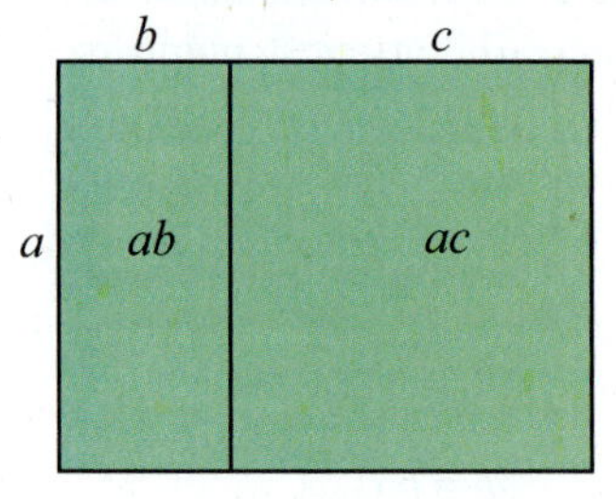

FIGURE 7

Example 13: Visualizing the Distributive Property

Consider the equation $a(b+c)=ab+ac$, which demonstrates the distributive property. We can represent the equation geometrically to understand why the distributive property works.

Figure 7 shows that the total area of the shaded region represents the product $a(b+c)$, and that total area is the sum of two smaller areas that represent the products ab and ac.

While the field properties are of fundamental importance to algebra, they imply further properties that are often of more immediate use.

PROPERTIES: Cancellation Properties

In the following properties, A, B, and C represent algebraic expressions. The symbol $\Leftrightarrow$ can be read as "if and only if" or "is equivalent to."

Property	Description
$A=B \Leftrightarrow A+C=B+C$	**Additive Cancellation** Adding the same quantity to both sides of an equation results in an equivalent equation.
For $C \neq 0$, $A=B \Leftrightarrow A \cdot C=B \cdot C$.	**Multiplicative Cancellation** Multiplying both sides of an equation by the same nonzero quantity results in an equivalent equation.

PROPERTIES: Zero-Factor Property

Let A and B represent algebraic expressions. If the product of A and B is 0, then at least one of A and B is itself 0. Using the symbol $\Rightarrow$ for "implies," we write

$$AB=0 \Rightarrow A=0 \text{ or } B=0.$$

Example 14: Using the Cancellation and Zero-Factor Properties

a.

$$y+12=18$$
$$y+12+(-12)=18+(-12)$$
$$y=6$$

Using additive cancellation, we add -12 to both sides, then simplify.

This shows that the equation $y+12=18$ is equivalent to the equation $y=6$.

b. $-6x = 30$

$$-6x\left(-\frac{1}{6}\right) = 30\left(-\frac{1}{6}\right)$$

$$x = -5$$

Using multiplicative cancellation, we multiply both sides by $-\frac{1}{6}$, then simplify.

Alternatively, dividing both sides by –6 yields the same result. Thus, the equation $-6x = 30$ is equivalent to equation $x = -5$. We can see how cancellation properties can help us *solve* equations for variables.

c. Multiplying both sides of the equation $x^2 - x = 2$ by 0 leads to the equation $0 = 0$ a true statement. However, these two equations are not equivalent!

While replacing x in the first equation by -1 or 2 leads to a true statement, any other value for x leads to a false statement. By contrast, the equation $0 = 0$ is true for all values of x, as there is no x in the equation to replace with a number. This example illustrates why we must multiply both sides of an equation by a nonzero quantity to apply multiplicative cancellation.

d. The equation $(x - y)(x + y) = 0$ means that $x - y = 0$ or $x + y = 0$, by the Zero-Factor Property. Remember that the only way for a product of two (or more) factors to be 0 is for *at least* one of the factors to be 0 itself. For instance, in this example it might be that *both* $x - y = 0$ and $x + y = 0$. If $x = 0$ and $y = 0$, this is indeed the case.

1.1 EXERCISES

PRACTICE

Which elements of the following sets are **a.** natural numbers, **b.** whole numbers, **c.** integers, **d.** rational numbers, **e.** irrational numbers, **f.** real numbers? See Example 1.

1. $\left\{19,\ -4.3,\ -\sqrt{3},\ \frac{0}{15},\ 2^5,\ -33\right\}$

2. $\left\{5\sqrt{7},\ 4\pi,\ \sqrt{16},\ 3.\bar{3},\ -1,\ \frac{22}{7},\ |-8|\right\}$

Plot the real numbers in the following sets on a number line. Choose the unit length appropriately for each set.

3. $\{-4.5, -1, 2.5\}$ **4.** $\{-24, 2, 15\}$ **5.** $\{5.1, 5.2, 5.8\}$ **6.** $\left\{0, \frac{1}{2}, \frac{5}{6}\right\}$

Select all of the symbols from the set $\{<, \leq, >, \geq\}$ that can be placed in the blank to make each statement true.

7. $12 \_\_\_ 14$ **8.** $-102 \_\_\_ 9$ **9.** $3 \_\_\_ 3$

10. $-50 \_\_\_ -45$ **11.** $-3.4 \_\_\_ -3.5$ **12.** $-\frac{1}{4} \_\_\_ -\frac{1}{3}$

Write each statement as an inequality, using the appropriate inequality symbol.

13. "$2a+b$ is strictly greater than c"

14. "2 is less than or equal to x"

15. "$2c$ is no more than $3d$"

16. "$6+x$ is greater than or equal to $4x$"

Describe each of the following sets using set-builder notation. There may be more than one correct way to do this. See Example 4.

17. $\{5,6,7,\ldots,105\}$

18. $\{2,3,5,7,11,13,17,\ldots\}$

19. $\{1,2,4,8,16,32,\ldots\}$

20. $\{-6,-3,0,3,6,9\}$

21. $\left\{\ldots,\frac{1}{3},\frac{1}{5},\frac{1}{7},\frac{1}{9},\ldots\right\}$

22. $\{0,1,2,3,4,5,\ldots\}$

Write each set as an interval using interval notation. See Example 5.

23. $\{x \mid -3 \le x < 19\}$

24. $\{x \mid x < 4\}$

25. $x < 15$

26. $-9 \le x \le 6$

27. The positive real numbers

28. The nonnegative real numbers

Graph the following intervals.

29. $[5,14)$

30. $[-9,-1]$

31. $(0,2)$

32. $(-3,18]$

33. $(-\infty,7]$

34. $(25,\infty)$

Simplify the following set expressions. See Example 6.

35. $[-7,7)\cup(2,5)$

36. $(-5,2]\cup(2,4]$

37. $(-5,2]\cap(2,4]$

38. $[3,5]\cap[2,4]$

39. $(-\infty,4]\cup(0,\infty)$

40. $(-\infty,\infty)\cap[-\pi,21)$

Simplify the following set expressions. See Example 7.

41. $\mathbb{Q}\cap\mathbb{Z}$

42. $\mathbb{N}\cup\mathbb{R}$

43. $\mathbb{N}\cup\mathbb{Z}\cap\mathbb{Q}$

44. $(-4.8,-3.5)\cap\mathbb{Z}$

Evaluate the absolute value expressions. See Examples 8 and 9.

45. $-|-11|$

46. $|3-7|$

47. $\left|\sqrt{3}-\sqrt{5}\right|$

48. $\left|-\sqrt{2}\right|$

49. $\frac{|-x|}{|x|}\ (x \ne 0)$

50. $-|4-9|$

51. $-\left|-4-|-11|\right|$

52. $-\left|-\sqrt{|-9|}-|-9|\right|$

Find the distance on the real number line between each pair of numbers given. See Example 8.

53. $a = 8, b = 3$

54. $a = 5, b = 5$

55. $a = 4, b = -2$

56. $a = -12, b = -1$

Identify the components of the algebraic expressions, as indicated. See Example 10.

57. Identify the terms in the expression $3x^2y^3 - 2\sqrt{x+y} + 7z$.

58. Identify the factors in the term $-2\sqrt{x+y}$.

59. Identify the coefficients in the expression $x^2 + 8.5x - 14y^3$.

60. Identify the factors in the term $8.5x$.

61. Identify the terms in the expression $\frac{-5x}{2yz} - 8x^5y^3 + 6.9z$.

62. Identify the coefficients in the expression $\frac{-5x}{2yz} - 8x^5y^3 + 6.9z$.

Evaluate the following algebraic expressions for the given values of the variables. See Example 11.

63. $3x^3 + 5x - 2$ for $x = -3$

64. $\sqrt{2x} + \frac{3x}{4}$ for $x = 8$

65. $-3\pi y + 8x + y^3$ for $x = 2$ and $y = -2$

66. $y\sqrt{x^3 - 2} + \sqrt{x - 2y} - 3y$ for $x = 3$ and $y = -\frac{1}{2}$

67. $|x - 9y| - (8z - 8)$ for $x = -3$, $y = 1$, and $z = 5$

68. $\frac{x^2y^3}{8z} - \frac{|2xy|}{8z}$ for $x = 2$, $y = -1$, and $z = 3$

Identify the property that justifies each of the following statements. If one of the cancellation properties is being used to transform an equation, identify the quantity that is being added to both sides or the quantity by which both sides are being multiplied. See Examples 12 and 14.

69. $(x - y)(z^2) = (z^2)(x - y)$

70. $3 - 7 = -7 + 3$

71. $4(y - 3) = 4y - 12$

72. $-3(4x^6z) = (-3)(4)(x^6z) = -12x^6z$

73. $4 + (-3 + x) = (4 - 3) + x = 1 + x$

74. $(x + y)\left(\frac{1}{x+y}\right) = 1$

75. $25x^3 = 10y \Leftrightarrow 5x^3 = 2y$

76. $-14y = 7 \Leftrightarrow y = -\frac{1}{2}$

77. $14 - x = 2x \Leftrightarrow 14 = 3x$

78. $(a + b)(x) = 0 \Rightarrow a + b = 0$ or $x = 0$

79. $\frac{x}{6}+\frac{y}{3}-2=0 \Leftrightarrow x+2y-12=0$

80. $x^2z=0 \Rightarrow x^2=0$ or $z=0$

81. $5+3x-y=2x-y \Leftrightarrow 5+x=0$

82. $(x-3)(x+2)=0 \Rightarrow x-3=0$ or $x+2=0$

APPLICATIONS

83. Jess, Stan, Nina, and Michele are in a marathon. Twenty-five minutes after beginning, Jess has run 3.4 miles, Stan has run 4 miles, Nina has run 2.25 miles, and Michele has walked 1.6 miles. Using 0 as the beginning point, plot each competitor's location on a real number line using an appropriate interval.

84. Freddie, Sarah, Elizabeth, JR, and Aubrey are trying to line up by height for a photo shoot. JR is the tallest and Elizabeth is the shortest. Freddie is taller than Sarah, and Sarah is taller than Aubrey. Express their lineup using appropriate inequality symbols.

85. Sue boards an eastbound train in Center Station at the same time Joy boards a westbound train in Center Station. After riding the Straight Line for 20 minutes, Sue's train has traveled 13 miles east, while Joy's train (also on the Straight Line) has traveled 7 miles west. Find the distance between the two trains at this time. (Assume the Straight Line is true to its name and that the tracks lie literally along a straight line.)

86. The admission prices at the local zoo are as follows.

Admission Prices

Children under 2	free
Children under 12	$3
Adults	$7
Seniors (65 and up)	$5

Express the age range for each of these prices in set-builder notation and interval notation.

87. A particular fudge recipe calls for at least 3 but no more than 4 cups of sugar and at least $\frac{1}{2}$ but no more than $\frac{2}{3}$ of a cup of walnuts. Express the amount of sugar and nuts needed in both set-builder and interval notation.

88. At the beginning of the month, your checking account contains $128. For your birthday, your mother deposits $50 and your grandmother deposits $25. After you write three checks for $17, $23, and $62, you make a deposit of $41. At the end of the month, your bank removes half of the balance to put in your savings account and then charges you a $5 fee for doing so. How much do you have remaining in your checking account?

89. A particular liquid boils at 268 °F. Given the formula $C=\frac{5}{9}(F-32)$ for converting temperatures from Celsius (C) to Fahrenheit (F), find the boiling point of this liquid in the Celsius scale. Round your answer to two decimal places.

90. Stephen received \$75 as a gift from his aunt. With this money, he decided to start saving to buy the newest gaming console, which costs \$398 after tax. After working two weeks at his part-time job, he got one check for \$123 and a second check for \$98. How much more does Stephen need to save to buy his gaming console?

91. Body mass index, abbreviated BMI, is one way doctors determine an adult's weight status. A BMI below 18.5 is considered underweight, the range 18.5–24.9 is normal, the range 25.0–29.9 is overweight, and a BMI above 30.0 indicates obesity. The formula used to determine BMI is $\text{BMI} = 703\left(\dfrac{\text{weight in pounds}}{(\text{height in inches})^2}\right)$.

Derek weighs 180 pounds and is 73 inches tall. Use this formula to determine Derek's BMI and weight status. Round your answer to one decimal place.

92. The Du Bois Method provides a formula used to estimate your body's surface area in meters squared: $\text{BSA} = 0.007184h^{0.725}w^{0.425}$ where h is height in centimeters and w is weight in kilograms. Assume Juan is 193 cm tall and weighs 88 kg. Use the Du Bois Method to estimate his body's surface area in square meters. Round your answer to two decimal places.

93. Samantha drops a tennis ball from the top of the mathematics building. If it takes the ball 3.42 seconds to hit the ground, use the formula $\text{distance} = \dfrac{1}{2}(\text{acceleration})(\text{time})^2$ to find the height of the building, which is equivalent to the distance the ball falls. Use the value of 32 ft/s^2 for the acceleration of a falling object. Round your answer to the nearest foot.

WRITING & THINKING

94. Choose a number. Multiply it by 3 and then add 4. Now multiply by 2 and subtract 8. Finally divide by 6. What do you notice about your final answer? Explain why you got this as a result.

95. After taking a poll in her town, Sally began grouping the citizens into various sets. One set contained all the citizens with brown hair and another set contained all the citizens with blue eyes. What do you know about the citizens who would be listed in the union of these two sets? What do you know about the citizens who would be listed in the intersection of these two sets?

96. Can a natural number be irrational? Explain.

97. Are all whole numbers also integers? Are all integers also whole numbers? Explain your answers.

98. In your own words, define absolute value.

99. Write a short paragraph explaining the similarities and differences between $>$ and $\geq$.

100. In your own words, explain the difference between a union and an intersection of two sets.

TECHNOLOGY

Select all of the symbols from the set $\{<, \leq, >, \geq\}$ that can be placed in the blank to make each statement true. Use a graphing utility to check your answers.

101. $-2.9 \_\_\_ -3.1$

102. $2.1 \_\_\_ -5.5$

103. $100 \_\_\_ -4$

104. $0.001 \_\_\_ -99.8$

105. $\frac{1}{3} \_\_\_ \frac{1}{4}$

106. $-\frac{1}{5} \_\_\_ -\frac{3}{4}$

Use a graphing utility to evaluate the following algebraic expressions.

107. $\sqrt{x^4y - z} + \frac{x - y^3}{z^2}$ for $x = -3$, $y = 2$, and $z = -2$

108. $\frac{(x - pq^2)^3}{2q^3}$ for $x = -5$, $p = 2$, and $q = -3$

109. $\frac{|x^2 - y^3| - 4x}{3y^5}$ for $x = 2$ and $y = 3$

110. $\sqrt{p^3q - q^3} - |p + q^2|$ for $p = -5$ and $q = 2$

1.2 PROPERTIES OF EXPONENTS AND RADICALS

TOPICS

1. Natural number and integer exponents
2. Properties of exponents
3. Scientific notation
4. Working with geometric formulas
5. Radical notation
6. Simplifying and combining radical expressions
7. Rational number exponents

TOPIC 1: Natural Number and Integer Exponents

As we progress in Chapter 1, we will encounter a variety of algebraic expressions. As discussed in Section 1.1, algebraic expressions are made up of constants and variables, combined by the operations of addition, subtraction, multiplication, division, exponentiation, and the taking of roots. In this section, we will explore the meaning of exponentiation and the properties of exponents.

We will begin with the most basic type of exponent: an exponent consisting of a natural number.

DEFINITION: Natural Number Exponents

If a is any real number and if n is any natural number, then $a^n = \underbrace{a \cdot a \cdot \cdots \cdot a}_{n \text{ factors}}$. That is, a^n is merely a shorter, and more precise, way of denoting the product of n factors of a. In the expression a^n, a is called the **base**, and n is the **exponent**. The process of multiplying n factors of a is called "raising a to the nth power," and the expression a^n may be referred to as "the nth power of a" or "a to the nth power." Note that $a^1 = a$.

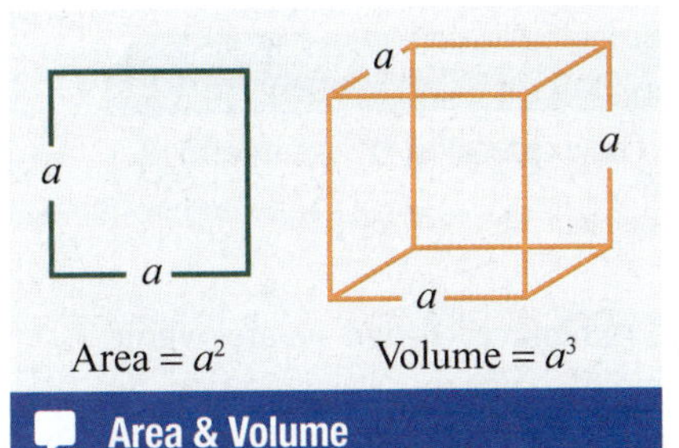

Area & Volume

Other phrases are also commonly used to denote the raising of something to a power, especially when the power is 2 or 3. For instance, a^2 is often referred to as "a squared" and a^3 is often referred to as "a cubed." These phrases have their basis in geometry, as the area of a square region with side length a is a^2 and the volume of a three-dimensional cube, again with side length a, is a^3.

If the base and exponent are known constants, the expression a^n may be evaluated and written as a simple number. Even if an expression with one or more natural number exponents contains a variable, some simplification may be possible based only on the above definition.

Example 1: Using Natural Number Exponents

TECHNOLOGY

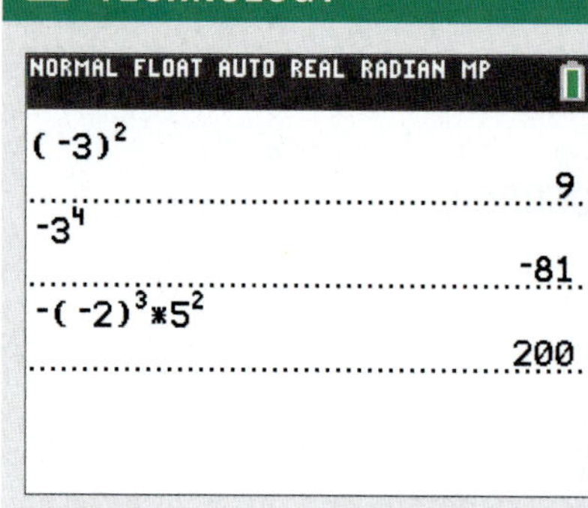

a. $4^3 = 4 \cdot 4 \cdot 4 = 64$

Thus, "four cubed is sixty-four."

b. $(-3)^2 = (-3)(-3) = 9$

Thus, "negative three squared is nine."

c. $-3^4 = -(3 \cdot 3 \cdot 3 \cdot 3) = -81$

Note that the exponent of 4 applies only to the number 3. After raising 3 to the 4th power, the result is multiplied by -1.

d. $-(-2)^3 \cdot 5^2 = -(-2)(-2)(-2)(5 \cdot 5) = -(-8)(25) = 200$

e. $x^3 \cdot x^4 = (x \cdot x \cdot x)(x \cdot x \cdot x \cdot x) = x^7$

Even though x is a variable, preventing us from writing the expression as simply a number, we can use the definition of natural number exponents to write the product in a simpler way.

f. $\dfrac{7^6}{7^4} = \dfrac{7 \cdot 7 \cdot \cancel{7} \cdot \cancel{7} \cdot \cancel{7} \cdot \cancel{7}}{\cancel{7} \cdot \cancel{7} \cdot \cancel{7} \cdot \cancel{7}} = 7 \cdot 7 = 49$

We can cancel four factors of 7 from the numerator and denominator of the original fraction, leaving us with 7^2, or 49.

The ultimate goal is to give meaning to the expression a^n for any real number n and to do so in such a way that certain properties of exponents hold consistently. For example, analysis of Example 1e leads to the observation that if n and m are natural numbers, then

$$a^n \cdot a^m = \underbrace{a \cdot a \cdot \cdots \cdot a}_{n \text{ factors}} \cdot \underbrace{a \cdot a \cdot \cdots \cdot a}_{m \text{ factors}} = \underbrace{a \cdot a \cdot \cdots \cdot a}_{n+m \text{ factors}} = a^{n+m}.$$

To extend the meaning of a^n to the case where $n = 0$, we might start by noting that we would like the following property to hold.

$$a^0 \cdot a^m = a^{0+m} = a^m$$

In order to make this specific property hold in the case when one exponent is 0, we want $a^0 \cdot a^m$ to be equal to a^m. This suggests the following definition.

DEFINITION: 0 as an Exponent

For any real number $a \neq 0$, we define $a^0 = 1$. The expression 0^0 is undefined, just as division by 0 is undefined.

Note that this definition means $a^n \cdot a^m = a^{n+m}$ whenever n and m are whole numbers. With this small extension of the meaning of exponents as inspiration, let us continue. Consider the following pattern.

The exponent is decreased by one at each step.

$$3^3 = 27$$
$$3^2 = 9$$
$$3^1 = 3$$
$$3^0 = 1$$
$$3^{-1} = ?$$
$$3^{-2} = ?$$
$$3^{-3} = ?$$

The result is $\frac{1}{3}$ of the result from the previous line.

In order to maintain the pattern that has begun to emerge, we are led to complete the pattern with $3^{-1} = \frac{1}{3}$, $3^{-2} = \frac{1}{9}$, and $3^{-3} = \frac{1}{27}$. In general, we define negative integer exponents as follows.

DEFINITION: Negative Integer Exponents

For any real number $a \neq 0$, and for any natural number n, $a^{-n} = \frac{1}{a^n}$. (We don't allow a to be 0 simply to avoid the possibility of division by 0.) Since any negative integer is the negative of a natural number, this defines exponentiation by negative integers.

Example 2: Simplifying Exponents

a. $\frac{y^2}{y^7} = \frac{\cancel{y} \cdot \cancel{y}}{y \cdot y \cdot y \cdot y \cdot y \cdot \cancel{y} \cdot \cancel{y}} = \frac{1}{y \cdot y \cdot y \cdot y \cdot y} = \frac{1}{y^5} = y^{-5}$

b. $\frac{6x^2}{-3x^2} = \frac{6}{-3} = -2$

Note that the variable x cancels out entirely. Recall, as in Example 1c, $-3x^2$ means -3 times x^2, not the quantity $(-3x)$ squared.

c. $5^0 5^{-3} = 5^{0-3} = 5^{-3} = \frac{1}{125}$

Note that $5^0 = 1$, as does a^0 for any $a \neq 0$.

d. $\frac{1}{t^{-3}} = \frac{1}{\frac{1}{t^3}} = 1 \cdot \frac{t^3}{1} = t^3$

e. $(x^2y)^3 = (x^2y)(x^2y)(x^2y) = x^2 \cdot x^2 \cdot x^2 \cdot y \cdot y \cdot y = x^6y^3$

TECHNOLOGY

```
NORMAL FLOAT AUTO REAL RADIAN MP
5⁰*5^(-3)
                              0.008
5^(-3)
                              0.008
```

TOPIC 2: Properties of Exponents

The following properties of exponents are used frequently in algebra. Most of these properties have been illustrated already in Examples 1 and 2. All of them can be readily demonstrated by applying the definition of integer exponents.

PROPERTIES: Properties of Exponents

In the following properties, a and b may be taken to represent constants, variables, or more complicated algebraic expressions. The letters n and m represent integers.

Property	Example(s)
1. $a^n \cdot a^m = a^{n+m}$	$3^3 \cdot 3^{-1} = 3^{3+(-1)} = 3^2 = 9$
2. $\dfrac{a^n}{a^m} = a^{n-m}$	$\dfrac{7^9}{7^{10}} = 7^{9-10} = 7^{-1}$
3. $a^{-n} = \dfrac{1}{a^n}$	$5^{-2} = \dfrac{1}{5^2} = \dfrac{1}{25}$ and $x^3 = \dfrac{1}{x^{-3}}$
4. $\left(a^n\right)^m = a^{nm}$	$\left(2^3\right)^2 = 2^{3\cdot 2} = 2^6 = 64$
5. $(ab)^n = a^n b^n$	$(7x)^3 = 7^3x^3 = 343x^3$ and $\left(-2x^5\right)^2 = (-2)^2\left(x^5\right)^2 = 4x^{10}$
6. $\left(\dfrac{a}{b}\right)^n = \dfrac{a^n}{b^n}$	$\left(\dfrac{3}{x}\right)^2 = \dfrac{3^2}{x^2} = \dfrac{9}{x^2}$ and $\left(\dfrac{1}{3z}\right)^2 = \dfrac{1^2}{(3z)^2} = \dfrac{1}{9z^2}$

Here and throughout this lesson, assume that every expression is defined. That is, if an exponent is 0, then the base is nonzero, and if an expression appears in the denominator of a fraction, then that expression is nonzero. Remember that $a^0 = 1$ for every $a \neq 0$.

Example 3: Using Properties of Exponents

Simplify the following expressions by using the properties of exponents. Write the final answers with only positive exponents.

a. $\left(17x^4 - 5x^2 + 2\right)^0$

b. $\dfrac{\left(x^2y^3\right)^{-1}z^{-2}}{x^3z^{-3}}$

c. $\dfrac{\left(-2x^3y^{-1}\right)^{-3}}{\left(18x^{-3}\right)^0(xy)^{-2}}$

d. $\left(7xz^{-2}\right)^2\left(5x^2y\right)^{-1}$

Solution

a. $\left(17x^4 - 5x^2 + 2\right)^0 = 1$

Any nonzero expression with an exponent of 0 is 1.

b. $$\frac{\left(x^2y^3\right)^{-1}z^{-2}}{x^3z^{-3}} = \frac{x^{-2}y^{-3}z^{-2}}{x^3z^{-3}} = \frac{z^3}{x^3x^2y^3z^2} = \frac{z}{x^5y^3}$$

Note that we have used several properties in this example. We could have used the applicable properties in many different orders to achieve the same result. Also note that the final answer contains only positive exponents. If we had not been told to write the answer in this way, we could have written the result as $zx^{-5}y^{-3}$.

c. $$\frac{\left(-2x^3y^{-1}\right)^{-3}}{\left(18x^{-3}\right)^0\left(xy\right)^{-2}} = \frac{(-2)^{-3}x^{-9}y^3}{x^{-2}y^{-2}}$$
$$= (-2)^{-3}x^{-9-(-2)}y^{3-(-2)}$$
$$= (-2)^{-3}x^{-7}y^5$$
$$= \frac{y^5}{-8x^7}$$
$$= -\frac{y^5}{8x^7}$$

We have chosen to apply the appropriate properties in a slightly different order than in the previous example, just to illustrate an alternative way to go about the task of simplifying such an expression.

d. $$\left(7xz^{-2}\right)^2\left(5x^2y\right)^{-1} = \frac{49x^2z^{-4}}{5x^2y}$$
$$= \frac{49}{5yz^4}$$

Note that the variable x disappeared entirely from the expression. If we had simplified the expression in a slightly different order, we would have obtained a factor of x^{2-2}, which is 1.

CAUTION

Many common errors result from forgetting the exact forms of the properties of exponents, as shown below.

Incorrect Statements	Corrected Statements
$x^2x^5 = x^{10}$	$x^2x^5 = x^{2+5} = x^7$
$2^42^3 = 4^7$	$2^42^3 = 2^{4+3} = 2^7$
$(3+4)^2 = 3^2 + 4^2$	$(3+4)^2 = 7^2$
$\left(x^2+3y\right)^{-1} = \frac{1}{x^2} + \frac{1}{3y}$	$\left(x^2+3y\right)^{-1} = \frac{1}{x^2+3y}$
$(3x)^2 = 3x^2$	$(3x)^2 = 3^2x^2 = 9x^2$
$\frac{x^5}{x^{-2}} = x^3$	$\frac{x^5}{x^{-2}} = x^{5-(-2)} = x^7$

TOPIC 3: Scientific Notation

Scientific notation is an important application of exponents. Very large and very small numbers arise naturally in a variety of situations, and working with them without scientific notation is an unwieldy and error-prone process. Scientific notation uses the properties of exponents to rewrite very large and very small numbers in a less clumsy form. It applies to those numbers which contain a long string of 0s before the decimal point (very large numbers) or to those which contain a long string of 0s between the decimal point and the nonzero digits (very small numbers).

DEFINITION: Scientific Notation

A number is in scientific notation when it is written in the form

$$a \times 10^n,$$

where $1 \le |a| < 10$ and n is an integer. If n is a positive integer, the number is large in magnitude; if n is a negative integer, the number is small in magnitude (close to 0). The number a itself can be either positive or negative, and the sign of a determines the sign of the number as a whole.

CAUTION

The sign of the exponent n in scientific notation does *not* determine the sign of the number as a whole. The sign of n only determines if the number is large (positive n) or small (negative n) in magnitude.

TECHNOLOGY

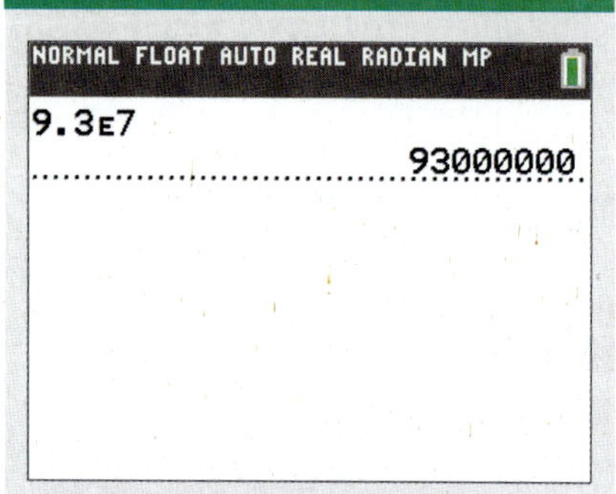

On a calculator, scientific notation is often expressed with an E. To represent this on a TI-84 Plus, press 2nd and ,.

Example 4: Using Scientific Notation

a. The distance from Earth to the sun is approximately 93,000,000 miles. Scientific notation takes advantage of the observation that multiplication of a number by 10 moves the decimal point one place to the right, and we can repeat this process as many times as necessary. Thus 9.3×10^7 is equal to 93,000,000, as 93,000,000 is obtained from 9.3 by moving the decimal point 7 places to the right.

$$9.3 \times 10^7 = 9\underbrace{3000000}_{\text{7 places}}$$

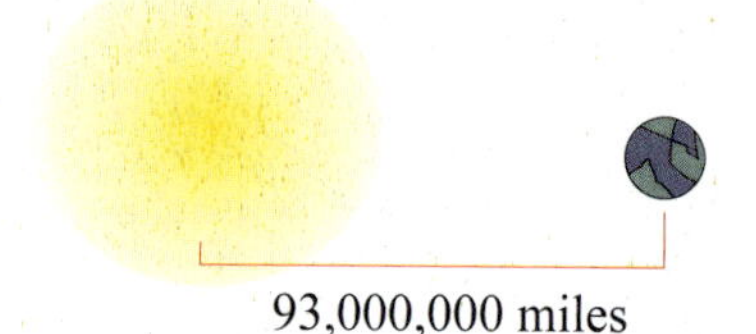

b. The mass of an electron, in kilograms, is approximately

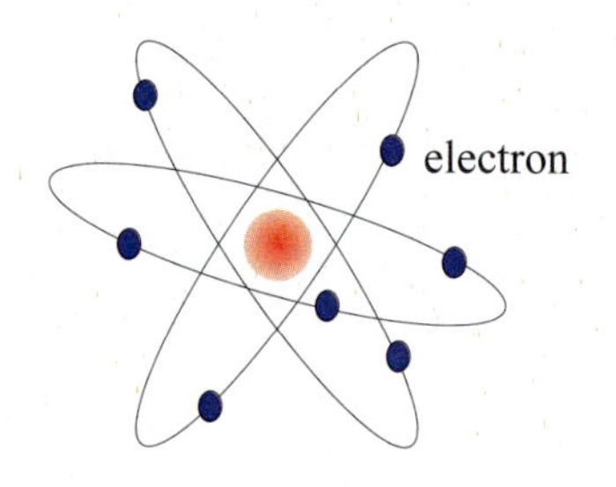

$$0.000000000000000000000000000000911,$$

clearly not a convenient number to work with. Scientific notation takes advantage of the observation that multiplication of a number by 10^{-1} (which is equivalent to division by 10) moves the decimal point one place to the left. Again, we can repeat this process as many times as necessary. Thus in scientific notation,

$$0.000000000000000000000000000000911 = 9.11 \times 10^{-31}.$$

c. The speed of light in a vacuum is approximately 3×10^8 meters per second. In standard (nonscientific) notation, this number is written as 300,000,000.

We can also use the properties of exponents to simplify computations involving two or more numbers that are large or small in magnitude, as illustrated by the next set of examples.

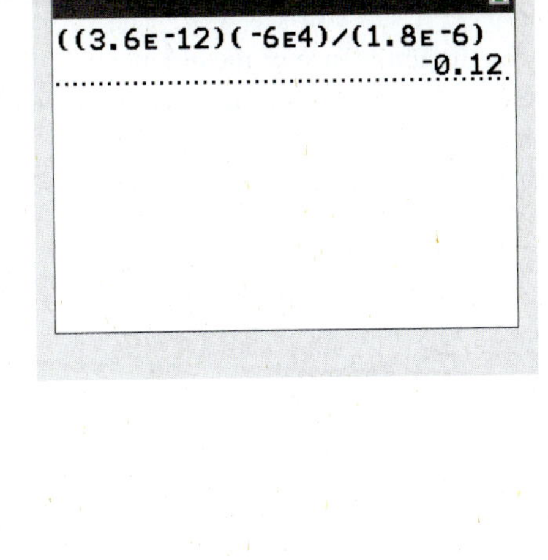

Example 5: Simplifying Expressions with Scientific Notation

Simplify the following expressions, writing your answer either in scientific or standard notation, as appropriate.

a. $\dfrac{(3.6\times10^{-12})(-6\times10^{4})}{1.8\times10^{-6}}$ **b.** $\dfrac{(7\times10^{34})(3\times10^{-12})}{6\times10^{-7}}$

Solution

a. $\dfrac{(3.6\times10^{-12})(-6\times10^{4})}{1.8\times10^{-6}}=\dfrac{(3.6)(-6)}{1.8}\times10^{-12+4-(-6)}$

$=-12\times10^{-2}$

$=-0.12$

We have written the final answer in standard notation (as it is not inconvenient to do so). Note that in scientific notation, it would be written as -1.2×10^{-1}.

b. $\dfrac{(7\times10^{34})(3\times10^{-12})}{6\times10^{-7}}=\dfrac{(7)(3)}{6}\times10^{34+(-12)-(-7)}$

$=3.5\times10^{29}$

This answer is best written in scientific notation.

TOPIC 4: Working with Geometric Formulas

Exponents occur in a very natural way when geometric formulas are considered. Some problems require nothing more than using a basic geometric formula, but others will require a bit more work. Often, the exact geometric formula that you need to solve a given problem can be derived from simpler formulas.

We will look at several examples of how a new geometric formula is built up from known formulas. The general rule of thumb in each case is to break down the task at hand into smaller pieces that can be easily handled.

Example 6: Using Geometric Formulas

Find formulas for each of the following:

a. The surface area of a box

b. The surface area of a soup can

c. The volume of a birdbath in the shape of half of a sphere

d. The volume of a gold ingot whose shape is a right trapezoidal cylinder

Solution

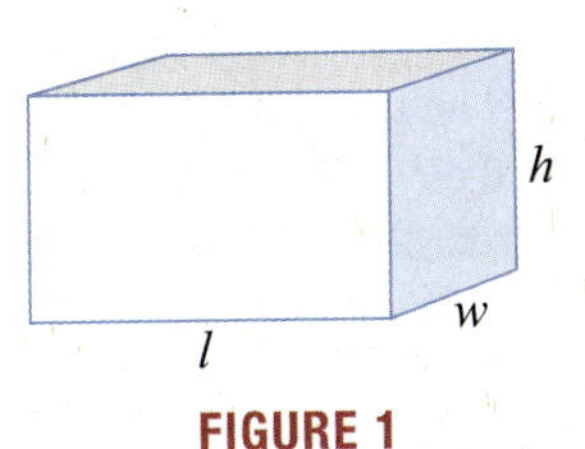

FIGURE 1

a. A box whose six faces are all rectangular is characterized by its length l, its width w, and its height h. The area of a rectangle is one of the basic formulas that you should be familiar with. The formula for the surface area of a box, then, is just the sum of the areas of the six rectangular sides. If we let S stand for the total surface area, we obtain the formula $S=lw+lw+lh+lh+hw+hw$ or $S=2lw+2lh+2hw$.

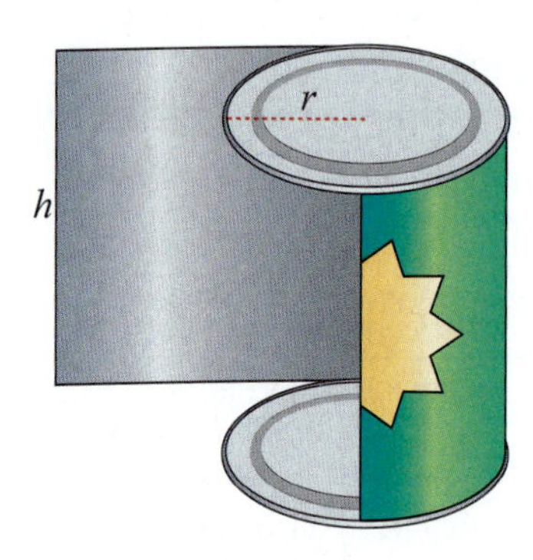

FIGURE 2

b. A soup can, an example of a right circular cylinder, is characterized by its height h and the radius r of the circle that makes up the base (or the top).

To determine the surface area of such a shape, imagine removing the top and bottom surfaces, cutting the soup can as shown in Figure 2, and flattening out the curved piece of metal making up the side. The flattened piece of metal would be a rectangle with height h and width $2\pi r$. Do you see why? The width of the rectangle would be the same as the circumference of the circular top and base, and the circumference of a circle is $2\pi r$. So the surface area of the curved side is $2\pi rh$. We also know that the area of a circle is πr^2, so if we let S stand for the surface area of the entire can, we have $S = \pi r^2 + \pi r^2 + 2\pi rh$, or $S = 2\pi r^2 + 2\pi rh$.

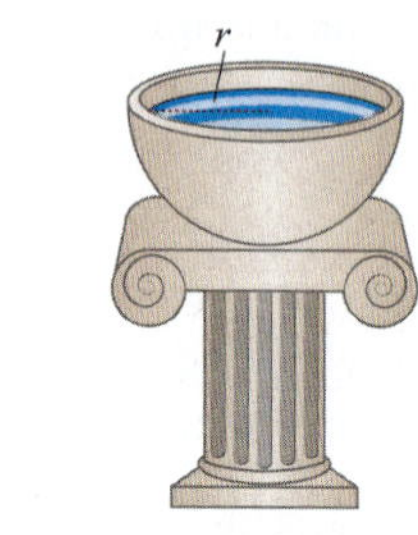

FIGURE 3

c. The volume of a sphere of radius r is $\frac{4}{3}\pi r^3$, and the birdbath of which we are to find the volume has the shape of half a sphere. So if we let V stand for the birdbath's volume,

$$V = \left(\frac{1}{2}\right)\left(\frac{4}{3}\pi r^3\right), \text{ or } V = \frac{2}{3}\pi r^3.$$

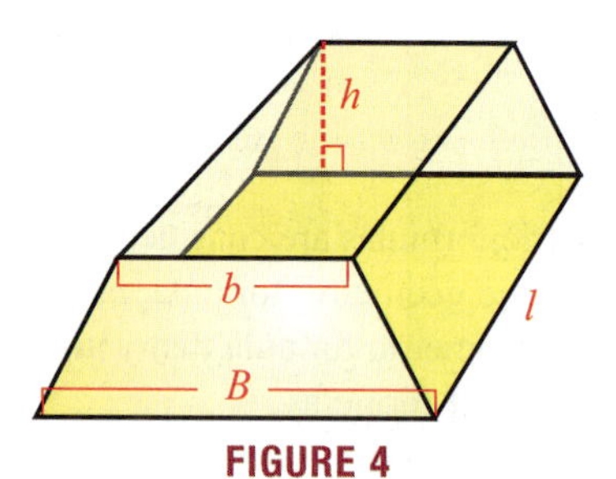

FIGURE 4

d. A *right cylinder* is the three-dimensional object generated by extending a plane region along an axis perpendicular to itself for a certain distance. (Such objects are often called prisms when the plane region is a polygon.) The volume of any right cylinder is the product of the area of the plane region and the distance that region has been extended perpendicular to itself. The gold ingot under consideration in this example is a right cylinder based on a trapezoid, as shown in Figure 4. The area of the trapezoid is $\frac{1}{2}(B+b)h$ and the ingot has length l, so its volume is $V = \frac{1}{2}(B+b)hl$. This could also be written as $V = \frac{(B+b)hl}{2}$.

TOPIC 5: Radical Notation

Taking the nth root of an expression (where n is a natural number) is the opposite operation of exponentiation by n. For example, the equation $4^3 = 64$ (which can be read as "four cubed is sixty-four") implies that the cube root of 64 is 4 (written $\sqrt[3]{64} = 4$). Similarly, the equation $2^4 = 16$ leads us to write $\sqrt[4]{16} = 2$ (read "the fourth root of sixteen is two").

At first glance, then, the definition of nth roots appears to be a simple matter, but there are some important difficulties to resolve. For instance, the statement $(-2)^4 = 16$ is also true, which might lead us to write $\sqrt[4]{16} = -2$. Can $\sqrt[4]{16}$ equal 2 and −2? Also, since any real number can be raised to any natural number power, we might be led to think that for any natural number n and any real number a, the nth root of a should be defined. What about $\sqrt[4]{-16}$? In order to evaluate $\sqrt[4]{-16}$ we seek a number whose fourth power is −16. Is there such a number?

The complete resolution of such issues will have to wait until Section 1.5, but we can begin now by defining $\sqrt[n]{a}$ in the cases where $\sqrt[n]{a}$ is a real number.

DEFINITION: Radical Notation

Case 1: n is an even natural number. If a is a nonnegative real number and n is an even natural number, $\sqrt[n]{a}$ is the nonnegative real number b with the property that $b^n = a$. That is, $\sqrt[n]{a} = b$ if and only if $a = b^n$. Note that $\left(\sqrt[n]{a}\right)^n = a$ and $\sqrt[n]{a^n} = a$.

Case 2: n is an odd natural number. If a is any real number and n is an odd natural number, $\sqrt[n]{a}$ is the real number b (whose sign will be the same as the sign of a) with the property that $b^n = a$. Again, $\sqrt[n]{a} = b$ if and only if $a = b^n$, $\left(\sqrt[n]{a}\right)^n = a$, and $\sqrt[n]{a^n} = a$.

The expression $\sqrt[n]{a}$ gives the nth root of a in **radical notation**. The natural number n is called the **index**, a is the **radicand**, and $\sqrt{}$ is called a **radical sign**. By convention, $\sqrt[2]{}$ is usually simply written as $\sqrt{}$.

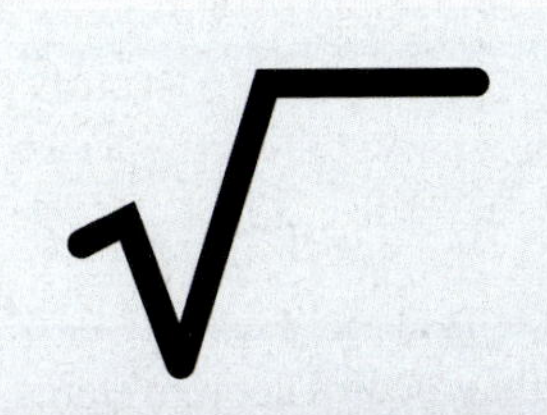

Radical History

The origin of the radical symbol $\sqrt{}$ is not entirely clear, though one school of thought speculates that its shape evolved from the letter "r," representing the first letter of the Latin word "radix" for "root." It was introduced in print by the German mathematician Christoff Rudolff in 1525, and later promoted by René Descartes.

The important distinction between the two cases is that when n is even, $\sqrt[n]{a}$ is defined only when a is nonnegative, whereas if n is odd, $\sqrt[n]{a}$ is defined for all real numbers a. This difference will be remedied in Section 1.5 by the introduction of complex numbers.

Mathematicians prevent any ambiguity in the meaning of $\sqrt[n]{a}$ when n is even and a is nonnegative by defining $\sqrt[n]{a}$ to be the *nonnegative number* whose n^{th} power is a. For example, $\sqrt{4}$ is equal to 2, not -2. Similarly, $\sqrt[4]{16}$ equals 2, not -2.

Note that just as in the case of powers, alternative phrases are commonly used when the index is 2 or 3. We usually read $\sqrt{a}$ as the "square root of a" and $\sqrt[3]{a}$ as the "cube root of a."

DEFINITION: Perfect Powers

A **perfect square** is an integer equal to the square of another integer. The square root of a perfect square is always an integer.

A **perfect cube** is an integer equal to the cube of another integer. The cube root of a perfect cube is always an integer.

Example 7: Using Radical Notation

a. $\sqrt[5]{-32} = -2$ because $(-2)^5 = -32$.

b. $\sqrt[4]{-16}$ is not a real number, as no real number raised to the fourth power is -16.

c. $-\sqrt[4]{16} = -2$. Note that the fourth root of 16 is a real number, which is then multiplied by -1.

d. $\sqrt{0} = 0$. In fact, $\sqrt[n]{0} = 0$ for any natural number n.

e. $\sqrt[n]{1} = 1$ for any natural number n. $\sqrt[n]{-1} = -1$ for any odd natural number n.

f. $\sqrt[3]{-\frac{27}{64}} = -\frac{3}{4}$ because $\left(-\frac{3}{4}\right)^3 = -\frac{27}{64}$.

g. $\sqrt[5]{-\pi^5} = -\pi$ because $(-\pi)^5 = (-1)^5 \pi^5 = -\pi^5$.

h. $\sqrt[4]{(-3)^4} = \sqrt[4]{81} = 3$. In general, if n is an even natural number, $\sqrt[n]{a^n} = |a|$ for any real number a. Remember, though, that $\sqrt[n]{a^n} = a$ if n is an odd natural number.

Example 8: Using Radical Notation

Given a right triangle with legs of length a and b, the Pythagorean Theorem states that the length of the hypotenuse c is given by $c = \sqrt{a^2 + b^2}$. In this formula from the Pythagorean Theorem, find the following:

a. The radicand

b. The index

c. The value of c if $a = 5$ and $b = 12$

d. The value of c if $a = 1$ and $b = 2$

Solution

a. The radicand is the quantity beneath the radical sign, $a^2 + b^2$.

b. Because no index is indicated, the index is 2.

c. $c = \sqrt{a^2 + b^2}$

$= \sqrt{(5)^2 + (12)^2}$ Substitute.

$= \sqrt{169}$ Since 169 is a perfect square, the solution is an integer.

$= 13$

d. $c = \sqrt{a^2 + b^2}$

$= \sqrt{(1)^2 + (2)^2}$ Substitute.

$= \sqrt{5}$ Since 5 is not a perfect square, the solution is not an integer.

TOPIC 6: Simplifying and Combining Radical Expressions

When solving equations that contain radical expressions, it is often helpful to simplify the expressions first. The word *simplify* arises frequently in mathematics, and its meaning depends on the context.

DEFINITION: Simplified Radical Form

A radical expression is in **simplified form** when the following conditions are met:

1. The radicand contains no factor with an exponent greater than or equal to the index of the radical.
2. The radicand contains no fractions.
3. The denominator, if there is one, contains no radical.
4. The greatest common factor of the index and any exponent occurring in the radicand is 1. That is, the index and any exponent in the radicand have no common factor other than 1.

These four conditions deserve some explanation. The first condition is a reflection of the fact that roots and powers undo one another.

$$\sqrt[n]{a^n} = \begin{cases} |a| & \text{if } n \text{ is even} \\ a & \text{if } n \text{ is odd} \end{cases}$$

So, if the radicand contains an exponent greater than or equal to the index, a factor can be brought out from under the radical.

The second and third conditions both aim to remove radicals from the denominators of fractions. The process of removing radicals from the denominator is called *rationalizing the denominator* and is useful when solving equations that contain radicals.

The fourth condition is again a reflection of the fact that roots and powers undo one another, and is perhaps best motivated by an example.

$$\sqrt[3]{a^6} = \sqrt[3]{a^2a^2a^2} = \sqrt[3]{\left(a^2\right)^3} = a^2$$

This simplification was possible because the exponent in the radicand and the index had a common factor of 3. This sort of simplification must be approached with caution, however, as will be seen in the following examples.

Before illustrating the process of simplifying radicals, a few useful properties of radicals will be listed. These properties can all be proved using nothing more than the definition of roots, but their validity will be more clear once we have discussed rational exponents later in this section.

PROPERTIES: Properties of Radicals

In the following properties, a and b may be taken to represent constants, variables, or more complicated algebraic expressions. The letters n and m represent natural numbers. Here and throughout this lesson, assume that all expressions are defined and are real numbers.

Property	Example
1. $\sqrt[n]{ab} = \sqrt[n]{a} \cdot \sqrt[n]{b}$	$\sqrt[3]{3x^6y^2} = \sqrt[3]{3} \cdot \sqrt[3]{x^3} \cdot \sqrt[3]{x^3} \cdot \sqrt[3]{y^2}$ $= \sqrt[3]{3} \cdot x \cdot x \cdot \sqrt[3]{y^2} = x^2\sqrt[3]{3y^2}$
2. $\sqrt[n]{\frac{a}{b}} = \frac{\sqrt[n]{a}}{\sqrt[n]{b}}$	$\sqrt[4]{\frac{x^4}{16}} = \frac{\sqrt[4]{x^4}}{\sqrt[4]{16}} = \frac{\lvert x \rvert}{2}$
3. $\sqrt[m]{\sqrt[n]{a}} = \sqrt[mn]{a}$	$\sqrt[3]{\sqrt{64}} = \sqrt[3]{\sqrt[2]{64}} = \sqrt[6]{64} = 2$

Example 9: Simplifying Radical Expressions

Simplify the following radical expressions.

a. $\sqrt[3]{-16x^8y^4}$ **b.** $\sqrt{8z^6}$ **c.** $\sqrt[3]{\frac{72x^2}{y^3}}$

Solution

a. $\sqrt[3]{-16x^8y^4} = \sqrt[3]{(-2)^3 \cdot 2 \cdot x^3 \cdot x^3 \cdot x^2 \cdot y^3 \cdot y}$
$= -2x^2y\sqrt[3]{2x^2y}$

Note that since the index is 3, we look for all of the *perfect cubes* in the radicand.

b. $\sqrt{8z^6} = \sqrt{2^2 \cdot 2 \cdot (z^3)^2}$
$= \lvert 2z^3 \rvert \sqrt{2}$
$= 2\lvert z^3 \rvert \sqrt{2}$

Remember that $\sqrt[n]{a^n} = \lvert a \rvert$ if n is even, so absolute value signs are necessary around the factor of z^3.

c. $\sqrt[3]{\frac{72x^2}{y^3}} = \frac{\sqrt[3]{8 \cdot 9 \cdot x^2}}{\sqrt[3]{y^3}}$
$= \frac{2\sqrt[3]{9x^2}}{y}$

All perfect cubes have been brought out from under the radical, and the denominator has been rationalized.

TECHNOLOGY

```
NORMAL FLOAT AUTO REAL RADIAN MP
√(9+16)
                                5
√(9)+√(16)
                                7
```

Notice that the expressions are not equal.

CAUTION

As with the properties of exponents, many mistakes arise from forgetting the properties of radicals. One common error, for instance, is to rewrite $\sqrt{a+b}$ as $\sqrt{a}+\sqrt{b}$. These two expressions are not equal! To convince yourself of this, evaluate the two expressions with actual constants in place of a and b. For example, note that $5=\sqrt{9+16}\neq\sqrt{9}+\sqrt{16}=7$.

Rationalizing denominators sometimes requires more effort than in Example 9c, and in fact is sometimes impossible. The following methods will, however, take care of two common cases.

Case 1: Denominator is a single term containing a root.

If the denominator is a single term containing a factor of $\sqrt[n]{a^m}$, we will take advantage of the fact that $\sqrt[n]{a^m}\cdot\sqrt[n]{a^{n-m}}=\sqrt[n]{a^m\cdot a^{n-m}}=\sqrt[n]{a^n}$ and this last expression is either a or $|a|$, depending on whether n is odd or even. Now, if we multiply the denominator by a factor of $\sqrt[n]{a^{n-m}}$, we must also multiply the numerator by the same factor. Thus, in this case we multiply the fraction by $\frac{\sqrt[n]{a^{n-m}}}{\sqrt[n]{a^{n-m}}}$, as shown in the following example.

$$\frac{1}{\sqrt{a}}=\frac{1}{\sqrt{a}}\cdot\frac{1}{1}=\frac{1}{\sqrt{a}}\cdot\frac{\sqrt{a}}{\sqrt{a}}=\frac{\sqrt{a}}{a}$$

Case 2: Denominator consists of two terms, one or both of which are square roots.

To discuss the method used in this case, let $A+B$ represent the denominator of the fraction under consideration, where at least one of A and B stands for a square root term. We will take advantage of the following fact.

$$(A+B)(A-B)=A(A-B)+B(A-B)=A^2-AB+BA-B^2=A^2-B^2$$

Note that the exponents of 2 in the end result negate the square root (or roots) initially in the denominator. Just as in Case 1, though, we can't multiply the denominator by $A-B$ unless we multiply the numerator by this same factor. The method is thus to multiply the fraction by $\frac{A-B}{A-B}$ as demonstrated in the equation below. The factor $A-B$ is called the **conjugate radical expression** of $A+B$.

$$\frac{1}{\sqrt{a}+\sqrt{b}}=\frac{1}{\sqrt{a}+\sqrt{b}}\cdot\frac{1}{1}=\frac{1}{\sqrt{a}+\sqrt{b}}\cdot\frac{\sqrt{a}-\sqrt{b}}{\sqrt{a}-\sqrt{b}}=\frac{\sqrt{a}-\sqrt{b}}{a-b}$$

Example 10: Rationalizing the Denominator

Simplify the following radical expressions.

a. $\sqrt[5]{\dfrac{-4x^6}{8y^2}}$ **b.** $\dfrac{4}{\sqrt{7}+\sqrt{3}}$ **c.** $\dfrac{-\sqrt{5x}}{5-\sqrt{x}}$

Solution

a.
$$\sqrt[5]{\frac{-4x^6}{8y^2}} = \frac{\sqrt[5]{-4x\cdot x^5}}{\sqrt[5]{8y^2}} = \frac{-x\sqrt[5]{4x}}{\sqrt[5]{2^3y^2}}\cdot\frac{\sqrt[5]{2^2y^3}}{\sqrt[5]{2^2y^3}} = \frac{-x\sqrt[5]{16xy^3}}{\sqrt[5]{2^5y^5}} = \frac{-x\sqrt[5]{16xy^3}}{2y}$$

The first step is to simplify the numerator and denominator and determine what factor the denominator must be multiplied by in order to eliminate the radical. Remember to multiply the numerator by the same factor.

b.
$$\frac{4}{\sqrt{7}+\sqrt{3}} = \left(\frac{4}{\sqrt{7}+\sqrt{3}}\right)\left(\frac{\sqrt{7}-\sqrt{3}}{\sqrt{7}-\sqrt{3}}\right) = \frac{4\left(\sqrt{7}-\sqrt{3}\right)}{7-3} = \frac{4\left(\sqrt{7}-\sqrt{3}\right)}{4} = \sqrt{7}-\sqrt{3}$$

Again, we have multiplied the fraction by 1, but this time we multiply the numerator and denominator by the *conjugate radical* of the original denominator.

c.
$$\frac{-\sqrt{5x}}{5-\sqrt{x}} = \left(\frac{-\sqrt{5x}}{5-\sqrt{x}}\right)\left(\frac{5+\sqrt{x}}{5+\sqrt{x}}\right) = \frac{-5\sqrt{5x}-\sqrt{5x^2}}{25-x} = \frac{-5\sqrt{5x}-x\sqrt{5}}{25-x}$$

In this example, the original denominator is the sum of 5 and $-\sqrt{x}$, so the conjugate radical of the denominator is $5+\sqrt{x}$. Note that in simplifying the term $-\sqrt{5x^2}$, we can simply write $-x\sqrt{5}$ instead of $-|x|\sqrt{5}$, since the original expression is not real if x is negative.

As an aside, there are occasions when rationalizing the numerator is desirable. For instance, some problems in calculus (which the author encourages all college students to take as a consciousness-raising experience!) are much easier to solve after rationalizing the numerator of a given fraction. This is accomplished by the same method, as seen in the next example.

Example 11: Rationalizing the Numerator

Rationalize the numerator of the fraction $\dfrac{\sqrt{4x}-\sqrt{6y}}{2x-3y}$.

Solution

$$\begin{aligned}\frac{\sqrt{4x}-\sqrt{6y}}{2x-3y} &= \left(\frac{\sqrt{4x}-\sqrt{6y}}{2x-3y}\right)\left(\frac{\sqrt{4x}+\sqrt{6y}}{\sqrt{4x}+\sqrt{6y}}\right)\\ &= \frac{4x-6y}{(2x-3y)\left(\sqrt{4x}+\sqrt{6y}\right)}\\ &= \frac{2(2x-3y)}{(2x-3y)\left(2\sqrt{x}+\sqrt{6y}\right)}\\ &= \frac{2}{2\sqrt{x}+\sqrt{6y}}\end{aligned}$$

We begin by multiplying the numerator and denominator by the conjugate radical of the *numerator*. Note that we could have simplified the term $\sqrt{4x}$ first, but this would not have altered the work in any substantial way, and the final answer would be the same.

Often, a sum of two or more radical expressions can be combined into one. This can be done if the radical expressions are **like radicals**, meaning that they have the same index and the same radicand. It is frequently necessary to simplify the radical expressions before it can be determined if they are like or not.

Example 12: Combining Radical Expressions

Combine the radical expressions, if possible.

a. $-3\sqrt{8x^5}+\sqrt{18x}$ **b.** $\sqrt[3]{54x^3}+\sqrt{50x^2}$ **c.** $\sqrt{\dfrac{1}{12}}-\sqrt{\dfrac{25}{48}}$

Solution

a. $$\begin{aligned}-3\sqrt{8x^5}+\sqrt{18x} &= -3\sqrt{2^2\cdot 2\cdot x^4\cdot x}+\sqrt{2\cdot 3^2\cdot x}\\ &= -6\left|x^2\right|\sqrt{2x}+3\sqrt{2x}\\ &= -6x^2\sqrt{2x}+3\sqrt{2x}\\ &= \left(-6x^2+3\right)\sqrt{2x}\end{aligned}$$

We begin by simplifying the radicals. Note that upon simplification, the two radicals have the same index and the same radicand. Also note that the absolute value bars around the factor of x^2 are unnecessary, since x^2 is always nonnegative.

b. $$\begin{aligned}\sqrt[3]{54x^3}+\sqrt{50x^2} &= \sqrt[3]{2\cdot 3^3\cdot x^3}+\sqrt{2\cdot 5^2\cdot x^2}\\ &= 3x\sqrt[3]{2}+5|x|\sqrt{2}\end{aligned}$$

Upon simplification, the radicands are the same, but the indices are not. We have written the radicals in simplest form, but they cannot be combined.

c. $\sqrt{\frac{1}{12}}-\sqrt{\frac{25}{48}}=\frac{1}{\sqrt{2^2\cdot 3}}-\frac{\sqrt{5^2}}{\sqrt{4^2\cdot 3}}$

$=\frac{1}{2\sqrt{3}}\cdot\frac{\sqrt{3}}{\sqrt{3}}-\frac{5}{4\sqrt{3}}\cdot\frac{\sqrt{3}}{\sqrt{3}}$

$=\frac{2\sqrt{3}}{4\cdot 3}-\frac{5\cdot\sqrt{3}}{4\cdot 3}$

$=-\frac{3\sqrt{3}}{12}=-\frac{\sqrt{3}}{4}$

The first step, again, is to simplify the radicals. Note that both denominators are being rationalized in the second step. The first fraction must be multiplied by $\frac{2}{2}$ in order to get a common denominator. After subtracting, the fraction can be reduced.

TOPIC 7: Rational Number Exponents

We can now return to the task of defining exponentiation, and give meaning to a^r when r is a rational number.

DEFINITION: Rational Number Exponents

Meaning of $a^{\frac{1}{n}}$**:** If n is a natural number and if $\sqrt[n]{a}$ is a real number, then $a^{\frac{1}{n}}=\sqrt[n]{a}$.

Meaning of $a^{\frac{m}{n}}$**:** If m and n are natural numbers with $n \neq 0$, if m and n have no common factors greater than 1, and if $\sqrt[n]{a}$ is a real number, then $a^{\frac{m}{n}}=\sqrt[n]{a^m}=\left(\sqrt[n]{a}\right)^m$. Either $\sqrt[n]{a^m}$ or $\left(\sqrt[n]{a}\right)^m$ can be used to evaluate $a^{\frac{m}{n}}$, as they are equal. $a^{\frac{-m}{n}}$ is defined to be $\frac{1}{a^{\frac{m}{n}}}$.

In addition to giving meaning to rational exponentiation, the above definition provides us with the means of converting between *radical notation* and *exponential notation*. In some cases, one notation or the other is more convenient, and the choice of notation should be based on which makes the task at hand easier.

Although originally stated only for integer exponents, the properties of exponents also hold for rational exponents (and for real exponents as well, though we don't require that fact at the moment). And now that rational exponentiation has been defined, we are better able to demonstrate the properties of radicals mentioned earlier.

$$\sqrt[m]{\sqrt[n]{a}}=\left(a^{\frac{1}{n}}\right)^{\frac{1}{m}}=a^{\frac{1}{n}\cdot\frac{1}{m}}=a^{\frac{1}{mn}}=\sqrt[mn]{a}$$

In fact, the desire to extend the properties of exponents from integers to rationals forces us to define rational exponents as we did above. For instance, since we want

$$\left(a^{\frac{1}{2}}\right)^2=a^{\frac{1}{2}\cdot 2}=a^1=a,$$

it must be the case that $a^{\frac{1}{2}}=\sqrt{a}$.

We close this section with a variety of examples illustrating radical notation, exponential notation, and the properties of both.

Example 13: Simplifying Expressions

Simplify each of the following expressions, writing your answer using the same notation as the original expression.

a. $27^{\frac{-2}{3}}$

b. $\sqrt[9]{-8x^6}$

c. $\left(5x^2+3\right)^{\frac{8}{3}}\left(5x^2+3\right)^{\frac{-2}{3}}$

d. $\sqrt[5]{\sqrt[3]{x^2}}$

e. $\dfrac{5x-y}{(5x-y)^{\frac{-1}{3}}}$

TECHNOLOGY

```
NORMAL FLOAT AUTO REAL RADIAN MP
27^(-2/3)▸Frac
                                1/9
27^(1/3)
                                  3
Ans^(-2)▸Frac
                                1/9
```

Solution

a. $27^{\frac{-2}{3}}=\left(27^{\frac{1}{3}}\right)^{-2}$
$=3^{-2}$
$=\dfrac{1}{3^2}$
$=\dfrac{1}{9}$

The only task here is to evaluate the expression. We could also have begun by noting that $27^{\frac{-2}{3}}=\left(27^{-2}\right)^{\frac{1}{3}}$, but this would have made the calculations much more tedious.

b. $\sqrt[9]{-8x^6}=-\sqrt[9]{2^3x^6}$
$=-2^{\frac{3}{9}}x^{\frac{6}{9}}$
$=-2^{\frac{1}{3}}x^{\frac{2}{3}}$
$=-\sqrt[3]{2x^2}$

To begin, the only factor that can be brought out from under the radical is −1, but we can still simplify the expression by reducing the exponents and index. Note that we switched to exponential form temporarily in doing this.

c. $\left(5x^2+3\right)^{\frac{8}{3}}\left(5x^2+3\right)^{\frac{-2}{3}}=\left(5x^2+3\right)^{\frac{8}{3}+\left(\frac{-2}{3}\right)}$
$=\left(5x^2+3\right)^{\frac{6}{3}}$
$=\left(5x^2+3\right)^2$

Since the bases are the same, we only have to apply the property $a^na^m=a^{n+m}$.

d. $\sqrt[5]{\sqrt[3]{x^2}}=\sqrt[15]{x^2}$

Use the property stating $\sqrt[m]{\sqrt[n]{a}}=\sqrt[mn]{a}$ to simplify this expression.

e. $\dfrac{5x-y}{(5x-y)^{\frac{-1}{3}}}=(5x-y)^{1-\left(\frac{-1}{3}\right)}$
$=(5x-y)^{\frac{4}{3}}$

We use the property $\dfrac{a^n}{a^m}=a^{n-m}$ to simplify this expression.

Simplifying the expressions in the next example requires a bit more work and/or caution.

Example 14: Simplifying Radical Expressions

a. Simplify the expression $\sqrt[4]{x^2}$.

b. Write $\sqrt[3]{2} \cdot \sqrt{3}$ as a single radical.

Solution

a.
$$\sqrt[4]{x^2} = \left(x^2\right)^{\frac{1}{4}} = |x|^{\frac{1}{2}} = \sqrt{|x|}$$

We might be tempted to write $\sqrt[4]{x^2}$ as simply $\sqrt{x}$, but note that $\sqrt[4]{x^2}$ is defined for *all* real numbers x, while $\sqrt{x}$ is defined only for nonnegative real numbers. We can rectify this disparity by first making sure the radicand is nonnegative.

b.
$$\sqrt[3]{2} \cdot \sqrt{3} = 2^{\frac{1}{3}} \cdot 3^{\frac{1}{2}} = 2^{\frac{2}{6}} \cdot 3^{\frac{3}{6}} = \left(2^2\right)^{\frac{1}{6}}\left(3^3\right)^{\frac{1}{6}} = 4^{\frac{1}{6}} \cdot 27^{\frac{1}{6}} = 108^{\frac{1}{6}} = \sqrt[6]{108}$$

We can make use of the property $a^n b^n = (ab)^n$ if we can first make the exponents equal. We do so by finding the least common denominator of $\frac{1}{3}$ and $\frac{1}{2}$, writing both fractions with this common denominator, and then making use of the property $a^{nm} = \left(a^n\right)^m$.

1.2 EXERCISES

PRACTICE

Simplify each of the following expressions, writing your answer with only positive exponents. See Examples 1 and 2.

1. $(-2)^4$ **2.** -2^4 **3** $3^2 \cdot 3^2$

4. $\dfrac{7^4}{7^5}$ **5.** $\dfrac{x^5}{x^2}$ **6.** $n^2 \cdot n^5$

Use the properties of exponents to simplify each of the following expressions, writing your answer with only positive exponents. See Examples 1, 2, and 3.

7. $\dfrac{3t^{-2}}{t^3}$ **8.** $9^0 x^3 y^0$ **9.** $\dfrac{2n^3}{n^{-5}}$

10. $\dfrac{x^7y^{-3}z^{12}}{x^{-1}z^9}$ **11.** $\dfrac{s^5y^{-5}z^{-11}}{s^8y^{-7}}$ **12.** $\dfrac{\left(3yz^{-2}\right)^0}{3y^2z}$

13. $\left[9m^2-\left(2n^2\right)^3\right]^{-1}$ **14.** $\dfrac{(-3a)^{-2}\left(bc^{-2}\right)^{-3}}{a^5c^4}$ **15.** $\left[\left(5m^4n^{-2}\right)^{-1}\right]^{-2}$

16. $\left[\left(4a^2b^{-5}\right)^{-1}\right]^{-3}$ **17.** $\left[\left(3^{-1}x^{-1}y\right)\left(x^2y\right)^{-1}\right]^{-3}$ **18.** $\left[\dfrac{100^0\left(x^{-1}y^3\right)^{-1}}{x^2y}\right]^{-3}$

19. $\left(5z^6-\left(3x^3\right)^4\right)^{-1}$ **20.** $\left[\dfrac{y^6\left(xy^2\right)^{-3}}{3x^{-3}z}\right]^{-2}$

Convert each number from scientific notation to standard notation, or vice versa, as indicated. See Example 4.

21. $-912{,}000{,}000$; convert to scientific **22.** 0.00000021; convert to scientific

23. 3.2×10^7; convert to standard **24.** 1.934×10^{-4}; convert to standard

25. There are approximately 31,536,000 seconds in a calendar year. Express the number of seconds in scientific notation.

26. Together, the 46 human chromosomes are estimated to contain some 3.0×10^9 base pairs of DNA. Express the number of pairs of DNA in standard notation.

27. A white blood cell is approximately 3.937×10^{-4} inches in diameter. Express this diameter in standard notation.

28. The probability of winning the lottery with one dollar is approximately 0.0000002605. Express this probability in scientific notation.

Evaluate each expression using the properties of exponents. Use a calculator only to check your final answer. See Example 5.

29. $\dfrac{\left(2\times10^3\right)\left(7\times10^{-2}\right)}{\left(5\times10^4\right)}$ **30.** $\dfrac{\left(8\times10^{-3}\right)\left(3\times10^{-2}\right)}{\left(2\times10^5\right)}$

31. $\left(2\times10^{-13}\right)\left(5.5\times10^{10}\right)\left(-1\times10^3\right)$ **32.** $\left(6\times10^{21}\right)\left(5\times10^{-19}\right)\left(5\times10^4\right)$

33. $\dfrac{4\times10^{-6}}{\left(5\times10^4\right)\left(8\times10^{-3}\right)}$ **34.** $\dfrac{\left(4.6\times10^{12}\right)\left(9\times10^3\right)}{\left(1.5\times10^8\right)\left(2.3\times10^{-5}\right)}$

Apply the definition of integer exponents to demonstrate the following properties.

35. $a^n\cdot a^m=a^{n+m}$ **36.** $\left(a^n\right)^m=a^{nm}$ **37.** $(ab)^n=a^nb^n$

Evaluate the following radical expressions. See Example 7.

38. $-\sqrt{9}$

39. $\sqrt[3]{-27}$

40. $\sqrt{-25}$

41. $\sqrt[3]{-\frac{27}{125}}$

42. $\sqrt{\frac{25}{121}}$

43. $-\sqrt[3]{-8}$

44. $\sqrt[4]{\sqrt{16}-\sqrt[3]{-27}+\sqrt{81}}$

45. $\sqrt{\frac{\sqrt[3]{-64}}{-\sqrt{144}-\sqrt{169}}}$

Simplify the following radical expressions. See Example 9.

46. $\sqrt[3]{-8x^6y^9}$

47. $\sqrt{2x^6y}$

48. $\sqrt[7]{x^{14}y^{49}z^{21}}$

49. $\sqrt{\frac{x^2}{4x^4y^6}}$

50. $\sqrt[3]{\frac{a^3b^{12}}{27c^6}}$

51. $\sqrt[4]{\frac{x^{12}y^8}{16}}$

52. $\sqrt[5]{\frac{y^{30}z^{25}}{32x^{35}}}$

53. $\sqrt[5]{32x^7y^{10}}$

Simplify the following radicals by rationalizing the denominators. See Example 10.

54. $\sqrt[3]{\frac{4x^2}{3y^4}}$

55. $\frac{-\sqrt{3a^3}}{\sqrt{6a}}$

56. $\frac{10}{\sqrt{7}-\sqrt{2}}$

57. $\frac{3}{\sqrt{6}-\sqrt{3}}$

58. $\frac{\sqrt{x}}{\sqrt{x}-\sqrt{2}}$

59. $\frac{\sqrt{x}+\sqrt{y}}{\sqrt{x}-\sqrt{y}}$

Rationalize the numerators of the following expressions. See Example 11.

60. $\frac{3+\sqrt{y}}{6}$

61. $\frac{\sqrt{13}+\sqrt{t}}{13-t}$

62. $\frac{2\sqrt{x}+\sqrt{y}}{\sqrt{x}-\sqrt{y}}$

Combine the radical expressions, if possible. See Example 12.

63. $\sqrt[3]{-16x^4}+5x\sqrt[3]{2x}$

64. $\sqrt{27xy^2}-4\sqrt{3xy^2}$

65. $\sqrt{7x}-\sqrt[3]{7x}$

66. $-x^2\sqrt[3]{54x}+3\sqrt[3]{2x^7}$

67. $\sqrt[5]{32x^{13}}+3x\sqrt[5]{x^8}$

68. $\sqrt[3]{-16z^4}+6z\sqrt[3]{2z}$

Simplify the following expressions, writing your answer using the same notation as the original expression. See Example 13.

69. $\sqrt[3]{\sqrt[4]{x^{36}}}$

70. $32^{-\frac{3}{5}}$

71. $(3x^2-4)^{\frac{1}{3}}(3x^2-4)^{\frac{5}{3}}$

72. $81^{\frac{3}{4}}$

73. $(-8)^{\frac{2}{3}}$

74. $625^{-\frac{3}{4}}$

75. $\sqrt[3]{\sqrt[5]{y^{25}}}$

76. $\frac{(a-b)^{-\frac{2}{3}}}{(a-b)^{-2}}$

Convert the following expressions from radical notation to exponential notation, or vice versa. Simplify each expression in the process, if possible.

77. $256^{-\frac{3}{4}}$ **78.** $\sqrt[12]{x^3}$ **79.** $\sqrt[6]{\frac{2}{72}}$ **80.** $\left(36n^4\right)^{\frac{5}{6}}$

Simplify the following expressions. See Example 14.

81. $\sqrt{5}\cdot\sqrt[4]{5}$ **82.** $\sqrt[3]{x^7}\cdot\sqrt[9]{x^6}$ **83.** $\sqrt[5]{y^{16}}\cdot\sqrt[25]{y^{20}}$ **84.** $\sqrt[4]{7}\cdot\sqrt[16]{7}$

Apply the definition of rational exponents to demonstrate the following properties.

85. $\sqrt[n]{ab}=\sqrt[n]{a}\cdot\sqrt[n]{b}$ **86.** $\sqrt[n]{\frac{a}{b}}=\frac{\sqrt[n]{a}}{\sqrt[n]{b}}$ **87.** $\sqrt[m]{\sqrt[n]{a}}=\sqrt[mn]{a}$

APPLICATIONS

88. The prism shown below is a right triangular cylinder, where the base is a right triangle. Find the volume of the prism in terms of b, h, and l.

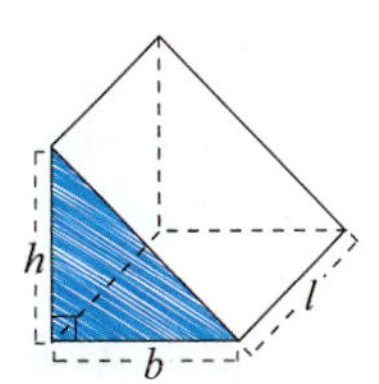

89. Determine the volume of the right circular cylinder shown, in terms of r and h.

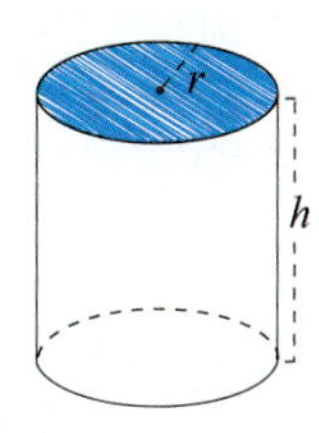

90. Matt wants to let people in the future know what life is like today, so he goes shopping for a time capsule. Capacity, along with price and quality, is an important consideration for him. One time capsule he looks at is a right circular cylinder with a hemisphere on each end. Find the volume of the time capsule, given that the length l of the cylinder is 16 inches and the radius r is 3 inches.

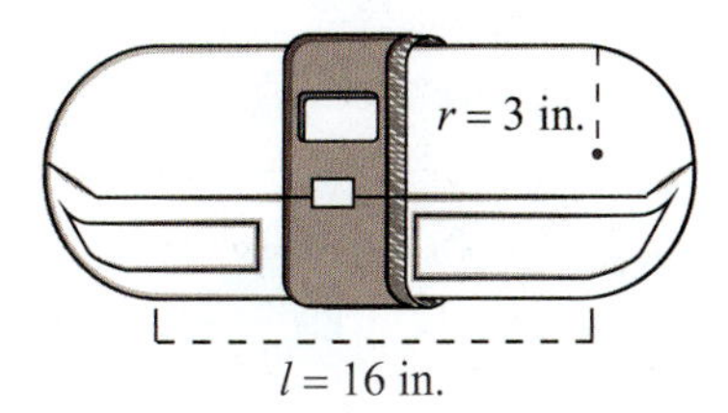

91. Bill and Dee are buying a new house. The house is a right cylinder based on a trapezoid atop a rectangular prism. The bases of the trapezoid are $B = 10$ m and $b = 8$ m, and the length of the house is $l = 15$ m. The height of the house up to the bottom of the roof is $H = 3$ m, and the height of the roof is $h = 1$ m. Find the volume of the house.

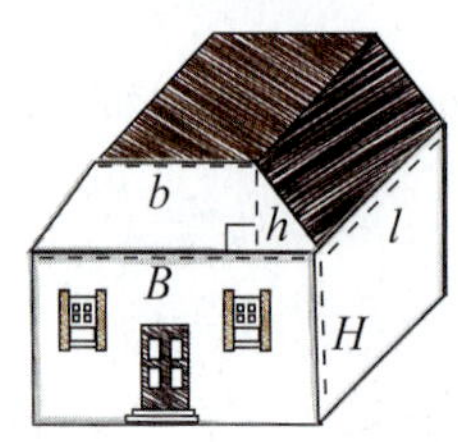

92. Determine the expression for the volume of water contained in an above-ground circular swimming pool that has a diameter of 18 feet, assuming the water has a uniform depth of d feet.

93. The floor of a rectangular bedroom measures N feet wide and M feet long. The height of the walls is 7 feet. Find an expression for the number of square feet of wallpaper needed to cover all the walls. (Ignore the presence of doors and windows.)

94. The interior surface of the birdbath in Example 6c needs to be painted with a waterproof (and nontoxic) coating. Determine the expression for the interior surface area.

95. The prism shown below is a right triangular cylinder, where the base is a right triangle. Find the surface area of the prism in terms of b, h, and l.

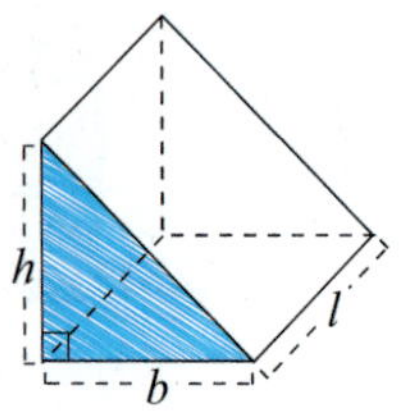

96. A jeweler decides to construct a pendant for a necklace by simply attaching equilateral triangles to each edge of a regular hexagon. The edge length of one of the points of the resulting star is $d = 0.8$ cm. Find the formula for the area of the star in terms of d and then evaluate for $d = 0.8$ cm (rounding to three decimal places). Use the fact that the area of an equilateral triangle of side length d is $A = \frac{d^2\sqrt{3}}{4}$.

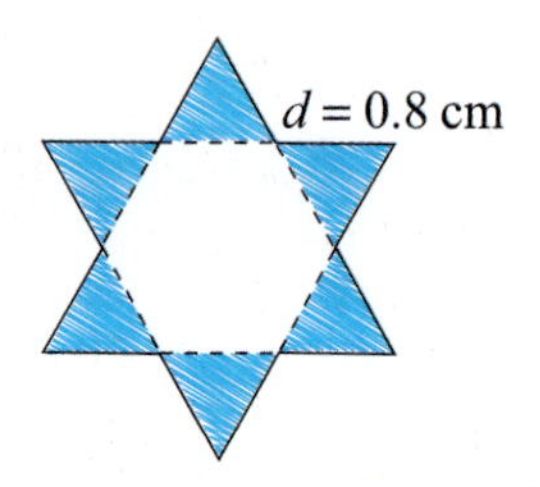

97. Ilyana has made a home for her pet guinea pig (Ralph) in the shape of a right triangular cylinder. Before she can put the new home in Ralph's cage, she must paint it with a nontoxic outer coat. If the front of the home has a base of 17.5 cm and a height of 15 cm and the length of the home is 25 cm, what is the surface area of Ralph's home, rounded to the nearest square centimeter? The small bottle of nontoxic coating will cover up to 1500 cm^2. Will the small bottle contain enough nontoxic coating to cover Ralph's home?

98. Einstein's Theory of Special Relativity tells us that $E = mc^2$, where E is energy (in joules, J), m is mass (in kilograms, kg), and c is the speed of light (in meters per second, m/s). This equation may also be written as $\sqrt{\frac{E}{m}} = c$. Assume you know $E = 418{,}400$ J and $m = 4.655 \times 10^{-12}$ kg. Use this information to estimate the speed of light.

WRITING & THINKING

99. Give a few examples of instances in which it would be more useful to use scientific notation rather than standard.

100. In February of 2006, the US national debt was approximately 8.2 trillion dollars. How is saying 8.2 trillion similar to scientific notation? How is it different?

101. In your own words, explain why $a^0 = 1$.

102. Explain, in your own words, why the square root of a negative number is not a real number.

103. Explain, in your own words, why exponents and roots are evaluated at the same time in the order of operations.

1.3 POLYNOMIALS AND FACTORING

TOPICS

1. The terminology of polynomial expressions
2. Basic operations with polynomials
3. Greatest common factor (GCF)
4. Factoring by grouping
5. Factoring special binomials
6. Factoring trinomials
7. Factoring expressions containing noninteger rational exponents

TOPIC 1: The Terminology of Polynomial Expressions

Polynomials through the Centuries

Diophantus of Alexandria, who lived in the 3rd century AD, was the author of a number of books that outlined methods for working with expressions that we now call polynomials. The French mathematician Pierre de Fermat was heavily influenced through his reading of some of Diophantus' works. In particular, in 1637 Fermat was inspired by a passage in Diophantus' book *Arithmetica* to state that the equation $a^n + b^n = c^n$ has no solutions in positive integers a, b, and c if n is any integer greater than or equal to 3 (a statement that came to be known as Fermat's Last Theorem).

Polynomials are a special class of algebraic expressions. Each term in a polynomial consists only of a number multiplied by variables (if it is multiplied by anything at all) raised to nonnegative integer exponents.

The number in any such term is called the **coefficient** of the term, and the sum of the exponents of the variables is the **degree of the term**. If a given term consists only of a nonzero number a (known as a **constant term**), its degree is defined to be 0, while the term 0 is not assigned any degree at all. The polynomial as a whole is also given a degree: the **degree of a simplified polynomial** is the largest of the degrees of the individual terms.

Every polynomial consists of some finite number of terms. Polynomials consisting of a single term are called **monomials**, those consisting of two terms are **binomials**, and those consisting of three terms are **trinomials**. The following example illustrates the use of the above terminology as applied to some specific polynomials.

Example 1: Polynomial Expressions

a. The expression $14x^6$ is a monomial in the variable x. The coefficient of the term is 14 and the degree of the term is 6, which is also the degree of the polynomial.

b. The polynomial $-3x^4y^2 + 5.4x^3y^4$ is a binomial in the two variables x and y. The degree of the first term is 6, and the degree of the second term is 7, so the degree of the polynomial as a whole is 7. The coefficient of the first term is –3 and the coefficient of the second term is 5.4.

c. The single number 5 can be considered a polynomial. In particular, it is a monomial of degree 0. The rationale for assigning degree 0 to nonzero constants such as 5 is that 5 can be thought of as $5x^0$ (or 5 times *any* variable raised to the 0 power). The coefficient of this monomial is itself: 5.

d. The polynomial $\frac{2}{3}x^3y^5 - z + y^{10} + 3$ has four terms and is a polynomial in three variables. If one of the terms of a polynomial consists only of a number, it is referred to as the constant term. The degree of this polynomial is 10, and the degrees of the individual terms are, from left to right, 8, 1, 10 and 0. The coefficients are $\frac{2}{3}$, -1, 1, and 3.

e. The expression $x^3 + 2x + 7x^{-1}$ is **not** a polynomial because it contains a term in which a variable is raised to a *negative* exponent.

f. The expression $\sqrt{x}$ is also **not** a polynomial, because it is equivalent to an expression with a variable raised to a *fractional* exponent, $x^{\frac{1}{2}}$.

The majority of the polynomials in this book are polynomials of a single variable and can be described generically as follows.

DEFINITION: Polynomials of a Single Variable

A polynomial in the variable x of degree n can be written in the form

$$a_nx^n + a_{n-1}x^{n-1} + \cdots + a_1x + a_0,$$

where $a_n, a_{n-1}, \ldots, a_1, a_0$ are real numbers, $a_n \neq 0$, and n is a nonnegative integer. This form is called **descending order**, because the powers descend from left to right. The **leading coefficient** of this polynomial is a_n.

TOPIC 2: Basic Operations with Polynomials

The variables in polynomials, as in all algebraic expressions, represent numbers, and consequently polynomials can be added, subtracted, multiplied and divided according to the field properties discussed in Section 1.1.

We will discuss addition (and thus subtraction) and multiplication of polynomials here, while division of polynomials will be covered in Chapter 5.

To add two or more polynomials, we use the field properties to combine **like terms** (also called **similar terms**). These are terms that have the same variables raised to the same powers. When subtracting one polynomial from another, we first distribute the minus sign over the terms of the polynomial being subtracted and then add.

Example 2: Adding and Subtracting Polynomials

Add or subtract the polynomials as indicated.

a. $(2x^3y - 3y + z^2x) + (3y + z^2 - 3xz^2 + 4)$ **b.** $(4ab^3 - b^3c) - (4 - b^3c)$

Solution

> **NOTE**
> The first step is always to identify and group like terms. The like terms are then combined using the distributive property.

a. $\left(2x^3y-3y+z^2x\right)+\left(3y+z^2-3xz^2+4\right)$

$=2x^3y+\left(-3y+3y\right)+\left(xz^2-3xz^2\right)+z^2+4$

$=2x^3y+y\left(-3+3\right)+xz^2\left(1-3\right)+z^2+4$

$=2x^3y-2xz^2+z^2+4$

Note that the terms z^2x and $-3xz^2$ are similar, since multiplication is commutative.

Combine like terms.

b. $\left(4ab^3-b^3c\right)-\left(4-b^3c\right)$

$=4ab^3-b^3c-4+b^3c$

$=4ab^3-4+\left(-b^3c+b^3c\right)$

$=4ab^3-4$

Begin by distributing the minus sign over all the terms in the second polynomial.

Combine like terms.

Polynomials are multiplied using the same properties. The distributive property is of particular importance in multiplying polynomials correctly. In general, the product of two polynomials, one with n terms and one with m terms, will consist initially of nm terms. If any of the resulting terms are similar, they are then combined in the last step.

Example 3: Multiplying Polynomials

Multiply the polynomials, as indicated.

a. $\left(2x^2y-z^3\right)\left(4+3z-3xy\right)$ **b.** $\left(3ab-a^2\right)\left(ab+a^2\right)$

Solution

> **NOTE**
> We use the distributive property first, and multiply each term of the first polynomial by each term of the second.

a. $\left(2x^2y-z^3\right)\left(4+3z-3xy\right)$

$=2x^2y\left(4+3z-3xy\right)-z^3\left(4+3z-3xy\right)$

$=8x^2y+6x^2yz-6x^3y^2-4z^3-3z^4+3xyz^3$

Multiply the second polynomial by each term of the first.

None of the resulting terms are similar, so the final answer is a polynomial of 6 terms.

b. $\left(3ab-a^2\right)\left(ab+a^2\right)$

$=3ab\left(ab+a^2\right)-a^2\left(ab+a^2\right)$

$=3a^2b^2+3a^3b-a^3b-a^4$

$=3a^2b^2+2a^3b-a^4$

Multiply the second polynomial by each term of the first.

We combine the two similar terms to obtain the final trinomial.

Example 4: The FOIL Method

When a binomial is multiplied by another binomial, as in Example 3b, the acronym FOIL is commonly used as a reminder of the four necessary products. Again consider the product $(3ab-a^2)(ab+a^2)$. Each letter represents a pair of terms to multiply.

First	$(3ab-a^2)(ab+a^2)$	$(3ab)(ab)=3a^2b^2$
Outer	$(3ab-a^2)(ab+a^2)$	$(3ab)(a^2)=3a^3b$
Inner	$(3ab-a^2)(ab+a^2)$	$(-a^2)(ab)=-a^3b$
Last	$(3ab-a^2)(ab+a^2)$	$(-a^2)(a^2)=-a^4$

Once again, we sum the resulting terms, combining like terms to find the same answer as before; $(3ab-a^2)(ab+a^2)=3a^2b^2+2a^3b-a^4$. While the FOIL acronym is a helpful tool, it is important to realize that it is nothing more than a special application of the distributive property.

Some products of binomial expressions are so common that their forms are worth remembering. In the three formulas that follow, A and B may represent simple polynomial terms but can also stand for more complicated expressions. Each of the formulas can be verified by applying the FOIL method to perform the indicated multiplication and then combining like terms.

f(x) FORMULA: Special Product Formulas

In the following equations, A and B represent algebraic expressions.

1. $(A-B)(A+B)=A^2-B^2$
2. $(A+B)^2=A^2+2AB+B^2$
3. $(A-B)^2=A^2-2AB+B^2$

Example 5: Using Special Product Formulas

Use a special product formula to perform the indicated operations.

a. $(2x-y)^2$

b. $(\sqrt{x}-\sqrt{y})(\sqrt{x}+\sqrt{y})$

Solution

a. The expression $(2x-y)^2$ is of the form $(A-B)^2$ with $A=2x$ and $B=y$, so we apply the third special product formula.

$$\begin{aligned}(2x-y)^2&=(2x)^2-2(2x)(y)+(y)^2\\&=4x^2-4xy+y^2\end{aligned}$$

b. Using the first special product formula, we have the following.

$$\left(\sqrt{x}-\sqrt{y}\right)\left(\sqrt{x}+\sqrt{y}\right)=\left(\sqrt{x}\right)^2-\left(\sqrt{y}\right)^2$$
$$=x-y$$

TOPIC 3: Greatest Common Factor (GCF)

Factoring, in general, means reversing the process of multiplication in order to find two or more expressions whose product is the original expression. For instance, the factored form of the polynomial x^2-3x+2 is $(x-2)(x-1)$, since if the product $(x-2)(x-1)$ is expanded we obtain x^2-3x+2 (as you should verify). Factoring is thus a logical topic to discuss immediately after multiplication, but the process of factoring is not as straightforward as multiplication.

Consequently, various methods have been devised to help. In some cases, more than one method may be used to factor a given algebraic expression, but all the methods will give the same result. In other cases, it may not be immediately obvious which method should be used, and the factoring process may require much trial and error.

We begin with some additional terminology. We say that a polynomial with integer coefficients is **factorable** if it can be written as a product of two or more polynomials, all of which also have integer coefficients. If this cannot be done, we say the polynomial is **irreducible** (over the integers), or **prime**. The goal of factoring a polynomial is to **completely factor** it: to write it as a product of prime polynomials. This means no factors can be factored any further.

Factoring out those factors common to all the terms in an expression is the simplest factoring method to apply and should be done first if possible. The **greatest common factor (GCF)** among all the terms is the product of all the factors common to each. For instance, $2x$ is a factor common to all the terms in the polynomial $12x^5-4x^2+8x^3z^3$, but $4x^2$ is the greatest common factor. This factoring method is a matter of applying the distributive property to "undistribute" the greatest common factor.

Example 6: Factoring Out the Greatest Common Factor

Factor each polynomial by factoring out the greatest common factor.

a. $12x^5-4x^2+8x^3z^3$

b. $-24ax^2+60a$

c. $\left(a^2-b\right)-3\left(a^2-b\right)$

d. $\left(x^2+y\right)^3+3\left(x^2+y\right)^2$

Solution

a. $12x^5 - 4x^2 + 8x^3z^3$
$= (4x^2)(3x^3) + (4x^2)(-1) + (4x^2)(2xz^3)$
$= (4x^2)(3x^3 - 1 + 2xz^3)$

As we noted above, $4x^2$ is the greatest common factor. Applying the distributive property in reverse leads to the factored form of this degree 6 trinomial.

b. $-24ax^2 + 60a$
$= (-12a)(2x^2) + (-12a)(-5)$
$= -12a(2x^2 - 5)$

An alternative form of the final answer is $12a(-2x^2 + 5)$. We would have obtained this answer if we had factored out $12a$ initially. These two answers are equivalent.

c. $(a^2 - b) - 3(a^2 - b)$
$= (a^2 - b)(1) + (a^2 - b)(-3)$
$= (a^2 - b)(1 - 3)$
$= -2(a^2 - b)$

In factoring out the greatest common factor $a^2 - b$, remember that it is being multiplied first by 1 and then by -3. One common source of error in factoring is to forget factors of 1.

d. $(x^2 + y)^3 + 3(x^2 + y)^2$
$= (x^2 + y)^2(x^2 + y) + 3(x^2 + y)^2$
$= (x^2 + y)^2(x^2 + y + 3)$

The greatest common factor is $(x^2 + y)^2$.

Now that a few examples have been presented, one more word on factoring is in order. As mentioned above, the goal in factoring a polynomial is to write it as a product of prime polynomials, and we continue the factoring process until *all* the factors are prime. The exception to this is that we generally do not factor monomials further. For instance, in Example 6a we do not write the factor $4x^2$ as $(2)(2)(x)(x)$.

TOPIC 4: Factoring by Grouping

Many polynomials have a GCF of 1, and the first factoring method is therefore not directly applicable. But if the terms of the polynomial are grouped in a suitable way, the GCF method may apply to each group, and a common factor might subsequently be found among the groups. **Factoring by grouping** is the name given to this process, and it is important to realize that this is a trial and error process. Your first attempt at grouping and factoring may not succeed, and you may have to try several different ways of grouping the terms.

Example 7: Factoring by Grouping

Factor each polynomial by grouping.

a. $6x^2 - y + 2x - 3xy$ **b.** $ax - ay - bx + by$ **c.** $4x - 2x^2 - 2x^3 + x^4$

Solution

a.
$$\begin{aligned} &6x^2 - y + 2x - 3xy \\ &= (6x^2 + 2x) + (-y - 3xy) \\ &= 2x(3x+1) + y(-1-3x) \\ &= 2x(3x+1) - y(3x+1) \\ &= (3x+1)(2x-y) \end{aligned}$$

The GCF of the four terms in the polynomial is 1, so the Greatest Common Factor method doesn't directly apply. The first and third terms have a GCF of $2x$, while the second and fourth have a GCF of y, so we group accordingly. After factoring the two groups, we notice that $3x + 1$ and $-1 - 3x$ differ only by a minus sign (and the order). This means $3x + 1$ can be factored out.

b.
$$\begin{aligned} &ax - ay - bx + by \\ &= a(x-y) - b(x-y) \\ &= (x-y)(a-b) \end{aligned}$$

The first two terms have a common factor, as do the last two, so we proceed accordingly. In this problem, we could also have grouped the first and third terms, and the second and fourth terms, and obtained the same result.

c.
$$\begin{aligned} &4x - 2x^2 - 2x^3 + x^4 \\ &= x(4 - 2x - 2x^2 + x^3) \\ &= x[2(2-x) - x^2(2-x)] \\ &= x[(2-x)(2-x^2)] \\ &= x(2-x)(2-x^2) \end{aligned}$$

We can begin by factoring out x from each of the four terms. The first two terms of the result then have a common factor of 2, and the second two terms have a common factor of $-x^2$. This allows us to factor out $(2-x)$ from the two groups.

CAUTION

One common error in factoring is to stop after groups within the original polynomial have been factored. For instance, while we have done some factoring to achieve the expression $2x(3x+1) + y(-1-3x)$ in Example 7a, this is *not* in factored form. An expression is only factored if it is written as a *product* of two or more factors. The expression $2x(3x+1) + y(-1-3x)$ is a sum of two smaller expressions.

TOPIC 5: Factoring Special Binomials

Three types of binomials can always be factored by following the patterns outlined below. You should verify these patterns by multiplying out the products on the right-hand side of each one.

$f(x)$ FORMULA: Factoring Special Binomials

In the following equations, A and B represent algebraic expressions.

Difference of Two Squares: $A^2 - B^2 = (A-B)(A+B)$

Difference of Two Cubes: $A^3 - B^3 = (A-B)(A^2 + AB + B^2)$

Sum of Two Cubes: $A^3 + B^3 = (A+B)(A^2 - AB + B^2)$

Note that there is no similar pattern for factoring $A^2 + B^2$.

Example 8: Factoring Special Binomials

Use the special factoring patterns to factor the following binomials.

a. $49x^2 - 9y^6$

b. $27a^6b^{12} + c^3$

c. $125y^3 - 8z^3$

d. $64 - (x+y)^3$

Solution

a. $49x^2 - 9y^6$

$$= (7x)^2 - (3y^3)^2$$
$$= (7x - 3y^3)(7x + 3y^3)$$

A difference of two squares
$A = 7x,\ B = 3y^3$
$A^2 - B^2$
$= (A-B)(A+B)$

b. $27a^6b^{12} + c^3$

$$= (3a^2b^4)^3 + (c)^3$$
$$= \Big(\underbrace{3a^2b^4}_{A} + \underbrace{c}_{B}\Big)\Big(\underbrace{(3a^2b^4)^2}_{A^2} - \underbrace{(3a^2b^4)(c)}_{AB} + \underbrace{(c)^2}_{B^2}\Big)$$
$$= (3a^2b^4 + c)(9a^4b^8 - 3a^2b^4c + c^2)$$

A sum of two cubes
$A = 3a^2b^4,\ B = c$
$A^3 + B^3$
$= (A+B)(A^2 - AB + B^2).$

c. $125y^3 - 8z^3$

$$= (5y)^3 - (2z)^3$$
$$= (5y - 2z)\big((5y)^2 + (5y)(2z) + (2z)^2\big)$$
$$= (5y - 2z)(25y^2 + 10yz + 4z^2)$$

A difference of two cubes
$A = 5y,\ B = 2z$
$A^3 - B^3$
$= (A-B)(A^2 + AB + B^2)$

d. $64 - (x+y)^3$

$$= 4^3 - (x+y)^3$$
$$= \big(4 - (x+y)\big)\big(4^2 + 4(x+y) + (x+y)^2\big)$$
$$= (4 - x - y)(16 + 4x + 4y + x^2 + 2xy + y^2)$$

In this difference of two cubes, the second cube is itself a binomial. But the factoring pattern still applies, leading to the final factored form of the original binomial.

TOPIC 6: Factoring Trinomials

In factoring a trinomial of the form $ax^2 + bx + c$, the goal is to find two binomials $px + q$ and $rx + s$ such that

$$ax^2 + bx + c = (px + q)(rx + s).$$

Since $(px+q)(rx+s) = prx^2 + (ps+qr)x + qs$ we seek p, q, r, and s such that $a = pr$, $b = ps + qr$, and $c = qs$.

$$ax^2 + bx + c = \underbrace{pr}_{a}x^2 + \underbrace{(ps+qr)}_{b}x + \underbrace{qs}_{c}$$

In general, this may require trial and error, but the following guidelines will help.

Case 1: Leading Coefficient is 1.

In this case, p and r must both be 1, so we only need q and s such that $x^2 + bx + c = x^2 + (q+s)x + qs$. That is, we need two integers whose sum is b, the coefficient of x, and whose product is c, the constant term.

Example 9: Factoring a Trinomial

To factor $x^2 + x - 12$ we can begin by writing $x^2 + x - 12 = (x + \boxed{?})(x + \boxed{?})$ and then try to find two integers to replace the question marks. The two integers we seek must have a product of -12, and the fact that the product is negative means that one integer must be positive and one negative. The only possibilities are $\{1,-12\}$, $\{-1,12\}$, $\{2,-6\}$, $\{-2,6\}$, $\{3,-4\}$, and $\{-3,4\}$, and when we add the requirement that the sum must be 1, we are left with $\{-3,4\}$. Thus $x^2 + x - 12 = (x-3)(x+4)$.

Case 2: Leading Coefficient is not 1.

In this case, trial and error may still be an effective way to factor the trinomial $ax^2 + bx + c$, especially if a, b, and c are relatively small in magnitude. If, however, trial and error seems to be taking too long, the following steps use factoring by grouping to minimize the amount of guessing required.

PROCEDURE: Factoring a Trinomial by Grouping

To factor the trinomial $ax^2 + bx + c$, perform the following steps.

Step 1: Multiply a and c.

Step 2: Factor ac into two integers whose sum is b. If no such factors exist, the trinomial is irreducible over the integers.

Step 3: Rewrite b in the trinomial with the sum found in step 2, and distribute. The resulting polynomial of four terms may now be factored by grouping.

Example 10: Factoring a Trinomial by Grouping

To factor the trinomial $6x^2 - x - 12$ by trial and error, we would begin by noting that if it can be factored, the factors must be of the form $(x+\boxed{?})(6x+\boxed{?})$ or $(2x+\boxed{?})(3x+\boxed{?})$. If we use the grouping method, we form the product $(6)(-12) = -72$ and then factor -72 into two integers whose sum is -1. The two numbers -9 and 8 work, so we write $6x^2 - x - 12 = 6x^2 + (-9+8)x - 12 = 6x^2 - 9x + 8x - 12$. Now proceed by grouping.

$$\begin{aligned} 6x^2 - 9x + 8x - 12 &= 3x(2x-3) + 4(2x-3) \\ &= (2x-3)(3x+4) \end{aligned}$$

Some trinomial expressions are known as "perfect square trinomials" because their factored form is the square of a binomial expression. For example, $x^2 - 6x + 9 = (x-3)^2$. (Either the trial and error method or factoring by grouping can give us this answer.) In general, such trinomials will have one of the following two forms.

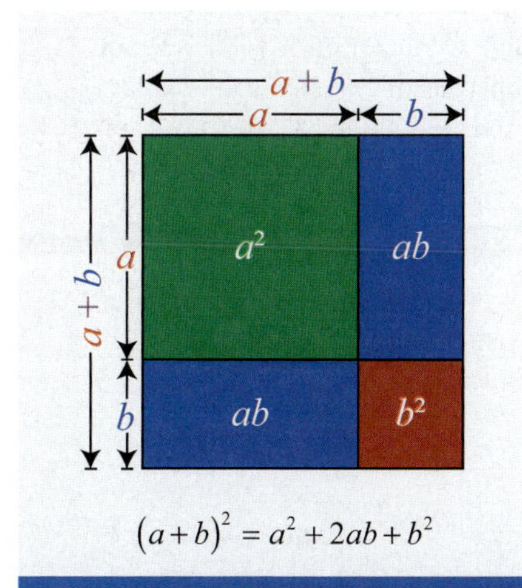

Perfect Square Trinomials

f(x) FORMULA: Perfect Square Trinomials

In the following, A and B represent algebraic expressions.

$$A^2 + 2AB + B^2 = (A+B)^2$$
$$A^2 - 2AB + B^2 = (A-B)^2$$

Example 11: Perfect Square Trinomials

Factor the algebraic expression $x^2 + 10x + 25$.

Solution

The expression appears to be in the form of a perfect square trinomial, but we need to check that the value of the middle term follows the above pattern. Taking $A = x$ and $B = 5$, we see that $2AB = 10x$, so the expression does match the perfect square trinomial form.

Thus the factored form of $x^2 + 10x + 25 = (x+5)^2$, and $x^2 + 10x + 25$ is a perfect square trinomial.

TOPIC 7: Factoring Expressions Containing Noninteger Rational Exponents

This last method does not apply to polynomials, as polynomials cannot have noninteger rational exponents. It will, however, be very useful in solving problems later in this book and in other math classes. The method applies to negative fractional exponents as well as positive.

To factor an algebraic expression that has fractional exponents, identify the smallest exponent among the terms, then factor out the variable raised to that smallest exponent from each of the terms. Factor out any other common factors and simplify if possible.

Example 12: Factoring Expressions with Noninteger Rational Exponents

Factor each of the following algebraic expressions.

a. $3x^{-\frac{2}{3}} - 6x^{\frac{1}{3}} + 3x^{\frac{4}{3}}$

b. $(x-1)^{\frac{1}{2}} - (x-1)^{-\frac{1}{2}}$

Solution

a. $3x^{-\frac{2}{3}} - 6x^{\frac{1}{3}} + 3x^{\frac{4}{3}}$

$= 3x^{-\frac{2}{3}}(1-2x+x^2)$

$= 3x^{-\frac{2}{3}}(x^2-2x+1)$

$= 3x^{-\frac{2}{3}}(x-1)(x-1)$

$= 3x^{-\frac{2}{3}}(x-1)^2$

Under the guidelines above, we factor out $3x^{-\frac{2}{3}}$. Note that we use the properties of exponents to obtain the terms in the second factor.

We notice that the second factor is a second-degree trinomial, which is itself factorable. In fact, it is an example of a perfect square trinomial.

b. $(x-1)^{\frac{1}{2}} - (x-1)^{-\frac{1}{2}}$

$= (x-1)^{-\frac{1}{2}}[(x-1)-1]$

$= (x-1)^{-\frac{1}{2}}(x-2)$

In this example we factor out $(x-1)^{-\frac{1}{2}}$ again using the properties of exponents to obtain the terms in the second factor.

1.3 EXERCISES

PRACTICE

Classify each of the following expressions as either a polynomial or not a polynomial. For those that are polynomials, identify the degree of the polynomial and the number of terms (use the words monomial, binomial, and trinomial if applicable). See Example 1.

1. $3x^{\frac{3}{2}} - 2x$

2. $17x^2y^5 + 2z^3 - 4$

3. $5x^{10} + 3x^3 - 2y^3z^8 + 9$

4. πx^3

5. 8

6. 0

7. $7^3xy^2 + 4y^4$

8. abc^2d^3

9. $4x^2 + 7xy + 5y^2$

10. $3n^4m^{-3} + n^2m$

11. $\frac{y^2z}{4} + 2yz^4$

12. $6x^4y + 3x^2y^2 + xy^5$

Write each of the following polynomials in descending order, and identify **a.** the degree of the polynomial, and **b.** the leading coefficient.

13. $-4x^{10} - x^{13} + 9 + 7x^{11}$

14. $9x^8 - 9x^{10}$

15. $4s^3 - 10s^5 + 2s^6$

16. $4-2x^5+x^2$ **17.** $9y^6-2+y-3y^5$ **18.** $4n+6n^2-3$

19. $8z^2+\pi z^5-2z+1$ **20.** $-6y^5-3y^7+12y^6$

Add or subtract the polynomials, as indicated. See Example 2.

21. $(-4x^3y+2xz-3y)-(2xz+3y+x^2z)$ **22.** $(4x^3-9x^2+1)+(-2x^3-8)$

23. $(x^2y-xy-6y)+(xy^2+xy+6x)$ **24.** $(5x^2-6x+2)-(4-6x-3x^2)$

25. $(a^2b+2ab+ab^2)-(ab^2+5ab+a^2b)$ **26.** $(x^4+2x^3-x+5)-(x^3-x-x^4)$

27. $(xy-4y+xy^2)+(3y-x^2y-xy)$ **28.** $(-8x^4+13-9x^2)-(8-2x^4)$

Multiply the polynomials, as indicated. See Examples 3 and 4.

29. $(3a^2b+2a-3b)(ab^2+7ab)$ **30.** $(x^2-2y)(x^2+y)$

31. $(3a+4b)(a-2b)$ **32.** $(x+xy+y)(x-y)$

33. $(6x-3y)(x+6y)$ **34.** $(5y+x)(4y-2x)$

35. $(7y^2+x)(y^2-5x)$ **36.** $(y^2+x)(3y^2-7x)$

37. $(6xy^2-3x+4y)(x^2y+6xy)$ **38.** $(2xy^2+4y-6x)(x^2y-5xy)$

Use a special product formula to perform the indicated operations. See Example 5.

39. $(3a+b)^2$ **40.** $(x-5y)^2$

41. $(2x-3y)(2x+3y)$ **42.** $(x-3y)^2$

43. $(-x-2y)^2$ **44.** $(\sqrt{2x}-\sqrt{3y})(\sqrt{2x}+\sqrt{3y})$

45. $\left(\frac{1}{x}-y\right)\left(\frac{1}{x}+y\right)$ **46.** $[(x-y)-z][(x-y)+z]$

Factor each polynomial by factoring out the greatest common factor. See Example 6.

47. $4m^2n+16m^3+7m$ **48.** $3a^2b+3a^3b-9a^2b^2$

49. $5(a-b^2)+(a-b^2)$ **50.** $3x^3y-9x^4y+12x^3y^2$

51. $2x^6-14x^3+8x$ **52.** $27x^7y+9x^6y-9x^4yz$

53. $(x^3-y)^2-(x^3-y)$ **54.** $6xy^3+9y^3-12xy^4$

55. $12y^6-8y^2-16y^5$ **56.** $(2x+y^2)^4-(2x+y^2)^6$

Factor each polynomial by grouping. See Example 7.

57. $a^3+ab-a^2b-b^2$ **58.** $ax-2bx-2ay+4by$

59. $z+z^2+z^3+z^4$ **60.** $x^2+3xy+3y+x$

61. $nx^2-2y-2x^2+ny$ **62.** $2ac-3bd+bc-6ad$

63. $ax-5bx+5ay-25by$ **64.** $3ac-5bd+bc-15ad$

Use the special factoring patterns to factor the following binomials. See Example 8.

65. $4x^2 - 121$

66. $64z^3 + 216$

67. $49a^2 - 144b^2$

68. $x^3 - 27y^3$

69. $25x^4y^2 - 9$

70. $27a^9 + 8b^{12}$

71. $x^3 - 1000y^3$

72. $64x^6 - 125y^3z^9$

73. $m^6 + 125n^9$

74. $49a^6 - 9b^2c^4$

75. $27x^6 - 8y^{12}z^3$

76. $(3x-6)^2 - (y-2x)^2$

77. $16z^2y^4 - 9x^8$

78. $512x^6 + 729y^3$

79. $343y^9 - 27x^3z^6$

80. $(2x+y^2)^2 - (y^2-3)^2$

Factor the following trinomials. See Examples 9, 10, and 11.

81. $x^2 + 2x - 15$

82. $x^2 + 6x + 9$

83. $x^2 - 2x + 1$

84. $x^2 - 5x + 6$

85. $x^2 - 4x + 4$

86. $x^2 + 5x + 4$

87. $y^2 + 14y + 49$

88. $x^2 - 3x - 18$

89. $x^2 + 13x + 22$

90. $y^2 + y - 42$

91. $y^2 - 9y + 8$

92. $6x^2 + 5x - 6$

93. $5a^2 - 37a - 24$

94. $25y^2 + 10y + 1$

95. $5x^2 + 27x - 18$

96. $6y^2 - 13y - 8$

97. $16y^2 - 25y + 9$

98. $10m^2 + 29m + 10$

99. $8a^2 - 2a - 3$

100. $20y^2 + 21y - 5$

101. $12y^2 - 19y + 5$

102. $10y^2 - 11y - 6$

Factor the following algebraic expressions. See Example 12.

103. $(2x-1)^{-\frac{3}{2}} + (2x-1)^{-\frac{1}{2}}$

104. $2x^{-2} + 3x^{-1}$

105. $7a^{-1} - 2a^{-3}b$

106. $(3z+2)^{\frac{5}{3}} - (3z+2)^{\frac{2}{3}}$

107. $10y^{-2} - 2y^{-5}x$

108. $4y^{-3} + 12y^{-4}$

109. $(5x+7)^{\frac{7}{3}} - (5x+7)^{\frac{4}{3}}$

110. $(8x+6)^{-\frac{7}{2}} - (8x+6)^{-\frac{1}{2}}$

111. $7y^{-1} + 5y^{-4}$

112. $5x^{-4} - 4x^{-5}y$

WRITING & THINKING

113. Pneumothorax is a disease in which air or gas collects between the lung and the chest wall, causing the lung to collapse. When this disease is evident, the following formula is used to determine the degree of collapse of the lungs, represented as a percent:

$$\text{Degree} = 100\left(1 - \frac{L^3}{H^3}\right)$$

In this formula, L is the diameter of one lung and H is the diameter of one hemithorax (or half the chest cavity). Is this formula a polynomial? If so, find its degree and the number of terms. If not, explain.

114. You are trying to find a formula for the area of a certain trapezoid. You know the height of the trapezoid is x^2 the bottom base is $2x^2+4$, and the top base is $6x+2$. Insert these values into the formula for the area of a trapezoid. Is the result a polynomial? If so, find the degree of the polynomial, the leading oefficient, and the number of terms in the polynomial. If not, explain.

115. **a.** Given a rectangular picture frame with sides of $2x+1$ and x^3+4, find the area of the picture frame. Is the result a polynomial? If so, find the degree of the polynomial, the leading coefficient, and the number of terms in the polynomial. If not, explain.

b. Now find the perimeter of the picture frame. Is this a polynomial? If so, find the degree of the polynomial, the leading coefficient, and the number of terms in the polynomial. If not, explain.

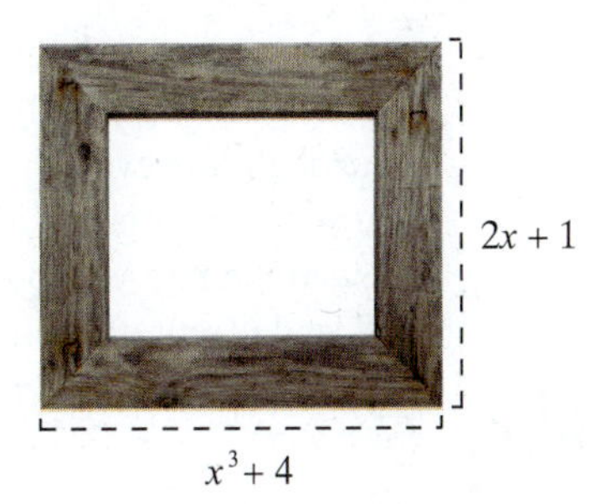

1.4 RATIONAL EXPRESSIONS

TOPICS

1. Simplifying rational expressions
2. Combining rational expressions
3. Simplifying complex rational expressions

TOPIC 1: Simplifying Rational Expressions

We will often encounter fractions in which the variable appears in the denominator. Some additional skills and experience in working with such fractions allows us to deal with them when, for instance, they appear in equations. In this section, we will learn how to work with a class of fractions called *rational expressions*.

DEFINITION: Rational Expressions

A **rational expression** is an expression that can be written as a *ratio* of two polynomials $\frac{P}{Q}$. Such a fraction is undefined for any value(s) of the variable(s) for which $Q = 0$. A given rational expression is *simplified* or *reduced* when P and Q contain no common factors (other than 1 and -1).

To simplify rational expressions, we factor the polynomials in the numerator and denominator completely and then cancel any common factors. It is important to remember, however, that the simplified rational expression may be defined for values of the variable (or variables) that the original (unsimplified) expression is not, and the two versions are equal only where they are both defined. That is, if A, B, and C represent algebraic expressions,

$$\frac{AC}{BC} = \frac{A}{B} \text{ only where } B \neq 0 \text{ and } C \neq 0.$$

TECHNOLOGY

X	Y1
-3	-2.333
-2	-2
-1	-3
0	ERROR
1	7
2	ERROR
3	6.3333
4	7
5	7.8
6	8.6667
7	9.5714

NORMAL FLOAT AUTO REAL RADIAN MP
PRESS [ENTER] TO EDIT

Y1=(X³-8)/(X²-2X)

Notice that x-coordinates of 0 and 2 result in an error. To arrive at this screen using a TI-84 Plus, press Y= and enter the equation. Next, press 2nd and graph to view the table.

Example 1: Simplifying Rational Expressions

Simplify the following rational expressions, and indicate values of the variable that must be excluded.

a. $\frac{x^3 - 8}{x^2 - 2x}$

b. $\frac{x^2 - x - 6}{3 - x}$

Solution

a. $\frac{x^3 - 8}{x^2 - 2x} = \frac{\cancel{(x-2)}\left(x^2 + 2x + 4\right)}{x\cancel{(x-2)}}$

$= \frac{x^2 + 2x + 4}{x}, \quad x \neq 0, 2$

After factoring both polynomials, we cancel the common factor of $x - 2$. Note that even though the final expression is defined when $x = 2$, the first and last expressions are equal only where both are defined.

b. $\dfrac{x^2-x-6}{3-x}=\dfrac{(x+2)\cancel{(x-3)}}{-\cancel{(x-3)}}$

$=\dfrac{x+2}{-1}$

$=-x-2, \quad x \neq 3$

The denominator is already factored, but we bring out a factor of -1 from the denominator in order to cancel a common factor of $x-3$. Note that the original and simplified versions are only equal for values of x not equal to 3.

CAUTION

Remember that only common *factors* can be canceled! A very common error is to think that common terms from the numerator and denominator can be canceled. For instance, the statement $\dfrac{\cancel{x}+4}{x^{\cancel{2}}}=\dfrac{4}{x}$ is **incorrect**. The expression $x+4$ is completely factored, so the x that appears in the numerator is not a factor that can be canceled with one of the x's in the denominator. The expression $\dfrac{x+4}{x^2}$ is already simplified as far as possible.

TOPIC 2: Combining Rational Expressions

Rational expressions are combined by the operations of addition, subtraction, multiplication, and division the same way that numerical fractions are. That is, in order to add or subtract two rational expressions, a common denominator must first be found. In order to multiply two rational expressions, the two numerators are multiplied and the two denominators are multiplied. Finally, in order to divide one rational expression by another, the first is multiplied by the reciprocal of the second. Remember to check to see if the resulting rational expression can be simplified in each case. Remember also that the final combined rational expression is only equal to the original given expression where both are defined; any values of the variable(s) that make the to-be-combined rational expressions *or* the final combined rational expression undefined must be excluded.

No matter which operation is being considered, it is generally best to factor all the numerators and denominators before combining rational expressions. This is the first step in finding the Least Common Denominator (LCD) of two rational expressions, and a common denominator is essential before adding or subtracting. Factoring first is also an efficient way to identify any common factors that can be canceled if the operation is multiplication or division.

Example 2: Combining Rational Expressions

Add or subtract the rational expressions, as indicated.

a. $\dfrac{2x-1}{x^2+x-2}-\dfrac{2x}{x^2-4}$

b. $\dfrac{x+1}{x+3}+\dfrac{x^2+x-2}{x^2-x-6}-\dfrac{x^2-2x+9}{x^2-9}$

Solution

a.
$$\frac{2x-1}{x^2+x-2}-\frac{2x}{x^2-4}$$
$$=\frac{2x-1}{(x+2)(x-1)}-\frac{2x}{(x+2)(x-2)}$$
$$=\frac{x-2}{x-2}\cdot\frac{2x-1}{(x+2)(x-1)}-\frac{x-1}{x-1}\cdot\frac{2x}{(x+2)(x-2)}$$
$$=\frac{2x^2-5x+2}{(x-2)(x+2)(x-1)}-\frac{2x^2-2x}{(x-1)(x+2)(x-2)}$$
$$=\frac{-3x+2}{(x-2)(x+2)(x-1)}$$

We begin by factoring both denominators, and note that the LCD is $(x-2)(x+2)(x-1)$. In the first fraction, we multiply the top and bottom by $x-2$. In the second fraction, we multiply the top and bottom by $x-1$.

After subtracting the second numerator from the first, we are done. Note that there are no common factors to cancel.

b.
$$\frac{x+1}{x+3}+\frac{x^2+x-2}{x^2-x-6}-\frac{x^2-2x+9}{x^2-9}$$
$$=\frac{x+1}{x+3}+\frac{\cancel{(x+2)}(x-1)}{(x-3)\cancel{(x+2)}}-\frac{x^2-2x+9}{(x-3)(x+3)}$$
$$=\frac{x-3}{x-3}\cdot\frac{x+1}{x+3}+\frac{x+3}{x+3}\cdot\frac{x-1}{x-3}-\frac{x^2-2x+9}{(x-3)(x+3)}$$
$$=\frac{x^2-2x-3+x^2+2x-3-x^2+2x-9}{(x-3)(x+3)}$$
$$=\frac{x^2+2x-15}{(x-3)(x+3)}$$
$$=\frac{(x+5)\cancel{(x-3)}}{\cancel{(x-3)}(x+3)}$$
$$=\frac{x+5}{x+3}$$

We again factor all the polynomials, and note that the second rational expression can be reduced. We do this before determining that the LCD is $(x-3)(x+3)$.

After multiplying each numerator and denominator by the required factors in order to obtain a common denominator, we combine the numerators and simplify.

After factoring the resulting numerator, there is a common factor that can be canceled.

Example 3: Combining Rational Expressions

Multiply or divide the rational expressions, as indicated.

a. $\frac{x^2+3x-10}{x+3}\cdot\frac{x-3}{x^2-x-2}$ **b.** $\frac{x^2+5x-14}{3x}\div\frac{x^2-4x+4}{9x^3}$

Solution

a.
$$\frac{x^2+3x-10}{x+3}\cdot\frac{x-3}{x^2-x-2}$$
$$=\frac{(x+5)(x-2)}{x+3}\cdot\frac{x-3}{(x-2)(x+1)}$$
$$=\frac{(x+5)\cancel{(x-2)}(x-3)}{(x+3)\cancel{(x-2)}(x+1)}$$
$$=\frac{(x+5)(x-3)}{(x+3)(x+1)}$$

We begin by factoring both numerators and denominators, and then write the product of the two rational expressions as a single fraction.

Since we have already factored the polynomials completely, any common factors that can be canceled are easily identified.

b. $\dfrac{x^2+5x-14}{3x} \div \dfrac{x^2-4x+4}{9x^3}$

$= \dfrac{(x+7)(x-2)}{3x} \cdot \dfrac{9x^3}{(x-2)^2}$

$= \dfrac{9x^3(x+7)(x-2)}{3x(x-2)^2}$

$= \dfrac{3x^2(x+7)}{x-2}$

We divide the first rational expression by the second by inverting the second fraction and multiplying. Note that we have factored all the polynomials and inverted the second fraction in one step.

Now we proceed to cancel common factors (including constant factors) to obtain the final answer.

TOPIC 3: Simplifying Complex Rational Expressions

A **complex rational expression** is a fraction in which the numerator or denominator (or both) contains at least one rational expression. Complex rational expressions can always be rewritten as simple rational expressions. One way to do this is to simplify the numerator and denominator individually and then divide the numerator by the denominator as in Example 3b. Another way, which is frequently faster, is to multiply the numerator and denominator by the LCD of all the fractions that make up the complex rational expression. This method will be illustrated in the next example.

Example 4: Simplifying Complex Rational Expressions

Simplify the complex rational expressions.

a. $\dfrac{\frac{1}{x+h}-\frac{1}{x}}{h}$

b. $\dfrac{x^{-1}-y^{-1}}{x^{-2}-y^{-2}}$

Solution

a. $\dfrac{\frac{1}{x+h}-\frac{1}{x}}{h} = \dfrac{\frac{1}{x+h}-\frac{1}{x}}{\frac{h}{1}} \cdot \dfrac{(x+h)(x)}{(x+h)(x)}$

$= \dfrac{x-(x+h)}{(h)(x+h)(x)}$

$= \dfrac{-h}{(h)(x+h)(x)}$

$= \dfrac{-1}{x(x+h)}$

This complex rational expression contains two rational expressions in the numerator. It may be helpful to write the denominator as a fraction, as we have done here, in order to determine that the LCD of all the fractions making up the overall expression is $(x+h)(x)$. We multiply the numerator and denominator by the LCD (so we are multiplying the overall expression by 1), then cancel the common factor of h to get the final answer.

b. $$\frac{x^{-1}-y^{-1}}{x^{-2}-y^{-2}}=\frac{\frac{1}{x}-\frac{1}{y}}{\frac{1}{x^2}-\frac{1}{y^2}}$$

This expression is a complex rational expression, a fact that is more clear once we rewrite the terms that have negative exponents as fractions.

$$=\frac{\frac{1}{x}-\frac{1}{y}}{\frac{1}{x^2}-\frac{1}{y^2}}\cdot\frac{x^2y^2}{x^2y^2}$$

The LCD in this case is x^2y^2, so we multiply top and bottom by this and factor the resulting polynomials.

$$=\frac{xy^2-x^2y}{y^2-x^2}$$

We cancel the common factor of $(y-x)$ to obtain the final simplified expression.

$$=\frac{xy\cancel{(y-x)}}{\cancel{(y-x)}(y+x)}$$

$$=\frac{xy}{y+x}$$

1.4 EXERCISES

PRACTICE

Simplify the following rational expressions, indicating which real values of the variable must be excluded. See Example 1.

1. $\frac{2x^2+7x+3}{x^2-2x-15}$ **2.** $\frac{x^2+5x-6}{x^3+2x^2-3x}$ **3.** $\frac{x^3+2x^2-3x}{x+3}$ **4.** $\frac{x^2-4x+4}{x^2-4}$

5. $\frac{x^2+5x-6}{x^2+4x-5}$ **6.** $\frac{2x^2+7x-15}{x^2+3x-10}$ **7.** $\frac{x+1}{x^3+1}$ **8.** $\frac{x^3+x}{3x^2+3}$

9. $\frac{2x^2+11x+5}{x+5}$ **10.** $\frac{x^4-x^3}{x^2-3x+2}$ **11.** $\frac{2x^2+11x-21}{x+7}$ **12.** $\frac{8x^3-27}{2x-3}$

Add or subtract the rational expressions, as indicated, and simplify your answer. See Example 2.

13. $\frac{x-3}{x+5}+\frac{x^2+3x+2}{x-3}$

14. $\frac{x^2-1}{x-2}-\frac{x-1}{x+1}$

15. $\frac{x+2}{x-3}-\frac{x-3}{x+5}-\frac{1}{x^2+2x-15}$

16. $\frac{x+1}{x-3}+\frac{x^2+3x+2}{x^2-x-6}-\frac{x^2-2x-3}{x^2-6x+9}$

17. $\frac{x^2+1}{x-3}+\frac{x-5}{x+3}$

18. $\frac{x-37}{(x+3)(x-7)}+\frac{3x+6}{(x-7)(x+2)}-\frac{3}{x+3}$

19. $\frac{x^2+2x-35}{x-5}+\frac{x-4}{x+3}$

20. $\frac{y+2}{y-2}+\frac{y-6}{y+4}+\frac{4}{y^2+2y-8}$

21. $\dfrac{x+2}{x-6}+\dfrac{x^2+5x+6}{x^2-3x-18}-\dfrac{x^2-4x-12}{x^2-12x+36}$

22. $\dfrac{y^2+2}{y+3}-\dfrac{y-4}{y-3}$

Multiply or divide the rational expressions, as indicated, and simplify your answer. See Example 3.

23. $\dfrac{y-2}{y+1}\cdot\dfrac{y^2-1}{y-2}$

24. $\dfrac{a^2-3a-4}{a-2}\div\dfrac{a^2-2a-8}{a-2}$

25. $\dfrac{2x^2-5x-12}{x-3}\cdot\dfrac{x^2-x-6}{x-4}$

26. $\dfrac{z^2+2z+1}{2z^2+3z+1}\cdot\dfrac{2z^2-5z-3}{z+1}$

27. $\dfrac{y^2-11y+24}{y+6}\div\dfrac{y^2+5y-24}{y+6}$

28. $\dfrac{y^2+8y+16}{5y^2+22y+8}\cdot\dfrac{5y^2-13y-6}{y+4}$

29. $\dfrac{5y^2-27y-18}{y-5}\cdot\dfrac{y^2-6y+5}{y-6}$

30. $\dfrac{4z^2+20z-56}{z^2-8z+12}\div\dfrac{5z^2+43z+56}{15z^2-66z-144}$

31. $\dfrac{3b^2+9b-84}{b^2-5b+4}\div\dfrac{5b^2+37b+14}{-10b^2+6b+4}$

32. $\dfrac{3x^2-x-10}{x-1}\cdot\dfrac{x^2-1}{6x^2+x-15}\div\dfrac{x^2-x-2}{2x^2+5x-12}$

Simplify the complex rational expressions. See Example 4.

33. $\dfrac{\frac{3}{x}+\frac{x}{3}}{2-\frac{1}{x}}$

34. $\dfrac{\frac{1}{x}-\frac{1}{y}}{\frac{1}{x}+\frac{1}{y}}$

35. $\dfrac{6x-6}{3-\frac{3}{x^2}}$

36. $\dfrac{x^{-2}-y^{-2}}{y-x}$

37. $\dfrac{\frac{1}{r}-\frac{1}{s}}{r+\frac{1}{r}}$

38. $\dfrac{\frac{1}{x^2}-\frac{1}{y^2}}{\frac{1}{y^3}-\frac{1}{xy^2}}$

39. $\dfrac{\frac{m}{n}-\frac{n}{m}}{m-n}$

40. $\dfrac{\frac{1}{y}-\frac{1}{x+3}}{\frac{1}{x}-\frac{y}{x^2+3x}}$

41. $\dfrac{x+y^{-1}}{x^{-1}+y}$

42. $\dfrac{1+xy}{x^{-2}-y^2}$

43. $\dfrac{x^2-y^2}{y^{-2}-x^{-2}}$

44. $\dfrac{xy^{-1}+\left(\frac{x}{y}\right)^{-1}}{x^{-2}+y^{-2}}$

45. $\dfrac{\frac{1}{7y}+\frac{1}{x-2}}{\frac{1}{11x}+\frac{7y}{11x^2-22x}}$

46. $\dfrac{8z+8}{2-\frac{2}{z^2}}$

47. $\dfrac{25x^{-2}-9z^{-2}}{\frac{5z+3x}{x^2}}$

48. $\dfrac{\frac{3y}{5}-\frac{5}{3y}}{3-\frac{5}{y}}$

Perform the indicated operations on the following rational expressions, and simplify your answer.

49. $\left(\frac{x^2-3x}{x^2+6x-27}-\frac{2}{x+9}\right)\cdot\frac{x+9}{x+2}$

50. $\frac{2y(y-1)}{y^2+6y-16}\div\frac{2}{y+8}-\frac{2}{y-2}$

51. $\left(\frac{z^2-17z+30}{z^2+2z-8}+\frac{6}{z-2}\right)\div\frac{1}{z^2-5z-36}$

52. $\frac{y+3}{2y+18}+\frac{y^2+2y+4}{y^2+3y-54}\cdot\frac{y-6}{y+3}$

53. $\frac{y^2+2y-15}{y+1}\cdot\left(\frac{y^2+3y+4}{y^2+3y-10}+\frac{y+4}{y+5}\right)\div\frac{y-3}{y-2}$

54. $\frac{y+6}{y-3}\left(\frac{y+5}{y-3}+\frac{y-3}{y+6}-\frac{y^2+4}{y^2+3y-18}\right)$

1.5 COMPLEX NUMBERS

TOPICS

1. The imaginary unit *i* and its properties
2. Basic operations with complex numbers
3. Roots and complex numbers

TOPIC 1: The Imaginary Unit *i* and Its Properties

Birth of Imaginary Numbers

Gerolamo Cardano was an Italian mathematician, physician, scientist, gambler, inventor, and philosopher famous for, in addition to having an enormous range of interests, being one of the first people to recognize the utility of what are now known as imaginary numbers. In fact, prior to Cardano's time, mathematicians were even reluctant to use negative numbers, but in his 1545 book *Ars Magna*, Cardano showed how negative numbers and square roots of negative numbers arise naturally in solving polynomial equations.

In Section 1.2 we encountered a problem with the real number system: there is a lack of symmetry in the definition of roots of real numbers. Recall that so far we have defined even roots only for nonnegative numbers, but we have defined odd roots for both positive and negative numbers (as well as 0).

This asymmetry is a reflection of the fact that the real number system is not *algebraically complete*. Roughly, this means that there are polynomial equations with real coefficients that have no real solutions! Consider the following question:

For a given nonzero real number a, how many solutions does the equation $x^2 = a$ *have?*

As we have noted previously, the equation has the two solutions $x = \sqrt{a}$ and $x = -\sqrt{a}$ if a is positive, but no (real) solutions if a is negative. The following definition changes this situation.

DEFINITION: The Imaginary Unit *i*

The **imaginary unit *i*** is defined as $i = \sqrt{-1}$. In other words, i has the property that its square is -1: $i^2 = -1$.

This allows us to immediately define square roots of negative numbers in general, as follows.

DEFINITION: Square Roots of Negative Numbers

If a is a positive real number, $\sqrt{-a} = i\sqrt{a}$. Note that by this definition, and by a logical extension of exponentiation, $\left(i\sqrt{a}\right)^2 = i^2\left(\sqrt{a}\right)^2 = -a$.

Example 1: The Number *i*

a. $\sqrt{-16} = i\sqrt{16} = i(4) = 4i$. As is customary, we write a constant such as 4 before letters in algebraic expressions, even if, as in this case, the letter is not a variable. Remember that i has a fixed meaning: i is the square root of -1.

b. $\sqrt{-8} = i\sqrt{8} = i\left(2\sqrt{2}\right) = 2i\sqrt{2}$. As is customary, again, we write the radical factor last. You should verify that $\left(2i\sqrt{2}\right)^2$ is indeed -8.

c. $i^3 = i^2 i = (-1)(i) = -i$, and $i^4 = i^2 i^2 = (-1)(-1) = 1$. The simple fact that $i^2 = -1$ allows us, by our extension of exponentiation, to determine i^n for any natural number n.

d. $(-i)^2 = (-1)^2 i^2 = i^2 = -1$. This shows that $-i$ also has the property that its square is -1.

The **powers of *i*** follow a pattern that repeats with every fourth power.

$$\begin{aligned} i^1 &= i & \qquad i^{4n+1} &= i \\ i^2 &= -1 & i^{4n+2} &= -1 \\ i^3 &= -i & i^{4n+3} &= -i \\ i^4 &= 1 & i^{4n} &= 1 \end{aligned}$$

Example 2: Powers of *i*

Compute the following powers of i.

a. i^9 **b.** i^{28} **c.** i^{102}

Solution

a. When we divide 9 by 4, we have a remainder of 1. This means this power of i takes the form i^{4n+1}, so $i^9 = i$.

b. When we divide 28 by 4, the remainder is 0. This means this power of i is of the form i^{4n}, so $i^{28} = 1$.

c. When we divide 102 by 4, the remainder is 2. This means that this power of i fits the form i^{4n+2}, so $i^{102} = -1$.

The definition of the imaginary unit i leads to the following definition of complex numbers.

DEFINITION: Complex Numbers

For any two real numbers a and b, the sum $a + bi$ is a **complex number**. The collection $\mathbb{C} = \{a + bi \mid a \text{ and } b \text{ are both real}\}$ is called the set of complex numbers and is another example of a field. The number a is called the **real part** of $a + bi$, and the number b is called the **imaginary part**. If the imaginary part of a given complex number is 0, the number is simply a real number. If the real part of a given complex number is 0, the number is a **pure imaginary number**.

Note that the set of real numbers is a subset of the complex numbers: every real number *is* a complex number with 0 as the imaginary part. The set of complex numbers is the largest set of numbers that will appear in this text.

Do not be misled by the names into thinking that complex numbers, with their possible imaginary parts, are unimportant or physically meaningless. In many applications, complex numbers, even pure imaginary numbers, arise naturally and have important implications. For instance, the fields of electrical engineering and fluid dynamics both rely on complex number arithmetic.

TOPIC 2: Basic Operations with Complex Numbers

The set of complex numbers is a field, so the field properties discussed in Section 1.1 will apply. In particular, every complex number has an additive inverse (its negative), and every nonzero complex number has a multiplicative inverse (its reciprocal). Further, sums and products (and hence differences and quotients) of complex numbers are complex numbers, and can be written in the standard form $a + bi$. Given several complex numbers combined by the operations of addition, subtraction, multiplication, or division, the goal is to *simplify* the expression into the standard form $a + bi$.

Sums, differences, and products of complex numbers are easily simplified by remembering the definition of i and by thinking of every complex number $a + bi$ as a binomial.

PROCEDURE: Simplifying Complex Expressions

Step 1: Add, subtract, or multiply the complex numbers, as required, by treating every complex number $a + bi$ as a polynomial expression. Remember, though, that i is not actually a variable. Treating $a + bi$ as a binomial in i is just a handy device.

Step 2: Complete the simplification by using the fact that $i^2 = -1$.

Example 3: Simplifying Complex Expressions

Simplify the following complex expressions.

a. $(4+3i)+(-5+7i)$

b. $(-2+3i)-(-5+3i)$

c. $(3+2i)(-2+3i)$

d. $(2-3i)^2$

NOTE

Remember that a complex number is not simplified until it has the form $a + bi$.

TECHNOLOGY

```
NORMAL FLOAT AUTO REAL RADIAN MP
(4+3i)+(-5+7i)
                        -1+10i
```

On a TI-84 Plus, the imaginary number i is accessed by pressing 2nd and then .

Solution

a. $$\begin{aligned}(4+3i)+(-5+7i)&=(4-5)+(3+7)i\\&=-1+10i\end{aligned}$$

As if adding polynomials, we combine the real parts, then the imaginary parts.

b. $$\begin{aligned}(-2+3i)-(-5+3i)&=(-2+3i)+(-(-5)-3i)\\&=(-2+5)+(3-3)i\\&=3+0i\\&=3\end{aligned}$$

Begin by distributing the minus sign over the second complex number.

c. $$\begin{aligned}(3+2i)(-2+3i)&=-6+9i-4i+6i^2\\&=-6+(9-4)i+6(-1)\\&=-6+5i-6\\&=-12+5i\end{aligned}$$

After multiplying, combine the two terms containing i and rewrite i^2 as -1.

d. $(2-3i)^2 = (2-3i)(2-3i)$
$= 4-6i-6i+9i^2$
$= 4-12i+9(-1)$
$= -5-12i$

Squaring this complex number also leads to four terms, which we simplify as in part c.

DEFINITION: Complex Conjugates

Given any complex number $a + bi$, the complex number $a - bi$ is called its **complex conjugate**.

A very useful property of the complex conjugate is demonstrated below; the product of any complex number and its complex conjugate is a *real* number.

$$(a+bi)(a-bi) = a^2 - abi + abi - b^2i^2 = a^2 + b^2$$

This fact is critical in dividing complex numbers. In order to simplify a quotient of complex numbers, we need to rewrite it in the standard form $a + bi$. We simplify the quotient of two complex numbers by multiplying the numerator and denominator of the fraction by the complex conjugate of the denominator. This multiplication leaves a real number in the denominator so that a straightforward simplification leads to the standard form.

NOTE

Always begin by finding the complex conjugate of the denominator, then multiply the numerator and denominator by this conjugate.

Note that this process is very similar to the process we used when rationalizing the denominator of a radical expression in Section 1.2.

Example 4: Dividing Complex Numbers

Simplify the following expressions.

a. $\dfrac{2+3i}{3-i}$ **b.** $(4-3i)^{-1}$ **c.** $\dfrac{1}{i}$

TECHNOLOGY

```
NORMAL FLOAT AUTO REAL RADIAN MP
(2+3i)/(3-i)▶Frac
                    3/10+11/10 i
```

On a TI-84 Plus, fraction conversion is accessed by pressing math and then selecting ▶Frac.

Solution

a. $\dfrac{2+3i}{3-i} = \left(\dfrac{2+3i}{3-i}\right)\left(\dfrac{3+i}{3+i}\right)$

$= \dfrac{(2+3i)(3+i)}{(3-i)(3+i)}$

$= \dfrac{6+2i+9i+3i^2}{9+3i-3i-i^2}$

$= \dfrac{3+11i}{10} = \dfrac{3}{10} + \dfrac{11}{10}i$

$3 + i$ is the complex conjugate of the denominator.

Multiply the numerator and the denominator by the complex conjugate.

We can often leave the answer in the form $\dfrac{3+11i}{10}$.

b. $(4-3i)^{-1} = \dfrac{1}{4-3i}$

$= \left(\dfrac{1}{4-3i}\right)\left(\dfrac{4+3i}{4+3i}\right)$

$= \dfrac{4+3i}{(4-3i)(4+3i)}$

$= \dfrac{4+3i}{16-9i^2}$

$= \dfrac{4+3i}{25} = \dfrac{4}{25} + \dfrac{3}{25}i$

Rewrite the original expression as a fraction.

Then multiply the top and bottom by the complex conjugate of the denominator and proceed as in part a.

c. $\dfrac{1}{i} = \left(\dfrac{1}{i}\right)\left(\dfrac{-i}{-i}\right)$

$= \dfrac{-i}{-i^2}$

$= \dfrac{-i}{1} = -i$

Here we write the reciprocal of the imaginary unit as a complex number. With this as a starting point, we could now calculate i^{-2}, i^{-3}, ...

TOPIC 3: Roots and Complex Numbers

We have now defined $\sqrt{a}$ without ambiguity: given a positive real number a, $\sqrt{a}$ is the positive real number whose square is a, and $\sqrt{-a}$ is defined to be $i\sqrt{a}$. These are called the **principal square roots**, to distinguish them from $-\sqrt{a}$ and $-i\sqrt{a}$ respectively. (Remember, both $\sqrt{a}$ and $-\sqrt{a}$ are square roots of a.)

⚠ CAUTION

In simplifying radical expressions, we have made frequent use of the properties that if $\sqrt{a}$ and $\sqrt{b}$ are real numbers, then

$$\sqrt{a}\sqrt{b} = \sqrt{ab} \text{ and } \frac{\sqrt{a}}{\sqrt{b}} = \sqrt{\frac{a}{b}}.$$

There is a subtle but important condition in the above statement: $\sqrt{a}$ and $\sqrt{b}$ must both be *real* numbers. If this condition is not met, these properties of radicals do not necessarily hold. For instance,

$$\sqrt{(-9)(-4)} = \sqrt{36} = 6, \text{ but } \sqrt{-9}\sqrt{-4} = (3i)(2i) = 6i^2 = -6.$$

In order to apply either of these two properties, then, first simplify any square roots of negative numbers by rewriting them as pure imaginary numbers.

Fallacious Proofs

Many "proofs" of obvious fallacies are based on misapplications of properties of radicals. As an example, consider the following alleged proof that 0 = 2.

$$\begin{aligned} 0 &= 1-1 \\ &= 1-\sqrt{1} \\ &= 1-\sqrt{(-1)(-1)} \\ &= 1-\sqrt{-1}\sqrt{-1} \\ &= 1-i^2 \\ &= 1-(-1) \\ &= 2 \end{aligned}$$

Example 5: Roots and Complex Numbers

Simplify the following expressions.

a. $\left(2-\sqrt{-3}\right)^2$

b. $\dfrac{\sqrt{4}}{\sqrt{-4}}$

Solution

a.
$$\begin{aligned}\left(2-\sqrt{-3}\right)^2 &= \left(2-\sqrt{-3}\right)\left(2-\sqrt{-3}\right)\\ &= 4-4\sqrt{-3}+\sqrt{-3}\sqrt{-3}\\ &= 4-4i\sqrt{3}+\left(i\sqrt{3}\right)^2\\ &= 4-4i\sqrt{3}-3\\ &= 1-4i\sqrt{3}\end{aligned}$$

Each $\sqrt{-3}$ is converted to $i\sqrt{3}$ before multiplying.

b.
$$\begin{aligned}\frac{\sqrt{4}}{\sqrt{-4}} &= \frac{2}{2i}\\ &= \frac{1}{i}\\ &= -i\end{aligned}$$

We simplify each radical before dividing.

We already simplified $\frac{1}{i}$ in Example 4c, so we quickly obtain the correct answer of $-i$.

1.5 EXERCISES

PRACTICE

Evaluate the following square root expressions. See Example 1.

1. $\sqrt{-25}$

2. $\sqrt{-12}$

3. $-\sqrt{-27}$

4. $-\sqrt{-100}$

5. $\sqrt{-32x}$, $x>0$

6. $\sqrt{-x^2}$

7. $\sqrt{-29}$

8. $(-i)^2\sqrt{-64}$

Simplify the following complex expressions. See Examples 2, 3, and 4.

9. $(4-2i)-(3+i)$

10. $(4-i)(2+i)$

11. $(3-i)^2$

12. i^7

13. $(7i-2)+(3i^2-i)$

14. $(3+i)(3-i)$

15. $(5-3i)^2$

16. $(5+i)(2-9i)$

17. i^{13}

18. $(9-4i)(9+4i)$

19. $11i^{314}$

20. i^{132}

21. $(7-3i)^2$

22. $(4-3i)(7+i)$

23. $(3i)^2$

24. $(1+i)+i$

25. $i(5-i)$

26. $i^{11}\left(\frac{6}{i^3}\right)$

27. $(10i^2-9i)+(9+5i)$

28. $(-5i)^3$

29. $i^7\left(\frac{49}{7i^2}\right)$

30. $\frac{1+2i}{1-2i}$

31. $\frac{10}{3-i}$

32. $\frac{i}{2+i}$

33. $\frac{1}{i^9}$

34. $(2+5i)^{-1}$

35. i^{-25}

36. $\frac{1}{i^{27}}$

37. $\frac{52}{5+i}$

38. $(2-3i)^{-1}$

39. $\frac{4i}{5+7i}$

40. i^{-4}

41. $\frac{5+i}{4+i}$

Simplify the following expressions. See Example 5.

42. $(3+\sqrt{-2})^2$

43. $(1+\sqrt{-6})^2$

44. $\frac{\sqrt{18}}{\sqrt{-2}}$

45. $(\sqrt{-32})(-\sqrt{-2})$

46. $(\sqrt{-9})(\sqrt{-2})$

47. $\frac{\sqrt{-98}}{3i\sqrt{-2}}$

48. $(\sqrt{-8})(\sqrt{-2})$

49. $(5+\sqrt{-3})^2$

50. $\frac{\sqrt{-72}}{5i\sqrt{-2}}$

APPLICATIONS

51. Electrical engineers often use j, rather than i, to represent imaginary numbers. This is to prevent confusion with their use of i, which often represents current. Under this convention, assume the impedance of a particular part of a series circuit is $4-3j$ ohms and the impedance of another part of the circuit is $2+6j$ ohms. Find the total impedance of the circuit. (Impedances in series are simply added.)

52. Consider the formula $V=IZ$, where V is voltage (in volts), I is current (in amps), and Z is impedance (in ohms). If you know the current of a circuit is $5-4j$ amps and the impedance is $8+2j$ ohms, find the voltage.

53. If you know the voltage of a circuit is $35+5j$ volts and the current is $3+j$ amps, find the impedance.

WRITING & THINKING

54. Explain why it may be useful to be able to use imaginary numbers in real-world math.

TECHNOLOGY

Use a graphing utility to simplify the following complex expressions.

55. $\dfrac{3-2i}{1+i}$

56. $(3-2i)^4$

57. $\dfrac{2500}{(3+i)^4}$

58. $(2-5i)(3+7i)(1-4i)$

59. $(1+i)^5(3-i)^2$

60. $\dfrac{3+7i}{(2-5i)(1+3i)}$

61. $\dfrac{6+3i}{2-4i}$

62. $(5-3i)^5$

63. $\dfrac{400}{(6+2i)^3}$

64. $(6-3i)(8+i)(7-4i)$

65. $(5-3i)^4(7+2i)^3$

66. $\dfrac{4+3i}{(7-2i)(5+4i)}$

1.6 LINEAR EQUATIONS IN ONE VARIABLE

TOPICS

1. Equations and the meaning of solutions
2. Solving linear equations
3. Solving linear absolute value equations
4. Solving linear equations for one variable
5. Applications of linear equations

TOPIC 1: Equations and the Meaning of Solutions

Truth in Mathematics

An identity may also be referred to as a *tautology*, a word with Greek roots. The ancient Greeks defined a tautology as a statement that is true merely by virtue of repeating itself, which has negative connotations. But beginning in the 19th century, "tautology" took on its modern meaning in the field of mathematical logic.

In mathematics, an **equation** is a statement that two expressions are equal. If the statement is always true for any allowable value(s) of the variable(s), then the equation is an **identity**. (Unless otherwise stated, it is assumed that the context for a given equation is the set of real numbers, and variables are to be replaced by values that result in real numbers.) If the statement is never true, it is a **contradiction**. The third (and last) possibility is that the equation is true for some values of the variable(s) and false for others. These equations are called **conditional**, and the goal in solving such equations is to discover the set of values that the variable(s) can be replaced by to make the equation true. This set is called the **solution set** of the equation. Any one element of the solution set is called a **solution** of the equation.

Example 1: Identifying Types of Equations

a. The equations $x^{\frac{1}{2}}(x+1)=x^{\frac{3}{2}}+x^{\frac{1}{2}}$ and $5+3=8$ are identities. Note that in the first equation the only allowable replacements for x are nonnegative real numbers, but for all such numbers the equation is true. There are no variables in the second equation, so any value for any variable automatically satisfies this true statement.

b. The equation $t+3=t$ is an example of a contradiction. The solution set of this equation is the empty set, $\varnothing$, since no value for t satisfies the equation.

c. The equation $x^2=9$ is conditional. The solution set of the equation is $\{-3,3\}$, as any other value for x results in a false statement.

The basic method for finding the solution set of an equation is to transform it into a simpler equation whose solution set is more clear. Two equations that have the same solution set are called **equivalent equations**, and transforming an equation into an equivalent but simpler equation constitutes much of the mechanical work of algebra. The field properties and the cancellation properties, discussed in Section 1.1, are the means by which equations are transformed into equivalent equations. Bear in mind, however, the strong possibility that more than one step may be required to transform the equation into an equivalent equation whose solution is apparent. Persistence and care are the key to successfully solving equations!

TOPIC 2: Solving Linear Equations

Linear equations in one variable are arguably the least complicated type of equation to solve.

> **DEFINITION:** Linear Equations in One Variable
>
> A **linear equation in one variable**, say the variable x, is an equation that can be transformed into the form $ax+b=0$, where a and b are real numbers and $a \neq 0$. Such equations are also called **first-degree equations**, as x appears to the first power.

A linear equation is an example of a polynomial equation, and the general method of solving polynomial equations will be discussed in Section 1.8.

Note that linear equations in one variable take very few steps to solve. Once the equation has been written in the form $ax+b=0$, the first cancellation property implies

$$ax+b=0 \Leftrightarrow ax=-b, \qquad \text{(add } -b \text{ to both sides)}$$

and the second cancellation property implies

$$ax=-b \Leftrightarrow x=-\frac{b}{a}. \qquad \text{(multiply both sides by } \frac{1}{a}\text{)}$$

A given linear equation does not have to be written in the form $ax+b=0$ initially, and it is likely that some of the field properties (such as the distributive and commutative properties) must be applied first.

TECHNOLOGY

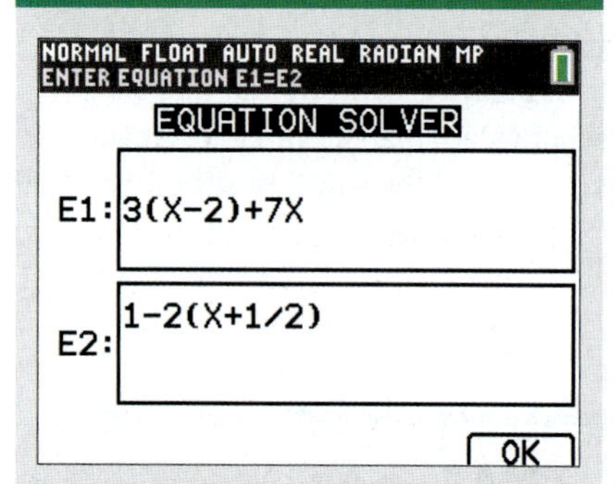

NORMAL FLOAT AUTO REAL RADIAN MP
SOLUTION IS MARKED ▪
3(X−2)+7X=1−2(X+1/2)
▪X=0.5
bound={−1E99,1E99}
▪E1−E2=0
SOLVE

To access the equation solver on a TI-84 Plus CE, press math and then press the up arrow key to select the last option. Press enter to open it.
For E1 enter the left-hand side of the equation, and for E2 enter the right-hand side of the equation. Press graph twice to select OK and then SOLVE.

Example 2: Solving Linear Equations in One Variable

Solve the following equations.

a. $3(x-2)+7x=1-2\left(x+\frac{1}{2}\right)$

b. $3x-7=3(x-2)$

c. $\frac{y}{6}+\frac{2y-1}{2}=\frac{y+1}{3}$

d. $0.25(x-3)+0.08=0.15x$

e. $5x+12=5(x+3)-3$

Solution

a.
$$\begin{aligned} 3(x-2)+7x&=1-2\left(x+\frac{1}{2}\right)\\ 3x-6+7x&=1-2x-1\\ 10x-6&=-2x\\ 12x-6&=0\\ 12x&=6\\ x&=\frac{1}{2}\end{aligned}$$

As is typical, we must apply some of the field properties discussed in Section 1.1 in order to solve the equation. The distributive property leads to the second equation. Combining like terms leads to the third.

The cancellation properties then allow us to complete the process and solve the equation.

b.
$$\begin{aligned} 3x-7&=3(x-2)\\ 3x-7&=3x-6\\ 3x-3x&=-6+7\\ 0&=1\end{aligned}$$

We first distribute, then add $-3x$ and 7 to both sides in order to combine like terms. In this problem, however, the variable cancels out and we are left with a false statement.

Thus, the equation is a contradiction and has no solutions. The solution set is the empty set, $\varnothing$.

c.
$$\frac{y}{6}+\frac{2y-1}{2}=\frac{y+1}{3}$$
$$6\left(\frac{y}{6}+\frac{2y-1}{2}\right)=6\left(\frac{y+1}{3}\right)$$
$$6\cdot\frac{y}{6}+6\cdot\frac{2y-1}{2}=6\cdot\frac{y+1}{3}$$
$$y+3(2y-1)=2(y+1)$$
$$y+6y-3=2y+2$$
$$y+6y-2y=2+3$$
$$5y=5$$
$$y=1$$

Although it is not a necessary step, many people prefer to get rid of any fractions that might appear by multiplying both sides of the equation by the least common denominator (LCD). Remember to multiply every term by the LCD.

Note the cancellation that has occurred. We have replaced $\frac{6}{6}$ with 1, $\frac{6}{2}$ with 3, and $\frac{6}{3}$ with 2. Like terms are then combined, and multiplication by $\frac{1}{5}$ leads to the final answer.

d.
$$0.25(x-3)+0.08=0.15x$$
$$25(x-3)+8=15x$$
$$25x-75+8=15x$$
$$10x=67$$
$$x=6.7$$

One approach to solving an equation with decimals is to multiply both sides by the appropriate power of 10 to eliminate the decimals. In this problem, multiplying both sides by 100 results in a simpler linear equation. Another approach is to retain the decimals throughout the solution process.

e.
$$5x+12=5(x+3)-3$$
$$5x+12=5x+15-3$$
$$5x-5x=15-3-12$$
$$0=0$$

In this problem, the variable cancels out and we are left with a true statement.

Thus, the equation is an identity. The solution set is all real numbers, $\mathbb{R}$.

TOPIC 3: Solving Linear Absolute Value Equations

Linear absolute value equations in one variable are closely related to linear equations and are solved in a similar fashion. The difference is that a linear absolute value equation contains at least one variable term inside absolute value symbols. If these symbols were removed, the equation would be linear.

Such equations are solved by recognizing that the absolute value of any quantity is either the original quantity or its negative, depending on whether the quantity is positive or negative to start with. For instance, $|ax+b|$ equals either $ax+b$ or $-(ax+b)$, depending on the sign of $ax+b$. Until the equation is solved, we don't know the sign of $ax+b$, and so we must consider both cases. This means that, in general, every occurrence of an absolute value term in an equation leads to *two* equations with the absolute value signs removed. As an example,

$$|ax+b|=c \text{ means } ax+b=c \text{ or } -(ax+b)=c.$$

CAUTION

A word of warning: the apparent solutions obtained by the above method may not solve the original absolute value equation! Absolute value equations are one class of equations (there are others, as we shall see) in which it is very important to check your final answer in the original equation. An apparent solution that does not solve the original problem is called an **extraneous solution**.

Example 3: Solving Absolute Value Equations

NOTE

Begin by rewriting each absolute value term as two new equations, one with a positive sign and one with a negative sign.

Solve the absolute value equations.

a. $|3x-2|=1$ **b.** $|x-4|=|2x+1|$ **c.** $|6x-7|+5=3$

Solution

a. $|3x-2|=1$ Rewrite the absolute value equation without absolute value bars.

$$\begin{aligned} 3x-2&=1 &\quad\text{or}\quad -(3x-2)&=1\\ 3x&=3 & 3x-2&=-1\\ x&=1 & 3x&=1\\ & & x&=\frac{1}{3}\end{aligned}$$

The result is two linear equations that can be solved using the method illustrated earlier in this section.

Now check each solution in the original equation: $|3(1)-2|=|3-2|=|1|=1$, and $\left|3\left(\frac{1}{3}\right)-2\right|=|1-2|=|-1|=1$, so the solution set is $x=\frac{1}{3},\ 1$.

TECHNOLOGY

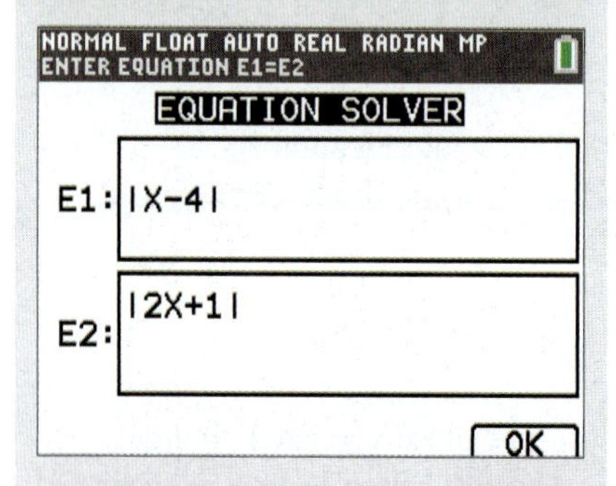

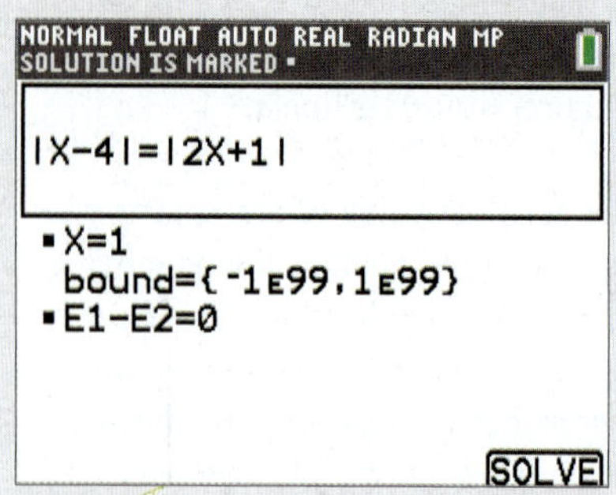

A word of caution—only one answer is generated here using the equation solver, but there are two solutions. Additional work must be done to find multiple solutions.

b. This equation has two absolute value terms, which leads to four linear equations when the absolute value bars are removed.

$$|x-4|=|2x+1| \Rightarrow \begin{cases} +(x-4)=+(2x+1)\\ +(x-4)=-(2x+1)\\ -(x-4)=+(2x+1)\\ -(x-4)=-(2x+1)\end{cases}$$

Note that the top and bottom equations are equivalent, as are the two middle equations. Thus, we have two linear equations to solve.

$$|x-4|=|2x+1|$$

$$\begin{aligned} x-4&=2x+1 &\quad\text{or}\quad -(x-4)&=2x+1\\ -x&=5 & -3x&=-3\\ x&=-5 & x&=1\end{aligned}$$

We proceed as before, applying the distributive property and combining like terms.

Finally, we check the apparent solutions in the original equation.

$$\begin{aligned} |(-5)-4|&=|2(-5)+1| &\qquad |(1)-4|&=|2(1)+1|\\ |-9|&=|-9| & |-3|&=|3|\\ 9&=9 & 3&=3\end{aligned}$$

Both apparent solutions are actual solutions, so the solution set is $x=-5,\ 1$.

c.

$$|6x-7|+5=3$$

$$|6x-7|=-2 \quad \text{Isolate the absolute value term.}$$

$$6x-7=-2 \quad \text{or} \quad -(6x-7)=-2 \quad \text{Again, we rewrite the original equation as two linear equations.}$$

$$6x=5 \qquad -6x=-9$$

$$x=\frac{5}{6} \qquad x=\frac{3}{2}$$

When we check the solutions this time, we find that neither apparent solution actually solves the equation!

$$\left|6\left(\frac{5}{6}\right)-7\right|+5=3 \qquad \left|6\left(\frac{3}{2}\right)-7\right|+5=3$$

$$|5-7|=-2 \qquad |9-7|=-2$$

$$|-2|=-2 \qquad |2|=-2$$

$$2\neq-2 \qquad 2\neq-2$$

Thus, the equation is a contradiction, and the solution set is $\varnothing$.

Example 3c illustrates the importance of checking the final answers in the original equation. If we had thought about the equation $|6x-7|+5=3$ before attempting to solve it, especially in the form $|6x-7|=-2$, we wouldn't have been surprised that there is no solution: any absolute value expression is automatically nonnegative, so there is no value of x for which $|6x-7|$ could be -2.

Before leaving the subject of absolute value equations, we will use a different technique to solve one last equation. In Section 1.1 we saw that $|x-a|$ means, geometrically, the distance between the real numbers x and a. The solution set of the equation $|x-5|=|x+3|$ thus consists of all real numbers x that are an equal distance from 5 and -3 (note that $|x+3|=|x-(-3)|$). A moment's thought tells us that the only solution is the number 1, the number halfway between -3 and 5.

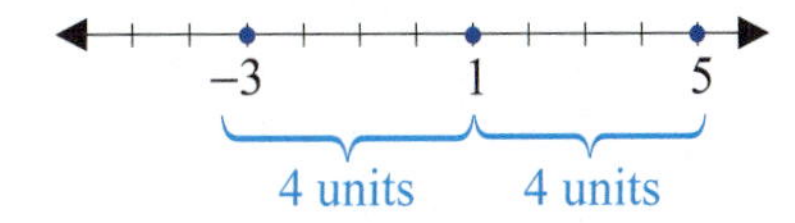

FIGURE 1: Solving $|x-5|=|x+3|$ Geometrically

Solve the equation $|x-5|=|x+3|$ algebraically, as in Example 3b, and verify that the solution set does indeed consist of only one number.

TOPIC 4: Solving Linear Equations for One Variable

One common task in applied mathematics is to solve a given equation in two or more variables for one of the variables. **Solving for a variable** means to transform the equation into an equivalent one in which the specified variable is isolated on one side of the equation. For linear equations we accomplish this by the same methods we have used in the previous examples.

Example 4: Solving Linear Equations for One Variable

Solve each of the following equations for the specified variable. All of the equations are formulas that arise in various applications, and they are linear in the specified variable.

a. $P = 2l + 2w$; solve for w

b. $A = P\left(1+\frac{r}{m}\right)^{mt}$; solve for P

c. $S = 2\pi r^2 + 2\pi rh$; solve for h

Solution

a.

$$P = 2l + 2w$$
$$P - 2l = 2w$$
$$\frac{P-2l}{2} = w$$
$$w = \frac{P-2l}{2}$$

This is the formula for the perimeter P of a rectangle of length l and width w. We first add $-2l$ to both sides, and then multiply by $\frac{1}{2}$.

The last equation is no different from the preceding one, but it is conventional to put the specified variable on the left side of the equation.

b.

$$A = P\left(1+\frac{r}{m}\right)^{mt}$$
$$\frac{A}{\left(1+\frac{r}{m}\right)^{mt}} = P$$
$$P = A\left(1+\frac{r}{m}\right)^{-mt}$$

This is the equation for compound interest. If principal P is invested at an annual rate r for t years, compounded m times a year, the value of the investment at time t is A. This formula is linear in the variables P and A, though not in m, t, or r. One step is all that is required to solve this equation for P, but the last equation uses one of the properties of exponents to make the result neater.

c.

$$S = 2\pi r^2 + 2\pi rh$$
$$S - 2\pi r^2 = 2\pi rh$$
$$\frac{S - 2\pi r^2}{2\pi r} = h$$
$$h = \frac{S - 2\pi r^2}{2\pi r}$$

This is the formula for the surface area of a right circular cylinder of radius r and height h. It is linear in the variables S and h, but not in r.

Two steps are all that are necessary to solve this formula for h. Solving this formula for r requires a technique that will be discussed in Section 1.8.

TOPIC 5: Applications of Linear Equations

Many applications lead to equations more complicated than those that we have studied so far, but good examples of linear equations arise from certain distance and simple interest problems. This is because the basic distance and simple interest formulas are linear in all of their variables.

Distance: $d = rt$, where d is the distance traveled at rate r for time t.

Simple Interest: $I = Prt$, where I is the interest earned on principal P invested at rate r for time t.

FIGURE 2

Example 5: Calculating Average Speed

The distance from Shreveport, LA to Austin, TX by one route is 325 miles. If Kevin made the trip in five and a half hours, what was his average speed?

Solution

We know d and t, and need to solve the linear equation $325 = \frac{11}{2}r$ for r (note that we have written five and a half as $\frac{11}{2}$). This is accomplished by multiplying both sides by $\frac{2}{11}$.

$$325 = \frac{11}{2}r$$
$$\frac{650}{11} = r$$
$$r \approx 59.1 \text{ mph}$$

Alternatively, the time can be expressed in decimal form.

$$325 = 5.5r$$
$$\frac{325}{5.5} = r$$
$$r \approx 59.1 \text{ mph}$$

Example 6: Calculating Average Interest Rate

Julie invested \$1500 in a risky high-tech stock on January 1st. On July 1st, her stock is worth \$2100. She knows that her investment is very volatile and that it does not earn interest at a constant rate, but she wants to determine her average annual rate of return at this point in the year. What effective annual rate of return has she earned so far?

Solution

The interest that Julie has earned in half a year is \$600 (or \$2100 − \$1500). Replacing P with 1500, t with $\frac{1}{2}$, and I with 600 in the formula $I = Prt$, we have the following equation.

$$600 = (1500)\left(\frac{1}{2}\right)r$$
$$\frac{1200}{1500} = r$$
$$r = 0.8$$
$$r = 80\% \text{ average rate of return per year}$$

Another class of problems, called mixture problems, also gives rise to linear equations, although an extra step may be necessary in order to arrive at an equation in just one variable.

Example 7: Solving a Mixture Problem

For a certain chemistry experiment, 500 milliliters of a 14% acid solution is required (a 14% acid solution means that 14% of any given quantity of solution is acid, while the remaining 86% of the solution is water). The supply room only has 10% solution and 20% solution made. How many milliliters of each must be mixed in order to obtain 500 milliliters of 14% solution?

Solution

Although we will ultimately construct a linear equation in a single variable, it's often easier to first organize the information in a mixture problem using two variables.

Accordingly, if we let x represent the number of milliliters of 10% solution we need and y the number of milliliters of 20% solution, the given information can be organized as follows.

Type of Solution	Amount of Solution	Amount of Acid
10%	x	$0.1x$
20%	y	$0.2y$
14%	500 ml	$(0.14)(500)$

TABLE 1

Since we want to wind up with 500 milliliters of solution, the second column of the table indicates that $x+y=500$. The third column lists the amount of acid present in the two existing solutions and the amount of acid that will be present in the 14% solution to be made. Adding up these amounts gives us the equation $0.1x+0.2y=(0.14)(500)$.

We can convert the second equation into one with a single variable by making use of the first equation in the form $y=500-x$. The result is then a linear equation in x that we can solve.

$$\begin{aligned} 0.1x+0.2(500-x) &= (0.14)(500) \\ 0.1x+100-0.2x &= 70 \\ -0.1x &= -30 \\ x &= 300 \end{aligned}$$

Now that we know $x=300$, we also know that $y=500-300=200$, so 300 milliliters of 10% solution mixed with 200 milliliters of 20% solution will give us the 500 milliliters of 14% solution that is required.

1.6 EXERCISES

PRACTICE

Solve the following linear equations. See Examples 1 and 2.

1. $-3(2t-4)=7(1-t)$
2. $5(2x-1)=3(1-x)+5x$
3. $\frac{y+5}{4}=\frac{1-5y}{6}$
4. $3x+5=3(x+3)-4$
5. $3w+5=2(w+3)-4$
6. $3x+5=3(x+3)-5$
7. $\frac{4s-3}{2}+\frac{7}{4}=\frac{8s+1}{4}$
8. $\frac{4x-3}{2}+\frac{3}{8}=\frac{7x+3}{4}$
9. $\frac{4z-3}{2}+\frac{3}{8}=\frac{8z+3}{4}$
10. $3(2w+13)=5w+w\left(7-\frac{3}{w}\right)$
11. $\frac{6}{7}(m-4)-\frac{11}{7}=1$
12. $0.08p+0.09=0.65$
13. $0.6x+0.08=2.3$
14. $0.9x+0.5=1.3x$
15. $0.73x+0.42(x-2)=0.35x$
16. $\frac{8y-2}{4}+\frac{6}{8}=\frac{16y+2}{8}$
17. $\frac{3}{7}(y-2)-\frac{14}{7}=-5$
18. $6(5w-5)=-31(3-w)$
19. $\frac{7x-5}{4}+\frac{14}{8}=\frac{14x+4}{8}$
20. $\frac{3}{11}(y-2)-\frac{33}{11}=-6$
21. $3z+3=3(z+4)-9$
22. $4y+9=4(y+4)-10$
23. $2.8x+1.2=3.2x$
24. $0.73z+0.34=9.1$
25. $0.24x+0.58(x-6)=0.82x-3.67$

Solve the following absolute value equations. See Example 3.

26. $|3x-2|=5$
27. $-|3y+5|+6=2$
28. $|4x+3|+2=0$
29. $|6x-2|=0$
30. $|-8x+2|=14$
31. $|2x-109|=731$
32. $|4x-4|-40=0$
33. $|5x-3|=7$
34. $|4x+15|=3$
35. $-|6x+1|=11$
36. $|-14y+3|+3=2$
37. $|3x-2|-1=|5-x|$

Solve the following absolute value equations geometrically and algebraically. See Figure 1.

38. $|x+3|=|x-7|$
39. $|x-3|-|x-7|=0$
40. $|2-x|=|2+x|$
41. $|x|=|x+1|$
42. $|x+97|=|x+101|$
43. $\left|x+\frac{1}{4}\right|=\left|x-\frac{3}{4}\right|$
44. $|z-51|-|z-5|=0$
45. $\left|x-\frac{5}{7}\right|=\left|x+\frac{3}{7}\right|$
46. $|6y-3|=|5y+5|$

Solve each of the following equations for the indicated variable. See Example 4.

47. Circumference of a circle: $C = 2\pi r$; solve for r

48. Ideal Gas Law: $PV = nRT$; solve for T

49. Velocity: $v^2 = v_0^{\,2} + 2ax$; solve for a

50. Area of a trapezoid: $A = \frac{1}{2}h(b+c)$; solve for h

51. Temperature conversions: $C = \frac{5}{9}(F-32)$; solve for F

52. Volume of a right circular cone: $V = \frac{1}{3}\pi r^2 h$; solve for h

53. Surface area of a rectangular prism: $A = 2lw + 2wh + 2hl$; solve for h

54. Distance: $d = rt_1 + rt_2$; solve for r

55. Kinetic energy of protons: $K = \frac{1}{2}mv^2$; solve for m

56. Finance: $A = P(1+rt)$; solve for t

APPLICATIONS

57. A riverboat leaves port and proceeds to travel downstream at an average speed of 15 miles per hour. How long will it take for the boat to arrive at the next port, 95 miles downstream?

58. Two trucks leave a warehouse at the same time. One travels due east at an average speed of 45 miles per hour, and the other travels due west at an average speed of 55 miles per hour. After how many hours will they be 450 miles apart?

59. Two cars leave a rest stop at the same time and proceed to travel down the highway in the same direction. One travels at an average rate of 62 miles per hour, and the other at an average rate of 59 miles per hour. How far apart are the two cars after four and a half hours?

60. Two trains are 630 miles apart, heading directly toward each other on parallel tracks. The first train is traveling at 95 mph, and the second train is traveling at 85 mph. How long will it be before the trains pass each other?

61. Two brothers, Rick and Tom, each inherit $10,000. Rick invests his inheritance in a savings account with an annual return of 2.25%, while Tom invests his in a CD paying 6.15% annually. How much more money does Tom have than Rick after 1 year?

62. Sarah, sister to Rick and Tom in the previous problem, also inherits $10,000, but she invests her inheritance in a global technology mutual fund. At the end of 1 year, her investment is worth $12,800. What has her effective annual rate of return been?

63. An industrial acid-etching procedure calls for 3 gallons of a 46% hydrofluoric acid solution, but the supplier currently only has 44% solution and 50% solution. How many gallons of each should be mixed for the procedure?

64. An agricultural stress test calls for soaking seeds in 8% saline solution. The scientist running the test wants to make use of 1 liter of 20% saline solution that is already made up. How much pure water should she add to the 20% solution to obtain an 8% solution?

65. A total of 39 tickets were sold for a puppet show, with child tickets selling for \$7.50 and adult tickets selling for \$10.00. The ticket sales raised \$330.00 in all. How many child tickets and how many adult tickets were sold?

66. Joe's Java Joint wants to make a blend of two coffees that can be sold for \$15 per pound. The first of the two types of coffee costs \$18 per pound, while the second costs \$13 per pound. How many pounds of each should be mixed to get 10 pounds of the desired blend?

67. Bob buys a large screen digital TV priced at \$9,500, but pays \$10,212.50 with tax. What is the rate of tax where Bob lives?

68. Will and Matt are brothers. Will is 6 feet, 4 inches tall, and Matt is 6 feet, 7 inches tall. How tall is Will as a percentage of Matt's height? How tall is Matt as a percentage of Will's height?

69. A farmer wants to fence in three square garden plots situated along a road, as shown, and he decides not to install fencing along the edge of the road. If he has 182 feet of fencing material total, what dimensions should he make each square plot?

70. Find three consecutive integers whose sum is 288. (**Hint:** If n represents the smallest of the three, then $n+1$ and $n+2$ represent the other two numbers.)

71. Find three consecutive odd integers whose sum is 165. (**Hint:** If n represents the smallest of the three, then $n+2$ and $n+4$ represent the other two numbers.)

72. Kathy buys last year's best-selling novel, in hardcover, for \$15.05. This is a 30% discount from the original price. What was the original price?

73. The highest point on Earth is the peak of Mount Everest. If you climbed to the top, you would be approximately 29,035 feet above sea level. Remembering that a mile is 5280 feet, what percentage of the height of the mountain would you have to climb to reach a point two miles above sea level?

TECHNOLOGY

Use a graphing utility to solve the following equations. Round your answers to two decimal places if necessary.

74. $453x = 95(34x+291)$

75. $-0.23 = 0.79x - 0.47(x+0.98)$

76. $254 + 0.98(x-124) = 0$

77. $323x - 1745 = 531(68x-887)$

1.7 LINEAR INEQUALITIES IN ONE VARIABLE

TOPICS

1. Solving linear inequalities
2. Solving double linear inequalities
3. Solving linear absolute value inequalities
4. Applications of linear inequalities

TOPIC 1: Solving Linear Inequalities

If the equality symbol in a linear equation is replaced with $<$, $\leq$, $>$, or $\geq$, the result is a **linear inequality**. One difference between linear equations and linear inequalities is the way in which the solutions are described. Typically, the solution of a linear inequality consists of some interval of real numbers; such solutions can be described graphically or with interval notation. The process of obtaining the solution, however, is much the same as the process for solving linear equations, with the one important difference discussed below.

When solving linear inequalities, the field properties outlined in Section 1.1 all still apply, and we often use the distributive and commutative properties in order to simplify one or both sides of an inequality. The additive version of the two cancellation properties is also used in the same way as in solving equations. The one difference lies in applying the multiplicative cancellation property. When dealing with linear *equations* we can multiply both sides by a positive or negative value and obtain an equivalent equation. When dealing with linear *inequalities*, some problems arise when multiplying both sides by a negative value.

Example 1: Multiplying Inequalities by Negative Numbers

Consider the following two inequalities: $-3 < 2$ and $x < 0$. Observe what happens if we multiply both sides of each inequality by -1.

1. The statement $-3 < 2$ is true, but if we multiply both sides by -1, we obtain the false statement $3 < -2$.

2. Now consider the inequality $x < 0$. If we multiply both sides by -1, we have the inequality $-x < 0$. But these two statements can't both be true!

These examples show that multiplicative cancellation must behave a bit differently for linear inequalities. Note that if we reverse the inequality sign in our results, we actually get true statements. This provides a clue to how we approach multiplicative cancellation in the case of linear inequalities.

PROPERTIES: Cancellation Properties for Inequalities

In the following properties, A, B, and C represent algebraic expressions and D represents a nonzero constant. Each of the properties is stated for the inequality symbol $<$, but they are also true when the symbol $<$ is replaced with $>$, $\leq$, or $\geq$ and the restrictions on D remain the same.

Property	Description
If $A<B$, then $A+C<B+C$.	Adding the same quantity to both sides of an inequality results in an equivalent inequality.
If $A<B$ and $D>0$, then $A \cdot D<B \cdot D$.	If both sides of an inequality are multiplied by a positive constant, the sense of the inequality is unchanged.
If $A<B$ and $D<0$, then $A \cdot D>B \cdot D$.	If both sides are multiplied by a negative constant, the sense of the inequality is reversed.

Keep in mind that multiplying (or dividing) both sides of an inequality by a negative quantity requires reversing, or "flipping" the inequality symbol. We will see this several times in the examples to follow.

Example 2: Solving Linear Inequalities

Solve the following inequalities, using interval notation to describe the solution set.

a. $5-2(x-3) \leq -(1-x)$

b. $\dfrac{3(a-2)}{2} < \dfrac{5a}{4}$

Solution

a. $5-2(x-3) \leq -(1-x)$ Begin by using the distributive property, then combine like terms.

$5-2x+6 \leq -1+x$

$-2x+11 \leq -1+x$

$-3x \leq -12$ Now, all we need to do is divide by −3.

$x \geq 4$ Note the reversal of the inequality symbol.

In interval notation, the solution is $[4, \infty)$.

b. $\frac{3(a-2)}{2}<\frac{5a}{4}$

$4\left(\frac{3(a-2)}{2}\right)<4\left(\frac{5a}{4}\right)$

$6(a-2)<5a$

$6a-12<5a$

$a<12$

Just as with equations, fractions in inequalities can be eliminated by multiplying both sides by the least common denominator.

Since we do not need to multiply or divide by a negative value, the sense of the inequality does not change.

Thus, in interval notation, the solution is $(-\infty,12)$.

The solutions in Example 2 were described using interval notation, but solutions can also be described by set-builder notation or by graphing. Graphing a solution to an inequality can lead to a better understanding of which real numbers solve the inequality.

The symbols used in this text for graphing intervals are the same as the symbols in interval notation. Parentheses are used to indicate excluded endpoints of intervals, and brackets are used when the endpoints are included in the interval. The portion of the number line that constitutes the interval is then shaded. (Other commonly used symbols in graphing are open circles for parentheses and filled-in circles for brackets.)

For example, the two solutions from Example 2, $[4,\infty)$ and $(-\infty,12)$, are graphed as follows.

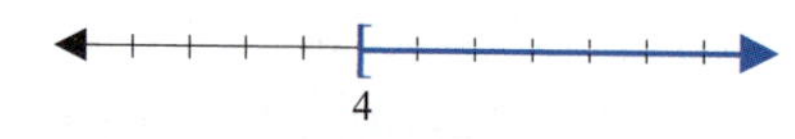

FIGURE 1: Graph of the Interval $[4,\infty)$

FIGURE 2: Graph of the Interval $(-\infty,12)$

Example 3: Graphing Intervals of Real Numbers

Graph the following intervals.

a. $[-3, 6]$ **b.** $(-\infty, 5]$ **c.** $[2, 9)$

Solution

a.

FIGURE 3

Both endpoints are included in the interval.

b.

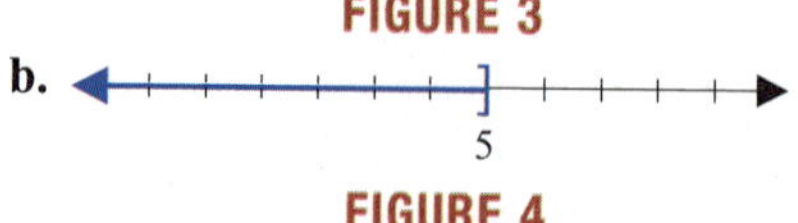

FIGURE 4

The left-hand side of the graph extends to negative infinity.

c.

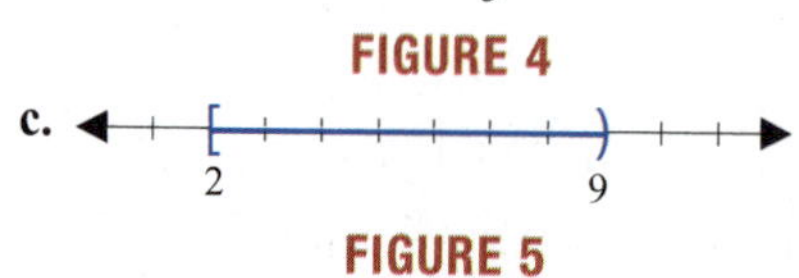

FIGURE 5

The left endpoint is included in the interval, while the right endpoint is excluded.

TOPIC 2: Solving Double Linear Inequalities

A **compound inequality** is a statement containing two or more distinct inequalities joined by either of the words "and" or "or." When two inequalities are joined by the word "and" (meaning both inequalities must be true), it is common to write the statement as a **double inequality**. We'll explore how to solve double inequalities with an application example you are likely to encounter at some point.

Example 4: Calculating Final Grades

The final grade in a class depends on the grades of 5 exams, each worth a maximum of 100 points. Suppose Janice's scores on the first four tests are 67, 82, 73, and 85. Assuming a grade of B corresponds to a numerical grade greater than or equal to 80 and less than 90, what scores can she make on the fifth test to get a B in the class?

NOTE

We follow the same process in solving double inequalities as we do for standard ones. The only difference is that operations are applied to all three "sides" of the statements.

Solution

If we let x represent the fifth test score, we need to solve the following double inequality.

$$80 \leq \frac{67 + 82 + 73 + 85 + x}{5} < 90$$

This could be solved by breaking it into the two inequalities

$$80 \leq \frac{67 + 82 + 73 + 85 + x}{5} \text{ and } \frac{67 + 82 + 73 + 85 + x}{5} < 90,$$

but it is more efficient to solve both at the same time as a double inequality.

$$80 \leq \frac{67+82+73+85+x}{5} < 90$$

First, combine like terms in the numerator of the fraction.

$$80 \leq \frac{307+x}{5} < 90$$

$$400 \leq 307 + x < 450$$

$$93 \leq x < 143$$

Just as if we were working with a single inequality, we begin by multiplying by 5, then subtract 307 from all three parts.

Thus, the mathematical solution to the double inequality is $[93,143)$. However, this is not the solution to our application problem!

Why not? There is an additional restriction on the solution set based on the context of the problem; each exam is worth a maximum of 100 points. This means that any value in the calculated solution set that is greater than 100 does not apply, making the actual solution $[93,100]$.

FIGURE 6

Example 5: Solving Double Linear Inequalities

Solve the following double inequalities.

a. $-1 < 3 - 2x \leq 5$ **b.** $2(2x-1) \leq 4x+2 \leq 4(x+1)$

Solution

a. $-1 < 3-2x \leq 5$ Begin by subtracting 3 from all three expressions.

$-4 < -2x \leq 2$ Since we divide each expression by –2, we must reverse each inequality symbol.The final double inequality is identical to the one before it, but has been written so that the smaller number appears first.

$2 > x \geq -1$

$-1 \leq x < 2$

In interval notation, the solution is $[-1,2)$.

b. $2(2x-1) \leq 4x+2 \leq 4(x+1)$ First, apply the distributive property.

$4x-2 \leq 4x+2 \leq 4x+4$

$-2 \leq 2 \leq 4$ The variable disappears from the inequality, and we are left to assess whether the statement is true.

Since we are left with a true statement, the double inequality is true for all values of the variable x, and the solution set is $(-\infty,\infty)$.

TOPIC 3: Solving Linear Absolute Value Inequalities

An **absolute value inequality** is an inequality in which some variable expression appears inside absolute value symbols. In the problems that we will study, the inequalities would be linear if the absolute value symbols were not there.

The Geometric Meaning of Absolute Value

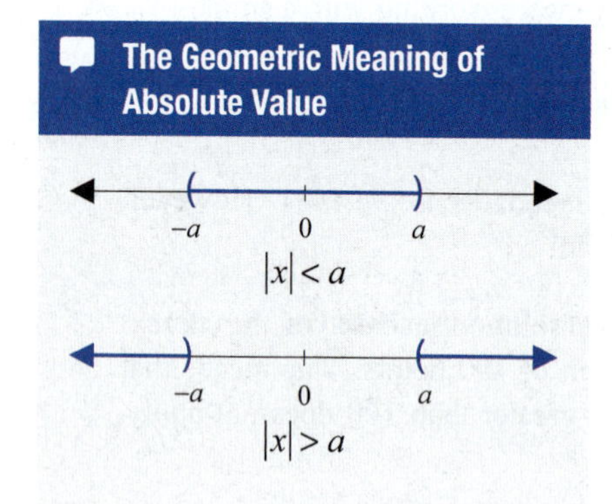

The geometric meaning of absolute value provides the method by which absolute value inequalities are solved. Recall that $|x|$ represents the distance between x and 0 on the real number line. If a is a positive real number, the inequality $|x| < a$ means that x is less than a units from 0, and the inequality $|x| > a$ means that x is greater than a units from 0 (similar interpretations hold for the symbols $\leq$ and $\geq$). This means that absolute value inequalities can be written without absolute values as follows.

$$|x| < a \quad \Leftrightarrow \quad -a < x < a$$

and

$$|x| > a \quad \Leftrightarrow \quad x < -a \text{ or } x > a$$

While most absolute value inequalities will be more complicated than the two above, they will serve as templates for the rewriting of such inequalities without the absolute values. Take note of the fact that the < symbol leads to a set of two inequalities that must *both* be true, while the > symbol leads to a solution in which *either* of two inequalities must hold.

Example 6: Solving Linear Absolute Value Inequalities

Solve the following absolute value inequalities.

a. $|4-2x|>6$

b. $2|3y-2|+3\leq 11$

c. $|5+2s|\leq -3$

d. $|5+2s|\geq -3$

Solution

a.

$$|4-2x|>6$$

$$\begin{aligned} 4-2x&<-6 & & & 4-2x&>6 \\ -2x&<-10 & \text{or} & & -2x&>2 \\ x&>5 & & & x&<-1 \end{aligned}$$

We can rewrite the inequality without absolute values and begin solving the two independent inequalities.

After dividing by −2, we need to reverse the sense of the inequality.

The solution is $(-\infty,-1)\cup(5,\infty)$.

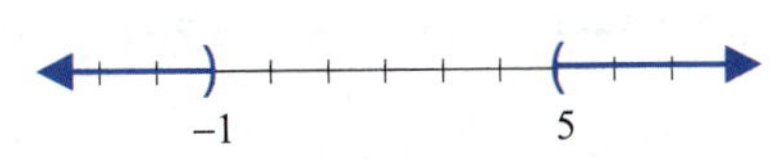

FIGURE 7

b.

$$\begin{aligned} 2|3y-2|+3&\leq 11 \\ |3y-2|&\leq 4 \\ -4\leq 3y-2&\leq 4 \\ -2\leq 3y&\leq 6 \\ -\frac{2}{3}\leq y&\leq 2 \end{aligned}$$

Isolate the absolute value expression by subtracting 3 from both sides, then dividing both sides by 2.

After rewriting the inequality as described earlier, we have a double inequality to solve.

Thus, the solution is $\left[-\frac{2}{3},2\right]$.

FIGURE 8

c. $|5+2s|\leq -3$

The solution is $\varnothing$.

Just as in Example 3c of Section 1.6, we conclude that the solution set is the empty set, as it is impossible for the absolute value of any expression to be negative.

d. $|5+2s|\geq -3$

The solution is $\mathbb{R}$.

Since every absolute value is greater than or equal to 0, the equation is true for all s.

TOPIC 4: Applications of Linear Inequalities

Many real-world applications leading to inequalities involve notions such as "is no greater than," "at least as large as," "does not exceed," and so on. Phrases such as these all have precise mathematical translations that use one of the four inequality symbols.

Let's look at how the phrases translate when variables are added.

Example 7: Translating Inequality Phrases

"x is no greater than y"

This means that x is not greater than y, which is the same as saying x is less than or equal to y, so this translates to $x \leq y$

"x is at least as large as y"

If x is at least as large as y, then it can either be as large as (equal to) y or larger (greater) than y, so this phrase translates to $x \geq y$.

"x does not exceed y"

Compare this to the first phrase; the words "is no" carry the same meaning as "does not," and "greater than" is a synonym for "exceed." The two phrases have the same meaning, and so "x does not exceed y" also translates to $x \leq y$.

While it is important to be able to reason out what inequality a particular phrase represents, it is also useful to have a reliable technique for doing these translations.

Given a statement like "x (phrase) y," one method is to ask whether the statement makes sense if x is less than y, if x is equal to y, and if x is greater than y. The answers to these three questions uniquely determine the appropriate inequality symbol.

Applying the process to the first two phrases from Example 7 and one new phrase, we have the results shown in Table 1.

Phrase	Can x be less than y?	Can x equal y?	Can x be greater than y?	Inequality
"x is no greater than y"	Yes	Yes	No	$x \leq y$
"x is at least as large as y"	No	Yes	Yes	$x \geq y$
"y exceeds x"	Yes	No	No	$x < y$

TABLE 1

Example 8: Applications of Inequalities

Express each of the following problems as an inequality, and then solve the inequality.

a. The average daily high temperature in Santa Fe, NM, over the course of three days exceeded 75. Given that the high on the first day was 72 and the high on the third day was 77, what can we say about the high temperature on the second day?

b. As a test for quality at a plant manufacturing silicon wafers for computer chips, a random sample of 10 batches of 1000 wafers each must not detect more than 5 defective wafers per batch on average. In the first 9 batches tested, the average number of defective wafers per batch is found to be 4.78 (to the nearest hundredth). What is the maximum number of defective wafers that can be found in the 10th batch for the plant to pass the quality test?

Solution

a. We'll begin building the inequality with an expression for calculating the average. Letting x represent the high temperature on the second day, we have the following.

$$\frac{72+x+77}{3}$$

What inequality symbol do we use? The problem states the average *exceeded* 75. To exceed is to be greater than, so we use the $>$ symbol.

$$\frac{72+x+77}{3}>75$$

We then proceed to solve the inequality.

$$\begin{aligned}\frac{72+x+77}{3}&>75\\149+x&>225\\x&>76\end{aligned}$$

Thus, the high temperature on the second day exceeded 76 degrees.

b. The phrase "must not detect more than 5 defective wafers per batch on average" means the average number must be less than or equal to 5. Let x denote the maximum number of defective wafers in the last batch.

$$\begin{aligned}\frac{(9)(4.78)+x}{10}&\leq 5\\43.02+x&\leq 50\\x&\leq 6.98\end{aligned}$$

The number of defective wafers found in the first 9 batches is $(9)(4.78)=43.02$.

Since it is not possible to have a fractional number of wafers, there must have been 43 defective wafers in the first 9 batches, so the maximum allowable number of defective wafers in the final batch is 7.

1.7 EXERCISES

PRACTICE

Determine which elements of $S=\{12,-9,3.14,-2.83,1,5.24,8,-3,4\}$ satisfy each inequality below.

1. $7y-33.6<-8.6+2y$
2. $-2.2y-18.8\geq 5.2(1-y)$
3. $-40<4y-8\leq 4$
4. $-4<-2(z-2)\leq 2$

Solve the following linear inequalities. Describe the solution set using interval notation and by graphing. See Examples 2 and 3.

5. $4+3t\leq t-2$
6. $x-7\geq 5+3x$
7. $5y-24<-9.6+2y$
8. $-\frac{v+2}{3}>\frac{5-v}{2}$
9. $4.2x-5.6<1.6+x$
10. $8.5y-3.5\geq 2.5(3-y)$
11. $-2(3-x)<-2x$
12. $\frac{1-x}{5}>\frac{-x}{10}$
13. $4w+7\leq -7w+4$
14. $-5(p-3)>19.8-p$
15. $\frac{6f-2}{5}<\frac{5f-3}{4}$
16. $\frac{u-6}{7}\geq\frac{2u-1}{3}$
17. $0.04n+1.7<0.13n-1.45$
18. $2k+\frac{3}{2}<5k-\frac{7}{3}$
19. $\frac{4x+4}{5}>\frac{3x+2.6}{4}$
20. $-1.4z-19.6\geq 4.4(1-z)$
21. $6m+\frac{7}{4}>\frac{4m+5.8}{5}$
22. $-3.9n-5.4\geq 6.2(2-3n)$

Solve the following double inequalities. Describe the solution set using interval notation and by graphing. See Examples 4 and 5.

23. $-4<3x-7\leq 8$
24. $5\leq 2m-3\leq 13$
25. $-36<3x-6\leq 12$
26. $2<3(x+2)\leq 21$
27. $-8\leq\frac{z}{2}-4<-5$
28. $6(x-1)<2(3x+5)\leq 6x+10$
29. $3<\frac{w+3}{8}\leq 9$
30. $4\leq\frac{p+7}{-2}<9$
31. $\frac{1}{3}<\frac{7}{6}(l-3)<\frac{2}{3}$
32. $-10<-2(4+y)\leq 9$
33. $\frac{1}{4}\leq\frac{g}{2}-3<5$
34. $-1.2\leq\frac{x+3}{-5}\leq 0.2$
35. $0.08<0.03c+0.13\leq 0.16$

Solve the following absolute value inequalities. Describe the solution set using interval notation and by graphing. See Example 6.

36. $|x-2| \geq 5$

37. $|4-2x| > 11$

38. $4+|3-2y| \leq 6$

39. $4+|3-2y| > 6$

40. $2|z+5| < 12$

41. $7-\left|\frac{q}{2}+3\right| \geq 12$

42. $4|z+3| \leq 28$

43. $-3|4-t| < -6$

44. $-3|4-t| > -6$

45. $3|4-t| < -6$

46. $7-|4-2y| \leq -5$

47. $11-\left|\frac{w}{4}+1\right| \geq 12$

48. $5.5+|x-7.2| \leq 3.5$

49. $6-5|x+2| \geq -4$

50. $|2x-1| < x+4$

51. $|3t+4| > -8$

52. $2 < |6w-2|+7$

The words "and" and "or" can appear explicitly between two inequalities, and their meaning in such cases is the same as in absolute value inequalities. If two inequalities are joined by the word "and," the solution set consists of all those real numbers that satisfy both inequalities; that is, the solution set overall is the intersection of the two individual solution sets. If the word "or" appears between two inequalities, the solution set consists of all those real numbers that satisfy at least one of the two inequalities; in other words, the solution set overall is the union of the two individual solution sets.

Guided by the above paragraph, solve the following inequality problems. Describe the solution set using interval notation and by graphing.

53. $t < 2t-3$ and $-3(t+4) > -57$

54. $7-\frac{3x}{5} < \frac{2}{5}$ or $2-3x \geq 5$

55. $-2(a-1) < 4$ and $6+a \leq 9$

56. $-2(a-1) < 4$ and $6-a \leq 9$

57. $-2(a-1) < 4$ or $6+a \leq 9$

58. $\frac{5n+6}{3} < -10$ and $-3(n-1) < -6$

59. $\frac{23x-3}{-7} \leq 7$ and $-x < -(4x-9)$

60. $7-\frac{x}{3} \leq 14+\frac{x}{2}$ or $-3x < 15$

APPLICATIONS

61. In a class in which the final course grade depends entirely on the average of four equally weighted 100-point tests, Cindy has scored 96, 94, and 97 on the first three. The professor has announced that there will be a 15-point bonus problem on the fourth test, and anyone who finishes the semester with an average of more than 100 will receive an A+. What interval of scores on the fourth test will give Cindy an A for the semester (an average between 90 and 100, inclusive), and what interval will give Cindy an A+?

62. In a series of 30 racquetball games played to date, Larry has won 10, giving him a winning average so far of 33.3% (to the nearest tenth of a percent). If he continues to play, what interval describes the number of games he must now win in a row to have an overall winning average greater than 50%?

63. Assume that the national average SAT score for high school seniors is 1020 out of 1600. A group of seven students receive their scores in the mail, and six of them look at their scores. Two students scored 1090, one got an 1120, two others each got a 910, and the sixth student received an 880. What interval of scores can the seventh student receive to pull the group's average above the national average?

64. The central bank of a certain country tries to keep the inflation rate below 5.0% on an annual basis. Assume that inflation rates for the first three quarters of a given year are as follows: 5.2%, 4.3%, and 4.7%. What interval of inflation rates for the final quarter would satisfy the government's goal?

1.8 POLYNOMIAL AND POLYNOMIAL-LIKE EQUATIONS IN ONE VARIABLE

TOPICS

1. Solving quadratic equations by factoring
2. Solving "perfect square" quadratic equations
3. Solving quadratic equations by completing the square
4. The quadratic formula
5. Applications of quadratic equations
6. Solving quadratic-like equations
7. Solving general polynomial equations by factoring
8. Solving polynomial-like equations by factoring

TOPIC 1: Solving Quadratic Equations by Factoring

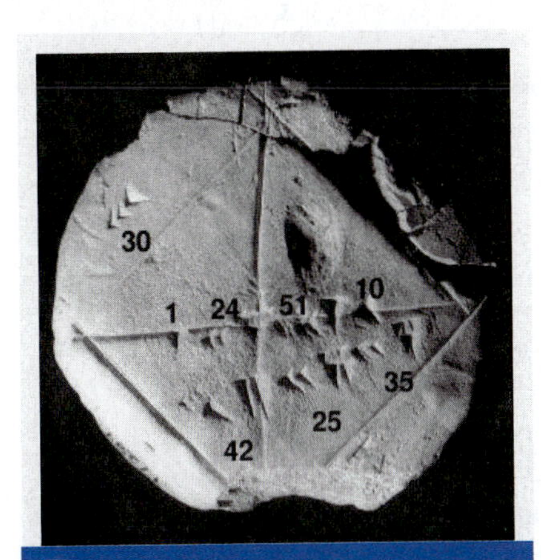

Quadratic Equations circa 2000 BC

Babylonian mathematicians, as far back as 2000 BC, had methods for solving certain area problems that today would take the form of quadratic equations. For instance, they had a well-developed method for finding values for x and y that would solve two equations of the form $x+y=p$ and $xy=q$ at the same time, which is equivalent to finding solutions to the quadratic equation $x^2-px+q=0$.

In Section 1.6, we studied first-degree polynomial equations in one variable. We will now expand the class of one-variable polynomial equations that we can solve to include **quadratic** (or **second-degree polynomial**) equations.

Recall that the method of solving linear equations was particularly straightforward, and that the method always works for *any* such equation. We will, by the end of this section, develop a method for solving one-variable second-degree equations that is also guaranteed to work. This is in contrast to polynomial equations in general. In fact, it can be shown that for polynomial equations of degree five and higher there is *no* method that always works.

Our development will begin with a formal definition of quadratic equations, and we will then proceed to study those quadratic equations that can be solved by applying the factoring skills learned in Section 1.3.

DEFINITION: Quadratic Equations

A **quadratic equation in one variable**, say the variable x, is an equation that can be transformed into the form

$$ax^2+bx+c=0,$$

where a, b, and c are real numbers and $a \neq 0$. Such equations are also called **second-degree** equations, as x appears to the second power. The name quadratic comes from the Latin word *quadrus*, meaning "square."

The key to using factoring to solve a quadratic equation, or indeed any polynomial equation, is to rewrite the equation so that 0 appears by itself on one side. This then allows us to use the Zero-Factor Property discussed in Section 1.1.

If the trinomial $ax^2 + bx + c$ can be factored, it can be written as a product of two linear factors A and B. The Zero-Factor Property then implies that the only way for $ax^2 + bx + c$ to be 0 is if one (or both) of A and B is 0. This is all we need to solve the equation.

Example 1: Solving Quadratic Equations by Factoring

Solve the quadratic equations by factoring.

a. $x^2 + \frac{5x}{2} = \frac{3}{2}$ **b.** $s^2 + 9 = 6s$ **c.** $5x^2 + 10x = 0$

Solution

a.

$$x^2 + \frac{5x}{2} = \frac{3}{2}$$

$$2x^2 + 5x = 3$$

To make the polynomial easier to factor, we multiply both sides by the LCD.

$$2x^2 + 5x - 3 = 0$$

$$(2x-1)(x+3) = 0$$

$$2x - 1 = 0 \quad \text{or} \quad x + 3 = 0$$

Although we could factor $2x^2 + 5x$, this would not do us any good. We must have 0 on one side in order to apply the Zero-Factor Property.

$$x = \frac{1}{2} \quad \text{or} \quad x = -3$$

After factoring, we have two linear equations to solve. The solution set is $\left\{\frac{1}{2}, -3\right\}$.

b.

$$s^2 + 9 = 6s$$

$$s^2 - 6s + 9 = 0$$

$$(s-3)^2 = 0$$

Again, we rewrite the equation with 0 on one side, and then factor the quadratic.

$$s - 3 = 0 \quad \text{or} \quad s - 3 = 0$$

$$s = 3$$

In this example, the two linear factors are the same. In such cases, the single solution is called a *double solution* or a *double root.*

c.

$$5x^2 + 10x = 0$$

$$5x(x+2) = 0$$

$$5x = 0 \quad \text{or} \quad x + 2 = 0$$

$$x = 0 \quad \text{or} \quad x = -2$$

An alternative approach in this example would be to divide both sides by 5 at the very beginning. This would lead to the equation $x(x+2) = 0$, which gives us the same solution set of $\{0, -2\}$.

TOPIC 2: Solving "Perfect Square" Quadratic Equations

The factoring method is fine when it works, but there are two potential problems with the method: (1) the second-degree polynomial in question might not factor over the integers, and (2) even if the polynomial does factor, the factored form may not be obvious.

In some cases where the factoring method is unsuitable, the solution can be obtained by using our knowledge of square roots. If A is an algebraic expression and if c is a constant, the equation $A^2 = c$ means $A = \sqrt{c}$ or $A = -\sqrt{c}$. We will find it convenient to summarize this as follows.

$$A^2 = c \quad \text{implies} \quad A = \pm\sqrt{c}$$

If a given quadratic equation can be written in the form $A^2 = c$, we can use the above observation to obtain two linear equations that can be easily solved.

Example 2: Perfect Square Quadratic Equations

Solve the quadratic equations by taking square roots.

a. $(2x+3)^2 = 8$ **b.** $(x-5)^2 + 4 = 0$

Solution

NOTE

In the factoring method, we move all terms to one side. Here, we isolate a term that is squared, ideally with only a constant on the other side.

a. $(2x+3)^2 = 8$

$2x+3 = \pm\sqrt{8}$

$2x+3 = \pm 2\sqrt{2}$

We begin by taking the square root of each side, keeping in mind that there are two numbers whose square is 8.

$2x = -3 \pm 2\sqrt{2}$

$x = \dfrac{-3 \pm 2\sqrt{2}}{2}$

We solve the two linear equations at once by subtracting 3 from both sides and then dividing both sides by 2. The solution set is $\left\{\dfrac{-3+2\sqrt{2}}{2}, \dfrac{-3-2\sqrt{2}}{2}\right\}$.

b. $(x-5)^2 + 4 = 0$

$(x-5)^2 = -4$

Before taking square roots, we isolate the perfect square algebraic expression on one side, and put the constant on the other.

$x - 5 = \pm\sqrt{-4}$

$x - 5 = \pm 2i$

$x = 5 \pm 2i$

In this example, taking square roots leads to two complex number solutions. (See Section 1.5 for a review of complex numbers.) The solution set is $\{5+2i, 5-2i\}$.

TOPIC 3: Solving Quadratic Equations by Completing the Square

There are potential pitfalls, once again, with the method just developed. If the quadratic equation under consideration appears in the form $A^2 = c$, the method works well (even if the ultimate solutions wind up being complex, as in Example 2b). But what if the equation doesn't have the form $A^2 = c$?

The method of **completing the square** allows us to write an arbitrary quadratic equation $ax^2+bx+c=0$ in the desired form. This method is outlined as follows.

PROCEDURE: Completing the Square

Step 1: Write the equation $ax^2+bx+c=0$ in the form $ax^2+bx=-c$.

Step 2: Divide by a, if $a\neq 1$, so that the coefficient of x^2 is 1: $x^2+\frac{b}{a}x=-\frac{c}{a}$.

Step 3: Divide the coefficient of x by 2, square the result, and add this to both sides: $x^2+\frac{b}{a}x+\left(\frac{b}{2a}\right)^2=-\frac{c}{a}+\left(\frac{b}{2a}\right)^2$.

Step 4: The trinomial on the left side will now be a perfect square trinomial. That is, it can be written as the square of a binomial.

At this point, the equation will have the form $A^2=c$ and can be solved by taking the square root of both sides.

Example 3: Completing the Square

Solve the quadratic equations by completing the square.

a. $x^2-2x-6=0$ **b.** $9x^2+3x=2$

Solution

a.
$$x^2-2x-6=0$$
$$x^2-2x=6$$
$$x^2-2x+1=6+1$$
$$(x-1)^2=7$$
$$x-1=\pm\sqrt{7}$$
$$x=1\pm\sqrt{7}$$

After moving the constant term to the right-hand side, we divide -2 (the coefficient of x) by 2 to get -1, and add $(-1)^2$ to both sides. The trinomial on the left can now be factored.

Taking square roots leads to two easily solved linear equations.

b.
$$9x^2+3x=2$$
$$x^2+\frac{1}{3}x=\frac{2}{9}$$
$$x^2+\frac{1}{3}x+\frac{1}{36}=\frac{2}{9}+\frac{1}{36}$$
$$\left(x+\frac{1}{6}\right)^2=\frac{1}{4}$$
$$x+\frac{1}{6}=\pm\frac{1}{2}$$
$$x=-\frac{1}{6}\pm\frac{1}{2}$$
$$x=\frac{1}{3},-\frac{2}{3}$$

The constant term is already isolated on the right-hand side, so our first step is to divide by 9 (and simplify the resulting fractions, if possible).

Half of the coefficient of x is $\frac{1}{6}$, and the square of this is $\frac{1}{36}$.

After simplifying the sum of fractions on the right, we take the square root of each side.

Since the answer of $-\frac{1}{6}\pm\frac{1}{2}$ can be simplified, we do so to obtain the final answer.

It is time to mention an incidental benefit of the method we have devised. The polynomial in Example 3a, x^2-2x-6, does not factor over the integers. That is, it cannot be written as a product of two first-degree polynomials with integer coefficients. Nevertheless, it can be factored as a product of two first-degree polynomials.

$$x^2-2x-6=\left(x-1-\sqrt{7}\right)\left(x-1+\sqrt{7}\right)$$

We know this because the two solutions of the equation $x^2-2x-6=0$ are $1+\sqrt{7}$ and $1-\sqrt{7}$, and we know that there is a close relationship between factors of a quadratic and the solutions of the equation in which that quadratic is equal to 0.

Specifically, if a quadratic can be factored as $(x-p)(x-q)$, then p and q solve the equation $(x-p)(x-q)=0$.

How does the quadratic $9x^2+3x-2$ factor? This quadratic comes from Example 3b, so we might guess factors of $x-\frac{1}{3}$ and $x+\frac{2}{3}$. But,

$$\left(x-\frac{1}{3}\right)\left(x+\frac{2}{3}\right)=x^2+\frac{1}{3}x-\frac{2}{9}.$$

It shouldn't be surprising that the product of these two factors has a leading coefficient of 1, since each of them individually has a leading coefficient of 1. To get the correct leading coefficient, we need to multiply by 9.

$$9\left(x-\frac{1}{3}\right)\left(x+\frac{2}{3}\right)=9\left(x^2+\frac{1}{3}x-\frac{2}{9}\right)=9x^2+3x-2$$

To make the factors look better, we can factor 9 into two factors of 3 and rearrange the products.

$$9\left(x-\frac{1}{3}\right)\left(x+\frac{2}{3}\right)=3\left(x-\frac{1}{3}\right)\cdot 3\left(x+\frac{2}{3}\right)=(3x-1)(3x+2)$$

TOPIC 4: The Quadratic Formula

The method of completing the square will always serve to solve any equation of the form $ax^2+bx+c=0$. But this begs the question: why not just solve $ax^2+bx+c=0$ once and for all? Since a, b, and c represent arbitrary constants, the ideal situation would be to find a formula for the solutions of $ax^2+bx+c=0$ based on a, b, and c. That is exactly what the quadratic formula is—a formula that gives the solution to *any* equation of the form $ax^2+bx+c=0$. We will derive the formula now, using what we have learned.

$$ax^2 + bx + c = 0$$

$$x^2 + \frac{b}{a}x = -\frac{c}{a}$$

We begin with Steps 1 and 2, moving the constant to the right-hand side and dividing by a.

$$x^2 + \frac{b}{a}x + \frac{b^2}{4a^2} = -\frac{c}{a} + \frac{b^2}{4a^2}$$

$$\left(x + \frac{b}{2a}\right)^2 = -\frac{4ac}{4a^2} + \frac{b^2}{4a^2}$$

$$\left(x + \frac{b}{2a}\right)^2 = \frac{b^2 - 4ac}{4a^2}$$

We next divide $\frac{b}{a}$ by 2 to get $\frac{b}{2a}$ and add $\left(\frac{b}{2a}\right)^2 = \frac{b^2}{4a^2}$ to both sides. Note that to add the fractions on the right, we need a common denominator of $4a^2$.

$$x + \frac{b}{2a} = \pm\frac{\sqrt{b^2 - 4ac}}{2a}$$

$$x = \frac{-b}{2a} \pm \frac{\sqrt{b^2 - 4ac}}{2a}$$

Taking square roots leads to two linear equations, which we then solve for x.

$$x = \frac{-b \pm \sqrt{b^2 - 4ac}}{2a}$$

Since the fractions have the same denominator, they are easily added to obtain the final formula.

f(x) FORMULA: The Quadratic Formula

The solutions of the general quadratic equation $ax^2 + bx + c = 0$, with $a \neq 0$, are given by the **quadratic formula**: $x = \frac{-b \pm \sqrt{b^2 - 4ac}}{2a}$.

The expression beneath the radical, $b^2 - 4ac$, is called the **discriminant**. Its value determines the number and type (real or complex) of solutions.

Discriminant	Number of Distinct Solutions	Type of Solutions	Notes
$b^2 - 4ac > 0$	2	Real	The solutions are always different.
$b^2 - 4ac = 0$	1	Real	This solution is a double root.
$b^2 - 4ac < 0$	2	Complex	The solutions are complex conjugates.

Example 4: Using the Quadratic Formula

Solve the quadratic equations using the quadratic formula.

a. $8x^2 - 4x = 1$

b. $t^2 + 6t + 13 = 0$

Solution

a.

$$8x^2 - 4x = 1$$

$$8x^2 - 4x - 1 = 0$$

$$a = 8, \quad b = -4, \quad c = -1$$

Before applying the quadratic formula, move all the terms to one side so a, b, and c can be identified correctly.

$$x = \frac{-(-4) \pm \sqrt{(-4)^2 - 4(8)(-1)}}{2(8)}$$

Apply the quadratic formula by making the appropriate replacements for a, b, and c.

$$x = \frac{4 \pm \sqrt{16+32}}{16}$$

$$x = \frac{4 \pm \sqrt{48}}{16}$$

The discriminant, 48, is positive.

$$x = \frac{4 \pm 4\sqrt{3}}{16}$$

$$x = \frac{1 \pm \sqrt{3}}{4}$$

We can cancel out the common factor of 4 in the numerator and denominator.

Thus, the solutions are $x = \frac{1+\sqrt{3}}{4}$ and $x = \frac{1-\sqrt{3}}{4}$; two unique, real solutions.

b. $t^2 + 6t + 13 = 0$

The equation is already in the proper form to apply the quadratic formula.

$$a = 1, \quad b = 6, \quad c = 13$$

$$t = \frac{-(6) \pm \sqrt{(6)^2 - 4(1)(13)}}{2(1)}$$

Substitute the values for a, b, and c.

$$t = \frac{-6 \pm \sqrt{36-52}}{2}$$

$$t = \frac{-6 \pm \sqrt{-16}}{2}$$

The discriminant is negative, so we know the solutions will be complex.

$$t = \frac{-6 \pm 4i}{2}$$

$$t = -3 \pm 2i$$

Again, we cancel out a common factor.

Thus, the solutions are $t = -3 + 2i$ and $t = -3 - 2i$; two complex conjugate solutions.

TECHNOLOGY

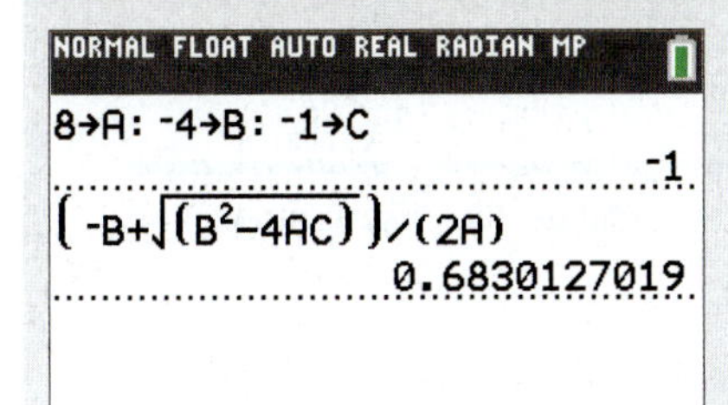

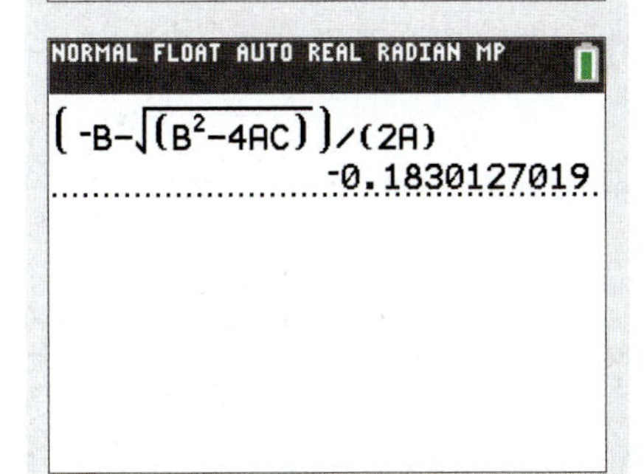

To use the quadratic formula on a TI-84 Plus, first define the variables *A*, *B*, and *C*. Enter the numerical value for *A*, press sto→, press alpha math for A, and press alpha . to enter a colon. Continue to store the values for *B* and *C*, and then press enter.

Once your variables are defined, enter the quadratic formula using alpha and the corresponding letters of the variables. Remember to use parentheses!

To obtain both solutions, enter the formula once with addition of the square root term in the numerator and once with subtraction of the square root term in the numerator.

Calculating the discriminant can be a very useful tool; it provides a quick check of whether solutions are reasonable, and later on we will find it helpful in classifying the graphs of quadratic equations.

Example 5: The Discriminant

For each of the following quadratic equations, calculate the discriminant and determine the number and type of solutions.

a. $-2x^2 + 12x - 18 = 0$ **b.** $5x^2 + 7x + 2 = 0$ **c.** $x^2 - 4x + 9 = 0$

Solution

We identify the values of a, b, and c, then calculate the discriminant $b^2 - 4ac$.

a. $-2x^2 + 12x - 18 = 0$

$a = -2, \quad b = 12, \quad c = -18$

We substitute, then calculate the discriminant.

$$b^2 - 4ac = (12)^2 - 4(-2)(-18) = 144 - 144 = 0$$

Since the discriminant is zero, we know there will be one real solution, also known as a double root.

b. $5x^2 + 7x + 2 = 0$

$a = 5, \quad b = 7, \quad c = 2$

$$b^2 - 4ac = (7)^2 - 4(5)(2) = 49 - 40 = 9$$

This time, the discriminant is positive, so there are two distinct real solutions.

c. $x^2 - 4x + 9 = 0$

$a = 1, \quad b = -4, \quad c = 9$

$$b^2 - 4ac = (-4)^2 - 4(1)(9) = 16 - 36 = -20$$

The discriminant is negative, so the equation has two complex conjugate solutions.

While we can solve any quadratic equation using the quadratic formula, it is not always the easiest or most efficient method of solution. For example; if an equation is already in factored form, it is much easier to read off the solutions by applying the Zero-Factor Property than to multiply out the factors and then apply the quadratic formula. Similarly, if an equation is already in the form $A^2 = c$, it's much easier to apply the square root method than to use the quadratic formula. In the following example, try to use the easiest, most efficient method to solve the different quadratic equations.

Example 6: Methods of Solving Quadratic Equations

Solve each of the following quadratic equations, identifying the most efficient method of solution.

a. $4x^2 - 25 = 0$

b. $(2x-3)^2 = 7$

c. $3x^2 - 11x - 4 = 0$

d. $3x^2 - 10x - 4 = 0$

Solution

a. The left-hand side is a difference of squares, so factoring is the easiest method.

$$4x^2 - 25 = 0$$
Factor the difference of squares.

$$(2x-5)(2x+5) = 0$$

$$2x - 5 = 0 \text{ or } 2x + 5 = 0$$
We need to solve two linear equations.

$$x = \frac{5}{2} \text{ or } x = -\frac{5}{2}$$
We have two unique real solutions.

b. The left-hand side is already a squared quantity, but the right-hand side is not zero, so we want to use the square root method.

$$(2x-3)^2 = 7$$
Take the square root of both sides.

$$2x - 3 = \pm\sqrt{7}$$
Simplify the linear equation.

$$2x = 3 \pm \sqrt{7}$$

$$x = \frac{3 \pm \sqrt{7}}{2}$$
We have two unique real solutions.

c. While the quadratic formula certainly works, this quadratic equation is factorable.

$$3x^2 - 11x - 4 = 0$$
Use trial and error or factoring by grouping to factor the trinomial into two binomials.

$$(3x+1)(x-4) = 0$$

$$3x + 1 = 0 \text{ or } x - 4 = 0$$
The Zero-Factor Property gives us two linear equations to solve.

$$x = -\frac{1}{3} \text{ or } x = 4$$

d. Here, the equation is not factorable, so we use the quadratic formula.

$$3x^2 - 10x - 4 = 0$$

$$a = 3, \quad b = -10, \quad c = -4$$
Identify the values of a, b, and c, then substitute into the quadratic formula.

$$x = \frac{-(-10) \pm \sqrt{(-10)^2 - 4(3)(-4)}}{2(3)}$$
All that remains is to simplify the solutions.

$$x = \frac{10 \pm \sqrt{148}}{6}$$

$$x = \frac{10 \pm 2\sqrt{37}}{6}$$
We have two unique real solutions.

$$x = \frac{5 \pm \sqrt{37}}{3}$$

TOPIC 5: Applications of Quadratic Equations

When an object near the surface of Earth is moving under the influence of gravity alone, its height above the surface is described by a quadratic polynomial in the variable t, where t stands for time and is usually measured in seconds. The phrase "moving under the influence of gravity alone" means that all other forces that could potentially affect the object's motion, such as air resistance or mechanical lifting forces, are either negligible or absent. The phrase "near the surface of Earth" means that we are considering objects that travel short vertical distances relative to the radius of Earth; the following formula doesn't apply, for instance, to rockets shot into orbit. As an example, think of someone throwing a baseball into the air on a windless day. After the ball is released, gravity is the only force acting on it.

If we let h represent the height at time t of such an object,

$$h = -\frac{1}{2}gt^2 + v_0 t + h_0,$$

where g, v_0, and h_0 are all constants: g is the force due to gravity, v_0 is the initial velocity which the object has when $t = 0$, and h_0 is the height of the object when $t = 0$ (we normally say that ground level corresponds to a height of 0). If t is measured in seconds and h in feet, g is 32 ft/s^2. If t is measured in seconds and h in meters, g is 9.8 m/s^2.

Many applications involving the above formula will result in a quadratic equation that must be solved for t. In some cases, one of the two solutions must be discarded as meaningless in the given problem.

Example 7: Gravity Problems

Robert stands on the topmost tier of seats in a baseball stadium, and throws a ball out onto the field with a vertical upward velocity of 60 ft/s. The ball is 50 feet above the ground at the moment he releases the ball. When does the ball land?

Solution

First, note that although the thrown ball has a horizontal velocity as well as a vertical velocity (otherwise it would go straight up and come straight back down), the horizontal velocity is irrelevant in such questions. All we are interested in is when the ball lands on the ground ($h = 0$). If we wanted to determine where in the field the ball lands, we would have to know the horizontal velocity as well.

We are given the following: $h_0 = 50 \text{ ft}$ and $v_0 = 60 \text{ ft/s}$. Since the units in the problem are feet and seconds, we know to use $g = 32 \text{ ft/s}^2$. What we are interested in determining is the time, t, when the height, h, of the ball is 0. That is, we need to solve the equation $0 = -16t^2 + 60t + 50$ for t.

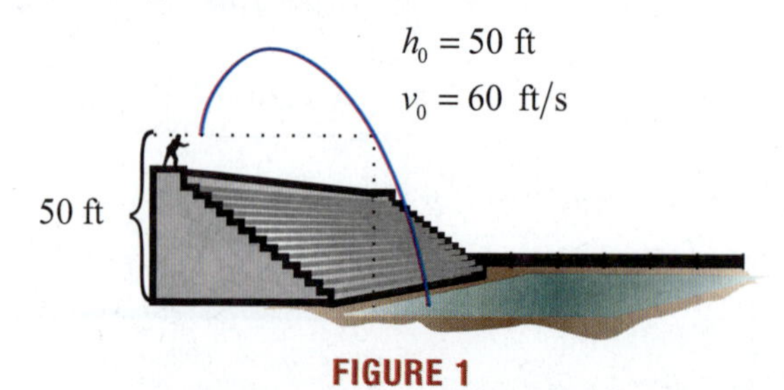

FIGURE 1

$$0 = -16t^2 + 60t + 50$$

$$0 = 8t^2 - 30t - 25$$

$$t = \frac{30 \pm \sqrt{900 + 800}}{16}$$

The quadratic polynomial in the equation doesn't factor over the integers, so the quadratic formula is a good method to use. To simplify the calculations, we can begin by dividing both sides of the equation by −2.

$$t = \frac{30 \pm 10\sqrt{17}}{16}$$

$$t = \frac{15 \pm 5\sqrt{17}}{8}$$

We then proceed to reduce the radical and simplify the resulting fraction (which actually describes two solutions).

$$t \approx \cancel{-0.70}, 4.45$$

These numbers are most meaningful in decimal form.

The negative solution is immaterial in this problem, as this represents a time before the ball is thrown, so it is discarded. The ball lands 4.45 seconds after being thrown.

TOPIC 6: Solving Quadratic-Like Equations

A polynomial equation of degree n in one variable, say x, is an equation that can be written in the form $a_n x^n + a_{n-1}x^{n-1} + \cdots + a_1 x + a_0 = 0$, where each a_i is a constant and $a_n \neq 0$. As we have seen, such equations can always be solved if $n = 1$ or $n = 2$, but in general there is no method for solving polynomial equations that is guaranteed to find all solutions. There are formulas, called the *cubic* and *quartic* formulas, that solve third- and fourth-degree polynomial equations, but there are no formulas to solve polynomial equations of degree five (or higher)! Moreover, many nonpolynomial equations have no solution method that is guaranteed to work.

However, even though no formula exists for solving these equations, it may still be possible to solve them! In this section, we'll use factoring, the Zero-Factor Property, and our knowledge of quadratic equations to solve higher-degree polynomial equations.

The Zero-Factor Property applies whenever a product of any finite number of factors is equal to 0; if $A_1 \cdot A_2 \cdot \cdots \cdot A_n = 0$, then at least one of the A_i's must equal 0. Recall that we used the Zero-Factor Property to solve quadratic equations. This means that if we can rewrite an equation in a quadratic form, we can use the Zero-Factor Property to solve this equation as well.

DEFINITION: Quadratic-Like Equations

An equation is **quadratic-like**, or **quadratic in form**, if it can be written in the form

$$aA^2 + bA + c = 0,$$

where a, b, and c are constants, $a \neq 0$, and A is an algebraic expression. Such equations can be solved by first solving for A and then solving for the variable in the expression A. This method of solution is called **substitution**.

Example 8: Quadratic-Like Equations

Solve the quadratic-like equations.

a. $(x^2+2x)^2-7(x^2+2x)-8=0$ **b.** $y^{\frac{2}{3}}+4y^{\frac{1}{3}}-5=0$

Solution

a.
$$(x^2+2x)^2-7(x^2+2x)-8=0$$
$$A^2-7A-8=0$$
$$(A-8)(A+1)=0$$
$$A=8,-1$$

Making the substitution $A=x^2+2x$ transforms the quadratic-like equation into a quadratic equation that can be solved by factoring.

$$A=8 \quad \text{or} \quad A=-1$$
$$x^2+2x=8 \qquad x^2+2x=-1$$
$$x^2+2x-8=0 \qquad x^2+2x+1=0$$
$$(x+4)(x-2)=0 \qquad (x+1)^2=0$$
$$x=-4 \text{ or } x=2 \text{ or } x=-1$$

Once we have solved for A, we replace A with x^2+2x and solve for x.

Note that -1 is a double root, while -4 and 2 are single roots.

While the substitution method does not necessarily introduce extraneous solutions, you should still check that each solution solves the original quadratic-like equation.

> **NOTE**
> The quadratic equations in Example 8a could also be solved using the quadratic formula with the help of a calculator. See Example 4a for instructions on using the quadratic formula with a TI-84 Plus.

b.
$$y^{\frac{2}{3}}+4y^{\frac{1}{3}}-5=0$$
$$\left(y^{\frac{1}{3}}\right)^2+4\left(y^{\frac{1}{3}}\right)-5=0$$
$$A^2+4A-5=0$$
$$(A+5)(A-1)=0$$
$$A=-5,1$$

Using properties of exponents, we can see that the substitution $A=y^{\frac{1}{3}}$ will make this equation quadratic.

$$A=-5 \quad \text{or} \quad A=1$$
$$y^{\frac{1}{3}}=-5 \qquad y^{\frac{1}{3}}=1$$
$$y=(-5)^3 \qquad y=(1)^3$$
$$y=-125 \quad \text{or} \quad y=1$$

Now that we've solved for A, we reverse the substitution and solve for y in each case.

Once again, you should confirm that both values do indeed solve the original equation $y^{\frac{2}{3}}+4y^{\frac{1}{3}}-5=0$.

TOPIC 7: Solving General Polynomial Equations by Factoring

If an equation consists of a polynomial on one side and 0 on the other, and if the polynomial can be factored completely, then the equation can be solved by using the Zero-Factor Property. If the coefficients in the polynomial are all real, the polynomial can, in principle, be factored into a product of first-degree and second-degree factors. In practice, however, this may be difficult to accomplish unless the degree of the polynomial is small or the polynomial is easily recognizable as a special product. For higher-degree polynomials, the GCF factoring method and factoring by grouping can be very effective. You may want to review some of the factoring techniques from Section 1.3 as you work through these problems.

Example 9: Solving Equations by Factoring

Solve the equations by factoring.

a. $x^4 = 9$ **b.** $y^3 + y^2 - 4y - 4 = 0$

c. $8t^3 - 27 = 0$ **d.** $3x^4 + 18x^3 - 21x^2 = 0$

NOTE

While checking solutions is always a good practice, solving by factoring does not produce any extraneous solutions.

TECHNOLOGY

```
NORMAL FLOAT AUTO REAL RADIAN MP
SOLUTION IS MARKED ▪
Y³+Y²-4Y-4=0
▪Y=2
 bound={⁻1E99,1E99}
▪E1-E2=0
SOLVE
```

A word of caution—only one answer is generated using the equation solver on a TI-84 Plus, but there are three solutions. Additional work must be done to find multiple solutions.

Solution

a.

$$x^4 = 9$$
$$x^4 - 9 = 0$$
$$(x^2 - 3)(x^2 + 3) = 0$$

After isolating 0 on one side, we see the polynomial is a difference of two squares, which can always be factored.

$$x^2 = 3 \quad \text{or } x^2 = -3$$
$$x = \pm\sqrt{3} \text{ or } x = \pm\sqrt{-3}$$
$$x = \pm\sqrt{3} \text{ or } x = \pm i\sqrt{3}$$

The Zero-Factor Property gives us two equations, both of which can be solved by taking square roots.

b.

$$y^3 + y^2 - 4y - 4 = 0$$
$$y^2(y+1) - 4(y+1) = 0$$

We factor the initial equation by grouping.

$$(y+1)(y^2 - 4) = 0$$
$$(y+1)(y-2)(y+2) = 0$$

We can factor the second term further, since it is a difference of squares.

$$y = -1 \text{ or } y = 2 \text{ or } y = -2$$

There are three solutions by the Zero-Factor Property.

c.

$$8t^3 - 27 = 0$$
$$(2t)^3 - 3^3 = 0$$
$$(2t-3)(4t^2 + 6t + 9) = 0$$

The polynomial in this case is a difference of two cubes, which can always be factored.

$$2t - 3 = 0 \quad \text{or} \quad 4t^2 + 6t + 9 = 0$$
$$2t = 3 \qquad t = \frac{-(6) \pm \sqrt{6^2 - 4(4)(9)}}{2(4)}$$
$$t = \frac{3}{2} \qquad t = \frac{-6 \pm \sqrt{36 - 144}}{8}$$
$$t = \frac{-6 \pm \sqrt{-108}}{8}$$
$$t = \frac{-6 \pm 6i\sqrt{3}}{8}$$
$$\text{or} \quad t = \frac{-3 \pm 3i\sqrt{3}}{4}$$

The Zero-Factor Property gives us two equations to solve. One of the equations is quadratic, and we can use the quadratic formula to solve it.

Thus, we have three solutions to the original equation.

d.

$$3x^4 + 18x^3 - 21x^2 = 0$$
$$3x^2(x^2 + 6x - 7) = 0$$

Factoring out the GCF of $3x^2$ yields a quadratic term that we can factor.

$$3x^2(x-1)(x+7) = 0$$
$$x^2(x-1)(x+7) = 0$$
$$x = 0, 1, \text{ or } -7$$

The Zero-Factor Property yields three solutions to the original equation.

TOPIC 8: Solving Polynomial-Like Equations by Factoring

The last type of equations that we will consider in this section are equations that are not polynomials but can be solved using the methods we have developed so far. We have already seen one such equation in Example 8b: the equation $y^{\frac{2}{3}}+4y^{\frac{1}{3}}-5=0$ is quadratic-like, and can be solved using polynomial methods. Like the equation in Example 8b, some polynomial-like equations can be solved by substitution, transforming them into polynomial equations. Other equations can be solved by rewriting the equation so that 0 appears on one side, factoring the equation, and then applying the Zero-Factor Property. Often, equations involving rational exponents can be solved by factoring out a common factor, as in the following examples.

Example 10: Solving Equations by Factoring

Solve the equations by factoring.

a. $x^{\frac{7}{3}}+x^{\frac{4}{3}}-2x^{\frac{1}{3}}=0$

b. $(x-1)^{\frac{1}{2}}-(x-1)^{-\frac{1}{2}}=0$

Solution

a.

$$x^{\frac{7}{3}}+x^{\frac{4}{3}}-2x^{\frac{1}{3}}=0$$
$$x^{\frac{1}{3}}\left(x^2+x-2\right)=0$$
$$x^{\frac{1}{3}}(x+2)(x-1)=0$$

Recall that in cases like this, we factor out x raised to the lowest exponent. In this case, the remaining factor is a factorable trinomial.

$$x^{\frac{1}{3}}=0 \quad \text{or} \quad x+2=0 \quad \text{or} \quad x-1=0$$
$$x=0 \quad \text{or} \quad x=-2 \quad \text{or} \quad x=1$$

The Zero-Factor Property leads to three simple equations.

b.

$$(x-1)^{\frac{1}{2}}-(x-1)^{-\frac{1}{2}}=0$$
$$(x-1)^{-\frac{1}{2}}\left((x-1)-1\right)=0$$
$$(x-1)^{-\frac{1}{2}}(x-2)=0$$

Again, we factor out the common algebraic expression raised to the lowest exponent.

$$(x-1)^{-\frac{1}{2}}=0 \quad \text{or} \quad x-2=0$$
$$\frac{1}{(x-1)^{\frac{1}{2}}}=0 \quad \text{or} \quad x=2$$

This equation leads to two equations, only one of which has a solution. (Note that there is no value for x which would solve the first of the two equations.)

$$x=2$$

The original equation has only one solution.

TECHNOLOGY

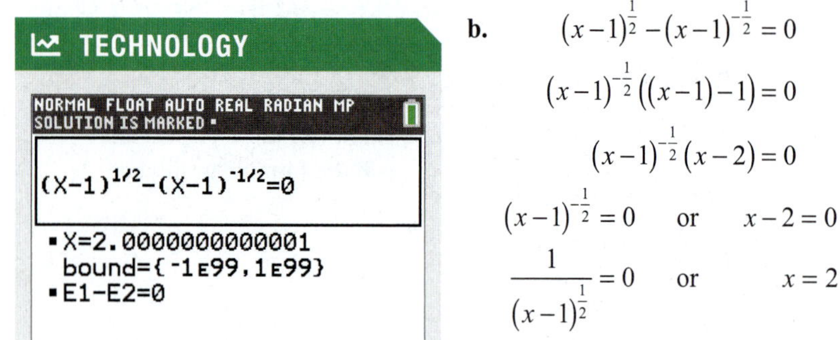

This screen was obtained using the equation solver on a TI-84 Plus CE.

We will end this section with one more example of an equation that can be solved using a technique we learned while studying quadratic equations.

Example 11: Solving Equations by Factoring

Solve the equation $x^{-2} - 7x^{-1} + 12 = 0$ by factoring.

Solution

While the given equation is not quadratic, it certainly bears a strong resemblance to one that is. In fact, if we make the substitution $y = x^{-1}$, we obtain the quadratic equation

$$y^2 - 7y + 12 = 0,$$

which we can solve by factoring the trinomial as $(y-3)(y-4)$. This gives us $y = 3$ or $y = 4$. Translating this back to the variable x, we have $x^{-1} = 3$ or $x^{-1} = 4$, so $x = \frac{1}{3}$ or $x = \frac{1}{4}$.

1.8 EXERCISES

PRACTICE

Solve the following quadratic equations by factoring. See Example 1.

1. $2x^2 - x = 3$

2. $x^2 - 14x + 49 = 0$

3. $y(2y+9) = -9$

4. $3x^2 + 33 = 2x^2 + 14x$

5. $5x^2 + 2x + 3 = 4x^2 + 6x - 1$

6. $(x-7)^2 = 16$

7. $4x^2 - 9 = 0$

Solve the following quadratic equations by taking square roots. See Example 2.

8. $(x-3)^2 = 9$

9. $(a-2)^2 = -5$

10. $(y-18)^2 - 1 = 0$

11. $9 = (3s+2)^2$

12. $(2x-1)^2 = 8$

13. $x^2 - 4x + 4 = 49$

Solve the following quadratic equations by completing the square. See Example 3.

14. $x^2 + 8x + 7 = -8$

15. $2x^2 + 6x - 10 = 10$

16. $2x^2 + 7x - 15 = 0$

17. $4x^2 - 4x - 63 = 0$

18. $4x^2 - 56x + 195 = 0$

19. $y^2 + 22y + 96 = 0$

Solve the following quadratic equations using the quadratic formula. See Example 4.

20. $3x^2 - 4 = -x$

21. $2.6z^2 - 0.9z + 2 = 0$

22. $a(a+2) = -1$

23. $3x^2 - 2x = 0$

24. $6x^2 + 5x - 4 = 3x - 2$

25. $4x^2 - 14x - 27 = 3$

Calculate the discriminant and use it to determine the number and type of solutions of the following quadratic equations. See Example 5.

26. $2x^2 - x + 5 = 0$

27. $x^2 + x - 5 = 0$

28. $-3x^2 - 2x + 2 = 0$

29. $2x^2 - 4x + 2 = 0$

Solve the following quadratic equations using any appropriate method. See Example 6.

30. $y^2 + 9y = -40.50$

31. $(z-11)^2 = 9$

32. $256t^2 - 324 = 0$

33. $(9y-6)^2 = 121y^2$

34. $2x^2 + 8x - 3 = 6x$

35. $4z^2 + 14z = 10z - 3$

36. $x^2 - 6x = 27$

37. $y^2 - 2y + 1 = -289$

38. $-3(b+5)^2 = -768$

39. $7x^2 - 42x = 0$

40. $5x^2 - 5x - 10 = 0$

41. $4w^2 + 10w + 5 = 3w^2 + 18w - 10$

42. $|x^2 - 3x| = 2$ (**Hint:** Replace $|x^2 - 3x|$ first with $x^2 - 3x$ and solve the resulting equation, then replace it with $-(x^2 - 3x)$ and solve the resulting equation.)

43. $|x^2 - x| = 2$

44. $|x^2 - 8| = 1$

Solve the following quadratic-like equations. See Example 8.

45. $(x-1)^2 + (x-1) - 12 = 0$

46. $(y-5)^2 - 11(y-5) + 24 = 0$

47. $(x^2+1)^2 + (x^2+1) - 12 = 0$

48. $(x^2-13)^2 + (x^2-13) - 12 = 0$

49. $(x^2-2x+1)^2 + (x^2-2x+1) - 12 = 0$

50. $2y^{\frac{2}{3}} + y^{\frac{1}{3}} - 1 = 0$

51. $2x^{\frac{2}{3}} - 7x^{\frac{1}{3}} + 3 = 0$

52. $(x^2-6x)^2 + 4(x^2-6x) - 5 = 0$

53. $(y^2-5)^2 + 5(y^2-5) - 36 = 0$

54. $(x^2+7)^2 + 8(x^2+7) + 12 = 0$

55. $(t^2-t)^2 - 8(t^2-t) + 12 = 0$

56. $2x^{\frac{1}{2}} - 5x^{\frac{1}{4}} + 2 = 0$

57. $3x^{\frac{2}{3}} - x^{\frac{1}{3}} - 2 = 0$

58. $5y^{\frac{2}{3}} + 33y^{\frac{1}{3}} + 18 = 0$

Solve the following polynomial equations by factoring. See Example 9.

59. $a^3 - 3a^2 = a - 3$

60. $2x^3 + x^2 + 2x + 1 = 0$

61. $x^4 + 5x^2 - 36 = 0$

62. $y^3 + 8 = 0$

63. $5s^3 + 6s^2 - 20s = 24$

64. $8a^3 - 27 = 0$

65. $16a^4 = 81$

66. $6x^3 + 8x^2 = 14x$

67. $14x^3 + 27x^2 - 20x = 0$

68. $5z^3 + 28z^2 = 49z$

69. $27x^3 + 64 = 0$

70. $x^3 - 4x^2 + x = 4$

Solve the following equations by factoring. See Examples 10 and 11.

71. $3x^{\frac{11}{3}} + 2x^{\frac{8}{3}} - 5x^{\frac{5}{3}} = 0$

72. $(y-6)^{\frac{-5}{2}} + 7(y-6)^{\frac{-3}{2}} = 0$

73. $y^{-2} - 2y^{-1} + 1 = 0$

74. $2x^{\frac{13}{5}} - 5x^{\frac{8}{5}} + 2x^{\frac{3}{5}} = 0$

75. $(2x-5)^{\frac{1}{3}} - 3(2x-5)^{\frac{-2}{3}} = 0$

76. $x^{-4} - 13x^{-2} + 36 = 0$

77. $y^{\frac{7}{2}} - 5y^{\frac{5}{2}} + 6y^{\frac{3}{2}} = 0$

78. $(t+4)^{\frac{2}{3}} + 2(t+4)^{\frac{8}{3}} = 0$

79. $y^{-2} - 2y^{-1} - 35 = 0$

80. $x^{\frac{11}{2}} - 6x^{\frac{9}{2}} + 9x^{\frac{7}{2}} = 0$

81. $5y^{\frac{11}{3}} + 3y^{\frac{8}{3}} - 2y^{\frac{5}{3}} = 0$

82. $(3x-3)^{\frac{-1}{3}} - 5(3x-3)^{\frac{-4}{3}} = 0$

83. $x^{-2} + 8x^{-1} + 15 = 0$

84. $(y+3)^{\frac{2}{5}} + 4(y+3)^{\frac{7}{5}} = 0$

APPLICATIONS

85. How long would it take for a ball dropped from the top of a 144-foot building to hit the ground?

86. Suppose that instead of being dropped, as in Exercise 85, a ball is thrown upward with a velocity of 40 feet per second from a height of 144 feet. Assuming it misses the building on the way back down, how long after being thrown will it hit the ground?

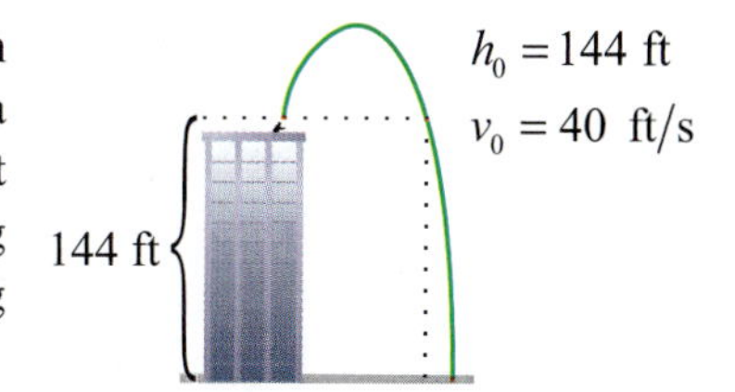

87. A slingshot is used to shoot a BB at a velocity of 96 feet per second straight up from ground level. When will the BB reach its maximum height of 144 feet?

88. A rock is thrown off a cliff with a velocity of 20 meters per second. It is thrown upward from a height of 24 meters and misses the cliff on the way back down. When will the rock be 7 meters from ground level? (Round your answer to one decimal place.)

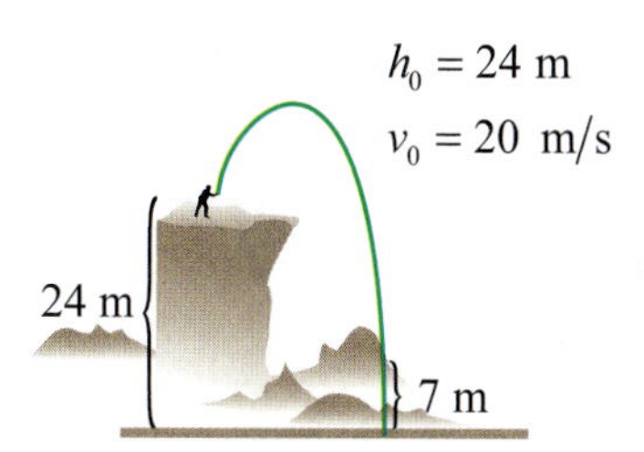

89. Luke, an experienced bungee jumper, leaps from a tall bridge and falls toward the river below. The bridge is 170 feet above the water and Luke's bungee cord is 110 feet long unstretched. When will Luke's cord begin to stretch? (Round your answer to one decimal place.)

WRITING & THINKING

90. Compare the answers to Exercises 85 and 87 and explain why they are the same.

Use the connection between solutions of polynomial equations and polynomial factoring to answer the following questions. See the discussion after Example 3.

91. Factor the quadratic $x^2 - 6x + 13$.

92. Factor the quadratic $9x^2 - 6x - 4$.

93. Factor the quadratic $4x^2 + 12x + 1$.

94. Factor the quadratic $25x^2 - 10x + 2$.

95. Determine b and c so that the equation $x^2 + bx + c = 0$ has the solution set $\{-3, 8\}$.

96. Find b, c, and d so the equation $x^3 + bx^2 + cx + d = 0$ has solutions of -3, -1, and 5.

97. Find b, c, and d so the equation $x^3 + bx^2 + cx + d = 0$ has solutions of -2, 0, and 6.

98. Find b and c so the equation $x^3 + bx^2 + cx = 0$ has solutions of 0, 1, and -7.

99. Find a, c, and d so the equation $ax^3 + 4x^2 + cx + d = 0$ has solutions of -4, 6, and -6.

100. Find a, b, and d so the equation $ax^3 + bx^2 + 3x + d = 0$ has solutions of -3, $-\frac{1}{2}$, and 0.

101. Find a, b, and c so the equation $ax^3 + bx^2 + cx + 6 = 0$ has solutions of $-\frac{3}{5}, \frac{2}{3}$, and 1.

TECHNOLOGY

Use a graphing utility to solve the following polynomial equations.

102. $5x^2 - 3x = 17$

103. $5x^2 - 3x = -17$

104. $(a+4)(4a-3) = 5$

105. $10\pi r + \pi r^2 = 107$

106. $4.8x^2 + 3.5x - 9.2 = 0$

107. $(3x-1)(3-x) = 2x + 5$

1.9 RATIONAL AND RADICAL EQUATIONS IN ONE VARIABLE

TOPICS

1. Solving rational equations
2. Applications of rational equations
3. Solving radical equations
4. Solving equations with positive rational exponents
5. Solving equations for one variable

TOPIC 1: Solving Rational Equations

When an equation contains a fraction in which the variable appears in the denominator, the solution process can be challenging. In this section, we will learn how to solve equations containing such expressions.

A **rational equation** is an equation that contains at least one rational expression, while any nonrational expressions are polynomials. Our general approach to solving such equations is to multiply each term in the equation by the LCD of all the rational expressions; this has the effect of converting rational equations into polynomial equations, which we have already learned how to solve.

There is one important difference between rational and polynomial equations: it is quite possible that one or more rational expressions in a rational equation are not defined for some values of the variable. These values cannot possibly be solutions of the equation, and must be excluded from the solution set. However, these excluded values may appear as solutions of the polynomial equation derived from the original rational equation. These are extraneous solutions! Errors can be avoided by keeping track of what values of the variable are disallowed and/or checking all solutions in the original equation.

Example 1: Solving Rational Equations

Solve the following rational equations.

a. $\dfrac{x^3+3x^2}{x^2-2x-15}=\dfrac{4x+5}{x-5}$

b. $\dfrac{3x^2}{5x-1}-1=0$

Solution

a.

$$\frac{x^3+3x^2}{x^2-2x-15}=\frac{4x+5}{x-5}$$

$$\frac{x^2\cancel{(x+3)}}{(x-5)\cancel{(x+3)}}=\frac{4x+5}{x-5}$$

$$(x-5)\cdot\frac{x^2}{(x-5)}=(x-5)\cdot\frac{4x+5}{x-5}$$

$$x^2=4x+5$$

$$x^2-4x-5=0$$

$$(x-5)(x+1)=0$$

$$x=\cancel{5},-1$$

$$x=-1$$

In order to cancel factors, and in order to determine the LCD, we begin by factoring all the numerators and denominators. This also tells us the very important fact that 5 and –3 cannot be solutions of the equation.

After multiplying both sides of the equation by the LCD, we have a second-degree polynomial equation that can be solved by factoring.

Note that we already determined that 5 cannot be a solution. It must be discarded.

b.

$$\frac{3x^2}{5x-1}-1=0$$

$$(5x-1)\cdot\frac{3x^2}{5x-1}-(5x-1)=0$$

$$3x^2-5x+1=0$$

$$x=\frac{5\pm\sqrt{25-12}}{6}$$

$$x=\frac{5\pm\sqrt{13}}{6}$$

There is no factoring possible in this problem, so we just note that $\frac{1}{5}$ is the one value for x that must be excluded.

After multiplying through by the LCD, we have a second-degree polynomial equation that can be solved by the quadratic formula. Since neither of the roots is $\frac{1}{5}$, both solve the original equation.

TECHNOLOGY

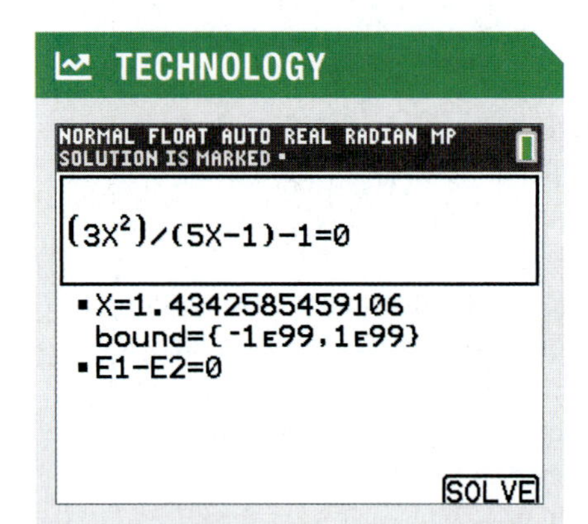

If one or more of the expressions in a rational equation appears inside absolute value symbols, we can use a variation on the solution process to solve the equation. We simply need to remember that each absolute value expression can represent two corresponding expressions without the absolute value symbols.

Example 2: Solving an Absolute Value Rational Equation

Solve the equation $|x-5|=\frac{7}{x+1}$.

Solution

The key step is to remember that $|x-5|$ can represent either $x-5$ or $-(x-5)$, depending on whether $x-5$ is positive or negative; that is, every occurrence of an absolute value expression in an equation leads to two equations without the absolute value symbols. In this case, we need to solve both

$$x-5=\frac{7}{x+1}\text{ and }-(x-5)=\frac{7}{x+1},$$

remembering to check for extraneous solutions at the end.

We will carry out the solutions of the two equations in parallel, as the steps are the same.

$$x-5=\frac{7}{x+1} \qquad\qquad -(x-5)=\frac{7}{x+1}$$

$$(x-5)(x+1)=7 \qquad\qquad (-x+5)(x+1)=7$$

$$x^2-4x-12=0 \qquad\qquad -x^2+4x-2=0$$

$$x=\frac{4\pm\sqrt{16+48}}{2} \qquad\qquad x=\frac{-4\pm\sqrt{16-8}}{-2}$$

$$x=-2,6 \qquad\qquad x=2\pm\sqrt{2}$$

Checking the two potential solutions obtained from the first equation, we see that it is indeed the case that $|6-5|$ is equal to $\frac{7}{6+1}$, as both expressions are equal to 1, so $x=6$ is a solution of the original equation. However, $|-2-5|=7$ while $\frac{7}{-2+1}=-7$, so $x=-2$ is extraneous and not a solution.

We check the two potential solutions obtained from the second equation the same way, remembering a technique for working with radical expressions in order to compare the two sides of the equation, as follows.

$$\left|2+\sqrt{2}-5\right|=\left|-3+\sqrt{2}\right|=3-\sqrt{2}$$

and

$$\frac{7}{2+\sqrt{2}+1}=\left(\frac{7}{3+\sqrt{2}}\right)\left(\frac{3-\sqrt{2}}{3-\sqrt{2}}\right)=\frac{7\left(3-\sqrt{2}\right)}{9-2}=3-\sqrt{2}$$

$$\left|2-\sqrt{2}-5\right|=\left|-3-\sqrt{2}\right|=3+\sqrt{2}$$

and

$$\frac{7}{2-\sqrt{2}+1}=\left(\frac{7}{3-\sqrt{2}}\right)\left(\frac{3+\sqrt{2}}{3+\sqrt{2}}\right)=\frac{7\left(3+\sqrt{2}\right)}{9-2}=3+\sqrt{2}$$

So both $x=2+\sqrt{2}$ and $x=2-\sqrt{2}$ are valid solutions of the original equation, in addition to $x=6$, and thus the complete solution set is $\left\{6,2\pm\sqrt{2}\right\}$.

TOPIC 2: Applications of Rational Equations

Many seemingly different applications fall into a class of problems known as work-rate problems. What these applications have in common is two or more "workers" acting in unison to complete a task. The workers can be, for example, employees on a job, machines manufacturing a part, or inlet and outlet pipes filling or draining a tank. Typically, each worker is capable of doing the task alone, and works at an individual rate regardless of whether others are involved.

The goal in a work-rate problem is usually to determine how fast the task at hand can be completed, either by the workers together or by one of the workers individually. There are two keys to solving a work-rate problem:

1. **The rate of work is the reciprocal of the time needed to complete the task.**
   If a given job can be done by a worker in x units of time, the worker works at a rate of $\frac{1}{x}$ jobs per unit of time. For instance, if Jane can overhaul an engine in 2 hours, her rate of work is $\frac{1}{2}$ of the job per hour. If a faucet can fill a sink in 5 minutes, its rate is $\frac{1}{5}$ of the sink per minute. This also means that the time needed to complete a task is the reciprocal of the rate of work: if a faucet fills a sink at the rate of $\frac{1}{5}$ of the sink per minute, it takes 5 minutes to fill the sink.

2. **Rates of work are "additive."**
   This means that, in the ideal situation, two workers working together on the same task will have a combined rate of work that is the sum of their individual rates. For instance, if Jane's rate in overhauling an engine is $\frac{1}{2}$ and Ted's rate is $\frac{1}{3}$, their combined rate is $\frac{1}{2}+\frac{1}{3}$, or $\frac{5}{6}$. That is, together they can overhaul an engine in $\frac{6}{5}$ hours, or one hour and twelve minutes.

Example 3: Filling a Pool

One hose can fill a swimming pool in 12 hours. The owner buys a second hose that can fill the pool at twice the rate of the first one. If both hoses are used together, how long does it take to fill the pool?

Solution

The rate of work of the first hose is $\frac{1}{12}$, so the rate of the second hose is $\frac{1}{6}$. If we let x denote the time needed to fill the pool when both hoses are used together, the sum of the two individual rates must equal $\frac{1}{x}$. So we need to solve the equation $\frac{1}{12}+\frac{1}{6}=\frac{1}{x}$.

$$\frac{1}{12}+\frac{1}{6}=\frac{1}{x}$$
$$x+2x=12$$
$$3x=12$$
$$x=4$$

As is typical, this work-rate problem leads to a rational equation which we can solve using the methods of this section.

After multiplying by the LCD, $12x$, we are left with a polynomial equation (linear in this case).

Thus, using both hoses, it will take 4 hours to fill the pool.

Example 4: Filling a Pool Poorly

The pool owner in Example 3 is a bit clumsy, and one day proceeds to fill his empty pool with the two hoses but accidentally turns on the pump that drains the pool also. Fortunately, the pump rate is slower than the combined rate of the two hoses, and the pool fills anyway, but it takes 10 hours to do so. How long does it take the pump to empty the pool?

Solution

First, let's translate the information in the problem into a work-rate rational equation. If we let x denote the time it takes the pump to empty the pool, we can say that the pump has a filling rate of $-\frac{1}{x}$ (since emptying is the opposite of filling). Since the two hoses can fill the pool in 4 hours, the combined filling rate of the two hoses is $\frac{1}{4}$. Finally, the total rate at which the pool is filled on this unfortunate day is $\frac{1}{10}$.

Again, the sum of the individual rates is equal to the combined rate, so the following rational equation reflects this situation.

$$\frac{1}{4}-\frac{1}{x}=\frac{1}{10}$$

Again, we multiply each term by the LCD, $20x$, to arrive at a polynomial equation to solve.

$$5x-20=2x$$

$$3x=20$$

$$x=\frac{20}{3}$$

Thus, working alone, the pump can empty the pool in $\frac{20}{3}$ hours, or 6 hours and 40 minutes.

TOPIC 3: Solving Radical Equations

The last one-variable equations we will discuss are those that contain radical expressions. A **radical equation** is an equation that has at least one radical expression containing a variable, while any nonradical expressions are polynomial terms. As with the rational equations discussed previously, we will develop a general method of solution that converts a given radical equation into a polynomial equation. We will see that just as with rational equations, we must check our potential solutions carefully to see if they actually solve the original radical equation.

Our method of solving rational equations involved multiplying both sides of the equation by an algebraic expression (the LCD of all the rational expressions), and in some cases potential solutions had to be discarded because they led to division by 0 in one or more of the rational expressions. Something similar can happen with radical equations.

Since our goal is to convert a given radical equation into a polynomial equation, one reasonable approach is to raise both sides of the equation to whatever power is necessary to "undo" the radical (or radicals). The problem is that this does *not* result in an equivalent equation; remember that we can only transform an equation into an equivalent equation by adding the same quantity to both sides or by multiplying both sides by a nonzero quantity. We won't *lose* any solutions by raising both sides of an equation to the same power, but we may *gain* some extraneous solutions. We identify these and discard them by checking all of our eventual solutions in the original equation. A simple example will make this clear.

Example 5: Causing Extraneous Solutions

Consider the equation $x = -3$.

Solution

This equation is so basic that it is its own solution. But for this demonstration, suppose we square both sides, obtaining the following equation.

$$x^2 = 9$$

This second-degree equation can be solved by factoring the polynomial $x^2 - 9$ or by taking the square root of both sides, and in either case we obtain the solution set $\{-3, 3\}$. That is, by squaring both sides of the original equation, we gained a second (extraneous) solution.

PROCEDURE: Solving Radical Equations

Step 1: Begin by isolating the radical expression on one side of the equation. If there is more than one radical expression, choose one to isolate on one side.

Step 2: Raise both sides of the equation by the power necessary to "undo" the isolated radical. That is, if the radical is an nth root, raise both sides to the nth power.

Step 3: If any radical expressions remain, simplify the equation if possible and then repeat steps 1 and 2 until the result is a polynomial equation. When a polynomial equation has been obtained, solve the equation using polynomial methods.

Step 4: Check your solutions in the original equation! Any extraneous solutions must be discarded.

If the equation contains many radical expressions, and especially if they have different indices, eliminating all the radicals may be a long process! The equations that we will solve will not require more than a few repetitions of steps 1 and 2.

Example 6: Radical Equations

Solve the radical equations.

a. $\sqrt{1-x}-1=x$ **b.** $\sqrt{x+1}+\sqrt{x+2}=1$ **c.** $\sqrt[4]{x^2+8x+7}-2=0$

Solution

a. $\sqrt{1-x}-1=x$

$\sqrt{1-x}=x+1$ Isolate the radical expression.

$\left(\sqrt{1-x}\right)^2=(x+1)^2$ Since we have square root, square both sides.

$1-x=x^2+2x+1$ The result is a second-degree polynomial

$0=x^2+3x$ equation that can be solved by factoring.

$0=x(x+3)$

$x=0,-3$ We have two apparent solutions to check.

Now we need to check each apparent solution in the original equation.

$$\begin{aligned}\sqrt{1-0}-1&=0\\\sqrt{1}-1&=0\\0&=0\end{aligned}\qquad\qquad\begin{aligned}\sqrt{1-(-3)}-1&=-3\\\sqrt{4}-1&=-3\\1&\neq-3\end{aligned}$$

Thus, -3 is an extraneous solution, so the solution set is $\{0\}$.

b. $\sqrt{x+1}+\sqrt{x+2}=1$ This equation has two radical expressions, so we isolate one of them initially.

$\sqrt{x+1}=1-\sqrt{x+2}$

$\left(\sqrt{x+1}\right)^2=\left(1-\sqrt{x+2}\right)^2$

$x+1=1-2\sqrt{x+2}+x+2$ We square both sides to eliminate the isolated radical, and then proceed to simplify and isolate the remaining radical.

$2\sqrt{x+2}=2$

$\sqrt{x+2}=1$

$x+2=1$ Finally, we square both sides again and solve the polynomial equation.

$x=-1$

We check the apparent solution in the original equation.

$$\begin{aligned}\sqrt{(-1)+1}+\sqrt{(-1)+2}&=1\\\sqrt{0}+\sqrt{1}&=1\\1&=1\end{aligned}$$

Thus, the solution set is $\{-1\}$.

c. $\sqrt[4]{x^2+8x+7}-2=0$

$\sqrt[4]{x^2+8x+7}=2$ — First isolate the radical.

$\left(\sqrt[4]{x^2+8x+7}\right)^4=2^4$ — In this case, the radical is a fourth root, so we raise both sides to the fourth power.

$x^2+8x+7=16$

$x^2+8x-9=0$

$(x+9)(x-1)=0$ — The resulting second-degree equation can again be solved by factoring.

$x=-9,1$

Once again, we check the apparent solutions in the original equation.

$$\begin{aligned}\sqrt[4]{(-9)^2+8(-9)+7}-2&=0\\ \sqrt[4]{16}-2&=0\\ 2-2&=0\\ 0&=0\end{aligned} \qquad \begin{aligned}\sqrt[4]{(1)^2+8(1)+7}-2&=0\\ \sqrt[4]{16}-2&=0\\ 2-2&=0\\ 0&=0\end{aligned}$$

Both apparent solutions are actual solutions, so the solution set is $\{-9,1\}$.

TOPIC 4: Solving Equations with Positive Rational Exponents

In Section 1.8, we encountered equations with rational exponents that we solved by factoring or by quadratic methods. Now that we have learned how to solve radical equations, we have another option for solving equations with *positive* rational exponents. Recall that we defined positive rational exponents in terms of radicals in Section 1.2.

DEFINITION: Positive Rational Number Exponents

Meaning of $a^{\frac{m}{n}}$: If m and n are natural numbers with $n \neq 0$, if m and n have no common factors greater than 1, and if $\sqrt[n]{a}$ is a real number, then $a^{\frac{m}{n}}=\sqrt[n]{a^m}=\left(\sqrt[n]{a}\right)^m$.

This means that an equation with positive rational exponents can be rewritten as one with radical terms, which we can then solve using the methods for solving radical equations.

Example 7: Rational Exponents

Solve the following equations with rational exponents.

a. $x^{\frac{2}{3}} - 9 = 0$

b. $\left(32x^2 - 32x + 17\right)^{\frac{1}{4}} = 3$

Solution

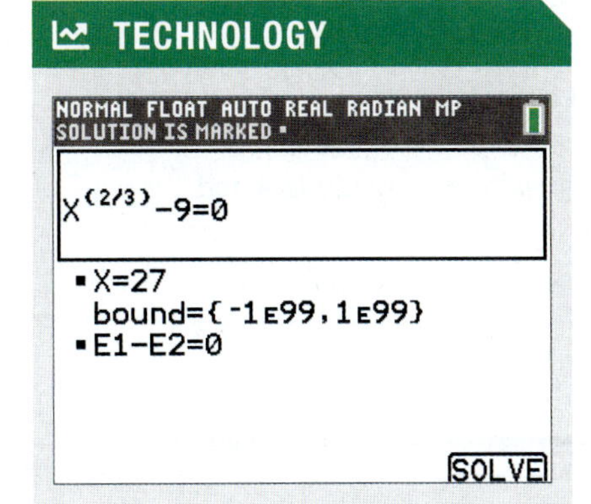

a. $x^{\frac{2}{3}} - 9 = 0$ — The term containing the rational exponent can be rewritten as a radical expression, so we will begin by isolating that term.

$x^{\frac{2}{3}} = 9$

$\sqrt[3]{x^2} = 9$ — Rewrite the left-hand side as a radical.

$x^2 = 9^3$ — Cubing both sides eliminates the cube root.

$x = \pm 9^{\frac{3}{2}}$ — Raising both sides to the $\frac{1}{2}$ power solves the equation for x, but we can evaluate the expression on the right-hand side.

$x = \left(\pm 9^{\frac{1}{2}}\right)^3$

$x = (\pm 3)^3$

$x = \pm 27$ — Note that both +3 and −3 must be considered.

Plugging the values in, we see that $(27)^{\frac{2}{3}} = 9$ and $(-27)^{\frac{2}{3}} = 9$, so both are solutions to the original equation.

b. $\left(32x^2 - 32x + 17\right)^{\frac{1}{4}} = 3$ — The exponent of $\frac{1}{4}$ indicates we will need to raise both sides to the fourth power.

$\sqrt[4]{32x^2 - 32x + 17} = 3$

$32x^2 - 32x + 17 = 3^4$

$32x^2 - 32x + 17 = 81$ — We are left with a second-degree polynomial equation that can be solved by factoring.

$32x^2 - 32x - 64 = 0$

$32\left(x^2 - x - 2\right) = 0$

$x^2 - x - 2 = 0$

$(x - 2)(x + 1) = 0$

$x = 2, -1$ — Note that both solutions again satisfy the original equation.

Confirm, by substituting the solutions into the original equation, that both apparent solutions truly solve the equation with rational exponents.

TOPIC 5: Solving Equations for One Variable

The procedure we use to solve radical equations can also be used to solve a given equation for a specified variable. We illustrate the process with one last example.

Example 8: Escape Speed

The speed required for an object to escape from the gravitational pull of a planet is called the **escape speed** of the planet. The escape speed is given by the equation $v_e = \sqrt{\frac{2GM}{r}}$, where v_e is the escape speed, G is the universal gravitation constant, M is the mass of the planet, and r is the radius of the planet. Solve this equation for r.

Solution

We follow the same procedure for solving radical equations.

$$v_e = \sqrt{\frac{2GM}{r}}$$ The radical expression is already isolated.

$$v_e^2 = \frac{2GM}{r}$$ Square both sides to eliminate the radical.

$$r = \frac{2GM}{v_e^2}$$ Solve for r.

1.9 EXERCISES

PRACTICE

Solve the following rational equations. See Examples 1 and 2.

1. $\frac{2x^3+4x^2}{x^2-4x-12} = \frac{-7x-6}{x-6}$

2. $\frac{-x^2}{x-1} - 3 = 0$

3. $\frac{3}{x-2} + \frac{2}{x+1} = 1$

4. $\frac{x}{x-1} + \frac{2}{x-3} = -\frac{2}{x^2-4x+3}$

5. $\frac{1}{t-3} + \frac{1}{t+2} = \frac{t}{t+3}$

6. $\frac{z}{6+z} + \frac{z-1}{6-z} = \frac{z}{6-z}$

7. $\frac{y}{y-1} + \frac{2}{y-3} = \frac{y^2}{y^2-4y+3}$

8. $\frac{2}{2x+1} - \frac{x}{x-4} = \frac{-3x^2+x-4}{2x^2-7x-4}$

9. $\frac{2}{2b+1} + \frac{2b^2-b+4}{2b^2-7b-4} = \frac{b}{b-4}$

10. $\frac{2}{n+3} + \frac{3}{n+2} = \frac{6}{n}$

11. $\frac{1}{x-3} + \frac{1}{x+3} = \frac{2x}{x^2-9}$

12. $\frac{3}{x-1} - \frac{3}{x+2} = \frac{9}{x^2+x-2}$

13. $\frac{1}{|x-3|} = 2$

14. $\frac{3}{|x+1|} = 1$

15. $\frac{1}{|x-3|} + \frac{1}{|x+1|} = 1$

16. $\frac{1}{x-2} + \frac{2}{|x-1|} = 2$

Solve the following radical equations. See Example 6.

17. $\sqrt{4-x}-x=2$

18. $\sqrt{3y+4}+\sqrt{5y+6}=2$

19. $\sqrt{3-3x}-3=\sqrt{3x+2}$

20. $\sqrt{x^2-4x+5}-x+2=0$

21. $\sqrt{x^2-4x+4}+2=3x$

22. $\sqrt{50+7s}-s=8$

23. $\sqrt[3]{3-2x}-\sqrt[3]{x+1}=0$

24. $\sqrt[4]{x^2-x}=\sqrt[4]{x-1}$

25. $\sqrt[4]{2x+3}=-1$

26. $\sqrt{11x+3}+4x=18$

27. $\sqrt{2b-1}+3=\sqrt{10b-6}$

28. $\sqrt{5x+5}=\sqrt{4x-7}+2$

29. $\sqrt{x+10}+1=x-1$

30. $\sqrt{x+1}+10=x-1$

31. $\sqrt{x^2-10}-1=x+1$

32. $\sqrt[3]{5x^2-14x}=-2$

33. $\sqrt[5]{7t^2+2t}=\sqrt[5]{5t^2+4}$

34. $\sqrt[3]{y^3-7y+2}=\sqrt[3]{2-3y}$

35. $\sqrt{14y^2-18y+4}+2=2y$

36. $\sqrt{9x+4}=\sqrt{7x+1}+1$

37. $\sqrt{4z+41}+3=z+2$

Solve the following equations. See Example 7.

38. $(x+3)^{\frac{1}{4}}+2=0$

39. $(2x-5)^{\frac{1}{4}}=(x-1)^{\frac{1}{4}}$

40. $(2x-1)^{\frac{2}{3}}=x^{\frac{1}{3}}$

41. $(3y^2+9y-5)^{\frac{1}{2}}=y+3$

42. $(3x-5)^{\frac{1}{5}}=(x+1)^{\frac{1}{5}}$

43. $w^{\frac{3}{5}}+8=0$

44. $z^{\frac{4}{3}}-\frac{16}{81}=0$

45. $x^{\frac{2}{3}}-\frac{25}{49}=0$

46. $(x^2+21)^{\frac{-3}{2}}=\frac{1}{125}$

47. $(x-2)^{\frac{2}{3}}=(14-x)^{\frac{1}{3}}$

48. $(x^2+7)^{\frac{-3}{2}}=\frac{1}{64}$

49. $(y-2)^{\frac{2}{3}}=(13y-66)^{\frac{1}{3}}$

Solve each of the following formulas for the indicated variable. See Example 8.

50. The formula $T=2\pi\sqrt{\frac{l}{g}}$ gives the period T of a pendulum of length l. Solve this formula for l.

51. The formula $c=\sqrt{a^2+b^2}$ gives the length of the hypotenuse c of a right triangle. Solve this formula for a.

52. Einstein's Theory of Relativity states that $E=mc^2$. Solve this equation for c.

53. The formula $\omega=\sqrt{\frac{k}{m}}$ gives the angular frequency ω of a mass m suspended from a spring of spring constant k. Solve this formula for m.

54. The formula $V = \frac{4}{3}\pi r^3$ gives the volume of a sphere with radius r. Solve the equation for r.

55. The formula $F = \frac{mv^2}{r}$ gives the force on an object in circular motion. Solve the equation for v.

56. The formula for lateral acceleration, used in automotives, is $a = \frac{1.227r}{t^2}$. Solve this equation for t.

57. According to one guideline regarding body mass index (BMI), a healthy mass for an adult male can be found using the formula $m = 23h^2$, where m is expressed in kilograms and h in meters. Solve this equation for h.

58. Kepler's Third Law is $T^2 = \frac{4\pi^2 r^3}{GM}$. It relates the period T of a planet to the radius r of its orbit and the sun's mass M. Solve this formula for r.

59. The equation $r = \frac{2gm}{c^2}$ is the Schwarzschild Radius Formula used to find the radius of a black hole in space. Solve the equation for c.

60. The total mechanical energy of an object with mass m at height h in a closed system can be written as $ME = \frac{1}{2}mv^2 + mgh$. Solve for v, the velocity of the object, in terms of the given quantities.

61. Recall, the formula for the Pythagorean Theorem states that $a^2 + b^2 = c^2$. Solve this formula for b.

62. In a circuit with an AC power source, the total impedance Z depends on the resistance R, the capacitance C, the inductance L, and the frequency of the current ω according to $Z = \sqrt{R^2 + \left(\omega L - \frac{1}{\omega C}\right)^2}$. Solve this equation for the inductance L.

63. The formula used to find the orbital period for circular Keplerian orbits is $P = \frac{2\pi}{\sqrt{\frac{u}{a^3}}}$. Solve this equation for a.

APPLICATIONS

64. If Joanne were to paint her living room alone, it would take 5 hours. Her sister Lisa could do the job in 7 hours. How long would it take them working together?

65. The hot water tap can fill a given sink in 4 minutes. If the cold water tap is turned on as well, the sink fills in 1 minute. How long would it take for the cold water tap to fill the sink alone?

66. The hull of Jack's yacht needs to be cleaned. He can clean it by himself in 5 hours, but he asks his friend Thomas to help him. If it takes 3 hours for the two men to clean the hull of the boat, how long would it have taken Thomas alone?

67. Two hoses, one of which has a flow rate three times the other, can together fill a tank in 3 hours. How long does it take each of the hoses individually to fill the tank?

68. Officials begin to release water from a full man-made lake at a rate that would empty the lake in 12 weeks, but a river that can fill the lake in 30 weeks is replenishing the lake at the same time. How long does it take to empty the lake?

69. In order to flush deposits from a radiator, a drain that can empty the entire radiator in 45 minutes is left open at the same time it is being filled at a rate that would fill it in 30 minutes. How long does it take for the radiator to fill?

70. Jimmy and Janice are picking strawberries. Janice can fill a bucket in a half hour, but Jimmy continues to eat the strawberries that Janice has picked at a rate of one bucket per 1.5 hours. How long does it take Janice to fill her bucket?

71. A farmer can plow a given field in 2 hours less time than it takes his son. If they acquire two tractors and work together, they can plow the field in 5 hours. How long does it take the father alone? Round your answer to one decimal place.

CHAPTER 1 PROJECT

Polynomials

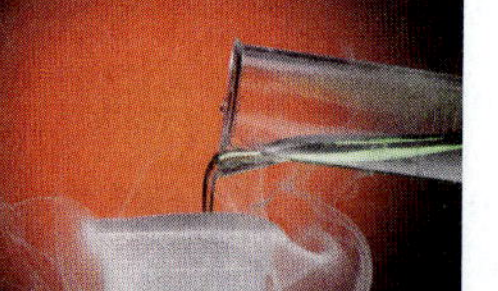

A chemistry professor calculates final grades for her class using the polynomial

$$A = 0.3f + 0.15h + 0.4t + 0.15p,$$

where A is the final grade, f is the final exam, h is the homework average, t is the chapter test average, and p is the semester project.

The following is a table containing the grades for various students in the class.

Name	Final Exam	Homework Average	Test Average	Project
Alex	77	95	79	85
Ashley	91	95	88	90
Barron	82	85	81	75
Elizabeth	75	100	84	80
Gabe	94	90	90	85
Lynn	88	85	80	75

1. Find the final grade for each student, rounded to one decimal place.

2. Who has the highest total score?

3. Why is the final grade raised more with a grade of 100 on the final exam than with a grade of 100 on the semester project?

4. Assume you are a student in this class. With one week until the final exam, you have a homework average of 85, a test average of 85, and a 95 on the semester project. What score must you make on the final exam to achieve at least a 90.0 overall? Round your answer to one decimal place.

CHAPTER 1 REVIEW EXERCISES

Section 1.1

Which elements of the following set are **a.** natural numbers, **b.** whole numbers, **c.** integers, **d.** rational numbers, **e.** irrational numbers, **f.** real numbers?

1. $\left\{\frac{3}{7}, -\sqrt{4}, 2^3, 5.3, |-2.1|, \sqrt{17}, 0\right\}$

Describe the following set using set-builder notation. There may be more than one correct way to do this.

2. $\left\{\frac{1}{2}, \frac{1}{4}, \frac{1}{6}, \frac{1}{8}, \frac{1}{10}, \ldots\right\}$

Write each set as an interval using interval notation.

3. $4 \le x < 17$

4. $\{x \mid -8 \le x \le -1\}$

Evaluate the absolute value expressions.

5. $-|-4-3|$

6. $-|11-2|$

7. $|\sqrt{9}-7|$

8. $|\sqrt{5}-\sqrt{11}|$

9. $-\frac{|x|}{|-x|}$

10. Liz, Monica, Peter, James, and Melissa are comparing their ages. Liz is older than Peter and Melissa is the youngest. James is the oldest and Peter is older than Monica. Order them from youngest to oldest.

Identify the components of the algebraic expressions, as indicated.

11. Identify the terms in the expression $\frac{x^2}{2y} + 12.1x - \sqrt{y+5}$.

12. Identify the coefficients in the expression $\frac{x^2}{2y} + 12.1x - \sqrt{y+5}$.

Evaluate the following algebraic expressions for the given values of the variables.

13. $7y^2 - \frac{1}{3}\pi xy + 8x^3$ for $x = -2$ and $y = 2$

14. $x^2z^3 + 5\sqrt{3x-2y}$ for $x = 2, y = 1,$ and $z = -1$

15. $|-3x + x^2y| - \frac{xy}{2}$ for $x = -3$ and $y = 4$

16. $3\sqrt{\frac{xy}{3}} - 2y^2$ for $x = 2$ and $y = 6$

Identify the property that justifies each of the following statements. If one of the cancellation properties is being used to transform an equation, identify the quantity that is being added to both sides or the quantity by which both sides are being multiplied.

17. $-4 + x = x - 4$

18. $12a^2 = 8b \Leftrightarrow 3a^2 = 2b$

19. $(x-3)(z-2) = 0 \Rightarrow x - 3 = 0 \text{ or } z - 2 = 0$

Simplify the following set expressions.

20. $(-4,8)\cup[5,13]$

21. $(-4,8)\cap[5,13]$

Section 1.2

Use the properties of exponents to simplify each of the following expressions, writing your answer with only positive exponents.

22. $\left(2^3a^{-2}b^4\right)^{-1}c^{-3}$

23. $\dfrac{-4t^0\left(s^2t^{-2}\right)^{-3}}{2^3st^{-3}}$

24. $\left[\left(3y^{-2}z\right)^{-1}\right]^{-3}$

25. $\dfrac{3^2x^{-4}\left(y^2z\right)^{-2}}{\left(2z^{-3}\right)^{-1}y^{-6}}$

Convert each number from scientific notation to standard notation, or vice versa, as indicated.

26. -3.005×10^{-4}; convert to standard

27. 69,520,000; convert to scientific

Evaluate each expression using the properties of exponents. Use a calculator only to check your final answer.

28. $\left(3.46\times10^8\right)\left(1.2\times10^4\right)$

29. $\dfrac{2.4\times10^{-12}}{(1.2)\times10^{-4}}$

30. Sam is making a piñata in the shape of a sphere and needs to know how much candy to buy to fill it. If the radius of the piñata is 10 inches, what is the volume of the piñata?

Evaluate the following radical expressions.

31. $\sqrt{3^2+4^2}$

32. $\dfrac{\sqrt[3]{\sqrt{15}}}{\sqrt{\sqrt[3]{5}}}$

Simplify the following radical expressions. Rationalize all denominators and use only positive exponents.

33. $\sqrt{25x^{20}}$

34. $\sqrt{16x^2}$

35. $\sqrt[3]{-64x^{-9}y^3}$

36. $\dfrac{\sqrt{3a^3}}{\sqrt{12a}}$

37. $\sqrt[3]{\dfrac{8x^2}{3y^{-4}}}$

38. $\sqrt[4]{\dfrac{a^9b^{-4}}{81}}$

39. $\dfrac{4}{\sqrt{2}-\sqrt{6}}$

40. $\dfrac{3}{\sqrt{x}+\sqrt{2}}$

Simplify the following expressions.

41. $\sqrt{18x^3y}-\sqrt[3]{16x^4y}$

42. $\left(2\sqrt{3}-5\sqrt{2}\right)^2$

Convert the following expressions from radical notation to exponential notation, or vice versa. Simplify each expression in the process, if possible.

43. $\sqrt{x^{-5}} \cdot \sqrt[4]{x^3}$

44. $\left(49x^4\right)^{\frac{1}{2}}\left(16x^{12}\right)^{\frac{3}{4}}$

Section 1.3

Add or subtract the polynomials, as indicated.

45. $\left(-4m^2 - 5m^3 + 4\right) + \left(m^4 + 7m^2 - 2\right)$

46. $(2xy + 3x) - \left(8x^2y - 6xy + 3x - y\right)$

Multiply the polynomials, as indicated.

47. $\left(x^2 + y\right)\left(3x - 4y^3\right)$

48. $(a + 5b)(5a - 7ab + 2b)$

Factor each of the following polynomials.

49. $x^2 - x - 12$

50. $2x^2 + x - 15$

51. $6a^2 - 7a - 5$

52. $4a^2 - 9b^4$

53. $36x^6 - y^2$

54. $nx + 3mx - 2ny - 6my$

55. $2x^2 + 6x - 5xy - 15y$

56. $8x^3y^2 + 4x^3y - 12xy^2$

Factor the following algebraic expressions.

57. $(3x - 2y)^{\frac{4}{3}} - (3x - 2y)^{\frac{2}{3}}$

58. $8x^{-2} + 5x^{-1}$

Section 1.4

Simplify the following rational expressions, indicating which real values of the variable must be excluded.

59. $\dfrac{x^3 + 6x^2 + 9x}{x^3 - 9x}$

60. $\dfrac{x^2 - 9}{x^3 - 27}$

Perform the indicated operations on the rational expressions and simplify your answer.

61. $\dfrac{1}{x} - \dfrac{3}{x+2} - \dfrac{6}{x^2 + 2x}$

62. $\dfrac{a^3 - 8}{a^2 - 4} \div \dfrac{a^3 + 2a^2 + 4a}{a^3 + 2a^2} \cdot \dfrac{1}{a^2 + a}$

Simplify the complex rational expressions.

63. $\dfrac{\dfrac{1}{2a} - \dfrac{1}{2b}}{\dfrac{2}{a} + \dfrac{2}{b}}$

64. $\dfrac{\dfrac{x}{3} - \dfrac{3}{x}}{-\dfrac{3}{x} + 1}$

65. $\dfrac{\dfrac{x}{y} - \dfrac{y}{x}}{x^{-1} - y^{-1}}$

Section 1.5

Evaluate the following square root expressions.

66. $-\sqrt{-8x}$

67. $i^3\sqrt{-9}$

Simplify the following expressions.

68. $(7-2i)+(9i-5)$

69. $(5-3i)-(-12i)$

70. $(3-i)(6i^2-4)$

71. $\frac{17}{4-i}$

72. $\frac{2i}{3-i}$

73. $\frac{3+4i}{3-4i}$

74. $(\sqrt{-3})(\sqrt{-16})$

75. $(8-\sqrt{-2})^2$

76. $\frac{2i\sqrt{-27}}{\sqrt{-16}}$

Section 1.6

Solve the following linear equations.

77. $2y-(1-y)=y+2(y-1)$

78. $\frac{x}{2}-\frac{1}{3}=x-\frac{1}{3}-\frac{x}{2}$

79. $-0.2x-0.5=-0.4x+0.75$

80. $-2(x-5)+1=3+(7x-2)$

Solve the following absolute value equations.

81. $|2x-7|=1$

82. $|2y-5|-1=|3-y|$

83. $|7z+5|+3=8$

84. $|w-5|=|3w+1|$

Solve the following absolute value equations geometrically and algebraically.

85. $|-2x+1|=7$

86. $|x+4|-|x-1|=0$

Solve each of the following equations for the indicated variable.

87. Area of a trapezoid: $A=\frac{1}{2}h(b+c)$; solve for c

88. Volume of a rectangular pyramid: $V=\frac{1}{3}lwh$; solve for l

89. Temperature conversions: $F=\frac{9}{5}C+32$; solve for C

90. Two trains leave the station at the same time in opposite directions. One travels at an average rate of 90 miles per hour, and the other at an average rate of 95 miles per hour. How far apart are the two trains after an hour and twenty minutes? Round your answer to one decimal place.

91. Two firefighters, Jake and Rose, each have \$5000 to invest. Jake invests his money in a money market account with an annual return of 3.25%, while Rose invests hers in a CD paying 4.95% annually. How much more money does Rose have than Jake after 1 year?

Section 1.7

Solve the following linear inequalities. Describe each solution set using interval notation and by graphing.

92. $-8x+3\geq -9x+10$

93. $4(2x-5)<-3(-3x+8)$

94. $\dfrac{-2(x-1)}{3}\leq\dfrac{-2x}{4}$

95. $3.1(2x-1)>7.2-4.1x$

96. $-5<3m+1<13$

97. $-14<-2(3+y)\leq 8$

98. $2<\dfrac{x+1}{4}\leq 7$

99. $-5|3+t|>-10$

100. $3+|2x-1|<1$

101. $-2|x-1|+|3x-3|\geq 7$

102. $6+\dfrac{x}{5}\leq\dfrac{4}{5}$ or $5+2x\geq x-2$

103. $\dfrac{8x-5}{9}\leq 3$ or $2(3x-16)\geq 4(x-3)$

104. $2.9x+1.8<3(1.3x+6)$ and $7x<5x+34$

Section 1.8

Solve the following quadratic equations.

105. $5x^2-13x-6=0$

106. $x^2=7$

107. $2(x-2)^2=-18$

108. $15x^2+3x+2=-8x$

109. $x^2-8x+14=0$

110. $3x^2-x+3=-7x$

111. $x^2=6x-16$

112. $-2x-7=-4x^2$

113. $2x^2+3x-10=10$

114. $x^2-7x-2=-12$

115. $1.7z^2-3.8z-2=0$

116. $2x^2+7x=x^2+2x-6$

Solve the following quadratic-like equations.

117. $(x^2+2)^2-7(x^2+2)+12=0$

118. $y^{\frac{2}{3}}+y^{\frac{1}{3}}-6=0$

119. $(t+2)^2-2(t+2)=24$

120. $x^4-13x^2+36=0$

Solve the following equations by factoring.

121. $x^3-4x^2-2x+8=0$

122. $2x^3+2x=5x^2$

123. $x^3-x^2+4x-4=0$

124. $x^4+7x^2-18=0$

125. $x^{\frac{7}{2}}-3x^{\frac{5}{2}}-4x^{\frac{3}{2}}=0$

126. $x^{\frac{7}{3}}+7x^{\frac{4}{3}}-8x^{\frac{1}{3}}=0$

127. $(x-2)^{\frac{3}{4}}+2(x-2)^{\frac{7}{4}}=0$

128. $(x-1)^{-\frac{1}{2}}+4(x-1)^{\frac{1}{2}}=0$

Use the connection between solutions of polynomial equations and polynomial factoring to answer the following questions.

129. Find b and c so the equation $x^3 + bx^2 + cx = 0$ has solutions of –2, 0, and 4.

130. Given that the equation $x^2 - 6x + m - 1 = 0$ has only one root, find m.

131. If the sum of the roots of the equation $x^2 + mx - 6 = 0$ is 5, then what is m?

Section 1.9

Solve the following rational equations.

132. $\frac{1}{x+2} + \frac{1}{x-3} - \frac{x}{x-3} = 0$

133. $\frac{1}{x-2} - \frac{x}{x+2} = \frac{2}{x^2-4}$

134. $\frac{y}{y-1} + \frac{1}{y-4} = \frac{y^2}{y^2-5y+4}$

135. $\frac{2}{x+1} - \frac{x}{x-3} = \frac{3x-21}{x^2-2x-3}$

136. Jim cleans a house in 6 hours. John cleans the same house in 8 hours. How long does it take for them to clean the house together?

Solve the following equations.

137. $\sqrt{-4-x} - 4 = x$

138. $\sqrt{5x-1} = 4 + \sqrt{x+3}$

139. $\sqrt{2x^2+8x+1} - x - 3 = 0$

140. $\sqrt{10x^2-14x+16} + 1 = 3x$

141. $x + 2 = \left(-x^2 + 11x + 19\right)^{\frac{1}{2}}$

142. $\left(2x^2 + 14x\right)^{\frac{1}{4}} = \left(-x^2 - 8\right)^{\frac{1}{4}}$

143. $(2x-5)^{\frac{1}{6}} = (x-2)^{\frac{1}{6}}$

144. $\left(x^2 + x - 16\right)^{\frac{1}{3}} = 2(x-1)^{\frac{1}{3}}$

145. The formula for the volume of a cone with radius r and height h is $V = \frac{1}{3}\pi r^2 h$. Solve the equation for r.

Visitors Overview
Apr 19
Apr 26
May 3
Visitors
2,958

CHAPTER 2

Equations and Inequalities in Two Variables

SECTIONS

2.1 The Cartesian Coordinate System

2.2 Circles

2.3 Linear Equations in Two Variables

2.4 Slope and Forms of Linear Equations

2.5 Parallel and Perpendicular Lines

2.6 Linear Inequalities in Two Variables

WHAT IF ...

What if, as a sales manager, you were asked to forecast next year's sales for your company? How could you use previous data to make an informed and accurate prediction?

By the end of this chapter, you'll be able to apply the skills regarding linear inequalities in two variables to applications concerning price, weight, volume, time, materials, and more. On page 176, you'll solve problems like the one given above. You'll master this type of problem using the formula for the slope of a line on page 166 and the equation of a line in two variables.

René Descartes

Introduction

This chapter introduces the Cartesian coordinate system, the two-dimensional framework that underlies most of the material throughout the remainder of this text.

Many problems that we seek to solve with mathematics are most naturally described with two variables. The existence of two variables in a problem leads naturally to the use of a two-dimensional system in which to work, especially when we attempt to depict the situation graphically, but it took many centuries for the ideas presented in this chapter to evolve. Some of the greatest accomplishments of the French mathematician and philosopher René Descartes (1596–1650) were his contributions to the then fledgling field of analytic geometry, the marriage of algebra and geometry. In *La Géométrie*, an appendix to a volume of scientific philosophy, Descartes laid out the basic principles by which algebraic problems could be construed as geometric problems, and the methods by which solutions to the geometric problems could be interpreted algebraically. As many later mathematicians expanded upon Descartes' work, analytic geometry came to be an indispensable tool in understanding and solving problems of both an algebraic and geometric nature.

In this chapter, we will use the Cartesian coordinate system, named in honor of Descartes, primarily to study linear equations and inequalities in two variables. As we will see, a graph is one of the best ways to describe the solutions of such problems. Sections 2.3 through 2.6 will introduce the basic means by which we can construct graphs and consequently shift between the algebraic and geometric views of a given linear equation or linear inequality.

The Cartesian coordinate system will continue to play a prominent role as we proceed to other topics, such as relations and functions, in later chapters. Mastery of the foundational concepts in this chapter will be essential to understanding these related ideas.

2.1 THE CARTESIAN COORDINATE SYSTEM

TOPICS

1. The Cartesian coordinate system
2. Graphing equations
3. The distance and midpoint formulas

TOPIC 1: The Cartesian Coordinate System

Descartes' Inspiration

There is a legend, possibly apocryphal in nature, that Descartes' inspiration for what we now call the Cartesian plane came from watching a fly crawl around on the ceiling of his bedroom. It occurred to him that he could completely describe the path of the fly if he knew at every moment the fly's distance from each of two walls. The two distances define an ordered pair of numbers uniquely associated with the fly's position.

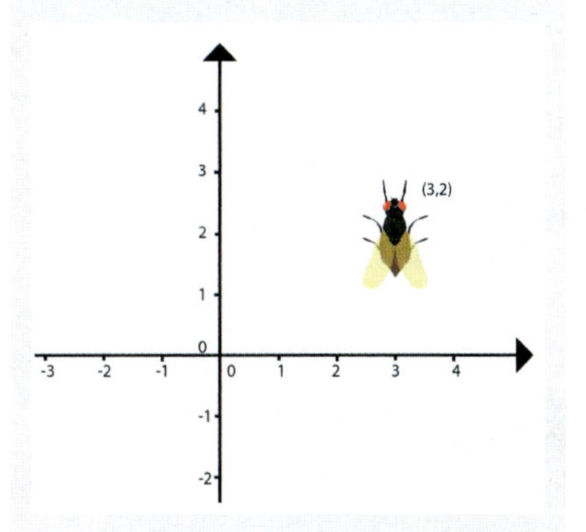

The last chapter dealt, essentially, with algebra of a single variable. Many of the methods learned in Chapter 1 revolve around solving equations and inequalities of one variable, and even in those equations with two or more variables we were interested in tasks such as solving the equation for one particular variable.

Our focus now shifts. While many important problems can be phrased, mathematically, as one-variable equations or inequalities, many more problems are most naturally expressed with two or more variables, and solving such a problem requires determining *all* of the values of *all* of the variables that make the equation or inequality true. This seemingly simple observation serves as the motivation for the material in this section. Consider, for example, an equation in the two variables x and y. A particular solution of the equation, if there is one, will consist of a value for x and a *corresponding* value for y; a solution of the equation cannot consist of a value for only one of the variables. The solution consists of a *pair* of ordered numbers (x, y), called an ordered pair.

DEFINITION: Ordered Pairs

An **ordered pair** (a, b) consists of two real numbers a and b. Unlike sets, the order of the elements in an ordered pair matters; that is, (a, b) is not equal to (b, a) unless $a = b$. In a given ordered pair (a, b), the number a is called the **first coordinate** and the number b is called the **second coordinate**.

This leads us naturally to the concept of a two-dimensional coordinate system. Just as the real number line, a one-dimensional coordinate system, is the natural arena in which to depict solutions of one-variable equations and inequalities, a two-dimensional coordinate system is the natural place to graph solutions of two-variable equations and inequalities. The coordinate system we use is named after René Descartes, the 17th-century French mathematician largely responsible for its development.

DEFINITION: The Cartesian Coordinate System

The **Cartesian coordinate system** (also called the **Cartesian plane**) consists of two perpendicular real number lines (each called an **axis**) intersecting at the 0 point of each line. The point of intersection is called the **origin** of the system, and the four quarters defined by the two lines are called the **quadrants** of the plane, numbered as indicated in Figure 1. Because the Cartesian plane consists of two crossed real lines, it is often given the symbol $\mathbb{R} \times \mathbb{R}$, or $\mathbb{R}^2$.

Each point P in the plane is identified by an ordered pair. The first coordinate indicates the horizontal displacement of the point from the origin, and the second coordinate indicates the vertical displacement. Figure 1 is an example of a Cartesian coordinate system and illustrates how several ordered pairs are **graphed**, or **plotted**.

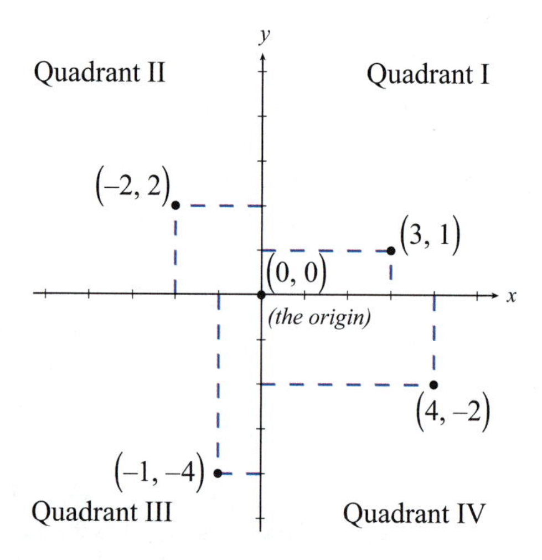

FIGURE 1: The Cartesian Plane

CAUTION

Unfortunately, mathematics uses parentheses to denote ordered pairs as well as open intervals, which sometimes leads to confusion. Context is the key to interpreting notation correctly. For instance, in the context of solving a one-variable inequality, the notation $(-2, 5)$ most likely refers to the open interval with endpoints at -2 and 5, while in the context of solving an equation in two variables, $(-2, 5)$ probably refers to a point in the Cartesian plane.

Example 1: Plotting Points in the Cartesian Coordinate System

Plot the following ordered pairs on the Cartesian plane, and identify which quadrant they lie in (or which axis they lie on).

a. $(2,3)$ **b.** $(-5,0)$ **c.** $(1,-3)$

d. $(-2,4)$ **e.** $(-6,-6)$ **f.** $(0,5)$

> **NOTE**
>
> There is no required method when plotting points, but it is helpful to establish a set pattern that you follow. For example, you may always count the horizontal value first, then the vertical displacement.

Solution

a.

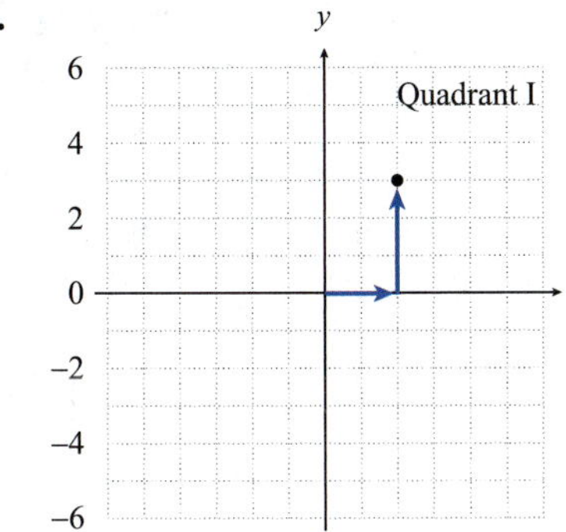

FIGURE 2

b.

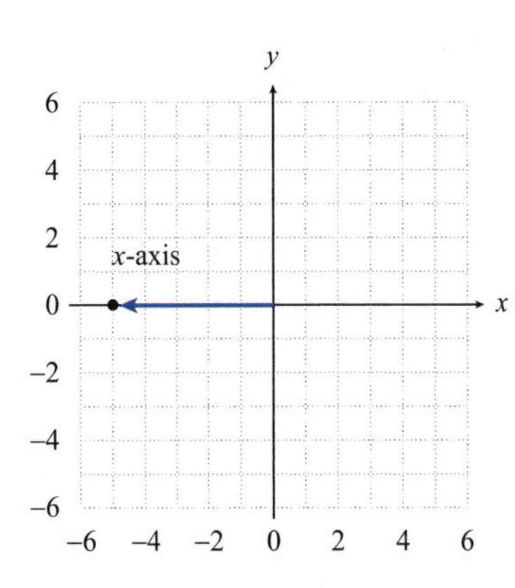

FIGURE 3

c.

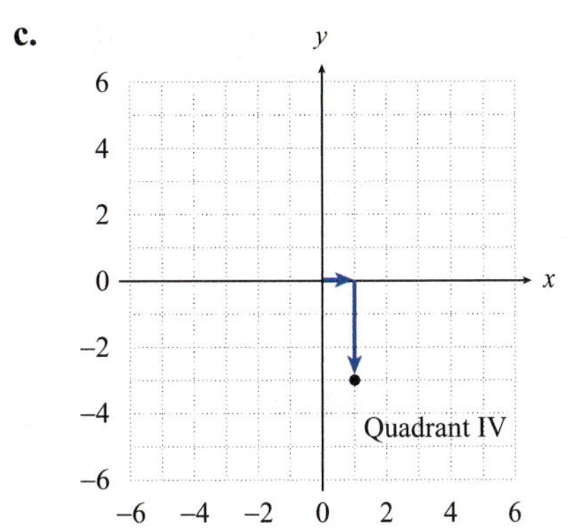

FIGURE 4

d.

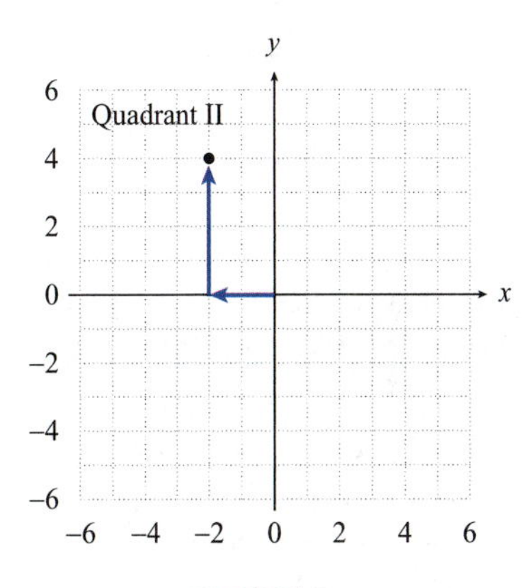

FIGURE 5

e.

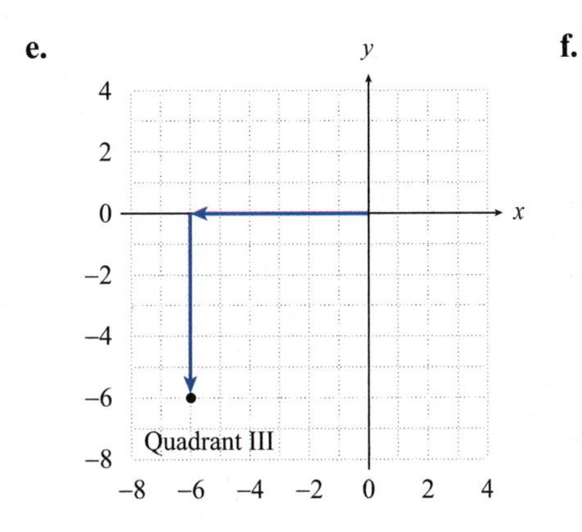

FIGURE 6

f.

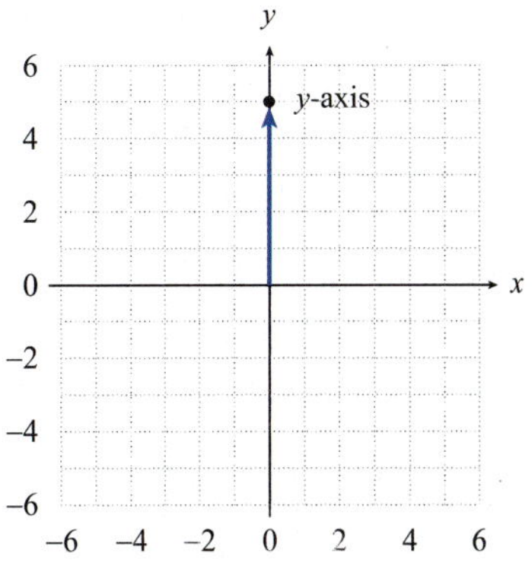

FIGURE 7

TOPIC 2: Graphing Equations

A solution of an equation in x and y must consist of a value a for x and a corresponding value b for y. It is natural to write such a solution as an ordered pair (a, b), and equally natural to graph it as a point in the plane whose coordinates are a and b. In this context we refer to the horizontal number line as the ***x*-axis**, the vertical number line as the ***y*-axis**, and the two coordinates of the ordered pair (a, b) as the ***x*-coordinate** and the ***y*-coordinate**.

As we will see, an equation in x and y usually consists of far more than one ordered pair (a, b). The **graph of an equation** is a plot in the Cartesian plane of *all* of the ordered pairs that make up the solution set of the equation.

We can make rough sketches of the graphs of many equations just by plotting enough solutions to give us a sense of the entire solution set. We can find individual ordered pair solutions of a given equation by selecting numbers that seem appropriate for one of the variables and then solving the equation for the other variable. This changes the task of solving a two-variable equation into that of solving a one-variable equation, and we have all the methods of Chapter 1 at our disposal to accomplish this. The process is illustrated in Example 2.

Example 2: Graphing Equations in Two Variables

Sketch graphs of the following equations by plotting points.

a. $2x - 5y = 10$ **b.** $x^2 + y^2 - 6x = 0$ **c.** $y = x^2 - 2x$

Solution

NOTE

Rather than solving a new equation each time you substitute a value, it is more efficient to solve the equation for one variable before making substitutions (see Section 1.6). This method is shown in Example 2b.

a.

x	y
-3	?
0	?
?	0
?	5
1	?

TABLE 1

$2x - 5y = 10$ →

x	y
-3	$-\frac{16}{5}$
0	-2
5	0
$\frac{35}{2}$	5
1	$-\frac{8}{5}$

TABLE 2

In each row in Table 1, we select a value for one of the two variables.

Once we substitute a value, the equation $2x - 5y = 10$ can be solved for the other variable. An example of this is shown below the tables.

$$\begin{aligned} 2(-3) - 5y &= 10 \\ -6 - 5y &= 10 \\ -5y &= 16 \\ y &= -\frac{16}{5} \end{aligned}$$

This gives us a list of 5 ordered pairs that can be plotted, though the ordered pair $\left(\frac{35}{2}, 5\right)$ is off the coordinate system we draw.

TECHNOLOGY

NORMAL FLOAT AUTO REAL RADIAN MP
PRESS + FOR △Tbl

X	Y1
-1	-2.4
0	-2
1	-1.6
2	-1.2
3	-0.8
4	-0.4
5	0
6	0.4
7	0.8
8	1.2
9	1.6

X=-1

You can view a table of specific points on the graph of an equation by selecting Y= and entering the equation, and then selecting 2nd and graph.

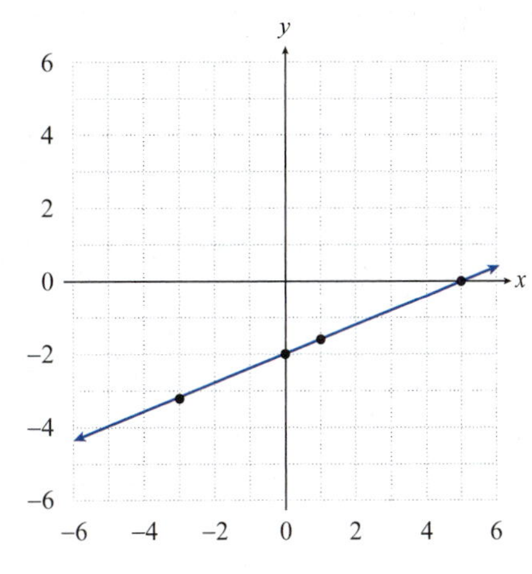

FIGURE 8

The four ordered pairs appear to lie on a straight line, and this is indeed the case. To gain more confidence in this fact, we could continue to plot more solutions of the equation, and we would find they all lie along the line that has been drawn through the four plotted ordered pairs. The infinite number of solutions of the equation are depicted by the line drawn through the plotted points.

b. For this example, we solve the original equation for y, then substitute several values for x to generate a series of y-values.

$$x^2 + y^2 - 6x = 0$$
$$y^2 = 6x - x^2$$
$$y = \pm\sqrt{6x - x^2}$$

x	y
0	?
1	?
2	?
3	?
4	?
5	?
6	?

TABLE 3

$x^2 + y^2 - 6x = 0$ →

x	y
0	0
1	$\pm\sqrt{5} \approx \pm 2.2$
2	$\pm 2\sqrt{2} \approx \pm 2.8$
3	± 3
4	$\pm 2\sqrt{2} \approx \pm 2.8$
5	$\pm\sqrt{5} \approx \pm 2.2$
6	0

TABLE 4

Again, we plot enough solutions to feel confident in sketching the entire solution set. Note that for $x < 0$ and $x > 6$, the corresponding y would be imaginary, and thus irrelevant when graphing the equation. Similarly, for $y < -3$ and for any $y > 3$, the corresponding x would be a complex number (you can use the quadratic formula to verify this).

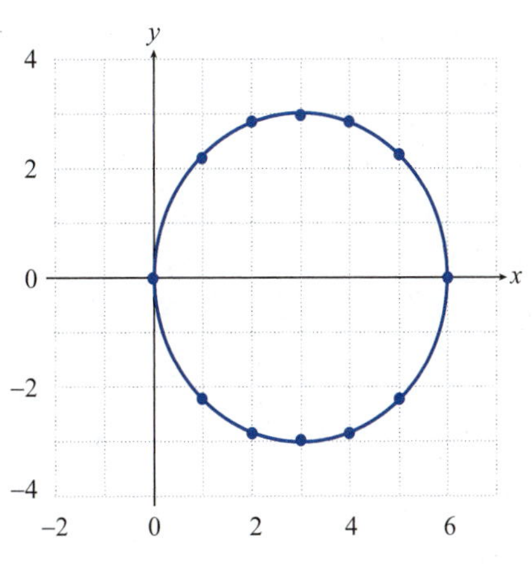

FIGURE 9

Once we plot enough solutions, the graph of the equation begins to take the shape of a circle. In fact, the graph is a circle, with center $(3, 0)$ and a radius of 3, but we will not be able to prove this claim until the next section.

c.

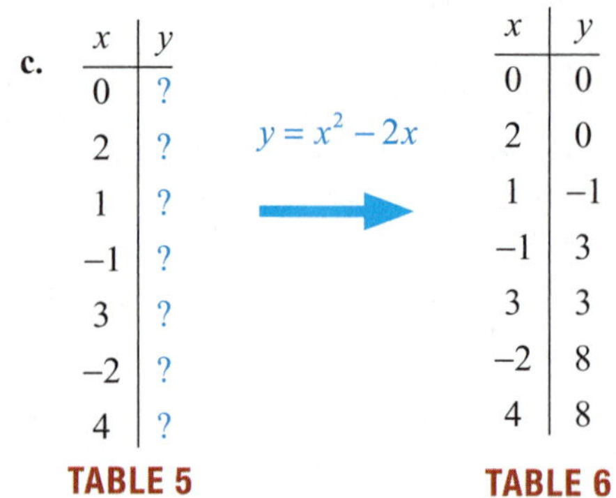

x	y
0	?
2	?
1	?
−1	?
3	?
−2	?
4	?

TABLE 5

x	y
0	0
2	0
1	−1
−1	3
3	3
−2	8
4	8

TABLE 6

Since this equation is already solved for y, we use a table of x-values and substitute them in the given equation. Again, enough points should be plotted to give some idea of the nature of the entire solution set of the equation.

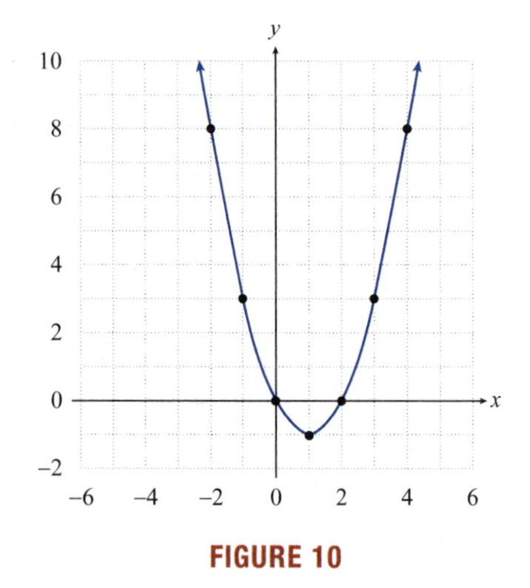

FIGURE 10

The graph of $y = x^2 - 2x$ is a shape known as a *parabola*. We will encounter these shapes again in Section 3.3, and we will be able, at that time, to prove that our rough sketch in Figure 10 is indeed the graph of the solution set of $y = x^2 - 2x$.

Plotting points is, for the most part, easily accomplished, but it is also a rather crude and tedious method, and there are a few concerns about graphing by plotting points:

Have we really plotted enough points to accurately "fill in" the gaps and sketch the entire solution set of each equation? Is filling in the gaps justified in the first place? What proof do we have that *all* the ordered pairs along our sketches actually solve the corresponding equation? Finally, is there a faster and more sophisticated way to determine the graph of an equation?

These concerns are not trivial. Throughout this chapter, we will address them for linear equations. Much of the rest of this textbook works to answer these questions for more complicated equations and graphs. Regardless, plotting points is an important skill since it is so useful when dealing with new or unfamiliar situations.

TECHNOLOGY: Graphing an Equation

If we were trying to sketch the graph of the equation $y = 0.1x^4 - 2.2x^2 + 2.4x + 4.5$, the method of plotting enough ordered pairs that solve the equation would not be very efficient, or accurate. We can use a TI-84 Plus to graph this equation.

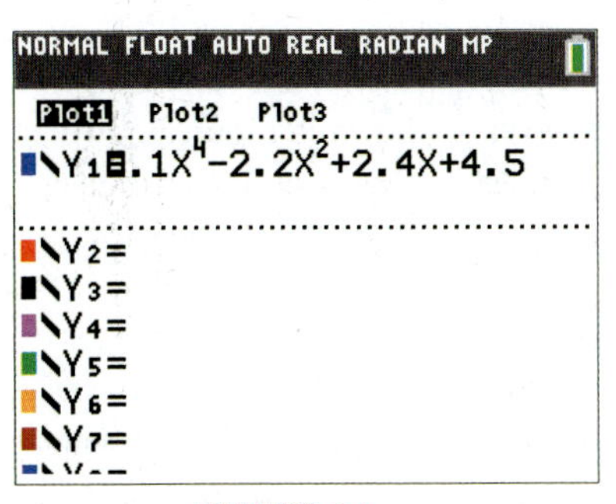

FIGURE 11

Press Y= and type in the equation next to Y1, as shown in Figure 11. To type in the variable x, press X,T,θ,n.

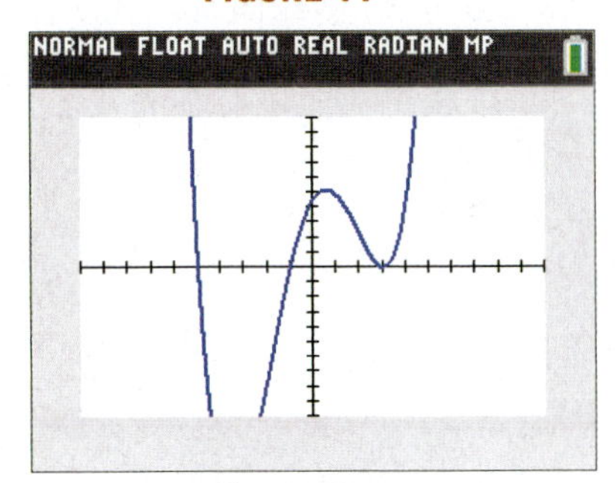

FIGURE 12

Press graph and the graph in Figure 12 should appear.

Notice that you can't see the very bottom of the curve. Oftentimes when we use a TI-84 Plus to graph equations, we have to adjust the viewing window to see the whole graph. To do so, press window. The default window displays the graph with x- and y-values ranging from −10 to 10. This window can be changed by changing the values for Xmin, Xmax, Ymin, and Ymax. Since the graph that appears descends below our viewing screen, we need to change the value for Ymin to something smaller, like −20, as shown in Figure 13.

Press graph again and the TI-84 Plus will display the graph again with the new window settings (see Figure 14).

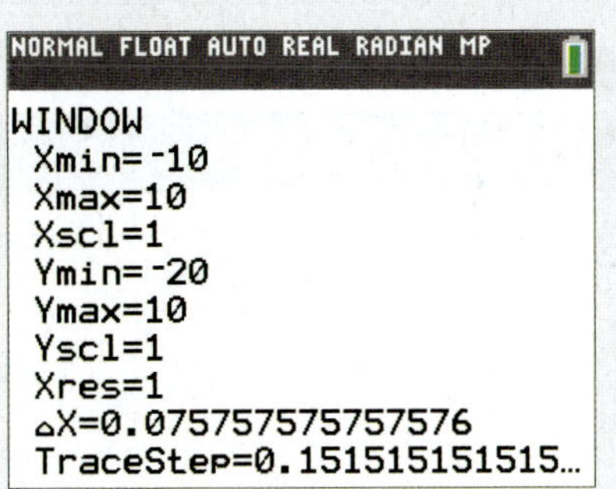

FIGURE 13

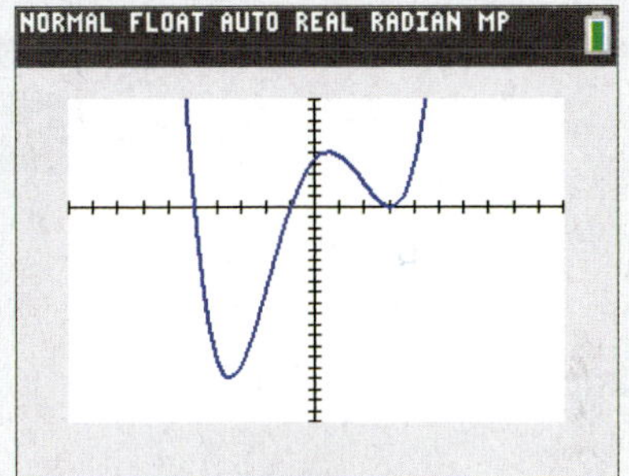

FIGURE 14

Keep in mind that the screen is not square: one unit on the x-axis looks longer than one unit on the y-axis, so the picture will not be an accurate representation unless the window is set to a ratio of about 3:2. One way to attain a window with this ratio is to press zoom and select ZSquare. This will change the values in the WINDOW screen.

Finally, notice that the equation we graphed has two variables, specifically x and y, and is solved for y. An equation must be in this form in order to graph it on a TI-84 Plus. For example, in order to graph the equation $4x + 2y = 1$, we would first have to solve the equation for y: $y = -2x + \frac{1}{2}$.

TOPIC 3: The Distance and Midpoint Formulas

Throughout the rest of this text, we will have reasons for wanting to know, on occasion, the *distance* between two points in the Cartesian plane. We already have the tools necessary to answer this question, and we will now derive a formula that we can apply whenever necessary.

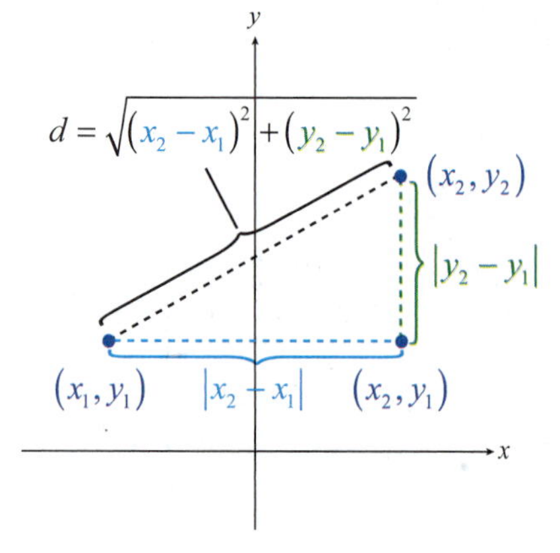

FIGURE 15: The Distance Formula

Let (x_1, y_1) and (x_2, y_2) be the coordinates of two points in the plane. By drawing the dotted lines parallel to the coordinate axes as shown in Figure 15, we can form a right triangle. Note that we are able to determine the coordinates of the vertex at the right angle from the other two vertices (x_1, y_1) and (x_2, y_2).

The lengths of the two legs of the triangle are easy to find, as they are just distances between numbers on real number lines. (The absolute value symbols are present since, in general, $x_2 - x_1$ and $y_2 - y_1$ could be positive or negative.) Recall that the Pythagorean Theorem states that for a right triangle the sum of the squares of its legs is equal to the square of the hypotenuse ($a^2 + b^2 = c^2$). We can apply the Pythagorean Theorem to determine the distance labeled in Figure 15.

$$d^2 = \left(|x_2 - x_1|\right)^2 + \left(|y_2 - y_1|\right)^2, \text{ so}$$

$$d = \sqrt{(x_2 - x_1)^2 + (y_2 - y_1)^2}$$

Notice that the absolute value symbols are not necessary in the final formula, as any quantity squared is automatically nonnegative.

f(x) FORMULA: Distance Formula

The **distance** between two points (x_1, y_1) and (x_2, y_2) in the Cartesian plane is given by the following formula.

$$d = \sqrt{(x_2 - x_1)^2 + (y_2 - y_1)^2}$$

Example 3: Using the Distance Formula

Calculate the distance between the following pairs of points.

a. $(-4, -2)$ and $(-7, 2)$

b. $(5, 1)$ and $(-1, 3)$

Solution

a. $d = \sqrt{((-4)-(-7))^2 + ((-2)-2)^2}$ — Substitute the coordinates of each point into the distance formula.

$= \sqrt{3^2 + (-4)^2}$

$= \sqrt{9 + 16}$ — Simplify.

$= \sqrt{25}$

$= 5$ — Note that we only take the positive square root, since we are calculating a distance.

TECHNOLOGY

```
NORMAL FLOAT AUTO REAL RADIAN MP
√((5-(-1))²+(1-3)²)
                    6.32455532
2√(10)
                    6.32455532
```

When entering the distance formula, be sure to include all the necessary parentheses.

b. $d = \sqrt{(5-(-1))^2 + (1-3)^2}$

$= \sqrt{36+4}$

$= \sqrt{40}$

$= 2\sqrt{10}$

Again, substitute the coordinates into the distance formula, then simplify.

Simplify the radical by factoring out $2^2 = 4$.

We will also want to be able to determine the midpoint of a line segment in the plane. That is, given two points (x_1, y_1) and (x_2, y_2), we want to know the coordinates of the point exactly halfway between the two given points.

Consider the points plotted in Figure 16. The x-coordinate of the midpoint is the average of the two x-coordinates of the given points, and the y-coordinate of the midpoint is the average of the two y-coordinates.

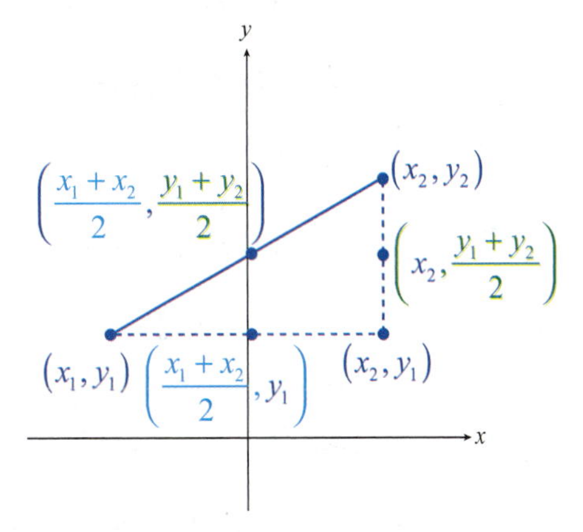

FIGURE 16: The Midpoint Formula

Since x_1 and x_2 are numbers on a real number line, and y_1 and y_2 are numbers on a (different) real number line, determining the averages of these two pairs of numbers is straightforward: the average of the x-coordinates is $\frac{x_1+x_2}{2}$ and the average of the y-coordinates is $\frac{y_1+y_2}{2}$. Putting these two coordinates together gives us the desired formula.

FORMULA: Midpoint Formula

The **midpoint** between two points (x_1, y_1) and (x_2, y_2) in the Cartesian plane has the following coordinates.

$$\left(\frac{x_1+x_2}{2}, \frac{y_1+y_2}{2}\right)$$

Example 4: Using the Midpoint Formula

Calculate the midpoint of the line connecting each pair of points.

a. $(5,1)$ and $(-1,3)$

b. $(3,0)$ and $(-6,11)$

Solution

a. $\left(\frac{5+(-1)}{2}, \frac{1+3}{2}\right) = (2, 2)$

b. $\left(\frac{3+(-6)}{2}, \frac{0+11}{2}\right) = \left(-\frac{3}{2}, \frac{11}{2}\right)$

In each case, we simply substitute the coordinates of each point into the midpoint formula. This has the effect of averaging both x-coordinates and both y-coordinates.

2.1 EXERCISES

PRACTICE

Plot the following sets of points in the Cartesian plane. See Example 1.

1. $\{(-3, 2), (5, -1), (0, -2), (3, 0)\}$

2. $\{(-4, 0), (0, -4), (-3, -3), (3, -3)\}$

3. $\{(3, 4), (-2, -1), (-1, -3), (-3, 0)\}$

4. $\{(2, 2), (0, 3), (4, -5), (-1, 3)\}$

5. $\{(0,5), (-3,2), (2,4), (1,1)\}$

6. $\{(8,3), (-3,4), (-4,-6), (3,-4)\}$

7. $\{(-5,-4), (3,2), (4,5), (-2,-1), (-4,-4), (1,1)\}$

8. $\{(-2,5), (0,1), (1,-1), (1,-3), (0,0), (-1,2), (0,-2)\}$

Identify the quadrant in which each point lies, if possible. If a point lies on an axis, specify which part (positive or negative) of which axis (x or y). See Example 1.

9. $(-2, -4)$ **10.** $(0, -12)$ **11.** $(4, -7)$ **12.** $(-2, 0)$

13. $(9, 0)$ **14.** $(3, 26)$ **15.** $(-4, -7)$ **16.** $(0, 1)$

17. $(17, -2)$ **18.** $(-\sqrt{2}, 4)$ **19.** $(-1, 1)$ **20.** $(-4, 0)$

21. $(3, -9)$ **22.** $(0, 0)$ **23.** $(4, 3)$ **24.** $(-3, -11)$

25. $(0, -97)$ **26.** $\left(\frac{1}{3}, 0\right)$

For each of the following equations, determine the value of the missing entries in the accompanying table of ordered pairs. Then plot the ordered pairs and sketch your guess of the complete graph of the equation. See Example 2.

27. $6x - 4y = 12$

x	y
0	?
?	0
3	?
?	3

28. $y = x^2 + 2x + 1$

x	y
?	0
1	?
?	1
2	?
−3	?

29. $x = y^2$

x	y
0	?
1	?
4	?
9	?
?	$-\sqrt{2}$

30. $5x - 2 = -y$

x	y
?	0
0	?
1	?
?	7
−2	?

31. $x^2 + y^2 = 9$

x	y
0	?
?	0
−1	?
1	?
?	2

32. $y = -x^2$

x	y
0	?
−1	?
1	?
−2	?
2	?

Determine **a.** the distance between the following pairs of points, and **b.** the midpoint of the line segment joining each pair of points. See Examples 3 and 4.

33. $(-2, 3)$ and $(-5, -2)$

34. $(-1, -2)$ and $(2, 2)$

35. $(0, 7)$ and $(3, 0)$

36. $\left(-\frac{1}{2}, 5\right)$ and $\left(\frac{9}{2}, -7\right)$

37. $(-2, 0)$ and $(0, -2)$

38. $(5, 6)$ and $(-3, -2)$

39. $(13, -14)$ and $(-7, -2)$

40. $(-8, 3)$ and $(2, 11)$

41. $(-3, -3)$ and $(5, -9)$

42. $(7, -7)$ and $(-7, -6)$

43. $(5, -4)$ and $(-1, 5)$

44. $(4, 6)$ and $(2, -7)$

45. $(8, 8)$ and $(-2, -2)$

46. $\left(3, \frac{26}{5}\right)$ and $\left(9, -\frac{14}{5}\right)$

47. Given $(10, 4)$ and $(x, -2)$, find x such that the distance between these two points is 10.

48. Given $(1, y)$ and $(13, -3)$, find y such that the distance between these two points is 15.

49. Given $(x, 3)$ and $(-6, y)$, find x and y such that the midpoint between these two points is $(2, 2)$.

Find the perimeter of the triangle whose vertices are the specified points in the plane.

50. $(-2, 3)$, $(-2, 1)$, and $(-5, -2)$

51. $(-1, -2)$, $(2, -2)$, and $(2, 2)$

52. $(6, -1)$, $(-6, 4)$, and $(9, 3)$

53. $(3, -4)$, $(-7, 0)$, and $(-2, -5)$

54. $(-3, 7)$, $(5, 1)$, and $(-3, -14)$

55. $(-12, -3)$, $(-7, 9)$, and $(9, -3)$

APPLICATIONS

56. Two college friends are taking a weekend road trip. Friday they leave home and drive 87 miles north for a night of dinner and dancing in the city. The next morning they drive 116 miles east to spend a day at the beach. If they drive straight home from the beach the next day, how far do they have to travel on Sunday?

57. Your backpacker's guide contains a grid map of Paris, with each unit on the grid representing 0.25 kilometers. If the Eiffel Tower is located at $(-8,-1)$ and the Arc de Triomphe is located at $(-8,4)$, what is the direct distance (not walking distance, which would have to account for bridges and roadways) between the two monuments in kilometers?

58. Your hotel, located at $(-1,-2)$ on the map from Exercise 57, is advertised as exactly halfway between the Eiffel Tower and Notre Dame. What are the grid coordinates of Notre Dame on your map? Find the direct distance from the Eiffel Tower to Notre Dame, rounded to the nearest hundredth of a kilometer.

59. The navigator of a submarine plots the position of the submarine and surrounding objects using a rectangular coordinate system, where each block is one square meter.

a. If his submarine is located at $(50,231)$ and the mobile base to which he is heading is located at $(83,478)$, how far is he from the mobile base?

b. Suppose there is another submarine located halfway between the first submarine and the mobile base. What is the position of the second sub?

60. At the entrance to Paradise Island Theme Park you are given a map of the park that is in the form of a grid, with the park entrance located at $(-5,-5)$. After walking past three rides and the restrooms, you arrive at the Tsunami Water Ride, which is located at $(-3,-1)$ on the grid. If you have traveled halfway along a straight line to your favorite ride, Thundering Tower, where on the grid is your favorite ride located? How far is Thundering Tower from the park entrance on the map?

WRITING & THINKING

61. Use the distance formula to prove that the triangle with vertices at the points $(1,1)$, $(-2,-5)$, and $(3,0)$ is a right triangle. Then determine the area of the triangle.

62. Use the distance formula to prove that the triangle with vertices at the points $(-2,2)$, $(1,-2)$, and $(2,5)$ is isosceles. Then determine the area of the triangle. (**Hint:** Make use of the midpoint formula.)

63. Use the distance formula to prove that the triangle with vertices at the points $(5,1)$, $(-3,7)$, and $(8,5)$ is a right triangle. Then determine the area of the triangle.

64. Use the distance formula to prove that the triangle with vertices at the points $(1,2)$, $(-2,0)$, and $(3,5)$ is isosceles. Then determine the area of the triangle. (**Hint:** Make use of the midpoint formula.)

65. Use the distance formula to prove that the triangle with vertices at the points $(2,2)$, $(6,3)$, and $(4,11)$ is a right triangle. Then determine the area of the triangle.

66. Use the distance formula to prove that the triangle with vertices at the points $(2,-1)$, $(4,3)$, and $(-2,-3)$ is isosceles. Then determine the area of the triangle. (**Hint:** Make use of the midpoint formula.)

67. Use the distance formula to prove that the polygon with vertices at the points $(-2,-1)$, $(6,5)$, $(-2,5)$, and $(6,-1)$ is a rectangle. Then determine the area of the rectangle. (**Hint:** It may help to plot the points before you begin.)

68. Plot the points $(-3,3)$, $(-5,-2)$, $(3,-2)$, and $(1,3)$ to demonstrate they are the vertices of a trapezoid. Then determine the area of the trapezoid.

TECHNOLOGY

Determine appropriate settings on a graphing utility so that each of the given points will lie within the viewing window. Answers will vary slightly.

69. $\{(-4,1),(2,8),(5,7)\}$

70. $\{(12,3),(5,-11),(-9,6)\}$

71. $\{(3,2),(-2,4),(5,-3)\}$

72. $\{(30,55),(40,25),(-80,-10)\}$

73. $\{(3.75,-8.5),(-5.25,6.0),(7.5,-2.25)\}$

74. $\{(63,99),(-87,34),(45,-22)\}$

2.2 CIRCLES

TOPICS

1. The standard form of the equation of a circle
2. Graphing circles

TOPIC 1: The Standard Form of the Equation of a Circle

The primary focus of this chapter will be the study of lines in the Cartesian plane, but we will first make use of a formula introduced in the last section to study another elementary two-dimensional shape.

Circles in the plane can be described mathematically with just two pieces of information: the circle's center and the circle's radius. To be concrete, suppose (h, k) is the location of the circle's center, and suppose the radius is given by the positive real number r. Our goal is to develop an equation in the two variables x and y so that every solution (x, y) of the equation corresponds to a point on the circle.

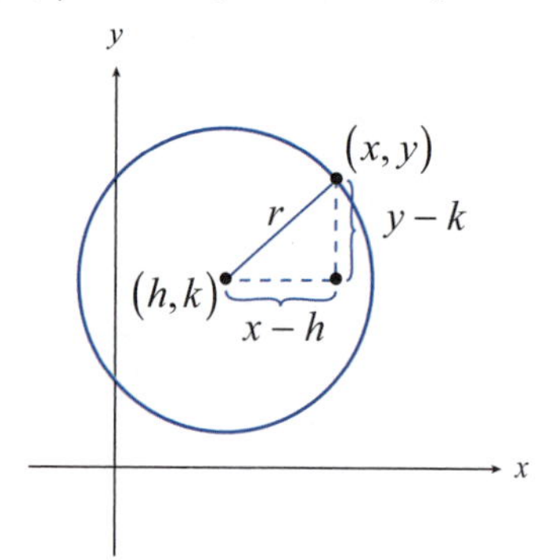

FIGURE 1: Graph of a Circle

The main tool that we need is the distance formula derived in Section 2.1. As we can see in Figure 1, every point (x, y) on the circle is at the same distance r from the circle's center (h, k), so the distance formula tells us that

$$r = \sqrt{(x-h)^2 + (y-k)^2}.$$

This form of the equation is in many ways the most natural, but the equations of circles are usually presented in the radical-free form that results from squaring both sides.

$$r^2 = (x-h)^2 + (y-k)^2$$

DEFINITION: Standard Form of a Circle

The **standard form** of the equation for a circle of radius r and center (h,k) is

$$(x-h)^2 + (y-k)^2 = r^2.$$

Example 1: Standard Form of a Circle

Find the standard form of the equation for the circle with radius 3 and center $(-2, 7)$.

Solution

We are given $h = -2$, $k = 7$, and $r = 3$, so we substitute these values into the standard form.

$$(x-(-2))^2 + (y-7)^2 = 3^2$$

This can be simplified as follows.

$$(x+2)^2 + (y-7)^2 = 9$$

Example 2: Standard Form of a Circle

Find the standard form of the equation for the circle with a diameter whose endpoints are $(-4, -1)$ and $(2, 5)$.

Solution

The midpoint of a diameter of a circle is the circle's center, so we can use the midpoint formula to find the center (h, k).

$$(h, k) = \left(\frac{-4+2}{2}, \frac{-1+5}{2}\right) = (-1, 2)$$

The distance from either diameter endpoint to the center gives us the circle's radius. Since we ultimately will want r^2, we can use a slight variation of the distance formula to find it.

$$r^2 = (-4-(-1))^2 + (-1-2)^2 = 9+9 = 18$$

Thus, the equation of the circle is

$$(x+1)^2 + (y-2)^2 = 18.$$

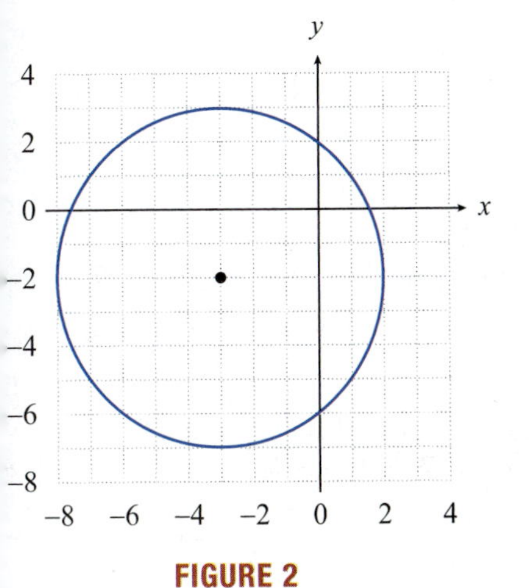

FIGURE 2

Example 3: Standard Form of a Circle

Find the standard form of the equation for the circle shown in Figure 2.

Solution

We can identify the center by looking at the graph; it lies at $(h,k) = (-3,-2)$. To find the radius, we could find any point on the circle, then use the distance formula. A convenient choice is the point $(-3,3)$ since it is directly up from the center. We can simply count off the distance to find that the radius is 5. Plugging our information into the standard form equation, we have

$$(x-(-3))^2 + (y-(-2))^2 = 5^2$$
$$(x+3)^2 + (y+2)^2 = 25.$$

TOPIC 2: Graphing Circles

We will often need to reverse the process illustrated in the first three examples. That is, given an equation for a circle, we will need to determine the circle's center and radius and possibly graph the circle. If the equation is given in standard form, this is straightforward since the standard form presents all the information needed to graph a circle.

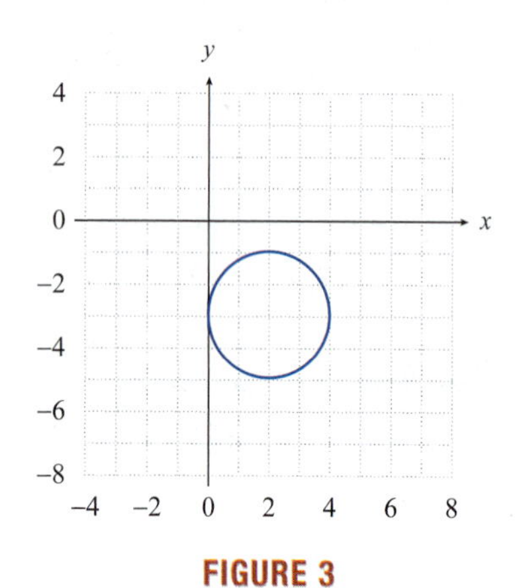

FIGURE 3

Example 4: Graphing Circles

Sketch the graph of the circle defined by $(x-2)^2+(y+3)^2=4$.

Solution

The only preliminary step is to slightly rewrite the equation in the form

$$(x-2)^2+(y-(-3))^2=2^2.$$

From this, we see that $(h,k)=(2,-3)$ and $r=2$. The graph of the equation is shown in Figure 3.

More often, the equation for a circle will not be given to us in quite so neat a fashion. We may have to apply some algebraic manipulation in order to determine that a given equation describes a circle and to determine the center and radius of that circle. Fortunately, the algebraic technique of *completing the square* is usually all that is required.

Example 5: Graphing Circles and Completing the Square

Sketch the graph of the equation $x^2+y^2+8x-2y=-1$.

Solution

We need to complete the square in the variable x and the variable y, and we do so as follows.

$$\begin{aligned} x^2+y^2+8x-2y&=-1 \\ (x^2+8x)+(y^2-2y)&=-1 \\ (x^2+8x+16)+(y^2-2y)&=-1+16 \\ (x+4)^2+(y^2-2y)&=15 \\ (x+4)^2+(y^2-2y+1)&=15+1 \\ (x+4)^2+(y-1)^2&=16 \end{aligned}$$

Begin by rearranging the equation so we can complete the square for each variable.

The coefficient on x is 8, so we add $4^2=16$ to both sides, then rewrite the x terms.

The coefficient on y is −2, so we add $(-1)^2=1$ to both sides, then rewrite the y terms.

We now see that the equation does indeed describe a circle, and that the center of the circle is $(-4,1)$ and the radius is 4. The graph is shown in Figure 4.

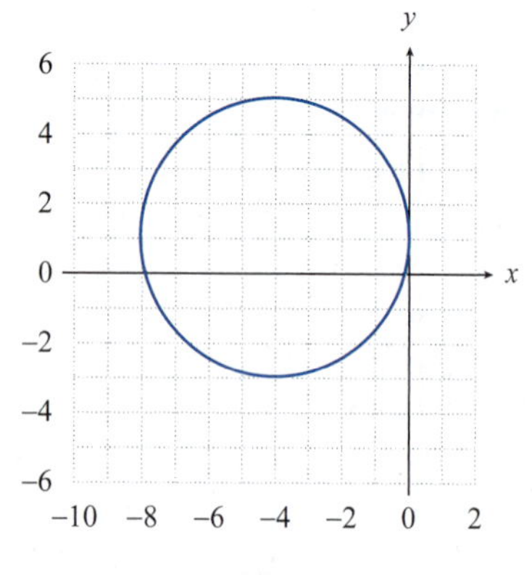

FIGURE 4

2.2 EXERCISES

PRACTICE

Find the standard form of the equation for the circle. See Examples 1 and 2.

1. Center $(-4,-3)$; radius 5
2. Center at origin; radius 3
3. Center $(7,-9)$; radius 3
4. Center $(-2, 2)$; radius 2
5. Center $(0,0)$; radius $\sqrt{6}$
6. Center $(6,3)$; radius 8
7. Center $(\sqrt{5}, \sqrt{3})$; radius 4
8. Center $\left(\frac{5}{3}, \frac{8}{5}\right)$; radius $\sqrt{8}$
9. Center $(7,2)$; passes through $(7,0)$
10. Center $(3,3)$; passes through $(1,3)$
11. Center $(-3,8)$; passes through $(-4,9)$
12. Center $(0,0)$; passes through $(2,10)$
13. Center $(4,8)$; passes through $(1,9)$
14. Center $(12,-4)$; passes through $(-9,5)$
15. Center at the origin; passes through $(6,-7)$
16. Center $(13,-2)$; passes through $(8,-3)$
17. Endpoints of a diameter are $(-8,6)$ and $(1,11)$
18. Endpoints of a diameter are $(5,3)$ and $(8,-3)$
19. Endpoints of a diameter are $(-7,-4)$ and $(-5,7)$
20. Endpoints of a diameter are $(2,3)$ and $(7,4)$
21. Endpoints of a diameter are $(0,0)$ and $(-13,-14)$
22. Endpoints of a diameter are $(4,10)$ and $(0,3)$
23. Endpoints of a diameter are $(0,6)$ and $(8,0)$
24. Endpoints of a diameter are $(6,9)$ and $(4,9)$

Find the standard form of the equation for the circle. See Example 3.

25.

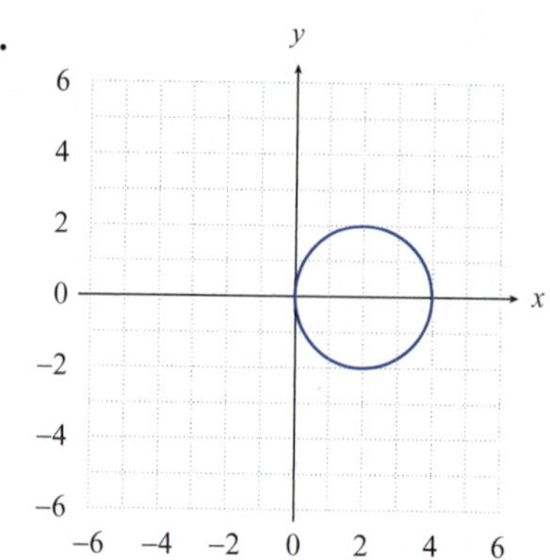

26.

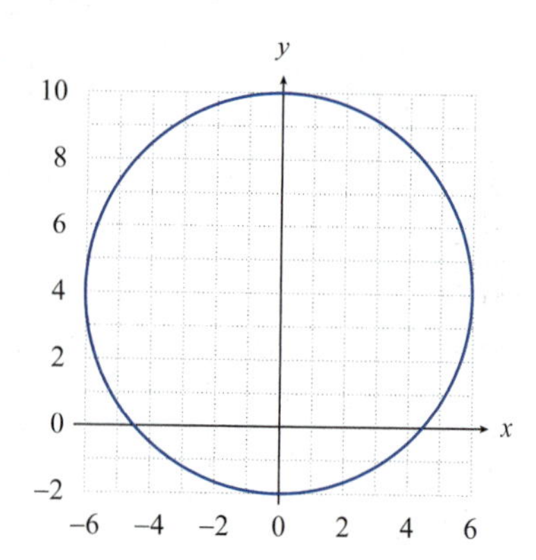

27.

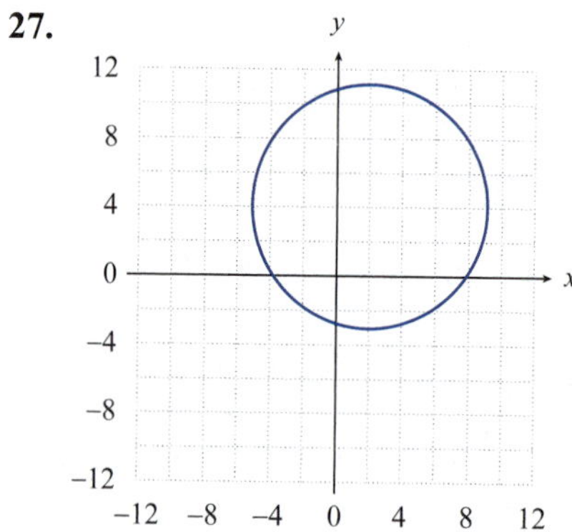

28.

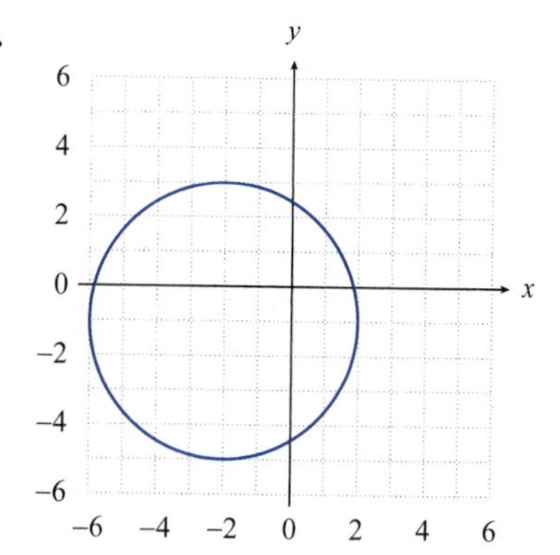

29.

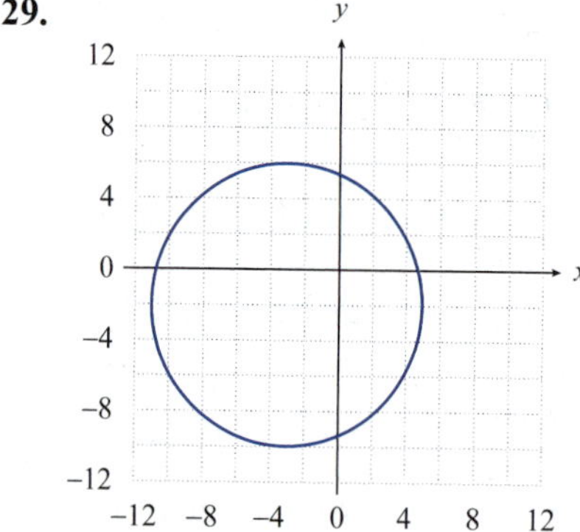

Sketch a graph of the equation and find the center and radius of each circle. See Examples 4 and 5.

30. $x^2 + y^2 = 25$

31. $x^2 + y^2 = 36$

32. $x^2 + (y-3)^2 = 16$

33. $x^2 + (y-8)^2 = 9$

34. $(x+2)^2 + y^2 = 49$

35. $(x-8)^2 + y^2 = 8$

36. $(x-9)^2 + (y-4)^2 = 49$

37. $(x+5)^2 + (y+4)^2 = 4$

38. $(x+2)^2+(y-7)^2=64$

39. $(x-5)^2+(y+5)^2=5$

40. $x^2+y^2-2x+10y+1=0$

41. $x^2+y^2-4x+4y-8=0$

42. $x^2+y^2+6x+5=0$

43. $x^2+y^2+10y+9=0$

44. $x^2+y^2-x-y=2$

45. $x^2+y^2+6y-2x=-2$

46. $(x-5)^2+y^2=225$

47. $4x^2+4y^2=256$

48. $(x-3)^2+(y+2)^2=81$

49. $x^2+y^2-6x+4y-3=0$

50. $(x+2)^2+(y-1)^2=61$

51. $(x-1)^2+y^2=9$

52. $x^2+(y+2)^2=49$

53. $x^2+y^2-4x+8y-16=0$

54. $x^2+y^2+8x=9$

55. $4x^2+4y^2-24x+24y=28$

2.3 LINEAR EQUATIONS IN TWO VARIABLES

TOPICS

1. Recognizing linear equations in two variables
2. Intercepts of the coordinate axes
3. Horizontal and vertical lines

TOPIC 1: Recognizing Linear Equations in Two Variables

Lines in Use

While straight lines may seem somewhat limited in scope compared to all the other sorts of curves that can be drawn in the plane, they have the virtue of being very easy to work with. This quality is so important that *linearization*, or the process of approximating a curve with a straight line at a given point, is common throughout fields of mathematics and in scientific and engineering applications.

Example 2a in Section 2.1 was our first encounter with a linear equation in two variables. Our first goal in this section is to recognize when an equation in two variables is linear; we want to know when the solution set of an equation is a straight line in the Cartesian plane.

DEFINITION: Linear Equations in Two Variables

A **linear equation in two variables**, say the variables x and y, is an equation that can be written in the form $ax + by = c$, where a, b, and c are constants and a and b are not both zero. This form of such an equation is called the **standard form**.

An equation may be linear but not appear in standard form. Some algebraic manipulation is often necessary in order to determine if a given equation is linear. We will see in the next section that there are other forms of linear equations that are useful in different situations. For now, we will focus on the standard form when identifying linear equations.

Example 1: Linear Equations in Two Variables

Determine if the following equations are linear.

a. $3x - (2 - 4y) = x - y + 1$

b. $3x + 2(x + 7) - 2y = 5x$

c. $\dfrac{x+2}{3} - y = \dfrac{y}{5}$

d. $7x - (4x - 2) + y = y + 3(x - 1)$

e. $4x^3 - 2y = 5x$

f. $x^2 - (x - 3)^2 = 3y$

Solution

a. $3x - (2 - 4y) = x - y + 1$

$3x - 2 + 4y = x - y + 1$ First, apply the distributive property.

$3x - x + 4y + y = 1 + 2$ Arrange the variables on one side.

$2x + 5y = 3$ Combine like terms. The equation is linear.

b. $3x+2(x+7)-2y=5x$

$3x+2x+14-2y=5x$ Begin, again, with the distributive property.

$5x-5x-2y=-14$ Move the variables to one side.

$-2y=-14$ Combine like terms. The x variable disappears, indicating a coefficient of 0, but the coefficient on y is nonzero, so the equation is still linear.

$y=7$

c. $\frac{x+2}{3}-y=\frac{y}{5}$ For this equation, we need to separate the fraction into a variable part and a constant part.

$\frac{1}{3}x+\frac{2}{3}-y=\frac{1}{5}y$

$\frac{1}{3}x-y-\frac{1}{5}y=-\frac{2}{3}$ Once again, we move all the variables to one side, then combine like terms.

$\frac{1}{3}x-\frac{6}{5}y=-\frac{2}{3}$ The equation is linear. Note that we could also have begun by clearing the fractions.

d. $7x-(4x-2)+y=y+3(x-1)$ After simplifying this equation, we see that coefficients on x and y are both 0. Thus, the equation is not linear.

$7x-4x+2+y=y+3x-3$

$3x-3x+y-y=-3-2$

$0=-5$ Further, the equation simplifies to a false statement, so it actually has no solutions!

e. $4x^3-2y=5x$ The presence of the cubed term in this already simplified equation makes it not linear.

f. $x^2-(x-3)^2=3y$ First, expand the squared binomial term.

$x^2-x^2+6x-9=3y$

$6x-3y=9$ In contrast to the last equation, when we simplify this equation the result is linear.

TOPIC 2: Intercepts of the Coordinate Axes

Often, the goal in working with a given linear equation is to graph its solution set. Since two points determine a line, all we need to do to graph the solution set of a linear equation is to find two different solutions.

If the equation under consideration is in the two variables x and y, it is natural to call the point where the graph crosses the x-axis the ***x*-intercept** and the point where it crosses the y-axis the ***y*-intercept**. If the line does indeed cross both axes, the two intercepts are easy to find: the y-coordinate of the x-intercept is 0, and the x-coordinate of the y-intercept is 0.

> **DEFINITION: The x- and y-Intercepts**
>
> Given a graph in the Cartesian plane, any point where the graph intersects the x-axis is called an ***x*-intercept**, and any point where the graph intersects the y-axis is called a ***y*-intercept**. All x-intercepts are of the form $(c, 0)$ and all y-intercepts are of the form $(0, c)$.

It should be noted that such phrases as "an x-intercept of 3" and "a y-intercept of 5" are common alternatives to, respectively, "an x-intercept at $(3, 0)$" and "a y-intercept at $(0, 5)$."

Example 2: Finding Intercepts and Graphing Linear Equations

Find the x- and y-intercepts of the given equations, and sketch their graphs.

a. $3x - 4y = 12$

b. $4x - (3 - x) + 2y = 7$

Solution

a. $$3x - 4y = 12$$

$$3(0) - 4y = 12 \qquad 3x - 4(0) = 12$$
$$y = -3 \qquad x = 4$$
y-intercept: $(0, -3)$ x-intercept: $(4, 0)$

To find the two intercepts, first set x equal to 0 and solve for y, then set y equal to 0 and solve for x.

This gives us the coordinates of the two intercepts, which we plot in Figure 1.

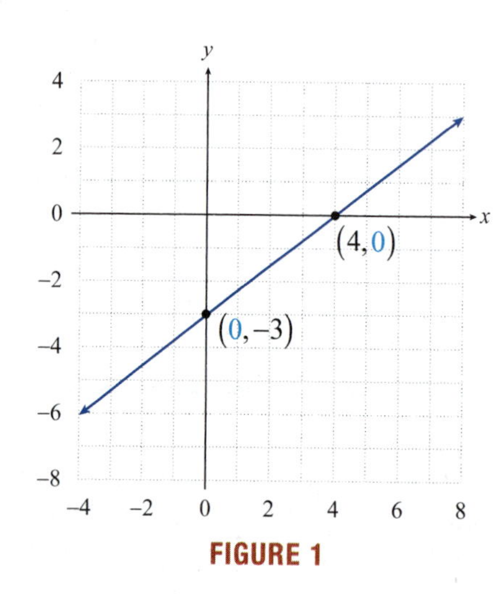

FIGURE 1

Once we have plotted the intercepts, drawing a straight line through them gives us the graph of the equation.

> **TECHNOLOGY**
>
> NORMAL FLOAT AUTO REAL RADIAN MP
> PRESS [ENTER] TO EDIT
>
> | X | Y₁ |
> |---|---|
> | -1 | -3.75 |
> | 0 | -3 |
> | 1 | -2.25 |
> | 2 | -1.5 |
> | 3 | -0.75 |
> | 4 | 0 |
> | 5 | 0.75 |
> | 6 | 1.5 |
> | 7 | 2.25 |
> | 8 | 3 |
> | 9 | 3.75 |
>
> Y1=3/4X−3
>
> To use a TI-84 Plus to generate a table of coordinates of points that lie on the graph of the equation, press Y= and enter the equation solved for y, $y = \frac{3}{4}x - 3$. Then press 2nd and graph to view the table. Notice that the y-value at $x = 0$ is -3 and the y-value at $x = 4$ is 0.

b. $$4x - (3 - x) + 2y = 7$$
$$5x + 2y = 10$$
$$5(0) + 2y = 10 \qquad 5x + 2(0) = 10$$
$$y = 5 \qquad x = 2$$
y-intercept: $(0, 5)$ x-intercept: $(2, 0)$

Again, find the two intercepts by setting the appropriate variables equal to 0.

Solving the resulting equations in one variable yields the intercept solutions.

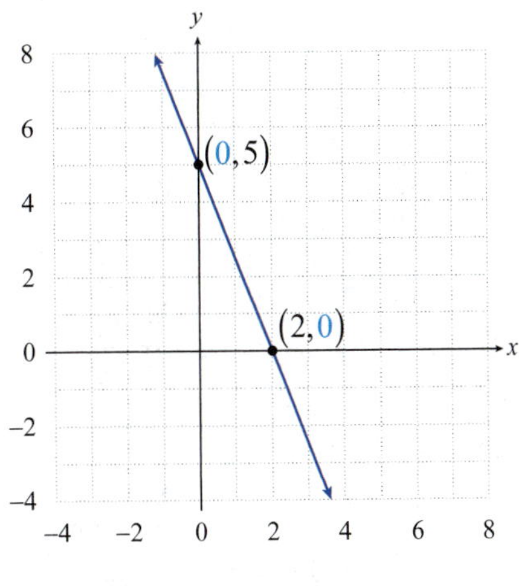

FIGURE 2

Plot the two intercepts, then draw the line passing through these two points. Note that in both graphs, the location of the origin has been chosen in order to conveniently plot the intercepts.

TECHNOLOGY: Finding Intercepts

Since all linear equations can be solved for the variable y, we can use a TI-84 Plus to graph them. Doing so enables us to use a TI-84 Plus to find the x- and y-intercepts. Suppose we've graphed the equation $y = 2x - 6$.

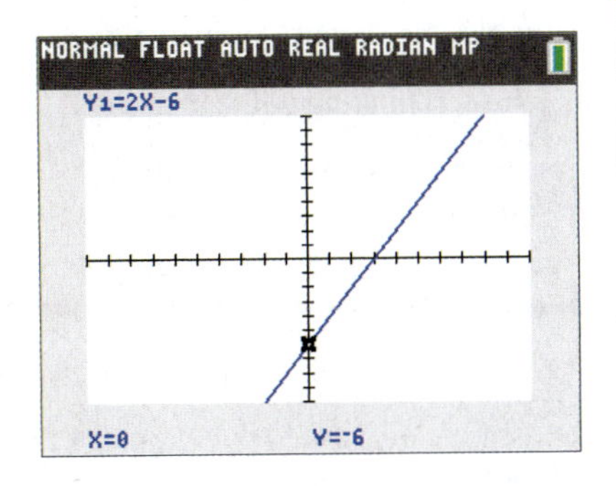

FIGURE 3

To find the y-intercept, press trace. Since the y-intercept occurs where $x = 0$, use the arrows to move the cursor along the line until the x-value is 0. Alternatively, just press 0 and enter to place the cursor at that point. The corresponding y-value, −6, is shown at the bottom of the screen (Figure 3). So the point $(0, -6)$ is the y-intercept.

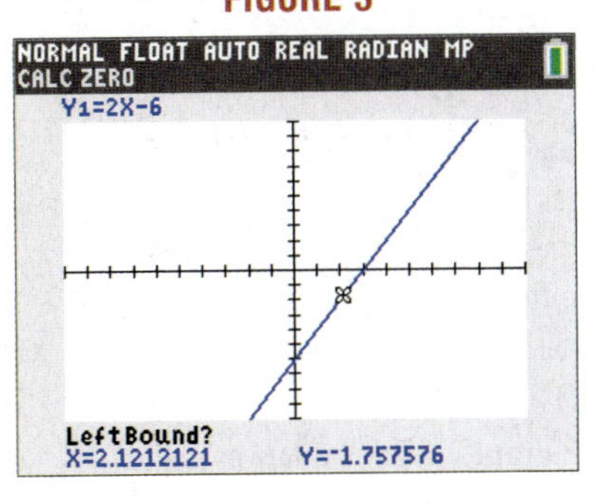

FIGURE 4

We will find the x-intercept using a different method. Press 2nd trace to access the CALCULATE menu and select zero and press enter. The screen should now display the graph with LeftBound? shown at the bottom (Figure 4). Use the arrows to move the cursor anywhere to the left of where the line crosses the x-axis and press enter. The screen should now say RightBound? (Figure 5). Use the right arrow to move the cursor to the right of where the line crosses the x-axis and press enter again. The text should now read Guess? (Figure 6). Press enter a third time and the x- and y-values of the x-intercept will appear at the bottom of the screen (Figure 7).

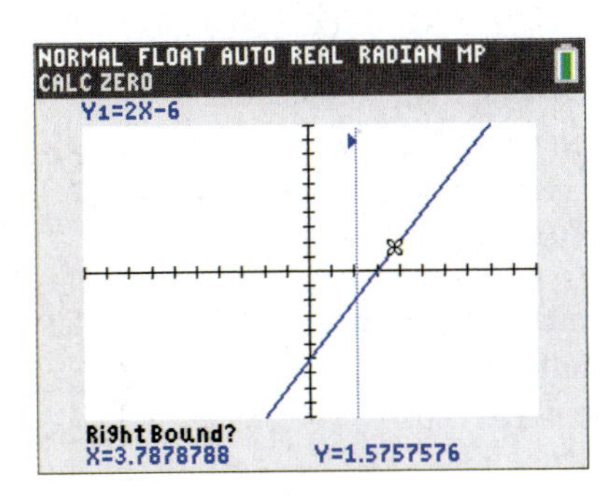

FIGURE 5

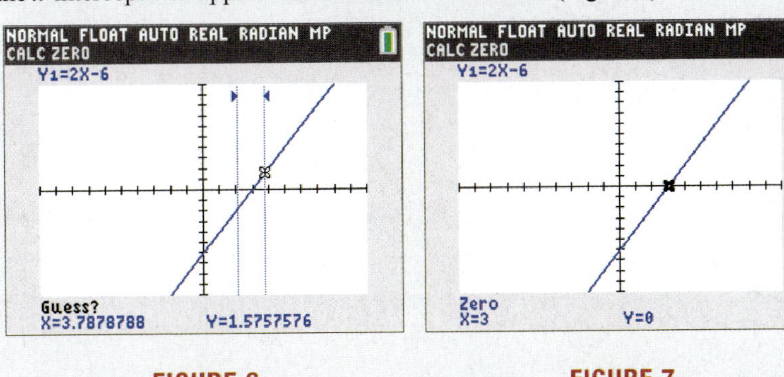

FIGURE 6

FIGURE 7

So the x-intercept is $(3, 0)$. Both of these techniques can be used to find the x- and y-intercepts of any equation graphed with a TI-84 Plus, not just linear equations.

TOPIC 3: Horizontal and Vertical Lines

If a linear equation doesn't have two intercepts, the graphing process in Example 2 does not work. When might this happen? One case is when the line passes through the origin, $(0,0)$. Then, the x-intercept and y-intercept are the same point instead of two distinct points. The other possibility is that the graph may be parallel to an axis (and thus not have one intercept); this happens when the equation is a horizontal or vertical line. In order to graph these equations, a second point (not an intercept) must be found in order to have two points to connect with a line.

Equations of horizontal or vertical lines are missing one of the two variables. In the absence of any other information, it is impossible to know if the solutions of equations like $x = 4$ or $y = -3$ consist of a point on the real number line or a line in the Cartesian plane. You must rely on the context of the problem to know how many variables should be considered. Throughout this chapter, all equations are assumed to be in two variables unless otherwise stated, so an equation of the form $ax = d$ (with $a \neq 0$) or $by = d$ (with $b \neq 0$) should be thought of as representing a line in the plane.

Consider an equation of the form $ax = d$. The variable y is absent, so *any* value for y will give a solution as long as we pair it with $x = \frac{d}{a}$. Thinking of the solution set as a set of ordered pairs, the solution consists of ordered pairs with a fixed first coordinate and arbitrary second coordinate. This describes, geometrically, a vertical line with an x-intercept of $\frac{d}{a}$. Similarly, the equation $by = d$ represents a horizontal line with y-intercept of $\frac{d}{b}$.

Example 3: Graphing Horizontal and Vertical Lines

Graph the following equations.

a. $5x = 0$ **b.** $2x - 2 = 3$ **c.** $3x + 2(x + 7) - 2y = 5x$

Solution

a. $5x = 0$

$x = 0$

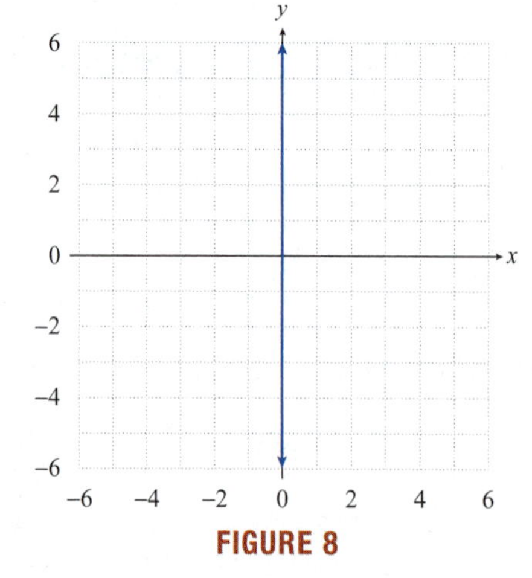

FIGURE 8

The first step is to divide both sides by 5, leaving the simple equation $x = 0$.

The graph of this equation is the y-axis, as all ordered pairs on the y-axis have an x-coordinate of 0.

This equation is unique in that it has an infinite number of y-intercepts (since each point on the graph is on the y-axis) and one x-intercept (the origin).

Similarly, the equation $y = 0$ has an infinite number of x-intercepts and one y-intercept.

b. $2x - 2 = 3$

$x = \frac{5}{2}$

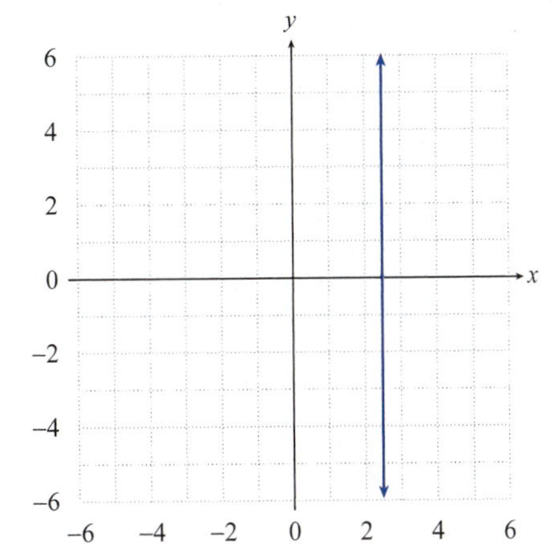

FIGURE 9

Upon simplifying, note that this equation also represents a vertical line, this time passing through $\frac{5}{2}$ on the x-axis.

c. $3x + 2(x + 7) - 2y = 5x$

$y = 7$

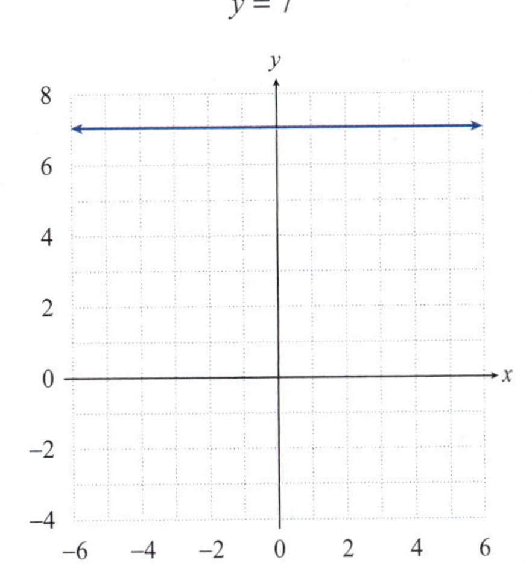

FIGURE 10

We encountered this equation in Example 1b and have already written it in standard form as shown.

The graph of this equation is the horizontal line consisting of all those ordered pairs whose y-coordinate is 7.

2.3 EXERCISES

PRACTICE

Determine if the following equations are linear. See Example 1.

1. $3x + 2(x - 4y) = 2x - y$

2. $9x + 4(y - x) = 3$

3. $9x^2 - (x + 1)^2 = y - 3$

4. $3x + xy = 2y$

5. $8 - 4xy = x - 2y$

6. $\frac{x - y}{2} + \frac{7y}{3} = 5$

7. $\frac{6}{x}-\frac{5}{y}=2$

8. $3x-3(x-2y)=y+1$

9. $2y-(x+y)=y+1$

10. $(3-y)^2-y^2=x+2$

11. $x^2-(x-1)^2=y$

12. $(x+y)^2-(x-y)^2=1$

13. $x(y+1)=16-y(1-x)$

14. $\frac{x-3}{2}=\frac{4+y}{5}$

15. $x-2x^2+3=\frac{x-7}{2}$

16. $x-3=\frac{4x+17}{5}$

17. $13x-17y=y(7-2x)$

18. $y^2-3y=(1+y)^2-2x$

19. $x-1=\frac{2y}{x}-x$

20. $3x-4=89(x-y)-y$

21. $x-x(1+x)=y-3x$

22. $x^2-2x=3-x^2+y$

23. $\frac{2y-5}{14}=\frac{x-3}{9}$

24. $16x=y(4+(x-3))-xy$

Find the x- and y-intercepts of the given equations, if possible, and then sketch their graphs. See Examples 2 and 3.

25. $4x-3y=12$

26. $y-3x=9$

27. $5-y=10x$

28. $y-2x=y-4$

29. $3y=9$

30. $2x-(x+y)=x+1$

31. $x+2y=7$

32. $y-x=x-y$

33. $y=-x$

34. $2x-3=1-4y$

35. $3y+7x=7(3+x)$

36. $4-2y=-2-6x$

37. $x+y=1+2y$

38. $3y+x=2x+3y+4$

39. $3(x+y)+1=x-5$

Match each equation to the correct graph.

40. $y=2x+3$

41. $2x+3y=4$

42. $2x-1=5$

43. $y+3-x=3$

44. $4y+3=11$

45. $5y-x-1=4y+3x+5$

a.

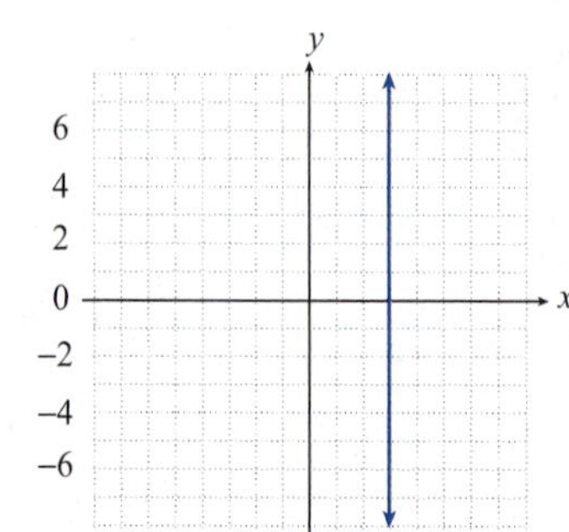

b.

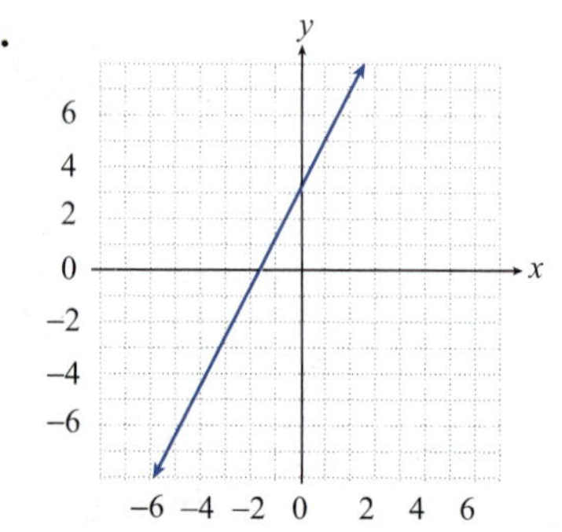

c.

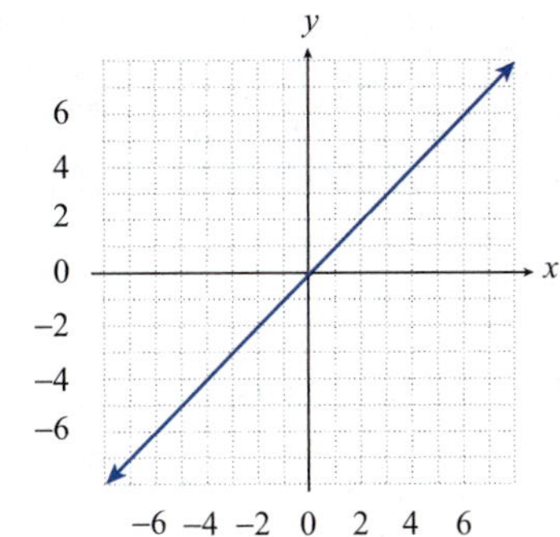

d.

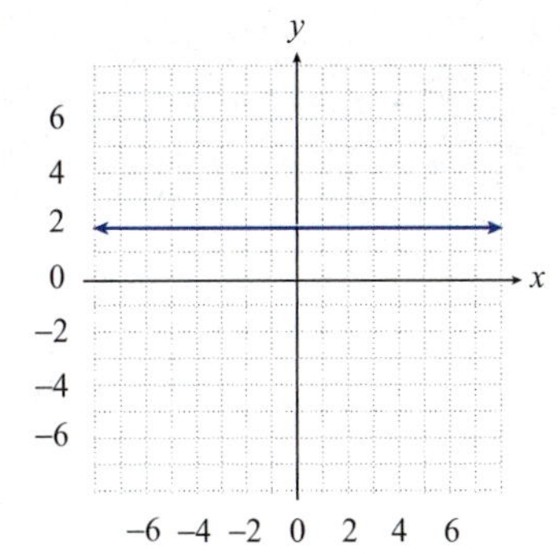

e.

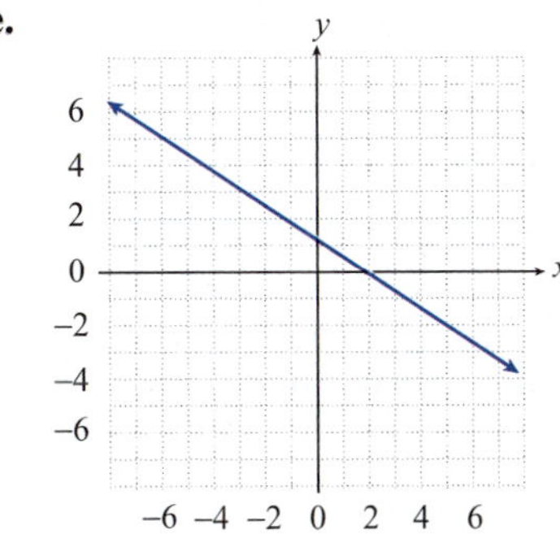

f.

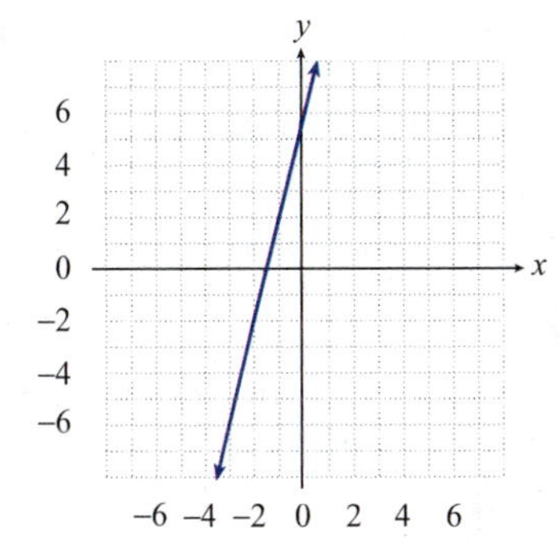

Solve each equation for the specified variable.

46. Standard form of a line: $ax + by = c$; solve for y

47. Perimeter of a triangle: $P = a + b + c$; solve for a

48. Surface area of a rectangular solid: $S = 2lw + 2wh + 2lh$; solve for w

APPLICATIONS

49. In your history class, you were told that the current population of Jamaica is approximately 24,000 more than 9 times the population of the Bahamas. Using j to represent the population of Jamaica and b to represent the population of the Bahamas, write this in the form of an equation. Then solve your equation for b to find an equation representing the population of the Bahamas. Are these equations linear?

50. The lowest point in the ocean, the bottom of the Mariana Trench, is about 1100 feet deeper than 26 times the depth of the lowest point on land, the Dead Sea. Find an equation to express the depth of the Mariana Trench, m, in terms of the depth of the Dead Sea, d. Then solve your equation for d to find the depth of the Dead Sea in terms of the depth of the Mariana Trench. Are these equations linear?

2.4 SLOPE AND FORMS OF LINEAR EQUATIONS

TOPICS

1. The slope of a line
2. The slope-intercept form of the equation of a line
3. The point-slope form of the equation of a line

TOPIC 1: The Slope of a Line

There are several ways to characterize a given line in a plane. We have already used one way repeatedly: two distinct points in the Cartesian plane determine a line. Another, often more useful, approach is to identify just one point on the line and to indicate how "steeply" the line is rising or falling as we scan the plane from left to right. It turns out that a single number is sufficient to convey this notion of "steepness."

DEFINITION: The Slope of a Line

Let L stand for a given line in the Cartesian plane, and let (x_1, y_1) and (x_2, y_2) be the coordinates of any two distinct points on L. The slope of the line L is the ratio $\frac{y_2 - y_1}{x_2 - x_1}$, which can be described in words as "change in y over change in x" or "rise over run."

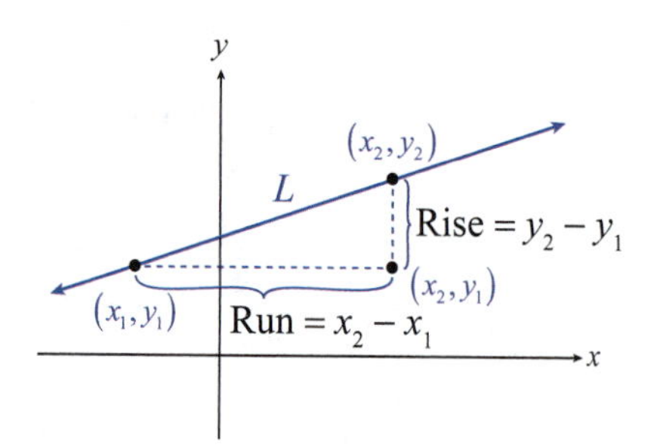

FIGURE 1: Rise and Run between Two Points

In the line drawn in Figure 1, the ratio $\frac{y_2 - y_1}{x_2 - x_1}$ is positive, the line rises from the lower left to the upper right, and we say that the line has a positive slope. If the rise and run have opposite signs, the slope of the line is negative and the line under consideration would fall from the upper left to the lower right.

CAUTION

It doesn't matter how you assign the labels (x_1, y_1) and (x_2, y_2) to the two points you are using to calculate slope, but it *is* important that you are consistent as you apply the formula. You cannot change the order in which you are subtracting as you determine the numerator and denominator in the slope formula.

Correct: $\dfrac{y_2 - y_1}{x_2 - x_1}$ or $\dfrac{y_1 - y_2}{x_1 - x_2}$ **Incorrect:** $\dfrac{y_1 - y_2}{x_2 - x_1}$ or $\dfrac{y_2 - y_1}{x_1 - x_2}$

Example 1: Calculating the Slope of a Line

Determine the slopes of the lines passing through the following pairs of points in $\mathbb{R}^2$.

a. $(-4, -3)$ and $(2, -5)$ **b.** $\left(\dfrac{3}{2}, 1\right)$ and $\left(1, -\dfrac{4}{3}\right)$ **c.** $(-2, 7)$ and $(1, 7)$

Solution

a. $\dfrac{-5-(-3)}{2-(-4)} = \dfrac{-2}{6} = -\dfrac{1}{3}$ We calculate the slope in two ways; first set $(x_1, y_1) = (-4, -3)$ and $(x_2, y_2) = (2, -5)$.

$\dfrac{-3-(-5)}{-4-2} = \dfrac{2}{-6} = -\dfrac{1}{3}$ We get the same result using $\dfrac{y_1 - y_2}{x_1 - x_2}$.

b. $\dfrac{1-\left(-\dfrac{4}{3}\right)}{\dfrac{3}{2}-1} = \dfrac{\dfrac{7}{3}}{\dfrac{1}{2}} = \dfrac{7}{3} \cdot \dfrac{2}{1} = \dfrac{14}{3}$ The final answer tells us that the line through the two points rises 14 units for every run of 3 units horizontally.

c. $\dfrac{7-7}{-2-1} = \dfrac{0}{-3} = 0$ These two points have the same y-coordinate and thus lie on a horizontal line.

TECHNOLOGY

The formula $\dfrac{y_2 - y_1}{x_2 - x_1}$ is only valid if $x_2 - x_1 \neq 0$, since division by zero is undefined. What lines have undefined slope? If $x_2 - x_1 = 0$, then the line has two points with the same x-coordinate, which defines a *vertical* line. The other extreme is that the numerator is 0 (and the denominator is nonzero), in which case the slope is 0. For what sorts of lines will this happen? If two points on a line have the same y-coordinate, that is if $y_1 = y_2$, the line must be *horizontal*.

PROPERTIES: Slopes of Horizontal and Vertical Lines

Horizontal lines, which can be written in the form $y = c$, have a **slope of 0**.

Vertical lines, which can be written in the form $x = c$, have an **undefined slope**.

NOTE

Intercepts are often good points to use in calculating the slope since they have at least one coordinate equal to zero.

TECHNOLOGY

NORMAL FLOAT AUTO REAL RADIAN MP
PRESS [ENTER] TO EDIT

X	Y1
-1	-5.333
0	-4
1	-2.667
2	-1.333
3	0
4	1.3333
5	2.6667
6	4
7	5.3333
8	6.6667
9	8

Y1=(4/3)X−4

To use a TI-84 Plus to generate a table of coordinates of points that lie on the graph of the equation, press Y= and enter the equation solved for y, $y=\frac{4}{3}x-4$. Then press 2nd and graph to view the table. Notice that the y-value at $x=0$ is -4 and the y-value at $x=3$ is 0.

Example 2: Calculating the Slope of a Line

Determine the slopes of the lines defined by the following equations.

a. $4x-3y=12$

b. $2x+7y=9$

c. $x=-\frac{3}{4}$

d. $y=9$

Solution

a. First, we find two points on the line by calculating the intercepts.

$$4x-3y=12$$

$$\begin{aligned} 4(0)-3y&=12 & 4x-3(0)&=12 \\ -3y&=12 & 4x&=12 \\ y&=-4 & x&=3 \end{aligned}$$

Recall that the x-intercept is found by setting y equal to 0 and solving for x, and vice versa for the y-intercept.

y-intercept: $(0,-4)$ x-intercept: $(3,0)$

$$\text{slope}=\frac{-4-0}{0-3}=\frac{-4}{-3}=\frac{4}{3}$$

Once we have two points, we apply the slope formula.

b. x-intercept: $\left(\frac{9}{2},0\right)$

second point on the line: $(1,1)$

$$\text{slope}=\frac{1-0}{1-\frac{9}{2}}=\frac{1}{-\frac{7}{2}}=-\frac{2}{7}$$

In this example, we have found the x-intercept. We do not have to find both intercepts; the point $(1,1)$ is on the line and is simple to use in calculation.

c. This equation is of the form $x=c$, and is a vertical line. Therefore, the slope is undefined.

d. This equation is of the form $y=c$, and is a horizontal line. Therefore, it has a slope of 0.

Note that the line in Example 2a has a positive slope and the line in Example 2b has a negative slope. Without graphing these lines, we know that the first line will rise from the lower left to the upper right part of the plane, while the second line will fall from the upper left to the lower right. You should practice your graphing skills and verify that these observations are indeed correct.

TOPIC 2: The Slope-Intercept Form of the Equation of a Line

Example 2 illustrates the most elementary way of determining the slope of a line from an equation. With a little work, we can develop a faster method for determining not only the slope of a line, but also the y-intercept.

Consider a nonvertical line in the plane. The variable y must appear in the linear equation that describes the line (otherwise the line would be vertical), so the equation can be solved for y. The result will be an equation of the form $y = mx + b$, where m and b are constants, and it turns out that these constants provide a lot of information about the graph of the line.

Suppose that (x_1, y_1) and (x_2, y_2) are two points that lie on the line $y = mx + b$. Then, it must be the case that $y_1 = mx_1 + b$ and $y_2 = mx_2 + b$. If we use these two points to determine the slope of the line, we obtain the following.

$$\text{slope} = \frac{y_2 - y_1}{x_2 - x_1} = \frac{(mx_2 + b) - (mx_1 + b)}{x_2 - x_1} = \frac{m\cancel{(x_2 - x_1)}}{\cancel{x_2 - x_1}} = m$$

Now, let's calculate the y-intercept of this line. As usual, we substitute 0 for x and then solve for y.

$$y = mx + b$$
$$y = m(0) + b$$
$$y = b$$

So, the y-intercept is $(0, b)$. Thus, the two constants m and b describe the slope and y-intercept of the line. As such, we call this the *slope-intercept* form of a linear equation.

DEFINITION: Slope-Intercept Form of a Line

If the equation of a nonvertical line in x and y is solved for y, the result is an equation in **slope-intercept form**.

$$y = mx + b$$

The constant m is the slope of the line, and the y-intercept of the line is $(0, b)$.

If the variable x does not appear in the equation, the slope is 0 and the equation is simply of the form $y = b$. Thus, its graph is a horizontal line.

We can make use of the slope-intercept form of a line to graph the line, as illustrated in the following example.

Example 3: Slope-Intercept Form of a Line

Use the slope-intercept form of the line to graph the equation $4x - 3y = 6$.

Solution

$$4x - 3y = 6$$
$$-3y = -4x + 6$$
$$y = \frac{4}{3}x - 2$$

Solving the equation for y puts it in slope-intercept form. Once we have done this, we know that the line has a slope of $\frac{4}{3}$ and crosses the y-axis at -2.

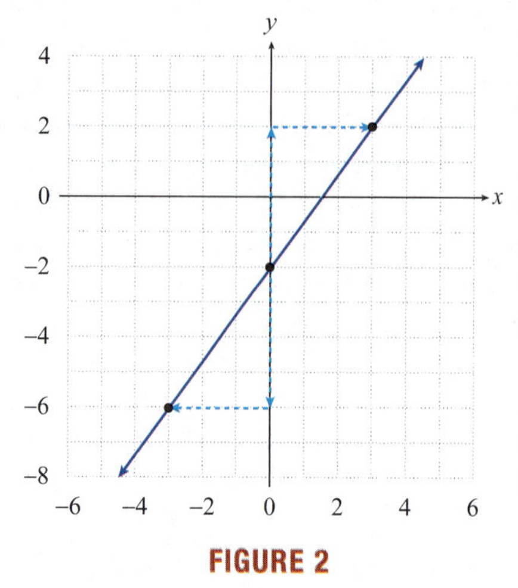

FIGURE 2

Immediately, we can plot the y-intercept.

A second point can now be found by using the fact that slope is "rise over run." This means a second point must lie 4 units up and 3 units to the right, i.e. at $(3,2)$.

Alternatively, we could locate a second point by moving down 4 units and moving to the left 3 units.

In some cases, we can also make use of the slope-intercept form to find the equation of a line that has certain properties.

Example 4: Slope-Intercept Form of a Line

Find the equation of the line that passes through the point $(0, 3)$ and has a slope of $-\frac{3}{5}$. Then graph the line.

NOTE

If a line is already in slope-intercept form, it's usually easiest to graph the line by plotting the y-intercept and then using the slope to find a second point.

Solution

First, we write the equation of this line in slope-intercept form. We are given the y-intercept of $(0,3)$ and the slope of $-\frac{3}{5}$.

$$y = mx + b$$

$$y = -\frac{3}{5}x + 3$$

We can immediately write down the equation since the only information we need is the y-intercept and the slope.

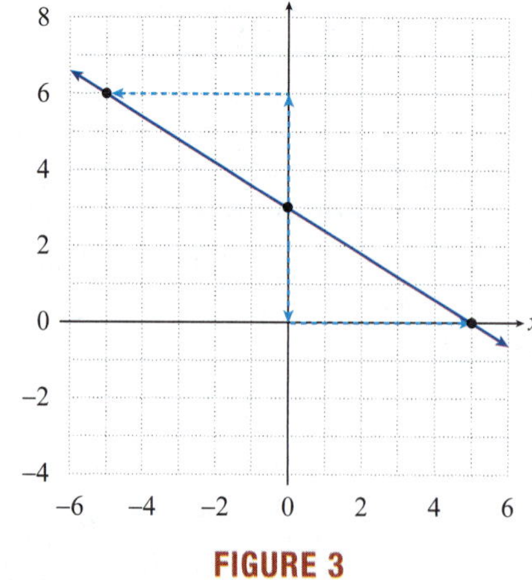

FIGURE 3

We first plot the y-intercept, then move down 3 units and to the right 5 units to find a second point.

Or, we could have plotted a second point by moving up 3 units and to the left 5 units. Both methods make use of the fact that the slope is $-\frac{3}{5}$.

TOPIC 3: The Point-Slope Form of the Equation of a Line

As in Example 4, we can easily find the slope-intercept form of a line given its slope and y-intercept. In the most general case, we would like to be able to construct an equation of a line given the slope of the line and *any* point on the line (not just the y-intercept). This would allow us to find the equation of a line given only two points on that line (since we can determine the slope from two points). The *point-slope* form of a line meets these requirements.

Suppose we know that a given line has slope m and passes through the point (x_1, y_1). If we plug any two points on the line into the slope formula, we must get a result of m. Choosing (x_1, y_1) and the generic point (x, y), the definition of slope states that

$$\frac{y - y_1}{x - x_1} = m.$$

A simple rearrangement of the slope equation leads to a linear equation defined by the slope and the coordinates of a single, arbitrary point.

$$y - y_1 = m(x - x_1)$$

This is the point-slope form.

DEFINITION: Point-Slope Form of a Line

The **point-slope form** of the equation for the line passing through the point (x_1, y_1) with slope m is

$$y - y_1 = m(x - x_1).$$

Note that m, x_1, and y_1 are all constants, while x and y are variables. Note also that since a line, by definition, has slope m, vertical lines cannot be described in this form.

Example 5: Using the Point-Slope Form of a Line

Find the equation, in slope-intercept form, of the line that passes through the point $(-2, 5)$ with slope 3.

NOTE

When asked for an equation in either standard form or slope-intercept form, it's frequently easiest to write the point-slope form and then convert the equation to the desired form.

Solution

Since we are given the slope of the line and a point on the line, we can substitute directly into the point-slope form, then solve for y to obtain the slope-intercept form.

$$y - y_1 = m(x - x_1) \quad \text{The point-slope form}$$
$$y - 5 = 3(x - (-2))$$
$$y - 5 = 3(x + 2) \quad \text{The slope-intercept form}$$
$$y - 5 = 3x + 6$$
$$y = 3x + 11$$

We know that two distinct points in the plane are sufficient to determine a line. With our knowledge of the point-slope form of a line, we can now deduce the equation for the line determined by two points.

Example 6: Using the Point-Slope Form of a Line

Find the equation, in slope-intercept form, of the line that passes through the two points $(-3,-2)$ and $(1,6)$.

Solution

We already have a point (actually, two) on the line, but we still need the slope to use the point-slope form. We can calculate this using the two points and the slope formula.

$$m = \frac{-2-6}{-3-1} = \frac{-8}{-4} = 2$$

Now we can substitute into the point-slope form, then solve for y to obtain the desired slope-intercept equation.

$$\begin{aligned} y - y_1 &= m(x - x_1) \\ y - 6 &= 2(x-1) \\ y - 6 &= 2x - 2 \\ y &= 2x + 4 \end{aligned}$$

Note that no matter which point we substitute into the point-slope form, the resulting slope-intercept equation is the same.

We close this section with a summary of the different forms of linear equations, what information we need to write them, and what they are each most useful for.

Standard Form: $ax + by = c$

Information Required: Typically, we arrive at the standard form when given a linear equation in another form.

Potential Uses: The standard form is most useful for easily calculating the x- and y-intercepts.

Slope-Intercept Form: $y = mx + b$

Information Required: The slope m and the y-intercept $(0,b)$.

Potential Uses: The slope-intercept form makes it very easy to find the y-intercept and slope, and therefore to graph the line.

Point-Slope Form: $y - y_1 = m(x - x_1)$

Information Required: The slope m and a point on the line (x_1, y_1) or two points on the line (x_1, y_1) and (x_2, y_2).

Potential Uses: The point-slope form allows us to find the equation for a line when the y-intercept is unknown.

2.4 EXERCISES

PRACTICE

Determine the slope of the line passing through the specified points. See Example 1.

1. $(0,-3)$ and $(-2,5)$
2. $(-3,2)$ and $(7,-10)$
3. $(4,5)$ and $(-1,5)$
4. $(3,-1)$ and $(-7,-1)$
5. $(3,-5)$ and $(3,2)$
6. $(0,0)$ and $(-2,5)$
7. $(-2,1)$ and $(-5,-1)$
8. $\left(\frac{1}{2},-7\right)$ and $\left(\frac{3}{4},-5\right)$
9. $\left(10,\frac{1}{5}\right)$ and $\left(4,-\frac{4}{5}\right)$
10. $(-2,4)$ and $(6,9)$
11. $(0,-21)$ and $(-3,0)$
12. $(-3,-5)$ and $(-2,8)$
13. $\left(\frac{1}{3},9\right)$ and $(2,4)$
14. $(29,-17)$ and $(31,-29)$
15. $(7,4)$ and $(-6,13)$

Determine the slopes of the lines defined by the following equations. See Example 2.

16. $8x-2y=11$
17. $2x+8y=11$
18. $12x-4y=-9$
19. $4y=13$
20. $\frac{x-y}{3}+2=4$
21. $7x=2$
22. $3y-2=\frac{x}{5}$
23. $3-y=2(5-x)$
24. $3(2y-1)=5(2-x)$
25. $\frac{x+2}{3}+2(1-y)=-2x$
26. $2y-7x=4y+5x$
27. $x-7=\frac{2y-1}{-5}$

Use the slope-intercept form to graph the equations. See Example 3.

28. $6x-2y=4$
29. $3y+2x-9=0$
30. $5y-15=0$
31. $x+4y=20$
32. $\frac{x-y}{2}=-1$
33. $3x+7y=8y-x$
34. $-4x-4y=8$
35. $-5x+3y+16=0$
36. $3x=3y-21$

Find the equation, in slope-intercept form, of the line with the given *y*-intercept and slope. See Example 4.

37. y-intercept $(0,-3)$; slope of $\frac{3}{4}$

38. y-intercept $(0,5)$; slope of -3

39. y-intercept $(0,-7)$; slope of $-\frac{5}{2}$

40. y-intercept $(0,6)$; slope of 4

41. y-intercept $(0,-9)$; slope of -5

42. y-intercept $(0,2)$; slope of $\frac{1}{2}$

Find the equation, in standard form, of the line passing through the given point with the given slope.

43. point $(-1,-3)$; slope of $\frac{3}{2}$

44. point $(6,0)$; slope of $\frac{5}{4}$

45. point $(-3,5)$; slope of 0

46. point $(-2,-13)$; undefined slope

47. point $(3,-1)$; slope of 10

48. point $(-1,3)$; slope of $-\frac{2}{7}$

49. point $(5,11)$; slope of -3

50. point $(5,-9)$; slope of $-\frac{1}{2}$

Find the equation, in standard form, of the line passing through the specified points.

51. $(-1,3)$ and $(2,-1)$

52. $(1,3)$ and $(-2,3)$

53. $(2,-2)$ and $(2,17)$

54. $(-9,2)$ and $(1,5)$

55. $(3,-1)$ and $(8,-1)$

56. $\left(\frac{4}{3},1\right)$ and $\left(\frac{2}{5},\frac{3}{7}\right)$

57. $(-2,8)$ and $(5,6)$

58. $(8,-10)$ and $(8,0)$

59. $(7,5)$ and $(-9,5)$

60. $(7,7)$ and $(9,-8)$

61. $\left(\frac{2}{3},\frac{5}{4}\right)$ and $\left(\frac{3}{5},\frac{9}{8}\right)$

62. $(-5,-5)$ and $(10,-11)$

Match each equation or description to the correct graph.

63. $-3x - 2y = 17$

64. $-4y + 10 = -4x$

65. $-6y + 9 = \dfrac{x}{-2}$

66. point $(-9, 7)$; slope $\dfrac{4}{3}$

67. point $(-2, 4)$; slope -2

68. point $(0, -5)$; slope -9

a.

b.

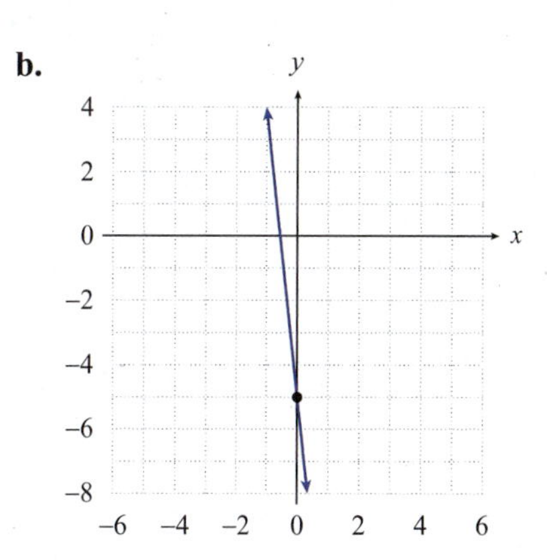

c.

d.

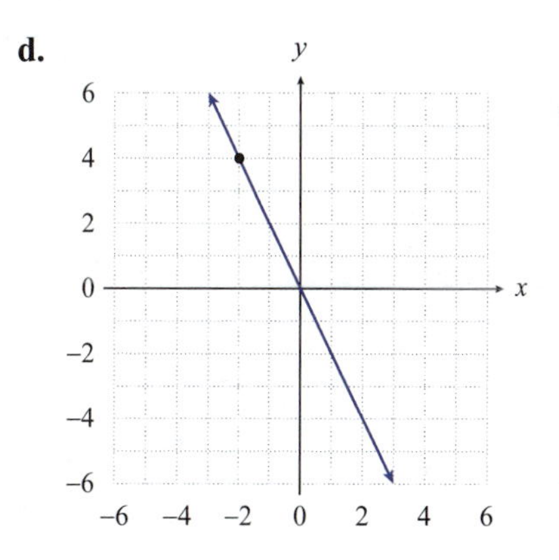

e.

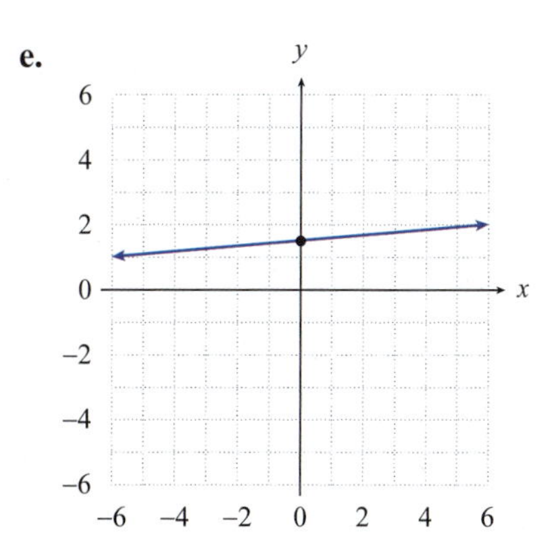

f.

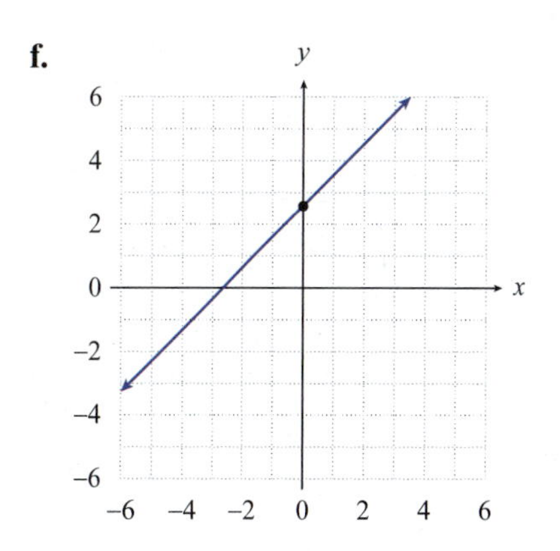

APPLICATIONS

69. A bottle manufacturer has determined that the total cost (C) in dollars of producing x bottles is $C = 0.25x + 2100$.

a. What is the cost of producing 500 bottles?
b. What are the fixed costs (costs incurred even when 0 bottles are produced)?
c. What is the increase in cost for each bottle produced?

70. Sales at Glover's Golf Emporium have been increasing linearly for the past couple of years. Last year, sales were $163,000. This year, sales were $215,000. If sales continue to increase at this linear rate, predict the sales for next year.

71. Amy owns stock in a company. If the stock had a value of $2500 in 2018 when she purchased it, what has been the average change in value per year if in 2020 the stock was worth $3150?

72. For tax and accounting purposes, businesses often have to depreciate equipment values over time. One method of depreciation is the straight-line method. Three years ago Hilde Construction purchased a bulldozer for $51,500. Using the straight-line method, the bulldozer has now depreciated to a value of $43,200. If V equals the value at the end of year t, write a linear equation expressing the value of the bulldozer over time. How many years from the purchase date will the value equal $0? Round your answer to two decimal places.

2.5 PARALLEL AND PERPENDICULAR LINES

TOPICS

1. Slopes of parallel lines
2. Slopes of perpendicular lines

TOPIC 1: Slopes of Parallel Lines

In this section, we will explore the relationship between slope and the geometric concepts of parallel and perpendicular lines. This will allow us to use algebra to construct lines parallel or perpendicular to a given line. We begin with parallel lines. Consider the following figure showing the "rise" and "run" of two parallel lines.

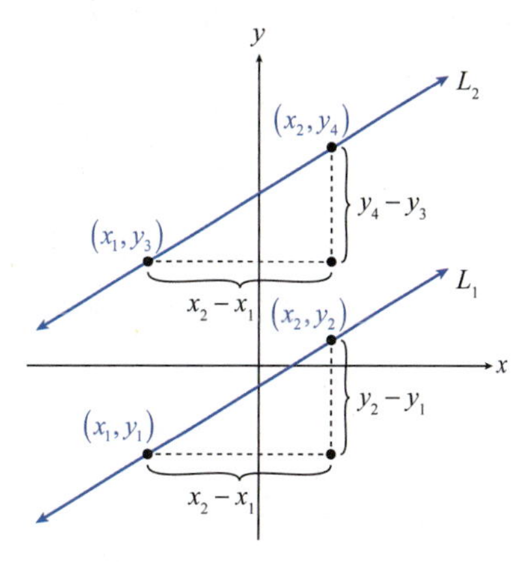

FIGURE 1: Slopes of Parallel Lines

In Figure 1, we can see that if we choose the right set of points on each line, the respective rises and runs are equal. More importantly, no matter which points we choose, the relative *ratio* of rise to run, which we have defined as the slope, is the same for two parallel lines. We thus have a simple algebraic characterization of parallel lines.

THEOREM: Slopes of Parallel Lines

Two nonvertical lines with slopes m_1 and m_2 are **parallel** if and only if $m_1 = m_2$. Also, two vertical lines (with undefined slopes) are always parallel to each other.

This fact gives us a straightforward way of finding the equations of lines parallel to a given line. First, we calculate the slope of the given line, and then construct new lines using that same slope. Usually, it will be easier to use the slope-intercept or point-slope form of a linear equation, since the slope appears directly.

Example 1: Finding Equations of Parallel Lines

Find equations for two lines parallel to each of the given lines.

a. $y = -\frac{2}{3}x + 4$ **b.** $10x - 2y = 14$

Solution

a. This line is already in slope-intercept form, so we immediately know the slope of the line is $-\frac{2}{3}$. Any line parallel to this one must also have a slope of $-\frac{2}{3}$. To find two parallel lines, we can simply change the value of the y-intercept. Two such examples are $y = -\frac{2}{3}x + 1$ and $y = -\frac{2}{3}x - 10$.

NOTE

When given a line in standard form, it is usually easiest to find its slope by rewriting it in slope-intercept form than to find two points on the line and calculate the slope directly.

b. This line is in standard form. Our first step is to rewrite it in slope-intercept form.

$$10x - 2y = 14$$

$$-2y = -10x + 14 \quad \text{Subtract } 10x \text{ from both sides.}$$

$$y = \frac{-10}{-2}x + \frac{14}{-2} \quad \text{Divide each term by } -2.$$

$$y = 5x - 7 \quad \text{Simplify.}$$

Again, once the line is in slope-intercept form, we can change the y-intercept to find two lines parallel to the original line.

Two such examples are $y = 5x$ and $y = 5x + 8$.

Example 2: Finding Equations of Parallel Lines

Find the equation, in slope-intercept form, for the line that is parallel to the line $3x + 5y = 23$ and passes through the point $(-2, 1)$.

Solution

Again, our first step is to write the initial equation in slope-intercept form.

$$3x + 5y = 23$$

$$5y = -3x + 23$$

$$y = -\frac{3}{5}x + \frac{23}{5}$$

This tells us that the slope of the line whose equation we seek is $-\frac{3}{5}$. We also know that the line is to pass through $(-2, 1)$, so we can use the point-slope form to obtain the desired equation.

TECHNOLOGY

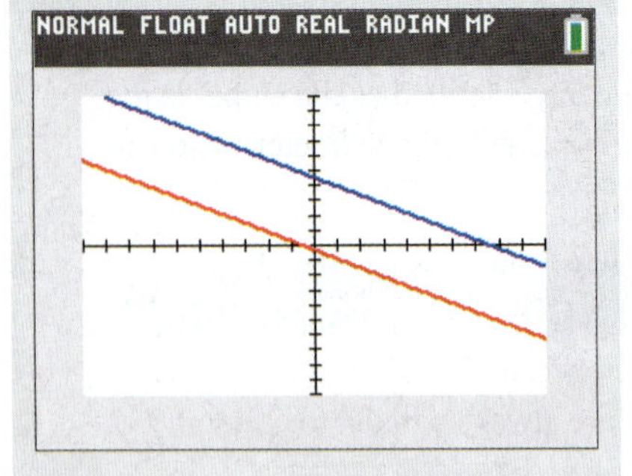

The screen above displays the graphs of both $3x + 5y = 23$ and the parallel line passing through $(-2,1)$. To view multiple graphs on a TI-84 Plus, press Y= and enter the equations in terms of y. Then press graph.

$$y - y_1 = m(x - x_1)$$

$$y - 1 = -\frac{3}{5}(x - (-2))$$ Begin by substituting our known information into the point-slope form: $m = -\frac{3}{5}$, $(x_1, y_1) = (-2, 1)$.

$$y - 1 = -\frac{3}{5}(x + 2)$$

$$y - 1 = -\frac{3}{5}x - \frac{6}{5}$$

$$y = -\frac{3}{5}x - \frac{1}{5}$$ The instructions asked for the equation in slope-intercept form, so we solve for y to obtain the final answer.

We can also use the knowledge that parallel lines have the same slope to answer questions that are more geometric in nature.

Example 3: Identifying a Quadrilateral

Determine if the quadrilateral (four-sided figure) in Figure 2 is a parallelogram (a quadrilateral in which both pairs of opposite sides are parallel).

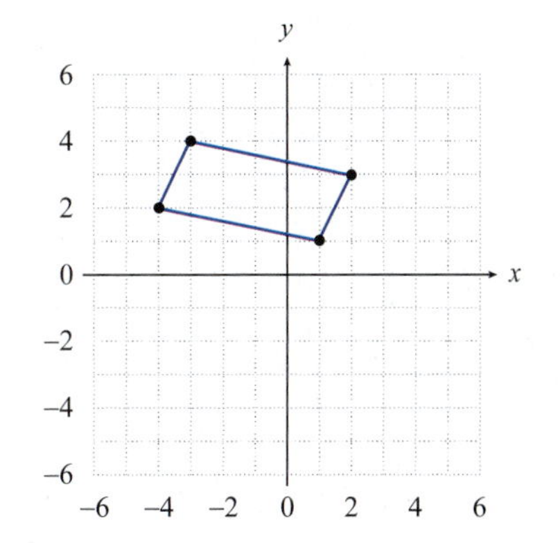

FIGURE 2

Solution

The four vertices are plotted in Figure 2, and the sides of the quadrilateral drawn. The figure is a parallelogram if the left and right sides are parallel and the top and bottom sides are parallel. The slopes of the left and right sides are, respectively,

$$\frac{4-2}{-3-(-4)} = 2 \text{ and } \frac{3-1}{2-1} = 2,$$

and the slopes of the top and bottom sides are, respectively,

$$\frac{4-3}{-3-2} = -\frac{1}{5} \text{ and } \frac{2-1}{-4-1} = -\frac{1}{5}.$$

Thus, the figure is indeed a parallelogram.

TOPIC 2: Slopes of Perpendicular Lines

The relationship between the slopes of perpendicular lines is a bit less obvious than that of parallel lines. Consider a nonvertical line L_1, and two points (x_1, y_1) and (x_2, y_2) on the line, as shown in Figure 3. These two points can be used to calculate the slope m_1 of L_1, with the result that $m_1 = \frac{a}{b}$, where $a = y_2 - y_1$ and $b = x_2 - x_1$.

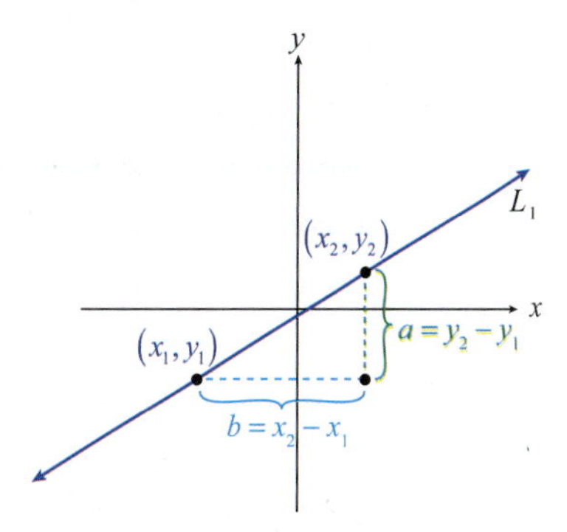

FIGURE 3: Definition of a and b

If we now draw a line L_2 perpendicular to L_1, we can use a and b to determine the slope m_2 of line L_2. There are an infinite number of lines that are perpendicular to L_1; one of them is drawn in Figure 4.

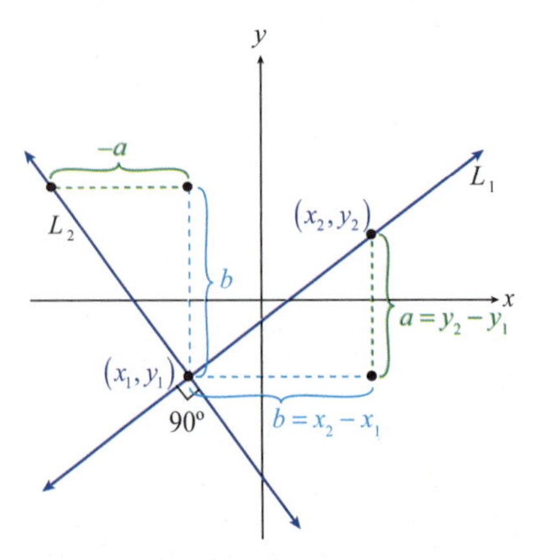

FIGURE 4: Perpendicular Lines

Note that in rotating the line L_1 by 90° to obtain L_2, we have also rotated the right triangle drawn with dashed lines, so the sides of the triangle are the same length. But to travel along the line L_2 from the point (x_1, y_1) to the second point drawn requires a positive rise and a negative run, whereas the rise and run between (x_1, y_1) and (x_2, y_2) are both positive. In other words, $m_2 = -\frac{b}{a}$, the negative reciprocal of the slope m_1.

This relationship always exists between the slopes of two perpendicular lines, assuming neither one is vertical. If one line is vertical, any line perpendicular to it will be horizontal with a slope of zero, while if one line is horizontal, any line perpendicular to it will be vertical with undefined slope. This is summarized next.

THEOREM: Slopes of Perpendicular Lines

Suppose m_1 and m_2 represent the slopes of two lines, neither of which is vertical. The two lines are perpendicular if and only if $m_1 = -\frac{1}{m_2}$ (equivalently, $m_2 = -\frac{1}{m_1}$ and $m_1 m_2 = -1$). If one of two perpendicular lines is vertical, the other is horizontal, and their slopes are, respectively, undefined and zero.

The following examples illustrate how we can use the relationship between slopes of perpendicular lines to solve problems.

Example 4: Finding Equations of Perpendicular Lines

For each line given, find the equation of a perpendicular line.

a. $y = -\frac{4}{9}x + 2$ **b.** The line passing through the points $(-1, 3)$ and $(4, 1)$.

Solution

a. This line is in slope-intercept form, so we can immediately identify the slope as $-\frac{4}{9}$. The slope of any perpendicular line must equal $\frac{9}{4}$, the negative reciprocal of the original slope. Thus, one solution is $y=\frac{9}{4}x$.

b. Since we only need the slope of the original line, there is no need to find its equation; we can calculate the slope directly from the given points.

$$m=\frac{1-3}{4-(-1)}=-\frac{2}{5}$$

Again, the slope of a line perpendicular to the line through the given points must have a slope equal to the negative reciprocal of $-\frac{2}{5}$, which is $\frac{5}{2}$. One such perpendicular line is $y=\frac{5}{2}x+6$.

TECHNOLOGY

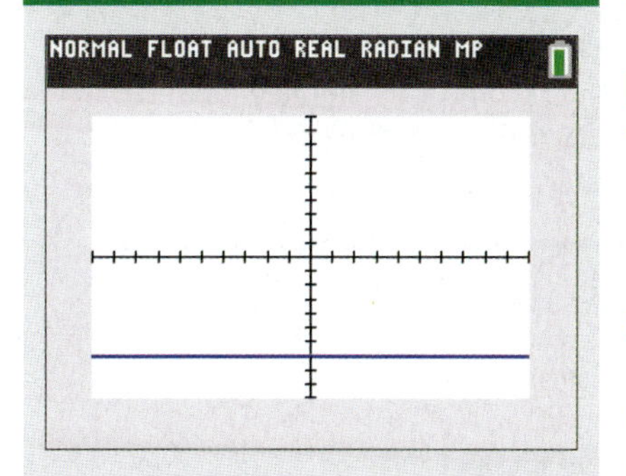

Example 5: Finding Equations of Perpendicular Lines

Find the equation, in standard form, of the line that passes through the point $(-3, 13)$ and is perpendicular to the line $y=-7$.

Solution

The line $y=-7$ is a horizontal line, and hence any line perpendicular to it must be a vertical line, having the form $x=c$. Since the perpendicular line must pass through the point $(-3, 13)$, the desired solution is $x=-3$.

NOTE

Remember that if you encounter a horizontal or vertical line, you cannot use the slope formulas to find a perpendicular line.

Given a pair of lines, we can use their slopes to determine if they are parallel, perpendicular, or neither. Note that the only information we need is the slope! The equations do not have to be written in the same form, and we do not need to know anything about their intercepts. Find the most efficient way to calculate the slope of each line to avoid any unnecessary work.

Example 6: Identifying Parallel and Perpendicular Lines

For each pair of lines, determine if the lines are parallel, perpendicular, or neither.

a. $3x-7y=12$ and $14x+6y=-5$

b. $y-\frac{263}{4}=9\left(x+\frac{77}{13}\right)$ and the line passing through the points $(0,4)$ and $(2,22)$

c. $y=\frac{3}{4}x+1$ and $y=\frac{4}{3}x-5$

TECHNOLOGY

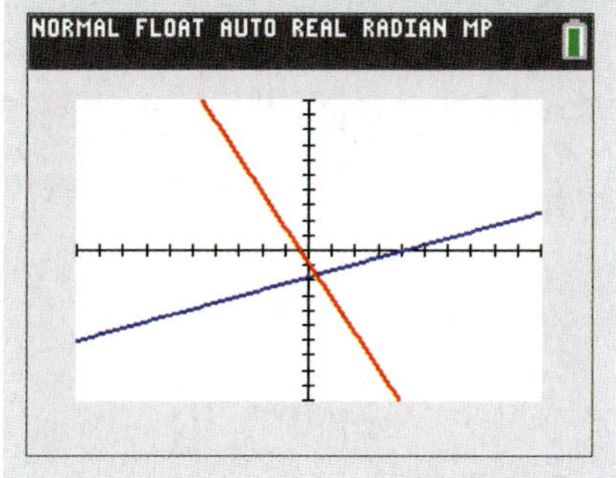

Solution

a. Both equations are in standard form, so our first step is to rewrite them in slope-intercept form to identify the slopes.

$$\begin{aligned} 3x-7y&=12 \\ -7y&=-3x+12 \\ y&=\frac{3}{7}x-\frac{12}{7} \end{aligned} \qquad \begin{aligned} 14x+6y&=-5 \\ 6y&=-14x-5 \\ y&=-\frac{7}{3}x-\frac{5}{6} \end{aligned}$$

NOTE

A pair of lines cannot be both parallel and perpendicular.

Are the lines parallel? No, the slopes are not equal.
Are the lines perpendicular? Yes, the slopes are negative reciprocals of each other.
Thus, the lines are perpendicular.

b. One line is in point-slope form, so we can see its slope is 9. We calculate the slope of the other line using the two points given.

$$m = \frac{22-4}{2-0} = \frac{18}{2} = 9$$

Are the lines parallel? Yes, the slopes are equal.
Thus, the lines are parallel. (Note that we didn't need to find the equation of the second line.)

c. Both lines are in slope-intercept form, so we can read off the slopes: $\frac{3}{4}$ and $\frac{4}{3}$.

Are the lines parallel? No, the slopes are not equal.
Are the lines perpendicular? No, the slopes are reciprocals, not *negative* reciprocals.
Thus, the lines are neither parallel nor perpendicular.

2.5 EXERCISES

PRACTICE

Find the equation, in slope-intercept form, for the line parallel to the given line and passing through the indicated point. See Examples 1 and 2.

1. $y-4x=7$; $(-1, 5)$

2. $6x+2y=19$; $(-6, -13)$

3. $3x+2y=3y-7$; $(3, -2)$

4. $2-\frac{y-3x}{3}=5$; $(0, -2)$

5. $y-4x=7-4x$; $(23, -9)$

6. $2(y-1)+\frac{x+3}{5}=-7$; $(-5, 0)$

7. $6y-4=-3(1-2x)$; $(-2, -2)$

8. $5-\frac{7y+5x}{2}=1$; $(4, 1)$

9. $2(y-1)-\frac{7x+1}{3}=-3$; $(1, 10)$

10. $8y-6=-3(4-x)$; $(11, -5)$

Each set of four ordered pairs defines the vertices, in counterclockwise order, of a quadrilateral. Determine if the quadrilateral is a parallelogram. See Example 3.

11. $\{(-2,2),(-5,-2),(2,-3),(5,1)\}$

12. $\{(-1,6),(-4,7),(-2,3),(1,1)\}$

13. $\{(-3,3),(-2,-2),(3,-1),(2,4)\}$

14. $\{(-2,-3),(-3,-6),(1,-2),(2,1)\}$

15. $\{(-6,-2),(-1,0),(-3,4),(-8,2)\}$

16. $\{(-3,-2),(3,-3),(5,2),(-1,3)\}$

17. $\{(-1,-1),(5,1),(3,5),(-2,3)\}$

18. $\{(0,1),(6,0),(7,4),(1,6)\}$

Determine if the two lines are parallel. See Example 6.

19. $y = 8x + 7$ and $y = -8x + 7$

20. $x - 5y = 2$ and $5x - y = 2$

21. $2x - 3y = (x - 1) - (y - x)$ and $-2y - x = 9$

22. $3 - (2y + x) = 7(x - y)$ and $\frac{5y + 1}{4} = 3 + 2x$

23. $6 = -12(x - y) + y$ and $13y = -12x + 3$

24. $\frac{2x - 3y}{3} = \frac{x - 1}{6}$ and $2y - x = 3$

25. $\frac{x - y}{2} = \frac{x + y}{3}$ and $\frac{2x + 3}{5} - 4y = 1 + 2y$

26. $5 - (4y + 3x) = 5(x - y)$ and $y + 4 = 5 + 8x$

27. $7x - 2(x + 3) = 5y - x$ and $-6x = 1 - 5y$

28. $\frac{2y + 11x}{3} = x + 1$ and $7x - 8y = 9x + 7$

29. $\frac{x - y}{5} = \frac{x + y}{3} - 1$ and $7 = -2(x - y) + 6y$

30. $2x + 5y = 14$ and the line passing through the points $(8, -5)$ and $(3, -3)$

Find the equation, in slope-intercept form, for the line perpendicular to the given line and passing through the indicated point. See Examples 4 and 5.

31. $3x + 2y = 3y - 7$; $(3, -2)$

32. $6y + 2x = 1$; $(-4, -12)$

33. $-y + 3x = 5 - y$; $(-2, 7)$

34. $x + y = 5$; origin

35. $x = \frac{1}{4}y - 3$; $(1, -1)$

36. $2(y + x) - 3(x - y) = -9$; $(2, 5)$

37. $4x + 8y = 4y - 3$; $(-2, 1)$

38. $\frac{3x - y}{4} = \frac{4x - 5}{2}$; $(8, 5)$

39. $4(y + x) - 8(x - y) = -1$; $(6, 10)$

40. $\frac{3x + 4}{3} - 3y = 1 - 4y$; $(2, -8)$

Determine if the two lines are perpendicular. See Example 6.

41. $x - 5y = 2$ and $5x - y = 2$

42. $y = 5x + 4$ and $y = -\frac{1}{5}x - 9$

43. $3x + y = 2$ and $x + 3y = 2$

44. $\frac{3x - y}{3} = x + 2$ and $x = 9$

45. $5x - 6(x + 1) = 2y - x$ and $2y - (x + y) = 4y + x$

46. $-6y+3x=7$ and $8x-3(x+1)=3y-x$

47. $-x=-\frac{2}{5}y+2$ and $5y=2x$

48. $\frac{7x-5y}{4}=x+2$ and $-3y-3x=2x+4$

49. $3(4-x)=6y+3$ and $-3y-2x=3-8x$

50. $\frac{x-1}{2}+\frac{3y+2}{3}=-9$ and $3y-5x=x+5$

51. $1-\frac{2y-5x}{2}=7x+4$ and $9x-2y=11$

52. $y-\frac{2}{3}=4\left(x+\frac{7}{11}\right)$ and the line passing through the points $(-2,4)$ and $(7,-14)$

Each set of four ordered pairs defines the vertices, in counterclockwise order, of a quadrilateral. Use the ideas in this section to determine if the quadrilateral is a rectangle.

53. $\{(-2,2),(-5,-2),(2,-3),(5,1)\}$

54. $\{(2,-1),(-2,1),(-3,-1),(1,-3)\}$

55. $\{(1,2),(3,-3),(9,-1),(7,4)\}$

56. $\{(5,-7),(1,-13),(28,-31),(32,-25)\}$

57. $\{(-5,-1),(0,-6),(5,-1),(0,4)\}$

58. $\{(-3,-3),(3,-2),(1,2),(-5,1)\}$

APPLICATIONS

59. A construction company is building a new suspension bridge that has support cables attached to a center tower at various heights. One cable is attached at a height of 30 feet and connects to the roadbed 50 feet from the base of the tower. If the support cables should run parallel to each other, how far from the base should the company attach a cable whose other end is connected to the tower at a height of 25 feet?

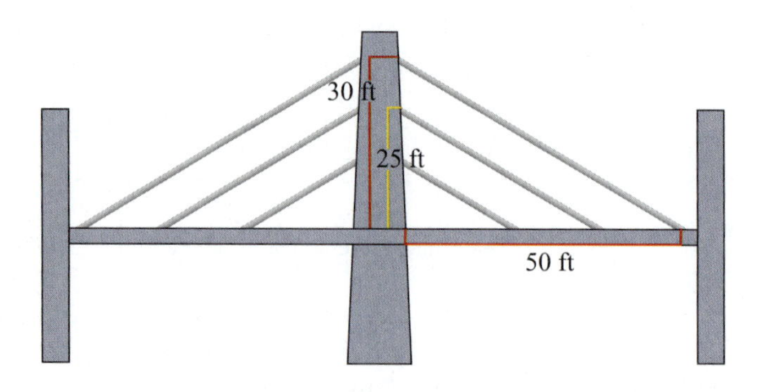

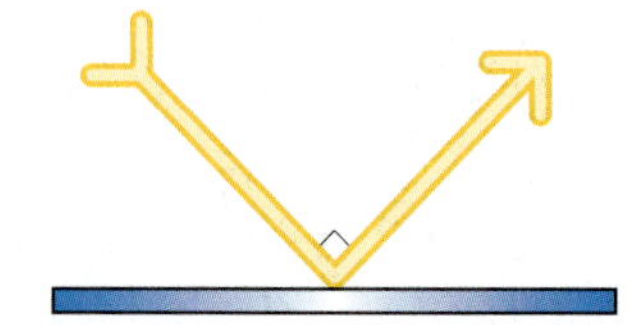

60. A light beam hits a mirror and is reflected off the mirror at a right angle. If the line formed by the original beam of light can be described by an equation of the form $y=-3.2x+b$ (for some constant b), write the form of an equation that describes the line of the reflected beam (use an arbitrary constant c in your answer).

2.6 LINEAR INEQUALITIES IN TWO VARIABLES

TOPICS

1. Graphing linear inequalities
2. Graphing linear inequalities joined by "and" or "or"
3. Graphing linear absolute value inequalities

TOPIC 1: Graphing Linear Inequalities

Just as in one variable, linear inequalities in two variables have much in common with linear equations in two variables. The first similarity is in the definition: if the equality symbol in a linear equation in two variables is replaced with <, >, ≤, or ≥, the result is a **linear inequality in two variables**. In other words, a linear inequality in the two variables x and y is an inequality that can be written in the form

$$ax + by < c,\ ax + by > c,$$
$$ax + by \leq c, \text{ or } ax + by \geq c,$$

where a, b, and c are constants and a and b are not both 0.

Another similarity lies in the solution process. The solution set of a linear inequality in two variables consists of all the ordered pairs in the Cartesian plane that lie on one side of a line in the plane, possibly including those points on the line. The first step in solving such an inequality then is to identify and graph this line. This line is simply the graph of the equation that results from replacing the inequality symbol in the original problem with an equality symbol.

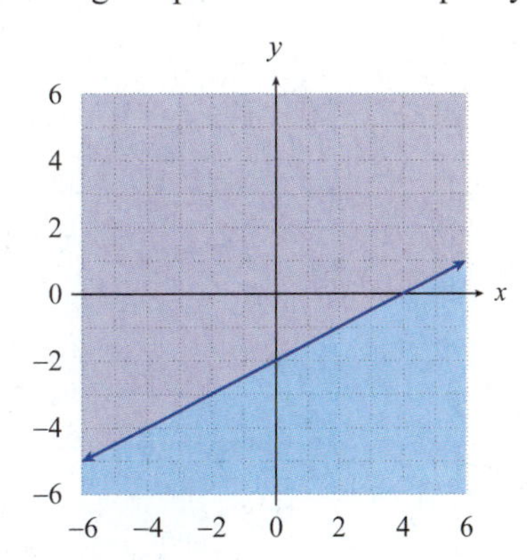

FIGURE 1: Dividing the Plane

Any line divides the plane into two **half-planes**; given a linear inequality, all of the points in one of the two half-planes will solve the inequality. In addition, the points on the **boundary line** also solve the inequality if the inequality symbol is ≤ or ≥, and this fact must be denoted graphically.

PROCEDURE: Solving Linear Inequalities in Two Variables

Step 1: Graph the line that results from replacing the inequality symbol with =.

Step 2: Make the line solid if the inequality symbol is ≤ or ≥ (not strict) and dashed if the symbol is < or > (strict). A solid line indicates that points on the line are included in the solution set while a dashed line indicates that points on the line are excluded from the solution set.

Step 3: Determine which of the half-planes defined by the boundary line solves the inequality by substituting a **test point** from one of the two half-planes into the inequality. If the resulting numerical statement is true, all the points in the same half-plane as the test point solve the inequality. Otherwise, the points in the other half-plane solve the inequality. Shade in the half-plane that solves the inequality.

Example 1: Solving Linear Inequalities

NOTE

Whenever possible, use the origin as a test point, since it makes for a very simple calculation.

Solve the following linear inequalities by graphing their solution sets.

a. $3x+2y<12$ **b.** $x-y\leq 0$ **c.** $x>3$

Solution

a.

$$3x+2y=12$$

$$3(0)+2y=12 \qquad 3x+2(0)=12$$

$$y=6 \qquad x=4$$

To graph the boundary line, replace the inequality symbol with an equal sign, then find the x- and y-intercepts.

x-intercept of boundary: $(4,0)$

y-intercept of boundary: $(0,6)$

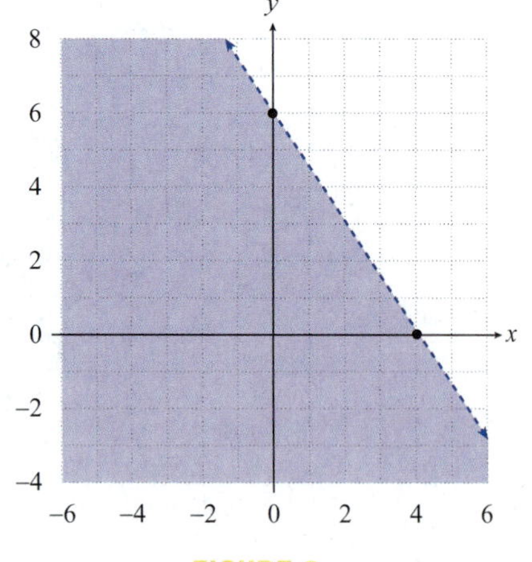

FIGURE 2

The graph of the equation is drawn with a dashed line since the inequality is strict.

The origin $(0,0)$ clearly lies on one side of the boundary line, so we can use it as our test point.

When both x and y in the inequality are replaced with 0, we obtain the true statement $0<12$. This tells us to shade the half-plane that contains $(0,0)$.

b. $x - y \leq 0$

$x - y = 0$

$y = x$

In slope-intercept form, the equation for the boundary is $y = x$, a line with a y-intercept of $(0,0)$ and a slope of 1.

Graph the boundary with a solid line since the inequality is not strict. Thus, points on the boundary do solve the inequality.

This time, the origin cannot be used as a test point, since it lies directly on the boundary line. The point $(1,-1)$ lies below the boundary, and if we substitute $x = 1$ and $y = -1$ into the inequality $x - y \leq 0$, we obtain the false statement $2 \leq 0$. Thus, we shade the half-plane that does not contain $(1,-1)$.

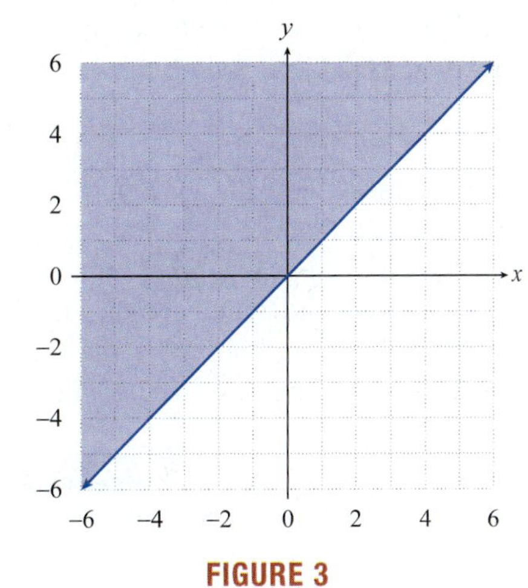

FIGURE 3

c. $x > 3$

In this example, the boundary line is a vertical line with x-intercept $(3,0)$.

Since the inequality is strict, we graph the boundary line as a dashed line.

The origin can be used as a convenient test point, and results in the false statement $0 > 3$, leading us to shade the half-plane to the right of the boundary.

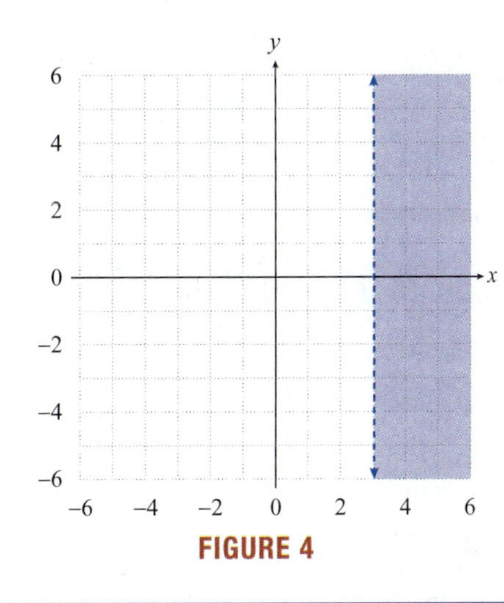

FIGURE 4

Another way to decide which half-plane to shade is to consider each inequality after solving for y (or for x if y is not present, as in Example 1c). The inequality in Example 1a is satisfied by all those ordered pairs for which y is less than $-\frac{3}{2}x + 6$. That is, if the line $y = -\frac{3}{2}x + 6$ is graphed (with a dashed line), the solution set of the inequality consists of all those ordered pairs whose y-coordinate is *less than* (lies below) $-\frac{3}{2}$ times the x-coordinate, plus 6 (those ordered pairs below the boundary line). Similarly, in Example 1b, we shade in those ordered pairs for which the y-coordinate is *greater than or equal to* the x-coordinate. Finally, in Example 1c, we shade in the ordered pairs for which the x-coordinate is *greater than* (lies to the right of) 3.

TECHNOLOGY: Graphing Inequalities

Just as we can use a graphing utility to graph linear equations, we can also graph linear inequalities like $y < -3x + 5$. On a TI-84 Plus, press Y= and type in the right-hand side of the inequality. Pressing graph would display the line $y = -3x + 5$, which is the boundary line of the solution set.

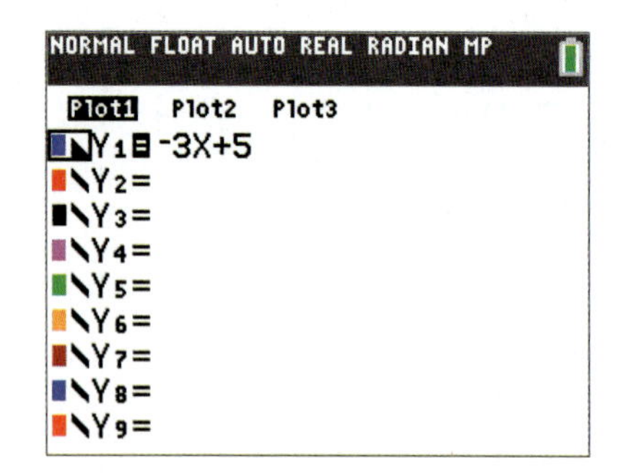

FIGURE 5

If we test the point $(0,0)$, we find that it is a solution to the inequality, so we know to shade the half-plane below the boundary line. In the Y= editor, use the left arrow to move the cursor to the left of the Y1, where there is a small diagonal line. Press enter and a new menu comes up on a TI-84 Plus CE. Here, use the down arrow and then the left arrow until the option selected for Line shows the shading below the boundary. Then press enter twice and the Y= editor should appear as in Figure 5.

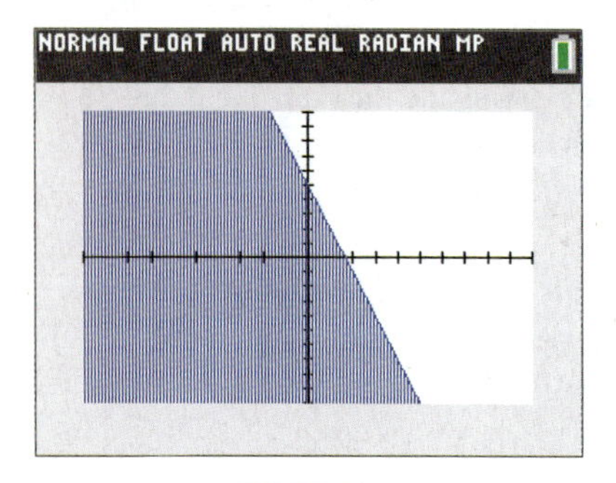

FIGURE 6

With that selection, pressing graph will display the boundary line with the half-plane below it shaded, which is the solution to the inequality (see Figure 6).

In order to shade the half-plane above the boundary line, which would be the solution to $y > -3x + 5$, we would need to return to the Y= editor and again place the cursor to the left of the Y1. Press enter and then use the arrow keys on a TI-84 Plus CE to select the line style showing the shading above the boundary. Press enter twice and the Y= editor should appear as in Figure 7. Then press graph to display the half-plane shaded above the boundary line as in Figure 8.

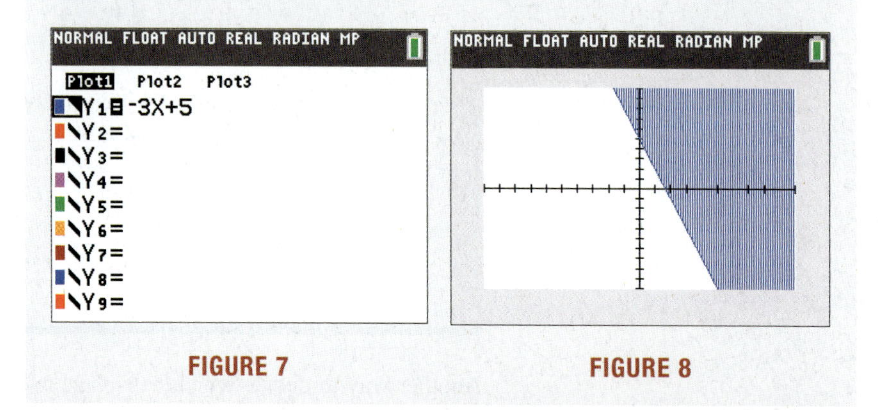

FIGURE 7 FIGURE 8

CAUTION

Notice that the boundary line appears in the calculator as a solid line. However, because we are finding the solution to a strict inequality, the boundary line is not included in the answer.

TOPIC 2: Graphing Linear Inequalities Joined by "And" or "Or"

Often, we need to identify those points in the Cartesian plane that satisfy more than one inequality. In the problems that we will examine, we will identify the portion of the plane that satisfies two or more linear inequalities joined by the word "and" or the word "or."

In Section 1.1, we defined the union of two sets A and B, denoted $A \cup B$, as the set containing all elements that are in set A **or** set B, and we defined the intersection of two sets A and B, denoted $A \cap B$, as the set containing all elements that are in both A **and** B.

If we let A denote the portion of the plane that solves one inequality and B the portion of the plane that solves a second inequality, then $A \cup B$ represents the solution set of the two inequalities joined by the word "or" and $A \cap B$ represents the solution set of the two inequalities joined by the word "and."

Visually, we can think of the union as the combining of two regions, while we can think of the intersection as the overlaps between two regions. Keep in mind that $A \cup B$ *contains* both of the sets A and B (and so is at least as large as either one individually), while $A \cap B$ is *contained in* both A and B (and so is no larger than either individual set).

To find the solution sets in the following problems, we will solve each linear inequality individually and then form the union or the intersection of the individual solutions, as appropriate.

Example 2: Solving Linear Inequalities

Graph the solution sets that satisfy the following inequalities.

a. $5x - 2y < 10$ and $y \le x$ **b.** $x + y < 4$ or $x \ge 4$

Solution

a. $5x - 2y < 10$ and $y \le x$

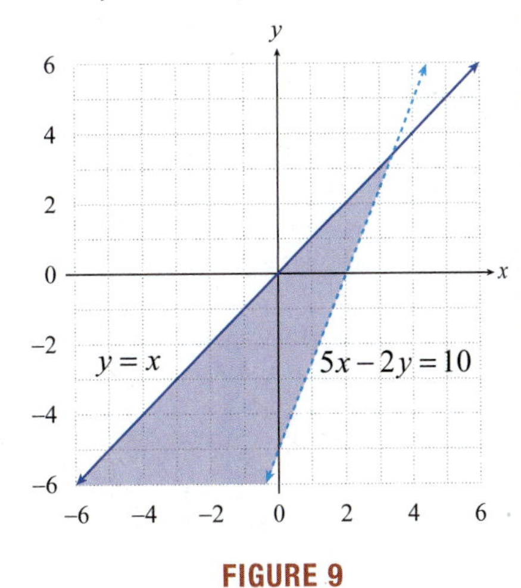

FIGURE 9

First graph the individual inequalities. We graph the line $5x - 2y = 10$ with a dashed line and the line $y = x$ with a solid line.

We then note that the half-plane lying above $5x - 2y = 10$ solves the first inequality (verify this by using the test point $(0,0)$) and that the half-plane lying below $y = x$ solves the second inequality (this can be verified using the test point $(1,-1)$).

Since the two inequalities are joined by the word "and", we shade in the intersection of the two half-planes, resulting in the graph shown in Figure 9.

b. $x + y < 4$ or $x \geq 4$

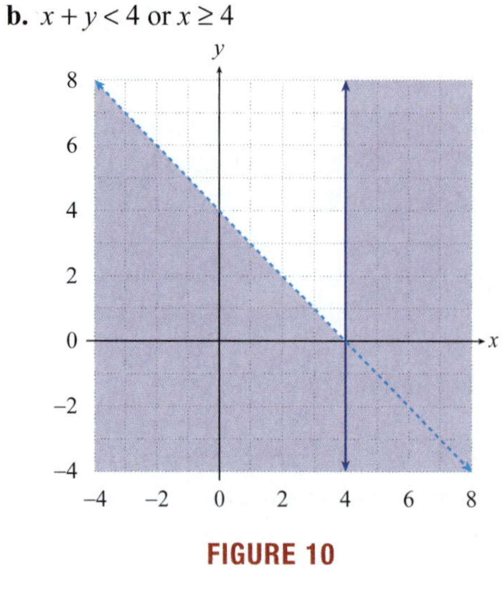

FIGURE 10

Again, begin by solving each inequality separately. We graph the line $x + y = 4$ with a dashed line, and the line $x = 4$ with a solid line, and note that the half-plane below $x + y = 4$ solves the first inequality and that the half-plane to the right of $x = 4$ solves the second.

Since the two inequalities are joined by "or," we shade the *union* of the two half-planes. Note that this results in a larger solution set than either of the two individual solutions. Any ordered pair in the shaded region will satisfy one or both of the inequalities.

TOPIC 3: Graphing Linear Absolute Value Inequalities

Unions and intersections of regions of the plane can also arise when solving inequalities involving absolute values. In Section 1.7, we saw that an inequality of the form $|x| < a$ can be rewritten as the joint condition $x > -a$ and $x < a$, which corresponds to the intersection of two sets. Similarly, an inequality of the form $|x| > a$ can be rewritten as $x < -a$ or $x > a$, a union of two sets.

In the Cartesian plane, solutions of such inequalities will be, respectively, intersections and unions of half-planes. The next example demonstrates how such solution sets can be found.

Example 3: Solving Absolute Value Linear Inequalities

Graph the solution set in $\mathbb{R}^2$ that satisfies the joint conditions $|x-3| > 1$ and $|y-2| \leq 3$.

Solution

First, we need to find the solution to each individual inequality.

$$|x-3| > 1$$

$$x-3 > 1 \text{ or } x-3 < -1$$

Begin by rewriting the absolute value inequality as two linear inequalities.

$$x > 4 \text{ or } \quad x < 2$$

Simplify. We have a union of solutions.

$$|y-2| \leq 3$$

$$y-2 \leq 3 \text{ and } y-2 \geq -3$$

Again, rewrite the inequality without absolute value signs.

$$y \leq 5 \text{ and } \quad y \geq -1$$

This time, we have an intersection of solutions.

Now we can graph the solutions to the individual conditions.

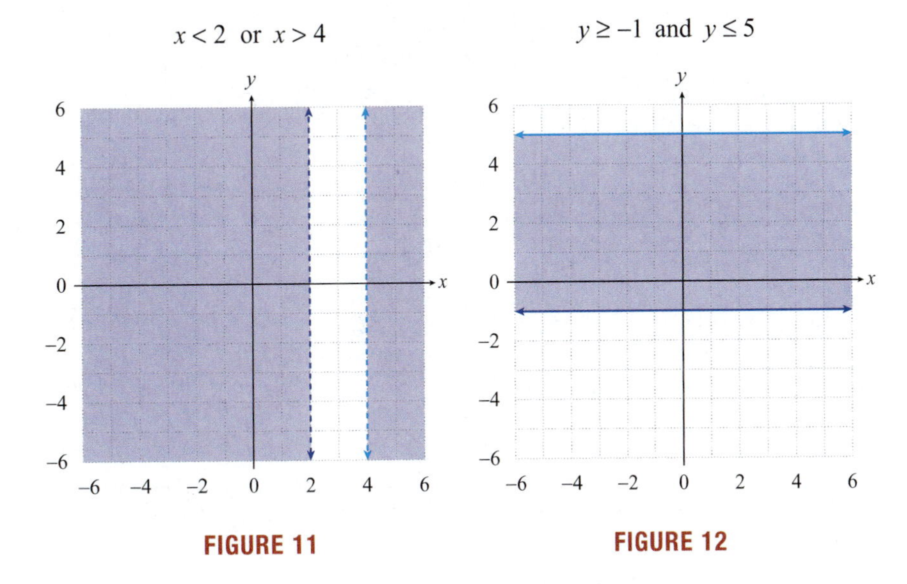

FIGURE 11

FIGURE 12

We now intersect the solution sets to obtain the final answer.

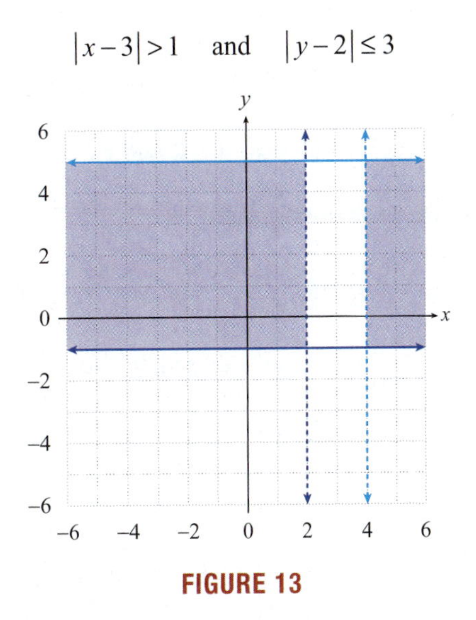

FIGURE 13

We will return to the notion of finding solutions to two or more inequalities when we study systems of linear inequalities and linear programming in Section 11.7.

2.6 EXERCISES

PRACTICE

Solve the following linear inequalities by graphing their solution sets. See Example 1.

1. $x-3y<6$
2. $y<2x-1$
3. $x>\frac{3}{4}y$
4. $x-3y\geq 6$
5. $3x-y\leq 2$
6. $\frac{2x-y}{4}>1$
7. $y<-2$
8. $x+1\geq 0$
9. $x+y<0$
10. $x+y>0$
11. $-(y-x)>-\frac{5}{2}-y$
12. $-2y\leq -x+4$
13. $5(y+1)\geq -x$
14. $3x-7y\geq 7(1-y)+2$
15. $x-y<2y+3$

Graph the solution sets that satisfy the following inequalities. See Example 2.

16. $y>-3x-6$ or $y\leq 2x-7$
17. $y\geq -2$ and $y>1$
18. $y\geq -2x-5$ and $y\leq -6x-9$
19. $y\leq 4x+4$ and $y>7x+7$
20. $x-3y\geq 6$ and $y>-4$
21. $x-3y\geq 6$ or $y>-4$
22. $3x-y\leq 2$ and $x+y>0$
23. $x>1$ and $y>2$
24. $x>1$ or $y>2$
25. $x+y>-2$ and $x+y<2$
26. $y>-2$ and $2y>-3x-4$
27. $3y>x+2$ or $4y\leq -x-2$
28. $y\leq -x$ and $2y+3x>-4$
29. $5x+6y<-30$ and $x\geq 2$
30. $6y-2x>-6$ or $y>6$
31. $x>-3$ or $y\geq 4$
32. $-2y<-3x-6$ or $-3y\geq -6x-18$
33. $x<6$ and $x\geq -5$

Graph the solution sets that satisfy the following linear absolute value inequalities. See Example 3.

34. $|x-3|<2$
35. $|x-3|>2$
36. $|3y-1|\leq 2$
37. $|2x-4|>2$
38. $1-|y+3|<-1$
39. $|x+1|<2$ and $|y-3|\leq 1$
40. $|x-3|\geq 1$ or $|y-2|\leq 1$
41. $|x-y|<1$
42. $|x+y|\geq 1$
43. $|4x-2y-3|\leq 5$
44. $|2x-3|\geq 1$ or $|2y+3|\geq 1$
45. $|y-3x|\leq 2$ and $|y|<2$

Match the following inequalities to the appropriate graph.

46. $-8y+5x \geq -8y+5$

47. $x < -2$ and $x \geq -5$

48. $|-7x-4y+23| \leq 16$

49. $y \leq 3x-6$ and $y > 2x-4$

50. $|3y-2x| > 17$ and $|y+6| \geq 1$

51. $4(y+2) < -x$

52. $-y < 6x+3$ or $4y \geq 3x-6$

53. $|7x+4| \leq 5$ or $|7y+4| \leq 5$

a.

b.

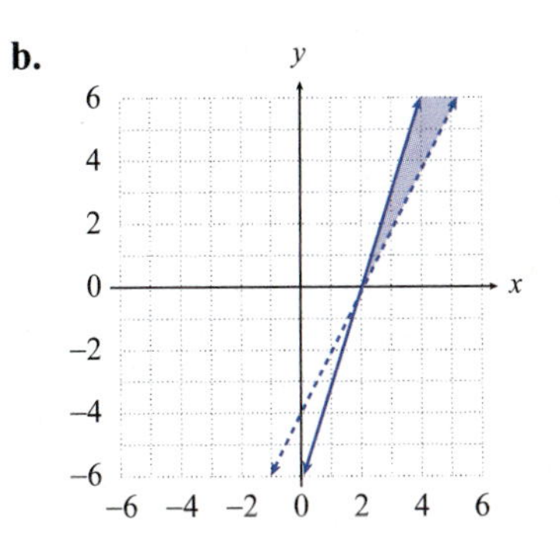

c.

d.

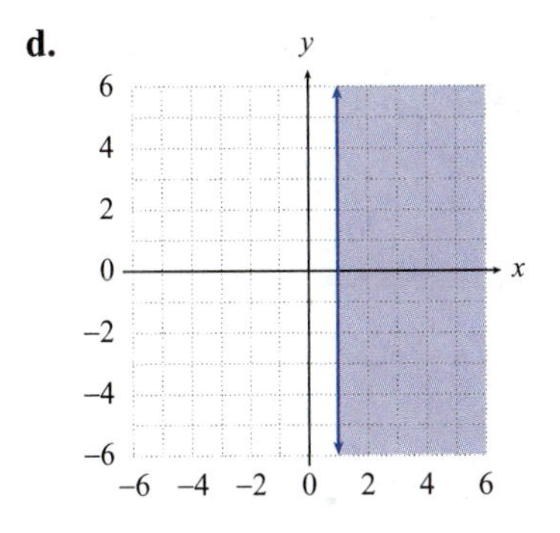

e.

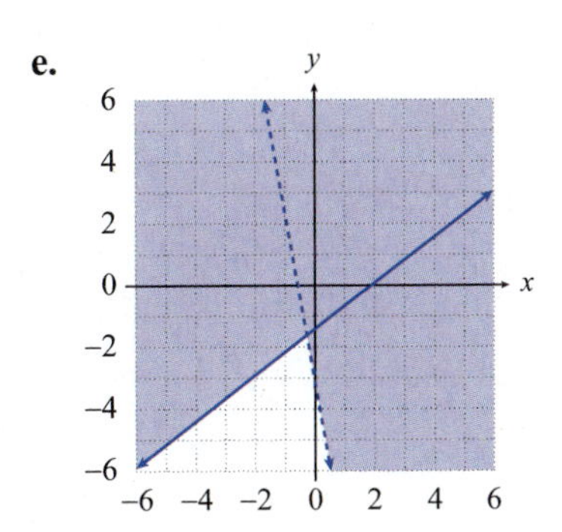

f.

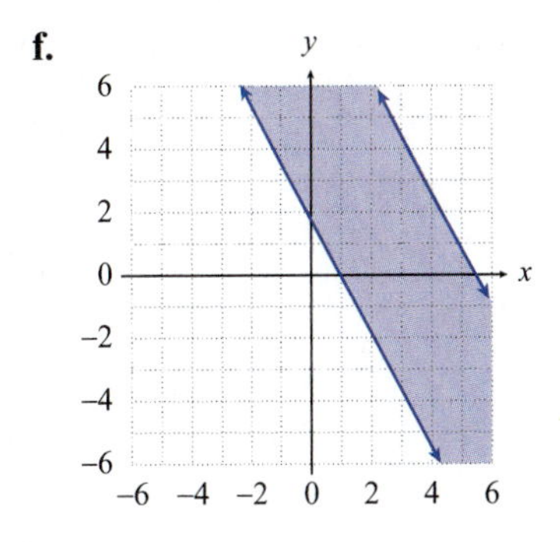

g.

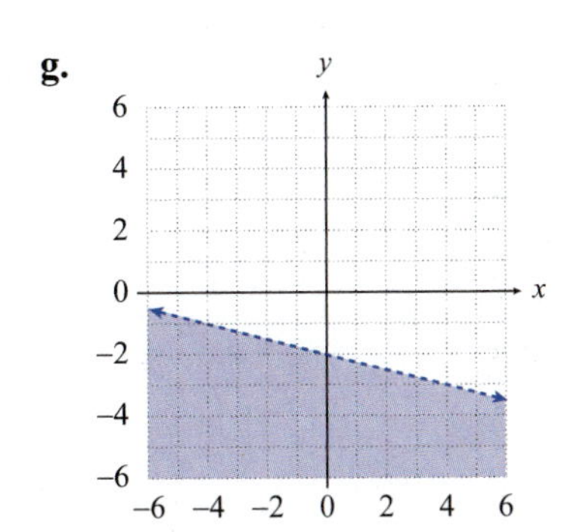

h.

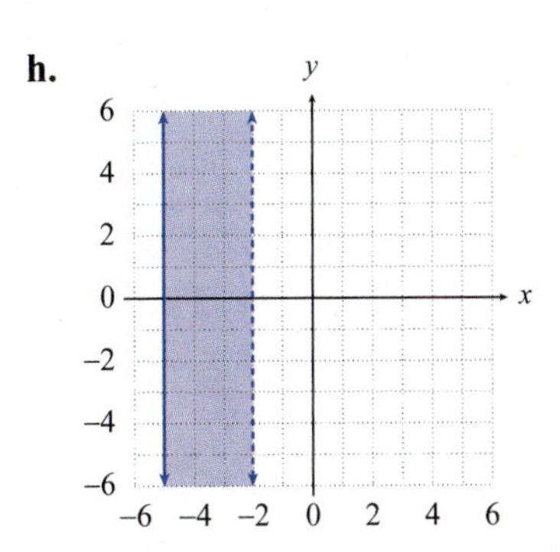

APPLICATIONS

54. It costs Happy Land Toys \$5.50 in variable costs per doll produced. If total costs must remain less than \$200, write a linear inequality describing the relationship between cost and dolls produced.

55. Trish is having a garden party where she wants to have several arrangements of lilies and orchids for decoration. The lily arrangements cost \$12 each and the orchids cost \$22 each. If Trish wants to spend less than \$150 on flowers, write a linear inequality describing the number of each arrangement she can purchase. Graph the inequality.

56. Rob has 300 feet of fencing he can use to enclose a small rectangular area of his yard for a garden. Assuming Rob may or may not use all the fencing, write a linear inequality describing the possible dimensions of his garden. Graph the inequality.

57. Flowertown Canoes produces two types of canoes. The two-person model costs \$73 to produce and the one-person model costs \$46 to produce. Write a linear inequality describing the number of each canoe the company can produce and keep costs under \$1750. Graph the inequality.

CHAPTER 2 PROJECT

Using the Pythagorean Theorem

Assume that a company that builds radio towers has hired you to supervise the installation of steel support cables for several newly built structures. Your task is to find the point at which the cables should be secured to the ground. Assume that the cables reach from ground level to the top of each tower. The cables have been precut by a subcontractor and have been labeled for each tower. Finding the correct distance from the base is necessary because each cable must be grounded before being attached to a tower to avoid damaging the equipment by electric shock.

The following is your work list for this week.

Tower Name	Tower Height	Cable Length	Distance from Base
Shelbyville Tower	58 ft	75 ft	______
Brockton Tower	100 ft	125 ft	______
Springfield Tower	77 ft	98 ft	______
Ogdenville Tower	130 ft	170 ft	______

1. Use the Pythagorean formula $\left(a^2 + b^2 = c^2\right)$ to determine how far from the base of the towers to attach the cables to the ground.

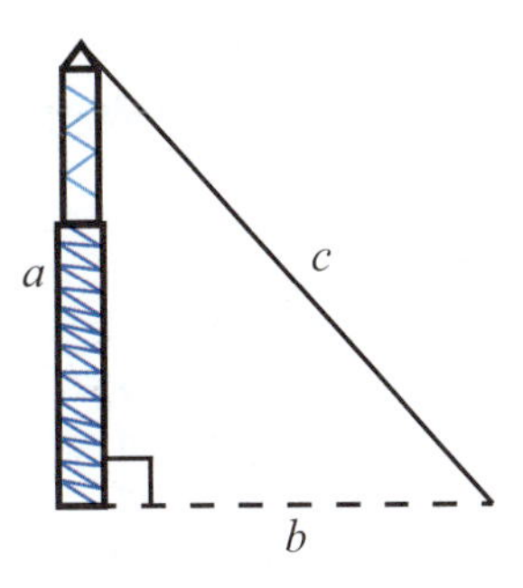

2. What length of cable would you need for a 150-foot tower if the grounding point has to be 100 feet from the base?

3. You have 400 feet of cable and wish to attach two lengths opposite one another to the top of a tower 200 feet tall, securing both lengths to the ground at a required distance of 75 feet from the tower's base. Do you have enough cable to do so? If so, how much is left after attaching the two cables? If not, you decide in advance to run one length to the top of the tower and use the remaining length of cable on the other side, attaching it to the highest point possible on the tower. How far down from the top would that attachment point be?

4. The tallest radio tower in the United States is in the Oro Valley near Tucson, Arizona. A cable from its top attached to the ground 260 feet from its base is 700 feet long. How tall is the radio tower?

CHAPTER 2 REVIEW EXERCISES

Section 2.1

Plot the following sets of points in the Cartesian plane.

1. $\{(7,3),(-2,4),(3,0),(-1,-6)\}$

2. $\{(4,-4),(-6,3),(-3,-1),(-4,2)\}$

3. $\{(2,1),(-4,5),(3,-7),(2,3)\}$

Identify the quadrant in which each point lies, if possible. If a point lies on an axis, specify which part (positive or negative) of which axis (*x* or *y*).

4. $(0,0)$

5. $(1,0)$

6. $(3,-2)$

For each of the following equations, determine the value of the missing entries in the accompanying table of ordered pairs. Then plot the ordered pairs and sketch your guess of the complete graph of the equation.

7. $3x-2y=6$

x	y
?	0
0	?
−1	?
?	−2
−2	?

8. $3x=y^2-4$

x	y
0	?
?	0
?	$-\sqrt{7}$
−1	?
?	3

Determine **a.** the distance between the following pairs of points, and **b.** the midpoint of the line segment joining each pair of points.

9. $(2,-6)$ and $(3,-7)$

10. $(-4,-3)$ and $(4,-9)$

11. $(-3,6)$ and $(-7,0)$

12. $(5,-1)$ and $(-4,3)$

13. Given $A(-4,2)$, $B(x,y)$, and $C(1,-1)$, find $x+y$ if C is the midpoint of the line segment $\overline{AB}$.

Find the perimeter of the triangle whose vertices are the specified points in the plane.

14. $(-3,2)$, $(-3,0)$, and $(-6,-3)$

15. $(8,-3)$, $(2,-3)$, and $(2,5)$

16. Use the distance formula to prove that the triangle with vertices at the points $(-2,2),(0,3)$, and $(4,-5)$ is a right triangle and determine the area of the triangle.

Section 2.2

Find the standard form of the equation for each circle described below.

17. Radius 4; center $\left(\sqrt{5}, -\sqrt{2}\right)$

18. Endpoints of a diameter are $(1,-3)$ and $(-5,3)$.

19. Center at $(2,-1)$; passes through $(4,3)$

20. Endpoints of a diameter are $(1,2)$ and $(-5,8)$.

21. What is the radius and center of the circle $(x+3)^2+(y-1)^2=8$?

22. Given that point $(a,4)$ is on the circle $x^2+y^2=25$, find a.

Sketch a graph of the circle defined by the given equation. Then state the radius and center of the circle.

23. $(x+5)^2+(y-2)^2=16$

24. $x^2+(y-3)^2=10$

25. $(x-1)^2+(y+4)^2=9$

26. $x^2+y^2+6x-10y=-5$

Section 2.3

Determine if the following equations are linear.

27. $3x+y(4-2x)=8$

28. $y-3(y-x)=8x$

29. $9x^2-(3x+1)^2=y-3$

30. $8x-3y=4(x-1)+y$

31. $2x(3y-1)=7$

32. $3x^2+2=(x+2)^2-1$

Find the *x*- and *y*-intercepts of the given equations, if possible, and then sketch their graphs.

33. $4y-12=8x$

34. $3(2y+1)=5y-4x+3$

35. $2x+y-2=2(3+x)$

36. $3y-4x=-2(3x-y)$

37. $2x+3y=18$

38. $4x+y=12+y$

Section 2.4

Determine the slope of the line passing through the specified points.

39. $(-2,5)$ and $(-3,-7)$

40. $(3,6)$ and $(7,-10)$

41. $(3,5)$ and $(3,-7)$

Use the slope-intercept form to graph the equations.

42. $6x-3y=9$ **43.** $2y+5x+9=0$ **44.** $15y-5x=0$

Find the equation, in standard form, of the line passing through the given point with the given slope.

45. point $(4,-1)$; slope of 1 **46.** point $(-2,3)$; slope of $\frac{3}{2}$

Find the equation, in slope-intercept form, of the line with the given *y*-intercept and slope.

47. *y*-intercept $(0,-2)$; slope of $\frac{5}{9}$ **48.** *y*-intercept $(0,9)$; slope of $-\frac{7}{3}$

Find the equation, in standard form, of the line passing through the specified points.

49. $(5,7)$ and $(3,-2)$ **50.** $\left(\frac{3}{2},1\right)$ and $\left(-3,\frac{5}{2}\right)$

51. A sales person receives a monthly salary of \$2800 plus a commission of 8% of sales. Write a linear equation for the sales person's monthly wage *W*, in terms of monthly sales, *s*.

Section 2.5

Determine if the two lines are perpendicular, parallel, or neither.

52. $x-4y=3$ and $4x-y=2$

53. $3x+y=2$ and $x-3y=25$

54. $\frac{3x-y}{3}=x+2$ and $\frac{y}{3}+x=9$

Find the equation, in slope-intercept form, for the line parallel to the given line and passing through the indicated point.

55. $y-3x=10$; $(-2,4)$ **56.** $3(y+1)=\frac{x-3}{2}$; $(-6,3)$

57. $y=2x+1$; $(1,-1)$ **58.** $3y-2=-5(2x-1)$; $(2,-5)$

Find the equation, in slope-intercept form, for the line perpendicular to the given line and passing through the indicated point.

59. $y=\frac{3}{4}x-1$; $(6,-2)$ **60.** $2(y-3)=\frac{2x+3}{3}$; $(-5,-4)$

61. $y=8$; $(7,1)$ **62.** $5x+7y-2=10$; $\left(\frac{2}{7},-1\right)$

Each set of four ordered pairs defines the vertices, in counterclockwise order, of a quadrilateral. Determine if the quadrilateral is a rectangle.

63. $\{(-2,1),(-1,-1),(3,1),(2,3)\}$ **64.** $\{(-2,2),(-3,-1),(2,-3),(2,1)\}$

Section 2.6

Solve the following linear inequalities by graphing their solution sets.

65. $x - 2y < 4$

66. $y < 3x + 2$

67. $\frac{4x + y}{3} \geq 2$

Graph the solution sets that satisfy the following inequalities.

68. $7x - 2y \geq 8$ and $y < 5$

69. $x - 4y \geq 6$ or $y > -2$

70. $y - x > 0$ and $x < 2$

Graph the solution sets that satisfy the following linear absolute value inequalities.

71. $|2x + 5| < 3$

72. $|2x - 1| < 5$

73. $|x - y| < 3$

74. $-5 + |x - 3| > -1$

75. $|2x + 1| < 3$ or $|y + 3| \geq 4$

76. $|x| > 4$ and $\left|\frac{2y - 1}{3}\right| < 3$

77. A candle store makes a \$3 profit for every novelty candle sold and a \$4 profit for every accompanying candle holder sold. Write a linear inequality describing the number of each type of item that needs to be sold in order to make a total profit of at least \$1500.

CHAPTER 3

Relations, Functions, and Their Graphs

SECTIONS

WHAT IF ...

What if you visited outer space in a space shuttle? Knowing how much you weigh on Earth, how much would you weigh 1000 miles above Earth?

By the end of this chapter, you'll be able to describe and manipulate relations, functions, and their graphs. To calculate your weight in outer space, you'll need to solve a variation problem like the one on page 271. You'll master this type of problem using the definition of inverse variation, found on page 266.

Gottfried Leibniz

German mathematician and philosopher Gottfried Leibniz is usually credited as one of the fathers of calculus.

Introduction

This chapter begins with a study of relations, which are generalizations of the equations in two variables discussed in Chapter 2, and then moves on to the more specialized topic of functions. As concepts, relations and functions are more abstract, but at the same time far more powerful and useful than the equations studied thus far in this text. Functions, in particular, lie at the heart of a great deal of the mathematics that you will encounter from this point on.

The history of the function concept serves as a good illustration of how mathematics develops. One of the first people to use the idea in a mathematical context was the German mathematician and philosopher Gottfried Leibniz (1646–1716), one of two people (along with Isaac Newton) usually credited with the development of calculus. Initially, Leibniz and other mathematicians tended to use the term to indicate that one quantity could be defined in terms of another by some sort of algebraic expression, and this (incomplete) definition of function is often encountered even today in elementary mathematics. As the problems that mathematicians were trying to solve increased in complexity, however, it became apparent that functional relations between quantities existed in situations where no algebraic expression defining the function was possible. One example came from the study of heat flow in materials, in which a description of the temperature at a given point at a given time was often given in terms of an infinite sum, not an algebraic expression.

The result of numerous refinements and revisions of the function concept is the definition that you will encounter in this chapter, and is essentially due to the German mathematician Lejeune Dirichlet (1805–1859). Dirichlet also refined our notion of what is meant by a variable, and gave us our modern understanding of dependent and independent variables, all of which you will soon encounter.

The proof of the power of functions lies in the multitude and diversity of their applications. The subtle and easily overlooked advantage of function notation also deserves special mention. As you work through Chapter 3, pay special attention to how function notation works. A solid understanding of what function notation means is essential to using functions.

3.1 RELATIONS AND FUNCTIONS

TOPICS

1. Relations, domain, and range
2. Functions and the vertical line test
3. Function notation and function evaluation
4. Implied domain of a function

TOPIC 1: Relations, Domain, and Range

In Chapter 2, we saw many examples of equations in two variables. Any such equation automatically defines a relation between the two variables present in the sense that each ordered pair on the graph of the equation relates a value for one variable (namely, the first coordinate of the ordered pair) to a value for the second variable (the second coordinate). Many applications of mathematics involve relating one variable to another, and this notion merits much further study.

We begin by generalizing the above observation and defining a relation to be simply a collection of ordered pairs; any collection of ordered pairs constitutes a relation, whether the ordered pairs correspond to the graph of an equation or not. This definition is repeated below, along with two related ideas.

Set Theory as a Foundation

Relations, like most objects in mathematics, are defined using the ideas and terminology of the branch of mathematics known as set theory. Since set theory serves such a foundational purpose, it may be surprising to learn that modern set theory is a relatively young area of study. It began in the 1870s principally with the work of two German mathematicians, Georg Cantor and Richard Dedekind. Their initial ideas sparked a flurry of activity that led relatively quickly to the formulation of the set theory axioms on which almost all of mathematics is now based.

DEFINITION: Relations, Domain, and Range

A **relation** is a set of ordered pairs. Any set of ordered pairs automatically relates the set of first coordinates to the set of second coordinates, and these sets have special names. The **domain** of a relation is the set of all the first coordinates, and the **range** of a relation is the set of all second coordinates.

It is important to understand that relations can be described in many different ways. We have already noted that an equation in two variables describes a relation, as the solution set of the equation is a collection of ordered pairs. Relations can also be described with a simple list of ordered pairs (if the list is not too long), with a picture in the Cartesian plane, and by many other means.

Example 1 demonstrates some of the common ways of describing relations and identifies the domain and range of each relation.

Example 1: Relations, Domain, and Range

a. The set $R = \{(-4,2),(6,-1),(0,0),(-4,0),(\pi,\pi^2)\}$ is a relation consisting of five ordered pairs. The domain of R is the set $\{-4,6,0,\pi\}$, as these four numbers appear as first coordinates in the relation. Note that it is not necessary to list the number –4 twice in the domain, even though it appears twice as a first coordinate in the relation. The range of R is the set $\{2,-1,0,\pi^2\}$, as these are the numbers that appear as second coordinates. Again, it is not necessary to list 0 twice in the range, even though it is used twice as a second coordinate in the relation. The graph of this relation is simply a picture of the five ordered pairs plotted in the Cartesian plane, as shown in Figure 1.

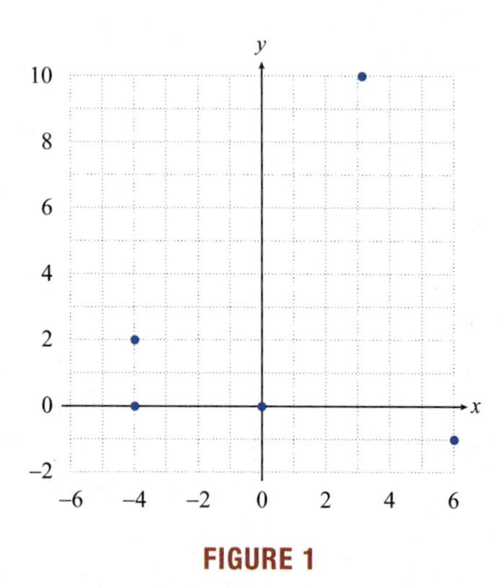

FIGURE 1

b. The equation $-3x+7y=13$ describes a relation. In contrast to part a., this relation consists of an infinite number of ordered pairs, so it is not possible to list them all as a set. As an example, one of the ordered pairs in the relation is $(-2,1)$, since $-3(-2)+7(1)=13$. Although it is not possible to list all the elements of this relation, it is possible to graph it; it is the graph of the solution set of the equation, as drawn in Figure 2. The domain and range of this relation are both the set of real numbers, as every real number appears as both a first coordinate and a second coordinate in the relation.

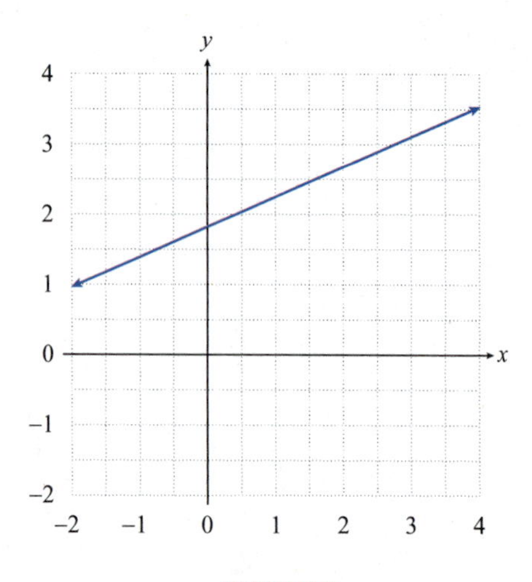

FIGURE 2

c. The rectangle in Figure 3 describes a relation. Some of the elements of the relation are $(-1,1),(-1,-2),(-0.3,2),(0,-2)$, and $(1,-0.758)$, but this is another example of a relation with an infinite number of elements so we cannot list all of them. Using the interval notation defined in Section 1.1, the domain of this relation is the closed interval $[-1,1]$ and the range is the closed interval $[-2,2]$.

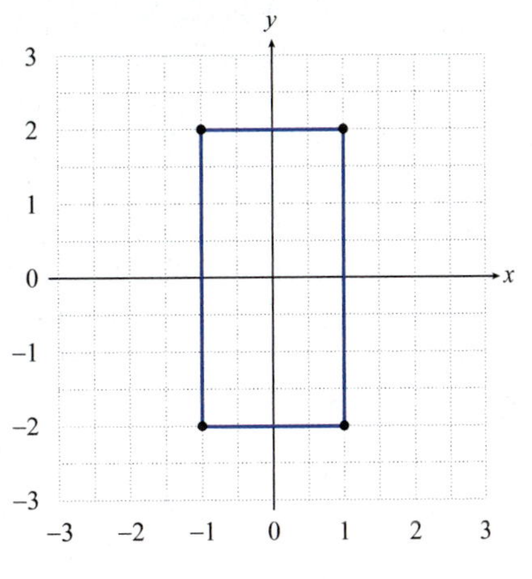

FIGURE 3

d. The rectangle in Figure 4 describes another relation, similar to the last but still different. The shading in the picture indicates that all ordered pairs lying inside the rectangle, as well as those actually on the rectangle, are elements of the relation. The domain is again the closed interval $[-1,1]$ and the range is again the closed interval $[-2,2]$, but this relation is not identical to the one in part c. For instance, the ordered pairs $(0,0)$ and $(0.2,1.5)$ are elements of this relation but are not elements of the relation in Example 1c.

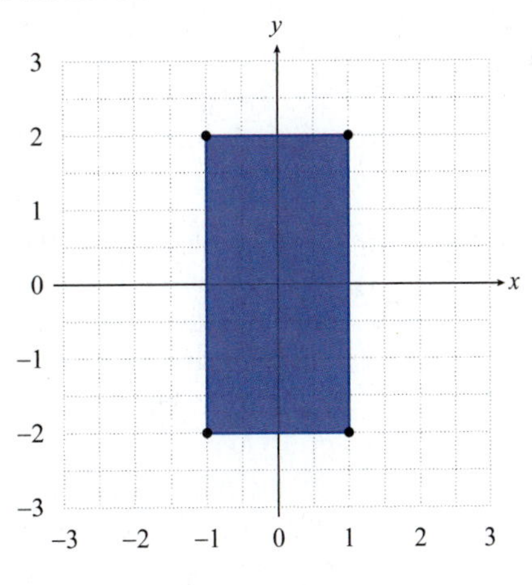

FIGURE 4

e. Although we will almost never encounter relations in this text that do not consist of ordered pairs of real numbers, there is nothing in our definition to prevent us from considering more exotic relations. For example, the set $S = \{(x, y) | x \text{ is the mother of } y\}$ is a relation among people. The domain of S is the set of all mothers, and the range of S is the set of all people. (Although advances in cloning are occurring rapidly, as of the writing of this text, no one has yet been born without a mother!)

TOPIC 2: Functions and the Vertical Line Test

As important as relations are in mathematics, a special type of relation, called a function, is of even greater use.

DEFINITION: Function

A **function** is a relation in which every element of the domain is paired with exactly one element of the range. Equivalently, a function is a relation in which no two distinct ordered pairs have the same first coordinate.

Note that there is a difference in the way domains and ranges are treated in the definition of a function: the definition allows for the two distinct ordered pairs to have the same second coordinate, as long as their first coordinates differ. A picture helps in understanding this distinction.

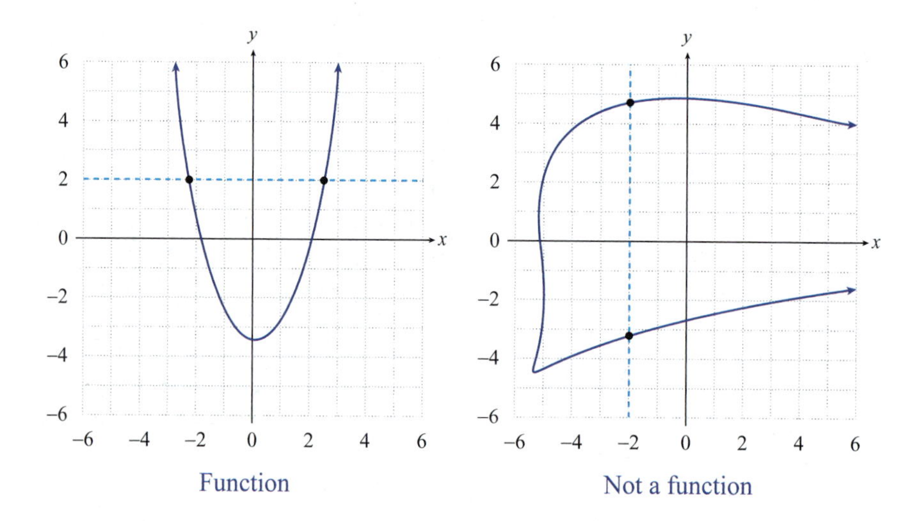

FIGURE 5: Definition of Functions

The relation on the left in Figure 5 has pairs of points that share the same y-value (one such pair is indicated on the graph where the dashed line intersects the curve). This means that some elements of the range are paired with more than one element of the domain. However, each element of the domain is paired with exactly one element of the range. Thus, this relation is a function.

On the other hand, the relation on the right in Figure 5 has pairs of points that share the same x-value (one such pair is indicated on the graph where the dashed line intersects the curve). This means that some elements of the domain are paired with more than one element of the range. This relation is not a function.

Example 2: Is the Relation a Function?

For each relation in Example 1, identify whether the relation is also a function.

Solution

a. The relation in Example 1a is not a function because the two ordered pairs $(-4, 2)$ and $(-4, 0)$ have the same first coordinate. If either one of these ordered pairs were deleted from the relation, the relation would be a function.

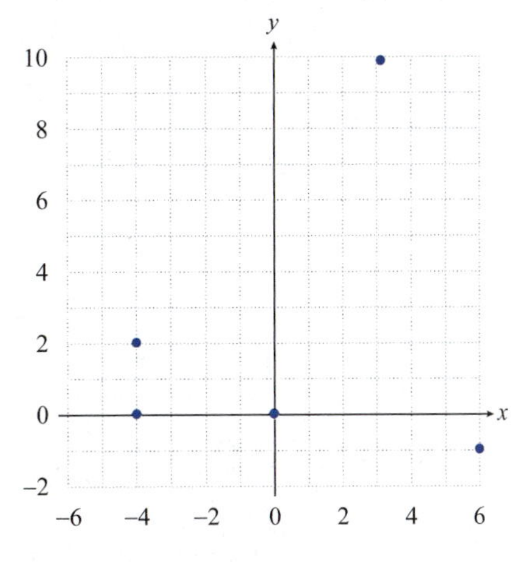

FIGURE 6

b. The relation in Example 1b is a function. Any two distinct ordered pairs that solve the equation $-3x+7y=13$ have different first coordinates. This can also be seen from the graph of the equation. If two ordered pairs have the same first coordinate, they must be aligned vertically, and no two ordered pairs on the graph of $-3x+7y=13$ have this property.

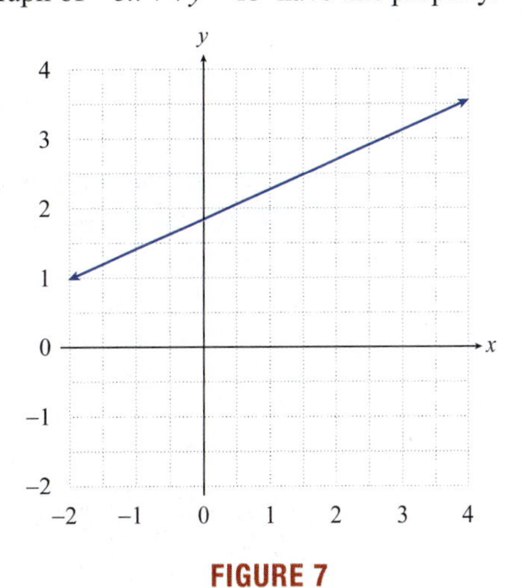

FIGURE 7

c. The relation in Example 1c is not a function. To prove that a relation is not a function, it is only necessary to find two ordered pairs with the same first coordinate, and the pairs $(0, 2)$ and $(0, -2)$ show that this relation fails to be a function.

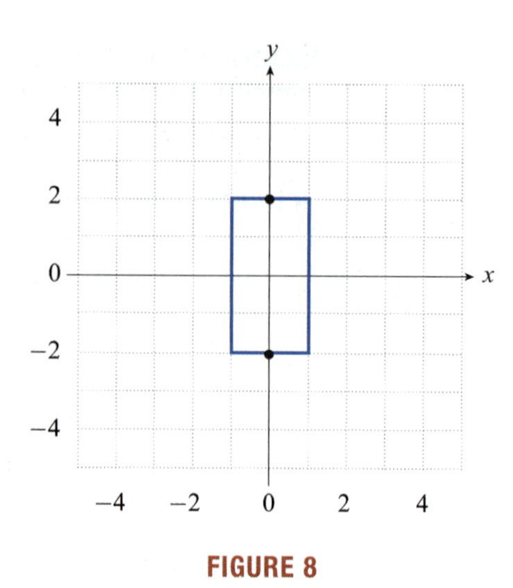

FIGURE 8

d. The relation in Example 1d is also not a function. In fact, we can use the same two ordered pairs as in the previous part to prove this fact.

e. Finally, the relation in Example 1e also fails to be a function. Think of two people who have the same mother; this gives us two ordered pairs with the same first coordinate: (mother, child 1) and (mother, child 2). Thus, the relation is not a function.

In Example 2b, we noted that two ordered pairs in the plane have the same first coordinate only if they are aligned vertically. We could also have used this criterion to determine that the relation in Example 1a is not a function, since the two ordered pairs $(-4, 2)$ and $(-4, 0)$ clearly lie on the same vertical line. This visual method of determining whether a relation is a function, called the **vertical line test**, is very useful when an accurate graph of the relation is available.

THEOREM: The Vertical Line Test

If a relation can be graphed in the Cartesian plane, the relation is a function if and only if no vertical line passes through the graph more than once. If even one vertical line intersects the graph of the relation two or more times, the relation fails to be a function.

CAUTION

Note that vertical lines that miss the graph of a relation entirely don't prevent the relation from being a function; it is only the presence of a vertical line that hits the graph two or more times that indicates the relation isn't a function.

The next example illustrates some more applications of the vertical line test.

Example 3: Functions and the Vertical Line Test

a. The relation $R = \{(-3,2),(-1,0),(0,2),(2,-4),(4,0)\}$, graphed in Figure 9, is a function. Any given vertical line in the plane either intersects the graph once or not at all.

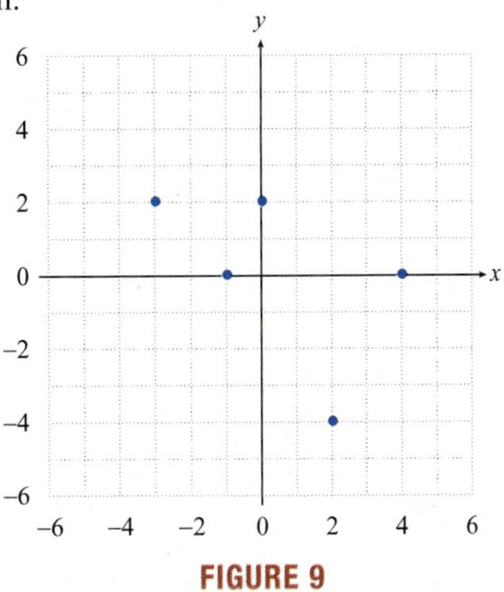

FIGURE 9

b. The relation graphed in Figure 10 is not a function, as there are many vertical lines that intersect the graph more than once. The dashed line is one such vertical line.

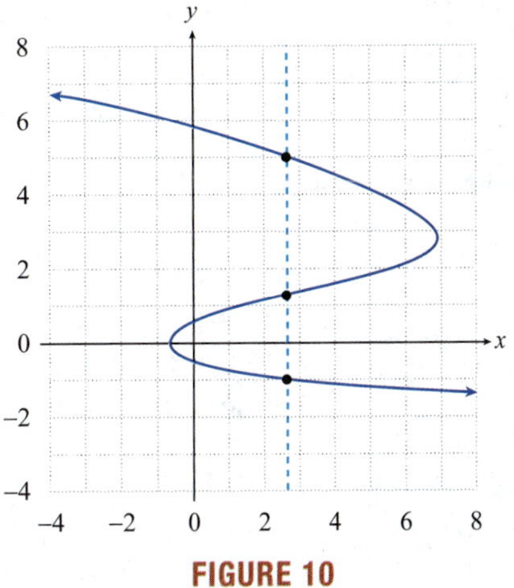

FIGURE 10

c. The relation graphed in Figure 11 is a function. In this case, every vertical line in the plane intersects the graph exactly once.

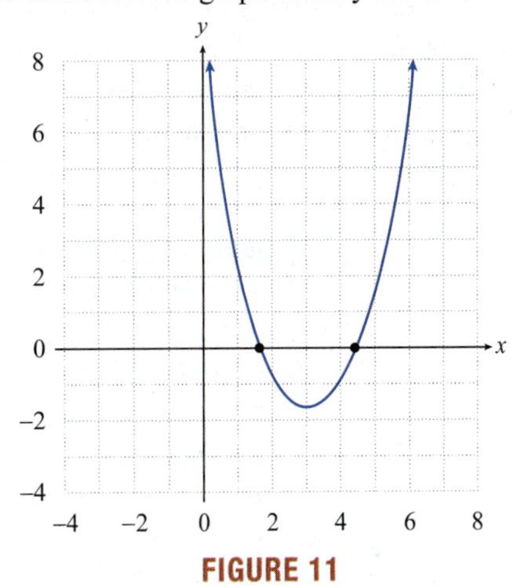

FIGURE 11

TOPIC 3: Function Notation and Function Evaluation

Functions are useful in mathematics for a variety of reasons, all of which ultimately depend on the property that differentiates functions from other relations. Specifically, given an element of the domain of a function, the function relates it to a unique element of the range; there is no ambiguity in which element of the range to associate with a given domain element. Relations that are not functions fail this criterion of uniqueness. For instance, the equation $x = y^2$ describes a perfectly good relation, but every positive value of x in the domain is associated with two elements of the range: for example, both $(9,3)$ and $(9,-3)$ lie on the graph of $x = y^2$.

Because functions assign a unique element of the range to each element of the domain, functions are often defined by means of a notation not yet seen in this text, called function notation.

Function notation accomplishes a number of things at once: it gives a name to the function being defined, it provides a formula to work with when using the function, and it makes the solving of many, many mathematical problems far easier.

LEONARD EULER.

The History of Function Notation

The prolific Swiss mathematician Leonhard Euler made many profound contributions, over the course of his seventy-six years, to diverse areas of mathematics. Along with brilliant insights and solutions to confounding problems, Euler is credited with popularizing such things as the use of i for the imaginary unit and, in the 1740s, the $f(x)$ function notation.

As an example of function notation, consider the function defined by the graph of $10x - 5y = 15$. We know this represents a function because if we graph $10x - 5y = 15$ we obtain a nonvertical line, and such a line always passes the vertical line test. If we need to know which specific element of the range corresponds to a given element of the domain, say $x = 4$, we can substitute 4 for x in the equation and solve for y. But if we have to do this for more than one domain element, it is foolish to repeat the process over and over. Why not just solve for y once and for all? If we do this, we obtain the equivalent equation $y = 2x - 3$. Now, given any element of the domain (that is, given any value for x), we can determine the corresponding range element (the value for y). In this context, x is called the **independent variable** and y is called the **dependent variable**, as the value for y depends on the value for x.

In function notation, the function defined by $10x - 5y = 15$ is written as $f(x) = 2x - 3$, where we have arbitrarily chosen to name the function f. The key point here is that we have used a mathematical formula to describe what the function named f does to whatever element of the domain it is given. The notation $f(x) = 2x - 3$ is read "f of x equals two times x, minus three" and indicates that when given a specific value for x, the function f returns two times that value, minus three. For instance, $f(x) = 2(4) - 3$, or $f(4) = 5$ (this last line is read "f of 4 equals 5"). Notice that the only difference between the equation $y = 2x - 3$ and the function $f(x) = 2x - 3$ is one of notation: we have replaced y with $f(x)$.

Example 4: Function Notation

Each of the following equations in x and y represents a function. Rewrite each one using function notation, and then evaluate each function at $x = -3$.

a. $y = \dfrac{3}{x} + 2$

b. $7x + 3 = 2y - 1$

c. $y - 5 = x^2$

d. $\sqrt{1-x} - 2y = 6$

integers) and the codomain is the entire set of real numbers. Since the range of j is the set $\{1,4,9,16,25,...\}$, which is not the same as the codomain, j is not onto.

TOPIC 4: Implied Domain of a Function

Sometimes, especially in applications, the domain of a function under consideration is defined at the same time as the function itself. Often, however, no mention is made explicitly of the values that may be "plugged into" the function, and it is up to us to determine what the domain is. In these cases, the domain of the function is implied by the formula used in defining the function. It is assumed that the domain of the function consists of all those real numbers at which the function can be evaluated to obtain a real number. This means, for instance, that any values for the argument that result in division by zero or an even root of a negative number must be excluded from the domain.

Example 8: Implied Domain of a Function

Determine the implied domain of each of the following functions.

a. $f(x)=5x-\sqrt{3-x}$

b. $g(x)=\dfrac{x-3}{x^2-1}$

Solution

a. Looking at the formula, we can identify what may cause the function to be undefined.

$f(x)=5x-\sqrt{3-x}$

We can always multiply a number by 5, but taking the square root of a negative number is undefined.

The square root term is defined as long as $3-x\geq 0$. Solving this inequality for x, we have $x\leq 3$.

Using interval notation, the domain of the function f is the interval $(-\infty,3]$.

b. Again, we first identify potential "dangers" in the formula for this function.

$g(x)=\dfrac{x-3}{x^2-1}$

We can safely substitute any value in the numerator, but we can't let the denominator equal zero.

The denominator will equal zero whenever $x^2-1=0$. This tells us that we must exclude $x=-1$ and $x=1$ from the domain.

In interval notation, the domain of g is $(-\infty,-1)\cup(-1,1)\cup(1,\infty)$.

3.1 EXERCISES

PRACTICE

For each of the following relations, determine the domain and range. See Example 1.

1. $R=\{(-2,5),(-2,3),(-2,0),(-2,-9)\}$

2. $S=\{(0,0),(-5,2),(3,3),(5,3)\}$

3. $A=\{(\pi,2),(-2\pi,4),(3,0),(1,7)\}$

4. $B=\{(3,3),(-4,3),(3,8),(3,-2)\}$

5. $T=\{(x,y)\mid x\in\mathbb{Z} \text{ and } y=2x\}$

6. $U=\{(\pi,y)\mid y\in\mathbb{Q}\}$

7. $C=\{(x,3x+4)\mid x\in\mathbb{Z}\}$

8. $D=\{(5x,3y)\mid x\in\mathbb{Z} \text{ and } y\in\mathbb{Z}\}$

9. $3x-4y=17$

10. $x+y=0$

11. $x=|y|$

12. $y=x^2$

13. $y=-1$

14. $x=3$

15. $x=4x$

16. $y=7\pi^2$

17.

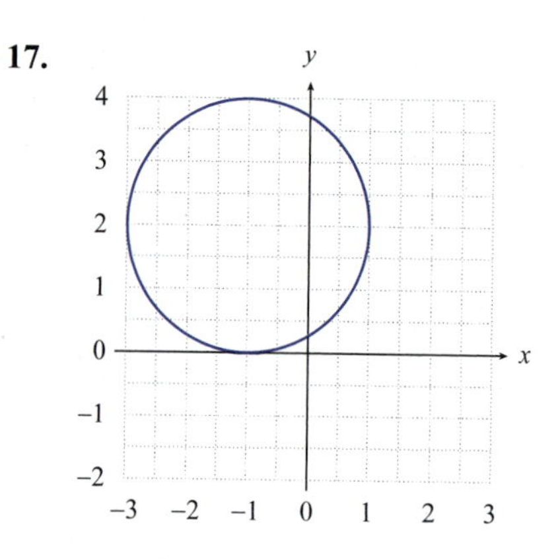

18.

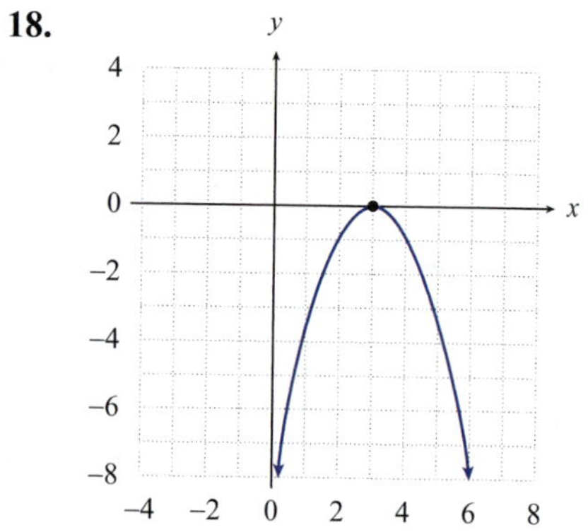

19.

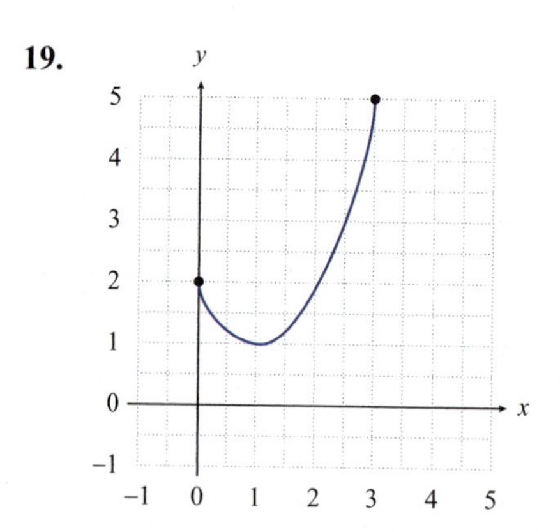

20.

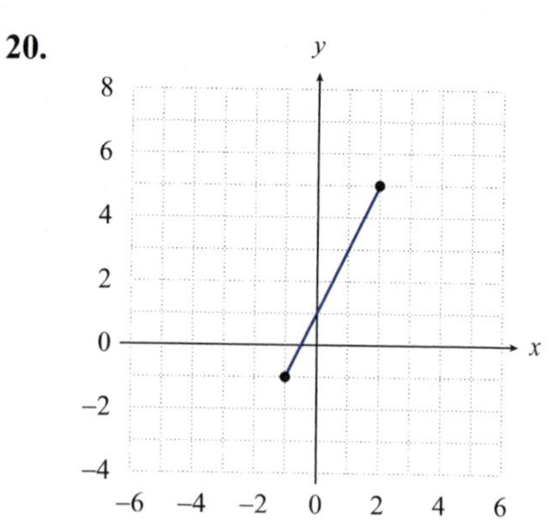

21.

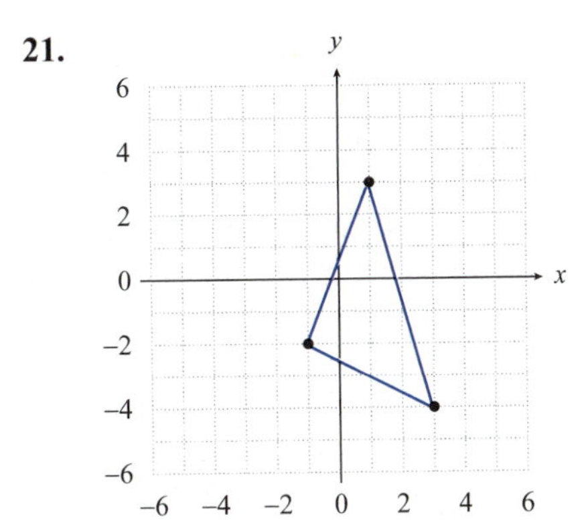

22.

23. $V = \{(x, y) \mid x \text{ is the brother of } y\}$

24. $W = \{(x, y) \mid y \text{ is the daughter of } x\}$

Determine which of the following relations is a function. For those that are not functions, identify two ordered pairs with the same first coordinate. See Examples 2 and 3.

25. $R = \{(-2, 5), (2, 4), (-2, 3), (3, -9)\}$

26. $S = \{(3, -2), (4, -2)\}$

27. $T = \{(-1, 2), (1, 1), (2, -1), (-3, 1)\}$

28. $U = \{(4, 5), (2, -3), (-2, 1), (4, -1)\}$

29. $V = \{(6, -1), (3, 2), (6, 4), (-1, 5)\}$

30. $W = \{(2, -3), (-2, 4), (-3, 2), (4, -2)\}$

31.

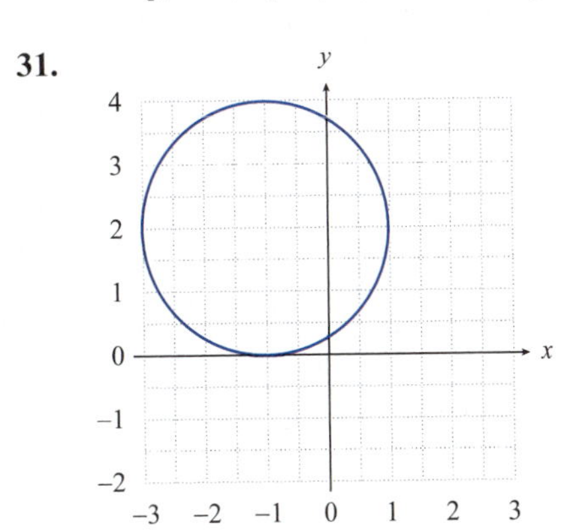

32.

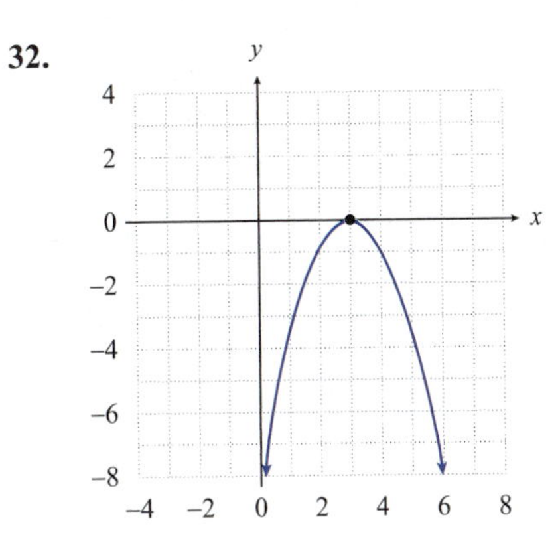

33.

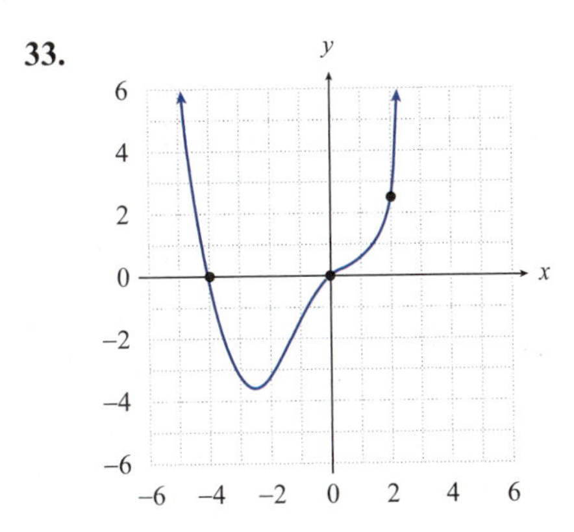

34.

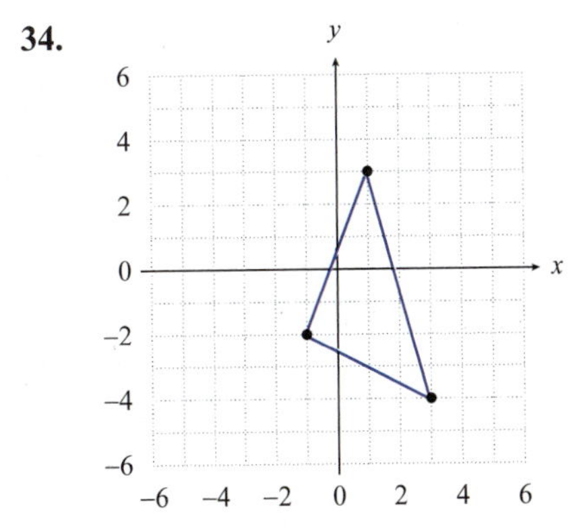

35.

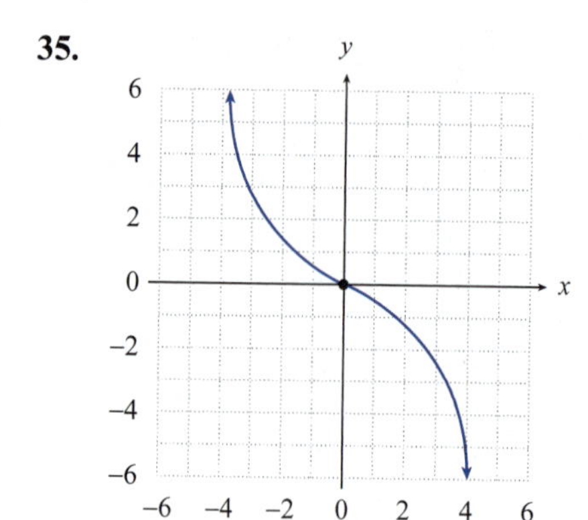

36.

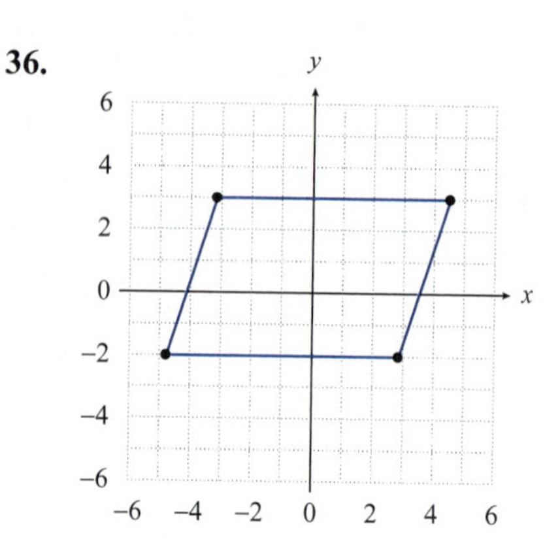

Identify which of the following relations is a function by determining whether there is a unique *y*-value related to every *x*-value in the relation's domain. For those that are not functions, identify two ordered pairs with the same first coordinate.

37. $y = \dfrac{1}{x}$ **38.** $x = y^2 - 1$ **39.** $x + y^2 = 0$ **40.** $y = 2x^2 - 4$

41. $y = \dfrac{x-1}{x+2}$ **42.** $x^2 + y^2 = 1$ **43.** $y = |x - 2|$ **44.** $y = x^3$

45. $y^2 - x^2 = 3$ **46.** $y = \sqrt{x} - 4$

Rewrite each of the following relations as a function of *x*. Then evaluate the function at $x = -1$. See Example 4.

47. $6x^2 - x + 3y = x + 2y$ **48.** $2y - \sqrt[3]{x} = x - (x-1)^2$

49. $\dfrac{x+3y}{5} = 2$ **50.** $x^2 + y = 3 - 4x^2 + 2y$

51. $y - 2x^2 = -2(x + x^2 + 5)$ **52.** $\dfrac{9y+2}{6} = \dfrac{3x-1}{2}$

Use the graph below of a function *f* to answer the following questions. See Example 5.

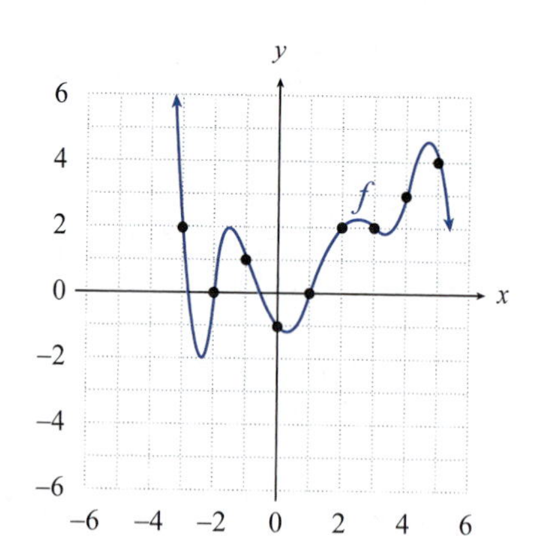

53. What is the value of $f(-1)$? **54.** What is the value of $f(0)$?

55. What is the value of $f(4)$?

56. For what integer value(s) of x is $f(x)=4$?

57. For what integer value(s) of x is $f(x)=2$?

58. For what integer value(s) of x is $f(x)=0$?

For each of the following functions, determine **a.** $f(2)$, **b.** $f(x-1)$, **c.** $f(x+a)-f(x)$, and **d.** $f(x^2)$. See Example 6.

59. $f(x)=x^2+3x$

60. $f(x)=\sqrt{x}$

61. $f(x)=3x+2$

62. $f(x)=-x^2-7$

63. $f(x)=2(5-3x)$

64. $f(x)=2x^2+\sqrt[4]{x}$

65. $f(x)=\sqrt{1-x}-3$

66. $f(x)=\dfrac{-\sqrt{1-x}+5}{2}$

Determine $\dfrac{f(x+h)-f(x)}{h}$ for each of the following functions. See Example 6c.

67. $f(x)=x^2-5x$

68. $f(x)=x^3+2$

69. $f(x)=\dfrac{1}{x+2}$

70. $f(x)=6x^2-7x+3$

71. $f(x)=5x^2$

72. $f(x)=(x+3)^2$

73. $f(x)=2x-7$

74. $f(x)=\sqrt{x}$

75. $f(x)=x^{\frac{1}{2}}-4$

76. $f(x)=\dfrac{3}{x}$

Identify the domain, the codomain, and the range of each of the following functions. See Example 7.

77. $f:\mathbb{R}\to\mathbb{R}$ by $f(x)=3x$

78. $g:\mathbb{Z}\to\mathbb{Z}$ by $g(x)=3x$

79. $f:\mathbb{Z}\to\mathbb{Z}$ by $f(x)=x+5$

80. $g:[0,\infty)\to\mathbb{R}$ by $g(x)=\sqrt{x}$

81. $h:\mathbb{N}\to\mathbb{N}$ by $h(x)=x+5$

82. $h:\mathbb{N}\to\mathbb{R}$ by $h(x)=\dfrac{x}{2}$

Determine the implied domain of each of the following functions. See Example 8.

83. $f(x)=\sqrt{x-1}$

84. $g(x)=\sqrt[5]{x+3}-2$

85. $h(x)=\dfrac{3x}{x^2-x-6}$

86. $f(x)=(2x+6)^{\frac{1}{2}}$

87. $g(x)=\sqrt[4]{2x^2+3}$

88. $h(x)=\dfrac{3x^2-6x}{x^2-6x+9}$

89. $s(x)=\dfrac{2x}{1-3x}$

90. $f(x)=(x^2-5x+6)^3$

91. $c(x)=\dfrac{x-1}{2-x}$

92. $g(x)=\dfrac{5}{\sqrt{3-x^2}}$

93. $f(x)=\sqrt{x+6}+1$

94. $g(x)=-5x^2-4x$

95. $h(x)=\dfrac{-3(-5+5x)}{x}$

96. $h(x)=\sqrt{3-x}$

WRITING & THINKING

97. Justify why the following statement is true: All functions are relations, but not all relations are functions.

TECHNOLOGY

Use a graphing utility to evaluate each of the following functions at the specified values of x.

98. $f(x)=\dfrac{7x^{\frac{5}{3}}-2x^{\frac{1}{3}}}{x^{\frac{1}{2}}}$; find $f(8)$ and $f(12)$

99. $g(x)=\sqrt{x^3-4x^2+2x+31}$; find $g(2)$ and $g(3)$

100. $f(x)=\dfrac{2x^5-9x^3+12}{4x^3-7x+6}$; find $f(-3)$ and $f(2)$

101. $g(x)=(5x^2-7x+1)^3$; find $g(-19)$ and $g(12)$

102. $f(x)=\dfrac{\sqrt{x^4+6x^3-4x+13}}{4x^3+2x^2-12}$; find $f(-4)$ and $f(6)$

103. $g(x)=\dfrac{(3x^3-2x+9)^4}{(7x^2-5x)^2}$; find $g(-5)$ and $g(4)$

3.2 LINEAR FUNCTIONS

TOPICS

1. Linear functions and their graphs
2. Linear regression

TOPIC 1: Linear Functions and Their Graphs

Much of the next several sections of this chapter will be devoted to gaining familiarity with some of the types of functions that commonly arise in mathematics. We will begin in this section by discussing the class of linear functions.

Recall that a linear equation in two variables is an equation whose graph consists of a straight line in the Cartesian plane. Similarly, a linear function (of one variable) is a function whose graph is a straight line. We can define such functions algebraically as follows.

DEFINITION: Linear Function

A **linear function** f in the variable x is any function that can be written in the form $f(x)=mx+b$, where m and b are real numbers. If $m \neq 0$, $f(x)=mx+b$ is also called a **first-degree polynomial function.**

Previously, we learned that a function defined by an equation in x and y can be written in function form by solving the equation for y and then replacing y with $f(x)$. This process can be reversed, so the linear function $f(x)=mx+b$ appears in equation form as $y=mx+b$, a linear equation written in slope-intercept form. Thus, the graph of a linear function is a straight line with slope m and y-intercept $(0,b)$.

NOTE

A function cannot represent a vertical line (since it fails the vertical line test). Vertical lines can represent the graphs of equations, but not functions.

As we noted in Section 3.1, the graph of a function is a plot of all the ordered pairs that make up the function; that is, the graph of a function f is the plot of all the ordered pairs in the set $\{(x,y) \mid f(x)=y\}$. We have a great deal of experience in plotting such sets if the ordered pairs are defined by an equation in x and y, but we have only plotted a few functions that have been defined with function notation. Any function of x defined with function notation can be written as an equation in x and y by replacing $f(x)$ with y, so the graph of a function f consists of a plot of the ordered pairs in the set $\{(x,f(x)) \mid x \in \text{domain of } f\}$. Consider the function $f(x)=-3x+5$. Figure 1 contains a table of four ordered pairs defined by the function and a graph of the function with the four ordered pairs noted.

x	$f(x)$
-1	8
0	5
1	2
2	-1

FIGURE 1: $f(x) = -3x + 5$

Again, note that every point on the graph of the function in Figure 1 is an ordered pair of the form $(x, f(x))$; we have simply highlighted four of them with dots.

We could have graphed the function $f(x) = -3x + 5$ by noting that it is a straight line with a slope of -3 and a y-intercept of 5. We use this approach in the following example.

Example 1: Graphing Linear Functions

Graph the following linear functions.

a. $f(x) = 3x + 2$

b. $g(x) = 3$

Solution

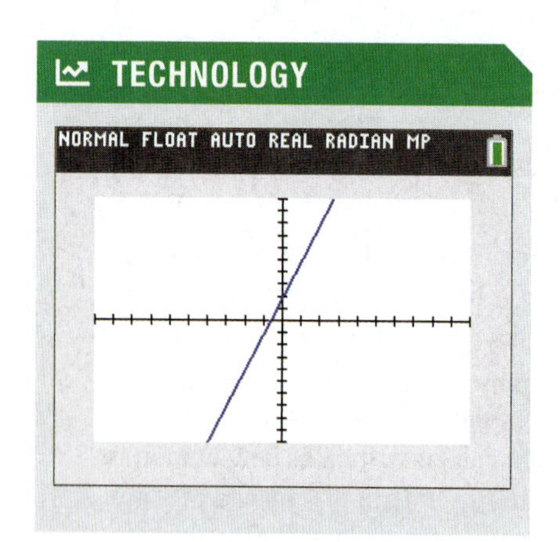

a.

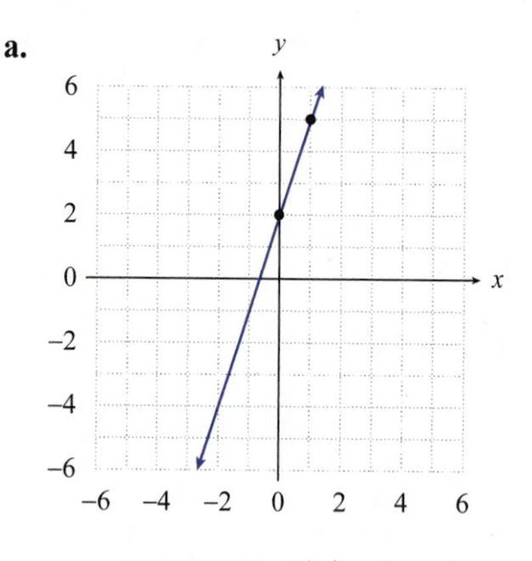

FIGURE 2: $f(x) = 3x + 2$

The function f is a line with a slope of 3 and a y-intercept of 2.

To graph the function, plot the ordered pair $(0, 2)$ and locate another point on the line by moving up 3 units and over to the right 1 unit, giving the ordered pair $(1, 5)$. Once these two points have been plotted, connecting them with a straight line completes the process.

b.

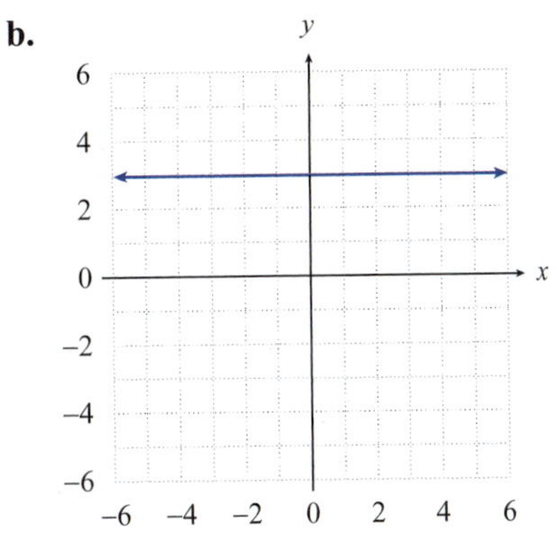

FIGURE 3: $g(x)=3$

The graph of the function g is a straight line with a slope of 0 and a y-intercept of 3.

A linear function with a slope of 0 is also called a **constant function**, as it turns any input into one fixed constant—in this case the number 3. The graph of a constant function is always a horizontal line.

Example 2: Finding a Linear Function Given Its Graph

Find a formula for the linear function whose graph is given in Figure 4.

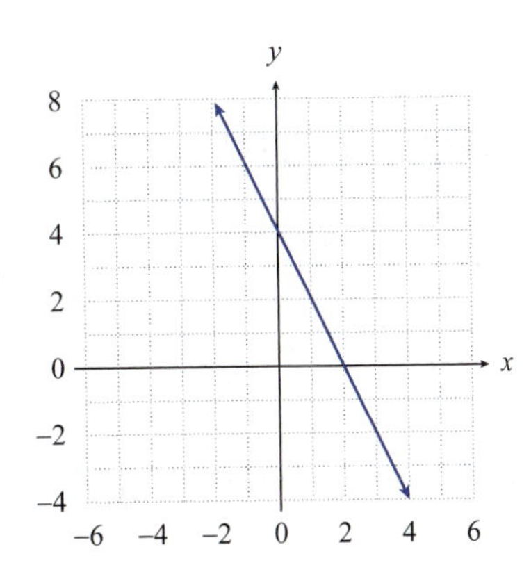

FIGURE 4

Solution

The graph shows a y-intercept of 4, giving us one of the two constants necessary to define a linear function. To determine the slope, we can use the fact that $(0,4)$ and $(2,0)$ both lie on the graph and hence

$$m=\frac{0-4}{2-0}=\frac{-4}{2}=-2.$$

If we name the function f, the formula for this function is thus

$$f(x)=-2x+4.$$

TOPIC 2: Linear Regression

We have seen that, given a linear equation in the form $y=mx+b$, we can easily determine any number of points lying on the line. We have also seen how to determine an equation for a line passing through any two distinct points (x_1,y_1) and (x_2,y_2). In many practical applications, the goal is to do just the opposite to the greatest extent possible. That is, given a number of points $(x_1,y_1),(x_2,y_2),\ldots,(x_n,y_n)$, we would like to find the equation $y=mx+b$ whose graph comes closest to "fitting" the points.

An example will best illustrate this goal. Figure 5 contains a plot of the growth of a new financial advising firm, with the vertical axis representing the number of clients acquired and the horizontal axis representing the number of weeks after the firm's opening. The actual data points are listed in Table 1.

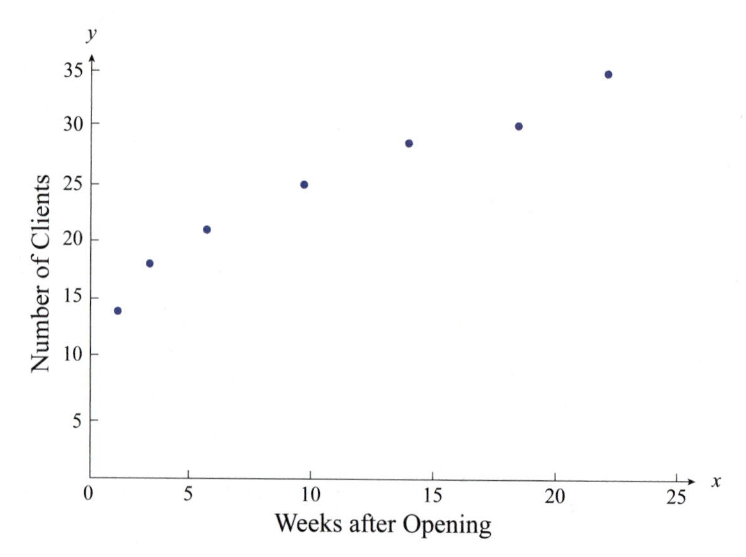

FIGURE 5: Growth of Number of Clients

x Weeks after Opening	y Number of Clients
1	14
3	18
6	21
10	25
14	28
18	30
22	35

TABLE 1

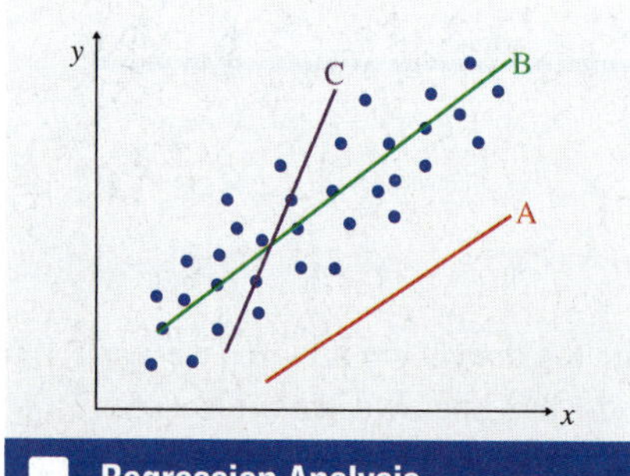

Regression Analysis

Linear regression is just one tool in a larger set known collectively as *regression analysis.* In broad terms, regression analysis is the art of approximating relationships between variables, whether that relationship is linear, as in the case of linear regression, or of some other form. Various methods exist for finding the relationships, but one of the most common is the least-squares method used in this text.

The owner of the business would like to be able to make a projection about the number of customers in the near future, based on the assumption that the growth can be approximately modeled by a straight line. A crude method would be to "eyeball" a straight line—that is, to literally draw a straight line on the graph coming as close as possible to the given data points. If the line were extended far enough to the right, a very approximate guess for the number of clients obtained by a particular week could be made.

This method's disadvantage is its imprecision. Different people are likely to draw different straight lines, and there is no resultant equation with which to work. Both of these flaws can be avoided by using **linear regression** to find the line that most closely fits given data.

Linear regression gives us the slope-intercept form for the line whose graph minimizes the sum of the squares of the deviations between the line and the actual data points. For this reason, the specific process we use is known as the **least-squares method**. The formulas in the method are easy enough to state, but their derivation is best left to a course in either calculus or linear algebra. We will illustrate the steps in the method by finding the best-fitting line for the seven data points in Figure 5 (with intermediate calculations rounded to six decimal places and final calculations rounded to two decimal places).

Step 1: Calculate $\bar{x}$ and $\bar{y}$.

We use $\bar{x}$ and $\bar{y}$ to denote the averages (or *means*) of the x-values and y-values, respectively, of all the given data points. For our example,

$$\bar{x} = \frac{1+3+6+10+14+18+22}{7} \approx 10.571429$$

and

$$\bar{y} = \frac{14+18+21+25+28+30+35}{7} \approx 24.428571.$$

Step 2: Calculate the values of Δx and Δy.

For each x-value there is a corresponding Δx, which represents the difference between the x-value and $\bar{x}$. Similarly, each y-value gives rise to a Δy. The Greek letter Δ ("delta") is often used to indicate difference, and these particular differences are read as "delta x" and "delta y." Tables 2 and 3 contain these calculations.

x	$\Delta x = x - \bar{x}$
1	−9.571429
3	−7.571429
6	−4.571429
10	−0.571429
14	3.428571
18	7.428571
22	11.428571

TABLE 2

y	$\Delta y = y - \bar{y}$
14	−10.428571
18	−6.428571
21	−3.428571
25	0.571429
28	3.571429
30	5.571429
35	10.571429

TABLE 3

Step 3: Calculate $\sum \Delta x \Delta y$ and $\sum (\Delta x)^2$.

The Greek letter $\sum$ ("sigma") is often used to indicate a sum of terms; you will encounter the symbol again and learn more about its use in Section 12.1. In this context, $\sum \Delta x \Delta y$ is just a convenient way to indicate that corresponding Δx's and Δy's from Tables 2 and 3 should be multiplied together, and the resulting products added. Similarly, $\sum (\Delta x)^2$ represents the sum of the squares of all the Δx's. Tables 4 and 5 show the computations of each quantity.

Δx	Δy	$\Delta x \Delta y$
−9.571429	−10.428571	99.816327
−7.571429	−6.428571	48.673469
−4.571429	−3.428571	15.673469
−0.571429	0.571429	−0.326531
3.428571	3.571429	12.244898
7.428571	5.571429	41.387756
11.428571	10.571429	120.816327
		$\sum \Delta x \Delta y = 338.285715$

TABLE 4

Δx	$(\Delta x)^2$
−9.571429	91.612253
−7.571429	57.326537
−4.571429	20.897963
−0.571429	0.326531
3.428571	11.755099
7.428571	55.183667
11.428571	130.612235
	$\sum (\Delta x)^2 = 367.714285$

TABLE 5

Step 4: Calculate the slope m and y-intercept b for the linear regression line of best fit.

The remaining tasks are easy. The slope for the linear regression line is

$$m = \frac{\sum \Delta x \Delta y}{\sum (\Delta x)^2} = \frac{338.285715}{367.714285} \approx 0.919969 \approx 0.920$$

and the y-intercept is

$$b = \overline{y} - m\overline{x} = 24.428571 - (0.919969)(10.571429) \approx 14.7.$$

So the equation for the linear regression line fitting the data in this example is $y = 0.920x + 14.7$. Figure 6 contains the original data points and the graph of the line we have just found.

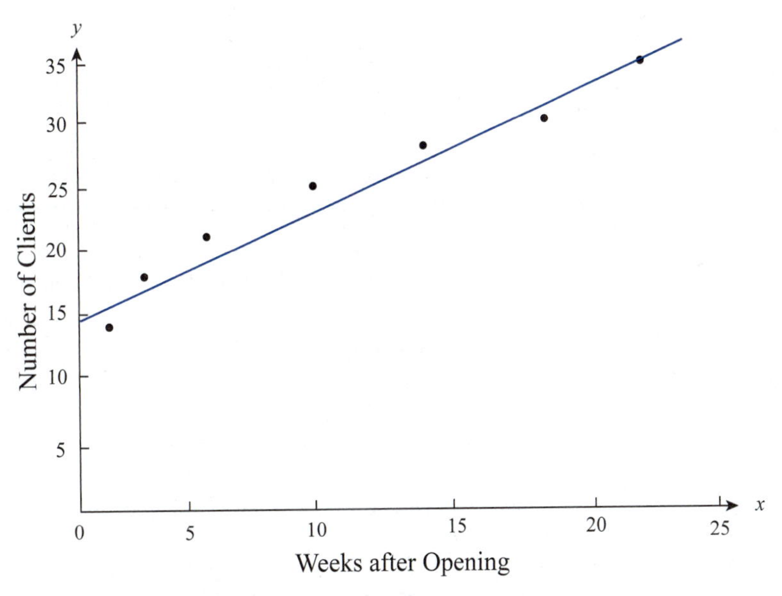

FIGURE 6: Linear Regression Line and Data

A concept closely related to linear regression is that of *correlation*. Given a collection of points $(x_1, y_1), (x_2, y_2), \ldots, (x_n, y_n)$, the question of whether the collection shows a linear dependence of y on x can be asked. The **Pearson correlation coefficient r** is a number that allows us to answer this question objectively, and its computation is similar to the computations used in linear regression. With the linear regression notation just introduced, the formula for the Pearson correlation coefficient is

$$r = \frac{\sum \Delta x \Delta y}{\sqrt{\sum (\Delta x)^2} \sqrt{\sum (\Delta y)^2}}.$$

The number r always satisfies $-1 \le r \le 1$, with $r = 0$ indicating no linear dependence of y on x and a value of $|r|$ close to 1 indicating a strong linear dependence. The sign of r is also meaningful, as a positive r corresponds to a positive slope in the line of best fit and a negative r corresponds to a negative slope in the line of best fit.

Continuing our analysis of the weekly growth of clients given the seven data points in Table 1, we already know

$$\sum \Delta x \Delta y = 338.285715 \quad \text{and} \quad \sum (\Delta x)^2 = 367.714285,$$

and a similar computation gives us the following.

$$\sum (\Delta y)^2 = 317.714285$$

Thus, the correlation coefficient for this data is

$$\begin{aligned} r &= \frac{\sum \Delta x \Delta y}{\sqrt{\sum (\Delta x)^2} \sqrt{\sum (\Delta y)^2}} \\ &= \frac{338.285715}{\sqrt{367.714285}\sqrt{317.714285}} \approx 0.990. \end{aligned}$$

This value close to 1 indicates a very strong (positive) linear dependence of y on x, which means that the line of best fit for the data that we previously found is a good means of predicting future growth.

Example 3: Finding the Line of Best Fit and Correlation Coefficient

Given the points graphed in Figure 7, **a.** use linear regression to find and graph the line of best fit, and **b.** find the Pearson correlation coefficient r.

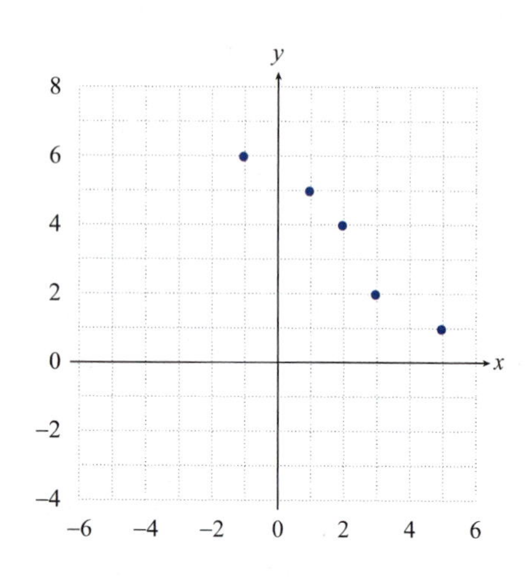

FIGURE 7

Solution

The coordinates of the five graphed points are

$$\{(-1,6),(1,5),(2,4),(3,2),(5,1)\}.$$

The averages of the x- and y-values are thus

$$\bar{x}=\frac{-1+1+2+3+5}{5}=2 \quad \text{and} \quad \bar{y}=\frac{6+5+4+2+1}{5}=3.6,$$

which we use to construct tables of the values of Δx and Δy.

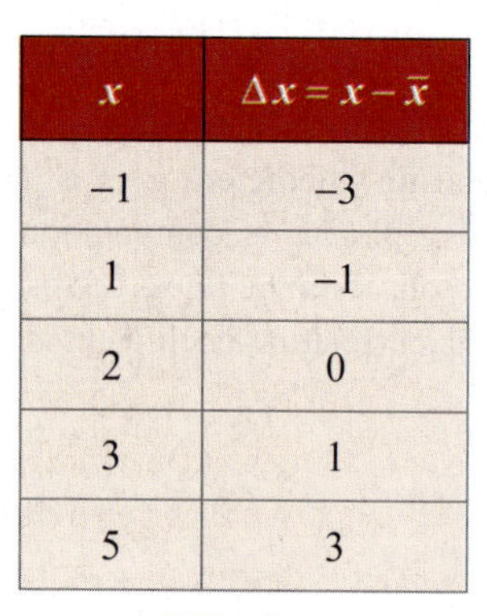

x	$\Delta x = x - \bar{x}$
−1	−3
1	−1
2	0
3	1
5	3

TABLE 6

y	$\Delta y = y - \bar{y}$
6	2.4
5	1.4
4	0.4
2	−1.6
1	−2.6

TABLE 7

We will need to know $\sum(\Delta x)^2$, $\sum \Delta x \Delta y$, and $\sum(\Delta y)^2$, so we make a third table leading to these sums.

Δx	Δy	$(\Delta x)^2$	$\Delta x \Delta y$	$(\Delta y)^2$
−3	2.4	9	−7.2	5.76
−1	1.4	1	−1.4	1.96
0	0.4	0	0	0.16
1	−1.6	1	−1.6	2.56
3	−2.6	9	−7.8	6.76
		$\sum(\Delta x)^2 = 20$	$\sum \Delta x \Delta y = -18$	$\sum(\Delta y)^2 = 17.2$

TABLE 8

a. We have all we need now to determine that

$$m = \frac{\sum \Delta x \Delta y}{\sum (\Delta x)^2} = \frac{-18}{20} = -0.9$$

and

$$b = \bar{y} - m\bar{x} = 3.6 - (-0.9)(2) = 5.4.$$

So, the line of best fit is $y = -0.9x + 5.4$. Figure 8 shows the graph of this line along with the given points.

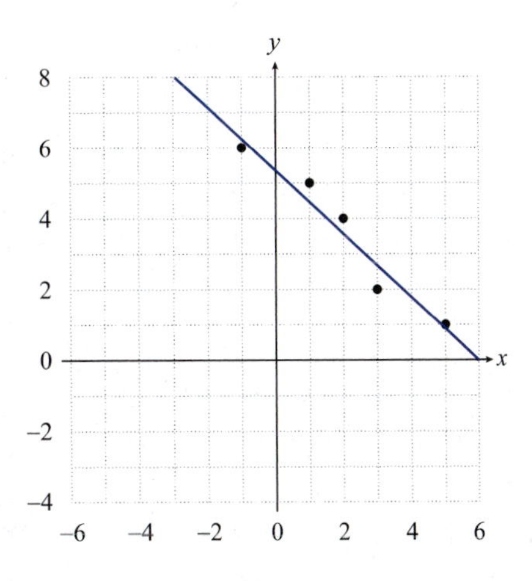

FIGURE 8

b. We also have all we need to determine the Pearson correlation coefficient r.

$$\begin{aligned} r &= \frac{\sum \Delta x \Delta y}{\sqrt{\sum (\Delta x)^2}\sqrt{\sum (\Delta y)^2}} \\ &= \frac{-18}{\sqrt{20}\sqrt{17.2}} \\ &\approx -0.970 \end{aligned}$$

The fact that r is negative reflects the fact that the slope of the best-fitting line is negative, and the fact that r is relatively close to 1 in magnitude means that this line of best fit is a reasonably good model of the behavior of the points.

Our next example illustrates a case in which a line is *not* a good model of the behavior of the given data.

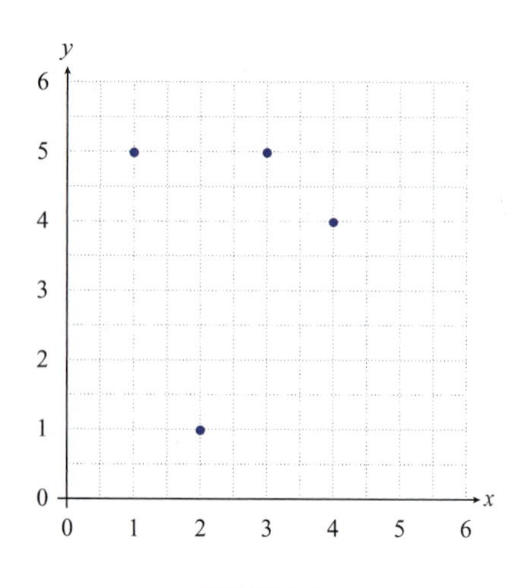

FIGURE 9

Example 4: Finding the Line of Best Fit and Correlation Coefficient

Given the points graphed in Figure 9, **a.** use linear regression to find and graph the line of best fit, and **b.** find the Pearson correlation coefficient r.

Solution

The coordinates of the four graphed points are

$$\{(1,5),(2,1),(3,5),(4,4)\}.$$

The averages of the x- and y-values are thus

$$\bar{x}=\frac{1+2+3+4}{4}=2.5 \quad \text{and} \quad \bar{y}=\frac{5+1+5+4}{4}=3.75,$$

which we use to construct tables of the values of Δx and Δy.

x	$\Delta x = x-\bar{x}$
1	−1.5
2	−0.5
3	0.5
4	1.5

TABLE 9

y	$\Delta y = y-\bar{y}$
5	1.25
1	−2.75
5	1.25
4	0.25

TABLE 10

We will need to know $\sum(\Delta x)^2$, $\sum \Delta x \Delta y$, and $\sum(\Delta y)^2$, so we make a third table leading to these sums.

Δx	Δy	$(\Delta x)^2$	$\Delta x\Delta y$	$(\Delta y)^2$
−1.5	1.25	2.25	−1.875	1.5625
−0.5	−2.75	0.25	1.375	7.5625
0.5	1.25	0.25	0.625	1.5625
1.5	0.25	2.25	0.375	0.0625
		$\sum(\Delta x)^2=5$	$\sum \Delta x\Delta y=0.5$	$\sum(\Delta y)^2=10.75$

TABLE 11

a. We have all we need now to determine that

$$m=\frac{\sum \Delta x\Delta y}{\sum(\Delta x)^2}=\frac{0.5}{5}=0.1$$

and

$$b=\bar{y}-m\bar{x}=3.75-(0.1)(2.5)=3.5.$$

So, the line of best fit is $y = 0.1x + 3.5$. Figure 10 shows the graph of this line along with the given points.

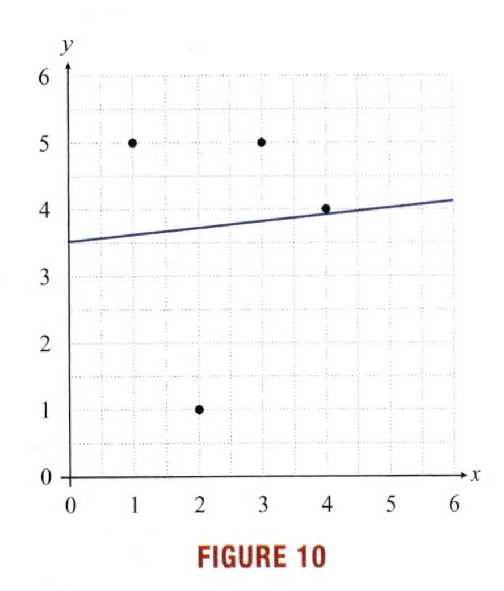

FIGURE 10

b. We also have all we need to determine the Pearson correlation coefficient r.

$$
\begin{aligned}
r &= \frac{\sum \Delta x \Delta y}{\sqrt{\sum (\Delta x)^2}\sqrt{\sum (\Delta y)^2}} \\
&= \frac{0.5}{\sqrt{5}\sqrt{10.75}} \\
&\approx 0.068
\end{aligned}
$$

The fact that r is positive reflects the fact that the slope of the best-fitting line is positive, but the fact that r is relatively close to 0 in magnitude means that this line is a poor model of the behavior of the data.

TECHNOLOGY: Graphing a Line of Best Fit

Graphing utilities can also be used to find a line of best fit, and can be used to check your understanding of the material in this chapter. Using a TI-84 Plus, you can find a line of best fit through the following steps.

First, you need to ensure DiagnosticOn is set in order to calculate the Pearson correlation coefficient r. Press 2nd and Ø to access the CATALOG. Scroll down to DiagnosticOn and press enter to paste the command into the home screen. Press enter again to set DiagnosticOn.

Now, press stat and select Edit. To clear any preexisting lists, move your cursor over the title bar and press clear, then enter. Enter your data vertically by variable (see Figure 11), then press 2nd and Y=.

NORMAL FLOAT AUTO REAL RADIAN MP

L1	L2	L3	L4	L5
1	14	------	------	------
3	18			
6	21			
10	25			
14	28			
18	30			
22	35			
------	------			

L2(7)=35

FIGURE 11

TECHNOLOGY: Graphing a Line of Best Fit (continued)

Select `1:` to get the screen for `Plot1`. Highlight the appropriate options, as shown in Figure 12. Make sure the names for `Xlist` and `Ylist` match the names of the data lists you created. To change the names shown, press `2nd` and `stat`, then select the appropriate name from the list. To plot the points, press `zoom` and select `ZoomStat`.

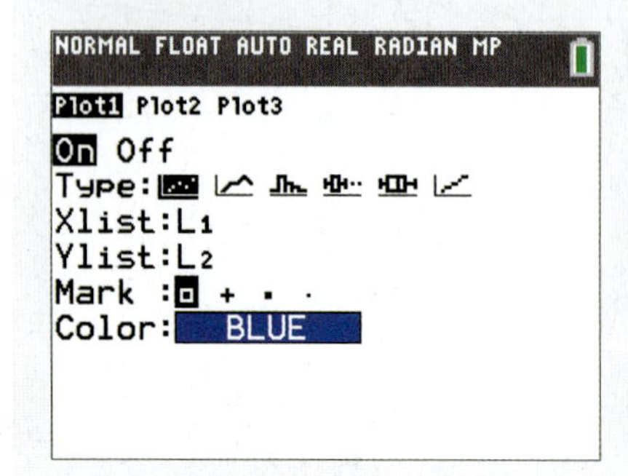

FIGURE 12

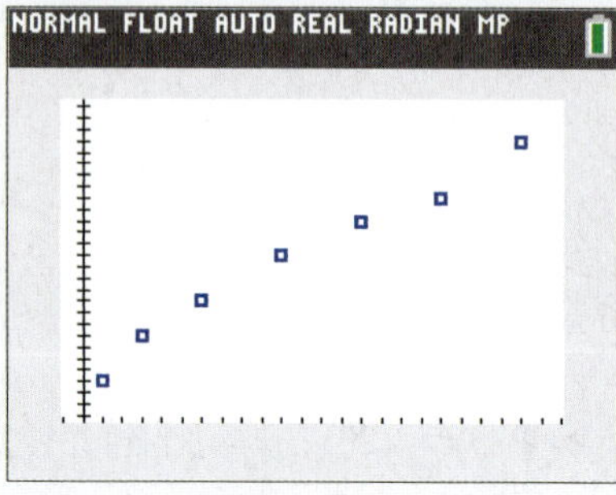

FIGURE 13

Next find and graph the equation of the line of best fit. The steps described to do so assume that the `STAT WIZARDS` feature is turned off. If your TI-84 Plus has this feature on, you will turn it off by pressing `mode`. Scroll down to `STAT WIZARDS` and select `OFF`.

Press `stat`, then scroll to the right to the `CALC` menu and select `LinReg(ax+b)`. This pastes the regression command into the home screen. Now, use `2nd` and `stat` to paste your list names into the home screen, separated by a `,`. Place another `,` and then press `vars`, scroll right to the `Y-VARS` menu, and select `Function`, then `Y1`. Press `enter` and you will see the values of the slope a and y-intercept b to plug into $y = ax + b$ to get the line of best fit (shown in Figure 14).

Now press `zoom` and select `ZoomStat` to see the data and regression line (shown in Figure 15).

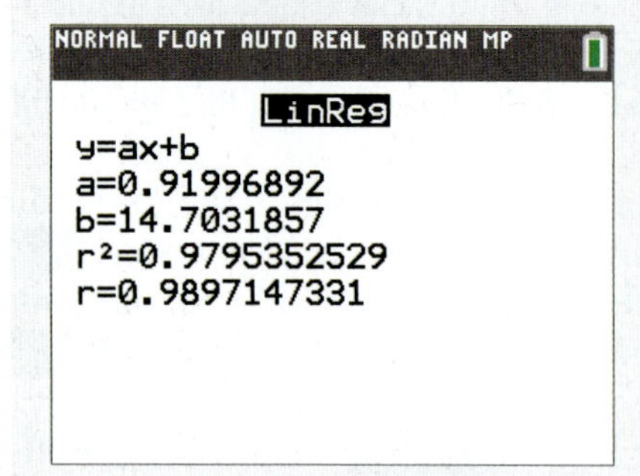

FIGURE 14

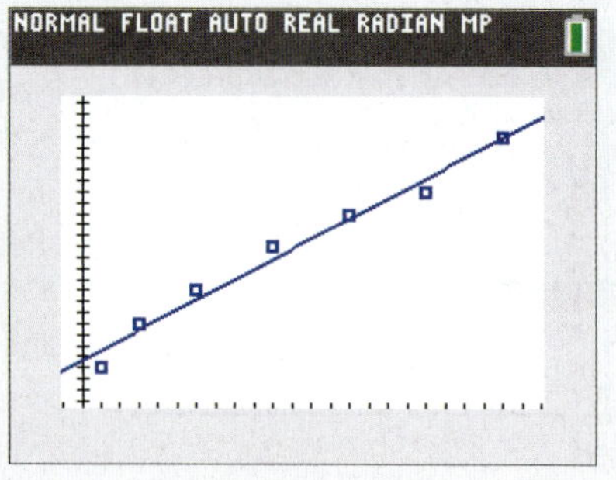

FIGURE 15

3.2 EXERCISES

PRACTICE

Graph the following linear functions. See Example 1.

1. $f(x)=-5x+2$
2. $g(x)=\frac{3x-2}{4}$
3. $h(x)=-x+2$
4. $p(x)=-2$
5. $g(x)=3-2x$
6. $r(x)=2-\frac{x}{5}$
7. $f(x)=-2(1-x)$
8. $a(x)=3\left(1-\frac{1}{3}x\right)+x$
9. $f(x)=2-4x$
10. $g(x)=\frac{2x-8}{4}$
11. $h(x)=5x-10$
12. $k(x)=3x-\frac{2+6x}{2}$
13. $m(x)=\frac{-x+25}{10}$
14. $q(x)=1.5x-1$
15. $w(x)=(x-2)-(2+x)$

Match the following functions with their graphs.

16. $f(x)=(8x-14)-(-17+2x)$
17. $f(x)=3x-\frac{7+8x}{3}$
18. $f(x)=\frac{6}{2}-\frac{2}{8}x$
19. 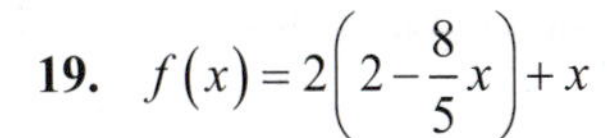 $f(x)=2\left(2-\frac{8}{5}x\right)+x$

a.

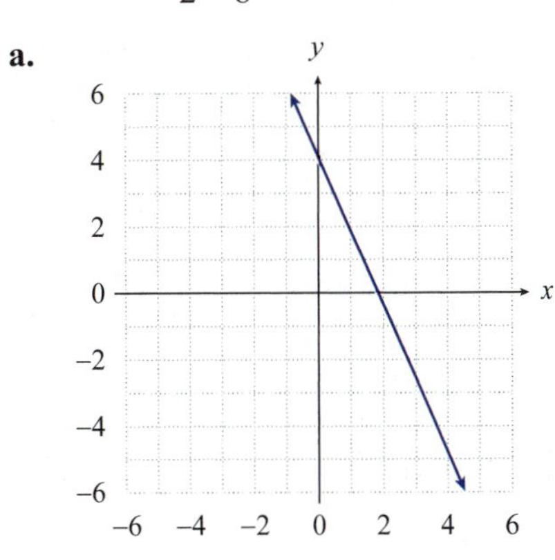

b.

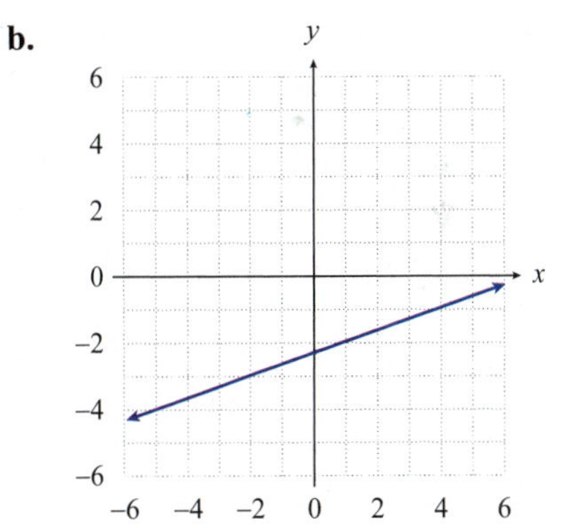

c.

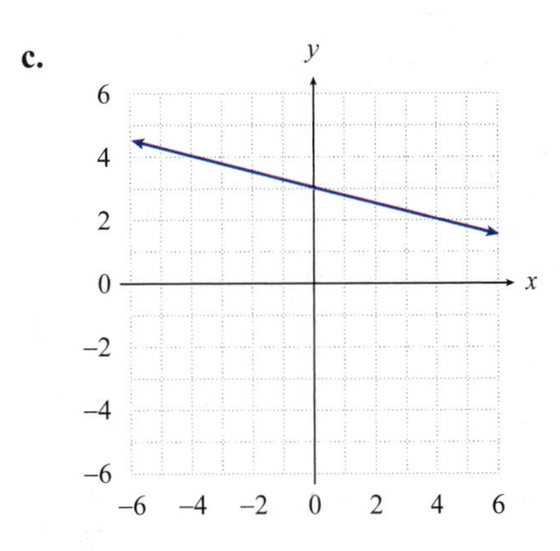

d.

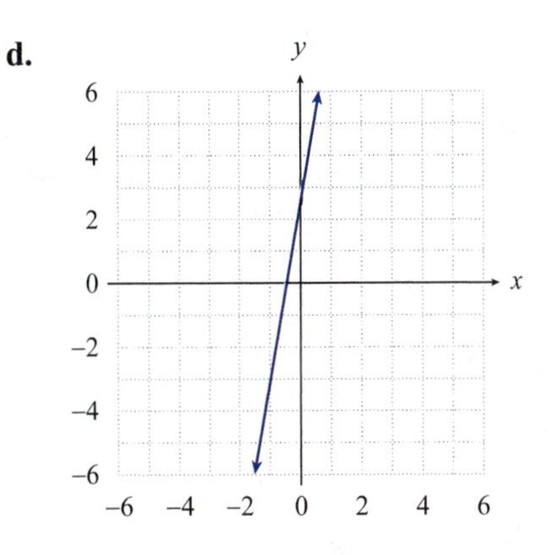

Find a formula for the linear function depicted in each of the following graphs. See Example 2.

20.

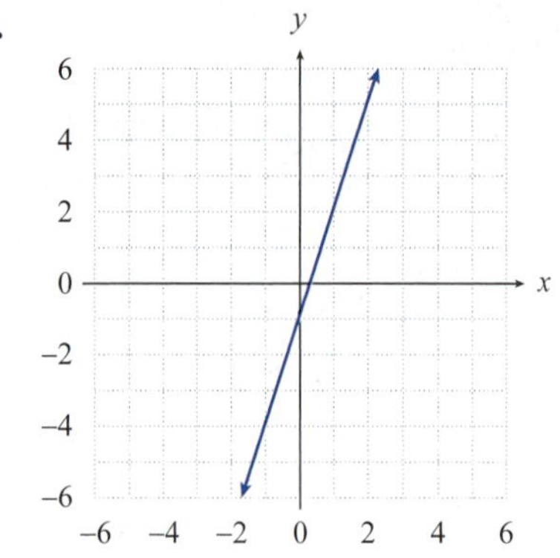

21.

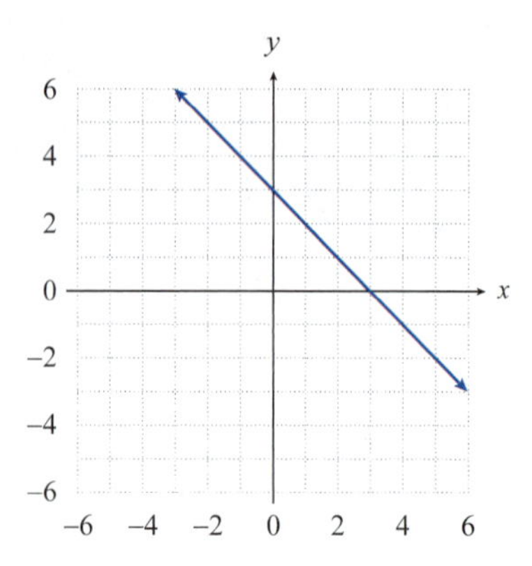

22.

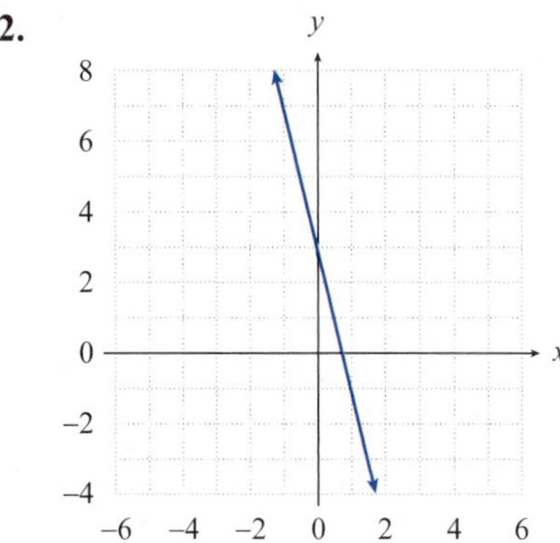

23.

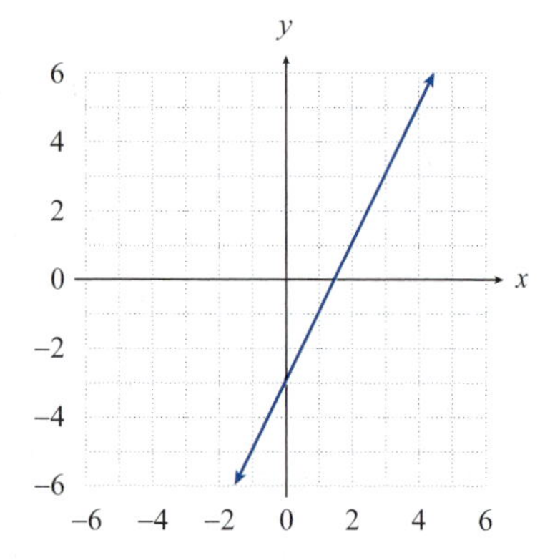

For the given points, **a.** use linear regression to find and graph the line of best fit along with the points, and **b.** find the Pearson correlation coefficient r. See Examples 3 and 4.

24. $(0,3),(1,5),(2,7),(3,8)(5,9),(6,9)$

25. $(1,10),(2,8),(3,7),(4,6),(5,5),(6,5),(7,4)$

26. $(1,9.6),(2,8.7),(3,7.7),(4,6.1),(5,5.0)$

27. $(1,5.2),(2,6.4),(3,8.1),(4,9.2),(5,10.6)$

28. $(1,5),(2,-1),(3,5),(4,0),(5,4)$

29. $(-2,2),(-1,2),(0,-1),(1,-1),(2,0)$

APPLICATIONS

30. An automobile company is running a new television commercial in five cities with approximately the same population. The following table shows the number of times the commercial is run on TV in each city and the number of car sales (in hundreds).

a. Find the linear regression line for the data given in the table.
b. Graph the data and the regression line on the same set of axes.
c. Find the Pearson correlation coefficient r.

Number of TV Commercials (x)	4	5	12	16	18
Car Sales (in hundreds) (y)	3	5	5	8	7

31. The following table shows the high school grade-point averages (HS-GPA) and the college grade-point averages (C-GPA) after 1 year of college for 10 students.

a. Find the linear regression line for the data given in the table.
b. Graph the data and the regression line on the same set of axes.
c. Find the Pearson correlation coefficient r.

HS-GPA (x)	2.0	2.0	2.2	2.2	2.7	3.2	3.2	3.3	3.5	3.7
C-GPA (y)	1.5	1.8	2.0	1.5	2.0	2.8	3.0	3.5	3.5	3.4

32. Keisha makes refrigerator magnets and has just started selling them along with other handcrafted items at craft shows. After four shows, she has collected the data below regarding price per magnet versus number of magnets sold.

a. Find the linear regression line for the data given in the table.
b. Graph the data and the regression line on the same set of axes.
c. Find the Pearson correlation coefficient r.

Number of Magnets Sold (x)	1	3	5	8
Price per Magnet (y)	\$14	\$11	\$10	\$6

WRITING & THINKING

33. The data in a set A has a Pearson correlation coefficient of 0.98 and the data in a set B has a Pearson correlation coefficient of –0.98. Which data set has the strongest correlation? Explain.

34. What does the sign of the Pearson correlation coefficient tell you about the line of best fit?

3.3 QUADRATIC FUNCTIONS

TOPICS

1. Quadratic functions and their graphs
2. Quadratic regression
3. Maximization/minimization problems

TOPIC 1: Quadratic Functions and Their Graphs

In Section 1.8, we learned how to solve quadratic equations in one variable. We will now study quadratic *functions* of one variable and relate this new material to what we already know.

DEFINITION: Quadratic Function

A **quadratic function** f in the variable x, also known as a **second-degree polynomial function**, is any function that can be written in the form $f(x) = ax^2 + bx + c$, where a, b, and c are real numbers and $a \neq 0$.

The graph of any quadratic function is a roughly U-shaped curve known as a **parabola**. We will study parabolas in more generality in Chapter 10, but in this section we will learn how to graph parabolas as they arise in the context of quadratic functions.

The graph in Figure 1 is an example of a parabola. Specifically, it is the graph of the quadratic function $f(x) = x^2$, and the table contains a few of the ordered pairs on the graph.

x	$f(x)$
-3	9
-1	1
0	0
2	4

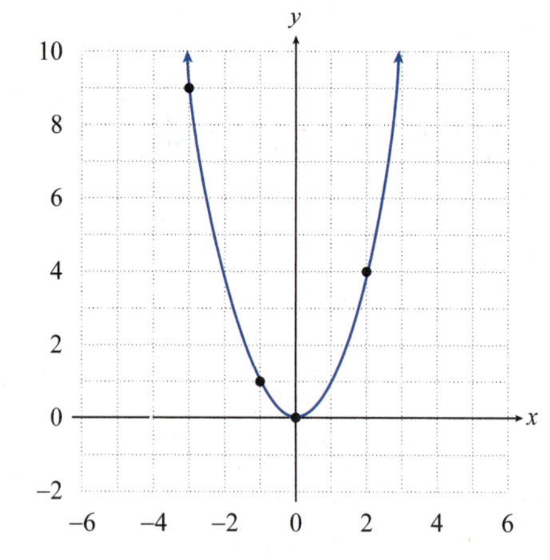

FIGURE 1: $f(x) = x^2$

As Figure 1 demonstrates, one key characteristic of a parabola is that it has one (and only one) point at which the graph "changes direction." This point is known as the **vertex**; scanning the plane from left to right, it is the point on a parabola where the graph stops going down and begins to go up (if the parabola opens upward) or stops going up and begins to go down (if the parabola opens downward). Another characteristic of a parabola is that it is symmetric with respect to its **axis**, a straight line passing through the vertex and dividing the parabola into two halves that are mirror images of each other. Every parabola that represents the graph of a quadratic function has a vertical axis, but we will see parabolas later in the text that have nonvertical axes. Finally, parabolas can be relatively skinny or relatively broad, meaning that the curve of the parabola at the vertex can range from very sharp to very flat.

Our knowledge of linear equations gives us a very satisfactory method for sketching the graphs of linear functions. Our goal now is to develop a method that allows us to sketch the graphs of quadratic functions. We *could*, given a quadratic function, simply plot a number of points and then connect them with a U-shaped curve, but this method has several drawbacks, chief among them being that it is slow and not very accurate. Ideally, we will develop a method that is fast and that pinpoints some of the important characteristics of a parabola (such as the location of its vertex) exactly.

We will develop our graphing method by working from the answer backward. We will first see what effects various mathematical operations have on the graphs of parabolas and then see how this knowledge lets us graph a general quadratic function.

To begin, the graph of the function $f(x)=x^2$, shown in Figure 1, is the basic parabola. We already know its characteristics: its vertex is at the origin, its axis is the *y*-axis, it opens upward, and the sharpness of the curve at its vertex will serve as a convenient reference when discussing other parabolas.

Now consider the function $g(x)=(x-3)^2$, obtained by replacing x in the formula for f with $x-3$. We know x^2 is equal to 0 when $x=0$. What value of x results in $(x-3)^2$ equaling 0? The answer is $x=3$. In other words, the point $(0,0)$ on the graph of f corresponds to the point $(3,0)$ on the graph of g. With this in mind, examine the table and graph in Figure 2.

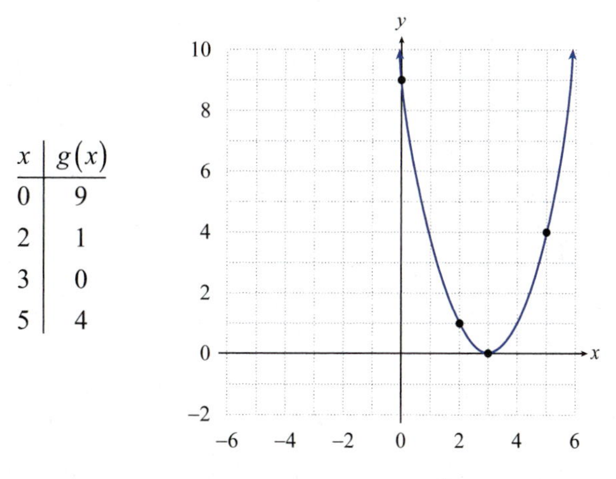

x	$g(x)$
0	9
2	1
3	0
5	4

FIGURE 2: $g(x)=(x-3)^2$

Notice that the shape of the graph of g is identical to that of f, but it has been shifted over to the right by 3 units. This is our first example of how we can manipulate graphs of functions, a topic we will fully explore in Section 4.1.

Now consider the function h obtained by replacing the x in x^2 with $x + 7$. As with the functions f and g, $h(x)=(x+7)^2$ is nonnegative for all values of x, and only one value for x will return a value of 0: $h(-7)=0$. Compare the table and graph in Figure 3 with those in Figures 1 and 2.

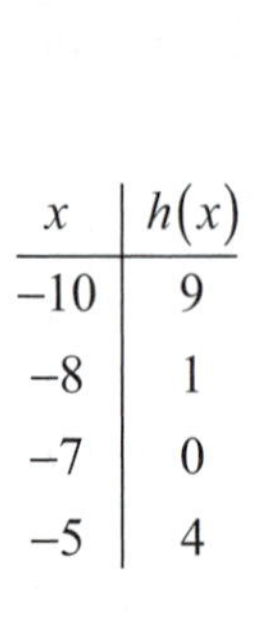

x	$h(x)$
−10	9
−8	1
−7	0
−5	4

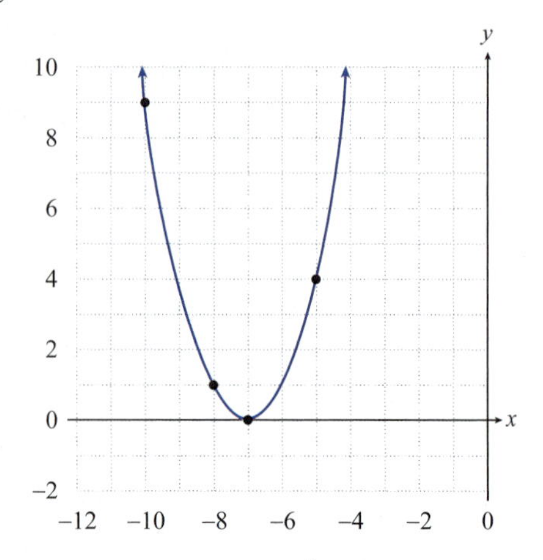

FIGURE 3: $h(x)=(x+7)^2$

So we have seen how to shift the basic parabola to the left and right: the graph of $g(x)=(x-h)^2$ has the same shape as the graph of $f(x)=x^2$, but it is shifted h units to the right if h is positive and h units to the left if h is negative.

How do we shift a parabola up and down? To move the graph of $f(x)=x^2$ up by a fixed number of units, we need to add that number of units to the second coordinate of each ordered pair. Similarly, to move the graph down we subtract the desired number of units from each second coordinate. To see this, consider the table and graphs for the two functions $j(x)=x^2+5$ and $k(x)=x^2-2$ in Figure 4.

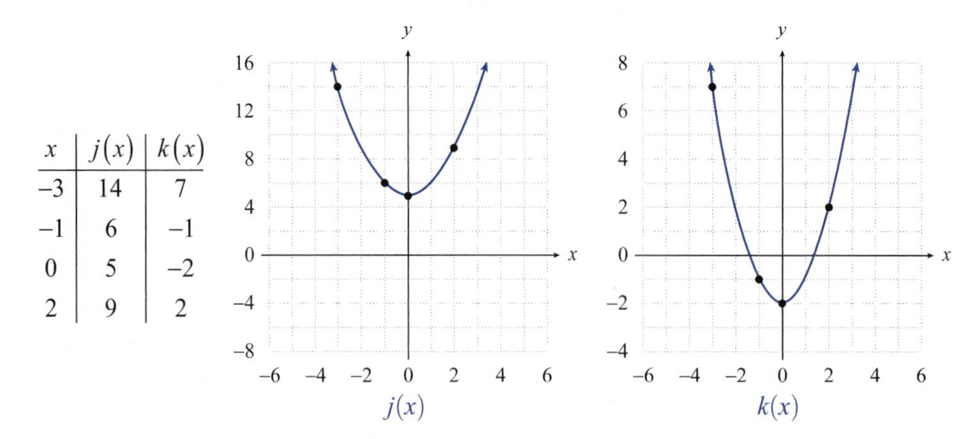

x	$j(x)$	$k(x)$
−3	14	7
−1	6	−1
0	5	−2
2	9	2

FIGURE 4: $j(x)=x^2+5$ and $k(x)=x^2-2$

Finally, how do we make a parabola skinnier or broader? To make the basic parabola skinnier (to make the curve at the vertex sharper), we need to stretch the graph vertically. We can do this by multiplying the formula x^2 by a constant a greater than 1 to obtain the formula ax^2. Multiplying the formula x^2 by a constant a that lies between 0 and 1 makes the parabola broader (it makes the curve at the vertex flatter). Finally, multiplying x^2 by a negative constant a turns all of the nonnegative outputs of f into nonpositive outputs, resulting in a parabola that opens downward instead of upward.

Compare the graphs of $l(x)=6x^2$ and $m(x)=-\frac{1}{2}x^2$ in Figure 5 to the basic parabola $f(x)=x^2$.

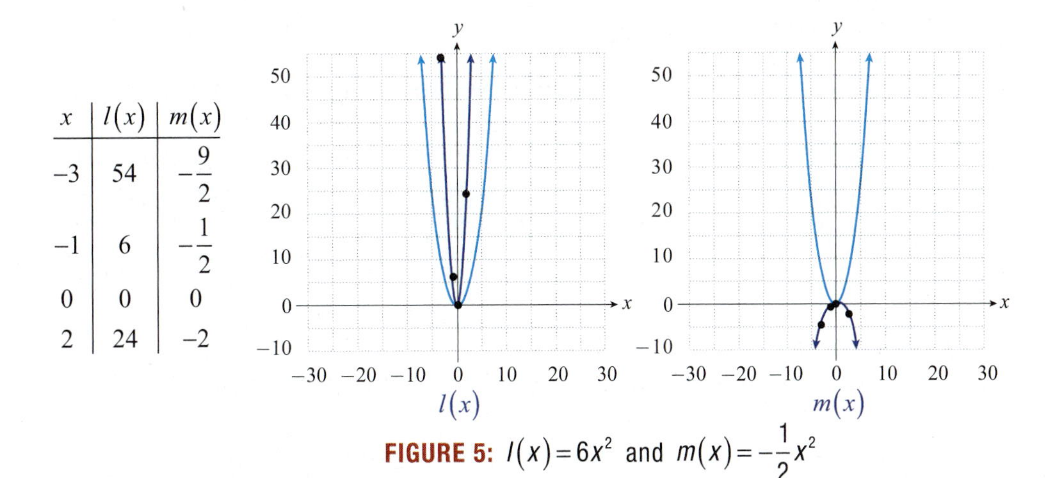

x	$l(x)$	$m(x)$
-3	54	$-\frac{9}{2}$
-1	6	$-\frac{1}{2}$
0	0	0
2	24	-2

FIGURE 5: $l(x)=6x^2$ and $m(x)=-\frac{1}{2}x^2$

The following form of a quadratic function brings together all of the above ways of altering the basic parabola $f(x)=x^2$.

DEFINITION: Vertex Form of a Quadratic Function

The graph of the function $g(x)=a(x-h)^2+k$, where a, h, and k are real numbers and $a\neq 0$, is a parabola whose vertex is at (h,k). The parabola is narrower than $f(x)=x^2$ if $|a|>1$ and is broader than $f(x)=x^2$ if $0<|a|<1$. The parabola opens upward if a is positive and downward if a is negative.

Where does this definition come from? Consider this construction of the vertex form.

$f(x)=x^2$	Begin with the basic parabola, with vertex $(0,0)$.
$g(x)=(x-h)^2$	This represents a horizontal shift of h.
$g(x)=(x-h)^2+k$	This step adds a vertical shift of k, making the new vertex (h,k).
$g(x)=a(x-h)^2+k$	Finally, apply the stretch/compress factor of a. If a is negative, the parabola opens downward.

The question now is this: given a quadratic function $f(x)=ax^2+bx+c$, how do we determine the location of its vertex, whether it opens upward or downward, and whether it is skinnier or broader than the basic parabola? All of this information is available if the equation is in vertex form, so we need to convert the formula ax^2+bx+c into the form $a(x-h)^2+k$. It turns out that we can *always* do this by completing the square on the first two terms of the expression.

NOTE

Finding and plotting the x-intercepts is a great way to see the shape of the function.

Example 1: Graphing Quadratic Functions

Sketch the graph of the function $f(x) = -x^2 - 2x + 3$. Locate the vertex and the x-intercepts.

Solution

First, identify the vertex of the function by completing the square as follows.

$f(x) = -x^2 - 2x + 3$

$= -(x^2 + 2x) + 3$ — First, factor out the leading coefficient of -1 from the first two terms.

$= -(x^2 + 2x + 1) + 1 + 3$ — Complete the square on the x^2 and $2x$ terms.

$= -(x+1)^2 + 4$ — Because of the -1 in front of the parentheses, this amounts to adding -1 to the function, so we compensate by adding 1 as well.

Completing the square places the equation in vertex form, and we rewrite the expression $-(x+1)^2 + 4$ as $-(x-(-1))^2 + 4$, so the vertex is $(-1, 4)$.

The instructions also ask us to identify the x-intercepts. An x-intercept of the function f is any point on the x-axis where $f(x) = 0$, so we need to solve the equation $-x^2 - 2x + 3 = 0$. This can be done by factoring.

$-x^2 - 2x + 3 = 0$ — First, divide each term by -1.

$x^2 + 2x - 3 = 0$

$(x+3)(x-1) = 0$ — Factor into two binomials.

$x = -3, 1$ — The Zero-Factor Property gives us the x-intercepts.

Therefore, the x-intercepts are located at $(-3, 0)$ and $(1, 0)$.

The vertex form of the function, $f(x) = -(x+1)^2 + 4$, tells us that this quadratic opens downward, has its vertex at $(-1, 4)$, and is neither skinnier nor broader than the basic parabola. We now also know that it crosses the x-axis at -3 and 1. Putting this all together, we obtain the following graph.

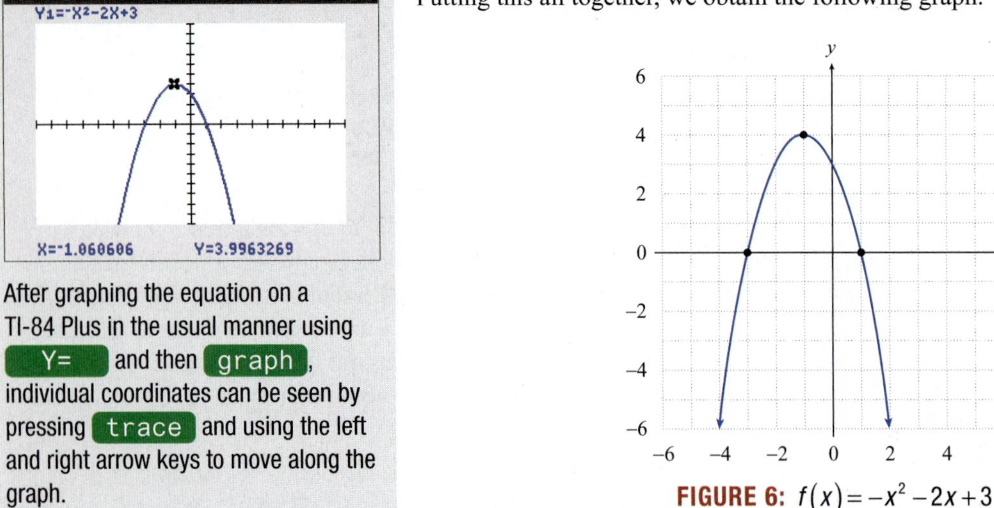

FIGURE 6: $f(x) = -x^2 - 2x + 3$

TECHNOLOGY

After graphing the equation on a TI-84 Plus in the usual manner using Y= and then graph, individual coordinates can be seen by pressing trace and using the left and right arrow keys to move along the graph.

In Section 1.8, we completed the square on the generic quadratic equation to develop the quadratic formula. We can use a similar approach to transform the standard form of a quadratic function into vertex form.

$$\begin{aligned} f(x) &= ax^2+bx+c \\ &= a\left(x^2+\frac{b}{a}x\right)+c \\ &= a\left(x^2+\frac{b}{a}x+\frac{b^2}{4a^2}\right)-a\left(\frac{b^2}{4a^2}\right)+c \\ &= a\left(x+\frac{b}{2a}\right)^2-\frac{b^2}{4a}+c \\ &= a\left(x+\frac{b}{2a}\right)^2+\frac{4ac-b^2}{4a} \end{aligned}$$

As always, begin by factoring the leading coefficient a from the first two terms.

To complete the square, add the square of half of $\frac{b}{a}$ inside the parentheses. We need to balance the equation by subtracting $a\left(\frac{b^2}{4a^2}\right)$ outside the parentheses, then simplify.

f(x) FORMULA: Vertex of a Quadratic Function

Given a quadratic function $f(x)=ax^2+bx+c$, the graph of f is a parabola with a vertex given by

$$\left(-\frac{b}{2a}, f\left(-\frac{b}{2a}\right)\right)=\left(-\frac{b}{2a}, \frac{4ac-b^2}{4a}\right).$$

Example 2: Using the Vertex Formula

Find the vertex of each of the following quadratic functions using the vertex formula.

a. $f(x)=x^2-4x+8$ **b.** $g(x)=3x^2+5x-1$

NOTE

If the x-coordinate of the vertex is simple, use substitution to find the y-coordinate.

If the x-coordinate is complicated, use the explicit formula (the right-hand form in the vertex formula).

Solution

a. Begin by using the formula to find the x-coordinate of the vertex.

$1x^2-4x+8$ Note that the value of a is 1.

$-\frac{b}{2a}=-\frac{-4}{2(1)}=2$ Substitute a and b into the formula and simplify.

At this point, we need to decide how to find the y-coordinate. Since the x-coordinate is an integer, substitute it directly into the original equation, finding $f\left(-\frac{b}{2a}\right)$.

$$\begin{aligned} f(2) &= 2^2-4(2)+8 \\ &= 4 \end{aligned}$$

Thus, the vertex of the graph of $f(x)$ is $(2,4)$.

b. Again, begin by finding the x-coordinate of the vertex.

$$3x^2+5x-1=0$$
$$-\frac{b}{2a}=-\frac{5}{2(3)}=-\frac{5}{6}$$

Substitute a and b into the formula and simplify.

Here, the x-coordinate is a fraction, so substituting it into the original equation leads to messy calculations. Instead, use the explicit formula to find the coordinate.

$$\begin{aligned}\frac{4ac-b^2}{4a}&=\frac{4(3)(-1)-(5)^2}{4(3)}\\&=\frac{-12-25}{12}\\&=-\frac{37}{12}\end{aligned}$$

Substitute a, b, and c into the formula and simplify.

Thus, the vertex of the graph of $g(x)$ is $\left(-\frac{5}{6},-\frac{37}{12}\right)$.

Example 3: Finding a Quadratic Function Given Its Graph

a. Find a formula for the quadratic function whose graph is given in Figure 7.

b. Use the formula to determine the coordinates of the parabola's vertex.

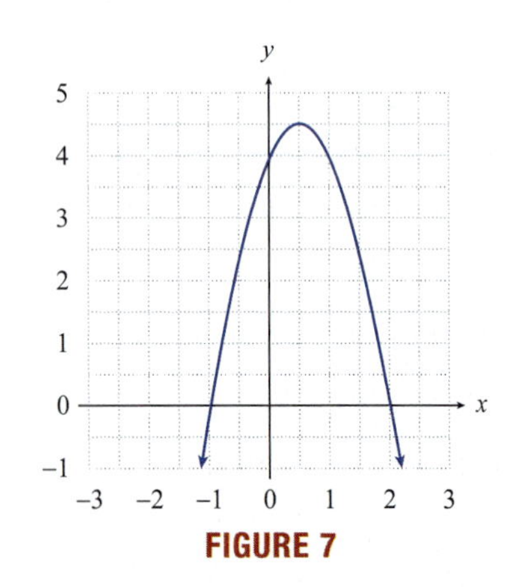

FIGURE 7

Solution

a. The graph indicates that the parabola has x-intercepts of -1 and 2, so we know that the quadratic function corresponding to this graph can be factored as follows.

$$f(x)=a(x+1)(x-2)$$

The fact that the parabola opens downward tells us that a must be negative, but we can determine it exactly by noting that $f(0)=4$. Substituting this into the above formula gives us an equation we can solve for a.

$$\begin{aligned}f(0)&=4\\a(0+1)(0-2)&=4\\-2a&=4\\a&=-2\end{aligned}$$

Hence, $f(x)=-2(x+1)(x-2)=-2x^2+2x+4$.

b. Given the quadratic function $f(x) = -2x^2 + 2x + 4$, we know the coefficients are $a = -2$, $b = 2$, and $c = 4$. We can now determine the coordinates of the vertex of the parabola as follows.

$$\left(-\frac{b}{2a}, \frac{4ac - b^2}{4a}\right) = \left(-\frac{2}{2(-2)}, \frac{4(-2)(4) - (2)^2}{4(-2)}\right)$$
$$= \left(\frac{1}{2}, \frac{9}{2}\right)$$

TOPIC 2: Quadratic Regression

In Section 3.2, we learned how to use the least-squares method to find the line that best fits a given set of points in the plane, a process called linear regression. If, instead of a line, we think that a parabola better reflects the behavior of the points, we can use the same least-squares method to find a quadratic function that best fits the points. With this goal in mind, the process is called **quadratic regression**; and, as you might expect, we will learn additional regression processes as we study additional classes of functions.

Just as technological aids are useful in linear regression, many graphing utilities can also determine quadratic regression curves. The specific commands vary with the technology.

Example 4: Quadratic Regression

Given the points graphed in Figure 8, use quadratic regression to find and graph the quadratic function of best fit.

TECHNOLOGY

To find and graph the quadratic function of best fit for a given set of points using a TI-84 Plus, perform the same steps as described in Section 3.2 for linear regression except select QuadReg from the CALC menu.

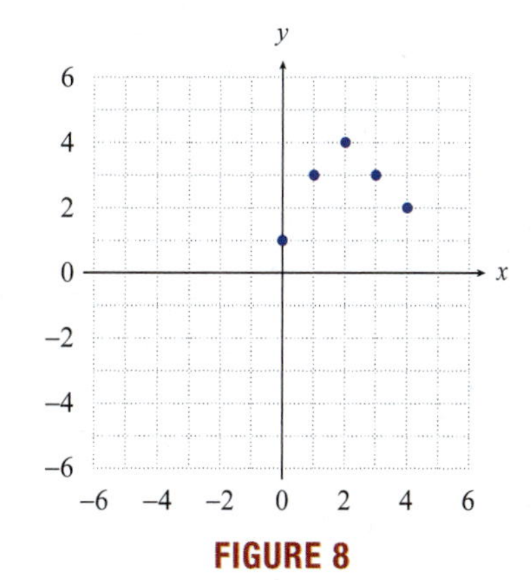

FIGURE 8

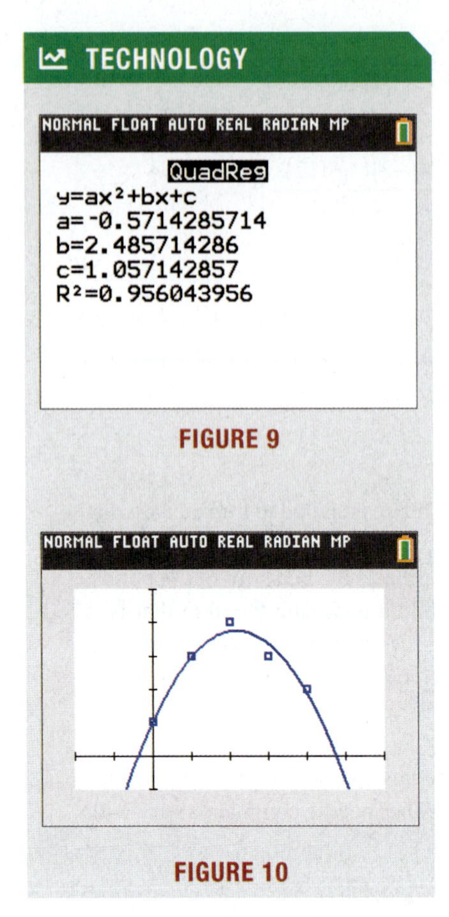

FIGURE 9

FIGURE 10

Solution

The coordinates of the five graphed points are $\{(0,1),(1,3),(2,4),(3,3),(4,2)\}$.

Using the least-squares method to find a quadratic function that best fits the points results in the following function.

$$f(x) = -0.571x^2 + 2.49x + 1.06$$

Figure 11 shows the graph of this function along with the given five points.

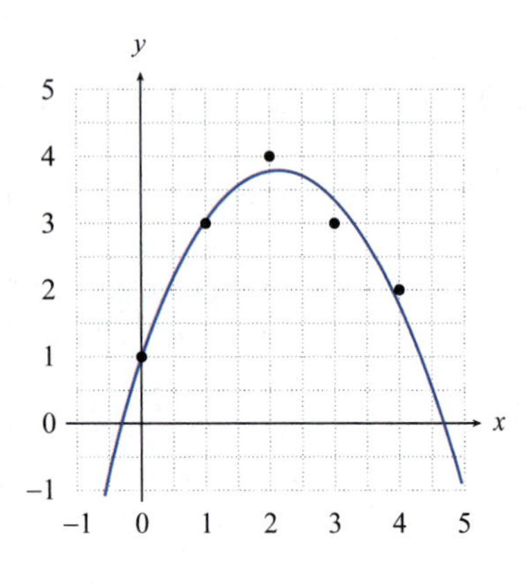

FIGURE 11

Example 5: Quadratic Regression

a. Find the quadratic function that best fits the points shown in Figure 12.

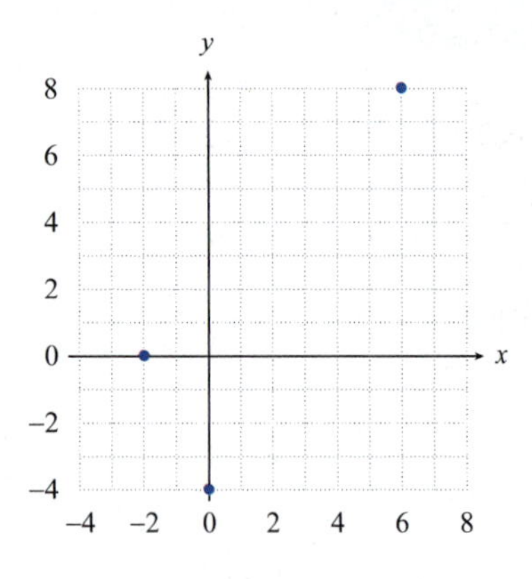

FIGURE 12

b. Use your result to determine the coordinates of the vertex of the best-fitting parabola.

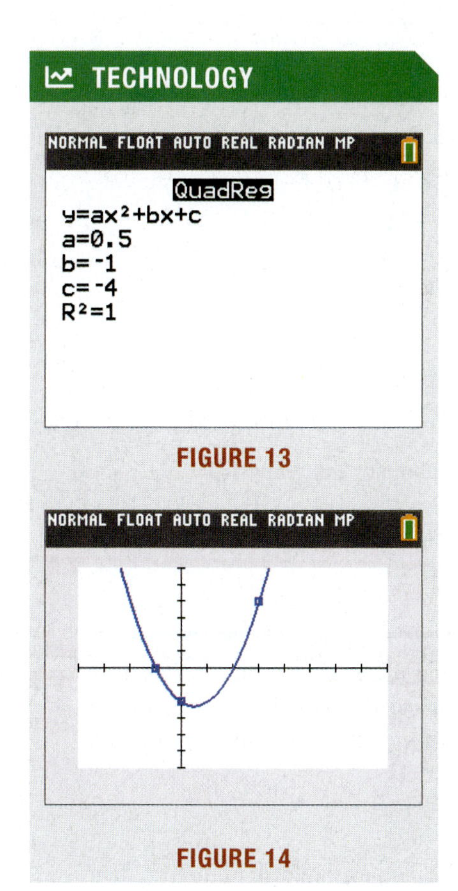

FIGURE 13

FIGURE 14

Solution

a. The coordinates of the three graphed points are $\{(-2,0),(0,-4),(6,8)\}$.

Using the least-squares method to find a quadratic function that best fits the points results in the function

$$f(x)=0.5x^2-x-4.$$

Figure 15 shows the graph of this function along with the given three points.

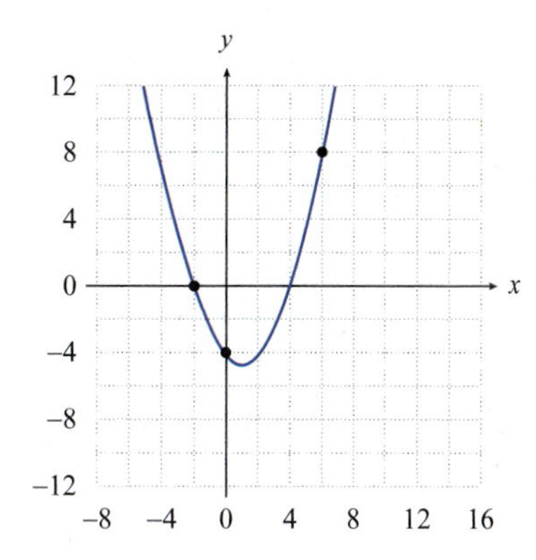

FIGURE 15

b. Given the quadratic function $f(x)=0.5x^2-x-4$, we know the coefficients are $a=0.5$, $b=-1$, and $c=-4$. We can now determine the coordinates of the vertex of the parabola as follows.

$$\left(-\frac{b}{2a},\frac{4ac-b^2}{4a}\right)=\left(-\frac{-1}{2(0.5)},\frac{4(0.5)(-4)-(-1)^2}{4(0.5)}\right)$$
$$=\left(1,-\frac{9}{2}\right)$$

TOPIC 3: Maximization/Minimization Problems

Many applications of mathematics involve determining the value (or values) of the variable x that returns either the maximum or minimum possible value of some function $f(x)$. Such problems are called max/min problems for short. Examples from business include minimizing cost functions and maximizing profit functions. Examples from physics include maximizing a function that measures the height of a rocket as a function of time and minimizing a function that measures the energy required by a particle accelerator.

If we have a max/min problem involving a quadratic function, we can solve it by finding the vertex. Recall that the vertex is the only point where the graph of a parabola "changes direction." This means it will be the minimum value of a function (if the parabola opens upward) or the maximum value (if the parabola opens downward).

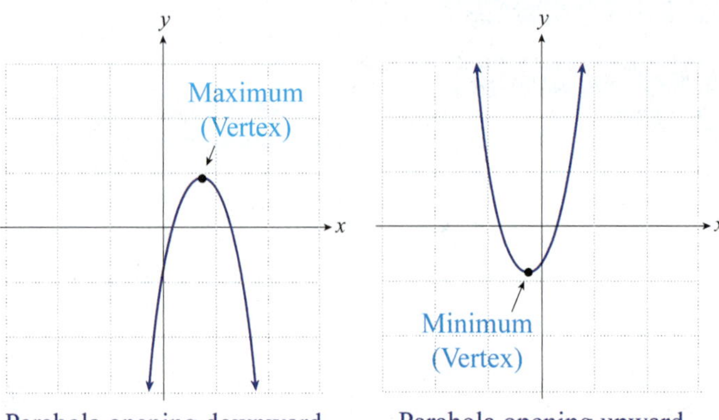

FIGURE 16: Maximum/Minimum Values of Quadratic Functions

Example 6: Fencing a Garden

A farmer plans to use 100 feet of spare fencing material to form a rectangular garden plot against the side of a long barn, using the barn as one side of the plot. How should he split up the fencing among the other three sides in order to maximize the area of the garden plot?

Solution

If we let x represent the length of one side of the plot, as shown in Figure 17, then the dimensions of the plot are x feet by $100-2x$ feet. A function representing the area of the plot is $A(x)=x(100-2x)$.

If we multiply out the formula for A, we recognize it as a quadratic function $A(x)=-2x^2+100x$. This is a parabola opening downward, so the vertex will be the maximum point on the graph of A.

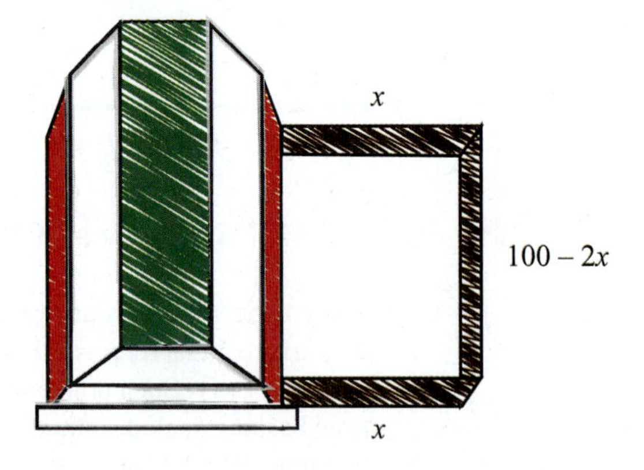

FIGURE 17

Using the vertex formula we know that the vertex of A is the ordered pair $\left(-\frac{100}{2(-2)}, A\left(-\frac{100}{2(-2)}\right)\right)$, or $(25, A(25))$. Thus, to maximize area, we should let $x=25$, and so $100-2x=50$. The resulting maximum possible area, $25 \cdot 50$, or 1250 square feet, is also the value $A(25)$.

Example 7: Quadratic Regression and Maximization

Suzanne occasionally plays guitar in a coffee shop and sells CDs of her music when she does. She's been experimenting with the price to charge per CD and has noted the relationship between price and number of CDs sold shown in Table 1, where the corresponding revenue (the product of price and number sold) is also shown. If she assumes the dependence of revenue on number of CDs sold is quadratic in nature, what number of CDs sold will maximize her revenue? What is the corresponding price per CD?

Price per CD	Number of CDs Sold	Revenue
\$16	2	\$32
\$7	11	\$77
\$3	15	\$45

TABLE 1

Solution

If we think of Revenue as a function of Number of CDs Sold, the coordinates of the three data points Suzanne has collected are $\{(2,32),(11,77),(15,45)\}$.

Using the least-squares method to find a quadratic function R that best fits the points results in the function $R(x) = -x^2 + 18x$.

Figure 18 shows the graph of this function along with the three data points.

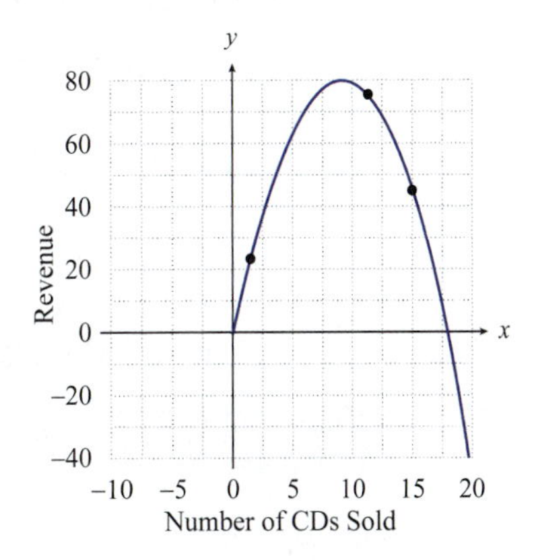

FIGURE 18

Given the formula $R(x) = -x^2 + 18x$, we know that $a = -1$, $b = 18$, and $c = 0$, so the vertex of the parabola has the following coordinates.

$$\left(-\frac{b}{2a}, \frac{4ac - b^2}{4a}\right) = \left(-\frac{18}{2(-1)}, \frac{4(-1)(0) - (18)^2}{4(-1)}\right)$$
$$= (9, 81)$$

So Suzanne should aim to sell 9 CDs to maximize her revenue at \$81, and the corresponding price per CD will be $\frac{\$81}{9} = \$9$.

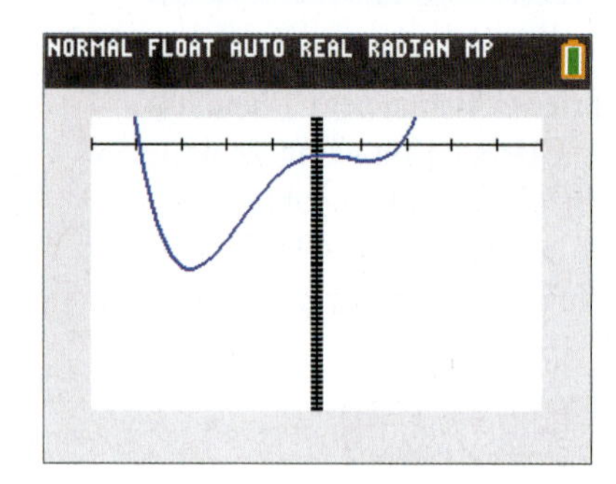

FIGURE 19

TECHNOLOGY: Finding the Maximum or Minimum of a Function

As we've seen, finding the maximum or minimum possible values of some function $f(x)$ can be extremely important, and we have a method for doing so when the function is quadratic. But what if we wanted to find the minimum of the function $f(x)=x^4+2x^3-7x^2+2x-4$? One way is to graph it on a TI-84 Plus, shown in Figure 19 with the following window settings: Xmin=-5, Xmax=5, Ymin=-100, Ymax=10.

To find the minimum, press 2nd trace to access the CALCULATE menu and select minimum. (If we were trying to find the maximum, we would select maximum.) The screen should now display the graph with LeftBound? shown at the bottom. Use the left arrow to move the cursor anywhere to the left of where the minimum appears to be and press enter. The screen should now say RightBound? at the bottom of the graph. Use the right arrow to move the cursor to the right of where the minimum appears to be and press enter again. The text at the bottom of the graph should now read Guess? (see Figure 20). Press enter a third time and the x- and y-values of the minimum will appear at the bottom of the screen (see Figure 21).

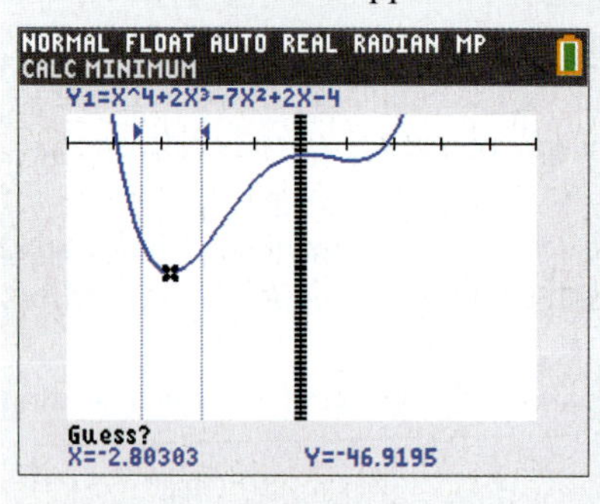

FIGURE 20

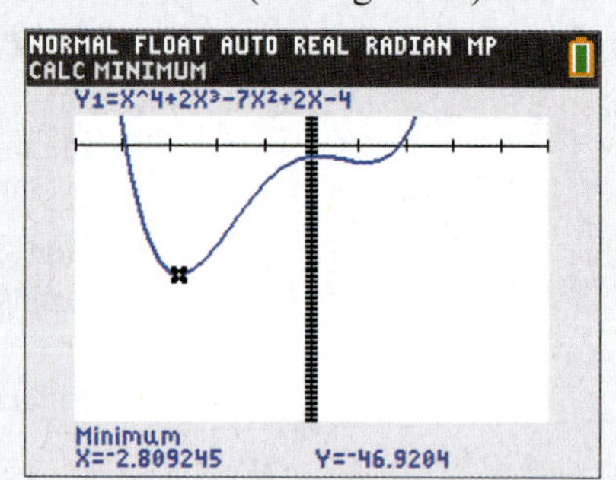

FIGURE 21

So the minimum is approximately $(-2.809,-46.920)$.

3.3 EXERCISES

PRACTICE

Graph the following quadratic functions, accurately locating the vertices and x-intercepts (if any). See Example 1.

1. $f(x)=(x-2)^2+3$
2. $g(x)=-(x+2)^2-1$
3. $h(x)=x^2+6x+7$
4. $F(x)=3x^2+2$
5. $G(x)=x^2-x-6$
6. $p(x)=-2x^2+2x+12$
7. $q(x)=2x^2+4x+3$
8. $r(x)=-3x^2-1$
9. $s(x)=\dfrac{(x-1)^2}{4}$
10. $m(x)=x^2+2x+4$
11. $n(x)=(x+2)(2-x)$
12. $p(x)=-x^2+2x-5$

13. $f(x) = 4x^2 - 6$

14. $k(x) = 2x^2 - 4x$

15. $q(x) = (x+10)(x-2)+36$

Match the following functions with their graphs.

16. $f(x) = -x^2 + 2x$

17. $f(x) = x^2 + 7x + 6$

18. $f(x) = \dfrac{x^2 - 8x + 16}{2}$

19. $f(x) = (x-5)(x+3)+16$

a.

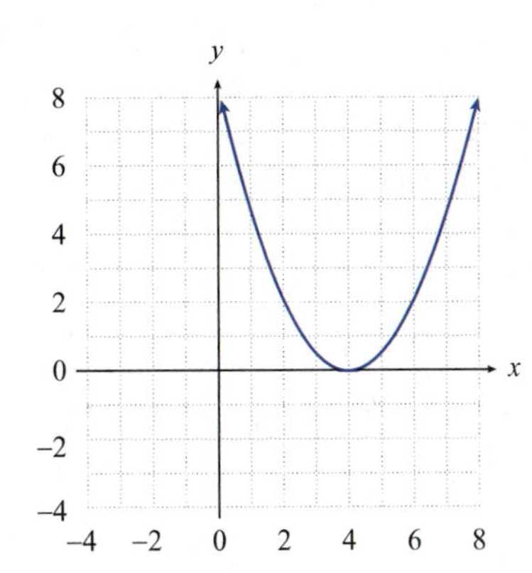

b.

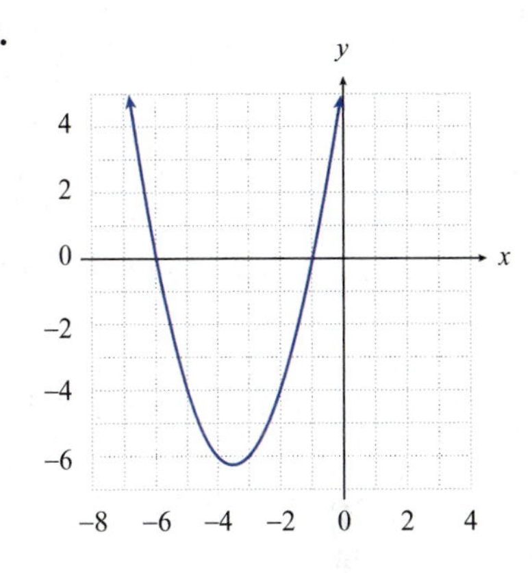

c.

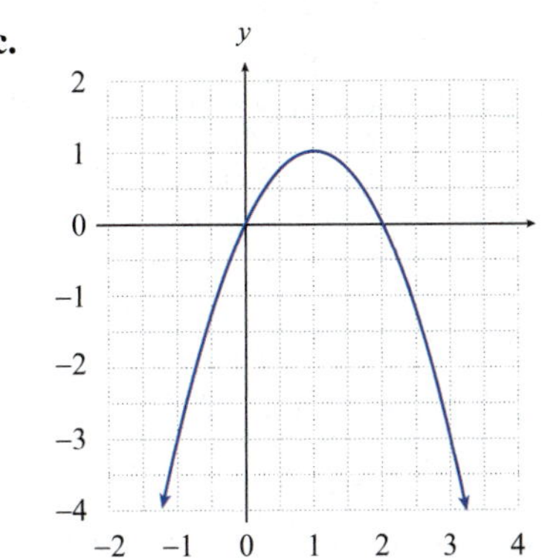

d.

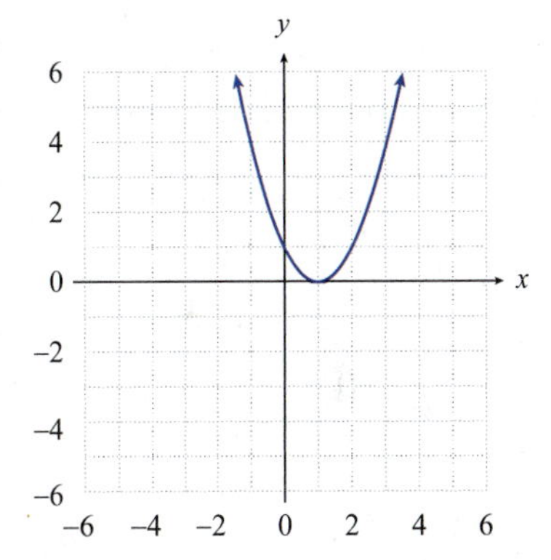

For each of the following parabolic graphs, **a.** find a formula for the corresponding quadratic function, and **b.** use the formula to determine the coordinates of the parabola's vertex. See Example 3.

20.

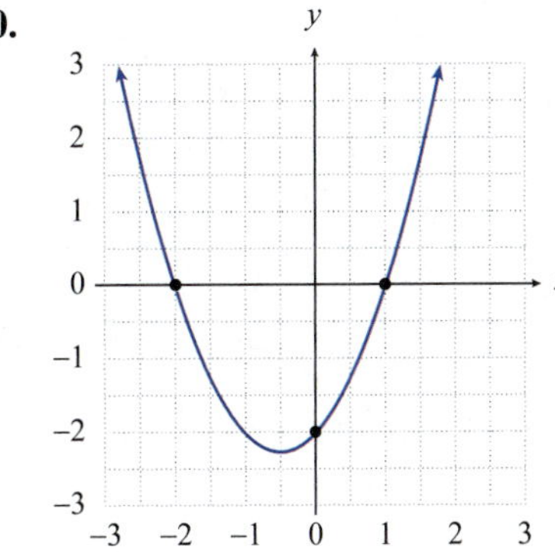

21.

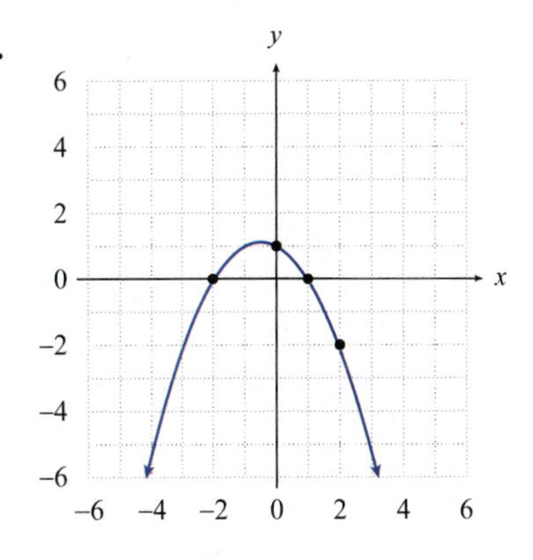

22.

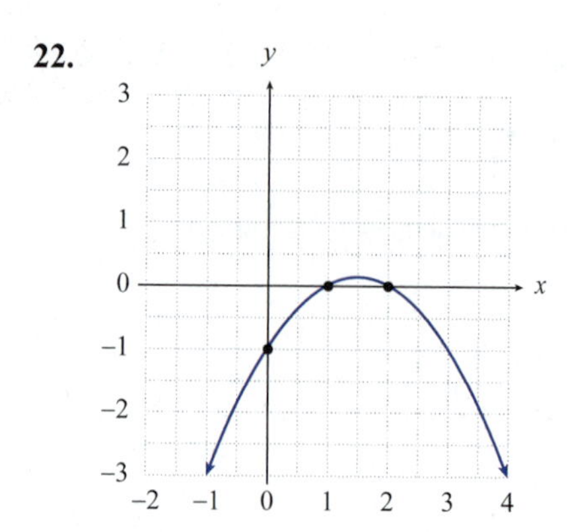

23.

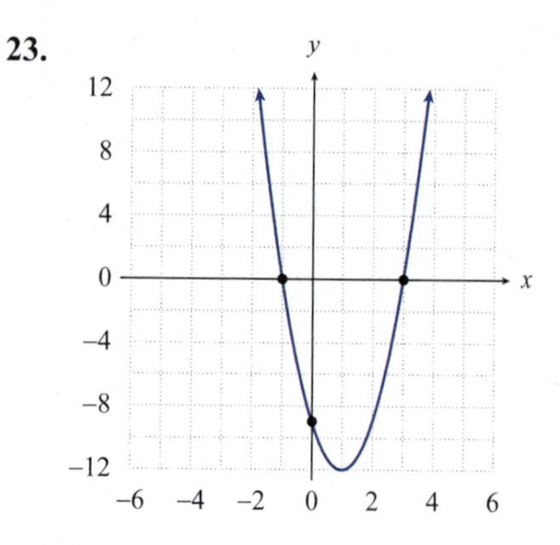

Given the points graphed in each of the following figures, use quadratic regression to find and graph each quadratic function of best fit. See Example 4.

24.

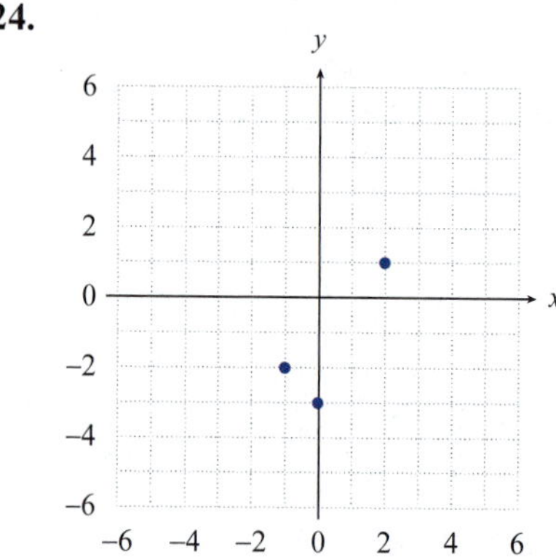

25.

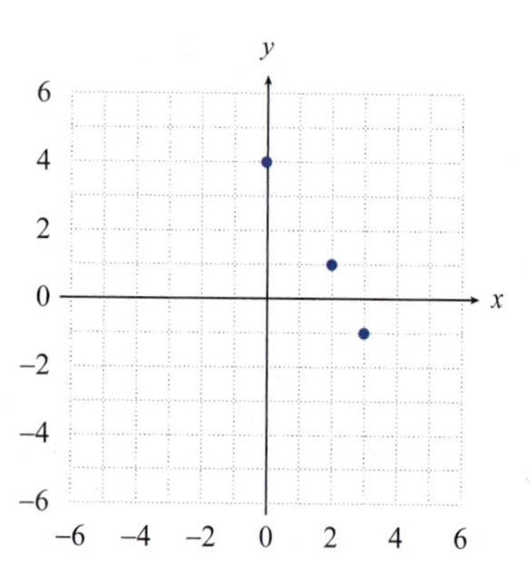

26.

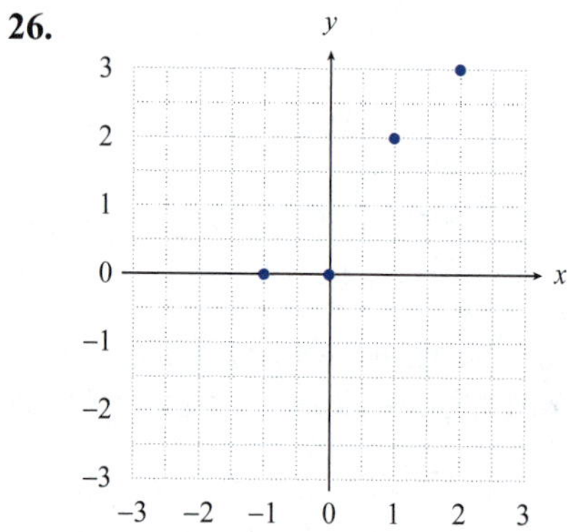

27.

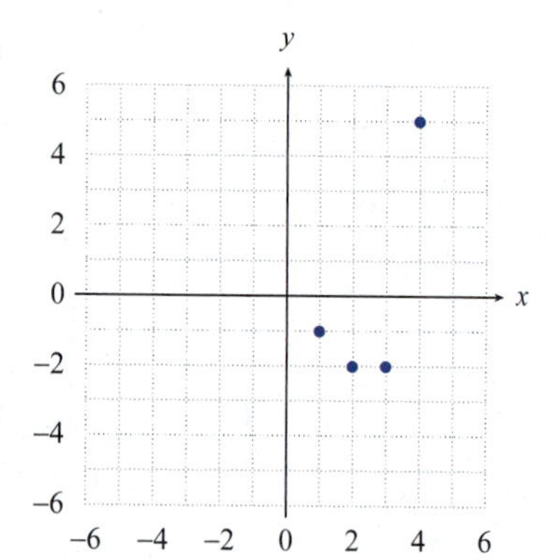

Given the points graphed in each of the following figures, **a.** find the quadratic function that best fits the points, and **b.** use your result to determine the coordinates of the vertex of the best-fitting parabola. See Example 5.

28.

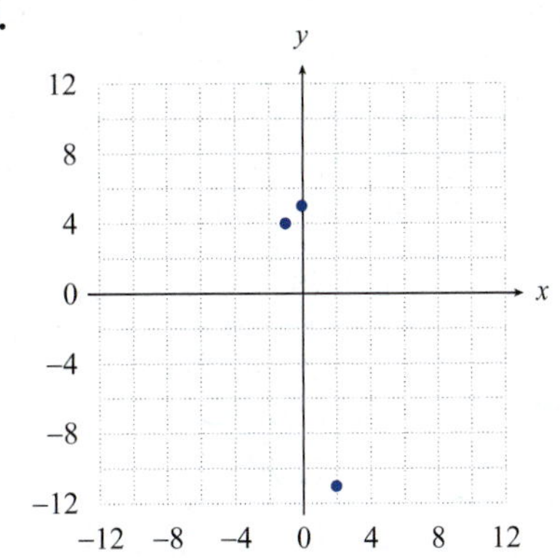

29.

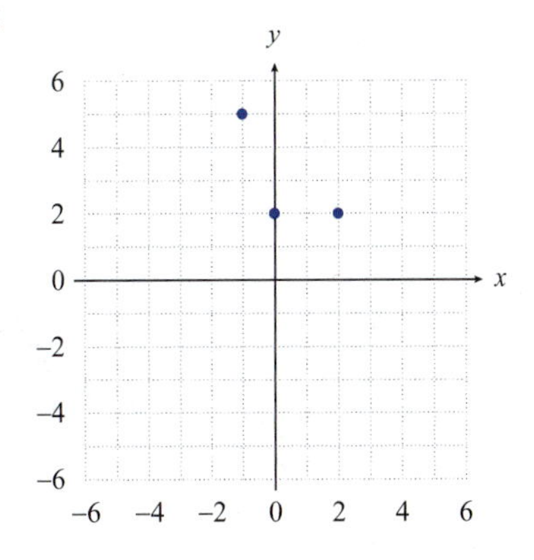

30.

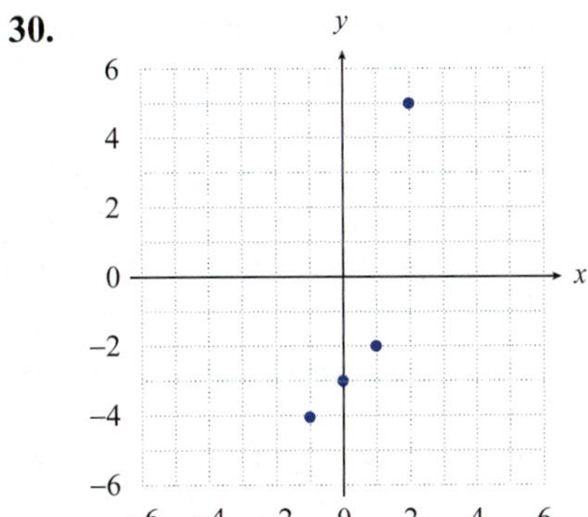

31.

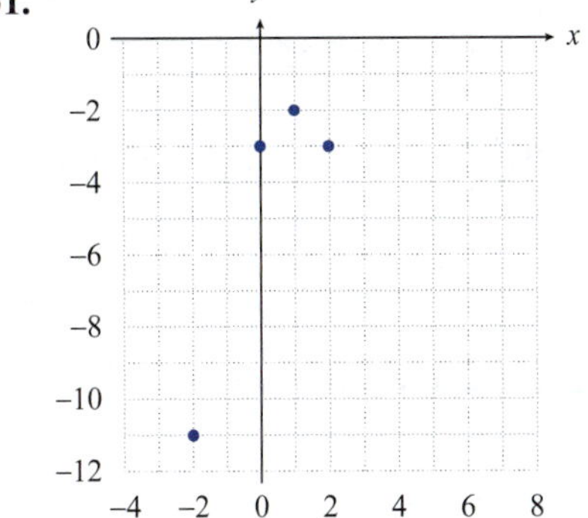

APPLICATIONS

32. Cindy wants to construct three rectangular dog-training arenas side by side, as shown, using a total of 400 feet of fencing. What should the overall length and width be in order to maximize the area of the three combined arenas? (**Hint:** Let x represent the width, as shown, and find an expression for the overall length in terms of x.)

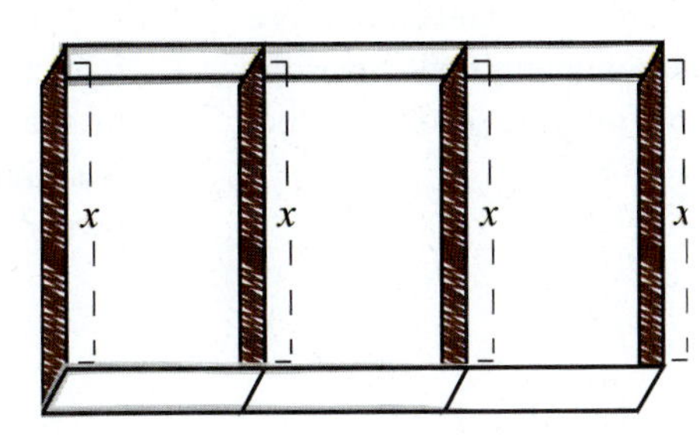

33. Among all the pairs of numbers with a sum of 10, find the pair whose product is maximum.

34. Among all rectangles that have a perimeter of 20, find the dimensions of the one whose area is largest.

35. Find the point on the line $2x + y = 5$ that is closest to the origin. (**Hint:** Instead of trying to minimize the distance between the origin and points on the line, minimize the square of the distance.)

36. Among all the pairs of numbers (x, y) such that $2x + y = 20$, find the pair for which the sum of the squares is minimum.

37. A rancher has a rectangular piece of sheet metal that is 20 inches wide by 10 feet long. He plans to fold the metal to create a narrow three-sided channel and weld two other sheets of metal to the ends to form a watering trough 10 feet long, as shown. How should he fold the metal in order to maximize the volume of the resulting trough?

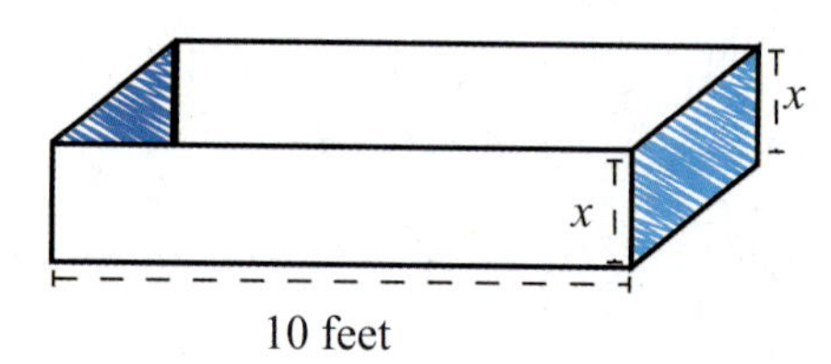

38. Find a pair of numbers whose product is maximum if the pair must have a sum of 16.

39. Search the Seas cruise ship has a conference room onboard that can hold up to 60 people. Companies can reserve the room for groups of 38 or more. If the group contains 38 people, the company pays \$60 per person. The cost per person is reduced by \$1 for each person in excess of 38. Find the size of the group that maximizes the income for the owners of the ship and find this income.

40. The back of George's property is a creek. George would like to enclose a rectangular area, using the creek as one side and fencing for the other three sides, to create a vegetable garden. If he has 300 feet of material, what is the maximum possible area of the garden?

41. Find a pair of numbers whose product is maximum if two times the first number plus the second number is 48.

42. The total revenue for Thompson's Studio Apartments is given by the function

$$R(x) = 100x - 0.1x^2,$$

where x is the number of rooms rented. What number of rooms rented produces the maximum revenue?

43. The total revenue of Tran's Machinery Rental is given by the function

$$R(x) = 300x - 0.4x^2,$$

where x is the number of units rented. What number of units rented produces the maximum revenue?

44. The total cost of producing a type of small car is given by

$$C(x) = 9000 - 135x + 0.045x^2,$$

where x is the number of cars produced. How many cars should be produced to incur minimum cost?

45. The total cost of manufacturing a set of golf clubs is given by

$$C(x) = 800 - 10x + 0.20x^2,$$

where x is the number of sets of golf clubs produced. How many sets of golf clubs should be manufactured to incur minimum cost?

46. The owner of a parking lot is going to enclose a rectangular area with fencing, using an existing fence as one of the sides. The owner has 220 feet of new fencing material (which is much less than the length of the existing fence). What is the maximum possible area that the owner can enclose?

In Exercises 47–49, use the formula $h(t) = -16t^2 + v_0t + h_0$ for the height at time t of an object thrown vertically upward with velocity v_0 (in feet per second) from an initial height of h_0 (in feet).

47. Sitting in a tree, 48 feet above ground level, Sue shoots a pebble straight up with a velocity of 64 feet per second. What is the maximum height attained by the pebble?

48. A ball is thrown upward with a velocity of 48 feet per second from the top of a 144-foot building. What is the maximum height of the ball?

49. A rock is thrown upward with a velocity of 80 feet per second from the top of a 64-foot-high cliff. What is the maximum height of the rock?

Use quadratic regression to answer the following questions. See Example 7.

50. Darlena has started taking photos at amateur dog racing events, later offering the photos for sale to the dog owners by email. The prices she has charged per photo at each of her first three events, and the corresponding number of photos sold and total revenue raised, appear in the following table. Treating revenue as a function of the number of photos sold, a graph of the three data points is also shown. If she uses quadratic regression to fit a curve to the data, what number of photos sold and what price per photo will maximize her revenue?

Price per Photo	Number of Photos Sold	Revenue
\$28	2	\$56
\$24	4	\$96
\$12	10	\$120

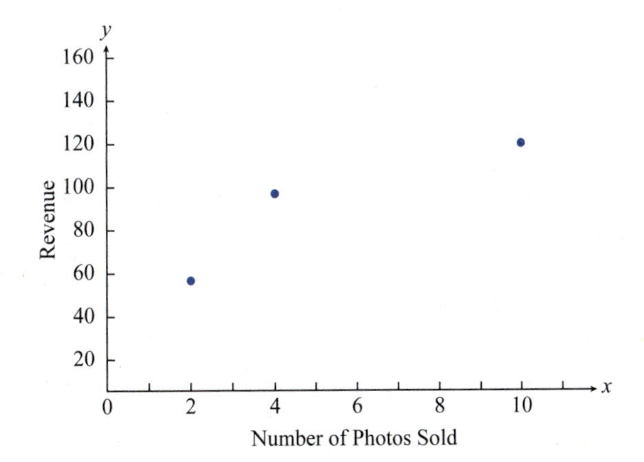

51. Joe makes a video of his friend Zach throwing a baseball as hard as he can straight up in the air. Looking at the video frame by frame later, they estimate that Zach released the ball, at a time they designate as $t = 0$ seconds, at a height of 7 feet. The ball appears to be the same height as the top of a nearby 20-foot-tall billboard at time $t = 0.23$ seconds on the way up and again at time $t = 3.56$ seconds on the way down. If they use quadratic regression to fit a curve to these three points, what maximum height did the baseball reach?

WRITING & THINKING

52. Without graphing, state the number of x-intercepts for each of the following functions and describe the location of the vertex in relation to the x-axis.

a. $y=(x-2)^2$

b. $y=(x-2)(x+2)$

c. $y=-(x-3)(x-1)$

d. $y=-(x-\sqrt{3})(x+\sqrt{3})$

e. $y=x(x+1)$

f. $y=-(x^2+1)$

TECHNOLOGY

Use a graphing utility to graph each of the following quadratic functions. Then determine the vertex and x-intercepts.

53. $f(x)=2x^2-16x+31$

54. $f(x)=-x^2-2x+3$

55. $f(x)=x^2-8x-20$

56. $f(x)=x^2-4x$

57. $f(x)=25-x^2$

58. $f(x)=3x^2+18x$

59. $f(x)=x^2+2x+1$

60. $f(x)=3x^2-8x+2$

61. $f(x)=-x^2+10x-4$

62. $f(x)=\frac{1}{2}x^2+x-1$

3.4 OTHER COMMON FUNCTIONS

TOPICS

1. Power functions of the form ax^n
2. Power functions of the form ax^{-n}
3. Power functions of the form $ax^{\frac{1}{n}}$
4. The absolute value function
5. The greatest integer function
6. Piecewise-defined functions

In Sections 3.2 and 3.3, we investigated the behavior of linear and quadratic functions, but these are just two types of commonly occurring functions; there are many other functions that arise naturally in solving various problems. In this section, we will explore several other classes of functions, building up a portfolio of functions to be familiar with.

TOPIC 1: Power Functions of the Form ax^n

We already know what the graph of any function of the form $f(x) = ax$ or $f(x) = ax^2$ looks like, as these are, respectively, simple linear and quadratic functions. What happens to the graphs as we increase the exponent and consider functions of the form $f(x) = ax^3$, $f(x) = ax^4$, etc.? To answer this, we must first define a new class of functions.

DEFINITION: Power Functions

A **power function** is a function of the form $f(x) = ax^r$, where a and r are real numbers.

The behavior of a function of the form $f(x) = ax^n$, where a is a real number and n is a natural number, falls into one of two categories. Consider the graphs in Figure 1.

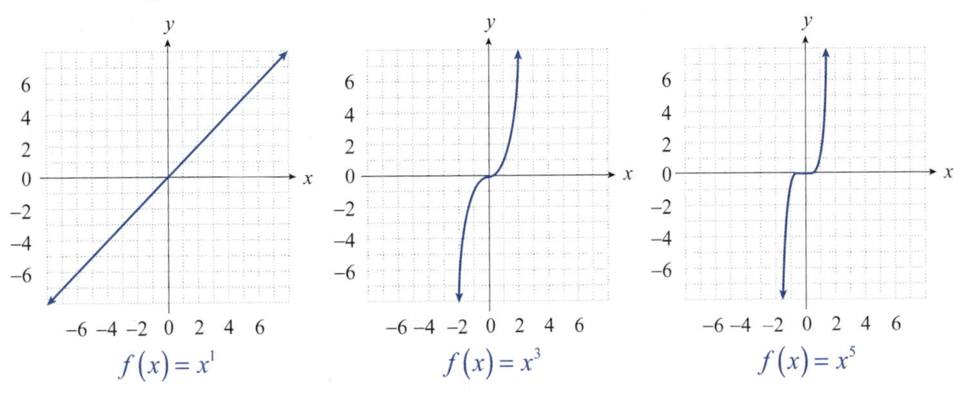

FIGURE 1: Odd Exponents

The three graphs in Figure 1 show the behavior of $f(x) = x^n$ for the first three odd exponents. Note that in each case, the domain and the range of the function are both the entire set of real numbers; the same is true for higher odd exponents as well. Now, consider the graphs in Figure 2.

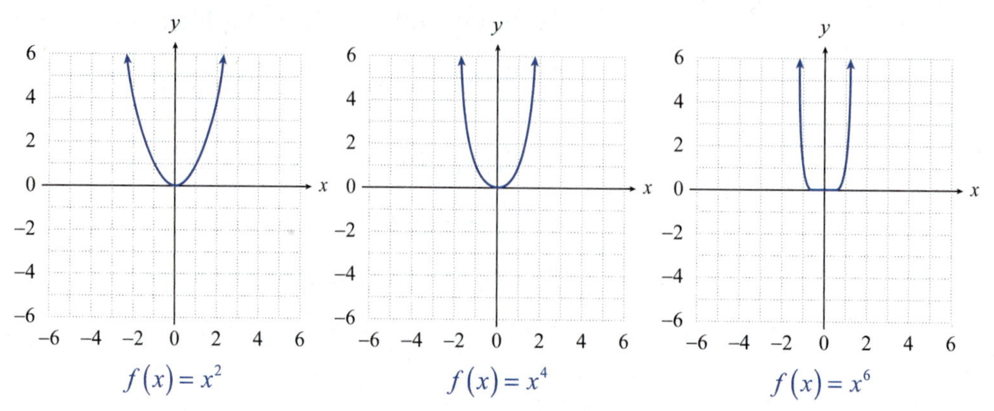

FIGURE 2: Even Exponents

These three functions are also similar to one another. The first one is the basic parabola we studied in Section 3.3. The other two bear some similarity to parabolas, but are flatter near the origin and rise more steeply for $|x| > 1$. For any function of the form $f(x) = x^n$ where n is an even natural number, the domain is the entire set of real numbers and the range is the interval $[0, \infty)$.

Multiplying a function of the form x^n by a constant a has the effect that we noticed in Section 3.3. If $|a| > 1$, the graph of the function is stretched vertically; if $0 < |a| > 1$, the graph is compressed vertically; and if $a < 0$, the graph is reflected with respect to the x-axis. We can use this knowledge, along with plotting a few specific points, to quickly sketch graphs of any function of the form $f(x) = ax^n$.

Example 1: Power Functions of the Form ax^n

Sketch the graphs of the following functions.

a. $f(x) = \dfrac{x^4}{5}$

b. $g(x) = -x^3$

Solution

a.

The graph of the function f will have the same basic shape as the function x^4, but compressed vertically because of the factor of $\frac{1}{5}$. To make the sketch more accurate, calculate the coordinates of a few points on the graph. Figure 3 illustrates that $f(-1) = \frac{1}{5}$ and that $f(2) = \frac{16}{5}$.

FIGURE 3: $f(x) = \dfrac{x^4}{5}$

TECHNOLOGY

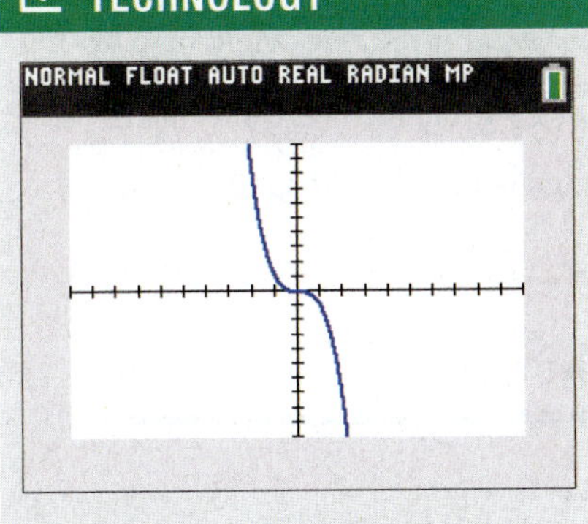

b.

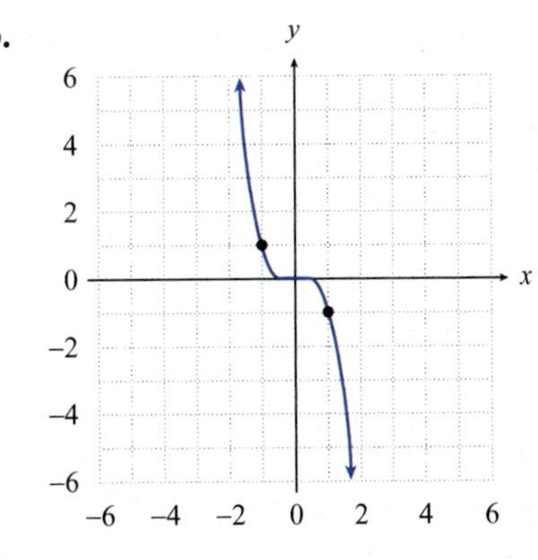

FIGURE 4: $g(x) = -x^3$

We know that the function g will have the same shape as the function x^3, but reflected over the x-axis because of the factor of -1. The graph in Figure 4 illustrates this.

We also plot a few points on the graph of g, namely $(-1,1)$ and $(1,-1)$, as a check.

TOPIC 2: Power Functions of the Form ax^{-n}

We could also describe the following functions as having the form $\frac{a}{x^n}$, where a is a real number and n is a natural number. Once again, the graphs of these functions fall roughly into two categories, as illustrated in Figures 5 and 6.

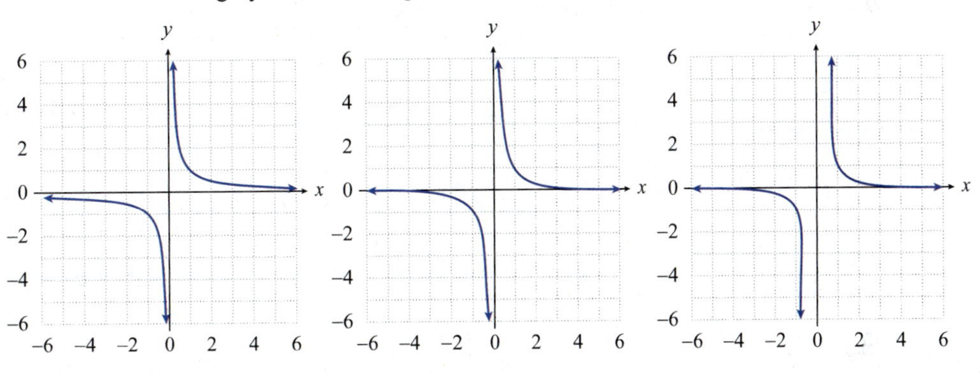

$f(x) = x^{-1} = \frac{1}{x}$ $f(x) = x^{-3} = \frac{1}{x^3}$ $f(x) = x^{-5} = \frac{1}{x^5}$

FIGURE 5: Odd Exponents

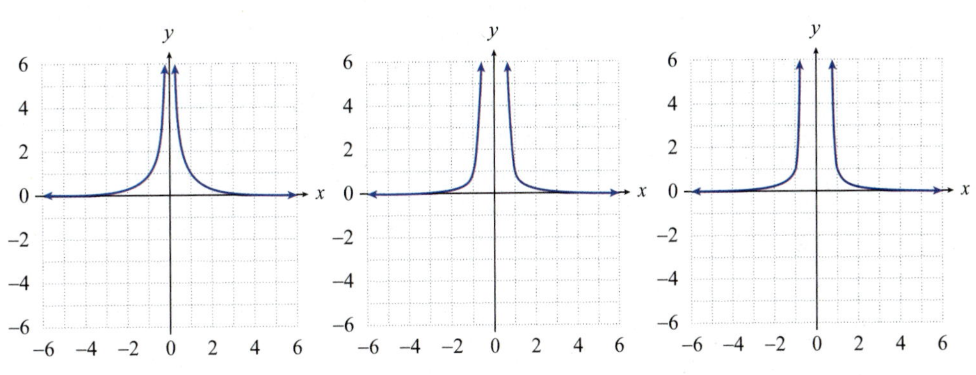

$f(x) = x^{-2} = \frac{1}{x^2}$ $f(x) = x^{-4} = \frac{1}{x^4}$ $f(x) = x^{-6} = \frac{1}{x^6}$

FIGURE 6: Even Exponents

As with functions of the form ax^n, increasing the exponent on functions of the form $\frac{a}{x^n}$ sharpens the curve of the graph near the origin. Note that the domain of any function of the form $f(x)=\frac{a}{x^n}$ is $(-\infty,0)\cup(0,\infty)$, but the range depends on whether n is even or odd. When n is odd, the range is also $(-\infty,0)\cup(0,\infty)$; and when n is even, the range is $(0,\infty)$.

Example 2: Power Functions of the Form ax^{-n}

Sketch the graph of the function $f(x)=-\frac{1}{4x}$.

TECHNOLOGY

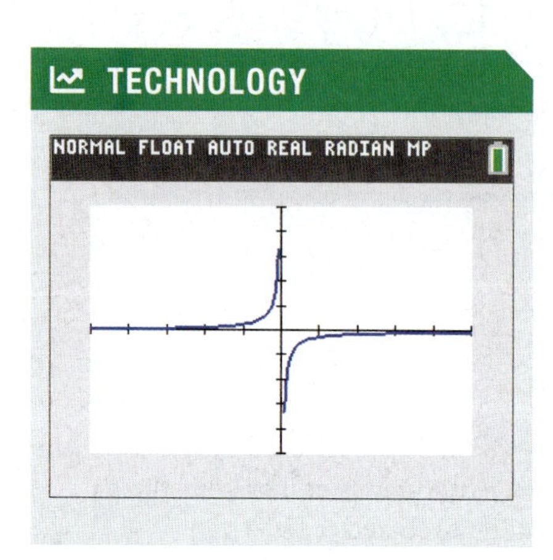

Solution

The graph of the function f is similar to that of the function $x^{-1}=\frac{1}{x}$, with two differences. We obtain the formula $-\frac{1}{4x}$ by multiplying $\frac{1}{x}$ by $-\frac{1}{4}$, a negative number between -1 and 1. So one difference is that the graph of f is the reflection of $\frac{1}{x}$ with respect to the x-axis. The other difference is that the graph of f is compressed vertically.

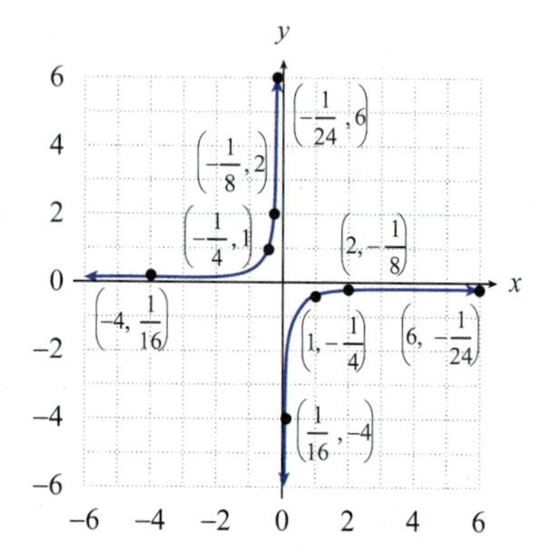

FIGURE 7: $f(x)=-\frac{1}{4x}$

With the above in mind, we can calculate the coordinates of a few points (such as $\left(-\frac{1}{4},1\right)$ and $\left(1,-\frac{1}{4}\right)$) and sketch the graph of f as shown in Figure 7.

TOPIC 3: Power Functions of the Form $ax^{\frac{1}{n}}$

Using radical notation, these are functions of the form $a\sqrt[n]{x}$, where a is again a real number and n is a natural number. Square root and cube root functions, in particular, are commonly seen in mathematics.

Functions of this form again fall into one of two categories, depending on whether n is odd or even. To begin with, note that the domain and range are both the entire set of real numbers when n is odd, and both are the interval $[0,\infty)$ when n is even. Figures 8 and 9 illustrate the two basic shapes of functions of this form.

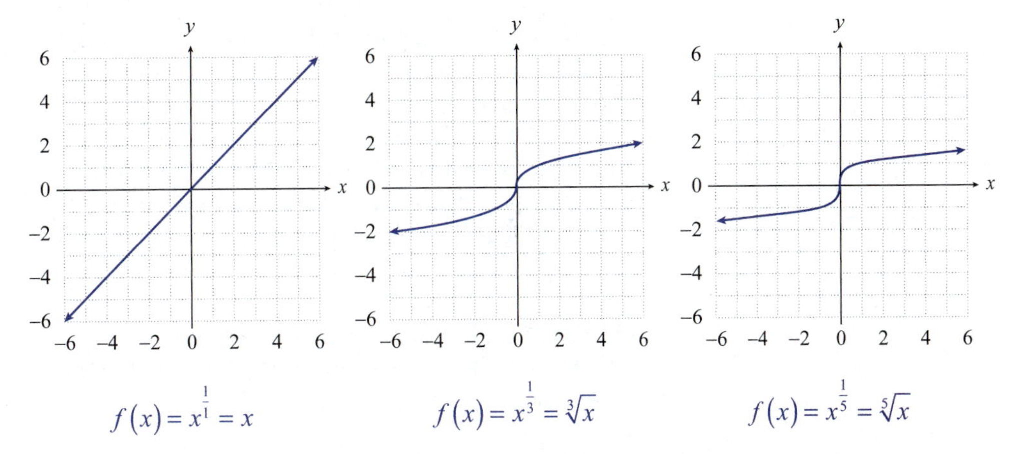

FIGURE 8: Odd Roots

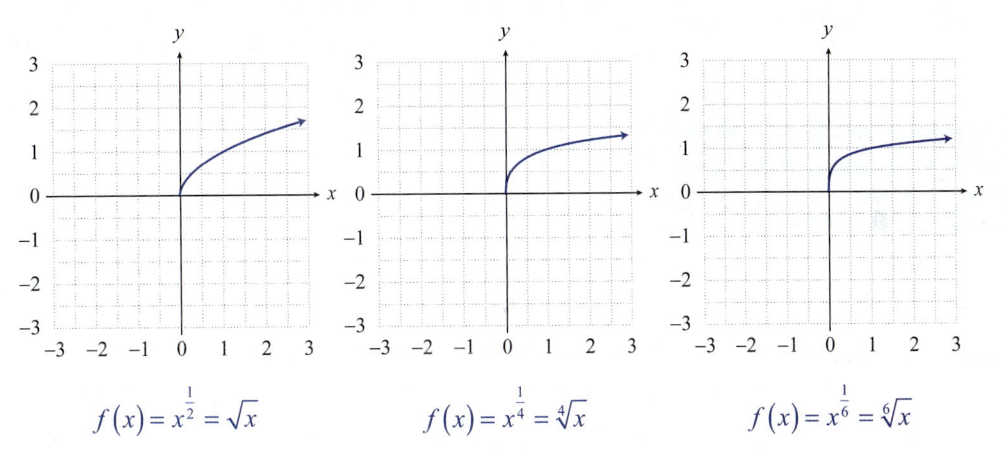

FIGURE 9: Even Roots

At this point, you may be thinking that the graphs in Figures 8 and 9 appear familiar. The shapes in Figure 8 are the same as those seen in Figure 1, but rotated by 90 degrees and reflected with respect to the x-axis. Similarly, the shapes in Figure 9 bear some resemblance to those in Figure 2, except that half of the graphs appear to have been erased. This resemblance is no accident, given that n^{th} roots undo n^{th} powers. We will explore this observation in much more detail in Section 4.4.

TOPIC 4: The Absolute Value Function

The basic absolute value function is $f(x) = |x|$. Note that for any value of x, $f(x)$ is nonnegative, so the graph of f should lie on or above the x-axis. One way to determine its exact shape is to review the following definition of absolute value.

$$|x| = \begin{cases} x & \text{if } x \geq 0 \\ -x & \text{if } x < 0 \end{cases}$$

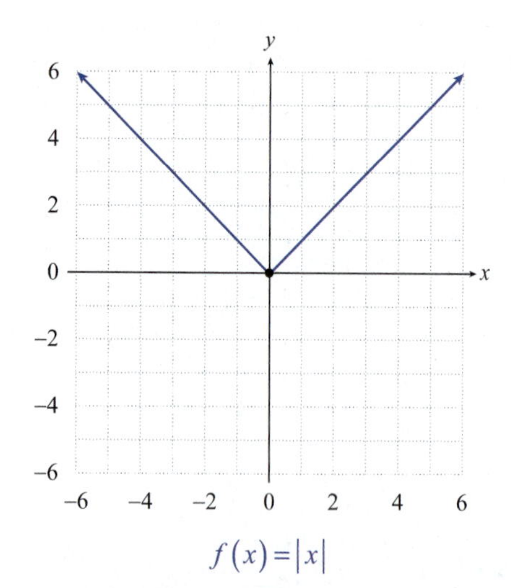

$f(x)=|x|$

FIGURE 10: The Absolute Value Function

This means that for nonnegative values of x, $f(x)$ is a linear function with a slope of 1, and for negative values of x, $f(x)$ is a linear function with a slope of −1. Both linear functions have a y-intercept of 0, so the complete graph of f is as shown in Figure 10.

The effect of multiplying $|x|$ by a real number a is what we have come to expect: if $|a|>1$, the graph is stretched vertically; if $0<|a|<1$ the graph is compressed vertically; and if a is negative, the graph is reflected with respect to the x-axis.

TECHNOLOGY

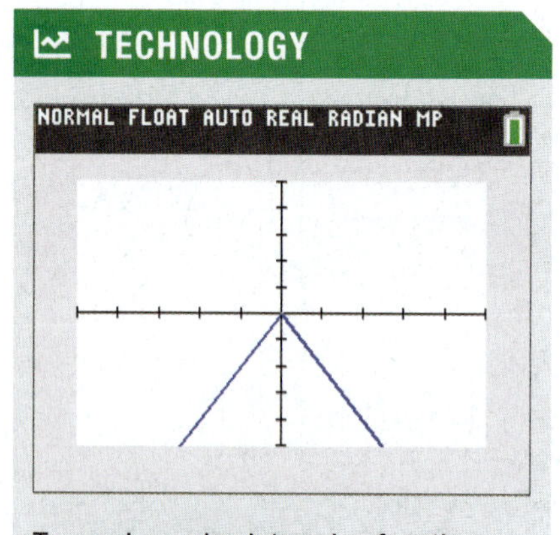

To graph an absolute value function, press Y= . To enter the absolute value, press math and then use the right arrow to select NUM. Select abs(. The absolute value symbol will appear in your equation.

Example 3: The Absolute Value Function

Sketch the graph of the function $f(x)=-2|x|$.

Solution

The graph of f will be a vertically stretched version of $y=|x|$ reflected over the x-axis. As always, we can plot a few points to verify that our reasoning is correct. In Figure 11, we have plotted the values of $f(-4)$ and $f(2)$.

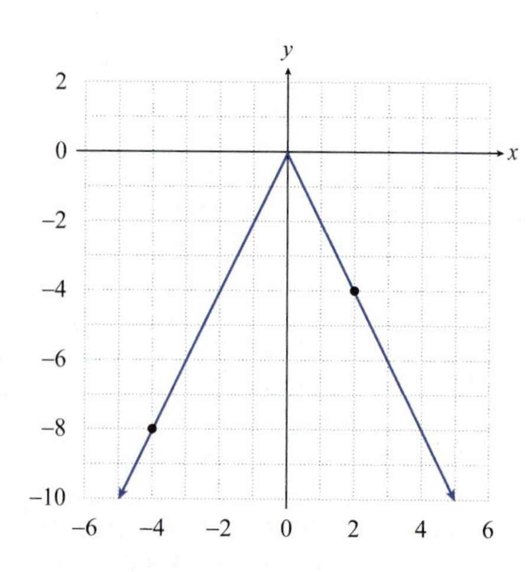

FIGURE 11: $f(x)=-2|x|$

TOPIC 5: The Greatest Integer Function

DEFINITION: The Greatest Integer Function

The greatest integer function, $f(x)=[\![x]\!]$, is a function commonly encountered in computer science applications. It is defined as follows: the **greatest integer of** x is the largest integer less than or equal to x. For instance, $[\![4.3]\!]=4$ and $[\![-2.9]\!]=-3$ (note that −3 is the largest integer to the left of −2.9 on the real number line).

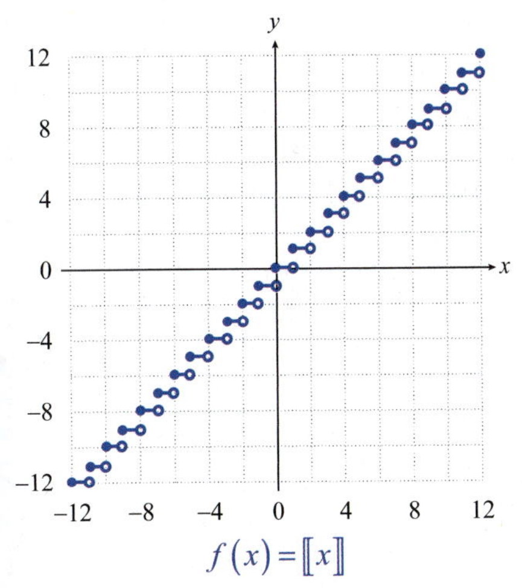

FIGURE 12: The Greatest Integer Function

Careful study of the greatest integer function reveals that its graph must consist of intervals where the function is constant, and that these portions of the graph must be separated by discrete "jumps," or breaks, in the graph. For instance, any value for x chosen from the interval $[1,2)$ results in $f(x)=1$, but $f(2)=2$. Similarly, any value for x chosen from the interval $[-3,-2)$ results in $f(x)=-3$, but $f(-2)=-2$.

Our graph of the greatest integer function must somehow indicate this repeated pattern of jumps. In cases like this, it is conventional to use an open circle on the graph to indicate that the function is either undefined at that point or is defined to be another value. Closed circles are used to emphasize that a certain point really does lie on the graph of the function. With these conventions in mind, the graph of the greatest integer function appears in Figure 12.

TOPIC 6: Piecewise-Defined Functions

There is no rule stating that a function needs to be defined by a single formula. In fact, we have worked with such a function already; in evaluating the absolute value of x, we use one formula if x is greater than or equal to 0 and a different formula if x is less than 0. Obviously, we can't have two rules govern the same input, but we can have multiple formulas on separate pieces of a function's domain.

NOTE

Always pay close attention to the boundary points of each interval. Remember that only one rule applies at each point.

DEFINITION: Piecewise-Defined Function

A **piecewise-defined function** is a function defined in terms of two or more formulas, each valid for its own unique portion of the real number line. In evaluating a piecewise-defined function f at a certain value of x, it is important to correctly identify which formula is valid for that particular value.

Example 4: Piecewise-Defined Function

Sketch the graph of the function $f(x)=\begin{cases}-2x-2 & \text{if } x\le -1\\ x^2 & \text{if } x>-1\end{cases}$.

Solution

The function f is a piecewise-defined function with different formulas for two separate intervals. To graph f, graph each portion separately, making sure that each formula is applied only on the appropriate interval.

The function f is a linear function on the interval $(-\infty,-1]$ and a quadratic function on the interval $(-1,\infty)$.

The complete graph appears in Figure 13, with the points $f(-4)=6$ and $f(2)=4$ noted in particular.

Note the use of a closed circle at $(-1,0)$ to emphasize that this point is part of the graph and the use of an open circle at $(-1,1)$ to indicate that this point is not part of the graph. That is, the value of $f(-1)$ is 0, not 1.

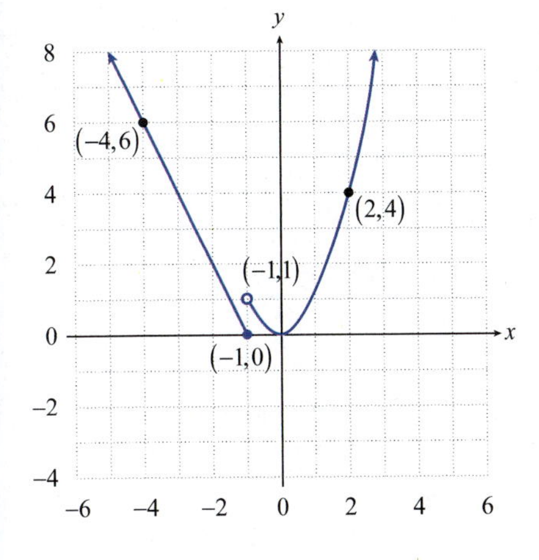

FIGURE 13: $y=f(x)$

3.4 EXERCISES

PRACTICE

Sketch the graphs of the following functions. Pay particular attention to intercepts, if any, and locate these accurately. See Examples 1 through 4.

1. $f(x)=-\frac{x}{2}$

2. $g(x)=2x^2$

3. $F(x)=x^{\frac{1}{2}}$

4. $h(x)=x^{-1}$

5. $p(x)=-\frac{2}{x}$

6. $q(x)=-\sqrt[3]{x}$

7. $G(x)=-|x|$

8. $k(x)=\frac{1}{x^3}$

9. $G(x)=\frac{\sqrt{x}}{2}$

10. $H(x)=0.5x^{\frac{1}{3}}$

11. $r(x)=3|x|$

12. $p(x)=-\frac{1}{x^2}$

13. $W(x)=\frac{x^4}{16}$

14. $k(x)=\frac{x^3}{9}$

15. $h(x)=2\sqrt[3]{x}$

16. $d(x)=2x^5$

17. $S(x)=4x^{-2}$

18. $f(x)=-x^2$

19. $r(x)=\frac{\sqrt[3]{x}}{3}$

20. $s(x)=\frac{|x|}{3}$

21. $t(x)=\frac{x^6}{4}$

22. $f(x)=2[\![x]\!]$

23. $P(x)=-[\![x]\!]$

24. $m(x)=\left[\!\!\left[\frac{x}{2}\right]\!\!\right]$

25. $f(x)=\begin{cases} 3-x & \text{if } x<-2 \\ x^{\frac{1}{3}} & \text{if } x\geq -2 \end{cases}$

26. $g(x)=\begin{cases} -x^2 & \text{if } x\leq 1 \\ x^2 & \text{if } x>1 \end{cases}$

27. $r(x)=\begin{cases} \frac{1}{x} & \text{if } x<1 \\ -x & \text{if } x>1 \end{cases}$

28. $p(x)=\begin{cases} x+1 & \text{if } x<-2 \\ x^3 & \text{if } -2\leq x<3 \\ -1-x & \text{if } x\geq 3 \end{cases}$

29. $q(x)=\begin{cases} -1 & \text{if } x\in\mathbb{Z} \\ 1 & \text{if } x\notin\mathbb{Z} \end{cases}$

30. $s(x)=\begin{cases} \frac{x^2}{3} & \text{if } x<0 \\ -\frac{x^2}{3} & \text{if } x\geq 0 \end{cases}$

31. $v(x)=\begin{cases} x^2 & \text{if } -1\leq x\leq 1 \\ |x| & \text{if } x<-1 \text{ or } x>1 \end{cases}$

32. $M(x)=\begin{cases} x & \text{if } x\in\mathbb{Z} \\ -x & \text{if } x\notin\mathbb{Z} \end{cases}$

33. $t(x)=\begin{cases} x^4 & \text{if } x\leq 1 \\ [\![x]\!] & \text{if } x>1 \end{cases}$

34. $N(x)=\begin{cases} x^2 & \text{if } x\in\mathbb{Z} \\ [\![x]\!] & \text{if } x\notin\mathbb{Z} \end{cases}$

35. $h(x)=\begin{cases}-|x| & \text{if } x<2\\ [\![x]\!] & \text{if } x\ge 2\end{cases}$

36. $u(x)=\begin{cases}[\![x]\!] & \text{if } x\le 1\\ 2x-2 & \text{if } x>1\end{cases}$

Match the following functions to their graphs.

37. $f(x)=-2x^4$

38. $f(x)=-\dfrac{7}{9x^4}$

39. $f(x)=-4\left[\!\left[\dfrac{x}{4}\right]\!\right]$

40. $f(x)=-\dfrac{7\sqrt[3]{x}}{3}$

41. $f(x)=-\dfrac{8}{9}|x|$

42. $f(x)=-4\sqrt{x}$

43. $f(x)=\dfrac{3}{7}|x|$

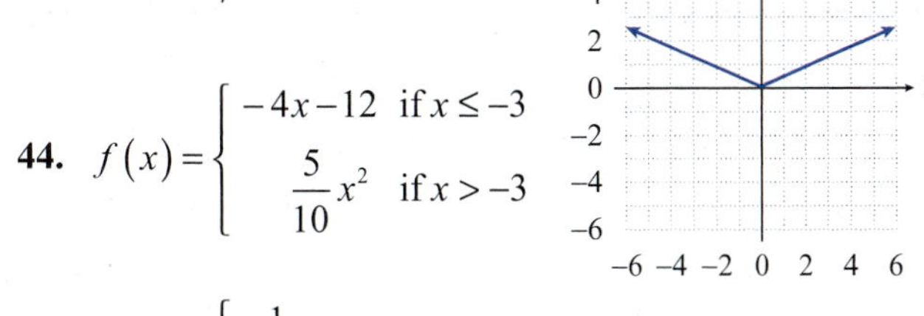

44. $f(x)=\begin{cases}-4x-12 & \text{if } x\le -3\\ \dfrac{5}{10}x^2 & \text{if } x>-3\end{cases}$

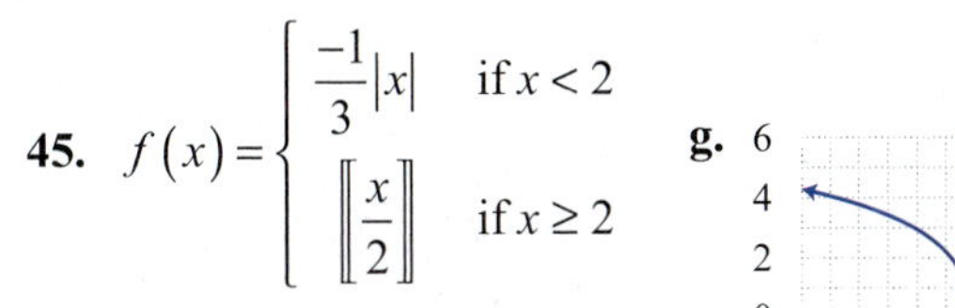

45. $f(x)=\begin{cases}\dfrac{-1}{3}|x| & \text{if } x<2\\ \left[\!\left[\dfrac{x}{2}\right]\!\right] & \text{if } x\ge 2\end{cases}$

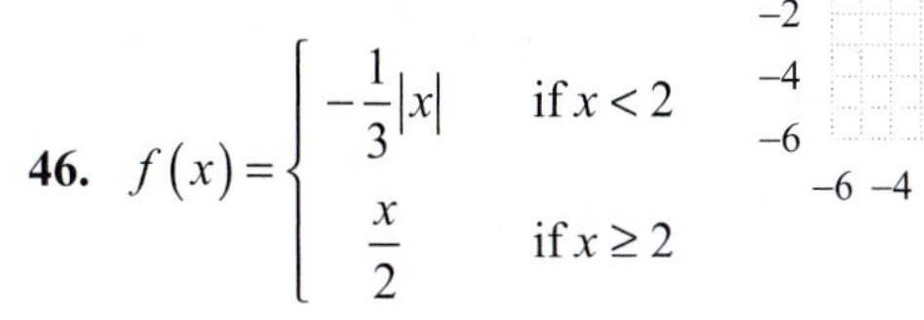

46. $f(x)=\begin{cases}-\dfrac{1}{3}|x| & \text{if } x<2\\ \dfrac{x}{2} & \text{if } x\ge 2\end{cases}$

a.

b.

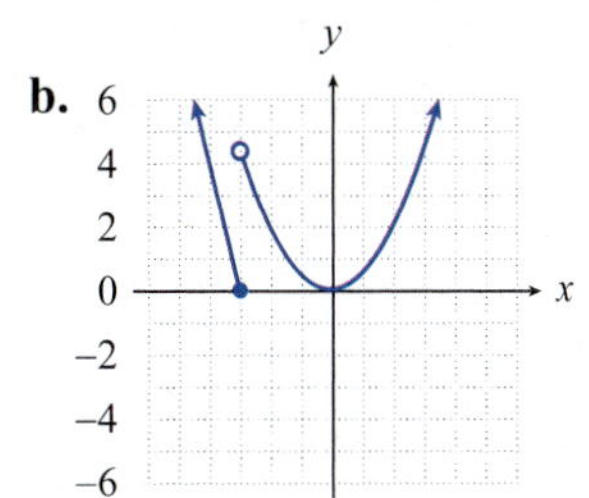

c.

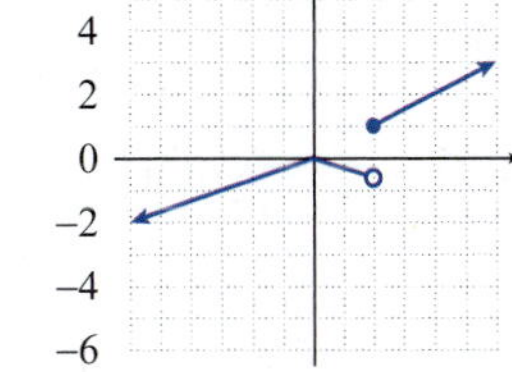

d.

e.

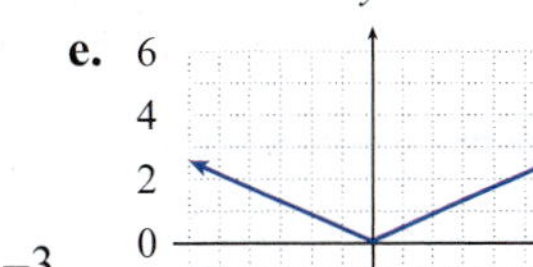

f.

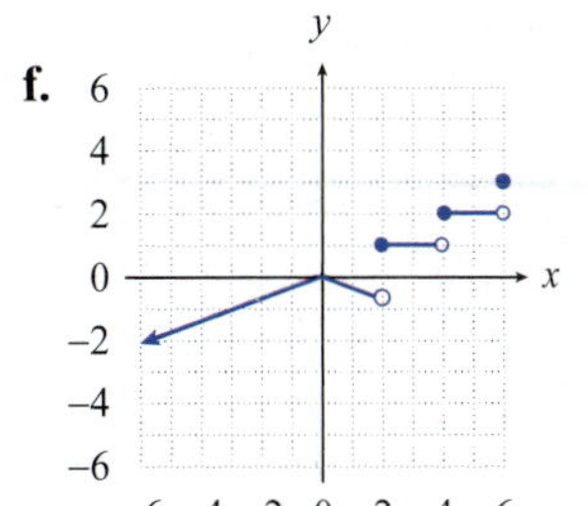

g.

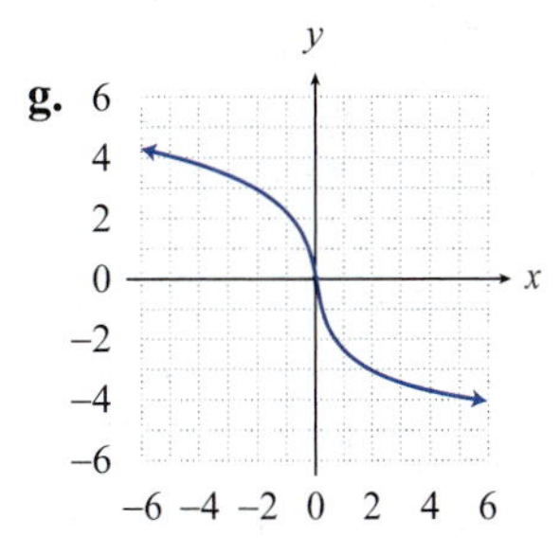

h.

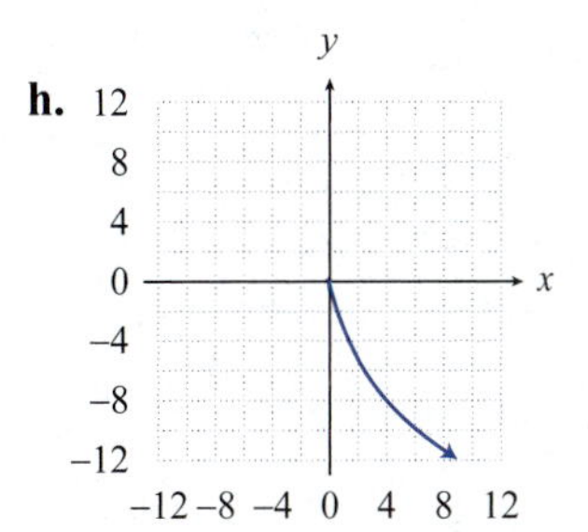

i.

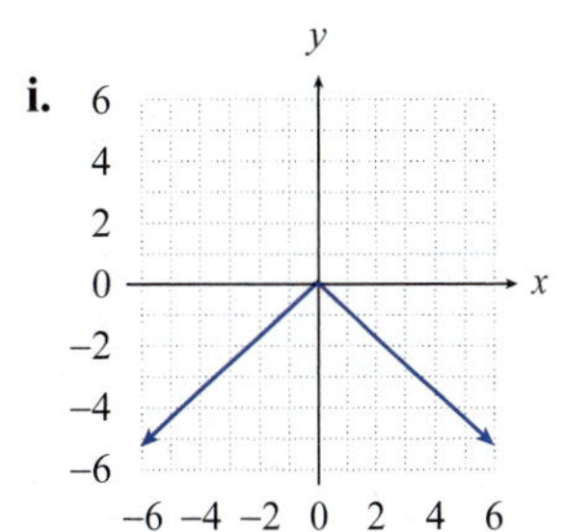

j.

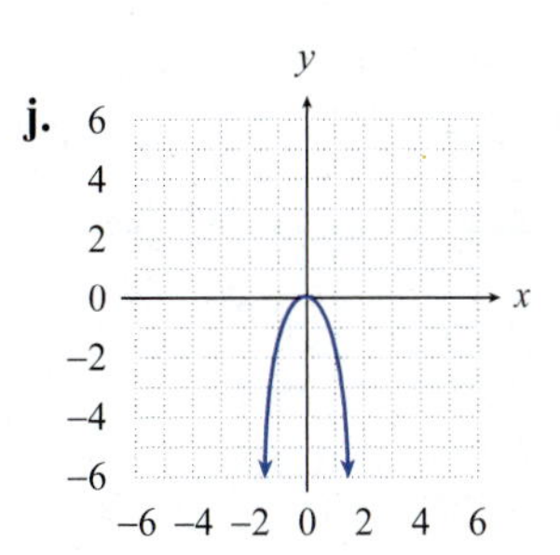

TECHNOLOGY

Use a graphing utility to graph the following functions. Experiment with different viewing windows until you obtain a sketch that seems to capture the meaningful parts of the graph.

47. $f(x)=10x^5-x^3$

48. $g(x)=x^5+x^2$

49. $f(x)=x^3-5x^2+x$

50. $g(x)=\sqrt{x}-x^2$

51. $f(x)=\sqrt{x}+3x-1$

52. $g(x)=x^4-3x^3+2$

3.5 VARIATION AND MULTIVARIABLE FUNCTIONS

TOPICS

1. Direct variation
2. Inverse variation
3. Joint variation
4. Multivariable functions

TOPIC 1: Direct Variation

A number of natural phenomena exhibit the mathematical property of variation: one quantity varies (or changes) as a result of a change in another quantity. One example is the electrostatic force of attraction between two oppositely charged particles, which varies in response to the distance between the particles. Another example is the distance traveled by a falling object, which varies as time increases. The principle underlying variation is that of functional dependence; in the first example, the force of attraction is a function of distance, and in the second example the distance traveled is a function of time.

We have now gained enough familiarity with functions that we can define the most common forms of variation.

DEFINITION: Direct Variation

We say that y **varies directly as** (or **is proportional to**) the nth power of x if there is a nonzero constant k (called the **constant of proportionality**) such that

$$y = kx^n.$$

Many variation problems involve determining what, exactly, the constant of proportionality is in a given situation. This can be easily done if enough information is given about how the various quantities in the problem vary with respect to one another; and once k is determined, many other questions can be answered. The following example illustrates the solution of a typical direct variation problem.

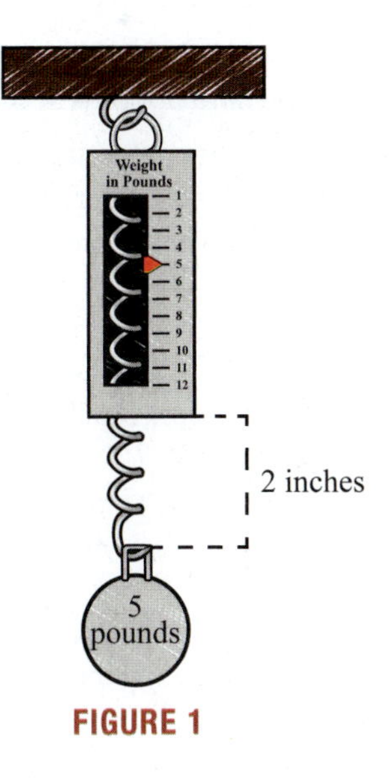

FIGURE 1

Example 1: Direct Variation

Hooke's Law says that the force exerted on a spring varies directly with the distance that the spring is stretched. If a 5-pound weight suspended on a spring scale stretches the spring 2 inches, how far will a 13-pound weight stretch it?

Solution

The first sentence tells us that $F = kx$, where F represents the force exerted by the spring and x represents the distance that the spring is stretched. When a weight is suspended on a spring scale (and is stationary), the force exerted upward by the spring must equal the force downward due to gravity, so the spring exerts a force of 5 pounds when a 5-pound weight is suspended from it. So the second sentence tells us that $5 = 2k$, or $k = \frac{5}{2}$. We can now answer the question.

$$13 = \frac{5}{2}x$$
$$\frac{26}{5} = x$$

So the spring stretches $\frac{26}{5} = 5.2$ inches when a 13-pound weight is suspended from it.

TOPIC 2: Inverse Variation

In many situations, an increase in one quantity results in a corresponding decrease in another quantity, and vice versa. Again, this is a natural illustration of a functional relationship between quantities, and an appropriate name for this type of relationship is *inverse variation*.

DEFINITION: Inverse Variation

We say that y **varies inversely as** (or **is inversely proportional to**) the nth power of x if there is a nonzero constant k such that

$$y = \frac{k}{x^n}.$$

The method of solving an inverse variation problem is identical to that seen in the first example. First, write an equation that expresses the nature of the relationship (including the as-yet-unknown constant of proportionality). Second, use the given information to determine the constant of proportionality. Third, use the knowledge gained to answer the question.

Example 2: Inverse Variation

The weight of a person, relative to Earth, is inversely proportional to the square of the person's distance from the center of Earth. Using a radius for Earth of 6370 kilometers, how much does a 180-pound man weigh when flying in a jet 9 kilometers above Earth's surface?

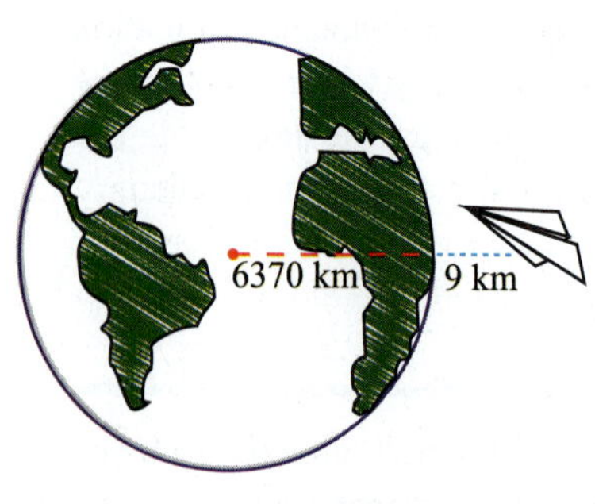

FIGURE 2

Solution

If we let W stand for the weight of a person and d the distance between the person and Earth's center, the first sentence tells us that

$$W = \frac{k}{d^2}.$$

The second sentence gives us enough information to determine k. Namely, we know that $W = 180$ (pounds) when $d = 6370$ (kilometers). Solving the equation for k and substituting in the values that we know, we obtain

$$k = Wd^2 = (180)(6370)^2 \approx 7.3 \times 10^9.$$

When the man is 9 kilometers above Earth's surface, we know $d = 6379$, so the man's weight while flying is

$$W = \frac{(180)(6370)^2}{(6379)^2}$$
$$\approx 179.49 \text{ pounds.}$$

Flying is not, therefore, a terribly effective way to lose weight.

TOPIC 3: Joint Variation

In more complicated situations, it may be necessary to identify more than two variables and to express how the variables relate to one another. And it may very well be the case that one quantity varies directly with respect to some variables and inversely with respect to others. For instance, Newton's Universal Law of Gravitation says that the force of gravitational attraction F between two bodies of mass m_1 and mass m_2 varies directly as the product of the masses and inversely as the square of the distance between the masses: $F = \frac{km_1m_2}{d^2}$.

When one quantity varies directly as two or more other quantities, the word *jointly* is often used.

DEFINITION: Joint Variation

We say that z **varies jointly as** (or **is jointly proportional to**) the nth power of x and the mth power of y if there is a nonzero constant k such that

$$z = kx^n y^m.$$

Example 3: Joint Variation

The volume of a right circular cylinder varies jointly as the height and the square of the radius. Express this relationship in equation form.

Solution

This simple problem merely asks for the form of the variation equation. If we let V stand for the volume of a right circular cylinder, r for its radius, and h for its height, we would write

$$V = kr^2h.$$

We are already familiar with this volume formula and know that the constant of proportionality is actually π, so we could provide more information and write

$$V = \pi r^2 h.$$

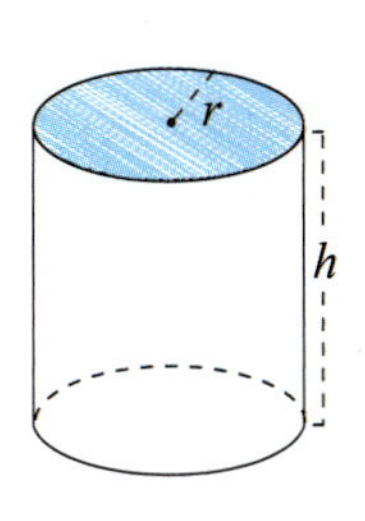

FIGURE 3

TOPIC 4: Multivariable Functions

The topic of variation provides an excellent opportunity to introduce functions that depend on two or more arguments. Examples abound in both pure and applied mathematics.

Consider again how the force of gravity F between two objects depends on the masses m_1 and m_2 of the objects and the distance d between them:

$$F = \frac{km_1m_2}{d^2}.$$

If we change any of the three quantities m_1, m_2, and d, the force F changes in response. A slight extension of our familiar function notation leads us then to express F as a function of m_1, m_2, and d and to write

$$F(m_1, m_2, d) = \frac{km_1m_2}{d^2}.$$

In fact, we can be a bit more precise and replace the constant of proportionality k with G, the universal gravitational constant. Through many measurements in many different experiments, G has been determined to be approximately $6.67\times10^{-11}\ \text{N}\cdot\text{m}^2/\text{kg}^2$ (N stands for the unit of force called the *newton*; 1 newton of force gives a mass of 1 kg an acceleration of $1\ \text{m/s}^2$). If we use this value for G, we must be sure to measure the masses of the objects in kilograms and the distance between them in meters.

The next example illustrates an application of this function to which all of us on Earth can relate.

Example 4: Finding the Force of Gravity

Determine the approximate force of gravitational attraction between Earth and the moon.

Solution

The mass of Earth is approximately 6.0×10^{24} kg and the mass of the moon is approximately 7.4×10^{22} kg. The distance between these two bodies varies, but on average it is 3.8×10^{8} m. Using function notation, we would write

$$\begin{aligned} F\left(6.0\times10^{24}, 7.4\times10^{22}, 3.8\times10^{8}\right) &= \frac{\left(6.67\times10^{-11}\right)\left(6.0\times10^{24}\right)\left(7.4\times10^{22}\right)}{\left(3.8\times10^{8}\right)^2} \\ &= 2.1\times10^{20}\ \text{N}. \end{aligned}$$

It is this force of mutual attraction that keeps the moon in orbit around Earth.

3.5 EXERCISES

PRACTICE

Mathematical modeling is the process of finding a function that describes how quantities or variables relate to one another. The function is called the mathematical model. (Mathematical modeling will be studied in far greater detail in Section 3.6.) Find the mathematical model for each of the following verbal statements.

1. A varies directly as the product of b and h.

2. V varies directly as the product of four-thirds and r cubed.

3. W varies inversely as d squared.

4. P varies inversely as V.

5. r varies inversely as t.

6. S varies directly as the product of four and r squared.

7. x varies jointly as the cube of y and the square of z.

8. a varies jointly as the square of b and inversely as c.

Find the mathematical model for each of the following verbal statements and then use it to solve for the unknown variable. See Examples 1, 2, and 3.

9. Suppose that y varies directly as the square root of x and that $y = 36$ when $x = 16$. What is y when $x = 20$?

10. Suppose that y varies inversely as the cube of x and that $y = 0.005$ when $x = 10$. What is y when $x = 5$?

11. Suppose that y varies directly as the cube root of x and that $y = 75$ when $y = 125$. What is y when $y = 128$?

12. Suppose that y is proportional to the 5th power of x and that $y = 96$ when $x = 2$. What is y when $x = 5$?

13. Suppose that y varies inversely as the square of x and that $y = 3$ when $x = 4$. What is y when $x = 8$?

14. Suppose that y varies inversely as the square of x and that $y = 8$ when $x = 6$. What is y when $x = 20$?

15. Suppose that y is inversely proportional to the 4th power of x and that $y = 15$ when $x = 4$. What is y when $x = 20$?

16. Suppose that z varies jointly as the square of x and the cube of y and that $z = 768$ when $x = 4$ and $y = 2$. What is z when $x = 3$ and $y = 2$?

17. Suppose that z is jointly proportional to x and y and that $z = 90$ when $x = 1.5$ and $y = 3$. What is z when $x = 0.8$ and $y = 7$?

18. Suppose that z is jointly proportional to x and the cube of y and that $z = 9828$ when $x = 13$ and $y = 6$. What is z when $x = 7$ and $y = 8$?

19. Suppose that z varies directly as the square of x and inversely as y. If $z = 36$ when $x = 6$ and $y = 7$, what value does z have when $x = 12$ and $y = 21$?

20. The quantity F is jointly proportional to a and b and varies inversely as c. If $F = 10$ when $a = 6$, $b = 5$, and $c = 2$, what is the value of F when $a = 12$, $b = 6$, and $c = 3$?

21. The variable a is proportional to $\sqrt{b}$. If $a = 15$ when $b = 9$, what is a when $b = 12$?

22. The variable a varies directly as b. If $a = 3$ when $b = 9$, what is a when $b = 7$?

23. The variable a varies directly as the square of b. If $a = 9$ when $b = 2$, what is a when $b = 4$?

24. The variable a is proportional to the square of b and varies inversely as the square root of c. If $a = 108$ when $b = 6$ and $c = 4$, what is a when $b = 4$ and $c = 9$?

25. The variable a varies jointly as b and c. If $a = 210$ when $b = 14$ and $c = 5$, what is the value of a when $b = 6$ and $c = 6$?

26. The variable a varies directly as the cube of b and inversely as c. If $a = 9$ when $b = 6$ and $c = 7$, what is the value of a when $b = 3$ and $c = 21$?

APPLICATIONS

27. The distance that an object falls from rest, when air resistance is negligible, varies directly as the square of the time. A stone dropped from rest travels 144 feet in the first 3 seconds. How far does it travel in the first 4 seconds?

28. A record store manager observes that the number of records sold seems to vary inversely as the price per record. If the store sells 840 records per week when the price per record is \$15.99, how many does he expect to sell if he lowers the price to \$14.99?

29. A person's Body Mass Index (BMI) is used by physicians to determine if a patient's weight falls within reasonable guidelines relative to the patient's height. The BMI varies directly as a person's weight in pounds and inversely as the square of a person's height in inches. Given that a 6-foot-tall man weighing 180 pounds has a BMI of 24.41, what is the BMI of a woman weighing 120 pounds with a height of 5 feet 4 inches?

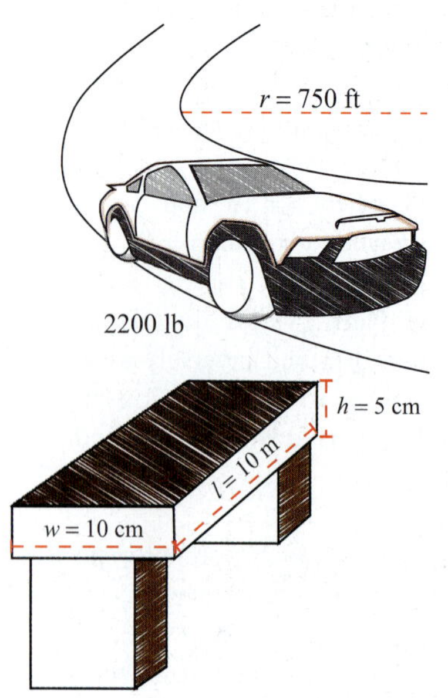

30. The force necessary to keep a car from skidding as it travels along a circular arc varies directly as the product of the weight of the car and the square of the car's speed, and inversely as the radius of the arc. If it takes 241 pounds of force to keep a 2200-pound car moving 35 miles per hour on an arc whose radius is 750 feet, how many pounds of force would be required if the car were to travel 40 miles per hour?

31. If a beam of width w, height h, and length l is supported at both ends, the maximum load that the beam can hold varies directly as the product of the width and the square of the height, and inversely as the length. A given beam 10 meters long with a width of 10 centimeters and a height of 5 centimeters can hold a load of 200 kilograms when the beam is supported at both ends. If the supports are moved inward so that the effective length of the beam is shorter, the beam can support more load. What should the distance between the supports be if the beam has to hold a load of 300 kilograms?

32. In a simple electric circuit connecting a battery and a light bulb, the current I varies directly as the voltage V but inversely as the resistance R. When a 1.5-volt battery is connected to a light bulb with resistance 0.3 ohms (Ω), the current that travels through the circuit is 5 amps. Find the current if the same light bulb is connected to a 6-volt battery.

33. The amount of time it takes for water to flow down a drainage pipe is inversely proportional to the square of the radius of the pipe. If a pipe of radius 1 cm can empty a sink in 25 seconds, find the radius of a pipe that would allow the sink to drain completely in 16 seconds.

34. The perimeter of a square varies directly as the length of the side of a square. If the perimeter of a square is 308 inches when one side is 77 inches, what is the perimeter of a square when the side is 133 inches?

35. The circumference of a circle varies directly as the diameter. A circular pizza slice has a length of 6.5 inches when the circumference of the pizza is 40.82 inches. What would the circumference of a pizza be if the pizza slice has a length of 5.5 inches?

36. The volume of a cylinder varies jointly as its height and the square of its radius. If a cylinder has the measurements $V = 301.44$ cubic inches, $r = 4$ inches, and $h = 6$ inches, what is the volume of a cylinder that has a radius of 6 inches and a height of 8 inches?

37. The surface area of a right circular cylinder varies directly as the sum of the radius times the height and the square of the radius. With a height of 18 in. and a radius of 7 in., the surface area of a right circular cylinder is 1099 in.^{2} What would the surface area be if the height equaled 5 in. and the radius equaled 3.2 in.?

38. The gravitational force, F, between an object and Earth is inversely proportional to the square of the distance from the object to the center of Earth. If an astronaut weighs 193 pounds on the surface of Earth, what will this astronaut weigh 1000 miles above Earth? Assume that the radius of Earth is 4000 miles.

39. In an electrical schematic, the voltage across a load is directly proportional to the power used by the load but inversely proportional to the current through the load. If a computer is connected to a wall outlet and the computer needs 18 volts to run and absorbs 54 watts of power, the current through the computer is 3 amps. Find the power absorbed by the computer if the same 18-volt computer is attached to a circuit with a loop current of 0.5 amps.

40. A hot dog vendor has determined that the number of hot dogs she sells a day is inversely proportional to the price she charges. The vendor wants to decide if increasing her price by 50 cents will drive away too many customers. On average, she sells 80 hot dogs a day at a price of \$3.50. How many hot dogs can she expect to sell if the price is increased by 50 cents?

41. The price of gasoline purchased varies directly with the number of gallons of gas purchased. If 16 gallons of gas are purchased for \$34.40, what is the price of purchasing 20 gallons?

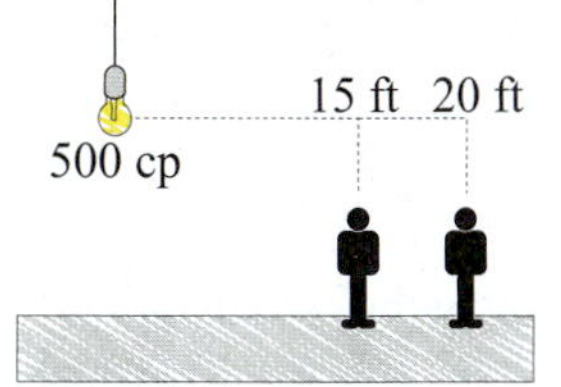

42. The illumination I of a light source varies directly as the intensity i and inversely as the square of the distance d. If a light source with an intensity of 500 cp (candlepower) has an illumination of 20 fc (foot-candles) at a distance of 15 feet, what is the illumination at a distance of 20 feet?

43. The resistance of a wire varies directly as its length and inversely as the square of the diameter. When a wire is 500 feet long and has a diameter of 0.015 in., it has a resistance of 20 ohms. What is the resistance of a wire that is 1200 feet long and has a diameter of 0.025 in.?

In Exercises 44–45, use Hooke's Law, which says that the force exerted on a spring varies directly with the distance that the spring is stretched.

44. A hanging spring will stretch 9 cm if a weight of 15 g is placed on the end of the spring. How far will the spring stretch if the weight is increased to 20 g?

45. If a 32-pound weight suspended on a spring scale stretches the spring 17 inches, how far will a 37-pound weight stretch the spring?

In Exercises 46–47, use Boyle's Law, which says that at a constant temperature, the volume of a gas in a container varies inversely as the pressure on the gas.

46. If the volume is 100 cubic centimeters under a pressure of 800 pascals, what would be the volume of the gas if the pressure was decreased to 400 pascals?

47. If a gas has a volume of 252 cubic inches under a pressure of 5 pounds per square inch, what will its volume be if the pressure is increased to 6 pounds per square inch?

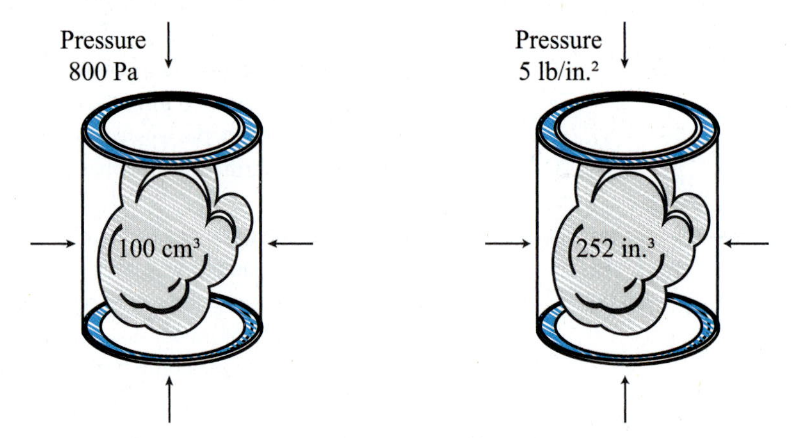

In Exercises 48–50, express the indicated quantities as functions of the other variables. See Example 4.

48. A person's body mass index (BMI) varies directly as a person's weight in pounds and inversely as the square of a person's height in inches. Given that a 6-foot-tall man weighing 180 pounds has a BMI of 24.41, express BMI as a function of weight (w) and height (h).

49. The electric pressure varies directly as the square of the surface charge density (σ) and inversely as the permittivity (ε). If the surface charge density is 6 coulombs per unit area and the free space permittivity equals 3, the pressure is equal to 6 N/m^2. Express the electric pressure as a function of surface charge density and permittivity.

50. The volume of a right circular cylinder varies directly as the radius squared times the height of the cylinder. If the radius is 7 and the height is 4, the volume is equal to 615.44. Determine the expression of the volume of a right circular cylinder.

3.6 MATHEMATICAL MODELS

TOPICS

1. Constructing mathematical models
2. Interpolation and extrapolation

TOPIC 1: Constructing Mathematical Models

Many applications of mathematics are motivated by the desire to understand how one variable relates to other variables—we studied a number of such relationships in the last section. For example, a common goal in ecology is to understand how the population of an organism depends on various environmental variables; and much effort in business and economics goes into finding relationships between cost, revenue, interest rates, and other quantities. The process of developing such relationships and then using and interpreting them is called *mathematical modeling*. A mathematical model for a given application may take the form of an equation involving the relevant variables or a function in which the dependence of one variable on others is more explicit. In this section, we will focus on how such models are constructed and used.

In general, the key steps in mathematical modeling are the following:

1. With a given question or goal in mind, identify and name the relevant variable quantities.
2. Translate verbal, graphical, or other knowledge about the variables into an equation or function. Determine the domain of each independent variable that makes sense in the context of the question.
3. Use the constructed equation or function to answer the given question. Frequently, the model is then also used for predictive purposes and/or the relationship between the variables is analyzed to provide better understanding of the original question.

We will illustrate these steps with several examples.

Example 1: Modeling the Value of Computers

A financial-services firm plans to purchase a number of computers, use them for five years, and then trade them in and replace them. The computers they want to buy cost \$2000 each and they've been told they'll get a 10% credit of the purchase price at trade-in time. For accounting purposes, they want to be able to assign a value to the computers at any given point in time after purchase, up to five years.

a. Assuming a linear depreciation in the value of the computers over the five-year time frame, how does the value of each computer vary as a function of time?

b. At what point will the computers have lost half of their initial value?

Solution

a. Since we are asking for the value of each computer over time, we will let V stand for *value* (measured in dollars) and t for *time* (measured in years). With the information we're given, V is a function of time, and time alone. Further, we are given that $V(0)=2000$, and that $V(5)=10\%$ of 2000, or $V(5)=200$. This means that each computer loses \$1800 of value over five years, so each year the loss in value is $\frac{\$1800}{5}=\$360$. Thus, at the end of one year, $V(1)=2000-(360)(1)$, at the end of two years, $V(2)=2000-(360)(2)$, and in general, at the end of t years, $V(t)=2000-360t$ for $0\le t\le 5$. The function $V(t)=2000-360t$ thus provides the answer to the first question, as we can use it to determine the value of each computer at any time t. To check our work, note that V is indeed a linear function of t and that $V(0)=2000-360(0)=2000$ and $V(5)=2000-360(5)=2000-1800=200$.

b. To answer the second question, we're looking for the value of t at which $V(t)=\frac{2000}{2}=1000$. Solving this equation for t, we obtain

$$\begin{aligned} 2000-360t &= 1000 \\ -360t &= -1000 \\ t &\approx 2.78 \text{ years.} \end{aligned}$$

Since $0.78\cdot 12=9.36$, the value of each computer will be halved a little more than 2 years and 9 months after purchase.

The construction of a mathematical model often calls for the use of well-known formulas or formulas that can be looked up, as seen in the next two examples.

Example 2: Modeling the Area of Garden Plots

Anne and Kay plan to use 600 feet of fencing to construct two large rectangular garden plots along a river, as shown in Figure 1, with the plots sharing a common side and no fence along the river. The overall dimensions can vary a bit, but they want to ensure that the total enclosed area is at least 20,000 ft^2.

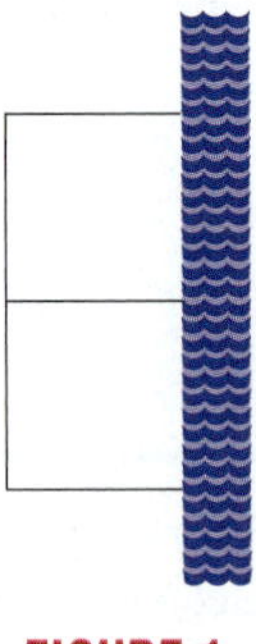

FIGURE 1

a. Can this be done? If so, what are the bounds on the dimensions of the plots that will guarantee at least this much enclosed area?

b. What is the maximum possible enclosed area?

Solution

Since Anne and Kay are concerned with the total enclosed area of the two plots, we will represent that quantity with an appropriate label, say A. Since A is the area of a rectangle (divided into two subrectangles), we are led to assign labels to the two dimensions of the rectangle; we will use x for one dimension and y for the other, as shown in Figure 2. Note that there will actually be three runs of fencing of length x, so the total amount of fencing used will be $3x+y$.

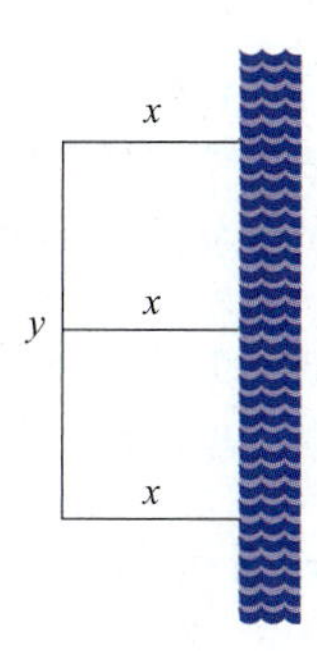

FIGURE 2

Our knowledge of the area of a rectangle tells us that $A = xy$. This is fine as far as it goes, but this relationship between A, x, and y doesn't take into account the fact that $3x + y$ can be no larger than 600. We can reduce the number of variables by letting $y = 600 - 3x$ (which means we will use all 600 feet of fencing) and substituting this expression for y. With this substitution, we can write A as a function of x alone to obtain

$$\begin{aligned} A(x) &= x(600 - 3x) \\ &= -3x^2 + 600x. \end{aligned}$$

Note that the domain of A (that is, the set of values for x that make sense in this problem) is the interval $[0, 200]$, with $A = 0$ at both endpoints of the domain.

a. To answer the first question, we now ask whether the inequality $-3x^2 + 600x \geq 20,000$ has a solution. We first use the quadratic formula to solve the equation $-3x^2 + 600x = 20,000$ to obtain the solutions

$$x = \frac{100(3 \pm \sqrt{3})}{3} \approx 42.3 \text{ ft and } 157.7 \text{ ft.}$$

It's also instructive to graph the function $A(x)$, as shown in Figure 3. This graph includes a horizontal line corresponding to an area of 20,000 ft^2, which illustrates that any choice of x between 42.3 ft and 157.7 ft will lead to an overall area of the combined plots of at least 20,000 ft^2.

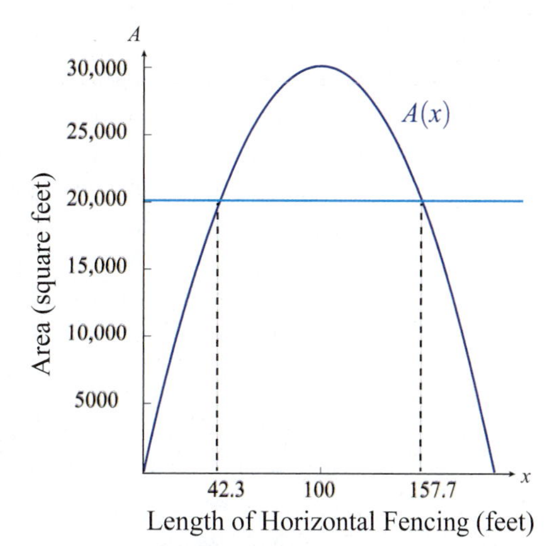

FIGURE 3

b. To answer the second question, the graph of A certainly hints that the maximum combined area is approximately 30,000 ft^2 and that this would be achieved by letting $x = 100$ (and consequently letting $y = 30,000$). We can prove that 30,000 ft^2 is indeed the maximum possible combined area by using the methods of Section 3.3 to identify $(100, 30,000)$ as the vertex of the downward-opening parabolic function $A(x)$.

Example 3: Modeling the Weight of a Person with a Given Mass

Consider the weight, or force of gravitational attraction, of a person with a given mass.

a. How does the force of gravitational attraction between a 75 kg person and Earth vary as a function of the person's height above the surface of Earth?

b. How does the force at sea level compare to the force when the person is in a jet at 30,000 ft?

c. What if the person were on the International Space Station?

d. How far above the surface of Earth would the person have to be for the force to be half as much as the force at sea level?

Solution

Unlike the formula for the area of a rectangle, it's likely that the force for gravitational attraction between two masses, and such information as the mass and radius of Earth and the height of the International Space Station, will need to be looked up. But all the details to model the requested force as a function of height are readily found, and indeed some have appeared already in this text. The force F in newtons between two masses m_1 and m_2 (in kilograms) whose centers of mass are a distance d meters apart is given by the formula

$$F = G\frac{m_1 m_2}{d^2},$$

where the universal gravitational constant G is approximately $6.67\times10^{-11}\ \mathrm{N}\cdot\mathrm{m}^2/\mathrm{kg}^2$. We also need the mass and radius of Earth, which we find to be approximately 5.97×10^{24} kg and 6.38×10^{6} m, respectively.

a. If we let m_1 be the mass of Earth, m_2 the mass of the person (given as 75 kg for the person in question), and h the distance between the person's center of mass and the surface of Earth, we have

$$F(h) = \left(6.67\times10^{-11}\right)\frac{\left(5.97\times10^{24}\right)(75)}{\left(6.38\times10^{6}+h\right)^2}.$$

Note that in writing this formula, we've used function notation to point out that F is now simply a function of the variable h—every other quantity is a fixed constant.

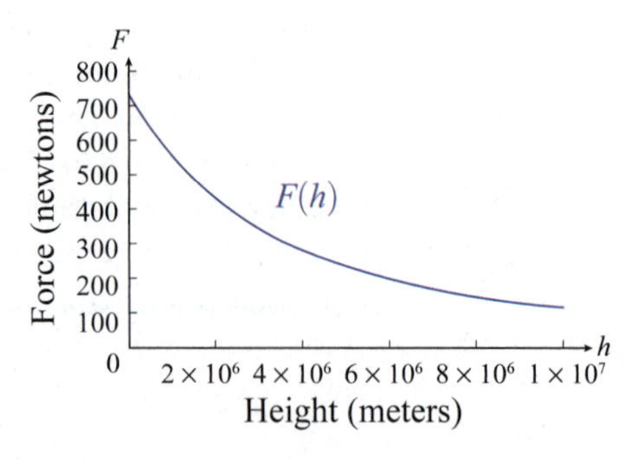

FIGURE 4

Before we proceed to answer the remaining specific questions we were asked, and to help us better understand the magnitude of the numbers involved, it's helpful to experiment with graphs of F as a function of h. The graph shown in Figure 4 indicates that we need to allow h to be quite large in order to see the overall shape of F. Here, we've graphed F over the h-interval from 0 meters to 10,000,000 meters (about 6214 miles).

b. Note that the force of attraction at sea level $(h = 0 \text{ m})$ is $F(0) \approx 734 \text{ N}$ and the force of attraction at 30,000 feet (about 9144 m) is $F(9144) \approx 732 \text{ N}$, or about 99.7% of the sea-level force.

c. The height of the International Space Station varies a bit, but we will use an average value of 248 miles, about 399,117 meters, to obtain $F(399{,}117) \approx 650 \text{ N}$, or about 88.6% of the sea-level force.

d. To answer the final question, we solve the equation $F(h) = \frac{734}{2} = 367$ as follows.

$$\begin{aligned}
(6.67 \times 10^{-11}) \frac{(5.97 \times 10^{24})(75)}{(6.38 \times 10^{6} + h)^2} &= 367 \\
\frac{(6.67 \times 10^{-11})(5.97 \times 10^{24})(75)}{367} &= (6.38 \times 10^{6} + h)^2 \\
8.14 \times 10^{13} &\approx (6.38 \times 10^{6} + h)^2 \\
9.02 \times 10^{6} &\approx 6.38 \times 10^{6} + h \\
h &\approx 2.64 \times 10^{6} \text{ m}
\end{aligned}$$

Note that we omitted the negative root in the second-to-last step on the left side of the equation since the height cannot be negative. Putting this into context, this distance is about 41% of the radius of Earth itself.

The basis of many business models is the relationship between the cost, revenue, and profit functions. If we let $C(x)$ represent the cost of producing x items of a certain product, and $R(x)$ the revenue in selling the x items, then the profit earned by the x items is revenue minus cost, or $P(x) = R(x) - C(x)$. Our last construction illustrates this important relationship.

Example 4: Modeling Profit Earned from Selling Tables

A furniture maker has been in the habit of making 10 copies of a particular style of small table during a production run, and has been able to sell them in a reasonable time frame at \$100 each. After carefully analyzing his market and the demand for the table, he has determined that he can gain one additional customer in the same time frame for every \$2 decrease in price, and conversely lose one customer with every \$2 increase in price. He has a fixed cost of \$450 for a production run of the tables, and each table costs an additional \$20 to make.

a. Assuming that he prices the tables in accordance with his market analysis, what profit will he earn by selling a given number of tables?

b. Is there a production interval that he needs to fall within in order to be profitable?

c. If there is a profitable interval, what is the ideal number of tables to produce?

d. What price corresponds to that number of tables?

Solution

The cost function is straightforward, as it is simply the fixed cost of \$450 plus another \$20 for every table made. So if we let x represent the number of tables made in a production run (and consequently sold during the reasonable time frame), then the cost function for making x tables is $C(x) = 450 + 20x$.

The revenue function in this situation requires a bit more work to determine. We can organize the given information and work toward the desired function by noting that when $x = 10$, the corresponding price is \$100 and the revenue taken in is $(10)(\$100) = \$1000$. Similar computations for prices above and below \$100 will help us identify a pattern.

x = Number of tables sold	Price per table	R = Total revenue
$10 + (-1) = 9$	$100 - 2(-1) = \$102$	$(9)(\$102) = \$918$
$10 + (0) = 10$	$100 - 2(0) = \$100$	$(10)(\$100) = \$1000$
$10 + (1) = 11$	$100 - 2(1) = \$98$	$(11)(\$98) = \$1078$

In general, if we let n be the number of \$2 increments above or below \$100 in the price, then a price of $100 - 2n$ corresponds to $x = 10 + n$. But there's no need to work with both variables x and n, as we can rewrite the equation as $n = x - 10$ and note that the price corresponding to x tables sold is

$$100 - 2n = 100 - 2(x - 10) = 120 - 2x,$$

which tells us that the total revenue raised by selling x tables is

$$R(x) = x(120 - 2x) = -2x^2 + 120x.$$

a. This means the profit function for this situation is

$$\begin{aligned} P(x) &= R(x) - C(x) \\ &= (-2x^2 + 120x) - (450 + 20x) \\ &= -2x^2 + 100x - 450, \end{aligned}$$

providing the answer to the first question.

b. We recognize the graph of this function as a downward-opening parabola, which should immediately alert us to the fact that, for small enough x and large enough x, $P(x)$ is guaranteed to be negative, an outcome to be avoided. Indeed, we know that in general the graph of a downward-opening parabola need not have any portion above the x-axis. But if there *is* a portion of the graph that lies above the x-axis, that is the portion we want to identify, as that is the profitable interval for this product.

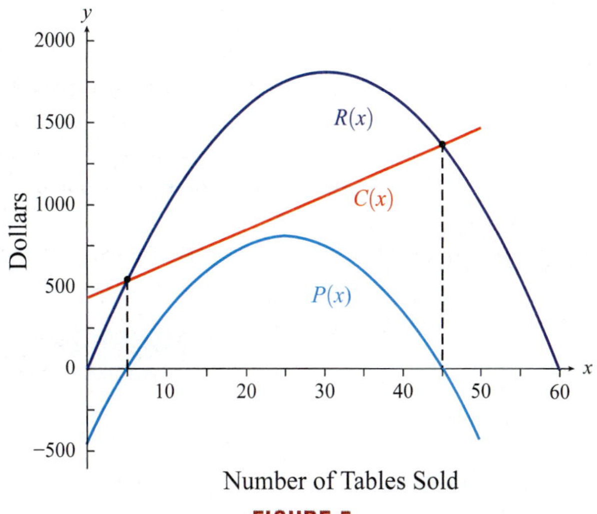

FIGURE 5

Figure 5 shows the graphs of all three functions, $C(x)$, $R(x)$, and $P(x)$, over the interval $[0,50]$. The graphs tell us many things, among them that $P(x)$ is indeed positive (so the product is profitable) over a certain production interval, which we will soon identify. The endpoints of the profitable interval are those values of x for which $C(x)=R(x)$, which are known as *break-even points* in this context. Also, as we would expect with a downward-opening parabola, the vertex represents the point at which maximum profit can be achieved, so we will want to identify that as well.

The profitable interval of production lies between the two x-intercepts of $P(x)$, which we can find by factoring.

$$\begin{aligned} P(x) &= -2x^2+100x-450 \\ &= -2(x^2-50x+225) \\ &= -2(x-5)(x-45) \end{aligned}$$

The solutions of $P(x)=0$ are thus $x=5$ and $x=45$, so the desired interval is $[5,45]$.

c. The methods of Section 3.3 for locating the vertex of a parabola tell us that the vertex of the graph of P is $(25, P(25))$, or $(25,800)$, so the ideal number of tables to produce is 25, which will maximize profit at \$800.

d. The price per table that corresponds to selling $x=25$ tables is $120-2(25)=\$70$.

TOPIC 2: Interpolation and Extrapolation

It's often the case that the basis for the construction of a mathematical model is simply a collection of data points. If we again have the goal of wanting to understand how one variable depends on another or if we want to use the data to infer or predict behavior, the set of data alone may be inadequate for the task. In such cases, we must make additional assumptions about the data, as we now discuss.

We will illustrate the key ideas with an example. Table 1 is a listing of the United States population in every census year from 1950 to 2010.

Year	Population
1950	151,325,798
1960	179,323,175
1970	203,302,031
1980	226,542,199
1990	248,709,873
2000	281,421,906
2010	308,745,538

Source: US Census Bureau

TABLE 1: US Population, 1950–2010

It's easy to imagine how this collection of data implicitly defines a function, say P, that returns the population of the country at year t, even for years that don't end in 0 or for years that precede 1950 or follow 2010. Just to emphasize the point, the table tells us, for instance, that $P(1950) = 151,325,798$ and $P(2010) = 308,745,538$. But to extend P to years between those in the table, or outside the interval $[1950, 2010]$, we need to make some estimates. The acts of estimating values of a function based on given data are called *interpolation* and *extrapolation*, and are perhaps best introduced with a picture. Figure 6 contains a plot of the seven data points from Table 1, along with a line that seems to approximate the general behavior of the points well.

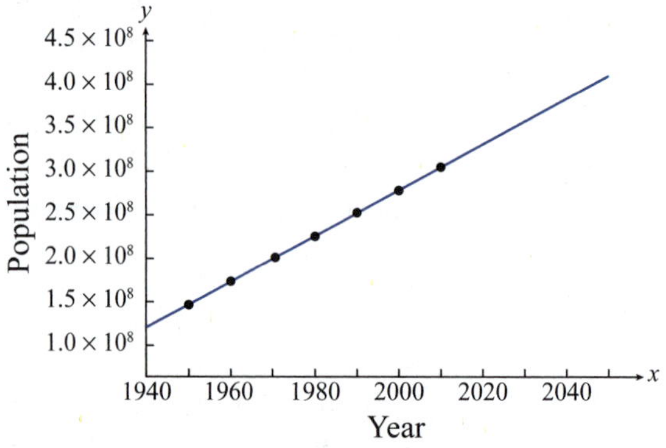

FIGURE 6: Graph of US Population Data, 1950–2010

Year	Population
1790	3,929,214
1800	5,308,483
1810	7,239,881
1820	9,638,453
1830	12,860,702
1840	17,063,353
1850	23,191,876
1860	31,443,321
1870	38,558,371
1880	50,189,209
1890	62,979,766
1900	76,212,168
1910	92,228,496
1920	106,021,537
1930	123,202,624
1940	132,164,569
1950	151,325,798
1960	179,323,175
1970	203,302,031
1980	226,542,199
1990	248,709,873
2000	281,421,906
2010	308,745,538

Source: US Census Bureau

TABLE 2: US Population, 1790–2010

To **interpolate** a value from a table or graph means to arrive at an estimate for the value of the implied function *between* two known data points, while to **extrapolate** means to guess at a value *beyond* the given points. In our example, we might interpolate a value for $P(1985)$ to be 240 million (2.4×10^8 in scientific notation) and extrapolate a value for $P(2060)$ to be a bit less than 440 million. However, this census-data example is also a good illustration of the limitations of tables and graphs. Figure 7 contains the plot of an expanded version of the same US population data (Table 2), along with a curve that approximates the behavior.

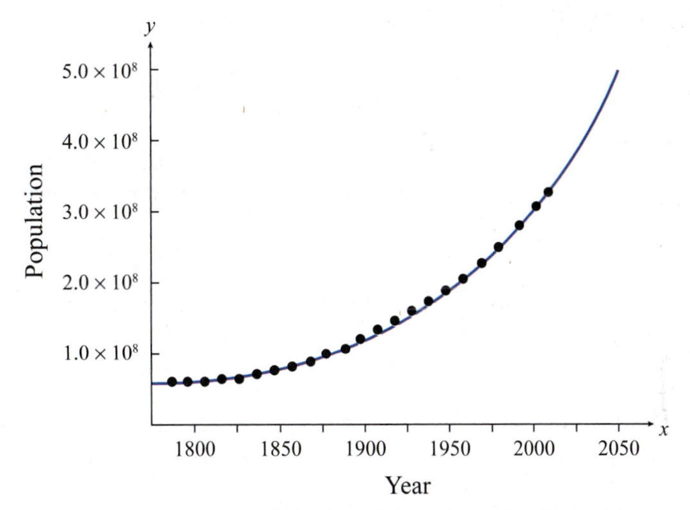

FIGURE 7: Graph of US Population Data, 1790–2010

If we were asked to extrapolate a value for $P(2060)$ on the basis of this expanded graph, we would probably guess a value a bit larger than 460 million, significantly larger than the value from our linear extrapolation.

The line and curve in Figures 6 and 7 are lines of best fit constructed using the least-squares method discussed in Sections 3.2 and 3.3. But as we learned in those sections, the method hinges upon making an assumption about the nature of the data. In Figure 6 we assumed the seven data points lie along a straight line, and the linear-regression algorithm gives us a line of best fit with the equation

$$y = 2.57809\times10^6 x - 4.87613\times10^9$$

which, when we let $x = 2060$, gives us a y-value of 4.35×10^8, or 435 million. In Figure 7, the underlying assumption was that the data points can be modeled by a quadratic function, and the curve is the graph of the equation

$$y = 6776.43x^2 - 2.43900\times10^7 x + 2.19523\times10^{10}.$$

When we let $x = 2060$ in this equation, we obtain a y-value of 4.65×10^8, or 30 million people more than the extrapolation using the linear approximation. Clearly, the discrepancy between the two results leaves some doubt as to a reasonable prediction for the population in the year 2060, but in fact the discrepancy points to even larger issues. Is either assumption of a linear relationship or quadratic relationship in the data really warranted? If we think about the nature of populations, neither seems like a realistic reflection of the way populations grow.

For instance, population growth over a long span of time always seems to show either an increase in the rate of growth or else a tapering off in the rate, neither of which is linear behavior. And the vertex that is present in the graph of any quadratic function also doesn't seem to be something typically seen in graphs of population growth. Is there a better assumption for the shape of the graph? The answer is yes, and we will return to the question of curve fitting several more times as we gain experience with more classes of functions.

3.6 EXERCISES

APPLICATIONS

Construct a mathematical model as appropriate for each of the following situations, and then use your model to answer the accompanying questions.

1. A tinsmith wants to make a small windowsill planter from a $20 \text{ cm} \times 60 \text{ cm}$ sheet of copper. She'll form it by cutting equally sized squares from each of the four corners of the sheet, folding up the resulting flaps to form the sides of the planter, and then soldering the four vertical edges.

   a. Construct a model for the volume of the planter based on the side length of the square cut from each corner, and determine the feasible domain for the model.
   b. Is it possible to construct such a planter with a volume of 2000 cm^3 or larger?
   c. What is the maximum possible volume, rounded to the nearest whole number?
   d. If she wants the ratio of the planter's width to height to be 2, what will be the ratio of its length to width?

2. Suppose the tinsmith of the previous problem wants to construct a similar planter (out of a sheet of copper with dimensions to be determined) but with ratios of height to width to length of $1:2:3$.

   a. Construct a model for the volume of such a planter based on the height.
   b. What must be the ratio of the width to length of the original sheet of copper?
   c. Rounding to the nearest whole number, what minimum height of such a planter will have a volume of 2000 cm^3 or larger?

3. A car dealership wants to try out a new leasing arrangement, which will allow a buyer to trade a car back in to the dealership for a certain amount of credit at any time throughout the first three years of ownership. For a car in good condition, the arrangement will value the car at $\frac{1}{3}$ the original price at the end of three years, and will depreciate the value of the car linearly over the course of the three years.

   a. Construct a model for the value at time t (in years) of a car with initial purchase price P, for $0 \leq t \leq 3$.
   b. At what point in time does a car have half its original value?
   c. What is the value of a car at the end of the first year?

4. Picture a person standing a certain distance away from a streetlamp at night, with the streetlamp casting a shadow of the person.

a. Find a model for the length s of the shadow of a person of height h cast by a 15-foot-tall streetlamp d feet away.
b. For a 5-foot-tall person, express the length s as a function of d.
c. How does the shadow of a 5-foot-tall person compare to that of a 6-foot-tall person when both are standing 20 feet from the streetlamp?
d. How far away does the 6-foot-tall person have to be for the person's shadow to be 6 feet long?

5. Consider a point on an arbitrary nonvertical line in the plane.

a. Find a model for the distance d between a point on the line $y = mx + b$ and the origin.
b. What form does the model take for lines that are horizontal?
c. What form does the model take for lines that pass through the origin?
d. What form does the model take for a line that passes through the origin and has slope $\sqrt{3}$?

6. a. Find a model for the product of two nonnegative numbers whose sum is 10.
b. What are the largest and smallest possible products?

7. a. Find a model for the sum of the cubes of two nonnegative numbers whose sum is 10.
b. Expressing the sum of the cubes as a function of one variable, graph the sum.
c. What are the largest and smallest possible sums?

8. a. Find a model for the Body Mass Index (BMI) of a person, given that BMI varies directly as a person's weight in pounds and inversely as the square of the person's height in inches.
b. With the additional information that a 6 ft tall person weighing 175 lb has a BMI of 23.73, what is the BMI of someone 5 ft, 6 in. tall weighing 140 lb?
c. How much weight would the 5 ft, 6 in. tall person from part b. need to gain or lose to have a BMI of 22?
d. What is the BMI for a person with the same weight but two inches taller than the 5 ft, 6 in. person?

9. a. Use the gravitational-attraction model of Example 3 to determine how far away from the surface of Earth a person would have to be to feel the force of attraction felt on the moon (about one-sixth that felt on Earth).
b. How does this distance compare to Earth's radius?
c. Why do astronauts appear to be weightless when they are in orbit so much closer to Earth?

10. a. Find a model for the area of sheet metal that must be used to make a cylindrical can, including both top and bottom, in terms of the can's radius r and height h.
b. Modify the model to be a function of r only, given that the volume of the can is to be 1000 cm^3 (1 liter).
c. Estimate the value of r that will minimize the area of metal required.

11. a. Find a model for the surface area of a cube of side length x in terms of the cube's volume.

b. What is the surface area of a cube that has a volume of 1000 mm^3?

12. a. Find a model for the surface area of a sphere of radius r in terms of the sphere's volume.

b. What is the surface area of a sphere that has a volume of $\frac{500}{3} \text{ mm}^3$?

13. Maria plans to build a rectangular dog run with an area of 1800 ft^2 at the edge of her property. She wants to use a solid fence that costs $\$6/\text{ft}$ for the side that will sit on the edge of her property, but is willing to use fencing that only costs $\$2/\text{ft}$ for the other three sides.

a. Find a model for the total cost of the fencing.

b. Estimate the length of solid fence that will minimize her total cost.

14. *Restaurante Caro* frequently offers a special prix fixe meal and has been charging \$150 per person for the event. At that price, they've been averaging 30 customers each time. Their marketing firm has convinced them that they'll gain a customer for every dollar they lower the cost of the event, and conversely lose a customer for every dollar they raise the cost. Their fixed cost per event is \$1500 and preparing each customer's meal costs an additional \$20.

a. Find the cost, revenue, and profit functions for these prix fixe events.

b. What are the break-even points in terms of customers served?

c. Is there a number of customers that will maximize their profit?

d. If so, what price per person should they charge?

15. A ceramicist made 6 bowls of a certain style and sold them for \$12 each, just breaking even at that price. Each time she starts up her kiln and makes any number of the bowls, she has a fixed cost of \$36, and there is an additional cost in materials to make each bowl. Polling the six customers who bought the bowls, she learned that only half would have bought the same bowl at a price of \$21.

a. Assuming a linear relationship between price per bowl and the number of bowls sold, and given the information she has, find the revenue, cost, and profit functions for selling a certain number of bowls.

b. Is there another break-even point for her product?

c. Is there an ideal number of bowls she should make in each production run?

d. If so, what should she charge per bowl and what maximum profit can she expect?

Interpolate and extrapolate, as appropriate, to answer the questions with the given data. Use a graphing utility to answer questions marked with [graphing icon].

16. Use the linear and quadratic functions of best fit modeling the US population data in Tables 1 and 2 of this section to answer the following questions.

a. How do the interpolated populations for 1955 compare in the two models?

b. What is the extrapolated linear-model population for the year 1800? How do you interpret this result?

c. What is the calculated quadratic-model population for the year 1800? How does this compare to the actual population in 1800 on which the quadratic model is based?

d. Imagine tracing the quadratic model population far back in time, before the US actually existed as a country. What is the extrapolated quadratic model population in the year 1000? How do you interpret this result?

17. [graphing icon] The table below shows the height at half-second intervals of a rock that breaks free from the top of a 200-foot-tall cliff and falls without obstruction to the river below.

Time t (in seconds)	Height (in feet)
0	200
0.5	196
1.0	184
1.5	164
2.0	136
2.5	100
3.0	56

a. Graph the heights (either by hand or with a graphing utility) and estimate the time the rock hits the water.

b. Find the linear function of best fit that models the height of the rock, and graph the function along with the given heights. By the linear model, what is the extrapolated time when the rock hits the water? What is the calculated linear-model height of the rock at time $t = 0$?

c. Find the quadratic function of best fit that models the height of the rock, and graph the function along with the given heights. By the quadratic model, what is the extrapolated time when the rock hits the water? What is the calculated quadratic-model height of the rock at time $t = 0$?

d. Which model appears to be more appropriate?

18. Karen bought a puppy and has been tracking its weight at the end of each month since it was born. She was told by the dog's breeder that the dog should have an adult weight somewhere between 40 and 45 pounds.

End of Month	Weight (in pounds)
1	3
2	5
3	8
4	11
5	17
6	22
7	28
8	32
9	35
10	38
11	40
12	42

a. Find the linear function of best fit that models the dog's weight, and graph the function along with the given weights. By the linear model, what is the extrapolated weight at the end of the second year? How do you interpret this result?

b. Find the quadratic function of best fit that models the dog's weight, and graph the function along with the given weights. By the quadratic model, what is the extrapolated weight at the end of the second year? How do you interpret this result?

CHAPTER 3 PROJECT

Demand Curves and Revenue

The graph of the relationship between the price p of a given product and the quantity x of that product sold is called the **demand curve** for the product. It typically reflects the fact that an increase in price results in lower sales and a decrease in price results in higher sales. The revenue realized by selling quantity x of the product at price p is then given by the function $R(x) = xp$.

1. One common relationship between p and x is an equation of the form

$$p + bx = a,$$

where a and b are positive constants.

a. With the horizontal axis representing x and the vertical axis representing p, sketch the general shape of the demand curve associated with this relationship. (**Hint:** It may help to initially pick specific values for a and b in order to identify the behavior.)

b. In words, how would you describe the dependence of p on x?

c. What value of x corresponds to the largest feasible value for p?

d. What value of p corresponds to the largest feasible value for x?

e. Find a formula for the revenue function $R(x)$ as a function of x alone.

f. What class of function is $R(x)$?

g. How would you describe the graph of $R(x)$?

2. Another common relationship between p and x is an equation of the form

$$px = a,$$

where again a is a positive constant.

a. With the horizontal axis representing x and the vertical axis representing p, sketch the general shape of the demand curve associated with this relationship. (It may again help to initially pick a specific value for a in order to identify the behavior.)

b. What is the smallest feasible value for x in this relationship?

c. What value of p corresponds to the smallest feasible value for x?

d. Is there a smallest feasible value for p in this relationship?

e. Find a formula for the revenue function $R(x)$.

f. What class of function is $R(x)$?

g. How would you describe the graph of $R(x)$?

CHAPTER 3 REVIEW EXERCISES

Section 3.1

For each of the following relations, determine the domain and range and determine whether the relation is a function.

1. $R = \{(-2,9),(-3,-3),(-2,2),(-2,-9)\}$

2. $f = \{(-3,0),(-1,4),(0,3),(3,3),(4,-1)\}$

3. $R = \{(x,2) \mid x \in \mathbb{R}\}$

4. $S = \{(x,4x) \mid x \in \mathbb{Z}\}$

5. $3x - 4y = 17$

6. $x = y^2 - 6$

7. $x = \sqrt{y-4}$

8. $y = -5$

9.

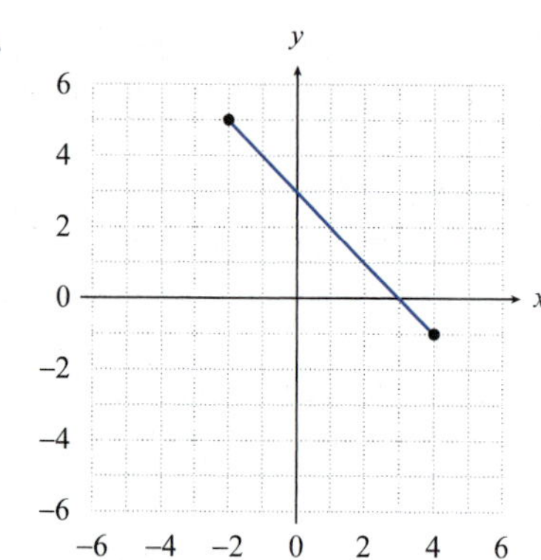

10.

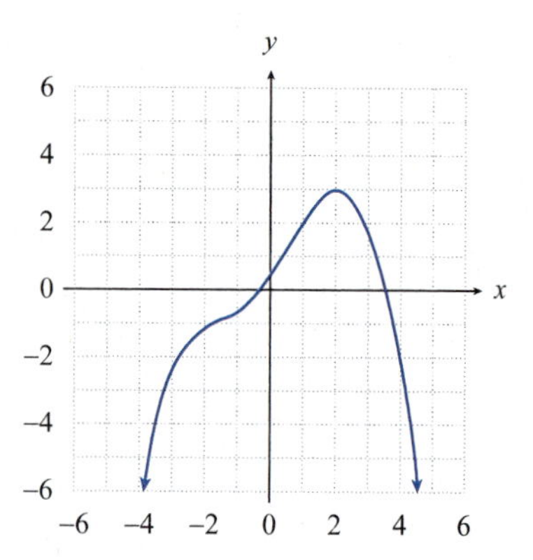

Rewrite each of the following relations as a function of *x*. Then evaluate the function at $x = -2$.

11. $\dfrac{y+4}{\sqrt{x+11}} - 3y = 3(1-y)$

12. $x^2 - 4x + 3y = x + 2y$

Use the graph below of the function *f* to answer each of the following questions.

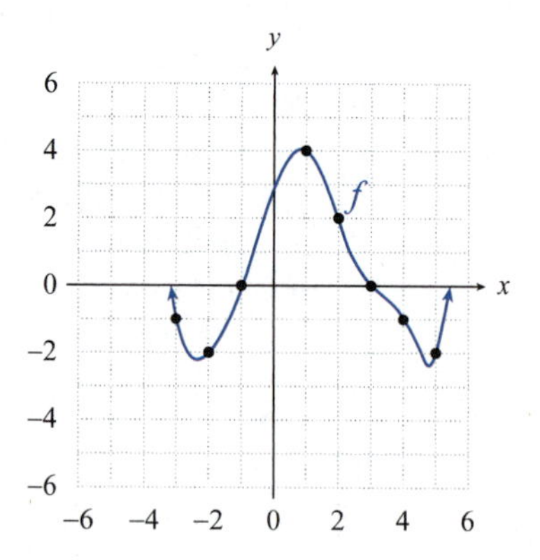

13. What is the value of $f(1)$?

14. What is the value of $f(3)$?

15. For what integer value(s) of *x* is $f(x) = 0$?

16. For what integer values(s) of *x* is $f(x) = -2$?

Given $f(x)=\sqrt{x}$ and $g(x)=\sqrt[3]{x^2}$, evaluate the following.

17. $f(x+h)$

18. $\dfrac{f(x+h)-f(x)}{x+h}$

19. $g(x+h)$

20. $\dfrac{g(x+h)-g(x)}{h}$

Identify the domain, the codomain, and the range of each of the following functions.

21. $g:\mathbb{N}\to\mathbb{R}$ by $g(x)=\dfrac{3x}{4}$

22. $h:\mathbb{R}\to\mathbb{R}$ by $h(x)=5x+1$

Determine the implied domain of each of the following functions.

23. $f(x)=\sqrt{|-x-3|}$

24. $f(x)=\dfrac{x}{1-x}$

Section 3.2

Graph the following linear functions.

25. $f(x)=7x-2$

26. $g(x)=\dfrac{2x-6}{3}$

27. $p(x)=-2x-3$

28. $h(x)=-5+\dfrac{x}{4}$

29. $k(x)=2(1-2x)+x$

30. $f(x)=\dfrac{4x+3}{2}-2x$

Find a formula for the linear function depicted in each of the following graphs.

31.

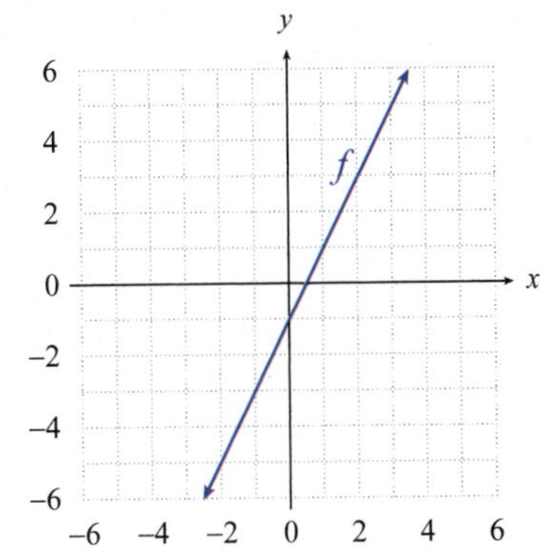

32.

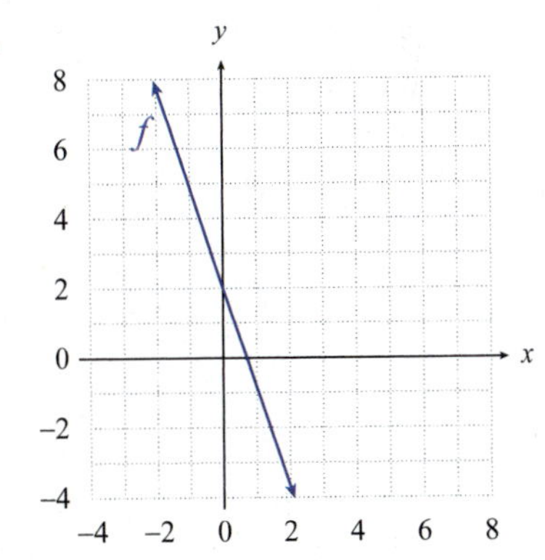

For the given points, **a.** use linear regression to find and graph the line of best fit along with the points, and **b.** find the Pearson correlation coefficient r.

33. $(1,2),(2,2),(3,4),(4,6),(5,5)$

Section 3.3

Graph the following quadratic functions, accurately locating the vertices and x-intercepts (if any).

34. $f(x)=(x-1)^2-1$

35. $g(x)=-(x+3)^2-2$

36. $p(x)=x^2-2$

37. $k(x)=-x^2+4x$

38. $h(x)=x^2+2x-3$

39. $f(x)=-x^2+5$

40. For the parabolic graph, **a.** find a formula for the corresponding quadratic function, and **b.** use the formula to determine the coordinates of the parabola's vertex.

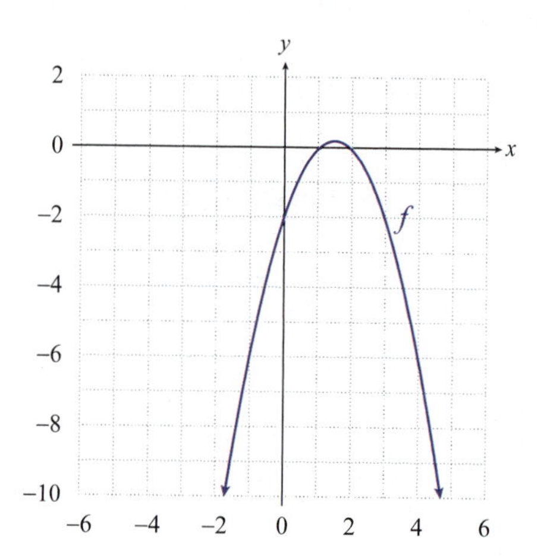

41. Given the points graphed in the following figure, **a.** find the quadratic function that best fits the points, and **b.** use your result to determine the coordinates of the vertex of the best-fitting parabola.

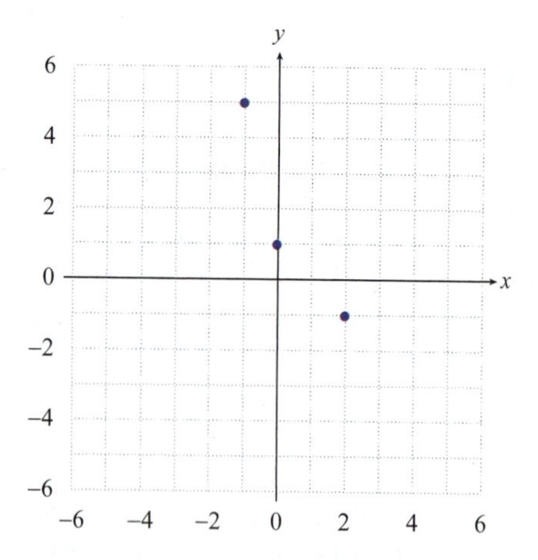

42. The total revenue for McDaniel's Storage Plus is given as the function

$$R(x) = -0.4x^2 + 100x - 5250,$$

where x is the number of storage units rented. What number of units rented produces the maximum revenue?

Section 3.4

Sketch the graphs of the following functions. Pay particular attention to intercepts, if any, and locate these accurately.

43. $f(x) = -4|x|$

44. $g(x) = 3\sqrt{x}$

45. $r(x) = \frac{1}{x^2}$

46. $p(x) = -2x^4$

47. $q(x) = -\frac{1}{x^3}$

48. $k(x) = \frac{\sqrt[3]{x}}{2}$

49. $f(x) = 4x^3$

50. $f(x) = -\frac{2}{x^2}$

51. $f(x) = \left[\!\left[\frac{2x}{3}\right]\!\right]$

52. $f(x) = \begin{cases} x^2 & \text{if } x < 1 \\ \frac{1}{x} & \text{if } x \geq 1 \end{cases}$

53. $g(x) = \begin{cases} (x+1)^2 - 1 & \text{if } x \leq 0 \\ \sqrt[3]{x} & \text{if } x > 0 \end{cases}$

54. $h(x) = \begin{cases} -|x| & \text{if } x < 3 \\ (x-4)^2 + 1 & \text{if } x \geq 3 \end{cases}$

55. $f(x) = \begin{cases} x^2 & \text{if } x \leq -2 \\ \frac{1}{x^2} & \text{if } x > -2 \end{cases}$

56. $q(x) = \begin{cases} 3x - 1 & \text{if } x < 1 \\ x^4 & \text{if } x \geq 1 \end{cases}$

57. $g(x) = \begin{cases} 2|x| & \text{if } x < 2 \\ \sqrt{x} & \text{if } x \geq 2 \end{cases}$

Section 3.5

Find the mathematical model for each of the following verbal statements.

58. The quantity V varies directly as the product of r squared and h.

59. The value of y varies directly as the cube of a and inversely as the square root of b.

Solve the following variation problems.

60. Suppose that y varies directly as the square of x and that $y = 567$ when $x = 9$. What is y when $x = 4$?

61. Suppose that y is inversely proportional to the square root of x and that $y = 45$ when $x = 64$. What is y when $x = 25$?

62. A video store manager observes that the number of videos rented seems to vary inversely as the price of a rental. If the store's customers rent 1050 videos per month when the price per rental is \$3.49, how many videos per month does he expect to rent if he lowers the price to \$2.99?

Section 3.6

63. Determine the approximate distance between Earth, which has a mass of approximately 6.4×10^{24} kg, and an object that has a mass of 6.42×10^{22} kg, if the gravitational force between them equals approximately 4.95×10^{21} N. Remember, $F = \frac{km_1m_2}{d^2}$ and the universal gravitational constant equals 6.67×10^{-11} N$\cdot$m^2/kg^2.

64. a. Find a model for the volume of a cylindrical can in terms of the can's radius r, given that the surface area of the rectangle used to make the cylindrical portion of the can is constrained to be 100 in^2.

b. Find the height of such a can if the volume is to be 150 in^3.

65. Robert is planning to build a fenced rectangular garden by the side of a road, with fencing that costs \$8/ft for the length along the road and fencing that costs \$5/ft for the other three sides. He wants the garden to have an area of 1200 ft^2.

a. Find a model for the total cost of the fencing.

b. Estimate the size of the garden that will minimize the total cost of the fencing and estimate that minimum cost.

66. Carlotta throws a baseball straight up as hard as she can. It reaches a maximum height of 30 meters, and the table below shows its height in quarter-second intervals from that point on.

a. Graph the heights (either by hand or with a graphing utility) and estimate the time the ball hits the ground.

b. Find the linear function of best fit that models the height of the ball, and graph the function along with the given heights. By the linear model, what is the extrapolated time when the ball hits the ground? What is the calculated linear-model height of the ball at time $t = 0$?

c. Find the quadratic function of best fit that models the height of the ball, and graph the function along with the given heights. By the quadratic model, what is the extrapolated time when the ball hits the ground? What is the calculated quadratic-model height of the ball at time $t = 0$?

Time t (in seconds)	Height (in meters)
0	30
0.25	29.4
0.5	27.6
0.75	24.5
1.0	20.2
1.25	14.7
1.5	8.0

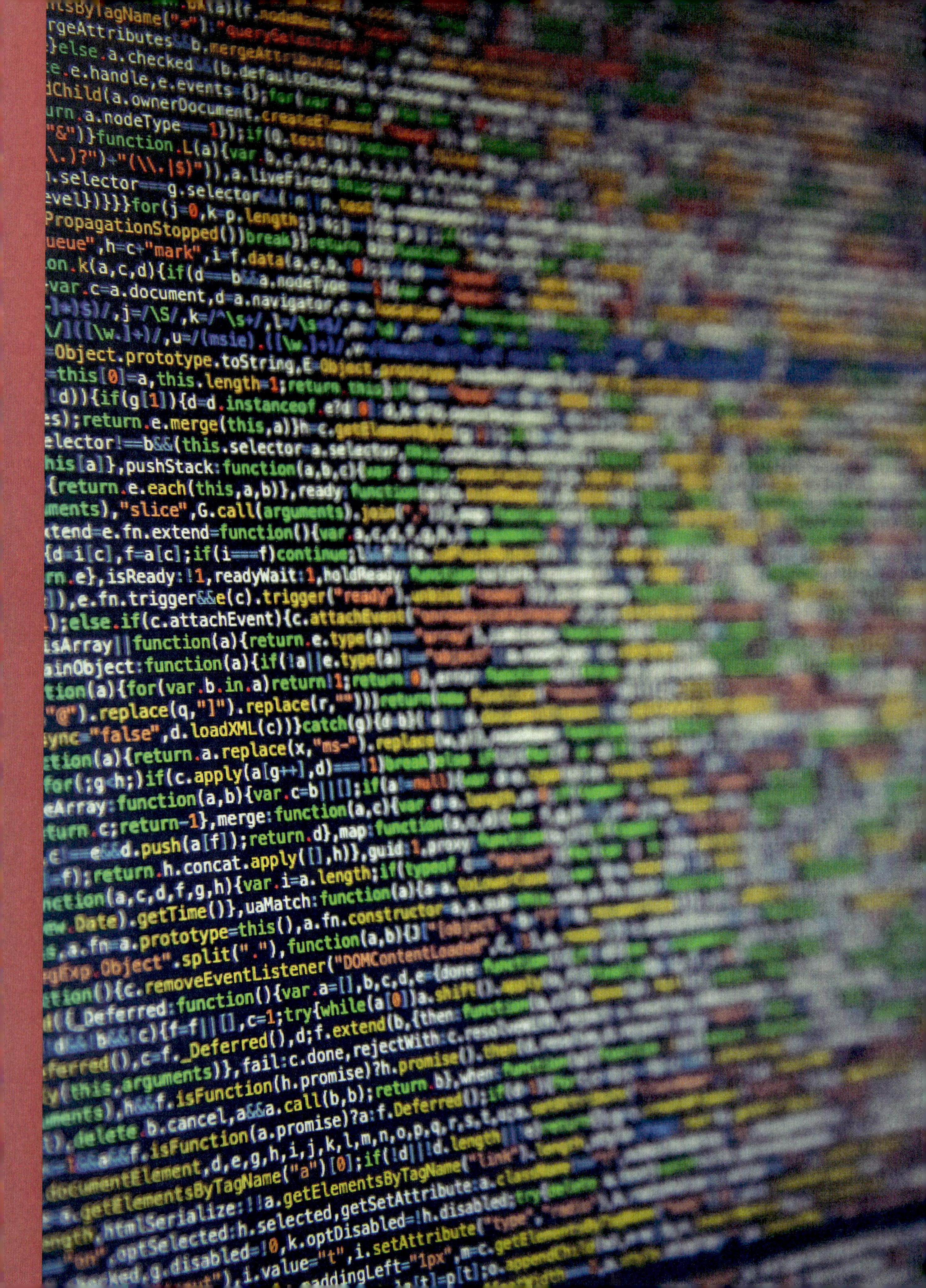

tsByTagName("*")
rgeAttributes
}else a.checked
e.handle,e.events
dChild(a.ownerDocument.createElement
rn a.nodeType===1));if(
"&")}function L(a){var
\.)?")+"(\\.|$)")),a.liveFired
.selector===g.selector
evel}}}}for(j=0,k=p.length;j<k;j
PropagationStopped())break}}return
ueue",h=c+"mark",i=f.data(a,e,h,
on k(a,c,d){if(d===b&&a.nodeType
var c=a.document,d=a.navigator,e
Object.prototype.toString,E=Object.prototype
this[0]=a,this.length=1;return this
d)){if(g[1]){d=d instanceof e?d[0]
s);return e.merge(this,a)}h=c.getElementById
elector!==b&&(this.selector=a.selector
his[a]},pushStack:function(a,b,c){var
{return e.each(this,a,b)},ready:function
ments),"slice",G.call(arguments).join(",")
tend=e.fn.extend=function(){var a,c,d,f,g,h,
{d=i[c],f=a[c];if(i===f)continue;
rn e},isReady:!1,readyWait:1,holdReady:function
),e.fn.trigger&&e(c).trigger("ready").unbind("ready")
);else if(c.attachEvent){c.attachEvent("onreadystatechange"
isArray||function(a){return e.type(a)==="array"}
ainObject:function(a){if(!a||e.type(a)!=="object"
tion(a){for(var b in a)return!1;return!0},error
"@").replace(q,"]").replace(r,""))){return
ync="false",d.loadXML(c)}catch(g){d=b}
tion(a){return a.replace(x,"ms-").replace
for(;g<h;)if(c.apply(a[g++],d)===!1)break}else
eArray:function(a,b){var c=b||[];if(a!=null)
turn c;return-1},merge:function(a,c){var
e.push(a[f]);return d},map:function(a,c,d){
f);return h.concat.apply([],h)},guid:1,proxy
nction(a,c,d,f,g,h){var i=a.length;if(typeof
w Date).getTime()},uaMatch:function(a){a=a.toLowerCase
a.fn=a.prototype=this(),a.fn.constructor=a,a.sub=this
gExp Object".split(" "),function(a,b){I["[object
tion(){c.removeEventListener("DOMContentLoaded"
Deferred:function(){var a=[],b,c,d,e={done:function
c){f=f||[],c=1;try{while(a[0])a.shift().apply
ferred(),c=f._Deferred(),d;f.extend(b,{then:function
(this,arguments)},fail:c.done,rejectWith:c.resolveWith
ments),h&&f.isFunction(h.promise)?h.promise().then
delete b.cancel,a&&a.call(b,b);return b},when:function
f.isFunction(a.promise)?a:f.Deferred();if(d
documentElement,d,e,g,h,i,j,k,l,m,n,o,p,q,r,s,t,u;
getElementsByTagName("a")[0];if(!d||!d.length
a.getElementsByTagName("link").length
htmlSerialize:!!a.getElementsByTagName
optSelected:h.selected,getSetAttribute:a.className
disabled=!0,k.optDisabled=!h.disabled;try{delete
),i.value="t",i.setAttribute("type","radio")
paddingLeft="1px",m=c.getElementById

CHAPTER 4

Working with Functions

SECTIONS

WHAT IF ...

What if you received an encoded message? Would you be able to read it? Could you figure out the code that was used and respond?

By the end of this chapter, you'll be able to use mathematical functions to encode and decode messages.

Benoit Mandelbrot

Introduction

In this chapter, we will learn a wide variety of techniques that allow us to build and use new functions given the portfolio of basic functions that we studied in the last chapter.

We start with techniques that are almost tangible in nature. In many situations, we'll need to work with functions that have the essential characteristics of one of the basic functions we already know, but in some modified form. In particular, we'll learn techniques that allow us to shift or stretch the graph of a basic function either in a vertical direction or in a horizontal direction, or possibly both. We'll also learn how to reflect the graph of a function with respect to one or both of the axes. To round out the discussion, we'll then define and use certain properties of functions that have their basis in characteristics of their graphs.

Another way in which we can construct new functions with desired characteristics is by combining two or more basic functions. Initially, we'll see how the elementary mathematical operations of addition, subtraction, multiplication, and division extend naturally to similar operations on functions. But functions can also be combined through a fifth operation known as *composition*, with results that open up entirely new realms of mathematics. Familiarity with and skill in using composition is essential, for instance, in the study of calculus and later courses in mathematics. Function composition is also a key concept in the creation of *fractals*, a term coined by the mathematician Benoit Mandelbrot (1924–2010) to describe highly complex objects that display infinite self-similarity. An example of mathematical art created through the operation of composition will appear in this chapter.

Finally, function composition leads to the idea of the inverse of a function, the topic with which we'll close the chapter. The ability to invert, or undo, the effect of a function is often central to the process of solving a problem, and thus is naturally of interest to us. But more generally, the concept of function inverses leads to completely new classes of functions and many deep and fundamental theorems in mathematics, some of which you will be introduced to in chapters to come.

4.1 TRANSFORMATIONS OF FUNCTIONS

TOPICS

1. Shifting graphs vertically and horizontally
2. Reflecting graphs
3. Stretching graphs vertically and horizontally
4. Order of transformations

TOPIC 1: Shifting Graphs Vertically and Horizontally

Much of the material in this section was introduced in Section 3.3, in our discussion of quadratic functions. You may want to review the ways in which the basic quadratic function $f(x)=x^2$ can be shifted, stretched, and reflected as you work through the more general ideas here.

THEOREM: Horizontal Shifting/Translation

Let $f(x)$ be a function, and let h be a fixed real number. If we replace x with $x-h$, we obtain a new function $g(x)=f(x-h)$. The graph of g has the same shape as the graph of f, but shifted h units to the right if $h>0$ and shifted h units to the left if $h<0$.

Example 1: Horizontal Shifting/Translation

Sketch the graphs of the following functions.

a. $f(x)=(x+2)^3$

b. $g(x)=|x-4|$

NOTE

Begin by identifying the underlying function that is being shifted.

Solution

a.

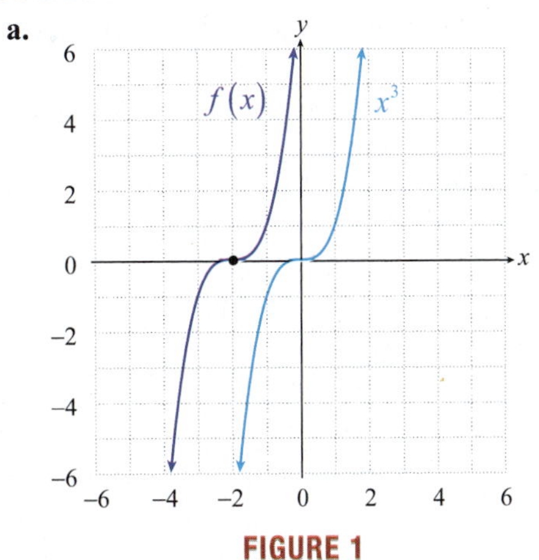

FIGURE 1

The basic function being shifted is x^3.

Begin by drawing the basic cubic shape (the shape of $y=x^3$).

Since x is replaced with $x+2$, the graph of $f(x)$ is the graph of x^3 shifted 2 units to the left.

Note, for example, that $(-2,0)$ is one point on the graph.

b.

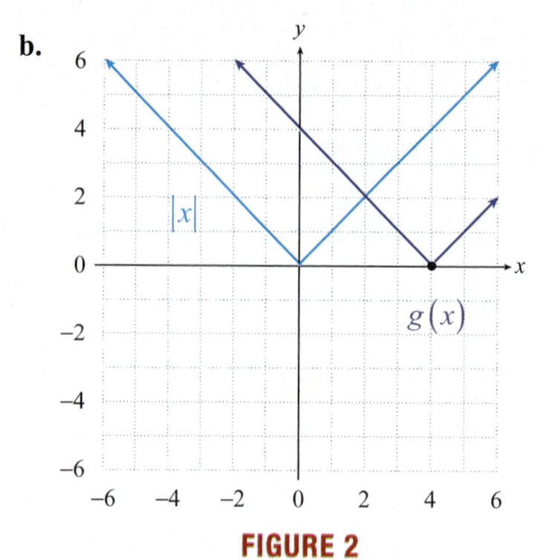

The basic function being shifted is $|x|$.

Start by graphing the basic absolute value function.

The graph of $g(x)=|x-4|$ has the same shape, but shifted 4 units to the right.

Note, for example, that $(4,0)$ lies on the graph of g.

FIGURE 2

> **CAUTION**
>
> The minus sign in the expression $x-h$ is critical. When you see an expression in the form $x+h$ you must think of it as $x-(-h)$.
>
> Consider a specific example: replacing x with $x-5$ shifts the graph 5 units to the *right*, since 5 is positive. Replacing x with $x+5$ shifts the graph 5 units to the *left*, since we have actually replaced x with $x-(-5)$.

> **THEOREM: Vertical Shifting/Translation**
>
> Let $f(x)$ be a function whose graph is known, and let k be a fixed real number. The graph of the function $g(x)=f(x)+k$ is the same shape as the graph of f, but shifted k units up if $k>0$ and k units down if $k<0$.

Example 2: Vertical Shifting/Translation

Sketch the graphs of the following functions.

a. $f(x)=\frac{1}{x}+3$ **b.** $g(x)=\sqrt[3]{x}-2$

> **NOTE**
>
> As before, begin by identifying the basic function being shifted.

Solution

a.

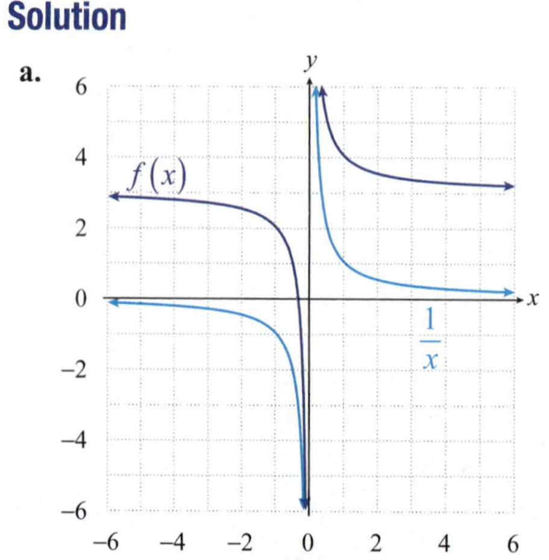

The basic function being shifted is $\frac{1}{x}$.

The graph of $f(x)=\frac{1}{x}+3$ is the graph of $y=\frac{1}{x}$ shifted 3 units up.

Note that this doesn't change the domain.

However, the range is affected; the range of f is $(-\infty,3)\cup(3,\infty)$.

FIGURE 3

b.

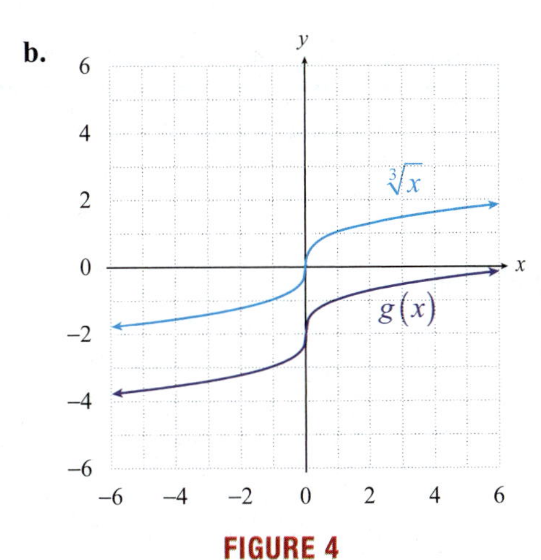

FIGURE 4

The basic function being shifted is $\sqrt[3]{x}$.

Begin by graphing the basic cube root shape.

To graph $g(x)=\sqrt[3]{x}-2$, the graph of $y=\sqrt[3]{x}$ is shifted 2 units down.

Example 3: Horizontal and Vertical Shifting

Sketch the graph of the function $f(x)=\sqrt{x+4}+1$.

Solution

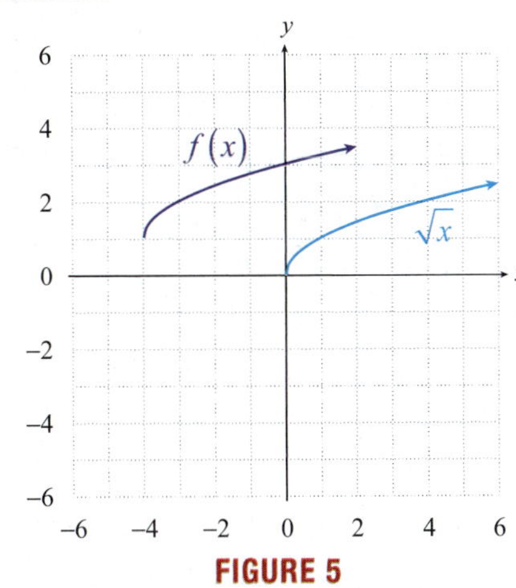

FIGURE 5

The basic function being shifted is $\sqrt{x}$.

Begin by graphing the basic square root shape.

In $f(x)$ we have replaced x with $x+4$, so shift the basic function 4 units to the left.

Then shift the resulting function 1 unit up.

NOTE

In this case, it doesn't matter which shift we apply first. However, when functions get more complicated, it is usually best to apply horizontal shifts before vertical shifts.

TOPIC 2: Reflecting Graphs

We next consider transformations that result in the graph of a function being reflected with respect to one axis or the other.

THEOREM: Reflecting with Respect to the Axes

Given a function $f(x)$,

1. the graph of the function $g(x)=-f(x)$ is the reflection of the graph of f with respect to the x-axis;
2. the graph of the function $g(x)=f(-x)$ is the reflection of the graph of f with respect to the y-axis.

In other words, a function is reflected with respect to the x-axis by multiplying the entire function by -1, and it is reflected with respect to the y-axis by replacing x with $-x$.

Example 4: Reflecting with Respect to the Axes

Sketch the graphs of the following functions.

a. $f(x) = -x^2$

b. $g(x) = \sqrt{-x}$

NOTE

We state that a function is reflected with respect to a particular axis. Visually, this means the function is reflected over (across) that axis.

Solution

a.

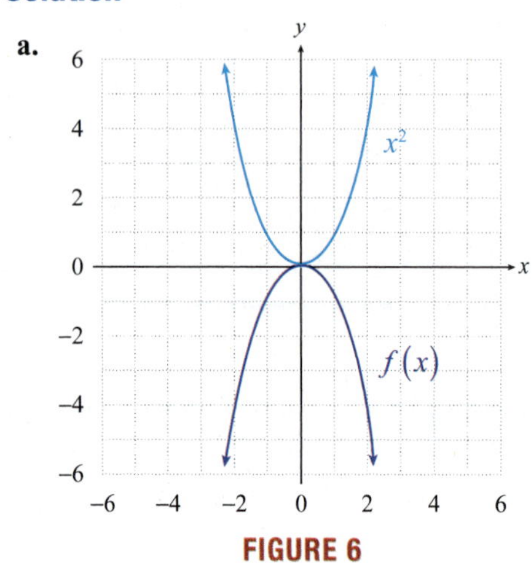

FIGURE 6

To graph $f(x) = -x^2$, begin with the graph of the basic parabola $y = x^2$.

The entire function is multiplied by -1, so reflect the graph over the x-axis, resulting in the original shape turned upside down.

Note that the domain is still the entire real line, but the range of f is the interval $(-\infty, 0]$.

b.

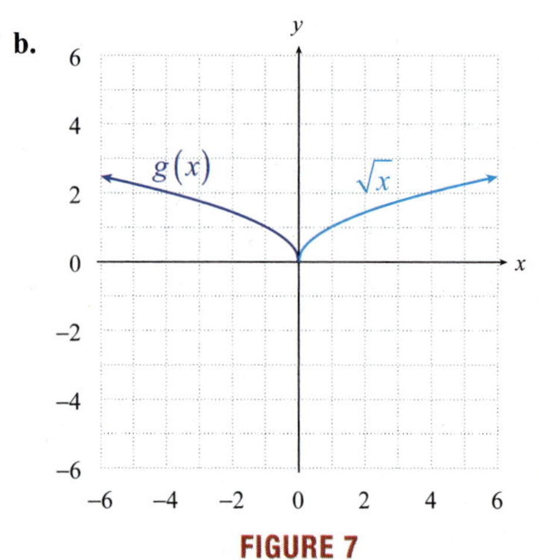

FIGURE 7

To graph $g(x) = \sqrt{-x}$, begin by graphing $y = \sqrt{x}$, the basic square root function.

In $g(x)$, x has been replaced with $-x$, so reflect the graph with respect to the y-axis.

Note that this changes the domain but not the range. The domain of g is the interval $(-\infty, 0]$ and the range is $[0, \infty)$.

TOPIC 3: Stretching Graphs Vertically and Horizontally

The graph of a function can be stretched or compressed either vertically or horizontally by two final transformations. We first consider vertical transformations.

THEOREM: Vertical Stretching and Compressing

Let $f(x)$ be a function and let a be a positive real number.

1. The graph of the function $g(x) = af(x)$ is stretched vertically compared to the graph of f by a factor of a if $a > 1$.
2. The graph of the function $g(x) = af(x)$ is compressed vertically compared to the graph of f by a factor of a if $0 < a < 1$.

Example 5: Vertical Stretching and Compressing

Sketch the graphs of the following functions.

a. $f(x)=\dfrac{\sqrt{x}}{10}$ **b.** $g(x)=5|x|$

Solution

> **NOTE**
> When graphing stretched or compressed functions, it may help to plot a few points of the new function.

a.

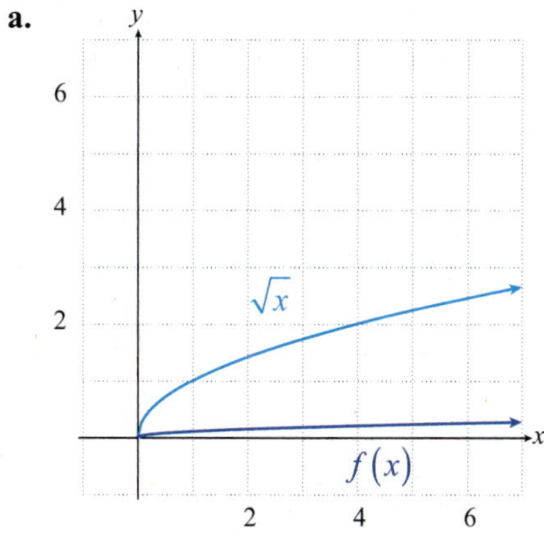

FIGURE 8

Begin with the graph of $\sqrt{x}$.

The shape of $f(x)$ is similar to the shape of $\sqrt{x}$, but all of the y-coordinates have been multiplied by the factor of $\dfrac{1}{10}$ and are consequently much smaller.

b.

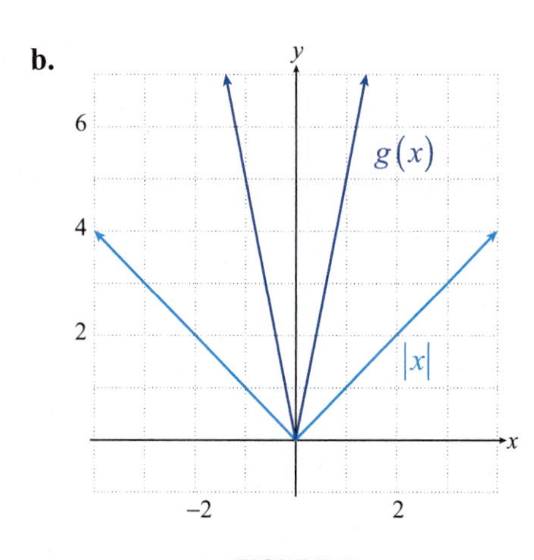

FIGURE 9

Begin with the graph of the absolute value function.

In contrast to the last example, the graph of $g(x)=5|x|$ is stretched compared to the standard absolute value function.

Every second coordinate is multiplied by a factor of 5.

We can stretch or compress the graph of a function horizontally by rescaling the argument used to define the function. For instance, if we replace x with ax in the definition of a given function f, then the point on the graph of $f(x)$ where $x=1$ will correspond to the point on the graph of $f(ax)$ where $x=\dfrac{1}{a}$, since $f\left(a\cdot\dfrac{1}{a}\right)=f(1)$. In similar fashion, the point on the graph of $f(x)$ where $x=c$ will correspond to the point on the graph of $f(ax)$ where $x=\dfrac{c}{a}$, since $f\left(a\cdot\dfrac{c}{a}\right)=f(c)$. We summarize this as follows.

THEOREM: Horizontal Stretching and Compressing

Let $f(x)$ be a function and let a be a positive real number.

1. The graph of the function $g(x)=f(ax)$ is stretched horizontally compared to the graph of f by a factor of $\frac{1}{a}$ if $0<a<1$.
2. The graph of the function $g(x)=f(ax)$ is compressed horizontally compared to the graph of f by a factor of $\frac{1}{a}$ if $a>1$.

Example 6: Horizontal Stretching and Compressing

Sketch the graph of the function $f(x)=\left(\frac{x}{3}\right)^3$.

Solution

We obtain the argument $\frac{x}{3}$ from x by multiplying by the constant $\frac{1}{3}$, which means the graph of $\left(\frac{x}{3}\right)^3$ is stretched out compared to the graph of x^3 by a factor of $\frac{1}{1/3}$, or 3. So, for instance, the point $(1,1)$ on the graph of x^3 corresponds to the point $(3,1)$ on the graph of f. Alternatively, we can rewrite the definition of f as $f(x)=\frac{x^3}{27}$, which we can interpret as the product of the constant $\frac{1}{27}$ and the function x^3. Thinking of f in this manner, the graph of f is compressed vertically by a factor of $\frac{1}{27}$ compared to the graph of x^3, with the same result.

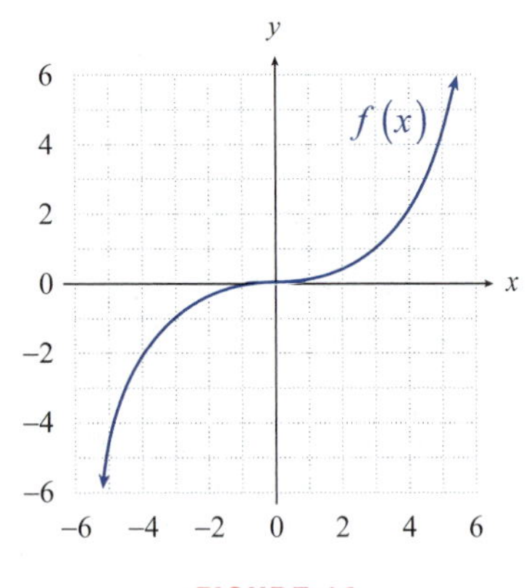

FIGURE 10

TOPIC 4: Order of Transformations

As we saw in Example 6, some transformations of functions can be interpreted in different but equivalent ways. And if several transformations are made to a basic function, it's even more likely that the result can be analyzed in several ways. In such a case, the following guidelines are useful in unraveling the sequence of transformations. Keep in mind, however, that for a complicated enough function there may be several different but equally correct ways to interpret the transformations.

PROCEDURE: Order of Transformations

If a function g has been constructed from a simpler function f through a number of transformations, g can be understood by looking for transformations in the following order.

1. Horizontal shifts
2. Horizontal and vertical stretching and compressing
3. Reflections
4. Vertical shifts

Example 7: Order of Transformations

Describe the transformations needed to construct the function

$$g(x) = -2\sqrt{2x+2} + 3$$

from the basic square root function using the order of transformations.

Solution

1. First, it's helpful to write $\sqrt{2x+2}$ as $\sqrt{2(x+1)}$ in order to separate the effects of the first two types of transformations. In this form, we see that the horizontal shift results from replacing x with $x + 1$. Figures 11 and 12 show the result of this transformation from $\sqrt{x}$ to $\sqrt{x+1}$.

2. Next, $\sqrt{x+1}$ has been transformed into $\sqrt{2(x+1)}$. That is, the argument of the square root function has been multiplied by 2, resulting in the horizontal compression by a factor of $\frac{1}{2}$ shown in Figure 13. A vertical stretching by a factor of 2 is also accomplished by transforming $\sqrt{2(x+1)}$ into $2\sqrt{2(x+1)}$, with the result shown in Figure 14. Note that the combined effect of these two transformations can also be viewed as a vertical stretching of the graph of $\sqrt{x+1}$ by a factor of $2\sqrt{2}$.

3. The next-to-last transformation is the reflection with respect to the x-axis achieved by multiplying the function by -1, as shown in Figure 15.

4. Finally, the graph is shifted 3 units up, as shown in Figure 16.

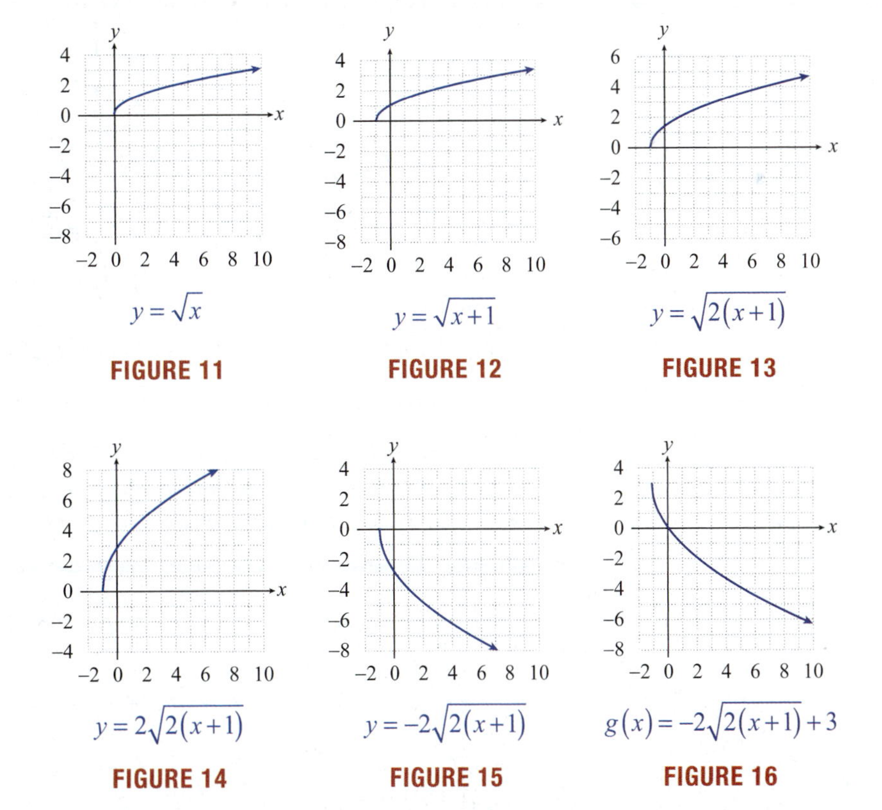

$y = \sqrt{x}$ **FIGURE 11**

$y = \sqrt{x+1}$ **FIGURE 12**

$y = \sqrt{2(x+1)}$ **FIGURE 13**

$y = 2\sqrt{2(x+1)}$ **FIGURE 14**

$y = -2\sqrt{2(x+1)}$ **FIGURE 15**

$g(x) = -2\sqrt{2(x+1)} + 3$ **FIGURE 16**

Example 8: Order of Transformations

Sketch the graph of the function $f(x)=\frac{1}{2-x}$.

Solution

The basic function that f is similar to is $\frac{1}{x}$. Following the order of transformations, we determine how to sketch the graph of f:

1. If we replace x with $x+2$ (shifting the graph 2 units to the left), we obtain the function $\frac{1}{x+2}$, which is closer to what we want.
2. There does not appear to be any stretching or compressing transformation.
3. If we replace x with $-x$, we have $\frac{1}{-x+2}=\frac{1}{2-x}$, which is equal to f. This reflects the graph of $\frac{1}{x+2}$ with respect to the y-axis.
4. Since we have already found f, we know there is no vertical shift.

The entire sequence of transformations is shown in Figure 17, ending with the graph of f.

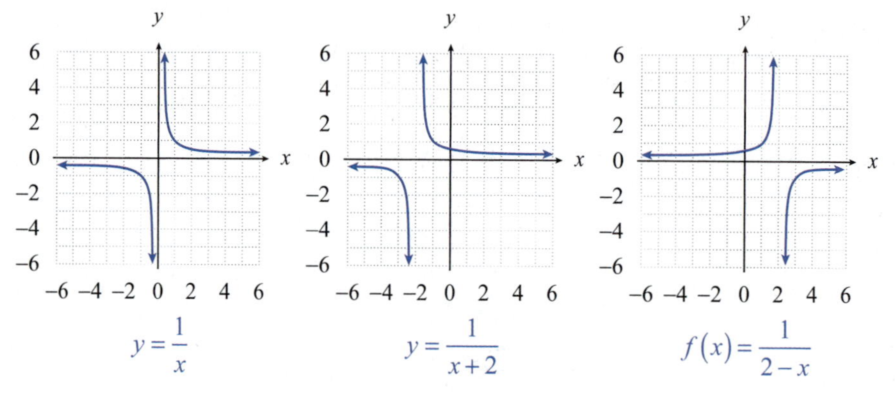

FIGURE 17

Note: An alternate approach to graphing $f(x)=\frac{1}{2-x}$ is to rewrite the function in the form $f(x)=-\frac{1}{x-2}$. In this form, the graph of f is the graph of $\frac{1}{x}$ shifted two units to the right, and then reflected with respect to the x-axis. The result is the same, as you should verify. Rewriting an equation in a different, equivalent form never changes its graph.

4.1 EXERCISES

PRACTICE

For each function or graph below, determine the basic function that has been shifted, reflected, stretched, or compressed.

1. $f(x) = -(1-x)^2 + 2$

2. $f(x) = \dfrac{1}{x-4} + 5$

3. $f(x) = \sqrt[3]{x+6} - 2$

4. $f(x) = -2 + 2|x-3|$

5. $f(x) = \sqrt{x+2} - 5$

6. $f(x) = [\![-2-x]\!]$

7. $f(x) = \dfrac{1}{(x+2)^2} + 1$

8. $f(x) = \dfrac{\sqrt{-x}}{2} + 4$

9. $f(x) = (x+6)^3$

10. $f(x) = (1-2x)^3$

11. $f(x) = 3\left|\dfrac{x}{2} - 1\right|$

12.

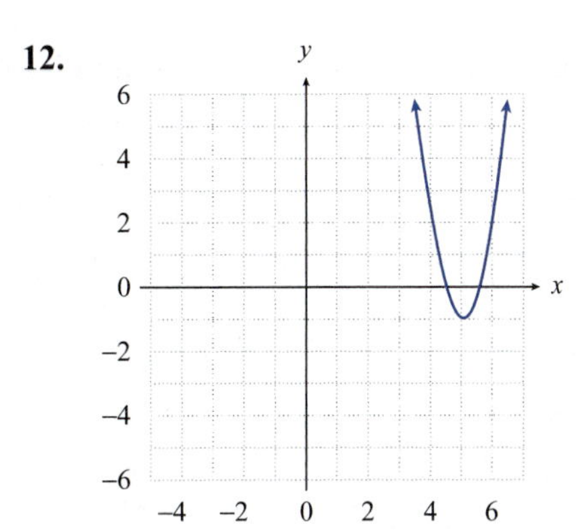

13.

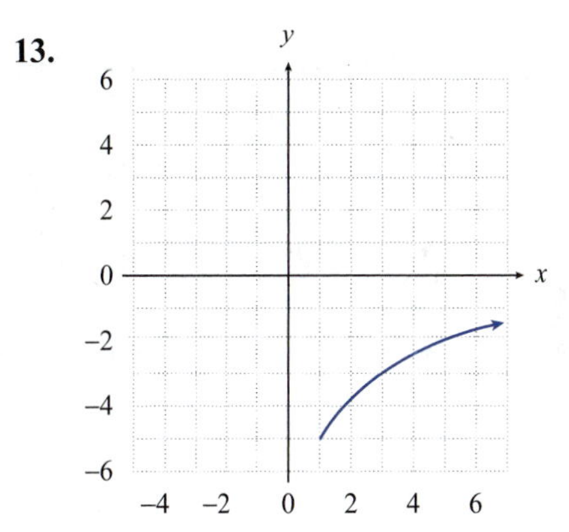

14.

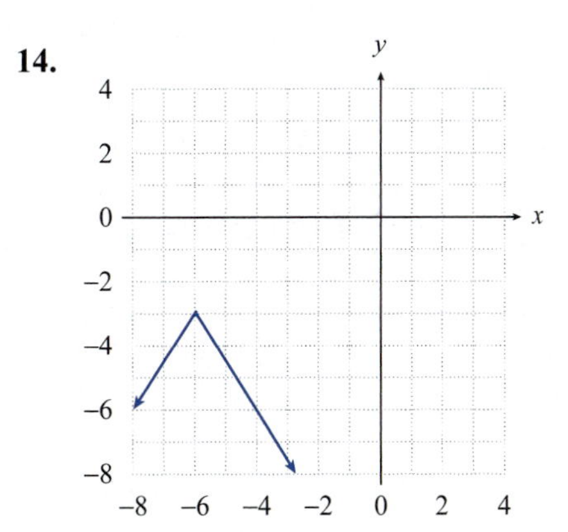

15.

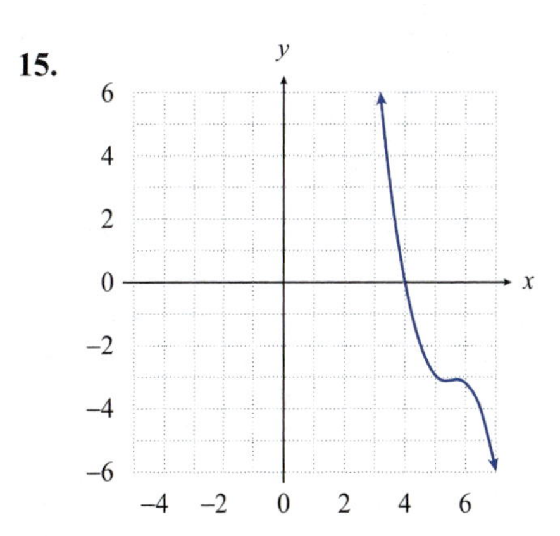

Sketch the graphs of the following functions by first identifying the more basic functions that have been shifted, reflected, stretched, or compressed. Then determine the domain and range of each function. See Examples 1 through 6.

16. $f(x)=(x+2)^3$

17. $G(x)=|x-4|$

18. $p(x)=-(x+1)^2+2$

19. $g(x)=\sqrt{x+3}-1$

20. $q(x)=(1-x)^2$

21. $r(x)=-\sqrt[3]{x}$

22. $s(x)=\sqrt{2-x}$

23. $F(x)=\dfrac{|x+2|}{3}+3$

24. $w(x)=\dfrac{1}{(x-3)^2}$

25. $v(x)=\dfrac{1}{3x}-2$

26. $f(x)=\dfrac{1}{2-x}$

27. $k(x)=\sqrt{-x}+2$

28. $b(x)=\sqrt[3]{x+2}-5$

29. $b(x)=[\![x-4]\!]+4$

30. $R(x)=4-2|x|$

31. $S(x)=(3-x)^3$

32. $g(x)=-\dfrac{1}{x+1}$

33. $h(x)=\dfrac{x^2}{2}-3$

34. $W(x)=1-|4-x|$

35. $W(x)=-\dfrac{|x-1|}{4}$

36. $S(x)=\dfrac{1}{x^2}+3$

37. $V(x)=-3\sqrt{x-1}+2$

38. $f(x)=\sqrt{2x-2}-2$

39. $g(x)=(2x-3)^2+1$

40. $f(x)=|1-2x|-1$

41. $g(x)=-(3x+3)^3$

42. $g(x)=x^2-6x+9$ (**Hint:** Find a better way to write the function.)

43. $h(x)=\dfrac{|x|}{x}$ (**Hint:** Evaluate h at a few points to understand its behavior.)

44. $W(x)=\dfrac{x-1}{|x-1|}$

45. $s(x)=[\![x-2]\!]$

Write a formula for each of the functions described.

46. Use the function $g(x)=x^2$. Move the function 3 units to the left and 4 units down.

47. Use the function $g(x)=x^2$. Move the function 4 units to the right and 2 units up.

48. Use the function $g(x)=x^2$. Reflect the function across the x-axis and move it 6 units up.

49. Use the function $g(x)=x^2$. Move the function 2 units to the right and reflect across the y-axis.

50. Use the function $g(x)=x^3$. Compress the function horizontally by a factor of 3 and move it 1 unit down.

51. Use the function $g(x)=x^3$. Move the function 1 unit to the left and reflect across the y-axis.

52. Use the function $g(x)=x^3$. Move the function 10 units to the right and 4 units up.

53. Use the function $g(x)=\sqrt{x}$. Move the function 5 units to the left and reflect across the x-axis.

54. Use the function $g(x)=\sqrt{x}$. Reflect the function across the y-axis and move it 3 units down.

55. Use the function $g(x)=\sqrt{x}$. Stretch the function horizontally by a factor of 2, reflect it with respect to the y-axis, and move it 3 units up.

56. Use the function $g(x)=|x|$. Move the function 7 units to the left, reflect across the x-axis, and reflect across the y-axis.

57. Use the function $g(x)=|x|$. Move the function 8 units to the right, 2 units up, and reflect across the x-axis.

Use your knowledge about transformations to find a possible formula for the function $f(x)$ given its graph.

58.

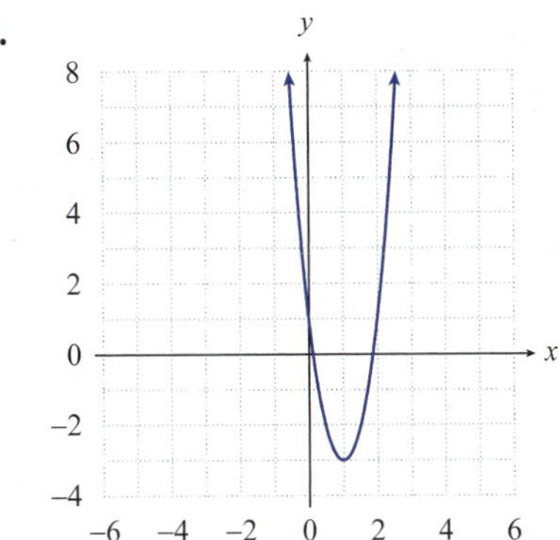

59.

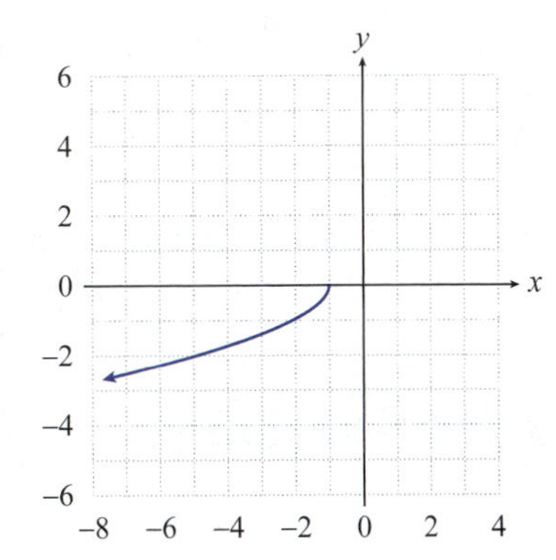

60.

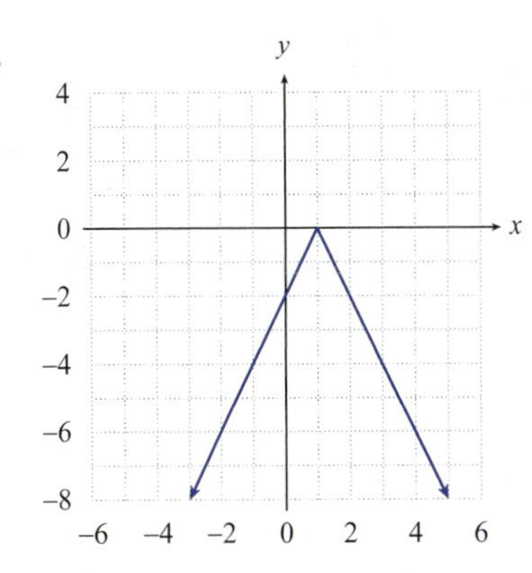

61.

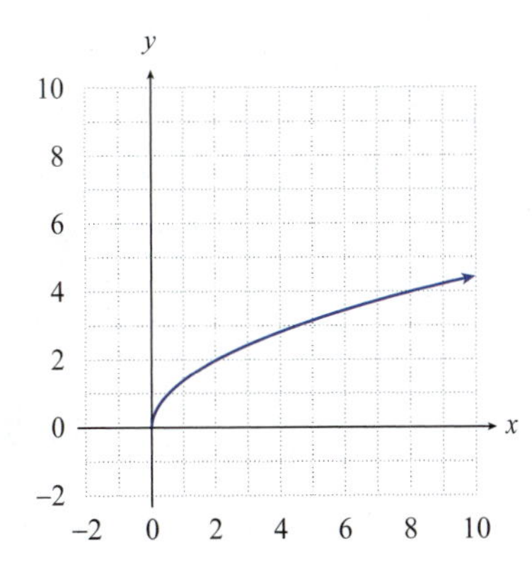

62.

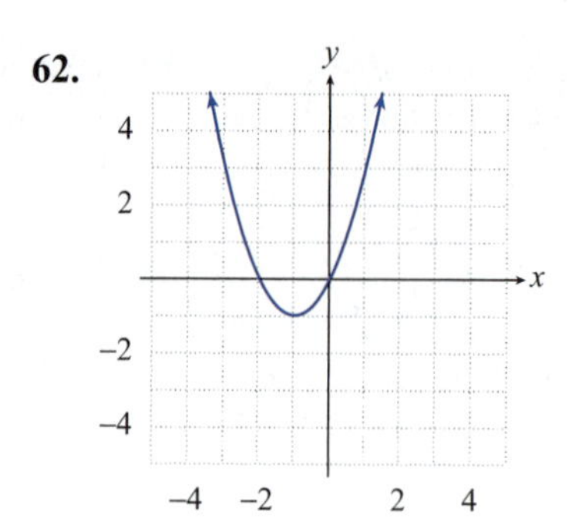

63.

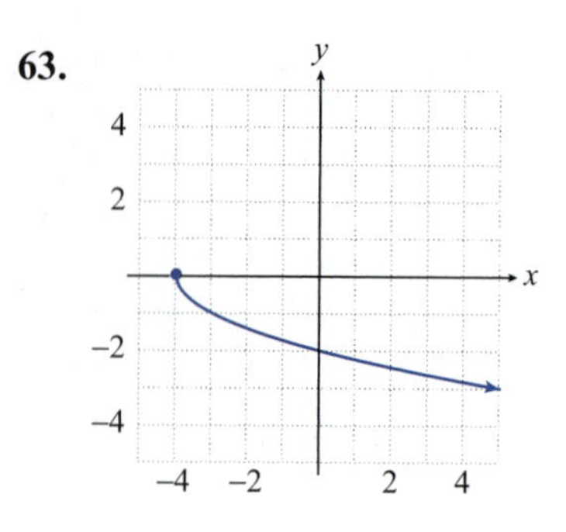

64.

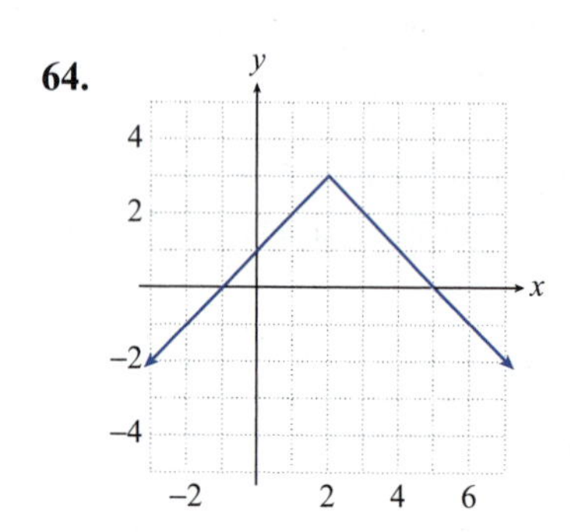

65.

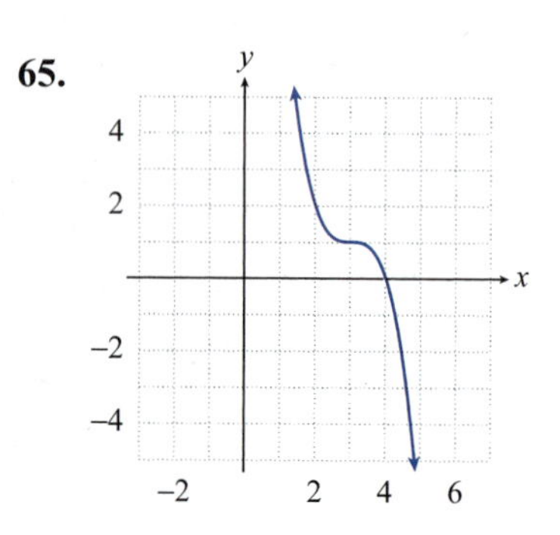

TECHNOLOGY

Mentally sketch the graph of the given function by identifying the basic shape that has been shifted, reflected, stretched, or compressed. Then use a graphing utility to graph the function and check your reasoning

66. $f(x) = -2(3-x)^3 + 5$

67. $f(x) = \dfrac{3}{x+5} - 1$

68. $f(x) = \dfrac{-1}{(x-2)^2} - 3$

69. $f(x) = -3|x+2| - 4$

70. $f(x) = -\sqrt{1-x} + 2$

71. $f(x) = \sqrt[3]{2+x} - 1$

Write a possible equation for the function depicted on the graphing utility. The function is shown in a $[-10,10]$ by $[-10,10]$ viewing window.

72.

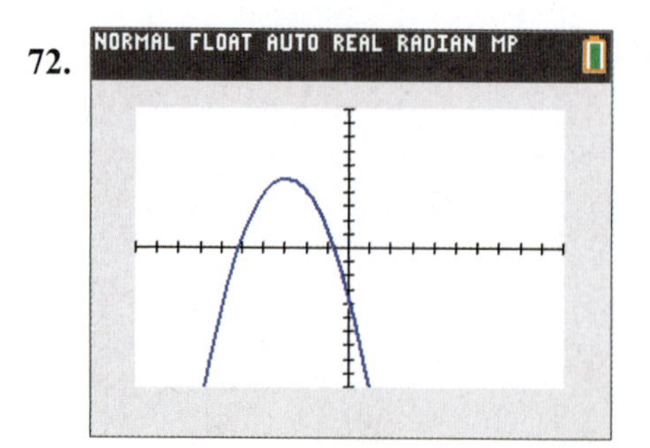

73.

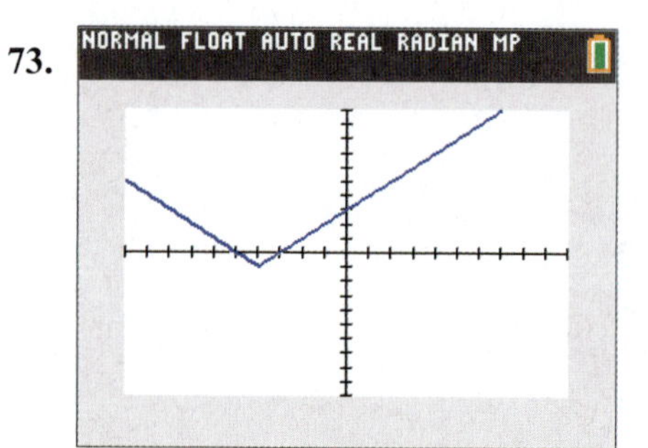

74.

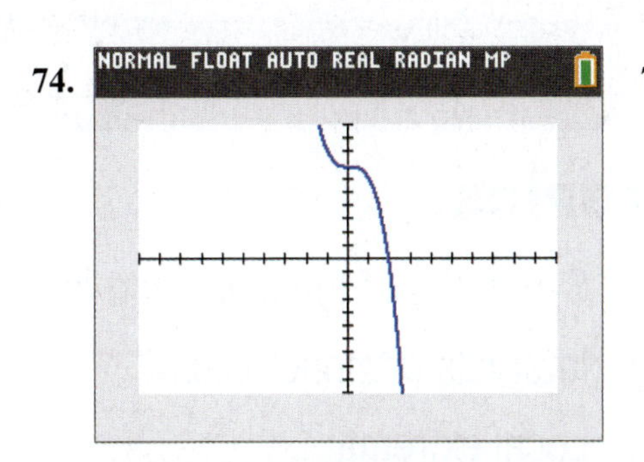
NORMAL FLOAT AUTO REAL RADIAN MP

75.

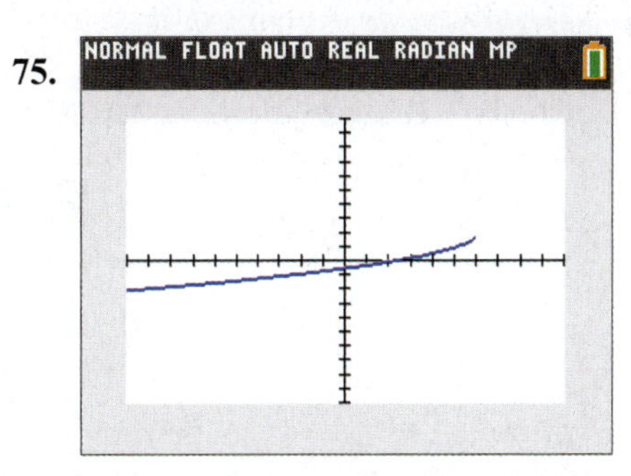
NORMAL FLOAT AUTO REAL RADIAN MP

76.

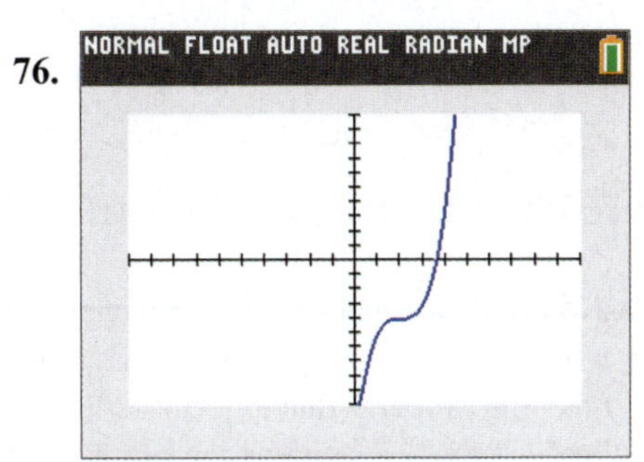
NORMAL FLOAT AUTO REAL RADIAN MP

77.

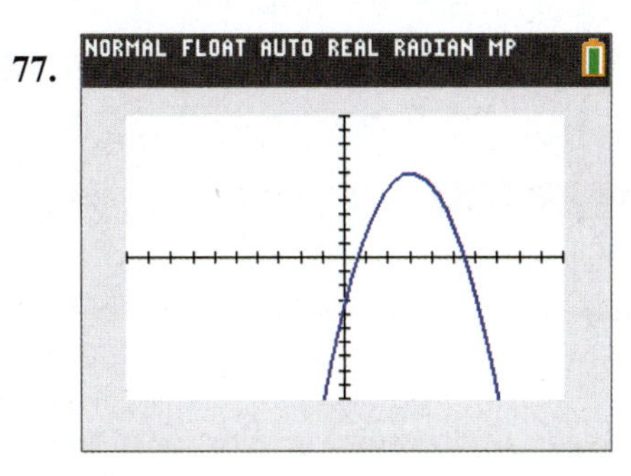
NORMAL FLOAT AUTO REAL RADIAN MP

4.2 PROPERTIES OF FUNCTIONS

TOPICS

1. Symmetry of functions and equations
2. Intervals of monotonicity
3. Local extrema
4. Average rate of change

TOPIC 1: Symmetry of Functions and Equations

In the last section, we learned that replacing x with $-x$ reflects the graph of a function with respect to the y-axis. It may be the case, however, that a given function f has the property that $f(-x)=f(x)$ for all x in the domain, meaning that the graph of f is the same as the graph of its reflection. This property of a graph is the subject of the following definition.

DEFINITION: *y*-Axis Symmetry

The graph of a function f has **y-axis symmetry**, or is **symmetric with respect to the *y*-axis**, if $f(-x)=f(x)$ for all x in the domain of f. Such functions are called **even functions**.

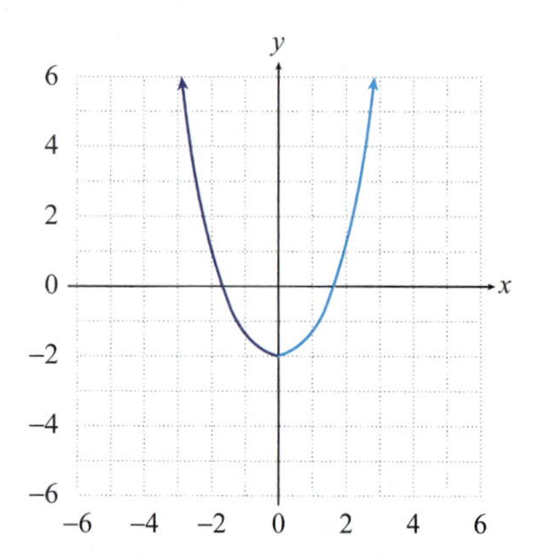

FIGURE 1: A Function with *y*-Axis Symmetry

Functions whose graphs have y-axis symmetry are called even functions because polynomial functions with only even exponents form one large class of functions with this property. Consider the function $f(x)=7x^8-5x^4+2x^2-3$. This function is a polynomial of four terms, all of which have even degree. If we replace x with $-x$ and simplify the result, we obtain the function f again.

$$\begin{aligned} f(-x) &= 7(-x)^8-5(-x)^4+2(-x)^2-3 \\ &= 7x^8-5x^4+2x^2-3 \\ &= f(x) \end{aligned}$$

Be aware, however, that such polynomial functions are not the only even functions. We will see many more examples as we proceed.

There is another class of functions for which replacing x with $-x$ results in the exact negative of the original function. That is, $f(-x) = -f(x)$ for all x in the domain, and this means changing the sign of the x-coordinate of a point on the graph also changes the sign of the y-coordinate.

What does this mean geometrically? Suppose f is such a function and $(x, f(x))$ is a point on the graph of f. If we change the signs of both coordinates, we obtain a new point that is the original point reflected through the origin (we can also think of this as reflected over the y-axis, then the x-axis).

For instance, if $(x, f(x))$ lies in the first quadrant, $(-x, -f(x))$ lies in the third; and if $(x, f(x))$ lies in the second quadrant, $(-x, -f(x))$ lies in the fourth. But since $f(-x) = -f(x)$, the point $(-x, -f(x))$ can be rewritten as $(-x, f(-x))$.

Written in this form, we know that $(-x, f(-x))$ is a point on the graph of f, since *any* point of the form $(?, f(?))$ lies on the graph of f. So a function with the property $f(-x) = -f(x)$ has a graph that is symmetric with respect to the origin.

DEFINITION: Origin Symmetry

The graph of a function f has **origin symmetry**, or is **symmetric with respect to the origin**, if $f(-x) = -f(x)$ for all x in the domain of f. Such functions are called **odd functions**.

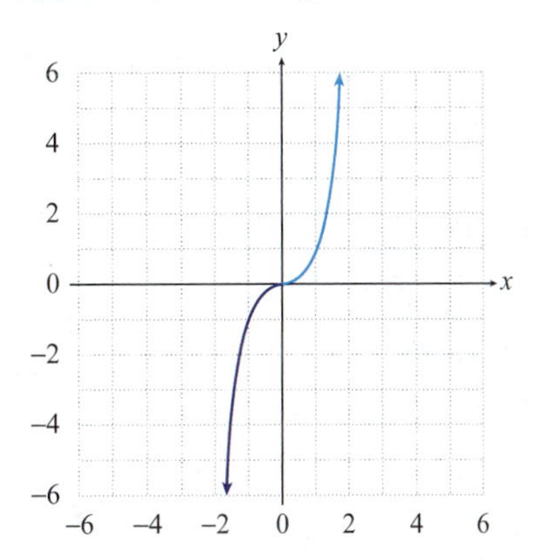

FIGURE 2: A Function with Origin Symmetry

As you might guess, such functions are called odd because polynomial functions with only odd exponents serve as simple examples. For instance, the function $f(x) = -2x^3 + 8x$ is odd.

$$\begin{aligned} f(-x) &= -2(-x)^3 + 8(-x) \\ &= -2(-x^3) + 8(-x) \\ &= 2x^3 - 8x \\ &= -f(x) \end{aligned}$$

As far as functions are concerned, y-axis and origin symmetry are the two principal types of symmetry. What about x-axis symmetry? It is certainly possible to draw a graph that displays x-axis symmetry; but unless the graph lies entirely on the x-axis, such a graph cannot represent a function. Why not? Draw a few graphs that are symmetric with respect to the x-axis; then apply the vertical line test to these graphs. In order to have x-axis symmetry, if (x, y) is a point on the graph, then $(x, -y)$ must also be on the graph, and thus the graph can not represent a function.

This brings us back to relations. Recall that any equation in x and y defines a relation between the two variables. There are three principal types of symmetry that equations can possess.

DEFINITION: Symmetry of Equations

We say that an equation in x and y is **symmetric with respect to**

1. the **y-axis** if replacing x with $-x$ results in an equivalent equation;
2. the **x-axis** if replacing y with $-y$ results in an equivalent equation;
3. the **origin** if replacing x with $-x$ and y with $-y$ results in an equivalent equation.

Knowing the symmetry of a function or an equation can serve as a useful aid in graphing. For instance, when graphing an even function it is only necessary to graph the part to the right of the y-axis, as the left half of the graph is the reflection of the right half with respect to the y-axis. Similarly, if a function is odd, the left half of its graph is the reflection of the right half through the origin.

Example 1: Using Symmetry to Graph Relations

Sketch the graphs of the following relations, making use of symmetry.

a. $f(x) = \frac{1}{x^2}$ **b.** $g(x) = x^3 - x$ **c.** $x = y^2$

NOTE

If you don't know where to begin when sketching a graph, plotting points often helps you understand the basic shape.

Solution

a.

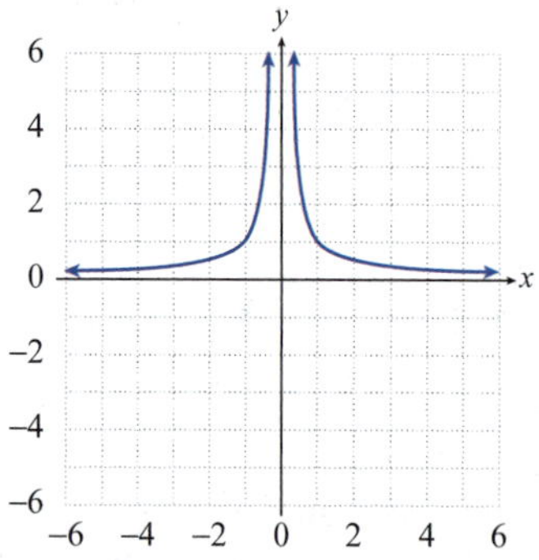

FIGURE 3

This is a relation that we previously graphed in Section 3.4. Note that it is indeed an even function and exhibits y-axis symmetry.

$$f(-x) = \frac{1}{(-x)^2} = \frac{1}{x^2} = f(x)$$

b.

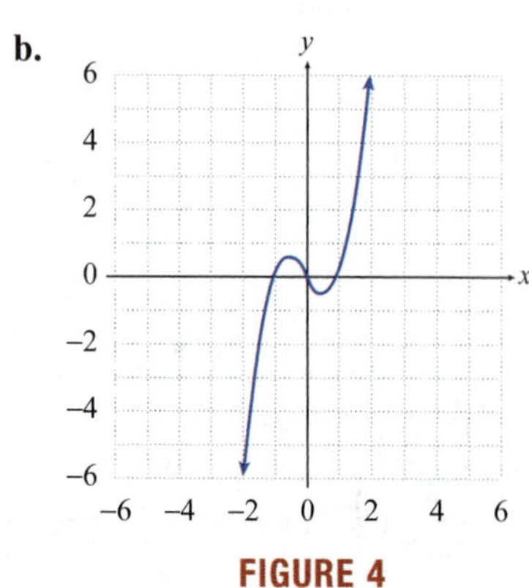

FIGURE 4

While we do not yet have the tools to graph general polynomial functions, we can obtain a good sketch of $g(x) = x^3 - x$.

First, g is odd: $g(-x) = -g(x)$ (verify this).

If we calculate a few values, such as $g(0) = 0$, $g\left(\frac{1}{2}\right) = -\frac{3}{8}$, $g(1) = 0$, and $g(2) = 6$, and then reflect these through the origin, we get a good idea of the shape of g.

c.

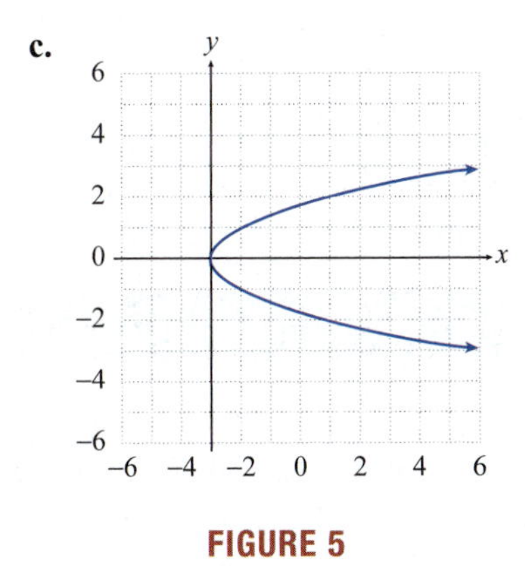

FIGURE 5

The equation $x = y^2$ is not a function, but it is a relation in x and y that has x-axis symmetry. If we replace y with $-y$ and simplify the result, we obtain the original equation.

$$x = (-y)^2$$
$$x = y^2$$

The upper half of the graph is the function $y = \sqrt{x}$, so drawing this and its reflection over the x-axis gives us the complete graph of $x = y^2$.

Summary of Symmetry

The first column in Table 1 summarizes the behavior of a graph in the Cartesian plane if it possesses any of the three types of symmetry we covered. If the graph is of an equation in x and y, the algebraic method in the second column can be used to identify the symmetry. The third column gives the algebraic method used to identify the type of symmetry if the graph is that of a function $f(x)$. Finally, the fourth column contains an example of each type of symmetry.

A graph is symmetric with respect to:	If the graph is of an equation in x and y, the equation is symmetric with respect to:	If the graph is of a function $f(x)$, the function is symmetric with respect to:	Example:
The y-axis if whenever the point (x, y) is on the graph, the point $(-x, y)$ is also on the graph.	The y-axis if replacing x with $-x$ results in an equivalent equation.	The y-axis if $f(-x) = f(x)$. We say the function is even.	$(-x, y)$ (x, y)
The x-axis if whenever the point (x, y) is on the graph, the point $(x, -y)$ is also on the graph.	The x-axis if replacing y with $-y$ results in an equivalent equation.	Not applicable (unless the graph consists only of points on the x-axis).	(x, y) $(x, -y)$
The origin if whenever the point (x, y) is on the graph, the point $(-x, -y)$ is also on the graph.	The origin if replacing x with $-x$ and y with $-y$ results in an equivalent equation.	The origin if $f(-x) = -f(x)$. We say the function is odd.	(x, y) $(-x, -y)$

TABLE 1: Summary of Symmetry

TOPIC 2: Intervals of Monotonicity

In many applications, it is useful to identify the intervals of the x-axis for which a function f is increasing in value, decreasing in value, or remaining constant. In this context, we say that these are intervals on which f is **monotone**.

DEFINITION: Increasing, Decreasing, and Constant

We say that a function f is

1. **increasing on an interval** if for any x_1 and x_2 in the interval with $x_1 < x_2$, it is the case that $f(x_1) < f(x_2)$;
2. **decreasing on an interval** if for any x_1 and x_2 in the interval with $x_1 < x_2$, it is the case that $f(x_1) > f(x_2)$;
3. **constant on an interval** if for any x_1 and x_2 in the interval, it is the case that $f(x_1) = f(x_2)$.

Based on this definition, an interval on which a particular function is monotone may be open, closed, or half-open. In practice, however, we are usually most interested in identifying the largest open intervals on which a given function is monotone. This often results in a subdivision of the domain of the function into disjoint open intervals separated by points at which the function changes monotonicity, as seen in Examples 2 and 3.

Determining the intervals of monotonicity of a function is a task that can be quite demanding, and in many cases is best tackled with the tools of calculus. We will look at some problems now in which algebra and our intuition will be sufficient.

Example 2: Determining Intervals of Monotonicity

Determine the open intervals of monotonicity of the function $f(x) = (x-2)^2 - 1$.

Solution

We know that the graph of f is the basic parabola shifted 2 units to the right and 1 unit down, as shown in Figure 6.

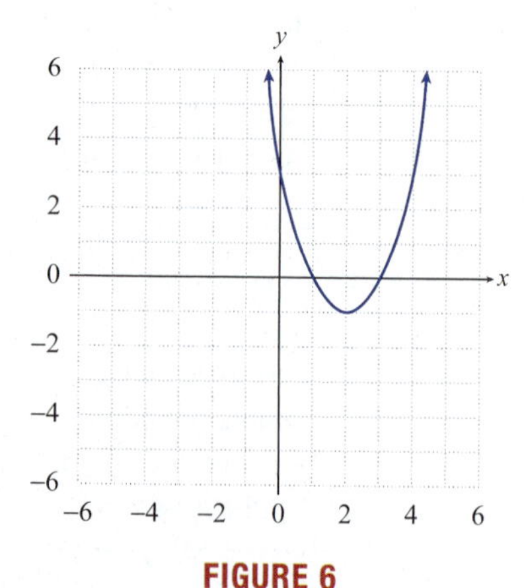

FIGURE 6

From the graph, we can see that f is decreasing on the interval $(-\infty, 2)$ and increasing on the interval $(2, \infty)$. Remember that these are intervals of the x-axis: if x_1 and x_2 are any two points in the interval $(-\infty, 2)$ with $x_1 < x_2$, then $f(x_1) > f(x_2)$. In other words, f is falling on this interval as we scan the graph from left to right. On the other hand, f is rising on the interval $(2, \infty)$ as we scan the graph from left to right.

Example 3: Modeling the Water Level of a River

The water level of a certain river varied over the course of a year as follows. In January, the level was 13 feet. From that level, the water increased linearly to a level of 18 feet in May. The water remained constant at that level until July, at which point it began to decrease linearly to a final level of 11 feet in December. Graph the water level as a function of time and determine the open intervals of monotonicity.

Solution

If we let 1 to 12 represent January through December, the intervals of monotonicity are as follows: increasing on $(1, 5)$, constant on $(5, 7)$, and decreasing on $(7, 12)$. The graph of the water level as a function of the month is shown in Figure 7.

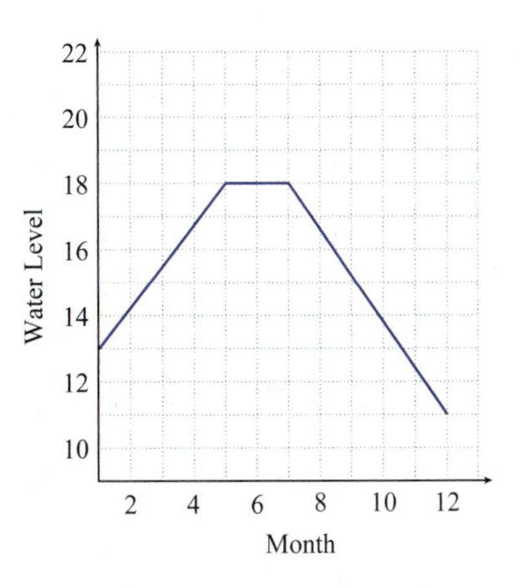

FIGURE 7

TOPIC 3: Local Extrema

In Section 3.3, we learned how to solve max/min problems involving quadratic functions by finding the vertex of the associated parabola. The vertex of an upward-opening parabola is the parabola's lowest point, and the vertex of a downward-opening parabola is its highest point. We now generalize this notion with the following definition.

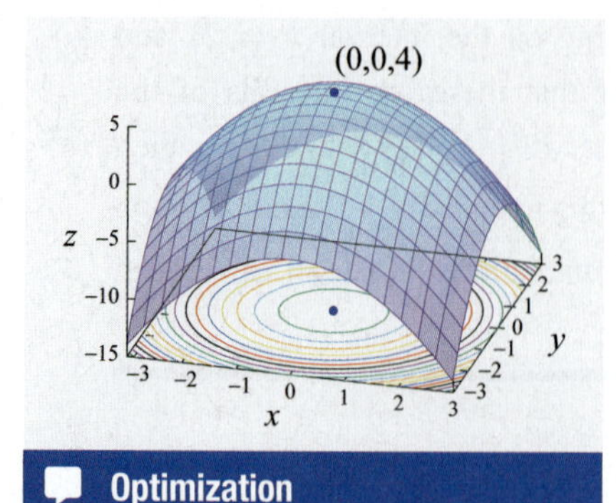

Optimization

The process of locating the local extrema of a function is often referred to as *optimization*, and in general the methods of calculus make the process much easier.

DEFINITION: Local Extrema

A function f has a **local maximum at c** if there is an open interval (a,b) containing c for which $f(x) \leq f(c)$ for all x in (a,b). In this case we say $f(c)$ is the **local maximum value** of f. Similarly, f has a **local minimum at c** if there is an open interval (a,b) containing c for which $f(x) \geq f(c)$ for all x in (a,b), and in this case we say $f(c)$ is the **local minimum value** of f. The local maxima and minima of a function are collectively referred to as **local extrema**.

Informally, the local maximum of a function is the highest of all nearby points on the graph of the function, and the local minimum is the lowest of all nearby points. And while local extrema may appear in cases where a function doesn't exhibit monotonicity, it's useful to note that *if* a given function f is decreasing on a half-open interval of the form $(a,c]$ and increasing on a half-open interval of the form $[c,b)$, then f has a local minimum at c and its value is $f(c)$ (see Figure 8). Similarly, if f is increasing on $(a,c]$ and decreasing on $[c,b)$, then f has a local maximum value of $f(c)$ at c (see Figure 9).

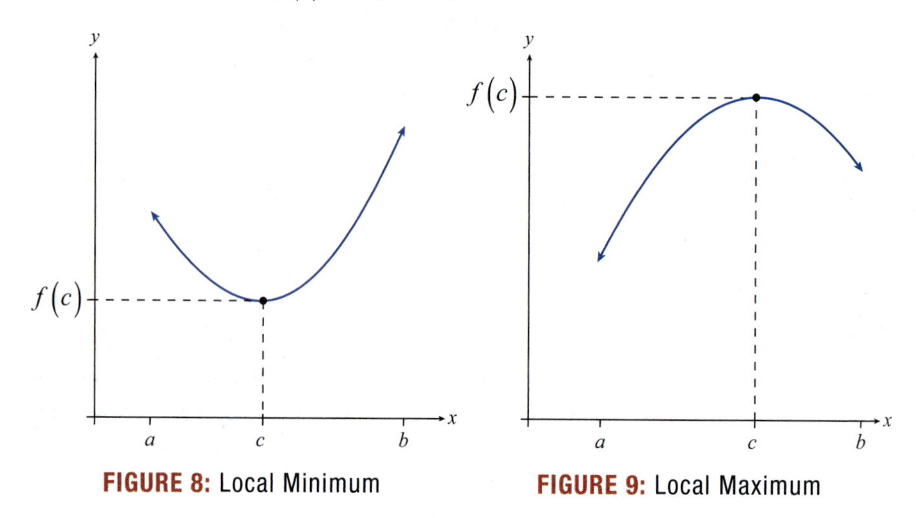

FIGURE 8: Local Minimum

FIGURE 9: Local Maximum

Example 4: Determining Local Extrema

Figure 10 shows the graph of a function f.

a. Determine the locations and types of the local extrema of f.

b. Determine the values of the local extrema of f.

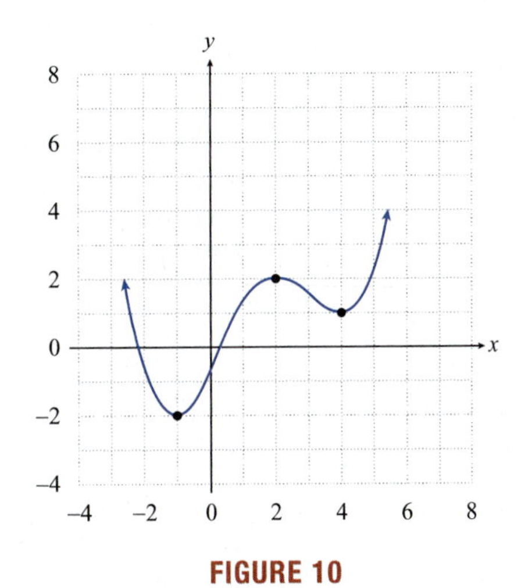

FIGURE 10

Solution

a. The function f has local minima at -1 and at 4, since for all x near -1 (for instance, within the interval $(-2,0)$) $f(x) \geq f(-1)$ and for all x near 4 (say, within $(3,5)$) $f(x) \geq f(4)$. The function f has a local maximum at 2, since $f(x) \leq f(2)$ for all x near 2.

b. Based on the graph, the value of the local minimum at -1 is -2 and the value of the local minimum at 4 is 1. The value of the local maximum at 2 is 2.

TOPIC 4: Average Rate of Change

We will close this section with one more way to analyze a function. The *average rate of change* of a function over a given interval provides an overall summary of the function's behavior over the interval. And in a sense to be made more precise soon, the average rate of change of a function, particularly over a small interval, can be thought of as a generalization of the idea of slope.

We will use a familiar situation to illustrate the concept. Suppose that you begin driving a car toward a certain destination and that the function $d(t)$ represents the miles traveled t hours after you start out. As you navigate the route, slowing down for curves, stopping at intersections or for other reasons, and speeding up as conditions allow, you would not expect $d(t)$ to be a linear function of t. But the idea of the average rate of travel is a natural one, and it can be calculated no matter how nonlinear d may be. For instance, if it takes 3 hours to reach your destination and you travel 150 miles in all, then the average rate of travel is $\frac{150 \text{ miles}}{3 \text{ hours}} = 50 \text{ mph}$. Stated in terms of the function d, and using the facts that $d(0) = 0$ and $d(3) = 150$, the average rate of travel over the entire 3 hours is

$$\frac{d(3) - d(0)}{3 - 0} = \frac{150 \text{ miles}}{3 \text{ hours}} = 50 \text{ mph}.$$

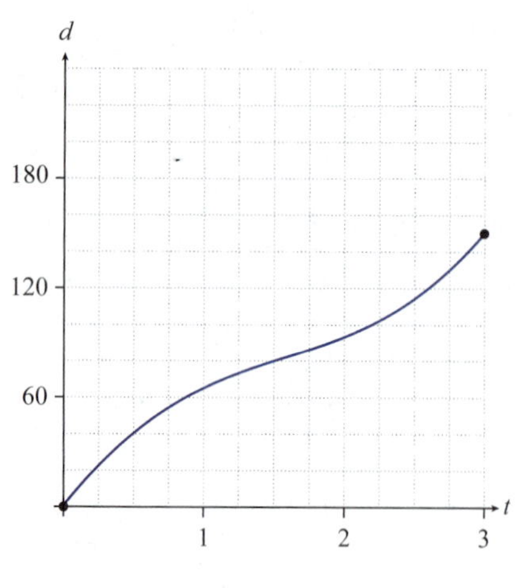

FIGURE 11

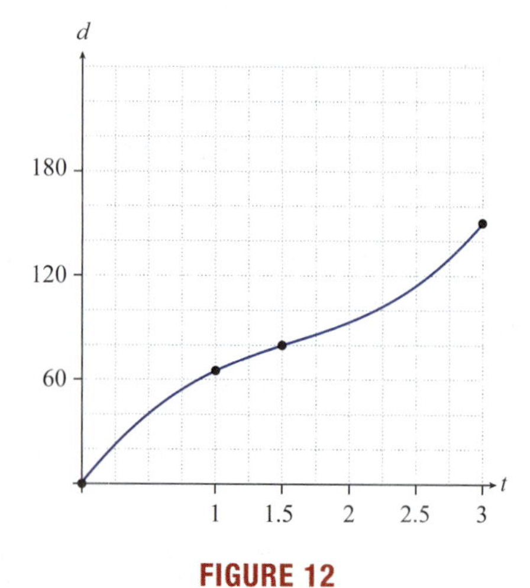

FIGURE 12

If we know more details about the distance function d, we can say more about the trip. Suppose the graph in Figure 11 represents the function $d(t)$ over the time interval $[0,3]$. It's clear that the distance traveled is not a linear function of t, which means that there are time intervals during which the car traveled faster or slower than at other times. We can use the same average rate of travel calculation to make this observation more precise. For instance, if we know that $d(1)=65$ and $d(1.5)=80$, as shown in Figure 12, then the average rate of travel over this half-hour interval was

$$\frac{d(1.5)-d(1)}{1.5-1}=\frac{80 \text{ miles}-65 \text{ miles}}{0.5 \text{ hours}}=\frac{15 \text{ miles}}{0.5 \text{ hours}}=30 \text{ mph}.$$

Figure 13 shows a close-up of the graph of d over the time interval $[1,1.5]$, and also shows a dashed line drawn between the two points on the graph at the endpoints of the interval. The dashed line is known as a *secant* line in this context, and its slope is 30, the average rate of travel over the interval. Because d is nearly a linear function over this interval, the graph of d lies close to the secant line; we say the secant line is a good approximation of d over this interval. In physical terms, this means that the car was traveling at a speed, or rate of travel, not too different from 30 mph over this particular half hour. Figure 14 shows again the graph of d over the whole 3-hour time interval, but with the secant line from $(0,0)$ to $(3,150)$ added. We've already determined that the slope of this secant line is 50, and Figure 14 shows that there are times along the trip where the speedometer of the car would indicate a number quite different from 50 mph.

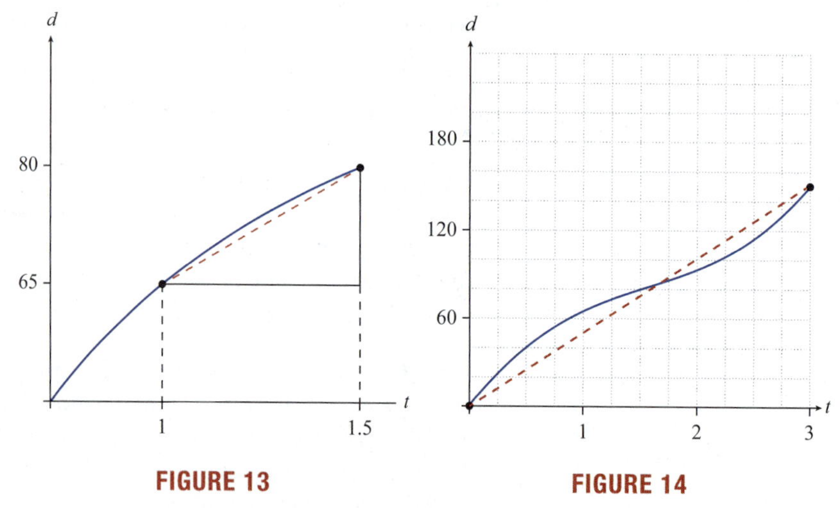

FIGURE 13 **FIGURE 14**

We now formalize the above ideas with a general definition.

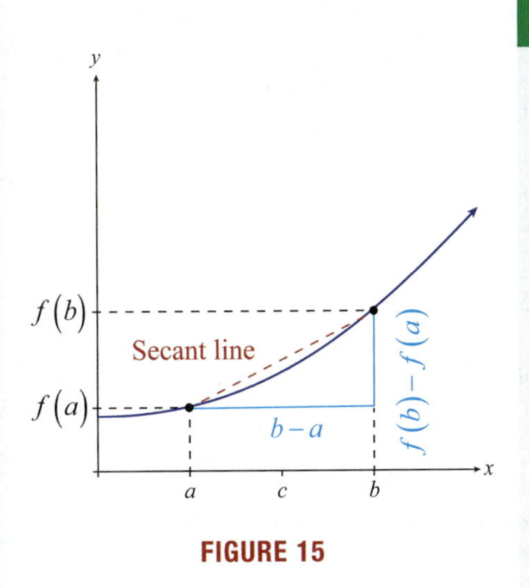

FIGURE 15

DEFINITION: Average Rate of Change

Given a function f defined on an interval $[a,b]$, $a \neq b$, the **average rate of change** of f over $[a,b]$ is $\dfrac{f(b)-f(a)}{b-a}$.

If $y = f(x)$, then any of the following expressions may be used to represent the average rate of change of f over $[a,b]$:

$$\frac{\text{change in } f}{\text{change in } x} = \frac{\text{change in } y}{\text{change in } x} = \frac{\Delta f}{\Delta x} = \frac{\Delta y}{\Delta x} = \frac{f(b)-f(a)}{b-a},$$

where the Greek letter Δ is used in this context to denote a change in the variable that follows it. The average rate of change of f over $[a,b]$ represents the slope of the **secant line** drawn between the points $(a, f(a))$ and $(b, f(b))$ on the graph of f, as shown in Figure 15.

Example 5: Determining Average Rate of Change

Given the function $f(x) = 3x^2 - 5x + 2$, determine the average rate of change over each of the following intervals.

a. $[1,3]$

b. $[-2,2]$

c. $[c, c+h]$, where $h \neq 0$

Solution

a. Using one of the equivalent choices of notation for average rate of change, we find the following.

$$\begin{aligned}\frac{\Delta f}{\Delta x} &= \frac{f(3)-f(1)}{3-1}\\ &= \frac{(3\cdot 3^2 - 5\cdot 3 + 2)-(3\cdot 1^2 - 5\cdot 1 + 2)}{3-1}\\ &= \frac{14-0}{2}\\ &= 7\end{aligned}$$

b. Over this interval, we calculate the average rate of change as follows.

$$\begin{aligned}\frac{\Delta f}{\Delta x} &= \frac{f(2)-f(-2)}{2-(-2)}\\ &= \frac{(3\cdot 2^2 - 5\cdot 2 + 2)-\left[3(-2)^2 - 5(-2) + 2\right]}{2-(-2)}\\ &= \frac{4-24}{4}\\ &= -5\end{aligned}$$

c. While neither c nor h are given as fixed numbers, we can find the average rate of change over the interval $[c,c+h]$ using the same process.

$$\begin{aligned}\frac{\Delta f}{\Delta x}&=\frac{f(c+h)-f(c)}{(c+h)-c}\\&=\frac{\left[3(c+h)^2-5(c+h)+2\right]-\left(3\cdot c^2-5\cdot c+2\right)}{(c+h)-c}\\&=\frac{\left(3c^2+6ch+3h^2-5c-5h+2\right)-\left(3c^2-5c+2\right)}{h}\\&=\frac{6ch+3h^2-5h}{h}\\&=6c+3h-5\end{aligned}$$

Differentiation and the Difference Quotient

In calculus, the act of determining the value of the difference quotient at a point c as h goes to 0 is called *differentiation*, and the resulting value (if there is one) is called the *derivative*.

The computation of the average rate of change of a function over an interval of the form $[c,c+h]$, as seen in Example 5c, occurs frequently in calculus, and the ratio

$$\frac{f(c+h)-f(c)}{h}$$

is known as the **difference quotient of f at c with increment h**. Its usefulness can be immediately seen by applying the result to parts a. and b. of Example 5:

a. For the interval $[1,3]$, $c=1$ and $h=2$, so $\frac{\Delta f}{\Delta x}=6\cdot1+3\cdot2-5=7$.

b. For the interval $[-2,2]$, $c=-2$ and $h=4$, so $\frac{\Delta f}{\Delta x}=6(-2)+3\cdot4-5=-5$.

In other words, once the difference quotient computation has been performed once, the average rate of change of a given function over any interval can be quickly found by identifying c and h and using the simplified difference quotient result. More importantly, the difference quotient is used in calculus to determine the behavior of functions over intervals with smaller and smaller increments h. Note also that the difference quotient is the same even if $h<0$, as long as f is defined on the interval $[c+h,c]$.

Example 6: Determining Intervals of Average Rate of Change

Use the given graph of the function f and the marked locations on the x-axis as endpoints to determine intervals over which the average rate of change of f is

a. positive, **b.** negative, **c.** zero.

Solution

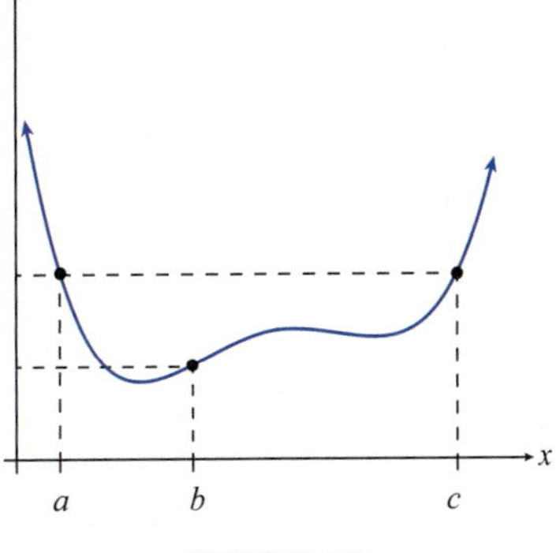

FIGURE 16

a. Since $f(c)-f(b)>0$ and $c-b>0$, $\dfrac{f(c)-f(b)}{c-b}>0$. That is, the average rate of change of f is positive on $[b,c]$.

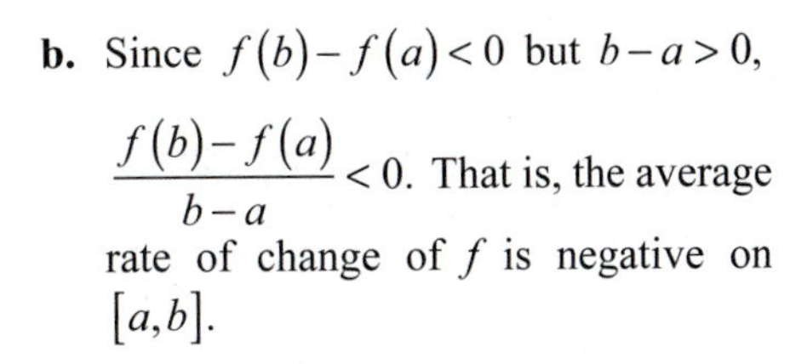

b. Since $f(b)-f(a)<0$ but $b-a>0$, $\dfrac{f(b)-f(a)}{b-a}<0$. That is, the average rate of change of f is negative on $[a,b]$.

c. Since $f(c)=f(a)$, $\dfrac{f(c)-f(a)}{c-a}=0$. That is, the average rate of change of f is zero on $[a,c]$.

4.2 EXERCISES

PRACTICE

Determine if each of the following relations is a function. If so, determine whether it is even, odd, or neither. Also determine if it has *y*-axis symmetry, *x*-axis symmetry, origin symmetry, or none of the above, and then sketch the graph of the relation. See Example 1.

1. $f(x)=|x|+3$ **2.** $g(x)=x^3$ **3.** $h(x)=x^3-1$

4. $w(x)=\sqrt[3]{x}$ **5.** $x=-y^2$ **6.** $3y-2x=1$

7. $x+y=1$ **8.** $F(x)=(x-1)^2$ **9.** $x=y^2+1$

10. $x=2|y|$ **11.** $g(x)=\dfrac{x^2}{5}-5$ **12.** $s(x)=\left[\!\left[x+\dfrac{1}{2}\right]\!\right]$

13. $m(x)=\sqrt[3]{x}-1$ **14.** $xy=2$ **15.** $x+y^2=3$

For each of the following functions, find the open intervals of monotonicity where the function is increasing, decreasing, or constant. See Examples 2 and 3.

16. $f(x)=(x+3)^2$ **17.** $g(x)=-|x-2|$ **18.** $h(x)=\dfrac{1}{x-1}$

19. $H(x)=\dfrac{1}{(x+3)^2}$ **20.** $G(x)=\sqrt{x+1}$ **21.** $F(x)=-2$

22. $p(x)=-30|x-1|$ **23.** $q(x)=(4-x)^2+1$

24. $r(x)=\dfrac{(x-7)^4}{-2}+4$ **25.** $P(x)=\begin{cases}(x+3)^2 & \text{if } x<-1\\ 1 & \text{if } x\geq -1\end{cases}$

26. $Q(x)=\begin{cases}|x-1| & \text{if } x\leq 3\\ 5-x & \text{if } x>3\end{cases}$

Using the graph of each of the following functions determine **a.** the locations and types of the local extrema, and **b.** the values of the local extrema. See Example 4.

27.

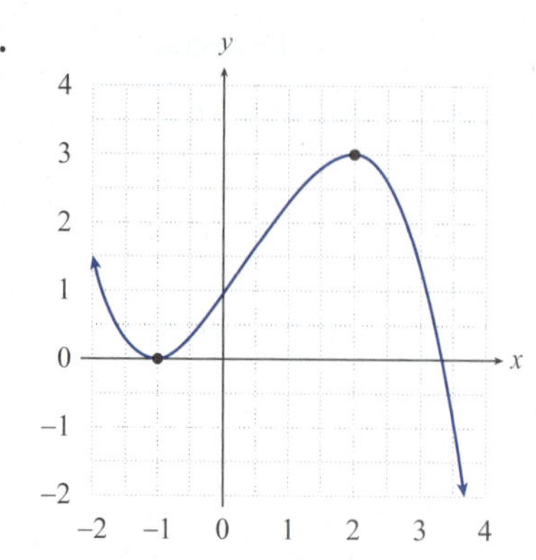

28.

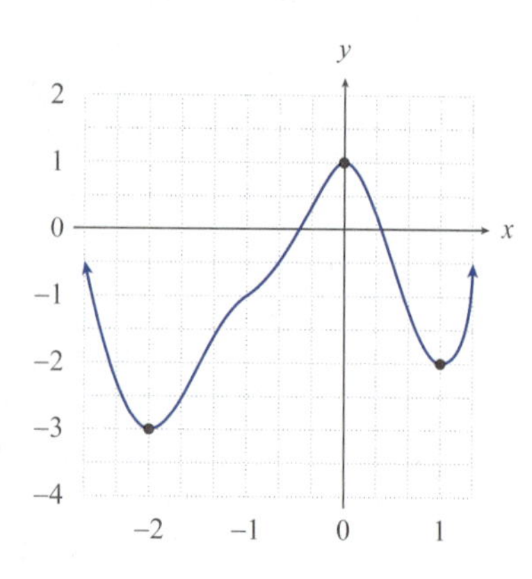

Using the graph and given formula for each of the following functions, determine **a.** the locations and types of the local extrema, and **b.** the values of the local extrema.

29. $f(x) = -2x^3 + 3x^2 + 12x - 5$

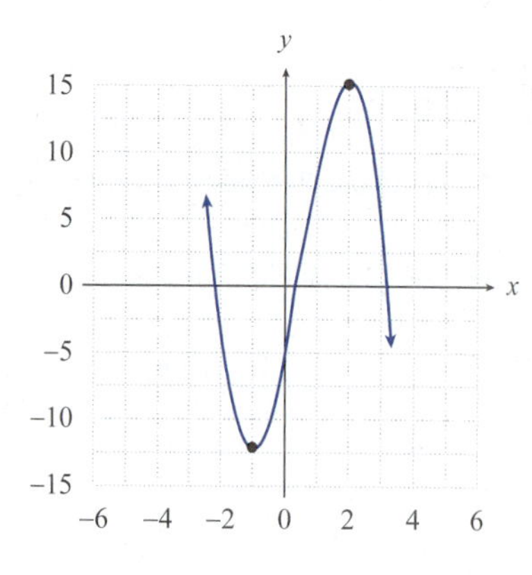

30. $f(x) = \frac{x^3}{3} - \frac{x^2}{2} - 6x$

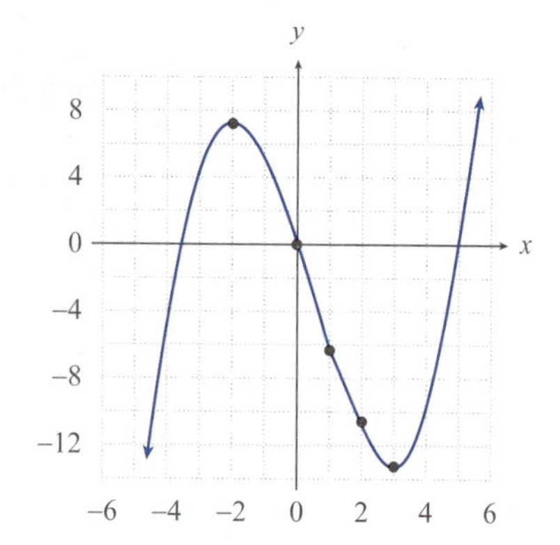

31. $f(x) = \frac{x^4}{4} - \frac{x^3}{3} - 3x^2 + 7$

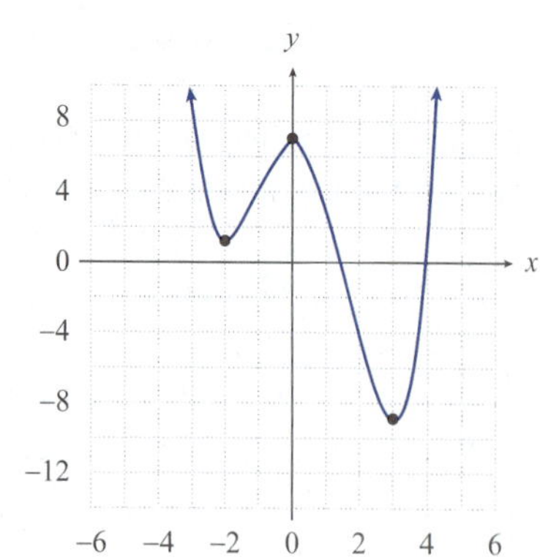

32. 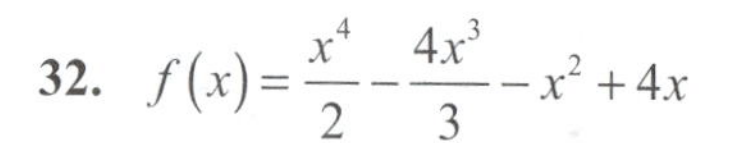 $f(x) = \frac{x^4}{2} - \frac{4x^3}{3} - x^2 + 4x$

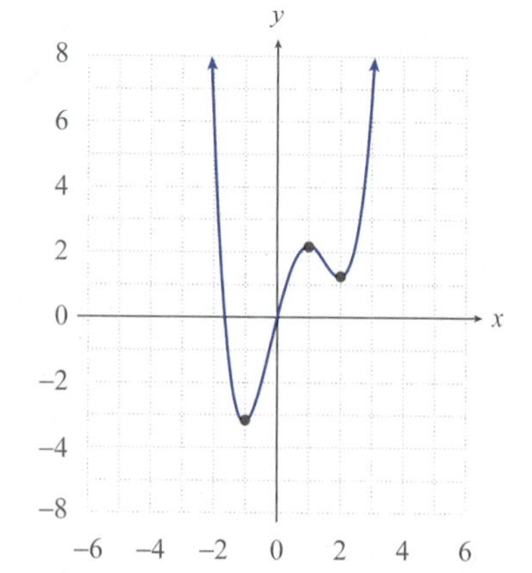

For each of the following functions, determine **a.** the locations and types of the local extrema, and **b.** the values of the local extrema. A sketch of the graph may be helpful.

33. $f(x)=(x-5)^2+2$

34. $f(x)=-(x+1)^2+3$

35. $f(x)=-x^2-4x-3$

36. $f(x)=x^2-10x+27$

37. $f(x)=5|x-3|-2$

38. $f(x)=-|x+1|+2$

For each given function and interval, determine the average rate of change of the function over the interval. See Example 5.

39. $f(x)=x^3-2x;\ [1,3]$

40. $f(x)=3x+17;\ [-2,0]$

41. $f(x)=x^2-5x+3;\ [2,5]$

42. $f(x)=-x^3+x^2-7;\ [-1,1]$

43. $f(x)=\sqrt{x};\ [2,4]$

44. $f(x)=-x^2+3x-1;\ [-2,1]$

45. $f(x)=x^2-3;\ [c,c+h]$

46. $f(x)=-3x^2+2x-5;\ [c,c+h]$

47. $f(x)=\frac{1}{x};\ [3,4]$

48. $f(x)=\frac{2}{x+1};\ [-3,-2]$

Use each given graph of a function and the marked locations on the x-axis as endpoints to determine intervals over which the average rate of change of the function is **a.** positive, **b.** negative, **c.** zero. See Example 6.

49.

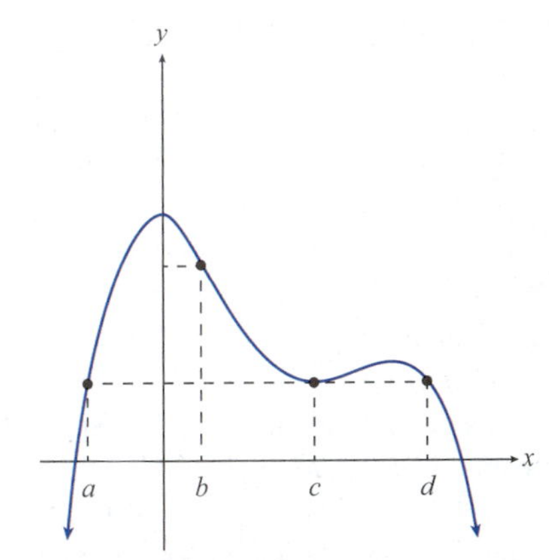

50.

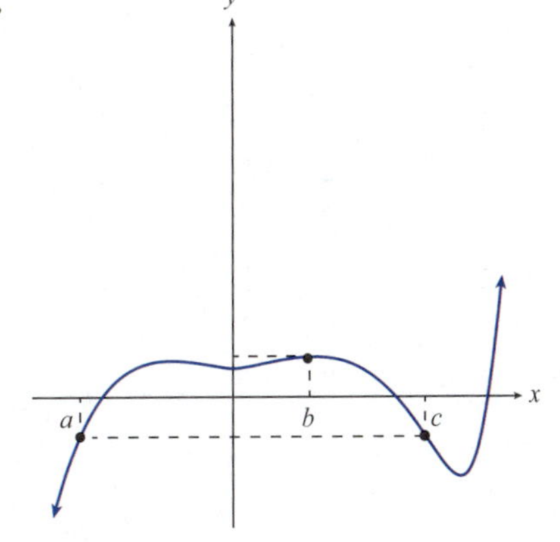

APPLICATIONS

51. During the summer months, the water level of a garden pool varies as water is added and as it evaporates. On May 1st the pool was 3.4 feet deep. After a steady and linear increase due to rain, the depth had increased to 4.9 feet on June 1st. By July 1st the water level had decreased linearly to 4.2 feet. Knowing that the pool would be covered for the winter, the owner filled the pool (in an essentially linear fashion) until it reached 5 feet on August 1st. Graph the water level as a function of time and determine the open intervals of monotonicity.

52. The profit made by a hot dog vendor is given by the function

$$P(x)=\begin{cases}2x-3 & \text{if } x\ge 0 \text{ and } x<7\\ \frac{1}{4}x^2 & \text{if } x\ge 7\end{cases}$$

where x is the number of hot dogs sold. Graph the profit function and determine the open intervals of monotonicity.

53. The cost incurred by a newspaper stand is given by the function

$$C(x)=\begin{cases}-2\sqrt{x}+8 & \text{if } x\ge 0 \text{ and } x<3\\ -x+8 & \text{if } x\ge 3\end{cases}$$

where x is the number of newspapers sold. Graph the cost function and determine the open intervals of monotonicity.

WRITING & THINKING

54. Determine the average rate of change of the function $f(x)=5x-19$ over each of the following intervals: $[-3,-2],[1,7],[c,c+h]$. What do you conclude from your calculations?

55. Let $f(x)=mx+b$, where m and b are both unspecified constants. Determine the average rate of change of f over several different intervals of your choice. What do you conclude from your calculations?

56. Let $f(x)=3x^2-7x+2$. Find the difference quotient of f at c with increment h. What happens to this difference quotient as the increment h becomes very small?

57. Let $f(x)=px^2+qx+r$, where p, q, and r are unspecified constants. Find the difference quotient of f at c with increment h. What happens to this difference quotient as the increment h becomes very small?

58. What can be deduced about the average rate of change of a function if the function is increasing over the interval?

59. What can be deduced about the monotonicity of a function over an interval if the function's average rate of change is positive over the interval?

60. What can be deduced about the monotonicity of a function over an interval if the function's average rate of change is zero over the interval?

4.3 COMBINING FUNCTIONS

TOPICS

1. Combining functions arithmetically
2. Composing functions
3. Decomposing functions
4. Recursive graphics

TOPIC 1: Combining Functions Arithmetically

In Section 4.1, we gained experience in building new functions from old ones by shifting, reflecting, and stretching the old functions. In this section, we will explore more ways of building functions.

We begin with four arithmetic ways of combining two or more functions to obtain new functions. The basic operations are very familiar to you: addition, subtraction, multiplication, and division. The difference is that we are applying these operations to functions, but as we will see, the arithmetic combination of functions is based entirely on the arithmetic combination of numbers.

DEFINITION: Addition, Subtraction, Multiplication, and Division of Functions

Let f and g be two functions. The **sum** $f+g$, **difference** $f-g$, **product** fg, and **quotient** $\frac{f}{g}$ are four new functions defined as follows.

1. $(f+g)(x)=f(x)+g(x)$
2. $(f-g)(x)=f(x)-g(x)$
3. $(fg)(x)=f(x)g(x)$
4. $\left(\frac{f}{g}\right)(x)=\frac{f(x)}{g(x)}$, provided that $g(x)\neq 0$

The domain of each of these new functions consists of the common elements (or the intersection of elements) of the domains of f and g individually, with the added condition that in the quotient function we have to omit those elements for which $g(x)=0$.

We have already combined functions arithmetically, though we didn't use the terminology of functions at the time, when we studied basic operations with polynomials in Section 1.3. With the above definition, we can determine the sum, difference, product, or quotient of two functions at one particular value for x, or find a formula for these new functions based on the formulas for f and g, if they are available.

Example 1: Combining Functions Arithmetically

Given that $f(-2)=5$ and $g(-2)=-3$, find $(f-g)(-2)$ and $\left(\frac{f}{g}\right)(-2)$.

Solution

By the definition of the difference and quotient of functions,

$$\begin{aligned}(f-g)(-2)&=f(-2)-g(-2)\\&=5-(-3)\\&=8,\end{aligned}$$

and

$$\begin{aligned}\left(\frac{f}{g}\right)(-2)&=\frac{f(-2)}{g(-2)}\\&=\frac{5}{-3}\\&=-\frac{5}{3}.\end{aligned}$$

Example 2: Combining Functions Arithmetically

Given the two functions $f(x)=4x^2-1$ and $g(x)=\sqrt{x}$, find $(f+g)(x)$ and $(fg)(x)$.

Solution

By the definition of the sum and product of functions,

$$\begin{aligned}(f+g)(x)&=f(x)+g(x)\\&=4x^2-1+\sqrt{x},\end{aligned}$$

and

$$\begin{aligned}(fg)(x)&=f(x)g(x)\\&=\left(4x^2-1\right)\left(\sqrt{x}\right)\\&=4x^{\frac{5}{2}}-x^{\frac{1}{2}}.\end{aligned}$$

What are the domains of $f+g$ and fg? We first need to find the domains of the individual functions f and g.

Domain of f: $(-\infty,\infty)$ since f is a quadratic function

Domain of g: $[0,\infty)$ since square roots of negative numbers are undefined

Since the domain of two functions combined arithmetically is the intersection of the individual domains, $f+g$ and fg both have a domain of $[0,\infty)$.

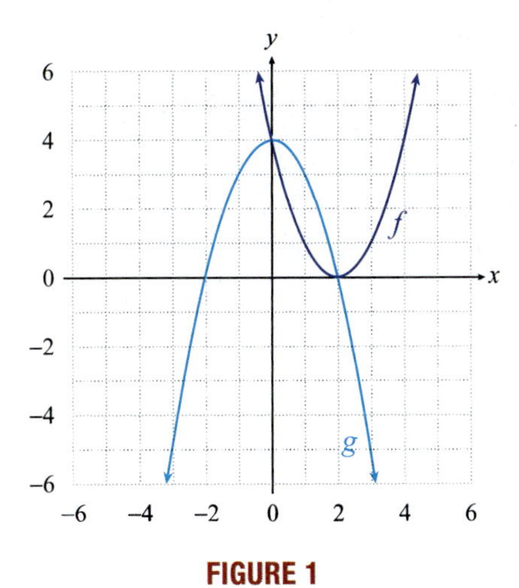

FIGURE 1

Example 3: Combining Functions Arithmetically

Given the graphs of f and g in Figure 1, determine the domain of $f+g$ and $\dfrac{f}{g}$ and evaluate $(f+g)(1)$ and $\left(\dfrac{f}{g}\right)(1)$.

Solution

From the graph, we can see that the domain of both f and g is the set of all real numbers $(-\infty,\infty)$. This means that the domain of $f+g$ is also $(-\infty,\infty)$. To find the domain of the quotient, we need to check where $g(x)=0$. The graph shows us that this occurs when $x=\pm 2$, so the domain of $\dfrac{f}{g}$ is all real numbers *except* 2 and −2.

$$(-\infty,-2)\cup(-2,2)\cup(2,\infty)$$

To evaluate the new functions, we need to find $f(1)$ and $g(1)$ using the graph.

We can see that $f(1)=1$ and $g(1)=3$, which means

$$(f+g)(1)=1+3=4 \text{ and } \left(\frac{f}{g}\right)(1)=\frac{1}{3}.$$

TOPIC 2: Composing Functions

A fifth way of combining functions is to form the *composition* of one function with another. Informally speaking, this means to apply one function to the output of another function. The symbol for composition is an open circle.

DEFINITION: Composing Functions

Let f and g be two functions. The **composition** of f and g, denoted $f\circ g$, is the function defined by $(f\circ g)(x)=f(g(x))$. The domain of $f\circ g$ consists of all x in the domain of g for which $g(x)$ is in turn in the domain of f. The function $f\circ g$ is read "f composed with g," or "f of g."

The diagram in Figure 2 is a schematic of the composition of two functions. To calculate $(f\circ g)(x)$ we first apply the function g, calculating $g(x)$, then apply the function f to the result, calculating $f(g(x))=(f\circ g)(x)$.

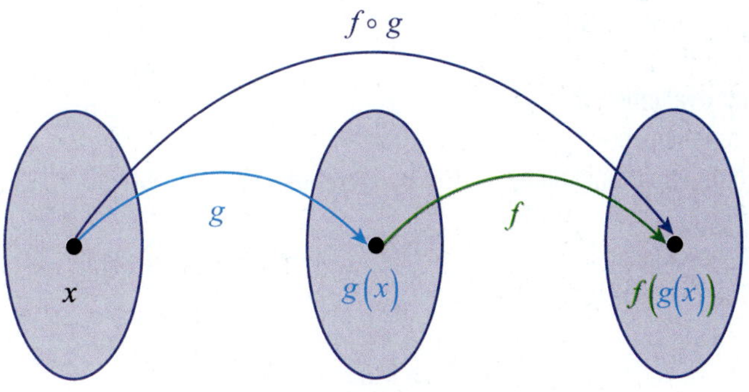

FIGURE 2: Composition of f and g

As with the four arithmetic ways of combining functions, we can evaluate the composition of two functions at a single point, or find a formula for the composition if we have been given formulas for the individual functions.

⚠ CAUTION

Note that the order of f and g is important. In general, we can expect the function $f \circ g$ to be *different* from the function $g \circ f$. In formal terms, the composition of two functions, unlike the sum and product of two functions, is not commutative.

Example 4: Composing Functions

Given $f(x) = x^2$ and $g(x) = x - 3$, find the following.

a. $(f \circ g)(6)$ **b.** $(g \circ f)(6)$

c. $(f \circ g)(x)$ **d.** $(g \circ f)(x)$

TECHNOLOGY

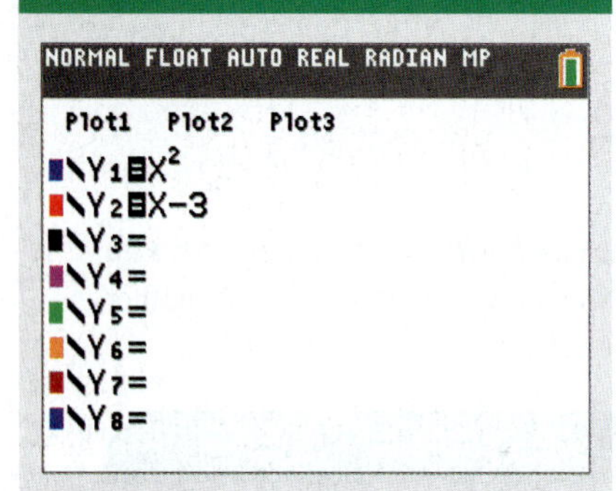

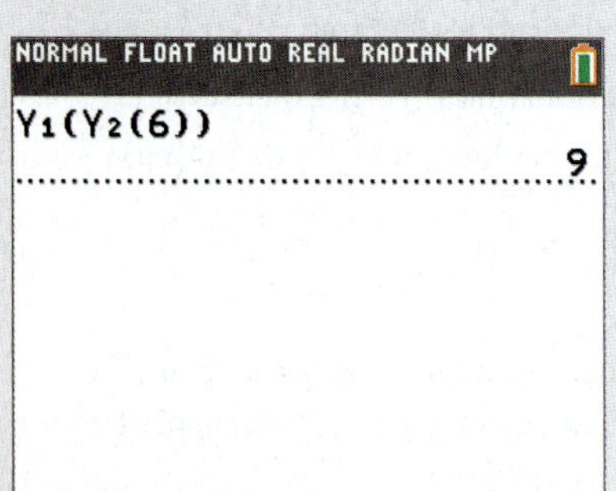

Use Y= to enter the two functions as Y1 and Y2. To input an expression for evaluating the composition of functions on the home screen, input each function by pressing vars, scrolling right to the Y-VARS menu, selecting Function, and then selecting either Y1 or Y2.

Solution

a. Since $(f \circ g)(6) = f(g(6))$, the first step is to calculate $g(6)$.

$$g(6) = 6 - 3 = 3$$

Then, apply f to the result.

$$(f \circ g)(6) = f(g(6)) = f(3) = 3^2 = 9$$

b. This time, we begin by finding $f(6)$.

$$f(6) = 6^2 = 36$$

Now, apply g to the result.

$$g(f(6)) = g(36) = 36 - 3 = 33$$

c. To find the formula for $f \circ g$ we apply the definition of composition, then simplify.

$(f \circ g)(x) = f(g(x))$	Write out the definition of composition.
$= f(x - 3)$	Substitute the formula for $g(x)$.
$= (x - 3)^2$	Apply the formula for $f(x)$.
$= x^2 - 6x + 9$	Simplify.

d. To find a formula for the function $g \circ f$ we follow the same process.

$(g \circ f)(x) = g(f(x))$	Write out the definition of composition.
$= g(x^2)$	Substitute the formula for $f(x)$.
$= x^2 - 3$	Apply the formula for $g(x)$; the result is already simplified.

Note that once we have found formulas $f \circ g$ and $g \circ f$ we can answer the first two parts by directly plugging into these formulas.

$$(f \circ g)(6) = 6^2 - 6(6) + 9 = 9$$
$$(g \circ f)(6) = 6^2 - 3 = 33$$

CAUTION

When evaluating the composition $(f \circ g)(x)$ at a point x, there are two reasons the value might be undefined:

1. If x is not in the domain of g, then $g(x)$ is undefined and we can't evaluate $f(g(x))$.
2. If $g(x)$ is not in the domain of f, then $f(g(x))$ is undefined and we can't evaluate it.

In either case, $(f \circ g)(x) = f(g(x))$ is undefined, and x is not in the domain of $(f \circ g)(x)$.

Example 5: Domains of Compositions of Functions

Let $f(x) = \sqrt{x-5}$ and $g(x) = \dfrac{2}{x+1}$. Evaluate the following.

a. $(f \circ g)(-1)$

b. $(f \circ g)(1)$

Solution

a. $(f \circ g)(-1) = f(g(-1))$

But, if we try to evaluate $g(-1)$, we see that it is undefined, so $(f \circ g)(-1)$ is also undefined.

b. $(f \circ g)(1) = f(g(1))$

First, we evaluate $g(1)$.

$$g(1) = \frac{2}{1+1} = \frac{2}{2} = 1$$

We plug this result into $f(x)$ but see that $\sqrt{1-5} = \sqrt{-4}$ is undefined. Thus, $(f \circ g)(1)$ is also undefined.

Example 6: Domains of Compositions of Functions

Let $f(x)=x^2-4$ and $g(x)=\sqrt{x}$. Find formulas and state the domains for the following.

a. $f \circ g$ **b.** $g \circ f$

Solution

a. $(f \circ g)(x) = f(g(x))$

$= f(\sqrt{x})$ Substitute the formula for $g(x)$ into $f(x)$.

$= (\sqrt{x})^2 - 4$ Simplify.

$= x - 4$

While the domain of $x-4$ is the set of all real numbers, the domain of $f \circ g$ is $[0,\infty)$ since only nonnegative numbers can be plugged into g.

b. $(g \circ f)(x) = g(f(x))$

$= g(x^2-4)$ Substitute the formula for $f(x)$ into $g(x)$.

$= \sqrt{x^2-4}$ The answer is already simplified.

The domain of $g \circ f$ consists of all x for which $x^2-4 \geq 0$, or $x^2 \geq 4$. We can write this in interval form as $(-\infty,-2] \cup [2,\infty)$.

TOPIC 3: Decomposing Functions

Often, functions can be best understood by recognizing them as a composition of two or more simpler functions. We have already seen instances of this: shifting, reflecting, stretching, and compressing can all be thought of as a composition of two or more functions. For example, the function $h(x)=(x-2)^3$ is a composition of the functions $f(x)=x^3$ and $g(x)=x-2$.

$$\begin{aligned} f(g(x)) &= f(x-2) \\ &= (x-2)^3 \\ &= h(x) \end{aligned}$$

To "decompose" a function into a composition of simpler functions, it is usually best to identify what the function does to its argument from the inside out. That is, identify the first thing that is done to the variable, then the second, and so on. Each action describes a less complex function, and can be identified as such. The composition of these functions, with the innermost function corresponding to the first action, the next innermost corresponding to the second action, and so on, is then equivalent to the original function.

Decomposition can often be done in several different ways. Consider, for example, the function $f(x)=\sqrt[3]{5x^2-1}$. Next we illustrate just a few of the ways f can be written as a composition of functions. Be sure you understand how each of the different compositions is equivalent to f.

1. $g(x)=\sqrt[3]{x}$
$h(x)=5x^2-1$

$$g(h(x))=g(5x^2-1)$$
$$=\sqrt[3]{5x^2-1}$$
$$=f(x)$$

2. $g(x)=\sqrt[3]{x-1}$
$h(x)=5x^2$

$$g(h(x))=g(5x^2)$$
$$=\sqrt[3]{5x^2-1}$$
$$=f(x)$$

3. $g(x)=\sqrt[3]{x}$
$h(x)=5x-1$
$j(x)=x^2$

$$g(h(j(x)))=g(h(x^2))$$
$$=g(5x^2-1)$$
$$=\sqrt[3]{5x^2-1}$$
$$=f(x)$$

Example 7: Decomposing Functions

Decompose the function $f(x)=|x^2-3|+2$ into the following.

a. a composition of two functions

b. a composition of three functions

NOTE

These are **not** the only possible solutions for the decompositions of $f(x)$.

Solution

a. $g(x)=|x|+2$
$h(x)=x^2-3$

$$g(h(x))=g(x^2-3)$$
$$=|x^2-3|+2$$
$$=f(x)$$

b. $g(x)=x+2$
$h(x)=|x-3|$
$j(x)=x^2$

$$g(h(j(x)))=g(h(x^2))$$
$$=g(|x^2-3|)$$
$$=|x^2-3|+2$$
$$=f(x)$$

TOPIC 4: Recursive Graphics

Recursion, in general, refers to using the output of a function as its input, and repeating the process a certain number of times. In other words, recursion refers to the composition of a function with itself, possibly many times. Recursion has many varied uses, one of which is a branch of mathematical art.

> So, Nat'ralists observe, a Flea
> Hath smaller Fleas that on him prey,
> And these have smaller yet to bite 'em,
> And so proceed *ad infinitum*
> —Jonathan Swift, *On Poetry: A Rapsody*

There is some special notation to describe recursion. If f is a function, $f^2(x)$ is used in this context to stand for $f(f(x))$, or $(f \circ f)(x)$, not $[f(x)]^2$! Similarly, $f^3(x)$ stands for $f(f(f(x)))$, or $(f \circ f \circ f)(x)$, and so on. The functions f^2, $f^3, \ldots$ are called **iterates** of f, with f^n being the $\boldsymbol{n}^{\text{th}}$ **iterate** of f.

Some of the most famous recursively generated mathematical art is based on functions whose inputs and outputs are complex numbers. Recall from Section 1.5 that every complex number can be expressed in the form $a + bi$, where a and b are real numbers and i is the imaginary unit. A one-dimensional coordinate system, such as the real number line, is insufficient to graph complex numbers, but complex numbers are easily graphed in a two-dimensional coordinate system.

To graph the number $a+bi$, we treat it as the ordered pair (a,b) and plot it as a point in the Cartesian plane, where the horizontal axis represents pure real numbers and the vertical axis represents pure imaginary numbers.

Benoit Mandelbrot used the function $f(z)=z^2+c$, where both z and c are variables representing complex numbers, to generate the image known as the Mandelbrot set in the 1970s. The basic idea is to evaluate the sequence of iterates $f(0)=0^2+c=c,\ f^2(0)=f(c)=c^2+c,\ f^3(0)=f(c^2+c)=(c^2+c)^2+c, \ldots$ for various complex numbers c and determine if the sequence of complex numbers stays close to the origin or not. Those complex numbers c that result in so-called "bounded" sequences are colored black, while those that lead to unbounded sequences are colored white. The author has used similar ideas to generate his own recursive art, as described below.

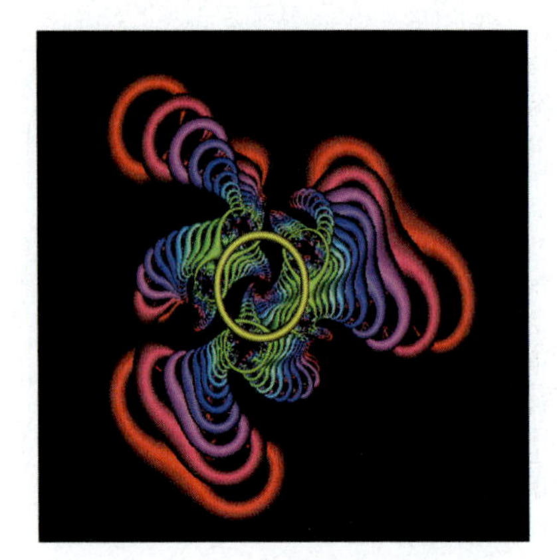

FIGURE 3

The image "*i* of the storm" reproduced in Figure 3 is based on the function $f(z)=\frac{(1-i)z^4+(7+i)z}{2z^5+6}$, where again z is a variable that will be replaced with complex numbers. The image is actually a picture of the complex plane, with the origin in the very center of the golden ring. The golden ring consists of those complex numbers that lie a distance between 0.9 and 1.1 units from the origin. The rules for coloring other complex numbers in the plane are as follows: given an initial complex number z not on the gold ring, $f(z)$ is calculated. If the complex number $f(z)$ lies somewhere on the gold ring, the original number z is colored the deepest shade of green. If not, the iterate $f^2(z)$ is calculated. If this result lies in the gold ring, the original z is colored a bluish shade of green. If not, the process continues up to the 12th iterate $f^{12}(z)$, using a different color each time. If $f^{12}(z)$, lies in the gold ring, z is colored red, and if not the process halts and z is colored black.

The idea of recursion can be used to generate any number of similar images, with the end result usually striking and often surprising even to the creator.

4.3 EXERCISES

PRACTICE

In each of the following exercises, use the information given to determine **a.** $(f+g)(-1)$, **b.** $(f-g)(-1)$, **c.** $(fg)(-1)$, and **d.** $\left(\frac{f}{g}\right)(-1)$. See Examples 1, 2, and 3.

1. $f(-1)=-3$ and $g(-1)=5$

2. $f(-1)=0$ and $g(-1)=-1$

3. $f(x)=x^2-3$ and $g(x)=x$

4. $f(x)=\sqrt[3]{x}$ and $g(x)=x-1$

5. $f(-1)=15$ and $g(-1)=-3$

6. $f(x)=\frac{x+5}{2}$ and $g(x)=6x$

7. $f(x)=x^4+1$ and $g(x)=x^{11}+2$

8. $f(x)=\frac{6-x}{2}$ and $g(x)=\sqrt{\frac{x}{-4}}$

9. $f=\{(5,2),(0,-1),(-1,3),(-2,4)\}$ and $g=\{(-1,3),(0,5)\}$

10. $f=\{(3,15),(2,-1),(-1,1)\}$ and $g(x)=-2$

11.

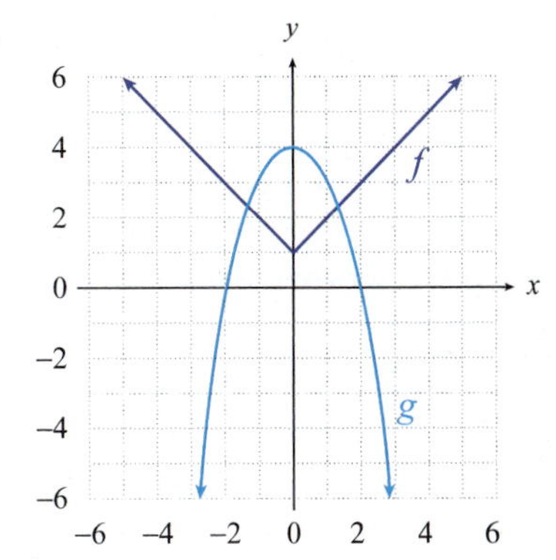

12.

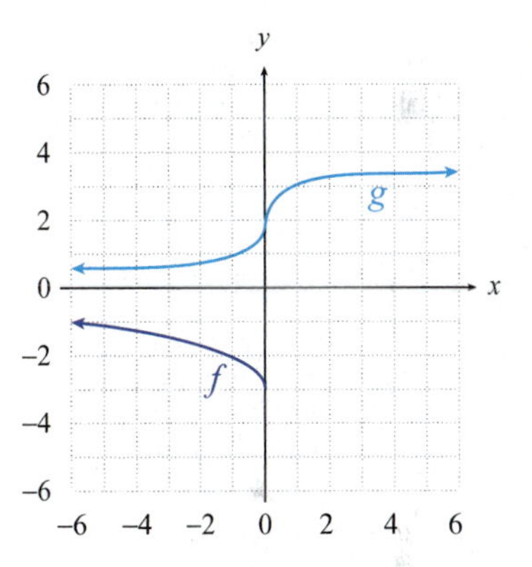

13.

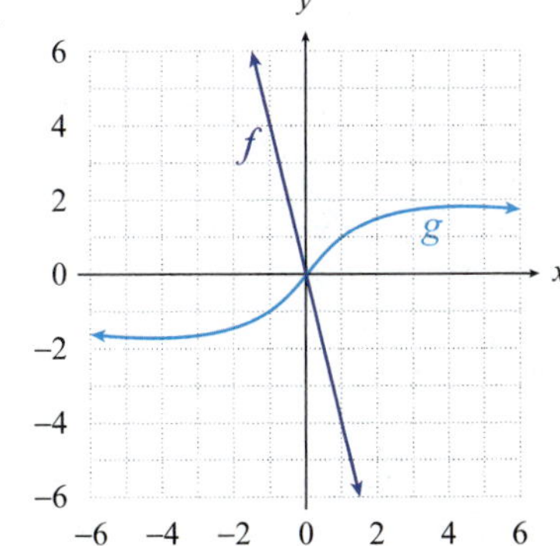

14.

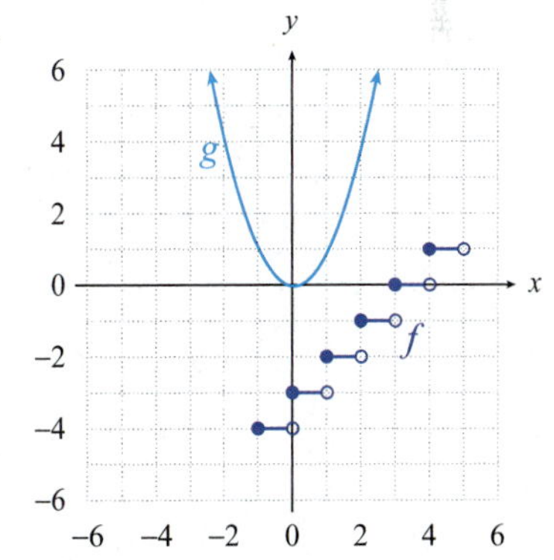

In each of the following exercises, find **a.** the formula and domain for $(f+g)$ and **b.** the formula and domain for $\frac{f}{g}$. See Examples 2 and 3.

15. $f(x)=|x|$ and $g(x)=\sqrt{x}$

16. $f(x)=x^2-1$ and $g(x)=\sqrt[3]{x}$

17. $f(x)=x-1$ and $g(x)=x^2-1$

18. $f(x)=x^{\frac{3}{2}}$ and $g(x)=x-3$

19. $f(x)=3x$ and $g(x)=x^3-8$

20. $f(x)=x^3+4$ and $g(x)=\sqrt{x-2}$

21. $f(x)=-2x^2$ and $g(x)=|x+4|$

22. $f(x)=6x-1$ and $g(x)=x^{\frac{2}{3}}$

In each of the following exercises, use the information given to determine $(f \circ g)(3)$. See Examples 4 and 5.

23. $f(-5)=2$ and $g(3)=-5$

24. $f(\pi)=\pi^2$ and $g(3)=\pi$

25. $f(x)=x^2-3$ and $g(x)=\sqrt{x}$

26. $f(x)=\sqrt{x^2-9}$ and $g(x)=1-2x$

27. $f(x)=2+\sqrt{x}$ and $g(x)=x^3+x^2$

28. $f(x)=x^{\frac{3}{2}}-3$ and $g(x)=\left|\frac{4x}{3}\right|$

29. $f(x)=\sqrt{x+6}$ and $g(x)=\sqrt{4x-3}$

30. $f(x)=\sqrt{\frac{3x}{14}}$ and $g(x)=x^4-x^3-x^2-x$

31.

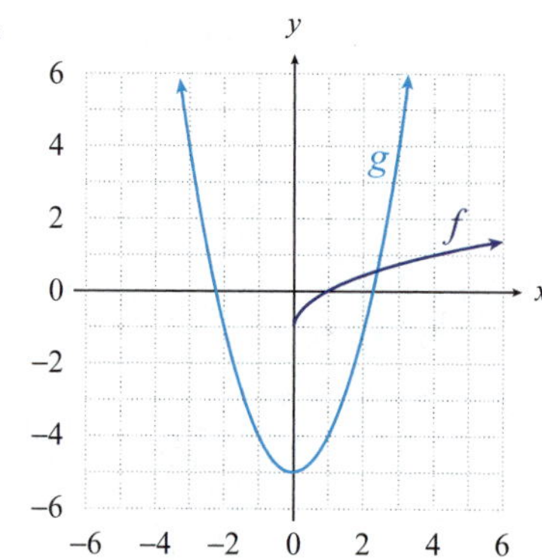

32.

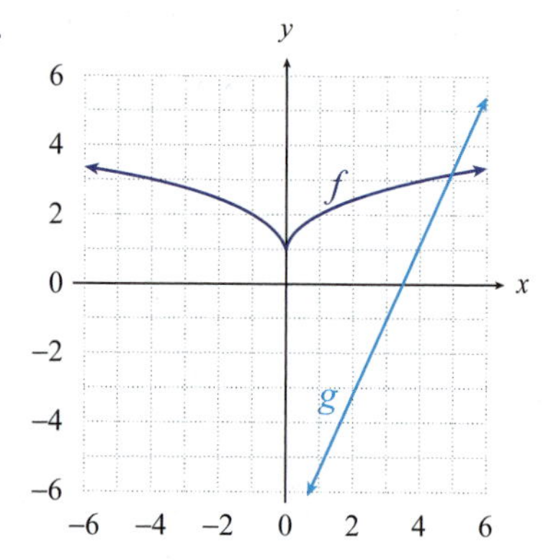

33.

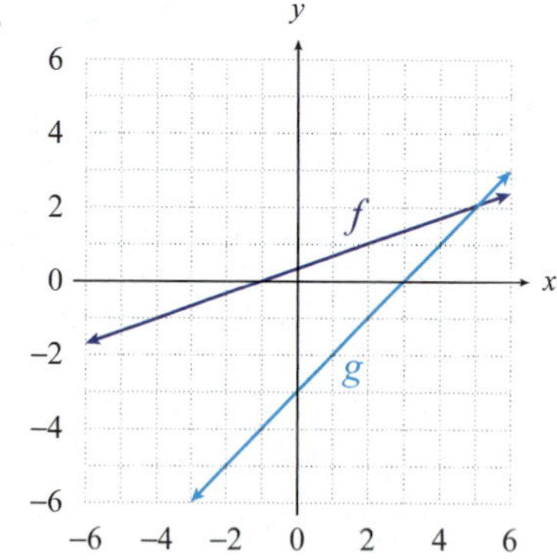

In each of the following exercises, find **a.** the formula and domain for $f \circ g$ and **b.** the formula and domain for $g \circ f$. See Example 6.

34. $f(x)=\frac{1}{x}$ and $g(x)=x-1$

35. $f(x)=\frac{4x-2}{3}$ and $g(x)=\frac{1}{x}$

36. $f(x)=1-x$ and $g(x)=\sqrt{x}$

37. $f(x)=|x-3|$ and $g(x)=x^3+1$

38. $f(x)=x^2+2x$ and $g(x)=x-3$

39. $f(x)=\sqrt{x-1}$ and $g(x)=\frac{x+1}{2}$

40. $f(x)=x^3+4x^2$ and $g(x)=|x|-1$

41. $f(x)=-3x+2$ and $g(x)=x^2+2$

42. $f(x)=x+2$ and $g(x)=\dfrac{x^2+3}{2}$

43. $f(x)=\sqrt{x-1}$ and $g(x)=x^2$

Write each of the following functions as a composition of two functions. Answers will vary. See Example 7.

44. $f(x)=\sqrt[3]{3x^2-1}$

45. $f(x)=\dfrac{2}{5x-1}$

46. $f(x)=|x-2|+3$

47. $f(x)=x+\sqrt{x+2}-5$

48. $f(x)=|x^3-5x|+7$

49. $f(x)=\dfrac{\sqrt{x-3}}{x^2-6x+9}$

50. $f(x)=\sqrt{2x^3-3}-4$

51. $f(x)=|x^2+3x|-3$

52. $f(x)=\dfrac{3}{4x-2}$

In each of the following exercises, use the information given to find $g(x)$.

53. $f(x)=|x+3|$ and $(f+g)(x)=|x+3|+\sqrt{x+5}$

54. $f(x)=x$ and $(f\circ g)(x)=\dfrac{x+12}{-3}$

55. $f(x)=x^2-3$ and $(f-g)(x)=x^3+x^2+4$

56. $f(x)=x^2$ and $(g\circ f)(x)=\sqrt{-x^2+5}+4$

APPLICATIONS

57. The volume of a right circular cylinder is given by the formula $V=\pi r^2 h$. If the height h is three times the radius r, show the volume V as a function of r.

58. The surface area S of a wind sock is given by the formula $S=\pi r\sqrt{r^2+h^2}$, where r is the radius of the base of the wind sock and h is the height of the wind sock. As the wind sock is being knitted by an automated knitter, the height h increases with time t according to the formula $h(t)=\frac{1}{4}t^2$. Find the surface area S of the wind sock as a function of time t and radius r.

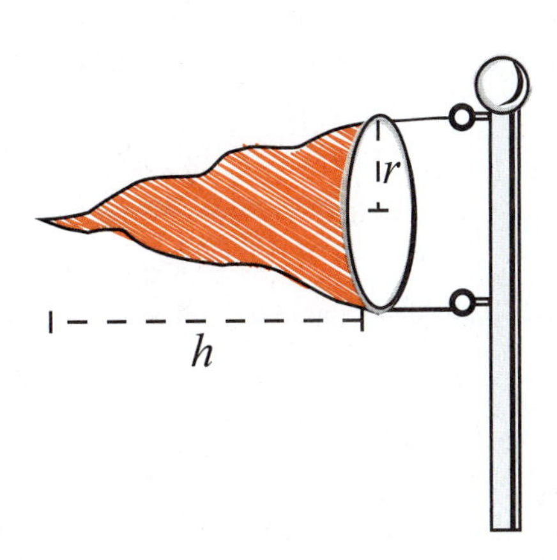

59. The volume V of the wind sock described in the previous exercise is given by the formula $V=\frac{1}{3}\pi r^2 h$ where r is the radius of the wind sock and h is the height of the wind sock. If the height h increases with time t according to the formula $h(t)=\frac{1}{4}t^2$, find the volume V of the wind sock as a function of time t and radius r.

60. A widget factory produces n widgets in t hours of a single day. The number of widgets the factory produces is given by the formula $n(t)=10{,}000t-25t^2$, $0\le t\le 9$. The cost c in dollars of producing n widgets is given by the formula $c(n)=2040+1.74n$. Find the cost c as a function of time t.

WRITING & THINKING

61. Given two odd functions f and g, show that $f\circ g$ is also odd. Verify this fact with the particular functions $f(x)=\sqrt[3]{x}$ and $g(x)=\frac{-x^3}{3x^2-9}$. Recall that a function is odd if $f(-x)=-f(x)$ for all x in the domain of f.

62. Given two even functions f and g, show that the product is also even. Verify this fact with the particular functions $f(x)=2x^4-x^2$ and $g(x)=\frac{1}{x^2}$. Recall that a function is even $f(-x)=f(x)$ for all x in the domain of f.

As mentioned in Topic 4, a given complex number c is said to be in the Mandelbrot set if, for the function $f(z)=z^2+c$, the sequence of iterates $f(0), f^2(0), f^3(0), \ldots$ stays close to the origin (which is the complex number $0+0i$). It can be shown that if any single iterate falls more than 2 units in distance (magnitude) from the origin, then the remaining iterates will grow larger and larger in magnitude. In practice, computer programs that generate the Mandelbrot set calculate the iterates up to a predecided point in the sequence, such as $f^{50}(0)$, and if no iterate up to this point exceeds 2 in magnitude, the number c is admitted to the set. The magnitude of a complex number $a+bi$ is the distance between the point (a,b) and the origin, so the formula for the magnitude of $a+bi$ is $\sqrt{a^2+b^2}$.

Use the above criterion to determine, without a calculator or computer, if the following complex numbers are in the Mandelbrot set or not.

63. $c=0$

64. $c=1$

65. $c=i$

66. $c=-1$

67. $c=1+i$

68. $c=-i$

69. $c=1-i$

70. $c=-1-i$

71. $c=2$

72. $c=-2$

4.4 INVERSES OF FUNCTIONS

TOPICS

1. Inverses of relations
2. Inverse functions and the horizontal line test
3. Finding the inverse of a function

TOPIC 1: Inverses of Relations

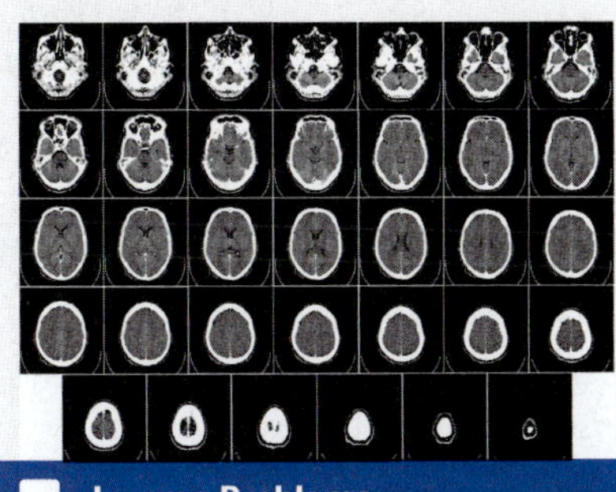

Inverse Problems

In science and engineering applications, an *inverse problem* is one in which the goal is to work backward from a known outcome in order to determine the inputs that could have produced it. Many of the hardest and most important problems in a broad array of industries, such as medical imaging, artificial intelligence, and signal processing, fall into the category of inverse problems.

In many problems, "undoing" one or more mathematical operations plays a critical role in the solution process. For instance, to solve the equation $3x+2=8$, the first step is to "undo" the addition of 2 on the left-hand side (by subtracting 2 from both sides) and the second step is to "undo" the multiplication by 3 (by dividing both sides by 3). In the context of more complex problems, the "undoing" process is often a matter of finding and applying the inverse of a function.

We begin with the more general idea of the inverse of a relation. Recall that a relation is just a set of ordered pairs; the inverse of a given relation is the set of these ordered pairs with the first and second coordinates of each exchanged.

DEFINITION: Inverse of a Relation

Let R be a relation. The **inverse of R**, denoted R^{-1}, is the relation defined by switching the first and second coordinates of each ordered pair that is an element of R.

$$R^{-1}=\{(b,a)|(a,b)\in R\}$$

Example 1: Finding the Inverse of a Relation

Determine the inverse of each of the following relations. Then graph each relation and its inverse, and determine the domain and range of both.

a. $R=\{(4,-1),(-3,2),(0,5)\}$ **b.** $y=x^2$

Solution

a. $R=\{(4,-1),(-3,2),(0,5)\}$
$R^{-1}=\{(-1,4),(2,-3),(5,0)\}$

For each ordered pair, switch the first and second coordinates (x- and y-coordinates).

Recall that the domain is the set of first coordinates, and the range is the set of second coordinates.

R: Domain $=\{4,-3,0\}$ Range $=\{-1,2,5\}$
R^{-1}: Domain $=\{-1,2,5\}$ Range $=\{4,-3,0\}$

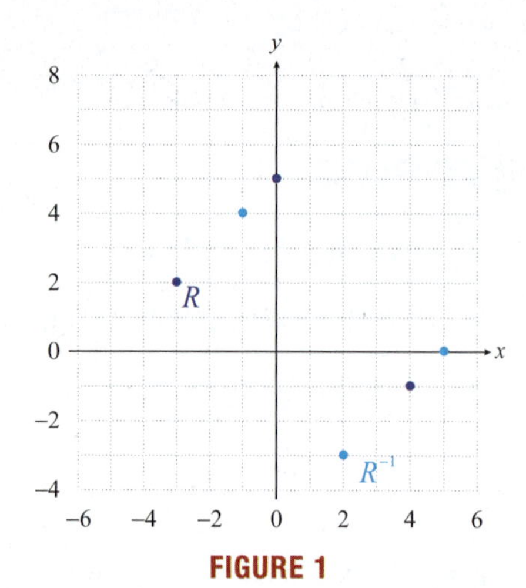

FIGURE 1

The relation R consists of three ordered pairs, and its inverse is simply these three ordered pairs with the coordinates exchanged. Note that the domain of R is the range of R^{-1} and vice versa.

b. $R = \{(x, y) \mid y = x^2\}$

$R^{-1} = \{(x, y) \mid x = y^2\}$

R: Domain $= \mathbb{R}$ Range $= [0, \infty)$

R^{-1}: Domain $= [0, \infty)$ Range $= \mathbb{R}$

In this problem, R is described by the given equation in x and y. The inverse relation is the set of ordered pairs in R with the coordinates exchanged, so we can describe the inverse relation by just exchanging x and y in the equation, as shown in Figure 2.

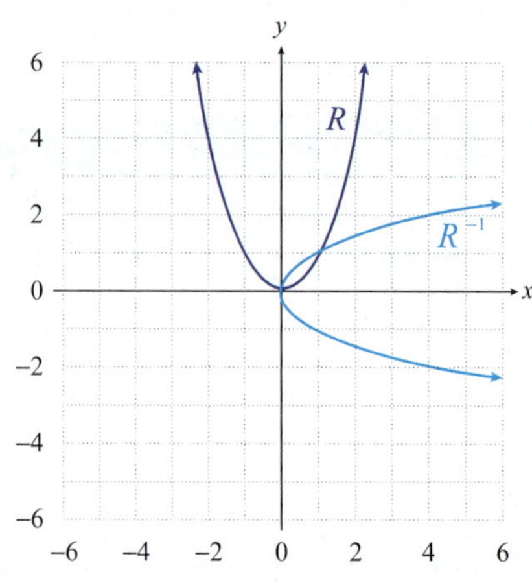

FIGURE 2

Note that the shapes of the graph of the relation and the graph of its inverse are essentially the same.

Consider the graphs of the two relations and their respective inverses in Example 1. By definition, an ordered pair (b, a) lies on the graph of a relation R^{-1} if and only if (a, b) lies on the graph of R, so it shouldn't be surprising that the graphs of a relation and its inverse bear some resemblance to one another. Specifically, they are mirror images of one another with respect to the line $y = x$. If you were to fold the Cartesian plane in half along the line $y = x$ in the two examples above, you would see that the points in R and R^{-1} coincide with one another.

The two relations in Example 1 illustrate another important point. Note that in both cases, R is a function, as its graph passes the vertical line test. By the same criterion, R^{-1} in Example 1a is also a function, but R^{-1} in Example 1b is not. The conclusion to be drawn is that even if a relation is a function, its inverse may or may not be a function.

TOPIC 2: Inverse Functions and the Horizontal Line Test

We have a convenient graphical test for determining when a relation is a function (the vertical line test); we would like to have a similar test for determining when the inverse of a relation is a function.

In practice, we will only be concerned with the question of when the inverse of a function *f*, denoted f^{-1}, is itself a function.

CAUTION

We are faced with another example of the reuse of notation. Note that f^{-1} does *not* stand for $\frac{1}{f}$ when *f* is a function! We use an exponent of -1 to indicate the reciprocal of a number or an algebraic expression, but when applied to a function or a relation it stands for the inverse relation.

Assume that *f* is a function. f^{-1} will only be a function itself if its graph passes the vertical line test; that is, only if each element of the domain of f^{-1} is paired with exactly one element of the range of f^{-1}. This is identical to saying that each element of the range of *f* is paired with exactly one element of the domain of *f*. In other words, every *horizontal* line in the plane must intersect the graph of *f* no more than once.

THEOREM: The Horizontal Line Test

Let *f* be a function. We say that the graph of *f* passes the **horizontal line test** if every horizontal line in the plane intersects the graph no more than once. If *f* passes the horizontal line test, then f^{-1} is also a function.

The horizontal line test is only useful if the graph of *f* is available to study. We can also phrase the above condition in a nongraphical manner. The inverse of *f* will only be a function if for every pair of distinct elements x_1 and x_2 in the domain of *f*, we have $f(x_1) \neq f(x_2)$. This criterion is important enough to merit a name.

DEFINITION: One-to-One Function

A function *f* is **one-to-one** if, for every pair of distinct elements x_1 and x_2 in the domain of *f*, we have $f(x_1) \neq f(x_2)$. This means that every element of the range of *f* is paired with exactly one element of the domain of *f*.

To sum up, the inverse f^{-1} of a function *f* is also a function if and only if *f* is one-to-one. If the graph of *f* is available for examination, *f* is one-to-one if and only if its graph passes the horizontal line test.

If we now examine Example 1 again, we see that the function *R* in Example 1a is one-to-one, and so we know that its inverse is also a function. On the other hand, the function *R* in Example 1b is not one-to-one (plenty of horizontal lines pass through the graph twice), so its inverse is not a function.

Example 2: Inverse Functions

Determine if the following functions have inverse functions.

a. $f(x)=|x|$ **b.** $g(x)=(x+2)^3$

> **NOTE**
>
> Even when a function f does not have an inverse *function*, it always has an inverse *relation*.

Solution

a. The function f does not have an inverse function, a fact demonstrated by showing that its graph does not pass the horizontal line test. We can also prove this algebraically: although $-3 \neq 3$, we have $f(-3)=f(3)$. Note that it only takes two ordered pairs to show that f does not have an inverse function.

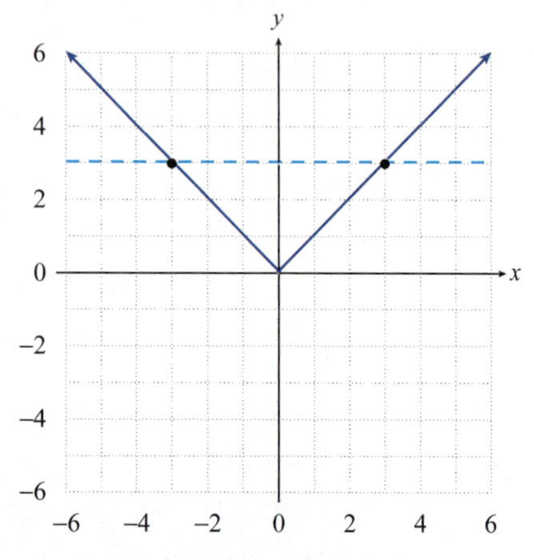

FIGURE 3: $f(x)=|x|$

b. The graph of g is the standard cubic shape shifted horizontally two units to the left. We can see this graph passes the horizontal line test, so g has an inverse function.

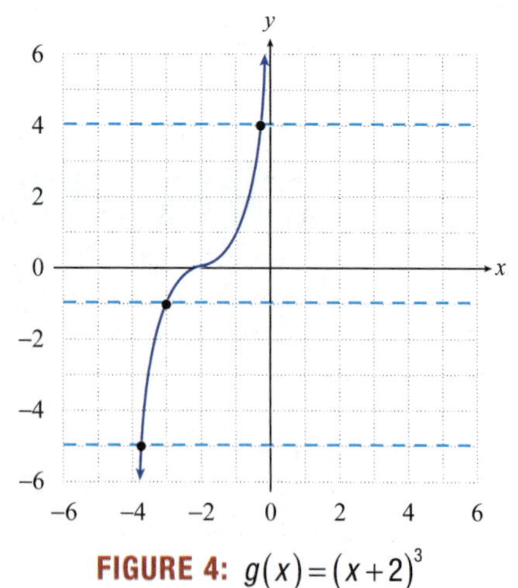

FIGURE 4: $g(x)=(x+2)^3$

Algebraically, any two distinct elements of the domain of g lead to different values when substituted into g, so g is one-to-one and hence has an inverse function.

Consider the function in Example 2a again. As we noted, the function $f(x)=|x|$ is not one-to-one, and so cannot have an inverse function. However, if we *restrict the domain* of *f* by specifying that the domain is the interval $[0,\infty)$, the new function, with this restricted domain, is one-to-one and has an inverse function. Of course, this **restriction of domain** changes the function; in this case the graph of the new function is the right-hand half of the graph of the absolute value function.

TOPIC 3: Finding the Inverse of a Function

In applying the notion of the inverse of a function, we will often begin with a formula for *f* and want to find a formula for f^{-1}. This will allow us, for instance, to transform equations of the form

$$f(x)=y \text{ into the form } x=f^{-1}(y).$$

Before we discuss the general algorithm for finding a formula for f^{-1}, consider the problem with which we began this section. If we define $f(x)=3x+2$, the equation $3x+2=8$ can be written as $f(x)=8$. Note that *f* is one-to-one, so f^{-1} is a function. If we can find a formula for f^{-1}, we can transform the equation into $x=f^{-1}(8)$. This is a complicated way to solve this equation, but it illustrates the use of inverse functions.

What should the formula for f^{-1} be? Consider what *f* does to its argument. The first action is to multiply *x* by 3, and the second is to add 2. To "undo" *f*, we need to negate these two actions in reverse order: subtract 2 and then divide the result by 3. So,

$$f^{-1}(x)=\frac{x-2}{3}.$$

Applying this to the problem at hand, we obtain

$$x=f^{-1}(8)=\frac{8-2}{3}=2.$$

This method of analyzing a function *f* and then finding a formula for f^{-1} by undoing the actions of *f* in reverse order is conceptually important and works for simple functions. For other functions, however, the following algorithm may be necessary as a standardized way to find the inverse formula.

PROCEDURE: Finding Formulas of Inverse Functions

Let *f* be a one-to-one function, and assume that *f* is defined by a formula. To find a formula for f^{-1}, perform the following steps.

Step 1: Replace $f(x)$ in the definition of *f* with the variable *y*. The result is an equation in *x* and *y* that is solved for *y* at this point.

Step 2: Solve the equation for *x*.

Step 3: Replace the *x* in the resulting equation with $f^{-1}(x)$ and replace each occurrence of *y* with *x*.

Example 3: Finding Formulas of Inverse Functions

Find the inverse of each of the following functions.

a. $f(x)=(x-1)^3+2$ **b.** $g(x)=\dfrac{x-3}{2x+1}$

Solution

a.

$f(x)=(x-1)^3+2$	
$y=(x-1)^3+2$	Replace $f(x)$ with y.
$y-2=(x-1)^3$	Subtract 2 from both sides.
$\sqrt[3]{y-2}=x-1$	Take the cube root of both sides.
$\sqrt[3]{y-2}+1=x$	Add 1 to both sides.
$f^{-1}(x)=\sqrt[3]{x-2}+1$	Replace x with $f^{-1}(x)$ and y with x.

b.

$g(x)=\dfrac{x-3}{2x+1}$	
$y=\dfrac{x-3}{2x+1}$	Replace $g(x)$ with y.
$y(2x+1)=x-3$	Multiply both sides by $2x+1$.
$2xy+y=x-3$	Distribute the product on the left.
$2xy-x=-y-3$	Collect all terms with x on the left.
$x(2y-1)=-y-3$	Factor out x.
$x=\dfrac{-y-3}{2y-1}$	Proceed to isolate x on one side.
$g^{-1}(x)=\dfrac{-x-3}{2x-1}$	Replace x with $g^{-1}(x)$ and y with x.

Example 4: Finding the Inverse of a Restricted-Domain Function

Find two suitable restrictions of the domain so that the function $f(x)=(x+1)^2-3$ has an inverse function, then find a formula for the inverse for each restricted function.

Solution

The graph of f is an upward-opening parabola with vertex at $(-1,-3)$, as shown in Figure 5. We can restrict its domain so that the resulting function is one-to-one by requiring either that $x\in(-\infty,-1]$ or that $x\in[-1,\infty)$.

Choice 1: If we choose $[-1,\infty)$ as the domain of the restricted function, then the expression $x+1$ is guaranteed to be nonnegative. We will use this fact as we work through the procedure to find a formula for the inverse function.

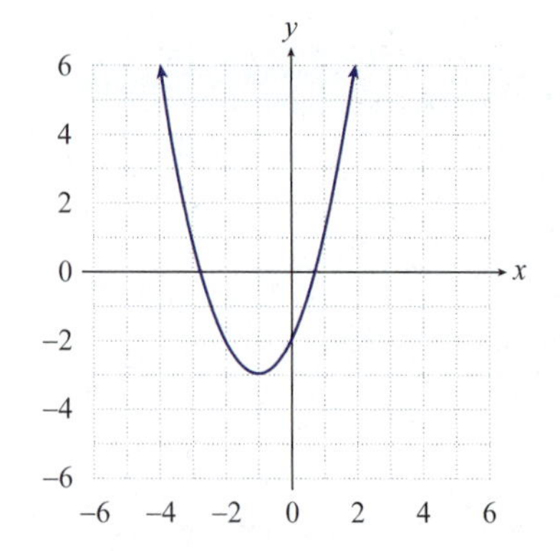

FIGURE 5: $f(x)=(x+1)^2-3$

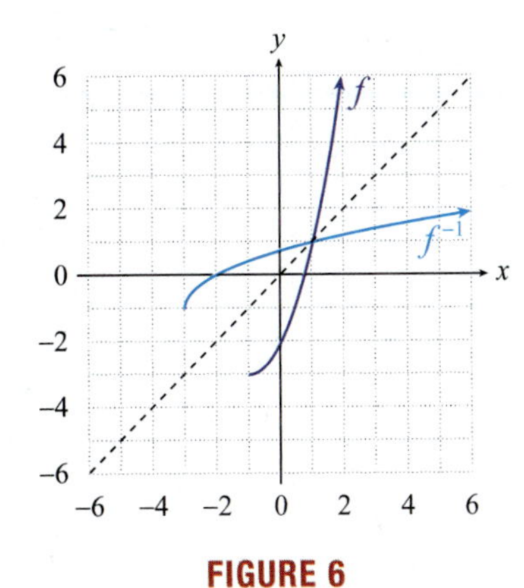

FIGURE 6

$$
\begin{aligned}
f(x) &= (x+1)^2 - 3 & \\
y &= (x+1)^2 - 3 & \text{Replace } f(x) \text{ with } y. \\
y+3 &= (x+1)^2 & \text{Add 3 to both sides.} \\
\sqrt{y+3} &= x+1 & \text{Since } x+1 \text{ is nonnegative, use the nonnegative root.} \\
\sqrt{y+3}-1 &= x & \text{Subtract 1 from both sides.} \\
f^{-1}(x) &= \sqrt{x+3}-1 & \text{Replace } x \text{ with } f^{-1}(x) \text{ and } y \text{ with } x.
\end{aligned}
$$

Figure 6 shows the graph of the function f restricted to the domain $[-1,\infty)$ along with a graph of the function's inverse.

Choice 2: If instead we choose $(-\infty,-1]$ as the domain of the restricted function, then the expression $x+1$ is guaranteed to be nonpositive. This fact will lead us to choose the other possible root when we take the square root in the following procedure.

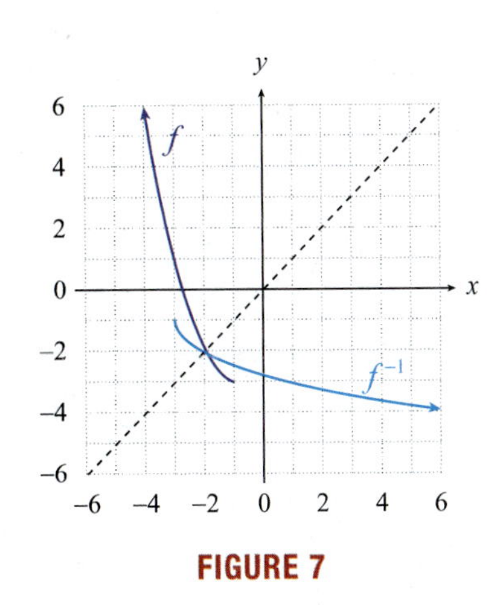

FIGURE 7

$$
\begin{aligned}
f(x) &= (x+1)^2 - 3 & \\
y &= (x+1)^2 - 3 & \text{Replace } f(x) \text{ with } y. \\
y+3 &= (x+1)^2 & \text{Add 3 to both sides.} \\
-\sqrt{y+3} &= x+1 & \text{Since } x+1 \text{ is nonpositive, use the nonpositive root.} \\
-\sqrt{y+3}-1 &= x & \text{Subtract 1 from both sides.} \\
f^{-1}(x) &= -\sqrt{x+3}-1 & \text{Replace } x \text{ with } f^{-1}(x) \text{ and } y \text{ with } x.
\end{aligned}
$$

Figure 7 shows the graph of f restricted to this domain, along with the inverse corresponding function.

As with relations and their inverses, the domain of the inverse function is the range of the original function, and the range of the inverse function is the domain of the original function. This fact is illustrated with Figures 6 and 7 in Example 4.

We can use the functions and their inverses from Example 3 to illustrate one last important point. The key characteristic of the inverse of a function is that it undoes the function. This means that if a function and its inverse are composed together, in either order, the resulting function has no effect on any allowable input!

THEOREM: Composition of Functions and Inverses

Given a function f and its inverse f^{-1}, the following statements are true:

$$f\left(f^{-1}(x)\right) = x \text{ for all } x \in \text{Dom}\left(f^{-1}\right), \text{ and}$$
$$f^{-1}\left(f(x)\right) = x \text{ for all } x \in \text{Dom}(f).$$

Example 5: Composition of Functions and Inverses

Use the functions f and g of Example 3 to demonstrate that the composition of a function and its inverse leaves any input unchanged.

a. $f(x)=(x-1)^3+2$ and $f^{-1}(x)=(x-2)^{\frac{1}{3}}+1$

b. $g(x)=\dfrac{x-3}{2x+1}$ and $g^{-1}(x)=\dfrac{-x-3}{2x-1}$

Solution

a.
$$\begin{aligned} f\left(f^{-1}(x)\right) &= f\left((x-2)^{\frac{1}{3}}+1\right) \\ &= \left((x-2)^{\frac{1}{3}}+1-1\right)^3+2 \\ &= \left((x-2)^{\frac{1}{3}}\right)^3+2 \\ &= x-2+2 \\ &= x \end{aligned}$$

A similar calculation shows that $f^{-1}\left(f(x)\right)=x,$ as you should verify.

b.
$$\begin{aligned} g^{-1}\left(g(x)\right) &= g^{-1}\left(\frac{x-3}{2x+1}\right) \\ &= \frac{-\dfrac{x-3}{2x+1}-3}{2\left(\dfrac{x-3}{2x+1}\right)-1} \\ &= \left(\frac{-\dfrac{x-3}{2x+1}-3}{2\left(\dfrac{x-3}{2x+1}\right)-1}\right)\left(\frac{2x+1}{2x+1}\right) \\ &= \frac{-x+3-6x-3}{2x-6-2x-1} \\ &= \frac{-7x}{-7} \\ &= x \end{aligned}$$

Similarly, $g\left(g^{-1}(x)\right)=x,$ as you should verify.

4.4 EXERCISES

PRACTICE

Graph the inverse of each of the following relations, and state its domain and range. See Example 1.

1. $R = \{(-4,2),(3,2),(0,-1),(3,-2)\}$

2. $S = \{(-3,-3),(-1,-1),(0,1),(4,4)\}$

3. $y = x^3$

4. $y = |x| + 2$

5. $x = |y|$

6. $x = -\sqrt{y}$

7. $y = \frac{1}{2}x - 3$

8. $y = -x + 1$

9. $y = \sqrt{x} + 2$

10. $T = \{(4,2),(3,-1),(-2,-1),(2,4)\}$

11. $x = y^2 - 2$

12. $y = 2\sqrt{x}$

Determine if each of the following functions is a one-to-one function. If so, graph the inverse of the function and state its domain and range.

13. $y = 2x + 3$

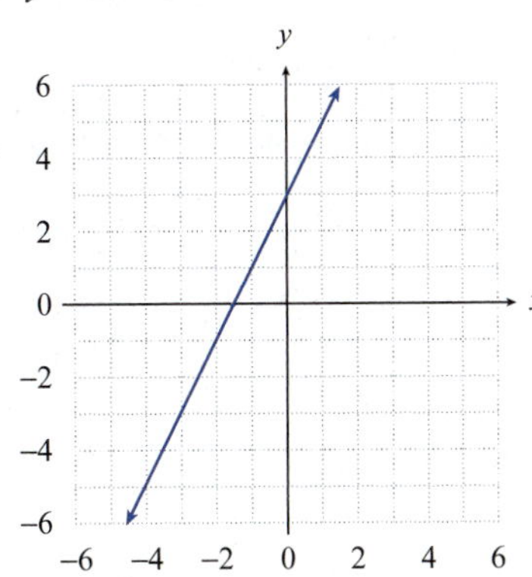

14. 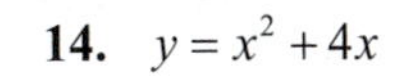 $y = x^2 + 4x$

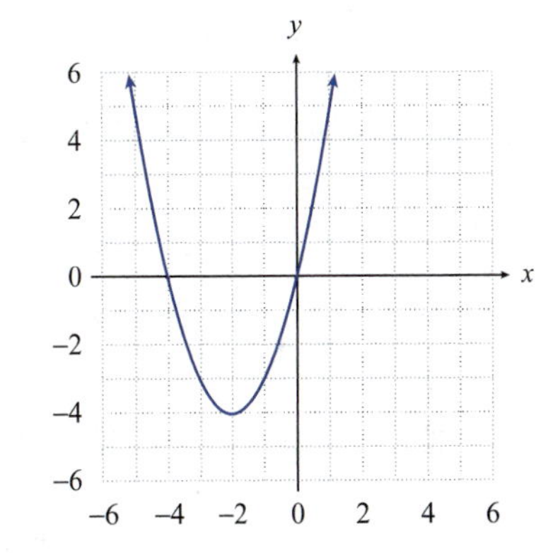

15. $y = \frac{1}{x^2}$

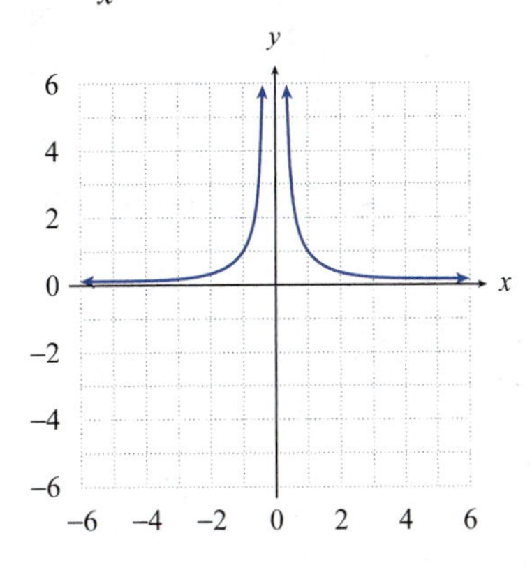

16. 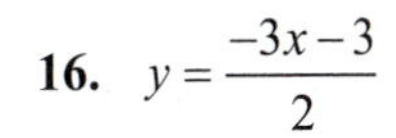 $y = \frac{-3x-3}{2}$

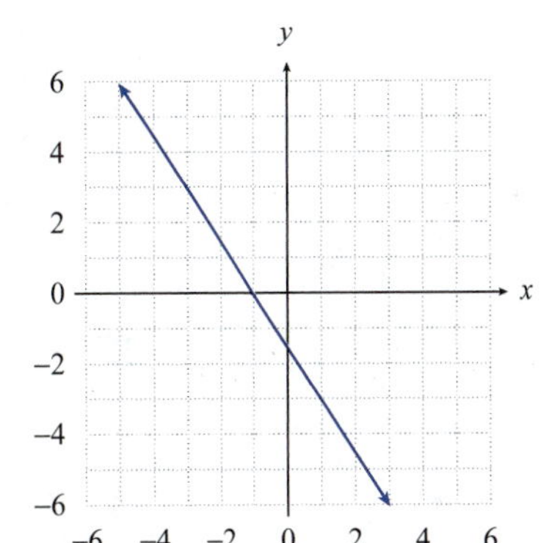

Determine if the following functions have inverse functions. If not, suggest a domain to restrict the function to so that it would have an inverse function (answers will vary). See Example 2.

17. $f(x)=x^2+1$

18. $g(x)=(x-2)^3-1$

19. $h(x)=\sqrt{x+3}$

20. $s(x)=\dfrac{1}{x^2}$

21. $G(x)=3x-5$

22. $F(x)=-x^2+5$

23. $r(x)=-\sqrt{x^3}$

24. $b(x)=\dfrac{1}{x}$

25. $f(x)=x^2-4x$

26. $m(x)=\dfrac{13x-2}{4}$

27. $H(x)=|x-12|$

28. $p(x)=10-x^2$

Find a formula for the inverse of each of the following functions. If necessary, first restrict the domain of the function. See Examples 3 and 4.

29. $f(x)=x^{\frac{1}{3}}-2$

30. $g(x)=4x-3$

31. $r(x)=\dfrac{x-1}{3x+2}$

32. $s(x)=\dfrac{1-x}{1+x}$

33. $F(x)=(x-5)^3+2$

34. $G(x)=\sqrt[3]{3x-1}$

35. $V(x)=\dfrac{x+5}{2}$

36. $W(x)=\dfrac{1}{x}$

37. $h(x)=x^{\frac{3}{5}}-2$

38. $A(x)=(x^3+1)^{\frac{1}{5}}$

39. $J(x)=\dfrac{2}{1-3x}$

40. $k(x)=\dfrac{x+4}{3-x}$

41. $h(x)=x^7+6$

42. $F(x)=\dfrac{3-x^5}{-9}$

43. $r(x)=\sqrt[5]{2x}$

44. $P(x)=(2+3x)^3$

45. $f(x)=3(2x)^{\frac{1}{3}}$

46. $q(x)=(x-2)^2+2$

47. $f(x)=(x-3)^2+2$

48. $f(x)=|x+2|+3$

49. $f(x)=(x+1)^4-2$

In each of the following exercises, verify that $f(f^{-1}(x))=x$ and that $f^{-1}(f(x))=x$. See Example 5.

50. $f(x)=\dfrac{3x-1}{5}$ and $f^{-1}(x)=\dfrac{5x+1}{3}$

51. $f(x)=\sqrt[3]{x+2}-1$ and $f^{-1}(x)=(x+1)^3-2$

52. $f(x)=\dfrac{2x+7}{x-1}$ and $f^{-1}(x)=\dfrac{x+7}{x-2}$

53. $f(x)=x^2,\ x\geq 0$ and $f^{-1}(x)=\sqrt{x}$

54. $f(x)=2x-3$ and $f^{-1}(x)=\dfrac{x+3}{2}$

55. $f(x)=\sqrt[3]{x+1}$ and $f^{-1}(x)=x^3-1$

56. $f(x)=\dfrac{1}{x}$ and $f^{-1}(x)=\dfrac{1}{x}$

57. $f(x)=\dfrac{x-5}{2x+3}$ and $f^{-1}(x)=\dfrac{3x+5}{1-2x}$

58. $f(x)=(x-2)^2,\ x\geq 2$ and $f^{-1}(x)=\sqrt{x}+2,\ x\geq 0$

59. $f(x)=\dfrac{1}{1+x}$ and $f^{-1}(x)=\dfrac{1-x}{x}$

Match the following functions with the graphs of the inverses of the functions.

60. $f(x)=x^3$

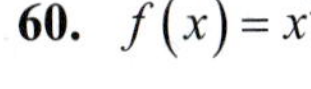

61. $f(x)=x-5$

62. $f(x)=\sqrt{x-4}$

63. $f(x)=x^2,\ x\geq 0$

64. $f(x)=\dfrac{x}{4}$

65. $f(x)=\sqrt[3]{x+1}$

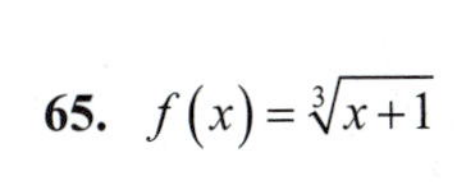

a.

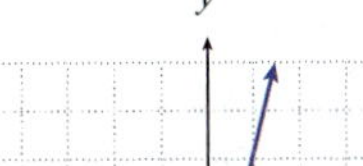

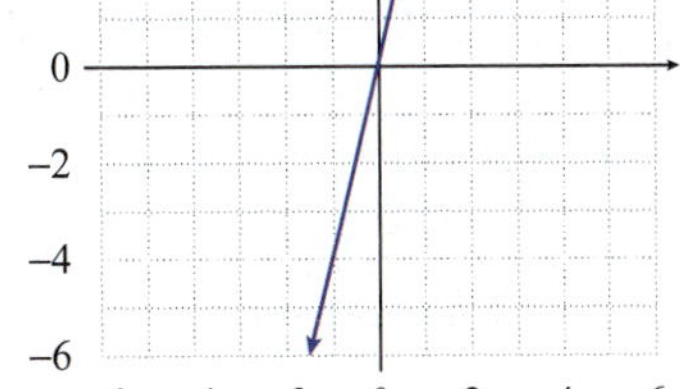

b.

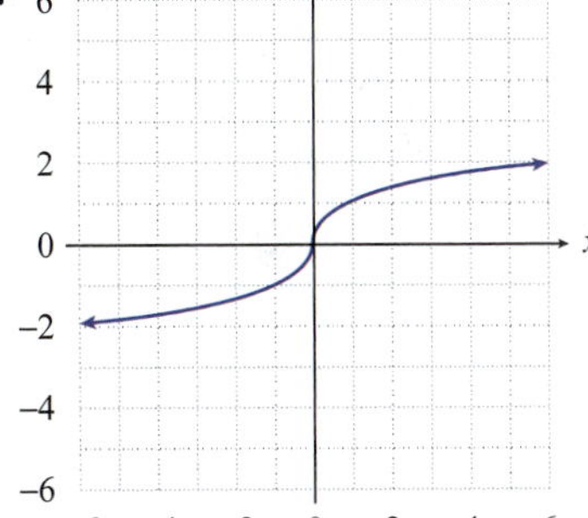
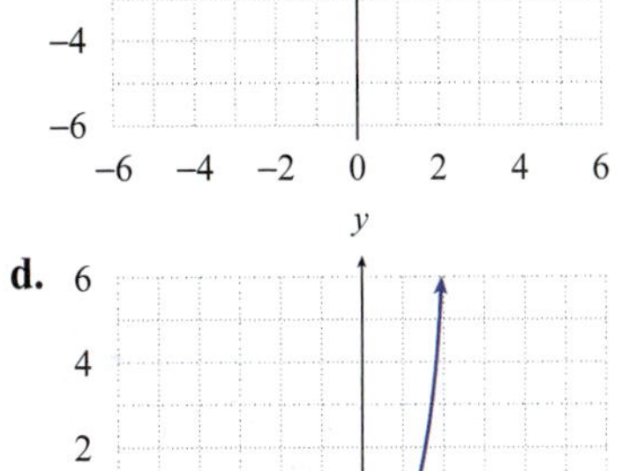

c.

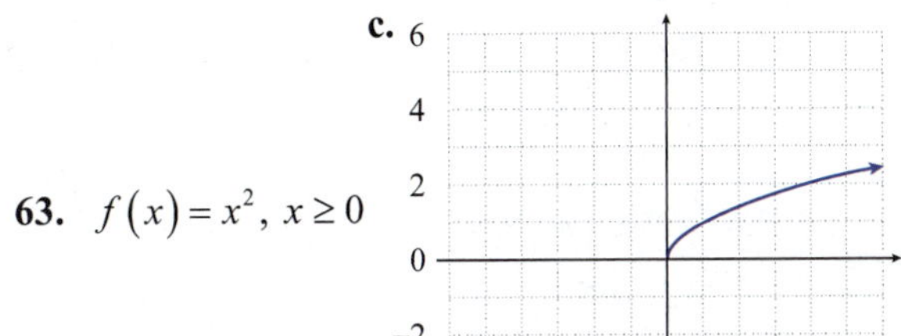
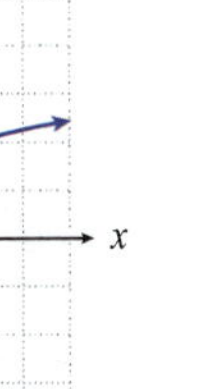
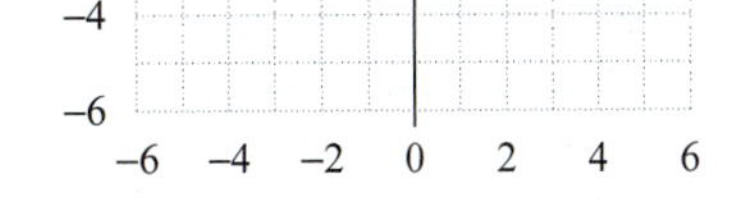

d.

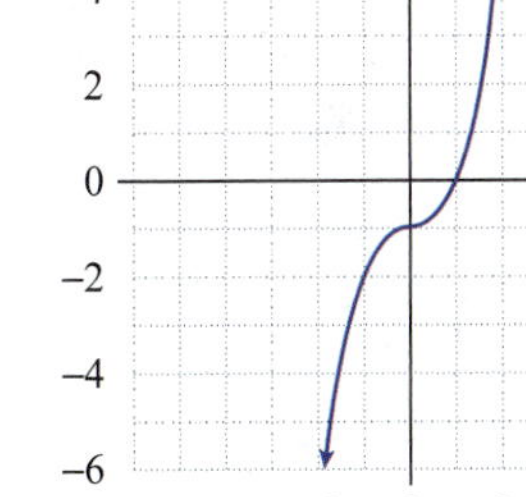
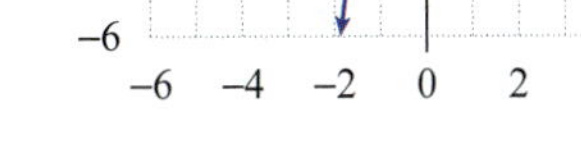

e.

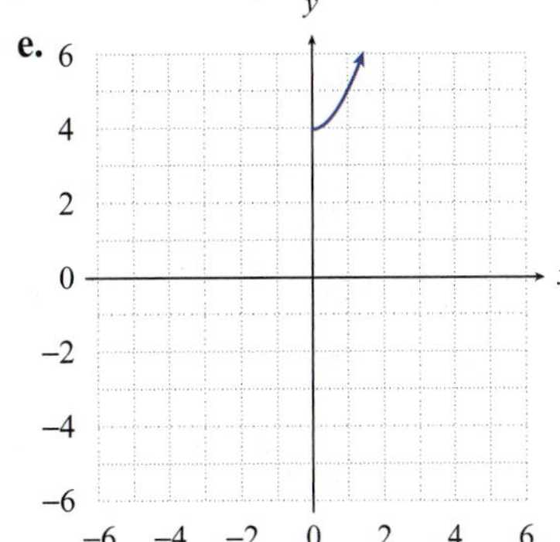

f.

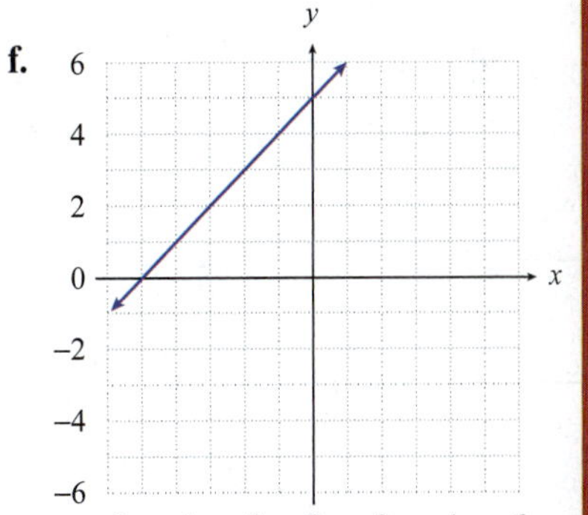

APPLICATIONS

An inverse function can be used to encode and decode words and sentences by assigning each letter of the alphabet a numerical value (A = 1, B = 2, C = 3, ..., Z = 26).
Example: Use the function $f(x)=x^2$ to encode the word PRECALCULUS. The encoded message would be 256 324 25 9 1 144 9 441 144 441 361. The word can then be decoded by using the inverse function $f^{-1}(x)=\sqrt{x}$. The inverse values are 16 18 5 3 1 12 3 21 12 21 19 which translates back to the word PRECALCULUS. Encode or decode the following words using the numerical values A = 1, B = 2, C = 3, ..., Z = 26.

66. Encode the message SANDY SHOES using the function $f(x)=4x-3$.

67. Encode the message WILL IT RAIN TODAY using the function $f(x)=8x$.

68. The following message was encoded using the function $f(x)=8x-7$. Decode the message.
41 137 65 145 9 33 33 169 113 89 89 33 193 9 1 89 89 1 105 25 57 113 137 145 33 145 57 113 33 145

69. The following message was encoded using the function $f(x)=5x+1$. Decode the message.
91 26 66 26 66 11 26 91 126 76 106 91 96 106 71 11 61 76 16 56

70. The following message was encoded using the function $f(x)=x^3$. Decode the message.
27 1 8000 27 512 1 12167 1 10648 125

71. The following message was encoded using the function $f(x)=-3-5x$. Decode the message.
−13 −28 −8 −18 −43 −33 −108 −73 −48 −73 −103 −43 −28 −98 −108 −73

TECHNOLOGY

A graphing utility can be used to verify the inverse of a function. Use a graphing utility to graph each of the following functions and its inverse in the same viewing window. Determine the domain and range of the inverse.

72. $f(x)=\sqrt{x+5}$ and $f^{-1}(x)=x^2-5$

73. $f(x)=x^3-1$ and $f^{-1}(x)=\sqrt[3]{x+1}$

74. $f(x)=\dfrac{2x+1}{x-1}$ and $f^{-1}(x)=\dfrac{x+1}{x-2}$

75. $f(x)=\dfrac{4}{\sqrt{x}}$ and $f^{-1}(x)=\dfrac{16}{x^2}$

76. $f(x)=-\sqrt{x^2-16},\ x\geq 4$ and $f^{-1}(x)=\sqrt{x^2+16}$

77. $f(x)=x^2+3,\ x\geq 0$ and $f^{-1}(x)=\sqrt{x-3}$

CHAPTER 4 PROJECT

The Ozone Layer

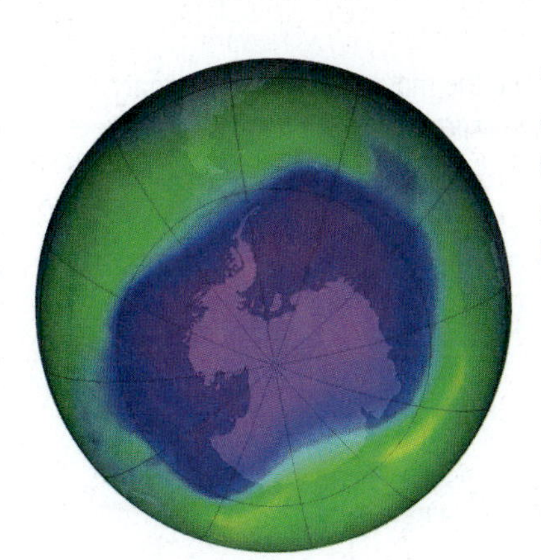

Photo Courtesy of NASA

The level of ozone in Earth's atmosphere has been closely watched since the 1970s, with the level of ozone depletion at the south pole a good indicator of global trends. One model of the region of polar depletion assumes the region is circular and that its radius, over a certain period of time, grows at a constant rate of 2.6 kilometers per hour.

1. Write the area of the circle as a function of the radius r.
2. Assuming that t is measured in hours, that $t = 0$ corresponds to the start of the annual growth of the hole, and that the radius of the hole is initially 0 kilometers, write the radius as a function of time t.
3. Write the area of the circle as a function of time t.
4. What is the radius after 3 hours?
5. What is the radius after 5.5 hours?
6. What is the area of the circle after 3 hours?
7. What is the area of the circle after 5.5 hours?
8. What is the average rate of change of the area from 3 hours to 5.5 hours?
9. What is the average rate of change of the area from 5.5 hours to 8 hours?
10. Is the average rate of change of the area increasing or decreasing as time passes?

CHAPTER 4 REVIEW EXERCISES

Section 4.1

Sketch the graphs of the following functions by first identifying the more basic functions that have been shifted, reflected, stretched, or compressed. Then determine the domain and range of each function.

1. $f(x)=(x-1)^3+2$

2. $G(x)=4|x+3|$

3. $m(x)=\dfrac{1}{(x+2)^2}$

4. $g(x)=-\sqrt[3]{x}+4$

5. $r(x)=\dfrac{1}{x-2}-3$

6. $f(x)=\sqrt{x-1}+3$

7. $g(x)=\sqrt{\dfrac{x}{2}}+1$

8. $f(x)=-\sqrt{-4x}$

Write a formula for each of the functions described below.

9. Use the function $g(x)=x^2$. Move the function 1 unit to the right and 2 units down.

10. Use the function $g(x)=|x|$. Move the functions 3 units to the right and reflect across the x-axis.

11. Use the function $g(x)=\sqrt{x}$. Reflect the function across the x-axis and move it 4 units up.

Section 4.2

Determine if each of the following relations is a function. If so, determine whether it is even, odd, or neither. Also determine if it has *y*-axis symmetry, *x*-axis symmetry, origin symmetry, or none of the above.

12. $y=|2x-1|$

13. $y=\dfrac{1}{x^2}+1$

14. $x=-5|y|$

For each of the following functions, find the open intervals of monotonicity where the function is increasing, decreasing, or constant.

15. $f(x)=(x-2)^4-6$

16. $R(x)=\begin{cases}(x+2)^2 & \text{if } x<-1\\ -x & \text{if } x\geq -1\end{cases}$

17. Given the following graph of a function determine, **a.** the locations and types of the local extrema, and **b.** the values of the local extrema.

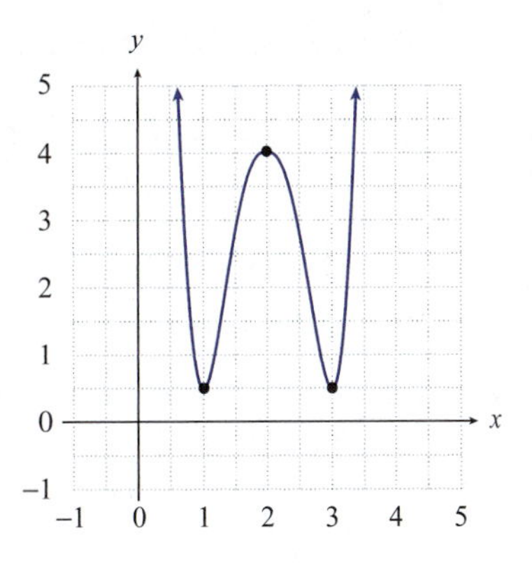

For each given function and interval, determine the average rate of change of the function over the interval.

18. $f(x)=x^2;\ [3,4]$

19. $f(x)=\dfrac{1}{x};\ [1,3]$

20. $f(x)=\sqrt{x};\ [1,4]$

21. $f(x)=x^2-x^3;\ [-1,2]$

Section 4.3

In each of the following exercises, use the information given to determine **a.** $(f+g)(2)$, **b.** $(f-g)(2)$, **c.** $(fg)(2)$, and **d.** $\left(\dfrac{f}{g}\right)(2)$.

22. $f(x)=-x^2+x$ and $g(x)=\dfrac{1}{x}$

23. $f(2)=4$ and $g(2)=-1$

24. $f(x)=\sqrt{2x}$ and $g(x)=x+3$

25. $f=\{(0,4),(2,8)\}$ and $g=\{(-2,2),(0,3),(2,-10)\}$

In each of the following exercises, find **a.** the formula and domain for $f+g$ and **b.** the formula and domain for $\frac{f}{g}$.

26. $f(x)=x^2$ and $g(x)=\sqrt{x}$

27. $f(x)=\frac{1}{x-2}$ and $g(x)=\sqrt[3]{x}$

28. $f(x)=3x$ and $g(x)=(x-1)^2$

29. $f(x)=x^2-4$ and $g(x)=\sqrt[3]{x}-1$

In each of the following exercises, use the information given to determine $(f\circ g)(3)$.

30. $f(x)=-x+1$ and $g(x)=-x-1$

31. $f(x)=\frac{x^{-1}}{18}-3$ and $g(x)=\frac{x-4}{x^3}$

32. $f(-3)=4$ and $g(3)=-3$

33. $f(x)=\frac{x}{3}$ and $g(x)=-\sqrt{x+1}$

In each of the following exercises, find **a.** the formula and domain for $f\circ g$ and **b.** the formula and domain for $g\circ f$.

34. $f(x)=4x-1$ and $g(x)=x^3+2$

35. $f(x)=\frac{1}{\sqrt{x-4}}$ and $g(x)=x+2$

36. $f(x)=2x^2+1$ and $g(x)=x-4$

37. $f(x)=3x$ and $g(x)=\sqrt{x-3}$

Write each of the following functions as a composition of two functions. Answers may vary.

38. $f(x)=\frac{3}{3x^2+1}$

39. $f(x)=\frac{\sqrt{x+2}}{x^2+4x+4}$

In each of the following exercises, use the information given to find $g(x)$.

40. $f(x)=6x-1$ and $(f\circ g)(x)=x+3$

41. $f(x)=\sqrt{x}+3$ and $(g\circ f)(x)=\frac{2}{\sqrt{x}+3}+1$

Section 4.4

Graph the inverse of each of the following relations, and state its domain and range.

42. $R=\{(3,4),(-1,5),(0,2),(-6,-1)\}$

43. $y=3x+1$

44. $y=\frac{\sqrt{x}}{2}$

Find a formula for the inverse of each of the following functions.

45. $r(x)=\dfrac{2}{7x-1}$

46. $g(x)=\dfrac{4x-3}{x}$

47. $f(x)=x^{\frac{1}{5}}-6$

48. $p(x)=2\sqrt{x-1}+3$

49. $f(x)=\dfrac{2x-3}{x+1}$

50. $f(x)=\sqrt[3]{x+2}-1$

51. $f(x)=8x+3$

52. $f(x)=(x-1)^2-3,\ x\geq 1$

Verify that $f\left(f^{-1}(x)\right)=x$ and that $f^{-1}\left(f(x)\right)=x$.

53. $f(x)=\dfrac{6x-7}{2-x}$ and $f^{-1}(x)=\dfrac{2x+7}{6+x}$

CHAPTER 5

Polynomial and Rational Functions

SECTIONS

WHAT IF ...

What if you owned a skateboard manufacturing company? How would you decide how many skateboards to produce in order to maximize your profits?

By the end of this chapter, you'll be able to apply the skills regarding polynomials to solve business application problems. You'll encounter the skateboard problem on page 371. You'll master this type of problem using techniques for solving polynomial inequalities, found on page 365.

Carl Friedrich Gauss

Introduction

In Chapters 3 and 4, we studied properties of functions in general and learned the nomenclature and notation commonly used when working with functions and their graphs. In this chapter, we narrow our focus and concentrate on polynomial functions.

We have already seen many examples of polynomial functions but have barely scratched the surface as far as obtaining a deep understanding of polynomials is concerned. We will soon see many mathematical methods that are peculiar to polynomials and make many observations that are relevant only to polynomials. Some of these methods include polynomial division and the Rational Zero Theorem, and many of the observations point out the strong connection between factors of polynomials, solutions of polynomial equations, and (when appropriate) graphs of polynomial functions. Our deeper understanding of polynomial functions will then make solving polynomial inequalities a far easier task than it would be otherwise.

The discussion of polynomials concludes with the Fundamental Theorem of Algebra, a theorem that makes a deceptively simple claim about polynomials, but one that nevertheless manages to tie together nearly all of the methods and observations of the chapter. The German mathematician Carl Friedrich Gauss (1777–1855), one of the towering figures in the history of mathematics, first proved this theorem in 1799. In doing so, he accomplished something that many brilliant people (among them Isaac Newton and Leonhard Euler) had attempted, and it is all the more remarkable that Gauss did so in his doctoral dissertation at the age of twenty-two! The Fundamental Theorem of Algebra operates on many levels: philosophically, it can be seen as one of the most elegant and fundamental arguments for the necessity of complex numbers, while pragmatically it is of great importance in solving polynomial equations. It thus ties together observations about polynomials that originated with Italian mathematicians of the 16th century, and points the way toward later work on polynomials by such people as Niels Abel and Évariste Galois.

The chapter ends with an opportunity to put to use many of the skills acquired thus far. An understanding of the subject of the last section, rational functions, depends not only on a knowledge of polynomials (of which rational functions are ratios), but also of x- and y-intercepts, factoring, and transformations of functions.

5.1 POLYNOMIAL FUNCTIONS AND POLYNOMIAL INEQUALITIES

TOPICS

1. Zeros of polynomial functions and solutions of polynomial equations
2. Graphing factored polynomial functions
3. Solving polynomial inequalities

TOPIC 1: Zeros of Polynomial Functions and Solutions of Polynomial Equations

The Long History of Polynomial Equations

Over the course of millennia, mathematicians gradually considered polynomial equations of higher degree and greater generality. Solution methods for certain cubic equations were known by Chinese and Babylonian scholars as long ago as 2000 BC. More than 3500 years later, in the 1600s AD, European mathematicians had begun to believe that a general n^{th}-degree polynomial equation had as many as, but no more than, n solutions, but the proof of this fact had to wait nearly another two centuries.

At this point, we have studied how linear and quadratic polynomial functions behave, and we have tools guaranteed to solve all linear and quadratic equations. We have also studied some elementary higher-degree polynomials (those of the form ax^n). In this section, we begin a more complete exploration of higher-degree polynomials.

Not surprisingly, the complexity of polynomial functions increases with the degree; higher-degree polynomials are usually more difficult to graph accurately, and we cannot necessarily expect to find exact solutions to higher-degree polynomial equations. Nevertheless, there is much that we can say about a given polynomial function.

In order to make our work as general as possible, we will refer to a generic n^{th}-degree polynomial function $p(x)=a_nx^n+a_{n-1}x^{n-1}+\cdots+a_1x+a_0$, where n is a nonnegative integer, $a_n, a_{n-1}, \ldots, a_1, a_0$ all represent constants (that may be real or complex), and $a_n \neq 0$. **Throughout this chapter, $p(x)$ will refer to this generic polynomial.** We begin by identifying which values of the variable make a polynomial function equal to zero.

DEFINITION: Zeros of a Polynomial

The number c (c may be a complex number) is a **zero** of the polynomial function $p(x)$ if $p(c)=0$. This is also expressed by saying that c is a **root** of the polynomial or a **solution** of the equation $p(x)=0$.

If $p(x)$ is a polynomial with real coefficients, and if c is a real number zero of $p(x)$, then the statement $p(x)=0$ means the graph of p crosses the x-axis at the point $(c,0)$. In this case, c may also be referred to as an x-intercept of $p(x)$.

The task of determining the zeros of a polynomial arises in many contexts, two of which are solving polynomial equations and graphing polynomials.

DEFINITION: Polynomial Equations

A **polynomial equation in one variable**, say the variable x, is an equation that can be written in the form $a_n x^n + a_{n-1}x^{n-1} + \cdots + a_1 x + a_0 = 0$, where n is a nonnegative integer and $a_n, a_{n-1}, \ldots, a_1, a_0$ are constants. Assuming $a_n \neq 0$, we say such an equation is of **degree** n and call a_n the **leading coefficient**.

Just as with linear and quadratic equations (which are polynomial equations of degree 1 and 2, respectively), a polynomial equation may not appear in the form of the definition. The first task is often to rewrite the equation so that one side is zero. Then, the zeros of the polynomial on the other side are the solutions of the equation.

Note that, given a polynomial equation in the form $p(x) = 0$, the zeros of p are precisely the solutions of the equation (and vice versa).

Example 1: Solutions of Polynomial Equations

Verify that the given values of x solve the corresponding polynomial equations.

a. $6x^2 - x^3 = 12 + 5x;\ x = 4$

b. $x^2 = 2x - 5;\ x = 1 + 2i$

c. $\dfrac{x}{1-i} = 3x^2;\ x = 0$

NOTE

When verifying a zero, there is no need to rewrite the equation in the form of the definition.

Solution

a.

$$6x^2 - x^3 = 12 + 5x$$

$$6(4)^2 - (4)^3 \stackrel{?}{=} 12 + 5(4)$$ Substitute $x = 4$ throughout the equation.

$$96 - 64 \stackrel{?}{=} 12 + 20$$ Simplify both sides.

$$32 = 32$$ This is a true statement, so 4 is a solution.

b.

$$x^2 = 2x - 5$$

$$(1+2i)^2 \stackrel{?}{=} 2(1+2i) - 5$$ Substitute $x = 1 + 2i$ in the equation.

$$1 + 4i + 4i^2 \stackrel{?}{=} 2 + 4i - 5$$ Expand both sides.

$$1 + 4i - 4 \stackrel{?}{=} 2 + 4i - 5$$ Simplify, using the fact that $i^2 = -1$.

$$-3 + 4i = -3 + 4i$$ This is a true statement, so $1 + 2i$ is a solution.

c.

$$\frac{x}{1-i} = 3x^2$$

$$\frac{0}{1-i} \stackrel{?}{=} 3(0)^2$$ Substitute 0 for x in the equation.

$$0 = 0$$ After simplifying, we arrive at a true statement. Thus, 0 is a solution to the polynomial equation.

TECHNOLOGY: Finding Zeros of Polynomials

In Chapter 2, we saw how to find the x-intercepts of a linear equation on a TI-84 Plus. The same method can be used to find the x-intercepts, or zeros, of any function graphed on a TI-84 Plus. The main difference is that with linear functions, there can be no more than one zero, but other functions might have more. Consider the graph of the function $f(x) = x^2 + 4x - 6$.

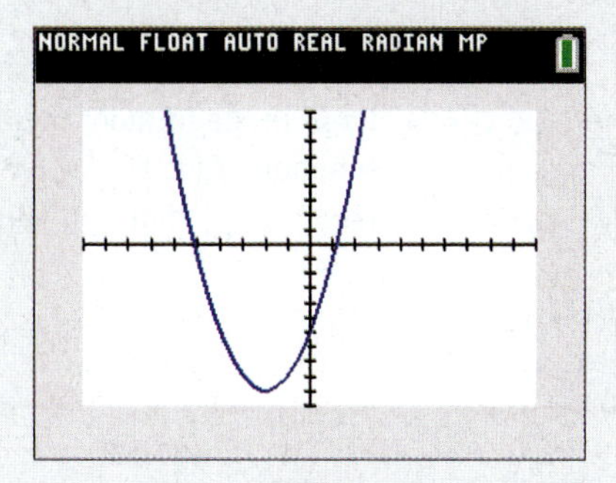

FIGURE 1

We can see that there are two zeros that appear to be located near $x = -5$ and $x = 1$. To check more accurately, press 2nd trace to access the CALC menu, then select zero. The screen should now display the graph with the words Left Bound? shown at the bottom. Choose which zero you want to find and use the arrows to move the cursor anywhere to the left of that intercept and press enter. The screen should now say Right Bound? Use the right arrow to move the cursor to the right of that same intercept and press enter again. (Be sure there is only one x-intercept between what you select as the left bound and what you select as the right bound.) The text should now read Guess? Press enter a third time and the x- and y-values of that x-intercept will appear at the bottom of the screen.

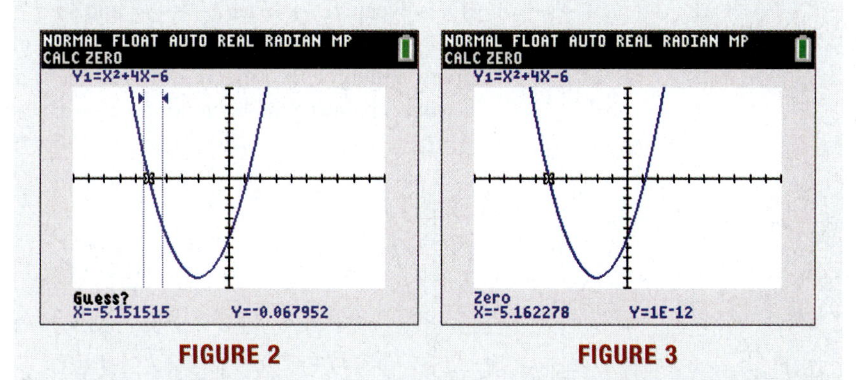

FIGURE 2 FIGURE 3

The leftmost zero occurs at approximately $x \approx -5.162$. Note that sometimes, as in Figure 3, the display will read a very small number rather than exactly zero. To find the other zero, repeat the process, this time focusing on the rightmost zero. We find that it occurs at $x \approx 1.162$.

TOPIC 2: Graphing Factored Polynomial Functions

Consider a generic polynomial function $p(x)$ with all real coefficients. Our goal is to be able to sketch the graph of such a function, paying particular attention to the behavior of p as $x \to -\infty$ and as $x \to \infty$, (where the symbol $\to$ is read "approaches") and to the x- and y-intercepts of p. We will begin by looking at the behavior of a polynomial function as $x \to \pm\infty$.

The graph of $p(x)$ is similar to the graph of $a_n x^n$ for values of x that are very large in magnitude; the leading term of $p(x)$ dominates the behavior. Take the function $f(x) = x^4 - 3x^3 - 5x^2 + 8$. The graph in Figure 4 shows that as x gets large, $f(x)$ takes on values very similar to x^4 and thus has a very similar graph.

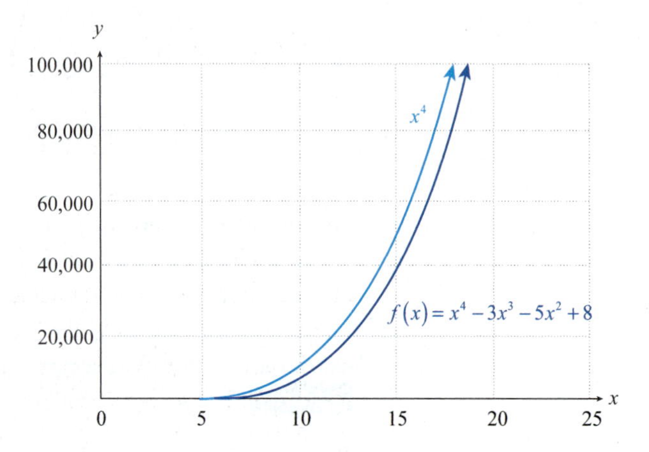

FIGURE 4: Similar Behavior as $x \to \infty$

This means we can understand the behavior of all polynomials as $x \to \pm\infty$ just by understanding how functions of the form $f(x) = a_n x^n$ behave. We know that if n is even, $x^n \to \infty$ as $x \to -\infty$ and as $x \to \infty$, and if n is odd, then $x^n \to -\infty$ as $x \to -\infty$ and $x^n \to \infty$ as $x \to \infty$. We also know that if a_n is negative, the graph undergoes a reflection across the x-axis; this reverses the sign of every y-value, changing the behavior as $x \to \pm\infty$. Figure 5 shows a few examples.

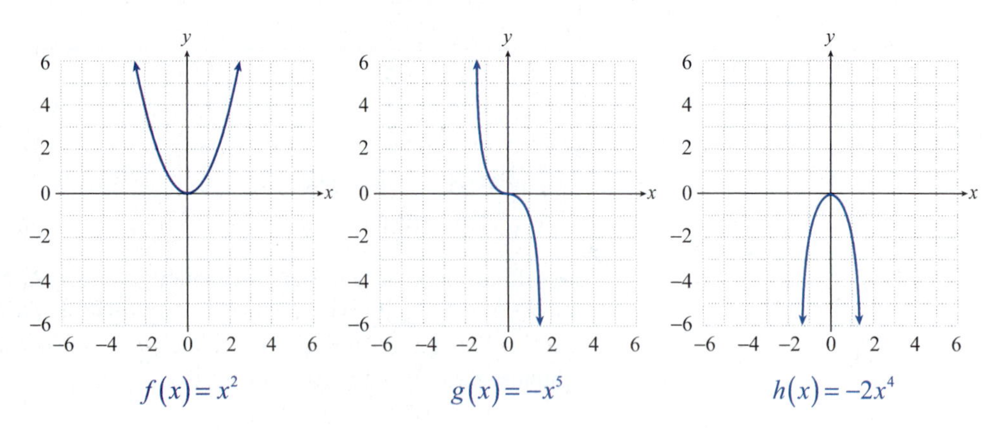

FIGURE 5: Examples of Behavior as $x \to \infty$

PROPERTIES: Behavior of Polynomials as $x \to \pm\infty$

Given a polynomial function $p(x)$ with degree n, the behavior of $p(x)$ as $x \to \pm\infty$ can be determined from the leading term $a_n x^n$.

	n is even	n is odd
a_n is positive	as $x \to -\infty$, $p(x) \to +\infty$ as $x \to +\infty$, $p(x) \to +\infty$ The graph rises to the left and rises to the right.	as $x \to -\infty$, $p(x) \to -\infty$ as $x \to +\infty$, $p(x) \to +\infty$ The graph falls to the left and rises to the right.
a_n is negative	as $x \to -\infty$, $p(x) \to -\infty$ as $x \to +\infty$, $p(x) \to -\infty$ The graph falls to the left and falls to the right.	as $x \to -\infty$, $p(x) \to +\infty$ as $x \to +\infty$, $p(x) \to -\infty$ The graph rises to the left and falls to the right.

Near the origin, however, the graph of $p(x)$ is likely to be quite different from the graph of $a_n x^n$. Recall our example from before: x^4 and $f(x) = x^4 - 3x^3 - 5x^2 + 8$ both behave the same as $x \to \pm\infty$. Observe how differently these functions behave near the origin.

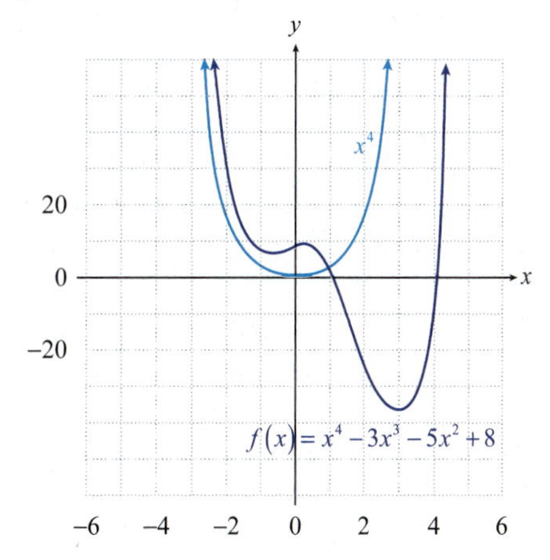

FIGURE 6: Different Behavior near the Origin

Finding the x- and y-intercepts of a polynomial function will help us sketch a more complete graph. Given a polynomial in the form $p(x) = a_n x^n + a_{n-1}x^{n-1} + \cdots + a_1 x + a_0$, finding the y-intercept is not difficult, as it simply requires evaluating $p(0)$.

$$\begin{aligned} p(0) &= a_n (0)^n + a_{n-1}(0)^{n-1} + \cdots + a_1(0) + a_0 \\ &= a_0 \end{aligned}$$

Thus, the y-intercept is $(0, a_0)$. On the other hand, finding the x-intercepts requires solving the polynomial equation $0 = a_n x^n + a_{n-1}x^{n-1} + \cdots + a_1 x + a_0$, which often takes more effort.

Writing the equation in a different form can make our task much easier. If we factor a given polynomial f into a product of linear factors, each linear factor with real coefficients corresponds to an x-intercept of the graph of f. To see why this is true, consider the polynomial function

$$f(x) = (3x-5)(x+2)(2x-6).$$

To determine the x-intercepts of this polynomial we need to solve the equation

$$0 = (3x-5)(x+2)(2x-6).$$

The Zero-Factor Property tells us that the only solutions are those values of x for which $3x-5=0$, $x+2=0$, or $2x-6=0$. Solving these three linear equations gives us the x-coordinates of the three x-intercepts of f: $\left\{\frac{5}{3}, -2, 3\right\}$.

Working with a polynomial in factored form almost always makes finding the x-intercepts easier, but we do have to adjust how we calculate the y-intercept and the behavior as $x \to \pm\infty$. If a polynomial is in factored form, we can not simply read off the value of the y-intercept. However, substituting $x = 0$ is not much more work.

$$\begin{aligned} f(0) &= (3(0)-5)((0)+2)(2(0)-6) \\ &= (-5)(2)(-6) \\ &= 60 \end{aligned}$$

Similarly, when a polynomial is in factored form, we can't directly see the term $a_n x^n$ to determine the end behavior. Instead of multiplying out f completely, we just determine how the leading x^3 term arises. The third-degree term comes from multiplying together the $3x$ from the first factor, the x from the second factor, and the $2x$ from the third factor. Thus, $a_n x^n = 6x^3$. Since the leading coefficient is positive and the degree is odd, we know that $f(x) \to -\infty$ as $x \to -\infty$ and $f(x) \to \infty$ as $x \to \infty$.

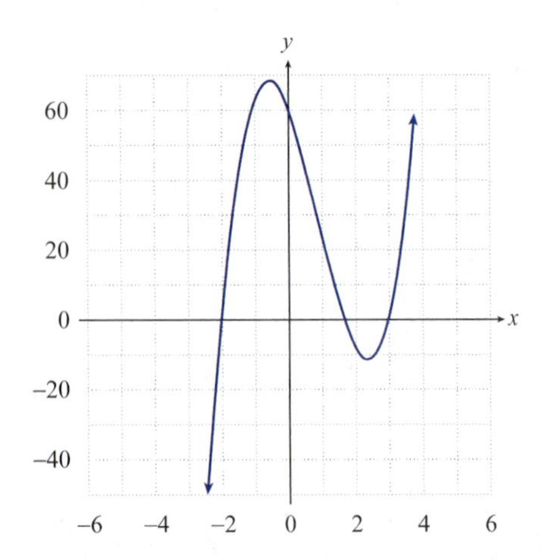

FIGURE 7: $f(x) = (3x-5)(x+2)(2x-6)$

Putting it all together, along with a few computed values of f, we obtain the sketch in Figure 7 (note the difference in the horizontal and vertical scales).

This method of graphing factored polynomials raises an important question: What do we do if we are given a polynomial in nonfactored form? In other words, how do we solve the generic n^{th}-degree polynomial equation $a_n x^n + a_{n-1}x^{n-1} + \cdots + a_1 x + a_0 = 0$? We'll study this question throughout the rest of this chapter, so for the remainder of this section all polynomials will either be given in factored form or be factorable with the tools we already possess.

Example 2: Graphing Polynomial Functions

Sketch the graphs of the following polynomial functions, paying particular attention to the x-intercept(s), the y-intercept, and the behavior as $x \to \pm\infty$.

a. $f(x) = -x(2x+1)(x-2)$

b. $g(x) = x^2 + 2x - 3$

c. $h(x) = x^4 - 1$

NOTE

As always, plotting additional points will help in sketching an accurate graph.

Solution

a. Begin with the x-intercepts. Using the Zero-Factor Property, we solve $f(x) = 0$.

$$-x(2x+1)(x-2) = 0$$
$$x = -\frac{1}{2}, 0, 2$$

Thus, the x-intercepts are $\left(-\frac{1}{2}, 0\right), (0,0)$, and $(2,0)$.

We could plug in $x = 0$ to find the y-intercept, but we already found it when calculating the x-intercepts! The y-intercept is the origin $(0,0)$.

All that remains is to determine the end behavior. Find the highest-degree term by multiplying $(-x)(2x)(x) = -2x^3$. The leading coefficient is negative, and the degree is odd, so $f(x) \to \infty$ as $x \to -\infty$ and $f(x) \to -\infty$ as $x \to \infty$.

Putting all this together, we obtain the following graph.

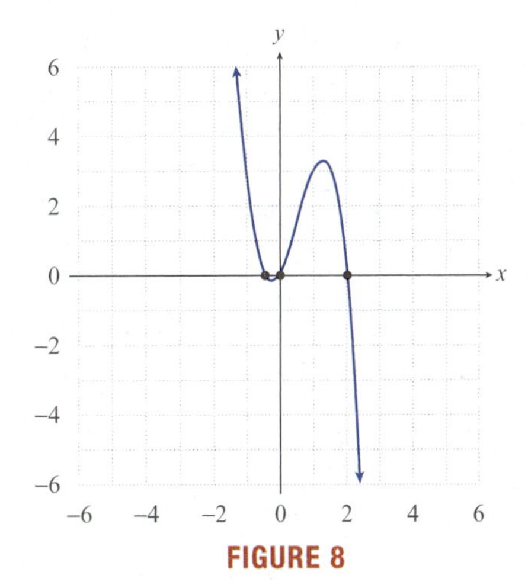

FIGURE 8

b. This polynomial is not in factored form. Before factoring, we can collect information about the y-intercept and behavior as $x \to \pm\infty$.

$g(x) = x^2 + 2x - 3$, so $a_0 = -3$ and the leading term is x^2. This means that the y-intercept is $(0,-3)$ and that $g(x) \to \infty$ as $x \to \pm\infty$.

Now, to find the x-intercepts, we factor the polynomial.

$$x^2 + 2x - 3 = 0$$
$$(x+3)(x-1) = 0$$
$$x = -3, 1$$

Thus, the x-intercepts are $(-3,0)$ and $(1,0)$.

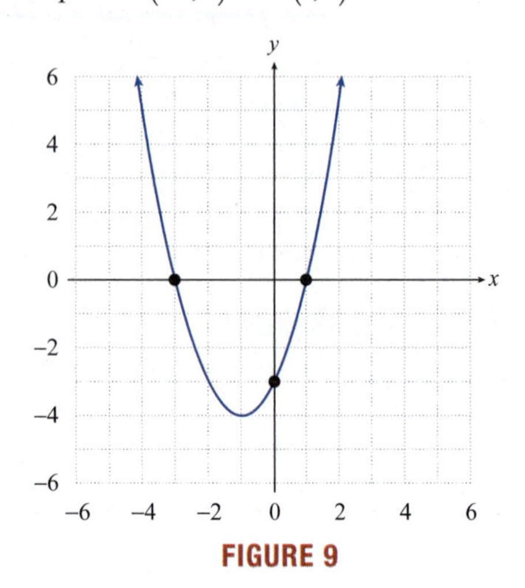

FIGURE 9

c. As with the previous example, we determine the y-intercept and end behavior before factoring.

$h(x)=x^4-1$, so $a_0=-1$ and the leading term is x^4. This tells us that the y-intercept is $(0,-1)$ and that $h(x)\to\infty$ as $x\to\pm\infty$.

Once again, we factor the polynomial to calculate the x-intercepts.

$$x^4-1=0 \quad \text{A difference of squares}$$
$$(x^2-1)(x^2+1)=0 \quad \text{Another difference of squares}$$
$$(x-1)(x+1)(x^2+1)=0$$
$$x=-1,1 \quad \text{Only the real solutions lead to } x\text{-intercepts.}$$

Here, the x-intercepts are $(-1,0)$ and $(1,0)$.

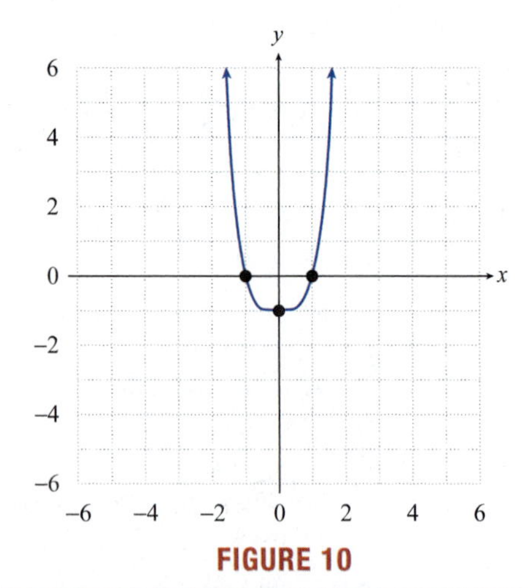

FIGURE 10

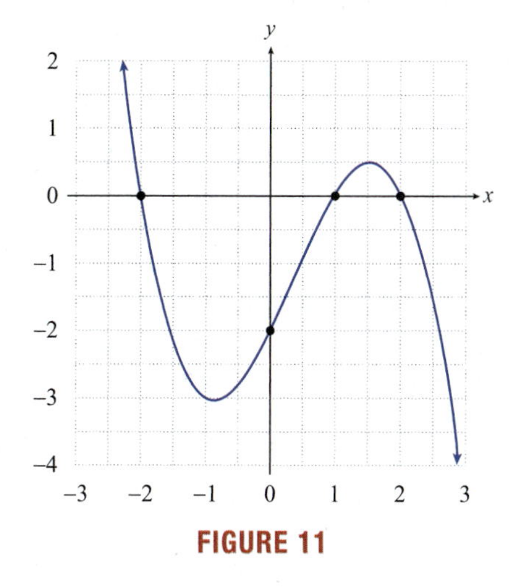

FIGURE 11

Example 3: Constructing a Polynomial Function from a Graph

Find the polynomial of lowest possible degree that corresponds to the graph in Figure 11.

Solution

Let $p(x)$ represent the polynomial to be found. The graph indicates that the zeros of p are -2, 1, and 2, so $p(x)$ must have factors of $(x+2)$, $(x-1)$, and $(x-2)$. In addition, $p(x)$ may have a constant factor a—in fact, we know a must be a negative number because the graph has the appearance of a cubic polynomial with a negative leading coefficient. So, we know that $p(x)=a(x+2)(x-1)(x-2)$, with a still to be determined. We can solve for a by noting that the graph has a y-intercept at $(0,-2)$, so

$$\begin{aligned} p(0) &= -2 \\ a(0+2)(0-1)(0-2) &= -2 \\ 4a &= -2 \\ a &= -\frac{1}{2}. \end{aligned}$$

Hence, $p(x)=\left(-\frac{1}{2}\right)(x+2)(x-1)(x-2)$ is the desired polynomial.

TOPIC 3: Solving Polynomial Inequalities

DEFINITION: Polynomial Inequalities

A **polynomial inequality** is any inequality that can be written in the form

$$p(x)<0,\ \ p(x)\le 0,\ \ p(x)>0, \text{ or } p(x)\ge 0,$$

where $p(x)$ is a polynomial function.

If we have an accurate graph of the function, solving a polynomial inequality is very straightforward; simply read where the function is positive and where it is negative.

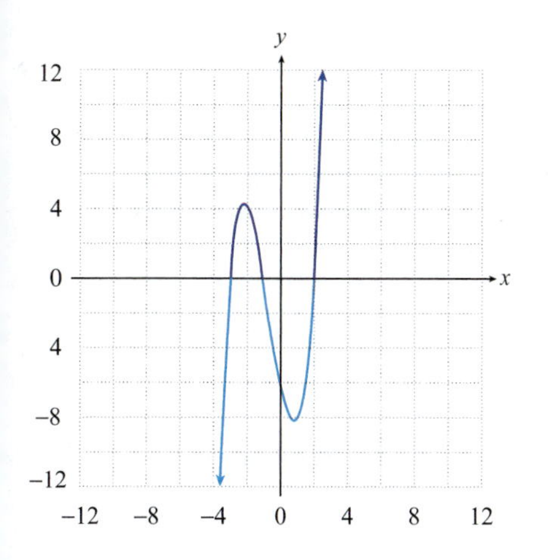

FIGURE 12: $f(x)=(x+3)(x+1)(x-2)$

Example 4: Solving Polynomial Inequalities Using a Graph

Solve the polynomial inequalities, given the graph of $f(x)=(x+3)(x+1)(x-2)$.

a. $(x+3)(x+1)(x-2)<0$ **b.** $(x+3)(x+1)(x-2)\ge 0$

Solution

a. From the graph, we see that $f(x)<0$ on the intervals $(-\infty,-3)$ and $(-1,2)$. Since the inequality is strict, we do not include the endpoints. Thus, the solution to this inequality is $(-\infty,-3)\cup(-1,2)$.

b. Similarly, the graph reveals that $f(x)>0$ on the intervals $(-3,-1)$ and $(2,\infty)$. This inequality is not strict, so we include the x-values where $f(x)=0$. Therefore, the solution to this inequality is $[-3,-1]\cup[2,\infty)$.

What if we do not have an accurate graph of the function? The graphing methods we have covered so far tell us if a polynomial $p(x)$ is positive or negative as $x \to \pm\infty$, but give no information about the sign of $p(x)$ near the origin.

As such, we need an algebraic method for solving polynomial inequalities when a detailed graph is not available. Having such a method will be a helpful tool in sketching graphs of polynomials.

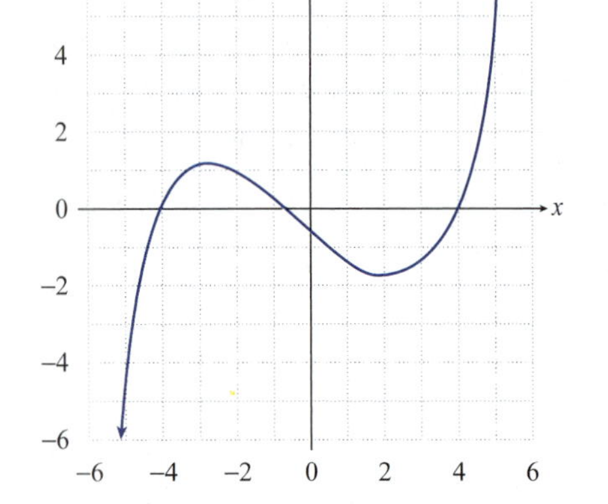

FIGURE 13: Continuous Function

Our method depends on a property of polynomials called **continuity**. While a more formal definition of continuity requires tools from calculus, intuitively, a continuous function has no "breaks" in it, in other words its graph can be drawn without lifting the pencil off the paper.

Why is continuity important? Look at the graphs in Figures 13 and 14. If a graph is not continuous, it can change sign without passing through the x-axis (thus, the value of the function "skips" zero). However, with a polynomial (or any continuous function), the only way for the graph to change sign is to pass through zero! This means that between each pair of consecutive zeros, the graph of $p(x)$ is always positive or always negative. Knowing this gives us a method for solving polynomial inequalities.

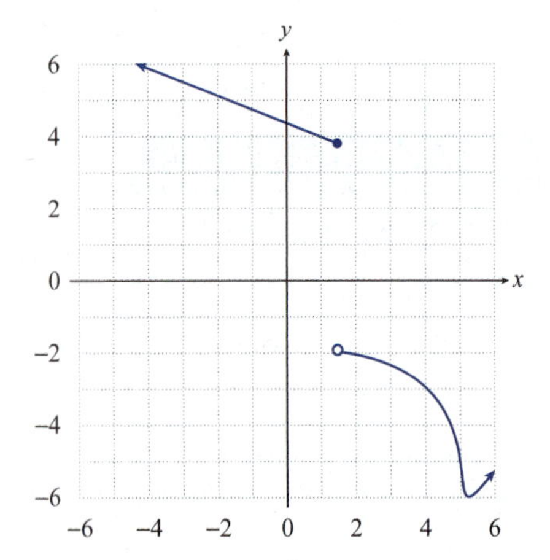

FIGURE 14: Discontinuous Function

PROCEDURE: Solving Polynomial Inequalities Using the Sign-Test Method

To solve a polynomial inequality $p(x) < 0$, $p(x) \leq 0$, $p(x) > 0$, or $p(x) \geq 0$, perform the following steps.

Step 1: Find the real zeros of $p(x)$. Equivalently, find the real solutions of $p(x) = 0$.

Step 2: Place the zeros on a number line, splitting it into intervals.

Step 3: Within each interval, select a **test point** and evaluate p at that number. If the result is positive, then $p(x) > 0$ for all x in the interval. If the result is negative, then $p(x) < 0$ for all x in the interval.

Step 4: Write the solution set, consisting of all of the intervals that satisfy the given inequality. If the inequality is not strict (uses $\leq$ or $\geq$), then the zeros are included in the solution set as well.

Example 5: Solving Polynomial Inequalities Using the Sign-Test Method

Solve the polynomial inequality $(x-2)(x+5)(x-4) > 0$.

NOTE

When choosing test points, integers (especially 0) are usually easiest to work with.

Solution

Follow the steps in the procedure for the Sign-Test Method.

Step 1: Find the zeros of $p(x) = (x-2)(x+5)(x-4)$. By the Zero-Factor Property, we can see that $p(x) = 0$ when $x = -5$, 2, or 4.

Step 2: Place the zeros on a number line.

FIGURE 15

This splits the number line into the intervals $(-\infty,-5)$, $(-5,2)$, $(2,4)$, and $(4,\infty)$.

Step 3: Evaluate $p(x)$ for a test point in each interval.

Interval	Test Point	Evaluate	Result
$(-\infty,-5)$	$x=-6$	$p(-6)=(-6-2)(-6+5)(-6-4)$ $=(-8)(-1)(-10)$ $=-80$	$p(x)<0$ on $(-\infty,-5)$ Negative
$(-5,2)$	$x=0$	$p(0)=(0-2)(0+5)(0-4)$ $=(-2)(5)(-4)$ $=40$	$p(x)>0$ on $(-5,2)$ Positive
$(2,4)$	$x=3$	$p(3)=(3-2)(3+5)(3-4)$ $=(1)(8)(-1)$ $=-8$	$p(x)<0$ on $(2,4)$ Negative
$(4,\infty)$	$x=6$	$p(6)=(6-2)(6+5)(6-4)$ $=(4)(11)(2)$ $=88$	$p(x)>0$ on $(4,\infty)$ Positive

TABLE 1

Step 4: Write the solution set to the original inequality, $p(x)>0$. From our table, we see that $p(x)>0$ on the intervals $(-5,2)$ and $(4,\infty)$. Since the inequality is strict, we do not include the zeros of $p(x)$ in our solution set. Thus, the solution to the inequality is $(-5,2)\cup(4,\infty)$.

Example 6: Solving Polynomial Inequalities

Solve the polynomial inequality $(x+2)(x-1)^2(x-3)\le 0$.

Solution

We will use a slight variation on our solution method which saves some computation time. The first two steps are the same, and since the zeros of the polynomial $p(x)=(x+2)(x-1)^2(x-3)$ are –2, 1, and 3, we divide the real number line at these points as shown in Figure 16.

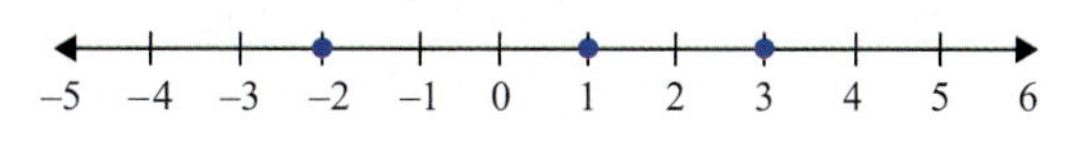

FIGURE 16

We also again select a test point within each of the four intervals defined by the zeros, but instead of computing the exact value of $p(x)$ at each test point, we simply determine its sign. For instance, using the test point $x=-3$ in the interval $(-\infty,-2)$, we know that $p(-3)$ is positive because $(-3+2)<0$, $(-3-1)^2>0$, $(-3-3)<0$ and the product of a negative factor, a positive factor, and a negative factor is positive. In Figure 17, a representative test point for each interval is shown, followed by a string of "+" and "–" signs indicating the sign of each factor of $p(x)$ at the test point. Finally, the sign of $p(x)$ overall on the interval is shown.

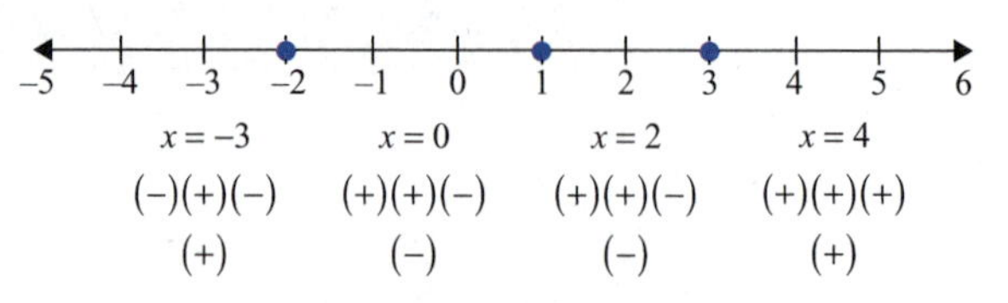

FIGURE 17

The last step of the method is to use the sign chart to identify the interval(s) on which $p(x)\le 0$. Since the inequality is not strict and the two intervals on which $p(x)$ is nonpositive overlap at $x=1$, the solution is $[-2,3]$.

5.1 EXERCISES

PRACTICE

Verify that the given values of *x* solve the corresponding polynomial equations. See Example 1.

1. $9x^2-4x=2x^3+15;\ x=-1$
2. $x^2-4x=-13;\ x=2-3i$
3. $x^2+13=4x;\ x=2+3i$
4. $3x^3+(5-3i)x^2=(2+5i)x-2i;\ x=i$
5. $9x^2-4x=2x^3+15;\ x=3$
6. $9x^2-4x=2x^3+15;\ x=\frac{5}{2}$
7. $3x^3+(5-3i)x^2=(2+5i)x-2i;\ x=-2$
8. $x^5-10x^4-80x^2=32-80x-40x^3;\ x=2$
9. $4x^5-8x^4-12x^3=16x^2-25x-69;\ x=3$
10. $x^2-4x-12=0;\ x=6$
11. $23x^7-12x^5=63x^4-3x^2;\ x=0$
12. $x^2+74=10x;\ x=5+7i$
13. $4x^2+32x+(8+i)x^3=-8;\ x=2i$
14. $8x-17=x^2;\ x=4-i$
15. $(5-3i)x-3x=4-6i;\ x=2$
16. $x^6-x^5+7x^4+x^3-9x=-1;\ x=1$
17. $6x^7-3x^5=3x^4-6x^2;\ x=-1$

Determine if the given values of x are zeros of the corresponding polynomial equations. See Example 1.

18. $16x = x^3 + x^2 + 20;\ x = -5$

19. $x^4 - 13x^2 + 12 = -x^3 + x;\ x = -1$

20. $x^4 - 3x^3 - 10x^2 = 0;\ x = 2$

21. $4x^5 - 216x^2 = 36x^3 - 24x^4;\ x = -6$

22. $x^3 - 8ix + 30 = 15x + 2x^2 + 16i;\ x = -i$

23. $x^3 - 7x^2 + 4x - 28 = 0;\ x = 2i$

Solve the following polynomial equations by factoring and/or using the quadratic formula, making sure to identify all the solutions. See Section 1.8 for review, if necessary.

24. $x^3 - x^2 - 6x = 0$

25. $x^2 - 2x + 5 = 0$

26. $x^4 + x^2 - 2 = 0$

27. $2x^2 + 5x = 3$

28. $9x^2 = 6x - 1$

29. $x^4 - 8x^2 + 15 = 0$

30. $x^3 - x^2 = 72x$

31. $x^2 + 5x = -\dfrac{25}{4}$

32. $2x^2 + 5 = 11x$

33. $x^4 - 8x^3 + 25x^2 = 0$

34. $x^4 - 13x^2 + 36 = 0$

35. $x^4 + 7x^2 = 8$

For each of the following polynomials, determine the degree and the leading coefficient; then determine the behavior of the graph as $x \to \pm\infty$.

36. $p(x) = 2x^4 - 3x^3 - 6x^2 - x - 23$

37. $j(x) = 4x^7 + 5x^5 + 12$

38. $r(x) = (3x+5)(x-2)(2x-1)(4x-7)$

39. $h(x) = -6x^5 + 2x^3 - 7x$

40. $g(x) = (x-5)^3(2x+1)(-x-1)$

41. $f(x) = -2(x+4)(x-4)(x^2)$

For each of the following polynomial functions, determine the behavior of its graph as $x \to \pm\infty$ and identify the x- and y-intercepts. Use this information to sketch the graph of each polynomial. See Example 2.

42. $f(x) = (x-3)(x+2)(x+4)$

43. $g(x) = (3-x)(x+2)(x+4)$

44. $f(x) = (x-2)^2(x+5)$

45. $h(x) = -(x+2)^3$

46. $r(x) = x^2 - 2x - 3$

47. $s(x) = x^3 + 3x^2 + 2x$

48. $f(x) = -(x-2)(x+1)^2(x+3)$

49. $g(x) = (x-3)^5$

Find the polynomial of lowest possible degree that corresponds to the given graph. See Example 3.

50.

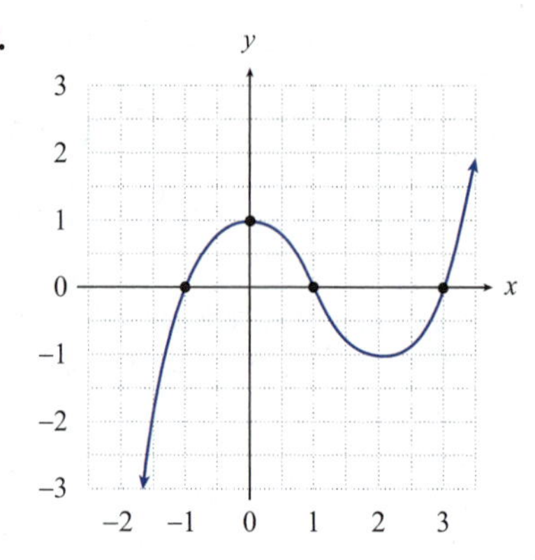

51.

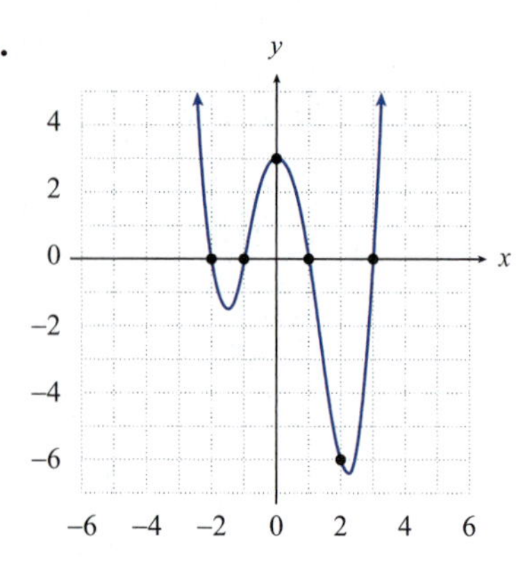

52.

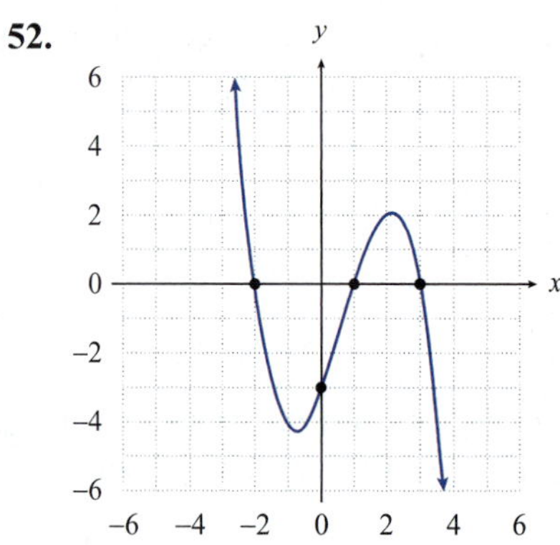

53.

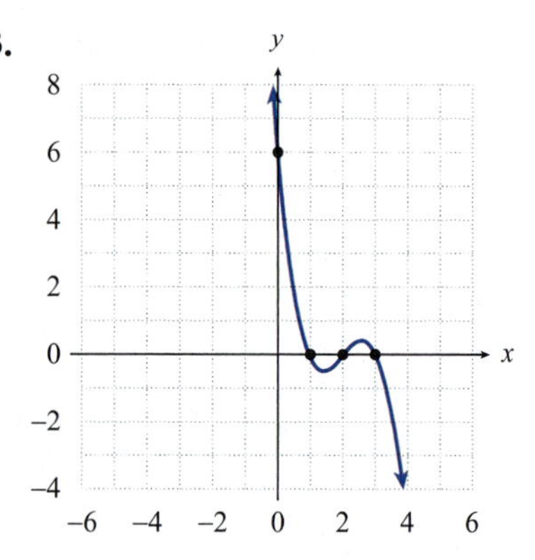

In Exercises 54 through 59, use the behavior as $x \to \pm\infty$ and the intercepts to match each polynomial with its graph.

54. $g(x) = (x+1)^2(x-3)^2$

55. $h(x) = 1-(x+2)^2$

56. $f(x) = (x-1)(x+2)(3-x)$

57. $r(x) = x^2 - x - 6$

58. $s(x) = (x-1)^3 - 2$

59. 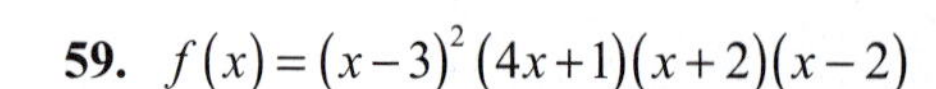 $f(x) = (x-3)^2(4x+1)(x+2)(x-2)$

a.

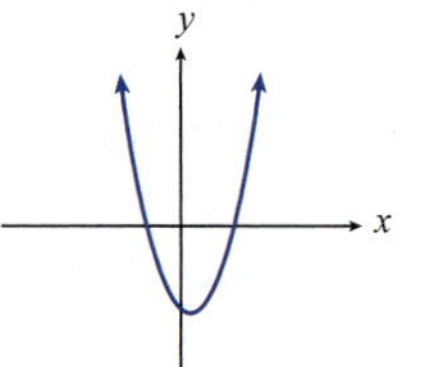

b.

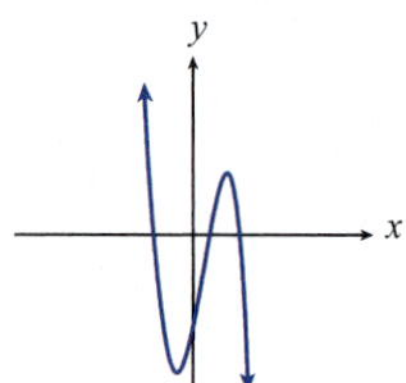

c.

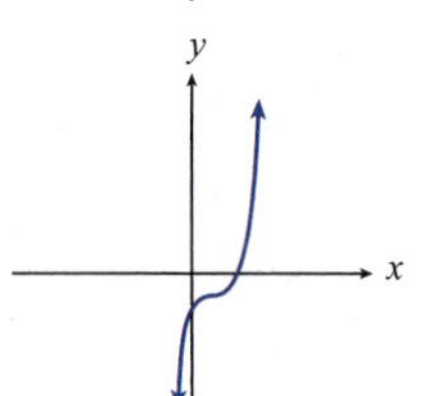

d.

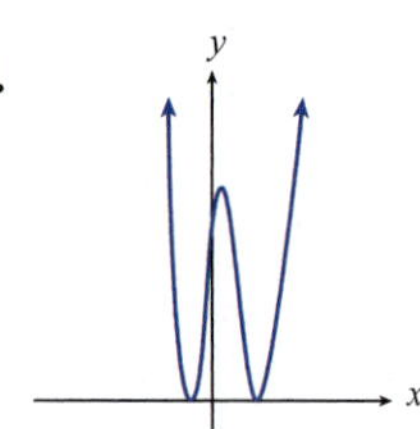

e.

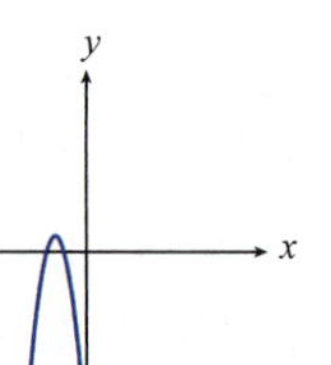

f.

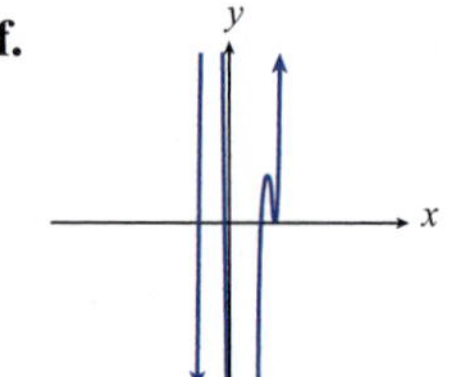

Match each of the following functions to the appropriate description.

60. $z(x) = (x-1)(x+2)(4-x)$

61. $r(x) = x^2 - 6x - 7$

62. $s(x) = x^3 + 3x^2 + 2x$

63. $g(x) = (x-1)(x+4)(3-x)$

64. $s(x) = x^3 + 5x^2 + 4x$

65. $s(x) = x^2 - 5x - 6$

a. cubic curve increasing as $x \to \infty$, has x-intercepts of 0, –1, and –2, and crosses the y-axis at 0.

b. parabola that opens up, has x-intercepts at 6 and –1, crosses the y-axis at –6.

c. cubic curve increasing as $x \to \infty$, has x-intercepts of 0, –1, and –4, and crosses the y-axis at 0.

d. parabola that opens up, has x-intercepts at 7 and –1, crosses the y-axis at –7.

e. cubic curve decreasing as $x \to \infty$, has x-intercepts at 1, 4, and –2, crosses the y-axis at –8.

f. cubic curve decreasing as $x \to \infty$, has x-intercepts of 1, 3, and –4, and crosses the y-axis at –12.

Solve the following polynomial inequalities. See Examples 5 and 6.

66. $x^2 - x - 6 \leq 0$

67. $x^2 > x + 6$

68. $(x+2)^2(x-1)^2 > 0$

69. $x^3 + 3x^2 + 2x < 0$

70. $(x-2)(x+1)(x+3) \geq 0$

71. $(x-1)(x+2)(3-x) \leq 0$

72. $-x^3 - x^2 + 30x > 0$

73. $(x^2-1)(x-4)(x+5) \leq 0$

74. $x^4 + x^2 > 0$

75. $4x^2 < 6x + 4$

76. $x^2(x+4)(x-3) > 0$

77. $(x-3)(x+4)(2-x) > 0$

APPLICATIONS

For Exercises 78–83, use the fact that profit is equal to revenue minus cost.

78. A small start-up skateboard company projects that the cost per month of manufacturing x skateboards will be $C(x) = 10x + 300$, and the revenue per month from selling x skateboards will be $r(x) = -x^2 + 50x$. For what value(s) of x will the company break even or make a profit?

79. A manufacturer has determined that the revenue from the sale of x cordless telephones is given by $r(x) = -x^2 + 15x$. The cost of producing x telephones is $C(x) = 135 - 17x$. For what value(s) of x will the company break even or make a profit?

80. The revenue from the sale of x fire extinguishers is estimated to be $r(x)=9-x^2$. The total cost of producing x fire extinguishers is $C(x)=209-33x$. For what value(s) of x will the company break even or make a profit?

81. A manufacturer has determined that the cost and revenue of producing and selling x telescopes are $C(x)=253-7x$ and $r(x)=27x-x^2$, respectively. For what value(s) of x will the company break even or make a profit?

82. A company that produces and sells compact refrigerators has found that the revenue from the sale of x compact refrigerators is $r(x)=-x^2+30x-370$. The cost function is given by $C(x)=6-25x$. For what value(s) of x will the company break even or make a profit?

83. An electronics company is deciding whether or not to begin producing phones. The company must determine if a profit can be made on the phones. The profit function is modeled by the equation $P(x)=x+0.27x^2-0.0015x^3-300$, where x is the number of phones produced in hundreds. Given this equation, how many phones must the company produce to make a profit?

84. The population of sea lions on an island is represented by the function $L(m)=110m^2-0.35m^4+750$, where m is the number of months the sea lions have been observed on the island. Given this information, how many more months will there be sea lions on the island?

85. The population of mosquitoes in a city in Florida is modeled by the function $M(w)=200w^2-0.01w^4+1200$, where w is the number of weeks since the town began spraying for mosquitoes. How many weeks will it take for all the mosquitoes to die?

5.2 POLYNOMIAL DIVISION AND THE DIVISION ALGORITHM

TOPICS

1. The Division Algorithm and the Remainder Theorem
2. Polynomial long division
3. Synthetic division
4. Constructing polynomials with given zeros

TOPIC 1: The Division Algorithm and the Remainder Theorem

In the last section, we made use of the factored form of a polynomial in graphing polynomials and in solving polynomial inequalities. In this section, we will formalize the observations we made and introduce additional techniques for working with polynomials.

Recall from arithmetic that if a natural number is divided by a smaller one (the divisor), the result can be expressed as a whole number (called the quotient) plus a remainder that is less than the divisor. The polynomial version of this concept is called the *Division Algorithm*, and it will prove to be very useful in helping us factor, evaluate, and graph polynomials.

THEOREM: The Division Algorithm

Let $p(x)$ and $d(x)$ be polynomials such that $d(x) \neq 0$ and with the degree of d less than or equal to the degree of p. Then there are unique polynomials $q(x)$ and $r(x)$, called the **quotient** and the **remainder**, respectively, such that

$$\underbrace{p(x)}_{\text{dividend}} = \underbrace{q(x)}_{\text{quotient}} \cdot \underbrace{d(x)}_{\text{divisor}} + \underbrace{r(x)}_{\text{remainder}}.$$

Either the degree of the remainder r is less than the degree of the **divisor** d, or the remainder is 0, in which case we say d **divides evenly** into the polynomial p. If the remainder is 0, the two polynomials q and d are **factors** of p.

If we divide every term in the equation $p(x) = q(x) \cdot d(x) + r(x)$ by the polynomial d, we obtain another form of the Division Algorithm.

$$\frac{p(x)}{d(x)} = q(x) + \frac{r(x)}{d(x)}$$

This version of the Division Algorithm will be very useful when we study rational functions in Section 5.5.

In many cases, we may need to divide a given polynomial p by a divisor of the form $d(x)=x-c$. The Division Algorithm tells us that the remainder is guaranteed to be either 0, or else a polynomial of degree 0 (since the degree of d is 1). In either case, the remainder polynomial is guaranteed to be simply a number, so $p(x)=q(x)(x-c)+r$ where r is a constant.

What does this mean if c happens to be a zero of the polynomial p? If this is the case, $p(c)=0$, so

$$\begin{aligned} 0 &= p(c) \\ &= q(c)(c-c)+r \\ &= r. \end{aligned}$$

The remainder is 0. Thus, if c is a zero of the polynomial p, $p(x)=q(x)(x-c)$. Conversely, if $x - c$ divides a given polynomial p evenly, then we know $p(x)=q(x)(x-c)$ and hence $p(c)=q(c)(c-c)=0$, so c is a zero of p. Together, these two observations constitute a major tool that we use in graphing polynomials and in solving polynomial equations, summarized below.

THEOREM: Zeros and Linear Factors

The number c is a **zero** of a polynomial $p(x)$ if and only if the linear polynomial $x - c$ is a factor of p. In this case, $p(x)=q(x)(x-c)$ for some quotient polynomial q. This also means that c is a solution of the polynomial equation $p(x)=0$, and if p is a polynomial with real coefficients and if c is a real number, then c is an x-intercept of p.

This reasoning also leads to a more general conclusion called the *Remainder Theorem*.

THEOREM: The Remainder Theorem

If the polynomial $p(x)$ is divided by $x - c$, the remainder is $p(c)$.

$$p(x)=q(x)(x-c)+p(c)$$

TOPIC 2: Polynomial Long Division

To make use of the Division Algorithm and the Remainder Theorem, we need to be able to actually divide one polynomial by another. Polynomial long division is the analog of numerical long division and provides the means for dividing any polynomial by another of equal or smaller degree.

We will begin looking at polynomial long division with an example, then write out a formal procedure. As you follow Example 1, notice the similarities between polynomial division and numerical division.

Example 1: Polynomial Long Division

Divide the polynomial $x^2+2x-24$ by the polynomial $x+6$.

Solution

$$x+6\overline{)x^2+2x-24}$$

Set up the division by arranging the terms of each polynomial in descending order of powers of x.

$$\begin{array}{r} x \\ x+6\overline{)x^2+2x-24} \end{array}$$

Divide the first term in the dividend, x^2, by the first term in the divisor, x.

$$\begin{array}{r} x \\ x+6\overline{)x^2+2x-24} \\ -\underline{(x^2+6x)} \end{array}$$

Multiply each term in the divisor by the result. We align like terms under those in the dividend. There is a minus sign because the next step is to subtract these terms.

$$\begin{array}{r} x \\ x+6\overline{)x^2+2x-24} \\ -\underline{(x^2+6x)} \\ -4x-24 \end{array}$$

Subtract x^2+6x from x^2+2x. Then, bring down -24 from the original dividend. This forms a new dividend to continue the process.

$$\begin{array}{r} x-4 \\ x+6\overline{)x^2+2x-24} \\ -\underline{(x^2+6x)} \\ -4x-24 \\ -\underline{(-4x-24)} \\ 0 \end{array}$$

We then apply the steps to the new dividend; divide, multiply, and subtract. At this point, there is nothing to bring down from the original dividend.

After subtracting, we are left with 0, which tells us that the remainder is 0.

Thus, the quotient is $x-4$ with a remainder of 0.

PROCEDURE: Polynomial Long Division

Step 1: Arrange the terms of each polynomial in descending order.

Step 2: Divide the first term in the dividend by the first term in the divisor. This gives the first term of the quotient.

Step 3: Multiply the entire divisor by the result (the first term of the quotient) and write this beneath the dividend so that like terms line up.

Step 4: Subtract the product from the dividend.

Step 5: Bring down the rest of the original dividend, forming a new dividend.

Step 6: Repeat the process with the new dividend. Continue until the degree of the remainder is less than the degree of the divisor.

Example 2: Polynomial Long Division

Divide the polynomial $6x^5 - 5x^4 + 10x^3 - 15x^2 - 19$ by the polynomial $2x^2 - x + 3$.

Solution

$$
\begin{array}{r@{}l}
 & \underline{\;3x^3 \qquad\qquad\qquad\qquad\qquad\qquad\quad} \\
2x^2 - x + 3 & \big)\,6x^5 - 5x^4 + 10x^3 - 15x^2 + 0x - 19 \\
 & \underline{-\left(6x^5 - 3x^4 + \;9x^3\right)} \\
 & \qquad\;\; -2x^4 + \quad x^3 - 15x^2 + 0x - 19
\end{array}
$$

When arranging the terms of the dividend, we insert a placeholder of $0x$. This makes it easier to keep like terms aligned, which prevents errors.

$$
\begin{array}{r@{}l}
 & \underline{\;3x^3 - \;x^2 \qquad\qquad\qquad\qquad\qquad\quad} \\
2x^2 - x + 3 & \big)\,6x^5 - 5x^4 + 10x^3 - 15x^2 + 0x - 19 \\
 & \underline{-\left(6x^5 - 3x^4 + \;9x^3\right)} \\
 & \qquad\;\; -2x^4 + \quad x^3 - 15x^2 + 0x - 19 \\
 & \qquad \underline{-\left(-2x^4 + \quad x^3 - \;3x^2\right)} \\
 & \qquad\qquad\qquad\qquad\;\; -12x^2 + 0x - 19
\end{array}
$$

Divide.

Multiply.

Subtract and bring down.

$$
\begin{array}{r@{}l}
 & \underline{\;3x^3 - \;x^2 \qquad\quad - \;6 \qquad\qquad\quad} \\
2x^2 - x + 3 & \big)\,6x^5 - 5x^4 + 10x^3 - 15x^2 + 0x - 19 \\
 & \underline{-\left(6x^5 - 3x^4 + 9x^3\right)} \\
 & \qquad\;\; -2x^4 + \quad x^3 - 15x^2 + 0x - 19 \\
 & \qquad \underline{-\left(-2x^4 + \quad x^3 - \;3x^2\right)} \\
 & \qquad\qquad\qquad\qquad\;\; -12x^2 + 0x - 19 \\
 & \qquad\qquad\qquad\quad \underline{-\left(-12x^2 + 6x - 18\right)} \\
 & \qquad\qquad\qquad\qquad\qquad\qquad -6x - \;1
\end{array}
$$

To complete the division, repeat the procedure one more time.

At this point, the process halts, as the degree of $-6x - 1$ is smaller than the degree of the divisor.

Thus, the solution is $3x^3 - x^2 - 6 + \dfrac{-6x - 1}{2x^2 - x + 3}$. We can also say the quotient is $3x^2 - x^2 - 6$ with a remainder of $-6x - 1$.

⚠ CAUTION

Although polynomial long division is a straightforward process, one common error is to forget to distribute the minus sign in each step as one polynomial is subtracted from the one above it. A good way to avoid this error is to put parentheses around the polynomial being subtracted, as shown in Examples 1 and 2.

Long division also works on polynomials with complex coefficients. When graphing polynomials, we work with those that have only real coefficients, but complex values can arise in intermediate steps of the graphing process. Further, in solving polynomial equations, we have seen (in some quadratic equations) that complex numbers may be the *only* solutions. Thus, we need to be able to handle division with complex numbers.

Example 3: Polynomial Long Division with Complex Numbers

Divide $p(x) = x^4 + 1$ by $d(x) = x^2 + i$.

Solution

$$x^2 + 0x + i \overline{\big)\, x^4 + 0x^3 + 0x^2 + 0x + 1}$$

Insert placeholders in both polynomials.

$$\begin{array}{r} x^2 \\ x^2 + 0x + i \overline{\big)\, x^4 + 0x^3 + 0x^2 + 0x + 1} \\ -\left(x^4 + 0x^3 + ix^2\right) \\ \hline -ix^2 + 0x + 1 \end{array}$$

The procedure is exactly the same as with all real coefficients.
Notice that we can use placeholders in the intermediate steps as well.

$$\begin{array}{r} x^2 - i \\ x^2 + 0x + i \overline{\big)\, x^4 + 0x^3 + 0x^2 + 0x + 1} \\ -\left(x^4 + 0x^3 + ix^2\right) \\ \hline -ix^2 + 0x + 1 \\ -\left(-ix^2 + 0x + 1\right) \\ \hline 0 \end{array}$$

When complex numbers are involved, we may need complex number arithmetic.

In the product step, use the fact that $(i)(-i) = -i^2 = 1$.

The remainder is zero, so we are finished.

Thus, the quotient is $x^2 - i$. There is no remainder, which tells us that the quotient is a factor of $p(x)$. In fact, we can write $x^4 + 1 = \left(x^2 + i\right)\left(x^2 - i\right)$.

TOPIC 3: Synthetic Division

Synthetic division is a shortened version of polynomial long division that can be used when the divisor is of the form $x - c$ for some constant c.

Synthetic division is more efficient because it omits the variables in the division process. Instead of various powers of the variable, synthetic division uses a tabular arrangement to keep track of the coefficients of the dividend and, ultimately, the coefficients of the quotient and the remainder. Consider the following long division of $-2x^3 + 8x^2 - 9x + 7$ by $x - 2$.

The Birth of Synthetic Division

The Italian mathematician Paolo Ruffini submitted the method of synthetic division for divisors of the form $x - c$ in 1804, in response to a challenge by the Italian Scientific Society; his method is consequently also known as Ruffini's rule. A more generalized form of synthetic division exists for divisors of higher degree.

$$\begin{array}{r}
-2x^2+4x-1 \\
x-2\overline{)-2x^3+8x^2-9x+7} \\
\underline{-(-2x^3+4x^2)} \\
4x^2-9x+7 \\
\underline{-(4x^2-8x)} \\
-x+7 \\
\underline{-(-x+2)} \\
5
\end{array}$$

First, we will place the constant c and the coefficients of the dividend in a row:

$$x-2\overline{)-2x^3+8x^2-9x+7} \quad \rightarrow \quad \underline{2|} \;\; -2 \;\; 8 \;\; -9 \;\; 7$$

Note that the number 2, which corresponds to c in the form $x - c$, appears without the minus sign. This is a very important and easily overlooked fact. When dividing by $x-c$ using synthetic division, the number that appears in the upper left is c.

Now, look at the key subtractions that occur in the long division.

$$\begin{array}{r}
-2x^2+4x-1 \\
x-2\overline{)-2x^3+8x^2-9x+7} \\
\underline{-(-2x^3+4x^2)} \\
4x^2-9x+7 \\
\underline{-(4x^2-8x)} \\
-1x+7 \\
\underline{-(-x+2)} \\
5
\end{array}$$

Because the x term of the divisor has a coefficient of 1, the result of each subtraction is the coefficient of the next term of the quotient. The result of the final subtraction is the remainder. Further, each subtraction begins with the coefficient of the dividend in the same "column." Thus, the result of our synthetic division should look like this.

$$\begin{array}{r|rrrr}
2 & -2 & 8 & -9 & 7 \\
\hline
 & -2 & 4 & -1 & 5
\end{array}$$

All that remains is to figure out how to get from the top row to the bottom row. Let's look one more time at the long division, except that we'll distribute the negative sign on each subtraction step, turning it into an addition.

$$
\begin{array}{r}
-2x^2+4x-1 \\
x-2\overline{)-2x^3+8x^2-9x+7} \\
\underline{+\left(2x^3-4x^2\right)} \\
4x^2-9x+7 \\
\underline{+\left(-4x^2+8x\right)} \\
-x+7 \\
\underline{+(x-2)} \\
5
\end{array}
$$

Note the following two observations:

1. The first coefficient of the quotient matches the first coefficient of the dividend.

2. To find the next coefficient of the quotient, add the product of c and the previous coefficient of the quotient to the corresponding coefficient of the dividend.

Figure 1 illustrates these calculations and then gives the completed table.

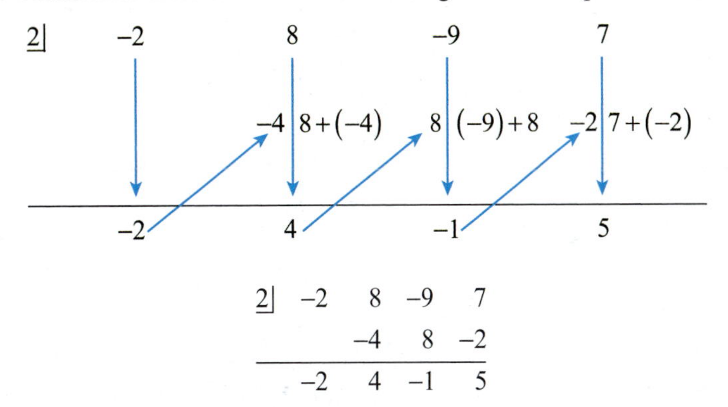

FIGURE 1: Synthetic Division

PROCEDURE: Synthetic Division

Step 1: If the divisor is $x-c$, write down c followed by the coefficients of the dividend.

Step 2: Write the leading coefficient of the dividend on the bottom row.

Step 3: Multiply c by the value placed on the bottom, and place the product in the next column, in the second row.

Step 4: Add the values in this column, giving a new value in the bottom row.

Step 5: Repeat this process until the table is complete.

Step 6: The numbers in the bottom row are the coefficients of the quotient, plus the remainder (which is the final value in this row). Note that the first term of the quotient will have degree one less than the first term of the dividend.

By the Remainder Theorem, synthetic division also provides a quick means of determining $p(c)$ for a given polynomial p, since $p(c)$ is the remainder when $p(c)$ is divided by $x-c$. If $p(c)=0$ then we know that c is a zero of $p(c)$ and that $x-c$ is a factor of $p(c)$. These facts are used in Example 4.

Example 4: Synthetic Division

For each polynomial p below, divide p by $x-c$ using synthetic division. Use the result to determine if the given c is a zero. If not, determine $p(c)$.

a. $p(x)=-2x^4+11x^3-5x^2-3x+15;\ c=5$

b. $p(x)=3x^8+9x^7-x^3-3x^2+x-1;\ c=-3$

NOTE

Remember, unlike in long division, we *add* the vertically aligned values of synthetic division.

Solution

a.

$$\begin{array}{r|rrrrr} 5 & -2 & 11 & -5 & -3 & 15 \\ & & & & & \\ \hline & -2 & & & & \end{array}$$

Place c in the upper-left corner, then write the coefficients of p on the top line.

$$\begin{array}{r|rrrrr} 5 & -2 & 11 & -5 & -3 & 15 \\ & & -10 & & & \\ \hline & -2 & 1 & & & \end{array}$$

Multiply $-2\cdot 5=-10$ and write down the result. Then add $11+(-10)$ to get 1, the next coefficient.

$$\begin{array}{r|rrrrr} 5 & -2 & 11 & -5 & -3 & 15 \\ & & -10 & 5 & 0 & -15 \\ \hline & -2 & 1 & 0 & -3 & 0 \end{array}$$

Continue the process, showing products and quotient coefficients. The last is the remainder of 0.

Because the remainder is 0, we know that $c=5$ is a zero of $p(x)$. With this information, we can factor $p(x)$ as follows.

$$-2x^4+11x^3-5x^2-3x+15=\left(-2x^3+x^2-3\right)(x-5)$$

b. While placeholders for missing terms are useful in long division, they are *necessary* when doing synthetic division.

$$\begin{array}{r|rrrrrrrrr} -3 & 3 & 9 & 0 & 0 & 0 & -1 & -3 & 1 & -1 \\ & & & & & & & & & \\ \hline & 3 & & & & & & & & \end{array}$$

Set up the synthetic division, including placeholders for x^6, x^5, and x^4.

$$\begin{array}{r|rrrrrrrrr} -3 & 3 & 9 & 0 & 0 & 0 & -1 & -3 & 1 & -1 \\ & & -9 & 0 & 0 & 0 & 0 & 3 & 0 & -3 \\ \hline & 3 & 0 & 0 & 0 & 0 & -1 & 0 & 1 & -4 \end{array}$$

Proceed with the synthetic division.

This time the remainder is not zero. This means that $c=-3$ is not a zero of $p(x)$. The Remainder Theorem tells us that $p(-3)=-4$.

Examine Example 4b again. The standard way to determine $p(-3)$ is to simplify $p(-3)=3(-3)^8+9(-3)^7-(-3)^3-3(-3)^2+(-3)-1$. This is a tedious calculation, but the result, $3(6561)+9(-2187)-(-27)-3(9)+(-3)-1$, does indeed equal -4. Compare these calculations with the far simpler synthetic division used in the example to see how useful synthetic division can be when evaluating polynomials.

Just as with long division, we can perform synthetic division on polynomials with complex coefficients (as long as the divisor is still first-degree).

Example 5: Synthetic Division (with Complex Numbers)

Compute $\dfrac{-3x^3+(5-2i)x^2+(-4+i)x+(1-i)}{x-1+i}$ using synthetic division.

Solution

$$\begin{array}{c|cccc} 1-i & -3 & 5-2i & -4+i & 1-i \\ & & & & \\ \hline & -3 & & & \end{array}$$

Set up the synthetic division. The dividend has several complex coefficients.

$$\begin{array}{c|cccc} 1-i & -3 & 5-2i & -4+i & 1-i \\ & & -3+3i & 3-i & -1+i \\ \hline & -3 & 2+i & -1 & 0 \end{array}$$

Throughout the division we need complex number arithmetic. In particular, $(1-i)(2+i)=2-i-i^2=3-i$.

Thus, the quotient is $-3x^2+(2+i)x-1$.

TOPIC 4: Constructing Polynomials with Given Zeros

The last topic in this section concerns reversing the division process. We now know the connection between a polynomial's zeros and factors: c is a zero of the polynomial $p(x)$ if and only if $x-c$ is a factor of $p(x)$. We can make use of this fact to construct polynomials that have certain desired properties, as illustrated in Example 6.

NOTE

When constructing polynomials from a set of factors, it is easier to keep the result in factored form until the last step. Often, it is fine to leave the polynomial in factored form.

Example 6: Constructing Polynomials

Construct a polynomial that has the given properties.

a. Third-degree, zeros of -3, 2, and 5, and goes to $-\infty$ as $x\to\infty$.

b. Fourth-degree, zeros of -5, -2, 1, and 3, and y-intercept at $(0,15)$.

Solution

a. We need $p(x)$ to have zeros -3, 2, and 5, so it must have linear factors of $(x+3)$, $(x-2)$, and $(x-5)$. Three linear factors gives us a third-degree polynomial, so there can be no more factors. Putting these together, we have the following.

$$p(x)=(x+3)(x-2)(x-5)$$

But does $p(x) \to -\infty$ as $x \to \infty$? No, if we multiply out, the leading term of this polynomial would be x^3, which has a positive leading coefficient. To fix this, we multiply the entire polynomial by -1.

$$\begin{aligned} p(x) &= -(x+3)(x-2)(x-5) \\ &= -x^3 + 4x^2 + 11x - 30 \end{aligned}$$

b. Once again, $p(x)$ is a product of linear factors, identified by the required zeros.

$$p(x) = (x+5)(x+2)(x-1)(x-3)$$

Our second condition is that the y-intercept must be $(0,15)$. If we substitute $x = 0$, we see that $p(0) = (5)(2)(-1)(-3) = 30$, so the y-intercept is $(0,30)$. To fix this, we might try subtracting 15 from the polynomial.

$$p(x) \stackrel{?}{=} (x+5)(x+2)(x-1)(x-3) - 15$$

But, we can not do this because it causes -5, -2, 1 and 3 to no longer be zeros! Instead, we multiply $p(x)$ by $\frac{1}{2}$.

$$\begin{aligned} p(x) &= \frac{1}{2}(x+5)(x+2)(x-1)(x-3) \\ &= \frac{1}{2}x^4 + \frac{3}{2}x^3 - \frac{15}{2}x^2 - \frac{19}{2}x + 15 \end{aligned}$$

5.2 EXERCISES

PRACTICE

Use polynomial long division to rewrite each of the following fractions in the form $q(x) + \frac{r(x)}{d(x)}$, where $d(x)$ is the denominator of the original fraction, $q(x)$ is the quotient, and $r(x)$ is the remainder. See Examples 1 through 3.

1. $\dfrac{6x^4 - 2x^3 + 8x^2 + 3x + 1}{2x^2 + 2}$

2. $\dfrac{5x^2 + 9x - 6}{x + 2}$

3. $\dfrac{x^3 - 6x^2 + 12x - 10}{x^2 - 4x + 4}$

4. $\dfrac{7x^5 - x^4 + 2x^3 - x^2}{x^2 + 1}$

5. $\dfrac{4x^3 - 6x^2 + x - 7}{x + 2}$

6. $\dfrac{x^3 + 2x^2 - 4x - 8}{x - 3}$

7. $\dfrac{3x^5 + 18x^4 - 7x^3 + 9x^2 + 4x}{3x^2 - 1}$

8. $\dfrac{9x^5 - 10x^4 + 18x^3 - 28x^2 + x + 3}{9x^2 - x - 1}$

9. $\dfrac{2x^5 - 5x^4 + 7x^3 - 10x^2 + 7x - 5}{x^2 - x + 1}$

10. $\dfrac{14x^5 - 2x^4 + 27x^3 - 3x^2 + 9x}{2x^3 + 3x}$

11. $\dfrac{x^4+x^2-20x-8}{x-3}$

12. $\dfrac{2x^5-3x^2+1}{x^2+1}$

13. $\dfrac{9x^3+2x}{3x-5}$

14. $\dfrac{-4x^5+8x^3-2}{2x^3+x}$

15. $\dfrac{2x^2+x-8}{x+3}$

16. $\dfrac{5x^5+x^4-13x^3-2x^2+6x}{x^3-2x}$

17. $\dfrac{2x^3-3ix^2+11x+(1-5i)}{2x-i}$

18. $\dfrac{9x^3-(18+9i)x^2+x+(-2-i)}{x-2-i}$

19. $\dfrac{3x^3+ix^2+9x+3i}{3x+i}$

20. $\dfrac{35x^4+(14-10i)x^3-(7+4i)x^2+2ix}{7x-2i}$

Use synthetic division to determine if the given value for c is a zero of the corresponding polynomial. If not, determine $p(c)$. See Example 4.

21. $p(x)=32x^5-80x^4+80x^3-40x^2+10x+2;\ c=1$

22. $p(x)=32x^5-80x^4+80x^3-40x^2+10x+2;\ c=\dfrac{1}{2}$

23. $p(x)=12x^4-7x^3-32x^2-7x+6;\ c=2$

24. $p(x)=12x^4-7x^3-32x^2-7x+6;\ c=1$

25. $p(x)=12x^4-7x^3-32x^2-7x+6;\ c=\dfrac{1}{3}$

26. $p(x)=2x^2-(3-5i)x+(3-9i);\ c=-2$

27. $p(x)=8x^4-2x+6;\ c=1$

28. $p(x)=x^4-1;\ c=1$

29. $p(x)=x^5+32;\ c=-2$

30. $p(x)=3x^5+9x^4+2x^2+5x-3;\ c=-3$

31. $p(x)=2x^2-(3-5i)x+(3-9i);\ c=-3i$

32. $p(x)=x^2-6x+13;\ c=2$

33. $p(x)=x^2-6x+13;\ c=3-2i$

34. $p(x)=3x^3-13x^2-28x-12;\ c=-2$

35. $p(x)=3x^3-13x^2-28x-12;\ c=6$

36. $p(x)=2x^3-8x^2-23x+63;\ c=2$

37. $p(x)=2x^3-8x^2-23x+63;\ c=5$

38. $p(x)=x^4-3x^3-3x^2+11x-6;\ c=1$

39. $p(x)=x^4-3x^3-3x^2+11x-6;\ c=-2$

40. $p(x)=x^4-3x^3-3x^2+11x-6;\ c=3$

Use synthetic division to rewrite each of the following fractions in the form $q(x)+\frac{r(x)}{d(x)}$, where $d(x)$ is the denominator of the original fraction, $q(x)$ is the quotient, and $r(x)$ is the remainder. See Example 5.

41. $\frac{x^3+x^2-18x+9}{x+5}$

42. $\frac{-2x^5+4x^4+3x^3-7x^2+3x-2}{x-2}$

43. $\frac{x^8+x^7-3x^3-3x^2+3}{x+1}$

44. $\frac{x^8-5x^7-3x^3+15x^2-2}{x-5}$

45. $\frac{4x^3-(16+4i)x^2+(14+4i)x+(-6-2i)}{x-3-i}$

46. $\frac{x^6-2x^5+2x^4+4x^2-8x+8}{x-1+i}$

47. $\frac{x^5-3x^4+x^3-5x^2+18}{x-2}$

48. $\frac{x^5-3x^4+x^3-5x^2+18}{x-3}$

49. $\frac{x^4+(i-1)x^3+(1-i)x^2+ix}{x+i}$

50. $\frac{x^6+8x^5+x^3+8x^2-14x-112}{x+8}$

51. $\frac{2x^3-10ix^2+5x+(8-3i)}{x-3i}$

52. $\frac{4x^5-6x^4+10x^3-4x^2-4x}{x-1}$

WRITING & THINKING

Construct a polynomial function with the stated properties. See Example 6.

53. Second-degree, zeros of -4 and 3, and goes to $-\infty$ as $x \to -\infty$.

54. Third-degree, zeros of -2, 1, and 3, and a y-intercept of -12.

55. Second-degree, zeros of $2-3i$ and $2+3i$, and a y-intercept of -13.

56. Third-degree, zeros of $1-i$, $2+i$, and -1, and a leading coefficient of -2.

57. Fourth-degree and a single x-intercept of 3.

58. Second-degree, zeros of $-\frac{3}{4}$ and 2, and a y-intercept of 6.

59. Fourth-degree, zeros of -3, -2, and 1, and a y-intercept of 18.

60. Third-degree, zeros of 1, 2, and 3, and passes through the point $(4, 12)$.

APPLICATIONS

61. A box company makes a variety of boxes, all with volume given by the formula $x^3+10x^2+31x+30$. If the height is given by $x + 3$, what is the formula for the surface area of the base?

5.3 LOCATING REAL ZEROS OF POLYNOMIAL FUNCTIONS

TOPICS

1. The Rational Zero Theorem
2. Descartes' Rule of Signs
3. Bounds of real zeros
4. The Intermediate Value Theorem

TOPIC 1: The Rational Zero Theorem

Abel and Galois

Two mathematicians are principally credited with proving that no formula like the quadratic formula can exist to solve polynomial equations of degree five or higher. The Norwegian mathematician Niels Henrik Abel proved the fact in 1824, improving on an earlier incomplete proof by Paolo Ruffini from 1799. Unfortunately, Abel suffered from tuberculosis and died at the age of 26 before his achievements were recognized. The impossibility of such a formula also follows from the deeper insights of the French mathematician Évariste Galois, who hurriedly scribbled the important details of what is known as Galois Theory in a letter to a friend in 1832, the night before he was fatally wounded in a duel over a love affair at the age of 20.

Given a polynomial function $p(x)$, we now know that c is a zero if and only if $x-c$ is a factor of p. Furthermore, if $x-c$ is a factor of p, we can use either polynomial long division or synthetic division to actually divide p by $x-c$ and find the quotient polynomial q, allowing us to write $p(x)=(x-c)q(x)$. This leaves us with a polynomial q of smaller degree and brings us one step closer to factoring p completely.

What we are lacking is a method for finding the zeros of p when it doesn't easily factor. The techniques from the last two sections cannot be put to use until we have some way of locating a zero c as a starting point.

Unfortunately, it can be proven that there is no formula that, like the quadratic formula, identifies all the zeros of a polynomial of degree five or higher. We do, however, have tools that give us hints about where to look for zeros of a given polynomial. In this section, we will study several such tools, beginning with the Rational Zero Theorem.

THEOREM: The Rational Zero Theorem

If $f(x) = a_n x^n + a_{n-1}x^{n-1} + \cdots + a_1 x + a_0$ is a polynomial with integer coefficients with $a_n \neq 0$, then any rational zero of f must be of the form $\frac{p}{q}$, where p is a factor of the constant term a_0 and q is a factor of the leading coefficient a_n.

CAUTION

Before applying the Rational Zero Theorem, we note two things it *doesn't* do:

1. The theorem doesn't necessarily find even a single zero of a polynomial; instead, it identifies a list of rational numbers that could *potentially* be zeros.

2. The theorem says nothing about irrational or complex zeros. If a polynomial has zeros that are either irrational or complex, we must resort to other means to find them.

Example 1: The Rational Zero Theorem

For each of the polynomials, list all of the potential rational zeros. Then write the polynomial in factored form and identify the actual zeros.

a. $f(x) = 2x^3 + 5x^2 - 4x - 3$ **b.** $g(x) = 27x^4 - 9x^3 - 33x^2 - x - 4$

NOTE

After generating the list of possible rational zeros, it's easiest to use synthetic division to check the zeros.

Solution

a. To apply the Rational Zero Theorem, find the factors of a_0 and a_3.

Factors of a_0: $\pm\{1, 3\}$

Factors of a_3: $\pm\{1, 2\}$

Note that we take both the positive and negative factors into consideration.

Possible rational zeros: $\pm\left\{1, 3, \frac{1}{2}, \frac{3}{2}\right\}$

If there are any rational zeros, they will come from this set of 8 numbers.

Now perform synthetic division on a trial-and-error basis with the potential rational zeros.

$$\begin{array}{r|rrrr} 1 & 2 & 5 & -4 & -3 \\ & & 2 & 7 & 3 \\ \hline & 2 & 7 & 3 & 0 \end{array}$$

Performing synthetic division with $c = 1$ gives a remainder of 0.

TECHNOLOGY

NORMAL FLOAT AUTO REAL RADIAN MP
PRESS + FOR △Tbl

X	Y1
-4	-35
-3	0
-2	9
-1	4
0	-3
1	0
2	25
3	84
4	189
5	352
6	585

X=1

Thus, we can factor $f(x)$ as follows:

$$f(x) = (x-1)(2x^2 + 7x + 3)$$

Use the result from synthetic division.

$$= (x-1)(2x+1)(x+3)$$

Factor the quadratic.

Actual zeros: $\left\{1, -\frac{1}{2}, -3\right\}$

Apply the Zero-Factor Property to determine the actual zeros.

b. Again, begin by listing the factors of the leading coefficient and the constant term.

Factors of a_0 : $\pm\{1, 2, 4\}$

Factors of a_4 : $\pm\{1, 3, 9, 27\}$

Possible rational zeros : $\pm\left\{1, 2, 4, \frac{1}{3}, \frac{2}{3}, \frac{4}{3}, \frac{1}{9}, \frac{2}{9}, \frac{4}{9}, \frac{1}{27}, \frac{2}{27}, \frac{4}{27}\right\}$

While it may be daunting to consider 24 potential rational zeros, appreciate the fact that the Rational Zero Theorem has eliminated all rational numbers except these 24! Before we begin trial-and-error, consider a few tips for choosing possible zeros.

1. Begin with integer values. The synthetic division and resulting quotient will usually be simpler than when trying fractions.

2. Begin with values near 1. This will keep the numbers in calculations smaller, allowing you to test values a bit more quickly.

$$\begin{array}{r|rrrrr} -1 & 27 & -9 & -33 & -1 & -4 \\ & & -27 & 36 & -3 & 4 \\ \hline & 27 & -36 & 3 & -4 & 0 \end{array}$$

Synthetic division with $c=-1$ tells us that $(x+1)$ is a factor of $g(x)$.

The resulting quotient is still a cubic, so we proceed with synthetic division. It also leads to another consideration when selecting potential zeros.

3. If the value c is a zero, do not forget to test c with the quotient, as $(x-c)$ can appear as a factor multiple times.

$$\begin{array}{r|rrrr} -1 & 27 & -36 & 3 & -4 \\ & & -27 & 63 & -66 \\ \hline & 27 & -63 & 66 & -70 \end{array}$$

In this case, the synthetic division fails to find a zero. We simply proceed with the next candidate.

$$\begin{array}{r|rrrr} \frac{4}{3} & 27 & -36 & 3 & -4 \\ & & 36 & 0 & 4 \\ \hline & 27 & 0 & 3 & 0 \end{array}$$

This time, the remainder is 0, so we have uncovered another factor, leaving us with a quadratic quotient.

We can then write the factorization using the result from synthetic division. The last step in factoring comes from solving a quadratic equation.

$$\begin{aligned} g(x) &= (x+1)\left(x-\frac{4}{3}\right)\left(27x^2+3\right) \\ &= 27(x+1)\left(x-\frac{4}{3}\right)\left(x-\frac{i}{3}\right)\left(x+\frac{i}{3}\right) \\ &= (x+1)(3x-4)(3x-i)(3x+i) \end{aligned}$$

By solving the equation $27x^2+3=0,$ we find $27x^2+3=27\left(x-\frac{i}{3}\right)\left(x+\frac{i}{3}\right).$

Actual zeros: $\left\{-1, \frac{4}{3}, \frac{i}{3}, -\frac{i}{3}\right\}$

TOPIC 2: Descartes' Rule of Signs

Descartes' Rule of Signs is another tool to aid us in our search for zeros. Unlike the Rational Zero Theorem, Descartes' Rule doesn't identify candidates; instead, it gives us guidelines on how many positive real zeros and how many negative real zeros we can expect a given polynomial to have. Again, this tool tells us nothing about complex zeros.

THEOREM: Descartes' Rule of Signs

Let $f(x)=a_nx^n+a_{n-1}x^{n-1}+\cdots+a_1x+a_0$ be a polynomial with real coefficients, and assume $a_n \neq 0$. A **variation in sign** of *f* is a change in the sign of one coefficient of *f* to the next, either from positive to negative or vice versa.

1. The number of **positive real zeros** of *f* is either the number of variations in sign of $f(x)$ or is less than this number by a positive even integer.

2. The number of **negative real zeros** of *f* is either the number of variations in sign of $f(-x)$ or is less than this number by a positive even integer.

Note that in order to apply Descartes' Rule of Signs, it is critical to first write the terms of the polynomial in descending order. Also, unless the number of variations in sign is 0 or 1, the rule does not give us a definitive answer for the number of zeros to expect. For instance, if the number of variations in sign of $f(x)$ is 4, we know only that there will be 4, 2, or 0 positive real zeros.

Example 2: Descartes' Rule of Signs

Use Descartes' Rule of Signs to determine the possible numbers of positive and negative real zeros of each of the following polynomials.

a. $f(x) = 2x^3 + 3x^2 - 14x - 21$ **b.** $g(x) = 3x^3 - 10x^2 + \frac{51}{4}x - \frac{13}{4}$

Solution

a. $f(x) = 2x^3 + 3x^2 \underset{\text{sign change}}{} -14x - 21$ One sign change in $f(x)$.

$f(-x) = 2(-x)^3 + 3(-x)^2 - 14(-x) - 21$ Plug in $-x$, then simplify.

$= -2x^3 \underset{\text{sign change}}{} +3x^2 + 14x \underset{\text{sign change}}{} -21$ Two sign changes in $f(-x)$.

By Descartes' Rule of Signs, there is exactly 1 positive real zero and either 2 or 0 negative real zeros of $f(x)$.

b. $g(x) = 3x^3 - 10x^2 + \frac{51}{4}x - \frac{13}{4}$ Three changes of sign in $g(x)$.

$g(-x) = 3(-x)^3 - 10(-x)^2 + \frac{51}{4}(-x) - \frac{13}{4}$ Plug in $-x$, then simplify.

$= -3x^3 - 10x^2 - \frac{51}{4}x - \frac{13}{4}$ There are no sign changes in $g(-x)$.

By Descartes' Rule of Signs, there are either 3 or 1 positive real zeros, but no negative real zeros of $g(x)$.

TOPIC 3: Bounds of Real Zeros

Although the Rational Zero Theorem and Descartes' Rule of Signs are useful for determining the zeros of a polynomial, Examples 1 and 2 show that more guidance would certainly be welcome, especially guidance that reduces the number of potential zeros that must be tested by trial and error. The following theorem does just that.

THEOREM: Upper and Lower Bounds of Zeros

Let $f(x)$ be a polynomial with real coefficients, a positive leading coefficient, and degree ≥ 1. Let a be a negative number and b be a positive number. Then:

1. No real zero of f is larger than b (we say b is an **upper bound** of the zeros of f) if the last row in the synthetic division of $f(x)$ by $x-b$ contains no negative numbers. That is, b is an upper bound of the zeros if the quotient and remainder have no negative coefficients when $f(x)$ is divided by $x-b$.

2. No real zero of f is smaller than a (we say a is a **lower bound** of the zeros of f) if the last row in the synthetic division of $f(x)$ by $x-a$ has entries that alternate in sign (0 can count as either positive *or* negative).

Example 3 revisits the polynomial $f(x)=2x^3+3x^2-14x-21$ that we studied in Example 2a and illustrates the use of the above theorem.

Example 3: Finding Bounds of Real Zeros

Use synthetic division to identify upper and lower bounds of the real zeros of the polynomial $f(x)=2x^3+3x^2-14x-21$.

NOTE

Finding the smallest upper bound and largest lower bound possible will help us eliminate as many potential zeros as possible.

Solution

Begin by testing any positive number as a potential upper bound.

2	2	3	−14	−21
		4	14	0
	2	7	0	−21

Synthetic division shows that 2 is not necessarily an upper bound, as the last row contains a negative number.

It is best to begin with a small value, then test progressively larger ones, as this will help in finding the smallest upper bound.

3	2	3	−14	−21
		6	27	39
	2	9	13	18

The number 3 is an upper bound according to the theorem, as all of the coefficients in the last row are nonnegative.

This tells us that all real zeros (including irrational zeros) of f are less than or equal to 3.

We continue by testing a value for the lower bound.

−3	2	3	−14	−21
		−6	9	—
	2	−3	−5	—

The synthetic division has not been completed, because as soon as the signs in the last row cease to alternate, we know −3 is not a lower bound.

NOTE

Remember that if a 0 appears in the last row, it can be counted as either positive or negative, whichever leads to a sequence of alternating signs.

Move on by testing a lower number.

−4	2	3	−14	−21
		−8	20	−24
	2	−5	6	−45

We find that −4 is a lower bound, as the signs in the last row alternate.

Thus, we see that –4 is a lower bound. Combined with the upper bound, we now know that all real zeros of f lie in the interval $[-4,3]$.

Example 4: Finding the Zeros of a Polynomial

Use the results of Example 3, in conjunction with the Rational Zero Theorem, to find the actual zeros of $f(x)=2x^3+3x^2-14x-21$.

Solution

Start by finding the potential rational zeros.

$$\text{Factors of } a_0\text{: } \pm\{1, 3, 7, 21\}$$

$$\text{Factors of } a_3\text{: } \pm\{1, 2\}$$

$$\text{Possible rational zeros: } \pm\left\{1, 3, 7, 21, \frac{1}{2}, \frac{3}{2}, \frac{7}{2}, \frac{21}{2}\right\}$$

Now, apply the lower and upper bounds. This allows us to eliminate any potential zeros greater than 3 or less than –4.

$$\text{Possible rational zeros: } \left\{1,-1,3,-3,\frac{1}{2},-\frac{1}{2},\frac{3}{2},-\frac{3}{2},-\frac{7}{2}\right\}$$

Now, use synthetic division to test potential rational zeros.

$$\begin{array}{r|rrrr} -\frac{3}{2} & 2 & 3 & -14 & -21 \\ & & -3 & 0 & 21 \\ \hline & 2 & 0 & -14 & 0 \end{array}$$

The quotient is $2x^2-14$. We can find the remaining two zeros by using the square root method.

$$\begin{aligned} 2x^2-14&=0 \\ 2x^2&=14 \\ x^2&=7 \\ x&=\pm\sqrt{7} \end{aligned}$$

Thus, the actual zeros of $f(x)$ are $\left\{-\frac{3}{2},\sqrt{7},-\sqrt{7}\right\}$.

CAUTION

Don't read more into the Upper and Lower Bounds Theorem than is actually there. For instance, –3 actually *is* a lower bound of the zeros of $f(x)=2x^3+3x^2-14x-21$, but the theorem is not powerful enough to indicate this. The work in Example 3 shows that -4 is a lower bound, but the theorem fails to spot the fact that –3 is a better lower bound. The trade-off for this weakness in the theorem is that it is quickly and easily applied.

TOPIC 4: The Intermediate Value Theorem

The last technique for locating zeros that we study makes use of a property of polynomials called *continuity*, which we briefly discussed when solving polynomial inequalities. Although a more complete discussion of continuity requires the use of limits, one consequence of continuity that we will use now is that the graph of a continuous function has no "breaks" in it. That is, assuming that the function can be graphed at all, it can be drawn without lifting your pencil.

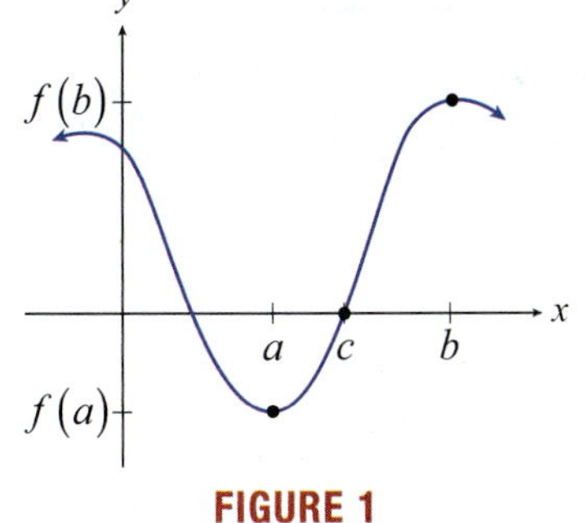

FIGURE 1

THEOREM: Intermediate Value Theorem

Assume that $f(x)$ is a polynomial with real coefficients, and that a and b are real numbers with $a < b$. If $f(a)$ and $f(b)$ differ in sign, then there is at least one point c such that $a < c < b$ and $f(c) = 0$. That is, at least one zero of f lies between a and b. See Figure 1.

CAUTION

The Intermediate Value Theorem can only tell us that there *is* a zero between two x-values, it can not prove that a zero *does not* exist between two values. If $f(a)$ and $f(b)$ do not differ in sign, there may still be one or more zeros between a and b.

We can use the Intermediate Value Theorem to prove that a zero of a given polynomial must lie in a particular interval. Repeated application of this process allows us to "home in" on the zero, generating a good approximation.

Example 5: Intermediate Value Theorem

a. Show that $f(x) = x^3 + 3x - 7$ has a zero between 1 and 2.

b. Find an approximation of the zero to the nearest tenth.

TECHNOLOGY

NORMAL FLOAT AUTO REAL RADIAN MP
PRESS [ENTER] TO EDIT

X	Y₁
-2	-21
-1	-11
0	-7
1	-3
2	7
3	29
4	69
5	133
6	227
7	357
8	529

$Y_1 = X^3+3X-7$

Solution

a. To use Intermediate Value Theorem, we need to calculate $f(1)$ and $f(2)$.

$$f(1) = (1)^3 + 3(1) - 7 = -3 \qquad f(1) \text{ is negative.}$$

$$f(2) = (2)^3 + 3(2) - 7 = 7 \qquad f(2) \text{ is positive.}$$

Because $f(1)$ and $f(2)$ differ in sign, the Intermediate Value Theorem states that f has a zero between 1 and 2.

b. To estimate this zero to the nearest tenth, we plug in more values to shrink the interval where the zero could potentially lie. We begin with 1.5.

$$f(1.5) = (1.5)^3 + 3(1.5) - 7 = 0.875$$

We see that $f(1.5)$ is positive, but small. Since $f(1)$ is negative, we might expect that the zero is slightly less than 1.5.

$$f(1.4) = (1.4)^3 + 3(1.4) - 7 = -0.056$$

Now the Intermediate Value Theorem tells us that the zero lies between 1.4 and 1.5. We need to test one more point to determine the zero to the nearest tenth.

$$f(1.45)=(1.45)^3+3(1.45)-7=0.398625$$

Once more, the Intermediate Value Theorem narrows the interval to $(1.4,1.45)$, which shows that the value of the zero, to the nearest tenth, is 1.4.

TECHNOLOGY: Finding Zeros of Polynomials

Graphing utilities are useful aids in finding real zeros of polynomials. They can be used, for example, to quickly eliminate some candidates for zeros given by the Rational Zero Theorem. The Zoom feature available on most of them is a convenient tool when applying the Intermediate Value Theorem (see your owner's manual for help with your specific utility). Figures 2, 3, and 4 are of the graph of $f(x)=x^3+3x-7$ as we zoom in on the approximate zero identified in Example 5.

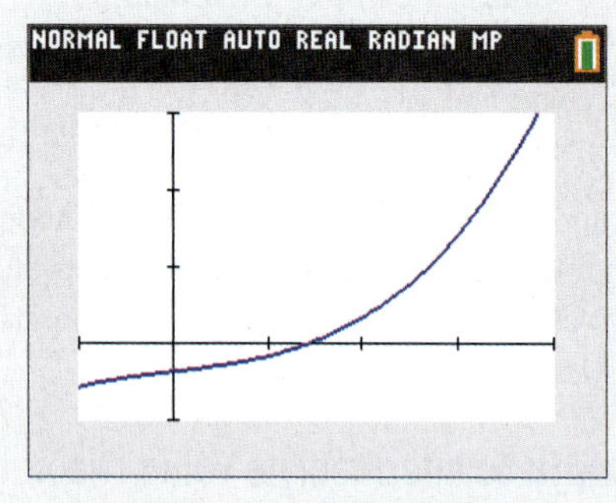

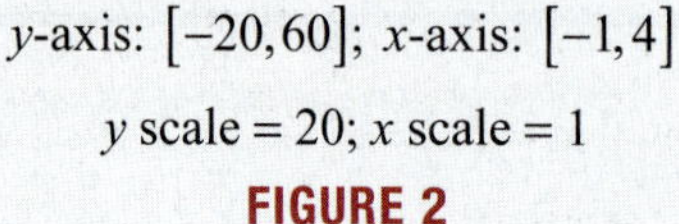

y-axis: $[-20,60]$; x-axis: $[-1,4]$

y scale $=20$; x scale $=1$

FIGURE 2

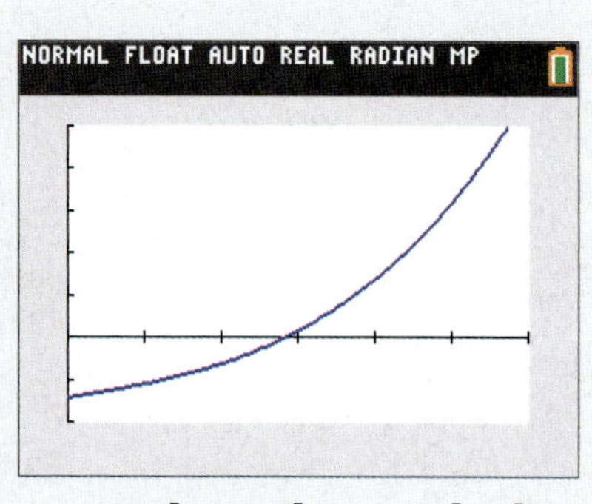

y-axis: $[-10,25]$; x-axis: $[0,3]$

y scale $=5$; x scale $=0.5$

FIGURE 3

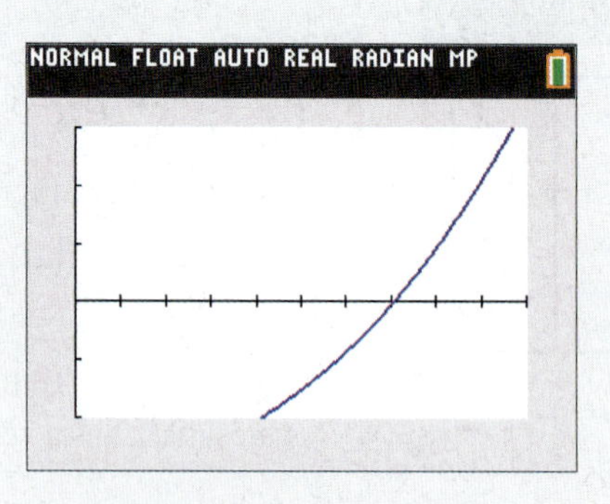

y-axis: $[-4,6]$; x-axis: $[0,2]$

y scale $=2$; x scale $=0.2$

FIGURE 4

5.3 EXERCISES

PRACTICE

List all of the potential rational zeros of the following polynomials. Then use polynomial division and the quadratic formula, if necessary, to identify the actual zeros. See Example 1.

1. $f(x) = 3x^3 + 5x^2 - 26x + 8$

2. $g(x) = -2x^3 + 11x^2 + x - 30$

3. $p(x) = x^4 - 5x^3 + 10x^2 - 20x + 24$

4. $h(x) = x^3 - 3x^2 + 9x + 13$

5. $q(x) = x^3 - 10x^2 + 23x - 14$

6. $r(x) = x^4 + x^3 + 23x^2 + 25x - 50$

7. $s(x) = 2x^3 - 9x^2 + 4x + 15$

8. $t(x) = x^3 - 6x^2 + 13x - 20$

9. $j(x) = 3x^4 - 3$

10. $k(x) = x^4 - 10x^2 + 24$

11. $m(x) = x^3 + 11x^2 - x - 11$

12. $g(x) = x^3 - 6x^2 - 5x + 30$

Using the Rational Zero Theorem or your answers to the preceding problems, solve the following polynomial equations.

13. $x^4 + x - 2 = -2x^4 + x + 1$

14. $x^4 + 10 = 10x^2 - 14$

15. $x^3 - 3x^2 + 9x + 13 = 0$

16. $3x^3 + 5x^2 = 26x - 8$

17. $x^4 + 10x^2 - 20x = 5x^3 - 24$

18. $-2x^3 + 11x^2 + x = 30$

19. $2x^3 - 12x^2 + 26x = 40$

20. $2x^3 + 9x^2 + 4x = 15$

21. $x^4 + x^3 + 23x^2 = 50 - 25x$

22. $x^3 + 23x = 10x^2 + 14$

23. $x^3 + 11x^2 = 11 + x$

24. $-6x^2 + x^3 = 5x - 30$

Use Descartes' Rule of Signs to determine the possible numbers of positive and negative real zeros of each of the following polynomials. See Example 2.

25. $f(x) = x^3 + 8x^2 + 17x + 10$

26. $g(x) = x^3 + 2x^2 - 5x - 6$

27. $f(x) = x^3 - 6x^2 + 3x + 10$

28. $g(x) = x^3 + 6x^2 + 11x + 6$

29. $f(x) = x^4 - 5x^3 - 2x^2 + 40x - 48$

30. $g(x) = x^3 + 3x^2 + 3x + 9$

31. $f(x) = x^4 - 25$

32. $g(x) = x^4 - 7x^3 + 5x^2 + 31x - 30$

33. $f(x) = 5x^5 - x^4 + 2x^3 + x - 9$

34. $g(x) = -6x^7 - x^5 - 7x^3 - 2x$

35. $f(x) = -5x^{11} - 14x^9 - 10x^7 - 15x^5$

36. $g(x) = 2x^4 + 7x^3 + 28x^2 + 112x - 64$

Use synthetic division to identify upper and lower bounds of the real zeros of the following polynomials. See Example 3.

37. $f(x) = x^3 + 4x^2 + x - 4$

38. $f(x) = 2x^3 - 3x^2 - 8x - 3$

39. $f(x) = x^3 - 6x^2 + 3x + 10$

40. $g(x) = x^3 + 6x^2 + 11x + 6$

41. $f(x) = x^4 - 5x^3 - 2x^2 + 40x - 48$

42. $g(x) = x^3 + 3x^2 + 3x + 9$

43. $f(x) = x^4 - 25$

44. $g(x) = x^4 - 7x^3 + 5x^2 + 31x - 30$

45. $f(x) = 2x^3 - 7x^2 - 28x - 12$

46. $g(x) = x^5 + x^4 - 9x^3 - x^2 + 20x - 12$

Using your answers to the preceding problems, polynomial division, and the quadratic formula, if necessary, find all of the zeros of the following polynomials.

47. $f(x) = x^3 + 4x^2 - x - 4$

48. $f(x) = 2x^3 - 3x^2 - 8x - 3$

49. $f(x) = x^3 - 6x^2 + 3x + 10$

50. $g(x) = x^3 + 6x^2 + 11x + 6$

51. $f(x) = x^4 - 5x^3 - 2x^2 + 40x - 48$

52. $g(x) = x^3 + 3x^2 + 3x + 9$

53. $f(x) = x^4 - 25$

54. $g(x) = x^4 - 7x^3 + 5x^2 + 31x - 30$

55. $f(x) = 2x^3 - 7x^2 - 28x - 12$

56. $g(x) = x^5 + x^4 - 9x^3 - x^2 + 20x - 12$

Use the Intermediate Value Theorem to show that each of the following polynomials has a real zero between the indicated values. See Example 5.

57. $f(x) = 5x^3 - 4x^2 - 31x - 6$; -3 and -1

58. $f(x) = x^4 - 9x^2 - 14$; 1 and 4

59. $f(x) = x^4 + 2x^3 - 10x^2 - 14x + 21$; 2 and 3

60. $f(x) = -x^3 + 2x^2 + 13x - 26$; -4 and -3

Show that each of the following equations must have a solution between the indicated real numbers.

61. $14x + 10x^2 = x^4 + 2x^3 + 21$; 2 and 3

62. $x^3 - 2x^2 = 13(x - 2)$; -4 and -3

Using any of the methods discussed in this section as guides, find all of the real zeros of the following functions.

63. $f(x) = 3x^3 - 18x^2 + 9x + 30$

64. $f(x) = -4x^3 - 19x^2 + 29x - 6$

65. $f(x) = 3x^5 + 7x^4 + 12x^3 + 28x^2 - 15x - 35$

66. $f(x) = 2x^4 + 5x^3 - 9x^2 - 15x + 9$

67. $f(x) = -15x^4 + 44x^3 + 15x^2 - 72x - 28$

68. $f(x) = 2x^4 + 13x^3 - 23x^2 - 32x + 20$

69. $f(x) = 3x^4 + 7x^3 - 25x^2 - 63x - 18$

70. $f(x) = x^5 + 7x^4 + 5x^3 - 43x^2 - 42x + 72$

71. $f(x) = 2x^5 - 3x^4 - 47x^3 + 103x^2 + 45x - 100$

72. $f(x) = x^6 - 125x^4 + 4804x^2 - 57{,}600$

Using any of the methods discussed in this section as guides, solve the following equations.

73. $x^3 + 6x^2 + 11x = -6$

74. $x^3 - 7x = 6(x^2 - 10)$

75. $x^3 + 9x^2 = 2x + 18$

76. $6x^3 + 14 = 41x^2 + 9x$

77. $4x^3 = 18x^2 + 106x + 48$

78. $3x^3 + 15x^2 - 6x = 72$

79. $8x^4 + 24 + 8x = 2x^3 + 38x^2$

80. $x^4 + 7x^2 = 3x^3 + 21x$

81. $6x^6 - 10x^5 - 9x^4 + 27x^3 = 20x^2 + 18x - 30$

82. $4x^5 - 5x^4 + 20x^2 = 6x^3 + 25x + 30$

WRITING & THINKING

83. Create a proof of the Rational Zero Theorem by following the suggested steps.

a. Assuming $\frac{p}{q}$ is a zero of the polynomial $f(x) = a_n x^n + a_{n-1}x^{n-1} + \cdots + a_1 x + a_0$, show that the equation $a_n\left(\frac{p}{q}\right)^n + a_{n-1}\left(\frac{p}{q}\right)^{n-1} + \cdots + a_1\left(\frac{p}{q}\right) + a_0 = 0$ can be written in the form $a_n p^n + a_{n-1}p^{n-1}q + \cdots + a_1 pq^{n-1} = -a_0 q^n$.

b. It can be assumed that $\frac{p}{q}$ is written in lowest terms (that is, the greatest common divisor of p and q is 1). By examining the left-hand side of the last equation above, show that p must be a divisor of the right-hand side, and hence a factor of a_0.

c. By rearranging the equation so that all terms with a factor of q are on one side, use a similar argument to show that q must be a factor of a_n.

5.4 THE FUNDAMENTAL THEOREM OF ALGEBRA

TOPICS

1. The Fundamental Theorem of Algebra
2. Multiple zeros and their geometric meaning
3. Conjugate pairs of zeros
4. Summary of polynomial methods

TOPIC 1: The Fundamental Theorem of Algebra

We are now ready to tie together all that we have learned about polynomials, and we begin with a powerful but deceptively simple-looking statement called the Fundamental Theorem of Algebra.

THEOREM: The Fundamental Theorem of Algebra

If p is a polynomial of degree n, with $n \geq 1$, then p has **at least one zero**. That is, the equation $p(x) = 0$ has at least one solution. It is important to note that the zero of p, and consequently the solution of $p(x) = 0$, may be a nonreal complex number.

A Recurring Theme

Gauss considered the Fundamental Theorem of Algebra so important that he returned to the topic repeatedly, publishing a total of four different proofs over his lifetime—the first in 1799, two in 1816, and the fourth in 1850.

Mathematicians began to suspect the truth of this statement in the first half of the 17th century, but a convincing proof did not appear until the German mathematician Carl Friedrich Gauss (1777–1855) provided one in his doctoral dissertation in 1799 at just 22 years of age!

Although the proof of the Fundamental Theorem of Algebra is beyond the scope of this text, we can use it to prove a consequence that summarizes much of the previous three sections. The following theorem has great implications in solving polynomial equations and in graphing real-coefficient polynomial functions. It tells us that our goal of factoring a polynomial completely is always at least theoretically possible.

THEOREM: The Linear Factors Theorem

Given the polynomial $p(x) = a_n x^n + a_{n-1}x^{n-1} + \cdots + a_1 x + a_0$, where $n \geq 1$ and $a_n \neq 0$, p can be factored as $p(x) = a_n(x - c_1)(x - c_2)\cdots(x - c_n)$, where $c_1, c_2, \ldots, c_n$ are constants (possibly nonreal complex constants and not necessarily distinct). In other words, **an nth-degree polynomial can be factored as a product of n linear factors.**

Proof

The Fundamental Theorem of Algebra tells us that $p(x)$ has at least one zero; call it c_1. Using the Division Algorithm, we know $(x-c_1)$ is a factor of p, and we can write

$$p(x)=(x-c_1)q_1(x),$$

where $q_1(x)$ is a polynomial of degree $n-1$. Note that the leading coefficient of q_1 must be a_n, since we divided p by $(x-c_1)$, a polynomial with leading coefficient of 1.

If the degree of q_1 is 0 (that is, if $n = 1$), then $q_1(x)=a_1$ and $p(x)=a_1(x-c_1)$. Otherwise, q_1 is of degree 1 or larger, and by the Fundamental Theorem of Algebra, q_1 itself has at least one zero; call it c_2. By the same reasoning, then, we can write

$$p(x)=(x-c_1)(x-c_2)q_2(x),$$

where q_2 is a polynomial of degree $n-2$, also with leading coefficient a_n. We can perform this process a total of n times (and no more), at which point we have the desired result:

$$p(x)=a_n(x-c_1)(x-c_2)\cdots(x-c_n).$$

CAUTION

The Linear Factors Theorem *does not* tell us the following things:

1. The theorem does not tell us that a polynomial has all real zeros. Some, or all, of the constants $c_1,c_2,\ldots,c_n$ may be nonreal complex numbers.
2. The theorem does not tell us that a polynomial has n *distinct* zeros. Some, or all, of the constants $c_1,c_2,\ldots,c_n$ may be identical.
3. The theorem does tell us that any polynomial can be written as a product of linear factors; it does not tell us *how to determine* the linear factors.

In the case where all of the coefficients of $p(x)=a_nx^n+a_{n-1}x^{n-1}+\cdots+a_1x+a_0$ are real, the Linear Factors Theorem tells us that the graph of p has *at most* n x-intercepts (and can only have *exactly* n x-intercepts if all n zeros are real and distinct). Indirectly, the theorem tells us something more: the graph of p can have at most $n-1$ *turning points*. A **turning point** of a graph is a point where the graph changes behavior from decreasing to increasing or vice versa. These facts are summarized below.

THEOREM: Interpreting the Linear Factors Theorem

The graph of an **n^{th}-degree polynomial function has at most n x-intercepts and at most $n-1$ turning points**. This also means that an nth-degree polynomial function has at most n zeros.

Figure 1 illustrates that the degree of a polynomial only gives us an upper bound on the number of x-intercepts and turning points. Note that the graph of the fourth-degree polynomial f has just two x-intercepts and only one turning point, while the graph of the third-degree polynomial g has three x-intercepts and two turning points.

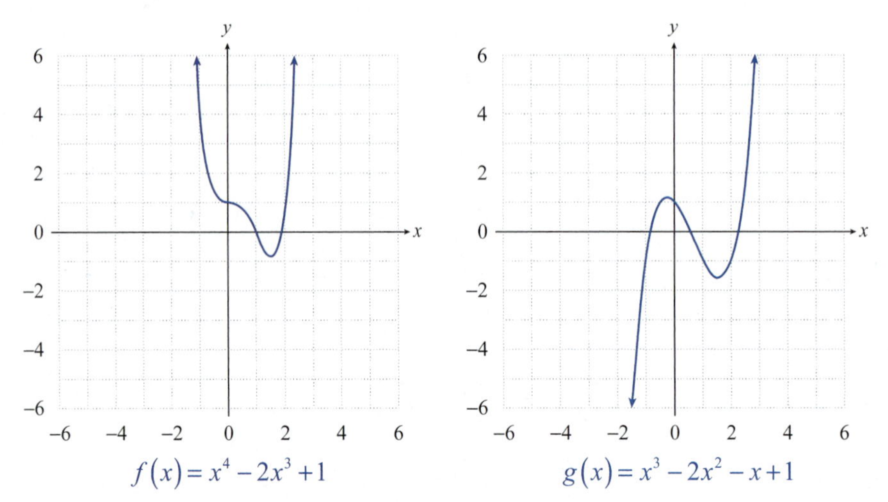

FIGURE 1: x-Intercepts and Turning Points

TOPIC 2: Multiple Zeros and Their Geometric Meaning

We know that, for example, the functions $(x-3)$, $(x-3)^2$, and $(x-3)^{15}$ are not the same and do not behave the same way. Yet, they have the same set of zeros: $\{3\}$. We need a way to classify functions in which a particular linear factor appears more than once.

DEFINITION: Multiplicity of Zeros

If the linear factor $(x-c)$ appears $k>0$ times in the factorization of a polynomial (or as $(x-c)^k$), we say the number c is a **zero of multiplicity *k*.**

If we are graphing a polynomial p for which c is a real zero of multiplicity k, then c is certainly an x-intercept of the graph of p, but the behavior of the graph near c depends on two characteristics:

1. Whether k is equal to or greater than 1.
2. Whether k is even or odd.

Before generalizing these concepts, let's look at a few examples.

Consider the function $f(x)=(x-1)^3$. We see that 1 is the only zero of f, and it is a zero of multiplicity 3. From our work with transformations of functions, we know the graph of f is the basic cubic shape shifted to the right by 1 unit, as shown in Figure 2. Note that the graph of f appears to "flatten out" near the zero.

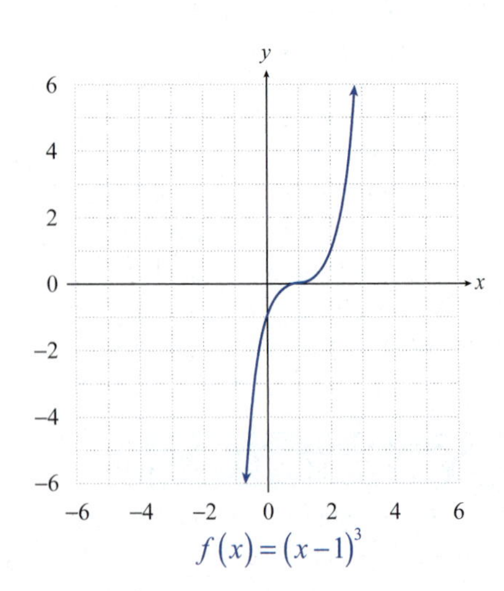

FIGURE 2: Zero of Multiplicity 3

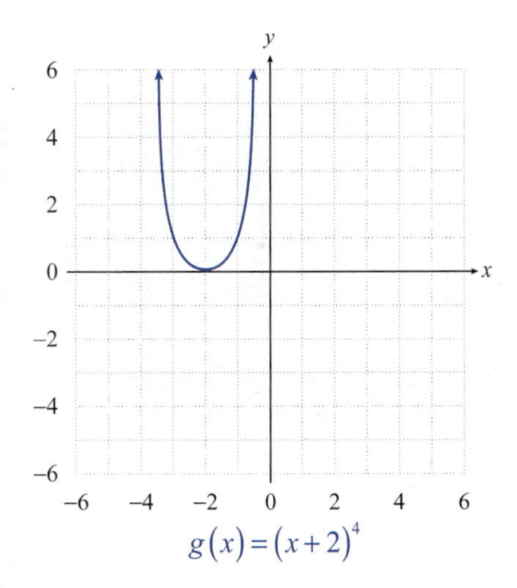

FIGURE 3: Zero of Multiplicity 4

Compare the behavior of $f(x)=(x-1)^3$ near its zero to the behavior of the function $g(x)=(x+2)^4$ near its own zero of −2, a zero of multiplicity 4. Figure 3 shows the graph of g. Again, the graph of g flattens out near the zero. Unlike the zero of multiplicity 3, in this case we see that the function has the same sign before and after the zero.

PROPERTIES: Geometric Meaning of Multiplicity

If c is a real zero of multiplicity k of a polynomial p (alternatively, if $(x-c)^k$ is a factor of p), the graph of p will touch the x-axis at $(c,0)$ and

- cross through the x-axis if k is odd, or
- stay on the same side of the x-axis if k is even.

Further, if $k>1$, the graph of p will "flatten out" near $(c,0)$.

With an understanding of how a zero's multiplicity affects a polynomial, constructing a reasonably accurate sketch of the graph becomes easier.

Example 1: Graphing Polynomial Functions

Sketch the graph of the polynomial $f(x)=(x+2)(x+1)^2(x-3)^3$.

Solution

We begin with the steps from before. Since f has even degree (6) and a positive leading coefficient (1), we know the end behavior: $f(x)\to\infty$ as $x\to\pm\infty$.

Then plug in $x=0$ to find the y-intercept.

$$\begin{aligned} f(0) &= (0+2)(0+1)^2(0-3)^3 \\ &= -54 \end{aligned}$$

Thus, f has its y-intercept at $(0,-54)$.

Using our knowledge of multiplicity, we can determine that f crosses the x-axis at −2 and 3, but not at −1, and that the graph of f flattens out near −1 and 3. Putting all of this together, we obtain the sketch in Figure 4.

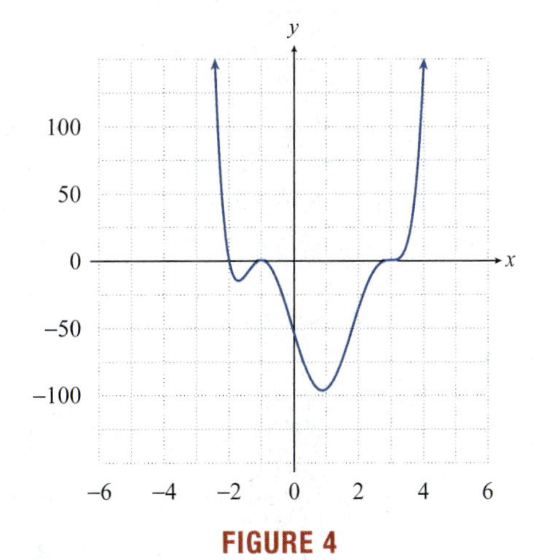

FIGURE 4

Note the extreme difference in the scales of the two axes.

To fill in portions of the graph between zeros accurately, we must still compute a few values of the function. For instance, $f(1)=-96$ and $f(2)=-36$. This also gives us a way to double-check our analysis of the behavior on either side of a zero.

TOPIC 3: Conjugate Pairs of Zeros

If all of the coefficients of a polynomial $p(x)$ are real, then p is a function that transforms real numbers into other real numbers, and consequently p can be graphed in the Cartesian plane. Nonetheless, it is very possible that some of the constants $c_1, c_2, \ldots, c_n$ in the factored form of p, $p(x) = a_n(x - c_1)(x - c_2)\cdots(x - c_n)$, might be nonreal complex numbers. For example, $x^2 + 1 = (x - i)(x + i)$. It turns out that such + roots must occur in pairs.

THEOREM: The Conjugate Roots Theorem

Let $p(x) = a_n x^n + a_{n-1}x^{n-1} + \cdots + a_1 x + a_0$ be a polynomial with only real coefficients. If the complex number $a + bi$ is a zero of p, then so is the complex number $a - bi$. In terms of the linear factors of p, this means that if $x - (a + bi)$ is a factor of p, then so is $x - (a - bi)$.

We can make use of this fact in several ways. For instance, if we are given one nonreal zero of a real-coefficient polynomial, we automatically know a second zero. The theorem is also useful when constructing polynomials with specified properties.

Another consequence of the Conjugate Roots Theorem is that every polynomial with real coefficients can be factored into a product of linear factors and irreducible quadratic factors with real coefficients. Each irreducible quadratic factor with real coefficients corresponds to the product of two factors of the form $x - (a + bi)$ and $x - (a - bi)$.

Example 2: Factoring Polynomials

Given that $4 - 3i$ is a zero of the polynomial $f(x) = x^4 - 8x^3 + 200x - 625$, factor f completely.

Solution

By the Conjugate Roots Theorem, since $4 - 3i$ is a zero of f, we know that $4 + 3i$ is a zero as well.

This gives us two ways to proceed:

1. We could divide f by $x - (4 - 3i)$ and then divide the result by $x - (4 + 3i)$ (most efficiently done with synthetic division).
2. Or, we could multiply $x - (4 - 3i)$ and $x - (4 + 3i)$ and divide f by their product (using polynomial long division).

In either case, we will be left with a quadratic polynomial that we know we can factor.

If we take the second approach, the first step is as follows:

$$\begin{aligned}(x - (4 - 3i))(x - (4 + 3i)) &= (x - 4 + 3i)(x - 4 - 3i) \\ &= x^2 - 4x \cancel{- 3ix} - 4x + 16 \cancel{+ 12i} \cancel{+ 3ix} \cancel{- 12i} - 9i^2 \\ &= x^2 - 8x + 25\end{aligned}$$

Now we divide f by this product:

$$\begin{array}{r} x^2 \qquad\quad -25 \\ x^2-8x+25\overline{\big)\,x^4-8x^3+\ 0x^2+200x-625} \\ -\left(x^4-8x^3+25x^2\right) \qquad\qquad\quad \\ \hline -25x^2+200x-625 \\ -\left(-25x^2+200x-625\right) \\ \hline 0 \end{array}$$

The quotient, x^2-25, is a difference of two squares and is easily factored, giving us our final result:

$$f(x)=(x-4+3i)(x-4-3i)(x-5)(x+5)$$

Example 3: Constructing Polynomials

Construct a fourth-degree real-coefficient polynomial function f with zeros of 2, –5, and $1+i$ such that $f(1)=12$.

Solution

Since $1+i$ is one of the zeros and f is to have only real coefficients, $1-i$ must be a zero as well by the Conjugate Roots Theorem. Based on this, f must be of the form

$$f(x)=a_4(x-(1+i))(x-(1-i))(x-2)(x+5)$$

for some real constant a_4. We must find a_4 so that $f(1)=12$. In order to do this, we begin by multiplying out $(x-(1+i))(x-(1-i))$:

$$\begin{aligned} (x-(1+i))(x-(1-i)) &= (x-1-i)(x-1+i) \\ &= x^2-2x+2 \end{aligned}$$

We then plug in $x=1$ and $f(1)=12$, then solve for a_4.

$$\begin{aligned} f(1) &= a_4\left(1^2-2(1)+2\right)(1-2)(1+5) && \text{Substitute } x=1. \\ 12 &= a_4(1)(-1)(6) && \text{Substitute } f(1)=12. \\ 12 &= -6a_4 \\ -2 &= a_4 && \text{Solve for } a_4. \end{aligned}$$

In factored form, the polynomial is $f(x)=-2(x-1-i)(x-1+i)(x-2)(x+5)$, which, if multiplied out, is $f(x)=-2x^4-2x^3+28x^2-52x+40$.

TOPIC 4: Summary of Polynomial Methods

All of the methods that you have learned in this chapter may be useful in solving a particular polynomial problem, whether it focuses on graphing a polynomial function, solving a polynomial equation, or solving a polynomial inequality. Now that all of the methods have been introduced, it makes sense to summarize them and see how they contribute to the big picture.

Recall that, in general, an nth-degree polynomial function has the form

$$p(x) = a_n x^n + a_{n-1}x^{n-1} + \cdots + a_1 x + a_0,$$

where $a_n \neq 0$ and any (or all) of the coefficients may be nonreal complex numbers. Keep in mind that it only makes sense to talk about graphing p in the Cartesian plane if all of the coefficients are real. Similarly, a polynomial inequality in which p appears on one side only makes sense if all the coefficients are real. For this reason, most of the polynomials in this text have only real coefficients.

Nevertheless, complex numbers often arise when working with polynomials, as some of the numbers $c_1, c_2, \ldots, c_n$ in the factored form of p,

$$p(x) = a_n(x - c_1)(x - c_2)\cdots(x - c_n),$$

may be nonreal even if all of $a_0, a_1, a_2, \ldots, a_n$ are real. The fact that p can, in principle, be factored is a direct consequence of the Fundamental Theorem of Algebra.

Factoring p into a product of linear factors as shown is the central point in solving a polynomial equation and (when the coefficients of p are real) in graphing a polynomial and solving a polynomial inequality. Specifically,

- The solutions of the polynomial equation $p(x) = 0$ are the numbers $c_1, c_2, \ldots, c_n$.
- When $a_0, a_1, a_2, \ldots, a_n$ are all real, the x-intercepts of the graph of p are the real numbers in the list $c_1, c_2, \ldots, c_n$. If a given c_i appears in the list k times, it is a *zero of multiplicity k*. If an x-intercept of p is of multiplicity k, the behavior of p near that x-intercept depends on whether k is even or odd. Any nonreal zeros in the list must appear in conjugate pairs.
- When $a_0, a_1, a_2, \ldots, a_n$ are all real, the solution of the polynomial inequality $p(x) > 0$ consists of all the open intervals on the x-axis where the graph of p lies strictly above the x-axis. The solution of $p(x) < 0$ consists of all the open intervals where the graph of p lies strictly below the x-axis. The solutions of $p(x) \geq 0$ and $p(x) \leq 0$ consist of closed intervals. Testing a single point in each interval suffices to determine the sign of p on that interval.

The remaining topics discussed in this chapter are observations and techniques that aid us in filling in the details of the big picture.

- The observation that the degree of a polynomial and the sign of its leading coefficient tell us how the graph of the polynomial behaves as $x \to -\infty$ and as $x \to \infty$.

- The observation that the graph of p crosses the y-axis at the easily computed point $(0, p(0))$.
- The technique of polynomial long division, useful in dividing one polynomial by another of the same or smaller degree.
- The technique of synthetic division, a shortcut that applies when dividing a polynomial by a polynomial of the form $x - c$. Recall that the remainder of this division is the value $p(c)$.
- The Rational Zero Theorem, which provides a list of potential rational zeros for polynomials with integer coefficients.
- Descartes' Rule of Signs, which provides guidance on the number of positive and negative real zeros that a real-coefficient polynomial might have.
- The Upper and Lower Bounds rule, which indicates an interval in which to search for all the zeros of a real-coefficient polynomial.
- The Intermediate Value Theorem, which can be used to "home in" on a real zero of a given polynomial.

As you solve various polynomial problems, try to keep the big picture in mind. Often, it is useful to literally keep a picture, namely the graph of the polynomial, in mind even if the problem does not specifically involve graphing.

5.4 EXERCISES

Throughout these exercises, a graphing utility may be helpful in identifying zeros and in checking your graphing, if permitted by your instructor.

PRACTICE

Sketch the graph of each factored polynomial. See Example 1.

1. $f(x) = (x+1)^4(x-2)^3(x-1)$

2. $g(x) = -x^3(x-1)(x+2)^2$

3. $f(x) = -x(x+2)(x-1)^2$

4. $g(x) = (x+2)(x-1)^3$

5. $f(x) = (x-1)^4(x-2)(x-3)$

6. $g(x) = (x+1)^2(x-2)^3$

7. $f(x) = (x-4)(x+2)^2(x-3)^3$

8. $g(x) = (x+3)(x-1)^5$

Use all available methods to factor each of the following polynomials completely, and then sketch the graph of each one. See Example 1.

9. $f(x) = x^5 + 4x^4 + x^3 - 10x^2 - 4x + 8$

10. $p(x) = 2x^3 - x^2 - 8x - 5$

11. $s(x) = -x^4 + 2x^3 + 8x^2 - 10x - 15$

12. $f(x) = -x^3 + 6x^2 - 12x + 8$

13. $H(x) = x^4 - x^3 - 5x^2 + 3x + 6$

14. $h(x) = x^5 - 11x^4 + 46x^3 - 90x^2 + 81x - 27$

15. $f(x) = 2x^3 + 11x^2 + 20x + 12$

16. $g(x) = x^4 + 3x^3 - 5x^2 - 21x - 14$

Use all available methods to solve each polynomial equation.

17. $x^5 + 4x^4 + x^3 = 10x^2 + 4x - 8$

18. $x^4 + 15 = 2x^3 + 8x^2 - 10x$

19. $x^4 + x^3 + 3x^2 + 5x - 10 = 0$

20. $x^3 - 9x^2 = 30 - 28x$

21. $x^5 + x^4 - x^3 + 7x^2 - 20x + 12 = 0$

22. $2x^4 - 5x^3 - 2x^2 + 15x = 0$

23. $x^5 + 15x^3 + 16 = x^4 + 15x^2 + 16x$

24. $x^3 - 5 = 5x^2 - 9x$

Use all available methods (in particular, the Conjugate Roots Theorem, if applicable) to factor each of the following polynomials completely, making use of the given zero if one is given. See Example 2.

25. $f(x) = x^4 - 9x^3 + 27x^2 - 15x - 52; 3 - 2i$ is a zero.

26. $g(x) = x^3 - (1 - i)x^2 - (8 - i)x + (12 - 6i); 2 - i$ is a zero.

27. $f(x) = x^3 - (2 + 3i)x^2 - (1 - 3i)x + (2 + 6i)$; 2 is a zero.

28. $p(x) = x^4 - 2x^3 + 14x^2 - 8x + 40; 2i$ is a zero.

29. $n(x) = x^4 - 4x^3 + 6x^2 + 28x - 91; 2 + 3i$ is a zero.

30. $G(x) = x^4 - 14x^3 + 98x^2 - 686x + 2401$; $7i$ is a zero.

31. $f(x) = x^4 - 3x^3 + 5x^2 - x - 10$

32. $g(x) = x^6 - 8x^5 + 25x^4 - 40x^3 + 40x^2 - 32x + 16$

33. $r(x) = x^4 + 7x^3 - 41x^2 + 33x$

34. $d(x) = x^5 - x^4 - 18x^3 + 18x^2 + 81x - 81$

35. $P(x) = x^3 - 6x^2 + 28x - 40$

36. $g(x) = x^6 - x^4 - 16x^2 + 16$

WRITING & THINKING

Construct polynomial functions with the stated properties. See Example 3.

37. Third-degree, only real coefficients, -1 and $5+i$ are two of the zeros, y-intercept is -52.

38. Fourth-degree, only real coefficients, $\sqrt{7}$ and $i\sqrt{5}$ are two of the zeros, y-intercept is -35.

39. Fifth-degree, 1 is a zero of multiplicity 3, -2 is the only other zero, leading coefficient is 2.

40. Fifth-degree, only real coefficients, 0 is the only real zero, $1+i$ is a zero of multiplicity 1, leading coefficient is 1.

41. Fourth-degree, only real coefficients, x-intercepts are 0 and 6, $-2i$ is a zero, leading coefficient is 3.

42. Fifth-degree, -2 is a zero of multiplicity 2, another integer is a zero of multiplicity 3, y-intercept is 108, leading coefficient is 1.

43. Third-degree, only real coefficients, -4 and $3 + i$ are two of the zeros, y-intercept is -40.

44. Fifth-degree, 1 is a zero of multiplicity 4, -2 is the only other zero, leading coefficient is 4.

45. Third-degree, only real coefficients, -4 and $4 + i$ are two of the zeros, y-intercept is -68.

46. Assume $f(x)$ is an nth-degree polynomial with real coefficients. Explain why the following statement is true: If n is even, the number of turning points is odd and if n is odd, the number of turning points is even.

APPLICATIONS

47. An open-top box is to be constructed from a 10 inch by 18 inch sheet of tin by cutting out squares from each corner as shown and then folding up the sides. Let $V(x)$ denote the volume of the resulting box.

a. Write $V(x)$ as a product of linear factors.
b. For which values of x is $V(x) = 0$?
c. Which answers from part b. are physically possible?

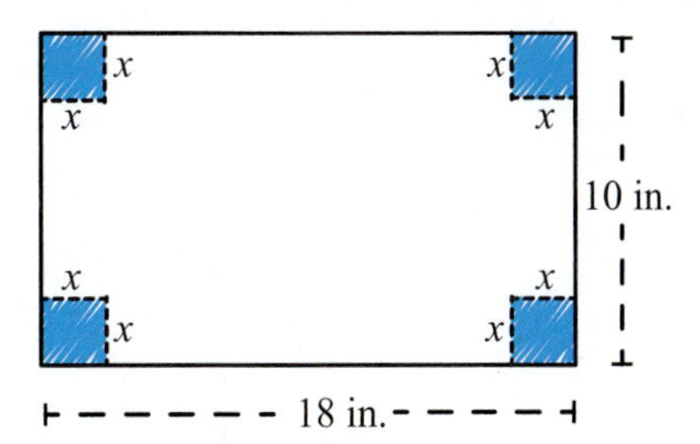

48. An open-top box is to be constructed from a 10 inch by 15 inch sheet of tin by cutting out squares from each corner and then folding up the sides. Let $V(x)$ denote the volume of the resulting box.

a. Write $V(x)$ as a product of linear factors.
b. For which values of x is $V(x) = 0$?
c. Which of your answers from part b. are physically possible?

49. An open-top box is to be constructed from a 9 inch by 17 inch sheet of tin by cutting out squares from each corner and then folding up the sides. Let $V(x)$ denote the volume of the resulting box.

a. Write $V(x)$ as a product of linear factors.
b. For which values of x is $V(x) = 0$?
c. Which of your answers from part b. are physically possible?

5.5 RATIONAL FUNCTIONS AND RATIONAL INEQUALITIES

TOPICS

1. Characteristics of rational functions
2. Vertical asymptotes
3. Horizontal and oblique asymptotes
4. Graphing rational functions
5. Solving rational inequalities

TOPIC 1: Characteristics of Rational Functions

The study of polynomials leads directly to a study of rational functions, which are ratios of polynomials. Since rational functions can have variables in the denominators of fractions, their behavior can be significantly more complex than that of polynomials.

DEFINITION: Rational Functions

A **rational function** is a function that can be written in the form

$$f(x) = \frac{p(x)}{q(x)},$$

where $p(x)$ and $q(x)$ are polynomial functions and $q(x) \neq 0$. Even though q is not allowed to be identically zero, there will often be values of x for which $q(x)$ is zero, and at these values the function is undefined. Consequently, the **domain of** f consists of all real numbers except those for which $q(x) = 0$.

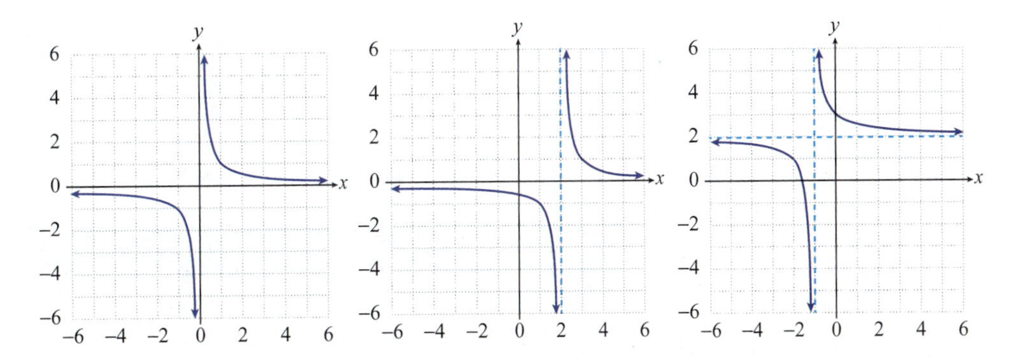

FIGURE 1: Graphs of Three Rational Functions

Each of the three graphs in Figure 1 has a new feature: vertical and horizontal dashed lines. These dashed lines are *not* part of the function, they are examples of *asymptotes*, and they serve as guides to understanding the function. Roughly speaking, an asymptote is an auxiliary line that the graph of a function approaches. Three kinds of asymptotes will appear in our study of rational functions: vertical, horizontal, and oblique.

DEFINITION: Vertical Asymptotes

The vertical line $x = c$ is a **vertical asymptote** of a function f if $f(x)$ increases in magnitude without bound as x approaches c. Examples of vertical asymptotes appear in Figure 2. The graph of a rational function cannot intersect a vertical asymptote.

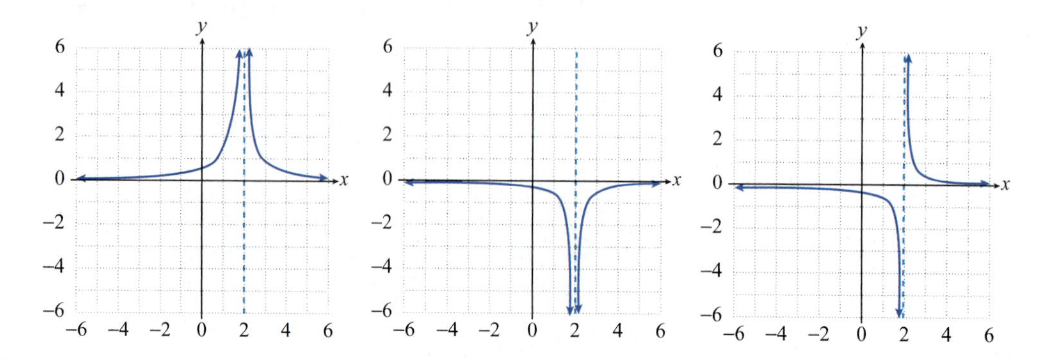

FIGURE 2: Vertical Asymptotes

To understand how vertical asymptotes arise, let's observe what happens to the function $f(x) = \frac{1}{x}$ as x gets closer to 0 from the left and from the right.

x	$f(x) = \frac{1}{x}$
−1	−1
−0.1	−10
−0.01	−100
−0.001	−1000
−0.0001	−10,000
−0.00001	−100,000

x	$f(x) = \frac{1}{x}$
1	1
0.1	10
0.01	100
0.001	1000
0.0001	10,000
0.00001	100,000

TABLE 1: Values of $f(x) = \frac{1}{x}$ as x Approaches 0

We can see that as x gets closer and closer to 0 (where the function is undefined), the value of f increases in magnitude without bound. The graph reflects this, as the curve gets steeper and steeper, never touching the line $x = 0$.

DEFINITION: Horizontal Asymptotes

The horizontal line $y = c$ is a **horizontal asymptote** of a function f if $f(x)$ approaches the value c as $x \to -\infty$ or as $x \to \infty$. Examples of horizontal asymptotes appear in Figure 3. The graph of a rational function may intersect a horizontal asymptote near the origin, but will eventually approach the asymptote from one side only as x increases in magnitude.

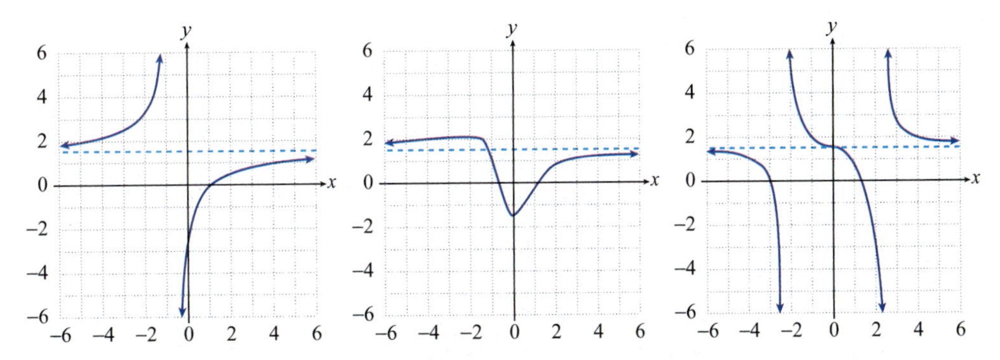

FIGURE 3: Horizontal Asymptotes

> **DEFINITION:** Oblique Asymptotes
>
> A nonvertical, nonhorizontal line may also be an asymptote of a function f. Examples of **oblique** (or **slant**) **asymptotes** appear in Figure 4. Again, the graph of a rational function may intersect an oblique asymptote near the origin, but will eventually approach the asymptote from one side only as $x \to \infty$ or $x \to -\infty$.

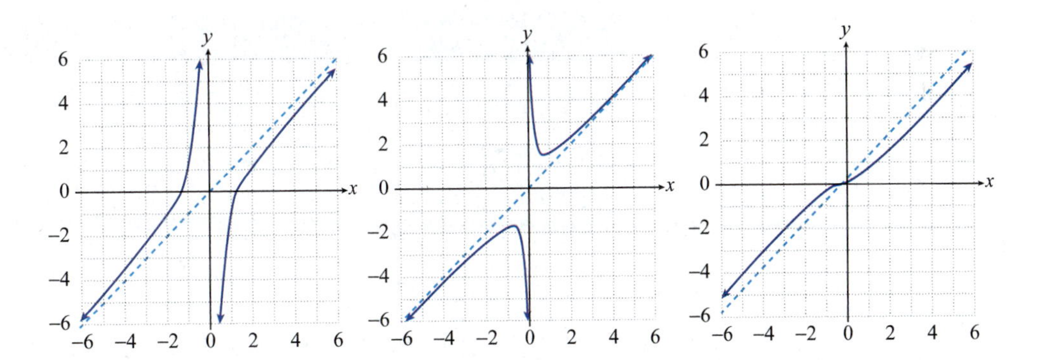

FIGURE 4: Oblique Asymptotes

As Figures 2, 3, and 4 illustrate, the behavior of rational functions with respect to asymptotes can vary considerably. In order to describe the behavior of a given rational function more easily, we have specific asymptote notation.

> **DEFINITION:** Asymptote Notation
>
> The notation $x \to c^-$ is used when describing the behavior of a graph as x approaches the value c from the left (the negative side). The notation $x \to c^+$ is used when describing behavior as x approaches c from the right (the positive side). The notation $x \to c$ is used when describing behavior that is the same on both sides of c.

Figure 5 illustrates how the above notation can be used to describe the behavior of functions.

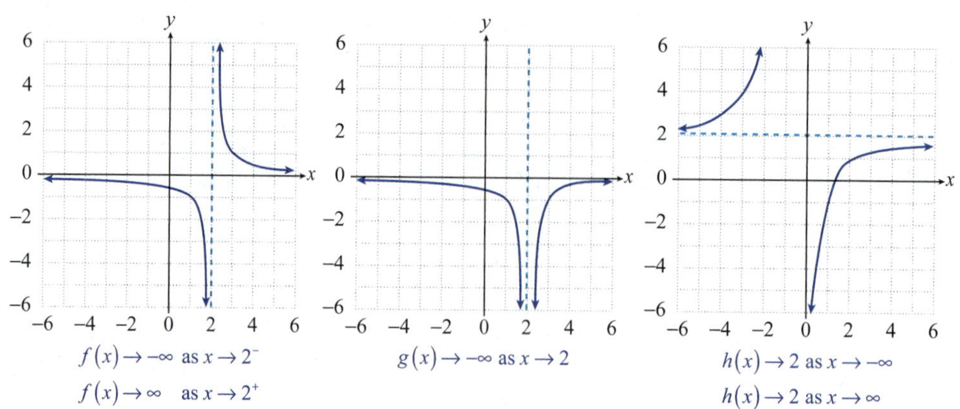

FIGURE 5: Asymptote Notation

TOPIC 2: Vertical Asymptotes

With the above notation and examples as background, we are ready to delve into the details of identifying asymptotes for rational functions.

THEOREM: Equations for Vertical Asymptotes

If the rational function $f(x)=\dfrac{p(x)}{q(x)}$ has been written in reduced form (so that p and q have no common factors), the vertical line $x=c$ is a **vertical asymptote** of f if and only if c is a zero of the polynomial q. In other words, f has vertical asymptotes at the x-intercepts of q.

Note that the numerator of a rational function is irrelevant in locating the vertical asymptotes, assuming that all common factors in the fraction have been canceled. However, if the numerator and denominator share a common factor of $(x-c)$, the value c will be out of the domain of the function, but the line $x=c$ will not be a vertical asymptote.

Example 1: Vertical Asymptotes

Find the domains and the equations for the vertical asymptotes of the following functions.

a. $f(x)=\dfrac{32}{x+2}$ **b.** $g(x)=\dfrac{x^2+1}{x^2+2x-15}$ **c.** $h(x)=\dfrac{x^2-x}{x-1}$

NOTE

We must always find the domain before canceling common factors. Even if a zero is removed from the denominator when finding the reduced form, that value is not part of the domain.

Solution

a. To answer both questions, we need to calculate the zeros of the denominator (which is already in factored form).

$$x+2=0$$
$$x=-2$$

Since the function f is in reduced form, this zero is the only point excluded from the domain. This means the domain of f is $(-\infty,-2)\cup(-2,\infty)$.

Further, we know that f has a vertical asymptote of $x=-2$.

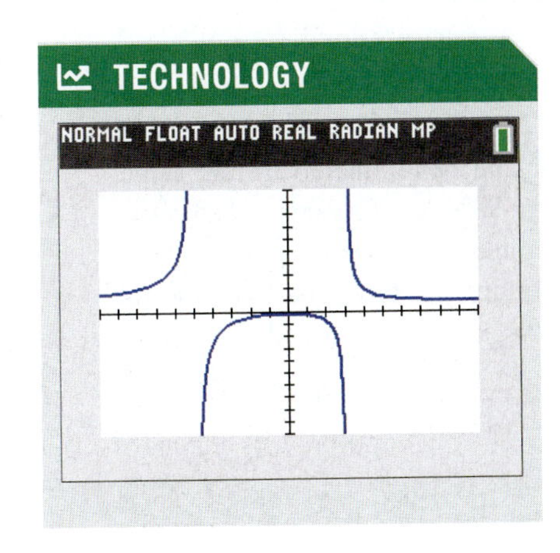

b. In order to determine the domain of the rational function we have to factor the denominator. After making note of the domain, we will look for common factors to cancel, so we may as well factor the numerator, if possible.

$$g(x)=\frac{x^2+1}{x^2+2x-15}=\frac{x^2+1}{(x+5)(x-3)}$$

Since the values -5 and 3 both make the denominator zero, the domain of g is $(-\infty,-5)\cup(-5,3)\cup(3,\infty)$.

The numerator is a sum of two squares and cannot be factored, so the rational function is already in reduced form. This means that the equations of the two vertical asymptotes are $x=-5$ and $x=3$.

c. The denominator is already in factored form, so we can see that its only zero is at $x=1$. Thus, the domain of h is $(-\infty,1)\cup(1,\infty)$.

Now that we have found the domain, we can look for common factors to cancel.

$$h(x)=\frac{x^2-x}{x-1}=\frac{x(x-1)}{x-1}$$
$$=x$$

By canceling the common factor of $(x-1)$, we have the reduced form $h(x)=x$, which applies only for values in the domain of h (all real numbers except for 1). Since the reduced form of h has no denominator, h has no vertical asymptotes.

TOPIC 3: Horizontal and Oblique Asymptotes

To determine horizontal and oblique asymptotes, we are interested in the behavior of a function $f(x)$ as $x\to-\infty$ and as $x\to\infty$. If f is a rational function, f is a ratio of two polynomials p and q, so we can begin by considering the effect p and q have on one another. There are many possibilities, but in order to develop some intuition about rational functions, consider the following examples:

- $f(x)=\dfrac{6x+1}{5x^2-2x+3}$

  Any polynomial of degree greater than or equal to 1 gets larger and larger in magnitude (without bound) as $x\to\pm\infty$, and that is true of the numerator and denominator of this function. The critical point, though, is that the degree of the denominator is larger than the degree of the numerator, and hence the behavior of the denominator dominates. As $x\to-\infty$ and as $x\to\infty$, the increasingly large values of the denominator overpower the large values of the numerator, and the fraction as a whole approaches 0. (Think about the result of dividing a fixed number, no matter how large, by a much larger number.) In symbols, $f(x)\to 0$ as $x\to\pm\infty$.

- $f(x)=\dfrac{6x^2-3x+2}{3x^2+5x-17}$

  In this case, the degrees of the numerator and denominator are the same, and neither dominates the other as $x \to \pm\infty$. We know, however, that in any given polynomial the highest power term dominates the rest far away from the origin. In this case, the numerator "looks" more and more like $6x^2$ and the denominator "looks" more and more like $3x^2$ as $x \to -\infty$ and as $x \to \infty$. As a whole, then, the fraction approaches $\dfrac{6x^2}{3x^2}=\dfrac{6}{3}=2$ far away from the origin. In symbols, $f(x) \to 2$ as $x \to \pm\infty$.

- $f(x)=\dfrac{7x^3+2x-1}{x^2+4x}$

  A comparison of the degrees indicates that the numerator of this fraction dominates the denominator far away from the origin. From our work in Section 5.2, we know that dividing a third-degree polynomial by a second-degree polynomial results in a first-degree quotient with a possible remainder. As we will see, the quotient (whose graph is a straight line since its degree is 1) is the oblique asymptote of *f*.

- $f(x)=\dfrac{3x^5-2x^3+7x^2-1}{4x^3+19x^2-3x+5}$

  If the degree of the numerator is larger than the degree of the denominator by more than 1 (as in this case), the rational function increases in magnitude without bound as $x \to \pm\infty$, but does not approach any straight line asymptote.

THEOREM: Equations for Horizontal and Oblique Asymptotes

Let $f(x)=\dfrac{p(x)}{q(x)}$ be a rational function, where p is an nth-degree polynomial with leading coefficient a_n, q is an mth-degree polynomial with leading coefficient b_m, and $p(x)$ and $q(x)$ have no common factors other than constants. Then the asymptotes of *f* are found as follows.

1. If $n<m$, the horizontal line $y=0$ (the *x*-axis) is the **horizontal asymptote** for *f*.
2. If $n=m$, the horizontal line $y=\dfrac{a_n}{b_m}$ is the **horizontal asymptote** for *f*.
3. If $n=m+1$, the line $y=g(x)$ is an **oblique asymptote** for *f*, where *g* is the quotient polynomial obtained by dividing *p* by *q*. (The remainder polynomial is irrelevant.)
4. If $n>m+1$, there is **no** straight line **horizontal** or **oblique asymptote** for *f*.

Example 2: Horizontal and Oblique Asymptotes

Find the equation for the horizontal or oblique asymptote of the following functions.

a. $f(x)=\dfrac{x^2+1}{x^2+2x-15}$

b. $g(x)=\dfrac{x^3+x^2+2x+2}{x^2+9}$

c. $h(x)=\dfrac{3x^4+10x-7}{x^6+x^5-x^2-1}$

d. $j(x)=\dfrac{2x^4-3x^2+8}{x^2-25}$

NOTE

Always begin by comparing the degrees of the numerator and the denominator.

Solution

a. First, note that the degree of the numerator and denominator of f are both equal to two. This means that the line $y=\dfrac{a_2}{b_2}=1$ is the horizontal asymptote of f.

b. Here, the numerator and denominator have no common factors and the degree of the numerator is one more than the degree of the denominator. So we know g has an oblique asymptote, equal to the quotient polynomial of the numerator and denominator. To find it, we need to perform polynomial division.

$$\begin{array}{r} x+1 \\ x^2+9\overline{\big)\,x^3+x^2+2x+2} \\ -\left(x^3+0x^2+9x\right) \\ \hline x^2-7x+2 \\ -\left(x^2+0x+9\right) \\ \hline -7x-7 \end{array}$$

This tells us that $g(x)=x+1+\dfrac{-7x-7}{x^2+9}$, but we only need the quotient, $x+1$, to find that the equation for the oblique asymptote is $y=x+1$.

c. In this case, the degree of the numerator of h is two *less* than the degree of the denominator. This means that the line $y=0$ is the horizontal asymptote of h.

d. For $j(x)$, the degree of the numerator is two *more* than the degree of the denominator. Thus, j has no horizontal or oblique asymptotes.

TOPIC 4: Graphing Rational Functions

Much of our experience in graph sketching will be useful as we graph rational functions. In addition to the standard steps of identifying the x-intercepts (if any) and y-intercept (if there is one), we will make use of asymptotes when graphing rational functions. The following is a list of suggested steps.

PROCEDURE: Graphing Rational Functions

Given a rational function f,

Step 1: Factor the denominator in order to determine the domain of f. Any points excluded from the domain correspond to holes (undefined points) or vertical asymptotes in the eventual graph.

Step 2: Factor the numerator as well and cancel any common factors. Zeros of the numerator and denominator arising from common linear factors are the x-coordinates of holes in the eventual graph.

Step 3: Examine the remaining linear factors in the denominator to determine the equations for any vertical asymptotes.

Step 4: Compare the degrees of the numerator and denominator to determine if there is a horizontal or oblique asymptote. If so, find its equation.

Step 5: Determine the y-intercept if 0 is in the domain of f.

Step 6: Determine the x-intercepts, if there are any, by setting the numerator of the reduced fraction equal to 0.

Step 7: Plot enough points to determine the behavior of f between x-intercepts and between vertical asymptotes.

Example 3: Graphing Rational Functions

Sketch the graphs of the following rational functions.

a. $f(x)=\dfrac{x^2-x}{x-1}$ **b.** $g(x)=\dfrac{x^2+1}{x^2+2x-15}$ **c.** $h(x)=\dfrac{x^3+x^2+2x+2}{x^2+9}$

Solution

a. The denominator of f is already factored, so we know that the domain of f consists of all real numbers except for $x=1$.

As in Example 1c, we factor the numerator to see if there are any common factors.

$$f(x)=\frac{x^2-x}{x-1}=\frac{x(x-1)}{x-1}$$
$$=x$$

This means that except for at $x=1$, where f is undefined, we have $f(x)=x$. We already know how to graph this function, so the remaining steps are unnecessary. The graph of f is the line $y=x$, excluding the point $(1,1)$, since $x=1$ is not in the domain of f. The result is that a "hole" appears in the graph f at $x=1$.

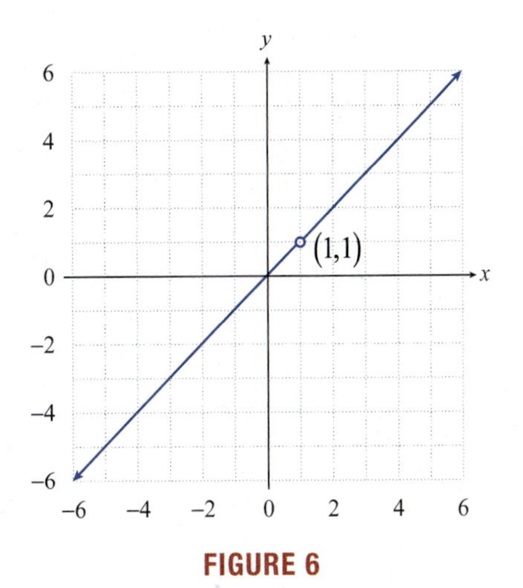

FIGURE 6

b. In Example 1b, we factored the denominator as follows.

$$g(x)=\frac{x^2+1}{x^2+2x-15}=\frac{x^2+1}{(x+5)(x-3)}$$

This means the domain of g excludes the values $x=-5$ and $x=3$. Since the numerator cannot be factored, we also know that the lines $x=-5$ and $x=3$ are vertical asymptotes of g.

Next, we look at the degrees of the numerator and denominator. As we saw in Example 2a, the degrees are the same, so the line $y=1$ is the horizontal asymptote. Plotting the asymptotes provides us a framework for graphing $g(x)$.

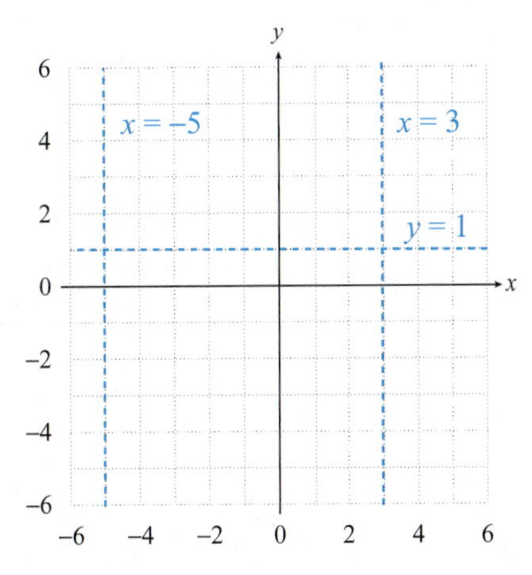

FIGURE 7

Setting $x=0$, we find that the y-intercept lies at $\left(0,-\frac{1}{15}\right)$. Since there is no real solution to the equation $x^2+1=0$, g has no x-intercepts.

Plotting a few points in each region between asymptotes gives us an idea of the general shape of the graph, shown in Figure 8.

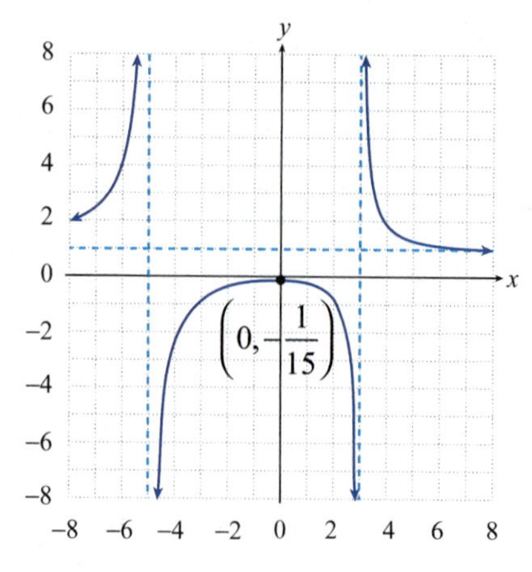

FIGURE 8

Incidentally, while the range of some rational functions, such as $f(x)=\frac{x^2-x}{x-1}$, is easy to determine (you should verify that $(-\infty,1)\cup(1,\infty)$ is the range of f), the range of g is deceptively difficult to determine without the tools of calculus. Although it's not apparent from the graph that we have just sketched, the range of g is $\left(-\infty,\frac{7-\sqrt{65}}{16}\right)\cup\left(\frac{7+\sqrt{65}}{16},\infty\right)$, or approximately $(-\infty,-0.066)\cup(0.941,\infty)$.

NOTE

There really isn't much we can say with honesty about the range of rational functions without the tools of calculus. For instance, even the range of the deceptively simple rational function in Example 3b isn't possible to determine without differentiation to find the critical points. The graphs in Figures 9 and 10 indicate subtle behavior near 0 and 16.

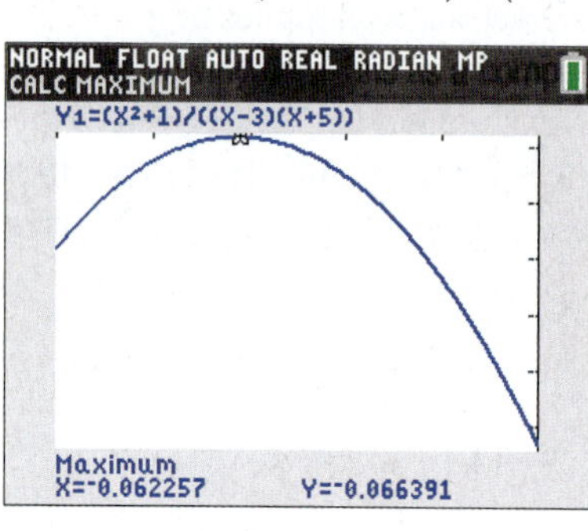

FIGURE 9

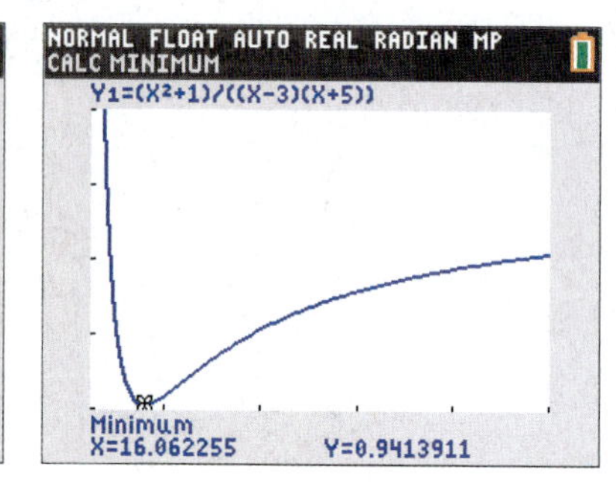

FIGURE 10

c. The denominator of this function cannot be factored, so there are no restrictions on the domain of h. Further, we saw in Example 2b that this function has an oblique asymptote of $y=x+1$.

As usual, we calculate the y-intercept by substituting $x=0$.

$$h(0)=\frac{0^3+0^2+2(0)+2}{0^2+9}=\frac{2}{9}$$

There are different approaches to finding the x-intercepts. Looking at the numerator, we might guess that -1 is a zero of the numerator. A quick calculation confirms this: $(-1)^3+(-1)^2+2(-1)+2=0$. This means we can factor the numerator. Using synthetic or long division, we have $(x+1)(x^2+2)$. Thus, $(-1,0)$ is the only x-intercept.

With the intercepts and a few other plotted points, we obtain the graph of h.

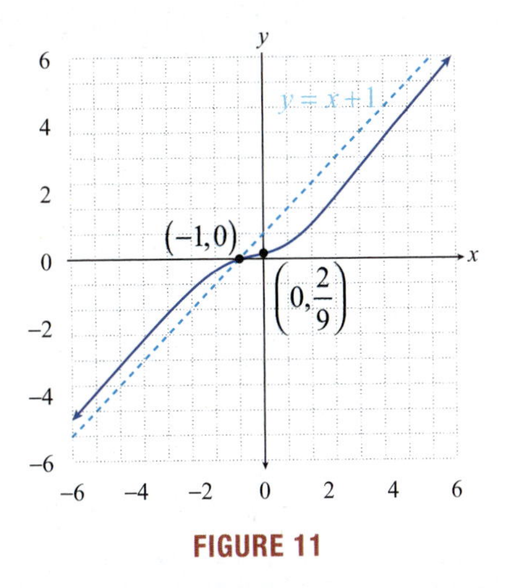

FIGURE 11

TOPIC 5: Solving Rational Inequalities

Now that we have discussed rational functions and equations, we can consider rational inequalities. We will see that solving rational inequalities involves similar work as solving polynomial inequalities. For this discussion, we will assume that all rational functions under consideration are in reduced form.

DEFINITION: Rational Inequalities

A **rational inequality** is any inequality that can be written in the form:

$$f(x) < 0, \ f(x) \leq 0, \ f(x) > 0, \text{ or } f(x) \geq 0,$$

where $f(x)$ is a rational function.

Just as with polynomial inequalities, solving rational inequalities is relatively simple if we have an accurate graph of the function. All we need to do is identify the intervals on which the function is positive and negative, then determine which intervals satisfy the inequality.

However, since graphing rational functions is even more difficult than graphing polynomial functions, we should depend on an algebraic method.

Recall that to solve a polynomial inequality, we made use of the fact that polynomial functions are continuous. This let us apply the *Sign-Test Method*, in which testing a single point on each interval (between x-intercepts) described the sign behavior of the entire function.

Although reduced rational functions are not always continuous, we know exactly where they are discontinuous: at their vertical asymptotes. This means that we can apply the Sign-Test Method, as long as we account for a possible sign change at each vertical asymptote.

PROCEDURE: Solving Rational Inequalities Using the Sign-Test Method

To solve a rational inequality $f(x)<0$, $f(x)\le 0$, $f(x)>0$, or $f(x)\ge 0$, where the rational function $f(x)=\dfrac{p(x)}{q(x)}$ is in reduced form:

Step 1: Find the real zeros of the numerator $p(x)$. These values are the **zeros** of *f*.

Step 2: Find the real zeros of the denominator $q(x)$. These values are the locations of the **vertical asymptotes** of *f*.

Step 3: Place the values from steps 1 and 2 on a number line, splitting it into intervals.

Step 4: Within each interval, select a **test point** and evaluate *f* at that number. If the result is positive, then $f(x)>0$ for all *x* in the interval. If the result is negative, then $f(x)<0$ for all *x* in the interval.

Step 5: Write the **solution set**, consisting of all of the intervals that satisfy the given inequality. If the inequality is not strict (uses $\le$ or $\ge$), then the zeros of *p* are included in the solution set as well. The zeros of *q* are never included in the solution set (as they are not in the domain of *f*).

Example 4: Solving Rational Inequalities

Solve the rational inequality $\dfrac{x^2+1}{x^2+2x-15}>0$.

Solution

Begin by finding the zeros of the numerator and denominator. As we've seen in previous examples, the rational expression can be factored as follows:

$$\frac{x^2+1}{x^2+2x-15}=\frac{x^2+1}{(x+5)(x-3)}$$

Thus, we can see that the numerator has no zeros, while the denominator has zeros at $x=-5$ and $x=3$.

Therefore, we place the values -5 and 3 on a number line.

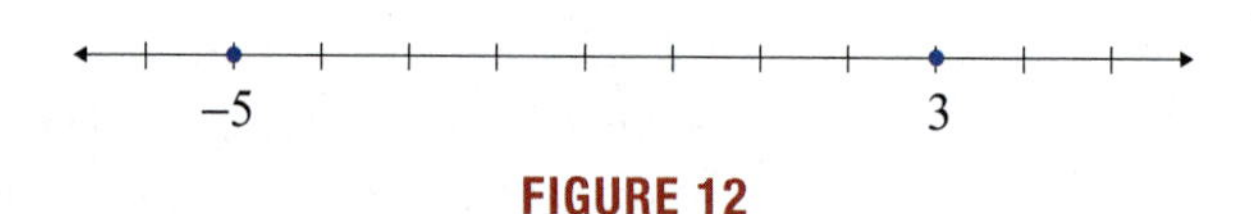

FIGURE 12

This splits the number line into the intervals $(-\infty,-5)$, $(-5,3)$, $(3,\infty)$. We then evaluate $f(x)$ for a test point in each interval.

Interval	Test Point	Evaluate	Result
$(-\infty,-5)$	$x=-6$	$f(-6)=\dfrac{(-6)^2+1}{(-6)^2+2(-6)-15}=\dfrac{37}{9}$	$f(x)>0$ on $(-\infty,-5)$ Positive
$(-5,3)$	$x=0$	$f(0)=\dfrac{(0)^2+1}{(0)^2+2(0)-15}=-\dfrac{1}{15}$	$f(x)<0$ on $(-5,3)$ Negative
$(3,\infty)$	$x=4$	$f(4)=\dfrac{(4)^2+1}{(4)^2+2(4)-15}=\dfrac{17}{9}$	$f(x)>0$ on $(3,\infty)$ Positive

TABLE 2

Then, the solution to the inequality consists of those intervals where $f(x)>0$, which is the set $(-\infty,-5)\cup(3,\infty)$.

CAUTION

It is very important to write rational inequalities in their proper form before solving them. Attempting to simplify them may cause us to neglect asymptotes. Suppose, for example, we tried to solve the inequality

$$\frac{x}{x+2}<3$$

by multiplying through by $x+2$, thus clearing the inequality of fractions. The result would be

$$x<3x+6,$$

a simple linear inequality. We can solve this to obtain $x>-3$, or the interval $(-3,\infty)$. But does this really work?

The number $-\frac{5}{2}$ is in this interval, but

$$\frac{-\frac{5}{2}}{-\frac{5}{2}+2}=5,$$

which is not less than 3. Also, note that -4 solves the inequality, but -4 is not in the interval $(-3,\infty)$.

What went wrong? By multiplying through by $x+2$, we made the assumption that $x+2$ was positive, since we did not worry about possibly reversing the inequality symbol. But we don't know beforehand if $x+2$ is positive or not, since we are trying to solve for the variable x.

To solve this inequality correctly, we have to write it in the standard form of a rational inequality, as shown in Example 5.

Example 5: Solving Rational Inequalities

Solve the rational inequalities $\frac{x}{x+2} < 3$ and $\frac{x}{x+2} \leq 3$.

Solution

Most of the work can be done for both inequalities at the same time.

We begin by subtracting 3 from both sides and then write the left-hand side as a single rational function.

$$\frac{x}{x+2} - 3 < 0 \quad \text{and} \quad \frac{x}{x+2} - 3 \leq 0$$

$$\frac{x}{x+2} - \frac{3(x+2)}{x+2} < 0 \qquad \frac{x}{x+2} - \frac{3(x+2)}{x+2} \leq 0$$

$$\frac{-2x-6}{x+2} < 0 \qquad \frac{-2x-6}{x+2} \leq 0$$

Then factor −2 from the numerator of the fraction and divide both sides by −2 (reversing the inequality symbol) to obtain the simpler inequalities.

$$\frac{x+3}{x+2} > 0 \quad \text{and} \quad \frac{x+3}{x+2} \geq 0$$

Now that the inequalities are in standard form, we can follow the procedure for solving rational inequalities.

The zeros of the numerator and denominator are −3 and −2, respectively. Placing these values on a number line,

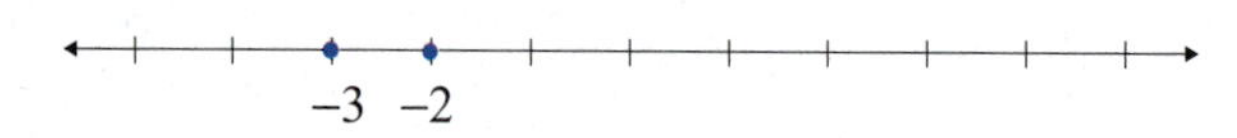

FIGURE 13

we have the intervals $(-\infty,-3)$, $(-3,-2)$, and $(-2,\infty)$.

Interval	Test Point	Evaluate	Result
$(-\infty,-3)$	$x=-4$	$f(-4)=\frac{(-4)+3}{(-4)+2}$ $=\frac{1}{2}$	$f(x)>0$ on $(-\infty,-3)$ Positive
$(-3,-2)$	$x=-2.5$	$f(-2.5)=\frac{(-2.5)+3}{(-2.5)+2}$ $=-1$	$f(x)<0$ on $(-3,-2)$ Negative
$(-2,\infty)$	$x=0$	$f(0)=\frac{(0)+3}{(0)+2}$ $=\frac{3}{2}$	$f(x)>0$ on $(-2,\infty)$ Positive

TABLE 3

The final step is to evaluate which intervals satisfy each inequality. The solution to the first inequality is the union of the two intervals where f is positive.

$$(-\infty,-3)\cup(-2,\infty)$$

For the second inequality, we have to decide which endpoints to include. We include $x=-3$, since this is a zero of the rational function, but we do not include $x=-2$, since the value is not in the domain of f. Thus, the solution to the second inequality is

$$(-\infty,-3]\cup(-2,\infty).$$

5.5 EXERCISES

PRACTICE

Find equations for the vertical asymptotes, if any, for each of the following rational functions. See Example 1.

1. $f(x)=\dfrac{5}{x-1}$
2. $f(x)=\dfrac{x^2+3}{x+3}$
3. $f(x)=\dfrac{x^2-4}{x+2}$
4. $f(x)=\dfrac{-3x+5}{x-2}$
5. $f(x)=\dfrac{3x^2+1}{x-2}$
6. $f(x)=\dfrac{x^2+2x}{x+1}$
7. $f(x)=\dfrac{x^2-4}{2x-x^2}$
8. $f(x)=\dfrac{x+2}{x^2-9}$
9. $f(x)=\dfrac{x^2-2x-3}{2x^2-5x-3}$
10. $f(x)=\dfrac{2x^2+2x-4}{x^2+2x+1}$
11. $f(x)=\dfrac{x^3-27}{x^2+5}$
12. $f(x)=\dfrac{x^2+5}{x^3-27}$
13. $f(x)=\dfrac{x^2-1}{x^2-8x+7}$
14. $f(x)=\dfrac{2x^2+7x-14}{2x^2+7x-15}$
15. $f(x)=\dfrac{x^3-6x^2+11x-6}{x^3+8}$
16. $f(x)=\dfrac{x^2-2x-15}{x-5}$
17. $f(x)=\dfrac{x^2-16}{x^2-4}$
18. $f(x)=\dfrac{x^2+4x+4}{x^2+x-2}$

Find equations for the horizontal or oblique asymptotes, if any, for each of the following rational functions. See Example 2.

19. $f(x)=\dfrac{5}{x-1}$
20. $f(x)=\dfrac{x^2+3}{x+3}$

21. $f(x)=\dfrac{x^4-4}{x^2+2}$

22. $f(x)=\dfrac{x^2-4}{2x-x^2}$

23. $f(x)=\dfrac{x+2}{x^2-9}$

24. $f(x)=\dfrac{x^2-2x-3}{2x^2-5x-3}$

25. $f(x)=\dfrac{2x^2+2x-4}{x^2+2x+1}$

26. $f(x)=\dfrac{-3x+5}{x-2}$

27. $f(x)=\dfrac{3x^2+1}{x-2}$

28. $f(x)=\dfrac{x^3-27}{x^2+5}$

29. $f(x)=\dfrac{x^2+5}{x^3-27}$

30. $f(x)=\dfrac{x^2+2x}{x+1}$

31. $f(x)=\dfrac{x^2-81}{x^3+7x-12}$

32. $f(x)=\dfrac{x^3-3x^2+2x}{x-7}$

33. $f(x)=\dfrac{x^2-9x+4}{x+2}$

34. $f(x)=\dfrac{-x^5+2x^2}{5x^5+3x^3-7}$

35. $f(x)=\dfrac{5x^2-x+12}{x-1}$

36. $f(x)=\dfrac{2x^2-5x+6}{x-3}$

Sketch the graphs of the following rational functions, making use of your work in the problems above and additional information about intercepts and any other points that may be useful. See Example 3.

37. $f(x)=\dfrac{5}{x-1}$

38. $f(x)=\dfrac{x^2+3}{x+3}$

39. $f(x)=\dfrac{x^2-4}{x+2}$

40. $f(x)=\dfrac{x^2-4}{2x-x^2}$

41. $f(x)=\dfrac{x+2}{x^2-9}$

42. $f(x)=\dfrac{x^2-2x-3}{2x^2-5x-3}$

43. $f(x)=\dfrac{2x^2+2x-4}{x^2+2x+1}$

44. $f(x)=\dfrac{-3x+5}{x-2}$

45. $f(x)=\dfrac{3x^2+1}{x-2}$

46. $f(x)=\dfrac{x^3-27}{x^2+5}$

47. $f(x)=\dfrac{x^2+5}{x^3-27}$

48. $f(x)=\dfrac{x^2+2x}{x+1}$

For each graph, find any **a.** vertical asymptotes, **b.** horizontal asymptotes, **c.** oblique asymptotes, **d.** visible x-intercepts, or **e.** visible y-intercepts.

49.

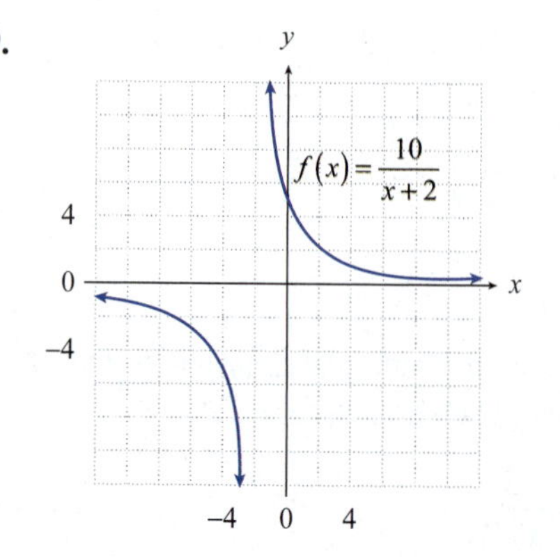

50.

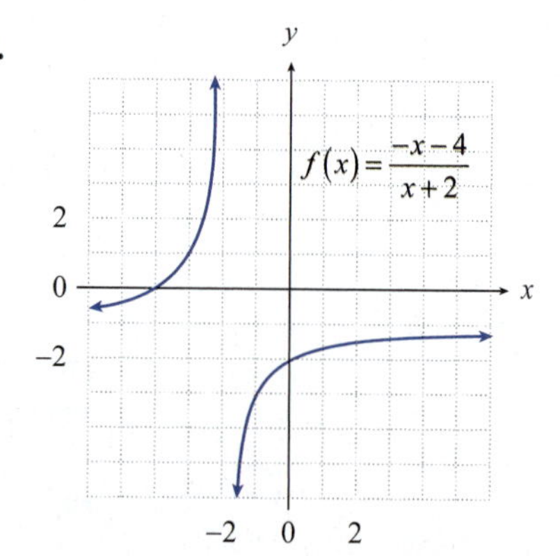

51.

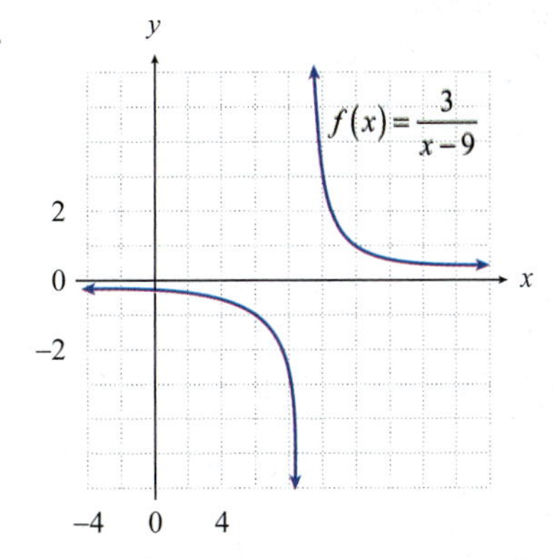

52.

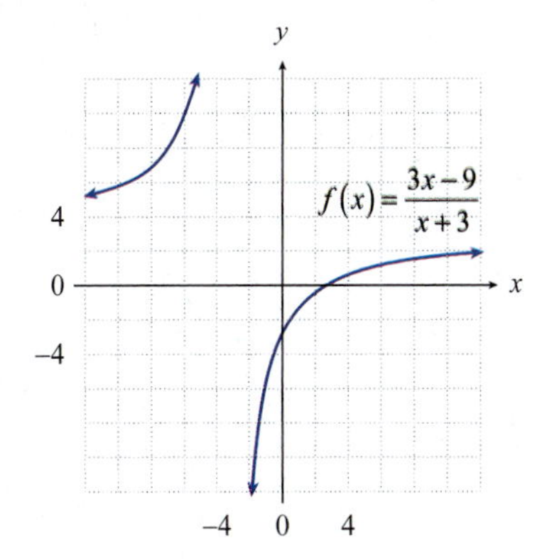

53.

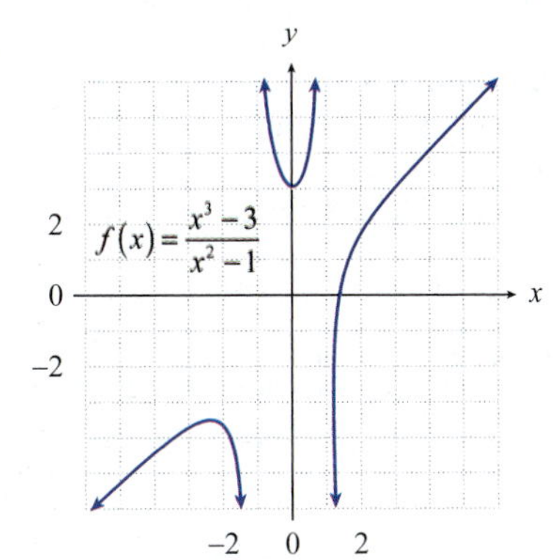

54.

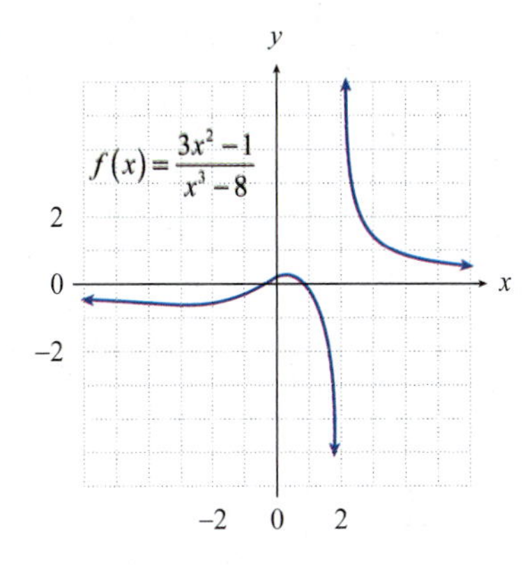

55.

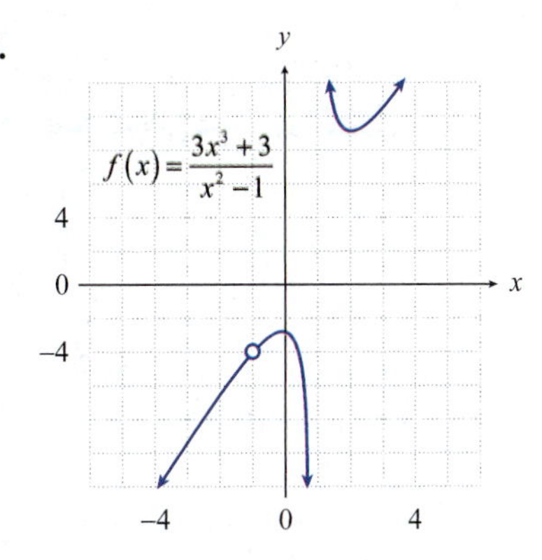

56.

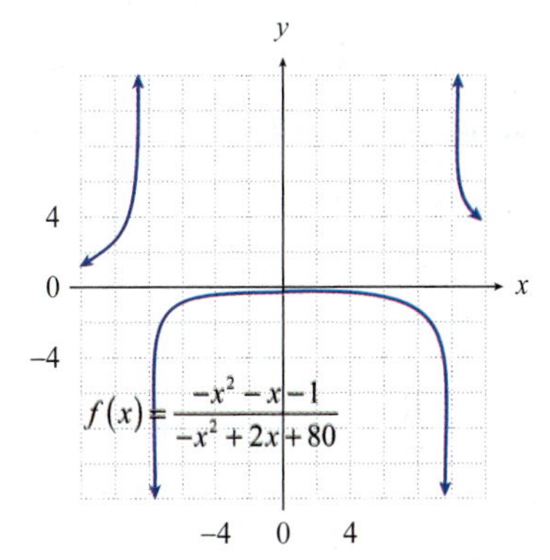

Solve the following rational inequalities. See Examples 4 and 5.

57. $2x < \frac{4}{x+1}$

58. $\frac{5}{x-2} \geq \frac{3x}{x-2}$

59. $\frac{5}{x-2} > \frac{3}{x+2}$

60. $\frac{x}{x^2-x-6} \leq \frac{-1}{x^2-x-6}$

61. $\frac{x}{x^2-x-6} \leq \frac{-2}{x^2-x-6}$

62. $x > \frac{1}{x}$

63. $\frac{4}{x-3} \leq \frac{4}{x}$

64. $\frac{x-7}{x-3} \geq \frac{x}{x-1}$

65. $\frac{x}{x^2+3x+2} > \frac{1}{x^2+3x+2}$

66. $\frac{1}{x-4} \geq \frac{1}{x+1}$

67. $\frac{x}{x+1} \geq \frac{x+1}{x}$

68. $\frac{x}{x^2-2x-3} > \frac{3}{x^2-2x-3}$

APPLICATIONS

69. April raises a species of aquarium fish, and the total number of fish she has follows the formula

$$p(t)=\frac{200t}{t+1},$$

where $t \geq 0$ represents the number of months since she began.

a. Sketch the graph of $p(t)$ for $t \geq 0$.

b. What happens to April's fish population in the long run?

70. If an object is placed a distance x from a lens with a focal length of f, the image of the object will appear a distance y on the opposite side of the lens, where x, f, and y are related by the equation $\frac{1}{x}+\frac{1}{y}=\frac{1}{f}$.

a. Express y as a function of x and f.

b. Graph your function for a lens with a focal length of 30 mm $(f = 30)$. What happens to y as the distance x increases?

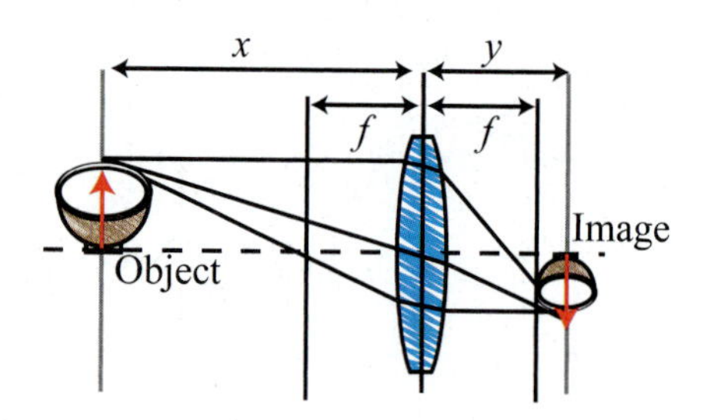

71. At t minutes after injection, the concentration (in mg/L) of a certain drug in the bloodstream of a patient is given by the formula

$$c(t)=\frac{20t}{t^2+1}.$$

a. Sketch the graph of $c(t)$ for $t \geq 0$.

b. What happens to the concentration of the drug in the long run?

CHAPTER 5 PROJECT

Polynomial Functions

Ace Electric Motorcycles is a new business that has just finished the design phase for its first product, and they're ready to begin production. The monthly costs for their staff, utilities, and plant lease is projected to be \$729,000, and the cost to produce each motorcycle will be \$4500. Their business plan projects that the revenue from selling x motorcycles will be $-3x^2 + 10{,}000x$.

1. Given that profit is equal to revenue minus cost, what is their profit for producing and selling x motorcycles per month?

2. What monthly interval of production will be profitable for the company?

3. Given that revenue is equal to the number of items sold times the price per item, what price range for the motorcycle corresponds to the profitable production interval?

4. What number of motorcycles produced and sold each month will maximize their profit?

5. Their executive team is prepared for monthly fixed costs to be as much as \$300,000 higher. In that worst case scenario, what does their monthly profitable production interval become?

CHAPTER 5 REVIEW EXERCISES

Section 5.1

Verify that the given values of x solve the corresponding polynomial equations.

1. $4x^3 - 5x^2 = -3x + 18;\ x = 2$
2. $x^2 - 6x = -13;\ x = 3 + 2i$
3. $x^3 + x = 6x^2 - 164;\ x = 5 - 4i$
4. $x^3 + (1+4i)x = (7-2i)x^2 - 2i + 36;\ x = -2i$

Solve the following polynomial equations by factoring and/or using the quadratic formula, making sure to identify all the solutions.

5. $x^4 - 7x^2 + 10 = 0$
6. $x^5 - x^3 - 2x = 0$
7. $x^4 + 4 = 4x^2$
8. $6x^2 + 8x = -x^3$
9. $x^4 + x^3 = x^2$
10. $x^2 + 4x + 7 = 0$

For each of the following polynomial functions, determine the behavior of its graph as $x \to \pm\infty$ and identify the x- and y-intercepts. Use this information to then sketch the graph of each polynomial.

11. $f(x) = (x+2)(x-1)(x-3)$
12. $f(x) = (x-2)^2(x+1)^2$
13. $g(x) = x^2 - 5x + 4$
14. $h(x) = -x^3 - 7x^2 - 10x$

Solve the following polynomial inequalities.

15. $2x^2 + 15 \le 11x$
16. $(x-3)^2(x+1)^2 > 0$
17. $(x-4)(x+2)(x^2-1) \le 0$
18. $x^3 - 2x^2 - 8x \ge 0$
19. $x^2(x-2)(1-x) < 0$
20. $-3x^2 + 7x - 2 > 0$
21. A manufacturer has determined that the revenue from the sale of x video games is given by $r(x) = -x^2 + 12x$. The cost of producing x video games is $C(x) = 120 - 22x$. Given that profit is revenue minus cost, what value(s) for x will give the company a nonnegative profit?

Section 5.2

Use polynomial long division to rewrite each of the following fractions in the form $q(x) + \frac{r(x)}{d(x)}$, where $d(x)$ is the denominator of the original fraction, $q(x)$ is the quotient, and $r(x)$ is the remainder.

22. $\dfrac{8x^4 - 6x^3 + 2x^2 + 3x + 4}{2x^2 - 1}$
23. $\dfrac{11x^2 + 2x - 5}{x - 3}$
24. $\dfrac{x^4 - 3x^2 + x - 8}{x^2 + 3x + 2}$
25. $\dfrac{2x^5 - 4x^3 - x^2 + x - 2}{x^2 - x}$
26. $\dfrac{2x^3 + ix^2 - 12x - 4 + i}{2x + i}$

Use synthetic division to determine if the given value for c is a zero of the corresponding polynomial. If not, determine $p(c)$.

27. $p(x) = 6x^5 - 23x^4 - 95x^3 + 70x^2 + 204x - 72;\ c = 1$

28. $p(x) = 48x^4 + 10x^3 - 51x^2 - 10x + 3;\ c = \frac{1}{6}$

29. $p(x) = 18x^5 - 87x^4 + 110x^3 - 28x^2 - 16x + 3;\ c = \frac{2}{3}$

Use synthetic division to rewrite each of the following fractions in the form $q(x) + \frac{r(x)}{d(x)}$, where $d(x)$ is the denominator of the original fraction, $q(x)$ is the quotient, and $r(x)$ is the remainder.

30. $\frac{x^4 - 2x^3 - x^2 + x - 21}{x - 3}$

31. $\frac{-x^4 - x^3 - x^2 + 2x + 69}{x + 3}$

32. $\frac{x^5 + 2x^4 + 3x^3 + 6x^2 - 5x + 13}{x + 2}$

33. $\frac{-x^4 + 8x^3 - 6x^2 - 4x + 2}{x - 1}$

34. $\frac{x^4 + (4 - 2i)x^3 - (1 + 8i)x^2 + (3 + 2i)x - 6i}{x - 2i}$

Construct a polynomial function with the stated properties.

35. Second-degree, zeros of -2 and 6, and goes to ∞ as $x \to \infty$.

36. Fourth-degree and a single x-intercept of -4 and y-intercept $(0, 128)$.

37. Third-degree, zeros of ± 2 and 3 and passing through the point $(4, 24)$.

Section 5.3

List all of the potential rational zeros of the following polynomials. Then use polynomial division and the quadratic formula, if necessary, to identify the actual zeros.

38. $f(x) = x^4 + 3x^3 - 3x^2 - 11x - 6$

39. $g(x) = 2x^3 - 11x^2 + 18x - 9$

40. $h(x) = 2x^3 + 2x^2 - 9x + 9$

41. $p(x) = x^4 + 8x^3 + 22x^2 + 24x + 9$

Using the Rational Zero Theorem or your answers to the preceding problems, solve the following polynomial equations.

42. $2x^4 - 6x^2 = -6x^3 + 22x + 12$

43. $2x^3 - 9x^2 + 18x = 9 + 2x^2$

44. $2x^3 + 9 = 9x - 2x^2$

45. $x^4 - x^5 = -x^5 - 8x^3 - 22x^2 - 24x - 9$

Use Descartes' Rule of Signs to determine the possible numbers of positive and negative real zeros of each of the following polynomials.

46. $f(x) = 2x^4 - 3x^3 - x^2 + 3x + 10$

47. $g(x) = x^6 - 4x^5 - 2x^4 + x^3 - 6x^2 - 11x + 6$

Use synthetic division to identify integer upper and lower bounds of the real zeros of the following polynomials.

48. $f(x) = 2x^3 - 11x^2 + 3x + 36$

49. $g(x) = 4x^3 - 16x^2 - 79x - 35$

Using your answers to the preceding problems, polynomial division, and the quadratic formula, if necessary, find all of the zeros of the following polynomials.

50. $f(x) = 2x^3 - 11x^2 + 3x + 36$

51. $g(x) = 4x^3 - 16x^2 - 79x - 35$

Use the Intermediate Value Theorem to show that each of the following polynomials has a real zero between the indicated values.

52. $f(x) = 2x^4 - 6x^3 + x - 5;\ -2$ and 0

53. $f(x) = -x^3 + 3x^2 + x - 3;\ 2$ and 4

Find all of the real zeros of the following functions.

54. $f(x) = x^4 - 5x^3 + 5x^2 + 5x - 6$

55. $g(x) = x^3 - 4x^2 + 9x - 36$

56. $f(x) = x^3 + 6x^2 + 11x + 6$

57. $f(x) = x^3 - 7x^2 + 13x - 3$

Solve the following equations.

58. $x^4 - 2x^3 + 10x^2 = 9(2x - 1)$

59. $2x^3 = 7x^2 - 4x - 4$

60. $-8 = 3x^3 + 4x^2 + 6x$

Section 5.4

Throughout these exercises, a graphing utility may be helpful in identifying zeros and in checking your graphs.

Sketch the graph of each factored polynomial.

61. $f(x) = (x+4)^2(x-1)$

62. $g(x) = x(x-3)(x+4)^3$

Factor each of the following polynomials completely, and then sketch the graph of each one.

63. $f(x) = x^3 - 3x^2 + x - 3$

64. $f(x) = x^5 - x^4 - 2x^3 - x^2 + x + 2$

Solve each polynomial equation.

65. $3x^5 + x^4 + 5x^3 = x^2 + 28x + 20$

66. $8x^5 + 12x^4 - 18x^3 - 35x^2 = 18x + 3$

67. $x^5 + 3x^4 + 3x^3 + 9x^2 = 4(x+3)$

Factor each of the following polynomials completely, making use of the given zero.

68. $f(x)=14x^4-109x^3+296x^2-321x+70;\ 2+i$ is a zero

69. $f(x)=x^4-5x^3+19x^2-125x-150;\ -5i$ is a zero

70. $f(x)=2x^4+3x^3-7x^2+8x+6;\ 1+i$ is a zero

71. $f(x)=4x^3+10x^2-x+15;\ -3$ is a zero

Construct polynomial functions with the stated properties.

72. Fourth-degree, only real coefficients, $\frac{1}{2}$ and $1+2i$ are two of the zeros, y-intercept is -30, leading coefficient is 2.

73. Fifth-degree, only real coefficients, -1 is a zero of multiplicity 3, $\sqrt{6}$ is a zero, y-intercept is -6, leading coefficient is 1.

74. Fifth-degree, only real coefficients, 1 is a zero of multiplicity 3, $\sqrt{3}$ is a zero, y-intercept is 3, leading coefficient is 1.

Section 5.5

Find equations for the vertical asymptotes, if any, for each of the following rational functions.

75. $f(x)=\dfrac{4}{2x-5}$

76. $f(x)=\dfrac{x^2-3x+2}{x-1}$

77. $f(x)=\dfrac{x^2-1}{x-x^2}$

78. $f(x)=\dfrac{x^2-x-6}{x^2-6x+9}$

Find equations for the horizontal or oblique asymptotes, if any, for each of the following rational functions.

79. $f(x)=\dfrac{2x^3+5x^2-1}{x^2-2x}$

80. $f(x)=\dfrac{x^2-x+8}{3x^2-7}$

81. $f(x)=\dfrac{x^2-9}{x+3}$

82. $f(x)=\dfrac{x^2+2x-3}{(x+1)^3}$

Sketch the graphs of the following rational functions.

83. $f(x)=\dfrac{2x}{x+1}$

84. $f(x)=\dfrac{4x^2}{x^2+3x}$

85. $f(x)=\dfrac{x^2+2}{x+2}$

86. $f(x)=\dfrac{x+1}{x^2-4}$

Solve the following rational inequalities.

87. $\dfrac{7}{x+3}\geq\dfrac{2x}{x+3}$

88. $\dfrac{x}{x^2-5x+6}\leq\dfrac{3}{x^2-5x+6}$

89. $\dfrac{x-4}{x+3}<\dfrac{x}{x-2}$

90. $\dfrac{x-2}{x+3}<2$

NIKE

CHAPTER 6

Exponential and Logarithmic Functions

SECTIONS

WHAT IF ...

What if you had to predict the amount of time it will take for the Earth's population to reach 20 billion (ignoring the effect of limited resources on population growth)? How would you perform this calculation?

By the end of this chapter, you'll be able to apply the skills regarding exponential and logarithmic functions to estimate the behavior of growing populations, interest-earning accounts, and decaying elements. On page 494 you'll encounter a problem like the population problem given above. You'll master this type of problem using tools such as the summary of logarithmic properties, found on page 485.

John Napier

Introduction

This chapter introduces two entirely new classes of functions, both of which are of enormous importance in many natural and man-made contexts. As we will see, the two classes of functions are inverses of one another, though historically exponential and logarithmic functions were developed independently and for unrelated reasons.

We will begin with a study of exponential functions. These are functions in which the variable appears in the exponent while the base is a constant, just the opposite of what we have seen so often in the individual terms of polynomials. As with many mathematical concepts, the argument can be made that exponential functions exist in the natural world independently of mankind and that consequently mathematicians have done nothing more than observe (and formalize) what there is to be seen. Exponential behavior is exhibited, for example, in the rate at which radioactive substances decay, in how the temperature of an object changes when placed in an environment held at a constant temperature, and in the fundamental principles of population growth. But exponential functions also arise in discussing such man-made phenomena as the growth of investment funds.

In fact, we will use the formula for compound interest to motivate the introduction of the most famous and useful base for exponential functions, the irrational constant e (the first few digits of which are $2.718281828459\ldots$). The Swiss mathematician Leonhard Euler (1707–1783), who identified many of this number's unique properties (such as the fact that $e=1+\frac{1}{1}+\frac{1}{1\cdot 2}+\frac{1}{1\cdot 2\cdot 3}+\cdots$), was one of the first to recognize the fundamental importance of e and in fact is responsible for the choice of the letter e as its symbol. The constant e also arises very naturally in the context of calculus, but that discussion must wait for a later course.

Logarithms are inverses, in the function sense, of exponentials, but historically their development was for very different reasons. Much of the development of logarithms is due to John Napier (1550–1617), a Scottish writer of political and religious tracts and an amateur mathematician and scientist. It was the goal of simplifying computations of large numbers that led him to devise early versions of logarithmic functions, to construct what today would be called tables of logarithms, and to design a prototype of what would eventually become a slide rule. Today it is not necessary to resort to logarithms in order to carry out difficult omputations, but the properties of logarithms make them invaluable in solving certain equations. Further, the fact that they are inverses of exponential functions means they have just as much a place in the natural world.

6.1 EXPONENTIAL FUNCTIONS AND THEIR GRAPHS

TOPICS

1. Characteristics of exponential functions
2. Graphing exponential functions
3. Solving elementary exponential equations

TOPIC 1: Characteristics of Exponential Functions

Transcendental Functions

Exponential functions are the first examples in this text of *transcendental* functions, so called because their definitions "transcend" or go beyond basic algebraic operations. The formal definition of a transcendental function is a function that does not satisfy a specific type of polynomial equation, while an *algebraic* function is one that does. Polynomial functions, root functions, and rational functions are all examples of algebraic functions, while logarithmic, trigonometric, and hyperbolic functions are further examples of transcendental functions.

We have studied many functions in which a variable is raised to a constant power, including polynomial functions such as $f(x) = x^3 + 2x^2 - 1$, radical functions like $g(x) = x^{\frac{1}{3}}$, and rational functions such as $h(x) = x^{-2}$.

An *exponential function* is a function in which a constant is raised to a variable power. Exponential functions are extremely important because of the large number of natural situations in which they arise. Examples include radioactive decay, population growth, compound interest, spread of epidemics, and rates of temperature change.

DEFINITION: Exponential Functions

Let a be a fixed, positive real number not equal to 1. The **exponential function with base a** is the function

$$f(x) = a^x.$$

Why do we have the restrictions $a > 0$, $a \neq 1$? The base of the exponent can't be negative, since a^x would not be real for many values of x. For example, if $a = -1$ and $x = \frac{1}{2}$, a^x is not real.

If we let $a = 1$, we don't have a problem producing real numbers. Instead, the "exponential" function turns out to be constant. Recall that for all values of x, $1^x = 1$. For this reason, 1^x is considered a constant function, not an exponential function, and should always be written in its simplified form, 1.

Note that for any positive constant a, a^x is defined for all real numbers x. Consequently, the domain of $f(x) = a^x$ is the set of real numbers. What about the range of f? Since a is positive, we know that a^x must be positive. We will see that the range of all exponential functions is the set of all positive real numbers.

Recall that if a is any nonzero number, a^0 is defined to be 1. This means the y-intercept of any exponential function, regardless of the base, is the point $(0,1)$. Beyond this, exponential functions fall into two classes, depending on whether a lies between 0 and 1 or if a is larger than 1.

Consider the calculations shown in Table 1 for two simple exponential functions.

x	$f(x)=\left(\frac{1}{3}\right)^x$	$g(x)=2^x$
-2	$f(-2)=9$	$g(-2)=\frac{1}{4}$
-1	$f(-1)=3$	$g(-1)=\frac{1}{2}$
0	$f(0)=1$	$g(0)=1$
1	$f(1)=\frac{1}{3}$	$g(1)=2$
2	$f(2)=\frac{1}{9}$	$g(2)=4$

TABLE 1

Note that the values of f decrease as x increases while the values of g do just the opposite. We say that f is an example of a *decreasing* function and g is an example of an *increasing* function. If we plot these points and then fill in the gaps with a smooth curve, we get the graphs of f and g that appear in Figure 1.

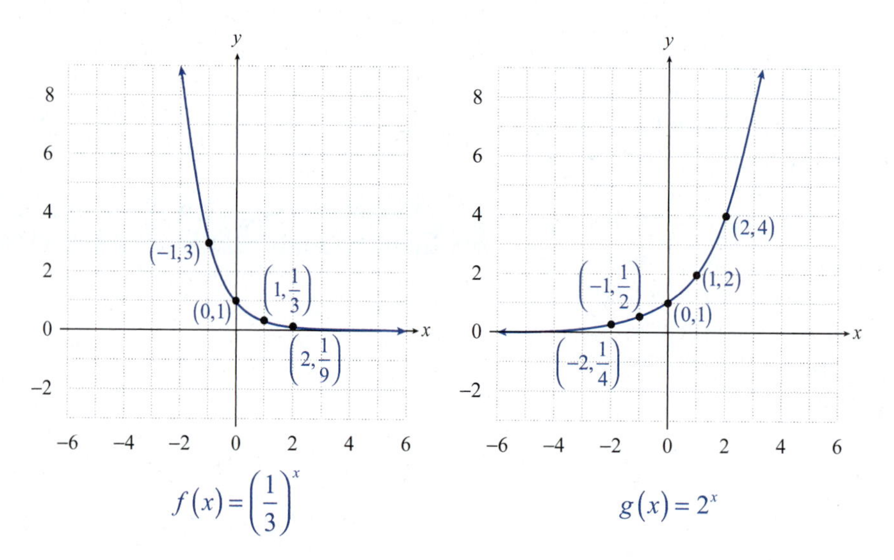

FIGURE 1: Two Exponential Functions

The graphs in Figure 1 suggest that the range of an exponential function is $(0,\infty)$, and that is indeed the case. Note that the base does not matter: the range of a^x is the positive real numbers for any allowable base (that is, for any positive a not equal to 1).

PROPERTIES: Behavior of Exponential Functions

Given a positive real number a not equal to 1, the function $f(x) = a^x$ is

- a **decreasing function** if $0 < a < 1$, with $f(x) \to \infty$ as $x \to -\infty$ and $f(x) \to 0$ as $x \to \infty$, and is
- an **increasing function** if $a > 1$, with $f(x) \to 0$ as $x \to -\infty$ and $f(x) \to \infty$ as $x \to \infty$.

In either case, the point $(0,1)$ lies on the graph of f, the domain of f is the set of real numbers, and the range of f is the set of positive real numbers.

TOPIC 2: Graphing Exponential Functions

Given that all exponential functions take one of two basic shapes (depending on whether a is less than 1 or greater than 1), they are relatively easy to graph. Plotting a few points, including the y-intercept of $(0,1)$, will provide an accurate sketch.

Example 1: Graphing Exponential Functions

Sketch the graphs of the following exponential functions.

a. $f(x) = 3^x$ **b.** $g(x) = \left(\frac{1}{2}\right)^x$

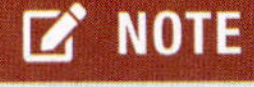

NOTE

Plugging in $x = -1$, $x = 0$, and $x = 1$ will produce a good idea of the shape of the graph.

Solution

In both cases, we plot 3 points by plugging in $x = -1$, $x = 0$, and $x = 1$. We then connect the points with a smooth curve.

a.

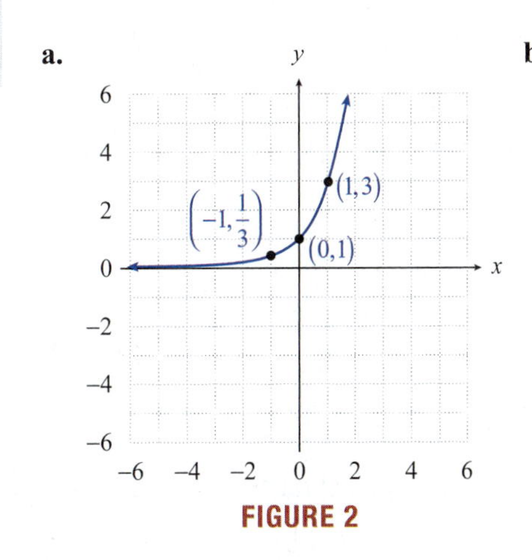

FIGURE 2

b.

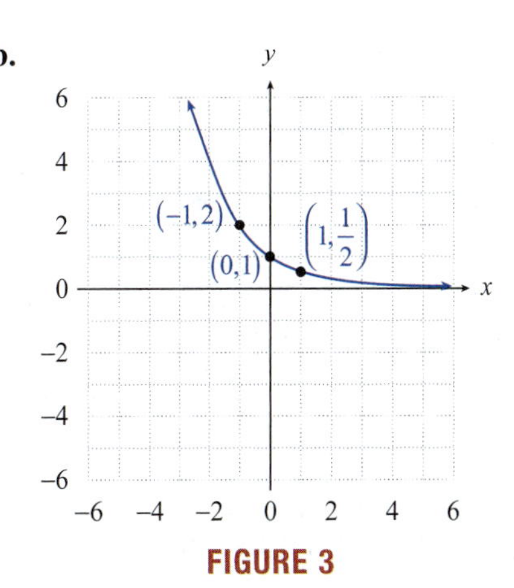

FIGURE 3

An exponential function, like any function, can be transformed in ways that result in the graph being shifted, reflected, stretched, or compressed. It often helps to graph the base function before trying to graph the transformed one.

Example 2: Graphing Exponential Functions

Sketch the graphs of each of the following functions. State their domain and range.

a. $f(x)=\left(\frac{1}{2}\right)^{x+3}$ **b.** $g(x)=-3^x+1$ **c.** $h(x)=2^{-x}$

NOTE

For each example, first graph the base function, then apply any transformations.

Solution

a.

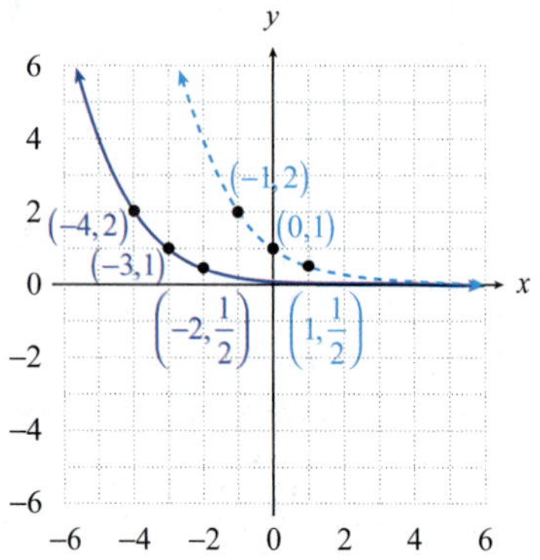

First, draw the graph of the function $\left(\frac{1}{2}\right)^x$, as in Example 1b.

Then, since x has been replaced by $x+3$, shift the graph to the left by 3 units.

FIGURE 4

Domain $=(-\infty,\infty)$, Range $=(0,\infty)$

b.

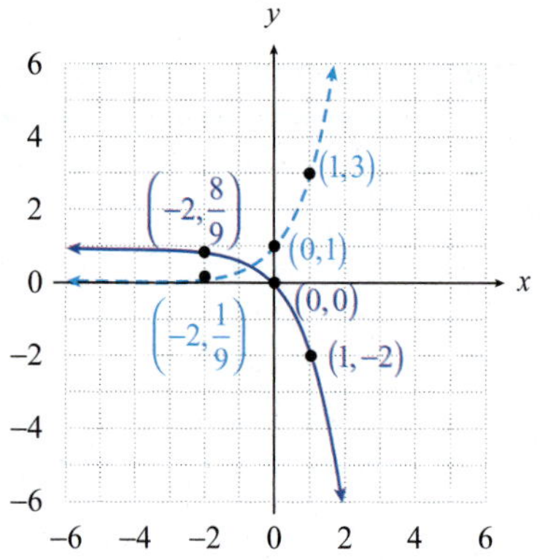

Begin with the graph of 3^x shown as a dashed curve.

The effect of multiplying a function by -1 is to reflect the graph with respect to the x-axis.

Following this, the second transformation of adding 1 to the function causes a vertical shift of the graph. The solid curve is the graph of $g(x)=-3^x+1$.

FIGURE 5

Domain $=(-\infty,\infty)$, Range $=(-\infty,1)$

c.

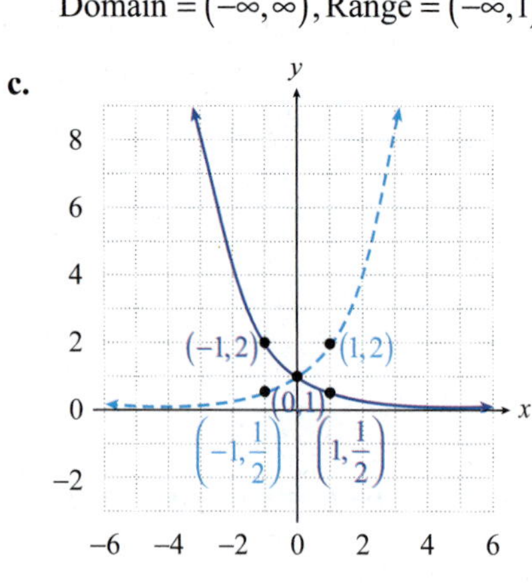
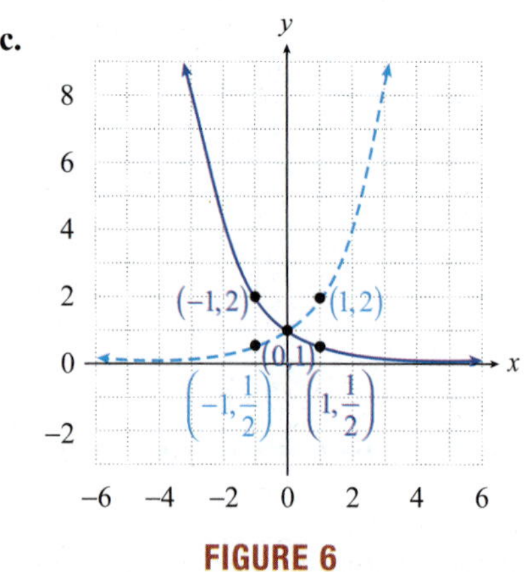

FIGURE 6

Domain $=(-\infty,\infty)$, Range $=(0,\infty)$

Begin by graphing the base function 2^x, shown as a dashed curve.

For h, x has been replaced by $-x$, so we reflect the graph of 2^x across the y-axis to obtain the graph of h, shown as a solid curve.

Note that this graph is also the graph of $\left(\frac{1}{2}\right)^x$. Using properties of exponents is another way to think about this problem:

$$2^{-x}=\left(2^{-1}\right)^x=\left(\frac{1}{2}\right)^x.$$

TOPIC 3: Solving Elementary Exponential Equations

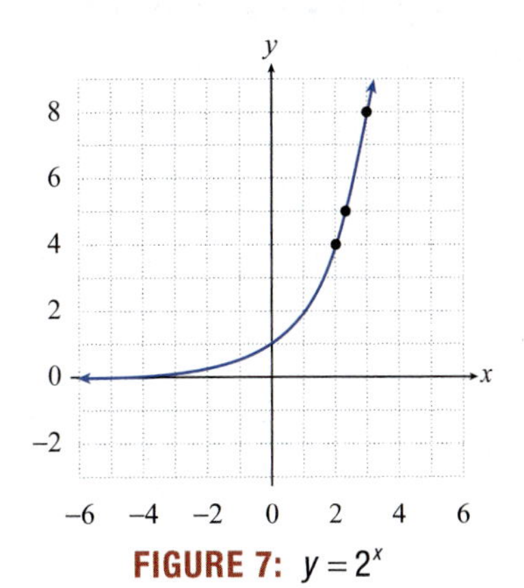

FIGURE 7: $y = 2^x$

As you might expect, an equation in which the variable appears as an exponent is called an *exponential equation*. We are not yet ready to tackle exponential equations in full generality. Even something as simple as

$$2^x = 5$$

currently stumps us. We know that $2^2 = 4$ and $2^3 = 8$, so the answer must be between 2 and 3, but beyond this we don't have a method to proceed. This must wait until we have discussed a class of functions called *logarithms* (which will happen in Section 6.3).

However, we *are* ready to solve exponential equations that can be written in the form

$$a^x = a^b,$$

where a is an exponential base (positive and not equal to 1) and b is a constant. You might guess, just from the form of the equation $a^x = a^b$, that the solution is $x = b$. This guess is correct, but we need to investigate why it is true. Note, for instance, that the solution of the (nonexponential) equation $1^x = 1^3$, is not simply $x = 3$; any real number x solves the equation because both sides reduce to 1.

The reason that the single value b is the solution of an exponential equation of the form $a^x = a^b$ is that the exponential function $f(x) = a^x$ is one-to-one (its graph passes the horizontal line test). Recall that if g is a one-to-one function, then the only way for $g(x_1)$ to equal $g(x_2)$ is if $x_1 = x_2$.

In the case of the function $f(x) = a^x$, the equation $a^x = a^b$ is equivalent to the statement $f(x) = f(b)$, and this implies $x = b$, since f is one-to-one.

An exponential equation may not appear in the simple form $a^x = a^b$ initially. Note the procedure you may need to take to solve an elementary exponential equation.

PROCEDURE: Solving Elementary Exponential Equations

To solve an elementary exponential equation perform the following steps.

Step 1: Isolate the exponential. Move the exponential containing x to one side of the equation and any constants or other variables in the expression to the other side. Simplify, if necessary.

Step 2: Find a base that can be used to rewrite both sides of the equation.

Step 3: Equate the powers, and solve the resulting equation.

Example 3: Solving Elementary Exponential Equations

Solve the following exponential equations.

a. $25^x - 125 = 0$ **b.** $8^{y-1} = \frac{1}{2}$ **c.** $\left(\frac{2}{3}\right)^x = \frac{9}{4}$

NOTE

As always, it is a good practice to check your solution in the original equation.

Solution

a. $25^x - 125 = 0$ Begin by isolating the term with the variable on one side.

$25^x = 125$

$\left(5^2\right)^x = 5^3$ We can write both sides with the same base since 25 and 125 are both powers of 5.

$5^{2x} = 5^3$ Simplify using properties of exponents.

$2x = 3$ We then equate the exponents, resulting in a linear equation, which we can easily solve.

$x = \frac{3}{2}$

b. $8^{y-1} = \frac{1}{2}$ Again, we need to rewrite both sides using the same base.

$\left(2^3\right)^{y-1} = 2^{-1}$ We see that 8 and $\frac{1}{2}$ can both be written as powers of 2.

$2^{3y-3} = 2^{-1}$ Simplify using properties of exponents.

$3y - 3 = -1$ Set the exponents equal to each other.

$3y = 2$ Solve the resulting linear equation.

$y = \frac{2}{3}$

c. $\left(\frac{2}{3}\right)^x = \frac{9}{4}$ Sometimes, the choice of base is not obvious.

$\left(\frac{2}{3}\right)^x = \left(\frac{3}{2}\right)^2$ Initially, we write the right-hand side as shown.

$\left(\frac{2}{3}\right)^x = \left(\frac{2}{3}\right)^{-2}$ Then, we can make the two bases equal by using properties of exponents.

$x = -2$ After making both bases the same, we equate the exponents to find the solution.

6.1 EXERCISES

PRACTICE

Sketch the graphs of the following functions. State their domain and range. See Examples 1 and 2.

1. $f(x) = 4^x$

2. $g(x) = (0.5)^x$

3. $s(x) = 3^{x-2}$

4. $f(x) = \left(\frac{1}{3}\right)^{x+1}$

5. $r(x) = 5^{x-2} + 3$

6. $h(x) = 1 - 2^{x+1}$

7. $f(x) = 2^{-x}$

8. $r(x) = 3^{2-x}$

9. $g(x) = 3\left(2^{-x}\right)$

10. $h(x) = 2^{2x}$

11. $s(x) = (0.2)^{-x}$

12. $f(x) = \frac{1}{2^x} + 1$

13. $g(x) = 3 - 2^{-x}$

14. $r(x) = \frac{1}{2^{3-x}}$

15. $h(x) = \left(\frac{1}{2}\right)^{5-x}$

16. $m(x) = 3^{2x+1}$

17. $p(x) = 2 - 4^{2-x}$

18. $q(x) = 5^{3-2x}$

19. $r(x) = \left(\frac{9}{2}\right)^{-x}$

20. $p(x) = \left(\frac{1}{3}\right)^{2-x}$

21. $r(x) = 1 - \left(\frac{15}{4}\right)^x$

Solve the following exponential equations. See Example 3.

22. $5^x = 125$

23. $3^{2x-1} = 27$

24. $9^{2x-5} = 27^{x-2}$

25. $10^x = 0.01$

26. $4^{-x} = 16$

27. $2^x = \left(\frac{1}{2}\right)^{13}$

28. $2^{x+1} = 64^3$

29. $\left(\frac{2}{3}\right)^{x+3} = \left(\frac{9}{4}\right)^{-x}$

30. $\left(\frac{1}{5}\right)^{x-4} = 625^{\frac{1}{2}}$

31. $4^{3x+2} = \left(\frac{1}{4}\right)^{-2x}$

32. $5^x = 0.2$

33. $7^{x^2+3x} = \frac{1}{49}$

34. $3^{x^2+4x} = 81^{-1}$

35. $\left(\frac{1}{2}\right)^{x-3} = \left(\frac{1}{4}\right)^{x-5}$

36. $64^{x+\frac{7}{6}} = 2$

37. $6^{2x} = 36^{2x-3}$

38. $4^{2x-5} = 8^{\frac{x}{2}}$

39. $\left(\frac{2}{5}\right)^{2x+4} = \left(\frac{4}{25}\right)^{11}$

40. $4^{4x-7} = \frac{1}{64}$

41. $-10^x = -0.001$

42. $3^x = 27^{x+4}$

43. $1000^{-x} = 10^{x-8}$

44. $1^{3x-7} = 4^{2-x}$

45. $5^{3x-1} = 625^x$

46. $3^{2x-7} = 81^{\frac{x}{2}}$

Match the graphs of the following functions to the appropriate equation.

47. $f(x) = 2^{3x}$

48. $h(x) = 5^x - 1$

49. $g(x) = 2\left(4^{x-1}\right)$

50. $p(x) = 1 - 2^{-x}$

51. $f(x) = 6^{4-x}$

52. $r(x) = \dfrac{1}{3^x}$

53. $m(x) = -2 + 2^{-3x}$

54. $g(x) = \left(\dfrac{1}{4}\right)^{1+x}$

55. $h(x) = 3^{\frac{1}{2}x}$

56. $s(x) = 1^x - 4$

a.

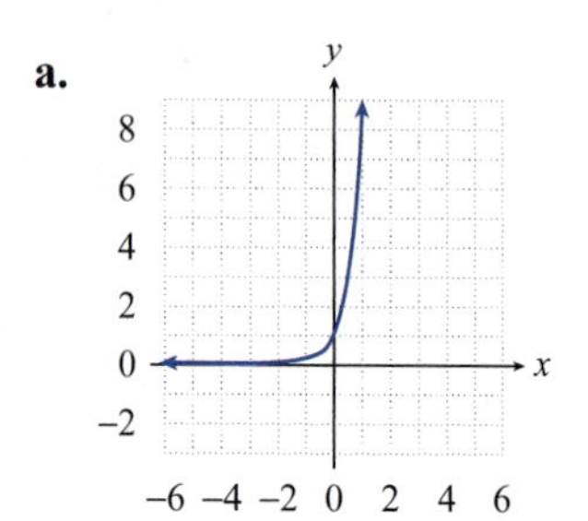

b.

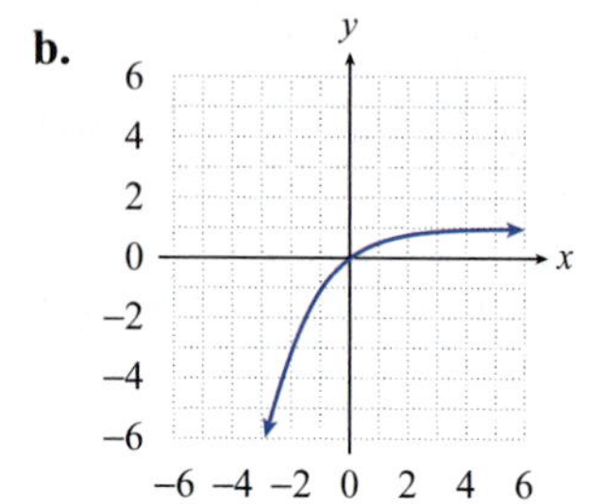

c.

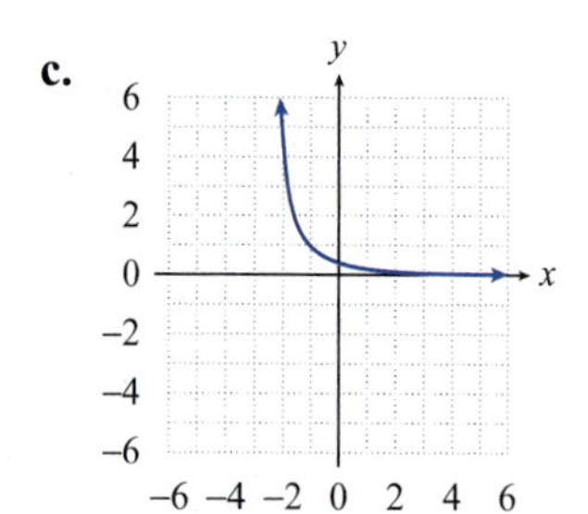

d.

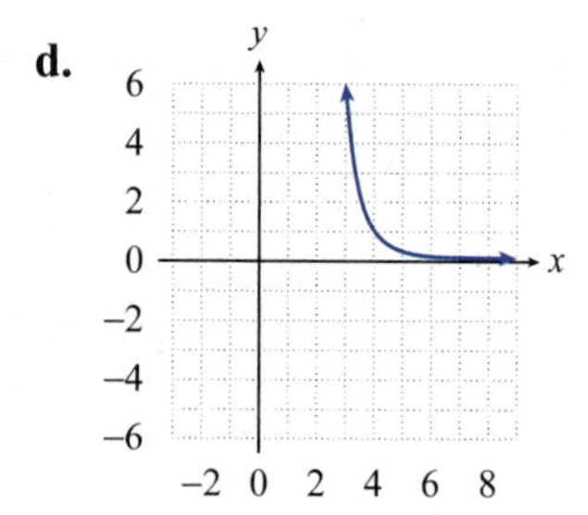

e.

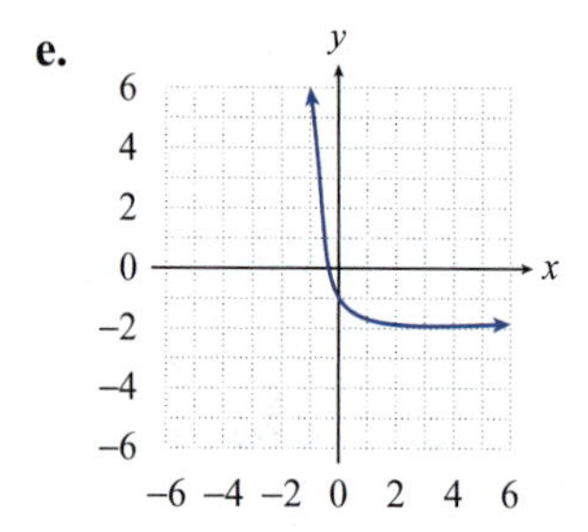

f.

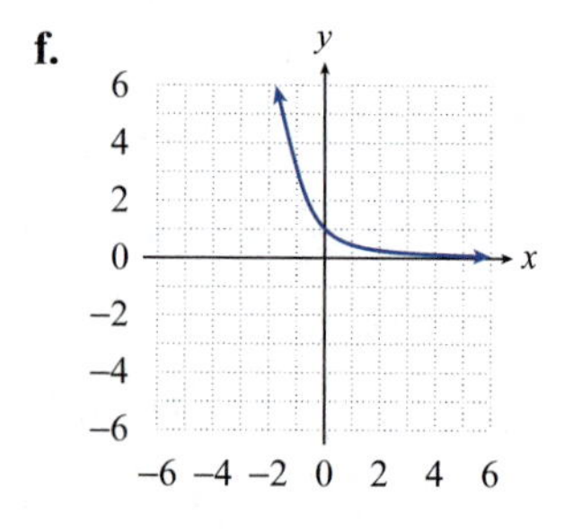

g.

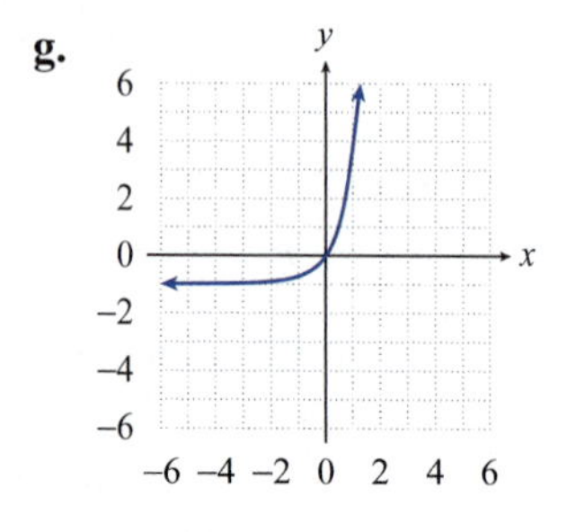

h.

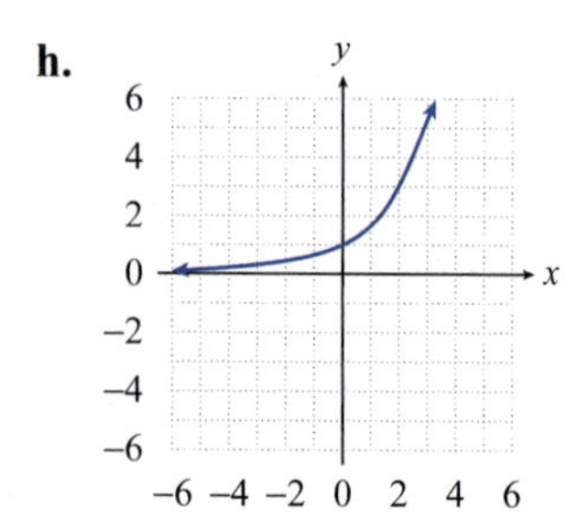

i.

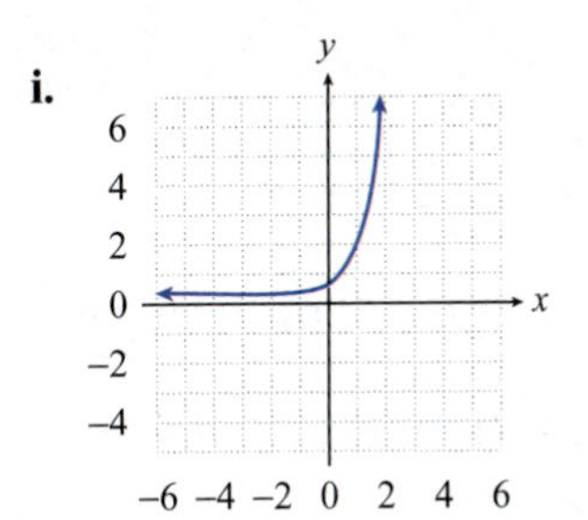

j.

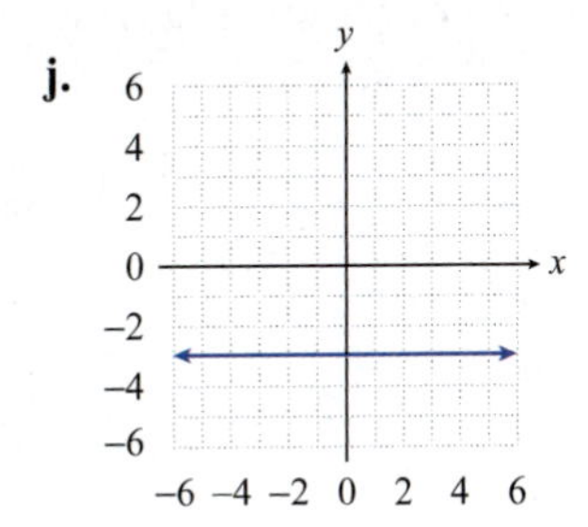

6.2 EXPONENTIAL MODELS

TOPICS

1. Models of population growth
2. Models of radioactive decay
3. Compound interest and the number e
4. Exponential regression

TOPIC 1: Models of Population Growth

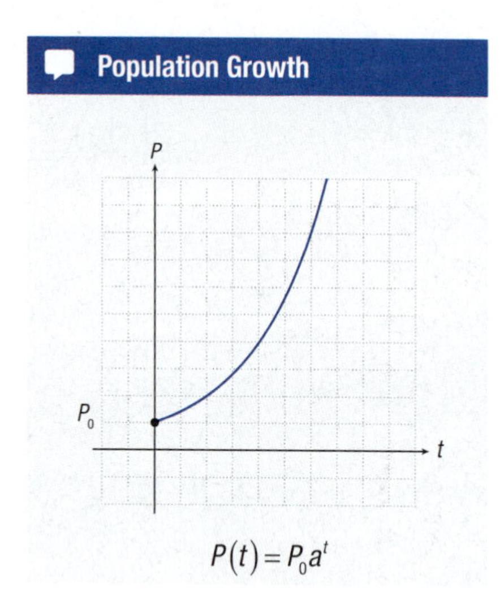

Exponential functions arise naturally in a wide array of situations. In this section, we will study a few such situations in some detail, beginning with population models.

Many people working in such areas as mathematics, biology, and sociology study mathematical models of population. The models represent many types of populations, such as people in a given city or country, wolves in a wildlife habitat, or number of bacteria in a Petri dish. While such models can be quite complex, depending on factors like availability of food, space constraints, and effects of disease and predation, at their core, many population models assume that population growth displays exponential behavior.

The reason for this is that the growth of a population usually depends to a large extent on the number of members capable of producing more members. This assumes an abundant food supply and no constraints on population from lack of space, but at least initially this is often the case. In any situation where the rate of growth of a population is proportional to the size of the population, the population will grow exponentially, so we can write the function

$$P(t)=P_0a^t.$$

This function tells us the size of a population $P(t)$ at time t. What do the other variables in the function represent?

Consider what happens if we substitute $t=0$. $P(0)=P_0a^0$, and since $a^0=1$ for all a, we have $P(0)=P_0$. Thus, P_0 is the *initial population* (population at time 0).

Recall from our work in graphing exponential equations that when $a>1$, the larger the value of a is, the faster the exponential function grows. In models of population growth, a is greater than 1 (otherwise we would have population decay). For this reason, the value of a is the *growth rate* of the population.

Often, we will need to use information about a population to determine the values of P_0 and a. Once we have the function for population growth, we can answer other questions about the population.

Example 1 illustrates this with a specific model of bacterial population growth.

Example 1: Population Growth

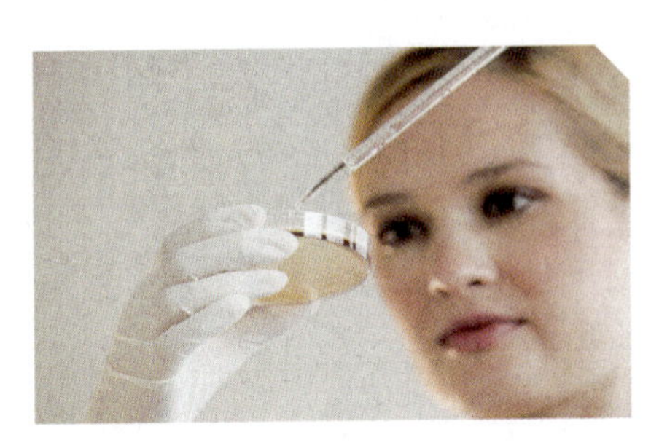

A biologist is culturing bacteria in a Petri dish. She begins with 1000 bacteria, and supplies sufficient food so that for the first five hours the bacteria population grows exponentially, doubling every hour.

a. Find a function that models the population growth of this bacteria culture.

b. Determine when the population reaches 16,000 bacteria.

c. Calculate the population two and a half hours after the scientist begins.

Solution

a. We know that we seek a function of the form $P(t) = P_0 a^t$. Since the scientist starts with 1000 bacteria, the initial population, $P_0 = 1000$.

To solve for a, we use the fact that the population doubles every hour.

$$2000 = P(1)$$
$$2000 = 1000a^1$$

Substitute, using the fact that the population after one hour equals 2000 bacteria.

$$2000 = 1000a$$
$$a = 2$$

Simplify, then solve for a.

Thus, a function that models the population growth of the bacteria culture is $P(t) = 1000(2)^t$, where t is measured in hours.

b. We wish to find the time t for which $P(t) = 16{,}000$.

$16{,}000 = 1000(2)^t$	Substitute the desired population value.
$16 = 2^t$	Divide both sides by 1000.
$2^4 = 2^t$	Rewrite both sides with the same base, 2.
$t = 4$	Equate the exponents and solve.

Thus, the bacteria culture reaches a population of 16,000 in 4 hours.

c. To calculate the population two and half hours after growth begins, we substitute $t = 2.5$ into the population model function.

$$P(2.5) = 1000(2)^{2.5}$$

While we could rewrite this using rational exponents, and try to find an exact value, frequently this will be very tedious or impossible, so we use a calculator to evaluate.

$$P(2.5) \approx 5657 \text{ bacteria}$$

TECHNOLOGY

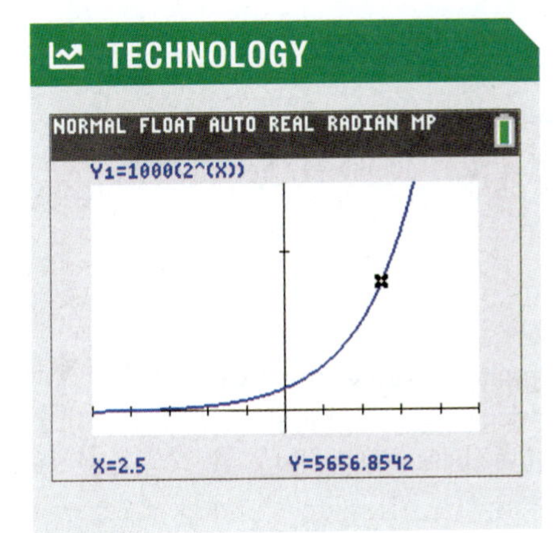

TOPIC 2: Models of Radioactive Decay

Radioactive Decay

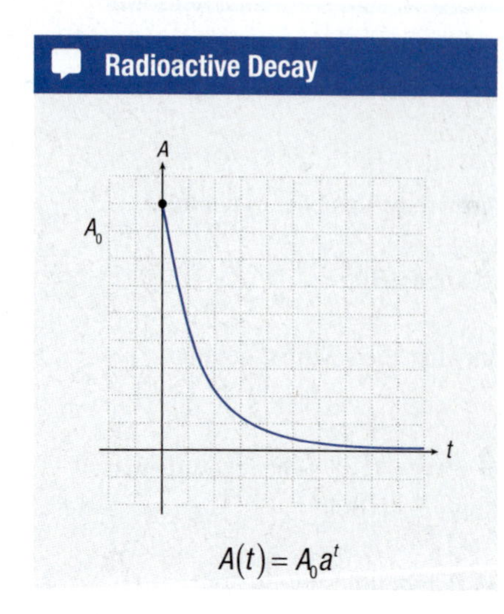

In contrast to populations (at least healthy populations), radioactive substances diminish with time. To be exact, the mass of a radioactive substance decreases over time as the substance decays into other elements. Since exponential functions with a base between 0 and 1 are decreasing functions, this suggests that radioactive decay is modeled by

$$A(t) = A_0 a^t,$$

where $A(t)$ represents the amount of a given substance at time t, A_0 is the amount at time $t = 0$, and a is a number between 0 and 1.

The fact of radioactive decay is important (indirectly) in using radioactivity for power generation, and important again in working out the details of storing spent radioactive fuel rods. Radioactive decay also has other uses, one of which goes by the name of *radiocarbon dating*. This is a technique used by archaeologists, anthropologists, and others to estimate how long ago an organism died. The method depends on the fact that living organisms constantly absorb molecules of the radioactive substance carbon-14 while alive, but the intake of carbon-14 ceases once the organism dies. It is believed that the percentage of carbon-14 on Earth (relative to other isotopes of carbon) has been relatively constant over time, so the first step in the method is to determine the percentage of carbon-14 in the remains of a given organism. By comparing this (smaller) percentage to the percentage found in living tissue, an estimate of when the organism died can then be made.

The mathematics of the age estimation depends on the fact that half of a given mass of carbon-14 decays over a period of 5730 years. This is known as the *half-life* of carbon-14; every radioactive substance has a half-life, and the half-life is usually an important feature when working with such substances. In the case of carbon-14, the half-life of 5730 years means that if, for instance, an organism contained 12 grams of carbon-14 at death, it would contain 6 grams after 5730 years, 3 grams after another 5730 years, and so on. (In exponential growth, there is the related concept of *doubling time*, which is the length of time needed for the function to double in value. The doubling time for the bacteria population in Example 1 is 1 hour.)

Example 2: Radioactive Decay

Determine the base a so that the function $A(t) = A_0 a^t$ accurately describes the decay of carbon-14 as a function of t years.

Solution

Note that $A(0) = A_0 a^0 = A_0$, so A_0 represents the amount of carbon-14 at time $t = 0$. Since we are seeking a general formula, we don't know what A_0 is specifically; that is, the value of A_0 will vary depending on the details of the situation. But we can still determine the value of the base constant a.

What we know is that half of the original amount of carbon-14 decays over a period of 5730 years, so $A(5730)$ will be half of A_0. This gives us the following equation.

$$A_0\left(a^{5730}\right) = \frac{A_0}{2}$$

To solve this for a, we can first divide both sides by A_0; the fact that A_0 then cancels from the equation just emphasizes that its exact value is irrelevant for the task at hand. Now we have the following equation.

$$a^{5730} = \frac{1}{2}$$

At this point, a calculator is called for, as we need to take the 5730th root of both sides. This gives us the value for a that we seek.

$$a = \left(\frac{1}{2}\right)^{\frac{1}{5730}} \approx 0.999879$$

TECHNOLOGY

```
NORMAL FLOAT AUTO REAL RADIAN MP
(1/2)^(1/5730)
                    0.9998790392
```

The function is thus $A(t) = A_0(0.999879)^t$ (using our approximate value for a). We can now verify that the function behaves as expected by evaluating A at various multiples of the half-life for carbon-14, as shown.

$$A(1 \cdot 5730) = A_0(0.999879)^{5730} \approx 0.5A_0$$
$$A(2 \cdot 5730) = A_0(0.999879)^{11,460} \approx 0.25A_0$$
$$A(3 \cdot 5730) = A_0(0.999879)^{17,190} \approx 0.125A_0$$

TOPIC 3: Compound Interest and the Number *e*

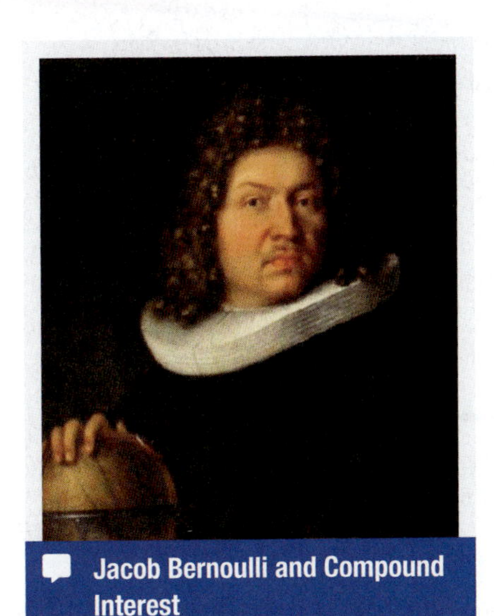

Jacob Bernoulli and Compound Interest

Questions regarding investing and interest have motivated research in mathematics for millennia. The Swiss mathematician Jacob Bernoulli, one member of the large mathematically-inclined Bernoulli family, defined the constant that came to be called *e* in 1683 while studying questions regarding compound interest.

One of the most commonly encountered applications of exponential functions is in compounding interest. We run into compound interest when earning money (by interest on a savings account or investment) and when spending money (on car loans, mortgages, and credit cards).

The basic compound interest formula can be understood by considering what happens when money is invested in a savings account. Typically, a savings account is set up to pay interest at an annual rate of r (which we will write in decimal form) compounded n times per year. For instance, a bank may offer an annual interest rate of 5% (meaning $r = 0.05$) compounded monthly (so $n = 12$).

Compounding is the act of calculating the interest earned on an investment and adding that amount to the investment. An investment in a monthly compounded account will have interest added to it twelve times over the course of a year, once each month.

Suppose an amount of P (for principal) dollars is invested in a savings account at an annual rate of r compounded n times per year. We want a formula for the amount of money $A(t)$ in the account after t years.

If we say that a period is the length of time between compoundings, interest is calculated at the rate of $\frac{r}{n}$ per period (for instance, if $r = 0.05$ and $n = 12$, interest is earned at a rate of $\frac{0.05}{12} \approx 0.00417$ per month). Table 1 illustrates how compounding increases the amount in the account over the course of several periods.

Period	Amount
0	$A = P$
1	$A = P\left(1+\frac{r}{n}\right)$
2	$A = P\left(1+\frac{r}{n}\right)\left(1+\frac{r}{n}\right) = P\left(1+\frac{r}{n}\right)^2$
3	$A = P\left(1+\frac{r}{n}\right)^2\left(1+\frac{r}{n}\right) = P\left(1+\frac{r}{n}\right)^3$
k	$A = P\left(1+\frac{r}{n}\right)^{k-1}\left(1+\frac{r}{n}\right) = P\left(1+\frac{r}{n}\right)^k$

TABLE 1: Effect of Compounding Interest on an Investment of P Dollars

Notice that the accumulation at the end of each period is the accumulation at the end of the preceding period multiplied by $\left(1+\frac{r}{n}\right)$; that is, the account has the amount from the end of the preceding period plus that amount multiplied by $\frac{r}{n}$. The amount A is then a function of t, the number of years the account stays active. Since the number of investment periods in t years is nt, we obtain the following formula.

f(x) FORMULA: Compound Interest Formula

An investment of P dollars, compounded n times per year at an annual interest rate of r, has a value after t years of

$$A(t) = P\left(1+\frac{r}{n}\right)^{nt}.$$

Example 3: Compound Interest Formula

Sandy invests \$10,000 in a savings account earning 4.5% annual interest compounded quarterly. What is the value of her investment after three and a half years?

TECHNOLOGY

```
NORMAL FLOAT AUTO REAL RADIAN MP
10000(1+0.045/4)^((4)(3.5))
                    11695.51862
```

Solution

We know that $P = 10,000$, $r = 0.045$ (remember to express the interest rate in decimal form), $n = 4$ (since the account is compounded four times per year), and $t = 3.5$. Now we substitute and evaluate.

$$\begin{aligned} A(3.5) &= 10,000\left(1+\frac{0.045}{4}\right)^{(4)(3.5)} \\ &= 10,000(1.01125)^{14} \\ &\approx \$11,695.52 \end{aligned}$$

Thus, after three and a half years, Sandy's investment grows to \$11,695.52.

The compound interest formula can also be used to determine the interest rate of an existing savings account, as shown in Example 4.

Example 4: Compound Interest Formula

Nine months after depositing \$520.00 in a monthly compounded savings account, Frank checks his balance and finds the account has \$528.84. Being the forgetful type, he can't remember what the annual interest rate for his account is, and sees the bank is advertising a rate of 2.5% for new accounts. Should he close out his existing account and open a new one?

Solution

As in Example 3, we will begin by identifying the known quantities in the compound interest formula: $P = 520$, $n = 12$ (12 compoundings per year), and $t = 0.75$ (nine months is three-quarters of a year).

Further, the amount in the account at this time is $A = \$528.84$. This gives us the equation

$$528.84 = 520\left(1+\frac{r}{12}\right)^{(12)(0.75)}$$

to solve for r, the annual interest rate.

$$528.84 = 520\left(1+\frac{r}{12}\right)^{9}$$ Simplify the exponent.

$$1.017 = \left(1+\frac{r}{12}\right)^{9}$$ Divide both sides by 520.

$$1.001875 \approx 1+\frac{r}{12}$$ Take the ninth root of both sides.

$$0.001875 \approx \frac{r}{12}$$ Simplify to solve for r.

$$0.0225 \approx r$$

Thus, Frank's current savings account is paying an annual interest rate of 2.25%, so he would gain a slight advantage by switching to a new account.

Even though the interest rate is divided by n, the number of periods per year, increasing the frequency of compounding always increases the total interest earned on the investment. This is because interest is always calculated on the current balance. For example, consider how much interest you earn in one year if you invest \$1000 in an account with a 5% annual interest rate. If the interest is compounded once per year, you earn \$50 in interest. However, if the interest is compounded *twice* per year, you earn 2.5%, or \$25, for the first period, then 2.5% *of the new balance* of \$1025 in the second period. This brings the total interest earned to \$50.63! Table 2 shows the interest earned in one year on an investment of \$1000 in an account with a 5% annual interest rate, compounded n times per year.

n	Calculation	Value after 1 Year
1 (annually)	$A = 1000(1+0.05)$	\$1050.00
2 (biannually)	$A = 1000\left(1+\frac{0.05}{2}\right)^2$	\$1050.63
4 (quarterly)	$A = 1000\left(1+\frac{0.05}{4}\right)^4$	\$1050.95
12 (monthly)	$A = 1000\left(1+\frac{0.05}{12}\right)^{12}$	\$1051.16
52 (weekly)	$A = 1000\left(1+\frac{0.05}{52}\right)^{52}$	\$1051.25
365 (daily)	$A = 1000\left(1+\frac{0.05}{365}\right)^{365}$	\$1051.27
8760 (hourly)	$A = 1000\left(1+\frac{0.05}{8760}\right)^{8760}$	\$1051.27

TABLE 2: Value of an Investment with Interest Compounded n Times per Year

Looking at Table 2, we see that the amount of interest keeps growing as we divide the year into more compounding periods, but that this growth slows dramatically. It looks as if there is a *limit* to how much interest we can earn, and this is in fact the case.

Let's examine the formula as $n \to \infty$. In order to do this we need to perform some algebraic manipulation of the formula.

$$A(t) = P\left(1+\frac{r}{n}\right)^{nt}$$

$$= P\left(1+\frac{1}{m}\right)^{rmt}$$ Substitute $m = \frac{n}{r}$, so $\frac{1}{m} = \frac{r}{n}$.

$$= P\left(\left(1+\frac{1}{m}\right)^{m}\right)^{rt}$$ Bring together the instances of the variable m using the properties of exponents.

Although the manipulation may appear strange, it has accomplished the important task of isolating the part of the formula that changes as $n \to \infty$. Looking at the change of variables $m = \frac{n}{r}$, we see that letting $n \to \infty$ means that $m \to \infty$ as well. Since every other quantity remains fixed, we only have to understand what happens to

$$\left(1+\frac{1}{m}\right)^{m}$$

as m grows without bound.

This is not a trivial undertaking. We might think that letting m get larger and larger would make the expression grow larger and larger, as m is the exponent in the expression and the base is larger than 1. But at the same time, the base approaches 1 as m increases without bound, and 1 raised to any power is simply 1. It turns out that these two effects balance one another out, as we can see in Table 3.

TECHNOLOGY

NORMAL FLOAT AUTO REAL RADIAN MP

L1	L2	L3	L4	L5
10	2.5937	------	------	------
100	2.7048			
1000	2.7169			
10000	2.7182			
100000	2.7183			
------	------			

L1(6)=

Press stat and select Edit... to enter the data for m as L1. To calculate and input the corresponding values in L2, highlight the L2 column name and enter

L2=(1+1/L1)^L1

(L1 can be selected by pressing 2nd and stat.)

m	$\left(1+\frac{1}{m}\right)^m$
10	2.59374
100	2.70481
1000	2.71692
10,000	2.71815
100,000	2.71827

TABLE 3: Values of $\left(1+\frac{1}{m}\right)^m$ as m Increases

As m gets larger and larger, we find that the expression approaches a fixed number.

$$\left(1+\frac{1}{m}\right)^m \to 2.718281828459$$

This value is a very important irrational number in mathematics, so important that it gets its own dedicated symbol, e, just as π is conventionally reserved for the irrational number obtained by dividing the circumference of a circle by its diameter.

DEFINITION: The Number e

The number e is defined as the value of $\left(1+\frac{1}{m}\right)^m$ as $m \to \infty$.

$$e \approx 2.71828182846$$

Let's look back at the compound interest formula from before.

$$A(t) = P\left(1+\frac{r}{n}\right)^{nt} = P\left(\left(1+\frac{1}{m}\right)^m\right)^{rt}$$

This means that if P dollars is invested in an account that is **compounded continuously**, that is, with $n \to \infty$, then the amount in the account is determined by the following formula, which results from substituting the definition of e.

FORMULA: Continuous Compounding Formula

An investment of P dollars, compounded continuously at an annual interest rate of r, has a value after t years of

$$A(t) = Pe^{rt}.$$

Example 5: Continuous Compounding Formula

If Sandy (last seen in Example 3) has the option of investing her \$10,000 in a continuously compounded account earning 4.5% annual interest, what will be the value of her account in three and a half years?

Solution

Again, the solution boils down to substituting the correct values and evaluating the result. Here, $P = 10,000$, $r = 0.045$, and $t = 3.5$.

$$\begin{aligned} A(3.5) &= 10,000e^{(0.045)(3.5)} \\ &= 10,000e^{0.1575} \\ &\approx \$11,705.81 \end{aligned}$$

This account earns \$10.29 more than the quarterly compounded account in Example 3.

As we will see in Section 6.4, all exponential functions can be expressed with the base e (or any other base, for that matter). The base e is so commonly used for exponential functions that it is often called the *natural base*. For instance, the formula for the radioactive decay of carbon-14, using the base e, is as follows.

$$A(t) = A_0 e^{-0.000121t}$$

You should verify that this version of the decay formula does indeed give the same values for $A(t)$ as the version derived in Example 2.

TOPIC 4: Exponential Regression

TECHNOLOGY

To find and graph the exponential function of best fit for a given set of points using a TI-84 Plus, perform the same steps as described in Section 3.2 for linear regression, except select ExpReg from the CALC menu.

Just as linear regression and quadratic regression can be used to fit, respectively, a straight line and a parabola to a given collection of points, **exponential regression** can be used to fit an exponential curve to points that we suspect exhibit exponential behavior. Many graphing utilities are able to determine exponential curves of best fit, with the specific commands varying with the technology used. Specifically, exponential regression is the process of determining constants a and b so that the graph of the function $f(x) = ab^x$ models the given data well. It should be noted, however, that the results of exponential regression may differ depending on the algorithm used by the technology.

Example 6: Exponential Regression

Given the points

$$\{(1,0.5),(2,1.0),(3,1.7),(4,2.6),(5,4.5)\},$$

a. use a TI-84 Plus to fit both an exponential curve and a parabola to the points, and

b. compare the results by calculating the sums of the squares of the differences between the y-values of the given points and the values of the regression functions at the corresponding x-values.

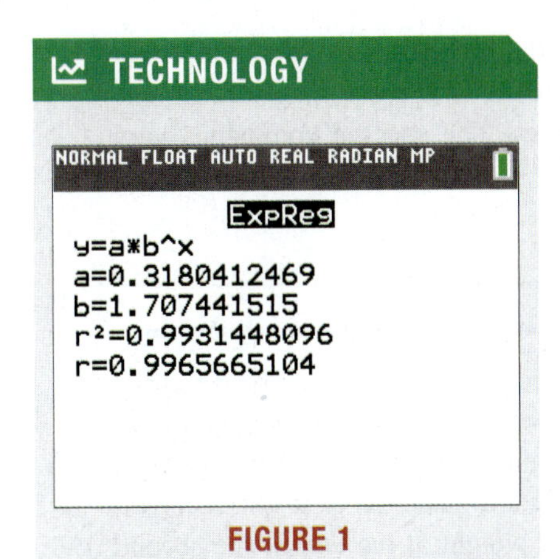

FIGURE 1

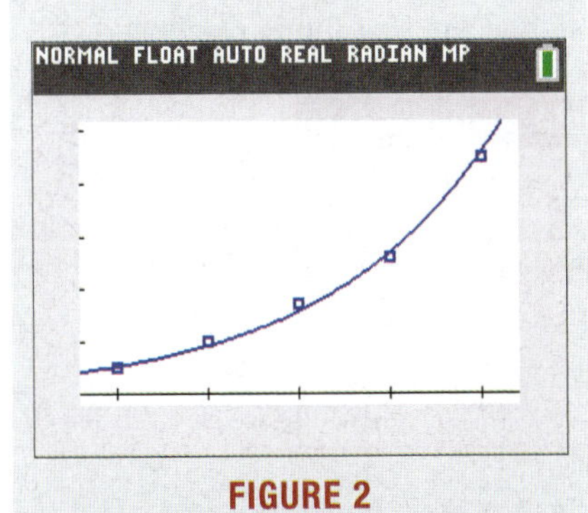

FIGURE 2

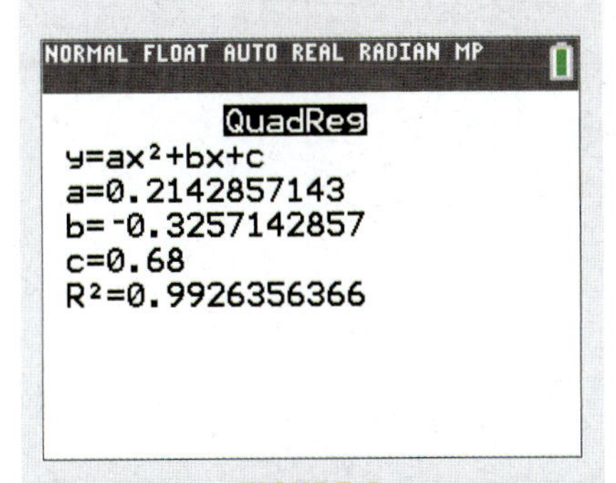

FIGURE 3

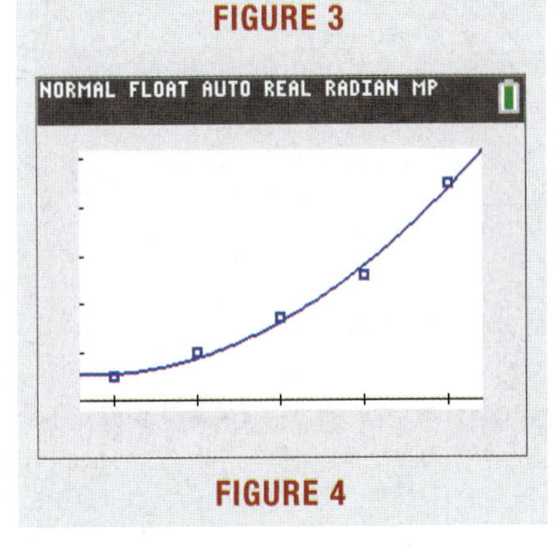

FIGURE 4

Solution

a. For the given five points in this example, a TI-84 Plus returns the following exponential function of best fit as shown in Figure 1.

$$f(x) = 0.3180(1.707)^x$$

A graph of f along with the given points is shown in Figure 2.

A graphing utility can also be used to find a quadratic function that best fits a given set of points. In this case, the equation of the best-fitting parabola, as shown in Figure 3, is the function

$$g(x) = 0.2143x^2 - 0.3257x + 0.6800,$$

and a graph of g with the given points appears in Figure 4.

b. The values of the exponential regression function f at the x-values of the given points are

$$f(1) \approx 0.5428,\ f(2) \approx 0.9266,\ f(3) \approx 1.5817,\ f(4) \approx 2.7000,$$
$$\text{and } f(5) \approx 4.6089.$$

The differences between these values and the respective y-values of the given points are

$$0.0428,\ -0.0734,\ -0.1183,\ 0.1000, \text{and } 0.1089;$$

and when we square each of these differences and add them we obtain the approximate least-squares measure of 0.0431. We use this as a way to measure the accuracy of the fit of the curve, since the least-squares method of curve-fitting seeks to make this sum as small as possible.

For the quadratic regression function g, the corresponding differences at the five x-values are

$$0.0686,\ -0.1142,\ -0.0684,\ 0.2060, \text{ and } -0.0910,$$

leading to an approximate sum-of-squares measure of 0.0731, about 1.7 times as large as the exponential least-squares measure.

Thus, the exponential regression function is a better model of the data.

Logistic curves are a family of curves based on exponential functions that are designed to model behavior often seen in biology, ecology, population studies, epidemiology, economics, and other disciplines. The graph of a logistic function, known as a *sigmoid curve*, is S-shaped. It consists of an initial portion displaying exponential growth, a middle portion that is more linear in nature, and then a final portion that displays an exponential tapering off toward a maximum value. More precisely, a logistic function can be written in the form

$$f(x) = \frac{c}{1 + ae^{-bx}},$$

where a, b, and c are positive constants.

TECHNOLOGY

To find and graph the logistic function of best fit for a given set of points using a TI-84 Plus, perform the same steps as described in Section 3.2 for linear regression, except select Logistic from the CALC menu.

Logistic regression is, as the name suggests, the process of using an algorithm to fit a logistic curve to a given collection of data. Many graphing utilities can compute equations of logistic regression curves. The specific commands vary with the technology.

Example 7: Logistic Regression

Exercise 18 of Section 3.6 contains a table (reproduced in Table 4) of the weight of a puppy at the end of each of its first 12 months of life. Karen, the puppy's owner, had been told to expect the dog to have an adult weight somewhere between 40 and 45 pounds. Use a graphing utility to fit a logistic curve to the recorded weights and use the result to predict the dog's weight at the end of the second year. Compare the logistic curve to the best-fitting line and best-fitting parabola.

End of Month	Weight (in pounds)
1	3
2	5
3	8
4	11
5	17
6	22
7	28
8	32
9	35
10	38
11	40
12	42

TABLE 4

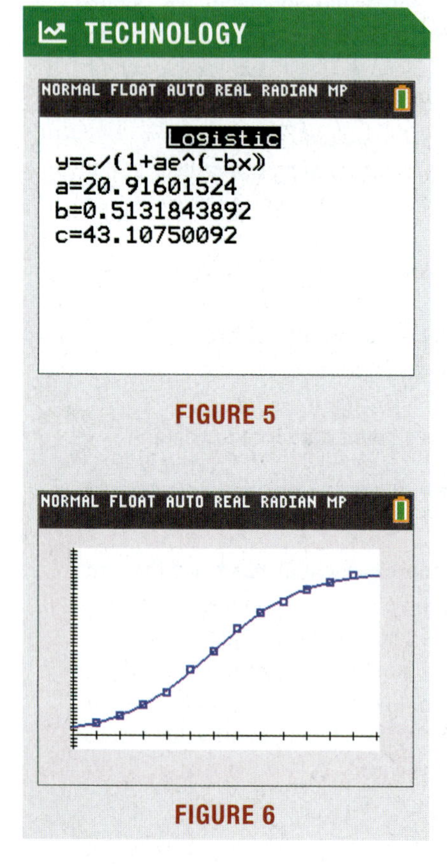

FIGURE 5

FIGURE 6

Solution

Using a graphing utility to perform a logistic regression, we find the following logistic function that best fits the given data.

$$f(x) = \frac{43.108}{1 + 20.916e^{-0.51318x}}$$

A graph of f, along with the given 12 data points, appears in Figure 6.

To predict the dog's weight at the end of the second year, we let $x = 24$ and find that

$$f(24) \approx 43.1 \text{ lb.}$$

To illustrate the utility of a logistic curve in a modeling situation such as this, Figure 7 contains the graphs of the best-fitting line (the red line A) and best-fitting parabola (the green curve B) for the given data, along with the logistic curve just found (the blue curve C). The graphs indicate that both linear and quadratic regression are inappropriate tools in this case, while the logistic regression function appears to be much more realistic. As a particular point of comparison, the weights predicted by the linear and quadratic models at the end of month 24 are, respectively, 92 pounds and 68 pounds, far outside the expected range of 40 to 45 pounds.

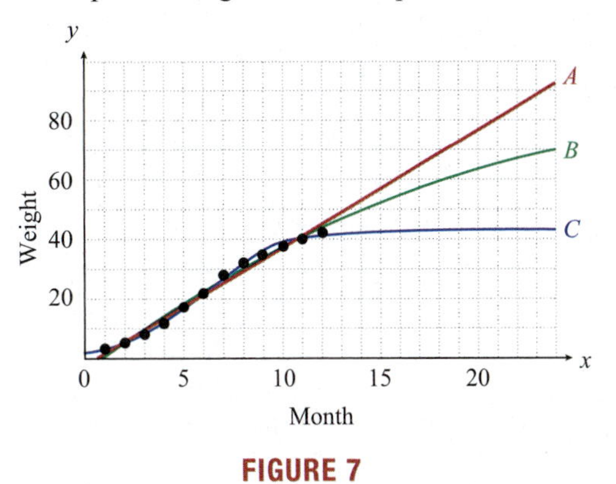

FIGURE 7

6.2 EXERCISES

APPLICATIONS

1. A new virus has broken out in isolated parts of Africa and is spreading exponentially through tribal villages. The growth of this new virus can be mapped using the following formula where V stands for the number of people in the village who are infected with the virus, P stands for the number of people in a village and d stands for the number of days since the virus first appeared. According to this equation, how many people in a village of 300 will be infected after 5 days?

$$V = P\left(1 - e^{-0.18d}\right)$$

2. A prototype for an electric motorcycle uses a battery whose energy capacity $C(d)$, in kilowatt-hours (kWh), is given by the formula $C(d) = 12e^{-0.02d}$, where d represents the number of days since receiving a full charge. What is the battery's energy capacity 30 days after being fully charged?

3. A young economics student has come across a very profitable investment scheme in which his money will accrue interest according to the equation listed below, where C represents the investment value after m months for an initial investment of I dollars. If this student invests \$1250 into this lucrative endeavor, how much money will he have after 24 months? I represents the investment and m represents the number of months the money has been invested for.

$$C = Ie^{0.08m}$$

4. A family releases a couple of pet rabbits into the wild. Upon being released the rabbits begin to reproduce at an exponential rate, as shown in the formula below. After 2 years how large is the rabbit population P, where n stands for the initial rabbit population (2) and m stands for the number of months?

$$P = ne^{0.5m}$$

5. Inside a business network, an email worm was downloaded by an employee. This worm goes through the infected computer's address book and sends itself to all the listed email addresses. This worm very rapidly works its way through the network following the equation below, where C is the number of computers in the network and W is the number of computers infected h hours after the worm is initially downloaded. After only 8 hours, how many computers has the worm infected if there are 150 computers in the network?

$$W = C\left(1 - e^{-0.12h}\right)$$

6. A construction crew has been assigned to build an apartment complex. The work of the crew can be modeled using the exponential formula below, where A is the total number of apartments to be built, w is the number of weeks, and F is the number of finished apartments. Out of a total of 100 apartments, how many apartments have been finished after 4 weeks of work?

$$F = A\left(1 - e^{-0.1w}\right)$$

7. The half-life of radium is approximately 1600 years.

   a. Determine a so that $A(t) = A_0a^t$ describes the amount of radium left after t years, where A_0 is the amount at time $t = 0$.
   b. How much of a 1-gram sample of radium would remain after 100 years?
   c. How much of a 1-gram sample of radium would remain after 1000 years?

8. The radioactive element polonium-210 has a relatively short half-life of 138 days, and one way to model the amount of polonium-210 remaining after t days is with the function $A(t) = A_0e^{-0.005023t}$, where A_0 is the mass at time $t = 0$ (note that $A(140) = \frac{A_0}{2}$.) What percentage of the original mass of a sample of polonium-210 remains after 365 days?

9. A certain species of fish is to be introduced into a new man-made lake, and wildlife experts estimate the population will grow according to $P(t) = (1000)2^{\frac{t}{3}}$, where t represents the number of years from the time of introduction.

   a. What is the doubling time for this population of fish?
   b. How long will it take for the population to reach 8000 fish, according to this model?

10. The population of a certain inner-city area is estimated to be declining according to the model $P(t) = 237{,}000e^{-0.018t}$, where t is the number of years from the present. What does this model predict the population will be in ten years?

11. In an effort to control vegetation overgrowth, 100 rabbits are released in an isolated area that is free of predators. After one year, it is estimated that the rabbit population has increased to 500. Assuming exponential population growth, what will the population be after another six months?

12. Assuming a current world population of 7.75 billion people, an annual growth rate of 1.9% per year, and a worst-case scenario of exponential growth, what will the world population be in **a.** 10 years? **b.** 50 years?

13. Madiha has \$3500 that she wants to invest in a simple savings account for two and a half years, at which time she plans to close out the account and use the money as a down payment on a car. She finds one local bank offering an annual interest rate of 2.75% compounded monthly, and another bank offering an annual interest rate of 2.7% compounded daily (365 times per year). Which bank should she choose?

14. Madiha, from the last problem, does some more searching and finds an online bank offering an annual rate of 2.75% compounded continuously. How much more money will she earn over two and a half years if she chooses this bank rather than the local bank offering the same rate compounded monthly?

15. Tom hopes to earn \$1000 in interest in three years time from \$10,000 that he has available to invest. To decide if it's feasible to do this by investing in a simple monthly compounded savings account, he needs to determine the annual interest rate such an account would have to offer for him to meet his goal. What would the annual rate of interest have to be?

16. An investment firm claims that its clients usually double their principal in five years time. What annual rate of interest would a savings account, compounded monthly, have to offer in order to match this claim?

17. The function $C(t) = C_0(1 + r)^t$ models the rise in the cost of a product that has a cost of C_0 today, subject to an average yearly inflation rate of r for t years. If the average annual rate of inflation over the next decade is assumed to be 3%, what will the inflation-adjusted cost of a \$100,000 house be in 10 years? Round your answer to the nearest dollar.

18. Given the inflation model $C(t) = C_0(1 + r)^t$ (see Exercise 17), and given that a loaf of bread that currently sells for \$3.60 sold for \$3.10 six years ago, what has the average annual rate of inflation been for the past six years?

19. The function $N(t) = \dfrac{10{,}000}{1 + 999e^{-t}}$ models the number of people in a small town who have caught the flu t weeks after the initial outbreak.

a. How many people were ill initially?

b. How many people have caught the flu after eight weeks?

c. Determine what happens to the function $N(t)$ as $t \to \infty$.

20. The concentration $C(t)$, in milligrams per liter, of a certain drug in the bloodstream after t minutes is given by the formula $C(t) = 0.05\left(1 - e^{-0.2t}\right)$. What is the concentration after 10 minutes?

21. Carbon-11 has a radioactive half-life of approximately 20 minutes, and is useful as a diagnostic tool in certain medical applications. Because of the relatively short half-life, time is a crucial factor when conducting experiments with this element.

a. Determine a so that $A(t) = A_0 a^t$ describes the amount of carbon-11 left after t minutes, where A_0 is the amount at time $t = 0$.

b. How much of a 2 kg sample of carbon-11 would be left after 30 minutes?

c. How many milligrams of a 2 kg sample of carbon-11 would be left after six hours?

22. Charles has recently inherited \$8000 that he wants to deposit into a savings account. He has determined that his two best bets are an account that compounds annually at a rate of 3.20% and an account that compounds continuously at an annual rate of 3.15%. Which account would pay Charles more interest?

23. Marshall invests \$1250 in a mutual fund which boasts a 5.7% annual return compounded semiannually (twice a year). After three and a half years, Marshall decides to withdraw his money.

a. How much is in his account?

b. How much has he made in interest from his investment?

24. Adam is working in a lab testing bacteria populations. After starting out with a population of 375 bacteria, he observes the change in population and notices that the population doubles every 27 minutes.

a. Find the equation for the population P in terms of time t in minutes, rounding a to the nearest thousandth.

b. Find the population after two hours.

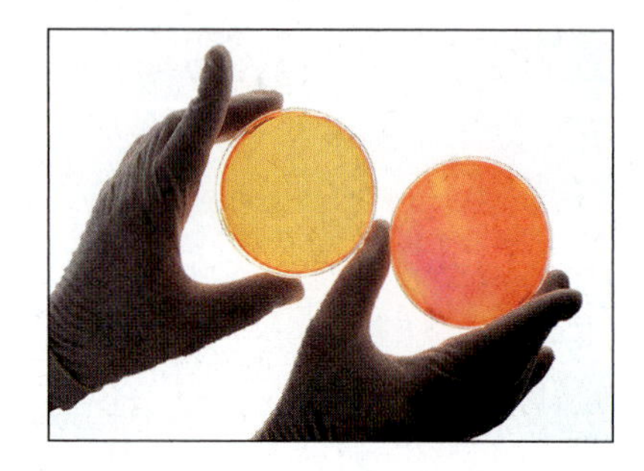

25. Your credit union offers a special interest rate of 10% compounded monthly for the first year for a student savings account opened in August if the student deposits \$5000 or more. You received a total of \$9000 for graduation, and you decide to deposit all of it in this special account. Assuming you open your account in August and make no withdrawals for the first year, how much money will you have in your account at the end of February (after six months)? How much will you have at the end of the following July (after one full year)?

26. You have a savings account of \$3000 with an interest rate of 6.8%.

a. How much interest would be earned in two years if the interest is compounded annually?

b. How much interest would be earned in two years if the interest is compounded semiannually?

c. In which case do you make more money on interest? Explain why this is so.

27. If $2500 is invested in a continuously compounded certificate of deposit with an annual interest rate of 4.2%, what would be the account balance at the end of three years?

28. The new furniture store in town boasts a special in which you can buy any set of furniture in their store and make no monthly payments for the first year. However, the fine print says that the interest rate of 7.25% is compounded quarterly beginning when you buy the furniture. You are considering buying a set of living room furniture for $4000 but know you cannot save up more than $4500 in one year's time. Can you fully pay off your furniture on the one year anniversary of having bought the furniture? If so, how much money will you have left over? If not, how much more money will you need?

29. When Nicole was born, her grandmother was so excited about her birth that she opened a certificate of deposit in Nicole's honor to help send her to college. Now at age 18, Nicole's account has $81,262.93. How much did her grandmother originally invest if the interest rate has been 8.1% compounded annually?

30. Inflation is a relative measure of your purchasing power over time. The formula for inflation is the same as the compound interest formula, but with $n = 1$. Given the current values below, what will the values of the following items be 10 years from now if inflation is at 6.4%?

a. an SUV: $38,000
b. a loaf of bread: $1.79
c. a gallon of milk: $3.40
d. your salary: $34,000

31. Depreciation is the decrease of an item's value and can be determined using a formula similar to that for compound interest:

$$V = P(1-r)^t,$$

where V is the new value.

If the particular car you buy upon graduation from college costs $17,500 and depreciates at a rate of 16% per year, what will the value of the car be in 5 years when you pay it off?

32. Assume the interest on your credit card is compounded continuously with an APR (annual percentage rate) of 19.8%. If you put your first term bill of $3984 on your credit card, but do not have to make payments until you graduate (4 years later), how much will you owe when you start making payments?

33. Suppose you deposit $5000 in an account for five years at an annual interest rate of 8.5%.

a. What would be the ending account balance if the interest is continuously compounded?
b. What would be the ending account balance if the interest is compounded daily?
c. Are these two answers similar? Why or why not?

Use the regression commands on a TI-84 Plus to fit the requested curves. See Examples 6 and 7.

34. Linda invested $10,000 in the stock market five years ago and has recorded the value of her investment annually since then, as shown in the table.

Year	Value
0	$10,000
1	$10,800
2	$11,400
3	$12,300
4	$13,200
5	$14,100

a. Find an exponential function that models the growth of her investment over the past five years.

b. Use your result to estimate the value of her investment after one more year, assuming she continues to earn the same effective annual interest rate.

35. The table contains US population data for the census years from 1850 to 1900.

Year	Population
1850	23,191,876
1860	31,443,321
1870	38,558,371
1880	50,189,209
1890	62,979,766
1900	76,212,168

Source: US Census Bureau

a. Use linear, quadratic, and exponential regression to fit curves to the data.

b. Use your results to extrapolate back in time to estimate the US population in the year 1800, and compare the estimates with the actual US Census population in that year of 5,308,483. What do you conclude about extrapolation of these regression models so far outside the known data?

36. A biologist conducts a six-month field study of a small plot of land, collecting data on, among other things, the population of a species of meadow mouse.

Month	Population
0	2
1	3
2	5
3	7
4	10
5	12
6	13

a. Fit an exponential curve to the data, and use your result to extrapolate the mouse population at the end of one year.

b. Fit a logistic curve to the data, and use your result to extrapolate the mouse population at the end of one year.

c. What do you conclude about these two regression models and extrapolation so far outside the known data?

TECHNOLOGY

Use a graphing utility to sketch the graphs of the following functions.

37. $m(x)=1-3e^{x}$

38. $p(x)=e^{4x}-2$

39. $b(x)=\frac{1}{e^{x-2}}$

40. $m(x)=e^{2x^{2}-3x+1}$

41. $g(x)=e^{x+3}-3$

42. $m(x)=6e^{2x}-2$

6.3 LOGARITHMIC FUNCTIONS AND THEIR GRAPHS

TOPICS

1. Characteristics of logarithmic functions
2. Graphing logarithmic functions
3. Evaluating elementary logarithmic expressions
4. Solving elementary logarithmic equations
5. Common and natural logarithms

TOPIC 1: Characteristics of Logarithmic Functions

Currently, we are only able to solve a small subset of possible exponential equations.

Solvable	Not Easily Solvable Yet
We can solve the equation $$2^x = 8$$ by writing 8 as 2^3 and equating exponents. This is an example of an elementary exponential equation we learned to solve in Section 6.1.	We cannot solve the equation $$2^x = 9$$ in the same way, although this equation is only slightly different. All we can say at the moment is that x must be a bit larger than 3.
If we know A, P, n, and t, we can solve the compound interest equation $$A = P\left(1 + \frac{r}{n}\right)^{nt}$$ for the annual interest rate r, as we did in Example 4 of Section 6.2.	If we know A, P, and t, it is not so easy to solve the continuously compounded interest equation $$A = Pe^{rt}$$ for the annual interest rate r.

TABLE 1

In both of the first two equations, the variable x appears in the exponent (making them exponential equations), and we aren't able to rewrite the equation with the variable outside the exponent. We can only solve the first equation because we can rewrite 8 as a power of 2, and even this doesn't remove the variable from the exponent.

In the second pair of equations, we are able to solve for r in the first case, but it is difficult in the continuous-compounding case: once again the variable is inconveniently "stuck" in the exponent. We need a way of undoing exponentiation.

This might seem familiar. We have "undone" functions before (whether we realized it at the time or not) by finding inverses of functions. Therefore, we need to find the inverse of the general exponential function $f(x) = a^x$. Fortunately, we know before we start that $f(x) = a^x$ has an inverse, because the graph of f passes the horizontal line test (regardless of whether $0 < a < 1$ or $a > 1$).

If we apply the algorithm for finding inverses of functions, we have the following.

$f(x) = a^x$	Begin with the function.
$y = a^x$	Replace $f(x)$ with y, and proceed to solve for x.
$? = x$	At this point, we are stuck again. What is x equal to?

No concept or notation that we have encountered up to this point allows us to solve the equation $y = a^x$ for the variable x, and this is the reason for introducing a new class of functions called *logarithms*.

DEFINITION: Logarithmic Functions

Let a be a fixed positive real number not equal to 1. The **logarithmic function with base *a*** is defined to be the inverse of the exponential function with base a, and is denoted $\log_a x$. In symbols, if $f(x) = a^x$, then $f^{-1}(x) = \log_a x$.

In equation form, the definition of logarithm means that the equations

$$x = a^y \text{ and } y = \log_a x$$

are equivalent. Note that a is the base in both equations: either the base of the exponential function or the base of the logarithmic function.

Example 1: Exponential and Logarithmic Equations

Use the definition of logarithmic functions to rewrite the following exponential equations as logarithmic equations.

a. $8 = 2^3$ **b.** $5^4 = 625$ **c.** $7^x = z$

Then rewrite the following logarithmic equations as exponential equations.

d. $\log_3 9 = 2$ **e.** $3 = \log_8 512$ **f.** $y = \log_2 x$

Solution

a. $3 = \log_2 8$

b. $4 = \log_5 625$

Note that in each case, the base of the exponential equation is the base of the logarithmic equation.

c. $x = \log_7 z$

d. $3^2 = 9$

e. $8^3 = 512$

f. $2^y = x$

TOPIC 2: Graphing Logarithmic Functions

Since logarithmic functions are inverses of exponential functions, we can learn a great deal about the graphs of logarithmic functions by recalling the graphs of exponential functions. For instance, the domain of $\log_a x$ (for any allowable a) is the positive real numbers, because the positive real numbers constitute the range of a^x (the domain of f^{-1} is the range of f). Similarly, the range of $\log_a x$ is the entire set of real numbers, because the real numbers make up the domain of a^x.

Recall that the graphs of a function and its inverse are reflections of one another with respect to the line $y = x$. Since exponential functions come in two forms based on the value of a $(0 < a < 1$ and $a > 1)$, logarithmic functions also fall into two categories. In both of the graphs in Figure 1, the dashed curve is the graph of an exponential function and the solid curve is the corresponding logarithmic function.

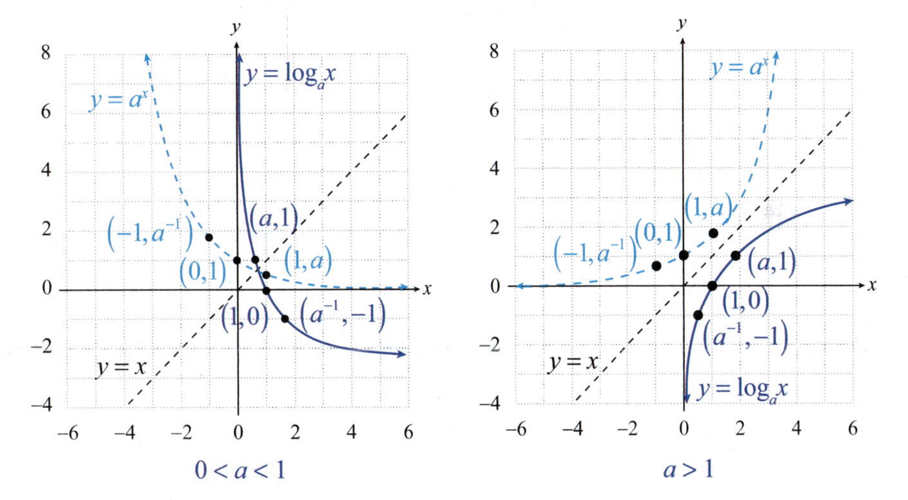

FIGURE 1: The Two Classes of Logarithmic Functions

In all cases, the points $(1,0)$, $(a,1)$, and $(a^{-1},-1)$ lie on the graph of a logarithmic function with base a. Frequently, these points will be enough to get a good idea of the shape of the graph.

Note that the domain of each of the logarithmic functions in Figure 1 is indeed $(0, \infty)$, and that the range in each case is $(-\infty, \infty)$. Also note that the y-axis is a vertical asymptote for both, and that neither has a horizontal asymptote.

⚠ CAUTION

While it may appear so on the graph, logarithmic functions do **not** have a horizontal asymptote. The reason that logarithmic functions often look like they have a horizontal asymptote is because they are among the slowest growing functions in mathematics! Consider the function $f(x) = \log_2 x$. We have $f(1024) = 10$, $f(2048) = 11$, and $f(4096) = 12$. While the function may increase its value very slowly as x increases, it never approaches an asymptote.

Example 2: Graphing Logarithmic Functions

Sketch the graphs of the following logarithmic functions.

a. $f(x) = \log_3 x$ **b.** $g(x) = \log_{\frac{1}{2}} x$

> **NOTE**
>
> Once again, plotting a few key points will provide a good idea of the shape of the function.

Solution

a.

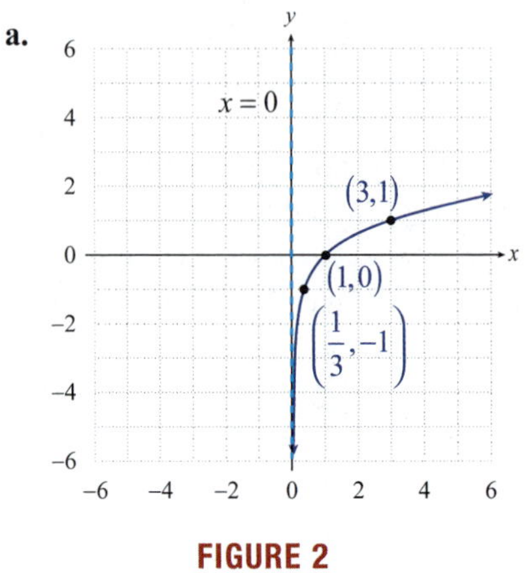

FIGURE 2

b.

y
x
x = 0
$\left(\frac{1}{2},1\right)$
(1,0)
(2,−1)
6
4
2
0
−2
−4
−6
−6 −4 −2 0 2 4 6

FIGURE 3

Example 3: Graphing Logarithmic Functions

Sketch the graph of the following functions. State their domain and range.

a. $f(x) = \log_3(x+2)+1$ **b.** $g(x) = \log_2(-x-1)$

c. $h(x) = \log_{\frac{1}{2}} x - 2$

> **NOTE**
>
> Begin by graphing the base function, then apply any transformations.

Solution

a.

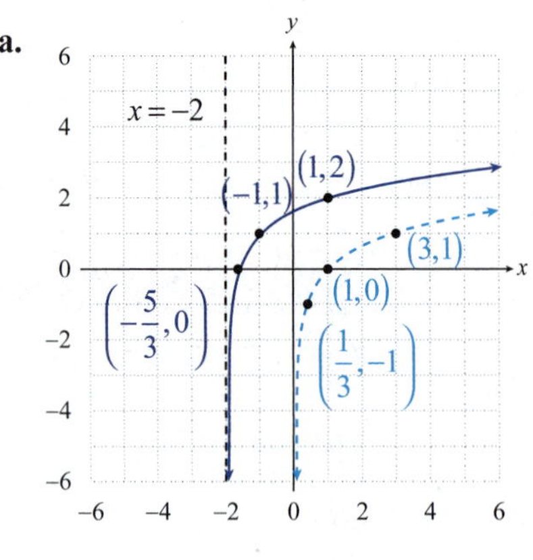

FIGURE 4

Begin by graphing the base function, which is $\log_3 x$. This is the dashed curve.

Since x has been replaced by $x+2$, we shift the graph 2 units to the left.

To find the graph of f, we shift the result up 1 unit, since 1 has been added to the function. This is the solid curve.

Note that the asymptote has also shifted to the left.

Domain $= (-2,\infty)$, Range $= (-\infty,\infty)$

b.

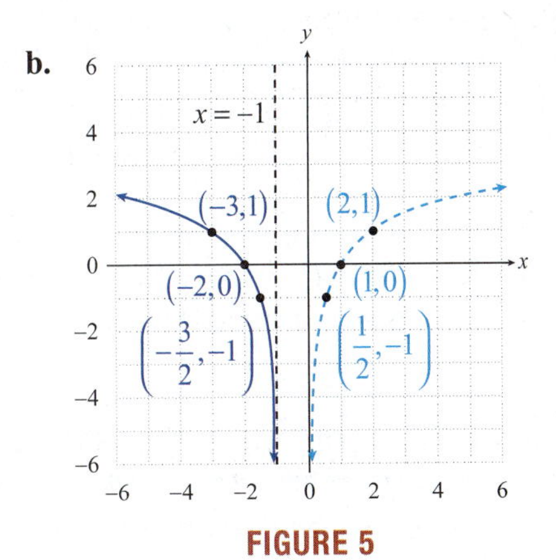

The basic shape of the graph of g is the same as the shape of $y = \log_2 x$, which is the dashed curve.

To obtain g from $\log_2 x$, the variable x is replaced with $x - 1$, which shifts the graph 1 unit to the right, and then x is replaced by $-x$, which reflects the graph with respect to the y-axis.

FIGURE 5

Domain $= (-\infty,-1)$, Range $= (-\infty,-\infty)$

c.

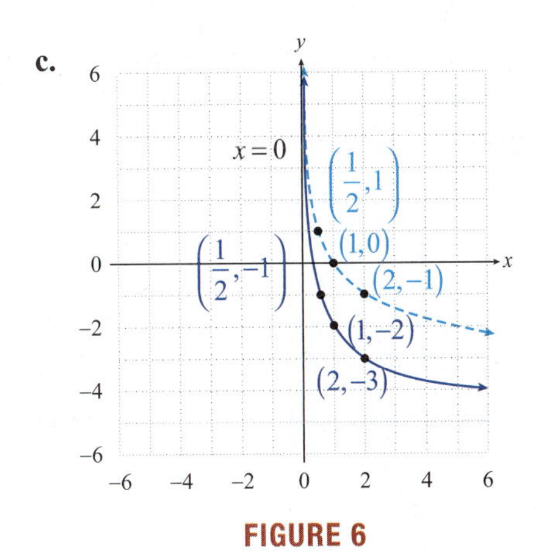

We begin with the dashed curve, which is the graph of $\log_{\frac{1}{2}} x$.

We then shift the graph 2 units

down to obtain the graph of the function h.

FIGURE 6

Domain $= (0,\infty)$, Range $= (-\infty,\infty)$

TOPIC 3: Evaluating Elementary Logarithmic Expressions

Now that we have graphed logarithmic functions, we can augment our understanding of their behavior with a few algebraic observations. These will enable us to evaluate some logarithmic expressions and solve some elementary logarithmic equations.

First, our work in Example 2 suggests that the point $(1,0)$ is always on the graph of $\log_a x$ for any allowable base a, and this is indeed the case. A similar observation is that $(a,1)$ is always on the graph of $\log_a x$. These two facts are actually just restatements of two corresponding facts about exponential functions, a consequence of the definition of logarithms.

$$\log_a 1 = 0, \text{ because } a^0 = 1$$

$$\log_a a = 1, \text{ because } a^1 = a$$

More generally, we can use the fact that the functions $\log_a x$ and a^x are inverses of one another to write the following.

$$\log_a\left(a^x\right) = x \text{ and } a^{\log_a x} = x$$

In Example 4, we use the first statement to evaluate logarithmic expressions. Note the similarity to solving exponential equations; we rewrite the argument of the logarithm as a power of the base, allowing us to simplify the expression.

Example 4: Logarithmic Expressions

Evaluate the following logarithmic expressions.

a. $\log_5 25$ **b.** $\log_{\frac{1}{2}} 2$ **c.** $\log_4 8$ **d.** $\log_\pi\left(\sqrt{\pi}\right)$

e. $\log_{17} 1$ **f.** $\log_{16} 4$ **g.** $\log_{\frac{1}{9}} 3$ **h.** $\log_{10}\left(\frac{1}{100}\right)$

Solution

NOTE

We write each of the equivalent exponential equations as a reference.

a. $\log_5 25 = \log_5\left(5^2\right)$ Rewrite 25 as a power of 5.

$= 2$ Equivalent exponential equation: $25 = 5^2$

b. $\log_{\frac{1}{2}} 2 = \log_{\frac{1}{2}}\left(\left(\frac{1}{2}\right)^{-1}\right)$ Rewrite 2 as a power of $\frac{1}{2}$.

$= -1$ Equivalent exponential equation: $2 = \left(\frac{1}{2}\right)^{-1}$

c. $\log_4 8 = \log_4\left(4^{\frac{3}{2}}\right)$ Rewrite 8 as a power of 4.

$= \frac{3}{2}$ Equivalent exponential equation: $8 = 4^{\frac{3}{2}}$

d. $\log_\pi\left(\sqrt{\pi}\right) = \log_\pi\left(\pi^{\frac{1}{2}}\right)$ Rewrite the radical as a rational exponent.

$= \frac{1}{2}$ Equivalent exponential equation: $\sqrt{\pi} = \pi^{\frac{1}{2}}$

e. $\log_{17} 1 = 0$ Equivalent exponential equation: $1 = 17^0$

f. $\log_{16} 4 = \log_{16}\left(16^{\frac{1}{2}}\right)$ Rewrite 4 as a power of 16.

$= \frac{1}{2}$ Equivalent exponential equation: $4 = 16^{\frac{1}{2}}$

g. $\log_{\frac{1}{9}} 3 = \log_{\frac{1}{9}}\left(\left(\frac{1}{9}\right)^{-\frac{1}{2}}\right)$ Rewrite 3 as a power of $\frac{1}{9}$.

$= -\frac{1}{2}$ Equivalent exponential equation: $3 = \left(\frac{1}{9}\right)^{-\frac{1}{2}}$

h. $\log_{10}\left(\frac{1}{100}\right) = \log_{10}\left(10^{-2}\right)$ Rewrite $\frac{1}{100}$ as a power of 10.

$= -2$ Equivalent exponential equation: $\frac{1}{100} = 10^{-2}$

TOPIC 4: Solving Elementary Logarithmic Equations

Although a given exponential equation can be written in an equivalent logarithmic form (and vice versa), it's commonly the case that one form or the other is easier to solve. We demonstrate this fact in Example 5, starting off in each case with an equation that is difficult to solve in its original form.

Example 5: Solving Logarithmic Equations

Use elementary properties of exponents and logarithms to solve the following equations.

a. $\log_6(2x) = -1$ **b.** $3^{\log_{3x} 2} = 2$ **c.** $\log_2 8^x = 5$

Solution

a. $\log_6(2x) = -1$

$2x = 6^{-1}$ Convert the equation to exponential form.

$2x = \frac{1}{6}$ Simplify the exponent, then solve for x.

$x = \frac{1}{12}$

b. $3^{\log_{3x} 2} = 2$ This time, we convert from the exponential form to the logarithmic form.

$\log_{3x} 2 = \log_3 2$

$3x = 3$ We can equate the bases, just like we equate the exponents in an elementary exponential equation.

$x = 1$

c. $\log_2(8^x) = 5$

$8^x = 2^5$ Rewrite the equation in exponential form.

$(2^3)^x = 2^5$ Rewrite 8 as a power of 2.

$2^{3x} = 2^5$ Simplify using properties of exponents.

$3x = 5$ Set the exponents equal to each other, then solve.

$x = \frac{5}{3}$

TOPIC 5: Common and Natural Logarithms

We have already mentioned the fact that the number e, called the natural base, plays a fundamental role in many important real-world situations and in higher mathematics, so it is not surprising that the logarithmic function with base e is worthy of special attention. For historical reasons, namely the fact that our number system is based on powers of 10, the logarithmic function with base 10 is also singled out.

DEFINITION: Common and Natural Logarithms

- The function $\log_{10} x$ is called the **common logarithm** and is usually written $\log x$.
- The function $\log_e x$ is called the **natural logarithm** and is usually written $\ln x$.

Another way in which these particular logarithms are special is that most calculators, if they are capable of calculating logarithms at all, are only equipped to evaluate common and natural logarithms. Such calculators normally have a button labeled "log" for the common logarithm and a button labeled "ln" for the natural logarithm.

TECHNOLOGY: Inputting Logarithms, Graphing Logarithmic Functions

Common and natural logarithms can be entered into a TI-84 Plus using the log and ln buttons, respectively. For instance, to graph the function $f(x)=\log(3-x)$, press Y= and log. The opening parenthesis appears automatically, so we just have to type in the argument, 3 − X,T,θ,n, and the right-hand parenthesis. When you press graph, the graph shown in Figure 8 should appear.

CAUTION

Notice that the graph appears to cut off around the point where $x = 3$. We know, however, that the graph has a vertical asymptote at $x = 3$, and approaches that asymptote even though it does not show on the graphing calculator.

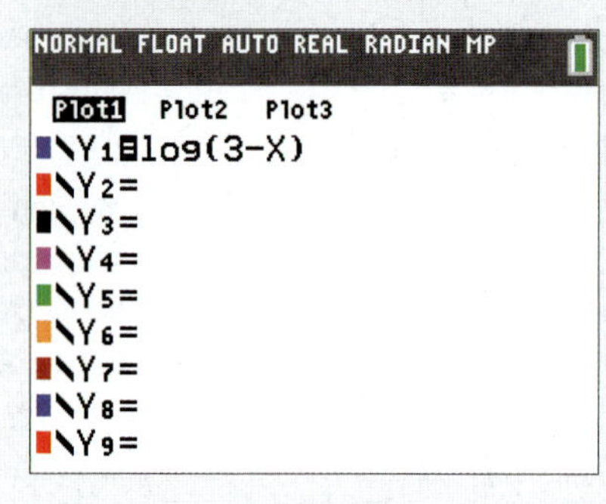

FIGURE 7

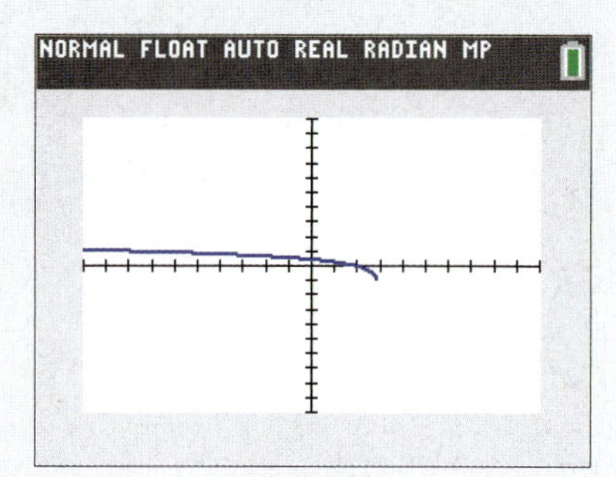

FIGURE 8

PROPERTIES: Properties of Natural Logarithms

$$\ln x = y \quad \Leftrightarrow \quad e^y = x$$

Properties	Reasons
1. $\ln 1 = 0$	Raise e to the power 0 to get 1.
2. $\ln e = 1$	Raise e to the power 1 to get e.
3. $\ln e^x = x$	Raise e to the power x to get e^x.
4. $e^{\ln x} = x$	$\ln x$ is the power to which e must be raised to get x.

Example 6: Evaluating Logarithmic Expressions

Evaluate the following logarithmic expressions.

a. $\ln\left(\sqrt[3]{e}\right)$

b. $\log 1000$

c. $\ln(4.78)$

d. $\log(10.5)$

Solution

a. $\ln\left(\sqrt[3]{e}\right) = \ln\left(e^{\frac{1}{3}}\right) = \frac{1}{3}$

No calculator is necessary for this problem, just an application of an elementary property of logarithms.

b. $\log 1000 = \log 10^3 = 3$

Again, no calculator is required. Equivalent exponential equation: $2 = \left(\frac{1}{2}\right)^{-1}$

c. $\ln(4.78) \approx 1.564$

This time, a calculator is needed, and only an approximate answer can be given. Be sure to use the correct logarithm.

d. $\log(10.5) \approx 1.021$

Again, we must use a calculator, though we can say beforehand that the answer should be only slightly larger than 1, as $\log 10 = 1$ and 10.5 is only slightly larger than 10.

6.3 EXERCISES

PRACTICE

Write the following equations in logarithmic terms.

1. $625 = 5^4$

2. $216 = 6^3$

3. $x^3 = 27$

4. $b^2 = 3.2$

5. $4.2^3 = C$

6. $1.3^2 = V$

7. $4^x = 31$

8. $16^{2x} = 215$

9. $(4x)^{\sqrt{3}} = 13$

10. $e^x = \pi$

11. $2^{e^x} = 11$

12. $4^e = N$

Write the following logarithmic equations as exponential equations.

13. $\log_3 81 = 4$

14. $\log_2\left(\frac{1}{8}\right) = -3$

15. $\log_b 4 = \frac{1}{2}$

16. $\log_y 9 = 2$

17. $\log_2 15 = b$

18. $\log_5 8 = d$

19. $\log_5 W = 12$

20. $\log_7 T = 6$

21. $\log_\pi (2x) = 4$

22. $\log_{\sqrt{3}}(2\pi) = x$

23. $\ln 2 = x$

24. $\ln(5x) = 3$

Sketch the graphs of the following functions. State their domain and range. See Examples 2 and 3.

25. $f(x) = \log_3(x-1)$

26. $g(x) = \log_5(x+2) - 1$

27. $r(x) = \log_{\frac{1}{2}}(x-3)$

28. $p(x) = 3 - \log_2(x+1)$

29. $q(x) = \log_3(2-x)$

30. $s(x) = \log_{\frac{1}{3}}(5-x)$

31. $h(x) = \log_7(x-3) + 3$

32. $m(x) = \log_{\frac{1}{2}}(1-x)$

33. $f(x) = \log_3(6-x)$

34. $p(x) = 4 - \log(x+3)$

35. $s(x) = -\log_{\frac{1}{3}}(-x)$

36. $g(x) = \log_5(2x) - 1$

Match the graph of the appropriate equation to the logarithmic function.

37. $f(x) = \log_2(x-1)$

38. $f(x) = \log_2(2-x)$

39. $f(x) = \log_2(-x)$

40. $f(x) = \log_2(x-3)$

41. $f(x) = 1 - \log_2 x$

42. $f(x) = -\log_2 x$

43. $f(x) = -\log_2(-x)$

44. $f(x) = \log_2 x$

45. $f(x) = \log_2(x+3)$

a.

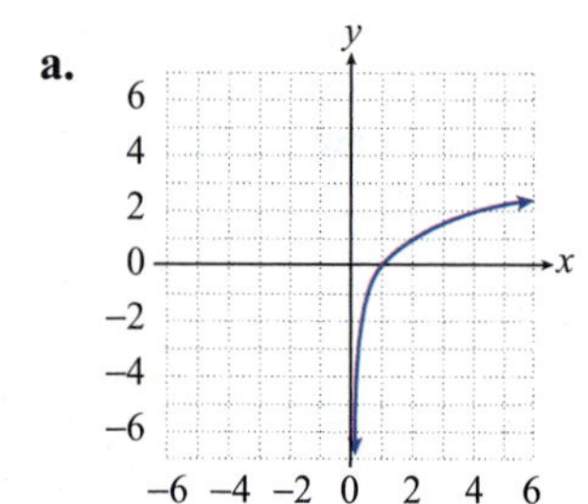

b.

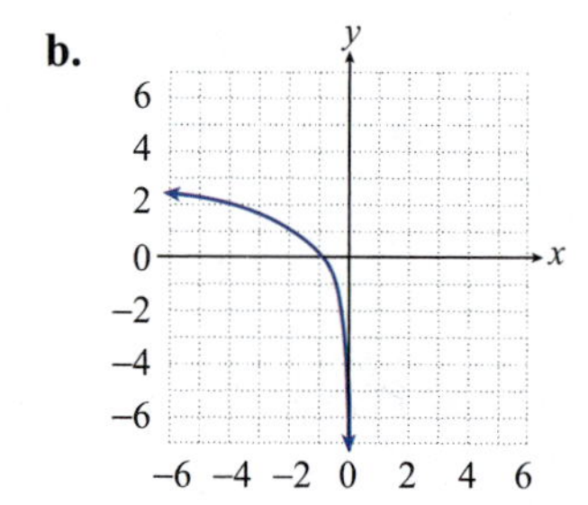

c.

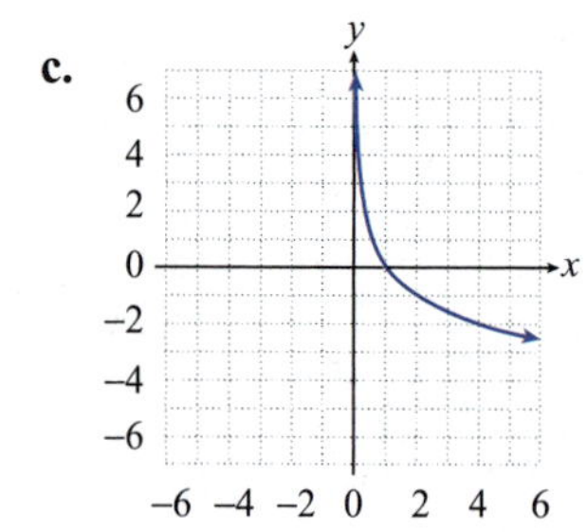

d.

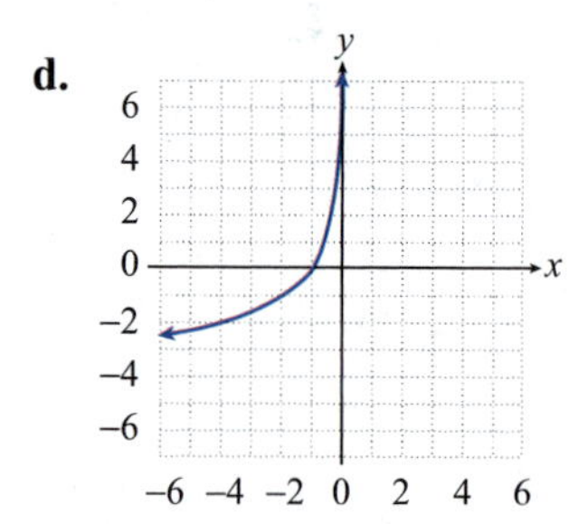

e.

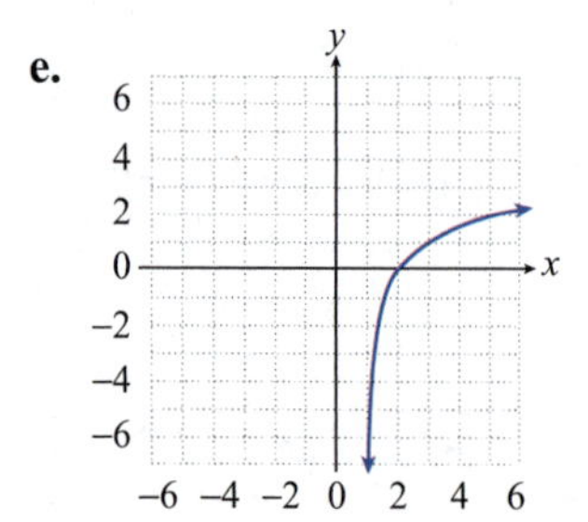

f.

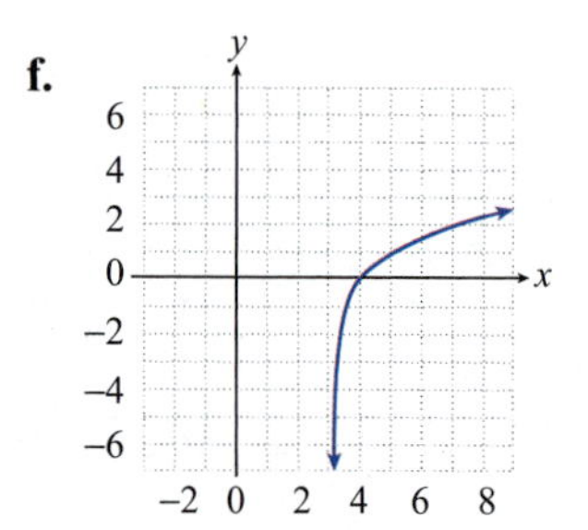

g.

h.

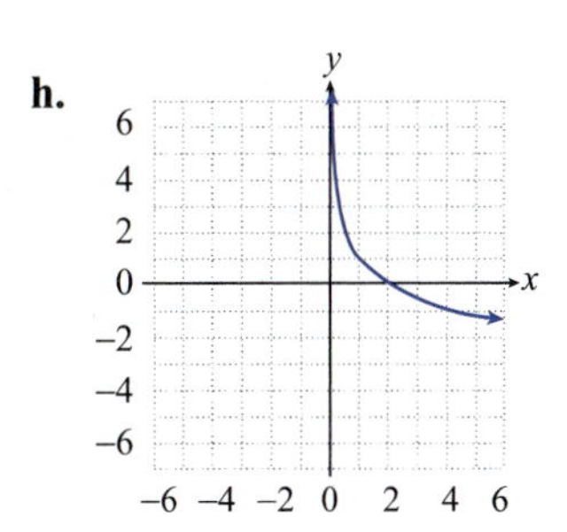

i.

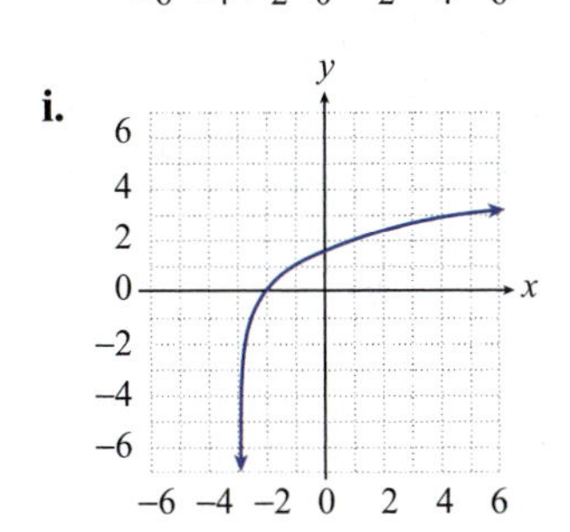

Evaluate the following logarithmic expressions without the use of a calculator. See Examples 4 and 6.

46. $\log_7\left(\sqrt{7}\right)$

47. $\log_{\frac{1}{2}} 4$

48. $\log_9\left(\frac{1}{81}\right)$

49. $\log_3 27$

50. $\log_{27} 3$

51. $\log_9\left(\frac{1}{3}\right)$

52. $\log_{27} 9$

53. $\log_{\frac{1}{16}}\left(\frac{1}{8}\right)$

54. $\log_3\left(\log_{27} 3\right)$

55. $\ln e^{2.89}$

56. $\log(0.0001)$

57. $\log_a\left(a^{\frac{5}{3}}\right)$

58. $\ln\left(\frac{1}{e}\right)$

59. $\log\left(\log\left(10^{10}\right)\right)$

60. $\log_3 1$

61. $\ln\left(\sqrt[5]{e}\right)$

62. $\log_{\frac{1}{16}} 4$

63. $\log_8 4^{\log 1000}$

Use the elementary properties of logarithms to solve the following equations. See Example 5.

64. $\log_{16} x = \frac{3}{4}$

65. $\log_{16}\left(x^{\frac{1}{2}}\right) = \frac{3}{4}$

66. $\log_{16} x = -\frac{3}{4}$

67. $\log_5\left(5^{\log_3 x}\right) = 2$

68. $\log_a\left(a^{\log_b x}\right) = 0$

69. $\log_3\left(9^{2x}\right) = -2$

70. $\log_{\frac{1}{3}}\left(3^x\right) = 2$

71. $\log_7(3x) = -1$

72. $4^{\log_3 x} = 0$

73. $\log\left(x^{10}\right) = 10$

74. $\log_x\left(\log_{\frac{1}{2}}\left(\frac{1}{4}\right)\right) = 1$

75. $6^{\log_x\left(e^2\right)} = e$

Hint: Note that $\log_a b = \log_{a^2} b^2$. This follows from the fact that

$\log_a b = y \Leftrightarrow b = a^y \Leftrightarrow b^2 = a^{2y} = \left(a^2\right)^y \Leftrightarrow \log_{a^2} b^2 = y.$

Solve the following logarithmic equations, using a calculator if necessary to evaluate the logarithms. See Examples 5 and 6. Express your answer either as a fraction or a decimal rounded to two decimal places.

76. $\log(3x) = 2.1$

77. $\log(x^2) = -2$

78. $\ln(x+1) = 3$

79. $\ln(2x) = -1$

80. $\ln(e^x) = 5.6$

81. $\ln(\ln(x^2)) = 0$

82. $\log 19 = 3x$

83. $\log(e^x) = 5.6$

84. $\log_9(2x-1) = 2$

85. $\log(\log(x-2)) = 1$

86. $\log(300^{\log x}) = 9$

6.4 LOGARITHMIC PROPERTIES AND MODELS

TOPICS

1. Properties of logarithms
2. The change of base formula
3. Applications of logarithmic functions
4. Logarithmic regression

TOPIC 1: Properties of Logarithms

In Section 6.3, we introduced logarithmic functions and studied some of their elementary properties. The motivation was our inability, at that time, to solve certain exponential equations. Let us reconsider the two sample problems that initiated our discussion of logarithms and see if we have made progress. We begin with the continuously compounding interest problem.

Example 1: Continuously Compounded Interest

Anne reads an ad in the paper for a new bank in town. The bank is advertising "continuously compounded savings accounts" in an attempt to attract customers, but fails to mention the annual interest rate. Curious, she goes to the bank and is told by an account agent that if she were to invest \$10,000 in an account, her money would grow to \$10,202.01 in one year's time. But strangely, the agent also refuses to divulge the yearly interest rate. What rate is the bank offering?

Solution

We need to solve the equation $A = Pe^{rt}$ for r, given that $A = 10,202.01$, $P = 10,000$, and $t = 1$.

$10,202.01 = 10,000e^{r(1)}$	Substitute the given values.
$1.020201 = e^{r}$	Divide both sides by 10,000.
$\ln(1.020201) = r$	Convert to logarithmic form.
$r \approx 0.02$	Evaluate using a calculator.

Note that we use the natural logarithm since the base of the exponential function is e. While we must use a calculator, we can now solve the equation for r.

Example 2: Solving Exponential Equations

Solve the equation $2^x = 9$.

Solution

We convert the equation to logarithmic form to obtain the solution $x = \log_2 9$. Unfortunately, this answer still doesn't tell us anything about x in decimal form, other than that it is bound to be slightly more than 3. Further, we can't use a calculator to evaluate $\log_2 9$ since the base is neither 10 nor e.

As Example 2 shows, our ability to work with logarithms is still incomplete. In this section we will derive some important properties of logarithms that allow us to solve more complicated equations, as well as provide a decimal approximation to the solution of $2^x = 9$.

The following properties of logarithmic functions are analogs of corresponding properties of exponential functions, a consequence of how logarithms are defined.

PROPERTIES: Properties of Logarithms

Let a (the logarithmic base) be a positive real number not equal to 1, let x and y be positive real numbers, and let r be any real number.

1. $\log_a(xy) = \log_a x + \log_a y$ ("the log of a product is the sum of the logs")
2. $\log_a\left(\frac{x}{y}\right) = \log_a x - \log_a y$ ("the log of a quotient is the difference of the logs")
3. $\log_a(x^r) = r\log_a x$ ("the log of something raised to a power is the power times the log")

We illustrate the link between these properties and the related properties of exponents by proving the first one. Try proving the second and third as further practice.

Proof

Let $m = \log_a x$ and $n = \log_a y$. The equivalent exponential forms of these two equations are $x = a^m$ and $y = a^n$.

Since we are interested in the product xy, note that

$$xy = a^m a^n = a^{m+n}.$$

The statement $xy = a^{m+n}$ can then be converted to logarithmic form, giving us

$$\log_a(xy) = m + n.$$

Referring back to the definition of m and n, we have $\log_a(xy) = \log_a x + \log_a y$.

If the properties of logarithms appear strange at first, remember that they are just the properties of exponents restated in logarithmic form.

CAUTION

Errors in working with logarithms often arise from incorrect recall of the logarithmic properties. The comparisons below highlight some common mistakes.

Incorrect Statements	Correct Statements
$\log_a(x+y)=\log_a x+\log_a y$	$\log_a(xy)=\log_a x+\log_a y$
$\log_a(xy)=(\log_a x)(\log_a y)$	$\log_a(xy)=\log_a x+\log_a y$
$\dfrac{\log_a x}{\log_a y}=\log_a x-\log_a y$	$\log_a\left(\dfrac{x}{y}\right)=\log_a x-\log_a y$
$\dfrac{\log_a x}{\log_a y}=\log_a\left(\dfrac{x}{y}\right)$	$\log_a\left(\dfrac{x}{y}\right)=\log_a x-\log_a y$
$\dfrac{\log_a(xz)}{\log_a(yz)}=\dfrac{\log_a x}{\log_a y}$	$\dfrac{\log_a(xz)}{\log_a(yz)}=\dfrac{\log_a x+\log_a z}{\log_a y+\log_a z}$

In some situations, we will find it useful to use properties of logarithms to decompose a complicated expression into a sum or difference of simpler expressions, while in other situations we will do the reverse, combining a sum or a difference of logarithms into one logarithm. Examples 3 and 4 illustrate these processes.

NOTE

As long as the base is the same for each term, its value does not affect the use of the properties.

Example 3: Expanding Logarithmic Expressions

Use the properties of logarithms to expand the following expressions as much as possible (that is, decompose the expressions into sums or differences of the simplest possible terms).

a. $\log_4\left(64x^3\sqrt{y}\right)$ **b.** $\log_a\left(\sqrt[3]{\dfrac{xy^2}{z^4}}\right)$ **c.** $\log\left(\dfrac{2.7\times10^4}{x^{-2}}\right)$

Solution

a.

$$\log_4\left(64x^3\sqrt{y}\right)=\log_4 64+\log_4\left(x^3\right)+\log_4\left(\sqrt{y}\right)$$

Use the first property to rewrite the expression as three terms.

$$=\log_4\left(4^3\right)+\log_4\left(x^3\right)+\log_4\left(y^{\frac{1}{2}}\right)$$

$$=3+3\log_4 x+\frac{1}{2}\log_4 y$$

We can evaluate the first term and rewrite the second and third terms using the third property.

b.

$$\log_a\left(\sqrt[3]{\frac{xy^2}{z^4}}\right)=\log_a\left(\left(\frac{xy^2}{z^4}\right)^{\frac{1}{3}}\right)$$

Rewrite the radical as an exponent.

$$=\frac{1}{3}\log_a\left(\frac{xy^2}{z^4}\right)$$

Bring the exponent in front of the logarithm using the third property.

$$=\frac{1}{3}\left(\log_a x+\log_a\left(y^2\right)-\log_a\left(z^4\right)\right)$$

Expand the expression using the first two properties.

$$=\frac{1}{3}\left(\log_a x+2\log_a y-4\log_a z\right)$$

Apply the third property to the terms that result.

c. Recall that if a base is not explicitly written, it is assumed to be 10. This base is convenient when working with numbers in scientific notation.

$$\log\left(\frac{2.7\times10^4}{x^{-2}}\right) = \log(2.7) + \log(10^4) - \log(x^{-2})$$ Expand using the first and second properties.

$$= \log(2.7) + 4 + 2\log x$$ Evaluate the first two terms and use the third property on the last term.

$$\approx 4.43 + 2\log x$$

It is appropriate to either evaluate $\log(2.7)$ or leave it in exact form. Use the context of the problem to decide which form is more convenient.

Example 4: Condensing Logarithmic Expressions

Use the properties of logarithms to condense the following expressions as much as possible (that is, rewrite the expressions as a sum or difference of as few logarithms as possible).

a. $2\log_3\left(\frac{x}{3}\right) - \log_3\left(\frac{1}{y}\right)$ **b.** $\ln(x^2) - \frac{1}{2}\ln y + \ln 2$ **c.** $\log_b 5 + 2\log_b(x^{-1})$

Solution

> **NOTE**
> Often, there will be multiple orders in which we can apply the properties to find the final result.

a. $$2\log_3\left(\frac{x}{3}\right) - \log_3\left(\frac{1}{y}\right) = \log_3\left(\frac{x}{3}\right)^2 + \log_3\left(\frac{1}{y}\right)^{-1}$$ Use the third property to make the coefficients appear as exponents.

$$= \log_3\left(\frac{x^2}{9}\right) + \log_3 y$$ Evaluate the exponents.

$$= \log_3\left(\frac{x^2y}{9}\right)$$ Combine terms using the first property.

b. $$\ln(x^2) - \frac{1}{2}\ln y + \ln 2 = \ln(x^2) - \ln\left(y^{\frac{1}{2}}\right) + \ln 2$$ Rewrite each term to have a coefficient of 1 or -1 using the third property.

$$= \ln\left(\frac{x^2}{y^{\frac{1}{2}}}\right) + \ln 2$$ We can then combine the terms using the second property.

$$= \ln\left(\frac{2x^2}{y^{\frac{1}{2}}}\right) \text{ or } \ln\left(\frac{2x^2}{\sqrt{y}}\right)$$ The final answer can be written in several different ways, two of which are shown.

c. $$\log_b 5 + 2\log_b(x^{-1}) = \log_b 5 + \log_b(x^{-2})$$ Rewrite the coefficient as an exponent, then combine terms.

$$= \log_b(5x^{-2}) \text{ or } \log_b\left(\frac{5}{x^2}\right)$$

TOPIC 2: The Change of Base Formula

The properties we just derived can be used to provide an answer to a question about logarithms that has been left unanswered thus far. A specific illustration of the question arose in Example 2: how do we determine the decimal form of a number like $\log_2 9$?

Surprisingly, to answer this question we will undo our work in Example 2. We assign a variable to the result $\log_2 9$, convert the resulting logarithmic equation into exponential form, take the natural logarithm of both sides of the equation, and then solve for the variable.

$x = \log_2 9$	Let x equal the result from before, $\log_2 9$.
$2^x = 9$	Convert the equation to exponential form.
$\ln(2^x) = \ln 9$	Take the natural logarithm of both sides.
$x\ln 2 = \ln 9$	Move the variable out of the exponent using the third property of logarithms.
$x = \dfrac{\ln 9}{\ln 2}$	Simplify.
$x \approx 3.17$	Evaluate with a calculator.

While using a logarithm with any base will give the correct solution to this problem, if a calculator is to be used to approximate the number $\log_2 9$, there are (for most calculators) only two good choices: the natural log and the common log. If we had done the work above with the common logarithm, the final answer would have been the same. That is,

$$\frac{\log 9}{\log 2} \approx 3.17.$$

And even though it would not be easy to evaluate,

$$\frac{\log_a 9}{\log_a 2} \approx 3.17$$

for any allowable logarithmic base a.

More generally, a logarithm with base b can be converted to a logarithm with base a through the same reasoning. This allows us to evaluate all logarithmic expressions.

f(x) FORMULA: Change of Base Formula

Let a and b both be positive real numbers, neither of them equal to 1, and let x be a positive real number. Then

$$\log_b x = \frac{\log_a x}{\log_a b}.$$

NOTE

Both the common and natural logarithms work in solving these problems.

TECHNOLOGY

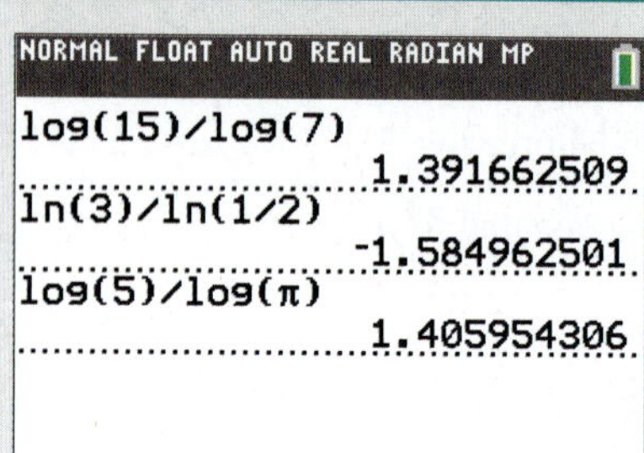

Example 5: Change of Base Formula

Evaluate the following logarithmic expressions, using the base of your choice.

a. $\log_7 15$ **b.** $\log_{\frac{1}{2}} 3$ **c.** $\log_\pi 5$

Solution

a. $\log_7 15 = \dfrac{\ln 15}{\ln 7}$ Apply the change of base formula.

≈ 1.392 Evaluate using a calculator.

b. $\log_{\frac{1}{2}} 3 = \dfrac{\log 3}{\log\left(\frac{1}{2}\right)}$ Apply the change of base formula. This time we use the common logarithm.

≈ -1.585 Since the base of the logarithm is a fraction, we should expect a negative answer.

c. $\log_\pi 5 = \dfrac{\log 5}{\log \pi}$ Once again, we apply the change of base formula, then evaluate using a calculator.

≈ 1.406

TOPIC 3: Applications of Logarithmic Functions

Logarithms appear in many different contexts and have a wide variety of uses. This is due partly to the fact that logarithmic functions are the inverses of exponential functions, and partly to the logarithmic properties we have discussed. In fact, the mathematician who can be most credited for "inventing" logarithms, John Napier (1550–1617) of Scotland, was inspired in his work by the convenience of what we now call logarithmic properties.

Computationally, logarithms are useful because they relocate exponents as coefficients, thus making them easier to work with. Consider, for example, a very large number such as 3×10^{17} or a very small number such as 6×10^{-9}. The common logarithm (used because it has a base of 10) expresses these numbers on a more comfortable scale.

$$\log\left(3\times10^{17}\right) = \log 3 + \log\left(10^{17}\right) = 17 + \log 3 \approx 17.477$$
$$\log\left(6\times10^{-9}\right) = \log 6 + \log\left(10^{-9}\right) = -9 + \log 6 \approx -8.222$$

Napier, working long before the advent of electronic calculating devices, devised logarithms in order to take advantage of this property.

In chemistry, the concentration of hydronium ions in a solution determines its acidity. Since concentrations are small numbers that vary over many orders of magnitude, it is convenient to express acidity in terms of the pH scale, as follows.

DEFINITION: The pH Scale

The **pH** of a solution is defined to be $-\log\left[H_3O^+\right]$, where $\left[H_3O^+\right]$ is the concentration of hydronium ions in units of moles/liter. Solutions with a pH less than 7 are said to be *acidic*, while those with a pH greater than 7 are *basic*.

FIGURE 1: pH of Common Substances

Example 6: The pH Scale

If a sample of orange juice is determined to have a $\left[H_3O^+\right]$ concentration of 1.58×10^{-4} moles/liter, what is its pH?

Solution

Applying the above formula (and using a calculator), the pH is equal to

$$\text{pH} = -\log\left(1.58\times10^{-4}\right) \approx -(-3.80) = 3.8.$$

After doing this calculation, the reason for the minus sign in the formula is more apparent. By multiplying the log of the concentration by -1, the pH of a solution is positive, which is convenient for comparative purposes.

The energy released during earthquakes can vary greatly, but logarithms provide a convenient way to analyze and compare the intensity of earthquakes.

DEFINITION: The Richter Scale

Earthquake intensity is measured on the **Richter scale** (named for the American seismologist Charles Richter, 1900–1985). In the original formula that follows, I_0 is the intensity of a just-discernible earthquake, I is the intensity of an earthquake being analyzed, and R is its ranking on the Richter scale.

$$R = \log\left(\frac{I}{I_0}\right)$$

By this measure, earthquakes range from a classification of minor ($R < 4$), to light ($4 \leq R < 5$), to moderate ($5 \leq R < 6$), to strong ($6 \leq R < 7$), to major ($7 \leq R < 8$), to great ($8 \leq R$).

The base 10 logarithm means that every increase of 1 unit on the Richter scale corresponds to an increase by a factor of 10 in the intensity. This is a characteristic of all logarithmic scales. Also, note that a barely discernible earthquake has a rank of 0, since $\log 1 = 0$.

Example 7: The Richter Scale

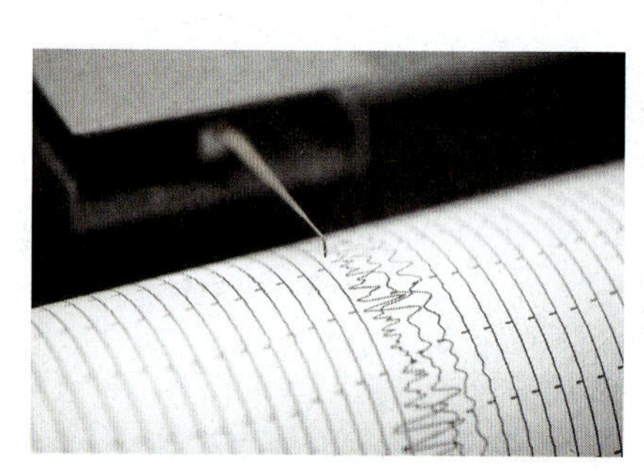

The January 2001 earthquake in the state of Gujarat in India was 7,940,000 times as intense as a 0-level earthquake. What was the Richter ranking of this devastating event?

Solution

If we let I denote the intensity of the Gujarat earthquake, then $I = 7{,}940{,}000 I_0$, so

$$\begin{aligned} R &= \log\left(\frac{7{,}940{,}000 I_0}{I_0}\right) \\ &= \log\left(7.94 \times 10^6\right) \\ &= \log(7.9) + \log\left(10^6\right) \\ &= \log(7.9) + 6 \\ &\approx 6.9. \end{aligned}$$

The Gujarat earthquake thus fell in the category of strong on the Richter scale.

Sound intensity is another quantity that varies greatly, and the measure of how the human ear perceives intensity, in units called decibels (dB), is very similar to the measure of earthquake intensity.

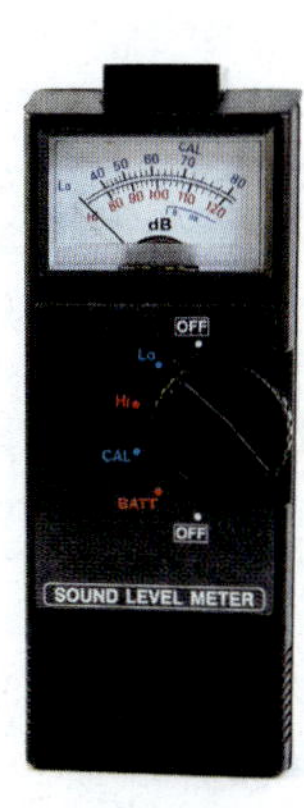

DEFINITION: The Decibel Scale

In the **decibel scale**, I_0 is the intensity of a just-discernible sound, I is the intensity of the sound being analyzed, and D is its decibel level.

$$D = 10\log\left(\frac{I}{I_0}\right)$$

Decibel levels range from 0 for a barely discernible sound, to 60 for the level of normal conversation, to 80 for heavy traffic, to 120 for a loud rock concert, and finally (as far as humans are concerned) to around 160, at which point the eardrum is likely to rupture.

Example 8: The Decibel Scale

Given that $I_0 = 10^{-12}$ watts/meter^2, what is the decibel level of jet airliner's engines at a distance of 45 meters, for which the sound intensity is 50 watts/meter^2?

TECHNOLOGY

```
NORMAL FLOAT AUTO REAL RADIAN MP
10(log(50/10^-12))
                        136.9897
```

Solution

$$\begin{aligned} D &= 10\log\left(\frac{50}{10^{-12}}\right) \\ &= 10\log\left(5 \times 10^{13}\right) \\ &= 10\left(\log 5 + 13\right) \\ &\approx 137 \end{aligned}$$

In other words, the sound level would probably not be literally earsplitting, but it would be very painful.

TOPIC 4: Logarithmic Regression

When a collection of data appears to display logarithmic behavior, **logarithmic regression** can be used to fit a logarithmic curve to the data. To be specific, logarithmic regression is the process of finding values for a and b so that the graph of the function $f(x) = a + b\ln x$ models the given data points as well as possible. Many graphing utilities have built-in functions to perform logarithmic regression. The specific commands vary with the technology, and it should be noted that the results of logarithmic regression may differ depending on the algorithm used by the technology.

TECHNOLOGY

To find and graph the logarithmic function of best fit for a given set of points using a TI-84 Plus, perform the same steps as described in Section 3.2 for linear regression except select LnReg from the CALC menu.

Example 9: Logarithmic Regression

A manufacturer of aviation instruments has designed a new pressure altimeter, a device that determines altitude as a function of atmospheric pressure. The data in Table 1 was collected in order to calibrate the device, where the pressure p (in pascals) was measured by the new device and its altitude h above sea level (in meters) was measured by another instrument of known accuracy. Use the data and logarithmic regression to find a logarithmic function that models altitude as a function of pressure and plot its graph along with the given points.

Pressure p (in pascals)	40,000	50,000	60,000	70,000	80,000	90,000	100,000
Altitude h (in meters)	7309	5670	4279	3064	1982	1005	113

TABLE 1

Solution

Entering the data into a TI-84 Plus and performing a logarithmic regression results in the function $h(p) = 90{,}678.9 - 7859.23 \ln p$. This is the logarithmic curve of best fit for the given data. The data is displayed in Figure 2, and the graph is shown in Figure 3.

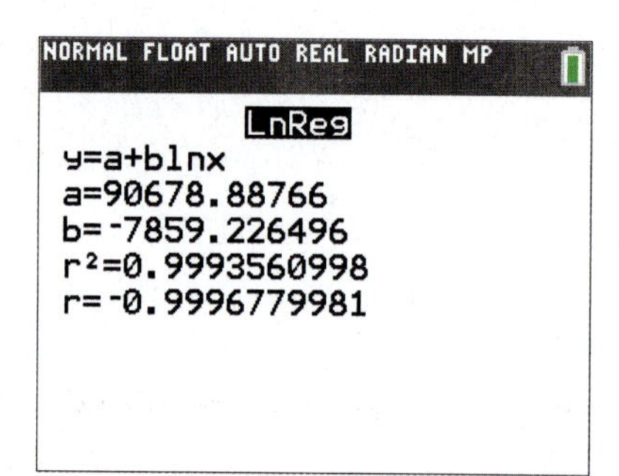

FIGURE 2

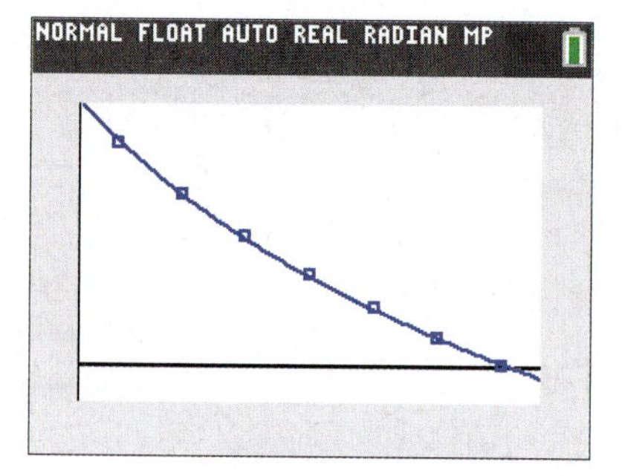

FIGURE 3

6.4 EXERCISES

PRACTICE

Use the properties of logarithms to expand the following expressions as much as possible. Simplify any numerical expressions that can be evaluated without a calculator. See Example 3.

1. $\log_5\left(125x^3\right)$

2. $\ln\left(\frac{x^2y}{3}\right)$

3. $\ln\left(\frac{e^2p}{q^3}\right)$

4. $\log(100x)$

5. $\log_9\left(9xy^{-3}\right)$

6. $\log_6\left(\sqrt[3]{\frac{p^2}{q}}\right)$

7. $\ln\left(\frac{\sqrt{x^3}pq^5}{e^7}\right)$

8. $\log_a\sqrt[5]{\frac{a^4b}{c^2}}$

9. $\log\left(\log\left(100x^3\right)\right)$

10. $\log_3(9x + 27y)$ **11.** $\log\left(\frac{10}{\sqrt{x+y}}\right)$ **12.** $\ln\left(\ln\left(e^{ex}\right)\right)$

13. $\log_2\left(\frac{y^2 + z}{16x^4}\right)$ **14.** $\log\left(\log\left(100{,}000^{2x}\right)\right)$ **15.** $\log_b\left(\sqrt{\frac{x^4 y}{z^2}}\right)$

16. $\ln\left(7x^2 - 42x + 63\right)$ **17.** $\log_b\left(ab^2c^b\right)$ **18.** $\ln\left(\ln\left(e^{e^x}\right)\right)$

Use the properties of logarithms to condense the following expressions as much as possible, writing each answer as a single term with a coefficient of 1. See Example 4.

19. $\log x - \log y$ **20.** $\log_5 x - 2\log_5 y$

21. $\log_5\left(x^2 - 25\right) - \log_5(x - 5)$ **22.** $\ln\left(x^2y\right) - \ln y - \ln x$

23. $\frac{1}{3}\log_2 x + \log_2(x + 3)$ **24.** $\frac{1}{5}\left(\log_7\left(x^2\right) - \log_7(pq)\right)$

25. $\ln 3 + \ln p - 2\ln q$ **26.** $2\left(\log_5\left(\sqrt{x}\right) - \log_5 y\right)$

27. $\log(x - 10) - \log x$ **28.** $2\log a^2b - \log\left(\frac{1}{b}\right) + \log\left(\frac{1}{a}\right)$

29. $3\left(\ln\left(\sqrt[3]{z^2}\right) - \ln(xy)\right)$ **30.** $\log_2(4x) - \log_2 x$

31. $\log_5 20 - \log_5 5$ **32.** $\log 30 - \log 2 - \log 5$

33. $\ln 15 + \ln 3$ **34.** $\ln 8 - \ln 4 + \ln 3$

35. $0.5\log_3 16 - \log_3 4$ **36.** $3\log_7 2 - 2\log_7 4$

37. $0.25\ln 81 + \ln 4$ **38.** $2(\log 4 - \log 1 + \log 2)$

39. $\log 11 + 0.5\log 9 - \log 3$ **40.** $3\log_4\left(x^2\right) + \log_4\left(x^6\right)$

41. $\log_8\left(2x^2 - 2y\right) - 0.25\log_8 16$ **42.** $\log_{3x} x^2 + \log_{3x} 18 - \log_{3x} 6$

Use the properties of logarithms to write each of the following as a single term that does not contain a logarithm.

43. $5^{2\log_5 x}$ **44.** $10^{\log y^2 - 3\log x}$ **45.** $e^{2 - \ln x + \ln p}$

46. $e^{5\left(\ln\sqrt[5]{3} + \ln x\right)}$ **47.** $10^{\log x^3 - 4\log y}$ **48.** $a^{\log_a b + 4\log_a \sqrt{a}}$

49. $10^{2\log x}$ **50.** $10^{4\log x - 2\log x}$ **51.** $\log_4 16 \cdot \log_x x^2$

52. $e^{\ln x + 2 + \ln x^2}$ **53.** $4^{\log_4(3x) + 0.5\log_4\left(16x^2\right)}$ **54.** $4^{2\log_2 6 - \log_2 9}$

Evaluate the following logarithmic expressions. See Example 5.

55. $\log_4 17$ **56.** $2\log_{\frac{1}{3}} 5$ **57.** $\log_9 8$

58. $\log_2 0.01$ **59.** $\log_{12} 10.5$ **60.** $\log(\ln 2)$

61. $\log_6 3^4$ **62.** $\log_7 14.3$ **63.** $\log_{\frac{1}{2}} \pi^{-2}$

64. $\log_{\frac{1}{5}} 626$

65. $\ln(\log 123)$

66. $\log_{17} 0.041$

67. $\log 16$

68. $\log_3 9$

69. $\log_5 20$

70. $\log_8 26$

71. $\log_4 0.25$

72. $\log_{1.8} 9$

73. $\log_{2.5} 34$

74. $\log_{0.5} 10$

75. $\log_4 2.9$

76. $\log_{0.4} 14$

77. $\log_{0.2} 17$

78. $\log_{0.16} 2.8$

Without using a calculator, evaluate the following expressions.

79. $\log_4 16$

80. $\log_5 25^3$

81. $\ln e^4 + \ln e^3$

82. $\log_4 \frac{1}{64}$

83. $\ln e^{1.5} - \log_4 2$

84. $\log_2 8^{(2\log_2 4 - \log_2 4)}$

Find the value of x in each of the following equations. Express your answer as exact as possible, or as a decimal rounded to two decimal places.

85. $\log_x 1024 = 4$

86. $\log_6 729 = x$

87. $\log_2 529 = x$

88. $\log_4 625 = x$

89. $\log_x 729 = 9$

90. $\log_4 x = 8$

91. $\log_{12} x = 1$

92. $\log_x 16{,}807 = 7$

93. $\log_4 x = 10$

APPLICATIONS

94. A certain brand of tomato juice has a $[H_3O^+]$ concentration of 6.31×10^{-5} moles/liter. What is the pH of this brand?

95. One type of detergent, when added to neutral water with a pH of 7, results in a solution with a $[H_3O^+]$ concentration that is 5.62×10^{-4} times weaker than that of the water. What is the pH of the solution?

96. What is the concentration of $[H_3O^+]$ in lemon juice with a pH of 2.1?

97. An earthquake in Chile in 2019 measured 6.7 on the Richter scale. What was the intensity, relative to a 0-level earthquake, of this event?

98. How much stronger was the 2001 Gujarat earthquake (6.9 on the Richter scale) than the 2019 earthquake described in Exercise 97?

99. A construction worker operating a jackhammer would experience noise with an intensity of 20 watts/meter^2 if it weren't for ear protection. Given that $I_0 = 10^{-12}$ watts/meter^2, what is the decibel level for such noise?

100. A microphone picks up the sound of a thunderclap and measures its decibel level as 105. Given that $I_0 = 10^{-12}$ watts/meter^2, with what sound intensity did the thunderclap reach the microphone?

101. Matt, a lifeguard, has to make sure that the pH of the swimming pool stays between 7.2 and 7.6. If the pH is out of this range, he has to add chemicals that alter the pH level of the pool. If Matt measures the $\left[H_3O^+\right]$ concentration in the swimming pool to be 2.40×10^{-8} moles/liter, what is the pH? Does he need to change the pH by adding chemicals to the water?

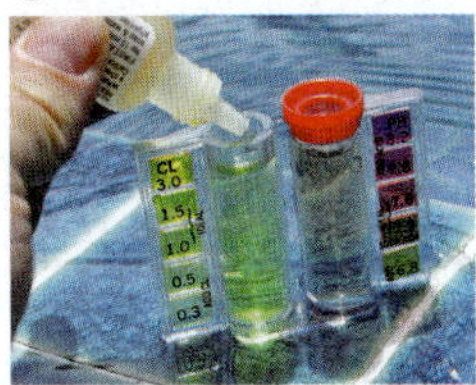

102. The intensity of a cat's soft purring is measured to be 2.19×10^{-11}. Given that $I_0=10^{-12}$ watts/meter^2, what is the decibel level of this noise?

103. Newton's Law of Cooling states that the rate at which an object cools is proportional to the difference between the temperature of the object and the surrounding temperature. If C denotes the surrounding temperature and T_0 denotes the temperature at time t = 0, the temperature of an object at time t is given by $T(t)=C+(T_0-C)e^{-kt}$, where k is a constant that depends on the particular object under discussion.

a. You are having friends over for tea and want to know how long after boiling the water it will be drinkable. If the temperature of your kitchen stays around 74 °F and you found online that the constant k for tea is approximately 0.049, how many minutes after boiling the water will the tea be drinkable (you prefer your tea no warmer than 140 °F)? Recall that water boils at 212 °F.

b. As you intern for your local crime scene investigation department, you are asked to determine at what time a victim died. If you are told k is approximately 0.1947 for a human body and the body's temperature was 72 °F at 1:00 a.m., and the body has been in a storage building at a constant 60 °F, approximately what time did the victim die? Recall the average temperature for a human body is 98.6 °F. Note in this situation, t is measured in hours.

c. When helping your father cook a turkey, you were told to remove the turkey when the thickest part had reached 180 °F. If you remove the turkey and place it on the table in a room that is 72 °F, and it cools to 155 °F in 20 minutes, what will the temperature of the turkey be at lunch time (an hour and 15 minutes after the turkey is removed from the oven)? Should you warm the turkey before eating?

TECHNOLOGY

Use a TI-84 Plus to find and graph a logarithmic function of best fit along with the given data. See Example 9.

104. The menu developers for a chain of coffee shops have conducted experiments to see how long customers take to make a drink choice, based on the number of drinks on the menu. Normalizing so that the average time needed to make a choice given just two drinks is 1 time unit, the average times needed to make a choice given n drinks on the menu are shown in the table below.

Number n of choices	2	3	4	5	6	7
Average time t to make a choice	1	1.4	1.7	2	2.2	2.3

105. The manufacturer of the new pressure altimeter in Example 9 expects the relative error in measured pressure to be larger for lower pressure values, so for pressures in the range of 25,000 pascals to 50,000 pascals altitude recordings were taken at 5000 pascal increments, as shown in the table below.

Pressure p (in pascals)	25,000	30,000	35,000	40,000	45,000	50,000
Altitude h (in meters)	10,541	9321	8257	7309	6452	5670

6.5 EXPONENTIAL AND LOGARITHMIC EQUATIONS

TOPICS

1. Converting between exponential and logarithmic forms
2. Applications of exponential and logarithmic equations
3. Analysis of a stock market investment

TOPIC 1: Converting between Exponential and Logarithmic Forms

At this point, we have all the tools we need to solve the most common sorts of exponential and logarithmic equations. All that is left is to develop our skill in using the tools.

We have already solved many exponential and logarithmic equations, using elementary facts about exponential and logarithmic functions to obtain solutions. However, many equations require a bit more work to solve. While there is no algorithm to follow in dealing with more complicated equations, if a given equation doesn't yield a solution easily, try converting it from exponential form to logarithmic form or vice versa.

All of the properties of exponents and their logarithmic counterparts are of great use as well. For reference, the logarithmic properties that we have noted throughout Sections 6.3 and 6.4 are restated here.

PROPERTIES: Summary of Logarithmic Properties

1. The equations $x = a^y$ and $y = \log_a x$ are equivalent, and are, respectively, the exponential form and the logarithmic form of the same statement.
2. The inverse of the function $f(x) = a^x$ is $f^{-1}(x) = \log_a x$, and vice versa.
3. A consequence of the last point is that $\log_a(a^x) = x$ and $a^{\log_a x} = x$. In particular, $\log_a 1 = 0$ and $\log_a a = 1$.
4. $\log_a(xy) = \log_a x + \log_a y$ ("the log of a product is the sum of the logs")
5. $\log_a\left(\frac{x}{y}\right) = \log_a x - \log_a y$ ("the log of a quotient is the difference of the logs")
6. $\log_a(x^r) = r\log_a x$ ("the log of something raised to a power is the power times the log")

The next several examples illustrate typical uses of the properties, and how converting between the exponential and logarithmic forms of an equation can lead to a solution.

Example 1: Solving Exponential Equations

Solve the equation $3^{2-5x} = 11$. Express the answer exactly and as a decimal approximation.

NOTE

This equation is not easily solved in exponential form, since the two sides of the equation do not have the same base.

Solution

There are two ways to convert the equation into logarithmic form. We will explore both, and see that they lead to the same answer.

The first method is to take the natural (or common) logarithm of both sides.

$$3^{2-5x} = 11$$

$$\ln\left(3^{2-5x}\right) = \ln 11$$ Take the natural logarithm of both sides.

$$(2-5x)\ln 3 = \ln 11$$ Use properties of logarithms to bring the variable out of the exponent.

$$2-5x = \frac{\ln 11}{\ln 3}$$ Divide both sides by $\ln 3$.

$$-5x = \frac{\ln 11}{\ln 3} - 2$$ Simplify.

$$x = -\frac{\ln 11}{5\ln 3} + \frac{2}{5}$$ An exact form of the answer.

The second method is to rewrite the equation using the definition of logarithms, then apply the change of base formula to work with natural (or common) logarithms.

$$3^{2-5x} = 11$$

$$2-5x = \log_3 11$$ Rewrite the equation using the definition of logarithms.

$$2-5x = \frac{\ln 11}{\ln 3}$$ Rewrite the logarithmic term using the change of base formula.

$$x = -\frac{\ln 11}{5\ln 3} + \frac{2}{5}$$ Applying the same algebra as above leads to the same exact answer.

The key step to finding the exact answer is to remove the variable from the exponent, which is achieved by converting to logarithmic form.

The key step to finding a decimal approximation is to change the base (of the exponent and logarithm) to either e or 10, allowing the use of a calculator.

$$x = -\frac{\ln 11}{5\ln 3} + \frac{2}{5} \approx -0.037$$ An approximate form of the answer

We can also use a calculator to verify this solution in the original equation.

Example 2: Solving Exponential Equations

Solve the equation $5^{3x-1} = 2^{x+3}$. Express the answer exactly and as a decimal approximation.

Solution

As in the first example, taking a logarithm of both sides is the key. We will use the common logarithm this time, but the natural logarithm would work just as well.

$$5^{3x-1} = 2^{x+3}$$

$$\log\left(5^{3x-1}\right) = \log\left(2^{x+3}\right)$$ Take the logarithm of both sides.

$$(3x-1)\log 5 = (x+3)\log 2$$

$$3x\log 5 - \log 5 = x\log 2 + 3\log 2$$ Bring the exponents down using a property of logarithms, then multiply the terms out.

$$3x\log 5 - x\log 2 = 3\log 2 + \log 5$$

$$x(3\log 5 - \log 2) = 3\log 2 + \log 5$$ Collect the terms with x on one side, then factor out x.

$$x = \frac{3\log 2 + \log 5}{3\log 5 - \log 2} \approx 0.892$$ Simplify and evaluate with a calculator.

The exact answer could appear in many different forms, depending on the base of the logarithm chosen and the order of logarithmic properties used in simplifying the answer. We could simplify it further as follows.

$$x = \frac{3\log 2 + \log 5}{3\log 5 - \log 2} = \frac{\log 8 + \log 5}{\log 125 - \log 2} = \frac{\log 40}{\log\left(\frac{125}{2}\right)}$$

Example 3: Solving Logarithmic Equations

NOTE

Since calculators can evaluate exponents of any base, it often does not matter what the base is when we convert logarithmic equations into their exponential forms.

Solve the equation $\log_7(3x - 2) = 2$.

Solution

Note that rewriting this equation using the change of base formula does not help, since the variable would still be trapped inside the logarithm. Instead, we use the definition of logarithms to rewrite the equation in exponential form.

$$\log_7(3x-2) = 2$$

$$3x - 2 = 7^2$$ Rewrite the equation in exponential form.

$$3x = 51$$ Simplify and solve for x.

$$x = 17$$

Example 4: Solving Logarithmic Equations

Solve the equation $\log_5 x = \log_5(2x+3) - \log_5(2x-3)$.

Solution

This is an example of a logarithmic equation that is not easily solved in logarithmic form. Once a few properties of logarithms have been utilized, the equation can be rewritten in a very familiar form.

$$\log_5 x = \log_5(2x+3) - \log_5(2x-3)$$

$$\log_5 x = \log_5\left(\frac{2x+3}{2x-3}\right)$$ Combine terms using a property of logarithms.

$$x = \frac{2x+3}{2x-3}$$ Equate the arguments since each term has the same base.

$$x(2x-3) = 2x+3$$ Multiply both sides by $2x-3$.

$$2x^2 - 3x = 2x+3$$ The result is a quadratic equation.

$$2x^2 - 5x - 3 = 0$$

$$(2x+1)(x-3) = 0$$ Rewrite the equation with 0 on one side and factor.

$$x = -\frac{1}{2}, 3$$ Solve using the Zero-Factor Property.

> **NOTE**
> We need to check for extraneous solutions when solving logarithmic equations. In particular, remember that logarithms of negative numbers are undefined.

A crucial step remains! While these two solutions definitely solve the quadratic equation, we must check that they solve the initial logarithmic equation, as the process of solving logarithmic equations can introduce extraneous solutions. If we check our two potential solutions in the original equation, we quickly discover that only one of them is valid.

$$\log_5\left(-\frac{1}{2}\right) = \log_5\left(2\left(-\frac{1}{2}\right)+3\right) - \log_5\left(2\left(-\frac{1}{2}\right)-3\right)$$

We can see that $-\frac{1}{2}$ is not a solution to the equation, because logarithms of negative numbers are undefined. We move on to the second solution.

$$\log_5 3 = \log_5(2(3)+3) - \log_5(2(3)-3)$$ Substitute into the equation.

$$\log_5 3 = \log_5 9 - \log_5 3$$ Simplify.

$$\log_5 3 = \log_5\left(\frac{9}{3}\right)$$ Combine the terms using a property of logarithms.

$$\log_5 3 = \log_5 3$$ A true statement

Thus, the solution to the equation is the single value $x = 3$.

TOPIC 2: Applications of Exponential and Logarithmic Equations

We will conclude our discussion of exponential and logarithmic equations by revisiting some important applications.

Example 5: Compounding Interest

Rita is saving up money for a down payment on a new car. She currently has \$5500 but she knows she can get a loan at a lower interest rate if she can put down \$6000. If she invests her \$5500 in a money market account that earns an annual interest rate of 4.8% compounded monthly, how long will it take her to accumulate the \$6000?

Solution

We need to solve the compound interest formula for t, the amount of time Rita invests her money. Given that $P = 5500$, $A(t) = 6000$, $r = 0.048$, and $n = 12$, we have

$$6000 = 5500\left(1 + \frac{0.048}{12}\right)^{12t}.$$

Our solution needs to be a decimal approximation to be of practical use, so we need to use the natural or common logarithm to rewrite the equation in logarithmic form.

$$6000 = 5500\left(1+\frac{0.048}{12}\right)^{12t}$$

$$\frac{6000}{5500} = \left(1+\frac{0.048}{12}\right)^{12t} \quad \text{Divide both sides by 5500.}$$

$$\ln\left(\frac{6000}{5500}\right) = \ln\left(\left(1+\frac{0.048}{12}\right)^{12t}\right) \quad \text{Take the natural logarithm of both sides.}$$

$$\ln\left(\frac{6000}{5500}\right) = 12t\ln(1.004) \quad \text{Bring the variable out of the exponent using a property of logarithms.}$$

$$\frac{\ln\left(\frac{6000}{5500}\right)}{12\ln(1.004)} = t \quad \text{Solve for } t.$$

$$t \approx 1.82$$

Since t is measured in years, the solution tells us that it will take a bit less than a year and 10 months for the \$5500 to grow to \$6000.

Example 6: Radiocarbon Dating

We have already discussed how the radioactive decay of carbon-14 can be used to arrive at age estimates of carbon-based fossils, and we constructed an exponential function describing the rate of decay. A much more common form of the function is

$$A(t) = A_0 e^{-0.000121t},$$

where $A(t)$ is the mass of carbon-14 remaining after t years, and A_0 is the mass of carbon-14 initially. Use this formula to determine

a. the half-life of carbon-14 and

b. the age of a fossilized organism containing 1.5 grams of carbon-14, given that a living organism of the same size contains 2.3 grams of carbon-14.

Solution

a. We are looking for the value of t for which $A(t)$ is half of A_0, or $\frac{A_0}{2}$.

$$\frac{A_0}{2} = A_0 e^{-0.000121t}$$

$$\frac{1}{2} = e^{-0.000121t}$$ Divide both sides by A_0.

$$\ln\left(\frac{1}{2}\right) = \ln\left(e^{-0.000121t}\right)$$ Since the base involved is e, using the natural logarithm makes calculation much simpler.

$$\ln\left(\frac{1}{2}\right) = -0.000121t$$ Simplify using a property of logarithms.

$$\frac{\ln\left(\frac{1}{2}\right)}{-0.000121} = t$$ Solve for t.

$$t \approx 5728 \text{ years}$$ Evaluate.

Thus, as we saw before, the half-life of carbon-14 is approximately 5728 years.

b. We plug the given information into the formula and solve for t.

$$1.5 = 2.3e^{-0.000121t}$$

$$\frac{1.5}{2.3} = e^{-0.000121t}$$ Divide both sides by 2.3.

$$\ln\left(\frac{1.5}{2.3}\right) = -0.000121t$$ Take the natural logarithm of both sides and simplify using a property of logarithms.

$$\frac{\ln\left(\frac{1.5}{2.3}\right)}{-0.000121} = t$$ Solve for t.

$$t \approx 3533 \text{ years}$$ Evaluate.

According to the radiocarbon dating, the fossil is about 3533 years old.

TOPIC 3: Analysis of a Stock Market Investment

Exponential growth is a recurring topic in any discussion of investing and in any long-term observation of stock and bond markets. Claims of the relevance of exponential functions as models of market behavior range from the fairly simple and defensible to the very complex and speculative. At this point, we have learned enough algebra to answer one simple but fundamental question that investors should periodically ask themselves: How well (or poorly) are my investments performing?

We will answer this question as it applies to an investment technique that many people currently use, called *dollar-cost averaging*. The technique calls for an investor to commit P dollars on a periodic basis in a chosen stock, mutual fund, bond, or other investment vehicle. The amount P and the period of investment are, of course, up to the investor, but once chosen should remain fixed. For instance, an investor might choose to invest \$100 on the first of every month in stock of a given company. The argument for dollar-cost averaging is that as the stock price fluctuates over time, the fixed \$100 per month investment buys fewer high-priced shares in the company and more lower-priced shares, thus hopefully increasing the investor's profits when he or she eventually decides to sell the shares.

One slight disadvantage of dollar-cost averaging is that it is not so easy to monitor the percentage gain (or loss) on the investment over time. The simplest sort of investment to analyze is a one-time investment; at any point it is possible to compare the current value of the investment to the purchase price and calculate the percentage change (even on an annual basis, if desired). And the performance of fixed periodic investments in a given stock is certainly harder to keep track of than, say, the performance of a savings account, which is often reported as an annual rate compounded monthly.

This points the way toward one possible solution. Suppose we invest P dollars on the first of January in Company XYZ, and suppose that one month later the value of our investment is A dollars. Then we could determine the annual rate r at which our investment has changed, assuming monthly compounding, by solving the equation

$$A = P\left(1 + \frac{r}{12}\right)$$

for r. We could then compare r to the rate for a typical monthly-compounded savings account to see how Company XYZ is doing. On the first of February we invest P dollars again. One month later our total investment has a new value A, and we need to solve the equation

$$A = \left(P\left(1 + \frac{r}{12}\right) + P\right)\left(1 + \frac{r}{12}\right) = P\left[\left(1 + \frac{r}{12}\right)^2 + \left(1 + \frac{r}{12}\right)\right]$$

for r if we want to compare the performance to a savings account again. In general, after n months (and just before our $(n+1)^{\text{th}}$ investment), we need to solve

$$A = P\left[\left(1 + \frac{r}{12}\right)^n + \left(1 + \frac{r}{12}\right)^{n-1} + \cdots + \left(1 + \frac{r}{12}\right)\right]$$

for r in order to keep track of our investment relative to a savings account.

This is a formidable task, even though at the moment we are only wondering if this can be done in theory, not in practice. Note that we are actually giving ourselves an nth-degree polynomial equation in r to solve. To make this clearer, let $x = 1 + \frac{r}{12}$. Then the equation becomes

$$A = P\left(x^n + x^{n-1} + \cdots + x\right).$$

If we can solve this equation for x, we can then determine r. But from Chapter 5, we know that an nth-degree polynomial equation has potentially n different solutions, only one of which has any meaning for us as far as this particular application is concerned. Is there any hope that we can identify that one solution?

Thankfully, yes. We will first make one more change to the equation, in order to eliminate the inconvenient dots. Notice that if we multiply both sides of the equation by $(x-1)$ and expand the polynomial product, most of the terms on the right-hand side cancel.

$$\begin{aligned} A\,(x-1) &= P\left(x^n + x^{n-1} + \cdots + x\right)(x-1) \\ Ax - A &= P\left[\left(x^{n+1} + x^n + \cdots + x^2\right) - \left(x^n + x^{n-1} + \cdots + x\right)\right] \\ Ax - A &= P\left(x^{n+1} - x\right) \end{aligned}$$

If we then put all the terms on one side, we obtain the final equation

$$x^{n+1} - \left(\frac{A}{P} + 1\right)x + \frac{A}{P} = 0.$$

This last equation has two changes of sign (note that $\frac{A}{P}$ is guaranteed to be positive), so Descartes' Rule of Signs tells us the equation has either 2 or 0 positive real solutions. Further, we already know a positive solution: $x = 1$ is certainly a solution, which shouldn't surprise us as we obtained the $(n+1)^{\text{th}}$-degree equation from the nth-degree equation by multiplying both sides by the factor $x - 1$. This means there must be exactly 2 positive real solutions, and the other positive solution is the one we care about.

Why do we only care about positive solutions? Note that if the r we are ultimately solving for is negative (an unpleasant but unfortunately real possibility), $x = 1 + \frac{r}{12}$ will be less than 1 but larger than 0. (The reason x can't be negative is for financial reasons, not mathematical reasons. The only way x could be negative is if r is so negative that the value A of our investment has actually fallen below 0, which can't happen.) If r is positive, then $x = 1 + \frac{r}{12}$ is larger than 1. In either event, the second positive real solution of our $(n+1)^{\text{th}}$-degree polynomial equation leads to a value for r and an understanding of how the stock is performing.

6.5 EXERCISES

PRACTICE

Solve the following exponential and logarithmic equations. Round your answer to two decimal places if necessary. See Examples 1 through 4.

1. $3e^{5x} = 11$

2. $4^{3-2x} = 7$

3. $11^{\frac{3}{x}} = 10$

4. $8^{3x+2} = 7^{2x+3}$

5. $e^{15-3x} = 28$

6. $10^{\frac{5}{x}} = 150$

7. $10^{2x+5} = e$

8. $e^{8x+e} = 8^{ex+8}$

9. $6^{x-7} = 7$

10. $2e^{3x} = 145$

11. $2^{6-x} = 10$

12. $e^{3x-6} = 10^{x+2}$

13. $e^{-4x-2} = 12$

14. $10^{x-9} = 2001$

15. $e^{x-4} = 4^{\frac{2x}{3}}$

16. $5^{x-2} = 20$

17. $8^{x^2+1} = 23$

18. $3^{\frac{4}{x}} = 15$

19. $6^{3x-4} = 36^{2x+4}$

20. $81^x = 3^{2x+16}$

21. $e^{2x} = 14$

22. $e^{4x} = e^{3x+14}$

23. $5^{5x-7} = 10^{2x}$

24. $10^{6x} = 3^{3x+4}$

25. $\log_5 x = 3$

26. $\log_2 x = 4$

27. $\log x + \log(4x) = 2$

28. $\log_4(x^2) - \log_4 x = 2$

29. $\ln(2x) - \ln 4 = 3$

30. $\ln(15x) - \ln 3 = 6$

31. $\log_4(x-3) + \log_4 2 = 3$

32. $\log_3 24 - \log_3 4 = x - 5$

33. $\log_5(8x) - \log_5 3 = 2$

34. $e^{2\ln x} = 4\log 10$

35. $9^{\log_3 x} = 16$

36. $3\log_8\left(512^{x^2}\right) = 36$

37. $\log_3(6x) - 2\log_3(6x) = 3$

38. $\ln(3e) = \log x$

39. $\ln\left(2^{4e^x}\right) = \ln\left(16^e\right)$

40. $\log(x-2) + \log(x+2) = 2$

41. $\log(x-3) + \log(x+3) = 4$

42. $\log_2(7x-4) = \log_2(16-3x)$

43. $\log_\pi(x-5) + \log_\pi(x+3) = \log_\pi(1-2x)$

44. $\log_3(x+3) + \log_3(x-5) = 2$

45. $\log x + \log(x-3) = 1$

46. $\log_7(3x+2) - \log_7 x = \log_7 4$

47. $\log_2 x + \log_2(x-7) = 3$

48. $\log_{12}(x-2) + \log_{12}(x-1) = 1$

49. $\log_3(x+1) - \log_3(x-4) = 2$

50. $\ln(x+1) + \ln(x-2) = \ln(x+6)$

51. $\log_4(x-3) + \log_4(x-2) = \log_4(x+1)$

52. $2\ln(x+3) = \ln(12x)$

53. $\log_5(x-1) + \log_5(x+4) = \log_5(x-5)$

54. $\log_{255}(2x+3) + \log_{255}(2x+1) = 1$

55. $\log_2(x-5) + \log_2(x+2) = 3$

56. $\log_6(x+1) + \log_6(x-4) = 2$

57. $\ln(x+2) + \ln(x) = 0$

58. $e^{2x} - 3e^x - 10 = 0$ (**Hint:** First solve for e^x.)

59. $2^{2x} - 12(2^x) + 32 = 0$ (**Hint:** First solve for 2^x.)

60. $e^{2x} + 2e^x - 8 = 0$

61. $3^{2x} - 12(3^x) + 27 = 0$

Using the properties of logarithmic functions, simplify the following functions as much as possible. Write each function as a single term with a coefficient of 1, if possible.

62. $f(x) = 0.5\ln(x^2)$

63. $f(x) = 0.25\log(16x^8)$

64. $f(x) = 4\ln(\sqrt{5x})$

65. $f(x) = 8\ln(\sqrt[4]{3x})$

66. $f(x) = 3\ln(e^x) - 3$

67. $f(x) = 10^{2x\log 16}$

68. $f(x) = 2\ln(x^3) + \ln(x^6)$

69. $f(x) = 2\ln(x^3) - \ln(x^6)$

70. $f(x) = \ln(x^2 + x) - \ln x$

71. $f(x) = 2\ln\left(5^{x\log_{20}(2\sqrt{5})}\right)$

72. $f(x) = e^{\ln(\log x^e - 1)}$

73. $f(x) = 2\ln(5^{\log_4 2})$

APPLICATIONS

74. Assuming that there are currently 8 billion people on Earth and a growth rate of 1.9% per year, how long will it take for Earth's population to reach 20 billion?

75. How long does it take for an investment to double in value if

a. The investment is in a monthly-compounded savings account earning 4% a year?
b. The investment is in a continuously-compounded account earning 7% a year?

76. Assuming a half-life of 5728 years, how long would it take for 3 grams of carbon-14 to decay to 1 gram?

77. Suppose a population of bacteria in a Petri dish has a doubling time of one and a half hours. How long will it take for an initial population of 10,000 bacteria to reach 100,000?

78. According to Newton's Law of Cooling, the temperature $T(t)$ of a hot object, at time t after being placed in an environment with a constant temperature C, is given by $T(t) = C + (T_0 - C)e^{-kt}$, where T_0 is the temperature of the object at time $t = 0$ and k is a constant that depends on the object.

If a hot cup of coffee, initially at 190 °F, cools to 125 °F in 5 minutes when placed in a room with a constant temperature of 75 °F, how long will it take for the coffee to reach 100 °F?

79. Wayne has $12,500 in a high interest savings account at 3.66% annual interest compounded monthly. Assuming he makes no deposits or withdrawals, how long will it take for his investment to grow to $15,000?

80. Ben and Casey both open money market accounts with 4.9% annual interest compounded continuously. Ben opens his account with $8700 while Casey opens her account with $3100.

a. How long will it take Ben's account to reach $10,000?
b. How long will it take Casey's account to reach $10,000?
c. How much money will be in Ben's account after the time found in part b.?

81. Cesium-137 has a half-life of approximately 30 years. How long would it take for 160 grams of cesium-137 to decay to 159 grams?

82. A chemist, running tests on an unknown sample from an illegal waste dump, isolates 50 grams of what he suspects is a radioactive element. In order to help identify the element, he would like to know its half-life. He determines that after 40 days only 44 grams of the original element remains. What is the half-life of this mystery element?

TECHNOLOGY

The dollar-cost averaging application in this section concluded with the equation

$$x^{n+1} - \left(\frac{A}{P} + 1\right)x + \frac{A}{P} = 0,$$

where A represents the total accumulation in an account in which amount P has been invested every month for n months. A graphing utility can be used to solve the equation for x when given known values for A, P, and n. Then the equivalent monthly-compounded annual rate of interest r can be found by solving the equation

$$x = 1 + \frac{r}{12}.$$

For instance, if a monthly investment of $P = 100$ dollars for $n = 12$ months results in an accumulation of $A = 1300$ dollars, the first step is to solve the following equation.

$$x^{12+1} - \left(\frac{1300}{100} + 1\right)x + \frac{1300}{100} = 0 \quad \Rightarrow \quad x^{13} - 14x + 13 = 0$$

As noted, $x = 1$ will always be one solution to the dollar-cost averaging equation, but the second positive solution is the one we seek. The second solution will lie between 0 and 1 if the equivalent rate r is negative and will be larger than 1 if r is positive. In this case, since $1200 has been invested and the total accumulation after 12 months is $1300, we know r is positive, and a graphing utility tells us that the second solution is indeed greater than 1; specifically, $x \approx 1.01225$. Solving $1.01225 = 1 + \frac{r}{12}$ for r gives us $r \approx 0.147$, or $r \approx 14.7\%$.

Use a graphing utility to determine the equivalent monthly-compounded interest rate r in each of the following scenarios, where P represents the amount invested each month and A is the accumulated value at the end of n months.

83. $n = 24$, $P = \$50.00$, and $A = \$1275.00$

84. $n = 120$, $P = \$100.00$, and $A = \$15{,}000.00$

85. $n = 12$, $P = \$50.00$, and $A = \$590.00$

86. $n = 24$, $P = \$75.00$, and $A = \$2000.00$

CHAPTER 6 PROJECT

Exponential Functions

Computer viruses cause enormous economic harm to businesses through costs associated with preventive measures, data recovery, and damaged equipment and reputation. The speed at which viruses can spread makes it difficult to manage an attack.

Suppose a new virus has been created and initially infects 100 computers in a large corporation through a single email. Let $t = 0$ be the time of the initial infection, and suppose the number of computers infected in the corporation doubles every 47 minutes. The infection is finally brought under control 6 hours later.

1. How many computers are infected at the 1 hour mark?

2. How many computers are infected after another 30 minutes?

3. How many computers are infected when the virus is finally brought under control?

4. A second virus is detected a week later, and the number of computers it infects t minutes after initial infection is estimated to be $50e^{\frac{t}{110}}$. To the nearest minute, what is the doubling time for this virus?

CHAPTER 6 REVIEW EXERCISES

Section 6.1

Sketch the graphs of the following functions. State their domain and range.

1. $f(x) = \left(\frac{1}{2}\right)^{x-1} + 3$

2. $r(x) = 2^{-x+4} - 2$

3. $h(x) = 3^x$

4. $f(x) = 1 - 2^{-x}$

5. $p(x) = \left(\frac{1}{4}\right)^x$

6. $s(x) = (0.2)^{x-2}$

7. $g(x) = 4 - 2^x$

8. $m(x) = \frac{1}{2^x} - 3$

9. $f(x) = \frac{1}{2^{4-x}}$

10. $r(x) = \left(\frac{9}{2}\right)^{3-x}$

Solve the following exponential equations.

11. $3^x = 243$

12. $2^{-x} = 16$

13. $0.5^x = 0.25$

14. $3^{3x-5} = 81$

15. $\left(\frac{2}{5}\right)^{-4x} = \left(\frac{25}{4}\right)^{x-1}$

16. $10{,}000^x = 10^{-2x-12}$

17. $9^{x-1} = 27^{-x+2}$

18. $\left(\frac{1}{3}\right)^{x-1} = 81^{\frac{1}{2}}$

19. $5^{3x-6} = 1$

Section 6.2

20. Melissa has recently inherited \$15,000 that she wants to deposit into a savings account for 10 years. She has determined that her two best bets are an account that compounds annually at a rate of 3.95% and an account that compounds continuously at an annual rate of 3.85%. Which account would pay Melissa more interest?

21. Bill has come upon a 37-gram sample of iodine-131. He isolates the sample and waits for 2 weeks. After this time period, only 11 grams of iodine-131 remain. What is the half-life of this isotope?

22. Katherine is working in a lab testing bacteria populations. Starting out with a population of 870 bacteria, she notices that the population doubles every 22 minutes. Find **a.** the equation for the population P in terms of time t in minutes, and **b.** the time it would take for the population to reach 7500 bacteria.

23. The number of fruit flies in an experimental population after t hours is given by $Q(t) = 20e^{0.03t},\ t \geq 0.$

a. How large is the population of fruit flies after 72 hours?
b. Find the initial number of fruit flies in the population.

24. In 1986, a nuclear reactor accident occurred in Chernobyl in what was then the Soviet Union. The explosion spread radioactive chemicals over hundreds of square miles and the government evacuated the city and surrounding areas. To see why the city is now uninhabited, consider the model $P = 10e^{-0.00002845t}$.

This model represents the amount of plutonium that remains (from an initial amount of 10 pounds) after t years. Sketch the graph of this function over the interval from $t = 0$ to $t = 100{,}000$. How much of the 10 pounds will remain after 100,000 years?

Section 6.3

Write the following equations in logarithmic terms.

25. $3^x = 8$ **26.** $(3a)^{\sqrt{2}} = 10$ **27.** $4^{3a} = 4096$

Write the following logarithmic equations as exponential equations.

28. $\log_4 64 = x$ **29.** $\log_3\left(\frac{1}{27}\right) = -3$ **30.** $\log_8(2A) = 3$

Sketch the graphs of the following functions. State their domain and range.

31. $f(x) = -\log_3(-x)$ **32.** $f(x) = \log_{\frac{1}{2}} x$ **33.** $m(x) = \log_{\frac{1}{2}}(x+2)$

Evaluate the following logarithmic expressions without the use of a calculator.

34. $\log_{27} 9^{\log 1000}$ **35.** $\log_{\frac{1}{3}} 9$ **36.** $\log_4\left(\frac{1}{64}\right)$

37. $\log_{\frac{1}{2}} 8$ **38.** $\log_{\sqrt{3}}\left(\frac{1}{3}\right)$ **39.** $\ln\left(\sqrt[3]{e^2}\right)$

Use the elementary properties of logarithms to solve the following equations.

40. $\log_6 6^{\log_5 x} = 3$ **41.** $\log_9 x^{\frac{1}{2}} = \frac{3}{4}$

42. $\log_x\left(\log_{\frac{1}{2}}\left(\frac{1}{16}\right)\right) = 2$ **43.** $\log_4(2x-1) = 2$

Solve the following logarithmic equations, using a calculator if necessary to evaluate the logarithms. Express your answer either as a fraction or a decimal rounded to two decimal places.

44. $\ln(4x) = 3.2$ **45.** $\ln(x-7) = 5$ **46.** $\log_7(4x-3) = 4$

Section 6.4

Use the properties of logarithms to expand the following expressions as much as possible. Simplify any numerical expressions that can be evaluated without a calculator.

47. $\log\sqrt{\dfrac{x^3}{4\pi^5}}$

48. $\ln\left(\dfrac{\sqrt{a^5}mn^2}{e^5}\right)$

49. $\log_3(27a^3)$

50. $\ln\left(\ln\left(e^{2ex}\right)\right)$

Use the properties of logarithms to condense the following expressions as much as possible, writing each answer as a single term with a coefficient of 1.

51. $\dfrac{1}{3}\left(\log_2(a^5) - \log_2(bc^3)\right)$

52. $\ln 4 - \ln(x^2) - 7\ln y$

53. $\log_2(x^2 - 9) - \log_2(x+3)$

54. $2\log a + 3\log b - \dfrac{1}{2}\log c - \log d$

55. $\log_3(x-2) + \log_3 x - \log_3(x^2+4)$

Use the properties of logarithms to write each of the following as a single term that does not contain a logarithm.

56. $6^{3\log_6 x}$

57. $5^{\log_5 x - 2\log_5 y}$

Evaluate the following logarithmic expressions.

58. $\log_3 17$

59. $\log_{1.4} 8$

60. $4\log_{\frac{1}{2}} 3$

Without using a calculator, evaluate the following expressions.

61. $\ln\left(\dfrac{1}{e^2}\right) + \ln e^2$

62. $\log_4(64^2)$

63. On the Richter scale, the magnitude R on a earthquake of intensity I is given by $R = \log\dfrac{I}{I_0}$, where $I_0 = 1$ is the minimum intensity used for comparison. Find the intensity per unit of area for the following values of R.

a. $R = 8.4$

b. $R = 6.85$

c. $R = 9.1$

Section 6.5

Solve the following exponential and logarithmic equations. When appropriate, write the answer as both an exact expression and as a decimal approximation. Round your answer to two decimal places if necessary.

64. $e^{8-5x} = 16$

65. $10^{\frac{6}{x}} = 321$

66. $7^{\frac{x}{3}-4} = 19$

67. $e^{4x} = 5^{3x+1}$

68. $24 = 3e^{x+2}$

69. $3^{2x-1} = 2^{2-x}$

70. $\ln(x+1)+\ln(x-1)=\ln(x+5)$

71. $\log_2(x+3)+\log_2(x+4)=\log_2(3x+8)$

72. $\log_5(8x-3)=3$

73. $\log_7(4x)-\log_7 6=2$

74. $\ln(5x+8)=\ln(40-3x)$

Using the properties of logarithmic functions, simplify the following functions as much as possible. Write each function as a single term with a coefficient of 1, if possible.

75. $f(x)=0.75\ln x^4$

76. $f(x)=6\log\sqrt{2x}$

77. $f(x)=4\log x^3-\log x^2$

78. $f(x)=0.5\ln(9x^6)$

79. $f(x)=2\log 7^{\log_9 3}$

80. $f(x)=2\ln 3^{\log_4 8}$

81. Rick puts \$6500 in a high interest money market account at 4.36% annual interest compounded monthly. Assuming he makes no deposits or withdrawals, how long will it take for his investment to grow to \$7000?

82. Sodium-24 has a half-life of approximately 15 hours. How long would it take for 350 grams of sodium-24 to decay to 12 grams?

CHAPTER 7

Trigonometric Functions

SECTIONS

WHAT IF ...

What if you were playing a game of pickup basketball with your friends? How would you model the path of the ball as you dribbled it in preparation for making a game-winning shot?

By the end of this chapter, you'll be able to evaluate and graph trigonometric functions. This type of function can be used to describe the behavior of many naturally occurring phenomena, such as waves and harmonic motion. On page 559, you will use a trigonometric function to model the displacement of a basketball. You will master this type of problem using the definition of simple harmonic motion and frequency on page 555.

Babylonian Numbers

Introduction

This chapter is the first of three dealing almost exclusively with an area of mathematics called *trigonometry*, an area which is big enough to be studied on its own (as it often is) but which is also intrinsically related to the algebraic and geometric concepts that make up the rest of this text. This close association will become even more apparent when you study calculus, so its inclusion in a book preparing you for calculus is appropriate.

The early history of trigonometry is not quite so well documented as that of geometry, but archaeologists have unearthed clay tablets indicating that Babylonian mathematicians around 2000 BC were already developing ideas that we would classify today as trigonometric. And that Babylonian heritage is of more than academic interest. We owe to the Babylonians of the 1st millennium BC our *degree* unit of angle measure. Although many competing arguments have been proposed as to why the Babylonians fixed on 360° as being the measure of one full rotation, there is no doubt that the convention began with them. (Some of the competing arguments are built around such things as connections between a full circle and the calendar, the fact that 360 can be factored many different ways, and the fact that Babylonians apparently divided the day into 12 "hours" of 30 parts each.)

The word "trigonometry" itself is Greek, and translates roughly as "measurement of triangles." From the start, trigonometry found important applications in astronomy, navigation, and surveying, and those applications have only grown in importance over time. Initially, trigonometry focused on ratios of side lengths of triangles, and that perspective is alive and strong still. But with the development of calculus in the 17th century, mathematicians also began to view the trigonometric relations as functions of real numbers. This secondary perspective lends itself to applications involving rotations or oscillations, and trigonometry quickly became an indispensable tool in engineering, explanations of wave propagation, and modern signal processing.

As with every topic in this text, try to keep the historical background in mind as you learn the material. Mathematics is not immune to societal and other pressures, and many aspects of trigonometry's history demonstrate this. The presentation in this text draws upon more than 2000 years of development, and would be very unfamiliar to early users of trigonometry. Relatively recent developments in technology have also had a profound effect on the way trigonometry is taught and learned—if you find yourself tiring at some point while studying this chapter, comfort yourself with the thought that dreaded "trig tables" are a thing of the past! Learning how to use them once constituted a large part of trigonometry, but calculators and computer software make such tedium unnecessary now. (If you have no idea what a "trig table" is, and want to subject yourself to a lecture on how kids today have it too easy, ask someone who learned trigonometry prior to the mid-1970s for an explanation.)

7.1 RADIAN AND DEGREE MEASURE

TOPICS

1. The unit circle and angle measure
2. Converting between degrees and radians
3. The Pythagorean Theorem and commonly encountered angles
4. Arc length and angular speed
5. Area of a circular sector

TOPIC 1: The Unit Circle and Angle Measure

Trigonometry is, at heart, the study of angles. Although much of trigonometry can be discussed without reference to angle measure (and in fact early Greek mathematicians did just that), we will find it very useful to have a method of describing the size of angles. As it turns out, there are two common ways to measure angles (as well as a number of less common ways). The method of measuring angles in terms of *degrees* is one that you are probably familiar with—references to degree measure occur in all sorts of nonmathematical contexts. But, as mentioned in the introduction to this chapter, the definition of degree has more of a cultural basis than a mathematical basis, so it shouldn't be too surprising that there is a way of measuring angles that makes more mathematical sense. That more mathematically useful way is called *radian measure.*

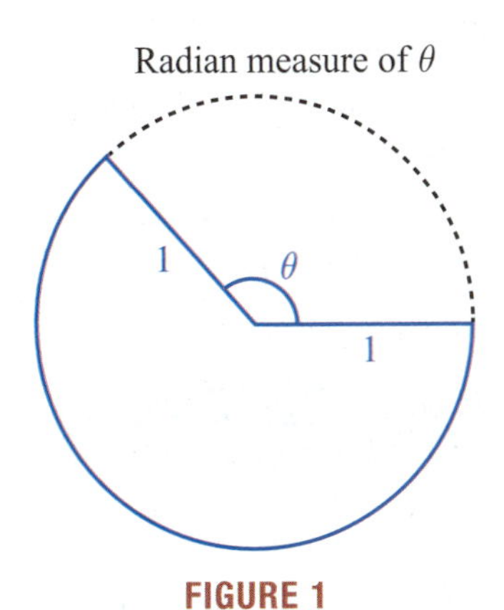

FIGURE 1

DEFINITION: Radian Measure

Let θ (the Greek letter *theta*) be an angle at the center of a circle of radius 1 (the unit circle), as shown in Figure 1. The measure of θ in **radians** (abbreviated as **rad**) is the length of that portion of the circle subtended by θ, which is the portion of the circumference shown as a dashed arc in Figure 1. (*Subtended*, in mathematics, can be taken to mean *opposite of, and stretching between*. An arc of a circle is subtended by its central angle and, conversely, a central angle is subtended by an arc of the circle.) Note that the unit of length measurement is immaterial. As long as the circle has a radius of 1 (unit), the length of the subtended portion of the circle (in the same units) is defined to be the radian measure of the angle.

In general, an angle θ is defined by any two rays R_1 and R_2 sharing a common origin, as shown in Figure 2. We can associate a sign with the measure of θ by designating one ray, say R_1, as the **initial side** and the other ray, R_2, as the **terminal side**. If θ is defined by a counterclockwise rotation from the initial side to the terminal side, we say θ has **positive measure**, and if θ is defined by a clockwise rotation we say it has **negative measure**. In Figure 2, the angle on the left has positive measure while the angle on the right is negative.

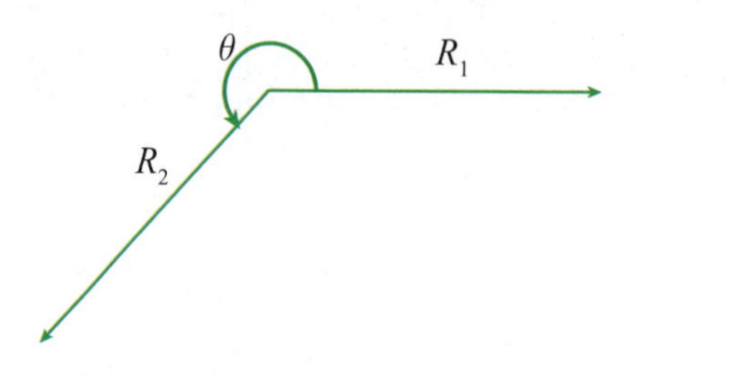

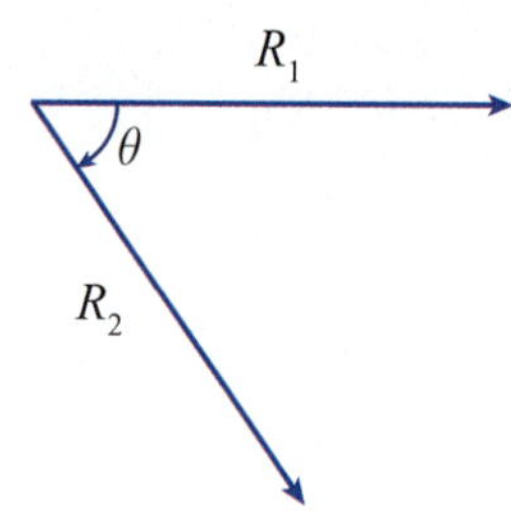

FIGURE 2

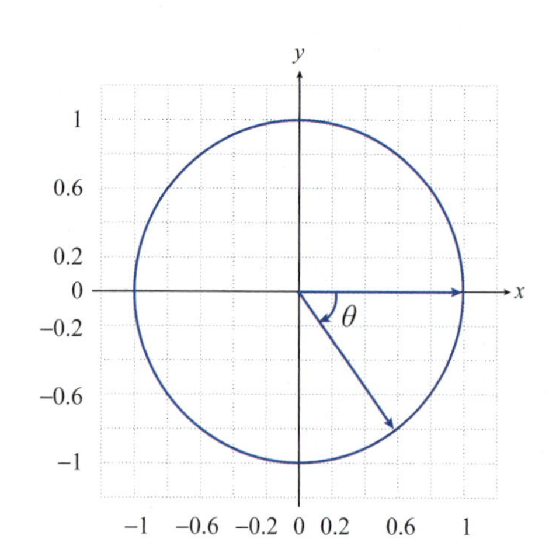

FIGURE 3: Standard Position of an Angle

In order to use our definition of radian to measure an angle defined by two rays, as in Figure 2, we place the vertex of the angle at the center of a **unit circle** (a circle of radius 1) and measure the length of the arc between the initial and terminal sides of the angle. Further, we say the angle is in **standard position** if its vertex is located at the origin of the Cartesian plane and its initial side lies along the positive x-axis. In this case, the unit circle is then the graph of the equation $x^2 + y^2 = 1$. Figure 3 illustrates the second angle from Figure 2 placed in standard position.

TOPIC 2: Converting between Degrees and Radians

Degrees	Radians
360°	2π
180°	π
90°	$\frac{\pi}{2}$

TABLE 1

Once radian measure has been defined, the next order of business is to acquire some familiarity with its use. To begin with, we want to be able to translate between degree measure and radian measure of an angle.

This is easily done if we recall the formula for the circumference C of a circle of radius r: $C = 2\pi r$. For the unit circle under discussion, $r = 1$ so $C = 2\pi$. If we think of the entire circumference as being the portion of the unit circle subtended by an angle of 360° (that is, an angle whose terminal side and initial side coincide), we have just determined that an angle of 360° corresponds to 2π radians. Using this as a starting point, we see that an angle of 180° corresponds to π radians (such an angle subtends half the circumference of the unit circle) and an angle of 90° corresponds to $\frac{\pi}{2}$ radians (that is, a right angle has a measure of $\frac{\pi}{2}$ radians). From the equation $180° = \pi$ rad, we can derive the following conversion formulas.

f(x) FORMULA: Angle Measurement Conversion

Since $180° = \pi\text{ rad}$, we know that $1° = \frac{\pi}{180}$ rad and $\left(\frac{180}{\pi}\right)^{\circ} = 1$ rad. Multiplying both sides of these equations by an arbitrary quantity x, we have

1. $x° = x\left(\frac{\pi}{180}\right)$ rad,
2. $x \text{ rad} = x\left(\frac{180}{\pi}\right)^{\circ}$.

In particular, note that 1 rad ≈ 57.296°, so an angle of 1 rad cuts off a bit less than one-sixth of a circle (an angle of 60° cuts off exactly one-sixth of a circle).

Example 1: Converting between Degrees and Radians

Convert the following angle measures as directed.

a. Express $\frac{\pi}{3}$ rad in degrees.

b. Express $270°$ in radians.

c. Express -2 rad in degrees.

Solution

a. $\frac{\pi}{3}\text{rad} = \left(\frac{\pi}{3}\right)\left(\frac{180}{\pi}\right)^{\circ} = 60°$

b. $270° = (270)\left(\frac{\pi}{180}\right)\text{rad} = \frac{3\pi}{2}\text{rad}$

c. $-2 \text{ rad} = (-2)\left(\frac{180}{\pi}\right)^{\circ} \approx -114.592°$

Before continuing, a note on terminology: Whenever an angle is measured in degrees, its measure will appear followed by the degree symbol (°); angles measured in radians will either appear with the abbreviation "rad" afterward or, more commonly, with no notation at all. It is a reflection of the importance of radian measure in mathematics that if no indication of the method of measurement appears, we are to assume the angle is measured in radians.

TOPIC 3: The Pythagorean Theorem and Commonly Encountered Angles

It is tempting, when teaching or learning a new area of mathematics, to restrict attention to examples that are artificially "nice." That is, examples in which complicated terms in the accompanying equations either never appear or else conveniently cancel, and examples in which the final answer is suspiciously missing any ugly fractions or approximations. With this in mind, you might dismiss the angles in the following discussion as unrealistically pleasant to work with. But the justification for studying the angles of 30°, 45°, and 60° is that, first, they actually do appear fairly frequently in real life and, second, they are undeniably useful in building an understanding of trigonometric functions. We will encounter them repeatedly in the sections that follow.

At the moment, we are primarily interested in determining the radian measures that correspond to these common angles, but that is a simple matter of applying the appropriate conversion formula. While we have them before us, therefore, we will also note how these angles relate to one another in the context of triangles. This knowledge will prove to be useful very soon.

The radian equivalents of 30°, 45°, 60°, and (for the sake of completeness) 90° are as follows.

$$30° = (30)\left(\frac{\pi}{180}\right) = \frac{\pi}{6}$$

$$45° = (45)\left(\frac{\pi}{180}\right) = \frac{\pi}{4}$$

$$60° = (60)\left(\frac{\pi}{180}\right) = \frac{\pi}{3}$$

$$90° = (90)\left(\frac{\pi}{180}\right) = \frac{\pi}{2}$$

NOTE

Remember that the absence of notation following an angle means the angle is measured in radians.

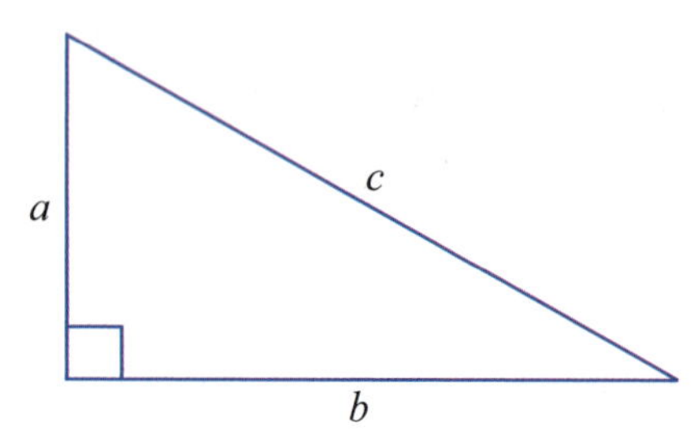

FIGURE 4: $a^2 + b^2 = c^2$

The triangle connection comes from an application of the Pythagorean Theorem. Recall that if a and b are the lengths of the two legs of a right triangle and if c is the length of the hypotenuse, then $a^2 + b^2 = c^2$ (see Figure 4). Recall also that the sum of the angles of a triangle is always 180°, or π radians. So a triangle with one angle of measure $\frac{\pi}{6}$ and a second angle of measure $\frac{\pi}{3}$ must have a right angle, and similarly for a triangle with two angles of measure $\frac{\pi}{4}$. These observations are illustrated in Figure 5.

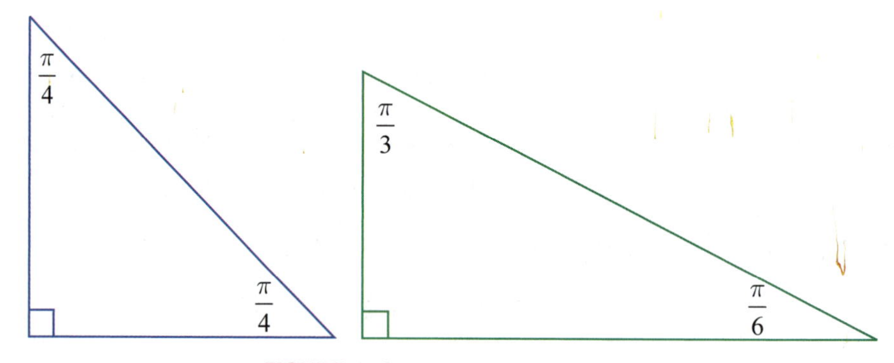

FIGURE 5: Common Right Triangles

Now, suppose the triangle on the left in Figure 5 has legs of length 1. The Pythagorean Theorem tells us then that the length of the hypotenuse is $\sqrt{1^2 + 1^2} = \sqrt{2}$. In general, any triangle with two angles of measure $\frac{\pi}{4}$ will be a right triangle with two legs of equal length, and the ratios of the lengths of the sides will be $1:1:\sqrt{2}$.

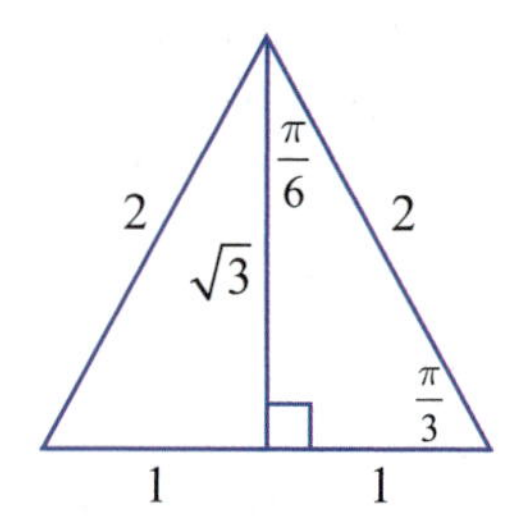

FIGURE 6: Doubling the $\frac{\pi}{6}$-$\frac{\pi}{3}$-$\frac{\pi}{2}$ Triangle

We have to work slightly harder to figure out the ratios of the lengths of the sides of the triangle on the right (the 30°-60°-90° triangle, in degree terms). Note that if we join the triangle with its mirror image, we obtain an equilateral triangle (since all the angles will measure 60°) as shown in Figure 6. This means that the length of the shorter leg of the original triangle must be half of the length of the hypotenuse. So if we assume the shorter leg has a length of 1, the hypotenuse has a length of 2 and the Pythagorean Theorem tells us that the longer leg has a length of $\sqrt{2^2 - 1^2} = \sqrt{3}$. In general, the ratio of the short leg to the long leg to the hypotenuse of such a triangle is $1:\sqrt{3}:2$.

Example 2: Determining the Measure of an Angle

Use the information in each diagram to determine the radian measure of the indicated angle.

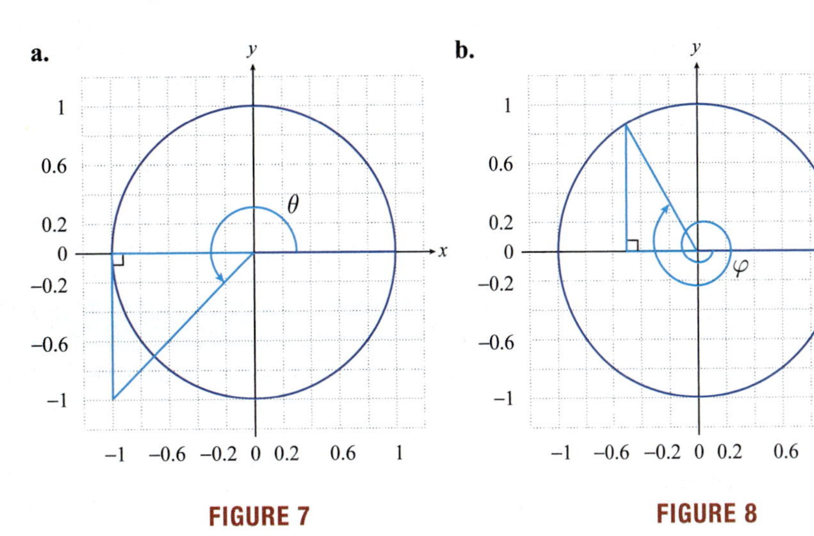

FIGURE 7 FIGURE 8

Solution

a. The angle θ is in standard position, and it is positive. It can also be seen that the measure of the angle is π radians plus a bit more, where the "bit more" comes from the angle whose initial side is the negative x-axis and whose terminal side contains the hypotenuse of the $\frac{\pi}{4}$-$\frac{\pi}{4}$-$\frac{\pi}{2}$ triangle shown in Figure 7. So the "bit more" must be $\frac{\pi}{4}$ radians, and the angle θ has measure $\pi+\frac{\pi}{4}=\frac{5\pi}{4}$.

b. The angle φ (the Greek letter *phi*) is also in standard position, but its measure is negative. It is defined by beginning at the positive x-axis and rotating -2π radians (that is, going full circle in the clockwise direction), continuing for another $-\pi$ radians (another half circle), and then continuing on for a bit more. This time, the angle corresponding to the "bit more" has its initial side on the negative x-axis and terminal side on the hypotenuse of a $\frac{\pi}{6}$-$\frac{\pi}{3}$-$\frac{\pi}{2}$ triangle. We know the triangle must be of this sort because its hypotenuse has length 1 (do you see why?) and its shorter leg has length $\frac{1}{2}$, so the ratio of the shorter leg to the hypotenuse is $1:2$. Hence the "bit more" must have measure $-\frac{\pi}{3}$ and altogether the measure of φ is $-2\pi-\pi-\frac{\pi}{3}=-\frac{10\pi}{3}$.

TOPIC 4 : Arc Length and Angular Speed

The advantages of radian measure over degree measure will appear repeatedly over the next several chapters (and later in calculus). The first advantage is actually just a restatement of the definition of radian. Recall that the radian measure of an angle is related to that portion of the unit circle cut off (or subtended) by the angle when the angle is placed at the center of the circle; the length of the subtended arc is an example of *arc length*. In other words, if θ is a central angle of a unit circle, then $\frac{\theta}{2\pi}$ is the fraction of the circle's circumference subtended by θ. More generally, if θ is a central angle of a circle of radius r, the length s of the portion of the circle subtended by θ is the same fraction multiplied by the circumference.

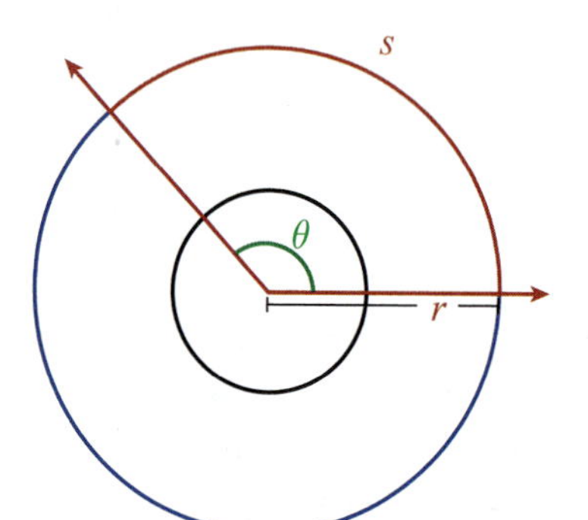

FIGURE 9

f(x) FORMULA: Arc Length

Given a circle of radius r, the length s of the arc subtended by a central angle θ (in radians) is given by the following formula.

$$\begin{aligned} s &= \left(\frac{\theta}{2\pi}\right)(2\pi r) \\ &= r\theta \end{aligned}$$

(In Figure 9, the smaller circle is the unit circle, while the larger circle has radius r.)

Example 3: Applying the Arc Length Formula

The Galapagos Islands lie almost exactly on the equator, and are located at 90° West longitude. Suppose a ship sails along the equator from the Galapagos to the International Date Line (at 180° longitude). How far does the ship travel? (Assume the radius of Earth is 6370 kilometers.)

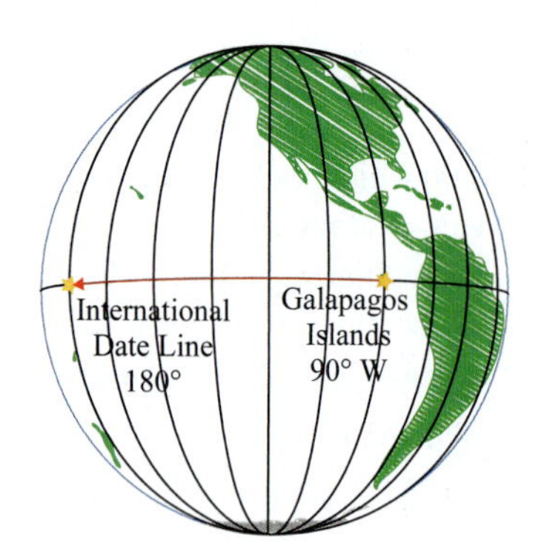

FIGURE 10

Solution

The critical observation is that the ship travels one-quarter of the circumference of Earth (from 90° West of 0° longitude to 180° from 0°), so the angle (at the center of Earth) described by the ship's path is $\frac{\pi}{2}$. Using the arc length formula, the distance traveled is

$$s = (6370)\left(\frac{\pi}{2}\right) \approx 10{,}006 \text{ km}.$$

Now that we have a convenient formula for arc length, we can easily determine the speed with which an object traverses a given arc. For instance, if we are told that the ship in Example 3 takes 15 days to make its journey, we can calculate that its average speed is

$$\frac{10{,}006}{(15)(24)} \approx 27.8\frac{\text{km}}{\text{h}}.$$

Often, information about rate of travel along a circle's circumference is given in terms of *angular speed*, which is a measure of the angle traversed over time.

DEFINITION: Angular Speed and Linear Speed

If an object moves along an arc of a circle defined by a central angle θ in time t, the object is said to have an **angular speed** ω (the Greek letter *omega*) given by

$$\omega = \frac{\theta}{t}.$$

If the circle has a radius of r, the distance traveled in time t is the arc length s, and the **linear speed** v is given by

$$v = \frac{s}{t} = \frac{r\theta}{t} = r\omega.$$

Example 4: Finding the Angular and Linear Speeds

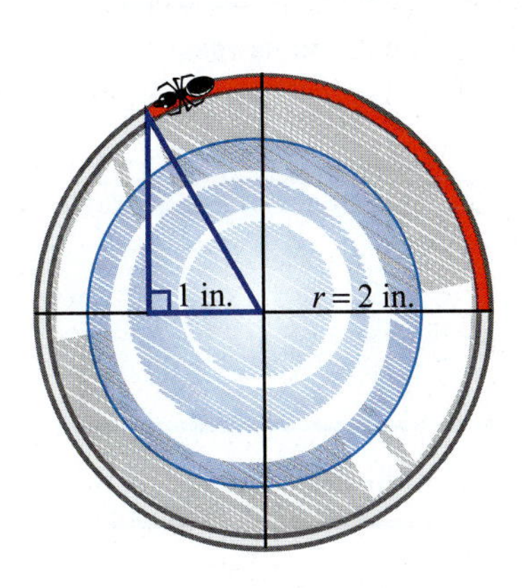

FIGURE 11

Suppose an ant crawls along the rim of a circular glass with radius 2 inches, and traverses the arc indicated in Figure 11 in 20 seconds. What are the angular and linear speeds of the ant, and how far does it travel?

Solution

As in Example 2b, we can determine that the triangle shown in Figure 11 is a $\frac{\pi}{6}$-$\frac{\pi}{3}$-$\frac{\pi}{2}$ right triangle, and the angle corresponding to its vertex at the rigin must have a measure of $\frac{\pi}{3}$. This means the ant describes an angle of

$$\theta = \frac{\pi}{2} + \frac{\pi}{6} = \frac{2\pi}{3}$$

as it crawls, so its angular speed is

$$\omega = \frac{\theta}{t} = \frac{\frac{2\pi}{3}}{20} = \frac{\pi}{30}\frac{\text{rad}}{\text{s}}.$$

Given that the radius of the glass is 2 inches, the linear speed of the ant is as follows.

$$v = r\omega = 2\left(\frac{\pi}{30}\right) = \frac{\pi}{15} \approx 0.21\frac{\text{in.}}{\text{s}}$$

Finally, the distance the ant travels is $s = r\theta = (2)\left(\frac{2\pi}{3}\right) = \frac{4\pi}{3}$ in. or approximately 4.19 in.

CAUTION

The arc length and angular speed formulas, as well as the area formula that follows, are only true for angles measured in radians. Equivalent but less convenient formulas can be derived for angles measured in terms of degrees.

TOPIC 5 : Area of a Circular Sector

We will close this section with one last example of the value of measuring angles in radians.

A **sector** of a circle is the portion of a circle between two radii. The area of a sector, then, can range from 0 to πr^2 square units, where the radius of the circle is assumed to be r units. The two radii defining a given sector can also be taken to be the initial and terminal sides of a central angle θ. Just as $\frac{\theta}{2\pi}$ represents the fraction of a circle's circumference subtended by θ, this same ratio represents the portion of a circle's area contained in the sector of angular size θ. This gives us the following formula.

f(x) FORMULA: Sector Area

The area A of a sector with a central angle of θ in a circle of radius r is

$$A = \left(\frac{\theta}{2\pi}\right)\left(\pi r^2\right) = \frac{r^2\theta}{2}.$$

Example 5: Finding the Area of a Sector

Determine the areas of the sectors defined by the given radii and angles.

a. Circle of radius 3 cm, central angle of 52°

b. Circle of radius $\frac{1}{2}$ ft, central angle of $\frac{4\pi}{3}$

Solution

a. In order to use the sector area formula, the first step is to convert the angle measure to radians.

$$52° = (52)\left(\frac{\pi}{180}\right) = \frac{13\pi}{45}$$

Now, the formula is easily applied.

$$A = \frac{\left(3^2\right)\left(\frac{13\pi}{45}\right)}{2} = \frac{13\pi}{10} \approx 4.08 \text{ cm}^2$$

b. Since the angle is given in radians, we immediately have the following.

$$A = \frac{\left(\frac{1}{2}\right)^2\left(\frac{4\pi}{3}\right)}{2} = \frac{\pi}{6} \approx 0.52 \text{ ft}^2$$

7.1 EXERCISES

PRACTICE

Convert the radian measure to degrees. See Example 1.

1. $\frac{5\pi}{4}$ **2.** $\frac{\pi}{180}$ **3.** $-\frac{3\pi}{8}$ **4.** $-\frac{7\pi}{6}$ **5.** $\frac{2\pi}{3}$

6. $\frac{7\pi}{20}$ **7.** $\frac{5\pi}{6}$ **8.** $\frac{11\pi}{10}$ **9.** $-\frac{9\pi}{4}$ **10.** $-\frac{5\pi}{3}$

Convert the degree measure to radians. See Example 1.

11. $47°$ **12.** $93°$ **13.** $132°$ **14.** $154°$ **15.** $148°$

16. $120°$ **17.** $480°$ **18.** $520°$ **19.** $125°$ **20.** $90°$

Convert each of the following angle measures as directed. See Example 1.

21. Express $\frac{3\pi}{2}$ in degrees.

22. Express $-\frac{9\pi}{4}$ in degrees.

23. Express 3π in degrees.

24. Express $\frac{\pi}{12}$ in degrees.

25. Express $-\frac{2\pi}{5}$ in degrees.

26. Express $\frac{2\pi}{3}$ in degrees.

27. Express $20°$ in radians.

28. Express $340°$ in radians.

29. Express $-144°$ in radians.

30. Express $66°$ in radians.

31. Express $30°$ in radians.

32. Express $180°$ in radians.

The unit circle shown below shows several angles in radians or degrees. Fill in the corresponding radian or degree measure for Exercises 33–44.

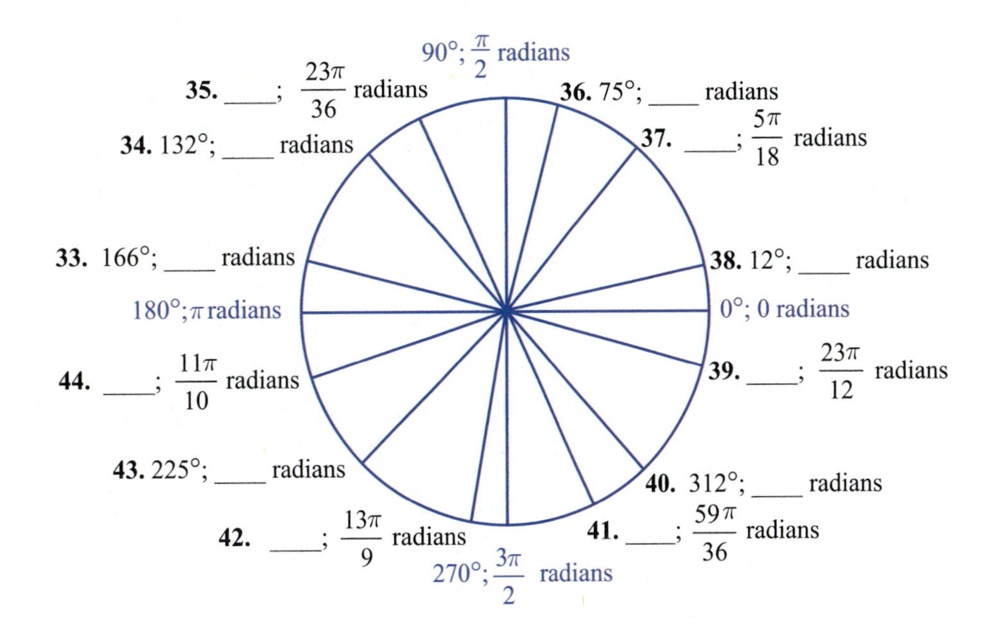

Use the information in each diagram to determine the radian measure of the indicated angle. See Example 2.

45.

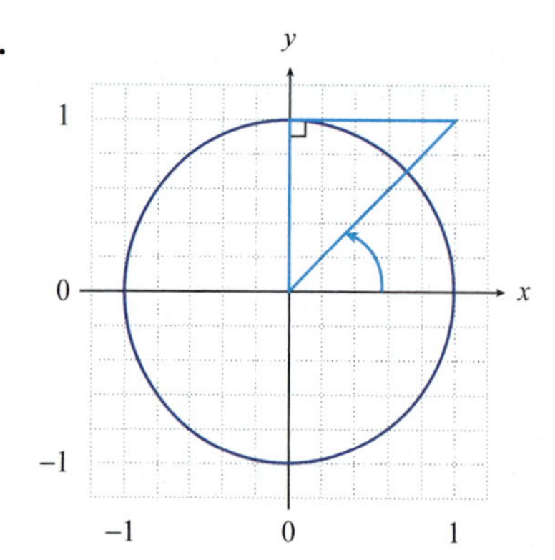

46.

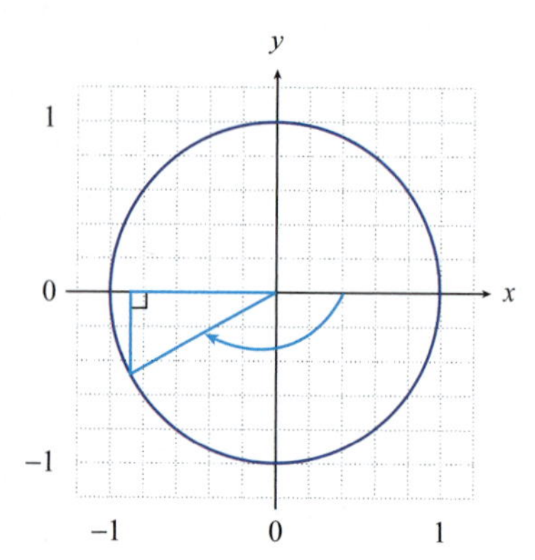

47.

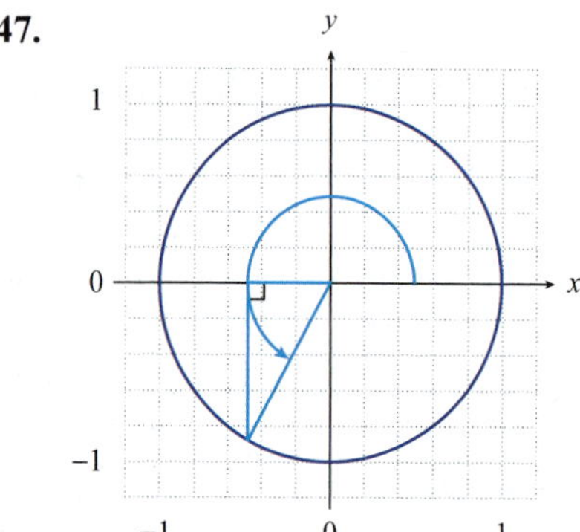

48.

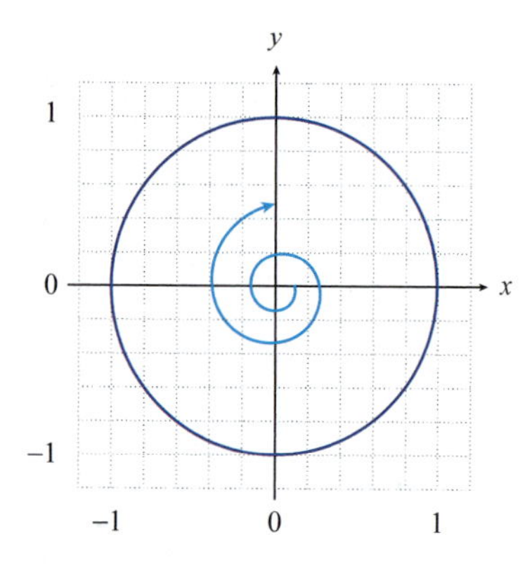

49.

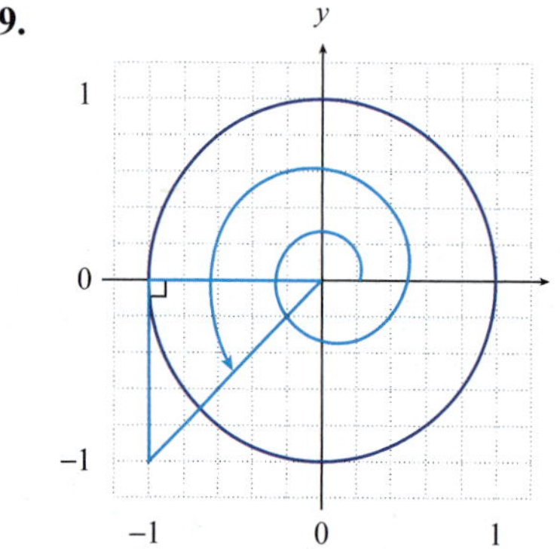

50.

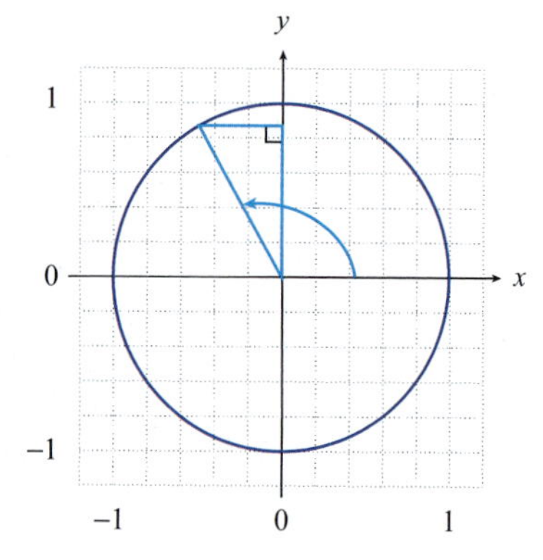

Sketch the indicated angles. See Example 2.

51. $\dfrac{5\pi}{2}$

52. $-60°$

53. $210°$

54. $-\dfrac{\pi}{3}$

55. $\dfrac{7\pi}{4}$

56. $120°$

Find the length of the arc subtended by the given central angle θ on a circle of radius r. Round your answers to two decimal places.

57. $r = 4$ in.; $\theta = 1$

58. $r = 9$ cm; $\theta = \dfrac{\pi}{2}$

59. $r = 15$ ft; $\theta = \dfrac{\pi}{4}$

60. $r = 80$ km; $\theta = 180°$

61. $r = 16.5$ m; $\theta = 30°$

62. $r = 7$ ft; $\theta = 90°$

Find the radian measure of the central angle θ given the radius r and the length s of the arc subtended by θ. Leave your answers in fraction form.

63. $r = 14$ ft; $s = 63$ ft

64. $r = 16$ in.; $s = 6$ in.

65. $r = 23.5$ dm; $s = 10.5$ dm

66. $r = 13$ cm; $s = 130$ cm

67. $r = 2$ km; $s = 22.5$ km

68. $r = 33$ ft; $s = 11$ ft

APPLICATIONS

Find the indicated arc length in each of the following problems. Round your answers to two decimal places. See Example 3.

69. Given a circle of radius 5 inches, find the length of the arc subtended by a central angle of 17° (**Hint:** Convert to radians first).

70. Given a circle of radius 22.5 cm, find the length of the arc subtended by a central angle of 3π.

71. Given a circle with a diameter of 6 feet, find the length of the arc subtended by a central angle of 68° (**Hint:** Convert to radians first).

72. Given a circle of radius 7 m, find the length of the arc subtended by a central angle of $\frac{7\pi}{8}$.

73. A fly walking around the edge of a circular table 6 feet in diameter subtends a central angle of 35°. What distance does the fly walk?

74. Assuming that Columbia, SC and Daytona Beach, FL have the same longitude (81° W), use a radius of 6370 km for Earth and the following to find the distance between the two cities.

City	Latitude
Columbia, SC	34° N
Daytona Beach, FL	29.25° N

75. Given that two cities on the equator are 100 miles apart and have the same latitude (that is, one is due west of the other), what is the difference in their longitudes? Use a value of 3960 miles for the radius of Earth.

76. Using a radius of 1.2 cm for the average eyeball, find the degree measure of the central angle formed to meet the edges of an iris (the colored portion of the eye) with an arc length of 9 mm.

77. Find the distance between Denver, CO and Roswell, NM which lie on the same longitude. The latitude of Denver is 39.75° N and the latitude of Roswell is 33.3° N. Use a radius of 3960 miles for Earth.

78. Find the distance between Atlanta, Georgia and Cincinnati, Ohio which lie on the same longitude. The latitude of Atlanta is 33.67° N and the latitude of Cincinnati is 39.17° N. Assume Earth's radius is 6370 km.

79. Find the distance between Greenwich, England and Valencia, Spain which lie on the same longitude. The latitude of Greenwich is 51.48° N and the latitude of Valencia is 39.47° N. Assume Earth's radius is 6370 km.

80. Find the distance between La Paz, Bolivia and Caracas, Venezuela which lie on the same longitude. The latitude of La Paz is 16.50° S and the latitude of Caracas is 10.52° N. Assume Earth's radius is 6370 km.

81. Find the distance between Bucharest, Romania and Johannesburg, South Africa which lie on the same longitude. The latitude of Bucharest is 44.43° N and the latitude of Johannesburg is 26.21° S. Assume Earth's radius is 6370 km.

The following problems ask you to determine the angular and/or linear speeds of various objects. Round your answers to two decimal places. See Example 4.

82. An industrial circular saw blade has a 10-inch radius and spins at 1000 rpm. Find **a.** the angular speed of a tooth of the blade in radians per minute and **b.** the linear speed of the tooth in feet per second.

83. Earth takes roughly 23 hours and 56 minutes to rotate once about its axis. Using a radius for Earth of 3960 miles, what is the linear speed in miles per hour (relative to the center of Earth) of a person standing on the equator? (Ignore, for the purposes of this problem, such motion as the rotation of Earth about the sun.)

84. A stationary exercise bike is ridden at a constant speed, causing the wheel to spin at a rate of 50 revolutions per minute. If a tack becomes lodged in the tire of radius 14 inches, find **a.** the angular speed of the tack in radians per minute and **b.** the linear speed of the tack in feet per minute.

85. A horse is tethered and urged to trot such that it completes a circular path every 5 seconds. If the rope which tethers it is 20 feet long, what is the linear speed of the horse in miles per hour?

86. The wheels of a certain bike are 28 inches in diameter. If the wheels are rotating at 210 revolutions per minute, how fast is the bicycle moving in miles per hour?

87. The floppy disk drive (FDD) was invented in 1967 to store information for computer users. The first floppy drive used an 8-inch disk and had a radius of 3.91 inches. The drive motor would spin at 300 rotations per minute (RPM).

a. Find the angular speed of the 8-inch disk in radians per second.
b. Find the linear speed of a particular point on the circumference of the 8-inch disk in inches per second.

88. The 8-inch floppy disk drive evolved into a smaller 5.25-inch disk that was used in the personal computers (PC) in the early 1980s. The 5.25-inch disk had a radius of 2.53 in. The usual drive motor for the 5.25-inch disk would spin at 360 rotations per minute.

a. Find the angular speed of the 5.25-inch disk in radians per second.
b. Find the linear speed of a particular point on the circumference of the 5.25-inch disk in inches per second.

Exercises 89–100 ask you to calculate the area of a sector of a circle. Round your answers to two decimal places. See Example 5.

89. Find the area of the shaded portion of the circle.

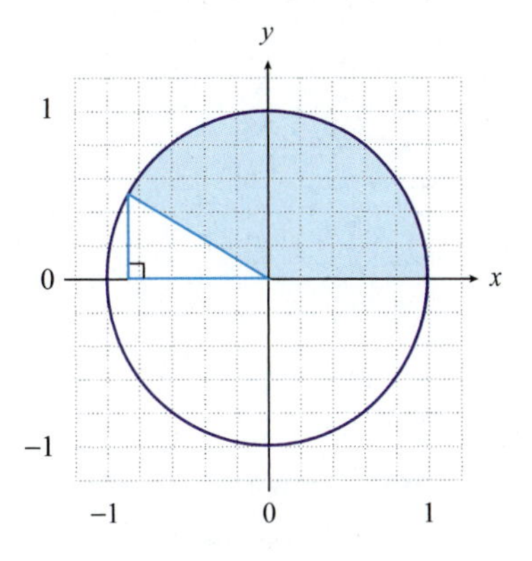

90. Find the area of the shaded portion of the circle.

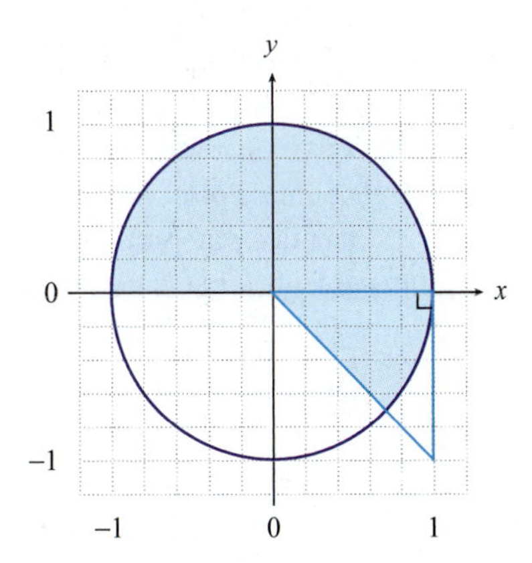

91. Find the area of the sector of a circle of radius 7 cm with a central angle of $70°$.

92. Find the area of the sector of a circle of radius 3.5 ft with a central angle of $27°$.

93. Find the area of the sector of a circle of radius 4 m with a central angle of $\frac{3\pi}{5}$.

94. Find the area of the sector of a circle of radius 16 in. with a central angle of $138°$.

95. Find the area of the sector of a circle of radius 20 ft with a central angle of $\frac{\pi}{2}$.

96. Find the area of the sector of a circle of radius 19 km with a central angle of $5.31°$.

97. A pie of radius 5 in. is cut into 8 equal pieces. What is the area of each piece?

98. The minute hand of a clock extends out to the edge of the clock's face, which is a circle of radius 2 in. What area does the minute hand sweep out between 9:05 and 9:25?

99. The circular spinner for a board game is divided into 6 equal wedges, each of a different color. If the radius is 5 cm, what area is encompassed by 2 wedges?

100. A lawn sprinkler throws water over a distance of 20 ft. If it rotates back and forth through an angle of 50°, what is the area of the region it waters?

101. Two gears are rotating to turn a conveyor belt. The smaller gear rotates 80° as the larger gear rotates 50°. If the larger gear has a radius of 18.7 in., what is the radius of the smaller gear?

102. Two water mills are on display at a local museum. The smaller water mill rotates counterclockwise and turns the larger water mill in a clockwise direction. If the smaller water mill has a radius of 5.23 ft and the larger water mill has a radius of 8.16 ft, what is the degree of rotation of the larger wheel when the smaller rotates 60°?

7.2 TRIGONOMETRIC FUNCTIONS AND RIGHT TRIANGLES

TOPICS

1. The trigonometric functions
2. Evaluating trigonometric functions
3. Applications of trigonometric functions

TOPIC 1: The Trigonometric Functions

Now that we are equipped with a useful way of measuring angles, we can proceed to define the basic trigonometric functions that will dominate the discussion in this and the next two chapters.

It is worthwhile to reflect again on the fact that the material we are studying was developed by countless individuals over the span of several thousand years, and that the cultures they lived in and the problems they hoped to solve varied greatly. Much of the early impetus in developing trigonometry came from astronomy, but by the time of the Renaissance the utility of trigonometry in navigation, surveying, and engineering was thoroughly well recognized. That utility has only increased with time, along with the fields in which trigonometric skill is necessary. We will find trigonometry particularly useful, for example, in solving plane geometry problems.

With the benefit of thousands of years of development behind us, we will take an approach to defining the six basic trigonometric functions that calls upon the work of many different eras. That is, our treatment of trigonometry would not be immediately recognizable to, say, early Greek mathematicians or to Italian mathematicians of the 15th and 16th centuries. Instead, our definitions will be motivated by the desire to make the trigonometric functions most readily useful in a wide variety of applications.

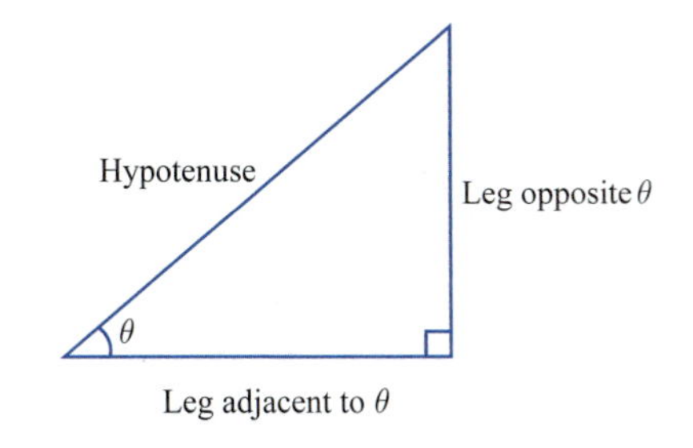

FIGURE 1: Legs Labeled Relative to θ

To begin, consider a right triangle such as the one shown in Figure 1. We will define six functions which are functions of the angle θ. In order to do so, we label the two legs of the triangle as **adjacent to** and **opposite** θ, as shown.

DEFINITION: Sine, Cosine, and Tangent

Assume θ is one of the acute (less than a right angle) angles in a right triangle, as in Figure 1, and let *adj* and *opp* stand for, respectively, the lengths of the legs adjacent to and opposite θ. Let *hyp* stand for the length of the hypotenuse of the right triangle. Then the **sine**, **cosine**, and **tangent** of θ, abbreviated $\sin\theta$, $\cos\theta$, and $\tan\theta$, are the ratios

$$\sin\theta = \frac{\text{opp}}{\text{hyp}}, \quad \cos\theta = \frac{\text{adj}}{\text{hyp}}, \quad \tan\theta = \frac{\text{opp}}{\text{adj}}.$$

Note that sine, cosine, and tangent are indeed functions of the angle θ, and to be consistent with our functional notation we would expect to see, for example, $\sin(\theta) = \frac{\text{opp}}{\text{hyp}}$. By convention, though, the parentheses around the argument are omitted unless called for to make the meaning clear.

Incidentally, the name sine appears to have evolved through a complicated history of abbreviations and mistranslations, beginning with an Arabic word for "half-chord." Our name for the ratio comes from the Latin word *sinus*, which means "bay," but the reference to water is entirely accidental. On the other hand, *cosine* and *tangent*, along with the three functions still to be defined, have meaningful names. More on the subject of names will appear soon.

The remaining three basic trigonometric functions are reciprocals of the first three, as follows.

DEFINITION: Cosecant, Secant, and Cotangent

Again, assume θ is one of the acute angles in a right triangle, as in Figure 1. Then the **cosecant**, **secant**, and **cotangent** of θ, abbreviated $\csc\theta$, $\sec\theta$, and $\cot\theta$, are the reciprocals, respectively, of $\sin\theta$, $\cos\theta$, and $\tan\theta$. That is,

$$\csc\theta = \frac{1}{\sin\theta} = \frac{\text{hyp}}{\text{opp}}, \quad \sec\theta = \frac{1}{\cos\theta} = \frac{\text{hyp}}{\text{adj}}, \quad \cot\theta = \frac{1}{\tan\theta} = \frac{\text{adj}}{\text{opp}}.$$

Example 1: Finding the Values of the Six Trigonometric Functions

Use the information contained in the two figures to determine the values of the six trigonometric functions of θ.

a.

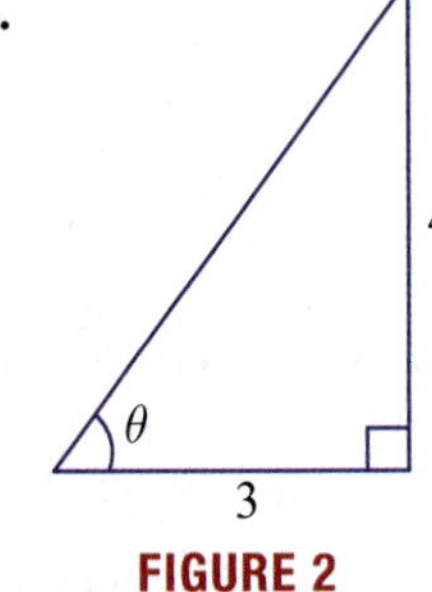

FIGURE 2

b.

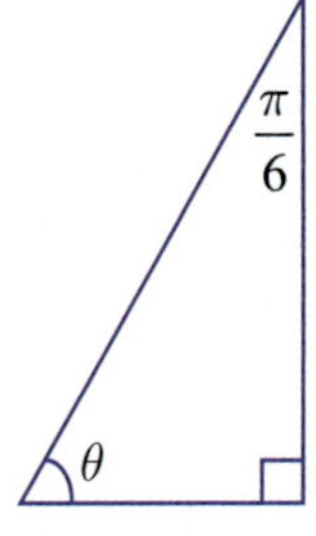

FIGURE 3

Solution

a. With the information given, we can determine $\tan\theta$ and $\cot\theta$ without effort (make sure you see why this is so). In order to evaluate the remaining four trigonometric functions at θ, all we need to do is determine the length of the hypotenuse. By the Pythagorean Theorem,

$$\begin{aligned}(\text{hyp})^2 &= (\text{adj})^2 + (\text{opp})^2 \\ &= 3^2 + 4^2 \\ &= 25\end{aligned}$$

so hyp $= 5$.

Thus,

$$\sin\theta = \frac{\text{opp}}{\text{hyp}} = \frac{4}{5}, \quad \cos\theta = \frac{\text{adj}}{\text{hyp}} = \frac{3}{5}, \quad \tan\theta = \frac{\text{opp}}{\text{adj}} = \frac{4}{3}$$

and

$$\csc\theta = \frac{1}{\sin\theta} = \frac{5}{4}, \quad \sec\theta = \frac{1}{\cos\theta} = \frac{5}{3}, \quad \cot\theta = \frac{1}{\tan\theta} = \frac{3}{4}.$$

b. Since the triangle pictured contains a right angle and an angle of $\frac{\pi}{6}$ radians, the angle θ must have measure $\frac{\pi}{3}$; in degree terms, this is a 30°-60°-90° triangle. Such triangles always have sides in the ratio $1:\sqrt{3}:2$, so even though we are not given the length of any side, we can still evaluate all six trigonometric functions at θ. For example, if the shorter leg (the leg adjacent to θ) is assumed to have a length of a, then the other leg must have a length of $a\sqrt{3}$ and the hypotenuse must have a length of $2a$. So,

$$\sin\theta = \frac{\text{opp}}{\text{hyp}} = \frac{a\sqrt{3}}{2a} = \frac{\sqrt{3}}{2}.$$

Similarly,

$$\cos\theta = \frac{\text{adj}}{\text{hyp}} = \frac{1}{2}, \quad \tan\theta = \frac{\text{opp}}{\text{adj}} = \sqrt{3},$$

$$\csc\theta = \frac{1}{\sin\theta} = \frac{2}{\sqrt{3}}, \quad \sec\theta = \frac{1}{\cos\theta} = 2, \quad \cot\theta = \frac{1}{\tan\theta} = \frac{1}{\sqrt{3}}.$$

TOPIC 2: Evaluating Trigonometric Functions

The last example introduces the sort of reasoning we can use to evaluate trigonometric functions of many angles, and we will employ similar methods often in what follows. In other cases we will want a numerical approximation of the value of some trigonometric function, and a calculator will prove to be very useful.

Example 2: Evaluating Trigonometric Functions

Evaluate the tangent and secant of $\theta = \frac{\pi}{4}$.

Solution

With practice, you'll be able to determine $\tan\left(\frac{\pi}{4}\right)$ and $\sec\left(\frac{\pi}{4}\right)$ mentally, but initially it's very useful to draw a picture in order to visualize what is being asked. Since we are working with an angle of $\frac{\pi}{4}$ radians (45°), the remaining angle of our right triangle must be the same size. We've already noted that the sides of such a triangle have lengths in the ratio $1:1:\sqrt{2}$, so we can draw a triangle as shown in Figure 4.

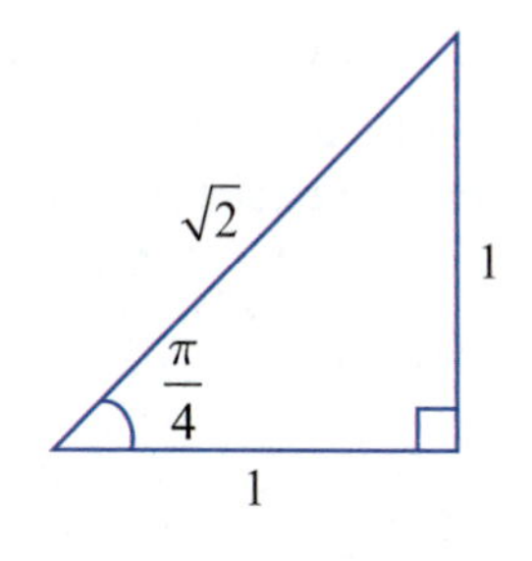

FIGURE 4

Note that the lengths of the sides have been arbitrarily chosen to be 1, 1, and $\sqrt{2}$. Any three numbers in the ratio of $1:1:\sqrt{2}$ could be used (the common factor will cancel out in the evaluation of any trigonometric function), so we may as well use these relatively simple numbers.

Now notice that

$$\tan\left(\frac{\pi}{4}\right)=\frac{1}{1}=1$$

and

$$\sec\left(\frac{\pi}{4}\right)=\frac{\sqrt{2}}{1}=\sqrt{2}.$$

Up to this point in this chapter, a calculator has not been needed to perform any of the evaluations. But if we are asked to evaluate an expression such as $\sin(56.4°)$ (we may encounter such an expression while solving a real-world application), some technological assistance is called for.

Fortunately, calculators (and computer software) that are equipped to handle trigonometric functions are readily available and easily used. However, it is important to remember that angles can be measured in terms of either degrees or radians, and *you* are responsible for putting the calculator in the correct mode (degree or radian) before performing the evaluation. This warning deserves to be repeated:

⚠ CAUTION

Before using a calculator to evaluate a given trigonometric expression, determine whether the angle in the expression is measured in degrees or radians. Then put the calculator in the appropriate mode prior to the evaluation.

Example 3: Evaluating Trigonometric Functions Using a Calculator

Use a calculator to evaluate the following expressions.

a. $\sin(56.4°)$ **b.** $\cot\left(\frac{5\pi}{11}\right)$

Solution

a. Refer to the user's manual to determine how to put your calculator in degree mode. Typically, there is a button labeled "mode" and pressing it leads to the option of choosing either "degree" or "radian." Once the calculator is in the correct mode, press the "sin" button, enter 56.4 on the number pad, and press the "=" or "Enter" button. The answer, rounded to 4 decimal places, is

$$\sin(56.4°)\approx 0.8329.$$

The exact number of digits on your display will depend on the calculator and its current settings. If your display reads –0.1481, your calculator is in the incorrect (radian) mode for this problem.

b. The absence of the degree symbol in the expression $\frac{5\pi}{11}$ tells us that the angle is measured in radians, so the first step is to place your calculator in radian mode. Next, recall that cotangent is the reciprocal of tangent. Most calculators don't have buttons specifically for the cotangent, secant, and cosecant functions; if you need to evaluate an expression containing one of these functions, take the reciprocal of, respectively, the tangent, cosine, or sine of the given angle.

In this case, use your calculator to confirm that

$$\tan\left(\frac{5\pi}{11}\right) \approx 6.9552,$$

and therefore

$$\cot\left(\frac{5\pi}{11}\right) = \frac{1}{\tan\left(\frac{5\pi}{11}\right)} \approx 0.1438.$$

As mentioned at the start of this section, trigonometry has been around for several thousand years. For all but the last few decades of that very long history, users of trigonometry did not have the option of being able to punch a few buttons on a calculator in order to perform their calculations. In the not-too-distant past, a large part of trigonometry consisted of teaching students how to use tables of predetermined evaluations (so-called "trig tables"). Thankfully, we are past the need for such instruction; however, one legacy of the precalculator days of trigonometry lives on and needs to be discussed. That legacy concerns notation.

Today, decimal notation most naturally suits the use of calculators. When calculations were done by hand, however, angles were more commonly expressed in the "degrees, minutes, seconds" (DMS) notation, and we still encounter this notation frequently in some contexts (surveying and astronomy, to name two). We need, therefore, to be able to convert from the DMS notation to decimal notation.

DEFINITION: Degree, Minute, and Second Notation

In the context of angle measure,

$$1' = \text{one minute} = \left(\frac{1}{60}\right)(1^\circ)$$

and

$$1'' = \text{one second} = \left(\frac{1}{60}\right)(1') = \left(\frac{1}{3600}\right)(1^\circ).$$

For instance, an angle given as 14°37′23″ ("14 degrees, 37 minutes, 23 seconds") can also be written in decimal form (rounded to four decimal places) as 14.6231°, since

$$14^\circ 37' 23'' = 14 + \frac{37}{60} + \frac{23}{3600} \approx 14.6231^\circ.$$

TOPIC 3: Applications of Trigonometric Functions

One way or another, most applications of trigonometry involve using given information about a triangle to determine something else about the triangle. At this point, the triangles we work with are all right triangles, and a basic knowledge of the trigonometric functions and the Pythagorean Theorem suffice as tools. The process of determining unknown angles and/or dimensions from known data in such cases is often termed **solving right triangles**. In later sections, we will consider arbitrary triangles and will enlarge our collection of tools with a variety of trigonometric identities and theorems.

Example 4: Using Trigonometric Functions

Before cutting down a dead tree in your yard, you very sensibly decide to determine its height. Backing up 40 feet from the tree (which rises straight up from level ground), you use a *theodolite* (a surveyor's instrument that accurately measures angles) and note that the angle between the ground and the top of the tree is $61°55'39''$. How tall is the tree?

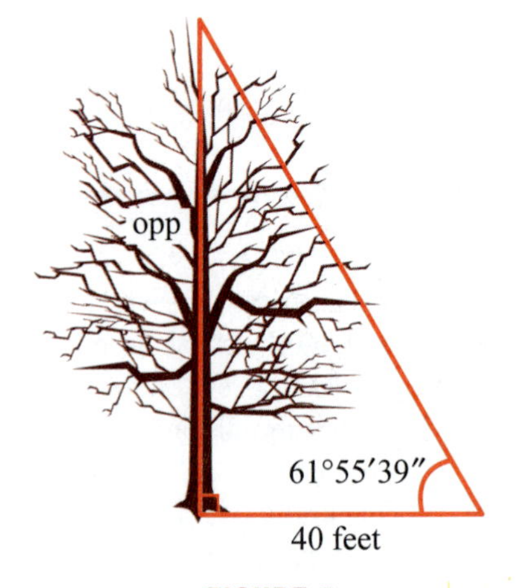

FIGURE 5

Solution

As in so many problems, a picture is of great help. Note that this problem indeed involves solving a right triangle: we know the measure of an angle and the length of the angle's adjacent leg, and we want the length of the opposite leg. This observation gives us the best clue as to which trigonometric function to use; since tangent and cotangent are the two that don't depend on the length of the hypotenuse, chances are one of these is a good choice.

Note also that we have to convert the theodolite's reading into decimal form before using a calculator.

$$61°55'39'' = 61 + \frac{55}{60} + \frac{39}{3600} = 61.9275°$$

Now we can use Figure 5 to see that

$$\tan(61.9275°) = \frac{\text{opp}}{40},$$

so

$$\text{opp} = 40\tan(61.9275°) \approx 75.$$

That is, the tree is approximately 75 feet tall.

Example 5: Using Trigonometric Functions

The manufacturer of a certain brand of 16-foot ladder recommends that, when in use, the angle between the ground and the ladder should equal 75°. What distance should the foot of the ladder be from the base of the wall it is leaning against?

FIGURE 6

Solution

Since we are given information about an angle, its adjacent leg, and the hypotenuse of a right triangle, cosine is the logical trigonometric function to use in solving this problem (equivalently, secant could be used, but calculators are equipped with a "cos" and not a "sec" button so our current technology tends to lead to the use of cosine).

We want to determine the length of the adjacent leg when the ladder is resting against the wall with its recommended angle of 75°, and we note that

$$\cos 75° = \frac{\text{adj}}{16}.$$

This gives us

$$\text{adj} = 16\cos 75° \approx 4.14 \text{ feet},$$

or a bit less than 4 feet, 2 inches.

In many surveying problems, it is frequently necessary to determine the height of some distant object when it is impossible or impractical to measure how far away the object is. One way to determine the height anyway begins with the diagram in Figure 7. In the diagram, assume that the distance d and angles α and β can be measured, but the distance x is unknown. How can we determine the height h?

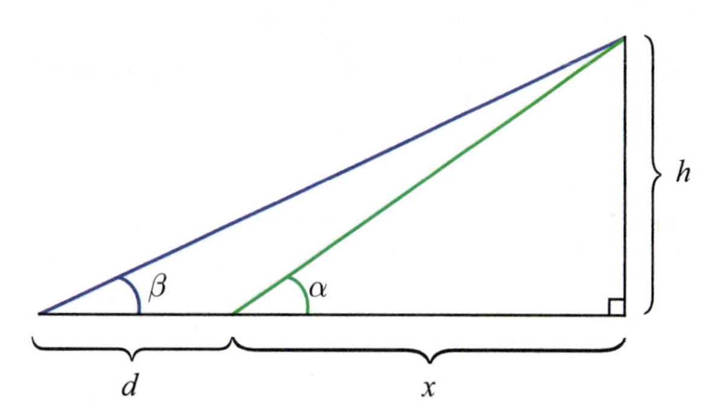

FIGURE 7: Determining h from Two Angles

There are two right triangles apparent in the diagram, and since we know or desire to know something about an angle and its opposite and adjacent legs in each triangle, the tangent function looks promising. We'll begin with the two trigonometric relations

$$\tan\alpha = \frac{h}{x}$$

and

$$\tan\beta = \frac{h}{x+d}.$$

The premise is that we want to find a formula for h in terms of α, β, and d alone, since we don't know the length x. We can start by solving the two equations above for h to get

$$h = x\tan\alpha \quad \text{and} \quad h = (x+d)\tan\beta.$$

These two equations actually are an example of a *system of two equations* in the two unknowns x and h, and we will study such systems in detail in Chapter 11. For our present purposes, we'll note that if we can solve one equation for x and use the result in the other equation, we'll obtain a single equation in which h is the only unknown (this is called the *method of substitution*). For instance, if we solve the first equation for x,

$$x = \frac{h}{\tan\alpha},$$

and make that substitution for x in the second equation, we obtain

$$h = \left(\frac{h}{\tan\alpha} + d\right)\tan\beta.$$

Now we can solve this equation for h.

$$h = \left(\frac{h}{\tan\alpha} + d\right)\tan\beta$$

$$h = \frac{h\tan\beta}{\tan\alpha} + d\tan\beta \qquad \text{Multiply through by } \tan\beta.$$

$$h\tan\alpha = h\tan\beta + d\tan\alpha\tan\beta \qquad \text{Clear fractions by multiplying by } \tan\alpha.$$

$$h\tan\alpha - h\tan\beta = d\tan\alpha\tan\beta \qquad \text{Isolate terms with } h \text{ on one side.}$$

$$h(\tan\alpha - \tan\beta) = d\tan\alpha\tan\beta \qquad \text{Factor out } h.$$

$$h = \frac{d\tan\alpha\tan\beta}{\tan\alpha - \tan\beta} \qquad \text{Divide to solve for } h.$$

This formula is well-known to surveyors, though it often appears in the slightly more appealing form obtained by dividing the numerator and denominator of the fraction on the right by $\tan\alpha\tan\beta$, as follows.

$$\begin{aligned} h &= \frac{d\tan\alpha\tan\beta}{\tan\alpha - \tan\beta} \\ &= \frac{d}{\frac{\tan\alpha}{\tan\alpha\tan\beta} - \frac{\tan\beta}{\tan\alpha\tan\beta}} \\ &= \frac{d}{\frac{1}{\tan\beta} - \frac{1}{\tan\alpha}} \\ &= \frac{d}{\cot\beta - \cot\alpha} \end{aligned}$$

Example 6: Using Trigonometric Functions

Approached from one direction, Mt. Baldy rises out of a perfectly level desert plain. A surveyor standing in the desert some distance from the mountain measures the angle of elevation between the desert floor and the top of the mountain to be $60°1'6''$. She then backs up 1000 feet and determines the new angle of elevation to be $56°3'23''$. How high above the desert plain does Mt. Baldy rise?

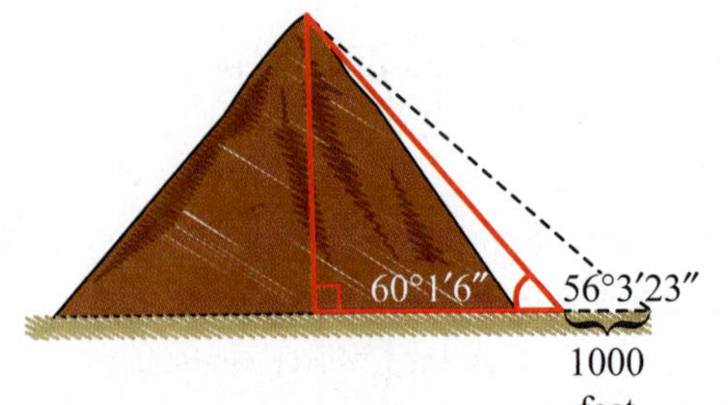

FIGURE 8

Solution

Using the notation of the derivation above, we are given $\alpha = 60°1'6''$, $\beta = 56°3'23''$, and $d = 1000$. Converting to decimal notation, $\alpha \approx 60.018333°$ and $\beta \approx 56.056389°$, so we calculate the approximate height of Mt. Baldy as follows.

$$\begin{aligned} h &= \frac{1000}{\cot(56.056389°) - \cot(60.018333°)} \\ &\approx \frac{1000}{0.673078 - 0.576924} \\ &\approx 10{,}400 \text{ feet} \end{aligned}$$

7.2 EXERCISES

Use the information contained in each figure to determine the values of the six trigonometric functions of θ. Rationalize all denominators in your answers. See Example 1.

1.

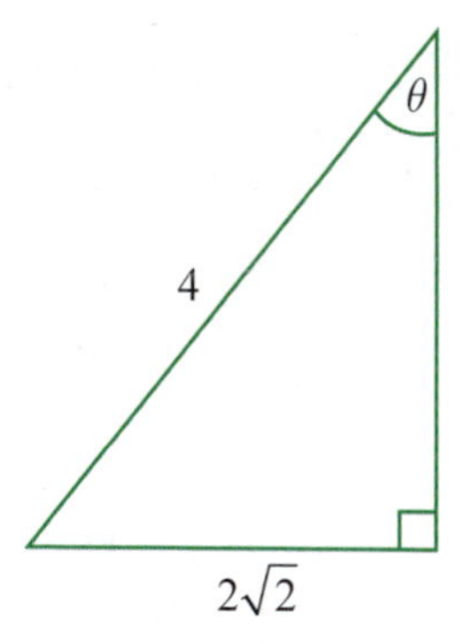

2.

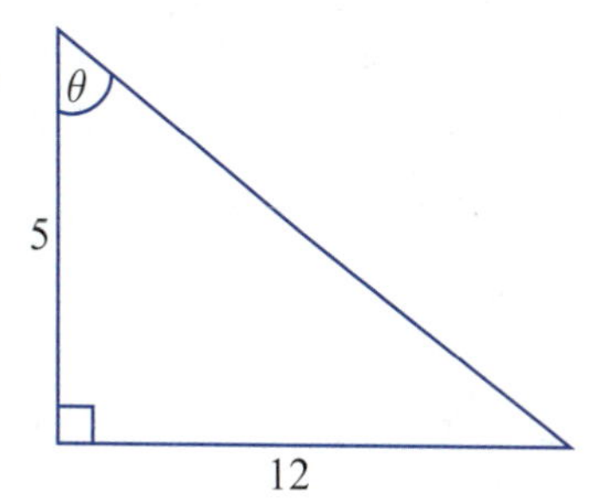

3.

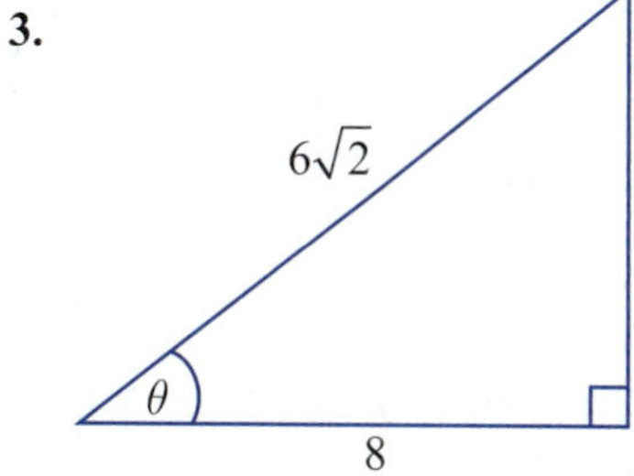

4.

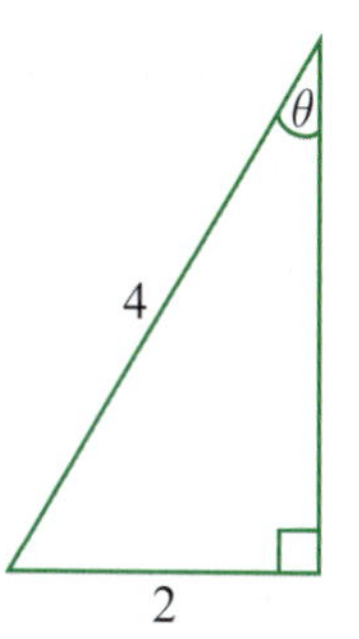

5.

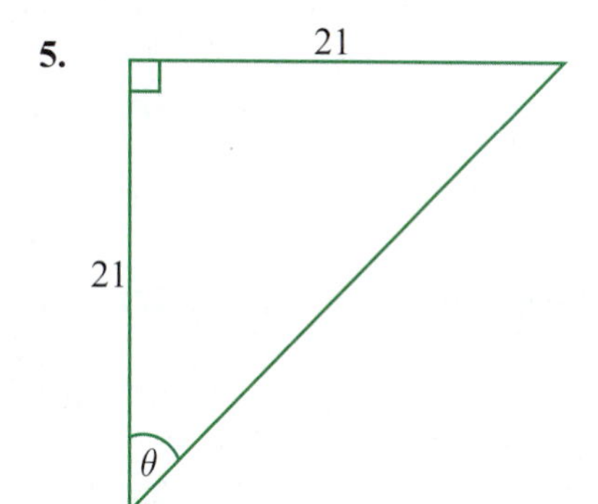

6.

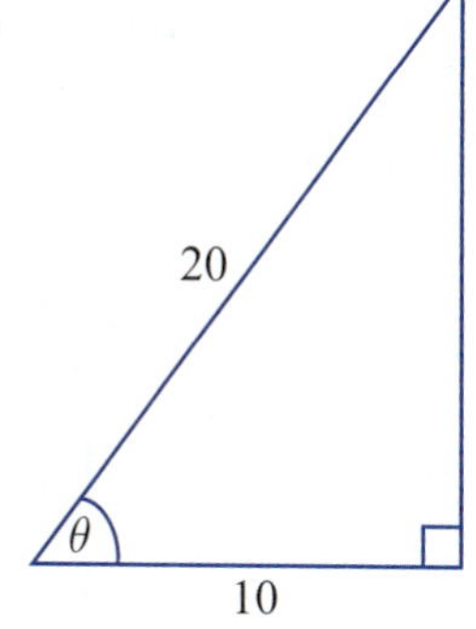

7.

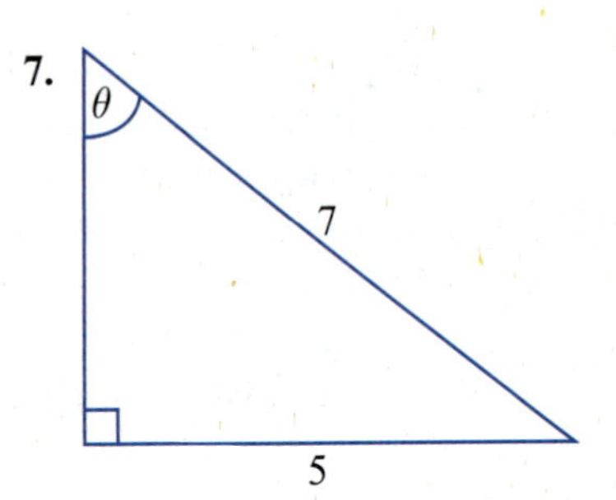

8.

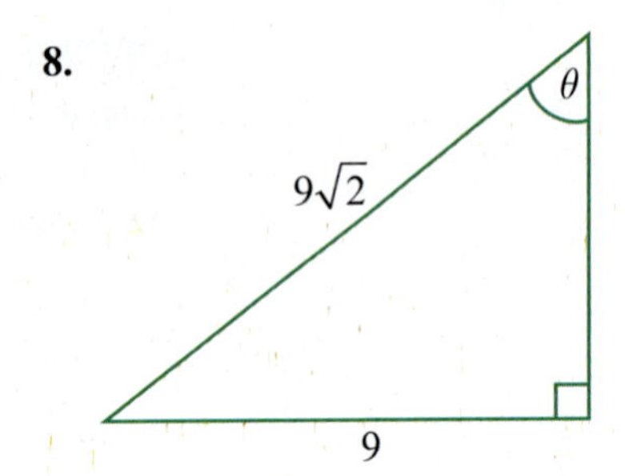

9.

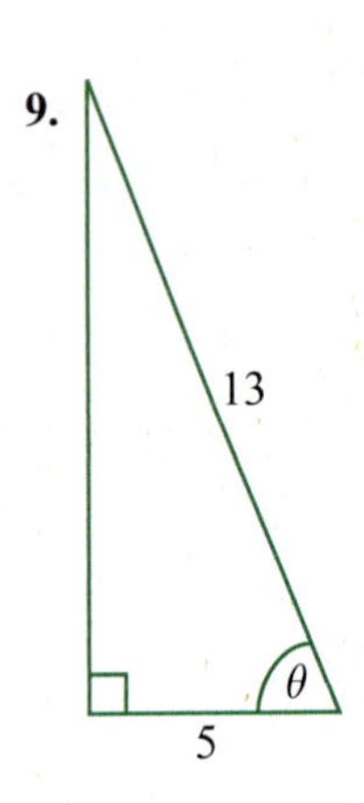

10.

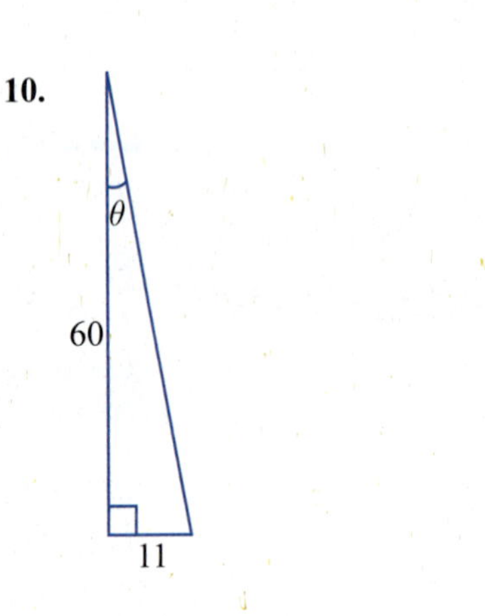

11.

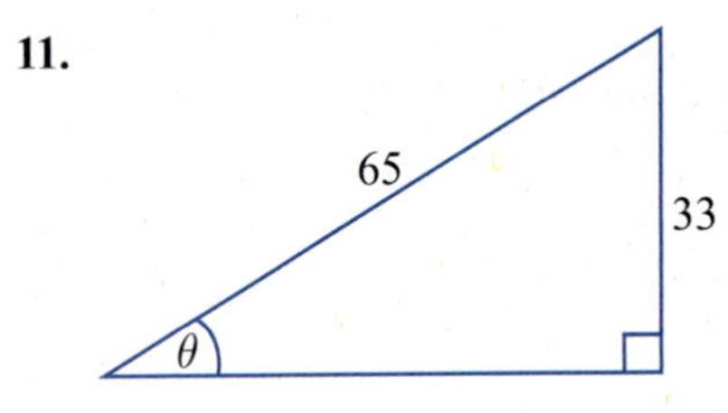

12.

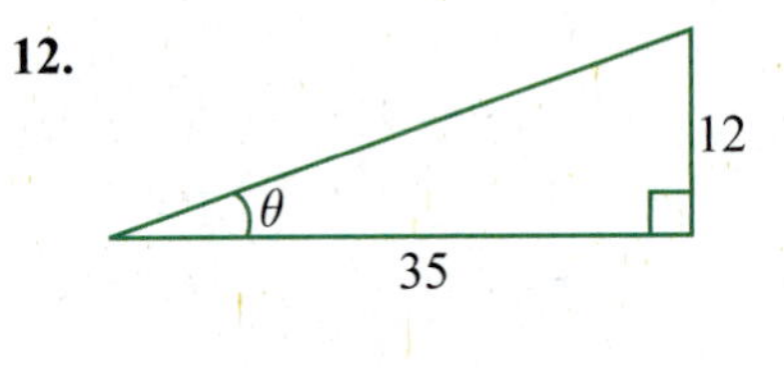

13.

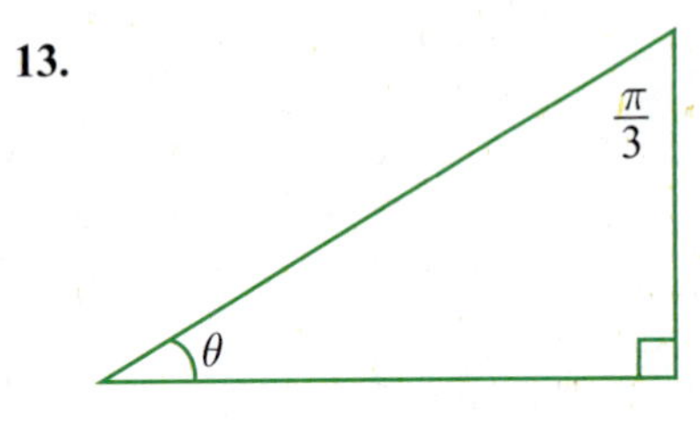

14.

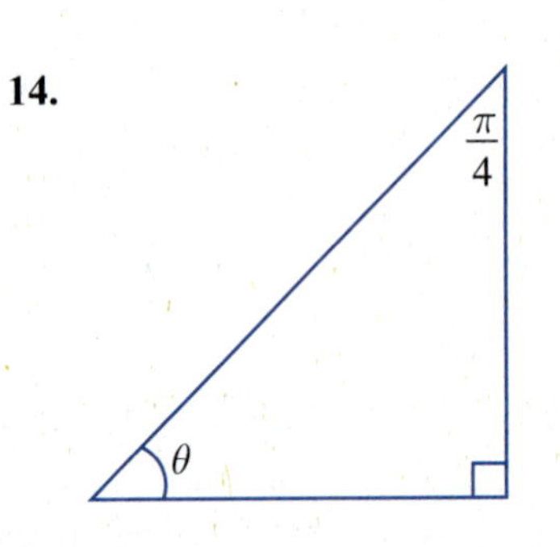

Evaluate the expressions, using a calculator if necessary. Rationalize all denominators in your answers. See Examples 2 and 3.

15. sine and cosecant of $\frac{\pi}{4}$

16. cosine and tangent of $\frac{\pi}{7}$

17. $\sec 60°$

18. $\tan 71°$ and $\cot 71°$

19. $\csc\left(\frac{\pi}{6}\right)$

20. sine of $\frac{3\pi}{7}$

21. secant and tangent of $5°$

22. cosine of $28.37°$

23. cotangent of $\frac{\pi}{3}$

24. $\sin\left(\frac{2\pi}{5}\right)$ and $\cos\left(\frac{2\pi}{5}\right)$

25. $\tan 87.2°$

26. $\csc 54°$

Use a calculator to evaluate each of the following expressions. Round your answers to four decimal places. See Example 3.

27. $\sin 84°$ **28.** $\cos 72°$ **29.** $\tan 46°$ **30.** $\csc 17°$

31. $\sec 88°$ **32.** $\cot 59°$ **33.** $\tan\left(\frac{2\pi}{5}\right)$ **34.** $\cos\left(\frac{\pi}{4}\right)$

35. $\sin\left(\frac{\pi}{8}\right)$ **36.** $\cot\left(\frac{2\pi}{7}\right)$ **37.** $\sec\left(\frac{\pi}{3}\right)$ **38.** $\csc\left(\frac{5\pi}{11}\right)$

Convert each expression from degrees, minutes, seconds (DMS) notation to decimal notation. Round your answers to four decimal places.

39. $38°54'19''$ **40.** $56°12'1''$ **41.** $25°18'90''$

42. $6°8'50''$ **43.** $21°39'56''$ **44.** $88°30'600''$

Determine the value of the given trigonometric expression given the value of another trigonometric expression. Round your answers to four decimal places.

45. Find $\sin\theta$ if $\csc\theta = 8.7$.

46. Find $\cos\theta$ if $\sec\theta = -\frac{7}{4}$.

47. Find $\tan\theta$ if $\cot\theta = \frac{\sqrt{15}}{3}$.

48. Find $\cot\theta$ if $\tan\theta = 2.5$.

49. Find $\sec\theta$ if $\cos\theta = 0.2$.

50. Find $\csc\theta$ if $\sin\theta = -\frac{1}{5}$.

Determine whether the following statements are true or false. Use a calculator when necessary.

51. If $\sin\theta = 0.8$, then $\csc\theta = 1.25$.

52. If $\cos\theta = 0.96$, then $\sec\theta = 1\frac{1}{24}$.

53. If $\tan\theta = 4\frac{4}{9}$, then $\cot\theta = 0.225$.

54. If $\sin\theta = 0.5625$, then $\csc\theta = 2.48$.

55. If $\cos\theta = 0.75$, then $\sec\theta = \frac{8}{3}$.

56. If $\tan\theta = 0.2540$, then $\cot\theta = 3.937$.

APPLICATIONS

Use an appropriate trigonometric function and a calculator if necessary to solve each of the following problems. Round your answers to two decimal places. See Examples 4 and 5.

57. A hang glider wants to determine if a certain vertical cliff is a suitable height for her liftoff. From a distance of 40 yards, she measures the angle from the ground to the cliff's tip as $88°55'24''$. How high is the cliff in feet?

58. A mahi-mahi is hooked on 70 feet of fishing line, 10 feet of which is above the surface of the water. The angle of depression from the water's surface to the line is $40°$. How deep is the fish?

59. A filing cabinet is 3 feet and 4 inches tall from the floor. If a piece of string is stretched from the top of the cabinet to a point on the floor, and the angle between the string and the floor is $11°$, what is the length of the string?

60. A tree being cut down makes a 70° angle with the ground when the tip of the tree is directly above a spot that is 40 feet from the base of the tree. Find the height of the tree.

61. Stephen is standing 15 yards from a stream, but instead of walking directly towards the stream, he decides to take a more scenic (though straight-line) path to the stream. If the angle between the scenic route and the stream is 18°, how far did Stephen walk?

62. The builder of a parking garage wants to build a ramp at an angle of 16° that covers a horizontal span of 40 feet. What is the vertical rise of the ramp?

63. A kitesurfer's lines are 20 m long and make an angle of 37° with the ocean while heading away from the beach under current wind conditions. How high above the water is the kite flying?

64. An anthropologist studying a tribe of indigenous people wants to know the dimensions of their stone-hewn temple. After walking 15 m from the structure, she measures the angle to its top to be 53°. What is the height of the temple?

65. A radio tower has a 64-foot shadow cast by the sun. If the angle from the tip of the shadow to the top of the tower is 78.5°, what is the height of the radio tower?

66. A ladder is propped up to a barn at an 80° angle. If the ladder is 22 feet long, what is the approximate height where the top of the ladder touches the barn?

67. The ramp of a moving truck touches the ground 12 feet away from the end of the truck. If the ramp makes an angle of 30° relative to the ground, what is the length of the ramp?

68. The angle of elevation of a flying kite is $61°7'21''$. If the other end of the 40-foot string attached to the kite is tied to the ground, what is the approximate height of the kite?

69. A length of rope is attached from the top of a dock to the rope tie device located on the underside of the boat at the water's surface. The rope is 33 feet in length and has an angle of elevation relative to the surface of the water of 12°. How high above the water does the dock sit?

In Exercises 70–74, use the formula from Example 6.

70. A surveyor wants to find the width of a river without crossing it. He sights an abandoned tire on the opposite bank (the banks are straight and parallel) and measures the angle from where he stands relative to the shore to be 31°. After walking precisely 15 feet away from the tire, he measures the same angle to be 13.5°. How wide is the river?

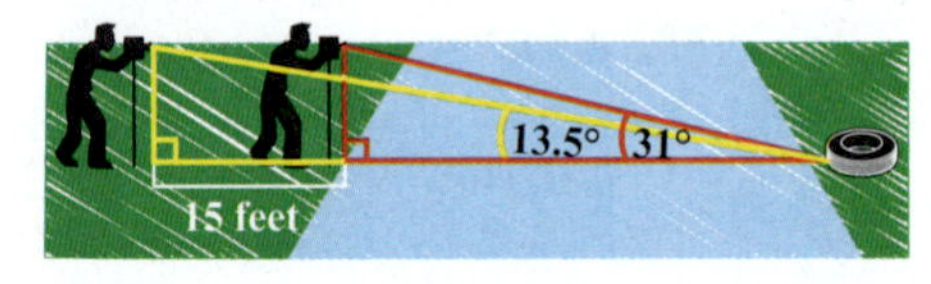

71. A drawbridge operator in a control room observes a sailboat approaching and finds the angle of depression to the boat to be 9°. Twenty minutes later, the angle to the same boat is 19°. If the sailboat has traveled 68.2 m, how high above water is the control room?

72. A birdwatcher discovers a hawk's nest in a tree some distance away. She wants to determine its height, so she measures the angle from the level ground to the nest at 40°. After approaching 25 feet closer to the tree, she finds the same angle to be 52.5°. How high does the nest sit, in feet?

73. A surveyor standing some distance from a plateau measures the angle of elevation from the ground to the top of the plateau to be $46°57'12''$. The surveyor then walks forward 800 feet and measures the angle of elevation to be $55°37'70''$. What is the height of the plateau?

74. A surveyor standing some distance from a hill measures the angle of elevation from the ground to the top of the hill to be $83°45'97''$. The surveyor then steps back 300 feet and measures the angle of elevation to be $75°44'16''$. What is the height of the hill?

WRITING & THINKING

75. To physically demonstrate how the trigonometric functions are defined in terms of ratios of sides of a triangle, Adam draws the triangle shown. Belinda draws another triangle, beginning with the same angle but then proceeding to make her triangle significantly larger, also shown. Nevertheless, when Adam measures the lengths of his triangle's sides and calculates the values of all six trigonometric functions for the angle, and Belinda does the same for her triangle, they obtain the same results. What explains this outcome?

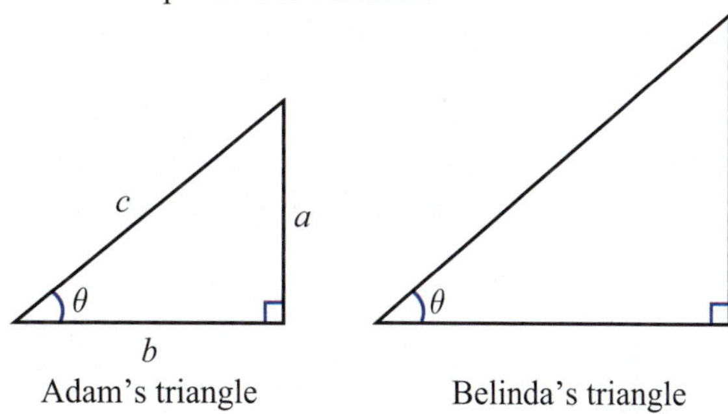

7.3 TRIGONOMETRIC FUNCTIONS AND THE UNIT CIRCLE

TOPICS

1. Extending the domains of the trigonometric functions
2. Evaluating trigonometric functions using reference angles
3. Relationships between trigonometric functions

TOPIC 1: Extending the Domains of the Trigonometric Functions

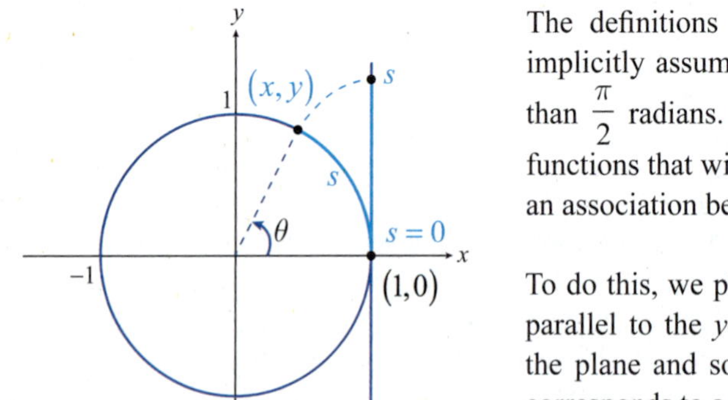

FIGURE 1: The Case $s > 0$

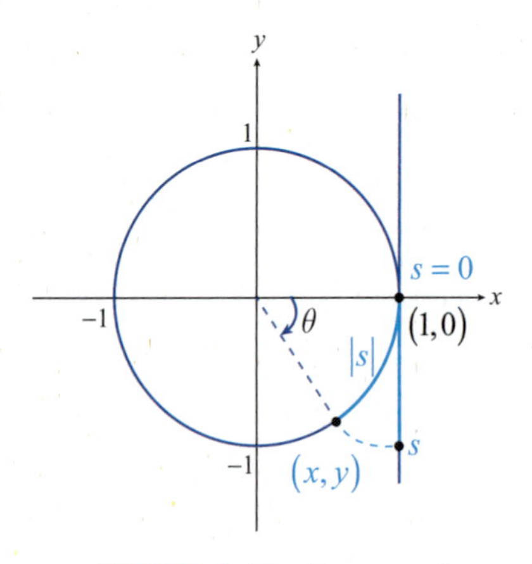

FIGURE 2: The Case $s < 0$

The definitions of the six trigonometric functions given in the last section implicitly assumed that the angle under discussion was greater than 0 and less than $\frac{\pi}{2}$ radians. In this section we'll take an alternative approach to defining the functions that will allow us to greatly extend their domains. We begin by making an association between any given real number and a point on the unit circle.

To do this, we place a third copy of the real number line in the Cartesian plane parallel to the y-axis, situated so that $s = 0$ coincides with the point $(1,0)$ in the plane and so that the positive numbers rise upward. A real number $s > 0$ corresponds to a line segment of length s on the new number line extending from the point 0 up to the point s, as shown in Figure 1, and a real number $s < 0$ corresponds to a line segment of length $|s|$ on the line extending from the point 0 down to the point s, as shown in Figure 2. We now imagine the line segment wrapping around the unit circle centered at the origin, counterclockwise if s is positive and clockwise if s is negative, again as shown in Figures 1 and 2. The endpoint of the wrapped line segment on the circle is the point (x, y) that we associate with the real number s.

One immediate fact to note is that if $|s|$ is larger than 2π, which is the circumference of the unit circle, then the line segment will wrap completely around the circle at least once and overlap itself. In fact, while any given number s gives rise to one and only one point (x, y), any given point (x, y) is associated with an infinite number of values of s. For instance, if $s = 1$ then the point (x, y) associated with 1 is also associated with $1+2\pi$, $1+4\pi$, $1+6\pi$, and so on, as well as $1-2\pi$, $1-4\pi$, etc. More succinctly, we say that (x, y) is associated with all real numbers of the form $1+2\pi n$, where n can be any integer.

The wrapping procedure we've described means that each real number s defines an arc of length $|s|$ on the unit circle, with the initial end of the arc located at $(1,0)$ and the arc proceeding counterclockwise to (x, y) if s is positive or clockwise to (x, y) if s is negative. If we now let θ be the angle corresponding to the arc length $|s|$, as shown in Figures 1 and 2, then the arc length formula tells us that $|s| = r|\theta|$. And using the facts that the sign of θ corresponds to the sign of s and that $r = 1$, we know that $s = \theta$, a fact we will use repeatedly.

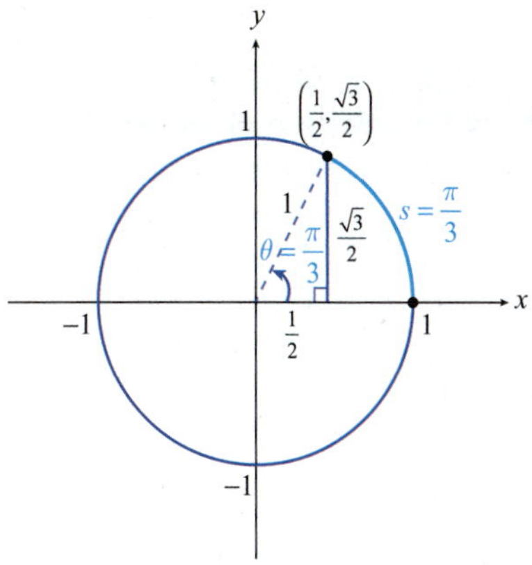

Example 1: Finding the Point on the Unit Circle Associated with a Real Number

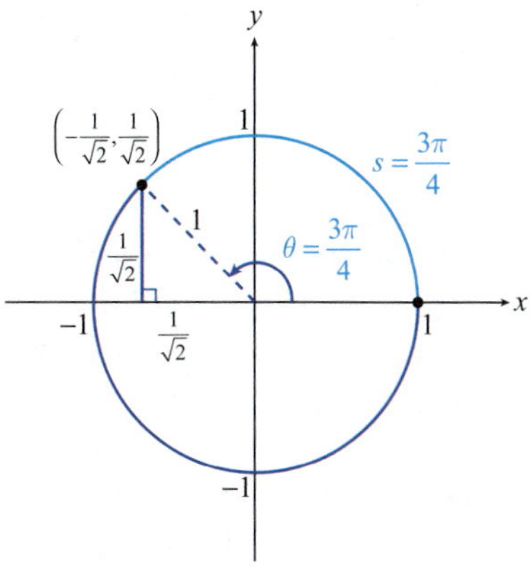

Determine the point (x, y) on the unit circle associated with each real number s.

a. $s = \dfrac{\pi}{3}$ **b.** $s = -\dfrac{5\pi}{2}$

Solution

a. Figure 3 shows an arc of length $\dfrac{\pi}{3}$ wrapped counterclockwise around the unit circle and shows that the angle θ associated with s is also $\dfrac{\pi}{3}$. By dropping a vertical line segment from the to-be-determined point (x, y) down to the x-axis, we see that the angle θ and the point (x, y) automatically determine a familiar right triangle, one of whose angles is $\dfrac{\pi}{3}$. The remaining angle must thus be $\dfrac{\pi}{6}$, and we now know that the ratio of the sides of the triangle must be $1:\sqrt{3}:2$. Since the hypotenuse has length 1, the leg adjacent to θ must have length $\dfrac{1}{2}$ and the leg opposite θ must have length $\dfrac{\sqrt{3}}{2}$. Thus, the point associated with s is $\left(\dfrac{1}{2}, \dfrac{\sqrt{3}}{2}\right)$.

FIGURE 3

b. Two important characteristics to note about the real number $-\dfrac{5\pi}{2}$ are that the number is negative and larger than 2π in magnitude. So the arc defined by this value of s will wrap around clockwise from the point $(1,0)$ and overlap itself by a length of $\left|-\dfrac{5\pi}{2}\right| - 2\pi$, or $\dfrac{\pi}{2}$ units. Expressed another way, since $-\dfrac{5\pi}{2} = -2\pi - \dfrac{\pi}{2}$, the point (x, y) associated with $s = -\dfrac{5\pi}{2}$ is also associated with $s = -\dfrac{\pi}{2}$, as shown in Figure 4. The terminal side of the angle $-\dfrac{\pi}{2}$ points straight downward, so the point (x, y) must be $(0, -1)$.

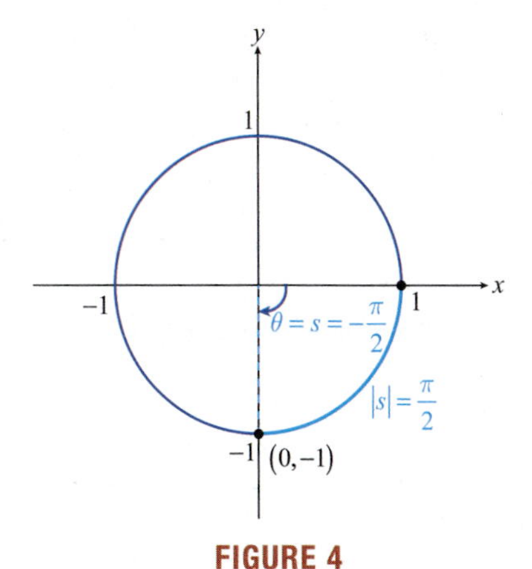

FIGURE 4

Example 2: Finding All Real Numbers Associated with a Point on the Unit Circle

Determine all real numbers s associated with the point $(x, y) = \left(-\dfrac{1}{\sqrt{2}}, \dfrac{1}{\sqrt{2}}\right)$ on the unit circle.

Solution

The given point $\left(-\dfrac{1}{\sqrt{2}}, \dfrac{1}{\sqrt{2}}\right)$ on the unit circle and the arc ending at that point are shown in Figure 5. Constructing a right triangle as we did in Example 1a, we see the legs are of equal length and recognize another of the commonly encountered triangles. From this we know that the angle θ must equal $\dfrac{3\pi}{4}$, and so the real number s must also be $\dfrac{3\pi}{4}$. But as we've seen, any other real number that differs from $\dfrac{3\pi}{4}$ by a multiple of 2π will also be associated with the point $\left(-\dfrac{1}{\sqrt{2}}, \dfrac{1}{\sqrt{2}}\right)$, so the complete set of real numbers associated with the point $\left(-\dfrac{1}{\sqrt{2}}, \dfrac{1}{\sqrt{2}}\right)$ are those of the form $\dfrac{3\pi}{4} + 2n\pi$, where n represents an arbitrary integer.

FIGURE 5

We will use the observation $s=\theta$ to show that, as we proceed to define sine and cosine for all real numbers s, we also automatically define them for arbitrary angles θ. Similarly, with noted exceptions, the remaining four trigonometric functions will also be defined for all real numbers and arbitrary angles.

DEFINITION: Extending the Domains of the Trigonometric Functions

Let s be a real number and let (x,y) be the point on the unit circle associated with s. We define the six trigonometric functions with argument s as follows.

$$\sin s = y \qquad \cos s = x \qquad \tan s = \frac{y}{x},\ x \neq 0$$

$$\csc s = \frac{1}{y},\ y \neq 0 \qquad \sec s = \frac{1}{x},\ x \neq 0 \qquad \cot s = \frac{x}{y},\ y \neq 0$$

To see how this definition of the six functions is consistent with and extends the right-triangle-based definition of Section 7.2, consider again the angle $\theta=\frac{\pi}{3}$ in Example 1a, and recall that the real number $s=\frac{\pi}{3}$ is associated with the point $\left(\frac{1}{2},\frac{\sqrt{3}}{2}\right)$. Looking at just one of the six trigonometric functions, say cosine, the right-triangle-based definition gives us

$$\cos\left(\frac{\pi}{3}\right)=\frac{\text{adj}}{\text{hyp}}=\frac{\frac{1}{2}}{1}=\frac{1}{2}$$

and our new extended definition also gives us

$$\cos\left(\frac{\pi}{3}\right)=x=\frac{1}{2}.$$

But the advantage of the new definition appears when we consider a nonacute angle, such as $\theta=\frac{3\pi}{4}$. In Example 2 we determined that the point $\left(-\frac{1}{\sqrt{2}},\frac{1}{\sqrt{2}}\right)$ on the unit circle is associated with the real number $s=\frac{3\pi}{4}$, so in this case, examining cosine again, we have

$$\cos\left(\frac{3\pi}{4}\right)=x=-\frac{1}{\sqrt{2}},$$

a result that wouldn't have been possible with the right-triangle-based definition. Further advantages of our extended definition are illustrated in Example 3.

Example 3: Finding the Values of the Six Trigonometric Functions

Determine the values of the six trigonometric functions of each angle θ.

a. $\theta=-\frac{5\pi}{2}$ **b.** $\theta=210°$

Solution

a. From Example 1b, we know that $(0,-1)$ is the point on the unit circle that is associated with $s=\theta=-\frac{5\pi}{2}$, so we can proceed to apply the extended definition of the trigonometric functions to obtain the following.

$$\sin\left(-\frac{5\pi}{2}\right) = y = -1 \quad \cos\left(-\frac{5\pi}{2}\right) = x = 0 \quad \tan\left(-\frac{5\pi}{2}\right) = \frac{y}{x} \text{ is undefined}$$

$$\csc\left(-\frac{5\pi}{2}\right) = \frac{1}{y} = -1 \quad \sec\left(-\frac{5\pi}{2}\right) = \frac{1}{x} \text{ is undefined} \quad \cot\left(-\frac{5\pi}{2}\right) = \frac{x}{y} = 0$$

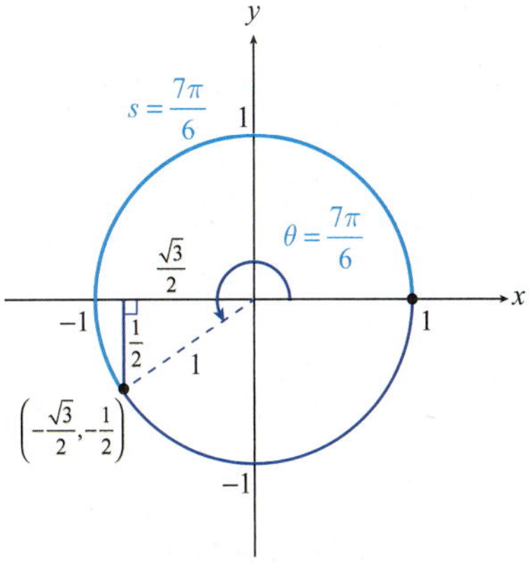

FIGURE 6

b. The angle $\theta = 210°$ is $30°$ more than $180°$, so we can construct a $30°$-$60°$-$90°$ triangle again and determine that the point on the unit circle associated with $210°$ (or $\frac{7\pi}{6}$ radians) is $\left(-\frac{\sqrt{3}}{2}, -\frac{1}{2}\right)$, as shown in Figure 6. So the values of the six trigonometric functions are as follows.

$$\sin 210° = -\frac{1}{2} \quad \cos 210° = -\frac{\sqrt{3}}{2} \quad \tan 210° = \frac{1}{\sqrt{3}}$$

$$\csc 210° = -2 \quad \sec 210° = -\frac{2}{\sqrt{3}} \quad \cot 210° = \sqrt{3}$$

As our extended definition indicates, and as Example 3 shows, some of the trigonometric functions are not defined for certain argument values. Specifically, if (x, y) is the point on the unit circle associated with the real number s, then $\tan s$ and $\sec s$ are not defined when $x = 0$, and $\csc s$ and $\cot s$ are not defined when $y = 0$. Relating real numbers s to their equivalent angles θ and thinking about what it means geometrically if $x = 0$ or $y = 0$, we can describe the domains of all six functions as follows, where n represents an arbitrary integer:

- sine and cosine are defined for all real numbers;
- tangent and secant are defined for all real numbers except for those of the form $\frac{\pi}{2} + n\pi$;
- cosecant and cotangent are defined for all real numbers except for those of the form $n\pi$.

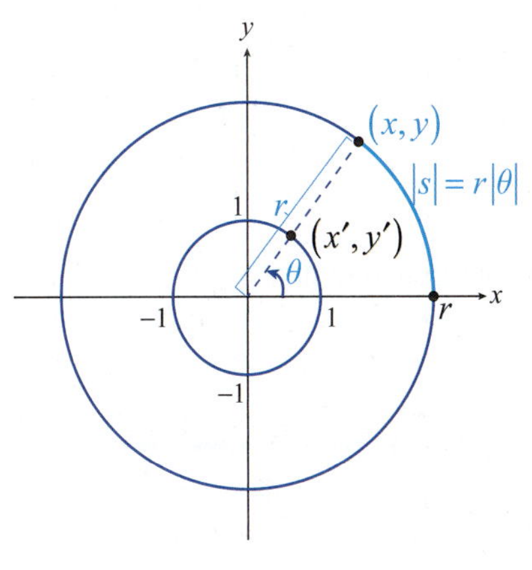

FIGURE 7

One last rephrasing of our extended definition is useful, as it allows us to evaluate the trigonometric functions at an angle θ without reference to the unit circle. Given an angle θ in standard position, let (x, y) be any point (other than the origin) on the terminal side of the angle, and let $r = \sqrt{x^2 + y^2}$. Let (x', y') be the point on the unit circle on the ray extending from the origin to (x, y), as shown in Figure 7. Then by similar triangles, we know

$$y' = \frac{y'}{1} = \frac{y}{r} \quad \text{and} \quad x' = \frac{x'}{1} = \frac{x}{r},$$

and hence

$$\frac{1}{y'} = \frac{r}{y}, \quad \frac{1}{x'} = \frac{r}{x}, \quad \frac{y'}{x'} = \frac{y}{x}, \quad \text{and} \quad \frac{x'}{y'} = \frac{x}{y}.$$

Using these observations, we restate the extended definition of the six trigonometric functions as follows.

DEFINITION: Trigonometric Functions Defined for an Arbitrary Angle

Let θ be an angle in standard position, let (x, y) be any point (other than the origin) on the terminal side of the angle θ, and let $r = \sqrt{x^2 + y^2}$. We define the six trigonometric functions with argument θ as follows.

$$\sin\theta = \frac{y}{r} \qquad \cos\theta = \frac{x}{r} \qquad \tan\theta = \frac{y}{x},\ x \neq 0$$

$$\csc\theta = \frac{r}{y},\ y \neq 0 \qquad \sec\theta = \frac{r}{x},\ x \neq 0 \qquad \cot\theta = \frac{x}{y},\ y \neq 0$$

TOPIC 2: Evaluating Trigonometric Functions Using Reference Angles

The last example hinted at the fact that the evaluation of a trigonometric function at a nonacute angle can be related to its evaluation at an angle in the interval $\left[0, \frac{\pi}{2}\right]$. Such angles are called *reference angles*, and the precise definition is as follows.

DEFINITION: Reference Angles

Given an angle θ in standard position, the **reference angle** θ' associated with it is the angle formed by the x-axis and the terminal side of θ. Reference angles are always greater than or equal to 0 and less than or equal to $\frac{\pi}{2}$ radians; that is, $0 \leq \theta' \leq \frac{\pi}{2}$.

Figure 8 illustrates four ways in which a given angle θ can relate to its reference angle θ'.

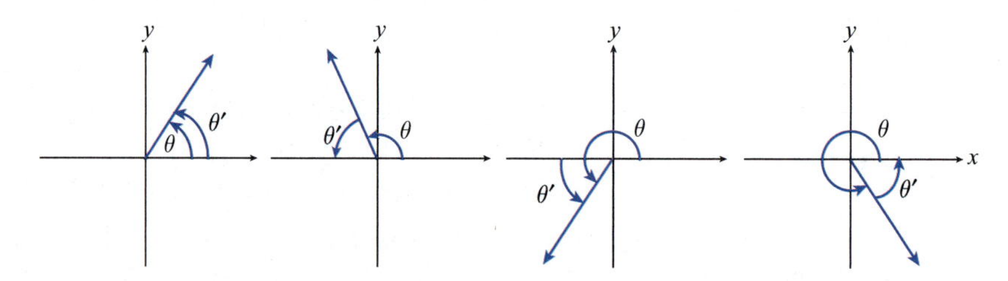

FIGURE 8: Angles and Associated Reference Angles

Example 4: Finding the Reference Angle

Find the reference angle associated with each of the following angles.

a. $\theta = \frac{9\pi}{8}$

b. $\varphi = -655°$

Solution

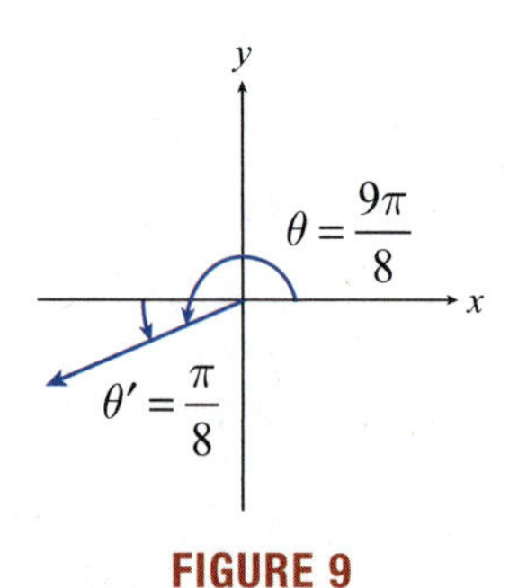

FIGURE 9

a. The terminal side of $\theta = \frac{9\pi}{8}$ lies in the third quadrant, so the reference angle is determined by it and the negative x-axis.

$$\theta' = \frac{9\pi}{8} - \pi = \frac{\pi}{8}$$

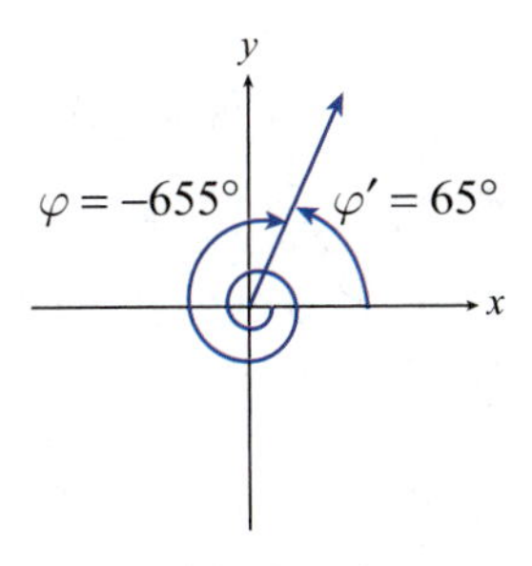

FIGURE 10

b. The terminal side of $\varphi = -655°$ lies in the first quadrant, since the angle describes one complete clockwise revolution and more than three-quarters of a second revolution starting from the positive x-axis. Since two full revolutions would bring us back to the positive x-axis, we find the reference angle between the positive x-axis and the terminal side of φ as follows.

$$\varphi' = 2(360°) - 655° = 65°$$

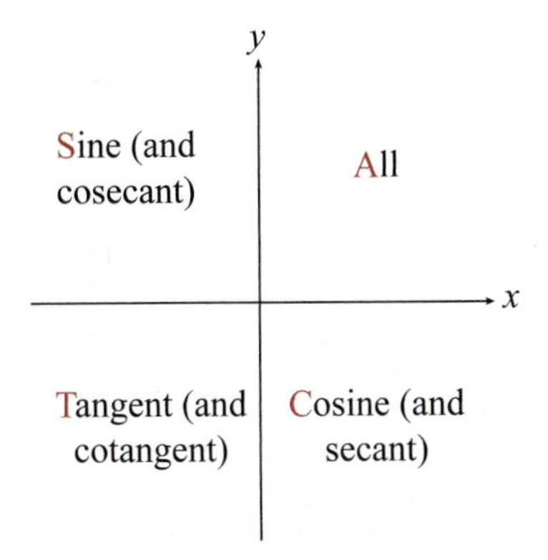

FIGURE 11

The value of reference angles lies in the fact that a trigonometric function evaluated at a given angle will be the same as the function evaluated at the reference angle, except possibly for sign. We implicitly used this fact in Example 3, and the reason why it's true is not hard to see. Since we have defined the trigonometric functions in terms of the coordinates of a point chosen on the terminal side of an angle, the value doesn't depend at all on how many revolutions around the origin (or in which direction) the angle describes. In Example 3b, for instance, the key step lay in determining that the reference angle for $210°$ is $30°$. This led to the construction of a $30°$-$60°$-$90°$ triangle and the easy evaluation of all six trigonometric functions.

Signs of the Trigonometric Functions

Quadrant	Positive	Negative
I	all	none
II	sin, csc	cos, sec, tan, cot
III	tan, cot	sin, csc, cos, sec
IV	cos, sec	sin, csc, tan, cot

TABLE 1

Sign is the one thing that may differ between the value of a trigonometric function at θ and the value of the same function at the reference angle θ'. By thinking about the signs of the x- and y-coordinates of points in the four quadrants, we see that all trig functions are positive in the first quadrant, that sine (and its reciprocal cosecant) are positive in the second quadrant, that tangent (and its reciprocal cotangent) are positive in the third quadrant, and that cosine (and its reciprocal secant) are positive in the fourth quadrant. A bit of propaganda has evolved as a mnemonic to help students remember this: "All Students Take Calculus" may remind you that, beginning in the first quadrant, the functions that are positive are as shown in Figure 11 and Table 1.

Example 5: Evaluating Trigonometric Functions

Evaluate the following.

a. $\cos\left(\frac{4\pi}{3}\right)$ **b.** $\tan(-225°)$

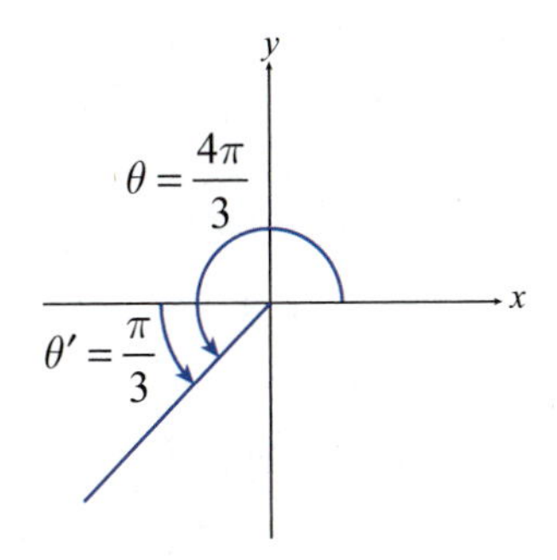

FIGURE 12

Solution

a. The terminal side of $\frac{4\pi}{3}$ lies in the third quadrant and its reference angle is $\frac{\pi}{3}$. We know that $\cos\left(\frac{\pi}{3}\right) = \frac{1}{2}$, but cosine is negative in the third quadrant, so $\cos\left(\frac{4\pi}{3}\right) = -\frac{1}{2}$.

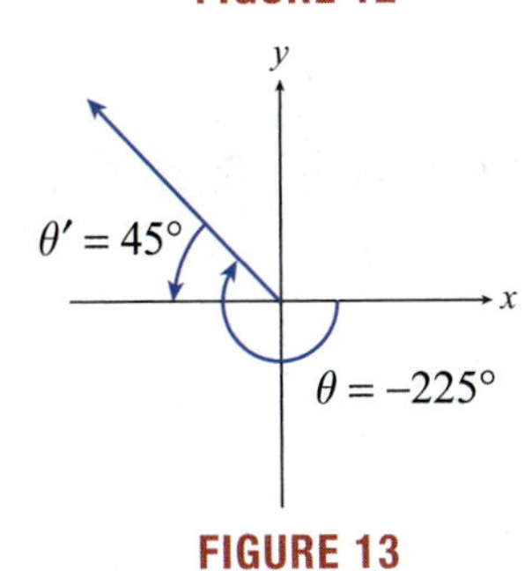

FIGURE 13

b. The terminal side of $-225°$ lies in the second quadrant and its reference angle is $45°$. Since tangent is negative in the second quadrant,

$$\tan(-225°) = -\tan 45° = -1.$$

TOPIC 3: Relationships between Trigonometric Functions

If you have been studying the definitions and examples in this chapter carefully, you may be starting to develop the feeling that there is a great deal of redundancy in trigonometry. This is no illusion; as specific examples, we know now that there are two ways to measure angles and that three of the trigonometric functions are just reciprocals of the other three. There are other redundancies that are a bit more subtle, but you may have developed a sense of them already. We will end this section with a discussion of the qualities of trigonometric functions that lead to this feeling.

The preceding paragraph should not be taken to mean, however, that some of what you are learning in this chapter is pointless. Take, for example, the fact that cosecant, secant, and cotangent are simply reciprocals of sine, cosine, and tangent, and therefore seem unnecessary. In calculus, you'll encounter problems that are more easily stated and solved in terms of, for example, secant rather than cosine. This argument (that a problem is easier to solve with the choice of one function over another) is potent, and can't be disregarded. Why discard something if its existence makes life easier?

Another reason, if not a justification, for the existence of all six trigonometric functions is historical. This history is worth spending a paragraph or two on just for the light it sheds on the nomenclature of trigonometry, but as it turns out the nomenclature in turn leads to some useful facts.

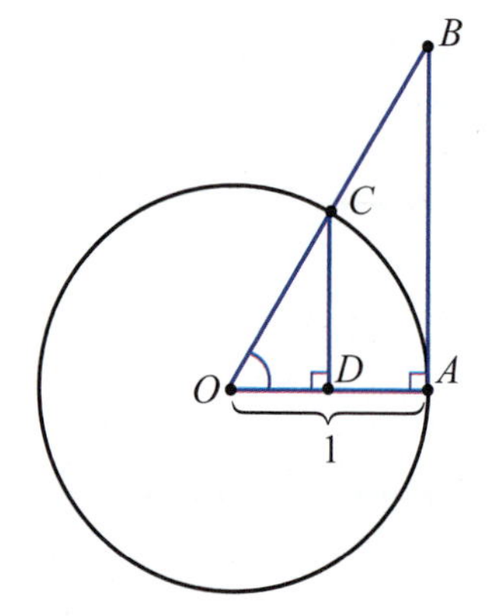

FIGURE 14: Etymology of the Trigonometric Functions

Starting with a circle of radius 1, construct the line segments as shown in Figure 14. The words *secant* and *tangent* are derived from Latin names for, respectively, the line segments $\overline{OB}$ and $\overline{AB}$ (we use "tangent" in everyday language to describe a situation where one object is just touching another). If the line segment passing through C and D were continued down until it intersected the circle again, it would form a *chord*; as it is, $\overline{CD}$ forms a half-chord and *sine* is the word that evolved to denote its length.

Make sure you see how the lengths of the line segments $\overline{OB}$, $\overline{AB}$, and $\overline{CD}$ represent, respectively, the secant, tangent, and sine of the central angle. (**Hint:** It's important to remember that $\overline{OA}$ and $\overline{OC}$ have length 1.) The next step is to see how these three functions lead to the remaining three trigonometric functions (cosine, cosecant, and cotangent.)

Complementary angles are two angles that, when combined, form a right angle. That is, the sum of their measures is $\frac{\pi}{2}$ (or 90° in terms of degree measure). The cosine, cosecant, and cotangent of an angle θ are, respectively, the sine, secant, and tangent of the *complement* of θ. In terms of formulas, this gives us our first set of identities.

IDENTITIES: Cofunction Identities

Given an angle θ (measured in radians), $\frac{\pi}{2}-\theta$ is the measure of its complement, so

$$\cos\theta = \sin\left(\frac{\pi}{2}-\theta\right), \quad \csc\theta = \sec\left(\frac{\pi}{2}-\theta\right), \quad \text{and} \quad \cot\theta = \tan\left(\frac{\pi}{2}-\theta\right).$$

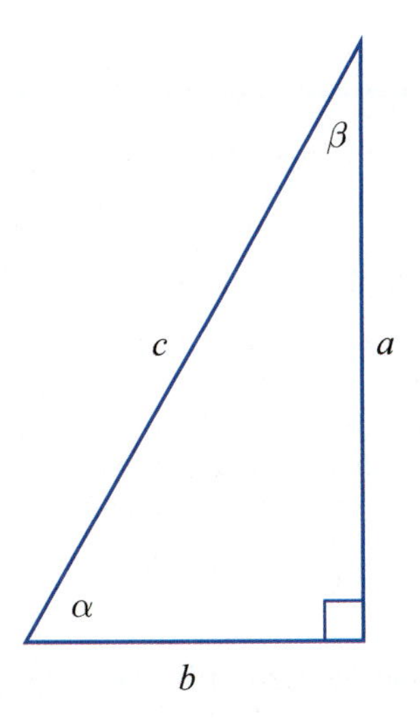

FIGURE 15

For angles between 0 and $\frac{\pi}{2}$, a good way to visualize the relationship between sine, secant, tangent, and their respective cofunctions is with a right triangle as in Figure 15.

Since the sum of the angles of any triangle is $\frac{\pi}{2}$ radians and a right angle has measure $\frac{\pi}{2}$, it must be the case that $\alpha+\beta=\frac{\pi}{2}$; that is, α and β are complementary angles. So using the right-triangle-based definitions, we obtain the following relationships.

$$\sin\alpha=\frac{a}{c}=\cos\beta$$

$$\sec\alpha=\frac{c}{b}=\csc\beta$$

$$\tan\alpha=\frac{a}{b}=\cot\beta$$

An alternative argument for the cofunction identities justifies their use for any angle θ; a sketch of the argument is as follows. If (x,y) is the point on the unit circle associated with a given angle θ, then the reflection of (x,y) with respect to the line $y=x$, namely the point (y,x), is the point associated with the angle $\frac{\pi}{2}-\theta$. By our extended definitions of the trigonometric functions, then,

$$\cos\theta=x \quad \text{and} \quad \sin\left(\frac{\pi}{2}-\theta\right)=x, \quad \text{so} \quad \cos\theta=\sin\left(\frac{\pi}{2}-\theta\right).$$

Similar reasoning leads to the remaining cofunction identities.

We will encounter many more identities in the sections to come. In general, trigonometric identities are equations that are useful in simplifying or evaluating expressions. For reference purposes, we list here another set of three identities that we have seen before.

IDENTITIES: Reciprocal Identities

For a given angle θ for which both sides of the equation are defined,

$$\csc\theta=\frac{1}{\sin\theta}, \quad \sec\theta=\frac{1}{\cos\theta}, \quad \text{and} \quad \cot\theta=\frac{1}{\tan\theta}.$$

We'll finish this initial list of identities with two that you may have already noted. Recall our definitions of tangent and cotangent in terms of a point (x,y) on the terminal side of an angle θ:

$$\tan\theta=\frac{y}{x} \quad \text{and} \quad \cot\theta=\frac{x}{y}.$$

Since $\sin\theta=\frac{y}{r}$ and $\cos\theta=\frac{x}{r}$, these equations lead to the following identities.

IDENTITIES: Quotient Identities

For a given angle θ for which both sides of the equation are defined,

$$\tan\theta=\frac{\sin\theta}{\cos\theta} \quad \text{and} \quad \cot\theta=\frac{\cos\theta}{\sin\theta}.$$

Example 6: Using Cofunction Identities

Express each of the following in terms of the appropriate cofunction, and verify the equivalence of the two expressions.

a. $\cos\left(-\frac{5\pi}{11}\right)$ **b.** $\cot 195°$

Solution

a. $\cos\left(-\frac{5\pi}{11}\right) = \sin\left(\frac{\pi}{2} - \left(-\frac{5\pi}{11}\right)\right)$ The cosine of an angle is equal to the sine of the complement of the angle.

$= \sin\left(\frac{11\pi}{22} + \frac{10\pi}{22}\right)$ Simplify the argument.

$= \sin\left(\frac{21\pi}{22}\right)$ Now use a calculator to verify that $\cos\left(-\frac{5\pi}{11}\right)$ and $\sin\left(\frac{21\pi}{22}\right)$ are both approximately 0.1423.

b. $\cot 195° = \tan(90° - 195°)$ The same relation between cotangent and tangent applies, though the angles are measured in degrees in this example.

$= \tan(-105°)$ Verify that both sides are approximately 3.7321.

We'll conclude this section with examples that illustrate another way to use the relationships between trigonometric functions.

Example 7: Using the Relationships between Trigonometric Functions

Given that $\cos\theta = -\frac{\sqrt{3}}{2}$ and $\tan\theta$ is negative, determine θ and $\tan\theta$.

Solution

Since $\cos\theta$ is negative, we know the terminal side of θ must lie in either the second or third quadrant (remember: "All Students Take Calculus" reminds us the cosine is only positive in the first and fourth quadrants). Given that $\tan\theta$ is also negative, the terminal side of θ must lie in the second quadrant (tangent is positive in the third). A diagram is helpful at this point.

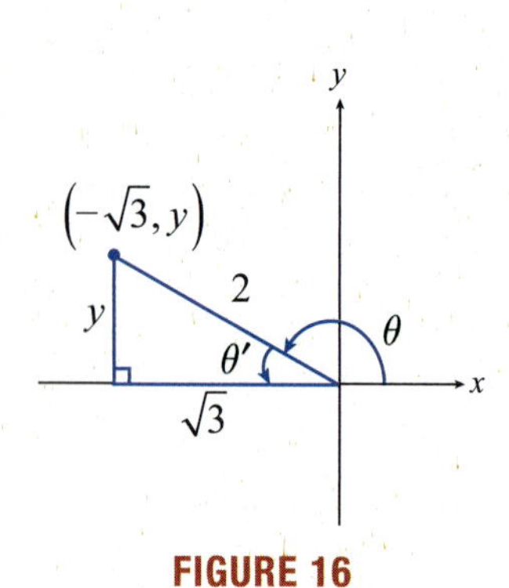

FIGURE 16

In Figure 16, we've drawn θ with its terminal side in quadrant II and we've drawn the reference angle θ'. We've also noted the relative magnitudes of the adjacent leg and hypotenuse of the right triangle, as indicated by the fact that $\cos\theta = -\frac{\sqrt{3}}{2}$. The actual lengths don't have to be $\sqrt{3}$ and 2, respectively, but they do have to be some multiple of $\sqrt{3}$ and 2, so we may as well use the simplest choice.

What we were not given initially is the length labeled y and the actual angle θ. But from Figure 16, we can now recognize that the triangle is a familiar one and that $\theta' = \frac{\pi}{6}$, so $\theta = \frac{5\pi}{6}$. Finally, it must be the case that $y = 1$, and therefore $\tan\theta = -\frac{1}{\sqrt{3}}$.

Example 8: Using the Relationships between Trigonometric Functions

Given that $\cot\theta = 0.4$ and θ lies in the first quadrant, determine $\sin\theta$.

Solution

All trigonometric functions are ratios, so it will probably be useful to express cotangent as a fraction. The result will help us construct a right triangle that relates to the given information. To that end, note that $\cot\theta = 0.4 = \frac{4}{10} = \frac{2}{5}$. If we take the numerator and denominator as the lengths of the adjacent and opposite sides of a right triangle, we are led to the diagram in Figure 17.

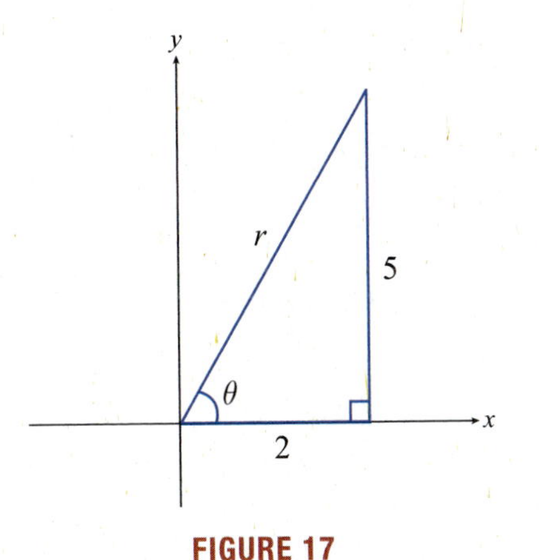

FIGURE 17

Now we can use the Pythagorean Theorem to determine that $r = \sqrt{4+25} = \sqrt{29}$, and so $\sin\theta = \frac{5}{\sqrt{29}}$.

7.3 EXERCISES

PRACTICE

Determine the point (x, y) on the unit circle associated with each real number s. See Example 1.

1. $s = \frac{\pi}{6}$

2. $s = 225°$

3. $s = -120°$

4. $s = -\frac{\pi}{4}$

5. $s = \pi$

6. $s = -\frac{8\pi}{3}$

7. $s = -750°$

8. $s = 855°$

9. $s = \frac{31\pi}{6}$

Determine all real numbers s associated with each point (x, y) on the unit circle. See Example 2.

10. $(x, y) = \left(-\frac{1}{\sqrt{2}}, -\frac{1}{\sqrt{2}}\right)$

11. $(x, y) = \left(\frac{1}{2}, -\frac{\sqrt{3}}{2}\right)$

12. $(x, y) = (0, -1)$

13. $(x, y) = \left(-\frac{\sqrt{3}}{2}, -\frac{1}{2}\right)$

Determine the values of the six trigonometric functions of each angle θ, using a calculator and rounding your answers to four decimal places if necessary. See Examples 3, 4, and 5.

14. $\theta = 45°$ **15.** $\theta = \frac{\pi}{2}$ **16.** $\theta = 60°$

17. $\theta = \frac{3\pi}{4}$ **18.** $\theta = \frac{5\pi}{2}$ **19.** $\theta = -520°$

20. $\theta = 305°$ **21.** $\theta = -1105°$ **22.** $\theta = 6\pi$

23. $\theta = 670°$ **24.** $\theta = \frac{3\pi}{2}$ **25.** $\theta = -215°$

26. $\theta = \frac{5\pi}{4}$ **27.** $\theta = 780°$ **28.** $\theta = -445°$

Determine the reference angle associated with the given angle. See Example 4.

29. $\theta = 98°$ **30.** $\theta = \frac{9\pi}{2}$ **31.** $\theta = -60°$

32. $\theta = \frac{5\pi}{4}$ **33.** $\theta = \frac{5\pi}{2}$ **34.** $\theta = 313°$

35. $\theta = \frac{7\pi}{6}$ **36.** $\theta = -168°$ **37.** $\theta = \frac{6\pi}{5}$

38. $\theta = 216°$ **39.** $\theta = \frac{3\pi}{2}$ **40.** $\theta = -330°$

41. $\theta = \frac{7\pi}{4}$ **42.** $\theta = 718°$ **43.** $\theta = 105°$

Determine the quadrant in which the terminal side of the angle θ is located.

44. $\sin\theta > 0$ and $\tan\theta < 0$ **45.** $\sin\theta < 0$ and $\cot\theta > 0$

46. $\tan\theta > 0$ and $\sec\theta > 0$ **47.** $\cos\theta > 0$ and $\cot\theta < 0$

48. $\sec\theta < 0$ and $\csc\theta < 0$ **49.** $\cot\theta > 0$ and $\csc\theta > 0$

50. $\cot\theta > 0$ and $\cos\theta < 0$ **51.** $\sin\theta > 0$ and $\sec\theta < 0$

In Exercises 52–61, match the given angle θ with the correct reference angle θ' among the answer choices **a.–c.** Answers will be used more than once.

a. $\theta' = 30°$ **b.** $\theta' = 45°$ **c.** $\theta' = 60°$

52. $\theta = 300°$ **53.** $\theta = 150°$ **54.** $\theta = -135°$ **55.** $\theta = 210°$

56. $\theta = -120°$ **57.** $\theta = 315°$ **58.** $\theta = 510°$ **59.** $\theta = 600°$

60. $\theta = 855°$ **61.** $\theta = 480°$

In Exercises 62–76, **a.** rewrite the expression in terms of the given angle's reference angle, and then **b.** evaluate the result, using a calculator and rounding your answers to four decimal places if necessary. See Example 5.

62. $\tan 98°$ **63.** $\sin\left(\frac{9\pi}{2}\right)$ **64.** $\cos(-60°)$

65. $\tan\left(\frac{5\pi}{4}\right)$ **66.** $\cos\left(\frac{5\pi}{2}\right)$ **67.** $\sin 313°$

68. $\cos\left(\frac{7\pi}{6}\right)$

69. $\tan(-168°)$

70. $\cos\left(\frac{6\pi}{5}\right)$

71. $\sin 216°$

72. $\tan\left(\frac{3\pi}{2}\right)$

73. $\cos(-330°)$

74. $\sin\left(\frac{7\pi}{4}\right)$

75. $\tan 718°$

76. $\sin 105°$

Use the appropriate identity to answer each of the following questions. Choose only one answer per question.

77. Which choice is equivalent to $\sin 18°$?

a. $\tan 72°$ **b.** $\cos 72°$ **c.** $\csc 72°$ **d.** $\sec 162°$ **e.** $\cos 162°$

78. Which choice is equivalent to $\sec\left(\frac{\pi}{6}\right)$?

a. $\csc\left(\frac{\pi}{3}\right)$ **b.** $\cos\left(\frac{\pi}{2}\right)$ **c.** $\sin\left(\frac{\pi}{6}\right)$ **d.** $\cos\left(\frac{\pi}{3}\right)$ **e.** $\tan\left(\frac{\pi}{6}\right)$

79. Which choice is equivalent to tan $\tan\left(\frac{\pi}{12}\right)$?

a. $\sin\left(\frac{\pi}{2}\right)$ **b.** $\cos\left(\frac{\pi}{12}\right)$ **c.** $\cot\left(\frac{\pi}{2}\right)$ **d.** $\cot\left(\frac{\pi}{12}\right)$ **e.** $\cot\left(\frac{5\pi}{12}\right)$

80. Which choice is equivalent to $\cos 87°$?

a. $\sin 93°$ **b.** $\cos 93°$ **c.** $\sin 273°$ **d.** $\sec 3°$ **e.** $\sin 3°$

Express each of the following in terms of the appropriate cofunction, and verify the equivalence of the two expressions, using a calculator and rounding your answers to four decimal places if necessary. See Example 6.

81. $\cot 135°$

82. $\sec\left(\frac{\pi}{2}\right)$

83. $\sin(-60°)$

84. $\cos\left(-\frac{3\pi}{4}\right)$

85. $\csc\left(\frac{5\pi}{6}\right)$

86. $\cot 313°$

87. $\cos\left(-\frac{3\pi}{6}\right)$

88. $\csc(-168°)$

89. $\sin\left(-\frac{4\pi}{5}\right)$

90. $\sec 216°$

91. $\csc\left(\frac{3\pi}{2}\right)$

92. $\cos(-15°)$

93. $\cot\left(\frac{\pi}{4}\right)$

94. $\tan(-105°)$

95. $\sec 105°$

Using a calculator, determine $\tan\theta$ and $\cot\theta$ for each of the following exercises. Round your answers to three decimal places.

96. $\sin\theta = 0.978$ and $\cos\theta = 0.208$

97. $\sin\theta = 0.588$ and $\cos\theta = -0.809$

98. $\sin\theta = -0.966$ and $\cos\theta = -0.259$

99. $\sin\theta = -0.866$ and $\cos\theta = -0.5$

100. $\sin\theta = -0.699$ and $\cos\theta = 0.743$

101. $\sin\theta = -0.995$ and $\cos\theta = -0.105$

Use the given information about each angle to evaluate the expressions, if possible. If no angle with the stated properties exists, determine the reason. See Examples 7 and 8.

102. Given that $\cos\theta = \frac{\sqrt{12}}{4}$ and $\tan\theta$ is negative, determine θ and $\tan\theta$.

103. Given that $\csc\theta = 1.25$ and the terminal side of θ lies in the second quadrant, determine $\cot\theta$.

104. Given that $\tan\theta = \frac{\sqrt{3}}{3}$ and $\sin\theta$ is positive, determine θ and $\sin\theta$.

105. Given that $\sin\theta = 2$ and θ is positive, determine $\tan\theta$.

106. Given that $\cot\theta = \frac{3}{4}$ and $\sin\theta$ is negative, determine $\sec\theta$.

107. Given that $\sin\theta = \frac{\sqrt{3}}{2}$ and θ lies in the second quadrant, determine θ and $\tan\theta$.

108. Given that $\sec\theta = 0.3$ and the terminal side of θ lies in the fourth quadrant, determine $\csc\theta$.

The three cofunction identities presented in this section have three companion identities, as follows:

$$\sin\theta = \cos\left(\frac{\pi}{2} - \theta\right), \quad \sec\theta = \csc\left(\frac{\pi}{2} - \theta\right), \quad \text{and} \quad \tan\theta = \cot\left(\frac{\pi}{2} - \theta\right).$$

Express each of the following in terms of the appropriate cofunction. Evaluate both the given expression and the cofunction expression with your calculator to verify that the expressions are equivalent.

109. $\sin\left(\frac{7\pi}{4}\right)$

110. $\csc\left(\frac{8\pi}{3}\right)$

111. $\cot\left(\frac{3\pi}{4}\right)$

112. $\cos\left(-\frac{5\pi}{3}\right)$

113. $\tan 15°$

114. $\sec(-315°)$

WRITING & THINKING

115. Prove the three cofunction identities given in the directions for Exercises 109–114. (**Hint:** For the first identity, begin with the observation that $\sin\theta = \sin\left(\frac{\pi}{2} - \left(\frac{\pi}{2} - \theta\right)\right)$ and then apply one of the three original cofunction identities.)

7.4 GRAPHS OF SINE AND COSINE FUNCTIONS

TOPICS

1. Graphing sine and cosine functions
2. Periodicity and symmetry
3. Amplitude, frequency, and phase shifts
4. Simple harmonic motion
5. Damped harmonic motion

TOPIC 1: Graphing Sine and Cosine Functions

In the preceding sections, you have been exposed to some of the highlights of more than 2000 years of thought regarding trigonometry and its uses. With the fundamentals out of the way, it's time to reflect on a subtle but important point.

Most of the applications of the trigonometric functions implicitly view them as either functions of (acute) angles or as functions of real numbers. (Incidentally, the trigonometric functions can be extended even further to be functions of complex numbers, but that discussion will have to wait for a later course.) We've had quite a bit of experience with applications of the first sort; chances are, if you sketch a triangle in the course of solving a problem, you're using the "angle" point of view. In this section, we will concentrate on the second point of view.

To emphasize the theme of this section and the next, we will now frequently use x instead of θ to represent the argument of a given trigonometric function, and x will be a variable standing for (nearly) any real number. To start with, we know from Section 7.3 that sine and cosine are functions of any real number x, so the first order of business will be to graph them as functions defined on the real line. The process will be familiar to you—we have taken similar steps to graph the functions we encountered in Chapters 3 through 6.

Recall that the most elementary approach to graphing a function for the first time is to plot points. That is, evaluate a given function $f(x)$ for a sufficient number of values x so that, when all the calculated points of the form $(x, f(x))$ are plotted, a reasonable guess for the graph of f can be constructed. We'll do this for the two functions sine and cosine at the same time.

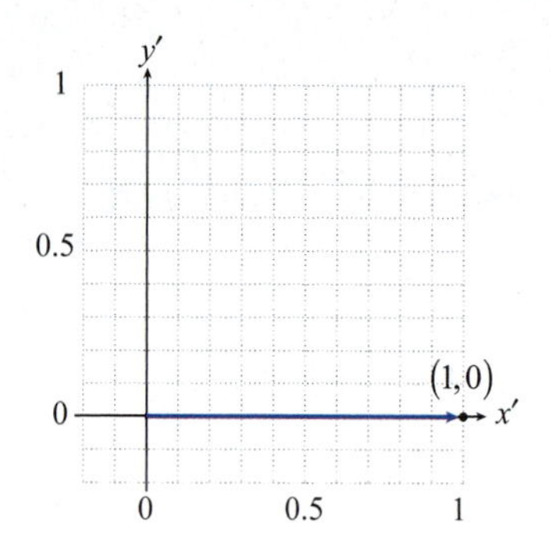

FIGURE 1

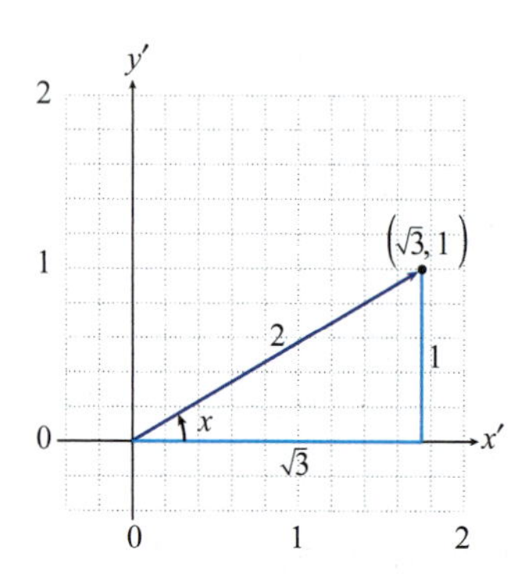

FIGURE 2

For instance, for $x = 0$ we know that $\sin 0 = 0$ and $\cos 0 = 1$. We could use a calculator to confirm these facts, but it's better to *understand* why the values are what they are. To do so, we use the method of the previous section and place the angle x in standard position with its vertex at the origin of the Cartesian plane, as shown in Figures 1 and 2. Note, however, that since we are now using x instead of θ to label the angle, we will temporarily label the horizontal and vertical axes of the plane with, respectively, x' and y' to avoid confusion. In this context, letting $x = 0$ tells us to consider an angle whose initial and terminal sides both lie along the positive x'-axis (see Figure 1), and the definitions of sine and cosine in Section 7.3 then lead to

$$\sin 0 = \frac{0}{1} = 0 \quad \text{and} \quad \cos 0 = \frac{1}{1} = 1.$$

Similarly, for $x = \frac{\pi}{6}$ (see Figure 2), we know

$$\sin\left(\frac{\pi}{6}\right) = \frac{1}{2} \quad \text{and} \quad \cos\left(\frac{\pi}{6}\right) = \frac{\sqrt{3}}{2}.$$

If we let x take on other values between 0 and 2π that are easy to work with, we obtain Table 1.

x	$\sin x$	$\cos x$
0	0	1
$\frac{\pi}{6}$	$\frac{1}{2}$	$\frac{\sqrt{3}}{2}$
$\frac{\pi}{4}$	$\frac{1}{\sqrt{2}}$	$\frac{1}{\sqrt{2}}$
$\frac{\pi}{3}$	$\frac{\sqrt{3}}{2}$	$\frac{1}{2}$
$\frac{\pi}{2}$	1	0
$\frac{2\pi}{3}$	$\frac{\sqrt{3}}{2}$	$-\frac{1}{2}$
$\frac{3\pi}{4}$	$\frac{1}{\sqrt{2}}$	$-\frac{1}{\sqrt{2}}$
$\frac{5\pi}{6}$	$\frac{1}{2}$	$-\frac{\sqrt{3}}{2}$

x	$\sin x$	$\cos x$
π	0	-1
$\frac{7\pi}{6}$	$-\frac{1}{2}$	$-\frac{\sqrt{3}}{2}$
$\frac{5\pi}{4}$	$-\frac{1}{\sqrt{2}}$	$-\frac{1}{\sqrt{2}}$
$\frac{4\pi}{3}$	$-\frac{\sqrt{3}}{2}$	$-\frac{1}{2}$
$\frac{3\pi}{2}$	-1	0
$\frac{5\pi}{3}$	$-\frac{\sqrt{3}}{2}$	$\frac{1}{2}$
$\frac{7\pi}{4}$	$-\frac{1}{\sqrt{2}}$	$\frac{1}{\sqrt{2}}$
$\frac{11\pi}{6}$	$-\frac{1}{2}$	$\frac{\sqrt{3}}{2}$

TABLE 1: Selected Values of Sine and Cosine

This table represents the values of sine and cosine as x assumes the values of convenient angles, beginning with $x = 0$ and rotating around counterclockwise for one full circle. There's no need to let x take on higher values, or negative numbers for that matter, since the table will simply repeat. For instance,

$\sin(2\pi) = \sin 0$ and $\sin\left(-\frac{\pi}{6}\right) = \sin\left(\frac{11\pi}{6}\right)$. This observation is a recognition of the *periodicity* of sine and cosine; we say that these two functions both have a period of 2π. We will discuss the periodicity of all trigonometric functions more thoroughly soon.

Figure 3 shows the result of plotting the calculated points for sine, with the curve drawn to smoothly pass through the points.

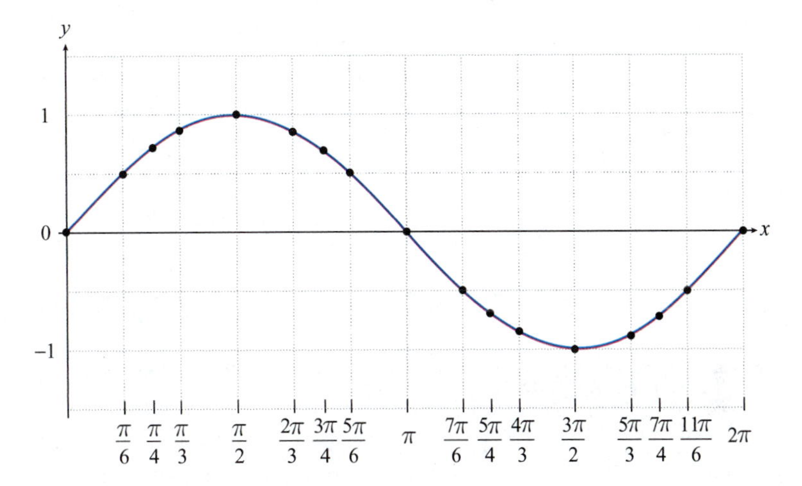

FIGURE 3: Graph of Sine between 0 and 2π

Similarly, Figure 4 contains the result of plotting the calculated points for cosine.

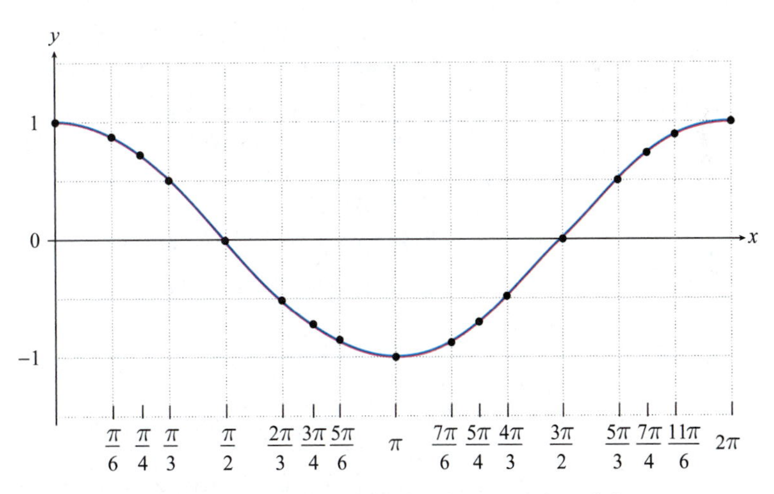

FIGURE 4: Graph of Cosine between 0 and 2π

One fact that leaps out from a glance at the graphs is that both sine and cosine take on values only between −1 and 1. This makes perfect sense since the length of the hypotenuse of a right triangle is always greater than or equal to the length of either leg, but the graphs drive this point home in a way that tables of figures don't. Another fact that is starting to appear is that sine and cosine seem to have similar shapes, one shifted horizontally with respect to the other. This is clearer if we extend the graphs to more of the real line, making use of each function's periodicity.

In Figure 5, both the sine and cosine functions are graphed over a longer interval of the x-axis.

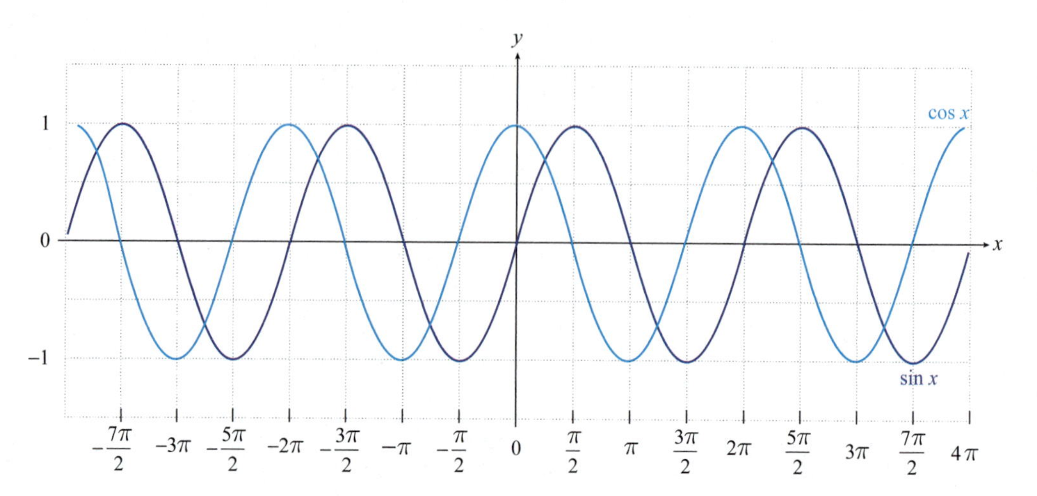

FIGURE 5: Graphs of Sine and Cosine

> **NOTE**
> The domain of both sine and cosine is $(-\infty,\infty)$ and the range of both sine and cosine is $[-1,1]$.

As we continue in this section and the next, we will use similar reasoning to construct or to recognize graphs of various trigonometric functions. As with nearly all of our graphing, however, our intent is not to compete with graphing utilities for the most accurate pictures. Instead, our goal is to understand the behavior of functions and be able to pick out important qualities like periodicity, symmetry, the existence of asymptotes, and so on.

Example 1: Transformation of Cosine, Part 1

Based on the graphs of sine and cosine in Figure 5 and the values of sine and cosine shown in Table 1, construct a transformation of cosine that is equal to sine.

Solution

Based on Figure 5, it appears that the graph of cosine is the same shape as the graph of sine, but shifted $\frac{\pi}{2}$ units to the left. So using the techniques of Section 4.1, the graph of $\cos\left(x-\frac{\pi}{2}\right)$ should coincide with the graph of sine, as the replacement of x with $x-\frac{\pi}{2}$ shifts the graph of cosine to the right by $\frac{\pi}{2}$ units. We can verify that this transformation works at a few convenient angles by making use of the calculations in Table 1, as follows.

$$\cos\left(\frac{3\pi}{4}-\frac{\pi}{2}\right)=\cos\left(\frac{\pi}{4}\right)=\frac{1}{\sqrt{2}}=\sin\left(\frac{3\pi}{4}\right)$$

$$\cos\left(\frac{4\pi}{3}-\frac{\pi}{2}\right)=\cos\left(\frac{5\pi}{6}\right)=-\frac{\sqrt{3}}{2}=\sin\left(\frac{4\pi}{3}\right)$$

$$\cos\left(\frac{11\pi}{6}-\frac{\pi}{2}\right)=\cos\left(\frac{4\pi}{3}\right)=-\frac{1}{2}=\sin\left(\frac{11\pi}{6}\right)$$

While these few calculations don't prove that $\cos\left(x-\frac{\pi}{2}\right)=\sin x$ for all x, they certainly suggest we're on the right track. We will soon return to this problem and prove that this transformation of cosine does indeed equal sine everywhere.

TOPIC 2: Periodicity and Symmetry

With a few graphs to look back on, we'll now formally define what we mean by periodicity.

DEFINITION: Period of a Function

A function f is said to be **periodic** if there is a positive number p such that

$$f(x+p)=f(x)$$

for all x in the domain of f. The smallest such number p is called the **period** of f.

For instance, we know that $\sin(x+2\pi)=\sin x$ and $\cos(x+2\pi)=\cos x$. It's also true that $\sin(x+2n\pi)=\sin x$ for any integer n, but 2π is the smallest positive constant p for which $\sin(x+p)=\sin x$, so the period of sine (and cosine and cosecant) is 2π.

Example 2: Finding the Period of Trigonometric Functions

Determine the periods of the secant, cosecant, tangent, and cotangent functions.

Solution

Since secant is the reciprocal of cosine, we might reasonably guess that the period of secant is the same as the period of cosine. We can prove this algebraically as follows:

$$\sec(x+2\pi)=\frac{1}{\cos(x+2\pi)}=\frac{1}{\cos x}=\sec x,$$

so the period of secant is no larger than 2π. And since cosine is the reciprocal of secant, if there were a smaller positive number p for which $\sec(x+p)=\sec x$, the same sort of argument would show that cosine would have the same period p, contradicting what we know about the period of cosine. Thus the period of secant is 2π. Note, however, that secant is not defined for all real numbers; the domain of secant is all real numbers except where cosine $=0$, namely $\pm\left\{\frac{\pi}{2},\frac{3\pi}{2},\frac{5\pi}{2},\ldots\right\}$.

The same reasoning shows that the period of cosecant, being the reciprocal of sine, is also 2π.

We'll have to work a bit harder to determine the period of tangent. A reasonable guess might again be 2π, since four of the trigonometric functions have period 2π, but we'll see that this time the guess is wrong. Recall the general definition of the tangent function: given an angle θ and any point (x,y) on the terminal side of θ (other than the origin),

$$\tan\theta=\frac{y}{x}.$$

FIGURE 6: Tangent of an Angle and the Angle plus π

If $\theta \in \left(0, \frac{\pi}{2}\right)$, $\tan\theta$ is positive because both y and x are positive. If $\theta \in \left(\pi, \frac{3\pi}{2}\right)$, $\tan\theta$ is again positive because y and x are both negative. If the terminal side of θ lies in the second or fourth quadrant, $\tan\theta$ is negative. This alone is enough to tell us that the period of tangent can't be less than π (do you see why?). But more precisely, consider what the tangent function does to a given angle θ and to $\theta + \pi$, as illustrated in Figure 6.

Since $\frac{-y}{-x} = \frac{y}{x}$ (with the restriction $x \neq 0$), we see that $\tan(\theta + \pi) = \tan\theta$. This, along with the observation in the preceding paragraph, tells us that the period of tangent is π, and consequently the period of cotangent is also π.

In Chapter 4 we defined the terms *even* and *odd* as applied to functions. Recall that a function is even if $f(-x) = f(x)$ for all x in the domain, and odd if $f(-x) = -f(x)$ for all x in the domain. Geometrically, this means that the graph of f is symmetric with respect to the y-axis or the origin, respectively. Since the graph of cosine is symmetric with respect to the y-axis, cosine is an even function, and since the graph of sine is symmetric with respect to the origin, sine is odd. Note that from these two facts, we can quickly determine the symmetry of the remaining four trigonometric functions. For example, the following equations show that the cosecant and tangent functions are both odd.

$$\csc(-x) = \frac{1}{\sin(-x)} = \frac{1}{-\sin x} = -\frac{1}{\sin x} = -\csc x$$

$$\tan(-x) = \frac{\sin(-x)}{\cos(-x)} = \frac{-\sin(x)}{\cos(x)} = -\frac{\sin x}{\cos x} = -\tan x$$

For reference, the symmetry classifications for all six trigonometric functions are summarized as the following identities.

IDENTITIES: Even/Odd Identities

$\sin(-x) = -\sin x$ $\quad\cos(-x) = \cos x$ $\quad\tan(-x) = -\tan x$

$\csc(-x) = -\csc x$ $\quad\sec(-x) = \sec x$ $\quad\cot(-x) = -\cot x$

Example 3: Transformation of Cosine, Part 2

Use a cofunction identity and an even/odd identity to prove the transformation statement $\cos\left(x - \frac{\pi}{2}\right) = \sin x$ for all x.

Solution

Given any x, $\left(\frac{\pi}{2}-x\right)+x=\frac{\pi}{2}$, so $\frac{\pi}{2}-x$ and x are complements of one another. Hence, by one of the cofunction identities,

$$\cos\left(\frac{\pi}{2}-x\right)=\sin x.$$

And since cosine is an even function,

$$\cos\left(x-\frac{\pi}{2}\right)=\cos\left(-\left(x-\frac{\pi}{2}\right)\right)=\cos\left(\frac{\pi}{2}-x\right)=\sin x.$$

TOPIC 3: Amplitude, Frequency, and Phase Shifts

In actual use, a trigonometric function is unlikely to appear in its basic form. That is, in order to be useful in solving a problem, a given function will probably have to be modified somewhat. Geometrically, this means that the graph of a function will appear stretched, compressed, reflected, or shifted relative to the graph of its basic form. We will finish out this section with a discussion of the more common transformations.

We have had quite a bit of experience with transformations of functions in general, but in the context of trigonometry some additional nomenclature is useful. We'll begin with a definition of one particularly useful term.

DEFINITION: Amplitude of Sine and Cosine Curves

Given a fixed real number a, the **amplitude** of the function $f(x)=a\sin x$ or the function $g(x)=a\cos x$ is the value $|a|$. As we know, the multiplication of $\sin x$ or $\cos x$ by a stretches (or compresses, if $-1<a<1$) the graph vertically by a factor of $|a|$, so the amplitude represents the distance between the x-axis and the maximum value of the function.

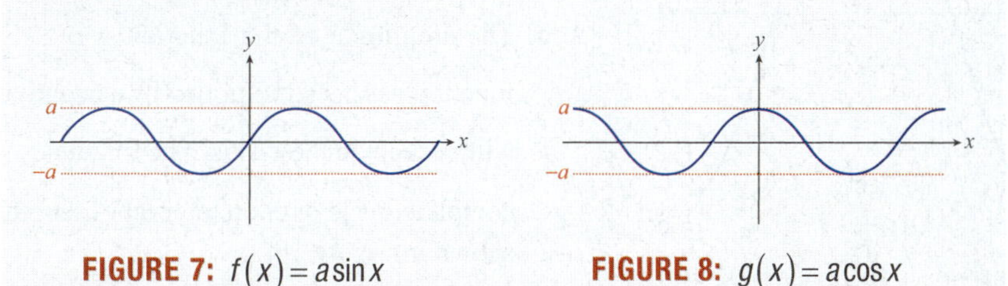

FIGURE 7: $f(x)=a\sin x$ **FIGURE 8:** $g(x)=a\cos x$

As we have seen, the shapes of the sine and cosine curves are identical; one is merely shifted horizontally with respect to the other. For this reason, both graphs are said to be *sinusoidal*, so we have now defined the amplitude of a sinusoidal curve.

Our next definition relates to a different type of transformation. As we saw in Section 4.1, replacing the argument x of a function with the argument bx, where b is a fixed real number, has the effect of either stretching out (if $0<|b|<1$) or compressing (if $|b|>1$) the graph of the function horizontally. When applied to sinusoidal graphs, this changes the number of times the graph goes through one complete cycle over an interval of length 1.

DEFINITION: Frequency of Sine and Cosine Curves

Given a fixed real number b, the **frequency** of the function $f(x)=\sin(bx)$ or the function $g(x)=\cos(bx)$ is the number $\frac{b}{2\pi}$. When the independent variable represents time, measured in seconds, the measurement of frequency is stated in terms of **cycles per second**, or **hertz (Hz)**.

Example 4: Finding Amplitude and Frequency

Determine the amplitude and frequency of each of the following functions. Then use your results to sketch the graph of one complete cycle of each function starting at $x=0$.

a. $f(x)=3\sin\left(\frac{x}{2}\right)$ **b.** $g(x)=\left(-\frac{1}{2}\right)\cos(2\pi x)$

Solution

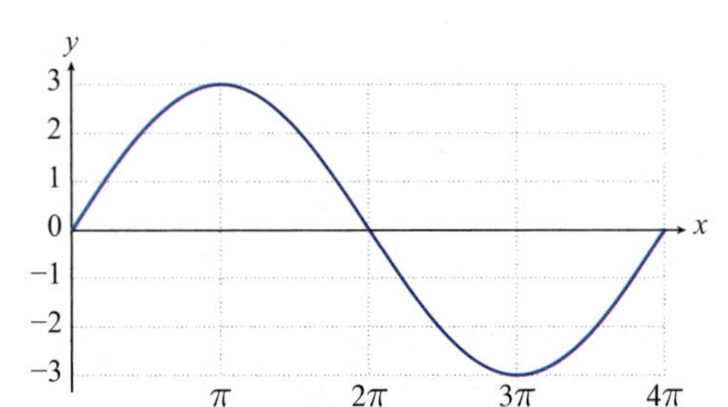

FIGURE 9: $f(x)=3\sin\left(\frac{x}{2}\right)$

a. Regardless of the argument, sine always oscillates between -1 and 1 in magnitude. Multiplying $\sin\left(\frac{x}{2}\right)$ by 3 then stretches the graph vertically by a factor of 3, so the function f has an amplitude of 3. The number b multiplying x is $\frac{1}{2}$, so the frequency of f is $\frac{\frac{1}{2}}{2\pi}=\frac{1}{4\pi}$. This means f goes through this fraction of a cycle over an interval of length 1 or, put another way, requires an interval of length 4π in order to go through one complete cycle. Using this fact and its amplitude, the graph of f is shown in Figure 9.

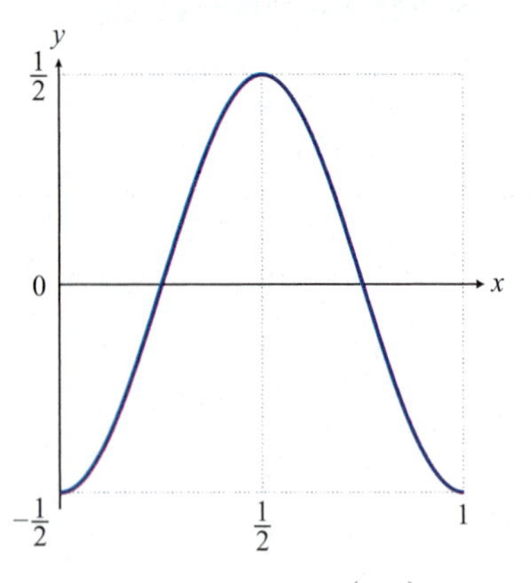

FIGURE 10: $g(x)=\left(-\frac{1}{2}\right)\cos(2\pi x)$

b. The amplitude of the function g is $\left|-\frac{1}{2}\right|=\frac{1}{2}$, but the fact that the cosine function has been multiplied by a negative number means the graph is reflected with respect to the x-axis. The frequency of g is $\frac{2\pi}{2\pi}=1$, so g goes through one complete cycle over an interval of length 1. These facts lead to the graph of g shown in Figure 10.

Using the reasoning of Example 4, we now know that the transformed functions $f(x)=\sin(bx)$ and $g(x)=\cos(bx)$ go through one complete cycle over an interval of length $\frac{2\pi}{b}$. Note that this is the reciprocal of the frequency of these functions. This makes sense, since the frequency measures how many cycles occur over an interval of length 1, while the period is the length of the interval required for one complete cycle. We use this observation to refine our definition of period in this context.

DEFINITION: Period Revisited

Given a fixed real number b, the **period** of the function $f(x)=\sin(bx)$ or the function $g(x)=\cos(bx)$ is the number $\frac{2\pi}{b}$. The period and frequency of a sinusoidal function are reciprocals of one another.

Example 5: Finding the Period Using Frequency

The A and D strings of a cello are tuned to vibrate at frequencies of 220 Hz and 146.8 Hz, respectively. What are the periods of the waves that these two strings produce?

Solution

A frequency of 220 Hz means that the A string produces a wave with 220 cycles per second, so its period (the length of time required for one cycle) is

$$\frac{1}{220} \approx 0.00455 \text{ s.}$$

Similarly, the period of the wave produced by the D string is

$$\frac{1}{146.8} \approx 0.00681 \text{ s.}$$

One more transformation leads to special nomenclature when applied to trigonometric functions. In general, we know that replacing the argument x with the argument $x-c$ shifts the graph of a function horizontally, to the right if c is positive and to the left if c is negative.

In the context of trigonometry, shifting a function to the left or right is called a **phase shift**, and in general it may occur in combination with a change in period and amplitude. For instance, the function $f(x)=3\sin(5\pi x-4\pi)$ has an amplitude of 3, a phase shift, and an altered period. One way to determine the phase shift and period precisely is to rewrite the function as follows.

$$\begin{aligned} f(x) &= 3\sin(5\pi x-4\pi) \\ &= 3\sin\left(5\pi\left(x-\frac{4}{5}\right)\right) \end{aligned}$$

The function $3\sin(5\pi x)$ has a period of $\frac{2}{5}$, but when x is replaced with $x-\frac{4}{5}$, we know the graph of $3\sin(5\pi x)$ gets shifted to the right by $\frac{4}{5}$ units.

Another way to determine details about the period and phase shift of $f(x)=3\sin(5\pi x-4\pi)$ is to relate the beginning and end of one cycle of the function to the beginning and end of one cycle of $\sin x$. That is, we know that a cycle of f will begin when its argument is equal to 0 and end when its argument is equal to 2π.

Solving these two equations for x gives us the following.

$$5\pi x - 4\pi = 0 \qquad 5\pi x - 4\pi = 2\pi$$
$$5\pi x = 4\pi \qquad 5\pi x = 6\pi$$
$$x = \frac{4}{5} \qquad x = \frac{6}{5}$$

So one complete cycle of f occurs over the interval between $x = \frac{4}{5}$ and $x = \frac{6}{5}$, telling us again that the period is $\frac{2}{5}$.

These observations are summarized for the functions sine and cosine. The same sort of analysis will prove to be just as useful for the remaining trigonometric functions, however.

DEFINITION: Amplitude, Period, and Phase Shift Combined

Given constants a, b (such that $b > 0$), and c, the functions

$$f(x) = a\sin(bx - c) \quad \text{and} \quad g(x) = a\cos(bx - c)$$

have **amplitude** $|a|$, **period** $\frac{2\pi}{b}$, and a **phase shift** of $\frac{c}{b}$. The left endpoint of one cycle of either function is $\frac{c}{b}$ and the right endpoint is $\frac{c}{b} + \frac{2\pi}{b}$.

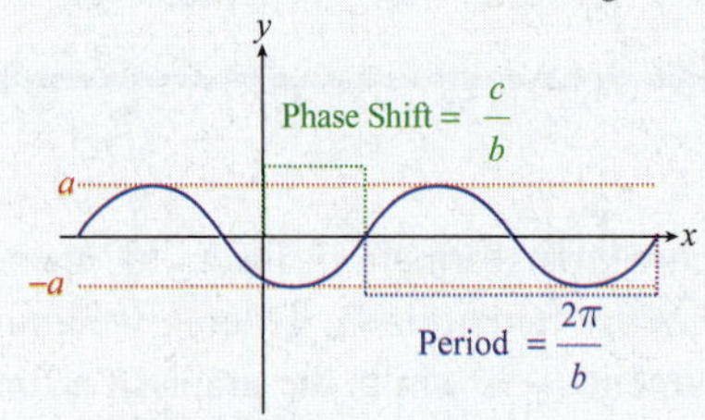

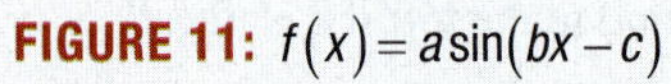
FIGURE 11: $f(x) = a\sin(bx - c)$

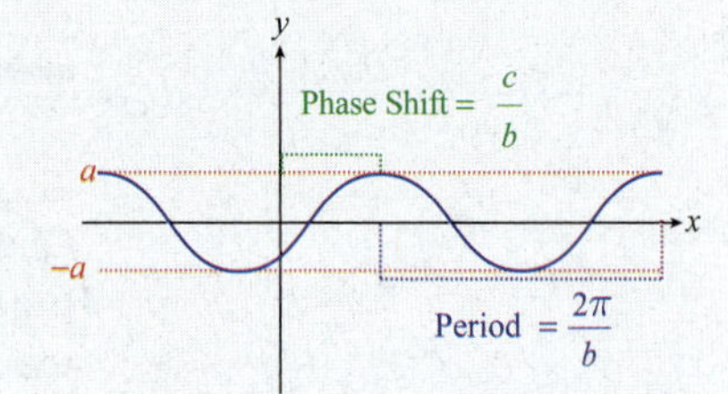

FIGURE 12: $g(x) = a\cos(bx - c)$

Example 6: Graphing Transformed Trigonometric Functions

Sketch the graph of $f(x) = -2\cos(\pi x - \pi)$.

Solution

The graph of the basic cosine function completes one cycle over an interval of length 2π, so we can determine the left and right endpoints of one cycle of f by solving $\pi x - \pi = 0$ and $\pi x - \pi = 2\pi$.

$$\pi x - \pi = 0 \qquad \pi x - \pi = 2\pi$$
$$\pi x = \pi \qquad \pi x = 3\pi$$
$$x = 1 \qquad x = 3$$

This tells us that f has a period of 2 and the cycle of cosine that occurs over the interval $[0, 2\pi]$ instead occurs for f over the interval $[1, 3]$. However, the -2 factor of f tells us that the amplitude of f is 2 and the graph of the cycle is reflected with respect to the x-axis. Putting this all together we get the graph shown in Figure 13, which shows two complete cycles of f.

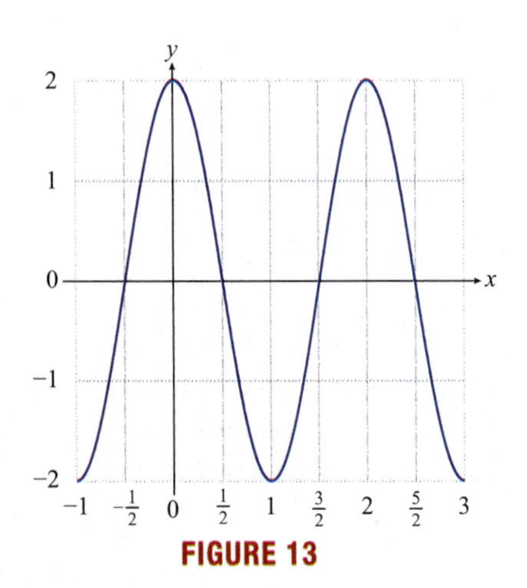

FIGURE 13

Example 7: Graphing Transformed Trigonometric Functions

Sketch the graph of $g(x) = 1 + \sin\left(x - \frac{\pi}{4}\right)$.

Solution

The graph of g is the graph of $\sin x$ shifted to the right by $\frac{\pi}{4}$ units and up by 1 (recall that adding a constant to a function merely shifts the graph up or down, according to whether the constant is positive or negative). Neither the amplitude nor period has changed, however. Our sketch is thus the following.

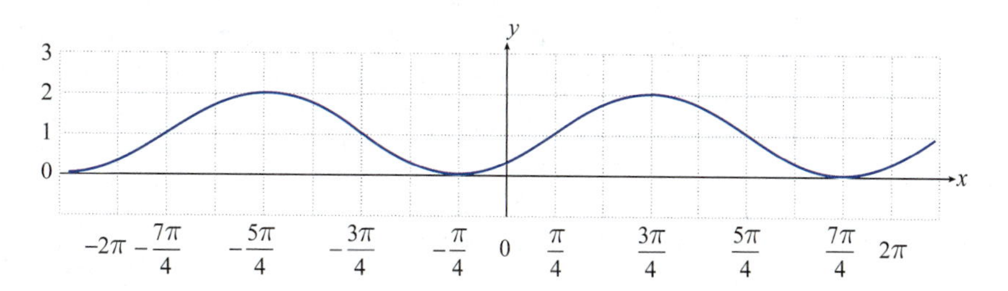

FIGURE 14

TOPIC 4: Simple Harmonic Motion

Sinusoidal curves arise in many physical situations, often when an object is displaying behavior that is the subject of the following definition.

DEFINITION: Simple Harmonic Motion

If an object is oscillating and its displacement from some midpoint at time t can be described by either $f(t) = a\sin(bt)$ or $g(t) = a\cos(bt)$, the object is said to be in **simple harmonic motion (SHM)**. In both cases, the maximum displacement of the object from its midpoint is the amplitude $|a|$ and its frequency of oscillation is $\frac{b}{2\pi}$.

Example 8: Using Simple Harmonic Motion

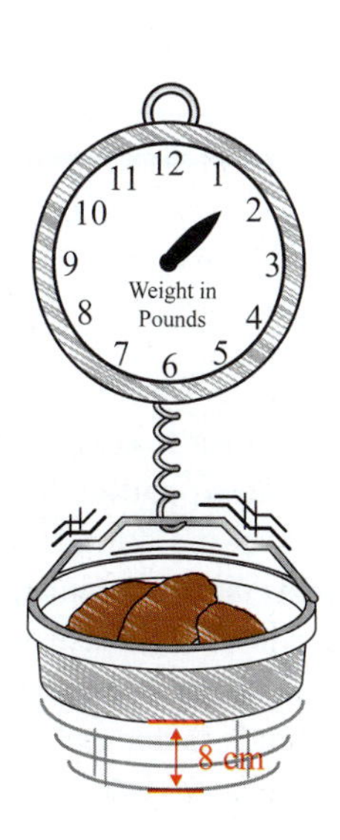

FIGURE 15

A first approximation to the motion of an object suspended at the end of a spring and set into vertical oscillation is given by $y = a\cos(bt)$, where y is the displacement of the object above or below its rest position. Suppose several potatoes are dumped into the basket of a grocer's scale, which then proceeds to bounce up and down with a frequency of 3 Hz. Given that the distance traveled between peaks of the basket's oscillation is 8 centimeters, find a mathematical model for the basket's motion.

Solution

From the given information, we know that our model must describe an object traveling 4 centimeters above and 4 centimeters below its rest position; in other words, the amplitude must be 4. We have a choice to make, however: we can define our coordinate system so that a positive displacement is either up or down. The most natural choice in this problem is probably to choose positive displacement as being in the upward direction, so we would like to arrange it so that $y = -4$ when $t = 0$. That is, a good model of the basket's oscillation will have the basket starting out 4 centimeters below its midpoint position at the moment $(t = 0)$ when the potatoes are dumped in. This means we want to set $a = -4$.

A frequency of 3 Hz means that the basket makes 3 complete up-and-down cycles every second. This means that $b = 6\pi$, and so our model is $y = -4\cos(6\pi t)$. To see how our model relates to the physical situation, consider the following graph.

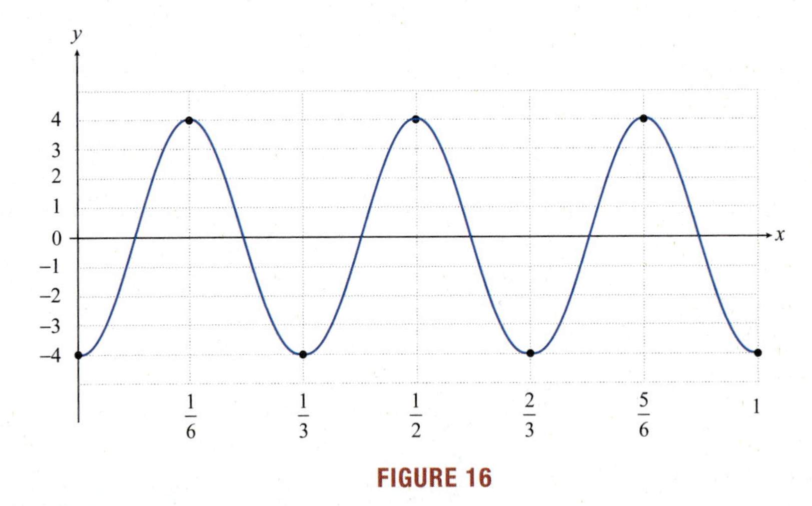

FIGURE 16

From the graph, it's clear that the basket really does make 3 complete cycles during the course of the first second, that it starts out 4 centimeters below the midpoint position, and that its maximum displacement above the midpoint is also 4 centimeters.

TOPIC 5: Damped Harmonic Motion

Our experience tells us that the simple model seen in Example 8 isn't very accurate over a long period of time—eventually, the basket settles down and ceases its oscillatory motion. If we want to more accurately describe the up-and-down motion of an object suspended from a spring, we need to account for the fact that oscillations are often *damped* as time progresses. That is, the amplitude of the oscillations decreases according to some rule. We can easily modify our sinusoidal wave in such a way by multiplying a sine or cosine function by an amplitude factor that is not constant. Remember: the fundamental behavior of a sinusoidal curve is to oscillate between values of -1 and 1. If we multiply such a wave by a decreasing amplitude, the result will be a wave whose oscillations diminish over time.

Example 9: Modeling Damped Harmonic Motion

Sketch the graph of $f(t) = -4e^{-t}\cos(6\pi t)$.

Solution

We've already graphed the function $f(t) = -4\cos(6\pi t)$ (this function was our simple model for the motion of the grocer's basket in Example 8). The factor of e^{-t} provides the desired damping effect. In Figure 17, the graphs of $4e^{-t}$ and $-4e^{-t}$ are included to show how they describe the "envelope" of amplitude modulation. The result is that the magnitude of the displacement of the grocer's basket decreases over time. Notice, however, the period is unaffected.

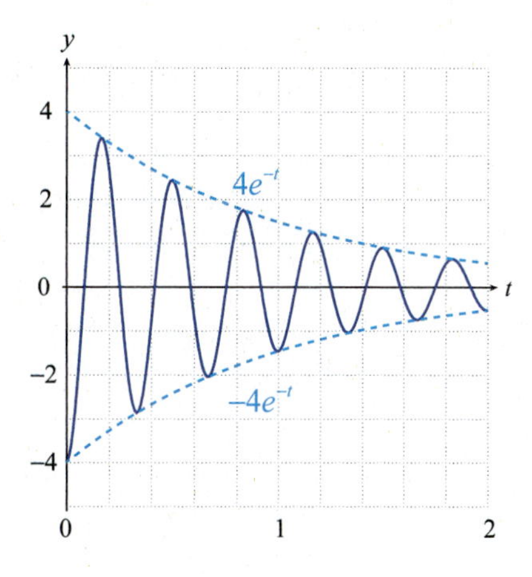

FIGURE 17

In the exercises to follow, you'll be asked to sketch graphs of similar products of damping factors and trigonometric functions.

7.4 EXERCISES

PRACTICE

Determine the amplitude and frequency of each of the following functions. Then sketch one complete cycle of each function starting at $x = 0$. See Example 4.

1. $f(x) = 2\cos(2x)$

2. $g(x) = -5\sin(\pi x)$

3. $g(x) = \dfrac{\sin(3x)}{2}$

4. $f(x) = \dfrac{\cos\left(\frac{\pi x}{2}\right)}{3}$

Determine the amplitude, period, and phase shift of each of the following functions. See Examples 6 and 7.

5. $f(x) = 2\cos x$

6. $g(x) = \frac{3}{2}\sin x$

7. $f(x) = 5 + 4\cos x$

8. $f(x) = \sin(x - 5)$

9. $h(x) = -\sin x$

10. $h(x) = \frac{\cos x}{2}$

11. $g(x) = -3\cos(x + 7)$

12. $f(x) = \frac{2}{3}\sin x$

13. $f(x) = 2\sin(2x)$

14. $h(x) = -3\cos\left(\frac{1}{2}x\right)$

15. $g(x) = \frac{3\sin(\pi\theta)}{2}$

16. $g(x) = \cos(3\pi\theta - 2)$

17. $f(x) = 0.5\sin(8x + 1)$

18. $h(x) = 7\cos\left(x \cdot \frac{\pi}{2} + \frac{3}{2}\right)$

19. $g(x) = \frac{8\cos(2\pi x + 4)}{5}$

20. $g(x) = 2 - \frac{3}{4}\sin(-3 + x)$

Sketch the graph of each of the following functions. See Examples 6 and 7.

21. $f(x) = \cos(\pi x)$

22. $g(x) = -2\sin(5x)$

23. $g(x) = 3\sin(x - 2\pi)$

24. $g(x) = \sin\left(x - \frac{\pi}{4}\right)$

25. $f(x) = 4\cos\left(\frac{3x}{2} + \frac{\pi}{2}\right)$

26. $g(x) = 2\cos(4x - 2)$

27. $f(x) = \cos(x - \pi)$

28. $g(x) = 3\sin(4x)$

29. $f(x) = -\sin(2\pi x)$

30. $g(x) = 1 + \sin(x - 2\pi)$

31. $f(x) = 2 - \cos(2\pi x)$

32. $g(x) = 5 - 2\sin\left(x - \frac{\pi}{2}\right)$

33. $f(x) = -3 + 5\cos x$

34. $g(x) = 2 - \sin\left(2x - \frac{\pi}{4}\right)$

35. $f(x) = \frac{1}{2} - 5\sin\left(\frac{1}{2}x - \frac{\pi}{2}\right)$

36. $g(x) = 1 - \frac{1}{4}\cos\left(\frac{1}{4}x - \frac{\pi}{2}\right)$

Sketch each of the following functions modeling damped harmonic motion. See Example 9.

37. $g(t) = -2e^{-t}\cos(5\pi t)$

38. $f(t) = e^{-t}\sin\left(\frac{3\pi}{4}t\right)$

39. $g(t) = e^{t}\sin\left(3t - \frac{\pi}{2}\right)$

40. $g(t) = 3e^{-t}\cos\left(5t - \frac{\pi}{2}\right)$

41. $f(t) = -3 + 5e^{-t}\cos t$

42. $f(t) = -5e^{t}\cos\left(\frac{3\pi}{2}t\right)$

43. $f(t) = \frac{1}{2}e^{-t}\sin\left(\frac{5}{6}t - 4\pi\right) + 2$

44. $g(t) = 2 + e^{-t}\sin\left(t - \frac{\pi}{4}\right)$

APPLICATIONS

In Exercises 45–46, use the relationship between frequency and period to answer the question. See Example 5.

45. Many grandfather clocks have a pendulum that swings with a period of two seconds. What is the frequency of such a pendulum?

46. A heart rate of 1200 beats per minute (bpm) is typical for a hummingbird. What is the length of the period, in seconds, of such a heart rate?

47. A baby is playing with a toy attached above his head on a coiled spring. The baby pulls the toy down a distance of 3 inches from its equilibrium position, and then releases it. The time for one oscillation is 2 seconds. Find the amplitude and period, then give the function for its displacement.

48. A pull cord for a lighted ceiling fan is swinging back and forth. The end of the cord swings a total distance of 4 inches from end to end at an average speed of 9 inches per second. Find the period of oscillation.

49. Marcel is bouncing a basketball at an average speed of 10 ft/s with the ball coming up to his waist on each bounce. The distance from the ground to his waist is approximately 3 feet. Find the amplitude and period, then give the function for its displacement.

50. A leaf floating on the water of a perfectly calm pond is suddenly disturbed by a series of waves caused by a landing duck. At time $t = 0$ seconds, the leaf initially bobs upward 5 centimeters, and then continues to oscillate up and down with a period of 2 seconds.

a. Find a function that models the simple harmonic motion described.

b. Assuming the amplitude of the waves diminishes over time by a factor of $e^{\frac{-t}{5}}$, modify your SHM model accordingly and graph the result.

WRITING & THINKING

51. **a.** Find a transformation of cosine that shifts its graph to the left so as to coincide with the graph of sine.

b. Using n to represent an arbitrary integer, find an expression that describes the infinite number of transformations of cosine that are equal to sine.

52. Prove the even and odd identities for the secant and cotangent functions.

7.5 GRAPHS OF OTHER TRIGONOMETRIC FUNCTIONS

TOPICS

1. Graphing tangent and cotangent functions
2. Graphing secant and cosecant functions

TOPIC 1: Graphing Tangent and Cotangent Functions

In this short section, we will use our familiarity with the graphs of sine and cosine, and the fact that the remaining four trigonometric functions can be defined in terms of sine and cosine, to quickly graph and analyze those four functions. We begin with the tangent and cotangent functions.

First, recall from Example 2 of Section 7.4 that both tangent and cotangent have a period of π. Recall also that we determined tangent and cotangent are odd functions, so their graphs are symmetric with respect to the origin. (We proved this for tangent by using the facts that $\tan x = \frac{\sin x}{\cos x}$, that sine is an odd function, and that cosine is an even function; we could use a similar argument for cotangent, or simply note that the reciprocal of an odd function is again odd.) Remember next that tangent is defined for all real numbers except for odd multiples of $\frac{\pi}{2}$. Finally, either by drawing some triangles or using previously calculated values for sine and cosine and again the fact that $\tan x = \frac{\sin x}{\cos x}$, we evaluate tangent at 0, $\frac{\pi}{6}$, $\frac{\pi}{4}$, and $\frac{\pi}{3}$ to obtain Table 1.

x	$\tan x$
0	0
$\frac{\pi}{6}$	$\frac{1}{\sqrt{3}}$
$\frac{\pi}{4}$	1
$\frac{\pi}{3}$	$\sqrt{3}$

TABLE 1

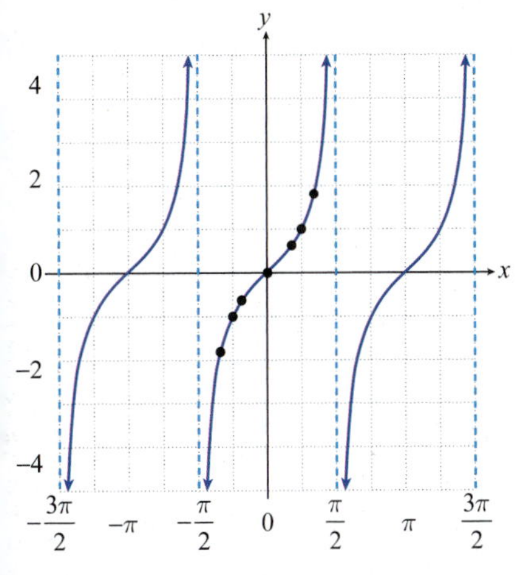

FIGURE 1: Graph of Tangent

Putting all these facts together and drawing a smooth curve through the calculated points and their reflections through the origin, we can sketch out the graph of tangent that appears in Figure 1, where we have shown three cycles of the function. In Exercise 15, you will be guided through three different ways of sketching out the graph of cotangent.

Note the vertical asymptotes that appear at odd multiples of $\frac{\pi}{2}$, a characteristic that neither sine nor cosine possesses. The graphs of cotangent, secant, and cosecant also all have periodic vertical asymptotes, and we can always remind ourselves of the locations of the asymptotes by recalling the following identities.

$$\tan x = \frac{\sin x}{\cos x}, \quad \cot x = \frac{\cos x}{\sin x}, \quad \sec x = \frac{1}{\cos x}, \quad \text{and} \quad \csc x = \frac{1}{\sin x}$$

This means that tangent and secant have vertical asymptotes where $\cos x = 0$ and cotangent and cosecant have vertical asymptotes where $\sin x = 0$.

As with sine and cosine, the other four trigonometric functions can be transformed in a number of ways, resulting in graphs that are shifted or stretched vertically or horizontally.

NOTE

The domain of tangent is the set $\left\{x \mid x \neq (2n+1)\frac{\pi}{2}, n \in \mathbb{Z}\right\}$, and the range of tangent is $(-\infty, \infty)$. As you will see in Exercise 15, the domain of cotangent is the set $\{x \mid x \neq n\pi, n \in \mathbb{Z}\}$, and the range of cotangent is again $(-\infty, \infty)$.

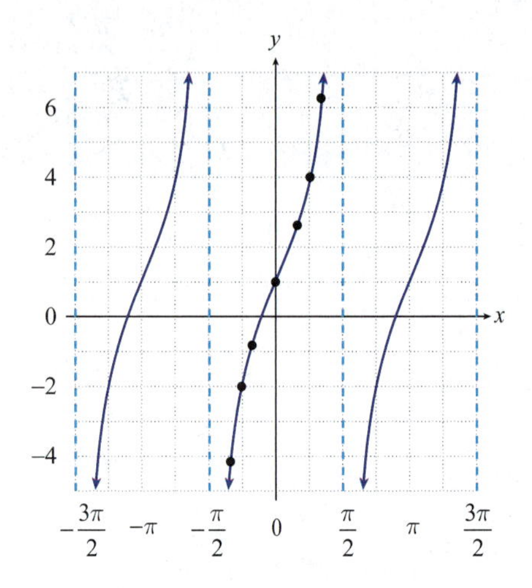

FIGURE 2: $f(x)=1+3\tan x$

Example 1: Graphing Transformed Trigonometric Functions

Sketch the graph of $f(x)=1+3\tan x$.

Solution

As always, multiplying a function by a constant has the effect of stretching or compressing the graph of the function vertically. In this case, multiplying $\tan x$ by 3 stretches the graph vertically. The effect of this is most easily seen at odd multiples of $\frac{\pi}{4}$, where the value of $3\tan x$ will be ± 3 instead of ± 1.

The effect of adding a constant, in this case 1, to a function simply shifts the entire graph upward by 1 unit. Incorporating these two transformations into the graph of the basic tangent function, we are led to the graph in Figure 2.

The next example demonstrates a horizontal stretching and shift of the graph of the tangent function, using a technique from Section 7.4 to determine intervals over which a complete cycle occurs.

Example 2: Graphing Transformed Trigonometric Functions

Sketch the graph of $f(x)=\tan\left(\frac{\pi x}{4}-\frac{\pi}{2}\right)$.

Solution

Since the graph of $\tan x$ has one complete cycle between the asymptotes $x=-\frac{\pi}{2}$ and $x=\frac{\pi}{2}$, the graph of $\tan\left(\frac{\pi x}{4}-\frac{\pi}{2}\right)$ has one complete cycle between asymptotes defined by $\frac{\pi x}{4}-\frac{\pi}{2}=-\frac{\pi}{2}$ and $\frac{\pi x}{4}-\frac{\pi}{2}=\frac{\pi}{2}$. Solving these two equations for x will give us an interval over which f has one complete cycle.

$$\frac{\pi x}{4}-\frac{\pi}{2}=-\frac{\pi}{2} \qquad \frac{\pi x}{4}-\frac{\pi}{2}=\frac{\pi}{2}$$
$$\frac{\pi x}{4}=0 \qquad \frac{\pi x}{4}=\pi$$
$$x=0 \qquad x=4$$

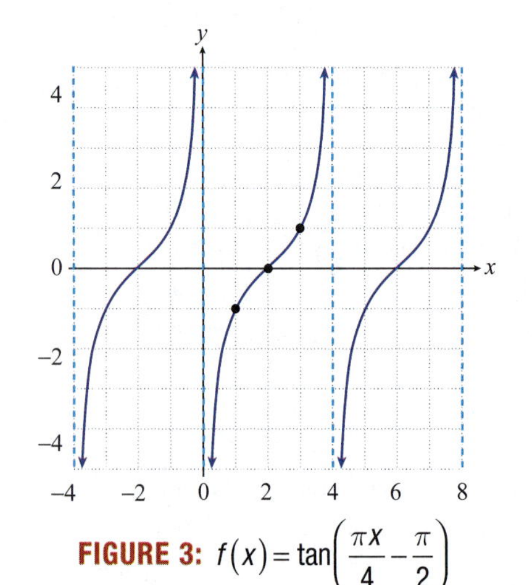

FIGURE 3: $f(x)=\tan\left(\frac{\pi x}{4}-\frac{\pi}{2}\right)$

This tells us that f has a period of 4 and vertical asymptotes at multiples of 4, with the graph of tangent being stretched out a bit horizontally accordingly. The graph appears in Figure 3, with values of f at $x=1$, $x=2$, and $x=3$ highlighted.

TOPIC 2: Graphing Secant and Cosecant Functions

We will work through the process of identifying the key features of cosecant here, culminating in a sketch of its graph. Similar work for secant will be left to you in Exercise 16.

Example 3: Graphing the Cosecant Function

Sketch the graph of the cosecant function.

Solution

Let's again begin with characteristics of cosecant that we already know. First, as the reciprocal of an odd function, cosecant is odd. Second, it has the same period as sine, namely 2π. Third, cosecant will have vertical asymptotes where $\sin x = 0$, which is every multiple of π.

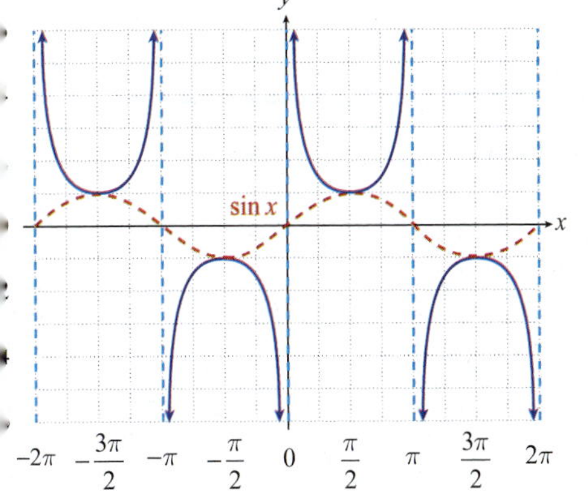

FIGURE 4: Graph of Cosecant

A few other observations about the graph of sine will help us to quickly visualize the graph of cosecant. Since sine always takes on values between -1 and 1 (inclusive), cosecant will always take on values greater than or equal to 1 in magnitude. Moreover, since the reciprocal of a number has the same sign as the number, cosecant and sine will be positive on the same (open) intervals and negative on the same (open) intervals. One last observation will be helpful: since the graph of sine is symmetric with respect to the lines $y = \pm\frac{\pi}{2}$, $y = \pm\frac{3\pi}{2}$, and so on, the graph of cosecant will also be symmetric with respect to these lines.

Putting this all together, we obtain the graph shown in Figure 4. The figure also shows the graph of sine as a dashed curve—be sure you understand how the graphs of the two functions relate to one another.

> **NOTE**
>
> The domain of cosecant is the set $\{x \mid x \neq n\pi, n \in \mathbb{Z}\}$, and the range of cosecant is $(-\infty,-1] \cup [1,\infty)$.
>
> As you will see in Exercise 16, the domain of secant is the set $\left\{x \mid x \neq (2n+1)\frac{\pi}{2}, n \in \mathbb{Z}\right\}$, and the range of secant is again $(-\infty,-1] \cup [1,\infty)$.

Example 4: Graphing Transformed Trigonometric Functions

Sketch the graph of $f(x) = -2\csc\left(\pi x + \frac{\pi}{2}\right)$.

Solution

We'll again first determine intervals over which complete cycles of f occur. A positive upward-opening half-cycle of cosecant occurs between 0 and π on the x-axis, so setting the argument of f equal to these numbers we obtain the following.

$$\begin{aligned} \pi x + \frac{\pi}{2} &= 0 \\ \pi x &= -\frac{\pi}{2} \\ x &= -\frac{1}{2} \end{aligned} \qquad \begin{aligned} \pi x + \frac{\pi}{2} &= \pi \\ \pi x &= \frac{\pi}{2} \\ x &= \frac{1}{2} \end{aligned}$$

However, note that the factor of -2 in the definition of f reflects the graph of cosecant and stretches it out vertically by a factor of 2, so the half-cycle between $-\frac{1}{2}$ and $\frac{1}{2}$ on the x-axis will actually be a downward-opening curve. By similar reasoning, the downward-opening half-cycle of cosecant that occurs between π and 2π will transform into a stretched upward-opening half-cycle of f between $\frac{1}{2}$ and $\frac{3}{2}$. Figure 5 shows a graph of 2 complete cycles of f.

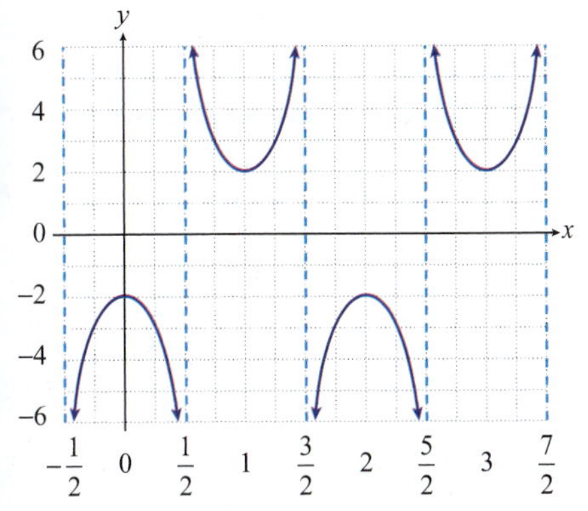

FIGURE 5: $f(x) = -2\csc\left(\pi x + \frac{\pi}{2}\right)$

We've now gained quite a lot of experience with the process of graphing transformed trigonometric functions. The periodic nature of trigonometric functions means, however, that the reverse process of identifying a function from its graph is not so clear cut, as our last example illustrates.

Example 5: Identifying a Trigonometric Graph

Find a function corresponding to the graph shown in Figure 6.

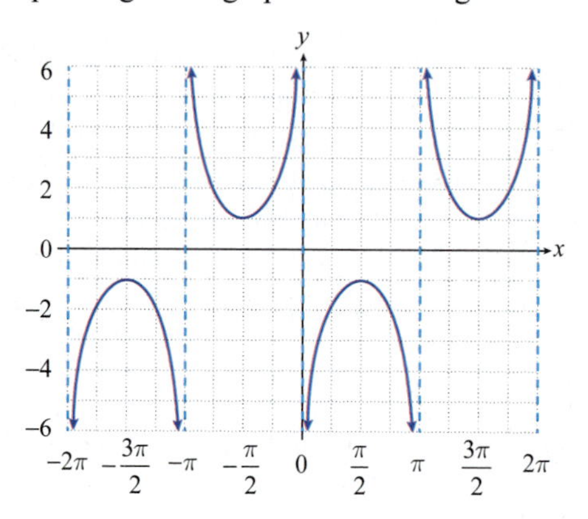

FIGURE 6

Solution

The graph shown in Figure 6 is certainly similar to the graph of cosecant in many ways, but after a transformation of some sort. One such transformation that may come to mind is that the graph appears to be the reflection of cosecant with respect to the y-axis, which corresponds to replacing x with $-x$ in the definition of the function. So the graph in Figure 6 could be of the function $f(x)=\csc(-x)$. However, someone else may look at the graph and see it as a horizontal shift to the left by π units of the graph of cosecant, which would be the function $g(x)=\csc(x+\pi)$. A third person may recognize it as the graph of cosecant reflected with respect to the x-axis, which would be the function $h(x)=-\csc x$. Once we recall that cosecant is an odd function, we remember that $\csc(-x)=-\csc x$, so we shouldn't be surprised that f and h are two possible functions corresponding to the given graph. In fact, f, g, and h are just three of an infinite number of functions all having the graph in Figure 6.

7.5 EXERCISES

PRACTICE

Sketch the graph of each of the following functions. See Examples 1 through 4.

1. $f(x)=\csc\left(\frac{3\pi}{4}x\right)$

2. $g(x)=\tan\left(3\pi x-\frac{\pi}{2}\right)$

3. $g(x)=\frac{1}{3}\sec(2x)$

4. $f(x)=-5\cot(\pi x)$

5. $g(x) = \csc\left(\frac{3\pi}{2}x - \frac{1}{2}\right)$

6. $g(x) = \cot\left(\frac{\pi x}{4}\right)$

7. $f(x) = 5\tan\left(3\pi - \frac{\pi}{2}x\right)$

8. $f(x) = 4 + \csc\left(1 - \frac{5\pi}{4}x\right)$

9. $f(x) = 1 - \cot\left(x - \frac{\pi}{2}\right)$

10. $g(x) = 1 + \tan\left(\pi x - \frac{\pi}{4}\right)$

11. $f(x) = 4 + \tan\left(x + \frac{3\pi}{2}\right)$

12. $f(x) = 1 - 2\sec(2\pi x)$

13. $g(x) = 2 + \frac{5}{6}\sec\left(\frac{1}{2}x - \pi x\right)$

14. $f(x) = \frac{1}{2}\tan\left(\frac{3}{4}x - 2\pi\right) + 3$

WRITING & THINKING

15. Sketch the graph of the cotangent function, using each of the following approaches and noting how they produce the same result.

a. Use the identity $\cot x = \frac{\cos x}{\sin x}$.

b. Use the identity $\cot x = \frac{1}{\tan x}$.

c. Use the identity $\cot x = \tan\left(\frac{\pi}{2} - x\right)$ and the fact that tangent is an odd function.

16. Sketch the graph of the secant function, using each of the following approaches and noting how they produce the same result.

a. Use the identity $\sec x = \frac{1}{\cos x}$.

b. Use the identity $\sec x = \csc\left(\frac{\pi}{2} - x\right)$ and the fact that cosecant is an odd function.

7.6 INVERSE TRIGONOMETRIC FUNCTIONS

TOPICS

1. The inverse trigonometric functions
2. Evaluating inverse trigonometric functions
3. Applications of inverse trigonometric functions

TOPIC 1: The Inverse Trigonometric Functions

The rationale for the inverse trigonometric functions is the rationale for inverses of functions in general. In many situations, we will want to find an angle having a certain property, and our method will be to "undo" the action of a given trigonometric function. As a simple example, suppose we need to find an acute angle θ for which $\sin\theta = \frac{1}{2}$. Our experience is sufficient for this task; we've worked with nice angles enough to recognize that $\sin\left(\frac{\pi}{6}\right) = \frac{1}{2}$, so it must be the case that $\theta = \frac{\pi}{6}$. But what if we seek an angle φ for which $\sin\varphi = 0.7$? The problem is similar, but we don't yet have a way to determine φ.

Recall from Chapter 4, however, that a function will have an inverse only if it is one-to-one. Recall also, if the graph of the function is available, this means the graph must pass the horizontal line test; this is something the trigonometric functions markedly fail to do. Fortunately, there is a way out. By restricting the domain of a trigonometric function wisely, we can make it one-to-one and thus invertible. We will go through the process step by step for the sine function and then briefly show how the other trigonometric functions are dealt with similarly.

There are many ways in which we could restrict the domain of sine in order to make it one-to-one, but we are guided also by the desire to not lose more than we have to in the restriction. For instance, we could specify that we will only define sine over the interval $\left[0, \frac{\pi}{2}\right]$, but by doing so we prevent the newly defined function from ever taking on a negative value (note that $0 \le \sin x \le 1$ for $x \in \left[0, \frac{\pi}{2}\right]$). Figure 1 indicates that $\left[-\frac{\pi}{2}, \frac{\pi}{2}\right]$ is the largest interval containing $\left[0, \frac{\pi}{2}\right]$ that we could choose for the restricted domain; the graph of sine is shown, where the solid curve is a portion of sine that is one-to-one and takes on all values between -1 and 1.

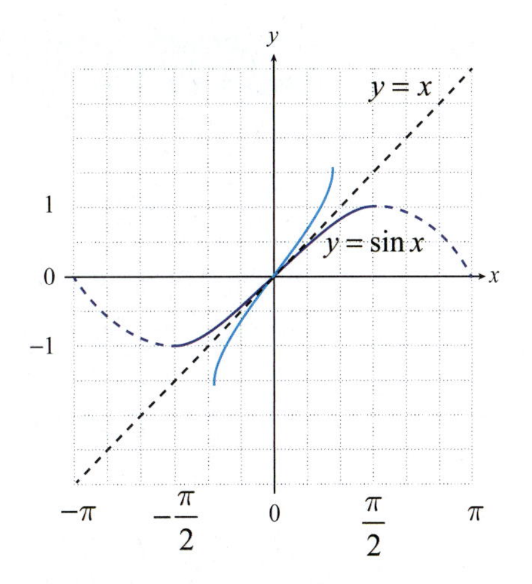

FIGURE 1: Restricting the Domain of Sine

In practice, context will tell us whether we want to think of sine as being defined on the entire real line or only over the interval $\left[-\frac{\pi}{2}, \frac{\pi}{2}\right]$, but the biggest hint will be whether we need to apply the inverse of the sine function. If so, the restricted domain for sine is called for.

Figure 1 also shows the reflection of the restricted graph of sine with respect to the line $y = x$. This reflection is the graph of the inverse of the sine function.

Two notations are commonly used for the inverse trigonometric functions. In the case of the (restricted) sine function, the statement $a = \sin b$ is equivalent to the statements

$$\sin^{-1} a = b \quad \text{and} \quad \arcsin a = b.$$

The arcsine notation is derived from the fact that $\arcsin y$ is the length of the arc (on the unit circle) corresponding to the angle x. The $\sin^{-1} y$ notation is in keeping with our use of f^{-1} to stand for the inverse of the function f.

CAUTION

When using the notation $\sin^{-1} x$, remember that

$$\sin^{-1} x \neq \frac{1}{\sin x}.$$

In order to avoid this possible source of confusion, some texts use only arcsine for the inverse sine function.

We summarize this with the following formal definition.

DEFINITION: Arcsine

Given $x \in [-1,1]$, **arcsine** is defined by either of the following:

$$\arcsin x = y \Leftrightarrow x = \sin y \quad \text{or} \quad \sin^{-1} x = y \Leftrightarrow x = \sin y$$

In words, $\arcsin x$ is the angle whose sine is x; that is, $\sin(\arcsin x) = x$. Since the restricted domain of sine is $\left[-\frac{\pi}{2}, \frac{\pi}{2}\right]$ and its range is $[-1,1]$, the domain of arcsine is $[-1,1]$ and its range is $\left[-\frac{\pi}{2}, \frac{\pi}{2}\right]$.

The solid portions of the curves in Figures 2 and 3 show similar restrictions of the cosine and tangent functions. The figures also show the reflections of those restricted graphs with respect to the line $y = x$. Again, there is a choice to be made when restricting both functions; the figures show a restriction of cosine to the interval $[0, \pi]$ and a restriction of tangent to the interval $\left(-\frac{\pi}{2}, \frac{\pi}{2}\right)$.

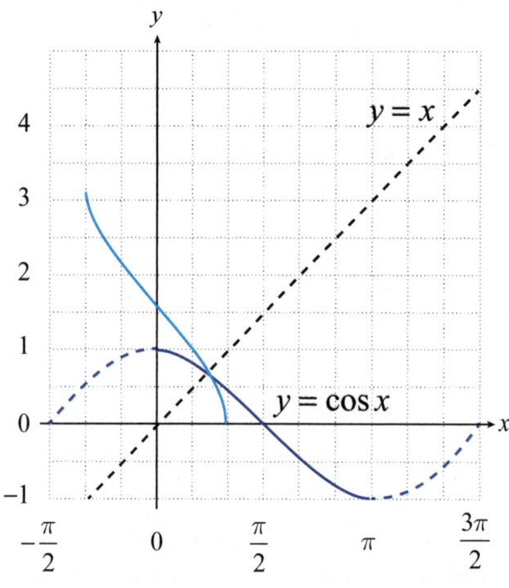

FIGURE 2: Restricting the Domain of Cosine

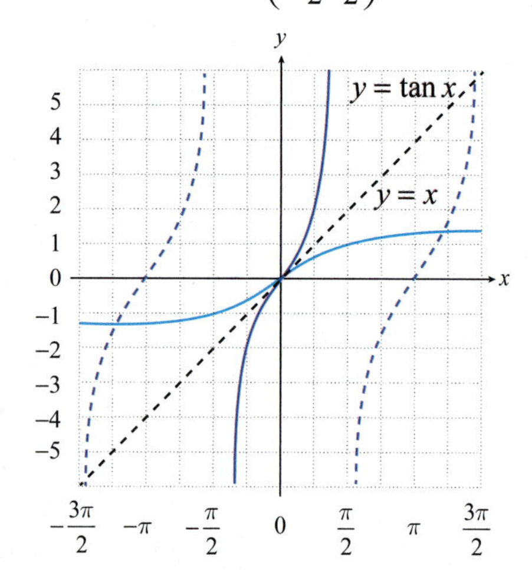

FIGURE 3: Restricting the Domain of Tangent

We will simply summarize the domains and ranges of the remaining three inverse trigonometric functions in the following definition, with graphs of all six inverse functions shown in Figure 4.

DEFINITION: Inverse Trigonometric Functions

Function	Domain	Range	Notation 1	Notation 2
Inverse Sine	$[-1,1]$	$\left[-\frac{\pi}{2},\frac{\pi}{2}\right]$	$\arcsin x = y \Leftrightarrow x = \sin y$	$\sin^{-1} x = y \Leftrightarrow x = \sin y$
Inverse Cosine	$[-1,1]$	$[0,\pi]$	$\arccos x = y \Leftrightarrow x = \cos y$	$\cos^{-1} x = y \Leftrightarrow x = \cos y$
Inverse Tangent	$(-\infty,\infty)$	$\left(-\frac{\pi}{2},\frac{\pi}{2}\right)$	$\arctan x = y \Leftrightarrow x = \tan y$	$\tan^{-1} x = y \Leftrightarrow x = \tan y$
Inverse Cosecant	$(-\infty,-1]\cup[1,\infty)$	$\left[-\frac{\pi}{2},0\right)\cup\left(0,\frac{\pi}{2}\right]$	$\operatorname{arccsc} x = y \Leftrightarrow x = \csc y$	$\csc^{-1} x = y \Leftrightarrow x = \csc y$
Inverse Secant	$(-\infty,-1]\cup[1,\infty)$	$\left[0,\frac{\pi}{2}\right)\cup\left(\frac{\pi}{2},\pi\right]$	$\operatorname{arcsec} x = y \Leftrightarrow x = \sec y$	$\sec^{-1} x = y \Leftrightarrow x = \sec y$
Inverse Cotangent	$(-\infty,\infty)$	$\left(-\frac{\pi}{2},0\right)\cup\left(0,\frac{\pi}{2}\right]$	$\operatorname{arccot} x = y \Leftrightarrow x = \cot y$	$\cot^{-1} x = y \Leftrightarrow x = \cot y$

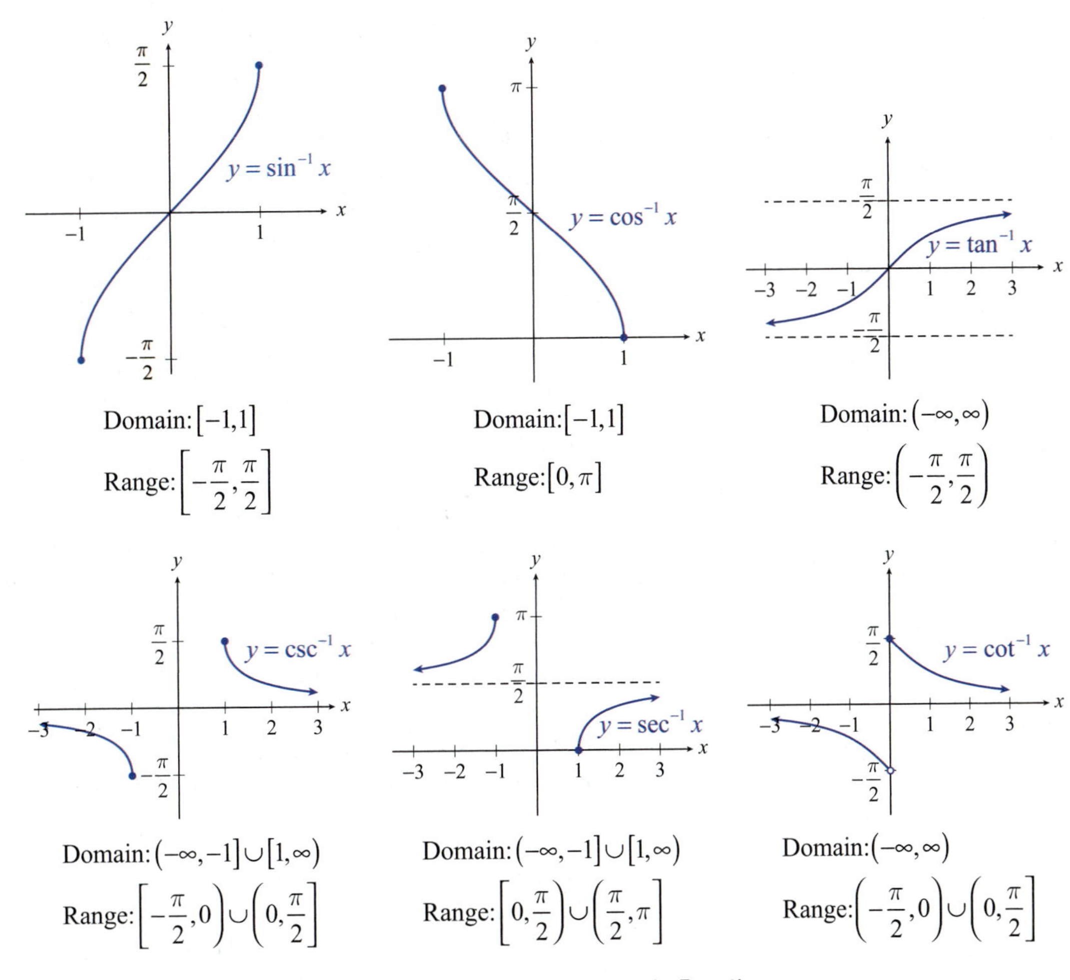

FIGURE 4: Inverse Trigonometric Functions

TOPIC 2: Evaluating Inverse Trigonometric Functions

The evaluation of inverse trigonometric functions can take several forms, depending on context. One meaning is the actual numerical evaluation of an expression containing an inverse trig function; this may or may not require the use of a calculator. Another meaning is the simplification of expressions containing inverse trig functions, using nothing more than our knowledge of how functions and their inverses behave relative to one another. We'll begin with some numerical examples.

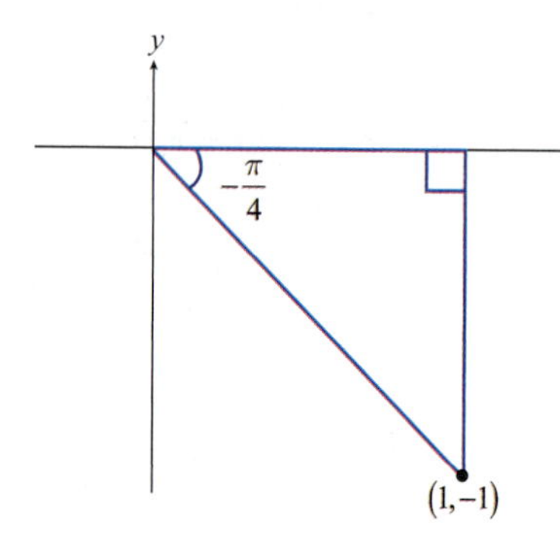

FIGURE 5

Example 1: Evaluating Inverse Trigonometric Functions

Evaluate the following expressions.

a. $\arctan(-1)$ **b.** $\csc^{-1} 2$ **c.** $\sin^{-1}(2.3)$

Solution

a. A glance at the graph of arctangent in Figure 4 tells us that $\arctan(-1)$ is defined and is a negative number apparently halfway between 0 and $-\frac{\pi}{2}$ (that is, a negative angle whose terminal side is in the fourth quadrant). By drawing the appropriate right triangle, we can verify that it is indeed the case that $\tan\left(-\frac{\pi}{4}\right) = -1$, so $\arctan(-1) = -\frac{\pi}{4}$.

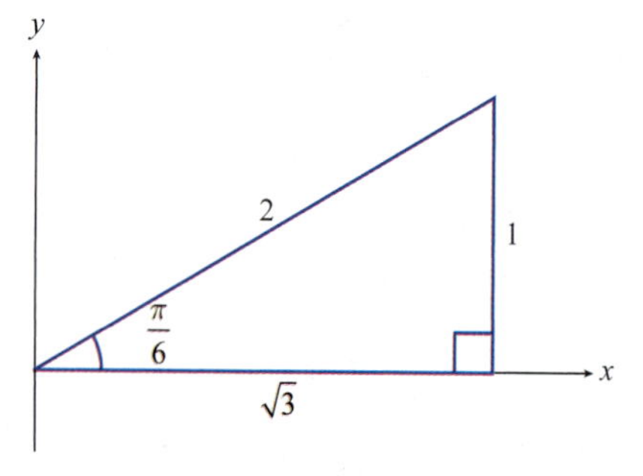

FIGURE 6

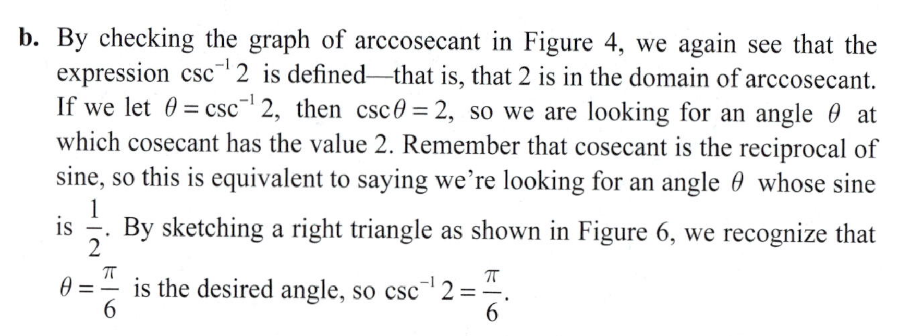

b. By checking the graph of arccosecant in Figure 4, we again see that the expression $\csc^{-1} 2$ is defined—that is, that 2 is in the domain of arccosecant. If we let $\theta = \csc^{-1} 2$, then $\csc\theta = 2$, so we are looking for an angle θ at which cosecant has the value 2. Remember that cosecant is the reciprocal of sine, so this is equivalent to saying we're looking for an angle θ whose sine is $\frac{1}{2}$. By sketching a right triangle as shown in Figure 6, we recognize that $\theta = \frac{\pi}{6}$ is the desired angle, so $\csc^{-1} 2 = \frac{\pi}{6}$.

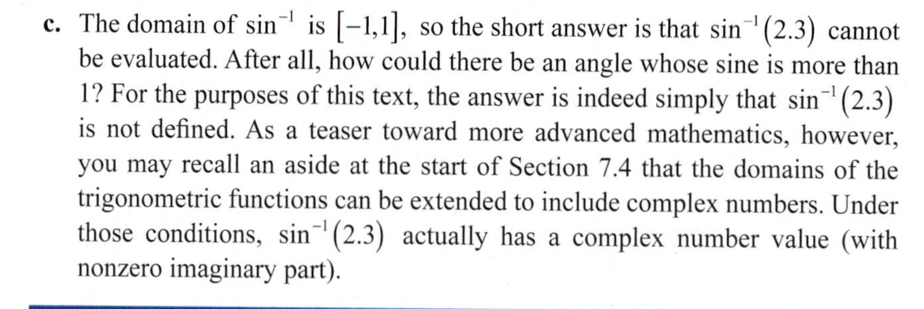

c. The domain of $\sin^{-1}$ is $[-1,1]$, so the short answer is that $\sin^{-1}(2.3)$ cannot be evaluated. After all, how could there be an angle whose sine is more than 1? For the purposes of this text, the answer is indeed simply that $\sin^{-1}(2.3)$ is not defined. As a teaser toward more advanced mathematics, however, you may recall an aside at the start of Section 7.4 that the domains of the trigonometric functions can be extended to include complex numbers. Under those conditions, $\sin^{-1}(2.3)$ actually has a complex number value (with nonzero imaginary part).

The single most important attribute of the inverse trigonometric functions is that they reverse the action of the functions they are associated with; remember that, in general, $f^{-1}(f(x)) = x$ and $f(f^{-1}(x)) = x$. But these statements are only true if all of the expressions contained in them make sense. A solid understanding of domains and ranges prevents potential errors, as shown in the next example.

Example 2: Evaluating Compositions of Trigonometric Functions

Evaluate the following expressions, if possible.

a. $\arcsin\left(\sin\left(\frac{3\pi}{4}\right)\right)$

b. $\cos\left(\cos^{-1}(-0.2)\right)$

c. $\tan^{-1}\left(\tan\left(\frac{7\pi}{6}\right)\right)$

Solution

a. The potential error in this problem is to assume that $\arcsin\left(\sin\left(\frac{3\pi}{4}\right)\right)=\frac{3\pi}{4}$, since arcsine and sine are inverse functions of one another. But $\frac{3\pi}{4}$ lies outside the range of arcsin, which is $\left[-\frac{\pi}{2},\frac{\pi}{2}\right]$, so we know this can't be the answer. The key is to evaluate the expressions individually.

$$\sin\left(\frac{3\pi}{4}\right)=\frac{1}{\sqrt{2}}$$

$$\arcsin\left(\frac{1}{\sqrt{2}}\right)=\frac{\pi}{4}$$

b. The number -0.2 lies in the domain of arccosine, and all real numbers lie in the domain of cosine, so all the parts of the expression $\cos\left(\cos^{-1}(-0.2)\right)$ make sense, and we are safe in stating $\cos\left(\cos^{-1}(-0.2)\right)=-0.2$. If we wanted to explore the expression a bit further, we could note that, from the graph of arccosine in Figure 4, $\cos^{-1}(-0.2)$, is some positive number, and further that $\cos^{-1}(-0.2)$ must be greater than $\frac{\pi}{2}$ since $\cos\left(\cos^{-1}(-0.2)\right)$ is negative. This is indeed the case: $\cos^{-1}(-0.2)$ is approximately 101.5°.

c. We run into the same problem with $\tan^{-1}\left(\tan\left(\frac{7\pi}{6}\right)\right)$ as with $\arcsin\left(\sin\left(\frac{3\pi}{4}\right)\right)$, but we will present a slightly different way of thinking about the resolution here. Instead of evaluating $\tan\left(\frac{7\pi}{6}\right)$ literally, consider only the steps involved in doing so. The first step is to determine that the reference angle for $\frac{7\pi}{6}$ is $\frac{\pi}{6}$, and the second step is to note that $\tan\left(\frac{7\pi}{6}\right)$ and $\tan\left(\frac{\pi}{6}\right)$ have the same sign (the terminal side of $\frac{7\pi}{6}$ is in the third quadrant, and tangent is positive there). So $\tan^{-1}\left(\tan\left(\frac{7\pi}{6}\right)\right)=\tan^{-1}\left(\tan\left(\frac{\pi}{6}\right)\right)=\frac{\pi}{6}$, as $\frac{\pi}{6}$ lies in the range of $\tan^{-1}$.

The last example demonstrated the evaluation of compositions of trig functions with their inverses, but other compositions are possible. In many cases, a picture aids greatly in the computation.

Example 3: Evaluating Compositions of Trigonometric Functions

Evaluate the following expressions.

a. $\tan\left(\sin^{-1}\left(-\frac{4}{5}\right)\right)$ **b.** $\cos\left(\arctan\left(0.4\right)\right)$

Solution

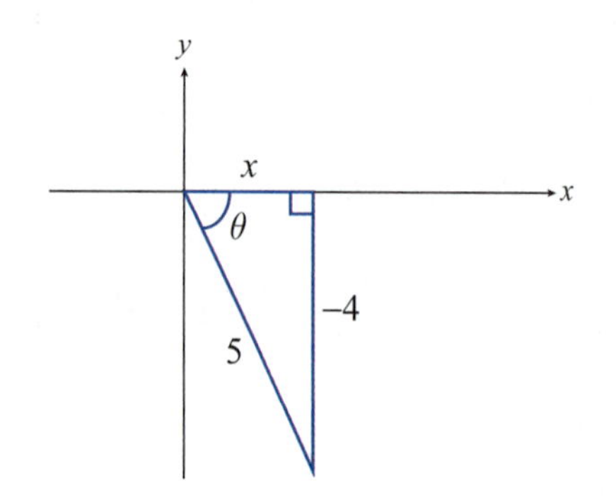

FIGURE 7

a. Remember that the range of arcsine is $\left[-\frac{\pi}{2},\frac{\pi}{2}\right]$, and in particular that $\sin^{-1}\left(-\frac{4}{5}\right)$ will lie between $-\frac{\pi}{2}$ and 0 (the graph in Figure 4 tells us that arcsine of a negative number is negative). If we let $\theta=\sin^{-1}\left(-\frac{4}{5}\right)$ then $\sin\theta=-\frac{4}{5}$, and we can sketch the triangle in Figure 7 to illustrate the relationship between θ and the given numbers.

The Pythagorean Theorem allows us to calculate x.

$$x=\sqrt{5^2-\left(-4\right)^2}=\sqrt{9}=3$$

Now we can see that $\tan\theta=-\frac{4}{3}$, so

$$\tan\left(\sin^{-1}\left(-\frac{4}{5}\right)\right)=-\frac{4}{3}.$$

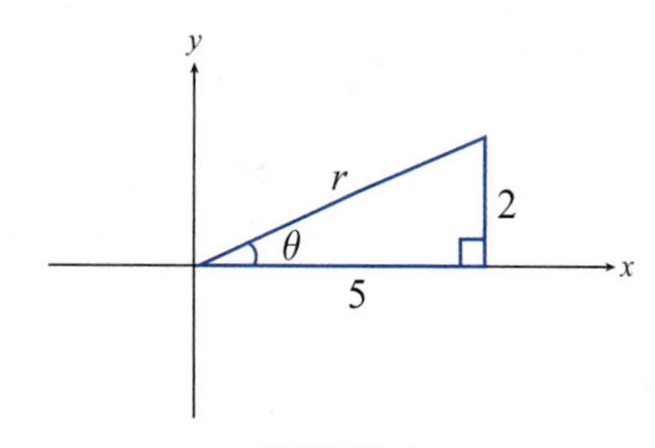

FIGURE 8

b. We can employ the same method and let $\theta=\arctan\left(0.4\right)$. This leads to

$$\tan\theta=0.4=\frac{4}{10}=\frac{2}{5}$$

and then to the triangle in Figure 8.

The Pythagorean Theorem gives us $r=\sqrt{5^2+2^2}=\sqrt{29}$, and so $\cos\left(\arctan\left(0.4\right)\right)=\frac{5}{\sqrt{29}}$.

TOPIC 3: Applications of Inverse Trigonometric Functions

Many applications calling for the use of inverse trigonometric functions are dynamic, and feature an angle that is changing over time; you will encounter many such problems in calculus. The first step is often to determine a formula for a given angle in terms of other quantities, as illustrated in the next examples.

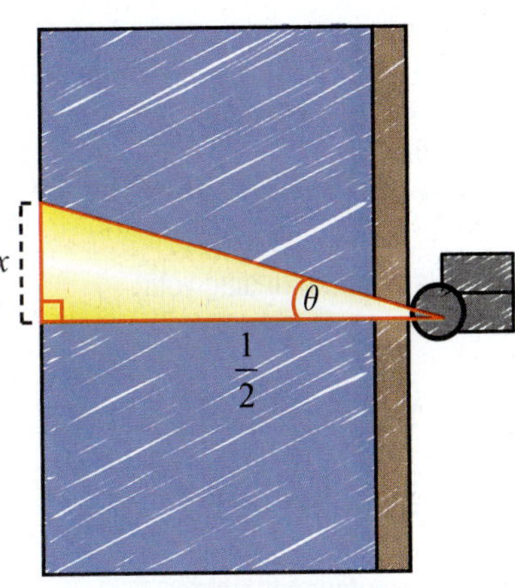

FIGURE 9

Example 4: Using Inverse Trigonometric Functions

A lighthouse is to be constructed half a mile from a long, straight reef, as shown in Figure 9. In order to ensure the light illuminates certain portions of the reef within specified lengths of time, the engineer needs a formula for θ in terms of x. Find such a formula.

Solution

From Figure 9, we see that $\tan\theta = \frac{x}{\frac{1}{2}} = 2x$, so the formula for θ is

$$\theta = \tan^{-1}(2x).$$

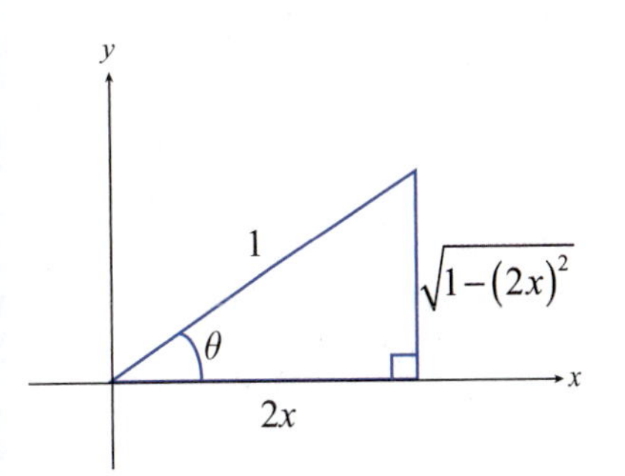

FIGURE 10

Example 5: Using Inverse Trigonometric Functions

Express $\sin\left(\cos^{-1}(2x)\right)$ as an algebraic function of x, assuming $-\frac{1}{2} \le x \le \frac{1}{2}$.

Solution

Let $\theta = \cos^{-1}(2x)$. Then $\cos\theta = 2x$, and we are led to consider a sketch like the one in Figure 10.

In the sketch, we have chosen the simplest lengths for the adjacent side and the hypotenuse that make $\cos\theta = 2x$, though any positive multiple of these lengths would also work. And as always, once the lengths of two sides of the right triangle have been determined, the Pythagorean Theorem provides the length of the third side. Now we can refer to the sketch to see that

$$\sin\left(\cos^{-1}(2x)\right) = \sin\theta = \frac{\sqrt{1-4x^2}}{1} = \sqrt{1-4x^2}.$$

7.6 EXERCISES

PRACTICE

Evaluate each of the following expressions without the use of a calculator. See Example 1.

1. $\sin^{-1}(-1)$ **2.** $\cos^{-1}\left(\frac{\sqrt{2}}{2}\right)$ **3.** $\tan^{-1}1$ **4.** $\cot^{-1}\left(-\frac{\sqrt{3}}{3}\right)$

5. $\sec^{-1}\left(\frac{2\sqrt{3}}{3}\right)$ **6.** $\csc^{-1}(-2)$ **7.** $\arcsin 0$ **8.** $\arccos(-1)$

9. $\arctan\left(-\sqrt{3}\right)$ **10.** $\text{arccot}\left(-\sqrt{3}\right)$ **11.** $\text{arcsec}\, 2$ **12.** $\text{arccsc}\left(\sqrt{2}\right)$

13. $\text{arccot}(-1)$ **14.** $\tan^{-1}\left(\frac{\sqrt{3}}{3}\right)$ **15.** $\cos^{-1}\left(-\frac{1}{2}\right)$ **16.** $\csc^{-1}2$

17. $\arcsin\left(-\frac{1}{2}\right)$ **18.** $\sec^{-1}(-1)$ **19.** $\operatorname{arccsc} 1$ **20.** $\arctan 0$

21. $\sin^{-1}\left(\frac{\sqrt{2}}{2}\right)$ **22.** $\arccos\left(-\frac{\sqrt{2}}{2}\right)$ **23.** $\operatorname{arcsec}(-2)$ **24.** $\cot^{-1}\left(-\sqrt{3}\right)$

Evaluate each of the following expressions, if possible, using a calculator and rounding your answers to four decimal places if necessary.

25. $\sin^{-1}(-0.2)$ **26.** $\cos^{-1} 4$ **27.** $\sin^{-1}(-0.9)$

28. $\tan^{-1} 5$ **29.** $\cos^{-1}(-0.4)$ **30.** $\tan^{-1}(0.8)$

Some calculators are not equipped with arccosecant, arcsecant, and arccotangent buttons, but expressions involving these functions can still be evaluated. To evaluate, for example, $\csc^{-1} x$, let $\theta = \csc^{-1} x$. Then

$$\csc\theta = x$$
$$\frac{1}{\sin\theta} = x$$
$$\sin\theta = \frac{1}{x}$$
$$\theta = \sin^{-1}\left(\frac{1}{x}\right).$$

This means that arccosecant, arcsecant, and arccotangent can all be evaluated using the following formulas.

$$\csc^{-1} x = \sin^{-1}\left(\frac{1}{x}\right)$$
$$\sec^{-1} x = \cos^{-1}\left(\frac{1}{x}\right)$$
$$\cot^{-1} x = \tan^{-1}\left(\frac{1}{x}\right), \text{ with } \cot^{-1} 0 = \frac{\pi}{2}$$

Use these formulas to evaluate each of the following expressions, if possible, using a calculator and rounding your answers to four decimal places if necessary.

31. $\csc^{-1} 5$ **32.** $\sec^{-1}(-0.5)$ **33.** $\cot^{-1} 150$

34. $\cot^{-1}(0.2)$ **35.** $\csc^{-1}(-8.9)$ **36.** $\sec^{-1} 2$

Evaluate each of the following expressions, if possible. See Example 2.

37. $\cos^{-1}\left(\cos\left(\frac{2\pi}{4}\right)\right)$ **38.** $\sin^{-1}\left(\sin\left(\frac{3\pi}{2}\right)\right)$ **39.** $\tan\left(\tan^{-1}(0.5)\right)$

40. $\sin^{-1}\left(\sin\left(\frac{7\pi}{6}\right)\right)$ **41.** $\cos\left(\cos^{-1}(-0.8)\right)$ **42.** $\tan^{-1}\left(\tan\left(\frac{5\pi}{4}\right)\right)$

Evaluate each of the following expressions, if possible, using a calculator and rounding your answers to four decimal places if necessary. See Example 3.

43. $\sin(\arctan(0.4))$ **44.** $\sin^{-1}\left(\cos\left(\frac{3\pi}{2}\right)\right)$ **45.** $\cos(\tan^{-1}(0.5))$

46. $\arcsin(\tan 1)$ **47.** $\tan(\cos^{-1}(-0.8))$ **48.** $\tan^{-1}(\cos 5)$

Find the value of each of the following expressions without using a calculator.

49. $\sin\left(\arctan\left(\sqrt{3}\right)\right)$ **50.** $\cos(\sec^{-1}(-2))$

51. $\tan(\operatorname{arccot} 1)$ **52.** $\csc\left(\arccos\left(-\frac{\sqrt{3}}{2}\right)\right)$

53. $\tan\left(\sin^{-1}\left(-\frac{\sqrt{2}}{2}\right)\right)$ **54.** $\sec\left(\csc^{-1}\left(\frac{2\sqrt{3}}{3}\right)\right)$

55. $\cos(\cot^{-1}(-1))$ **56.** $\sec\left(\arcsin\left(-\frac{1}{2}\right)\right)$

57. $\cot\left(\operatorname{arcsec}\left(\sqrt{2}\right)\right)$ **58.** $\cot(\operatorname{arccsc}(-2))$

59. $\sin\left(\cos^{-1}\left(\frac{\sqrt{2}}{2}\right)\right)$ **60.** $\sec\left(\tan^{-1}\left(-\frac{\sqrt{3}}{3}\right)\right)$

61. $\sec\left(\arccos\left(-\frac{\sqrt{2}}{2}\right)\right)$ **62.** $\tan(\csc^{-1}(-2))$

63. $\sin\left(\operatorname{arcsec}\left(\sqrt{2}\right)\right)$ **64.** $\csc\left(\cot^{-1}\left(\sqrt{3}\right)\right)$

65. $\cot\left(\sin^{-1}\left(-\frac{\sqrt{2}}{2}\right)\right)$ **66.** $\cos\left(\arctan\left(-\frac{\sqrt{3}}{3}\right)\right)$

Express each of the following functions as a purely algebraic function. See Example 5.

67. $\tan(\cos^{-1} x)$ **68.** $\cot\left(\sin^{-1}\left(\frac{2}{x}\right)\right)$ **69.** $\sec(\tan^{-1}(3x))$

70. $\tan\left(\sin^{-1}\left(\frac{x}{\sqrt{x^2+3}}\right)\right)$ **71.** $\sin(\sec^{-1} x)$ **72.** $\cos\left(\tan^{-1}\left(\frac{x}{4}\right)\right)$

Using a calculator, find the value of θ in degrees. Remember to make sure your calculator is in the correct mode.

73. $\theta = \sin^{-1}(0.74184113)$ **74.** $\theta = \arctan(-0.258416)$

75. $\theta = \operatorname{arccsc}(1.847526)$ **76.** $\theta = \sec^{-1}(-1.1224539)$

77. $\theta = \cot^{-1}(0.57496998)$

Using a calculator, find the value of θ in radians. Remember to make sure your calculator is in the correct mode.

78. $\theta = \arccos(-0.1115598)$

79. $\theta = \text{arccot}(1.547773)$

80. $\theta = \tan^{-1}(5.999999)$

81. $\theta = \csc^{-1}(-1.333333)$

82. $\theta = \arcsin(0.65937229)$

Sketch the graph of each of the following functions. Then graph the function using a graphing utility to check your answer.

83. $f(x) = \sin^{-1}(x-3)$

84. $f(x) = \sec^{-1}(2x)$

85. $f(x) = \arctan\left(\frac{x}{2}\right)$

86. $f(x) = 2\arccos x$

APPLICATIONS

87. Kim is watching a space shuttle launch from an observation spot two miles away from the launch pad. Find the angle of elevation to the shuttle for each of the following heights. Round your answers to four decimal places.

a. 0.5 miles **b.** 2 miles **c.** 2.8 miles

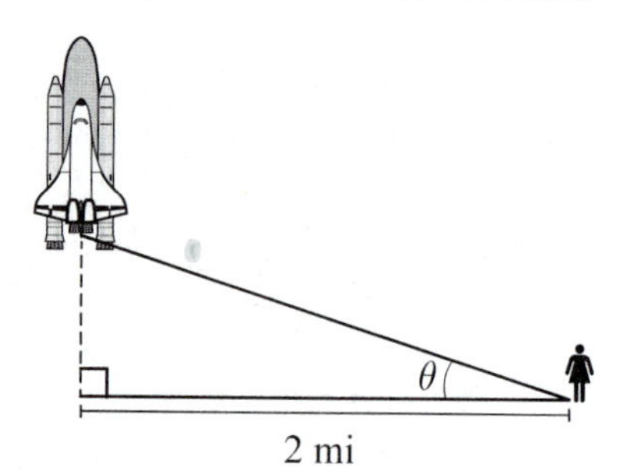

88. Jesse is rowing in the men's singles race. The length of the oar from the side of the shell to the water is 7 feet. At what angle is the oar from the side of the boat when the blade is at the following distance from the boat? Round your answers to four decimal places.

a. 2 feet **b.** 3 feet **c.** 5 feet

CHAPTER 7 PROJECT

Trigonometric Applications

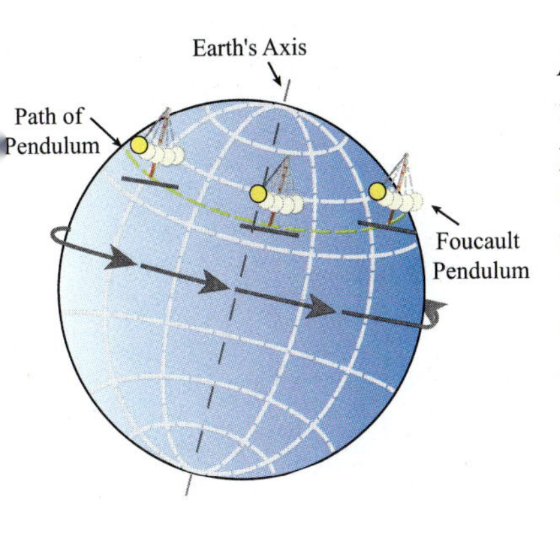

At the Paris Observatory in 1851, Jean Foucault used a long pendulum to prove that Earth is rotating. As it swings, the pendulum appears to change its path. However, it is not the pendulum that changes path, but the room rotating underneath it. At the North Pole, Earth revolves 360° underneath the pendulum over 24 hours. The path of a pendulum at the equator does not revolve at all; instead, the pendulum travels in a huge circle while Earth spins. At points between the two, the pendulum cannot show how far it travels, but it can show how much planet Earth is revolving underneath it. To calculate how much our planet revolves in a particular location, use the following equation.

$$\text{degrees of revolution} = 360° \sin\left(\text{latitude of location}\right)$$

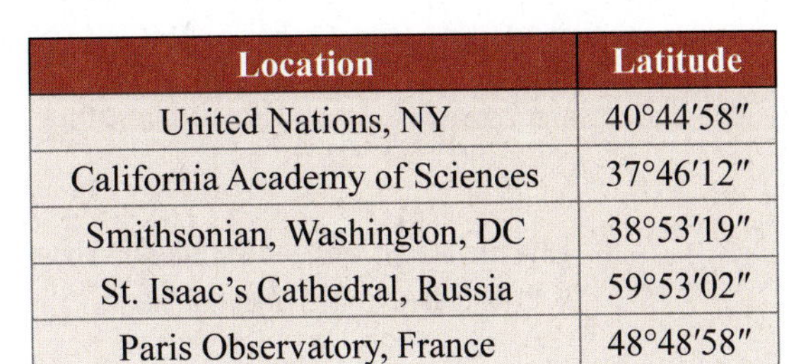

Location	Latitude
United Nations, NY	40°44′58″
California Academy of Sciences	37°46′12″
Smithsonian, Washington, DC	38°53′19″
St. Isaac's Cathedral, Russia	59°53′02″
Paris Observatory, France	48°48′58″

1. In a 24-hour period, how many degrees does Earth revolve at the locations specified in the table?

Your university has decided to install a 50-foot Foucault Pendulum in your science building and asked you to make sure that there is enough room.

2. If the pendulum swings a total of 16°, how long is the arc traced in the air by the tip of the pendulum during one swing?
3. The school plans to build a small wall encircling the swinging pendulum. What should the diameter of the circle be if they want the tip of the pendulum to come within 6 inches of the wall?
4. When the pendulum reaches the farthest point from the center, how much higher will the tip be compared to when it is at the center?
5. If the science center only has room for a circular wall of diameter 12 feet, how many degrees can the pendulum swing and still stay 6 inches from the wall?
6. The Foucault Pendulum in the United Nations building has a length of 75 feet and a period of 10 seconds. Assuming simple harmonic motion and that at $t = 0$ the pendulum is at its farthest distance away (6 feet from the center of the circle), what function models the motion of the pendulum? Graph this function.

CHAPTER 7 REVIEW EXERCISES

Section 7.1

Convert each of the following angle measures as directed.

1. Express $\frac{\pi}{45}$ in degrees.
2. Express $\frac{6\pi}{5}$ in degrees.
3. Express $-\frac{7\pi}{4}$ in degrees.
4. Express $-\frac{3\pi}{10}$ in degrees.
5. Express 42° in radians.
6. Express 60° in radians.
7. Express –79° in radians.
8. Express –25° in radians.

Sketch the indicated angles.

9. 300°
10. $-\frac{9\pi}{4}$

Find the length of the arc subtended by the given central angle θ on a circle of radius r. Round your answers to two decimal places.

11. $r = 5$ ft; $\theta = 180°$
12. $r = 8$ km; $\theta = \frac{3\pi}{4}$

Find the indicated arc length in each of the following problems. Round your answers to two decimal places.

13. Given a circle of radius 16.3 meters, find the length of the arc subtended by a central angle of $\frac{3\pi}{5}$.
14. Given a circle of radius 8 inches, find the length of the arc subtended by a central angle of 72° (**Hint:** Convert to radians first).

15. Find the distance between Vancouver, British Columbia and San Francisco, California which have the same longitude. The latitude of Vancouver is 49.25° N and the latitude of San Francisco is 37.77° N. Assume the Earth's radius is 6470 km.

Find the radian measure of the central angle θ given the radius r and the length s of the arc subtended by θ. Leave your answers in fraction form.

16. $r = 18$ ft; $s = 52$ ft
17. $r = 6.4$ dm; $s = 19.2$ dm

Solve the following problems. Round your answers to two decimal places.

18. A gear in a machine has a radius of 14 cm and is rotating at a constant speed of 600 rpm. Find **a.** the angular speed of a tooth of the gear in radians per minute and **b.** the linear speed of the tooth in meters per second.

19. Find the area of the sector of a circle of radius 18 ft with a central angle of $\frac{2\pi}{3}$.

Section 7.2

Use the information contained in each figure to determine the values of the six trigonometric functions of θ. Rationalize all denominators in your answers.

20. 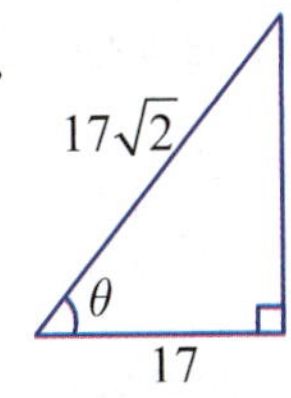

21. 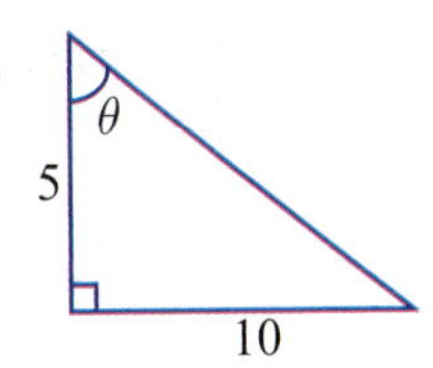

Use a calculator to evaluate each of the following expressions. Round your answers to four decimal places.

22. $\sin 82°$

23. $\cot 14°$

24. $\csc\left(\frac{5\pi}{12}\right)$

25. $\cos\left(\frac{3\pi}{7}\right)$

Convert each expression from degrees, minutes, seconds (DMS) notation to decimal notation. Round your answers to four decimal places.

26. $36°56'14''$

27. $15°12'73''$

Determine whether the following statements are true or false. Use a calculator when necessary.

28. If $\tan\theta = 1.6$, then $\cot\theta = 0.625$.

29. If $\csc\theta = 3.4$, then $\sin\theta = 1.7$.

Use an appropriate trigonometric function and a calculator if necessary to solve each of the following problems. Round your answer to two decimal places.

30. A wheelchair ramp touches the ground 15 feet away from the top of the steps. If the ramp makes an angle of 30° relative to the ground, how long is the ramp?

31. A building is 83 feet tall and a cable is stretched from the top of the building to the ground. If the angle between the cable and the ground is 40°, how long is the cable?

Section 7.3

Determine the values of the six trigonometric functions of each angle θ, using a calculator and rounding your answers to four decimal places if necessary.

32. $\theta = 90°$

33. $\theta = -460°$

34. $\theta = \frac{\pi}{4}$

35. $\theta = \frac{7\pi}{3}$

Determine the reference angle associated with the given angle.

36. $\theta = 86°$

37. $\theta = -143°$

38. $\theta = \frac{3\pi}{2}$

39. $\theta = \frac{11\pi}{4}$

Determine the quadrant in which the terminal side of the angle θ is located.

40. $\csc\theta > 0$ and $\tan\theta > 0$

41. $\sec\theta < 0$ and $\cot\theta > 0$

In Exercises 42 and 43, **a.** rewrite the expression in terms of the given angle's reference angle, and then **b.** evaluate the result, using a calculator and rounding your answer to four decimal places if necessary.

42. $\sin 290°$

43. $\tan\left(\frac{4\pi}{3}\right)$

Express each of the following in terms of the appropriate cofunction, and verify the equivalence of the two expressions, using a calculator and rounding your answer to four decimal places if necessary.

44. $\csc 193°$

45. $\sin(-42°)$

46. $\cot\left(\frac{3\pi}{4}\right)$

47. $\cos\left(\frac{5\pi}{4}\right)$

Use the given information about each angle to evaluate the expression, if possible. If no angle with the stated properties exists, determine the reason.

48. Given that $\sin\theta = \frac{\sqrt{2}}{2}$ and $\tan\theta$ is negative, determine θ and $\tan\theta$.

49. Given that $\csc\theta = \frac{13}{12}$ and θ lies in the first quadrant, determine $\sec\theta$.

Section 7.4

Determine the amplitude, period, and phase shift of each of the following functions.

50. $f(x) = 3\cos(4x)$

51. $h(x) = 10 + 6\cos x$

52. $g(x) = 6 - \frac{1}{2}\sin(3\theta - \pi)$

53. $f(x) = -3 + 9\sin(2\theta + 2\pi)$

Sketch the graph of each of the following functions.

54. $f(x) = \frac{1}{2}\cos(3\pi x)$

55. $g(x) = 4\sin(2x - 5)$

56. $f(\theta) = 5\cos\left(\theta - \frac{\pi}{3}\right)$

57. $h(x) = 2 + \sin(x - \pi)$

58. $g(x) = 1 - \frac{1}{2}\sin\left(\frac{1}{2}x + \frac{\pi}{4}\right)$

59. $f(t) = \frac{1}{2}e^{-t}\cos(t + 2\pi) - 1$

Section 7.5

Sketch the graph of each of the following functions.

60. $f(x) = 1 - \tan\left(x - \frac{\pi}{2}\right)$

61. $f(x) = \cot\left(\frac{\pi x}{4} + \frac{\pi}{4}\right)$

62. $f(x) = \frac{1}{2}\sec(2\pi x)$

63. $f(x) = -3\csc\left(\frac{x}{2} + \frac{\pi}{2}\right) + 1$

Section 7.6

Evaluate each of the following expressions without the use of a calculator.

64. $\cos^{-1}\left(\frac{\sqrt{3}}{2}\right)$

65. $\cos^{-1} 0$

66. $\arctan(-1)$

67. $\csc^{-1}\left(\frac{2\sqrt{3}}{3}\right)$

Evaluate each of the following expressions, if possible, using a calculator and rounding your answer to four decimal places if necessary.

68. $\sin^{-1} 2$

69. $\tan^{-1}(0.5)$

Evaluate each of the following expressions, if possible.

70. $\arccos(\sin \pi)$

71. $\cos\left(\cos^{-1}(0.9)\right)$

72. $\tan\left(\tan^{-1}(0.75)\right)$

73. $\sin^{-1}\left(\cos\left(\frac{3\pi}{4}\right)\right)$

Find the value of each of the following expressions without using a calculator.

74. $\sin\left(\arctan\left(\frac{\sqrt{3}}{3}\right)\right)$

75. $\cot\left(\sec^{-1} 2\right)$

76. $\cos\left(\arcsin\left(\frac{-\sqrt{2}}{2}\right)\right)$

77. $\sec\left(\sin^{-1}\left(\frac{\sqrt{3}}{2}\right)\right)$

Using a calculator, find the value of θ in radians. Remember to make sure your calculator is in the correct mode.

78. $\theta = \cos^{-1}(0.3492581)$

79. $\theta = \tan^{-1}(-4.18249588)$

Using a calculator, find the value of θ in degrees. Remember to make sure your calculator is in the correct mode.

80. $\theta = \arcsin(-0.66666667)$

81. $\theta = \arccos(0.56894372)$

Express the following function as a purely algebraic function.

82. $\tan\left(\sin^{-1}\left(\frac{x}{\sqrt{x^2+4}}\right)\right)$

CHAPTER 8

Trigonometric Identities and Equations

SECTIONS

WHAT IF ...

What if you were walking up to bat in a baseball game? At what angle from the ground should you hit the ball to make sure that it clears the back fence, giving you a home run?

By the end of this chapter, you will have been introduced to and know how to solve equations using several fundamental trigonometric identities. These identities are very useful for simplifying trigonometric expressions. On page 621, you will find a baseball problem similar to the one given above. You will need to use identities such as the double-angle identities on page 604 and inverse trigonometric functions to find the angle to hit the ball in order to achieve the needed range.

François Viéte

Introduction

The last chapter introduced the trigonometric functions and their basic properties and uses. This chapter now delves deeper into how the functions relate to one another; this deeper understanding will allow us to simplify unwieldy expressions, solve trigonometric equations, and extend our grasp of trigonometry yet further.

The relationships between the trigonometric functions fall into a category of equations called *identities*, the formal definition of which was given in Chapter 1. Briefly, a trigonometric identity (also sometimes referred to as a trigonometric formula) is an equation that is always true. As such, there is no need to solve an identity—instead, identities are used as tools in accomplishing other tasks. Identities began to be used soon after the appearance of trigonometric functions. By the second century AD, Ptolemy, the Greek astronomer and geographer, was using identities in his work with the chords of a circle, the accepted form of trigonometry at the time. For his early work entitled *Almagest*, he used the sum and difference formulas and something that resembled the modern day half angle formula in order to update the current chord tables to an accuracy of three decimal places.

Identities must either be discovered or verified, and both processes call for the application of algebra. Further, the use of an identity usually calls for algebraic skill; some of the uses we will explore include simplifying expressions, evaluating trigonometric functions exactly, and solving *conditional* equations (which are equations that are *not* always true). In fact, the underlying theme of this chapter is the marriage of trigonometry and algebra. Whereas the last chapter focused almost exclusively on the basics of trigonometry, this chapter and the next bring algebra back into the discussion. The union of trigonometry and algebra culminates in the last section of the chapter, where you will see that such algebraic methods as factoring and solving for specific terms are critical in being able to solve trigonometric equations.

The concept of mixing algebra and trigonometry did not happen overnight. For centuries mathematicians separated the realms of real numbers and geometric ideas. It was not until the sixteenth century that mathematicians finally began to use algebra for abstract quantities rather than simply for concrete values. This transition was thanks in part to François Viéte (1540–1603), who was the first to consistently apply algebraic methods to his work in trigonometry. He introduced the sum-to-product formulas, the law of tangents, and a recurrence formula that allows $\cos(nx)$ to be presented in terms of the cosine of lower multiples of x. Although Viéte's breakthrough application of algebra to trigonometry may seem remarkably basic, it is good to remember that many important mathematical discoveries come from examining a problem in a new, unexpected, and sometimes simple way.

8.1 FUNDAMENTAL TRIGONOMETRIC IDENTITIES

TOPICS

1. Fundamental trigonometric identities
2. Simplifying trigonometric expressions
3. Verifying trigonometric identities
4. Trigonometric substitutions

TOPIC 1: Fundamental Trigonometric Identities

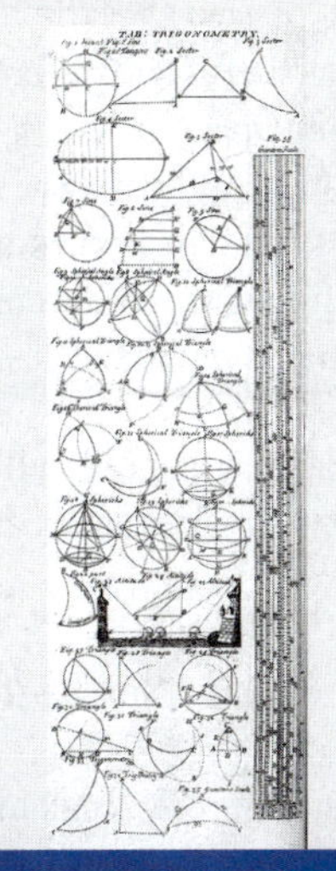

Naming the Cofunctions

As mentioned in the last chapter, cosine, cosecant, and cotangent are all so-named because they are, respectively, the sine, secant, and tangent of the *co*mplement of a given angle. The names "cosine" and "cotangent" were coined by the English clergyman, astronomer, and mathematician Edmund Gunter in the early 1600s, although the ratios they refer to were used under different names prior to that time. The first known use of the name "cosecant" was in the early 1700s.

Chapter 7 presented the foundations of trigonometry, and concentrated largely on the geometric and functional aspects of the material. This chapter focuses more on the relationships between trigonometric functions and on the marriage of algebra and trigonometry. It does so by introducing and then using statements of equality known as *trigonometric identities*. We begin with a review of the identities already seen, though three of the equations below have only been alluded to in passing.

IDENTITIES: Identities Already Seen

Reciprocal Identities

$$\csc x = \frac{1}{\sin x} \qquad \sec x = \frac{1}{\cos x} \qquad \cot x = \frac{1}{\tan x}$$

$$\sin x = \frac{1}{\csc x} \qquad \cos x = \frac{1}{\sec x} \qquad \tan x = \frac{1}{\cot x}$$

Quotient Identities

$$\tan x = \frac{\sin x}{\cos x} \qquad \cot x = \frac{\cos x}{\sin x}$$

Cofunction Identities

$$\cos x = \sin\left(\frac{\pi}{2} - x\right) \qquad \csc x = \sec\left(\frac{\pi}{2} - x\right) \qquad \cot x = \tan\left(\frac{\pi}{2} - x\right)$$

$$\sin x = \cos\left(\frac{\pi}{2} - x\right) \qquad \sec x = \csc\left(\frac{\pi}{2} - x\right) \qquad \tan x = \cot\left(\frac{\pi}{2} - x\right)$$

Period Identities

$$\sin(x + 2\pi) = \sin x \qquad \cos(x + 2\pi) = \cos x$$

$$\csc(x + 2\pi) = \csc x \qquad \sec(x + 2\pi) = \sec x$$

$$\tan(x + \pi) = \tan x \qquad \cot(x + \pi) = \cot x$$

IDENTITIES: Identities Already Seen (continued)

Even/Odd Identities

$\sin(-x) = -\sin x$ $\cos(-x) = \cos x$ $\tan(-x) = -\tan x$

$\csc(-x) = -\csc x$ $\sec(-x) = \sec x$ $\cot(-x) = -\cot x$

Pythagorean Identities

$\sin^2 x + \cos^2 x = 1$ $\tan^2 x + 1 = \sec^2 x$ $1 + \cot^2 x = \csc^2 x$

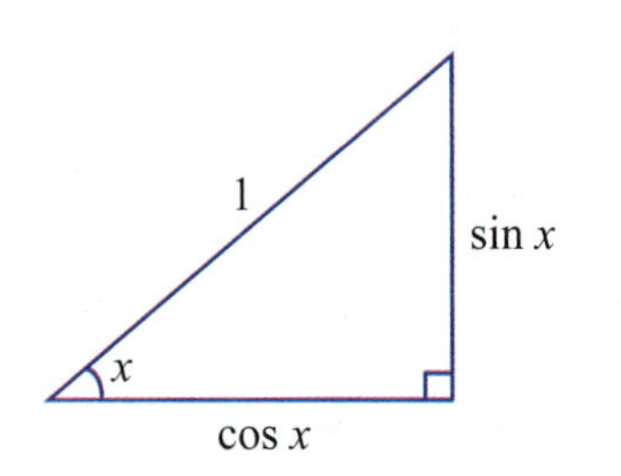

FIGURE 1: Derivation of $\sin^2 x + \cos^2 x = 1$

The Pythagorean identities have not explicitly appeared yet, but they are based on an idea we have used frequently. Their name alludes to the familiar Pythagorean Theorem, and the first identity follows from consideration of a diagram such as that in Figure 1.

The diagram assumes that x is a real number between 0 and $\frac{\pi}{2}$, but a similar diagram can be drawn given any real number (recall that the use of reference angles makes this possible). If a right triangle with a hypotenuse of length 1 is drawn, then the legs of the right triangle must be of length $\sin x$ and $\cos x$ (be sure you see how this follows from the definitions of sine and cosine). The Pythagorean Theorem then leads to the first Pythagorean identity; dividing through by $\cos^2 x$ and $\sin^2 x$ leads to, respectively, the second and third Pythagorean identities.

The ideas behind the identities can lead to statements that appear, superficially, to be different. For instance, we know that the 2π-periodicity of sine makes all of the following statements true.

$$\sin(x - 6\pi) = \sin x, \quad \sin(x + 4\pi) = \sin(x + 2\pi), \text{ and } \sin(x + 2\pi) = \sin(x - 2\pi)$$

Deeper contemplation of the graph of sine, as illustrated in Figure 2, leads to the similar identity

$$\sin x = -\sin(x + \pi).$$

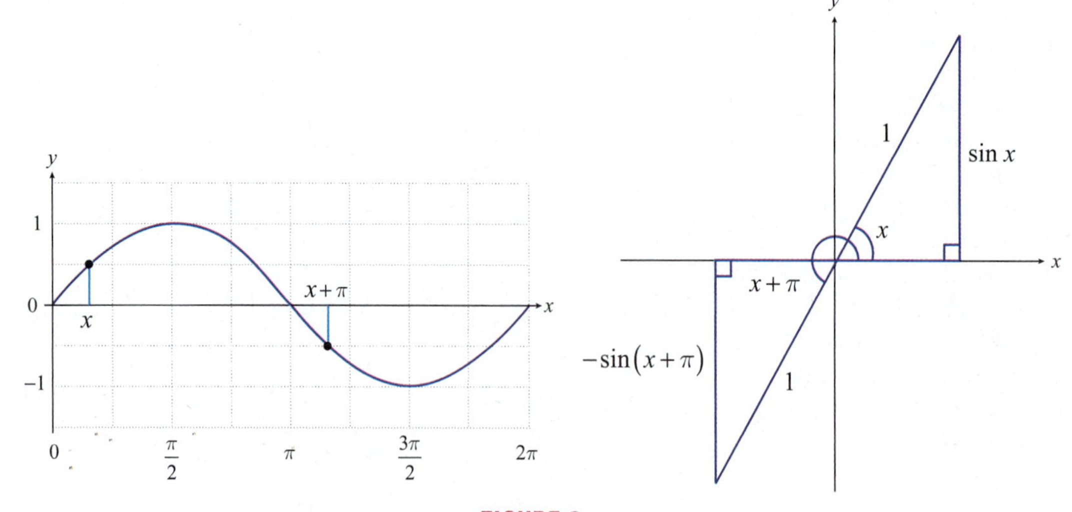

FIGURE 2

Similar statements based on periodicity can be deduced for the other trigonometric functions.

TOPIC 2: Simplifying Trigonometric Expressions

One common use of trigonometric identities is in simplifying expressions. Frequently, the first answer obtained in solving a trigonometry-related problem is unnecessarily complicated, and the judicious use of identities leads to a simpler form of the answer. This happens often, for instance, in calculus problems.

Example 1: Simplifying Trigonometric Expressions

Simplify the expression $\cos\theta + \sin\theta\tan\theta$.

Solution

Skill in using identities to simplify trigonometric expressions only comes with practice, but the guiding principle is to rewrite the expression in such a way that the use of one or more identities becomes apparent. In the case of the expression $\cos\theta + \sin\theta\tan\theta$, a good way to begin is to rewrite it in terms of only sine and cosine, since three distinct functions are unnecessary.

$$\begin{aligned}\cos\theta + \sin\theta\tan\theta &= \cos\theta + \sin\theta\left(\frac{\sin\theta}{\cos\theta}\right)\\ &= \cos\theta + \frac{\sin^2\theta}{\cos\theta}\end{aligned}$$

Now, the presence of $\sin^2\theta$ should remind you of the first Pythagorean identity (one that you'll find yourself using frequently). Remember that $\sin^2\theta + \cos^2\theta = 1$, so if some other term can be rewritten in such a way that $\sin^2\theta$ is added to $\cos^2\theta$, there is a good chance that the identity applies. In the problem at hand,

$$\begin{aligned}\cos\theta + \frac{\sin^2\theta}{\cos\theta} &= \frac{\cos^2\theta}{\cos\theta} + \frac{\sin^2\theta}{\cos\theta}\\ &= \left(\frac{1}{\cos\theta}\right)\left(\cos^2\theta + \sin^2\theta\right)\\ &= \frac{1}{\cos\theta}\\ &= \sec\theta.\end{aligned}$$

The fact that $\cos\theta + \sin\theta\tan\theta = \sec\theta$ is not at all obvious initially, but is also not at all surprising. Such equivalences between relatively complicated and relatively simple expressions occur frequently in trigonometry, and it's usually worth spending some amount of time to see if a complicated expression can be simplified. After all, if the expression $\cos\theta + \sin\theta\tan\theta$ were the answer to a real-world problem, and if the next step was to evaluate the expression for numerous values of θ, it would certainly be easier to simply evaluate $\sec\theta$ instead.

Example 2: Simplifying Trigonometric Expressions

Simplify the expression $\cot\alpha + \dfrac{\sin\alpha}{1+\cos\alpha}$.

Solution

As in Example 1, rewriting the expression in terms of only sine and cosine is a good way to start. And as in so many problems, finding a common denominator and combining the resulting fractions is a promising way to proceed. After that, the remaining steps suggest themselves clearly.

$$\begin{aligned}
\cot\alpha + \frac{\sin\alpha}{1+\cos\alpha} &= \frac{\cos\alpha}{\sin\alpha} + \frac{\sin\alpha}{1+\cos\alpha} \\
&= \left(\frac{\cos\alpha}{\sin\alpha}\right)\left(\frac{1+\cos\alpha}{1+\cos\alpha}\right) + \left(\frac{\sin\alpha}{1+\cos\alpha}\right)\left(\frac{\sin\alpha}{\sin\alpha}\right) \\
&= \frac{\cos\alpha + \cos^2\alpha + \sin^2\alpha}{\sin\alpha(1+\cos\alpha)} \\
&= \frac{\cos\alpha + 1}{\sin\alpha(1+\cos\alpha)} \\
&= \frac{1}{\sin\alpha} \\
&= \csc\alpha
\end{aligned}$$

Multiply by appropriate factors to achieve a common denominator.

Apply the first Pythagorean identity.

Cancel the common factors.

TOPIC 3: Verifying Trigonometric Identities

The identities we have seen to this point are all fundamental, and in fact some are merely restatements of definitions. Nevertheless, they are also very useful and serve as good examples of the class of equations known as identities. Recall from Section 1.6 that an identity is an equation that is true for any (allowable) value(s) of the variable(s); this would be a good time to review the list of trigonometric identities at the start of this section and verify that those equations indeed fit this description.

Most equations that we encounter in algebra are not identities, and in fact the goal of much of our work is to determine exactly which values of the variable(s) make the equation true. In Section 1.6 equations that were not identities or contradictions were labeled *conditional*, but in practice the adjective is usually dropped since most equations we encounter are of this sort. We have already solved many simple conditional trigonometric equations, but we will study such equations in much greater depth in Section 8.4. At that time, we will see that trigonometric identities are very useful in determining the solutions of trigonometric equations.

Before we get to that point, though, we need to build up our repertoire of identities, and one step in doing so is to verify that a proposed identity really is true for all values of the variable. This is called *verifying an identity,* and the process is often similar to using identities to simplify expressions. While there is no guaranteed method to use in such verification, there are some general guidelines to follow.

PROCEDURE: Guidelines for Verifying Trigonometric Identities

1. **Work with one side at a time.** Choose one side of the equation to work with and simplify it. The more complicated side is usually the best choice. The goal is to transform it into the other side.
2. **Apply trigonometric identities as appropriate.** To do so, it will probably be necessary to combine fractions, add or subtract terms, factor expressions, or use other algebraic manipulations.
3. **Rewrite in terms of sine and cosine if necessary.** If you are stuck, expressing everything in terms of sine and cosine often leads to inspiration.

Example 3: Verifying Trigonometric Identities

Verify the identity $2\csc^2 x = \dfrac{1}{1-\cos x} + \dfrac{1}{1+\cos x}$.

Solution

The right-hand side is more complicated, so we'll begin with it. Combining the two fractions is a good way to begin, especially if we notice that a denominator of $1-\cos^2 x$ will eventually appear.

$$\frac{1}{1-\cos x} + \frac{1}{1+\cos x} = \frac{1+\cos x + 1 - \cos x}{(1-\cos x)(1+\cos x)}$$

Modify each fraction in order to obtain a common denominator, then combine.

$$= \frac{2}{1-\cos^2 x}$$

Multiply out the denominator.

$$= \frac{2}{\sin^2 x}$$

Apply the first Pythagorean identity.

$$= 2\csc^2 x$$

In verifying some identities, it may be easiest to simplify both sides of the equation individually. The goal in this case is to achieve a single simpler expression that is equivalent to both sides of the original equation.

Example 4: Verifying Trigonometric Identities

Verify the identity $\dfrac{\tan^2 x}{1+\sec x} = \dfrac{1-\cos x}{\cos x}$.

Solution

It's not immediately clear that either side is more complicated than the other, so we can try simplifying both. Beginning with the left-hand side, we can use another of the Pythagorean identities as follows.

$$\frac{\tan^2 x}{1+\sec x} = \frac{\sec^2 x - 1}{1+\sec x}$$

Use the second Pythagorean identity to rewrite the numerator.

$$= \frac{(\sec x - 1)(\sec x + 1)}{1+\sec x}$$

Factor the difference of two squares.

$$= \sec x - 1$$

Cancel the common factors.

The right-hand side is more easily dealt with.

$$\frac{1-\cos x}{\cos x} = \frac{1}{\cos x} - \frac{\cos x}{\cos x}$$
$$= \sec x - 1$$

Break the single fraction into two and simplify.

Since both sides of the original equation are equivalent to $\sec x - 1$, the identity is true.

Example 5: Verifying Trigonometric Identities

Verify the identity $\frac{\cos\varphi \cot\varphi}{1-\sin\varphi} - 1 = \csc\varphi$.

Solution

The left-hand side is clearly more complicated, and there are several ways of beginning the process of simplifying it. First, rewriting the numerator of the fraction in terms of sine and cosine looks promising, as a factor of $\cos^2\varphi$ would then appear. Second, obtaining a denominator of $1-\sin^2\varphi$ in the fraction would be easily done and would allow one of the Pythagorean identities to apply. We'll try both ideas.

$$\frac{\cos\varphi\cot\varphi}{1-\sin\varphi} - 1 = \frac{\cos\varphi\left(\frac{\cos\varphi}{\sin\varphi}\right)}{1-\sin\varphi} - 1$$

Rewrite cotangent in terms of sine and cosine.

$$= \left(\frac{\cos^2\varphi}{\sin\varphi(1-\sin\varphi)}\right)\left(\frac{1+\sin\varphi}{1+\sin\varphi}\right) - 1$$

Multiply appropriately in order to obtain $1-\sin^2\varphi$ in the denominator.

$$= \left(\frac{\cos^2\varphi}{\sin\varphi(1-\sin^2\varphi)}\right)(1+\sin\varphi) - 1$$

$$= \left(\frac{\cos^2\varphi}{\sin\varphi\cos^2\varphi}\right)(1+\sin\varphi) - 1$$

Apply the first Pythagorean identity.

$$= \frac{1+\sin\varphi}{\sin\varphi} - 1$$

Cancel the common factors.

$$= \frac{1}{\sin\varphi} + 1 - 1$$
$$= \csc\varphi$$

Break apart the fraction and simplify.

TOPIC 4: Trigonometric Substitutions

In several classes of calculus problems, it is very convenient to be able to replace certain algebraic expressions with trigonometric expressions. Most often, these replacements depend on one of the Pythagorean identities, and the work involved is reminiscent of that in earlier examples in this section. The next example illustrates a typical trigonometric substitution.

Example 6: Using Trigonometric Substitutions

Use the substitution $\sin\theta = \frac{x}{2}$ to write $\sqrt{4-x^2}$ as a trigonometric expression. Assume $0 \le \theta \le \frac{\pi}{2}$.

Solution

Although it is not necessary for the task at hand, a diagram motivating the substitution may be helpful. The triangle in Figure 3 illustrates the geometric relation between θ and the various algebraic expressions.

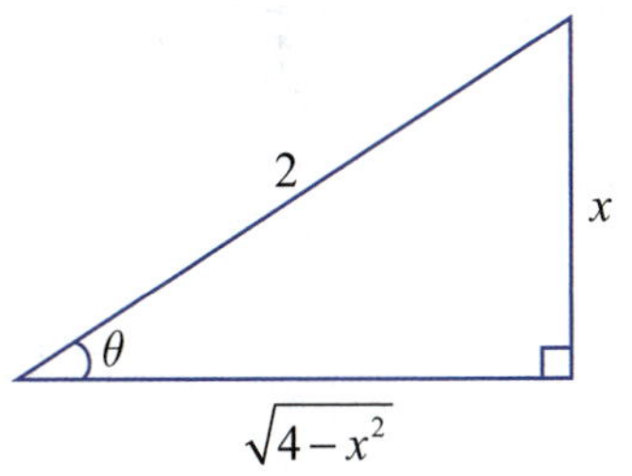

FIGURE 3

The suggested substitution can be rewritten as $x = 2\sin\theta$, and so we obtain the following.

$$\begin{aligned} \sqrt{4-x^2} &= \sqrt{4-(2\sin\theta)^2} \\ &= \sqrt{4-4\sin^2\theta} \\ &= 2\sqrt{1-\sin^2\theta} \\ &= 2\sqrt{\cos^2\theta} \\ &= 2\cos\theta \end{aligned}$$

We can write $2\cos\theta$ instead of $2|\cos\theta|$ since the restriction $0 \le \theta \le \frac{\pi}{2}$ means $\cos\theta \ge 0$.

8.1 EXERCISES

PRACTICE

Use trigonometric identities to simplify the expressions. There may be more than one correct answer. See Examples 1 and 2.

1. $\tan x \csc x$
2. $\dfrac{1}{\tan^2\theta + 1}$
3. $\dfrac{\tan t}{\sec t}$
4. $\cot^2 x - \cot^2 x \cos^2 x$
5. $\sin(-x)\tan x$
6. $\dfrac{1}{\sec^2 x} + \sin x \cos\left(\dfrac{\pi}{2} - x\right)$
7. $\sin(\alpha + 2\pi)\sec\alpha$
8. $\sin t(\csc t - \sin t)$
9. $\cos y(1 + \tan^2 y)$
10. $\dfrac{1}{\cos x \csc(-x)}$
11. $\dfrac{1 - \tan^2 x}{\cot^2 x - 1}$
12. $\dfrac{\sin\beta \tan\left(\dfrac{\pi}{2} - \beta\right)}{\cos\beta}$

Use the suggested substitution to rewrite the given expression as a trigonometric expression. See Example 6. Assume $0 \le \theta \le \frac{\pi}{2}$.

13. $\sqrt{x^2+1}, \quad x = \tan\theta$

14. $\sqrt{x^2-16}, \quad x = 4\sec\theta$

15. $\sqrt{9-x^2}, \quad \cos\theta = \frac{x}{3}$

16. $\sqrt{4x^2+100}, \quad \cot\theta = \frac{x}{5}$

17. $\sqrt{64-x^2}, \quad x = 8\sin\theta$

18. $\sqrt{x^2-4}, \quad x = 2\csc\theta$

19. $\sqrt{x^2+25}, \quad \tan\theta = \frac{x}{5}$

20. $\sqrt{144-9x^2}, \quad x = 4\cos\theta$

WRITING & THINKING

Verify the following trigonometric identities. See Examples 3, 4, and 5.

21. $(1-\cos\theta)(1+\cos\theta) = \sin^2\theta$

22. $\csc x - \sin x = \cos x \cot x$

23. $\sec^2 y - \tan^2 y = \sec y \cos(-y)$

24. $(1-\sin\beta)(\sec\beta\tan\beta) = \frac{\sin\beta}{1+\sin\beta}$

25. $\dfrac{\sin\left(\frac{\pi}{2}-x\right)}{\cos\left(\frac{\pi}{2}-x\right)} = \cot x$

26. $\dfrac{\sec^2\theta}{\tan\theta} = \sec\theta\csc\theta$

27. $\dfrac{1}{\tan x} + \tan x = \dfrac{\sec^2 x}{\tan x}$

28. $\sin^2 t + \sin^2\left(\frac{\pi}{2}-t\right) = 1$

29. $\dfrac{1}{\sin(\theta+2\pi)+1} + \dfrac{1}{\csc(\theta+2\pi)+1} = 1$

30. $3 + \cot^2\alpha = 2 + \csc^2\alpha$

31. $\sin^2 x - \sin^4 x = \cos^2(-x) - \cos^4(-x)$

32. $\cot\left(\frac{\pi}{2}-\beta\right)\cot\beta = 1$

33. $\dfrac{\cos\left(\frac{\pi}{2}-\alpha\right)}{\csc\alpha} - 1 = \sin\alpha\cot(-\alpha)\cos(-\alpha)$

Show how the identities below follow from the first Pythagorean identity.

34. $\tan^2 x + 1 = \sec^2 x$

35. $1 + \cot^2 x = \csc^2 x$

8.2 SUM AND DIFFERENCE IDENTITIES

TOPICS

1. Sum and difference identities
2. Using sum and difference identities for exact evaluation
3. Using sum and difference identities for verification and simplification

TOPIC 1: Sum and Difference Identities

This section and the next will introduce new identities, selected proofs of the identities, and many examples of how the identities can be used. The first group of identities concerns functions acting on sums or differences of angles.

IDENTITIES: Sum and Difference Identities

Sine Identities

$\sin(u+v)=\sin u\cos v+\cos u\sin v$ $\qquad$ $\sin(u-v)=\sin u\cos v-\cos u\sin v$

Cosine Identities

$\cos(u+v)=\cos u\cos v-\sin u\sin v$ $\qquad$ $\cos(u-v)=\cos u\cos v+\sin u\sin v$

Tangent Identities

$\tan(u+v)=\dfrac{\tan u+\tan v}{1-\tan u\tan v}$ $\qquad$ $\tan(u-v)=\dfrac{\tan u-\tan v}{1+\tan u\tan v}$

We will outline the proof of two of these identities here, and indicate in the exercises how the other four identities can then be derived.

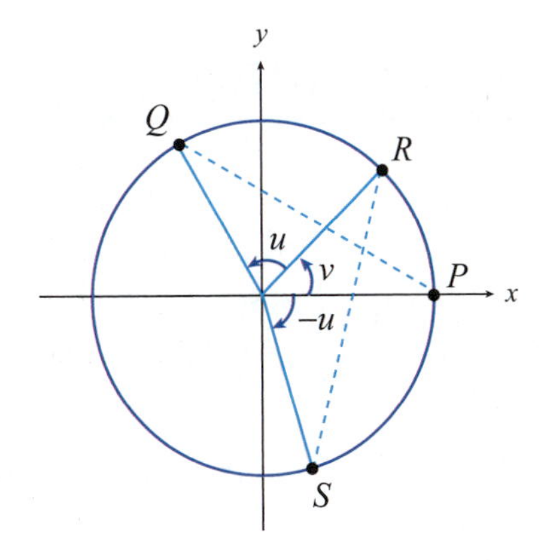

FIGURE 1: Derivation of Cosine Sum Identity

Consider the unit circle in Figure 1. In the figure, angle v has been placed in standard position and angle u has been positioned in such a way that angle $u+v$ is in standard position. In addition, the negative of angle u has also been drawn in standard position.

Using the fact that the radius of the circle is 1, we can identify the coordinates of the points P, Q, R, and S.

$$P=(1,0), Q=(\cos(u+v),\sin(u+v)), R=(\cos v,\sin v), \text{ and } S=(\cos u,-\sin u)$$

Note that for point S, we have used the fact that $\sin(-u)=-\sin u$ and $\cos(-u)=\cos u$. Note also that the chord $\overline{PQ}$ has the same length as the chord $\overline{RS}$, since the subtended angles both have magnitude $u+v$. Since we know the coordinates of the endpoints of the chords, we can use the distance formula to obtain the following equation.

$$\sqrt{(\cos(u+v)-1)^2+(\sin(u+v)-0)^2}=\sqrt{(\cos v-\cos u)^2+(\sin v+\sin u)^2}$$

The square roots are easily eliminated by squaring both sides, and we can then proceed to expand the squared terms.

$$\cos^2(u+v) - 2\cos(u+v) + 1 + \sin^2(u+v) =$$
$$\cos^2 v - 2\cos u \cos v + \cos^2 u + \sin^2 v + 2\sin u \sin v + \sin^2 u$$

By now, you should be attuned to the presence of $\sin^2(\ )$ and $\cos^2(\ )$ terms; every pair of these having the same argument can be replaced with 1. Making that replacement three times in the preceding equation gives us

$$2 - 2\cos(u+v) = 2 - 2\cos u \cos v + 2\sin u \sin v,$$

and subtracting 2 from both sides and then dividing by –2 yields

$$\cos(u+v) = \cos u \cos v - \sin u \sin v.$$

With this identity at our disposal, the rest follow quickly. In particular, the difference identity for cosine is very easily proved by replacing v with $-v$ in the sum identity and making use of the fact that cosine is an even function and sine is odd.

$$\begin{aligned}\cos(u-v) &= \cos u \cos(-v) - \sin u \sin(-v) \\ &= \cos u \cos v + \sin u \sin v\end{aligned}$$

TOPIC 2: Using Sum and Difference Identities for Exact Evaluation

It was fairly easy, back in Chapter 7, to determine the exact values of the trigonometric functions acting on the angles $\frac{\pi}{6}$, $\frac{\pi}{4}$, and $\frac{\pi}{3}$. It may seem odd, therefore, that these are still the only (acute) angles for which we can perform exact evaluation. To this point, for instance, our only option for evaluating $\sin 75°$ has been to use a calculator and note that $\sin 75° \approx 0.9659$. The sum and difference identities extend our ability to obtain exact values greatly, as seen in the next example.

Example 1: Using the Sum and Difference Identities for Exact Evaluation

Determine the exact value of $\sin 75°$.

Solution

All problems of this sort will call for us to express the given angle in terms of angles about which we know more. In this case, we'll use the fact that $75° = 45° + 30°$.

$$\begin{aligned}\sin 75° &= \sin(45° + 30°) \\ &= \sin 45° \cos 30° + \cos 45° \sin 30° \\ &= \left(\frac{1}{\sqrt{2}}\right)\left(\frac{\sqrt{3}}{2}\right) + \left(\frac{1}{\sqrt{2}}\right)\left(\frac{1}{2}\right) \\ &= \frac{\sqrt{3}+1}{2\sqrt{2}}\end{aligned}$$

You can use a calculator to verify that this exact value is approximately 0.9659.

Example 2: Using the Sum and Difference Identities for Exact Evaluation

Determine the exact value of $\cos 75°$.

Solution

The purpose of this example is twofold. The first is to point out that we are starting to build up a significant collection of tools and knowledge. The second is to emphasize that two identical answers may appear, superficially, to be different. We could certainly use the sum identity for cosine to evaluate $\cos 75°$, taking steps very similar to those in Example 1. Using this approach, we would obtain

$$\cos 75° = \frac{\sqrt{6}-\sqrt{2}}{4}.$$

But, coming immediately after the evaluation of $\sin 75°$, it would also make sense to use the identity $\cos^2 x + \sin^2 x = 1$ to obtain

$$\begin{aligned}\cos 75° &= \sqrt{1-\sin^2 75°}\\ &= \sqrt{1-\left(\frac{\sqrt{3}+1}{2\sqrt{2}}\right)^2}\\ &= \sqrt{1-\frac{4+2\sqrt{3}}{8}}\\ &= \sqrt{\frac{2-\sqrt{3}}{4}}\\ &= \frac{\sqrt{2-\sqrt{3}}}{2}.\end{aligned}$$

Are these two answers actually the same?

You can use a calculator to determine that both are approximately 0.2588, but it is more convincing to prove the equivalence mathematically.

First, note that since $\sqrt{2} < \sqrt{6}$ and $\sqrt{3} < \sqrt{4} = 2,$ we know that both

$$\frac{\sqrt{6}-\sqrt{2}}{4} > 0 \text{ and } \frac{\sqrt{2-\sqrt{3}}}{2} > 0.$$

Next, note that

$$\left(\frac{\sqrt{6}-\sqrt{2}}{4}\right)^2 = \frac{1}{16}\left(6-2\sqrt{12}+2\right) = \frac{1}{16}\left(8-4\sqrt{3}\right) = \frac{1}{4}\left(2-\sqrt{3}\right)$$

and

$$\left(\frac{\sqrt{2-\sqrt{3}}}{2}\right)^2 = \frac{1}{4}\left(2-\sqrt{3}\right),$$

so

$$\left(\frac{\sqrt{6}-\sqrt{2}}{4}\right)^2 = \left(\frac{\sqrt{2-\sqrt{3}}}{2}\right)^2.$$

Since the expressions being squared are both positive, taking the square root of both sides gives us the result

$$\frac{\sqrt{6}-\sqrt{2}}{4} = \frac{\sqrt{2-\sqrt{3}}}{2}.$$

Example 3: Using the Sum and Difference Identities for Exact Evaluation

Determine the exact value of $\tan\left(\frac{\pi}{12}\right)$.

Solution

To make use of previous results, we need to note that $\frac{\pi}{12} = \frac{\pi}{3} - \frac{\pi}{4}$.

$$\begin{aligned}\tan\left(\frac{\pi}{12}\right) &= \tan\left(\frac{\pi}{3} - \frac{\pi}{4}\right)\\ &= \frac{\tan\left(\frac{\pi}{3}\right) - \tan\left(\frac{\pi}{4}\right)}{1 + \tan\left(\frac{\pi}{3}\right)\tan\left(\frac{\pi}{4}\right)}\\ &= \frac{\sqrt{3} - 1}{1 + \left(\sqrt{3}\right)(1)}\\ &= \frac{\sqrt{3} - 1}{\sqrt{3} + 1}\end{aligned}$$

Although the need arises less often, the sum and difference identities can also be used in reverse.

Example 4: Using the Sum and Difference Identities for Exact Evaluation

Determine the exact value of $\sin 80° \cos 20° - \cos 80° \sin 20°$.

Solution

The key step is to recognize the expression as the difference identity for sine. We can then proceed as follows.

$$\begin{aligned}\sin 80° \cos 20° - \cos 80° \sin 20° &= \sin\left(80° - 20°\right)\\ &= \sin 60°\\ &= \frac{\sqrt{3}}{2}\end{aligned}$$

TOPIC 3: Using Sum and Difference Identities for Verification and Simplification

The truth of the cofunction identities can be easily seen through the use of a unit circle diagram, as in the argument at the start of this section. But the difference identities can also be used to furnish a very quick algebraic verification, as in the next example.

Example 5: Using the Sum and Difference Identities

Use a difference identity to verify that $\sin\left(\frac{\pi}{2}-x\right)=\cos x$.

Solution

$$\begin{aligned}\sin\left(\frac{\pi}{2}-x\right)&=\sin\left(\frac{\pi}{2}\right)\cos x-\cos\left(\frac{\pi}{2}\right)\sin x\\&=(1)(\cos x)-(0)(\sin x)\\&=\cos x\end{aligned}$$

The identities in this section are useful in simplifying trigonometric functions of sums or differences no matter how unwieldy they may initially appear to be.

Example 6: Using the Sum and Difference Identities

Evaluate the expression $\sin\left(\cos^{-1}\left(\frac{3}{5}\right)-\tan^{-1}\left(\frac{12}{5}\right)\right)$.

Solution

From our previous experience with compositions of trigonometric functions and inverse trigonometric functions, we know the general approach to take; the only additional complication here is that the argument of sine is a difference. If we let $u=\cos^{-1}\left(\frac{3}{5}\right)$ and $v=\tan^{-1}\left(\frac{12}{5}\right)$, then $\cos u=\frac{3}{5}$ and $\tan v=\frac{12}{5}$ and we are led to construct two right triangles as shown in Figure 2 and Figure 3.

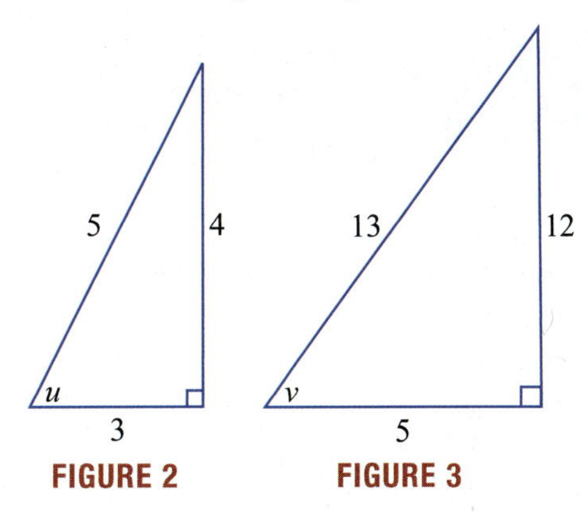

FIGURE 2 FIGURE 3

We can now proceed to apply a difference identity and evaluate, referring to the triangles we've constructed.

$$\begin{aligned}\sin\left(\cos^{-1}\left(\frac{3}{5}\right)-\tan^{-1}\left(\frac{12}{5}\right)\right)&=\sin(u-v)\\&=\sin u\cos v-\cos u\sin v\\&=\left(\frac{4}{5}\right)\left(\frac{5}{13}\right)-\left(\frac{3}{5}\right)\left(\frac{12}{13}\right)\\&=\frac{20}{65}-\frac{36}{65}\\&=-\frac{16}{65}\end{aligned}$$

Example 7: Using the Sum and Difference Identities

Express $\sin\left(\tan^{-1} x + \cos^{-1} x\right)$ as an algebraic function of x.

Solution

We follow the procedure that we used in Example 6—the fact that the expression contains a variable x instead of constants doesn't alter the approach. If we let $u = \tan^{-1} x$ and $v = \cos^{-1} x$, then $\tan u = x$ and $\cos v = x$ and we are led to consider two right triangles such as those in Figure 4 and Figure 5.

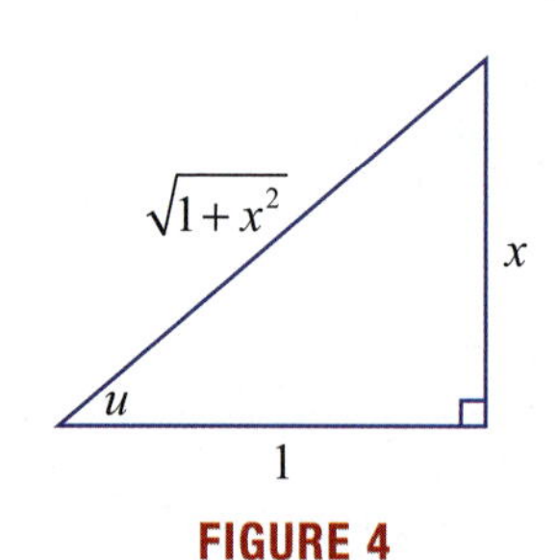

FIGURE 4

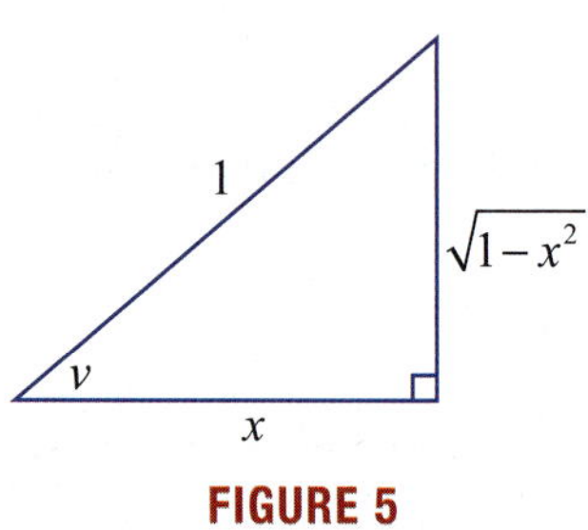

FIGURE 5

We can now apply the appropriate sum identity.

$$\begin{aligned}
\sin\left(\tan^{-1} x + \cos^{-1} x\right) &= \sin(u+v) \\
&= \sin u \cos v + \cos u \sin v \\
&= \left(\frac{x}{\sqrt{1+x^2}}\right)(x) + \left(\frac{1}{\sqrt{1+x^2}}\right)\left(\sqrt{1-x^2}\right) \\
&= \frac{x^2}{\sqrt{1+x^2}} + \frac{\sqrt{1-x^2}}{\sqrt{1+x^2}} \\
&= \frac{x^2 + \sqrt{1-x^2}}{\sqrt{1+x^2}}
\end{aligned}$$

A Trigonometric Difference Quotient

The left-hand side of the identity in Example 8 is one form of the difference quotient of the sine function. In order to find the slope of the line tangent to the graph of sine at the point (x, sin x), the limit of the difference quotient as $h \to 0$ must be considered. But, as is common, this cannot be done directly—in this case, a trigonometric identity must be employed to rewrite the difference quotient in a more useful form.

Example 8: Using the Sum and Difference Identities

Show that $\dfrac{\sin(x+h) - \sin x}{h} = \sin x\left(\dfrac{\cos h - 1}{h}\right) + \cos x\left(\dfrac{\sin h}{h}\right)$.

Solution

$$\begin{aligned}
\frac{\sin(x+h) - \sin x}{h} &= \frac{\sin x \cos h + \cos x \sin h - \sin x}{h} \\
&= \frac{\sin x \cos h - \sin x}{h} + \frac{\cos x \sin h}{h} \\
&= \sin x\left(\frac{\cos h - 1}{h}\right) + \cos x\left(\frac{\sin h}{h}\right)
\end{aligned}$$

The last example of an application of the sum/difference identities leads to a surprising conclusion. As you know, the graph of cosine has the same shape as the graph of sine shifted to the left by $\frac{\pi}{2}$ (one of the cofunction identities is the algebraic form of this statement). But the connection between cosine and sine goes deeper; as it turns out, a *sum* of a sine term and a cosine term can be rephrased in terms of sine or cosine alone. This is true even if the sine and cosine functions have different amplitudes.

To see this, consider the expression $A\sin x + B\cos x$, a sum of a sine curve with amplitude A and a cosine curve with amplitude B. This has some resemblance to the right-hand side of the sum identity for sine, but only if A can be replaced with a cosine term and B with a sine term. That is the inspiration for the following:

$$A\sin x + B\cos x = \sqrt{A^2+B^2}\left(\frac{A}{\sqrt{A^2+B^2}}\sin x + \frac{B}{\sqrt{A^2+B^2}}\cos x\right).$$

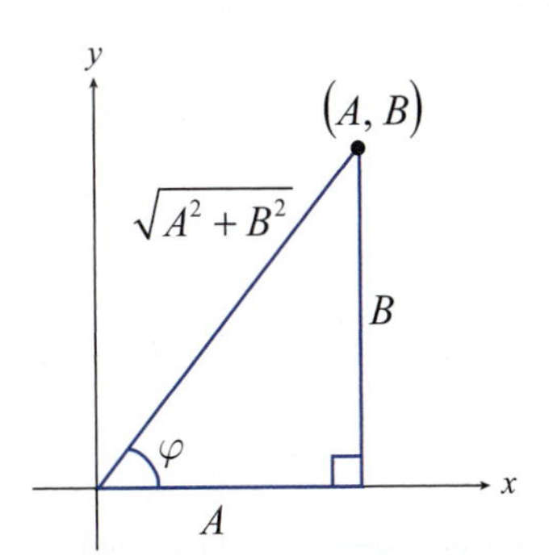

FIGURE 6: Determining the Phase Shift φ

If an angle φ exists for which

$$\cos\varphi = \frac{A}{\sqrt{A^2+B^2}} \quad \text{and} \quad \sin\varphi = \frac{B}{\sqrt{A^2+B^2}},$$

then the sum identity gives us

$$\begin{aligned} A\sin x + B\cos x &= \sqrt{A^2+B^2}\,(\cos\varphi\sin x + \sin\varphi\cos x) \\ &= \sqrt{A^2+B^2}\,\sin(x+\varphi). \end{aligned}$$

Does such a φ exist? Figure 6 makes it clear that φ can be determined.

THEOREM: Sum of Sines and Cosines

$$\begin{aligned} A\sin x + B\cos x &= \sqrt{A^2+B^2}\left(\frac{A}{\sqrt{A^2+B^2}}\sin x + \frac{B}{\sqrt{A^2+B^2}}\cos x\right) \\ &= \sqrt{A^2+B^2}\,\sin(x+\varphi) \end{aligned}$$

$$\text{where } \cos\varphi = \frac{A}{\sqrt{A^2+B^2}} \quad \text{and} \quad \sin\varphi = \frac{B}{\sqrt{A^2+B^2}}.$$

Example 9: Using the Sum of Sines and Cosines Theorem

Express the function $f(x) = \sin x - \sqrt{3}\cos x$ in terms of a single sine function, and graph the result.

Solution

Since $A = 1$ and $B = -\sqrt{3}$, $\sqrt{A^2+B^2} = 2$ and φ must satisfy

$$\cos\varphi = \frac{1}{2} \quad \text{and} \quad \sin\varphi = \frac{-\sqrt{3}}{2}.$$

This means φ lies in the fourth quadrant, and as a negative angle it can be expressed as $\varphi = -\frac{\pi}{3}$ (we could also write $\varphi = \frac{5\pi}{3}$). Using the above derivation,

$$f(x) = 2\sin\left(x - \frac{\pi}{3}\right),$$

and the graph of this function is shown in Figure 7.

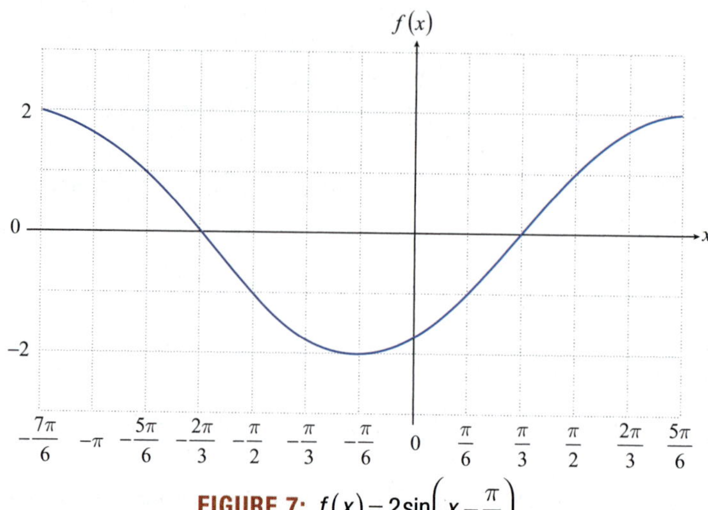

FIGURE 7: $f(x) = 2\sin\left(x - \frac{\pi}{3}\right)$

8.2 EXERCISES

PRACTICE

Use the sum and difference identities to determine the exact value of each of the following expressions. See Examples 1, 2, and 3.

1. $\cos\left(\frac{\pi}{4} + \frac{\pi}{3}\right)$

2. $\sin\left(\frac{\pi}{6} + \frac{3\pi}{4}\right)$

3. $\tan\left(\frac{4\pi}{3} + \frac{5\pi}{4}\right)$

4. $\sin\left(\frac{2\pi}{3} + \frac{\pi}{4}\right)$

5. $\tan\left(\frac{\pi}{3} - \frac{3\pi}{4}\right)$

6. $\tan\left(\frac{4\pi}{3} - \frac{5\pi}{4}\right)$

7. $\cos\left(\frac{\pi}{4} - \frac{\pi}{6}\right)$

8. $\sin\left(\frac{5\pi}{4} - \frac{\pi}{3}\right)$

9. $\sin\left(\frac{7\pi}{4} + \frac{2\pi}{3}\right)$

10. $\cos\left(\frac{\pi}{3} - \frac{\pi}{4}\right)$

11. $\tan 75°$

12. $\tan 15°$

13. $\sin 165°$

14. $\cos(-15°)$

15. $\tan 255°$

16. $\cos 195°$

17. $\cos 165°$

18. $\sin\left(\frac{\pi}{12}\right)$

19. $\tan\left(\frac{5\pi}{12}\right)$

20. $\cos\left(\frac{7\pi}{12}\right)$

21. $\cos\left(\frac{25\pi}{12}\right)$

22. $\sin\left(\frac{13\pi}{12}\right)$

23. $\sin\left(\frac{11\pi}{12}\right)$

24. $\tan\left(\frac{7\pi}{12}\right)$

25. $\sin\left(\frac{5\pi}{12}\right)$

26. $\tan\left(\frac{\pi}{12}\right)$

27. Suppose that $\sin\alpha = \frac{4}{5}$ and $\sin\beta = \frac{5}{13}$ and both α and β are in quadrant I. Find $\cos(\alpha - \beta)$.

28. Suppose that $\sin\alpha = -\frac{15}{17}$ and $\cos\beta = -\frac{3}{5}$ and the terminal sides of both α and β are in quadrant III. Find $\sin(\alpha - \beta)$.

29. Suppose that $\cos\alpha = -\frac{15}{17}$ and $\cos\beta = -\frac{3}{5}$, the terminal side of α is in quadrant II, and the terminal side of β is in quadrant III. Find $\sin(\alpha + \beta)$.

30. Suppose that $\cos\alpha = -\frac{24}{25}$ and $\sin\beta = \frac{5}{13}$, the terminal side of α is in quadrant III, and β is in quadrant I. Find $\cos(\alpha + \beta)$.

31. Suppose that $\cos\alpha = \frac{2}{5}$ and $\cos\beta = \frac{1}{5}$ and both α and β are in quadrant I. Find $\sin(\beta - \alpha)$.

32. Suppose that $\cos\alpha = -\frac{2}{3}$ and $\sin\beta = -\frac{2\sqrt{2}}{3}$, the terminal side of α is in quadrant III, and the terminal side of β is in quadrant IV. Find $\tan(\alpha + \beta)$.

Use the sum and difference identities to rewrite each of the following expressions as a trigonometric function of one angle, and then evaluate the result. See Example 4.

33. $\sin 15° \cos 30° + \cos 15° \sin 30°$

34. $\cos\left(\frac{5\pi}{12}\right)\cos\left(\frac{2\pi}{3}\right) + \sin\left(\frac{5\pi}{12}\right)\sin\left(\frac{2\pi}{3}\right)$

35. $\dfrac{\tan 100° + \tan 35°}{1 - \tan 100° \tan 35°}$

36. $\sin 125° \cos 35° - \cos 125° \sin 35°$

37. $\dfrac{\tan\left(\frac{5\pi}{16}\right) - \tan\left(\frac{\pi}{16}\right)}{1 + \tan\left(\frac{5\pi}{16}\right)\tan\left(\frac{\pi}{16}\right)}$

38. $\cos 15° \cos 15° - \sin 15° \sin 15°$

39. $\sin 70° \cos 80° + \cos 70° \sin 80°$

40. $\cos\left(\frac{\pi}{5}\right)\cos\left(\frac{3\pi}{10}\right) - \sin\left(\frac{\pi}{5}\right)\sin\left(\frac{3\pi}{10}\right)$

41. $\cos 182° \cos 47° + \sin 182° \sin 47°$

42. $\dfrac{\tan\left(\frac{5\pi}{12}\right) + \tan\left(\frac{3\pi}{4}\right)}{1 - \tan\left(\frac{5\pi}{12}\right)\tan\left(\frac{3\pi}{4}\right)}$

43. $\dfrac{\tan 70° - \tan 10°}{1 + \tan 70° \tan 10°}$

44. $\sin\left(\frac{5\pi}{12}\right)\cos\left(\frac{\pi}{12}\right)-\cos\left(\frac{5\pi}{12}\right)\sin\left(\frac{\pi}{12}\right)$

Evaluate each of the following expressions. See Example 6.

45. $\tan\left(\cos^{-1}\left(\frac{3}{5}\right)+\sin^{-1}\left(\frac{12}{13}\right)\right)$

46. $\cos\left(\arctan 1+\arccos\left(\frac{1}{2}\right)\right)$

47. $\sin\left(\arctan \sqrt{3}+\arctan\left(\frac{4}{3}\right)\right)$

48. $\tan\left(\sin^{-1}\left(\frac{1}{\sqrt{2}}\right)-\cos^{-1}\left(\frac{5}{13}\right)\right)$

Express each of the following as an algebraic function of *x*. See Example 7.

49. $\sin\left(\sin^{-1}(2x)+\cos^{-1}(2x)\right)$

50. $\sin\left(\arctan(2x)-\arccos(2x)\right)$

51. $\cos\left(\arctan(2x)-\arcsin x\right)$

52. $\cos\left(\cos^{-1}x-\sin^{-1}x\right)$

53. $\cos\left(\arccos x+\arcsin(2x)\right)$

54. $\sin\left(\sin^{-1}x-\cos^{-1}x\right)$

Express each of the following functions in terms of a single sine function, and graph the result. See Example 9.

55. $f(x)=\sin x+\cos x$

56. $g(x)=\sin x+\sqrt{3}\cos x$

57. $h(\beta)=\sin(2\beta)-\cos(2\beta)$

58. $f(\theta)=-\sqrt{3}\sin\theta+\cos\theta$

59. $g(u)=5\sin(5u)+12\cos(5u)$

60. $h(v)=8\cos\left(\frac{v}{2}\right)+6\sin\left(\frac{v}{2}\right)$

WRITING & THINKING

Use the sum and difference identities to verify the following identities. See Example 5.

61. $\tan\left(\frac{\pi}{2}-\theta\right)=\cot\theta$ (**Hint:** Use sine and cosine.)

62. $\cos^2 u-\sin^2 v=\cos(u+v)\cos(u-v)$

63. $\cos\left(\frac{3\pi}{2}-\alpha\right)=-\sin\alpha$

64. $\sin(\beta-\theta)+\sin(\beta+\theta)=2\sin\beta\cos\theta$

65. $\tan\left(\alpha-\frac{5\pi}{4}\right)=\frac{\tan\alpha-1}{1+\tan\alpha}$

66. $\sec\left(\frac{\pi}{2}-u\right)=\csc u$

67. $\tan(\pi+2\pi)=0$

68. $\sin\left(\frac{5\pi}{6}+\theta\right)=\frac{1}{2}\left(\cos\theta-\sqrt{3}\sin\theta\right)$

69. $\sin(u+v)\sin(u-v)=\sin^2 u-\sin^2 u$

70. $\cos\left(\frac{7\pi}{4}-\beta\right)=\frac{\sqrt{2}}{2}(\cos\beta-\sin\beta)$

71. Use a cofunction identity to prove the sum and difference identities for sine. (**Hint:** Note that $\sin(u+v)=\cos\left(\frac{\pi}{2}-(u+v)\right)=\cos\left(\left(\frac{\pi}{2}-u\right)-v\right)$ and apply the difference identity for cosine.)

72. Given sum identities for sine and cosine, prove the sum identity for tangent.

73. Prove or disprove that $\sin(u+v)+\sin(u-v)=2\sin u\cos v$.

74. Prove or disprove that $\dfrac{\cos(u+v)}{\cos u \cos v}=\tan u+\tan v$.

75. Prove or disprove that $\dfrac{\cos(u-v)}{\cos(u+v)}=2\tan u \tan v$.

76. Prove or disprove that $\dfrac{\sin(u+v)}{\sin(u-v)}=\dfrac{\tan u+\tan v}{\tan u-\tan v}$.

77. Use the sine and cosine difference formulas to prove $\tan(u-v)=\dfrac{\tan u-\tan v}{1+\tan u \tan v}$.

TECHNOLOGY

Using a graphing utility, determine whether the following identities are true or false. (**Hint:** Graph both expressions on each side of the equality separately and determine if the graphs coincide.)

78. $\sin\left(\dfrac{\pi}{2}-\theta\right)=\cos\theta$

79. $\cos\left(\theta-\dfrac{3\pi}{2}\right)=-\sin\theta$

80. $\cot(\pi+\theta)=-\tan\theta$

81. $\sec\left(\dfrac{\pi}{2}-\theta\right)=\csc\theta$

82. $\sin\left(\dfrac{\pi}{6}+\theta\right)=\dfrac{1}{2}\left(\cos\theta+\sqrt{3}\sin\theta\right)$

83. $\dfrac{1+\tan\theta}{1-\tan\theta}=-\tan\theta$

8.3 PRODUCT-SUM IDENTITIES

TOPICS

1. Double-angle identities
2. Power-reducing identities
3. Half-angle identities
4. Product-to-sum and sum-to-product identities

TOPIC 1: Double-Angle Identities

We will enlarge our repertoire greatly in this section, introducing and using four useful classes of identities. We will also prove a few selected identities in order to demonstrate typical methods of proof.

Our first class of identities contains those in which the argument of a given trigonometric function is twice an angle, and the identities are thus commonly called *double-angle identities.*

IDENTITIES: Double-Angle Identities

Sine Identity: $\sin(2u) = 2\sin u \cos u$

Cosine Identities:

$$\begin{aligned}\cos(2u) &= \cos^2 u - \sin^2 u \\ &= 2\cos^2 u - 1 \\ &= 1 - 2\sin^2 u\end{aligned}$$

Tangent Identity: $\tan(2u) = \dfrac{2\tan u}{1 - \tan^2 u}$

These identities are easily verified using the sum identities of the previous section. We will prove two of the cosine identities here, and leave the remainder of the proofs as exercises. For the first, note that

$$\begin{aligned}\cos(2u) &= \cos(u + u) \\ &= \cos u \cos u - \sin u \sin u \\ &= \cos^2 u - \sin^2 u.\end{aligned}$$

And for the second cosine identity, we can apply a Pythagorean identity:

$$\begin{aligned}\cos(2u) &= \cos^2 u - \sin^2 u \\ &= \cos^2 u - \left(1 - \cos^2 u\right) \\ &= 2\cos^2 u - 1.\end{aligned}$$

Example 1: Using the Double-Angle Identities

Given that $\cos x = -\frac{2}{\sqrt{5}}$ and that $\sin x$ is positive, determine $\cos(2x)$, $\sin(2x)$, and $\tan(2x)$.

Solution

First, note that since $\cos x$ is negative and $\sin x$ is positive, we know that x lies in quadrant II. The Pythagorean Theorem (or the first Pythagorean identity) then tells us that $\sin x = \frac{1}{\sqrt{5}}$, as follows:

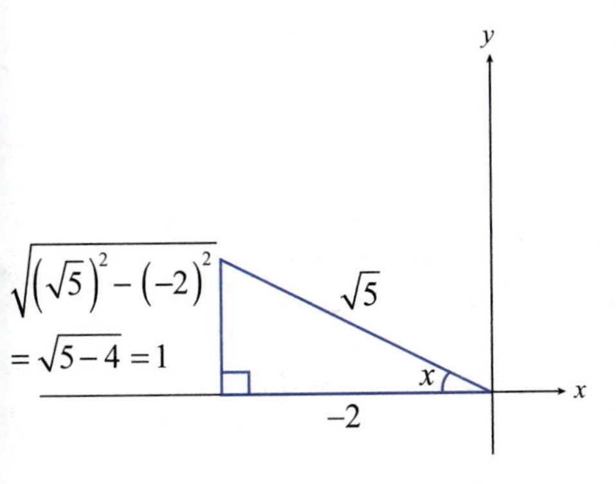

FIGURE 1

$$\sin x = \sqrt{1-\left(-\frac{2}{\sqrt{5}}\right)^2} = \sqrt{1-\frac{4}{5}} = \frac{1}{\sqrt{5}}.$$

We can now easily apply two of the double-angle identities:

$$\cos(2x) = \cos^2 x - \sin^2 x = \left(-\frac{2}{\sqrt{5}}\right)^2 - \left(\frac{1}{\sqrt{5}}\right)^2 = \frac{4}{5} - \frac{1}{5} = \frac{3}{5}$$

and

$$\sin(2x) = 2\sin x\cos x = 2\left(\frac{1}{\sqrt{5}}\right)\left(-\frac{2}{\sqrt{5}}\right) = -\frac{4}{5}.$$

We can now determine $\tan(2x)$ by noting that $\tan(2x) = \frac{\sin(2x)}{\cos(2x)} = -\frac{4}{3}$, but we could also have determined that $\tan x = -\frac{1}{2}$ and used the tangent double-angle identity as follows:

$$\tan(2x) = \frac{2\tan x}{1-\tan^2 x} = \frac{2\left(-\frac{1}{2}\right)}{1-\left(-\frac{1}{2}\right)^2} = \frac{-1}{\frac{3}{4}} = -\frac{4}{3}.$$

The double-angle identities (in conjunction with others) can be used to prove new identities, as shown in the next example.

Example 2: Using Trigonometric Identities

Prove that $\sin(3x) = 3\sin x - 4\sin^3 x$.

Solution

$\sin 3x = \sin(2x + x)$	Rewrite the argument as a sum.
$= \sin(2x)\cos x + \cos(2x)\sin x$	Apply the sum identity for sine.
$= 2\sin x\cos x\cos x + (1-2\sin^2 x)\sin x$	Apply two of the double-angle identities.
$= 2\sin x\cos^2 x + \sin x - 2\sin^3 x$	
$= 2\sin x(1-\sin^2 x) + \sin x - 2\sin^3 x$	Apply the first Pythagorean identity.
$= 2\sin x - 2\sin^3 x + \sin x - 2\sin^3 x$	
$= 3\sin x - 4\sin^3 x$	

Not surprisingly, the same sort of argument can be used on any expression of the form $\sin(nx)$ or $\cos(nx)$, where n is a positive integer. In fact, $\cos(nx)$ can be rewritten as an nth-degree polynomial in $\cos x$ (the polynomials are called *Chebyshev polynomials*, after the 19th-century Russian Pafnuty Lvovich Chebyshev, a mathematician famous for, among many other things, the variety of spellings of his last name). A similar statement is almost true for $\sin(nx)$; see the exercises for more on this topic.

TOPIC 2: Power-Reducing Identities

The double-angle identities just introduced are useful when $\sin x$, $\cos x$, or $\tan x$ is known and an expression containing $\sin(nx)$, $\cos(nx)$, or $\tan(nx)$ (for some positive integer n) must be dealt with. But they are also useful when faced with an expression containing $\sin^2 x$, $\cos^2 x$, or $\tan^2 x$. Quite often in calculus, for instance, it's very helpful to be able to lower the power of a trigonometric expression—that is, to be able to replace, say, $\cos^2 x$, with an expression containing only cosine to the first power. The following three identities follow quickly from the double-angle identities and accomplish the power-reducing task.

IDENTITIES: Power-Reducing Identities

Sine Identity: $$\sin^2 x = \frac{1-\cos(2x)}{2}$$

Cosine Identity: $$\cos^2 x = \frac{1+\cos(2x)}{2}$$

Tangent Identity: $$\tan^2 x = \frac{1-\cos(2x)}{1+\cos(2x)}$$

The first identity is easily proved by beginning with one of the double-angle identities for cosine.

$$\cos(2x) = 1 - 2\sin^2 x$$
$$2\sin^2 x = 1 - \cos(2x)$$
$$\sin^2 x = \frac{1-\cos(2x)}{2}$$

We begin with the one double-angle identity that contains $\sin^2 x$ and no other squared terms, and then proceed to solve for $\sin^2 x$.

The power-reducing identities can also be used to reduce powers other than 2, as shown in the next example.

Example 3: Using the Power-Reducing Identities

Express $\sin^5 x$ in terms containing only first powers of sine and cosine.

Solution

The procedure is to break the expression down into pieces to which the power-reducing identities apply, repeating the process if necessary.

$$\begin{aligned}
\sin^5 x &= \sin^2 x \sin^2 x \sin x \\
&= \left(\frac{1-\cos(2x)}{2}\right)\left(\frac{1-\cos(2x)}{2}\right)\sin x \\
&= \left(\frac{1-2\cos(2x)+\cos^2(2x)}{4}\right)\sin x \\
&= \frac{1}{4}\left(\sin x - 2\cos(2x)\sin x + \cos^2(2x)\sin x\right) \\
&= \frac{1}{4}\left(\sin x - 2\cos(2x)\sin x + \left(\frac{1+\cos(4x)}{2}\right)\sin x\right) \\
&= \frac{1}{4}\left(\sin x - 2\cos(2x)\sin x + \frac{\sin x}{2} + \frac{\cos(4x)\sin x}{2}\right) \\
&= \frac{1}{4}\left(\frac{3\sin x}{2} - 2\cos(2x)\sin x + \frac{\cos(4x)\sin x}{2}\right)
\end{aligned}$$

TOPIC 3: Half-Angle Identities

We have reached the point where we have enough identities to accomplish most of the common trigonometric tasks that may present themselves. For instance, if we needed the exact evaluation of $\sin\left(\frac{\pi}{8}\right)$, $\cos\left(\frac{\pi}{8}\right)$, or $\tan\left(\frac{\pi}{8}\right)$, we could start with an appropriate power-reducing identity and make use of the fact that twice $\frac{\pi}{8}$ is $\frac{\pi}{4}$, one of the "nice" angles. The following box contains general identities relating sine, cosine, and tangent of half an angle to expressions involving the whole angle. After proving one of the identities, we will proceed to determine the exact values of $\sin\left(\frac{\pi}{8}\right)$, $\cos\left(\frac{\pi}{8}\right)$, and $\tan\left(\frac{\pi}{8}\right)$.

IDENTITIES: Half-Angle Identities

Sine Identity: $\sin\left(\frac{x}{2}\right) = \pm\sqrt{\frac{1-\cos x}{2}}$

Cosine Identity: $\cos\left(\frac{x}{2}\right) = \pm\sqrt{\frac{1+\cos x}{2}}$

Tangent Identities: $\tan\left(\frac{x}{2}\right) = \frac{1-\cos x}{\sin x} = \frac{\sin x}{1+\cos x}$

The proofs of the first two half-angle identities follow from the corresponding power-reducing identities, with the sign of the result being determined by the quadrant in which $\frac{x}{2}$ lies (see the exercises for guidance in the proofs). For the tangent half-angle identity, we begin by replacing x with $\frac{x}{2}$ in the power-reducing identity and proceed as follows.

$$\tan^2\left(\frac{x}{2}\right)=\frac{1-\cos\left(2\left(\frac{x}{2}\right)\right)}{1+\cos\left(2\left(\frac{x}{2}\right)\right)}$$ Apply the tangent power-reducing identity with an argument of $\frac{x}{2}$.

$$\tan\left(\frac{x}{2}\right)=\pm\sqrt{\frac{1-\cos x}{1+\cos x}}$$ Simplify the right-hand side and take the square root of both sides.

$$\tan\left(\frac{x}{2}\right)=\pm\sqrt{\left(\frac{1-\cos x}{1+\cos x}\right)\left(\frac{1-\cos x}{1-\cos x}\right)}$$ Multiply the numerator and denominator by $1-\cos x$.

$$\tan\left(\frac{x}{2}\right)=\pm\sqrt{\frac{(1-\cos x)^2}{1-\cos^2 x}}$$ Expand the denominator of the right-hand side and apply the Pythagorean identity.

$$\tan\left(\frac{x}{2}\right)=\pm\frac{|1-\cos x|}{|\sin x|}$$ Simplify the radical.

All that remains is to simplify the formula a bit. First, note that $1-\cos x$ is always nonnegative, so the absolute value in the numerator is unnecessary. The next step is to note that $\tan\left(\frac{x}{2}\right)$ and $\sin x$ have the same sign for any value of x; the verification of this is left to the reader (consider individually the four cases of $0\le\frac{x}{2}<\frac{\pi}{2}$, $\frac{\pi}{2}<\frac{x}{2}\le\pi$, $-\frac{\pi}{2}<\frac{x}{2}\le 0$, and $-\pi\le\frac{x}{2}<-\frac{\pi}{2}$). This means that if the absolute value in the denominator is removed, then the $\pm$ in the formula can also be discarded, since its only purpose is to remind us to choose the correct sign depending on the quadrant in which $\frac{x}{2}$ lies. We are thus led to the desired result; proving the second formula in the tangent half-angle identity is left as an exercise.

Example 4: Using the Half-Angle Identities

Determine the exact values of $\sin\left(\frac{\pi}{8}\right)$, $\cos\left(\frac{\pi}{8}\right)$, and $\tan\left(\frac{\pi}{8}\right)$.

Solution

Since $\cos\left(\frac{\pi}{4}\right)=\sin\left(\frac{\pi}{4}\right)=\frac{\sqrt{2}}{2}$, we can easily determine that

$$\sin\left(\frac{\pi}{8}\right)=\sqrt{\frac{1-\cos\left(\frac{\pi}{4}\right)}{2}}=\sqrt{\frac{1-\frac{\sqrt{2}}{2}}{2}}=\sqrt{\frac{2-\sqrt{2}}{4}}=\frac{1}{2}\sqrt{2-\sqrt{2}},$$

$$\cos\left(\frac{\pi}{8}\right)=\sqrt{\frac{1+\cos\left(\frac{\pi}{4}\right)}{2}}=\sqrt{\frac{1+\frac{\sqrt{2}}{2}}{2}}=\sqrt{\frac{2+\sqrt{2}}{4}}=\frac{1}{2}\sqrt{2+\sqrt{2}},$$

and

$$\tan\left(\frac{\pi}{8}\right)=\frac{1-\cos\left(\frac{\pi}{4}\right)}{\sin\left(\frac{\pi}{4}\right)}=\frac{1-\frac{\sqrt{2}}{2}}{\frac{\sqrt{2}}{2}}=\frac{2\sqrt{2}-2}{2}=\sqrt{2}-1.$$

We could also have used the fact that $\tan x = \dfrac{\sin x}{\cos x}$ for the last calculation. Verifying that the two seemingly different answers obtained are actually equivalent is left to the reader.

Example 5: Using the Half-Angle Identities for Exact Evaluation

Determine the exact value of $\cos 105°$.

Solution

Once we note that $105°$ is half of $210°$, and recall that the reference angle of $210°$ is $30°$, the calculation is straightforward. We must also be careful, however, to insert a minus sign before the radical, as $105°$ lies in the second quadrant and $\cos 105°$ is therefore negative.

$$\cos 105° = -\sqrt{\frac{1+\cos 210°}{2}} = -\sqrt{\frac{1-\frac{\sqrt{3}}{2}}{2}} = -\frac{1}{2}\sqrt{2-\sqrt{3}}$$

TOPIC 4: Product-to-Sum and Sum-to-Product Identities

The final set of identities in this section is easily verified using the sum and difference identities, and the proofs are left to the reader. The following box contains identities that allow us to rewrite certain products of trigonometric functions as sums or differences of functions.

IDENTITIES: Product-to-Sum Identities

$$\sin x \cos y = \frac{1}{2}\left[\sin(x+y)+\sin(x-y)\right]$$

$$\cos x \sin y = \frac{1}{2}\left[\sin(x+y)-\sin(x-y)\right]$$

$$\sin x \sin y = \frac{1}{2}\left[\cos(x-y)-\cos(x+y)\right]$$

$$\cos x \cos y = \frac{1}{2}\left[\cos(x+y)+\cos(x-y)\right]$$

Example 6: Using the Product-to-Sum Identities

Express $\sin^5 x$ in terms containing only first powers of sine.

Solution

In Example 3, we used a power-reducing identity to determine that

$$\sin^5 x = \frac{1}{4}\left(\frac{3\sin x}{2} - 2\cos(2x)\sin x + \frac{\cos(4x)\sin x}{2}\right).$$

We can now use one of the product-to-sum identities to rewrite two of the terms in this expression, as follows.

$$\begin{aligned}\sin^5 x &= \frac{1}{4}\left[\frac{3\sin x}{2} - 2\cos(2x)\sin x + \frac{\cos(4x)\sin x}{2}\right]\\ &= \frac{1}{4}\left[\frac{3\sin x}{2} - 2\left(\frac{1}{2}(\sin(3x) - \sin x)\right) + \frac{1}{2}\left(\frac{1}{2}(\sin(5x) - \sin(3x))\right)\right]\\ &= \frac{3}{8}\sin x - \frac{1}{4}\sin(3x) + \frac{1}{4}\sin x + \frac{1}{16}\sin(5x) - \frac{1}{16}\sin(3x)\\ &= \frac{5}{8}\sin x - \frac{5}{16}\sin(3x) + \frac{1}{16}\sin(5x)\end{aligned}$$

Consider, now, the first product-to-sum identity, with u and v in place of x and y:

$$\sin u \cos v = \frac{1}{2}\left[\sin(u+v) + \sin(u-v)\right].$$

If we replace u with $\frac{x+y}{2}$ and v with $\frac{x-y}{2}$, we obtain

$$\sin\left(\frac{x+y}{2}\right)\cos\left(\frac{x-y}{2}\right) = \frac{1}{2}(\sin x + \sin y).$$

Multiplying this last equation through by 2 gives us an identity that allows us to rewrite a sum of sine functions as a product of a sine and cosine function, and it is the first of the four sum-to-product identities in the following box.

IDENTITIES: Sum-to-Product Identities

$$\sin x + \sin y = 2\sin\left(\frac{x+y}{2}\right)\cos\left(\frac{x-y}{2}\right)$$

$$\sin x - \sin y = 2\cos\left(\frac{x+y}{2}\right)\sin\left(\frac{x-y}{2}\right)$$

$$\cos x + \cos y = 2\cos\left(\frac{x+y}{2}\right)\cos\left(\frac{x-y}{2}\right)$$

$$\cos x - \cos y = -2\sin\left(\frac{x+y}{2}\right)\sin\left(\frac{x-y}{2}\right)$$

Example 7: Using the Sum-to-Product Identities

Verify the identity $\frac{\sin(2x)+\sin(4x)}{\cos(2x)+\cos(4x)} = \tan(3x)$.

Solution

We can apply two of the sum-to-product identities to verify the statement.

$$\frac{\sin(2x)+\sin(4x)}{\cos(2x)+\cos(4x)} = \frac{2\sin\left(\frac{6x}{2}\right)\cos\left(\frac{-2x}{2}\right)}{2\cos\left(\frac{6x}{2}\right)\cos\left(\frac{-2x}{2}\right)} = \frac{\sin(3x)}{\cos(3x)} = \tan(3x)$$

8.3 EXERCISES

PRACTICE

Use the given information to determine $\cos(2x)$, $\sin(2x)$, and $\tan(2x)$, if possible. See Example 1.

1. $\sin x = \frac{3}{5}$ and $\cos x$ is positive
2. $\tan x = -4$ and $\sin x$ is negative
3. $\cos x = -\frac{2}{\sqrt{6}}$ and $\sin x$ is positive
4. $\sin x = \frac{1}{\sqrt{5}}$ and $\tan x$ is positive
5. $\tan x = \frac{1}{\sqrt{3}}$ and $\cos x$ is negative
6. $\cos x = -3$ and $\tan x$ is negative

Use a power-reducing identity to rewrite the given expression as directed. See Example 3.

7. Rewrite $\sin^3 x$ in terms containing only first powers of sine and cosine.
8. Rewrite $\sin^4 x$ in terms containing only first powers of cosine.
9. Rewrite $\sin^4 x \cos^2 x$ in terms containing only first powers of cosine.
10. Rewrite $\cos^3 x \sin^2 x$ in terms containing only first powers of cosine.
11. Rewrite $\tan^4 x \sin x$ in terms containing only first powers of sine and cosine.
12. Rewrite $\sin^8 x$ in terms containing only first powers of cosine.

Determine the exact value of each of the following expressions. See Examples 4 and 5.

13. $\sin\left(\frac{3\pi}{8}\right)$
14. $\tan(112°30')$
15. $\cos\left(-\frac{\pi}{12}\right)$
16. $\tan\left(\frac{7\pi}{12}\right)$
17. $\sin 75°$
18. $\cos 165°$

Use the product-to-sum identities to rewrite the given expression as a sum or difference. See Example 6.

19. $\sin(3x)\cos(3x)$
20. $\cos\left(\frac{\pi}{4}\right)\sin\left(\frac{\pi}{4}\right)$
21. $5 \cos 105° \sin 15°$
22. $2 \cos 75° \cos 45°$
23. $\sin(x+y)\sin(x-y)$
24. $\sin\left(\frac{5\pi}{6}\right)\cos\left(\frac{\pi}{6}\right)$
25. $\sin\left(\frac{5\pi}{4}\right)\sin\left(\frac{2\pi}{3}\right)$
26. $\cos\beta \cos(3\beta)$
27. $2\cos\left(\frac{\pi}{3}\right)\sin\left(\frac{\pi}{6}\right)$

Use the sum-to-product identities to rewrite the given expression as a product. See Example 7.

28. $\sin\left(\frac{\pi}{4}\right)+\sin\left(\frac{3\pi}{4}\right)$

29. $\sin(6x)+\sin(2x)$

30. $\cos 60° + \cos 30°$

31. $\cos(3\beta)-\cos\beta$

32. $\sin\pi-\sin\left(\frac{\pi}{2}\right)$

33. $\sin 135° - \sin 15°$

34. $\cos(6x)-\cos(2x)$

35. $\cos\left(\frac{7\pi}{6}\right)-\cos\left(\frac{\pi}{4}\right)$

36. $\sin(\pi+\theta)+\sin(\pi-\theta)$

WRITING & THINKING

Verify the following trigonometric identities. See Example 2.

37. $\tan(3x)=\frac{3\tan x-\tan^3 x}{1-3\tan^2 x}$

38. $\sin(2x)=\frac{2\tan x}{1+\tan^2 x}$

39. $\frac{\sin(4x)-\sin(2x)}{\cos(4x)+\cos(2x)}=\tan x$

40. $\frac{\sin(3x)}{\sin x}=3-4\sin^2 x$

41. $2\sin^2(3x)=1-\cos(6x)$

42. $\sin(3x)=3\sin x\cos^2 x-\sin^3 x$

43. Two of the double-angle identities were proved in this section. Prove the remaining three double-angle identities.

44. The power-reducing identity for sine was proved in this section. Prove the remaining two power-reducing identities.

45. Prove the half-angle identities for sine and cosine by replacing x with $\frac{x}{2}$ in an appropriately chosen identity.

46. One of the formulas in the tangent half-angle identity was proved in this section. Prove the second formula in the tangent half-angle identity.

47. As mentioned in this section, $\cos(nx)$ can be expressed as a polynomial of degree n in $\cos x$; such polynomials are called Chebyshev polynomials. For $\sin(nx)$, the equivalent rewriting is a product of $\sin x$ and a polynomial of degree $n-1$ in $\cos x$. Expand $\sin(nx)$ and $\cos(nx)$ for $n = 2$, 3, and 4 and compare the results.

8.4 TRIGONOMETRIC EQUATIONS

TOPICS

1. Solving trigonometric equations using algebraic techniques
2. Solving trigonometric equations using inverse trigonometric functions

TOPIC 1: Solving Trigonometric Equations Using Algebraic Techniques

The majority of the trigonometric equations we have seen thus far in this chapter have been identities, and the goal has been to either verify the identity or use it in simplifying a trigonometric expression. In this section, we will study equations that are conditional—that is, equations which are not true for all allowable values of the variable. The goal, of course, is to identify exactly those values of the variable that *do* make a given equation true.

The identities we have learned in the first three sections of this chapter will prove to be very useful again, especially when combined with standard algebraic techniques covered in previous chapters. It is important to remember, however, that the solution to an equation consists of *all* values that make the equation true. Since trigonometric functions are periodic, it is not at all unusual for a trigonometric equation to have an infinite number of solutions, and the goal is to find and describe all of them.

One common approach in solving trigonometric equations, especially if only one trigonometric function is involved, is to isolate the trigonometric expression on one side of the equation. Our knowledge of the behavior of trigonometric functions will then often allow us to describe the complete solution set.

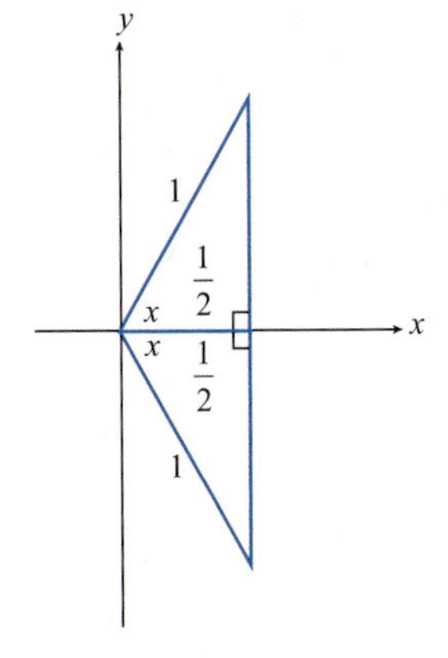

FIGURE 1

Example 1: Solving Equations by Isolating the Trigonometric Function

Solve the equation $3-6\cos x=0$.

Solution

Isolating $\cos x$ is easily accomplished.

$$3-6\cos x=0$$
$$3=6\cos x$$
$$\cos x=\frac{1}{2}$$

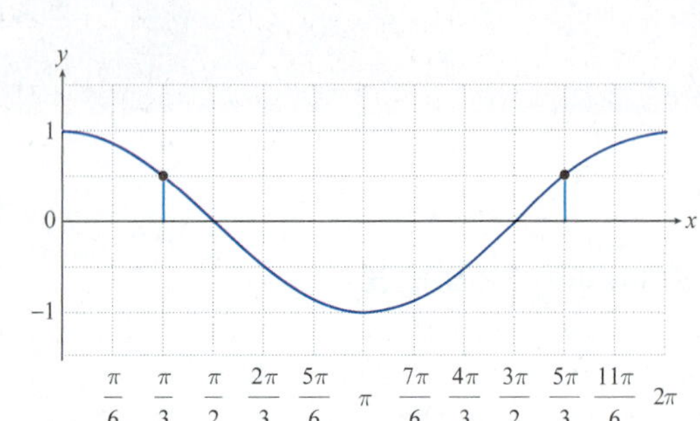

FIGURE 2

This tells us several things. First, the fact that $\cos x$ is positive means that x lies in the first or fourth quadrant. This means that $0 \le x \le \frac{\pi}{2}$ or $-\frac{\pi}{2} \le x \le 0$ (we could also use the range $\frac{3\pi}{2} \le x \le 2\pi$ to describe fourth-quadrant angles). We can then sketch a triangle or two, if necessary, to remind ourselves that $\cos\left(\frac{\pi}{3}\right) = \frac{1}{2}$ and $\cos\left(-\frac{\pi}{3}\right) = \frac{1}{2}$.

Then, since cosine is 2π-periodic, we can describe the complete solution set as

$$x = \frac{\pi}{3} + 2n\pi \quad \text{or} \quad x = -\frac{\pi}{3} + 2n\pi,$$

where n is any integer. Note that this set could be described, less precisely, as follows.

$$\left\{\ldots, -\frac{7\pi}{3}, -\frac{5\pi}{3}, -\frac{\pi}{3}, \frac{\pi}{3}, \frac{5\pi}{3}, \frac{7\pi}{3}, \ldots\right\}$$

Example 2: Solving Equations by Isolating the Trigonometric Function

Solve the equation $\tan^2 x - 1 = 2$.

Solution

The trigonometric function can again be isolated fairly easily.

$$\begin{aligned} \tan^2 x - 1 &= 2 \\ \tan^2 x &= 3 \\ \tan x &= \pm\sqrt{3} \end{aligned}$$

As in Example 1, the task now is to identify *all* the angles that solve the last equation (which, because of the plus-or-minus symbol, is actually two equations). Since tangent has a period of π, we'll be done if we find all the solutions in an interval of length π and add to them all integer multiples of π. For instance, on the interval $\left[-\frac{\pi}{2}, \frac{\pi}{2}\right]$, $\tan x = \sqrt{3}$ when $x = \frac{\pi}{3}$ and $\tan x = -\sqrt{3}$ when $x = -\frac{\pi}{3}$. So the complete solution set is described as follows.

$$x = \frac{\pi}{3} + n\pi \quad \text{or} \quad x = -\frac{\pi}{3} + n\pi$$

The last example included a squared trigonometric term, but the equation was easily solved by taking the square root of both sides. More complicated quadratic-like trigonometric equations can be solved by factoring, by completing the square, or by the quadratic formula.

Example 3: Solving Trigonometric Equations by Factoring

Solve the equation $\sin^2 x - \sin x = \sin x + 3$.

Solution

The equation $\sin^2 x - \sin x = \sin x + 3$ is quadratic in sin x; that is, if sin x were replaced by u, it would simply be a quadratic equation. This observation provides the key to its solution: we will use our previous experience with quadratic equations to solve for sin x and then use our knowledge of trigonometry to solve for x.

$$\sin^2 x - \sin x = \sin x + 3$$
$$\sin^2 x - 2\sin x - 3 = 0$$
$$(\sin x - 3)(\sin x + 1) = 0$$

Remember that the first step in solving quadratic equations by factoring or by the quadratic formula is to collect all the terms on one side. In this problem, we can easily factor the resulting left-hand side.

As in any quadratic equation solved by factoring, the two factors now lead to two equations.

$\sin x = 3$ or $\sin x = -1$

no real solution $\qquad x = \frac{3\pi}{2} + 2n\pi$

The first equation has no real number solution, as 3 lies outside the range of the sine function. The second equation has an infinite number of solutions.

The examples in this section so far have involved only one trigonometric function. If an equation initially appears with two or more functions, one approach to solving it is to use an identity to obtain a new equation in only one function.

Example 4: Solving Equations Using Trigonometric Identities

Solve the equation $-2\cos^2 x + 1 = \sin x$.

Solution

The presence of a $\cos^2 x$ or $\sin^2 x$ term is almost always an indication that the first Pythagorean identity may be useful.

$$-2\cos^2 x + 1 = \sin x$$
$$-2(1 - \sin^2 x) + 1 = \sin x$$

Use the first Pythagorean identity to rewrite the equation in terms of sine alone.

$$2\sin^2 x - \sin x - 1 = 0$$
$$(2\sin x + 1)(\sin x - 1) = 0$$

Solve the resulting quadratic in sine.

As in the last example, we have two equations to solve.

$$2\sin x + 1 = 0 \quad \text{or} \quad \sin x - 1 = 0$$
$$\sin x = -\frac{1}{2} \qquad \sin x = 1$$
$$x = -\frac{5\pi}{6} + 2n\pi, \quad -\frac{\pi}{6} + 2n\pi \qquad x = \frac{\pi}{2} + 2n\pi$$

The solutions to the first equation are angles lying in the third and fourth quadrants, while the solution to the second equation is a right angle. Three equations are thus necessary to describe the solution set.

There is another way to consider the equation $-2\cos^2 x + 1 = \sin x$ that is very instructive, though usually not sufficient to clearly identify the solutions. The equation is satisfied at those values of x for which the graphs of the functions $f(x) = -2\cos^2 x + 1$ and $g(x) = \sin x$ coincide. Figure 3 contains the graphs of these two functions over the interval $[-\pi, \pi]$; note that the two graphs intersect at exactly the points found in Example 4.

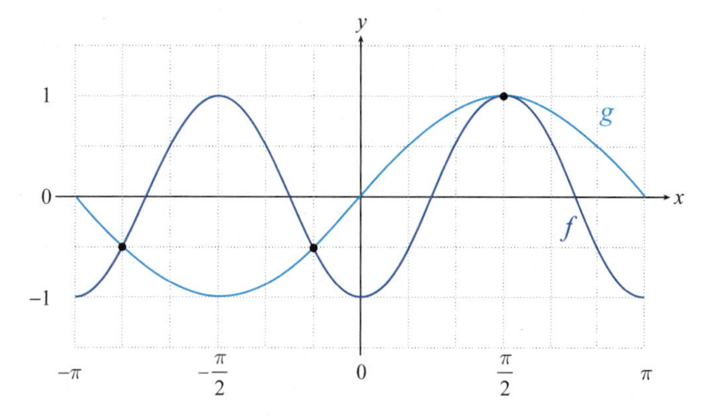

FIGURE 3: Graphical Representation of Solution to $-2\cos^2 x + 1 = \sin x$

Example 5: Solving Trigonometric Equations by Graphing

Solve the equation $\sin x = \cos x$ on the interval $[0, 2\pi)$.

Solution

We have used the graphs of sine and cosine to such an extent that the solutions to this equation should come as no surprise. Graphically, we are looking for those x's at which sine and cosine coincide. The graph in Figure 4 illustrates the two points on the interval $[0, 2\pi)$ where this happens.

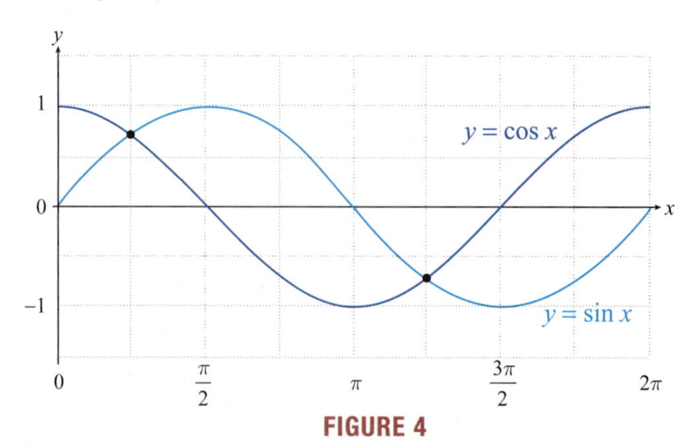

FIGURE 4

From the graph, it appears that the solution of the equation is the set $\left\{\frac{\pi}{4}, \frac{5\pi}{4}\right\}$, and this is indeed the case. Algebraically, we could also solve the equation $\sin x = \cos x$ by dividing both sides by $\cos x$ and making use of our knowledge of the tangent function. Note, though, that we can only do this after we determine that we won't lose any solutions in the process—dividing by $\cos x$ is only a legitimate step if $\cos x$ is nonzero. But when cosine *is* zero, sine is not, so we can safely discard those values of x as they can't possibly be part of the solution set. After the division, we have the equation $\tan x = 1$, and the solution set of this equation on the interval $[0, 2\pi)$ is $\left\{\frac{\pi}{4}, \frac{5\pi}{4}\right\}$.

In some nontrigonometric equations, such as those involving radicals or fractional exponents, raising both sides to a power is a useful technique. The same is true for some trigonometric equations. But just as in the nontrigonometric case, we must check our answers at the end to guard against extraneous solutions; remember that squaring both sides of an equation, for instance, often introduces solutions that do not solve the original problem.

Example 6: Solving Equations Using Trigonometric Identities

Solve the equation $\sin x - 1 = \cos x$ on the interval $[0, 2\pi)$.

Solution

$$\begin{aligned} \sin x - 1 &= \cos x \\ (\sin x - 1)^2 &= \cos^2 x \\ \sin^2 x - 2\sin x + 1 - \cos^2 x &= 0 \\ \sin^2 x - 2\sin x + 1 - (1 - \sin^2 x) &= 0 \\ 2\sin^2 x - 2\sin x &= 0 \\ 2(\sin x)(\sin x - 1) &= 0 \end{aligned}$$

By squaring both sides of the equation, we can apply an identity and rewrite the equation in terms of a single trigonometric function.

The new equation is quadratic in sine and can be solved by factoring.

$$\sin x = 0 \quad \text{or} \quad \sin x = 1$$

$$x = 0, \pi \quad \text{or} \quad x = \frac{\pi}{2}$$

At this point, there are three potential solutions that we must test in the original equation.

$$\sin 0 - 1 \stackrel{?}{=} \cos 0 \qquad \sin\left(\frac{\pi}{2}\right) - 1 \stackrel{?}{=} \cos\left(\frac{\pi}{2}\right) \qquad \sin \pi - 1 \stackrel{?}{=} \cos \pi$$

$$-1 \neq 1 \qquad 0 = 0 \qquad -1 = -1$$

Checking the potential solutions reveals that the solution set is actually $\left\{\frac{\pi}{2}, \pi\right\}$.

Many of the identities we have encountered contain arguments that are multiples of angles, and such arguments can appear in conditional equations as well.

Example 7: Solving Equations by Isolating the Trigonometric Function

Solve the equation $\frac{\sec(3t)}{2} - 1 = 0$.

Solution

$$\frac{\sec(3t)}{2} - 1 = 0$$
$$\sec(3t) = 2$$

Begin by isolating the single trigonometric term in the equation.

$$3t = \frac{\pi}{3} + 2n\pi \quad \text{or} \quad 3t = -\frac{\pi}{3} + 2n\pi$$

There are two solutions of the equation in any interval of length 2π.

$$t = \frac{\pi}{9} + \frac{2n\pi}{3} \quad \text{or} \quad t = -\frac{\pi}{9} + \frac{2n\pi}{3}$$

Divide both sides (including the $2n\pi$ term) by 3.

TOPIC 2: Solving Trigonometric Equations Using Inverse Trigonometric Functions

It is, of course, no accident that the examples so far in this section have had "nice" solutions based on the commonly encountered angles of $\frac{\pi}{6}$, $\frac{\pi}{4}$, $\frac{\pi}{3}$, and $\frac{\pi}{2}$. The preceding examples have been undeniably convenient for demonstrating the sorts of techniques used in solving trigonometric equations. But real-world problems do not, necessarily, give rise to equations with such easily described solutions. It is in solving these more unwieldy problems that inverse trigonometric functions are most useful.

Example 8: Solving Equations Using Inverse Trigonometric Functions

Solve the equation $\tan^2 x + 2\tan x = 3$ on the interval $\left(-\frac{\pi}{2}, \frac{\pi}{2}\right)$.

Solution

$$\tan^2 x + 2\tan x = 3$$
$$\tan^2 x + 2\tan x - 3 = 0$$

The equation is quadratic in tangent and can be solved by factoring.

$$(\tan x + 3)(\tan x - 1) = 0$$
$$\tan x = -3 \quad \text{or} \quad \tan x = 1$$
$$x = \tan^{-1}(-3) \quad \text{or} \quad x = \frac{\pi}{4}$$

At this point, one of the two resulting equations can be solved easily. The other solution is $\tan^{-1}(-3)$, which is approximately -1.2490.

Example 9: Solving Equations Using Inverse Trigonometric Functions

Solve the equation $6\sin x - 2 = \sin x$ on the interval $[0, 2\pi)$.

Solution

$$6\sin x - 2 = \sin x$$
$$5\sin x = 2$$
$$\sin x = 0.4$$

The trigonometric function can be easily isolated.

$x = \sin^{-1}(0.4)$ or $x = \pi - \sin^{-1}(0.4)$

Note that since $\sin x$ is positive, x lies in the first or second quadrant. The solution in the first quadrant is $\sin^{-1}(0.4)$, and the solution in the second quadrant has $\sin^{-1}(0.4)$ as its reference angle.

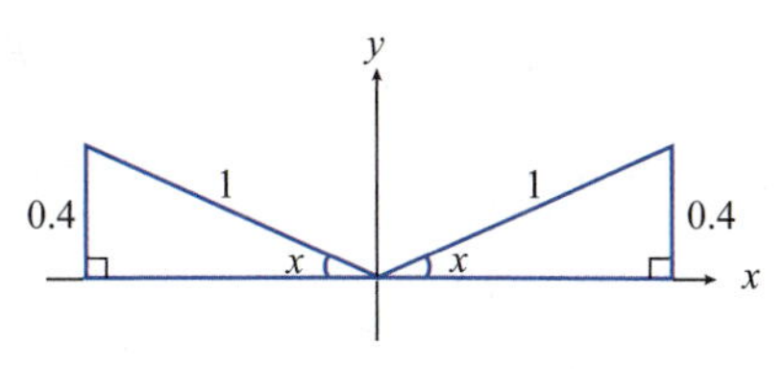

FIGURE 5

8.4 EXERCISES

PRACTICE

Use trigonometric identities and algebraic methods, as necessary, to solve the following trigonometric equations. See Examples 1 through 7.

1. $2\sin x + 1 = 0$

2. $4\sin^2 x + 2 = 3$

3. $\sqrt{2} - 2\cos x = 0$

4. $4 \cos x = 2$

5. $2\cos x - \sqrt{3} = 0$

6. $\sin(2x) = \sqrt{3}\cos(2x)$

7. $-\frac{1}{\sqrt{48}\sin x} = \frac{1}{8}$

8. $\sin^2 x - \sin x = 2 \sin x - 2$

9. $\sqrt{3}\tan x + 1 = -2$

10. $(3\tan^2 x - 1)(\tan^2 x - 3) = 0$

11. $\sec^2 x - 1 = 0$

12. $\sin^2 x = \sin^2 x + \cos^2 x$

13. $\sec x + \tan x = 1$

14. $\cos x + \sin x \tan x = 2$

15. $\sin^2 x + \cos^2 x + \tan^2 x = 0$

16. $\cos^2 x = 3 \sin^2 x$

17. $2 \cos^2 x - 3 = 5 \cos x$

18. $\sin^3 x = \sin x$

19. $\frac{\cot(2x)}{\sqrt{3}} = -1$

20. $\cos x - 1 = \sin x$

Use trigonometric identities, algebraic methods, and inverse trigonometric functions, as necessary, to solve the following trigonometric equations on the interval $[0,2\pi)$. See Examples 8 and 9.

21. $2\sin^2 x + 7\sin x = 4$

22. $\tan^2 x = \tan x + 6$

23. $2\cos^2 x - 1 = 0$

24. $\sec^2 x - 3 = -\tan x$

25. $0.05\sin^3 x = 0.1\sin x$

26. $12\sin x - 1 = 8\sin x$

27. $2\cos^2 x + 11\cos x = -5$

28. $10\sin x - 16 = -11 + 18\sin x$

29. $\sec^2 x - 2 + 8\tan x = 42$

30. $\sin(-x) = 5\sin x - 3$

Verify that the *x*-values given are solutions to the given equation.

31. $2\cos x + 1 = 0$

a. $x = \frac{2\pi}{3}$ **b.** $x = \frac{4\pi}{3}$

32. $3\sec^2 x - 4 = 0$

a. $x = \frac{\pi}{6}$ **b.** $x = \frac{5\pi}{6}$

33. $2\sin^2 x - \sin x - 1 = 0$

a. $x = \frac{\pi}{2}$ **b.** $x = \frac{7\pi}{6}$

34. $\tan^2(3x) = 3$

a. $x = \frac{\pi}{9}$ **b.** $x = \frac{2\pi}{9}$

35. $\csc^4 x - 4\csc^2 x = 0$

a. $x = \frac{\pi}{6}$ **b.** $x = \frac{5\pi}{6}$

36. $3\cot^2 x - 1 = 0$

a. $x = \frac{\pi}{3}$ **b.** $x = \frac{2\pi}{3}$

37. $2\cot x + 1 = -1$

a. $x = \frac{3\pi}{4}$ **b.** $x = \frac{7\pi}{4}$

38. $\csc^2 x = 2\cot x$

a. $x = \frac{\pi}{4}$ **b.** $x = \frac{5\pi}{4}$

39. $2\sec x + 1 = \sec x + 3$

a. $x = \frac{\pi}{3}$ **b.** $x = \frac{5\pi}{3}$

40. $2\sin^2(2x) = 1$

a. $x = \frac{\pi}{8}$ **b.** $x = \frac{3\pi}{8}$

Determine if the value given is a solution to the trigonometric equation. If the value of *x* is not a solution, give all solutions to the equation.

41. $2\cos x = -1;\quad x = \frac{4\pi}{3} + 2n\pi$

42. $\tan(3x)(\tan x - 1) = 0;\quad x = \frac{\pi}{4} + n\pi$

43. $3\sec^2 x = 4;\quad x = \frac{\pi}{6} + n\pi$

44. $\sin^2 x - 3\cos^2 x = 0;\quad x = \frac{\pi}{3} + 2n\pi$

45. $\sqrt{3}\csc x = 2;\quad x = \frac{2\pi}{3} + 2n\pi$

46. $2\sin^2 x - 1 = 0;\quad x = \frac{\pi}{4} + n\pi$

47. $\tan x = -\sqrt{3};\quad x = \frac{\pi}{6} + n\pi$

48. $\tan^2(3x) - 3 = 0;\quad x = \frac{2\pi}{9} + \frac{n\pi}{3}$

49. $3\cot^2 x = 1;\quad x = \frac{\pi}{3} + n\pi$

50. $\cos(2x)(2\cos x + 1) = 0;\quad x = \frac{5\pi}{6} + n\pi$

Use trigonometric identities and algebra, as necessary, to rewrite the following equations to be quadratic-like, and then solve each on the interval $[0,2\pi)$.

51. $2\sin^2 x - \sin x - 1 = 0$

52. $2\sin^2 x + 3\cos x - 3 = 0$

53. $\sin x - \cos x - 1 = 0$

54. $\tan x + \sqrt{3} = \sec x$

55. $\cos(2x) - \cos x = 0$

56. $2\cos^2 x - \sqrt{3}\cos x = 0$

57. $\csc^2 x - 2\cot x = 0$

Solve the following equations on the interval $[0°,360°)$. Give the exact answers when appropriate; otherwise, round your answers to one decimal place.

58. $\sin^2 x \cos x - \cos x = 0$

59. $\cos^2 x = \sin^2 x$

60. $\tan x = \cot x$

61. $2 \sin x = \csc x + 1$

62. $\sec^2 x - 2 \tan x = 4$

63. $\sin^2 x = 2 \sin x - 3$

64. $2 \cos^2 x - 1 = -2 \cos x$

65. $2\sin x \cot x + \sqrt{3}\cot x - 2\sqrt{3}\sin x - 3 = 0$

Solve the algebraic and trigonometric equations given. Restrict the solutions of the trigonometric equations to the interval $[0,2\pi)$. Give the exact answers for s; round your answers for t to four decimal places.

66. $6s^2 - 13s + 6 = 0;\quad 6\cos^2 t - 13\cos t + 6 = 0$

67. $s^2 + s - 12 = 0;\quad \sin^2 t + \sin t - 12 = 0$

68. $2s^2 + 7s - 15 = 0;\quad 2\tan^2 t + 7\tan t - 15 = 0$

69. $4s^2 - 4s - 1 = 0;\quad 4\cos^2 t - 4\cos t - 1 = 0$

APPLICATIONS

70. If an arrow is shot by an archer with an initial velocity v_0 at an angle of θ in reference to the horizontal, then its range, the horizontal distance it travels, is given by $r = \frac{1}{32}v_0^2 \sin(2\theta)$. If the initial velocity is $v_0 = 100$ feet per second and the arrow hits a target 300 feet from where the archer is standing, what is the value, in degrees, of the angle θ? Round your answer to one decimal place.

71. A baseball leaves a bat at an angle of θ in reference to the horizontal. The initial velocity is $v_0 = 95$ feet per second. The ball is caught 160 feet from where it is hit. What is the value, in degrees, of the angle θ if the range, the horizontal distance traveled by the ball, is given by $r = \frac{1}{32}v_0^2 \sin(2\theta)$? Round your answer to one decimal place.

TECHNOLOGY

Use a graphing utility to approximate the solutions of the given equation on the interval $[0,2\pi)$. Round your answers to four decimal places.

72. $x\tan x - 3 = 0$

73. $2\sin x + \cos x = 0$

74. $2\cos^2 x - \sin x = 0$

75. $\cot^2 x - \sec^2 x = 0$

76. $2\sin x - \csc^2 x = 0$

77. $2\sin x = 1 - 2\cos x$

78. $\log x = -\sin x$

79. $\sin\left(\frac{x}{2}\right) = 2\cos(2x)$

Use a graphing utility to solve the following equations on the interval $[0°,360°)$. Remember to change the mode to degrees.

80. $2\sin(2x) = \sqrt{3}$

81. $\sin(3x) - \frac{1}{2} = 0$

82. $2\sin(4x) - 1 = 0$

WRITING & THINKING

83. While working Exercises 66–69, what did you observe as the maximum number of real solutions the algebraic equations can have?

84. While working Exercises 66–69, what did you observe as the maximum number of real solutions the trigonometric equations can have on the interval $[0, 2\pi)$?

85. While working Exercises 80–82, what did you observe about the solutions to the equations of the form $y = \sin(ax)$ on the interval $[0°, 360°)$?

CHAPTER 8 PROJECT

Trigonometric Identities

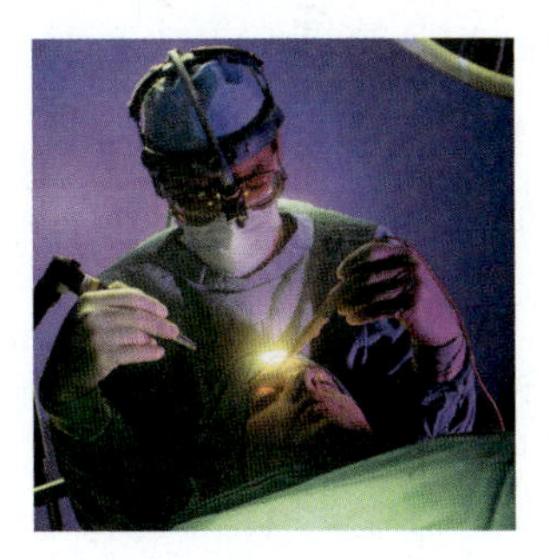

Lasers are used in such varied applications as video players, checkout counter scanners, surgical operations, and weaponry. A beam of light from a flashlight shines for a couple hundred yards, but a laser's narrow band of light can be reflected off the moon and detected on Earth. A laser has this ability because its light is *coherent*. Coherent light means that each light wave has exactly the same amplitude, direction, and phase. Coherence reflects the superposition principle which states that when combining two waves, the resulting wave is the sum of the two individual waves.

Let's examine how the superposition principle works.

1. Consider two waves with a difference in displacement of $\frac{\pi}{2}$.

$$y_1 = 2\sin(kx - \omega t)$$
$$y_2 = 2\sin\left(kx - \omega t + \frac{\pi}{2}\right)$$

Using a trigonometric identity, add these two waves to find the equation of their superposition. What is the amplitude of the resulting wave?

2. The following equations are given.

$$y_1 = A\sin(kx - \omega t)$$
$$y_2 = A\sin(kx - \omega t + \delta)$$

a. For what values of δ would the amplitude be the largest? (This happens when two waves are coherent and it is called constructive interference.)

b. What is the smallest amplitude possible for $y_1 + y_2$?

c. For what values of δ would the smallest amplitude occur? (This is called destructive interference.)

3. Graph the following equations.

$$y_1 = 3\sin t$$
$$y_2 = 3\sin\left(t + \frac{\pi}{3}\right)$$

Now graph $y_1 + y_2$. Discuss the relationship between the three graphs.

CHAPTER 8 REVIEW EXERCISES

Section 8.1

Use trigonometric identities to simplify the expressions. There may be more than one correct answer.

1. $\cot x \sec x$

2. $\left(\csc^2 x-1\right)\cos^2\left(\frac{\pi}{2}-x\right)$

3. $\frac{\tan(-y)}{\cot(\pi+y)}$

4. $\frac{\tan^2\alpha}{\csc^2\left(\frac{\pi}{2}-\alpha\right)}+\frac{\cos\alpha}{\sec(\alpha+2\pi)}$

5. $\sin^2\theta\sec^2\theta-\csc^2\left(\frac{\pi}{2}-\theta\right)+\sin(-\theta)$

6. $\sin\left(\frac{\pi}{2}-x\right)\cos(x+2\pi)+\sin(-x)\sec\left(\frac{\pi}{2}-x\right)$

Verify the following trigonometric identities.

7. $(\tan x+\sec x)(\sec x-\tan x)=1$

8. $\cos^2 x\tan^2 x=1-\frac{1}{\sec^2 x}$

9. $\frac{\cos\left(\frac{\pi}{2}-t\right)}{\tan(-t)}=-\cos t$

10. $5+\tan^2 y=4+\sec^2 y$

11. $\tan(\theta+\pi)=-\frac{\sec(\theta+2\pi)}{\csc(-\theta)}$

12. $-\tan\left(\frac{\pi}{2}-x\right)\tan(-x)-\tan^2 x\sin^2\left(\frac{\pi}{2}-x\right)=\cos^2 x$

Use the suggested substitution to rewrite the given expression as a trigonometric expression. Assume $0\leq\theta\leq\frac{\pi}{2}$.

13. $\sqrt{16+x^2},\quad \tan\theta=\frac{x}{4}$

14. $\sqrt{64-16x^2},\quad 2\sin\theta=x$

15. $\sqrt{25x^2-100},\quad \csc\theta=\frac{x}{2}$

16. $\sqrt{9x^2+36},\quad x=2\tan\theta$

Section 8.2

Use the sum and difference identities to determine the exact value of each of the following expressions.

17. $\cos\left(\frac{\pi}{2}+\frac{5\pi}{3}\right)$

18. $\cos 255°$

19. $\sin(-15°)$

20. $\sin\left(\frac{5\pi}{4}+\frac{\pi}{6}\right)$

21. $\tan\left(\pi-\frac{2\pi}{3}\right)$

22. $\tan 105°$

23. Suppose that $\sin\alpha = \frac{8}{17}$ and $\sin\beta = \frac{3}{5}$ and the terminal sides of both α and β are in quadrant II. Find $\tan(\alpha - \beta)$.

24. Suppose that $\sin\alpha = \frac{5}{13}$ and $\cos\beta = \frac{4}{5}$, the terminal side of α is in quadrant II, and the terminal side of β is in quadrant IV. Find $\sin(\alpha - \beta)$.

Use the sum and difference identities to rewrite each of the following expressions as a trigonometric function of one angle, and then evaluate the result.

25. $\sin 175° \cos 35° + \cos 175° \sin 35°$

26. $\dfrac{\tan\left(\frac{9\pi}{8}\right) - \tan\left(\frac{3\pi}{8}\right)}{1 + \tan\left(\frac{9\pi}{8}\right)\tan\left(\frac{3\pi}{8}\right)}$

Use the sum and difference identities to verify the following identities.

27. $\sin x = \cos\left(\frac{\pi}{2} - x\right)$

28. $\cos(\alpha + \beta) - \cos(\alpha - \beta) = -2\sin\alpha\sin\beta$

Express each of the following as an algebraic function of *x*.

29. $\cos\left(\sin^{-1} x + \tan^{-1} x\right)$

30. $\cos\left(\cos^{-1}(2x) + \tan^{-1}(2x)\right)$

Express each of the following functions in terms of a single sine function.

31. $f(x) = \sqrt{2}\sin x - \sqrt{2}\cos x$

32. $h(\alpha) = \sqrt{3}\sin(4\alpha) - \cos(4\alpha)$

Section 8.3

Use the given information to determine $\cos(2x)$, $\sin(2x)$, and $\tan(2x)$ if possible.

33. $\tan x = \frac{4}{3}$ and $\sin x$ is positive

34. $\sin x = \frac{-1}{\sqrt{10}}$ and $\tan x$ is positive

Verify the following trigonometric identities.

35. $\cos(3x) = 4\cos^3 x - 3\cos x$

36. $\dfrac{\sin(4x)}{4} = \sin x \cos x - 2\sin^3 x \cos x$

Use a power-reducing identity to rewrite the given expression as directed.

37. Rewrite $\sin^3 x \cos^2 x$ in terms containing only the first powers of sine and cosine.

38. Rewrite $\tan^2 x \sin^3 x$ in terms containing only the first powers of sine and cosine.

Determine the exact value of each of the following expressions.

39. $\tan\left(\frac{5\pi}{12}\right)$

40. $\cos(157.5°)$

41. $\tan 15°$

42. $\sin\left(-\frac{5\pi}{8}\right)$

Use the product-to-sum identities to rewrite the given expression as a sum or difference.

43. $\cos(x+y)\sin(x-y)$

44. $\cos\left(\frac{3\pi}{4}\right)\cos\left(\frac{\pi}{6}\right)$

45. $\sin 165^\circ \cos 15^\circ$

46. $\sin(4x)\sin(3x)$

Use the sum-to-product identities to rewrite the given expression as a product.

47. $\sin(5\alpha)-\sin(3\alpha)$

48. $\cos 225^\circ + \cos 15^\circ$

49. $\cos\left(\frac{3\pi}{4}\right)-\cos\left(\frac{2\pi}{3}\right)$

50. $\sin\left(\frac{5\pi}{6}\right)+\sin\left(\frac{\pi}{3}\right)$

Section 8.4

Use the trigonometric identities and algebraic methods, as necessary, to solve the following trigonometric equations.

51. $8\cos^2 x+1=7$

52. $2\sin^2 x=\sin x$

53. $-\sin^2 x+4\cos x+1=0$

54. $\tan^3 x=\tan x$

55. $-2\sin^2 x=-\cos x-1$

56. $\sin x+\cos x\cot x=-2$

Use trigonometric identities, algebraic methods, and inverse trigonometric functions, as necessary, to solve the following trigonometric equations on the interval $[0,2\pi)$.

57. $3\tan^2 x+9=10$

58. $\sin^2 x=3-2\sin x$

Determine if the value given is a solution to the trigonometric equation. If the value of x is not a solution, give all solutions to the equation.

59. $4\sin^2 x=3;\quad x=\frac{5\pi}{3}+2n\pi$

60. $\frac{1}{2}\csc x+1=2;\quad x=\frac{3\pi}{4}+2n\pi$

61. $\tan(2x)\cos x=-\frac{\sqrt{3}}{2};\quad x=\frac{\pi}{6}+2n\pi$

62. $\sin x+\cos(2x)=1;\quad x=\frac{5\pi}{6}+2n\pi$

Solve the following equations on the interval $[0^\circ,360^\circ)$. Give exact answers when appropriate; otherwise, round your answers to one decimal place.

63. $\cos^2 x\sin x=\sin x$

64. $2\cos^2 x+7\cos x=4$

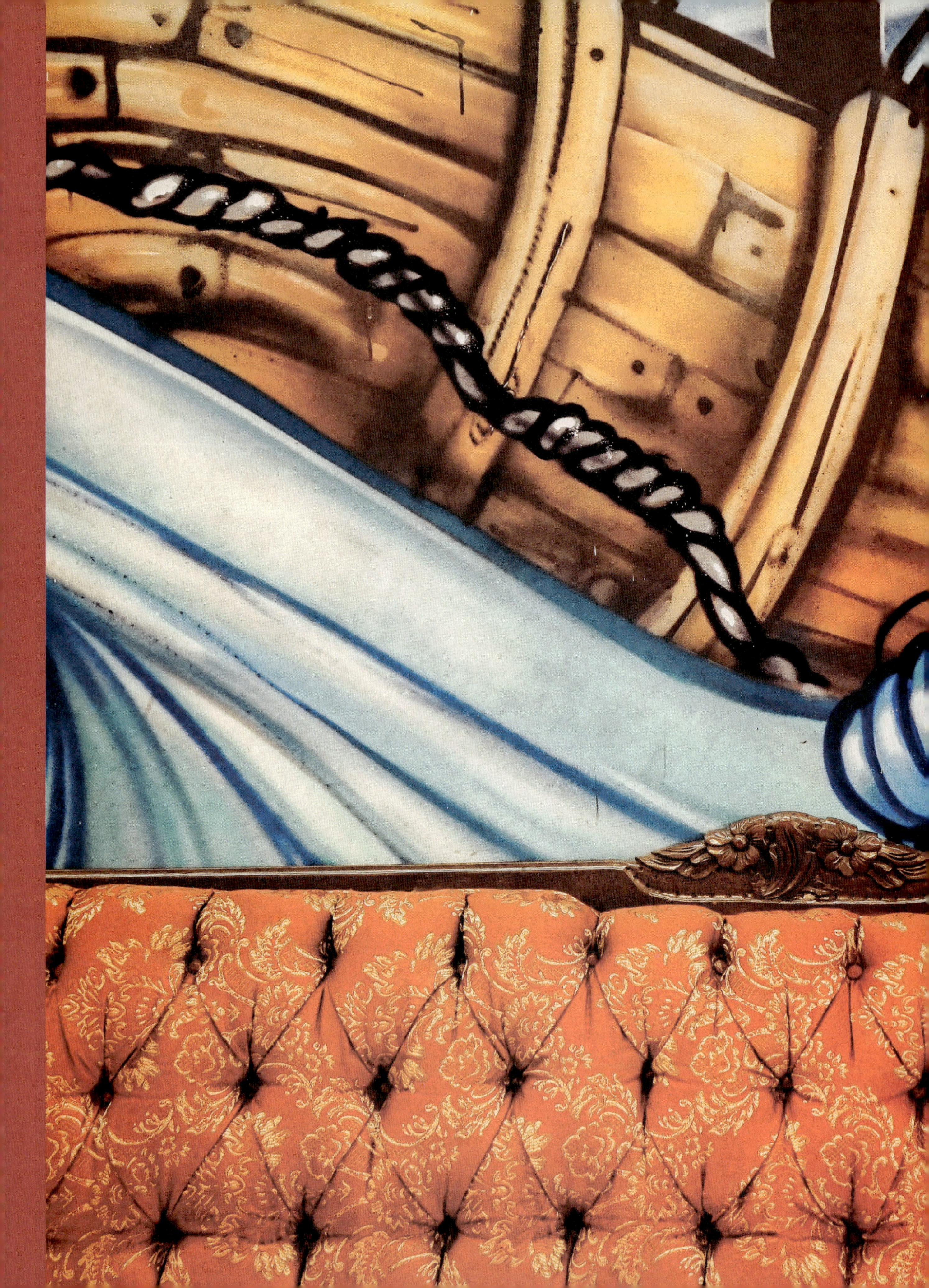

CHAPTER 9

Additional Topics in Trigonometry

SECTIONS

WHAT IF ...

What if you were moving to a new apartment and needed to move your couch out of your living room? If you knew the weight of your couch, could you determine the force necessary to slide it from your living room to the front door?

By the end of this chapter, you will be able to perform many operations with vectors, including addition, scalar multiplication, and a special kind of vector multiplication called the dot product. Vectors have many uses in the physical sciences due to their ability to concisely describe and perform calculations about the forces acting around us. On page 690, you will solve problems similar to the one given above. You will master these types of problems using tools such as vector operations and magnitude using components on page 684.

William Rowan Hamilton

Introduction

This is the last of three chapters devoted to the introduction and use of trigonometric concepts. Chapter 7 introduced the nomenclature of trigonometry and the six fundamental trigonometric functions, and Chapter 8 focused on a deeper understanding of how those functions behave. The goal of this chapter is to show how trigonometry relates to a surprisingly large number of topics seen elsewhere in this text, even in situations which would at first seem to have nothing in common with trigonometry.

The applications of trigonometry seen in this chapter vary widely. As a preview, here is a cursory list of the uses that will be developed:

- new theorems and relations that are exceedingly useful in surveying, navigation, and other practical concerns
- new ways to measure areas of triangles
- a second planar coordinate system that simplifies the graphing and solving of some kinds of equations
- a second way of describing curves in the plane that allows us to answer questions that have been awkward up to this point
- a second way of representing complex numbers that simplifies multiplication and division, and offers an easy method for the calculation and visualization of roots
- a very powerful method of representing *directed magnitudes*, allowing us to mathematically describe quantities in which the direction that a magnitude is applied is of critical importance
- a new mathematical operation that greatly simplifies the solving of problems involving, for example, force and work

As usual, the concepts presented come from many different eras and cultures. One of the formulas for triangular area is named for a 1st-century mathematician from Alexandria, but was probably known by mathematicians from several different areas a century or two before. The method of representing directed magnitudes owes much to the work of the 19th-century Irish mathematician William Rowan Hamilton. And the use of trigonometry in handling complex numbers developed over the course of several centuries and through the work of many mathematicians working toward many widely differing goals.

9.1 THE LAW OF SINES

TOPICS

1. The Law of Sines
2. Applications of the Law of Sines
3. The Law of Sines and the area of a triangle

TOPIC 1: The Law of Sines

Chapter 7 introduced the trigonometric functions and demonstrated how they are used to solve a variety of problems involving triangles. Chapter 8 then explored the properties of the trigonometric functions and developed relationships and identities that greatly extended their usefulness. But the triangles analyzed to this point have all been right triangles, and even our more general treatment of the trigonometric functions has drawn, explicitly or implicitly, upon properties of right triangles. In this section we will expand the class of triangles we can analyze to include *oblique* triangles—those that have no right angle.

Most of the problems in this section contain a step in which a triangle must be *solved*; that is, given some information about a triangle's sides and/or angles, we will need to determine something else about the triangle. Note that there are six fundamental quantities that can be measured for any triangle: the lengths of the three sides and the sizes of the three angles. We will see that if we know the length of any one side and any two of the other five quantities, we can determine the remaining three quantities.

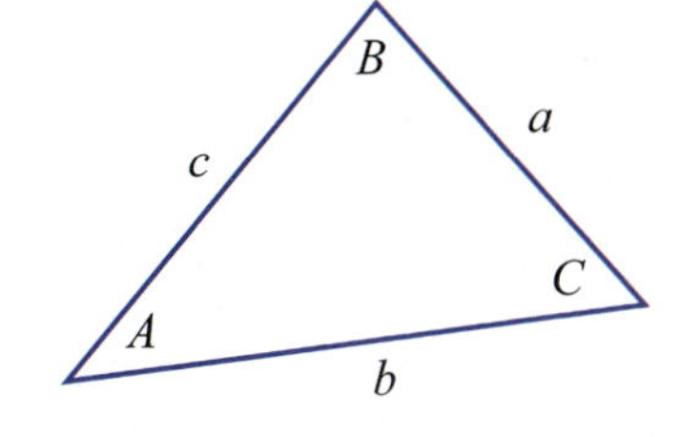

FIGURE 1: Oblique Triangle with Sides and Angles Labeled

In keeping with tradition, we will let A, B, and C denote the measures of the three angles of a given triangle and a, b, and c the lengths of the sides, with the side of length a opposite the angle of measure A and similarly for b and c. We will also follow the convention of letting a letter stand for both a quantity and the side or angle it measures. Figure 1 illustrates a typical oblique triangle and the labeling of its sides and angles.

In describing what we know about a triangle, we will let S represent knowledge about a side and A knowledge about an angle, so for instance the notation SAS is shorthand for a situation in which we know the lengths of two sides and the measure of the angle between them. Two laws allow us to solve triangles about which we know some minimal amount of information regarding sides and angles. The two laws, known as the Law of Sines and the Law of Cosines, apply to different cases as summarized in Table 1. We will study and learn how to apply the Law of Sines in this section and then discuss the Law of Cosines in the next section.

Law of Sines	Law of Cosines
Two angles and a side (AAS or ASA)	Two sides and the included angle (SAS)
Two sides and a nonincluded angle (SSA)	Three sides (SSS)

TABLE 1: Use of the Law of Sines and the Law of Cosines

We are now ready to state the Law of Sines.

THEOREM: The Law of Sines

Given a triangle with sides and angles labeled according to the convention in Figure 1, the following equation represents the **Law of Sines.**

$$\frac{\sin A}{a} = \frac{\sin B}{b} = \frac{\sin C}{c}$$

Note that since A, B, and C represent measures of angles in a triangle, all three must lie between 0 and π radians; hence, $\sin A$, $\sin B$, and $\sin C$ are all nonzero and the Law of Sines can also be written using the reciprocals.

$$\frac{a}{\sin A} = \frac{b}{\sin B} = \frac{c}{\sin C}$$

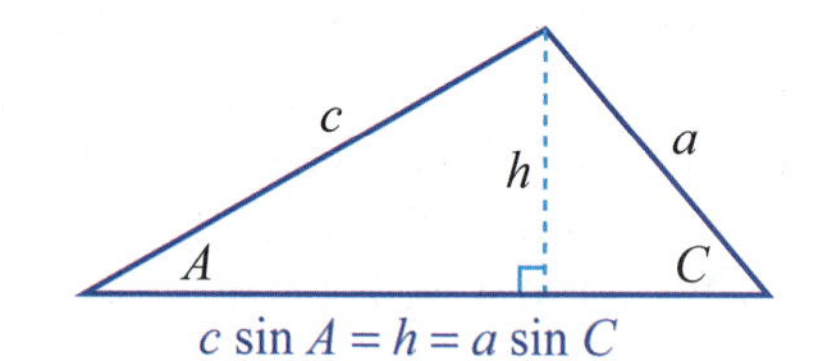

FIGURE 2: The Law of Sines (Acute Case)

While the Law of Sines is typically applied to oblique triangles, its truth relies on decomposing a triangle into two right triangles. For instance, if angles A and C are acute, we can construct the altitude h and determine its value two different ways, as shown in Figure 2.

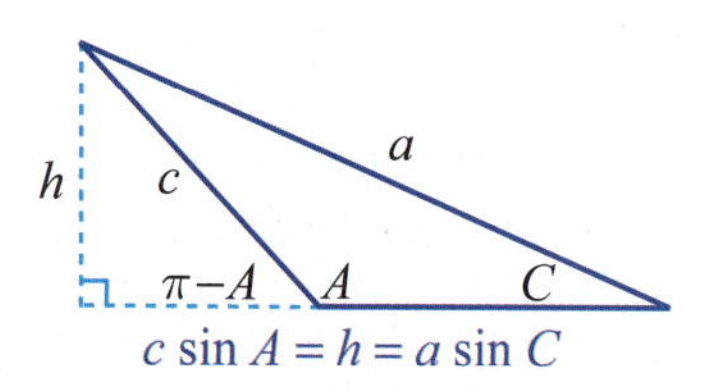

FIGURE 3: The Law of Sines (Obtuse Case)

The exact same relation is true if one of the angles, say A, is obtuse, but this realization depends on recalling that $\sin A = \sin(\pi - A)$ (that is, the sine of an angle in the second quadrant is equal to the sine of the reference angle). Figure 3 illustrates this case.

In either case, dividing both sides of the equation $c \sin A = a \sin C$ by ac leads to the Law of Sines as it relates to A, C, a, and c. Similar diagrams provide the rest of the law.

TOPIC 2: Applications of the Law of Sines

We will proceed with illustrations of how the Law of Sines is applied. Each of the following examples will be labeled according to the information we are given or can infer about its sides and angles.

Example 1: Using the Law of Sines in an AAS Situation

Sarah is piloting a hot-air balloon, and finds herself becalmed directly above a long straight road. She notices mile markers on the road, and determines the angle of depression to the two markers as shown in Figure 4. How far is she from marker A? What is her altitude?

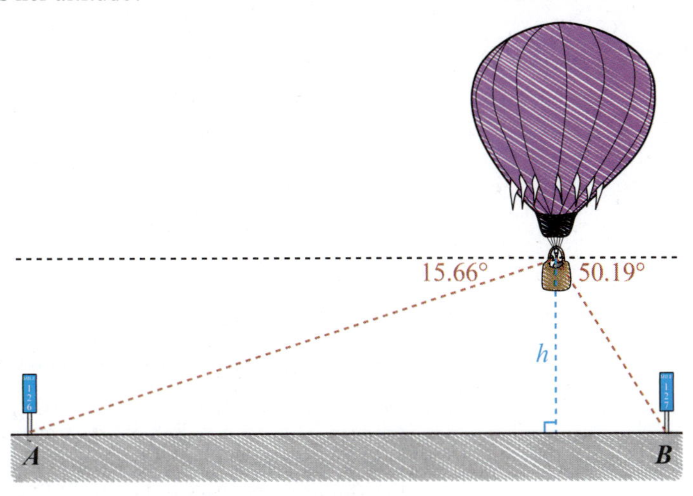

FIGURE 4

Solution

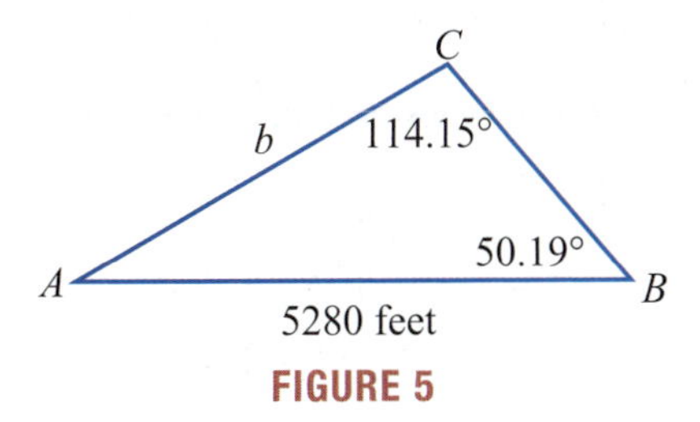

FIGURE 5

To begin, note that the angle at mile marker B also has measure $50.19°$ (this follows from the equality of opposite angles formed by a line cutting two parallel lines). Further, the angle at Sarah's position has measure $180° - 15.66° - 50.19° = 114.15°$. Expressing the one length that we know in feet, we have the information shown in Figure 5.

We can now use the Law of Sines to obtain the equation

$$\frac{b}{\sin(50.19°)} = \frac{5280}{\sin(114.15°)},$$

which we can solve for length b to get

$$b = [\sin(50.19°)]\frac{5280}{\sin(114.15°)} \approx 4445 \text{ feet.}$$

To determine Sarah's altitude h, we can use the fact that the measure of the angle at marker A is $15.66°$ and observe that

$$\sin(15.66°) = \frac{h}{4445}.$$

Solving this for h, we obtain $h \approx 1200$ feet.

Example 2: Using the Law of Sines in an ASA Situation

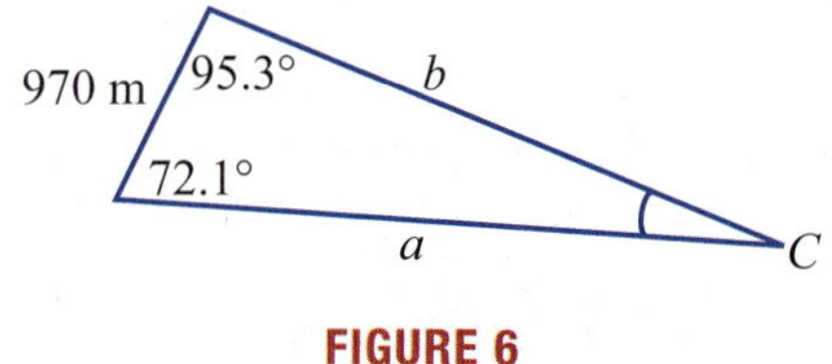

FIGURE 6

A surveyor has the task of determining the dimensions of the triangular plot of land shown in Figure 6. He has already measured the length of the short side of the plot, and has also determined the measure of two of the angles. What are the lengths of the other two sides?

Solution

The first step is to determine the measure of the third angle. Since the sum of the angles in a triangle is 180°, the third angle has measure $180° - 95.3° - 72.1°$, or 12.6°. The Law of Sines now tells us

$$\frac{970}{\sin(12.6°)} = \frac{a}{\sin(95.3°)} \text{ and } \frac{970}{\sin(12.6°)} = \frac{b}{\sin(72.1°)}.$$

Solving for a and b, we obtain $a \approx 4428$ meters and $b \approx 4231$ meters.

As Examples 1 and 2 demonstrated, problems in which two angles and a side are known are readily dealt with. Usually, the application of some simple facts about triangles and angles, followed by the Law of Sines, allows us to completely determine the dimensions of a triangle in the AAS and ASA cases. Unfortunately, the same is not true for the SSA case. Given two sides a and b and the angle A opposite a, there may be no triangle, exactly one triangle, or two triangles that satisfy the conditions. To see why this is so, consider the four possibilities that may occur when A is acute:

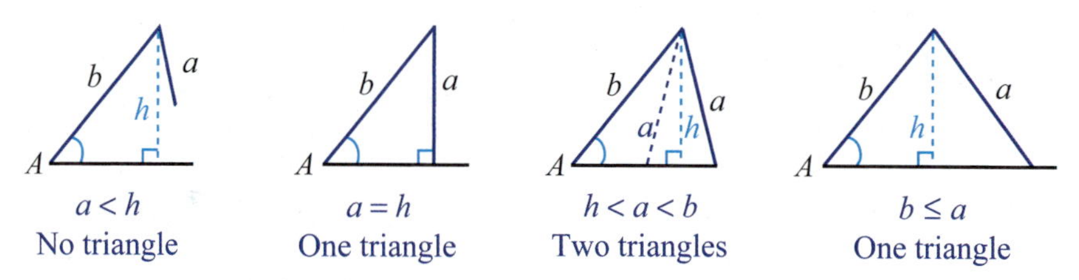

FIGURE 7: SSA Case with Acute Angle *A*

As Figure 7 shows, the size of a in relation to h and b, where $h = b \sin A$, determines whether a triangle exists which fits the given information, and whether the triangle is unique if it does exist. There are fewer possibilities if A is obtuse, but the ambiguity is still present. Figure 8 illustrates the obtuse case.

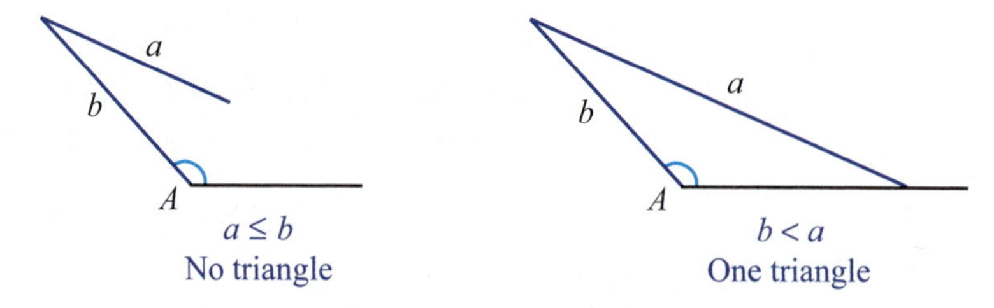

FIGURE 8: SSA Case with Obtuse Angle *A*

Because of the uncertainty over whether a unique triangle exists in the SSA case, it is sometimes called the **ambiguous case**. In practice, the ambiguity is resolved through consideration of additional information.

Example 3: Using the Law of Sines in an SSA Situation with Two Solutions

Create a triangle, if possible, for which $A = \frac{\pi}{6}$, $b = 12$ cm, and $a = 7$ cm.

Solution

A sketch is often useful in order to get an initial idea of whether such a triangle is possible. In this case, a rough sketch might look something like Figure 9.

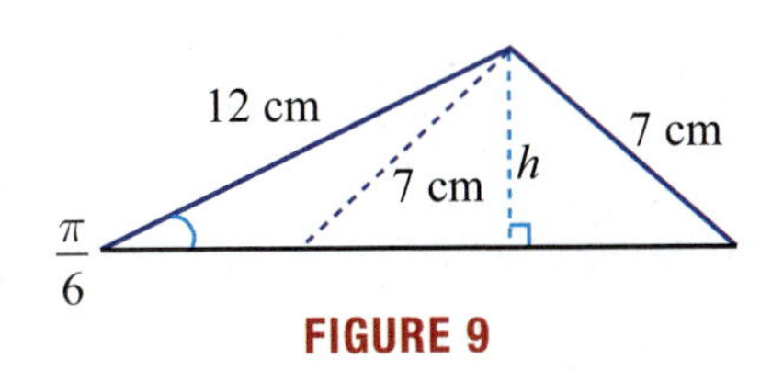

FIGURE 9

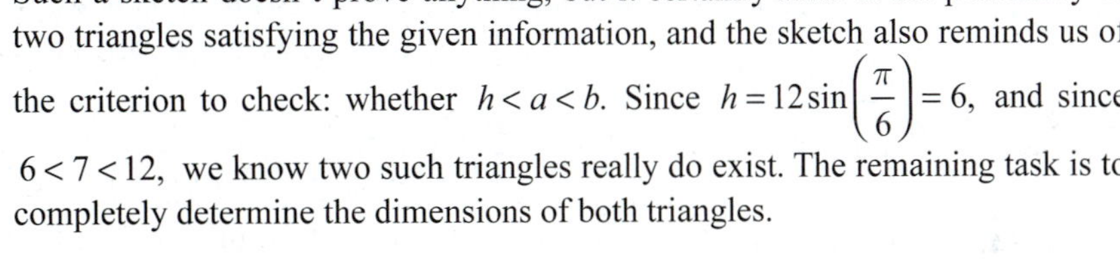

Such a sketch doesn't prove anything, but it certainly hints at the possibility of two triangles satisfying the given information, and the sketch also reminds us of the criterion to check: whether $h < a < b$. Since $h = 12\sin\left(\frac{\pi}{6}\right) = 6$, and since $6 < 7 < 12$, we know two such triangles really do exist. The remaining task is to completely determine the dimensions of both triangles.

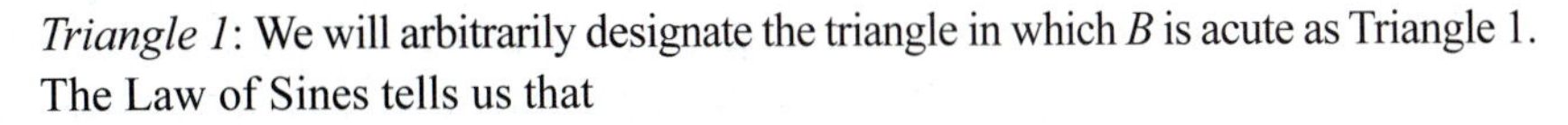

Triangle 1: We will arbitrarily designate the triangle in which B is acute as Triangle 1. The Law of Sines tells us that

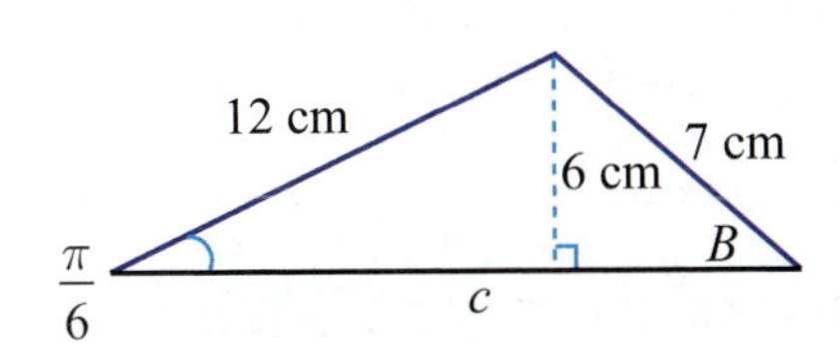

FIGURE 10: Triangle 1

$$\frac{\sin\left(\frac{\pi}{6}\right)}{7} = \frac{\sin B}{12},$$

and so $\sin B = \frac{6}{7}$. Thus, $B = \sin^{-1}\left(\frac{6}{7}\right) \approx 1.03$ (remember that this is in radians). This means that angle C has measure $\pi - \frac{\pi}{6} - \sin^{-1}\left(\frac{6}{7}\right)$, or approximately 1.59, and a second application of the Law of Sines gives us

$$\frac{c}{\sin(1.59)} = \frac{7}{\sin\left(\frac{\pi}{6}\right)},$$

so $c \approx 14.0$ cm.

Triangle 2: In the second triangle, B is obtuse and we have $\sin B = \sin(\pi - B) = \frac{6}{7}$. From our work in Triangle 1, we know that the acute angle $\pi - B \approx 1.03$. Therefore, $B \approx \pi - 1.03 \approx 2.11$. (a quick check with a calculator will verify that $\sin(2.11) \approx \frac{6}{7}$). In this case, $C \approx \pi - \frac{\pi}{6} - 2.11 = 0.51$, and hence

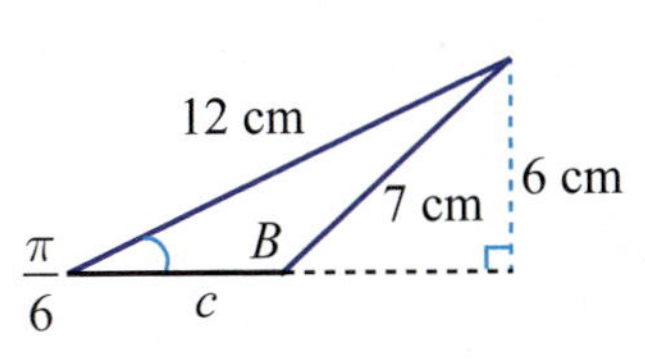

FIGURE 11: Triangle 2

$$\frac{c}{\sin(0.51)} = \frac{7}{\sin\left(\frac{\pi}{6}\right)}.$$

Solving this for c gives us $c \approx 6.8$ cm.

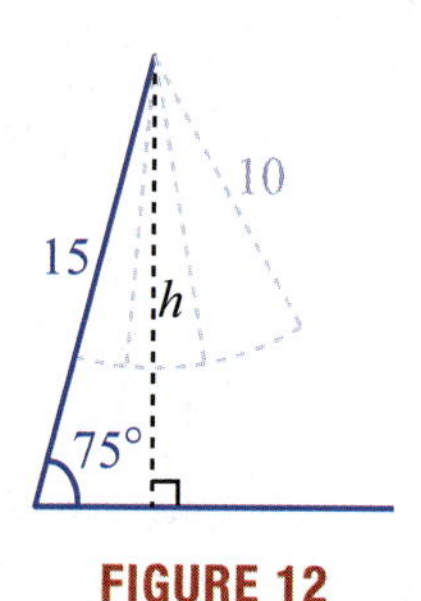

FIGURE 12

Example 4: Using the Law of Sines in an SSA Situation with No Solution

Create a triangle, if possible, for which $A = 75°$, $b = 15$ units, and $a = 10$ units.

Solution

A sufficiently accurate sketch may lead you to conclude that no such triangle is possible. The proof of this fact follows from noting that $h = 15\sin 75° \approx 14.5$, and that $a < h$. In other words, a is too short to reach side c and form a triangle, so no triangle satisfies the given conditions.

TOPIC 3: The Law of Sines and the Area of a Triangle

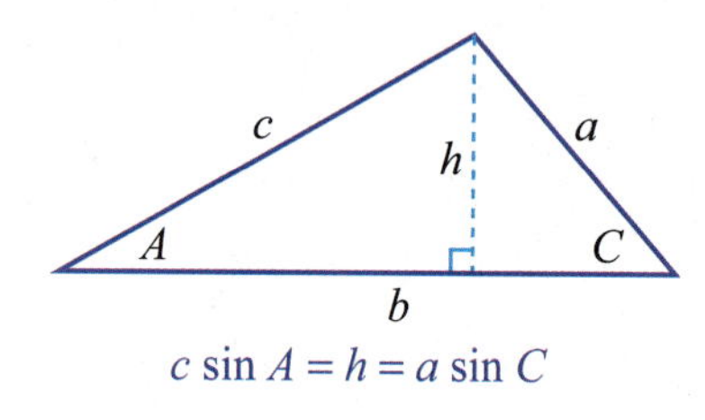

$c \sin A = h = a \sin C$

FIGURE 13: Two Formulas for Height

Look again at the proof of the Law of Sines. In that discussion, we determined the height of a given triangle two different ways, both of which involved the sine of an angle. The two formulas in one particular case are depicted again for reference in Figure 13.

Recall that the area of a triangle is given by the formula $\left(\frac{1}{2}\right)(\text{base})(\text{height})$; in Figure 13, the length of the base is b, and we have two ways of determining the height. However, multiplying the base b by either expression for height results in a product of the lengths of two sides and the sine of the included angle. This same sort of product occurs if we instead consider one of the other sides of the triangle to be the base. In summary, we have derived the following formula.

THEOREM: Area of a Triangle (Sine Formula)

The area of a triangle is one-half the product of the lengths of any two sides and the sine of their included angle. That is,

$$\text{Area} = \frac{1}{2}ab\sin C = \frac{1}{2}bc\sin A = \frac{1}{2}ac\sin B.$$

Example 5: Using the Sine Formula to Find the Area of a Triangle

A businessman has an opportunity to buy a triangular plot of land in a part of town known colloquially as "Five Points," as five roads intersect there to form five equal angles. The property has 147 feet of frontage on one road, and 207 feet of frontage on the other. What is the square footage of the property?

Solution

Since the roads intersect to form five equal angles, the included angle of the plot must be $360° \div 5 = 72°$. The Sine Formula then quickly gives us the area of the triangular plot of land.

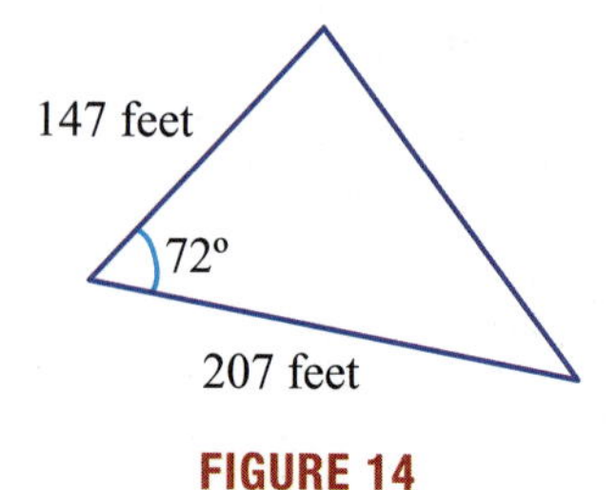

FIGURE 14

$$\text{Area} = \frac{1}{2}(147)(207)(\sin 72°)$$
$$\approx 14{,}470 \text{ ft}^2$$

9.1 EXERCISES

PRACTICE

Solve for the remaining angle and sides of the triangles. See Examples 1 and 2.

1. $A = 30°, B = 45°, a = 3$
2. $A = 60°, B = 40°, a = 2$
3. $A = 70°, B = 50°, b = 4$
4. $A = 100°, B = 20°, b = 1$
5. $B = 70°, C = 30°, c = 2$
6. $B = 120°, C = 40°, b = 6$
7. $A = 20°, B = 10°, a = 2$
8. $B = 100°, C = 30°, a = 3$

Create a triangle, if possible, using the given information and the Law of Sines. See Examples 3 and 4.

9. $A = 40°, a = 2, b = 4$
10. $A = 40°, a = 4, b = 4$
11. $C = 45°, a = 2, c = 4$
12. $A = 32°, a = 4, b = 7$
13. $C = 140°, b = 1, c = 9$
14. $A = 60°, a = 5, c = 6$
15. $B = 80°, a = 2, b = 6$
16. $B = 50°, b = 2, c = 5$
17. $B = 110°, a = 1, b = 8$
18. $A = 60°, a = 10, b = 6$
19. $C = 42°, b = 9, c = 3$
20. $B = 13.2°, A = 63.7°, b = 21.2$
21. $A = 6°23', B = 64°15', c = 2.5$
22. $C = 100°, a = 18.1, c = 20.4$
23. $A = 108°, a = 9, b = 8.9$
24. $C = 24°, b = 2.4, c = 1.5$
25. $B = 16.9°, A = 29.7°, b = 17.8$
26. $A = 46°53', B = 74°13', c = 3.1$
27. $C = 116°, a = 24.1, c = 25$
28. $A = 10°, a = 2, b = 5$
29. $A = 30°, a = 15, b = 13$
30. $C = 74°, b = 4.5, c = 23$

Find the area of the triangle using the given information. See Example 5.

31. $A = 131°, b = 10, c = 25$
32. $B = 60°7', c = 18, a = 6$
33. $C = 103°, a = 10, b = 2$
34. $B = 54°, a = 10, c = 7$
35. $A = 67°49', c = 4.2, b = 9.5$
36. $C = 46°, b = 20, a = 19$
37. $A = 86°, b = 24, c = 28$

APPLICATIONS

38. A plane flies 730 miles from Charleston, SC to Cleveland, OH with a bearing of N 30° W (30° West of North). The plane then flies from Cleveland to Dallas, TX at a S 42° W bearing (42° West of South). How far is Dallas from Charleston (assume Dallas and Charleston are at the same latitude)?

39. Jack wants to build a tree house. His parents worry that he is building it too high. If Jack's dad is looking at the tree house location from a 70° angle and then moves back 10 feet so he can see it at a 50° angle, how high is the tree house location?

40. A telephone pole was recently hit by a car and now leans 6° from the vertical. A point 40 feet away from the base of the pole has an angle of elevation of 36° to the top of the pole. How tall is the pole?

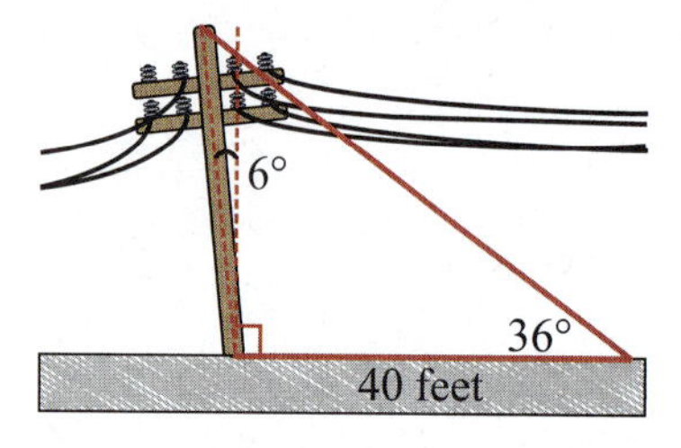

41. As Brandy prepares to land her plane on the runway, she is descending at a 10° angle from the horizontal. Behind her is a marker on the ground that is 500 feet from the runway, and at the marker the angle between the ground and Brandy's plane is 50°. What is the actual distance (not ground distance) between Brandy's plane and the runway?

42. A surveyor sets up two positions A and B 500 yards apart as a baseline on a beach. From position A, he measures an angle of 75° between the baseline and a buoy offshore. From position B he measures an angle of 50° between the baseline and the buoy. How far is the buoy from each of the two positions?

43. Kristin is playing miniature golf. She hits the ball and it bounces off a brick, making a 110° angle. Her ball comes to a stop 8 feet away at a 20° angle from where it started. How far did the ball travel?

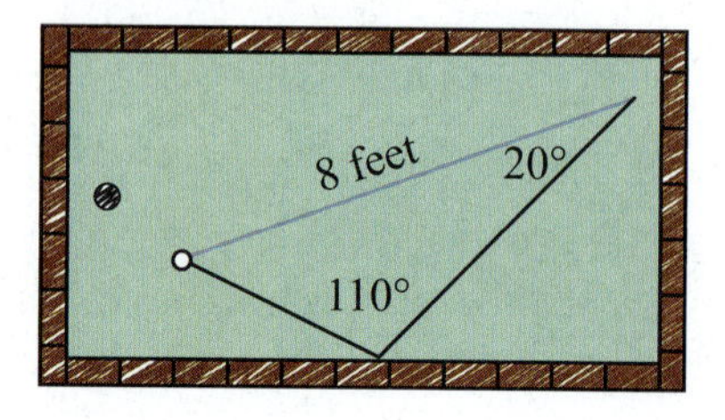

44. Janet is racing her friend Susan. Susan runs 10° away from Janet for 2000 feet. If she has to turn at a 50° angle to get back to Janet's path, how much shorter was Janet's run?

45. Two pieces of mail blew out into the yard. When Bob went to pick them up, he walked 10 feet to the first piece, turned 40° and walked to the next piece, and finally turned 150° to walk back to where he started. How far did Bob walk?

46. A horizontal bridge is suspended over a gorge, with the bridge and the sides of the gorge forming a downward-pointing isosceles triangle. If the bridge makes an 80° angle with the side of the gorge and the bridge is 1000 feet long, how deep is the gorge?

47. Brittany and Jim are playing catch. They are standing 30 feet away from each other. Ryan wants to join them and stands at a 50° angle away from Jim and at a 70° angle away from Brittany. How far away is Ryan from Jim and Ryan from Brittany?

48. Alan is golfing and sets up for a long drive. He slices it and hits a tree 80 feet away. The ball ricochets off of it at a 110° angle and comes to a stop 20° away from the direction he hit it. How far from Alan did the ball land?

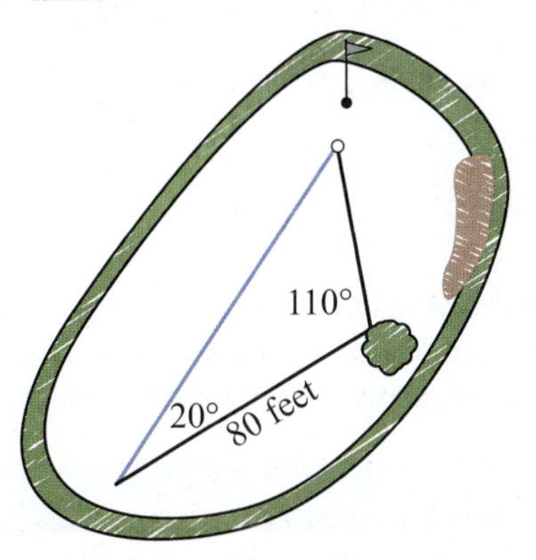

49. An airplane has to fly between 3 airports. The trip from the 1st to the 2nd is 120 miles. After landing at the 2nd airport, the airplane must turn 140° to head toward the 3rd airport. At the 3rd airport it must turn 100° to head to the first airport. How far does the airplane have to travel from the 2nd airport to the 3rd?

50. A ping pong net has become bent at a 70° angle instead of a 90° angle. The bottom of the net is 4.5 feet away from the end of the table. If the top of the net is 4.35 feet away from the end of the table, how high is the net?

51. A gymnast bends over backward until her hands touch the ground, at which point there is an angle of 60° between an imaginary line from her waist to her feet and an imaginary line from her waist to her hands. If the distance from her feet to her waist is 3 feet and the distance from her feet to her hands is 3.3 feet, what is the distance between her waist and hands?

52. Nancy wants to plant wildflowers between the two intersecting paths in her garden. If the paths intersect at a 72° angle and she wants the flowers to extend 12 feet down one path and 15 feet down the other, how large is the area she wants to plant?

53. An A-frame house overlooking the Atlantic Ocean has windows entirely covering one end. If the roof intersects at a 54° angle and the roof is 21 feet long from peak to ground, how much area do the windows cover?

9.2 THE LAW OF COSINES

TOPICS

1. The Law of Cosines
2. Applications of the Law of Cosines
3. Heron's Formula and the area of a triangle

TOPIC 1: The Law of Cosines

The Law of Sines is of no use if we are given information about three sides of a triangle or two sides and the included angle of a triangle. Fortunately, the Law of Cosines handles the SSS and SAS cases easily.

THEOREM: The Law of Cosines

Given a triangle ABC, with sides labeled conventionally, the following equations are all true. These equations represent the **Law of Cosines**.

$$a^2 = b^2 + c^2 - 2bc\cos A$$
$$b^2 = a^2 + c^2 - 2ac\cos B$$
$$c^2 = a^2 + b^2 - 2ab\cos C$$

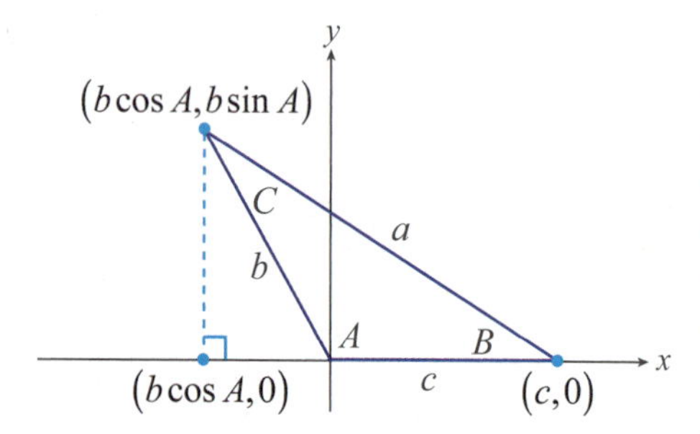

FIGURE 1: Proving the Law of Cosines

The similarity of these statements to the Pythagorean Theorem is no coincidence, and you should verify that the Law of Cosines, when applied to a right triangle, reduces to the Pythagorean Theorem. The Law of Cosines is thus an extension of the simpler theorem, but its proof depends on constructing an auxiliary right triangle to which we can apply the Pythagorean Theorem. To prove the first statement, for example, position the triangle under consideration so that angle A is in standard position in the plane, as shown in Figure 1.

Applying the Pythagorean Theorem to the right triangle formed by the points $(c, 0)$, $(b\cos A, b\sin A)$, and $(b\cos A, 0)$, we obtain

$$\begin{aligned} a^2 &= (c - b\cos A)^2 + (b\sin A)^2 \\ &= c^2 - 2bc\cos A + b^2\cos^2 A + b^2\sin^2 A \\ &= b^2(\cos^2 A + \sin^2 A) + c^2 - 2bc\cos A \\ &= b^2 + c^2 - 2bc\cos A. \end{aligned}$$

The other two parts of the Law of Cosines are proved similarly.

TOPIC 2: Applications of the Law of Cosines

In solving triangles, it's useful to remember a simple fact: the longest side of a triangle is opposite the largest angle. This is important especially when working backward to determine an angle from the sine of the angle, as we will see in Example 1. And to avoid possible confusion when applying the Law of Cosines to solve for an angle, it's a good idea to solve for the largest angle first. If the largest angle is obtuse, the other two angles must be acute (a triangle can have at most one obtuse angle). If the largest angle is acute, this means the other two angles (which are by definition smaller) must also be acute.

Example 1: Using the Law of Cosines in an SSS Situation

Determine the three angles for a triangle in which $a = 3$ inches, $b = 5$ inches, and $c = 7$ inches.

Solution

Guided by the previous observations, we'll solve for C, the largest angle, first. By the Law of Cosines,

$$7^2 = 3^2 + 5^2 - 2(3)(5)\cos C$$
$$49 = 34 - 30\cos C$$

and so $\cos C = \frac{(49-34)}{(-30)} = -\frac{1}{2}$. Since $\cos C$ is negative, we know C is obtuse, and a calculator tells us that $C = \cos^{-1}(-0.5) = 120°$. (We can also determine C without a calculator by first determining C's reference angle; since $\cos 60° = 0.5$, we know that $C = 180° - 60° = 120°$.)

We can now use either law to determine another angle. Using the Law of Sines to determine A, we have

$$\frac{\sin A}{3} = \frac{\sin 120°}{7}$$

and so $\sin A \approx 0.37$. Using a calculator, this means $A \approx 21.79°$ (and we know that A is not $180° - 21.79°$ because we have already determined that A is acute). We can now determine that $B \approx 180° - 120° - 21.79° = 38.21°$.

Example 2: Using the Law of Cosines in an SAS Situation

The course of a sailboat race instructs the sailors to head due east 11 kilometers to the first buoy. They are then to veer off to port (to the left) by 20° and proceed for another 15 kilometers, at which point they should find the second buoy. What bearing would a sailor take to go from the starting point directly to the second buoy?

Solution

A picture is definitely in order.

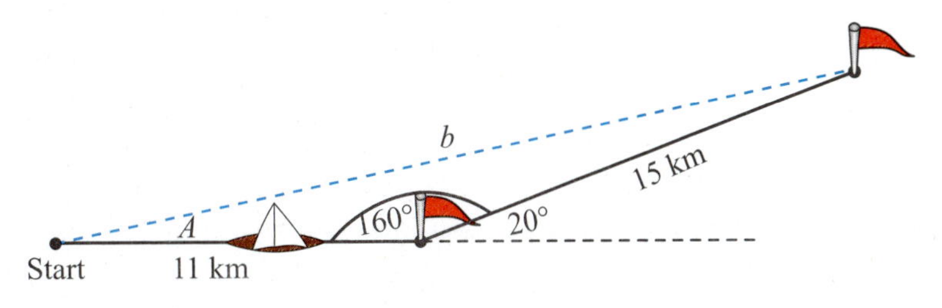

FIGURE 2

We wish to determine A, but neither law allows us to do so directly. However, the Law of Cosines will allow us to determine the third side, which we have labeled b in the diagram.

$$\begin{aligned} b^2 &= 11^2 + 15^2 - 2(11)(15)\cos 160° \\ &= 121 + 225 - 330\cos 160° \\ &\approx 656.10 \end{aligned}$$

This gives us $b \approx 25.61$ kilometers, and we can now apply the Law of Sines to determine $\sin A$.

$$\frac{\sin A}{15} = \frac{\sin 160°}{25.61}$$

This gives us $\sin A \approx 0.20$ and thus $A \approx 11.54°$ North of East.

TOPIC 3: Heron's Formula and the Area of a Triangle

Recall that the area of a triangle can be found with any one of the versions of the Sine Formula.

$$\text{Area } = \frac{1}{2}ab\sin C = \frac{1}{2}bc\sin A = \frac{1}{2}ac\sin B$$

Remarkably, there is also a formula for triangular area that does not depend on knowing *any* of the three angles. The formula is usually called Heron's Formula (Heron was an Alexandrian mathematician of the first century AD), but there is evidence that Archimedes (287 BC–212 BC) and mathematicians of other cultures had also discovered it.

THEOREM: Area of a Triangle (Heron's Formula)

Given a triangle with sides a, b, and c, let $s = \frac{a+b+c}{2}$. Then the following is true.

$$\text{Area} = \sqrt{s(s-a)(s-b)(s-c)}$$

The Sine Formula for the area of a triangle and the Law of Cosines allow us to derive Heron's Formula. Beginning with the Sine Formula for area, followed by one of the Pythagorean identities and some factoring, we know

$$\begin{aligned} \text{Area}^2 &= \frac{1}{4}a^2b^2\sin^2 C \\ &= \frac{1}{4}a^2b^2\left(1-\cos^2 C\right) \\ &= \frac{1}{4}a^2b^2(1-\cos C)(1+\cos C). \end{aligned}$$

By the Law of Cosines, we can rewrite the last two factors as follows.

$$1-\cos C = 1-\frac{a^2+b^2-c^2}{2ab} \qquad 1+\cos C = 1+\frac{a^2+b^2-c^2}{2ab}$$

$$= \frac{2ab-a^2-b^2+c^2}{2ab} \qquad = \frac{2ab+a^2+b^2-c^2}{2ab}$$

$$= \frac{c^2-\left(a^2-2ab+b^2\right)}{2ab} \qquad = \frac{\left(a^2+2ab+b^2\right)-c^2}{2ab}$$

$$= \frac{c^2-(a-b)^2}{2ab} \qquad = \frac{(a+b)^2-c^2}{2ab}$$

$$= \frac{(c+a-b)(c-a+b)}{2ab} \qquad = \frac{(a+b+c)(a+b-c)}{2ab}$$

Replacing the above expressions for the original factors, we now have

$$\begin{aligned}
\text{Area}^2 &= \frac{1}{4}a^2b^2\left(\frac{(c+a-b)(c-a+b)}{2ab}\right)\left(\frac{(a+b+c)(a+b-c)}{2ab}\right)\\
&= \frac{1}{16}(a+b+c)(c-a+b)(c+a-b)(a+b-c)\\
&= \left(\frac{a+b+c}{2}\right)\left(\frac{-a+b+c}{2}\right)\left(\frac{a-b+c}{2}\right)\left(\frac{a+b-c}{2}\right)\\
&= s(s-a)(s-b)(s-c).
\end{aligned}$$

Taking the square root of both sides completes the process.

Example 3: Using Heron's Formula to Find the Area of a Triangle

A set designer is putting together a backdrop for a play, and one element of the scene is a large triangular piece of wood. The edges of the triangle are of lengths 4 meters, 7 meters, and 9 meters. She wants to know the square area of the triangle in order to estimate the amount of paint needed to cover it.

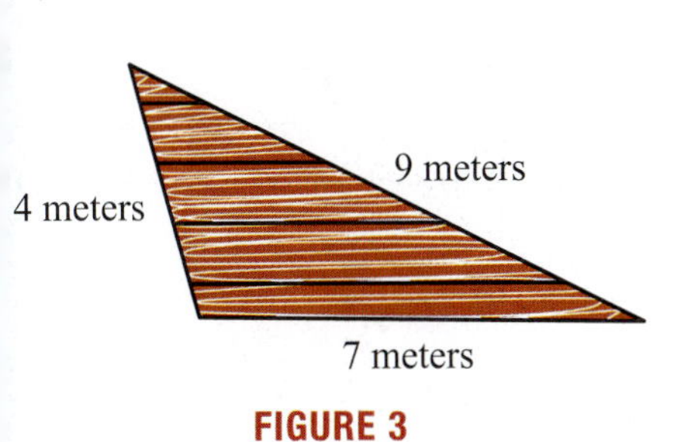

FIGURE 3

Solution

Heron's Formula is applied as follows:

$$s = \frac{4+7+9}{2} = 10$$

and so

$$\begin{aligned}
\text{Area} &= \sqrt{(10)(10-4)(10-7)(10-9)}\\
&= \sqrt{(10)(6)(3)(1)}\\
&= \sqrt{180}\\
&= 6\sqrt{5}\\
&\approx 13.4 \text{ m}^2.
\end{aligned}$$

9.2 EXERCISES

PRACTICE

Solve for the remaining angles and side of the triangles. See Example 2.

1. $A = 60°, b = 3, c = 7$

2. $A = 40°, b = 2, c = 3$

3. $B = 50°, a = 4, c = 6$

4. $B = 45°, a = 5, c = 4$

5. $C = 30°, a = 8, b = 6$

6. $A = 110°, b = 2, c = 1$

7. $C = 70°, a = 5, b = 7$

8. $B = 100°, a = 1, c = 3$

Solve for the angles of the given triangles. See Example 1.

9. $a = 3, b = 4, c = 2$

10. $a = 5, b = 2, c = 6$

11. $a = 8, b = 6, c = 3$

12. $a = 9, b = 4, c = 7$

13. $a = 5, b = 5, c = 5$

14. $a = 6, b = 4, c = 7$

15. $a = 5, b = 3, c = 4$

16. $a = 7, b = 2, c = 8$

Create a triangle, if possible, using the given information and the Law of Cosines. See Examples 1 and 2.

17. $A = 65°, c = 13, b = 7$

18. $C = 35°, b = 12, a = 14$

19. $B = 24.2°, a = 13.3, c = 21.2$

20. $C = 46°7', a = 27.8, b = 19.4$

21. $A = 103°, c = 8, b = 6.3$

22. $C = 75°4', b = 15.4, a = 16.8$

23. $b = 12, c = 9, a = 15$

24. $c = 4.78, b = 16.46, a = 16.54$

25. $b = 4.2, a = 7.6, c = 9.2$

26. $b = 6.84, c = 10.87, a = 7.37$

27. $a = 76.45, b = 94.45, c = 84.42$

28. $a = 5, b = 10, c = 7$

Find the area of the triangle using the given information. See Example 3.

29. $b = 12, c = 18, a = 15$

30. $a = 3, b = 7, c = 8$

31. $a = 5.45, b = 4.83, c = 9$

32. $a = 4.2, b = 9.1, c = 11.5$

APPLICATIONS

33. A log is seen floating down a stream. The log is first spotted 10 feet away. Ten seconds later the log is 70 feet away, making a 60° angle between the two sightings. How far did the log travel?

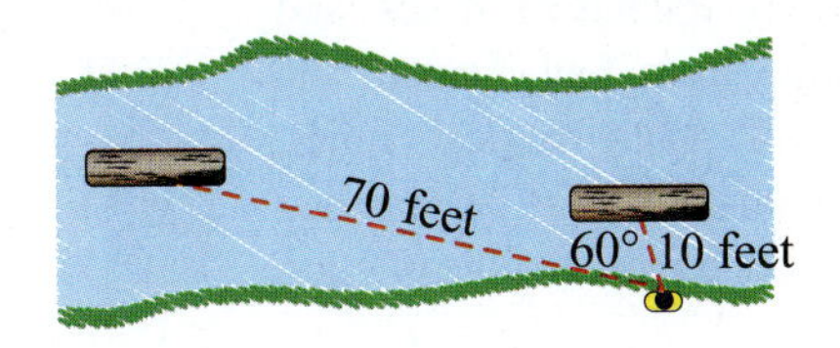

34. A bullet is fired and ricochets off a metal sign 100 feet away, making an 80° angle as it speeds toward a tree where it embeds itself. If the sign and tree are 60 feet apart, how far did the bullet stop from where it was fired?

35. Astronomers once thought the sun revolved around Earth. The sun is 9.3×10^7 miles away and moves 15° across the sky in an hour. Assuming the sun travels in a straight line, how far would it have had to travel?

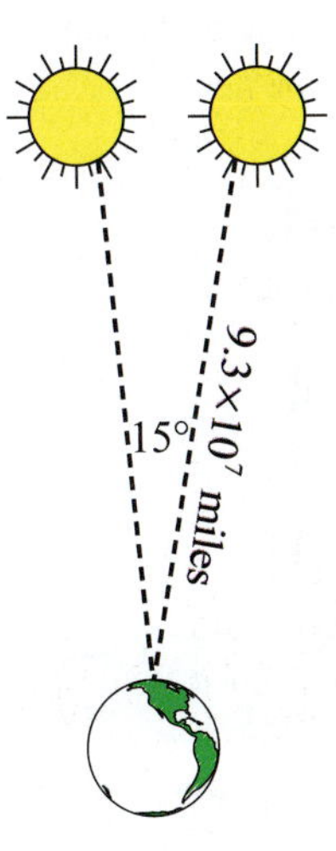

36. A pitcher 60 feet away throws a baseball to Joey. Joey bunts the ball at a 20° angle away from the pitcher. If the ball travels 50 feet, how far does the pitcher have to run to pick up the ball?

37. Nick is surfing a wave that carries him for 20 feet. He executes a sharp turn making a 100° angle. He rides the wave for 5 feet more before he topples into the water. How far is Nick from where he started?

38. A farmer puts a piece of fence across an inside corner of his barn to make a pen for his chickens. The lengths of the sides of the pen are 7 feet, 5 feet, and 8 feet. What are the respective angles?

39. Teresa wants to make a picture frame with two 5-inch and two 12-inch pieces of wood. If the diagonal length is 13 inches, what do the inside angles have to be for the two imaginary triangles?

40. Brian is up to bat. He hits the ball straight at the pitcher 60 feet away. The ball ricochets off the pitcher's shoulder at a 100° angle and comes to rest 40 feet away from the pitcher. How far did the ball travel away from Brian?

41. A plane took off and ascended for 1000 feet before leveling off. Once level, the plane flew for 500 feet, which put it 1480 feet directly away from where it started. After leveling off, what is the angle between the plane's current horizontal flight path and its ascending flight path?

42. Bob wants to build an ice skating rink in his backyard, but his wife says he can only use the part beyond the wood-chipped path running through their yard. How large would his rink be if it is triangular with sides of length 20 feet, 23 feet, and 32 feet?

43. The USS *Cyclops* mysteriously disappeared somewhere in the Bermuda Triangle in 1918. Miami, Florida; San Juan, Puerto Rico; and the Bermudas are generally accepted as the three points of the triangle. The distances from Miami to San Juan and from Miami to the Bermudas are both 908.2 nautical miles and the distance from San Juan to the Bermudas is 839.1 nautical miles. How large an area must be searched to look for the remains of the missing ship?

44. Brian just bought a used sailboat, but it needs a new triangular sail. The dimensions of the sail are 11 feet × 12 feet × 7 feet.

a. What are the measures of the three angles of the sail?
b. How much fabric would a sail of this size require?
c. Suppose he plans to make the sail three different colors by dividing the largest angle so the 12-foot side is split into three sections of 5 feet (blue), 4 feet (yellow), and 3 feet (red), respectively. How much fabric of each color would he need?

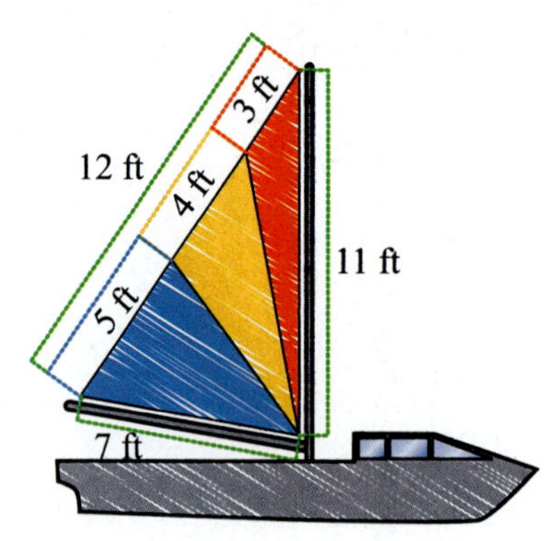

45. Any regular (all sides are equal) n-sided polygon can be divided into n identical triangles by drawing a line from each vertex to the center of the polygon. A pentagon would be divided as shown in the figure.

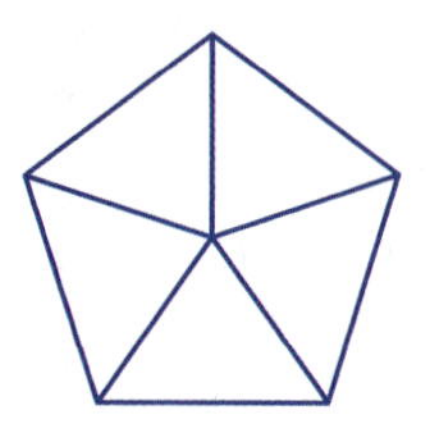

a. If each side has a length of 6 inches, what would the area of the given pentagon be?
b. Using a similar method, what would be the area of an octagon with sides of length 11 inches?
c. What would be the area of a five-pointed star where each line segment has a length of 8 inches?

9.3 POLAR COORDINATES AND POLAR EQUATIONS

TOPICS

1. The polar coordinate system
2. Converting between polar and Cartesian coordinates
3. The form of polar equations
4. Graphing polar equations

TOPIC 1: The Polar Coordinate System

The Cartesian coordinate system has served us well to this point. But there are many situations for which a rectangular coordinate system is not the most natural choice. Some planar images and some equations in two variables have a symmetry that is awkward to express in terms of the familiar x- and y-coordinates. Polar coordinates provide an alternative framework for these cases.

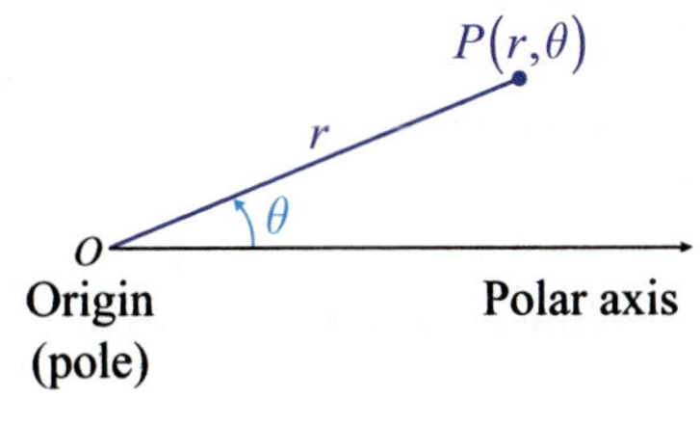

FIGURE 1

The **polar coordinate system**, like the Cartesian coordinate system, serves as a means of locating points in the plane, and both systems are centered at a point O called the **origin**, sometimes referred to as the **pole** in the polar system. Starting from the origin O, a ray (or half-line) called the **polar axis** is drawn; in practice, the polar axis is usually drawn extending horizontally to the right, so that it corresponds with the positive x-axis in the Cartesian system. Now, given any point P in the plane other than the origin, the line segment $\overline{OP}$ has a unique positive length; we will label this length r (as in *radius*). Finally, we let θ denote the angle, measured counterclockwise, between the polar axis and the segment $\overline{OP}$, and we say (r,θ) are **polar coordinates** of the point P (see Figure 1). The origin is the unique point for which $r=0$ and for which the angle θ is irrelevant; the coordinates $(0,\theta)$ refer to O for any angle θ. Figure 2 illustrates the process of determining r and θ for several points.

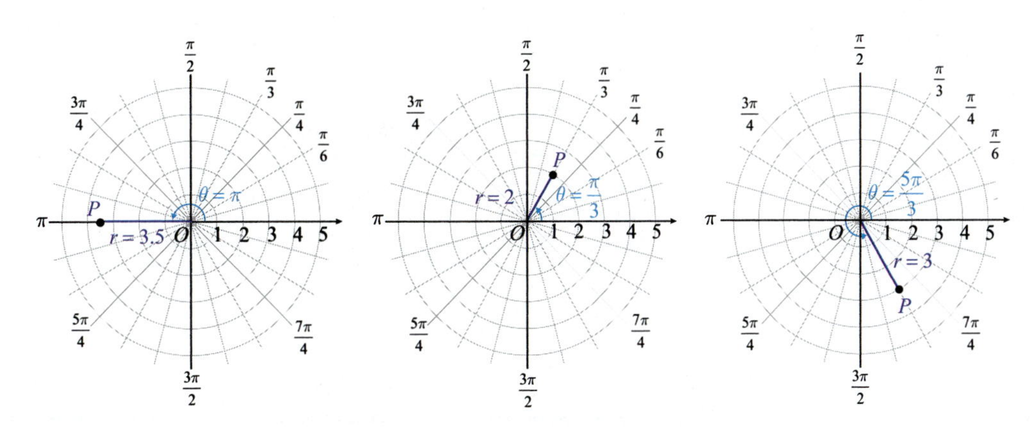

FIGURE 2: The Polar Coordinate System

Before proceeding further, it is important to recognize one very critical difference between Cartesian and polar coordinates. A given point P corresponds to unique Cartesian coordinates (x, y), but the polar coordinates (r, θ) of P as described above are only one of an infinite number of ways of specifying P in the polar system. Our familiarity with trigonometric functions indicates one reason for this: $(r, \theta + 2n\pi)$ also represents P, since θ and $\theta + 2n\pi$ have the same terminal sides for any integer n. Further, $(-r, \theta + (2n+1)\pi)$ also represents P for any integer n, given the interpretation that $-r$ indicates travel in the opposite direction through the origin. These observations are illustrated in Figure 3 with alternate descriptions of the points from Figure 2.

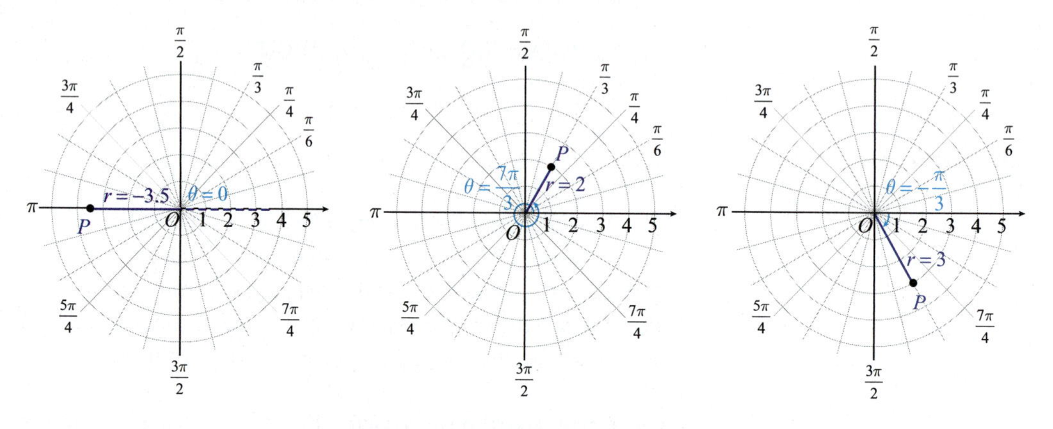

FIGURE 3: Alternate Polar Coordinates

Example 1: Plotting in Polar Coordinates

Plot the following points given by the polar coordinates:

$$\left(2, \frac{3\pi}{4}\right), \left(3.5, -\frac{5\pi}{2}\right), \left(-1, \frac{4\pi}{3}\right)$$

Solution

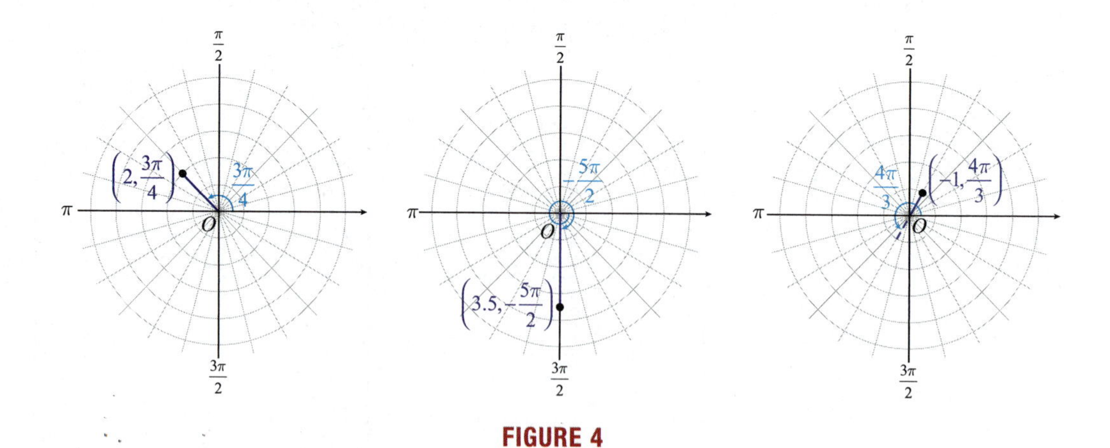

FIGURE 4

TOPIC 2: Converting between Polar and Cartesian Coordinates

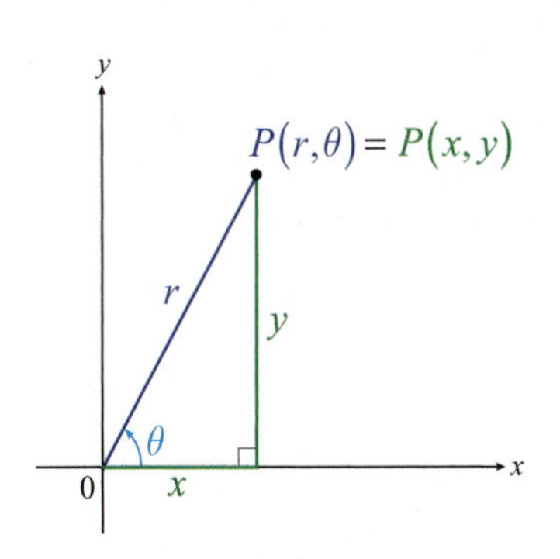

FIGURE 5:

Converting between Polar and Cartesian Coordinates

Since we have two systems by which to specify points in the plane, it should come as no surprise that we will occasionally need to be able to translate information from one system to the other. As we will soon see, this will be especially useful when faced with an equation that is awkward to graph in one coordinate system, but straightforward in the other. Fortunately, converting from Cartesian coordinates to polar coordinates, and vice versa, is easily accomplished. To do so, we will assume that the polar axis is aligned with the positive x-axis and that a fixed point P has Cartesian coordinates (x, y) and polar coordinates (r, θ). Then, as seen in Figure 5, $r^2 = x^2 + y^2$ and

$$\cos\theta = \frac{x}{r}, \quad \sin\theta = \frac{y}{r}, \quad \text{and} \quad \tan\theta = \frac{y}{x}.$$

These relations, and the diagram in Figure 5, should seem familiar—we encountered them first in defining the trigonometric functions. They are restated here as conversion formulas.

f(x) FORMULA: Coordinate Conversion

Converting from Polar to Cartesian Coordinates

Given (r, θ), x and y are defined as follows.

$$x = r\cos\theta, \quad y = r\sin\theta$$

Converting from Cartesian to Polar Coordinates

Given (x, y), r and θ are defined as follows.

$$r^2 = x^2 + y^2, \quad \tan\theta = \frac{y}{x} \quad (x \neq 0)$$

Make note of the quadrant of the original Cartesian coordinates when converting, to be sure the polar coordinates fall in the same quadrant (see Example 3).

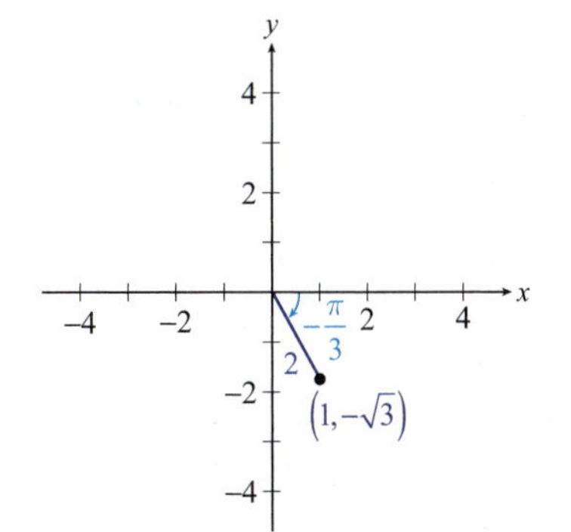

FIGURE 6

Example 2: Converting from Polar to Cartesian Coordinates

Convert the following points from polar to Cartesian coordinates.

a. $\left(2, -\frac{\pi}{3}\right)$ **b.** $\left(-3, \frac{\pi}{4}\right)$

Solution

a. We calculate $x = 2\cos\left(-\frac{\pi}{3}\right) = 1$ and $y = 2\sin\left(-\frac{\pi}{3}\right) = -\sqrt{3}$, so the Cartesian coordinates are $\left(1, -\sqrt{3}\right)$ (see Figure 6).

b. We calculate $x = -3\cos\left(\frac{\pi}{4}\right) = -\frac{3}{\sqrt{2}}$ and $y = -3\sin\left(\frac{\pi}{4}\right) = -\frac{3}{\sqrt{2}}$, so the Cartesian coordinates are $\left(-\frac{3}{\sqrt{2}}, -\frac{3}{\sqrt{2}}\right)$ (see Figure 7).

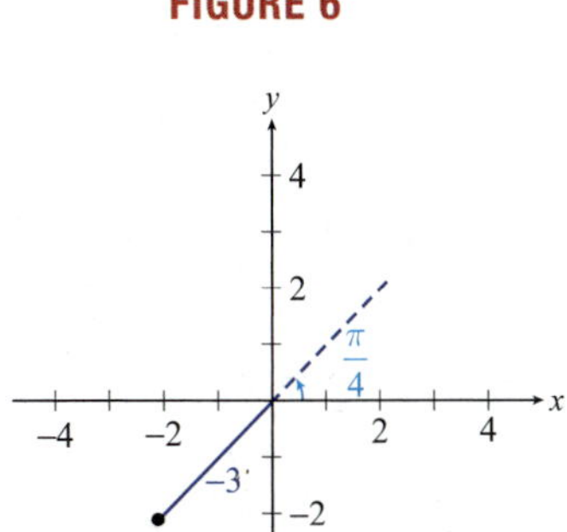

FIGURE 7

Example 3: Converting from Cartesian to Polar Coordinates

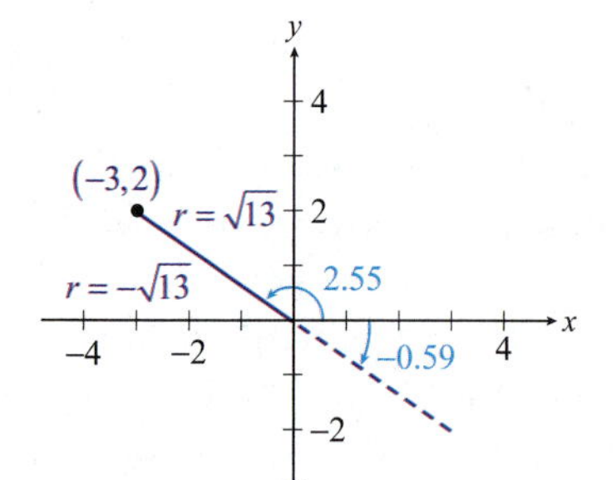

FIGURE 8

Convert the following points from Cartesian to polar coordinates.

a. $(-3,2)$ **b.** $(\sqrt{3},-1)$

Solution

a. To avoid making an error, be sure you have some rough idea of what the answer should be before doing any calculations. Since $(-3,2)$ is in the second quadrant, one possible conversion will lead to $\frac{\pi}{2}<\theta<\pi$. To get the exact angle, we use the fact that $\tan\theta=-\frac{2}{3}$, so $\theta=\tan^{-1}\left(-\frac{2}{3}\right)\approx 2.55$ (remember, this is in radians). Depending on your calculator, though, you may have found $\tan^{-1}\left(-\frac{2}{3}\right)\approx -0.59$, an angle in the fourth quadrant; if so, it is up to you to remember to either add π in order to get an angle in the second quadrant, or to use a negative value for r. The radius is more easily determined: $r=\sqrt{(-3)^2+(2)^2}=\sqrt{13}$. The two answers we have found are thus $(\sqrt{13},2.55)$ and $(-\sqrt{13},-0.59)$ (see Figure 8).

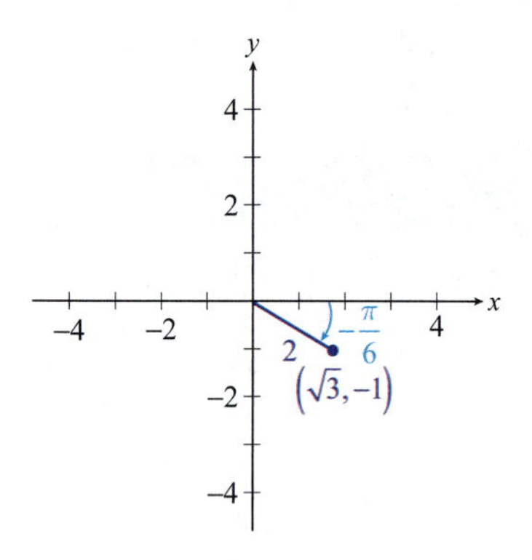

FIGURE 9

b. The polar coordinates of $(\sqrt{3},-1)$ are more easily determined. The point lies in the fourth quadrant, so the two most obvious conversions will lead to either $\frac{3\pi}{2}<\theta<2\pi$ or $-\frac{\pi}{2}<\theta<0$. The coordinates $(\sqrt{3},-1)$ give rise to a familiar triangle with convenient angles, and it's probably easiest to use $\theta=-\frac{\pi}{6}$. The radius is $r=2$, so one set of polar coordinates for the point is $\left(2,-\frac{\pi}{6}\right)$ (see Figure 9).

TOPIC 3: The Form of Polar Equations

In the most general terms, a *polar equation* is an equation in the variables r and θ that defines a relationship between these polar coordinates (just as equations in x and y define a relationship between rectangular coordinates). A solution to a polar equation is thus an ordered pair (r,θ) that makes the equation true, and the graph of a polar equation consists of all such ordered pairs. Many polar equations can be written in the form $r=f(\theta)$; in such a form, the distance from the origin is expressed as a function of the angle θ and the graph of the equation is usually fairly easy to determine. We will soon study many examples of polar equations, but first we will examine the process of translating an equation from one coordinate system to another.

The basic tools for such translation are the formulas for coordinate conversion used in Examples 2 and 3. Translating from rectangular coordinates to polar coordinates is particularly straightforward: simply replace every occurrence of x with $r\cos\theta$ and every occurrence of y with $r\sin\theta$. Translation in the opposite direction may require significantly more effort.

Example 4: Rewriting an Equation in Polar Form

Rewrite the equation $x^2 - 2x + y^2 = 0$ in polar form.

Solution

Making the appropriate substitutions in the equation $x^2 - 2x + y^2 = 0$ and simplifying, we obtain the following.

$$\begin{aligned}(r\cos\theta)^2 - 2(r\cos\theta) + (r\sin\theta)^2 &= 0\\ r^2\cos^2\theta - 2r\cos\theta + r^2\sin^2\theta &= 0\\ r^2(\cos^2\theta + \sin^2\theta) - 2r\cos\theta &= 0\\ r^2 - 2r\cos\theta &= 0\end{aligned}$$

Example 5: Rewriting an Equation in Rectangular Form

Rewrite the equation $2r = \sec\theta$ in rectangular form.

Solution

One good way to begin is to rewrite the equation as follows.

$$\begin{aligned}2r &= \sec\theta\\ 2r &= \frac{1}{\cos\theta}\\ r\cos\theta &= \frac{1}{2}\end{aligned}$$

Now we recognize the term on the left-hand side as x, and we see that the equation is $x = \frac{1}{2}$.

TOPIC 4: Graphing Polar Equations

Gaining a sense of confidence in your ability to graph polar equations calls for nothing more than familiarity with a number of examples. As with equations in rectangular coordinates, we will begin with some simple, but illustrative, examples.

Example 6: Graphing Polar Equations

Sketch the graphs of the following polar equations, and then convert the equations to rectangular coordinates.

a. $r = 3$ **b.** $\theta = \frac{2\pi}{3}$

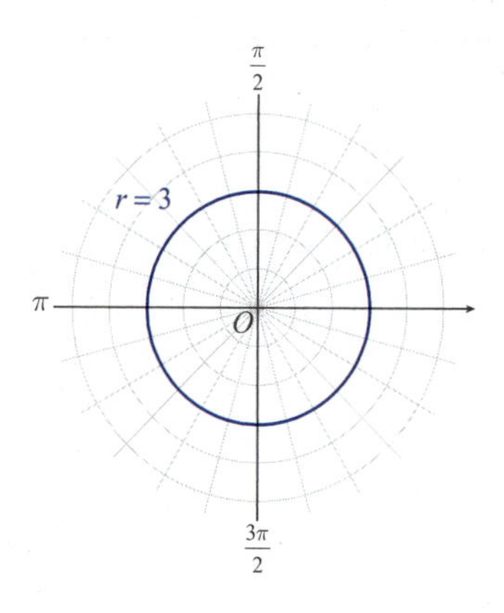

FIGURE 10

Solution

a. Since θ doesn't appear in the equation $r = 3$, θ is allowed to take on any value. The equation thus describes all points (r, θ) for which $r = 3$; that is, a circle of radius 3, as shown in Figure 10. The equation in rectangular coordinates is $x^2 + y^2 = 9$.

b. In this equation, r is allowed to take on any value. But every point (r, θ) that satisfies the equation must have $\theta = \frac{2\pi}{3}$. The graph is thus a straight line passing through the origin (see Figure 11), and $\frac{y}{x} = \tan\theta = \tan\left(\frac{2\pi}{3}\right) = -\sqrt{3}$, so $y = -\sqrt{3}x$.

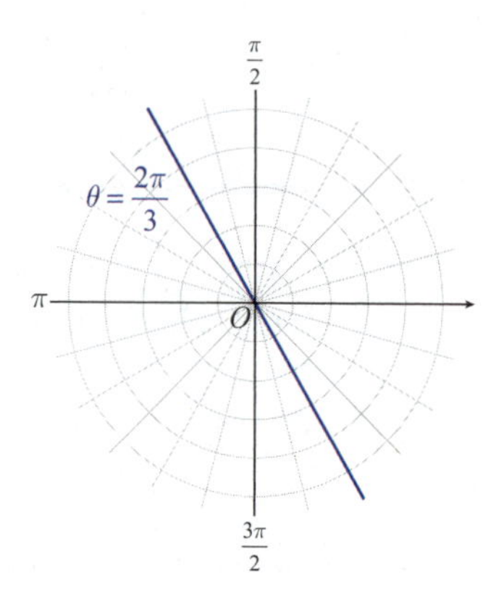

FIGURE 11

In general we expect an equation in polar coordinates to contain both r's and θ's. As with rectangular coordinates, the most basic approach to sketching the graph of a polar equation is to plot some representative points and to connect the points as seems appropriate. This method, applied judiciously and perhaps in combination with some algebra, will take us far.

Example 7: Graphing Polar Equations

Sketch the graph of the equation $r = 2\cos\theta$.

Solution

We can begin by calculating some values of r for given values of θ.

θ	0	$\frac{\pi}{6}$	$\frac{\pi}{4}$	$\frac{\pi}{3}$	$\frac{\pi}{2}$	$\frac{2\pi}{3}$	$\frac{3\pi}{4}$	$\frac{5\pi}{6}$
r	2	$\sqrt{3}$	$\sqrt{2}$	1	0	-1	$-\sqrt{2}$	$-\sqrt{3}$

TABLE 1

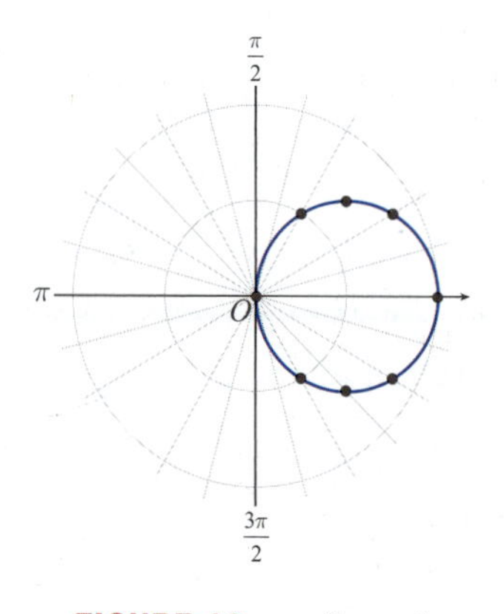

FIGURE 12: $r = 2\cos\theta$

Now if we plot the pairs (r, θ) from Table 1, we obtain the points shown in Figure 12. These certainly appear to lie along the circumference of a circle, so the points have been connected with a curve. If we convert the equation into rectangular coordinates, we see that the graph indeed is a circle of radius 1 centered at $x = 1, y = 0$.

$$r = 2\cos\theta$$

$$r^2 = 2r\cos\theta \quad \text{Multiply both sides by } r.$$

$$x^2 + y^2 = 2x \quad \text{Convert to rectangular coordinates.}$$

$$x^2 - 2x + y^2 = 0$$

$$x^2 - 2x + 1 + y^2 = 1 \quad \text{Complete the square.}$$

$$(x-1)^2 + y^2 = 1$$

We have used symmetry in the past as an aid in graphing functions and equations, and the concept is no less useful in polar coordinates. Consider the three types of symmetry illustrated in Figure 13.

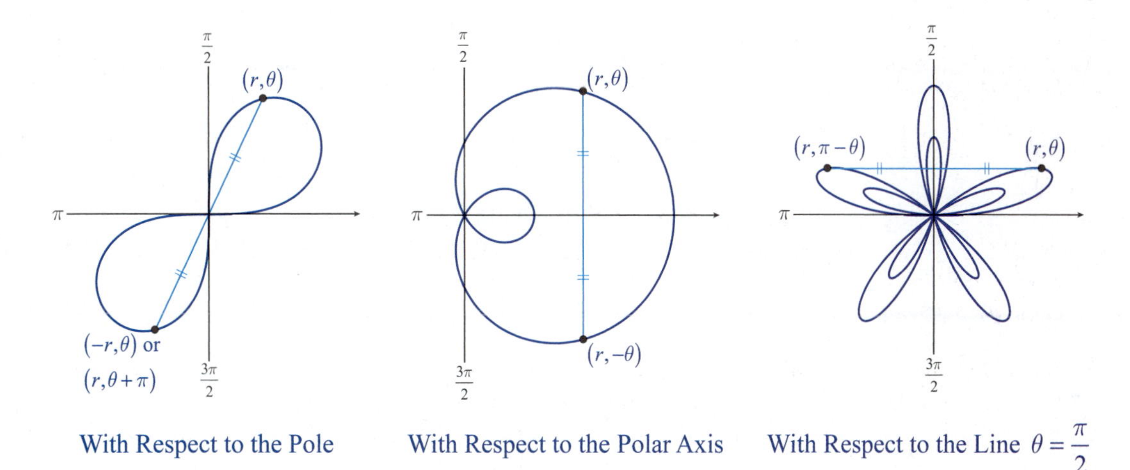

FIGURE 13: Symmetry in Polar Coordinates

Algebraically, symmetry can be recognized with the following tests.

DEFINITION: Symmetry of Polar Equations

An equation in r and θ is **symmetric with respect to**

1. the **pole** if replacing r with $-r$ (or replacing θ with $\theta + \pi$) results in an equivalent equation;
2. the **polar axis** if replacing θ with $-\theta$ results in an equivalent equation;
3. the **vertical line** $\theta = \frac{\pi}{2}$ if replacing θ with $\pi - \theta$ results in an equivalent equation.

We will conclude this section with a catalog of some polar equations that arise frequently enough to have been given names. Exploring a few of these further will give us the opportunity to apply the symmetry tests and gain more familiarity with graphing in polar coordinates.

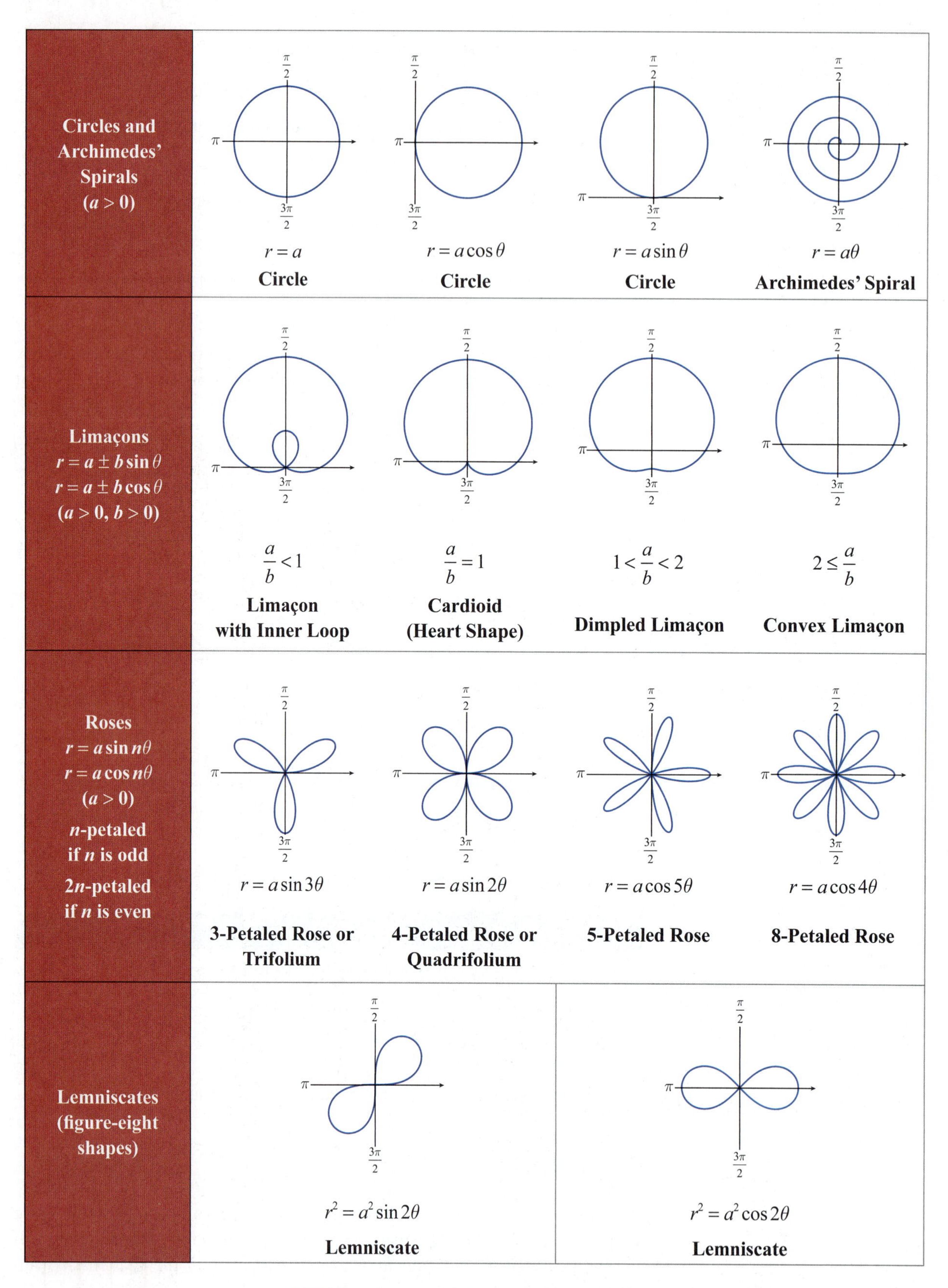

TABLE 2: Common Polar Equations and Graphs

Example 8: Graphing Common Polar Equations Using Symmetry

Use symmetry and the table of common polar equations and graphs to sketch the graph of $r = 3\cos 4\theta$.

Solution

The equation $r = 3\cos 4\theta$ possesses all three kinds of symmetry, and reference to Table 2 indicates that the graph is an 8-petaled rose. Several aspects of the graph can be determined easily, such as the fact that the maximum distance between the origin and points on the graph is 3; this follows from the fact that $-1 \le \cos 4\theta \le 1$ for all θ. Some points can be easily calculated by hand, but a graphing utility is very useful in generating a large number of points. The polar coordinates of points on the graph for $0 \le \theta \le 2\pi$ in increments of $\frac{\pi}{16}$, are shown in Table 3.

θ	0	$\frac{\pi}{16}$	$\frac{\pi}{8}$	$\frac{3\pi}{16}$	$\frac{\pi}{4}$	$\frac{5\pi}{16}$	$\frac{3\pi}{8}$	$\frac{7\pi}{16}$	$\frac{\pi}{2}$	$\frac{9\pi}{16}$	$\frac{5\pi}{8}$
r	3	$\frac{3}{\sqrt{2}}$	0	$-\frac{3}{\sqrt{2}}$	-3	$-\frac{3}{\sqrt{2}}$	0	$\frac{3}{\sqrt{2}}$	3	$\frac{3}{\sqrt{2}}$	0
θ	$\frac{11\pi}{16}$	$\frac{3\pi}{4}$	$\frac{13\pi}{16}$	$\frac{7\pi}{8}$	$\frac{15\pi}{16}$	π	$\frac{17\pi}{16}$	$\frac{9\pi}{8}$	$\frac{19\pi}{16}$	$\frac{5\pi}{4}$	$\frac{21\pi}{16}$
r	$-\frac{3}{\sqrt{2}}$	-3	$-\frac{3}{\sqrt{2}}$	0	$\frac{3}{\sqrt{2}}$	3	$\frac{3}{\sqrt{2}}$	0	$-\frac{3}{\sqrt{2}}$	-3	$-\frac{3}{\sqrt{2}}$
θ	$\frac{11\pi}{8}$	$\frac{23\pi}{16}$	$\frac{3\pi}{2}$	$\frac{25\pi}{16}$	$\frac{13\pi}{8}$	$\frac{27\pi}{16}$	$\frac{7\pi}{4}$	$\frac{29\pi}{16}$	$\frac{15\pi}{8}$	$\frac{31\pi}{16}$	2π
r	0	$\frac{3}{\sqrt{2}}$	3	$\frac{3}{\sqrt{2}}$	0	$-\frac{3}{\sqrt{2}}$	-3	$-\frac{3}{\sqrt{2}}$	0	$\frac{3}{\sqrt{2}}$	3

TABLE 3

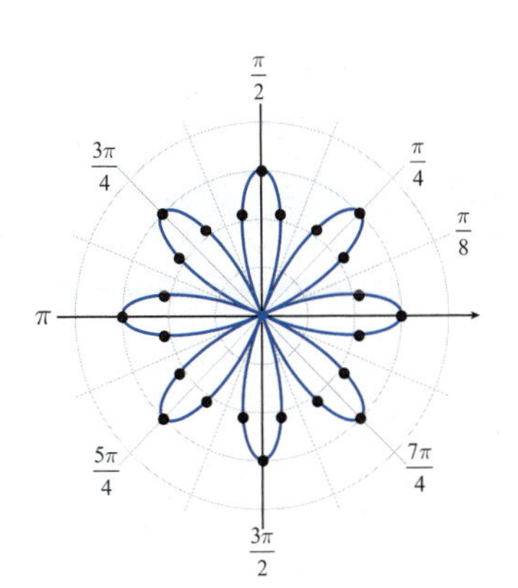

FIGURE 14: $r = 3\cos 4\theta$

These points, and the rest of the graph, are plotted in Figure 14.

Example 9: Graphing Common Polar Equations Using Symmetry

Use symmetry and the table of common polar equations and graphs to sketch the graph of $r = 1 + 2\cos\theta$.

Solution

The graph of $r = 1 + 2\cos\theta$ is a limaçon with an inner loop. Since cosine is an even function,

$$1 + 2\cos\theta = 1 + 2\cos(-\theta)$$

and the graph is symmetric with respect to the polar axis. Some points on the graph are as follows.

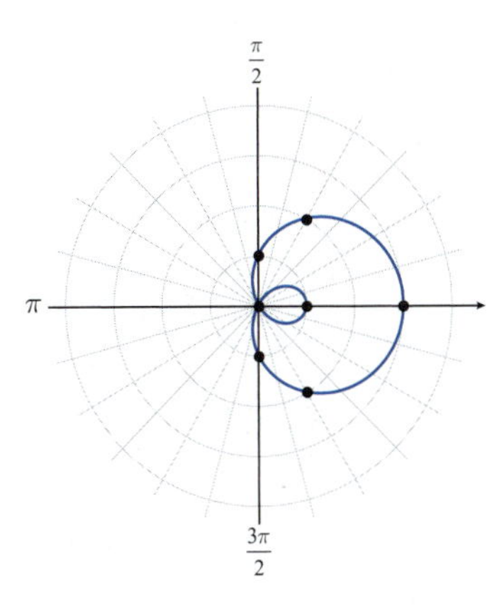

FIGURE 15: $r = 1 + 2\cos\theta$

θ	0	$\frac{\pi}{3}$	$\frac{\pi}{2}$	$\frac{2\pi}{3}$	π	$\frac{4\pi}{3}$	$\frac{3\pi}{2}$	$\frac{5\pi}{3}$	2π
r	3	2	1	0	−1	0	1	2	3

TABLE 4

The graph of $r = 1 + 2\cos\theta$ is shown in Figure 15.

TECHNOLOGY: Graphing Polar Equations

Graphing utilities can be very useful in creating accurate graphs of polar equations, though their capabilities are limited and vary considerably. Figures 16 and 17 show the graphs of a cardioid and a lemniscate on a TI-84 Plus. To generate these graphs, the selected graphing mode needs to be POLAR and the polar equations are entered in the Y = editor. Note that to enter θ, we press X,T,θ,n.

Often, a polar equation must be solved for r in order to use the graphing feature. Note that in order to graph the lemniscate shown, we solved the equation $r^2 = 4\sin 2\theta$ for r and then typed in both the positive and negative square roots that resulted.

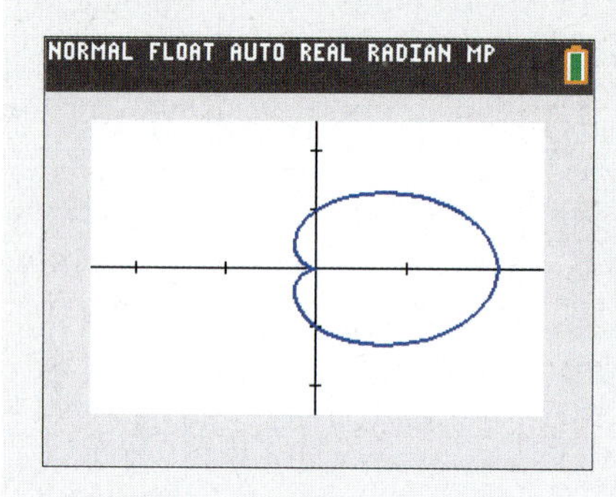

$[-2.5, 2.5]$ by $[-2.5, 2.5]$

FIGURE 16: $r = 1 + \cos\theta$

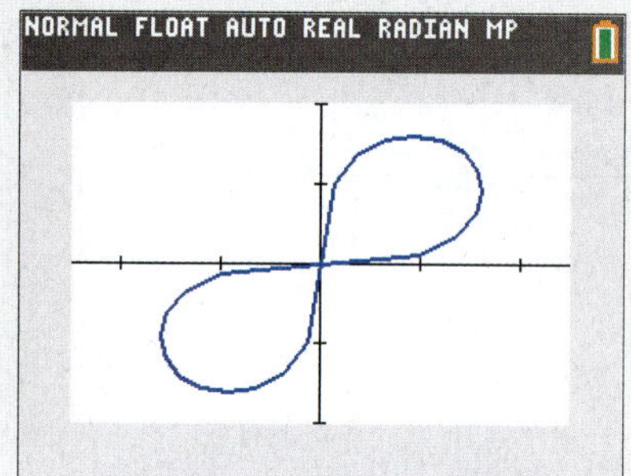

$[-2.5, 2.5]$ by $[-2, 2]$

FIGURE 17: $r^2 = 4\sin 2\theta$

9.3 EXERCISES

PRACTICE

Plot the point given by the polar coordinates. See Example 1.

1. $\left(-1, \frac{5\pi}{4}\right)$

2. $\left(-5, \frac{3\pi}{2}\right)$

3. $\left(\frac{1}{4}, -\frac{7\pi}{6}\right)$

4. $\left(\sqrt{3}, -\frac{\pi}{3}\right)$

5. $\left(\frac{44}{9}, -\pi\right)$

6. $\left(\frac{7}{\sqrt{2}}, \frac{\pi}{2}\right)$

Convert the point from polar to Cartesian coordinates. See Example 2.

7. $\left(5, \frac{7\pi}{4}\right)$

8. $(0, 2\pi)$

9. $\left(6.25, -\frac{3\pi}{4}\right)$

10. $\left(-2.25, \frac{\pi}{4}\right)$

11. $\left(3, -\frac{5\pi}{6}\right)$

12. $\left(-11, \frac{5\pi}{6}\right)$

Convert the point from Cartesian to polar coordinates. See Example 3.

13. $(-3, 0)$

14. $\left(-6, \sqrt{3}\right)$

15. $(12, -1)$

16. $(8, 0)$

17. $\left(-\sqrt{3}, 9\right)$

18. $(-5, -5)$

Rewrite the rectangular equation in polar form. See Example 4.

19. $x^2 + y^2 = 25$

20. $x^2 + y^2 = 81$

21. $x = 12$

22. $y = 16$

23. $y = x$

24. $y = b$

25. $x = 16a$

26. $x^2 + y^2 = a$

27. $x^2 + y^2 = 4ax$

28. $x^2 + y^2 = 4ay$

29. $y^2 - 4 = 4x$

30. $x^2 + y^2 = 36a^2$

Rewrite the polar equation in rectangular form. See Example 5.

31. $r = 5\cos\theta$

32. $r = 8\sin\theta$

33. $r = 7$

34. $\theta = \frac{\pi}{6}$

35. $18r = 9\csc\theta$

36. $r = 2\sec\theta$

37. $r^2 = \sin 2\theta$

38. $r = \frac{2}{1 - \cos\theta}$

39. $r = \frac{12}{4\sin\theta + 7\cos\theta}$

40. $r = \frac{16}{4 + 4\sin\theta}$

Rewrite the polar equation in rectangular form; then sketch the graph. See Examples 6 and 7.

41. $r=2$

42. $r=6$

43. $\theta=\frac{5\pi}{6}$

44. $\theta=\frac{\pi}{4}$

45. $r=7\sec\theta$

46. $r=2\csc\theta$

Sketch a graph of the given polar equation. See Examples 8 and 9.

47. $r=4$

48. $r=5$

49. $\theta=\frac{4\pi}{3}$

50. $\theta=\frac{-\pi}{3}$

51. $r=6\cos\theta$

52. $r=2\sin\theta$

53. $r=3-3\sin\theta$

54. $r=6+5\cos\theta$

55. $r=7(1+\cos\theta)$

56. $r=2(1-2\sin\theta)$

57. $r=4-3\sin\theta$

58. $r=3+4\sin\theta$

59. $r=3\sin 3\theta$

60. $r=5\sin 3\theta$

61. $r=2\sin 2\theta$

62. $r=4\sin 2\theta$

63. $r=5\cos 5\theta$

64. $r=4\cos 5\theta$

65. $r=4\cos 4\theta$

66. $r=3\cos 4\theta$

67. $r^2=16\sin 2\theta$

68. $r^2=9\cos 2\theta$

Find all points of intersection of the given polar curves.

69. $r=\sin\theta,\quad r=\cos\theta$

70. $r=\sin 2\theta,\quad r=\cos\theta$

71. $r=1-\cos\theta,\quad r=1+\sin\theta$

72. $r^2=4\sin\theta,\quad r=1-\sin\theta$

WRITING & THINKING

73. For a fixed real number a, explain in geometric terms how the graphs of $f(\theta)$ and $f(\theta-a)$ are related. (**Hint:** For guidance, recall the rectangular analogue.)

74. a. Describe the graph of $r=\sec\left(\theta-\frac{\pi}{4}\right)$.

b. How are the graphs of $r=k\sec\left(\theta-\frac{\pi}{4}\right)$ related as k ranges over nonzero values? (Do not use graphing technology.)

TECHNOLOGY

Use a graphing utility to sketch each of the given curves. Whenever applicable, explore how different values of the parameter(s) affect the shape of the graph. Experiment with both integer and noninteger parameters.

75. $r = \cos k\theta$

76. $r = 1 - k_1 \sin k_2\theta$

77. $r = \dfrac{1 + k\sin\theta}{1 - k\sin\theta}$

78. $r = \theta\cos\theta, \quad -2\pi \le \theta \le 2\pi$ (Garfield curve)

79. $r = 1 + 2\sin\left(\dfrac{\theta}{2}\right)$ (nephroid of Freeth)

80. $r = k_1 + k_2\theta$

81. $r = 1 - k_1 \cos k_2\theta$

9.4 PARAMETRIC EQUATIONS

TOPICS

1. Applications of parametric equations
2. Graphing parametric equations by eliminating the parameter
3. Constructing parametric equations to describe a graph

TOPIC 1: Applications of Parametric Equations

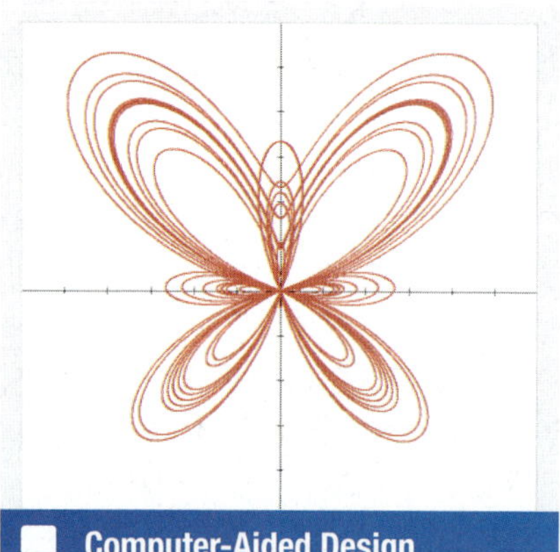

Computer-Aided Design

Parametric equations are often used to depict curves and surfaces in the field of computer-aided design, as this form of representation makes such actions as rotations, translations, and scaling much easier.

Many curves in the plane are most naturally defined in terms of a variable called a **parameter**, a variable different from those representing the coordinates. For instance, if (x, y) denotes the position of a thrown object, both x and y may be thought of as functions of a third variable t, representing time. In fact, thinking of the position (x, y) as a function of time t provides more information than a simple equation relating x and y: by calculating (x, y) for various values of t, we can determine not just the shape of the object's flight, but *where* the object is at any given time. If we have determined exactly how x and y depend on t, so that

$$x = f(t) \quad \text{and} \quad y = g(t)$$

for some functions f and g, we have determined **parametric equations** for the curve traced out by (x, y).

The example provided by a thrown object is important enough to warrant further study, and a thorough analysis will lead to our first illustration of how parametric curves are sketched. In Section 1.8, we saw that the height of an object, moving only under the influence of the force of gravity g, is given by the formula

$$h(t) = -\frac{1}{2}gt^2 + v_0 t + h_0,$$

where v_0 is the initial (vertical) velocity and h_0 is the initial height of the object. In this formula, t represents the variable time; in the derivation that follows, t will retain this meaning, but we will now also consider t as the parameter describing the object's travel.

Only strictly vertical travel was considered in Section 1.8, but we now have the tools to analyze more realistic scenarios. We will set up our coordinate system so that the object starts off at position $(0, h_0)$ at time $t = 0$, and we will assume that it has an initial velocity of v_0 at an angle θ, as shown in Figure 1.

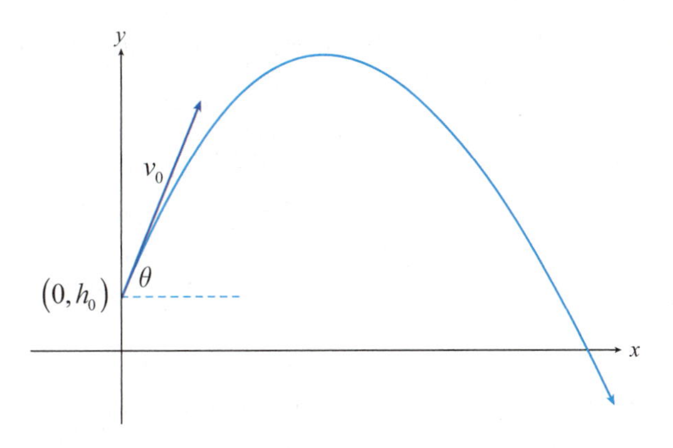

FIGURE 1: Describing an Object's Travel Parametrically

It is important to realize that v_0 is the size of the initial velocity in the direction θ, and that therefore the initial *vertical* velocity is $v_0 \sin\theta$ and the initial *horizontal* velocity is $v_0 \cos\theta$. If we let (x, y) stand for the position of the object at time t, the formula from Section 1.8 tells us

$$y = -\frac{1}{2}gt^2 + (v_0 \sin\theta)t + h_0,$$

so we have succeeded in writing y as a function of t. The situation in the horizontal direction is different. Because there is no force affecting the object's horizontal velocity over time (we are assuming there is no frictional force present), the horizontal displacement is simply the product of the horizontal velocity and time (remember, $\text{distance} = \text{rate} \times \text{time}$). That is,

$$x = (v_0 \cos\theta)t.$$

We now have parametric equations defining the path of the object's travel. A specific application of these general equations follows.

Example 1: Determining Parametric Equations and Their Graphs

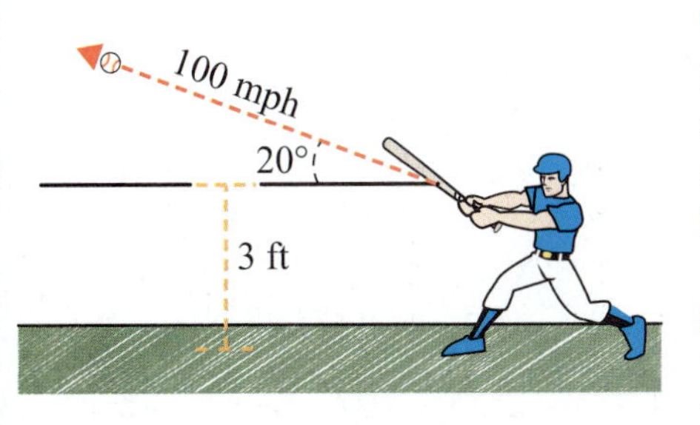

FIGURE 2

Suppose a baseball is hit 3 feet above the ground, and that it leaves the bat at a speed of 100 miles an hour at an angle of 20° from the horizontal. Construct a set of parametric equations describing the ball's flight, and sketch a graph of the ball's travel.

Solution

With the information given, it's appropriate to work in units of feet and seconds. Recall that $g = 32 \text{ ft/s}^2$, and that all the quantities must be expressed in the same units. Hence,

$$h_0 = 3 \text{ feet}$$

and

$$v_0 = 100\ \frac{\text{miles}}{\text{hour}} = \left(\frac{100 \text{ miles}}{\text{hour}}\right)\left(\frac{5280 \text{ feet}}{1 \text{ mile}}\right)\left(\frac{1 \text{ hour}}{3600 \text{ seconds}}\right) \approx 146.7\ \frac{\text{feet}}{\text{second}}.$$

We can now determine the parametric equations:

$$y = -\frac{1}{2}(32)t^2 + (146.7)(\sin 20°)t + 3$$
$$\approx -16t^2 + 50.2t + 3$$

and

$$x = (146.7)(\cos 20°)t$$
$$\approx 137.9t.$$

For the purposes of this application, the part of the curve we are interested in begins with $t = 0$, the moment the bat hits the ball. Note that at this time, $x = 0$ and $y = 3$; that is, the ball starts off at the point $(0,3)$, as we intended. As time moves on, the x-coordinate increases and the y-coordinate at first increases and then decreases. Considering the physical situation giving rise to the equations, we are probably only interested in the curve up to the point where $y = 0$ (when the ball hits the ground). If we calculate (x, y) for values of t from 0 to 3.5, in half-second increments, we obtain the following table:

t	0	0.5	1.0	1.5
(x, y)	$(0,3)$	$(69.0, 24.1)$	$(137.9, 37.2)$	$(206.9, 42.3)$

t	2.0	2.5	3.0	3.5
(x, y)	$(275.8, 39.4)$	$(344.8, 28.5)$	$(413.7, 9.6)$	$(482.7, -17.3)$

TABLE 1

We can now plot these points and connect them, as usual, with a smooth curve. The resulting graph is shown in Figure 3.

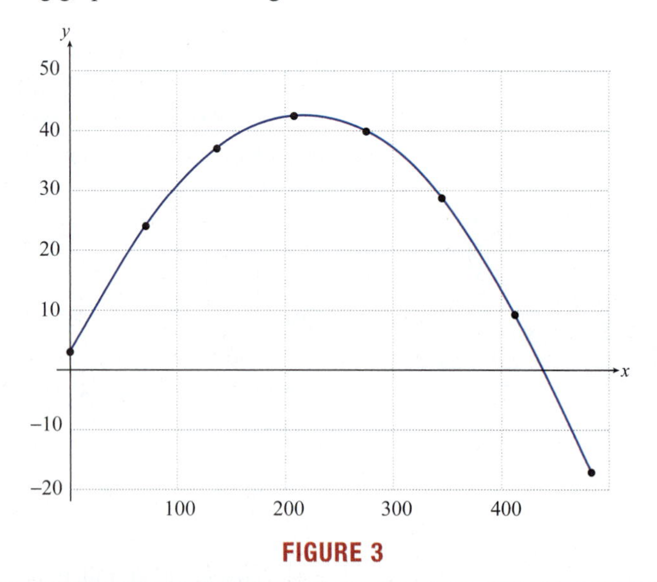

FIGURE 3

We have completed the tasks set in Example 1, but there are still some advantages to parametric curves to point out. First, curves described by parametric equations automatically possess an **orientation**, which is the direction traveled along the curve as the parameter increases. If the parameter represents time, as in Example 1, the orientation has an obvious physical meaning: it is the direction the ball travels as it traces out the curve. Second, the parametric description of the curve allows many questions to be easily answered. Consider, for example, the following question.

Example 2: Using Parametric Equations

Suppose the ball in Example 1 has been hit toward a 10-foot-high fence that is 400 feet from home plate. Will the ball clear the fence?

Solution

The graph of the ball's flight makes it appear that the answer may be yes, but pictures can be deceptive and a graph is certainly not a proof. However, we can easily resolve the situation in two steps: first, determine the time when the ball is 400 feet from home plate, and then calculate the height of the ball at that time. We determine the time by setting the equation for x equal to 400 and solving for t.

$$\begin{aligned} 137.9t &= 400 \\ t &= 2.9 \text{ seconds} \end{aligned}$$

Then we evaluate the height y at this time.

$$\begin{aligned} y &= -16(2.9)^2 + 50.2(2.9) + 3 \\ &= 14.0 \text{ feet} \end{aligned}$$

So the ball does indeed clear the fence.

TOPIC 2: Graphing Parametric Equations by Eliminating the Parameter

While the parametric form of a curve is very useful in answering some questions and for some graphing purposes, it should not be assumed that one form or the other is better for *all* purposes. In some cases, it is convenient to convert parametric equations into a single equation relating x and y; we will call such equations rectangular equations, to make the distinction with parametric equations.

The process of converting from parametric form to rectangular form is called, appropriately enough, **eliminating the parameter**. The most common method is to solve one of the equations for the parameter and to substitute the result for the parameter in the other equation.

Example 3: Eliminating the Parameter

Sketch the graph described by the parametric equations $x = \frac{1}{\sqrt{t-1}} + 1$ and $y = \frac{1}{t-1}$ by eliminating the parameter.

Solution

If we solve the second equation for t, the result is as follows.

$$\begin{aligned} y &= \frac{1}{t-1} \\ t-1 &= \frac{1}{y} \\ t &= \frac{1}{y} + 1 \end{aligned}$$

Substituting this into the first equation and simplifying, we obtain an equation in x and y.

$$\begin{aligned} x &= \frac{1}{\sqrt{\left(\frac{1}{y}+1\right)-1}} + 1 \\ &= \frac{1}{\sqrt{\frac{1}{y}}} + 1 \\ &= \sqrt{y} + 1 \end{aligned}$$

We are more accustomed to graphing such equations in x and y if they are solved for y, and doing so in this case leads to a curve we know well.

$$\begin{aligned} x &= \sqrt{y} + 1 \\ x-1 &= \sqrt{y} \\ (x-1)^2 &= y \end{aligned}$$

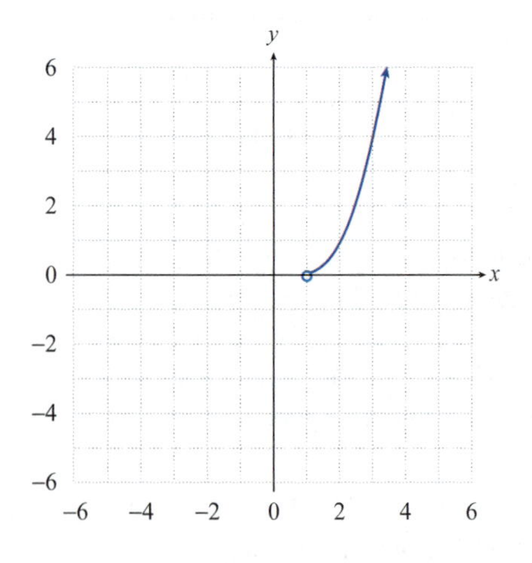

FIGURE 4

As we know, the graph of this last equation is an upward-opening parabola with vertex at $(1,0)$. But before sketching this curve and proclaiming it as the answer, look again at the original parametric equations. The formulas in t are only valid for $t > 1$ (because of the presence of $\sqrt{t-1}$), and this restriction on t leads to corresponding restrictions on x and y. If we consider all possible values of t bigger than 1, we see that the restriction on y is simply $y > 0$. However, as $t \to \infty$, $x \to 1$ and as $t \to 1$, $x \to \infty$. In particular, note that there is no value of t for which $x = 1$ (or anything smaller than 1). The graph defined by the original parametric equations is thus only half of the parabola, as shown in Figure 4.

In addition to time t, angle measure θ is often used as a parameter. In fact, we have encountered parametric equations in θ already, though this was not clear at the time. Consider the following familiar example.

Example 4: Eliminating the Parameter

Sketch the graph described by the parametric equations $x = 5\cos\theta$ and $y = 5\sin\theta$ for $0 \le \theta \le 2\pi$.

Solution

We have many ways of determining the shape of this curve. One way is to recognize the equations $x = 5\cos\theta$ and $y = 5\sin\theta$ as specific instances of the formulas $x = r\cos\theta$ and $y = r\sin\theta$ that relate rectangular and polar coordinates. If we make this connection, we see immediately that $r = 5$ and the curve is thus a circle of radius 5 centered at the origin.

But we don't have to rely on this insight in order to obtain the answer. Another approach is to eliminate the parameter as in the last example, and again we have several options. One is to solve for θ in one equation and substitute the result in the other equation, using the techniques of Section 7.6 to simplify the expression. The process for $0 < \theta < \frac{\pi}{2}$ is as follows:

$$y = 5\sin\theta$$
$$\sin\theta = \frac{y}{5}$$
$$\theta = \sin^{-1}\left(\frac{y}{5}\right)$$

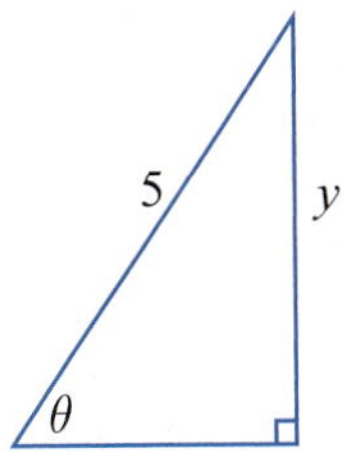

FIGURE 5

and so

$$x = 5\cos\left(\sin^{-1}\left(\frac{y}{5}\right)\right)$$
$$= 5\left(\frac{\sqrt{25-y^2}}{5}\right)$$
$$= \sqrt{25-y^2}.$$

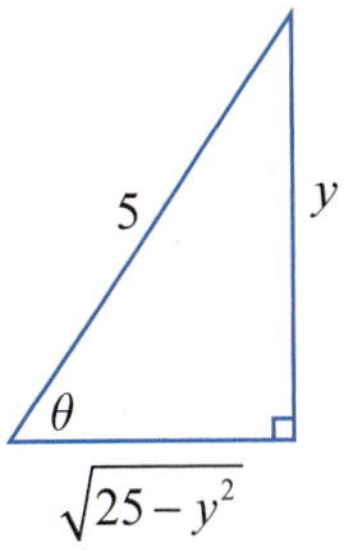

FIGURE 6

This last equation is more meaningful if we square both sides and rearrange terms slightly.

$$x = \sqrt{25-y^2}$$
$$x^2 = 25 - y^2$$
$$x^2 + y^2 = 25$$

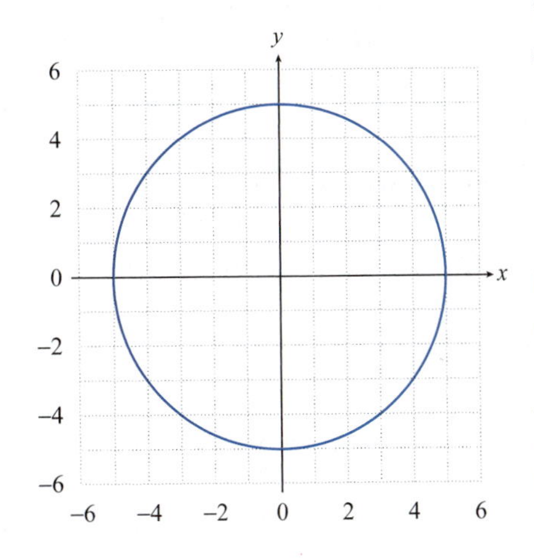

FIGURE 7

Using this method, we arrive at the rectangular form for the equation of a circle of radius 5 centered at the origin.

Another method of eliminating the parameter is indirect, but very useful. If we square both sides of each of the parametric equations, we can apply the Pythagorean identity $\sin^2\theta + \cos^2\theta = 1$ as follows:

$$x^2 = 25\cos^2\theta \quad \text{and} \quad y^2 = 25\sin^2\theta$$

so

$$\begin{aligned} x^2 + y^2 &= 25\cos^2\theta + 25\sin^2\theta \\ &= 25\left(\cos^2\theta + \sin^2\theta\right) \\ &= 25. \end{aligned}$$

Using any method, we arrive at the graph shown in Figure 7.

TOPIC 3: Constructing Parametric Equations to Describe a Graph

As we saw in Examples 1 and 2, the parametric form of an equation can prove to be very useful in answering some questions. This naturally leads us to the question of *how* parametric equations describing a given curve can be constructed. We will consider several methods.

Example 5: Constructing Parametric Equations

Construct parametric equations describing the graph of $y = x^2 + 3$.

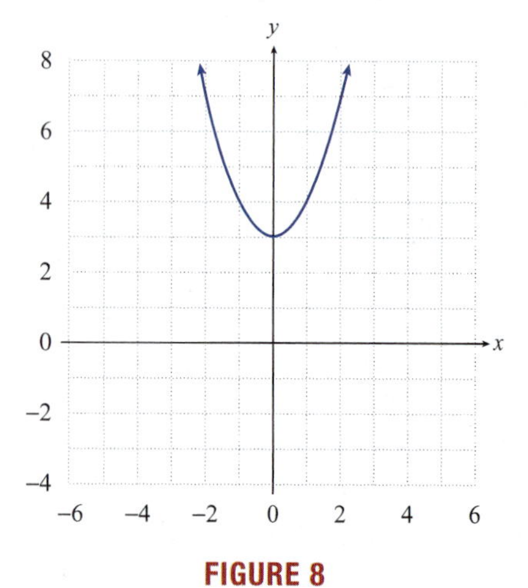

FIGURE 8

Solution

One hallmark of parametric descriptions of graphs is that they are not unique. Geometrically, the reason for this is that a given curve can be traced out by two parameters at different "speeds" or even in opposite directions. We will illustrate this point with two solutions to the problem in this example.

Solution 1: One parameterization of the graph is achieved by letting $x = t$ and thereby defining $y = t^2 + 3$. As t increases, the parabola defined by the parametric equations is traced out from left to right. Plot a few points to verify this claim.

Solution 2: One of an infinite number of alternative solutions, and one that traces out the graph in the opposite direction, is achieved by letting $x = 5 - t$ (there is no special significance to the number 5 here). Now, as t increases, x decreases and so the points on the parabola will be traced out from right to left. We know this to be true even before determining that $y = x^2 + 3 = (5 - t)^2 + 3 = t^2 - 10t + 28$. Again, plot a few points to verify the claim.

Example 6: Constructing Parametric Equations

A **cycloid** is the curve traced out by a point on a circle as it rolls along a straight line (see Figure 9). Fix a point P on a circle of radius a, and assume that P lies initially at the origin and that the circle rolls to the right, as shown. Find parametric equations describing the cycloid traced out by P.

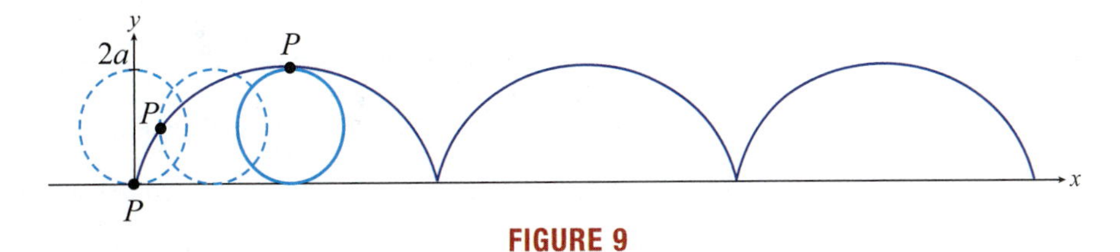

FIGURE 9

Solution

The first step is to make a wise choice of parameter. To this end, let θ measure the angle between the ray extending straight down from the circle's center and the ray extending from the circle's center through the point P. Then when the circle is in the initial position, the two rays coincide and $\theta = 0$. As the circle rolls to the right, θ increases; when the circle is in the position of the second circle shown in Figure 9, $\theta = \frac{\pi}{2}$, and $\theta = \pi$ when the point P reaches its topmost position the first time (the third circle in Figure 9).

Consider the enlarged diagram Figure 10. If we let (x, y) denote the coordinates of the point P, our goal is to write x and y as functions of θ.

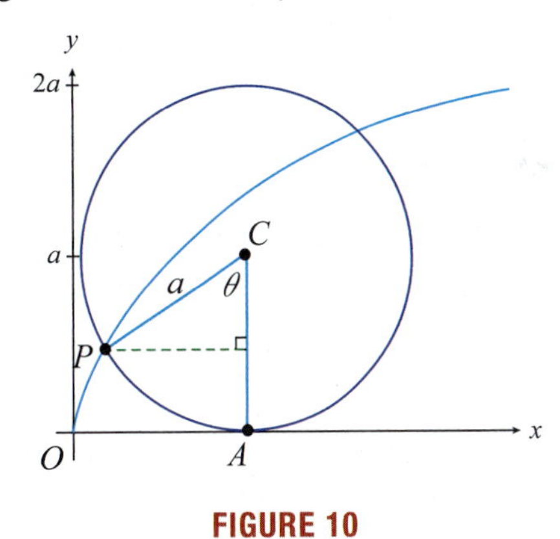

FIGURE 10

Our first observation is that the line segment $\overline{OA}$ must have the same length as the arc $\widehat{PA}$, since the circle rolls without slipping. The arc-length formula tells us the length of $\widehat{PA}$ is $a\theta$, so this is the length of $\overline{OA}$ as well. This means the coordinates of point C are $(a\theta, a)$ and that the coordinates of point A are $(a\theta, 0)$. The dashed line in the diagram has length $a\sin\theta$, and the other leg of the right triangle has length $a\cos\theta$. Putting this all together, we have

$$x + a\sin\theta = a\theta \quad \text{and} \quad y + a\cos\theta = a.$$

Solving for x and y, we obtain the desired parametric equations (verifying that the equations above remain true as θ takes on larger values is left to the reader).

$$x = a(\theta - \sin\theta) \quad \text{and} \quad y = a(1 - \cos\theta)$$

9.4 EXERCISES

PRACTICE

1. Given the parametric equations $x = 5 + t$ and $y = \frac{\sqrt{t}}{(t-2)}$, construct a table of the points (x, y) that result from integer t-values from 0 to 6, and then sketch the curve.

2. Given the parametric equations $x = \frac{\tan\theta}{2}$ and $y = \cos^2\theta + 3$, construct a table of the points (x, y) that result from the values $\theta = 0, \frac{\pi}{6}, \frac{\pi}{3}, \frac{\pi}{2}, \frac{2\pi}{3}, \frac{5\pi}{6}$, and π. Using these points, sketch the graph of the equations.

Sketch the graphs of the following parametric equations by eliminating the parameter. See Examples 3 and 4.

3. $x = 3(t+1)$ and $y = 2t$

4. $x = \sqrt{t-2}$ and $y = 3t - 2$

5. $x = 1 + t$ and $y = \frac{t-3}{2}$

6. $x = |t+3|$ and $y = t - 5$

7. $x = \frac{t}{4}$ and $y = t^2$

8. $x = \frac{t}{t+2}$ and $y = \sqrt{t}$

9. $x = \sqrt{t+3}$ and $y = t + 3$

10. $x = \frac{2}{|t-3|}$ and $y = 2t - 1$

11. $x = \cos\theta$ and $y = 2\sin\theta$

12. $x = 3\sin\theta - 1$ and $y = \frac{\cos\theta}{2}$

13. $x = 1 - \sin\theta$ and $y = \sin\theta - 1$

14. $x = 2\cos\theta$ and $y = 3\cos\theta$

15. $x = 2\sin\theta + 2$ and $y = 2\cos\theta + 2$

16. $x = \sin\theta$ and $y = 4 - 3\cos\theta$

Construct parametric equations describing the graphs of the following equations. See Example 5.

17. $y = (x+1)^2$

18. $y = 5x - 2$

19. $y = -x^2 - 5$

20. $x^2 + \frac{y^2}{4} = 1$

21. $x = y^2 + 4$

22. $y = x^2 + 1$

23. $y = \frac{1}{x}$

24. $x = 4y - 6$

25. $y = |x - 1|$

26. $x = 2(y-3)$

27. $y^2 = 1 - x^2$

28. $x = \frac{1}{3y}$

29. $y = x^2 - x - 6$

Construct parametric equations for the line with the given attributes. (Answers will vary.)

30. Slope -2, passing through $(-5, -2)$

31. Slope $\frac{1}{4}$, passing through $(10, 12)$

32. Slope 3, passing through $(7, 2)$

33. Passing through $(0, 0)$ and $(7, 4)$

34. Passing through $(6, -3)$ and $(2, 3)$

35. Passing through $(12, 3)$ and $(-4, -5)$

Using the given values for *x*, construct parametric equations describing the graph of each of the following equations.

36. $y = 3x + 1$, given that $x = 2 + t$

37. $y = 2 - |x|$, given that $x = t - 5$

38. $y = 5 - x^2$, given that $x = t + 1$

39. $5 = 2y + x$, given that $x = 4t$

40. $\frac{x^2}{2} + 6 = y$, given that $x = t - 4$

41. $(x-2)^2 = y$, given that $x = 5t + 1$

Construct parametric equations for the circle with *x* the given attributes. (Answers will vary.)

42. Center $(0,0)$, radius 1

43. Center $(-4,2)$, radius 3

44. Center $(7,-5)$, radius 4

45. Center $(0,-2)$, radius 6

APPLICATIONS

46. François shoots a basketball at an angle of 48° from the horizontal. It leaves his hands 7 ft from the ground with a velocity of 21 ft/s.

a. Construct parametric equations representing the path of the ball.
b. Sketch a graph of the basketball's flight.
c. If the basket is 15 ft away and 11 ft high, will he make the shot?

47. Suppose that a circus performer is shot from a cannon at a rate of 80 mph, at an angle of 60° from the horizontal. The cannon sits on a platform 10 feet above the ground.

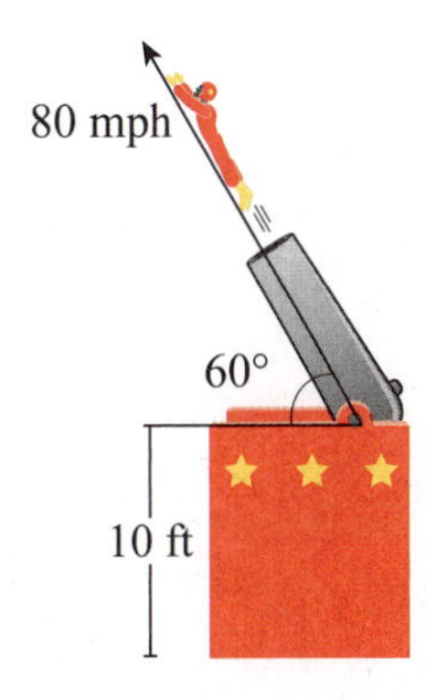

a. Construct parametric equations representing the performer's path as he flies through the air.
b. Sketch a graph of his flight.
c. How high is the acrobat 1.5 seconds after leaving the cannon?
d. How far from the cannon should a landing net be placed, if it is placed at ground level?
e. At what time *t* will the performer land in the net?
f. If a 12-foot-high wall of flames is placed 70 feet from the cannon, will he clear it unharmed?

48. On his morning paper route, John throws a newspaper from his car window 3.5 ft from the ground. The paper has an initial velocity of 10 ft/s and is tossed at an angle of $10°$ from the horizontal.

a. Construct parametric equations modeling the path of the newspaper.
b. Sketch a graph of the paper's path.

49. A wheel of radius 12 inches rolls along a flat surface in a straight line. There is a fixed point P that initially lies at the point $(0,0)$. Find parametric equations defining the cycloid traced out by P.

50. A ball is rolled on the floor in a straight line from one person to another person. The ball has a radius of 3 cm and there is a fixed point P located on the ball. Let the person rolling the ball represent the origin. Find parametric equations defining the cycloid traced out by P.

9.5 TRIGONOMETRIC FORM OF COMPLEX NUMBERS

TOPICS

1. The complex plane
2. Complex numbers in trigonometric form
3. Multiplying and dividing complex numbers
4. Powers of complex numbers
5. Roots of complex numbers

TOPIC 1: The Complex Plane

Complex numbers were introduced in Section 1.5, and at that point the most basic arithmetic operations were extended from the field of real numbers to the field of complex numbers. Since then you have been exposed to many algebraic and trigonometric concepts, some of which can now be put to use to enlarge our understanding of complex numbers and expand our ability to work with them.

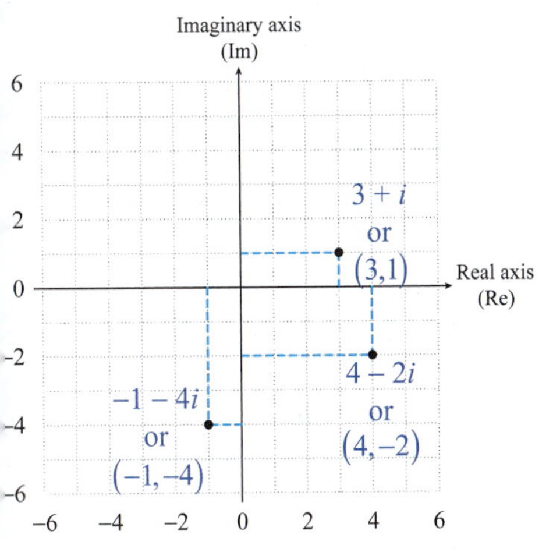

FIGURE 1: The Complex Plane

The first idea we can put to use is that of a two-dimensional plane. We will use the plane as a way to visualize complex numbers, exactly as we use the real line to visualize real numbers; this use was foreshadowed in Topic 4 on recursive graphics in Section 4.3. We need a two-dimensional framework in order to plot complex numbers since complex numbers consist of two independent components—the real part and the imaginary part. In fact, for the purposes of graphing, we identify a given complex number $z = a + bi$ with the ordered pair (a,b). In this context, we label the horizontal axis of a rectangular coordinate system the **real axis**, and we label the vertical axis the **imaginary axis**; the plane as a whole is called the **complex plane**. Figure 1 illustrates how several complex numbers are graphed.

The association of a complex number $z = a + bi$ with the ordered pair (a,b) in the plane gives rise immediately to a way to measure the size of z.

DEFINITION: Magnitude of a Complex Number

We say the **magnitude** of a complex number $z = a + bi$, also known as the **modulus** or **absolute value** of z, is the real number

$$|z| = \sqrt{a^2 + b^2},$$

and we use $|z|$ to denote this nonnegative quantity.

The formula for $|z|$ is nothing more than the formula for the distance between (a,b) and the origin of the plane, so the magnitude of a complex number is a measure of how distant it is from the complex number $0 + 0i$ (which we usually just write as 0). Note also that if z is real, $|z|$ has the same meaning as the absolute value of a real number.

Example 1: Finding the Magnitude

Determine the magnitudes of the following complex numbers.

a. $-2+5i$ b. $1-3i$

Solution

a. $|-2+5i| = \sqrt{(-2)^2+(5)^2} = \sqrt{29}$

b. $|1-3i| = \sqrt{(1)^2+(-3)^2} = \sqrt{10}$

Example 2: Graphing Regions in the Complex Plane

Graph the regions of the complex plane defined by the following.

a. $\{z \mid |z| \leq 1\}$ b. $\{z = a+bi \mid |z| \leq 1 \text{ and } b \geq 0\}$

Solution

a. In words, $\{z \mid |z| \leq 1\}$ is the set of all complex numbers with magnitude less than or equal to 1. The region appears in Figure 2.

b. The region $\{z = a+bi \mid |z| \leq 1 \text{ and } b \geq 0\}$ consists of only those complex numbers with magnitude less than or equal to 1 and nonnegative coefficients of i. The graph of the region is a half circle as shown in Figure 3.

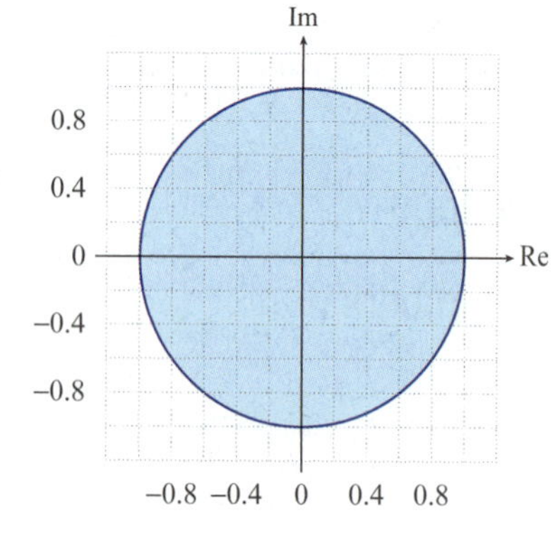

FIGURE 2

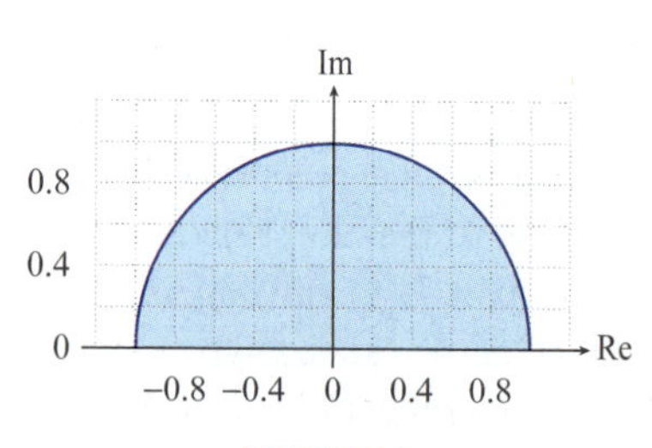

FIGURE 3

TOPIC 2: Complex Numbers in Trigonometric Form

We can also put our knowledge of trigonometry to use in describing complex numbers, with the result that many operations, such as multiplication, division, and the taking of roots, are made much easier.

To see how trigonometry applies, consider how the graph of a complex number gives rise to an angle θ in a natural way, as shown in Figure 4. Given $z = a+bi$, we know

$$\sin\theta = \frac{b}{|z|} \text{ and } \cos\theta = \frac{a}{|z|},$$

so $a = |z|\cos\theta$ and $b = |z|\sin\theta$. This leads to the following definition.

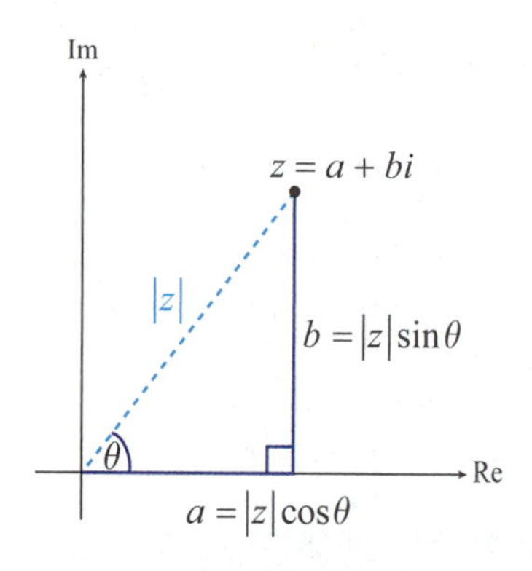

FIGURE 4

DEFINITION: Trigonometric Form of $z = a + bi$

Given $z = a+bi$, the **trigonometric form** of z is given by

$$\begin{aligned} z &= |z|\cos\theta + i|z|\sin\theta \\ &= |z|(\cos\theta + i\sin\theta), \end{aligned}$$

where $|z| = \sqrt{a^2+b^2}$ and θ satisfies $\tan\theta = \frac{b}{a}$. The angle θ is called the **argument** of z.

In this way, we also relate the graphing of complex numbers to the polar coordinates discussed in Section 9.3. And as mentioned in that section, the angle θ corresponding to a specific point in the plane is not unique; however, any two arguments for a given complex number will differ by a multiple of 2π.

Example 3: Writing Complex Numbers in Trigonometric Form

Write each of the following complex numbers in trigonometric form.

a. $-1+i$ **b.** $3-4i$

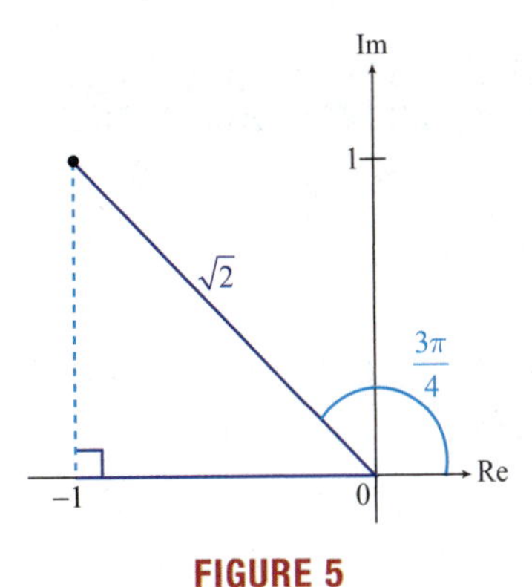

FIGURE 5

Solution

a. $|-1+i| = \sqrt{1+1} = \sqrt{2}$ and the argument of $-1+i$ is $\theta = \tan^{-1}(-1) = \frac{3\pi}{4}$. Note that $-1+i$ lies in the second quadrant of the complex plane, as shown in Figure 5. In trigonometric form, $-1+i$ is written $\sqrt{2}\left(\cos\left(\frac{3\pi}{4}\right)+i\sin\left(\frac{3\pi}{4}\right)\right)$.

b. The magnitude of $3-4i$ is 5, but the argument is not so neatly expressed. In degrees, $\theta = \tan^{-1}\left(-\frac{4}{3}\right) \approx -53.1°$. So $3-4i \approx 5(\cos(-53.1°)+i\sin(-53.1°))$.

Example 4: Writing Complex Numbers in Standard Form

Write the complex number $2\left(\cos\left(\frac{\pi}{6}\right)+i\sin\left(\frac{\pi}{6}\right)\right)$ in standard form.

Solution

$$2\left(\cos\left(\frac{\pi}{6}\right)+i\sin\left(\frac{\pi}{6}\right)\right) = 2\left(\frac{\sqrt{3}}{2}+i\frac{1}{2}\right) = \sqrt{3}+i$$

Although it requires calculus to prove, there is a famous and useful identity that relates exponential functions and arguments of complex numbers. The identity is called **Euler's Formula**, and states $e^{i\theta} = \cos\theta + i\sin\theta$. In the context of the trigonometric form of complex numbers, we can use this to write

$$z = |z|(\cos\theta + i\sin\theta) = |z|e^{i\theta}.$$

We will return to this elegant formula soon.

TOPIC 3: Multiplying and Dividing Complex Numbers

The remainder of this section will show how, as promised, the trigonometric form of complex numbers simplifies certain computations. We start with multiplication and division of complex numbers.

Given $z_1 = |z_1|(\cos\theta_1 + i\sin\theta_1)$ and $z_2 = |z_2|(\cos\theta_2 + i\sin\theta_2)$, we can use our basic knowledge of complex multiplication to write the product as

$$\begin{aligned} z_1 z_2 &= |z_1|(\cos\theta_1 + i\sin\theta_1)|z_2|(\cos\theta_2 + i\sin\theta_2) \\ &= |z_1||z_2|\left[(\cos\theta_1\cos\theta_2 - \sin\theta_1\sin\theta_2) + i(\sin\theta_1\cos\theta_2 + \cos\theta_1\sin\theta_2)\right]. \end{aligned}$$

The combinations of sines and cosines that result should look familiar: they appear in the sum and difference trigonometric identities of Section 8.2. Applying these, we can simplify the previous product to obtain

$$z_1 z_2 = |z_1||z_2|\left[\cos(\theta_1 + \theta_2) + i\sin(\theta_1 + \theta_2)\right].$$

A similar result can be obtained for the division of complex numbers, but the verification will be left to the reader. The following box summarizes the observations.

f(x) FORMULA: Product and Quotient of Complex Numbers

Given $z_1 = |z_1|(\cos\theta_1 + i\sin\theta_1)$ and $z_2 = |z_2|(\cos\theta_2 + i\sin\theta_2)$,

$$z_1 z_2 = |z_1||z_2|\left[\cos(\theta_1 + \theta_2) + i\sin(\theta_1 + \theta_2)\right]$$

and

$$\frac{z_1}{z_2} = \frac{|z_1|}{|z_2|}\left[\cos(\theta_1 - \theta_2) + i\sin(\theta_1 - \theta_2)\right] \quad (z_2 \neq 0).$$

Example 5: Multiplying Complex Numbers

Use the product formula to find the product of $z_1 = 3\left(\cos\left(\frac{5\pi}{6}\right) + i\sin\left(\frac{5\pi}{6}\right)\right)$ and $z_2 = 2\left(\cos\left(\frac{2\pi}{3}\right) + i\sin\left(\frac{2\pi}{3}\right)\right)$.

Solution

$$\begin{aligned} z_1 z_2 &= (3)(2)\left(\cos\left(\frac{5\pi}{6} + \frac{2\pi}{3}\right) + i\sin\left(\frac{5\pi}{6} + \frac{2\pi}{3}\right)\right) \\ &= 6\left(\cos\left(\frac{9\pi}{6}\right) + i\sin\left(\frac{9\pi}{6}\right)\right) \\ &= 6\left(\cos\left(\frac{3\pi}{2}\right) + i\sin\left(\frac{3\pi}{2}\right)\right) \\ &= -6i \end{aligned}$$

Example 6: Dividing Complex Numbers

Use the quotient formula to divide $z_1 = 6(\cos 117° + i\sin 117°)$ by $z_2 = 2(\cos 72° + i\sin 72°)$.

Solution

$$\begin{aligned} \frac{z_1}{z_2} &= \frac{6(\cos 117° + i\sin 117°)}{2(\cos 72° + i\sin 72°)} \\ &= 3(\cos(117° - 72°) + i\sin(117° - 72°)) \\ &= 3(\cos 45° + i\sin 45°) \\ &= \frac{3}{\sqrt{2}} + i\frac{3}{\sqrt{2}} \end{aligned}$$

To remember the product and quotient formulas just derived, it's useful to note that multiplication of complex numbers corresponds to addition of the arguments, while division of complex numbers corresponds to subtraction of the arguments (along with a respective product or quotient of the magnitudes). This correspondence should seem familiar: it's exactly the behavior of multiplication and division of exponential functions. This similarity is no accident, as we see if we rephrase multiplication and division using Euler's Formula as follows.

$$z_1 z_2 = \left(|z_1| e^{i\theta_1}\right)\left(|z_2| e^{i\theta_2}\right) = |z_1||z_2| e^{i\theta_1} e^{i\theta_2} = |z_1||z_2| e^{i(\theta_1+\theta_2)}$$

and

$$\frac{z_1}{z_2} = \frac{|z_1| e^{i\theta_1}}{|z_2| e^{i\theta_2}} = \left(\frac{|z_1|}{|z_2|}\right)\left(\frac{e^{i\theta_1}}{e^{i\theta_2}}\right) = \frac{|z_1|}{|z_2|} e^{i(\theta_1-\theta_2)}.$$

This behavior of multiplication and division of complex numbers also manifests itself graphically, as seen in Example 7.

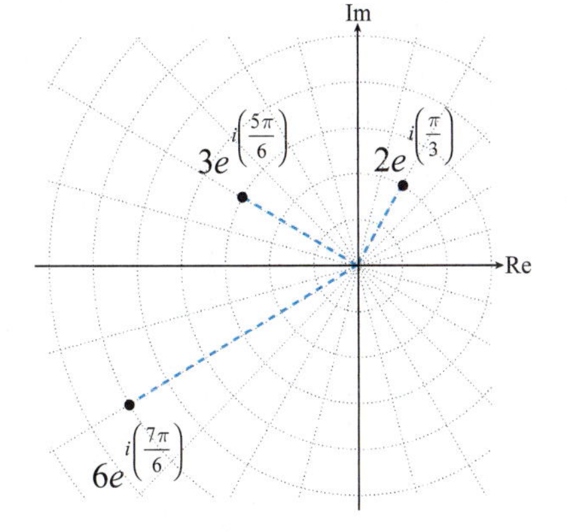

FIGURE 6

Example 7: Graphing Complex Numbers and Their Product

Plot the two complex numbers $2e^{i\left(\frac{\pi}{3}\right)}$ and $3e^{i\left(\frac{5\pi}{6}\right)}$ and then, on the same graph, plot their product.

Solution

First note that, by the previous formula, $\left(2e^{i\left(\frac{\pi}{3}\right)}\right)\left(3e^{i\left(\frac{5\pi}{6}\right)}\right) = 6e^{i\left(\frac{7\pi}{6}\right)}$. We can now use the magnitudes and arguments of the three complex numbers as their polar coordinates and construct the graph in Figure 6. Note in particular how the argument of the product is the sum of the arguments of the two given complex numbers.

TOPIC 4: Powers of Complex Numbers

The product formula can be used to greatly speed up the process of calculating powers of complex numbers. For instance, if $z = |z|(\cos\theta + i\sin\theta)$, then

$$\begin{aligned} z^2 &= \left[|z|(\cos\theta + i\sin\theta)\right]^2 \\ &= |z|^2\left(\cos(2\theta) + i\sin(2\theta)\right). \end{aligned}$$

Similarly,

$$\begin{aligned} z^3 &= z^2 z \\ &= |z|^2\left(\cos(2\theta) + i\sin(2\theta)\right)|z|(\cos\theta + i\sin\theta) \\ &= |z|^3\left(\cos(3\theta) + i\sin(3\theta)\right). \end{aligned}$$

The pattern that is developing is even more apparent if we use Euler's Formula.

$$\begin{aligned} z^n &= \left(|z| e^{i\theta}\right)^n \\ &= |z|^n e^{in\theta} \end{aligned}$$

These observations are named after the early 18th-century mathematician Abraham De Moivre.

$f(x)$ FORMULA: Powers of Complex Numbers (De Moivre's Theorem)

Given $z = |z|(\cos\theta + i\sin\theta)$ and a positive integer n,

$$z^n = |z|^n\left(\cos(n\theta) + i\sin(n\theta)\right).$$

Using Euler's Formula, this appears in the form

$$z^n = |z|^n e^{in\theta}.$$

Example 8: Using De Moivre's Theorem

Use De Moivre's Theorem to calculate $\left(\sqrt{3} - i\right)^7$.

Solution

In one form,

$$\begin{aligned}\left(\sqrt{3} - i\right)^7 &= 2^7 e^{7i\left(-\frac{\pi}{6}\right)} \\ &= 128e^{i\left(-\frac{7\pi}{6}\right)} \\ &= 128e^{i\left(\frac{5\pi}{6}\right)}.\end{aligned}$$

In the alternative form, $\left(\sqrt{3} - i\right)^7 = 128\left(\cos\left(\frac{5\pi}{6}\right) + i\sin\left(\frac{5\pi}{6}\right)\right)$.

TOPIC 5: Roots of Complex Numbers

De Moivre's Theorem can be used to determine roots of complex numbers. The first step is to realize that if $z = |z|(\cos\theta + i\sin\theta)$ is a nonzero complex number and if n is a positive integer, z has n distinct n^{th} roots. This follows from an application of the Fundamental Theorem of Algebra: the nth-degree polynomial equation $w^n = z$ has n solutions (here, w is a complex variable).

One n^{th} root of z is easily determined. If we let $w_0 = |z|^{\frac{1}{n}}\left(\cos\left(\frac{\theta}{n}\right) + i\sin\left(\frac{\theta}{n}\right)\right)$, then by De Moivre's Theorem

$$\begin{aligned}w_0^{\,n} &= \left(|z|^{\frac{1}{n}}\right)^n\left(\cos\left(n\frac{\theta}{n}\right) + i\sin\left(n\frac{\theta}{n}\right)\right) \\ &= |z|(\cos\theta + i\sin\theta) \\ &= z,\end{aligned}$$

so w_0 is indeed an nth root. But as we know, replacing θ with $\theta + 2k\pi$ results in an equivalent way of writing z. This observation leads to the following.

$f(x)$ FORMULA: Roots of Complex Numbers

Given the nonzero complex number $z = |z|(\cos\theta + i\sin\theta)$ and the positive integer n,

$$w_k = |z|^{\frac{1}{n}}\left[\cos\left(\frac{\theta + 2k\pi}{n}\right) + i\sin\left(\frac{\theta + 2k\pi}{n}\right)\right]$$

defines n distinct n^{th} roots of z for $k = 0, 1, \ldots, n-1$. Alternatively,

$$w_k = |z|^{\frac{1}{n}} e^{i\left(\frac{\theta + 2k\pi}{n}\right)}.$$

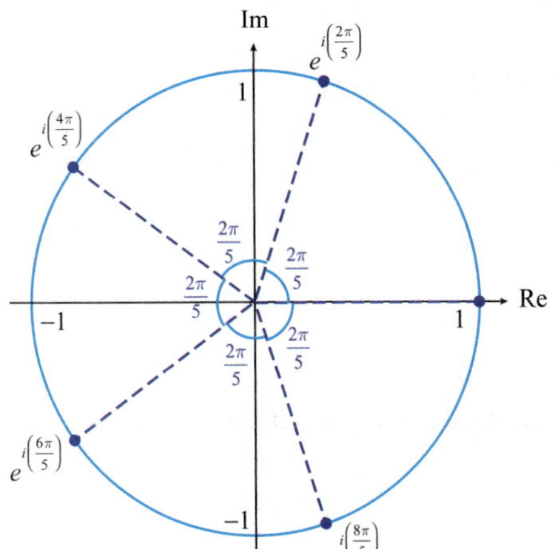

FIGURE 7

Example 9: Finding Roots of Complex Numbers

Find all the fifth roots of 1, and graph their locations in the complex plane.

Solution

The easiest way to describe the five fifth roots uses the Euler formulation. Since 1 can be written as e^{0i} we know $\theta = 0$. Thus we have

$$\left\{(1)^{\frac{1}{5}} e^{i\left(\frac{0+2(0)\pi}{5}\right)}, (1)^{\frac{1}{5}} e^{i\left(\frac{0+2(1)\pi}{5}\right)}, (1)^{\frac{1}{5}} e^{i\left(\frac{0+2(2)\pi}{5}\right)}, (1)^{\frac{1}{5}} e^{i\left(\frac{0+2(3)\pi}{5}\right)}, (1)^{\frac{1}{5}} e^{i\left(\frac{0+2(4)\pi}{5}\right)}\right\}$$

which simplifies to

$$\left\{1, e^{i\left(\frac{2\pi}{5}\right)}, e^{i\left(\frac{4\pi}{5}\right)}, e^{i\left(\frac{6\pi}{5}\right)}, e^{i\left(\frac{8\pi}{5}\right)}\right\}.$$

The graph of the five roots appears in Figure 7. Note that they are evenly spaced around a circle of radius 1.

Example 10: Finding Roots of Complex Numbers

Find all the fourth roots of $-1 - i\sqrt{3}$.

Solution

Again, the roots are most easily written using Euler's Formula, and we begin by writing $-1 - i\sqrt{3}$ as $2e^{i\left(\frac{4\pi}{3}\right)}$. Then the four fourth roots are

$$\left\{2^{\frac{1}{4}} e^{i\left(\frac{\pi}{3}\right)}, 2^{\frac{1}{4}} e^{i\left(\frac{5\pi}{6}\right)}, 2^{\frac{1}{4}} e^{i\left(\frac{4\pi}{3}\right)}, 2^{\frac{1}{4}} e^{i\left(\frac{11\pi}{6}\right)}\right\}.$$

9.5 EXERCISES

PRACTICE

Graph and determine the magnitudes of the following complex numbers. See Example 1.

1. $3+5i$
2. $-1+3i$
3. $2-4i$
4. $-6-i$
5. $4+4i$
6. $5+2i$

Sketch z_1, z_2, z_1+z_2, and z_1z_2 on the same complex plane.

7. $z_1=7+2i, z_2=-2+3i$
8. $z_1=-1-2i, z_2=4+4i$
9. $z_1=3+i, z_2=5-i$
10. $z_1=5-2i, z_1=-1+5i$

Graph the regions of the complex plane defined by the following. See Example 2.

11. $\{z \mid |z|<3\}$
12. $\{z \mid 1\le|z|\le 4\}$
13. $\{z \mid |z|\ge 1\}$
14. $\{z=a+bi \mid a>1, b>2\}$
15. $\{z=a+bi \mid a\ge b\}$
16. $\{z \mid |z|=4\}$

Write each of the following complex numbers in trigonometric form. See Example 3.

17. $-3-i$
18. $5-3i$
19. $1+2i$
20. $3+\sqrt{3}i$
21. $4+2i$
22. $5-\sqrt{2}i$
23. $\sqrt{2}-\sqrt{2}i$
24. $2\sqrt{3}-2i$
25. $3+4i$
26. $1+i$
27. $4-4\sqrt{3}i$
28. $2\sqrt{2}-i$

Write each of the following complex numbers in standard form. See Example 4.

29. $3\left(\cos\left(\frac{5\pi}{6}\right)+i\sin\left(\frac{5\pi}{6}\right)\right)$
30. $\cos\left(\frac{\pi}{3}\right)+i\sin\left(\frac{\pi}{3}\right)$
31. $2\left(\cos\left(\frac{4\pi}{3}\right)+i\sin\left(\frac{4\pi}{3}\right)\right)$
32. $6(\cos\pi+i\sin\pi)$
33. $5\left(\cos\left(\frac{3\pi}{4}\right)+i\sin\left(\frac{3\pi}{4}\right)\right)$
34. $\cos\left(\frac{5\pi}{4}\right)+i\sin\left(\frac{5\pi}{4}\right)$
35. $\frac{3}{2}(\cos 150°+i\sin 150°)$
36. $4(\cos 210°+i\sin 210°)$
37. $5[\cos(78°20')+i\sin(78°20')]$
38. $3[\cos(121°40')+i\sin(121°40')]$

Perform the following operations and show the answer in both trigonometric form and standard form. See Examples 5 and 6.

39. $[4(\cos 60° + i\sin 60°)][4(\cos 330° + i\sin 330°)]$

40. $[3(\cos 180° + i\sin 180°)][4(\cos 30° + i\sin 30°)]$

41. $\left[\sqrt{2}\left(\cos\left(\frac{5\pi}{4}\right) + i\sin\left(\frac{5\pi}{4}\right)\right)\right]\left[3\sqrt{3}\left(\cos\left(\frac{\pi}{6}\right) + i\sin\left(\frac{\pi}{6}\right)\right)\right]$

42. $\left[10\left(\cos\left(\frac{5\pi}{6}\right) + i\sin\left(\frac{5\pi}{6}\right)\right)\right]\left[6\left(\cos\left(\frac{2\pi}{3}\right) + i\sin\left(\frac{2\pi}{3}\right)\right)\right]$

43. $(-1+3i)(\sqrt{3}+i)$

44. $2i(4+5i)$

45. $\dfrac{6(\cos 225° + i\sin 225°)}{3(\cos 45° + i\sin 45°)}$

46. $\dfrac{8(\cos 135° + i\sin 135°)}{4(\cos 30° + i\sin 30°)}$

47. $\dfrac{10\left(\cos\left(\frac{5\pi}{3}\right) + i\sin\left(\frac{5\pi}{3}\right)\right)}{3\left(\cos\left(\frac{7\pi}{6}\right) + i\sin\left(\frac{7\pi}{6}\right)\right)}$

48. $\dfrac{12\left(\cos\left(\frac{10\pi}{3}\right) + i\sin\left(\frac{10\pi}{3}\right)\right)}{6(\cos(2\pi) + i\sin(2\pi))}$

49. $\dfrac{-i}{1+i}$

50. $\dfrac{-2-2i}{4+3i}$

51. $\dfrac{2e^{\frac{2\pi}{3}i}}{e^{\frac{\pi}{4}i}}$

52. $(e^{210° i})(e^{90° i})$

Plot both complex numbers and, on the same graph, plot their product. See Example 7.

53. $\left[4\left(\cos\left(\frac{5\pi}{6}\right) + i\sin\left(\frac{5\pi}{6}\right)\right)\right][2(\cos\pi + i\sin\pi)]$

54. $\left[5\left(\cos\left(\frac{3\pi}{2}\right) + i\sin\left(\frac{3\pi}{2}\right)\right)\right]\left[3\left(\cos\left(\frac{5\pi}{6}\right) + i\sin\left(\frac{5\pi}{6}\right)\right)\right]$

55. $(2-5i)(\sqrt{2}+2i)$

56. $(-4-2i)(-\sqrt{3}+4i)$

57. $\left(2e^{\frac{\pi}{3}i}\right)\left(3e^{\frac{5\pi}{4}i}\right)$

58. $\left(5e^{\frac{5\pi}{3}i}\right)(e^{\pi i})$

Use De Moivre's Theorem to calculate the following. See Example 8.

59. $(1-\sqrt{3}i)^5$

60. $\left(\frac{\sqrt{2}}{2} + \frac{\sqrt{2}}{2}i\right)^{22}$

61. $(5+3i)^{17}$

62. $(-\sqrt{3}+i)^{13}$

63. $\left(\cos\left(\frac{\pi}{4}\right) + i\sin\left(\frac{\pi}{4}\right)\right)^8$

64. $[2(\cos 135° + i\sin 135°)]^4$

Find the indicated roots of the following and graphically represent each set in the complex plane. See Examples 9 and 10.

65. The fourth roots of -1.

66. The cube roots of $64i$.

67. The square roots of $2\sqrt{3}+2i$.

68. The fourth roots of $-1-i$.

69. The fourth roots of 256.

70. The fourth roots of $16\left(\cos\left(\frac{4\pi}{3}\right)+i\sin\left(\frac{4\pi}{3}\right)\right)$.

71. The square roots of $4(\cos 120° + i\sin 120°)$.

Solve the following equations. See Examples 9 and 10.

72. $z^3 - i = 0$

73. $z^2 - 4\sqrt{3} - 4i = 0$

74. $z^4 + 81i = 0$

75. $z^5 + 32 = 0$

76. $z^3 + 4\sqrt{2} - i = 0$

77. $z^2 + 25i = 0$

9.6 VECTORS IN THE CARTESIAN PLANE

TOPICS

1. Vector terminology
2. Basic vector operations
3. Component form of a vector
4. Applications of vectors

TOPIC 1: Vector Terminology

Many quantities are defined by their size. For example, length, area, mass, price, and temperature are fully determined by a single number; such numbers, representing only magnitude, are called **scalars**. Other quantities, however, cannot be adequately described by a single number. Force and velocity, for instance, possess both a *magnitude* and a *direction*, and a complete description of these quantities must somehow include both. Such *directed magnitudes* are called **vectors**.

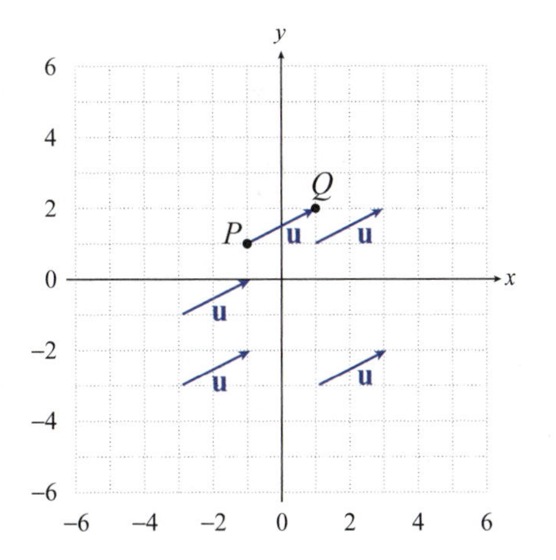

FIGURE 1: Five Depictions of **u**

We will introduce vectors in the setting of the two-dimensional plane, but the study of vectors in general constitutes an enormous area of mathematics, and vectors are easily extended into spaces of any dimension. In the plane, vectors are often represented as directed line segments (informally known as "arrows"). Such a directed line segment begins at a point P called the **initial point** and ends at a point Q called the **terminal point**, and the notation $\overrightarrow{PQ}$ is used to refer to the directed line segment. A subtle but very important point, though, is that a vector is characterized *entirely* by its direction and its magnitude, not by its initial and terminal points. That is, for a specific pair of points P and Q, $\overrightarrow{PQ}$ is only one way of depicting the vector it represents. We will use bold lowercase letters to denote vectors in general, and an expression such as $\mathbf{u} = \overrightarrow{PQ}$ means that **u** is a vector whose length is the same as the length of the line segment $\overline{PQ}$ and whose direction is defined by the initial point P and the terminal point Q. To make this important point clear, Figure 1 illustrates five different ways of depicting the one vector **u**.

Since vectors are characterized entirely by length and direction, we say that two vectors **u** and **v** with the same length and the same direction are **equal** (no matter where they might actually be depicted, if they are graphed) and we write $\mathbf{u} = \mathbf{v}$. The **length** of a vector may also be referred to as its **magnitude** or **norm**, and is denoted $\|\mathbf{u}\|$. In Figure 1, the length of **u** is the same as the length of the line segment from $(-1,1)$ to $(1,2)$, and the familiar distance formula tells us the following.

$$\|\mathbf{u}\| = \left\|\overrightarrow{PQ}\right\| = \sqrt{(1-(-1))^2 + (2-1)^2} = \sqrt{5}$$

TOPIC 2: Basic Vector Operations

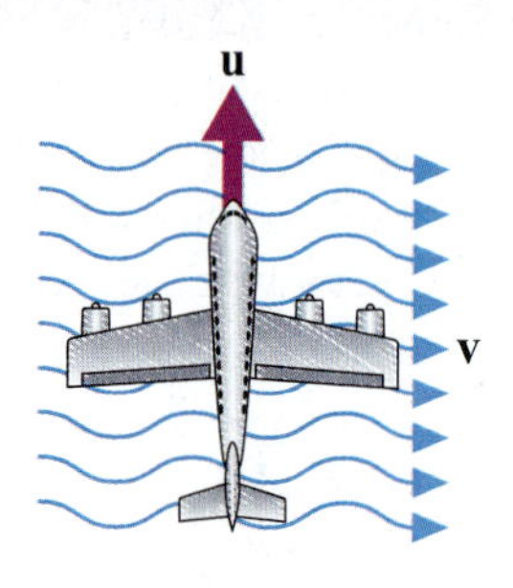

FIGURE 2

The applications of vectors motivate the basic ways that vectors can be combined. Consider the following scenario: wind with a velocity of **v** is affecting the flight of a plane that would be flying, if the wind were absent, with a velocity of **u**. What is the actual velocity of the plane given the presence of the wind? Both the plane and the wind possess, individually, a speed and a direction; the combination of the plane's velocity and the wind's velocity results in a third velocity that represents the actual progress of the plane.

The use of vectors facilitates such calculations. **Vector addition** is the operation called for in this scenario; the plane's actual velocity is given by $\mathbf{u} + \mathbf{v}$, and this expression is referred to as the **sum** of the two vectors. Graphically, $\mathbf{u} + \mathbf{v}$ can be represented by depicting **u**, placing **v** so that its initial point coincides with the terminal point of **u**, and then drawing a directed line segment from the initial point of **u** to the terminal point of **v**.

Example 1: Graphing the Sum of Two Vectors

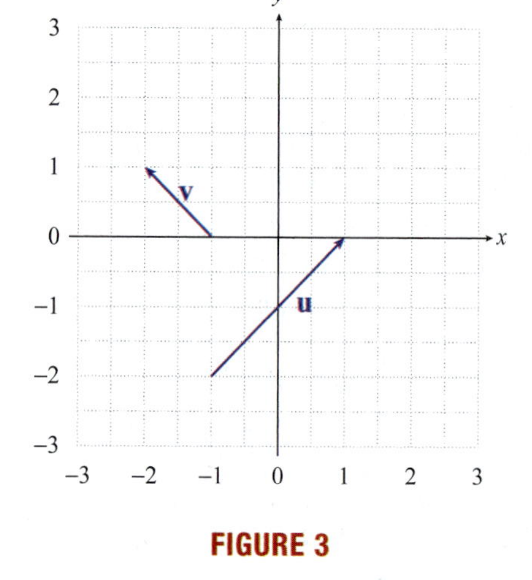

FIGURE 3

Given the two vectors **u** and **v** depicted in Figure 3, construct a graphical representation of $\mathbf{u} + \mathbf{v}$.

Solution

The simplest way to proceed is to leave **u** alone and to redraw **v** so that its initial point is the terminal point of **u**. If we then construct a directed line segment from **u**'s initial point to **v**'s terminal point, the result will depict $\mathbf{u} + \mathbf{v}$. This has been done in Figure 4.

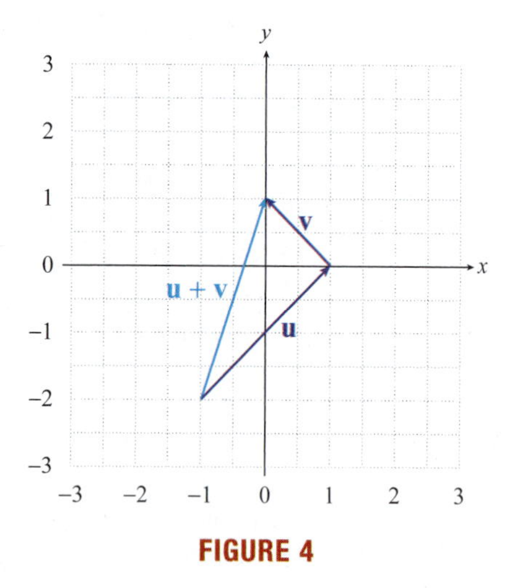

FIGURE 4

Now consider a second scenario. A person is attempting to paddle a kayak upstream, and her velocity would be **u** if the water weren't flowing. But suppose the water *is* flowing, with a magnitude equal to the kayaker's but the exact opposite direction. If we let **v** denote the water's velocity, then $\mathbf{u} + \mathbf{v} = \mathbf{0}$, where $\mathbf{0}$ is the **zero vector**. The zero vector is the unique vector which has a magnitude of 0, and in the context of vectors its function is equivalent to the number 0 on the real line (or the number 0 on the complex plane). Further, it makes sense to write $\mathbf{v} = -\mathbf{u}$ and to say that $-\mathbf{u}$ is the *additive inverse* of the vector **u**.

Consider one last scenario. Suppose you are walking along a hallway with a velocity **u**, and realize you are late for a meeting. You break into a slow run, moving three times faster than you were before. Your direction is unchanged, but your magnitude is three times as large; a convenient way to denote your new velocity is 3**u**, i.e., the scalar 3 times the vector **u**. Negative numbers can multiply vectors too: $(-1)\mathbf{u}$ means the same as $-\mathbf{u}$ and represents a vector with the same magnitude as **u** but the opposite direction.

In all, the set of vectors we are describing possesses properties that are very familiar to us from our study of real and complex numbers. The complete set of such properties appears below.

PROPERTIES: Properties of Vector Addition and Scalar Multiplication

Assume **u**, **v**, and **w** represent vectors, while a and b represent scalars. Then the following hold.

Vector Addition Properties	Scalar Multiplication Properties
$\mathbf{u}+\mathbf{v}=\mathbf{v}+\mathbf{u}$	$a(\mathbf{u}+\mathbf{v})=a\mathbf{u}+a\mathbf{v}$
$\mathbf{u}+(\mathbf{v}+\mathbf{w})=(\mathbf{u}+\mathbf{v})+\mathbf{w}$	$(a+b)\mathbf{u}=a\mathbf{u}+b\mathbf{u}$
$\mathbf{u}+\mathbf{0}=\mathbf{u}$	$(ab)\mathbf{u}=a(b\mathbf{u})=b(a\mathbf{u})$
$\mathbf{u}+(-\mathbf{u})=\mathbf{0}$	$1\mathbf{u}=\mathbf{u}$, $0\mathbf{u}=\mathbf{0}$, and $a\mathbf{0}=\mathbf{0}$
	$\lVert a\mathbf{u}\rVert=\lvert a\rvert\cdot\lVert\mathbf{u}\rVert$

Example 2: Graphing Vectors

Given the two vectors **u** and **v** depicted in Figure 5, construct graphical representations of $-2\mathbf{u}$, $3\mathbf{u}+\mathbf{v}$, and $\mathbf{u}-\mathbf{v}$.

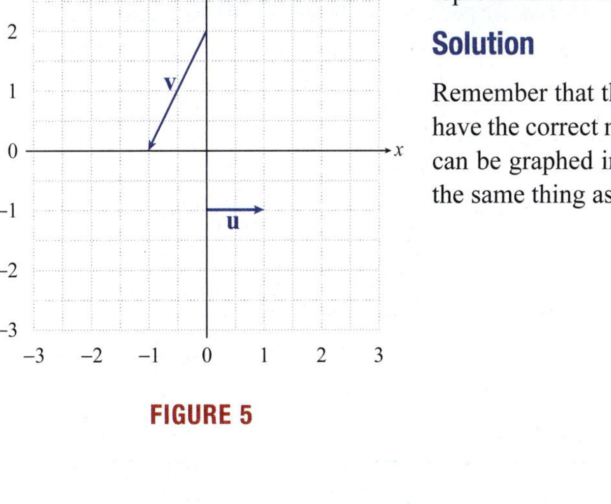

FIGURE 5

Solution

Remember that the vectors can be placed anywhere on the plane, as long as they have the correct magnitude and direction. With a bit of planning, all three answers can be graphed in the same image, as shown in Figure 6. Note that $\mathbf{u}-\mathbf{v}$ means the same thing as $\mathbf{u}+(-\mathbf{v})$.

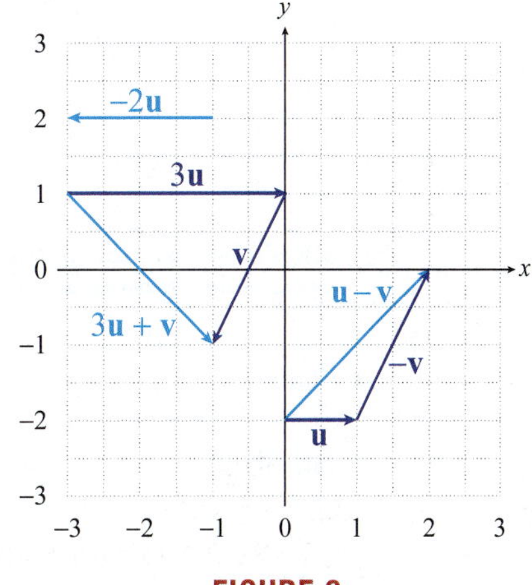

FIGURE 6

TOPIC 3: Component Form of a Vector

Our graphical work with vectors to this point has illustrated the important differences between scalars and vectors, and it has also begun to show how vectors can be used to solve certain problems. In order for vectors to be really useful, we need a way to treat them in a more analytical fashion; that is, we need a form that can be used in equations.

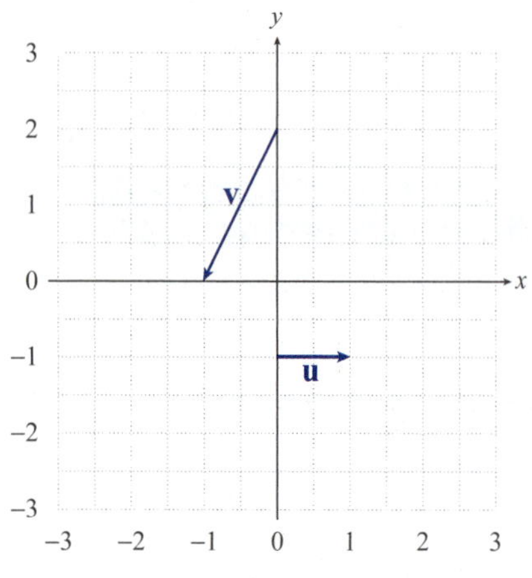

FIGURE 7

The **component form** of a vector is one such form. Consider again the two vectors **u** and **v** from Example 2. One way of completely characterizing **u** is to note that the horizontal displacement between the initial and terminal points is 1, while the vertical displacement is 0. The notation for this observation is $\mathbf{u} = \langle 1,0 \rangle$; in a similar fashion, we would write $\mathbf{v} = \langle -1,-2 \rangle$. The similarity between this notation and ordered pair notation is no accident, as the order in which the components appear is critical in both cases. Note, however, the very important distinction between vectors and points in the plane: $(-1,-2)$ refers to the one single point that is 1 unit to the left and 2 units down from the origin, while $\langle -1,-2 \rangle$ refers to the vector whose horizontal displacement is −1 and whose vertical displacement is −2. A depiction of $\mathbf{v} = \langle -1,-2 \rangle$ can be placed anywhere in the plane, as long as the displacement between the initial point and the terminal point is correct.

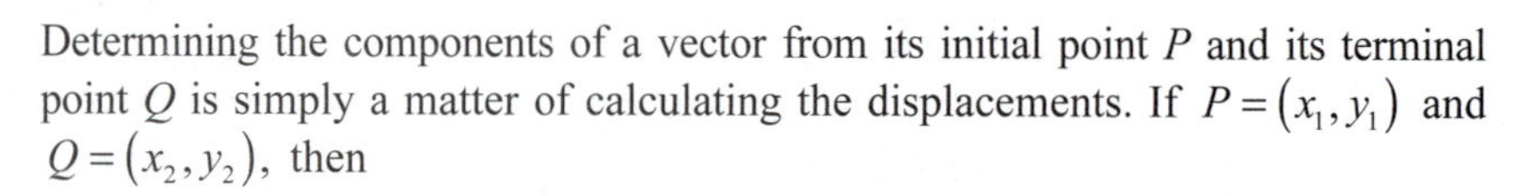

Determining the components of a vector from its initial point P and its terminal point Q is simply a matter of calculating the displacements. If $P = (x_1, y_1)$ and $Q = (x_2, y_2)$, then

$$\overrightarrow{PQ} = \langle x_2 - x_1, y_2 - y_1 \rangle.$$

The component form of a vector provides a simple and precise way of performing vector operations.

f(x) FORMULA: Vector Operations and Magnitude Using Components

Given two vectors $\mathbf{u} = \langle u_1, u_2 \rangle$ and $\mathbf{v} = \langle v_1, v_2 \rangle$ and a scalar a, we have the following:

$$\mathbf{u} + \mathbf{v} = \langle u_1 + v_1, u_2 + v_2 \rangle$$

$$a\mathbf{u} = \langle au_1, au_2 \rangle$$

$$\|\mathbf{u}\| = \sqrt{u_1^2 + u_2^2}$$

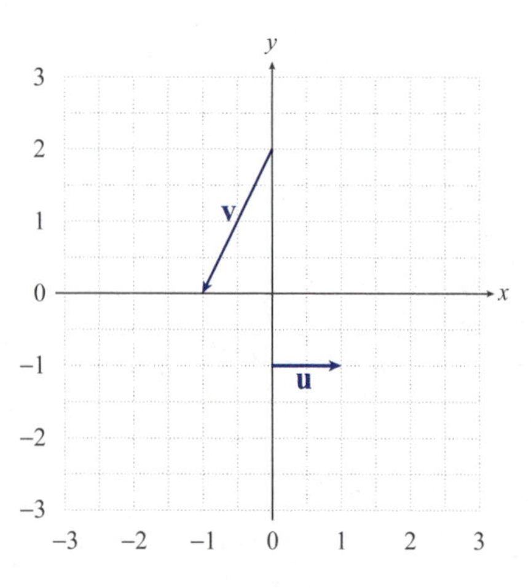

FIGURE 8

Example 3: Using Vector Operations

Given the two vectors **u** and **v** depicted in Figure 8, determine the component forms of $-2\mathbf{u}$, $3\mathbf{u} + \mathbf{v}$, and $\mathbf{u} - \mathbf{v}$, and then find the magnitudes of **u** and **v**.

Solution

In the previous discussion, we determined that $\mathbf{u} = \langle 1,0 \rangle$ and $\mathbf{v} = \langle -1,-2 \rangle$; note how these components can be determined by identifying $(0,-1)$ as the initial point of **u**, $(1,-1)$ as the terminal point of **u**, and $(0,2)$ and $(-1,0)$ as the initial and terminal points of **v**.

The remainder of the work is straightforward.

$$\begin{aligned}
-2\mathbf{u} &= \langle (-2)(1),(-2)(0) \rangle = \langle -2,0 \rangle, \\
3\mathbf{u}+\mathbf{v} &= \langle 3,0 \rangle + \langle -1,-2 \rangle = \langle 2,-2 \rangle, \\
\mathbf{u}-\mathbf{v} &= \langle 1,0 \rangle - \langle -1,-2 \rangle = \langle 2,2 \rangle, \\
\|\mathbf{u}\| &= \sqrt{1^2+0^2} = 1, \\
\|\mathbf{v}\| &= \sqrt{(-1)^2+(-2)^2} = \sqrt{5}
\end{aligned}$$

The vector **u** in the preceding two examples is just one example of a **unit vector**, which is any vector with a length of 1. There are an infinite number of unit vectors (think of any arrow of length 1 pointing in an arbitrary direction), but two of them are given special names.

$$\mathbf{i} = \langle 1,0 \rangle \quad \text{and} \quad \mathbf{j} = \langle 0,1 \rangle$$

In other words, **i** and **j** are the two unit vectors parallel to the coordinate axes. These two unit vectors are useful in some contexts because *any* vector in the plane can be written in terms of **i** and **j**, as follows.

$$\langle a,b \rangle = a\langle 1,0 \rangle + b\langle 0,1 \rangle = a\mathbf{i} + b\mathbf{j}$$

The last expression in the equation above is an example of a **linear combination**; any sum of scalar multiples of vectors constitutes a linear combination of the vectors, and the properties of linear combinations turn out to be very important in the study of vectors.

Example 4: Finding the Unit Vector and Linear Combination of a Vector

Let $\mathbf{u} = \langle -5,3 \rangle$. Find **a.** a unit vector pointing in the same direction as **u**, and **b.** the linear combination of **i** and **j** that is equivalent to **u**.

Solution

a. The key in constructing a unit vector with the same direction as **u** is to multiply **u** by an appropriate scalar a such that $\|a\mathbf{u}\| = 1$. We can solve this for a as follows.

$$\begin{aligned}
\|a\mathbf{u}\| &= 1 \\
|a|\|\mathbf{u}\| &= 1 \\
a &= \frac{1}{\|\mathbf{u}\|}
\end{aligned}$$

Note that the absolute value sign around a has been dropped in the third equation, since we know in advance that we want to multiply **u** by a positive number. In our case, we have $\|\mathbf{u}\| = \sqrt{25+9} = \sqrt{34}$, so the desired unit vector is

$$\left(\frac{1}{\sqrt{34}}\right)\langle -5,3 \rangle = \left\langle -\frac{5}{\sqrt{34}}, \frac{3}{\sqrt{34}} \right\rangle.$$

b. The second task is fairly straightforward: $\mathbf{u} = \langle -5,3 \rangle = -5\mathbf{i} + 3\mathbf{j}$.

In some cases, information about a vector's magnitude and its angle with respect to the positive x-axis may be given to us. If we need to work instead with the components of the vector, our knowledge of trigonometry comes to the rescue. The right triangle in Figure 9 is familiar from our study of sine and cosine.

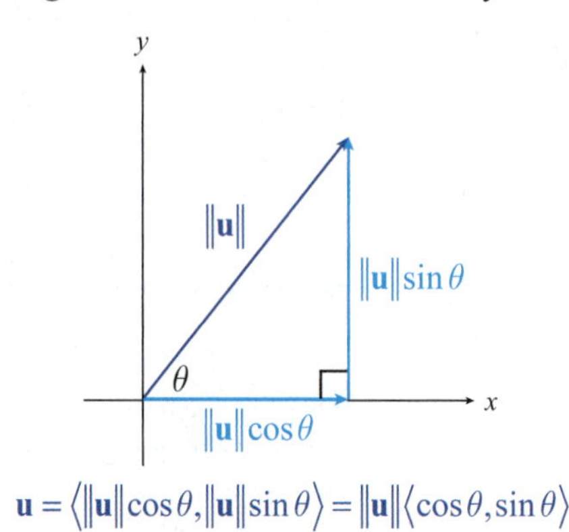

$$\mathbf{u} = \langle \|\mathbf{u}\|\cos\theta, \|\mathbf{u}\|\sin\theta \rangle = \|\mathbf{u}\|\langle \cos\theta, \sin\theta \rangle$$

FIGURE 9: Components Derived from Magnitude and Angle

Example 5: Finding the Vector Form of the Velocity

A baseball is hit and leaves the bat at a speed of 100 mph at an angle of $20°$ from the horizontal. Express this velocity in vector form.

Solution

The speed is the vector's magnitude, and $\theta = 20°$. So the ball's velocity as it leaves the bat is the vector

$$100\langle \cos 20°, \sin 20° \rangle \approx \langle 93.97, 34.20 \rangle.$$

TOPIC 4: Applications of Vectors

Let us return to an application that served as the inspiration for vector addition.

Example 6: Applying Vector Operations

An airplane is flying at a speed of 200 miles per hour at a bearing of N 33° E when it encounters wind with a velocity of 35 miles per hour at a bearing of N 47° W. What is the resultant true velocity of the airplane?

Solution

The first step is to express the plane's velocity and the wind's velocity as vectors. A bearing of N 33° E means 33° East of North, but it is more convenient to think of this as 57° North of East (in keeping with our convention of measuring angles relative to the positive x-axis). Similarly, the wind's bearing of N 47° W equates to a bearing of 137° as measured from due East. If we let $\mathbf{p}$ denote the plane's velocity and $\mathbf{w}$ the wind's velocity, we have the following.

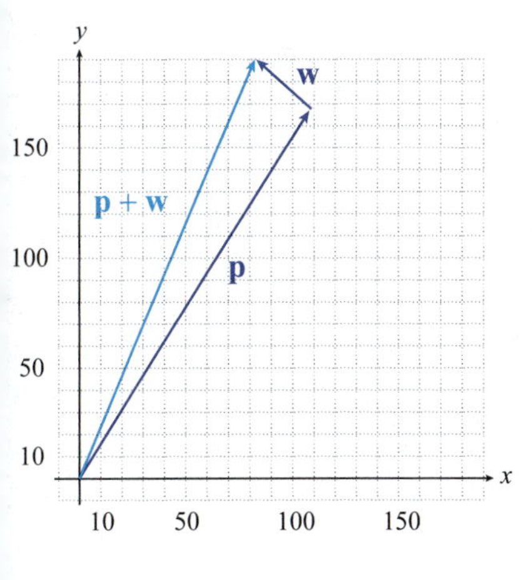

FIGURE 10

$$\mathbf{p} = 200\langle \cos 57°, \sin 57° \rangle \approx \langle 108.9, 167.7 \rangle$$

$$\mathbf{w} = 35\langle \cos 137°, \sin 137° \rangle \approx \langle -25.6, 23.9 \rangle$$

The plane's true velocity is now $\mathbf{p} + \mathbf{w} = \langle 83.3, 191.6 \rangle$. It may also be useful to determine that the speed of the plane is now $\|\mathbf{p} + \mathbf{w}\| = \sqrt{(83.3)^2 + (191.6)^2} \approx 208.9$ miles per hour, and that its bearing is 66.5° North of East (which, using conventional bearing notation, would be written as N 23.5° E). This last angle is derived from the fact that $\tan\theta = \dfrac{191.6}{83.3}$, so $\theta = \tan^{-1}\left(\dfrac{191.6}{83.3}\right) \approx 66.5°$. Figure 10 illustrates the three vectors in this problem.

Example 7: Applying Vector Operations

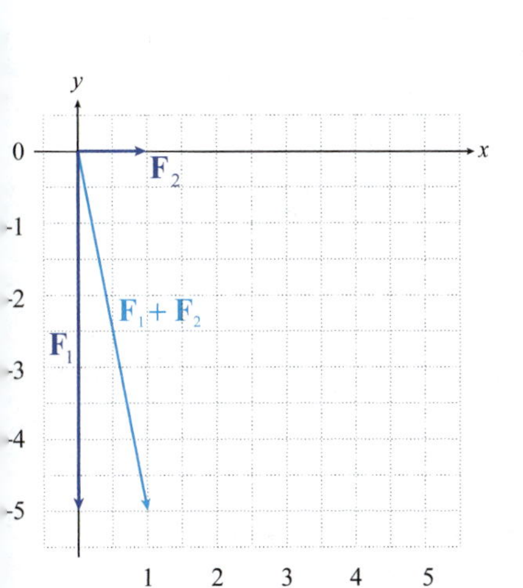

FIGURE 11

A cat is slowly pushing a 5-pound plant across a table, with the intention of knocking it off the edge (determining why cats feel the need to do so is beyond the scope of this text). The cat is pushing with a force of 1 pound. What is the total force being applied to the plant?

Solution

Weight is itself a force—it is the force due to gravity that Earth exerts on an object. Forces exerted on an object are added as vectors, and the result is the total applied force. If we let $\mathbf{F}_1$ denote the weight of the plant and $\mathbf{F}_2$ the force exerted by the cat, the force $\mathbf{F}$ on the plant is

$$\mathbf{F} = \mathbf{F}_1 + \mathbf{F}_2 = \langle 0, -5 \rangle + \langle 1, 0 \rangle = \langle 1, -5 \rangle.$$

The magnitude of $\mathbf{F}$ is $\sqrt{1+25} \approx 5.1$ pounds, and Figure 11 illustrates the situation.

9.6 EXERCISES

PRACTICE

Use the figure to sketch a graph for the specified vector. See Examples 1 and 2.

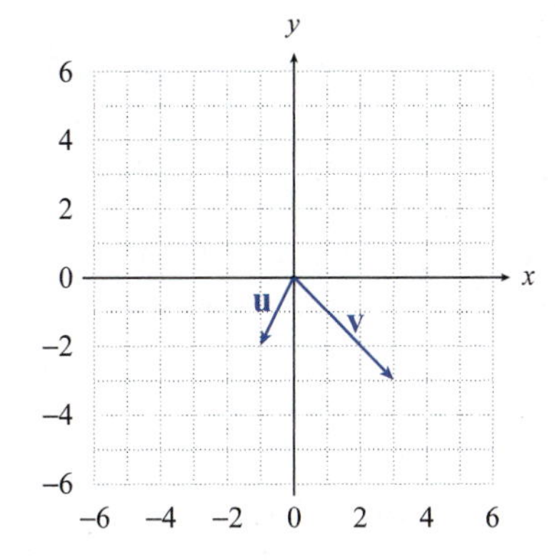

1. $-\mathbf{u}$

2. $2\mathbf{u} + \mathbf{v}$

3. $3\mathbf{v}$

4. $-\dfrac{1}{2}\mathbf{u} - \mathbf{v}$

5. $2\mathbf{u} - 2\mathbf{v}$

6. $\mathbf{u} + 3\mathbf{v}$

Find the component form and the magnitude of vector **v** for each of the following. See Example 3.

7.

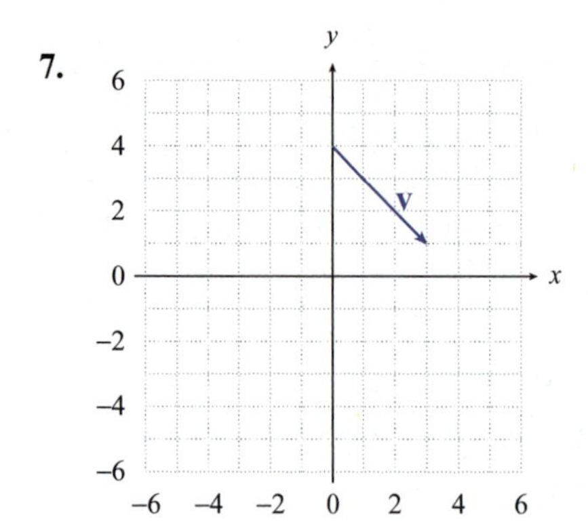

8.

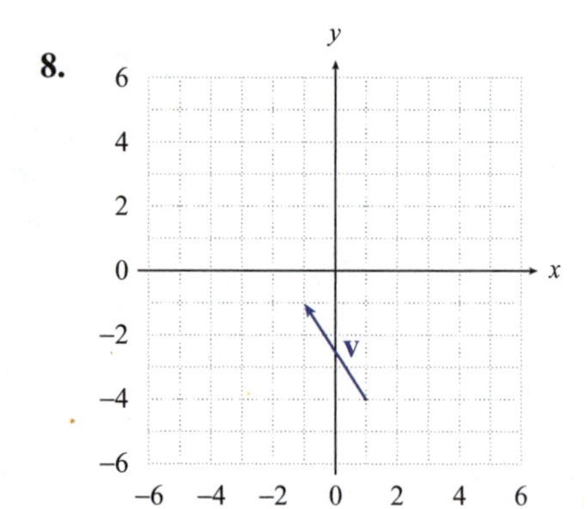

9.

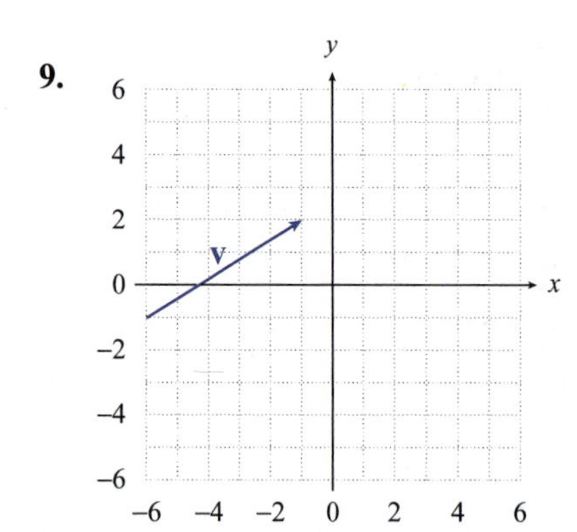

10.

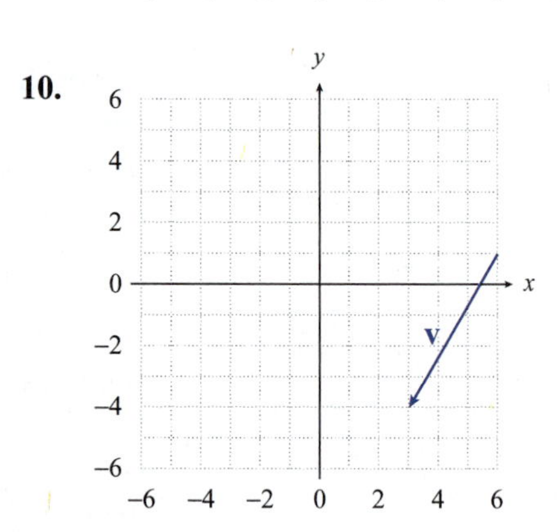

Find the component form and the magnitude of a vector **v** defined by the given points. Assume the first point given is the initial point and the second point given is the terminal point. See Example 3.

11. $(-2,4),(3,3)$

12. $(1,6),(2,3)$

13. $(5,-2),(-2,5)$

14. $(4,0),(-1,7)$

15. $(3,4),(-1,-2)$

16. $(1,-6),(0,0)$

For each of the following, calculate **a.** $2\mathbf{u}+\mathbf{v}$, **b.** $-\mathbf{u}+3\mathbf{v}$, and **c.** $-2\mathbf{v}$. See Example 3.

17. $\mathbf{u}=\langle -2,4\rangle, \mathbf{v}=\langle 2,0\rangle$

18. $\mathbf{u}=\langle 4,1\rangle, \mathbf{v}=\langle 2,5\rangle$

19. $\mathbf{u}=\langle 2,0\rangle, \mathbf{v}=\langle -3,4\rangle$

20. $\mathbf{u}=\langle 1,3\rangle, \mathbf{v}=\langle 4,4\rangle$

21. $\mathbf{u}=\langle -1,-4\rangle, \mathbf{v}=\langle -3,-2\rangle$

22. $\mathbf{u}=\langle 0,-5\rangle, \mathbf{v}=\langle -1,2\rangle$

For each of the following graphs, determine the component forms of $-\mathbf{u}$, $2\mathbf{u} - \mathbf{v}$, and $\mathbf{u} + \mathbf{v}$ and find the magnitudes of $\mathbf{u}$ and $\mathbf{v}$. See Example 3.

23.

24.

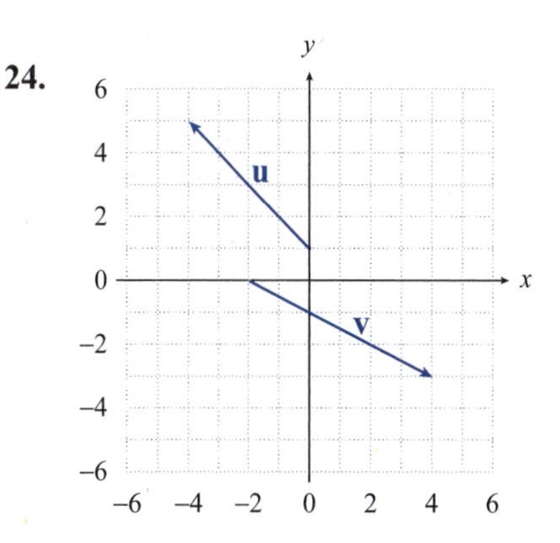

25.

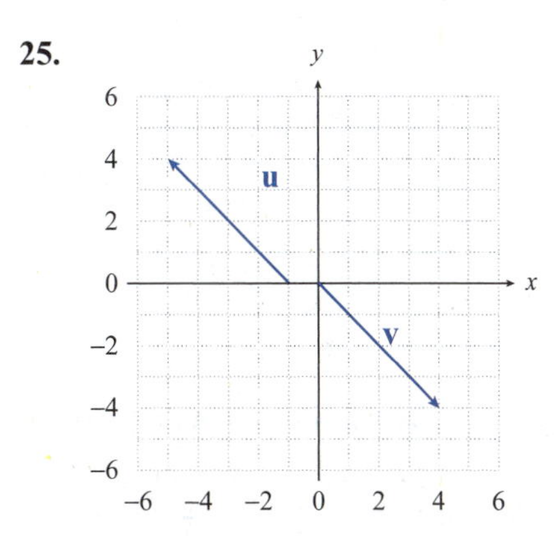

26.

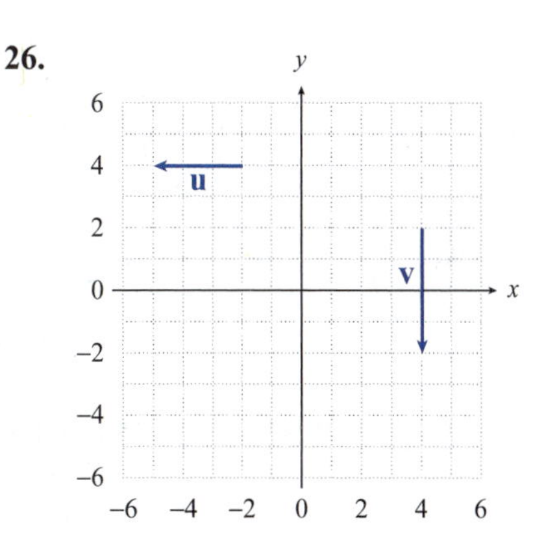

Given the vector $\mathbf{u}$, find **a.** a unit vector pointing in the same direction as $\mathbf{u}$, and **b.** the linear combination of $\mathbf{i}$ and $\mathbf{j}$ that is equivalent to $\mathbf{u}$. See Example 4.

27. $\mathbf{u} = \langle 6, -3 \rangle$

28. $\mathbf{u} = \langle 1, 4 \rangle$

29. $\mathbf{u} = \langle -5, -1 \rangle$

30. $\mathbf{u} = \langle -4, 3 \rangle$

31. $\mathbf{u} = \langle 2, 3 \rangle$

32. $\mathbf{u} = \langle 5, 2 \rangle$

Find the magnitude and direction angle of the vector $\mathbf{v}$.

33. $\mathbf{v} = 5(\cos 30°\mathbf{i} + \sin 30°\mathbf{j})$

34. $\mathbf{v} = 7(\cos 45°\mathbf{i} + \sin 45°\mathbf{j})$

35. $\mathbf{v} = 4\mathbf{i} + 3\mathbf{j}$

36. $\mathbf{v} = -2\mathbf{i} - 2\mathbf{j}$

Find the component form of $\mathbf{v}$ given its magnitude and the angle it makes with the positive x-axis. See Example 5.

37. $\|\mathbf{v}\| = 6,\ \theta = 30°$

38. $\|\mathbf{v}\| = \frac{5}{2},\ \theta = 0°$

39. $\|\mathbf{v}\| = 18,\ \theta = 135°$

40. $\|\mathbf{v}\| = 3\sqrt{3},\ \theta = 90°$

41. $\|\mathbf{v}\| = 1,\ \theta = 120°$

42. $\|\mathbf{v}\| = 4\sqrt{2},\ \theta = 45°$

43. $\|\mathbf{v}\| = 4,\ \mathbf{v}$ in the direction of $2\mathbf{i} + 3\mathbf{j}$

44. $\|\mathbf{v}\| = 7,\ \mathbf{v}$ in the direction of $\mathbf{i} + 4\mathbf{j}$

APPLICATIONS

45. A paper airplane is launched into the air at a speed of 4 ft/s and at an angle of 30° from the horizontal. Express this velocity in vector form.

46. A golf ball is driven into the air at a speed of 75 miles per hour and at an angle of 50° from the horizontal. Express this velocity in vector form.

47. A sailboat is traveling at a speed of 45 miles per hour with a bearing of N 59° W, when it encounters a front with winds blowing at 15 miles per hour with a bearing of S 3° E. What is the resultant true velocity of the sailboat?

48. An underwater missile is traveling at a speed of 350 miles per hour and bearing of S 17° W, when it meets a current traveling at 44 miles per hour in the direction of N 61° W. What is the resultant true velocity of the underwater missile?

49. Prometheus is slowly pushing a 1235-pound boulder across a flat plain with a force of 150 pounds. What is the total force **F** being applied to the boulder, and what is the magnitude of **F**?

50. A boy is pushing a toy truck across the floor. If the toy weighs 3 pounds and the boy is exerting half a pound of pressure on the toy, what is the total force **F** being applied to the toy truck, and what is the magnitude of **F**?

9.7 THE DOT PRODUCT

TOPICS

1. Properties of the dot product
2. Projections of vectors
3. Applications of the dot product

TOPIC 1: Properties of the Dot Product

First Uses of the Dot Product

The American mathematical physicist Josiah Gibbs developed and promoted the use of the dot product (and other vector operations) in the 1880s, inspired by the earlier work of the German mathematician Hermann Grassmann. Gibbs was motivated by the practical applications of vectors, as was the British mathematical physicist Oliver Heaviside, who independently conducted similar work around the same time.

This section introduces a vector operation that is distinct from the vector addition and scalar multiplication that you learned in Section 9.6. The reason for introducing this third operation is simply its usefulness in performing a variety of tasks.

The operation goes by a variety of names; it is often referred to as the **dot product** of two vectors, due to the notation used, but it is also often called the **scalar product** because it is an operation that turns two vectors into a scalar. More advanced texts often refer to the operation as the **inner product**, to differentiate it from a number of other products that can be defined on vectors (including one called, as you might have guessed, the *outer product*).

DEFINITION: The Dot Product in the Plane

Given two vectors $\mathbf{u} = \langle u_1, u_2 \rangle$ and $\mathbf{v} = \langle v_1, v_2 \rangle$, the **dot product** $\mathbf{u} \cdot \mathbf{v}$ of the two vectors is the scalar defined by

$$\mathbf{u} \cdot \mathbf{v} = u_1 v_1 + u_2 v_2.$$

Note that the dot product of two vectors is, indeed, a scalar, and that it can be positive, 0, or negative. Its calculation is simple, assuming the components of the two vectors are known.

Example 1: Calculating the Dot Product

Calculate each of the following dot products.

a. $\langle -5, 2 \rangle \cdot \langle 3, -1 \rangle$ **b.** $\langle -5, 2 \rangle \cdot \langle 2, 5 \rangle$ **c.** $\langle -5, 2 \rangle \cdot \langle -5, 2 \rangle$

Solution

a. $\langle -5, 2 \rangle \cdot \langle 3, -1 \rangle = (-5)(3) + (2)(-1) = -15 - 2 = -17$

b. $\langle -5, 2 \rangle \cdot \langle 2, 5 \rangle = (-5)(2) + (2)(5) = -10 + 10 = 0$

c. $\langle -5, 2 \rangle \cdot \langle -5, 2 \rangle = (-5)(-5) + (2)(2) = 25 + 4 = 29$

A number of properties of the dot product can be proven. The verifications of the following statements are left to the reader.

PROPERTIES: Elementary Properties of the Dot Product

Given two vectors $\mathbf{u} = \langle u_1, u_2 \rangle$ and $\mathbf{v} = \langle v_1, v_2 \rangle$ and a scalar a, the following hold.

1. $\mathbf{u} \cdot \mathbf{v} = \mathbf{v} \cdot \mathbf{u}$
2. $\mathbf{0} \cdot \mathbf{u} = 0$
3. $\mathbf{u} \cdot (\mathbf{v} + \mathbf{w}) = \mathbf{u} \cdot \mathbf{v} + \mathbf{u} \cdot \mathbf{w}$
4. $a(\mathbf{u} \cdot \mathbf{v}) = (a\mathbf{u}) \cdot \mathbf{v} = \mathbf{u} \cdot (a\mathbf{v})$
5. $\mathbf{u} \cdot \mathbf{u} = \|\mathbf{u}\|^2$

Example 2: Using the Properties of the Dot Product

Given that the dot product of a vector **u** with itself is 8, determine the magnitude of **u**.

Solution

Since $\|\mathbf{u}\|^2 = \mathbf{u} \cdot \mathbf{u} = 8,$ it follows that $\|\mathbf{u}\| = 2\sqrt{2}.$

There is another property, however, whose truth is not so immediately clear. It often goes by the name of the Dot Product Theorem.

THEOREM: The Dot Product Theorem

Let two nonzero vectors **u** and **v** be depicted so that their initial points coincide, and let θ be the smaller of the two angles formed by **u** and **v** (so $0 \le \theta \le \pi$). Then

$$\mathbf{u} \cdot \mathbf{v} = \|\mathbf{u}\| \|\mathbf{v}\| \cos\theta.$$

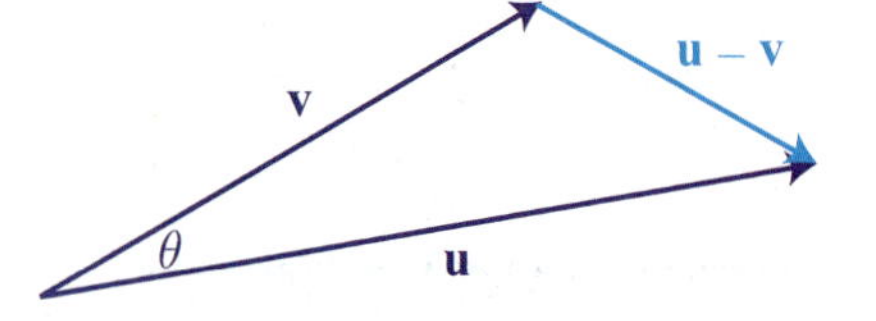

FIGURE 1: Proving the Dot Product Theorem

The proof of this statement makes good use of the Law of Cosines and several of the elementary properties listed previously. Starting with the two vectors **u** and **v** drawn as directed, the third side of a triangle is formed by **u** − **v** as shown in Figure 1.

By the Law of Cosines, then,

$$\|\mathbf{u} - \mathbf{v}\|^2 = \|\mathbf{u}\|^2 + \|\mathbf{v}\|^2 - 2\|\mathbf{u}\| \|\mathbf{v}\| \cos\theta.$$

But we have a second way of expressing $\|\mathbf{u} - \mathbf{v}\|^2.$

$$\begin{aligned} \|\mathbf{u} - \mathbf{v}\|^2 &= (\mathbf{u} - \mathbf{v}) \cdot (\mathbf{u} - \mathbf{v}) \\ &= \mathbf{u} \cdot \mathbf{u} - \mathbf{u} \cdot \mathbf{v} - \mathbf{v} \cdot \mathbf{u} + \mathbf{v} \cdot \mathbf{v} \\ &= \|\mathbf{u}\|^2 - 2(\mathbf{u} \cdot \mathbf{v}) + \|\mathbf{v}\|^2 \end{aligned}$$

Equating these two different ways of expressing $\|\mathbf{u}-\mathbf{v}\|^2$ and canceling like terms, we obtain the following.

$$-2\|\mathbf{u}\|\|\mathbf{v}\|\cos\theta = -2(\mathbf{u}\cdot\mathbf{v})$$

Dividing both sides by −2 then gives us the desired result.

While the Dot Product Theorem is occasionally useful in calculating the dot product of two given vectors, it is more often used to determine the angle between two vectors.

Example 3: Applying the Dot Product Theorem

Find the angle between the vector $\mathbf{u} = \langle -3, 1\rangle$ and the vector $\mathbf{v} = \langle 2, 5\rangle$.

Solution

By the Dot Product Theorem,

$$\cos\theta = \frac{\mathbf{u}\cdot\mathbf{v}}{\|\mathbf{u}\|\|\mathbf{v}\|} = \frac{\langle -3, 1\rangle \cdot \langle 2, 5\rangle}{\left(\sqrt{10}\right)\left(\sqrt{29}\right)} = \frac{-1}{\sqrt{290}}$$

and so

$$\theta = \cos^{-1}\left(-\frac{1}{\sqrt{290}}\right) \approx 93.4°.$$

The Dot Product Theorem also makes it clear that pairs of vectors for which θ is a right angle are easily identified by the dot product. If $\theta = \frac{\pi}{2}$ for two given vectors $\mathbf{u}$ and $\mathbf{v}$, then $\cos\theta = 0$ and hence the dot product is 0. We turn this into a definition as follows:

DEFINITION: Orthogonal Vectors

Two vectors $\mathbf{u}$ and $\mathbf{v}$ are **orthogonal** (or **perpendicular**) if

$$\mathbf{u}\cdot\mathbf{v} = 0.$$

To accompany the notion of orthogonality, it is appropriate to also define a related concept.

DEFINITION: Parallel Vectors

Two vectors $\mathbf{u}$ and $\mathbf{v}$ are **parallel** if each is a scalar multiple of the other. That is, $\mathbf{u} = k\mathbf{v}$ for some real number k (and consequently, $\mathbf{v} = \frac{1}{k}\mathbf{u}$).

Geometrically, two vectors are parallel if either they point in the same direction or exact opposite directions.

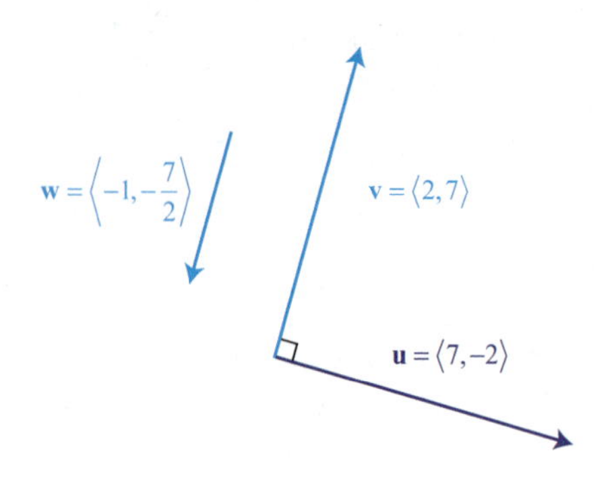

FIGURE 2

Example 4: Finding an Orthogonal Vector

Find a vector orthogonal to the vector $\mathbf{u} = \langle 7, -2 \rangle$.

Solution

There are an infinite number of vectors $\mathbf{v}$ for which $\mathbf{u} \cdot \mathbf{v} = 0$, and hence an infinite number of correct answers. One such answer is $\mathbf{v} = \langle 2, 7 \rangle$, and another is $\mathbf{w} = \left\langle -1, -\frac{7}{2} \right\rangle$, as shown in Figure 2. (Note that $\mathbf{u}$ and $\mathbf{v}$ have been placed so that their initial points coincide, while $\mathbf{w}$ is shown elsewhere in the plane.) Any multiple of $\langle 2, 7 \rangle$ is orthogonal to $\mathbf{u}$.

TOPIC 2: Projections of Vectors

The solution of many problems depends on being able to decompose a vector into a sum of two vectors in a specific way. Usually, the task amounts to writing a vector $\mathbf{u}$ as a sum $\mathbf{w}_1 + \mathbf{w}_2$ where $\mathbf{w}_1$ is parallel to a second vector $\mathbf{v}$ and $\mathbf{w}_2$ is orthogonal to $\mathbf{v}$.

In the language of vectors, we say $\mathbf{w}_1$ is the **projection of u onto v**, and write $\mathbf{w}_1 = \text{proj}_\mathbf{v}\mathbf{u}$. If we can determine $\text{proj}_\mathbf{v}\mathbf{u}$, then we can easily find $\mathbf{w}_2$ by noting that

$$\mathbf{w}_2 = \mathbf{u} - \text{proj}_\mathbf{v}\mathbf{u},$$

and since $\mathbf{w}_2$ is perpendicular to $\mathbf{v}$, some books refer to $\mathbf{w}_2$ as $\text{perp}_\mathbf{v}\mathbf{u}$.

The first step is to find a formula for $\text{proj}_\mathbf{v}\mathbf{u}$ in terms of $\mathbf{u}$ and $\mathbf{v}$. From our study of trigonometry and the right triangle in Figure 3, we know that

$$\cos\theta = \frac{\|\mathbf{w}_1\|}{\|\mathbf{u}\|},$$

so

$$\|\mathbf{w}_1\| = \|\mathbf{u}\|\cos\theta.$$

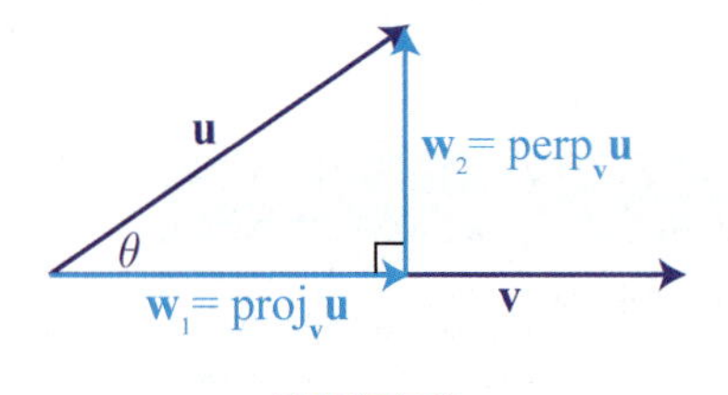

FIGURE 3

We can now apply the Dot Product Theorem and write

$$\|\mathbf{w}_1\| = \|\mathbf{u}\|\cos\theta = \|\mathbf{u}\|\left(\frac{\mathbf{u} \cdot \mathbf{v}}{\|\mathbf{u}\|\|\mathbf{v}\|}\right) = \frac{\mathbf{u} \cdot \mathbf{v}}{\|\mathbf{v}\|}.$$

This, however, is only the *magnitude* of the projection of $\mathbf{u}$ onto $\mathbf{v}$. In order to construct the actual vector $\mathbf{w}_1$, we need to multiply this magnitude by a unit vector pointing in the same direction as $\mathbf{v}$.

$$\mathbf{w}_1 = \left(\frac{\mathbf{u} \cdot \mathbf{v}}{\|\mathbf{v}\|}\right)\left(\frac{\mathbf{v}}{\|\mathbf{v}\|}\right) = \left(\frac{\mathbf{u} \cdot \mathbf{v}}{\|\mathbf{v}\|^2}\right)\mathbf{v}$$

Note that $\mathbf{w}_1$ is indeed parallel to $\mathbf{v}$ since $\mathbf{w}_1$ is a scalar multiple of $\mathbf{v}$.

DEFINITION: Projection of u onto v

Let $\mathbf{u}$ and $\mathbf{v}$ be nonzero vectors. The **projection of u onto v** is the vector

$$\text{proj}_{\mathbf{v}}\mathbf{u} = \left(\frac{\mathbf{u}\cdot\mathbf{v}}{\|\mathbf{v}\|^2}\right)\mathbf{v}.$$

Example 5: Finding the Projection of u onto v

Find the projection of $\mathbf{u} = \langle 2, 4\rangle$ onto $\mathbf{v} = \langle 7, -1\rangle$, and then write $\mathbf{u}$ as a sum of two orthogonal vectors, one of which is $\text{proj}_{\mathbf{v}}\mathbf{u}$.

Solution

First,

$$\begin{aligned}\text{proj}_{\mathbf{v}}\mathbf{u} &= \left(\frac{\langle 2, 4\rangle\cdot\langle 7, -1\rangle}{\|\langle 7, -1\rangle\|^2}\right)\langle 7, -1\rangle \\ &= \left(\frac{14-4}{\left(\sqrt{49+1}\right)^2}\right)\langle 7, -1\rangle \\ &= \left(\frac{1}{5}\right)\langle 7, -1\rangle \\ &= \left\langle \frac{7}{5}, -\frac{1}{5}\right\rangle.\end{aligned}$$

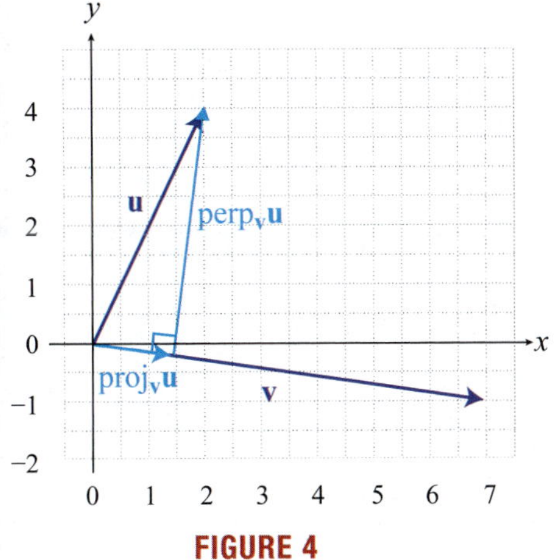

FIGURE 4

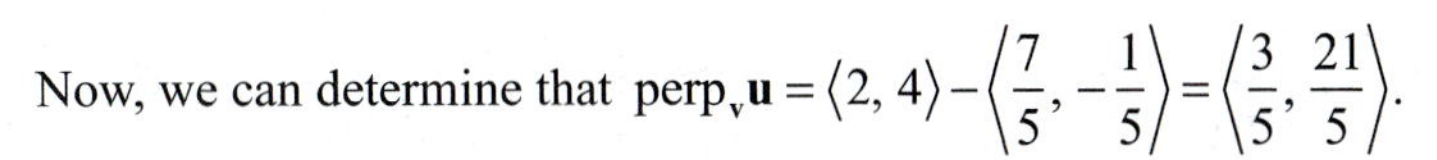
Now, we can determine that $\text{perp}_{\mathbf{v}}\mathbf{u} = \langle 2, 4\rangle - \left\langle \frac{7}{5}, -\frac{1}{5}\right\rangle = \left\langle \frac{3}{5}, \frac{21}{5}\right\rangle.$

So, we can write $\mathbf{u}$ as the sum of the two orthogonal vectors $\text{proj}_{\mathbf{v}}\mathbf{u}$ and $\text{perp}_{\mathbf{v}}\mathbf{u}$.

$$\mathbf{u} = \text{proj}_{\mathbf{v}}\mathbf{u} + \text{perp}_{\mathbf{v}}\mathbf{u} = \left\langle \frac{7}{5}, -\frac{1}{5}\right\rangle + \left\langle \frac{3}{5}, \frac{21}{5}\right\rangle = \langle 2, 4\rangle = \mathbf{u}$$

Figure 4 shows the graphs of $\mathbf{u}$ and $\mathbf{v}$ as well as the orthogonal decompositions of $\mathbf{u}$, labeled $\text{proj}_{\mathbf{v}}\mathbf{u}$ and $\text{perp}_{\mathbf{v}}\mathbf{u}$.

TOPIC 3: Applications of the Dot Product

The dot product and associated concepts are very useful in answering some common questions. We will close this section with a few examples.

Example 6: Applying the Dot Product

A boat and trailer, which together weigh 650 pounds, are to be pulled up a boat ramp that has an incline of 30°. What force is required to merely prevent the boat and trailer from rolling down the ramp?

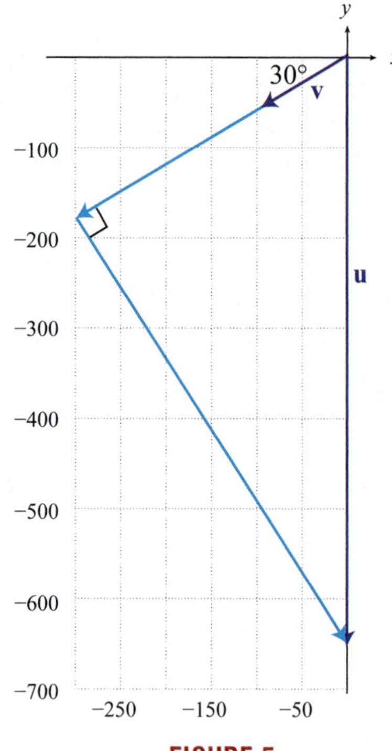

FIGURE 5

Solution

As seen in Example 7 of Section 9.6, weight is simply an example of force: it is the force that Earth exerts on an object through gravity. To gain some insight into this problem, suppose the question were "What is the force needed to lift the boat and trailer straight up?" The answer would be that 650 pounds of force is needed to counter the pull of gravity, so anything greater than 650 pounds of force would result in the boat and trailer being lifted.

But the actual situation isn't so extreme. Only the portion of the combined weight that is parallel to the boat ramp must be countered. That is, we are looking for the component of the 650 pounds that is directed at an angle of 30° from the horizontal. The picture at right illustrates the vectors under discussion.

In Figure 5, the long vector of magnitude 650 pointing straight down represents the weight of the boat and trailer, and the vector labeled **v** points down the boat ramp. We seek the component of the weight vector in the direction of the boat ramp. The weight vector is simply $\mathbf{u} = \langle 0, -650 \rangle$. A vector of any length pointing down the boat ramp can be used; one that is easy to work with is $\mathbf{v} = \langle -\sqrt{3}, -1 \rangle$ (note that these vector components correspond to the adjacent and opposite sides of the familiar 30-60-90 right triangle).

Now,

$$\begin{aligned}
\text{proj}_{\mathbf{v}}\mathbf{u} &= \left(\frac{\langle 0, -650 \rangle \cdot \langle -\sqrt{3}, -1 \rangle}{\left\| \langle -\sqrt{3}, -1 \rangle \right\|^2} \right) \langle -\sqrt{3}, -1 \rangle \\
&= \left(\frac{(-650)(-1)}{\left(\sqrt{3+1}\right)^2} \right) \langle -\sqrt{3}, -1 \rangle \\
&= \left(\frac{650}{4} \right) \langle -\sqrt{3}, -1 \rangle \\
&= \left\langle -\frac{325\sqrt{3}}{2}, \frac{-325}{2} \right\rangle .
\end{aligned}$$

The magnitude of this projection, which is 325 pounds, is thus the force that must be exerted to keep the boat and trailer stationary. Anything more than this will allow the boat and trailer to be pulled up the ramp.

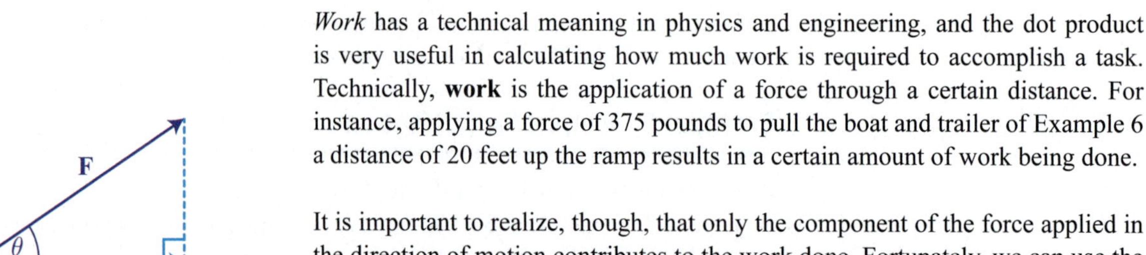

Work has a technical meaning in physics and engineering, and the dot product is very useful in calculating how much work is required to accomplish a task. Technically, **work** is the application of a force through a certain distance. For instance, applying a force of 375 pounds to pull the boat and trailer of Example 6 a distance of 20 feet up the ramp results in a certain amount of work being done.

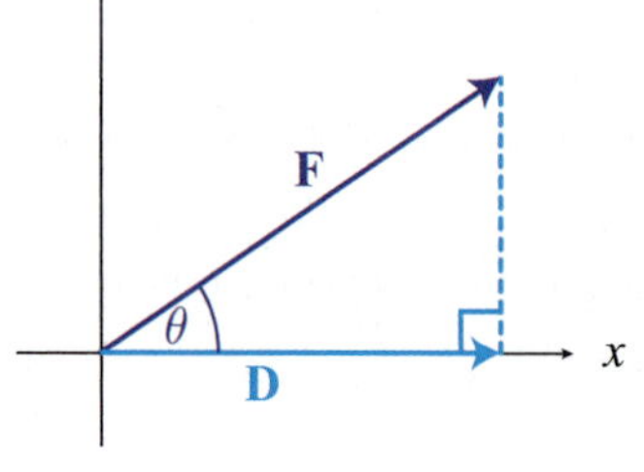

FIGURE 6

It is important to realize, though, that only the component of the force applied in the direction of motion contributes to the work done. Fortunately, we can use the dot product to make this calculation. If we let **D** represent the vector of the motion (so that $\|\mathbf{D}\|$ is the distance traveled) and **F** the force applied (not necessarily in the same direction as **D**), then the component of the force in the direction of motion is $\|\mathbf{F}\|\cos\theta$, where θ is, as usual, the angle between the two vectors. The work done is then the product of $\|\mathbf{D}\|$ and $\|\mathbf{F}\|\cos\theta$, that is, $\mathbf{W} = \|\mathbf{D}\|\,\|\mathbf{F}\|\cos\theta$. But this last expression should look familiar—it appears in the Dot Product Theorem, allowing us to write the simpler formula below.

$$\mathbf{W} = \mathbf{F} \cdot \mathbf{D}$$

Example 7: Applying the Dot Product

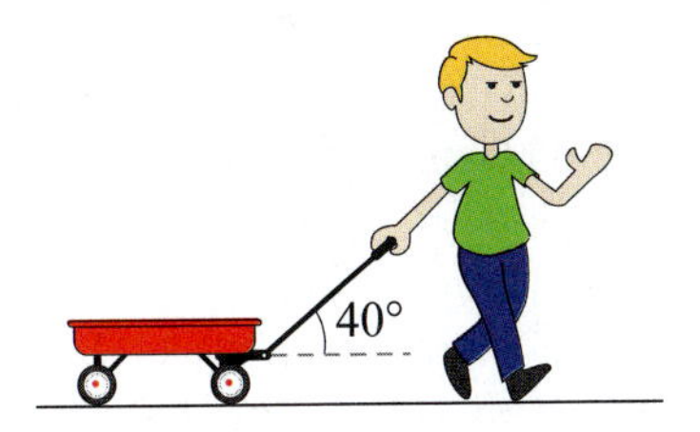

FIGURE 7

A child pulls a wagon along a sidewalk, exerting a force of 15 pounds on the handle of the wagon. The handle is at an angle of 40° to the horizontal. If the child pulls the wagon a distance of 50 feet, what work has been done?

Solution

We start by defining the force and distance vectors.

$$\mathbf{F} = 15\langle \cos 40°, \sin 40° \rangle \text{ and } \mathbf{D} = \langle 50, 0 \rangle$$

The calculation is now straightforward.

$$\begin{aligned} \mathbf{W} &= 15\langle \cos 40°, \sin 40° \rangle \cdot \langle 50, 0 \rangle \\ &\approx (15)(0.766)(50) + (15)(0.643)(0) \\ &= 574.5 \text{ foot-pounds} \end{aligned}$$

9.7 EXERCISES

PRACTICE

Calculate each of the following dot products. See Example 1.

1. $\langle 4, 3 \rangle \cdot \langle 5, -1 \rangle$ **2.** $\langle 2, 4 \rangle \cdot \langle -1, -1 \rangle$ **3.** $\langle 3, 5 \rangle \cdot \langle 2, 0 \rangle$

4. $\langle -1, 6 \rangle \cdot \langle 6, 1 \rangle$ **5.** $\langle 2, 2 \rangle \cdot \langle 2, 2 \rangle$ **6.** $\langle 1, 2 \rangle \cdot \langle 3, 4 \rangle$

7. $\langle -4, 3 \rangle \cdot \langle 2, 3 \rangle$ **8.** $\langle -2, -4 \rangle \cdot \langle 6, 2 \rangle$

9. $\mathbf{u} = 5\mathbf{i} + \mathbf{j},\ \mathbf{v} = -2\mathbf{i} + 3\mathbf{j}$ **10.** $\mathbf{u} = \mathbf{i} - 5\mathbf{j},\ \mathbf{v} = -2\mathbf{i} - 4\mathbf{j}$

Find the indicated quantity given $\mathbf{u} = \langle -2, 3 \rangle$ and $\mathbf{v} = \langle 4, 4 \rangle$. See Example 1.

11. $\mathbf{v} \cdot \mathbf{v}$ **12.** $4\mathbf{u} \cdot \mathbf{v}$ **13.** $(\mathbf{u} \cdot \mathbf{u})\mathbf{u}$ **14.** $(\mathbf{u} \cdot \mathbf{v})2\mathbf{v}$

Find the magnitude of **u** using the dot product. See Example 2.

15. $\mathbf{u} = \langle 6, -1 \rangle$ **16.** $\mathbf{u} = \langle 10, 3 \rangle$ **17.** $\mathbf{u} = 2\mathbf{i} + 7\mathbf{j}$ **18.** $\mathbf{u} = -3\mathbf{i} + 4\mathbf{j}$

Find the angle between the given vectors. See Example 3.

19. $\mathbf{u} = \langle -2, 3 \rangle,\ \mathbf{v} = \langle 1, 0 \rangle$ **20.** $\mathbf{u} = \langle 5, 4 \rangle,\ \mathbf{v} = \langle 3, 2 \rangle$

21. $\mathbf{u} = \langle 3, 5 \rangle$, $\mathbf{v} = \langle 4, 4 \rangle$

22. $\mathbf{u} = \langle -4, 2 \rangle$, $\mathbf{v} = \langle 1, 5 \rangle$

23. $\mathbf{u} = -\mathbf{i} + 2\mathbf{j}$, $\mathbf{v} = 3\mathbf{i} - 3\mathbf{j}$

24. $\mathbf{u} = 5\mathbf{i} + 2\mathbf{j}$, $\mathbf{v} = 4\mathbf{i} + \mathbf{j}$

25. $\mathbf{u} = \cos\left(\frac{3\pi}{4}\right)\mathbf{i} + \sin\left(\frac{3\pi}{4}\right)\mathbf{j}$, $\mathbf{v} = \cos\left(\frac{\pi}{2}\right)\mathbf{i} + \sin\left(\frac{\pi}{2}\right)\mathbf{j}$

26. $\mathbf{u} = \cos\left(\frac{\pi}{4}\right)\mathbf{i} + \sin\left(\frac{\pi}{4}\right)\mathbf{j}$, $\mathbf{v} = \cos\left(\frac{5\pi}{6}\right)\mathbf{i} + \sin\left(\frac{5\pi}{6}\right)\mathbf{j}$

Use vectors to find the interior angles of the triangles given the following sets of vertices.

27. $(3, 3), (4, 2), (-1, -6)$

28. $(0, 0), (0, 5), (3, 6)$

29. $(-2, -1), (2, 4), (-4, 5)$

30. $(6, 3), (-5, 2), (-6, 1)$

Find **u** · **v** where θ is the angle between **u** and **v**. See Example 3.

31. $\|\mathbf{u}\| = 25$, $\|\mathbf{v}\| = 5$, $\theta = 120°$

32. $\|\mathbf{u}\| = 4$, $\|\mathbf{v}\| = 64$, $\theta = \frac{\pi}{6}$

33. $\|\mathbf{u}\| = 16$, $\|\mathbf{v}\| = 4$, $\theta = \frac{3\pi}{4}$

34. $\|\mathbf{u}\| = 9$, $\|\mathbf{v}\| = 10$, $\theta = \frac{2\pi}{3}$

Find two vectors orthogonal to the given vector. See Example 4. Answers may vary.

35. $\mathbf{u} = \langle 3, -3 \rangle$

36. $\mathbf{u} = \langle 4, 1 \rangle$

37. $\mathbf{u} = \langle 2, -6 \rangle$

38. $\mathbf{u} = \langle 5, 4 \rangle$

Determine whether **u** and **v** are orthogonal, parallel, or neither. See Example 4.

39. $\mathbf{u} = \langle 2, -3 \rangle$, $\mathbf{v} = \langle 1, 6 \rangle$

40. $\mathbf{u} = \langle -12, 30 \rangle$, $\mathbf{v} = \left\langle \frac{1}{2}, -\frac{5}{4} \right\rangle$

41. $\mathbf{u} = 2\mathbf{i} - 2\mathbf{j}$, $\mathbf{v} = -\mathbf{i} - \mathbf{j}$

42. $\mathbf{u} = \mathbf{i}$, $\mathbf{v} = -2\mathbf{i} + 2\mathbf{j}$

Find the projection of **u** onto **v**, and then write **u** as a sum of two orthogonal vectors, one of which is $\text{proj}_{\mathbf{v}}\mathbf{u}$. See Example 5.

43. $\mathbf{u} = \langle 1, 3 \rangle$, $\mathbf{v} = \langle 4, 2 \rangle$

44. $\mathbf{u} = \langle 2, 2 \rangle$, $\mathbf{v} = \langle 1, -7 \rangle$

45. $\mathbf{u} = \langle 3, -5 \rangle$, $\mathbf{v} = \langle 6, 2 \rangle$

46. $\mathbf{u} = \langle 0, 3 \rangle$, $\mathbf{v} = \langle 2, 6 \rangle$

47. $\mathbf{u} = \langle -3, -3 \rangle$, $\mathbf{v} = \langle -4, -1 \rangle$

48. $\mathbf{u} = \langle 4, 2 \rangle$, $\mathbf{v} = \langle 1, 5 \rangle$

Find the work done on a particle moving from J to K if the magnitude and direction of the force are given by **F**. See Example 7.

49. $J = (1, 4)$, $K = (5, 6)$, $\mathbf{F} = \langle 2, 3 \rangle$

50. $J = (-3, 2)$, $K = (0, 5)$, $\mathbf{F} = \langle 4, 2 \rangle$

51. $J = (3, 0)$, $K = (-4, -2)$, $\mathbf{F} = -\mathbf{i} + 2\mathbf{j}$

52. $J = (3, -3)$, $K = (5, 1)$, $\mathbf{F} = 6\mathbf{i} - 3\mathbf{j}$

APPLICATIONS

53. A truck with a gross weight of 25,000 pounds is parked on an 8° slope. What force is required to prevent the truck from rolling down the hill?

54. A child sits in his go-cart at the start position of a race atop a hill. If the hill has a slope of 3°, and the child and go-cart have a total weight of 250 pounds, what force is required to keep them stationary at the start position?

55. A woman on skis holds herself stationary, with the use of her ski poles, on a slope that is 45° from the horizontal. If the woman and her skis have a total weight of 155 pounds, what is the force required to prevent her from sliding down the slope?

56. A loaded furniture dolly is being pulled up a 15° ramp by a mover. When he pauses to rest, he has to exert 103.53 pounds of force just to keep the dolly stationary. How much does the loaded dolly weigh?

57. A child pulls a sled over the snow, exerting a force of 25 pounds on the attached rope. The rope is 35° from the horizontal. If the child pulls the sled a distance of 80 feet, what work has been done?

58. The world's strongest man pulls a log 200 feet, and the tension in the cable connecting the man and log is 3000 pounds. What is the work being done if the cable is being held 15° from the horizontal?

59. A recreational vehicle pulls a passenger car behind it, exerting 1250 pounds on the attachment point. The angle of attachment is 30° from the horizontal. If the RV pulls the car a distance of 2 miles, what work has been done?

9.8 HYPERBOLIC FUNCTIONS

TOPICS

1. The hyperbolic functions
2. Hyperbolic identities
3. The inverse hyperbolic functions

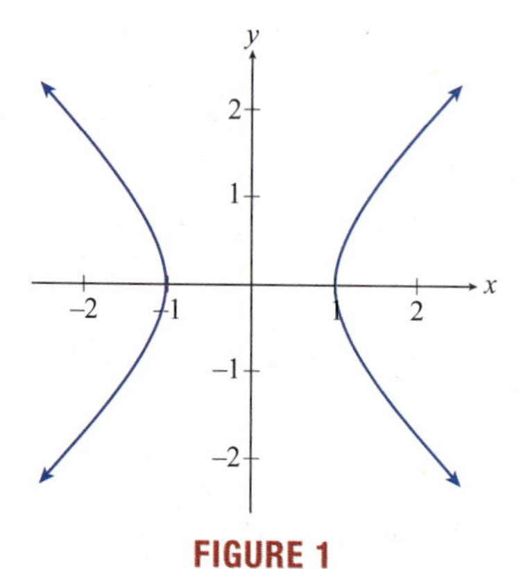

FIGURE 1

The graph of the equation $x^2 - y^2 = 1$, shown in Figure 1, is an example of a *hyperbola*. We will study hyperbolas in greater generality in Section 10.3, but for the moment we are concerned only with this one specific example, as it serves as the basis for the family of so-called *hyperbolic functions*.

The six hyperbolic functions, along with their inverses, have the same relationship to hyperbolas that the more familiar trigonometric functions have to circles. Although the original reasons for their introduction have faded in relevance, their usefulness in solving certain types of equations and in describing important physical phenomena has given them continued life.

TOPIC 1: The Hyperbolic Functions

We will introduce the first two hyperbolic functions via a technique that is useful in its own right. And while this method of introduction doesn't reflect their historical development, we will quickly be able to show how the hyperbolic functions come by their name.

Recall from Section 4.2 that a function f is said to be even if $f(-x) = f(x)$ and odd if $f(-x) = -f(x)$, in each case for all x in the domain. Because these are fairly restrictive properties, it may be surprising to realize that *any* function defined on an interval centered at the origin can be written as the sum of an even function and an odd function. The decomposition for any such function f is described as follows.

$$f(x) = \underbrace{\frac{f(x) + f(-x)}{2}}_{\text{even part}} + \underbrace{\frac{f(x) - f(-x)}{2}}_{\text{odd part}}$$

Since the function e^x is defined on the entire real line (and hence its domain can be said to be centered at 0), we can apply the decomposition and label the even and odd parts as follows.

$$e^x = \underbrace{\frac{e^x + e^{-x}}{2}}_{\cosh x} + \underbrace{\frac{e^x - e^{-x}}{2}}_{\sinh x}$$

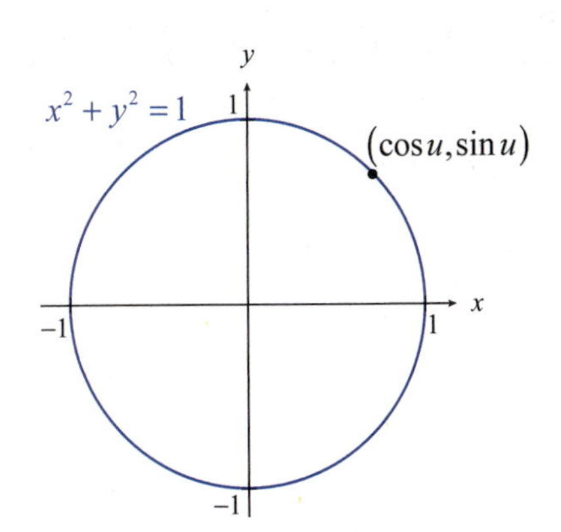

FIGURE 2: Circular Cosine and Sine Functions

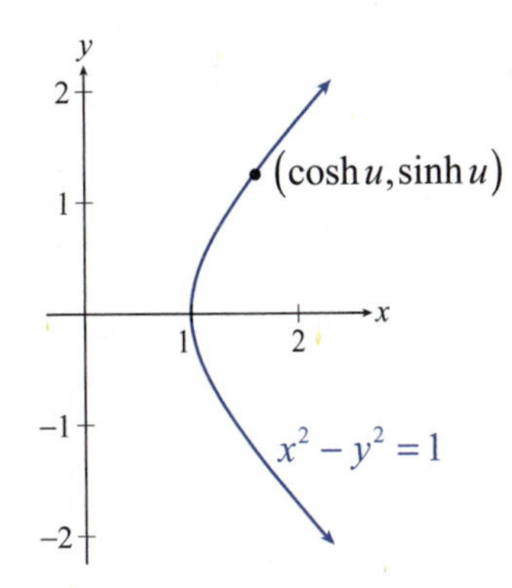

FIGURE 3: Hyperbolic Cosine and Sine Functions

The notation *cosh* stands for *hyperbolic cosine* and *sinh* stands for *hyperbolic sine*; they are pronounced, respectively, as "kosh" (rhymes with "gosh") and "cinch" (rhyming with "pinch"). By the previous derivation, we know that cosh is an even function and that sinh is odd. But their names come from the fact that, as u takes on different real values, $\cosh u$ and $\sinh u$ trace out the right-hand branch of the hyperbola $x^2 - y^2 = 1$ in the same way that $\cos u$ and $\sin u$ trace out the circle $x^2 + y^2 = 1$. In fact, some early treatises on trigonometry made equal reference to *circular trigonometric* functions and *hyperbolic trigonometric* functions and developed their theory in parallel.

Figures 2 and 3 illustrate how $(\cos u, \sin u)$ and $(\cosh u, \sinh u)$ trace out their respective curves. Note that in the case of the circular trigonometric functions, u represents the radian measure between the positive x-axis and the ray passing through $(\cos u, \sin u)$. The parameter u doesn't have any similar physical meaning in the hyperbolic case, aside from the fact that if u is positive, $(\cosh u, \sinh u)$ is a point on the upper half of the hyperbola, while a negative u corresponds to a point $(\cosh u, \sinh u)$ on the lower half of the hyperbola. (You should verify that, given the definitions $\cosh u = \frac{e^u + e^{-u}}{2}$ and $\sinh u = \frac{e^u - e^{-u}}{2}$, it is indeed the case that $\cosh^2 u - \sinh^2 u = 1$.)

Once the first two hyperbolic trigonometric functions have been defined, the remaining four follow; all six are formally defined in Table 1. For each one, the graph of the function is shown with its asymptotic curves, if appropriate, as dashed lines (note how the asymptotic curves relate to the definitions of the functions).

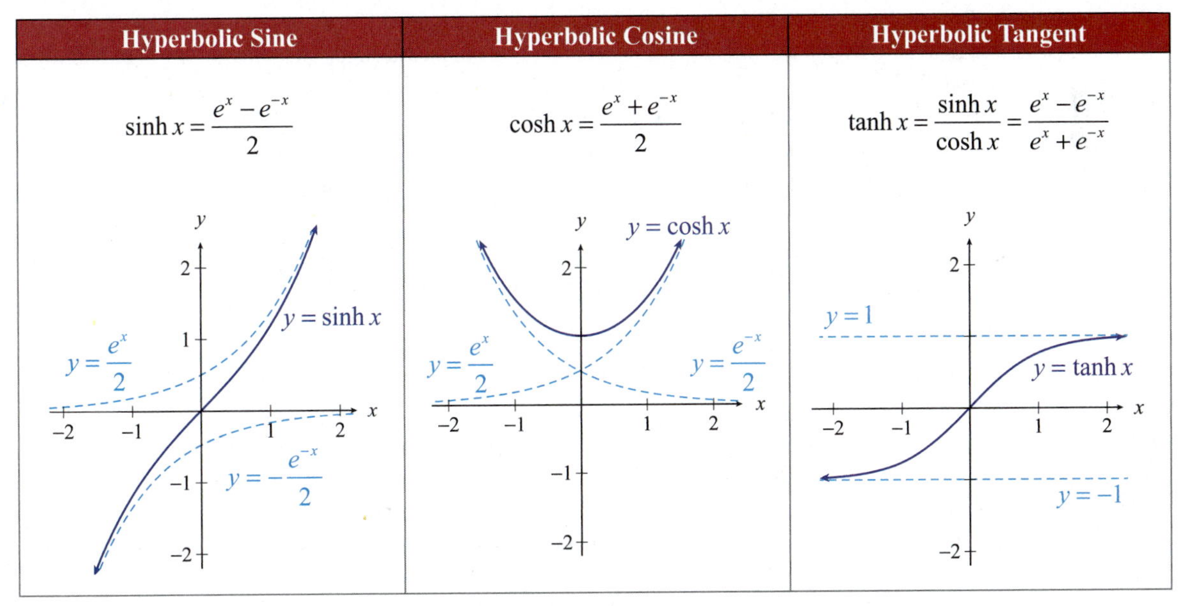

Hyperbolic Sine	Hyperbolic Cosine	Hyperbolic Tangent
$\sinh x = \frac{e^x - e^{-x}}{2}$	$\cosh x = \frac{e^x + e^{-x}}{2}$	$\tanh x = \frac{\sinh x}{\cosh x} = \frac{e^x - e^{-x}}{e^x + e^{-x}}$

TABLE 1: Hyperbolic Functions

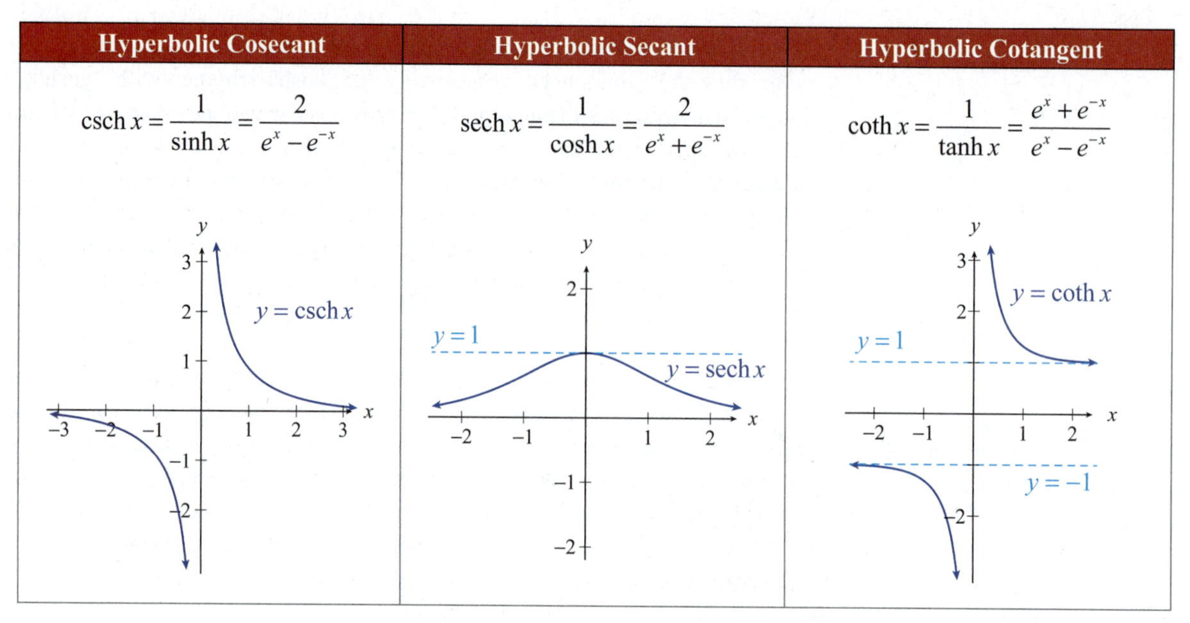

Hyperbolic Cosecant	Hyperbolic Secant	Hyperbolic Cotangent
$\operatorname{csch} x = \dfrac{1}{\sinh x} = \dfrac{2}{e^x - e^{-x}}$	$\operatorname{sech} x = \dfrac{1}{\cosh x} = \dfrac{2}{e^x + e^{-x}}$	$\coth x = \dfrac{1}{\tanh x} = \dfrac{e^x + e^{-x}}{e^x - e^{-x}}$

TABLE 1: Hyperbolic Functions (continued)

The fact that the hyperbolic functions are defined in terms of exponential functions gives us a straightforward means of evaluating the functions, as shown in Example 1.

Example 1: Evaluating Hyperbolic Functions

Evaluate each of the following expressions. If necessary, use a calculator and round your answer to two decimal places.

a. $\cosh 0$ **b.** $\tanh(\ln 3)$ **c.** $\sinh(-2)$

Solution

a. $$\cosh 0 = \frac{e^0 + e^{-(0)}}{2} = \frac{1+1}{2} = 1$$

b. $$\tanh(\ln 3) = \frac{e^{\ln 3} - e^{-\ln 3}}{e^{\ln 3} + e^{-\ln 3}} = \frac{3 - e^{\ln\left(\frac{1}{3}\right)}}{3 + e^{\ln\left(\frac{1}{3}\right)}} = \frac{3 - \frac{1}{3}}{3 + \frac{1}{3}} = \frac{\frac{8}{3}}{\frac{10}{3}} = \frac{4}{5}$$

c. $$\sinh(-2) = \frac{e^{-2} - e^{-(-2)}}{2} = \frac{\frac{1}{e^2} - e^2}{2} \approx -3.63$$

TOPIC 2: Hyperbolic Identities

Like the trigonometric functions, hyperbolic functions give rise to identities that are useful in simplifying expressions. The following are a few of the more common identities, the proofs of which are all very similar (see Exercises 16–35).

IDENTITIES: Elementary Hyperbolic Identities

$\cosh^2 x - \sinh^2 x = 1$ $\quad\quad$ $\tanh^2 x = 1 - \operatorname{sech}^2 x$ $\quad\quad$ $\coth^2 x = 1 + \operatorname{csch}^2 x$

$\sinh(x+y) = \sinh x \cosh y + \cosh x \sinh y$

$\cosh(x+y) = \cosh x \cosh y + \sinh x \sinh y$

$\sinh(2x) = 2\sinh x \cosh x$ $\quad\quad$ $\cosh(2x) = \cosh^2 x + \sinh^2 x$

$\sinh^2 x = \dfrac{\cosh(2x) - 1}{2}$ $\quad\quad$ $\cosh^2 x = \dfrac{\cosh(2x) + 1}{2}$

You have already verified the first identity—it is the one that gives rise to the name of these functions. We will verify two others in Example 2.

Example 2: Verification of Identities

Verify each of the following identities.

a. $\tanh^2 x = 1 - \operatorname{sech}^2 x$ $\quad\quad$ **b.** $\sinh^2 x = \dfrac{\cosh(2x) - 1}{2}$

Solution

a. $\cosh^2 x - \sinh^2 x = 1$

$$1 - \frac{\sinh^2 x}{\cosh^2 x} = \frac{1}{\cosh^2 x}$$

$$1 - \tanh^2 x = \operatorname{sech}^2 x$$

$$\tanh^2 x = 1 - \operatorname{sech}^2 x$$

b.
$$\begin{aligned}
\sinh^2 x &= \left(\frac{e^x - e^{-x}}{2}\right)\left(\frac{e^x - e^{-x}}{2}\right) \\
&= \frac{1}{4}\left(e^{2x} - 2 + e^{-2x}\right) \\
&= \frac{1}{2}\left(\frac{e^{2x} + e^{-2x}}{2} - 1\right) \\
&= \frac{\cosh(2x) - 1}{2}
\end{aligned}$$

TOPIC 3: The Inverse Hyperbolic Functions

Just as the six trigonometric functions have inverses, the six hyperbolic functions also have inverses. The notation we use to denote inverse hyperbolic functions follows the standard practice of denoting the inverse of the function f by f^{-1}. In each of the following graphs, a hyperbolic function (or a restricted portion, if necessary) is shown, with the inverse hyperbolic function being its reflection across the line $y = x$.

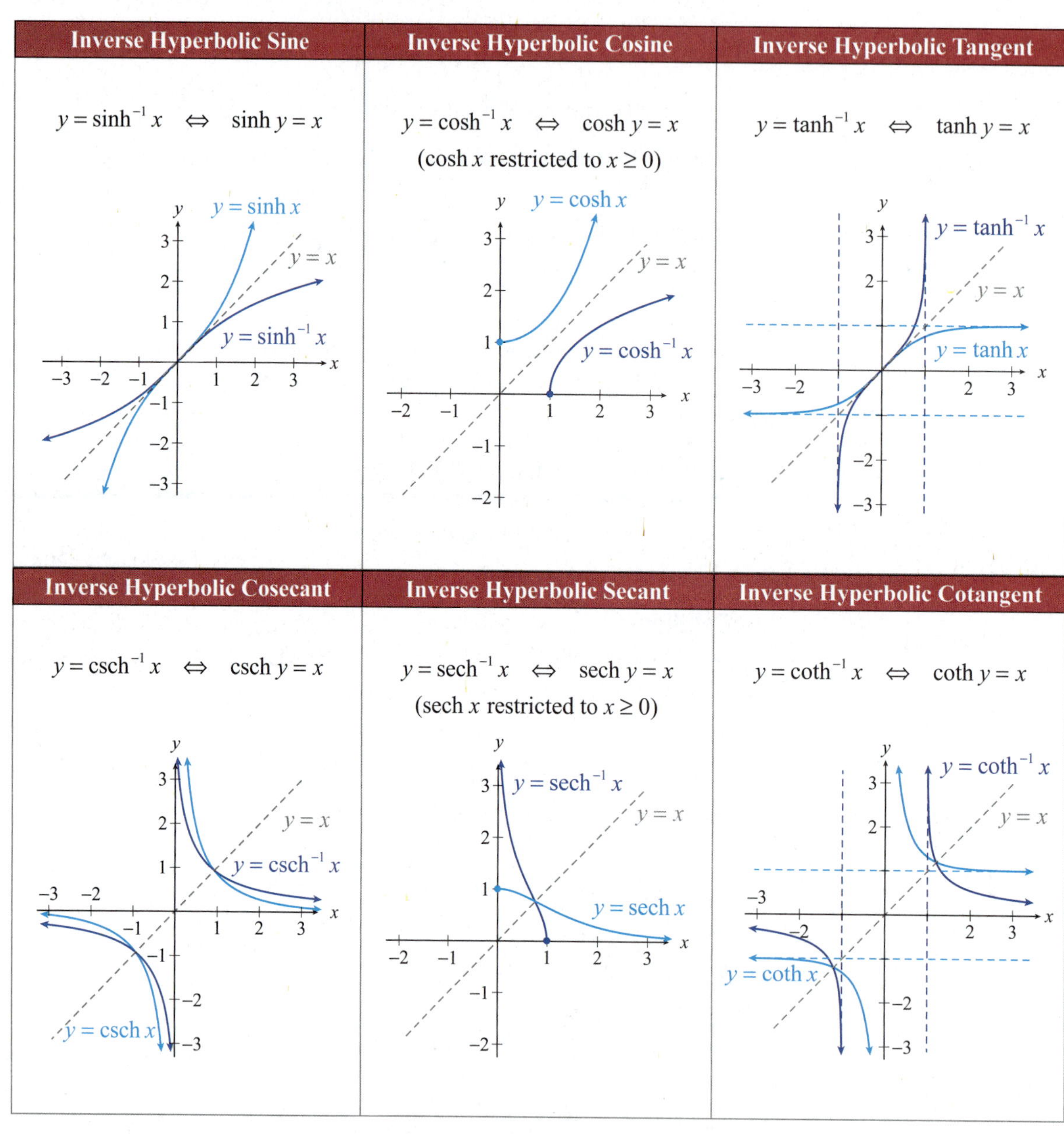

TABLE 2: Inverse Hyperbolic Functions

As with the inverse trigonometric functions, there is an alternative way of denoting the inverse hyperbolic functions. However, the meaning of the terms *arsinh* and *arcosh* (synonyms for $\sinh^{-1}$ and $\cosh^{-1}$, respectively) has become clouded and somewhat confused over time. The alternatives *arcsine* and *arccosine* for $\sin^{-1}$ and $\cos^{-1}$ reflect the fact that, given a point (x, y) in the first quadrant of the unit circle centered at the origin, $\cos^{-1}x$ and $\sin^{-1}y$ both represent the radian measure u shown in Figure 4, and hence also the *arc* length s. In similar fashion, $\cosh^{-1}x$ and $\sinh^{-1}y$ are related to the *area* of the hyperbolic sector shown in Figure 5, and the *ar* of *arsinh* and *arcosh* consequently arises from the word "area." As it turns out, the parameter u is equal to twice the area A, a fact that can be proved using calculus.

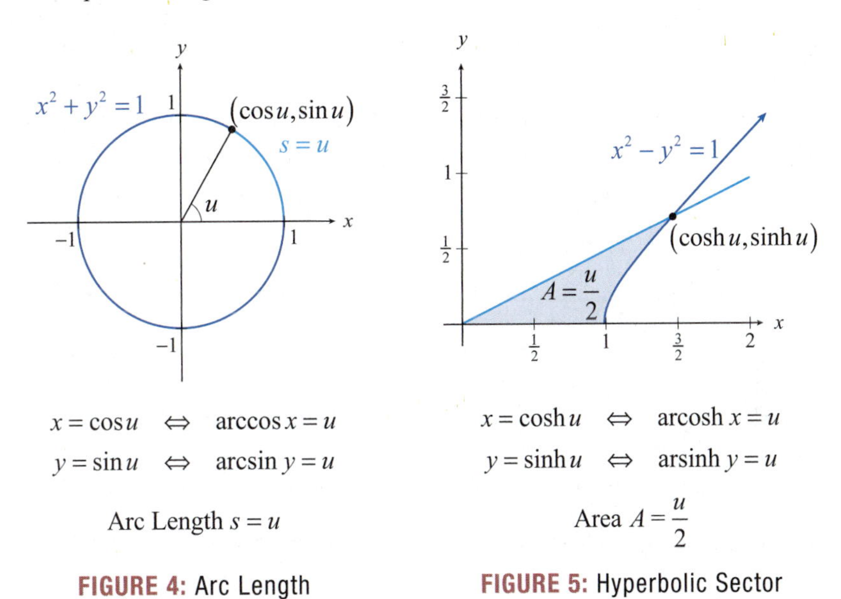

FIGURE 4: Arc Length

FIGURE 5: Hyperbolic Sector

Unfortunately, the similarity between *ar* and *arc* led, over time, to the frequent use of *arcsinh* and *arccosh* instead of *arsinh* and *arcosh*—these names are often seen in software packages and programming languages.

In a few cases, inverse hyperbolic functions can be evaluated using the sort of reasoning illustrated in Example 3. More generally, the functions can be evaluated through the use of formulas that you will develop in Exercises 39–43.

Example 3: Evaluating Inverse Hyperbolic Functions

Evaluate $\sinh^{-1} 0$.

Solution

Let $\sinh^{-1} 0 = x$. Then $0 = \sinh x$, and hence $0 = \frac{e^x - e^{-x}}{2}$. This can only happen if $e^x - e^{-x} = 0$, or $e^x = e^{-x}$. Examining the graph of e^x (or indeed any exponential function), you can see that the only x for which e^x will have the same value as e^{-x} is $x = 0$. Alternatively, multiplying the equation $e^x = e^{-x}$ through by e^x results in the equation $e^{2x} = 1$, which has the single solution $2x = 0$, and hence $x = 0$ (this method is similar to the approach you will use in Exercises 39 and 40). By either argument, $x = 0$ and so $\sinh^{-1} 0 = 0$.

9.8 EXERCISES

PRACTICE

Evaluate each of the following expressions. Round your answers to two decimal places if necessary. See Example 1.

1. $\sinh 0$
2. $\tanh 0$
3. $\operatorname{sech} 0$
4. $\sinh(\ln 2)$
5. $\tanh(\ln 2)$
6. $\cosh(-2)$
7. $\coth(-2)$
8. $\operatorname{csch}(-2)$
9. $\cosh 5$

Use each of the given equations to classify the hyperbolic function as even or odd. Then use the definition of the function to prove your assertion.

10. $\sinh(-x) = -\sinh x$
11. $\cosh(-x) = \cosh x$
12. $\tanh(-x) = -\tanh x$
13. $\coth(-x) = -\coth x$
14. $\operatorname{sech}(-x) = \operatorname{sech} x$
15. $\operatorname{csch}(-x) = -\operatorname{csch} x$

Verify each of the following identities. See Example 2.

16. $e^{kx} = \cosh(kx) + \sinh(kx)$
17. $e^{-kx} = \cosh(kx) - \sinh(kx)$
18. $\sinh(x+y) = \sinh x \cosh y + \cosh x \sinh y$
19. $\sinh(x-y) = \sinh x \cosh y - \cosh x \sinh y$
20. $\cosh(x+y) = \cosh x \cosh y + \sinh x \sinh y$
21. $\cosh(x-y) = \cosh x \cosh y - \sinh x \sinh y$
22. $\sinh(2x) = 2\sinh x \cosh x$
23. $\cosh(2x) = \cosh^2 x + \sinh^2 x$
24. $\cosh^2 x = \dfrac{\cosh(2x)+1}{2}$
25. $\coth^2 x = 1 + \operatorname{csch}^2 x$
26. $\tanh(x+y) = \dfrac{\tanh x + \tanh y}{1 + \tanh x \tanh y}$
27. $\tanh(x-y) = \dfrac{\tanh x - \tanh y}{1 - \tanh x \tanh y}$
28. $\tanh(2x) = \dfrac{2\tanh x}{1 + \tanh^2 x}$

29. $\coth(2x) = \dfrac{1+\coth^2 x}{2\coth x}$

30. $\sinh\left(\dfrac{x}{2}\right) = \pm\sqrt{\dfrac{\cosh x - 1}{2}}$

31. $\cosh\left(\dfrac{x}{2}\right) = \sqrt{\dfrac{\cosh x + 1}{2}}$

32. $\tanh\left(\dfrac{x}{2}\right) = \pm\sqrt{\dfrac{\cosh x - 1}{\cosh x + 1}}$

33. $(\cosh x + \sinh x)^2 = \cosh(2x) + \sinh(2x)$

34. $(\cosh x + \sinh x)^n = \cosh(nx) + \sinh(nx), \quad n \in \mathbb{N}$

35. $(\cosh x - \sinh x)^n = \cosh(nx) - \sinh(nx), \quad n \in \mathbb{N}$

Evaluate each of the following expressions. See Example 3.

36. $\cosh^{-1} 1$

37. $\tanh^{-1} 0$

38. $\operatorname{sech}^{-1} 1$

WRITING & THINKING

Given that the hyperbolic functions can be expressed in terms of exponential functions, it's not surprising that their inverses can be expressed in terms of logarithms. For example, if we let $y = \sinh^{-1} x$, then $x = \sinh y$ and we have the following.

$$x = \frac{e^y - e^{-y}}{2}$$

$$2x = e^y - e^{-y}$$

$$e^y - 2x - e^{-y} = 0$$

$$e^{2y} - 2xe^y - 1 = 0 \qquad \text{Multiply through by } e^y.$$

$$\left(e^y\right)^2 - 2xe^y - 1 = 0 \qquad \text{Express as a quadratic in } e^y.$$

$$e^y = \frac{2x \pm \sqrt{4x^2+4}}{2} \qquad \text{Solve for } e^y.$$

$$e^y = x + \sqrt{x^2+1} \qquad x - \sqrt{x^2+1} < 0 \text{ but } e^y > 0, \text{ so discard } x - \sqrt{x^2+1}.$$

$$y = \ln\left(x + \sqrt{x^2+1}\right) \qquad \text{Take the natural logarithm of both sides.}$$

Use this procedure to verify each of the following identities.

39. $\cosh^{-1} x = \ln\left(x + \sqrt{x^2 - 1}\right), \quad x \geq 1$

40. $\tanh^{-1} x = \dfrac{1}{2}\ln\left(\dfrac{1+x}{1-x}\right), \quad -1 < x < 1$

(**Hint:** Begin by setting $y = \tanh^{-1} x$; then write the equation as $\tanh y = x$, square both sides, and apply the identity $\tanh^2 y = 1 - \operatorname{sech}^2 y$. Solve the result for $\cosh y$, apply $\cosh^{-1}$ to both sides, and apply the result of the previous exercise. Then apply some logarithmic properties.)

Given a function f, let $\frac{1}{f}$ denote its reciprocal—that is, $\left(\frac{1}{f}\right)(x)=\frac{1}{f(x)}$. The following is a useful relationship between the functions f^{-1} and $\left(\frac{1}{f}\right)^{-1}$, assuming both of these inverse functions exist.

$$\left(\frac{1}{f}\right)\left(f^{-1}\left(\frac{1}{x}\right)\right)=\frac{1}{f\left(f^{-1}\left(\frac{1}{x}\right)\right)}=\frac{1}{\frac{1}{x}}=x$$

$$\text{so}\quad \left(\frac{1}{f}\right)^{-1}(x)=f^{-1}\left(\frac{1}{x}\right).$$

Applied to hyperbolic functions, this fact indicates the following.

$$\operatorname{csch}^{-1}x=\sinh^{-1}\frac{1}{x},\qquad \operatorname{sech}^{-1}x=\cosh^{-1}\frac{1}{x},\qquad \coth^{-1}x=\tanh^{-1}\frac{1}{x}$$

Use these relationships to verify each of the following identities.

41. $\operatorname{csch}^{-1}x=\ln\left(\frac{1}{x}+\frac{\sqrt{1+x^2}}{|x|}\right),\quad x\neq 0$

42. $\operatorname{sech}^{-1}x=\ln\left(\frac{1+\sqrt{1-x^2}}{x}\right),\quad 0<x\leq 1$

43. $\coth^{-1}x=\frac{1}{2}\ln\left(\frac{x+1}{x-1}\right),\quad |x|>1$

Evaluate each of the following expressions using the formulas from Exercises 39–43. Round your answers to two decimal places.

44. $\sinh^{-1}2$

45. $\operatorname{csch}^{-1}(-3)$

46. $\cosh^{-1}5$

47. $\operatorname{sech}^{-1}(0.8)$

48. $\tanh^{-1}(-0.3)$

49. $\coth^{-1}3$

CHAPTER 9 PROJECT

Trigonometric Applications

Built by King Khufu from 2589–2566 BC to serve as his tomb, the Great Pyramid of Giza covers 13 acres and weighs more than 6.5 million tons. The Great Pyramid is the oldest and the only remaining wonder of the 7 Wonders of the Ancient World. Even using modern technology, engineers in the 21st century would have difficulties recreating this impressive structure. Today, we can only put forth theories as to how this amazing structure was created.

When first built, the Egyptians called the Great Pyramid Ikhet (which means Glorious Light) because the sides of the pyramid were covered in highly polished limestone that would have shone brightly under the hot Egyptian Sun.

1. When built, the length of each side of the base was 754 ft and the distance from each corner of the base to the peak was 718 ft. What was the surface area of the four sides?

The Egyptians quarried most of the stone locally, but they also floated huge granite blocks down the Nile River from Aswan.

2. Your barge with the latest shipment of granite for King Khufu is quickly approaching Giza. The river is flowing at 195 yards per minute. You command your oarsmen to start rowing towards shore at a 65° angle from the direction of the current. They can row 260 yards per minute.

    a. What is the resultant true velocity of the barge?

    b. The Nile River is 840 yards wide near Giza. If the boat is in the center of the river and the dock is 750 yards ahead, will they hit the bank before or after it? By how much will they miss it? (**Hint:** (velocity)(time) = distance.)

3. Once the boat reaches the shore, the granite stones need to be moved into place.

    a. If a granite block weighs 8300 pounds and 12 of your men are each pulling on it (in the same direction) with a force of 115 pounds, what is the total force being applied?

    b. In order to get the stone to the necessary spot, the stone must be pulled up a ramp that has been built around the pyramid. If the ramp has a 9° grade, what force is necessary to keep the block from sliding?

    c. If the top of the pyramid is currently 320 feet above the desert, how long does the ramp have to be?

    d. How much work is it for the 12 men to drag the stone to the top of the pyramid? (Remember, in this case work is done in both the horizontal and vertical directions.)

CHAPTER 9 REVIEW EXERCISES

Section 9.1

Create a triangle, if possible, using the given information.

1. $A = 30°, B = 45°, b = 4$
2. $a = 15, c = 13, C = 57°$
3. $A = 74°20', C = 37°, c = 23$
4. $b = 8, c = 13, C = 78°$
5. Find the area of a triangle for which $a = 3$, $b = 7$, and $C = 75°$.

Section 9.2

Create a triangle, if possible, using the given information and the Law of Cosines.

6. $A = 62°, b = 8, c = 10$
7. $B = 94°\ 7', a = 6, c = 14$
8. $a = 9, b = 2.5, c = 7.3$
9. $a = 10.8, b = 13.4, c = 6$
10. The base of a 25 ft ladder is positioned 7 ft away from an office building situated on a slight hill, and the ladder and ground form a 62° angle. At what angle and at what height does the ladder touch the building?
11. Find the area of a triangle for which $a = 5$, $b = 11$, and $c = 9$.

Section 9.3

Convert the point from polar to Cartesian coordinates.

12. $\left(-3.45, \frac{\pi}{3}\right)$
13. $\left(7, \frac{7\pi}{6}\right)$

Convert the point from Cartesian to polar coordinates.

14. $\left(-\sqrt{3}, -1\right)$
15. $(10, 12)$

Rewrite the rectangular equation in polar form.

16. $x^2 + y^2 = 16a^2$
17. $x^2 + y^2 = 9ax$

Rewrite the polar equation in rectangular form.

18. $r = 2\cos\theta$
19. $r = \dfrac{16}{4\cos\theta + 4\sin\theta}$

Sketch a graph of the given polar equation.

20. $r = 4\sin(3\theta)$
21. $r^2 = 25\cos(2\theta)$

Section 9.4

Sketch the graphs of the following parametric equations by eliminating the parameter.

22. $x = \frac{1}{36t}$ and $y = t^2$

23. $x = t + 5$ and $y = |t - 2|$

24. $x = \frac{3}{4t - 2}$ and $y = 2t - 2$

25. $x = 4\sin\theta$ and $y = \cos\theta + 1$

Construct parametric equations describing the graphs of the following equations.

26. $y^2 = x^2 + 4$

27. $6x = 2 - y$

Construct parametric equations for the line or conic with the given attributes.

28. Line: passing through $(14, 4)$ and $(-3, -8)$

29. Circle: center $(1, 1)$, radius 1

Section 9.5

Graph and determine the magnitudes of the following complex numbers.

30. $4 + 5i$

31. $-3 + 3i$

Sketch z_1, z_2, $z_1 + z_2$, and z_1z_2 on the same complex plane.

32. $z_1 = -2 - 3i, z_2 = 6 + 3i$

33. $z_1 = 4 + 2i, z_2 = -5 + i$

Graph the regions of the complex plane defined by the following.

34. $\{z \mid 2 \leq |z| \leq 3\}$

35. $\{z = a + bi \mid a > 2, b > 3\}$

Write each of the following complex numbers in trigonometric form.

36. $2\sqrt{3} - 3i$

37. $1 + 4i$

Write each of the following complex numbers in standard form.

38. $4\left(\cos\left(\frac{7\pi}{4}\right) + i\sin\left(\frac{7\pi}{4}\right)\right)$

39. $3(\cos 60° + i\sin 60°)$

Perform the following operations and show the answer in both trigonometric form and standard form.

40. $\left[\sqrt{3}\left(\cos\left(\frac{2\pi}{3}\right) + i\sin\left(\frac{2\pi}{3}\right)\right)\right]\left[4\sqrt{3}\left(\cos\left(\frac{7\pi}{6}\right) + i\sin\left(\frac{7\pi}{6}\right)\right)\right]$

41. $\frac{5(\cos 240° + i\sin 240°)}{(\cos 120° + i\sin 120°)}$

42. $\frac{-\sqrt{3} + i}{1 - i\sqrt{3}}$

43. $\left(12e^{35°i}\right)\left(2e^{280°i}\right)$

Use De Moivre's Theorem to calculate the following.

44. $\left(1+\sqrt{3}i\right)^6$

45. $\left[3\left(\cos 240° + i\sin 240°\right)\right]^{11}$

Find the indicated roots of the following and graphically represent each set in the complex plane.

46. The square roots of $-144i$.

47. The cube roots of $125\left(\cos\left(\frac{7\pi}{4}\right)+i\sin\left(\frac{7\pi}{4}\right)\right)$.

Solve the following equations.

48. $z^4 - 1 + i = 0$

49. $z^3 + 4\sqrt{2} - 4i\sqrt{2} = 0$

Section 9.6

Find the component form and the magnitude of a vector **v** defined by the given points. Assume the first point given is the initial point and the second point given is the terminal point.

50. $(-1, 0), (4, -5)$

51. $(6, 5), (-4, -1)$

For each of the following, calculate and graph **a.** 2**u** + **v**, **b.** –**u** + 3**v**, and **c.** –2**v**.

52. $\mathbf{u} = \langle 1,3\rangle, \mathbf{v} = \langle -5,2\rangle$

53. $\mathbf{u} = \langle 1,-1\rangle, \mathbf{v} = \langle 4,-3\rangle$

For each of the following graphs, determine the component forms of –**u**, 2**u** – **v**, and **u** + **v** and find the magnitudes of **u** and **v**.

54.

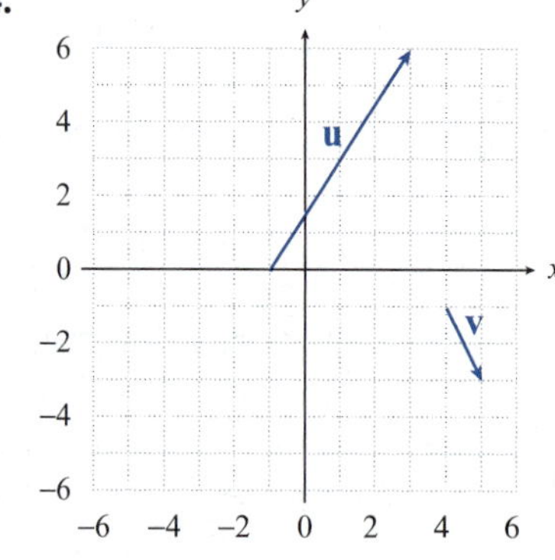

55.

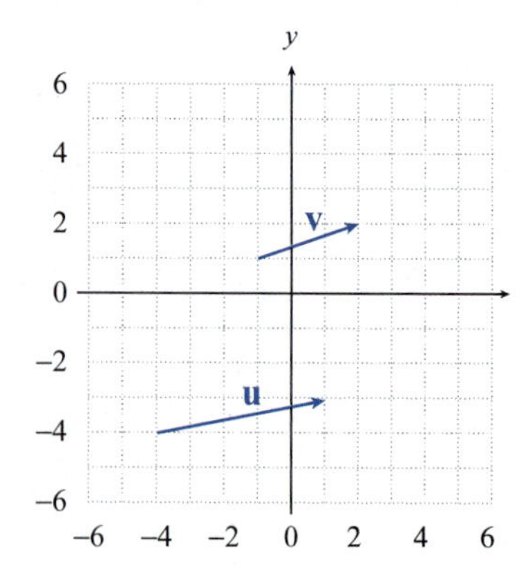

Given the vector **u**, find **a.** a unit vector pointing in the same direction as **u**, and **b.** the linear combination of **i** and **j** that is equivalent to **u**.

56. $\mathbf{u} = \langle -4,5\rangle$

57. $\mathbf{u} = \langle 6,3\rangle$

Find the magnitude and direction angle of the vector **v**.

58. $\mathbf{v} = 4\left(\cos 135°\mathbf{i} + \sin 135°\mathbf{j}\right)$

59. $\mathbf{v} = 5\mathbf{i} - \mathbf{j}$

Find the component form of **v** given its magnitude and the angle it makes with the positive *x*-axis.

60. $\|\mathbf{v}\| = 2\sqrt{2}, \theta = 60°$

61. $\|\mathbf{v}\| = 6,$ $\mathbf{v}$ in the direction of $3\mathbf{i} - 4\mathbf{j}$

62. A golf ball is driven into the air at a speed of 90 miles per hour and an angle of 45° from the horizontal. Express this velocity in vector form.

63. A sailboat is traveling at a speed of 55 miles per hour with a bearing of N 24° W, when it encounters a front with winds blowing at 20 miles per hour with a bearing of S 10° W. What is the resultant true velocity of the sailboat?

Section 9.7

Find the indicated quantity given $\mathbf{u} = \langle 1, -4 \rangle$ and $\mathbf{v} = \langle 2, 5 \rangle$.

64. $3\mathbf{u} \cdot \mathbf{v}$

65. $(\mathbf{u} \cdot \mathbf{v})3\mathbf{v}$

Find the magnitude of **u** using the dot product.

66. $\mathbf{u} = \langle -2, -3 \rangle$

67. $\mathbf{u} = -\mathbf{i} - 3\mathbf{j}$

Find the angle between the given vectors.

68. $\mathbf{u} = \langle 5, -5 \rangle, \mathbf{v} = \langle 1, 4 \rangle$

69. $\mathbf{u} = \cos\left(\frac{\pi}{4}\right)\mathbf{i} + \sin\left(\frac{\pi}{4}\right)\mathbf{j}, \mathbf{v} = \cos\left(\frac{2\pi}{3}\right)\mathbf{i} + \sin\left(\frac{2\pi}{3}\right)\mathbf{j}$

Find **u** · **v** where θ is the angle between **u** and **v**.

70. $\|\mathbf{u}\| = 16, \|\mathbf{v}\| = 2, \theta = 60°$

71. $\|\mathbf{u}\| = 8, \|\mathbf{v}\| = 9, \theta = \frac{2\pi}{3}$

Find the projection of **u** onto **v**, and then write **u** as a sum of two orthogonal vectors, one of which is $\text{proj}_{\mathbf{v}}\mathbf{u}$.

72. $\mathbf{u} = \langle 2, 3 \rangle, \mathbf{v} = \langle -1, 5 \rangle$

73. $\mathbf{u} = \langle 4, -1 \rangle, \mathbf{v} = \langle 2, 2 \rangle$

Find the work done in a particle moving from *J* to *K* if the magnitude and direction of the force are given by **F**.

74. $J = (2, 4), K = (3, 6), \mathbf{F} = \langle 1, 3 \rangle$

75. $J = (-5, 3), K = (0, 4), \mathbf{F} = \langle 5, 6 \rangle$

76. A truck with a gross weight of 33,000 pounds is parked on a 6° slope. What force is required to prevent the truck from rolling down the hill?

77. The world's strongest man pulls a log 160 feet, and the tension in the cable connecting the man and log is 2650 pounds. What is the work being done if the cable is being held 10° from the horizontal?

Section 9.8

Evaluate each of the following expressions. Round your answers to two decimal places if necessary.

78. $\cosh(\ln 2)$

79. $\sinh(\ln 4)$

80. $\operatorname{sech}(-2)$

81. $\coth 1$

82. Verify the identity $\cosh x + \cosh y = 2\cosh\left(\frac{x+y}{2}\right)\cosh\left(\frac{x-y}{2}\right)$.

tex
tex
tex
tex
tex
tex
tex
tex
CAPITAL

CHAPTER 10

Conic Sections

SECTIONS

WHAT IF ...

What if you were sailing a ship and didn't know your exact location in the sea but could receive signals from two radio transmitters? How would you determine your location?

By the end of this chapter, you'll be able to graph conic sections and describe them with equations. On page 751, you'll find that the answer to the ship lost at sea involves a hyperbola. You'll master this type of problem using tools such as the standard form of the equation of a hyperbola found on page 742.

Johannes Kepler

Introduction

This chapter is tightly focused on a family of plane curves called *conic sections*. We have actually studied simple examples of such curves already (circles and parabolas are conic sections), but the more comprehensive study presented in this chapter makes use of the advanced tools we now possess.

The study of conic sections, the curves obtained by intersecting a plane with a cone, has a history extending back to Greek mathematics of the third century BC and continuing very much to the present. In fact, few branches of mathematics display such a long-lived vitality. Such famous early mathematical figures as Euclid and Archimedes studied conics and discovered many of their properties, while later mathematicians like Isaac Newton, René Descartes, and Carl Friedrich Gauss continued the tradition and made significant contributions of their own. But perhaps the most interesting aspect of this history is the long span of time that lies between the early (almost purely intellectual) formulation of the theory and its modern (thoroughly pragmatic) applications.

The Greek mathematician Apollonius (c. 262–190 BC) is largely remembered today for his eight volume book entitled *Conic Sections*. Apollonius improved upon the slightly earlier work of Euclid and Archimedes and proceeded to develop almost all of the theory of conic sections you will encounter in this chapter. Apparently, he is also responsible for the names of the three varieties of conic sections: ellipses, parabolas, and hyperbolas. The names adhere to the Pythagorean tradition of identifying mathematical objects by their geometric properties, and the Greek words refer to the behavior of certain projections of the curves. (Incidentally, the three figures of speech known as *ellipsis*, *parabole*, and *hyperbole* have a similar root.) And while the mathematicians of Apollonius' time did have a few worldly uses for their knowledge of conics, it must be stressed that their driving force was the joy of intellectual accomplishment.

Contrast this with the applications of the theory that were discovered several millennia later. The German astronomer and mathematician Johannes Kepler (1571–1630) made the observation that planets orbit the Sun in elliptical paths, and proceeded to quantify those paths. The unique focusing property of parabolas leads to their use in modern lighting, in the design of satellite dishes, and in telescopes. And an analysis of the problems of long-range navigation led naturally to the use of hyperbolas in the navigational system known as LORAN. All of these applications, along with many others, make use of geometric properties identified by people with no foreknowledge of their eventual use.

10.1 ELLIPSES

TOPICS

1. Overview of conic sections
2. The standard form of the equation of an ellipse
3. Applications of ellipses

TOPIC 1: Overview of Conic Sections

The three types of conic sections—ellipses, parabolas, and hyperbolas—are so named because all three types of curves arise from intersecting a plane with a double circular cone. As shown in Figure 1, an **ellipse** is a closed curve resulting from the intersection of a cone with a plane that intersects only one *nappe* of the cone (the part of the cone on one side of the vertex). A **parabola** results from intersecting a cone with a plane that is parallel to a line on the surface of the cone passing through the vertex (a parabola also intersects only one nappe). Finally, a **hyperbola** is the intersection of a cone with a plane that intersects both nappes. In each case, we specify that the intersecting plane does not contain the vertex. A figure that results when the plane *does* contain the vertex is called a *degenerate* conic section and is a point, a line, or a pair of intersecting lines.

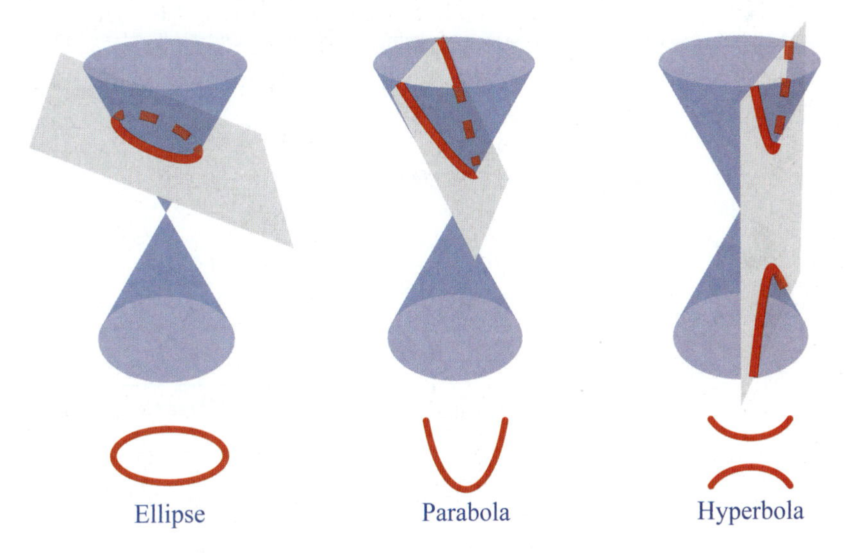

FIGURE 1: Conic Sections

These conic sections share similarities in their algebraic definitions as well. As curves in the Cartesian plane, every conic section is the graph of an equation that can be written in the form $Ax^2 + Bxy + Cy^2 + Dx + Ey + F = 0$, where A, B, C, D, E, and F are real constants. Initially, we study those conic sections for which $B = 0$; an analysis of the effect of allowing B to be nonzero requires trigonometry and appears in Section 10.4.

DEFINITION: Algebraic Definition of Conics

A conic section described by an equation of the form

$$Ax^2 + Cy^2 + Dx + Ey + F = 0,$$

where at least one of the two coefficients A and C is not equal to 0, is one of the following:

1. An **ellipse** if the product AC is positive
2. A **parabola** if the product AC is 0
3. A **hyperbola** if the product AC is negative

Finally, the three conic sections are characterized by certain geometric properties that are similar in nature. We will make use of these properties in the work to follow.

DEFINITION: Plane Geometric Definition of Conics

1. An ellipse consists of the set of points in the plane for which the sum of the distances d_1 and d_2 to two **foci** (plural of **focus**) is a fixed constant. See the first diagram in Figure 2.
2. A parabola consists of the set of points that are the same distance d from a line (called the **directrix**), and a point not on the line (called the **focus**). See the second diagram in Figure 2.
3. A hyperbola consists of the set of points for which the magnitude of the difference of the distances d_1 and d_2 to two **foci** is fixed. See the third diagram in Figure 2.

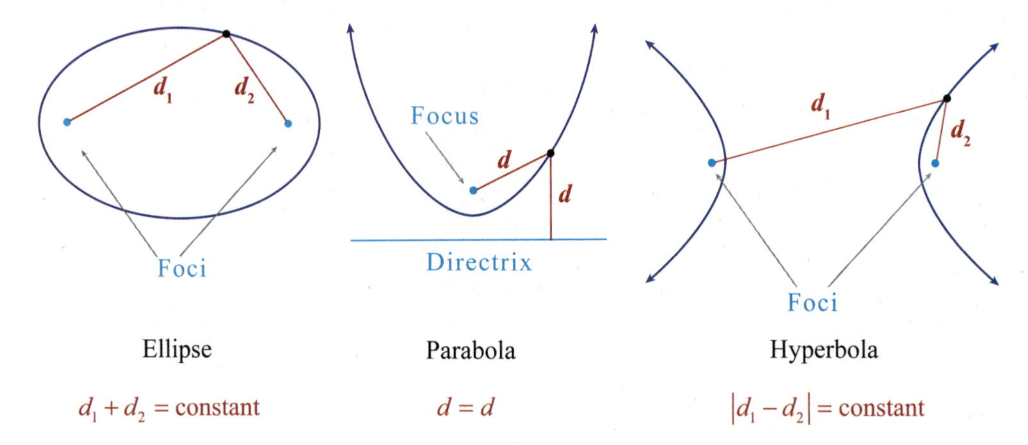

FIGURE 2: Plane Geometric Definition of Conics

TOPIC 2: The Standard Form of the Equation of an Ellipse

We would like to use the properties of the conic sections to derive useful forms of their equations. Specifically, we want to be able to construct the graph of a conic section from its equation and, reversing the process, be able to find the equation for a conic section with known properties. We begin with the ellipse.

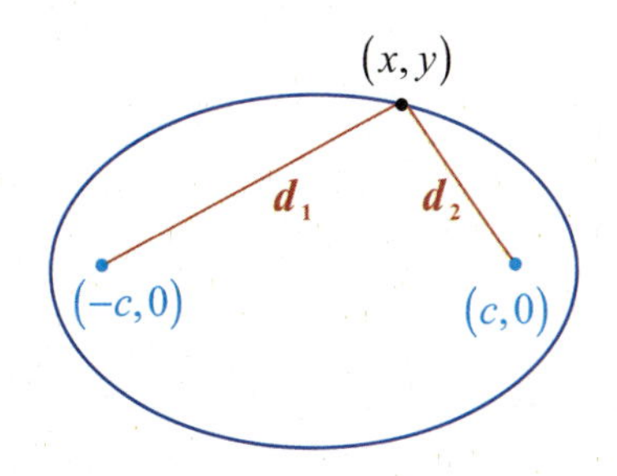

FIGURE 3: The Geometric Property of the Ellipse

Initially, assume we have an ellipse centered at the origin with foci at $(-c, 0)$ and $(c, 0)$. We want an equation in x and y that identifies all points (x, y) on the ellipse. The only property we need is that the sum $d_1 + d_2$ shown in Figure 3 is fixed. To make our algebra easier, we will denote this sum as $2a$; that is, $d_1 + d_2 = 2a$.

Using the distance formula to find d_1 and d_2, we know that

$$\sqrt{(x+c)^2 + y^2} + \sqrt{(x-c)^2 + y^2} = 2a.$$

While technically we have found an equation in x and y that describes all the points on the ellipse, it does not appear particularly easy to work with. We can make the equation easier to interpret by eliminating the radicals.

$$\left(\sqrt{(x+c)^2 + y^2}\right)^2 = \left(2a - \sqrt{(x-c)^2 + y^2}\right)^2$$

$$(x+c)^2 + y^2 = 4a^2 - 4a\sqrt{(x-c)^2 + y^2} + (x-c)^2 + y^2$$

$$\cancel{x^2} + 2cx + \cancel{c^2} + \cancel{y^2} = 4a^2 - 4a\sqrt{(x-c)^2 + y^2} + \cancel{x^2} - 2cx + \cancel{c^2} + \cancel{y^2}$$

There is still one radical left, so we isolate it and repeat the process:

$$4cx - 4a^2 = -4a\sqrt{(x-c)^2 + y^2}$$

$$a^2 - cx = a\sqrt{(x-c)^2 + y^2}$$

$$\left(a^2 - cx\right)^2 = a^2\left((x-c)^2 + y^2\right)$$

$$a^4 - \cancel{2a^2cx} + c^2x^2 = a^2x^2 - \cancel{2a^2cx} + a^2c^2 + a^2y^2$$

$$a^4 - a^2c^2 = a^2x^2 - c^2x^2 + a^2y^2$$

A change of variables will improve the appearance even more. From the diagram in Figure 3, we can see that $2a$ is larger than $2c$. Even in the extreme case of very skinny ellipses, the sum $d_1 + d_2$ (which we have called $2a$) will approach $2c$ but will always be larger than $2c$.

We can conclude that $a > c$, so $a^2 - c^2$ is a positive number and we can safely rename $a^2 - c^2$ as b^2. Making this substitution, we simplify further.

$$a^4 - a^2c^2 = a^2x^2 - c^2x^2 + a^2y^2$$

$$a^2\left(a^2 - c^2\right) = \left(a^2 - c^2\right)x^2 + a^2y^2$$

$$a^2b^2 = b^2x^2 + a^2y^2$$

One last cosmetic adjustment is to divide the equation by a^2b^2:

$$b^2x^2 + a^2y^2 = a^2b^2$$

$$\frac{x^2}{a^2} + \frac{y^2}{b^2} = 1$$

Note that this equation satisfies the algebraic definition for an ellipse. If we solve the equation for zero, we have

$$\frac{1}{a^2}x^2 + \frac{1}{b^2}y^2 - 1 = 0.$$

Highlighting the coefficients A and C of the general conic section form, we calculate $AC = \frac{1}{a^2b^2}$, which must be positive. This tells us that the equation indeed describes an ellipse in the plane.

Example 1: Graphing an Ellipse Centered at the Origin

Determine the foci of the ellipse $\frac{x^2}{16} + \frac{y^2}{9} = 1$, and then graph the ellipse.

Solution

The given equation is of the form $\frac{x^2}{a^2} + \frac{y^2}{b^2} = 1$, so it represents an ellipse centered at the origin with foci on the x-axis.

We see that $a = 4$ and $b = 3$, so using the fact that $a^2 - c^2 = b^2$ (by definition), we can calculate the coordinates of the foci.

$$\begin{aligned} a^2 - c^2 &= b^2 \\ c^2 &= a^2 - b^2 \\ c^2 &= 16 - 9 = 7 \\ c &= \pm\sqrt{7} \end{aligned}$$

Thus, the foci are located at $\left(-\sqrt{7}, 0\right)$ and $\left(\sqrt{7}, 0\right)$.

Again, we know the graph of this equation is an ellipse. To graph it, we will compute the x- and y-intercepts.

Setting x and y equal to zero, we have

$$\begin{aligned} \frac{(0)^2}{16} + \frac{y^2}{9} &= 1 \\ \frac{y^2}{9} &= 1 \\ y^2 &= 9 \\ y &= \pm 3 \end{aligned} \qquad \text{and} \qquad \begin{aligned} \frac{x^2}{16} + \frac{(0)^2}{9} &= 1 \\ \frac{x^2}{16} &= 1 \\ x^2 &= 16 \\ x &= \pm 4 \end{aligned}$$

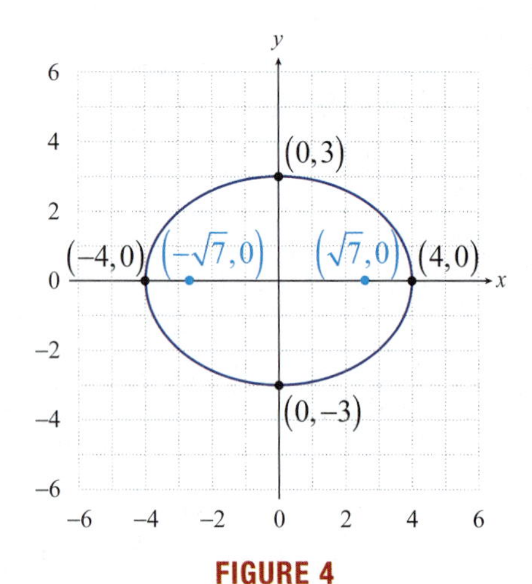

FIGURE 4

This shows that the x-intercepts are $(-4, 0)$ and $(4, 0)$ and the y-intercepts are $(0, -3)$ and $(0, 3)$. Note that these four points are extremes of the graph, as any x-value larger than 4 in magnitude leads to imaginary values for y, and any y-value larger than 3 in magnitude leads to imaginary values for x.

Connecting the four extreme points with an elliptically shaped curve, we obtain the graph in Figure 4.

TECHNOLOGY: Graphing an Ellipse

To graph the ellipse in Example 1, $\frac{x^2}{16}+\frac{y^2}{9}=1$, using a TI-84 Plus, first press apps. Select Conics, then select ELLIPSE. Now select 1: to enter the values from the equation, as shown in Figure 5. To view the graph, press graph. See Figure 6.

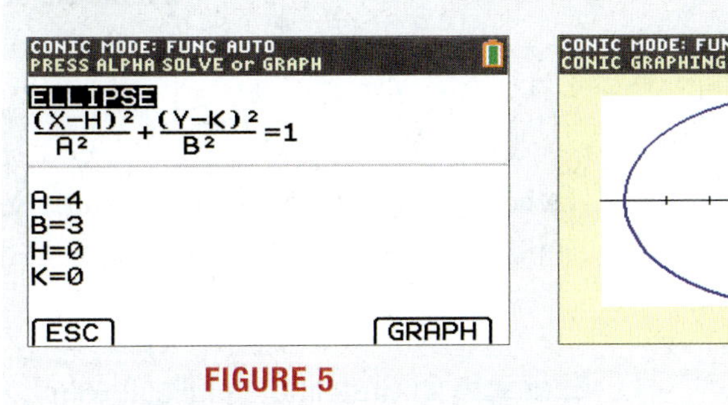

FIGURE 5

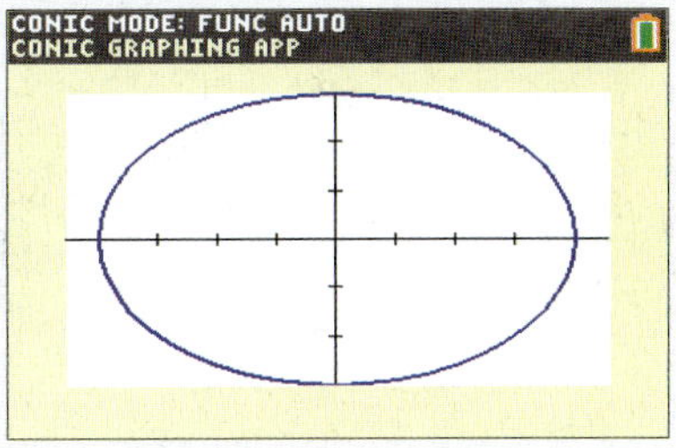

FIGURE 6

At this point, we know that

$$\frac{x^2}{a^2}+\frac{y^2}{b^2}=1$$

describes an ellipse centered at the origin that is elongated horizontally. If the foci of an origin-centered ellipse are instead at $(0,c)$ and $(0,-c)$, so that the ellipse is elongated vertically, then the equation is

$$\frac{x^2}{b^2}+\frac{y^2}{a^2}=1,$$

where $a>b>0$ and $c^2=a^2-b^2$.

The last step is to come up with the equation for an ellipse that is centered at the point (h,k). We already know how to do this: replacing x with $x-h$ in an equation shifts the graph of the equation h units horizontally, and replacing y with $y-k$ shifts the graph k units vertically. We introduce a few terms before writing the standard form.

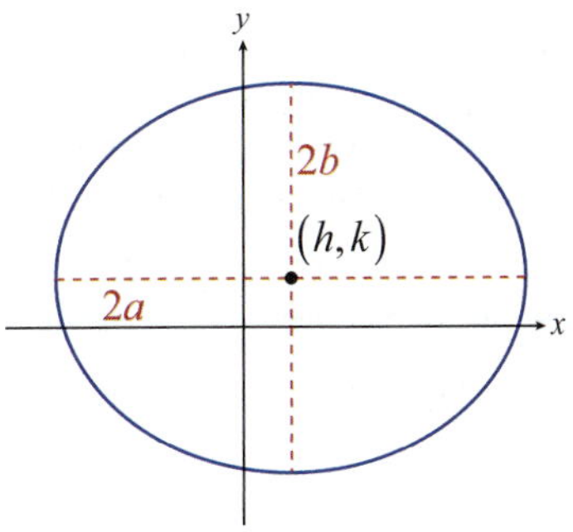

y $2b$ (h,k) $2a$ x

FIGURE 7

DEFINITION: Axes and Vertices of an Ellipse

The **major axis** of an ellipse is the line segment extending from one extreme point of the ellipse to the other and passing through the two foci and the center. The major axis has a length of $2a$. The two ends of the major axis are the **vertices** of the ellipse.

The **minor axis** is the line segment that also passes through the center and is perpendicular to the major axis; it spans the distance between the other two extremes of the ellipse and has a length of $2b$.

DEFINITION: Standard Form of the Equation of an Ellipse

Let a and b be positive constants with $a > b$. The **standard form of the equation of an ellipse** centered at (h,k) with major axis of length $2a$ and minor axis of length $2b$ is

- $\dfrac{(x-h)^2}{a^2}+\dfrac{(y-k)^2}{b^2}=1$ if the major axis is horizontal, and
- $\dfrac{(x-h)^2}{b^2}+\dfrac{(y-k)^2}{a^2}=1$ if the major axis is vertical.

In both cases, the two foci of the ellipse are located on the major axis c units away from the center of the ellipse, where $c^2=a^2-b^2$.

Example 2: Graphing an Ellipse with Equation in Standard Form

Graph the ellipse $\dfrac{(x-1)^2}{25}+\dfrac{(y+4)^2}{9}=1.$ Include the foci on the graph.

Solution

First, we note that the denominator on the x term is larger than the denominator on the y term. This means the major axis is horizontal. We can now collect information about the ellipse from the standard form equation:

$$\frac{(x-1)^2}{5^2}+\frac{(y-(-4))^2}{3^2}=1$$

The ellipse is centered at $(1,-4)$, and we have $a=5$ and $b=3$. This means the major axis extends from $(-4,-4)$ to $(6,-4)$ and the minor axis runs from $(1,-7)$ to $(1,-1)$. Connecting these points with an elliptical curve will provide a good graph.

Before graphing, we locate the foci.

$$c^2=a^2-b^2$$
$$c^2=25-9=16$$
$$c=\pm 4$$

Thus, the foci are located 4 units from the center, along the major axis. This places them at $(-3,-4)$ and $(5,-4)$. Putting all of this together, we graph the ellipse shown in Figure 8.

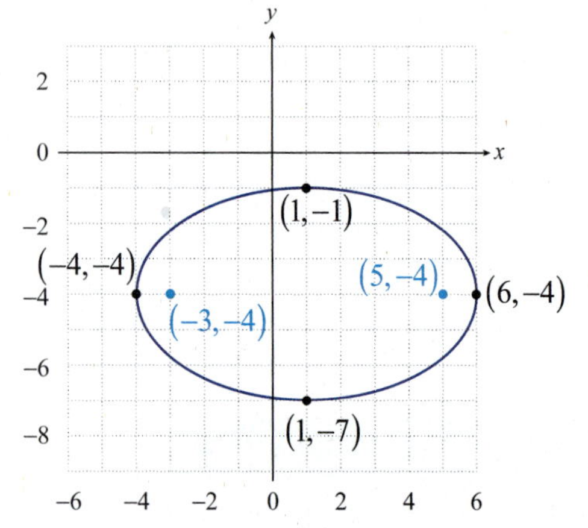

FIGURE 8

Example 3: Graphing an Ellipse with Equation Not in Standard Form

Graph the ellipse $25x^2 + 4y^2 + 100x - 24y + 36 = 0$.

Solution

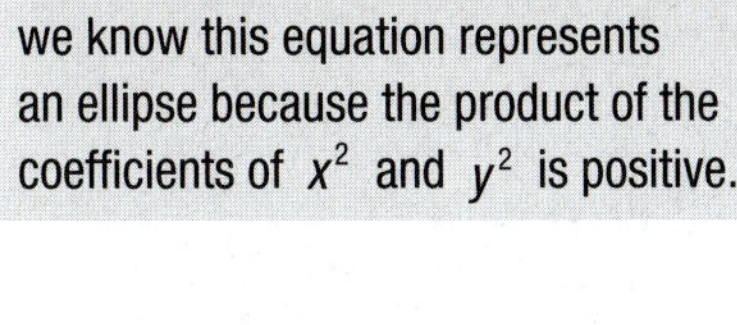

NOTE
Even though it is not in standard form, we know this equation represents an ellipse because the product of the coefficients of x^2 and y^2 is positive.

As written, the equation doesn't provide any useful information about the ellipse. To rewrite it in standard form, we need to complete the square for both variables.

$$25x^2 + 4y^2 + 100x - 24y + 36 = 0$$
Arrange the terms to pair like variables.

$$25x^2 + 100x + 4y^2 - 24y = -36$$

$$25\left(x^2 + 4x\right) + 4\left(y^2 - 6y\right) = -36$$
Factor to obtain a coefficient of 1 on each of the squared terms.

$$25\left(x^2 + 4x + 4\right) + 4\left(y^2 - 6y + 9\right) = -36 + 100 + 36$$
Complete the square. Note that we have to balance the equation.

$$25(x+2)^2 + 4(y-3)^2 = 100$$
Rewrite the perfect square trinomials as squared binomials.

$$\frac{(x+2)^2}{4} + \frac{(y-3)^2}{25} = 1$$
Divide both sides by 100.

At this point, we have an equivalent equation that is in the standard form for an ellipse. Using this form, we can find all the information we need to construct a graph.

$$\frac{(x-(-2))^2}{2^2} + \frac{(y-3)^2}{5^2} = 1$$

Reading the equation, we have the center at $(-2,3)$ with $a = 5$ and $b = 2$. Note that a^2 appears in the y term, so this ellipse is elongated vertically. This means that the extreme points of the ellipse lie at $(-4,3)$, $(0,3)$, $(-2,-2)$ and $(-2,8)$.

Using $c^2 = a^2 - b^2 = 21$, we have $c = \pm\sqrt{21} \approx \pm 4.6$. Thus, the foci lie about 4.6 units below and above the center of the ellipse, at $(-2,-1.6)$ and $(-2,7.6)$. Putting all this information together, we have the graph in Figure 9.

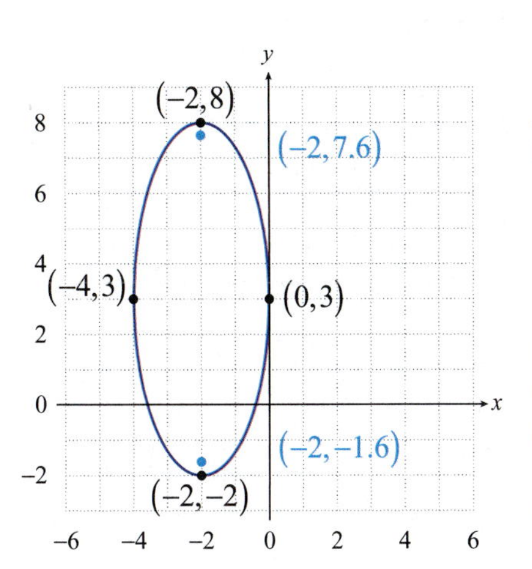

FIGURE 9

TECHNOLOGY: Graphing an Ellipse with a Vertical Major Axis

To graph the ellipse in Example 3 using a TI-84 Plus, first press `apps`. Select `Conics`, then select `ELLIPSE`. Now select `2:` to enter the values from the equation in standard form, $\frac{(x+2)^2}{4} + \frac{(y-3)^2}{25} = 1$, as shown in Figure 10. To view the graph, press `graph`. See Figure 11.

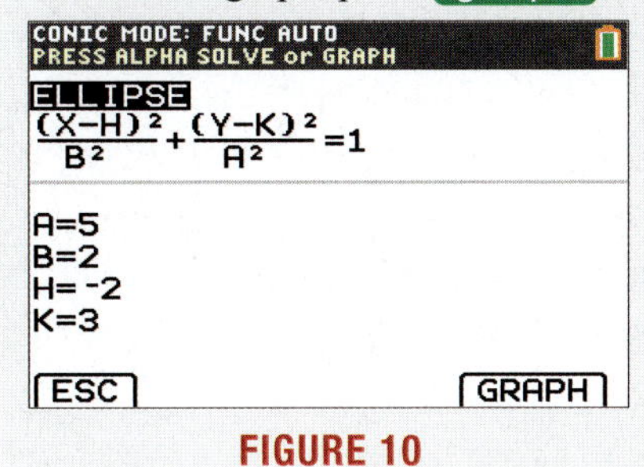

FIGURE 10

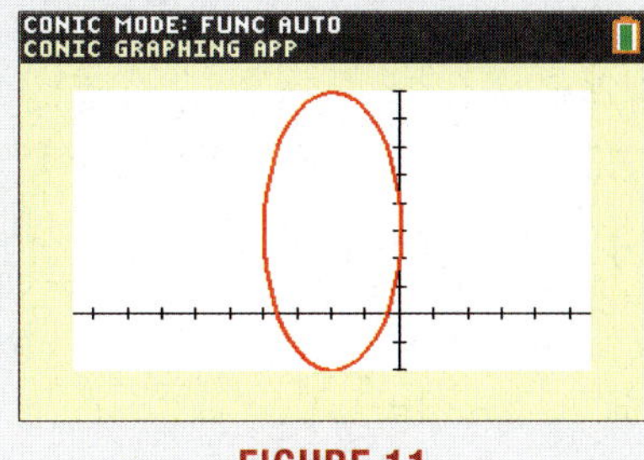

FIGURE 11

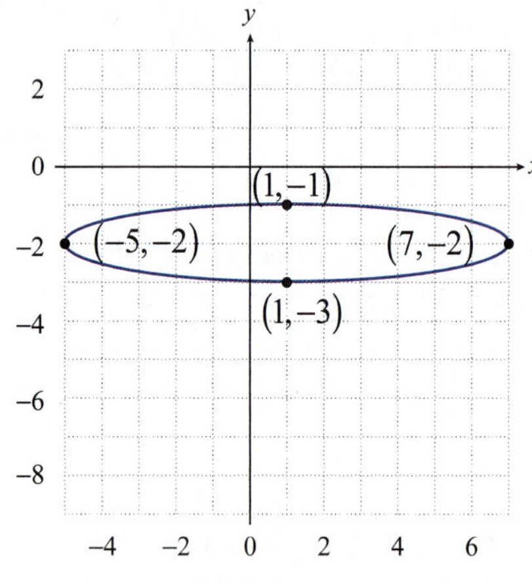

FIGURE 12

Example 4: Constructing the Equation for an Ellipse

Construct the standard form of the equation of the ellipse in Figure 12.

Solution

In order to construct the standard form equation, we need four pieces of information: the center of the ellipse, whether the major axis is horizontal or vertical, and the values of a and b.

Examining the graph, we see that the center is $(1,-2)$, the major axis is horizontal with length 12, and the minor axis has length 2. This gives us $a=6$ and $b=1$. Plugging this information into the standard form, we arrive at the equation for this ellipse:

$$\frac{(x-1)^2}{6^2}+\frac{(y-(-2))^2}{1^2}=1,$$

which we rewrite as $\frac{(x-1)^2}{36}+(y+2)^2=1.$

Note that the technique of graphing ellipses is similar to that of graphing circles, including the need to complete the square to obtain standard form, as in Example 3. In fact, circles are a special type of ellipse. Consider what happens if the numbers a and b in the standard form of an ellipse happen to be the same. If $a=b=r$, then the standard form is

$$\frac{(x-h)^2}{r^2}+\frac{(y-k)^2}{r^2}=1.$$

If we multiply through by r^2, we obtain $(x-h)^2+(y-k)^2=r^2$, the standard form of a circle. So a circle is an ellipse whose major and minor axes are the same length.

A circle, viewed as a type of ellipse, has only one focus, and that focus coincides with the center of the circle. Ellipses that are nearly circular have two foci, but they are relatively close to the center, while narrow ellipses have foci far away from the center. This gives us a convenient way to measure the relative "skinniness" of an ellipse.

DEFINITION: Eccentricity of an Ellipse

Given an ellipse with major and minor axes of lengths $2a$ and $2b$, respectively, the **eccentricity** of the ellipse, denoted by the symbol e, is defined by

$$e=\frac{c}{a}=\frac{\sqrt{a^2-b^2}}{a}.$$

If $e=0$, then $c=0$ and the ellipse is a circle. At the other extreme, c may be close to (but cannot equal) a, in which case e is close to 1. An eccentricity close to 1 indicates a relatively skinny ellipse.

TOPIC 3: Applications of Ellipses

Orbital Mechanics

Katherine Johnson was an American mathematician whose knowledge of analytic geometry and ability to quickly and accurately calculate trajectories led to an honored reputation over her 35-year career in the National Aeronautics and Space Administration. With the support of her family, she overcame racist and sexist prejudice to obtain her college education in the 1930s, and began work at NASA in the 1950s. She calculated the trajectories and other mission parameters for Alan Shepard's space flight in 1961 and John Glenn's first orbit of Earth in 1962. She was awarded the Presidential Medal of Freedom in 2015 and the Congressional Gold Medal in 2019.

Johannes Kepler (1571–1630) was the German-born astronomer and mathematician who first demonstrated that the planets in our solar system follow elliptical orbits. He did this by laboring over a period of twenty-one years to mathematically model the astronomical observations of his predecessor Tycho Brahe. The ultimate result was Kepler's Three Laws of Planetary Motion.

THEOREM: Kepler's Laws of Planetary Motion

1. The planets orbit the sun in elliptical paths, with the sun at one focus of each orbit.
2. A line segment between the sun and a given planet sweeps out equal areas of space in equal intervals of time.
3. The square of the time needed for a planet to complete a full revolution about the sun is proportional to the cube of the orbit's semimajor axis (half of the major axis).

The shapes of the various planetary orbits in our solar system vary widely, and these differences greatly influence seasons on the planets. For instance, Mars has a much more eccentric orbit than Earth, in both the informal and formal meaning of the term. The eccentricity of Mars' orbit is approximately 0.093, while the eccentricity of Earth's orbit is approximately 0.017. As a result, seasonal climate differences on Mars are much more dramatic than on Earth, including a large change in atmospheric pressure. Orbital eccentricities can be easily calculated from astronomical data, and we can use the eccentricity of a planet's orbit to answer particular questions.

Example 5: Planetary Orbits

Given that the furthest Earth gets from the sun is approximately 94.56 million miles, and that the eccentricity of Earth's orbit is approximately 0.017, estimate the closest approach of Earth to the sun.

Solution

Kepler's First Law states that each planet follows an elliptical orbit, and that the sun is positioned at one focus of the ellipse. Thus, Earth is furthest from the sun when it is at the end of the ellipse's major axis on the other side of the center from the sun.

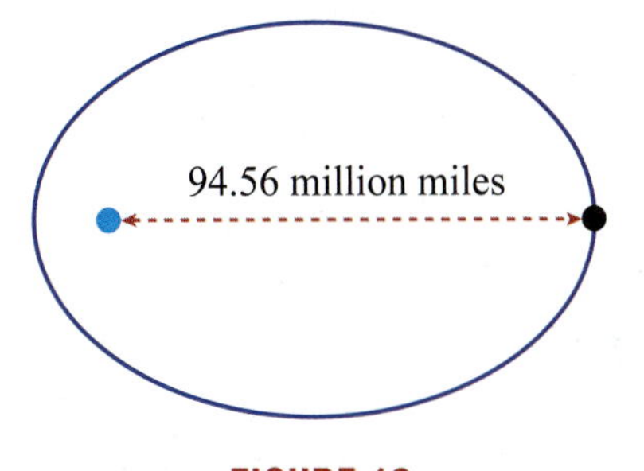

FIGURE 13

Using our standard terminology, and units of millions of miles, this means that

$$a + c = 94.56.$$

By the definition of eccentricity, we also know that

$$0.017 = \frac{c}{a}.$$

We can use these two equations to find the values of a and c.

$$a + c = 94.56$$

$$a + 0.017a = 94.56 \qquad \text{Substitute } c = 0.017a.$$

$$1.017a = 94.56 \qquad \text{Simplify.}$$

$$a \approx 92.98$$

Substituting back into our first equation, we have $c \approx 94.56 - 92.98$, so $c \approx 1.58$.

The closest approach to the sun must be a distance of $a - c$, which is approximately $92.98 - 1.58 = 91.40$ million miles.

10.1 EXERCISES

PRACTICE

Find the center, foci, and vertices of the ellipse that each equation describes. See Examples 1, 2 and 3.

1. $\dfrac{(x-5)^2}{4} + \dfrac{(y-2)^2}{25} = 1$

2. $\dfrac{(x+3)^2}{9} + \dfrac{(y+1)^2}{16} = 1$

3. $(x+2)^2 + 3(y+5)^2 = 9$

4. $4(x-4)^2 + (y-2)^2 = 8$

5. $x^2 + 6x + 2y^2 - 8y + 13 = 0$

6. $2x^2 + y^2 - 4x + 4y - 10 = 0$

7. $4x^2 + y^2 + 40x - 2y + 85 = 0$

8. $x^2 + 2y^2 - 6x + 16y + 37 = 0$

9. $x^2 + 3y^2 + 8x - 12y + 1 = 0$

10. $4x^2 + 3y^2 - 8x + 18y + 19 = 0$

11. $x^2 - 4x + 5y^2 - 1 = 0$

12. $x^2 + 4y^2 + 24y + 28 = 0$

Match the following equations to their graphs.

13. $\dfrac{(x-1)^2}{4} + \dfrac{y^2}{81} = 1$

14. $\dfrac{(x-3)^2}{49} + \dfrac{(y-2)^2}{25} = 1$

15. $\dfrac{(x+4)^2}{9} + \dfrac{(y-3)^2}{4} = 1$

16. $\dfrac{(x+3)^2}{36} + \dfrac{(y-1)^2}{64} = 1$

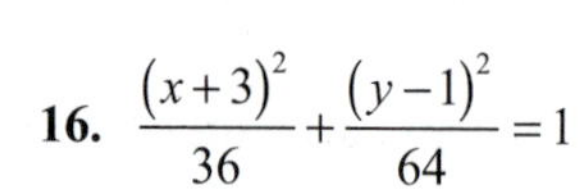

17. $\dfrac{(x+3)^2}{4} + \dfrac{(y+3)^2}{9} = 1$

18. $\dfrac{x^2}{16} + \dfrac{(y+5)^2}{4} = 1$

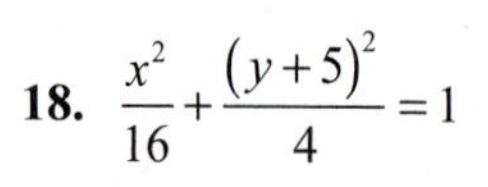

a.

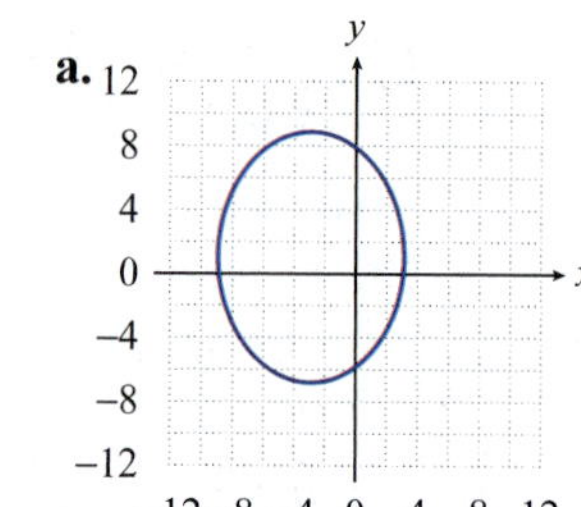

b.

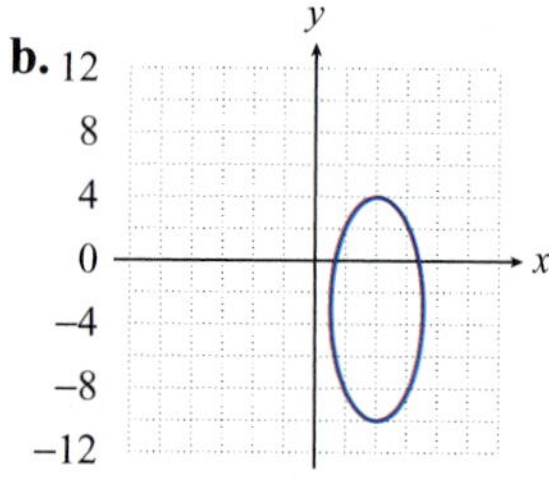

c.

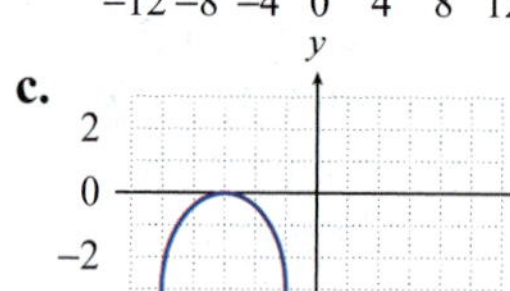

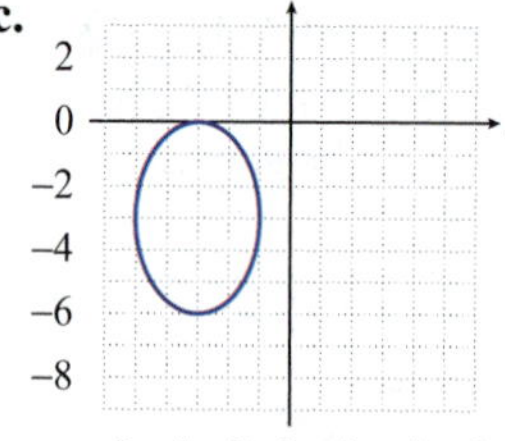

d.

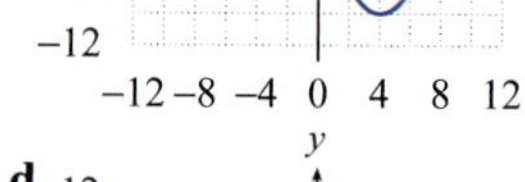

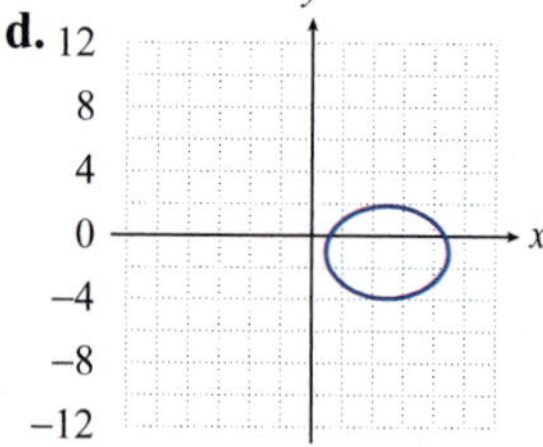

e.

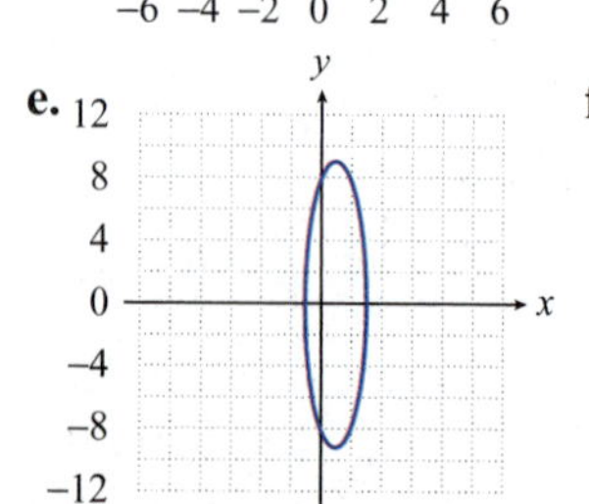

f.

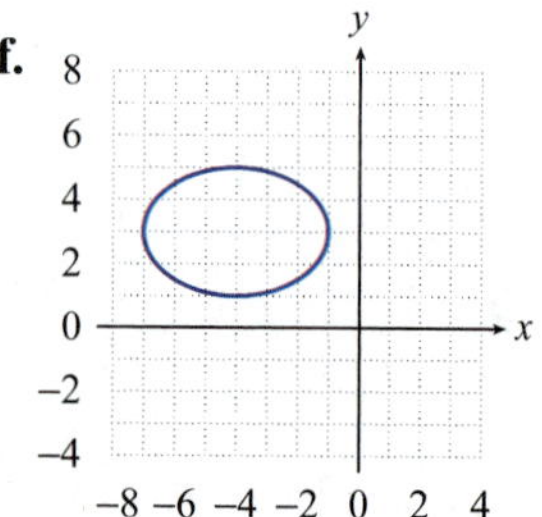

19. $\dfrac{(x-4)^2}{9}+\dfrac{(y+3)^2}{49}=1$

20. $\dfrac{(x-5)^2}{16}+\dfrac{(y+1)^2}{9}=1$

g.

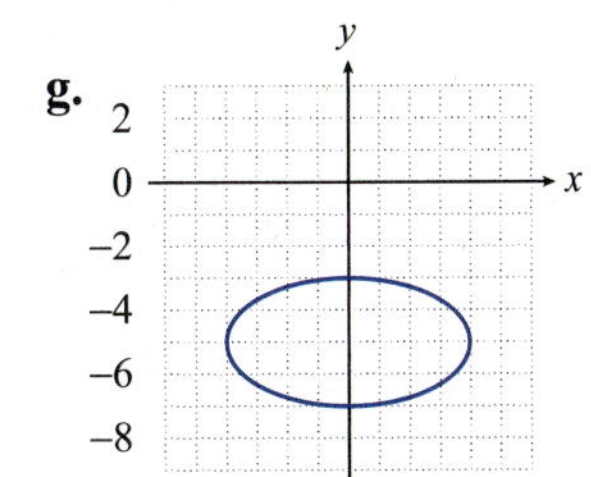

h.

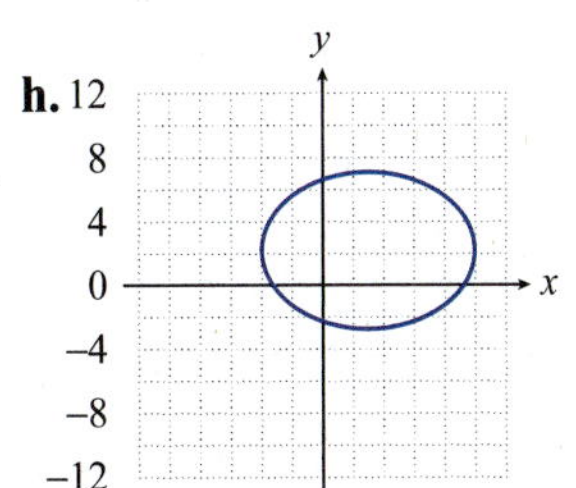

Sketch the graphs of the following ellipses and determine the coordinates of the foci. See Examples 1, 2, and 3.

21. $\dfrac{(x-3)^2}{9}+\dfrac{(y+1)^2}{1}=1$

22. $\dfrac{(x+5)^2}{4}+\dfrac{(y+2)^2}{16}=1$

23. $\dfrac{(x-3)^2}{9}+\dfrac{(y-4)^2}{4}=1$

24. $\dfrac{x^2}{25}+\dfrac{(y-3)^2}{16}=1$

25. $(x-1)^2+\dfrac{(y-4)^2}{4}=1$

26. $\dfrac{(x-4)^2}{16}+\dfrac{(y-4)^2}{4}=1$

27. $\dfrac{(x+1)^2}{25}+\dfrac{(y+5)^2}{4}=1$

28. $\dfrac{(x-2)^2}{9}+\dfrac{(y+1)^2}{9}=1$

29. $\dfrac{(x+2)^2}{16}+\dfrac{(y+1)^2}{9}=1$

30. $\dfrac{x^2}{25}+(y+2)^2=1$

31. $9x^2+16y^2+18x-64y=71$

32. $9x^2+4y^2-36x-24y+36=0$

33. $16x^2+y^2+160x-6y=-393$

34. $25x^2+4y^2-100x+8y+4=0$

35. $4x^2+9y^2+40x+90y+289=0$

36. $16x^2+y^2-64x+6y+57=0$

37. $4x^2+y^2+4y=0$

38. $9x^2+4y^2+108x-32y=-352$

In each of the following exercises, an ellipse is described by either a picture or by the properties it possesses. Find the equation, in standard form, for each ellipse. See Example 4.

39. Center at the origin, major axis of length 10 on the y-axis, foci 3 units from the center.

40. Center at $(-2,3)$, major axis of length 8 oriented horizontally, minor axis of length 4.

41. Vertices at $(1,4)$ and $(1,-2)$, foci $2\sqrt{2}$ units from the center.

42. Vertices at $(5,-1)$ and $(1,-1)$, minor axis of length 2.

43. Foci at $(0,0)$ and $(6,0)$, $e=\dfrac{1}{2}$.

44. Vertices at $(-1,4)$ and $(-1,0)$, $e=0$.

45. Vertices at $(-2,-1)$ and $(-2,-5)$, minor axis of length 2.

46. Vertices at $(-4,6)$ and $(-14,6)$, $e=\frac{2}{5}$.

47. Vertices at $(1,3)$ and $(9,3)$, one of the foci at $(6,3)$.

48. Foci at $(2,-4)$ and $(2,-8)$, minor axis of length 6.

49.

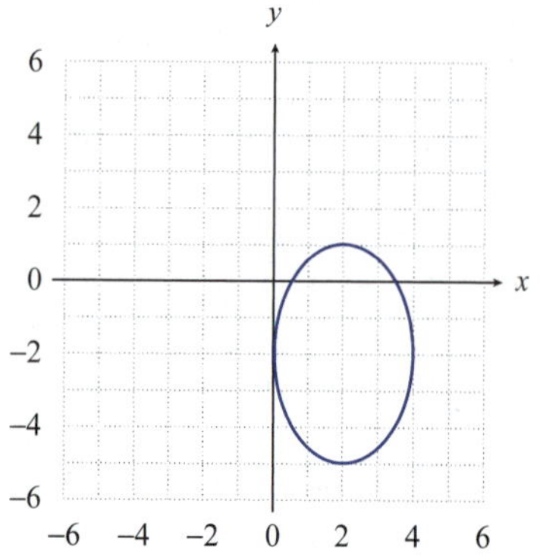

50.

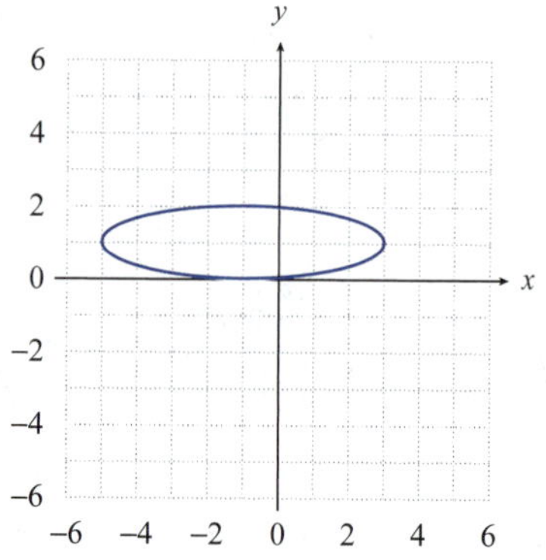

51.

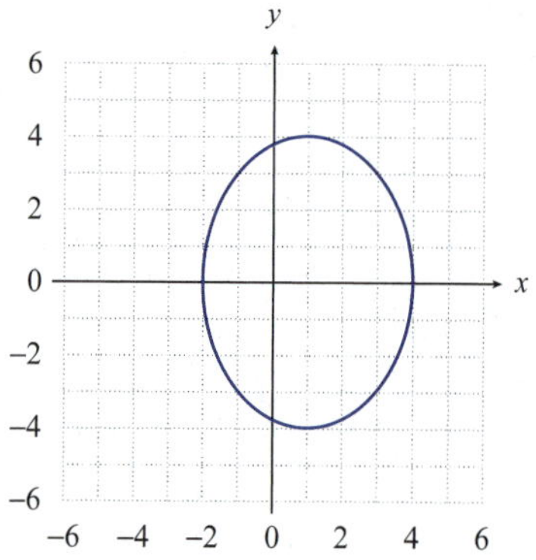

52.

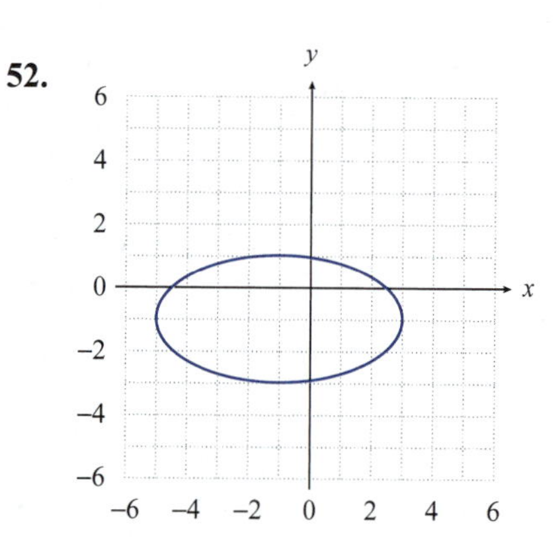

Find the eccentricity and the lengths of the minor and major axes of the following ellipses.

53. $\frac{x^2}{100}+\frac{y^2}{144}=1$

54. $\frac{x^2}{64}+\frac{y^2}{9}=1$

55. $x^2+9y^2=36$

56. $25x^2+4y^2=100$

57. $4x^2+16y^2=16$

58. $5x^2+8y^2=40$

59. $20x^2+10y^2=40$

60. $\frac{1}{4}x^2+\frac{1}{12}y^2=\frac{1}{2}$

61. $x^2=49-7y^2$

APPLICATIONS

62. The orbit of Halley's Comet is an ellipse with an eccentricity of 0.967. Its closest approach to the sun is approximately 54,591,000 miles. What is the farthest Halley's Comet ever gets from the sun?

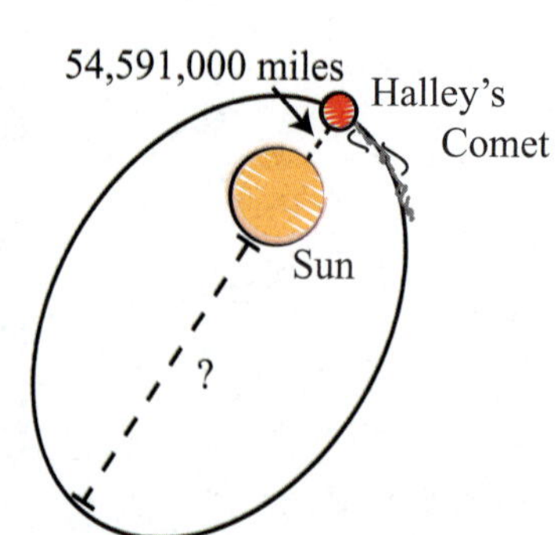

63. Pluto's closest approach to the sun is approximately 4.43×10^9 kilometers, and its maximum distance from the sun is approximately 7.37×10^9 kilometers. What is the eccentricity of Pluto's orbit?

64. Use the information given in Example 5 to determine the length of the minor axis of the ellipse formed by Earth's orbit around the sun.

65. The archway supporting a bridge over a river is in the shape of half an ellipse. The archway is 60 feet wide and is 15 feet tall at the middle. A boat is 10 feet wide and 14 feet, 9 inches tall. Is the boat capable of passing under the archway?

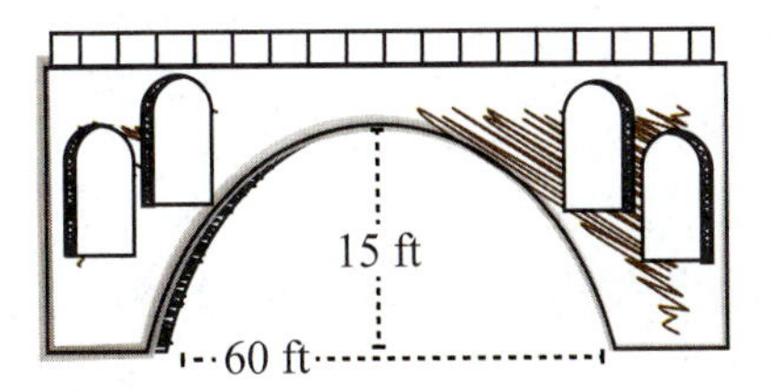

66. *The Whispering Gallery* in Chicago's Museum of Science and Industry is a giant ellipsoid that transmits the slightest whisper from one focus to the other focus. This giant ellipse is known to have a length of about 568 inches and a width of about 162 inches. Find the eccentricity of *The Whispering Gallery*. About how far apart are two whisperers when communicating in this gallery? Round your answers to four decimal places.

WRITING & THINKING

67. Since the sum of the distances from each of the two foci to any point on an ellipse is constant, we can draw an ellipse using the following method. Tack the ends of a length of string at two points (the foci) and, keeping the string taut by pulling outward with the tip of a pencil, trace around the foci to form an ellipse (the total length of the string remains constant). If you want to create an ellipse with a major axis of length 5 cm and a minor axis of length 3 cm, how long should your string be and how far apart should you place the tacks? Use the relationships of distances and formulas that you have learned in this section.

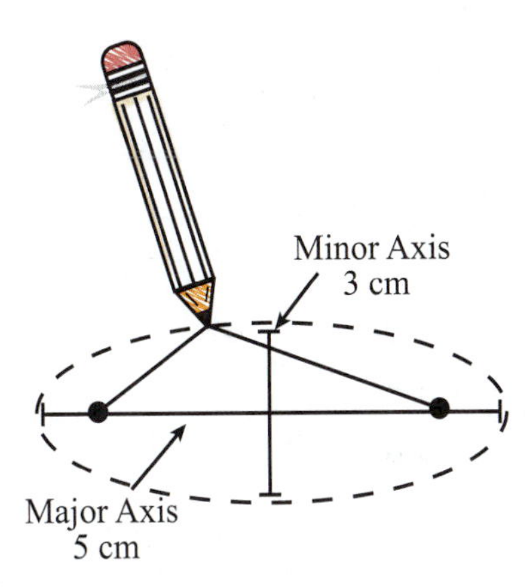

68. Using the method described in Exercise 67, describe the change in your ellipse when you move the two foci closer together. What happens when you move them farther apart?

TECHNOLOGY

Use a graphing utility to graph the following equations.

69. $15x^2 + 9y^2 + 150x - 36y = -276$

70. $5x^2 + 12y^2 - 20x + 144y + 392 = 0$

71. $3x^2 + 2y^2 = 3 - 18x$

72. $2x^2 + 5y^2 = 70y - 205$

10.2 PARABOLAS

TOPICS

1. The standard form of the equation of a parabola (as a conic section)
2. Applications of parabolas

TOPIC 1: The Standard Form of the Equation of a Parabola (as a Conic Section)

We have already studied parabolas in the context of quadratic functions. Specifically, any function of the form $f(x)=a(x-h)^2+k$ describes a vertically oriented parabola in the plane (opening upward or downward, depending on the sign of a) whose vertex is at (h,k).

The material in this section does not replace what we have learned. Instead, viewing parabolas as conic sections broadens our understanding of them, and allows us to work with parabolic curves that are not defined by functions.

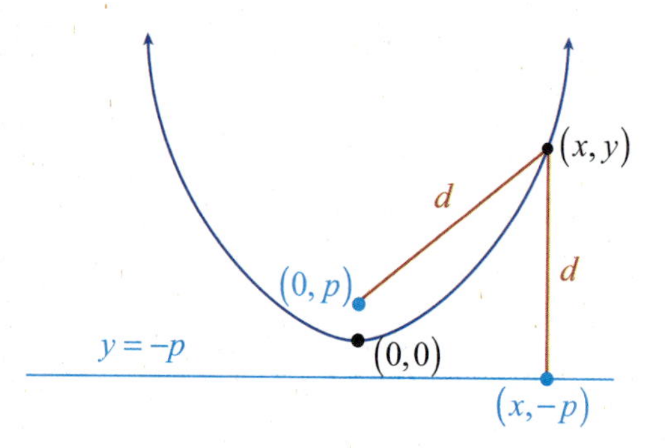

FIGURE 1: The Geometric Property of the Parabola

Just as with ellipses, we want to derive a useful form of the equation that describes parabolas from their characteristic geometric property. Recall that the plane geometric definition of a parabola is that each point on a given parabola is equidistant from a fixed point called the focus and a fixed line called the directrix. Let p denote the distance between the vertex of the parabola and the focus, and hence also the distance between the vertex and the directrix, as shown in Figure 1.

We begin our derivation by assuming the parabola is oriented vertically and that the vertex is at the origin. As we did with ellipses, we will generalize this basic result later.

Given this, we know that the focus of our parabola is at $(0,p)$ and the equation for the directrix is $y=-p$. The following equation uses the distance formula to characterize all those points that are of equal distance d from the focus and the directrix.

$$\sqrt{x^2+(y-p)^2}=\sqrt{(y+p)^2}$$

Note that the x-coordinate does not appear in the distance formula on the right side of the equation. This is because the shortest path from this parabola to the directrix is always a vertical line.

One way to make the equation look nicer is to square both sides. We then proceed to eliminate some terms and combine others to arrive at a useful form.

$$\begin{aligned} x^2+(y-p)^2 &= (y+p)^2 \\ x^2+y^2-2py+p^2 &= y^2+2py+p^2 \\ x^2 &= 4py \end{aligned}$$

The conclusion is that any equation that can be written in the above form describes a parabola with vertex at the origin, focus at $(0, p)$ and directrix the line $y = -p$. This is true even if p is negative; in this case the parabola opens downward.

To generalize, recall that replacing x with y and vice versa reflects a graph about the line $y = x$, so the equation $y^2 = 4px$ describes a horizontally oriented parabola with vertex at $(0, 0)$. Swapping x and y reflects the focus and directrix as well, so the focus of $y^2 = 4px$ is at $(p, 0)$ and the directrix is the vertical line $x = -p$. As always, replacing x with $x - h$ and y with $y - k$ shifts a graph h units horizontally and k units vertically.

Summarizing all of this information, we arrive at the standard form.

DEFINITION: Standard Form of the Equation of a Parabola

Let p be a nonzero real constant. The **standard form of the equation of a parabola** with vertex at (h, k) is

- $(x - h)^2 = 4p(y - k)$ if the parabola is vertically oriented. In this case, the focus is at $(h, k + p)$ and the equation of the directrix is $y = k - p$.
- $(y - k)^2 = 4p(x - h)$ if the parabola is horizontally oriented. In this case, the focus is at $(h + p, k)$ and the equation of the directrix is $x = h - p$.

As we have seen, parabolas can be relatively flat or relatively skinny. With quadratic functions, we saw that the coefficient a governed the flatness of a parabola. In standard form, the parameter p determines the level of flatness. To see how this effect works, consider the standard form for a vertically opening parabola with vertex at the origin.

$$x^2 = 4py$$

Solving this equation for y, we return to the quadratic equation form, with $a = \frac{1}{4p}$.

$$y = \frac{1}{4p}x^2$$

Thus, we see that if the magnitude of p increases, the magnitude of a decreases, producing a flatter parabola, while if the magnitude of p decreases, the magnitude of a increases, producing a skinnier parabola. This relationship holds even if the vertex moves or if the parabola opens horizontally.

Example 1: Graphing a Parabola with Equation in Standard Form

Graph the parabola $(y + 2)^2 = 8(x - 4)$ and determine its focus and directrix.

Solution

This equation is in standard form, so we can quickly determine the vertex, focus and directrix.

$$(y - (-2))^2 = 8(x - 4)$$

We can see the vertex is at $(4, -2)$. To find the focus and directrix we first calculate p: $8 = 4p$, so $p = 2$. This means the focus lies at $(6, -2)$ and the directrix is the vertical line $x = 2$.

To get an idea of the shape of the parabola, we need to find a few more points on its graph. Here we make use of the geometric property of the parabola.

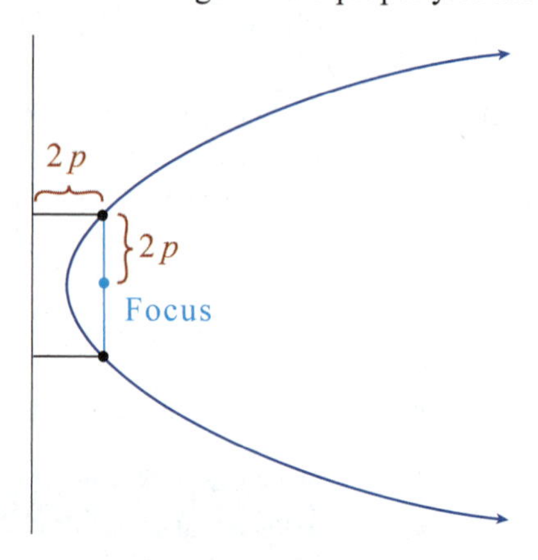

FIGURE 2: Two Easily Plotted Points

Recall that every point on the parabola must be equidistant from the focus and directrix. With our horizontally opening parabola, we know the focus is a distance of $2p$ from the directrix. Drawing the line parallel to the directrix that passes through the focus, we see that there are two points on the parabola that are $2p$ from the focus. By definition, these points must lie $2p$ above and below the focus.

Thus, two points on our parabola are $(6,-2-4)=(6,-6)$ and $(6,-2+4)=(6,2)$. Plotting these along with our vertex, we have the graph in Figure 3.

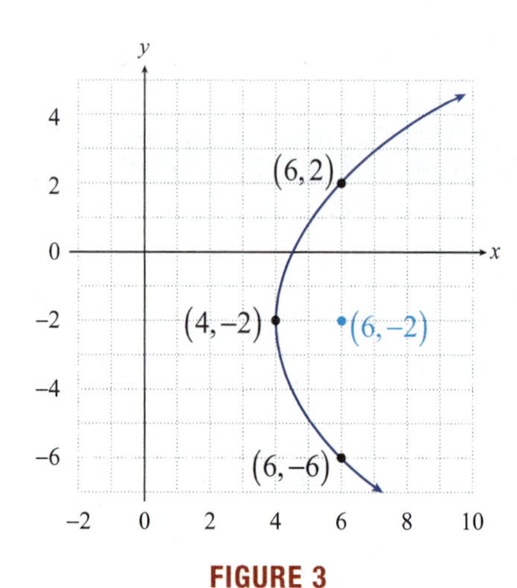

FIGURE 3

Example 2: Graphing a Parabola with Equation Not in Standard Form

Graph the parabola $-y^2+2x+2y+5=0$ and determine its focus and directrix.

> **NOTE**
> Even though it is not in standard form, we know this equation represents a parabola because the product of the coefficients of x^2 and y^2 is zero.

Solution

Written in this form, it is difficult to graph the parabola. We can rewrite the equation in standard form by completing the square with respect to y (the squared variable).

$$-y^2+2x+2y+5=0$$

$$y^2-2y=2x+5$$

Begin by rearranging the terms to obtain a coefficient of 1 on the y^2 term.

$$y^2-2y+1=2x+5+1$$

In order to complete the square, we have to add 1 to both sides.

$$(y-1)^2=2(x+3)$$

Rewrite the trinomial as a binomial squared.

$$(y-1)^2=4\left(\frac{1}{2}\right)(x+3)$$

Put the right-hand side into the form $4p(x-h)$ to make the value of p easy to see.

Now that the equation is in standard form, by inspection we can tell that the vertex is at $(-3,1)$ and that $p=\frac{1}{2}$. Since the focus is p units to the right of the vertex

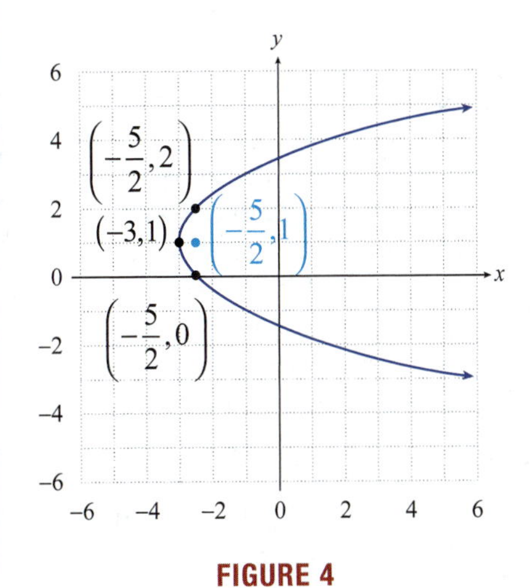

FIGURE 4

and the directrix is a vertical line p units to the left of the vertex, we obtain the following.

$$\text{focus: } \left(-\frac{5}{2},1\right), \quad \text{directrix: } x=-\frac{7}{2}$$

To sketch the graph, we begin by plotting the vertex at $(-3,1)$. Using the same process as in Example 1, note that the two points with x-coordinates of $-\frac{5}{2}$ are 1 unit away from the directrix; therefore they must be 1 unit away from the focus as well. Thus, $\left(-\frac{5}{2},0\right)$ and $\left(-\frac{5}{2},2\right)$ must also lie on the parabola. This gives us some idea of the "flatness" of the parabola.

TECHNOLOGY: Graphing a Horizontal Parabola

To graph the parabola in Example 2 using a TI-84 Plus, first press apps. Select Conics, then select PARABOLA. Now select 1: to enter a horizontal parabola and the values from the equation in standard form, $(y-1)^2=4\left(\frac{1}{2}\right)(x+3)$, as shown in Figure 5. To view the graph, press graph. See Figure 6.

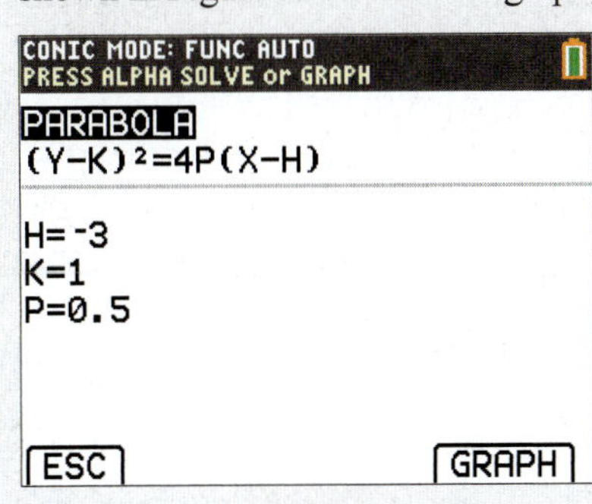

FIGURE 5

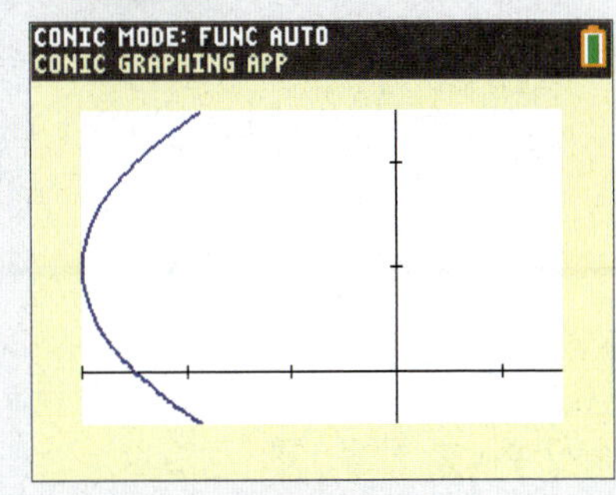

FIGURE 6

Example 3: Standard Form Equation of a Parabola

Given that the directrix of a parabola is the line $y=2$, that $p=-1$, and that the parabola is symmetric with respect to the line $x=2$, find its standard form equation.

Solution

Because the directrix $y=2$ is a horizontal line, the parabola opens vertically. This means we need to use the form

$$(x-h)^2=4p(y-k).$$

From the equation for the line of symmetry, we know the x-coordinate of the vertex (and the focus) must be 2. Moving down 1 unit from the directrix (since p is negative) puts the vertex at $(2,1)$. Plugging this information into the standard form, we have the following.

$$(x-2)^2=4(-1)(y-1)$$
$$(x-2)^2=-4(y-1)$$

We can verify our equation with a graph of the parabola.

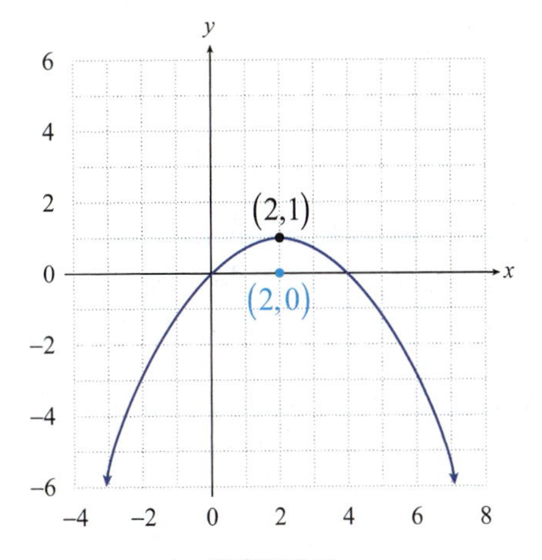

FIGURE 7

TOPIC 2: Applications of Parabolas

We have already seen an important algebraic application of the parabola: the path of an object thrown under the influence of gravity is a parabola. Algebraically, we know that the height $h(t)$ of an object with initial velocity v_0 and initial height h_0 is

$$h(t) = -\frac{1}{2}gt^2 + v_0 t + h_0,$$

where g, a constant, is the acceleration due to gravity (see Section 1.8 for details). This is a quadratic function, and we experience the fact that a parabola is the shape of its graph whenever we see, for example, a thrown baseball.

In some applications, the geometric properties of parabolas are the key issue, and the standard form may be more helpful than a quadratic equation. One useful geometric property of the parabola is that if rays of light emanate from the focus of the parabola, they reflect from the inner surface of the parabola in such a way that the reflected rays are parallel to the axis of symmetry, as shown in Figure 8.

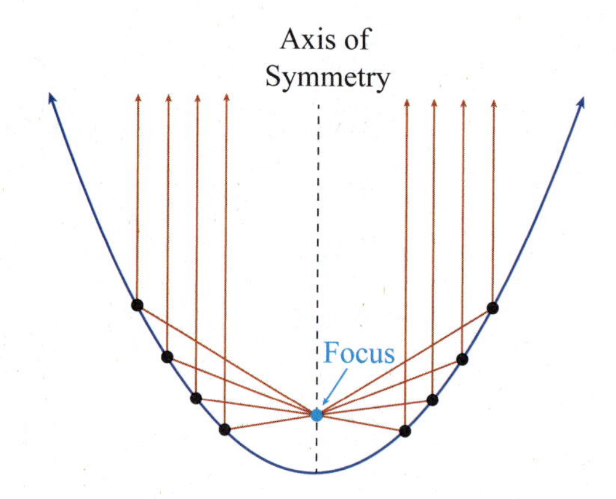

FIGURE 8: Parallel Rays of Reflection in a Parabola

This property is used in a variety of real-world applications. If a parabola is rotated about its axis of symmetry, a three-dimensional shape called a *paraboloid* is the result. A *parabolic mirror* is then made by coating the inner surface of a paraboloid with a reflecting material. Parabolic mirrors are the basis of searchlights and vehicle headlights, with a light source placed at the focus of the paraboloid. They are also the basis of one design of telescope, in which incoming (parallel) starlight is reflected to an eyepiece at the focus.

Example 4: Parabolic Mirrors

The Hale Telescope at the Mount Palomar observatory in California is a very large reflecting telescope. The paraboloid is the top surface of a large cylinder of Pyrex glass 200 inches in diameter. Along the outer rim, the cylinder is 26.8 inches thick, while at the center, it is 23 inches thick. Where is the focus of the parabolic mirror located?

NOTE

The same concept can be used to focus sunlight, intensely heating a small area at the focus. This is called a parabolic furnace.

Solution

First, we need to draw a picture of the situation. In order to make the math as easy as possible, we can locate the origin of our coordinate system at the vertex of a parabolic cross section of the mirror, and we can assume the parabola opens upward.

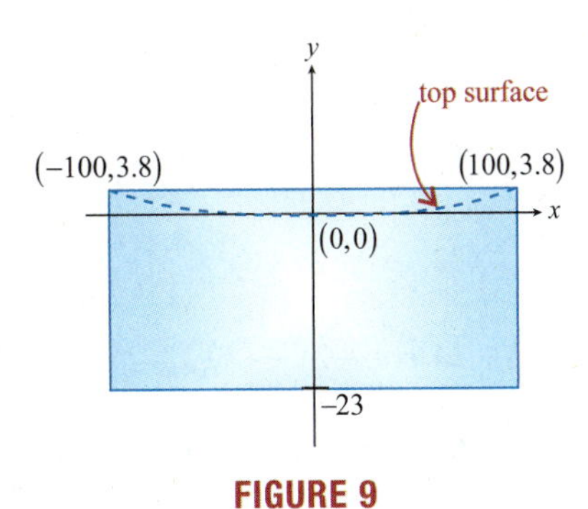

FIGURE 9

Since we placed the vertex at $(0,0)$, we know the equation $x^2 = 4py$ describes the shape of the cross section for some value p. If we can determine p, we can find the focus of the parabola.

To find p, we need the coordinates of another point on the parabola. The difference in thickness of the mirror between the center and the outer rim is 3.8 inches, and the mirror has a diameter of 200 inches, so the two points $(-100,3.8)$ and $(100,3.8)$ must lie on the graph. Plugging a point into the equation $x^2 = 4py$, we can solve for p.

$$\begin{aligned}(100)^2 &= 4p(3.8)\\ 10000 &= 15.2p\\ p &\approx 657.9 \text{ inches}\\ p &\approx 54.8 \text{ feet}\end{aligned}$$

We know that the focus of a parabola is p units from the vertex, so the focus of the Hale Telescope is nearly 55 feet from the mirror.

10.2 EXERCISES

PRACTICE

Graph the following parabolas and determine the focus and directrix of each. See Examples 1 and 2.

1. $(x+1)^2 = 4(y-3)$

2. $y^2 - 4y = 8x + 4$

3. $(y-4)^2 = -2(x-1)$

4. $(y-1)^2 = 8(x+3)$

5. $(x-2)^2 = 4(y+1)$

6. $(y+1)^2 = -12(x+1)$

7. $y^2 = 6x$

8. $x^2 = 2y$

9. $x^2 = 7y$

10. $x^2 = -5y$

11. $y = -12x^2$

12. $x = -4y^2$

13. $x = \frac{1}{6}y^2$

14. $\frac{1}{5}x = -y^2$

15. $y^2 + 16x = 0$

16. $-6x - 2y^2 = 0$

17. $4y + 2x^2 = 4$

18. $2y^2 - 10x = 10$

19. $y^2 + 2y + 12x + 37 = 0$

20. $x^2 - 8y = 6x - 1$

21. $x^2 + 6x + 8y = -17$

22. $x^2 + 2x + 8y = 31$

23. $y^2 + 6y - 2x + 13 = 0$

24. $y^2 - 2y - 4x + 13 = 0$

Match the following equations to their graphs.

25. $(x+2)^2 = 3(y-1)$

26. $(y-1)^2 = 2(x+2)$

27. $y^2 = 4(x+1)$

28. $x^2 = 2(y+1)$

29. $(x-1)^2 = -(y-2)$

30. $(y+2)^2 = 3x$

31. $(x-2)^2 = 4y$

32. $y^2 = -2(x+1)$

a.

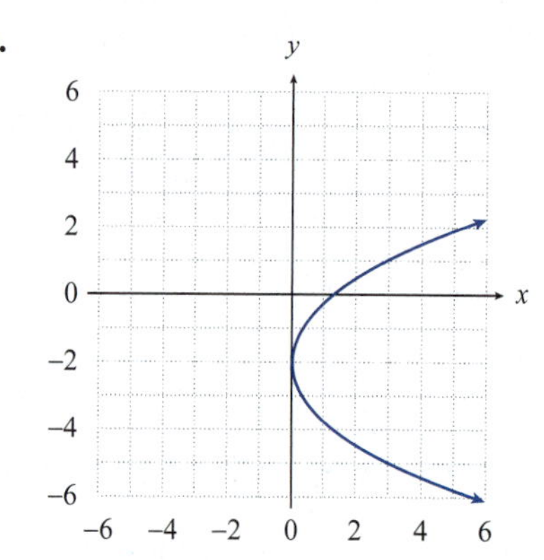

b.

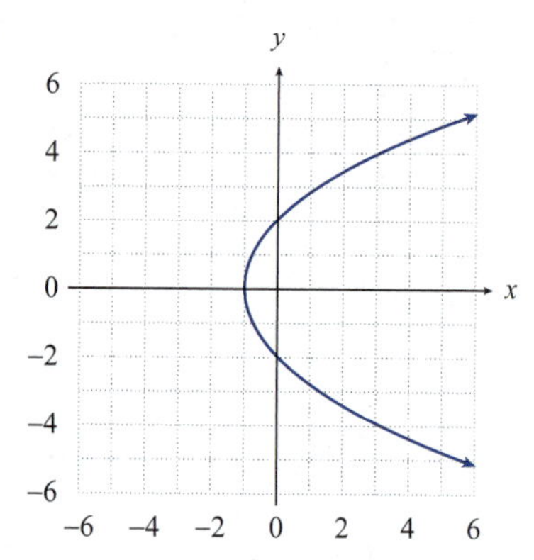

c.

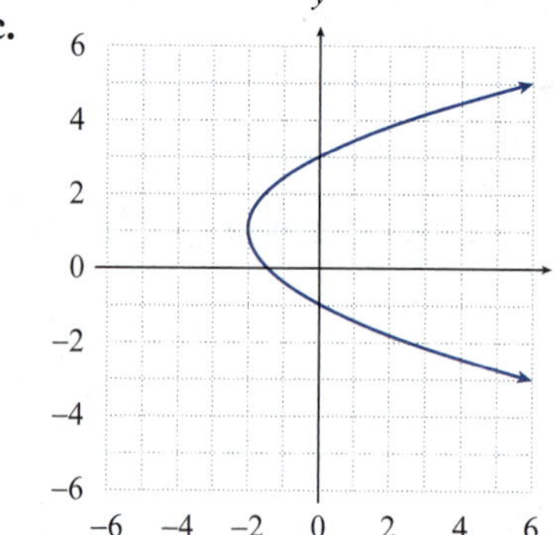

d.

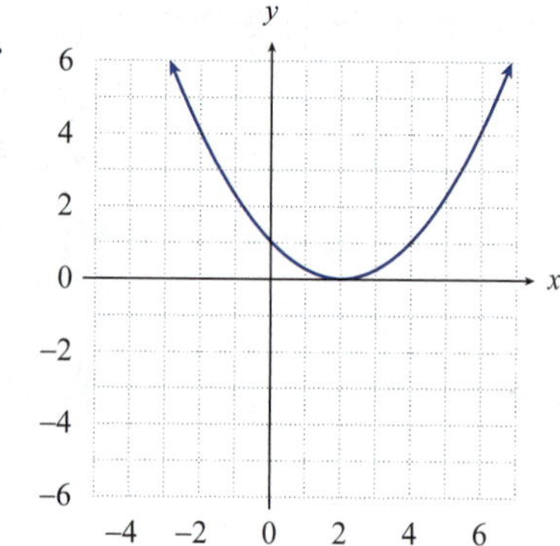

e.

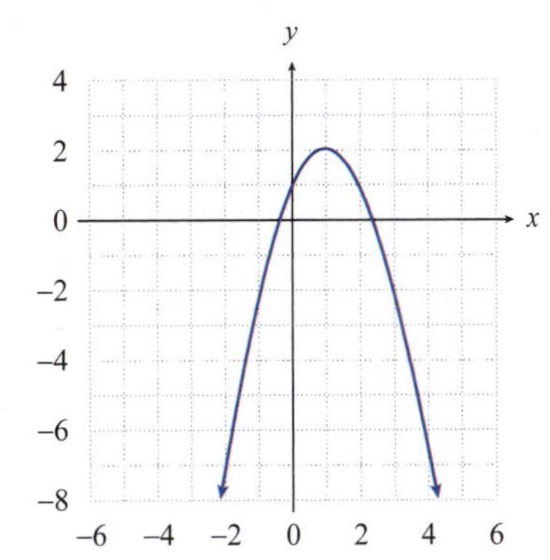

f.

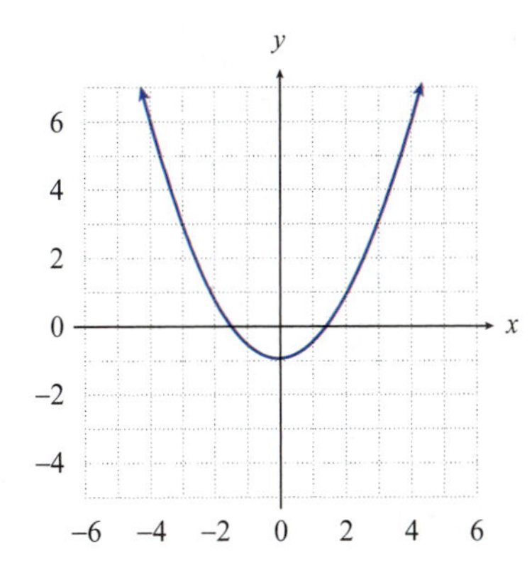

g.

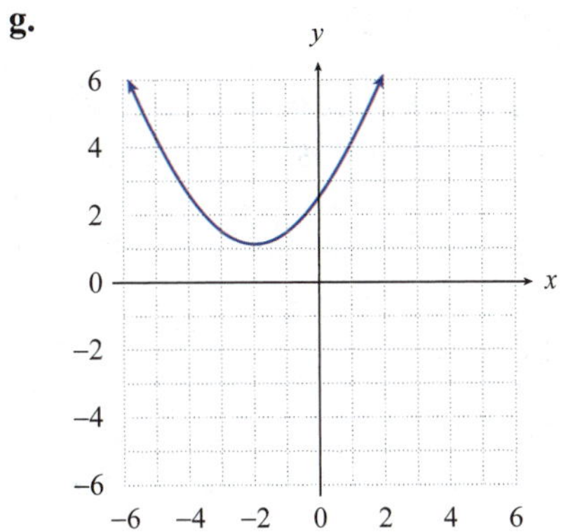

h.

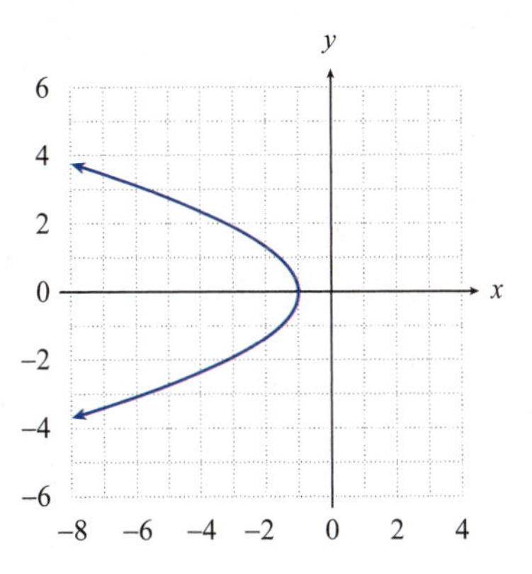

Find the equation, in standard form, for the parabola with the given properties or with the given graph. See Example 3.

33. Focus at $(-2,1)$, directrix is the y-axis.

34. Focus at $(-2,1)$, directrix is the x-axis.

35. Vertex at $(3,-1)$, focus at $(3,1)$.

36. Symmetric with respect to the line $y = 1$, directrix is the line $x = 2$, and $p = -3$.

37. Vertex at $(3,-2)$, directrix is the line $x = -3$.

38. Vertex at $(7,8)$, directrix is the line $x = \frac{27}{4}$.

39. Focus at $\left(-3, -\frac{3}{2}\right)$, directrix is the line $y = -\frac{1}{2}$.

40. Vertex at $(3,16)$, focus at $(3,11)$.

41. Vertex at $(-4,3)$, focus at $\left(-\frac{3}{2}, 3\right)$.

42. Symmetric with respect to the x-axis, focus at $(-3,0)$, and $p = 2$.

43.

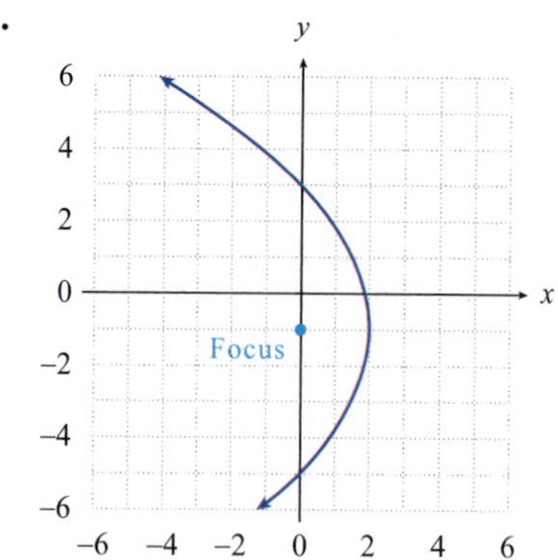

44.

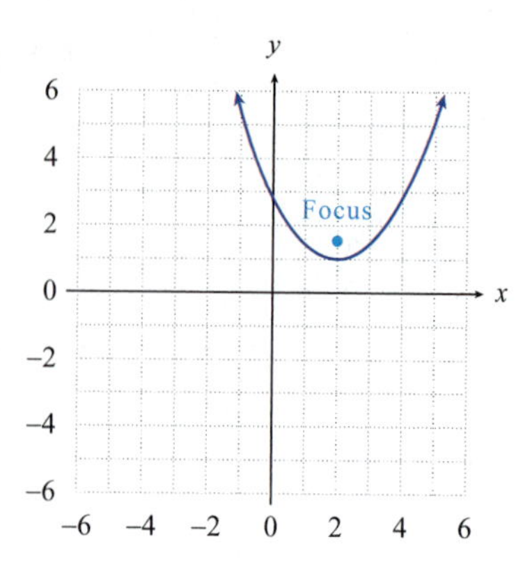

APPLICATIONS

45. One design for a solar furnace is based on the paraboloid formed by rotating the parabola $x^2 = 8y$ around its axis of symmetry. The object to be heated in the furnace is then placed at the focus of the paraboloid (assume that x and y are in units of feet). How far from the vertex of the paraboloid is the hottest part of the furnace?

46. A certain brand of satellite dish antenna is a paraboloid with a diameter of 6 feet and a depth of 1 foot. How far from the vertex of the dish should the receiver of the antenna be placed given that the receiver should be located at the focus of the paraboloid?

47. A spotlight is made by placing a strong lightbulb inside a reflective paraboloid formed by rotating the parabola $x^2 = 6y$ around its axis of symmetry (assume that x and y are in units of inches). In order to have the brightest, most concentrated light beam, how far from the vertex should the bulb be placed?

TECHNOLOGY

Use a graphing utility to graph the following equations.

48. $3x^2 - 4y + 24x = -56$

49. $y^2 + 2y = 8x - 41$

50. $y^2 - 6y + 4x = -17$

51. $x^2 - 6x + 12y + 21 = 0$

10.3 HYPERBOLAS

TOPICS

1. The standard form of the equation of a hyperbola
2. Applications of hyperbolas

TOPIC 1: The Standard Form of the Equation of a Hyperbola

In studying hyperbolas, we once again begin with the characteristic geometric property and use this to derive a useful form of the equation for a hyperbola which we call the standard form.

The characteristic geometric property of hyperbolas is similar to the one for ellipses. Recall that the points of an ellipse are those for which the sum of the distances to two foci is a fixed constant. For the points on a hyperbola, the magnitude of the *difference* of the distances to two foci is a fixed constant. This results in two disjoint pieces, called *branches*, of a hyperbola. The point halfway between the two branches is the center of the hyperbola, and the two points on the hyperbola closest to the center are the two vertices. Figure 1 illustrates how these parts of a hyperbola relate to one another.

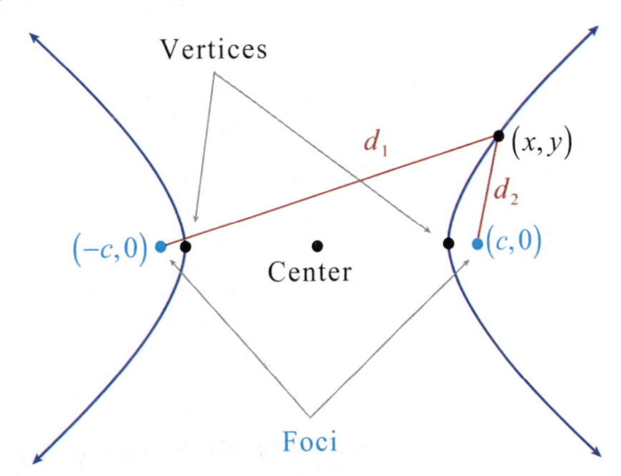

FIGURE 1: The Geometric Property of the Hyperbola

To begin, we will work with a hyperbola oriented as in Figure 1, with the center at the origin and the two foci at $(-c,0)$ and $(c,0)$.

We know that for every point (x,y) on the hyperbola, the quantity $|d_1 - d_2|$ is fixed; in order to simplify the algebra, let $2a$ denote this fixed quantity. Using the distance formula to express d_1 and d_2, we have:

$$\left|\sqrt{(x+c)^2+(y-0)^2}-\sqrt{(x-c)^2+(y-0)^2}\right|=2a$$

Note that if we find $|d_1 - d_2|$ for one of the vertices, it is equal to the distance between the two vertices, which is twice the distance between each vertex and the center. This means that the distance from the vertex to the center is equal to a.

We wish to manipulate this equation into a more useful form. Because of the absolute value symbols, this one equation actually represents two equations. The left-hand side represents either $d_1 - d_2$ or $d_2 - d_1$ depending on which quantity is nonnegative.

Fortunately, the two cases result in the same equation after we isolate one of the radicals and square both sides. Here, d_1 has been isolated and both sides squared:

$$\left(\sqrt{(x+c)^2+(y-0)^2}\right)^2 = \left(2a+\sqrt{(x-c)^2+(y-0)^2}\right)^2$$

$$(x+c)^2+y^2 = 4a^2+4a\sqrt{(x-c)^2+y^2}+(x-c)^2+y^2$$

This work is familiar, as it closely parallels the derivation of the standard form of the equation of an ellipse. In fact, if we follow the same simplification steps, we arrive at the form

$$\frac{x^2}{a^2}-\frac{y^2}{c^2-a^2}=1,$$

and as with ellipses we make a change of variables to improve the appearance. Note that a (the distance of either vertex from the center) is less than c (the distance of either focus from the center), so c^2-a^2 is positive and we can rename c^2-a^2 as b^2. This changes the above equation to

$$\frac{x^2}{a^2}-\frac{y^2}{b^2}=1,$$

with $b^2=c^2-a^2$ (we will write this as $c^2=a^2+b^2$ later so that we can solve for c).

To generalize this equation, we make the same observations as with ellipses and parabolas. Swapping x and y reflects the graph in Figure 1 with respect to the line $y=x$ (giving us a hyperbola with foci on the y-axis) and replacing x with $x-h$ and y with $y-k$ moves the center to (h,k).

DEFINITION: Standard Form of the Equation of a Hyperbola

Let a and b be positive constants. The **standard form of the equation of a hyperbola** with center at (h,k) is:

- $\dfrac{(x-h)^2}{a^2}-\dfrac{(y-k)^2}{b^2}=1$ if the foci are aligned horizontally (where the y-values of the foci are equal).
- $\dfrac{(y-k)^2}{a^2}-\dfrac{(x-h)^2}{b^2}=1$ if the foci are aligned vertically (where the x-values of the foci are equal).

In either case, the foci are located c units away from the center, where $c^2=a^2+b^2$, and the vertices are located a units away from the center.

The standard form of the equation of a hyperbola is useful, as it tells us by inspection where the center is and hence where the two vertices are (as well as the foci if we wish). Unfortunately, this knowledge alone leaves a lot of uncertainty about the shape of the hyperbola.

We need a reliable way to understand the "flatness" of the branches of a hyperbola, and it turns out that the branches of a hyperbola approach two *oblique asymptotes* far away from the center.

Consider a hyperbola centered at the origin with foci aligned horizontally, as in Figure 1. Then for some pair of constants a and b, the hyperbola is described by

$$\frac{x^2}{a^2} - \frac{y^2}{b^2} = 1.$$

To understand how this hyperbola behaves far away from the center, we can solve the equation for y:

$$y^2 = b^2\left(\frac{x^2}{a^2} - 1\right)$$

We are wondering how y behaves when x is large in magnitude, but the answer is not clear from the equation in this form. As is often the case, a little algebraic manipulation sheds light on the issue. Factoring out the fraction from the parentheses gives us the equation

$$y^2 = \frac{b^2x^2}{a^2}\left(1 - \frac{a^2}{x^2}\right),$$

and taking the square root of both sides leads to

$$y = \pm\frac{b}{a}x\sqrt{1 - \frac{a^2}{x^2}}.$$

The advantage of this last form is that as x goes to ∞ or $-\infty$, the radicand approaches 1. This means that y gets closer and closer to the value

$$\frac{b}{a}x \quad \text{or} \quad -\frac{b}{a}x$$

for values of x that are large in magnitude. In other words, the two straight lines

$$y = \frac{b}{a}x \quad \text{and} \quad y = -\frac{b}{a}x$$

(which intersect at the center) are the asymptotes of the hyperbola.

For hyperbolas whose foci are aligned vertically, the equations for the asymptotes are

$$x = \frac{b}{a}y \quad \text{and} \quad x = -\frac{b}{a}y;$$

that is, x and y exchange places. If we solve these last two equations for y, and also consider hyperbolas not centered at the origin by adding translations, we obtain the equations for asymptotes of all hyperbolas. This information allows us to sketch actual graphs of hyperbolas.

f(x) FORMULA: Asymptotes of Hyperbolas

- The asymptotes of the hyperbola $\frac{(x-h)^2}{a^2}-\frac{(y-k)^2}{b^2}=1$ are the two lines $y-k=\frac{b}{a}(x-h)$ and $y-k=-\frac{b}{a}(x-h)$.
- The asymptotes of the hyperbola $\frac{(y-k)^2}{a^2}-\frac{(x-h)^2}{b^2}=1$ are the two lines $y-k=\frac{a}{b}(x-h)$ and $y-k=-\frac{a}{b}(x-h)$.

As shown in Figure 2, these asymptotes are the diagonals of a rectangle centered at the center of the hyperbola with sides of length $2a$ and $2b$.

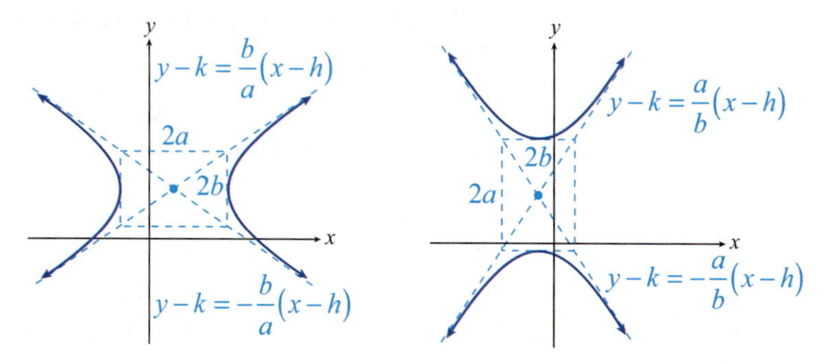

FIGURE 2: Asymptotes of Horizontally and Vertically Oriented Hyperbolas

Example 1: Graphing a Hyperbola with Equation in Standard Form

Graph the hyperbola $\frac{(x-2)^2}{25}-\frac{(y-1)^2}{4}=1$, indicating its asymptotes.

Solution

Since the hyperbola is written in standard form, the first step is to gather all of the information contained in the equation.

$$\frac{(x-2)^2}{25}-\frac{(y-1)^2}{4}=1$$

The positive term contains the x variable, so the foci of this hyperbola are aligned horizontally. We can also read that the center is $(2,1)$, $a=5$, and $b=2$. Using the center and the value of a, we know the vertices must lie at

$$(2-5,1)=(-3,1) \quad \text{and} \quad (2+5,1)=(7,1).$$

Using $c^2=a^2+b^2$, we know that $c=\sqrt{29}$, so we know the foci must lie at

$$\left(2-\sqrt{29},1\right)\approx(-3.4,1) \quad \text{and} \quad \left(2+\sqrt{29},1\right)\approx(7.4,1).$$

The best tool for graphing the hyperbola is its set of asymptotes. According to our formula, the asymptotes are the lines $y-1=\frac{2}{5}(x-2)$ and $y-1=-\frac{2}{5}(x-2)$. Using these lines and our vertices, we plot the graph of the hyperbola.

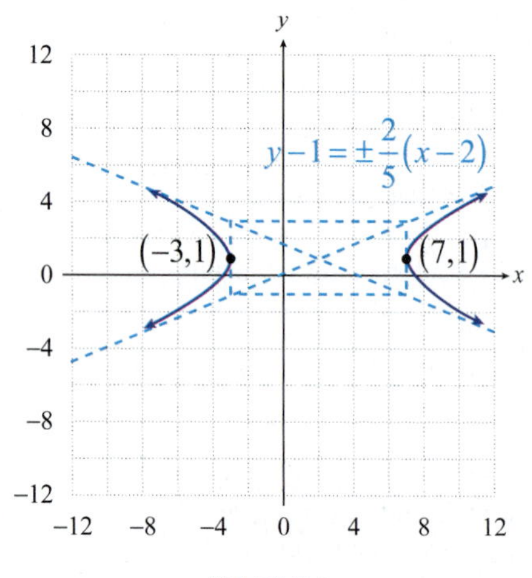

FIGURE 3

Example 2: Graphing a Hyperbola with Equation Not in Standard Form

> **NOTE**
>
> Even though it is not in standard form, we know this equation represents a hyperbola because the product of the coefficients of x^2 and y^2 is negative.

Graph the hyperbola $-16x^2+9y^2+96x+18y=279$, indicating its asymptotes.

Solution

In its current form, the equation tells us nothing about the graph of the hyperbola it represents. We need to complete the square (twice) to find the standard form.

$$-16x^2+9y^2+96x+18y=279$$

$$-16\left(x^2-6x\right)+9\left(y^2+2y\right)=279$$

Factor to obtain a coefficient of 1 on each of the squared terms.

$$-16\left(x^2-6x+9\right)+9\left(y^2+2y+1\right)=279-144+9$$

$$-16(x-3)^2+9(y+1)^2=144$$

Add the needed constant to complete each perfect square trinomial, and compensate by adding the appropriate numbers to the right-hand side as well.

$$\frac{(y+1)^2}{16}-\frac{(x-3)^2}{9}=1$$

Finally, divide by 144 to arrive at the standard form.

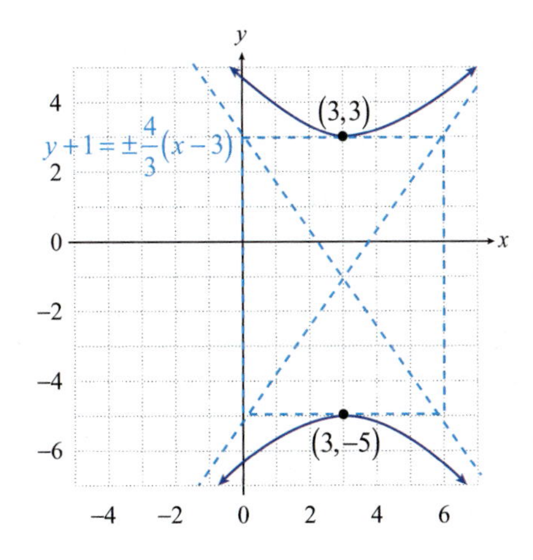

FIGURE 4

With the equation in standard form, we now know that the foci of the hyperbola are aligned vertically (since the positive fraction on the left is in the variable y), that $a=4$ and $b=3$, and that the center of the hyperbola is at $(3,-1)$. This tells us that the two vertices of the hyperbola must lie at

$$(3,-1-4)=(3,-5) \quad \text{and} \quad (3,-1+4)=(3,3).$$

Using $c^2=a^2+b^2=25$, we know that $c=5$, so the two foci are at

$$(3,-1-5)=(3,-6) \quad \text{and} \quad (3,-1+5)=(3,4).$$

For this example, we will use a shortcut to graph the asymptotes. We know the asymptotes pass through the center of the hyperbola, and we know that one has a slope of $\frac{4}{3}$ and the other a slope of $-\frac{4}{3}$. The rectangle drawn in Figure 4 is centered at $(3,-1)$, with the bottom and top edges 4 units away from the center and the left and right edges 3 units away. The asymptotes are the diagonals of this rectangle.

TECHNOLOGY: Graphing a Vertically Oriented Hyperbola

To graph the hyperbola in Example 2 using a TI-84 Plus, first press apps. Select Conics, then select HYPERBOLA. Now select 2: to enter the values from the equation in standard form, $\frac{(y+1)^2}{16}-\frac{(x-3)^2}{9}=1$, as shown in Figure 5. To view the graph, press graph. See Figure 6.

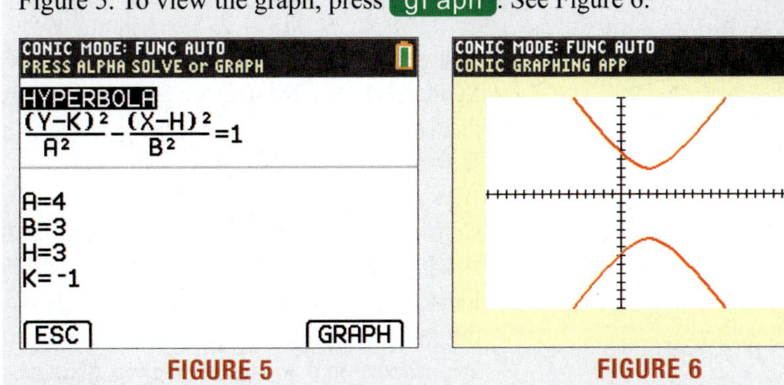

FIGURE 5 FIGURE 6

Example 3: Standard Form of the Equation of a Hyperbola

Given that the asymptotes of a hyperbola have slopes of $\frac{1}{2}$ and $-\frac{1}{2}$ and that the vertices of the hyperbola are at $(-1,0)$ and $(7,0)$, find its standard form equation.

Solution

The fact that the vertices (and hence the foci) are aligned horizontally tells us that we need to construct an equation of the form

$$\frac{(x-h)^2}{a^2}-\frac{(y-k)^2}{b^2}=1.$$

We need to determine h, k, a, and b. Since the vertices are 8 units apart, we know that $2a=8$, or $a=4$. We also know the center lies halfway between the vertices, at the point $(3,0)$. All that remains is to determine b.

Since the foci are aligned horizontally, we know the asymptotes are of the form

$$y-k=\frac{b}{a}(x-h) \text{ and } y-k=-\frac{b}{a}(x-h),$$

and we already know a, h, and k. Specifically, we know that the two given slopes must correspond to the fractions $\frac{b}{a}$ and $-\frac{b}{a}$. That is, $\frac{1}{2}=\frac{b}{a}$, so $\frac{1}{2}=\frac{b}{4}$, and so $b=2$. This means that the equation of the hyperbola is

$$\frac{(x-3)^2}{16}-\frac{y^2}{4}=1,$$

and the equations for the asymptotes are

$$y=\pm\frac{1}{2}(x-3).$$

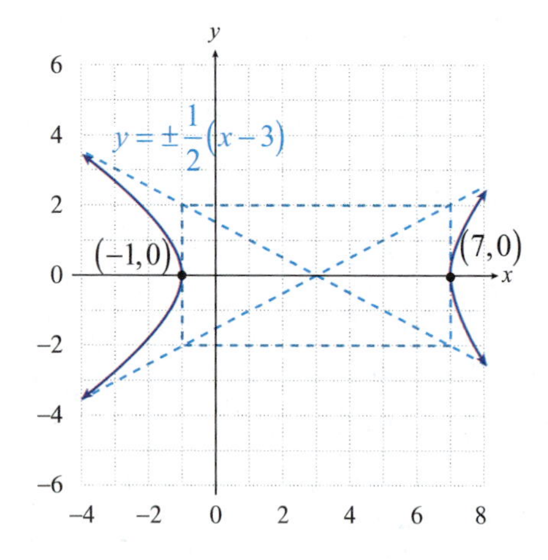

FIGURE 7

A sketch of the hyperbola, along with its asymptotes, appears in Figure 7.

TOPIC 2: Applications of Hyperbolas

Hyperbolas, like ellipses and parabolas, arise in a wide variety of contexts. Hyperbolas are seen in architecture and structural engineering (for example, the shape of a nuclear power plant's cooling towers) and in astronomy (comets that make a single pass through our solar system don't have elliptical orbits, but instead trace one branch of a hyperbola).

One important application of hyperbolas concerns guidance systems, such as LORAN (**LO**ng **RA**nge **N**avigation). LORAN is a radio-communication system that can be used to determine the location of a ship at sea, and the basis of LORAN is an understanding of hyperbolic curves.

Consider a situation in which two land-based radio transmitters, located at sites A and B in Figure 8, send out a signal simultaneously. A receiver on a ship, located at C, would receive the two signals at slightly different times due to the difference in the distances the signals must travel. Since the times for signal travel are proportional to the respective distances d_1 and d_2, the difference in time between receipt of the two signals is proportional to $|d_1 - d_2|$. In other words, a person on the ship can determine $|d_1 - d_2|$ by measuring the time difference in receiving the two signals.

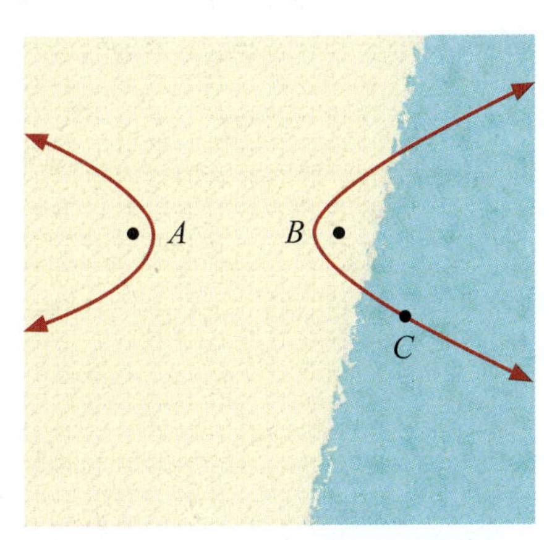

FIGURE 8: Possible Ship Locations

As Figure 8 indicates, knowing $|d_1 - d_2|$ and the locations of the radio transmitters defines a hyperbola of possible locations for the ship. This hyperbola provides a convenient means of navigation; by maintaining the same time difference between the signals, the captain ensures that the ship stays on a known hyperbolic path. Often LORAN stations are placed very close to the shore; the resulting hyperbolas have their vertices on the shore as well. Thus, different time differences of the signals correspond to different landing locations on the shore.

Example 4: LORAN

A ship measures a time difference of 0.000108 seconds between the signals of two LORAN signals sent from stations 100 miles apart along a coastline. If the ship maintains this time difference, where will it land on the coastline? Assume that the signal moves at the speed of light $(186,000$ miles per second$)$.

NOTE

As usual, setting up a convenient system of coordinates can make the resulting equations simpler.

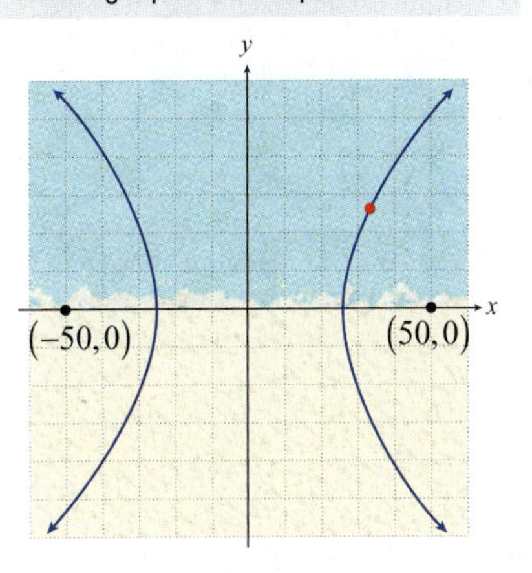

FIGURE 9

Solution

Begin by graphing the situation in the Cartesian plane. We place the two stations (which are the foci of the hyperbola) at $(-50,0)$ and $(50,0)$. The ship must lie on the hyperbola (and hopefully is in the water).

Note that the hyperbola intersects the shore at its vertices. Thus, if the ship maintains its path on the hyperbola, it will land at a vertex. We use the time difference to calculate the position of the vertex using the rate equation $d = rt$.

Substituting $r = 186,000\dfrac{\text{miles}}{\text{second}}$ and $t = 0.000108$ seconds, we have

$$d = \left(186,000\frac{\text{miles}}{\text{second}}\right)(0.000108 \text{ seconds}) \approx 20 \text{ miles}.$$

Recall from before that this distance $|d_1 - d_2| = 2a$, where a is the distance of the vertex from the center of the hyperbola. Thus, we know that the ship will reach shore 10 miles from the origin of our coordinate system.

Whether the ship is heading towards $(-10,0)$ or $(10,0)$ depends on which signal reached the ship's receiver first.

LORAN can actually determine the location of the ship by performing the same computations for another pair of simultaneous signals sent out from two additional transmitters, located at A' and B'. This defines a second hyperbola, and the ship must be at a point where the two hyperbolas intersect, as shown in Figure 10.

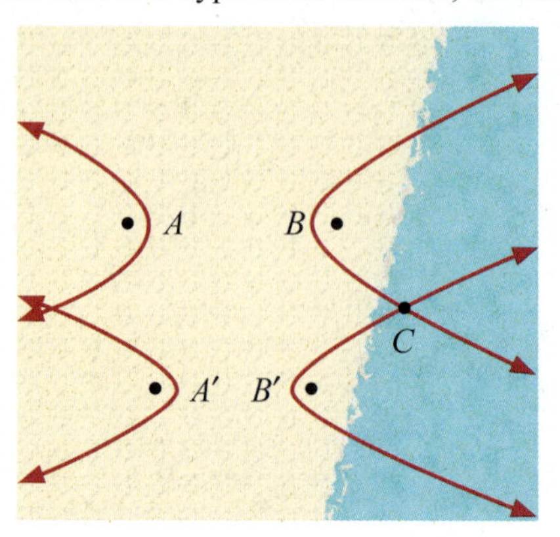

FIGURE 10: Two Sets of Transmitters

10.3 EXERCISES

PRACTICE

Sketch the graphs of the following hyperbolas, using asymptotes as guides. Determine the coordinates of the foci in each case. See Examples 1 and 2.

1. $\dfrac{(x+3)^2}{4}-\dfrac{(y+1)^2}{9}=1$
2. $\dfrac{(y-2)^2}{25}-\dfrac{(x+2)^2}{9}=1$
3. $4y^2-x^2-24y+2x=-19$
4. $x^2-9y^2+4x+18y-14=0$
5. $9x^2-25y^2=18x-50y+241$
6. $9x^2-16y^2+116=36x+64y$
7. $\dfrac{x^2}{16}-\dfrac{(y-2)^2}{4}=1$
8. $\dfrac{(y-1)^2}{9}-(x+3)^2=1$
9. $9y^2-25x^2-36y-100x=289$
10. $9x^2+18x=4y^2+27$
11. $9x^2-16y^2-36x+32y-124=0$
12. $x^2-y^2+6x-6y=4$
13. $\dfrac{(y-2)^2}{64}-\dfrac{(x+7)^2}{49}=1$
14. $\dfrac{(y-4)^2}{49}-\dfrac{(x+2)^2}{16}=1$
15. $\dfrac{(x+1)^2}{64}-\dfrac{(y+7)^2}{4}=1$
16. $\dfrac{(x+10)^2}{16}-\dfrac{(y+8)^2}{25}=1$

Find the center, foci, and vertices of the hyperbola that each equation describes.

17. $\dfrac{(x+3)^2}{4}-\dfrac{(y-2)^2}{9}=1$
18. $\dfrac{(y-2)^2}{16}-\dfrac{(x+1)^2}{9}=1$
19. $3(x-1)^2-(y+4)^2=9$
20. $(y-2)^2-2(x-4)^2=4$
21. $(x+2)^2-5(y-1)^2=25$
22. $6(y+2)^2-(x+1)^2=12$

23. $2x^2+12x-y^2-2y+9=0$

24. $y^2-9x^2+6y+72x-144=0$

25. $x^2-4y^2-2x=0$

26. $4y^2-x^2+32y+2x+47=0$

27. $4x^2-y^2-64x+10y+167=0$

28. $4x^2-9y^2-36y-72=0$

Match the following equations to their graphs.

29. $\frac{x^2}{9}-y^2=1$

30. $y^2-\frac{x^2}{4}=1$

31. $x^2-\frac{(y-3)^2}{4}=1$

32. $\frac{(x-3)^2}{4}-\frac{(y+1)^2}{9}=1$

33. $(y+2)^2-\frac{(x-2)^2}{4}=1$

34. $\frac{x^2}{9}-\frac{(y+2)^2}{4}=1$

35. $\frac{y^2}{4}-(x-1)^2=1$

36. $x^2-y^2=1$

a.

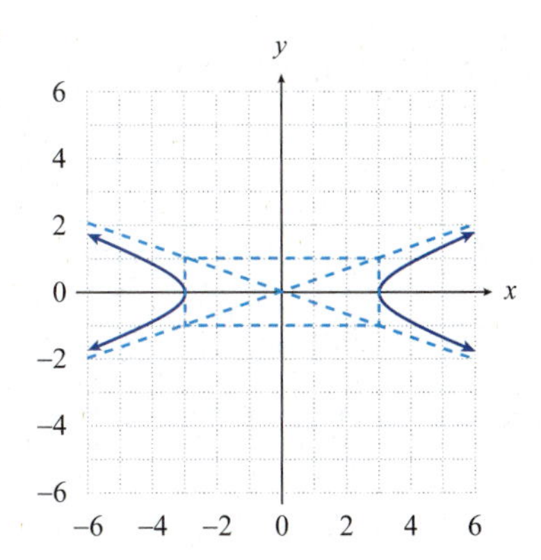

b.

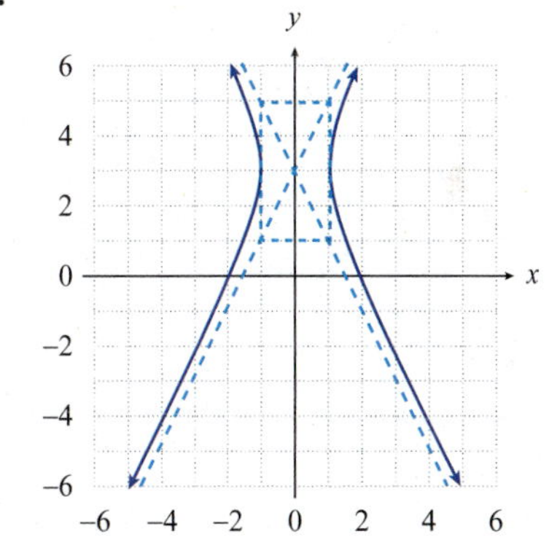

c.

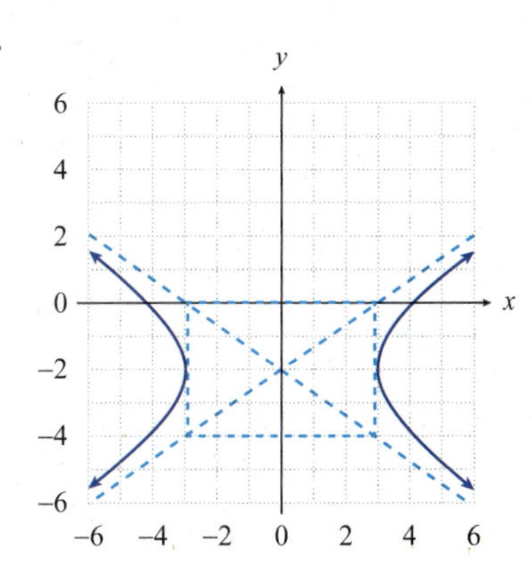

d.

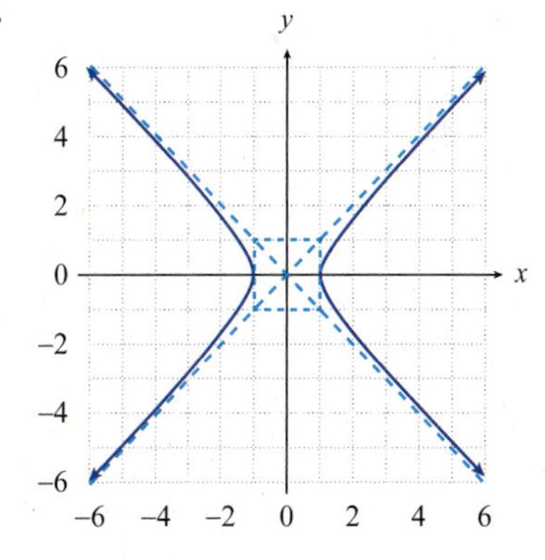

e.

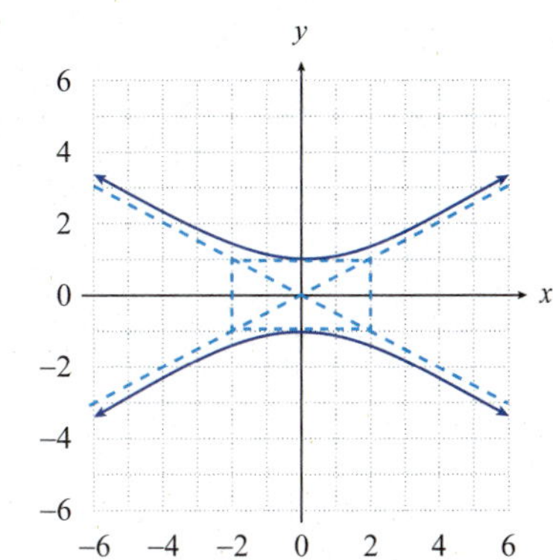

f.

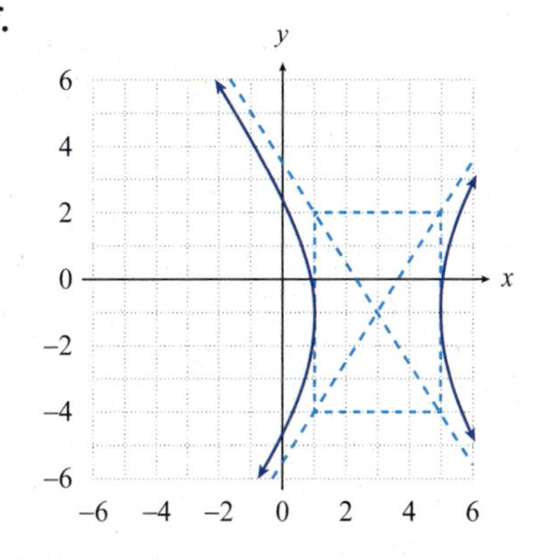

g.

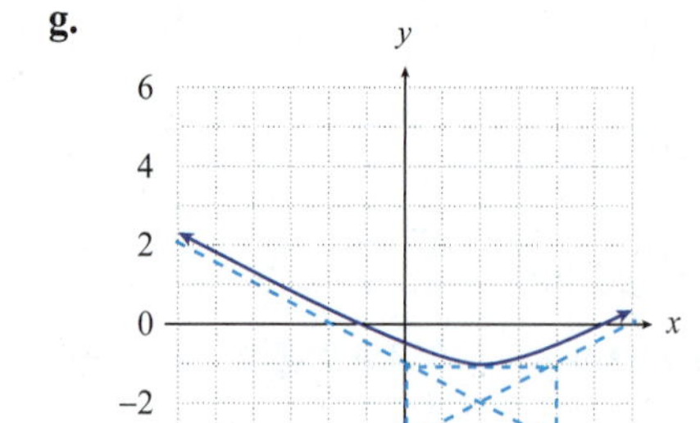

h.

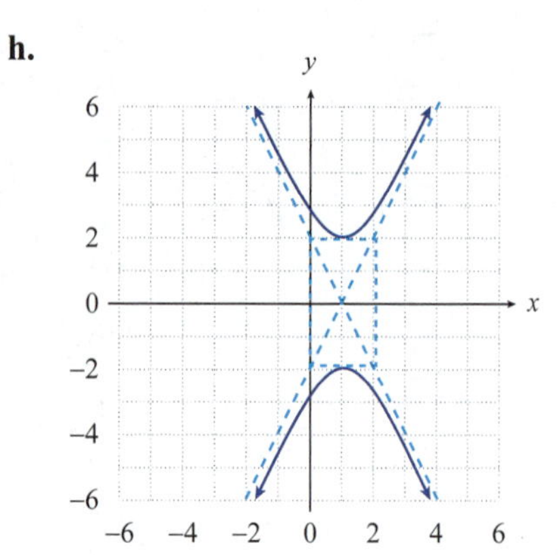

Find the equation, in standard form, for the hyperbola with the given properties or with the given graph. See Example 3.

37. Foci at $(-3,0)$ and $(3,0)$ and vertices at $(-2,0)$ and $(2,0)$.

38. Foci at $(1,5)$ and $(1,-1)$ and vertices at $(1,3)$ and $(1,1)$.

39. Asymptotes of $y=\pm 2x$ and vertices at $(0,-1)$ and $(0,1)$.

40. Asymptotes of $y=\pm(x-2)+1$ and vertices at $(-1,1)$ and $(5,1)$.

41. Foci at $(2,4)$ and $(-2,4)$ and asymptotes of $y=\pm 3x+4$.

42. Foci at $(-1,3)$ and $(-1,-1)$ and asymptotes of $y=\pm(x+1)+1$.

43. Foci at $(2,5)$ and $(10,5)$ and vertices at $(3,5)$ and $(9,5)$.

44. Foci at $(7,4)$ and $(7,-4)$ and vertices at $(7,1)$ and $(7,-1)$.

45. Asymptotes of $y=\pm(2x+8)+3$ and vertices at $(-6,3)$ and $(-2,3)$.

46. Asymptotes of $y=\pm\frac{4}{3}x-3$ and vertices at $(0,-7)$ and $(0,1)$.

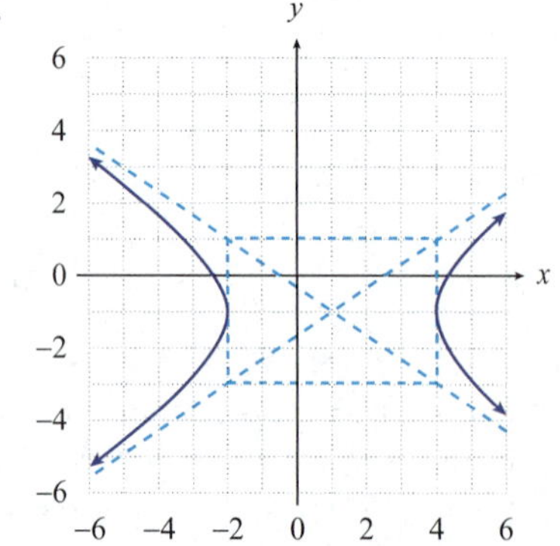

48.

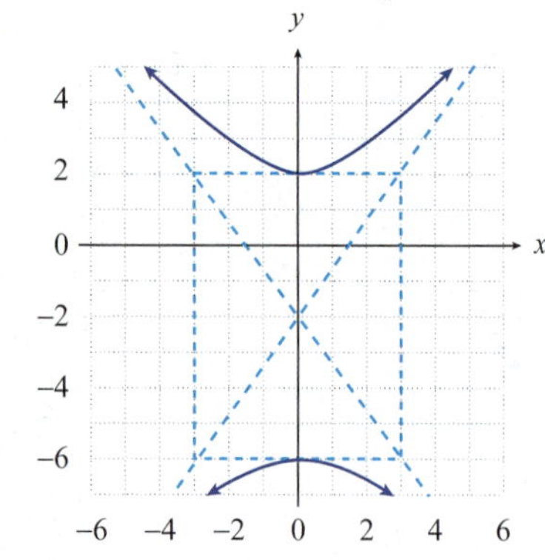

49.

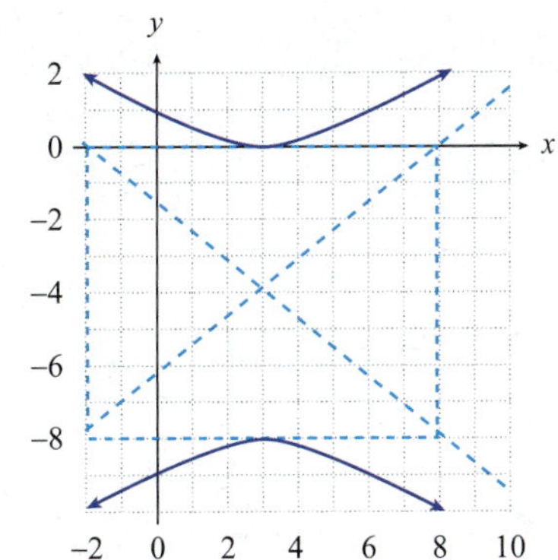

50.

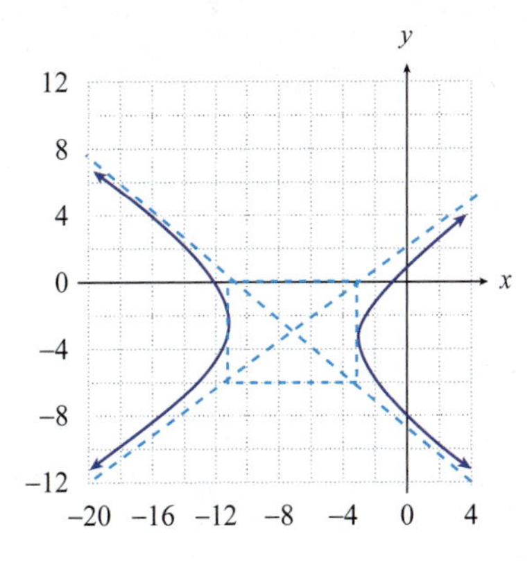

APPLICATIONS

51. As mentioned in this section, some comets trace one branch of a hyperbola through the solar system, with the sun at one focus. Suppose a comet is spotted that appears to be headed straight for Earth, as shown in the figure. As the comet gets closer, however, it becomes apparent that it will pass between Earth, which lies at the center of the hyperbolic path of the comet, and the sun. In the end, the closest the comet comes to Earth is 60,000,000 miles. Using an estimate of 94,000,000 miles for the distance from Earth to the sun, and positioning Earth at the origin of a coordinate system, find the equation for the path of the comet.

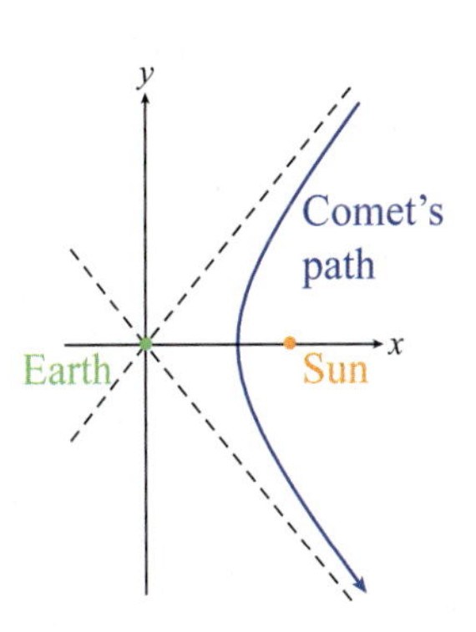

52. Suppose two LORAN radio transmitters are 26 miles apart. A ship at sea receives signals sent simultaneously from the two transmitters and is able to determine that the difference between the distances from the ship to each of the transmitters is 24 miles. By positioning the two transmitters on the y-axis, each 13 miles from the origin, find the equation for the hyperbola that describes the set of possible locations for the ship.

TECHNOLOGY

Use a graphing utility to graph the following equations.

53. $x^2 - 6y^2 = 15$

54. $4y^2 - 9x^2 = 18$

55. $3y^2 - 18x^2 = 36$

56. $x^2 - 6 = 3y^2$

57. $(y+2)^2 - 20 = 5x^2$

58. $(x+5)^2 = 3(y-2)^2 + 15$

59. $x^2 - 2y^2 = 4x + 12y + 26$

60. $2x^2 - y^2 + 12x + 2y = 3$

61. $5x^2 - y^2 + 20x = 10y + 25$

62. $x^2 - 5y^2 = 14x + 20y - 4$

10.4 ROTATION OF CONIC SECTIONS

TOPICS

1. Rotation relations
2. Rotation invariants

TOPIC 1: Rotation Relations

In the first three sections of this chapter, we studied equations of the form

$$Ax^2 + Cy^2 + Dx + Ey + F = 0.$$

The choice of letters for the coefficients is traditional, and certainly hints at a missing term. As mentioned briefly in Section 10.1, the most general form of a conic in Cartesian coordinates is

$$Ax^2 + Bxy + Cy^2 + Dx + Ey + F = 0,$$

and we are now ready to explore the consequences of adding a nonzero *xy*-term to the equation.

Geometrically, the condition $B \neq 0$ leads to a conic that is not oriented vertically or horizontally. The ellipses, parabolas, and hyperbolas graphed so far have all been aligned so that the axis (or axes in the case of ellipses and hyperbolas) of symmetry is parallel to the *x*-axis or *y*-axis. If $B \neq 0$, the graph is a conic section that has been rotated through some angle θ. We can exploit this fact to make the graphing of a rotated conic section relatively easy. Our method will be to introduce a new set of coordinate axes (i.e., a second Cartesian plane) rotated by an acute angle θ with respect to the original coordinate axes, and then to graph the conic in the new coordinate system. Algebraically, the goal is to begin with an equation of the form

$$Ax^2 + Bxy + Cy^2 + Dx + Ey + F = 0$$

and define a new set of coordinate axes x' and y' in which the equation has the form

$$A'x'^2 + C'y'^2 + D'x' + E'y' + F' = 0.$$

That is, the coefficient B' in the new coordinate system is 0, and hence the graphing techniques of the first three sections of this chapter apply.

We begin with a picture. Figure 1 is an illustration of two rectangular coordinate systems, with the new system rotated by an acute angle θ with respect to the original.

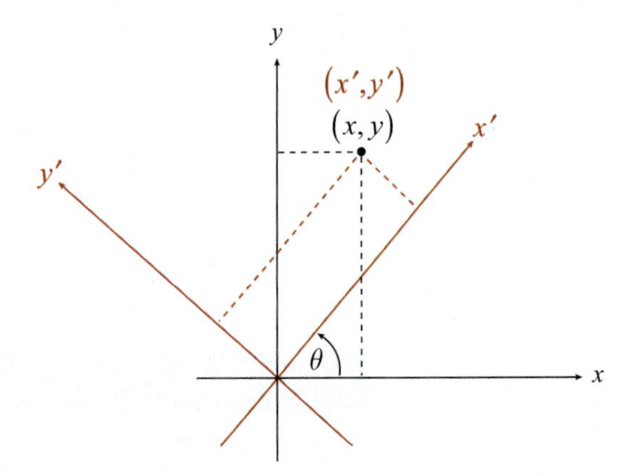

FIGURE 1: Two Coordinate Systems

The point in Figure 1 has two sets of coordinates, corresponding to the two coordinate planes. However, the distance r between the origin and the point is the same in both, and this fact and the introduction of the angle θ' shown in Figure 2 allow us to begin to relate the two sets of coordinates.

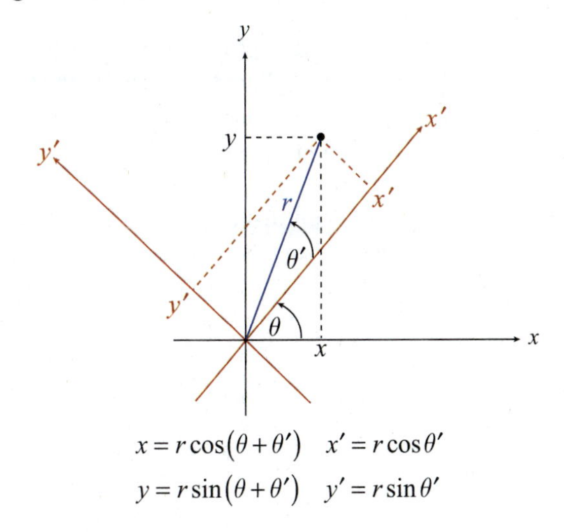

$$x = r\cos(\theta + \theta') \quad x' = r\cos\theta'$$
$$y = r\sin(\theta + \theta') \quad y' = r\sin\theta'$$

FIGURE 2: Relation between $x', y', x,$ and y

We can now apply one of the trigonometric identities from Chapter 8.

$$\begin{aligned} x &= r\cos(\theta + \theta') \\ &= r(\cos\theta\cos\theta' - \sin\theta\sin\theta') \\ &= r\cos\theta'\cos\theta - r\sin\theta'\sin\theta \\ &= x'\cos\theta - y'\sin\theta \end{aligned}$$

Similarly, $y = x'\sin\theta + y'\cos\theta$. We will also need to be able to express x' and y' in terms of x and y, and we can do so through clever application of another trigonometric identity. By multiplying the two equations we have just derived by $\cos\theta$ and $\sin\theta$, respectively, and then adding the results, we obtain the following.

$$\begin{array}{r} \left(x\cos\theta = x'\cos^2\theta - y'\sin\theta\cos\theta\right) \\ +\left(y\sin\theta = x'\sin^2\theta + y'\sin\theta\cos\theta\right) \\ \hline x\cos\theta + y\sin\theta = x'\left(\cos^2\theta + \sin^2\theta\right) \\ = x' \end{array}$$

A similar manipulation allows us to express y' in terms of x and y. All four relations are summarized in the following definition.

FORMULA: Rotation Relations

Given a rectangular coordinate system with axes x' and y' rotated by angle θ with respect to axes x and y, as in Figure 1, the two sets of coordinates (x', y') and (x, y) for the same point are related by the following.

$$x = x'\cos\theta - y'\sin\theta \qquad x' = x\cos\theta + y\sin\theta$$
$$y = x'\sin\theta + y'\cos\theta \qquad y' = -x\sin\theta + y\cos\theta$$

Example 1: Finding $x'y'$-Coordinates

Given that $\theta = \frac{\pi}{6}$, find the $x'y'$-coordinates of the point with xy-coordinates $(-1,5)$.

Solution

Using the rotation relations, we have that

$$\begin{aligned} x' &= (-1)\cos\frac{\pi}{6} + 5\sin\frac{\pi}{6} \\ &= (-1)\left(\frac{\sqrt{3}}{2}\right) + (5)\left(\frac{1}{2}\right) \\ &= \frac{5-\sqrt{3}}{2} \end{aligned}$$

and

$$\begin{aligned} y' &= -(-1)\sin\frac{\pi}{6} + 5\cos\frac{\pi}{6} \\ &= \left(\frac{1}{2}\right) + (5)\left(\frac{\sqrt{3}}{2}\right) \\ &= \frac{1+5\sqrt{3}}{2}. \end{aligned}$$

The $x'y'$-coordinates are thus $\left(\frac{5-\sqrt{3}}{2}, \frac{1+5\sqrt{3}}{2}\right)$.

We are familiar with the graph of $y = \frac{1}{x}$ from our work in Chapters 3, 4, and 5. The next example gives us another perspective on this equation, written in the form $xy = 1$.

Example 2: Using the Rotation Relations

Use the rotation $\theta = 45°$ to show that the graph of $xy = 1$ is a hyperbola.

Solution

Using the angle $\theta = 45°$ in the rotation relations, we convert the equation as follows.

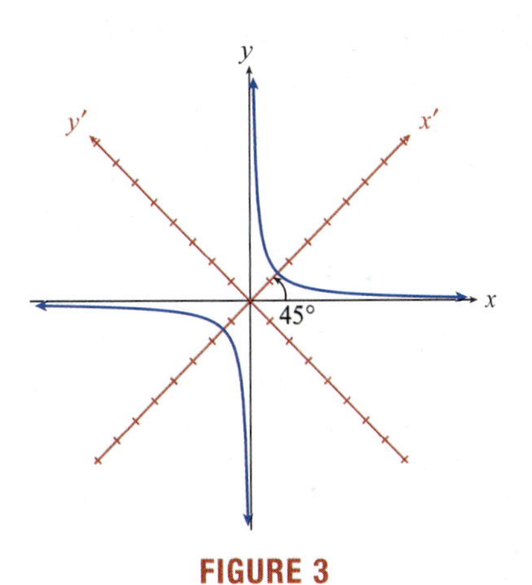

FIGURE 3

$$
\begin{aligned}
xy &= 1 \\
(x'\cos 45° - y'\sin 45°)(x'\sin 45° + y'\cos 45°) &= 1 \\
\left(\frac{x'}{\sqrt{2}} - \frac{y'}{\sqrt{2}}\right)\left(\frac{x'}{\sqrt{2}} + \frac{y'}{\sqrt{2}}\right) &= 1 \\
\frac{x'^2}{2} - \frac{x'y'}{2} + \frac{x'y'}{2} - \frac{y'^2}{2} &= 1 \\
\frac{x'^2}{2} - \frac{y'^2}{2} &= 1
\end{aligned}
$$

We recognize this last equation as a hyperbola in the $x'y'$-plane, with center at the origin and vertices $\sqrt{2}$ away. The asymptotes are $y' = \pm x'$, which correspond to the x- and y-axes.

Remember that the goal in general is to determine θ so that the conversion of the equation

$$Ax^2 + Bxy + Cy^2 + Dx + Ey + F = 0$$

has no $x'y'$-term. Example 2 will serve as the inspiration; by replacing x and y with the corresponding x' and y' expressions and simplifying the result, we can derive a formula for the appropriate angle θ. We begin with the following replacements.

$$
\begin{aligned}
&A(x'\cos\theta - y'\sin\theta)^2 + B(x'\cos\theta - y'\sin\theta)(x'\sin\theta + y'\cos\theta) \\
&+C(x'\sin\theta + y'\cos\theta)^2 + D(x'\cos\theta - y'\sin\theta) \\
&+E(x'\sin\theta + y'\cos\theta) + F = 0
\end{aligned}
$$

When the left-hand side of this equation is expanded and like terms collected, the result is an equation of the form

$$A'x'^2 + B'x'y' + C'y'^2 + D'x' + E'y' + F' = 0,$$

where

$$A' = A\cos^2\theta + B\cos\theta\sin\theta + C\sin^2\theta$$
$$B' = 2(C - A)\cos\theta\sin\theta + B(\cos^2\theta - \sin^2\theta)$$
$$C' = A\sin^2\theta - B\cos\theta\sin\theta + C\cos^2\theta$$
$$D' = D\cos\theta + E\sin\theta$$
$$E' = -D\sin\theta + E\cos\theta$$
$$F' = F.$$

Since we want $B' = 0$, this gives us an equation in θ to solve. To do so, we will use the double-angle formulas for both sine and cosine in reverse.

$$\begin{aligned} 2(C - A)\cos\theta\sin\theta + B(\cos^2\theta - \sin^2\theta) &= 0 \\ (C - A)\sin(2\theta) + B\cos(2\theta) &= 0 \\ B\cos(2\theta) &= (A - C)\sin(2\theta) \\ \frac{\cos(2\theta)}{\sin(2\theta)} &= \frac{A - C}{B} \\ \cot(2\theta) &= \frac{A - C}{B} \end{aligned}$$

This result is summarized in the following theorem.

THEOREM: Elimination of the *xy*-Term

The graph of the equation $Ax^2 + Bxy + Cy^2 + Dx + Ey + F = 0$ in the *xy*-plane is the same as the graph of the equation $A'x'^2 + C'y'^2 + D'x' + E'y' + F' = 0$ in the $x'y'$-plane, where the angle of rotation θ between the two coordinate systems satisfies the equation

$$\cot(2\theta) = \frac{A - C}{B}.$$

Although formulas relating the primed coefficients to the unprimed coefficients were derived in the preceding discussion, in practice it is easier to simply determine θ and use the rotation relations to convert equations, as shown in this next example.

Example 3: Eliminating the *xy*-Term

Graph the conic section $x^2 + 2\sqrt{3}xy + 3y^2 + \sqrt{3}x - y = 0$ by first determining the appropriate angle θ by which to rotate the axes.

Solution

Using the formula for eliminating the *xy*-term,

$$\cot(2\theta) = \frac{1 - 3}{2\sqrt{3}} = -\frac{1}{\sqrt{3}}.$$

Since the angle θ is to be acute, it must be the case that $2\theta = \frac{2\pi}{3}$ and hence $\theta = \frac{\pi}{3}$. By the rotation relations, then,

$$\begin{aligned} x &= x'\cos\left(\frac{\pi}{3}\right) - y'\sin\left(\frac{\pi}{3}\right) \\ &= \frac{1}{2}x' - \frac{\sqrt{3}}{2}y' \end{aligned}$$

and

$$\begin{aligned} y &= x'\sin\left(\frac{\pi}{3}\right) + y'\cos\left(\frac{\pi}{3}\right) \\ &= \frac{\sqrt{3}}{2}x' + \frac{1}{2}y'. \end{aligned}$$

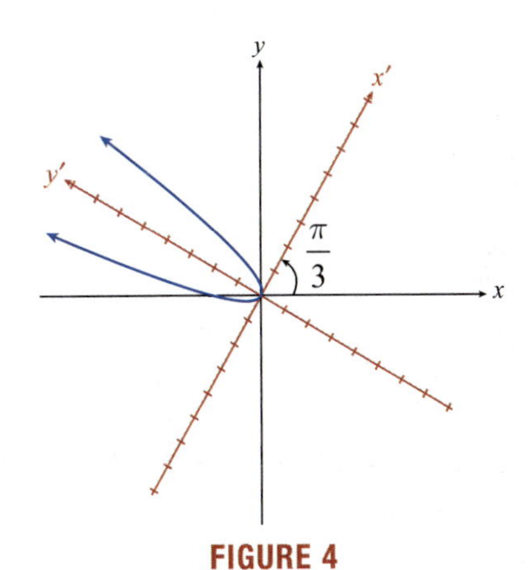

FIGURE 4

Making these substitutions into the equation, we obtain the following.

$$\left(\frac{1}{2}x' - \frac{\sqrt{3}}{2}y'\right)^2 + 2\sqrt{3}\left(\frac{1}{2}x' - \frac{\sqrt{3}}{2}y'\right)\left(\frac{\sqrt{3}}{2}x' + \frac{1}{2}y'\right) + 3\left(\frac{\sqrt{3}}{2}x' + \frac{1}{2}y'\right)^2$$
$$+\sqrt{3}\left(\frac{1}{2}x' - \frac{\sqrt{3}}{2}y'\right) - \left(\frac{\sqrt{3}}{2}x' + \frac{1}{2}y'\right) = 0$$

Multiplying out and collecting like terms in this equation results in

$$y' = 2x'^2,$$

a much simpler equation. We recognize this as a parabola with vertex at the origin. The graph of this equation in the $x'y'$-plane is shown in Figure 4.

TOPIC 2: Rotation Invariants

The relationships we derived between A, B, C, D, E, F and A', B', C', D', E', and F' lead to some interesting observations, one of which we have an immediate use for. Since $F' = F$, we say the constant term is *invariant* under rotation. Slightly less obviously, $A' + C' = A + C$, so $A + C$ is also said to be an invariant.

Less obvious still is the very important fact that $B'^2 - 4A'C' = B^2 - 4AC$. This invariant is called the **discriminant** of the conic section, and its sign (except in the case of degenerate conics) identifies the graph of the equation as an ellipse, a parabola, or a hyperbola. The discriminant thus generalizes the role that the product AC played in Sections 10.1, 10.2, and 10.3.

DEFINITION: Classifying Conics

Assuming the graph of the equation $Ax^2 + Bxy + Cy^2 + Dx + Ey + F = 0$ is a nondegenerate conic section, it is classified by its discriminant as follows:

1. **Ellipse** if $B^2 - 4AC < 0$
2. **Parabola** if $B^2 - 4AC = 0$
3. **Hyperbola** if $B^2 - 4AC > 0$

Example 4: Classifying and Graphing Conic Sections

Classify the conic section $13x^2+10xy+13y^2+42\sqrt{2}x-6\sqrt{2}y+18=0$ and then sketch its graph.

Solution

The discriminant is $10^2-4(13)(13)=-576$. Since this result is negative, the conic section is an ellipse. The next step is to determine the rotation angle θ.

$$\cot(2\theta)=\frac{13-13}{10}=0$$
$$2\theta=\frac{\pi}{2}$$
$$\theta=\frac{\pi}{4}$$

The rotation relations are thus

$$x=x'\cos\left(\frac{\pi}{4}\right)-y'\sin\left(\frac{\pi}{4}\right)$$
$$=\frac{x'}{\sqrt{2}}-\frac{y'}{\sqrt{2}}$$

and

$$y=x'\sin\left(\frac{\pi}{4}\right)+y'\cos\left(\frac{\pi}{4}\right)$$
$$=\frac{x'}{\sqrt{2}}+\frac{y'}{\sqrt{2}}.$$

Making these substitutions in the original equation gives us

$$13\left(\frac{x'}{\sqrt{2}}-\frac{y'}{\sqrt{2}}\right)^2+10\left(\frac{x'}{\sqrt{2}}-\frac{y'}{\sqrt{2}}\right)\left(\frac{x'}{\sqrt{2}}+\frac{y'}{\sqrt{2}}\right)+13\left(\frac{x'}{\sqrt{2}}+\frac{y'}{\sqrt{2}}\right)^2$$
$$+42\sqrt{2}\left(\frac{x'}{\sqrt{2}}-\frac{y'}{\sqrt{2}}\right)-6\sqrt{2}\left(\frac{x'}{\sqrt{2}}+\frac{y'}{\sqrt{2}}\right)+18=0,$$

which, when multiplied out and simplified, reduces to

$$9x'^2+4y'^2+18x'-24y'+9=0.$$

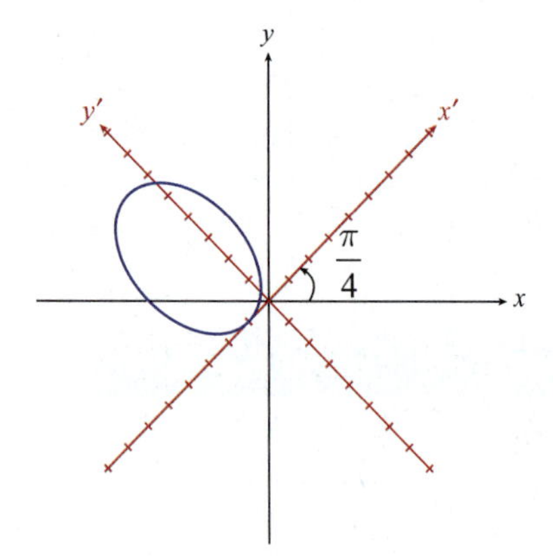

FIGURE 5

In order to easily graph this ellipse in the $x'y'$-plane, we follow the usual completing-the-square process.

$$9\left(x'^2+2x'\right)+4\left(y'^2-6y'\right)=-9$$
$$9\left(x'^2+2x'+1\right)+4\left(y'^2-6y'+9\right)=-9+9+36$$
$$9\left(x'+1\right)^2+4\left(y'-3\right)^2=36$$
$$\frac{\left(x'+1\right)^2}{4}+\frac{\left(y'-3\right)^2}{9}=1$$

We can now easily graph this ellipse whose center, in the $x'y'$-plane, is $(-1,3)$ and whose minor and major axes have lengths 4 and 6, respectively.

10.4 EXERCISES

PRACTICE

Find the $x'y'$-coordinates of each point for the given rotation angle θ. See Example 1.

1. $(8,6),\quad \theta=30°$
2. $(-5,1),\quad \theta=\frac{\pi}{3}$
3. $\left(\frac{-1}{2},\frac{-1}{8}\right),\quad \theta=\frac{\pi}{4}$
4. $(2.7,5),\quad \theta=15°$
5. $(13,-4),\quad \theta=78°$
6. $\left(\frac{12}{\sqrt{18}},\frac{240}{\sqrt{1152}}\right),\quad \theta=45°$
7. $\left(3.65,\frac{3}{8}\right),\quad \theta=\frac{\pi}{6}$
8. $\left(\frac{3+\sqrt{48}}{-\left(2\sqrt{12}+3\right)},\frac{\sqrt{4096}}{8\sqrt{25}\cdot\frac{1}{5}\sqrt{64}}\right),\quad \theta=\frac{\pi}{2}$

Use the discriminant to determine whether the equation of the given conic represents an ellipse, a parabola, or a hyperbola. See Example 4.

9. $2x^2-3xy+2y^2-2x=0$
10. $3x^2+7xy+5y^2-6x+7y+15=0$
11. $3x^2+8xy+4y^2-7=0$
12. $5x^2+6xy-3y^2-9=0$
13. $-2x^2-8xy+2y^2+2y+5=0$
14. $3x^2-6xy+3y^2+3x-9=0$
15. $x^2-xy+4y^2+2x-3y+1=0$
16. $x^2-4xy+4y^2+2x+3y-1=0$

Use the discriminant to classify each of the following conic sections. Then determine the angle θ that will allow you to convert the equation and eliminate the *xy*-term. Finally, sketch the graph of the conic section. See Examples 2, 3, and 4.

17. $xy=2$
18. $xy-4=0$
19. $x^2+2xy+y^2-x+y=0$
20. $7x^2+5\sqrt{3}xy+2y^2=14$
21. $22x^2+6\sqrt{3}xy+16y^2-49=276$
22. $2\sqrt{3}x^2-6xy+\sqrt{3}x-9y=0$
23. $34x^2+8\sqrt{3}xy+42y^2=1380$
24. $xy+x-4y=6$

Sketch the graphs of the following conic sections. See Examples 2, 3, and 4.

25. $x^2+6xy+y^2=18$
26. $x^2-4xy+3y^2=12$
27. $9x^2+14xy-9y^2=15$
28. $36x^2-19xy+8y^2=72$
29. $40x^2+20xy+10y^2+\left(2\sqrt{2}-6\right)x-\left(4\sqrt{2}+8\right)y=90$
30. $72x^2+19xy+4y^2=20$
31. $48x^2+15xy+7y^2=28$
32. $72x^2+18xy-9y^2=14$

Match the equation with its corresponding graph.

33. $3x^2 + 2xy + y^2 - 10 = 0$

34. $x^2 - 4xy + 4y^2 + 5\sqrt{5}y + 1 = 0$

35. $xy - 1 = 0$

36. $x^2 + y^2 - 16x + 39 = 0$

37. $x^2 - y^2 - 16 = 0$

38. $3x^2 + 8xy + 4y^2 - 7 = 0$

39. $4xy - 9 = 0$

40. $x^2 - 6xy + 9y^2 - 2y + 1 = 0$

a.

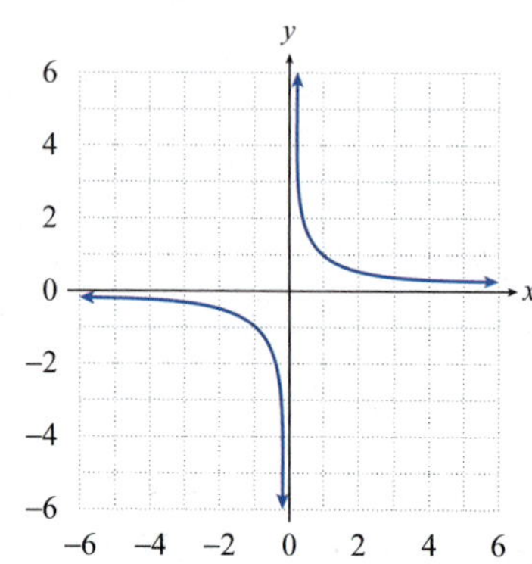

b.

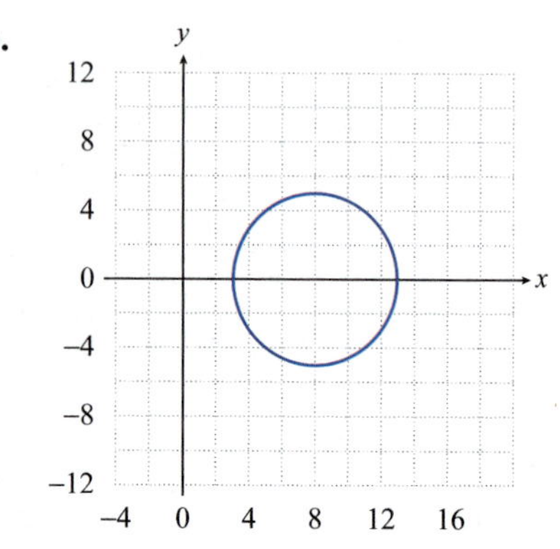

c.

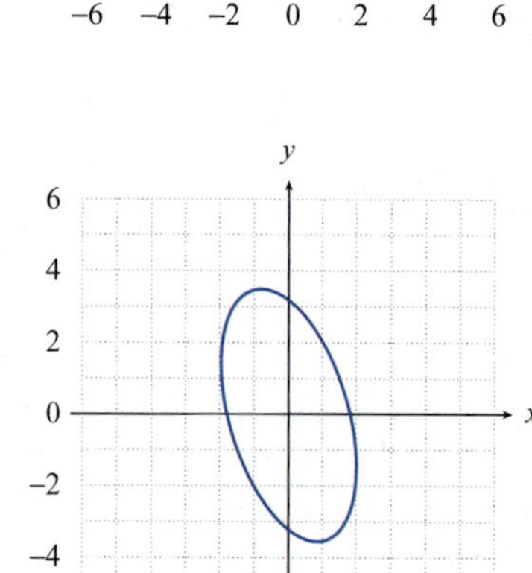

d.

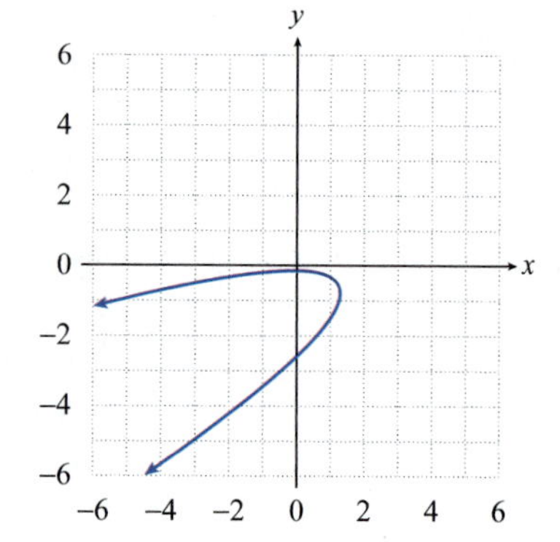

e.

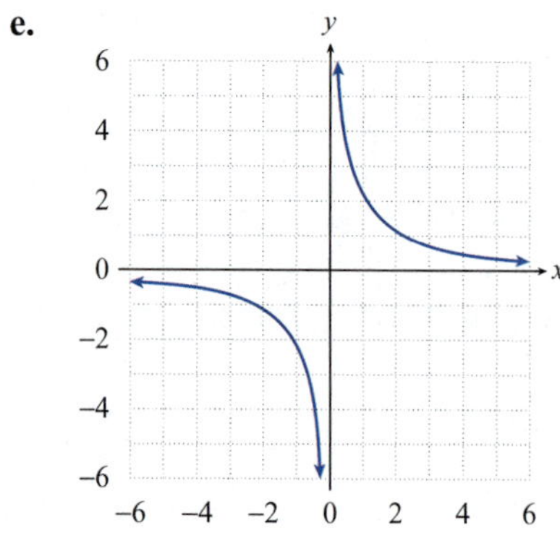

f.

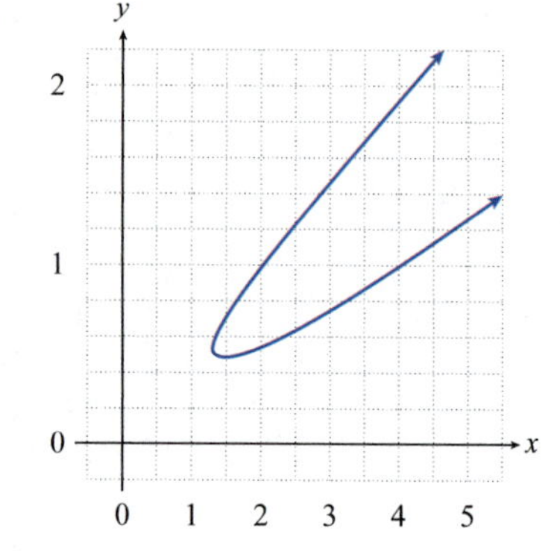

g.

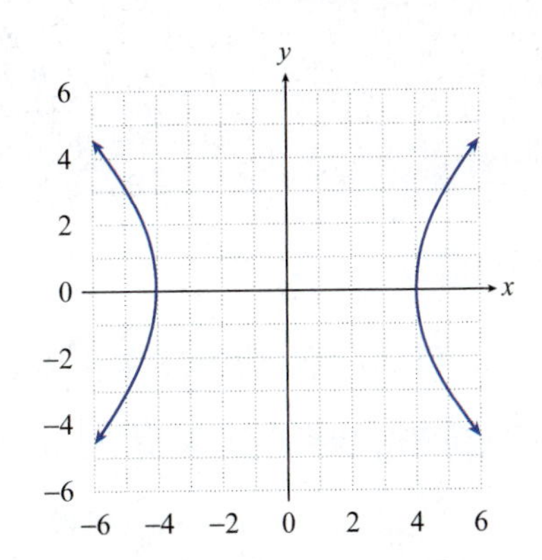

h.

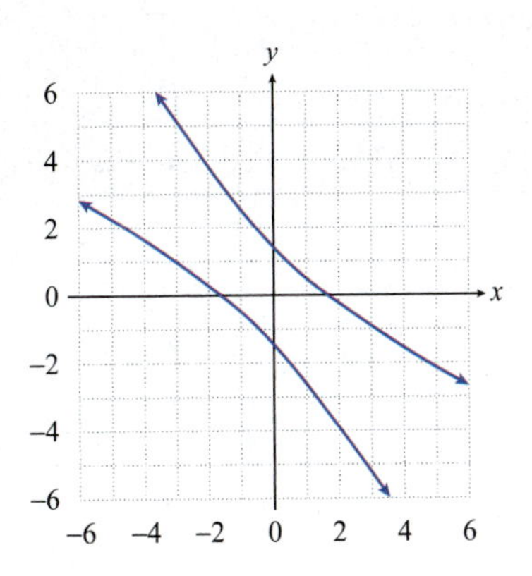

WRITING & THINKING

41. You have just used the rotation of axes to rotate the x- and y-axes until they were parallel to the axes of the conic. The resulting equation in the $x'y'$-plane is of the form

$$A'x'^2 + Bx'y' + E'y' + F' = 0.$$

What is wrong with the resulting equation?

42. What must the angle of rotation θ be if the coefficients of x^2 and y^2 are equal and $B \neq 0$? Support your answer.

43. Using the equation $7x^2 - 6\sqrt{3}xy + 13y^2 - 16 = 0$,

a. show that the rotation invariant $F = F'$ is true,
b. show that the rotation invariant $A + C = A' + C'$ is true, and
c. show that the rotation invariant $B^2 - 4AC = B'^2 - 4A'C'$ is true.

10.5 POLAR EQUATIONS OF CONIC SECTIONS

TOPICS

1. The focus and directrix of a conic section
2. Graphing polar equations of conic sections

The first three sections of this chapter dealt with the three varieties of conic sections individually, describing their geometric properties and the formulation of their associated equations in rectangular coordinates. This individual attention is useful when introducing conic sections, and rectangular coordinates are undeniably useful in working with conics to solve certain applications. But polar coordinates provide us with an alternative approach to the study of conic sections, one which possesses important advantages of its own.

TOPIC 1: The Focus and Directrix of a Conic Section

Probably the most striking virtue of the polar coordinate formulation of conics is that the equations for the three types of conics all have the same form. The equations are easily identified, and the magnitude of a parameter e, called the eccentricity, determines whether the conic is an ellipse, a parabola, or a hyperbola. You first encountered eccentricity in the discussion of ellipses—the use of e in this section is an extension of the original use. An additional characteristic of the polar form of conics is that all three varieties are defined in terms of a focus and a directrix; previously, the directrix only made an appearance in the discussion of parabolas.

Throughout this section, we will assume a point F, called the focus, lies at the origin of the plane and that a line L, called the directrix, lies d units away from the focus. Until we discuss rotated conics in polar form, the directrix will be oriented either vertically or horizontally, so the equation for the directrix will be $x=-d$, $x=d$, $y=-d$, or $y=d$. We will let $D(P,F)$ denote the distance between a variable point P and the fixed focus F, and we will let $D(P,L)$ denote the shortest distance between a variable point P and the fixed directrix L. The eccentricity e will be a fixed positive number for any given conic, and we will make frequent use of the polar coordinates (r,θ) of points in the plane.

DEFINITION: Focus and Directrix Description of Conics

Let $D(P,F)$ denote the distance between a variable point P and the fixed focus F, and let $D(P,L)$ denote the shortest distance between P and the fixed directrix L. A conic section consists of all points P in the plane that satisfy the equation

$$\frac{D(P,F)}{D(P,L)}=e,$$

where e is a fixed positive constant.

The conic is

- an ellipse if $0<e<1$,
- a parabola if $e=1$, and
- a hyperbola if $e>1$.

In words, a conic section consists of all those points for which the ratio of the distance to the focus and the distance to the directrix is a fixed constant e. This common definition for all three types of conic sections is pleasantly unified. Figure 1 illustrates three conic sections that share the directrix $x = -1$ but with three different eccentricities. The directrix and the focus are included in each of the three graphs.

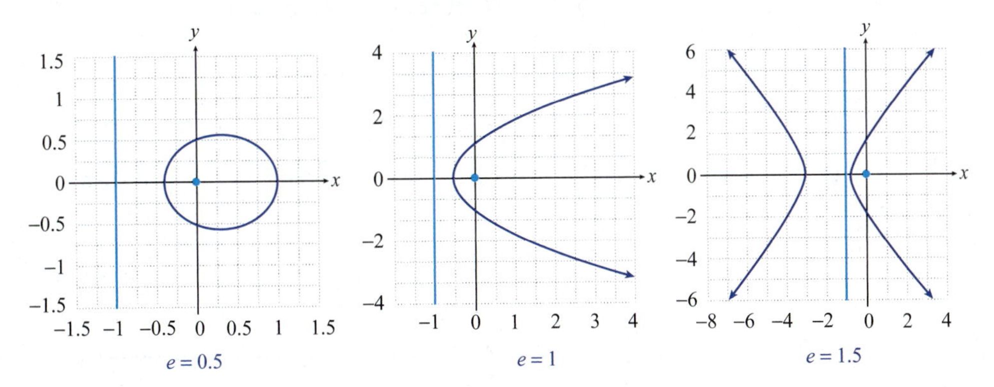

FIGURE 1: Common Directrix and Focus, Three Eccentricities

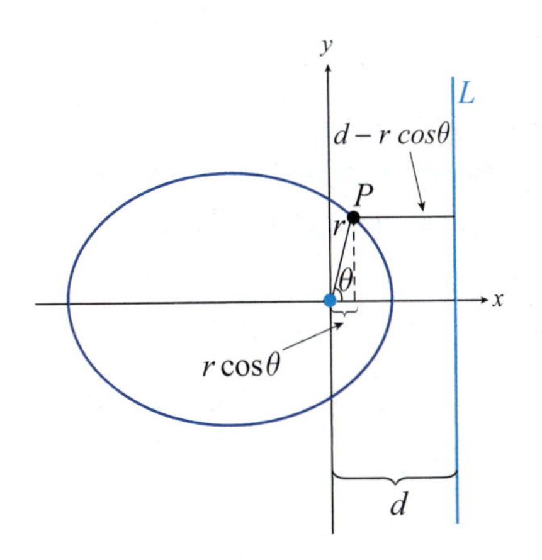

FIGURE 2: Determining $D(P, L)$

The next step is to use the focus/directrix description to develop the form of the polar equations for conics. If we let (r,θ) denote a point P on a given conic section, then $D(P,F) = r$. To determine $D(P,L)$ consider the diagram in Figure 2, which depicts an ellipse and a directrix L of the form $x = d$.

From Figure 2, we see that $D(P,L) = d - r\cos\theta$, so the equation in polar coordinates is

$$\frac{r}{d - r\cos\theta} = e.$$

This equation will be more useful if we solve for r.

$$\begin{aligned} r &= e(d - r\cos\theta) \\ r &= ed - er\cos\theta \\ r + er\cos\theta &= ed \\ r(1 + e\cos\theta) &= ed \\ r &= \frac{ed}{1 + e\cos\theta} \end{aligned}$$

The derivations of the equations for a directrix of the form $x = -d$, $y = -d$, or $y = d$ are very similar, and will be left to the reader. The four possible forms and their geometric meanings are described in the following box.

$f(x)$ FORMULA: Polar Equations of Conic Sections

In polar coordinates, the equation of a conic section with its focus at the origin and either a horizontal or vertical directrix has one of the following forms.

Equation of Conic Section	Directrix
$r=\dfrac{ed}{1+e\cos\theta}$	Vertical directrix $x=d$
$r=\dfrac{ed}{1-e\cos\theta}$	Vertical directrix $x=-d$
$r=\dfrac{ed}{1+e\sin\theta}$	Horizontal directrix $y=d$
$r=\dfrac{ed}{1-e\sin\theta}$	Horizontal directrix $y=-d$

Example 1: Identifying the Type and Directrix of a Conic

Identify the variety of each of the following conics, and determine the equation for the directrix in each case.

a. $r=\dfrac{15}{5-3\cos\theta}$ **b.** $r=\dfrac{2}{3+5\sin\theta}$

Solution

a. All we need to do is determine the two constants e and d, and we can accomplish this with algebraic manipulation. In order to have the correct form, the constant term in the denominator must be a 1.

$$\begin{aligned} r &= \frac{15}{5-3\cos\theta} \\ &= \frac{15\left(\frac{1}{5}\right)}{(5-3\cos\theta)\left(\frac{1}{5}\right)} \\ &= \frac{3}{1-\left(\frac{3}{5}\right)\cos\theta} \end{aligned}$$

This tells us that $e=\frac{3}{5}$, so the conic is an ellipse. To determine the constant d, we need to write the numerator as a product of e and d. Since the numerator is 3 and since we now know $e=\frac{3}{5}$, we can do this as follows.

$$r = \frac{3}{1 - \left(\frac{3}{5}\right)\cos\theta}$$

$$= \frac{\left(\frac{3}{5}\right)(5)}{1 - \left(\frac{3}{5}\right)\cos\theta}$$

Hence, $d = 5$. The last observation is that the trigonometric function in the denominator is cosine and that the sign between the two terms in the denominator is negative. By the guidelines for conics in polar form, this tells us that $x = -5$ is the equation for the directrix.

b. We use the same methods to determine first e and then d.

$$r = \frac{2}{3 + 5\sin\theta}$$

$$= \frac{\frac{2}{3}}{1 + \left(\frac{5}{3}\right)\sin\theta}$$

$$= \frac{\left(\frac{5}{3}\right)\left(\frac{2}{5}\right)}{1 + \left(\frac{5}{3}\right)\sin\theta}$$

From this form, we can see that $e = \frac{5}{3}$ and that the directrix is $y = \frac{2}{5}$. Thus, the conic section is a hyperbola and the directrix is horizontal and above the x-axis.

Example 2: Constructing a Polar Equation

Construct a polar equation for a leftward-opening parabola with focus at the origin and directrix 2 units from the focus.

Solution

Since we are discussing a parabola, $e = 1$. If the parabola is to open to the left, the directrix must be oriented vertically and must lie to the right of the focus, so $d = 2$ and the trigonometric function in the denominator must be cosine. Thus, the equation is

$$r = \frac{2}{1 + \cos\theta}.$$

The graph of the parabola appears in Figure 3, along with the focus and directrix.

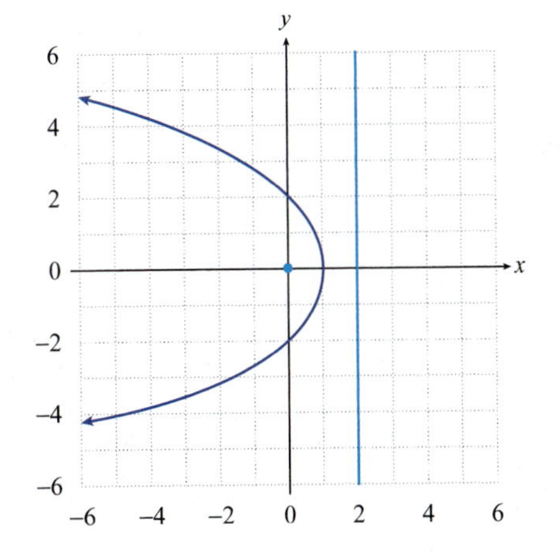

FIGURE 3

TOPIC 2: Graphing Polar Equations of Conic Sections

Example 1 illustrated how the two constants e and d can be easily determined from the equation for a conic in polar form. In order to get an even better understanding of a given conic section, it may be useful to determine the constants a, b, and c as well. These constants have the same meaning in this context as they did in Sections 10.1 and 10.3, that is, $2a$ is the distance between vertices of an ellipse or a hyperbola, $2b$ is the length of the minor axis of an ellipse or the width of the auxiliary rectangle that aids in sketching the asymptotes of a hyperbola, and $2c$ is the distance between foci of an ellipse or a hyperbola.

We begin with a. In Section 10.1, eccentricity was defined as $e=\frac{c}{a}$, and this relation still holds for the expanded definition of eccentricity that we now have. In originally discussing ellipses, we noted that $b^2=a^2-c^2$, and in originally discussing hyperbolas we noted that $b^2=c^2-a^2$; we can combine these two statements by noting that b is nonnegative and that $b^2=\left|a^2-c^2\right|$. Note that if $e=1$, then $a=c$, and hence $b=0$. This should not be surprising, as b played no role in the original discussion of parabolas.

When graphing an ellipse or a hyperbola described in polar form, it is usually easiest to first determine a. Using our observations, it is then easy to determine c and b. This process is illustrated in the next several examples.

Example 3: Graphing Conic Sections in Polar Form

Sketch the graph of the conic section $r=\frac{15}{5-3\cos\theta}$.

Solution

This is the first equation from Example 1, and we have already seen that the equation can be written in the form

$$r=\frac{\left(\frac{3}{5}\right)(5)}{1-\left(\frac{3}{5}\right)\cos\theta}.$$

So $e=\frac{3}{5}$, $d=5$, and the directrix is the line $x=-5$. This tells us that the graph is an ellipse and that the major axis is oriented horizontally (perpendicular to the directrix). The entire graph is traced out as θ increases from 0 to 2π, with the right vertex corresponding to $\theta=0$, and the left vertex corresponding to $\theta=\pi$ (halfway around the ellipse). When $\theta=0$, $r=\frac{15}{2}$ and when $\theta=\pi$, $r=\frac{15}{8}$. In rectangular coordinates, the coordinates of the right vertex are $\left(\frac{15}{2},0\right)$ and the coordinates of the left vertex are $\left(-\frac{15}{8},0\right)$ (remember that $\theta=\pi$ and a positive r corresponds to a point left of the origin).

Now that we know the coordinates of the two vertices, we can determine that

$$2a=\frac{15}{2}+\frac{15}{8}=\frac{75}{8}$$

and so $a=\frac{75}{16}$. Since $e=\frac{3}{5}$ and $c=ea$, this means

$$c = \left(\frac{3}{5}\right)\left(\frac{75}{16}\right) = \frac{45}{16}.$$

Finally, from the relation $b^2 = \left|a^2 - c^2\right|$, we can determine that $b = \frac{15}{4}$.

With this knowledge of a and b, we can sketch the graph in Figure 4.

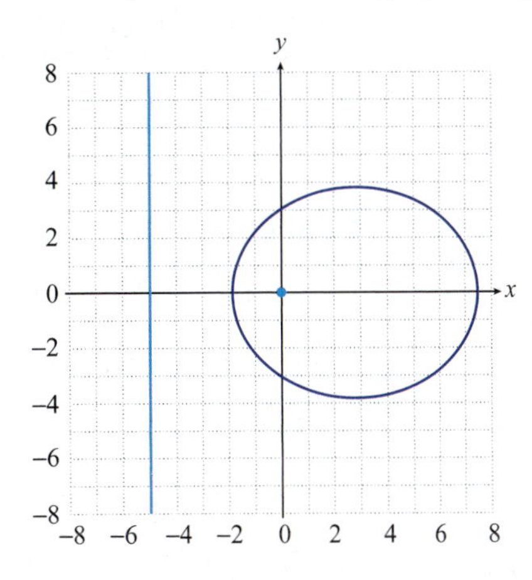

FIGURE 4

Example 4: Graphing Conic Sections in Polar Form

Sketch the graph of the conic section $r = \frac{2}{3 + 5\sin\theta}$.

Solution

This equation is the second from Example 1, and we already know $e = \frac{5}{3}$ and that $y = \frac{2}{5}$ is the equation for the directrix. This tells us that the graph is a hyperbola and that the vertices are aligned vertically (perpendicular again to the directrix). One vertex corresponds to $\theta = \frac{\pi}{2}$ while the other vertex corresponds to $\theta = \frac{3\pi}{2}$ (it might also be useful to note that the lower branch of the hyperbola crosses the x-axis at $\theta = 0$ and $\theta = \pi$). When $\theta = \frac{\pi}{2}$, $r = \frac{1}{4}$ and when $\theta = \frac{3\pi}{2}$, $r = -1$ (make note of the negative sign on r). This means that the rectangular coordinates of the two vertices are $\left(0, \frac{1}{4}\right)$ and $(0,1)$, so $2a = \frac{3}{4}$ and $a = \frac{3}{8}$. From this, $c = ea = \frac{5}{8}$ and the relation $b^2 = \left|a^2 - c^2\right|$ leads to $b = \frac{1}{2}$.

Our work in Section 10.3 tells us the equations for the asymptotes of the hyperbola are $y - \frac{5}{8} = \pm\frac{3}{4}x$. Now, using the vertices and asymptotes we calculated, we can sketch the graph in Figure 5.

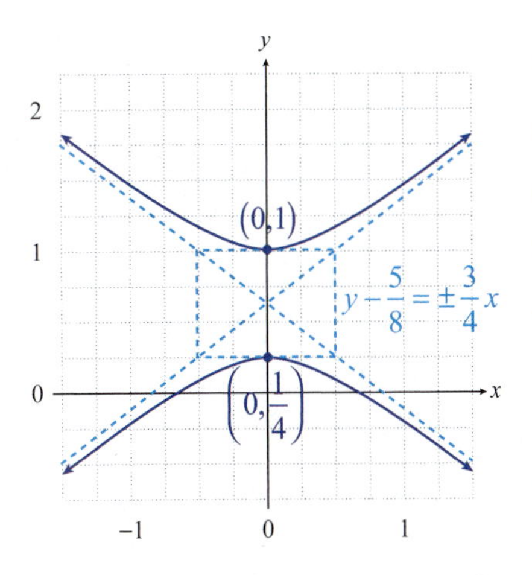

FIGURE 5

We will conclude this section with an example of rotation of conics in polar form. In general, the graph of an equation $r = f(\theta - \varphi)$ is the rotation of the graph of $r = f(\theta)$ by the angle φ counterclockwise. This makes rotation in polar coordinates particularly easy to handle.

Example 5: Graphing Conic Sections Using Rotation in Polar Coordinates

Sketch the graph of the conic section $r = \dfrac{2}{1+\cos\left(\theta - \frac{\pi}{6}\right)}$.

Solution

We constructed the equation $r = \dfrac{2}{1+\cos\theta}$ in Example 2, so we know its graph is a parabola opening to the left with directrix $x = 2$. The graph of $r = \dfrac{2}{1+\cos\theta}$ is repeated in Figure 6.

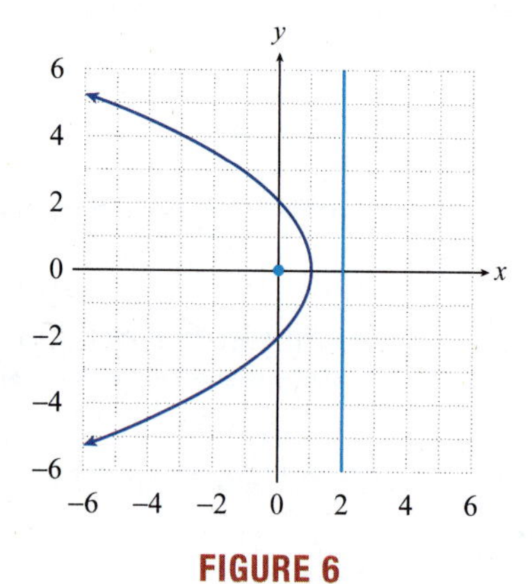

FIGURE 6

The graph of $r = \dfrac{2}{1+\cos\left(\theta - \frac{\pi}{6}\right)}$ is the same shape rotated $\dfrac{\pi}{6}$ radians counterclockwise.

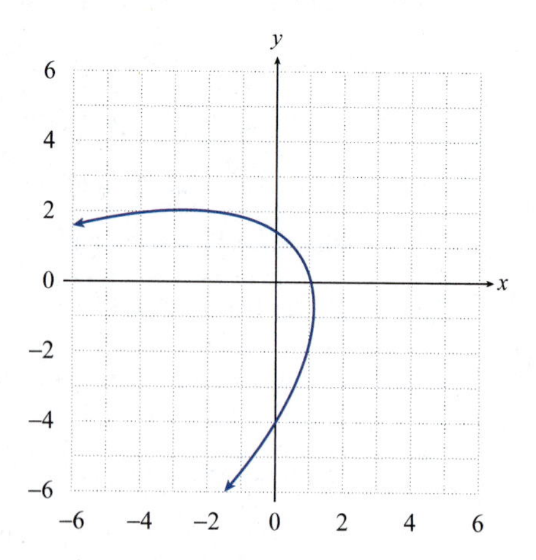

FIGURE 7

10.5 EXERCISES

PRACTICE

Match the polar equation with its corresponding graph. See Examples 2 and 3.

1. $r = \dfrac{3}{4 - \cos\theta}$

2. $r = \dfrac{9}{6 - 2\sin\theta}$

3. $r = \dfrac{3}{3 + 4\sin\theta}$

4. $r = \dfrac{1}{2 + 2\cos\theta}$

5. $r = \dfrac{6}{1 + 3\sin\theta}$

6. $r = \dfrac{6}{1 + 3\cos\theta}$

a.

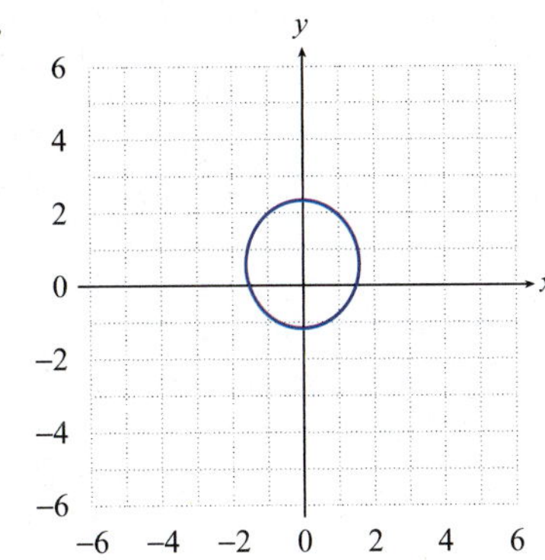

b.

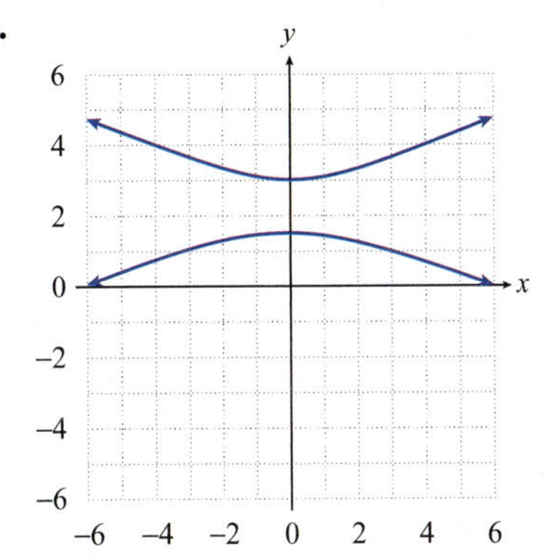

c.

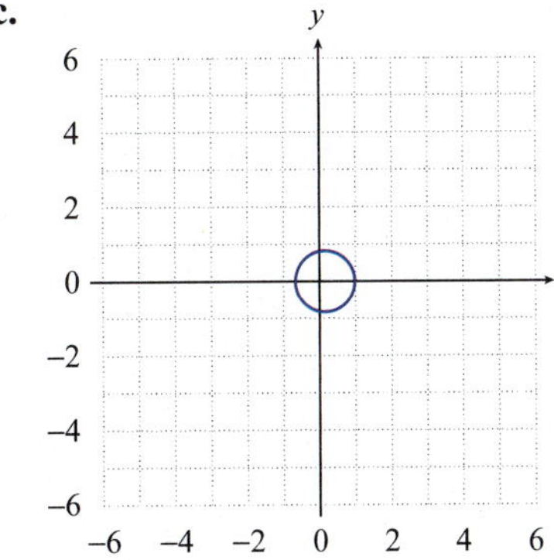

d.

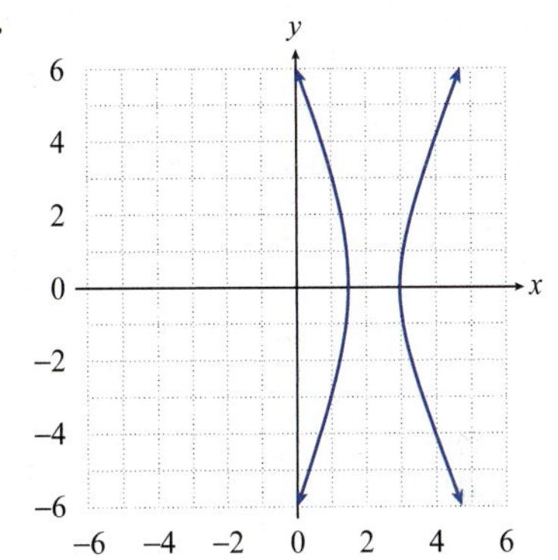

e.

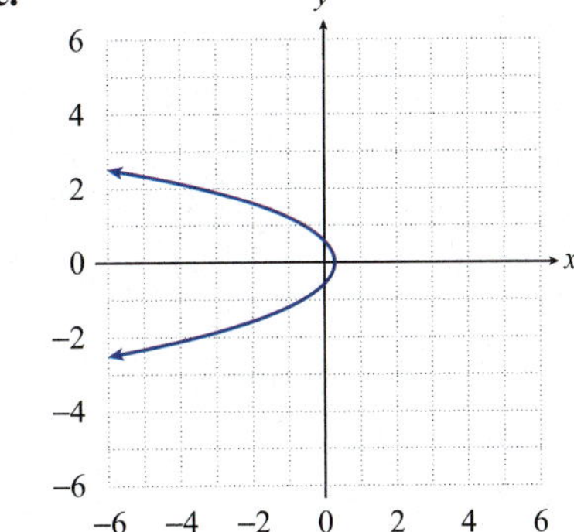

f.

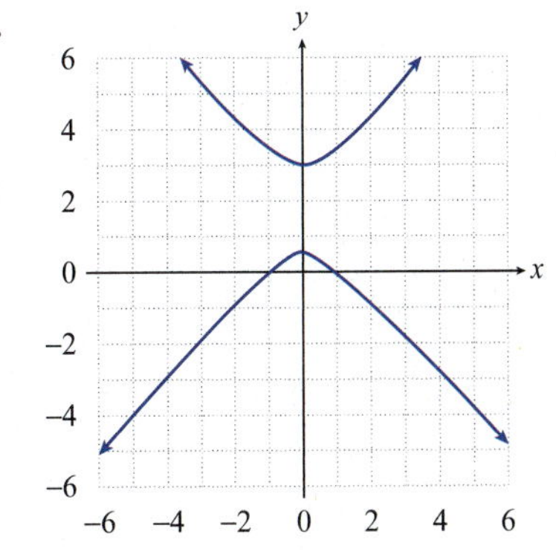

Identify each conic section and find the equation for its directrix. See Example 1.

7. $r = \dfrac{7}{1+6\sin\theta}$

8. $r = \dfrac{2}{1-\sin\theta}$

9. $r = \dfrac{3}{4-\cos\theta}$

10. $r = \dfrac{4}{2-2\cos\theta}$

11. $r = \dfrac{1}{1+3\cos\theta}$

12. $r = \dfrac{7}{3+2\sin\theta}$

13. $r = \dfrac{5}{2+\cos\theta}$

14. $r = \dfrac{3}{4-3\sin\theta}$

15. $r = \dfrac{6}{3-5\cos\theta}$

16. $r = \dfrac{8}{5-6\sin\theta}$

17. $r = \dfrac{3}{2+2\sin\theta}$

18. $r = \dfrac{-1}{3+4\cos\theta}$

19. $r = \dfrac{4}{6-7\cos\theta}$

20. $r = \dfrac{9}{5-4\sin\theta}$

Construct a polar equation for each conic section with the focus at the origin and the given eccentricity and directrix. See Example 2.

	Conic	Eccentricity	Directrix
21.	Parabola	$e=1$	$x=-2$
22.	Hyperbola	$e=2$	$x=-3$
23.	Hyperbola	$e=4$	$y=-\dfrac{3}{4}$
24.	Parabola	$e=1$	$x=2$
25.	Ellipse	$e=\dfrac{1}{4}$	$x=12$
26.	Ellipse	$e=\dfrac{1}{2}$	$y=8$

Sketch the graphs of the following conic sections. See Examples 3, 4, and 5.

27. $r = \dfrac{5}{1+3\cos\theta}$

28. $r = \dfrac{3}{2+\sin\theta}$

29. $r = \dfrac{4}{1-2\sin\theta}$

30. $r = \dfrac{6}{2-4\cos\theta}$

31. $r = \dfrac{9}{3-2\cos\theta}$

32. $r = \dfrac{5}{3+\sin\theta}$

33. $r = \dfrac{4}{1+2\cos\theta}$

34. $r = \dfrac{4}{2+2\sin\theta}$

35. $r = \dfrac{-3}{4-9\cos\theta}$

36. $r = \dfrac{9}{-4+\dfrac{3}{2}\sin\theta}$

37. $r = \dfrac{-11}{3-\cos\theta}$

38. $r = \dfrac{2}{10+4\sin\theta}$

39. $r = \dfrac{3}{7+3\cos\theta}$

40. $r = \dfrac{2}{2+3\cos\left(\theta-\dfrac{\pi}{4}\right)}$

41. $r = \dfrac{-7}{5 + 3\sin\left(\theta - \dfrac{\pi}{6}\right)}$

42. $r = \dfrac{5}{-2 - 4\sin\left(\theta + \dfrac{2\pi}{3}\right)}$

43. $r = \dfrac{4}{-3 - 2\cos\left(\theta + \dfrac{\pi}{3}\right)}$

44. $r = \dfrac{1}{1 + 4\sin\left(\theta + \dfrac{\pi}{6}\right)}$

45. $r = \dfrac{2}{1 + \cos\left(\theta - \dfrac{\pi}{4}\right)}$

46. $r = \dfrac{4}{2 + 2\sin\left(\theta - \dfrac{\pi}{3}\right)}$

APPLICATIONS

47. The planets of our solar system follow elliptical orbits with the sun located at one of the foci. If we assume the sun is located at the pole and the major axes of these elliptical orbits lie along the polar axis, the orbits of the planets can be expressed by the polar equation

$$r = \frac{\left(1 - e^2\right)a}{1 - e\cos\theta},$$

where e is the eccentricity. Verify the above equation.

48. Using the equation from Exercise 47, answer the following:

a. Show that the shortest distance from the sun to a planet, called the *perihelion*, is $r = a(1 - e)$.

b. Show that the longest distance from the sun to a planet, called the *aphelion*, is $r = a(1 + e)$.

c. Uranus is approximately 2.74×10^9 km away from the sun at perihelion and 3.00×10^9 km at aphelion. Find the eccentricity of Uranus' orbit.

d. The eccentricity of Neptune's path is 0.0113 and $a = 4.495 \times 10^9$ km. Determine the perihelion and aphelion distances for Neptune.

CHAPTER 10 PROJECT

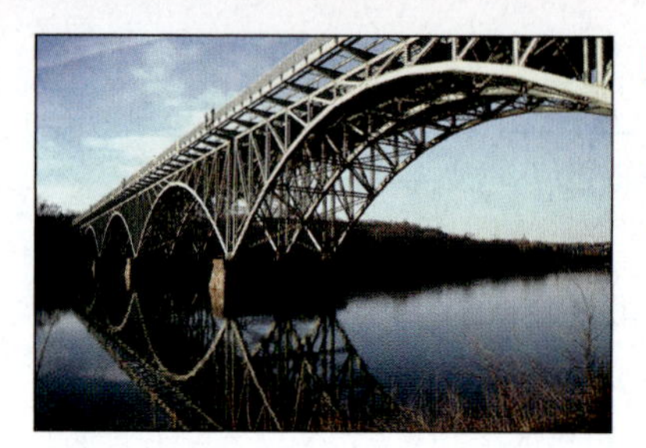

Constructing a Bridge

Plans are in process to develop an uninhabited coastal island into a new resort. Before development can begin, a bridge must be constructed joining the island to the mainland.

Two possibilities are being considered for the support structure of the bridge. The archway could be built as a parabola, or in the shape of a semiellipse.

Assume all measurements that follow refer to dimensions at high tide. The county building inspector has deemed that in order to establish a solid foundation, the space between supports must be at most 300 feet and the height at the center of the arch should be 80 feet. There is a commercial fishing dock located on the mainland whose fishing vessels travel constantly along this intracoastal waterway. The tallest of these ships requires 60 feet of clearance to pass comfortably beneath the bridge. With these restrictions, the width of a channel with a minimum height of 60 feet has to be determined for both possible shapes of the bridge to confirm that it will be suitable for the water traffic beneath it.

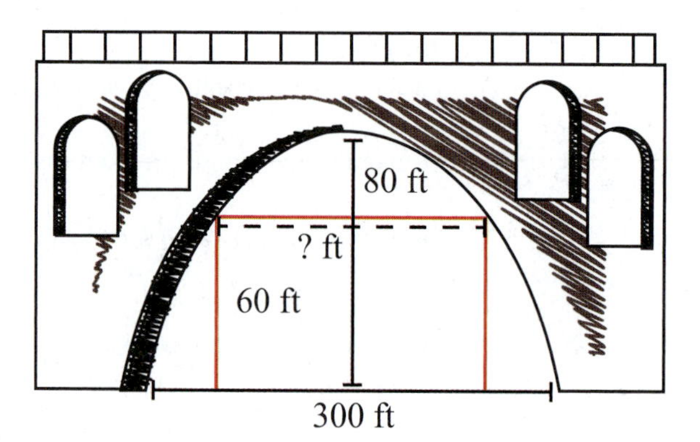

1. Find the equation of a parabola that will fit these constraints.
2. How wide is the channel with a minimum 60-foot vertical clearance for the parabola in question 1?
3. Find the equation of a semiellipse that will fit these constraints.
4. How wide is the channel with a minimum 60-foot vertical clearance for the semiellipse in question 3?
5. Which of these bridge designs would you choose, and why?
6. Suppose the tallest fishing ship installs a new antenna which raises the center height by 12 feet. How far off of center (to the left or right) can the ship now travel and still pass under the bridge without damage to the antenna

   a. for the parabola?
   b. for the semiellipse?

CHAPTER 10 REVIEW EXERCISES

Section 10.1

Find the center, foci, and vertices of the ellipse that each equation describes.

1. $(x-3)^2+4(y+1)^2=16$

2. $9x^2+4y^2+18x-16y+9=0$

Sketch the graphs of the following ellipses and determine the coordinates of the foci.

3. $\dfrac{(x+1)^2}{16}+\dfrac{(y-2)^2}{9}=1$

4. $x^2+9y^2-6x+18y=-9$

5. $3x^2+y^2=27$

6. $25x^2+4y^2-200x+300=0$

In each of the following exercises, an ellipse is described by either a picture or by the properties it possesses. Find the equation, in standard form, for each ellipse.

7. Center at $(-1,4)$, major axis is vertical and of length 8, foci $\sqrt{7}$ units from the center.

8. Foci at $(1,2)$ and $(7,2)$, $e=\dfrac{1}{2}$.

9. Vertices at $\left(\dfrac{7}{2},-1\right)$ and $\left(\dfrac{1}{2},-1\right)$, $e=0$.

10. Vertices at $(1,-8)$ and $(1,2)$, minor axis of length 6.

11. Foci at $(0,0)$ and $(4,0)$, major axis of length 8.

12. Center at $(0,4)$, $a=2c$, and vertices at $(-4,4)$ and $(4,4)$.

13.

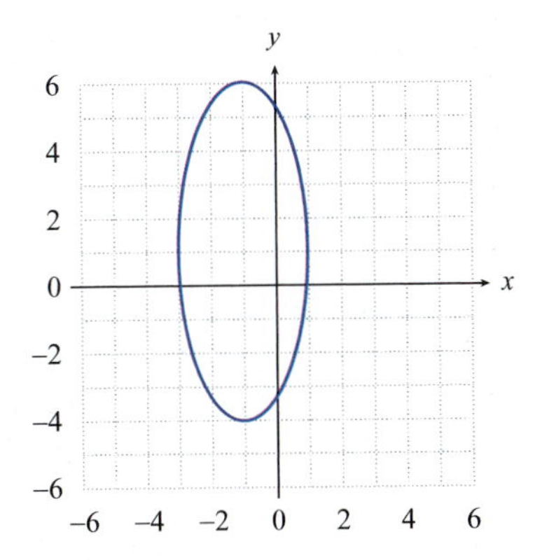

14.

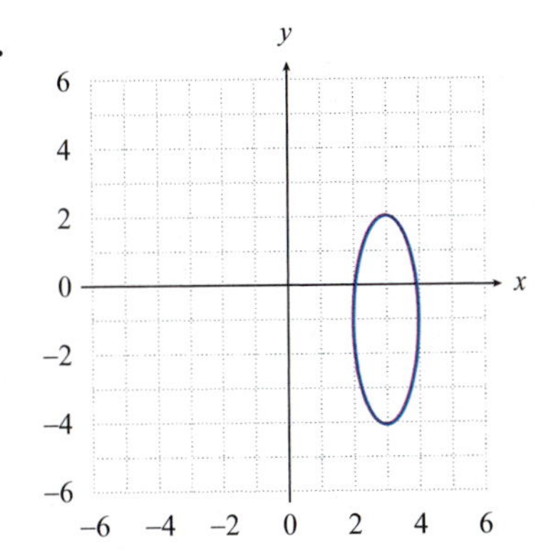

For exercises 15 and 16, use the fact that the area A of the ellipse $\dfrac{x^2}{a^2}+\dfrac{y^2}{b^2}=1$ is $A=\pi\cdot a\cdot b$ and $a+b=30$.

15. Write the area of the ellipse as a function of a.

16. Find the equation of an ellipse with an area of 200π square inches.

Section 10.2

Graph the following parabolas and determine the focus and directrix of each.

17. $(y+1)^2 = -12(x+3)$

18. $y^2 - 8y + 2x + 14 = 0$

19. $y^2 + 2y = 4x - 1$

20. $x + \frac{1}{4}y^2 = 0$

21. $2y + 4x^2 = 8$

22. $y^2 - 4y + 2x + 24 = 0$

Find the equation, in standard form, for the parabola with the given properties or with the given graph.

23. Vertex at $(-2,3)$, directrix is the line $y = 2$.

24. Vertex at $(5,-3)$, focus at $(5,1)$.

25. Focus at $(3,-1)$, directrix is the line $x = 2$.

26. Focus at $(1,-2)$, directrix is the x-axis.

27. Vertex at $(2,-1)$, directrix is the line $x = -2$.

28. Symmetric with respect to the x-axis, focus at $(-3,0)$, and $p = 4$.

29.

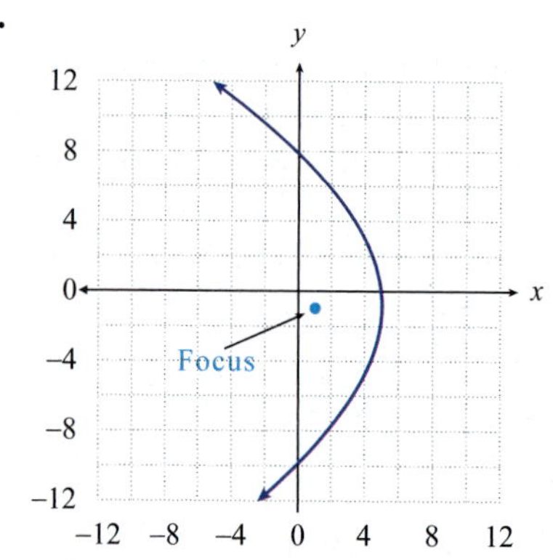

30.

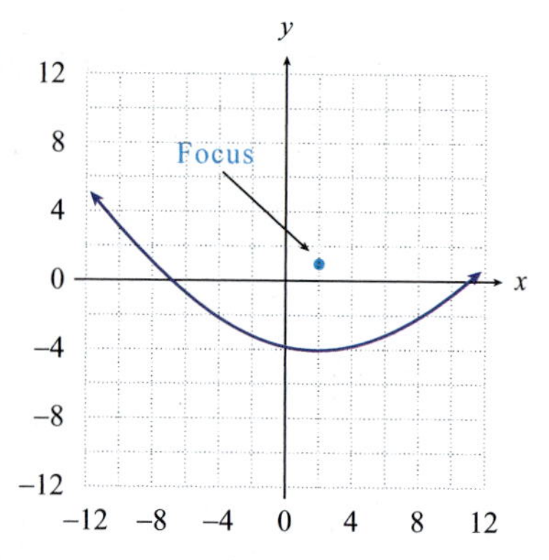

31. A motorcycle headlight is made by placing a strong light bulb inside a reflective paraboloid formed by rotating the parabola $x^2 = 5y$ around its axis of symmetry (assume that x and y are in units of inches). In order to have the brightest, most concentrated light beam, how far from the vertex should the bulb be placed?

Section 10.3

Sketch the graphs of the following hyperbolas, using asymptotes as guides. Determine the coordinates of the foci in each case.

32. $\frac{(y+2)^2}{9} - \frac{(x-2)^2}{16} = 1$

33. $9x^2 - 4y^2 + 54x - 8y + 41 = 0$

34. $x^2 - y^2 = 1$

35. $\frac{y^2}{25} - \frac{x^2}{144} = 1$

Find the center, foci, and vertices of the hyperbola that each equation describes.

36. $(x+1)^2 - 4(y-2)^2 = 36$

37. $x^2 - 9y^2 + 36y - 72 = 0$

38. $y^2 - 4x^2 - 2y - 32x = 67$

39. $\dfrac{(y-3)^2}{4} - \dfrac{(x-3)^2}{49} = 1$

Find the equation, in standard form, for the hyperbola with the given properties or with the given graph.

40. Vertices at $(4,-1)$ and $(-2,-1)$ and foci at $(5,-1)$ and $(-3,-1)$.

41. Asymptotes of $y = \pm\dfrac{5}{2}(x+1) - 2$ and vertices at $(-3,-2)$ and $(1,-2)$.

42. Foci at $(-1,-2)$ and $(-1,8)$ and asymptotes of $y = \pm\left(\dfrac{3}{4}x + \dfrac{3}{4}\right) + 3$.

43. Asymptotes of $y = \pm(3x-6) + 2$ and vertices at $(2,-1)$ and $(2,5)$.

44. Vertices at $(\pm 3, 0)$ and foci at $(\pm 5, 0)$.

45. Foci at $\left(-1, 7 \pm \sqrt{13}\right)$ and asymptotes of $y = \pm\dfrac{2}{3}(x+1) + 7$.

46.

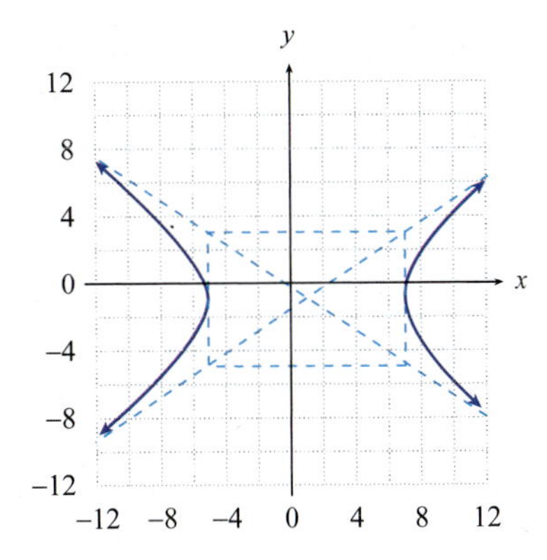

47.

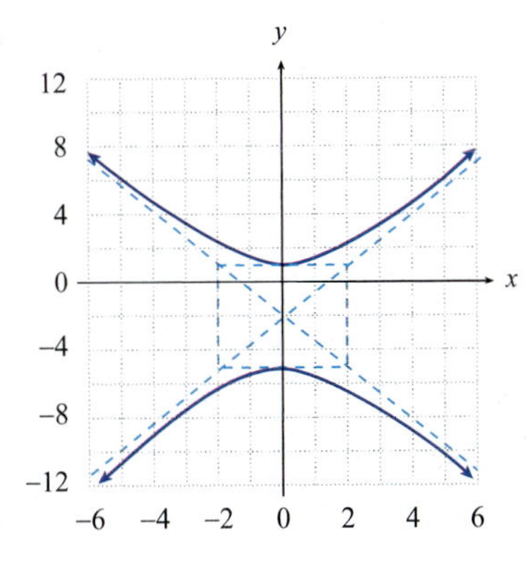

Section 10.4

Find the $x'y'$-coordinates of each point for the given rotation angle θ.

48. $(-8,7), \quad \theta = \dfrac{\pi}{4}$

49. $(22,86), \quad \theta = \dfrac{\pi}{3}$

50. $(4.6,-8.9), \quad \theta = 53°$

51. $\left(2\sqrt{3}, 6\sqrt{3}\right), \quad \theta = 30°$

Use the discriminant to classify each of the following conic sections. Then determine the angle θ that will allow you to convert the equation and eliminate the xy-term. Finally, sketch the graph of the conic section.

52. $xy - 6 = 0$

53. $10x^2 + 2\sqrt{3}xy + 12y^2 - 100y = 0$

54. $10\sqrt{3}x^2 + 42xy - 4\sqrt{3}y^2 = 187\sqrt{3}$

55. $x^2 + 2xy + y^2 + x - y = 0$

Section 10.5

Identify each conic section and find the equation for its directrix.

56. $r = \dfrac{5}{4 - 8\sin\theta}$

57. $r = \dfrac{7}{4 + 4\sin\theta}$

58. $r = \dfrac{4}{6 - 3\cos\theta}$

59. $r = \dfrac{7}{5 + 2\cos\theta}$

Construct a polar equation for each conic section with the focus at the origin and the given eccentricity and directrix.

Conic	Eccentricity	Directrix
60. Hyperbola	$e = 4$	$y = 3$
61. Ellipse	$e = \dfrac{1}{4}$	$x = 16$
62. Parabola	$e = 1$	$y = -7$
63. Hyperbola	$e = 9$	$x = \dfrac{1}{3}$

CHAPTER 11

Systems of Equations and Inequalities

SECTIONS

WHAT IF ...

What if you were given a large handful of nickels and pennies and were told their total value and the number of coins? Without counting, would you know how many of each coin you have?

By the end of this chapter, you'll be able to solve both linear and nonlinear systems of equations. Given two or more equations in the same variables, you will find solutions for the variables that solve all the equations simultaneously. You'll encounter the coin problem on page 792. You'll master this type of problem using tools such as the method for solving systems of linear equations by elimination, found on page 783.

Carl Friedrich Gauss

Introduction

In this chapter, we return to the study of linear equations, arguably the simplest class of equations. But as we will soon see, our understanding of linear equations has room for growth in many directions. In fact, the material in this chapter serves as a good illustration of how, given incentive and opportunity, mathematicians extend ideas and techniques beyond the familiar.

In the particular case of linear equations, this extension has taken the form of

1. considering equations containing more variables (we have already seen the first step in this process when we moved from linear equations in one variable to linear equations in two);
2. trying to find solutions that satisfy more than one linear equation at a time; such sets of equations are called *systems*;
3. making use of elementary methods to solve systems if possible, and developing entirely new methods if necessary or desirable.

The third point above constitutes the bulk of this chapter. We start off, in our quest to solve systems of linear equations, by using some intuitive methods that work admirably if the equations contain only a few variables and if the system has only a few equations. Several centuries ago, however, mathematicians began to realize the limitations of such simple methods. As the problems that people wanted to solve led to systems of many variables and equations, refinements were made to the elementary methods of solution and, eventually, entirely new techniques were developed. The two refinements we will study are called *Gaussian elimination* and *Gauss-Jordan elimination.* Some of the new techniques that were developed for solving large systems of equations make use of *determinants of matrices* and *matrix inverses*, two concepts that can serve as an introduction to higher mathematics.

The notion of the inverse of a matrix (a *matrix* is a rectangular array of numbers) closes out our discussion of solution methods, but in fact it brings us full circle and allows us to write systems of linear equations as single matrix equations. This illustrates how mathematicians, in extending our reach in terms of problem solving and in constructing new mathematics, are guided by the achievements of the past.

11.1 SOLVING SYSTEMS OF LINEAR EQUATIONS BY SUBSTITUTION AND ELIMINATION

TOPICS

1. Systems of linear equations
2. Solving systems of linear equations by substitution
3. Solving systems of linear equations by elimination
4. Systems of linear equations in three or more variables
5. Applications of systems of linear equations

TOPIC 1: Systems of Linear Equations

Many problems in mathematics can be described by two or more equations in two or more variables. When the equations are all linear, such a collection of equations is called a *system of linear equations*, or sometimes *simultaneous linear equations*. The word *simultaneous* refers to the goal of identifying the values of the variables (if there are any) that solve all of the equations simultaneously.

In the case of two linear equations in two variables, it turns out that there are only three possible configurations of solutions, which we see in Figure 1.

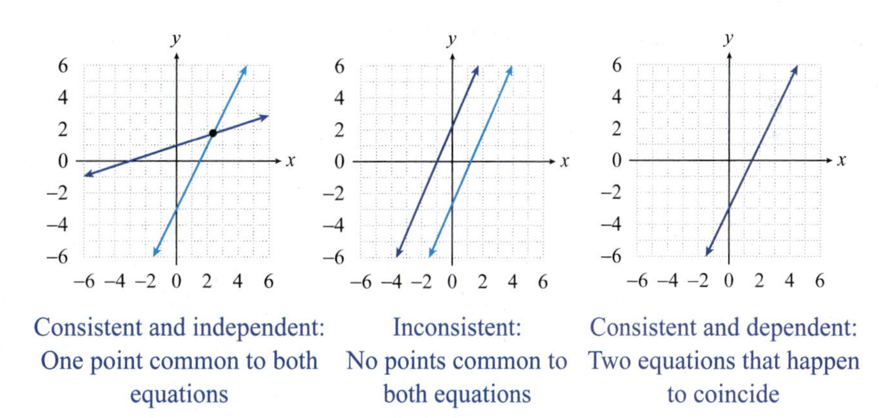

FIGURE 1: Solutions to Systems of Two Linear Equations

Now, each equation individually has an infinite number of solutions. What we are interested in are the points that solve both equations. These are *solutions to the system of equations*.

In the first graph of Figure 1, the two lines intersect in exactly one point. In the second graph, the two lines are parallel, so the system of equations has no solution since there is no point lying on both lines. In the final graph, the two lines actually coincide and appear as one, meaning that any ordered pair that solves one of the equations in the system solves the other as well, so the system has an infinite number of solutions.

We will encounter systems consisting of more than two equations and/or more than two variables, but larger systems have one important similarity to the two-variable, two-equation systems in Figure 1: every system of linear equations will have exactly one solution, no solution, or an infinite number of solutions.

DEFINITION: Solutions to Systems of Linear Equations

- A system of linear equations with no solution is called **inconsistent**.
- A system of linear equations that has at least one solution is called **consistent**.
  1. A consistent system with exactly one solution is called **independent**.
  2. A consistent system with more than one solution must have an infinite number of solutions and is called **dependent**.

Our goal is to develop systematic and effective methods of solving systems of linear equations. Because such systems arise in so many different contexts and are of great importance for both theoretical and practical reasons, many solution methods have been devised. We will learn two of them, the method of substitution and the method of elimination, in this section.

TOPIC 2: Solving Systems of Linear Equations by Substitution

The solution method of substitution hinges on solving one equation in a system for one of the variables, and substituting the result for that variable in the remaining equations. This can be a time-consuming process if the system is large (meaning more than a few equations and more than a few variables), and in fact the task may have to be repeated many times. But it is a very natural method to use for some small systems, as illustrated in Examples 1 and 2.

Example 1: Solving an Independent System by Substitution

Use the method of substitution to solve the system $\begin{cases} 2x - y = 1 \\ x + y = 5 \end{cases}.$

NOTE

Solving for a variable with a coefficient of 1 will make the process of substitution a bit easier.

Solution

Either equation can be solved for either variable, and the choice doesn't affect the final answer. We will solve the second equation for x, and then substitute the result in the first equation, giving us one equation in the variable y. Once we have solved for y, we can substitute the value of y into either original equation to find x.

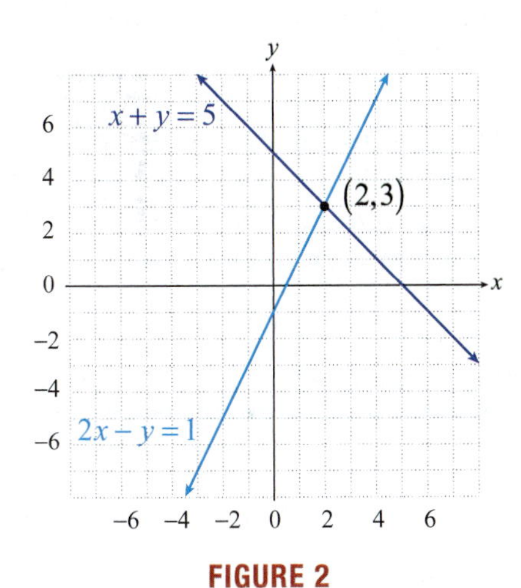

FIGURE 2

$x = -y + 5$	Solve the second equation for x.
$2(-y+5) - y = 1$	Substitute the result in the first equation.
$-2y + 10 - y = 1$	Simplify and solve for y.
$-3y = -9$	
$y = 3$	

We then substitute our answer for y into the original second equation.

$x + 3 = 5$	Substitute $y = 3$ in the second equation.
$x = 2$	Solve for x.

Thus, the solution to the system of equations is $(2,3)$.

Note how the graph in Figure 2 corresponds to the system and the ordered pair solution that we have found.

Example 2: Solving a Dependent System by Substitution

Use the method of substitution to solve the system $\begin{cases} -2x + 6y = 6 \\ x + 3 = 3y \end{cases}$.

Solution

$x = 3y - 3$	Solve the second equation for x.
$-2(3y-3) + 6y = 6$	Substitute the result in the first equation.
$-6y + 6 + 6y = 6$	Simplify.
$0 = 0$	

The resulting equation is always true. This means that for any value of y, letting $x = 3y - 3$ results in an ordered pair $(x, y) = (3y - 3, y)$ that solves both equations. Since there are an infinite number of solutions, the system is dependent.

Graphically, this means the graphs of the two equations are exactly the same. In fact, the first equation is simply a rearranged multiple of the second equation.

Algebraically, we can describe the solution set as $\{(3y-3, y) | y \in \mathbb{R}\}$. If we had solved either equation for y instead of x, we would have obtained the alternative but equivalent solution $\left\{\left(x, \frac{x+3}{3}\right) \middle| x \in \mathbb{R}\right\}$.

TOPIC 3: Solving Systems of Linear Equations by Elimination

In some systems, the expressions obtained by solving one equation for one variable are difficult to work with, no matter which variable we choose. In such cases, the solution method of elimination may be a more efficient choice. The elimination method is also often a better choice for larger systems.

The method of elimination is based on the goal of eliminating one variable in one equation by adding two equations together, and in fact, the elimination method is often called the *addition method.*

The method works by making sure that the resulting equation has fewer variables than either of the original two, simplifying the system. In fact, if the system is of the two-variable, two-equation variety, the new equation is ready to be solved for its remaining variable, and the solution of the system is then straightforward to find.

Example 3: Solving an Independent System by Elimination

Use the method of elimination to solve the system $\begin{cases} 5x+3y=-7 \\ 7x-6y=-20 \end{cases}.$

Solution

The coefficient of y in the second equation is –6, while the coefficient of y in the first equation is 3. This means that if we multiply all the terms in the first equation by 2, the coefficients of y will be negatives of one another, so adding the two equations will eliminate the y variable.

In order to keep track of these steps, we annotate our work with labeled arrows. When we modify the system, we are not changing the solutions, we are just writing an equivalent system that is easier to solve. The ultimate goal is to rewrite the system so that we can "read off" the answer.

The notation above the arrow indicates that we have modified the system by multiplying each term in equation 1 by the constant 2.

$$\begin{cases} 5x+3y=-7 \\ 7x-6y=-20 \end{cases} \xrightarrow{2E_1} \begin{cases} 10x+6y=-14 \\ 7x-6y=-20 \end{cases}$$

$$17x=-34 \quad \text{Add the equations.}$$

$$x=-2 \quad \text{Solve for } x.$$

We can then substitute $x=-2$ into either of the original equations to determine y. Here, we substitute into the first equation of the original system.

$$5(-2)+3y=-7$$
$$3y=3$$
$$y=1$$

The ordered pair $(-2,1)$ is thus the solution of the system. Note that using the second equation gives the same y-value and is a good way to check our work.

Example 4: Solving an Inconsistent System by Elimination

Use the method of elimination to solve the system $\begin{cases} 2x-3y=3 \\ 3x-\frac{9}{2}y=5 \end{cases}.$

Solution

Eliminate the variable x by multiplying both equations by a constant.

$$\begin{cases} 2x - 3y = 3 \\ 3x - \frac{9}{2}y = 5 \end{cases} \xrightarrow[-2E_2]{3E_1} \begin{array}{r} \begin{cases} 6x - 9y = 9 \\ -6x + 9y = -10 \end{cases} \\ \hline 0 = -1 \end{array}$$

Although the intent was to obtain coefficients of x that were negatives of one another, we have achieved the same thing for y.

When we add the equations, the result is $0 = -1$, a false statement. This means that no ordered pair solves both equations, and the system is inconsistent. Graphically, the two lines defined by the equations are parallel.

TECHNOLOGY: Solving Systems of Equations

The solution to a consistent pair of linear equations is the point common to both equations. Graphically speaking, the solution is the point where the graphs of the two equations intersect. We can use a graphing utility to find this point. Consider the following system of equations: $\begin{cases} 2x - 3y = -13 \\ x = y - 6 \end{cases}$. One way to solve this system using a TI-84 Plus is to graph each equation. Remember to solve for y before entering the equation in the Y= editor. After both equations are entered, press graph.

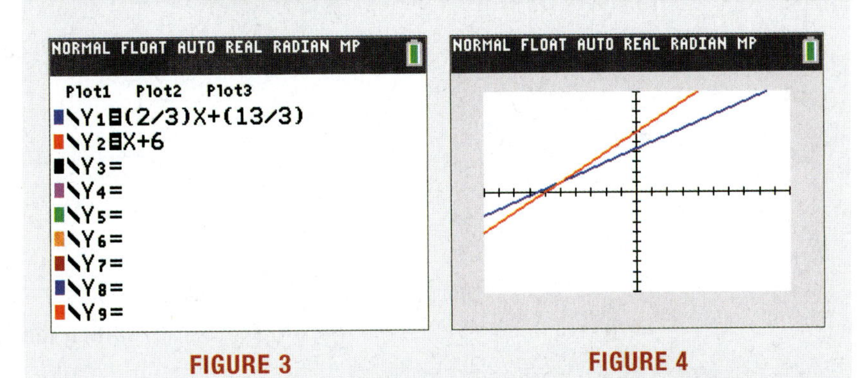

FIGURE 3 FIGURE 4

Once the graph of the two lines is displayed, press 2nd trace to access the CALCULATE menu and select intersect. The phrase First curve? should appear. Use the arrows to move the cursor along the first line to where it appears to intersect the other line and press enter. When the phrase Second curve? appears, press enter again (as the cursor should now be on the second line, still near the point of intersection). Now the word Guess? should appear. Press enter a final time and the x- and y-values of the point of intersection will appear at the bottom (see Figure 5).

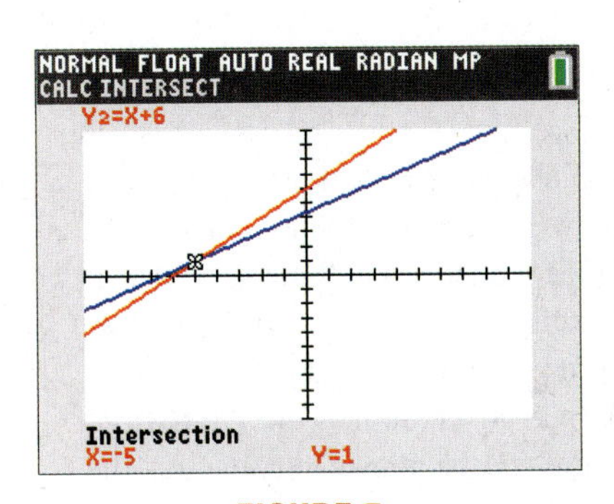

FIGURE 5

So the point where the lines intersect, and thus the solution to this system of equations, is $(-5,1)$. This method works with any system of equations that can be graphed on a calculator, not just linear ones.

TOPIC 4: Systems of Linear Equations in Three or More Variables

Algebraically, larger systems of equations can be dealt with in the same way as the two-variable, two-equation systems that we have studied (though the number of steps needed to obtain a solution might increase). Geometrically, however, larger systems can mean something quite different.

For example, if an equation contains three variables, say x, y, and z, a given solution of the equation must consist of an *ordered triple* of numbers, not an ordered pair. A graphical representation of the ordered triple requires three coordinate axes, as opposed to two. This leads to the concept of three-dimensional space, and a coordinate system with three axes. Figure 6 is an illustration of the way in which the positive x, y, and z axes are typically represented on a two-dimensional surface, such as a piece of paper, a computer monitor, or a blackboard. The negative portion of each of the axes is not drawn, and the three axes meet at the origin (the point with $(0,0,0)$ as its coordinates) at right angles. As an illustration of how ordered triples appear plotted in Cartesian *space*, the point $(1,2,4)$ is plotted (the thin colored lines are drawn merely for reference and are not part of the plot).

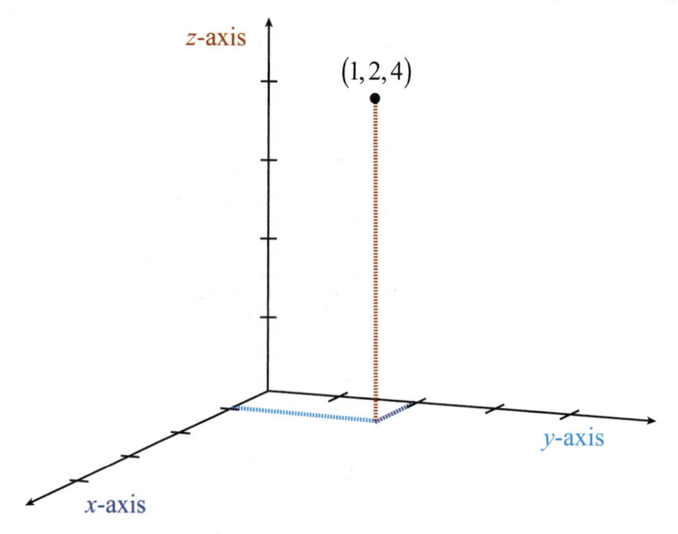

FIGURE 6: Plotting the Point $(1,2,4)$

We are interested in the graph of a linear equation in three variables. It turns out that any equation of the form

$$Ax + By + Cz = D,$$

where not all of A, B, and C are 0, depicts a plane in three-dimensional space. A system of linear equations in three variables will thus describe a collection of planes, one per equation. In a system of three linear equations in three variables, it is possible for the three planes to intersect in exactly one point. It is also possible, however, for two of the planes to intersect in a line while the third plane contains no point of that line, for the three planes to intersect in a common line, for two or three of the planes to be parallel, or for two or three of the planes to coincide. The addition of another variable and another equation leads to many more possibilities than we have seen thus far. However, it is still the case that a linear system has no solution, exactly one solution, or an infinite number of solutions. Figure 7 illustrates some of the possible configurations of a three-variable, three-equation linear system.

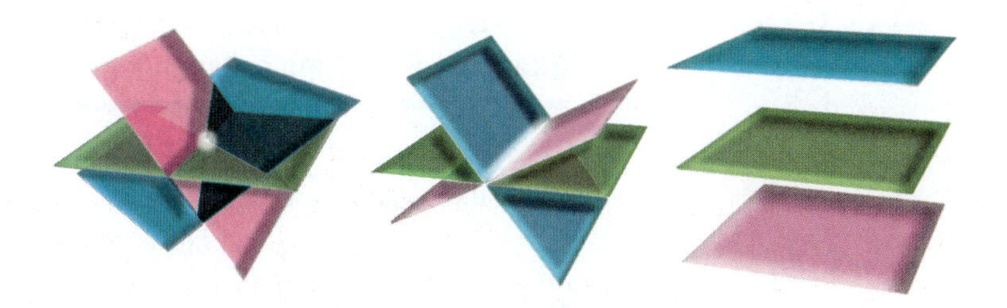

FIGURE 7: Three-Variable, Three-Equation Systems

Example 5: Solving an Independent System by Elimination

Solve the system $\begin{cases} 2x+y+z=6 \\ 2x+3y-z=-2. \\ -3x+2y-z=5 \end{cases}$

Solution

We will follow the same type of approach we used in Example 4. If we add the first equation to the second equation or the third equation, we will eliminate z, resulting in a two-equation system in the variables x and y.

$$\begin{array}{r} \text{Equation 1:} \\ \text{Equation 2:} \end{array} \begin{cases} 2x+y+z=6 \\ 2x+3y-z=-2 \end{cases} \qquad \begin{array}{r} \text{Equation 1:} \\ \text{Equation 3:} \end{array} \begin{cases} 2x+y+z=6 \\ -3x+2y-z=5 \end{cases}$$

$$4x+4y=4 \qquad\qquad -x+3y=11$$

Putting these two equations together, we have the following system.

$$\begin{cases} 4x+4y=4 \\ -x+3y=11 \end{cases}$$

We use elimination once more, multiplying the second equation by 4 to eliminate x.

$$\begin{cases} 4x+4y=4 \\ -x+3y=11 \end{cases} \xrightarrow{4E_2} \begin{cases} 4x+4y=4 \\ -4x+12y=44 \end{cases}$$

$$\begin{aligned} 16y&=48 \\ y&=3 \end{aligned}$$

Now that we have solved for y, we can plug this back into one of the equations from the two variable system to solve for x.

$$\begin{aligned} -x+3(3)&=11 \\ -x+9&=11 \\ x&=-2 \end{aligned}$$

Finally, we substitute both x and y into one of the equations from the original system.

$$\begin{aligned} 2(-2)+3+z&=6 \\ -4+3+z&=6 \\ z&=7 \end{aligned}$$

Thus, the solution to the system of equations is the ordered triple $(-2,3,7)$.

Example 6: Solving a Dependent System by Elimination

Solve the system $\begin{cases} 3x-5y+z=-10 \\ -x+2y-3z=-7 \\ x-y-5z=-24 \end{cases}$.

Solution

In general, a good approach to solving a large system of equations is to try to eliminate a variable and obtain a smaller system.

There are many possible ways to proceed. One option is to use the second equation (or a multiple of it) to eliminate x when we add it to the first and third equations. The result will be a two equation system in the variables y and z.

$$\begin{array}{l} \text{Equation 1:} \\ \text{Equation 2:} \end{array} \begin{cases} 3x-5y+z=-10 \\ -x+2y-3z=-7 \end{cases} \xrightarrow{3E_2} \begin{cases} 3x-5y+z=-10 \\ -3x+6y-9z=-21 \end{cases} \qquad \overline{\quad y-8z=-31}$$

$$\begin{array}{l} \text{Equation 2:} \\ \text{Equation 3:} \end{array} \begin{cases} -x+2y-3z=-7 \\ x-y-5z=-24 \end{cases} \qquad \overline{\quad y-8z=-31}$$

The two resulting equations are identical, and tell us that $y=8z-31$. We can now use any equation that contains x to determine the relation between x and z. For instance, the third equation in the system tells us that $x=y+5z-24$, or

$$x=(8z-31)+5z-24=13z-55.$$

One description of the solution set is thus $\{(13z-55, 8z-31, z) \mid z \in \mathbb{R}\}$. Geometrically, the three planes described by the equations of the system intersect in a line.

TOPIC 5: Applications of Systems of Linear Equations

Many applications that we have previously analyzed using a single equation are more naturally stated in terms of two or more equations. Consider, for example, the following mixture problem.

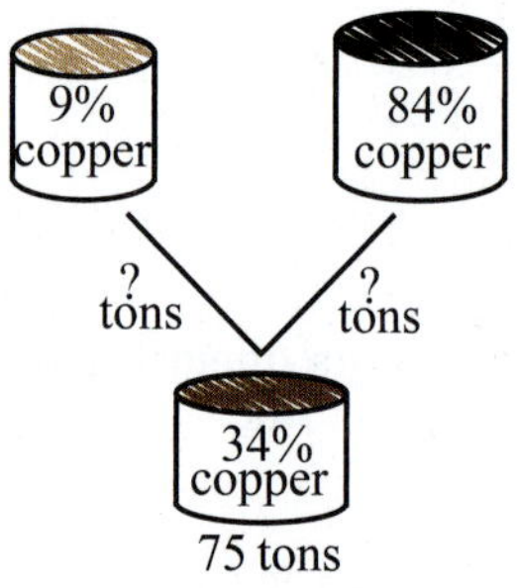

9% copper
84% copper
? tons
? tons
34% copper
75 tons

FIGURE 8

Example 7: Mixing Alloys

A foundry needs to produce 75 tons of an alloy that is 34% copper. It has supplies of 9% copper alloy and 84% copper alloy. How many tons of each alloy must be mixed to obtain the desired result?

Solution

Let x represent the number of tons of 9% copper alloy needed, and y the number of tons of 84% copper alloy needed. We have two variables, so we will need two equations to find the solution.

Since we need 75 total tons of alloy, one equation is $x+y=75$. We also know that 9% of the x tons and 84% of the y tons represent the total amount of copper, and this amount must equal 34% of 75 tons. The second equation is thus $0.09x+0.84y=0.34(75)$. This gives us a system that can be solved by elimination.

$$\begin{cases} x+y=75 \\ 0.09x+0.84y=0.34(75) \end{cases} \xrightarrow[100E_2]{-9E_1} \begin{cases} -9x-9y=-675 \\ 9x+84y=2550 \end{cases}$$
$$75y=1875$$
$$y=25$$

Substituting back, we have $x+25=75$, so $x=50$. Thus, 50 tons of 9% alloy and 25 tons of 84% alloy are needed.

Example 8: Determining Ages

If the ages of three girls, Xenia, Yolanda, and Zsa Zsa, are added, the result is 30. The sum of Xenia's and Yolanda's ages is Zsa Zsa's age, while Xenia's age subtracted from Yolanda's is half of Zsa Zsa's age a year ago. How old is each girl?

Solution

Let x, y, and z represent Xenia's, Yolanda's, and Zsa Zsa's ages, respectively. The first sentence tells us that $x+y+z=30$. The second sentence tells us that $x+y=z$, and that $y-x=\frac{z-1}{2}$. To make the work easier, these equations can be rewritten as follows.

$$\begin{cases} x+y+z=30 \\ x+y=z \\ y-x=\frac{z-1}{2} \end{cases} \xrightarrow{2E_3} \begin{cases} x+y+z=30 \\ x+y-z=0 \\ -2x+2y-z=-1 \end{cases}$$

From this, we see that the sum of the first and second equations results in $2x+2y=30$, or $x+y=15$, and the sum of the first and third equations is $-x+3y=29$. The method of elimination is the best choice for solving the new system

$$\begin{cases} x+y=15 \\ -x+3y=29 \end{cases}$$

as the sum of the two equations gives us $4y=44$, or $y=11$. We can use this value to determine that $x=4$, and then use these two values to determine that $z=15$. Geometrically, it means that the ordered triple $(4,11,15)$ is the point of intersection of the three planes described by these equations. In context, it means that Xenia is 4, Yolanda is 11, and Zsa Zsa is 15.

11.1 EXERCISES

PRACTICE

Use the method of substitution to solve the following systems of equations. If a system is dependent, express the solution set in terms of one of the variables. See Examples 1 and 2.

1. $\begin{cases} 2x - y = -12 \\ 3x + y = -13 \end{cases}$

2. $\begin{cases} 2x - 4y = -6 \\ 3x - y = -4 \end{cases}$

3. $\begin{cases} 3y = 9 \\ x + 2y = 11 \end{cases}$

4. $\begin{cases} -3x - y = 2 \\ 9x + 3y = -6 \end{cases}$

5. $\begin{cases} 2x + y = -2 \\ -4x - 2y = 5 \end{cases}$

6. $\begin{cases} 5x - y = -21 \\ 9x + 2y = -34 \end{cases}$

7. $\begin{cases} 2x - y = -3 \\ -4x + 2y = 6 \end{cases}$

8. $\begin{cases} 3x + 6y = -12 \\ 2x + 4y = -8 \end{cases}$

9. $\begin{cases} 2x + 5y = 33 \\ 3x = -3 \end{cases}$

10. $\begin{cases} 5x + 2y = 8 \\ 2x + y = 6 \end{cases}$

11. $\begin{cases} -2x + y = 5 \\ 9x - 2y = 5 \end{cases}$

12. $\begin{cases} 3x + y = 4 \\ -2x + 3y = 1 \end{cases}$

13. $\begin{cases} 4x - y = -1 \\ -8x + 2y = 2 \end{cases}$

14. $\begin{cases} 4x - 2y = 3 \\ -2x + y = -7 \end{cases}$

15. $\begin{cases} 9x - y = -1 \\ 3x + 2y = 44 \end{cases}$

Use the method of elimination to solve the following systems of equations. If a system is dependent, express the solution set in terms of one of the variables. See Examples 3 and 4.

16. $\begin{cases} 2x - 3y = 8 \\ 8x + 5y = -2 \end{cases}$

17. $\begin{cases} -2x + 3y = 13 \\ 4x + 2y = -18 \end{cases}$

18. $\begin{cases} 5x + 7y = 1 \\ -2x + 3y = -12 \end{cases}$

19. $\begin{cases} x + 2y = 17 \\ 3x + 4y = 39 \end{cases}$

20. $\begin{cases} 5x - 10y = 9 \\ -x + 2y = -3 \end{cases}$

21. $\begin{cases} -2x - 2y = 4 \\ 3x + 3y = -6 \end{cases}$

22. $\begin{cases} 4x + y = 11 \\ 3x - 2y = 0 \end{cases}$

23. $\begin{cases} 7x + 8y = -3 \\ -5x - 4y = 9 \end{cases}$

24. $\begin{cases} -2x - y = 9 \\ 4x + 2y = 1 \end{cases}$

25. $\begin{cases} -2x + 4y = 6 \\ 3x - y = -4 \end{cases}$

26. $\begin{cases} 5x - 6y = -1 \\ -4x + 3y = -10 \end{cases}$

27. $\begin{cases} \frac{2}{3}x + y = -3 \\ 3x + \frac{5}{2}y = -\frac{7}{2} \end{cases}$

28. $\begin{cases} \frac{x}{5} - y = -\frac{11}{5} \\ \frac{x}{4} + y = 4 \end{cases}$

29. $\begin{cases} \frac{2}{3}x + 2y = 1 \\ x + 3y = 0 \end{cases}$

30. $\begin{cases} -x - 5y = -6 \\ \frac{3}{5}x + 3y = 1 \end{cases}$

Use any convenient method to solve the following systems of equations. If a system is dependent, express the solution set in terms of one or more of the variables, as appropriate. See Examples 5 and 6.

31. $\begin{cases} x - y + 4z = -4 \\ 4x + y - 2z = -1 \\ -y + 2z = -3 \end{cases}$

32. $\begin{cases} x + 2y = -1 \\ y + 3z = 7 \\ 2x + 5z = 21 \end{cases}$

33. $\begin{cases} x + y = 4 \\ y + 3z = -1 \\ 2x - 2y + 5z = -5 \end{cases}$

34. $\begin{cases} 2x - y = 0 \\ 5x - 3y - 3z = 5 \\ 2x + 6z = -10 \end{cases}$

35. $\begin{cases} 3x - y + z = 2 \\ -6x + 2y - 2z = -4 \\ -3x + y - z = -2 \end{cases}$

36. $\begin{cases} 2x - 3y = -2 \\ x - 4y + 3z = 0 \\ -2x + 7y - 5z = 0 \end{cases}$

37. $\begin{cases} 3x - y + z = 2 \\ -6x + 2y - 2z = 1 \\ 5x + 2y - 3z = 2 \end{cases}$

38. $\begin{cases} 4x - y + 5z = 6 \\ 4x - 3y - 5z = -14 \\ -2x - 5z = -8 \end{cases}$

39. $\begin{cases} 3x + 8z = 3 \\ -3x + y - 7z = -2 \\ x + 2y + 3z = 3 \end{cases}$

40. $\begin{cases} x + 2y + z = 8 \\ 2x - 3y - 4z = -16 \\ x - 5y + 5z = 6 \end{cases}$

41. $\begin{cases} 2x - 7y - 4z = 7 \\ -x + 4y + 2z = -3 \\ 3y - 4z = -1 \end{cases}$

42. $\begin{cases} 4x + 4y - 2z = 6 \\ x - 5y + 3z = -2 \\ -2x - 2y + z = 3 \end{cases}$

43. $\begin{cases} 2x + 3y + 4z = 1 \\ 3x - 4y + 5z = -5 \\ 4x + 5y + 6z = 5 \end{cases}$

44. $\begin{cases} x - 4y + 2z = -1 \\ 2x + y - 3z = 10 \\ -3x + 12y - 6z = 3 \end{cases}$

45. $\begin{cases} x + 2y + 3z = 29 \\ 2x - y - z = -2 \\ 3x + 2y - 6z = -8 \end{cases}$

46. $\begin{cases} 5x - 2y + z = 14 \\ 8x + 4y = 12 \\ 9x = 18 \end{cases}$

47. $\begin{cases} 2x + 5y = 6 \\ 3y + 8z = -6 \\ x + 4y = -5 \end{cases}$

48. $\begin{cases} 4x + 3y + 4z = 5 \\ 5x - 6y - 2z = -12 \\ 5z = 20 \end{cases}$

49. $\begin{cases} 9x + 4y - 8z = -4 \\ -6x + 3y - 9z = -9 \\ 8y - 3z = 18 \end{cases}$

50. $\begin{cases} 21x - 7y + 51z = 141 \\ 13x + 9y - 5z = -19 \\ 19x - 8y + 23z = 30 \end{cases}$

APPLICATIONS

51. Karen empties out her purse and finds 45 loose coins, consisting entirely of nickels and pennies. If the total value of the coins is \$1.37, how many nickels and how many pennies does she have?

52. What choice of a, b, and c will force the graph of the polynomial $f(x) = ax^2 + bx + c$ to have a y-intercept of 5 and to pass through the points $(1,3)$ and $(2,0)$?

53. A tour organizer is planning on taking a group of 40 people to a musical. Balcony tickets cost \$29.95 and regular tickets cost \$19.95. The organizer collects a total of \$1048.00 from her group to buy the tickets. How many people chose to sit in the balcony?

54. How many ounces each of a 12% alcohol solution and a 30% alcohol solution must be combined to obtain 60 ounces of an 18% solution?

55. Eliza's mother is 20 years older than Eliza, but 3 years younger than Eliza's father. Eliza's father is 7 years younger than three times Eliza's age. How old is Eliza?

56. An investor decides at the beginning of the year to invest some of his cash in an account paying 8% annual interest, and to put the rest in a stock fund that ends up earning 15% over the course of the year. He puts \$2000 more in the first account than in the stock fund, and at the end of the year he finds he has earned \$1310 in interest. How much money was invested at each of the two rates?

57. Jack and Tyler went shopping for summer clothes. Shirts were \$12.47 each, including tax, and shorts were \$17.23 per pair, including tax. Jack and Tyler spent a total of \$156.21 on 11 items. How many shirts and pairs of shorts did they buy?

58. Three years ago, Bob was twice as old as Marla. Fifteen years ago, Bob was three times as old as Marla. How old is Bob?

59. Deyanira empties her pockets and finds 42 coins consisting of quarters, dimes, and pennies. There are twice as many pennies as dimes and quarters total. If the total value of the coins is \$2.13, how many coins of each denomination does she have?

60. If an investor has invested \$1000 in stocks and bonds, how much has he invested in stocks if he invested four times more in stocks than in bonds?

61. Twelve years ago, Jim was twice as old as Kristin. Sixteen years ago, Jim was three times older than Kristin. How old is Jim?

62. A movie brought in \$740 in ticket sales in one day. Tickets during the day cost \$5 and tickets at night cost \$7. If 120 tickets were sold, how many were sold during the day?

63. A computer has 24 screws in its case. If there are 7 times more slotted screws than thumb screws, how many thumb screws are in the computer?

64. Jael has \$10,000 she would like to invest. She has narrowed her options down to a certificate of deposit paying 5% annually, bonds paying 4% annually, and stocks with an expected annual rate of return of 13.5%. If she wants to invest twice as much in the stocks as in the certificate of deposit and she wants to earn \$1000 in interest by the end of the year, how much should she invest in each type of investment?

65. Lea ordered fruit baskets for three of her coworkers. One contained 5 apples, 2 oranges, and 1 mango and cost \$6.81. Another contained 2 mangos, 8 oranges, and 3 apples and cost \$11.88. The third contained 4 apples, 4 oranges, and 4 mangos and cost \$11.04. How much did each type of fruit cost?

TECHNOLOGY

Solve each of the following systems of equations using a graphing utility.

66. $\begin{cases} 98x+43y-82z=-784 \\ -65x+34y=3032 \\ 28y-13z=966 \end{cases}$

67. $\begin{cases} 7.5x+5.2y-9.3z=-23.971 \\ -6.8x+4.4y=2.708 \\ 0.9x-1.88y=-2.0194 \end{cases}$

68. $\begin{cases} -5x+2y-20z=14 \\ 2x-3y+10z=-19 \\ 7x+4y-7z=-7 \end{cases}$

69. $\begin{cases} -5.5x+2.2y-5.1z=11.29 \\ 1.8x+4.9y-0.5z=7.066 \\ 3.9x-2.6y+6.3z=-3.698 \end{cases}$

70. $\begin{cases} 5x-10y+11z=19 \\ 27x+9y+7z=-44 \\ 2x+19y-4z=-3 \end{cases}$

71. $\begin{cases} -23x+17y-7z=-51 \\ -13x+25y-11z=45 \\ 51x-21y-28z=-58 \end{cases}$

11.2 MATRIX NOTATION AND GAUSS-JORDAN ELIMINATION

TOPICS

1. Systems of linear equations, matrices, and augmented matrices
2. Gaussian elimination and row echelon form
3. Gauss-Jordan elimination and reduced row echelon form

TOPIC 1: Systems of Linear Equations, Matrices, and Augmented Matrices

Infinite Systems

Some systems of linear equations go beyond the merely large and enter the realm of the infinite. Anna Johnson Pell Wheeler was an American mathematician who worked with such systems in the early 1900s, when the area of study was known as *infinite dimensional linear algebra*—such study is now a part of *functional analysis*. Wheeler became a professor at Bryn Mawr College in 1918, teaching and serving there, with some interruption, until her retirement in 1948. Along with her research achievements, she was instrumental in helping the eminent mathematician Emmy Noether leave Nazi Germany in 1933 to also become a professor at Bryn Mawr.

Many applications of systems of linear equations, such as the scheduling of planes and passengers at an airport, involve thousands of variables and equations. The solution methods of substitution and elimination learned in Section 11.1 are inadequate for solving such large systems. In this section, we introduce the concept of a *matrix* and show how it can be used to both describe and solve a system of linear equations.

DEFINITION: Matrices and Matrix Notation

A **matrix** is a rectangular array of numbers, called **elements** or **entries** of the matrix. As the numbers are in a rectangular array, they naturally form **rows** and **columns**. It is often important to determine the size of a given matrix; we say that a matrix with m rows and n columns is an $m \times n$ matrix (read "m by n"), or of **order** $m \times n$. By convention, the number of rows is always stated first.

columns ↓ ↓ ↓ ↓

rows →

$$\begin{bmatrix} 4 & 0 & 1 & -3 \\ 12 & -5 & -5 & 8 \\ 0 & -7 & 25 & 2 \end{bmatrix}$$

This matrix has 3 rows and 4 columns, so it is a 3×4 matrix.

Matrices are often labeled with capital letters. The same letter in lowercase, with a pair of subscripts attached, is usually used to refer to its individual elements. For instance, if A is a matrix, a_{ij} refers to the element in the ith row and the jth column.

If we call the above matrix A, then we have $a_{23} = -5$ and $a_{32} = -7$.

Example 1: Matrices and Matrix Notation

Given the matrix $A = \begin{bmatrix} -27 & 0 & 1 \\ 5 & -\pi & 13 \end{bmatrix}$, determine

a. the order of A,

b. the value of a_{13},

c. the value of a_{21}.

Solution

a. A has 2 rows and 3 columns, and is thus a 2×3 matrix.

b. The value of the entry in the first row and third column is $a_{13} = 1$.

c. The value of the entry in the second row and first column is $a_{21} = 5$.

DEFINITION: Standard Form of a System of Linear Equations

A system of linear equations is in **standard form** when each equation has been simplified with its variables on the left-hand side and its constant term on the right-hand side. Each equation should have its variable terms listed in the same order.

For example, the system $\begin{cases} 3x + 4 = 7y \\ -2x + 8y = 18 \end{cases}$ is written in standard form as $\begin{cases} 3x - 7y = -4 \\ -x + 4y = 9 \end{cases}$.

Now that we have a standard way to organize the terms in a system of equations, we can put the matrix to use as a way to represent a system of equations concisely.

DEFINITION: Augmented Matrices

The **augmented matrix** of a system of linear equations written in standard form is a matrix consisting of the coefficients of the variables listed in their relative positions with an adjoined column of the constants of the system. The matrix of coefficients and the column of constants are customarily separated by a vertical bar.

For example, the augmented matrix for the system $\begin{cases} 3x - 7y = -4 \\ -x + 4y = 9 \end{cases}$ is $\left[\begin{array}{cc|c} 3 & -7 & -4 \\ -1 & 4 & 9 \end{array}\right]$.

The augmented matrix will have as many rows as there are equations in the system, and one more column than there are variables.

Example 2: Augmented Matrices

Construct the augmented matrix for the linear system $\begin{cases} \dfrac{2x-6y}{2}=3-z \\ z-x+5y=12 \\ x+3y-2=2z \end{cases}$.

Solution

Write each equation in standard form, then read off the coefficients and constants to construct the augmented matrix.

Linear System	Standard Form	Augmented Matrix
$\dfrac{2x-6y}{2}=3-z$	$x-3y+z=3$	
$z-x+5y=12$	$-x+5y+z=12$	$\left[\begin{array}{ccc\|c} 1 & -3 & 1 & 3 \\ -1 & 5 & 1 & 12 \\ 1 & 3 & -2 & 2 \end{array}\right]$
$x+3y-2=2z$	$x+3y-2z=2$	

TOPIC 2: Gaussian Elimination and Row Echelon Form

Consider the following augmented matrix:

$$\left[\begin{array}{ccc|c} 1 & 2 & -2 & 11 \\ 0 & 1 & -1 & 3 \\ 0 & 0 & 1 & -1 \end{array}\right]$$

We can translate this back into system form to obtain

$$\begin{cases} x+2y-2z=11 \\ \quad\quad y-z=3 \\ \quad\quad\quad z=-1 \end{cases},$$

and we note that this system can be solved when we substitute the last equation $z=-1$ into the second equation to obtain $y-(-1)=3$, or $y=2$, and then substitute again in the first equation to obtain $x+2(2)-2(-1)=11$, or $x=5$.

Typically, systems of equations aren't so straightforward to solve, at least not at first. Recall that the methods of substitution and elimination made systems of equations simpler to solve by reducing the number of equations and/or variables present.

Matrices, a powerful organizational tool, make it easier to transform complicated systems into ones like the example above. The process of *Gaussian elimination* (named after the mathematician Carl Friedrich Gauss) can transform any augmented matrix into a form like the one above, and the solution of the corresponding system then solves the original system as well. The technical name for an augmented matrix of this form is *row echelon form*.

DEFINITION: Row Echelon Form

A matrix is in **row echelon form** if each of the following is true.

1. The first nonzero entry in each row is 1. We call this a **leading 1**.
2. Every entry below a leading 1 is 0, and each leading 1 appears farther to the right than the leading 1s in the rows above it.
3. All rows consisting entirely of 0s (if there are any) appear at the bottom.

Example 3: Row Echelon Form

Determine if the following matrices are in row echelon form. If not, explain why not.

a. $\left[\begin{array}{ccc|c} 2 & 0 & 13 & -1 \\ 0 & 5 & 1 & 1 \\ 0 & 0 & 1 & -4 \end{array}\right]$

b. $\left[\begin{array}{ccc|c} 1 & 2 & -8 & 0 \\ 0 & 1 & 1 & 3 \\ 0 & 0 & 1 & 10 \end{array}\right]$

c. $\left[\begin{array}{ccc|c} 1 & 0 & 4 & -2 \\ 1 & 2 & 0 & 7 \\ 1 & -5 & -3 & 6 \end{array}\right]$

d. $\left[\begin{array}{ccc|c} 1 & -3 & 1 & 6 \\ 0 & 0 & 0 & 0 \\ 0 & 1 & -8 & -10 \end{array}\right]$

Solution

a. This matrix fails the first condition, as the first nonzero entry is not always 1.

$$\left[\begin{array}{ccc|c} 2 & 0 & 13 & -1 \\ 0 & 5 & 1 & 1 \\ 0 & 0 & 1 & -4 \end{array}\right]$$

b. This matrix meets each of the three conditions, and is in row echelon form.

c. This matrix fails the second condition, as nonzero values are present beneath a leading 1.

$$\left[\begin{array}{ccc|c} 1 & 0 & 4 & -2 \\ 1 & 2 & 0 & 7 \\ 1 & -5 & -3 & 6 \end{array}\right]$$

d. This matrix fails the third condition, since there is a row of all zeros that lies above a row with nonzero entries.

$$\left[\begin{array}{ccc|c} 1 & -3 & 1 & 6 \\ 0 & 0 & 0 & 0 \\ 0 & 1 & -8 & -10 \end{array}\right]$$

Since the real goal of Gaussian elimination is to solve systems of equations that are represented by augmented matrices, we need to make sure any changes we apply to the matrix do not change the solution of the system.

These "legal" operations are called *elementary row operations*. There are three such matrix operations, based on the fact that the following actions have no effect on the solutions of a system of equations.

- Interchanging the order of two equations in a system
- Multiplying all the terms in an equation by a nonzero constant
- Adding one equation to another (or a nonzero multiple of an equation)

DEFINITION: Elementary Row Operations

Let A be an augmented matrix corresponding to a system of equations. Each of the following operations on A results in the augmented matrix of an equivalent system. In the notation, R_i refers to row i of the matrix A.

1. Rows i and j can be interchanged. (Denoted as $R_i \leftrightarrow R_j$.)
2. Each entry in row i can be multiplied by a nonzero constant c. (Denoted as cR_i.)
3. Row j can be replaced with the sum of itself and a constant multiple of row i. (Denoted as $cR_i + R_j$.)

Typically, we write that an operation has been performed by connecting the original matrix to the new matrix with an arrow, writing the type of operation above it. For example, the matrix from Example 3d can be placed in row echelon form as follows.

$$\left[\begin{array}{ccc|c} 1 & -3 & 1 & 6 \\ 0 & 0 & 0 & 0 \\ 0 & 1 & -8 & -10 \end{array}\right] \xrightarrow{R_2 \leftrightarrow R_3} \left[\begin{array}{ccc|c} 1 & -3 & 1 & 6 \\ 0 & 1 & -8 & -10 \\ 0 & 0 & 0 & 0 \end{array}\right]$$

Example 4: Elementary Row Operations

Perform the indicated elementary row operation.

$$\left[\begin{array}{ccc|c} 1 & 2 & 3 & 0 \\ 5 & 4 & 1 & -7 \\ 6 & 1 & 8 & -9 \end{array}\right] \xrightarrow{-5R_1 + R_2} ?$$

Solution

The row operation indicates that we should add -5 times Row 1 to Row 2.

$$\left[\begin{array}{ccc|c} 1 & 2 & 3 & 0 \\ 5 & 4 & 1 & -7 \\ 6 & 1 & 8 & -9 \end{array}\right] \xrightarrow{-5R_1 + R_2} \left[\begin{array}{ccc|c} 1 & 2 & 3 & 0 \\ -5(1)+5 & -5(2)+4 & -5(3)+1 & -5(0)+(-7) \\ 6 & 1 & 8 & -9 \end{array}\right]$$

$$= \left[\begin{array}{ccc|c} 1 & 2 & 3 & 0 \\ 0 & -6 & -14 & -7 \\ 6 & 1 & 8 & -9 \end{array}\right]$$

Note that the first row now begins with 1 and the second row now begins with 0. The row operation has begun to change this matrix towards row echelon form.

In summary, Gaussian elimination refines the method of elimination by removing the need to write variables at each step and by providing a framework for systematic application of row operations. Just as with substitution and elimination, there is no one correct method to apply Gaussian elimination. With the upcoming examples, we will see some general guidelines to follow.

Example 5: Gaussian Elimination

Use Gaussian elimination to solve the system $\begin{cases} -2x+y-5z=-6 \\ x+2y-z=-8. \\ 3x-y+2z=2 \end{cases}$

Solution

First, we read off the augmented matrix corresponding to this system.

$$\left[\begin{array}{ccc|c} -2 & 1 & -5 & -6 \\ 1 & 2 & -1 & -8 \\ 3 & -1 & 2 & 2 \end{array}\right]$$

Now, we transform it into row echelon form. It is usually easiest to work one column at a time. After getting a leading 1 in the first row, use row operations to obtains 0s below it. Repeat this process with each successive column.

$$\left[\begin{array}{ccc|c} -2 & 1 & -5 & -6 \\ 1 & 2 & -1 & -8 \\ 3 & -1 & 2 & 2 \end{array}\right] \xrightarrow{R_1 \leftrightarrow R_2} \left[\begin{array}{ccc|c} 1 & 2 & -1 & -8 \\ -2 & 1 & -5 & -6 \\ 3 & -1 & 2 & 2 \end{array}\right]$$

Exchange Rows 1 and 2 to make 1 the first entry of the first row.

$$\xrightarrow{2R_1+R_2} \left[\begin{array}{ccc|c} 1 & 2 & -1 & -8 \\ 0 & 5 & -7 & -22 \\ 3 & -1 & 2 & 2 \end{array}\right]$$

Add 2 times Row 1 to Row 2 to get a 0 as the first entry of Row 2.

$$\xrightarrow{-3R_1+R_3} \left[\begin{array}{ccc|c} 1 & 2 & -1 & -8 \\ 0 & 5 & -7 & -22 \\ 0 & -7 & 5 & 26 \end{array}\right]$$

Add −3 times Row 1 to Row 3 to get a 0 as the first entry of Row 3.

$$\xrightarrow{\frac{1}{5}R_2} \left[\begin{array}{ccc|c} 1 & 2 & -1 & -8 \\ 0 & 1 & -\frac{7}{5} & -\frac{22}{5} \\ 0 & -7 & 5 & 26 \end{array}\right]$$

Multiply Row 2 by $\frac{1}{5}$ to make its first nonzero entry 1.

$$\xrightarrow{7R_2+R_3} \left[\begin{array}{ccc|c} 1 & 2 & -1 & -8 \\ 0 & 1 & -\frac{7}{5} & -\frac{22}{5} \\ 0 & 0 & -\frac{24}{5} & -\frac{24}{5} \end{array}\right]$$

Add 7 times Row 2 to Row 3 to get a 0 as the second entry of Row 3.

$$\xrightarrow{-\frac{5}{24}R_3} \left[\begin{array}{ccc|c} 1 & 2 & -1 & -8 \\ 0 & 1 & -\frac{7}{5} & -\frac{22}{5} \\ 0 & 0 & 1 & 1 \end{array}\right]$$

Multiply Row 3 by $-\frac{5}{24}$ to make its first nonzero entry a 1.

The third row in the last matrix tells us that $z = 1$. When we substitute this in the second equation, we obtain

$$y - \frac{7}{5}(1) = -\frac{22}{5},$$

so $y = -3$. From the first row equation, we then obtain

$$x + 2(-3) - 1(1) = -8,$$

or $x = -1$. Thus, the ordered triple $(-1, -3, 1)$ solves the system of equations.

TOPIC 3: Gauss-Jordan Elimination and Reduced Row Echelon Form

Once a matrix has been put in row echelon form, we can back-substitute, beginning with the last equation and working backward to solve the system of equations. We reached the row echelon form with matrix operations, and it turns out that we can perform the remaining steps in matrix form as well.

Just as Gaussian elimination is a refinement of the method of elimination, Gauss-Jordan elimination, co-named for Wilhelm Jordan (1842–1899), is a refinement of Gaussian elimination. The goal in the method is to put a given matrix into *reduced row echelon form.*

DEFINITION: Reduced Row Echelon Form

A matrix is said to be in **reduced row echelon form** if

1. it is in row echelon form and
2. each entry *above* a leading 1 is also 0.

Consider, for instance, the last matrix obtained in Example 5. If we begin with that matrix and eliminate entries (that is, convert to 0) above the leading 1 in the third column, and then do the same in the second column, we obtain a matrix in reduced row echelon form:

$$\left[\begin{array}{ccc|c} 1 & 2 & -1 & -8 \\ 0 & 1 & -\frac{7}{5} & -\frac{22}{5} \\ 0 & 0 & 1 & 1 \end{array}\right] \xrightarrow[R_3+R_1]{\frac{7}{5}R_3+R_2} \left[\begin{array}{ccc|c} 1 & 2 & 0 & -7 \\ 0 & 1 & 0 & -3 \\ 0 & 0 & 1 & 1 \end{array}\right] \xrightarrow{-2R_2+R_1} \left[\begin{array}{ccc|c} 1 & 0 & 0 & -1 \\ 0 & 1 & 0 & -3 \\ 0 & 0 & 1 & 1 \end{array}\right]$$

If we now write this in system form, we have

$$\begin{cases} x = -1 \\ y = -3 \\ z = 1 \end{cases}$$

which is equivalent to the original system, but in a form that directly tells us the solution of the system.

Example 6: Reduced Row Echelon Form

Use Gauss-Jordan elimination to solve the system $\begin{cases} x-2y+3z=-5 \\ 2x+3y-z=1 \\ -x-5y+4z=-6 \end{cases}$.

Solution

$$\left[\begin{array}{ccc|c} 1 & -2 & 3 & -5 \\ 2 & 3 & -1 & 1 \\ -1 & -5 & 4 & -6 \end{array}\right] \xrightarrow[R_1+R_3]{-2R_1+R_2} \left[\begin{array}{ccc|c} 1 & -2 & 3 & -5 \\ 0 & 7 & -7 & 11 \\ 0 & -7 & 7 & -11 \end{array}\right]$$

$$\xrightarrow{R_2+R_3} \left[\begin{array}{ccc|c} 1 & -2 & 3 & -5 \\ 0 & 7 & -7 & 11 \\ 0 & 0 & 0 & 0 \end{array}\right]$$

Note that the row of 0s at the bottom corresponds to the true statement 0 = 0, indicating that this system has an infinite number of solutions. In order to describe the solution set algebraically, we continue the row operations until the matrix is in reduced row echelon form.

$$\left[\begin{array}{ccc|c} 1 & -2 & 3 & -5 \\ 0 & 7 & -7 & 11 \\ 0 & 0 & 0 & 0 \end{array}\right] \xrightarrow{\frac{1}{7}R_2} \left[\begin{array}{ccc|c} 1 & -2 & 3 & -5 \\ 0 & 1 & -1 & \frac{11}{7} \\ 0 & 0 & 0 & 0 \end{array}\right]$$

$$\xrightarrow{2R_2+R_1} \left[\begin{array}{ccc|c} 1 & 0 & 1 & -\frac{13}{7} \\ 0 & 1 & -1 & \frac{11}{7} \\ 0 & 0 & 0 & 0 \end{array}\right]$$

The last matrix is in reduced row echelon form, as every leading 1 has 0s both below and above it. The corresponding system of equations is

$$\begin{cases} x+z=-\frac{13}{7} \\ y-z=\frac{11}{7} \end{cases}, \text{ which is equivalent to } \begin{cases} x=-z-\frac{13}{7} \\ y=z+\frac{11}{7} \end{cases}.$$

Thus, we can describe the solution set as $\left\{\left(-z-\frac{13}{7}, z+\frac{11}{7}, z\right) \middle| z \in \mathbb{R}\right\}$.

TECHNOLOGY: Matrices

Operations with matrices can also be done with a TI-84 Plus. Suppose we wanted to find the row reduced form of the matrix $\begin{bmatrix} 8 & -1 & 2 & 30 \\ -3 & 4 & 5 & -20 \\ 1 & 2 & -1 & -6 \end{bmatrix}$.

First, we need to enter in this matrix by pressing 2nd x^{-1} to access the matrix menu and use the arrows to highlight EDIT. Select [A] for matrix A and enter the dimensions, 3×4, and the values for each entry. The double subscripts appear at the bottom of the display as each value is entered. See Figure 1.

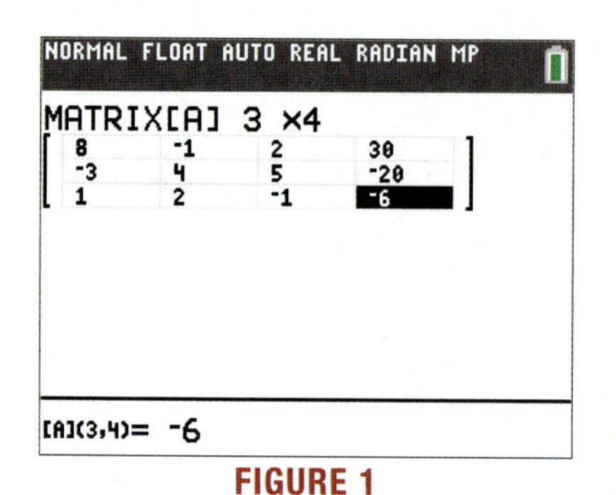

FIGURE 1

Return to the home screen by pressing 2nd mode. Now that the matrix is stored in the calculator as matrix A, we can perform operations. Access the matrix menu again, but this time select MATH. Scroll to select ref(and press enter.

Now we need to select which matrix we want to put in row echelon form, matrix A. To do this, press 2nd x^{-1} once more. NAMES and [A] should already be highlighted, so press enter. Add the right-hand parenthesis and press enter. See Figure 2.

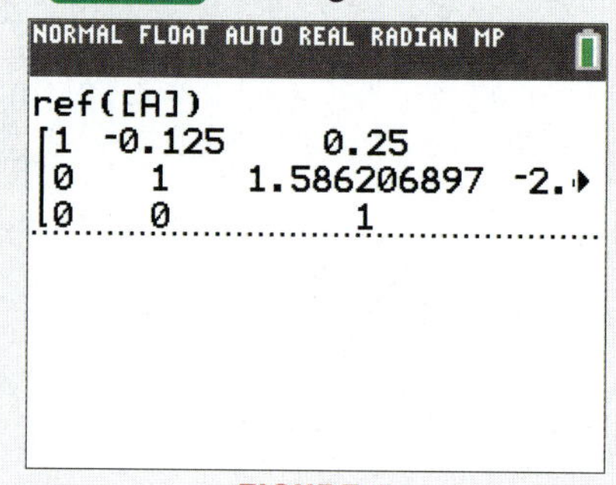

FIGURE 2

To view this matrix with fractional entries, press math and select ▸Frac. Press enter again on the main screen. See Figure 3.

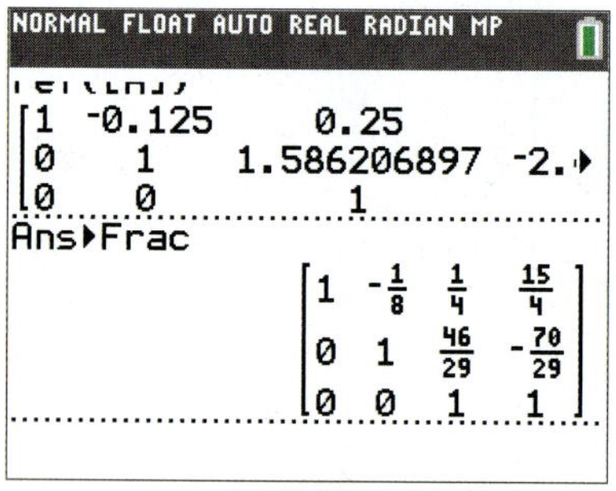

FIGURE 3

Note: If the entire matrix does not fit on the screen, you can use the arrow keys to view the right side of the matrix.

Finding the row-reduced echelon form of this matrix is similar. Access the matrix menu and select MATH, but this time highlight rref(and press enter. Select matrix A again, add the right-hand parenthesis, and press enter. See Figure 4.

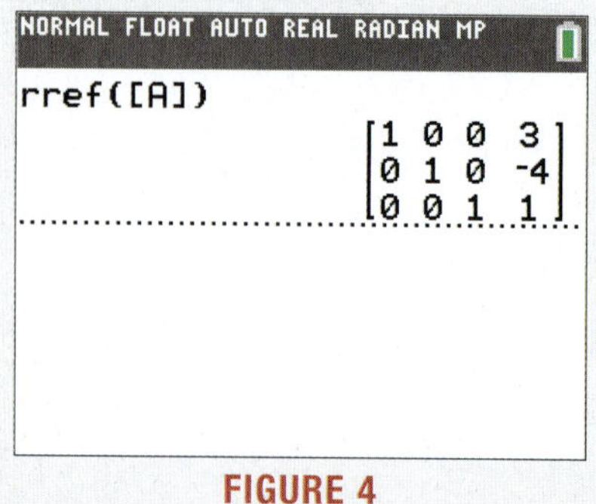

FIGURE 4

11.2 EXERCISES

PRACTICE

1. Let $A = \begin{bmatrix} 4 & -1 \\ 0 & 3 \\ 9 & -5 \end{bmatrix}$. Determine the following, if possible:

a. The order of A **b.** The value of a_{12} **c.** The value of a_{23}

2. Let $B = \begin{bmatrix} -7 & 2 & 11 \end{bmatrix}$. Determine the following, if possible:

a. The order of B **b.** The value of b_{12} **c.** The value of b_{31}

3. Let $C = \begin{bmatrix} 1 & 0 \\ 5 & -3 \\ 2 & 9 \\ \pi & e \\ 10 & -7 \end{bmatrix}$. Determine the following, if possible:

a. The order of C **b.** The value of c_{23} **c.** The value of c_{51}

4. Let $D = \begin{bmatrix} -8 & 13 & -1 \\ 0 & 6 & 3 \\ 0 & -9 & 0 \end{bmatrix}$. Determine the following, if possible:

a. The order of D **b.** The value of d_{23} **c.** The value of d_{33}

5. Let $E = \begin{bmatrix} -443 & 951 & 165 & 274 \\ 286 & -653 & 812 & -330 \\ 909 & 377 & 429 & -298 \end{bmatrix}$. Determine the following, if possible:

a. The order of E **b.** The value of e_{42} **c.** The value of e_{21}

6. Let $A = \begin{bmatrix} 9 & 5 & 0 \\ 7 & 4 & 2 \end{bmatrix}$. Determine the following, if possible:

a. The order of A **b.** The value of a_{22} **c.** The value of a_{13}

7. Let $B = \begin{bmatrix} 8 & 1 \\ 3 & 0 \\ 6 & 7 \end{bmatrix}$. Determine the following, if possible:

a. The order of B **b.** The value of b_{12} **c.** The value of b_{13}

8. Let $C = \begin{bmatrix} 65 & 32 & 91 & 45 \\ 23 & 18 & 75 & 47 \\ 8 & 63 & 28 & 31 \end{bmatrix}$. Determine the following, if possible:

a. The order of C **b.** The value of c_{43} **c.** The value of c_{23}

9. Let $D = \begin{bmatrix} 4 & 9 & 7 & 1 & 8 \\ 5 & 3 & 0 & 2 & 6 \end{bmatrix}$. Determine the following, if possible:

a. The order of D **b.** The value of d_{21} **c.** The value of d_{24}

Construct the augmented matrix that corresponds to each of the following systems of equations. See Example 2. (Answers may appear in slightly different, but equivalent, form.)

10. $\begin{cases} 4x+5y-3z=8 \\ 7x-2y+9=3 \\ 5x-6y+3z=0 \end{cases}$

11. $\begin{cases} y-2z+4=3x \\ \dfrac{x}{2}-4y-1=z \\ 3(-y+z)-1=0 \end{cases}$

12. $\begin{cases} 5x+\dfrac{y-z}{2}=3 \\ 7(z-x)+y-2=0 \\ x-(4-z)=y \end{cases}$

13. $\begin{cases} \dfrac{2-3x}{2}=y \\ 3z+2(x+y)=0 \\ 2x-y=2(x-3z) \end{cases}$

14. $\begin{cases} 2(z+3)-x+y=z \\ -3(x-2y)-1=5z \\ \dfrac{x}{3}-(y-2z)=x \end{cases}$

15. $\begin{cases} \dfrac{12x-1}{5}+\dfrac{y}{2}=\dfrac{3z}{2} \\ y-(x+3z)=-(1-y) \\ 2x-2-z-2y=7x \end{cases}$

16. $\begin{cases} \dfrac{3x+4y}{2}-3z=6 \\ 3(x-2y+9z)=0 \\ 2x+6y=3-z \end{cases}$

17. $\begin{cases} \dfrac{2x-4y}{3}=2z \\ 8x=2(y-3z)+7 \\ 3x=2y \end{cases}$

18. $\begin{cases} \dfrac{2(2x-y)}{3}+z=7 \\ 4=\dfrac{3}{-x+y+3z} \\ 4x-8y+4=9x \end{cases}$

19. $\begin{cases} 0.5x-14y=\dfrac{z}{4}-8 \\ \dfrac{x}{5}-y+\dfrac{z}{4}=\dfrac{y}{6}-3 \\ \dfrac{2}{3}\left(\dfrac{4}{y-x-1}\right)=\dfrac{5}{z} \end{cases}$

Construct the system of equations that corresponds to each of the following matrices.

20. $\left[\begin{array}{cc|c} 5 & 3 & 9 \\ 1 & 4 & 12 \end{array}\right]$

21. $\left[\begin{array}{cc|c} 1 & 0 & 8 \\ 0 & 1 & 3 \end{array}\right]$

22. $\left[\begin{array}{ccc|c} 14 & 0 & 1 & 16 \\ 3 & 6 & 4 & 0 \\ 8 & 2 & 5 & 21 \end{array}\right]$

23. $\left[\begin{array}{ccc|c} 1 & 3 & 6 & 16 \\ 0 & 1 & 2 & 9 \\ 0 & 0 & 1 & 4 \end{array}\right]$

24. $\left[\begin{array}{ccc|c} 2 & 1 & 1 & 22 \\ 1 & 3 & 1 & 17 \\ 1 & 1 & 4 & 8 \end{array}\right]$

25. $\left[\begin{array}{ccc|c} 0 & 9 & 13 & 27 \\ 2 & 0 & 21 & 19 \\ 7 & 18 & 0 & 32 \end{array}\right]$

Fill in the blanks by performing the indicated row operations. See Example 4.

26. $\left[\begin{array}{cc|c} 3 & 2 & -7 \\ 1 & 3 & 5 \end{array}\right] \xrightarrow{-3R_2+R_1} \underline{?}$

27. $\left[\begin{array}{cc|c} 2 & -5 & 3 \\ -4 & 3 & -1 \end{array}\right] \xrightarrow{2R_1+R_2} \underline{?}$

28. $\left[\begin{array}{cc|c} 4 & 2 & -8 \\ 3 & -9 & 0 \end{array}\right] \xrightarrow[\frac{1}{3}R_2]{\frac{1}{2}R_1} \underline{?}$

29. $\left[\begin{array}{cc|c} 9 & -2 & 7 \\ 1 & 3 & -2 \end{array}\right] \xrightarrow{R_1 \leftrightarrow R_2} \underline{?}$

30. $\left[\begin{array}{cc|c} 4 & 1 & 5 \\ 3 & 6 & 0 \end{array}\right] \xrightarrow{2R_1} \underline{?}$

31. $\left[\begin{array}{cc|c} 8 & -2 & -4 \\ 3 & -1 & 7 \end{array}\right] \xrightarrow{-2R_2} \underline{?}$

32. $\left[\begin{array}{cc|c} 9 & 12 & -6 \\ 15 & -3 & 0 \end{array}\right] \xrightarrow{-\frac{1}{3}R_1} \underline{?}$

33. $\left[\begin{array}{cc|c} 4 & 12 & -6 \\ 7 & 3 & 9 \end{array}\right] \xrightarrow{\frac{1}{2}R_1+R_2} \underline{?}$

34. $\left[\begin{array}{cc|c} 3 & 0 & 1 \\ 5 & 7 & -2 \end{array}\right] \xrightarrow{3R_1+R_2} \underline{?}$

35. $\left[\begin{array}{cc|c} 8 & -2 & 10 \\ 9 & -3 & 0 \end{array}\right] \xrightarrow[-\frac{2}{3}R_2]{\frac{1}{2}R_1} \underline{?}$

36. $\left[\begin{array}{ccc|c} 5 & 2 & 9 & 7 \\ 1 & 3 & -5 & 0 \\ 2 & -4 & 1 & 8 \end{array}\right] \xrightarrow[-R_1+R_3]{2R_2} \underline{?}$

37. $\left[\begin{array}{ccc|c} 6 & -2 & 5 & 14 \\ -7 & 19 & 2 & 3 \\ -9 & 11 & -4 & 7 \end{array}\right] \xrightarrow[0.5R_3]{3R_1} \underline{?}$

38. $\left[\begin{array}{ccc|c} 5 & 3 & 13 & 15 \\ 17 & 9 & -8 & -14 \\ 4 & -11 & 19 & 8 \end{array}\right] \xrightarrow{-2R_2+R_3} \underline{?}$

39. $\left[\begin{array}{ccc|c} 8 & 11 & 18 & 2 \\ 14 & 33 & -3 & -5 \\ -9 & 21 & 12 & 9 \end{array}\right] \xrightarrow[-2R_3+R_2]{\frac{1}{3}R_3+R_1} \underline{?}$

40. $\left[\begin{array}{ccc|c} 1 & 3 & -2 & 4 \\ 3 & -1 & 8 & 2 \\ -5 & 0 & 2 & 7 \end{array}\right] \xrightarrow[5R_1+R_3]{-3R_1+R_2} \underline{?}$

41. $\left[\begin{array}{ccc|c} 2 & 3 & -3 & 5 \\ 1 & 1 & 3 & 4 \\ 3 & 3 & 9 & 12 \end{array}\right] \xrightarrow[-3R_2+R_3]{-2R_2+R_1} \underline{?}$

42. $\left[\begin{array}{cc|c} -3 & 2 & 2 \\ 5 & -4 & 1 \end{array}\right] \xrightarrow{2R_1+R_2} \underline{?}$

43. $\left[\begin{array}{cc|c} -5 & 20 & -15 \\ 2 & -12 & 5 \end{array}\right] \xrightarrow[\frac{1}{2}R_2]{\frac{1}{5}R_1} \underline{?}$

44. $\left[\begin{array}{ccc|c} 2 & 2 & 3 & 7 \\ -3 & 2 & 8 & -2 \\ 1 & 5 & 2 & 6 \end{array}\right] \xrightarrow[3R_3+R_2]{-2R_3+R_1} \underline{?}$

45. $\left[\begin{array}{ccc|c} 1 & 5 & -9 & 11 \\ 1 & 4 & -1 & 4 \\ 4 & 3 & 5 & 45 \end{array}\right] \xrightarrow[-4R_1+R_3]{-R_1+R_2} \underline{?}$

For each matrix, determine if it is in row echelon form, reduced row echelon form, or neither.

46. $\left[\begin{array}{cc|c} 1 & 5 & 4 \\ 0 & 1 & 3 \end{array}\right]$

47. $\left[\begin{array}{ccc|c} 1 & 2 & 0 & 9 \\ 0 & 1 & 3 & 4 \\ 0 & 1 & 1 & 12 \end{array}\right]$

48. $\left[\begin{array}{ccc|c} 1 & 0 & 0 & 4 \\ 0 & 1 & 0 & 1 \\ 0 & 0 & 1 & 8 \end{array}\right]$

49. $\left[\begin{array}{ccc|c} 1 & 0 & 0 & 7 \\ 5 & 1 & 0 & 14 \\ 3 & 4 & 1 & -16 \end{array}\right]$

50. $\left[\begin{array}{ccc|c} 1 & 2 & 5 & 0 \\ 0 & 1 & 9 & 3 \\ 0 & 0 & 0 & 1 \end{array}\right]$

51. $\left[\begin{array}{cc|c} 0 & 1 & 3 \\ 1 & 0 & 6 \end{array}\right]$

Use Gaussian elimination and back-substitution to solve the following systems of equations. See Example 5.

52. $\begin{cases} 2x-4y=-6 \\ 3x-y=-4 \end{cases}$ **53.** $\begin{cases} 2x-5y=11 \\ 3x+2y=7 \end{cases}$ **54.** $\begin{cases} 5x-y=-21 \\ 9x+2y=-34 \end{cases}$

55. $\begin{cases} x-4y=-11 \\ 7x-y=4 \end{cases}$ **56.** $\begin{cases} x+2y=17 \\ 3x+4y=39 \end{cases}$ **57.** $\begin{cases} 2x+6y=4 \\ -4x-7y=7 \end{cases}$

58. $\begin{cases} 3x-2y=5 \\ -5x+4y=-3 \end{cases}$ **59.** $\begin{cases} 2x+y=-2 \\ -4x-2y=5 \end{cases}$ **60.** $\begin{cases} 6x-16y=10 \\ -3x+8y=4 \end{cases}$

61. $\begin{cases} 2x-3y=0 \\ 5x+y=17 \end{cases}$ **62.** $\begin{cases} 6x+3y=3 \\ x+y=3 \end{cases}$ **63.** $\begin{cases} 3x+6y=-12 \\ 2x+4y=-8 \end{cases}$

64. $\begin{cases} 4x+5y=9 \\ 8x+3y=-17 \end{cases}$ **65.** $\begin{cases} \frac{2}{3}x+2y=1 \\ x+3y=0 \end{cases}$ **66.** $\begin{cases} 13x-17y=-3 \\ -19x+15y=-35 \end{cases}$

67. $\begin{cases} 3x-9y-7z=-9 \\ 5x+11y-z=17 \\ -4x-8y+7z=5 \end{cases}$ **68.** $\begin{cases} 8x-y+5z=-8 \\ 11x-2y+9z=-9 \\ 7x-3y+13z=4 \end{cases}$ **69.** $\begin{cases} 17x+13y+8z=46 \\ -12x+3y+28z=-19 \\ 14x+5y-15z=-15 \end{cases}$

Use Gauss-Jordan elimination to solve the following systems of equations. See Example 6.

70. $\begin{cases} 2x-3y=8 \\ 8x+5y=-2 \end{cases}$ **71.** $\begin{cases} \frac{2}{3}x+y=-3 \\ 3x+\frac{5}{2}y=-\frac{7}{2} \end{cases}$ **72.** $\begin{cases} 3y=9 \\ x+2y=11 \end{cases}$

73. $\begin{cases} 6x+2y=-4 \\ -9x-3y=6 \end{cases}$ **74.** $\begin{cases} 3y=6 \\ 5x+2y=4 \end{cases}$ **75.** $\begin{cases} 3x+8y=-4 \\ x+2y=-2 \end{cases}$

76. $\begin{cases} -3x+2y=5 \\ 5x-2y=1 \end{cases}$ **77.** $\begin{cases} 9x-11y=10 \\ -4x+3y=-12 \end{cases}$ **78.** $\begin{cases} 9x-15y=-6 \\ -3x+11y=-10 \end{cases}$

79. $\begin{cases} 3x-8y=7 \\ 18x-35y=-23 \end{cases}$ **80.** $\begin{cases} 4x+y-3z=-9 \\ 2x-3z=-19 \\ 7x-y-4z=-29 \end{cases}$ **81.** $\begin{cases} -5x+9y+3z=1 \\ 3x+2y-6z=9 \\ x+4y-z=16 \end{cases}$

82. $\begin{cases} 2x-y=0 \\ 5x-3y-3z=5 \\ 2x+6z=-10 \end{cases}$ **83.** $\begin{cases} x+y=4 \\ y+3z=-1 \\ 2x-2y+5z=-5 \end{cases}$ **84.** $\begin{cases} 2x-3y=-2 \\ x-4y+3z=0 \\ -2x+7y-5z=0 \end{cases}$

85. $\begin{cases} 3x+8z=3 \\ -3x-7z=-3 \\ x+3z=1 \end{cases}$ **86.** $\begin{cases} 3x-y+z=2 \\ -6x+2y-2z=1 \\ 5x+2y-3z=2 \end{cases}$ **87.** $\begin{cases} x+2y=-1 \\ y+3z=7 \\ 2x+5z=21 \end{cases}$

88. $\begin{cases} 2x+8y-z=-5 \\ -5x+3y+4z=-6 \\ x-4y-5z=-8 \end{cases}$

89. $\begin{cases} 7x-8y+2z=-2 \\ 5x-3y-z=-3 \\ 8x+y-3z=7 \end{cases}$

90. $\begin{cases} 8x+14y-3z=3 \\ -6x+2y+7z=-13 \\ 8x+19y+3z=11 \end{cases}$

91. $\begin{cases} 8x+5y+3z=-2 \\ 12x-y-18z=1 \\ 7x+6y+10z=19 \end{cases}$

92. $\begin{cases} 4x+8y+7z=27 \\ -2x+9y-8z=-15 \\ 9x+13y+7z=-33 \end{cases}$

93. $\begin{cases} w-x+2z=9 \\ 2w+3y=-1 \\ -2w-5y-z=0 \\ x+2y=-4 \end{cases}$

94. $\begin{cases} 3w-x+5y+3z=2 \\ -4w-10y-2z=10 \\ w-x+2z=7 \\ 4w-2x+5y+5z=9 \end{cases}$

APPLICATIONS

95. The sum of three integers is 155. The first integer is sixteen more than the second. The third integer is seven less than the sum of the first integer and twice the second. What are the three integers?

96. Mario bought a pound of bacon, a dozen eggs, and a loaf of bread to make breakfast for his family. The total cost was \$7.42. The bacon cost \$0.03 more than twice the price of the bread and the eggs cost \$0.03 less than half the price of the bread. Find the price of each item.

97. The Pizza House sells three sizes of pizzas: small, medium, large. The prices of the pizzas are \$9.00, \$12.00, and \$15.00, respectively. In one day, they sold 82 pizzas for a total of \$1098.00. If the number of large pizzas sold was twice the number of medium pizzas sold, how many of each size pizza did the Pizza House sell?

11.3 DETERMINANTS AND CRAMER'S RULE

TOPICS

1. Evaluating determinants
2. Solving systems of linear equations using Cramer's Rule

TOPIC 1: Evaluating Determinants

The Meaning of Matrix

Although arrays of numbers were employed to solve what we now call systems of linear equations long before, the word "matrix" for such an array was coined by the English mathematician J. J. Sylvester in 1850. The word is derived from the Latin *mater*, meaning *mother*, as Sylvester wanted to indicate that a matrix of numbers gives rise to a number of determinants known as minors.

While Gaussian elimination and Gauss-Jordan elimination both provide significant improvements in ease and speed when solving systems of equations, they are not truly new methods. Each simply uses the matrix as a powerful, agile organizer to make the substitution and elimination methods simpler. Cramer's Rule, named after the Swiss mathematician Gabriel Cramer (1704–1752), is a solution method that truly brings something new to the discussion.

Cramer's Rule relies on the computation of a number, called the *determinant*, associated with every *square* matrix (a matrix with the same number of rows as columns).

DEFINITION: Determinant of a 2×2 Matrix

The **determinant** of the matrix $A = \begin{bmatrix} a_{11} & a_{12} \\ a_{21} & a_{22} \end{bmatrix}$, denoted $|A|$, is $a_{11}a_{22} - a_{21}a_{12}$.

CAUTION

If A is a matrix, $|A|$ stands for the determinant of A, not the absolute value of A. In fact, the concept of absolute value does not apply to matrices.

Unfortunately, this only defines determinants for matrices of one size, and the simplicity of the definition does not carry over to larger sizes. The good news, however, is that determinants of 3×3, 4×4, and larger matrices can all be related, ultimately, to determinants of 2×2 matrices. We will do a few calculations before considering larger matrices.

Example 1: Determinant of a 2×2 Matrix

Evaluate the determinants of each of the following matrices:

a. $A = \begin{bmatrix} -2 & 3 \\ -1 & 3 \end{bmatrix}$ **b.** $B = \begin{bmatrix} 2 & -3 \\ -4 & 6 \end{bmatrix}$

Solution

a. $|A| = a_{11}a_{22} - a_{21}a_{12}$
$= (-2)(3) - (-1)(3) = -6 + 3 = -3$

b. $|B| = b_{11}b_{22} - b_{21}b_{12}$
$= (2)(6) - (-4)(-3) = 12 - 12 = 0$

Calculating the determinant of a 3×3 matrix involves calculating three specific 2×2 determinants. Similarly, finding the determinant of a 4×4 matrix requires evaluating four specific 3×3 determinants.

These specific smaller determinants are called the *cofactors* of a matrix.

DEFINITION: Minors and Cofactors

Let A be an $n\times n$ matrix, and let i and j be numbers between 1 and n, so that a_{ij} is an element of A.

- The **minor** of the element a_{ij} is the determinant of the $(n-1)\times(n-1)$ matrix formed from A by deleting its ith row and j^{th} column.
- The **cofactor** of the element a_{ij} is $(-1)^{i+j}$ times the minor of a_{ij}. Thus, if $i+j$ is even, the cofactor of a_{ij} is the same as the minor of a_{ij}; if $i+j$ is odd, the cofactor is the negative of the minor.

$$\begin{bmatrix} + & - & + & - & \cdots \\ - & + & - & + & \cdots \\ + & - & + & - & \cdots \\ - & + & - & + & \cdots \\ \vdots & \vdots & \vdots & \vdots & \ddots \end{bmatrix}$$

FIGURE 1: The Cofactor Sign Matrix

The matrix of signs in Figure 1 may help you remember which minors to multiply by -1: if a_{ij} is in a position occupied by a $-$ sign, change the sign of the minor of a_{ij} to obtain the cofactor.

Example 2: Minors and Cofactors

Consider the matrix $A = \begin{bmatrix} -5 & 3 & 2 \\ 1 & 0 & -1 \\ -3 & 1 & 0 \end{bmatrix}$.

a. Evaluate the minor of a_{12}.

b. Evaluate the cofactor of a_{23}.

Solution

a. Finding the minor of a_{12} requires deleting the first row and second column of the matrix A. This gives us the following.

$$\begin{bmatrix} -5 & 3 & 2 \\ 1 & 0 & -1 \\ -3 & 1 & 0 \end{bmatrix}$$

Thus, the minor of $a_{12} = \begin{vmatrix} 1 & -1 \\ -3 & 0 \end{vmatrix} = (1)(0) - (-3)(-1) = 0 - 3 = -3.$

b. The cofactor of a_{23} is the minor of a_{23} multiplied by -1, since $2 + 3$ is odd.

$$a_{23} = (-1)^{2+3}\begin{vmatrix} -5 & 3 \\ -3 & 1 \end{vmatrix} = (-1)^5\left[(-5)(1) - (-3)(3)\right]$$
$$= (-1)(-5+9) = (-1)(4) = -4$$

DEFINITION: Determinant of an $n \times n$ Matrix

Evaluation of an $n \times n$ determinant is accomplished by **expansion** along a fixed row or column. The result does not depend on which row or column is chosen.

- To expand along the ith row, each element of that row is multiplied by its cofactor, and the n products are then added.
- To expand along the jth column, each element of that column is multiplied by its cofactor, and the n products are then added.

For example, if we expand along the first column of a 3×3 matrix, we get the following. Note the minus sign in front of a_{21}.

$$\begin{vmatrix} a_{11} & a_{12} & a_{13} \\ a_{21} & a_{22} & a_{23} \\ a_{31} & a_{32} & a_{33} \end{vmatrix} = a_{11}\begin{vmatrix} a_{22} & a_{23} \\ a_{32} & a_{33} \end{vmatrix} - a_{21}\begin{vmatrix} a_{12} & a_{13} \\ a_{32} & a_{33} \end{vmatrix} + a_{31}\begin{vmatrix} a_{12} & a_{13} \\ a_{22} & a_{23} \end{vmatrix}$$

We could expand along a row or a different column in a similar manner.

Example 3: Determinant of an $n \times n$ Matrix

Evaluate the determinant of the matrix $A = \begin{bmatrix} -1 & 3 & 2 \\ -2 & 0 & 0 \\ 4 & 1 & 5 \end{bmatrix}$.

Solution

NOTE

Minimize the number of computations by choosing which row or column to expand along carefully.

First, we decide which row or column to expand along. A row or column with many zeros is generally a good choice, since it makes the multiplication much easier. In this case, Row 2 has the most zeros, so expand along it.

$$\begin{aligned} \begin{vmatrix} -1 & 3 & 2 \\ -2 & 0 & 0 \\ 4 & 1 & 5 \end{vmatrix} &= -(-2)\begin{vmatrix} 3 & 2 \\ 1 & 5 \end{vmatrix} + (0)\begin{vmatrix} -1 & 2 \\ 4 & 5 \end{vmatrix} - (0)\begin{vmatrix} -1 & 3 \\ 4 & 1 \end{vmatrix} \\ &= -(-2)(13) + 0 - 0 \\ &= 26 \end{aligned}$$

Thus, $|A| = 26$. We get the same answer if we expand along a different row or column.

$$\begin{aligned} \begin{vmatrix} -1 & 3 & 2 \\ -2 & 0 & 0 \\ 4 & 1 & 5 \end{vmatrix} &= (-1)\begin{vmatrix} 0 & 0 \\ 1 & 5 \end{vmatrix} - (-2)\begin{vmatrix} 3 & 2 \\ 1 & 5 \end{vmatrix} + (4)\begin{vmatrix} 3 & 2 \\ 0 & 0 \end{vmatrix} \\ &= (-1)(0) - (-2)(13) + (4)(0) \\ &= 26 \end{aligned}$$

As we saw in the previous example, even 3×3 determinants can involve a large number of calculations, but by taking advantage of zeros, we are able to reduce the amount of work. We can take this a step farther by applying a few properties of determinants.

PROPERTIES: Properties of Determinants

1. A constant can be factored out of each of the terms in a given row or column when computing determinants.

$$\begin{vmatrix} 2 & -1 \\ 15 & 5 \end{vmatrix} = 5\begin{vmatrix} 2 & -1 \\ 3 & 1 \end{vmatrix} \quad \text{and} \quad \begin{vmatrix} 4 & 7 \\ 12 & 9 \end{vmatrix} = 4\begin{vmatrix} 1 & 7 \\ 3 & 9 \end{vmatrix}$$

2. Interchanging two rows or two columns changes the determinant by a factor of -1.

$$\begin{vmatrix} 2 & -1 \\ 15 & 5 \end{vmatrix} = -\begin{vmatrix} 15 & 5 \\ 2 & -1 \end{vmatrix} \quad \text{and} \quad \begin{vmatrix} 3 & -2 \\ 7 & 1 \end{vmatrix} = -\begin{vmatrix} -2 & 3 \\ 1 & 7 \end{vmatrix}$$

3. The determinant is unchanged by adding a multiple of one row (or column) to another row (or column).

$$\begin{vmatrix} 3 & -2 \\ 1 & -1 \end{vmatrix} \overset{-3R_2+R_1}{=} \begin{vmatrix} 0 & 1 \\ 1 & -1 \end{vmatrix}$$

Example 4: Properties of Determinants

Evaluate the determinant of the matrix $B = \begin{bmatrix} 4 & -2 & 3 & 0 \\ 2 & 1 & -1 & 3 \\ 3 & 0 & 1 & 1 \\ 2 & -2 & 0 & 0 \end{bmatrix}$.

Solution

Use the properties of determinants to try to obtain rows or columns with as many zeros as possible.

$$\begin{vmatrix} 4 & -2 & 3 & 0 \\ 2 & 1 & -1 & 3 \\ 3 & 0 & 1 & 1 \\ 2 & -2 & 0 & 0 \end{vmatrix} \overset{-3R_3+R_2}{=} \begin{vmatrix} 4 & -2 & 3 & 0 \\ -7 & 1 & -4 & 0 \\ 3 & 0 & 1 & 1 \\ 2 & -2 & 0 & 0 \end{vmatrix}$$

Applying the third property makes the fourth column have only one nonzero entry.

Now expand along the fourth column (remembering that the minus sign is part of the cofactor):

$$\begin{vmatrix} 4 & -2 & 3 & 0 \\ -7 & 1 & -4 & 0 \\ 3 & 0 & 1 & 1 \\ 2 & -2 & 0 & 0 \end{vmatrix} = -(1)\begin{vmatrix} 4 & -2 & 3 \\ -7 & 1 & -4 \\ 2 & -2 & 0 \end{vmatrix}$$

We can continue to apply the third property of determinants to simplify the evaluation of the 3×3 determinant.

$$|B| = -\begin{vmatrix} 4 & -2 & 3 \\ -7 & 1 & -4 \\ 2 & -2 & 0 \end{vmatrix} \overset{C_1+C_2}{=} -\begin{vmatrix} 4 & 2 & 3 \\ -7 & -6 & -4 \\ 2 & 0 & 0 \end{vmatrix}$$

Add the first column to the second.

$$= -(2)\begin{vmatrix} 2 & 3 \\ -6 & -4 \end{vmatrix}$$

Expand along the third row, which now has two zeros.

$$= (-2)(10)$$
$$= -20$$

Notice that we have reduced the work to the evaluation of only one 3×3 determinant, which in turn involved evaluating only one 2×2 determinant.

TECHNOLOGY: Evaluating Determinants

To find the determinant of a square matrix using a TI-84 Plus, first input the matrix $\begin{bmatrix} 2 & 4 & -8 \\ 1 & 3 & 6 \\ -7 & 5 & 1 \end{bmatrix}$. See Figure 2.

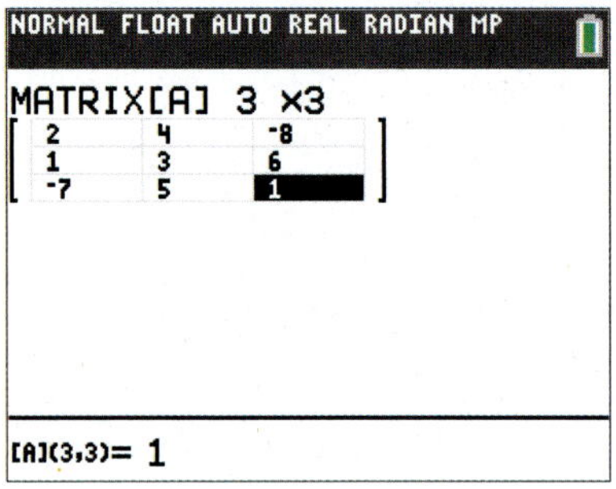

NORMAL FLOAT AUTO REAL RADIAN MP
MATRIX[A] 3 ×3
[2 4 -8]
[1 3 6]
[-7 5 1]
[A](3,3)= 1

FIGURE 2

Press 2nd and x^{-1} to open the matrix menu, select MATH and then det(. Next, select the name of the matrix from the matrix menu under NAMES. Add the closing parenthesis and press enter to calculate the determinant. See Figure 3.

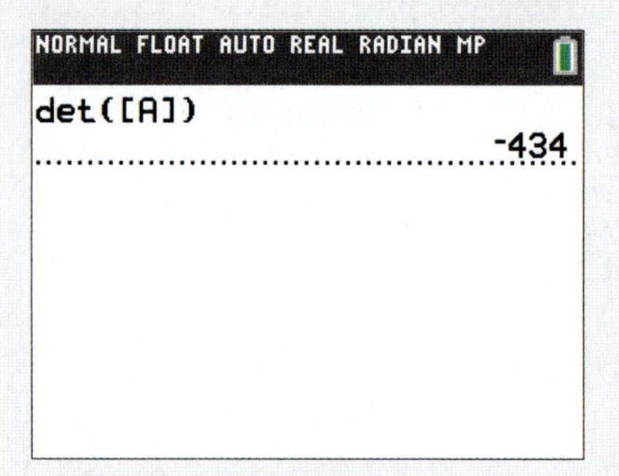

FIGURE 3

Remember that the definition of determinant only applies to square matrices; matrices that are not square do not have determinants. Notice that if we try to find the determinant of a 3×4 matrix, for instance, we get the error message shown in Figure 4.

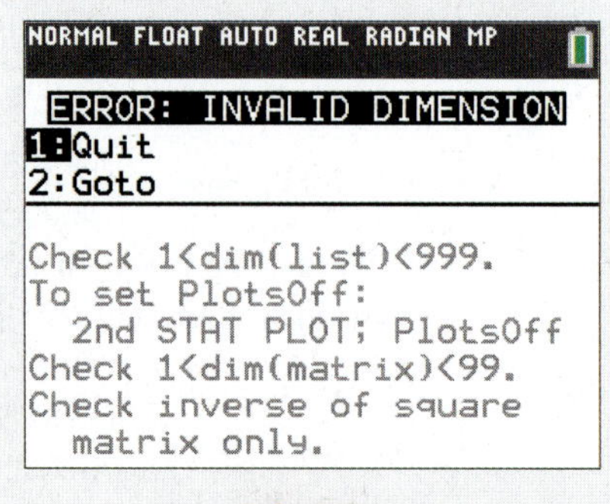

FIGURE 4

TOPIC 2: Solving Systems of Linear Equations Using Cramer's Rule

To understand the form of Cramer's Rule, we will solve the general two-variable, two-equation linear system by elimination. To do this, we note that any such system can be put into the form

$$\begin{cases} ax+by=e \\ cx+dy=f \end{cases}$$

where a, b, c, d, e, and f are all constants. If we can solve this system for x and y, then we will have a formula for the solution of any such system.

Using the method of elimination, we can obtain an equation in x alone by multiplying the first equation by d and the second equation by $-b$.

$$\begin{cases} ax+by=e \\ cx+dy=f \end{cases} \xrightarrow[-bE_2]{dE_1} \begin{cases} adx+bdy=ed \\ -bcx-bdy=-bf \end{cases}$$
$$(ad-bc)x=ed-bf$$

This equation can then be solved for x to obtain

$$x=\frac{ed-bf}{ad-bc}.$$

Similarly, the system can be solved for y to obtain

$$y=\frac{af-ce}{ad-bc}.$$

Note that these formulas only make sense if the denominator is not zero. We will deal with this possibility shortly.

These formulas are worthwhile on their own, but they are a bit complex and would be difficult to memorize. Note that each term in the two fractions appears in the form of a 2×2 determinant. In fact, these formulas are equivalent to the following equations.

$$x=\frac{\begin{vmatrix} e & b \\ f & d \end{vmatrix}}{\begin{vmatrix} a & b \\ c & d \end{vmatrix}} \quad \text{and} \quad y=\frac{\begin{vmatrix} a & e \\ c & f \end{vmatrix}}{\begin{vmatrix} a & b \\ c & d \end{vmatrix}}$$

The denominator D in both formulas is the determinant of the coefficient matrix, the square matrix consisting of the coefficients of the variables. If we let D_x and D_y represent the numerators of the formulas for x and y, respectively, then D_x is the determinant of the coefficient matrix with the first column (the x-column) replaced by the column of constants, and D_y is the determinant of the coefficient matrix with the second column replaced by the column of constants. Putting these observations together, we obtain Cramer's Rule for the two-variable, two-equation case.

THEOREM: Cramer's Rule for Two-Variable, Two-Equation Linear Systems

The solution of a two-variable, two-equation linear system $\begin{cases} ax+by=e \\ cx+dy=f \end{cases}$ is given by

$$x=\frac{D_x}{D} \quad \text{and} \quad y=\frac{D_y}{D},$$

where D is the determinant of the coefficient matrix $\begin{vmatrix} a & b \\ c & d \end{vmatrix}$, D_x is the determinant of the matrix formed by replacing the column of x-coefficients with the column of constant terms $\begin{vmatrix} e & b \\ f & d \end{vmatrix}$, and D_y is the determinant of the matrix formed by replacing the column of y-coefficients with the column of constant terms $\begin{vmatrix} a & e \\ c & f \end{vmatrix}$.

CAUTION

Whenever a fraction appears in our work, we need to ask if the expression in the denominator can ever be zero, and what it means if this happens. In Cramer's Rule, the determinant D can equal 0, which prevents us from using the given formulas.

If $D=0$, the system is either dependent or inconsistent. If both D_x and D_y are also zero, the system is dependent. If at least one of D_x and D_y is nonzero, the system has no solution.

Example 5: Cramer's Rule

Use Cramer's Rule to solve the following systems.

a. $\begin{cases} 4x-5y=3 \\ -3x+7y=1 \end{cases}$ **b.** $\begin{cases} -x+2y=-1 \\ 3x-6y=3 \end{cases}$

Solution

a. $D=\begin{vmatrix} 4 & -5 \\ -3 & 7 \end{vmatrix}=28-15=13$ Calculate D first. Since $D \neq 0$, we know there is a single solution to the system.

$D_x=\begin{vmatrix} 3 & -5 \\ 1 & 7 \end{vmatrix}=21-(-5)=26$ Calculate D_x and D_y.

$D_y=\begin{vmatrix} 4 & 3 \\ -3 & 1 \end{vmatrix}=4-(-9)=13$

Applying Cramer's Rule, we have $x=\frac{D_x}{D}=\frac{26}{13}=2$ and $y=\frac{D_y}{D}=\frac{13}{13}=1$, so the solution is $(2,1)$.

b. $D=\begin{vmatrix}-1 & 2\\ 3 & -6\end{vmatrix}=6-6=0$

$D_x=\begin{vmatrix}-1 & 2\\ 3 & -6\end{vmatrix}=6-6=0$

$D_y=\begin{vmatrix}-1 & -1\\ 3 & 3\end{vmatrix}=-3-(-3)=0$

Again we calculate D first. Since $D=0$ either the system has no solution or it has an infinite number of solutions.

Since D_x and D_y both equal zero, the system is dependent.

The solution set can be found by solving either equation for either variable: $\{(2y+1,y)\mid y\in\mathbb{R}\}$.

Cramer's Rule can be extended to solve any system of n linear equations in n variables. While Cramer's Rule is an extremely powerful method and remarkable for its succinctness, keep in mind that using Cramer's Rule to solve an n-equation, n-variable system entails calculating $(n+1)$ $n\times n$ determinants, so it is important to make use of the labor saving properties of determinants.

THEOREM: Cramer's Rule

A system of n linear equations in the n variables $x_1,x_2,\ldots,x_n$ can be written in the following form.

$$\begin{cases}a_{11}x_1+a_{12}x_2+\cdots+a_{1n}x_n=b_1\\ a_{21}x_1+a_{22}x_2+\cdots+a_{2n}x_n=b_2\\ \qquad\vdots\\ a_{n1}x_1+a_{n2}x_2+\cdots+a_{nn}x_n=b_n\end{cases}$$

The solution of the system is given by the formulas $x_1=\frac{D_{x_1}}{D}$, $x_2=\frac{D_{x_2}}{D}$, …, $x_n=\frac{D_{x_n}}{D}$, where D is the determinant of the coefficient matrix and D_{x_i} is the determinant of the same matrix with the ith column replaced by the column of constants $b_1,b_2,\ldots,b_n$.

If $D=0$ and if each $D_{x_i}=0$ as well, the system is dependent and has an infinite number of solutions. If $D=0$ and D_{x_i} is nonzero for at least one value of i, the system has no solution.

Example 6: Cramer's Rule

Use Cramer's Rule to solve the system $\begin{cases}3x-2y-2z=-1\\ 3y+z=-7\\ x+y+2z=0\end{cases}$.

Solution

Note how the properties of determinants are used to simplify each calculation. In each case, the row or column used for expansion is used for expansion is highlighted.

$$D=\begin{vmatrix}3 & -2 & -2\\ 0 & 3 & 1\\ 1 & 1 & 2\end{vmatrix}\overset{-3R_3+R_1}{=}\begin{vmatrix}0 & -5 & -8\\ 0 & 3 & 1\\ 1 & 1 & 2\end{vmatrix}=(1)\begin{vmatrix}-5 & -8\\ 3 & 1\end{vmatrix}=19$$

Since $D \neq 0$, the system has a unique solution.

$$D_x = \begin{vmatrix} -1 & -2 & -2 \\ -7 & 3 & 1 \\ 0 & 1 & 2 \end{vmatrix} \overset{-2C_2+C_3}{=} \begin{vmatrix} -1 & -2 & 2 \\ -7 & 3 & -5 \\ 0 & 1 & 0 \end{vmatrix} = -(1)\begin{vmatrix} -1 & 2 \\ -7 & -5 \end{vmatrix} = -19$$

$$D_y = \begin{vmatrix} 3 & -1 & -2 \\ 0 & -7 & 1 \\ 1 & 0 & 2 \end{vmatrix} \overset{-2C_1+C_3}{=} \begin{vmatrix} 3 & -1 & -8 \\ 0 & -7 & 1 \\ 1 & 0 & 0 \end{vmatrix} = (1)\begin{vmatrix} -1 & -8 \\ -7 & 1 \end{vmatrix} = -57$$

$$D_z = \begin{vmatrix} 3 & -2 & -1 \\ 0 & 3 & -7 \\ 1 & 1 & 0 \end{vmatrix} \overset{-C_1+C_2}{=} \begin{vmatrix} 3 & -5 & -1 \\ 0 & 3 & -7 \\ 1 & 0 & 0 \end{vmatrix} = (1)\begin{vmatrix} -5 & -1 \\ 3 & -7 \end{vmatrix} = 38$$

After evaluating D_x, D_y, and D_z, we know the solution is the single ordered triple

$$(x, y, z) = \left(\frac{D_x}{D}, \frac{D_y}{D}, \frac{D_z}{D}\right) = \left(\frac{-19}{19}, \frac{-57}{19}, \frac{38}{19}\right) = (-1, -3, 2).$$

Why does Cramer's Rule work? Once we have the formulas, we can use the properties of determinants (specifically, the first and third properties) to verify that they work. Consider again the following generic n-equation, n-variable system.

$$\begin{cases} a_{11}x_1 + a_{12}x_2 + \cdots + a_{1n}x_n = b_1 \\ a_{21}x_1 + a_{22}x_2 + \cdots + a_{2n}x_n = b_2 \\ \vdots \\ a_{n1}x_1 + a_{n2}x_2 + \cdots + a_{nn}x_n = b_n \end{cases}$$

If we assume that the system has a unique solution, which we will denote $(x_1, x_2, \ldots, x_n)$, then we can rewrite D_{x_1} as follows:

$$D_{x_1} = \begin{vmatrix} b_1 & a_{12} & \cdots & a_{1n} \\ b_2 & a_{22} & \cdots & a_{2n} \\ \vdots & \vdots & \cdots & \vdots \\ b_n & a_{n2} & \cdots & a_{nn} \end{vmatrix} = \begin{vmatrix} a_{11}x_1 + a_{12}x_2 + \cdots + a_{1n}x_n & a_{12} & \cdots & a_{1n} \\ a_{21}x_1 + a_{22}x_2 + \cdots + a_{2n}x_n & a_{22} & \cdots & a_{2n} \\ \vdots & \vdots & \cdots & \vdots \\ a_{n1}x_1 + a_{n2}x_2 + \cdots + a_{nn}x_n & a_{n2} & \cdots & a_{nn} \end{vmatrix}$$

We can now use the third property of determinants $n - 1$ times to simplify the first column. We do this by multiplying the second column by $-x_2$ and adding it to the first, multiplying the third column by $-x_3$ and adding it to the first, and so on. After doing this, we obtain

$$D_{x_1} = \begin{vmatrix} a_{11}x_1 & a_{12} & \cdots & a_{1n} \\ a_{21}x_1 & a_{22} & \cdots & a_{2n} \\ \vdots & \vdots & \cdots & \vdots \\ a_{n1}x_1 & a_{n2} & \cdots & a_{nn} \end{vmatrix} = x_1 \begin{vmatrix} a_{11} & a_{12} & \cdots & a_{1n} \\ a_{21} & a_{22} & \cdots & a_{2n} \\ \vdots & \vdots & \cdots & \vdots \\ a_{n1} & a_{n2} & \cdots & a_{nn} \end{vmatrix} = x_1 D.$$

Note that we have used the first property of determinants to factor x_1 out of the determinant. We can now solve this equation for x_1 (since $D \neq 0$) to obtain

$$x_1 = \frac{D_{x_1}}{D},$$

and we can then repeat the process for the remaining variables $x_2, x_3, \ldots, x_n$.

11.3 EXERCISES

PRACTICE

Evaluate each of the following determinants. See Example 1.

1. $\begin{vmatrix} 4 & -3 \\ 1 & 2 \end{vmatrix}$ **2.** $\begin{vmatrix} 5 & -2 \\ 5 & -2 \end{vmatrix}$ **3.** $\begin{vmatrix} 0 & 3 \\ -5 & 2 \end{vmatrix}$ **4.** $\begin{vmatrix} 34 & -2 \\ 17 & -1 \end{vmatrix}$

5. $\begin{vmatrix} a & x \\ x & b \end{vmatrix}$ **6.** $\begin{vmatrix} 5x & 2 \\ -x & 1 \end{vmatrix}$ **7.** $\begin{vmatrix} -2 & 2 \\ -2 & -2 \end{vmatrix}$ **8.** $\begin{vmatrix} ac & 2ad \\ bc & db \end{vmatrix}$

9. $\begin{vmatrix} -1 & 2 \\ 3 & 4 \end{vmatrix}$ **10.** $\begin{vmatrix} w & x \\ y & z \end{vmatrix}$ **11.** $\begin{vmatrix} -2 & 9 \\ 5 & -3 \end{vmatrix}$ **12.** $\begin{vmatrix} 2y & 3x \\ y-1 & x^2 \end{vmatrix}$

Solve for x by calculating the determinant.

13. $\begin{vmatrix} x-2 & 2 \\ 2 & x+1 \end{vmatrix} = 0$ **14.** $\begin{vmatrix} x+7 & -2 \\ 9 & x-2 \end{vmatrix} = 0$ **15.** $\begin{vmatrix} x+1 & 8 \\ 1 & x+3 \end{vmatrix} = 0$

16. $\begin{vmatrix} x-8 & 11 \\ -2 & x+5 \end{vmatrix} = 0$ **17.** $\begin{vmatrix} x+6 & 2 \\ -1 & x+3 \end{vmatrix} = 0$ **18.** $\begin{vmatrix} x-4 & -4 \\ 3 & x+9 \end{vmatrix} = 0$

19. $\begin{vmatrix} x+5 & 3 \\ 3 & x-3 \end{vmatrix} = 0$ **20.** $\begin{vmatrix} x+3 & 6 \\ 5 & x+7 \end{vmatrix} = 0$ **21.** $\begin{vmatrix} x-3 & 2 \\ 1 & x-4 \end{vmatrix} = 0$

Use the matrix $A = \begin{bmatrix} 2 & -1 & 5 \\ 0 & 1 & 3 \\ 1 & 0 & -2 \end{bmatrix}$ to evaluate the following. See Example 2.

22. The minor of a_{12}

23. The cofactor of a_{12}

24. The minor of a_{22}

25. The cofactor of a_{22}

26. The cofactor of a_{32}

27. The cofactor of a_{33}

28. The minor of a_{13}

29. The cofactor of a_{21}

30. The cofactor of a_{31}

Find the determinant by the method of expansion by cofactors along the given row or column. See Example 3.

31. $\begin{vmatrix} 4 & 5 & 3 \\ -1 & 2 & 7 \\ 11 & 6 & 2 \end{vmatrix}$ Expand along Row 3

32. $\begin{vmatrix} 8 & 2 & 0 \\ 3 & 4 & 7 \\ 1 & 0 & 2 \end{vmatrix}$ Expand along Column 3

33. $\begin{vmatrix} 5 & 8 & 5 \\ 0 & -6 & 3 \\ 2 & 4 & -1 \end{vmatrix}$ Expand along Row 2

34. $\begin{vmatrix} -4 & 2 & 1 \\ 9 & 12 & 8 \\ 0 & 6 & -3 \end{vmatrix}$ Expand along Column 1

35. $\begin{vmatrix} 13 & 0 & -7 \\ 4 & 2 & 3 \\ 1 & 4 & 0 \end{vmatrix}$ Expand along Row 2

36. $\begin{vmatrix} 7 & 0 & 1 \\ 2 & 5 & 3 \\ 8 & 6 & 2 \end{vmatrix}$ Expand along Column 3

37. $\begin{vmatrix} 8 & 0 & -7 & 5 \\ 4 & -2 & 3 & 3 \\ -1 & 1 & 0 & 2 \\ 2 & 0 & 6 & 0 \end{vmatrix}$ Expand along Row 4

38. $\begin{vmatrix} 4 & -2 & 9 & 2 \\ 7 & 0 & 1 & 7 \\ -6 & 3 & 0 & 1 \\ 3 & 1 & 2 & 0 \end{vmatrix}$ Expand along Column 2

Evaluate each of the following determinants. In each case, minimize the required number of computations by carefully choosing a row or column to expand along, and use the properties of determinants to simplify the process. See Examples 3 and 4.

39. $\begin{vmatrix} 2 & 0 & 1 \\ -5 & 1 & 0 \\ 3 & -1 & 1 \end{vmatrix}$

40. $\begin{vmatrix} 12 & 3 & 1 \\ 1 & 1 & -1 \\ 0 & 2 & 0 \end{vmatrix}$

41. $\begin{vmatrix} 12 & 3 & 6 \\ 2 & 2 & -4 \\ 0 & 2 & 0 \end{vmatrix}$

42. $\begin{vmatrix} 1 & 2 & 3 \\ 4 & 5 & 6 \\ 7 & 8 & 9 \end{vmatrix}$

43. $\begin{vmatrix} 2 & 1 & -3 & 0 \\ 1 & -2 & 1 & 0 \\ 0 & 1 & 0 & 1 \\ 2 & 0 & 1 & 1 \end{vmatrix}$

44. $\begin{vmatrix} x & 0 & 0 & 0 \\ 0 & x & 0 & 0 \\ 0 & 0 & x & 0 \\ 0 & 0 & 0 & x \end{vmatrix}$

45. $\begin{vmatrix} x & x & x & x \\ 0 & x & x & x \\ 0 & 0 & x & x \\ 0 & 0 & 0 & x \end{vmatrix}$

46. $\begin{vmatrix} 0 & 2 & 0 & 0 \\ -2 & -4 & 5 & 9 \\ 1 & 3 & -1 & 1 \\ 0 & 7 & 0 & 2 \end{vmatrix}$

47. $\begin{vmatrix} x & x & 0 & 0 \\ yz & x^3 & z & x^4 \\ z & xy & x & 0 \\ x^2 & 0 & 0 & 0 \end{vmatrix}$

Use Cramer's Rule to solve each system of equations. See Examples 5 and 6.

48. $\begin{cases} 2x-3y=8 \\ 8x+5y=-2 \end{cases}$

49. $\begin{cases} 5x+7y=9 \\ 2x+3y=-7 \end{cases}$

50. $\begin{cases} 5x-10y=9 \\ -x+2y=-3 \end{cases}$

51. $\begin{cases} -2x-2y=4 \\ 3x+3y=-6 \end{cases}$

52. $\begin{cases} \frac{2}{3}x+y=-3 \\ 3x+\frac{5}{2}y=-\frac{7}{2} \end{cases}$

53. $\begin{cases} \frac{2}{3}x+2y=1 \\ x+3y=0 \end{cases}$

54. $\begin{cases} x+2y=-1 \\ y+3z=7 \\ 2x+5z=21 \end{cases}$

55. $\begin{cases} 2x-y=0 \\ 5x-3y-3z=5 \\ 2x+6z=-10 \end{cases}$

56. $\begin{cases} 3x+8z=3 \\ -3x-7z=-3 \\ x+3z=1 \end{cases}$

57. $\begin{cases} 3w-x+5y+3z=2 \\ -4w-10y-2z=10 \\ w-x+2z=7 \\ 4w-2x+5y+5z=9 \end{cases}$

58. $\begin{cases} 2w+x-3y=3 \\ w-2x+y=1 \\ x+z=-2 \\ y+z=0 \end{cases}$

59. $\begin{cases} 3w-2x+y-5z=-1 \\ w+x-y+4z=2 \\ 4w-x-z=1 \\ 5w-x=9 \end{cases}$

60. $\begin{cases} -4x+y=1 \\ 7x+2y=407 \end{cases}$

61. $\begin{cases} 5x-4y=-49 \\ 24x-19y=179 \end{cases}$

62. $\begin{cases} 2w-3x+4y-z=21 \\ w+5x=2 \\ -2x+3y+z=12 \\ -3w+4z=-5 \end{cases}$

63. $\begin{cases} -5x+10y=3 \\ \frac{7}{2}x-7y=20 \end{cases}$

64. $\begin{cases} 23x+21y=-4 \\ x-3y=-8 \end{cases}$

65. $\begin{cases} w-x+y-z=2 \\ 2w-x+3y=-5 \\ x-2z=7 \\ 3w+4x=-13 \end{cases}$

APPLICATIONS

66. The three sides of a triangle are related as follows: the perimeter is 43 feet, the second side is 5 feet more than twice the first side, and the third side is 3 feet less than the sum of the other two sides. Find the lengths of the three sides of the triangle.

67. Eric's favorite candy bar and ice cream flavor have fat and calorie contents as follows: each candy bar has 5 grams of fat and 280 calories; each serving of ice cream has 10 grams of fat and 150 calories. How many candy bars and servings of ice cream did he eat during the weekend he consumed 85 grams of fat and 2300 calories from these two treats?

68. A farmer plants soybeans, corn, and wheat and rotates the planting each year on her 500-acre farm. In a particular year, the profits from her crops were $120 per acre of soybeans, $100 per acre of corn, and $80 per acre of wheat. She planted twice as many acres of corn as soybeans. How many acres did she plant with each crop that year if she made a total profit of $51,800?

TECHNOLOGY

Using a graphing utility find the determinant of the matrix.

69. $\begin{vmatrix} 0.1 & 0.4 & -0.7 \\ 0.3 & -0.1 & 0.2 \\ 0.5 & -0.2 & 0.3 \end{vmatrix}$

70. $\begin{vmatrix} 0.1 & 0.3 & 0.1 \\ 0.2 & -0.2 & -0.1 \\ -0.1 & -0.4 & 0.5 \end{vmatrix}$

71. $\begin{vmatrix} 2.2 & 0.3 & -1.7 \\ 0.4 & -0.2 & 0.1 \\ 0.2 & 0.3 & -1.6 \end{vmatrix}$

72. $\begin{vmatrix} 3.1 & 0.6 & -1.1 \\ 1.2 & 5.2 & -7.3 \\ -0.1 & -4.1 & 6.5 \end{vmatrix}$

73. $\begin{vmatrix} 13 & 23 & -21 \\ 17 & -32 & 14 \\ 15 & 12 & -16 \end{vmatrix}$

74. $\begin{vmatrix} 25 & 32 & 17 \\ -13 & 14 & -24 \\ 16 & 26 & 36 \end{vmatrix}$

Use a graphing utility and Cramer's Rule to solve each system of equations.

75. $\begin{cases} x-2y+3z=9 \\ -x+3y=-4 \\ 2x-5y+5z=17 \end{cases}$

76. $\begin{cases} 2x+4y+z=1 \\ x-2y-3z=2 \\ x+y-z=-1 \end{cases}$

77. $\begin{cases} w+x+y+z=6 \\ -w+2x+3y=0 \\ 2w-3x+4y+z=4 \\ w+x+2y-z=0 \end{cases}$

11.4 BASIC MATRIX OPERATIONS

TOPICS

1. Matrix addition
2. Scalar multiplication
3. Matrix multiplication
4. Transition matrices

TOPIC 1: Matrix Addition

We've already seen once in this book that the mathematical operations of addition, subtraction, multiplication and division can be defined on objects besides real numbers. Specifically, we saw how these operations can be defined for real-valued functions. In this section and in 11.5, we will explore how to define these operations for matrices.

DEFINITION: Matrix Addition

Two matrices A and B can be added to form the new matrix $A+B$ only if A and B are of the same order. The addition is performed by adding corresponding entries of the two matrices together; that is, the element in the ith row and the jth column of $A+B$ is given by $a_{ij}+b_{ij}$.

CAUTION

It is very important to note the restriction on the order of the two matrices in the above definition; a matrix with m rows and n columns can only be added to another matrix with m rows and n columns. In all aspects of matrix algebra, the orders of the matrices involved must be considered.

DEFINITION: Matrix Equality

Two matrices A and B are **equal**, denoted $A = B$, if they are of the same order and all corresponding entries of A and B are equal.

Example 1: Matrix Addition

Perform the indicated addition, if possible.

a. $\begin{bmatrix} -3 & 2 \\ 0 & -5 \\ 11 & -9 \end{bmatrix} + \begin{bmatrix} 3 & 17 \\ 5 & 4 \\ -10 & 4 \end{bmatrix}$

b. $\begin{bmatrix} 2 & -5 \\ 1 & 0 \\ 0 & 3 \\ -7 & 10 \end{bmatrix} + \begin{bmatrix} 2 & 1 & 0 & -7 \\ -5 & 0 & 3 & 10 \end{bmatrix}$

Solution

a. Both matrices are 3×2, so the sum is defined and

$$\begin{bmatrix} -3 & 2 \\ 0 & -5 \\ 11 & -9 \end{bmatrix} + \begin{bmatrix} 3 & 17 \\ 5 & 4 \\ -10 & 4 \end{bmatrix} = \begin{bmatrix} 0 & 19 \\ 5 & -1 \\ 1 & -5 \end{bmatrix}$$

Each entry in the first matrix is added to its corresponding entry in the second matrix.

b. The first matrix is 4×2 and the second is 2×4, so the addition cannot be performed.

The entries in matrices do not all have to be constants. In many applications of matrices, it is convenient to represent some of the entries initially as variables, with the intent of eventually solving for the variables. We can solve some examples of such *matrix equations* now, using only matrix addition and matrix equality.

Example 2: Matrix Equations

Determine the values of the variables that will make each of the following statements true.

a. $\begin{bmatrix} -3 & a & b \\ -2 & a+b & 5 \end{bmatrix} = \begin{bmatrix} c & 3 & 7 \\ -2 & d & 5 \end{bmatrix}$

b. $\begin{bmatrix} 3x \\ 4 \end{bmatrix} + \begin{bmatrix} -y \\ 2x \end{bmatrix} = \begin{bmatrix} 13 \\ 7y \end{bmatrix}$

Solution

a. Solving for the four variables a, b, c, and d is just a matter of comparing the entries one-by-one.

$$\begin{bmatrix} -3 & a & b \\ -2 & a+b & 5 \end{bmatrix} = \begin{bmatrix} c & 3 & 7 \\ -2 & d & 5 \end{bmatrix}$$

Comparing the top rows tells us that $a = 3$, $b = 7$, and $c = -3$.

In the bottom rows, note the -2 in the left corner of each matrix and the 5 in each right corner. If these constants were not equal we would have a contradiction, and there would be no way to make the matrix equation true.

The only variable left to solve for is d, which we see is equal to $a+b$. So, $d = 3+7 = 10$.

b. Each matrix consists of only two entries, and after performing the matrix addition on the left and comparing corresponding entries, we arrive at the following system of equations.

$$\begin{cases} 3x - y = 13 \\ 4 + 2x = 7y \end{cases}$$

We know a number of ways of solving such a system. If we choose to use Cramer's Rule, we first rewrite the system as follows.

$$\begin{cases} 3x - y = 13 \\ 2x - 7y = -4 \end{cases}$$

From here we can determine

$$D = \begin{vmatrix} 3 & -1 \\ 2 & -7 \end{vmatrix} = (3)(-7) - (2)(-1) = -19,$$

$$D_x = \begin{vmatrix} 13 & -1 \\ -4 & -7 \end{vmatrix} = (13)(-7) - (-4)(-1) = -95, \text{ and}$$

$$D_y = \begin{vmatrix} 3 & 13 \\ 2 & -4 \end{vmatrix} = (3)(-4) - (2)(13) = -38,$$

so $x = \dfrac{D_x}{D} = \dfrac{-95}{-19} = 5$ and $y = \dfrac{D_y}{D} = \dfrac{-38}{-19} = 2.$

TOPIC 2: Scalar Multiplication

In the context of matrix algebra, a *scalar* is a real number, and *scalar multiplication* refers to the product of a real number and a matrix. It may seem strange to combine two very different sorts of objects (a scalar and a matrix), but consider the following matrix sum.

$$\begin{bmatrix} -5 & 2 \\ 1 & -3 \\ -2 & 7 \end{bmatrix} + \begin{bmatrix} -5 & 2 \\ 1 & -3 \\ -2 & 7 \end{bmatrix} = \begin{bmatrix} -10 & 4 \\ 2 & -6 \\ -4 & 14 \end{bmatrix}$$

Since the result of adding a matrix to itself has the effect of doubling each entry, it makes sense to write

$$2\begin{bmatrix} -5 & 2 \\ 1 & -3 \\ -2 & 7 \end{bmatrix} = \begin{bmatrix} -10 & 4 \\ 2 & -6 \\ -4 & 14 \end{bmatrix}.$$

Extending the idea to all scalars (real numbers) leads to the following definition.

DEFINITION: Scalar Multiplication

If A is an $m \times n$ matrix and c is a scalar, cA stands for the $m \times n$ matrix for which each entry is c times the corresponding entry of A. In other words, the entry in the ith row and j^{th} column of cA is ca_{ij}.

Example 3: Scalar Multiplication

Given the matrices $A=\begin{bmatrix}-1 & 6 & 2\\-8 & 0 & 1\end{bmatrix}$ and $B=\begin{bmatrix}0 & -3 & 4\\1 & -2 & 6\end{bmatrix}$, write $-3A+2B$ as a single matrix.

Solution

Before calculating, note that the operation can be performed since both matrices are of the same order, 2×3.

This problem has two steps. First, multiply each entry of A and B by its corresponding scalar. Second, add the resulting matrices together.

$$-3A+2B=-3\begin{bmatrix}-1 & 6 & 2\\-8 & 0 & 1\end{bmatrix}+2\begin{bmatrix}0 & -3 & 4\\1 & -2 & 6\end{bmatrix}$$ Multiply each matrix by its scalar.

$$=\begin{bmatrix}3 & -18 & -6\\24 & 0 & -3\end{bmatrix}+\begin{bmatrix}0 & -6 & 8\\2 & -4 & 12\end{bmatrix}$$ Add the resulting matrices.

$$=\begin{bmatrix}3 & -24 & 2\\26 & -4 & 9\end{bmatrix}$$

TECHNOLOGY

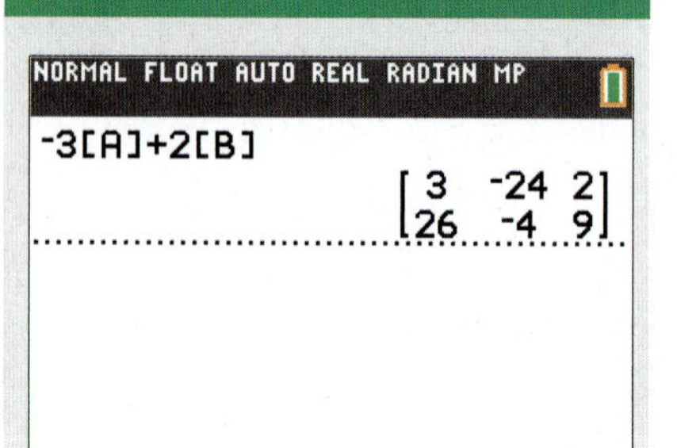

In the matrix menu, select EDIT to define matrices *A* and *B*. Enter the expression on the home screen, selecting each matrix as needed through the matrix menu.

As we try to understand algebraic operations on matrices, we can use real numbers as a model. Subtraction of real numbers is defined in terms of addition, and we define matrix subtraction in the same way.

DEFINITION: Matrix Subtraction

Let A and B be two matrices of the same order. The difference $A-B$ is defined by

$$A-B=A+(-B).$$

Example 4: Matrix Subtraction

Perform the indicated subtraction: $[3 \quad -5 \quad 2]-[-2 \quad -5 \quad 3]$.

Solution

Since both matrices are of order 1×3, we know the subtraction is possible. Subtract each entry in the second matrix from the corresponding entry in the first.

$$\begin{aligned}[3 \quad -5 \quad 2]-[-2 \quad -5 \quad 3]&=[3-(-2) \quad -5-(-5) \quad 2-3]\\&=[5 \quad 0 \quad -1]\end{aligned}$$

TOPIC 3: Matrix Multiplication

Early Matrix Multiplication

The first person to describe what we now know as matrix multiplication was the French mathematician Jacques Philippe Marie Binet in 1812. His description is based on the fact that matrices represent linear functions and that matrix multiplication corresponds to function composition, but his work and terminology actually predates the use of the word "matrix."

The definition of matrix multiplication is not as straightforward as that of matrix addition or subtraction. Simply put, matrix multiplication does not refer to multiplying the corresponding entries of two matrices together. To understand the process of matrix multiplication, it helps to first think about matrices as functions.

Note that the system of equations

$$\begin{cases} x' = ax + by \\ y' = cx + dy \end{cases}$$

can be thought of as a function that transforms the ordered pair (x, y) into the ordered pair (x', y'), and that the function is characterized entirely by the matrix

$$A = \begin{bmatrix} a & b \\ c & d \end{bmatrix}.$$

Similarly, the system

$$\begin{cases} x' = ex + fy \\ y' = gx + hy \end{cases}$$

is a function that transforms ordered pairs into ordered pairs, and it is characterized by the matrix

$$B = \begin{bmatrix} e & f \\ g & h \end{bmatrix}.$$

Now we can look at the result of plugging the output of the first function into the second function (*composing* the two functions so that B acts on the result of A).

If we let (x', y') denote the output of the first function, we obtain the new ordered pair (x'', y'') given by

$$\begin{cases} x'' = ex' + fy' \\ y'' = gx' + hy' \end{cases} \quad \text{or} \quad \begin{cases} x'' = e(ax + by) + f(cx + dy) \\ y'' = g(ax + by) + h(cx + dy) \end{cases}.$$

We can change the way the last system is written to obtain

$$\begin{cases} x'' = (ea + fc)x + (eb + fd)y \\ y'' = (ga + hc)x + (gb + hd)y \end{cases}$$

so the composition of the two functions, in the order BA, is characterized by the matrix

$$BA = \begin{bmatrix} ea + fc & eb + fd \\ ga + hc & gb + hd \end{bmatrix}.$$

Note that the entries of this last matrix are the sums of products, where the products in each sum are between elements of the rows of B with elements of the columns of A. This pattern is the basis of our formal definition of matrix multiplication.

DEFINITION: Matrix Multiplication

Two matrices A and B can be multiplied together, resulting in a new matrix denoted AB, only if the number of columns of A (the matrix on the left) is the same as the number of rows of B (the matrix on the right). Thus, if A is of order $m \times n$, the product AB is only defined if B is of order $n \times p$. The order of AB will be $m \times p$.

If we let c_{ij} denote the entry in the ith row and j^{th} column of AB, c_{ij} is obtained from the ith row of A and the jth column of B by the formula

$$c_{ij} = a_{i1}b_{1j} + a_{i2}b_{2j} + \cdots + a_{in}b_{nj}.$$

In words, c_{ij} equals the product of the first element of row i of matrix A and the first element of column j of matrix B, plus the product of the second element of row i and the second element of column j, and so on.

CAUTION

Unlike numerical multiplication, matrix multiplication is not commutative. That is, given two matrices A and B, AB *in general* is not equal to BA. As an illustration of this fact, suppose A is a 3×4 matrix and B is a 4×2 matrix. Then AB is defined (and is of order 3×2), but BA doesn't even exist. Even when both AB and BA are defined, they are generally not equal.

Example 5: Matrix Multiplication

Given the matrices $A = \begin{bmatrix} 2 & 0 \\ -5 & 1 \end{bmatrix}$ and $B = \begin{bmatrix} 7 & -2 \\ 3 & 1 \end{bmatrix}$, find AB.

Solution

Since both matrices are 2×2, they can be multiplied, and the result will also be a 2×2 matrix. For this example, we will follow the procedure one entry at a time.

$$AB = \begin{bmatrix} 2 & 0 \\ -5 & 1 \end{bmatrix}\begin{bmatrix} 7 & -2 \\ 3 & 1 \end{bmatrix} = \begin{bmatrix} 2(7)+0(3) & AB_{12} \\ AB_{21} & AB_{22} \end{bmatrix}$$

Sum the product of entries in the first row of A and first column of B.

$$= \begin{bmatrix} 2 & 0 \\ -5 & 1 \end{bmatrix}\begin{bmatrix} 7 & -2 \\ 3 & 1 \end{bmatrix} = \begin{bmatrix} 14 & 2(-2)+0(1) \\ AB_{21} & AB_{22} \end{bmatrix}$$

Then the first row of A and second column of B

$$= \begin{bmatrix} 2 & 0 \\ -5 & 1 \end{bmatrix}\begin{bmatrix} 7 & -2 \\ 3 & 1 \end{bmatrix} = \begin{bmatrix} 14 & -4 \\ -5(7)+1(3) & AB_{22} \end{bmatrix}$$

Then the second row of A and first column of B

$$= \begin{bmatrix} 2 & 0 \\ -5 & 1 \end{bmatrix}\begin{bmatrix} 7 & -2 \\ 3 & 1 \end{bmatrix} = \begin{bmatrix} 14 & -4 \\ -32 & -5(-2)+1(1) \end{bmatrix}$$

Finally, the second row of A and second column of B

$$= \begin{bmatrix} 14 & -4 \\ -32 & 11 \end{bmatrix}$$

Example 6: Matrix Multiplication

Given the matrices $A=\begin{bmatrix}2 & -3\\ 4 & -1\\ 1 & 0\end{bmatrix}$ and $B=\begin{bmatrix}5 & 0 & -2\\ -4 & 1 & 3\end{bmatrix}$, find the following products.

a. AB **b.** BA

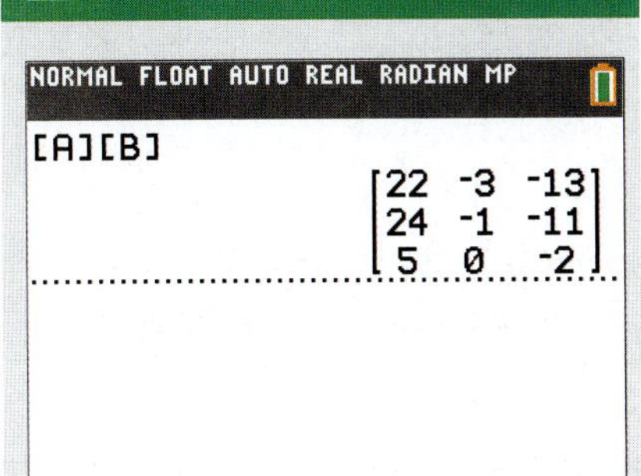

Solution

a. A is of order 3×2 and B is of order 2×3, so AB is defined and is of order 3×3. Each entry of AB is formed from a row of A and a column of B.

$$AB=\begin{bmatrix}2 & -3\\ 4 & -1\\ 1 & 0\end{bmatrix}\begin{bmatrix}5 & 0 & -2\\ -4 & 1 & 3\end{bmatrix}=\begin{bmatrix}2(5)+(-3)(-4) & 2(0)+(-3)(1) & 2(-2)+(-3)(3)\\ 4(5)+(-1)(-4) & 4(0)+(-1)(1) & 4(-2)+(-1)(3)\\ 1(5)+0(-4) & 1(0)+0(1) & 1(-2)+0(3)\end{bmatrix}$$

$$=\begin{bmatrix}22 & -3 & -13\\ 24 & -1 & -11\\ 5 & 0 & -2\end{bmatrix}$$

b. From the orders of the two matrices, we know BA exists and will be a 2×2 matrix. Note that each entry of BA is a sum of three products.

$$BA=\begin{bmatrix}5 & 0 & -2\\ -4 & 1 & 3\end{bmatrix}\begin{bmatrix}2 & -3\\ 4 & -1\\ 1 & 0\end{bmatrix}=\begin{bmatrix}5(2)+0(4)+(-2)(1) & 5(-3)+0(-1)+(-2)(0)\\ (-4)(2)+1(4)+3(1) & (-4)(-3)+1(-1)+3(0)\end{bmatrix}$$

$$=\begin{bmatrix}8 & -15\\ -1 & 11\end{bmatrix}$$

Now that matrix multiplication has been defined, it can be used in another way to illustrate the fact that matrices can be considered as functions. As already mentioned, the matrix

$$A=\begin{bmatrix}a & b\\ c & d\end{bmatrix}$$

characterizes the function that transforms (x, y) into (x', y') in the system

$$\begin{cases}x' = ax+by\\ y' = cx+dy\end{cases}.$$

This is even more clear if we associate the ordered pair (x, y) with the 2×1 matrix

$$\begin{bmatrix}x\\ y\end{bmatrix}.$$

Then the matrix product

$$\begin{bmatrix} a & b \\ c & d \end{bmatrix}\begin{bmatrix} x \\ y \end{bmatrix},$$

which has the "look" of a function acting on its argument, gives us the expressions in the system. Verify for yourself that

$$\begin{bmatrix} a & b \\ c & d \end{bmatrix}\begin{bmatrix} x \\ y \end{bmatrix} = \begin{bmatrix} ax+by \\ cx+dy \end{bmatrix}.$$

TOPIC 4: Transition Matrices

In a variety of important applications, matrices can be used to model how the "state" of a situation changes over time. To illustrate the idea, we will consider one such situation in some detail.

Suppose a new grocery store opens in a small town, with the intent of competing with the one existing store. The new store, which we will call store A, begins an aggressive advertising campaign, with the result that every month 45% of the customers of the existing store (store B) decide to start shopping at store A. Store B, however, responds with a strong appeal to win back customers and every month 30% of the customers of store A return to store B. In this highly dynamic situation, a number of questions may arise in the minds of the managers of the two stores, among them:

1. Given the known number of customers one month, how many customers can be expected the following month?

2. Can a certain percentage of the total number of the town's shoppers be expected in the long run? If so, what is that percentage?

To answer the first question, let a denote the number of customers of store A in a given month, and let b denote the number of customers of store B in the same month. In this simple example, we will assume that $a + b$, the total number of grocery store customers in the town, remains fixed over time.

In the following month, store A gains $0.45b$ customers (45% of its competitor's customers) and retains $0.7a$ of its existing customers. At the same time, store B gains $0.3a$ new customers (the 30% of A's clientele that switch back to store B), and retains $0.55b$ existing customers. In equation form, the number of customers of each store the following month are given by the following two expressions.

$$\begin{cases} 0.7a+0.45b \\ 0.3a+0.55b \end{cases}$$

Matrix multiplication allows us to write this information as

$$\begin{bmatrix} 0.7 & 0.45 \\ 0.3 & 0.55 \end{bmatrix}\begin{bmatrix} a \\ b \end{bmatrix},$$

and if we let P stand for the 2×2 matrix of percentages, we can think of P as a function that transforms one month's distribution of customers into the next month's. For instance, if store A has 360 customers and store B has 640 customers one month, then the following month their respective number of customers will be 540 and 460, as

$$\begin{bmatrix} 0.7 & 0.45 \\ 0.3 & 0.55 \end{bmatrix} \begin{bmatrix} 360 \\ 640 \end{bmatrix} = \begin{bmatrix} 540 \\ 460 \end{bmatrix}.$$

In such situations, a matrix like P is called a **transition matrix**, as it characterizes the transition of the system from one state to the next. Transition matrices are characterized by the fact that all of their entries are positive, and the sum of the entries in each column is 1.

To answer the second question, we might continue with this specific example and ask how the 540 and 460 customers of the respective stores realign themselves one month later. Note that

$$\begin{bmatrix} 0.7 & 0.45 \\ 0.3 & 0.55 \end{bmatrix} \begin{bmatrix} 540 \\ 460 \end{bmatrix} = \begin{bmatrix} 585 \\ 415 \end{bmatrix},$$

meaning that at the end of the second month, store A has 585 customers and store B has 415. But we can obtain the same result by applying the matrix P^2 to the original distribution of 360 and 640 customers, as

$$P^2 \begin{bmatrix} 360 \\ 640 \end{bmatrix} = P\left(P \begin{bmatrix} 360 \\ 640 \end{bmatrix} \right) = P \begin{bmatrix} 540 \\ 460 \end{bmatrix}.$$

In other words, we can form the composition of P with itself to obtain a function that transforms one month's distribution of customers into the distribution two months later. Note that

$$P^2 = \begin{bmatrix} 0.625 & 0.5625 \\ 0.375 & 0.4375 \end{bmatrix}$$

and that

$$\begin{bmatrix} 0.625 & 0.5625 \\ 0.375 & 0.4375 \end{bmatrix} \begin{bmatrix} 360 \\ 640 \end{bmatrix} = \begin{bmatrix} 585 \\ 415 \end{bmatrix}.$$

At this point, a calculator or computer software with matrix capability may be useful. We can use either aid to calculate high powers of P to see if there is a long-term trend in the distribution of the customers. For instance, to six decimal places,

$$P^5 = \begin{bmatrix} 0.600391 & 0.599414 \\ 0.399609 & 0.400586 \end{bmatrix},$$

and after a certain point calculators and software will round off the entries and give us a result of

$$P^n = \begin{bmatrix} 0.6 & 0.6 \\ 0.4 & 0.4 \end{bmatrix}$$

for large n (the value for n at which this happens will vary depending on the technology used). This means that, after a few months, store A can count on roughly 60% of the town's customers and store B can count on roughly 40% (the actual identities of the customers will keep changing from month to month, but the relative proportions will have stabilized). We can verify that the situation is stable by applying the transition matrix to an assumed 1000 customers split 60:40.

$$\begin{bmatrix} 0.7 & 0.45 \\ 0.3 & 0.55 \end{bmatrix}\begin{bmatrix} 600 \\ 400 \end{bmatrix} = \begin{bmatrix} 600 \\ 400 \end{bmatrix}$$

You will see another approach to determining the stable long-term state in the exercises.

11.4 EXERCISES

PRACTICE

Given $A = \begin{bmatrix} 3 & -2 \\ 1 & 0 \\ 0 & 5 \end{bmatrix}$, $B = \begin{bmatrix} 4 & -5 \\ 3 & 0 \\ -2 & 2 \end{bmatrix}$, $C = \begin{bmatrix} 2 & -1 \\ 6 & 10 \\ -3 & 7 \end{bmatrix}$, and $D = \begin{bmatrix} 3 & 2 & 5 \\ -2 & -4 & 1 \end{bmatrix}$,

determine the following, if possible. See Examples 1, 3, and 4.

1. $3A - B$

2. $B - 2D$

3. $3C$

4. $\frac{1}{2}D$

5. $3D + C$

6. $A + B + C$

7. $2A + 2B$

8. $\frac{3}{2}B + \frac{1}{2}C$

9. $C - 3A$

10. $3C - A$

11. $4A - 3D$

12. $2(A - 3B)$

Determine values of the variables that will make the following equations true, if possible. See Examples 1–4.

13. $\begin{bmatrix} 2a & b & 3 \\ -5 & 9 & 7 \end{bmatrix} = \begin{bmatrix} 6 & -1 & 3 \\ -5 & 9 & c-3 \end{bmatrix}$

14. $\begin{bmatrix} x \\ -9 \\ -1+z \end{bmatrix} = \begin{bmatrix} 8 \\ 3y \\ 5 \end{bmatrix}$

15. $\begin{bmatrix} a & 2b & c \end{bmatrix} + 3\begin{bmatrix} a & 2 & -c \end{bmatrix} = \begin{bmatrix} 8 & 2 & 2 \end{bmatrix}$

16. $\begin{bmatrix} w & 5x \\ 2y & z \end{bmatrix} - 5\begin{bmatrix} w & x \\ y & -z \end{bmatrix} = \begin{bmatrix} w+5 & 0 \\ 6 & 1 \end{bmatrix}$

17. $\begin{bmatrix} 3x \\ 2y \end{bmatrix} + \begin{bmatrix} x \\ -y \\ z \end{bmatrix} = \begin{bmatrix} 4 \\ 0 \\ 2 \end{bmatrix}$

18. $\begin{bmatrix} 2a & 3b & c \end{bmatrix} = \begin{bmatrix} 4 \\ 3 \\ 0 \end{bmatrix}$

19. $\begin{bmatrix} x \\ 3x \end{bmatrix} - \begin{bmatrix} y \\ 2y \end{bmatrix} = \begin{bmatrix} 5 \\ 20 \end{bmatrix}$

20. $7\begin{bmatrix} -1 \\ y \end{bmatrix} = \begin{bmatrix} 2x \\ 5x \end{bmatrix} + 3\begin{bmatrix} y \\ 1 \end{bmatrix}$

21. $2\begin{bmatrix} x \\ 2y \end{bmatrix} - 3\begin{bmatrix} 5y \\ -3x \end{bmatrix} = \begin{bmatrix} -9 \\ 31 \end{bmatrix}$

22. $2\begin{bmatrix} 3r & s & 2t \end{bmatrix} - \begin{bmatrix} r & s & t \end{bmatrix} = \begin{bmatrix} 15 & 3 & 9 \end{bmatrix}$

23. $2\begin{bmatrix} 2x^2 & x \\ 7x & 4 \end{bmatrix} - \begin{bmatrix} 5x \\ x-2 \end{bmatrix} = \begin{bmatrix} 2x & 0 \\ 6 & x^2 \end{bmatrix}$

24. $\begin{bmatrix} -x \\ 3 \end{bmatrix} - 5\begin{bmatrix} 2 \\ y \end{bmatrix} = \begin{bmatrix} -2y \\ 3x \end{bmatrix}$

25. $3\begin{bmatrix} 2a \\ -a \end{bmatrix} - 3\begin{bmatrix} 3b \\ 2b \end{bmatrix} = \begin{bmatrix} 3 \\ -54 \end{bmatrix}$

26. $2\begin{bmatrix} -s \\ -7 \end{bmatrix} + 2\begin{bmatrix} -2r \\ r \end{bmatrix} = -2\begin{bmatrix} 8 \\ s \end{bmatrix}$

Evaluate the following matrix products, if possible. See Examples 5 and 6.

27. $\begin{bmatrix} 3 & -2 & 1 \end{bmatrix}\begin{bmatrix} 5 & -1 \\ 0 & 3 \\ 9 & 4 \end{bmatrix}$

28. $\begin{bmatrix} 0 & -8 \\ 5 & 6 \end{bmatrix}\begin{bmatrix} 3 & 7 \end{bmatrix}$

29. $\begin{bmatrix} 3 & 7 \end{bmatrix}\begin{bmatrix} 0 & -8 \\ 5 & 6 \end{bmatrix}$

30. $\begin{bmatrix} 5 & 0 & -3 \end{bmatrix}\begin{bmatrix} 4 \\ 2 \\ -6 \end{bmatrix}$

31. $\begin{bmatrix} 3 & 9 & -4 \\ 0 & 0 & 2 \\ 5 & -2 & 7 \end{bmatrix}\begin{bmatrix} 3 & 2 \\ 2 & 1 \end{bmatrix}$

32. $\begin{bmatrix} 4 \\ 2 \\ -6 \end{bmatrix}\begin{bmatrix} 5 & 0 & -3 \end{bmatrix}$

33. $\begin{bmatrix} -3 & -6 & -3 \end{bmatrix}\begin{bmatrix} 6 & 9 \\ 6 & -8 \\ -8 & 8 \end{bmatrix}$

34. $\begin{bmatrix} 4 & -5 \\ 7 & -9 \end{bmatrix}\begin{bmatrix} -8 & 3 \end{bmatrix}$

35. $\begin{bmatrix} -3 \\ -5 \\ -6 \end{bmatrix}\begin{bmatrix} -5 & 1 & 8 \end{bmatrix}$

Given $A = \begin{bmatrix} -3 & 1 \\ 2 & 3 \end{bmatrix}$, $B = \begin{bmatrix} 8 & -5 \end{bmatrix}$, $C = \begin{bmatrix} 4 \\ 7 \\ -2 \end{bmatrix}$, and $D = \begin{bmatrix} -5 & 4 \\ -1 & -1 \end{bmatrix}$, determine the following, if possible. See Examples 5 and 6.

36. AB **37.** BA **38.** $BA + B$ **39.** A^2

40. C^2 **41.** CB **42.** D^2 **43.** $CD + C$

44. DA **45.** AD **46.** DB **47.** $(BD)A$

APPLICATIONS

48. Suppose that each month 20% of store B's customers switch to store A, and 10% of store A's customers switch back to store B. At the start of January, store A has 300 customers and store B has 700. How many customers can each store expect at the start of February? At the start of March?

49. Given the percentages stated in the last problem, what long-term proportion of the town's customers can each store expect? (A graphing utility may be used to compute high powers of the transition matrix, or you can use the method described in the following exercise.)

WRITING & THINKING

50. Suppose P is a 2×2 transition matrix, and we want to determine the effect of applying high powers of P to the matrix

$$\begin{bmatrix} x \\ y \end{bmatrix},$$

where $x + y$ is a fixed constant, say c. (In our competing store situation, $x + y = 1000$.) If the long-term behavior approaches a steady state, as in our two-store example, then there is some value for x and some value for y such that $x + y = c$ and

$$P\begin{bmatrix} x \\ y \end{bmatrix} = \begin{bmatrix} x \\ y \end{bmatrix}.$$

In other words, once the steady state has been reached, applying the matrix P to it has no effect on the state.

We can use this fact to actually solve for x and y as follows. Given the matrix

$$P = \begin{bmatrix} 0.7 & 0.45 \\ 0.3 & 0.55 \end{bmatrix},$$

write the equation

$$P\begin{bmatrix} x \\ y \end{bmatrix} = \begin{bmatrix} x \\ y \end{bmatrix}$$

in system form. You should find that the two equations that result are actually identical. But if we now also use the fact that $x + y = 1000$, we can solve for the variables and find that $x = 600$ and $y = 400$. Verify that this is indeed the case.

51. Your friend Jared is having trouble with matrices, so you offer to study with him. Check his solution of the following problem. If the solution is incorrect, explain the error that has been made.

$$2\begin{bmatrix} -8 & -9 & -1 \\ -8 & 1 & 5 \end{bmatrix} - 2\begin{bmatrix} -2 & 3 \\ 5 & -7 \\ -8 & -1 \end{bmatrix}$$

$$= \begin{bmatrix} -16 & -18 & -2 \\ -16 & 2 & 10 \end{bmatrix} + \begin{bmatrix} 4 & -6 \\ -10 & 14 \\ 16 & 2 \end{bmatrix}$$

$$= \begin{bmatrix} -16+4-18-10-2+16 & -16-6-18+14-2+2 \\ -16+4+2-10+10+2 & -16-6+2+14+10+2 \end{bmatrix}$$

$$= \begin{bmatrix} -26 & -26 \\ -8 & 6 \end{bmatrix}$$

TECHNOLOGY

Given $A = \begin{bmatrix} 3.8 & -1.2 & 4.6 \end{bmatrix}$, $B = \begin{bmatrix} -8.2 & -4.9 \\ 7.4 & -1.3 \\ 3.5 & -2.1 \end{bmatrix}$, $C = \begin{bmatrix} 6.3 \\ 5.7 \end{bmatrix}$, and $D = \begin{bmatrix} 2.8 & -7.1 \\ -5.4 & 6.6 \end{bmatrix}$, use a graphing utility to determine the following, if possible.

52. BD

53. CA

54. D^2

55. AB

56. DC

57. BC

11.5 INVERSES OF MATRICES

TOPICS

1. The matrix form of a system of linear equations
2. Finding the inverse of a matrix
3. Solving systems of linear equations using matrix inverses

TOPIC 1: The Matrix Form of a System of Linear Equations

As we saw in Section 11.4, if we express the ordered pair (x, y) as a 2×1 matrix, then the linear system

$$\begin{cases} ax+by=e \\ cx+dy=f \end{cases}$$

can be written as

$$\begin{bmatrix} a & b \\ c & d \end{bmatrix}\begin{bmatrix} x \\ y \end{bmatrix}=\begin{bmatrix} e \\ f \end{bmatrix}.$$

The fact that the matrix equation is equivalent to the system of equations above it is a great leap in efficiency: it converts a system of any number of equations into a single matrix equation. More importantly, the function interpretation of a matrix allows us to express a *whole system* of equations in a form like that of a *single linear equation* of a single variable.

Since the generic linear equation $ax=b$ can be solved by dividing both sides by a,

$$ax=b \quad \Leftrightarrow \quad x=\frac{b}{a} \ (\text{assuming } a\neq 0),$$

it is tempting to solve

$$\begin{bmatrix} a & b \\ c & d \end{bmatrix}\begin{bmatrix} x \\ y \end{bmatrix}=\begin{bmatrix} e \\ f \end{bmatrix}$$

by "dividing" both sides by the 2×2 matrix of coefficients. Unfortunately, we don't yet have a way of making sense of "matrix division."

We will return to this thought soon, but first we will see how some specific linear systems appear in matrix form.

Example 1: Matrix Equation

Write each linear system as a matrix equation.

a. $\begin{cases} -3x+5y=2 \\ x-4y=-1 \end{cases}$

b. $\begin{cases} 3y-x=-2 \\ 4-z+y=5 \\ z-3x+3=y-x \end{cases}$

Solution

a. Since the system is in standard form, we can just read off the coefficients of x and y to form the equation

$$\begin{bmatrix} -3 & 5 \\ 1 & -4 \end{bmatrix}\begin{bmatrix} x \\ y \end{bmatrix}=\begin{bmatrix} 2 \\ -1 \end{bmatrix}.$$

b. First, we write each equation in standard form.

$$\begin{cases} 3y-x=-2 \\ 4-z+y=5 \\ z-3x+3=y-x \end{cases} \longrightarrow \begin{cases} -x+3y=-2 \\ y-z=1 \\ -2x-y+z=-3 \end{cases}$$

Now we can read off the coefficients to form the matrix equation

$$\begin{bmatrix} -1 & 3 & 0 \\ 0 & 1 & -1 \\ -2 & -1 & 1 \end{bmatrix}\begin{bmatrix} x \\ y \\ z \end{bmatrix}=\begin{bmatrix} -2 \\ 1 \\ -3 \end{bmatrix}.$$

TOPIC 2: Finding the Inverse of a Matrix

In order to solve matrix equations like the two we obtained in Example 1, we need a way to "undo" the matrix of coefficients that appears in front of the column of variables.

In order to figure out how to "undo" a matrix, we will first need to understand how to do *nothing* to a matrix. Consider the following matrix products.

$$\begin{bmatrix} 1 & 0 \\ 0 & 1 \end{bmatrix}\begin{bmatrix} x \\ y \end{bmatrix} \quad \text{and} \quad \begin{bmatrix} 1 & 0 & 0 \\ 0 & 1 & 0 \\ 0 & 0 & 1 \end{bmatrix}\begin{bmatrix} x \\ y \\ z \end{bmatrix}$$

Evaluating these products, we have

$$\begin{bmatrix} 1 & 0 \\ 0 & 1 \end{bmatrix}\begin{bmatrix} x \\ y \end{bmatrix}=\begin{bmatrix} 1x+0y \\ 0x+1y \end{bmatrix}=\begin{bmatrix} x \\ y \end{bmatrix}$$

and

$$\begin{bmatrix} 1 & 0 & 0 \\ 0 & 1 & 0 \\ 0 & 0 & 1 \end{bmatrix}\begin{bmatrix} x \\ y \\ z \end{bmatrix}=\begin{bmatrix} 1x+0y+0z \\ 0x+1y+0z \\ 0x+0y+1z \end{bmatrix}=\begin{bmatrix} x \\ y \\ z \end{bmatrix}.$$

If these matrix products appear as the left-hand side of matrix equations, the equations would correspond to solutions of linear systems:

$$\begin{bmatrix} 1 & 0 \\ 0 & 1 \end{bmatrix}\begin{bmatrix} x \\ y \end{bmatrix} = \begin{bmatrix} a \\ b \end{bmatrix} \text{ corresponds to } \begin{cases} x = a \\ y = b \end{cases}$$

and

$$\begin{bmatrix} 1 & 0 & 0 \\ 0 & 1 & 0 \\ 0 & 0 & 1 \end{bmatrix}\begin{bmatrix} x \\ y \\ z \end{bmatrix} = \begin{bmatrix} a \\ b \\ c \end{bmatrix} \text{ corresponds to } \begin{cases} x = a \\ y = b. \\ z = c \end{cases}$$

When we multiply any matrix by one of these matrices, the original matrix is *unchanged.* This fact is very useful in solving matrix equations.

DEFINITION: Identity Matrices

The $n \times n$ **identity matrix**, denoted I_n (just I when there is no possibility of confusion), is the $n \times n$ matrix consisting of 1s on the *main diagonal* and 0s everywhere else. The **main diagonal** consists of those entries in the first row-first column, the second row-second column, and so on down to the nth row-n^{th} column. Every identity matrix has the form

$$I = \begin{bmatrix} 1 & 0 & 0 & \cdots & 0 \\ 0 & 1 & 0 & \cdots & 0 \\ 0 & 0 & 1 & \cdots & 0 \\ \vdots & \vdots & \vdots & \ddots & \vdots \\ 0 & 0 & 0 & \cdots & 1 \end{bmatrix}.$$

If the matrices A and B have appropriate order, so that the matrix products are defined, then $AI = A$ and $IB = B$. Thus, the identity matrix serves as the multiplicative identity on the set of appropriately sized matrices. In this sense, I serves the same purpose as the number 1 in the set of real numbers.

We know that a linear system of n equations and n variables can be expressed as a matrix equation $AX = B$, where A is an $n \times n$ matrix of coefficients, X is an $n \times 1$ matrix containing the n variables, and B is an $n \times 1$ matrix of the constants from the right-hand sides of the equations. If we could find a matrix, which we call A^{-1}, with the property that $A^{-1}A = I$, then we could use A^{-1} to "undo" the matrix A. We call the matrix A^{-1}, if it exists, the *inverse* of A. This is analogous to the fact that $\frac{1}{a}$, sometimes denoted a^{-1}, is the (multiplicative) inverse of the real number a.

DEFINITION: The Inverse of a Matrix

Let A be an $n \times n$ matrix. If there exists an $n \times n$ matrix A^{-1} such that

$$A^{-1}A = I_n \text{ and } AA^{-1} = I_n,$$

we call A^{-1} the **inverse** of A.

Example 2: Finding the Inverse of a Matrix

Find the inverse of the matrix $A=\begin{bmatrix}2 & -3\\ -1 & 2\end{bmatrix}$.

Solution

If we let $A^{-1}=\begin{bmatrix}w & x\\ y & z\end{bmatrix}$, we can use the equation $AA^{-1}=I$ to find w, x, y, and z.

$$\begin{bmatrix}2 & -3\\ -1 & 2\end{bmatrix}\begin{bmatrix}w & x\\ y & z\end{bmatrix}=\begin{bmatrix}1 & 0\\ 0 & 1\end{bmatrix}$$

Multiplying the left-hand side out, we see that we need to solve the equation

$$\begin{bmatrix}2w-3y & 2x-3z\\ -w+2y & -x+2z\end{bmatrix}=\begin{bmatrix}1 & 0\\ 0 & 1\end{bmatrix}$$

which, if we equate columns on each side, means we need to solve the two linear systems

$$\begin{cases}2w-3y=1\\ -w+2y=0\end{cases}\quad\text{and}\quad\begin{cases}2x-3z=0\\ -x+2z=1\end{cases}.$$

We have covered many methods for solving such systems. If we write the augmented matrix for each system, we get

$$\left[\begin{array}{cc|c}2 & -3 & 1\\ -1 & 2 & 0\end{array}\right]\quad\text{and}\quad\left[\begin{array}{cc|c}2 & -3 & 0\\ -1 & 2 & 1\end{array}\right].$$

Note that the left-hand sides of these matrices are the same. This allows us to combine them into a new kind of augmented matrix so we can use Gauss-Jordan elimination to solve the systems at the same time. Combining the matrices, we get

$$\left[\begin{array}{cc|cc}2 & -3 & 1 & 0\\ -1 & 2 & 0 & 1\end{array}\right].$$

When we change this new matrix into reduced row echelon form, we will have solved the first system with the numbers in the third column and the second system with the numbers in the fourth column.

$$\left[\begin{array}{cc|cc}2 & -3 & 1 & 0\\ -1 & 2 & 0 & 1\end{array}\right]\xrightarrow{R_1\leftrightarrow R_2}\left[\begin{array}{cc|cc}-1 & 2 & 0 & 1\\ 2 & -3 & 1 & 0\end{array}\right]\xrightarrow{2R_1+R_2}\left[\begin{array}{cc|cc}-1 & 2 & 0 & 1\\ 0 & 1 & 1 & 2\end{array}\right]$$

$$\xrightarrow{-R_1}\left[\begin{array}{cc|cc}1 & -2 & 0 & -1\\ 0 & 1 & 1 & 2\end{array}\right]\xrightarrow{2R_2+R_1}\left[\begin{array}{cc|cc}1 & 0 & 2 & 3\\ 0 & 1 & 1 & 2\end{array}\right]$$

This tells us that $w=2$ and $y=1$ (from the third column) and $x=3$ and $z=2$ (from the fourth column). So

$$A^{-1}=\begin{bmatrix}2 & 3\\ 1 & 2\end{bmatrix}.$$

We can now verify that

$$\begin{bmatrix} 2 & -3 \\ -1 & 2 \end{bmatrix}\begin{bmatrix} 2 & 3 \\ 1 & 2 \end{bmatrix}=\begin{bmatrix} 1 & 0 \\ 0 & 1 \end{bmatrix} \text{ and also } \begin{bmatrix} 2 & 3 \\ 1 & 2 \end{bmatrix}\begin{bmatrix} 2 & -3 \\ -1 & 2 \end{bmatrix}=\begin{bmatrix} 1 & 0 \\ 0 & 1 \end{bmatrix},$$

so we have indeed found A^{-1}.

Note how, during the solution process, the identity matrix passed from the right side of the matrix to the left, resulting in reduced row echelon form. With this observation, we can omit the intermediate step of constructing the systems of equations, and skip to the process of putting the appropriate augmented matrix into reduced row echelon form.

PROCEDURE: Finding the Inverse of a Matrix

Let A be an $n \times n$ matrix. The inverse of A can be found by

Step 1: Forming the augmented matrix $[A \mid I]$, where I is the $n \times n$ identity matrix.

Step 2: Using Gauss-Jordan elimination to put $[A \mid I]$ into the form $[I \mid B]$, if possible.

Step 3: Defining A^{-1} to be B.

If it is not possible to put $[A \mid I]$ into reduced row echelon form, then A doesn't have an inverse, and we say A is **not invertible.**

If the coefficient matrix of a system of equations is not invertible, it means that the system either has an infinite number of solutions or has no solution.

There is a shortcut for finding inverses of 2×2 matrices that can save you some time. This shortcut also quickly identifies those 2×2 matrices that are not invertible.

FORMULA: Inverse of a 2×2 Matrix

Let $A=\begin{bmatrix} a & b \\ c & d \end{bmatrix}$. Then $A^{-1}=\frac{1}{|A|}\begin{bmatrix} d & -b \\ -c & a \end{bmatrix}$, where $|A|=ad-bc$ is the determinant of A. Since $|A|$ appears in the denominator of a fraction, A^{-1} fails to exist if $|A|=0$.

In fact, we can extend this last observation to all square matrices.

THEOREM: Invertible Matrices

A square matrix A is **invertible** if and only if $|A| \neq 0$.

Example 3: Finding the Inverse of a Matrix

Find the inverses of the following matrices, if possible.

a. $A=\begin{bmatrix}3 & -5\\ 2 & 1\end{bmatrix}$ **b.** $B=\begin{bmatrix}2 & 4 & 2\\ -1 & 5 & -1\\ 3 & 1 & 3\end{bmatrix}$

Solution

a. Since A is 2×2, we can use the shortcut,

$$A^{-1}=\frac{1}{3-(-10)}\begin{bmatrix}1 & 5\\ -2 & 3\end{bmatrix}=\begin{bmatrix}\frac{1}{13} & \frac{5}{13}\\ \frac{-2}{13} & \frac{3}{13}\end{bmatrix}.$$

We can verify our work as follows:

$$\begin{bmatrix}3 & -5\\ 2 & 1\end{bmatrix}\begin{bmatrix}\frac{1}{13} & \frac{5}{13}\\ \frac{-2}{13} & \frac{3}{13}\end{bmatrix}=\begin{bmatrix}1 & 0\\ 0 & 1\end{bmatrix}=\begin{bmatrix}\frac{1}{13} & \frac{5}{13}\\ \frac{-2}{13} & \frac{3}{13}\end{bmatrix}\begin{bmatrix}3 & -5\\ 2 & 1\end{bmatrix}$$

b. Since B is a 3×3 matrix, we use the augmented identity matrix process.

$$\left[\begin{array}{ccc|ccc}2 & 4 & 2 & 1 & 0 & 0\\ -1 & 5 & -1 & 0 & 1 & 0\\ 3 & 1 & 3 & 0 & 0 & 1\end{array}\right]\xrightarrow[3R_2+R_3]{2R_2+R_1}\left[\begin{array}{ccc|ccc}0 & 14 & 0 & 1 & 2 & 0\\ -1 & 5 & -1 & 0 & 1 & 0\\ 0 & 16 & 0 & 0 & 3 & 1\end{array}\right]$$

$$\xrightarrow{R_1\leftrightarrow R_2}\left[\begin{array}{ccc|ccc}-1 & 5 & -1 & 0 & 1 & 0\\ 0 & 14 & 0 & 1 & 2 & 0\\ 0 & 16 & 0 & 0 & 3 & 1\end{array}\right]\xrightarrow[\frac{1}{14}R_2]{-R_1}\left[\begin{array}{ccc|ccc}1 & -5 & 1 & 0 & -1 & 0\\ 0 & 1 & 0 & \frac{1}{14} & \frac{1}{7} & 0\\ 0 & 16 & 0 & 0 & 3 & 1\end{array}\right]$$

$$\xrightarrow{-16R_2+R_3}\left[\begin{array}{ccc|ccc}1 & -5 & 1 & 0 & -1 & 0\\ 0 & 1 & 0 & \frac{1}{14} & \frac{1}{7} & 0\\ 0 & 0 & 0 & -\frac{8}{7} & \frac{5}{7} & 1\end{array}\right]$$

At this point, we can stop. Once the first three entries of any row are 0 in the 3×3 matrix, there is no way to put the matrix into reduced row echelon form. Thus, B has no inverse.

We could have seen this fact earlier by considering the determinant of B, which is 0, meaning B has no inverse. To quickly see that $|B|=0$, note that B has two identical columns. Recall that switching two columns changes the sign of the determinant, but if we switch identical columns, the determinant must remain the same! The only way this is possible is if $|B|=0$.

TOPIC 3: Solving Systems of Linear Equations Using Matrix Inverses

We have now assembled all the tools we need to solve linear systems by the inverse matrix method.

PROCEDURE: The Inverse Matrix Method

To solve a linear system of n equations in n variables

Step 1: Write the system in matrix form as $AX = B$, where A is the $n \times n$ matrix of coefficients, X is the $n \times 1$ matrix of variables, and B is the $n \times 1$ matrix of constants.

Step 2: Calculate A^{-1}, if it exists. If A^{-1} does not exist, the system either has an infinite number of solutions, or no solution, and a different method is required.

Step 3: Multiply both sides of the equation $AX = B$ by A^{-1}. We obtain

$$\begin{aligned} A^{-1}AX &= A^{-1}B \\ IX &= A^{-1}B \\ X &= A^{-1}B \end{aligned}$$

Step 4: The entries in the $n \times 1$ matrix $A^{-1}B$ are the solutions for the variables listed in the $n \times 1$ matrix X.

CAUTION

It is crucial to multiply both sides of the equation $AX = B$ by A^{-1} on the *left-hand side*. Recall that matrix multiplication is not commutative, so failing to do this can result in an incorrect answer, and often the multiplication will not even be defined.

Example 4: Inverse Matrix Method

Solve the following systems by the inverse matrix method.

a. $\begin{cases} 4x - 5y = 3 \\ -3x + 7y = 1 \end{cases}$

b. $\begin{cases} -x + 2y = 3 \\ 3x - 6y = -5 \end{cases}$

Solution

a. $\begin{cases} 4x - 5y = 3 \\ -3x + 7y = 1 \end{cases}$

$$\begin{bmatrix} 4 & -5 \\ -3 & 7 \end{bmatrix}\begin{bmatrix} x \\ y \end{bmatrix} = \begin{bmatrix} 3 \\ 1 \end{bmatrix}$$

Write the system in matrix form.

$$\begin{bmatrix} 4 & -5 \\ -3 & 7 \end{bmatrix}^{-1} = \frac{1}{28-15}\begin{bmatrix} 7 & 5 \\ 3 & 4 \end{bmatrix} = \begin{bmatrix} \frac{7}{13} & \frac{5}{13} \\ \frac{3}{13} & \frac{4}{13} \end{bmatrix}$$

Find the inverse of the coefficient matrix. Since the matrix is of order 2×2, we can use the shortcut to obtain the inverse quickly.

$$\begin{bmatrix} x \\ y \end{bmatrix} = \begin{bmatrix} \frac{7}{13} & \frac{5}{13} \\ \frac{3}{13} & \frac{4}{13} \end{bmatrix}\begin{bmatrix} 3 \\ 1 \end{bmatrix} = \begin{bmatrix} \frac{21}{13}+\frac{5}{13} \\ \frac{9}{13}+\frac{4}{13} \end{bmatrix} = \begin{bmatrix} 2 \\ 1 \end{bmatrix}$$

It is a good idea to check your work at this stage by making sure that the product of the original matrix and its inverse is the identity matrix.

Rewrite the equation $AX = B$ in the form $X = A^{-1}B$ and carry out the matrix multiplication.

Thus, the solution to the system is $(2,1)$.

b. $\begin{cases} -x+2y=3 \\ 3x-6y=-5 \end{cases}$

$$\begin{bmatrix} -1 & 2 \\ 3 & -6 \end{bmatrix}\begin{bmatrix} x \\ y \end{bmatrix} = \begin{bmatrix} 3 \\ -5 \end{bmatrix}$$

Again, write the system in matrix form.

$$\begin{bmatrix} -1 & 2 \\ 3 & -6 \end{bmatrix}^{-1} = \frac{1}{6-6}\begin{bmatrix} -6 & -2 \\ -3 & -1 \end{bmatrix}$$

The coefficient matrix has no inverse (since its determinant is 0), so the system has no solution or an infinite number of solutions.

$$D_x = \begin{vmatrix} 3 & 2 \\ -5 & -6 \end{vmatrix} = -18-(-10) \neq 0$$

Since $D_x \neq 0$, Cramer's Rule tells us the system has no solution.

This system has no solution.

We have now reached the end of our list of solution methods for linear systems. One important question has not yet been addressed. How should we choose a method of solution, given a system of equations?

Unfortunately, there is no simple answer. If the system is small, or looks especially simple, then the methods of substitution or elimination may be the quickest route to a solution. Generally, the larger and/or the more complicated the system, the more likely we are to prefer a matrix method of solution, such as Gauss-Jordan elimination, Cramer's Rule, or the inverse matrix method. If you find that your first choice leads to a computational headache, don't be reluctant to stop and try another method.

In practical applications, it is often the case that a number of closely related systems must be solved, and the only difference in the systems lies in the constants on the right-hand sides of the equations. That is, the coefficient matrix is the same for all the systems. In this case, the inverse matrix method is almost certainly the best choice, as illustrated in Example 5.

Example 5: Using the Inverse Matrix Method

Solve the following three linear systems.

$$\begin{cases} x-4y+z=-6 \\ -2x+y=5 \\ 3y-z=3 \end{cases} \qquad \begin{cases} x-4y+z=-17 \\ -2x+y=2 \\ 3y-z=14 \end{cases} \qquad \begin{cases} x-4y+z=3 \\ -2x+y=5 \\ 3y-z=-5 \end{cases}$$

Solution

The coefficient matrix is the same for the three systems, and this means we can avoid unnecessary computation if we use the inverse matrix method. We begin by finding the inverse of the coefficient matrix.

$$\left[\begin{array}{ccc|ccc} 1 & -4 & 1 & 1 & 0 & 0 \\ -2 & 1 & 0 & 0 & 1 & 0 \\ 0 & 3 & -1 & 0 & 0 & 1 \end{array}\right] \xrightarrow{2R_1+R_2} \left[\begin{array}{ccc|ccc} 1 & -4 & 1 & 1 & 0 & 0 \\ 0 & -7 & 2 & 2 & 1 & 0 \\ 0 & 3 & -1 & 0 & 0 & 1 \end{array}\right]$$

$$\xrightarrow{\frac{1}{3}R_3} \left[\begin{array}{ccc|ccc} 1 & -4 & 1 & 1 & 0 & 0 \\ 0 & -7 & 2 & 2 & 1 & 0 \\ 0 & 1 & -\frac{1}{3} & 0 & 0 & \frac{1}{3} \end{array}\right] \xrightarrow[7R_3+R_2]{4R_3+R_1} \left[\begin{array}{ccc|ccc} 1 & 0 & -\frac{1}{3} & 1 & 0 & \frac{4}{3} \\ 0 & 0 & -\frac{1}{3} & 2 & 1 & \frac{7}{3} \\ 0 & 1 & -\frac{1}{3} & 0 & 0 & \frac{1}{3} \end{array}\right]$$

$$\xrightarrow{-3R_2} \left[\begin{array}{ccc|ccc} 1 & 0 & -\frac{1}{3} & 1 & 0 & \frac{4}{3} \\ 0 & 0 & 1 & -6 & -3 & -7 \\ 0 & 1 & -\frac{1}{3} & 0 & 0 & \frac{1}{3} \end{array}\right] \xrightarrow{R_2 \leftrightarrow R_3} \left[\begin{array}{ccc|ccc} 1 & 0 & -\frac{1}{3} & 1 & 0 & \frac{4}{3} \\ 0 & 1 & -\frac{1}{3} & 0 & 0 & \frac{1}{3} \\ 0 & 0 & 1 & -6 & -3 & -7 \end{array}\right]$$

$$\xrightarrow[\frac{1}{3}R_3+R_2]{\frac{1}{3}R_3+R_1} \left[\begin{array}{ccc|ccc} 1 & 0 & 0 & -1 & -1 & -1 \\ 0 & 1 & 0 & -2 & -1 & -2 \\ 0 & 0 & 1 & -6 & -3 & -7 \end{array}\right]$$

$$A^{-1} = \begin{bmatrix} -1 & -1 & -1 \\ -2 & -1 & -2 \\ -6 & -3 & -7 \end{bmatrix}$$

The majority of the work has now been done, and we can solve the three systems quickly by multiplying by the three matrices consisting of the right-hand-side constants:

$$\begin{bmatrix} x \\ y \\ z \end{bmatrix} = \begin{bmatrix} -1 & -1 & -1 \\ -2 & -1 & -2 \\ -6 & -3 & -7 \end{bmatrix} \begin{bmatrix} -6 \\ 5 \\ 3 \end{bmatrix} = \begin{bmatrix} -2 \\ 1 \\ 0 \end{bmatrix}$$

$$\begin{bmatrix} x \\ y \\ z \end{bmatrix} = \begin{bmatrix} -1 & -1 & -1 \\ -2 & -1 & -2 \\ -6 & -3 & -7 \end{bmatrix} \begin{bmatrix} -17 \\ 2 \\ 14 \end{bmatrix} = \begin{bmatrix} 1 \\ 4 \\ -2 \end{bmatrix}$$

$$\begin{bmatrix} x \\ y \\ z \end{bmatrix} = \begin{bmatrix} -1 & -1 & -1 \\ -2 & -1 & -2 \\ -6 & -3 & -7 \end{bmatrix} \begin{bmatrix} 3 \\ 5 \\ -5 \end{bmatrix} = \begin{bmatrix} -3 \\ -1 \\ 2 \end{bmatrix}$$

TECHNOLOGY: Inverting Matrices

We can also use a graphing utility to find the inverse of a matrix. When using a TI-84 Plus, first define the matrix whose inverse we want to find. Then, enter the matrix on the home screen, press x^{-1}, and press enter.

To show the answer in fraction form, press math and select ▸Frac. If we defined matrix A to be $\begin{bmatrix} 7 & 4 \\ 1 & 2 \end{bmatrix}$, we would find its inverse as shown in Figure 1.

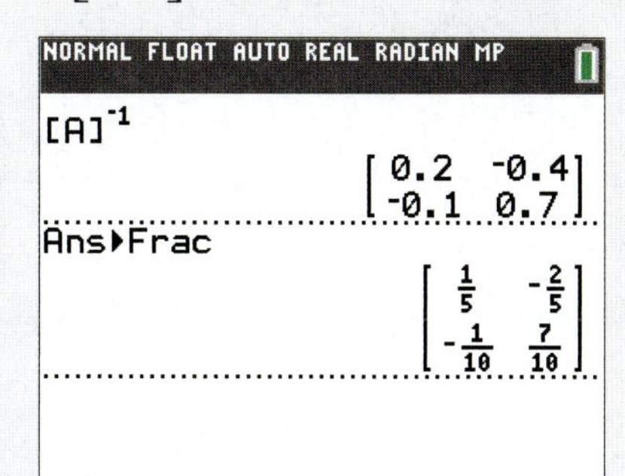

FIGURE 1

11.5 EXERCISES

PRACTICE

Write each of the following systems of equations as a single matrix equation. See Example 1.

1. $\begin{cases} 14x - 5y = 7 \\ x + 9y = 2 \end{cases}$

2. $\begin{cases} x - 5 = 9y \\ 3y - 2x = 8 \end{cases}$

3. $\begin{cases} -6 - 2y = x \\ 9x + 14 = 3y \end{cases}$

4. $\begin{cases} x - y = 5 \\ 2 - z = x \\ z - 3y = 4 \end{cases}$

5. $\begin{cases} 3x_1 - 7x_2 + x_3 = -4 \\ x_1 - x_2 = 2 \\ 8x_2 + 5x_3 = -3 \end{cases}$

6. $\begin{cases} x_3 = x_2 \\ x_2 = x_1 \\ x_1 = x_3 \end{cases}$

7. $\begin{cases} \dfrac{3x - 8y}{5} = 2 \\ y - 2 = 0 \end{cases}$

8. $\begin{cases} x - 7y = 5 \\ \dfrac{6 + x}{2} = 3y - 2 \end{cases}$

9. $\begin{cases} 4x = 3y - 9 \\ 13 - 2x = -4y \end{cases}$

10. $\begin{cases} -\frac{7}{3}y = \frac{5-x}{6} \\ x-5(y-3) = -2 \end{cases}$ **11.** $\begin{cases} 2x - y = -3z \\ y - x = 17 \\ 2 + z + 4x = 5y \end{cases}$ **12.** $\begin{cases} 2x_1 - 3x_3 = 7 \\ x_2 - 10x_3 = 0 \\ 2x_1 - x_2 + x_3 = 1 \end{cases}$

Find the inverse of each of the following matrices, if possible. See Examples 2 and 3.

13. $\begin{bmatrix} 0 & 4 \\ -5 & -1 \end{bmatrix}$ **14.** $\begin{bmatrix} -2 & -2 \\ -1 & 2 \end{bmatrix}$ **15.** $\begin{bmatrix} 3 & 4 \\ -4 & -5 \end{bmatrix}$

16. $\begin{bmatrix} -1 & -1 \\ -\frac{1}{4} & -\frac{1}{2} \end{bmatrix}$ **17.** $\begin{bmatrix} -\frac{1}{5} & 0 \\ \frac{1}{5} & \frac{1}{2} \end{bmatrix}$ **18.** $\begin{bmatrix} -7 & 2 \\ 7 & -2 \end{bmatrix}$

19. $\begin{bmatrix} -2 & -4 & -2 \\ 1 & -4 & 1 \\ 4 & -3 & 4 \end{bmatrix}$ **20.** $\begin{bmatrix} -3 & 0 & -4 \\ 2 & 5 & 4 \\ 1 & -5 & -2 \end{bmatrix}$ **21.** $\begin{bmatrix} -\frac{5}{11} & -\frac{8}{11} & 1 \\ \frac{13}{11} & \frac{12}{11} & -2 \\ -\frac{2}{11} & -\frac{1}{11} & 0 \end{bmatrix}$

22. $-\frac{1}{31}\begin{bmatrix} 17 & -8 & -2 \\ 1 & 5 & 9 \\ -6 & 1 & 8 \end{bmatrix}$ **23.** $\begin{bmatrix} -1 & 2 & -1 \\ 0 & 3 & -1 \\ 0 & 4 & -1 \end{bmatrix}$ **24.** $\begin{bmatrix} -1 & 0 & -1 \\ -\frac{3}{2} & \frac{1}{2} & -\frac{3}{2} \\ -\frac{1}{2} & 0 & -\frac{1}{4} \end{bmatrix}$

25. $\begin{bmatrix} -\frac{6}{5} & -\frac{2}{5} & -1 \\ \frac{3}{5} & \frac{1}{5} & 1 \\ 1 & 0 & 1 \end{bmatrix}$ **26.** $\begin{bmatrix} 2 & -2 & 1 \\ -2 & 2 & -3 \\ 1 & 0 & 2 \end{bmatrix}$ **27.** $\begin{bmatrix} 0 & 1 & 1 \\ 1 & 1 & 0 \\ 0 & 1 & 2 \end{bmatrix}$

28. $\begin{bmatrix} 9 & 8 & 7 \\ 6 & 5 & 4 \\ 3 & 2 & 1 \end{bmatrix}$ **29.** $\begin{bmatrix} \frac{2}{3} & \frac{8}{9} & \frac{1}{9} \\ -\frac{1}{3} & \frac{2}{9} & -\frac{2}{9} \\ -\frac{1}{3} & -\frac{7}{9} & -\frac{2}{9} \end{bmatrix}$ **30.** $\begin{bmatrix} -3 & -3 & -4 \\ 0 & \frac{1}{4} & \frac{1}{2} \\ 2 & 2 & 3 \end{bmatrix}$

For each pair of matrices, determine if either matrix is the inverse of the other.

31. $\begin{bmatrix} -5 & -2 \\ -7 & 4 \end{bmatrix}, \begin{bmatrix} 10 & 4 \\ 14 & -8 \end{bmatrix}$ **32.** $\begin{bmatrix} 9 & -18 \\ 3 & 12 \end{bmatrix}, \begin{bmatrix} -3 & -6 \\ -1 & -4 \end{bmatrix}$

33. $\begin{bmatrix} -6 & -1 & 1 \\ 4 & -1 & -2 \\ 1 & -1 & -1 \end{bmatrix}, \begin{bmatrix} -1 & -2 & 3 \\ 2 & 5 & -8 \\ -3 & -7 & 10 \end{bmatrix}$ **34.** $\begin{bmatrix} -1 & 4 & 5 \\ 3 & -11 & -17 \\ 4 & -17 & -19 \end{bmatrix}, \begin{bmatrix} -80 & -9 & -13 \\ -11 & -1 & -2 \\ -7 & -1 & -1 \end{bmatrix}$

35. $\begin{bmatrix} 2 & 0 & -1 \\ 3 & 4 & 2 \\ 1 & 1 & -3 \end{bmatrix}, \begin{bmatrix} 4 & 0 & -2 \\ 6 & 8 & 4 \\ 2 & 2 & -6 \end{bmatrix}$ **36.** $\begin{bmatrix} -7 & 0 & -2 \\ -10 & -1 & -2 \\ -7 & -1 & -1 \end{bmatrix}, \begin{bmatrix} -1 & 2 & -2 \\ 4 & -7 & 6 \\ 3 & -7 & 7 \end{bmatrix}$

Solve the following systems by the inverse matrix method, if possible. If the inverse matrix method doesn't apply, use any other method to determine if the system is inconsistent or dependent. See Example 4.

37. $\begin{cases} -2x-2y=9 \\ -x+2y=-3 \end{cases}$ **38.** $\begin{cases} 3x+4y=-2 \\ -4x-5y=9 \end{cases}$ **39.** $\begin{cases} -2x+3y=1 \\ 4x-6y=-2 \end{cases}$

40. $\begin{cases} -2x+4y=5 \\ x-4y=-3 \end{cases}$ **41.** $\begin{cases} -5x=10 \\ 2x+2y=-4 \end{cases}$ **42.** $\begin{cases} -3x+y=2 \\ 9x-3y=5 \end{cases}$

43. $\begin{cases} 8x+2y=26 \\ -16x-2y=-90 \end{cases}$ **44.** $\begin{cases} 3x-7y=-2 \\ -6x+14y=4 \end{cases}$ **45.** $\begin{cases} 3y=15 \\ 8x+4y=20 \end{cases}$

46. $\begin{cases} 4y+3z=-254 \\ 2x-2y-z=100 \\ -x+y-2z=155 \end{cases}$ **47.** $\begin{cases} 2x-y-3z=-10 \\ 2y-z=11 \\ -x+4z=0 \end{cases}$ **48.** $\begin{cases} 3y-4z=15 \\ x+2y-3z=9 \\ -x-y+2z=-5 \end{cases}$

Solve the following sets of systems by the inverse matrix method. See Example 5.

49. $\begin{cases} x+2y-z=2 \\ 3x+3y-z=-5 \\ 4x+4y-z=1 \end{cases}$ $\begin{cases} x+2y-z=1 \\ 3x+3y-z=1 \\ 4x+4y-z=1 \end{cases}$ $\begin{cases} x+2y-z=0 \\ 3x+3y-z=1 \\ 4x+4y-z=1 \end{cases}$

50. $\begin{cases} -x-y-2z=4 \\ x+3y+3z=0 \\ -3y-2z=9 \end{cases}$ $\begin{cases} -x-y-2z=1 \\ x+3y+3z=0 \\ -3y-2z=0 \end{cases}$ $\begin{cases} -x-y-2z=-2 \\ x+3y+3z=-3 \\ -3y-2z=1 \end{cases}$

51. $\begin{cases} -x+z=6 \\ -x+3y+2z=-11 \\ 2x-4y-3z=13 \end{cases}$ $\begin{cases} -x+z=-2 \\ -x+3y+2z=2 \\ 2x-4y-3z=-1 \end{cases}$ $\begin{cases} -x+z=-4 \\ -x+3y+2z=2 \\ 2x-4y-3z=0 \end{cases}$

TECHNOLOGY

Using a graphing utility, find the inverse of each of the following matrices, if possible. Round your answers to three decimal places if necessary.

52. $\begin{bmatrix} -7 & 3 \\ -1 & 2 \end{bmatrix}$ **53.** $\begin{bmatrix} -6 & 2 \\ -5 & 5 \end{bmatrix}$ **54.** $\begin{bmatrix} -2 & 0 & 2 \\ 2 & -3 & 1 \\ 1 & -2 & 3 \end{bmatrix}$

55. $\begin{bmatrix} 2.3 & 7.8 \\ -3.4 & 1.6 \end{bmatrix}$ **56.** $\begin{bmatrix} 4.5 & -9.4 & 6.9 \\ 8.6 & -2.8 & 1.2 \\ 3.1 & 0.3 & -7.0 \end{bmatrix}$ **57.** $\begin{bmatrix} 38 & -44 & 72 \\ -93 & 16 & 29 \\ 65 & 23 & -19 \end{bmatrix}$

11.6 PARTIAL FRACTION DECOMPOSITION

TOPICS

1. The pattern of partial fraction decompositions
2. Completing the partial fraction decomposition process

TOPIC 1: The Pattern of Partial Fraction Decompositions

Throughout this text, we have frequently found it useful or necessary to combine fractions. You have done this so often by now, and for such a variety of reasons, that you may not even be consciously aware of the process—the act of finding a common denominator and combining fractions may be second nature. There are occasions, however, when it is helpful to be able to reverse the process. In performing certain operations in calculus, for instance, the ability to write a single fraction as a sum of simpler fractions comes in very handy. The process of doing so is called **partial fraction decomposition**, and we will find the methods of this chapter useful in the execution of the process.

To be specific, the fractions we will want to decompose are proper rational functions; that is, fractions of the form

$$f(x) = \frac{p(x)}{q(x)}$$

where p and q are polynomials and the degree of p is less than the degree of q (recall that we already know how to perform polynomial division on fractions where the degree of p is greater than or equal to the degree of q). As a consequence of the Fundamental Theorem of Algebra and the Conjugate Roots Theorem (Chapter 5), if $q(x)$ has only real coefficients then it can be written as a product of factors of the form $(ax+b)^m$ and $(ax^2+bx+c)^n$, where m and n are positive integers, a, b, and c are real numbers, and ax^2+bx+c cannot be factored further without resorting to complex coefficients (we say ax^2+bx+c is irreducible). The appearance of such factors tells us how the rational function can be decomposed.

PROCEDURE: Decomposition Pattern

Given the proper rational function $f(x)=\frac{p(x)}{q(x)}$, assume $q(x)$ has been completely factored as a product of factors of the form $(ax+b)^m$ and $(ax^2+bx+c)^n$, where a, b, and c are real numbers, m and n are positive integers, and ax^2+bx+c is irreducible over the real numbers. Then $f(x)$ can be decomposed as a sum of simpler rational functions, which are determined based on the factors of $q(x)$, as follows.

1. Each factor of $q(x)$ of the form $(ax+b)^m$ leads to a sum of the following form.

$$\frac{A_1}{ax+b}+\frac{A_2}{(ax+b)^2}+\cdots+\frac{A_m}{(ax+b)^m}$$

2. Each factor of $q(x)$ of the form $(ax^2+bx+c)^n$ leads to a sum of the following form.

$$\frac{A_1x+B_1}{ax^2+bx+c}+\frac{A_2x+B_2}{(ax^2+bx+c)^2}+\cdots+\frac{A_nx+B_n}{(ax^2+bx+c)^n}$$

Example 1: Finding the Partial Fraction Decomposition of a Function

Write the form of the partial fraction decomposition of the rational function.

$$f(x)=\frac{p(x)}{x^3+6x^2+12x+8}$$

Assume that the degree of p is 2 or smaller.

Solution

The primary task in this problem is to factor the denominator. The Rational Zero Theorem (Section 5.3) tells us that the potential rational zeros of the denominator are $\pm\{1,2,4,8\}$ (remember that the potential rational zeros of a polynomial are the factors of the constant term divided by the factors of the leading coefficient). Synthetic division or long division can then be used to test these potential zeros one by one; the work following uses synthetic division to show that –2 is a zero.

$$\begin{array}{r|rrrr} -2 & 1 & 6 & 12 & 8 \\ & & -2 & -8 & -8 \\ \hline & 1 & 4 & 4 & 0 \end{array}$$

From this, we conclude that $x^3+6x^2+12x+8=(x+2)(x^2+4x+4)=(x+2)^3$. Following the guidelines of the decomposition pattern, we now know that $f(x)$ can be written as a sum of rational functions of the form

$$f(x)=\frac{A_1}{x+2}+\frac{A_2}{(x+2)^2}+\frac{A_3}{(x+2)^3}.$$

We can go no further with Example 1 since we were not given a specific polynomial in the numerator. Note, though, that the numerator played no role in our work—any rational function with the denominator $(x+2)^3$ can be decomposed as in Example 1, as long as the degree of the numerator is 2 or smaller.

Example 2: Finding the Partial Fraction Decomposition of a Function

Write the form of the partial fraction decomposition of the rational function.

$$f(x)=\frac{p(x)}{(x-5)^2(x^2+4)^2}$$

Assume that the degree of p is 5 or smaller.

Solution

In this example, the denominator is already appropriately factored; while it is true that x^2+4 can be factored as $(x-2i)(x+2i)$, partial fraction decomposition calls for leaving irreducible quadratics such as x^2+4 in their unfactored state. We now use the decomposition guidelines to rewrite $f(x)$.

$$f(x)=\frac{A_1}{x-5}+\frac{A_2}{(x-5)^2}+\frac{B_1x+C_1}{x^2+4}+\frac{B_2x+C_2}{(x^2+4)^2}$$

Note the choice of letters in the numerators of the decomposition. The names of the unknown coefficients do not matter, but it is important to realize that there are a total of six such coefficients which require six different symbols to denote them.

CAUTION

The fact that the form of the partial fraction decomposition of a proper rational function such as $\frac{p(x)}{(x-3)(x+3)}$ is $\frac{A_1}{x-3}+\frac{A_2}{x+3}$ may lead one to think that the form of the partial fraction decomposition of the similar function $\frac{p(x)}{(x-3)^2}$ is $\frac{A_1}{x-3}+\frac{A_2}{x-3}$. But $\frac{A_1}{x-3}+\frac{A_2}{x-3}$ is equal to $\frac{A_1+A_2}{x-3}$, so this couldn't possibly be equivalent to $\frac{p(x)}{(x-3)^2}$ if, for instance, $p(x)$ is a polynomial of degree 1. Instead, the partial fraction decomposition of $\frac{p(x)}{(x-3)^2}$ must be of the form $\frac{A_1}{x-3}+\frac{A_2}{(x-3)^2}$.

TOPIC 2: Completing the Partial Decomposition Process

Assuming the degree of the numerator is smaller than the degree of the denominator, the numerator plays no role in determining the form of the partial fraction decomposition of a given rational function. But it must be considered when we need to actually solve for the unknown coefficients appearing in the decomposition. It is at this stage that one of the methods of solving systems of equations may be called for.

Example 3: Finding the Partial Fraction Decomposition of a Function

Find the partial fraction decomposition of the rational function.

$$f(x)=\frac{-2x+14}{x^2+2x-3}$$

Solution

Since $x^2+2x-3=(x+3)(x-1)$, we know the following.

$$\frac{-2x+14}{x^2+2x-3}=\frac{A_1}{x+3}+\frac{A_2}{x-1}$$

There are many ways we can go about solving for A_1 and A_2, but most begin with the step of eliminating the fractions. Multiplying through by $(x+3)(x-1)$ leads immediately, upon canceling common factors, to the following equation.

$$-2x+14=A_1(x-1)+A_2(x+3)$$

This particular partial fraction decomposition can be accomplished simply—no advanced methods of solving systems of equations will be necessary. The key observation is that the equation above is true for *all* values of x; thus, we can substitute well-chosen values for x into the equation and quickly solve for the two coefficients. The values of x that are useful are those that make either $x-1$ or $x+3$ zero.

Let $x=1$ and solve for A_2.

$$\begin{aligned}-2(1)+14&=A_1(1-1)+A_2(1+3)\\12&=4A_2\\A_2&=3\end{aligned}$$

Let $x=-3$ and solve for A_1.

$$\begin{aligned}-2(-3)+14&=A_1(-3-1)+A_2(-3+3)\\20&=-4A_1\\A_1&=-5\end{aligned}$$

Thus, $\frac{-2x+14}{x^2+2x-3}=\frac{-5}{x+3}+\frac{3}{x-1}$.

Our answer can be checked by combining these two fractions.

The method shown in Example 3 will not always be sufficient, by itself, to solve for the unknown coefficients. Consider the next example.

Example 4: Finding the Partial Fraction Decomposition of a Function

Find the partial fraction decomposition of the rational function.

$$f(x)=\frac{3x+1}{(x+2)^3}$$

Solution

We already considered rational functions with this particular denominator in Example 1, and we know the following is true.

$$\frac{3x+1}{(x+2)^3}=\frac{A_1}{x+2}+\frac{A_2}{(x+2)^2}+\frac{A_3}{(x+2)^3}$$

Begin by eliminating fractions.

$$3x+1=A_1(x+2)^2+A_2(x+2)+A_3$$

One convenient value for x is -2, and making this substitution allows us to quickly solve for A_3 as follows.

$$\begin{aligned}3(-2)+1&=A_1(-2+2)^2+A_2(-2+2)+A_3\\-5&=A_3\end{aligned}$$

However, there are no other values for x that allow us to immediately determine A_1 and A_2. There are still several other ways we can complete the decomposition, though. One is to simply substitute two other values for x to obtain two equations in two unknowns. Since we now know that $A_3=-5$, we will make the substitutions for x in the following equation.

$$3x+6=A_1(x+2)^2+A_2(x+2)$$

In the work below, we chose to use $x=0$ and $x=-1$ because they make the calculations easy:

$$\begin{aligned}3(0)+6&=A_1(0+2)^2+A_2(0+2)\\6&=4A_1+2A_2\end{aligned}$$

and

$$\begin{aligned}3(-1)+6&=A_1(-1+2)^2+A_2(-1+2)\\3&=A_1+A_2.\end{aligned}$$

We recognize this as a linear system in the two variables A_1 and A_2. While we can use any of the methods of this chapter to solve it, we chose to use the elimination method.

$$\begin{cases} 4A_1 + 2A_2 = 6 \\ A_1 + A_2 = 3 \end{cases} \xrightarrow{-2E_2} \begin{cases} 4A_1 + 2A_2 = 6 \\ -2A_1 - 2A_2 = -6 \end{cases}$$
$$2A_1 \quad = 0$$

So $A_1 = 0$ and hence $A_2 = 3$. This gives us the following partial fraction decomposition.

$$\frac{3x+1}{(x+2)^3} = \frac{3}{(x+2)^2} + \frac{-5}{(x+2)^3}$$

There is another way of thinking about the decomposition process that is very important conceptually. After eliminating fractions, the decomposition in Example 4 has the following form.

$$3x+1 = A_1(x+2)^2 + A_2(x+2) + A_3$$

This is an equation relating two polynomials, a fact that becomes clear if we multiply out the right-hand side and collect the powers of x.

$$3x+1 = A_1x^2 + (4A_1 + A_2)x + (4A_1 + 2A_2 + A_3)$$

As we know, two polynomials are equal only if corresponding coefficients are equal. Equating the coefficients in this example gives us the following three equations.

$$\begin{cases} 0 = A_1 \\ 3 = 4A_1 + A_2 \\ 1 = 4A_1 + 2A_2 + A_3 \end{cases}$$

Note that there is no x^2 term on the left.

This is a linear system of three equations in three variables, and again any method from this chapter could be used to solve it. The verification that the solution of this system leads to the answer in Example 4 is left to the reader.

We conclude this section with an example that calls for the application of several skills learned in this text.

Example 5: Finding the Partial Fraction Decomposition of a Function

Find the partial fraction decomposition of the rational function.

$$f(x) = \frac{3x^5 + 6x^4 + 9x^3 + 7x^2 + 4x - 1}{(x^2 + x + 1)^2}$$

Solution

First, note that this rational function is improper—the degree of the numerator is greater than the degree of the denominator, so we must perform some polynomial division before the partial fraction decomposition. To do so, we first expand the denominator as follows.

$$(x^2 + x + 1)^2 = x^4 + 2x^3 + 3x^2 + 2x + 1$$

Polynomial long division (Section 5.2) then gives us the following result.

$$\frac{3x^5+6x^4+9x^3+7x^2+4x-1}{\left(x^2+x+1\right)^2}=3x+\frac{x^2+x-1}{\left(x^2+x+1\right)^2}$$

Now we need to decompose the fractional part.

Note that x^2+x+1 is irreducible. This follows from the use of the quadratic formula to solve the equation $x^2+x+1=0$; since the solutions of this equation are complex numbers, the linear factors of x^2+x+1 contain complex coefficients. The partial fraction decomposition thus has the following form.

$$\frac{x^2+x-1}{\left(x^2+x+1\right)^2}=\frac{A_1x+B_1}{x^2+x+1}+\frac{A_2x+B_2}{\left(x^2+x+1\right)^2}$$

Clearing the equation of fractions gives us the following.

$$x^2+x-1=\left(A_1x+B_1\right)\left(x^2+x+1\right)+\left(A_2x+B_2\right)$$
$$x^2+x-1=A_1x^3+\left(A_1+B_1\right)x^2+\left(A_1+B_1+A_2\right)x+\left(B_1+B_2\right)$$

From this polynomial equation we derive the following system.

$$\begin{cases} 0=A_1 \\ 1=A_1+B_1 \\ 1=A_1+B_1+A_2 \\ -1=B_1+B_2 \end{cases}$$

This system is easily solved, considering the equations in the order presented, and gives us $A_1=0, B_1=1, A_2=0,$ and $B_2=-2$. The following answer results.

$$\frac{3x^5+6x^4+9x^3+7x^2+4x-1}{\left(x^2+x+1\right)^2}=3x+\frac{1}{x^2+x+1}+\frac{-2}{\left(x^2+x+1\right)^2}$$

11.6 EXERCISES

PRACTICE

Write the form of the partial fraction decomposition of each of the following rational functions. In each case, assume the degree of the numerator is less than the degree of the denominator. See Examples 1 and 2.

1. $f(x)=\frac{p(x)}{x^2-x-6}$

2. $f(x)=\frac{p(x)}{x^2-2x-24}$

3. $f(x)=\frac{p(x)}{x^3+11x^2+40x+48}$

4. $f(x)=\frac{p(x)}{(x+5)^2\left(x^2+3\right)^2}$

5. $f(x)=\dfrac{p(x)}{(x+3)(x^2-4)}$

6. $f(x)=\dfrac{p(x)}{(x^2+5)(x^2+3x-4)}$

Match each rational expression with the form of its decomposition. The decompositions are labeled a.–h.

7. $\dfrac{2x-1}{(x+2)^3(x-2)}$

8. $\dfrac{2x-1}{x^2(x^2-4)}$

9. $\dfrac{2x-1}{x^3-4x^2+4x}$

10. $\dfrac{2x-1}{x^4-16}$

11. $\dfrac{2x-1}{x^3(x-2)^2}$

12. $\dfrac{2x-1}{x^5+2x^4+4x^3+8x^2}$

13. $\dfrac{2x-1}{x^3(x-2)}$

14. $\dfrac{2x-1}{x^5+6x^4+12x^3+8x^2}$

a. $\dfrac{A}{x}+\dfrac{B}{x^2}+\dfrac{C}{x^3}+\dfrac{D}{x-2}+\dfrac{E}{(x-2)^2}$

b. $\dfrac{A}{x}+\dfrac{B}{x^2}+\dfrac{C}{x+2}+\dfrac{D}{x-2}$

c. $\dfrac{A}{x}+\dfrac{B}{x^2}+\dfrac{C}{x^3}+\dfrac{D}{x-2}$

d. $\dfrac{A}{x+2}+\dfrac{B}{(x+2)^2}+\dfrac{C}{(x+2)^3}+\dfrac{D}{x-2}$

e. $\dfrac{Ax+B}{x^2+4}+\dfrac{C}{x+2}+\dfrac{D}{x-2}$

f. $\dfrac{A}{x}+\dfrac{B}{x^2}+\dfrac{C}{x+2}+\dfrac{Dx+E}{x^2+4}$

g. $\dfrac{A}{x+2}+\dfrac{B}{(x+2)^2}+\dfrac{C}{(x+2)^3}+\dfrac{D}{x}+\dfrac{E}{x^2}$

h. $\dfrac{A}{x}+\dfrac{B}{x-2}+\dfrac{C}{(x-2)^2}$

Find the partial fraction decomposition of each of the following rational functions. See Examples 3, 4, and 5.

15. $f(x)=\dfrac{3x^2+4}{x^3-4x}$

16. $f(x)=\dfrac{2x}{x^3+7x^2-6x-72}$

17. $f(x)=\dfrac{4x+2}{(x^3+8x)(x^2+2x-8)}$

18. $f(x)=\dfrac{5}{x^2+3x-4}$

19. $f(x)=\dfrac{5x}{x^2-6x+8}$

20. $f(x)=\dfrac{6x^2-4}{(x^2+3)(x+6)(x+5)}$

21. $f(x)=\dfrac{6x}{x^3+8x^2+9x-18}$

22. $f(x)=\dfrac{12x^2+x-1}{x^4+7x^3+5x^2-31x-30}$

23. $f(x)=\dfrac{1}{x^2-1}$

24. $f(x)=\dfrac{x+3}{(x^2+3)(x^2+x-6)}$

25. $f(x)=\dfrac{x^2-4}{(x^4-16)(x^2+2x-8)}$

26. $f(x)=\dfrac{x+1}{x^3-x}$

27. $f(x)=\dfrac{x+3}{x^2-4}$

28. $f(x)=\dfrac{x}{x^3+6x^2+11x+6}$

29. $f(x)=\dfrac{x}{x^4-16}$

30. $f(x)=\dfrac{5}{x^2-6x+8}$

31. $f(x)=\dfrac{2x+3}{(x^2-9)(x^2+4x-12)}$

32. $f(x)=\dfrac{x}{x^2-7x+12}$

33. $f(x)=\dfrac{x^2}{x^3+5x^2+3x-9}$

34. $f(x)=\dfrac{2x}{(x+4)(x^2-2x-3)}$

35. $f(x)=\dfrac{2x}{x^2-9}$

36. $f(x)=\dfrac{4x+3}{(x^2-9)(x^2-2x-24)}$

37. $f(x)=\dfrac{x^2}{x^3+4x^2-12x}$

38. $f(x)=\dfrac{2}{x^3+7x^2-8x}$

Write the partial fraction decomposition for the rational expression. You may check your answer by assigning a value to the constant *a* and graphing the result.

39. $\dfrac{1}{x(x+a)}$

40. $\dfrac{1}{x(a-x)}$

41. $\dfrac{1}{a^2-x^2}$

42. $\dfrac{1}{(x+1)(a-x)}$

43. $\dfrac{1}{(x+a)(x+1)}$

TECHNOLOGY

Using a graphing utility, determine whether the partial fraction decomposition is true or false by graphing the left and right side of the equation on the same coordinate plane.

44. $\dfrac{x+7}{x^2-x-6}=\dfrac{2}{x-3}-\dfrac{1}{x+2}$

45. $\dfrac{5x^2+20x+6}{x^3+2x^2+x}=\dfrac{6}{x}-\dfrac{1}{x+1}+\dfrac{9}{(x+1)^2}$

46. $\dfrac{4x^2-3x-4}{x^3+x^2-2x}=\dfrac{2}{x}-\dfrac{1}{x-1}+\dfrac{3}{x+2}$

47. $\dfrac{2x+4}{x^3-2x^2}=-\dfrac{2}{x}-\dfrac{2}{x^2}+\dfrac{2}{x-2}$

48. $\dfrac{1}{x^2+x-2}=\dfrac{1}{x-1}+\dfrac{1}{x+2}$

49. $\dfrac{6x^2+2}{x^2(x-3)^3}=\dfrac{4}{x}+\dfrac{2}{x^2}+\dfrac{3x}{x-3}+\dfrac{4}{(x-3)^2}+\dfrac{6x}{(x-3)^3}$

11.7 SYSTEMS OF LINEAR INEQUALITIES AND LINEAR PROGRAMMING

TOPICS

1. Systems of linear inequalities
2. Planar feasible regions
3. Linear programming in two variables

TOPIC 1: Systems of Linear Inequalities

Just as the solution of a system of linear equations consists of all points that solve each of the equations simultaneously, the solution of a system of linear inequalities consists of all points that solve each of the inequalities simultaneously. We saw similar systems in Section 2.6 when we graphed the solutions of linear inequalities joined by the word "and". Recall that, geometrically, the solution of two or more inequalities joined by "and" consists of the intersection of the individual solution regions. For systems in the plane, we find the intersecting solution region by graphing the boundary of each individual region, identifying which side of the boundary satisfies the given inequality, and then determining the portion of the plane that satisfies all the inequalities at once.

Example 1: Graphing the Solution of a System of Linear Inequalities

Graph the solution set to the following system of linear inequalities.

$$\begin{cases} x < 2 \\ y \le \dfrac{x}{2} + 1 \\ x + y \ge -3 \end{cases}$$

Solution

The boundary of the solution set of the first inequality is the vertical line $x = 2$, which we draw as a dashed line, as shown in Figure 1, to refect the fact that the inequality is strict. For this inequality, the region to the left of the boundary constitutes the solution set, since the x-coordinate of any solution point must be less than 2.

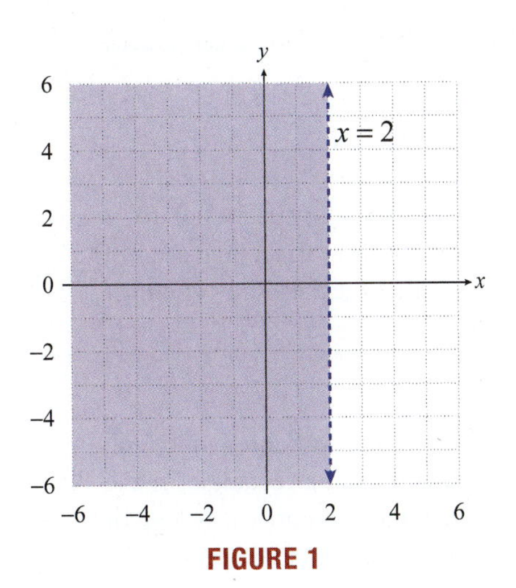

FIGURE 1

We similarly draw the lines $y = \frac{x}{2} + 1$ and $x + y = -3$ for the boundaries of the second and third inequalities, respectively, this time using solid lines, as the inequalities are not strict. Figures 2 and 3 show the solution sets of the two inequalities. Recall that one way to determine which side of a boundary constitutes the solution set of a given inequality is to use a convenient test point on one side or the other. For both of these inequalities, the test point $(0,0)$ is convenient.

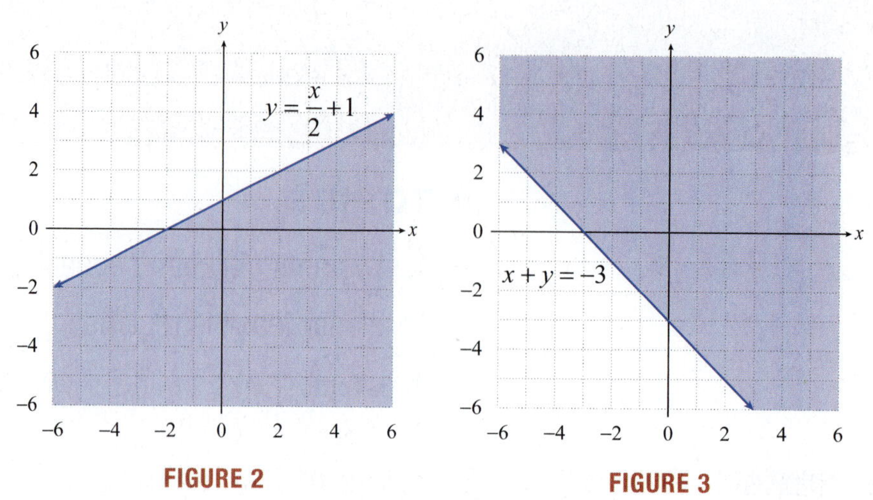

FIGURE 2 FIGURE 3

Alternatively, the inequality $y \leq \frac{x}{2} + 1$ explicitly states that the y-coordinate of any solution point must lie on or below the line $y = \frac{x}{2} + 1$; and the inequality $x + y \geq -3$, when rewritten as $y \geq -x - 3$, says that the y-coordinate of any solution point must lie on or above the boundary line. Figure 4 is the solution region for the system of inequalities since it satisfies all of them at once.

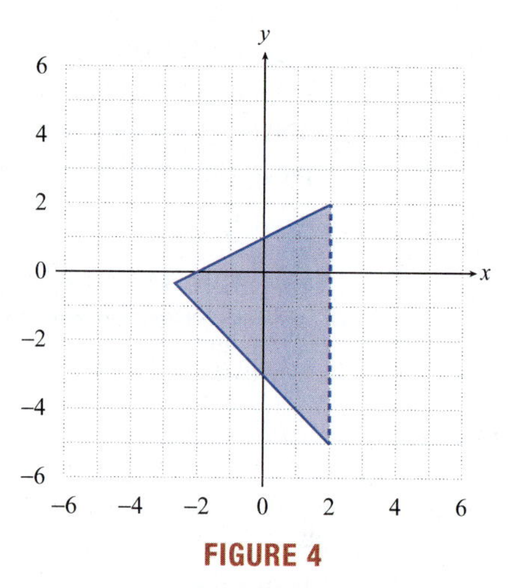

FIGURE 4

TOPIC 2: Planar Feasible Regions

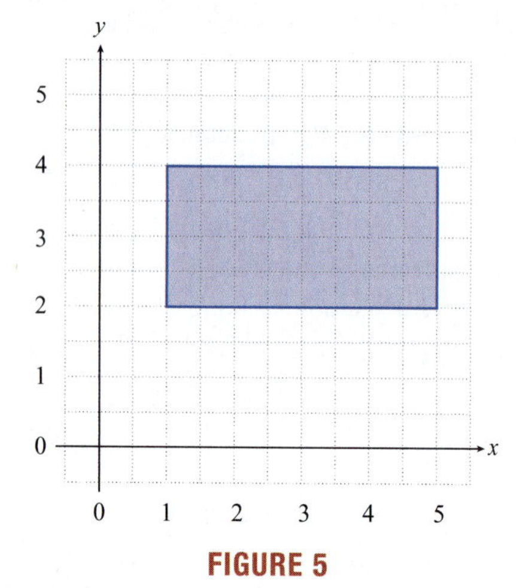

FIGURE 5

Many applications of mathematics require identifying the allowed values for a set of variables. If an application imposes limitations on the variables, the set of all allowable values forms a portion of the plane called the **feasible region** or **region of constraint**. For instance, the constraints $1 \leq x \leq 5$ and $2 \leq y \leq 4$ define the rectangular feasible region shown in Figure 5.

More often, constraint inequalities that involve both variables and the corresponding feasible region are a bit more interesting than a rectangle. If the constraint inequalities are all linear and taken together, they comprise a system of linear inequalities, and the techniques just reviewed can be used to determine the feasible region.

When feasible regions arise in applications of the sort we study in this section, they are typically not strict. For this reason, all examples from this point on will have feasible regions with solid boundary lines.

Example 2: Feasible Regions

A family orchard sells peaches and nectarines. To prevent a pest infestation, the number of nectarine trees in the orchard cannot exceed the number of peach trees. Also, because of the space requirements of each type of tree, the number of nectarine trees plus twice the number of peach trees cannot exceed 100 trees. Construct the constraints and graph the feasible region for this situation.

Solution

Let p be the number of peach trees and n be the number of nectarine trees. The constraints given in the problem can now be written in terms of p and n.

The second sentence tells us that $n \leq p$, and the third sentence tells us that $n + 2p \leq 100$. Since we can't have a negative number of trees, we also know that $n \geq 0$ and $p \geq 0$.

Letting the horizontal axis represent p and the vertical axis represent n, we can graph the feasible region defined by these constraints.

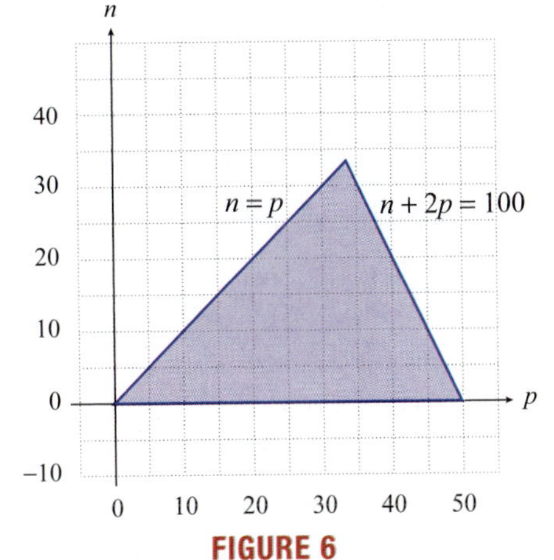

FIGURE 6

Any point (with integer coordinates) in the shaded feasible region corresponds to a certain number of peach trees and nectarine trees, and all such points meet the stated conditions.

TOPIC 3: Linear Programming in Two Variables

Often, given a feasible region, we want to find the point in the region such that the associated values of the variables maximize or minimize some quantity. Two important examples come from business: profit is a quantity that people usually wish to maximize, and cost is a quantity best kept to a minimum. When the quantity to be optimized (that is, either maximized or minimized) is a linear function of the variables concerned, the problem is known as a **linear programming** problem. Linear programming is used in a wide variety of applications, some involving thousands of variables, and a complete study of its techniques is beyond the scope of this text. However, the basic concepts are simple and easily applied, especially when working with applications involving just two variables.

DEFINITION: Bounded and Unbounded Regions

A **bounded** region is one in which all the points lie within some finite distance of the origin. An **unbounded** region has points that are arbitrarily far away from the origin.

PROCEDURE: Linear Programming

Step 1: Identify the variables to be considered in the problem, determine all the constraints on the variables imposed by the problem, and sketch the feasible region described by the constraints.

Step 2: Determine the function that is to be either maximized or minimized. This function is called the **objective function**.

Step 3: Evaluate the function at each of the vertices of the feasible region and compare the values found. If the feasible region is bounded, the optimum value of the function will occur at a vertex. If the feasible region is unbounded, the optimum value of the function may not exist, but if it does exist, it will occur at a vertex.

Example 3: Linear Programming

Find the maximum and minimum values of $f(x,y)=3x+2y$ subject to the following constraints.

$$\begin{cases} x \geq 0 \\ y \geq 0 \\ y \geq 3-3x \\ 2x+y \leq 10 \\ x+y \leq 7 \\ y \geq x-2 \end{cases}$$

Solution

The two variables are given to us, as are the constraints; all we must do to complete the first step is sketch the region defined by the constraints. Figure 7 shows the sketch, with selected boundaries identified by their corresponding equations.

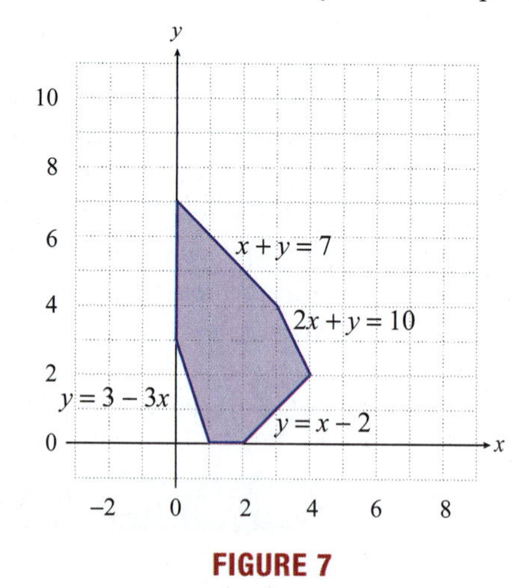

FIGURE 7

The function to be optimized is also given to us in this problem, so step two is complete. According to the linear programming method, the only remaining task is to evaluate the function at the vertices of the feasible region. Why should we expect the maximum and minimum values of *f* to occur at vertices, and not in the interior of the region?

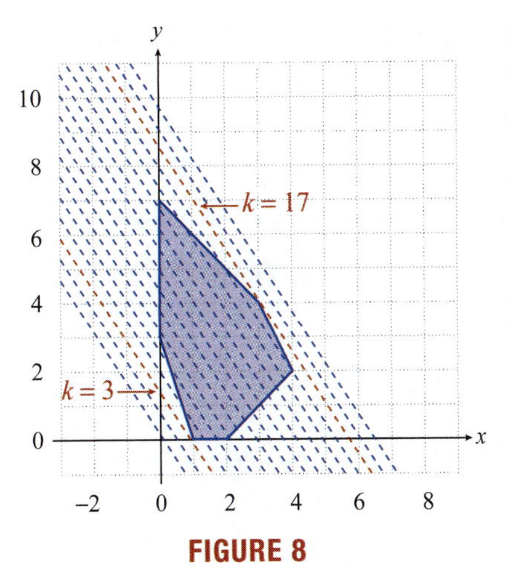

FIGURE 8

Another diagram will answer this question. The function $f(x,y)$ takes on a variety of values at different ordered pairs (x,y). But for any particular value, say k, there will be a line of ordered pairs in the plane such that $f(x,y)=k$. Such lines for different values of k will be parallel. Two particular values for k are labeled in Figure 8, and these values are the minimum and maximum values of f.

The extreme values will thus occur where a line of constant value *just touches* the region. This will be at a vertex or one of the edges of the region, if the edge is parallel to the line of constant value. Thus, evaluating the function at the vertices of the feasible region will suffice to identify the extreme values.

Each vertex of a feasible region is the intersection of two edges, so the coordinates of each vertex are found by solving a system of two equations. The six vertices of the region in this problem are determined by the following six systems:

$$\begin{cases} x+y=7 \\ 2x+y=10 \end{cases} \quad \begin{cases} 2x+y=10 \\ y=x-2 \end{cases} \quad \begin{cases} y=x-2 \\ y=0 \end{cases} \quad \begin{cases} y=0 \\ y=3-3x \end{cases} \quad \begin{cases} y=3-3x \\ x=0 \end{cases} \quad \begin{cases} x=0 \\ x+y=7 \end{cases}$$

Solving each of these systems, we find the vertices to be $(3,4)$, $(4,2)$, $(2,0)$, $(1,0)$, $(0,3)$, and $(0,7)$. Substituting each of these ordered pairs into $f(x,y)=3x+2y$, we find the maximum value of 17 occurs at the vertex $(3,4)$, and the minimum value of 3 occurs at the vertex $(1,0)$.

Example 4: Linear Programming

Find the maximum and minimum values of $f(x,y)=x+2y$ subject to the following constraints.

$$\begin{cases} x\geq 0 \\ y\geq 3-x \\ x-y\leq 1 \\ y\leq 2x+4 \end{cases}$$

Solution

The feasible region defined by the constraints is unbounded; a portion of its sketch appears in Figure 9.

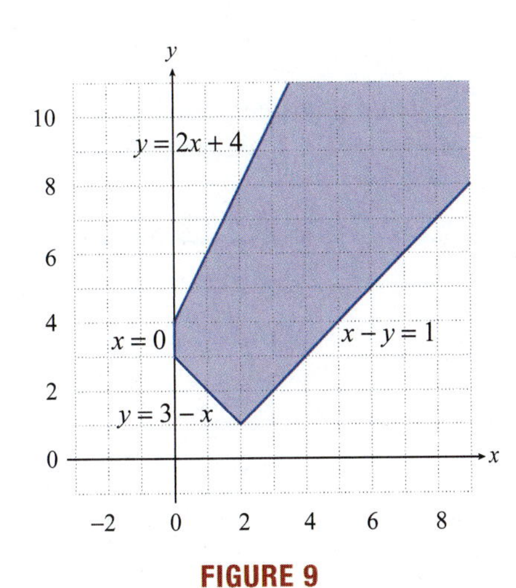

FIGURE 9

Because the region is unbounded, the function f may not have a maximum or a minimum value over the region. In this case, f has no maximum value. One way to see this is to note that ordered pairs of the form (x,x) lie in the region for all $x\geq\frac{3}{2}$, and $f(x,x)=x+2x=3x$ grows without bound as x increases.

On the other hand, f does have a minimum value over the feasible region. The three vertices are $(0,4)$, $(0,3)$, and $(2,1)$. Evaluating the function at these points, we obtain $f(0,4)=8$, $f(0,3)=6$, and $f(2,1)=4$, so the minimum value is 4.

Example 5 illustrates how linear programming can be applied in a typical business application.

Example 5: Linear Programming

A manufacturer makes two models of a certain electronic device. Model A requires 3 minutes to assemble, 4 minutes to test, and 1 minute to package. Model B requires 4 minutes to assemble, 3 minutes to test, and 6 minutes to package. The manufacturer can allot 7400 minutes total for assembly, 8000 minutes for testing, and 9000 minutes for packaging. Model A generates a profit of $7.00 and Model B generates a profit of $8.00. How many of each model should be made in order to maximize profit?

Solution

If we let x denote the number of Model A devices made and y the number of Model B devices made, the constraint inequalities are as follows.

$$\begin{cases} x \geq 0, y \geq 0 \\ 3x + 4y \leq 7400 \\ 4x + 3y \leq 8000 \\ x + 6y \leq 9000 \end{cases}$$

The sketch of the feasible region appears in Figure 10.

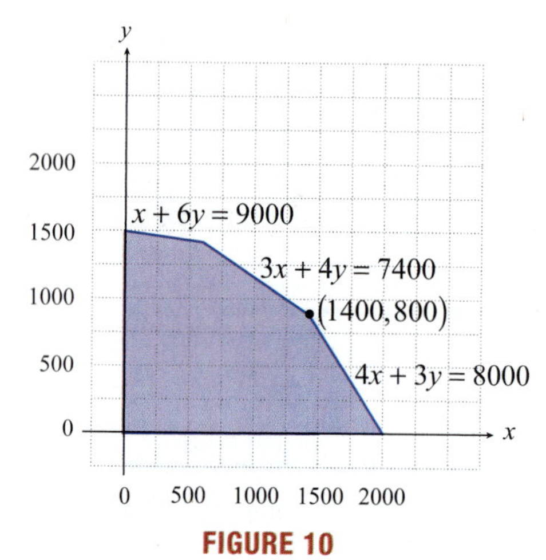

FIGURE 10

The profit function in this situation is $f(x, y) = 7x + 8y$, and our goal is to maximize this function. The vertices of the feasible region are $(0,1500)$, $(600,1400)$, $(1400,800)$, $(2000,0)$, and $(0,0)$. Manufacturing 0 units of Model A and 0 units of Model B is not going to be profitable, so we really only need to evaluate f at four vertices:

$$f(0,1500) = 12,000$$
$$f(600,1400) = 15,400$$
$$f(1400,800) = 16,200$$
$$f(2000,0) = 14,000$$

From these calculations, we conclude that the maximum profit is generated from making 1400 units of Model A and 800 units of Model B. The maximum profit would be $16,200.

11.7 EXERCISES

PRACTICE

Graph the solution set of each of the following systems of inequalities. See Example 1.

1. $\begin{cases} y \geq -2 \\ y > 1 \end{cases}$

2. $\begin{cases} y \geq -2x - 5 \\ y \leq -6x - 9 \end{cases}$

3. $\begin{cases} y \leq 4x + 4 \\ y > 7x + 7 \end{cases}$

4. $\begin{cases} x - 3y \geq 6 \\ y > -4 \end{cases}$

5. $\begin{cases} 3x - y \leq 2 \\ x + y > 0 \end{cases}$

6. $\begin{cases} x > 1 \\ y > 2 \end{cases}$

7. $\begin{cases} x + y > -2 \\ x + y < 2 \end{cases}$

8. $\begin{cases} y > -2 \\ 2y > -3x - 4 \end{cases}$

9. $\begin{cases} y \leq -x \\ 2y + 3x > -4 \end{cases}$

10. $\begin{cases} 5x + 6y < -30 \\ x \geq 2 \end{cases}$

11. $\begin{cases} x < 6 \\ x \geq -5 \end{cases}$

12. $\begin{cases} |x + 1| < 2 \\ |y - 3| \leq 1 \end{cases}$

13. $\begin{cases} |y - 3x| \leq 2 \\ |y| < 2 \end{cases}$

Find the minimum and maximum values of the given functions, subject to the given constraints. See Examples 3 and 4.

14. Objective Function:
$f(x, y) = 2x + 3y$
Constraints:
$\begin{cases} x \geq 0, y \geq 0 \\ x + y \leq 7 \end{cases}$

15. Objective Function:
$f(x, y) = 4x + y$
Constraints:
$\begin{cases} x \geq 0, y \geq 0 \\ x + y \leq 3 \end{cases}$

16. Objective Function:
$f(x, y) = 2x + 5y$
Constraints:
$\begin{cases} x \geq 0, y \geq 0 \\ x + y \leq 7 \end{cases}$

17. Objective Function:
$f(x, y) = 7x + 4y$
Constraints:
$\begin{cases} x \geq 0, y \geq 0 \\ 3x + y \leq 3 \end{cases}$

18. Objective Function:
$f(x, y) = 5x + 6y$
Constraints:
$\begin{cases} 0 \leq x \leq 7 \\ 0 \leq y \leq 10 \\ 8x + 5y \geq 40 \end{cases}$

19. Objective Function:
$f(x, y) = 9x + 7y$
Constraints:
$\begin{cases} 0 \leq x \leq 20 \\ 0 \leq y \leq 10 \\ 6x + 12y \geq 140 \end{cases}$

20. Objective Function:
$f(x,y) = 6x + 4y$
Constraints:
$$\begin{cases} 0 \le x \le 4 \\ 0 \le y \le 5 \\ 4x + 3y \ge 10 \end{cases}$$

21. Objective Function:
$f(x,y) = 3x + 7y$
Constraints:
$$\begin{cases} 0 \le x \le 8 \\ 0 \le y \le 6 \\ 7x + 10y \ge 50 \end{cases}$$

22. Objective Function:
$f(x,y) = 6x + 8y$
Constraints:
$$\begin{cases} x \ge 0, y \ge 0 \\ 4x + y \le 16 \\ x + 3y \le 15 \end{cases}$$

23. Objective Function:
$f(x,y) = x + 2y$
Constraints:
$$\begin{cases} x \ge 0, y \ge 0 \\ 3x + y \le 45 \\ x + 3y \le 24 \end{cases}$$

24. Objective Function:
$f(x,y) = 6x + y$
Constraints:
$$\begin{cases} x \ge 0, y \ge 0 \\ 3x + 4y \ge 24 \\ 3x + 4y \le 48 \end{cases}$$

25. Objective Function:
$f(x,y) = 15x + 30y$
Constraints:
$$\begin{cases} x \ge 0, y \ge 0 \\ 5x + 7y \ge 70 \\ 5x + 7y \le 140 \end{cases}$$

26. Objective Function:
$f(x,y) = 3x + 10y$
Constraints:
$$\begin{cases} x \ge 0 \\ 2x + 4y \ge 8 \\ 5x - y \le 10 \\ x + 3y \le 40 \end{cases}$$

27. Objective Function:
$f(x,y) = 20x + 30y$
Constraints:
$$\begin{cases} x \ge 0 \\ 12x + 6y \ge 120 \\ 9x - 6y \le 144 \\ x + 4y \le 12 \end{cases}$$

APPLICATIONS

28. A plane carrying relief food and water can carry a maximum of 50,000 pounds and is limited in space to carrying no more than 6000 cubic feet. Each container of water weighs 60 pounds and takes up 1 cubic foot, and each container of food weighs 50 pounds and takes up 10 cubic feet. What is the region of constraint for the numbers of containers of food and water that the plane can carry?

29. A furniture company makes two kinds of sofas, the Standard model and the Deluxe model. The Standard model requires 40 hours of labor to build, and the Deluxe model requires 60 hours of labor to build. The finish of the Deluxe model uses both teak and fabric, while the Standard uses only fabric, with the result that each Deluxe sofa requires 5 square yards of fabric and each Standard sofa requires 8 square yards of fabric. Given that the company can use 200 hours of labor and 25 square yards of fabric per week building sofas, what is the region of constraint for the numbers of Deluxe and Standard sofas the company can make per week?

30. Sarah is looking through a clothing catalog, and she is willing to spend up to \$80 on clothes and \$10 for shipping. Shirts cost \$12 each plus \$2 shipping, and a pair of pants costs \$32 plus \$3 shipping. What is the region of constraint for the numbers of shirts and pairs of pants Sarah can buy?

31. Suppose you inherit \$75,000 from a previously unknown (and highly eccentric) uncle and that the inheritance comes with certain stipulations regarding investments. First, the dollar amount invested in bonds must not exceed the dollar amount invested in stocks. Second, a minimum of \$10,000 must be invested in stocks, and a minimum of \$5000 must be invested in bonds. Finally, a maximum of \$40,000 can be invested in stocks. What is the region of constraint for the dollar amounts that can be invested in the two categories of stocks and bonds?

32. A manufacturer produces two models of computers. The times (in hours) required for assembling, testing, and packaging each model are listed in the following table.

Process	Model X	Model Y
Assemble	2.5	3
Test	2	1
Package	0.75	1.25

The total times available for assembling, testing, and packaging are 4000 hours, 2500 hours, and 1500 hours, respectively. The profits per unit are \$50 for Model X and \$52 for Model Y. How many of each type should be produced to maximize profit? What is the maximum profit?

33. A manufacturer produces two types of fans. The times (in minutes) required for assembling, packaging, and shipping each type are listed in the following table.

Process	Type X	Type Y
Assemble	20	25
Package	40	10
Ship	10	7.5

The total times available for assembling, packaging, and shipping are 4000 minutes, 4800 minutes, and 1500 minutes, respectively. The profits per unit are \$4.50 for Type X and \$3.75 for Type Y. How many of each type should be produced to maximize profit? What is the maximum profit?

34. Ashley is making a set of patchwork curtains for her apartment. She needs a minimum of 16 yards of the solid material, at least 5 yards of the striped material, and at least 20 yards of the flowered material. She can choose between two sets of precut bundles. The olive-based bundle costs \$10 per bundle and contains 8 yards of the solid material, 1 yard of the striped material, and 2 yards of the flowered material. The cranberry-based bundle costs \$20 per bundle and includes 2 yards of the solid material, 1 yard of the striped material, and 7 yards of the flowered material. How many of each bundle should Ashley buy to minimize her cost and yet buy enough material to complete the curtains? What is her minimum cost?

35. A volunteer has been asked to drop off some supplies at a facility housing victims of a hurricane evacuation. The volunteer would like to bring at least 60 bottles of water, 45 first aid kits, and 30 security blankets on his visit. The relief organization has a standing agreement with two companies that provide victim packages. Company A can provide packages of 5 water bottles, 3 first aid kits, and 4 security blankets at a cost of $1.50. Company B can provide packages of 2 water bottles, 2 first aid kits, and 1 security blanket at a cost of $1.00. How many of each package should the volunteer pick up to minimize the cost? What total amount does the relief organization pay?

36. On your birthday your grandmother gave you $25,000, but told you she would like you to invest the money for 10 years before you use any of it. Since you wish to respect your grandmother's wishes, you seek out the advice of a financial adviser. She suggests you invest at least $15,000 in municipal bonds yielding 6% and no more than $5000 in Treasury bills yielding 9%. How much should be placed in each investment so that income is maximized?

37. A boutique cell phone manufacturer produces two models: a retro model flip phone and a smart phone. The manufacturer's quota per day is to produce at least 100 flip phones and 80 smart phones. No more than 200 flip phones and 170 smart phones can be produced per day due to limitations on production. A total of at least 200 phones must be shipped every day.

a. If the production costs are $5 for a flip phone and $7 for a smart phone, how many of each model should be produced on a daily basis to minimize cost and what would that cost be?

b. If each flip phone results in a $2 loss but each smart phone results in a $5 gain, how many of each model should be manufactured daily to maximize profit? What is the maximum profit if this number of phones is produced?

11.8 SYSTEMS OF NONLINEAR EQUATIONS AND INEQUALITIES

TOPICS

1. Solving systems of nonlinear equations by graphing
2. Solving systems of nonlinear equations algebraically
3. Solving systems of nonlinear inequalities

TOPIC 1: Solving Systems of Nonlinear Equations by Graphing

In this last section of the chapter, we deal with systems of equations in which one or more of the equations are nonlinear. Because at least one equation is nonlinear, the matrix-based methods that we have learned *do not apply*. The theory of systems of nonlinear equations is much more complex than that of linear systems, so we will only use general guiding principles to solve them.

Since we have learned how to graph many different classes of equations, we can put our graphing knowledge to use in solving systems of nonlinear equations. Keep in mind, however, that graphing will only help us identify the *real* solutions; as we will see, some systems of nonlinear equations also have complex number solutions (with nonzero imaginary parts).

Example 1: Solving Systems of Nonlinear Equations by Graphing

Use graphing to approximate the real solution(s) of the following system, and then verify that your answer is correct.

$$\begin{cases} x+y=2 \\ x^2+y^2=4 \end{cases}$$

Solution

The first equation describes a line in the plane with an x-intercept at $(2,0)$ and a y-intercept of $(0,2)$. The second equation describes a circle of radius 2 centered at the origin.

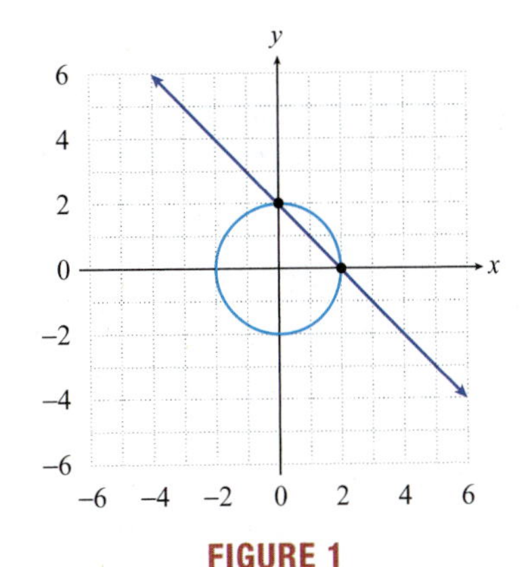

FIGURE 1

We are interested in whether any ordered pairs of numbers satisfy both equations simultaneously. Graphically, we are looking for the points of intersection of the two graphs.

The graph in Figure 1 suggests that $(2,0)$ and $(0,2)$ make up the real solutions of the system. We already know these two points lie on the line, and it can be verified that they also solve the second equation.

$$2^2+0^2=4 \quad \text{and} \quad 0^2+2^2=4$$

This example points out another very significant difference between systems of linear and systems of nonlinear equations. Recall that a system of linear equations is guaranteed to have no solution, one solution, or an infinite number of solutions. Systems of nonlinear equations may have any number of solutions.

Example 2: Solving Systems of Nonlinear Equations by Graphing

Use graphing to approximate the real solution(s) of the following system, and then verify that your answer is correct.

$$\begin{cases} x+2=y^2 \\ x^2+y^2=4 \end{cases}$$

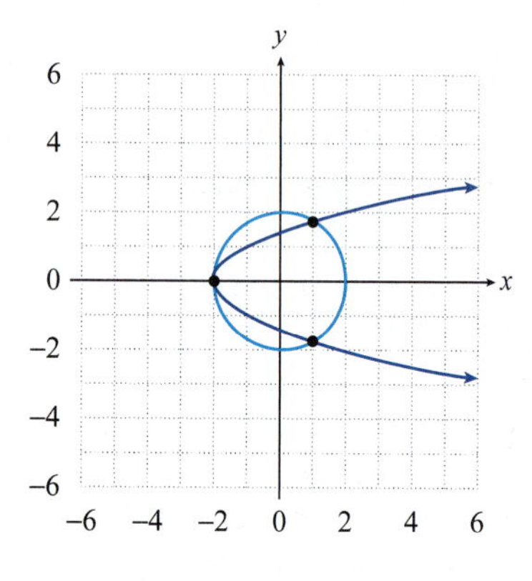

FIGURE 2

Solution

The first equation represents a horizontally oriented parabola opening to the right, with vertex at $(-2,0)$. The second equation is again a circle of radius 2 centered at the origin.

This time, the system appears to have three real solutions. The point $(-2,0)$ is easily verified to be one of them, but the other two are harder to determine just by looking at the graph in Figure 2. Noting that the x-coordinate of the other two points appears to be 1, we can replace x with 1 in the two equations to obtain the equation $y^2=3$ in both cases. Solving for y yields $y=\pm\sqrt{3}$, which means that $(1,\sqrt{3})$ and $(1,-\sqrt{3})$ are the other two solutions of the system. Checking these solutions verifies that both ordered pairs solve both equations.

TOPIC 2: Solving Systems of Nonlinear Equations Algebraically

In Example 2, we needed to combine graphical knowledge with some algebra in order to solve the system. The picture corresponding to the system suggested that there were three real solutions, and we can be fairly confident that we found all solutions to the system. But graphs can be misleading, especially if they are not drawn accurately. Consider, for example, the following very similar system.

Example 3: Solving Systems of Nonlinear Equations Algebraically

Solve the following system of nonlinear equations.

$$\begin{cases} x+4=y^2 \\ x^2+y^2=4 \end{cases}$$

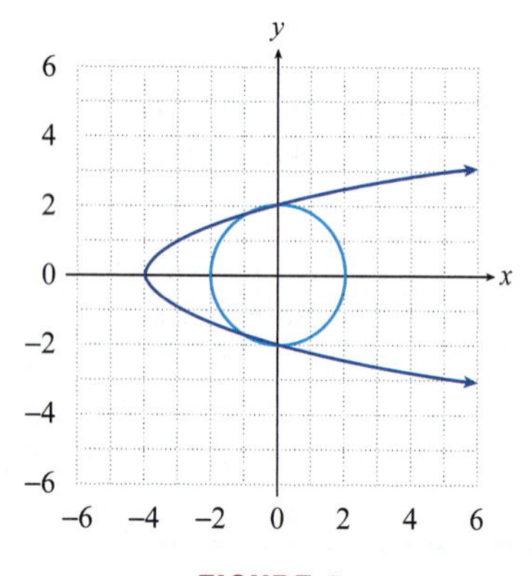

FIGURE 3

Solution

The only difference between this system and the one in Example 2 is that the vertex of the parabola is at $(-4,0)$. The graph of the system is shown in Figure 3.

Looking at Figure 3, it isn't clear exactly how many intersections exist or where those intersections are. A reasonable approximation might be the two points

$(0,2)$ and $(0,-2)$, and it can be verified that these two points do indeed lie on both equations. But if we were to conclude that these are the only two solutions, we would be wrong.

We need a more reliable method for this example, and algebra comes to the rescue. While we can't apply the matrix-based solution methods to nonlinear systems, we can use the methods of substitution and elimination. For the system

$$\begin{cases} x+4=y^2 \\ x^2+y^2=4 \end{cases}$$

the method of substitution works well if we solve both equations for y^2.

$$\begin{cases} x+4=y^2 \\ y^2=4-x^2 \end{cases}$$

The two expressions equal to y^2 both contain only the variable x, so if we equate the expressions we will have a single equation in one variable. The resulting equation is solved as shown.

$$\begin{aligned} x+4&=4-x^2 \\ x&=-x^2 \\ x^2+x&=0 \\ x(x+1)&=0 \\ x&=0,\ -1 \end{aligned}$$

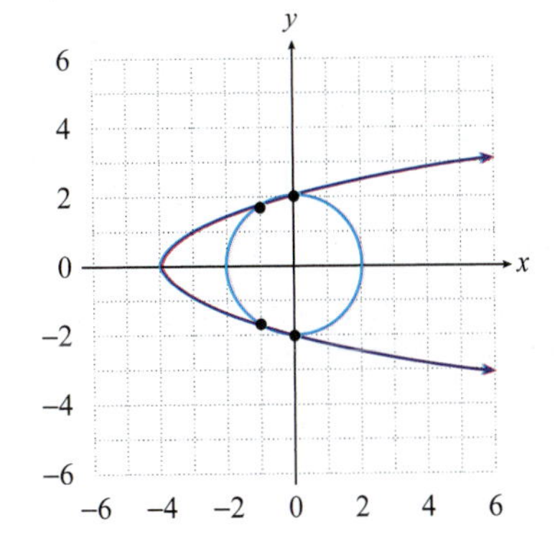

FIGURE 4

We already know that 0 is one possible value for x, as we have verified that $(0,2)$ and $(0,-2)$ are solutions of the system. But the value of -1 is new, and if we substitute $x=-1$ into either equation, we find that $(-1, \sqrt{3})$ and $(-1, -\sqrt{3})$ are also solutions. So the system has a total of four solutions, which we now indicate on the graph in Figure 4.

The next example of a system of nonlinear equations can be solved by the method of elimination.

Example 4: Solving Systems of Nonlinear Equations Algebraically

Solve the following system of nonlinear equations.

$$\begin{cases} y+1=(x-2)^2 \\ 3-y=(x-1)^2 \end{cases}$$

Solution

Before solving the system algebraically, you may want to graph these two vertically oriented parabolas to get some idea of what to expect.

We can eliminate the variable y by adding the two equations, leaving the quadratic equation $4=(x-2)^2+(x-1)^2$.

We can now solve the single-variable equation.

$$4=(x-2)^2+(x-1)^2$$
$$4=x^2-4x+4+x^2-2x+1$$
$$0=2x^2-6x+1$$
$$x=\frac{6\pm\sqrt{36-8}}{4}$$
$$x=\frac{3\pm\sqrt{7}}{2}$$

For each of the two values of x, we need to determine the corresponding values of y, and we can use either equation to do so. For this example, we will use the first equation.

For $x=\frac{3+\sqrt{7}}{2}$:

$$y=\left(\frac{3+\sqrt{7}}{2}-2\right)^2-1$$
$$y=\left(\frac{-1+\sqrt{7}}{2}\right)^2-1$$
$$y=\frac{8-2\sqrt{7}}{4}-1$$
$$y=2-\frac{\sqrt{7}}{2}-1$$
$$y=1-\frac{\sqrt{7}}{2}$$

For $x=\frac{3-\sqrt{7}}{2}$:

$$y=\left(\frac{3-\sqrt{7}}{2}-2\right)^2-1$$
$$y=\left(\frac{-1-\sqrt{7}}{2}\right)^2-1$$
$$y=\frac{8+2\sqrt{7}}{4}-1$$
$$y=2+\frac{\sqrt{7}}{2}-1$$
$$y=1+\frac{\sqrt{7}}{2}$$

Thus, the two solutions of the system are as follows.

$$\left(\frac{3+\sqrt{7}}{2},1-\frac{\sqrt{7}}{2}\right) \quad \text{and} \quad \left(\frac{3-\sqrt{7}}{2},1+\frac{\sqrt{7}}{2}\right)$$

Example 5: Solving Systems of Nonlinear Equations Algebraically

Solve the following system of nonlinear equations.

$$\begin{cases} \dfrac{(x+1)^2}{9}+y^2=1 \\ y^2=x+2 \end{cases}$$

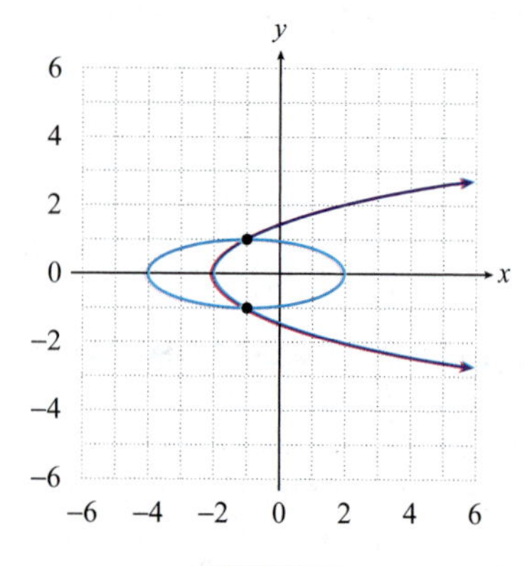

FIGURE 5

Solution

The first equation describes an ellipse, and the second describes a parabola opening to the right. The graph of the system in Figure 5 indicates that we should expect two real solutions.

If we solve this system algebraically, we discover that there is more to this problem.

Since the second equation is already solved for y^2, the method of substitution is a good option. Substituting in the first equation, we obtain

$$\frac{(x+1)^2}{9}+x+2=1.$$

We solve this second-degree equation in x the usual way.

$$\begin{aligned}\frac{(x+1)^2}{9}+x+2&=1\\ \frac{(x+1)^2}{9}+x+1&=0\\ (x+1)^2+9x+9&=0\\ x^2+11x+10&=0\\ (x+10)(x+1)&=0\\ x&=-10,\ -1\end{aligned}$$

The solution $x=-1$ isn't surprising, since our graph certainly suggested this value, but the solution $x=-10$ doesn't seem to relate to the picture at all. But remember, these are only the x-coordinates of the ordered pairs that solve the system, and we still need to determine the corresponding y-coordinates. We will use the second equation to do this.

For $x=-10$:

$$\begin{aligned}y^2&=-10+2\\ y^2&=-8\\ y&=\pm 2i\sqrt{2}\end{aligned}$$

For $x=-1$:

$$\begin{aligned}y^2&=-1+2\\ y^2&=1\\ y&=\pm 1\end{aligned}$$

Now we know that the two real solutions are $(-1,-1)$ and $(-1,1)$, as was suggested by our graph, and that the two ordered pairs $(-10,-2i\sqrt{2})$ and $(-10,2i\sqrt{2})$ also solve the system. These last two solutions don't appear on the graph because of their nonreal second coordinates.

TOPIC 3: Solving Systems of Nonlinear Inequalities

The feasible regions of Section 11.7 are portions of the plane that simultaneously satisfy a collection of linear inequalities, and thus represent solutions of systems of linear inequalities. Combining the skills learned there with our newly acquired knowledge of how to solve systems of nonlinear equations, we can now graph solutions of systems of nonlinear inequalities.

Recall that a linear inequality in two variables divides the Cartesian plane into two regions separated by a line: one region consists of all those points that satisfy the inequality, and the other region consists of those that do not. The boundary line is either a part of the solution region or not, depending on whether the inequality is nonstrict or strict. In general, a nonlinear inequality in two variables also separates the plane into regions, with the difference (as we would expect) that the boundary between the regions is not linear. To illustrate, the graphs in Figure 6 represent the solutions of, respectively, the inequalities $x^2+y^2<1$ and $y\geq x^2+1$.

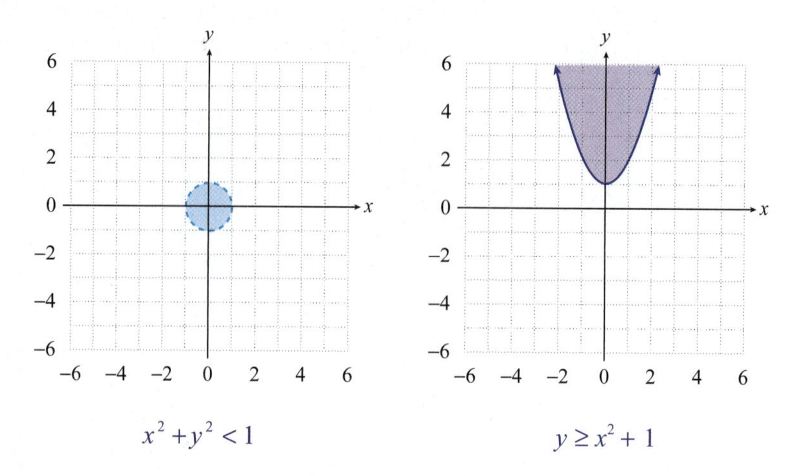

FIGURE 6: Solutions of Nonlinear Inequalities

As with linear inequalities, the solution regions can often be identified by considering the geometric meaning of the given nonlinear inequality (e.g., the solutions of $x^2 + y^2 < 1$ are those points lying within a radius of 1 of the origin, and the solutions of $y \geq x^2 + 1$ are those points for which the y-coordinate lies on or above the parabola $y = x^2 + 1$). Alternatively, a "test point" within a given region can be substituted into the inequality; if the point satisfies the inequality, that region is part of the solution set. All that remains in order to solve systems of nonlinear inequalities is to note that the solution of a system consists of the intersection of the solutions of the individual inequalities. We will illustrate this point with a variation of Example 5.

Example 6: Solving Systems of Nonlinear Inequalities by Graphing

Graph the solution of the following system of nonlinear inequalities.

$$\begin{cases} \dfrac{(x+1)^2}{9} + y^2 < 1 \\ y^2 \leq x + 2 \end{cases}$$

Solution

The region solving the first inequality is the interior of an ellipse (with the boundary excluded), while the region solving the second inequality consists of those points on and to the right of a rightward-opening parabola. Thus, the solution of the system consists of all those points in the plane lying in both regions; that is, the intersection of the regions. The graph of this intersection appears shaded in Figure 7 (note the dashed line for the portion of the boundary defined by the ellipse and the solid line for the portion defined by the parabola).

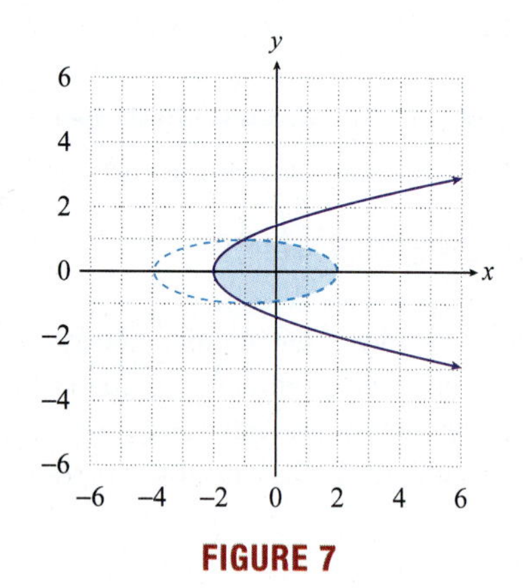

FIGURE 7

11.8 EXERCISES

PRACTICE

Use graphing to approximate the real solution(s) of the following systems, and then verify that your answers are correct. See Examples 1 and 2.

1. $\begin{cases} 3x-2y=6 \\ \dfrac{x^2}{4}+\dfrac{y^2}{9}=1 \end{cases}$

2. $\begin{cases} x+2y=2 \\ \dfrac{x^2}{4}+y^2=1 \end{cases}$

3. $\begin{cases} x^2+4y^2=5 \\ x^2+y^2=2 \end{cases}$

4. $\begin{cases} 4x^2+y^2=5 \\ 4(x-2)^2+y^2=5 \end{cases}$

5. $\begin{cases} y=x^2 \\ 2-y=x^2 \end{cases}$

6. $\begin{cases} x^2+(y-1)^2=4 \\ (x-3)^2+(y-1)^2=1 \end{cases}$

7. $\begin{cases} y-x^2=1 \\ y+2=4x^2 \end{cases}$

8. $\begin{cases} x^2+y^2=10 \\ x^2+y=-2 \end{cases}$

9. $\begin{cases} x=y^2-3 \\ x^2+4y^2=4 \end{cases}$

10. $\begin{cases} (x-1)^2+(y-6)^2=9 \\ (x-1)^2+(y+1)^2=16 \end{cases}$

11. $\begin{cases} (x+1)^2+(y-1)^2=4 \\ (x+1)^2+4(y-1)^2=4 \end{cases}$

12. $\begin{cases} x^2+y^2=1 \\ y=x^2-1 \end{cases}$

13. $\begin{cases} (x-2)^2+y^2=4 \\ (x+2)^2+y^2=4 \end{cases}$

14. $\begin{cases} x^2+y^2=9 \\ x^2+y^2-2x-3=1 \end{cases}$

15. $\begin{cases} x^2+y^2=9 \\ \dfrac{x^2}{9}+\dfrac{y^2}{25}=1 \end{cases}$

16. $\begin{cases} x^2+y^2=4 \\ -x^2=2y-1 \end{cases}$

17. $\begin{cases} x^2+\dfrac{y^2}{4}=1 \\ y=0 \end{cases}$

18. $\begin{cases} y=x^2+1 \\ y-1=x^3 \end{cases}$

19. $\begin{cases} 2y^2-3x^2=6 \\ 2y^2+x^2=22 \end{cases}$

20. $\begin{cases} y=x^3 \\ y=\sqrt[3]{x} \end{cases}$

21. $\begin{cases} x=y^2-4 \\ x+13=6y \end{cases}$

22. $\begin{cases} y=2x^2-3 \\ y=-x^2 \end{cases}$

23. $\begin{cases} 2y=x^2-4 \\ x^2+y^2=4 \end{cases}$

24. $\begin{cases} x^2-y^2=5 \\ \dfrac{x^2}{25}+\dfrac{4y^2}{25}=1 \end{cases}$

Solve the following systems of nonlinear equations. Be sure to check for nonreal solutions. See Examples 3, 4, and 5.

25. $\begin{cases} x^2 + y^2 = 30 \\ x^2 = y \end{cases}$

26. $\begin{cases} 3x^2 + 2y^2 = 12 \\ x^2 + 2y^2 = 4 \end{cases}$

27. $\begin{cases} x^2 - 1 = y \\ 4x + y = -5 \end{cases}$

28. $\begin{cases} x^2 + y^2 = 4 \\ 3x^2 + 4y^2 = 24 \end{cases}$

29. $\begin{cases} y = \dfrac{4}{x} \\ 2x^2 + y^2 = 18 \end{cases}$

30. $\begin{cases} xy = 5 \\ x^2 + y^2 = 10 \end{cases}$

31. $\begin{cases} y - x^2 = 4 \\ x^2 + y^2 = 16 \end{cases}$

32. $\begin{cases} y - x^2 = 6x \\ y = 4x \end{cases}$

33. $\begin{cases} 2x^2 + 3y^2 = 6 \\ x^2 + 3y^2 = 3 \end{cases}$

34. $\begin{cases} x^2 + y^2 = 4 \\ \dfrac{x^2}{4} - \dfrac{y^2}{8} = 1 \end{cases}$

35. $\begin{cases} 3x^2 - y = 3 \\ 9x^2 + y^2 = 27 \end{cases}$

36. $\begin{cases} \dfrac{1}{x} + \dfrac{1}{y} = 5 \\ \dfrac{1}{x} - \dfrac{1}{y} = -3 \end{cases}$

37. $\begin{cases} x + y^2 = 2 \\ 2x^2 - y^2 = 1 \end{cases}$

38. $\begin{cases} y - 2 = (x+3)^2 \\ \dfrac{1}{3}y = (x-1)^2 \end{cases}$

39. $\begin{cases} y - 2 = (x-2)^2 \\ y + 2 = (x-1)^2 \end{cases}$

40. $\begin{cases} y^2 + 2 = 2x^2 \\ y^2 = x^2 - 6 \end{cases}$

41. $\begin{cases} (x+1)^2 + y^2 = 10 \\ \dfrac{(x-2)^2}{4} + y^2 = 1 \end{cases}$

42. $\begin{cases} x^2 + y^2 = 10 \\ x^2 + y = 8 \end{cases}$

43. $\begin{cases} 2x = y - 1 \\ \dfrac{x^2}{25} + y^2 = 1 \end{cases}$

44. $\begin{cases} 2x^2 + y^2 = 4 \\ 2(x-1)^2 + y^2 = 3 \end{cases}$

45. $\begin{cases} x^2 + 7y^2 = 14 \\ x^2 + y^2 = 3 \end{cases}$

46. $\begin{cases} x^2 + y^2 = 25 \\ y^2 = x - 5 \end{cases}$

47. $\begin{cases} y = x^3 + 8x^2 + 17x + 10 \\ -y = x^3 + 8x^2 + 17x + 10 \end{cases}$

48. $\begin{cases} \dfrac{x^2}{25} + \dfrac{y^2}{16} = 1 \\ x^2 + y^2 = 16 \end{cases}$

49. $\begin{cases} xy = 6 \\ (x-2)^2 + (y-2)^2 = 1 \end{cases}$

50. $\begin{cases} y^2 = x+1 \\ \dfrac{x^2}{5} + \dfrac{y^2}{6} = 1 \end{cases}$

51. $\begin{cases} y = x^3 - 1 \\ 3y = 2x - 3 \end{cases}$

52. $\begin{cases} y+5 = (x+1)^2 \\ y-3 = (x-3)^2 \end{cases}$

53. $\begin{cases} xy - y = 4 \\ (x-1)^2 + y^2 = 10 \end{cases}$

54. $\begin{cases} 2x^2 + 5y^2 = 16 \\ 4x^2 + 3y^2 = 4 \end{cases}$

55. $\begin{cases} y = \sqrt{x-4} + 1 \\ (x-3)^2 + (y-1)^2 = 1 \end{cases}$

56. $\begin{cases} y = \sqrt[3]{x} \\ \sqrt{y} = x \end{cases}$

57. $\begin{cases} y^2 - y - 12 = x - x^2 \\ y - 1 + \dfrac{2x-12}{y} = 0 \end{cases}$

58. $\begin{cases} y = 7x^2 + 1 \\ x^2 + y^2 = 1 \end{cases}$

59. $\begin{cases} \dfrac{(y+2)^2}{(x+y)} = 1 \\ x = y^2 + 5y + 4 \end{cases}$

60. $\begin{cases} x = \sqrt{6y+1} \\ y = \sqrt{\dfrac{x^2+7}{2}} \end{cases}$

61. $\begin{cases} \dfrac{-2}{x^2} + \dfrac{1}{y^2} = 8 \\ \dfrac{9}{x^2} - \dfrac{2}{y^2} = 4 \end{cases}$

62. $\begin{cases} x^2 + 3x - 2y^2 = 5 \\ -4x^2 + 6y^2 = 3 \end{cases}$

Draw the graph and determine whether the ordered pairs are solutions to the system of inequalities.

63. $\begin{cases} x \geq 3 \\ y > 4 \end{cases}$ **a.** $(2,5)$ **b.** $(7,8)$ **c.** $(5,0)$ **d.** $(3,4)$

64. $\begin{cases} y \leq 2x+1 \\ y < 4 \\ y > x \end{cases}$ **a.** $(1,2)$ **b.** $(3,4)$ **c.** $(-1,-1)$ **d.** $(3,3)$

65. $\begin{cases} y \geq x^2 \\ y < x^3 \\ y \leq 4x \end{cases}$ **a.** $(2,2)$ **b.** $(2,4)$ **c.** $(2,8)$ **d.** $(3,9)$

66. $\begin{cases} y \geq x^2 - 2 \\ y \leq (x-2)^2 \\ 3y > 2x + 12 \end{cases}$ **a.** $(2,5)$ **b.** $(7,8)$ **c.** $(5,0)$ **d.** $(3,4)$

67. $\begin{cases} x < 4 \\ y \geq \sqrt{x} \\ 2y > -x \end{cases}$ **a.** $(2,5)$ **b.** $(7,8)$ **c.** $(5,0)$ **d.** $(3,4)$

Graph the following systems of inequalities. See Example 6.

68. $\begin{cases} y < 2x \\ y > x^2 \end{cases}$

69. $\begin{cases} y \leq 2x+3 \\ y \geq 0 \\ x \geq 0 \end{cases}$

70. $\begin{cases} y > x^2 \\ -3y \leq x-9 \end{cases}$

71. $\begin{cases} y \leq x \\ 2y > -x \\ x < 4 \end{cases}$

72. $\begin{cases} y \leq \sqrt{x} \\ 2y > (x-1)^2 - 4 \end{cases}$

73. $\begin{cases} y > x^3 \\ y \leq \sqrt[3]{x} \\ y > 0 \end{cases}$

74. $\begin{cases} y \geq x^3 \\ y \geq -x^3 \\ y < 2(x+1) \end{cases}$

75. $\begin{cases} x^2 + y^2 < 9 \\ -4y \geq x-12 \end{cases}$

76. $\begin{cases} y \leq \sin x \\ x \geq 0 \end{cases}$

APPLICATIONS

77. The area of a certain rectangle is 45 square inches, and its perimeter is 28 inches. Find the dimensions of the rectangle.

78. The product of two positive integers is 88, and their sum is 19. What are the integers?

79. Jack takes half an hour longer than his wife does to make the 210-mile drive between two cities. His wife drives 10 miles an hour faster. How fast do the two drive?

80. To construct the two garden beds shown below, 48.5 meters of fencing are needed. The combined area of the beds is 95 square meters. There are two possibilities for the overall dimensions of the two beds. What are they?

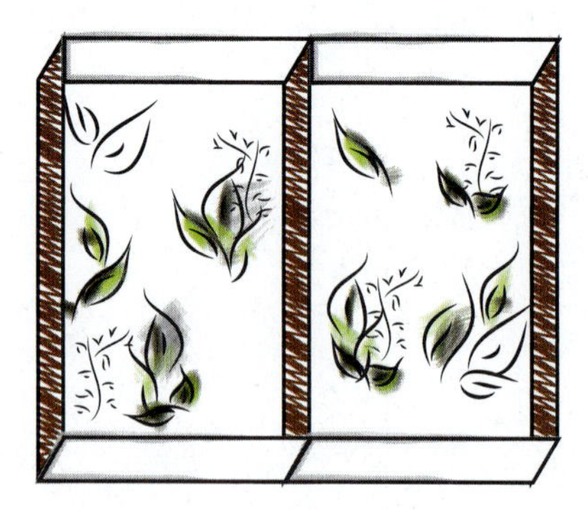

81. The product of two integers is –84, and their sum is –5. What are the integers?

82. Paul and Maria were driving the same 24-mile route, and they departed at the same time. After 20 minutes, Maria was 4 miles ahead of Paul. If it took Paul 10 minutes longer to reach their destination, how fast were they each driving?

83. The surface area of a certain right circular cylinder is $54\pi\ \text{cm}^2$ and the volume is $54\pi\ \text{cm}^3$. Find the height h and the radius r of this cylinder. (**Hint:** The formulas for the volume and surface area of a right circular cylinder are as follows: $V = \pi r^2 h$ and $SA = 2\pi rh + 2\pi r^2$.)

CHAPTER 11 PROJECT

Market Share Matrix

Assume you are the sales and marketing director for Joe's Java, a coffee shop located on a crowded city street corner. There are two competing coffee shops on this block—Buck's Café and Tweak's Coffee. The management has asked you to develop a marketing campaign to increase your market share from 25% to at least 35% within 6 months. With the resulting plan to meet this goal, you predict that each month

a. you will retain 93% of your customers, 4% will go to Buck's Café, and 3% will go to Tweak's Coffee;

b. Buck's Café will retain 91% of their customers, 6% will come to Joe's Java, and 3% will go to Tweak's Coffee; and

c. Tweak's Coffee will retain 92% of their customers, 3% will come to Joe's Java and 5% will go to Buck's Café.

The current percentage of the market is shown in this matrix:

$$x_0 = \begin{bmatrix} 0.25 \\ 0.45 \\ 0.30 \end{bmatrix} \begin{matrix} \textit{Joe's} \\ \textit{Buck's} \\ \textit{Tweak's} \end{matrix}$$

After one month the shares of the coffee shops are

$$x_1 = Px_0 = \begin{bmatrix} 0.93 & 0.06 & 0.03 \\ 0.04 & 0.91 & 0.05 \\ 0.03 & 0.03 & 0.92 \end{bmatrix} \begin{bmatrix} 0.25 \\ 0.45 \\ 0.30 \end{bmatrix} = \begin{bmatrix} 0.2685 \\ 0.4345 \\ 0.2970 \end{bmatrix}$$

1. Construct a table that lists the market share for all of the coffee shops at the end of each of the first 6 months.

2. Will your campaign be successful based on this model? (Will you reach 35% market share in 6 months?)

3. What actions do you think Buck's Café and Tweak's Coffee will take as your market share changes?

4. What effect could their actions have on the market?

CHAPTER 11 REVIEW EXERCISES

Section 11.1

Use any convenient method to solve the following systems of equations. If a system is dependent, express the solution set in terms of one or more of the variables, as appropriate.

1. $\begin{cases} 3x - y + z = 2 \\ -x + y - 2z = -4 \\ -6x + 2y - 2z = -7 \end{cases}$

2. $\begin{cases} 2x - y = 13 \\ 5x - 2y - z = 25 \\ 7x - 6z = -2 \end{cases}$

3. $\begin{cases} x + y - z = 1 \\ 3x - 4y - 5z = -1 \\ 6x - 3y + z = 20 \end{cases}$

4. $\begin{cases} 6x - 5y = 17 \\ -4x + 9y = -17 \end{cases}$

5. $\begin{cases} 3x - y = 2 \\ -6x + 2y = 5 \end{cases}$

6. $\begin{cases} 3x - 2y = -10 \\ x + 2y = 2 \end{cases}$

7. $\begin{cases} \dfrac{x}{3} + y - 1 = 0 \\ x + 3y = 3 \end{cases}$

8. $\begin{cases} \dfrac{x}{3} - \dfrac{y+1}{2} = 1 \\ \dfrac{x}{2} - \dfrac{y}{4} = \dfrac{3}{4} \end{cases}$

9. $\begin{cases} x + y = 5 \\ 2x - y = 4 \\ 5x + y = 17 \end{cases}$

10. $\begin{cases} 2x + 3y - 4z = -7 \\ x - y + 4z = 6 \\ x + y + z = 2 \end{cases}$

11. $\begin{cases} 3x - 2y + z = 10 \\ x + y + z = 30 \\ 2x - y - z = -6 \end{cases}$

12. $\begin{cases} 3x - 2y - 2z = -8 \\ x - y - z = -5 \\ x + y + z = -3 \end{cases}$

13. Find the equation of a parabola $y = ax^2 + bx + c$, passing through the points $(1,0)$, $(-4,9)$, and $(-1,2)$.

Section 11.2

14. Let $A = \begin{bmatrix} 2 & -8 & 9 \\ 7 & 3 & 0 \\ 11 & 6 & 1 \end{bmatrix}$. Determine the following, if possible:

a. The order of A **b.** The value of a_{12} **c.** The value of a_{21}

15. Let $B = \begin{bmatrix} 13 & 8 & 20 & 5 \end{bmatrix}$. Determine the following, if possible:

a. The order of B **b.** The value of b_{12} **c.** The value of b_{31}

Construct the augmented matrix that corresponds to each of the following systems of equations. (Answers may appear in slightly different, but equivalent, forms.)

16. $\begin{cases} 2x+(y-z)=3 \\ 2(y-x)+y-2=z \\ 3x-\dfrac{3-z}{2}=4y \end{cases}$

17. $\begin{cases} z-4x=5y \\ 14z+7(x+3y)=21 \\ 8x-y=-2(x-3z) \end{cases}$

Construct the system of equations that corresponds to each of the following matrices.

18. $\left[\begin{array}{cc|c} 8 & -2 & 2 \\ -1 & 5 & 3 \end{array}\right]$

19. $\left[\begin{array}{ccc|c} 8 & 0 & 7 & 5 \\ 0 & -3 & 4 & 16 \\ 16 & -2 & 1 & 2 \end{array}\right]$

20. $\left[\begin{array}{ccc|c} 3 & -7 & 6 & 9 \\ -11 & 0 & 3 & -14 \\ 0 & 0 & 8 & 2 \end{array}\right]$

Fill in the blanks by performing the indicated elementary row operations.

21. $\left[\begin{array}{cc|c} 3 & 1 & -2 \\ 1 & 2 & 3 \end{array}\right] \xrightarrow{-3R_2+R_1} \underline{?}$

22. $\left[\begin{array}{cc|c} 2 & 3 & 5 \\ -4 & -1 & 2 \end{array}\right] \xrightarrow{2R_1+R_2} \underline{?}$

23. $\left[\begin{array}{cc|c} 1 & -4 & -4 \\ 3 & -1 & 3 \end{array}\right] \xrightarrow{-2R_1+R_2} \underline{?}$

24. $\left[\begin{array}{ccc|c} -1 & 0 & 2 & -6 \\ 1 & -3 & 4 & 1 \\ -2 & -1 & -3 & 0 \end{array}\right] \xrightarrow[-R_1+R_3]{2R_2} \underline{?}$

Use Gaussian elimination and back-substitution to solve the following systems of equations.

25. $\begin{cases} 3x-y=7 \\ x-4y=6 \end{cases}$

26. $\begin{cases} \dfrac{x}{5}-\dfrac{y}{3}=2 \\ -6x+5y=20 \end{cases}$

Use Gauss-Jordan elimination to solve the following systems of equations.

27. $\begin{cases} 5x-4y=35 \\ 25x-18y=165 \end{cases}$

28. $\begin{cases} x-3y-4z=-5 \\ -x+7y+8z=17 \\ 2x-10y-12z=-10 \end{cases}$

Section 11.3

Evaluate the following determinants.

29. $\begin{vmatrix} x^3 & -x^2 \\ x^2 & x \end{vmatrix}$

30. $\begin{vmatrix} -1 & 3 & 1 \\ 1 & -4 & 0 \\ 0 & 2 & 3 \end{vmatrix}$

31. $\begin{vmatrix} -2 & -1 & -3 & 0 \\ 3 & 3 & 1 & 5 \\ 4 & 0 & 0 & 1 \\ 2 & 0 & 1 & 0 \end{vmatrix}$

32. $\begin{vmatrix} x^4 & x & x & 2x \\ 0 & x & x^3 & x \\ 0 & 0 & x & x \\ 0 & 0 & 0 & x^2 \end{vmatrix}$

Use the matrix $A = \begin{bmatrix} 0 & -3 & 1 \\ 2 & 0 & 5 \\ -1 & 3 & 2 \end{bmatrix}$ to evaluate the minor and cofactor of the following elements.

33. a_{12}

34. a_{31}

Use Cramer's Rule to solve each system of equations.

35. $\begin{cases} x + 6y = 2 \\ 3x - y = -13 \end{cases}$

36. $\begin{cases} x + 2y - 3z = -3 \\ -5x - y + 4z = -5 \\ 3x + y + z = 6 \end{cases}$

37. $\begin{cases} -4x + 2y = 3 \\ 2x - y = 4 \end{cases}$

38. $\begin{cases} x - 2y = 0 \\ x + y + z = 6 \\ 3x - y - 4z = 10 \end{cases}$

Section 11.4

Given $A = \begin{bmatrix} 2 & -8 & 3 \\ -1 & 0 & 5 \end{bmatrix}$, $B = \begin{bmatrix} 2 & 0 \\ -1 & 3 \end{bmatrix}$, $C = \begin{bmatrix} 3 & 1 & -6 \\ 8 & -3 & -7 \end{bmatrix}$, and $D = \begin{bmatrix} 0 & 4 \\ -3 & 11 \\ 7 & 1 \end{bmatrix}$, determine the following, if possible.

39. BA

40. B^2

41. $CD + C$

42. BD

43. $3A + C$

44. $AD + B$

Determine values of the variables that will make the following equations true, if possible.

45. $\begin{bmatrix} w & 5x \\ 2y & z \end{bmatrix} - 3\begin{bmatrix} w & x \\ 2 & -z \end{bmatrix} = \begin{bmatrix} 4 & 2 \\ y-3 & -16 \end{bmatrix}$

46. $\begin{bmatrix} 4x & 2y^2 & z \end{bmatrix} = \begin{bmatrix} 12 \\ 18 \\ -2 \end{bmatrix}$

47. $2\begin{bmatrix} x \\ -3y \end{bmatrix} - \begin{bmatrix} y \\ 2x \end{bmatrix} = \begin{bmatrix} 7 \\ 14 \end{bmatrix}$

48. $\begin{bmatrix} 3x \\ 5y \end{bmatrix} - \begin{bmatrix} y \\ -2x \end{bmatrix} = \begin{bmatrix} 4 \\ -3 \end{bmatrix}$

Evaluate the following matrix products, if possible.

49. $\begin{bmatrix} 7 & 1 & -1 \end{bmatrix}\begin{bmatrix} 1 & 6 \\ 2 & 1 \\ -3 & -3 \end{bmatrix}$

50. $\begin{bmatrix} 4 \\ 5 \\ 6 \end{bmatrix}\begin{bmatrix} -3 & 2 & 3 \end{bmatrix}$

Section 11.5

Write each of the following systems of equations as a single matrix equation.

51. $\begin{cases} x_1 - x_2 + 2x_3 = -4 \\ 2x_1 - 3x_2 - x_3 = 1 \\ -3x_1 + 6x_3 = 5 \end{cases}$

52. $\begin{cases} 3x - y + z = 4 \\ 2x - 5z = 1 \\ 4x + 3y - 6 = 0 \end{cases}$

Find the inverse of each of the following matrices, if possible.

53. $\begin{bmatrix} 4 & -2 \\ 2 & 3 \end{bmatrix}$

54. $\begin{bmatrix} 2 & 2 \\ \frac{1}{2} & 1 \end{bmatrix}$

55. $\begin{bmatrix} 4 & 12 \\ 3 & 9 \end{bmatrix}$

56. $\begin{bmatrix} -1 & 2 & 3 \\ 1 & -1 & 4 \\ 2 & 0 & -2 \end{bmatrix}$

For each pair of matrices, determine if either matrix is the inverse of the other.

57. $\begin{bmatrix} 3 & 12 \\ 2 & 9 \end{bmatrix}, \begin{bmatrix} 1 & 4 \\ \frac{2}{3} & 3 \end{bmatrix}$

58. $\begin{bmatrix} 1 & -2 \\ -3 & 4 \end{bmatrix}, \begin{bmatrix} -2 & -1 \\ -\frac{3}{2} & -\frac{1}{2} \end{bmatrix}$

59. $\begin{bmatrix} -2 & 4 & -3 \\ 0 & 6 & -3 \\ 0 & 8 & -3 \end{bmatrix}, \begin{bmatrix} -\frac{1}{2} & 1 & -\frac{1}{2} \\ 0 & -\frac{1}{2} & \frac{1}{2} \\ 0 & -\frac{4}{3} & 1 \end{bmatrix}$

60. $\begin{bmatrix} 5 & -3 & 7 \\ 6 & 0 & 2 \\ -9 & 1 & 0 \end{bmatrix}, \begin{bmatrix} -9 & 2 & 1 \\ 7 & 0 & -3 \\ 1 & 8 & 2 \end{bmatrix}$

Solve the following systems by the inverse matrix method, if possible. If the inverse matrix method doesn't apply, use any other method to determine if the system is inconsistent or dependent.

61. $\begin{cases} 5x + 9y = 2 \\ -2x - 3y = -1 \end{cases}$

62. $\begin{cases} 2y + 3z = 3 \\ -2x = 0 \\ 8x + 4y + 5z = -1 \end{cases}$

Solve the following set of systems by the inverse matrix method.

63. $\begin{cases} 2x - z = 3 \\ x + 4y + 2z = -1 \\ x + y = 5 \end{cases}$ $\begin{cases} 2x - z = 0 \\ x + 4y + 2z = 2 \\ x + y = 1 \end{cases}$ $\begin{cases} 2x - z = -1 \\ x + 4y + 2z = 1 \\ x + y = 2 \end{cases}$

Section 11.6

Write the form of the partial fraction decomposition of each of the following rational functions. In each case, assume the degree of the numerator is less than the degree of the denominator.

64. $f(x) = \frac{p(x)}{9x^4 - 6x^3 + x^2}$

65. $f(x) = \frac{p(x)}{x^2 + 3x - 4}$

Find the partial fraction decomposition of each of the following rational functions.

66. $f(x) = \frac{2x}{(x-1)(x+1)}$

67. $f(x) = \frac{x-4}{(2x-5)^2}$

68. $f(x) = \frac{2x^2 + x + 8}{(x^2+4)^2}$

Section 11.7

Graph the solution set of each of the following systems of inequalities.

69. $\begin{cases} 7x - 2y \ge 8 \\ y < 5 \end{cases}$

70. $\begin{cases} y - x > 0 \\ x < 2 \end{cases}$

Construct the constraints and graph the feasible regions for the following situations.

71. Each bag of nuts contains peanuts and cashews. The total number of nuts in the bag cannot exceed 60. There must be at least 20 peanuts and 10 cashews per bag. There can be no more than 40 peanuts or 40 cashews per bag. What is the region of constraint for the number of nuts per bag?

72. You wish to study at least 15 hours (over a 4-day span) for your upcoming statistics and biology tests. You need to study a minimum of 6 hours for each test. The maximum you wish to study for statistics is 10 hours and for biology is 8 hours. What is the region of constraint for the numbers of hours you should study for each test?

Find the minimum and maximum values of the given functions, subject to the given constraints.

73. Objective Function:

$f(x, y) = 6x + 10y$

Constraints:

$\begin{cases} x \ge 0,\ y \ge 0 \\ 2x + 5y \le 10 \end{cases}$

74. Objective Function:

$f(x, y) = 5x + 2y$

Constraints:

$\begin{cases} x \ge 0,\ y \ge 0 \\ x + y \le 10 \\ x + 2y \ge 10 \\ 2x + y \ge 10 \end{cases}$

75. Objective Function:

$f(x, y) = 5x + 4y$

Constraints:

$\begin{cases} x \ge 0,\ y \ge 0 \\ 2x + 3y \le 12 \\ 3x + y \le 12 \\ x + y \ge 2 \end{cases}$

76. Objective Function:

$f(x, y) = 70x + 82y$

Constraints:

$\begin{cases} x \ge 0,\ y \ge 0 \\ x \le 10,\ y \le 20 \\ x + y \ge 5 \\ x + 2y \le 18 \end{cases}$

77. Krueger's Pottery manufactures two kinds of hand-painted pottery: a vase and a pitcher. There are three processes to create the pottery: throwing (forming the pottery on the potter's wheel), baking, and painting. No more than 90 hours are available per day for throwing, only 120 hours are available per day for baking, and no more than 60 hours per day are available for painting. The vase requires 3 hours for throwing, 6 hours for baking, and 2 hours for painting. The pitcher requires 3 hours for throwing, 4 hours for baking, and 3 hours for painting. The profit for each vase is \$25 and the profit for each pitcher is \$30. How many of each piece of pottery should be produced a day to maximize profit? What would the maximum profit be if Krueger's produced this amount?

78. Pranas produces bionic arms and legs for those that are missing a limb. Pranas can produce at least 20, but no more than 60 arms in a week due to the lab limitations. They can produce at least 15, but no more than 40 legs in a week. To keep their research grant, the company must produce at least 50 limbs per week. It costs $450 to produce the bionic arm and $550 to produce the bionic leg. How many of each should be produced per week to minimize the cost? What would the minimum cost be if Pranas produced this amount?

Section 11.8

Use graphing to approximate the real solution(s) of the following systems, and then verify that your answers are correct.

79. $\begin{cases} (x-2)^2 + y = 2 \\ x - y = 2 \end{cases}$

80. $\begin{cases} x^2 + y^2 = 25 \\ -x - y = 5 \end{cases}$

Solve the following systems of nonlinear equations. Be sure to check for nonreal solutions.

81. $\begin{cases} x^2 + 2y^2 = 1 \\ x^2 = y \end{cases}$

82. $\begin{cases} x^2 + y^2 = 25 \\ 2x^2 - y^2 = 23 \end{cases}$

83. $\begin{cases} y = (x-1)^2 \\ y + 8 = (x+1)^2 \end{cases}$

Draw the graph and determine whether the ordered pairs are solutions to the system of inequalities.

84. $\begin{cases} y^2 \le 9 - x^2 \\ y < |x| \\ y > -|x| \end{cases}$ **a.** $(2,5)$ **b.** $(7,8)$ **c.** $(5,0)$ **d.** $(3,4)$

Graph the following systems of inequalities.

85. $\begin{cases} y \le \sin x \\ y > -\sin x \end{cases}$

86. $\begin{cases} y \le \sqrt{x+1} \\ y > x^2 - 1 \end{cases}$

87. $\begin{cases} x^2 y \le 1 \\ 2y \le x^2 + 2 \\ y < 16x^2 \end{cases}$

88. The product of two positive integers is 144, and their sum is 25. What are the integers?

89. Stephen and Scott were driving the same 72-mile route, and they departed at the same time. After 30 minutes, Stephen was 6 miles ahead of Scott. If it took Scott one more hour than Stephen to reach their destination, how fast were they each driving?

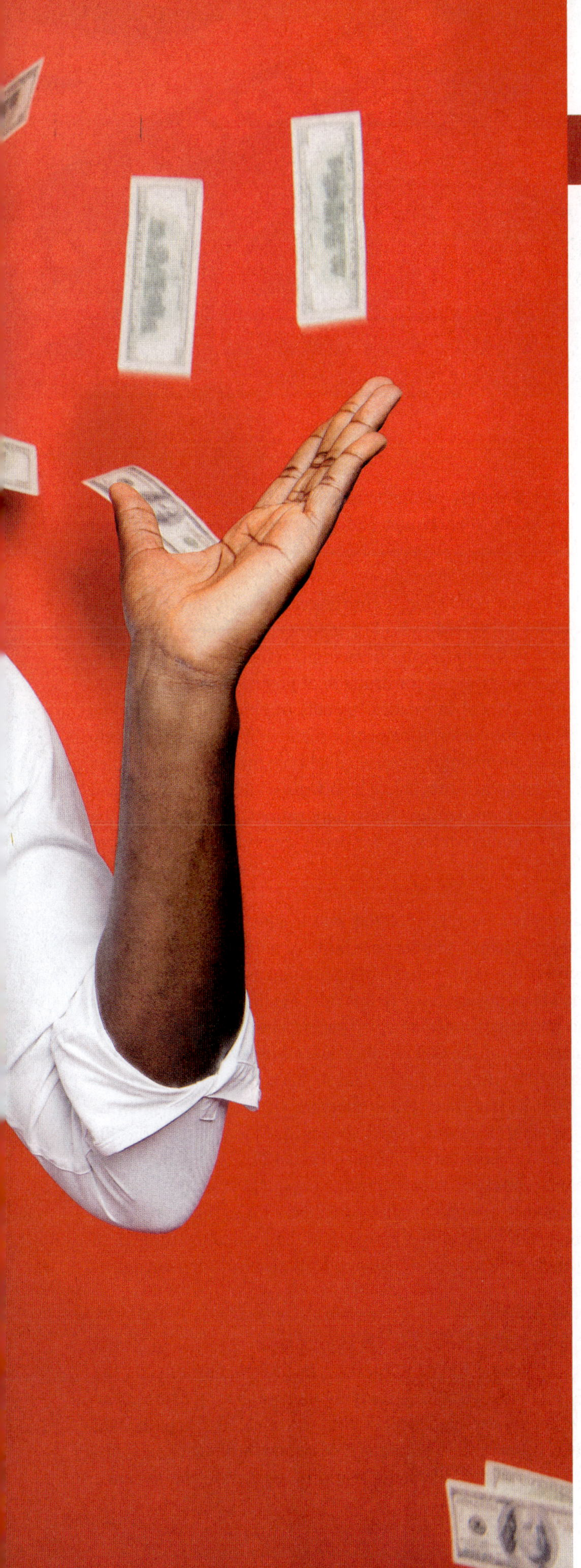

CHAPTER 12

Sequences, Series, Combinatorics, and Probability

SECTIONS

WHAT IF ...

What if your boss offered you a \$1 raise this year, a \$2 raise next year, a \$4 raise the following year, and promised to continue doubling in this manner. How much of a raise would you receive after 20 years? (**Hint:** Your boss is not likely to agree to this deal!)

By the end of this chapter, you'll be able to find individual terms in sequences and series as well as their general formulas. You'll solve application problems similar to the one above, including one involving a king, grains of wheat, and a chess board on page 922. You'll master this type of problem using tools such as the formula for the partial sum of a geometric sequence, found on page 914.

Blaise Pascal

Introduction

This chapter introduces a variety of mathematical topics that, depending on your major, you are likely to encounter and study in greater detail in later courses.

Nearly everyone will, at some point, find it convenient to know the basic facts about sequences and series presented here. For instance, sequences and series arise in the financial world in the guise of retirement planning, portfolio analysis, and loan repayment; in sociology in the construction of population models; in the study of computer algorithms; and in medical models of epidemic spread. The counting techniques of *combinatorics*, a branch of mathematics, are immensely useful in a large number of surprisingly diverse situations; the fields of statistics and computer science, to name just two, make heavy use of the methods studied here. Finally, the science of *probability* finds applications ranging from the deadly serious (actuarial mortality tables) to the purely entertaining (the modern "gaming" industry).

Probability, in particular, has a long, convoluted, and fascinating history. People have pondered questions of chance, and the possibility of analyzing or predicting fate, for all of recorded history (and longer). Mathematicians, as a subset of the general population, have been no exception to this rule, but their understanding of probability has had no particularly mathematical aspect until relatively recently. While Italian mathematicians of the Renaissance made some small progress in the area, mathematical probability is usually said to have been born in France in the year 1654. It was in that year that a member of the French nobility, Antoine Gombaud, Chevalier de Méré, posed several questions born from his gambling experience to the mathematician Blaise Pascal (1623–1662). Pascal, intrigued by the questions, consulted with Pierre de Fermat (1601–1665), another mathematician famous today largely for his so-called "Last Theorem." In a series of letters back and forth, Pascal and Fermat laid the foundation of modern mathematical probability, a field with profound applications inside and outside of the world of gambling.

The material in this chapter can be viewed as a bridge between algebra and further study in mathematics. All of the topics introduced in this chapter rely on the mathematical maturity you have gained by studying algebra, but there are many more areas of math that are not explored at all in this book. Calculus, in particular, is one such area; many of you will soon be taking at least one semester of calculus, a field of mathematics that extends the static world of algebra to the dynamic world in which we live.

12.1 SEQUENCES AND SERIES

TOPICS

1. Recursively and explicitly defined sequences
2. Summation notation and formulas
3. Partial sums and series
4. Fibonacci sequences

TOPIC 1: Recursively and Explicitly Defined Sequences

Many natural and mathematical phenomena can be described with a simple list of numbers. For instance, the growth of an isolated population of rabbits over the course of many months might be described with a list of the number of rabbits born each month. When we write out a list of the sum of the first n odd numbers, we can recognize it as another familiar list, the squares of the integers:

$$1=1$$
$$1+3=4$$
$$1+3+5=9$$
$$1+3+5+7=16$$
$$1+3+5+7+9=25$$

These lists are called *sequences*, and we can think of them as a special type of function.

DEFINITION: Sequences

An **infinite sequence** (or just **sequence**) is a function whose domain is the set of natural numbers $\{1,2,3,\ldots\}$. Instead of giving the function a name such as f and referring to the values of the sequence as $f(1), f(2), f(3),\ldots,$ we commonly use subscripts and refer to the **terms** of the sequence as $a_1, a_2, a_3, \ldots$. In other words, a_n stands for the nth term in the sequence.

Finite sequences are defined similarly, but the domain of a finite sequence is a set of the form $\{1,2,3,\ldots,k\}$ for some positive integer k.

A consequence of sequence notation is that we don't have a label, such as f, to attach to a sequence; instead, a given sequence is normally identified by some formula for determining a_n, the nth term of the sequence, and the entire sequence may be referred to as $\{a_n\}$. If it is necessary to discuss several sequences at once, use different letters to denote the different terms. For instance, a problem may refer to two sequences $\{a_n\}$ and $\{b_n\}$.

Example 1: Sequences

Determine the first five terms of the following sequences.

a. $a_n = 5n - 2$ **b.** $b_n = \frac{(-1)^n + 1}{2}$ **c.** $c_n = \frac{n}{n+1}$

NOTE

Sequences often have distinct patterns. As you write out the terms for each sequence, see if you can predict the next term before calculating it.

Solution

a. Replacing n in the formula for the general term with the first five positive integers, we obtain

$$a_1 = 5(1)-2, \quad a_2 = 5(2)-2, \quad a_3 = 5(3)-2, \quad a_4 = 5(4)-2, \quad a_5 = 5(5)-2,$$

so the sequence starts out as $3, 8, 13, 18, 23, \ldots$.

b. Again, we replace n with the first five positive integers to determine the first five terms of the sequence.

$$b_1 = \frac{(-1)^1+1}{2} = \frac{0}{2} = 0$$
$$b_2 = \frac{(-1)^2+1}{2} = \frac{2}{2} = 1$$
$$b_3 = \frac{(-1)^3+1}{2} = \frac{0}{2} = 0$$
$$b_4 = \frac{(-1)^4+1}{2} = \frac{2}{2} = 1$$
$$b_5 = \frac{(-1)^5+1}{2} = \frac{0}{2} = 0$$

Note that $(-1)^n$ is equal to -1 if n is odd and is equal to 1 if n is even.

When we add 1 to $(-1)^n$ and divide the result by 2, we obtain either 0 or 1, so the sequence begins $0, 1, 0, 1, 0, \ldots$.

This example illustrates the fact that a given value may appear more than once in a sequence.

c. Replacing n in the formula for the general term with the first five positive integers, we obtain

$$c_1 = \frac{1}{1+1}, \quad c_2 = \frac{2}{2+1}, \quad c_3 = \frac{3}{3+1}, \quad c_4 = \frac{4}{4+1}, \quad c_5 = \frac{5}{5+1},$$

so the sequence starts out as $\frac{1}{2}, \frac{2}{3}, \frac{3}{4}, \frac{4}{5}, \frac{5}{6}, \ldots$.

The formulas for the general n^{th} terms in Example 1 are all examples of **explicit** formulas, named because they provide a direct rule for calculating any term in the sequence. In many cases, though, an explicit formula for the general term cannot be found, or is not as easily determined.

In such cases, the terms of a sequence are often defined *recursively*. A **recursive** formula is one that refers to one or more of the terms preceding a_n in the definition for a_n. For instance, if the first term of a sequence is -5 and if it is known that each of the remaining terms of the sequence is 7 more than the term preceding it, the sequence can be defined by the rules $a_1 = -5$ and $a_n = a_{n-1} + 7$ for $n \geq 2$.

Example 2: Recursively Defined Sequences

Determine the first five terms of the following recursively defined sequences.

a. $a_1 = 3$ and $a_n = a_{n-1} + 5$ for $n \geq 2$ **b.** $a_1 = 2$ and $a_n = 3a_{n-1} + 1$ for $n \geq 2$

Solution

a. We find the first five terms by replacing n with the first five positive integers, just as in Example 1. Note that in using the recursive definition we must determine the elements of the sequence in order; that is, to determine a_5, we need to first know a_4. And to determine a_4, we need to first know a_3, and so on back to a_1.

$$\begin{aligned} a_1 &= 3 \\ a_2 &= a_1 + 5 = 3 + 5 = 8 \\ a_3 &= a_2 + 5 = 8 + 5 = 13 \\ a_4 &= a_3 + 5 = 13 + 5 = 18 \\ a_5 &= a_4 + 5 = 18 + 5 = 23 \end{aligned}$$

Thus, the first five terms of the sequence are $3, 8, 13, 18, 23, \ldots$.

The sequence defined by this recursive definition appears to be the same as the first sequence in Example 1, defined explicitly by $a_n = 5n - 2$. This illustrates the fact that a given sequence can often be defined several different ways.

b. Using the recursive formula $a_1 = 2$ and $a_n = 3a_{n-1} + 1$ for $n \geq 2$, we obtain

$$\begin{aligned} a_1 &= 2 \\ a_2 &= 3a_1 + 1 = 3(2) + 1 = 7 \\ a_3 &= 3a_2 + 1 = 3(7) + 1 = 22 \\ a_4 &= 3a_3 + 1 = 3(22) + 1 = 67 \\ a_5 &= 3a_4 + 1 = 3(67) + 1 = 202 \end{aligned}$$

Thus, the first five terms of the sequence are $2, 7, 22, 67, 202, \ldots$.

In general, an explicit formula is more useful when finding the terms of a sequence than a recursive formula. Consider the task of calculating a_{100} based on the formula $a_n = 5n - 2$, versus the same task given the formula $a_1 = 3$ and $a_n = a_{n-1} + 5$ for $n \geq 2$.

To an extent, the problems in Examples 1 and 2 can be turned around, so that the question is finding a formula for the general n^{th} term of a sequence given its first few terms. This is often the challenge in modeling situations, where the goal is to extrapolate the behavior of a sequence of numbers by finding a formula that produces the first few terms of the sequence. The catch is that there is always more than one formula that will produce identical terms of a sequence up to a certain point and then differ beyond that. Consider the following two explicit formulas:

$$a_n = 3n \text{ and } b_n = 3n + (n-1)(n-2)(n-3)(n-4)(n-5)n^2$$

These two formulas will produce identical results for the first five terms, but different results from then on. For this reason, the instructions in Example 3 ask for *possible* formulas for the given sequences.

Example 3: Finding the Formula for a Sequence

Find a possible formula for the general n^{th} term of the sequences that begin as follows.

a. $-3, 9, -27, 81, -243, \ldots$ **b.** $1, 3, 6, 10, 15, \ldots$

Solution

a. There is no general method to find a formula for the terms of a sequence. Observation is usually the best tool. If a formula for a given sequence does not come to mind quickly, it may help to associate each term of the sequence with its place in the sequence, as shown below.

$$\begin{array}{cccccc} 1 & 2 & 3 & 4 & 5 & \ldots \\ \updownarrow & \updownarrow & \updownarrow & \updownarrow & \updownarrow & \updownarrow \\ -3 & 9 & -27 & 81 & -243 & \ldots \end{array}$$

If a pattern is still not apparent, try rewriting the terms of the sequence in different ways. In this case, factoring the terms leads to the following.

$$\begin{array}{cccccc} 1 & 2 & 3 & 4 & 5 & \ldots \\ \updownarrow & \updownarrow & \updownarrow & \updownarrow & \updownarrow & \updownarrow \\ -3^1 & 3^2 & -3^3 & 3^4 & -3^5 & \ldots \end{array}$$

The nth term of the sequence is the nth power of 3, multiplied by -1 if n is odd. One way to express alternating signs in a sequence is to multiply the nth term by $(-1)^n$ (if the odd terms are negative) or by $(-1)^{n+1}$ (if the even terms are negative). In this case, a possible formula for the general n^{th} term is $a_n = (-1)^n (3)^n$, or $a_n = (-3)^n$.

Note that we might also have come up with the recursive formula $a_1 = -3$ and $a_n = -3a_{n-1}$ for $n \geq 2$.

b. Associate each term with its place in the sequence, as follows.

$$\begin{array}{cccccc} 1 & 2 & 3 & 4 & 5 & \ldots \\ \updownarrow & \updownarrow & \updownarrow & \updownarrow & \updownarrow & \updownarrow \\ 1 & 3 & 6 & 10 & 15 & \ldots \end{array}$$

In this case, factoring the terms does not seem to help in identifying a pattern, but thinking about the difference between successive terms does.

$$\begin{array}{cccccc} 1 & 2 & 3 & 4 & 5 & \ldots \\ \updownarrow & \updownarrow & \updownarrow & \updownarrow & \updownarrow & \updownarrow \\ 1 & 3 = 1 + 2 & 6 = 3 + 3 & 10 = 6 + 4 & 15 = 10 + 5 & \ldots \end{array}$$

This observation leads to the recursive formula $a_1 = 1$ and $a_n = a_{n-1} + n$ for $n \geq 2$.

TOPIC 2: Summation Notation and Formulas

First Use of $\sum$

The prolific 18th century Swiss mathematician Leonhard Euler was the first to introduce many of the math symbols and notations that have become customary, and the use of a capital Greek sigma for a sum, whether finite or infinite, is an example.

One very common use of sequence notation is to define terms that are to be added together. If the first n terms of a given sequence are to be added, we can write the sum as $a_1 + a_2 + \cdots + a_n$, but this can be confusing and unwieldy. Further, this notation doesn't describe the terms being added.

Summation notation (also known as *sigma notation*) provides a better option. Summation notation borrows the capital Greek letter Σ ("sigma") to denote the operation of summation, as described below.

DEFINITION: Summation Notation

The sum $a_1 + a_2 + \cdots + a_n$ is expressed in **summation notation** as $\sum_{i=1}^{n} a_i$.

When this notation is used, the letter i is called the **index of summation**, and a_i often appears as the formula for the ith term of a sequence. In the sum above, all the terms of the sequence beginning with a_1 and ending with a_n are to be added. The notation can be modified to indicate a different first or last term of the sum.

Example 4: Evaluating Sums

Rewrite the following sums in expanded form, then evaluate them.

a. $\sum_{i=1}^{4} (3i - 2)$ **b.** $\sum_{i=3}^{5} i^2$

Solution

a.
$$\begin{aligned}\sum_{i=1}^{4} (3i-2) &= (3\cdot 1 - 2) + (3\cdot 2 - 2) + (3\cdot 3 - 2) + (3\cdot 4 - 2)\\ &= 1 + 4 + 7 + 10\\ &= 22\end{aligned}$$

Replace the index i with the numbers 1 through 4.

b.
$$\begin{aligned}\sum_{i=3}^{5} i^2 &= 3^2 + 4^2 + 5^2\\ &= 9 + 16 + 25\\ &= 50\end{aligned}$$

Replace the index i with the numbers 3, 4, and 5.

Keep in mind that sigma notation is just a more concise way of representing addition. The following properties of sigma notation are just restatements of familiar properties of addition.

PROPERTIES: Properties of Sigma Notation

Let $\{a_n\}$ and $\{b_n\}$ be two sequences, and let c be a constant.

1. $\sum_{i=1}^{n}(a_i + b_i) = \sum_{i=1}^{n} a_i + \sum_{i=1}^{n} b_i$ (the terms of a sum can be rearranged)

2. $\sum_{i=1}^{n} ca_i = c\sum_{i=1}^{n} a_i$ (constants can be factored out of a sum)

3. $\sum_{i=1}^{n} a_i = \sum_{i=1}^{k} a_i + \sum_{i=k+1}^{n} a_i$ for any $1 \le k \le n-1$, (a sum can be broken apart into two smaller sums)

CAUTION

There are many statements that look similar to these properties, but are not true. For instance, it is not generally true that $\sum_{i=1}^{n} a_i b_i = \sum_{i=1}^{n} a_i \cdot \sum_{i=1}^{n} b_i$.

In addition to the arithmetic properties listed above, formulas for sums that occur frequently can be very useful when calculating more complicated sums. We will cover four such formulas in this section and prove two of them. A proof of the third formula will be covered in Section 12.4.

FORMULA: Four Summation Formulas

1. $\sum_{i=1}^{n} 1 = n$

2. $\sum_{i=1}^{n} i = \frac{n(n+1)}{2}$

3. $\sum_{i=1}^{n} i^2 = \frac{n(n+1)(2n+1)}{6}$

4. $\sum_{i=1}^{n} i^3 = \frac{n^2(n+1)^2}{4}$

The first formula really requires no proof at all. We can see that it is true just by writing the sum in expanded form.

$$\sum_{i=1}^{n} 1 = \underbrace{1+1+\cdots+1}_{n \text{ terms}} = n$$

One proof of the second formula begins by letting S stand for the sum.

$$S = \sum_{i=1}^{n} i = 1+2+\cdots+n$$

Since the addition can be performed in any order, the following is also true.

$$S = n+(n-1)+\cdots+1$$

Note the result of adding these two equations:

$$\begin{array}{rcllll} S & = & 1 & +2 & +\cdots & +n \\ S & = & n & +(n-1) & +\cdots & +1 \\ \hline 2S & = & (n+1) & +(n+1) & +\cdots & +(n+1) \end{array}$$

Since the term $n + 1$ appears n times on the right-hand side of the bottom equation, we have $2S = n(n+1)$ or, after dividing both sides by 2,

$$\sum_{i=1}^{n} i = \frac{n(n+1)}{2}.$$

Note that this provides an explicit formula for the nth term of the sequence in Example 3b. Since the nth term is the sum $1+2+\cdots+n$, the formula is

$$a_n = \frac{n(n+1)}{2}.$$

Example 5: Evaluating Sums

Use the above properties and formulas to evaluate the following sums.

a. $\sum_{i=1}^{9}(7i-3)$ **b.** $\sum_{i=4}^{6}3i^2$

Solution

a. $\sum_{i=1}^{9}(7i-3) = \sum_{i=1}^{9}7i - \sum_{i=1}^{9}3$ Apply the first property to split the given sum into two sums.

$= 7\sum_{i=1}^{9}i - 3\sum_{i=1}^{9}1$ Apply the second property to factor the constants out of each sum.

$= 7\left(\frac{9\cdot 10}{2}\right) - 3\cdot 9$ Use the summation formulas.

$= 315 - 27$ Simplify.

$= 288$

b. $\sum_{i=4}^{6}3i^2 = 3\sum_{i=4}^{6}i^2$ Use the second property to factor out the 3.

$= 3\sum_{i=1}^{6}i^2 - 3\sum_{i=1}^{3}i^2$ Apply the third property of sigma notation to obtain two sums that begin with $i = 1$.

$= 3\left(\frac{6\cdot 7\cdot 13}{6}\right) - 3\left(\frac{3\cdot 4\cdot 7}{6}\right)$ Use a summation formula.

$= 273 - 42$ Simplify.

$= 231$

NOTE

A good strategy is to break the sum into the simplest parts possible using the properties of sigma notation, then apply the known formulas.

TOPIC 3: Partial Sums and Series

Given a sequence $\{a_n\}$, we may need to calculate sums of the form $a_1 + a_2 + \cdots + a_n$ or the sum of *all* the terms, which we might express as $a_1 + a_2 + a_3 + \cdots$. The following definition formalizes these two ideas.

DEFINITION: Partial Sums and Series

Given an infinite sequence $\{a_n\}$, the ***n*^th^ partial sum** is $S_n = \sum_{i=1}^{n} a_i$. S_n is an example of a **finite series**. The **infinite series** associated with $\{a_n\}$ is the sum $\sum_{i=1}^{\infty} a_i$. Note that the adjective *infinite* refers to the infinite number of terms that appear in the sum and does not imply that the sum itself is infinite.

FIGURE 1: Visualizing an Infinite Series

It may seem that the sum of an infinite number of terms can't possibly be a finite number, especially if all of the terms are positive. But in fact, many simple examples of infinite series with finite sums exist. Consider the fractions of an inch marked on a typical ruler.

If half of an inch and a quarter of an inch are added, the result is three-quarters of an inch. If an additional eighth of an inch is added, the result is seven-eighths of an inch. If the ruler is of sufficient precision, a sixteenth of an inch can be added, resulting in a total of fifteen-sixteenths of an inch.

In principle, we can continue this process indefinitely, with the result that the sum of the fractions at every stage is closer to, but never exceeds, one inch. In the language of partial sums and series, we can say

$$S_n = \frac{1}{2} + \frac{1}{4} + \cdots + \frac{1}{2^n} = \frac{2^n - 1}{2^n}$$

and

$$S = \frac{1}{2} + \frac{1}{4} + \frac{1}{8} + \cdots = \sum_{i=1}^{\infty} \frac{1}{2^i} = 1.$$

An infinite series $\sum_{i=1}^{\infty} a_i$ **converges** if the sequence of partial sums $S_n = \sum_{i=1}^{n} a_i$ approaches a fixed real number S, and in this case we write $S = \sum_{i=1}^{\infty} a_i$.

If the sequence of partial sums does not approach some fixed real number (either by getting larger and larger in magnitude or by "bouncing around"), we say the series **diverges**. The next example illustrates both possible outcomes.

Example 6: Evaluating Partial Sums and Series

Examine the partial sums associated with each infinite series to determine if the series converges or diverges.

a. $\sum_{i=1}^{\infty}\left(\frac{1}{i}-\frac{1}{i+1}\right)$

b. $\sum_{i=1}^{\infty}(-1)^i$

Solution

a. $S_1=1-\frac{1}{2}=1-\frac{1}{1+1}$

Begin by evaluating the first few partial sums.

$S_2=\left(1-\frac{1}{2}\right)+\left(\frac{1}{2}-\frac{1}{3}\right)$

$=1-\frac{1}{2}+\frac{1}{2}-\frac{1}{3}=1-\frac{1}{3}=1-\frac{1}{2+1}$

$S_3=\left(1-\frac{1}{2}\right)+\left(\frac{1}{2}-\frac{1}{3}\right)+\left(\frac{1}{3}-\frac{1}{4}\right)$

$=1-\frac{1}{2}+\frac{1}{2}-\frac{1}{3}+\frac{1}{3}-\frac{1}{4}=1-\frac{1}{4}=1-\frac{1}{3+1}$

$\vdots$

Writing the terms out, we see that all of the terms except the first and last one cancel out. Since the length of the expression collapses down, this is an example of a *telescoping series*.

$S_n=1-\frac{1}{n+1}$

We can write a formula for the nth partial sum by observing the pattern.

The partial sums approach 1 as n gets larger and larger, therefore the series converges. In fact, $\sum_{i=1}^{\infty}\left(\frac{1}{i}-\frac{1}{i+1}\right)=1$.

b. $S_1=(-1)^1=-1$

Again, begin by calculating the first few partial sums to look for a pattern.

$S_2=(-1)^1+(-1)^2$

$=-1+1=0$

$S_3=(-1)^1+(-1)^2+(-1)^3$

$=-1+1+(-1)=-1$

$S_4=(-1)^1+(-1)^2+(-1)^3+(-1)^4$

$=-1+1+(-1)+1=0$

$\vdots$

$$S_n=\begin{cases}-1 & \text{if } n \text{ is odd}\\ 0 & \text{if } n \text{ is even}\end{cases}$$

A pattern emerges based on whether n is even or odd.

The partial sums of this series oscillate between -1 and 0 and do not approach a fixed number as n gets large. Therefore, this series diverges.

TECHNOLOGY: Sequences and Sums

If we are given an explicit formula for a sequence, like $a_n = 3n^2 - 8$, we can use a TI-84 Plus to find specific terms in that sequence. First, press Y= and enter the formula for the sequence as Y1, using X,T,θ,n for n. See Figure 2.

NORMAL FLOAT AUTO REAL RADIAN MP
Plot1 Plot2 Plot3
\Y1=3X²−8
\Y2=
\Y3=
\Y4=
\Y5=
\Y6=
\Y7=

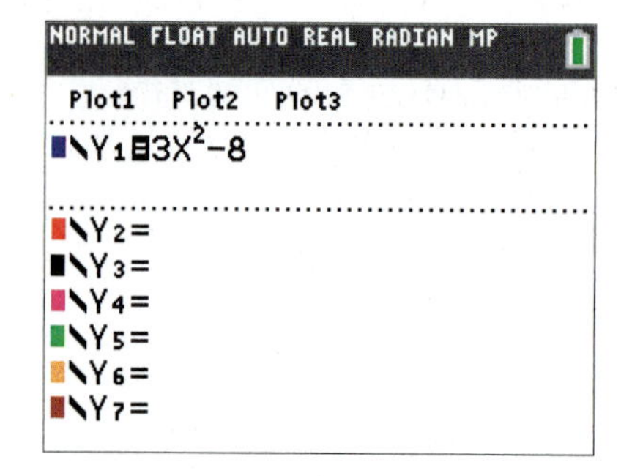

FIGURE 2

QUIT by pressing 2nd mode, and then press vars. Use the right arrow to highlight Y-VARS and press enter to select Function. Since we entered the sequence as Y1, press enter to select Y1. Then, use parentheses and enter the desired value for n. See Figure 3.

NORMAL FLOAT AUTO REAL RADIAN MP
Y1(19)
1075

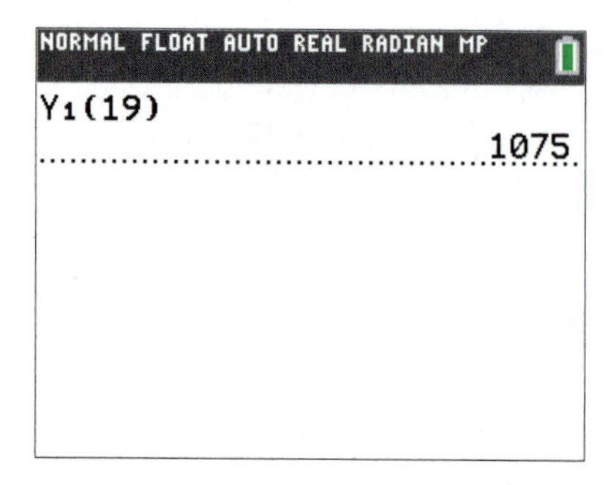

FIGURE 3

This tells us that the 19th term of the sequence, a_{19}, is 1075.

We can use a TI-84 Plus to evaluate partial sums as well as to find terms in a sequence. Suppose we wanted to evaluate $\sum_{i=2}^{7}\left(3i^2 - 8\right)$.

To find a sum, press 2nd stat and use the arrow key to highlight MATH. Then select sum(and press enter. See Figure 4.

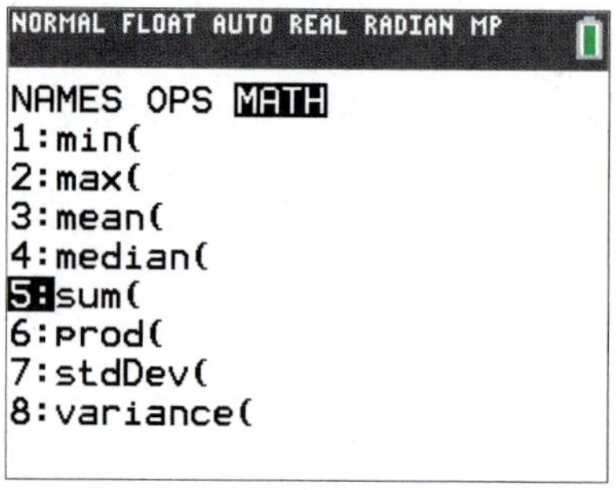

NORMAL FLOAT AUTO REAL RADIAN MP
NAMES OPS MATH
1:min(
2:max(
3:mean(
4:median(
5:sum(
6:prod(
7:stdDev(
8:variance(

FIGURE 4

In the parentheses for "sum" we need to enter the sequence $3i^2 - 8$. To do so, press 2nd stat, but this time highlight OPS, select seq(, and then press enter. See Figure 5.

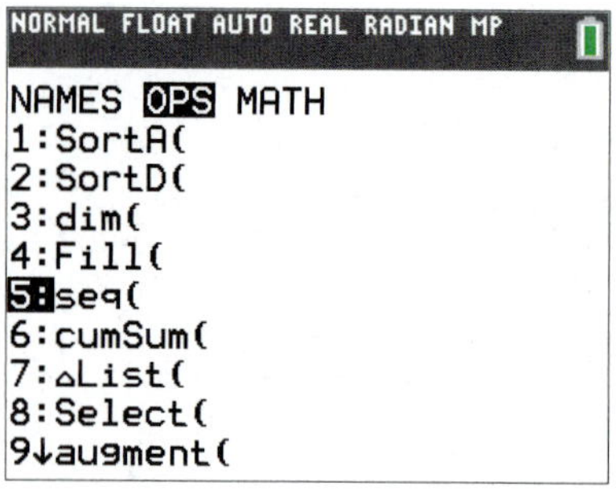

NORMAL FLOAT AUTO REAL RADIAN MP
NAMES OPS MATH
1:SortA(
2:SortD(
3:dim(
4:Fill(
5:seq(
6:cumSum(
7:ΔList(
8:Select(
9↓augment(

FIGURE 5

In the seq screen, type the sequence using X,T,θ,n in the Expr: line. The name of the variable is X, so press X,T,θ,n to enter X in the Variable line. Enter 2 for the start and 7 for the end. Type 1 for the step. Then select Paste and press enter. This will transfer our sequence into a single line command line. Press enter again to execute the command. See Figure 6.

NORMAL FLOAT AUTO REAL RADIAN MP
sum(seq(3X²−8,X,2,7,1)
369

FIGURE 6

So $\sum_{i=2}^{7}\left(3i^2 - 8\right) = 369$.

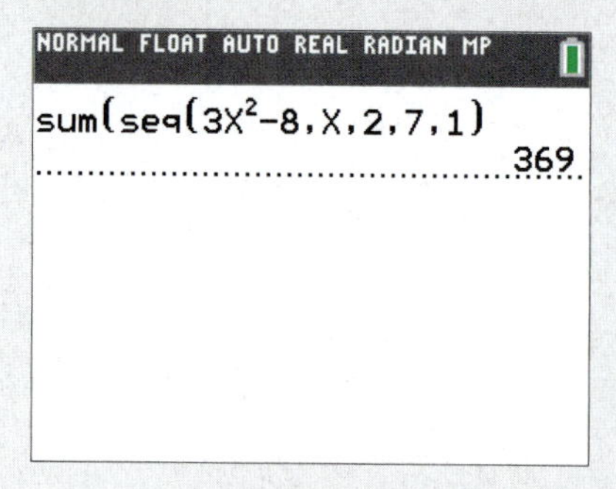

TOPIC 4: Fibonacci Sequences

One of the most famous sequences in mathematics has a long and colorful history. Leonardo Fibonacci (meaning "Leonardo, son of Bonaccio") was born in Pisa, in approximately the year 1175, and was one of the first of many famous Italian mathematicians of the Middle Ages. As a child and young adult, he traveled with his father (a merchant) to ports in northern Africa and the Middle East where Arab scholars had collected, preserved, and expanded the mathematics of many different cultures. Among other achievements, Leonardo is known for the book he wrote after returning to Italy, the *Liber abaci*, which exposed European scholars to Hindu-Arabic notation and the mathematics he had learned, as well as to his own contributions.

One of the problems in the *Liber abaci* gives rise to what we now call the **Fibonacci sequence**: $1,1,2,3,5,8,13,21,\ldots$. The problem, loosely translated, asks

How many pairs of rabbits can a single pair produce, if every month each pair of rabbits can beget a new pair and if each pair becomes productive in the second month of existence?

We can solve this problem using the sequence notation from this lesson. In the first month, we have just the initial pair of rabbits, so

$$a_1 = 1.$$

In the second month, this pair is still too young to reproduce, so again there is just one pair, meaning

$$a_1 = 1,\ a_2 = 1.$$

In the third month, the pair produces a new pair of rabbits, giving a total of 2:

$$a_1 = 1,\ a_2 = 1,\ a_3 = 2.$$

In the fourth month, we have 2 existing pairs, and one new pair (since the rabbits born in the third month are not yet old enough to reproduce), thus

$$a_1 = 1,\ a_2 = 1,\ a_3 = 2,\ a_4 = 3.$$

Can we now deduce a recursive formula for the number of pairs in the next month? For the fifth month ($n = 5$), the number of pairs will be equal to the number of existing pairs, which is equal to a_4, plus the number of new pairs. Since every pair that is at least 2 months old produces 1 new pair, this is equal to a_3. Thus, we have

$$\begin{aligned} a_5 &= a_4 + a_3 \\ &= 3 + 2 = 5 \end{aligned}$$

This logic applies to any month beginning with the third, so we can define the Fibonacci sequence recursively.

$$a_1 = 1,\ a_2 = 1 \text{ and } a_n = a_{n-1} + a_{n-2} \text{ for } n \geq 3$$

Generalized Fibonacci sequences are defined similarly, with the first two terms given and each successive term defined as the sum of the previous two.

12.1 EXERCISES

PRACTICE

Determine if each sequence is finite or infinite.

1. The sequence of odd numbers
2. $2, 4, 6, 8, 10, 12, \ldots$
3. $1, 3, 5, 7, 9, 11$
4. $3, 9, 27, 81, 243, \ldots$
5. The sequence of the days of the week
6. $1, 0, 1, 0, 1, 0, 1, 0, \ldots$
7. $1, 2, 3, 4, 5, 6, 7, \ldots$
8. The sequence of letters in the alphabet
9. The sequence of the number of ants in a colony recorded daily

Determine the first five terms of each sequence whose n^{th} term is defined as follows. See Examples 1 and 2.

10. $a_n = 7n - 3$
11. $a_n = -3n + 5$
12. $a_n = (-2)^n$
13. $a_n = \frac{3n}{n+2}$
14. $a_n = \frac{(-1)^n}{n^2}$
15. $a_n = \frac{(-1)^{n+1} 2^n}{3^n}$
16. $a_n = \left(-\frac{1}{3}\right)^{n-1}$
17. $a_n = \frac{n^2}{n+1}$
18. $a_n = \frac{(n-1)^2}{(n+1)^2}$
19. $a_n = \frac{n(n+1)}{2}\cos(n\pi)$
20. $a_n = (-2)^n + n$
21. $a_n = (-n+4)^3 - 1$
22. $a_n = \frac{2n^2}{3n-2}$
23. $a_n = (-1)^n \sqrt{n}$
24. $a_n = \frac{2^n}{n^2}$
25. $a_n = 4n - 3$
26. $a_n = -5n + 15$
27. $a_n = 2^{n-2}$
28. $a_n = 3^{-n-2}$
29. $a_n = (3n)^n$
30. $a_n = \sqrt[2^n]{64}$
31. $a_n = \frac{5n}{n+3}$
32. $a_n = \frac{n^2}{n+2}$
33. $a_n = \frac{n^2+n}{2}$
34. $a_n = (-1)^n n$
35. $a_n = \frac{(n+1)^2}{(n-1)^2}$
36. $a_n = n^2 + n$
37. $a_n = \frac{2n-1}{3n}$
38. $a_n = \sqrt{3n} + 1$
39. $a_n = -(n-1)^2$
40. $a_n = (n-1)(n+2)(n-3)$
41. $a_1 = 2$ and $a_n = (a_{n-1})^2$ for $n \geq 2$
42. $a_1 = -2$ and $a_n = 7a_{n-1} + 3$ for $n \geq 2$
43. $a_1 = 1$ and $a_n = na_{n-1}$ for $n \geq 2$
44. $a_1 = -1$ and $a_n = -a_{n-1} - 1$ for $n \geq 2$

45. $a_1 = 2$ and $a_n = \sqrt{(a_{n-1})^2 + 1}$ for $n \geq 2$

46. $a_n = n\sin\left(\frac{n\pi}{2}\right)$ **47.** $a_n = n^3 \sin\left(\frac{n\pi}{2}\right)$ **48.** $a_n = 2^n \cos(n\pi)$

Find a possible formula for the general n^{th} term of each sequence. Answers may vary. See Example 3.

49. $5, 12, 19, 26, 33, \ldots$

50. $-2, 4, -8, 16, -32, \ldots$

51. $-1, 2, -6, 24, -120, \ldots$

52. $\frac{1}{3}, \frac{2}{4}, \frac{3}{5}, \frac{4}{6}, \frac{5}{7}, \ldots$

53. $1, \frac{1}{4}, \frac{1}{9}, \frac{1}{16}, \frac{1}{25}, \ldots$

54. $1, \frac{1}{2}, \frac{1}{6}, \frac{1}{24}, \frac{1}{120}, \ldots$

55. $-34, -25, -16, -7, 2, \ldots$

56. $\frac{3}{14}, \frac{2}{15}, \frac{1}{16}, 0, -\frac{1}{18}, \ldots$

57. $\frac{1}{4}, \frac{1}{2}, 1, 2, 4, \ldots$

58. $-1, -6, -11, -16, -21, \ldots$

59. $\frac{1}{2}, \frac{1}{2}, \frac{3}{8}, \frac{1}{4}, \frac{5}{32}, \ldots$

60. $1, 4, 15, 64, 325, \ldots$

Translate each expanded sum that follows into summation notation, and vice versa. Then evaluate the sum. See Examples 4 and 5.

61. $\sum_{i=1}^{7}(3i - 5)$ **62.** $\sum_{i=1}^{5} -3i^2$ **63.** $1 + 8 + 27 + \cdots + 216$

64. $1 + 4 + 7 + \cdots + 22$ **65.** $\sum_{i=3}^{10} 5i^2$ **66.** $9 + 16 + 25 + \cdots + 81$

67. $\sum_{i=1}^{6} -3(2)^i$ **68.** $\sum_{i=6}^{13}(i+3)(i-10)$ **69.** $9 + 27 + 81 + \cdots + 19{,}683$

Find a formula for the nth partial sum S_n of each series. If the series is finite, determine the sum. If the series is infinite, determine if it converges or diverges, and if it converges, determine the sum. See Example 6.

70. $\sum_{i=1}^{100}\left(\frac{1}{i+3} - \frac{1}{i+4}\right)$ **71.** $\sum_{i=1}^{\infty}\left(\frac{1}{i+3} - \frac{1}{i+4}\right)$ **72.** $\sum_{i=1}^{\infty}\left(2^i - 2^{i-1}\right)$

73. $\sum_{i=1}^{15}\left(2^i - 2^{i-1}\right)$ **74.** $\sum_{i=1}^{49}\left(\frac{1}{2i} - \frac{1}{2i+2}\right)$ **75.** $\sum_{i=1}^{\infty}\left(\frac{1}{2i} - \frac{1}{2i+2}\right)$

76. $\sum_{i=1}^{100} \ln\left(\frac{i}{i+1}\right)$ (**Hint:** Make use of a property of logarithms to rewrite the sum.)

77. $\sum_{i=1}^{\infty} \ln\left(\frac{i}{i+1}\right)$ (**Hint:** Make use of a property of logarithms to rewrite the sum.)

78. $\sum_{i=1}^{30}\left(\frac{1}{2i+5} - \frac{1}{2i+7}\right)$ **79.** $\sum_{i=1}^{\infty}\left(\frac{1}{3i+1} - \frac{1}{3i+4}\right)$ **80.** $\sum_{i=1}^{65} \ln\left(\frac{i}{i+1}\right)$

Determine the first five terms of each generalized Fibonacci sequence.

81. $a_1 = 4$, $a_2 = 7$, and $a_n = a_{n-2} + a_{n-1}$ for $n \geq 3$

82. $a_1 = -9$, $a_2 = 1$, and $a_n = a_{n-2} + a_{n-1}$ for $n \geq 3$

83. $a_1 = 10$, $a_2 = 20$, and $a_n = a_{n-2} + a_{n-1}$ for $n \geq 3$

84. $a_1 = -17$, $a_2 = 13$, and $a_n = a_{n-2} + a_{n-1}$ for $n \geq 3$

85. $a_1 = 13$, $a_2 = -17$, and $a_n = a_{n-2} + a_{n-1}$ for $n \geq 3$

Determine the first five terms of each recursively defined sequence.

86. $a_1 = 2$, $a_2 = -3$, and $a_n = 3a_{n-1} + a_{n-2}$ for $n \geq 3$

87. $a_1 = 1$, $a_2 = -3$, and $a_n = a_{n-1}a_{n-2}$ for $n \geq 3$

88. $a_1 = 3$, $a_2 = 1$, and $a_n = \left(a_{n-2}\right)^{a_{n-1}}$ for $n \geq 3$

APPLICATIONS

89. Suppose you buy one cow and a number of bulls. In year one, your cow gives birth to a female calf and continues to bear another female calf every year for the rest of her life. Assuming that every calf born is female, that each cow begins calving in her third year (at age two), and that your cows never die, determine the number of cows (do not count the bulls) you will have at the end of the 14th year.

90. You borrow \$638 to buy a new car stereo. You plan to pay this sum back with monthly payments of \$74. The interest rate on your loan is 6% compounded monthly (recall that's 0.5% per month). Let A_n be the amount you owe at the end of the nth month. Find a recursive sequence to represent A_n. Use this sequence to find the amount owed after 4 months and the amount owed after 6 months. How many months will it take to pay off your loan?

WRITING & THINKING

91. Beginning with yourself, create a sequence describing the number of biological predecessors you have in each of the past 7 generations of your family.

92. The Fibonacci sequence is quite prevalent in nature. Do some research on your own to find an occurrence in nature (other than population growth) of the Fibonacci sequence.

12.2 ARITHMETIC SEQUENCES AND SERIES

TOPICS

1. Arithmetic sequences and series
2. The formula for the general term of an arithmetic sequence
3. Evaluating partial sums of arithmetic sequences

TOPIC 1: Arithmetic Sequences and Series

Suppose that the parents of a ten-year-old child decide to increase her \$1.00 / week allowance by \$0.50 / week with the start of each new year. The sequence describing her weekly allowance, beginning with age ten, is then

$$1.00, 1.50, 2.00, 2.50, 3.00, 3.50, \ldots$$

This type of sequence, in which the difference between any two consecutive terms is constant, is called an *arithmetic sequence.*

DEFINITION: Arithmetic Sequences

A sequence $\{a_n\}$ is an **arithmetic sequence** (also called an **arithmetic progression**) if there is a constant d such that $a_{n+1} - a_n = d$ for each $n = 1, 2, 3, \ldots$. The constant d is called the **common difference** of the sequence.

Since every sequence $\{a_n\}$ can be used to determine an associated series $a_1 + a_2 + a_3 + \cdots$, arithmetic series follow naturally from arithmetic sequences. We can prove that any nontrivial arithmetic series diverges.

If we denote the first term of the sequence a_1, the second term is then $a_1 + d$ (where d is the common difference), the third term is $(a_1 + d) + d = a_1 + 2d$, and so on.

So if we add up the first n terms of the sequence (that is, if we find the nth partial sum), we have

$$\begin{aligned} S_n &= a_1 + (a_1 + d) + (a_1 + 2d) + \cdots + (a_1 + (n-1)d) \\ &= \sum_{i=1}^{n} (a_1 + (i-1)d) \\ &= \sum_{i=1}^{n} a_1 + \sum_{i=1}^{n} (i-1)d \end{aligned}$$

Note the use of a property of Σ.

This may be enough to convince you that the partial sums are getting larger and larger in magnitude as n grows, since S_n consists of a_1 added to itself n times, plus a sum of multiples of d. However, having an explicit formula for the nth partial sum is useful, so we will simplify further.

$$S_n = \sum_{i=1}^{n} a_1 + \sum_{i=1}^{n} (i-1)d$$

$$= a_1 \sum_{i=1}^{n} 1 + d \sum_{i=1}^{n} (i-1)$$ Factor out constants using a property of sigma notation.

$$= na_1 + d\left(0+1+2+\cdots+(n-1)\right)$$ Evaluate the first term using a summation formula. Writing out the second term, we see it can be written using a different index of summation.

$$= na_1 + d\sum_{i=1}^{n-1} i$$

$$= na_1 + d\left(\frac{(n-1)n}{2}\right)$$ Apply another summation formula.

It is clear that the sequence of partial sums does not approach a fixed number S, except in the trivial case when $a_1 = 0$ and $d = 0$. Thus, except for the series $0+0+0+\cdots$, every arithmetic series diverges.

TOPIC 2: The Formula for the General Term of an Arithmetic Sequence

As with all sequences, an explicit formula for the general n^{th} term of an arithmetic sequence is very useful. We have already noted that if a_1 is the first term of an arithmetic sequence, and if the common difference is d, then $a_1 + d$ is the second term, $a_1 + 2d$ is the third term, and so on. This pattern is summarized below.

f(x) FORMULA: General Term of an Arithmetic Sequence

The explicit formula for the **general n^{th} term of an arithmetic sequence** is

$$a_n = a_1 + (n-1)d,$$

where d is the common difference for the sequence.

Example 1: General Term of an Arithmetic Sequence

Find the formula for the general n^{th} term of each arithmetic sequence.

a. $-3, 2, 7, 12, \ldots$

b. $a_1 = \frac{1}{3}$ and $a_4 = \frac{11}{6}$

c. $a_7 = -8$ and $d = -3$

NOTE

An arithmetic sequence can be defined by two terms, or by one term and the common difference.

Solution

a. First, find d, the difference, by calculating the difference between any two consecutive terms.

$$d = 2 - (-3) = 5$$

Since a_1 is listed in the sequence ($a_1 = -3$), we have all the information we need.

$$a_n = a_1 + (n-1)d$$

$$= -3 + (n-1)(5)$$ Substitute the known values.

$$= 5n - 8$$ Simplify.

b. Here, two nonconsecutive terms are given. Since $a_1 = \frac{1}{3}$, we can use the formula for the nth term to find d.

$$a_4 = a_1 + (4-1)d \quad \text{Write the formula for } n = 4.$$
$$\frac{11}{6} = \frac{1}{3} + 3d \quad \text{Substitute the known values.}$$
$$\frac{3}{2} = 3d \quad \text{Simplify.}$$
$$d = \frac{1}{2}$$

Now, substitute back in to the formula to find the general n^{th} term.

$$a_n = \frac{1}{3} + (n-1)\left(\frac{1}{2}\right) \quad \text{Substitute the known values.}$$
$$= \frac{1}{2}n - \frac{1}{6} \quad \text{Simplify.}$$

c. Similarly, we have enough information to use the formula to find a_1.

$$a_7 = a_1 + (7-1)d$$
$$-8 = a_1 + 6(-3) \quad \text{Substitute the known values.}$$
$$-8 = a_1 - 18 \quad \text{Simplify.}$$
$$a_1 = 10$$

Substitute back in to the formula to find a_n.

$$a_n = 10 + (n-1)(-3)$$
$$= -3n + 13$$

Example 2: Modeling Population Growth

A demographer models the population growth of a small town as an arithmetic progression. He knows that the population in 2012 was 12,790 and that in 2015 the population was 13,150. He wants to treat the population in 2010 as the first term of the arithmetic progression. What is the sought-after formula?

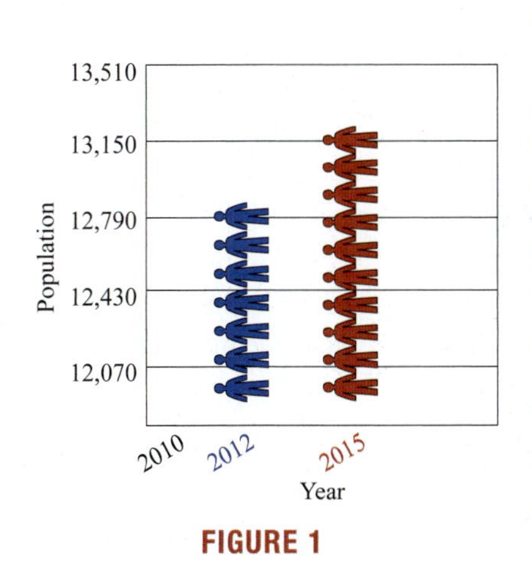

FIGURE 1

Solution

We know that a_1 represents the population in 2010. Similarly, a_2 represents the population in 2011 and a_3 represents the population in 2012.

The known information gives us $a_3 = 12{,}790$ and $a_6 = 13{,}150$ (since the population in 2015 is 13,150). If we let d represent the common difference, the population increases by $3d$ between 2012 and 2015:

$$3d = 13{,}150 - 12{,}790 = 360,$$
$$\text{so } d = 120.$$

Now, we can use the fact that $a_3 = 12,790$ to determine a_1,

$$\begin{aligned} a_3 &= a_1 + 2d \\ 12,790 &= a_1 + 2(120) \\ 12,790 - 240 &= a_1 \\ a_1 &= 12,550 \end{aligned}$$

Putting this information together, we have the desired formula:

$$\begin{aligned} a_n &= 12,550 + 120(n-1) \\ &= 12,430 + 120n \end{aligned}$$

Note that to apply the formula, we need to remember $n = 1$ corresponds to 2010, $n = 2$ corresponds to 2011, and so on.

TOPIC 3: Evaluating Partial Sums of Arithmetic Sequences

To begin our study of partial sums of arithmetic sequences, we will return to the ten-year-old girl whose weekly allowance grows in an arithmetic progression.

Example 3: Partial Sums of Arithmetic Sequences

Melinda, a ten-year-old, knows that her parents plan to increase her weekly allowance by \$0.50/week with the start of each new year, and she currently receives \$1.00/week. Assuming that Melinda receives an allowance every week from the current year up through the year she turns 18, what is her total take?

Solution

In the current year, Melinda will receive

$$(52)(\$1.00) = \$52.00.$$

The next year, when Melinda is 11, she will receive

$$(52)(\$1.50) = \$78.00.$$

The sum of these two amounts, \$130.00, is her total allowance during her 10th and 11th years. Extending this reasoning through Melinda's 18th year, we make use of the partial sum formula we already derived:

$$S_n = na_1 + d\left(\frac{(n-1)n}{2}\right)$$

We know that $a_1 = 52$ since her total allowance during the first year is \$52.00. Since her weekly allowance increases \$0.50 every year, Melinda's yearly allowance increases by $52(\$0.50) = \$26.00$ each year, so $d = 26$.

We then use the formula to calculate S_9, since Melinda's 18th year will be her 9th year of receiving allowance.

$$S_9 = 9 \cdot 52 + (26)\left(\frac{(9-1)\cdot 9}{2}\right) = 468 + 936 = 1404$$

Melinda will thus receive a total of $1404 in allowance money from age 10 through age 18.

We can rewrite the partial sum formula in a way that is useful when d has not yet been found. Recall that the formula for the general term of an arithmetic sequence is

$$a_n = a_1 + (n-1)d.$$

Rewriting this, we have $(n-1)d = a_n - a_1$, and we can use this fact in our partial sum formula to eliminate d, as follows:

$S_n = na_1 + d\left(\frac{(n-1)n}{2}\right)$	The known partial sum formula.
$= na_1 + \left(\frac{n}{2}\right)(n-1)d$	Rewrite the second term to isolate $(n-1)d$.
$= na_1 + \left(\frac{n}{2}\right)(a_n - a_1)$	Substitute.
$= \left(\frac{n}{2}\right)(a_1 + a_n)$	Simplify.

f(x) FORMULA: Partial Sums of Arithmetic Sequences

Let $\{a_n\}$ be an arithmetic sequence with common difference d. The **sum of the first *n* terms** (the nth partial sum) is given by both

$$S_n = na_1 + d\left(\frac{(n-1)n}{2}\right)$$

and

$$S_n = \left(\frac{n}{2}\right)(a_1 + a_n).$$

In practice, the choice of formula is based on whether it is easier to determine d or the last term, a_n, in the sum to be evaluated.

Example 4: Partial Sums of Arithmetic Sequences

Determine the sum of the first 100 positive odd integers.

Solution

Listing the first few positive odd integers, 1, 3, 5, 7, …, we can see that the nth odd number is equal to $2n-1$.

TECHNOLOGY

```
NORMAL FLOAT AUTO REAL RADIAN MP
sum(seq(2X-1,X,1,100,1)
                          10000
```

Thus, we have $a_1 = 1$, $d = 2$, $n = 100$, and $a_n = a_{100} = 2(100) - 1 = 199$. Given this information, we can use either partial sum formula to find the answer.

Using the first formula, $S_{100} = 100(1) + 2\left(\frac{(100-1)(100)}{2}\right)$

$$= 100 + 2\left(\frac{99 \cdot 100}{2}\right) = 10{,}000.$$

Using the second formula, $S_{100} = \left(\frac{100}{2}\right)(1 + 199) = 50 \cdot 200 = 10{,}000.$

12.2 EXERCISES

PRACTICE

Find the explicit formula for the general n^{th} term of each arithmetic sequence. See Example 1.

1. $-2, 1, 4, 7, 10, \ldots$

2. $5, 7, 9, 11, 13, \ldots$

3. $7, 5, 3, 1, -1, \ldots$

4. $a_2 = 14$ and $a_3 = 19$

5. $a_1 = 5$ and $a_5 = 41$

6. $a_2 = 13$ and $a_4 = 21$

7. $a_3 = -9$ and $d = -6$

8. $a_{12} = 43$ and $d = 3$

9. $a_5 = 100$ and $d = 19$

10. $-37, -20, -3, 14, 31, \ldots$

11. $\frac{7}{2}, \frac{9}{2}, \frac{11}{2}, \frac{13}{2}, \frac{15}{2}, \ldots$

12. $15, 11, 7, 3, -1, \ldots$

13. $a_1 = 12$ and $a_3 = -7$

14. $a_{73} = 224$ and $a_{75} = 230$

15. $a_1 = -1$ and $a_6 = -11$

16. $a_5 = -\frac{5}{2}$ and $d = \frac{3}{2}$

17. $a_4 = 17$ and $d = -4$

18. $a_{34} = -71$ and $d = -2$

Determine if each of the following sequences is arithmetic. If so, find the common difference.

19. The sequence of even numbers

20. $1, 2, 4, 7, 11, 16, \ldots$

21. $1, 2, 3, 4, 5, 6, 7, \ldots$

22. The Fibonacci sequence

23. $1, 2, 4, 8, 16, 32, \ldots$

24. $42, 38, 34, 30, 26, 22, \ldots$

25. $0, 1, 0, 2, 0, 3, 0, 4, \ldots$

26. $12, 12, 12, 12, 12, \ldots$

Given the initial term and the common difference, find the value of the 7th term of each of the arithmetic sequences.

27. $a_1 = 1$ and $d = 2$

28. $a_1 = 4$ and $d = -3$

29. $a_1 = 0$ and $d = \frac{1}{3}$

30. $a_1 = 3$ and $d = \pi$

31. $a_1 = 8$ and $d = -1$

32. $a_1 = \frac{1}{2}$ and $d = 3$

Given two terms, find the common difference and the first five terms of each of the arithmetic sequences.

33. $a_1 = 5$ and $a_2 = 7.5$

34. $a_6 = 27$ and $a_9 = 42$

35. $a_7 = 49$ and $a_{11} = 77$

36. $a_4 = 76$ and $a_8 = 156$

37. $a_5 = -26$ and $a_9 = 10$

38. $a_8 = 45$ and $a_{10} = 53$

Find the common difference of each of the following sequences. See Example 1.

39. $\{5n-3\}$

40. $\left\{3n - \frac{1}{2}\right\}$

41. $\{n+6\}$

42. $\{1-4n\}$

43. $\{\sqrt{2} - 2n\}$

44. $\{n\sqrt{3} + 5\}$

Use the given information about each arithmetic sequence to answer the question.

45. Given that $a_1 = -3$ and $a_5 = 5$, what is a_{100}?

46. In the sequence $24, 43, 62, \ldots$, which term is 955?

47. In the sequence $1, \frac{4}{3}, \frac{5}{3}, \ldots$, which term is 25?

48. Given that $a_5 = -\frac{5}{3}$ and $a_9 = 1$, what is a_{62}?

49. In the sequence $-16, -9, -2, \ldots$, what is a_{20}?

50. In the sequence $\frac{1}{4}, \frac{7}{16}, \frac{5}{8}, \ldots$, which term is $\frac{35}{8}$?

51. In the sequence $2, 5, 8, 11, \ldots$, what is the 9th term?

52. In the sequence $1, 3, 5, 7, \ldots$, what is the 6th term?

53. In the sequence $16, 12, 8, 4, \ldots$, what is the 7th term?

54. In the sequence $\frac{1}{2}, 2, \frac{7}{2}, 5, \ldots$, what is the 8th term?

55. In the sequence $-2, 1, 4, 7, \ldots$, what is the 6th term?

56. In the sequence $9, 6, 3, 0, \ldots$, what is the 10th term?

57. In the sequence $5, 10, 15, 20, \ldots$, what is the 11th term?

58. In the sequence $2\sqrt{2}, 4\sqrt{2}, 6\sqrt{2}, 8\sqrt{2}, \ldots$, what is the 7th term?

Find the value of the partial sum of each arithmetic sequence. See Example 4.

59. $\sum_{i=1}^{100}(3i-8)$ **60.** $\sum_{i=1}^{50}(-2i+5)$ **61.** $\sum_{i=5}^{90}(4i+9)$

62. $3+11+\cdots+795$ **63.** $25+18+\cdots+(-143)$ **64.** $-12+2+\cdots+674$

65. $\sum_{i=1}^{37}\left(-\frac{3}{5}i-6\right)$ **66.** $\sum_{i=100}^{200}(3i+57)$ **67.** $\sum_{i=2}^{42}(2i-22)$

68. $-90+(-77)+\cdots+92$ **69.** $7+3+\cdots+(-101)$ **70.** $4+\frac{81}{20}+\cdots+900$

APPLICATIONS

71. Cynthia borrows $21,000, interest-free, from her parents to help pay for her college education, and promises that upon graduation she will pay back the sum beginning with $1000 the first year and increasing the amount by $1000 with each successive year. How many years will it take for her to repay the entire $21,000?

72. A certain theatre is shaped so that the first row has 30 seats, and, moving toward the back, each successive row has two seats more than the previous one. If there are 40 rows, how many seats does the last row contain? How many seats are there altogether?

73. A brick mason spends a morning moving a pile of bricks from his truck to the work site by wheelbarrow. Each brick weighs two pounds, and on his first trip he transports 100 pounds. On each successive trip, as he tires, he decides to move one less brick. How many pounds of bricks has he transported after 20 trips?

74. The manager of a grocery store decides to create a display of soup cans by placing cans in a row on the floor and then stacking successive rows so that each level of the tower has one less can than the one below it. The manager wants the top row to have 5 cans, and the store has 290 cans that can be used for the display. If all of the cans are used, how many rows will the display have?

75. A man decides to lease a car and is told that his payment to the car dealership will be $50 in the first month. He is also told that every month thereafter, for the next 60 months, his payments will increase by $25. How much is his monthly payment after two years? How much has he paid in total after the first two years?

76. Your grandmother doesn't trust banks, so she decided to save for your college education by periodically adding money to a mason jar buried in her flower garden. She began the practice with \$65 and added \$15 every time she got her monthly paycheck. If she continued this routine for 18 years, how much money did she manage to save for you?

TECHNOLOGY

Use a graphing utility to evaluate each of the following sums.

77. $1.2+2.8+4.4+6.0+7.6+\cdots+28.4$

78. $5+\frac{5}{2}+0+\frac{-5}{2}+\cdots+\frac{-45}{2}$

79. $\sum_{i=1}^{25} 89.47-7.35i$

80. $\sum_{i=1}^{32} 4.12i+17.54$

81. $\sum_{i=49}^{83} (4.37i+8.21)$

82. $\sum_{i=23}^{79} \left(\frac{256i}{397}+\frac{57}{481}\right)$

12.3 GEOMETRIC SEQUENCES AND SERIES

TOPICS

1. Geometric sequences
2. The formula for the general term of a geometric sequence
3. Evaluating partial sums of geometric sequences
4. Evaluating infinite geometric series
5. Zeno's paradoxes

TOPIC 1: Geometric Sequences

Suppose Marilyn is nearing the end of a job hunt, and has a choice of two positions. Both offer a starting salary of $40,000 per year, but they differ in their projected future salaries. Employer A offers a yearly salary increase of $1250 per year. Employer B offers an increase of 3% per year for all employees. Assuming the positions are equally desirable in all other respects, which one should Marilyn choose?

This question doesn't have a simple answer. One important consideration is the number of years Marilyn anticipates working for her next employer. A table comparing the two projected salaries over the next decade will help us determine an answer. To construct this table, we need a formula for the salary of each company each year.

We know how to find such a formula for Employer A, as the sequence of yearly salaries forms an arithmetic progression with a common difference of $1250; this means the nth term of the sequence is given by

$$a_n = 40,000 + (n-1)1250.$$

The nth term of the sequence of salaries for Employer B is of a different form. Since each successive salary is 3% greater than the preceding one, the recursive formula

$$b_1 = 40,000 \text{ and } b_n = b_{n-1} + (0.03)b_{n-1} \text{ for } n \geq 2$$

describes the sequence of salaries. We can then simplify the formula by writing $b_n = (1.03)b_{n-1}$ for $n \geq 2$. With these two formulas, we can construct the table comparing salaries seen in Table 1 (all salaries are rounded to the nearest dollar).

	1	2	3	4	5	6	7	8	9	10
a_n	40,000	41,250	42,500	43,750	45,000	46,250	47,500	48,750	50,000	51,250
b_n	40,000	41,200	42,436	43,709	45,020	46,371	47,762	49,195	50,671	52,191

TABLE 1: Comparison of Salaries of Employers A and B in Year n

The first thing we notice is that Employer A offers a higher yearly salary up through the fourth year, but that thereafter Employer B pays more. So if Marilyn anticipates staying in her next job for five years or more, she may want to go with Employer B.

This reasoning doesn't consider the *accumulated* salary through year n; for a more accurate comparison of the two jobs, we need to compare the *partial sums*. We will return to the question of partial sums later in this section.

The salary sequence for Employer B is an example of a *geometric sequence*, and its identifying characteristic is that the *ratio* of consecutive terms in the sequence is a fixed constant. As we noted, the ratio of any term in the sequence to the preceding term is 1.03. (Remember that for arithmetic sequences, the *difference* between consecutive terms is fixed.)

DEFINITION: Geometric Sequences

A sequence $\{a_n\}$ is a **geometric sequence** (also called a **geometric progression**) if there is a constant $r \neq 0$ so that $\frac{a_{n+1}}{a_n} = r$ for each $n = 1, 2, 3, \ldots$. The constant r is called the **common ratio** of the sequence.

TOPIC 2: The Formula for the General Term of a Geometric Sequence

We can develop an explicit formula for the general n^{th} term of a geometric sequence by relating each term to a_1, the first term of the sequence, just as we did for arithmetic sequences.

Since the sequence is geometric, there is some fixed number r such that $\frac{a_{n+1}}{a_n} = r$ for each $n = 1, 2, 3, \ldots$. In particular,

$$\frac{a_2}{a_1} = r, \text{ or } a_2 = a_1 r.$$

Similarly,

$$\frac{a_3}{a_2} = r, \text{ so } a_3 = a_2 r = (a_1 r) r = a_1 r^2.$$

Extending this pattern, we arrive at an explicit formula for a_n.

FORMULA: General Term of a Geometric Sequence

The explicit formula for the **general n^{th} term of a geometric sequence** is

$$a_n = a_1 r^{n-1},$$

where r is the common ratio for the sequence.

Given this information, we now know that the yearly salary offered by Employer B is given by the formula

$$b_n = (40{,}000)(1.03)^{n-1}.$$

Example 1: General Term of a Geometric Sequence

Find the formula for the general n^{th} term of each geometric sequence.

a. $\frac{1}{3}, \frac{1}{9}, \frac{1}{27}, \ldots$ **b.** $3, -6, 12, \ldots$ **c.** $a_4 = \frac{5}{16}$ and $r = \frac{1}{2}$

Solution

a. First, use any two consecutive terms to determine the common ratio r.

$$r = \frac{\left(\frac{1}{9}\right)}{\left(\frac{1}{3}\right)} = \frac{3}{9} = \frac{1}{3}$$

Since we know that $a_1 = \frac{1}{3}$, the explicit formula is as follows.

$$a_n = a_1 r^{n-1}$$

$$= \frac{1}{3}\left(\frac{1}{3}\right)^{n-1} = \left(\frac{1}{3}\right)^n \qquad \text{Substitute the known values and simplify.}$$

b. Using the same process as before, we know $a_1 = 3$ and can calculate r.

$$r = \frac{-6}{3} = -2$$

Applying the formula, we have the following.

$$a_n = 3(-2)^{n-1}$$

c. Here we are given r and another term. We can use the formula to find a_1.

$$a_4 = a_1(r)^{4-1} \qquad \text{Write out the explicit formula for } n = 4.$$

$$\frac{5}{16} = a_1\left(\frac{1}{2}\right)^3 \qquad \text{Substitute the known values.}$$

$$\frac{5}{16} = a_1\left(\frac{1}{8}\right)$$

$$a_1 = \frac{5}{2} \qquad \text{Simplify and solve for } a_1.$$

We then substitute a_1 and r to find the explicit formula.

$$a_n = \frac{5}{2}\left(\frac{1}{2}\right)^{n-1}$$

Example 2: General Term of a Geometric Sequence

Given that the second term of a geometric sequence is –6 and the fifth term is 162, what is the fourth term?

Solution

Given that the sequence is geometric, we know that

$$-6 = a_1 r \text{ and } 162 = a_1 r^4.$$

We can eliminate a_1 from this pair of equations by dividing.

$$\frac{162}{-6} = \frac{a_1 r^4}{a_1 r}, \text{ so } r^3 = -27 \text{ and thus } r = -3$$

Given r, we can find a_1 using one of the equations above.

$$\begin{aligned} -6 &= a_1 r \\ a_1 &= \frac{-6}{-3} = 2 \end{aligned}$$

So the general term is given by $a_n = 2(-3)^{n-1}$, and we find the fourth term by plugging in $n = 4$.

$$a_4 = 2(-3)^{4-1} = -54$$

TOPIC 3: Evaluating Partial Sums of Geometric Sequences

Given a geometric sequence, many applications require us to compute the partial sum $S_n = a_1 + a_2 + \cdots + a_n$ for some value of n. We will now derive an explicit formula for the nth partial sum.

First, we use the formula for the nth term of a geometric sequence to rewrite S_n.

$$\begin{aligned} S_n &= a_1 + a_2 + \cdots + a_n \\ &= a_1 + a_1 r + a_1 r^2 + \cdots + a_1 r^{n-1} \end{aligned}$$

If we multiply both sides of this equation by r, we obtain

$$rS_n = a_1 r + a_1 r^2 + a_1 r^3 + \cdots + a_1 r^n.$$

If we now subtract the second equation from the first, most of the terms from the right-hand sides of the equations cancel.

$$\begin{aligned} S_n - rS_n &= \left(a_1 + \cancel{a_1 r} + \cancel{\cdots} + \cancel{a_1 r^{n-1}}\right) - \left(\cancel{a_1 r} + \cancel{a_1 r^2} + \cancel{\cdots} + a_1 r^n\right) \\ &= a_1 - a_1 r^n \end{aligned}$$

The goal is to solve for S_n, and we can do this by factoring the left and right sides of the equation and dividing.

$$S_n(1-r) = a_1(1-r^n)$$

$$S_n = \frac{a_1(1-r^n)}{1-r}$$

f(x) FORMULA: Partial Sum of a Geometric Sequence

Let $\{a_n\}$ be a geometric sequence with common ratio r, and assume r is neither 0 nor 1. Then **the nth partial sum of the sequence,** $S_n = a_1 + a_1r + a_1r^2 + \cdots + a_1r^{n-1}$, is given by

$$S_n = \frac{a_1(1-r^n)}{1-r}.$$

The condition $r \neq 1$ is necessary to prevent division by zero in the formula, but it isn't difficult to determine the partial sum without the formula if r happens to be 1. In this case, $S_n = a_1 + a_1 + \cdots + a_1 = na_1$.

With this formula in hand, along with one of the two partial sum formulas for arithmetic sequences, let us return to the question with which we began this section.

Example 3: Comparing Salary Offers

Recall that Marilyn has the option of taking a job with Employer A, who offers a yearly salary increase of \$1250, or Employer B, who offers a yearly salary increase of 3%. Given that the starting salary is \$40,000 for both, in what year does the accumulated salary paid by Employer B overtake the accumulated salary paid by Employer A?

Solution

We have already determined that Employer B pays a higher salary beginning with the fifth year of employment, but we need to compare the partial sums of the salaries paid by the two employers up through year n to answer the question above. Recall that the sequence of yearly salaries paid by Employer A is defined by

$$a_n = 40{,}000 + (n-1)1250$$

and that the sequence of yearly salaries paid by Employer B is defined by

$$b_n = 40{,}000(1.03)^{n-1}.$$

Recall also that one partial sum formula for an arithmetic sequence is

$$S_n = na_1 + d\left(\frac{(n-1)n}{2}\right),$$

where d is the common difference for the sequence.

In order to keep the partial sums straight, let A_n denote the sum of the salaries paid by Employer A through year n, and let B_n be the same for Employer B. Then

$$A_n = n(40,000) + 1250\left(\frac{(n-1)n}{2}\right) = 40,000n + 625(n^2 - n) = 625n^2 + 39,375n$$

and

$$B_n = \frac{40,000(1-1.03^n)}{1-1.03} = \left(\frac{40,000}{0.03}\right)(1.03^n - 1).$$

We can now use these formulas to compute the accumulated salaries paid by the two employers up through year n, as shown in Table 2.

	1	2	3	4	5	6	7	8
A_n	40,000	81,250	123,750	167,500	212,500	258,750	306,250	355,000
B_n	40,000	81,200	123,636	167,345	212,365	258,736	306,498	355,693

TABLE 2: Accumulated Salaries through Year 8

Comparing values in the table indicates that the accumulated salary paid by Employer B overtakes that of Employer A in the seventh year.

We can also use the geometric partial sum formula to evaluate certain expressions defined with the sigma notation, as shown in Example 4.

Example 4: Partial Sum of a Geometric Sequence

Evaluate $\sum_{i=2}^{7} 5\left(-\frac{1}{2}\right)^i$.

Solution

This is a partial sum of a geometric sequence, but as it is written, the first term and the common ratio of the sequence are not apparent. A good way to begin is to write the sum in expanded form.

$$\begin{aligned}\sum_{i=2}^{7} 5\left(-\frac{1}{2}\right)^i &= 5\left(-\frac{1}{2}\right)^2 + 5\left(-\frac{1}{2}\right)^3 + \cdots + 5\left(-\frac{1}{2}\right)^7 \\ &= \frac{5}{4} - \frac{5}{8} + \cdots - \frac{5}{128}\end{aligned}$$

We can see that $a_1 = \frac{5}{4}$ and that the common ratio is $r = -\frac{1}{2}$. Since there are six terms in the sum, we let $n = 6$ in the partial sum formula. Putting this all together, we have the following.

TECHNOLOGY

```
NORMAL FLOAT AUTO REAL RADIAN MP
sum(seq(5(-1/2)^X,X,2,7,1)
                    0.8203125
Ans▸Frac
                      105/128
```

This screen was obtained using the summation and sequence commands on a TI-84 Plus, as described in Section 12.1. To find the answer in fraction form, press math and select ▸Frac.

$$S_6 = \frac{a_1\left(1-r^6\right)}{1-r}$$ Write the formula for the 6th partial sum.

$$= \frac{\left(\frac{5}{4}\right)\left(1-\left(-\frac{1}{2}\right)^6\right)}{1-\left(-\frac{1}{2}\right)}$$ Substitute the known values.

$$= \frac{\left(\frac{5}{4}\right)\left(1-\frac{1}{64}\right)}{\left(\frac{3}{2}\right)}$$ Simplify.

$$= \left(\frac{5}{4}\right)\left(\frac{63}{64}\right)\left(\frac{2}{3}\right) = \frac{105}{128}$$

TOPIC 4: Evaluating Infinite Geometric Series

In Section 12.1, we used an intuitive approach to think about the result of adding half an inch to a quarter of an inch to an eighth of an inch, and so on. We came to the conclusion that if the process were continued indefinitely, the sum of the fractions would be 1. That is,

$$\frac{1}{2}+\frac{1}{4}+\frac{1}{8}+\cdots = 1, \text{ or } \sum_{i=1}^{\infty}\left(\frac{1}{2}\right)^i = 1.$$

Our reasoning was that the sequence of partial sums of the sequence

$$\left\{\left(\frac{1}{2}\right)^n\right\}$$

never exceeded 1, yet got closer and closer to 1 as more terms were added.

We now have the machinery to analyze infinite series like the one above more rigorously. For one thing, we know that the sequence

$$\frac{1}{2}, \frac{1}{4}, \frac{1}{8}, \ldots$$

is a geometric sequence, and that both the first term and the common ratio are $\frac{1}{2}$. We can use the partial sum formula, substituting $a_1 = \frac{1}{2}$ and $r = \frac{1}{2}$.

$$S_n = \frac{1}{2}+\frac{1}{4}+\cdots+\frac{1}{2^n} = \frac{\left(\frac{1}{2}\right)\left(1-\left(\frac{1}{2}\right)^n\right)}{\left(1-\frac{1}{2}\right)} = 1-\left(\frac{1}{2}\right)^n$$

If we write S_n as a single fraction, we obtain

$$S_n = \frac{2^n - 1}{2^n},$$

the formula we intuitively derived in Section 12.1. In either form, we can see that S_n approaches 1 as $n \to \infty$, but we can now generalize this observation for all convergent geometric series.

f(x) FORMULA: Sum of an Infinite Geometric Series

If $|r| < 1$, the infinite geometric series

$$\sum_{n=1}^{\infty} a_1 r^{n-1} = a_1 + a_1 r + a_1 r^2 + a_1 r^3 + \cdots$$

converges, and the sum of the series is given by $S = \frac{a_1}{1-r}$. We can use sigma notation with the index beginning at 0 to write the same series; in this form,

$$\sum_{n=0}^{\infty} a_1 r^n = a_1 + a_1 r + a_1 r^2 + a_1 r^3 + \cdots = \frac{a_1}{1-r}.$$

The proof of this result follows from the fact that if $|r| < 1$, then $r^n \to 0$ as $n \to \infty$, so $S_n = \frac{a_1\left(1 - r^n\right)}{1-r} \to \frac{a_1}{1-r}$ as $n \to \infty$.

In the case of the series $\frac{1}{2} + \frac{1}{4} + \frac{1}{8} + \cdots$, where $a_1 = \frac{1}{2}$ and $r = \frac{1}{2}$, the sum is

$$S = \frac{\frac{1}{2}}{1 - \frac{1}{2}} = \frac{\frac{1}{2}}{\frac{1}{2}} = 1.$$

Example 5: Infinite Geometric Series

Find the sums of the following series.

a. $\sum_{n=1}^{\infty} 5\left(-\frac{1}{2}\right)^{n-1}$ **b.** $\sum_{n=1}^{\infty} 3\left(\frac{1}{10}\right)^{n}$

Solution

a. First, we identify the values of a_1 and r. Plugging in $n = 1$, we see that $a_1 = 5$, and from the summation formula, we know $r = -\frac{1}{2}$.

Since the common ratio r is less than 1 in magnitude, we can apply the formula.

$$S = \frac{a_1}{1-r}$$

$$= \frac{5}{1 - \left(-\frac{1}{2}\right)} = \frac{5}{\left(\frac{3}{2}\right)} = \frac{10}{3}$$

Substitute the known values, then simplify.

b. It is important to note exactly how a series is written. The formula $S = \frac{a_1}{1-r}$ applies directly to two forms of summation notation:

$$\sum_{n=1}^{\infty} a_1 r^{n-1} \quad \text{or} \quad \sum_{n=0}^{\infty} a_1 r^n$$

In this case, the form of the series doesn't quite match either of these forms, so we write out the first few terms to identify the first term of the series and the common ratio.

$$\sum_{n=1}^{\infty} 3\left(\frac{1}{10}\right)^n = \frac{3}{10} + \frac{3}{100} + \frac{3}{1000} + \cdots$$

From this, we can see that $a_1 = \frac{3}{10}$ and $r = \frac{1}{10}$. Plugging these values in, we have the following.

$$\begin{aligned} S &= \frac{a_1}{1-r} \\ &= \frac{\left(\frac{3}{10}\right)}{1-\frac{1}{10}} = \frac{\left(\frac{3}{10}\right)}{\left(\frac{9}{10}\right)} = \frac{1}{3} \end{aligned}$$

Example 5 illustrates an important relation between geometric series and the decimal system we use to write real numbers. Note that we have shown that

$$\sum_{n=1}^{\infty} 3\left(\frac{1}{10}\right)^n = \frac{3}{10} + \frac{3}{100} + \frac{3}{1000} + \cdots = 0.33\overline{3}$$

is the decimal representation of the fraction $\frac{1}{3}$. Similarly,

$$0.\overline{52} = \frac{52}{100} + \frac{52}{10{,}000} + \frac{52}{1{,}000{,}000} + \cdots = \sum_{n=1}^{\infty} 52\left(\frac{1}{100}\right)^n$$

can be written in fractional form by noting that for this series $a_1 = \frac{52}{100}$ and $r = \frac{1}{100}$, so

$$0.\overline{52} = \frac{\frac{52}{100}}{1-\frac{1}{100}} = \frac{52}{99}.$$

TOPIC 5: Zeno's Paradoxes

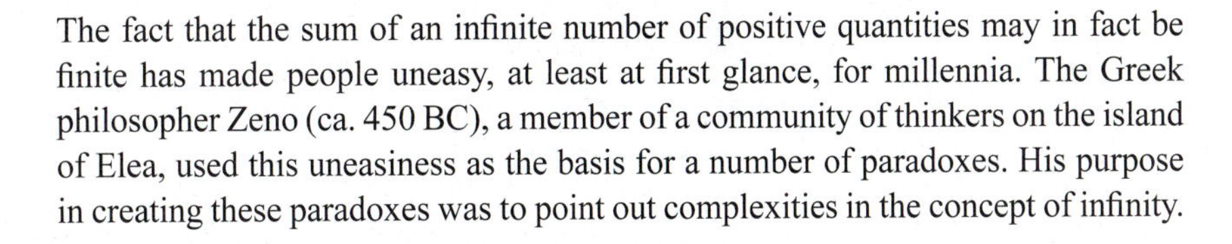

The fact that the sum of an infinite number of positive quantities may in fact be finite has made people uneasy, at least at first glance, for millennia. The Greek philosopher Zeno (ca. 450 BC), a member of a community of thinkers on the island of Elea, used this uneasiness as the basis for a number of paradoxes. His purpose in creating these paradoxes was to point out complexities in the concept of infinity.

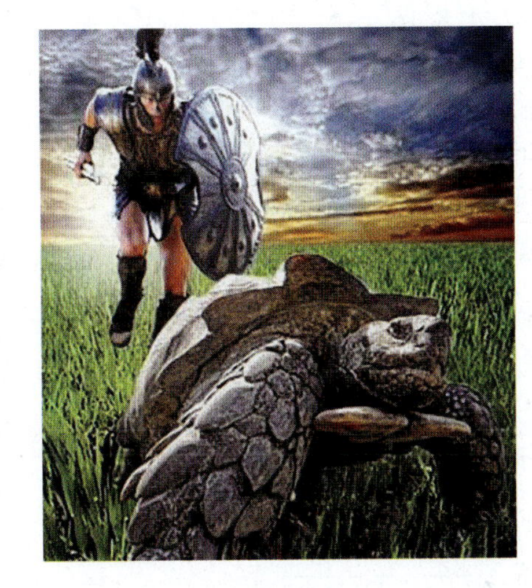

One of the most famous of Zeno's paradoxes concerns a race between the Greek hero Achilles and a tortoise. Because Achilles is so much faster, the tortoise is allowed to begin closer to the finish line. The paradox is that, once the race has begun, it will take a certain amount of time for Achilles to reach the tortoise's starting place, and that in this time the tortoise will have been able to move ahead some (small) distance to a new point.

By the time Achilles reaches this new point, the tortoise will have again been able to move ahead, and Achilles will have to repeat the process. Because this argument can be continued indefinitely, Zeno argued, Achilles can never actually catch the tortoise, let alone get ahead of him.

Our knowledge of infinite geometric series now allows us to resolve the paradox. To be specific, let us suppose that Achilles can run 100 feet per second (remember, he's a mythical hero), and that the tortoise begins with a 1000 foot head start. Let us also suppose Achilles can run 10 times as fast as the tortoise (this means the tortoise runs at 10 feet per second, so it's fairly heroic, too). Once the race starts, it takes Achilles 10 seconds to get to the tortoise's starting place, and in this 10 seconds the tortoise has managed to crawl another 100 feet. It takes Achilles an additional 1 second to cover this 100 feet, but during that 1 second the tortoise is able to crawl another 10 feet. If we continue the process, the total amount of time required for Achilles to catch the tortoise is then

$$T = 10 + 1 + \frac{1}{10} + \frac{1}{100} + \cdots$$

This sum is a geometric series with a first term of 10 and a common ratio of $\frac{1}{10}$, so

$$T = \frac{10}{1 - \frac{1}{10}} = \frac{100}{9} = 11.\bar{1} \text{ seconds.}$$

We can also determine that Achilles has run $1111.\bar{1}$ feet in this time, so the tortoise actually only manages to run $111.\bar{1}$ feet before being caught.

12.3 EXERCISES

PRACTICE

Find the explicit formula for the general n^{th} term of each geometric sequence. See Examples 1 and 2.

1. $-3,-6,-12,-24,-48,\ldots$

2. $7,\frac{7}{2},\frac{7}{4},\frac{7}{8},\frac{7}{16},\ldots$

3. $2,-\frac{2}{3},\frac{2}{9},-\frac{2}{27},\frac{2}{81},\ldots$

4. $a_1=5$ and $a_4=40$

5. $a_2=-\frac{1}{4}$ and $a_5=\frac{1}{256}$

6. $a_1=1$ and $a_4=-0.001$

7. $a_2=\frac{1}{7}$ and $r=\frac{1}{7}$

8. $a_3=\frac{9}{16}$ and $r=-\frac{3}{4}$

9. $a_3=9,\ a_5=81,$ and $r<0$

10. $-3,9,-27,81,-243,\ldots$

11. $3,2,\frac{4}{3},\frac{8}{9},\frac{16}{27},\ldots$

12. $-5,\frac{5}{4},-\frac{5}{16},\frac{5}{64},-\frac{5}{256},\ldots$

13. $a_3=28$ and $a_6=-224$

14. $a_2=-24$ and $a_5=-81$

15. $a_5=1$ and $a_6=2$

16. $a_4=\frac{343}{3}$ and $r=7$

17. $a_2=\frac{13}{17}$ and $r=\frac{4}{3}$

18. $a_4=8,\ a_8=128,$ and $r>0$

Determine if each of the following sequences is geometric. If so, find the common ratio.

19. The sequence of odd numbers

20. $4,4,4,4,4,4,\ldots$

21. $100,50,25,12.5,6.25,\ldots$

22. $2,5,11,23,47,\ldots$

23. $\frac{7}{8},\frac{7}{4},\frac{7}{2},7,14,\ldots$

24. The sequence of numbers called out at a Bingo game

25. $7,49,343,2401,\ldots$

26. $10,15,22.5,33.75,\ldots$

Given the two terms of a geometric sequence, find the common ratio and first five terms of the sequence.

27. $a_1=8$ and $a_2=24$

28. $a_6=\frac{1}{2}$ and $a_9=\frac{1}{54}$

29. $a_7=16$ and $a_{11}=256$

30. $a_4=108$ and $a_8=8748$

31. $a_5=100$ and $a_9=\frac{4}{25}$

32. $a_8=100$ and $a_{10}=1$

Use the given information about each geometric sequence to answer the question.

33. Given that $a_2 = -\frac{5}{2}$ and $a_5 = \frac{5}{16}$, what is a_{15}?

34. Given that $a_1 = 1$ and $a_4 = \frac{8}{27}$, what is the common ratio r?

35. Given that $a_3 = -2$ and $a_4 = -16$, what is a_{13}?

36. Given that $a_2 = 24$ and $a_5 = 375$, what is the common ratio r?

37. Given that $a_1 = -1$ and $a_3 = -4$, what is the common ratio r?

38. Given that $a_3 = 108$ and $a_4 = -648$, what is the common ratio r?

39. Given that $a_3 = -\frac{4}{25}$ and $a_7 = -\frac{4}{15{,}625}$, what is the common ratio r?

Each of the following sums is a partial sum of a geometric sequence. Use this fact to evaluate the sums. See Examples 3 and 4.

40. $\sum_{i=1}^{10} 3\left(-\frac{1}{2}\right)^i$

41. $\sum_{i=5}^{20} 5\left(\frac{3}{2}\right)^i$

42. $\sum_{i=10}^{40} 2^i$

43. $1 - \frac{1}{2} + \cdots + \frac{1}{16{,}384}$

44. $2 + 6 + \cdots + 39{,}366$

45. $5 - \frac{5}{3} + \cdots - \frac{5}{19{,}683}$

46. $1 - 3 + \cdots + 59{,}049$

47. $\sum_{i=4}^{15} 5(-2)^i$

48. $1 + \frac{3}{5} + \cdots + \frac{243}{3125}$

Determine if each of the following infinite geometric series converges. If so, find the sum. See Example 5.

49. $\sum_{i=0}^{\infty} -\frac{1}{2}\left(\frac{2}{3}\right)^i$

50. $\sum_{i=1}^{\infty} \left(\frac{4}{5}\right)^i$

51. $\sum_{i=0}^{\infty} \left(-\frac{9}{8}\right)^i$

52. $\sum_{i=0}^{\infty} \left(-\frac{8}{9}\right)^i$

53. $\sum_{i=5}^{\infty} \left(\frac{19}{20}\right)^i$

54. $\sum_{i=0}^{\infty} (-1)^i$

55. $\sum_{i=1}^{\infty} \frac{1}{3}(2)^{i-1}$

56. $\sum_{i=0}^{\infty} 5\left(\frac{6}{11}\right)^i$

57. $\sum_{i=4}^{\infty} \left(\frac{13}{24}\right)^i$

Write each of the following repeating decimal numbers as a fraction. See Example 5b.

58. $1.\overline{65}$

59. $0.\overline{123}$

60. $-0.\overline{5}$

61. $-3.\overline{8}$

62. $0.\overline{029}$

63. $9.\overline{98}$

APPLICATIONS

64. A rubber ball is dropped from a height of 10 feet, and on each bounce it rebounds up to 80% of its previous height. How far has it traveled vertically at the moment when it hits the ground for the tenth time? If we assume it bounces an infinite number of times, what is the total vertical distance traveled?

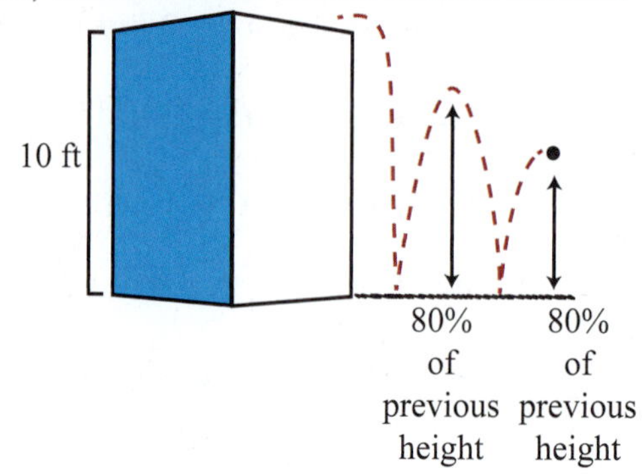

65. If \$10,000 is invested in a simple savings account with an annual interest rate of 4% compounded once a year, what is the value of the account after ten years?

66. If \$10,000 is invested in a simple savings account with an annual interest rate of 4% compounded once a month, what is the value of the account after ten years?

67. An ancient story about the game of chess tells of a king who offered to grant the inventor of the game a wish. The inventor replied, "Place a grain of wheat on the first square of the board, 2 grains on the second square, 4 grains on the third, and so on. The wheat will be my reward." How many grains of wheat would the king have had to come up with? (There are 64 squares on a chessboard.)

68. An isosceles right triangle is divided into two similar triangles, one of the new triangles is divided into two similar triangles, and this process is continued without end. If the shading pattern seen below is continued indefinitely, what fraction of the original triangle is shaded?

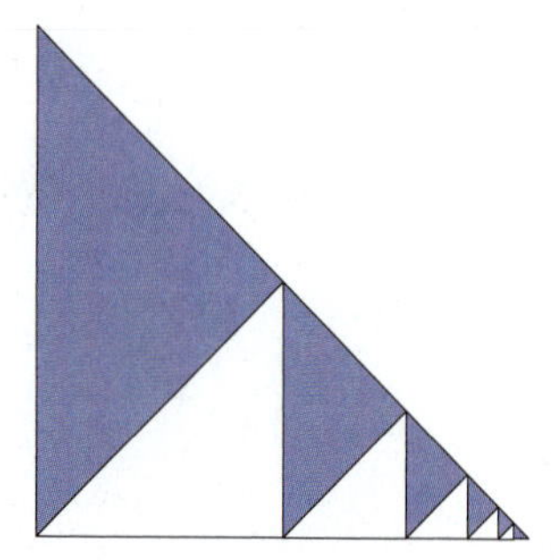

69. Each year the university admissions committee accepts 3% more students than they accepted in the previous year. If 2130 students were admitted in the first year of this trend, how many total students will have been admitted after 6 years?

70. On the day you were born, your parents deposited \$15,000 in a simple savings account for your college education. If the annual interest rate is 6.8%, compounded quarterly, how much money will be in the account when you begin college at the age of 18? What is the common ratio of this series?

71. Spamway, an internet advertising agency, uses e-mail forwards to collect addresses to which they send advertisements. They begin with an e-mail chain letter that they send to 10 people. According to the letter, each of those 10 people has to forward the e-mail to 10 more people or they will have 7 years of bad luck. Assuming the e-mails are received and forwarded only once a day and all the recipients are superstitious and follow the rules, how many e-mail addresses will Spamway have collected after 30 days? Is this series geometric? If so, find the common ratio.

72. Last summer you were a camp counselor for Sunny Days day camp. An arts and crafts project required you to distribute pieces of string (of different lengths) among the children. If you began with a piece of string 18 feet long and gave each child exactly half of your remaining string as you distributed, how long (in inches) was the seventh child's string?

WRITING & THINKING

73. Is it possible for a geometric sequence to also be an arithmetic sequence? If so, give an example and if not, explain your reasoning.

TECHNOLOGY

Use a graphing utility to evaluate each of the following sums.

74. $1.2+1.44+1.728+2.0736+\cdots+38.3376$

75. $5+\frac{5}{2}+\frac{5}{4}+\frac{5}{8}+\cdots+\frac{5}{256}$

76. $\sum_{i=53}^{92}(4.21)^i$

77. $\sum_{i=13}^{54}\left(\frac{19}{436}\right)^i$

78. $\sum_{i=23}^{79}\frac{25}{81}\left(\frac{57}{397}\right)^i$

79. $\sum_{i=16}^{71}3.42(5.26)^i$

12.4 MATHEMATICAL INDUCTION

TOPICS

1. The role of induction
2. Proofs by mathematical induction

TOPIC 1: The Role of Induction

Induction's Long History

The informal use of basic inductive reasoning can be found in the work of many scholars ranging geographically from China to Greece and going back more than 2000 years. Over time, proofs by induction became more rigorous and formal, with versions of what we call the Principle of Mathematical Induction appearing in Europe in the 1800s.

The first three sections of this chapter focused on concepts built on the natural numbers: sequences are functions whose domain is the natural numbers, and series are sums based on sequences. This section is similarly focused. While the word "induction" has many meanings in everyday language, the *Principle of Mathematical Induction* has a precise mathematical meaning that depends on the properties of natural numbers.

Mathematical induction is most often used to prove statements involving natural numbers. More precisely, if $P(n)$ is a statement about each natural number n, mathematical induction can be used to prove that the statement is true for all n. As an example, consider the following statement.

$$P(n)\text{: The sum of the first } n \text{ positive integers is } \frac{n(n+1)}{2}.$$

We already know this statement is true for each n, as we proved it in Section 12.1. However, induction provides a more powerful and more general method of proof, and we will see how it can be used to prove similar statements. In particular, we can use induction to prove the other summation formulas from Section 12.1.

It is important to see how induction relates to some of the other concepts in this chapter. We have seen several examples of recursive formulas; for instance, $a_1 = 1$ and $a_n = a_{n-1} + n$ for $n \geq 2$ is a recursive formula that generates the sequence of sums $1+2+\cdots+n$. As we know, this recursive formula is inconvenient for calculating, say, a_{100}, as we would need to first calculate a_{99}, which would in turn require us to calculate a_{98}, and so on back to a_1. In contrast, the explicit formula

$$a_n = \frac{n(n+1)}{2}$$

allows us to quickly determine a_{100} or any other such sum. As we will see, the Principle of Mathematical Induction is a useful tool for making the transition from a recursive formula to an explicit formula.

TOPIC 2: Proofs by Mathematical Induction

Consider again the following statement.

$$P(n)\text{: The sum of the first } n \text{ positive integers is } \frac{n(n+1)}{2}.$$

Suppose we demonstrate that **if** the statement is true for an integer k, **then** the statement is also true for the next integer, $k+1$. We can easily show that $P(1)$ is true, as the sum of the first 1 positive integers, namely 1, is indeed $\frac{(1)(1+1)}{2}=1$. Since $P(1)$ is true, we can conclude from our demonstration that $P(2)$ is also true. And since $P(2)$ is true, it follows that $P(3)$ is true. We don't have to explicitly show that $P(2)$ being true implies that $P(3)$ is true, because we've already demonstrated that, in general, the validity of $P(k)$ implies the validity of $P(k+1)$. This is the essence of the Principle of Mathematical Induction.

THEOREM: Principle of Mathematical Induction

Assume that $P(n)$ is a statement about each natural number n. Suppose that the following two conditions hold:

1. $P(1)$ is true.
2. For each natural number k, if $P(k)$ is true, then $P(k+1)$, is true.

Then the statement $P(n)$ is true for every natural number n.

An inductive proof thus consists of two steps. The first task, which we will call the **basic step**, is generally very easy; we simply verify that $P(1)$ is true. The second task, called the **induction step**, may require more work and algebraic manipulation.

It is important to realize that the induction step begins with a very powerful assumption: under the *assumption* that $P(k)$ is true, our goal is to prove that $P(k+1)$, is also true. The single most important step, and the step that is most unfamiliar initially, is the one in which the inductive assumption is used to simplify part of the statement $P(k+1)$ in order to prove that $P(k+1)$ is true. Be sure to note how the inductive assumption is used in the following examples.

Example 1: Proof by Mathematical Induction

Prove that for each natural number n, $1+3+5+\cdots+(2n-1)=n^2$.

Solution

Basic Step: The statement $P(1)$ is the equation $1=1^2$, which is clearly true.

Induction Step: As always, we begin this step with the assumption that $P(k)$ is true.

That is, $1+3+5+\cdots+(2k-1)=k^2$. We want to prove that $P(k+1)$ is true, so we need to show that $1+3+5+\cdots+(2(k+1)-1)=(k+1)^2$. To do this, we rewrite the left-hand side using the inductive assumption.

$$1+3+5+\cdots+(2(k+1)-1)$$

$$=1+3+5+\cdots+(2k-1)+(2(k+1)-1)$$

$$=k^2+(2(k+1)-1) \quad \text{Use the inductive assumption to rewrite all but the last term.}$$

$$=k^2+2k+1 \quad \text{Simplify.}$$

$$=(k+1)^2 \quad \text{Factor.}$$

Since we've completed the basic step and the induction step, the proof is complete.

Induction can be used to prove statements other than equations. The next two examples prove an inequality and a divisibility fact.

Example 2: Proof by Mathematical Induction

Prove that for each natural number n, $2n \le 2^n$.

Solution

Basic Step: The statement $P(1)$ is the inequality $2(1) \le 2^1$, or $2 \le 2$. This is true.

Induction Step: Assume that $2k \le 2^k$. Then we have the following.

$$2(k+1) = 2k+2 \quad \text{Begin by writing the left-hand side of } P(k+1).$$

$$\le 2^k+2 \quad \text{Use the inductive assumption.}$$

$$\le 2^k+2^k \quad \text{Use the fact that } k \ge 1 \text{ implies } 2 \le 2^k.$$

$$=2(2^k)$$

$$=2^{k+1} \quad \text{Simplify until the right-hand side of } P(k+1) \text{ is reached.}$$

Example 3: Proof by Mathematical Induction

Prove that for each natural number n, $8^n - 3^n$ is divisible by 5.

Solution

Basic Step: Since $8^1 - 3^1 = 5$ is divisible by 5, $P(1)$ is true.

Induction Step: Assume that $8^k - 3^k$ is divisible by 5. Then there is some integer p for which $8^k - 3^k = 5p$.

$$8^{k+1}-3^{k+1} = 8^k \cdot 8 - 3^k \cdot 3 \quad \text{Rewrite using properties of exponents.}$$

$$=8^k \cdot (5+3) - 3^k \cdot 3 \quad \text{Rewrite 8 as 5 + 3.}$$

$$=5 \cdot 8^k + 3 \cdot 8^k - 3 \cdot 3^k \quad \text{Apply the distributive property.}$$

$$=5 \cdot 8^k + 3 \cdot (8^k - 3^k) \quad \text{Factor the last two terms.}$$

$$=5 \cdot 8^k + 3 \cdot 5p \quad \text{Apply the inductive assumption.}$$

$$=5(8^k + 3p) \quad \text{Factor once more.}$$

This shows that $8^{k+1}-3^{k+1}$ is a product of 5 and the integer 8^k+3p, so we have shown that $8^{k+1}-3^{k+1}$ is divisible by 5, completing the inductive step, and thus the proof.

Finally, we will prove the formula for the summation of the first n squares. The final summation formula from Section 12.1 is left as an exercise (Exercise 21).

Example 4: Proof by Mathematical Induction

Prove that for each natural number n, $\sum_{i=1}^{n} i^2 = \frac{n(n+1)(2n+1)}{6}$.

Solution

Basic Step: Note that $P(1)$ is the statement $1^2 = \frac{(1)(1+1)(2+1)}{6}$, which is true.

Induction Step: The inductive assumption is $1^2+2^2+\cdots+k^2 = \frac{k(k+1)(2k+1)}{6}$.

Consider the left-hand side of $P(k+1)$.

$$1^2+2^2+\cdots+k^2+(k+1)^2$$

To show that $P(k+1)$ is true, we need to demonstrate that this sum is equal to the right-hand side of $P(k+1)$, namely $\frac{(k+1)((k+1)+1)(2(k+1)+1)}{6}$, better written as $\frac{(k+1)(k+2)(2k+3)}{6}$.

We use the inductive assumption to rewrite all but the last term of the left-hand side; some algebraic manipulation then leads to the desired result.

$$
\begin{aligned}
1^2+2^2+\cdots+(k+1)^2 &= 1^2+2^2+\cdots+k^2+(k+1)^2 && \text{Use the inductive assumption to rewrite all but the last term.}\\
&= \frac{k(k+1)(2k+1)}{6}+(k+1)^2 \\
&= \frac{k(k+1)(2k+1)}{6}+\frac{6(k+1)^2}{6} && \text{Rewrite the second term using the LCD.}\\
&= \frac{(k+1)}{6}\left[k(2k+1)+6(k+1)\right] && \text{Factor out } \frac{k+1}{6}.\\
&= \frac{(k+1)}{6}\left[2k^2+k+6k+6\right] && \text{Expand the remaining terms.}\\
&= \frac{(k+1)}{6}\left[2k^2+7k+6\right] && \text{Combine like terms.}\\
&= \frac{(k+1)}{6}\left[(k+2)(2k+3)\right] && \text{Factor the resulting quadratic.}\\
&= \frac{(k+1)(k+2)(2k+3)}{6} && \text{Rewrite in the desired form.}
\end{aligned}
$$

This completes the induction step, and thus the entire proof.

12.4 EXERCISES

PRACTICE

Find S_{k+1} for the given S_k.

1. $S_k = \dfrac{1}{3(k+2)}$

2. $S_k = \dfrac{k^2}{k(k-1)}$

3. $S_k = \dfrac{k(k+1)(2k+1)}{4}$

4. $S_k = \dfrac{1}{(2k-1)(2k+1)}$

Use the Principle of Mathematical Induction to prove the following statements.

5. $1+2+3+4+\cdots+n = \dfrac{n(n+1)}{2}$

6. $\dfrac{1}{2}+\dfrac{1}{2^2}+\dfrac{1}{2^3}+\cdots+\dfrac{1}{2^n} = 1-\dfrac{1}{2^n}$

7. $2+4+6+8+\cdots+2n = n(n+1)$

8. $\displaystyle\sum_{i=1}^{n} \frac{1}{(2i-1)(2i+1)} = \frac{n}{2n+1}$

9. $4^0+4^1+4^2+\cdots+4^{n-1} = \dfrac{4^n-1}{3}$

10. $2^n > n^2$ for all $n \geq 5$.

11. $\dfrac{1}{1\cdot 4}+\dfrac{1}{4\cdot 7}+\dfrac{1}{7\cdot 10}+\cdots+\dfrac{1}{(3n-2)(3n+1)} = \dfrac{n}{3n+1}$

12. $5^0 + 5^1 + 5^2 + \cdots + 5^{n-1} = \dfrac{5^n - 1}{4}$

13. $5 + 10 + 15 + \cdots + 5n = \dfrac{5n(n+1)}{2}$

14. $n^2 \geq 100n$ for all $n \geq 100$.

15. $\left(1+\dfrac{1}{1}\right)\left(1+\dfrac{1}{2}\right)\left(1+\dfrac{1}{3}\right)\cdots\left(1+\dfrac{1}{n}\right) = n+1$

16. $3 + 5 + 7 + \cdots + (2n+1) = n(n+2)$

17. $1 + 4 + 7 + 10 + \cdots + (3n-2) = \dfrac{n}{2}(3n-1)$

18. $-2 - 3 - 4 - \cdots - (n+1) = -\dfrac{1}{2}(n^2 + 3n)$

19. $3^n > 2n + 1$ for all $n \geq 2$.

20. $2^n > n$, for all $n \geq 1$

21. $1^3+2^3+3^3+4^3+\cdots+n^3=\dfrac{n^2(n+1)^2}{4}$

22. $1\cdot 2+2\cdot 3+3\cdot 4+\cdots+n(n+1)=\dfrac{n(n+1)(n+2)}{3}$

23. If $a>1$, then $a^n>1$.

24. $2^n>4n$ for all $n\geq 5$.

25. $1^4+2^4+3^4+4^4+\cdots+n^4=\dfrac{n(n+1)(2n+1)(3n^2+3n-1)}{30}$

26. $1^5+2^5+3^5+4^5+\cdots+n^5=\dfrac{n^2(n+1)^2(2n^2+2n-1)}{12}$

27. $\dfrac{1}{\sqrt{1}}+\dfrac{1}{\sqrt{2}}+\dfrac{1}{\sqrt{3}}+\cdots+\dfrac{1}{\sqrt{n}}>\sqrt{n}$ for all $n \geq 2$.

28. $1+3+5+7+\cdots+(2n-1)=n^2$

Use the Principle of Mathematical Induction to prove the given properties. (Assume *m* and *n* are natural numbers and *a*, *b*, and *x* are real numbers.)

29. $(ab)^n = a^n b^n$ (Assume a and b are constants.)

30. $(a^m)^n = a^{mn}$ (Assume a and m are constants.)

31. If $x_1 > 0, x_2 > 0, \ldots, x_n > 0$, then
$\ln(x_1 \cdot x_2 \cdot x_3 \cdot \cdots \cdot x_n) = \ln x_1 + \ln x_2 + \ln x_3 + \cdots + \ln x_n$.

32. 5 is a factor of $(2^{2n-1}+3^{2n-1})$.

33. 64 is a factor of $(9^n - 8n - 1)$ for all $n \geq 2$.

34. 3 is a factor of $(n^3 + 3n^2 + 2n)$.

35. $n^3 - n + 3$ is divisible by 3.

36. $5^n - 1$ is divisible by 4.

37. $n(n+1)(n+2)$ is divisible by 6.

APPLICATIONS

38. In the 19th century a mathematical puzzle was published telling of a mythical monastery in Benares, India with three crystal towers holding 64 disks made of gold. The disks are each of a different size and have holes in the middle so that they slide over the towers and sit in a stack with the largest on the bottom and the smallest on the top. The monks of the monastery were instructed to move all of the disks to the third tower following these three rules:

- Each disk sits over a tower except when it is being moved.
- No disk may ever rest on a smaller disk.
- Only one disk at a time may be moved.

According to the puzzle, when the monks complete their task, the world would end! To move n disks requires $H(n) = 2^n - 1$ moves. Prove this is true through mathematical induction.

39. If there are n people in a room, and every person shakes hands with every other person exactly once, then exactly $\frac{n(n-1)}{2}$ handshakes will occur. Prove this is true through mathematical induction.

40. Any monetary value of 4 cents or higher can be composed of twopence (a British two-cent coin) and nickels. Your basic step would be

$$4 \text{ cents} = \text{twopence} + \text{twopence}.$$

Use the fact that $k = 2t + 5n$ where k is the total monetary value, t is the number of twopence, and n is the number of nickels, to prove $P(k+1)$. (**Hint:** There are 3 induction steps to prove.)

WRITING & THINKING

41. What is wrong with this "proof" by induction?

Proposition: All horses are the same color. (In any set of n horses, all horses are the same color.)

Basic Step: If you have only one horse in a group, then all of the horses in that group have the same color.

Induction Step: Assume that in any group of n horses, all horses are the same color. Now take any group of $n + 1$ horses. Remove the first horse from this group and the remaining n horses must be of the same color because of the hypothesis. Now replace the first horse and remove the last horse. Once again, the remaining n horses must be the same color because of the hypothesis. Since the two groups overlap, all $n + 1$ horses must be the same color.

Thus by induction, any group of n horses are the same color.

12.5 COMBINATORICS

TOPICS

1. The Multiplication Principle of Counting
2. Permutations
3. Combinations
4. The Binomial Theorem
5. The Multinomial Theorem

TOPIC 1: The Multiplication Principle of Counting

Combinatorics can be informally defined as "the science of counting." Such a simple definition is somewhat misleading, however, and runs the risk of trivializing the topic. A better—though less pithy—definition of combinatorics is "the study of techniques used to determine the sizes, or cardinalities, of sets." Even this may sound deceptively simple until a few examples are studied. As we will see, there are many occasions when the cardinality (number of elements) of a perfectly well-defined set may be difficult to determine at first glance. We will also see that in many cases the size of a set is of more importance than the actual elements of the set.

One of the joys of combinatorics is that there are many problems that can be solved with nothing more than a few fairly intuitive ideas, the most basic of which is the *Multiplication Principle of Counting*. Before formally stating the principle, we will use it to solve a problem.

Example 1: Counting Phone Numbers

In the United States, telephone numbers consist of a 3-digit area code followed by a 7-digit local number. Neither the first digit of the area code nor the first digit of the local number can be 0 or 1. How many such phone numbers are there if

a. there is a further restriction that the middle digit of the area code must be 0 or 1?

b. there are no further restrictions?

Solution

a. Notice that this problem, in mathematical terms, concerns finding the cardinality of the set of all phone numbers of a certain form, but the words "set" and "cardinality" don't appear in the statement of the problem. This is very typical. Most of the combinatorics problems we will see use informal language to define a set, and then ask us, with equally informal language, to determine its size.

The method we will use to solve this problem is also very typical. We will count all the possible phone numbers by considering how we could go about constructing them. Every such phone number consists of a string of ten digits, with the restrictions that the first and fourth digits (reading from left to right) can't be either 0 or 1, but the second digit must be either 0 or 1. Since there are ten digits in all (0 through 9), this means there are eight possible ways to choose a digit for the first and fourth "slots", only two possible ways to choose a digit for the second slot, and ten possible ways to choose digits for all the remaining slots. This is illustrated in Figure 1.

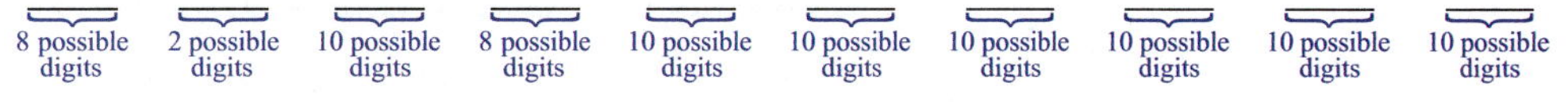

FIGURE 1: Constructing Allowable Phone Numbers

For the moment, consider how just the first two slots can be filled. Any of the eight allowable digits for the first slot can be paired up with either of the two allowable digits for the second slot, meaning that there are 8×2 ways of filling the first two slots altogether. In fact, all of the sixteen possible choices for the first two slots can be easily listed:

$$\underbrace{\underline{2}\underline{0}, \underline{3}\underline{0}, \underline{4}\underline{0}, \ldots, \underline{9}\underline{0}}_{8 \text{ ending in } 0}, \quad \underbrace{\underline{2}\underline{1}, \underline{3}\underline{1}, \underline{4}\underline{1}, \ldots, \underline{9}\underline{1}}_{8 \text{ ending in } 1}$$

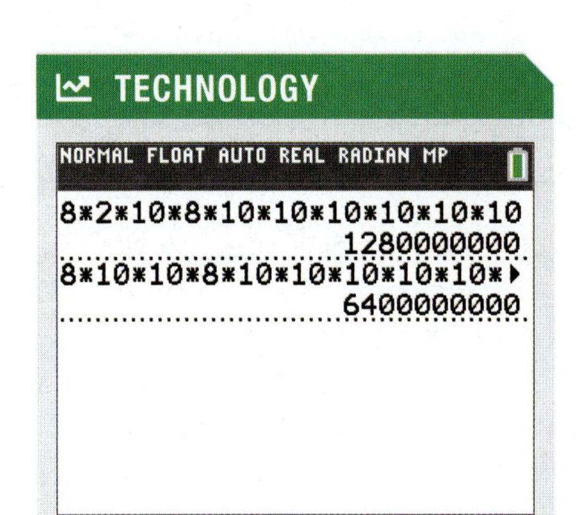

Now, any of these sixteen possible choices for the first two slots can be matched with any of the ten possible digits for the third slot, giving us a total of 160 ways of filling the first three slots. This pattern continues, so that the total number of phone numbers of the required form is

$$8 \times 2 \times 10 \times 8 \times 10 \times 10 \times 10 \times 10 \times 10 \times 10 = 1,280,000,000.$$

b. The same sort of argument applies if there is no restriction on the middle digit of the area code. In this case, the total number of allowable phone numbers is

$$8 \times 10 \times 10 \times 8 \times 10 \times 10 \times 10 \times 10 \times 10 \times 10 = 6,400,000,000.$$

The generalization of the reasoning we used in Example 1 is often stated in terms of a sequence of events.

THEOREM: The Multiplication Principle of Counting

Suppose $E_1, E_2, \ldots, E_n$ is a sequence of events, each of which has a certain number of possible outcomes. Suppose event E_1 has m_1 possible outcomes, and that after event E_1 has occurred, event E_2 has m_2 possible outcomes. Similarly, after event E_2, event E_3 has m_3 possible outcomes, and so on. Then the total number of ways that all n events can occur is

$$m_1 \cdot m_2 \cdot \cdots \cdot m_n.$$

An alternative interpretation of the principle is to think of the events as a sequence of tasks to be completed. In Example 1, for instance, each task consists of selecting a digit for a given slot. There are ten tasks in all, and the product of the number of ways each task can be performed gives us the total number of phone numbers.

Example 2: Using the Multiplication Principle of Counting

A certain state specifies that all non-personalized license plates consist of two letters followed by four digits, and that the letter O (which could be mistaken for the digit 0) cannot be used. How many such license plates are there?

Solution

Generating such a license plate is a matter of choosing six characters, the first two of which can be any of 25 letters and the last four of which can be any of 10 digits. The total number of such license plates is thus $25 \times 25 \times 10 \times 10 \times 10 \times 10 = 6,250,000$.

TOPIC 2: Permutations

A **permutation** of a set of objects is a linear arrangement of the objects. In any such arrangement, one of the objects is first, another is second, and so on, so the construction of a permutation of n objects consists of "filling in" n slots. We can thus use the Multiplication Principle of Counting to determine the number of permutations of a given set of objects.

Example 3: Calculating Permutations

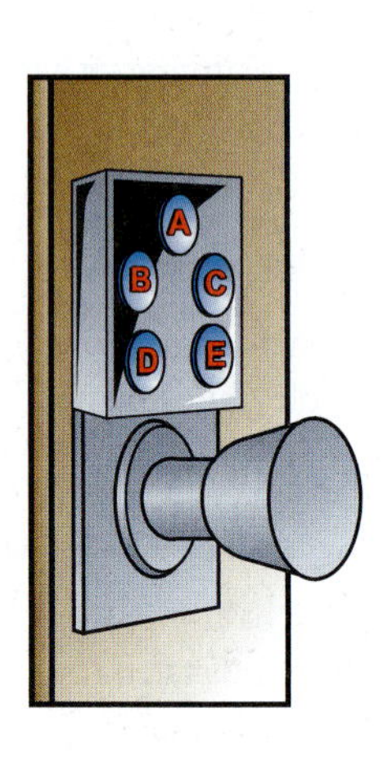

One brand of combination lock for a door consists of five buttons labeled A, B, C, D, and E. To unlock the door, each button is pressed once and only once. Given that the installer of the lock can set the combination to be any permutation of the five letters, how many such permutations are there?

Solution

The difference between this problem and Examples 1 and 2 is that once a letter has been chosen for a given slot, it can't be reused. So in constructing a combination code, there are five choices for the first letter but only four choices for the second letter, since whatever letter was used first cannot be used again.

Similarly, there are only three choices for the third slot and only two choices for the fourth slot. Finally, whichever of the five letters is left *must* be used for the fifth slot, so there is only one choice. Figure 2 illustrates the slot-filling process.

FIGURE 2: Constructing a Permutation of Five Characters

We can now use the Multiplication Principle of Counting to determine that there are $5 \times 4 \times 3 \times 2 \times 1 = 120$ such combinations.

Products of the form $n \times (n-1) \times (n-2) \times \cdots \times 2 \times 1$ occur so frequently that it makes sense to assign some shorthand notation to them, as follows.

DEFINITION: Factorial Notation

If n is a positive integer, we use the notation $n!$ (which is read "***n* factorial**") to stand for the product of all the integers from 1 to n. That is,

$$n! = n \times (n-1) \times (n-2) \times \cdots \times 2 \times 1$$

In addition, we define $0! = 1$.

CAUTION

Note that the factorial operation is defined only for positive integers and, as a special case, 0. You may encounter, in a later math class, a related function called the gamma function that extends the factorial idea to all positive real numbers.

If all permutation problems involved nothing more than a straightforward application of the Multiplication Principle of Counting, as in Example 3, there would be little left to say. But typically, the solution of a permutation problem requires counting the number of ways that a linear arrangement of k objects can be made from a collection of n objects, where $k \leq n$. Consider the following problem.

Example 4: Calculating Permutations

How many different five-letter combination codes are possible if every letter of the alphabet can be used, but no letter may be repeated?

Solution

The difference between this problem and the one in Example 3 is that there are now 26 choices for the first letter, 25 choices for the second, and so on. The corresponding "slot diagram" describing the number of ways each slot can be filled appears in Figure 3.

26 choices 25 choices 24 choices 23 choices 22 choices

FIGURE 3: Five-Letter Permutations Chosen from 26 Characters

The total number of such combination codes is thus

$$26 \times 25 \times 24 \times 23 \times 22 = 7{,}893{,}600.$$

Products like the one in Example 4 are also very common, and deserve special mention. To motivate the following formula, note that we can state the answer to Example 4 in terms of factorials as follows:

$$26 \times 25 \times 24 \times 23 \times 22 = \frac{(26 \times 25 \times 24 \times 23 \times 22) \times 21!}{21!} = \frac{26!}{21!}$$

If we generalize this, we obtain a formula for the number of permutations of length k that can be formed from a collection of n objects, usually expressed as the number of permutations of n objects taken k at a time.

f(x) FORMULA: Permutation Formula

The number of permutations of n objects taken k at a time is

$$_nP_k = \frac{n!}{(n-k)!}.$$

This is a conveniently short way of expressing the product

$$_nP_k = n \times (n-1) \times \cdots \times (n-k+1).$$

The formula for $_nP_k$ is especially useful when the number of factors to be multiplied is large, as in Example 5, assuming you have access to a calculator or computer with the factorial function.

TECHNOLOGY

To find the number of permutations using a TI-84 Plus, enter the first number, n (in this case 52), press math, and use the right arrow to highlight PROB. Then select nPr and enter the second number, r (in this case 20) and press enter.

Example 5: Using the Permutation Formula

A magician is preparing to demonstrate a card trick that involves 20 cards chosen at random from a standard deck of cards. Once chosen, the 20 cards are arranged in a row, and the order of the cards plays a role in the trick. How many such orderings are possible?

Solution

A standard deck of cards contains 52 distinct cards (13 cards in each of 4 suits), and this problem asks for the number of permutations of 52 cards taken 20 at a time. One way to determine this would be to evaluate $52 \times 51 \times \cdots \times 33$ (note that this is a product of 20 numbers), but a far faster way is to use the permutation formula:

$$_{52}P_{20} = \frac{52!}{(52-20)!} = \frac{52!}{32!} \approx \frac{8.07 \times 10^{67}}{2.63 \times 10^{35}} \approx 3.07 \times 10^{32}$$

The two factorials that appear in the formula above have been determined by a calculator. Note that $52!$, $32!$, and the final answer are all very large numbers, so it is convenient to use scientific notation.

TOPIC 3: Combinations

In contrast to permutations, where the order of the objects under consideration is of great importance, combinations are simply collections of objects. To be specific, a **combination** of n objects taken k at a time is one of the ways of forming a subset of size k from a set of size n, where again $k \leq n$. Combination problems typically ask us to determine the number of different subsets of size k that can be formed.

To emphasize the difference between permutations and combinations, and to begin to understand the combination formula that we will soon derive, consider the following problem.

Example 6: Calculating Permutations and Combinations

Let $S = \{a, b, c, d\}$.

a. How many permutations of size 3 can be formed from the set *S*?

b. How many combinations of size 3 can be formed from the set *S*?

TECHNOLOGY

To enter the factorial sign, press math, highlight PROB, and select !.

```
NORMAL FLOAT AUTO REAL RADIAN MP
4!/(4-3)!
                                24
```

Solution

a. The number of permutations of 4 objects taken 3 at a time is

$$_4P_3 = \frac{4!}{(4-3)!} = \frac{4!}{1!} = 24.$$

b. We have already determined there are 24 permutations of size 3 that can be formed. For the purpose of determining the corresponding number of combinations, it might be useful to actually list all the permutations:

abc	acb	bac	bca	cab	cba
abd	adb	bad	bda	dab	dba
acd	adc	cad	cda	dac	dca
bcd	bdc	cbd	cdb	dbc	dcb

If we now view these collections of objects as sets, all six permutations in the first row describe the single set $\{a, b, c\}$. Similarly, the six permutations in the second row are simply six different ways of describing the set $\{a, b, d\}$. The third row describes the set $\{a, c, d\}$ and the fourth row describes the set $\{b, c, d\}$. In all, there are only four combinations of 4 objects taken 3 at a time.

Although we don't yet have the formula we seek, let $_nC_k$ denote the number of combinations of *n* objects taken *k* at a time. Taking a cue from Example 6, we might work toward the desired formula by noting that each of the size *k* combinations formed from *S* gives rise to $k!$ permutations of size *k*, and we know that there are $_nP_k$ permutations taken *k* at a time overall. This means that

$$(k!)(_nC_k) = {_nP_k} \quad \text{or} \quad _nC_k = \frac{_nP_k}{k!}.$$

If we now replace $_nP_k$ with $\frac{n!}{(n-k)!}$, we have the following formula.

FORMULA: Combination Formula

The number of combinations of *n* objects taken *k* at a time is

$$_nC_k = \frac{n!}{(k!)(n-k)!}.$$

At this point, we have seen all of the counting techniques we will need. There is much more to the subject of combinatorics, but the three ideas we have discussed—the Multiplication Principle of Counting, the permutation formula, and the combination formula—are sufficient to answer many, many questions. This is especially true when the techniques are combined in various ways, as the next few examples show.

Example 7: The Combination Formula and Forming Committees

Senate Committee
11 Democrats | 10 Republicans | 1 Independent

Suppose a Senate committee consists of 11 Democrats, 10 Republicans, and 1 Independent. The chair of the committee wants to form a sub-committee to be charged with researching a particular issue, and decides to appoint 3 Democrats, 2 Republicans, and the 1 Independent to the sub-committee. How many different sub-committees are possible?

Solution

This is in large part a combination problem, as the chair needs to form a subset of size 3 from the 11 Democrats, a subset of size 2 from the 10 Republicans, and a subset of size 1 from the 1 Independent. (Note that the order of those chosen is irrelevant, which is why this is a combination problem and not a permutation problem.) The combination formula tells us that these tasks can be done in, respectively, ${}_{11}C_3$, ${}_{10}C_2$, and ${}_1C_1$ ways (there is only 1 way to choose 1 member from a set of 1). These numbers are

$${}_{11}C_3 = \frac{11!}{3!8!} = \frac{11\times10\times9\times\cancel{8!}}{3!\cancel{8!}} = \frac{990}{6} = 165$$

$${}_{10}C_2 = \frac{10!}{2!8!} = \frac{10\times9\times\cancel{8!}}{2!\cancel{8!}} = \frac{90}{2} = 45$$

$${}_1C_1 = \frac{1!}{1!0!} = \frac{1}{1} = 1 \text{ (remember that } 0! = 1\text{)}$$

Once the appropriate number of people from each party have been chosen, any of the 165 possible groups of 3 Democrats can be matched up with any of the 45 possible groups of 2 Republicans, and then further matched up with the 1 Independent. The Multiplication Principle of Counting thus tells us there are $165\times45\times1 = 7425$ ways of forming the desired sub-committee (any one of which, inevitably, is bound to offend one party or another).

TECHNOLOGY

To find the number of combinations using a TI-84 Plus, enter the first number, *n* (in this case 11), press `math`, and use the right arrow to highlight PROB. Then select nCr and enter the second number, *r* (in this case 3) and press `enter`.

```
NORMAL FLOAT AUTO REAL RADIAN MP
11C3
                              165
10C2
                               45
165*45*1
                             7425
```

Example 8: Counting Rules and Forming "Words"

How many different arrangements are there of the letters in the following words?

a. STIPEND **b.** SALAAM **c.** MISSISSIPPI

Solution

a. The goal is to count the total number of "words" that can be formed from the letters in the word STIPEND, using each letter once and only once (most of the arrangements will not actually be legitimate English words). Since the order of the letters is important, and since STIPEND contains 7 distinct letters, the answer is

$$ {}_7P_7 = \frac{7!}{0!} = 7! = 5040. $$

b. The word SALAAM contains 6 letters, but only 4 distinct letters. That is, the 3 A's are indistinguishable, so the answer is not simply ${}_6P_6$. In fact, 6! overcounts the total number of arrangements by a factor of 3!, because any one arrangement of the 6 letters in SALAAM is equivalent to 5 more arrangements. This is because the 3 A's can be permuted in 3! = 6 ways. Consider the arrangement MAALSA rewritten as MA*A*LSa so that the six different arrangements are more readily distinguishable:

MA*A*LSa, MAaLS*A*, M*A*ALSa, M*A*aLSA, MaALS*A*, Ma*A*LSA

This means the total number of arrangements is actually

$$ \frac{6!}{3!} = \frac{720}{6} = 120. $$

c. MISSISSIPPI contains 11 characters, but 4 of them are S's, 4 of them are I's, 2 of them are P's, and the remaining 1 character is M. If the 11 characters were all distinct, the total number of arrangements of the letters would be 11!, but we need to divide this number by 4! to account for the indistinguishable S's, and then divide again by 4! to account for the I's, and then again by 2! to account for the P's. This gives us a total of

$$ \frac{11!}{4!4!2!} = \frac{39,916,800}{(24)(24)(2)} = 34,650. $$

TOPIC 4: The Binomial Theorem

Consider the expressions $(x+y)^7$ and $(a+b+c)^3$. The first is a binomial raised to a power, and the second is a trinomial raised to a power. Expressions like these (or worse) can occur in the course of solving an algebra problem, and we often need to expand them in order to work toward the solution of the problem. We will use our knowledge of combinatorics to do so efficiently, tackling binomial expansions first and then expansions of powers of three or more terms as the final topic of this section.

We could use only elementary methods to expand such expressions. For example, we could expand $(x+y)^7$ as follows:

$$ \begin{aligned} (x+y)^7 &= (x+y)(x+y)(x+y)^5 \\ &= (x^2+2xy+y^2)(x+y)^5 \\ &= (x^2+2xy+y^2)(x+y)(x+y)^4 \\ &= \cdots \end{aligned} $$

But such work can be extremely tedious and error-prone if the exponent is larger than, say, 3 or 4. We can use the combinatorics methods we have seen to develop two formulas that greatly simplify the process.

When we expand an expression like $(x+y)^7$, we are really making sure we account for all the possible products that result from taking one term from each factor. For instance, when we multiply the terms that are boxed below, we obtain x^3y^4.

$$(\boxed{x}+y)(x+\boxed{y})(x+\boxed{y})(\boxed{x}+y)(x+\boxed{y})(\boxed{x}+y)(x+\boxed{y})$$

But there are many other choices of terms that also lead to x^3y^4, such as

$$(x+\boxed{y})(x+\boxed{y})(\boxed{x}+y)(\boxed{x}+y)(\boxed{x}+y)(x+\boxed{y})(x+\boxed{y}).$$

In all, there are 35 different ways of boxing 3 x's and 4 y's, meaning that there is a coefficient of 35 in front of x^3y^4 in the expansion of $(x+y)^7$.

Where did this number of 35 come from? We can relate the expansion of binomials (or, more generally, multinomials) to the "word" problems we studied in Example 8. For instance, the coefficient of x^3y^4 is the number of different arrangements of the letters *xxxyyyy*. Since these 7 letters consist of 3 x's and 4 y's, the total number of such arrangements is

$$\frac{7!}{3!4!} = \frac{5040}{(6)(24)} = 35.$$

(Note that the boxed letters above correspond to the "words" *xyyxyxy* and *yyxxxyy*.)

There is some special notation and language for the type of work we have just done.

DEFINITION: Binomial Coefficients

Given nonnegative integers n and k, with $k \le n$, we define

$$\binom{n}{k} = \frac{n!}{k!(n-k)!}.$$

The expression $\binom{n}{k}$ is called a **binomial coefficient**, as it corresponds to the coefficient of $x^k y^{n-k}$ in the expansion of $(x+y)^n$.

We have already seen the formula for binomial coefficients in another context.

$$\binom{n}{k} = {}_nC_k$$

The last step in expanding a binomial, now that we know how to find the coefficients, is to put all the pieces together. Consider again the expression $(x+y)^7$. When this is expanded, there will be terms containing y^7, xy^6, x^2y^5, x^3y^4, x^4y^3, x^5y^2, x^6y, and x^7 each multiplied by the appropriate binomial coefficient. Note that the sum of the exponents in each term is 7, as this is the total number of x's and y's in each term.

Example 9: Constructing a Binomial Expansion

Expand the expression $(x+y)^7$.

Solution

Based on the above reasoning, and the binomial coefficient formula, we have the following.

$$\begin{aligned}(x+y)^7 &= \binom{7}{0}x^0y^7 + \binom{7}{1}x^1y^6 + \binom{7}{2}x^2y^5 + \binom{7}{3}x^3y^4 + \binom{7}{4}x^4y^3 \\ &\quad + \binom{7}{5}x^5y^2 + \binom{7}{6}x^6y^1 + \binom{7}{7}x^7y^0 \\ &= y^7 + 7xy^6 + 21x^2y^5 + 35x^3y^4 + 35x^4y^3 + 21x^5y^2 + 7x^6y + x^7\end{aligned}$$

The generalization of our work is called the Binomial Theorem.

THEOREM: The Binomial Theorem

Given the positive integer n, and any two expressions A and B,

$$\begin{aligned}(A+B)^n &= \sum_{k=0}^{n}\binom{n}{k}A^kB^{n-k} \\ &= \binom{n}{0}A^0B^n + \binom{n}{1}A^1B^{n-1} + \binom{n}{2}A^2B^{n-2} + \cdots + \binom{n}{n-1}A^{n-1}B^1 + \binom{n}{n}A^nB^0.\end{aligned}$$

Note that A and B can be more complicated than we have considered so far.

Example 10: Using the Binomial Theorem

Expand the expression $(2x-y)^4$.

Solution

To use the Binomial Theorem, note that $A = 2x$ and $B = -y$. The theorem thus tells us the following.

$$\begin{aligned}(2x-y)^4 &= \binom{4}{0}(2x)^0(-y)^4 + \binom{4}{1}(2x)^1(-y)^3 + \binom{4}{2}(2x)^2(-y)^2 \\ &\quad + \binom{4}{3}(2x)^3(-y)^1 + \binom{4}{4}(2x)^4(-y)^0 \\ &= (-y)^4 + 4(2x)(-y)^3 + 6(2x)^2(-y)^2 + 4(2x)^3(-y) + (2x)^4 \\ &= y^4 - 8xy^3 + 24x^2y^2 - 32x^3y + 16x^4\end{aligned}$$

TOPIC 5: The Multinomial Theorem

We will finish with one last definition and formula. To expand an expression like $(a+b+c)^3$ we need to determine the coefficients of terms of the form $a^{k_1}b^{k_2}c^{k_3}$, where $k_1+k_2+k_3=3$. Reasoning similar to that preceding the Binomial Theorem leads to the following definition.

DEFINITION: Multinomial Coefficients

Let n be a positive integer, and let $k_1,k_2,\ldots,k_r$ be nonnegative integers such that $k_1+k_2+\cdots+k_r=n$. We define

$$\binom{n}{k_1,k_2,\ldots,k_r}=\frac{n!}{k_1!k_2!\cdots k_r!}.$$

Such an expression is called a **multinomial coefficient**. It is the coefficient of the term $A_1^{k_1}A_2^{k_2}\cdots A_r^{k_r}$ in the expansion of $(A_1+A_2+\cdots+A_r)^n$.

To expand something of the form $(A_1+A_2+\cdots+A_r)^n$, we identify all the terms of the form $A_1^{k_1}A_2^{k_2}\cdots A_r^{k_r}$, multiply each one by its corresponding multinomial coefficient, and add them up. This is called the Multinomial Theorem.

THEOREM: The Multinomial Theorem

Given the positive integer n and expressions $A_1,A_2,\ldots,A_r$,

$$(A_1+A_2+\cdots+A_r)^n=\sum_{k_1+k_2+\cdots+k_r=n}\binom{n}{k_1,k_2,\ldots,k_r}A_1^{k_1}A_2^{k_2}\cdots A_r^{k_r}.$$

Example 11: Using the Multinomial Theorem

Expand the expression $(a+b+c)^3$.

Solution

We will start by noting that the expansion will have terms containing a^3, a^2b, a^2c, ab^2, ac^2, abc, b^3, b^2c, bc^2, and c^3. (Note that these are all the ways that the three non-negative integer exponents can add up to 3.) Each of these terms must be multiplied by its corresponding multinomial coefficient. For instance, the coefficient of a^2c is

$$\binom{3}{2,0,1}=\frac{3!}{2!0!1!}=\frac{6}{2}=3.$$

Similarly, the coefficient of b^3 is

$$\binom{3}{0,3,0}=\frac{3!}{0!3!0!}=\frac{6}{6}=1.$$

Altogether, we obtain

$$(a+b+c)^3=a^3+3a^2b+3a^2c+3ab^2+3ac^2+6abc+b^3+3b^2c+3bc^2+c^3.$$

12.5 EXERCISES

PRACTICE

Consider each of the following situations and determine if each is a combination or permutation.

1. double scoop options from 29 ice cream flavors
2. a poker hand from a standard deck
3. a board committee chosen from 15 candidates
4. a seating chart for 24 students

Evaluate the following permutations. See Examples 5 and 6.

5. ${}_4P_2$
6. ${}_{15}P_2$
7. ${}_6P_5$
8. ${}_{19}P_{17}$

Evaluate the following combinations. See Example 7.

9. ${}_6C_4$
10. ${}_4C_2$
11. ${}_{12}C_5$
12. ${}_{21}C_{14}$

Determine how many different arrangements there are of the letters in each of the following words. See Example 8.

13. ABYSS
14. BANANA
15. COLLEGE
16. ALGEBRA
17. MATHEMATICS
18. FIBONACCI

Use the Binomial and Multinomial Theorems to expand each of the following expressions. See Examples 9, 10, and 11.

19. $(3x+y)^5$
20. $(x-2y)^7$
21. $(x-3)^4$
22. $(x^2-y^3)^4$
23. $(6x^2+y)^5$
24. $(4x+5y^2)^6$
25. $(7x^2+8y^2)^4$
26. $(x^3-y^2)^5$
27. $(x+y+z)^2$
28. $(a-2b+c)^3$
29. $(2x+5)^6$
30. $(2x+3y-z)^3$
31. What is the coefficient of the term containing x^3y in the expansion of $(2x+y)^4$?
32. What is the coefficient of the term containing x^4y^3 in the expansion of $(x^2-2y)^5$?
33. Find the first four terms in the expansion of $(x+3y)^{16}$.

34. Find the first three terms in the expansion of $(2x+3)^{13}$.

35. Find the first two terms in the expansion of $\left(3x^{\frac{1}{4}}+5y\right)^{17}$.

36. Find the 11th term in the expansion of $(x+2)^{24}$.

37. Find the 17th term in the expansion of $(2x+1)^{21}$.

38. Find the 9th term in the expansion of $(x-6y)^{12}$.

APPLICATIONS

Use the Multiplication Principle of Counting to answer the following questions. See Examples 1 and 2.

39. Suppose you write down someone's phone number on a piece of paper, but then accidentally wash it along with your laundry. Upon drying the paper, all you can make out of the number is 42? – 3?7?. How many different phone numbers fit this pattern?

40. How many different 7-digit phone numbers contain no odd digits? (Ignore the fact that certain 7-digit sequences are disallowed as phone numbers.)

41. How many different 7-digit phone numbers do not contain the digit 9? (Ignore the fact that certain 7-digit sequences are disallowed as phone numbers.)

42. A certain combination lock allows the buyer to set any combination of five letters, with repetition allowed, but each of the letters must be A, B, C, D, E, or F. How many combinations are possible?

43. In how many different orders can 15 runners finish a race, assuming there are no ties?

44. How many different 4-letter radio-station names can be made, assuming the first letter must be a K or a W? Assume repetition of letters is allowed.

45. How many different 4-letter radio-station names can be made from the call-letters K, N, I, T, assuming the letter K must appear first? Each of the four letters can be used only once.

46. Three men and three women line up in a row for a photograph, and decide men and women should alternate. In how many different ways can this be done? (Don't forget that the left-most person can be a man or a woman.)

47. How many different ways can a 10-question multiple choice test be answered, assuming every question has 5 possible answers?

48. How many different ways can your 12 favorite novels be arranged in a row?

49. How many different 6-character license plates can be formed if all 26 letters and 10 numerical digits can be used with repetition?

50. How many different 6-character license plates can be formed if all 26 letters and 10 numerical digits can be used without repetition?

51. How many different 6-character license plates can be formed if the first 3 places must be letters and the last 3 places must be numerical digits? (Assume repetition is not allowed.)

52. A box of crayons comes with 8 different colored crayons arranged in a single row. How many different ways can the crayons be ordered in the box?

Express the answer to the following permutation problems using permutation notation $\left({}_nP_k\right)$ and numerically. See Examples 3, 4, and 5.

53. Suppose you have a collection of 30 cherished math books. How many different ways can you choose 12 of them to arrange in a row?

54. In how many different ways can first-place, second-place, and third-place be decided in a 15-person race?

55. Suppose you need to select a user-ID for a computer account, and the system administrator requires that each ID consist of 8 characters with no repetition allowed. The characters you may choose from are the 26 letters of the alphabet (with no distinction between uppercase and lowercase) and the 10 digits. How many choices for a user-ID do you have?

56. How many different 5-letter "words" (they don't have to be actual English words) can be formed from the letters in the word PLASTIC?

57. Seven children rush into a room in which six chairs are lined up in a row. How many different ways can six of the seven children choose a chair to sit in? (The seventh remains standing.) How does the answer differ if there are seven chairs in the room?

58. At a meeting of 17 people, a president, vice president, secretary, and treasurer are to be chosen. How many different ways can these positions be filled?

59. Given 26 building blocks, each with a different letter of the alphabet printed on it, how many different 3-letter "words" can be formed?

Express the answer to the following combination problems using combination notation $\left({}_nC_k\right)$ and numerically. See Example 7.

60. In many countries, it is not uncommon for quite a few political parties to have their representatives in power. Suppose a committee composed of 10 Conservatives, 13 Liberals, 6 Greens, and 4 Socialists decides to form a sub-committee consisting of 3 Conservatives, 4 Liberals, 2 Greens and 1 Socialist. How many different such sub-committees can be formed?

61. A trade-union asks its members to select 3 people, from a slate of 7, to serve as representatives at a national meeting. How many different sets of 3 can be chosen?

62. Many lottery games are set up so that players select a subset of numbers from a larger set and the winner is the person whose selection matches that chosen by some random mechanism. The order of the numbers is irrelevant. How many choices of six numbers can be made from the numbers 1 through 49?

63. How many different lines can be drawn through a set of nine points in the plane, assuming that no three of the points are collinear? (Points are said to be *collinear* if a single line containing them can be drawn.)

64. Suppose you are taking a 10-question True-False test, and you are guessing that the professor has arranged it so that five of the answers are True and five are False. How many different ways are there of marking the test with five True answers and five False answers?

65. A caller in a Bingo game draws 5 marked ping pong balls from a basket of 75 and calls the numbers out to the players. How many different combinations are possible assuming that the order is irrelevant?

Use the techniques seen in this section, to answer the following questions.

66. How many different ways are there of choosing five cards from a standard 52-card deck and arranging them in a row? How many different five-card hands can be dealt from a standard 52-card deck?

67. Suppose you have 10 Physics texts, 8 Computer Science texts, and 13 Math texts. How many different ways can you select 4 of each to take with you on vacation?

68. Suppose you have 10 Physics texts, 8 Computer Science texts, and 13 Math texts. How many different ways can you select 4 of each and then arrange them in a row on a shelf, so that the books are grouped by discipline?

69. A certain ice cream store has four different kinds of cones and 28 different flavors of ice cream. How many different single-scoop ice cream cones is it possible to order at this ice cream store?

70. If a local pizza shop has three different types of crust, two different kinds of sauce, and five different toppings, how many different one topping pizzas can be ordered?

71. A man has 8 different shirts, 4 different shorts, and 3 different pairs of shoes. How many different outfits can the man choose from?

72. A couple wants to have three children. They want to know the different possible gender outcomes for birth order. How many different birth orders are possible?

73. A student has to make out his schedule for classes next fall. He has to take a math class, a science class, an elective, a history class, and an English class. There are three math classes to choose from, two science classes, four electives, three history classes, and four English classes. How many different schedules could the student have?

74. A basketball team has 12 different people on the team. The team consists of three point guards, two shooting guards, one weak forward, three power forwards, and three centers. How many different starting line-ups are possible? (The starting line-up will consist of one player in each of the 5 positions.)

75. How many 5-letter strings can be formed using the letters V, W, X, Y, and Z, if the same letter cannot be repeated?

76. If at the racetrack nine greyhounds are racing against each other, how many different first, second, and third place finishes are possible?

77. A basketball tryout has four distinct positions available on the team. If 25 people show up for tryouts, how many different ways can the four positions be filled?

78. If a trumpet player is practicing eight different pieces of music, in how many different orders can he play his pieces of music?

79. A pizza place has 12 total toppings to choose from. How many different 4-topping pizzas can be ordered?

80. How many different 5-digit numbers can be formed using each of the numbers 6, 8, 1, 9, and 4?

81. If eight cards are chosen randomly from a deck of 52, how many possible groups of eight can be chosen?

82. A baseball team has 15 players on the roster and abatting line-up consists of 9 players. How many different batting line-ups are possible?

83. How many different ways can two red balls, one orange ball, one black ball, and three yellow balls be arranged?

WRITING & THINKING

Pascal's triangle is a triangular arrangement of binomial coefficients, the first few rows of which appear as follows:

$$\begin{array}{c} 1 \\ 1 \quad 1 \\ 1 \quad 2 \quad 1 \\ 1 \quad 3 \quad 3 \quad 1 \\ 1 \quad 4 \quad 6 \quad 4 \quad 1 \\ \vdots \end{array}$$

Each number (aside from those on the perimeter of the triangle) is the sum of the two numbers diagonally adjacent to it in the previous row. Pascal's triangle is a useful way of generating binomial coefficients, with the nth row containing the coefficients of a binomial raised to the $(n-1)^{\text{st}}$ power. It can also be used to suggest useful relationships between binomial coefficients. Prove each of the following such relationships algebraically.

84. $\binom{n}{k}=\binom{n-1}{k-1}+\binom{n-1}{k}$ (Note that this is a restatement of how Pascal's triangle is formed.)

85. $\binom{n}{k}=\binom{n}{n-k}$

86. $\binom{n}{0}=\binom{n}{n}=1$

87. $\binom{n}{0}+\binom{n}{1}+\cdots+\binom{n}{n}=2^n$

(**Hint:** Use the Binomial Theorem on $(x+y)^n$ for a convenient choice of x and y.)

12.6 PROBABILITY

TOPICS

1. The language of probability
2. Computing probabilities using combinatorics
3. Unions, intersections, and independent events

TOPIC 1: The Language of Probability

The Problem of Points

As mentioned in the chapter introduction, the inspiration for modern mathematical probability began with a gambling quandary. The specific question posed to Pascal in 1654 was known as the Problem of Points, and it had been argued over for at least several centuries before Pascal and Fermat provided a well-reasoned solution. The problem supposes that two players of a game of chance have equal chances of winning and have agreed that the winner of a majority of games in a sequence will take the prize pot. If the sequence of games is interrupted before the end, how should the pot be fairly divided between the two players, based on the number of games each has won to that point?

We make use of the concept of *probability* nearly every day in a wide variety of ways. Most of the time, the use is informal; we might wonder what the probability of rain is during a game of tennis, or what the chances are of striking it rich by winning the state lottery. In mathematics, probability is much more rigorously defined, and the results of probabilistic analysis are used in many important applications.

DEFINITION: Terminology of Probability

An **experiment** is any activity that results in well-defined **outcomes**. In a given problem, we are usually concerned with finding the probability that one or more of the outcomes will occur, and we use the word **event** to refer to a set of outcomes.

The set of all possible outcomes of a given experiment is called the **sample space** of the experiment. This means that in the language of sets, an event is any subset of the sample space, including the empty set and the entire sample space.

For example, rolling a standard die is an experiment that has outcomes of the numbers 1, 2, 3, 4, 5, and 6. The sample space of this experiment is the set $S = \{1, 2, 3, 4, 5, 6\}$. The events consist of all the possible subsets of S. We might, for instance, be interested in the event E defined as "the number rolled is even." In terms of sets, $E = \{2, 4, 6\}$. Our intuition tells us that, if the die is a fair one, the probability of E occurring is one-half, as half the numbers are even and half are odd. We denote this by writing $P(E) = \frac{1}{2}$, which is read "the probability of E is one-half."

This intuition points the way toward the main tool we have for calculating probabilities.

FORMULA: Probabilities when Outcomes Are Equally Likely

We say that the outcomes of an experiment are *equally likely* if they all have the same probability of occurring. If E is an event of such an experiment, and if S is the sample space of the experiment, then the **probability of E** is given by

$$P(E) = \frac{n(E)}{n(S)}$$

where $n(E)$ and $n(S)$ are, respectively, the cardinalities of the sets E and S.

The formula tells us that probabilities are always going to be real numbers between 0 and 1, inclusive. The probability of an event E is 0 if E is the empty set, and 1 if E is the entire sample space S. In all other cases, the size of E is going to be between 0 and the size of S, so the fraction will yield a number between 0 and 1.

⚠ CAUTION

This formula has two important restrictions. First, it only applies in equally likely situations, so we can't use it to analyze weighted coins, crooked roulette tables, tampered decks of cards, and so on. Secondly, and more subtly, it assumes that the size of the sample space is finite (otherwise, the formula makes no sense), so we can't use it to analyze experiments based on, say, choosing a real number at random.

The complement of an event E, denoted E^C, is the set of all outcomes in the sample space that are not in E. Thus the probability of the complement of an event can be derived as follows.

$$P\left(E^C\right)=\frac{n(S)-n(E)}{n(S)}=\frac{n(S)}{n(S)}-\frac{n(E)}{n(S)}=1-P(E)$$

Example 1: Probabilities when Outcomes Are Equally Likely

A (fair) die is rolled once. Find the probability that the number rolled is

a. prime **b.** divisible by 5 **c.** 7

Solution

The sample space $S=\{1,2,3,4,5,6\}$ is the same for all three questions, and we can see that $n(S)=6$.

a. Let E be the event that the number is prime. Then $E=\{2,3,5\}$ since these are the the prime numbers between 1 and 6, so $n(E)=3$.
Then, $P(E)=\frac{n(E)}{n(S)}=\frac{3}{6}=\frac{1}{2}$.

b. Let F be the event that the number is divisible by 5. Then $F=\{5\}$, as this is the only integer from 1 to 6 that is divisible by 5.
So $P(F)=\frac{n(F)}{n(S)}=\frac{1}{6}$.

c. Let G be the event that the number is 7. In this case, $G=\varnothing$, the empty set, as there is no way for the top face to show a 7. This means $P(G)=0$.

TOPIC 2: Computing Probabilities Using Combinatorics

Most probability questions are not as basic as those in Example 1, and what makes a probability problem more complex is finding the size of an event and/or the sample space. This is a major application of the combinatorics techniques we covered in Section 12.5.

Example 2: Computing Probabilities

A pair of dice is rolled, and the sum of the top faces noted. What is the probability that the sum is

a. 2 **b.** 5 **c.** 7 or 11

Solution

The size of the sample space is the same for all three questions, so it makes sense to determine this first. In order to use the one probability formula we have, we need to make sure that all outcomes of the sample space are equally likely.

For this reason, we do *not* want to define the sample space to be all integers between 2 and 12. This is because these sums are not all equally likely. For instance, there is only one way for the sum to be 2: both top faces must show a 1. But there are two ways for the sum to be 3: one die (call it die A) shows a 1 and the other (die B) shows a 2, or else die A shows a 2 and die B shows a 1. Similarly, there are *three* ways for the sum of the top faces to be 4.

In order to define the sample space properly, we create a table of ordered pairs. In Table 1, the first number in each pair corresponds to the number showing on die A and the second number corresponds to the number showing on die B.

	1	2	3	4	5	6
1	(1,1)	(1,2)	(1,3)	(1,4)	(1,5)	(1,6)
2	(2,1)	(2,2)	(2,3)	(2,4)	(2,5)	(2,6)
3	(3,1)	(3,2)	(3,3)	(3,4)	(3,5)	(3,6)
4	(4,1)	(4,2)	(4,3)	(4,4)	(4,5)	(4,6)
5	(5,1)	(5,2)	(5,3)	(5,4)	(5,5)	(5,6)
6	(6,1)	(6,2)	(6,3)	(6,4)	(6,5)	(6,6)

TABLE 1: Sample Space for Rolling a Pair of Dice

Each of these ordered pairs is equally likely to come up, as any of the numbers 1 through 6 are equally likely for the first slot (die A) and similarly for the second slot (die B). Referring to the positions in the ordered pairs as "slots" points to a quick way of determining the size of the sample space. Since there are 6 choices for each slot, the Multiplication Principle of Counting tells us there are 36 possible outcomes of this experiment.

We can now proceed to answer the three specific questions.

a. There is only one ordered pair corresponding to a sum of 2, namely $(1,1)$, so the probability of this event is $\frac{1}{36}$.

b. There are four ordered pairs corresponding to a sum of 5:

$$\{(1,4),(2,3),(3,2),(4,1)\},$$

so the probability of this event is $\frac{4}{36}$, or $\frac{1}{9}$.

c. A sum of 7 or 11 comes from any of the following ordered pairs.

$$\{(1,6),(2,5),(3,4),(4,3),(5,2),(6,1),(5,6),(6,5)\}$$

Since there are eight elements in this event, the probability of rolling a sum of 7 or 11 is $\frac{8}{36}$, or $\frac{2}{9}$.

Example 3: Computing Probabilities

Suppose you are taking a 10-question True-False test, and you are completely unprepared for it. If you decide to guess on each question, what is the probability of getting 8 or more questions right?

Solution

The sample space for this problem consists of all the possible sequences of 10 answers, each of which is True or False. Combinatorics tells us that there are $2^{10} = 1024$ such sequences (2 choices for each of 10 "slots"), and these 1024 possible sequences are equally likely if your choice of True or False on each question is random.

The probability of getting 8 or more questions right can be broken up into three possibilities: getting 8 right, 9 right, or all 10 right.

There is only one sequence of 10 answers that are all correct. There are more ways of correctly answering exactly 9 questions, and we can use another tool from combinatorics to find out exactly how many ways. We need to select 9 "objects" from a set of 10 to be correctly answered questions, and the number of ways to do this is

$$_{10}C_9 = \frac{10!}{9!1!} = 10.$$

Similarly, there are $_{10}C_8$ ways of selecting 8 of the 10 questions, so there are

$$_{10}C_8 = \frac{10!}{8!2!} = \frac{10 \cdot 9}{2} = 45$$

ways of getting exactly 8 questions right.

Altogether, there are 45 + 10 + 1 = 56 ways of getting 8, 9, or 10 questions right, so the probability of this happening is

$$\frac{56}{1024} \approx 0.055.$$

This means you only have a 5.5% chance of scoring 80% or better by guessing, so studying ahead of time is clearly advantageous!

TOPIC 3: Unions, Intersections, and Independent Events

Since sample spaces and events are defined in terms of sets, it makes sense to use the terms *union* and *intersection* to describe ways of combining events. The notion of independence of events, on the other hand, is unique to probability. We will look first at probabilities of unions and intersections of events. For a review of set operations like union and intersection, see Section 1.1.

If E and F are two subsets of the same sample space S, then $E \cup F$ and $E \cap F$ are also subsets and constitute events in their own right. It makes sense, then, to talk about the probability of the events $E \cup F$ and $E \cap F$, and it's reasonable to suspect that they bear some relation to $P(E)$ and $P(F)$.

First, note that if all the outcomes in S are equally likely (as is the case with the problems we study in this section), then

$$P(E \cup F) = \frac{n(E \cup F)}{n(S)} \quad \text{and} \quad P(E \cap F) = \frac{n(E \cap F)}{n(S)},$$

so what we are really after are the relations between $n(E \cup F)$, $n(E \cap F)$, $n(E)$, and $n(F)$.

Consider again the experiment of rolling a die. Let E be the event "the number rolled is divisible by 2" and let F be the event "the number rolled is divisible by 3." For this small experiment, we can list the elements of each of the four events we are interested in, and then determine each event's cardinality:

$$E = \{2,4,6\};\ n(E) = 3$$
$$F = \{3,6\};\ n(F) = 2$$
$$E \cup F = \{2,3,4,6\};\ n(E \cup F) = 4$$
$$E \cap F = \{6\};\ n(E \cap F) = 1$$

Note that even though union is in some ways the set equivalent of numerical addition, the cardinality of the union of E and F in this example is not the sum of the cardinalities of E and F individually. The reason: both E and F contain the element 6, so if we simply add the sizes of E and F together to get the size of $E \cup F$, we wind up counting 6 twice. This happens in general when we try to determine the cardinality of a union of two sets, and we remedy the situation by subtracting the cardinality of the intersection from the sum of the individual cardinalities. By doing this, we count those elements that lie in both sets just once.

THEOREM: Cardinality of a Union of Sets

Let E and F be two finite sets. Then

$$n(E \cup F) = n(E) + n(F) - n(E \cap F).$$

Note that if E and F are *disjoint*, meaning they have no elements in common, then $n(E \cap F) = 0$, so $n(E \cup F) = n(E) + n(F)$.

We can use this fact to find a formula for the probability of the union of two events. Assuming E and F are two subsets of the same sample space S,

$$\begin{aligned} P(E \cup F) &= \frac{n(E \cup F)}{n(S)} \\ &= \frac{n(E)+n(F)-n(E \cap F)}{n(S)} \\ &= \frac{n(E)}{n(S)}+\frac{n(F)}{n(S)}-\frac{n(E \cap F)}{n(S)} \\ &= P(E)+P(F)-P(E \cap F). \end{aligned}$$

THEOREM: Probability of a Union of Two Events

Let E and F be two subsets of the same sample space. Then the **probability of the event "*E* or *F*,"** denoted $P(E \cup F)$, is given by the formula

$$P(E \cup F) = P(E)+P(F)-P(E \cap F).$$

The term $P(E \cap F)$ represents the probability of both events E and F happening. If, as sets, events E and F are disjoint (so $E \cap F = \varnothing$), then E and F are said to be *mutually exclusive*. In this case, $E \cap F = \varnothing$ and $P(E \cup F) = P(E)+P(F)$.

Example 4: Probability of a Union of Two Events

Assume a single die has been rolled and the number showing on top noted. Let E be the event "the number is divisible by 2" and let F be the event "the number is divisible by 3." Find the following probabilities.

a. $P(E \cap F)$ **b.** $P(E \cup F)$

Solution

We have already determined the sizes of all the relevant sets, so we are ready to apply the appropriate formulas.

a. Since $n(E \cap F) = 1,$ and since the sample space has six elements altogether,

$$P(E \cap F) = \frac{1}{6}.$$

b. Since $n(E) = 3$ and $n(F) = 2,$

$$\begin{aligned} P(E \cup F) &= P(E)+P(F)-P(E \cap F) \\ &= \frac{3}{6}+\frac{2}{6}-\frac{1}{6} \\ &= \frac{2}{3}. \end{aligned}$$

The formula for the probability of the union of three or more events can be complicated in general, and will be left for a later course. But if no two events have any elements in common, the formula is less complex. We say that events

$E_1, E_2, \ldots, E_n$ are **pairwise disjoint** if $E_i \cap E_j = \varnothing$ whenever $i \neq j$. Another way of saying this is that every possible pair of events in the collection $E_1, E_2, \ldots, E_n$ is a mutually exclusive pair.

THEOREM: Unions of Mutually Exclusive Events

If $E_1, E_2, \ldots, E_n$ are pairwise disjoint, then

$$P(E_1 \cup E_2 \cup \cdots \cup E_n) = P(E_1) + P(E_2) + \cdots + P(E_n).$$

Example 5: Unions of Mutually Exclusive Events

Suppose Jim has chosen a PIN of 8736 for his bank's ATM, and that all PINs at the bank consist of four digits from 0 to 9. Using the ATM one day, he senses someone looking over his shoulder as he enters the first two digits, and he decides to cancel the operation and leave. The next day Jim discovers his ATM card is missing. In the worst case scenario, a stranger now has Jim's ATM card and the first two digits of his PIN. As he calls the bank to cancel the card, he wonders what the chances are the unknown someone can guess the remaining digits in the three tries the ATM allows. What is the probability of this event?

Solution

If the first two digits are indeed known, the stranger has the task of filling in the last two digits, and there are 10 choices for each:

$$\underline{8} \quad \underline{7} \quad \underbrace{}_{10 \text{ choices}} \quad \underbrace{}_{10 \text{ choices}}$$

The size of the sample space is thus 100, and any guess at completing the PIN correctly has a probability equal to $\frac{1}{100}$ of being correct. Assuming the stranger tries three different ways of completing the PIN (so that the three events are pairwise disjoint), the probability of Jim's account being broken into is

$$\frac{1}{100} + \frac{1}{100} + \frac{1}{100} = \frac{3}{100}.$$

CAUTION

A common error often made in finding the answer to Example 5 gives an answer of $\frac{1}{100} + \frac{1}{99} + \frac{1}{98} = \frac{14{,}701}{485{,}100} \approx 0.0303$. This is often due to confusion about an area of statistics called conditional probability. The probability of guessing right on the second guess is not $\frac{1}{99}$, but is actually $\left(\frac{1}{99}\right)\left(\frac{99}{100}\right)$, where $\frac{99}{100}$ is the probability that the first guess is wrong. Note that $\left(\frac{1}{99}\right)\left(\frac{99}{100}\right) = \frac{1}{100}$.

The last probability idea we will consider is that of *independence*. Informally, we say that two events are independent if the occurrence of one of them has no effect on the occurrence of the other. Formally, independence of events is related to the probability of their intersection, as follows.

DEFINITION: Independent Events

Given two events E_1 and E_2 in the same sample space, we say E_1 and E_2 are **independent** if $P(E_1 \cap E_2) = P(E_1)P(E_2)$.

More generally, a collection of events $E_1, E_2, \ldots, E_n$ is **independent** if for any subcollection $E_{n_1}, E_{n_2}, \ldots, E_{n_k}$ of $E_1, E_2, \ldots, E_n$, it is true that

$$P(E_{n_1} \cap E_{n_2} \cap \cdots \cap E_{n_k}) = P(E_{n_1})P(E_{n_2})\cdots P(E_{n_k}).$$

Example 6: Independent Events

If a coin is flipped three times, what is the probability of it coming up tails all three times?

Solution

We actually have two good ways of answering this question, one using the notion of independence and one not.

Let E_i be the event "the coin comes up tails on the ith flip." We are interested, then, in the probability of $E_1 \cap E_2 \cap E_3$, the probability that we get tails each time. Since $P(E_i) = \frac{1}{2}$ for each $i = 1, 2, 3$ (remember, the coin is assumed to be fair), then

$$\begin{aligned} P(E_1 \cap E_2 \cap E_3) &= P(E_1)P(E_2)P(E_3) \\ &= \frac{1}{2} \cdot \frac{1}{2} \cdot \frac{1}{2} \\ &= \frac{1}{8}. \end{aligned}$$

The other way of obtaining the same answer is to consider the sample space made up of all possible three-toss sequences. Since each flip of the coin results in one of two possibilities, the Multiplication Principle of Counting tells us that there are $2^3 = 8$ possible sequences. Only one of these is the sequence consisting of three tails, so the probability of this event is $\frac{1}{8}$.

12.6 EXERCISES

PRACTICE

Below is the given probability that an event will occur; find the probability that it will not occur.

1. $P(E)=\frac{2}{5}$

2. $P(E)=0.72$

3. $P(E)=\frac{4}{13}$

4. $P(E)=0.15$

5. $P(E)=\frac{2}{3}$

6. $P(E)=0.49$

Apply the formulas for the probabilities of intersection and union to the following sets and determine **a.** $P(E\cap F)$ and **b.** $P(E\cup F)$. Let $n(S)$ equal the size of the sample space.

7. $n(S)=8, E=\{2,5\}, F=\{3,7,9\}$

8. $n(S)=10, E=\{1,2,5\}, F=\{1,2,3,5\}$

9. $n(S)=5, E=\{4,B\}, F=\{3\}$

10. $n(S)=8, E=\{A\}, F=\{B,C,D,E\}$

11. $n(S)=4, E=\{1,\beta\}, F=\{\alpha,2\}$

12. $n(S)=12, E=\{A,C,g,5,n,7,8,t,L\}, F=\{n,6\}$

13. $n(S)=16, E=\{1,2,A,m,13,Y,8\}, F=\{1,9,11,m\}$

14. $n(S)=11, E=\{m,7,D,4,\theta\}, F=\{\phi,D,3,7,m,\Sigma\}$

Determine the sample space of each of the following experiments.

15. A coin is flipped four times and the result recorded after each flip.

16. A card is drawn at random from the 13 hearts.

17. A coin is flipped and a card is drawn at random from the 13 hearts.

18. A quadrant of the Cartesian plane is chosen at random.

19. A slot machine lever is pulled; there are 3 slots, each of which can hold 6 different values.

20. An individual die is rolled twice and each of the two results is recorded.

21. At a casino, a roulette wheel spins until a ball comes to rest in one of the 38 pockets.

22. A lottery drawing consists of 6 randomly drawn numbers from 1 to 20; the order of the numbers matters in this case, and repetition is possible.

APPLICATIONS

23. An ordinary die is rolled. Find the probability of rolling

a. a 3 or higher.
b. an even composite number.

24. A card is drawn from a standard 52-card deck. Find the probability of drawing

a. a face card (jack, queen, or king) in the suit of hearts.
b. anything but an ace.
c. a black (clubs or spades) card that is not a face card.

25. A coin is flipped three times. Find the probability of getting

a. Heads exactly twice.
b. the sequence Heads, Tails, Heads.
c. two or more Heads.

26. A state lottery game is won by choosing the same six numbers (without repetition) as those selected by a mechanical device. The numbers are picked from the set $\{1,2,\ldots,49\}$, and the order of the numbers chosen is immaterial. What is the probability of winning?

27. What is the probability that a four-digit ATM PIN chosen at random ends in 7, 8, or 9?

28. Assume the probability of a newborn being male is one-half. What is the probability that a family with five children has exactly three boys?

29. What is the probability that a 9-digit driver's license number chosen at random will not have an 8 as a digit?

30. A roulette wheel in a casino has 38 pockets: 18 red, 18 black, and 2 green. Spinning the wheel causes a small ball to randomly drop into one of the pockets. All of the pockets are equally likely. The wheel is spun twice. Find the probability of getting

a. green both times.
b. black at least once.
c. red exactly once.

31. There is a 25% chance of rain for each of the next 2 days. What is the probability that it will rain on one of the days but not the other?

32. A pair of dice is rolled. Find the probability that the sum of the top faces is

a. seven.
b. seven or eleven.
c. an even number or a number divisible by three.
d. ten or higher.

33. A card is drawn from a standard 52-card deck. Find the probability of drawing

a. a face card or a diamond.
b. a face card but not a diamond.
c. a red face card or a king.

34. A state lottery game is won by choosing the same six numbers (without repetition) as those selected by a mechanical device. The numbers are picked from the set $\{1,2,\ldots,49\}$, and the order of the numbers chosen is immaterial. What is the probability of winning if someone buys 1000 tickets? (No two tickets have the same set of six numbers.) How many tickets would have to be bought to raise the probability of winning to one-half?

35. Two cards are drawn at random from a standard 52-card deck. What is the probability of them both being aces if

a. the first card is drawn, looked at, placed back in the deck, and the deck is then shuffled before the second card is drawn?
b. the two cards are drawn at the same time?

36. What is the probability of being dealt a five-card hand (from a standard 52-card deck) that has four cards of the same rank?

37. The probability of rain today is 75%, and the probability that Bob will forget to put the top up on his convertible is 25%. What is the probability of the inside of his car getting wet?

38. What is the probability of drawing 3 face cards in a row, without replacement, from a 52-card deck?

39. Two dice are rolled, and the difference is calculated by subtracting the smaller value from the larger value. Therefore, the difference may range from 0 to 5. Find the probability of each of the following differences:

a. 0
b. 1
c. 4

40. A letter is randomly chosen from the word MISSISSIPPI. What is the probability of the letter being an S?

41. Mike works in a company of 100 employees. If this year five people in the company are going to be randomly laid off, what is the probability that Mike will get laid off?

42. A pack of M&M's contains 10 yellow, 15 green, and 20 red pieces. What is the probability of choosing a green M&M out of the pack?

43. A jar of cookies has 3 sugar cookies, 4 chocolate chip cookies, and 2 peanut butter cookies. What is the probability of randomly choosing a peanut butter cookie out of the jar?

44. A big box of crayons contains 4 different blues, 3 different reds, 5 different greens, and 2 different yellows. What is the probability of randomly choosing a yellow crayon out of the box?

45. A bag of marbles contains 3 blue marbles, 2 red marbles, and 5 orange marbles. What is the probability of randomly picking a blue marble out of the bag?

46. Every week a teacher of a class of 25 randomly chooses a student to wash the blackboards. If there are 15 girls in the class, what is the probability that the student selected will be a boy?

47. If in a raffle 135 tickets are sold, how many tickets must be purchased for an individual to have a 20% chance of winning?

48. Zach is running for student council. At Zach's school the student council is chosen randomly from all qualified candidates. If there are 42 candidates running, including Zach, and a total of three positions, what is the probability that Zach will be selected for the council?

CHAPTER 12 PROJECT

Probability

You may be familiar with the casino game of roulette. But have you ever tried to compute the probability of winning on a given bet?

The roulette wheel has 38 total slots. The wheel turns in one direction and a ball is rolled in the opposite direction around the wheel until it comes to rest in one of the 38 slots. The slots are numbered 00, and 0–36. Eighteen of the slots between 1 and 36 are colored black and eighteen are colored red. The 0 and 00 slots are colored green and are considered neither even nor odd, and neither red nor black. These slots are the key to the house's advantage.

The following are some common bets in roulette:
A gambler may bet that the ball will land on a particular number, or a red slot, or a black slot, or an odd number, or an even number (not including 0 or 00). He or she could wager instead that the ball will land on a column (one of 12 specific numbers between 1 and 36), or on a street (one of 3 specific numbers between 1 and 36).

The payoffs for winning bets are

1 to 1 on odd, even, red, and black
2 to 1 on a column
11 to 1 on a street
35 to 1 any one number

1. Compute the probability of the ball landing on

a. a red slot.
b. an odd number.
c. the number 0.
d. a street (any of 3 specific numbers).
e. the number 2.

2. Based on playing each of the scenarios above (**a.–e.**) compute the winnings for each bet individually, if $5 is bet each time and all 5 scenarios lead to winnings.

3. If $1 is bet on hitting just one number, what would be the expected payoff? (**Hint:** Expected payoff is [(*probability of winning*) · (*payment for a win*)] − [(*probability of losing*) · (*payout for a loss*)].)

4. Given the information in question 3, would you like to play roulette on a regular basis? Why or why not? Why will the casino acquire more money in the long run?

CHAPTER 12 REVIEW EXERCISES

Section 12.1

Determine the first five terms of each sequence whose n^{th} term is defined as follows.

1. $a_n = (-3)^n$

2. $a_n = (-1)^n \sqrt[3]{n}$

3. $a_1 = -3, a_{n-1} = a_n + 1$ for $n \geq 2$

4. $a_n = \dfrac{n!}{n^n}$

Find a possible formula for the general n^{th} term of each sequence. Answers may vary.

5. $-7, -1, 5, 11, 17, \ldots$

6. $\dfrac{1}{2}, \dfrac{3}{4}, \dfrac{9}{8}, \dfrac{27}{16}, \dfrac{81}{32}, \ldots$

7. $0, 3, 8, 15, 24, 35, \ldots$

8. $\dfrac{3}{2}, \dfrac{5}{3}, \dfrac{7}{4}, \dfrac{9}{5}, \dfrac{11}{6}, \ldots$

9. $-2, -4, -12, -48, -240, \ldots$

10. $2, 6, 12, 20, 30, \ldots$

Translate each expanded sum that follows into summation notation, and vice versa. Then evaluate the sum.

11. $\sum_{i=3}^{8} (-2i + 3)$

12. $\sum_{i=2}^{7} (-2)^{i-1}$

13. $8 + 27 + 64 + \cdots + 343$

14. $\sum_{i=1}^{6} (2i - 3)$

15. $\sum_{i=1}^{5} -4(2^i)$

16. $8 + 18 + 32 + \cdots + 200$

Find a formula for the nth partial sum S_n of each series. If the series is finite, determine the sum. If the series is infinite, determine if it converges or diverges, and if it converges, determine the sum.

17. $\sum_{i=1}^{80} \left(\dfrac{1}{i+1} - \dfrac{1}{i+2} \right)$

18. $\sum_{i=1}^{\infty} \left(\dfrac{1}{i+1} - \dfrac{1}{i+2} \right)$

19. $\sum_{i=1}^{\infty} \left(3^i - 3^{i+1} \right)$

Determine the first five terms of each generalized Fibonacci sequence.

20. $a_1 = -2,\ a_2 = 5,$ and $a_n = a_{n-2} + a_{n-1}$ for $n \geq 3$

21. $a_1 = -10,\ a_2 = -12,$ and $a_n = a_{n-2} + a_{n-1}$ for $n \geq 3$

Section 12.2

Find the explicit formula for the general n^{th} term of each arithmetic sequence.

22. $5, 2, -1, -4, -7, \ldots$

23. $a_2 = 14$ and $a_4 = 19$

24. $a_7 = -43$ and $d = -9$

25. $a_1 = 2, a_4 = 11$

26. $a_9 = \dfrac{13}{2}, d = \dfrac{3}{4}$

27. $-5, 4, 13, 22, 31, \ldots$

Use the given information about each arithmetic sequence to answer the question.

28. Given that $a_1 = -2$ and $a_4 = -20$, what is a_{25}?

29. Given that $a_3 = 17$ and $a_7 = 29$, what is a_{89}?

30. In the sequence $8, 19, 30, \ldots$, which term is 668?

31. In the sequence $6, 1, -4, \ldots$, which term is -169?

32. In the sequence $\frac{8}{3}, \frac{10}{3}, 4, \ldots$, which term is $\frac{56}{3}$?

Find the value of the partial sum of each arithmetic sequence.

33. $\sum_{i=1}^{97} (2i - 7)$

34. $\sum_{i=1}^{60} (-4i + 3)$

35. Sylvia suspects that she has an ant infestation in her apartment. The first day she noticed them, she saw 10 ants in her kitchen. Each day she notices 4 more ants than on the previous day. If she doesn't call an exterminator, how many ants would she see on the fifteenth day?

Section 12.3

Find the explicit formula for the general n^{th} term of each geometric sequence.

36. $2, 8, 32, 128, 512, \ldots$

37. $3, \frac{3}{5}, \frac{3}{25}, \frac{3}{125}, \frac{3}{625}, \ldots$

38. $18, -6, 2, -\frac{2}{3}, \frac{2}{9}, \ldots$

39. $a_1 = 6$ and $a_4 = 384$

40. $a_2 = 20$ and $a_6 = 320$

41. $a_1 = 8$ and $a_4 = \frac{1}{8}$

Given the two terms of a geometric sequence, find the common ratio and first five terms of the sequence.

42. $a_1 = 4$ and $a_4 = 108$

43. $a_4 = \frac{5}{3}$ and $a_6 = \frac{20}{27}$

Use the given information about each geometric sequence to answer the question.

44. Given that $a_2 = \frac{3}{5}$ and $a_4 = \frac{1}{15}$, what is the common ratio r?

45. Given that $a_1 = 3$ and $a_4 = -24$, what is the common ratio r?

46. Given that $a_5 = -16$ and $a_6 = -4$, what is a_{11}?

Each of the following sums is a partial sum of a geometric sequence. Use this fact to evaluate the sums.

47. $\sum_{i=3}^{9} 3\left(\frac{1}{2}\right)^i$

48. $5 + 10 + \cdots + 20{,}480$

Determine if each of the following infinite geometric series converges. If so, find the sum.

49. $\sum_{i=0}^{\infty} -3\left(\frac{3}{4}\right)^i$

50. $\sum_{i=1}^{\infty}\left(-\frac{5}{4}\right)^i$

51. $\sum_{i=1}^{\infty}\frac{2}{5}\left(\frac{5}{7}\right)^i$

Section 12.4

Use the Principle of Mathematical Induction to prove the following statements.

52. $\frac{1}{1\cdot 3}+\frac{1}{3\cdot 5}+\frac{1}{5\cdot 7}+\cdots+\frac{1}{(2n-1)(2n+1)}=\frac{n}{2n+1}$

53. $5+8+11+\cdots+(3n+2)=\frac{n(3n+7)}{2}$

54. $1\cdot 3+2\cdot 4+3\cdot 5+\cdots+n(n+2)=\frac{n(n+1)(2n+7)}{6}$

55. For all natural numbers n, 11^n-7^n is divisible by 4.

56. For all natural numbers n, 7^n-1 is divisible by 3.

Section 12.5

Use the Multiplication Principle of Counting and the permutation and combination formulas to answer the following questions.

57. A license plate must contain 4 numerical digits followed by 3 letters. If the first digit cannot be 0 or 1, how many different license plates can be created?

58. How many different 7-digit phone numbers do not contain the digits 6 or 7?

59. In how many different orders can the letters in the word "aardvark" be arranged?

60. In how many different ways can first place, second place, and third place be awarded in a 10-person shot put competition?

61. At a meeting of 21 people, a president, vice president, secretary, treasurer, and recruitment officer are to be chosen. How many different ways can these positions be filled?

62. A college admissions committee selects 4 out of 12 scholarship finalists to receive merit-based financial aid. How many different sets of 4 recipients can be chosen?

Use the Binomial and Multinomial Theorems to expand each of the following expressions.

63. $(1-2y)^5$

64. $(x+2)^7$

65. $(5x^2-2y)^5$

66. $(x+2y+z)^3$

Section 12.6

Apply the formulas for the probabilities of intersection and union to the following sets and determine **a.** $P(E \cap F)$ and **b.** $P(E \cup F)$. Let $n(S)$ equal the size of the sample space.

67. $n(S)=9,\ E=\{3,5,7\},\ F=\{1,2,3,4\}$

68. $n(S)=6,\ E=\{A,B\},\ F=\{X,Y,Z\}$

69. $n(S)=7,\ E=\{\alpha,13\},\ F=\{\alpha,\beta,13,14\}$

70. $n(S)=8,\ E=\{a,4,m,7\},\ F=\{m,3,s\}$

71. $n(S)=10,\ E=\{3,4,X,Y,5,Z\},\ F=\{5,6,7\}$

72. A card is drawn from a standard 52-card deck. Find the probability of drawing

a. a seven or a club.

b. a face card but not a red queen.

c. a black three or a spade.

73. What is the probability of being dealt a five-card hand (from a standard 52-card deck) that contains only face cards?

74. There is a 10% chance of rain each individual day for an entire week. What is the probability that it will rain at least once during this seven- day period?

CHAPTER 13

An Introduction to Limits, Continuity, and the Derivative

SECTIONS

WHAT IF ...

What if you're given a function that tells you the position of an object over time, such as a ball that has been dropped? Can you translate knowledge of the ball's position into knowledge of its speed?

By the end of this chapter, you'll know how the mathematical concept of the *derivative* is used to answer this question and many others, some that are clearly related and some that appear initially to have little in common.

Archimedes of Syracuse

Introduction

This chapter opens with a discussion of two broadly defined problems whose solutions turn out to have much in common. Mathematically, the task of determining the instantaneous velocity of an object and the task of finding the line tangent to the graph of a function both depend on the concept of *limit*. Much of this chapter revolves around developing an intuitive sense as well as a rigorous definition of the concept.

The limit idea inherently involves motion and reflects a major difference between the relatively static world of algebra and the dynamic world of calculus. As with so many concepts in mathematics, the evolution of the idea spans cultures and ages. Some early thinkers, such as the philosopher Zeno of Elea (ca. 495–ca. 430 BC) and the mathematician Archimedes of Syracuse (ca. 287–ca. 212 BC), developed an appreciation of the power and depth of the concept centuries before later mathematicians overcame the difficulties of rigorously defining and using limits.

Zeno, in fact, is chiefly remembered today for the many paradoxes he devised, which illustrate the danger of naïve assumptions about limits and infinity—a variation of one of these paradoxes appears as Example 2 in Section 13.2. Archimedes, whose name usually appears on any list of the greatest mathematicians of all time, used methods we would now classify as belonging to calculus to achieve results that were centuries ahead of their time. Formulas for the area under a parabola and for the volumes and surface areas of certain three-dimensional objects are just a few examples.

Much later, in the seventeenth century AD, the ideas and methods involving limits were brought together and ultimately characterized as "calculus." But even after calculus came to be recognized as an especially rich branch of mathematics, the limit concept retained its somewhat dangerous reputation and continued to trick unwary thinkers. Great discoveries and advances were made using calculus throughout the 1600s and 1700s by a long list of famous mathematicians, but rigorous definitions of "limit" and related ideas didn't appear until the 1800s. The definition we use today is essentially due to the French mathematician Augustin-Louis Cauchy (1789–1857), who also refined and made rigorous the idea of *continuity* as it applies to functions.

The two problems that open the chapter serve as motivation for the mathematics that follows, but by the end of the chapter it will be apparent that they are only representatives of the many problems that can be solved by means of the *derivative* of a function.

13.1 RATES OF CHANGE AND TANGENTS

TOPICS

1. The difference quotient
2. The velocity problem
3. The tangent problem

TOPIC 1: The Difference Quotient

We first encountered the difference quotient in Section 4.2, while discussing the average rate of change of functions. In this section, we'll revisit the idea and see that it lies at the heart of two specific problems and, in fact, an entire branch of mathematics called *differential calculus*. We begin with a reminder of the definition of the difference quotient and related terminology.

DEFINITION: The Difference Quotient and Secant Line

Given a real number c, a nonzero real number h (which may be positive or negative), and a function f defined on an interval containing the points c and $c+h$, the **difference quotient of f at c with increment h** is the ratio

$$\frac{f(c+h)-f(c)}{h}.$$

If $y=f(x)$, any of the following expressions can be used to represent the average rate of change of f over the interval with endpoints c and $c+h$:

$$\frac{\text{change in } f}{\text{change in } x}=\frac{\text{change in } y}{\text{change in } x}=\frac{\Delta f}{\Delta x}=\frac{\Delta y}{\Delta x}=\frac{f(c+h)-f(c)}{h},$$

where the Greek letter Δ denotes a change in the variable that follows it. The average rate of change of f over the interval with endpoints c and $c+h$ represents the slope of the **secant line** drawn between the points $(c, f(c))$ and $(c+h, f(c+h))$ on the graph of f, as shown in Figure 1.

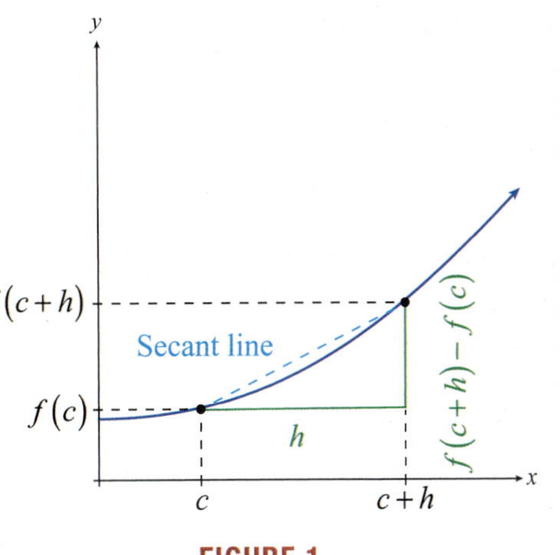

FIGURE 1

Example 1: Constructing and Simplifying the Difference Quotient

For each given function, construct and simplify the difference quotient at c with increment h.

a. $f(x)=-x^2+3x+2$ **b.** $g(x)=\frac{3}{x}$ **c.** $s(x)=\sqrt{x}$

Solution

a. The difference quotient of the function $f(x)=-x^2+3x+2$ at c with increment h is

$$\begin{aligned}\frac{f(c+h)-f(c)}{h} &= \frac{-(c+h)^2+3(c+h)+2-(-c^2+3c+2)}{h} \\ &= \frac{-c^2-2ch-h^2+3c+3h+2+c^2-3c-2}{h} \\ &= \frac{-2ch-h^2+3h}{h} \\ &= -2c-h+3.\end{aligned}$$

b. The difference quotient of the function $g(x)=\frac{3}{x}$ at c with increment h is

$$\begin{aligned}\frac{g(c+h)-g(c)}{h} &= \frac{\frac{3}{c+h}-\frac{3}{c}}{h} \\ &= \left(\frac{1}{h}\right)\left[\left(\frac{3}{c+h}\right)\left(\frac{c}{c}\right)-\left(\frac{3}{c}\right)\left(\frac{c+h}{c+h}\right)\right] \\ &= \left(\frac{1}{h}\right)\left(\frac{3c-3c-3h}{c^2+ch}\right) \\ &= \frac{-3}{c^2+ch}.\end{aligned}$$

c. The difference quotient of the function $s(x)=\sqrt{x}$ at c with increment h is

$$\frac{s(c+h)-s(c)}{h}=\frac{\sqrt{c+h}-\sqrt{c}}{h}.$$

For the moment, this solution is perfectly acceptable. However, we will see in Section 13.6 that there is good reason to "rationalize the numerator" and rewrite the expression. We can accomplish this with the following additional steps.

$$\begin{aligned}\frac{\sqrt{c+h}-\sqrt{c}}{h} &= \left(\frac{\sqrt{c+h}-\sqrt{c}}{h}\right)\left(\frac{\sqrt{c+h}+\sqrt{c}}{\sqrt{c+h}+\sqrt{c}}\right) \\ &= \frac{c+h-c}{h\left(\sqrt{c+h}+\sqrt{c}\right)} \\ &= \frac{1}{\sqrt{c+h}+\sqrt{c}}\end{aligned}$$

Example 2: Using the Difference Quotient to Determine the Average Rate of Change

Use the difference quotient of the function $f(x)=-x^2+3x+2$ at $c=3$ to determine the average rate of change of f over each of the given intervals.

a. $[2,3]$ **b.** $[2.9,3]$ **c.** $[3,3.0001]$

Solution

We've already found in Example 1 that the difference quotient of $f(x) = -x^2 + 3x + 2$ at c with increment h is $-2c - h + 3$, so we will use this result with $c = 3$ and appropriate values of h.

a. For the interval $[2,3]$, $h = -1$ so

$$-2c - h + 3 = -2(3) - (-1) + 3 = -6 + 1 + 3 = -2.$$

b. For the interval $[2.9,3]$, $h = -0.1$ so

$$-2c - h + 3 = -2(3) - (-0.1) + 3 = -6 + 0.1 + 3 = -2.9.$$

c. For the interval $[3,3.0001]$, $h = 0.0001$ so

$$-2c - h + 3 = -2(3) - (0.0001) + 3 = -6 - 0.0001 + 3 = -3.0001.$$

TOPIC 2: The Velocity Problem

We have already been introduced to the velocity problem in Section 4.2, when we discussed average rate of change. We are now ready to analyze the problem more fully and explore the deeper meaning of its solution.

The simple relationship *distance* = *rate* × *time*, or $d = rt$ is often one of the first applications of basic algebra to be studied. This equation can be taken as the definition of average rate r of travel for an object moving distance d over time t (that is, the **average rate** is the *ratio* of d over t), or it can be used to solve for any one of the variables if the other two quantities are known. It is undeniably useful in situations where an object's average rate is all that is desired (especially so if the object is moving at a constant rate), but it is just as undeniably limited in scope. Most of our everyday experiences involve objects that move at varying speeds, and we don't have to progress very far in mathematics before needing a way to handle more than just *average* rates of travel.

By convention, we often use the word *velocity* or *speed* when discussing a rate of travel that is not necessarily constant (later, we will further refine our usage and differentiate between speed and velocity). To emphasize the distinction with average rate even more, the word *instantaneous* sometimes appears before velocity. One of the best examples of instantaneous velocity is the number you read off your speedometer as you drive—at any given moment, the gauge indicates your velocity *at that instant.*

One form of the "velocity problem" can be put in this context: Suppose you are given the task of calculating a car's instantaneous velocity from perfect knowledge about its distance traveled over various lengths of time but without benefit of the speedometer. How would you do it?

To give us something to visualize and work with, suppose that the car is traveling along a straight road and $d(t)$ represents its location, relative to its starting point, at time t; we can set $t = 0$ as the moment when the car starts moving, so $d(0)$ is its initial location. Units of measurement are immaterial at the moment so they will not be specified—it is important only to remember that $d(t)$ represents distance with respect to time t. Let's suppose further that we want to determine the car's velocity at time $t = 10$. Knowing nothing more than the relationship $d = rt$, we can easily compute some average rates over time intervals that contain $t = 10$. For example, the average rate of travel from $t = 9$ to $t = 11$ is calculated as follows.

$$\frac{d(11)-d(9)}{11-9}=\frac{d(11)-d(9)}{2}$$

If we suspect that the instantaneous velocity of the car varies quite a bit over the time interval $[9,11]$ and that the average rate is not a good reflection of the velocity at exactly $t = 10$, we can easily compute the average rate over a shorter interval, say $[9.5,10.5]$.

$$\frac{d(10.5)-d(9.5)}{10.5-9.5}=d(10.5)-d(9.5)$$

And there's no need to stop there. Our intuition tells us that as the time intervals get shorter and shorter, the average rate of travel over those time intervals should get closer and closer to the instantaneous velocity at $t = 10$.

With a bit more thought, our intuition may tell us something else. Namely, it really shouldn't matter much if the small intervals we're considering are centered at $t = 10$ or not, as long as the intervals all contain that point in time. After all, the average rate of travel that we are computing is an approximation to the instantaneous velocity at *every* point of the interval, not just the midpoint. In practice, we may find it easier to use intervals that have our point of interest at an endpoint, rather than the midpoint. The difference quotient concept, with the point of interest as the common interval endpoint, is ideally suited for this purpose.

Example 3: Using Average Velocities to Estimate Instantaneous Velocity

A piece of rock falls into a 256-foot-deep canyon from its rim. If we ignore air resistance, its distance below the rim at time t is $d(t)=16t^2$, where d is measured in feet and t is measured in seconds. Estimate

a. the instantaneous velocity of the rock after 2 seconds.

b. the velocity of impact.

Solution

a. We will start by calculating the average velocity of the rock from $t = 2$ to $t = 3$. Note that the elapsed time is $\Delta t = 3 - 2 = 1$, while the distance traveled is $\Delta d = d(3)-d(2)$, so we can set up the following difference quotient to calculate the desired average velocity over the interval $[2,3]$.

$$\begin{aligned}\frac{\Delta d}{\Delta t}&=\frac{d(3)-d(2)}{3-2}\\&=\frac{16\cdot 3^2-16\cdot 2^2}{1}\\&=16\cdot 9-16\cdot 4=80\text{ ft/s}\end{aligned}$$

However, this is not expected to be an accurate reflection of the instantaneous velocity at $t = 2$, since the rock is accelerating significantly between $t = 2$ and $t = 3$. We obtain a more accurate guess if we work over a much shorter time interval, say the one from $t = 2$ to $t = 2.1$.

$$\begin{aligned}\frac{\Delta d}{\Delta t} &= \frac{d(2.1)-d(2)}{2.1-2}\\ &= \frac{16\cdot(2.1)^2-16\cdot 2^2}{0.1}\\ &= \frac{16\cdot(4.41)-16\cdot 4}{0.1} = 65.6 \text{ ft/s}\end{aligned}$$

This is a much better guess, but there is no need to stop here. Table 1 summarizes the average velocities over time intervals of decreasing lengths, all the way down to a length of 0.01 seconds.

Time Interval	$[2,2.09]$	$[2,2.08]$	$[2,2.07]$	$[2,2.06]$	$[2,2.05]$	$[2,2.04]$	$[2,2.03]$	$[2,2.02]$	$[2,2.01]$
Average Velocity $\Delta d/\Delta t$	65.44	65.28	65.12	64.96	64.8	64.64	64.48	64.32	64.16

TABLE 1

Since the velocity numbers in the table seem to be approaching 64 ft/s as the intervals are becoming shorter, we might guess that number to be the instantaneous velocity of the rock at $t = 2$. Notice also that in Table 1, 2 serves as the left endpoint of all of the intervals. Intuitively, we expect the same instantaneous velocity to be obtained if $t = 2$ were the right endpoint, or if it were in the interior of each interval, as long as the interval's length is decreasing and approaching 0. In our next table, we chose $t = 2$ to be the right endpoint of each time interval.

Time Interval	$[1.91,2]$	$[1.92,2]$	$[1.93,2]$	$[1.94,2]$	$[1.95,2]$	$[1.96,2]$	$[1.97,2]$	$[1.98,2]$	$[1.99,2]$
Average Velocity $\Delta d/\Delta t$	62.56	62.72	68.88	63.04	63.2	63.36	63.52	63.68	63.84

TABLE 2

Notice that the average velocities here are all less than in Table 1, but are increasing as the time interval is getting shorter, "shrinking" onto the point $t = 2$. This is precisely what should be expected because of the steady acceleration of the rock. In conclusion, both tables support our guess that the rock's instantaneous velocity at $t = 2$ is 64 ft/s downward. As a final confirmation, let us calculate the difference quotient over the very short interval $[1.999,2]$.

$$\begin{aligned}\frac{\Delta d}{\Delta t} &= \frac{d(2)-d(1.999)}{2-1.999}\\ &= \frac{16\cdot 2^2-16\cdot(1.999)^2}{0.001}\\ &= 63.984 \text{ ft/s}\end{aligned}$$

b. To estimate the velocity of impact, we first need to find when exactly impact occurs. Being in possession of the position function and knowing that at impact the position of the rock is exactly the depth of 256 feet, we can find the time of impact by solving

$$d(t) = 256$$
$$16t^2 = 256$$
$$t^2 = 16,$$

which gives us $t = 4$, since we can safely discard the negative root of $t = -4$.

To find the velocity of impact, we use the same strategy as in part a. but this time using $t = 4$ as the right endpoint of our intervals for our difference quotients. Table 3 summarizes our calculations.

Time Interval	$[3.91,4]$	$[3.92,4]$	$[3.93,4]$	$[3.94,4]$	$[3.95,4]$	$[3.96,4]$	$[3.97,4]$	$[3.98,4]$	$[3.99,4]$
Average Velocity $\Delta d/\Delta t$	126.56	126.72	126.88	127.04	127.2	127.36	127.52	127.68	127.84

TABLE 3

As before, we examine the table and conclude that the velocity of impact is approximately 128 ft/s (you may want to further confirm our finding by calculating difference quotients over even shorter time intervals).

TOPIC 3: The Tangent Problem

Our second problem will be presented in a purely mathematical context, but, as we will see, it is closely related to the first.

The word "tangent" comes from the Latin word *tangens*, and translates as "touching" (and while there is a connection with the trigonometric function that goes by the name tangent, that connection is not especially instructive at the moment). In mathematics, a line is tangent to a given curve if it "just touches" the curve. In some cases, it is easy to identify such lines; in Figure 2, each of the two solid lines is tangent to its associated curve, while the dashed lines are not. In these two examples, the tangent line touches the curve at exactly one point and doesn't cross the curve.

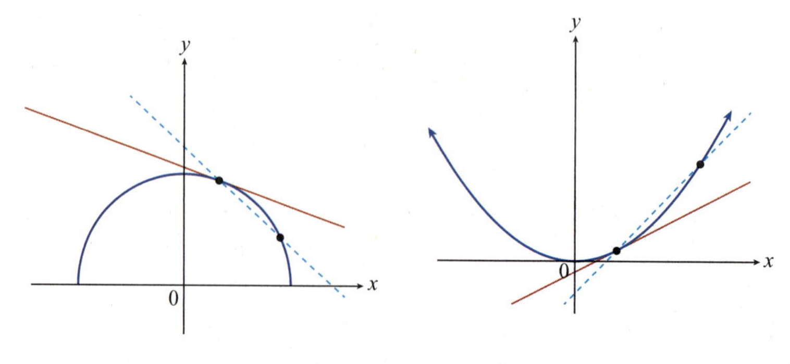

FIGURE 2: Tangent and Nontangent Lines

In many cases, however, it is not at all clear how to construct the line tangent to a curve at a given point, and, in any event, we will soon need a precise method of construction that results in an equation for a line (as opposed to just a rough sketch). How should we proceed?

Our two simple examples in Figure 2 are perhaps deceptive. If we think a bit harder about what it means for a line to be tangent to a curve, we might be led to the idea that a tangent line should represent the relative rise or fall of the curve at a particular point—in other words, the tangent line should have the same "slope" or trend as the curve at that point. With this interpretation, it's clear that sometimes a tangent line will intersect the curve at more than one point, and that sometimes it is not possible for a unique line to capture the trend at all. Figure 3 illustrates these possibilities.

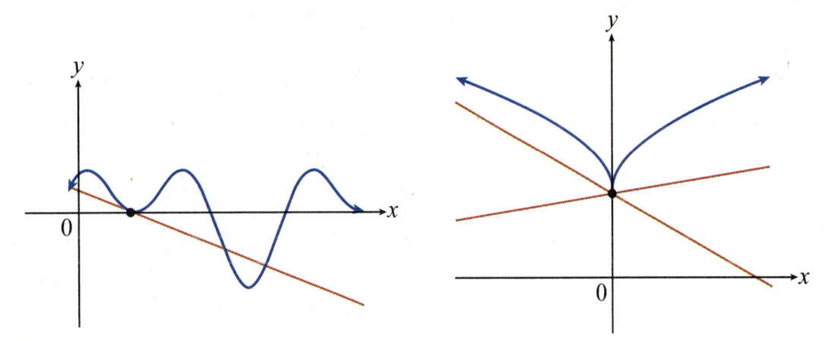

FIGURE 3: More Tangent and Nontangent Lines

To nail down precisely what we mean by "tangent," and to work toward an equation form for the tangent line, we can follow an approach similar to what we used for the velocity problem. Namely, we will use a process that allows us to construct approximations to the tangent line that are as accurate as we desire.

To set the stage, let's assume that we have a function f for which we want to construct the tangent line at $x = c$, that is, at the point $(c, f(c))$ on the graph of f. Note that all we really need is the slope of the tangent line, since we already know that we want the tangent to pass through the point $(c, f(c))$. While we don't yet know the slope of the tangent line, we *do* know the slope of any line that passes through $(c, f(c))$ and a point on the graph of f slightly removed from $(c, f(c))$; we have already defined such a line as a secant line (again, not to be confused with the trigonometric function secant). An easy way to denote a "slightly removed" point on the graph is to let $x = c + h$, where h is as close (but not equal) to 0 as we wish it to be. The "slightly removed" point is then $(c+h, f(c+h))$, and the slope of the associated secant line is as follows.

$$\frac{f(c+h)-f(c)}{(c+h)-c} = \frac{f(c+h)-f(c)}{h}$$

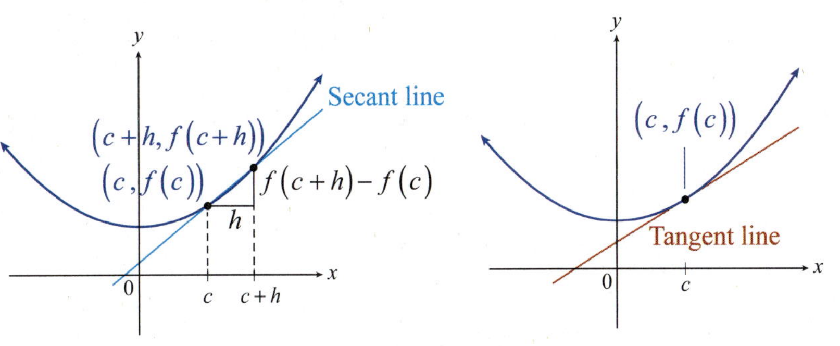

FIGURE 4

Figure 4 illustrates the connection between the tangent problem and the velocity problem as we have just constructed a difference quotient again. To make the connection even more apparent, note that in both problems the key concept is "rate of change." In the first, we seek the instantaneous rate of change of the position of an object, and we label the result "velocity." In the second, we seek the rate of change of the graph of a function at a particular point, and we interpret the result as the slope of the line tangent to the graph at that point. The next few examples illustrate how we can actually construct such tangent lines.

Example 4: Using Difference Quotients to Approximate the Slope of a Tangent Line

Construct difference quotients to approximate the slope of the line tangent to the graph of $f(x)=0.2x^3-1.8x^2+3.6x$ at the point $(1,2)$.

Solution

Note that in this problem, $c=1$. Just as in Example 3, we will construct difference quotients using several h-values that are decreasing in magnitude. To start off, letting $h=1$ we obtain the following.

$$\begin{aligned}\frac{f(1+1)-f(1)}{1}&=f(2)-f(1)\\&=0.2\cdot 2^3-1.8\cdot 2^2+3.6\cdot 2-\left(0.2\cdot 1^3-1.8\cdot 1^2+3.6\cdot 1\right)\\&=-0.4\end{aligned}$$

This means that -0.4 is the slope of the secant line corresponding to $h=1$. However, given that $c=1$, our choice of $h=1$ means that the x-coordinate of the "slightly removed" point in this case is $x=2$, which is a rather big step away from c, so we don't yet expect the slope of -0.4 to reflect the trend of the graph in any meaningful way. Let us check what happens if $h=0.5$.

$$\begin{aligned}\frac{f(1+0.5)-f(1)}{0.5}&=2\cdot\left[f(1.5)-f(1)\right]\\&=2\cdot\left[0.2\cdot(1.5)^3-1.8\cdot(1.5)^2+3.6\cdot 1.5-\left(0.2\cdot 1^3-1.8\cdot 1^2+3.6\cdot 1\right)\right]\\&=0.05\end{aligned}$$

This shows us that the associated secant line is now nearly horizontal, and its slope went from negative to positive. Decreasing h even further, we evaluate several more difference quotients, and Table 4 displays our results.

h	0.3	0.1	0.05	0.01	0.001
m_{sec}	0.258	0.482	0.5405	0.58802	0.59880

TABLE 4

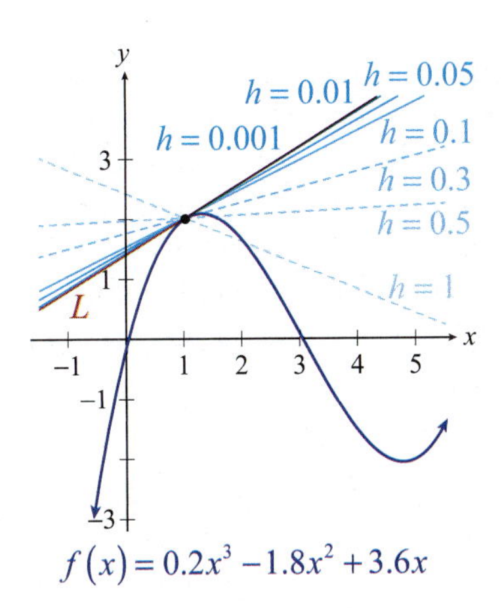

$f(x) = 0.2x^3 - 1.8x^2 + 3.6x$

FIGURE 5

In the second row of Table 4, the slope of the secant associated with each corresponding h-value is denoted by m_{sec}, and we conclude that the slope of the actual tangent line approaches $m = 0.6$. (Later we will learn how to confirm that it actually is equal to 0.6.)

Figure 5 shows the graph of $f(x)$ and the calculated secant lines, as well as the predicted tangent. Note how the secant lines move closer and closer to the tangent as h decreases. Also, the secant line corresponding to $h = 0.001$ is virtually indistinguishable from the tangent line L shown in red.

Example 5: Using Difference Quotients to Show that a Point has No Tangent Line

Show that the slope of a tangent line of the absolute value function $A(x) = |x|$ at $(0,0)$ can't be inferred from a table of difference quotients.

Solution

In this example, $c = 0$, so using the formula for the difference quotient with, say, $h = 0.5$ we obtain the following.

$$\frac{A(c+h) - A(c)}{h} = \frac{|0+h| - |0|}{h}$$
$$= \frac{|0.5|}{0.5} = 1$$

On the other hand, if h is negative, say $h = -0.5$, the difference quotient is as follows.

$$\frac{A(c+h) - A(c)}{h} = \frac{|0+h| - |0|}{h}$$
$$= \frac{|-0.5|}{-0.5}$$
$$= \frac{0.5}{-0.5} = -1$$

In fact, if we construct a table like before, but this time alternating between positive and negative h-values, something curious happens.

h	0.1	−0.1	0.05	−0.05	0.01	−0.01
m_{sec}	1	−1	1	−1	1	−1

TABLE 5

First of all, it seems that ± 1 are the only possible values for the slope of the secant, and second, it seems to be $+1$ when $h > 0$ and -1 when $h < 0$. Notice that this is an observation we can prove fairly easily.

If $h > 0$, since in this case $|h| = h$, the difference quotient becomes

$$\frac{|0+h|}{h} = \frac{|h|}{h} = \frac{h}{h} = 1,$$

regardless of the value of h. On the other hand, if $h < 0$, then $|h| = -h$, so

$$\frac{|0+h|}{h} = \frac{|h|}{h} = \frac{-h}{h} = -1,$$

and again, h can be any negative real number.

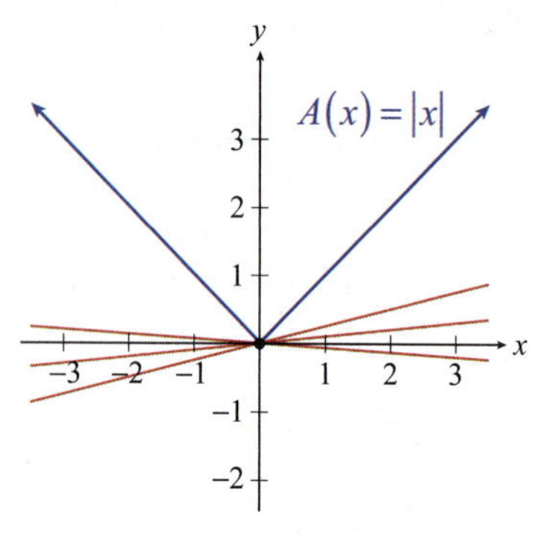

FIGURE 6

Notice that our results indicate that any secant line corresponding to a positive h-value will coincide with the right branch of the graph, while all secant lines obtained from $h < 0$ coincide with the left branch.

What does all this mean? There is no single real number being approached by the difference quotients or slopes as h is getting smaller. Graphically, this means that there is no tangent, no single line that would "best capture" the trend of the graph, none that would best align itself to the graph of $A(x)$ at $(0,0)$. The graph has a "sharp turn" there, making it impossible for a tangent line to exist, as shown in Figure 6. We also express this fact by saying that the graph is not "smooth" at $x = 0$.

13.1 EXERCISES

PRACTICE

Construct and simplify the difference quotient at c with increment h. See Example 1.

1. $f(x) = -7x + 5$

2. $g(x) = 2x + 13$

3. $p(x) = x^2 - 5x + 3$

4. $g(x) = 9x^2 - 17$

5. $f(x) = \dfrac{1}{6-x}$

6. $f(x) = \dfrac{-x}{3x+1}$

7. $g(x) = \ln x$

8. $p(x) = \dfrac{x^2 - 1}{x+3}$

9. $f(x) = 3\sqrt{x-2}$

Use the difference quotient of each function for one appropriate value of c to determine the average rate of change of the function over each of the given intervals. See Example 2.

10. $f(x) = -7x + 5$ **a.** $[5,7]$ **b.** $[5,5.5]$ **c.** $[4.9,5]$

11. $g(x) = 2x + 13$ **a.** $[-3,-1]$ **b.** $[-1,0]$ **c.** $[-1.0001,-1]$

12. $p(x) = x^2 - 5x + 3$ **a.** $[1,2]$ **b.** $[0.9,1]$ **c.** $[1,1.0001]$

13. $f(x)=\dfrac{1}{6-x}$ **a.** $[2,3]$ **b.** $[3,3.1]$ **c.** $[2.999,3]$

14. $f(x)=3\sqrt{x-2}$ **a.** $[3,4]$ **b.** $[2.9,3]$ **c.** $[3,3.0001]$

Estimate the slope of the tangent line shown in the given graph.

15.

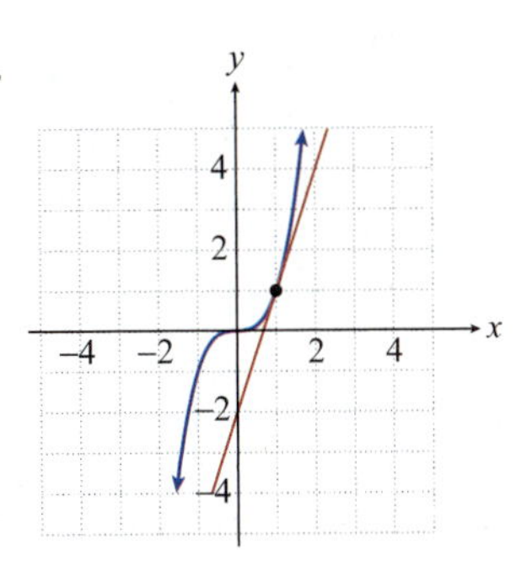

16.

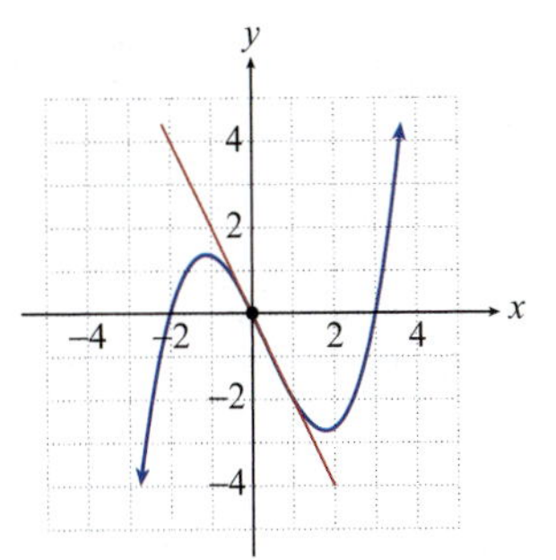

17.

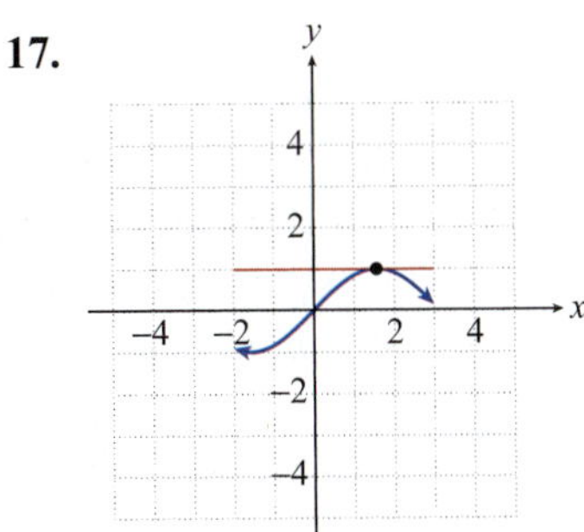

18.

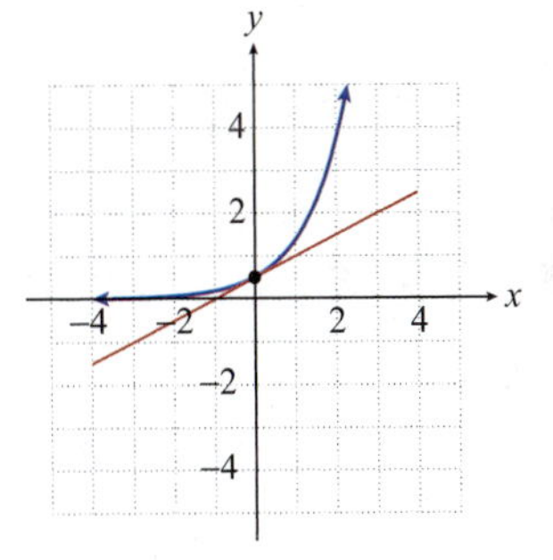

19.

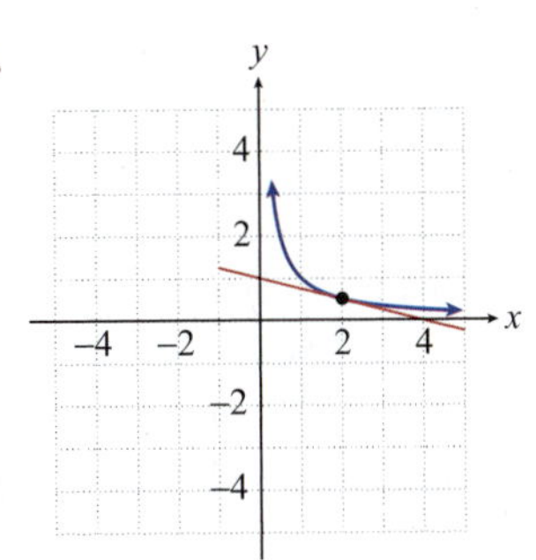

20.

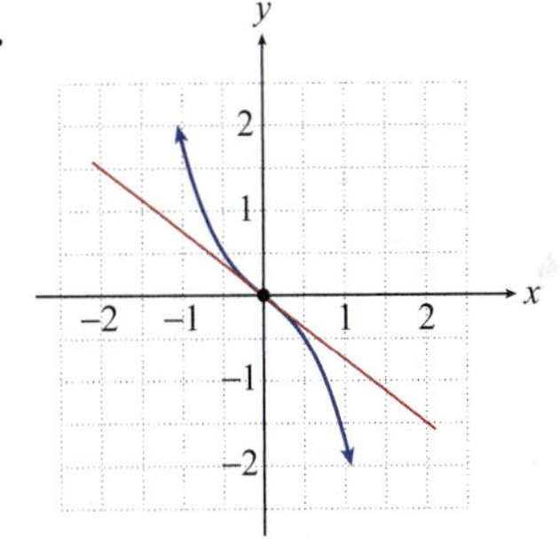

Use difference quotients to approximate the slope of the tangent to the graph of the function at the given point. Use at least five different *h*-values that are decreasing in magnitude. (Answers will vary.)

21. $f(x)=1-2x;\ (1,-1)$

22. $g(x)=\dfrac{5}{4}x-8;\ (8,2)$

23. $h(x)=\dfrac{1}{3}x^2-1;\ (3,2)$

24. $F(x)=3+x-\dfrac{x^2}{2};\ (4,-1)$

25. $G(x)=\dfrac{1}{4}x^3-x+1;\ (-2,1)$

26. $k(x)=10-x^{\frac{3}{2}};\ (4,2)$

27. $H(x)=\ln x+1;\ (e,2)$

28. $u(x)=\cos x;\ \left(\dfrac{\pi}{2},0\right)$

29. $v(x)=\log(2x)-1;\ (5,0)$

30. $w(x)=\tan x;\ (0,0)$

31. $p(x)=-x^4+1;\ (1,0)$

32. $q(x)=x^5-x+3;\ (0,3)$

APPLICATIONS

33. An arrow is shot into the air and its height in feet after t seconds is given by the function $f(t) = -16t^2 + 80t$. The graph of the curve $y = f(t)$ is shown.

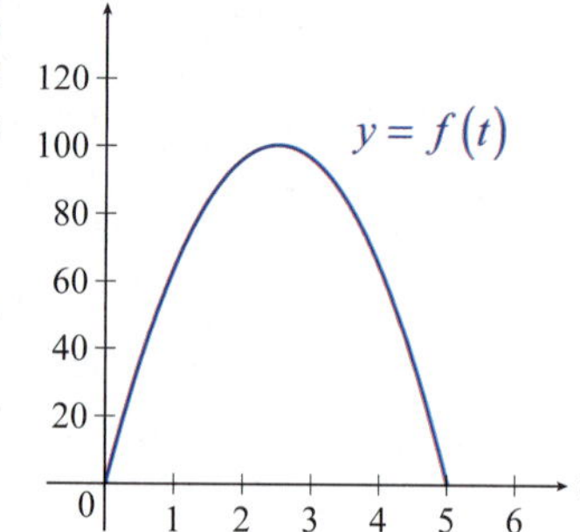

a. Find the height of the arrow when $t = 2$ seconds.
b. Find the instantaneous velocity of the arrow when $t = 2$ seconds.
c. Find the slope of the line tangent to the curve at $t = 2$ seconds.
d. Find the time it takes the arrow to reach its peak.

34. Suppose that a sailboat is observed, over a period of 5 minutes, to travel a distance from a starting point according to the function $s(t) = t^3 + 60t$, where t is time in minutes and s is the distance traveled in meters.

a. How far will it travel during the first 6 seconds?
b. What is the average velocity during the first 6 seconds?
c. Estimate how fast the boat is moving at the starting point.
d. Estimate how fast the boat is moving at the end of 3 minutes.

35. A model rocket is fired vertically upward. The height after t seconds is $h(t) = 192t - 16t^2$ feet.

a. What will be its height at the end of the first second?
b. What is the average velocity of the rocket during the first second?
c. Estimate the instantaneous velocity at $t = 0$ seconds.
d. Estimate the instantaneous velocity at $t = 4$ seconds.
e. When will the velocity be 0? (**Hint:** You may want to start with the initial velocity you found in part a. and use the fact that under the influence of gravity, when air resistance is ignored, vertical upward velocity decreases by 32 ft/s every second. Once you have a guess, test it by a table of difference quotients.)

36. A particle moving in a straight line is at a distance of $s(t) = 2.5t^2 + 18t$ feet from its starting point after t seconds, where $0 \le t \le 12$. Estimate the instantaneous velocity at **a.** $t = 6$ seconds and **b.** $t = 9$ seconds.

37. The distance, in meters, traveled by a moving particle in t seconds is given by $d(t) = 3t(t+1)$. Estimate the instantaneous velocity at **a.** $t = 0$ seconds, **b.** $t = 2$ seconds, and **c.** at time t_0. (**Hint:** Write the difference quotient corresponding to $t = t_0$, simplify, and try to find the value being approached by the expression as h decreases.)

38. The distance, in meters, traveled by a moving particle in t seconds is given by $d(t) = t^2 - 3t$. Estimate the instantaneous velocity at **a.** $t = 0$ seconds, **b.** $t = 4$ seconds, and **c.** at time t_0. (See the hint given in part c. of the previous problem.)

39. After start, on a straight stretch of the track, a race car's velocity changes according to the function $v(t) = -1.8t^2 + 18t$, when $0 \le t \le 10$, t is measured in seconds, and $v(t)$ is measured in meters per second.

a. When does peak velocity occur and what is it? (**Hint:** The graph of $v(t)$ may be helpful.)
b. When does peak deceleration occur?
c. Use difference quotients to estimate peak deceleration. Approximately what multiple of $g \approx 9.81\ \text{m/s}^2$ have you obtained?

40. If we ignore air resistance, a falling body will fall $16t^2$ feet in t seconds.

a. How far will it fall between $t = 2$ and $t = 2.1$?
b. What is its average velocity between $t = 2$ and $t = 2.1$?
c. Estimate its instantaneous velocity at $t = 2$.

TECHNOLOGY

Use a graphing utility to graph $f(x)$ along with three secant lines at the indicated x-value, corresponding to the difference quotients with h-values of 0.2, 0.1, and 0.01, respectively. Can you come up with a possible equation for the tangent? Use technology to test your conjecture.

41. $f(x) = x^2;\quad x = 2$

42. $f(x) = -x^3 + x + 1;\quad x = \frac{\sqrt{3}}{3}$

43. $f(x) = \sin x + \cos x;\quad x = 0$

44. $f(x) = 3\sqrt{x};\quad x = 4$

Use a graphing utility to graph the given function $f(x)$ along with $D(x) = \frac{f(x + 0.001) - f(x)}{0.001}$ in the same coordinate system. Explain how the function values of $D(x)$ are reflected on the graph of $f(x)$.

45. $f(x) = x^4$

46. $f(x) = x(3 - x)$

47. $f(x) = \sin x$

48. $f(x) = \ln x$

Use a graphing utility to find the x-values at which the graph of $f(x)$ does not have a tangent line. Explain.

49. $f(x) = -|x - 1| + 1$

50. $f(x) = |x^2 - 4|$

51. $f(x) = |\ln x|$

52. $f(x) = (x - 1)^{\frac{2}{3}}$

13.2 LIMITS IN THE PLANE

TOPICS

1. Limits in verbal, numerical, and visual forms
2. Vertical asymptotes and one-sided limits
3. Horizontal asymptotes and limits at infinity

TOPIC 1: Limits in Verbal, Numerical, and Visual Forms

Ghosts of Departed Quantities

The mathematicians of the 1600s and 1700s who developed the fundamentals of modern calculus at first used vaguely-defined small quantities called infinitesimals in place of rigorous limit arguments. This practice was famously criticized by the Irish philosopher George Berkeley in a treatise published in 1734, where he cuttingly referred to infinitesimals as the "ghosts of departed quantities." Infinitesimals were shown to be unnecessary by later mathematicians in the 1800s, so Berkeley's criticisms ceased to be relevant. Interestingly, however, the German-born mathematician Abraham Robinson showed in the early 1960s that infinitesimals could be rigorously defined and used in an alternate development of calculus, a framework called *nonstandard analysis.*

The need to exactly determine "limiting" behavior is a characteristic of calculus and other advanced mathematics courses. We encounter the notion of limiting behavior in many different contexts, mathematical and otherwise. If we consider the velocity of a car or a falling rock, we might be interested in calculating the average rate of travel over shorter and shorter time periods in order to estimate the *instantaneous* velocity at a point in time. Similarly, if we examine the graph of a nonlinear function, we might be interested in calculating the slope of the secant line over smaller and smaller intervals in order to estimate the slope of the tangent line at a point.

Every limit scenario, when described verbally, contains phrases of a dynamic nature. Examples of such phrases are "as h approaches 0," "as x goes to 3," and "as N grows without bound." In a less mathematical setting, the phrases might be something like "mortgage rates are expected to approach 5 percent" or "assuming the national debt keeps getting bigger and bigger." We might also see tables of data that hint at some limiting behavior.

In order to progress mathematically, we need notation that precisely describes whatever limit situation is under discussion; every such mathematical scenario can be expressed in terms of a function and a variable whose value is approaching a given point. We will begin with an informal introduction of limit notation here and follow up with a formal definition in Section 13.3.

For the purpose of discussion, let us suppose we have a function $f(x)$ defined on an open interval containing c, except possibly at c itself (the reason for this exception will be explained shortly). We say that the limit of $f(x)$ as x approaches c is L, and write

$$\lim_{x \to c} f(x) = L,$$

if the value of $f(x)$ is closer and closer to L as x takes on values closer and closer (but not equal) to c. That is, as the difference between x and c approaches 0, the difference between $f(x)$ and L also approaches 0.

Note the small but important caveat: the limit of a function $f(x)$ as x approaches c does not depend *at all* on the value of f at c; in fact, the limit may exist when f is not even defined at c. This technical detail highlights one distinction between the mathematical meaning of "limit" and casual usage of the word—in everyday language, the boundary between limiting behavior and behavior at the limit point is often blurred. The distinction is important mathematically because, as we will see repeatedly, the behavior of a function near the point c, but not at c, is the key.

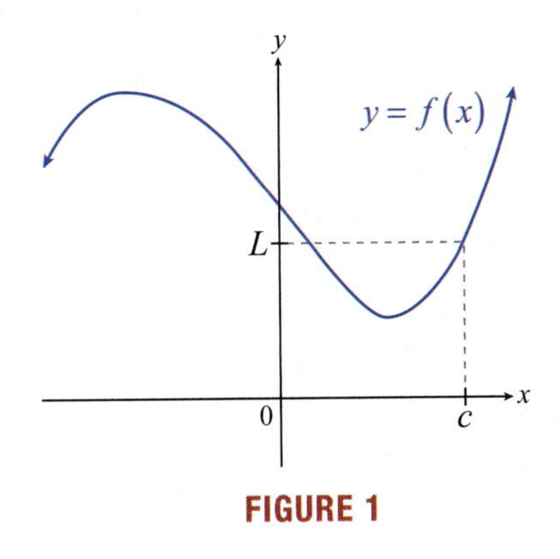

FIGURE 1

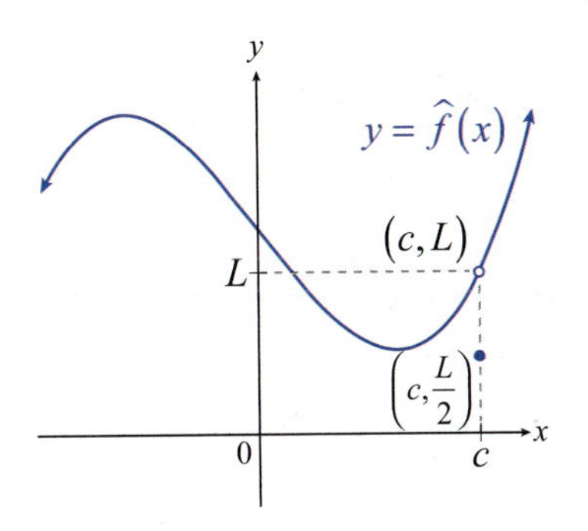

FIGURE 2

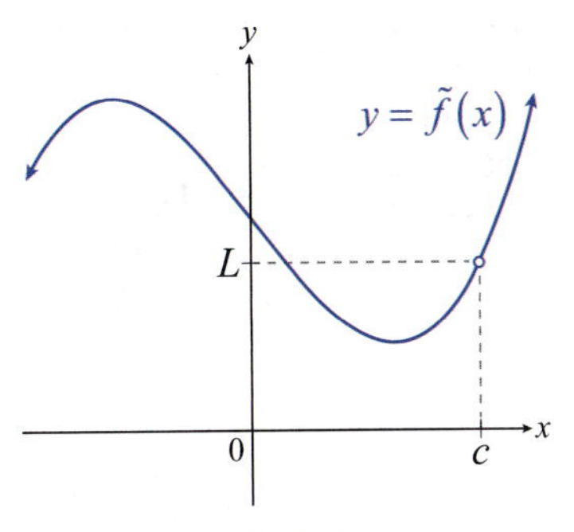

FIGURE 3

Example 1: Finding Limits Given Graphs

Suppose the graph shown in Figure 1 is that of the function $y = f(x)$. Using your previous experience with graphs of functions, it is reasonable to conclude that the function value at $x = c$ is L, that is, $f(c) = L$. Also, since the graph is a nice, smooth, unbroken curve, we expect $f(x)$ to be close to L if x is close to c. In other words, as x approaches c from either side along the x-axis, $f(x)$ will be approaching L, a fact we can confirm visually. As before, we can express this by writing

$$\lim_{x \to c} f(x) = L.$$

Next, suppose we redefine $f(c)$ to be $\frac{L}{2}$ (or some other number), but do not change anything else about f (let's call the resulting function $\hat{f}$). Graphically, this means that the point $\left(c, \hat{f}(c)\right)$ "jumps out" of the graph, and moves down to the $\frac{L}{2}$ level, as shown in Figure 2. In other words, by redefining a single function value, we have relocated *one point* $\left(c, f(c)\right)$ from the graph of f, but we haven't changed a bit the behavior of the function *near* $\left(c, f(c)\right)$. If x takes on values closer and closer to c, but not equal to c, the function value $\hat{f}(x)$ will still move closer and closer to L, which is the second coordinate of the empty circle (i.e., its altitude over the x-axis). In other words, we can still write

$$\lim_{x \to c} \hat{f}(x) = L.$$

Finally, we will consider one last scenario. Suppose that we delete the function value corresponding to $x = c$ altogether, thus making the function undefined at $x = c$. We will call the resulting function $\tilde{f}$. The empty circle in Figure 3 indicates that one point is "missing" from the graph; that is, the vertical line $x = c$ doesn't meet the graph at all. However, notice that this still doesn't have any influence on the function values *near* $x = c$. As before, when x approaches c, but is not equal to c, the function values $\tilde{f}(x)$ approach L, that is,

$$\lim_{x \to c} \tilde{f}(x) = L.$$

The conclusion to be drawn is that $\lim_{x \to c} f(x)$ may exist without the actual function value $f(c)$ being equal to it or, indeed, even defined.

Example 2: Applying Limits to a Paradox

According to a classic paradox, you can never eat an apple for the following reason. A certain amount of time is needed to consume the first half of the apple, and at that point half the apple remains. Additional time is needed to eat half of the remaining half, which constitutes one-quarter of the original apple. After that, more time is needed to consume half of the remaining quarter, or one-eighth of the apple, and so on. By this reasoning, the process of eating the apple never ends.

Of course, this runs counter to our experience, so where does the problem lie? If you were challenged by an argument such as the one above, could you provide a mathematically correct resolution of the paradox?

Solution

As we shall see, the concept of limits can help resolve the apparent flaw in the argument. Let us assume that a math student performs the experiment, and holding a stopwatch in her left hand, eats an apple in exactly 4 minutes. We will make the realistic assumption that the time necessary to eat a certain portion of the apple is directly proportional to its size, and we will monitor the portion of the fruit having been consumed. Proceeding as in the above argument, we first note that half of the apple is gone in the first 2 minutes. At the end of the third minute, another quarter is gone, so we stand at $\frac{1}{2}+\frac{1}{4}=\frac{3}{4}$ of the apple consumed, with 1 minute left on the clock. At 3 and a half minutes, $\frac{3}{4}+\frac{1}{8}=\frac{7}{8}$ is gone, and at 3 minutes and 45 seconds, $\frac{7}{8}+\frac{1}{16}=\frac{15}{16}$ is gone. With half of the remaining time left, that is, at 3 minutes and 52.5 seconds, $\frac{15}{16}+\frac{1}{32}=\frac{31}{32}$ of the apple is gone, and so on. If we examine the fractions we obtain, it is easy to see that they get closer and closer to 1 as time progresses toward 4 minutes. Mathematically, we say that the limit of these values is 1; that is, the student indeed ate the entire apple in 4 minutes, in accordance with what we would expect from our own experience.

As important as it is to understand what it means when we say that a certain limit exists, it is just as important to understand what it means when a limit fails to exist. We will again defer the technical definition of this notion to Section 13.3, but we can develop our intuitive grasp of the idea and introduce some more notation now.

Assume again that we have a function f defined on an open interval containing the point c, except possibly at c itself. The limit of $f(x)$ as x approaches c then fails to exist if the values of $f(x)$ do not approach any particular real number L. This could occur for various reasons: $f(x)$ might be getting larger and larger (in absolute value), the values of $f(x)$ may "jump" as x passes from one side of c to the other, or the values of $f(x)$ may simply not "settle down" to any particular value no matter how narrowly we focus on the region around c. Figure 4 illustrates these three possibilities.

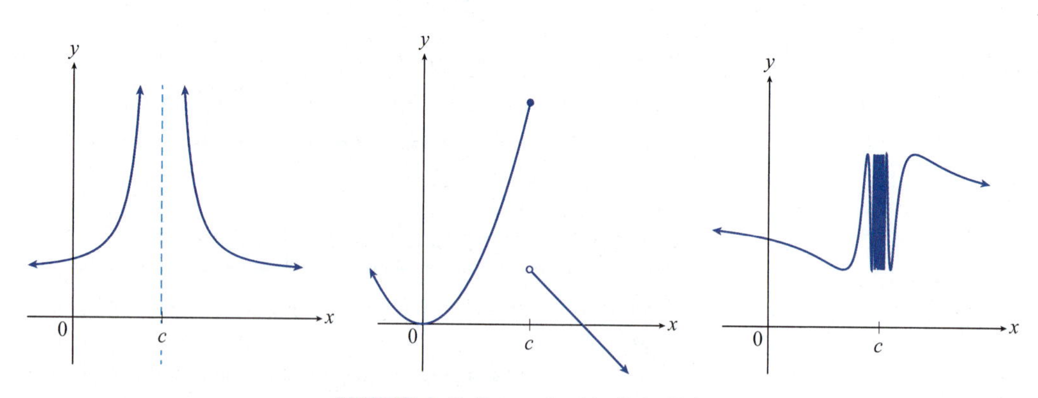

FIGURE 4: Failure of a Limit to Exist

While potentially confusing, we use the notation

$$\lim_{x\to c} f(x) = \infty \quad \text{or} \quad \lim_{x\to c} f(x) = -\infty$$

to indicate the nonexistence of a limit when it fails to exist because the function f either increases without bound (approaches positive infinity) or decreases without bound (approaches negative infinity) as $x \to c$. It should be remembered, however, that in either scenario the limit does not exist since f is not approaching a fixed real number. The notation simply describes the particular *way* in which the limit fails to exist.

TOPIC 2: Vertical Asymptotes and One-Sided Limits

Our familiarity with vertical asymptotes provides us many examples of the use of the limit concept and limit notation. And, conversely, limit notation can be used to make our understanding of asymptotic behavior more rigorous.

A distinguishing feature of a vertical asymptote is that, at least from one side, the function under discussion grows in magnitude without bound. Our first task is to introduce notation that succinctly and accurately describes such behavior. Assume the function f is defined on an open interval with right endpoint c. We write

$$\lim_{x\to c^-} f(x) = \infty$$

if the values of $f(x)$ increase without bound as x approaches the point c from the left. Similarly, we write

$$\lim_{x\to c^-} f(x) = -\infty$$

if the values of $f(x)$ decrease without bound (approach negative infinity) as x approaches the point c from the left. Figure 5 illustrates these two scenarios.

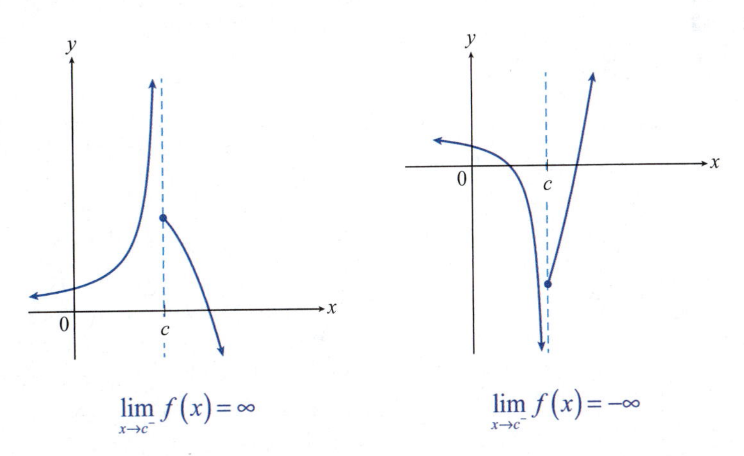

FIGURE 5: Asymptotic Behavior from the Left

Similarly, we use the notation $\lim_{x\to c^+} f(x)$ in discussing the limit of the function f as x approaches the point c from the right, with the implicit assumption that f is defined on an open interval whose *left* endpoint is c. And as noted before, we need to be careful in interpreting the four statements $\lim_{x\to c^-} f(x) = \infty$, $\lim_{x\to c^-} f(x) = -\infty$, $\lim_{x\to c^+} f(x) = \infty$, and $\lim_{x\to c^+} f(x) = -\infty$; all four indicate that these *one-sided* limits of f at c do not exist, since f is either increasing or decreasing without bound.

You may be wondering why the clause "defined on an open interval" keeps appearing in these limit discussions. The reason is that in order for us to study the limit of a function as x approaches a point c, we need to know that f is defined at all points near c (but not necessarily at c). The exact meaning of *near* will vary from one application to another, but by stipulating that f is defined on an open interval around c, or to one side or the other in the case of one-sided limits, we can be sure that $f(x)$ always represents a real number as x approaches c.

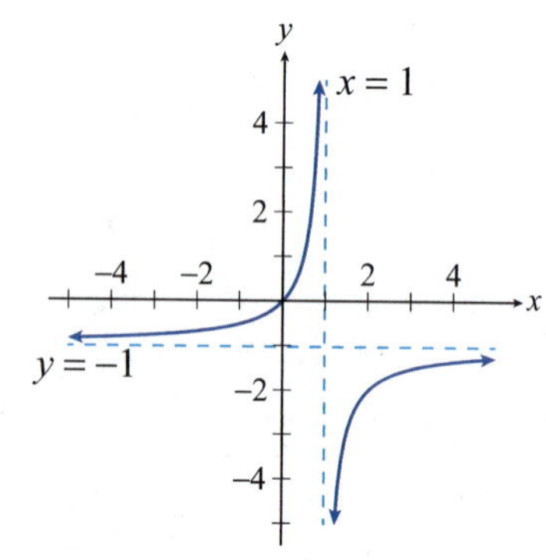

FIGURE 6: $f(x)=\dfrac{-x}{x-1}$

Example 3: Finding One-Sided Limits near Vertical Asymptotes

Use one-sided limit notation to describe the behavior of the following functions near their vertical asymptote(s).

a. $f(x)=\dfrac{-x}{x-1}$ **b.** $g(x)=\dfrac{x^2+1}{x^2-x-6}$ **c.** $h(x)=\dfrac{1}{(x-1)^2}$

Solution

a. As we can see from the graph in Figure 6, the sole vertical asymptote occurs at $x=1$, a fact we can deduce from the formula as well. A further examination of the graph reveals that as x approaches 1 from the left, the function values increase without bound, a fact we can express as follows:

$$\lim_{x\to1^-} f(x)=\infty$$

On the other hand, as x approaches 1 from the right, the function values decrease:

$$\lim_{x\to1^+} f(x)=-\infty$$

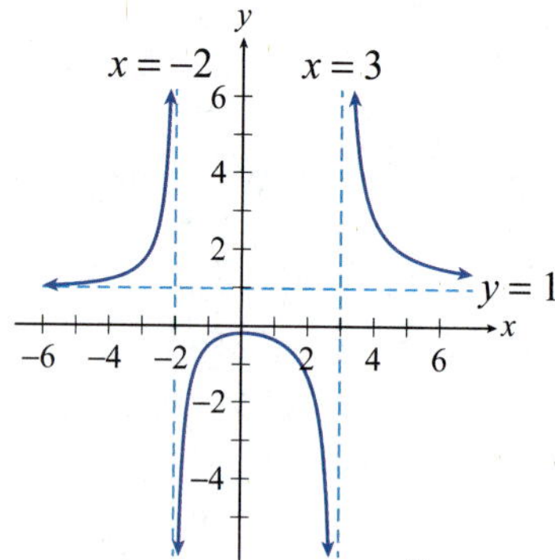

FIGURE 7: $g(x)=\dfrac{x^2+1}{x^2-x-6}$

b. As seen from the graph in Figure 7, $g(x)$ has two vertical asymptotes, at $x=-2$ and $x=3$, a fact we can confirm algebraically by factoring the denominator of $g(x)$. The function values increase without bound as x approaches -2 from the left and decrease in a similar manner as x approaches -2 from the right. Therefore, we can write

$$\lim_{x\to-2^-} g(x)=\infty \quad \text{and} \quad \lim_{x\to-2^+} g(x)=-\infty.$$

Examining the graph on both sides of the vertical asymptote $x=3$ in a similar manner, we conclude that

$$\lim_{x\to3^-} g(x)=-\infty \quad \text{and} \quad \lim_{x\to3^+} g(x)=\infty.$$

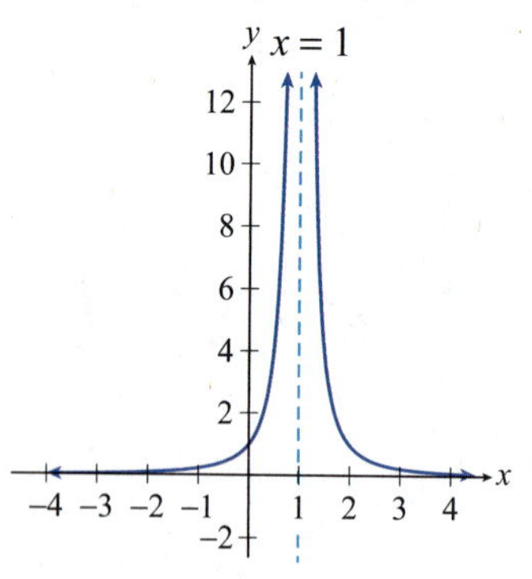

FIGURE 8: $h(x)=\dfrac{1}{(x-1)^2}$

c. Examining the graph of $h(x)$ in Figure 8, we see that the behavior is the same on both sides of the vertical asymptote $x=1$; namely, that of the function values increasing without bound as x approaches 1. This makes it unnecessary to distinguish between the two one-sided limits in this case; instead, we can simply write $\lim_{x\to1} h(x)=\infty$.

Example 4: Finding One-Sided Limits near Vertical Asymptotes

Deduce the one-sided infinite limits of $f(x)=\frac{1}{3-x}$ at its vertical asymptote without graphing the function.

Solution

We see from the formula that the sole vertical asymptote of $f(x)$ occurs at $x=3$ (which is where the denominator is 0). We will examine the two one-sided limits separately. Let us assume first that x is approaching 3 from the left. This means that the x-values are moving closer and closer to $x=3$ through values less than 3 (though approaching it in magnitude). For example, when $x=2.99$,

$$f(2.99)=\frac{1}{3-2.99}$$
$$=\frac{1}{0.01}=100.$$

Throughout this process of x approaching 3, $f(x)$ stays positive, but its denominator decreases; thus the function values themselves increase in magnitude.

A couple of easy calculations underscore this observation:

$$f(2.999)=\frac{1}{3-2.999}=\frac{1}{0.001}=1000,$$
$$f(2.9999)=\frac{1}{3-2.9999}=\frac{1}{0.0001}=10,000,$$

and so on. The closer x is to 3, the greater the function value $f(x)$ is in magnitude (remember, x never actually reaches the value of 3).

We can now summarize our observations by writing

$$\lim_{x\to 3^-}\frac{1}{3-x}=\infty.$$

If, on the other hand, x approaches 3 from the right, the denominator and thus the function value becomes negative. As an example,

$$f(3.01)=\frac{1}{3-3.01}$$
$$=\frac{1}{-0.01}=-100.$$

It is still true that the closer x is to 3, the greater the function values are in magnitude, but this time, they are all negative. For example, as you can easily check,

$$f(3.001)=-1000,$$
$$f(3.0001)=-10,000,$$

and so on. As before, we can express this by writing

$$\lim_{x\to 3^+}f(x)=-\infty.$$

We can briefly summarize the above as follows: the presence of a vertical asymptote $x = 3$ indicates that the values of $f(x)$ increase or decrease without bound near $x = 3$.

In general, these tendencies may or may not be different on the two sides of the asymptote, that is, for $x < 3$ and $x > 3$, respectively. We were, however, able to find out whether the respective one-sided limits were $+\infty$ or $-\infty$ from the sign of the expression as x approached 3 from a fixed side, even though a graph was not readily available.

With two-sided and one-sided limit notation in hand, we can now provide a more formal definition of vertical asymptote.

DEFINITION: Vertical Asymptote

We say that a function f has a **vertical asymptote** at the point c if at least one of the following statements holds:

$$\lim_{x \to c^-} f(x) = \infty \qquad \lim_{x \to c^+} f(x) = \infty \qquad \lim_{x \to c} f(x) = \infty$$

$$\lim_{x \to c^-} f(x) = -\infty \qquad \lim_{x \to c^+} f(x) = -\infty \qquad \lim_{x \to c} f(x) = -\infty$$

And now that one-sided limit notation has been introduced, it is easy to extend its meaning. We write

$$\lim_{x \to c^-} f(x) = L \quad \text{or} \quad \lim_{x \to c^+} f(x) = L$$

if the values of $f(x)$ get closer and closer to L as x approaches c from, respectively, the left or right. The next example illustrates the use of this notation.

Example 5: Finding One-Sided Limits Given Graphs

Use the graph of the function h in Figure 9 to decide whether the given one-sided limits exist. For those that do, find their values.

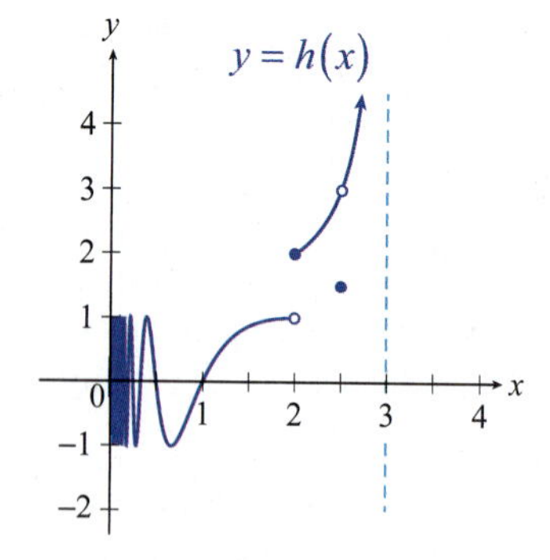

FIGURE 9

a. $\lim_{x \to 0^+} h(x)$ **b.** $\lim_{x \to 2^-} h(x)$ **c.** $\lim_{x \to 2^+} h(x)$

d. $\lim_{x \to 2.5^-} h(x)$ **e.** $\lim_{x \to 2.5^+} h(x)$ **f.** $\lim_{x \to 3^-} h(x)$

Solution

a. We see from the graph that $h(x)$ oscillates wildly between $+1$ and -1 on the right-hand side of $x = 0$. Moreover, the closer x moves to 0, the more oscillations we have; in fact, there are infinitely many oscillations near 0. (Our technology is incapable of graphing infinitely many waves, which is the reason we see a "shaded rectangle" immediately to the right of the y-axis.) Therefore, there is no single value being approached by the values of $h(x)$, so we conclude that $\lim_{x \to 0^+} h(x)$ does not exist.

b. As x approaches 2 from the left, the function values are approaching 1, so we conclude

$$\lim_{x \to 2^-} h(x) = 1.$$

Notice, however, that the actual function value is $h(2) = 2$. In general, we cannot expect the limit of a function at c to always equal the function value at c. In other words, as we previously observed in Example 1, what a function does *near* c may not be the same as what it does *at* c.

c. Examining the tendency of the function values as x approaches 2 from the right, we see that

$$\lim_{x \to 2^+} h(x) = h(2) = 2.$$

In other words, the one-sided limit and the function value agree in this case.

d–e. The graph shows that the function values $h(x)$ approach 3 as x approaches 2.5 from either direction, so we can write

$$\lim_{x \to 2.5^-} h(x) = \lim_{x \to 2.5^+} h(x) = 3.$$

Notice that just like in Example 1, since both one-sided limits exist and are equal, we conclude that the *two-sided* limit exists at 2.5, and

$$\lim_{x \to 2.5} h(x) = 3.$$

The last important observation we wish to make here is that the function value at $x = 2.5$ differs from the limit, so once again

$$\lim_{x \to 2.5} h(x) \neq h(2.5).$$

f. Because $x = 3$ is a vertical asymptote, $\lim_{x \to 3^-} h(x)$ does not exist, but we can write

$$\lim_{x \to 3^-} h(x) = \infty.$$

TOPIC 3: Horizontal Asymptotes and Limits at Infinity

We have just seen examples of when it is appropriate to say that a given limit (two-sided or one-sided) *is* $\pm\infty$; we will now study limits *at* $\pm\infty$.

Just as an unbounded limit at a point c corresponds to a vertical asymptote at that point, there is a connection between horizontal asymptotes and limits at infinity. Recall that a horizontal line passing through L on the y-axis is a horizontal asymptote for the function f if the graph of f approaches the line as x is allowed to either increase or decrease without bound. A given function can have either 0, 1, or 2 horizontal asymptotes, as illustrated in Figure 10; unlike vertical asymptotes, a function can cross a horizontal asymptote any number of times (or not at all).

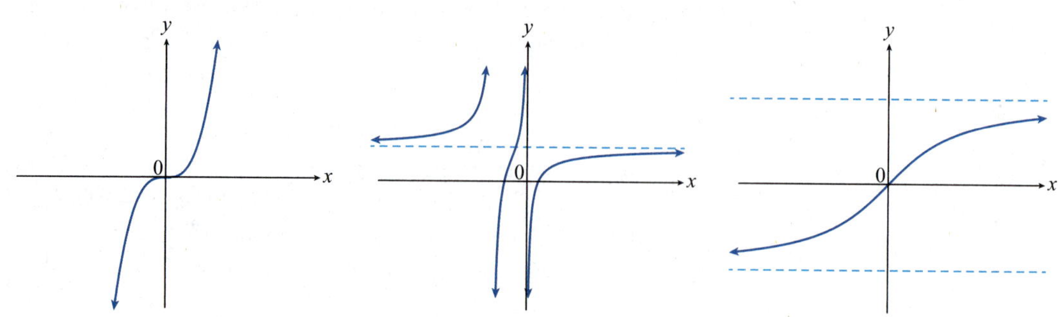

FIGURE 10: Limits at Infinity

And as with vertical asymptotes, the notion of limit is exactly what is needed to make the definition of horizontal asymptote more precise.

> **DEFINITION: Horizontal Asymptote**
>
> We say that the line $y = L$ is a **horizontal asymptote** for the function f if either of the following statements is true.
>
> $$\lim_{x\to\infty} f(x) = L \quad \text{or} \quad \lim_{x\to-\infty} f(x) = L$$

Example 6: Using Limits at Infinity to Identify Horizontal Asymptotes

Identify the horizontal asymptotes of the following functions using their graphs shown in Figures 11 and 12, respectively.

a. $f(x) = \frac{4}{\pi}\arctan x$ **b.** $g(x) = e^{-x}\cos(5x)$

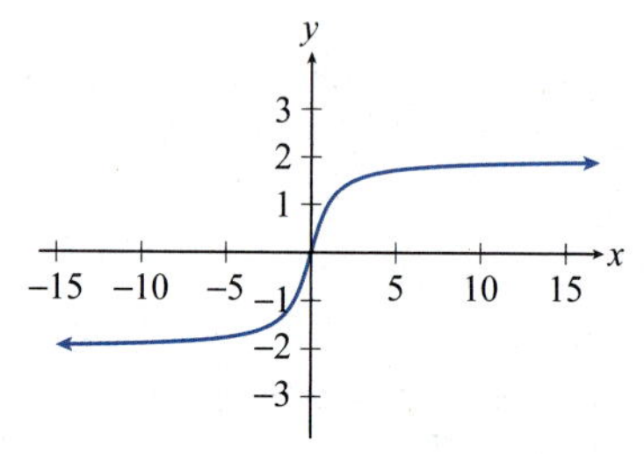

FIGURE 11: $f(x) = \frac{4}{\pi}\arctan x$

Solution

a. As shown by the graph in Figure 11, the values of $f(x)$ approach the value of 2 as x increases without bound, and they will be as close to 2 as we wish if we make sure x is large enough. In other words, we can write

$$\lim_{x\to\infty} f(x) = 2$$

and conclude that the line $y = 2$ is a horizontal asymptote for f. Similarly, we conclude that

$$\lim_{x\to-\infty} f(x) = -2,$$

and thus the line $y = -2$ is another horizontal asymptote for f.

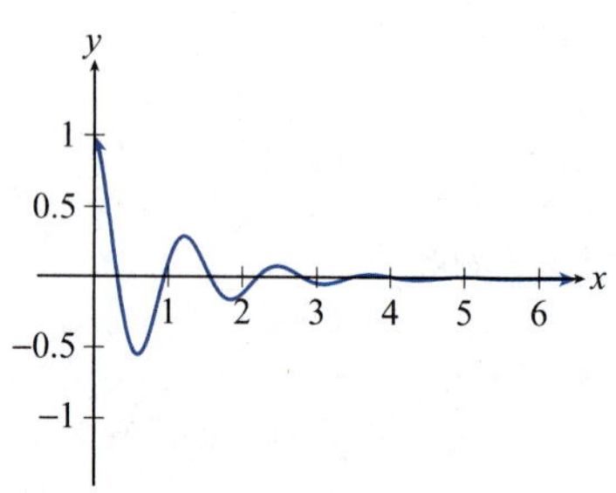

FIGURE 12: $g(x) = e^{-x}\cos(5x)$

b. The graph of g, though oscillating, is approaching the x-axis more and more as x increases (see Figure 12). In other words, as we will soon be able to prove, the function values of $g(x)$ will be as small in absolute value as we wish, if we merely choose x large enough. This is precisely what we need to conclude that the x-axis or $y = 0$ is a horizontal asymptote for g. Using limit notation, this means

$$\lim_{x\to\infty} g(x) = 0.$$

As a final remark, since the cosine factor causes $g(x)$ to oscillate, the graph actually crosses its horizontal asymptote infinitely often.

Many functions tend to increase in magnitude without bound as x goes to ∞ or $-\infty$; the first graph in Figure 10 is an example. In such cases, it is natural to use limit notation to provide more information on the behavior of the function. We write

$$\lim_{x\to\infty} f(x) = \infty$$

if the values of f increase without bound as $x \to \infty$, with similar meanings attached to the remaining three possibilities,

$$\lim_{x\to\infty} f(x) = -\infty, \quad \lim_{x\to-\infty} f(x) = \infty, \quad \text{and} \quad \lim_{x\to-\infty} f(x) = -\infty.$$

Example 7: Using Limit Notation to Describe Unbounded Behavior

Decide whether the given functions increase or decrease without bound as x goes to ∞ and $-\infty$. Use limit notation to express your conclusion.

a. $f(x) = -x^2$ **b.** $g(x) = x^3 - 7$ **c.** $h(x) = x^5 - 6x^2 + 3$

Solution

a. First, we will examine the function values $f(x)$ as x goes to ∞. Our first observation is that for any positive real number x, if $x > 1$ then $x^2 > x$, so if x tends to ∞, then x^2 will certainly grow in magnitude, without bound. Therefore, $f(x) = -x^2$ will *decrease* without bound as x tends to ∞. Using limit notation,

$$\lim_{x\to\infty} f(x) = -\infty.$$

Next, let us assume x goes to $-\infty$. Like before, x^2 will grow without bound, and since x^2 is positive for any nonzero x, we conclude that $f(x) = -x^2$ again decreases without bound. Thus,

$$\lim_{x\to-\infty} f(x) = -\infty.$$

b. We again let x go to ∞ first. Then by an argument similar to the one given in part a., x^3 grows without bound. This tendency is not influenced by subtracting 7 from x^3, so

$$\lim_{x\to\infty} g(x) = \infty.$$

If on the other hand, x goes to $-\infty$, $x^3 - 7$ will decrease without bound, since the cube of a negative number is negative. Thus we conclude that

$$\lim_{x\to-\infty} g(x) = -\infty.$$

c. As for the function h, we will first argue that its values grow without bound as x tends toward ∞. The reasoning is as follows. It is clear that the first term, x^5, increases without bound as x goes to ∞. And as x grows, the fifth-degree term increases so much faster than the quadratic term that their difference, $x^5 - 6x^2$, still increases. As an illustration, when $x = 3$, we have

$$x^5 - 6x^2 = 3^5 - 6 - 9 = 243 - 54 = 189,$$

but when, say, $x = 10$, we have

$$x^5 - 6x^2 = 10^5 - 6 - 10^2 = 100{,}000 - 600 = 99{,}400.$$

Adding 3 to the previous expression (or adding or subtracting any constant, for that matter) will not change the increasing tendency of the function values, so we conclude that

$$\lim_{x\to\infty} h(x) = \infty.$$

We can argue in a similar fashion in the case of x going to $-\infty$, but since the fifth power of a negative number is negative, x^5 *decreases* without bound in this case. This time, however, subtracting $6x^2$ only helps (but like before, the quadratic term would not be able to change the process, even if we added it to x^5). Again, the constant term is immaterial from the perspective of the tendency of the function values, so we conclude that

$$\lim_{x\to-\infty} h(x) = -\infty.$$

TECHNOLOGY: Estimating Limits Using a Graphing Utility

To this point, we have relied upon graphs, tables, intuition, and elementary reasoning in order to arrive at guesses for limits that we feel confident about. In Sections 13.3 and 13.4 we will learn a number of mathematical facts that allow us to accurately determine limits of functions without the need for guesswork. But we will occasionally make use of graphing utilities in order to guide our explorations.

Consider the function $f(x) = \dfrac{\sqrt{x^2+4}-2}{x^2}$.

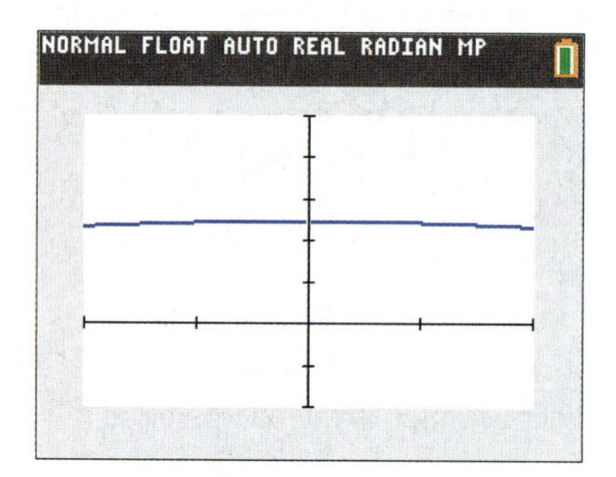

$[-1,1]$ by $[-0.2,0.5]$

FIGURE 13

If we needed to determine the behavior of this function as x goes to 0, without the benefit of the mathematical facts and techniques we will learn in the next two sections, we might note that the denominator of the fraction goes to 0 as x goes to 0, so we might guess that the function "blows up," that is, increases in magnitude without bound. That guess would be wrong—our analysis so far misses an important fact. A closer examination of the fraction reveals that the numerator also approaches 0 as x goes to 0, and this common situation can lead to a variety of possible outcomes.

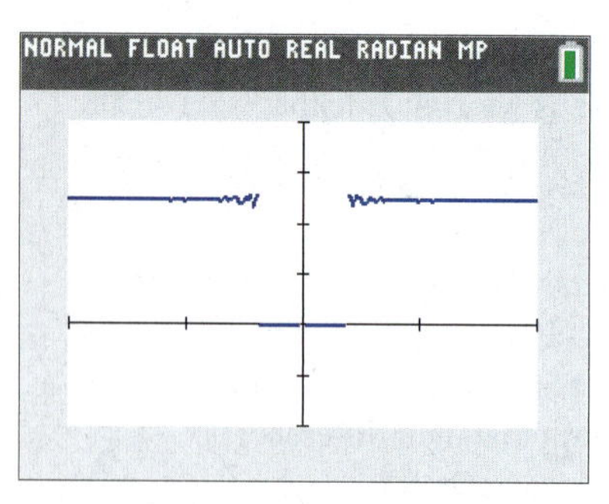

$[-0.00001,0.00001]$ by $[-0.2,0.4]$

FIGURE 14

Another option would be to use a graphing utility to graph the function near zero. Figures 13 and 14 show $f(x)$ graphed on a TI-84 Plus using different viewing windows. As you can see, these graphs may not lead to a correct guess since the function looks different in the different viewing windows.

As Figures 13 and 14 illustrate, zooming in on our graph can lead to more questions than answers. A prolonged exploration and a weighing of all the results would probably lead us to eventually guess that the limit of this function as x approaches 0 is $\frac{1}{4}$, but we might still be left with a shadow of doubt. Some graphing utilities also provide tools for determining limits, but software has its limitations and can be fooled. Fortunately, the limit theorems we will learn later will provide us the means to remove all doubt in such situations.

13.2 EXERCISES

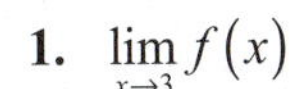

Use the graph of the function to find the indicated limit (if it exists).

1. $\lim_{x\to 3} f(x)$

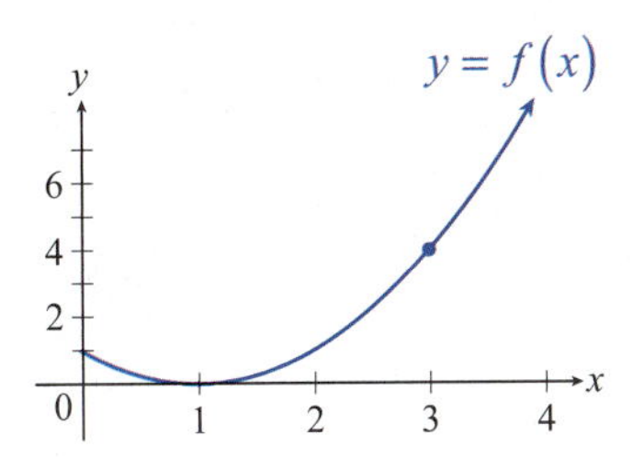

2. $\lim_{x\to 1} g(x)$

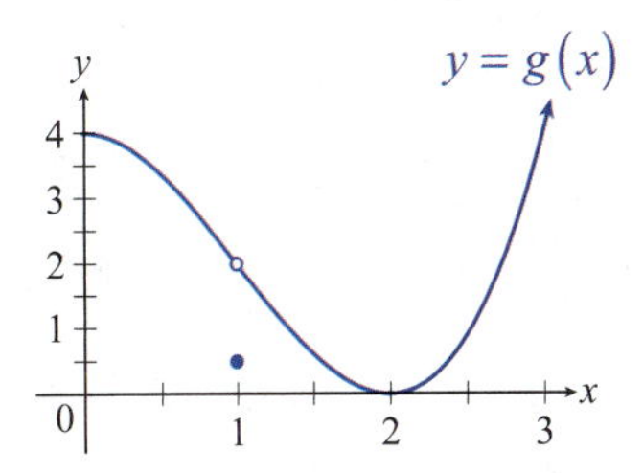

3. $\lim_{x\to 0} h(x)$

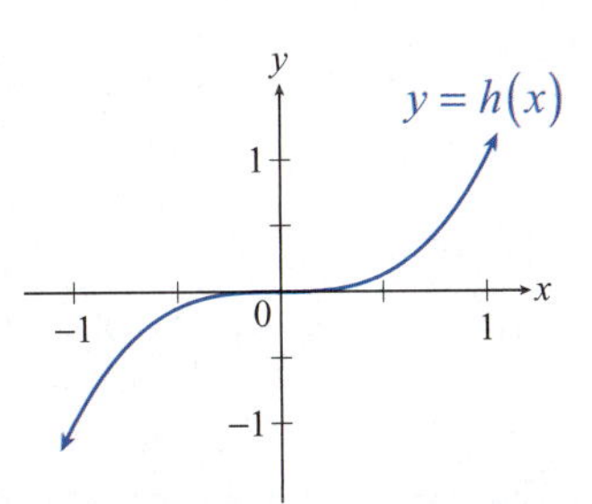

4. $\lim_{x\to 1} k(x)$

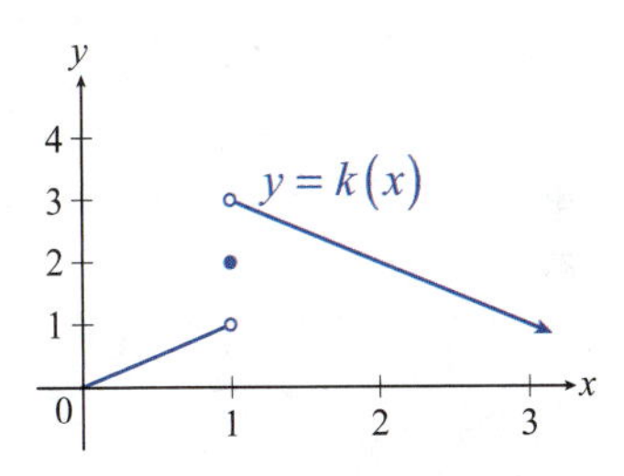

Create a table of values to estimate the value of the indicated limit without graphing the function. Choose the last *x*-value so that it is no more than 0.001 units from the given *c*-value.

5. $\lim_{x\to\sqrt{2}} \dfrac{x^2-2}{x-\sqrt{2}}$

6. $\lim_{x\to 3} \dfrac{x^3-9x^2+27x-27}{x-3}$

7. $\lim_{x\to 1} \dfrac{x^{10}-1}{x-1}$

8. $\lim_{x\to 0} \dfrac{4\sin x}{3x}$

9. $\lim_{x\to \pi} \dfrac{2\cos x-1}{1-\sin x}$

10. $\lim_{x\to 7^-} \dfrac{x^2-49}{x-7}$

11. $\lim_{x\to 7^+} \dfrac{x^2+49}{x-7}$

12. $\lim_{x\to 0^+} \dfrac{\sqrt{4+x}}{x}$

Use one-sided limit notation to describe the behavior of the function near its vertical asymptote(s).

13.

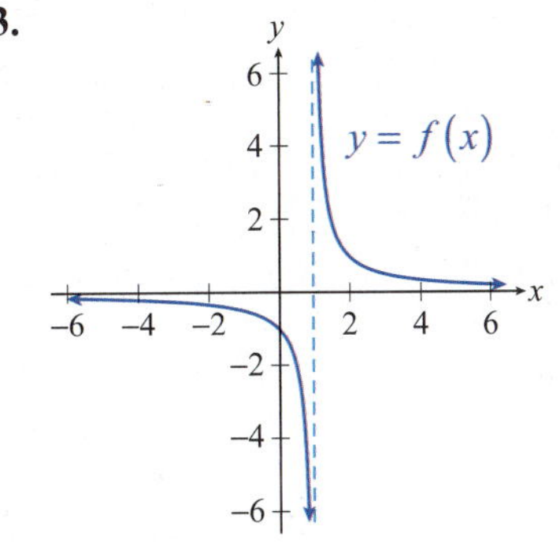

14.

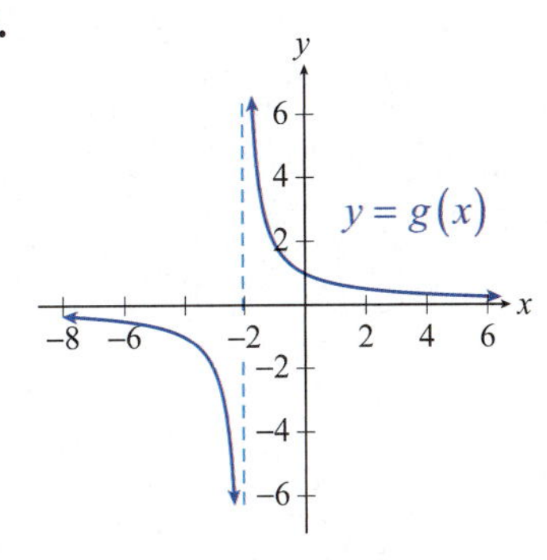

15.

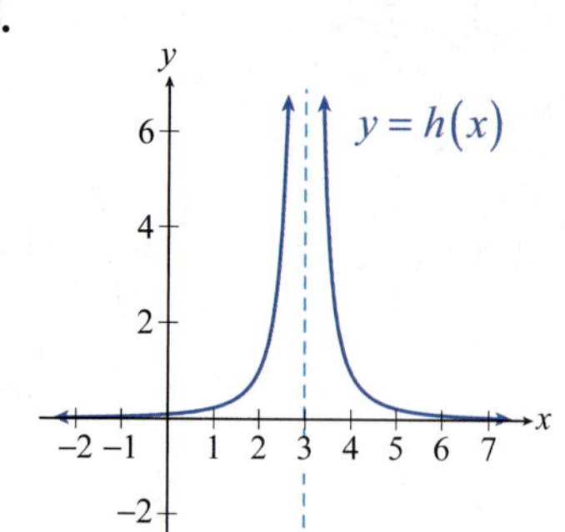

16.

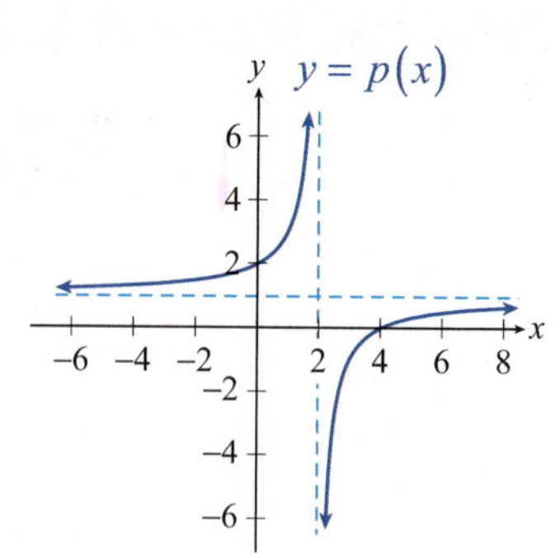

17.

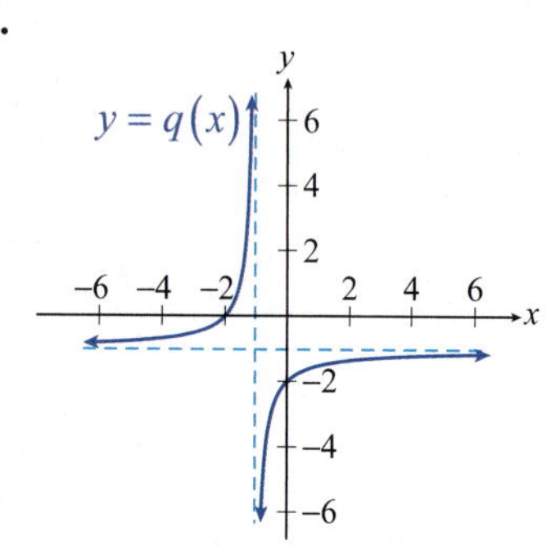

18.

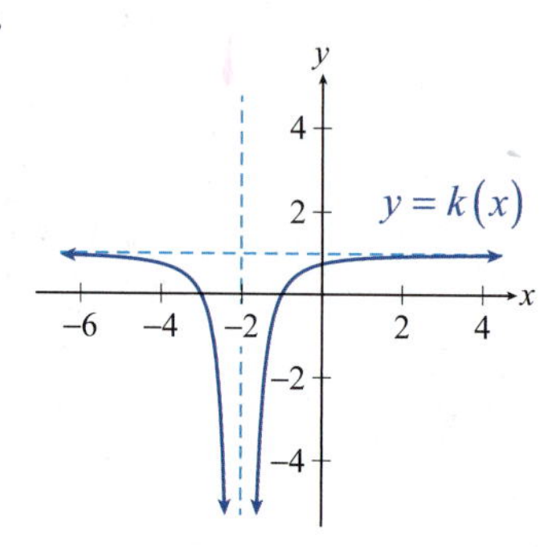

19.

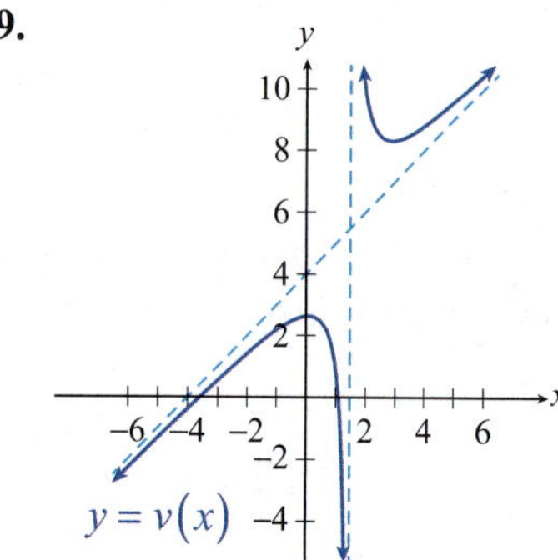

20.

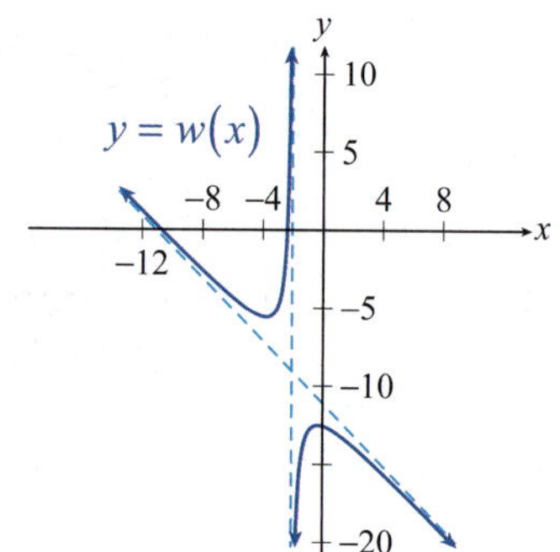

21.

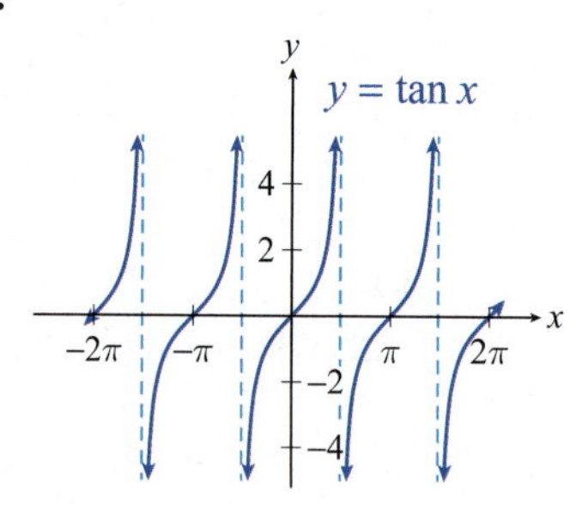

22.

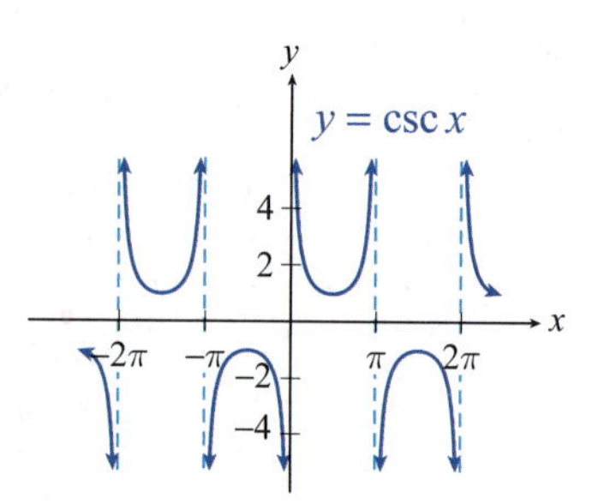

23.

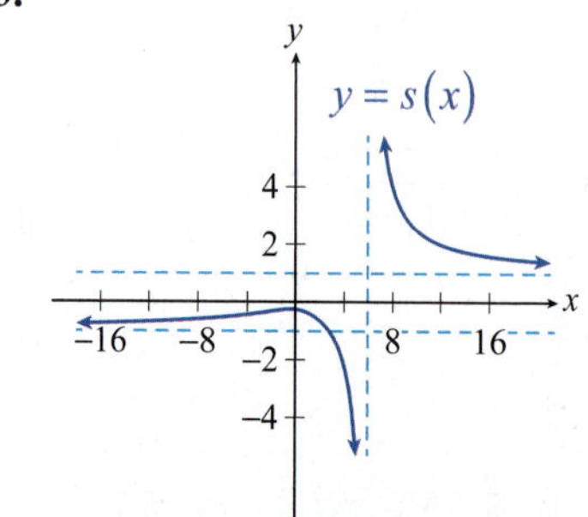

24.

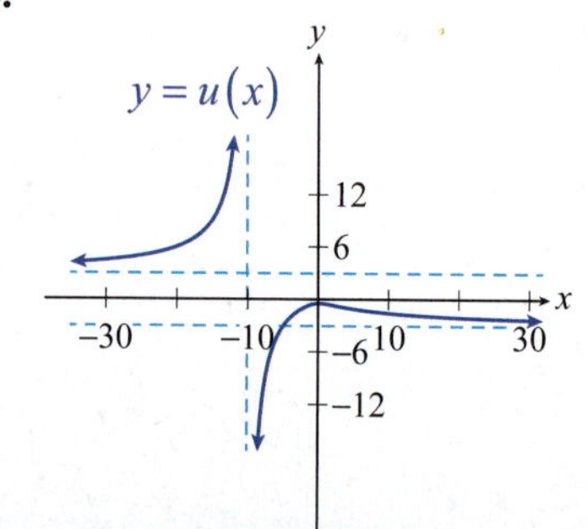

25–36. Consider the functions given in Exercises 13–24. Find their limits at ∞ and $-\infty$ (if they exist). When applicable, use the horizontal asymptote(s) as a guide.

Use the graph to find the indicated one-sided limits, if they exist.

37. a. $\lim\limits_{x\to 2^-} f(x)$ **b.** $\lim\limits_{x\to 2^+} f(x)$

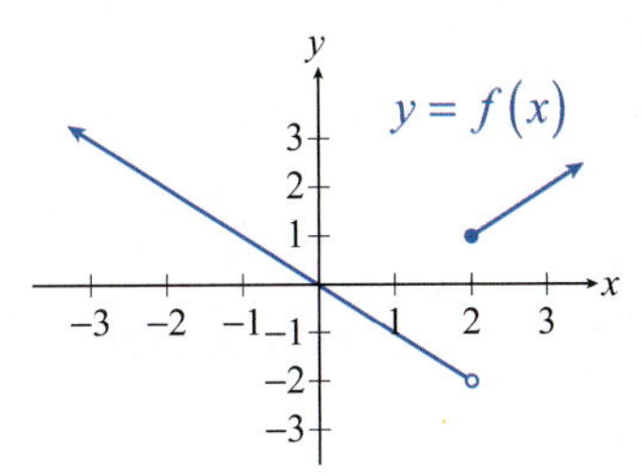

38. a. $\lim\limits_{x\to 1^-} g(x)$ **b.** $\lim\limits_{x\to 1^+} g(x)$

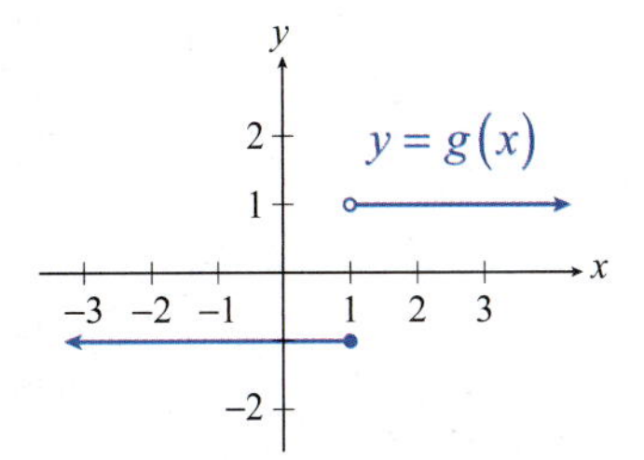

39. a. $\lim\limits_{x\to 1^-} h(x)$ **b.** $\lim\limits_{x\to 1^+} h(x)$

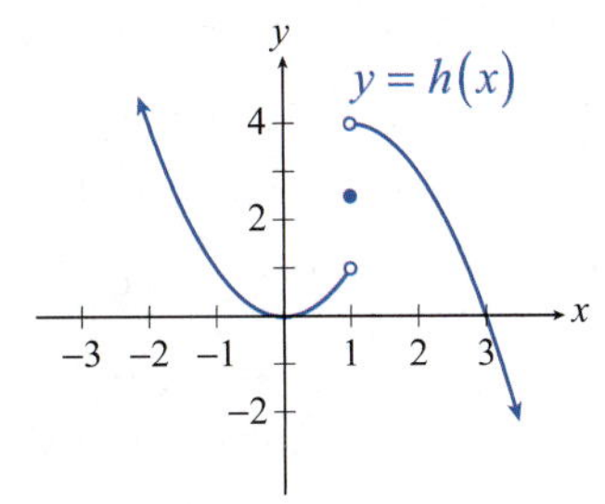

40. a. $\lim\limits_{x\to\left(-\frac{\pi}{2}\right)^-} F(x)$ **b.** $\lim\limits_{x\to\left(-\frac{\pi}{2}\right)^+} F(x)$

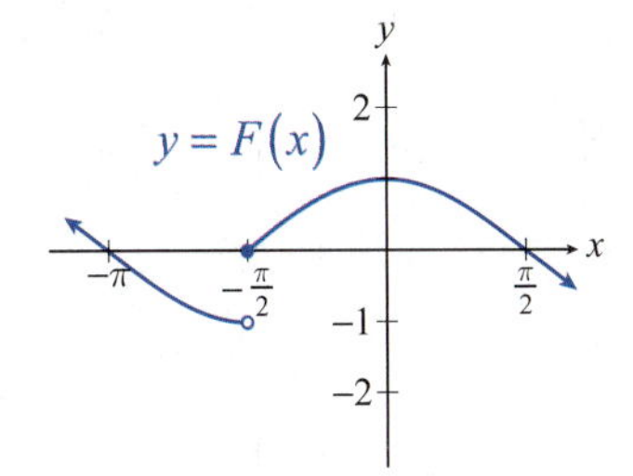

41. a. $\lim\limits_{x\to 0^-} G(x)$ **b.** $\lim\limits_{x\to 0^+} G(x)$

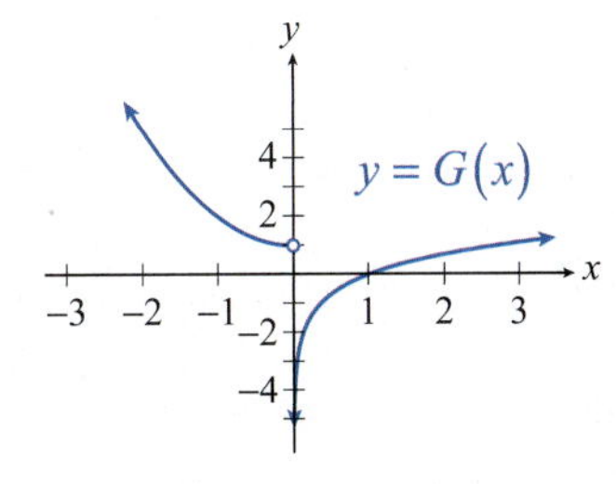

42. a. $\lim\limits_{x\to -1^-} k(x)$ **b.** $\lim\limits_{x\to -1^+} k(x)$

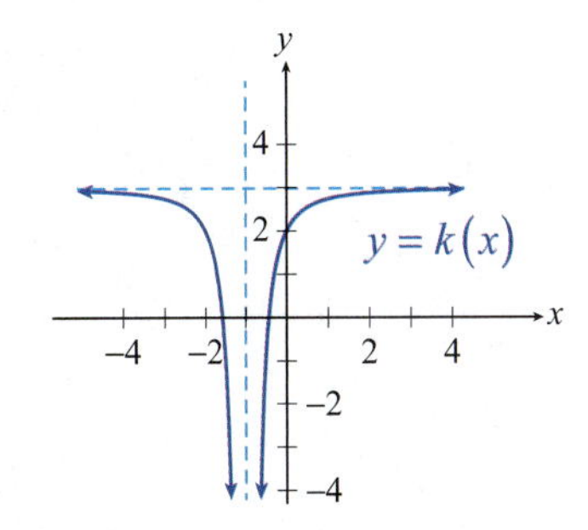

43. a. $\lim\limits_{x\to 0^-} H(x)$ **b.** $\lim\limits_{x\to 0^+} H(x)$

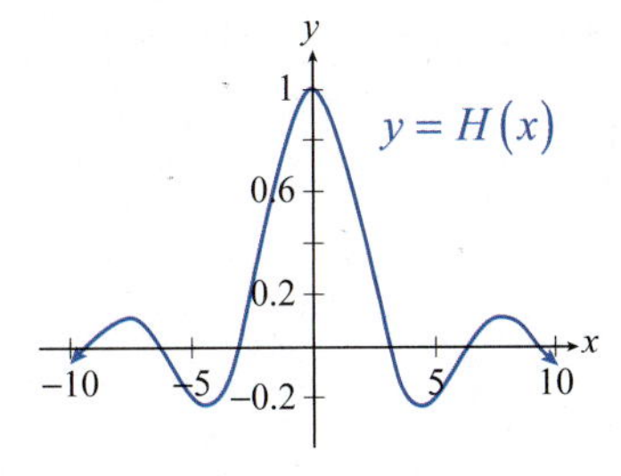

44. a. $\lim\limits_{x\to 2^-} u(x)$ **b.** $\lim\limits_{x\to 2^+} u(x)$

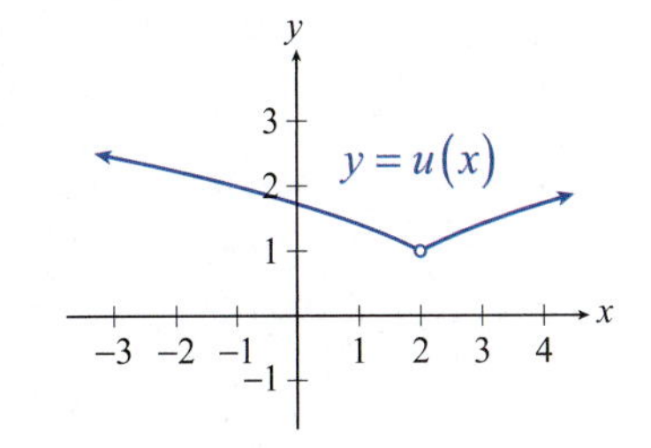

45. **a.** $\lim_{x\to 0^-} v(x)$ **b.** $\lim_{x\to 0^+} v(x)$

46. **a.** $\lim_{x\to 0^-} w(x)$ **b.** $\lim_{x\to 0^+} w(x)$

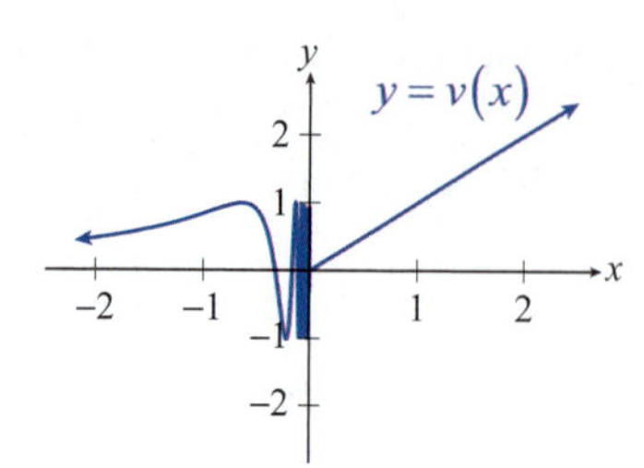

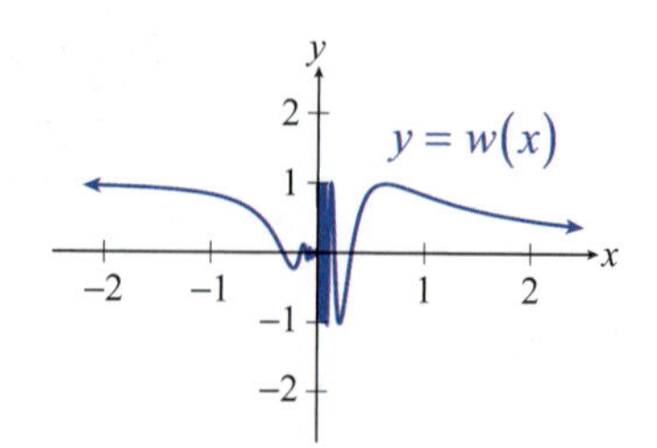

Use limit notation to describe the unbounded behavior of the given function as x approaches ∞ and/or $-\infty$.

47. $f(x)=x^3$

48. $g(x)=x^2+2.1x-1$

49. $h(x)=-x^4+0.2x^3$

50. $k(x)=-0.35x^5+x+1.35$

51. $F(x)=\sqrt{x+2}$

52. $G(x)=\sqrt[3]{x+1}-2.3$

53. $H(x)=|x+2|$

54. $K(x)=-|x+2|-1$

55. $u(x)=|x-1|+|x+2|$

56. $v(x)=e^{x+2}$

57. $s(x)=-10^{-x}+1$

58. $t(x)=\ln x-1$

WRITING & THINKING

Determine whether the given statement is true or false. If false explain or provide a counterexample.

59. If $\lim_{x\to c} f(x)$ does not exist, then $f(x)$ is undefined at $x=c$.

60. If $f(x)$ is undefined at $x=c$, then $\lim_{x\to c} f(x)$ does not exist.

61. If $f(x)$ is defined on $(0,\infty)$ and $y=0$ is a horizontal asymptote for $f(x)$, then there exists a number M such that if $x>M$ then $f(x)<\frac{1}{10^6}$.

62. If $f(x)$ has a vertical asymptote at $x=c$, then either $\lim_{x\to c} f(x)=\infty$ or $\lim_{x\to c} f(x)=-\infty$.

63. If $\lim_{x\to c} f(x)$ does not exist, then $\lim_{x\to c^-} f(x)\neq\lim_{x\to c^+} f(x)$ or at least one of $\lim_{x\to c^-} f(x)$ or $\lim_{x\to c^+} f(x)$ does not exist.

TECHNOLOGY

Use a graphing utility to decide whether the given limit exists by evaluating the function at several *x*-values approaching the indicated *c*-value. Then graph the function to confirm your findings. Do you obtain misleading graphs when choosing small viewing windows?

64. $\lim_{x \to 2} \frac{x^2 - 5x + 6}{x - 2}$

65. $\lim_{x \to 3} \frac{x - 2}{x^2 - 5x + 6}$

66. $\lim_{x \to -1.5} \frac{x^2 - 2.25}{x + 1.5}$

67. $\lim_{x \to 1} \frac{x - 1}{\sqrt{x} - 1}$

68. $\lim_{x \to 3} \frac{\sqrt{x+1} - 2}{x - 3}$

69. $\lim_{x \to 0} \frac{\sin(3x)}{2x}$

70. $\lim_{x \to 0^+} \cos\left(\frac{1}{x}\right)$

71. $\lim_{x \to 0^+} \left[x \cos\left(\frac{1}{x}\right)\right]$

72. Evaluate the function $f(x) = \left(1 + \frac{1}{x}\right)^x$ for several consecutive positive integers, and try to observe a tendency. Then use a graphing utility to graph $f(x)$ in a large viewing window and try to guess $\lim_{x \to \infty} f(x)$. Have you seen that number before?

13.3 THE MATHEMATICAL DEFINITION OF LIMIT

TOPICS

1. The formal definition of a limit
2. Proving a limit exists
3. Proving a limit does not exist

TOPIC 1: The Formal Definition of a Limit

We still have only an informal notion of what the statement $\lim_{x \to c} f(x) = L$ means. Now that our limit intuition has begun to develop, we can refine our informal notion and make the definition precise and mathematically useful. To be exact, we need to capture the idea that all the values of $f(x)$ become as close to L as we care to specify as x gets close enough to c.

The two parts of that previous sentence that need more work are "as close to L as we care to specify" and "as x gets close enough to c." The most common definition of limit refines those two phrases through the use of two Greek letters, epsilon (ε) and delta (δ), and the following is often referred to as the *epsilon-delta* definition of limit.

DEFINITION: Formal Definition of Limit

Let f be a function defined on an open interval containing c, except possibly at c itself. We say that the **limit of $f(x)$ as x approaches c is L**, and write $\lim_{x \to c} f(x) = L$, if for every number $\varepsilon > 0$ there is a number $\delta > 0$ such that $|f(x) - L| < \varepsilon$ whenever x satisfies $0 < |x - c| < \delta$.

The use of the letters ε and δ originated with the French mathematician Augustin-Louis Cauchy (1789–1857). His choice was deliberate, and served to remind the reader of the words *error* and *difference* (the French spellings are similar); you may find this correspondence useful as well. We may now interpret the statement $\lim_{x \to c} f(x) = L$ in this way: if we wish to guarantee that the *error* between $f(x)$ and L is less than the amount ε, it suffices to make sure that the *difference* between x and c is less than the amount δ (but remember that we do not allow x to equal c).

Figure 1 shows the relationship between L, c, ε, and δ for a given function f. The successive images illustrate the way in which δ depends on ε: as we specify smaller values for the error ε, the difference δ correspondingly shrinks. But in each image, any x chosen between $c - \delta$ and $c + \delta$ (except c itself) results in a value for $f(x)$ that is between $L - \varepsilon$ and $L + \varepsilon$.

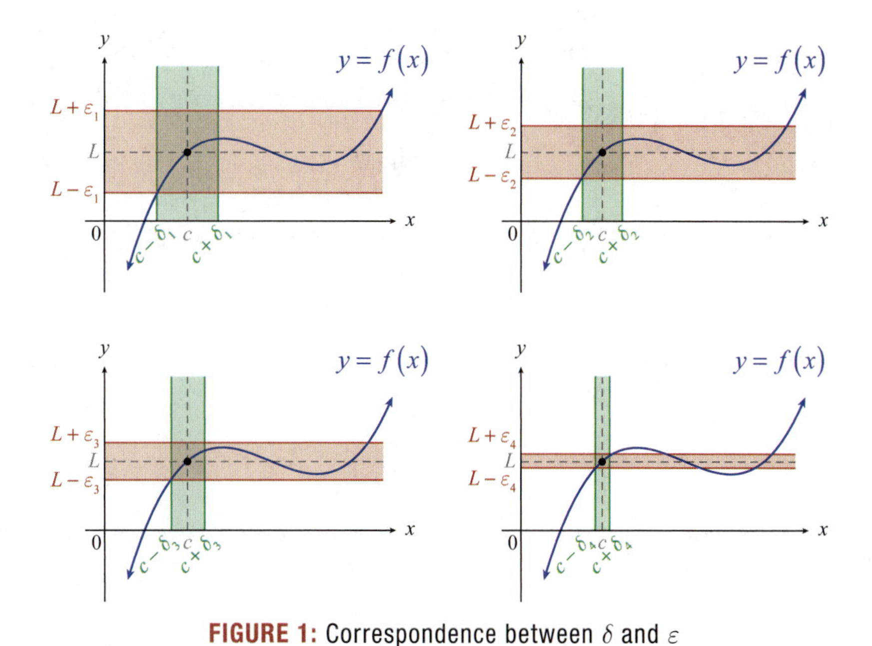

FIGURE 1: Correspondence between δ and ε

The true value of the formal limit definition is that it allows us to prove limit claims absolutely. In fact, you might consider the correspondence between ε and δ as a series of challenges and responses. If a skeptic wished to challenge the claim that $\lim_{x \to c} f(x) = L$ for the function f shown in Figure 1, each of the four ε's might be presented in turn with the demand "tell me a condition that guarantees $f(x)$ is within ε of L." The response, for each ε, would be the corresponding δ: if $x \neq c$ is chosen between $c - \delta$ and $c + \delta$ then $f(x)$ is guaranteed to fall between $L - \varepsilon$ and $L + \varepsilon$.

Note that the δ in each image is actually the largest possible choice—any positive number smaller than the pictured δ would also serve to guarantee $|f(x) - L| < \varepsilon$. Also note that, for this function and for the four ε's shown, the largest possible δ is dictated by where the line $y = L - \varepsilon$ intersects the graph of f. In the first and second images, in particular, it is clear that x could be selected from a longer interval to the right of c and still force $|f(x) - L| < \varepsilon$. Any larger δ than that shown would not suffice, however, for points selected to the left of c.

Example 1: Using Graphs to Estimate δ Given ε

In our first example, we will use the graph of $f(x) = -0.9x^2 + 2$ with $c = 0.5$ to estimate the value of δ corresponding to **a.** $\varepsilon = 1$, **b.** $\varepsilon = 0.5$, **c.** $\varepsilon = 0.25$, and **d.** $\varepsilon = 0.1$.

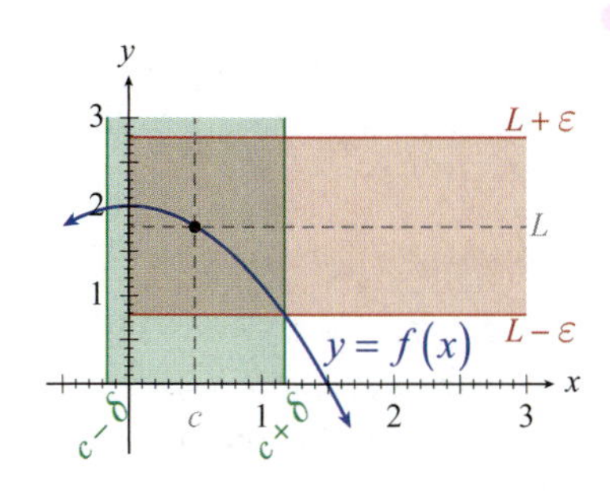

FIGURE 2

Solution

a. Like in our discussion before, we see that for $\varepsilon = 1$, the value of δ is dictated by where the line $y = L - \varepsilon$ intersects the graph of f (see Figure 2). If, for example, we choose $\delta = 0.65$, we can ensure that for any $x \neq c$ between $c - \delta$ and $c + \delta$, $f(x)$ will fall between $L - \varepsilon$ and $L + \varepsilon$.

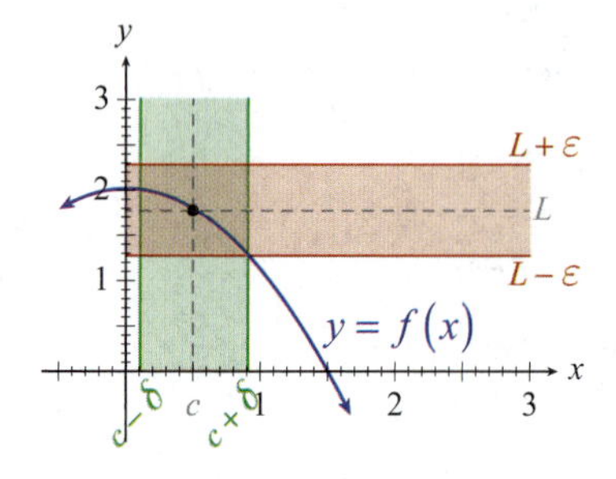

FIGURE 3

b. As for the case of $\varepsilon = 0.5$, using the illustration in Figure 3, it is clear that if $x \neq c$ falls between, say, 0.1 and 0.9, then $L - \varepsilon < f(x) < L + \varepsilon$. Thus $\delta = 0.4$ (or any smaller number) is a good choice.

c. As before, visually inspecting the graph in Figure 4 leads to the conclusion that when $\varepsilon = 0.25$, the value of $\delta = 0.2$ (or any smaller number) is a good choice, since 0.2 is less than the distance between c and either of the green vertical lines.

d. Again, a visual inspection reveals that $\varepsilon = 0.1$ will require a much smaller δ; $\delta = 0.1$ (or any smaller number) works well (see Figure 5).

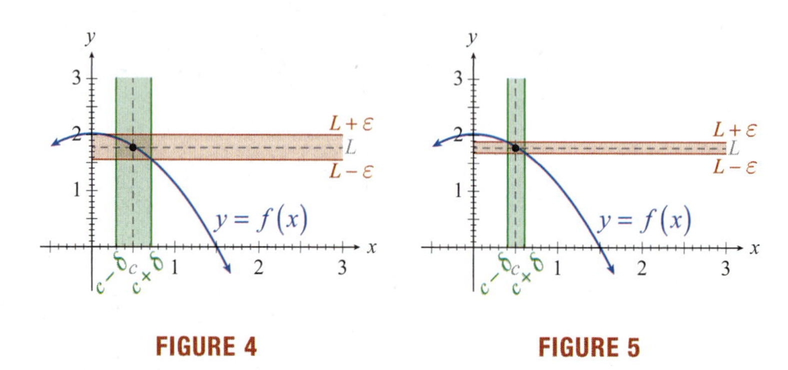

FIGURE 4 FIGURE 5

Example 2: Using Graphs to Determine Limits

Consider the following function:

$$g(x) = \begin{cases} x^3 & \text{if } x \neq 0.7 \\ 1 & \text{if } x = 0.7 \end{cases}$$

The open and closed circles in the graph in Figure 6 reflect the fact that x^3 has been redefined at $x = 0.7$: the coordinates of the "hole" are $\left(0.7, (0.7)^3\right) = (0.7, 0.343)$, while $g(0.7) = 1$.

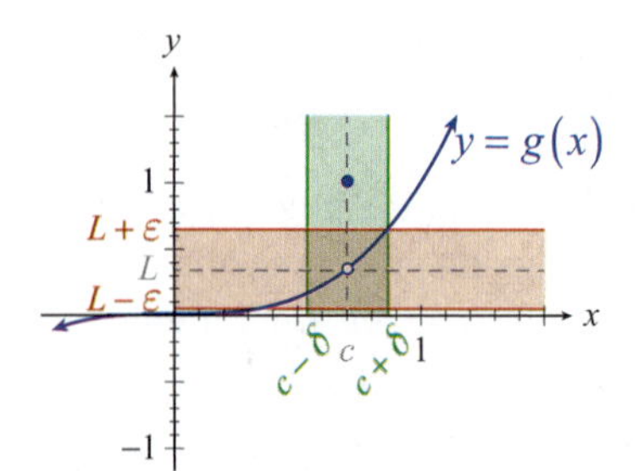

FIGURE 6

In addition, we can also learn from Figure 6 that there is a $\delta > 0$ corresponding to $\varepsilon = 0.3$ such that if we choose any $x \neq c$ from the interval $(c - \delta, c + \delta)$, the corresponding value of g will fall within ε of $(0.7)^3 = 0.343$.

Figures 7, 8, and 9 are illustrations of the existence of δ-values corresponding to smaller and smaller ε's:

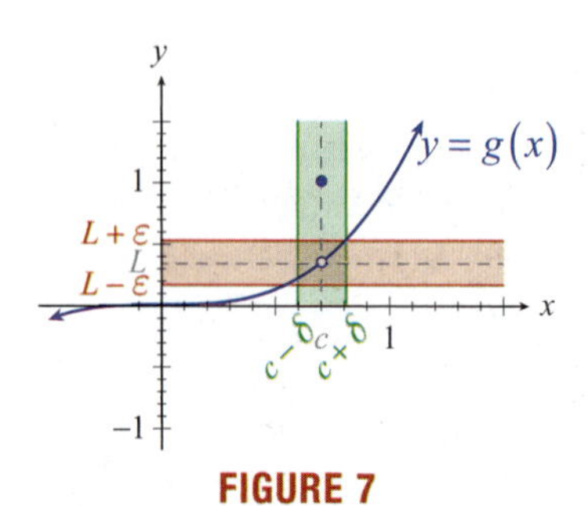

FIGURE 7

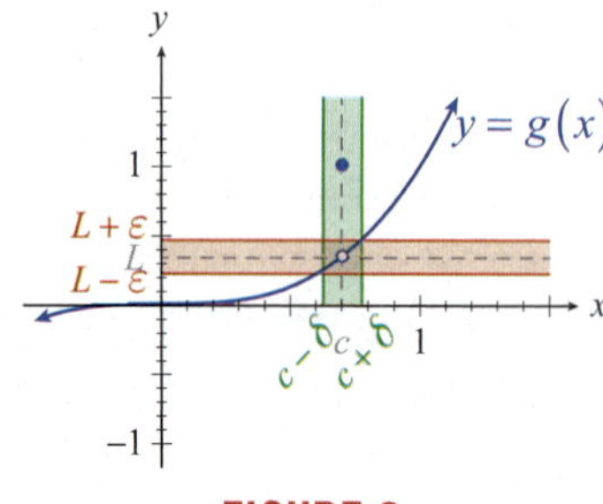

FIGURE 8

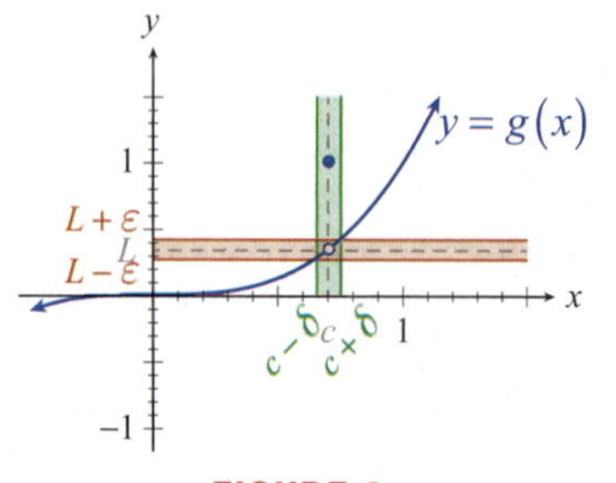

FIGURE 9

From these illustrations, it is at least visually clear that

$$\lim_{x\to 0.7} g(x) = (0.7)^3 = 0.343,$$

even though the function value $g(0.7) = 1$. In other words,

$$\lim_{x\to 0.7} g(x) \neq g(0.7);$$

the limit and the function value at $c = 0.7$ are not equal. More is actually true: we could define $g(0.7)$ to be *any* real number, without affecting the limit. In general, the function value $g(c)$ and the limit $\lim\limits_{x\to c} g(x)$ are entirely independent of each other; that is, the existence and/or value of one does not affect that of the other.

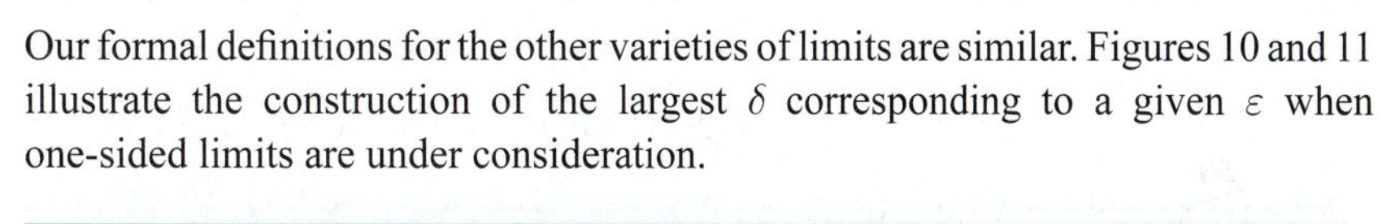

Our formal definitions for the other varieties of limits are similar. Figures 10 and 11 illustrate the construction of the largest δ corresponding to a given ε when one-sided limits are under consideration.

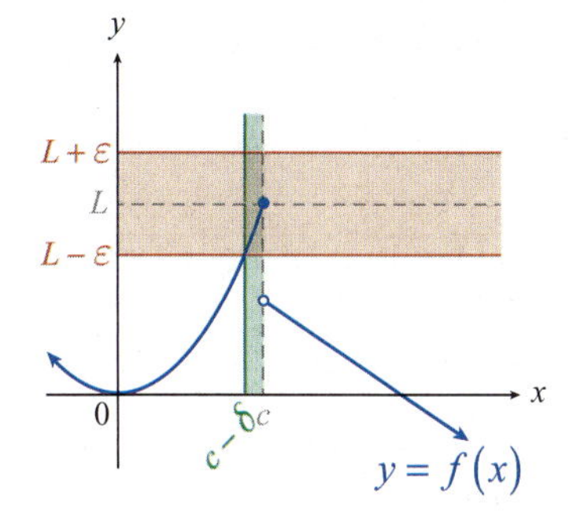

FIGURE 10: Left-Hand Limit

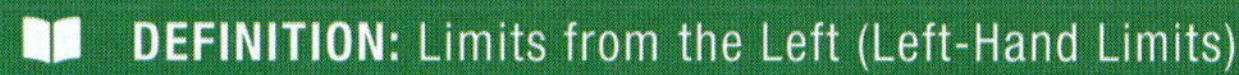

DEFINITION: Limits from the Left (Left-Hand Limits)

Let f be a function defined on an open interval whose right endpoint is c. We say that the **limit of $f(x)$ as x approaches c from the left is L**, and write $\lim\limits_{x\to c^-} f(x) = L$, if for every number $\varepsilon > 0$ there is a number $\delta > 0$ such that $|f(x) - L| < \varepsilon$ whenever x satisfies $c - \delta < x < c$.

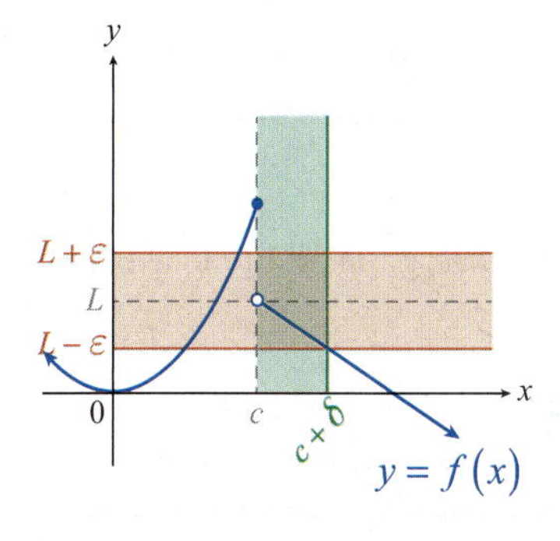

FIGURE 11: Right-Hand Limit

DEFINITION: Limits from the Right (Right-Hand Limits)

Let f be a function defined on an open interval whose left endpoint is c. We say that the **limit of $f(x)$ as x approaches c from the right is L**, and write $\lim\limits_{x\to c^+} f(x) = L$, if for every number $\varepsilon > 0$ there is a number $\delta > 0$ such that $|f(x) - L| < \varepsilon$ whenever x satisfies $c < x < c + \delta$.

We will wrap up the formal limit definitions with cases in which the symbol ∞ makes an appearance.

DEFINITION: Infinite Limits

Let f be a function defined on an open interval containing c, except possibly at c itself. We say that the **limit of $f(x)$ as x approaches c is positive infinity**, and write $\lim\limits_{x\to c} f(x) = \infty$, if for every positive number M there is a number $\delta > 0$ such that $f(x) > M$ whenever x satisfies $0 < |x - c| < \delta$.

Similarly, we say that the **limit of $f(x)$ as x approaches c is negative infinity**, and write $\lim\limits_{x\to c} f(x) = -\infty$, if for every negative number N there is a number $\delta > 0$ such that $f(x) < N$ whenever x satisfies $0 < |x - c| < \delta$. **Infinite one-sided limits** are defined in a manner analogous to finite one-sided limits.

If we wish to prove that a given function has an infinite limit at a point c, our task is to prove that given some (presumably large) number M, there is an interval $(c-\delta, c+\delta)$ on which the function is larger than M (except possibly at the point c itself). Figure 12 shows how an appropriate δ can be chosen for the given value of M.

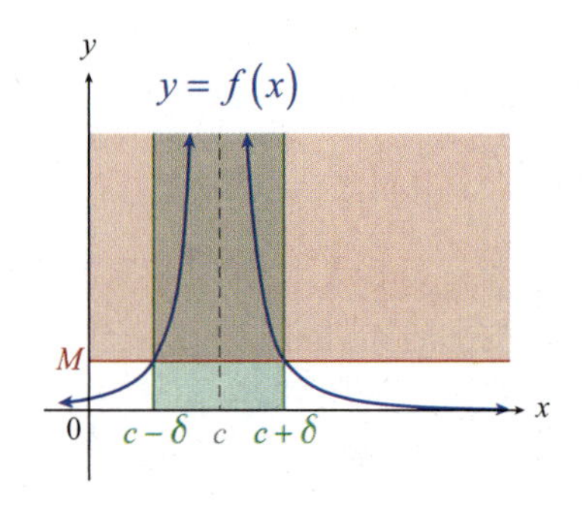

FIGURE 12: Correspondence between M and δ for an Infinite Limit

Finally, limits at infinity are defined formally as follows.

DEFINITION: Limits at Infinity

Let f be a function defined on some interval (a,∞). We say the **limit of f at infinity is L**, and write $\lim_{x\to\infty} f(x) = L$, if for every number $\varepsilon > 0$ there is a number N such that $|f(x) - L| < \varepsilon$ for all $x > N$.

Similarly, for a function defined on some interval $(-\infty, b)$ we say the **limit of f at negative infinity is L**, and write $\lim_{x\to-\infty} f(x) = L$, if for every number $\varepsilon > 0$ there is a number N such that $|f(x) - L| < \varepsilon$ for all $x < N$.

Be sure you understand how these formal definitions of limits at infinity relate to horizontal asymptotes.

TOPIC 2: Proving a Limit Exists

Although the formal definition of limit gives us the ability to prove limit claims beyond a shadow of a doubt, it does not immediately give us the means by which to determine the value of a given limit in the first place. In other words, we need to know what L is before we can prove that $\lim_{x\to c} f(x) = L$. Fortunately, as we will soon see, our limit definition is the basis for a number of theorems that will make determination of limits much easier.

As a stepping-stone toward those theorems, we will develop our ability to use the limit definition to prove some example claims.

Example 3: Using the ε-δ Definition of Limit to Prove a Limit Exists

Use the ε-δ definition of limit to prove that $\lim_{x\to 3}(2x-1) = 5$.

Solution

Suppose an $\varepsilon > 0$ is given. We need to find a $\delta > 0$ such that

$$0 < |x-3| < \delta \quad \Rightarrow \quad |(2x-1)-5| < \varepsilon.$$

Notice that this latter inequality is equivalent to the following:

$$\begin{aligned} |2x-6| < \varepsilon \quad &\Leftrightarrow \quad |2(x-3)| < \varepsilon \\ &\Leftrightarrow \quad |2|\cdot|x-3| < \varepsilon \\ &\Leftrightarrow \quad 2|x-3| < \varepsilon \\ &\Leftrightarrow \quad |x-3| < \frac{\varepsilon}{2} \end{aligned}$$

Reading this chain of equivalent inequalities from the bottom up, this means that $\delta = \frac{\varepsilon}{2}$ (or any smaller number) works. In other words, given $\varepsilon > 0$, if we choose $\delta = \frac{\varepsilon}{2}$, then $0 < |x-3| < \delta$ implies

$$|(2x-1)-5| = |2x-6| = |2(x-3)| = 2|x-3| < 2\delta = 2 \cdot \frac{\varepsilon}{2} = \varepsilon,$$

just as we needed to show.

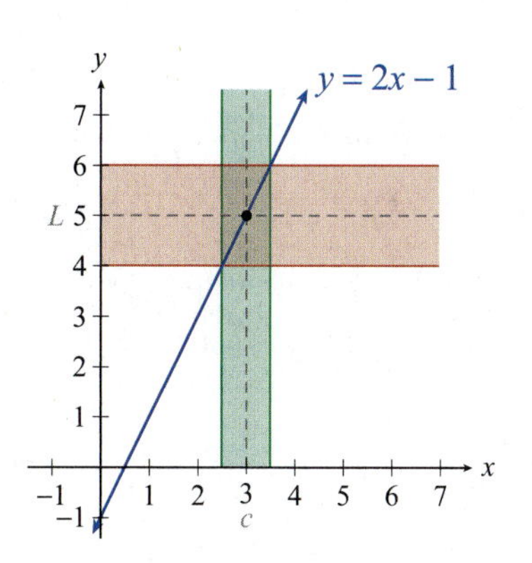

FIGURE 13

What all this means is that if a skeptic were to challenge you with a small $\varepsilon > 0$ of his or her choosing, you could always respond, for each ε, by picking $\delta = \frac{\varepsilon}{2}$: if $x \neq c$ is chosen between $3 - \delta$ and $3 + \delta$, then $2x - 1$ is guaranteed to fall between $5 - \varepsilon$ and $5 + \varepsilon$. For example, given $\varepsilon = 0.1$, let $\delta = 0.05$. For the reason that x needs to be twice as close to 3 as the skeptic's challenge, examine the graph in Figure 13, and note that the slope of the line is 2.

Example 4: Using the ε-δ Definition of Limit to Prove a Limit Exists

Use the ε-δ definition of limit to prove that $\lim_{x \to 4} x^2 = 16$.

Solution

Again, supposing that $\varepsilon > 0$ is given, we need to find a $\delta > 0$ such that

$$0 < |x-4| < \delta \quad \Rightarrow \quad |x^2 - 16| < \varepsilon.$$

Our first observation is that

$$|x^2 - 16| = |(x+4)(x-4)| = |x+4| \cdot |x-4|.$$

Next, we will agree to choose a $\delta > 0$ while also making sure that $\delta < 1$. Note that we can do this, since once a successful $\delta > 0$ is chosen, any smaller number works just fine.

Since $|x-4| < \delta$ translates into $4 - \delta < x < 4 + \delta$, which implies $3 < x < 5$ if $\delta < 1$, we have $|x+4| = x + 4 < 5 + 4 = 9$ and thus

$$|x^2 - 16| = |x+4| \cdot |x-4| < 9|x-4|.$$

So for the given $\varepsilon > 0$, let $\delta > 0$ be chosen so that it is less than the smaller of the numbers 1 and $\frac{\varepsilon}{9}$.

Then if $|x-4| < \delta$,

$$|x^2 - 16| = |x+4| \cdot |x-4| < 9|x-4| < 9\delta < 9 \cdot \frac{\varepsilon}{9} = \varepsilon,$$

this latter inequality finishing our proof.

Example 5: Using the ε-δ Definition of Limit to Prove a Limit Exists

Use the ε-δ definition of limit to prove that $\lim_{x\to\infty}\frac{x^2+1}{x^2}=1$.

Solution

As before, let $\varepsilon>0$ be given. We need to find a number N such that

$$x>N \quad\Rightarrow\quad \left|\frac{x^2+1}{x^2}-1\right|<\varepsilon.$$

Notice that since $\frac{x^2+1}{x^2}=1+\frac{1}{x^2}$, this latter inequality is equivalent to

$$\left|\left(1+\frac{1}{x^2}\right)-1\right|<\varepsilon \quad\Leftrightarrow\quad \left|\frac{1}{x^2}\right|<\varepsilon$$

$$\Leftrightarrow\quad \frac{1}{x^2}<\varepsilon,$$

since x^2 is nonnegative. Also observe that if we choose N large enough so that $\frac{1}{N^2}<\varepsilon$, then $x>N$ will imply $\frac{1}{x^2}<\frac{1}{N^2}<\varepsilon$. The above choice is always possible, by simply making sure $N>\frac{1}{\sqrt{\varepsilon}}$.

We can summarize our observations as follows. For a given $\varepsilon>0$, choose and fix a positive number N satisfying $N>\frac{1}{\sqrt{\varepsilon}}$. Then if $x>N$,

$$\left|\frac{x^2+1}{x^2}-1\right|=\left|1+\frac{1}{x^2}-1\right|=\left|\frac{1}{x^2}\right|=\frac{1}{x^2}<\frac{1}{N^2}<\frac{1}{\left(\frac{1}{\sqrt{\varepsilon}}\right)^2}=\frac{1}{\frac{1}{\varepsilon}}=\varepsilon,$$

and this latter chain of inequalities finishes our argument.

Example 6: Proving a Limit Does Not Exist

Prove that $\lim_{x\to2}\frac{1}{(x-2)^2}=\infty$.

Solution

For an arbitrary, but fixed, positive number M, we need to exhibit a number $\delta>0$ such that $\frac{1}{(x-2)^2}>M$ whenever x satisfies $0<|x-2|<\delta$. To that end, let us assume that $M>0$ is fixed, and choose a $\delta>0$ small enough so that $\frac{1}{\delta^2}\geq M$. Note that this is possible; for example, $\delta=\frac{1}{\sqrt{M}}$ (or any smaller number) works: if $0<|x-2|<\delta$, then

$$\frac{1}{(x-2)^2}>\frac{1}{\delta^2}=\frac{1}{\left(\frac{1}{\sqrt{M}}\right)^2}=\frac{1}{\frac{1}{M}}=M,$$

which is precisely the inequality we wanted.

As a final remark, we wish to emphasize that $\lim_{x \to 2} \frac{1}{(x-2)^2}$ actually does not exist, since ∞ is not a number. What is happening is that the limit fails to exist because the function $f(x) = \frac{1}{(x-2)^2}$ grows without bound as x approaches 2. We have established this using techniques as in the previous examples, and expressed the same fact using limit notation and the ∞ symbol.

TOPIC 3: Proving a Limit Does Not Exist

We will close this section with an illustration of how we can use our limit definition to prove that a given function does *not* have a limit at a given point. There are occasions when the ability to prove that something does not happen is just as useful as the ability to prove that it does.

Example 7: Proving a Limit Does Not Exist

Prove that for the function $f(x) = \cos\left(\frac{1}{x}\right)$, $\lim_{x \to 0} f(x)$ does not exist by showing that the function does not satisfy the ε-δ definition for any possible L at $c = 0$.

Solution

We will first need to think carefully about what it means for the definition to *fail* at $c = 0$. This will happen precisely when no number L works as the limit of the function at 0, in other words, when x can move closer to 0 than *any* given $\delta > 0$ without the corresponding function values $f(x)$ approaching any number. You can think about this, too, as a challenge game. Suppose a skeptic challenges you with a very small number δ, and you are able to show that the function values $f(x)$, corresponding to nonzero x-values with $-\delta < x < \delta$, do not approach any number L. If you are always able to respond to the challenge, no matter how small $\delta > 0$ becomes, you will have proved that $\lim_{x \to 0} f(x)$ does not exist. This is exactly what we endeavor to do in this example.

Suppose that $\delta > 0$ is given. Recall that $\cos(2k\pi) = 1$ and $\cos((2k+1)\pi) = -1$ for all $k \in \mathbb{Z}$. Choose and fix a big enough positive integer k such that $\frac{1}{2k\pi} < \delta$ (since δ is already fixed, you can achieve this by merely choosing k big enough). Next, we will use the real numbers $x_1 = \frac{1}{2k\pi}$ and $x_2 = \frac{1}{(2k+1)\pi}$ as follows. Note that both are less than δ in magnitude, but

$$\begin{aligned} f(x_1) &= \cos\left(\frac{1}{x_1}\right) \\ &= \cos\left(\frac{1}{\frac{1}{2k\pi}}\right) \\ &= \cos(2k\pi) = 1, \end{aligned}$$

while

$$\begin{aligned} f(x_2) &= \cos\left(\frac{1}{x_2}\right) \\ &= \cos\left(\frac{1}{\frac{1}{(2k+1)\pi}}\right) \\ &= \cos((2k+1)\pi) = -1. \end{aligned}$$

What this means is that we were able to find x-values less than δ in magnitude, namely x_1 and x_2, for which the corresponding function values $f(x_1)$ and $f(x_2)$ are a full 2 units apart. More importantly, we can do the same, no matter how small a $\delta > 0$ is specified. Thus the values of $f(x)$ cannot be approaching any limit L at all.

To put our argument on a more precise footing, we will show that the definition for the existence of $\lim_{x\to 0} f(x)$ fails, by showing that there is an $\varepsilon > 0$ for which no $\delta > 0$ exists to satisfy the definition.

Let us pick, say, $\varepsilon = 1$, and suppose that there is a $\delta > 0$ such that for some L,

$$|f(x) - L| < 1,$$

whenever $0 < |x - 0| < \delta$. Then proceed to find x_1, x_2 as above. Since both $|x_1| < \delta$ and $|x_2| < \delta$, by assumption we have first of all the inequality

$$|f(x_1) - L| = |1 - L| < 1.$$

What this means is that L is less than 1 unit away from 1; in particular, L is positive.

On the other hand, because of the choice of x_2, we also have the inequality

$$|f(x_2) - L| = |-1 - L| < 1;$$

in other words, L is less than 1 unit from -1 and therefore is negative.

Since such a number does not exist, $\lim_{x\to 0} f(x)$ cannot exist either.

13.3 EXERCISES

PRACTICE

Use the graph to estimate δ corresponding to the given ε satisfying the ε-δ definition of $\lim_{x \to c} f(x) = L$. See Example 1.

1.

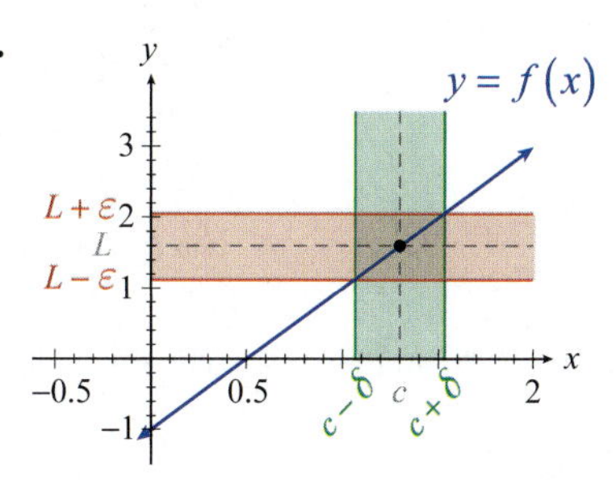

2.

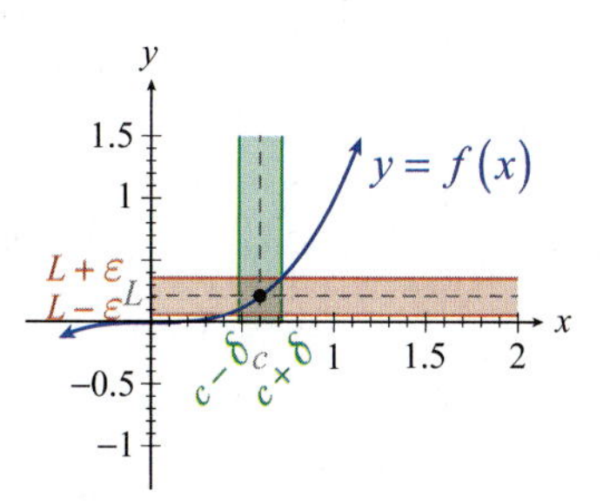

3.

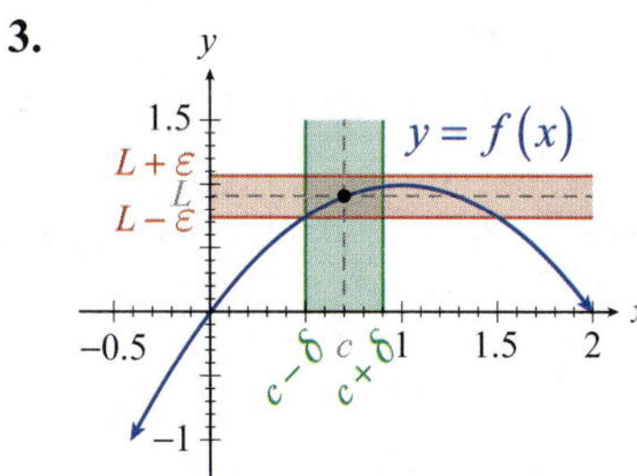

4.

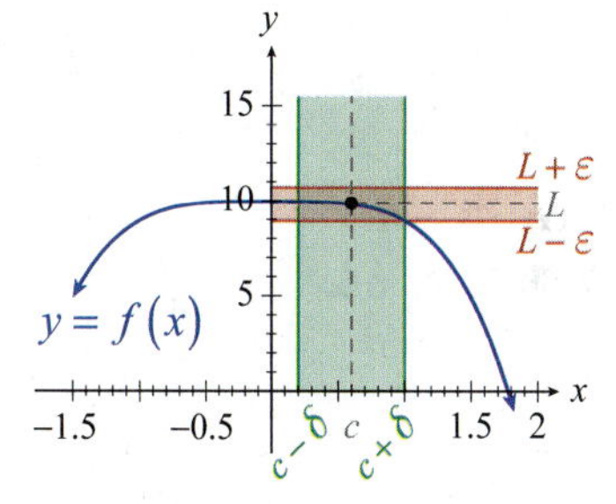

Find a $\delta > 0$ that satisfies the limit claim corresponding to $\varepsilon = 0.1$; that is, such that $0 < |x - c| < \delta$ would imply $|f(x) - L| < 0.1$.

5. $\lim_{x \to 2}(5x - 1) = 9$

6. $\lim_{x \to 1}(3x + 1) = 4$

7. $\lim_{x \to -1}(-x + 2) = 3$

8. $\lim_{x \to 6}\left(4 - \frac{x}{2}\right) = 1$

9. $\lim_{x \to 0} x^2 = 0$

10. $\lim_{x \to 8} \sqrt[3]{x} = 2$

11. $\lim_{x \to 1} \frac{1}{x} = 1$

12. $\lim_{x \to 0} e^x = 1$

13. $\lim_{x \to 1} \ln x = 0$

14. $\lim_{x \to 0} \cos x = 1$

Find a number N that satisfies the limit claim corresponding to $\varepsilon = 0.1$; that is, such that $x > N$ (or $x < N$, as appropriate) would imply $|f(x) - L| < 0.1$.

15. $\lim_{x \to \infty} \frac{1}{x} = 0$

16. $\lim_{x \to \infty} \frac{x^2 - 1}{x^2 + 2x} = 1$

17. $\lim_{x \to -\infty} \frac{x + 1}{x} = 1$

18. $\lim_{x \to \infty} \frac{2x}{\sqrt{x^2 + x}} = 2$

19. $\lim_{x \to -\infty} e^x = 0$

20. $\lim_{x \to \infty} \arctan x = \frac{\pi}{2}$

For the given function $f(x)$, find a $\delta > 0$ corresponding to $M = 100$; that is, such that $0 < |x - c| < \delta$ would imply $f(x) > 100$ (let $N = -100$ if the limit is $-\infty$, in which case $0 < |x - c| < \delta$ should imply $f(x) < -100$).

21. $\lim_{x\to 0} \frac{1}{x^2} = \infty$

22. $\lim_{x\to 0^+} \frac{1}{x} = \infty$

23. $\lim_{x\to -1} \frac{-1}{(x+1)^2} = -\infty$

24. $\lim_{x\to 0^+} \ln x = -\infty$

25. $\lim_{x\to \frac{\pi}{2}^-} \tan x = \infty$

26. $\lim_{x\to 0^+} \csc x = \infty$

Use the ε-δ definition to prove the limit claim. See Examples 3 and 4.

27. $\lim_{x\to 1} (2x+3) = 5$

28. $\lim_{x\to 7} x = 7$

29. $\lim_{x\to c} a = a$

30. $\lim_{x\to 4} \left(\frac{1}{4}x + 1\right) = 2$

31. $\lim_{x\to 0} \left(\frac{1}{2} - 4x\right) = \frac{1}{2}$

32. $\lim_{x\to 9} \left(5 - \frac{x}{3}\right) = 2$

33. $\lim_{x\to 1} x^3 = 1$

34. $\lim_{x\to 0} x^2 = 0$

35. $\lim_{x\to 0} \left(\frac{1}{2}|x|\right) = 0$

36. $\lim_{x\to -2} |x+2| = 0$

37. $\lim_{x\to 1^+} \sqrt{x-1} = 0$

38. $\lim_{x\to 0^-} \left(\sqrt[3]{x} + 1\right) = 1$

39. $\lim_{x\to 1} (x^2 + x) = 2$

40. $\lim_{x\to 3} (3x^2 - 9x + 5) = 5$

Give the formal definition of the limit claim. Then use the definition to prove the claim. See Examples 5 and 6.

41. $\lim_{x\to \infty} \frac{1+x}{x} = 1$

42. $\lim_{x\to -\infty} \frac{2}{x^2} = 0$

43. $\lim_{x\to \infty} \frac{1}{\sqrt{x}} = 0$

44. $\lim_{x\to \infty} \frac{1+3x^3}{x^3} = 3$

45. $\lim_{x\to -\infty} 2^x = 0$

46. $\lim_{x\to \infty} (e^{-x} - 1) = -1$

47. $\lim_{x\to \infty} \frac{\sin x}{x} = 0$

48. $\lim_{x\to \infty} (2\arctan x) = \pi$

49. $\lim_{x\to 0^+} \frac{1}{x} = \infty$

50. $\lim_{x\to 0} \frac{1}{x^4} = \infty$

51. $\lim_{x\to -1} \frac{-1}{(x+1)^2} = -\infty$

52. $\lim_{x\to 0^+} \log x = -\infty$

53. $\lim_{x\to \frac{\pi}{2}^+} \tan x = -\infty$

54. $\lim_{x\to 0^+} \csc x = \infty$

55. $\lim_{x\to 2} \frac{-3}{(x-2)^2} = -\infty$

56. $\lim_{x\to -2^-} \frac{x+3}{x+2} = -\infty$

Determine whether the limit exists. Prove your conclusion. See Example 7.

57. $\lim_{x\to 0^+} \sin\left(\frac{\pi}{x}\right)$

58. $\lim_{x\to 0^+} \left(x^2 \cos\left(\frac{1}{x}\right)\right)$

59. $\lim_{x\to 0} \frac{|x|}{x}$

60. $\lim_{x\to 1} f(x)$, where $f(x) = \begin{cases} 1 & \text{if } x \text{ is rational} \\ 0 & \text{if } x \text{ is irrational} \end{cases}$

61. $\lim_{x\to 0} g(x)$, where $g(x) = \begin{cases} x & \text{if } x \text{ is rational} \\ 0 & \text{if } x \text{ is irrational} \end{cases}$

APPLICATIONS

62. A piston is manufactured to fit into the cylinder of a certain automobile engine. Suppose that the diameter of the cylinder is 82 mm and that the cross-sectional area of the piston is not allowed to be less than 99.89% of that of the cylinder. If both are perfectly round, what does this mean in terms of maximum tolerance for the clearance between the piston and the cylinder wall? (Be sure to identify which function and data take the roles of $f(x)$, c, ε, and δ in this problem.)

63. The tension in a stretched steel wire (in newtons, N) is calculated by the formula $F = E\frac{\Delta L}{L_0}A$, where $E = 2\times 10^{11}\ \text{N/m}^2$ is the elastic modulus (or Young's modulus) of steel, ΔL is the elongation, L_0 the original length, and A the cross-sectional area (in m^2). Suppose a 1-meter-long steel string of radius 1 millimeter is stretched by 2 millimeters when tuning a musical instrument.

a. Calculate the tension in the string caused by the above tightening.

b. If we are not allowed to overload the string by more than 100 N, what is the tolerance in the amount of stretching? (Be sure to identify the function and data taking the roles of c, ε, and δ in this problem.)

WRITING & THINKING

64. Use ε and δ to state what $\lim_{x\to c} f(x) \neq L$ means.

Students gave the following definitions for the existence of a limit of $f(x)$ at c. Find and correct any errors.

65. "$\lim_{x\to c} f(x)$ exists if for any $\varepsilon > 0$ and real number L there is a $\delta > 0$ such that $0 < |x - L| < \delta$ implies $|f(x) - c| < \varepsilon$."

66. "$\lim_{x\to c} f(x)$ exists and equals L if for any $\varepsilon > 0$ and $\delta > 0$ whenever $0 < |x - c| < \varepsilon$, we have $|f(x) - L| < \delta$."

67. "If there is a real number L such that for an $\varepsilon > 0$ there is a $\delta > 0$ such that whenever $|x - c| < \delta$ and $x \neq c$, we have $|f(x) - L| < \varepsilon$, we say that the limit of the function at c is L."

68. "We say that $\lim_{x\to c} f(x) = L,$ if for any $\varepsilon > 0$ there is a $\delta > 0$ such that $|x-c| < \delta \Rightarrow |f(x) - L| < \varepsilon.$"

69. "We say that $\lim_{x\to c} f(x) = L,$ if for any $\varepsilon > 0$ there is a $\delta > 0$ such that $0 \le |x-c| \le \delta \Rightarrow |f(x) - L| \le \varepsilon.$"

70. "If the real number L is such that for any $\varepsilon > 0$ there is a $\delta > 0$ such that $0 < |x-c| < \delta \Rightarrow |f(x) - L| < \varepsilon,$ we say that $\lim_{x\to c} f(x) = L.$"

Determine whether the given statement is true or false. If false, explain or provide a counterexample.

71. If $f(c) = L,$ then as x approaches $c,$ $\lim_{x\to c} f(x) = L.$

72. If $\lim_{x\to c} f(x)$ exists and equals $L,$ then $f(c) = L.$

73. If $f(x) < g(x)$ for all $x \ne c,$ and both $\lim_{x\to c} f(x)$ and $\lim_{x\to c} g(x)$ exist, then $\lim_{x\to c} f(x) < \lim_{x\to c} g(x).$

TECHNOLOGY

Use a graphing utility to estimate the given limit. By zooming in appropriately, find δ-values that correspond to $\varepsilon = 0.1$.

74. $\lim_{x\to 5} \dfrac{x^2 - 5x + 6}{x - 2}$

75. $\lim_{x\to 0} \dfrac{\sqrt{x+5} - \sqrt{5}}{x}$

76. $\lim_{x\to 3.5} \dfrac{x^2 - 6.25}{x + 2.5}$

77. $\lim_{x\to 0} \dfrac{x - 1}{\sqrt{x} - 1}$

78. $\lim_{x\to 0} \dfrac{\sin 3x}{2x}$

79. $\lim_{x\to -\infty} \dfrac{2x + 3}{\sqrt{x^2 + 1}}$

80. $\lim_{x\to \infty} \dfrac{\sqrt{9x^2 + 1}}{x - 2}$

81. $\lim_{x\to -\infty} \dfrac{2x^2 + 1.5x - 7}{\sqrt{x^4 + 1}}$

82. $\lim_{x\to \infty} \left(\sqrt{x^2 + 3x + 5} - \sqrt{x^2 + 2x + 1}\right)$

83. $\lim_{x\to \infty} \left(1 + \dfrac{1}{x}\right)^x$

Use a graphing utility to locate a vertical asymptote of the given function. Then for such an asymptote $x = c$ find an appropriate value $\delta > 0$ such that $|x - c| < \delta \Rightarrow |f(x)| > 10.$ (Answers will vary.)

84. $f(x) = \dfrac{x^2 - 7}{x^3 + x + 1}$

85. $f(x) = \dfrac{3x + 1}{2x^4 + x - 5}$

86. $f(x) = \ln\left(\dfrac{x^2}{x^2 + 1}\right)$

87. $f(x) = \tan\left(\dfrac{1}{2}x + 3\right)$

88. $f(x) = \csc(2x + 1)$

89. $f(x) = \cot\left(\dfrac{1}{2}\cos x\right)$

13.4 DETERMINING LIMITS OF FUNCTIONS

TOPICS

1. Limit laws
2. Limit determination techniques

TOPIC 1: Limit Laws

We now have a formal definition of *limit* with which to work, and some experience in proving limit claims with epsilon-delta arguments. What we are lacking is a collection of tools allowing us to determine limits in the first place. The theorems in this section will help us do just that and, at the same time, provide the necessary proof that the resulting limits are correct.

We will begin with the basic limit laws and show how they can be used both in immediate applications and in deriving more powerful laws.

THEOREM: Basic Limit Laws

Let f and g be two functions such that both $\lim_{x \to c} f(x)$ and $\lim_{x \to c} g(x)$ exist, and let k be a fixed real number. Then the following laws hold.

Sum Law $\lim_{x \to c}\left[f(x)+g(x)\right]=\lim_{x \to c} f(x)+\lim_{x \to c} g(x)$

Difference Law $\lim_{x \to c}\left[f(x)-g(x)\right]=\lim_{x \to c} f(x)-\lim_{x \to c} g(x)$

Constant Multiple Law $\lim_{x \to c}\left[kf(x)\right]=k\lim_{x \to c} f(x)$

Product Law $\lim_{x \to c}\left[f(x)g(x)\right]=\lim_{x \to c} f(x)\cdot\lim_{x \to c} g(x)$

Quotient Law $\lim_{x \to c}\frac{f(x)}{g(x)}=\frac{\lim_{x \to c} f(x)}{\lim_{x \to c} g(x)}$, provided $\lim_{x \to c} g(x)\neq 0$

Proof

We will prove the Sum Law here. The proofs of the remaining laws are similar in nature but will be left for a Calculus course.

For ease of exposition, let $L=\lim_{x \to c} f(x)$ and $M=\lim_{x \to c} g(x)$; we need to show that $\lim_{x \to c}\left[f(x)+g(x)\right]=L+M$ using our ε-δ definition of limit. To this end, assume $\varepsilon>0$ is given. Then by assumption there exist $\delta_1>0$ and $\delta_2>0$ such that

$$0<|x-c|<\delta_1 \quad \Rightarrow \quad |f(x)-L|<\frac{\varepsilon}{2}$$

and

$$0<|x-c|<\delta_2 \quad \Rightarrow \quad |g(x)-M|<\frac{\varepsilon}{2}.$$

Note that we have found δ_1 and δ_2 so that the differences between the functions and their respective limits are smaller than $\frac{\varepsilon}{2}$. We did so in order to obtain the following consequence for all $x \neq c$ chosen within δ of c, where δ is the smaller of δ_1 and δ_2.

$$\begin{aligned} |f(x)+g(x)-(L+M)| &= |f(x)-L+g(x)-M| \\ &\leq |f(x)-L|+|g(x)-M| && \text{Triangle Inequality: } |a+b| \leq |a|+|b| \\ &< \frac{\varepsilon}{2}+\frac{\varepsilon}{2}=\varepsilon && \delta \leq \delta_1 \text{ and } \delta \leq \delta_2 \end{aligned}$$

Since we have demonstrated that the function $f(x)+g(x)$ is within ε of $L+M$ for all x such that $0<|x-c|<\delta$, our proof is complete.

Intuitively, the limit laws seem reasonable. For instance, if $\lim_{x\to c} f(x)=L$ and $\lim_{x\to c} g(x)=M$, the Product Law points out (in a precise manner) that for values of x close to c, the function $f \cdot g$ assumes values close to $L \cdot M$. As always, though, it is important to remember that statements about the limit at a point c describe behavior *near* c, not *at* c.

Example 1: Using Graphs and Limit Laws to Determine Limits

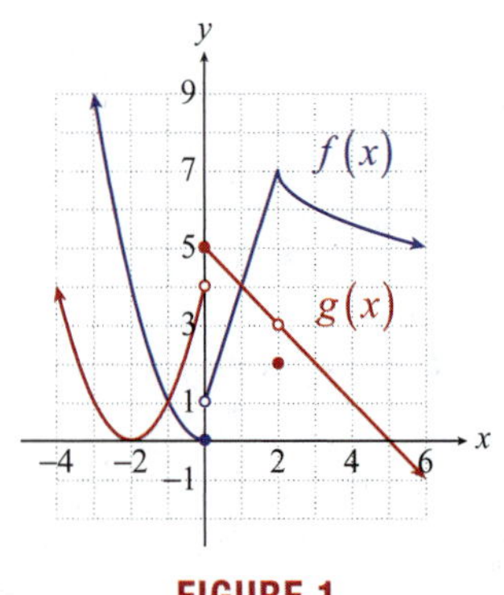

FIGURE 1

For the functions f and g graphed in Figure 1, determine if the following limits exist. If a particular limit exists, evaluate it. If not, give reasons why it fails to exist.

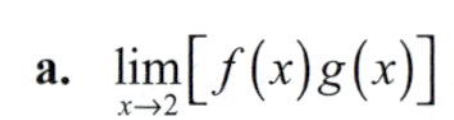

a. $\lim_{x\to 2}[f(x)g(x)]$ **b.** $\lim_{x\to 2}\left[\frac{3}{7}f(x)+g(x)\right]$

c. $\lim_{x\to 0}[g(x)-f(x)]$ **d.** $\lim_{x\to 0}[f(x)g(x)]$

Solution

a. As we can see from the figure, the limits of both f and g exist at $c=2$. In the case of g, we note once again that the actual function value at 2 has no bearing on the value of the limit.

$$\lim_{x\to 2} f(x)=7 \quad \text{and} \quad \lim_{x\to 2} g(x)=3$$

Now we evaluate the limit using the Product Law.

$$\lim_{x\to 2}[f(x)g(x)]=\lim_{x\to 2} f(x)\cdot \lim_{x\to 2} g(x)=7\cdot 3=21$$

b. Here we are going to use the Sum Law and the Constant Multiple Law.

$$\begin{aligned} \lim_{x\to 2}\left[\frac{3}{7}f(x)+g(x)\right] &= \lim_{x\to 2}\left[\frac{3}{7}f(x)\right]+\lim_{x\to 2} g(x)=\frac{3}{7}\lim_{x\to 2} f(x)+\lim_{x\to 2} g(x) \\ &= \frac{3}{7}\cdot 7+3=3+3=6 \end{aligned}$$

c. We will first examine the behavior of f near 0. It is clear from the graph that both one-sided limits exist, but they don't agree; thus the two-sided limit of f cannot exist at $x = 0$. In fact, $\lim_{x \to 0^-} f(x) = 0$ while $\lim_{x \to 0^+} f(x) = 1$, and so $\lim_{x \to 0} f(x)$ does not exist.

Similarly for g, $\lim_{x \to 0^-} g(x) = 4$ while $\lim_{x \to 0^+} g(x) = 5$, and thus $\lim_{x \to 0} g(x)$ cannot exist.

In words, since the one-sided limits are unequal, the two-sided limit of g cannot exist at 0. However, the limit laws do apply to one-sided limits, so we can determine those for $\left[g(x) - f(x)\right]$.

$$\lim_{x \to 0^-}\left[g(x) - f(x)\right] = \lim_{x \to 0^-} g(x) - \lim_{x \to 0^-} f(x) = 4 - 0 = 4$$

$$\lim_{x \to 0^+}\left[g(x) - f(x)\right] = \lim_{x \to 0^+} g(x) - \lim_{x \to 0^+} f(x) = 5 - 1 = 4$$

Notice from our findings that something interesting is actually going on here. Even though the two-sided limits of f and g do not exist individually, the one-sided limits of $g - f$ exist and agree at $x = 0$. What this means is that the limit of $g - f$ at 0 actually exists.

$$\lim_{x \to 0}\left[g(x) - f(x)\right] = 4$$

d. Using the Product Law for one-sided limits, we obtain

$$\lim_{x \to 0^-}\left[f(x)g(x)\right] = \lim_{x \to 0^-} f(x) \cdot \lim_{x \to 0^-} g(x) = 0 \cdot 4 = 0, \text{ while}$$

$$\lim_{x \to 0^+}\left[f(x)g(x)\right] = \lim_{x \to 0^+} f(x) \cdot \lim_{x \to 0^+} g(x) = 1 \cdot 5 = 5.$$

Since the one-sided limits of the product function $f \cdot g$ are unequal, we conclude that $\lim_{x \to 0}\left[f(x)g(x)\right]$ does not exist.

Example 2: Using Limit Laws to Determine Limits

Let $f(x) = \begin{cases} \dfrac{|x|}{x} & \text{if } x \neq 0 \\ 0 & \text{if } x = 0 \end{cases}$ and $g(x) = \cos x$. Determine whether $\lim_{x \to 0}\left[f(x)g(x)\right]$ exists.

Solution

We will first examine the behavior of $f(x)$ near 0. If x approaches 0 from the left, then since $x < 0$, we have

$$f(x) = \frac{|x|}{x} = \frac{-x}{x} = -1,$$

so f is the constant -1 for all negative x-values.

Therefore,

$$\lim_{x\to 0^-} f(x) = -1.$$

On the other hand, for all positive x-values $|x| = x$, so

$$f(x) = \frac{|x|}{x} = \frac{x}{x} = 1,$$

thus

$$\lim_{x\to 0^+} f(x) = 1.$$

Since the one-sided limits are unequal, $\lim_{x\to 0} f(x)$ does not exist. Consequently, the Product Law is not applicable, but since it does apply to one-sided limits, we can proceed to examine those separately. Recalling the well-known fact that $\lim_{x\to 0} \cos x = 1$, we obtain

$$\lim_{x\to 0^-}\left[f(x)g(x)\right] = \lim_{x\to 0^-} f(x)\cdot \lim_{x\to 0^-} g(x) = (-1)\cdot 1 = -1, \text{ while}$$

$$\lim_{x\to 0^+}\left[f(x)g(x)\right] = \lim_{x\to 0^+} f(x)\cdot \lim_{x\to 0^+} g(x) = 1\cdot 1 = 1.$$

Thus, we conclude that the two-sided limit, $\lim_{x\to 0}\left[f(x)g(x)\right]$, does not exist.

The five limit laws that we have listed so far can be quickly extended to a much larger collection when combined with one another and with a few easily proved statements. For instance, if the Product Law is applied to the product of a function with itself, we obtain the statement

$$\lim_{x\to c}\left[f(x)f(x)\right] = \lim_{x\to c} f(x)\cdot \lim_{x\to c} f(x) = \left[\lim_{x\to c} f(x)\right]^2$$

(assuming $\lim_{x\to c} f(x)$ exists). And the same law can now be applied to the product of $f(x)$ and $\left[f(x)\right]^2$ to reach a similar conclusion about the limit of the function $\left[f(x)\right]^3$. In general, repeated application of the Product Law results in the Positive Integer Power Law.

THEOREM: Positive Integer Power Law

Let f be a function for which $\lim_{x\to c} f(x)$ exists, and let m be a fixed positive integer. Then

$$\lim_{x\to c}\left[f(x)\right]^m = \left[\lim_{x\to c} f(x)\right]^m.$$

The two limit statements

$$\lim_{x \to c} 1 = 1 \quad \text{and} \quad \lim_{x \to c} x = c$$

are certainly reasonable and easily proved, and allow us to extend our list of limit laws to large classes of functions. In particular, we can now state the following two laws, which are examples of what are sometimes referred to as Direct Substitution Laws.

THEOREM: Polynomial Substitution Law

Let p be a polynomial function. Then

$$\lim_{x \to c} p(x) = p(c).$$

THEOREM: Rational Function Substitution Law

Let p and q be polynomial functions. Then if $q(c) \neq 0$,

$$\lim_{x \to c} \frac{p(x)}{q(x)} = \frac{p(c)}{q(c)}.$$

In Exercises 65–68, you will be guided through the proofs of the above statements.

Example 3: Using Limit Laws to Determine Limits

Find the following limits.

a. $\lim_{x \to 2} \left(4x^3 - 5x^2 + 1\right)$

b. $\lim_{x \to -1} \left(2x^4 + x^3 - 3x^2 + 7.5\right)$

c. $\lim_{x \to 3} \frac{x^2 + 2}{3x - 1}$

d. $\lim_{x \to 1} \frac{2x^3 - 5x + 1}{2x - x^2}$

Solution

a. First using the Sum and Difference Laws, we obtain

$$\begin{aligned} \lim_{x \to 2} \left(4x^3 - 5x^2 + 1\right) &= \lim_{x \to 2} \left(4x^3\right) - \lim_{x \to 2} \left(5x^2\right) + \lim_{x \to 2} (1) \\ &= 4 \lim_{x \to 2} x^3 - 5 \lim_{x \to 2} x^2 + 1, \end{aligned}$$

where in the last step we used the Constant Multiple Law. Next, the Positive Integer Power Law is applied.

$$4 \lim_{x \to 2} x^3 - 5 \lim_{x \to 2} x^2 + 1 = 4\left(\lim_{x \to 2} x\right)^3 - 5\left(\lim_{x \to 2} x\right)^2 + 1 = 4(2)^3 - 5(2)^2 + 1 = 13$$

Notice that our repeated application of the various rules eventually led us to finding the limit by simply substituting $x = 2$ into the polynomial. This is exactly what the Polynomial Substitution Law allows us to do; so henceforth we don't even have to go through the above, somewhat lengthy, process when finding the limits of polynomials.

b. In this case, we will simply use the Polynomial Substitution Law.

$$\lim_{x \to -1} \left(2x^4 + x^3 - 3x^2 + 7.5\right) = 2(-1)^4 + (-1)^3 - 3(-1)^2 + 7.5 = 5.5$$

c. Notice that our third limit is that of a rational function. Since both the numerator and denominator are polynomials, we can apply the Quotient Law combined with a repeated application of the Polynomial Substitution Law as follows.

$$\lim_{x \to 3} \frac{x^2 + 2}{3x - 1} = \frac{\lim_{x \to 3}\left(x^2 + 2\right)}{\lim_{x \to 3}(3x - 1)} = \frac{3^2 + 2}{3 \cdot 3 - 1} = \frac{9 + 2}{9 - 1} = \frac{11}{8}$$

Notice, however, that what we did was in effect substituting $x = 3$ into the given rational function. This is exactly what the Rational Function Substitution Law says we can always do unless the process results in a 0 denominator.

d. This time, we will simply refer to the Rational Function Substitution Law.

$$\lim_{x \to 1} \frac{2x^3 - 5x + 1}{2x - x^2} = \frac{2(1)^3 - 5 \cdot 1 + 1}{2 \cdot 1 - 1^2} = \frac{-2}{1} = -2$$

In summary, we wish to emphasize that when evaluating limits of a polynomial or rational function, our theorems allow for the calculation to be reduced to a simple matter of evaluating the function at the limit point. The only exception is for those c-values that cause the denominator of a given rational limit to equal 0.

We can gain another significant extension with the addition of the following law, the proof of which is a consequence of the Intermediate Value Theorem which we will see in Section 13.5.

THEOREM: Positive Integer Root Law

Let f be a function for which $\lim_{x \to c} f(x)$ exists, and let n be a fixed positive integer. Then

$$\lim_{x \to c} \sqrt[n]{f(x)} = \sqrt[n]{\lim_{x \to c} f(x)},$$

with the assumption that $\lim_{x \to c} f(x)$ is nonnegative if n is even.

Exercise 69 shows how the Positive Integer Power Law and the Positive Integer Root Law we have (and, if necessary, the Quotient Law) can be combined to yield the following.

THEOREM: Rational Power Law

Let f be a function for which $\lim_{x \to c} f(x)$ exists, and let m and n be fixed nonzero integers with no common factor. Then

$$\lim_{x \to c}\left[f(x)\right]^{\frac{m}{n}} = \left[\lim_{x \to c} f(x)\right]^{\frac{m}{n}},$$

with the assumption that $\lim_{x \to c} f(x)$ is nonnegative if n is even.

Example 4: Using Limit Laws to Determine Limits

Use the Rational Power Law to evaluate the following limits.

a. $\lim_{x \to 1}\left(\frac{5x+3}{x^2-2x+2}\right)^{\frac{7}{3}}$ **b.** $\lim_{x \to 2}\sqrt{\left(x^4+2x^2+4\right)^3}$ **c.** $\lim_{x \to 0}\sqrt[3]{4\cos^2 x}$

Solution

a. First of all, we claim that for $f(x) = \frac{5x+3}{x^2-2x+2}$, $\lim_{x \to 1} f(x)$ exists. In fact, by direct substitution we have the following.

$$\lim_{x \to 1} f(x) = \lim_{x \to 1}\frac{5x+3}{x^2-2x+2} = \frac{5\cdot 1+3}{1^2-2\cdot 1+2} = \frac{8}{1-2+2} = 8$$

It follows that the Rational Power Law applies.

$$\lim_{x \to 1}\left(\frac{5x+3}{x^2-2x+2}\right)^{\frac{7}{3}} = \left(\lim_{x \to 1}\frac{5x+3}{x^2-2x+2}\right)^{\frac{7}{3}} = 8^{\frac{7}{3}} = 2^7 = 128$$

b. To start us off, notice that $\lim_{x \to 2}\left(x^4+2x^2+4\right) = 28$, a fact easily verified by direct substitution. Also, since the above limit is positive and

$$\sqrt{\left(x^4+2x^2+4\right)^3} = \left(x^4+2x^2+4\right)^{\frac{3}{2}},$$

the Rational Power Law applies.

$$\lim_{x \to 2}\sqrt{\left(x^4+2x^2+4\right)^3} = \lim_{x \to 2}\left(x^4+2x^2+4\right)^{\frac{3}{2}} = \left[\lim_{x \to 2}\left(x^4+2x^2+4\right)\right]^{\frac{3}{2}} = (28)^{\frac{3}{2}}$$

$$= \left(\sqrt{28}\right)^3 = \left(2\sqrt{7}\right)^3 = 2^3\left(\sqrt{7}\right)^3 = 8\cdot 7\cdot\sqrt{7} = 56\sqrt{7}$$

c. Since $4\cos^2 x = (2\cos x)^2$, and since $\lim_{x \to 0}(2\cos x) = 2$, the Rational Power Law once again applies.

$$\lim_{x \to 0}\sqrt[3]{4\cos^2 x} = \lim_{x \to 0}(2\cos x)^{\frac{2}{3}} = \left[\lim_{x \to 0}(2\cos x)\right]^{\frac{2}{3}} = 2^{\frac{2}{3}} = \sqrt[3]{4}$$

TOPIC 2: Limit Determination Techniques

The limit laws we have stated greatly simplify the determination of many limits and also show that limits of polynomial and rational functions can be found simply by evaluating the function at the limit point (remember that in the case of rational functions the limit point must be in the domain of the function). Functions possessing this property, which we will refer to as the *Direct Substitution Property*, are called *continuous*; much more will be said about these well-behaved functions in the next section. But we will close this section with a discussion of techniques to use when the limit laws and/or direct substitution do not immediately apply.

Many of our techniques make use of the fact that the limit of a function f at a point c is determined entirely by its behavior near, but not at, c. If we can find a function g that is identical to f near c, and if the limit of g at the point c is easy to determine, then we are done. The following examples illustrate the steps typically taken in this process.

Example 5: Using Algebraic Techniques to Determine Limits

Find $\lim\limits_{x\to -3}\dfrac{x^2-9}{x+3}$.

Solution

First of all we note that we cannot use direct substitution to evaluate this limit, for the denominator equals 0 at $x=-3$. In fact, $f(x)=\dfrac{x^2-9}{x+3}$ is not even defined at $x=-3$. However, as far as the limit is concerned, that is not a problem at all (recall that the existence and/or value of $f(c)$ and $\lim\limits_{x\to c}f(x)$ have no bearing on each other). The limit not only exists, but finding it is surprisingly easy, using a bit of algebra.

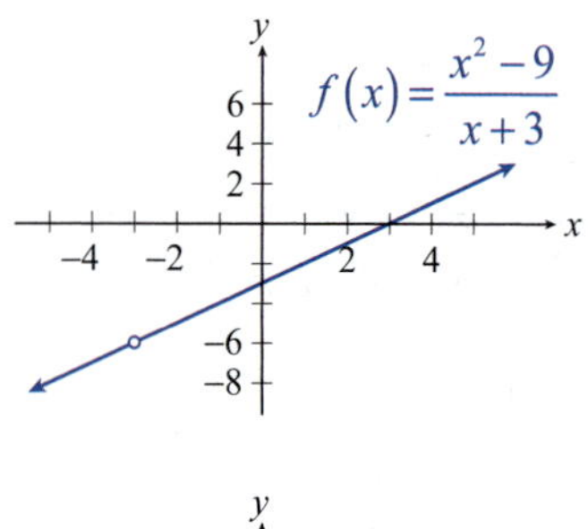

$$\lim_{x\to -3}\frac{x^2-9}{x+3}=\lim_{x\to -3}\frac{(x+3)(x-3)}{x+3}=\lim_{x\to -3}(x-3)=-3-3=-6$$

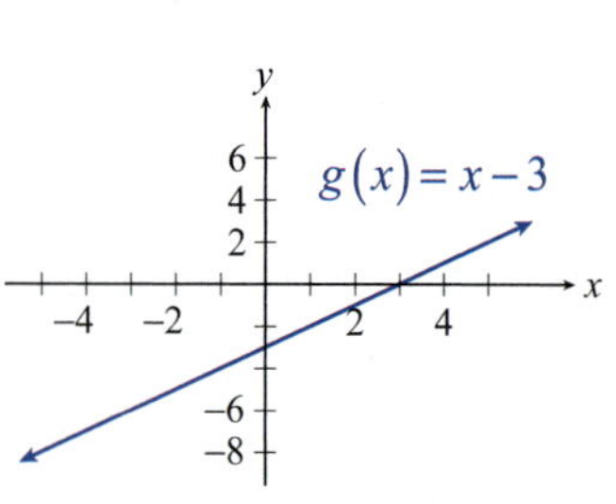

FIGURE 2

A few very important remarks are in order. First of all, canceling the factor of $(x+3)$ is legitimate, since even though x is approaching -3, it never is actually equal to -3 throughout the limit process, so we have not divided by 0. Second, the function $g(x)=x-3$ is actually different from $f(x)$. In fact, they agree everywhere but at $x=-3$; at which point $f(x)$ is undefined, but $g(-3)=-6$ (see Figure 2). However, the behaviors of f and g are identical as x approaches -3, and therefore, so are their limits. (Examining the chain of equalities above, note that we never actually stated the equality of f and g, but merely the fact that their limits at -3 were equal.)

Algebraic techniques similar to the one from Example 5 are common when evaluating limits analytically. We provide further illustrations in the next two examples.

Example 6: Using Algebraic Techniques to Determine Limits

Use algebra to evaluate $\lim\limits_{x\to 1}\dfrac{\sqrt{x+3}-2}{x-1}$.

Solution

The trick with a limit such as this is to use the "conjugate product rule" $(a+b)(a-b)=a^2-b^2$. If we multiply both the numerator and denominator by the "conjugate" of the expression containing the radical (in this case, $\sqrt{x+3}+2$), we obtain the following.

$$\lim_{x\to 1}\frac{\sqrt{x+3}-2}{x-1}=\lim_{x\to 1}\left(\frac{\sqrt{x+3}-2}{x-1}\cdot\frac{\sqrt{x+3}+2}{\sqrt{x+3}+2}\right)=\lim_{x\to 1}\frac{\left(\sqrt{x+3}\right)^2-2^2}{(x-1)\left(\sqrt{x+3}+2\right)}$$

$$=\lim_{x\to 1}\frac{x+3-4}{(x-1)\left(\sqrt{x+3}+2\right)}=\lim_{x\to 1}\frac{x-1}{(x-1)\left(\sqrt{x+3}+2\right)}$$

$$=\lim_{x\to 1}\frac{1}{\sqrt{x+3}+2}=\frac{1}{\sqrt{1+3}+2}=\frac{1}{2+2}=\frac{1}{4}$$

As in the previous example, the cancellation was legitimate, and even though in the process we did change the function, we did not change the value of the limit.

Example 7: Using Algebraic Techniques to Determine Limits

Find $\lim\limits_{h\to 0}\dfrac{(h+2)^2-4}{h}$.

Solution

We will take an algebraic approach very similar to the previous examples, but keeping in mind the fact that in the current problem, the variable is denoted by h, rather than the usual x. This, however, should not cause any difficulties.

$$\lim_{h\to 0}\frac{(h+2)^2-4}{h}=\lim_{h\to 0}\frac{\left(h^2+4h+4\right)-4}{h}=\lim_{h\to 0}\frac{h^2+4h}{h}$$

$$=\lim_{h\to 0}\frac{h(h+4)}{h}=\lim_{h\to 0}(h+4)=4$$

Again, canceling h is legitimate, since h never actually assumes the value 0.

Another technique for determining $\lim\limits_{x\to c}f(x)$ calls for finding two other functions, g and h, such that f is "squeezed between" g and h and for which $\lim\limits_{x\to c}g(x)$ and $\lim\limits_{x\to c}h(x)$ are easier to determine.

The following statement, which we will call the Squeeze Theorem, also goes by such names as the Sandwich Theorem and the Pinching Theorem.

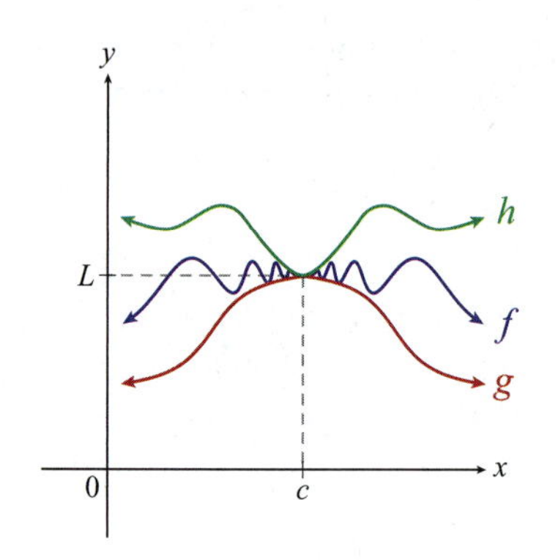

FIGURE 3: The Squeeze Theorem

THEOREM: The Squeeze Theorem

If $g(x) \le f(x) \le h(x)$ for all x in some open interval containing c, except possibly at c itself, and if $\lim_{x\to c} g(x) = \lim_{x\to c} h(x) = L$, then $\lim_{x\to c} f(x) = L$ as well.

The statement also holds for limits at infinity, that is, for $c = -\infty$ or $c = \infty$.

Example 8: Using the Squeeze Theorem to Determine Limits

Use the Squeeze Theorem to prove $\lim_{x\to 0}\left(x\sin\left(\frac{1}{x}\right)\right) = 0$.

Solution

As a first observation, we recall that the sine value of any angle is never greater than 1 or less than −1; that is, we have the well-known inequalities

$$-1 \le \sin\alpha \le 1$$

for any α. This certainly means that for any nonzero x, choosing $\alpha = \frac{1}{x}$,

$$-1 \le \sin\left(\frac{1}{x}\right) \le 1,$$

a fact you have seen before. Multiplying all sides of this chain of inequalities by $|x|$, we obtain the following.

$$-|x| \le x\sin\left(\frac{1}{x}\right) \le |x|$$

In words, we can say that while the sine function oscillates between −1 and 1, $x\sin\left(\frac{1}{x}\right)$ will oscillate between $-|x|$ and $|x|$. The previous inequality coupled with the fact that $\lim_{x\to 0}(-|x|) = \lim_{x\to 0}|x| = 0$ means that the functions $f(x) = x\sin\left(\frac{1}{x}\right)$, $g(x) = -|x|$, and $h(x) = |x|$ satisfy the hypotheses of the Squeeze Theorem. Thus, we can simply invoke the theorem, which ensures that the claim

$$\lim_{x\to 0}\left(x\sin\left(\frac{1}{x}\right)\right) = 0$$

is now proven.

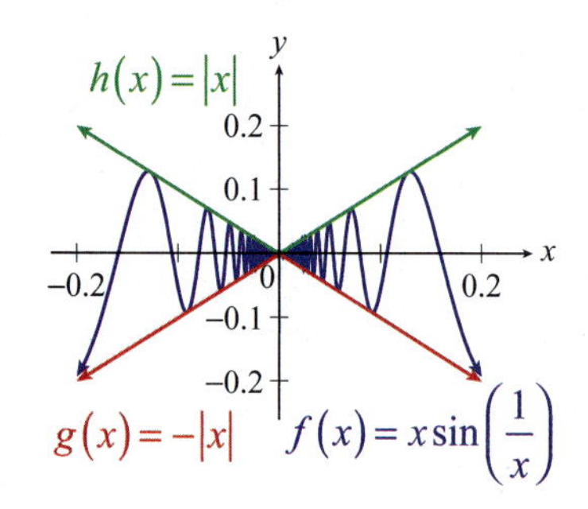

FIGURE 4

Figure 4 shows the Squeeze Theorem at work. Notice how f is "squeezed between" g and h near the origin.

We will end with one last limit theorem that will prove useful in some derivations to follow.

THEOREM: Upper Bound Theorem

If $f(x) \le g(x)$ for all x in some open interval containing c, except possibly at c itself, and if the limits of f and g both exist at c, then

$$\lim_{x\to c} f(x) \le \lim_{x\to c} g(x).$$

13.4 EXERCISES

PRACTICE

Use the graph to find the given limit. See Example 1.

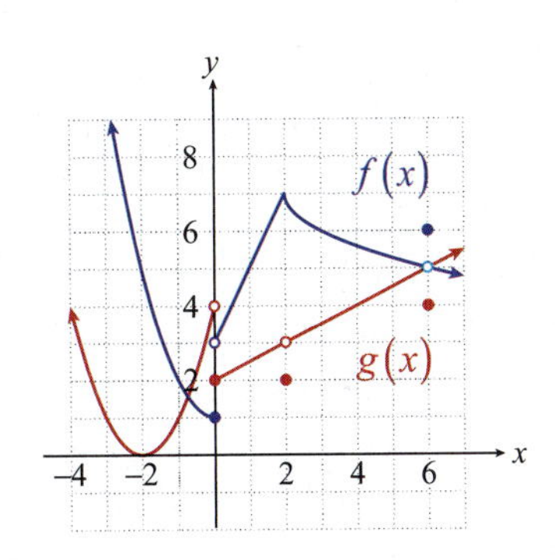

1. a. $\lim_{x\to 0^+}\left[g(x)-2f(x)\right]$ b. $\lim_{x\to 2^+}\left[g(x)f(x)\right]$

2. a. $\lim_{x\to 6}\left[g(x)+f(x)\right]$ b. $\lim_{x\to 0^-}\dfrac{f(x)}{2g(x)}$

Use appropriate limit laws to evaluate the given limit. See Examples 3 and 4.

3. $\lim_{x\to 4}5$

4. $\lim_{x\to 4}(5x)$

5. $\lim_{x\to 3}(2x+1)$

6. $\lim_{x\to \frac{1}{2}}(3-4x)$

7. $\lim_{x\to -3}x^2$

8. $\lim_{x\to -2}\left(-x^5\right)$

9. $\lim_{x\to 3}\left(2x^2-x+7\right)$

10. $\lim_{x\to -1}\left(3+x-\dfrac{5}{2}x^2\right)$

11. $\lim_{x\to \frac{1}{2}}\left(2x^3-3x^2+x-4\right)$

12. $\lim_{x\to -2}\left(3x^3-x^5\right)$

13. $\lim_{x\to 1}\dfrac{3x-7}{x+1}$

14. $\lim_{x\to -1}\dfrac{5x+3}{x^2-x}$

15. $\lim_{x\to 3}\left(\dfrac{4x}{11x-x^3}\right)^{\frac{1}{3}}$

16. $\lim_{t\to 1}\left(\dfrac{2t+t^3}{3t^2+1}\right)^{\frac{3}{2}}$

17. $\lim_{x\to -2}\sqrt[3]{5x^4-x^3+3x^2+2x+4}$

18. $\lim_{x\to 4}\sqrt{x^4+2x^2+1}$

19. $\lim_{x\to -3}\left(\dfrac{x^4-5x}{x^3+2x^2-4x}\right)^{\frac{4}{5}}$

20. $\lim_{x\to -5}\sqrt[3]{\left(x^4+2x^3+x^2\right)^2}$

Use algebra to evaluate the given limit. See Examples 5, 6, and 7.

21. $\lim_{x\to 6}\dfrac{x^2-36}{x-6}$

22. $\lim_{x\to -7}\dfrac{x+7}{x^2-49}$

23. $\lim_{x\to 3}\frac{3-13x+4x^2}{x-3}$

24. $\lim_{x\to 4}\frac{x^4-256}{x^2-16}$

25. $\lim_{x\to 5}\frac{2x^3-7x^2-14x-5}{x^2-25}$

26. $\lim_{x\to 2}\frac{x^3-8}{x^3-2x^2+2x-4}$

27. $\lim_{x\to 7}\frac{\sqrt{x+2}-3}{x-7}$

28. $\lim_{x\to 9}\frac{3-\sqrt{x}}{x-9}$

29. $\lim_{x\to 0}\frac{\sqrt{x+5}-\sqrt{5}}{x}$

30. $\lim_{x\to 0}\frac{\frac{1}{4+x}-\frac{1}{4}}{x}$

31. $\lim_{x\to 2}\frac{\frac{1}{3}-\frac{1}{1+x}}{x-2}$

32. If $f(x)=x^2$, find $\lim_{x\to 2}\frac{f(x)-f(2)}{x-2}$.

33. If $g(x)=x^2-2$, find $\lim_{h\to 0}\frac{g(3+h)-g(3)}{h}$.

34. If $k(x)=1-x+x^2$, find $\lim_{h\to 0}\frac{k(2-h)-k(2)}{h}$.

35. If $p(x)=x^3+x$, find $\lim_{x\to 1}\frac{p(x)-p(1)}{x-1}$.

36. If $F(x)=\frac{1}{x}$, find $\lim_{x\to\frac{1}{2}}\frac{F(x)-F\left(\frac{1}{2}\right)}{x-\frac{1}{2}}$.

37. $\lim_{h\to 0}\frac{\sqrt{x+h}-\sqrt{x}}{h}$

38. $\lim_{x\to 3}\frac{2x^2-3x-9}{x^2-9}$

39. $\lim_{x\to 2}\frac{x^4-16}{3x^2-5x-2}$

40. $\lim_{x\to -3}\frac{\frac{1}{3}+\frac{1}{x}}{x^3+27}$

41. $\lim_{x\to 2}\frac{3-\sqrt{x+7}}{x-2}$

42. $\lim_{x\to 8}\frac{8-x}{\sqrt[3]{x}-2}$

43. $\lim_{y\to 0}\left(\frac{1}{y}+\frac{1}{y^2-y}\right)$

44. $\lim_{x\to 1}\frac{\sqrt{x}-1}{\sqrt[3]{x}-1}$

Use $\lim_{x\to c}f(x)=3$ and $\lim_{x\to c}g(x)=-2$ to find the limit.

45. $\lim_{x\to c}\left[2f(x)-g(x)\right]$

46. $\lim_{x\to c}\frac{4f(x)+3g(x)}{f(x)-\frac{1}{2}g(x)}$

47. $\lim_{x\to c}\sqrt{\left[f(x)\right]^4+10\left[g(x)\right]^2}$

48. $\lim_{x\to c}\left(\left[f(x)-1\right]^2\sqrt[3]{g(x)}\right)$

49. $\lim_{x\to c}\left(\left[f(x)\right]^2+(x-2)g(x)\right)$

50. $\lim_{x\to c}\left(\frac{f(x)+g(x)}{\left[g(x)\right]^2}\right)^{\frac{3}{2}}$

Use the limit laws to find the one-sided limit.

51. $\lim_{x\to 0^+} \frac{x}{|x|}$

52. $\lim_{x\to 0^-} (\operatorname{sgn} x \cos x)$, where $\operatorname{sgn} x = \begin{cases} 1 & \text{if } x > 0 \\ 0 & \text{if } x = 0 \\ -1 & \text{if } x < 0 \end{cases}$

53. $\lim_{x\to 2^+} \frac{\sqrt{x-2}}{3x+1}$

54. $\lim_{x\to 1^-} \sqrt{1-x^2}$

55. $\lim_{x\to\left(\frac{1}{3}\right)^-} \frac{\sqrt{1-3x}}{6x+5}$

56. $\lim_{x\to 1^+} ([\![x]\!] - x)$ (See the definition of $f(x) = [\![x]\!]$ in Section 3.4.)

57. $\lim_{x\to 2^+} ([\![x]\!] e^x)$

58. $\lim_{x\to -1^-} \frac{[\![x]\!](2x^2+1)}{x+3}$

Use the Squeeze Theorem to prove the limit claim. See Example 8.

59. $\lim_{x\to 0}\left(x^2 \sin\left(\frac{1}{x}\right)\right) = 0$

60. $\lim_{x\to 0}(|x|\cos x) = 0$

61. $\lim_{x\to\infty} \frac{\cos x}{x} = 0$

62. $\lim_{x\to -\infty}(e^x \sin x) = 0$

63. $\lim_{x\to 0^+}\left(x^{\frac{3}{2}} e^{\cos\left(\frac{1}{x}\right)}\right) = 0$

64. $\lim_{x\to\infty} \frac{\sin^2 x + 1}{2+x} = 0$

WRITING & THINKING

65. Provide a rigorous proof of the limit claim $\lim_{x\to c} 1 = 1$. (**Hint:** Use the fact that for the constant 1 function, $f(x) = 1$ for all x, so in particular, if an $\varepsilon > 0$ is given, $|f(x) - 1| = |1 - 1| = 0$, which makes the choice of δ "easy.")

66. Provide a rigorous proof of the limit claim $\lim_{x\to c} x = c$. (**Hint:** Since $f(x) = x$ in this problem, for a given $\varepsilon > 0$ we need to ensure that $|f(x) - c| = |x - c| < \varepsilon$ as long as $0 < |x - c| < \delta$. This observation makes the choice of δ obvious.)

67. Use Exercise 66 and the basic limit laws to prove the Polynomial Substitution Law. (**Hint:** From Exercise 66 and a repeated application of the Product Law it follows that $\lim_{x\to c} x^k = c^k$. As a next step, from the Constant Multiple Law we can conclude that if $a \in \mathbb{R}$, $\lim_{x\to c}(ax^k) = ac^k$. From the above claim, a repeated application of the Sum Law will yield the result for a general polynomial.)

68. Use Exercise 67 and the basic limit laws to prove the Rational Function Substitution Law.

69. Combine the Positive Integer Power Law and the Positive Integer Root Law to prove the Rational Power Law. (**Hint:** Assuming first that both m and n are positive, we can write $\lim\limits_{x\to c}\left[f(x)\right]^{\frac{m}{n}} = \lim\limits_{x\to c}\left[\left[f(x)\right]^{\frac{1}{n}}\right]^m = \lim\limits_{x\to c}\left[\sqrt[n]{f(x)}\right]^m$. Now use the Positive Integer Power Law followed by the Positive Integer Root Law to obtain that the above limit is equal to $\left[\lim\limits_{x\to c}\sqrt[n]{f(x)}\right]^m = \left[\sqrt[n]{\lim\limits_{x\to c} f(x)}\right]^m$, from which the result follows. If m is negative, note that $\left[f(x)\right]^{\frac{m}{n}} = \dfrac{1}{\left[f(x)\right]^{-\frac{m}{n}}}$, where $-m$ is positive. Thus if we use the Quotient Law along with the previous argument, we obtain

$$\begin{aligned}\lim_{x\to c}\left[f(x)\right]^{\frac{m}{n}} &= \lim_{x\to c}\frac{1}{\left[f(x)\right]^{-\frac{m}{n}}}\\ &= \frac{1}{\lim\limits_{x\to c}\left[f(x)\right]^{-\frac{m}{n}}}\\ &= \frac{1}{\left[\lim\limits_{x\to c} f(x)\right]^{-\frac{m}{n}}}\end{aligned}$$

from which the result easily follows.)

70. Let $D(x)=\begin{cases}0 & \text{if } x \text{ is rational}\\ 1 & \text{if } x \text{ is irrational}\end{cases}$. Does $\lim\limits_{x\to 0} D(x)$ exist? Prove your answer.

71. Let $F(x)=\begin{cases}0 & \text{if } x \text{ is rational}\\ x^2 & \text{if } x \text{ is irrational}\end{cases}$. Does $\lim\limits_{x\to 0} F(x)$ exist? Prove your answer.

72. Prove that if $\lim\limits_{x\to c} f(x) = L$ and $\lim\limits_{x\to c} f(x) = K$, then $L = K$. In words, prove that if the limit of f exists at c, then the limit is unique.

73. Prove that if n and m are positive integers, then $\lim\limits_{x\to 1}\dfrac{x^n - 1}{x^m - 1} = \dfrac{n}{m}$.

74. Prove that if $\lim\limits_{x\to c} f(x) = 0$, then $\lim\limits_{x\to c}\left|f(x)\right| = 0$.

75. Prove that if $\lim\limits_{x\to c} f(x) = 0$ and $g(x)$ is such that $\left|g(x)\right| \le M$ for some number M (such functions are called bounded), then $\lim\limits_{x\to c}\left[f(x)g(x)\right] = 0$.

76. Prove that in Exercise 75, it is sufficient to require the boundedness of g only on an interval around c (except at c itself).

77. By finding functions f and g such that $\lim\limits_{x\to c} f(x) = 0$ but $\lim\limits_{x\to c}\left[f(x)g(x)\right] \ne 0$, show that it is necessary to impose a boundedness condition on g in Exercise 75.

78. Give examples of f and g to show that

a. the existence of $\lim_{x\to c}\left[f(x)+g(x)\right]$ does not imply the existence of $\lim_{x\to c} f(x)$ and

b. the existence of $\lim_{x\to c}\left[f(x)g(x)\right]$ does not imply the existence of $\lim_{x\to c} f(x)$.

APPLICATIONS

79. A *concave spherical mirror* is a part of the inside of a sphere, silvered to form a reflective surface. The radius r of the sphere is called the mirror's *radius of curvature*. If the size of such a mirror is small relative to its radius of curvature, light rays parallel to its principal axis are reflected through approximately a single point, called *focus*. In the following illustration, C denotes the center, F_d is the focus, while d is the distance between the incoming ray and the principal axis. Note that according to the Law of Reflection, the incoming and reflected rays make the same size angle α with the radius $\overline{CR}$ (this radius is called *normal* to the mirror surface). One way to determine the *focal length* (the distance between the mirror and the focus) is to find the limiting position of F_d as $d \to 0$. Noting that the triangle $\triangle CRF_d$ is isosceles, by similarity we obtain $\dfrac{CF_d}{\frac{r}{2}} = \dfrac{r}{\sqrt{r^2-d^2}}$. Use this observation to express CF_d and determine the focal length of the spherical mirror by taking the limit as $d \to 0$.

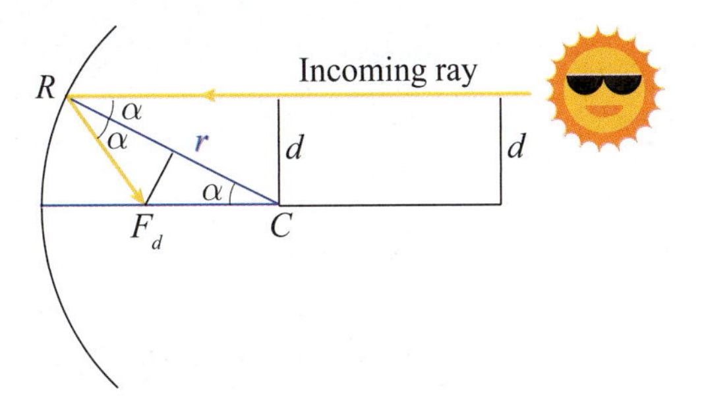

13.5 CONTINUITY

TOPICS

1. Continuity and discontinuity at a point
2. Continuous functions
3. Properties of continuity

TOPIC 1: Continuity and Discontinuity at a Point

Continuity of a function f at a point c was defined informally in Section 13.4 as synonymous with the Direct Substitution Property. That is, if $\lim_{x \to c} f(x) = f(c)$, then we say that f is continuous at c. Continuity, as it turns out, is one of those fundamental concepts that serves as the cornerstone for a great deal of other mathematical ideas. We will devote this section to a more thorough and rigorous study of the notion.

As is often the case with deep, fundamental concepts, a solid understanding of continuity was not easily achieved, and it long evaded mathematicians. A once-common and imprecise definition, still often seen today, is that a continuous function is one whose graph can be drawn without lifting pen or pencil off the paper. This coincides with the actual definition of continuity in the case of sufficiently simple functions but is inadequate for the purposes of calculus.

DEFINITION: Continuity at a Point

Given a function f defined on an open interval containing c, we say f is **continuous at c** if

$$\lim_{x \to c} f(x) = f(c).$$

If the domain of f is an interval containing c either as a left or right endpoint, we also define f to be **right-continuous** or **left-continuous at c** if, respectively,

$$\lim_{x \to c^+} f(x) = f(c) \quad \text{or} \quad \lim_{x \to c^-} f(x) = f(c).$$

In usage, continuity refers to either the first or the second definition depending on the context.

This definition actually requires f to possess three properties, and it's instructive to break the definition down into these three components. Specifically, in order for f to be continuous at the point c,

1. f must be defined at c;
2. the limit of f at c must exist (one-sided limit when c is an endpoint);
3. the value of the limit must equal $f(c)$.

If any one of the three properties fails to hold, then the function f is **discontinuous** at the point c and we call c a **point of discontinuity** of f. Examples 1 and 2 illustrate the identification of points of continuity and discontinuity.

Example 1: Finding Points of Continuity and Discontinuity

Find all points of continuity as well as all points of discontinuity for the function $f(x)$ given by its graph in Figure 1. For any discontinuities, identify those from the three properties on the previous page that fail to hold.

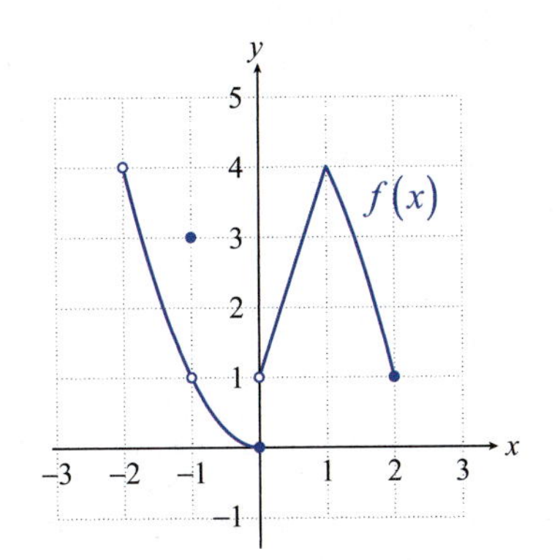

FIGURE 1

Solution

A careful examination of the graph, paying attention to the empty and full circles, reveals that the domain for this function is $(-2,2]$. Recalling our studies of limits from the previous two sections, we see that the only point where $\lim_{x \to c} f(x)$ fails to exist is $c = 0$; though there is a discrepancy between the limit and the function value at $c = -1$: $\lim_{x \to -1} f(x) = 1$, but $f(-1) = 3$.

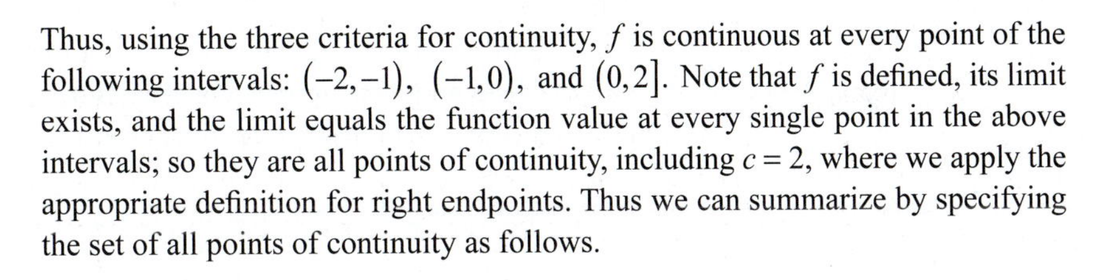

Thus, using the three criteria for continuity, f is continuous at every point of the following intervals: $(-2,-1)$, $(-1,0)$, and $(0,2]$. Note that f is defined, its limit exists, and the limit equals the function value at every single point in the above intervals; so they are all points of continuity, including $c = 2$, where we apply the appropriate definition for right endpoints. Thus we can summarize by specifying the set of all points of continuity as follows.

$$(-2,-1) \cup (-1,0) \cup (0,2]$$

To find the points of discontinuity, we first note that $c = -2$ is certainly one; for f is not even defined at $c = -2$ (criterion 1 fails). Therefore, while f is continuous at the right endpoint of its domain, at $c = 2$, it is discontinuous at the left endpoint, namely at $c = -2$. Next, for $c = -1$, as we mentioned above, $\lim_{x \to -1} f(x) \neq f(-1)$, so it is a point of discontinuity (criterion 3 fails). Note, however, that the definition $f(-1) = 1$ will make f continuous; that is, appropriately redefining the function will remove this discontinuity. In contrast, at $c = 0$ the one-sided limits are unequal, so $\lim_{x \to 0} f(x)$ does not even exist. This is not only a point of discontinuity (by virtue of criterion 2 failing), but one that cannot be removed by redefining the function. Such discontinuities are called *jump discontinuities*; the reason for the name should be clear from the graph.

Lastly, we note that f is certainly discontinuous everywhere outside of $[-2,2]$, for it is undefined at those points (again, criterion 1 fails). Thus, we can summarize the points of discontinuity of f as follows.

$$(-\infty,-2] \cup \{-1\} \cup \{0\} \cup (2,\infty)$$

Be sure you understand the use of different types of grouping symbols here.

Example 2: Finding Points of Discontinuity

Identify and examine the discontinuities of the following functions.

a. $f(x) = \dfrac{x^2 + x}{x}$ **b.** $g(x) = \dfrac{1}{x^2}$ **c.** $h(x) = \sin\left(\dfrac{1}{x}\right)$

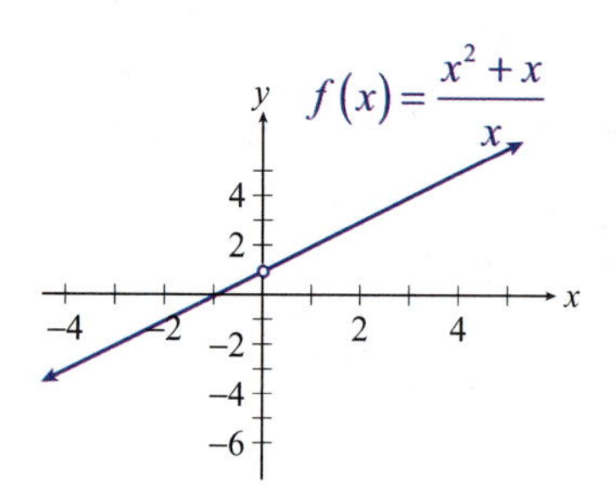

FIGURE 2

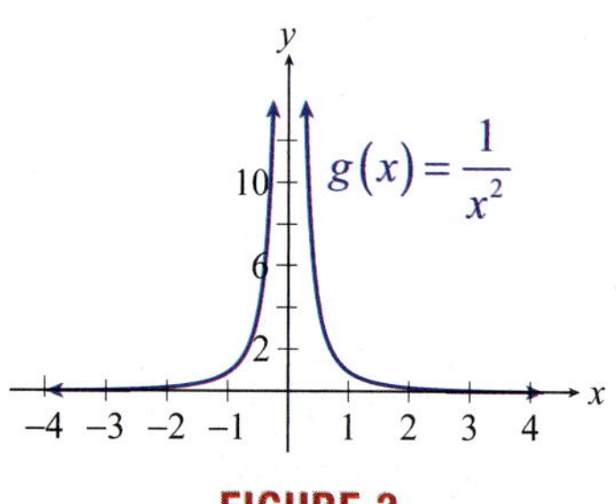

FIGURE 3

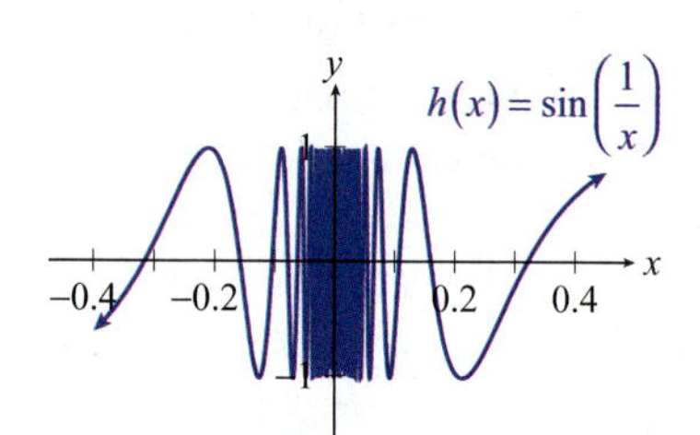

FIGURE 4

Solution

a. The function f is a rational function, and as such it possesses the Direct Substitution Property and thus is continuous at any point where the denominator is nonzero; that is, $f(x)$ is continuous at every $x \neq 0$. The point $x = 0$ is a point of discontinuity since f is not defined there; however, using a bit of algebra we can actually say more. Since

$$\lim_{x\to 0} f(x) = \lim_{x\to 0}\frac{x^2+x}{x} = \lim_{x\to 0}\frac{x(x+1)}{x} = \lim_{x\to 0}(x+1) = 1,$$

our conclusion is that $\lim_{x\to 0} f(x)$ actually exists, and that the graph of f agrees with that of $y = x + 1$ except for the point corresponding to $x = 0$, where f is undefined. In other words, you can think of the discontinuity of f arising at $x = 0$ as one caused by a single point "missing" from the graph, that is, the graph being the line $y = x + 1$ "punctured" at $(0,1)$, as shown in Figure 2. This is another example of a discontinuity that can be removed; the definition $f(0) = 1$ will make f continuous at $x = 0$.

b. Note that g is undefined and has a vertical asymptote at $x = 0$; moreover, $\lim_{x\to 0} g(x) = \infty$ (see Figure 3). We can argue, as we did in part a., that g is continuous at every $x \neq 0$, everywhere on its domain. The discontinuity at $x = 0$, however, is very different from that of $f(x)$ in part a., since $\lim_{x\to 0} g(x)$ doesn't exist by virtue of the function values approaching infinity. In other words, no definition will make g continuous at 0. This type of discontinuity is called an *infinite discontinuity*.

c. By an argument almost exactly like the one we gave in Example 7 of Section 13.3, one can show not only that h is undefined for $x = 0$, but that $\lim_{x\to 0} h(x)$ does not exist either. It follows from that argument that the reason for the nonexistence of the limit is wild oscillation: actually, infinitely many oscillations are squeezed into arbitrarily small neighborhoods of 0 (see Figure 4). Even good technology cannot do full justice to what is actually going on. It is quite clear, however, that we cannot define the value of $h(0)$ so as to make h continuous. We call this an *oscillating discontinuity*.

The points of discontinuity in Examples 1 and 2 illustrate the variety of ways in which one or more of the three properties of continuity can fail. The examples also illustrate that some discontinuities can be remedied by simply redefining the function at the point of discontinuity.

DEFINITION: Removable Discontinuity

If a function f has a point of discontinuity at c but $\lim_{x\to c} f(x)$ exists, c is called a **removable discontinuity** of f. The function can be made continuous at c by redefining f at c so that

$$f(c) = \lim_{x\to c} f(x).$$

If c is an endpoint of an interval on which f is defined, replace $\lim_{x\to c} f(x)$ with the appropriate one-sided limit.

If a given point of discontinuity is not removable, it is called **nonremovable**.

Example 3: Classifying Removable and Nonremovable Points of Discontinuity

Identify the discontinuities in Examples 1 and 2 as removable or nonremovable.

Solution

As we discussed in Example 1, f has a removable discontinuity at $c = -1$, since $\lim_{x \to -1} f(x)$ exists. This not being the case at $c = 0$, the latter is a nonremovable discontinuity. Also, since the right-hand limit of f exists at -2, it has a removable discontinuity there, which can be removed by defining $f(-2) = 4$.

As for the functions in Example 2, we have seen that f is the only one with a removable discontinuity, which occurs at $c = 0$.

The discontinuities of g and h (both at $c = 0$) are nonremovable. The reason in the case of g is that the function values approach infinity near 0, while h has infinitely many oscillations near its point of discontinuity.

If we rephrase the definition of continuity using the epsilon-delta definition of limit, we obtain the following.

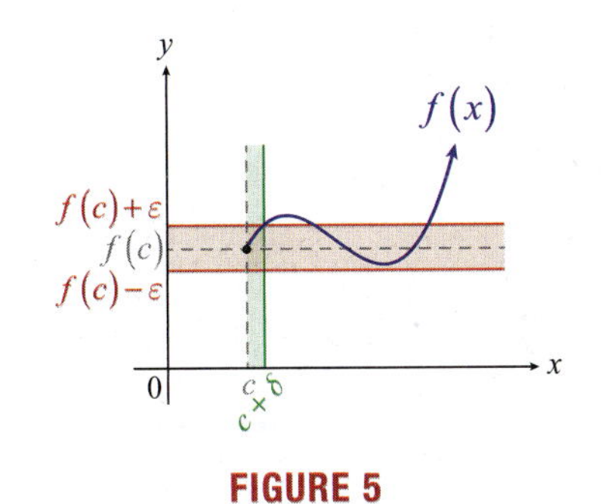

FIGURE 5

DEFINITION: Epsilon-Delta Continuity at a Point

Given a function f and a point c in the domain of f, we say f is continuous at c if for every number $\varepsilon > 0$ there is a number $\delta > 0$ such that $|f(x) - f(c)| < \varepsilon$ for all x in the domain of f satisfying $|x - c| < \delta$.

This alternate version of the definition makes it clear that continuity of a function at an endpoint of an interval is really the same idea as continuity elsewhere. The key idea is that f must be defined at c and that all values of $f(x)$ must be close to (meaning within ε of) the value $f(c)$ whenever x lies in the domain of f and is sufficiently close to (within δ of) the point c (see Figure 5).

TOPIC 2: Continuous Functions

Once continuity at a point has been defined, it is natural to extend the meaning of continuity to larger sets. For example, we say that a function is **continuous on an interval** if it is continuous at every point of that interval. For easily graphed functions defined on an interval, this extension agrees well with the intuitive sense that continuity corresponds to the ability to construct a graph without lifting pen from paper.

Our last extension is very similar, but it contains a few subtleties that merit careful consideration.

DEFINITION: Continuity of a Function

A function f is said to be **continuous** (or **continuous on its domain**) if it is continuous at every point of its domain.

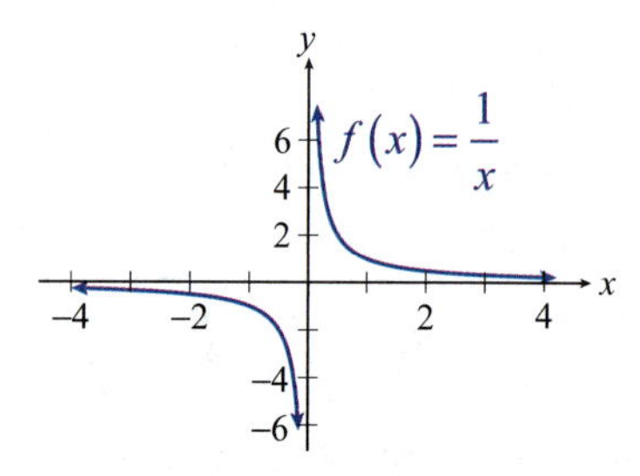

FIGURE 6

The first subtlety to be aware of is that a continuous function can have points of discontinuity. This seemingly paradoxical statement is entirely consistent with our definitions of continuity at a point and of continuity of a function. Consider, for example, the function $f(x)=\frac{1}{x}$ (see Figure 6). The domain of this function is $(-\infty,0)\cup(0,\infty)$, and f is indeed continuous at each point of its domain (you should verify for yourself that $\lim_{x\to c} f(x)=f(c)$ for every point c in the domain).

So f is a continuous function, but it is clear that 0 is a point of discontinuity for f since $\lim_{x\to 0} f(x)$ does not exist. Further, it should be noted that f is a continuous function that is not necessarily continuous on every interval of real numbers—for instance, f is not continuous on $[-3,3]$.

Example 4: Finding Points of Continuity and Discontinuity

Describe the continuity of $f(x)=\frac{x^2-1}{x-1}$. If applicable, identify all points of discontinuity.

Solution

Since f is a rational function, we know that it satisfies the Direct Substitution Property and is thus continuous everywhere except at $c=1$, where the denominator is 0. Note that $f(1)$ is undefined, making $c=1$ the only point of discontinuity for f. So once again, our seemingly paradoxical, but correct, conclusion is that f is a continuous function, and its only point of discontinuity is at $c=1$.

As a final remark, since $f(x)=\frac{x^2-1}{x-1}=\frac{(x-1)(x+1)}{x-1}=x+1$ for all $x\neq 1$, we identify the above discontinuity as removable, since the definition $f(1)=2$ makes f continuous on the entire real line.

Example 5: Using the ε-δ Definition of Continuity

Recall $g(x)=\frac{1}{x^2}$ from Example 2. As we have mentioned, its only point of discontinuity is at $c=0$, where it is undefined, but g is continuous everywhere on its domain. In this example, we will give a rigorous proof of this latter claim, using the epsilon-delta continuity definition.

Choose and fix an arbitrary, nonzero $c\in\mathbb{R}$. We will proceed to prove that $g(x)=\frac{1}{x^2}$ is continuous at c. Note that we may assume $c>0$ (in the negative case, the argument is similar, or one may take advantage of the fact that g is an even function).

Our epsilon-delta argument will be similar to those given in earlier examples to prove limit claims (this shouldn't come as a surprise, since our epsilon-delta continuity definition is itself closely related to the formal definition of limit). To start us off, suppose that $\varepsilon>0$ is given. We need to find a $\delta>0$ such that

$$|x-c|<\delta \;\Rightarrow\; \left|\frac{1}{x^2}-\frac{1}{c^2}\right|<\varepsilon.$$

We will first try to bound the quantity $\left|\frac{1}{x^2}-\frac{1}{c^2}\right|$ in order to get an idea of just how small δ needs to be to satisfy the above inequality. Observe that

$$\left|\frac{1}{x^2}-\frac{1}{c^2}\right| = \left|\frac{c^2-x^2}{x^2c^2}\right| = \frac{\left|(c-x)(c+x)\right|}{x^2c^2} = |x-c|\frac{x+c}{x^2c^2}.$$

Here we are safe to have omitted the absolute value symbols from the last factor, since we can assume that x is close enough to c so that $x+c$ is positive. In fact, we will insist that $\frac{1}{2}c < x < \frac{3}{2}c$ throughout this argument. Note that this is equivalent to requiring that $\delta < \frac{1}{2}c,$ which is not a loss of generality since once a successful $\delta > 0$ is chosen any smaller number works just fine.

Next, let's examine the quantity $\frac{x+c}{x^2c^2}$. Since x is in the interval specified above, $x+c < \frac{3}{2}c+c = \frac{5}{2}c,$ and $x^2c^2 > \left(\frac{1}{2}c\right)^2 c^2 = \frac{1}{4}c^4$. Thus, we have

$$|x-c|\frac{x+c}{x^2c^2} < |x-c|\frac{\frac{5}{2}c}{\frac{1}{4}c^4} = |x-c|\frac{10}{c^3}.$$

Note that since c is fixed throughout this argument, so is $\frac{10}{c^3}$. Putting the above two chains of equalities and inequalities together, we can summarize what we have obtained so far. If x is sufficiently close to c,

$$\left|\frac{1}{x^2}-\frac{1}{c^2}\right| < |x-c|\frac{10}{c^3}.$$

Now it is time to pick a $\delta > 0$ for the given ε. Let us choose it to be less than the smaller of the numbers $\frac{c}{2}$ and $\left(\frac{c^3}{10}\right)\varepsilon$. Then if $|x-c| < \delta,$

$$\left|\frac{1}{x^2}-\frac{1}{c^2}\right| < |x-c|\frac{10}{c^3} < \delta\frac{10}{c^3} < \frac{c^3}{10}\varepsilon\frac{10}{c^3} = \varepsilon,$$

and our proof is complete.

TOPIC 3: Properties of Continuity

Just as the limit laws greatly simplify the task of determining whether a function has a limit at a given point, we have a number of properties of continuity that allow us to quickly answer many questions regarding continuity of functions.

THEOREM: Properties of Continuous Functions

Let f and g be two functions both continuous at the point c, and let k be a fixed real number. Then the following combinations of f and g are also continuous at c.

Sums	$f+g$	"A sum of continuous functions is continuous."
Differences	$f-g$	"A difference of continuous functions is continuous."
Constant Multiples	$k \cdot f$	"A multiple of a continuous function is continuous."
Products	$f \cdot g$	"A product of continuous functions is continuous."
Quotients	$\frac{f}{g}$, provided $g(c) \neq 0$	"A quotient of continuous functions is continuous."

These properties follow immediately from the corresponding limit properties. As an example, we will prove the fourth property of continuous functions.

Proof

We assume that f and g are both continuous at c.

$$\begin{aligned}
\lim_{x \to c}(f \cdot g)(x) &= \lim_{x \to c}\left[f(x)g(x)\right] \\
&= \lim_{x \to c} f(x) \cdot \lim_{x \to c} g(x) && \text{The limit of a product is the product of the limits.} \\
&= f(c) \cdot g(c) && \text{Both } f \text{ and } g \text{ are continuous at } c. \\
&= (f \cdot g)(c)
\end{aligned}$$

As a consequence of these properties, polynomial and rational functions can be immediately classified as continuous functions. That is, every polynomial function and every rational function is continuous at every point of its domain.

Example 6: Understanding Rational Functions and Continuity

To illustrate the fact that a rational function is continuous everywhere on its domain, everywhere except for the zeros of the denominator, let us consider

$$r(x) = \frac{x^2 - 2x}{x^2 - 3x + 2}.$$

Since the denominator factors as $x^2 - 3x + 2 = (x-2)(x-1)$, we see that the domain of r is the set $D = \{x \in \mathbb{R} \mid x \neq 1, 2\}$. If $c \in D$, that is, c is an arbitrary real number other than 1 or 2, the Direct Substitution Property applies, and $r(c) = \lim_{x \to c} r(x)$, showing that r is continuous at every point of its domain. Using our limit determination techniques, we can even find out what happens at the two points of discontinuity. Since

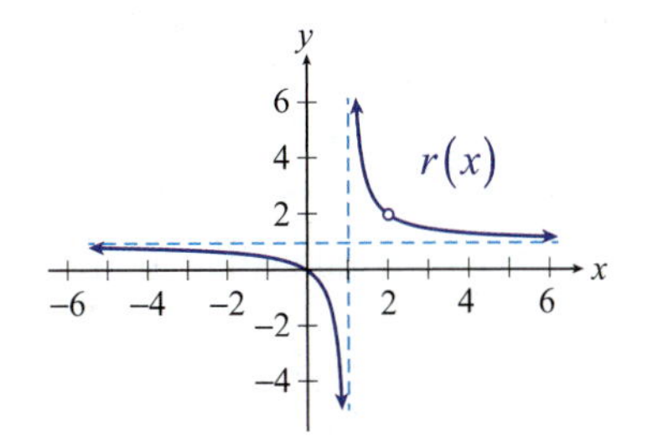

FIGURE 7

$$r(x)=\frac{x^2-2x}{x^2-3x+2}=\frac{x(x-2)}{(x-1)(x-2)}=\frac{x}{x-1},$$

it follows that $x=2$ is a removable discontinuity, and the graph of r agrees with that of $g(x)=\frac{x}{x-1}$ for all $x\neq 2$. The discontinuity at $x=1$, however, is nonremovable. Since the numerator is bounded near 1, while the denominator nears 0, r has a vertical asymptote at $x=1$. Notice how the graph in Figure 7 supports all of our findings.

Since sums, differences, products, and quotients of continuous functions are continuous, it is natural to ask if continuity is preserved under other combinations of functions. The following theorem is a bridge toward an important answer to that question.

THEOREM: "Limits Pass through a Continuous Function"

Suppose $\lim_{x\to c} g(x)=a$ and f is continuous at the point a. Then

$$\lim_{x\to c} f(g(x))=f\left(\lim_{x\to c} g(x)\right)=f(a).$$

In words, we say the limit operation passes inside the continuous function f.

The proof of this theorem can typically be found in a Calculus text—it is a nice example of the use of the epsilon-delta definitions of both limit and continuity. At the moment, though, we are more interested in the following application of the theorem.

THEOREM: "A Composition of Continuous Functions Is Continuous"

If g is continuous at the point c, and if f is continuous at $g(c)$, then the composite function $f\circ g$ is continuous at c.

Proof

Because g is continuous at c, we know that $\lim_{x\to c} g(x)=g(c)$, and so by replacing a with $g(c)$ in the previous theorem, it follows that

$$\lim_{x\to c} f(g(x))=f\left(\lim_{x\to c} g(x)\right)=f(g(c)).$$

Using composition notation, we have

$$\lim_{x\to c}(f\circ g)(x)=(f\circ g)(c).$$

Another useful theorem tells us that inverses of continuous functions are continuous.

THEOREM: "The Inverse of a Continuous Function Is Continuous"

If f is one-to-one and continuous on the interval (a,b), then f^{-1} is also a continuous function.

With these theorems in hand, the set of functions that we can classify as continuous instantly expands considerably.

Example 7: Using Theorems to Determine Continuity of Functions

Use the above theorems to describe the continuity of the following functions.

a. $F(x)=\sqrt[3]{\dfrac{2x+1}{x^2+5}}$ **b.** $G(x)=\sqrt{\dfrac{4x^5-3x^2+7}{9x^3+3x}}$

c. $H(x)=\sin\left(\dfrac{2x^3+x^2-3}{x^2-2x-8}\right)$ **d.** $K(x)=\arcsin\left(\sqrt{1-x^2}\right)$

Solution

a. Notice that F is a composite function; you can think of it as $F(x)=f(g(x))$, where $g(x)=\dfrac{2x+1}{x^2+5}$, and $f(x)=\sqrt[3]{x}$. Furthermore, g is continuous on all of $\mathbb{R}$, since its denominator is never 0. The same can be said about f, namely that it is continuous on the entire real line. Note that you can see this in at least two ways. Either apply the theorem preceding Example 7, since f is the inverse of $k(x)=x^3$, which is a continuous function mapping $\mathbb{R}$ onto $\mathbb{R}$; or by referring to the Positive Integer Root Law you can provide a one-line proof of the continuity of f as follows. If $c\in\mathbb{R}$ is arbitrary,

$$\lim_{x\to c} f(x)=\lim_{x\to c}\sqrt[3]{x}=\sqrt[3]{\lim_{x\to c} x}=\sqrt[3]{c}.$$

Now a simple application of the earlier theorem about continuity of compositions establishes the fact that $F(x)=f(g(x))$ is continuous on $\mathbb{R}$.

b. As in part a., we will use the "continuity of compositions" theorem, but while the conditions of the theorem were readily satisfied by both the inner and outer functions on all of $\mathbb{R}$ in the previous problem, we will need to pay a little extra attention here. To start off, we once again use the notation $G(x)=f(g(x))$; however, this time $g(x)=\dfrac{4x^5-3x^2+7}{9x^3+3x}$ (see Figure 8), while $f(x)=\sqrt{x}$. According to our theorem, $G=f\circ g$ will be continuous at all points of continuity c of $g(x)$ provided that f is also continuous at $g(c)$. A factoring argument like the one given in part a. shows that g is undefined and has a vertical asymptote at $x=0$; it is, however, continuous everywhere else. The outer function f, on the other hand, is only defined for $x\geq 0$, but it is continuous on its domain. This latter claim can be established by an argument similar to the one given for the cube root function in part a. In fact, notice that it is easy to generalize that argument and conclude that all root functions are continuous as a result of the previous theorem and the continuity of the functions x^n.

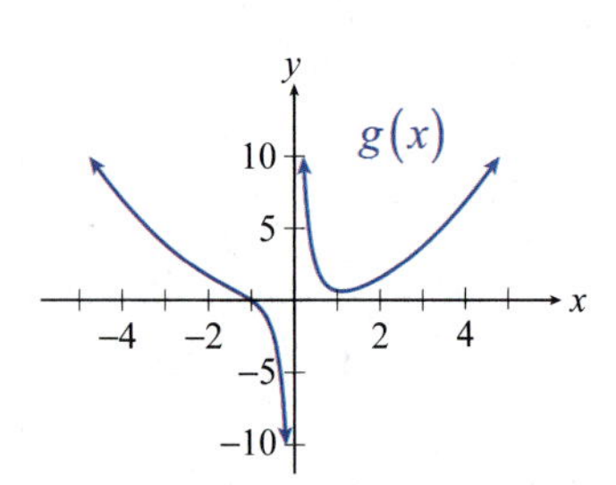

FIGURE 8

Summarizing our observations up to this point, and using the fact that the composition of continuous functions is continuous, we see that $G(x)=f(g(x))$ will be continuous at all nonzero real numbers c such that $g(c)\geq 0$.

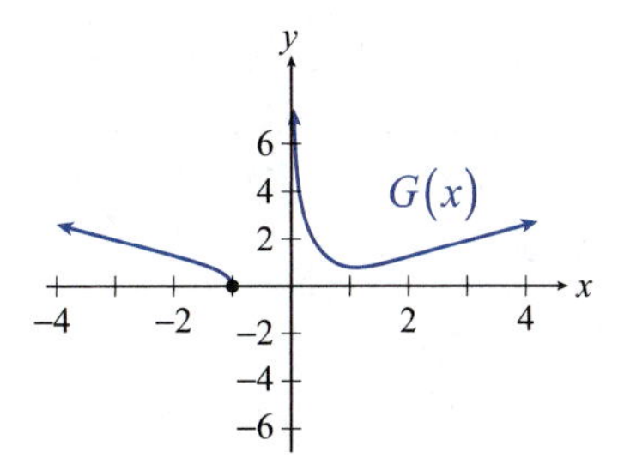

FIGURE 9

Our next observation is that $g(x) < 0$ on the interval $(-1,0)$, but is nonnegative everywhere else on its domain; that is, $g(x) \geq 0$ on $(-\infty,-1] \cup (0,\infty)$. (You can algebraically convince yourself of this fact, or simply examine the graph in Figure 8.)

Therefore, we conclude that $G(x) = f(g(x))$ is continuous on the set $(-\infty,-1] \cup (0,\infty)$. The graph of G supports our conclusion (see Figure 9).

c. Factoring the denominator easily shows that the inner function of this problem $g(x) = \dfrac{2x^3 + x^2 - 3}{x^2 - 2x - 8}$ has two vertical asymptotes, at $x = -2$ and $x = 4$, respectively. However, being a rational function, g is continuous everywhere else on $\mathbb{R}$. As we will show in Example 8, $f(x) = \sin x$ is continuous at every real number, so our conclusion is that $H(x)$ is continuous everywhere except at $x = -2$ and $x = 4$.

d. First of all, we observe that $g(x) = \sqrt{1 - x^2}$ is continuous everywhere on its domain of $[-1,1]$. Furthermore, the range of g is the interval $[0,1]$ (note that the graph of g is the upper semicircle of radius 1, centered at the origin). Using the theorem about the continuity of inverse functions, we conclude that $f(x) = \arcsin x$ is continuous at every point of the range of g and hence $K(x) = f(g(x))$ is continuous on its domain of $[-1,1]$.

In fact, most familiar functions (those that we bother to give names to) are continuous (though many have points of discontinuity—don't forget the subtle distinction). Trigonometric functions are continuous, a fact that is most easily proven through the use of an alternate formulation of continuity:

$$f \text{ is continuous at the point } c \text{ if and only if } \lim_{h \to 0} f(c+h) = f(c).$$

Example 8: Using Alternate Formulations to Prove Functions are Continuous

Use the alternate formulation of continuity to prove that $f(x) = \sin x$ is continuous on $\mathbb{R}$.

Solution

Our goal is to show that for any fixed real number c, $\lim_{h \to 0} \sin(c+h) = \sin c$.

Using the appropriate sum identity known from trigonometry along with the relevant limit laws,

$$\begin{aligned}
\lim_{h \to 0} \sin(c+h) &= \lim_{h \to 0} (\sin c \cos h + \cos c \sin h) \\
&= \lim_{h \to 0} (\sin c \cos h) + \lim_{h \to 0} (\cos c \sin h) \\
&= \sin c \lim_{h \to 0} \cos h + \cos c \lim_{h \to 0} \sin h,
\end{aligned}$$

where we also used the fact that during the limit process $\sin c$ and $\cos c$ are constants, so the Constant Multiple Law applies.

Our next step is to recall that as h approaches 0, $\lim_{h\to 0}\cos h = 1$ and $\lim_{h\to 0}\sin h = 0$. (You may recall these facts from the unit-circle definition of sine and cosine for small angles, or simply from the graphs of the sine and cosine functions.) Using these latter two limits we obtain

$$\begin{aligned}\sin c \lim_{h\to 0}\cos h + \cos c \lim_{h\to 0}\sin h &= \sin c \cdot 1 + \cos c \cdot 0 \\ &= \sin c,\end{aligned}$$

which is what we needed to prove.

Finally, we note that the continuity of $\cos x$ can be established by a similar argument, or by using the continuity of the sine function along with the identity $\cos x = \sin\left(\frac{\pi}{2} - x\right)$. Then using the properties of continuous functions, the continuity of the remaining four trigonometric functions $\tan x$, $\cot x$, $\sec x$, and $\csc x$ will follow.

Exponential functions are continuous, a fact largely due to design. For a given base a, the expression $a^{\frac{m}{n}}$ is easily understood for every rational number $\frac{m}{n}$, and a common way to extend the meaning of a^x to irrational numbers is to define a^x to be the limit of terms of the form a^{r_n}, where $\{r_n\}$ is a sequence of rational numbers approaching x (some work is required in order to show such a definition is unambiguous). And because inverse functions of continuous functions are continuous, the continuity of logarithms and inverse trigonometric functions is assured.

Example 9: Using Theorems to Determine Continuity of Functions

Use the above theorems to identify where the following functions are continuous.

a. $f(x) = \dfrac{\arctan(\ln x)}{x^2 + 1}$ **b.** $g(x) = \dfrac{xe^{\cos x}}{\sqrt{x^2 - 1} - 3}$

c. $h(x) = \dfrac{\ln x - \sin^{-1} x}{x^2 + x - 2}$

Solution

a. Being the inverses of continuous functions, $\ln x$ is continuous on $(0, \infty)$, and $\arctan x$ is continuous on the entire real line. Thus the composition $\arctan(\ln x)$ is continuous on $(0, \infty)$. Since the denominator of f, $x^2 + 1$, is never 0 on $\mathbb{R}$ (check this), it follows that the quotient $f(x)$ is continuous on $(0, \infty)$.

b. Since all three functions $\cos x$, e^x, and x are continuous on the entire real line, it follows from our theorems that both the composition $e^{\cos x}$ and the product $xe^{\cos x}$ will be continuous on all of $\mathbb{R}$. Next, just as we did in Example 6, we would like to use the theorem stating the continuity of the quotient of continuous functions, but we need to carefully check the continuity of the denominator of g. First, the square root function requires that the inequality $x^2 - 1 \geq 0$ be true for x. This means, we need $x^2 \geq 1$ to hold; that is, $x^2 \geq 1$ or $x^2 \leq -1$. Since the composition of continuous functions is continuous, we conclude that $\sqrt{x^2 - 1}$ is continuous on $(-\infty, -1] \cup [1, \infty)$. Last, but not least, in order for g to be continuous, its denominator cannot be 0, so we have to exclude the solutions of the equation $\sqrt{x^2-1}=3$. By squaring both sides and adding 1 we obtain $x^2=10$, that is, the points $x=\pm\sqrt{10}$ have to be excluded from the domain.

Summarizing our findings, we conclude that $g(x)$ is continuous on the following set.

$$\left(-\infty,-\sqrt{10}\right)\cup\left(-\sqrt{10},-1\right]\cup\left[1,\sqrt{10}\right)\cup\left(\sqrt{10},\infty\right)$$

c. Arguing as we did in part a., $\ln x$ and $\sin^{-1}x$ are continuous, since they are inverses of continuous functions. The domain of $\ln x$ is $(0,\infty)$, while $\sin^{-1}x$ is only defined on $[-1,1]$. However, notice that $x=1$ is a zero of the denominator, so by the properties of continuous functions, we conclude that $h(x)$ is continuous on the open interval $(0,1)$.

Collectively, the theorems in this section allow us to identify a large number of continuous functions very quickly. It might even be possible to do something about the occasional point of discontinuity. Recall that a discontinuity is termed *removable* if the limit at the point exists; in such cases, we can define the **continuous extension** of a function as in Example 10.

Example 10: Defining Continuous Extensions of Functions

Identify the removable discontinuities and define the continuous extension of $r(x)=\dfrac{x^2-5x}{x^3-3x^2-10x}$.

Solution

Factoring and canceling yields

$$r(x)=\frac{x^2-5x}{x^3-3x^2-10x}=\frac{x(x-5)}{x(x-5)(x+2)}=\frac{1}{x+2},$$

where the previous equality holds for all x except for $x=0$ and $x=5$. However, as we learned in Section 13.4, both $\lim\limits_{x\to 0} r(x)$ and $\lim\limits_{x\to 5} r(x)$ still exist.

$$\lim_{x\to 0} r(x)=\frac{1}{0+2}=\frac{1}{2},\quad \text{while}\quad \lim_{x\to 5} r(x)=\frac{1}{5+2}=\frac{1}{7}$$

Therefore, r has two removable discontinuities, at $x=0$ and $x=5$. Now we can define $\tilde{r}(x)$, the continuous extension of r, as follows.

$$\tilde{r}(x)=\begin{cases}\dfrac{x^2-5x}{x^3-3x^2-10x} & \text{if } x\neq 0 \text{ or } 5\\[2ex] \dfrac{1}{2} & \text{if } x=0\\[2ex] \dfrac{1}{7} & \text{if } x=5\end{cases}$$

Let us note, however, that although correct and instructive, it is not necessary in this case to use a piecewise defined function to define the continuous extension of r. Alternatively, it is much shorter to define it simply as $\tilde{r}(x)=\dfrac{1}{x+2}$. Note that this is equivalent to our piecewise definition above. Note also that both r and its continuous extension have nonremovable discontinuities at $x=-2$.

We will close this section with one last theorem that makes a connection between our rigorous definition of continuity and the intuitive "graph without interruptions" interpretation of continuity.

THEOREM: The Intermediate Value Theorem

If f is a continuous function defined on the closed interval $[a,b]$, then f takes on every value between $f(a)$ and $f(b)$. That is, if L is a real number between $f(a)$ and $f(b)$, then there is a c in the interval $[a,b]$ such that $f(c)=L$.

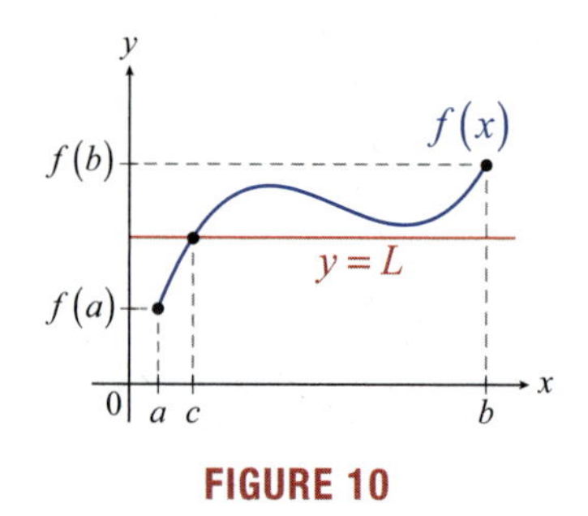

FIGURE 10

The proof of the Intermediate Value Theorem (IVT) relies upon a property of the real numbers called *completeness*; this concept and the proof of the theorem are presented in courses like Advanced Calculus and Introductory Real Analysis. But the implications of the Intermediate Value Theorem are of interest to us at the moment. Informally, the IVT says that the graph of a continuous function f defined on an interval $[a,b]$ cannot avoid intersecting any horizontal line $y=L$ if L is a value between $f(a)$ and $f(b)$ (see Figure 10). In other words, such a function f has no "breaks" or "jumps" that would allow it to skip over the value L. This property, also referred to as the *Intermediate Value Property*, is often used to prove that an equation of the form $f(x)=L$ must have at least one solution in the interval $[a,b]$. Note that the content of the IVT is precisely the fact that a continuous function f on a closed interval $[a,b]$ possesses the Intermediate Value Property.

Example 11: Using the Intermediate Value Theorem

Use the Intermediate Value Theorem to show that the equation $x^5+9x-4=0$ has a solution between 0 and 1.

Solution

Introducing the notation $f(x)=x^5+9x-4$, notice that we are attempting to prove the existence of a solution, or root, of the equation $f(x)=0$ on the interval $[0,1]$. The existence of such root c would mean that f takes on the value $L=0$ for some c in $[0,1]$. Does the Intermediate Value Theorem guarantee that? Notice first of all that f is certainly continuous on the closed interval $[0,1]$ (actually, it is continuous on the entire real line, but for the purposes of this problem we can safely ignore what happens outside of $[0,1]$). Next, since 0 and 1 play the roles of a and b in the theorem, respectively, we proceed to examine the values $f(0)$ and $f(1)$.

$$f(0)=0^5+9\cdot 0-4=-4<0$$
$$f(1)=1^5+9\cdot 1-4=6>0$$

In other words, a change of sign occurs on $[0,1]$; that is, f goes from negative to positive on $[0,1]$. The statement of the Intermediate Value Theorem is precisely the fact that a continuous function cannot do this and avoid intersecting the horizontal line $y=0$ (the x-axis). More precisely, since $L=0$ we have $-4<L<6$; that is, since 0 is a real number between $f(0)=-4$ and $f(1)=6$, by the Intermediate Value Theorem there is a c in $[0,1]$ such that $f(c)=L=0$. In other words, we have

$$c^5+9c-4=0.$$

Thus we proved the existence of a root between 0 and 1 for the given equation.

NOTE

The Intermediate Value Theorem is an *existence theorem*; in other words, we proved the *existence* of a solution between 0 and 1 without actually specifying what it is.

13.5 EXERCISES

PRACTICE

Find all points of continuity as well as all points of discontinuity for the given function. For any discontinuities, identify those from the three continuity criteria that fail to hold. See Examples 1 and 2.

1.

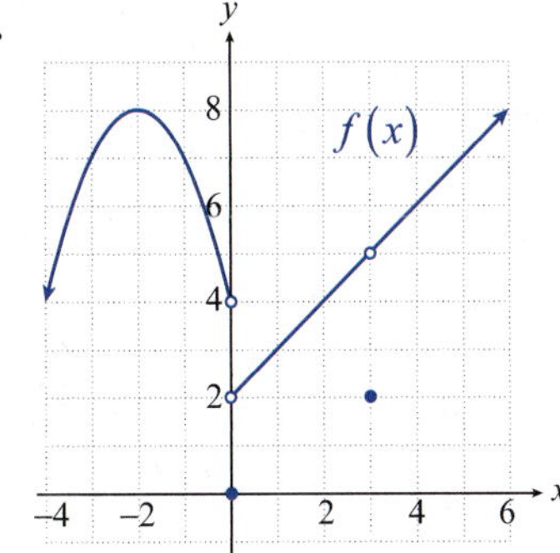

2.

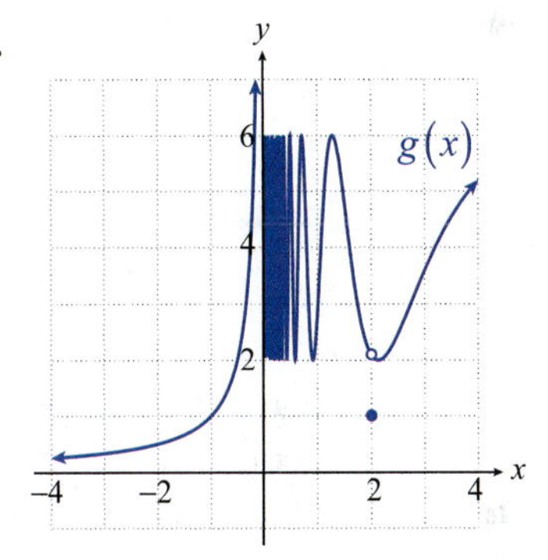

3. Sketch a graph of a function (a formula is not necessary) that has a removable discontinuity at $x=-1$, a jump discontinuity at $x=2$, but is right-continuous at $x=2$.

4. Sketch a graph of a function that has an infinite discontinuity at $x=0$ and an oscillating discontinuity at $x=5$ so that it is still left-continuous at $x=5$.

Find and classify the discontinuities (if any) of the function as removable or nonremovable. See Example 3.

5. $f(x)=\frac{1}{x}$

6. $g(x)=\frac{-2}{x-3}$

7. $h(x)=\frac{x^2-9}{x-3}$

8. $k(x)=\frac{x^2-2x}{x^2+5x-14}$

9. $u(x)=\frac{x^2-9}{x-2}$

10. $v(x)=\frac{x-1}{x^2+2x-3}$

11. $w(x)=\begin{cases} x+1 & \text{if } x\le 0 \\ \frac{1}{2}x^2+1 & \text{if } x>0 \end{cases}$

12. $f(x)=\begin{cases} \frac{1}{2}x-2 & \text{if } x\le 4 \\ x^3+1 & \text{if } x>4 \end{cases}$

13. $g(x)=\begin{cases} \tan x & \text{if } x<\frac{\pi}{2} \\ \cos x & \text{if } x\ge\frac{\pi}{2} \end{cases}$

14. $h(x)=\begin{cases} \cos x & \text{if } x\le 0 \\ \tan x+1 & \text{if } x>0 \end{cases}$

15. $F(u)=\frac{u-4}{\sqrt{u}-2},\quad u\ge 0$

16. $G(s)=\frac{s}{\sqrt{s+4}-2},\quad s\ge -4$

17. $H(t)=\frac{t}{\sqrt{t^2+2}}$

18. $u(x)=\cos\left(\frac{1-x^2}{1-x}\right)$

19. $v(t)=|\sin t|$

20. $K(x)=|x+2|+|x-1|$

21. $F(t)=\frac{t}{t^2-1}$

22. $G(t)=\frac{t}{t^2+1}$

23. $H(x)=|x+2|$

24. $k(x)=\frac{|x-4|}{x-4}$

25. $s(x)=[\![x+2]\!]$

26. $t(x)=4-[\![x]\!]$

27. $u(z)=[\![z^2]\!]$

28. $v(x)=x[\![x]\!]$

29. $w(x)=x\left[\!\left[\frac{1}{x}\right]\!\right]$

Use the ε-δ definition to prove that the function is continuous. See Example 5.

30. $f(x)=\frac{1}{x}$

31. $g(x)=3x-2$

32. $F(x)=x^3$

33. $G(x)=\sqrt{x}$

Use the theorems of this section to describe the continuity of the function. See Example 7.

34. $F(x)=\sqrt{\frac{x}{x^2+7x+12}}$

35. $G(x)=\sqrt{\frac{x^4-x^3-11x^2+9x+18}{2x^3+x}}$

36. $H(x)=\cos\left(\frac{2\ln(x-3)+1}{\sqrt[3]{x^2-2x-15}}\right)$

37. $f(x)=\arctan\left(\frac{x}{\sqrt{3-x^2}}\right)$

38. $g(x)=\ln(\arcsin(\pi x+1))$

39. $h(x)=\frac{\csc(\pi x+1)}{\sin(\pi e^{x+2})}$

Use the alternate formulation of continuity to prove that the function is continuous. See Example 8.

40. $f(x)=3x-5x^2$

41. $g(x)=\cos x$

42. $h(x)=\tan x$

43. $k(x)=e^x$

Identify the removable discontinuities and define the continuous extension of the function. See Example 10.

44. $f(x)=\dfrac{x^2+x-12}{x-3}$

45. $g(x)=\dfrac{x^3-2x^2-x+2}{x^2-3x+2}$

46. $h(x)=\dfrac{x-1}{\sqrt{x}-1}$

47. $F(x)=\dfrac{\sqrt{x+1}-2}{x-3}$

48. $G(x)=2^{-\frac{1}{x^2}}$

49. $H(x)=x\cos\left(\dfrac{\pi}{x}\right)$

Describe the continuity of the function on the given closed interval.

50. $S(x)=\sqrt{16-x^2}$ on $[-4,4]$

51. $T(x)=\left[\!\left[\dfrac{x}{3}\right]\!\right]$ on $[0,3]$

52. $U(x)=\begin{cases}\dfrac{1}{x^2-9} & \text{if } |x|<3\\ 0 & \text{if } |x|=3\end{cases}$ on $[-3,3]$

53. $V(x)=\begin{cases}x^2\sin\left(\dfrac{1}{x}\right) & \text{if } x\neq 0\\ 0 & \text{if } x=0\end{cases}$ on $\left[0,\dfrac{1}{\pi}\right]$

Find the value of *a* (or the values of *a* and *b*, where applicable) such that *f* is continuous on the entire real line.

54. $f(x)=\begin{cases}0 & \text{if } x\leq 0\\ ax & \text{if } 0<x<1\\ 2x+3 & \text{if } x\geq 1\end{cases}$

55. $f(x)=\begin{cases}x^3 & \text{if } x\leq 3\\ ax^2 & \text{if } x>3\end{cases}$

56. $f(x)=\begin{cases}-x^2 & \text{if } x<1\\ ax+b & \text{if } 1\leq x\leq 3\\ (x-3)^2+2 & \text{if } x>3\end{cases}$

57. $f(x)=\begin{cases}\cos x & \text{if } x\leq 0\\ -(x-a)^2+b & \text{if } 0<x<2\\ \dfrac{1}{2}x-2 & \text{if } x\geq 2\end{cases}$

Decide whether the Intermediate Value Theorem applies to the given function on the indicated interval. If so, find *c* as guaranteed by the theorem. If not, find the reason. See Example 11.

58. $f(x)=-x^2+x+3$ on $[0,3]$; $f(c)=1$

59. $g(x)=2x^3-x^2-1$ on $[-1,2]$; $g(c)=0$

60. $h(x)=\dfrac{x}{x+2}$ on $[0,4]$; $h(c)=\dfrac{1}{2}$

61. $F(x)=\dfrac{2x}{x-1}$ on $[0,2]$; $F(c)=2$

62. $G(x)=[\![x-2]\!]$ on $[-2,2]$; $G(c)=-\dfrac{1}{2}$

63. $H(x)=\sin\left(\dfrac{3x+2}{2}\right)$ on $\left[-\dfrac{2}{3},\dfrac{\pi-2}{3}\right]$; $H(c)=\dfrac{1}{2}$

Use the Intermediate Value Theorem to prove that the given equation has a solution on the indicated interval. See Example 11.

64. $x^3 - 7.5x^2 + 1.2x + 1 = 0$ on $[-1, 0]$

65. $2x^3 + x + 10 = 0$ on $[-2, 1]$

66. $\cos x = x^2$ on $[0, \pi]$

67. $\ln x - \sqrt{x-2} = 0$ on $[2, 5]$

68. $\dfrac{5}{x^2 + 2} = 1$ on $[-3, -1]$

69. $\cot\left(\dfrac{\pi x}{4}\right) - \dfrac{x}{x+2} = -\dfrac{1}{2}$ on $[1, 3]$

APPLICATIONS

70. Suppose that the outside temperature in Columbia, SC on a summer morning at 7:00 a.m. is 74 °F, and it shoots up to 98 °F by 1:00 p.m. Assuming that temperature changes continuously, prove that sometime between 7:00 a.m. and 1:00 p.m. the temperature was exactly 88.35 °F.

71. A hermit leaves his hut at the foot of a mountain one day at 6:00 a.m. and sets out to climb all the way to the top. He arrives at 6:00 p.m. and realizes that it is too late to go back, so he sets up camp for the night. At 6:00 a.m. the following day, he starts hiking back to his hut, taking the exact same route as the day before. This time, however, it is mostly downhill, so he makes much better time and arrives home at 2:00 p.m. Prove that there is a point along the hermit's route that he passed at exactly the same time on both days. (**Hint:** Apply the Intermediate Value Theorem or the Fixed Point Theorem. See Exercise 83.)

72. A long-distance phone company charges 31 cents for the first minute and 10 cents for each additional minute or any fraction thereof. Graph the cost as a function of time, find a formula for it, and describe the significance of its discontinuities. (**Hint:** Use the greatest integer function to construct your answer.)

73. If Δt denotes the length of the time interval between two events as measured by an observer on a spaceship moving at speed v, and ΔT is the length of the same time interval as measured from Earth, then the formula relating the two quantities is given by

$$\Delta T = \frac{\Delta t}{\sqrt{1 - \dfrac{v^2}{c^2}}},$$

where c is the speed of light. This phenomenon is called *time dilation*, and it follows from the theory of relativity. In essence, it says that a clock moving at speed v relative to an observer is perceived by the same observer to run slower.

a. Explain why we don't normally notice the time dilation effect in everyday life.

b. What is the significance of the discontinuity of ΔT (as a function of v)?

WRITING & THINKING

74. Prove the alternate formulation of continuity; that is, the statement that a function f is continuous at the point c if and only if $\lim_{h \to 0} f(c+h) = f(c)$.

75. Prove that if $f(x)$ is continuous and $f(c) > 0$, then there is a $\delta > 0$ such that $f(x) > 0$ for all $x \neq c$ in the interval $(c - \delta, c + \delta)$.

76. Prove that the Dirichlet function $\xi(x)=\begin{cases}0 & \text{if } x \text{ is rational}\\ 1 & \text{if } x \text{ is irrational}\end{cases}$ is discontinuous at every real number.

77. Prove that the function $f(x)=\begin{cases}0 & \text{if } x \text{ is rational}\\ x^2 & \text{if } x \text{ is irrational}\end{cases}$ is continuous only at the single point $c=0$.

78. Prove that if the functions f and g are both continuous on $\mathbb{R}$ and they agree on the rationals (i.e., $f(x)=g(x)$ for all $x \in \mathbb{Q}$), then $f=g$.

79. Prove that if f is continuous and never 0 on the interval $[a,b]$, then either $f(x)>0$ for every x in $[a,b]$, or $f(x)<0$ for every x in $[a,b]$.

80. (Existence of nth roots) Prove that if b is a positive real number and n a positive integer, then there is a positive real number c such that $c^n=b$. (**Hint:** Consider the continuous function $f(x)=x^n$ on the interval $[0,b+1]$.)

81. Prove that a circle of diameter d has a chord of length c for every number c between 0 and d.

82. Use the function

$$f(x)=\begin{cases}\sin\left(\dfrac{\pi}{x}\right) & \text{if } x\neq 0\\ 0 & \text{if } x=0\end{cases}$$

to prove that the converse of the Intermediate Value Theorem is false; in other words, a function may possess the Intermediate Value Property without being continuous.

83. (Fixed Point Theorem) Prove that if the function $f:[a,b]\to[a,b]$ is continuous, then there is a number c in $[a,b]$ with $f(c)=c$ (i.e., c is "fixed," or "not being moved," by f).

Determine whether the given statement is true or false. If false, explain or provide a counterexample.

84. If f is both left- and right-continuous at c, then f is continuous at c.

85. Any function f has an interval (a,b) on which it is continuous.

86. If c is a discontinuity of f, but f does not have a vertical asymptote at c, then c is a removable or jump discontinuity.

87. If $\lim_{x\to c} f(x)=L$, and $f(c)=L$, then f is continuous at c.

88. If c is a discontinuity of f, but $\lim_{x\to c^+} f(x)$ exists, then c is a removable or jump discontinuity.

TECHNOLOGY

89–94. Use a graphing utility to solve the equations given in Exercises 64–69 to four decimal places.

95–100. Use a graphing utility to graph the functions of Exercises 34–39 and explain how the graphs support your discussions of continuity in these exercises.

13.6 THE DERIVATIVE

TOPICS

1. The velocity and tangent problems revisited
2. The derivative as a function

TOPIC 1: The Velocity and Tangent Problems Revisited

We will wrap up this chapter by revisiting the two motivational problems that we began with, making use of the expertise in working with limits that we have since acquired. Our last topic of discussion will be an indication of how the concepts that have been introduced generalize and apply to a wealth of situations.

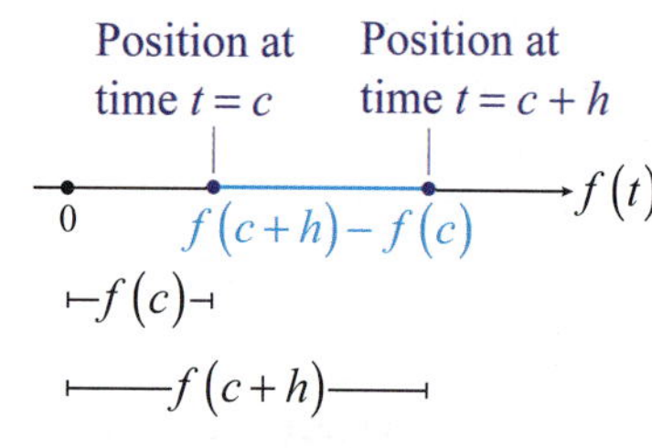

FIGURE 1

In Section 13.1, we developed a method for determining instantaneous velocity at a particular point c in time by considering the behavior of the fraction

$$\frac{f(c+h)-f(c)}{h}$$

for smaller and smaller values of h. The function $f(t)$ in this setting describes the position of an object at time t moving along a straight line, and the fraction was defined to be the **difference quotient of f at c with increment h**. Although we lacked the terminology and notation at the time, we can now succinctly define the (**instantaneous**) **velocity of the object at time c** to be

$$\lim_{h \to 0} \frac{f(c+h)-f(c)}{h}.$$

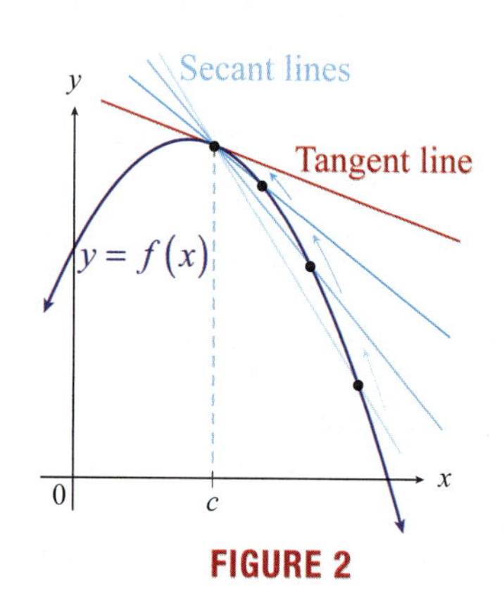

FIGURE 2

Similarly, we developed a way to arrive at the slope of the line tangent to the function $f(x)$ at the point c by considering the slopes of secant lines that approached the tangent line. Ultimately, we saw that the tangent line slope (if the tangent line exists) is also the limiting behavior of the difference quotient at c, so we can define the **slope of the line tangent to f at c** to also be

$$\lim_{h \to 0} \frac{f(c+h)-f(c)}{h}.$$

Given this concurrence, it likely comes as no surprise that limits of difference quotients will occupy our attention for some time, and it is appropriate to now introduce some additional notation.

DEFINITION: The Derivative at a Point

The **derivative of the function f at the point c**, denoted $f'(c)$, is

$$f'(c) = \lim_{h \to 0} \frac{f(c+h)-f(c)}{h},$$

provided the limit exists. ($f'(c)$ is read "f prime of c.")

Recall that in Section 13.1 we saw some alternate ways of expressing the same idea. In some contexts, it may be more natural to refer to the point $c + h$ as x, and to express the difference quotient as follows:

$$\frac{f(c+h)-f(c)}{h}=\frac{f(x)-f(c)}{x-c}$$

In this case, the derivative of f at c is obtained by letting x approach c, so an alternate definition of the derivative is

$$f'(c)=\lim_{x\to c}\frac{f(x)-f(c)}{x-c}.$$

One last variation of this idea comes from referring to $f(x)-f(c)$ as the "change in f" or "change in y" (denoted Δf or Δy) and to $x - c$ as the "change in x" (denoted Δx). This is commonly seen when the context makes it natural to refer to the equation $y = f(x)$ and to emphasize the fact that the numerator of the difference quotient represents a vertical difference while the denominator represents a horizontal difference. Using this notation, and with the implicit understanding that c is fixed, the derivative would be denoted as

$$f'(c)=\lim_{\Delta x\to 0}\frac{\Delta f}{\Delta x}=\lim_{\Delta x\to 0}\frac{\Delta y}{\Delta x}.$$

It is important to realize, however, that all these variations represent the same fundamental concept. Regardless of the difference quotient notation used, and regardless of whether we are seeking the instantaneous velocity of an object or the slope of a tangent line, $f'(c)$ indicates the **rate of change of the function f at the point c.**

Example 1: Finding the Equation of a Tangent Line

Find the slope of the line tangent to the graph of $f(x)=x^2-1$ at the point $(2,3)$. Then use this information to derive the equation of the tangent line.

Solution

As we have discussed above, the slope of the tangent line at the point $(2,3)$ will be the limit of the difference quotient at $c = 2$, or using our new terminology, the derivative of f at the point $c = 2$. Recall that we denote this number by $f'(c)$, and it is calculated as

$$f'(c)=\lim_{h\to 0}\frac{f(c+h)-f(c)}{h}.$$

We now proceed to evaluate the above derivative for the given $f(x)=x^2-1$ at $c = 2$:

$$\begin{aligned}f'(2)&=\lim_{h\to 0}\frac{f(2+h)-f(2)}{h}=\lim_{h\to 0}\frac{\left[(2+h)^2-1\right]-\left(2^2-1\right)}{h}\\&=\lim_{h\to 0}\frac{\left[\left(2^2+4h+h^2\right)-1\right]-\left(2^2-1\right)}{h}\end{aligned}$$

$$f'(2) = \lim_{h \to 0} \frac{4+4h+h^2-1-4+1}{h} = \lim_{h \to 0} \frac{4h+h^2}{h}$$

$$= \lim_{h \to 0} \frac{(4+h)h}{h} = \lim_{h \to 0} (4+h) = 4$$

So we conclude that the slope of the requested tangent is $m = f'(2) = 4$.

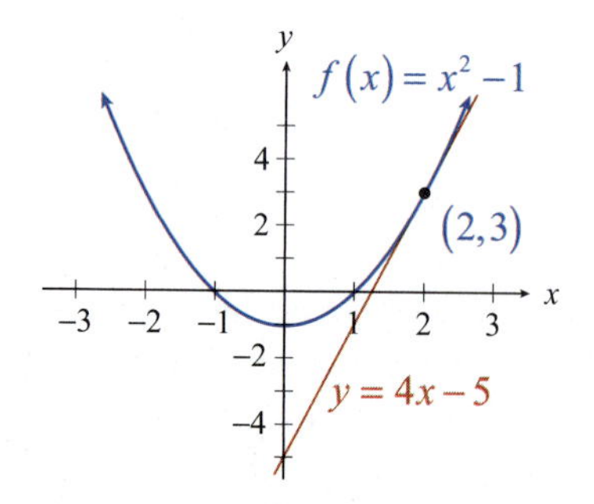

FIGURE 3

Notice that we now have all the information necessary to obtain the equation of the tangent line. It is a line with a slope of 4 that passes through the point $(2,3)$. Recall the point-slope form of the equation of a line with slope m that passes through the point (x_0, y_0) is given by

$$y - y_0 = m(x - x_0).$$

Applying this formula to our situation at hand, we obtain the equation

$$y - 3 = 4(x-2), \quad \text{or equivalently} \quad y = 4x - 5.$$

The graph of $f(x)$ along with the tangent line are shown in Figure 3.

Example 2: Finding Instantaneous Velocity

A particle is moving along a straight line so that its distance from the start is given by $d(t) = 5t - \frac{1}{2}t^2$, where d is measured in feet and t in seconds. What is the particle's instantaneous velocity at $t = 2$ seconds?

Solution

As we have seen, the instantaneous velocity at time $t = c$ of an object moving along a straight line is the derivative of its position function at $t = c$. Using this observation with our given function d and $t = c = 2$ seconds, we obtain the following:

$$d'(c) = d'(2) = \lim_{h \to 0} \frac{d(2+h) - d(2)}{h}$$

$$= \lim_{h \to 0} \frac{\left[5(2+h) - \frac{1}{2}(2+h)^2\right] - \left[5(2) - \frac{1}{2}(2^2)\right]}{h}$$

$$= \lim_{h \to 0} \frac{\left[10 + 5h - \frac{1}{2}(4 + 4h + h^2)\right] - (10-2)}{h}$$

$$= \lim_{h \to 0} \frac{\left(10 + 5h - 2 - 2h - \frac{1}{2}h^2\right) - 8}{h} = \lim_{h \to 0} \frac{3h - \frac{1}{2}h^2}{h}$$

$$= \lim_{h \to 0} \frac{\left(3 - \frac{1}{2}h\right)h}{h} = \lim_{h \to 0} \left(3 - \frac{1}{2}h\right) = 3$$

Thus we conclude that the instantaneous velocity of the particle at time 2 seconds is 3 ft/s.

Example 3: Finding Instantaneous Velocity

A piece of rock is dropped from a height of 196 feet, and its height above ground level is given by the function $h(t)=196-16t^2$ feet, where t is measured in seconds. Find the instantaneous velocity of the rock at **a.** $t=1$ second, **b.** $t=2$ seconds, and **c.** $t=2.5$ seconds.

What is the velocity of impact?

Solution

a. As before, the instantaneous velocity at time $t=c$ can be found by finding the derivative of the position function at c.

$$
\begin{aligned}
h'(c) &= \lim_{h\to 0}\frac{\left[196-16(c+h)^2\right]-\left(196-16c^2\right)}{h} \\
&= \lim_{h\to 0}\frac{196-16\left(c^2+2ch+h^2\right)-196+16c^2}{h} \\
&= \lim_{h\to 0}\frac{-32ch-16h^2}{h}=\lim_{h\to 0}\frac{h(-32c-16h)}{h} \\
&= \lim_{h\to 0}(-32c-16h)=-32c
\end{aligned}
$$

We now obtain the instantaneous velocity at $t=c=1$ by a simple substitution.

$$h'(1)=-32(1)=-32 \text{ ft/s}$$

Notice that unlike in our solution of the previous example, rather than substituting $c=1$ at the outset, we carried c through the computation to obtain a formula for the instantaneous velocity, which we subsequently evaluated at $c=1$. The advantage of this method lies in the fact that now we can use the formula to answer both parts b. and c.

b. $h'(2)=-32(2)=-64$ ft/s

c. $h'(2.5)=-32(2.5)=-80$ ft/s

Finally, we proceed to find the velocity of impact. The fundamental observation is that impact happens precisely when the position of the rock becomes 0. This helps us determine exactly how many seconds after being dropped the rock hits the ground. Thus we solve the equation $h(t)=0$.

$$
\begin{aligned}
196-16t^2 &= 0 \\
16t^2 &= 196 \\
t^2 &= 12.25 \\
t &= \pm 3.5
\end{aligned}
$$

Since time is always positive, we will only consider the solution $t=3.5$ s.

Now we can determine the velocity of impact, which is precisely the instantaneous velocity at the time of impact, by evaluating h' at $c=3.5$. Notice that we can once again use the formula we derived before:

$$h'(3.5) = -32(3.5) = -112 \text{ ft/s},$$

and we conclude that the rock hits the ground with a velocity of -112 ft/s.

In conclusion, we note that the negative signs in the answers mean that the direction of velocity is pointing downward throughout the motion. Also, while our answer is close to what we would measure in an actual experiment, it is important to remember that we ignored air resistance in this problem.

TOPIC 2: The Derivative as a Function

It will frequently be the case that we need to determine the derivative of a function at more than one point (as in Example 3) or even at every point of a given interval. In such cases, it is often no more difficult to determine the derivative at a general point than at a specific point. If we are successful in doing so, then what we have achieved is a definition of another function—this new function, having been *derived* from our original function, is referred to as the *derivative* of the function.

DEFINITION: The Derivative of a Function

The **derivative of f**, denoted f', is the function whose value at the point x is

$$f'(x) = \lim_{h \to 0} \frac{f(x+h) - f(x)}{h},$$

provided the limit exists.

This definition is exactly the same as the definition of the derivative at a point—the only difference is that we are extending the definition to all points at which the limit exists.

In usage, names other than f and f' may be more appropriate for the functions under consideration. For instance, in a problem describing the position of an object, labels such as $x(t)$ (if the object is moving along a horizontal line) or $h(t)$ (if the height of a thrown object is being discussed) may be used for the position function. In such cases, it is common to name the derivative of the position function $v(t)$ for velocity. In other words, $v(t)$ represents the instantaneous velocity of the object at time t. It is now appropriate to make a distinction between the ideas of *velocity* and *speed*: if $v(t)$ represents the velocity of an object at time t, $s(t) = |v(t)|$ represents its **speed** at time t.

Example 4: Finding Speed

A soccer player kicks a ball vertically upward so that its position relative to ground level is $h(t) = -4.9t^2 + 19.6t + 1$ meters, where t is measured in seconds. Find the velocity and speed of the soccer ball at **a.** $t = 1$ second, **b.** $t = 2.5$ seconds, and **c.** $t = 3.5$ seconds.

What is this ball's speed of impact upon its return? (Ignore air resistance.)

Solution

a. Note first of all that the position function realistically reflects the fact that the ball does not start from ground level. Its initial height can be found by substituting $t = 0$ into the position function.

$$h(0) = -4.9(0)^2 + 19.6(0) + 1 = 1 \text{ meter}$$

In order to answer the questions, next we proceed to find $v(t)$, the velocity function of the ball, which is the derivative of $h(t)$. Notice that we are using the alternate notation of Δt instead of h for the increment, so as not to cause confusion with the name of the position function.

$$\begin{aligned}
v(t) = h'(t) &= \lim_{\Delta t \to 0} \frac{h(t+\Delta t) - h(t)}{\Delta t} \\
&= \lim_{\Delta t \to 0} \frac{-4.9(t+\Delta t)^2 + 19.6(t+\Delta t) + 1 - \left(-4.9t^2 + 19.6t + 1\right)}{\Delta t} \\
&= \lim_{\Delta t \to 0} \frac{-4.9t^2 - 9.8t\Delta t - 4.9(\Delta t)^2 + 19.6t + 19.6\Delta t + 1 + 4.9t^2 - 19.6t - 1}{\Delta t} \\
&= \lim_{\Delta t \to 0} \frac{\Delta t(-9.8t - 4.9\Delta t + 19.6)}{\Delta t} \\
&= \lim_{\Delta t \to 0} (-9.8t - 4.9\Delta t + 19.6) \\
&= -9.8t + 19.6
\end{aligned}$$

Already in possession of the velocity function, it is now easy to find the ball's instantaneous velocity at $t = 1$.

$$v(1) = -9.8(1) + 19.6 = 9.8 \text{ m/s}$$

Note that the speed equals the velocity this time, since the latter is positive.

$$s(1) = |v(1)| = |9.8| = 9.8 \text{ m/s}$$

Notice also that, as one would expect, our function indicates that the ball is slowing down significantly from its initial velocity, which can be found by evaluating

$$v(0) = -9.8(0) + 19.6 = 19.6 \text{ m/s}.$$

b. Substituting again in the velocity function, we find

$$v(2.5) = -9.8(2.5) + 19.6 = -4.9 \text{ m/s}.$$

The negative sign shows that the ball turned back and is now traveling downward. Its speed, however, being independent of direction, is still positive.

$$s(2.5) = |v(2.5)| = |-4.9| = 4.9 \text{ m/s}$$

c. We expect the ball to accelerate as it falls, and that is exactly what we find upon substituting $t = 3.5$ into the velocity function.

$$v(3.5) = -9.8(3.5) + 19.6 = -14.7 \text{ m/s}$$

The negative sign shows that the direction of velocity is still downward, but its absolute value, the ball's speed, has increased.

$$s(3.5) = |v(3.5)| = |-14.7| = 14.7 \text{ m/s}$$

In fact, as it is with any free-falling body when air resistance is negligible, gravity increases the ball's speed by approximately 9.8 m/s every second. (Can you see this from our results?)

To answer the final question regarding the speed of impact, we first need to know when it happens, and then we substitute the appropriate t-value into the velocity function. Since height is 0 upon impact, we can find t by solving $h(t) = 0$, that is,

$$-4.9t^2 + 19.6t + 1 = 0.$$

An application of the quadratic formula yields the positive solution $t \approx 4.05$ seconds. Therefore, the speed of impact is approximately

$$s(4.05) = |v(4.05)| = |-9.8(4.05) + 19.6| = |-20.09| = 20.09 \text{ m/s}.$$

Example 5: Finding Derivatives and Tangent Lines to Graphs

Find the derivative of $f(x) = \sqrt{x}$, and use it to determine equations of the tangent lines to the graph of f at the points $(1, 1)$ and $(2, \sqrt{2})$.

Solution

We will start with the definition of the derivative, and make use of some of the limit-determination techniques learned in Section 13.4.

$$\begin{aligned}
f'(x) &= \lim_{h \to 0} \frac{f(x+h) - f(x)}{h} \\
&= \lim_{h \to 0} \frac{\sqrt{x+h} - \sqrt{x}}{h} \\
&= \lim_{h \to 0} \frac{\left(\sqrt{x+h} - \sqrt{x}\right)\left(\sqrt{x+h} + \sqrt{x}\right)}{h\left(\sqrt{x+h} + \sqrt{x}\right)} \\
&= \lim_{h \to 0} \frac{(x+h) - x}{h\left(\sqrt{x+h} + \sqrt{x}\right)} = \lim_{h \to 0} \frac{h}{h\left(\sqrt{x+h} + \sqrt{x}\right)} \\
&= \lim_{h \to 0} \frac{1}{\sqrt{x+h} + \sqrt{x}} = \frac{1}{\sqrt{x} + \sqrt{x}} = \frac{1}{2\sqrt{x}}
\end{aligned}$$

To find the equation of the tangent line at $(1,1)$, we first determine its slope by evaluating the derivative $f'(x) = \dfrac{1}{2\sqrt{x}}$ at $x = 1$.

$$m = f'(1) = \frac{1}{2\sqrt{1}} = \frac{1}{2}$$

Now the point-slope equation comes to bear, and we obtain

$$y-1=\frac{1}{2}(x-1), \quad \text{or} \quad y=\frac{1}{2}x+\frac{1}{2}.$$

We deal with the equation of the tangent line at $\left(2,\sqrt{2}\right)$ in a similar manner. Since

$$f'(2)=\frac{1}{2\sqrt{2}}=\frac{\sqrt{2}}{4},$$

the requested equation is

$$y-\sqrt{2}=\frac{\sqrt{2}}{4}(x-2), \quad \text{or} \quad y=\frac{\sqrt{2}}{4}x+\frac{\sqrt{2}}{2}.$$

Notice that we can use f' to find the slope of the tangent at *any* given point $x=c$, if it exists, by simply evaluating $f'(c)$.

Recall that in business, $C(x)$ is often used to represent the total cost of producing x items of a certain product, and is referred to as the cost function. Similarly, $R(x)$ stands for the total revenue when x items are sold, this latter being called the revenue function. Consequently, the profit function $P(x)$ satisfies $P(x)=R(x)-C(x)$; or in words, profit equals total revenue minus total cost. The corresponding derivatives, or rates of change, the functions $C'(x)$, $R'(x)$, and $P'(x)$, are called **marginal cost**, **marginal revenue**, and **marginal profit**, respectively. As an application, the following example involves some of the these functions.

Example 6: Finding Break-Even Points and Marginal Profit

If a table manufacturer has a cost function of $C(x)=225+2x^2$ along with its revenue function of $R(x)=117x-\frac{1}{4}x^2$, find **a.** all break-even points and **b.** the marginal profit when $x=10$, $x=20$, and $x=30$.

Solution

a. Break-even points occur when $C(x)=R(x)$, in other words, when revenue levels equal the total cost invested. This leads to the equation

$$225+2x^2=117x-\frac{1}{4}x^2.$$

After rearranging terms and factoring, this leads to

$$\frac{9}{4}(x-2)(x-50)=0,$$

thus we conclude that break-even points occur when $x=2$ and $x=50$ (see Figure 4).

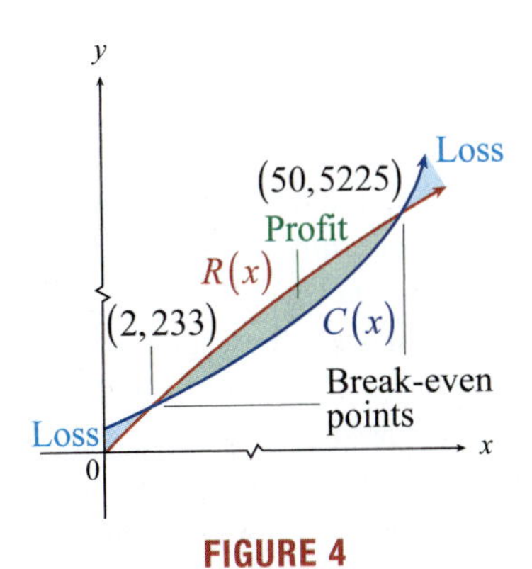

FIGURE 4

b. Since marginal profit is the rate of change of profit, we must first find the profit function and then calculate its derivative.

$$\begin{aligned} P(x) &= R(x) - C(x) \\ &= \left(117x - \frac{1}{4}x^2\right) - \left(225 + 2x^2\right) \\ &= 117x - \frac{1}{4}x^2 - 225 - 2x^2 \\ &= -\frac{9}{4}x^2 + 117x - 225 \end{aligned}$$

Next, we will use the definition of the derivative to obtain the marginal profit function.

$$\begin{aligned} P'(x) &= \lim_{h\to 0}\frac{P(x+h)-P(x)}{h} \\ &= \lim_{h\to 0}\frac{-\frac{9}{4}(x+h)^2 + 117(x+h) - 225 - \left(-\frac{9}{4}x^2 + 117x - 225\right)}{h} \end{aligned}$$

After expanding, rearranging terms, and canceling, this reduces to

$$\begin{aligned} P'(x) &= \lim_{h\to 0}\frac{h\left(-\frac{9}{2}x + 117 - \frac{9}{4}h\right)}{h} \\ &= \lim_{h\to 0}\left(-\frac{9}{2}x + 117 - \frac{9}{4}h\right) \\ &= -\frac{9}{2}x + 117. \end{aligned}$$

Now we can calculate the various marginal profits.

$$P'(10) = -\frac{9}{2}(10) + 117 = -45 + 117 = \$72$$

$$P'(20) = -\frac{9}{2}(20) + 117 = -90 + 117 = \$27$$

$$P'(30) = -\frac{9}{2}(30) + 117 = -135 + 117 = -\$18$$

In conclusion, an important remark is in order. Since the number of tables produced is always an integer, the increment h in any difference quotient should be thought of as an integer, so the smallest nonzero h possible is $h = 1$. This explains the following interpretation of marginal profit: $P'(x)$ is an estimate for the increase in $P(x)$ if x is increased by 1. Note also that in this example, $P'(x)$ is actually getting smaller as x gets larger. This shows that the rate of growth of profit per table is actually decreasing at a rate of \$18 per table when 30 tables are manufactured and sold. This does not necessarily mean there is a loss, but as production increases, the profit per table is growing less because costs are increasing faster than revenue.

13.6 EXERCISES

PRACTICE

Use the graph to estimate the derivative at the given points.

1. a. $x_1 = -1$ **b.** $x_2 = 1$

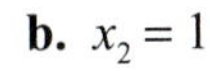

2. a. $x_1 = -2$ **b.** $x_2 = 0$

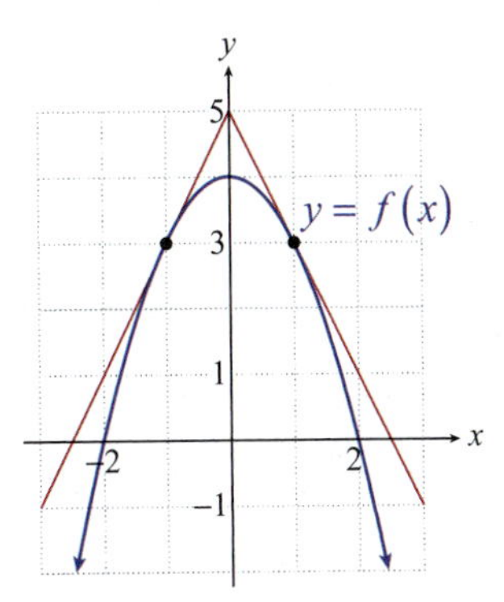

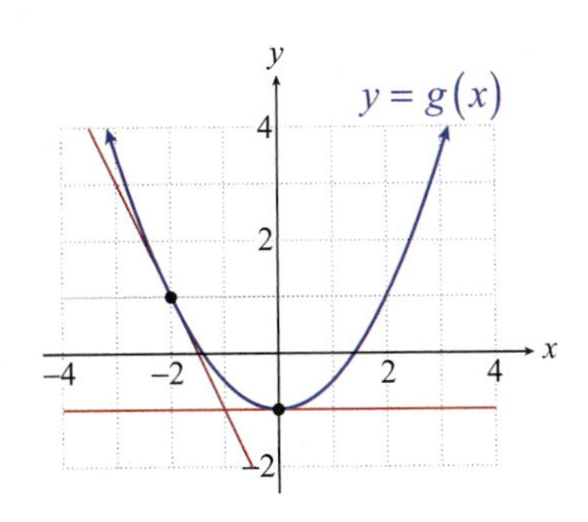

Find the equation of the tangent line to the graph of $f(x)$ at the given point. See Example 1.

3. $f(x) = x^2 - 2;\ (2,2)$

4. $f(x) = 3x - 2x^2;\ (-1,-5)$

5. $f(x) = \frac{1}{2}x + 4;\ (2,5)$

6. $f(x) = 1 - 5x;\ (0,1)$

7. $f(x) = x^3;\ (2,8)$

8. $f(x) = 5x - 2x^3;\ (-1,-3)$

9. $f(x) = \sqrt{x+1};\ (0,1)$

10. $f(x) = 2\sqrt{1-3x};\ (-1,4)$

11. $f(x) = \frac{1}{x};\ \left(\frac{1}{2},2\right)$

12. $f(x) = \frac{5}{1-2x};\ \left(-1,\frac{5}{3}\right)$

13. $f(x) = \frac{1}{\sqrt{x}};\ \left(4,\frac{1}{2}\right)$

14. $f(x) = \frac{2}{\sqrt{x+1}};\ (3,1)$

Use the definition (also called the *limit process*) to find the derivative function f' of the given function f. Find all x-values (if any) where the tangent line is horizontal. See Example 5.

15. $f(x) = 2$

16. $f(x) = 2x$

17. $f(x) = 4x + 5$

18. $f(x) = 3 - \frac{2}{5}x$

19. $f(x) = 3x^2$

20. $f(x) = 4 - 2x^2$

21. $f(x) = \frac{1}{2}x^2 + 5x - 7$

22. $f(x) = x - \frac{1}{3}x^2$

23. $f(x) = x^3 + x$

24. $f(x) = 7 + x - 3x^2 + x^3$

25. $f(x) = x^4$

26. $f(x) = \frac{1}{2x}$

27. $f(x) = \frac{5}{2x-4}$

28. $f(x) = \frac{x-2}{x+2}$

29. $f(x)=\frac{2x+1}{x-3}$

30. $f(x)=\frac{2}{x^2}$

31. $f(x)=\frac{1}{x^2+1}$

32. $f(x)=\frac{2}{x^2-2x}$

33. $f(x)=\sqrt{5x}$

34. $f(x)=\frac{1}{\sqrt{5x}}$

35. $f(x)=\sqrt{2x+1}$

36. $f(x)=\frac{1}{\sqrt{x-2}}$

37. $f(x)=\sqrt{x^2+1}$

38. $f(x)=\frac{1}{\sqrt{x^2+1}}$

Find the equation of a tangent line to the graph of the function that is parallel to the given line.

39. $f(x)=x^2+3;\quad y-6x+1=0$

40. $g(x)=2x-x^2;\quad y-5=4x$

41. $h(x)=\frac{1}{2x};\quad x+2y=3$

42. $F(x)=\frac{1}{x-3};\quad y+4x+7=0$

43. $G(x)=\frac{1}{\sqrt{x}};\quad 54y+x=1$

44. $H(x)=\frac{1}{\sqrt{x^2-7}};\quad 27y+4x-2=0$

Use the alternate form of the definition of the derivative $f'(c)=\lim\limits_{x\to c}\frac{f(x)-f(c)}{x-c}$ to evaluate the given slope.

45. $f(x)=5-\frac{1}{4}x;\quad f'(3.6)$

46. $g(x)=x^2+1;\quad g'(-1)$

47. $h(x)=(x+2)^2;\quad h'(3)$

48. $F(t)=\frac{1}{t-3};\quad F'(2)$

49. $G(x)=\frac{2}{5-x};\quad G'(7)$

50. $k(t)=\sqrt{t+5};\quad k'(11)$

51. $u(x)=2\sqrt{1-x};\quad u'(-3)$

52. $v(x)=\frac{1}{x^2+1};\quad v'(0)$

53. $w(s)=\frac{1}{\sqrt{s+4}};\quad w'(5)$

54. $F(t)=t^3-t;\quad F'(1)$

55. $G(s)=s^4;\quad G'(-2)$

56. $H(x)=\frac{2}{\sqrt{x^2+1}};\quad H'(0)$

Match the graph of f with the graph of its derivative f' (labeled a–d).

57.

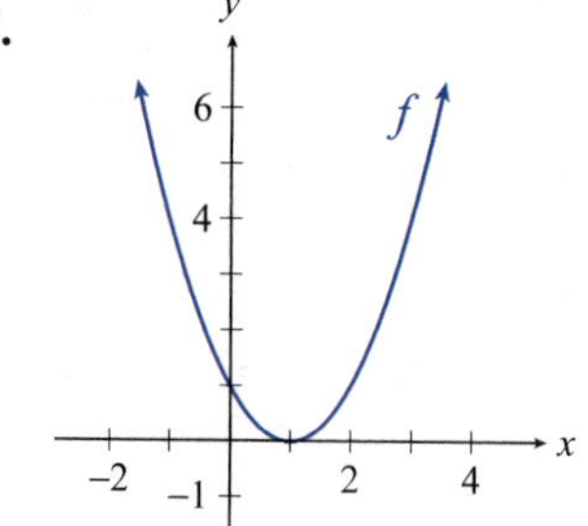

58.

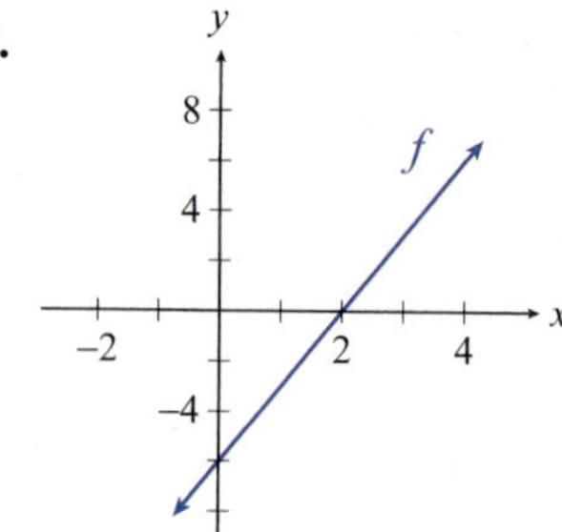

59.

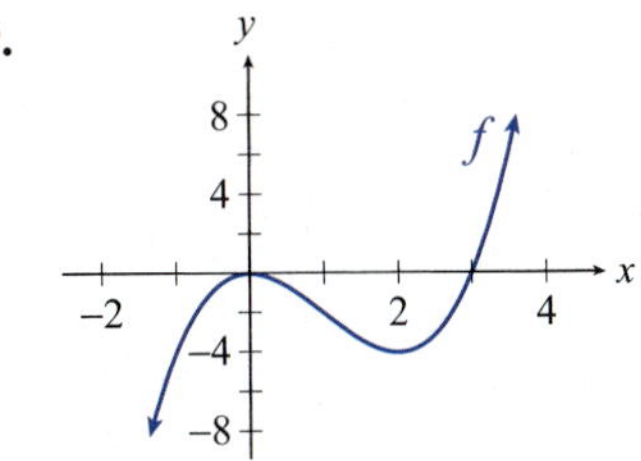

60.

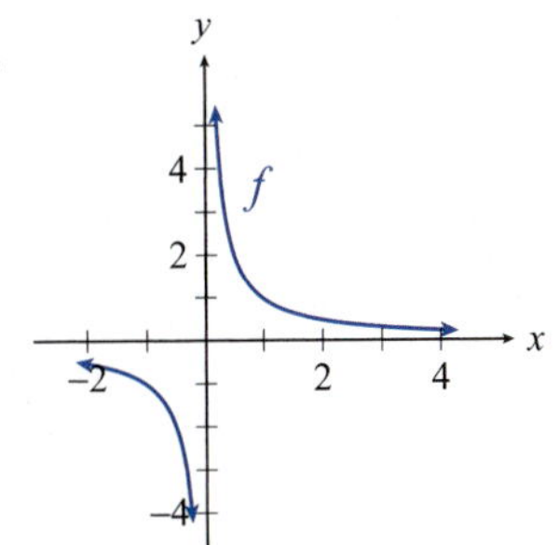

a.

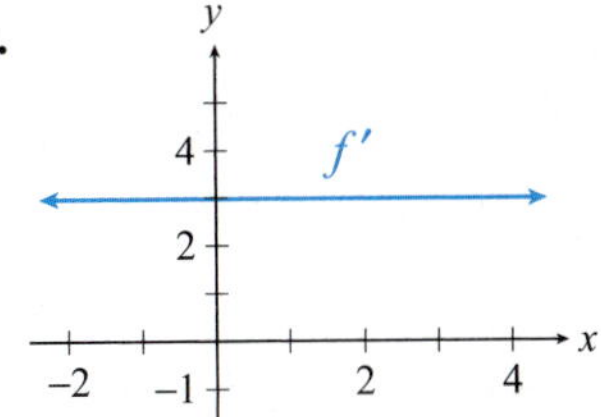

b.

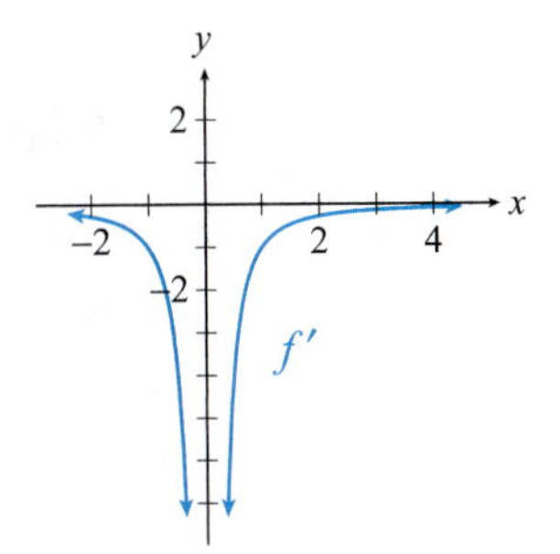

c.

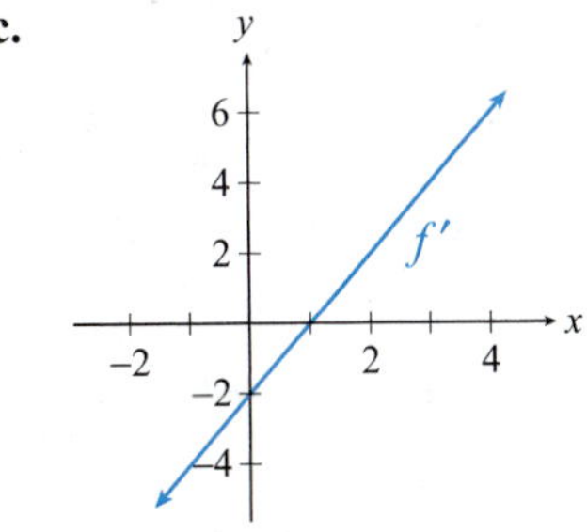

d.

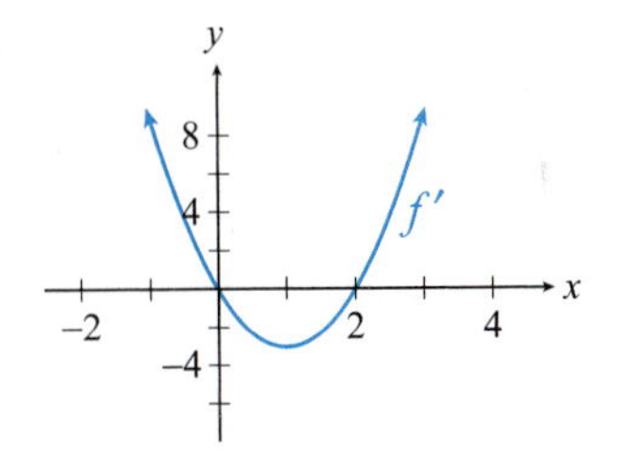

WRITING & THINKING

For Exercises 61–65, sketch the graph of a function *f* possessing the given characteristics. (A formula is useful, but not necessary.)

61. $f(0)=1,\ f'(0)=0,\ f'(x)<0$ for $x<0$, $f'(x)>0$ for $x>0$

62. $f(1)=0,\ f'(1)=0,\ f'(x)\geq 0$ on the entire real line

63. $f(x)>0$ on the entire real line, $f'(x)<0$ on the entire real line

64. $f(1)=1,\ f'(1)=-1,\ f'$ is nonzero on the entire real line

65. $f(1)=5,\ f'(x)=5$ on the entire real line

66. Prove that if $f(x)=c$ (a constant function), then $f'(x)=0$.

67. Use the definition of the derivative to prove that if $f(x)=x$, then $f'(x)=1$.

68. Generalize Exercise 67 to prove that if $f(x)$ is a linear function, then $f'(x)$ is constant.

69. Use the definition of the derivative to prove that if $f(x) = x^n$ for a positive integer n, then $f'(x) = nx^{n-1}$.

70. Recall from Section 4.2 that a function f is even if $f(-x) = f(x)$ and odd if $f(-x) = -f(x)$ throughout its domain. Prove that the derivative of an even function is odd and, vice versa, an odd function has an even derivative.

71. Find the equation of the line tangent to the graph of $f(x) = \frac{1}{x}$ at the point $(c, f(c))$. Prove that the area of the triangle bounded by the tangent line and the coordinate axes is the same for all $c \neq 0$.

APPLICATIONS

72. The position function of a moving particle is given by $p(t) = t^2 - 3t + 1$ feet at t seconds. Find all points in time where the particle's speed is 1 ft/s. When does it come to a momentary stop?

73. Repeat Exercise 72 with the position function $p(t) = \frac{1}{9}t^3 - t^2 + \frac{8}{3}t$.

74. A baseball is hit vertically upward with an initial speed of 80 ft/s. When does it slow down to 32 ft/s? How high does it go and how long is it aloft? (**Hint:** Use the position function $h(t) = -16t^2 + 80t$. Ignore air resistance.)

75. A rock is thrown upward from the edge of a 150 ft high cliff with an initial velocity of 48 ft/s.

a. Calculate the velocity and speed of the rock when it is exactly 32 ft above the person's hand.
b. How high does it go and when does it reach the bottom of the cliff?
c. What is the velocity of impact?

(**Hint:** Use $h(t) = -16t^2 + 48t + 150$ as the position function, where h is in feet, t in seconds. Ignore air resistance.)

76. A package is dropped from a small airplane 122.5 meters above the Earth. If we ignore air resistance, how much time does the package need to reach the ground and what is the speed of impact? (**Hint:** The position function is $h(t) = -4.9t^2 + 122.5$ meters, where t is measured in seconds. Use $g \approx 9.81 \text{ m/s}^2$.)

77. The following graph is a position function of a student's car relative to her home as she drove to class one morning. From the graph, recreate a possible story of her trip, mentioning distance, velocity, speed, and so forth.

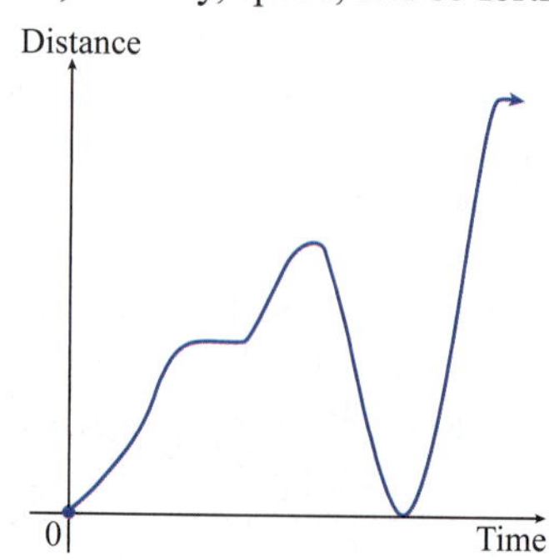

78. A manufacturer has determined that the revenue from the sale of x cordless telephones is given by $R(x) = 94x - 0.03x^2$ dollars. The cost of producing x telephones is $C(x) = 10{,}800 + 34x$ dollars.

a. Find the profit function $P(x)$ and any break-even points.

b. Find $P(200)$, $P(400)$, and $P(600)$.

c. Find the marginal profit function $P'(x)$.

d. Find $P'(200)$, $P'(400)$, and $P'(600)$.

79. The owner of a leather retailer has determined that he can sell x attaché cases if the price is $p = D(x) = 46 + 0.25x$ dollars ($D(x)$ is often called the demand function). The total cost for these cases is $C(x) = 0.15x^2 + 6x + 190$ dollars.

a. Find the profit function $P(x)$. (**Hint:** Find the revenue function $R(x)$ first.)

b. Find any break-even points.

c. Find $P(25)$, $P(30)$, and $P(40)$.

d. Find the marginal profit function $P'(x)$.

e. Find $P'(25)$, $P'(30)$, and $P'(40)$.

80. The average cost $\overline{C}(x)$ of manufacturing x units of a certain product is $\frac{C(x)}{x}$, where $C(x)$ is the total cost function.

a. Find the average cost function if $C(x) = 30 + 2x + 0.003x^2$.

b. What is the rate of change of average cost?

c. What value of x results in a minimum average cost? (**Hint:** Use the fact that when average cost is a minimum, its rate of change is 0. Alternatively, use a graphing utility to graph $C(x)$ for $x \geq 0$ and zoom in on the lowest point.)

81. The average manufacturing cost function of a product is given by $\overline{C}(x) = 20x^{-1} + 3$. Determine the cost function and the marginal cost function for the product. (**Hint:** See Exercise 80.)

TECHNOLOGY

82–105. Referring back to the functions given in Exercises 15–38, use a graphing utility to sketch the graph of f along with that of f' in the same viewing window. Compare the graphs and describe their relationship.

CHAPTER 13 PROJECT

Rideshare Function

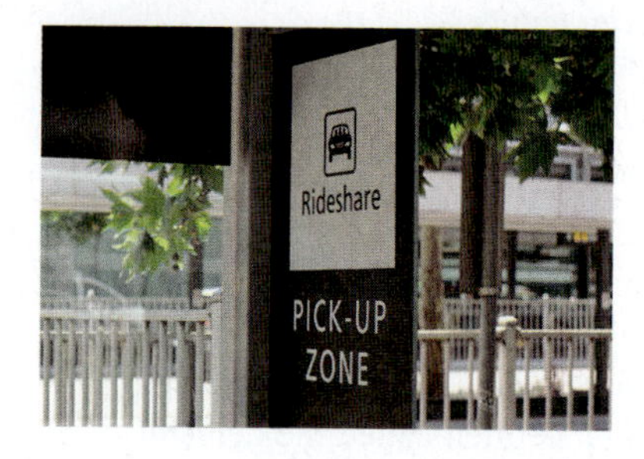

A certain rideshare company charges \$3.00 to pick up a passenger and an additional \$0.90 for each mile or fraction of a mile that the passenger is driven.

1. Graph a function $C(x)$ that reflects the cost to a passenger for a ride of x miles over the x-interval $[0,5]$.

2. Find and classify all points of discontinuity of the function C.

3. If it exists, determine $\lim_{x\to1.7} C(x)$. If the limit does not exist, determine the one-sided limits, if possible.

4. If it exists, determine $\lim_{x\to2} C(x)$. If the limit does not exist, determine the one-sided limits, if possible.

5. If it exists, determine $C'(1.7)$.

6. Is $C'(x)$ defined for every nonnegative x? If not, where is $C'(x)$ not defined?

7. What is the first interval of distance over which a passenger can ride and pay their fare in dollars with no change necessary? How much will the fare be?

CHAPTER 13 REVIEW EXERCISES

Section 13.1

Construct and simplify the difference quotient at c with increment h.

1. $f(x) = 5x - 7$

2. $g(x) = 3x^2 + x$

3. $h(x) = \dfrac{1}{2x+2}$

4. $f(x) = 9x^2 + 5$

Use the difference quotient of each function for one appropriate value of c to determine the average rate of change of the function over each of the given intervals.

5. $f(x) = 5x - 7$, interval $= [1, 1.1]$

6. $g(x) = 3x^2 + x$, interval $= [0.9, 1]$

7. $h(x) = \dfrac{1}{2x+2}$, interval $= [-0.01, 0]$

8. $f(x) = 9x^2 + 5$, interval $= [2, 2.01]$

Use difference quotients to approximate the slope of the tangent to the graph of the function at the given point. Use at least five different h-values that are decreasing in magnitude. (Answers will vary.)

9. $f(x) = 3x + 2;\ (0, 2)$

10. $g(x) = 2 - x^2;\ (1, 1)$

11. $h(x) = \sqrt{x-1};\ (2, 1)$

12. $k(x) = \sin x;\ (0, 0)$

13. A pellet is shot vertically upward from an initial height of 6 feet. Its height after t seconds is given by $h(t) = 6 + 608t - 16t^2$ feet. Use difference quotients to answer the questions below.

a. What will be the pellet's height at the end of the first second?

b. What is the average velocity of the pellet during the first two seconds?

c. Estimate the instantaneous velocity at $t = 0$ seconds.

d. Estimate the instantaneous velocity at $t = 2$ seconds.

e. When will the velocity be 0?

Section 13.2

Use the graph of the function to find the indicated (possibly one-sided) limits, if they exist.

14. $\lim\limits_{x \to 1} f(x)$

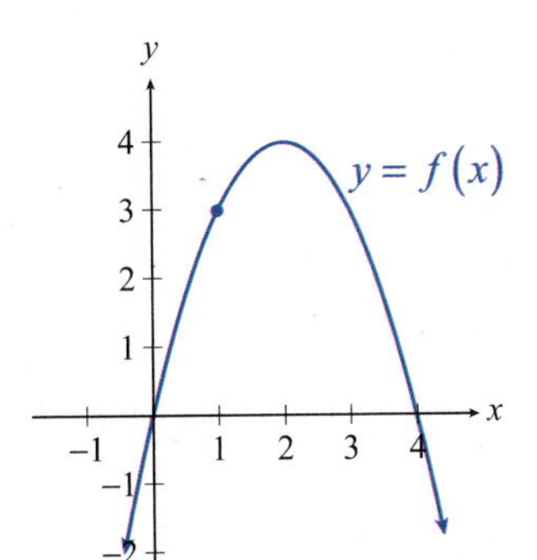

15. $\lim\limits_{x \to 1} g(x)$

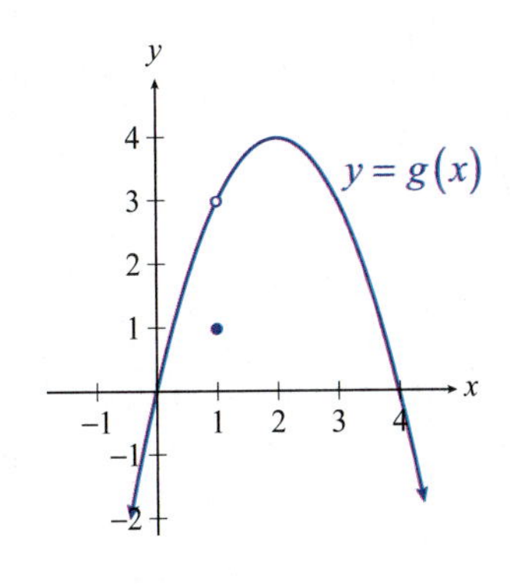

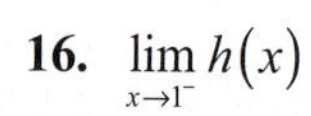

16. $\lim_{x\to1^-} h(x)$

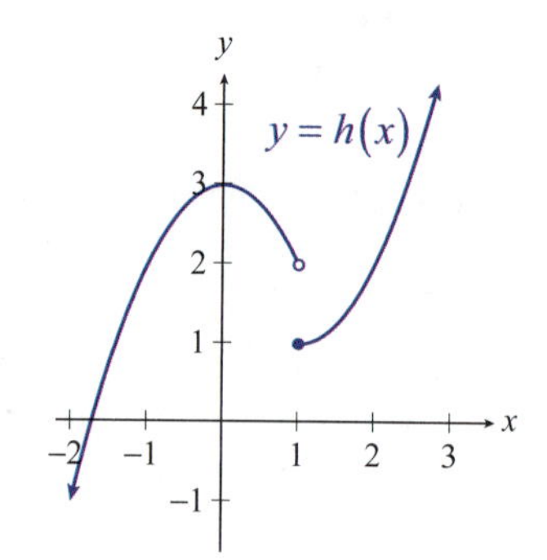

17. $\lim_{x\to-1^+} k(x)$

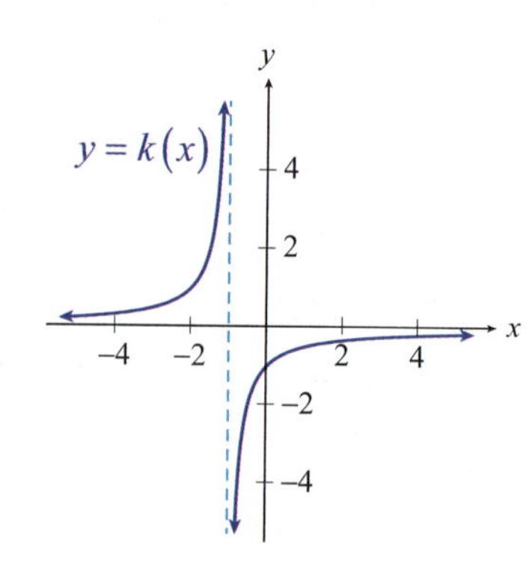

Create a table of values to estimate the value of the indicated limit without graphing the function. Choose the last *x*-value so that it is no more than 0.001 units from the given *c*-value.

18. $\lim_{x\to1} \dfrac{x^3-1}{x-1}$

19. $\lim_{x\to0} x^x$

20. $\lim_{x\to0} \dfrac{\sin(2x)}{4x}$

21. $\lim_{x\to0}\left(2x\sin\left(\dfrac{1}{4x}\right)\right)$

22. Use one-sided limit notation to describe the behavior of $f(x) = \dfrac{1}{x-1}$ near $x = 1$.

Section 13.3

Find a $\delta > 0$ that satisfies the limit claim corresponding to $\varepsilon = 0.01$; that is, such that $0 < |x - c| < \delta$ would imply $|f(x) - L| < 0.01$.

23. $\lim_{x\to0}(3-2x) = 3$

24. $\lim_{x\to4}\sqrt{x} = 2$

Give the formal definition of the limit claim. Then use the definition to prove the claim.

25. $\lim_{x\to1}(3x+1) = 4$

26. $\lim_{x\to1} x^2 = 1$

27. $\lim_{x\to1}\sqrt{x} = 1$

28. $\lim_{x\to2}\dfrac{2}{x} = 1$

Section 13.4

Use algebra and/or appropriate limit laws to evaluate the given limit (one-sided limit where indicated). If the limit is unbounded, use the symbol ∞ or $-\infty$ in your answer.

29. $\lim_{x\to3}(2x^2-3x+5)$

30. $\lim_{x\to-2}\left(\dfrac{x^3}{4}+2x^2-x+1\right)$

31. $\lim_{x\to3}\sqrt{x^3+2x^2+4}$

32. $\lim_{x\to-2}\dfrac{2x+1}{x^2-x}$

33. $\lim_{t\to1}\left(\dfrac{3t+5t^3}{t^2+1}\right)^{\frac{3}{2}}$

34. $\lim_{x\to4}\dfrac{x^2-16}{x-4}$

35. $\lim_{x\to-5}\dfrac{x+5}{x^2-25}$

36. $\lim_{x\to5^-}\dfrac{x+5}{x^2-25}$

37. $\lim_{x\to1^+}\frac{x^2+1}{x^4-1}$

38. $\lim_{x\to1}\frac{x^2-1}{x^4-1}$

39. $\lim_{x\to3}\frac{\sqrt{x+1}-2}{x-3}$

40. $\lim_{x\to0}\frac{\frac{1}{2+x}-\frac{1}{2}}{x}$

41. $\lim_{x\to0^-}\frac{2|x|}{x}$

42. $\lim_{x\to-2^+}\sqrt{4-x^2}$

43. $\lim_{x\to2^-}\left([\![x]\!]+2x\right)$

44. $\lim_{x\to1^+}\left([\![x]\!]x\right)$

45. If $f(x)=x^2$, find $\lim_{h\to0}\frac{f(x+h)-f(x)}{h}$.

Use the Squeeze Theorem to prove the limit claim.

46. $\lim_{x\to0}\left(x\cos\left(\frac{1}{x}\right)\right)=0$

47. $\lim_{x\to\infty}\frac{\sin x}{\ln x}=0$

Section 13.5

48. Sketch a graph of a function (a formula is not necessary) that is not continuous at $x=0$ from either direction, but both of its one-sided limits exist at $x=0$.

49. Sketch a graph of a function that is left-continuous at $x=0$, but its right-hand limit at $x=0$ doesn't exist.

Find and classify the discontinuities (if any) of the function as removable or nonremovable.

50. $f(x)=\frac{x-9}{\sqrt{x}-3},\quad x\ge0$

51. $g(x)=\frac{\sqrt{x}+2}{x-4},\quad x\ge0$

52. $h(x)=\frac{1}{\sqrt{x^2+1}}$

53. $t(x)=2+2[\![x]\!]$

54. $G(x)=\frac{x}{\sqrt{x+1}-1},\quad x\ge-1$

55. $k(x)=|x-3|+|x+1|$

Use the ε-δ definition to prove that the function is continuous.

56. $f(x)=3x-1$

57. $g(x)=2x^2$

58. Find the values of a and b such that f is continuous on the entire real line.

$$f(x)=\begin{cases}-1 & \text{if } x\le-3\\ ax+b & \text{if } -3<x<2\\ x^2 & \text{if } x\ge2\end{cases}$$

59. Use the Intermediate Value Theorem to prove that the equation $2x^5 + x + 1 = 0$ has a solution on the interval $[-1,1]$.

60. Use the Intermediate Value Theorem to show that the graphs of $f(x) = x^3$ and $g(x) = e^{-x}$ intersect.

Section 13.6

Find the equation of the tangent line to the graph of $f(x)$ at the given point.

61. $f(x) = x^2 + x;\ \ (1,2)$

62. $f(x) = \sqrt{x};\ \ (4,2)$

Use the definition (also called the *limit process*) to find the derivative function f' of the given function f. Find all x-values (if any) where the tangent line is horizontal.

63. $f(x) = 2x - x^2$

64. $f(x) = \dfrac{3}{x-2}$

For Exercises 65–66, sketch the graph of a function f possessing the given characteristics. (A formula is useful, but not necessary.)

65. f is continuous at 0, $f(0) = 1$, $f'(x) < 0$ for $x < 0$, $f'(x) > 0$ for $x > 0$, and $f'(0)$ does not exist

66. $g(1) < 0$, $g'(1) > 0$, and $g(2) > 0$, but $g'(2) < 0$

67. Prove that if $f(x)$ is a quadratic function, then $f'(x)$ is linear.

68. A small object is thrown upward with an initial velocity of $12\ \text{m/s}$ from the top of a 15 m high building.

a. How high does it go and when does it reach the ground?
b. What is the speed of impact?

(Hint: Use $h(t) = -5t^2 + 12t + 15$ as the position function, where h is in meters, t in seconds.)

69. The owner of a small toy manufacturer has determined that he can sell x toys if the price is $p = D(x) = 0.2x + 30$ dollars. The total cost as a function of x is given by $C(x) = 0.1x^2 + 15x + 247.5$ dollars.

a. Find the profit function $P(x)$.
b. Find any break-even points.
c. Find the marginal profit function.

Answer Key

Chapter 1: Algebraic Expressions, Equations, and Inequalities

1.1 Exercises

1. a. $19, 2^5$ **b.** $19, \frac{0}{15}, 2^5$

c. $19, \frac{0}{15}, 2^5, -33$

d. $19, -4.3, \frac{0}{15}, 2^5, -33$

e. $-\sqrt{3}$ **f.** All

3.  −4.5 −1 2.5

5. 5.1 5.2 5.8

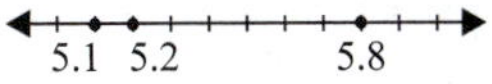

7. $<, \leq$ **9.** $\leq, \geq$

11. $>, \geq$ **13.** $2a+b>c$

15. $2c \leq 3d$

17. $\{n \mid n \text{ is an integer and } 5 \leq n \leq 105\}$

19. $\{2^n \mid n \text{ is a whole number}\}$

21. $\left\{\frac{1}{n} \middle| n \text{ is an odd integer}\right\}$

23. $[-3, 19)$ **25.** $(-\infty, 15)$

27. $(0, \infty)$

29.  5 14

31. 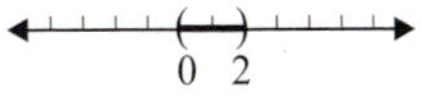 0 2

33. 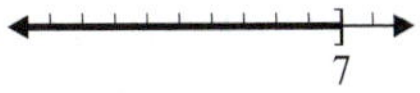 7

35. $[-7, 7)$ **37.** $\varnothing$

39. $(-\infty, \infty)$ **41.** $\mathbb{Z}$

43. $\mathbb{Z}$ **45.** -11

47. $\sqrt{5}-\sqrt{3}$ **49.** 1

51. -15 **53.** 5

55. 6

57. $3x^2y^3, -2\sqrt{x+y}, 7z$

59. $1, 8.5, -14$

61. $\frac{-5x}{2yz}, -8x^5y^3, 6.9z$

63. -98 **65.** $6\pi+8$

67. -20 **69.** Commutative

71. Distributive **73.** Associative

75. Multiplicative cancellation; $\frac{1}{5}$

77. Additive cancellation; x

79. Multiplicative cancellation; 6

81. Additive cancellation; $-2x+y$

83. 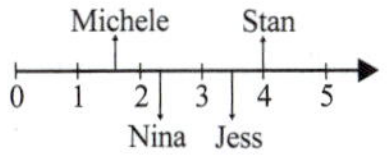 Michele Stan 0 1 2 3 4 5 Nina Jess

85. 20 miles

87. If sugar $= s$,

$\{s \mid 3 \leq s \leq 4\} = [3, 4]$

If walnuts $= w$,

$\left\{w \middle| \frac{1}{2} \leq w \leq \frac{2}{3}\right\} = \left[\frac{1}{2}, \frac{2}{3}\right]$

89. 131.11 °C

91. Derek's BMI is 23.7 and his weight status is normal.

93. 187 ft

95. The union contains the citizens with brown hair or blue eyes and the intersection contains the citizens with brown hair and blue eyes.

97. Yes, all whole numbers are also integers, but the negative integers are not whole numbers, so all integers are not whole numbers.

1.2 Exercises

1. 16 **3.** 81

5. x^3 **7.** $\frac{3}{t^5}$

9. $2n^8$ **11.** $\frac{y^2}{s^3z^{11}}$

13. $\frac{1}{9m^2-8n^6}$ **15.** $\frac{25m^8}{n^4}$

17. $27x^9$ **19.** $\frac{1}{5z^6-81x^{12}}$

21. -9.12×10^8 **23.** 32,000,000

25. 3.1536×10^7 **27.** 0.0003937

29. 2.8×10^{-3} **31.** -11

33. 1×10^{-8} **39.** -3

41. $-\frac{3}{5}$ **43.** 2

45. $\frac{2}{5}$ **47.** $|x^3|\sqrt{2y}$

49. $\frac{1}{2|xy^3|}$ **51.** $\frac{|x^3|y^2}{2}$

53. $2xy^2\sqrt[5]{x^2}$ **55.** $-\frac{|a|\sqrt{2}}{2}$

57. $\sqrt{6}+\sqrt{3}$ **59.** $\frac{x+2\sqrt{xy}+y}{x-y}$

61. $\frac{1}{\sqrt{13}-\sqrt{t}}$ **63.** $3x\sqrt[3]{2x}$

65. Not possible **67.** $5x^2\sqrt[5]{x^3}$

69. $|x^3|$ **71.** $(3x^2-4)^2$

73. 4 **75.** $y\sqrt[3]{y^2}$

77. $\frac{1}{64}$ **79.** $6^{-\frac{1}{3}}$

81. $\sqrt[4]{125}$ **83.** y^4

89. $\pi r^2 h$ **91.** 585 m^3

93. $14(N+M)$ ft^2

95. $bh+bl+hl+l\sqrt{b^2+h^2}$

97. 1651 cm^2; no

103. Because a root is the same as a fractional exponent.

1.3 Exercises

1. Not a polynomial

3. Degree 11; polynomial of four terms

5. Degree 0 monomial

7. Degree 4 binomial

9. Degree 2 trinomial

11. Degree 5 binomial

13. $-x^{13}+7x^{11}-4x^{10}+9$

a. 13 **b.** -1

15. $2s^6-10s^5+4s^3$ **a.** 6 **b.** 2

17. $9y^6-3y^5+y-2$ **a.** 6 **b.** 9

19. πz^5+8z^2-2z+1 **a.** 5 **b.** π

21. $-4x^3y-6y-x^2z$

23. $x^2y+xy^2+6x-6y$

25. $-3ab$

27. xy^2-x^2y-y

29. $3a^3b^3+21a^3b^2+2a^2b^2+14a^2b-3ab^3-21ab^2$

31. $3a^2-2ab-8b^2$

33. $6x^2+33xy-18y^2$

35. $7y^4-34xy^2-5x^2$

37. $6x^3y^3-3x^3y+36x^2y^3+4x^2y^2-18x^2y+24xy^2$

39. $9a^2+6ab+b^2$

41. $4x^2-9y^2$

43. $x^2+4xy+4y^2$

45. $\frac{1}{x^2}-y^2$

47. $m(4mn+16m^2+7)$

49. $6(a-b^2)$

51. $2x(x^5-7x^2+4)$

53. $(x^3-y)(x^3-y-1)$

55. $4y^2(3y^4-2-4y^3)$

57. $(a^2+b)(a-b)$

59. $z(1+z)(1+z^2)$

61. $(n-2)(x^2+y)$

63. $(a-5b)(x+5y)$

65. $(2x-11)(2x+11)$

67. $(7a-12b)(7a+12b)$

69. $(5x^2y-3)(5x^2y+3)$

71. $(x-10y)(x^2+10xy+100y^2)$

73. $(m^2+5n^3)(m^4-5m^2n^3+25n^6)$

75. $(3x^2-2y^4z)\times(9x^4+6x^2y^4z+4y^8z^2)$

77. $(4y^2z-3x^4)(4y^2z+3x^4)$

79. $(7y^3+3xz^2)\times(49y^6+21xy^3z^2+9x^2z^4)$

81. $(x+5)(x-3)$

83. $(x-1)^2$

85. $(x-2)^2$

87. $(y+7)^2$

89. $(x+11)(x+2)$

91. $(y-8)(y-1)$

93. $(5a+3)(a-8)$

95. $(x+6)(5x-3)$

97. $(16y-9)(y-1)$

99. $(4a-3)(2a+1)$

101. $(4y-5)(3y-1)$

103. $2x(2x-1)^{\frac{-3}{2}}$

105. $a^{-3}(7a^2-2b)$

107. $2y^{-5}(5y^3-x)$

109. $(5x+7)^{\frac{4}{3}}(5x+6)$

111. $y^{-4}(7y^3+5)$

113. No; a variable in the denominator is equivalent to a variable with a negative exponent.

115. **a.** Yes; degree = 4; leading coefficient = 2; terms = 4

b. Yes; degree = 3; leading coefficient = 2; terms = 3

1.4 Exercises

1. $\frac{2x+1}{x-5}; x\neq -3, 5$

3. $x(x-1); x\neq -3$

5. $\frac{x+6}{x+5}; x\neq -5, 1$

7. $\frac{1}{x^2-x+1}; x\neq -1$

9. $2x+1; x\neq -5$

11. $2x-3; x\neq -7$

13. $\frac{x^3+9x^2+11x+19}{(x-3)(x+5)}$

15. $\frac{13x}{(x-3)(x+5)}$

17. $\frac{x^3+4x^2-7x+18}{(x+3)(x-3)}$

19. $\frac{x^2+11x+17}{x+3}$

21. $\frac{x+2}{x-6}$

23. $y-1$

25. $(x+2)(2x+3)$

27. $\frac{y-8}{y+8}$

29. $5y^2-2y-3$

31. -6

33. $\frac{x^2+9}{6x-3}$

35. $\frac{2x^2}{x+1}$

37. $\frac{s-r}{r^2s+s}$

39. $\frac{m+n}{mn}$

41. $\frac{x}{y}$

43. x^2y^2

45. $\frac{11x}{7y}$

47. $\frac{5z-3x}{z^2}$

49. $\frac{x-2}{x+2}$

51. $\frac{(z^2-11z+54)(z-9)}{(z-2)}$

53. $\frac{2y^2+5y-4}{y+1}$

1.5 Exercises

1. $5i$

3. $-3i\sqrt{3}$

5. $4i\sqrt{2x}$

7. $i\sqrt{29}$

9. $1-3i$

11. $8-6i$

13. $-5+6i$

15. $16-30i$

17. i

19. -11

21. $40-42i$

23. -9

25. $1+5i$

27. $-1-4i$

29. $7i$

31. $3+i$

33. $-i$

35. $-i$

37. $10-2i$

39. $\frac{14}{37}+\frac{10}{37}i$

41. $\frac{21}{17}-\frac{1}{17}i$

43. $-5+2i\sqrt{6}$

45. 8

47. $-\frac{7}{3}i$

49. $22+10i\sqrt{3}$

51. $6+3j$ ohms

53. $11-2j$ ohms

1.6 Exercises

1. $t=-5$

3. $y=-1$

5. $w=-3$

7. $\mathbb{R}$ (Identity)

9. $\varnothing$ (Contradiction)

11. $m=7$

13. $x=3.7$

15. $x=1.05$

17. $y=-5$

19. $\mathbb{R}$ (Identity) **21.** $\mathbb{R}$ (Identity)

23. $x = 3$

25. $\varnothing$ (Contradiction)

27. $y = -\frac{1}{3}, -3$ **29.** $x = \frac{1}{3}$

31. $x = -311, 420$

33. $x = -\frac{4}{5}, 2$

35. $\varnothing$ (Contradiction)

37. $x = -2, 2$ **39.** $x = 5$

41. $x = -\frac{1}{2}$ **43.** $x = \frac{1}{4}$

45. $x = \frac{1}{7}$ **47.** $r = \frac{C}{2\pi}$

49. $a = \frac{v^2 - v_0^{\,2}}{2x}$

51. $F = \frac{9}{5}C + 32$

53. $h = \frac{A - 2lw}{2w + 2l}$

55. $m = \frac{2K}{v^2}$

57. $\frac{19}{3}$ hours, or 6 hours and 20 minutes

59. 13.5 miles **61.** \$390

63. 2 gallons 44%, 1 gallon 50%

65. 24 child tickets, 15 adult tickets

67. 7.5%

69. 26 feet by 26 feet

71. 53, 55, and 57

73. 36.4% **75.** $x \approx 0.72$

77. $x \approx 13.11$

1.7 Exercises

1. $\{-9, 3.14, -2.83, 1, -3, 4\}$

3. $\{-2.83, 1, -3\}$

5. $(-\infty, -3]$ (number line: -3)

7. $(-\infty, 4.8)$ (number line: 4.8)

9. $(-\infty, 2.25)$ (number line: 2.25)

11. $\left(-\infty, \frac{3}{2}\right)$ (number line: $\frac{3}{2}$)

13. $\left(-\infty, -\frac{3}{11}\right]$ (number line: $-\frac{3}{11}$)

15. $(7, \infty)$ (number line: 7)

17. $(35, \infty)$ (number line: 35)

19. $(-3, \infty)$ (number line: -3)

21. $(-0.11, \infty)$ (number line: -0.11)

23. $(1, 5]$ (number line: 1, 5)

25. $(-10, 6]$ (number line: -10, 6)

27. $[-8, -2)$ (number line: -8, -2)

29. $(21, 69]$ (number line: 21, 69)

31. $\left(\frac{23}{7}, \frac{25}{7}\right)$ (number line: $\frac{23}{7}$, $\frac{25}{7}$)

33. $\left[\frac{13}{2}, 16\right)$ (number line: $\frac{13}{2}$, 16)

35. $\left(-\frac{5}{3}, 1\right]$ (number line: $-\frac{5}{3}$, 1)

37. $\left(-\infty, -\frac{7}{2}\right) \cup \left(\frac{15}{2}, \infty\right)$ (number line: $-\frac{7}{2}$, $\frac{15}{2}$)

39. $\left(-\infty, \frac{1}{2}\right) \cup \left(\frac{5}{2}, \infty\right)$ (number line: $\frac{1}{2}$, $\frac{5}{2}$)

41. $\varnothing$

43. $(-\infty, 2) \cup (6, \infty)$ (number line: 2, 6)

45. $\varnothing$

47. $\varnothing$

49. $[-4, 0]$ (number line: -4, 0)

51. $(-\infty, \infty)$

53. $(3, 15)$ (number line: 3, 15)

55. $(-1, 3]$ (number line: -1, 3)

57. $(-\infty, \infty)$

59. $[-2, 3)$ (number line: -2, 3)

61. $[73, 113]$ for an A, $(113, 115)$ for an A+.

63. $(1140, 1600]$

1.8 Exercises

1. $\left\{\frac{3}{2}, -1\right\}$ **3.** $\left\{\frac{-3}{2}, -3\right\}$

5. $\{2\}$ **7.** $\left\{\pm\frac{3}{2}\right\}$

9. $\left\{2 \pm i\sqrt{5}\right\}$ **11.** $\left\{-\frac{5}{3}, \frac{1}{3}\right\}$

13. $\{-5, 9\}$ **15.** $\{-5, 2\}$

17. $\left\{-\frac{7}{2}, \frac{9}{2}\right\}$ **19.** $\{-16, -6\}$

21. $\{0.17 \pm 0.86i\}$

23. $\left\{0, \frac{2}{3}\right\}$ **25.** $\left\{-\frac{3}{2}, 5\right\}$

27. Two real solutions

29. One real solution

31. $\{8, 14\}$ **33.** $\left\{-3, \frac{3}{10}\right\}$

35. $\left\{\frac{-1 \pm i\sqrt{2}}{2}\right\}$ **37.** $\{1 \pm 17i\}$

39. $\{0, 6\}$ **41.** $\{3, 5\}$

43. $\left\{-1, 2, \frac{1 \pm i\sqrt{7}}{2}\right\}$

45. $\{-3, 4\}$ **47.** $\left\{\pm\sqrt{2}, \pm i\sqrt{5}\right\}$

49. $\left\{1 \pm 2i, 1 \pm \sqrt{3}\right\}$

51. $\left\{\frac{1}{8}, 27\right\}$ **53.** $\{\pm 2i, \pm 3\}$

55. $\{-1, \pm 2, 3\}$ **57.** $\left\{1, -\frac{8}{27}\right\}$

59. $\{\pm 1, 3\}$ **61.** $\{\pm 2, \pm 3i\}$

63. $\left\{\pm 2, -\frac{6}{5}\right\}$ **65.** $\left\{\pm\frac{3}{2}, \pm\frac{3i}{2}\right\}$

67. $\left\{-\frac{5}{2}, 0, \frac{4}{7}\right\}$

69. $\left\{-\frac{4}{3}, \frac{2 \pm 2i\sqrt{3}}{3}\right\}$

71. $\left\{-\frac{5}{3}, 0, 1\right\}$ **73.** $\{1\}$

75. $\{4\}$ **77.** $\{0, 2, 3\}$

79. $\left\{-\frac{1}{5}, \frac{1}{7}\right\}$ **81.** $\left\{-1, 0, \frac{2}{5}\right\}$

83. $\left\{-\frac{1}{3}, -\frac{1}{5}\right\}$ **85.** 3 seconds

87. 3 seconds **89.** 2.6 seconds

91. $(x - 3 - 2i)(x - 3 + 2i)$

93. $\left(2x + 3 - 2\sqrt{2}\right)\left(2x + 3 + 2\sqrt{2}\right)$

95. $b = -5$ and $c = -24$

97. $b = -4$, $c = -12$, and $d = 0$

99. $a = 1$, $c = -36$, and $d = -144$

101. $a = 15$, $b = -16$, and $c = -5$

1.9 Exercises

1. $\left\{-2, -\frac{3}{2}\right\}$ **3.** $\{3 \pm \sqrt{10}\}$

5. $\{-3 \pm \sqrt{6}\}$ **7.** $\{-2\}$

9. $\varnothing$

11. $(-\infty, -3) \cup (-3, 3) \cup (3, \infty)$

13. $\left\{\frac{5}{2}, \frac{7}{2}\right\}$

15. $\{1, -\sqrt{5}, 2 + \sqrt{5}\}$

17. $\{0\}$ **19.** $\varnothing$

21. $\{1\}$ **23.** $\left\{\frac{2}{3}\right\}$

25. $\varnothing$ **27.** $\left\{\frac{29}{8}\right\}$

29. $\{6\}$ **31.** $\varnothing$

33. $\{-2, 1\}$ **35.** $\{1\}$

37. $\{10\}$ **39.** $\{4\}$

41. $\{2\}$ **43.** $\{-32\}$

45. $\left\{\pm\frac{125}{343}\right\}$ **47.** $\{-2, 5\}$

49. $\{7, 10\}$

51. $a = \pm\sqrt{c^2 - b^2}$

53. $m = \frac{k}{\omega^2}$ **55.** $v = \pm\sqrt{\frac{Fr}{m}}$

57. $h = \pm\sqrt{\frac{m}{23}}$ **59.** $c = \pm\sqrt{\frac{2gm}{r}}$

61. $b = \pm\sqrt{c^2 - a^2}$

63. $a = \sqrt[3]{\frac{uP^2}{4\pi^2}}$

65. $\frac{4}{3}$ minutes, or 1 minute and 20 seconds

67. 4 hours and 12 hours

69. 90 minutes **71.** 9.1 hours

Chapter 1 Project

1. Alex: 81.7; Ashley: 90.3; Barron: 81.0; Elizabeth: 83.1; Gabe: 90.5; Lynn: 82.4

3. The final exam contributes a higher percentage of points to the final grade than does the semester project.

Chapter 1 Review Exercises

1. a. 2^3 **b.** $2^3, 0$

c. $-\sqrt{4}, 2^3, 0$

d. All except $\sqrt{17}$

e. $\sqrt{17}$ **f.** All

3. $[4, 17)$ **5.** -7

7. 4 **9.** -1

11. $\frac{x^2}{2y}$, $12.1x$, $-\sqrt{y+5}$

13. $\frac{4\pi}{3} - 36$ **15.** 51

17. Commutative Property

19. Zero-Factor Property

21. $[5, 8)$ **23.** $-\frac{t^9}{2s^7}$

25. $\frac{18y^2}{x^4 z^5}$ **27.** 6.952×10^7

29. 2.0×10^{-8} **31.** 5

33. $5x^{10}$ **35.** $-\frac{4y}{x^3}$

37. $\frac{2y\sqrt[3]{9x^2 y}}{3}$ **39.** $-\sqrt{2} - \sqrt{6}$

41. $3|x|\sqrt{2xy} - 2x\sqrt[3]{2xy}$

43. $\frac{1}{x^{\frac{7}{4}}}$

45. $m^4 - 5m^3 + 3m^2 + 2$

47. $3x^3 - 4x^2 y^3 + 3xy - 4y^4$

49. $(x + 3)(x - 4)$

51. $(2a + 1)(3a - 5)$

53. $(6x^3 + y)(6x^3 - y)$

55. $(2x - 5y)(x + 3)$

57. $(3x - 2y)^{\frac{2}{3}}\left[(3x - 2y)^{\frac{2}{3}} - 1\right]$

59. $\frac{x+3}{x-3}$, $x \neq 0, \pm 3$

61. $\frac{-2}{x}$ **63.** $\frac{b-a}{4a+4b}$

65. $-x - y$ **67.** 3

69. $5 + 9i$ **71.** $4 + i$

73. $-\frac{7}{25} + \frac{24}{25}i$ **75.** $62 - 16i\sqrt{2}$

77. $\varnothing$ (Contradiction)

79. $x = 6.25$ **81.** $x = 3, 4$

83. $z = -\frac{10}{7}, 0$ **85.** $x = -3, 4$

87. $c = \frac{2A}{h} - b$ **89.** $C = \frac{5}{9}(F - 32)$

91. \$85

93. $(4, \infty)$

95. $(1, \infty)$

97. $[-7, 4)$

99. $(-5, -1)$

101. $(-\infty, -6] \cup [8, \infty)$

103. $(-\infty, 4] \cup [10, \infty)$

105. $\left\{-\frac{2}{5}, 3\right\}$ **107.** $\{2 \pm 3i\}$

109. $\{4 \pm \sqrt{2}\}$ **111.** $\{3 \pm \sqrt{7}i\}$

113. $\left\{-4, \frac{5}{2}\right\}$ **115.** $\left\{\frac{19 \pm \sqrt{701}}{17}\right\}$

117. $\{\pm 1, \pm\sqrt{2}\}$ **119.** $\{-6, 4\}$

121. $\{\pm\sqrt{2}, 4\}$ **123.** $\{1, \pm 2i\}$

125. $\{-1, 0, 4\}$ **127.** $\left\{\frac{3}{2}, 2\right\}$

129. $b = -2$ and $c = -8$

131. -5 **133.** $\{0, 3\}$

135. $\{-5\}$ **137.** $\{-4\}$

139. $\{2\}$ **141.** $\left\{-\frac{3}{2}, 5\right\}$

143. $\{3\}$ **145.** $r = \sqrt{\frac{3V}{\pi h}}$

Chapter 2: Equations and Inequalities in Two Variables

2.1 Exercises

1.

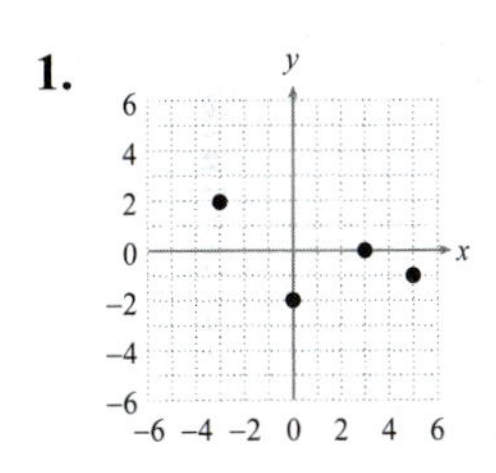

3.

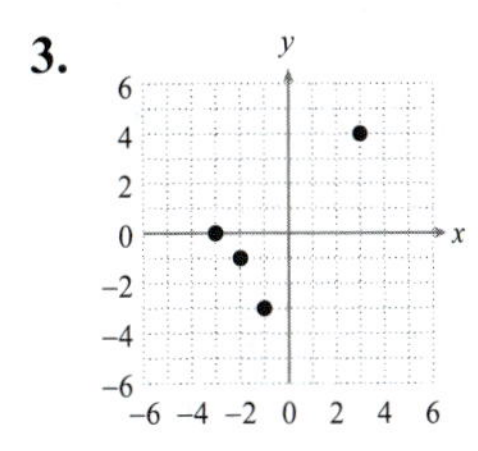

5.

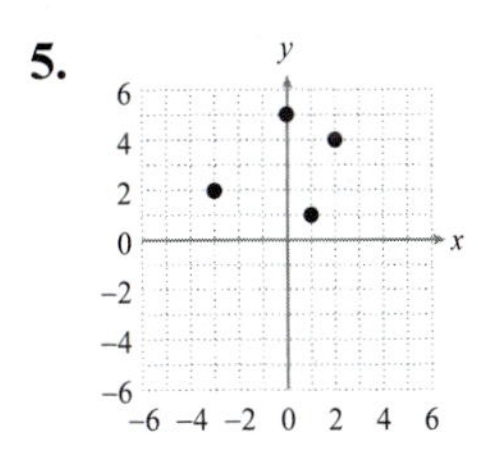

7.

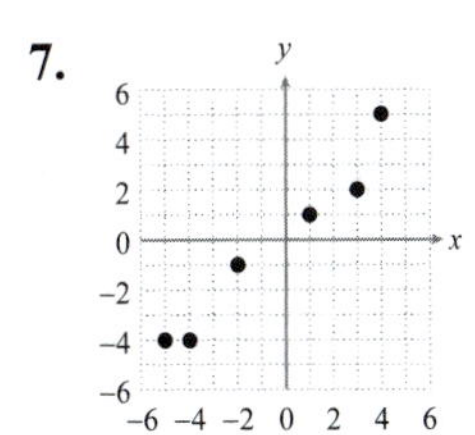

9. III **11.** IV

13. Positive x-axis

15. III **17.** IV

19. II **21.** IV

23. I

25. Negative y-axis

27. $\left\{(0,-3),(2,0),\left(3,\frac{3}{2}\right),(4,3)\right\}$

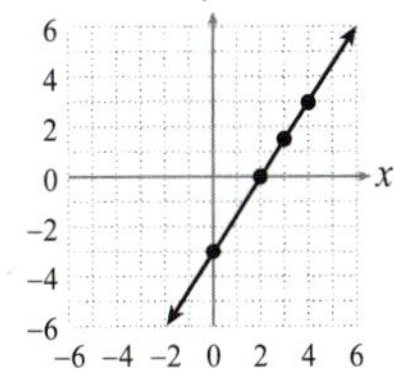

29. $\left\{(0,0),(1,\pm1),(4,\pm2),(9,\pm3),\left(2,-\sqrt{2}\right)\right\}$

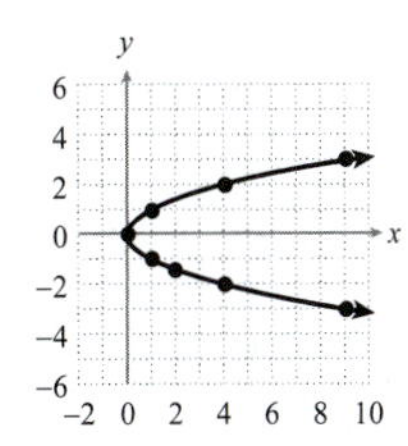

31. $\left\{(0,\pm3),(\pm3,0),\left(-1,\pm2\sqrt{2}\right),\left(1,\pm2\sqrt{2}\right),\left(\pm\sqrt{5},2\right)\right\}$

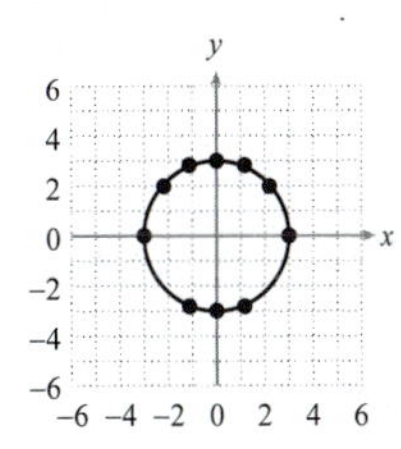

33. $\sqrt{34},\ \left(\frac{-7}{2},\frac{1}{2}\right)$

35. $\sqrt{58},\ \left(\frac{3}{2},\frac{7}{2}\right)$

37. $2\sqrt{2},(-1,-1)$

39. $4\sqrt{34},(3,-8)$

41. $10,\ (1,-6)$

43. $3\sqrt{13},\ \left(2,\frac{1}{2}\right)$

45. $10\sqrt{2},\ (3,3)$

47. $x=2$ or 18

49. $x=10,\ y=1$

51. 12

53. $2\sqrt{29}+\sqrt{26}+5\sqrt{2}$

55. 54

57. 1.25 kilometers

59. **a.** 249.19 meters

b. $\left(\frac{133}{2},\frac{709}{2}\right)$

61. area $=\frac{15}{2}$ **63.** area $=25$

65. area $=17$ **67.** area $=48$

69. $x=[-5,6];y=[-8,9]$

71. $x=[-3,6];y=[-4,5]$

73. $x=[-6,8];y=[-9,7]$

2.2 Exercises

1. $(x+4)^2+(y+3)^2=25$

3. $(x-7)^2+(y+9)^2=9$

5. $x^2+y^2=6$

7. $\left(x-\sqrt{5}\right)^2+\left(y-\sqrt{3}\right)^2=16$

9. $(x-7)^2+(y-2)^2=4$

11. $(x+3)^2+(y-8)^2=2$

13. $(x-4)^2+(y-8)^2=10$

15. $x^2+y^2=85$

17. $\left(x+\frac{7}{2}\right)^2+\left(y-\frac{17}{2}\right)^2=\frac{53}{2}$

19. $(x+6)^2+\left(y-\frac{3}{2}\right)^2=\frac{125}{4}$

21. $\left(x+\frac{13}{2}\right)^2+(y+7)^2=\frac{365}{4}$

23. $(x-4)^2+(y-3)^2=25$

25. $(x-2)^2+y^2=4$

27. $(x-2)^2+(y-4)^2=49$

29. $(x+3)^2+(y+2)^2=64$

31.

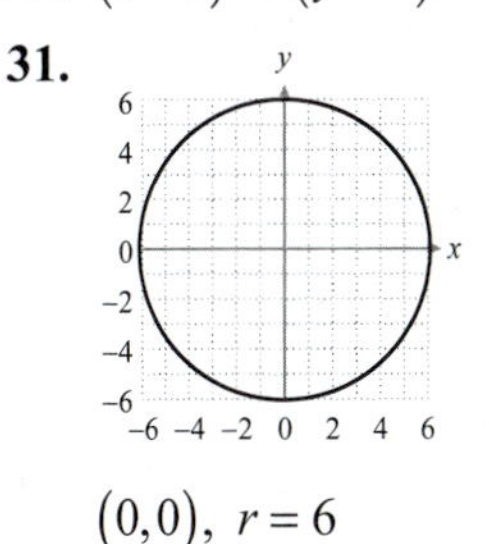

$(0,0),\ r=6$

33.

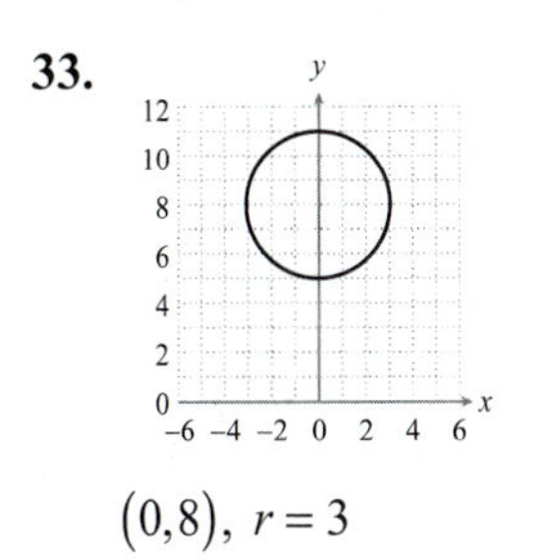

$(0,8),\ r=3$

35. 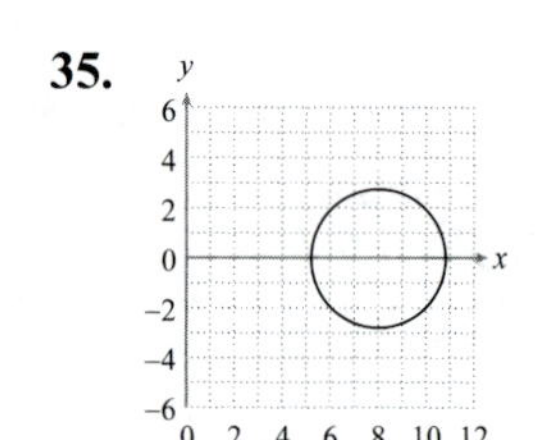

$(8,0),\ r=2\sqrt{2}$

37. 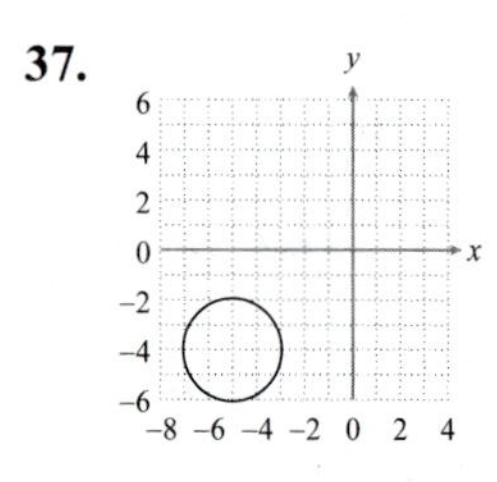

$(-5,-4),\ r=2$

39. 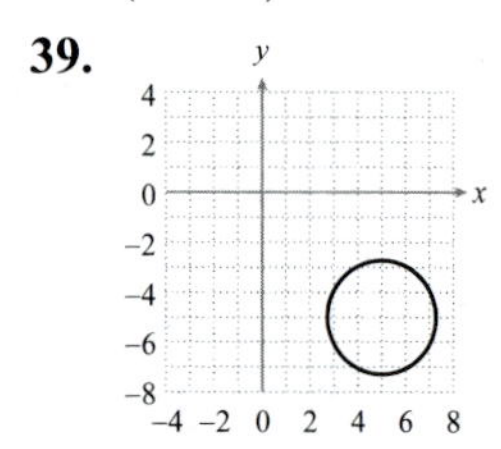

$(5,-5),\ r=\sqrt{5}$

41. 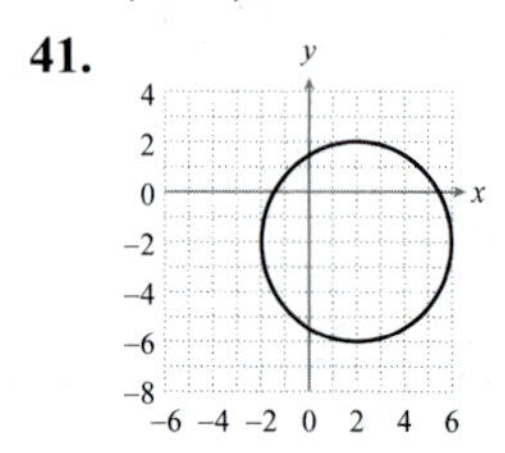

$(2,-2),\ r=4$

43. 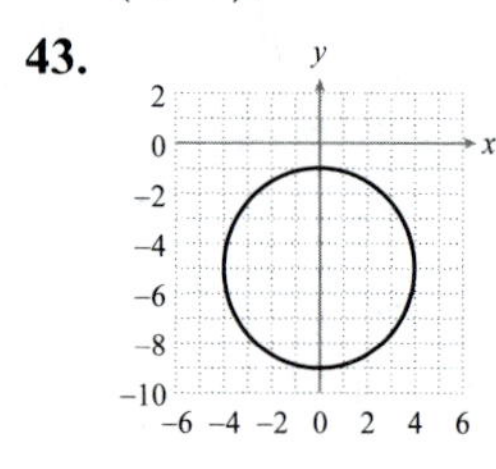

$(0,-5),\ r=4$

45. 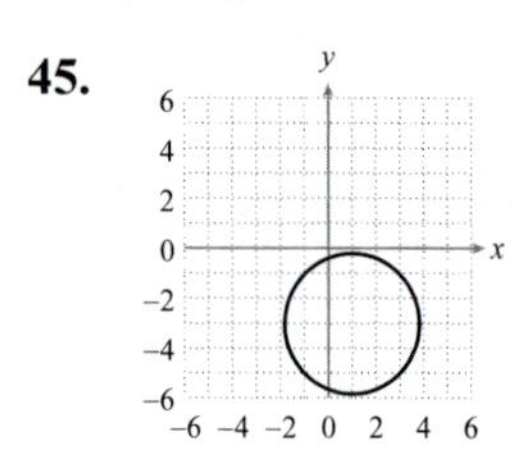

$(1,-3),\ r=2\sqrt{2}$

47. 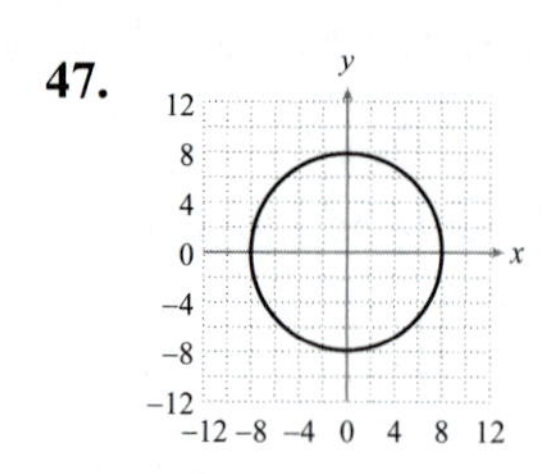

$(0,0),\ r=8$

49. 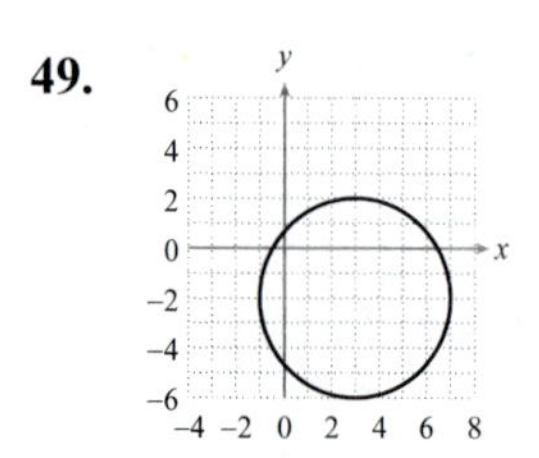

$(3,-2),\ r=4$

51. 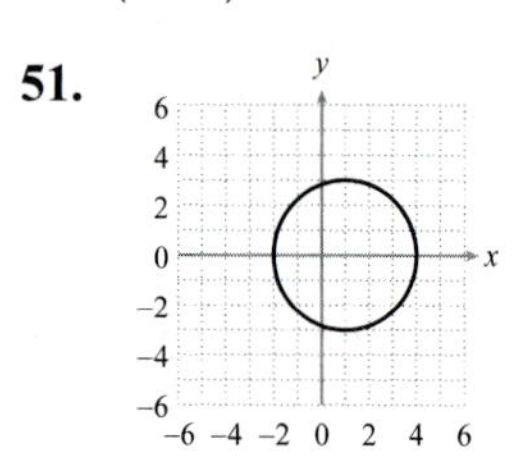

$(1,0),\ r=3$

53. 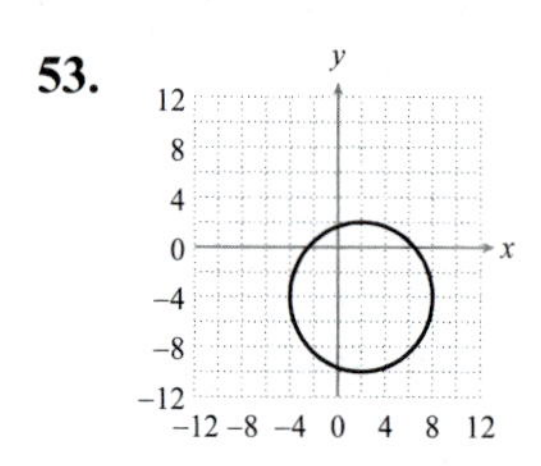

$(2,-4),\ r=6$

55. 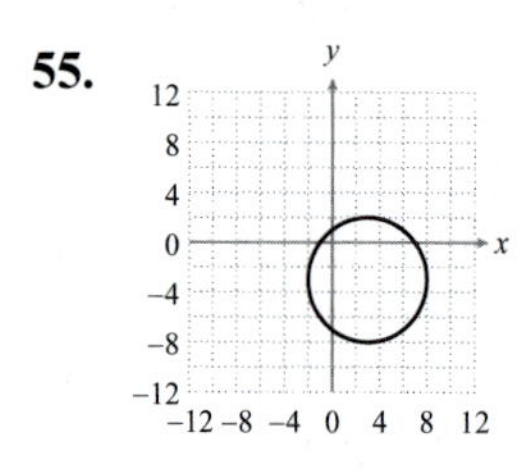

$(3,-3),\ r=5$

2.3 Exercises

1. Yes	**3.** No	**5.** No
7. No	**9.** Yes	**11.** Yes
13. Yes	**15.** No	**17.** No
19. No	**21.** No	**23.** Yes

25. 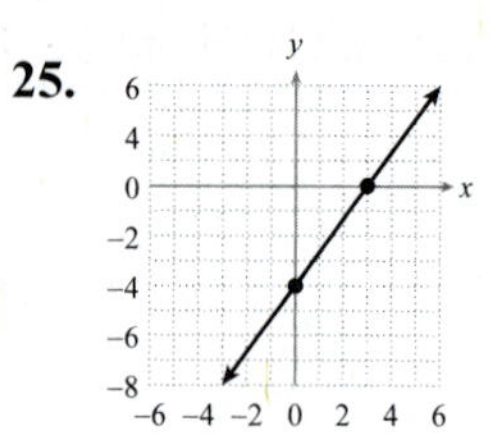

27. 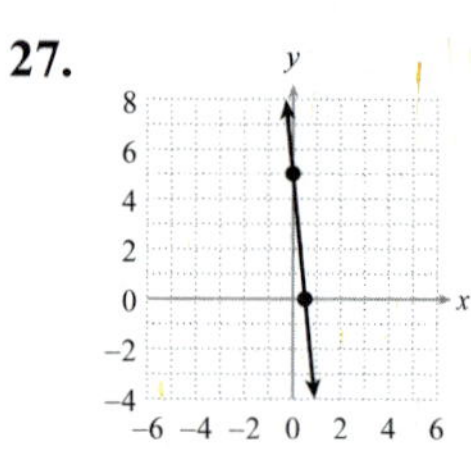

29. 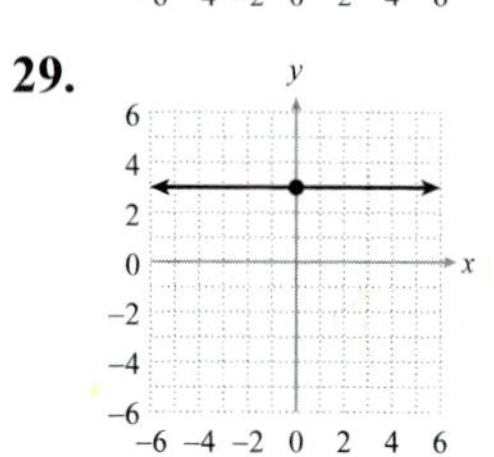

31. 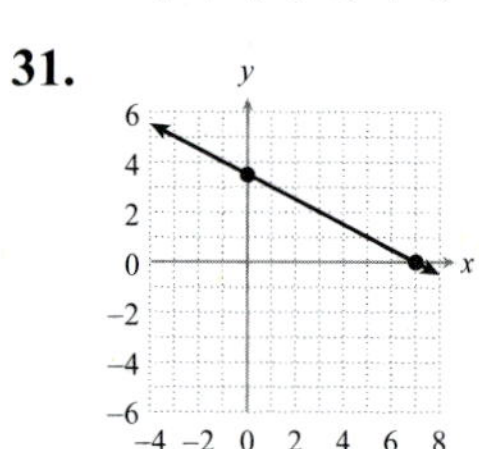

33. 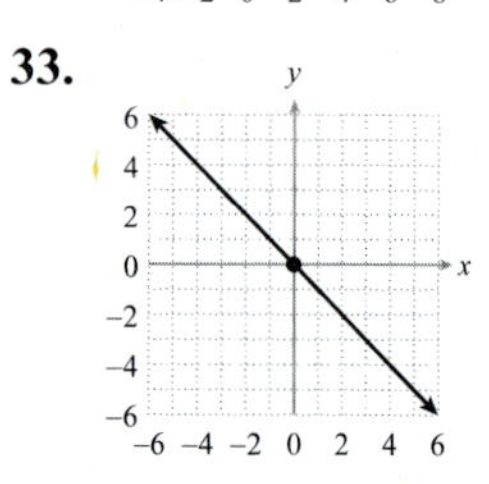

35. 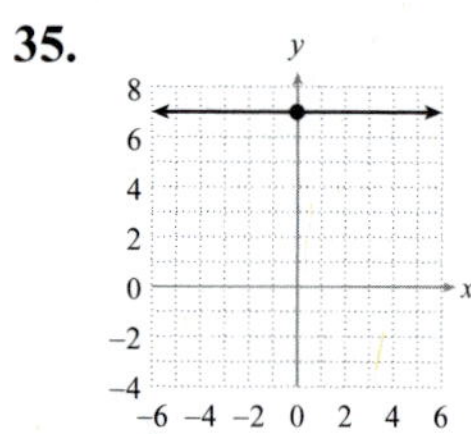

37. 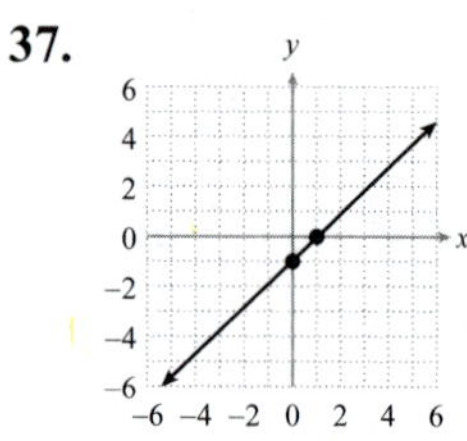

39. 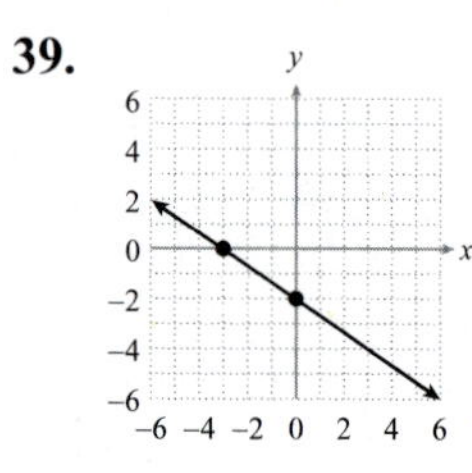

41. e **43.** c **45.** f

47. $a = P - b - c$

49. $j = 24{,}000 + 9b$; $b = \dfrac{j - 24{,}000}{9}$; Yes

2.4 Exercises

1. –4 **3.** 0

5. Undefined **7.** $\frac{2}{3}$

9. $\frac{1}{6}$ **11.** –7

13. –3 **15.** $-\frac{9}{13}$

17. $-\frac{1}{4}$ **19.** 0

21. Undefined **23.** 2

25. $\frac{7}{6}$ **27.** $-\frac{5}{2}$

29.

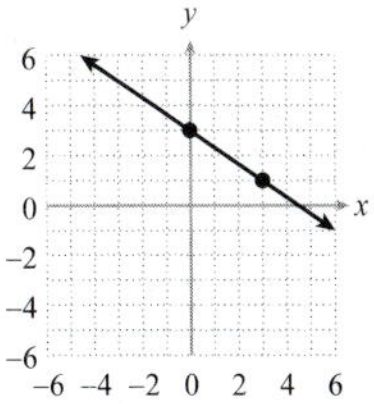

31.

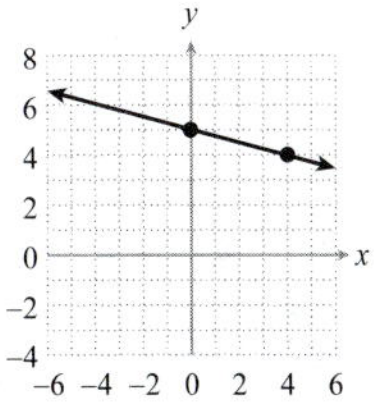

33.

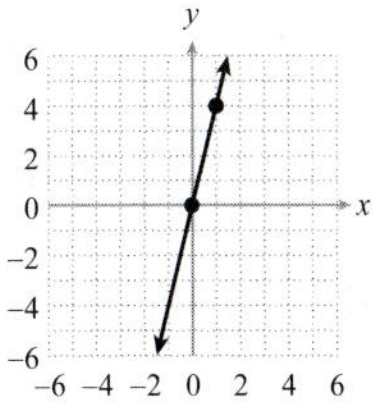

35.

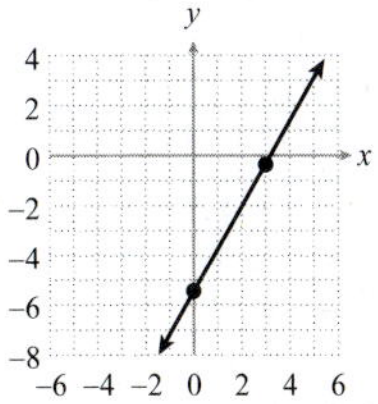

37. $y = \frac{3}{4}x - 3$ **39.** $y = -\frac{5}{2}x - 7$

41. $y = -5x - 9$ **43.** $3x - 2y = 3$

45. $y = 5$ **47.** $10x - y = 31$

49. $3x + y = 26$ **51.** $4x + 3y = 5$

53. $x = 2$ **55.** $y = -1$

57. $2x + 7y = 52$

59. $y = 5$ **61.** $15x - 8y = 0$

63. c **65.** e **67.** d

69. a. \$2225 **b.** \$2100 **c.** \$0.25

71. \$325

2.5 Exercises

1. $y = 4x + 9$ **3.** $y = 3x - 11$

5. $y = -9$ **7.** $y = x$

9. $y = \frac{7}{6}x + \frac{53}{6}$

11. Yes **13.** Yes **15.** Yes

17. No **19.** No **21.** No

23. No **25.** No **27.** Yes

29. No

31. $y = -\frac{1}{3}x - 1$ **33.** $y = 7$

35. $y = -\frac{1}{4}x - \frac{3}{4}$

37. $y = x + 3$ **39.** $y = -3x + 28$

41. No **43.** No **45.** No

47. No **49.** Yes **51.** No

53. No **55.** No **57.** Yes

59. $41\frac{2}{3}$ ft

2.6 Exercises

1.

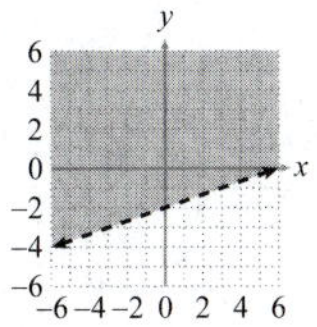

3.

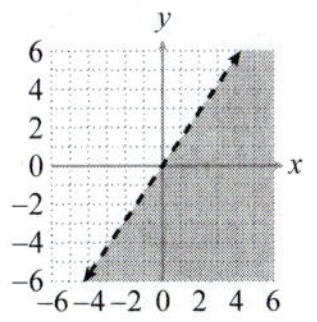

5.

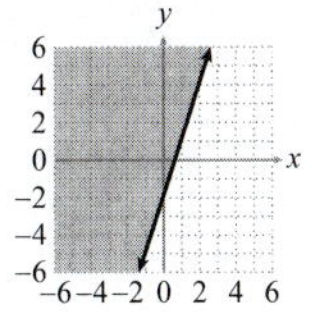

7.

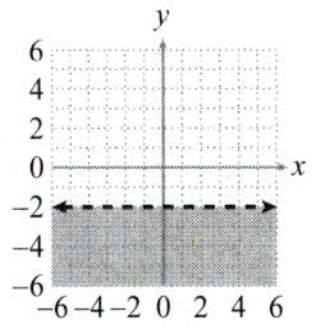

9.

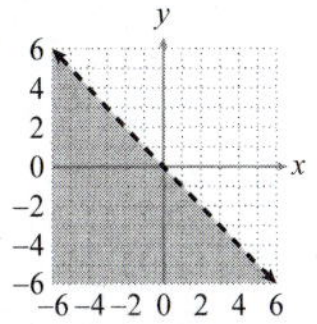

11.

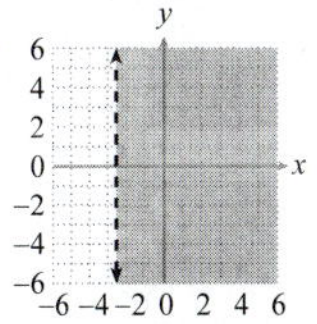

13.

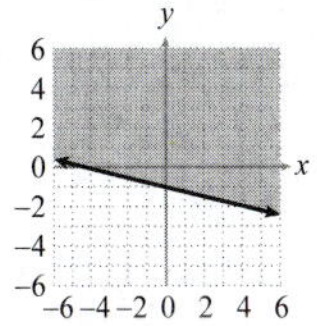

15.

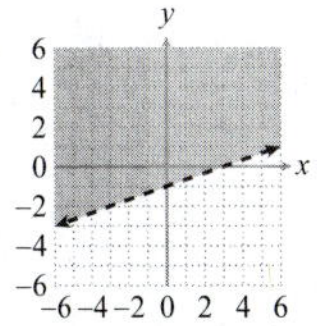

17.

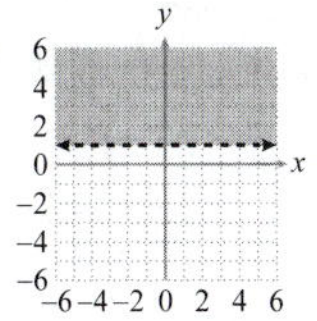

19.

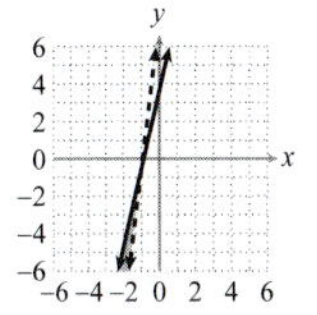

21.

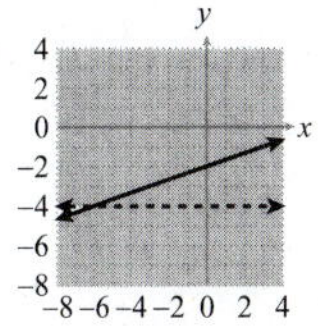

23.

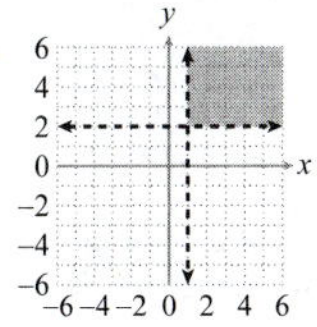

25.

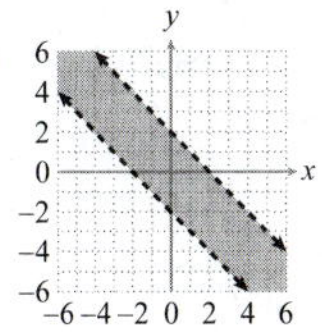

27.

29.

31.

33.

35.

37.

39.

41.

43.

45.

47. h

49. b

51. g

53. c

55. $12x + 22y < 150$

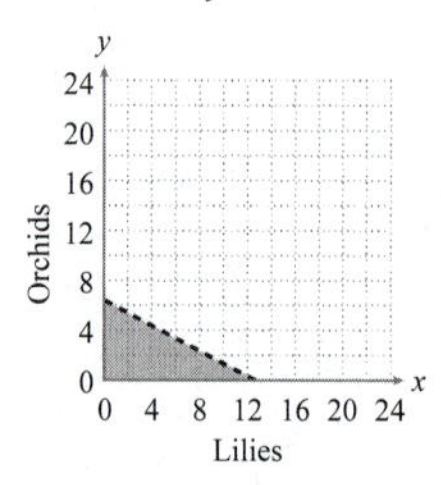

57. $73x + 46y < 1750$

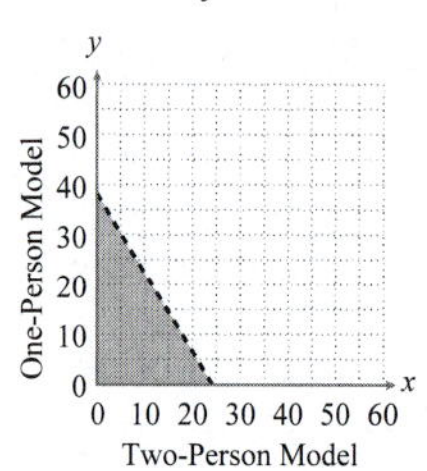

Chapter 2 Project

1. Shelbyville Tower: approx. 47.5 ft; Brockton Tower: 75 ft; Springfield Tower: approx. 60.6 ft; Ogdenville Tower: approx. 109.5 ft

3. No; attachment point: 29.4 ft from the top

Chapter 2 Review Exercises

1.

3.

5. Positive x-axis

7.

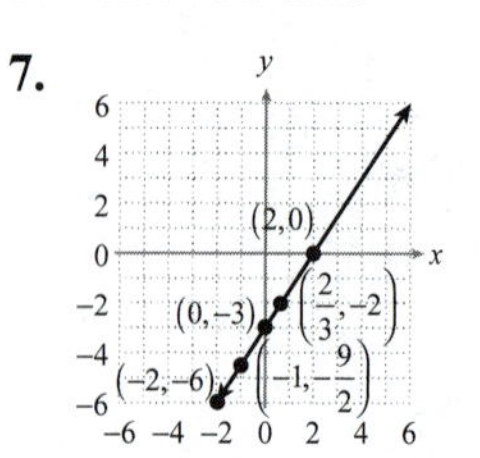

$(2,0), (0,-3), \left(-1,-\frac{9}{2}\right),$
$\left(\frac{2}{3},-2\right), (-2,-6)$

9. **a.** $\sqrt{2}$ **b.** $\left(\frac{5}{2},-\frac{13}{2}\right)$

11. **a.** $2\sqrt{13}$ **b.** $(-5,3)$

13. 2

15. 24

17. $\left(x-\sqrt{5}\right)^2 + \left(y+\sqrt{2}\right)^2 = 16$

19. $(x-2)^2 + (y+1)^2 = 20$

21. Center: $(-3,1)$; Radius: $2\sqrt{2}$

23.

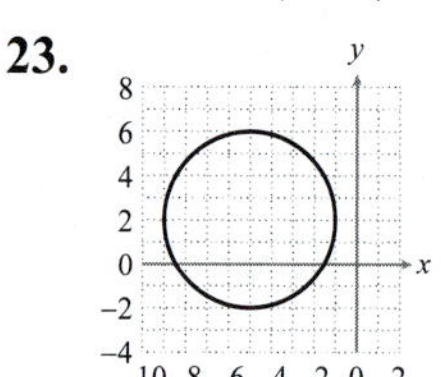

$r = 4;\ (h,k) = (-5,2)$

25.

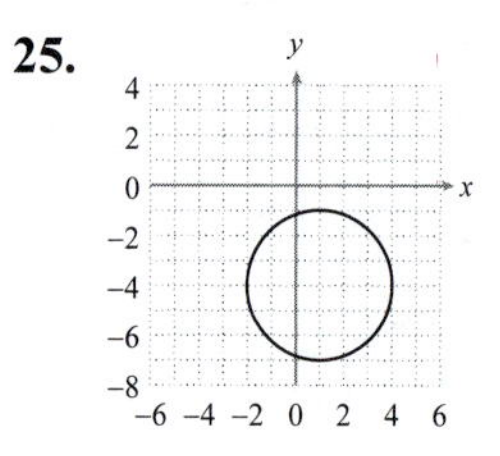

$r = 3;\ (h,k) = (1,-4)$

27. No

29. Yes

31. No

33.

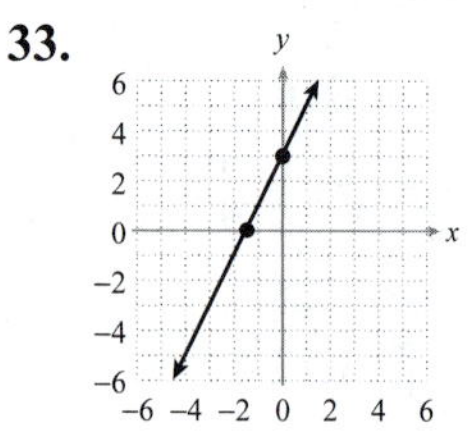

35.

37.

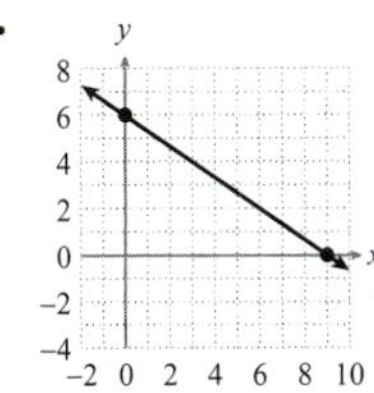

39. 12

41. Undefined

43.

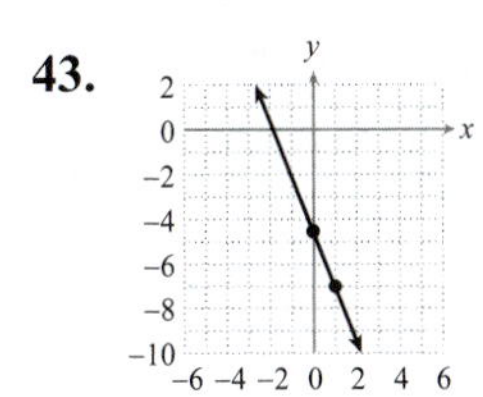

45. $x - y = 5$

47. $y = \frac{5}{9}x - 2$

49. $9x - 2y = 31$

51. $W = 0.08s + 2800$

53. Perpendicular

55. $y = 3x + 10$

57. $y = 2x - 3$

59. $y = -\frac{4}{3}x + 6$

61. $x = 7$

63. Yes

65.

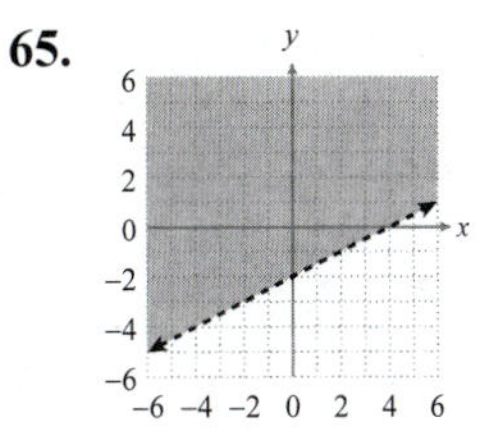

67.

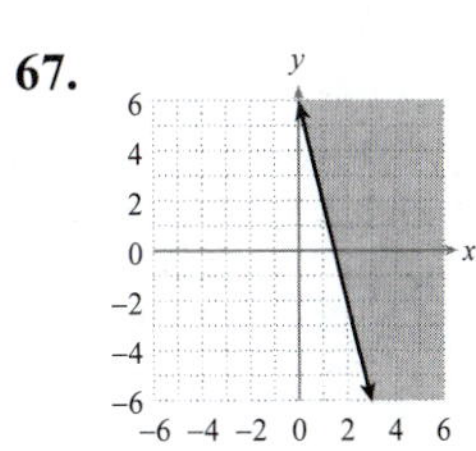

69.

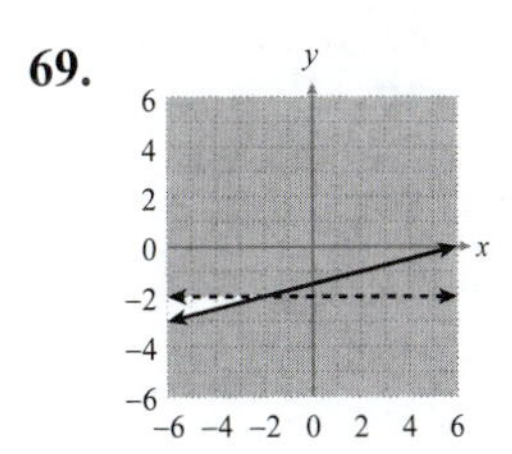

71.

73.

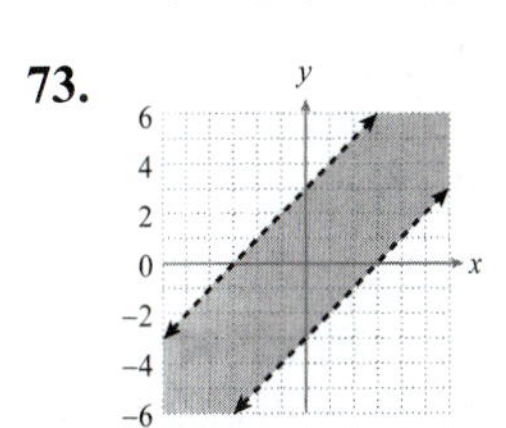

75.

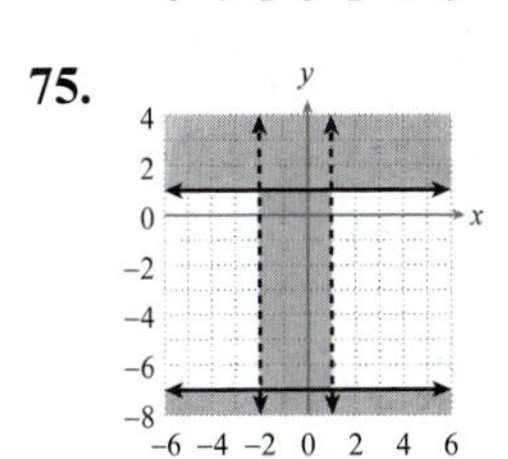

77. $3x + 4y \geq 1500$

Chapter 3: Relations, Functions, and Their Graphs

3.1 Exercises

1. Dom = $\{-2\}$, Ran = $\{5,3,0,-9\}$

3. Dom = $\{\pi,-2\pi,3,1\}$, Ran = $\{2,4,0,7\}$

5. Dom = $\mathbb{Z}$, Ran = even integers

7. Dom = $\mathbb{Z}$, Ran = $\{\ldots,-2,1,4,\ldots\}$

9. Dom = Ran = $\mathbb{R}$

11. Dom = $[0,\infty)$, Ran = $\mathbb{R}$

13. Dom = $\mathbb{R}$, Ran = $\{-1\}$

15. Dom = $\{0\}$, Ran = $\mathbb{R}$

17. Dom = $[-3,1]$, Ran = $[0,4]$

19. Dom = $[0,3]$, Ran = $[1,5]$

21. Dom = $[-1,3]$, Ran = $[-4,3]$

23. Dom = All males with siblings, Ran = All people who have brothers

25. Not a function; $(-2,5)$ and $(-2,3)$

27. Function

29. Not a function; $(6,-1)$ and $(6,4)$

31. Not a function; $(-1,0)$ and $(-1,4)$

33. Function **35.** Function

37. Function

39. Not a function; $(-1,-1)$ and $(-1,1)$

41. Function **43.** Function

45. Not a function; $(1,-2)$ and $(1,2)$

47. $f(x) = -6x^2 + 2x, f(-1) = -8$

49. $f(x) = \frac{-x+10}{3}, f(-1) = \frac{11}{3}$

51. $f(x) = -2x - 10, f(-1) = -8$

53. 1

55. 3

57. $x = -3, 2, 3$

59. **a.** 10 **b.** $x^2 + x - 2$ **c.** $2ax + 3a + a^2$ **d.** $x^4 + 3x^2$

61. **a.** 8 **b.** $3x - 1$ **c.** $3a$ **d.** $3x^2 + 2$

63. **a.** -2 **b.** $-6x + 16$ **c.** $-6a$ **d.** $-6x^2 + 10$

65. **a.** $i - 3$ **b.** $\sqrt{2-x} - 3$ **c.** $\sqrt{1-x-a} - \sqrt{1-x}$ **d.** $\sqrt{1-x^2} - 3$

67. $2x + h - 5$

69. $\frac{-1}{(x+h+2)(x+2)}$

71. $5(2x+h)$

73. 2

75. $\frac{\sqrt{x+h} - \sqrt{x}}{h}$

77. Dom = Cod = Ran = $\mathbb{R}$

79. Dom = Cod = Ran = $\mathbb{Z}$

81. Dom = Cod = $\mathbb{N}$, Ran = $\{6,7,8,\ldots\}$

83. $[1,\infty)$

85. $(-\infty,-2)\cup(-2,3)\cup(3,\infty)$

87. $\mathbb{R}$

89. $\left(-\infty,\frac{1}{3}\right)\cup\left(\frac{1}{3},\infty\right)$

91. $(-\infty,2)\cup(2,\infty)$

93. $[-6,\infty)$

95. $(-\infty,0)\cup(0,\infty)$

97. A function is a special relation in which every element of the domain is paired with exactly one element of the range.

3.2 Exercises

1.

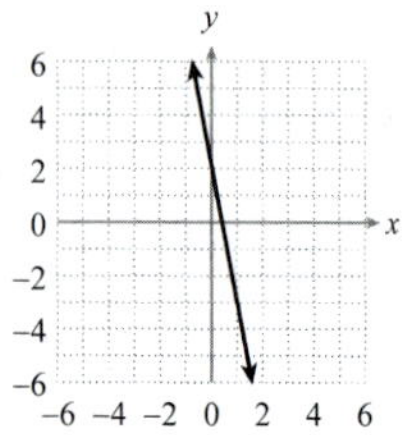

3.

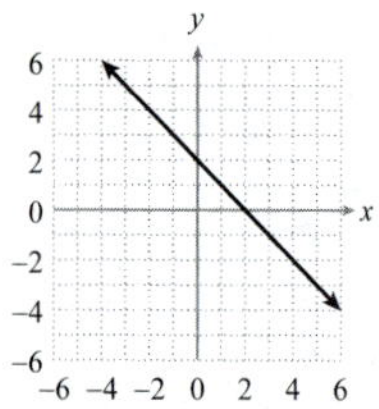

5.

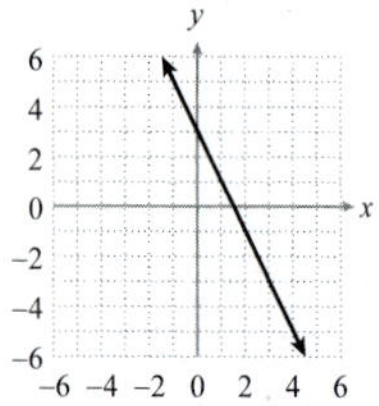

7.

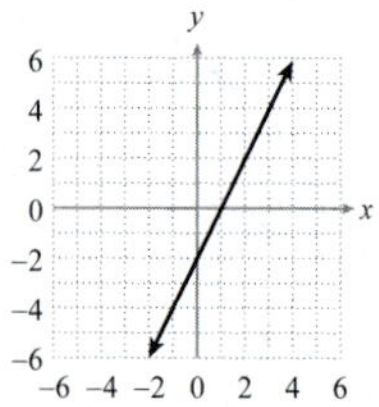

9.

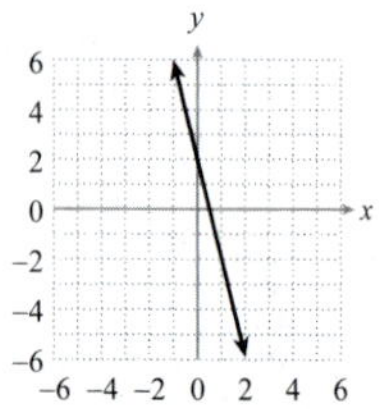

11.

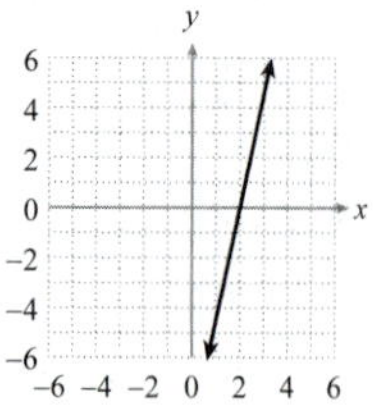

13.

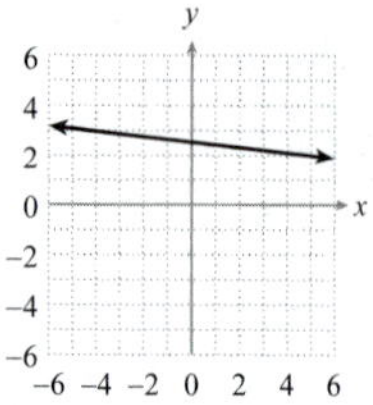

15.

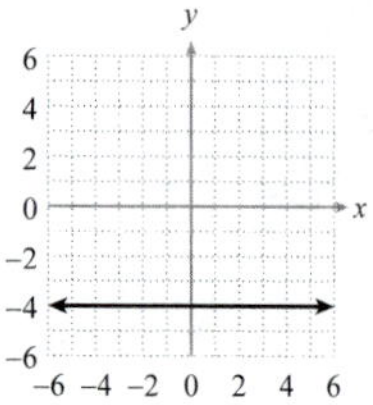

17. b **19.** a

21. $f(x)=-x+3$

23. $f(x)=2x-3$

25. a. $y=-0.93x+10.14$

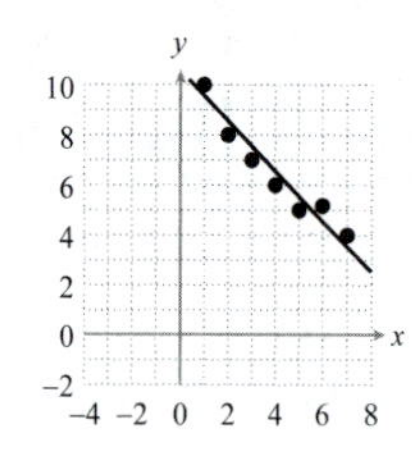

b. $r\approx-0.969$

27. a. $y=1.36x+3.82$

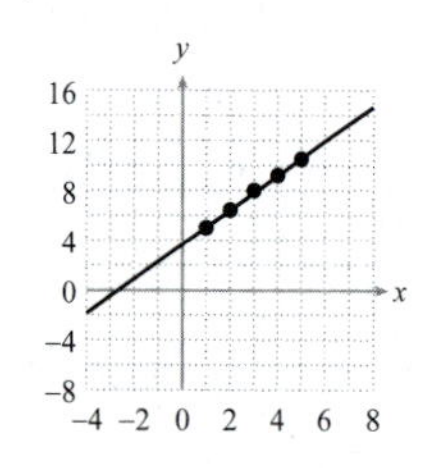

b. $r\approx0.998$

29. a. $y=-0.7x+0.4$

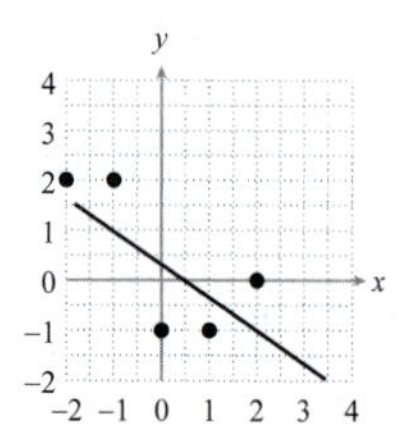

b. $r\approx-0.730$

31. a. $y=1.20x-0.85$

b.

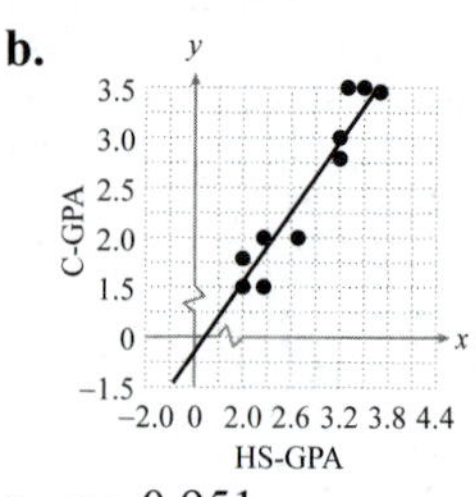

c. $r\approx0.951$

33. Neither. They are equal because the correlation is the absolute value.

3.3 Exercises

1. Vertex: $(2,3)$; no x-int.

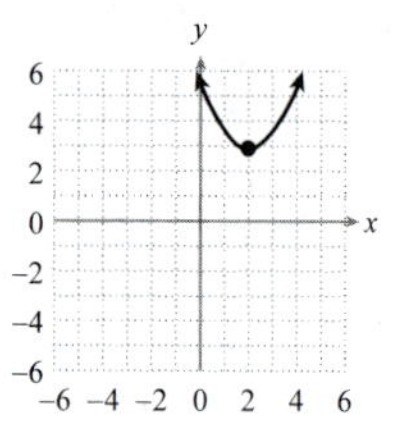

3. Vertex: $(-3,-2)$;

x-int.: $x=-3\pm\sqrt{2}$

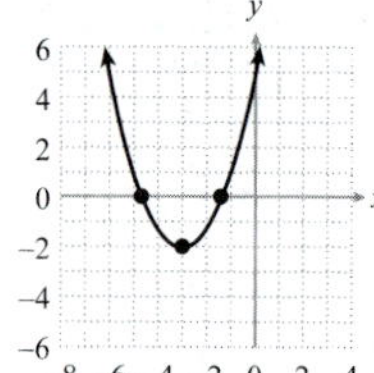

5. Vertex: $\left(\frac{1}{2},-\frac{25}{4}\right)$;

x-int.: $x=-2,3$

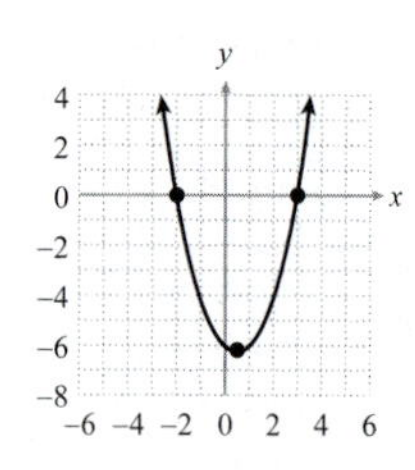

7. Vertex: $(-1,1)$; no x-int.

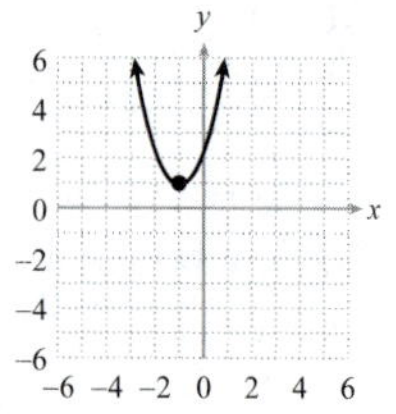

9. Vertex: $(1,0)$; x-int.: $x=1$

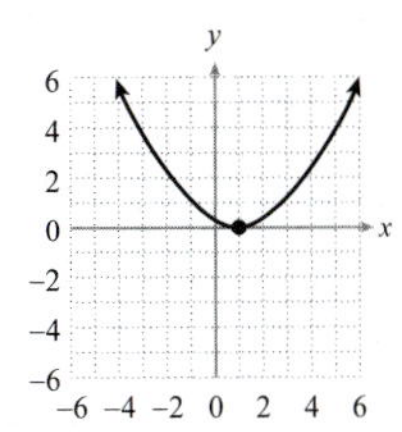

11. Vertex: $(0,4)$; x-int.: $x=-2,2$

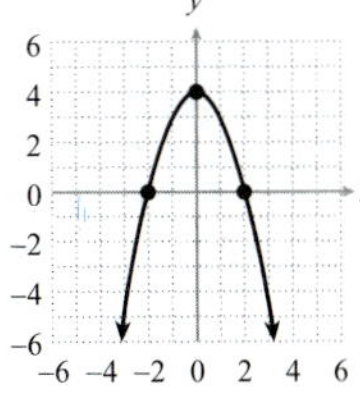

13. Vertex: $(0,-6)$; x-int.: $x=\pm\dfrac{\sqrt{6}}{2}$

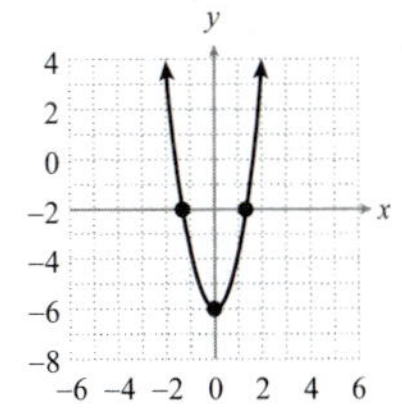

15. Vertex: $(-4,0)$; x-int.: $x=-4$

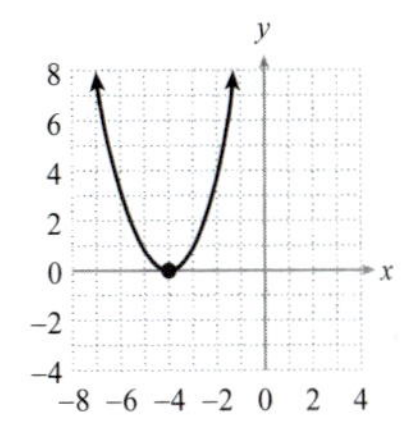

17. b **19.** d

21. a. $f(x)=-\frac{1}{2}(x+2)(x-1)$

b. $\left(-\frac{1}{2},\frac{9}{8}\right)$

23. a. $f(x)=3(x+1)(x-3)$

b. $(1,-12)$

25. $f(x)=-0.167x^2-1.17x+4$

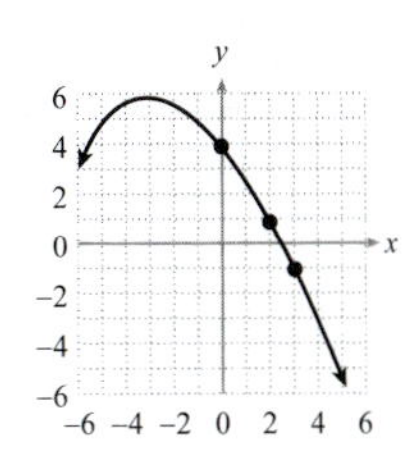

27. $f(x)=2x^2-8.2x+5.5$

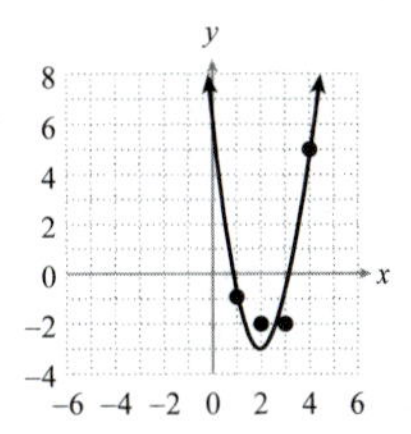

29. a. $f(x)=x^2-2x+2$

b. $(1,1)$

31. a. $f(x)=-x^2+2x-3$

b. $(1,-2)$

33. 5 and 5 **35.** $(2,1)$

37. The dimensions should be 5 inches by 10 inches by 10 feet

39. 49 people; $2401

41. 12 and 24 **43.** 375 units

45. 25 sets of golf clubs

47. 112 feet **49.** 164 feet

51. $h(t)=-15.88t^2+60.17t+7$; the maximum height of the baseball was approximately 64 feet.

53. Vertex: $(4,-1)$;

x-int.: $x=\dfrac{8\pm\sqrt{2}}{2}$

55. Vertex: $(4,-36)$;
x-int.: $x=-2,10$

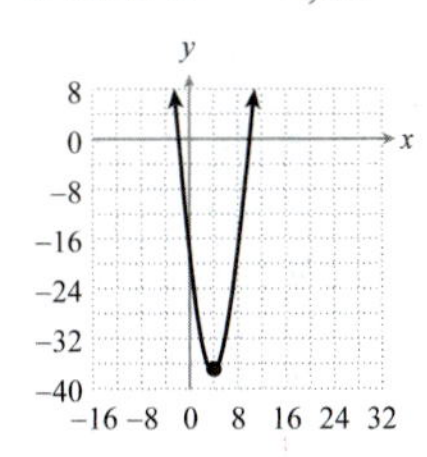

57. Vertex: $(0,25)$; x-int.: $x=-5,5$

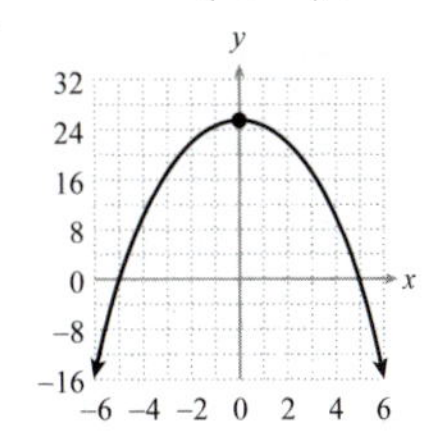

59. Vertex: $(-1,0)$; x-int.: $x=-1$

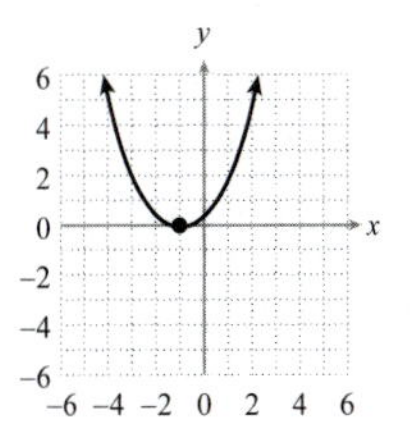

61. Vertex: $(5,21)$;
x-int.: $x=5\pm\sqrt{21}$

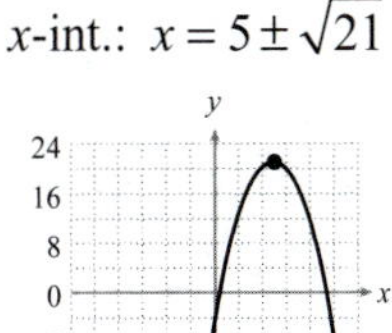

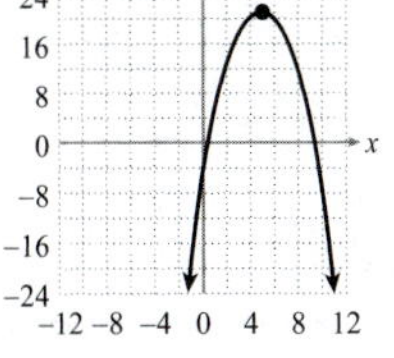

3.4 Exercises

1.

3.

5.

7.

9.

11.
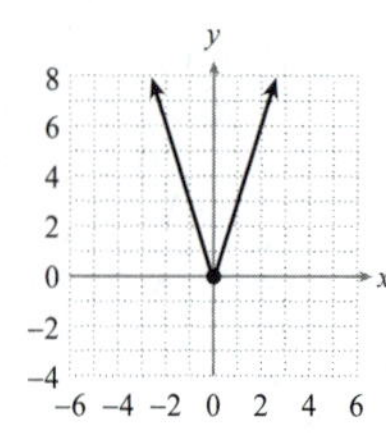

13.
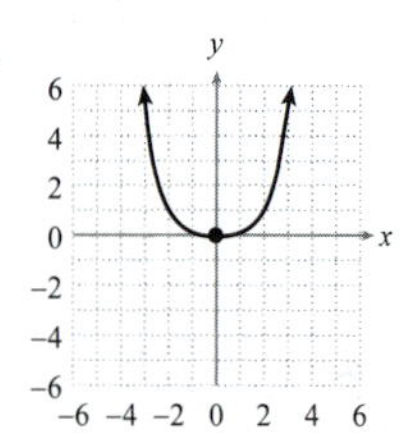

15.
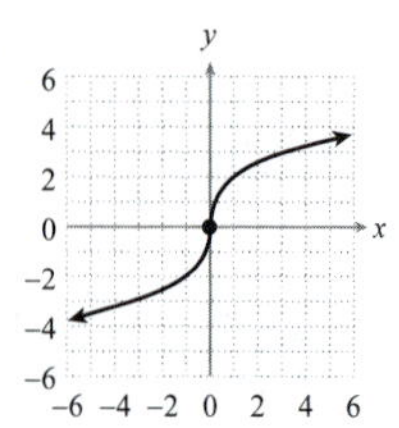

17.
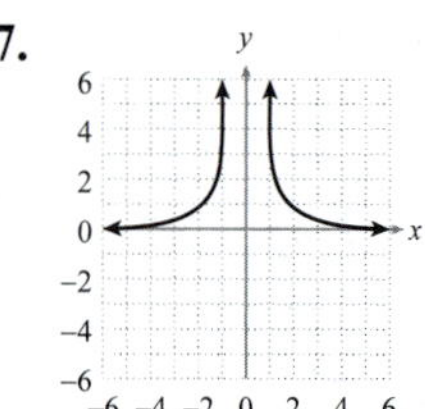

19.
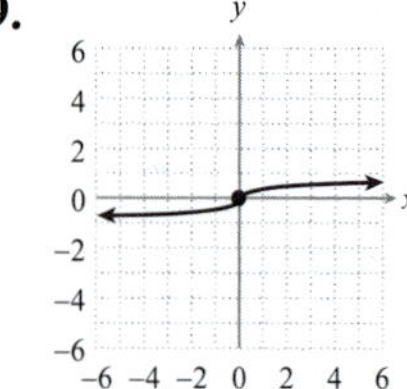

21.
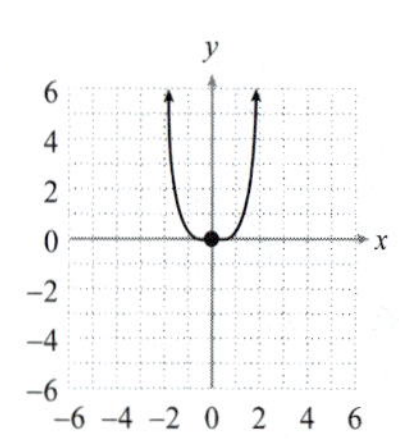

23.
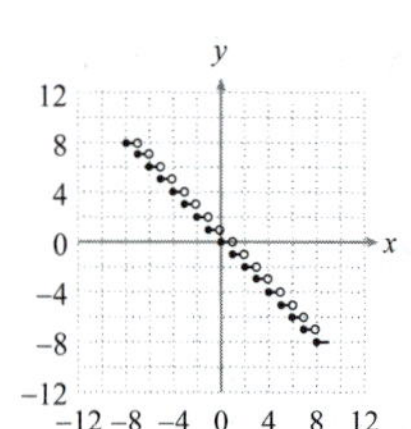

25.
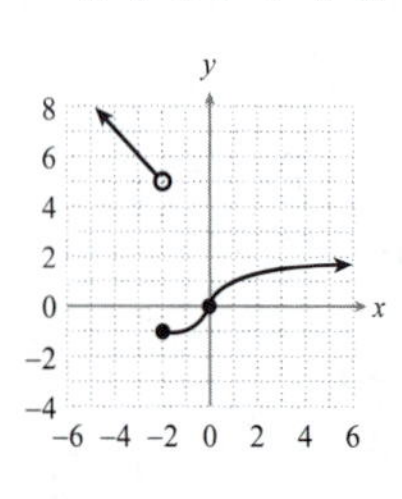

27.
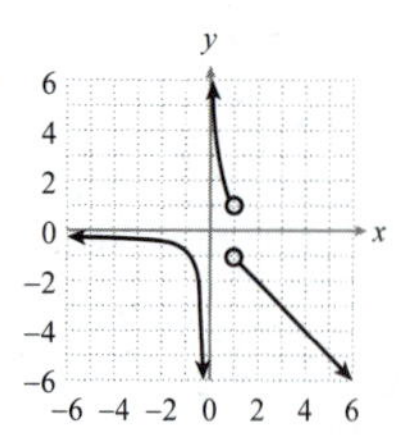

29.
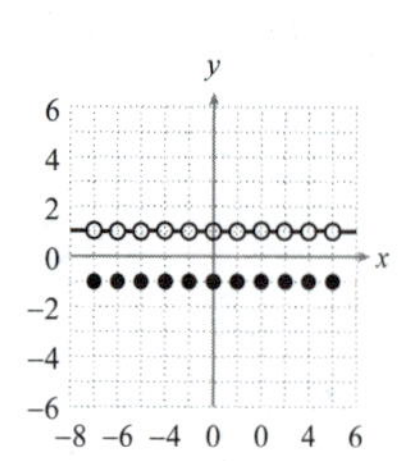

31.
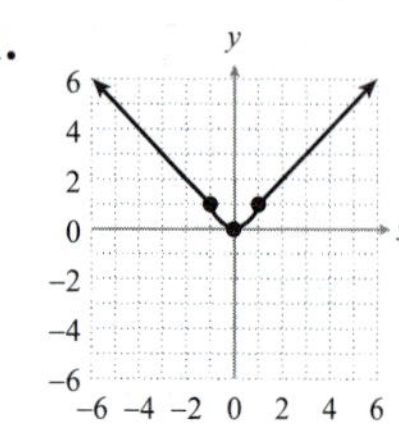

33.
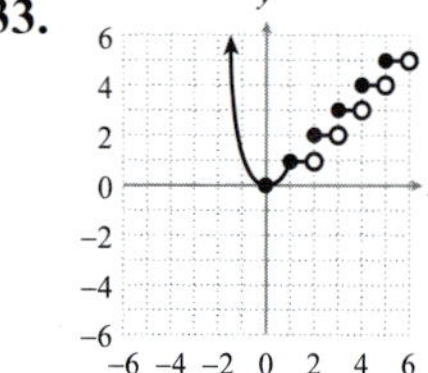

35.
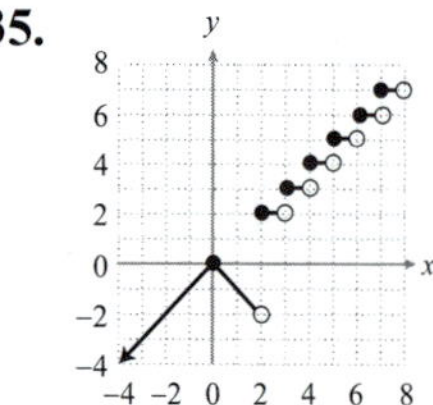

37. j **39.** a **41.** i

43. e **45.** f

3.5 Exercises

1. $A = kbh$ **3.** $W = \dfrac{k}{d^2}$

5. $r = \dfrac{k}{t}$ **7.** $x = ky^3z^2$

9. $y = 18\sqrt{5}$ **11.** $y = 60\sqrt[3]{2}$

13. $y = 0.75$ **15.** $y = 0.0024$

17. $z = 112$ **19.** $z = 48$

21. $a = 10\sqrt{3}$ **23.** $a = 36$

25. $a = 108$ **27.** 256 feet

29. 20.60 **31.** 6.7 meters

33. 1.25 centimeters

35. 34.54 inches **37.** 164.7872 in.^{2}

39. 9 watts **41.** $43

43. 17.28 ohms **45.** 19.66 inches

47. 210 cubic inches

49. $P(\sigma,\varepsilon) = \dfrac{\sigma^2}{2\varepsilon}$

3.6 Exercises

1. a. If x represents the side length of each square cut from a corner,
$$V(x) = (60-2x)(20-2x)x = 4x^3 - 160x^2 + 1200x$$
for $0 \le x \le 10$.

b. Yes; for example, $V(4) = 2496 \text{ cm}^3$.

c. Approximately 2525 cm^3, obtained by letting $x = 4.5$ cm.

d. A height of 5 cm will lead to a width of 10 cm and consequently a length of 50 cm, so the ratio of length to width will be 5:1.

3. a. $V(t) = P\left(1 - \dfrac{2t}{9}\right)$ for $0 \le t \le 3$

b. $\dfrac{9}{4}$ years, or 2 years and 3 months

c. $V(1) = \dfrac{7}{9}P$

5. a. $d = \sqrt{(1+m^2)x^2 + 2bmx + b^2}$

b. $d = \sqrt{x^2 + b^2}$

c. $d = \sqrt{(1+m^2)}\ |x|$

d. $d = 2|x|$

7. a. $s(x) = 30x^2 - 300x + 1000$ for $0 \le x \le 10$

b.
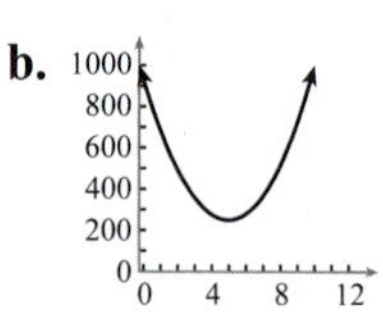

c. 1000, 250

9. a. 9.25×10^6

b. 1.45 times, or 145%

c. They aren't weightless, but instead in free fall. Astronauts fall toward Earth at the same rate as the craft they're in, making it appear as though they're floating.

11. a. $A=6V^{\frac{2}{3}}$

b. $A=600\text{ mm}^2$

13. a. $C=8x+4y=8x+\dfrac{7200}{x}$, where x is the length of fence along the boundary

b. 30 ft

15. a. $R(x)=x(30-3x)$,
$C(x)=36+6x$,
$P(x)=-3x^2+24x-36$

b. 2 **c.** 4

d. \$18 for \$12 maximum profit

17. a.

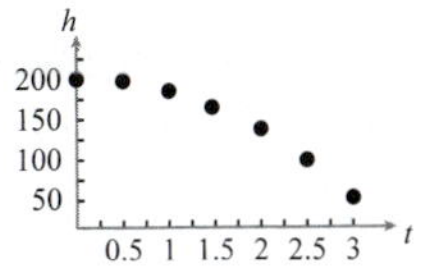

b.

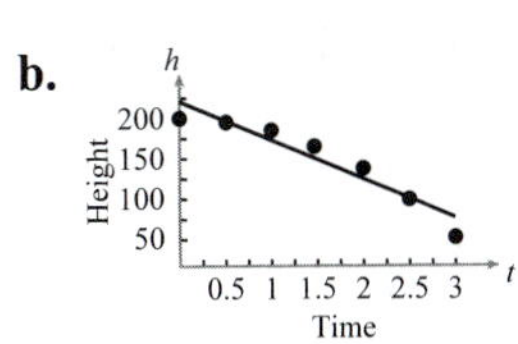

$h(t)=-48t+220$; approximately 4.58 s; 220 ft

c.

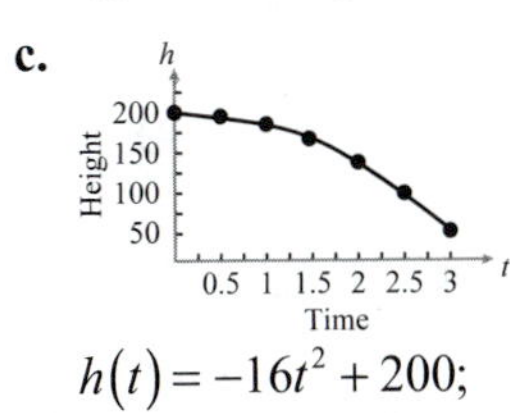

$h(t)=-16t^2+200$; approximately 3.54 s; 200 ft

d. The quadratic model is more accurate.

Chapter 3 Project

1. a.

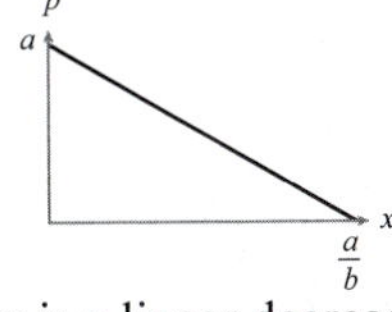

b. p is a linear decreasing function of x.

c. $x=0$ corresponds to $p=a$, so a price of a corresponds to no sales.

d. $p=0$ corresponds to $x=\dfrac{a}{b}$, which is the maximum feasible sales figure.

e. $R(x)=xp=x(a-bx)$
$=-bx^2+ax$

f. R is a quadratic function.

g. The graph of R is a downward-opening parabola.

Chapter 3 Review Exercises

1. Dom $=\{-2,-3\}$,
Ran $=\{-9,-3,2,9\}$; No

3. Dom $=\mathbb{R}$, Ran $=\{2\}$; Yes

5. Dom $=\mathbb{R}$, Ran $=\mathbb{R}$; Yes

7. Dom $=[0,\infty)$, Ran $=[4,\infty)$; Yes

9. Dom $=\{-2,4\}$,
Ran $=\{-1,5\}$; Yes

11. $f(x)=3\sqrt{x+11}-4$;
$f(-2)=5$

13. 4

15. $x=-1,\quad x=3$

17. $\sqrt{x+h}$

19. $\sqrt[3]{(x+h)^2}$

21. Dom $=\mathbb{N}$, Cod $=\mathbb{R}$,
Ran $=\left\{\dfrac{3}{4},\dfrac{3}{2},\dfrac{9}{4},\ldots\right\}$

23. $\mathbb{R}$

25.

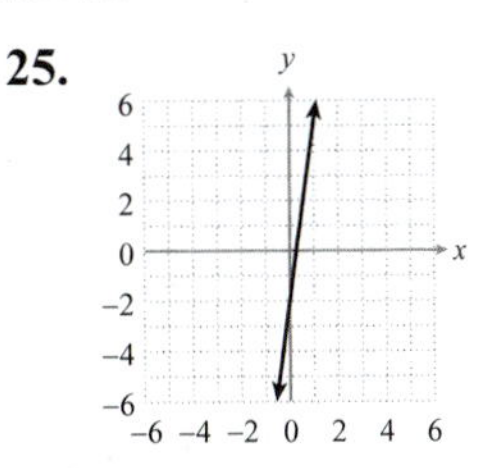

27.

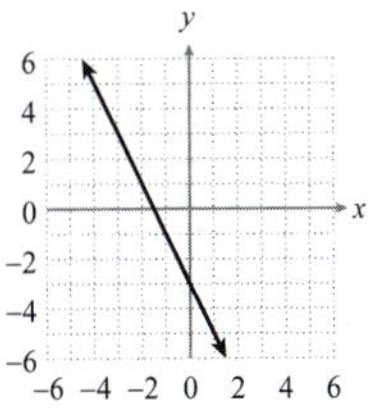

29.

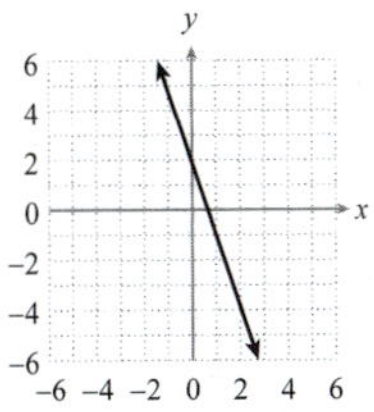

31. $f(x)=2x-1$

33. a. $y=x+0.8$

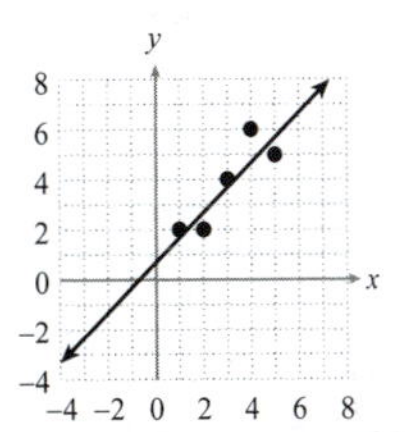

b. $r\approx0.884$

35.

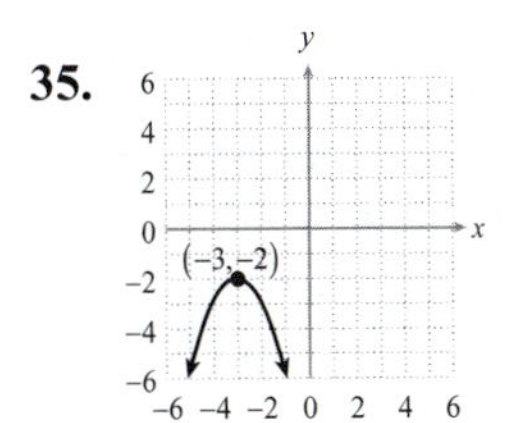

37.

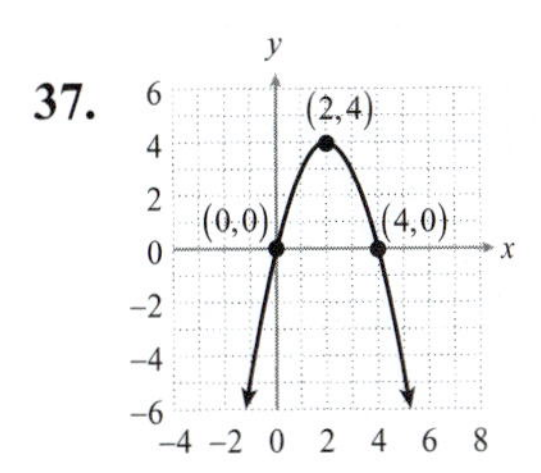

39.

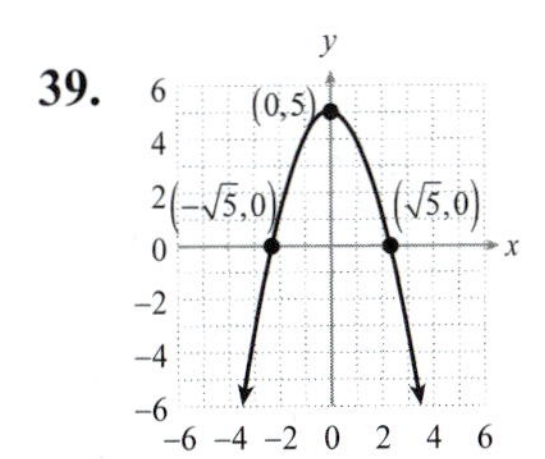

41. a. $f(x)=x^2-3x+1$

b. $(1.5,-1.25)$

43.

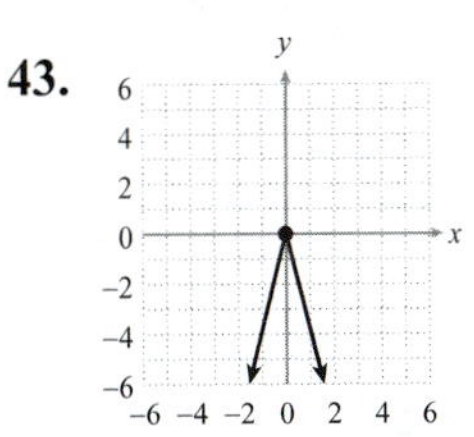

45.

47.

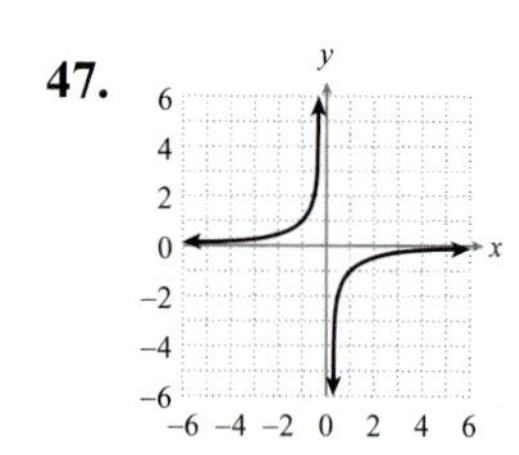

49.

51.

53.

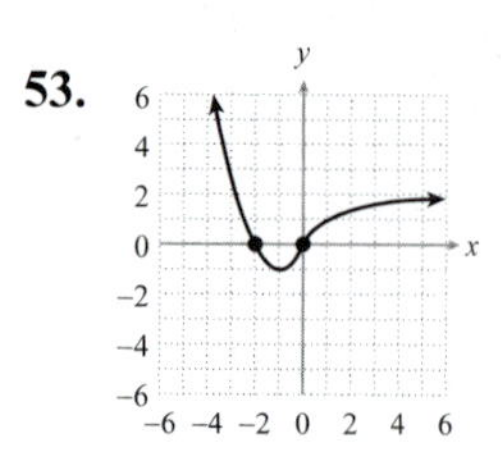

55.

57.

59. $y = \dfrac{ka^3}{\sqrt{b}}$

61. $y = 72$

63. 7.44×10^7 m

65. a. $C(x) = \dfrac{12{,}000}{x} + 13x$, where x is the length of fence along the road.

b. Approximately 30.38 ft (along the road) by 39.50 ft (not along the road); approximately $789.94

Chapter 4: Working with Functions

4.1 Exercises

1. $f(x) = x^2$ **3.** $f(x) = \sqrt[3]{x}$

5. $f(x) = \sqrt{x}$ **7.** $f(x) = \dfrac{1}{x^2}$

9. $f(x) = x^3$ **11.** $f(x) = |x|$

13. $f(x) = \sqrt{x}$ **15.** $f(x) = x^3$

17.

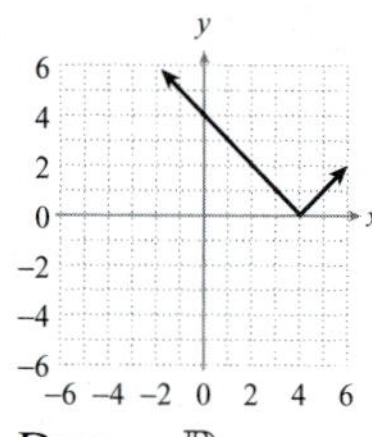

Dom = $\mathbb{R}$,

Ran = $[0, \infty)$

19.

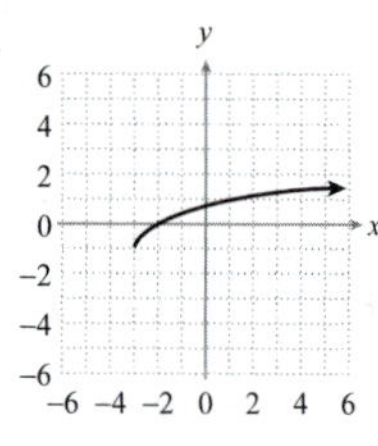

Dom = $[-3, \infty)$,

Ran = $[-1, \infty)$

21.

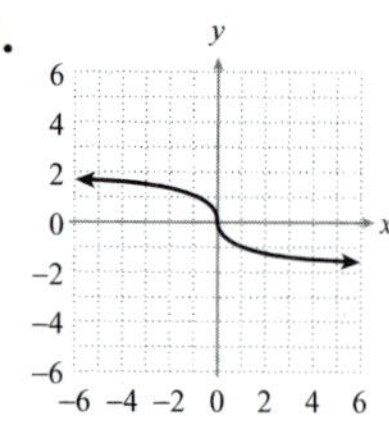

Dom = Ran = $\mathbb{R}$

23.

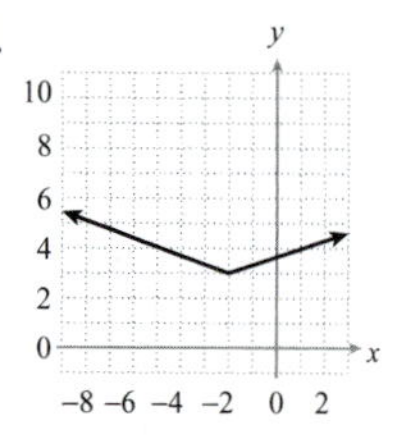

Dom = $\mathbb{R}$,

Ran = $[3, \infty)$

25.

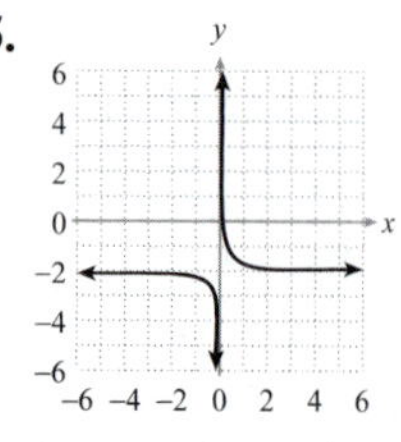

Dom = $(-\infty, 0) \cup (0, \infty)$,

Ran = $(-\infty, -2) \cup (-2, \infty)$

27.

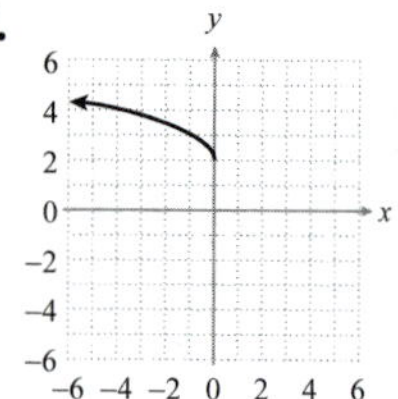

Dom = $(-\infty, 0]$,

Ran = $[2, \infty)$

29.

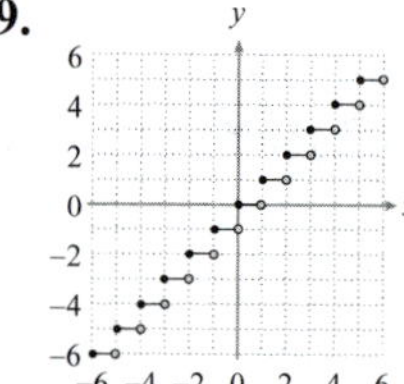

Dom = $\mathbb{R}$, Ran = $\mathbb{Z}$

31.

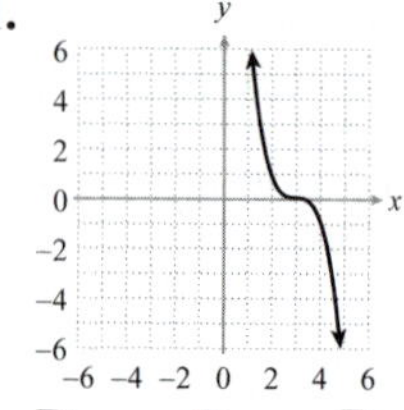

Dom = Ran = $\mathbb{R}$

33.

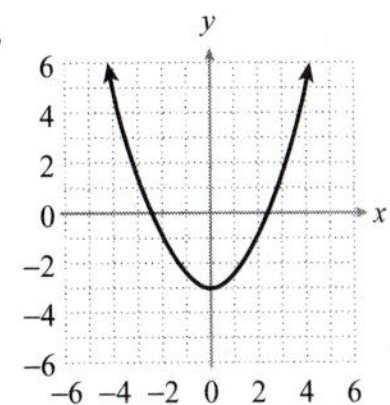

Dom = $\mathbb{R}$, Ran = $[-3,\infty)$

35.

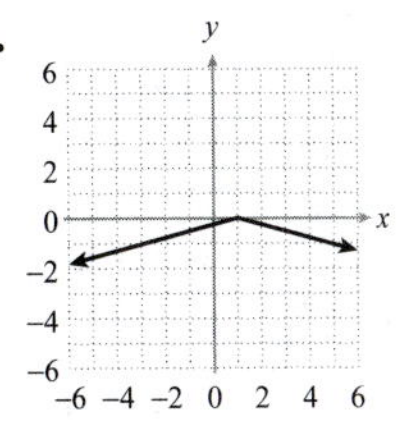

Dom = $\mathbb{R}$, Ran = $(-\infty,0]$

37.

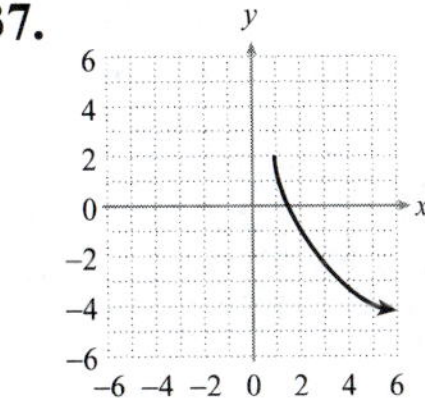

Dom = $[1,\infty)$, Ran = $(-\infty,2]$

39.

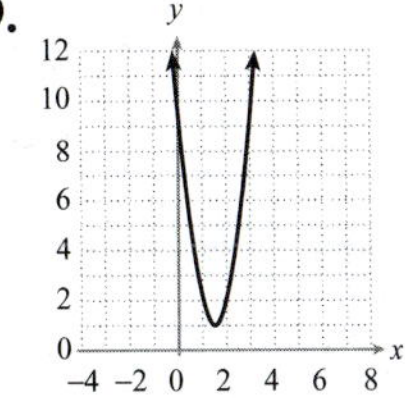

Dom = $\mathbb{R}$, Ran = $[1,\infty)$

41.

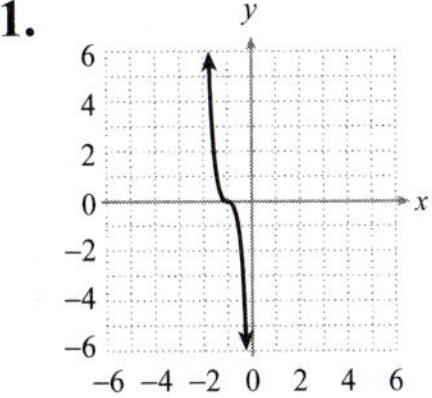

Dom = $\mathbb{R}$, Ran = $\mathbb{R}$

43.

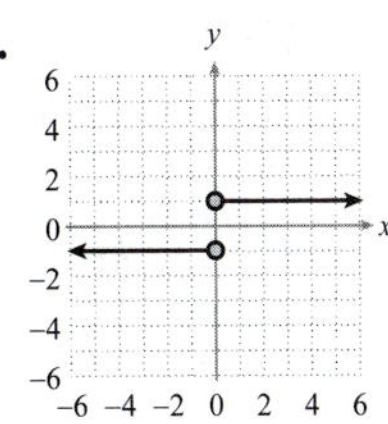

Dom = $(-\infty,0)\cup(0,\infty)$,

Ran = $\{-1,1\}$

45.

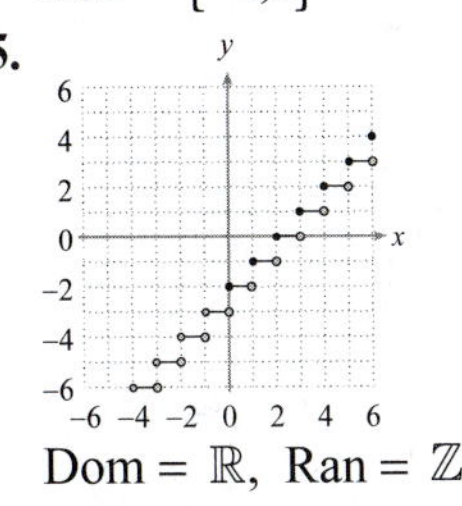

Dom = $\mathbb{R}$, Ran = $\mathbb{Z}$

47. $f(x)=(x-4)^2+2$

49. $f(x)=(-x-2)^2=(x+2)^2$

51. $f(x)=(-x+1)^3$

53. $f(x)=-\sqrt{x+5}$

55. $f(x)=\sqrt{-\dfrac{x}{2}+3}$

57. $f(x)=-|x-8|-2$

59. $f(x)=-\sqrt{-(x+1)}$

61. $f(x)=\sqrt{2x}$

63. $f(x)=-\sqrt{x+4}$

65. $f(x)=1-(x-3)^3$

73. $f(x)=|x+4|-1$

75. $f(x)=-\sqrt{6-x}+2$

77. $f(x)=6-(x-3)^2$

4.2 Exercises

1.

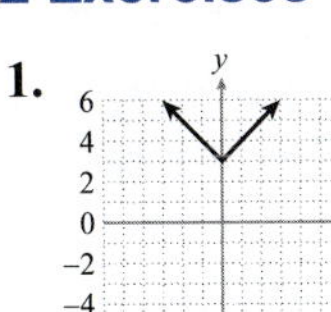

Even function; y-axis symmetry

3.

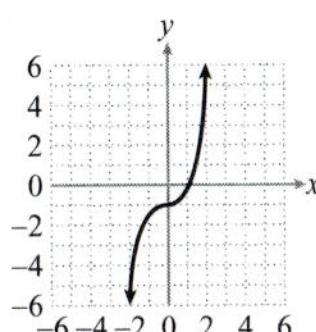

Neither; no symmetry

5. Not a function; x-axis symmetry

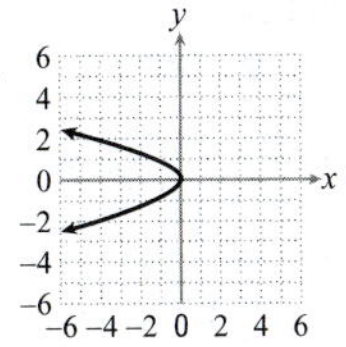

7. None of the above

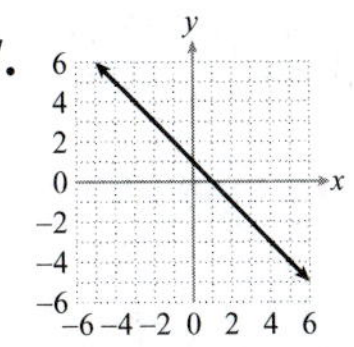

9. Not a function; x-axis symmetry

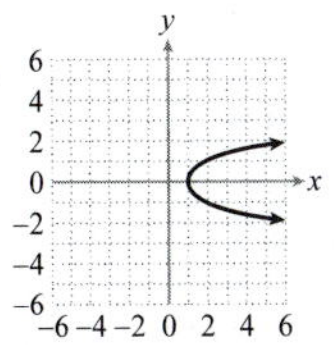

11. Even function; y-axis symmetry

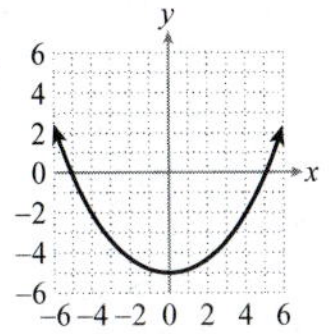

13. None of the above

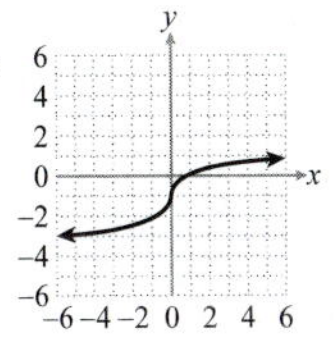

15.

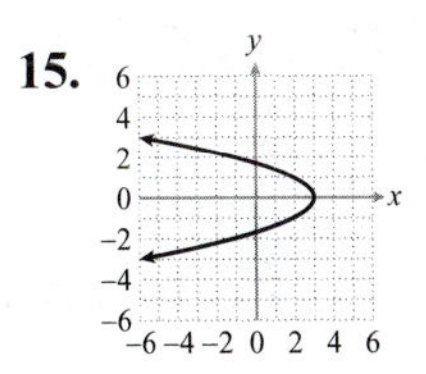

Not a function; x-axis symmetry

17. Inc. on $(-\infty,2)$,
Dec. on $(2,\infty)$

19. Inc. on $(-\infty,-3)$,
Dec. on $(-3,\infty)$

21. Constant on $\mathbb{R}$

23. Dec. on $(-\infty,4)$,
Inc. on $(4,\infty)$

25. Dec. on $(-\infty,-3)$,
Inc. on $(-3,-1)$,
Constant on $(-1,\infty)$

27. a. local min at -1, local max at 2

b. value at -1 is 0, value at 2 is 3

29. a. local min at -1, local max at 2

b. value at -1 is -12,
value at 2 is 15

31. a. local min at -2,
local max at 0, local min at 3

b. value at -2 is $\dfrac{5}{3}$, value at 0 is 7, value at 3 is $-\dfrac{35}{4}$

33. a. local min at 5

b. value at 5 is 2

35. a. local max at –2

b. value at –2 is 1

37. a. local min at 3

b. value at 3 is –2

39. 11 **41.** 2

43. $\dfrac{2-\sqrt{2}}{2}$

45. $2c+h$ **47.** $-\dfrac{1}{12}$

49. a. $[a,b]$ **b.** $[b,c],[b,d]$

c. $[a,c],[a,d],[c,d]$

51.

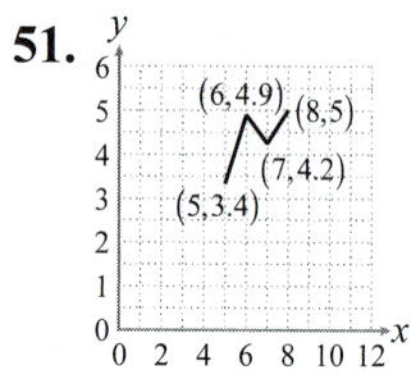

Inc. on $(5,6)$ and $(7,8)$

Dec. on $(6,7)$

53.

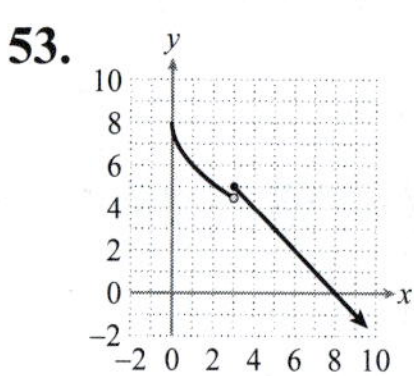

Dec. on $(0,3)\cup(3,\infty)$

55. The average rate of change is m for any interval.

57. $2cp+hp+q$; The difference quotient approaches $2cp+q$ as h gets very small.

59. Nothing; Exercise 49a illustrates this conclusion.

4.3 Exercises

1. a. 2 **b.** –8 **c.** –15 **d.** $-\dfrac{3}{5}$

3. a. –3 **b.** –1 **c.** 2 **d.** 2

5. a. 12 **b.** 18 **c.** –45 **d.** –5

7. a. 3 **b.** 1 **c.** 2 **d.** 2

9. a. 6 **b.** 0 **c.** 9 **d.** 1

11. a. 5 **b.** –1 **c.** 6 **d.** $\dfrac{2}{3}$

13. a. 3 **b.** 5 **c.** –4 **d.** –4

15. a. $|x|+\sqrt{x}$, Dom $=[0,\infty)$

b. $\dfrac{|x|}{\sqrt{x}}$, Dom $=(0,\infty)$

17. a. x^2+x-2, Dom $=\mathbb{R}$

b. $\dfrac{1}{x+1}$,

Dom: $(-\infty,-1)\cup(-1,1)\cup(1,\infty)$

19. a. x^3+3x-8, Dom $=\mathbb{R}$

b. $\dfrac{3x}{x^3-8}$,

Dom $=(-\infty,2)\cup(2,\infty)$

21. a. $-2x^2+|x+4|$, Dom $=\mathbb{R}$

b. $\dfrac{-2x^2}{|x+4|}$,

Dom $=(-\infty,-4)\cup(-4,\infty)$

23. 2 **25.** 0 **27.** 8

29. 3 **31.** 1 **33.** $\dfrac{1}{3}$

35. a. $\dfrac{4-2x}{3x}$,

Dom $=(-\infty,0)\cup(0,\infty)$

b. $\dfrac{3}{4x-2}$,

Dom $=\left(-\infty,\dfrac{1}{2}\right)\cup\left(\dfrac{1}{2},\infty\right)$

37. a. $|x^3-2|$, Dom $=\mathbb{R}$

b. $|x-3|^3+1$, Dom $=\mathbb{R}$

39. a. $\sqrt{\dfrac{x-1}{2}}$, Dom $=[1,\infty)$

b. $\dfrac{\sqrt{x-1}+1}{2}$, Dom $=[1,\infty)$

41. a. $-3x^2-4$, Dom $=\mathbb{R}$

b. $9x^2-12x+6$, Dom $=\mathbb{R}$

43. a. $\sqrt{x^2-1}$,

Dom $=(-\infty,-1]\cup[1,\infty)$

b. $x-1$, Dom $=\mathbb{R}$

45. $g(x)=\dfrac{2}{x}$, $h(x)=5x-1$,

$f(x)=g(h(x))$

47. $g(x)=x+\sqrt{x}-5$,

$h(x)=x+2$, $f(x)=g(h(x))$

49. $g(x)=\dfrac{\sqrt{x}}{x^2}$, $h(x)=x-3$,

$f(x)=g(h(x))$

51. $g(x)=x-3$, $h(x)=|x^2+3x|$,

$f(x)=g(h(x))$

53. $g(x)=\sqrt{x+5}$

55. $g(x)=-x^3-7$

57. $V=3\pi r^3$

59. $V=\dfrac{1}{12}\pi r^2t^2$

61. $(f\circ g)(x)=\sqrt[3]{\dfrac{-x^3}{3x^2-9}}$,

$(f\circ g)(-x)=\sqrt[3]{\dfrac{x^3}{3x^2-9}}$

$=-(f\circ g)(x)$

63. Yes **65.** Yes **67.** No

69. No **71.** No

4.4 Exercises

1.

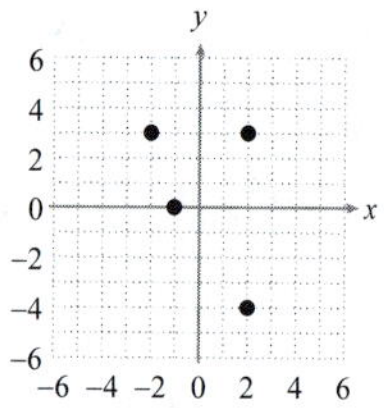

Dom $=\{2,-1,-2\}$,

Ran $=\{-4,3,0\}$

3.

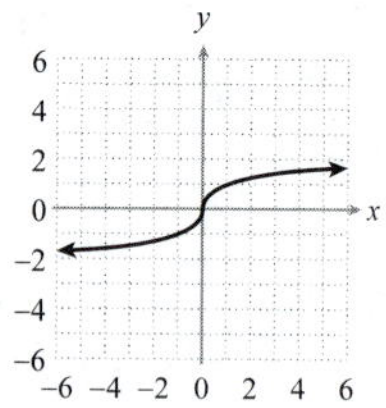

Dom $=$ Ran $=\mathbb{R}$

5.

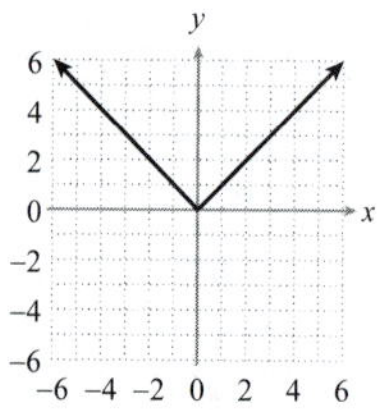

Dom $=\mathbb{R}$, Ran $=[0,\infty)$

7. 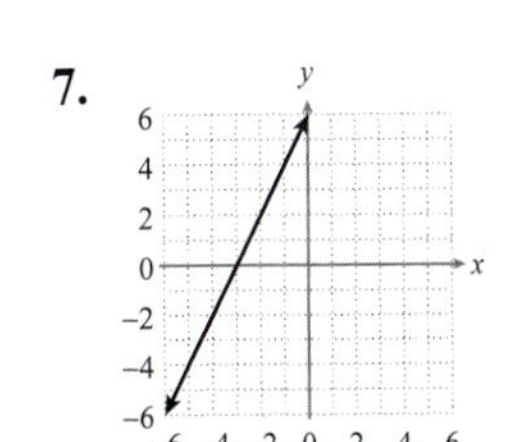

Dom = ℝ, Ran = ℝ

9. 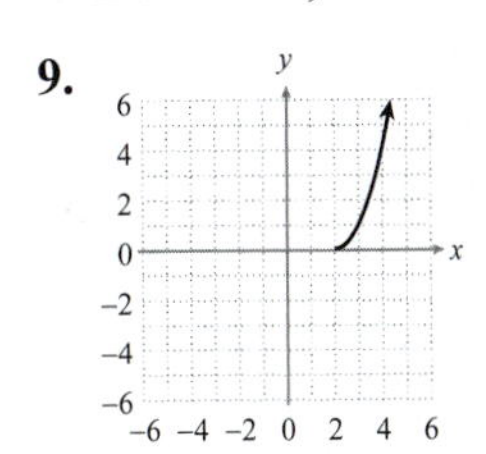

Dom = $[2,\infty)$, Ran = $[0,\infty)$

11. 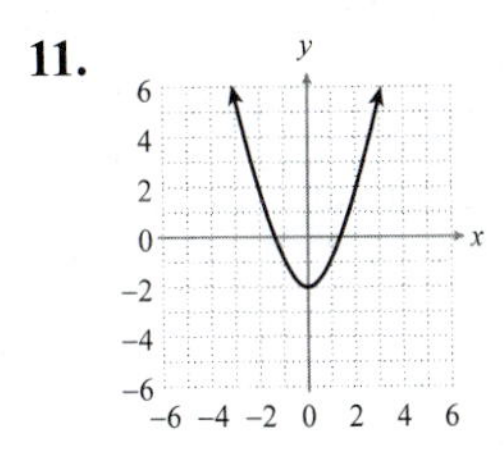

Dom = ℝ, Ran = $[-2,\infty)$

13. 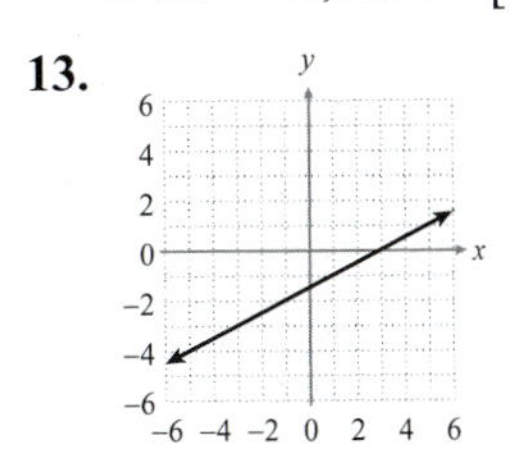

Dom = ℝ, Ran = ℝ

15. Not a one-to-one function $f(-1)=f(1)=1$

17. Restrict to $[0,\infty)$

19. Inverse exists

21. Inverse exists

23. Inverse exists

25. Restrict to $[2,\infty)$

27. Restrict to $[12,\infty)$

29. $f^{-1}(x)=(x+2)^3$

31. $r^{-1}(x)=\dfrac{-2x-1}{3x-1}$

33. $F^{-1}(x)=(x-2)^{\frac{1}{3}}+5$

35. $V^{-1}(x)=2x-5$

37. $h^{-1}(x)=(x+2)^{\frac{5}{3}}$

39. $J^{-1}(x)=\dfrac{x-2}{3x}$

41. $h^{-1}(x)=(x-6)^{\frac{1}{7}}$

43. $r^{-1}(x)=\dfrac{x^5}{2}$ **45.** $f^{-1}(x)=\dfrac{x^3}{54}$

47. $f^{-1}(x)=\sqrt{x-2}+3$ or $f^{-1}(x)=-\sqrt{x-2}+3$

49. $f^{-1}(x)=\sqrt[4]{x+2}-1$ or $f^{-1}(x)=-\sqrt[4]{x+2}-1$

61. f **63.** c **65.** d

67. 184 72 96 96 72 160 144 8 72 112 160 120 32 8 200

69. REMEMBER YOUR SUNBLOCK

71. BEACH FUN IN THE SUN

73. 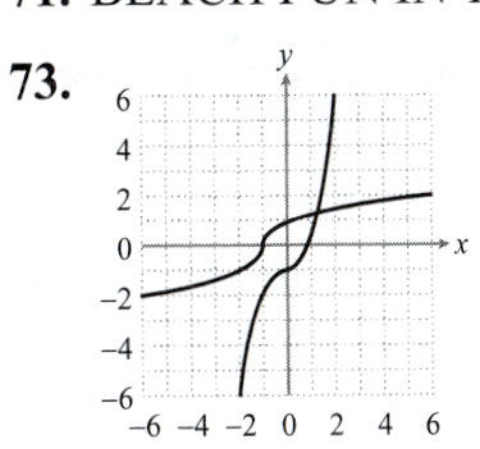

Dom = $(-\infty,\infty)$, Ran = $(-\infty,\infty)$

75. 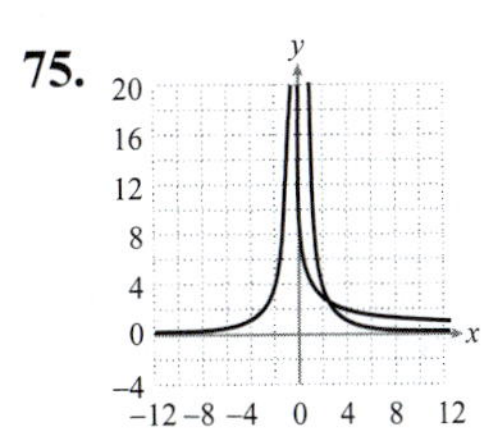

Dom = $(0,\infty)$, Ran = $(0,\infty)$

77. 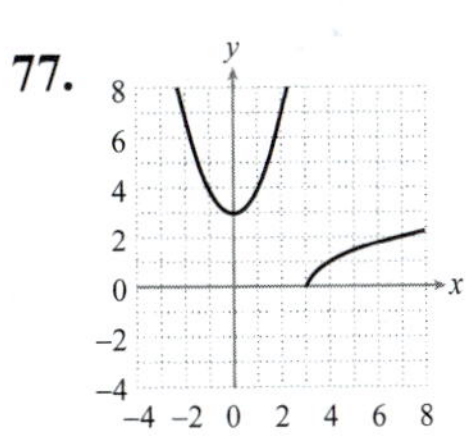

Dom = $[3,\infty)$, Ran = $[0,\infty)$

Chapter 4 Project

1. $A(r)=\pi r^2$

3. $A(t)=6.76\pi t^2$

5. $r(5.5)=14.3$ km

7. $A(5.5)\approx 642.4$ km^2

9. Approximately 286.7 km^2/hr

Chapter 4 Review Exercises

1. 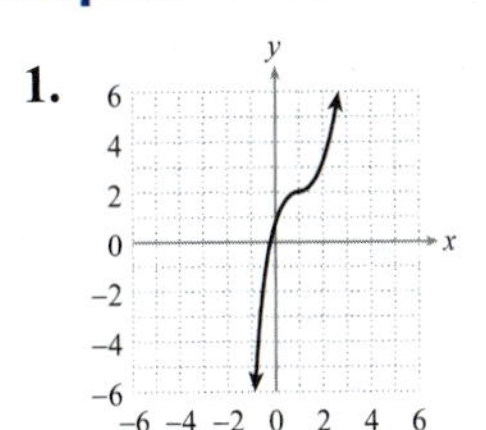

Dom = Ran = ℝ

3. 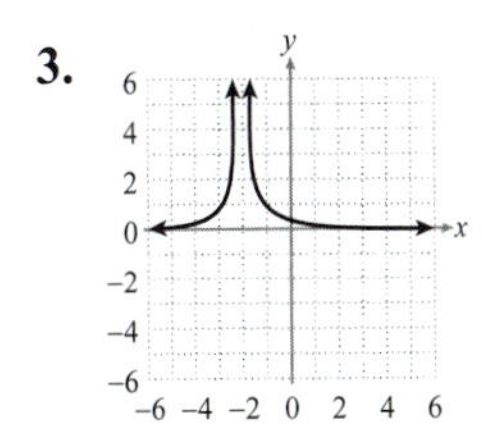

Dom = $(-\infty,-2)\cup(-2,\infty)$, Ran = $(0,\infty)$

5. 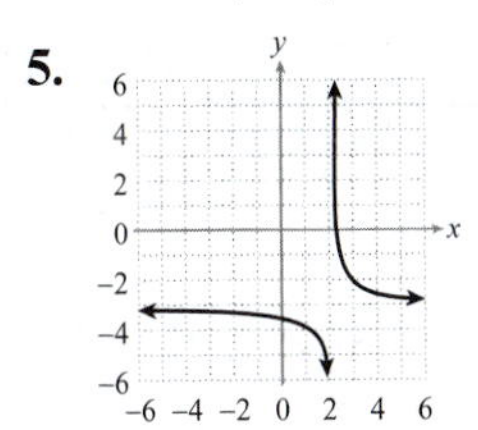

Dom = $(-\infty,2)\cup(2,\infty)$, Ran = $(-\infty,-3)\cup(-3,\infty)$

7. 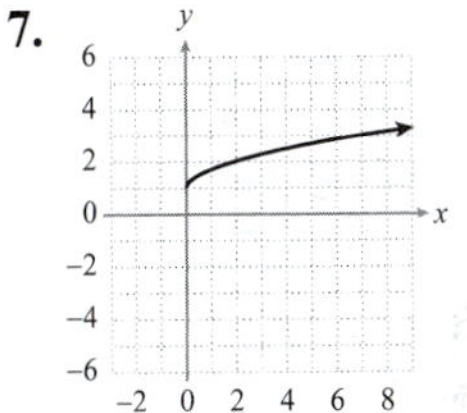

Dom = $[0,\infty)$, Ran = $[1,\infty)$

9. $f(x)=(x-1)^2-2$

11. $f(x)=-\sqrt{x}+4$

13. Even function; y-axis symmetry

15. Dec. on $(-\infty,2)$, Inc. on $(2,\infty)$

17. a. Local min at 1, local max at 2, local min at 3

b. Value at 1 is 1, value at 2 is 4, value at 3 is 1

19. $-\dfrac{1}{3}$ **21.** -2

23. a. 3 **b.** 5 **c.** -4 **d.** -4

25. a. -2 **b.** 18 **c.** -80 **d.** $-\dfrac{4}{5}$

27. a. $\frac{1}{x-2}+\sqrt[3]{x}$, Dom $=(-\infty,2)\cup(2,\infty)$

b. $\frac{1}{\sqrt[3]{x}(x-2)}$, Dom $=(-\infty,0)\cup(0,2)\cup(2,\infty)$

29. a. $x^2+\sqrt[3]{x}-5$, Dom $=\mathbb{R}$

b. $\frac{x^2-4}{\sqrt[3]{x}-1}$, Dom $=(-\infty,1)\cup(1,\infty)$

31. $-\frac{9}{2}$ **33.** $-\frac{2}{3}$

35. a. $\frac{1}{\sqrt{x-2}}$, Dom $=(2,\infty)$

b. $\frac{1}{\sqrt{x-4}}+2$, Dom $=(4,\infty)$

37. a. $3\sqrt{x-3}$, Dom $=[3,\infty)$

b. $\sqrt{3x-3}$, Dom $=[1,\infty)$

39. $g(x)=\frac{\sqrt{x}}{x^2}$, $h(x)=x+2$, $f(x)=g(h(x))$

41. $g(x)=\frac{2}{x}+1$

43.

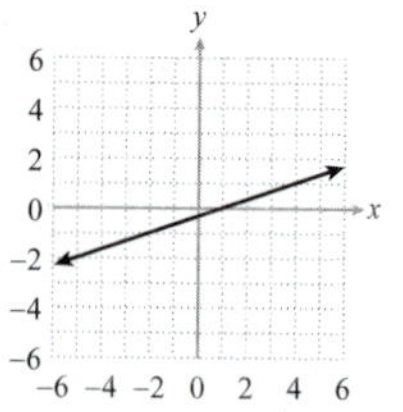

Dom = Ran = $\mathbb{R}$

45. $r^{-1}(x)=\frac{x+2}{7x}$

47. $f^{-1}(x)=(x+6)^5$

49. $f^{-1}(x)=\frac{-x-3}{x-2}$

51. $f^{-1}(x)=\frac{x-3}{8}$

Chapter 5: Polynomial and Rational Functions

5.1 Exercises

19. Yes **21.** Yes **23.** Yes

25. $1\pm 2i$ **27.** $-3, \frac{1}{2}$

29. $\pm\sqrt{3}, \pm\sqrt{5}$ **31.** $-\frac{5}{2}$

33. $0, 4\pm 3i$ **35.** $\pm 1, \pm 2i\sqrt{2}$

37. 7th-degree; lead coef. = 4; $j(x)\to-\infty$ as $x\to-\infty$; $j(x)\to\infty$ as $x\to\infty$

39. 5th-degree; lead coef. = −6; $h(x)\to\infty$ as $x\to-\infty$; $h(x)\to-\infty$ as $x\to\infty$

41. 4th-degree; lead coef. = −2; $f(x)\to-\infty$ as $x\to-\infty$ and $x\to\infty$

43.

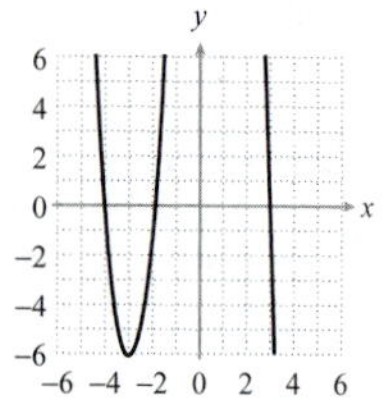

$g(x)\to\infty$ as $x\to-\infty$; $g(x)\to-\infty$ as $x\to\infty$

x-int: $(-4,0),(-2,0),(3,0)$; y-int: $(0,24)$

45.

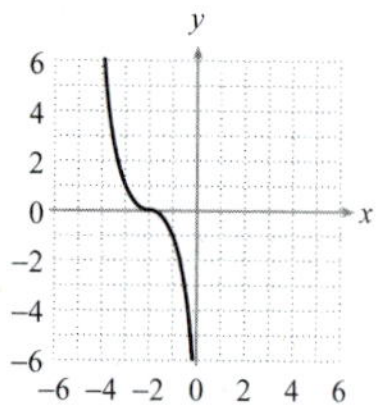

$h(x)\to\infty$ as $x\to-\infty$; $h(x)\to-\infty$ as $x\to\infty$

x-int: $(-2,0)$; y-int: $(0,-8)$

47.

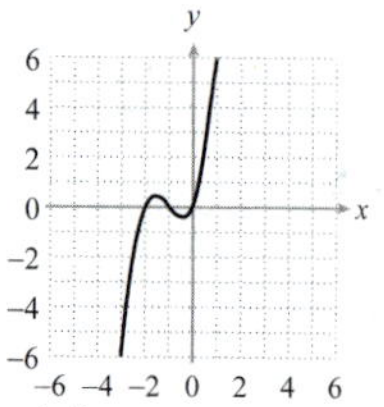

$s(x)\to-\infty$ as $x\to-\infty$; $s(x)\to\infty$ as $x\to\infty$

x-int: $(-2,0),(-1,0),(0,0)$; y-int: $(0,0)$

49.

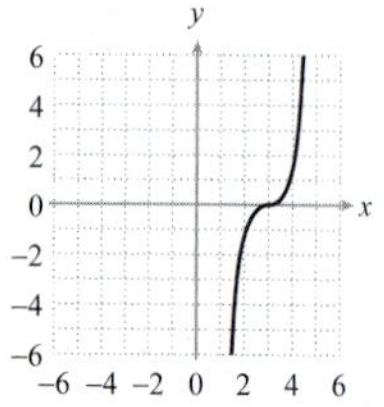

$g(x)\to-\infty$ as $x\to-\infty$; $g(x)\to\infty$ as $x\to\infty$

x-int: $(3,0)$; y-int: $(0,-243)$

51. $p(x)=\left(\frac{1}{2}\right)(x+2)\times(x+1)(x-1)(x-3)$

53. $p(x)=-(x-1)(x-2)(x-3)$

55. e **57.** a **59.** f

61. d **63.** f **65.** b

67. $(-\infty,-2)\cup(3,\infty)$

69. $(-\infty,-2)\cup(-1,0)$

71. $[-2,1]\cup[3,\infty)$

73. $[-5,-1]\cup[1,4]$

75. $\left(-\frac{1}{2},2\right)$

77. $(-\infty,-4)\cup(2,3)$

79. All integers between 5 and 27, inclusive

81. All integers between 11 and 23, inclusive

83. Between 3490 and 17,740 phones, inclusive.

85. About 141.4 weeks

5.2 Exercises

1. $3x^2-x+1+\frac{5x-1}{2x^2+2}$

3. $x-2+\frac{-2}{x^2-4x+4}$

5. $4x^2-14x+29+\frac{-65}{x+2}$

7. $x^3+6x^2-2x+5+\frac{2x+5}{3x^2-1}$

9. $2x^3-3x^2+2x-5$

11. $x^3+3x^2+10x+10+\frac{22}{x-3}$

13. $3x^2+5x+9+\frac{45}{3x-5}$

15. $2x-5+\frac{7}{x+3}$

17. $x^2-ix+6+\frac{1+i}{2x-i}$

19. x^2+3

21. $p(1)=4$

23. c is a zero

25. c is a zero

27. $p(1)=12$

29. c is a zero

31. c is a zero

33. c is a zero

35. c is a zero

37. $p(5)=-2$

39. c is a zero

41. $x^2-4x+2+\frac{-1}{x+5}$

43. $x^7-3x^2+\frac{3}{x+1}$

45. $4x^2-4x+2$

47. $x^4-x^3-x^2-7x-14-\frac{10}{x-2}$

49. x^3-x^2+x

51. $2x^2-4ix+17+\frac{8+48i}{x-3i}$

53. $f(x)=-x^2-x+12$

55. $f(x)=-x^2+4x-13$

57. $f(x)=x^4-12x^3+54x^2-108x+81$

59. $f(x)=3x^4+9x^3-9x^2-21x+18$

61. $SA=(x+5)(x+2)=x^2+7x+10$

5.3 Exercises

1. $\pm\left\{\frac{1}{3},\frac{2}{3},1,\frac{4}{3},2,\frac{8}{3},4,8\right\},\left\{-4,\frac{1}{3},2\right\}$

3. $\pm\{1,2,3,4,6,8,12,24\},\{\pm 2i,2,3\}$

5. $\pm\{1,2,7,14\},\{1,2,7\}$

7. $\pm\left\{\frac{1}{2},1,\frac{3}{2},\frac{5}{2},3,5,\frac{15}{2},15\right\},\left\{-1,\frac{5}{2},3\right\}$

9. $\pm\left\{\frac{1}{3},1,3\right\},\{-1,1,-i,i\}$

11. $\pm\{1,11\},\{-11,-1,1\}$

13. $\{-1,1,-i,i\}$

15. $\{-1,2-3i,2+3i\}$

17. $\{-2i,2i,2,3\}$

19. $\{4,1-2i,1+2i\}$

21. $\{-5i,5i,-2,1\}$

23. $\{-11,-1,1\}$

25. 0 pos., 3 or 1 neg.

27. 2 or 0 pos., 1 neg.

29. 3 or 1 pos., 1 neg.

31. 1 pos., 1 neg.

33. 3 or 1 pos., 0 neg.

35. 0 pos., 0 neg.

37. $[-5,1]$

39. $[-1,6]$

41. $[-3,6]$

43. $[-3,3]$

45. $[-3,6]$

47. $\{-4,-1,1\}$

49. $\{-1,2,5\}$

51. $\{2,3,\pm 2\sqrt{2}\}$

53. $\pm\{\sqrt{5},i\sqrt{5}\}$

55. $\left\{-2,-\frac{1}{2},6\right\}$

57. $f(-3)=-84,\ f(-1)=16$

59. $f(2)=-15,\ f(3)=24$

61. $f(2)=15,\ f(3)=-24$

63. $\{-1,2,5\}$

65. $\left\{-\frac{7}{3},\pm 1\right\}$

67. $\left\{-1,-\frac{2}{5},2,\frac{7}{3}\right\}$

69. $\left\{\pm 3,-2,-\frac{1}{3}\right\}$

71. $\left\{-5,\pm 1,\frac{5}{2},4\right\}$

73. $\{-3,-2,-1\}$

75. $\{-9,\pm\sqrt{2}\}$

77. $\left\{-3,-\frac{1}{2},8\right\}$

79. $\left\{\pm 2,-\frac{3}{4},1\right\}$

81. $\left\{\sqrt[3]{-2},\pm\frac{\sqrt{6}}{2},\frac{5}{3}\right\}$

5.4 Exercises

1.

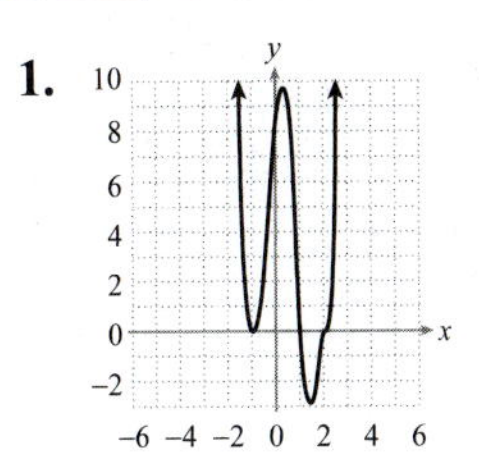

3.

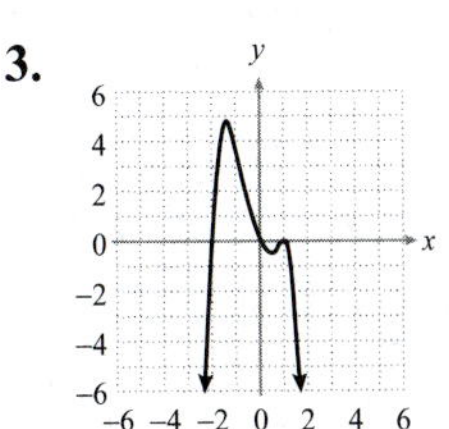

5.

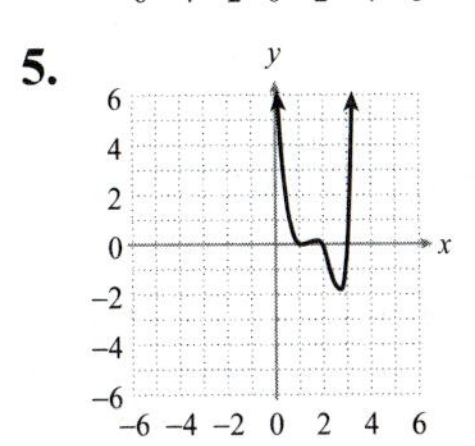

7.

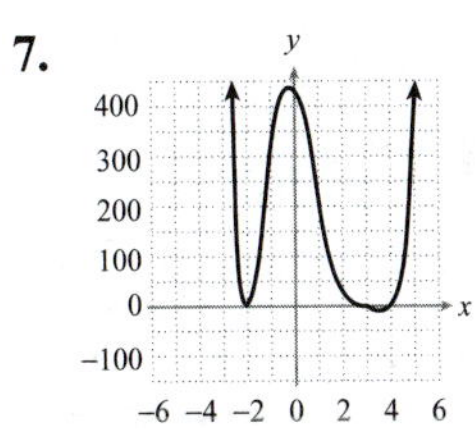

9.

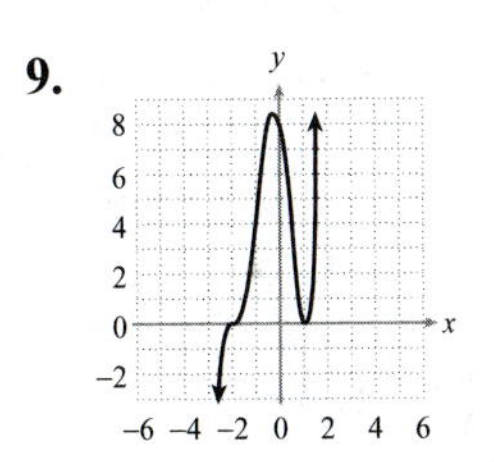

$f(x)=(x+2)^3(x-1)^2$

11.

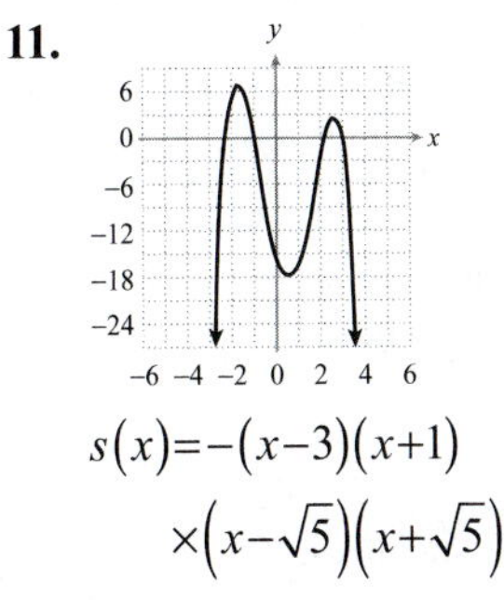

$s(x)=-(x-3)(x+1)\times(x-\sqrt{5})(x+\sqrt{5})$

13.

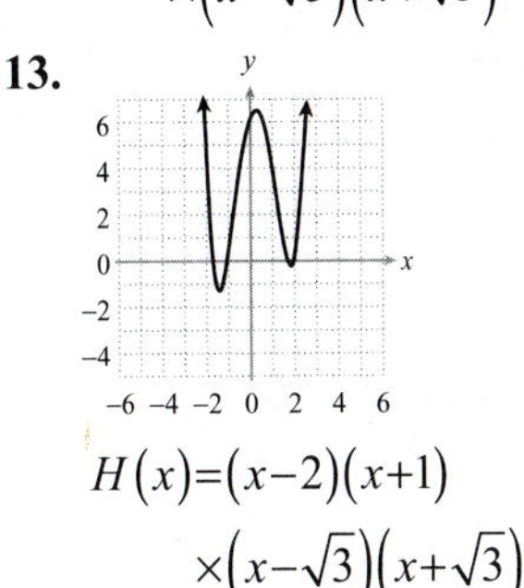

$H(x)=(x-2)(x+1)\times(x-\sqrt{3})(x+\sqrt{3})$

15. 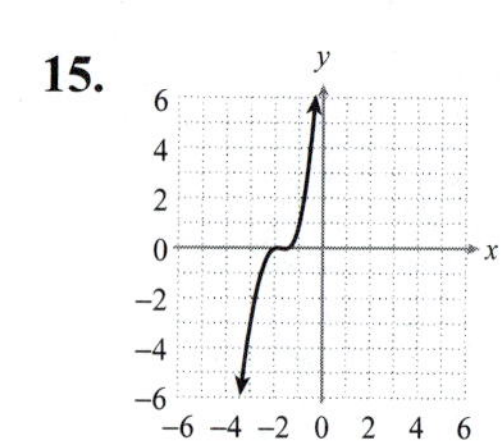

$f(x)=(x+2)^2(2x+3)$

17. $\{-2,1\}$ 19. $\{-2,1,\pm i\sqrt{5}\}$

21. $\{-3,1,\pm 2i\}$ 23. $\{\pm 1,\pm 4i\}$

25. $(x-3+2i)(x-3-2i)$
$\times(x-4)(x+1)$

27. $(x-1-3i)(x-2)(x+1)$

29. $(x+\sqrt{7})(x-\sqrt{7})(x-2+3i)$
$\times(x-2-3i)$

31. $(x-2)(x+1)(x-1+2i)$
$\times(x-1-2i)$

33. $x(x-3)(x-1)(x+11)$

35. $(x-2)(x-2+4i)(x-2-4i)$

37. $f(x)=-2x^3+18x^2-32x-52$

39. $f(x)=2x^5+2x^4-10x^3-2x^2$
$+16x-8$

41. $f(x)=3x^4-18x^3+12x^2-72x$

43. $f(x)=-x^3+2x^2+14x-40$

45. $f(x)=-x^3+4x^2+15x-68$

47. a. $V(x)=4x(5-x)(9-x)$
b. $x=0, x=5, x=9$
c. $x=0$ and $x=5$

49. a. $V(x)=x(17-2x)(9-2x)$
b. $x=0, x=\frac{9}{2}, x=\frac{17}{2}$
c. $x=0$ and $x=\frac{9}{2}$

5.5 Exercises

1. $x=1$

3. No vertical asymptote

5. $x=2$ 7. $x=0$

9. $x=-\frac{1}{2}$

11. No vertical asymptote

13. $x=7$ 15. $x=-2$

17. $x=-2, x=2$ 19. $y=0$

21. No horizontal or oblique asymptote

23. $y=0$ 25. $y=2$

27. $y=3x+6$ 29. $y=0$

31. $y=0$ 33. $y=x-11$

35. $y=5x+4$

37. 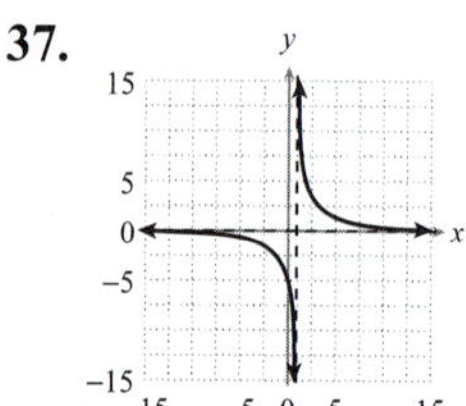

39. 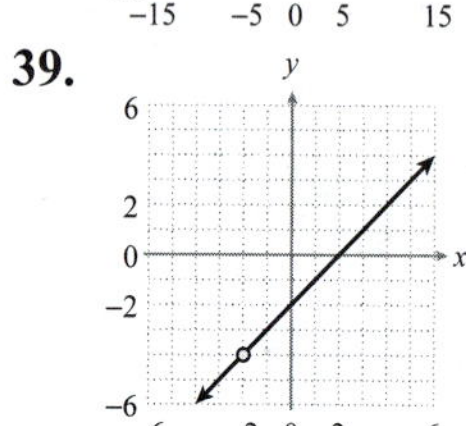

41. 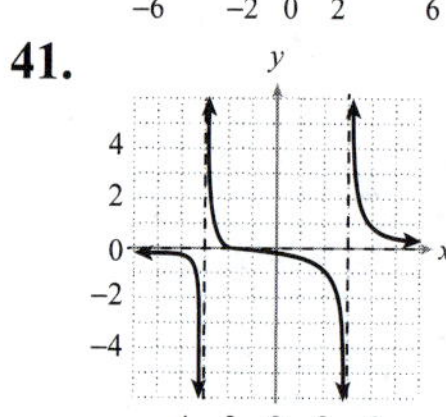

43. 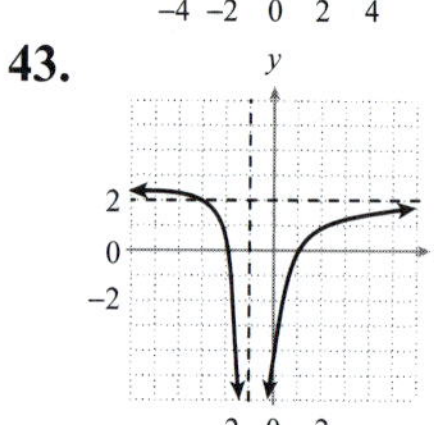

45. 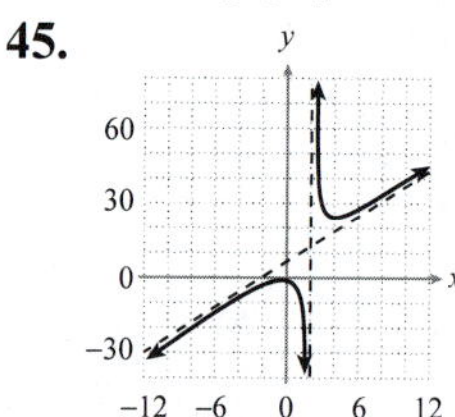

47. 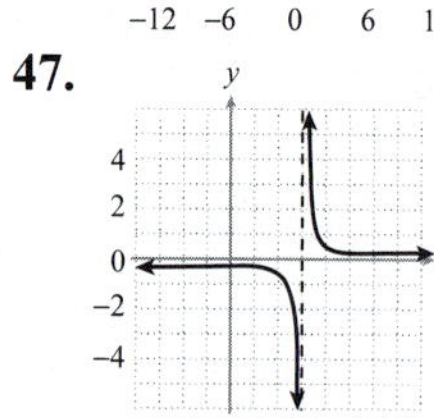

49. a. $x=-2$ b. $y=0$
c. None d. None e. $(0,5)$

51. a. $x=9$ b. $y=0$
c. None d. None
e. $\left(0,-\frac{1}{3}\right)$

53. a. $x=-1, x=1$ b. None
c. $y=x$ d. $(\sqrt[3]{3},0)$
e. $(0,3)$

55. a. $x=1$ b. None c. $y=3x$
d. None e. $(0,-3)$

57. $(-\infty,-2)\cup(-1,1)$

59. $(-8,-2)\cup(2,\infty)$

61. $(-\infty,-2)\cup(-2,3)$

63. $(0,3)$

65. $(-2,-1)\cup(1,\infty)$

67. $(-\infty,-1)\cup\left[-\frac{1}{2},0\right)$

69. a. 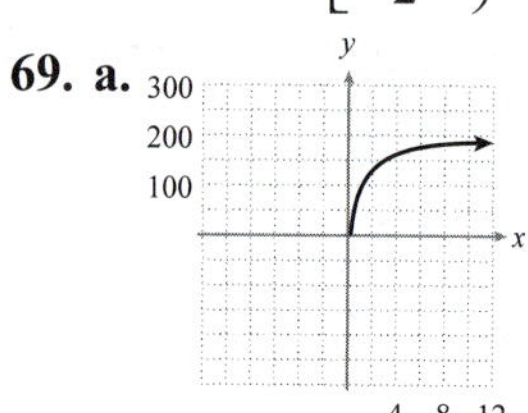

b. April's fish population approaches a maximum of 200 fish.

71. a. 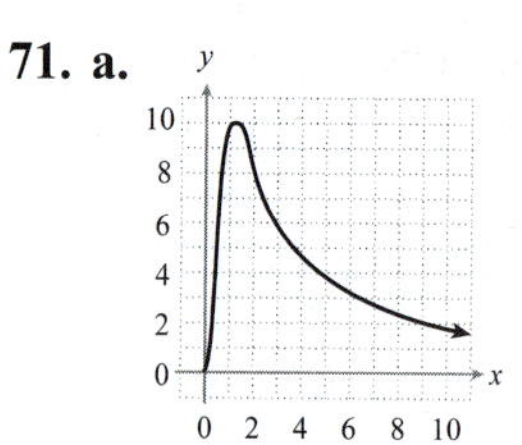

b. The concentration of the drug disappears in the long run.

Chapter 5 Project

1. $P(x)=-3x^2+5500x-729{,}000$

3. \$4933 for $x=1689$ to \$9568 for $x=144$

5. $[212,1621]$

Chapter 5 Review Exercises

5. $\pm\sqrt{2},\pm\sqrt{5}$ 7. $\pm\sqrt{2}$

9. $0,\frac{-1\pm\sqrt{5}}{2}$

11. x-int: $-2, 1, 3$; y-int: 6

$f(x) \to -\infty$ as $x \to -\infty$;

$f(x) \to \infty$ as $x \to \infty$

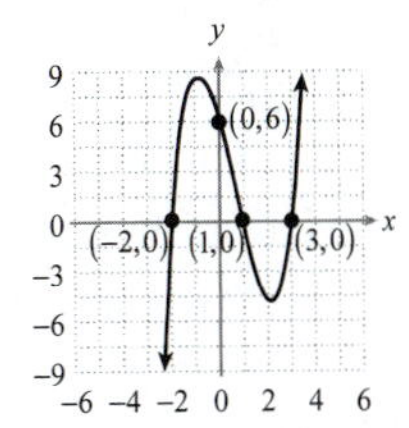

13. x-int: 1, 4; y-int: 4

$g(x) \to \infty$ as $x \to \pm\infty$

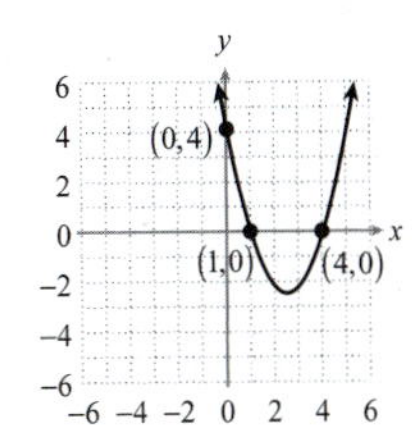

15. $\left[\frac{5}{2}, 3\right]$

17. $[-2,-1] \cup [1,4]$

19. $(-\infty,0) \cup (0,1) \cup (2,\infty)$

21. All integers between 4 and 30, inclusive

23. $11x + 35 + \dfrac{100}{x-3}$

25. $2x^3 + 2x^2 - 2x - 3 + \dfrac{-2x-2}{x^2 - x}$

27. $p(1) = 90$ **29.** $p\left(\frac{2}{3}\right) = -\frac{7}{3}$

31. $-x^3 + 2x^2 - 7x + 23$

33. $-x^3 + 7x^2 + x - 3 + \dfrac{-1}{x-1}$

35. $f(x) = x^2 - 4x - 12$

37. $f(x) = 2(x^2 - 4)(x - 3)$

39. $\pm\left\{\frac{1}{2}, 1, \frac{3}{2}, 3, \frac{9}{2}, 9\right\}, \left\{1, \frac{3}{2}, 3\right\}$

41. $\pm\{1,3,9\}, \{-3,-1\}$

43. $\left\{1, \frac{3}{2}, 3\right\}$ **45.** $\{-3,-1\}$

47. 4, 2 or 0 pos., 2 or 0 neg.

49. $[-3,7]$ **51.** $\left\{-\frac{5}{2}, -\frac{1}{2}, 7\right\}$

53. $f(2) = 3; f(4) = -15$

55. $\{\pm 3i, 4\}$ **57.** $\{3, 2 \pm \sqrt{3}\}$

59. $\left\{-\frac{1}{2}, 2\right\}$

61.

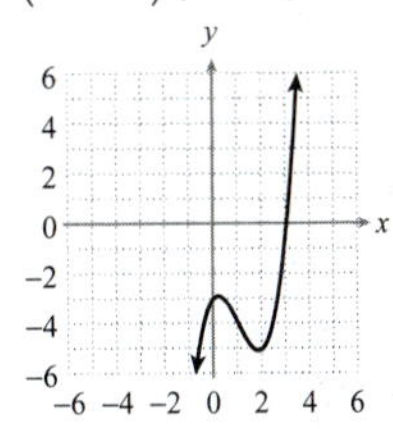

63. $(x^2 + 1)(x - 3)$

65. $\left\{-1, \frac{5}{3}, \pm 2i\right\}$

67. $\{\pm 2i, -3, \pm 1\}$

69. $(x - 5i)(x + 5i)(x - 6)(x + 1)$

71. $(x+3)\left(x - \dfrac{1 + i\sqrt{19}}{4}\right) \times \left(x - \dfrac{1 - i\sqrt{19}}{4}\right)$

73. $f(x) = x^5 + 3x^4 - 3x^3 - 17x^2 - 18x - 6$

75. $x = \frac{5}{2}$ **77.** $x = 0$

79. $y = 2x + 9$

81. No horizontal or oblique asymptote

83.

85.

87. $\left(-3, \frac{7}{2}\right]$

89. $\left(-3, \frac{8}{9}\right) \cup (2, \infty)$

Chapter 6: Exponential and Logarithmic Functions

6.1 Exercises

1.

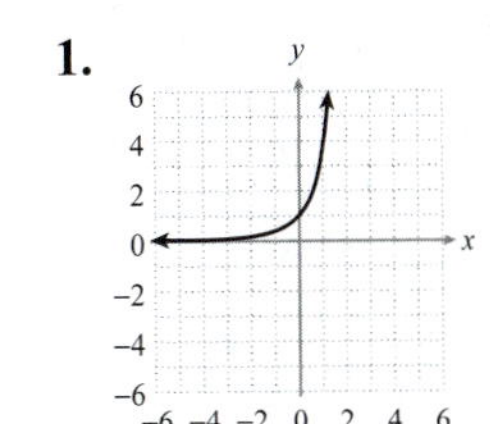

Dom $= (-\infty, \infty)$, Ran $= (0, \infty)$

3.

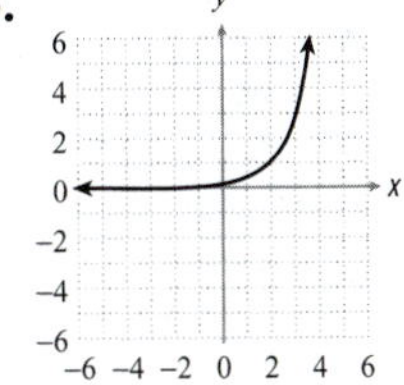

Dom $= (-\infty, \infty)$, Ran $= (0, \infty)$

5.

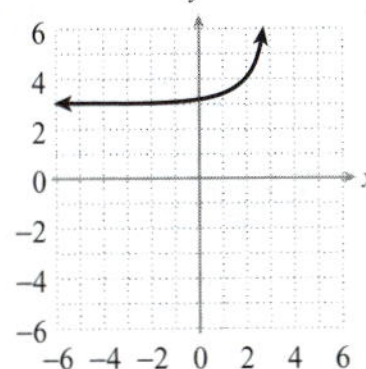

Dom $= (-\infty, \infty)$, Ran $= (3, \infty)$

7. Dom $= (-\infty, \infty)$, Ran $= (0, \infty)$

9. Dom $= (-\infty, \infty)$, Ran $= (0, \infty)$

11. Dom $= (-\infty, \infty)$, Ran $= (0, \infty)$

13. Dom $= (-\infty, \infty)$, Ran $= (-\infty, 3)$

15. Dom $= (-\infty, \infty)$, Ran $= (0, \infty)$

17. Dom $= (-\infty, \infty)$, Ran $= (-\infty, 2)$

19. Dom $= (-\infty, \infty)$, Ran $= (0, \infty)$

21. Dom $= (-\infty, \infty)$, Ran $= (-\infty, 1)$

23. $\{2\}$ **25.** $\{-2\}$ **27.** $\{-13\}$

29. $\{3\}$ **31.** $\{-2\}$

33. $\{-2, -1\}$

35. $\{7\}$ **37.** $\{3\}$ **39.** $\{9\}$

41. $\{-3\}$ **43.** $\{2\}$ **45.** $\{-1\}$

47. a **49.** i **51.** d

53. e **55.** h

6.2 Exercises

1. $V \approx 178$ people

3. $C \approx \$8526.20$

5. $W \approx 93$ computers

7. a. $a \approx 0.999567$

b. $A \approx 0.958$ grams

c. $A \approx 0.648$ grams

9. a. 3 years **b.** 9 years

11. 1118 rabbits

13. The bank offering 2.75% and monthly compounding.

15. Approximately 3.18%

17. \$134,392

19. a. 10 **b.** 7490 people

c. The function approaches 10,000 as time goes on.

21. a. $a \approx 0.965936$

b. $A \approx 0.707$ kg

c. $A \approx 7.628$ mg

23. a. \$1521.74 **b.** \$271.74

25. \$9459.48; \$9942.41

27. \$2835.71

29. \$20,000

31. \$7318.71

33. a. \$7647.95 **b.** \$7647.57

c. Yes; daily compounding is a frequency close enough to continuous compounding to make little difference at the hundredths place.

35. a. Linear:

$y = 1.06098 \times 10^6 x - 1.94223 \times 10^9$

Quadratic:

$y = 8501.22x^2 - 3.08186 \times 10^7 x + 2.79424 \times 10^{10}$

Exponential:

$y = 2.172414 \times 10^{-12} (1.023986)^x$

b. Linear: −32,466,000

Quadratic: 12,872,800

Exponential: 7,348,129

Actual: 5,308,483

None of these regression models appear to be very accurate this far from known data, though the population estimates given by the exponential model are closer to the actual population.

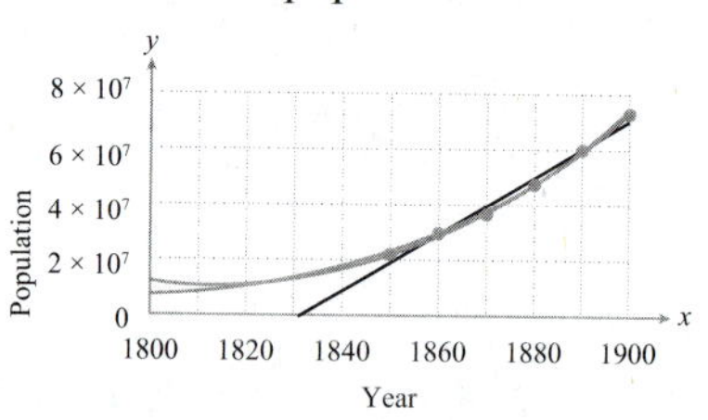

6.3 Exercises

1. $4 = \log_5 625$ **3.** $3 = \log_x 27$

5. $3 = \log_{4.2} C$ **7.** $x = \log_4 31$

9. $\sqrt{3} = \log_{4x} 13$ **11.** $e^x = \log_2 11$

13. $81 = 3^4$ **15.** $4 = b^{\frac{1}{2}}$

17. $15 = 2^b$ **19.** $W = 5^{12}$

21. $2x = \pi^4$ **23.** $e^x = 2$

25.

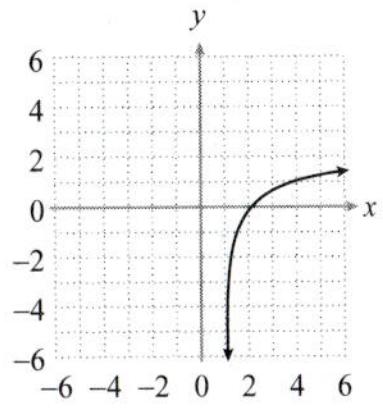

Dom: $(1,\infty)$, Ran: $(-\infty,\infty)$

27.

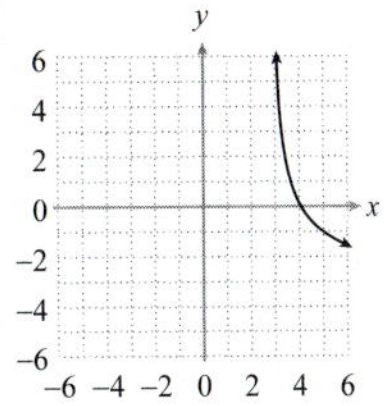

Dom: $(3,\infty)$, Ran: $(-\infty,\infty)$

29.

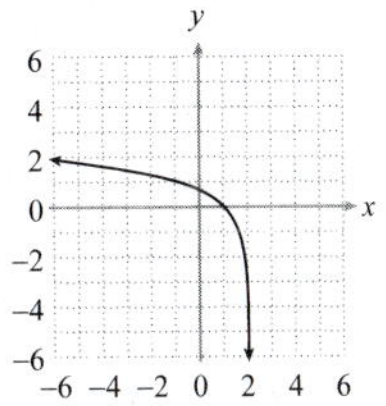

Dom: $(\infty,2)$, Ran: $(-\infty,\infty)$

31.

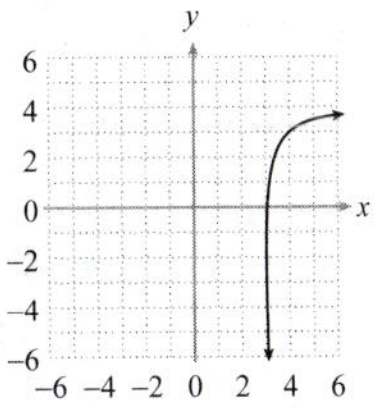

Dom: $(3,\infty)$, Ran: $(-\infty,\infty)$

33.

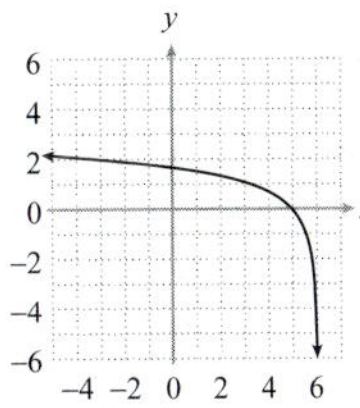

Dom: $(-\infty,6)$, Ran: $(-\infty,\infty)$

35.

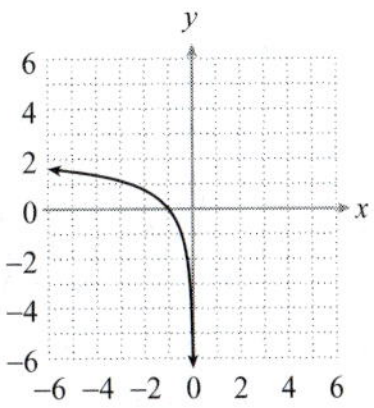

Dom: $(-\infty,0)$, Ran: $(-\infty,\infty)$

37. e **39.** b **41.** h

43. d **45.** i **47.** -2

49. 3 **51.** $-\frac{1}{2}$ **53.** $\frac{3}{4}$

55. 2.89 **57.** $\frac{5}{3}$ **59.** 1

61. $\frac{1}{5}$ **63.** 2 **65.** $\{64\}$

67. $\{9\}$ **69.** $\left\{-\frac{1}{2}\right\}$ **71.** $\left\{\frac{1}{21}\right\}$

73. $\{10\}$ **75.** $\{36\}$

77. $\left\{\pm\frac{1}{10}\right\}$ **79.** $\{0.18\}$

81. $\{\pm\sqrt{e}\}$ or $\{\pm 1.65\}$

83. $\{12.89\}$

85. $\{10,000,000,002\}$

6.4 Exercises

1. $3+3\log_5 x$

3. $2+\ln p-3\ln q$

5. $1+\log_9 x-3\log_9 y$

7. $\frac{3}{2}\ln x+\ln p+5\ln q-7$

9. $\log(2+3\log x)$

11. $1-\frac{1}{2}\log(x+y)$

13. $\log_2(y^2+z)-4\log_2 x-4$

15. $2\log_b x+\frac{1}{2}\log_b y-\log_b z$

17. $2+\log_b a+b\log_b c$

19. $\log\frac{x}{y}$

21. $\log_5(x+5)$

23. $\log_2\left(x^{\frac{4}{3}}+3x^{\frac{1}{3}}\right)$

25. $\ln\left(\frac{3p}{q^2}\right)$ **27.** $\log\left(\frac{x-10}{x}\right)$

29. $\ln\left(\frac{z^2}{x^3y^3}\right)$ **31.** $\log_5 4$

33. $\ln 45$ **35.** $\log_3 1=0$

37. $\ln 12$ **39.** $\log 11$

41. $\log_8(x^2-y)$ **43.** x^2

45. $\frac{e^2p}{x}$ **47.** $\frac{x^3}{y^4}$ **49.** x^2

51. 4 **53.** $12x^2$ **55.** 2.04

57. 0.95 **59.** 0.95 **61.** 2.45

63. 3.30 **65.** 0.74 **67.** 1.20

69. 1.86 **71.** -1 **73.** 3.85

75. 0.77 **77.** -1.76 **79.** 2

81. 7 **83.** 1

85. $4\sqrt{2}\approx 5.66$ **87.** 9.05

89. 2.08 **91.** 12

93. 1,048,576 **95.** 10.25

97. 5,011,872 times stronger

99. 133 decibels

101. 7.62; yes

103. **a.** 15.05 minutes

b. 7:00 p.m.

c. 112 °F; no

105. $h(p)=81,751.7-7027.82\ln p$

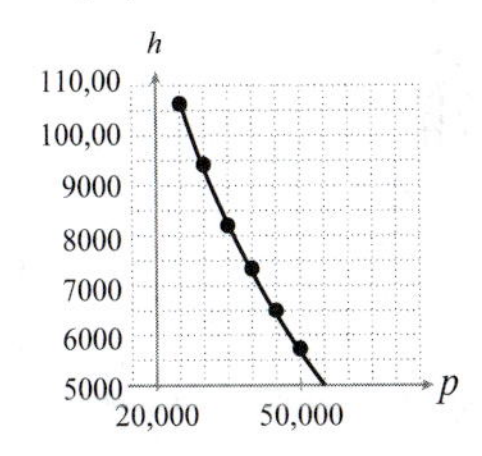

6.5 Exercises

1. $x\approx 0.26$ **3.** $x\approx 3.12$

5. $x\approx 3.89$ **7.** $x\approx -2.28$

9. $x\approx 8.09$ **11.** $x\approx 2.68$

13. $x\approx -1.12$ **15.** $x\approx 52.77$

17. $x\approx \pm 0.71$ **19.** $x=-12$

21. $x\approx 1.32$ **23.** $x\approx 3.27$

25. $x=125$ **27.** $x=5$

29. $x\approx 40.17$ **31.** $x=35$

33. $x\approx 9.38$ **35.** $x=4$

37. $x=\frac{1}{162}$ **39.** $x=1$

41. $x\approx 100.04$ **43.** No solution

45. $x=5$ **47.** $x=8$

49. $x=\frac{37}{8}$ **51.** $x=5$

53. No solution **55.** $x=6$

57. $x=\sqrt{2}-1$ **59.** $x=2,3$

61. $x=1,2$ **63.** $f(x)=\log 2x^2$

65. $f(x) = \ln 9x^2$

67. $f(x) = 256x$

69. $f(x) = \ln 1 = 0$

71. $f(x) = \ln 5^x$ **73.** $f(x) = \ln 5$

75. **a.** 17.36 years

b. 9.90 years

77. 4.98 hours **79.** 4.99 years

81. 0.271 years (about 99 days)

Chapter 6 Project

1. 242 **3.** 20,219

Chapter 6 Review Exercises

1.

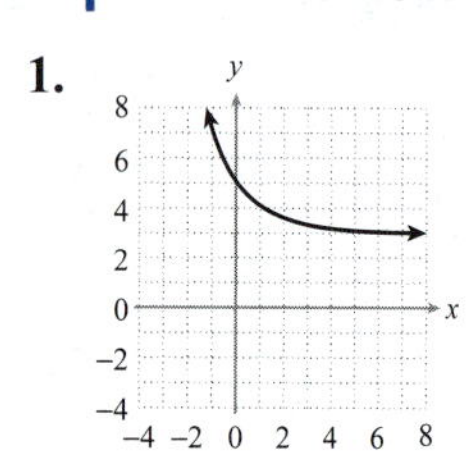

$\text{Dom} = (-\infty, \infty), \text{Ran} = (3, \infty)$

3.

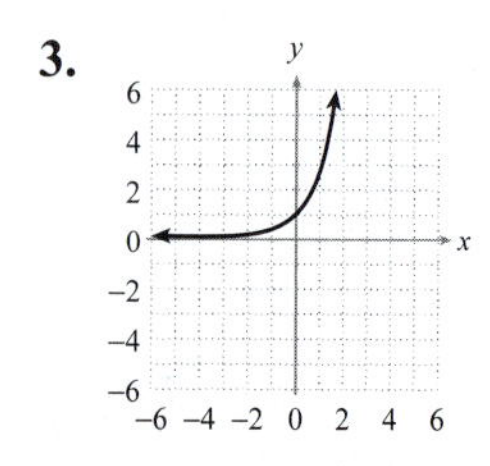

$\text{Dom} = (-\infty, \infty), \text{Ran} = (0, \infty)$

5.

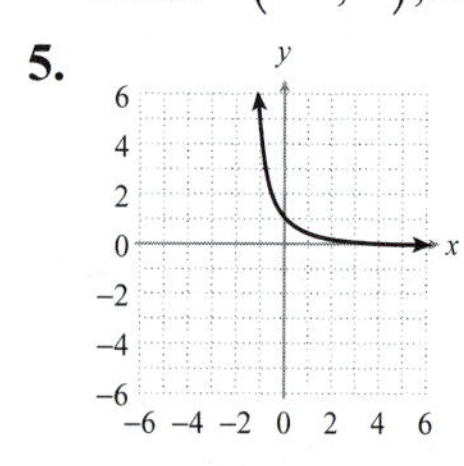

$\text{Dom} = (-\infty, \infty), \text{Ran} = (0, \infty)$

7.

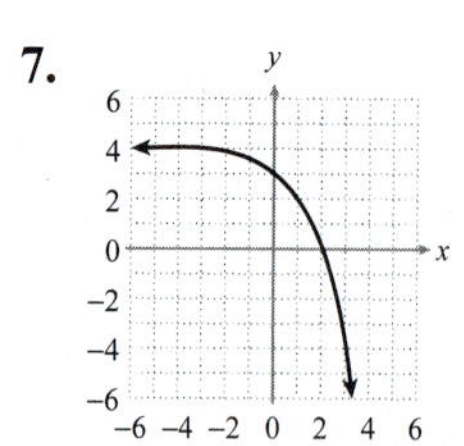

$\text{Dom} = (-\infty, \infty), \text{Ran} = (-\infty, 4)$

9.

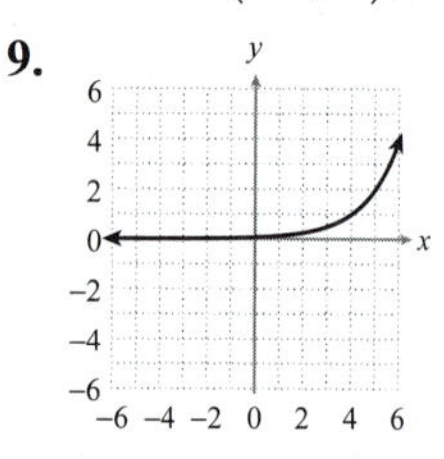

$\text{Dom} = (-\infty, \infty), \text{Ran} = (0, \infty)$

11. $x = 5$ **13.** $x = 2$

15. $x = -1$ **17.** $x = \dfrac{8}{5}$

19. $x = 2$ **21.** 8 days

23. **a.** 173 flies **b.** 20 flies

25. $x = \log_3 8$ **27.** $\log_4 4096 = 3a$

29. $3^{-3} = \dfrac{1}{27}$

31.

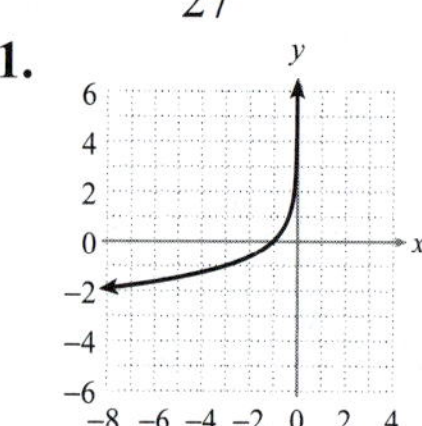

$\text{Dom} = (-\infty, 0), \text{Ran} = (-\infty, \infty)$

33.

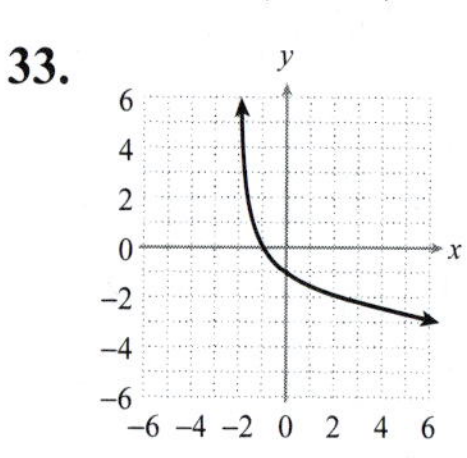

$\text{Dom} = (-2, \infty), \text{Ran} = (-\infty, \infty)$

35. −2 **37.** −3 **39.** $\dfrac{2}{3}$

41. $x = 27$ **43.** $x = \dfrac{17}{2}$

45. $x \approx 155.41$

47. $\dfrac{3}{2}\log x - \dfrac{5}{2}\log \pi - \log 2$

49. $3 + 3\log_3 a$ **51.** $\log_2\left(\dfrac{a^{\frac{5}{3}}}{b^{\frac{1}{3}}c}\right)$

53. $\log_2(x-3)$ **55.** $\log_3 \dfrac{x^2 - 2x}{x^2 + 4}$

57. $\dfrac{x}{y^2}$ **59.** 6.18 **61.** 0

63. **a.** 251,188,643

b. 7,079,458

c. 1,258,925,412

65. $\dfrac{6}{\log 321} \approx 2.39$

67. $\dfrac{\ln 5}{4 - 3\ln 5} \approx -1.94$

69. $\dfrac{\log 12}{\log 18} \approx 0.86$

71. −2 **73.** 73.5

75. $f(x) = \ln x^3$ **77.** $f(x) = \log x^{10}$

79. $f(x) = \log 7$

81. 20.4 months (1.7 years)

Chapter 7: Trigonometric Functions

7.1 Exercises

1. 225° **3.** −67.5° **5.** 120°

7. 150° **9.** −405° **11.** $\dfrac{47\pi}{180}$

13. $\dfrac{11\pi}{15}$ **15.** $\dfrac{37\pi}{45}$ **17.** $\dfrac{8\pi}{3}$

19. $\dfrac{25\pi}{36}$ **21.** 270° **23.** 540°

25. −72° **27.** $\dfrac{\pi}{9}$ **29.** $-\dfrac{4\pi}{5}$

31. $\dfrac{\pi}{6}$ **33.** $\dfrac{83\pi}{90}$ **35.** 115°

37. 50° **39.** 345° **41.** 295°

43. $\dfrac{5\pi}{4}$ **45.** $\dfrac{\pi}{4}$ **47.** $\dfrac{4\pi}{3}$

49. $\dfrac{13\pi}{4}$

51.

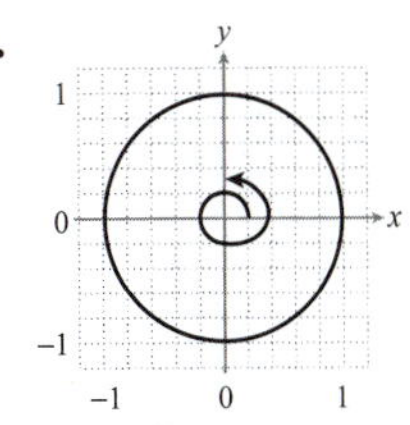

53.

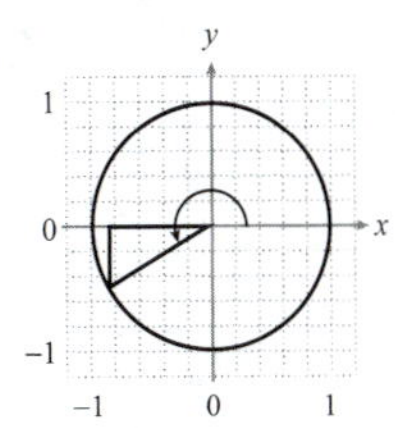

55.

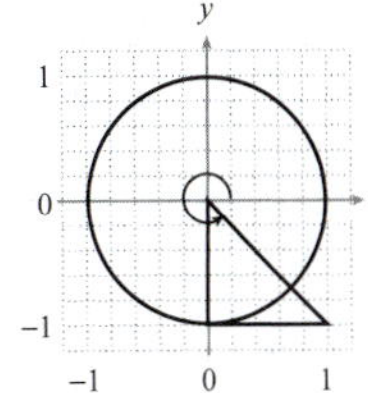

57. 4 in. **59.** 11.78 ft

61. 8.64 m **63.** $\frac{9}{2}$

65. $\frac{21}{47}$ **67.** $\frac{45}{4}$

69. 1.48 inches **71.** 3.56 feet

73. 1.83 feet **75.** 1.45°

77. 445.79 miles **79.** 1335.24 km

81. 7853.58 km **83.** 1039.61 mph

85. 17.14 mph

87. **a.** 10π rad/s **b.** 122.84 in./s

89. $\frac{5\pi}{12} \approx 1.31$ **91.** 29.93 cm^2

93. 15.08 m^2 **95.** 314.16 ft^2

97. 9.82 in.^{2} **99.** 26.18 cm

101. 11.7 in.

7.2 Exercises

1. $\sin\theta = \cos\theta = \frac{\sqrt{2}}{2}$, $\tan\theta = \cot\theta = 1$, $\csc\theta = \sec\theta = \sqrt{2}$

3. $\sin\theta = \frac{1}{3}$, $\cos\theta = \frac{2\sqrt{2}}{3}$, $\tan\theta = \frac{\sqrt{2}}{4}$, $\csc\theta = 3$, $\sec\theta = \frac{3\sqrt{2}}{4}$, $\cot\theta = 2\sqrt{2}$

5. $\sin\theta = \cos\theta = \frac{\sqrt{2}}{2}$, $\tan\theta = \cot\theta = 1$, $\csc\theta = \sec\theta = \sqrt{2}$

7. $\sin\theta = \frac{5}{7}$, $\cos\theta = \frac{2\sqrt{6}}{7}$, $\tan\theta = \frac{5\sqrt{6}}{12}$, $\csc\theta = \frac{7}{5}$, $\sec\theta = \frac{7\sqrt{6}}{12}$, $\cot\theta = \frac{2\sqrt{6}}{5}$

9. $\sin\theta = \frac{12}{13}$, $\cos\theta = \frac{5}{13}$, $\tan\theta = \frac{12}{5}$, $\csc\theta = \frac{13}{12}$, $\sec\theta = \frac{13}{5}$, $\cot\theta = \frac{5}{12}$

11. $\sin\theta = \frac{33}{65}$, $\cos\theta = \frac{56}{65}$, $\tan\theta = \frac{33}{56}$, $\csc\theta = \frac{65}{33}$, $\sec\theta = \frac{65}{56}$, $\cot\theta = \frac{56}{33}$

13. $\sin\theta = \frac{1}{2}$, $\cos\theta = \frac{\sqrt{3}}{2}$, $\tan\theta = \frac{\sqrt{3}}{3}$, $\csc\theta = 2$, $\sec\theta = \frac{2\sqrt{3}}{3}$, $\cot\theta = \sqrt{3}$

15. $\sin\left(\frac{\pi}{4}\right) = \frac{\sqrt{2}}{2}$, $\csc\left(\frac{\pi}{4}\right) = \sqrt{2}$

17. $\sec 60° = 2$

19. $\csc\left(\frac{\pi}{6}\right) = 2$

21. $\sec 5° \approx 1.0038$, $\tan 5° \approx 0.0875$

23. $\cot\left(\frac{\pi}{3}\right) = \frac{\sqrt{3}}{3}$

25. $\tan(87.2°) \approx 20.4465$

27. 0.9945 **29.** 1.0355

31. 28.6537 **33.** 3.0777

35. 0.3827 **37.** 2

39. 38.9053° **41.** 25.325°

43. 21.6656° **45.** 0.1149

47. 0.7746 **49.** 5

51. True **53.** True

55. False **57.** 751.19 feet

59. 17.47 feet **61.** 48.54 yards

63. 12.04 m **65.** 314.57 feet

67. 13.86 feet **69.** 6.86 feet

71. 20 m **73.** 3196.80 feet

75. The two triangles are similar, meaning that the lengths of the sides of Belinda's triangle are scaled by some common factor k relative to the corresponding sides of Adam's triangle. So, for instance, Adam would find that $\sin\theta = \frac{a}{c}$ and Belinda would find that $\sin\theta = \frac{ka}{kc} = \frac{a}{c}$.

7.3 Exercises

1. $\left(\frac{\sqrt{3}}{2}, \frac{1}{2}\right)$ **3.** $\left(-\frac{1}{2}, -\frac{\sqrt{3}}{2}\right)$

5. $(-1, 0)$ **7.** $\left(\frac{\sqrt{3}}{2}, -\frac{1}{2}\right)$

9. $\left(-\frac{\sqrt{3}}{2}, -\frac{1}{2}\right)$

11. $\frac{5\pi}{3} + 2n\pi, n \in \mathbb{Z}$

13. $\frac{7\pi}{6} + 2n\pi, n \in \mathbb{Z}$

15. $\sin\left(\frac{\pi}{2}\right) = \csc\left(\frac{\pi}{2}\right) = 1$, $\cos\left(\frac{\pi}{2}\right) = \cot\left(\frac{\pi}{2}\right) = 0$, $\tan\left(\frac{\pi}{2}\right) = \sec\left(\frac{\pi}{2}\right) = \text{undefined}$

17. $\sin\left(\frac{3\pi}{4}\right) = \frac{\sqrt{2}}{2}$, $\cos\left(\frac{3\pi}{4}\right) = -\frac{\sqrt{2}}{2}$, $\tan\left(\frac{3\pi}{4}\right) = \cot\left(\frac{3\pi}{4}\right) = -1$, $\csc\left(\frac{3\pi}{4}\right) = \sqrt{2}$, $\sec\left(\frac{3\pi}{4}\right) = -\sqrt{2}$

19. $\sin(-520°) \approx -0.3420$, $\cos(-520°) \approx -0.9397$, $\tan(-520°) \approx -0.3640$, $\csc(-520°) \approx -2.9238$, $\sec(-520°) \approx -1.0642$, $\cot(-520°) \approx -2.7475$

21. $\sin(-1105°) \approx -0.4226$,
$\cos(-1105°) \approx 0.9063$,
$\tan(-1105°) \approx -0.4663$,
$\csc(-1105°) \approx -2.3662$,
$\sec(-1105°) \approx 1.1034$,
$\cot(-1105°) \approx -2.1445$

23. $\sin 670° \approx -0.7660$,
$\cos 670° \approx 0.6428$,
$\tan 670° \approx -1.1918$,
$\csc 670° \approx -1.3054$,
$\sec 670° \approx 1.5557$,
$\cot 670° \approx -0.8391$

25. $\sin(-215°) \approx 0.5736$,
$\cos(-215°) \approx -0.8192$,
$\tan(-215°) \approx -0.7002$,
$\csc(-215°) \approx 1.7434$,
$\sec(-215°) \approx -1.2208$,
$\cot(-215°) \approx -1.4281$

27. $\sin 780° = \frac{\sqrt{3}}{2}$, $\cos 780° = \frac{1}{2}$,
$\tan 780° = \sqrt{3}$, $\csc 780° = \frac{2\sqrt{3}}{3}$,
$\sec 780° = 2$, $\cot 780° = \frac{\sqrt{3}}{3}$

29. $\theta' = 82°$ **31.** $\theta' = 60°$

33. $\theta' = \frac{\pi}{2}$ **35.** $\theta' = \frac{\pi}{6}$

37. $\theta' = \frac{\pi}{5}$ **39.** $\theta' = \frac{\pi}{2}$

41. $\theta' = \frac{\pi}{4}$ **43.** $\theta' = 75°$

45. IV **47.** IV **49.** I

51. II **53.** a **55.** a

57. b **59.** c **61.** c

63. $\sin\left(\frac{9\pi}{2}\right) = \sin\left(\frac{\pi}{2}\right) = 1$

65. $\tan\left(\frac{5\pi}{4}\right) = \tan\left(\frac{\pi}{4}\right) = 1$

67. $\sin 313° = -\sin 47° \approx -0.7314$

69. $\tan(-168°) = \tan 12° \approx 0.2126$

71. $\sin 216° = -\sin 36° \approx -0.5878$

73. $\cos(-330°) = \cos 30° = \frac{\sqrt{3}}{2}$

75. $\tan 718° = -\tan 2° \approx -0.0349$

77. b **79.** e

81. $\cot 135° = \cos(-45°) = -1$

83. $\sin(-60°) = \cos 150° = -\frac{\sqrt{3}}{2}$

85. $\csc\left(\frac{5\pi}{6}\right) = \sec\left(-\frac{\pi}{3}\right) = 2$

87. $\cos\left(-\frac{3\pi}{6}\right) = \sin\pi = 0$

89. $\sin\left(-\frac{4\pi}{5}\right) = \cos\left(\frac{13\pi}{10}\right)$
≈ -0.5878

91. $\csc\left(\frac{3\pi}{2}\right) = \sec(-\pi) = -1$

93. $\cot\left(\frac{\pi}{4}\right) = \tan\left(\frac{\pi}{4}\right) = 1$

95. $\sec 105° = \csc(-15°) \approx -3.8637$

97. $\tan\theta \approx -0.727$, $\cot\theta \approx -1.376$

99. $\tan\theta = 1.732$, $\cot\theta \approx 0.577$

101. $\tan\theta \approx 9.476$, $\cot\theta \approx 0.106$

103. $\csc\theta = \frac{125}{100} = \frac{5}{4}$, $\cot\theta = -\frac{3}{4}$

105. No such angle exists, as this would require the opposite side to be longer than the hypotenuse.

107. $\theta = -\frac{2\pi}{3}$, $\tan\theta = -\sqrt{3}$

109. $\cos\left(-\frac{5\pi}{4}\right)$, -0.7071

111. $\tan\left(-\frac{\pi}{4}\right)$, -1

113. $\cot 75°$, 0.2679

7.4 Exercises

1. Amplitude = 2,
frequency = $\frac{1}{\pi}$

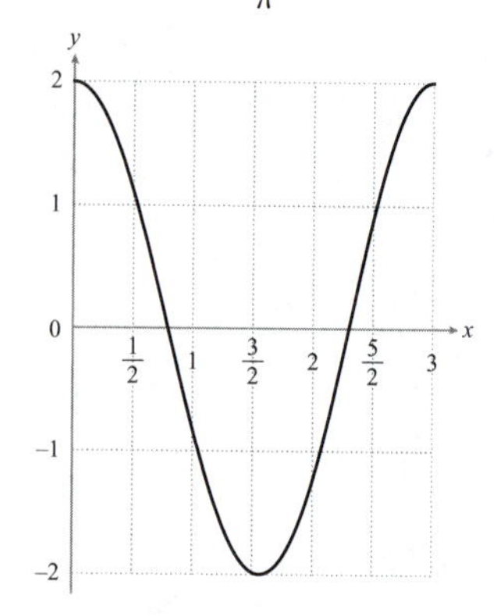

3. Amplitude = $\frac{1}{2}$,
frequency = $\frac{3}{2\pi}$

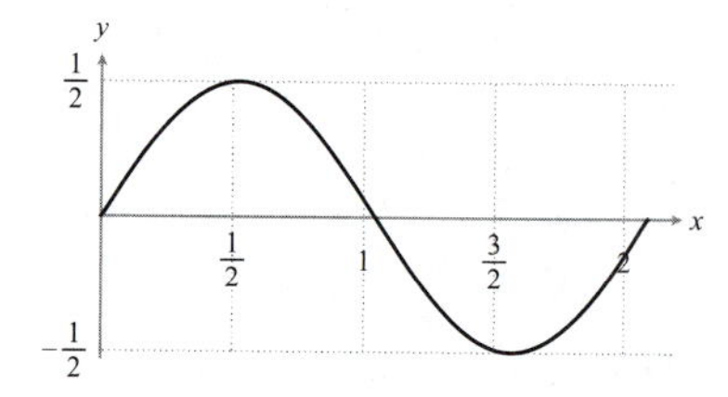

5. A = 2, P = 2π, no phase shift

7. A = 4, P = 2π, no phase shift

9. A = 1, P = 2π, no phase shift

11. A = 3, P = 2π, shifted left 7 units

13. A = 2, P = π, no phase shift

15. A = $\frac{3}{2}$, P = 2, no phase shift

17. A = 0.5, P = $\frac{\pi}{4}$, shifted left $\frac{1}{8}$ units

19. A = $\frac{8}{5}$, P = 1, shifted left $\frac{2}{\pi}$ units

21.

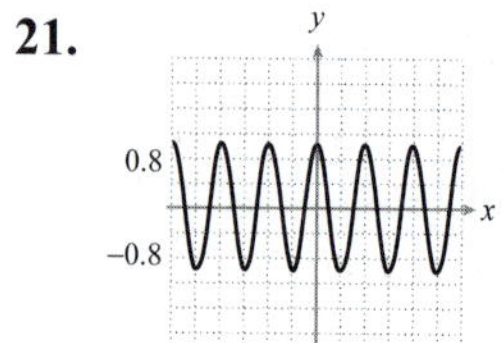

23.
y
4
2
0
−2
−4
x
−3π −π 0 π 3π

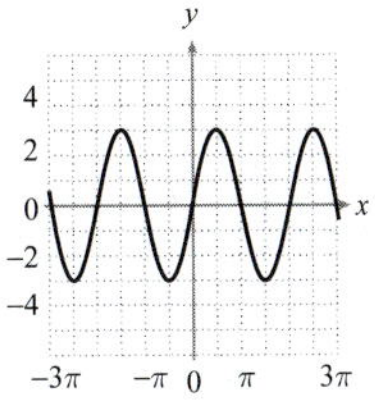

25.
y
2
−2
x
−5 5

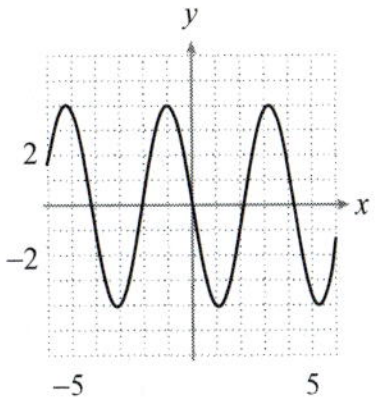

27.
y
0.8
−0.8
x
−5 5

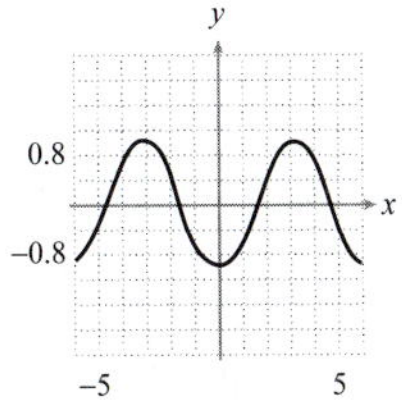

29.
y
0.8
−0.8
x
−1 1

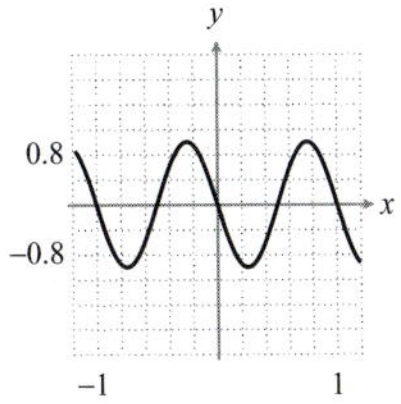

31.
y
3
1
−1
−3
x
−2 2

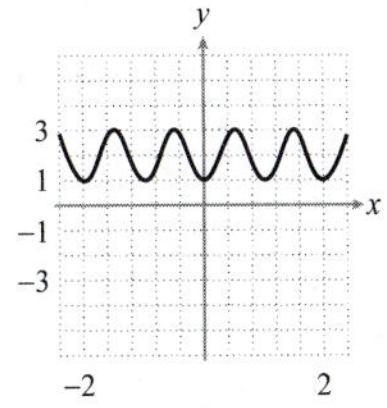

33.
y
10
−10
x
−10 10

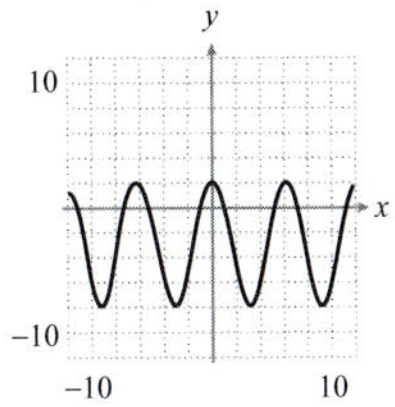

35.
y
10
−10
x
−10 10

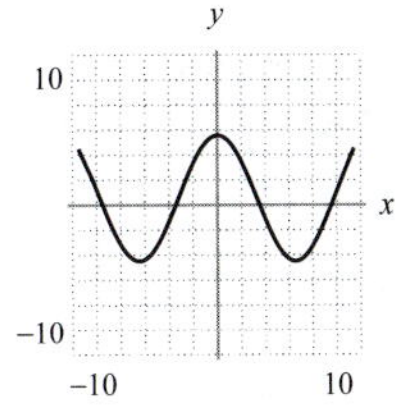

37.
g(t)
5
0
−5
t
1

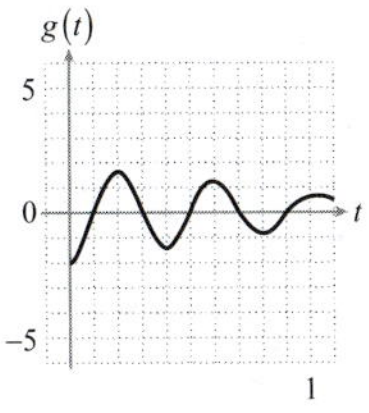

39.
g(t)
25
0
−25
t
4

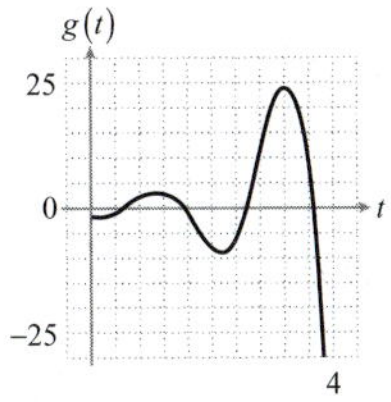

41.
f(t)
5
0
−5
t
10

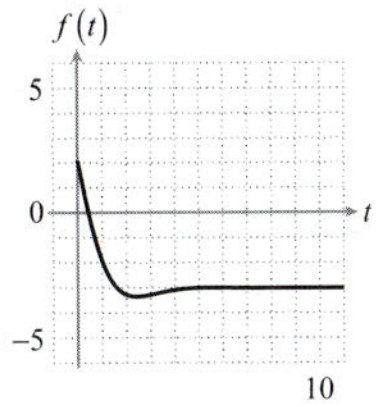

43.
f(t)
5
0
−5
t
5

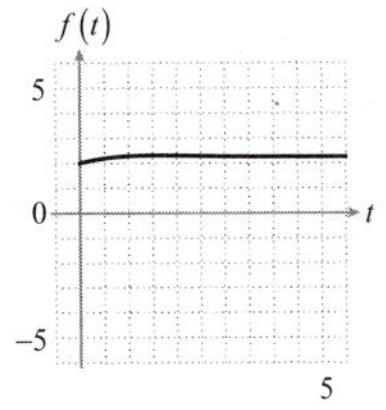

45. Frequency = 0.5 Hz

47. A = 3 inches, P = 2 seconds, $g(t) = 3\cos(\pi t)$

49. A = 1.5 ft, P = $\frac{3}{5}$ seconds, $g(t)=1.5\cos\left(\frac{10\pi}{3}t\right)$

51. a. $\cos\left(x+\frac{3\pi}{2}\right)$

b. $\cos\left(x-\frac{(4n+1)\pi}{2}\right)$

7.5 Exercises

1.
y 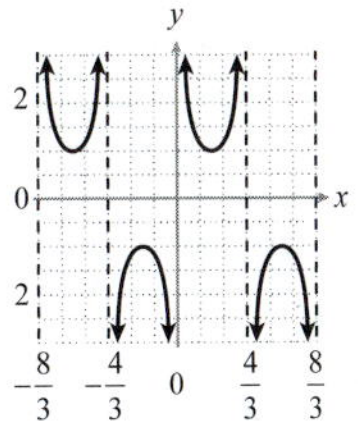
2
0
2
x
$-\frac{8}{3}$ $-\frac{4}{3}$ 0 $\frac{4}{3}$ $\frac{8}{3}$

3.
y
1
$\frac{1}{3}$
$-\frac{1}{3}$
−1
x
$-\frac{\pi}{4}$ 0 $\frac{\pi}{4}$ $\frac{3\pi}{4}$

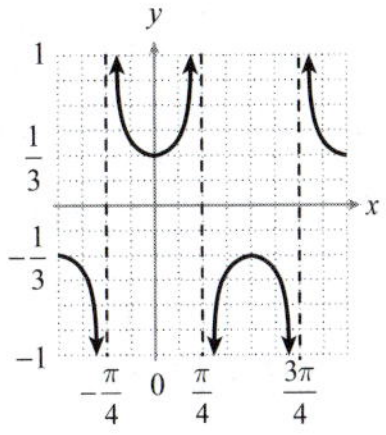

5.
y
5
0
−5
x
−1 1

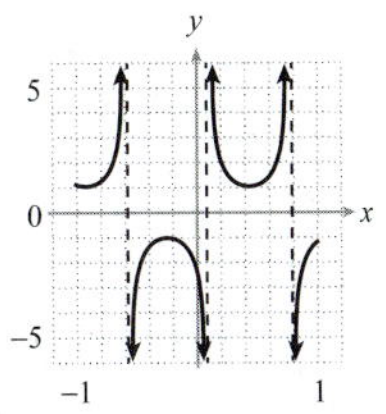

7.
y
15
5
−5
−15
x
−3 −1 1 3

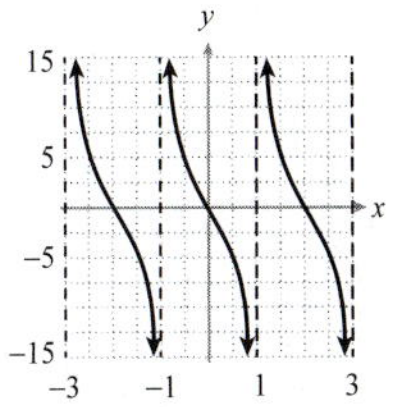

9.
y
6
4
2
0
−2
−4
−6
x
$-\frac{\pi}{2}$ $\frac{\pi}{2}$ π $\frac{3\pi}{2}$ 2π $\frac{5\pi}{2}$

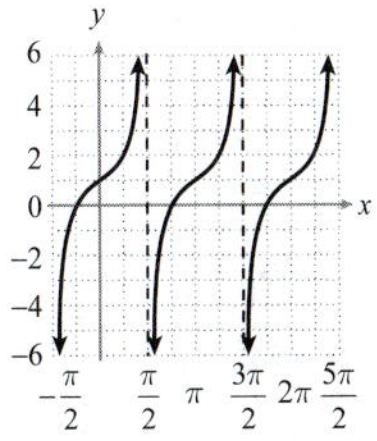

11.
y
10
−10
x
−5 5

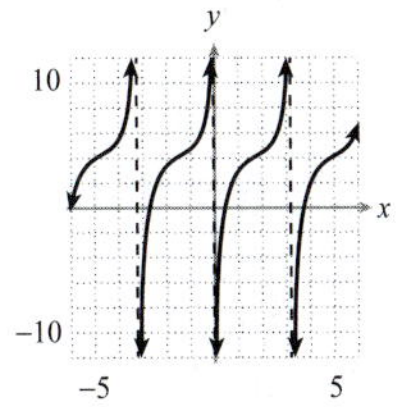

13.
y
2
−2
x
−2 2

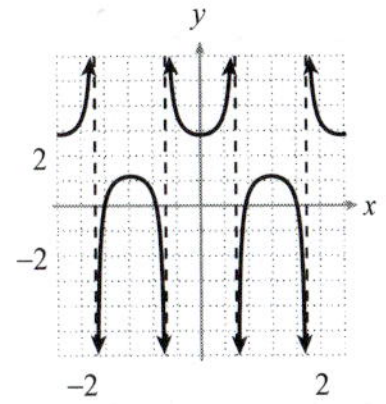

7.6 Exercises

1. $-\frac{\pi}{2}$ **3.** $\frac{\pi}{4}$ **5.** $\frac{\pi}{6}$

7. 0 **9.** $-\frac{\pi}{3}$ **11.** $\frac{\pi}{3}$

13. $-\frac{\pi}{4}$ **15.** $\frac{2\pi}{3}$ **17.** $-\frac{\pi}{6}$

19. $\frac{\pi}{2}$ **21.** $\frac{\pi}{4}$ **23.** $\frac{2\pi}{3}$

25. −0.2014 **27.** −1.1198

29. 1.9823 **31.** 0.2014

33. 0.0067 **35.** −0.1126

37. $\frac{\pi}{2}$ **39.** $\frac{1}{2}$ **41.** −0.8

43. 0.3714

45. 0.8944 **47.** 0.75

49. $\frac{\sqrt{3}}{2}$ **51.** 1 **53.** −1

55. $\frac{\sqrt{2}}{2}$ **57.** 1 **59.** $\frac{\sqrt{2}}{2}$

61. $-\sqrt{2}$ **63.** $\frac{\sqrt{2}}{2}$ **65.** -1

67. $\frac{\sqrt{1-x^2}}{x}$

69. $\sqrt{9x^2+1}$ **71.** $\left|\frac{1}{x}\right|\sqrt{x^2-1}$

73. $\theta = 47.88848865$

75. $\theta = 32.76975166$

77. $\theta = 60.10239082$

79. $\theta = 0.5736213262$

81. $\theta = -0.8480623625$

83.

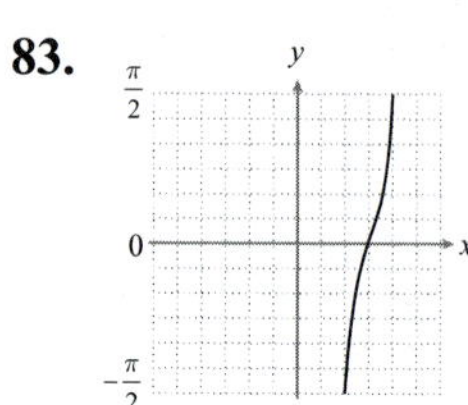

85.

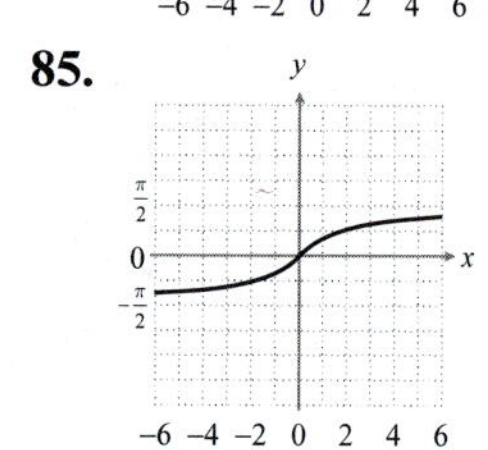

87. **a.** 0.2450

b. 0.7854

c. 0.9505

Chapter 7 Project

1. United Nations, NY: approx. 235.0°; California Academy of Sciences: approx. 220.5°; Smithsonian, Washington, DC: approx. 226.0°; St. Isaac's Cathedral, Russia: approx. 311.4°; Paris Observatory, France: approx. 270.9°

3. Approx. 14.92 ft

5. Approx. 12.6°

Chapter 7 Review Exercises

1. 4° **3.** –315°

5. $\frac{7\pi}{30}$ **7.** $-\frac{79\pi}{180}$

9.

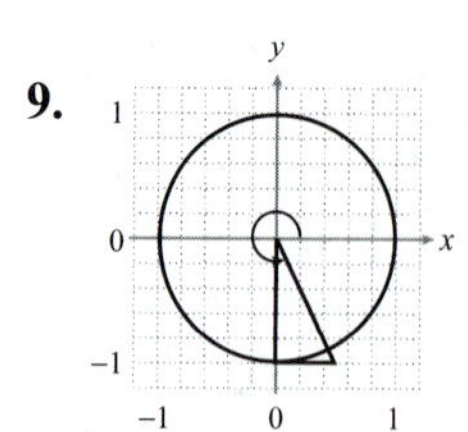

11. 15.71 ft **13.** 30.72 meters

15. 1296.35 km **17.** 3 rad

19. $108\pi\ \text{ft}^2$ or $339.29\ \text{ft}^2$

21. $\sin\theta = \frac{2\sqrt{5}}{5}, \cos\theta = \frac{\sqrt{5}}{5},$
$\tan\theta = 2, \csc\theta = \frac{\sqrt{5}}{2},$
$\sec\theta = \sqrt{5}, \cot\theta = \frac{1}{2}$

23. 4.0108 **25.** 0.2225

27. 15.2203° **29.** False

31. 129.13 feet

33. $\sin(-460°) \approx -0.9848,$
$\cos(-460°) \approx -0.1736,$
$\tan(-460°) \approx 5.6713,$
$\csc(-460°) \approx -1.0154,$
$\sec(-460°) \approx -5.7588,$
$\cot(-460°) \approx 0.1763$

35. $\sin\left(\frac{7\pi}{3}\right) = \frac{\sqrt{3}}{2}, \cos\left(\frac{7\pi}{3}\right) = \frac{1}{2},$
$\tan\left(\frac{7\pi}{3}\right) = \sqrt{3}, \csc\left(\frac{7\pi}{3}\right) = \frac{2\sqrt{3}}{3},$
$\sec\left(\frac{7\pi}{3}\right) = 2, \cot\left(\frac{7\pi}{3}\right) = \frac{\sqrt{3}}{3}$

37. $\theta' = 37°$ **39.** $\theta' = \frac{\pi}{4}$

41. III

43. $\tan\left(\frac{4\pi}{3}\right) = \tan\left(\frac{\pi}{3}\right) = \sqrt{3}$

45. $\sin(-42°) = \cos 132° \approx -0.6691$

47. $\cos\left(\frac{5\pi}{4}\right) = \sin\left(-\frac{3\pi}{4}\right) = \frac{-\sqrt{2}}{2}$

49. $\sin\theta = \frac{13}{5}$

51. $A = 6,\ P = 2\pi,$ no phase shift

53. $A = 9,\ P = \pi,$ shifted left π units

55.

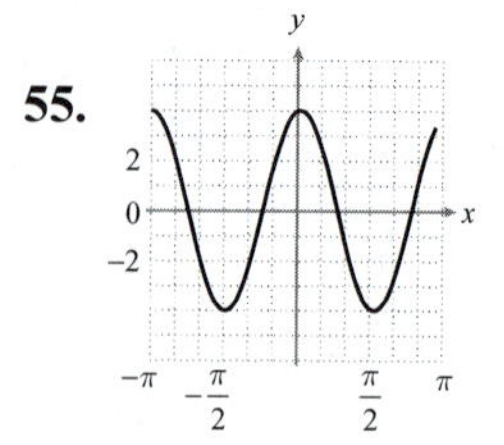

57.

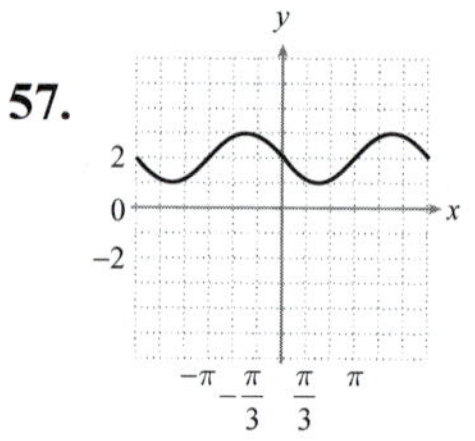

59.

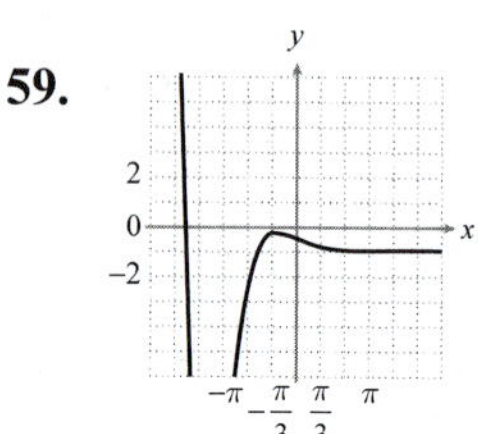

61.

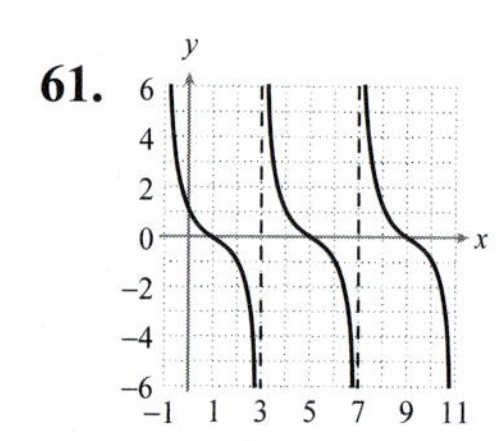

63.

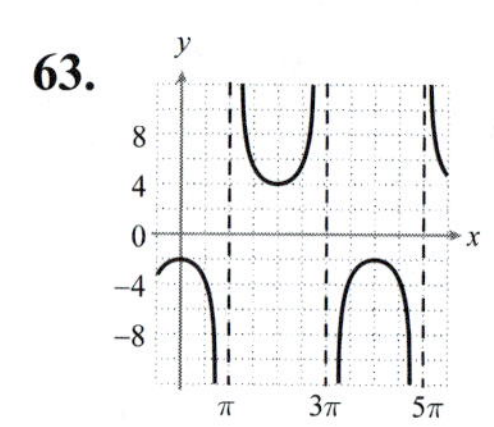

65. $\frac{\pi}{2}$ **67.** $\frac{\pi}{3}$

69. 0.4636 rad or 26.5651 deg

71. 0.9 **73.** $-\frac{\pi}{4}$

75. $\frac{\sqrt{3}}{3}$ **77.** 2

79. –1.336110366

81. 55.32339906

Chapter 8: Trigonometric Identities and Equations

8.1 Exercises

1. $\sec x$
3. $\sin t$
5. $\cos x - \sec x$
7. $\tan \alpha$
9. $\sec y$
11. $\tan^2 x$
13. $\sec \theta$
15. $3 \sin \theta$
17. $8 \cos \theta$
19. $5 \sec \theta$

8.2 Exercises

1. $\dfrac{-\sqrt{6}+\sqrt{2}}{4}$
3. $-\sqrt{3}-2$
5. $-\sqrt{3}-2$
7. $\dfrac{\sqrt{6}+\sqrt{2}}{4}$
9. $\dfrac{\sqrt{2}+\sqrt{6}}{4}$
11. $\sqrt{3}+2$
13. $\dfrac{\sqrt{6}-\sqrt{2}}{4}$
15. $\sqrt{3}+2$
17. $\dfrac{-\sqrt{6}-\sqrt{2}}{4}$
19. $\sqrt{3}+2$
21. $\dfrac{\sqrt{6}+\sqrt{2}}{4}$
23. $\dfrac{\sqrt{6}-\sqrt{2}}{4}$
25. $\dfrac{\sqrt{6}+\sqrt{2}}{4}$
27. $\dfrac{56}{65}$
29. $\dfrac{36}{85}$
31. $\dfrac{4\sqrt{6}-\sqrt{21}}{25}$
33. $\sin 45° = \dfrac{\sqrt{2}}{2}$
35. $\tan 135° = -1$
37. $\tan\left(\dfrac{\pi}{4}\right) = 1$
39. $\sin 150° = \dfrac{1}{2}$
41. $\cos 135° = -\dfrac{\sqrt{2}}{2}$
43. $\tan 60° = \sqrt{3}$
45. $-\dfrac{56}{33}$
47. $\dfrac{4+3\sqrt{3}}{10}$
49. 1
51. $\dfrac{2x^2+\sqrt{1-x^2}}{\sqrt{1+4x^2}}$
53. $x\left(\sqrt{1-4x^2} - 2\sqrt{1-x^2}\right)$
55. $\sqrt{2}\sin\left(x+\dfrac{\pi}{4}\right)$

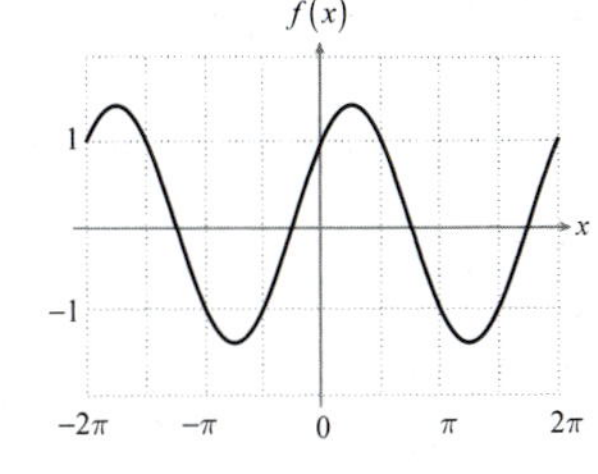

57. $\sqrt{2}\sin\left(2\beta-\dfrac{\pi}{4}\right)$

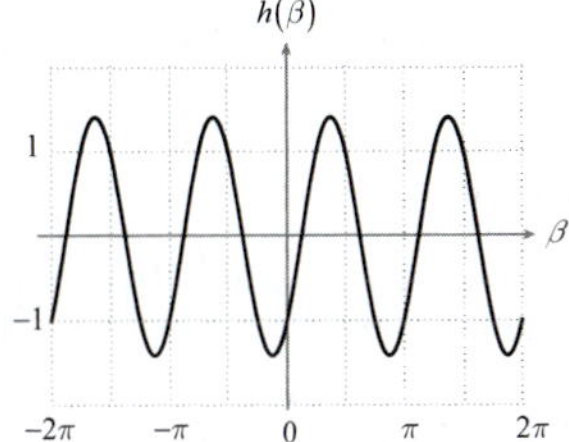

59. $13\sin(5u+1.1760)$

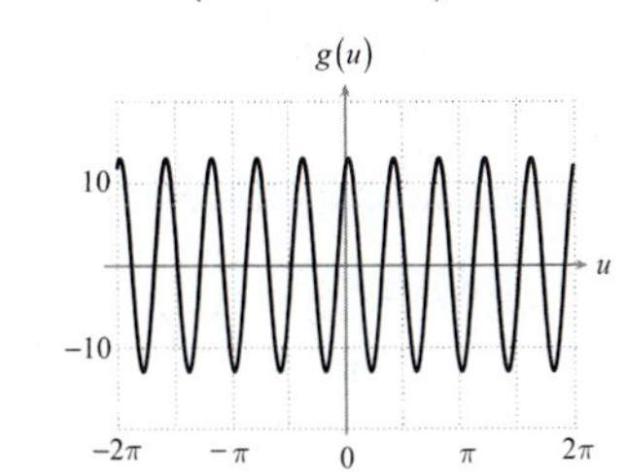

73. True
75. False
79. True
81. True
83. False

8.3 Exercises

1. $\cos(2x) = \dfrac{7}{25}$; $\sin(2x) = \dfrac{24}{25}$; $\tan(2x) = \dfrac{24}{7}$
3. $\cos(2x) = \dfrac{1}{3}$; $\sin(2x) = -\dfrac{2\sqrt{2}}{3}$; $\tan(2x) = -2\sqrt{2}$
5. $\cos(2x) = \dfrac{1}{2}$; $\sin(2x) = \dfrac{\sqrt{3}}{2}$; $\tan(2x) = \sqrt{3}$
7. $\dfrac{\sin x - \sin x\cos(2x)}{2}$
9. $\dfrac{1-\cos(2x)-\cos(4x)}{16} + \dfrac{\cos(2x)\cos(4x)}{16}$
11. $\dfrac{3\sin x - 4\sin x\cos(2x)}{3+4\cos(2x)+\cos(4x)} + \dfrac{\sin x\cos(4x)}{3+4\cos(2x)+\cos(4x)}$
13. $\dfrac{\sqrt{2+\sqrt{2}}}{2}$
15. $\dfrac{\sqrt{2+\sqrt{3}}}{2}$
17. $\dfrac{\sqrt{2+\sqrt{3}}}{2}$
19. $\dfrac{\sin(6x)+\sin 0}{2}$
21. $\dfrac{5(\sin 120° - \sin 90°)}{2}$
23. $\dfrac{\cos(2y)-\cos(2x)}{2}$
25. $\dfrac{\cos\left(\frac{7\pi}{12}\right)-\cos\left(\frac{23\pi}{12}\right)}{2}$
27. $\sin\left(\dfrac{\pi}{2}\right)-\sin\left(\dfrac{\pi}{6}\right)$
29. $2\sin(4x)\cos(2x)$
31. $-2\sin(2\beta)\sin\beta$
33. $2\cos 75°\sin 60°$
35. $-2\sin\left(\dfrac{17\pi}{24}\right)\sin\left(\dfrac{11\pi}{24}\right)$

47. $\cos(2x) = 2\cos^2 x - 1$

$\cos(3x) = 4\cos^3 x - 3\cos x$

$\cos(4x) = 8\cos^4 x - 8\cos^2 x + 1$

$\sin(2x) = 2\sin x \cos x$

$\sin(3x) = \sin x(4\cos^2 x - 1)$

$\sin(4x) = \sin x(8\cos^3 x - 4\cos x)$

8.4 Exercises

1. $x = \frac{7\pi}{6} + 2n\pi,\ x = \frac{11\pi}{6} + 2n\pi$

3. $x = \frac{\pi}{4} + 2n\pi,\ x = \frac{7\pi}{4} + 2n\pi$

5. $x = \frac{\pi}{6} + 2n\pi,\ x = \frac{11\pi}{6} + 2n\pi$

7. No solution

9. $x = \frac{2\pi}{3} + n\pi$

11. $x = n\pi$ **13.** $x = 2n\pi$

15. No solution

17. $x = \frac{2\pi}{3} + 2n\pi,\ x = \frac{4\pi}{3} + 2n\pi$

19. $x = \frac{5\pi}{12} + \frac{n\pi}{2}$

21. $x = \frac{\pi}{6}, \frac{5\pi}{6}$

23. $x = \frac{\pi}{4}, \frac{3\pi}{4}, \frac{5\pi}{4}, \frac{7\pi}{4}$

25. $x = 0, \pi, 2\pi$ **27.** $x = \frac{2\pi}{3}, \frac{4\pi}{3}$

29. $x = \tan^{-1}(-4 + \sqrt{59})$, $\tan^{-1}(-4 - \sqrt{59})$

41. True **43.** True

45. True **47.** $x = \frac{2\pi}{3} + n\pi$

49. True

51. $x = \frac{\pi}{2}, \frac{7\pi}{6}, \frac{11\pi}{6}$

53. $x = \frac{\pi}{2}, \pi$ **55.** $x = 0, \frac{2\pi}{3}, \frac{4\pi}{3}$

57. $x = \frac{\pi}{4}, \frac{5\pi}{4}$

59. $x = 45°, 135°, 225°, 315°$

61. $x = 90°, 210°, 330°$

63. No solution

65. $x = 30°, 210°, 240°, 300°$

67. $s = -4, 3$; t has no solution.

69. $s = \frac{1 \pm \sqrt{2}}{2}$

$t = 1.7794, 4.5038$

71. 17.3°

73. 2.6779 and 5.8195

75. 0.6662, 2.4754, 3.8078, and 5.6169

77. 1.9948 and 5.8592

79. 0.6993, 2.6078, 3.6754, and 5.5839

81. $x = 10°, 50°, 130°, 170°, 250°, 290°$

83. They can have at most two real solutions.

85. They may have as many as $2a$ solutions from 0° to 360°.

Chapter 8 Project

1. $y_1 + y_2 = 2\sqrt{2}\sin\left(kx - \omega t + \frac{\pi}{4}\right)$,

Amplitude: $2\sqrt{2}$

3.

We can suggest that the addition of two waves with equal amplitudes averages their displacements. (Answers will vary.)

Chapter 8 Review Exercises

1. $\csc x$ **3.** $-\tan^2 y$

5. $-1 - \sin\theta$ **13.** $4\sec\theta$

15. $10\cot\theta$ **17.** $\frac{\sqrt{3}}{2}$

19. $\frac{\sqrt{2} - \sqrt{6}}{4}$ **21.** $\sqrt{3}$

23. $\frac{13}{84}$ **25.** $\sin 210°; -\frac{1}{2}$

29. $\frac{\sqrt{1 - x^2} - x^2}{\sqrt{1 + x^2}}$

31. $2\sin\left(x + \frac{7\pi}{4}\right)$

33. $\cos(2x) = -\frac{7}{25}, \sin(2x) = \frac{24}{25}$, $\tan(2x) = -\frac{24}{7}$

37. $\frac{\sin x - \sin x\cos(4x)}{8}$

39. $2 + \sqrt{3}$

41. $2 - \sqrt{3}$

43. $\frac{\sin(2x) - \sin(2y)}{2}$

45. $\frac{1}{2}(\sin 180° + \sin 150°)$

47. $2\cos(4\alpha)\sin\alpha$

49. $-2\sin\left(\frac{17\pi}{24}\right)\sin\left(\frac{\pi}{24}\right)$

51. $x = \frac{\pi}{6} + 2n\pi,\ x = -\frac{\pi}{6} + 2n\pi$

53. $x = \frac{\pi}{2} + 2n\pi,\ x = -\frac{\pi}{2} + 2n\pi$

55. $x = \frac{\pi}{3} + 2n\pi,\ x = \frac{5\pi}{3} + 2n\pi$, $x = \pi + 2n\pi$

57. $x = \frac{\pi}{6}, \frac{5\pi}{6}, \frac{7\pi}{6}, \frac{11\pi}{6}$

59. True

61. $x = \frac{\pi}{3} + 2n\pi,\ x = \frac{2\pi}{3} + 2n\pi$

63. $x = 0°, 180°$

Chapter 9: Additional Topics in Trigonometry

9.1 Exercises

1. $C = 105°$, $b \approx 4.2426$, $c \approx 5.7956$
3. $C = 60°$, $a \approx 4.9067$, $c \approx 4.5221$
5. $A = 80°$, $a \approx 3.9392$, $b \approx 3.7588$
7. $C = 150°$, $b \approx 1.0154$, $c \approx 2.9238$
9. No triangle
11. $A \approx 20.7048°$, $B \approx 114.2952°$, $b \approx 5.1559$
13. $A \approx 35.9044°$, $B \approx 4.0956°$, $a \approx 8.2110$
15. $A \approx 19.1638°$, $C \approx 80.8362°$, $c \approx 6.0148$
17. $A \approx 6.7456°$, $C \approx 63.2544°$, $c \approx 7.6026$
19. $h \approx 6.02$, $c < h$, so a triangle cannot be created.
21. $C = 109°22'$, $a \approx 0.29$, $b \approx 2.39$

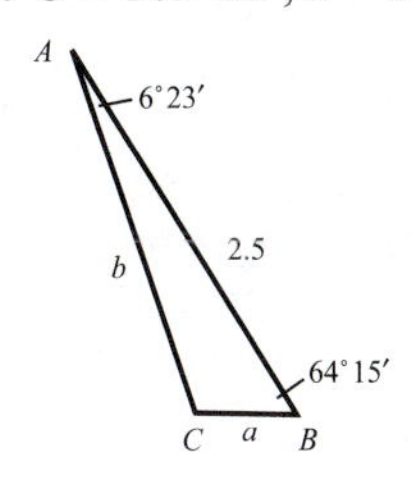

23. $c \approx 0.31$, $B \approx 70.13°$, $C \approx 1.87°$

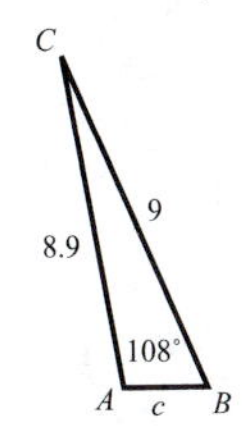

25. $C = 133.4°$, $a \approx 30.34$, $c \approx 44.49$

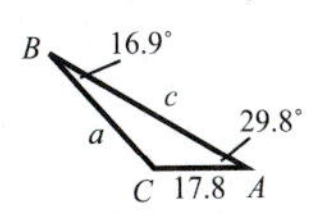

27. $b \approx 1.92$, $A \approx 60.05°$, $B \approx 3.95°$

A, 25, b, 116°, C, 24.1, B

29. $c \approx 24.78$, $C \approx 124.32°$, $B \approx 25.68°$

C, 13, 15, 30°, A, c, B

31. 94.34
33. 9.74
35. 18.47
37. 335.18
39. 21 feet
41. 442.3 feet
43. 9.4 feet
45. 26.3 feet
47. 32.6 feet, 26.5 feet
49. 105.5 miles
51. 3.5 feet
53. 178.3882 ft^2

9.2 Exercises

1. $a = \sqrt{37}$, $B \approx 25.2850°$, $C \approx 94.7150°$
3. $b \approx 4.5985$, $A \approx 41.7854°$, $C \approx 88.2146°$
5. $c \approx 4.1063$, $A \approx 103.0643°$, $B \approx 46.9357°$
7. $c \approx 7.0752$, $A \approx 41.6113°$, $B \approx 68.3887°$
9. $A \approx 46.5675°$, $B \approx 104.4775°$, $C \approx 28.9550°$
11. $A \approx 121.8554°$, $B \approx 39.5712°$, $C \approx 18.5734°$
13. $A = B = C = 60°$
15. $A = 90°$, $B \approx 36.8699°$, $C \approx 53.1301°$
17. $a \approx 11.88$, $B \approx 32.28°$, $C \approx 82.72°$

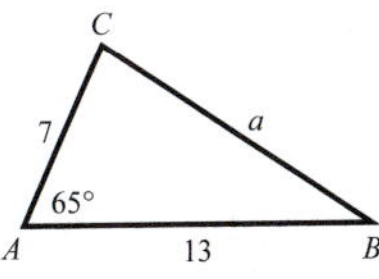

19. $b \approx 10.58$, $A \approx 31.01°$, $C \approx 124.79°$

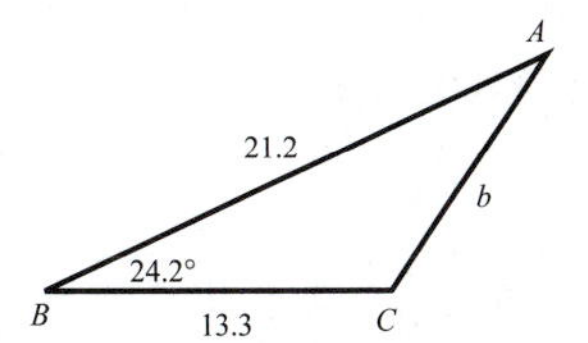

21. $a \approx 11.24$, $B \approx 33.1°$, $C \approx 43.9°$

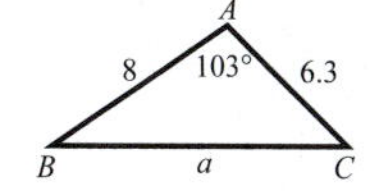

23. $A \approx 90°$, $B \approx 53.13°$, $C \approx 36.87°$

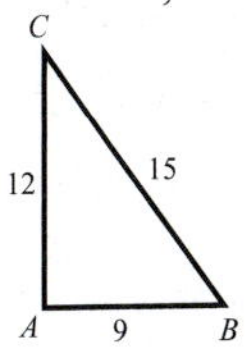

25. $A \approx 54.82°$, $B \approx 26.85°$, $C \approx 98.33°$

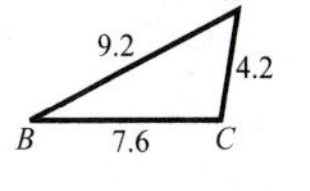

27. $A \approx 50.22°$, $B \approx 71.71°$, $C \approx 58.07°$

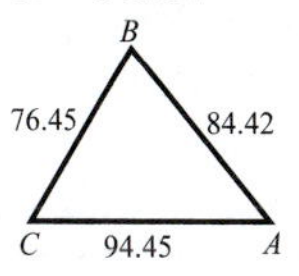

29. 89.29
31. 11.15
33. 65.5744 feet
35. 2.4×10^7 miles
37. 21.4413 feet
39. 22.6199°, 67.3801°
41. 160.1188°
43. 337,940.9589 square nautical miles
45. a. 61.9372 $in.^2$
    b. 584.2397 $in.^2$
    c. 136.1041 $in.^2$

9.3 Exercises

1.

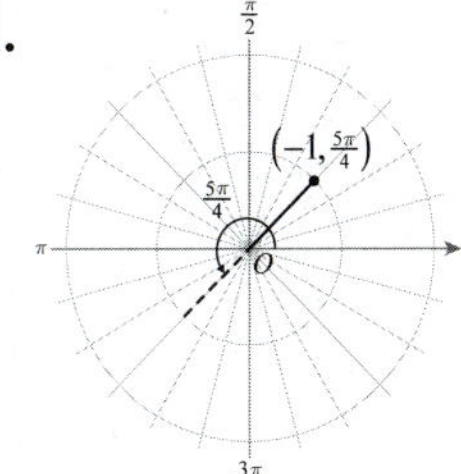

3.

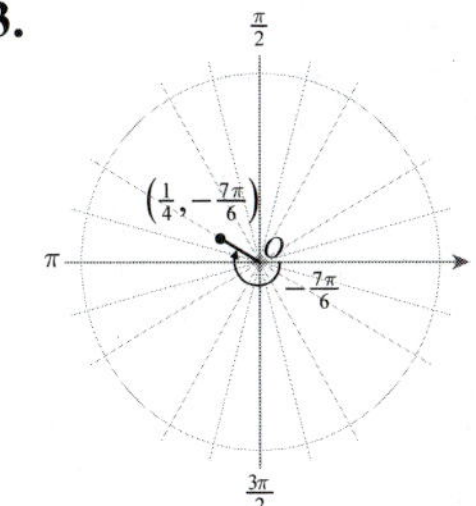

5.

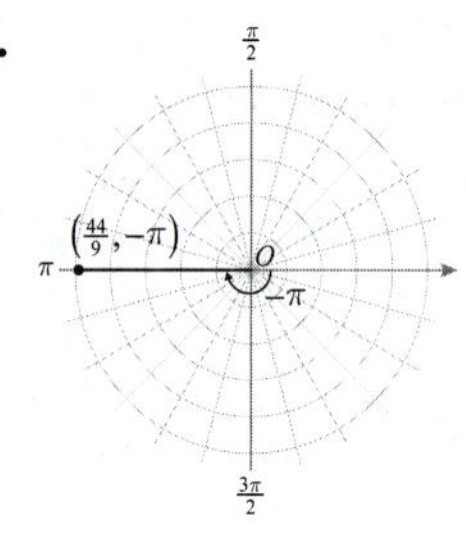

7. $(3.54, -3.54)$

9. $(-4.42, -4.42)$

11. $(-2.60, -1.50)$

13. $(-3, 0)$ and $(3, \pi)$

15. $\left(\sqrt{145}, -0.08\right)$ and $\left(-\sqrt{145}, 3.06\right)$

17. $\left(2\sqrt{21}, 1.76\right)$ and $\left(-2\sqrt{21}, -1.38\right)$

19. $r^2 = 25$

21. $r\cos\theta = 12$

23. $\sin\theta = \cos\theta$

25. $r\cos\theta = 16a$

27. $r^2 - 4ar\cos\theta = 0$

29. $r^2\sin^2\theta - 4r\cos\theta - 4 = 0$

31. $x^2 + y^2 = 5x$

33. $x^2 + y^2 = 49$

35. $y = \dfrac{1}{2}$

37. $x^4 + y^4 + 2x^2y^2 = 2xy$

39. $4y + 7x = 12$

41. $x^2 + y^2 = 4$

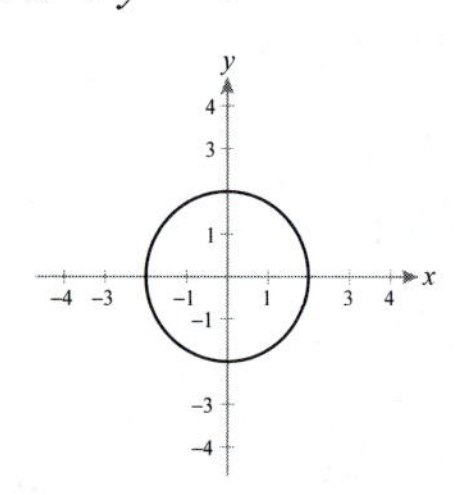

43. $y = -\dfrac{x}{\sqrt{3}}$

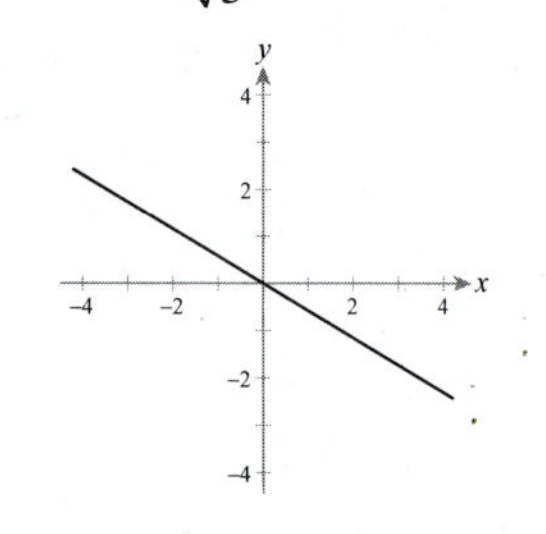

45. $x = 7$

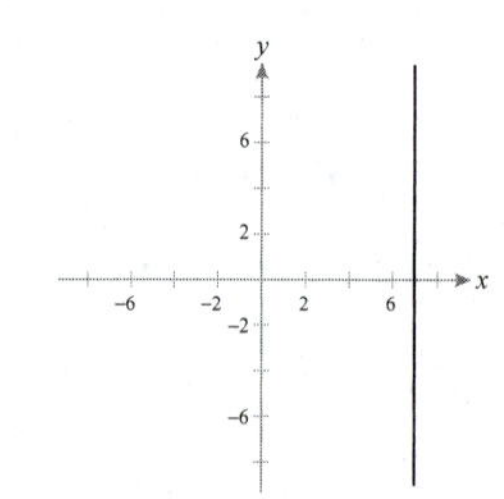

47.

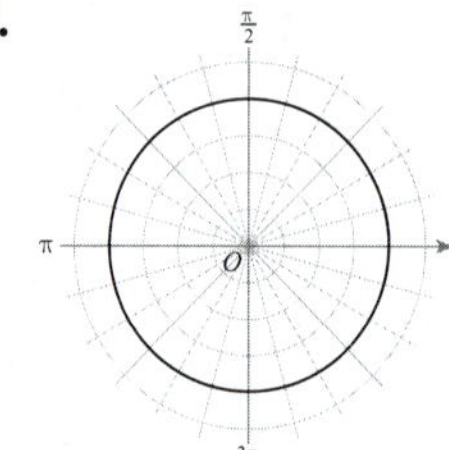

49.

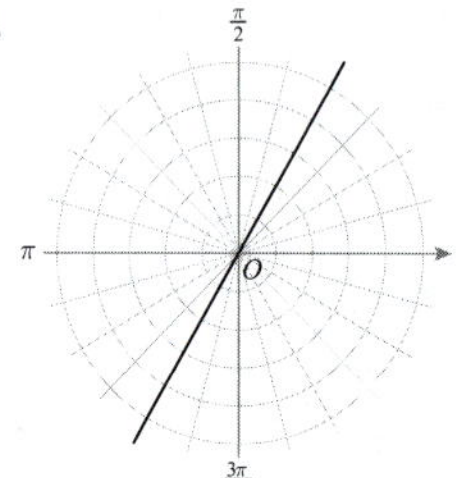

51.

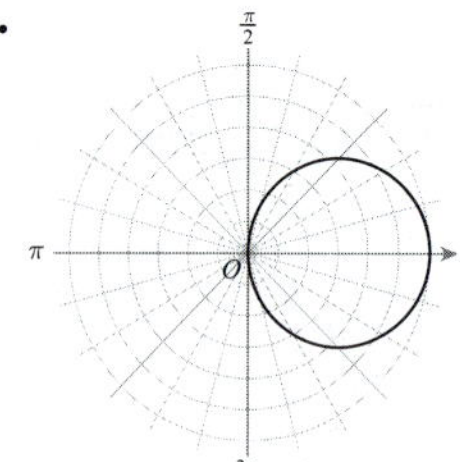

53.

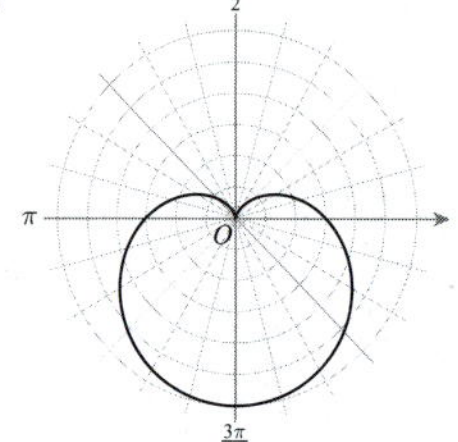

55.

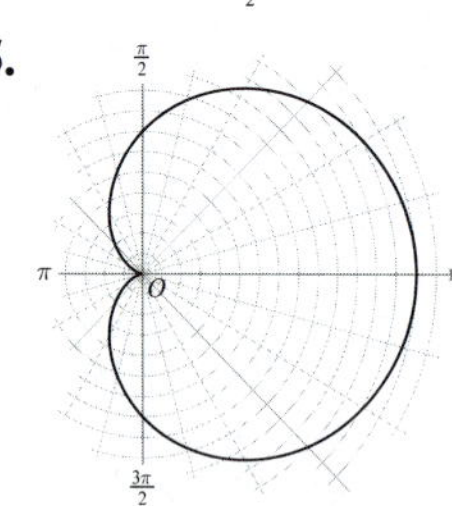

57.

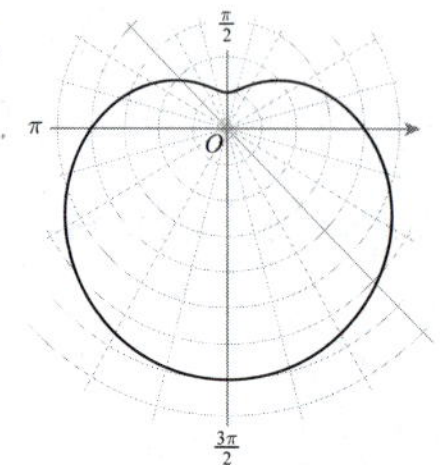

59.

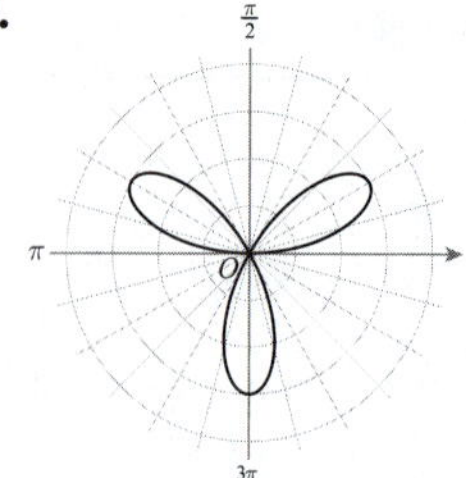

61.

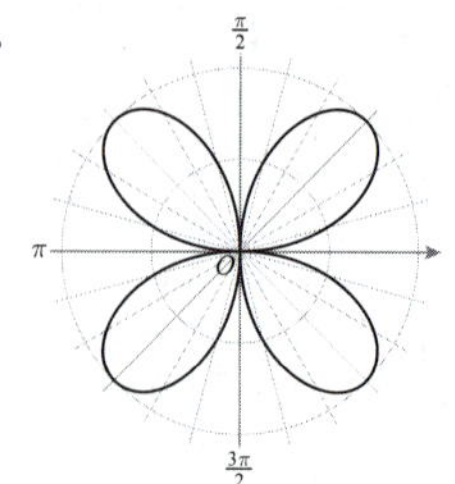

63.

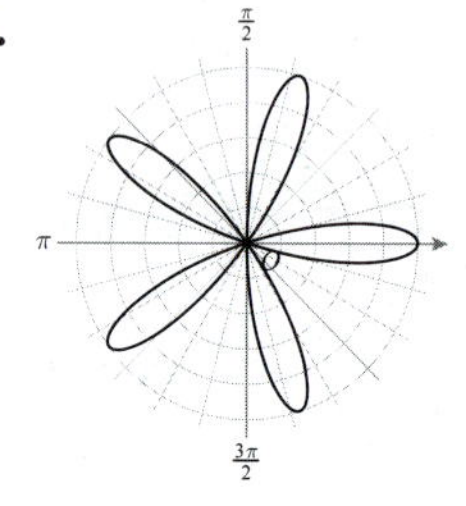

65.

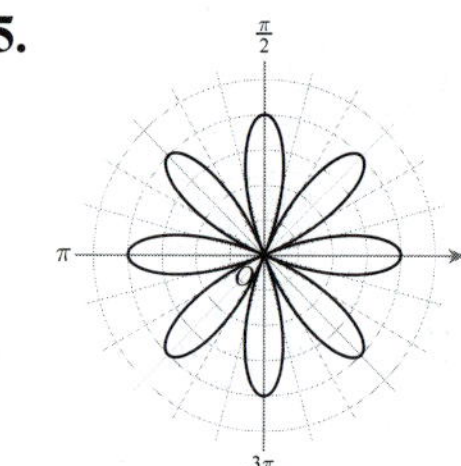

67.

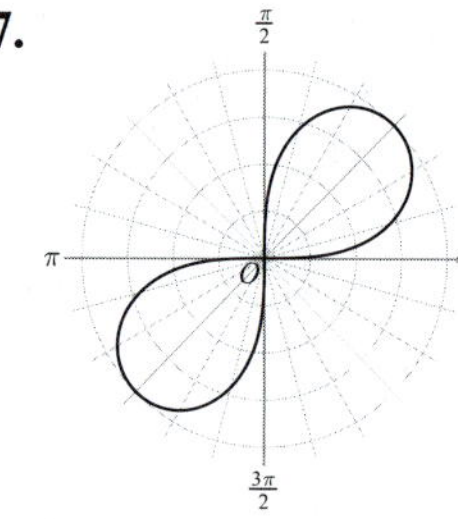

69. $(0,0), \left(\dfrac{\sqrt{2}}{2}, \dfrac{\pi}{4}\right)$

71. $(0,0), \left(1 + \dfrac{\sqrt{2}}{2}, \dfrac{3\pi}{4}\right),$

$\left(1 - \dfrac{\sqrt{2}}{2}, -\dfrac{\pi}{4}\right)$

73. The graph of $f(\theta - a)$ is that of $f(\theta)$ rotated about the origin by a radians.

9.4 Exercises

1.

t	x	y
0	5	0
1	6	1
2	7	undefined
3	8	$\sqrt{3}$
4	9	1
5	10	$\frac{\sqrt{5}}{3}$
6	11	$\frac{\sqrt{6}}{4}$

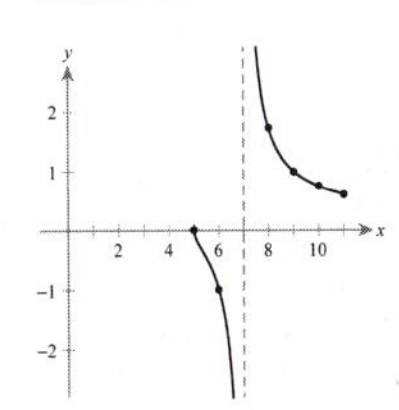

3. $2x = 3y + 6$

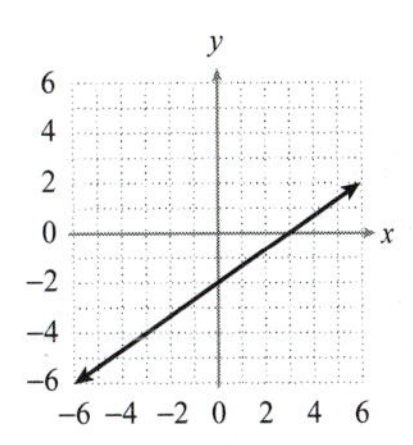

5. $x = 2y + 4$

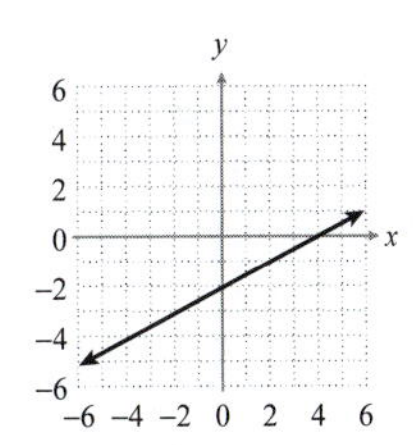

7. $y = 16x^2$

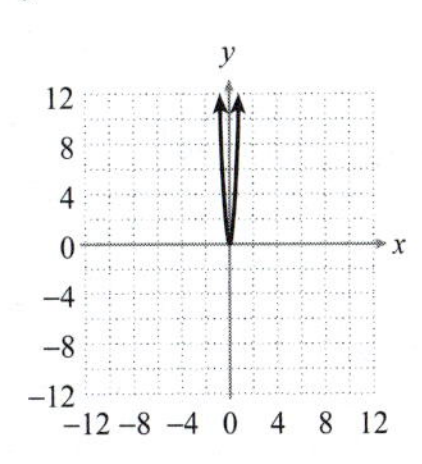

9. $x = \sqrt{y}$

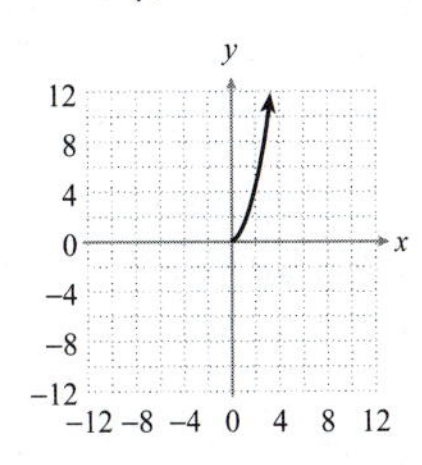

11. $y = \pm 2\sqrt{1 - x^2}$

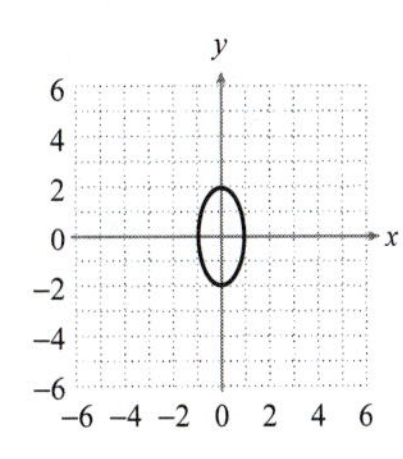

13. $y = -x, 0 \le x \le 2$

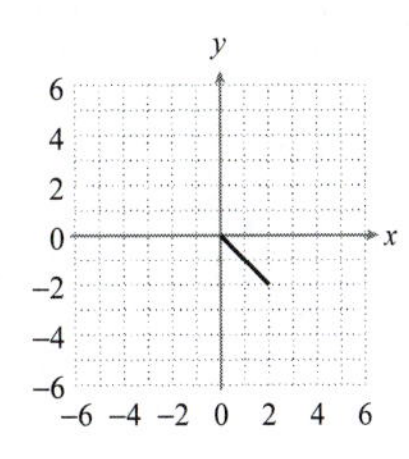

15. $y = \pm\sqrt{4x - x^2} + 2$

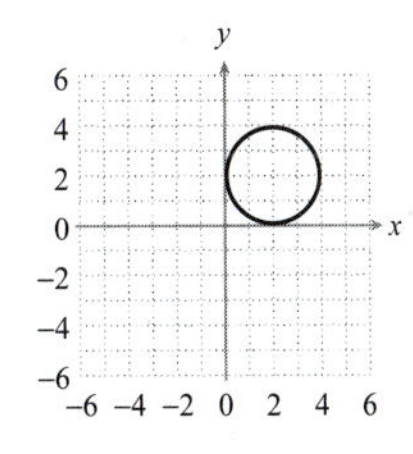

17. $x = t, y = t^2 + 2t + 1$

19. $x = t, y = -t^2 - 5$

21. $x = t, y = \pm\sqrt{t - 4}$

23. $x = t, y = \frac{1}{t}$

25. $x = t, y = |t - 1|$

27. $x = t, y = \pm\sqrt{1 - t^2}$

29. $x = t, y = t^2 - t - 6$

31. $x = t, y = \frac{1}{4}t + \frac{19}{2}$

33. $x = t, y = \frac{4}{7}t$

35. $x = t, y = \frac{1}{2}t - 3$

37. $x = t - 5, y = 2 - |t - 5|$

39. $x = 4t, y = -2t + \frac{5}{2}$

41. $x = 5t + 1, y = 25t^2 - 10t + 1$

43. $x = -4 + 3\cos\theta, y = 2 + 3\sin\theta$

45. $x = 6\cos\theta, y = -2 + 6\sin\theta$

47. a. $x \approx 58.67t,$
$y \approx -16t^2 + 101.61t + 10$

b.

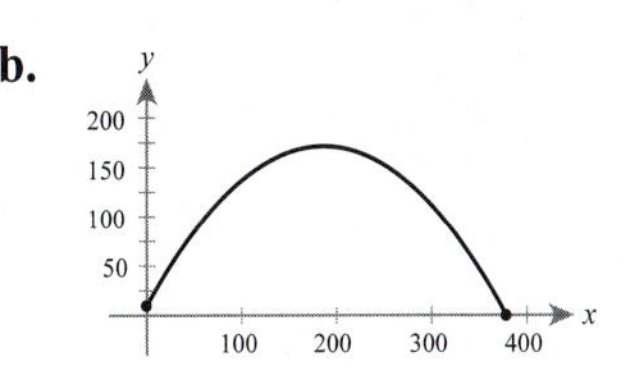

c. Approx. 126.42 ft

d. Approx. 378.42 ft

e. $t \approx 6.45$ s

f. Yes

49. $x = 12(\theta - \sin\theta),$
$y = 12(1 - \cos\theta)$

9.5 Exercises

1. $\sqrt{34}$

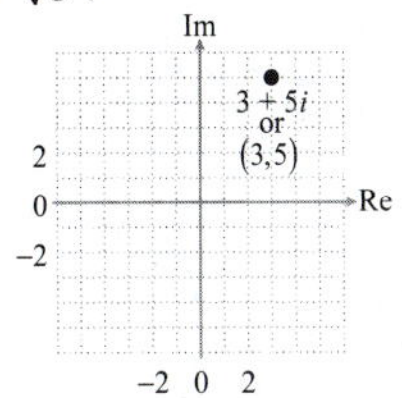

3. $\sqrt{20} = 2\sqrt{5}$

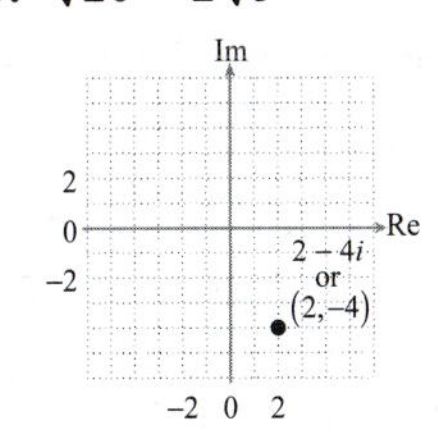

5. $\sqrt{32} = 4\sqrt{2}$

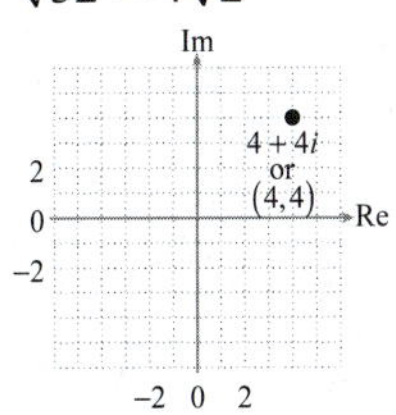

7.

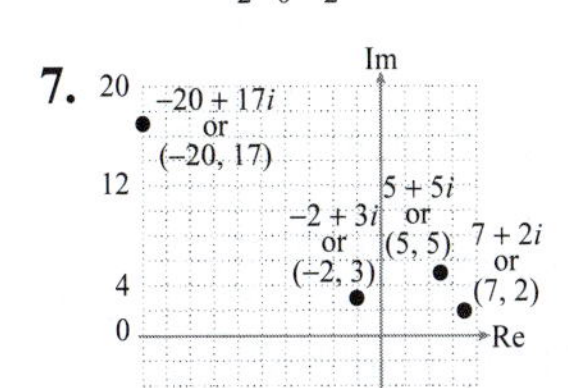

9.

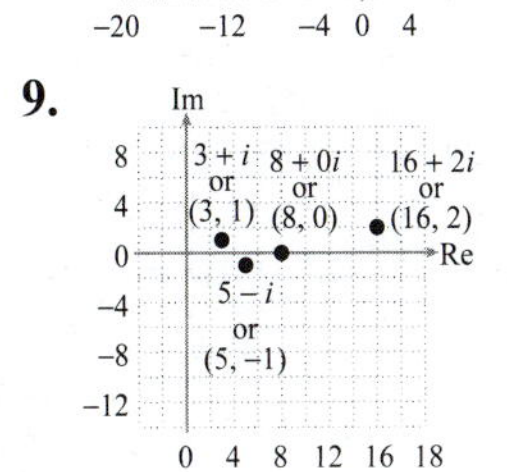

11.

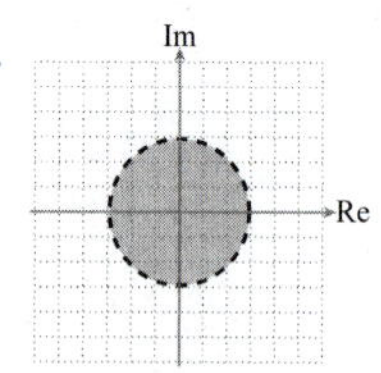

13.

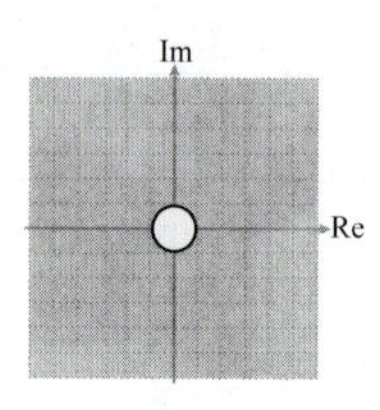

15.

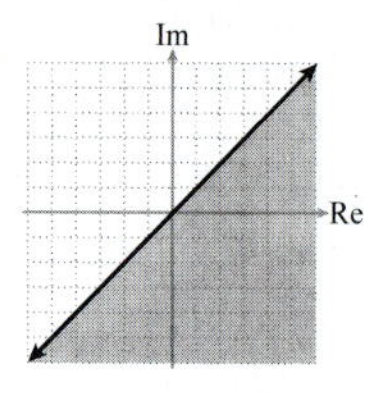

17. $\sqrt{10}\left(\cos(3.46)+i\sin(3.46)\right)$

19. $\sqrt{5}\left(\cos(1.11)+i\sin(1.11)\right)$

21. $2\sqrt{5}\left(\cos(0.46)+i\sin(0.46)\right)$

23. $2\left(\cos\left(-\frac{\pi}{4}\right)+i\sin\left(-\frac{\pi}{4}\right)\right)$

25. $5\left(\cos(0.93)+i\sin(0.93)\right)$

27. $8\left(\cos\left(-\frac{\pi}{3}\right)+i\sin\left(-\frac{\pi}{3}\right)\right)$

29. $\frac{-3\sqrt{3}}{2}+\frac{3i}{2}$

31. $-1-i\sqrt{3}$ **33.** $-\frac{5}{\sqrt{2}}+\frac{5i}{\sqrt{2}}$

35. $\frac{-3\sqrt{3}}{4}+\frac{3i}{4}$ **37.** $1.01+4.9i$

39. $16\left(\cos 30°+i\sin 30°\right)=8\sqrt{3}+8i$

41. $3\sqrt{6}\left(\cos\left(\frac{17\pi}{12}\right)+i\sin\left(\frac{17\pi}{12}\right)\right)$

$=-1.9-7.1i$

43. $2\sqrt{10}\left(\cos(2.42)+i\sin(2.42)\right)$

$=\left(-3-\sqrt{3}\right)+\left(3\sqrt{3}-1\right)i$

45. $2\left(\cos 180°+i\sin 180°\right)=-2$

47. $\frac{10}{3}\left(\cos\left(\frac{\pi}{2}\right)+i\sin\left(\frac{\pi}{2}\right)\right)=\frac{10i}{3}$

49. $\frac{1}{\sqrt{2}}\left(\cos\left(-\frac{3\pi}{4}\right)+i\sin\left(-\frac{3\pi}{4}\right)\right)$

$=-\frac{1}{2}-\frac{i}{2}$

51. $2\left(\cos\left(\frac{5\pi}{12}\right)+i\sin\left(\frac{5\pi}{12}\right)\right)$

$=0.52+1.93i$

53. $a=4\left(\cos\left(\frac{5\pi}{6}\right)+i\sin\left(\frac{5\pi}{6}\right)\right)$,

$b=2\left(\cos\pi+i\sin\pi\right)$,

$c=8\left(\cos\left(\frac{11\pi}{6}\right)+i\sin\left(\frac{11\pi}{6}\right)\right)$

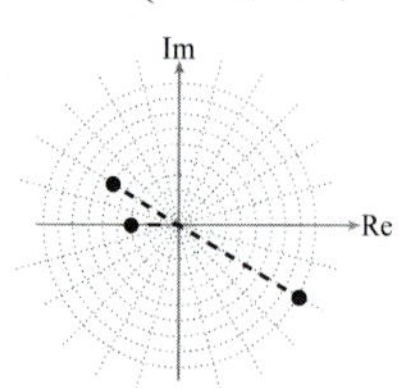

55. $a=2-5i, b=\sqrt{2}+2i$,

$c=10+2\sqrt{2}+i\left(4-5\sqrt{2}\right)$

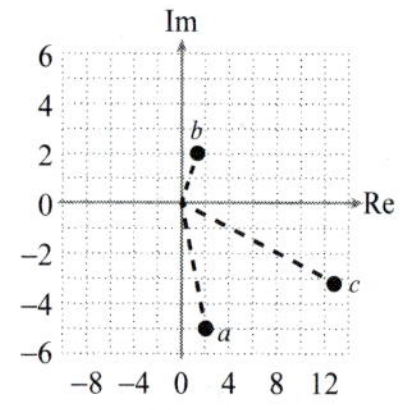

57. $a=2e^{\frac{\pi}{3}i}, b=3e^{\frac{5\pi}{4}i}, c=6e^{\frac{19\pi}{12}i}$

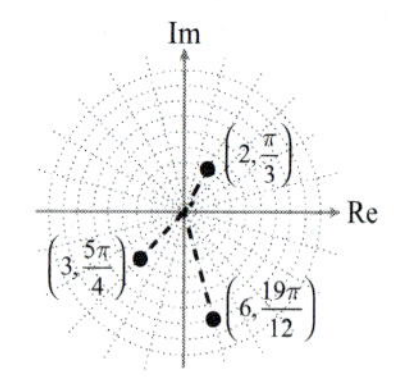

59. $32e^{\frac{\pi}{3}i}$

61. $1.04\times10^{13}e^{2.9i}$

63. $e^{2\pi i}$

65. $a=e^{\frac{\pi i}{4}}, b=e^{\frac{3\pi i}{4}}, c=e^{\frac{5\pi i}{4}}, d=e^{\frac{7\pi i}{4}}$

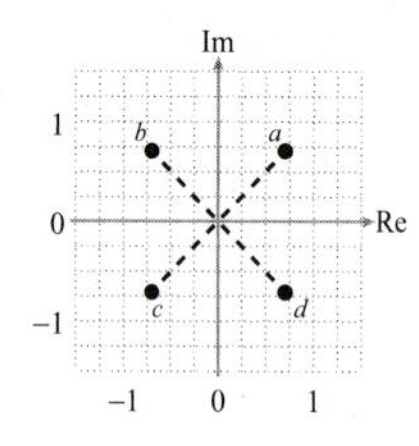

67. $a=2e^{\frac{\pi i}{12}}, b=2e^{\frac{13\pi i}{12}}$

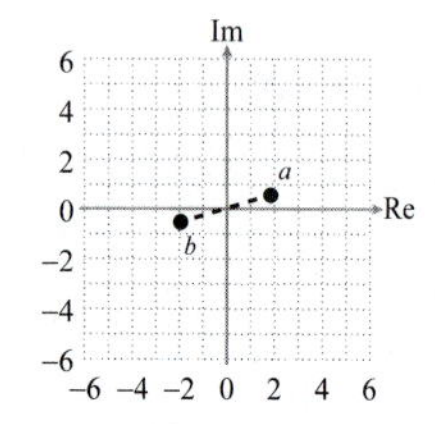

69. $a=4, b=4e^{\frac{\pi i}{2}}, c=4e^{\pi i}, d=4e^{\frac{3\pi i}{2}}$

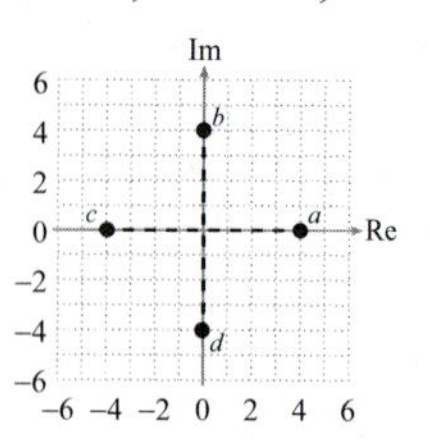

71. $a=2e^{60°i}, b=2e^{240°i}$

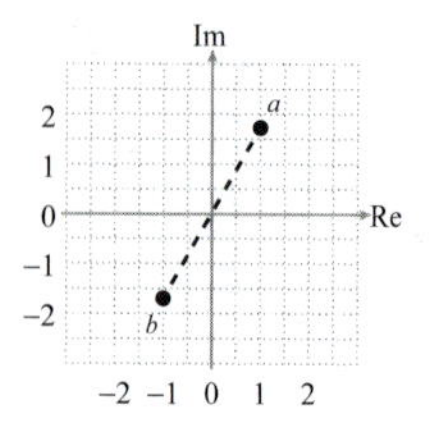

73. $2\sqrt{2}e^{\frac{\pi i}{12}}, 2\sqrt{2}e^{\frac{13\pi i}{12}}$

75. $2e^{\frac{\pi i}{5}}, 2e^{\frac{3\pi i}{5}}, 2e^{\pi i}, 2e^{\frac{7\pi i}{5}}, 2e^{\frac{9\pi i}{5}}$

77. $5e^{\frac{\pi i}{4}}, 5e^{\frac{3\pi i}{4}}$

9.6 Exercises

1.

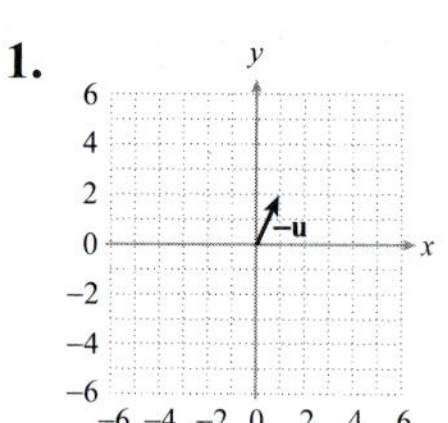

3.

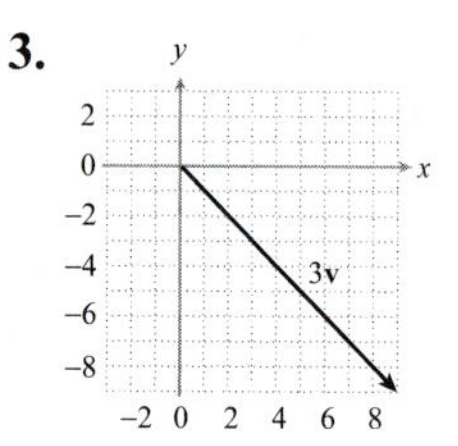

5.

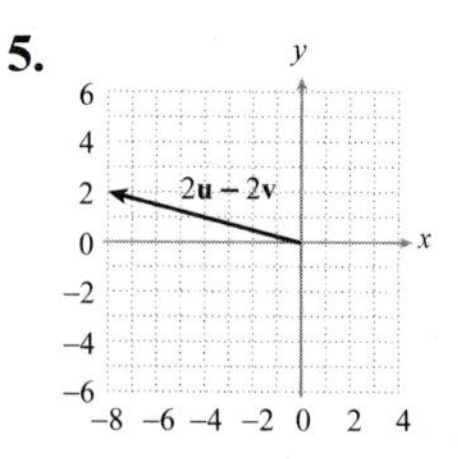

7. $\mathbf{v}=\langle 3,-3\rangle, \|\mathbf{v}\|=3\sqrt{2}$

9. $\mathbf{v}=\langle 5,3\rangle, \|\mathbf{v}\|=\sqrt{34}$

11. $\mathbf{v}=\langle 5,-1\rangle, \|\mathbf{v}\|=\sqrt{26}$

13. $\mathbf{v}=\langle -7,7\rangle, \|\mathbf{v}\|=7\sqrt{2}$

15. $\mathbf{v}=\langle -4,-6\rangle, \|\mathbf{v}\|=2\sqrt{13}$

17. a. $\langle -2,8\rangle$ **b.** $\langle 8,-4\rangle$
c. $\langle -4,0\rangle$

19. a. $\langle 1,4\rangle$ **b.** $\langle -11,12\rangle$
c. $\langle 6,-8\rangle$

21. a. $\langle -5,-10\rangle$ **b.** $\langle -8,-2\rangle$
c. $\langle 6,4\rangle$

23. $-\mathbf{u}=\langle -1,-1\rangle$,
$2\mathbf{u}-\mathbf{v}=\langle -1,5\rangle$,
$\mathbf{u}+\mathbf{v}=\langle 4,-2\rangle$,
$\|\mathbf{u}\|=\sqrt{2}, \|\mathbf{v}\|=3\sqrt{2}$

25. $-\mathbf{u}=\langle 4,-4\rangle$,
$2\mathbf{u}-\mathbf{v}=\langle -12,12\rangle$,
$\mathbf{u}+\mathbf{v}=\langle 0,0\rangle$,
$\|\mathbf{u}\|=4\sqrt{2}, \|\mathbf{v}\|=4\sqrt{2}$

27. a. $\left\langle \frac{2}{\sqrt{5}},-\frac{1}{\sqrt{5}}\right\rangle$
b. $\mathbf{u}=6\mathbf{i}-3\mathbf{j}$

29. a. $\left\langle \frac{-5}{\sqrt{26}},\frac{1}{\sqrt{26}}\right\rangle$
b. $\mathbf{u}=5\mathbf{i}-\mathbf{j}$

31. a. $\left\langle \frac{2}{\sqrt{13}},\frac{3}{\sqrt{13}}\right\rangle$
b. $\mathbf{u}=3\mathbf{i}+3\mathbf{j}$

33. $\|\mathbf{v}\|=5, \theta=30°$

35. $\|\mathbf{v}\|=5, \theta=36,9°$

37. $\langle 3\sqrt{3},3\rangle$

39. $\langle -9\sqrt{2},9\sqrt{2}\rangle$

41. $\left\langle -\frac{1}{2},\frac{\sqrt{3}}{2}\right\rangle$

43. $\left\langle \frac{8}{\sqrt{13}},\frac{12}{\sqrt{13}}\right\rangle$

45. $\langle 2\sqrt{3},2\rangle$

47. 38.67 mph, N 77.76° W

49. $\mathbf{F}=\langle 150,-1235\rangle$,
$|\mathbf{F}|=1244.08$ pounds

9.7 Exercises

1. 17 **3.** 6 **5.** 8
7. 1 **9.** −7 **11.** 32
13. $\langle -26,39\rangle$ **15.** $\sqrt{37}$
17. $\sqrt{53}$ **19.** 123.7°
21. 14.0° **23.** 161.6°
25. $\frac{\pi}{4}$ **27.** 8°, 69°, 103°
29. 57.1°, 60.8°, 62.1°
31. −62.5 **33.** $-32\sqrt{2}$
35. $\langle 1,1\rangle,\langle 5,5\rangle$
37. $\langle 3,1\rangle,\langle -6,-2\rangle$
39. Neither
41. Orthogonal
43. $\text{proj}_{\mathbf{v}}\mathbf{u}=\langle 2,1\rangle$,
$\text{perp}_{\mathbf{v}}\mathbf{u}=\langle -1,2\rangle$
45. $\text{proj}_{\mathbf{v}}\mathbf{u}=\left\langle \frac{6}{5},\frac{2}{5}\right\rangle$,
$\text{perp}_{\mathbf{v}}\mathbf{u}=\left\langle \frac{9}{5},-\frac{27}{5}\right\rangle$
47. $\text{proj}_{\mathbf{v}}\mathbf{u}=\left\langle -\frac{60}{17},-\frac{15}{17}\right\rangle$,
$\text{perp}_{\mathbf{v}}\mathbf{u}=\left\langle \frac{9}{17},-\frac{36}{17}\right\rangle$
49. 14 **51.** 3
53. 3479.3 pounds
55. 109.6 pounds
57. 1638.3 ft-lb
59. 11,431,535.3 ft-lb

9.8 Exercises

1. 0 **3.** 1 **5.** $\frac{3}{5}$
7. −1.04 **9.** 74.21 **11.** Even
13. Odd **15.** Odd **37.** 0
45. −0.33 **47.** 0.69 **49.** 0.35

Chapter 9 Project

1. Approx. 921,479 ft²
3. a. Approx. 8414 pounds
b. Approx. 1298.4 pounds
c. Approx. 2045.6 feet
d. 2,822,928 foot-pounds

Chapter 9 Review Exercises

1. $C=105°, c\approx 5.46, a\approx 2.83$

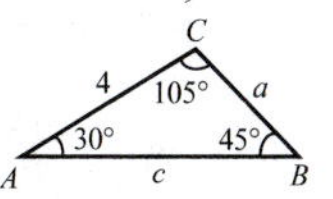

3. $B=68°40', a\approx 36.80, b\approx 35.60$

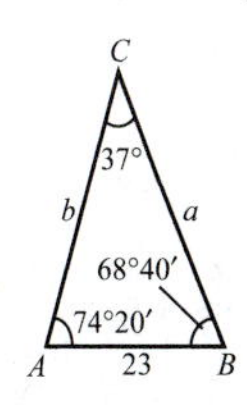

5. 10.14

7. $A\approx 22°31', C\approx 63°22', b\approx 15.62$

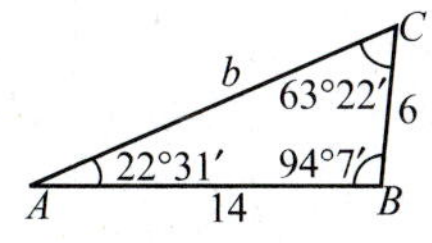

9. $A\approx 52.04°, B\approx 101.99°$,
$C\approx 25.98°$

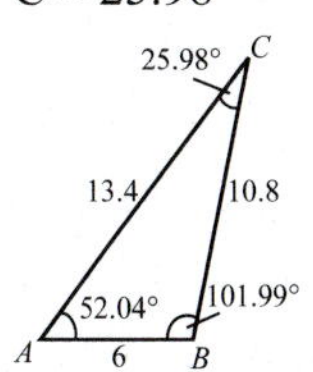

11. 22.19

13. $(-6.06,-3.5)$

15. $(15.62,0.88)$ and $(-15.62,-2.26)$

17. $r^2-9ar\cos\theta=0$

19. $x+y=4$

21.

$\frac{\pi}{2}$ π $\frac{3\pi}{2}$

23. $y=|x-7|$

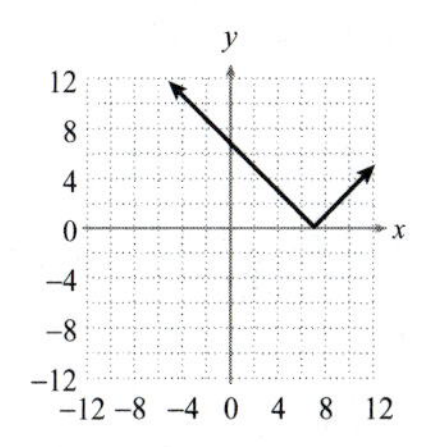

25. $\frac{x^2}{16}+(y-1)^2=1$

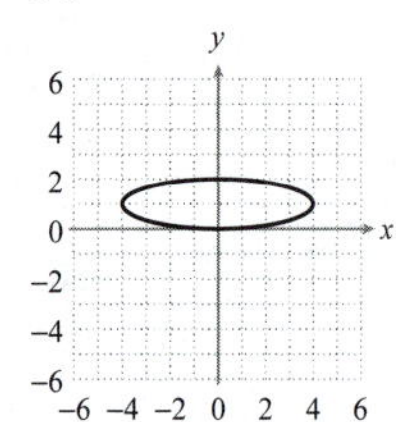

27. $x=t,\ y=2-6t$

29. $x=1+\cos\theta,\ y=1+\sin\theta$

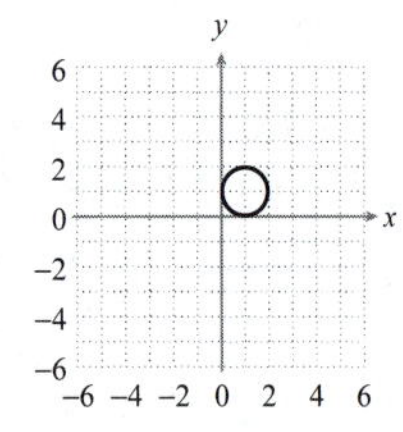

31. $3\sqrt{2}$

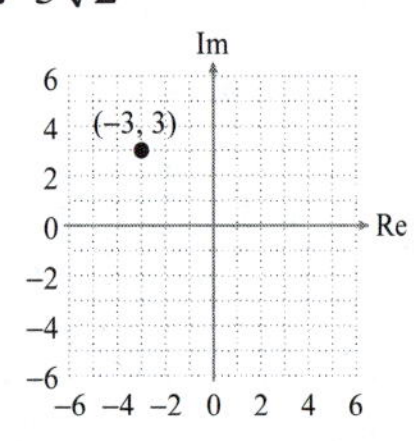

33.

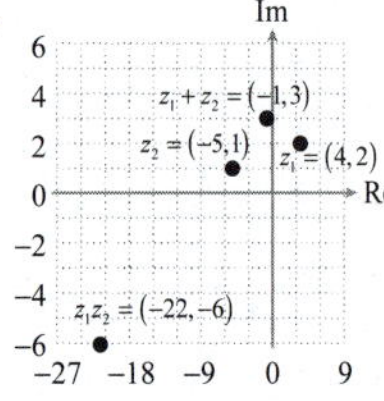

35.

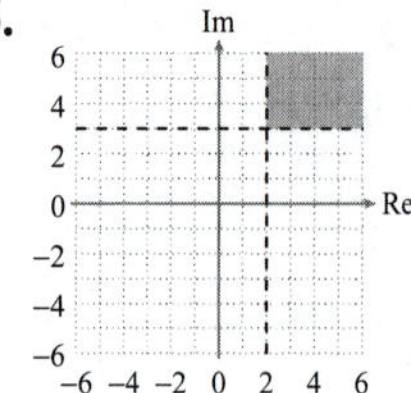

37. $\sqrt{17}\left(\cos(1.33)+i\sin(1.33)\right)$

39. $\frac{3}{2}+\frac{3i\sqrt{3}}{2}$

41. $5(\cos 120°+i\sin 120°)$, $-\frac{5}{2}+\frac{5i\sqrt{3}}{2}$

43. $24(\cos 315°+i\sin 315°)$, $12\sqrt{2}-12i\sqrt{2}$

45. $177{,}147e^{120°i}$

47. $a=5e^{i\left(\frac{7\pi}{12}\right)}, b=5e^{i\left(\frac{5\pi}{4}\right)}$, $c=5e^{i\left(\frac{23\pi}{12}\right)}$

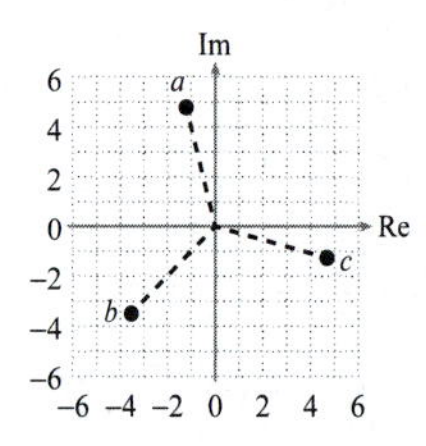

49. $2e^{i\left(\frac{\pi}{4}\right)}, 2e^{i\left(\frac{11\pi}{12}\right)}, 2e^{i\left(\frac{19\pi}{12}\right)}$

51. $\mathbf{v}=\langle -10,-6\rangle, \|\mathbf{v}\|=2\sqrt{34}$

53.

55. $-\mathbf{u}=\langle -5,-1\rangle, 2\mathbf{u}-\mathbf{v}=\langle 7,1\rangle$, $\mathbf{u}+\mathbf{v}=\langle 8,2\rangle, \|\mathbf{u}\|=\sqrt{26}$, $\|\mathbf{v}\|=\sqrt{10}$

57. a. $\left\langle \frac{2}{\sqrt{5}},\frac{1}{\sqrt{5}}\right\rangle$

b. $6\mathbf{i}+3\mathbf{j}$

59. $\|\mathbf{v}\|=\sqrt{26}, \theta=-11.3°$

61. $\langle 3.6,-4.8\rangle$

63. 40.01 mph, N 40.23° W

65. $\langle -108,-270\rangle$

67. $\sqrt{10}$ **69.** $\frac{5\pi}{12}$ **71.** −36

73. $\text{proj}_{\mathbf{v}}\mathbf{u}=\left\langle \frac{3}{2},\frac{3}{2}\right\rangle$, $\text{proj}_{\mathbf{v}}\mathbf{u}=\left\langle \frac{5}{2},\frac{-5}{2}\right\rangle$

75. 31

77. 417,558.5 ft-lb

79. $\frac{15}{8}$ **81.** 1.31

Chapter 10: Conic Sections

10.1 Exercises

1. Center: $(5,2)$
Foci: $(5,2\pm\sqrt{21})$
Vertices: $(5,7),(5,-3)$

3. Center: $(-2,-5)$
Foci: $(-2\pm\sqrt{6},-5)$
Vertices: $(1,-5),(-5,-5)$

5. Center: $(-3,2)$
Foci: $(-3\pm\sqrt{2},2)$
Vertices: $(-1,2),(-5,2)$

7. Center: $(-5,1)$
Foci: $(-5,1\pm 2\sqrt{3})$
Vertices: $(-5,5),(-5,-3)$

9. Center: $(-4,2)$
Foci: $(-4\pm 3\sqrt{2},2)$
Vertices: $(-4\pm 3\sqrt{3},2)$

11. Center: $(2,0)$
Foci: $(4,0),(0,0)$
Vertices: $(2\pm\sqrt{5},0)$

13. e **15.** f

17. c **19.** b

21.

$(3\pm 2\sqrt{2},-1)$

23.

$(3\pm\sqrt{5},4)$

25. $\left(1, 4 \pm \sqrt{3}\right)$

27. $\left(-1 \pm \sqrt{21}, -5\right)$

29. $\left(-2 \pm \sqrt{7}, -1\right)$

31. $\left(-1 \pm \sqrt{7}, 2\right)$

33. $\left(-5, 3 \pm \sqrt{15}\right)$

35. $\left(-5 \pm \sqrt{5}, -5\right)$

37. $\left(0, -2 \pm \sqrt{3}\right)$

39. $\dfrac{x^2}{16} + \dfrac{y^2}{25} = 1$

41. $(x-1)^2 + \dfrac{(y-1)^2}{9} = 1$

43. $\dfrac{(x-3)^2}{36} + \dfrac{y^2}{27} = 1$

45. $(x+2)^2 + \dfrac{(y+3)^2}{4} = 1$

47. $\dfrac{(x-5)^2}{16} + \dfrac{(y-3)^2}{15} = 1$

49. $\dfrac{(x-2)^2}{4} + \dfrac{(y+2)^2}{9} = 1$

51. $\dfrac{(x-1)^2}{9} + \dfrac{y^2}{16} = 1$

53. $e = \dfrac{\sqrt{11}}{6}$; major $= 24$; minor $= 20$

55. $e = \dfrac{2\sqrt{2}}{3}$; major $= 12$; minor $= 4$

57. $e = \dfrac{\sqrt{3}}{2}$; major $= 4$; minor $= 2$

59. $e = \dfrac{\sqrt{2}}{2}$; major $= 4$; minor $= 2\sqrt{2}$

61. $e = \dfrac{\sqrt{42}}{7}$; major $= 14$
minor $= 2\sqrt{7}$

63. $e \approx 0.249$

65. Yes, just barely, if the boat is centered on the river.

67. The string should be 5 cm long, and the tacks should be 4 cm apart.

10.2 Exercises

1.

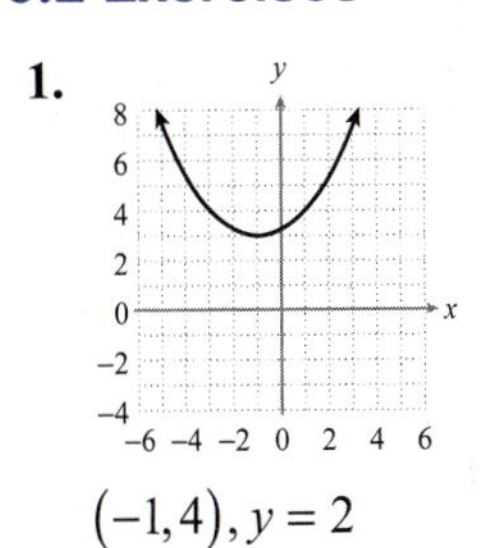

$(-1, 4),\ y = 2$

3.

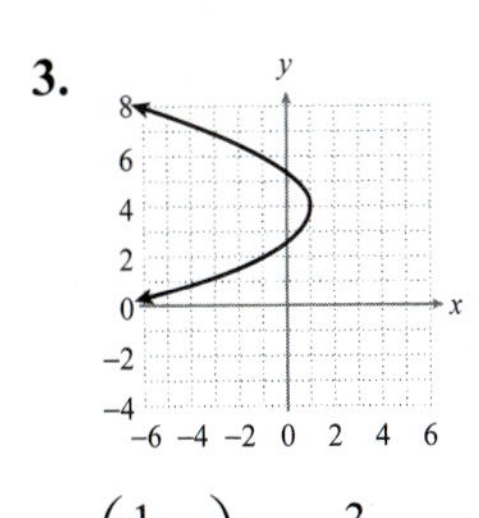

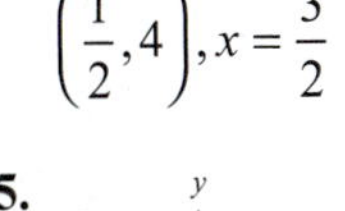

$\left(\dfrac{1}{2}, 4\right),\ x = \dfrac{3}{2}$

5.

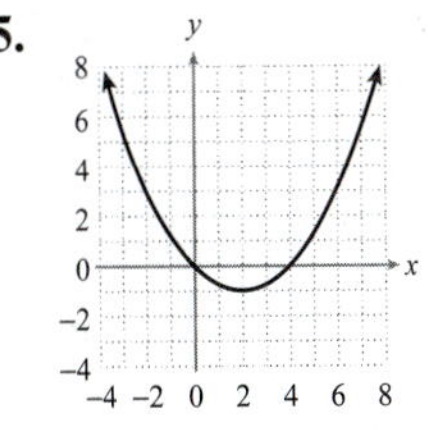

$(2, 0),\ y = -2$

7.

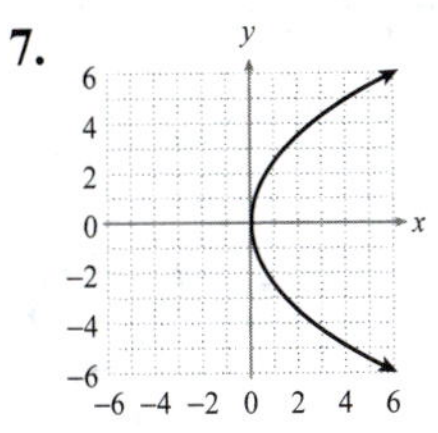

$\left(\dfrac{3}{2}, 0\right),\ x = -\dfrac{3}{2}$

9.

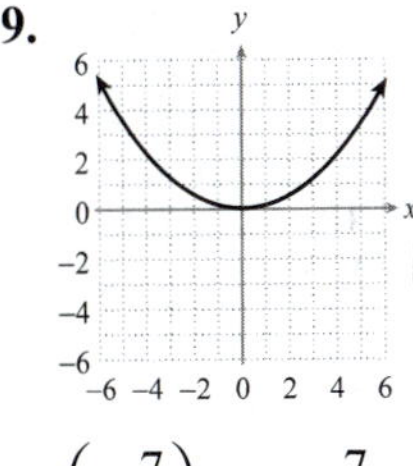

$\left(0, \dfrac{7}{4}\right),\ y = -\dfrac{7}{4}$

11.

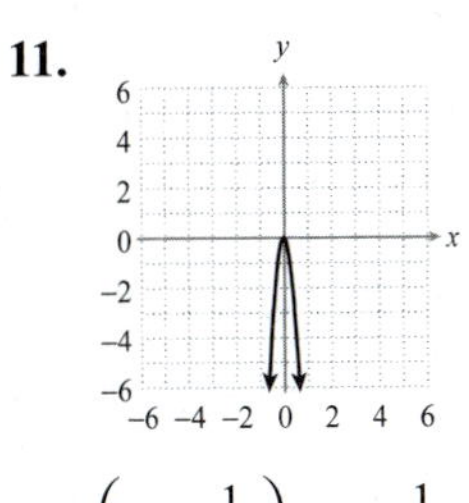

$\left(0, -\dfrac{1}{48}\right),\ y = \dfrac{1}{48}$

13.

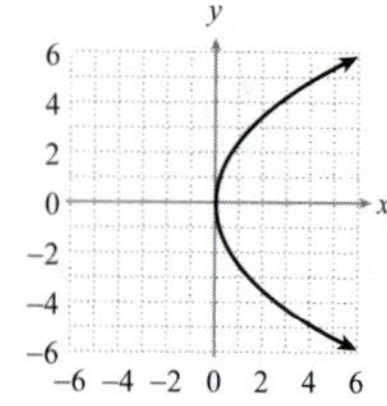

$\left(\frac{3}{2},0\right), x=-\frac{3}{2}$

15.

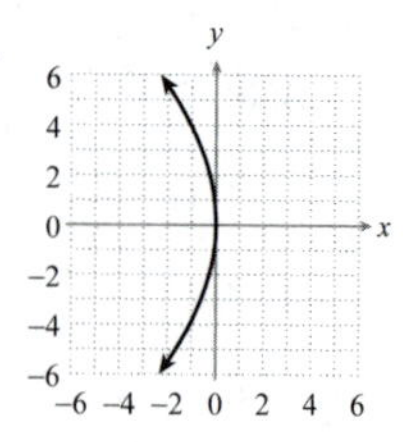

$(-4,0), x=4$

17.

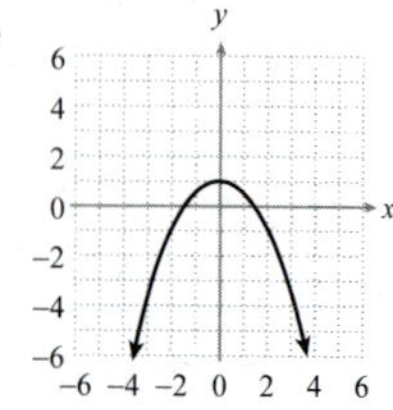

$\left(0,\frac{1}{2}\right), y=\frac{3}{2}$

19.

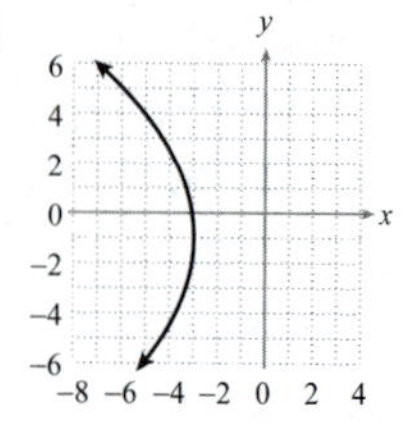

$(-6,-1), x=0$

21.

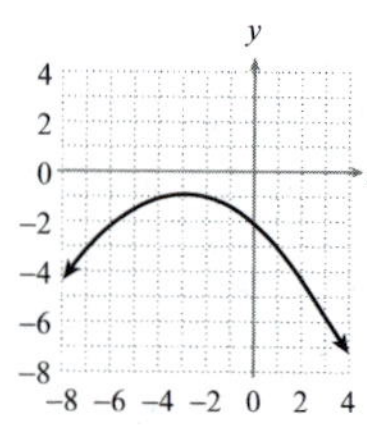

$(-3,-3), y=1$

23.

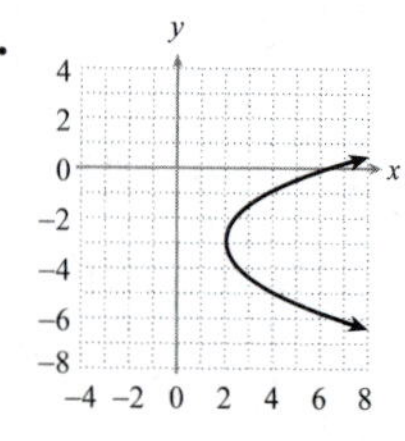

$\left(\frac{5}{2},-3\right), x=\frac{3}{2}$

25. g **27.** b

29. e **31.** d

33. $(y-1)^2=-4(x+1)$

35. $(x-3)^2=8(y+1)$

37. $(y+2)^2=24(x-3)$

39. $(x+3)^2=-2(y+1)$

41. $(y-3)^2=10(x+4)$

43. $(y+1)^2=-8(x-2)$

45. 2 feet **47.** 1.5 inches

10.3 Exercises

1.

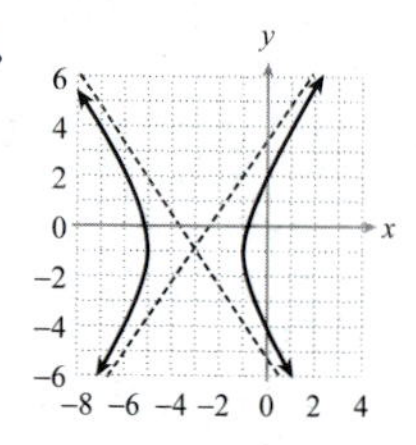

$\left(-3\pm\sqrt{13},-1\right)$

3.

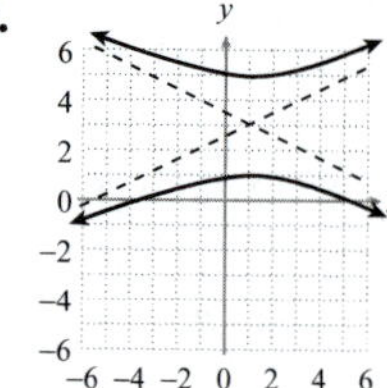

$\left(1, 3\pm2\sqrt{5}\right)$

5.

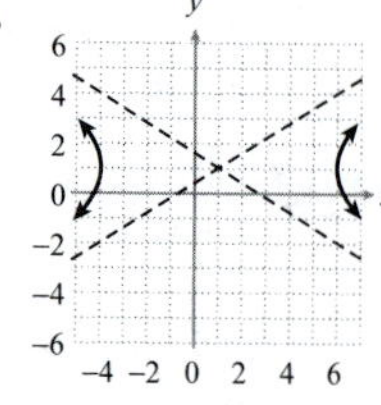

$\left(1\pm\sqrt{34},1\right)$

7.

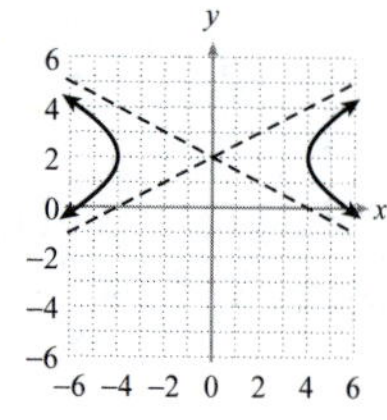

$\left(\pm2\sqrt{5}, 2\right)$

9.

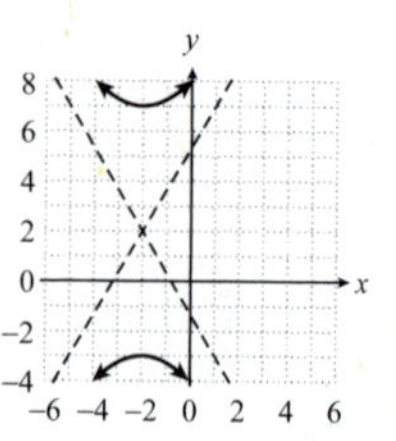

$\left(-2, 2\pm\sqrt{34}\right)$

11.

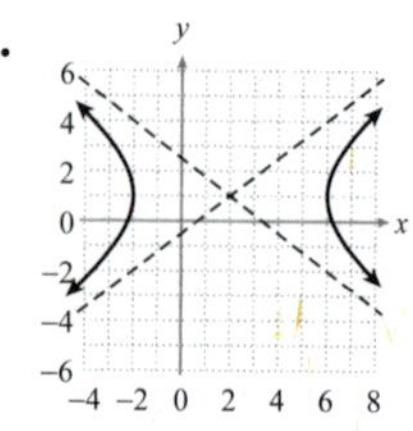

$(7,1),(-3,1)$

13.

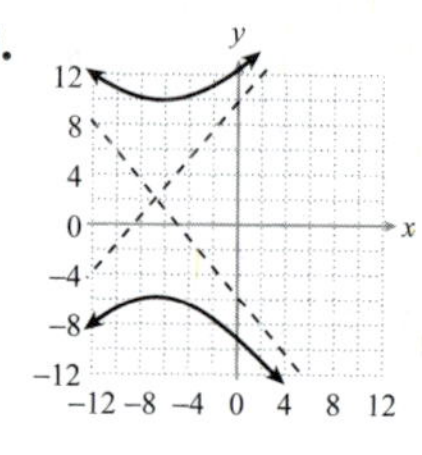

$\left(-7,2\pm\sqrt{113}\right)$

15.

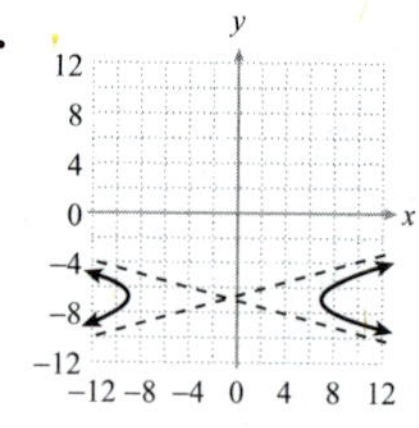

$\left(-1\pm2\sqrt{17},-7\right)$

17. Center: $(-3,2)$
Foci: $\left(-3\pm\sqrt{13}, 2\right)$
Vertices: $(-1,2),(-5,2)$

19. Center: $(1,-4)$
Foci: $\left(1\pm2\sqrt{3}, -4\right)$
Vertices: $\left(1\pm\sqrt{3}, -4\right)$

21. Center: $(-2,1)$
Foci: $\left(-2\pm\sqrt{30}, 1\right)$
Vertices: $(3,1),(-7,1)$

23. Center: $(-3,-1)$
Foci: $\left(-3\pm2\sqrt{3}, -1\right)$
Vertices: $(-1,-1),(-5,-1)$

25. Center: $(1,0)$

Foci: $\left(1 \pm \frac{\sqrt{5}}{2}, 0\right)$

Vertices: $(2,0),(0,0)$

27. Center: $(8,5)$

Foci: $(8 \pm 4\sqrt{5}, 5)$

Vertices: $(12,5),(4,5)$

29. a **31.** b

33. g **35.** h

37. $\frac{x^2}{4} - \frac{y^2}{5} = 1$

39. $y^2 - 4x^2 = 1$

41. $\frac{5x^2}{2} - \frac{5(y-4)^2}{18} = 1$

43. $\frac{(x-6)^2}{9} - \frac{(y-5)^2}{7} = 1$

45. $\frac{(x+4)^2}{4} - \frac{(y-3)^2}{16} = 1$

47. $\frac{(x-1)^2}{9} - \frac{(y+1)^2}{4} = 1$

49. $\frac{(y+4)^2}{16} - \frac{(x-3)^2}{25} = 1$

51. $\frac{x^2}{(6\times 10^7)^2} - \frac{y^2}{(7.2\times 10^7)^2} = 1$

10.4 Exercises

1. $(4\sqrt{3}+3, -4+3\sqrt{3})$

3. $\left(\frac{-5\sqrt{2}}{16}, \frac{3\sqrt{2}}{16}\right)$

5. $(-1.2097, -13.5476)$

7. $(3.3485, -1.5002)$

9. Ellipse **11.** Hyperbola

13. Hyperbola **15.** Ellipse

17. Hyperbola

$\theta = \frac{\pi}{4}, x'^2 - y'^2 = 4$

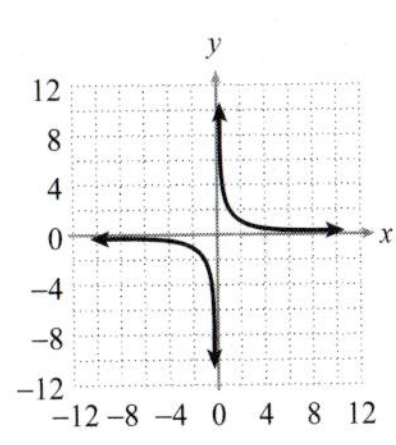

19. Parabola

$\theta = \frac{\pi}{4}, y' = -\sqrt{2}x'^2$

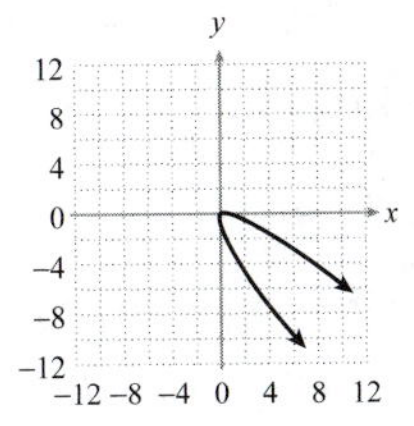

21. Ellipse

$\theta = \frac{\pi}{6}, \frac{x'^2}{13} + \frac{y'^2}{25} = 1$

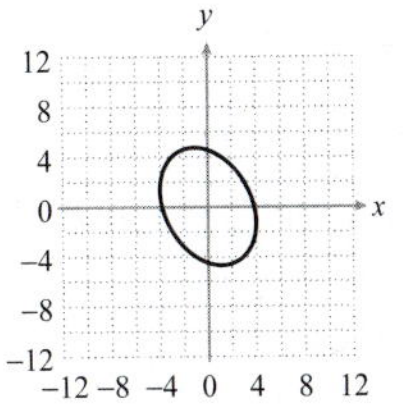

23. Ellipse

$\theta = \frac{\pi}{3}, \frac{x'^2}{30} + \frac{y'^2}{46} = 1$

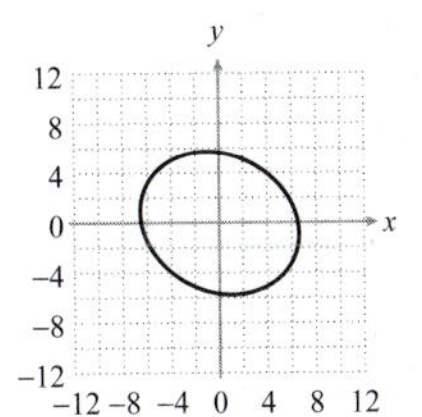

25.

27.

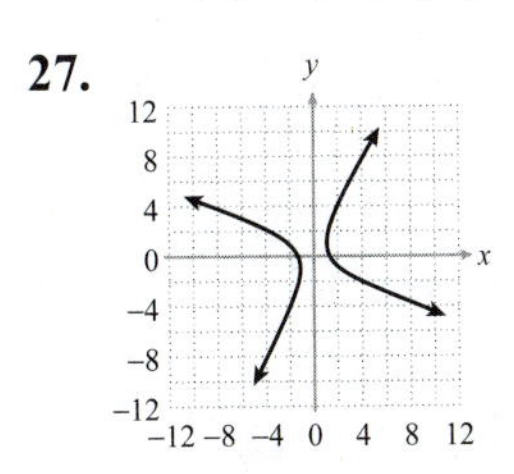

29.

31.

33. c **35.** a

37. g **39.** e

41. The objective of the rotation of axes is to eliminate the $x'y'$-term. If your final equation contains an $x'y'$-term, you know that a mistake has occurred.

43. a. Use the rotation of axes procedure to obtain the equation $4x'^2 + 16y'^2 - 16 = 0$. Now we know $F = -16$ and $F' = -16$. We can plug these values in $F = F'$ and obtain $-16 = -16$, which is true.

b. Use the rotation of axes procedure to obtain the equation $4x'^2 + 16y'^2 - 16 = 0$. Now we know $A = 7$, $C = 13$, $A' = 4$, and $C' = 16$. We can plug these values in $A + C = A' + C'$ and obtain $7 + 13 = 4 + 16$, or $20 = 20$, which is true.

c. Use the rotation of axes procedure to obtain the equation $4x'^2 + 16y'^2 - 16 = 0$. Now we know $A = 7, B = -6\sqrt{3}, C = 13,$ $A' = 4, B' = 0,$ and $C' = 16$. We can plug these values in $B^2 - 4AC = B'^2 - 4A'C'$ and obtain

$$(-6\sqrt{3})^2 - 4(7)(13) = (0)^2 - 4(4)(16)$$

or $-256 = -256$, which is true.

10.5 Exercises

1. c **3.** f **5.** b

7. Hyperbola, $y = \frac{7}{6}$

9. Ellipse, $x = -3$

11. Hyperbola, $x=\frac{1}{3}$

13. Ellipse, $x=5$

15. Hyperbola, $x=-\frac{6}{5}$

17. Parabola, $y=\frac{3}{2}$

19. Hyperbola, $x=-\frac{4}{7}$

21. $r=\frac{2}{1-\cos\theta}$

23. $r=\frac{3}{1-4\sin\theta}$

25. $r=\frac{3}{1+\frac{1}{4}\cos\theta}$

27.

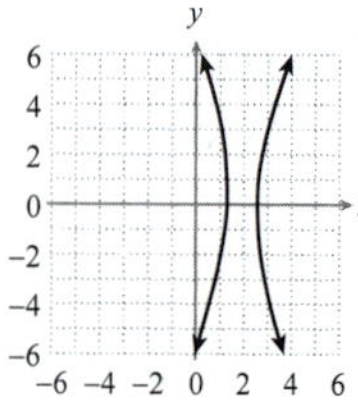

29.

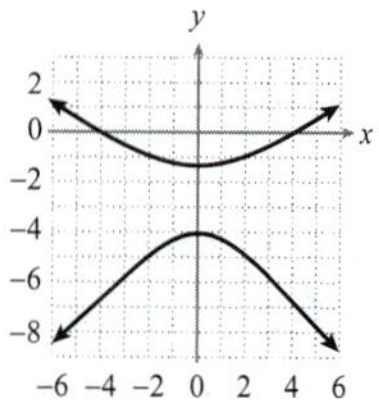

31.

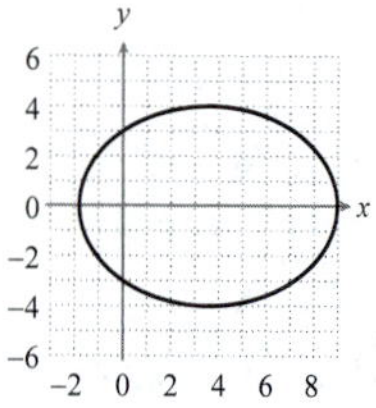

33.

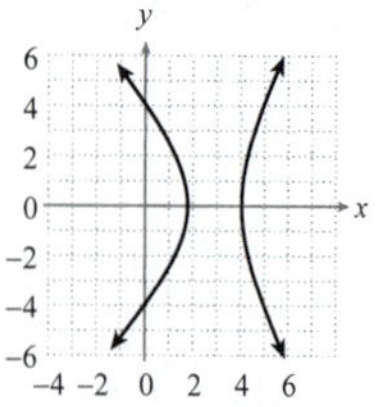

35.

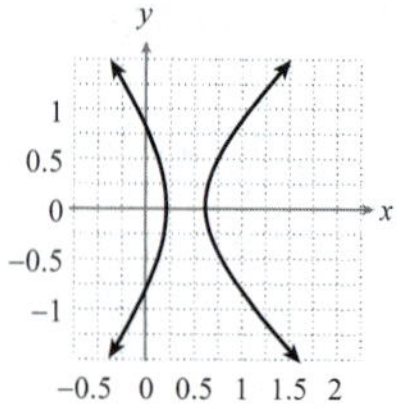

37.

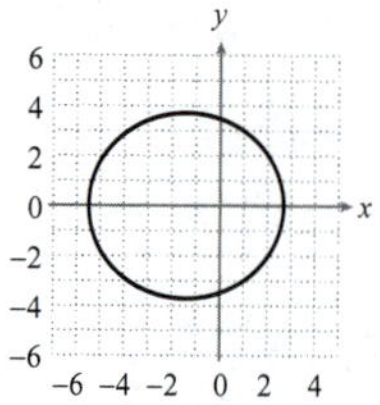

39.

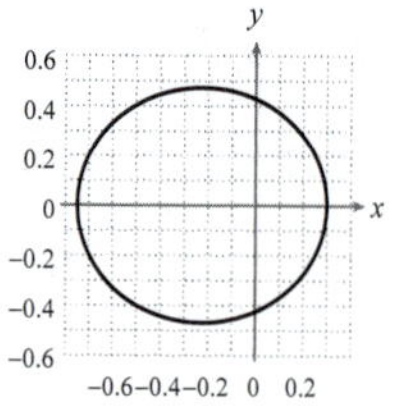

41.

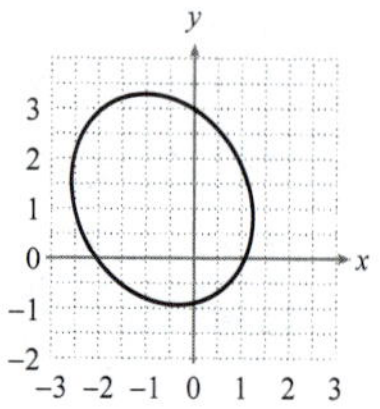

43.

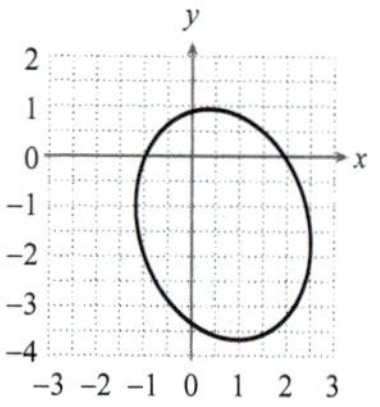

45.

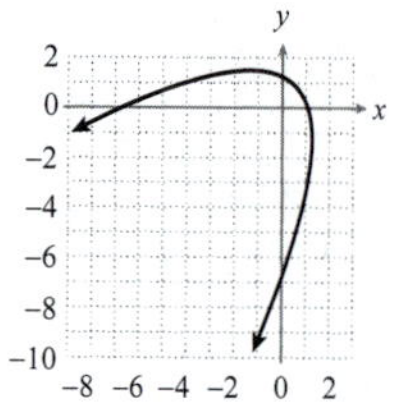

Chapter 10 Project

1. $x^2=-\frac{1125}{4}(y-80)$

3. $\frac{x^2}{22{,}500}+\frac{y^2}{6400}=1, y\geq 0$

5. Answers may vary.
The semiellipse design,
since it gives more space for ships
to pass through.

Chapter 10 Review Exercises

1. Center: $(3,-1)$

Vertices: $(7,-1),(-1,-1)$

Foci: $\left(3\pm 2\sqrt{3},-1\right)$

3.

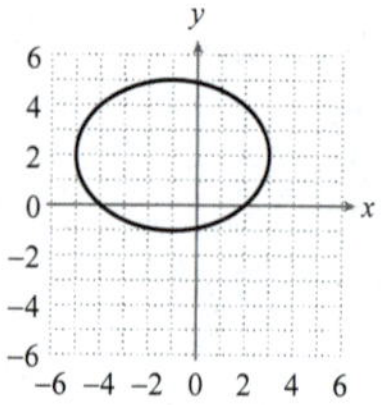

$\left(-1\pm\sqrt{7},2\right)$

5.

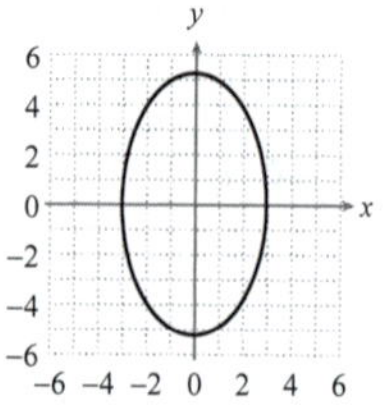

$\left(0,\pm 3\sqrt{2}\right)$

7. $\frac{(x+1)^2}{9}+\frac{(y-4)^2}{16}=1$

9. $\frac{4(x-2)^2}{9}+\frac{4(y+1)^2}{9}=1$

11. $\frac{(x-2)^2}{16}+\frac{y^2}{12}=1$

13. $\frac{(x+1)^2}{4}+\frac{(y-1)^2}{25}=1$

15. $30\pi a-\pi a^2$

17.

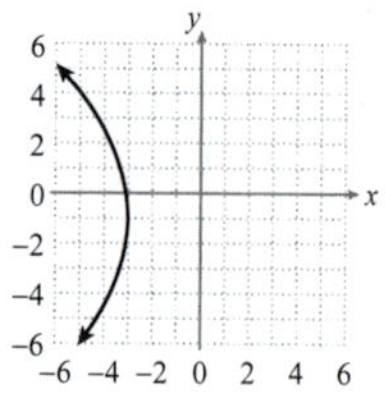

$(-6,-1), x=0$

19.

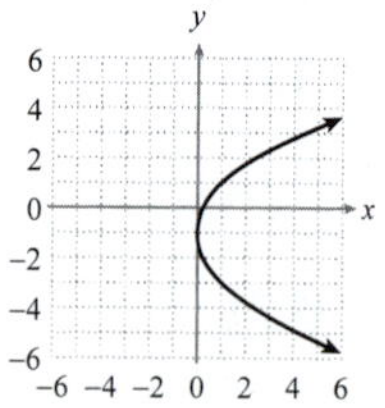

$(1,-1),\ x=-1$

21.

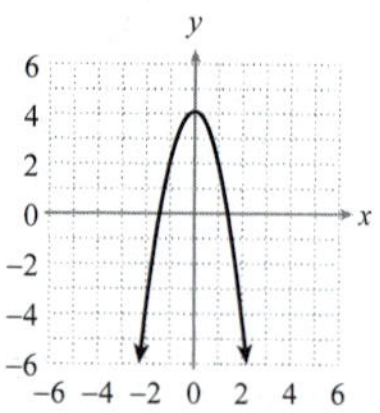

$\left(0,\frac{31}{8}\right),\ y=\frac{33}{8}$

23. $(x+2)^2 = 4(y-3)$

25. $(y+1)^2 = 2\left(x-\dfrac{5}{2}\right)$

27. $(y+1)^2 = 16(x-2)$

29. $(y+1)^2 = -16(x-5)$

31. $\dfrac{5}{4}$ inches

33.

$\left(-3\pm\sqrt{13},-1\right)$

35.

$(0,\pm 13)$

37. Center: $(0,2)$
Foci: $\left(\pm 2\sqrt{10},2\right)$
Vertices: $(\pm 6,2)$

39. Center: $(3,3)$
Foci: $\left(3,3\pm\sqrt{53}\right)$
Vertices: $(3,1),(3,5)$

41. $\dfrac{(x+1)^2}{4}-\dfrac{(y+2)^2}{25}=1$

43. $\dfrac{(y-2)^2}{9}-\dfrac{(x-2)^2}{1}=1$

45. $\dfrac{(y-7)^2}{4}-\dfrac{(x+1)^2}{9}=1$

47. $\dfrac{(y+2)^2}{9}-\dfrac{x^2}{4}=1$

49. $\left(11+43\sqrt{3},-11\sqrt{3}+43\right)$

51. $\left(3+3\sqrt{3},9-\sqrt{3}\right)$

53. Ellipse,
$\theta=\dfrac{\pi}{3}$,

$$\frac{\left(x'-\dfrac{25\sqrt{3}}{13}\right)^2}{\dfrac{25{,}000}{1521}}+\frac{\left(y'-\dfrac{25}{9}\right)^2}{\dfrac{25{,}000}{1053}}=1$$

55. Parabola,
$\theta=\dfrac{\pi}{4}, y'=\sqrt{2}x'^2$

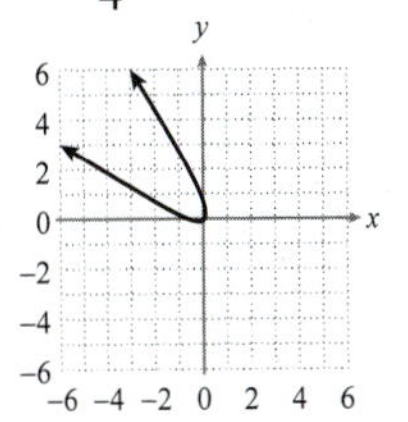

57. Parabola, $y=\dfrac{7}{4}$

59. Ellipse, $x=\dfrac{7}{2}$

61. $r=\dfrac{4}{1+\dfrac{1}{4}\cos\theta}$

63. $r=\dfrac{3}{1+9\cos\theta}$

Chapter 11: Systems of Equations and Inequalities

11.1 Exercises

1. $(-5,2)$ **3.** $(5,3)$

5. $\varnothing$

7. $\left\{\left(\dfrac{y-3}{2},y\right)\middle|\, y\in\mathbb{R}\right\}$

9. $(-1,7)$ **11.** $(3,11)$

13. $\left\{(x,4x+1)\,|\,x\in\mathbb{R}\right\}$

15. $(2,19)$ **17.** $(-5,1)$

19. $(5,6)$

21. $\left\{(-y-2,\ y)\,|\,y\in\mathbb{R}\right\}$

23. $(-5,4)$ **25.** $(-1,1)$

27. $(3,-5)$ **29.** $\varnothing$

31. $(-1,3,0)$ **33.** $(2,2,-1)$

35. $\left\{\left(\dfrac{y-z+2}{3},y,z\right)\middle|\, y\in\mathbb{R}, z\in\mathbb{R}\right\}$

37. $\varnothing$ **39.** $(1,1,0)$

41. $(9,1,1)$ **43.** $(3,1,-2)$

45. $(4,5,5)$ **47.** $\left(\dfrac{49}{3},\dfrac{-16}{3},\dfrac{5}{4}\right)$

49. $(0,3,2)$

51. 22 pennies, 23 nickels

53. 25 people

55. Eliza is 15 years old.

57. 7 shirts and 4 pairs of shorts

59. 3 quarters, 11 dimes, and 28 pennies

61. Jim is 28 years old.

63. 3 thumb screws

65. Apples: \$0.78, Oranges: \$0.93, Mangoes: \$1.05

67. $(0.43,1.28,3.64)$

69. $(-3.42,2.98,2.76)$

71. $(6,8,7)$

11.2 Exercises

1. a. 3×2 **b.** -1 **c.** None

3. a. 5×2 **b.** None **c.** 10

5. a. 3×4 **b.** None **c.** 286

7. a. 3×2 **b.** 1 **c.** None

9. a. 2×5 **b.** 5 **c.** 2

11. $\left[\begin{array}{ccc|c} -3 & 1 & -2 & -4 \\ \frac{1}{2} & -4 & -1 & 1 \\ 0 & -3 & 3 & 1 \end{array}\right]$

13. $\left[\begin{array}{ccc|c} -\frac{3}{2} & -1 & 0 & -1 \\ 2 & 2 & 3 & 0 \\ 0 & -1 & 6 & 0 \end{array}\right]$

15. $\left[\begin{array}{ccc|c} \frac{12}{5} & \frac{1}{2} & -\frac{3}{2} & \frac{1}{5} \\ 1 & 0 & 3 & 1 \\ 5 & 2 & 1 & -2 \end{array}\right]$

17. $\left[\begin{array}{ccc|c} \frac{2}{3} & -\frac{4}{3} & -2 & 0 \\ 8 & -2 & 6 & 7 \\ 3 & -2 & 0 & 0 \end{array}\right]$

19. $\left[\begin{array}{ccc|c} \frac{1}{2} & -14 & -\frac{1}{4} & -8 \\ \frac{1}{5} & -\frac{7}{6} & \frac{1}{4} & -3 \\ 5 & -5 & \frac{8}{3} & -5 \end{array}\right]$

21. $\begin{cases} x = 8 \\ y = 3 \end{cases}$

23. $\begin{cases} x + 3y + 6z = 16 \\ y + 2z = 9 \\ z = 4 \end{cases}$

25. $\begin{cases} 9y + 13z = 27 \\ 2x + 21z = 19 \\ 7x + 18y = 32 \end{cases}$

27. $\left[\begin{array}{cc|c} 2 & -5 & 3 \\ 0 & -7 & 5 \end{array}\right]$

29. $\left[\begin{array}{cc|c} 1 & 3 & -2 \\ 9 & -2 & 7 \end{array}\right]$

31. $\left[\begin{array}{cc|c} 8 & -2 & -4 \\ -6 & 2 & -14 \end{array}\right]$

33. $\left[\begin{array}{cc|c} 4 & 12 & -6 \\ 9 & 9 & 6 \end{array}\right]$

35. $\left[\begin{array}{cc|c} 4 & -1 & 5 \\ -6 & 2 & 0 \end{array}\right]$

37. $\left[\begin{array}{ccc|c} 18 & -6 & 15 & 42 \\ -7 & 19 & 2 & 3 \\ -4.5 & 5.5 & -2 & 3.5 \end{array}\right]$

39. $\left[\begin{array}{ccc|c} 5 & 18 & 22 & 5 \\ 32 & -9 & -27 & -23 \\ -9 & 21 & 12 & 9 \end{array}\right]$

41. $\left[\begin{array}{ccc|c} 0 & 1 & -9 & -3 \\ 1 & 1 & 3 & 4 \\ 0 & 0 & 0 & 0 \end{array}\right]$

43. $\left[\begin{array}{cc|c} -1 & 4 & -3 \\ 1 & -6 & \frac{5}{2} \end{array}\right]$

45. $\left[\begin{array}{ccc|c} 1 & 5 & -9 & 11 \\ 0 & -1 & 8 & -7 \\ 0 & -17 & 41 & 1 \end{array}\right]$

47. Neither **49.** Neither

51. Neither **53.** $(3,-1)$

55. $(1,3)$ **57.** $(-7,3)$

59. $\varnothing$ **61.** $(3,2)$

63. $\{(-2y-4, y) \mid y \in \mathbb{R}\}$

65. $\varnothing$ **67.** $(4,0,3)$

69. $(15,-21,8)$ **71.** $(3,-5)$

73. $\{(x, -3x-2) \mid x \in \mathbb{R}\}$

75. $(-4,1)$ **77.** $(6,4)$

79. $(-11,-5)$ **81.** $(7,3,3)$

83. $(2,2,-1)$ **85.** $\{(1,y,0) \mid y \in \mathbb{R}\}$

87. $(3,-2,3)$ **89.** $(2,3,4)$

91. $(9,-19,7)$ **93.** $(1,-2,-1,3)$

95. 42, 26, 87

97. Small: 10, Medium: 24, Large: 48

11.3 Exercises

1. 11 **3.** 15

5. $ab - x^2$ **7.** 8

9. -10 **11.** -39

13. $\{-2,3\}$ **15.** $\{-5,1\}$

17. $\{-5,-4\}$ **19.** $\{-6,4\}$

21. $\{2,5\}$ **23.** 3

25. -9 **27.** 2

29. -2 **31.** 159

33. 78 **35.** -254

36. -84 **37.** 404

39. 4 **41.** 120

43. 10 **45.** x^4

47. x^8 **49.** $(76,-53)$

51. $\{(-y-2, y) \mid y \in \mathbb{R}\}$

53. $\varnothing$ **54.** $(3,-2,3)$

55. $\{(-3z-5, -6z-10, z) \mid z \in \mathbb{R}\}$

57. $\left\{\left(\frac{-5y-z-5}{2}, \frac{-5y+3z-19}{2}, y, z\right) \middle| y \in \mathbb{R}, z \in \mathbb{R}\right\}$

59. $\left\{\left(-z+8, -5z+31, -2z+37, z\right) \middle| z \in \mathbb{R}\right\}$

61. $(1647, 2071)$

63. $\varnothing$

65. $(-3,-1,0,-4)$

67. Candy bars: 5, Ice cream: 6

69. 0.012 **71.** 0.564

73. 1194 **75.** $(1,-1,2)$

77. $(2,1,0,3)$

11.4 Exercises

1. $\begin{bmatrix} 5 & -1 \\ 0 & 0 \\ 2 & 13 \end{bmatrix}$ **3.** $\begin{bmatrix} 6 & -3 \\ 18 & 30 \\ -9 & 21 \end{bmatrix}$

5. Not possible **7.** $\begin{bmatrix} 14 & -14 \\ 8 & 0 \\ -4 & 14 \end{bmatrix}$

9. $\begin{bmatrix} -7 & 5 \\ 3 & 10 \\ -3 & -8 \end{bmatrix}$ **11.** Not possible

13. $a = 3, b = -1, c = 10$

15. $a = 2, b = -2, c = -1$

17. Not possible **19.** $x = 10, y = 5$

21. $x = 3, y = 1$ **23.** Not possible

25. $a = 8, b = 5$ **27.** $\begin{bmatrix} 24 & -5 \end{bmatrix}$

29. $\begin{bmatrix} 35 & 18 \end{bmatrix}$ **31.** Not possible

33. $\begin{bmatrix} -30 & -3 \end{bmatrix}$

35. $\begin{bmatrix} 15 & -3 & -24 \\ 25 & -5 & -40 \\ 30 & -6 & -48 \end{bmatrix}$

37. $\begin{bmatrix} -34 & -7 \end{bmatrix}$

39. $\begin{bmatrix} 11 & 0 \\ 0 & 11 \end{bmatrix}$

41. $\begin{bmatrix} 32 & -20 \\ 56 & -35 \\ -16 & 10 \end{bmatrix}$

43. Not possible

45. $\begin{bmatrix} 14 & -13 \\ -13 & 5 \end{bmatrix}$

47. $\begin{bmatrix} 179 & 76 \end{bmatrix}$

49. $\frac{2}{3}$ for store A; $\frac{1}{3}$ for store B

51. Solution is incorrect. Explanations may vary.

53. $\begin{bmatrix} 23.94 & -7.56 & 28.98 \\ 21.66 & -6.84 & 26.22 \end{bmatrix}$

55. $\begin{bmatrix} -23.94 & -26.72 \end{bmatrix}$

57. $\begin{bmatrix} -79.59 \\ 39.21 \\ 10.08 \end{bmatrix}$

11.5 Exercises

1. $\begin{bmatrix} 14 & -5 \\ 1 & 9 \end{bmatrix}\begin{bmatrix} x \\ y \end{bmatrix} = \begin{bmatrix} 7 \\ 2 \end{bmatrix}$

3. $\begin{bmatrix} 1 & 2 \\ 9 & -3 \end{bmatrix}\begin{bmatrix} x \\ y \end{bmatrix} = \begin{bmatrix} -6 \\ -14 \end{bmatrix}$

5. $\begin{bmatrix} 3 & -7 & 1 \\ 1 & -1 & 0 \\ 0 & 8 & 5 \end{bmatrix}\begin{bmatrix} x_1 \\ x_2 \\ x_3 \end{bmatrix} = \begin{bmatrix} -4 \\ 2 \\ -3 \end{bmatrix}$

7. $\begin{bmatrix} \frac{3}{5} & -\frac{8}{5} \\ 0 & 1 \end{bmatrix}\begin{bmatrix} x \\ y \end{bmatrix} = \begin{bmatrix} 2 \\ 2 \end{bmatrix}$

9. $\begin{bmatrix} 4 & -3 \\ 2 & -4 \end{bmatrix}\begin{bmatrix} x \\ y \end{bmatrix} = \begin{bmatrix} -9 \\ 13 \end{bmatrix}$

11. $\begin{bmatrix} 2 & -1 & 3 \\ -1 & 1 & 0 \\ 4 & -5 & 1 \end{bmatrix}\begin{bmatrix} x \\ y \\ z \end{bmatrix} = \begin{bmatrix} 0 \\ 17 \\ -2 \end{bmatrix}$

13. $\begin{bmatrix} -\frac{1}{20} & -\frac{1}{5} \\ \frac{1}{4} & 0 \end{bmatrix}$

15. $\begin{bmatrix} -5 & -4 \\ 4 & 3 \end{bmatrix}$

17. $\begin{bmatrix} -5 & 0 \\ 2 & 2 \end{bmatrix}$

19. Not invertible

21. $\begin{bmatrix} 2 & 1 & -4 \\ -4 & -2 & -3 \\ -1 & -1 & -4 \end{bmatrix}$

23. $\begin{bmatrix} -1 & 2 & -1 \\ 0 & -1 & 1 \\ 0 & -4 & 3 \end{bmatrix}$

25. $\begin{bmatrix} -1 & -2 & 1 \\ -2 & 1 & -3 \\ 1 & 2 & 0 \end{bmatrix}$

27. $\begin{bmatrix} -2 & 1 & 1 \\ 2 & 0 & -1 \\ -1 & 0 & 1 \end{bmatrix}$

29. $\begin{bmatrix} 2 & -1 & 2 \\ 0 & 1 & -1 \\ -3 & -2 & -4 \end{bmatrix}$

31. No **33.** Yes

35. No

37. $\left(-2, -\frac{5}{2}\right)$

39. $\left\{\left(\frac{3y-1}{2}, y\right) \middle| y \in \mathbb{R}\right\}$

41. $(-2,0)$ **43.** $(8,-19)$

45. $(0,5)$ **47.** $(-4,5,-1)$

49. $(-13,19,23)$; $(0,0,-1)$; $(1,-1,-1)$

51. $(1,-8,7)$; $(3,1,1)$; $(4,2,0)$

53. $\begin{bmatrix} \frac{-1}{4} & \frac{1}{10} \\ \frac{-1}{4} & \frac{3}{10} \end{bmatrix}$

55. $\begin{bmatrix} 0.053 & -0.258 \\ 0.113 & 0.076 \end{bmatrix}$

57. $\begin{bmatrix} 0.004 & -0.003 & 0.009 \\ 0 & 0.020 & 0.029 \\ 0.012 & 0.014 & 0.013 \end{bmatrix}$

11.6 Exercises

1. $\frac{A_1}{x-3} + \frac{A_2}{x+3}$

3. $\frac{A_1}{x+3} + \frac{A_2}{x+4} + \frac{A_3}{(x+4)^2}$

5. $\frac{A_1}{x+3} + \frac{A_2}{x-2} + \frac{A_3}{x+2}$

7. d **9.** h

11. a **13.** c

15. $-\frac{1}{x} + \frac{2}{x-2} + \frac{2}{x+2}$

17. $\frac{5}{72(x-2)} - \frac{1}{32x} - \frac{7}{288(x+4)} - \frac{x+17}{72(x^2+8)}$

19. $\frac{10}{x-4} - \frac{5}{x-2}$

21. $\frac{3}{2(x+3)} + \frac{3}{14(x-1)} - \frac{12}{7(x+6)}$

23. $\frac{1}{2(x-1)} - \frac{1}{2(x+1)}$

25. $\frac{1}{48(x-2)} - \frac{x}{80(x^2+4)} - \frac{3}{40(x^2+4)} - \frac{1}{120(x+4)}$

27. $\frac{5}{4(x-2)} - \frac{1}{4(x+2)}$

29. $\frac{1}{16(x+2)} + \frac{1}{16(x-2)} - \frac{x}{8(x^2+4)}$

31. $\frac{1}{24(x+6)} + \frac{1}{30(x+3)} - \frac{7}{40(x-2)} + \frac{1}{6(x-3)}$

33. $\frac{15}{16(x+3)} + \frac{1}{16(x-1)} - \frac{9}{4(x+3)^2}$

35. $\dfrac{1}{x+3}+\dfrac{1}{x-3}$

37. $\dfrac{1}{4(x-2)}+\dfrac{3}{4(x+6)}$

39. $\dfrac{1}{a}\left(\dfrac{1}{x}-\dfrac{1}{x+a}\right)$

41. $\dfrac{1}{2a}\left(\dfrac{1}{a+x}+\dfrac{1}{a-x}\right)$

43. $\dfrac{1}{a-1}\left(\dfrac{1}{x+1}+\dfrac{1}{x+a}\right)$

45. True

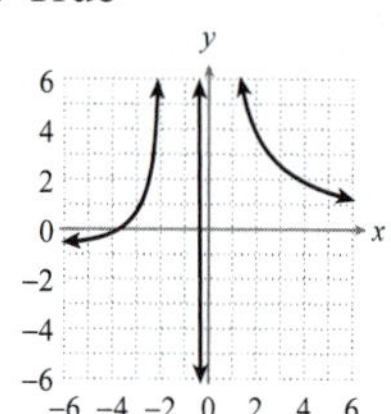

47. True

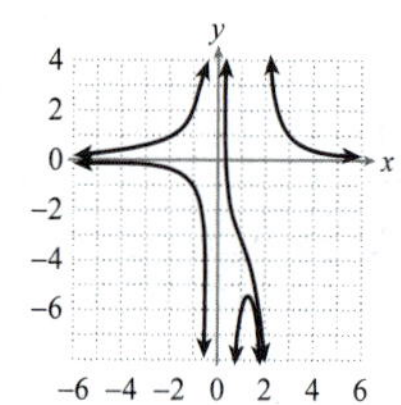

49. False

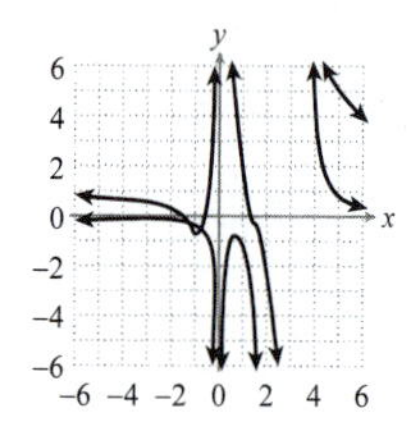

11.7 Exercises

1.

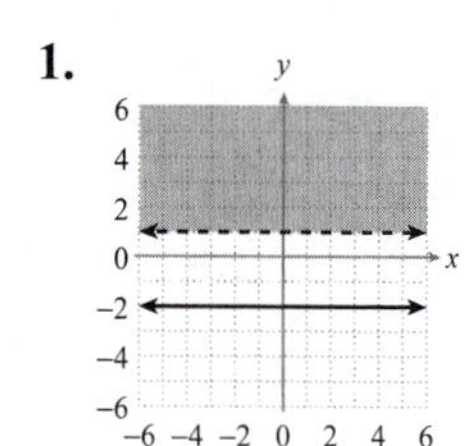

3.

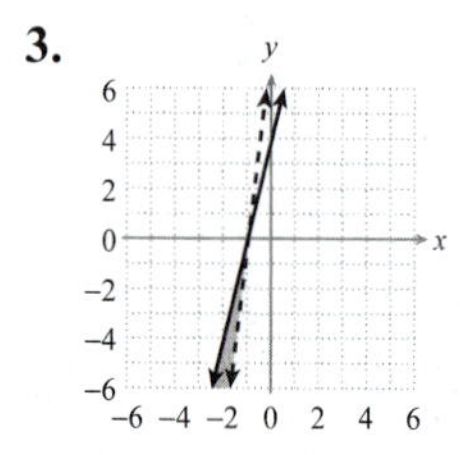

5.

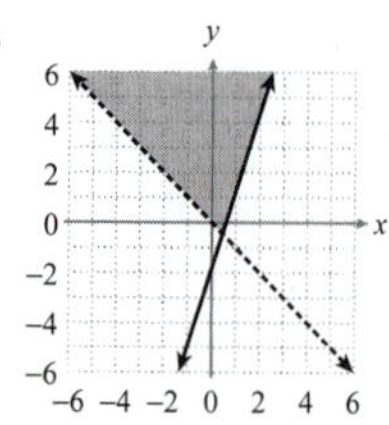

7.

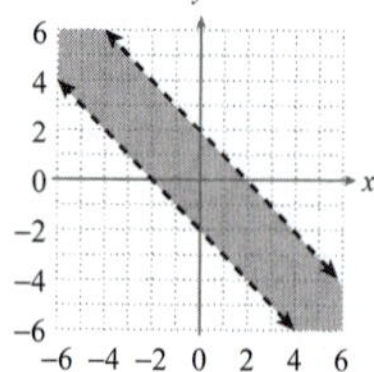

9.

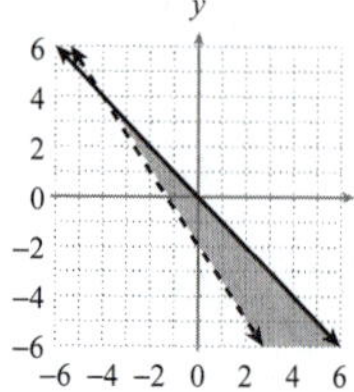

11.

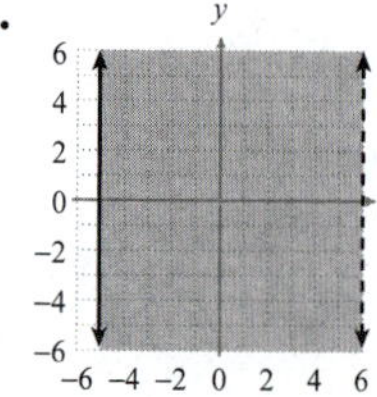

13.

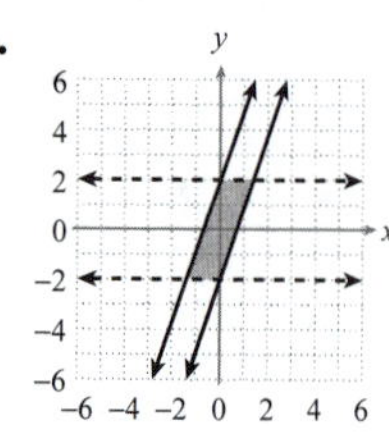

15. Min $= 0$ at $(0,0)$;
Max $= 12$ at $(3,0)$

17. Min $= 0$ at $(0,0)$;
Max $= 12$ at $(0,3)$

19. Min $= 100$ at $\left(\dfrac{10}{3},10\right)$;
Max $= 250$ at $(20,10)$

21. Min $= \dfrac{150}{7}$ at $\left(\dfrac{50}{7},0\right)$;
Max $= 66$ at $(8,6)$

23. Min $= 0$ at $(0,0)$;
Max $= \dfrac{165}{8}$ at $\left(\dfrac{111}{8},\dfrac{27}{8}\right)$

25. Min $= 210$ at $(14,0)$;
Max $= 600$ at $(0,20)$

27. Min $= \dfrac{680}{7}$ at $\left(\dfrac{88}{7},\dfrac{-36}{7}\right)$;
Max $= \dfrac{1980}{7}$ at $\left(\dfrac{108}{7},\dfrac{-6}{7}\right)$

29.
$40x+60y\le 200$;
$8x+5y\le 25$;
$x\ge 0$; $y\ge 0$

31.
$x+y\le 75{,}000$;
$x-y\ge 0$;
$x\le 40{,}000$;
$x\ge 10{,}000$;
$y\ge 5000$

33. Type X: 75 units;
Type Y: 100 units;
Maximum profit: $712.50

35. The volunteer could choose any of the following points: $(9,9),(11,6),(13,3),(15,0)$. In each of these points, the first coordinate represents the number of packages from Company A and the second coordinate represents the number of packages from Company B. The minimum cost is $22.50.

37. **a.** 120 flip phones;
80 smart phones;
Minimum cost: $1160

b. 100 flip phones;
170 smart phones;
Maximum profit: $650

11.8 Exercises

1. $\{(0,-3),(2,0)\}$

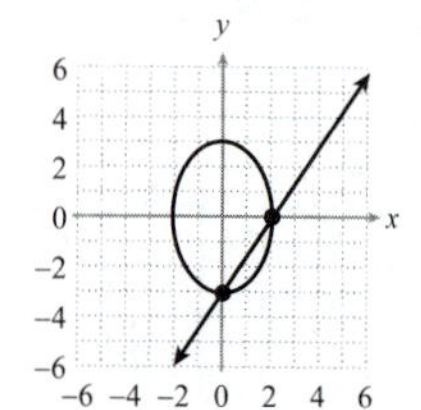

3. $\left\{\begin{matrix}(-1,-1),(1,1)\\(-1,1),(1,-1)\end{matrix}\right\}$

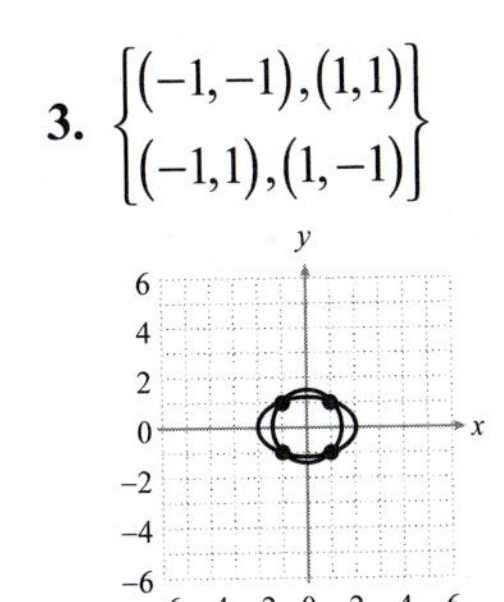

5. $\{(-1,1),(1,1)\}$

7. $\{(-1,2),(1,2)\}$

9. No solution

11. $\{(-3,1),(1,1)\}$

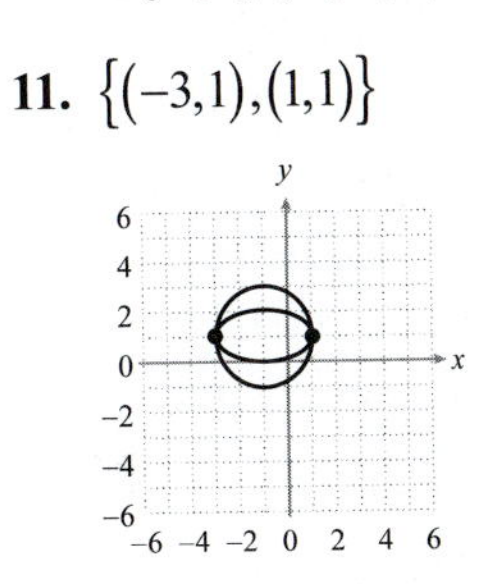

13. $\{(0,0)\}$

15. $\{(\pm 3,0)\}$

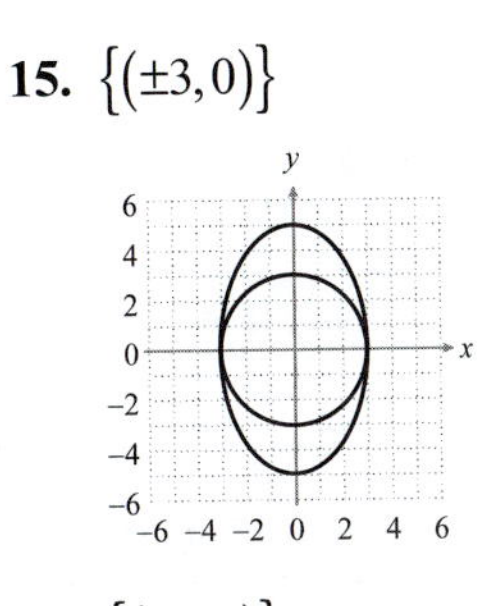

17. $\{(\pm 1,0)\}$

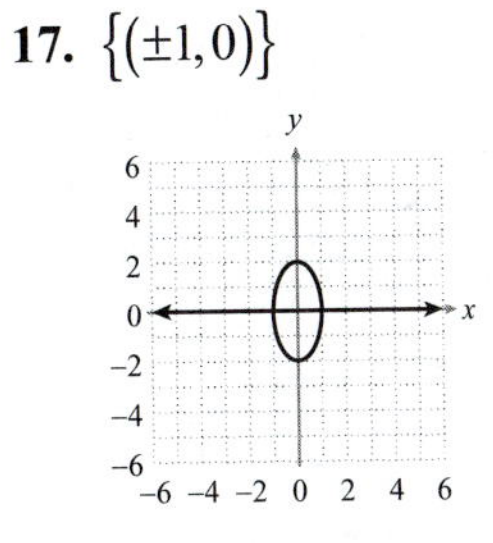

19. $\{(2,\pm 3),(-2,\pm 3)\}$

21. $\{(5,3)\}$

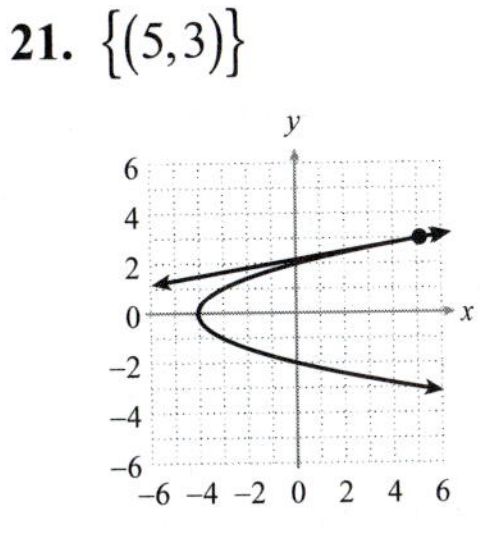

23. $\{(\pm 2,0),(0,-2)\}$

25. $\left\{\begin{matrix}(-i\sqrt{6},-6),(i\sqrt{6},-6),\\(-\sqrt{5},5),(\sqrt{5},5)\end{matrix}\right\}$

27. $\{(-2,3)\}$

29. $\left\{\begin{matrix}(-1,-4),(1,4),(-2\sqrt{2},-\sqrt{2}),\\(2\sqrt{2},\sqrt{2})\end{matrix}\right\}$

31. $\{(-3i,-5),(3i,-5),(0,4)\}$

33. $\{(-\sqrt{3},0),(\sqrt{3},0)\}$

35. $\{(-i,-6),(i,-6),(-\sqrt{2},3),(\sqrt{2},3)\}$

37. $\left\{\begin{matrix}(1,1),\left(-\frac{3}{2},\frac{\sqrt{14}}{2}\right),\\(1,-1),\left(-\frac{3}{2},-\frac{\sqrt{14}}{2}\right)\end{matrix}\right\}$

39. $\left\{\left(\frac{7}{2},\frac{17}{4}\right)\right\}$

41. $\left\{\begin{matrix}(-6,-i\sqrt{15}),(2,1),\\(-6,i\sqrt{15}),(2,-1)\end{matrix}\right\}$

43. $\left\{(0,1),\left(-\frac{100}{101},-\frac{99}{101}\right)\right\}$

45. $\left\{\begin{matrix}\left(-\frac{\sqrt{42}}{6},-\frac{\sqrt{66}}{6}\right),\\\left(-\frac{\sqrt{42}}{6},\frac{\sqrt{66}}{6}\right),\\\left(\frac{\sqrt{42}}{6},-\frac{\sqrt{66}}{6}\right),\\\left(\frac{\sqrt{42}}{6},\frac{\sqrt{66}}{6}\right)\end{matrix}\right\}$

47. $\{(-5,0),(-2,0),(-1,0)\}$

49. $\left\{\begin{matrix}(3,2),(2,3),\\\left(\frac{-1\pm i\sqrt{23}}{2},\frac{-1\pm i\sqrt{23}}{2}\right)\end{matrix}\right\}$

51. $\left\{(0,-1),\left(\frac{\sqrt{6}}{3},\frac{2\sqrt{6}}{9}-1\right)\right\}$

53. $\left\{\begin{matrix}(\sqrt{2}+1,2\sqrt{2}),(-\sqrt{2}+1,-2\sqrt{2}),\\(2\sqrt{2}+1,\sqrt{2}),(-2\sqrt{2}+1,-\sqrt{2})\end{matrix}\right\}$

55. $\{(4,1),(1,1+i\sqrt{3})\}$

57. $\{(0,4),(0,-3),(3,-2),(3,3)\}$

59. $\{(4,0)\}$

61. $\left\{\left(\pm\frac{1}{2},\pm\frac{1}{4}\right)\right\}$

63. b

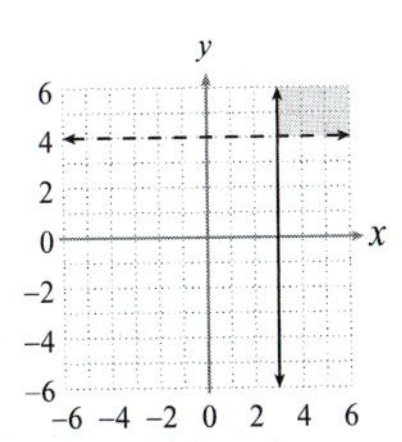

65. b, d

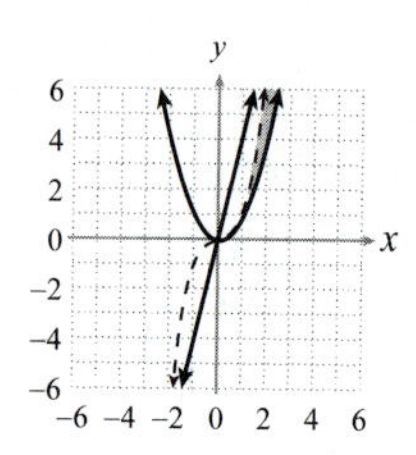

67. a, d

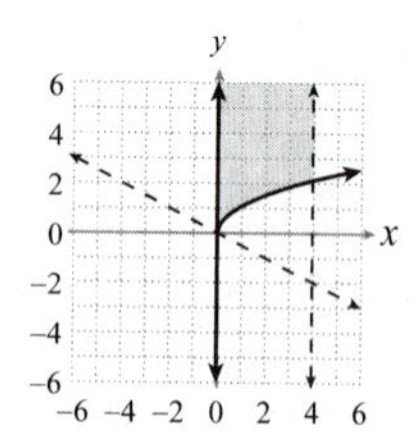

69.

71.

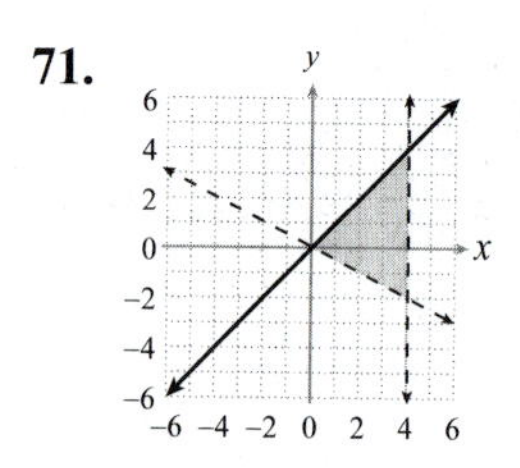

73.

75.

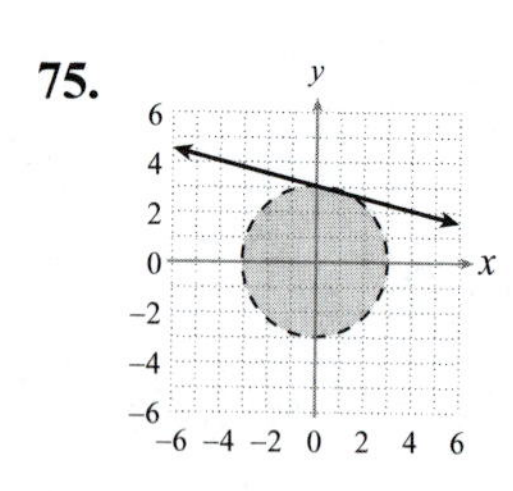

77. 9 inches by 5 inches

79. 60 mph and 70 mph

81. −12 and 7

83. $h = 6$ cm, $r = 3$ cm

Chapter 11 Project

1.

The number of months passed	Shop Name		
	Joe's Java	Buck's Café	Tweak's Coffee
1	26.85%	43.45%	29.70%
2	28.47%	42.10%	29.43%
3	29.88%	40.92%	29.20%
4	31.12%	39.89%	28.98%
5	32.21%	39.00%	28.80%
6	33.16%	38.21%	28.63%

3. As our market share increases, the shares of both Buck's Café and Tweak's Coffee decrease. As such, they will have to adjust their marketing strategies to stop the decrease and potentially increase their shares.

Chapter 11 Review Exercises

1. $\varnothing$ **3.** $(3,0,2)$

5. $\varnothing$

7. $\{(3-3y, y) \mid y \in \mathbb{R}\}$

9. $(3,2)$ **11.** $(8,12,10)$

13. $y = \frac{4}{15}x^2 - x + \frac{11}{15}$

15. a. 1×4 **b.** 8 **c.** None

17. $\left[\begin{array}{ccc|c} 4 & 5 & -1 & 0 \\ 1 & 3 & 2 & 3 \\ 10 & -1 & -6 & 0 \end{array}\right]$

19. $\begin{cases} 8x + 7z = 5 \\ -3y + 4z = 16 \\ 16x - 2y + z = 2 \end{cases}$

21. $\left[\begin{array}{cc|c} 0 & -5 & -11 \\ 1 & 2 & 3 \end{array}\right]$

23. $\left[\begin{array}{cc|c} 1 & -4 & -4 \\ 1 & 7 & 11 \end{array}\right]$

25. $(2,-1)$ **27.** $(3,-5)$

29. $2x^4$ **31.** 7

33. 9, −9 **35.** $(-4,1)$

37. $\varnothing$

39. $\begin{bmatrix} 4 & -16 & 4 \\ -5 & 8 & 12 \end{bmatrix}$

41. Not possible **43.** $\begin{bmatrix} 9 & -23 & 3 \\ 5 & -3 & 8 \end{bmatrix}$

45. $w = -2, x = 1, y = 3, z = -4$

47. $x = 2, y = -3$

49. $\begin{bmatrix} 12 & 46 \end{bmatrix}$

51. $\begin{bmatrix} 1 & -1 & 2 \\ 2 & -3 & -1 \\ -3 & 0 & 6 \end{bmatrix} \begin{bmatrix} x_1 \\ x_2 \\ x_3 \end{bmatrix} = \begin{bmatrix} -4 \\ 1 \\ 5 \end{bmatrix}$

53. $\begin{bmatrix} \frac{3}{16} & \frac{1}{8} \\ -\frac{1}{8} & \frac{1}{4} \end{bmatrix}$

55. Not possible

57. No **59.** Yes

61. $\left(1, -\frac{1}{3}\right)$

63. $(-15,20,-33), (-2,3,-4), (-9,11,-17)$

65. $\frac{A_1}{x+4} + \frac{A_2}{x-1}$

67. $-\frac{3}{2(2x-5)^2} + \frac{1}{2(2x-5)}$

69.

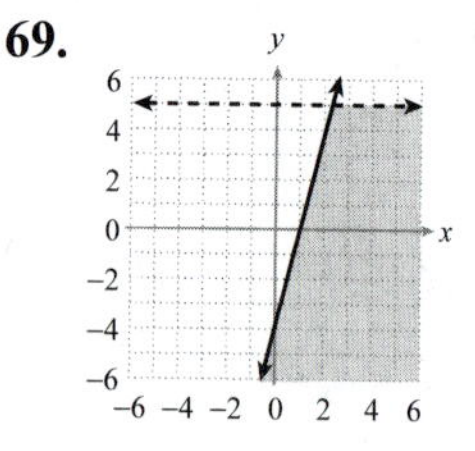

71. $x \geq 20,\ y \geq 10,\ x \leq 40,$ $y \leq 40,\ x + y \leq 60$

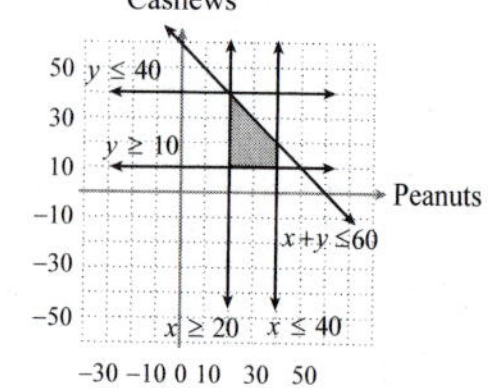

73. Min $= 0$ at $(0,0)$
Max $= 30$ at $(5,0)$

75. Min $= 8$ at $(0,2)$
Max $= 24$ at $\left(\frac{24}{7},\frac{12}{7}\right)$

77. 12 vases should be produced, 12 pitchers should be produced, Max profit: $660

79. $\{(0,-2),(3,1)\}$

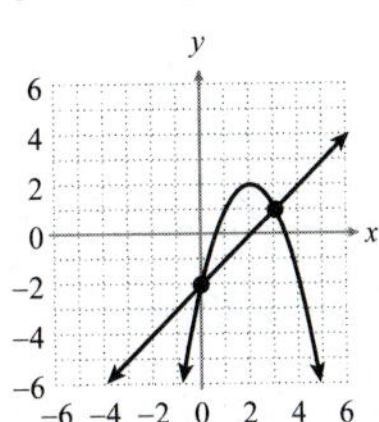

81. $\left\{(i,-1),(-i,-1),\left(\frac{\sqrt{2}}{2},\frac{1}{2}\right),\left(-\frac{\sqrt{2}}{2},\frac{1}{2}\right)\right\}$

83. $\{(2,1)\}$

85.

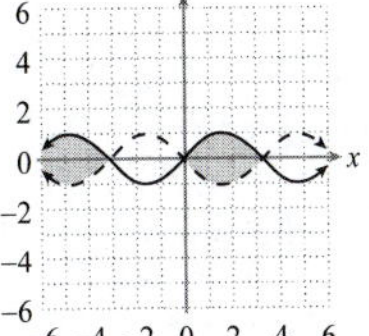

87.

89. 36 mph and 24 mph

Chapter 12: Sequences, Series, Combinatorics, and Probability

12.1 Exercises

1. Infinite

3. Finite **5.** Finite

7. Infinite **9.** Infinite

11. $2,-1,-4,-7,-10$

13. $1,\frac{3}{2},\frac{9}{5},2,\frac{15}{7}$

15. $\frac{2}{3},-\frac{4}{9},\frac{8}{27},-\frac{16}{81},\frac{32}{243}$

17. $\frac{1}{2},\frac{4}{3},\frac{9}{4},\frac{16}{5},\frac{25}{6}$

19. $-1,3,-6,10,-15$

21. $26,7,0,-1,-2$

23. $-1,\sqrt{2},-\sqrt{3},2,-\sqrt{5}$

25. $1,5,9,13,17$

27. $\frac{1}{2},1,2,4,8$

29. $3,36,729,20,736,759,375$

31. $\frac{5}{4},2,\frac{5}{2},\frac{20}{7},\frac{25}{8}$

33. $1,3,6,10,15$

35. Undefined, 9, 4, $\frac{25}{9},\frac{9}{4}$

37. $\frac{1}{3},\frac{1}{2},\frac{5}{9},\frac{7}{12},\frac{3}{5}$

39. $0,-1,-4,-9,-16$

41. $2,4,16,256,65,536$

43. $1,2,6,24,120$

45. $2,\sqrt{5},\sqrt{6},\sqrt{7},2\sqrt{2}$

47. 1, 0, -27, 0, 125

49. $a_n = 7n-2$

51. $a_1=-1, a_n=-na_{n-1}, n\ge 2$

53. $a_n=\left(\frac{1}{n}\right)^2$

55. $a_n = 9n-43$

57. $a_n = 2^{n-3}$

59. $a_n=\frac{n}{2^n}$

61. $-2+1+4+7+10+13+16=49$

63. $\sum_{i=1}^{6} i^3 = 441$

65. $45+80+125+180+245+320+405+500=1900$

67. $-6-12-24-48-96-192=-378$

69. $\sum_{i=2}^{9} 3^i = 29{,}520$

71. $S_n=\frac{n}{4(n+4)}, S=\frac{1}{4}$

73. $S_n=2^n-1, S_{15}=32{,}767$

75. $S_n=\frac{n}{2(n+1)}, S=\frac{1}{2}$

77. $S_n=-\ln(n+1)$, series diverges

79. $S_n=\frac{3n}{4(3n+4)}, S=\frac{1}{4}$

81. $4,7,11,18,29$

83. $10,20,30,50,80$

85. $13,-17,-4,-21,-25$

87. $1,-3,-3,9,-27$

89. 987

91. 1, 2, 4, 8, 16, 32, 64, 128

12.2 Exercises

1. $a_n=3n-5$

3. $a_n=-2n+9$

5. $a_n=9n-4$

7. $a_n=-6n+9$

9. $a_n=19n+5$

11. $a_n=n+\frac{5}{2}$

13. $a_n=-\frac{19}{2}n+\frac{43}{2}$

15. $a_n=-2n+1$

17. $a_n=-4n+33$

19. $d=2$ **21.** $d=1$ **23.** No

25. No **27.** 13 **29.** 2

31. 2

33. $d=2.5$; 5, 7.5, 10, 12.5, 15

35. $d=7$; 7, 14, 21, 28, 35

37. $d=9$; -62, -53, -44, -35, -26

39. $d=5$ **41.** $d=1$ **43.** $d=-2$

45. 195 **47.** a_{73} **49.** 117

51. 26 **53.** −8 **55.** 13

57. 55 **59.** 14,350

61. 17,114 **63.** −1475

65. $-\frac{3219}{5}$ **67.** 902

69. −1316 **71.** 6 years

73. 1620 pounds

75. \$625; \$8100

77. 266.4

79. −152

81. 10,382.05

12.3 Exercises

1. $a_n = -3(2)^{n-1}$

3. $a_n = 2\left(-\frac{1}{3}\right)^{n-1}$

5. $a_n = \left(-\frac{1}{4}\right)^{n-1}$

7. $a_n = \left(\frac{1}{7}\right)^{n-1}$

9. $a_n = (-3)^{n-1}$

11. $a_n = 3\left(\frac{2}{3}\right)^{n-1}$

13. $a_n = 7(-2)^{n-1}$

15. $a_n = \frac{1}{16}(2)^{n-1}$

17. $a_n = \frac{39}{68}\left(\frac{4}{3}\right)^{n-1}$

19. No **21.** $r = \frac{1}{2}$ **23.** $r = 2$

25. $r = 7$

27. $r = 3$; 8, 24, 72, 216, 648

29. $r = 2$; $\frac{1}{4}, \frac{1}{2}, 1, 2, 4$

31. $r = \frac{1}{5}$;
62,500, 12,500, 2500, 500, 100

33. $\frac{5}{16{,}384}$

35. −2,147,483,648

37. $r = \pm 2$ **39.** $r = \pm\frac{1}{5}$

41. $\frac{52{,}222{,}139{,}775}{1{,}048{,}576} \approx 49{,}802.9$

43. $\frac{10{,}923}{16{,}384} \approx 0.666687$

45. $\frac{73{,}810}{19{,}683} \approx 3.749936$

47. −109,200

49. $-\frac{3}{2}$ **51.** Series diverges

53. $\frac{2{,}476{,}099}{160{,}000} \approx 15.475619$

55. Series diverges

57. $\frac{28{,}561}{152{,}064} \approx 0.187822$

59. $\frac{123}{999}$ **61.** $-\frac{35}{9}$ **63.** $\frac{989}{99}$

65. \$14,802.44 **67.** 1.845×10^{19}

69. Approximately 13,778 students

71. $S_{30} = 1.1 \times 10^{30}; r = 10$

73. Yes; explanations will vary (any example such that $r = 1$ and $d = 0$).

75. 9.98

77. 2.137×10^{-18}

79. 6.54×10^{51}

12.4 Exercises

1. $S_{k+1} = \frac{1}{3k+9}$

3. $S_{k+1} = \frac{(k+1)(k+2)(2k+3)}{4}$

5. Basic Step:

$n = 1, 1 = 1$ and $\frac{1(1+1)}{2} = 1$;

Induction Step:

If $1+2+3+\cdots+k = \frac{k(k+1)}{2}$,

then $(1+2+3+\cdots+k)+(k+1)$

$= \frac{k(k+1)}{2} + (k+1)$

$= \frac{k^2+k+2k+2}{2}$

$= \frac{(k+1)(k+2)}{2}$

7. Basic Step:

$n = 1, 2(1) = 2$ and $1(1+1) = 2$;

Induction Step:

If $2+4+6+\cdots+2k = k(k+1)$,

then

$(2+4+6+\cdots+2k)+2(k+1)$

$= k^2+k+2k+2$

$= (k+1)(k+2)$

9. Basic Step:

$n = 1, 4^{1-1} = 1$ and $\frac{4^1-1}{3} = 1$;

Induction Step:

If $4^0+4^1+4^2+\cdots+4^{k-1} = \frac{4^k-1}{3}$,

then $4^0+4^1+4^2+\cdots+4^{k-1}+4^{k+1-1}$

$= \frac{4^k-1}{3}+4^k = \frac{4^k-1+3\cdot 4^k}{3}$

$= \frac{4\cdot 4^k-1}{3} = \frac{4^{k+1}-1}{3}$

11. Basic Step:

$n = 1, \frac{1}{(3(1)-2)(3(1)+1)} = \frac{1}{4}$

and $\frac{1}{3(1)+1} = \frac{1}{4}$;

Induction Step:

If $\frac{1}{1\cdot 4}+\frac{1}{4\cdot 7}+\frac{1}{7\cdot 10}$

$+\cdots+\frac{1}{(3k-2)(3k+1)} = \frac{k}{3k+1}$,

then $\frac{1}{1\cdot 4}+\frac{1}{4\cdot 7}+\frac{1}{7\cdot 10}$

$+\cdots+\frac{1}{(3k-2)(3k+1)}$

$+\frac{1}{(3(k+1)-2)(3(k+1)+1)}$

$= \left[\frac{1}{1\cdot 4}+\frac{1}{4\cdot 7}+\frac{1}{7\cdot 10} +\cdots+\frac{1}{(3k-2)(3k+1)}\right]$

$+\frac{1}{(3k+1)(3k+4)}$

$= \frac{k}{3k+1}+\frac{1}{(3k+1)(3k+4)}$

$= \frac{3k^2+4k+1}{(3k+1)(3k+4)}$

$= \frac{(3k+1)(k+1)}{(3k+1)(3k+4)} = \frac{(k+1)}{(3(k+1)+1)}$

13.
Basic Step:
$n=1, 5(1)=5$ and $\frac{5(1)(1+1)}{2}=5$;
Induction Step:
If $5+10+15+\cdots+5k=\frac{5k(k+1)}{2}$,
then $5+10+15+\cdots+5k+5(k+1)$
$=(5+10+15+\cdots+5k)+5k+5$
$=\frac{5k(k+1)}{2}+5k+5$
$=\frac{5k^2+15k+10}{2}$
$=\frac{5(k+1)(k+2)}{2}$
$=\frac{5(k+1)[(k+1)+1]}{2}$

15.
Basic Step:
$n=1,\ 1+\frac{1}{1}=2$ and $1+1=2$
Induction Step:
If $\left(1+\frac{1}{1}\right)\left(1+\frac{1}{2}\right)\left(1+\frac{1}{3}\right)\cdots\left(1+\frac{1}{k}\right)=k+1$,
then
$\left(1+\frac{1}{1}\right)\left(1+\frac{1}{2}\right)\left(1+\frac{1}{3}\right)\cdots\left(1+\frac{1}{k}\right)\left(1+\frac{1}{k+1}\right)$
$=(k+1)\left(1+\frac{1}{k+1}\right)$
$=k+1+\frac{k+1}{k+1}=(k+1)+1$

17.
Basic Step:
$n=1, 3(1)-2=1$ and $\frac{1}{2}(3(1)-1)=1$;
Induction Step:
If $1+4+7+10+\cdots+(3k-2)=\frac{k}{2}(3k-1)$,
then
$[1+4+7+10+\cdots+(3k-2)]$
$\quad+[3(k+1)-2]$
$=\frac{k}{2}(3k-1)+(3k+1)$
$=\frac{k(3k-1)+2(3k+1)}{2}$
$=\frac{3k^2+5k+2}{2}$
$=\frac{(k+1)(3k+2)}{2}$
$=\frac{k+1}{2}(3(k+1)-1)$

19. Basic Step:
$n=2, 3^2=9$ and $2(2)+1=5$,
so $3^2>2(2)+1$;
Induction Step:
If $3^k>2k+1$, then
$3^{k+1}=3^1\cdot 3^k>3(2k+1)$
$=6k+3>2k+3$
$=2k+2+1=2(k+1)+1$

21.
Basic Step:
$n=1, 1^3=1$ and $\frac{1^2(1+1)^2}{4}=1$;
Induction Step:
If $1^3+2^3+3^3+4^3+\cdots+k^3=\frac{k^2(k+1)^2}{4}$,
then
$(1^3+2^3+3^3+4^3+\cdots+k^3)+(k+1)^3$
$=\frac{k^2(k+1)^2}{4}+(k+1)^3$
$=\frac{k^2(k+1)^2+4(k+1)^3}{4}$
$=\frac{(k+1)^2(k+2)^2}{4}=\frac{(k+1)^2((k+1)+1)^2}{4}$

23. Basic Step:
$n=1, a^1=a$ so $a^1>1$, when $a>1$;
Induction Step:
If $a^k>1$, then
$a^{k+1}=a^k\cdot a^1>1\cdot a=a>1$

25.
Basic Step:
$n=1, 1^4=1$ and
$\frac{1(1+1)(2(1)+1)(3(1)^2+3(1)-1)}{30}=1$;
Induction Step:
If $1^4+2^4+3^4+\cdots+k^4$
$=\frac{k(k+1)(2k+1)(3k^2+3k-1)}{30}$,
then
$(1^4+2^4+3^4+\cdots+k^4)+(k+1)^4$
$=\frac{k(k+1)(2k+1)(3k^2+3k-1)}{30}+(k+1)^4$
$=\frac{6k^5+45k^4+130k^3+180k^2+119k+30}{30}$
$=\frac{(k+1)(k+2)(2k+3)(3k^2+9k+5)}{30}$
$=\frac{\left[\begin{array}{c}(k+1)(k+2)(2(k+1)+1)\\ \times(3(k+1)^2+3(k+1)-1)\end{array}\right]}{30}$

27.
Basic Step:
$n\ge 2, \frac{1}{\sqrt{1}}+\frac{1}{\sqrt{2}}=1+\frac{\sqrt{2}}{2}$
and $1+\frac{\sqrt{2}}{2}>\sqrt{2}$;
Induction Step:
If $\frac{1}{\sqrt{1}}+\frac{1}{\sqrt{2}}+\frac{1}{\sqrt{3}}+\cdots+\frac{1}{\sqrt{k}}>\sqrt{k}$,
then
$\left[\frac{1}{\sqrt{1}}+\frac{1}{\sqrt{2}}+\frac{1}{\sqrt{3}}+\cdots+\frac{1}{\sqrt{k}}\right]+\frac{1}{\sqrt{k+1}}$
$>\sqrt{k}+\frac{1}{\sqrt{k+1}}=\frac{\sqrt{k}\left(\sqrt{k+1}\right)+1}{\sqrt{k+1}}$
$=\frac{\sqrt{k^2+k}+1}{\sqrt{k+1}}>\frac{\sqrt{k^2}+1}{\sqrt{k+1}}$
$=\frac{k+1}{\sqrt{k+1}}=\sqrt{k+1}$

29. Basic Step:
$n=1, (ab)^1=ab$ and $a^1b^1=ab$;
Induction Step:
If $(ab)^k=a^kb^k$, then
$(ab)^{k+1}=(ab)^k\cdot(ab)$
$=a^kb^k\cdot ab=(a\cdot a^k)(b\cdot b^k)$
$=a^{k+1}b^{k+1}$

31.
Basic Step:
$n=1, \ln(x_1)=\ln x_1$;
Induction Step:
If $\ln(x_1\cdot x_2\cdot x_3\cdot\dots\cdot x_k)$
$=\ln x_1+\ln x_2+\ln x_3+\cdots+\ln x_k$
when $x_1>0, x_2>0,\ldots,x_n>0$,
then $\ln(x_1\cdot x_2\cdot x_3\cdot\dots\cdot x_k\cdot x_{k+1})$
$=\ln(x_1\cdot x_2\cdot x_3\cdot\dots\cdot x_k)+\ln(x_{k+1})$
$=(\ln x_1+\ln x_2+\ln x_3+\cdots+\ln x_k)+\ln x_{k+1}$

33.
Basic Step: $n=2$,
$(9^2-8(2)-1)=64$ of which 64 is a factor;
Induction Step:
If $(9^k-8k-1)=64p$ for some integer p,
then $(9^{k+1}-8(k+1)-1)$
$=9\cdot 9^k-8k-9$
$=9\cdot 9^k-9\cdot 8k+8\cdot 8k-9$
$=9(9^k-8k-1)+64k$
$=9(64p)+64k=64(9p+k)$

35.

Basic Step: $n = 1$,

$(1^3 - 1 + 3) = 3$, which is divisible by 3;

Induction Step:

If $\frac{k^3 - k + 3}{3} = p$

or $k^3 - k + 3 = 3p$ for some integer p,

then $(k+1)^3 - (k+1) + 3$

$= k^3 + 3k^2 + 2k + 3$

$= (k^3 - k + 3) + (3k^2 + 3k)$

$= 3p + 3(k^2 + k) = 3(p + k^2 + k)$

37.

Basic Step: $n = 1$,

$1(1+1)(1+2) = 6$, which is divisible by 6;

Induction Step:

If $\frac{k(k+1)(k+2)}{6} = p$

or $k(k+1)(k+2) = 6p$

for some integer p, then

$(k+1)(k+2)(k+3)$

$= k^3 + 6k^2 + 11k + 6$

$= (k^3 + 3k^2 + 2k) + (3k^2 + 9k + 6)$

$= k(k+1)(k+2) + 3(k+1)(k+2)$

$= 6p + 3(k+1)(k+2)$.

$6p$ is clearly divisible by 6.

In order for $3(k+1)(k+2)$ to be divisible by 6, it must be divisible by 2 and 3. It is clearly divisible by 3.

If k is odd, then the term $(k+1)$ must be even, making it divisible by 2.

If k is even, then the term $(k+2)$ is even, making it divisible by 2.

Therefore, $3(k+1)(k+2)$ is divisible by 6.

39.

$0 + 1 + 2 + 3 + \cdots + (n-1) = \frac{n(n-1)}{2}$;

Basic Step:

$n = 1, (1-1) = 0$ and $\frac{1(1-1)}{2} = 0$;

Induction Step:

If $0 + 1 + 2 + \cdots + (k-1) = \frac{k(k-1)}{2}$,

then

$[0 + 1 + 2 + \cdots + (k-1)] + (k + 1 - 1)$

$= \frac{k(k-1)}{2} + k = \frac{k^2 - k + 2k}{2}$

$= \frac{k(k+1)}{2} = \frac{(k+1)((k+1)-1)}{2}$

41. The induction step does not work for $n = 1$. In the case of $n = 1$, $n + 1 = 2$ and the groups formed by removing the first horse and then the last horse do not overlap.

12.5 Exercises

1. combination

3. combination

5. 12 **7.** 720

9. 15 **11.** 792

13. $\frac{5!}{2!} = 60$

15. $\frac{7!}{2!2!} = 1260$

17. $\frac{11!}{2!2!2!} = 4{,}989{,}600$

19. $243x^5 + 405x^4y + 270x^3y^2 + 90x^2y^3 + 15xy^4 + y^5$

21. $x^4 - 12x^3 + 54x^2 - 108x + 81$

23. $7776x^{10} + 6480x^8y + 2160x^6y^2 + 360x^4y^3 + 30x^2y^4 + y^5$

25. $2401x^8 + 10{,}976x^6y^2 + 18{,}816x^4y^4 + 14{,}336x^2y^6 + 4096y^8$

27. $x^2 + 2xy + 2xz + y^2 + 2yz + z^2$

29. $64x^6 + 960x^5 + 6000x^4 + 20{,}000x^3 + 37{,}500x^2 + 37{,}500x + 15{,}625$

31. 32

33. $x^{16} + 48x^{15}y + 1080x^{14}y^2 + 15{,}120x^{13}y^3$

35. $129{,}140{,}163x^{\frac{17}{4}} + 3{,}658{,}971{,}285x^4y$

37. $651{,}168x^5$

39. $10^3 = 1000$

41. $9^7 = 4{,}782{,}969$

43. $15! \approx 1.308 \times 10^{12}$

45. $3! = 6$

47. $5^{10} = 9{,}765{,}625$

49. $36^6 = 2{,}176{,}782{,}336$

51. $26 \cdot 25 \cdot 24 \cdot 10 \cdot 9 \cdot 8 = 11{,}232{,}000$

53. ${}_{30}P_{12} \approx 4.143 \times 10^{16}$

55. ${}_{36}P_8 \approx 1.220 \times 10^{12}$

57. ${}_7P_6 = 5040$

${}_7P_7 = 5040$ as well. (Having a child remain standing is numerically equivalent to putting a seventh chair in the room.)

59. ${}_{26}P_3 = 15{,}600$

61. ${}_7C_3 = 35$

63. ${}_9C_2 = 36$

65. ${}_{75}C_5 = 17{,}259{,}390$

67. ${}_{10}C_4 \cdot {}_8C_4 \cdot {}_{13}C_4 = 10{,}510{,}500$

69. 112 cones

71. 96 outfits

73. 288 schedules

75. 120 5-letter strings

77. 303,600 ways

79. 495 pizzas

81. 752,538,150 groups

83. 420 ways

85. $\binom{n}{n-k} = \frac{n!}{(n-k)!(n-(n-k))!}$

$= \frac{n!}{(n-k)!(n-n+k)!}$

$= \frac{n!}{(n-k)!k!} = \binom{n}{k}$

87. $2^n = (1+1)^n$

$= \sum_{k=0}^{n} \binom{n}{k} (1)^k (1)^{n-k} = \sum_{k=0}^{n} \binom{n}{k}$

$= \binom{n}{0} + \binom{n}{1} + \cdots + \binom{n}{n}$

12.6 Exercises

1. $\frac{3}{5}$ **3.** $\frac{9}{13}$

5. $\frac{1}{3}$

7. a. 0 **b.** $\frac{5}{8}$

9. a. 0 **b.** $\frac{3}{5}$

11. a. 0 **b.** 1

13. a. $\frac{1}{8}$ **b.** $\frac{9}{16}$

15. The set of all ordered 4-tuples made up of H's and T's. There are 16 such 4-tuples.

17. The set of all ordered pairs that have either an H or a T in the first slot and one of the 13 hearts in the second slot. There are 26 such ordered pairs.

19. The set of all ordered triples with any of the 6 values in each slot. There are 216 such triples.

21. The set of the 38 pockets.

23. a. $\frac{2}{3}$ **b.** $\frac{1}{3}$

25. a. $\frac{3}{8}$ **b.** $\frac{1}{8}$ **c.** $\frac{1}{2}$

27. $\frac{3}{10}$

29. $\frac{387,420,489}{1,000,000,000} \approx 0.3874$

31. $\frac{3}{8}$

33. a. $\frac{11}{26}$ **b.** $\frac{9}{52}$ **c.** $\frac{2}{13}$

35. a. $\frac{1}{169}$ **b.** $\frac{1}{221}$

37. 18.75%

39. a. $\frac{1}{6}$ **b.** $\frac{5}{18}$ **c.** $\frac{1}{9}$

41. $\frac{1}{20}$ **43.** $\frac{2}{9}$ **45.** $\frac{3}{10}$

47. 27 tickets

Chapter 12 Project

1. a. $\frac{9}{19}$ **b.** $\frac{9}{19}$ **c.** $\frac{1}{38}$

d. $\frac{3}{38}$ **e.** $\frac{1}{38}$

3. Approximately –$0.05 (or the person betting will lose about 5 cents per play on average)

Chapter 12 Review Exercises

1. –3, 9, –27, 81, –243

3. –3, –4, –5, –6, –7

5. $a_n = 6n - 13$ **7.** $a_n = n^2 - 1$

9. $a_1 = -2, a_n = n(a_{n-1})$ for $n \geq 2$

11. $-3-5-7-9-11-13=-48$

13. $\sum_{i=2}^{7} i^3 = 783$

15. $-8-16-32-64-128=-248$

17. $S_n = \frac{n}{2(n+2)}$, $S_{80} = \frac{20}{41}$

19. $S_n = 3 - 3^{n+1}$, series diverges

21. –10, –12, –22, –34, –56

23. $a_n = \frac{5}{2}n + 9$

25. $a_n = 3n - 1$ **27.** $a_n = 9n - 14$

29. 275 **31.** a_{36} **33.** 8827

35. 66

37. $a_n = 3\left(\frac{1}{5}\right)^{n-1}$

39. $a_n = 6(4)^{n-1}$

41. $a_n = 8\left(\frac{1}{4}\right)^{n-1}$

43. $r = \pm\frac{2}{3}; \pm\frac{45}{8}, \frac{15}{4}, \pm\frac{5}{2}, \frac{5}{3}, \pm\frac{10}{9}$

45. –2 **47.** $\frac{381}{512}$ **49.** –12

51. 1

53.

Basic Step: $n = 1, (3(1)+2) = 5$ and $\frac{1(3(1)+7)}{2} = 5$;

Induction Step:

If $5+8+11+\cdots+(3k+2) = \frac{k(3k+7)}{2}$,

then

$$5+8+11+\cdots+(3k+2)+(3(k+1)+2)$$
$$= \frac{k(3k+7)}{2} + (3k+5)$$
$$= \frac{3k^2+7k+6k+10}{2}$$
$$= \frac{(k+1)(3k+10)}{2}$$
$$= \frac{(k+1)(3(k+1)+7)}{2}$$

55. Basic Step: $n = 1$, $11^1 - 7^1 = 4$, which is divisible by 4;

Induction Step:

If $\frac{11^k - 7^k}{4} = p$ or $11^k - 7^k = 4p$ for some integer p, then

$$11^{k+1} - 7^{k+1} = 11 \cdot 11^k - 7 \cdot 7^k$$
$$= 4 \cdot 11^k + 7 \cdot 11^k - 7 \cdot 7^k$$
$$= 4 \cdot 11^k + 7(11^k - 7^k)$$
$$= 4 \cdot 11^k + 7(4p) = 4(11^k + 7p)$$

57. $8 \cdot 10^3 \cdot 26^3 = 140,608,000$

59. $\frac{8!}{3!2!} = 3360$

61. ${}_{21}P_5 = 2,441,880$

63. $-32y^5 + 80y^4 - 80y^3 + 40y^2 - 10y + 1$

65. $3125x^{10} - 6250x^8 y + 5000x^6 y^2 - 2000x^4 y^3 + 400x^2 y^4 - 32y^5$

67. a. $\frac{1}{9}$ **b.** $\frac{2}{3}$

69. a. $\frac{2}{7}$ **b.** $\frac{4}{7}$

71. a. $\frac{1}{10}$ **b.** $\frac{4}{5}$

73. $\frac{33}{108,290}$

Chapter 13: An Introduction to Limits, Continuity, and the Derivative

13.1 Exercises

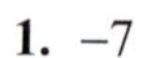

1. -7 **3.** $2c-5+h$

5. $\dfrac{1}{c^2-12c+ch-6h+36}$

7. $\dfrac{\ln(c+h)-\ln c}{h}$

9. $\dfrac{3\left(\sqrt{c-2+h}-\sqrt{c-2}\right)}{h}$ or $\dfrac{3}{\sqrt{c-2}+\sqrt{c-2+h}}$

11. $c=-1$

a. 2 **b.** 2 **c.** 2

13. $c=3$

a. 0.0833 **b.** 0.1149

c. 0.1111

15. 3 **17.** 0 **19.** $-\dfrac{1}{4}$

21. The exact answer is -2.

23. The exact answer is 2.

25. The exact answer is 2.

27. The exact answer is $\dfrac{1}{e}$ or approx. 0.3679.

29. The exact answer is $\dfrac{1}{5\ln 10}$ or approx. 0.0869.

31. The exact answer is -4.

33. a. 96 ft **b.** 16 ft/s

c. 16 **d.** 2.5 s

35. a. 176 ft **b.** 176 ft/s

c. 192 ft/s **d.** 64 ft/s **e.** 6 s

37. a. 3 m/s **b.** 15 m/s

c. $6t_0+3$

39. a. 5 s; 45 m/s **b.** 10 s

c. $-1.83g$

41.

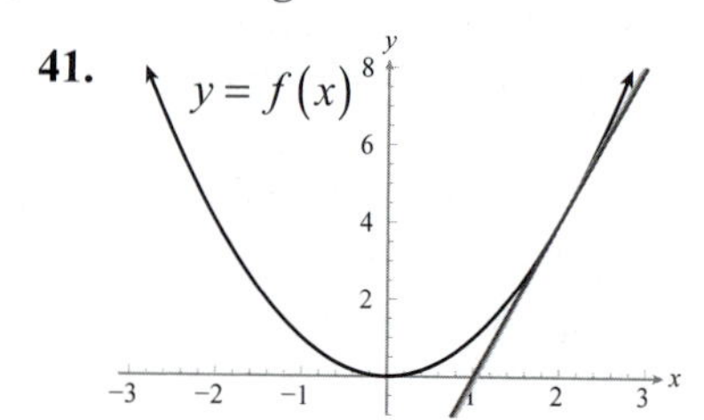

43.

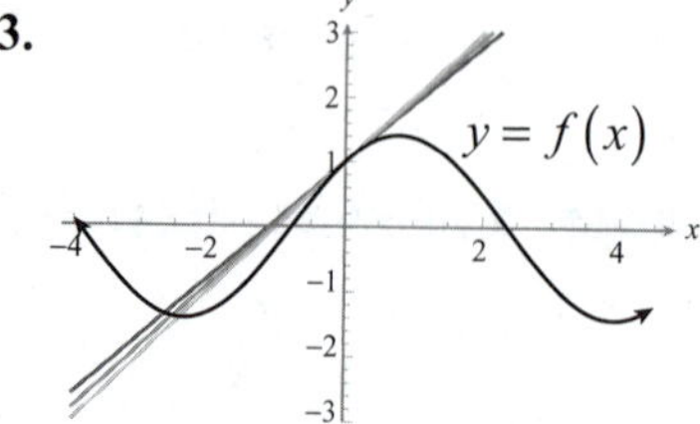

45.

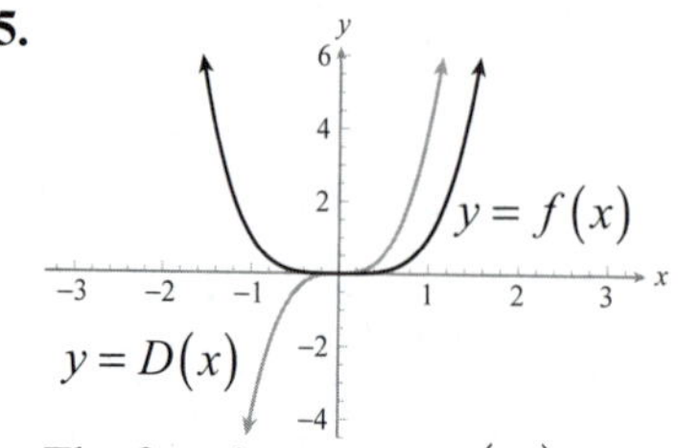

The function value $D(x_0)$ at any given x_0 is approximately equal to the slope of the graph of f at x_0.

47.

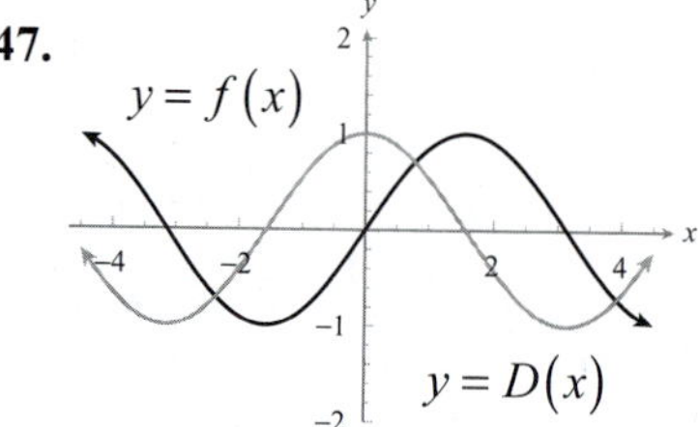

The function value $D(x_0)$ at any given x_0 is approximately equal to the slope of the graph of f at x_0.

49. $x=1$ **51.** $x=1$

13.2 Exercises

1. 4 **3.** 0

5.

x	y
1	2.414
1.4	2.814
1.41	2.824
1.414	2.828

The table points to a limit of $2\sqrt{2}\approx 2.828$.

7.

x	y
1.5	113.33
1.1	15.94
1.01	10.46
1.001	10.045

The table points to a limit of 10.

9.

x	y
3	−3.4696
3.14	−3.0048
3.141	−3.0018
3.1415	−3.0003

The table points to a limit of -3.

11.

x	y
7.5	210.5
7.1	994.1
7.01	9814.0
7.001	98014.0

The table points to an undefined limit.

13. $\lim_{x\to 1^-} f(x)=-\infty$; $\lim_{x\to 1^+} f(x)=\infty$

15. $\lim_{x\to 3^-} h(x)=\infty$; $\lim_{x\to 3^+} h(x)=\infty$

17. $\lim_{x\to -1^-} q(x)=\infty$; $\lim_{x\to -1^+} q(x)=-\infty$

19. $\lim_{x\to 1.5^-} v(x)=-\infty$; $\lim_{x\to 1.5^+} v(x)=\infty$

21. $\lim_{x\to\left(\frac{\pi}{2}+k\pi\right)^-} \tan x=\infty$;

$\lim_{x\to\left(\frac{\pi}{2}+k\pi\right)^+} \tan x=-\infty \quad (k\in\mathbb{Z})$

23. $\lim_{x\to 6^-} s(x)=-\infty$; $\lim_{x\to 6^+} s(x)=\infty$

25. $\lim_{x\to\infty} f(x)=0$; $\lim_{x\to-\infty} f(x)=0$

27. $\lim_{x\to\infty} h(x)=0$; $\lim_{x\to-\infty} h(x)=0$

29. $\lim_{x\to\infty} q(x)=-1$; $\lim_{x\to-\infty} q(x)=-1$

31. $\lim_{x\to\infty} v(x)=\infty$; $\lim_{x\to-\infty} v(x)=-\infty$

33. $\lim_{x\to\infty}\tan x$ does not exist;

$\lim_{x\to-\infty}\tan x$ does not exist

35. $\lim_{x\to\infty} s(x)=1$; $\lim_{x\to-\infty} s(x)=-1$

37. a. -2 **b.** 1

39. a. 1 **b.** 4

41. a. 1 **b.** $-\infty$

43. a. 1 **b.** 1

45. a. Does not exist **b.** 0

47. $\lim_{x\to\infty} f(x)=\infty$; $\lim_{x\to-\infty} f(x)=-\infty$

49. $\lim_{x\to\infty} h(x)=-\infty$; $\lim_{x\to-\infty} h(x)=-\infty$

51. $\lim_{x\to\infty} F(x) = \infty$

53. $\lim_{x\to\infty} H(x) = \infty$; $\lim_{x\to-\infty} H(x) = \infty$

55. $\lim_{x\to\infty} u(x) = \infty$; $\lim_{x\to-\infty} u(x) = \infty$

57. $\lim_{x\to-\infty} s(x) = -\infty$

59. False; see $h(x)$ of Example 5 at $x = 2$.

61. True

63. True

65. Does not exist

67. 2 **69.** $\frac{3}{2}$

71. 0

13.3 Exercises

1. $\delta \approx 0.23$ or smaller

3. $\delta \approx 0.2$ or smaller

5. $\delta = 0.02$ or smaller

7. $\delta = 0.1$ or smaller

9. $\delta = \sqrt{0.1}$ or smaller

11. $\delta = 0.0\overline{9}$ or smaller

13. $\delta = e^{0.1} - 1 \approx 0.1052$ or smaller

15. $N = 10$ or larger

17. $N = -10$ or smaller

19. $N = \ln 0.1$ or smaller

21. $\delta = 0.1$ or smaller

23. $\delta = 0.1$ or smaller

25. $\delta = \frac{\pi}{2} - \arctan 100 \approx 0.0099997$ or smaller

57. The limit does not exist.

59. The limit does not exist.

61. The limit is 0.

63. **a.** 1256.64 N

b. Approx. 0.16 mm

65. L is incorrectly quantified and switched with c.

67. ε is incorrectly quantified.

69. The inequality $0 \le |x - c|$ is incorrect.

71. False; the function value and limit at c need not be equal.

73. False; consider $f(x) = -x^2$ and $g(x) = x^2$ at $c = 0$.

85. Vert. asym.: $x \approx -1.3340$, $x \approx 1.1759$

87. Vert. asym.: $x = (2n+1)\pi - 6$, $n \in \mathbb{Z}$

89. Vert. asym.: $x = \frac{(2n-1)\pi}{2}$, $n \in \mathbb{Z}$, $n \ne 0$

13.4 Exercises

1. **a.** -4 **b.** 21

3. 5 **5.** 7 **7.** 9

9. 22 **11.** -4 **13.** -2

15. $\sqrt[3]{2}$ **17.** $\sqrt[3]{100}$ **19.** 16

21. 12 **23.** 11 **25.** $\frac{33}{5}$

27. $\frac{1}{6}$ **29.** $\frac{1}{2\sqrt{5}}$ **31.** $\frac{1}{9}$

33. 6 **35.** 4 **37.** $\frac{1}{2\sqrt{x}}$

39. $\frac{32}{7}$ **41.** $-\frac{1}{6}$ **43.** -1

45. 8 **47.** 11

49. $9 - 2(c - 2)$

51. 1 **53.** 0 **55.** 0

57. $-2e^2$

71. The limit is 0.

79. $\frac{r}{2}$

13.5 Exercises

1. Points of continuity: $(-\infty, 0) \cup (0, 3) \cup (3, \infty)$
Points of discontinuity: $c = 0$, $\lim_{x\to 0} f(x)$ does not exist; $c = 3$, $\lim_{x\to 3} f(x) \ne f(3)$

5. $c = 0$, nonremovable

7. $c = 3$, removable

9. $c = 2$, nonremovable

11. None

13. $c = -\frac{(2n+1)\pi}{2}$ for $n \in \{-1, 0, 1, 2, \ldots\}$, nonremovable

15. $c = 4$, removable

17. None

19. None

21. $c = \pm 1$, nonremovable

23. None

25. All integers, nonremovable

27. $c = \pm\sqrt{n}$, n a positive integer, nonremovable

29. $c = \frac{1}{n}$, n a nonzero integer, nonremovable

35. Continuous on $[-3, -1] \cup (0, 2] \cup [3, \infty)$

37. Continuous on $(-\sqrt{3}, \sqrt{3})$

39. Continuous except on $\left\{\frac{n-1}{\pi} \middle| n \in \mathbb{Z}\right\} \cup \{\ln k - 2 \mid k \in \mathbb{N}\}$

45. $g(1) = 2$ and $g(2) = 3$ will make g continuous.

47. $F(3) = \frac{1}{4}$ will make F continuous.

49. $H(0) = 0$ will make H continuous.

51. Not continuous at 3

53. Continuous on $\left[0, \frac{1}{\pi}\right]$

55. $a = 3$

57. $a = 0.5, b = 1.25$

59. Yes; $c = 1$

61. No; discontinuity at 1

63. Yes; $c = \frac{\pi}{9} - \frac{2}{3}$ or $c = \frac{5\pi}{9} - \frac{2}{3}$

73. **a.** Because speeds of everyday objects are smaller than c by orders of magnitude, the denominator of ΔT is approximately 1.

b. No moving object can reach the speed of light.

Index

D

E

F

G

H

I

J

K

L

M

N

Q

R

S

T

U

V

W

X

Y

Z

DETERMINANT OF A 2×2 MATRIX 11.3

The determinant of a 2×2 matrix $A=\begin{bmatrix} a_{11} & a_{12} \\ a_{21} & a_{22} \end{bmatrix}$ is given by the formula

$$|A| = a_{11}a_{22} - a_{21}a_{12}.$$

DETERMINANT OF AN $n\times n$ MATRIX 11.3

The minor of the element a_{ij} is the determinant of the $(n-1)\times(n-1)$ matrix formed from A by deleting the ith row and the jth column.

The cofactor of the element a_{ij} is $(-1)^{i+j}$ times the minor of a_{ij}.

Find the determinant of an $n\times n$ matrix by expanding along a fixed row or column.

- To expand along the ith row, each element of that row is multiplied by its cofactor and the n products are then added.
- To expand along the jth column, each element of that column is multiplied by its cofactor and the n products are then added.

CRAMER'S RULE 11.3

A system of n linear equations in the n variables $x_1, x_2, \ldots, x_n$ can be written in the following form.

$$\begin{cases} a_{11}x_1 + a_{12}x_2 + \cdots + a_{1n}x_n = b_1 \\ a_{21}x_1 + a_{22}x_2 + \cdots + a_{2n}x_n = b_2 \\ \quad\vdots \\ a_{n1}x_1 + a_{n2}x_2 + \cdots + a_{nn}x_n = b_n \end{cases}$$

The solution of the system is given by the n formulas

$$x_1 = \frac{D_{x_1}}{D}, x_2 = \frac{D_{x_2}}{D}, \ldots, x_n = \frac{D_{x_n}}{D},$$

where D is the determinant of the coefficient matrix and D_{x_i} is the determinant of the same matrix with the ith column of constants replaced by the column of constants $b_1, b_2, \ldots, b_n$.

MATRIX ADDITION 11.4

$A+B=$ the matrix such that $c_{ij}=a_{ij}+b_{ij}$ (c_{ij} is the element in the ith row and j^{th} column of $A+B$).

SCALAR MULTIPLICATION 11.4

$cA=$ the matrix such that the element in the ith row and j^{th} column is equal to ca_{ij}.

MATRIX MULTIPLICATION 11.4

$AB=$ the matrix such that $c_{ij}=a_{i1}b_{1j}+a_{i2}b_{2j}+\cdots+a_{in}b_{nj}$. (The length of each row in A must be the same as the length of each column of B.)

PROPERTIES OF SIGMA NOTATION 12.1

For sequences $\{a_n\}$ and $\{b_n\}$ and a constant c:

$$\sum_{i=1}^{n}(a_i+b_i)=\sum_{i=1}^{n}a_i+\sum_{i=1}^{n}b_i \qquad \sum_{i=1}^{n}ca_i=c\sum_{i=1}^{n}a_i$$

$$\sum_{i=1}^{n}a_i=\sum_{i=1}^{k}a_i+\sum_{i=k+1}^{n}a_i \text{ (for any } 1\le k\le n-1)$$

SUMMATION FORMULAS 12.1

$$\sum_{i=1}^{n}1=n \qquad \sum_{i=1}^{n}i=\frac{n(n+1)}{2}$$

$$\sum_{i=1}^{n}i^2=\frac{n(n+1)(2n+1)}{6} \qquad \sum_{i=1}^{n}i^3=\frac{n^2(n+1)^2}{4}$$

SEQUENCES AND SERIES

Arithmetic 12.2

Let $\{a_n\}$ be an arithmetic sequence with common difference d.

General term: $a_n=a_1+(n-1)d$

Partial sum: $S_n=na_1+d\left(\frac{(n-1)n}{2}\right)=\left(\frac{n}{2}\right)(a_1+a_n)$

Geometric 12.3

Let $\{a_n\}$ be a geometric sequence with common ratio r.

General term: $a_n=a_1r^{n-1}$

Partial sum: $S_n=\frac{a_1(1-r^n)}{1-r}$, if $r\neq 0,1$

Infinite sum: $S=\sum_{n=0}^{\infty}a_1r^n=\frac{a_1}{1-r}$, if $|r|<1$

PERMUTATION FORMULA 12.5

$${}_nP_k=\frac{n!}{(n-k)!}$$

COMBINATION FORMULA 12.5

$${}_nC_k=\binom{n}{k}=\frac{n!}{k!(n-k)!}$$

BINOMIAL THEOREM 12.5

$$(A+B)^n=\sum_{k=0}^{n}\binom{n}{k}A^kB^{n-k}$$

MULTINOMIAL COEFFICIENTS 12.5

$$\binom{n}{k_1,k_2,\ldots,k_r}=\frac{n!}{k_1!k_2!\cdots k_r!}$$

MULTINOMIAL THEOREM 12.5

$$(A_1+A_2+\cdots+A_r)^n=\sum_{k_1+k_2+\cdots+k_r=n}\binom{n}{k_1,k_2,\ldots,k_r}A_1^{k_1}A_2^{k_2}\cdots A_r^{k_r}$$

COMPLEX NUMBERS AND DE MOIVRE'S THEOREM 9.5

$z = a + bi$ $\qquad |z| = \sqrt{a^2 + b^2}$ $\qquad \tan\theta = \dfrac{b}{a}$

$$z = |z|(\cos\theta + i\sin\theta) = |z|e^{i\theta}$$

$$z^n = |z|^n\left(\cos(n\theta) + i\sin(n\theta)\right) = |z|^n e^{in\theta}$$

Distinct n^{th} roots of z for $k = 0, 1, \ldots, n-1$:

$$w_k = |z|^{\frac{1}{n}}\left[\cos\left(\frac{\theta + 2k\pi}{n}\right) + i\sin\left(\frac{\theta + 2k\pi}{n}\right)\right] = |z|^{\frac{1}{n}} e^{i\left(\frac{\theta + 2k\pi}{n}\right)}$$

VECTOR OPERATIONS AND MAGNITUDE 9.6

Given two vectors $\mathbf{u} = \langle u_1, u_2\rangle$ and $\mathbf{v} = \langle v_1, v_2\rangle$ and a scalar a, $\mathbf{u} + \mathbf{v} = \langle u_1 + v_1, u_2 + v_2\rangle$, $a\mathbf{u} = \langle au_1, au_2\rangle$, and $\|\mathbf{u}\| = \sqrt{{u_1}^2 + {u_2}^2}$.

PROPERTIES OF VECTOR OPERATIONS 9.6

For vectors **u**, **v**, and **w** and scalars a and b:

Vector Addition	Scalar Multiplication
$\mathbf{u} + \mathbf{v} = \mathbf{v} + \mathbf{u}$	$a(\mathbf{u} + \mathbf{v}) = a\mathbf{u} + a\mathbf{v}$
$\mathbf{u} + (\mathbf{v} + \mathbf{w}) = (\mathbf{u} + \mathbf{v}) + \mathbf{w}$	$(a + b)\mathbf{u} = a\mathbf{u} + b\mathbf{u}$
$\mathbf{u} + \mathbf{0} = \mathbf{u}$	$(ab)\mathbf{u} = a(b\mathbf{u}) = b(a\mathbf{u})$
$\mathbf{u} + (-\mathbf{u}) = \mathbf{0}$	$1\mathbf{u} = \mathbf{u}$, $0\mathbf{u} = \mathbf{0}$, and $a\mathbf{0} = \mathbf{0}$
	$\|a\mathbf{u}\| = \lvert a\rvert\|\mathbf{u}\|$

DOT PRODUCT 9.7

Given two vectors $\mathbf{u} = \langle u_1, u_2\rangle$ and $\mathbf{v} = \langle v_1, v_2\rangle$,

$$\mathbf{u}\cdot\mathbf{v} = u_1v_1 + u_2v_2.$$

ELEMENTARY PROPERTIES OF THE DOT PRODUCT 9.7

Given vectors **u**, **v**, and **w** and a scalar a:

$\mathbf{u}\cdot\mathbf{v} = \mathbf{v}\cdot\mathbf{u}$ $\qquad \mathbf{0}\cdot\mathbf{u} = 0$

$\mathbf{u}\cdot(\mathbf{v} + \mathbf{w}) = \mathbf{u}\cdot\mathbf{v} + \mathbf{u}\cdot\mathbf{w}$ $\qquad a(\mathbf{u}\cdot\mathbf{v}) = (a\mathbf{u})\cdot\mathbf{v} = \mathbf{u}\cdot(a\mathbf{v})$

$\mathbf{u}\cdot\mathbf{u} = \|\mathbf{u}\|^2$

THE DOT PRODUCT THEOREM 9.7

Let θ be the smaller of the two angles formed by nonzero vectors **u** and **v** (so $0 \le \theta \le \pi$). Then $\mathbf{u}\cdot\mathbf{v} = \|\mathbf{u}\|\|\mathbf{v}\|\cos\theta$.

ORTHOGONAL VECTORS 9.7

Two nonzero vectors **u** and **v** are said to be orthogonal (or perpendicular) if $\mathbf{u}\cdot\mathbf{v} = 0$.

PROJECTION OF u ONTO v 9.7

Let **u** and **v** be nonzero vectors. The projection of **u** onto **v** is the vector $\text{proj}_{\mathbf{v}}\mathbf{u} = \left(\dfrac{\mathbf{u}\cdot\mathbf{v}}{\|\mathbf{v}\|^2}\right)\mathbf{v}$.

ROTATION OF AXES 10.4

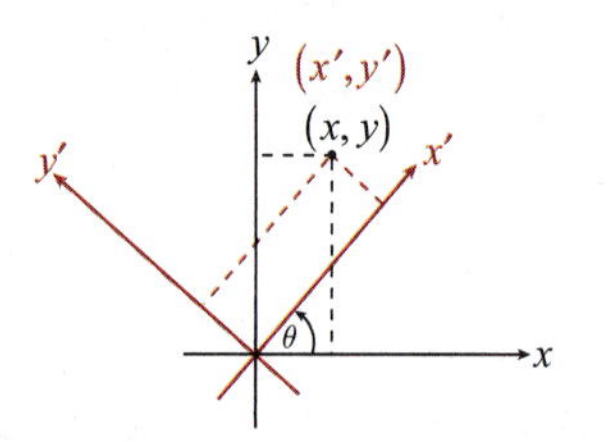

ROTATION RELATIONS 10.4

$x = x'\cos\theta - y'\sin\theta$ $\qquad x' = x\cos\theta + y\sin\theta$

$y = x'\sin\theta + y'\cos\theta$ $\qquad y' = -x\sin\theta + y\cos\theta$

ELIMINATION OF THE *xy*-TERM 10.4

The graph of the equation $Ax^2 + Bxy + Cy^2 + Dx + Ey + F = 0$ in the xy-plane is the same as the graph of the equation $A'x'^2 + C'y'^2 + D'x' + E'y' + F' = 0$ in the $x'y'$-plane, where the angle of rotation θ between the two coordinate systems satisfies $\cot(2\theta) = \dfrac{A - C}{B}$.

CLASSIFYING CONICS 10.4

Assuming the graph of the equation $Ax^2 + Bxy + Cy^2 + Dx + Ey + F = 0$ is a nondegenerate conic section, it is classified by its discriminant as follows:

- Ellipse if $B^2 - 4AC < 0$
- Parabola if $B^2 - 4AC = 0$
- Hyperbola if $B^2 - 4AC > 0$

POLAR EQUATIONS OF CONIC SECTIONS 10.5

A conic section consists of all points P in the plane that satisfy the equation

$$\frac{D(P, F)}{D(P, L)} = e,$$

where e is a fixed positive constant. The conic is

- an ellipse if $0 < e < 1$,
- a parabola if $e = 1$, and
- a hyperbola if $e > 1$.

Equation of Conic Section	Directrix
$r = \dfrac{ed}{1 + e\cos\theta}$	Vertical directrix $x = d$
$r = \dfrac{ed}{1 - e\cos\theta}$	Vertical directrix $x = -d$
$r = \dfrac{ed}{1 + e\sin\theta}$	Horizontal directrix $y = d$
$r = \dfrac{ed}{1 - e\sin\theta}$	Horizontal directrix $y = -d$

RECIPROCAL IDENTITIES 8.1

$$\csc x = \frac{1}{\sin x} \qquad \sec x = \frac{1}{\cos x} \qquad \cot x = \frac{1}{\tan x}$$

$$\sin x = \frac{1}{\csc x} \qquad \cos x = \frac{1}{\sec x} \qquad \tan x = \frac{1}{\cot x}$$

COFUNCTION IDENTITIES 8.1

$$\cos x = \sin\left(\frac{\pi}{2} - x\right) \qquad \sin x = \cos\left(\frac{\pi}{2} - x\right)$$

$$\csc x = \sec\left(\frac{\pi}{2} - x\right) \qquad \sec x = \csc\left(\frac{\pi}{2} - x\right)$$

$$\cot x = \sin\left(\frac{\pi}{2} - x\right) \qquad \tan x = \cot\left(\frac{\pi}{2} - x\right)$$

QUOTIENT IDENTITIES 8.1

$$\tan x = \frac{\sin x}{\cos x} \qquad \cot x = \frac{\cos x}{\sin x}$$

PERIOD IDENTITIES 8.1

$$\sin(x + 2\pi) = \sin x \qquad \csc(x + 2\pi) = \csc x$$

$$\cos(x + 2\pi) = \cos x \qquad \sec(x + 2\pi) = \sec x$$

$$\tan(x + \pi) = \tan x \qquad \cot(x + \pi) = \cot x$$

EVEN/ODD IDENTITIES 8.1

$$\sin(-x) = -\sin x \quad \cos(-x) = \cos x \quad \tan(-x) = -\tan x$$

$$\csc(-x) = -\csc x \quad \sec(-x) = \sec x \quad \cot(-x) = -\cot x$$

PYTHAGOREAN IDENTITIES 8.1

$$\sin^2 x + \cos^2 x = 1 \quad \tan^2 x + 1 = \sec^2 x \quad 1 + \cot^2 x = \csc^2 x$$

SUM AND DIFFERENCE IDENTITIES 8.2

$$\sin(u + v) = \sin u \cos v + \cos u \sin v$$

$$\sin(u - v) = \sin u \cos v - \cos u \sin v$$

$$\cos(u + v) = \cos u \cos v - \sin u \sin v$$

$$\cos(u - v) = \cos u \cos v + \sin u \sin v$$

$$\tan(u + v) = \frac{\tan u + \tan v}{1 - \tan u \tan v}$$

$$\tan(u - v) = \frac{\tan u - \tan v}{1 + \tan u \tan v}$$

DOUBLE-ANGLE IDENTITIES 8.3

$$\sin(2u) = 2\sin u \cos u$$

$$\tan(2u) = \frac{2\tan u}{1 - \tan^2 u}$$

$$\cos(2u) = \cos^2 u - \sin^2 u = 2\cos^2 u - 1 = 1 - 2\sin^2 u$$

POWER-REDUCING IDENTITIES 8.3

$$\sin^2 x = \frac{1 - \cos(2x)}{2} \qquad \cos^2 x = \frac{1 + \cos(2x)}{2}$$

$$\tan^2 x = \frac{1 - \cos(2x)}{1 + \cos(2x)}$$

HALF-ANGLE IDENTITIES 8.3

$$\sin\frac{x}{2} = \pm\sqrt{\frac{1 - \cos x}{2}} \qquad \cos\frac{x}{2} = \pm\sqrt{\frac{1 + \cos x}{2}}$$

$$\tan\frac{x}{2} = \frac{1 - \cos x}{\sin x} = \frac{\sin x}{1 + \cos x}$$

PRODUCT-TO-SUM IDENTITIES 8.3

$$\sin x \cos y = \frac{1}{2}\left[\sin(x + y) + \sin(x - y)\right]$$

$$\cos x \sin y = \frac{1}{2}\left[\sin(x + y) - \sin(x - y)\right]$$

$$\sin x \sin y = \frac{1}{2}\left[\cos(x - y) - \cos(x + y)\right]$$

$$\cos x \cos y = \frac{1}{2}\left[\cos(x + y) + \cos(x - y)\right]$$

SUM-TO-PRODUCT IDENTITIES 8.3

$$\sin x + \sin y = 2\sin\left(\frac{x + y}{2}\right)\cos\left(\frac{x - y}{2}\right)$$

$$\sin x - \sin y = 2\cos\left(\frac{x + y}{2}\right)\sin\left(\frac{x - y}{2}\right)$$

$$\cos x + \cos y = 2\cos\left(\frac{x + y}{2}\right)\cos\left(\frac{x - y}{2}\right)$$

$$\cos x - \cos y = -2\sin\left(\frac{x + y}{2}\right)\sin\left(\frac{x - y}{2}\right)$$

LAWS OF SINES AND COSINES

Law of Sines 9.1

$$\frac{\sin A}{a} = \frac{\sin B}{b} = \frac{\sin C}{c}$$

Law of Cosines 9.2

$$a^2 = b^2 + c^2 - 2bc\cos A$$

$$b^2 = a^2 + c^2 - 2ac\cos B$$

$$c^2 = a^2 + b^2 - 2ab\cos C$$

Area of a Triangle (Sine Formula) 9.1

$$\text{Area} = \frac{1}{2}ab\sin C = \frac{1}{2}bc\sin A = \frac{1}{2}ac\sin B$$

POLAR COORDINATES 9.3

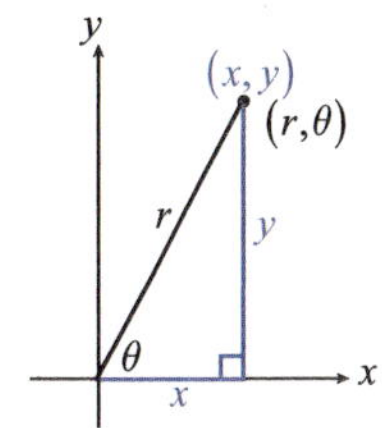

$$x = r\cos\theta$$

$$y = r\sin\theta$$

$$r^2 = x^2 + y^2$$

$$\tan\theta = \frac{y}{x}\ (x \neq 0)$$

CONVERTING BETWEEN RADIAN AND DEGREE MEASURE 7.1

$180° = \pi$ rad

$1° = \frac{\pi}{180}$ rad $\qquad \left(\frac{180}{\pi}\right)^{\circ} = 1$ rad

$x° = x\left(\frac{\pi}{180}\right)$ rad $\qquad x$ rad $= x\left(\frac{180}{\pi}\right)^{\circ}$

ARC LENGTH, AREA OF A SECTOR, ANGULAR SPEED, AND LINEAR SPEED 7.1

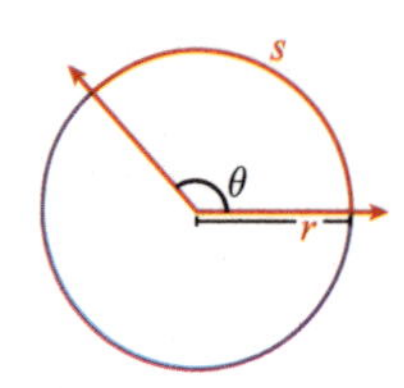

$s = \left(\frac{\theta}{2\pi}\right)(2\pi r) = r\theta \qquad A = \left(\frac{\theta}{2\pi}\right)(\pi r^2) = \frac{r^2\theta}{2}$

$\omega = \frac{\theta}{t} \qquad v = \frac{s}{t} = \frac{r\theta}{t} = r\omega$

TRIGONOMETRIC FUNCTIONS AND RIGHT TRIANGLES 7.2

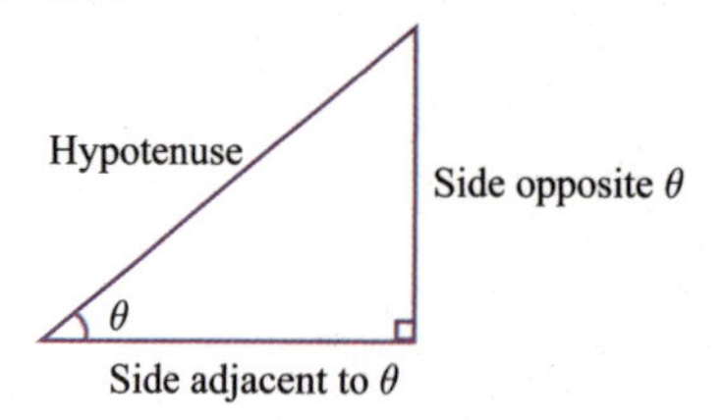

$\sin\theta = \frac{\text{opp}}{\text{hyp}} \qquad \csc\theta = \frac{1}{\sin\theta} = \frac{\text{hyp}}{\text{opp}}$

$\cos\theta = \frac{\text{adj}}{\text{hyp}} \qquad \sec\theta = \frac{1}{\cos\theta} = \frac{\text{hyp}}{\text{adj}}$

$\tan\theta = \frac{\text{opp}}{\text{adj}} \qquad \cot\theta = \frac{1}{\tan\theta} = \frac{\text{adj}}{\text{opp}}$

TRIGONOMETRIC FUNCTIONS OF ARBITRARY ANGLES 7.3

$\sin\theta = \frac{y}{r} \qquad \csc\theta = \frac{r}{y}$ (for $y \neq 0$)

$\cos\theta = \frac{x}{r} \qquad \sec\theta = \frac{r}{x}$ (for $x \neq 0$)

$\tan\theta = \frac{y}{x}$ (for $x \neq 0$) $\qquad \cot\theta = \frac{x}{y}$ (for $y \neq 0$)

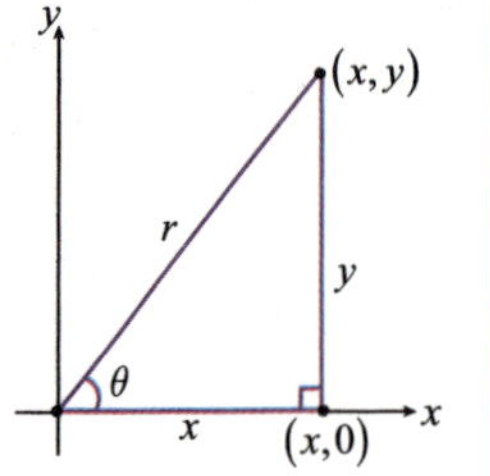

TRIGONOMETRIC FUNCTIONS AND THE UNIT CIRCLE 7.3

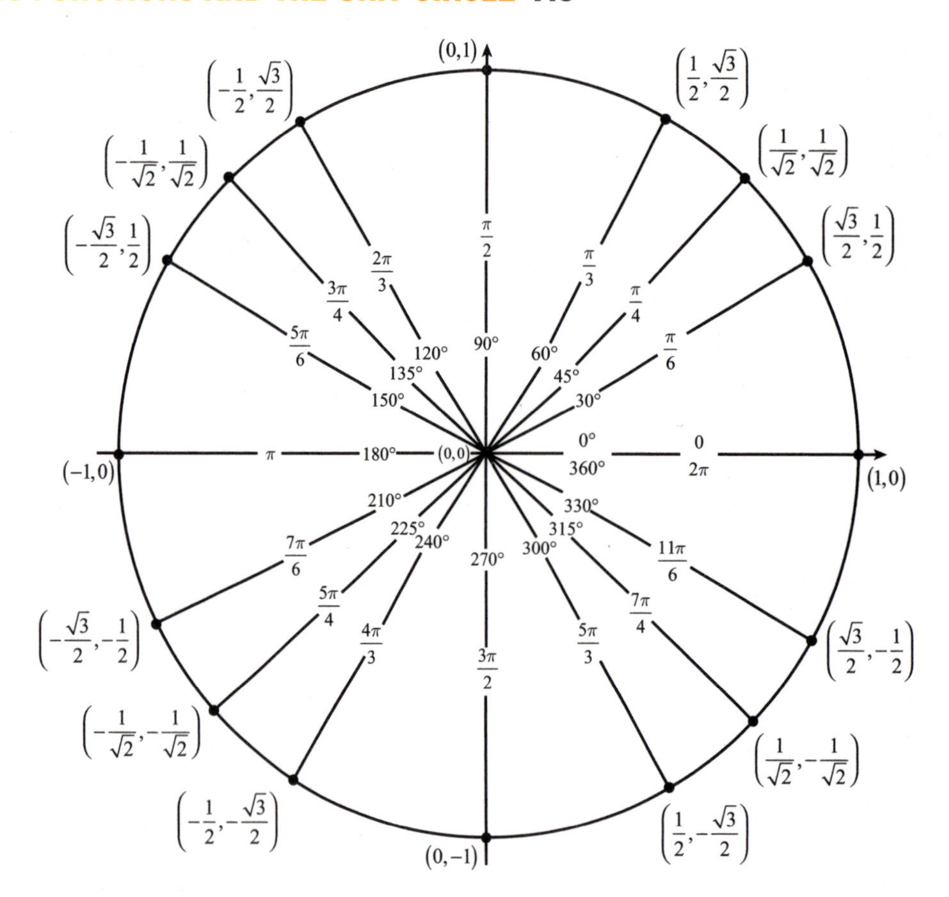

Each ordered pair represents $(\cos\theta, \sin\theta)$, and $\tan\theta = \frac{\sin\theta}{\cos\theta}$.